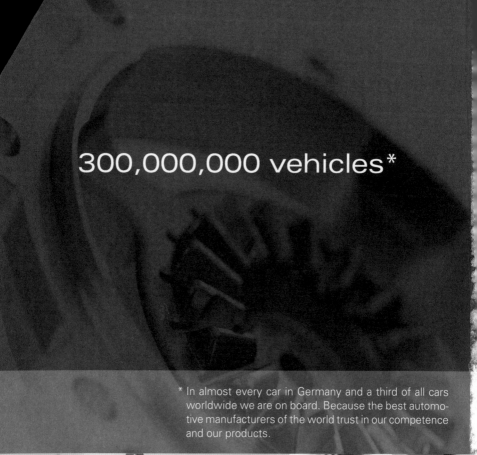

Automotive Handbook 9th Edition

Automotive Handbook

Imprint

Published by:
© Robert Bosch GmbH, 2014
Postfach 410960
D-76225 Karlsruhe.
Business Division
Automotive Aftermarket.

Scientific advisor and editor
Prof. Dr.-Ing. Konrad Reif,
Duale Hochschule Baden-Württemberg, Ravensburg, Campus Friedrichshafen,
Academic Program Director
Automotive Electronics and Mechatronic Systems.

Editorial staff
Dipl.-Ing. Karl-Heinz Dietsche.

Type setting
Dipl.-Ing. Karl-Heinz Dietsche;
Ann-Kathrin Bauer;
Detlev Sack, sacdesign, Stuttgart.

Translation
STAR Deutschland GmbH
Member of STAR Group

9th Edition, revised and extended, September 2014.

Printed in Germany.
Imprimé en Allemagne.

Distribution
John Wiley & Sons Ltd,
The Atrium,
Southern Gate, Chichester,
West Sussex PO19 8SQ, England
Telephone (+44) 1243 779777
Internet: www.wiley.com or www.wileyeurope.com
Email: cs-books@wiley.co.uk

ISBN 978-1-119-03294-6

Authors
Prof. Dr.-Ing. Konrad Reif,
Dipl.-Ing. Karl-Heinz Dietsche
and approx. 200 authors from industry and the university and college sector.

Technical graphics
Bauer & Partner, Gesellschaft für technische Grafik mbH, Stuttgart;
Schwarz Technische Grafik, Leonberg.

The following companies kindly placed text and picture matter and diagrams at our disposal:

Automotive Lighting Reutlingen GmbH.

Bosch Mahle Turbo Systems
GmbH & Co. KG, Stuttgart.

Daimler AG, Stuttgart.

Daimler AG, Sindelfingen.

Dr. Ing. h.c. F. Porsche AG, Weissach.

ETAS GmbH, Stuttgart.

Gates GmbH, Aachen.

IAV GmbH, Berlin.

IWIS Ketten, Joh. Winklhofer & Söhne
GmbH & Co. KG, München.

J. Eberspächer GmbH & Co. KG,
Esslingen.

Kiekert AG, Heiligenhaus.

Knorr-Bremse SfN, Schwieberdingen.

KS Gleitlager GmbH.

MAHLE Behr GmbH & Co. KG, Stuttgart.

MAN Nutzfahrzeuge Group.

MANN+HUMMEL GmbH, Ludwigsburg.

Michelin Reifenwerke AG & Co.KGaA,
Karlsruhe.

Robert Bosch Battery Systems GmbH,
Stuttgart

Saint-Gobain Sekurit International,
Herzogenrath.

Johnson Controls Autobatterie GmbH &
Co. KGaA, Hannover.

Volkswagen AG, Wolfsburg.

ZF Friedrichshafen AG, Friedrichshafen.

ZF Lenksysteme GmbH,
Schwäbisch Gmünd.

Duale Hochschule Baden-Württemberg,
Ravensburg, Campus Friedrichshafen.

Friedrich-Alexander-Universität
Erlangen-Nürnberg.

Hochschule Esslingen.

Hochschule Karlsruhe – Technik und
Wirtschaft.

Hochschule München.

Reinhold-Würth-Hochschule, Künzelsau.

Rheinisch-Westfälische
Technische Hochschule (RWTH) Aachen.

Foreword to the 9th Edition

Automotive engineering has become an extremely complex field over the last few years and decades. It is becoming increasingly more difficult to command an overview of the entire field and to maintain constant access to the subjects which are significant to automotive engineering. Many of these new subjects have in the meantime been covered in great detail in the wealth of available specialist literature. However, for those readers who wish to approach one of these new subjects for the first time, the available literature is neither easily manageable nor is it readable within the available timeframe.

This is where the Automotive Handbook comes in useful. It is structured in such a way as to be easily accessible even to those readers who are new to any individual subject. The most important subjects relevant to automotive engineering have been compiled in a compact, easily understandable, and practically relevant form. This is because all the content has been written by experts at Bosch and other vehicle manufacturers and suppliers who work in the very fields covered in the Handbook.

Many subjects have been newly incorporated into the 9th Edition, including driver-assistance systems, which are increasingly appearing in motor vehicles. In order to clearly bring out the essential character of the Automotive Handbook, the chapters on the basic scientific principles of automotive engineering have been revised and adapted to the latest automotive-engineering requirements. New additions include the chapter on Chemistry, and the chapter on Mathematics and all the subjects relating to materials have been greatly expanded. The Handbook has however also been revised, expanded and updated in many other places to make the contents even more understandable to the reader. In spite of the large number of contributing authors, every effort has been made to provide a uniform presentation and to maintain consistent classifications and nomenclature within its pages.

Over the eight decades in which the Automotive Handbook has been supporting the reader the format has been enlarged several times. In spite of this the designation Handbook has been retained. The format has once again been enlarged for the new edition to make it even easier to read. Not only has the format changed, but also the number of pages has increased. Around 280 new pages have been added.

This latest edition could not have been completed without the outstanding support of many individuals. Firstly, we would like to thank the authors of the individual articles, who, with great care and patience, have succeeded in delivering on schedule chapters of the highest substance and quality. Finally, we would like to thank all those readers who have provided useful suggestions and advice on corrections.

Friedrichshafen and Karlsruhe, April 2014,

The scientific advisor and editor
and the editorial staff.

Contents

Authors

Unless otherwise stated, the authors are employees of Robert Bosch GmbH.

Basic principles
Quantities and units
Prof. Dr. rer. nat. Susanne Schandl, Duale Hochschule Baden-Württemberg, Ravensburg, Campus Friedrichshafen.

Basic principles of mechanics
Prof. Dr.-Ing. Horst Haberhauer, Hochschule Esslingen.

Vibrations and oscillations
Dipl.-Ing. Sebastian Loos, Rheinisch-Westfälische Technische Hochschule (RWTH) Aachen.

Acoustics
Dipl.-Ing. Hans-Martin Gerhard, Dr. Ing. h.c. F. Porsche AG, Weissach.

Optical technology
Dipl.-Phys. Stefanie Mayer.

Hydrostatics, fluid mechanics
Prof. Dr.-Ing. Horst Haberhauer, Hochschule Esslingen.

Thermodynamics
Dr.-Ing. Ingo Stotz.

Electrical engineering
Dr.-Ing. Hans Roßmanith, Friedrich-Alexander-Universität Erlangen-Nürnberg.

Electronics
Prof. Dr.-Ing. Klemens Gintner, Hochschule Karlsruhe – Technik und Wirtschaft;
Dr. rer. nat. Ulrich Schaefer.

Chemistry
Dr. rer. nat. Jörg Ullmann.

Electrochemistry
Prof. Dr.-Ing. Matthias E. Rebhan, Hochschule München.

Mathematics and methods
Mathematics
Prof. Dr.-Ing. Matthias E. Rebhan, Hochschule München.

Finite-element method
Prof. Dipl.-Ing. Peter Groth, Hochschule Esslingen.

Control engineering
Dr.-Ing. Wolf-Dieter Gruhle, ZF Friedrichshafen AG, Friedrichshafen.

Materials
Materials
Dr.-Ing. Hagen Kuckert;
Dr. rer. nat. Jörg Ullmann;
Dr. rer. nat. Witold Pieper;
Dr. rer. nat. Waldemar Draxler;
Dipl.-Ing. Angelika Schubert;
Dipl.-Ing. Gert Lindemann;
Dr.-Ing. Carsten Tüchert;
Dr.-Ing. Sven Robert Raisch;
Dr.-Ing. Reiner Lützeler;
Dr. rer. nat. Jörg Bettenhausen;
Dr.-Ing. Gerrit Hülder;
Dipl.-Ing. Cornelius Gaida;
Dipl.-Phys. Klaus-Volker Schütt.

Heat treatment of metallic materials
Dr.-Ing. Jochen Schwarzer.

Corrosion and corrosion protection
Dipl.-Ing. (FH) Thomas Jäger.

Deposits and coatings
Dr. rer. nat. Manfred Rössler;
Dr. rer. nat. Ullrich Kraatz;
Dr. rer. nat. Christoph Treutler;
Dipl.-Ing. (FH) Hellmut Schmid.

Operating fluids
Lubricants
Dr. rer. nat. Gerd Dornhöfer.

Fuels
Dr. rer. nat. Jörg Ullmann.

Brake fluids, coolants
Dipl.-Ing. (FH) Lieselotte Häbe-Rapf.

Machine parts
Springs
Prof. Dr.-Ing. Horst Haberhauer,
Hochschule Esslingen.

Sliding bearings
Wolfgang Bickle,
KS Gleitlager GmbH.

Roller bearings
Dr.-Ing. Zhenhuan Wu.

Gears and tooth systems
Dipl.-Ing. Uwe von Ehrenwall.

Belt drives
Dipl.-Ing. Wolfgang Körfer,
Gates GmbH, Aachen.

Chain drives
Dr.-Ing. Thomas Fink,
IWIS Ketten, Joh. Winklhofer & Söhne
GmbH & Co. KG, München.

Joining and bonding techniques
Detachable connections
Prof. Dr.-Ing. Horst Haberhauer,
Hochschule Esslingen;
Dipl.-Ing. Rolf Bald.

Permanent joints
Dr.-Ing. Knud Nörenberg,
Volkswagen AG, Wolfsburg;
Dr. rer. nat. Patrick Stihler.

Internal-combustion engines
Internal-combustion engines, mixture formation, charge cycle and supercharging, reciprocating-piston engine
Dipl.-Ing. Heijo Oelschlegel,
Daimler AG, Stuttgart;
Prof. Dr. sc. techn. Thomas Koch,
Karlsruher Institut für Technologie (KIT);
Prof. Dr.-Ing. Klaus Binder,
Daimler AG, Stuttgart;
Dr.-Ing. Otmar Scharrer,
MAHLE Behr GmbH & Co. KG, Stuttgart.

Engine cooling
Dipl.-Ing. (FH) Ralf-Holger Schink,
MAHLE Behr GmbH & Co. KG, Stuttgart;
Dr.-Ing. Otmar Scharrer,
MAHLE Behr GmbH & Co. KG, Stuttgart.

Engine lubrication
Dipl.-Ing. Markus Kolczyk,
MANN+HUMMEL GmbH, Ludwigsburg;

Air-intake systems and intake manifolds
Dipl.-Ing. Andreas Weber,
MANN+HUMMEL GmbH, Ludwigsburg;
Dipl.-Ing. Andreas Pelz,
MANN+HUMMEL GmbH, Ludwigsburg;
Dipl.-Ing. Markus Kolczyk,
MANN+HUMMEL GmbH, Ludwigsburg;
Dipl. Ing. (FH) Alexander Korn,
MANN+HUMMEL GmbH, Ludwigsburg;
Dipl. Ing. (FH) Matthias Alex,
MANN+HUMMEL GmbH, Ludwigsburg;
Dr.-Ing. Matthias Teschner,
MANN+HUMMEL GmbH, Ludwigsburg;
Dipl. Ing. (FH) Mario Rieger,
MANN+HUMMEL GmbH, Ludwigsburg;
Dipl. Ing. Christof Mangold,
MANN+HUMMEL GmbH, Ludwigsburg;
Dipl.-Ing. Hedwig Schick,
MANN+HUMMEL GmbH, Ludwigsburg.

Turbochargers and superchargers
Dr.-Ing. Stefan Münz,
Bosch Mahle Turbo Systems
GmbH & Co. KG, Stuttgart.

Exhaust-gas system
Dr. rer. nat. Rolf Jebasinski,
J. Eberspächer GmbH & Co. KG,
Esslingen.

Emission-control and diagnosis legislation
Emission-control Legislation
Dr.-Ing. Matthias Tappe;
Dipl.-Ing. Michael Bender.

Exhaust-gas measuring techniques
Dipl.-Phys. Martin-Andreas Drühe;
Dipl.-Ing. Andreas Kreh;
Dipl.-Ing. Bernd Hinner;
Dr.-Ing. Matthias Tappe.

Diagnostics
Dr.-Ing. Markus Willimowski;
Dr.-Ing. Günter Driedger;
Dr. rer. nat. Hauke Wendt;
Dr.-Ing. Michael Hackner;
Dorothee Amann;
B. Eng. Varun Suri.

Management for spark-ignition engines
Management for spark-ignition engines
Dipl.-Ing. Armin Hassdenteufel.

Cylinder charge
Dr.-Ing. Martin Brandt.

Fuel supply system
Dipl.-Ing. Timm Hollmann;
Dipl.-Ing. Karsten Scholz;
Dipl.-Ing. Jens Wolber;
Dr.-Ing. Thomas Kaiser;
Dipl.-Ing. Uwe Müller;
Dipl.-Ing. (FH) Horst Kirschner.

Mixture formation
Dipl.-Ing. Andreas Binder;
Dipl.-Ing. Markus Gesk;
Dipl.-Ing. Andreas Glaser;
Dr.-Ing. Tilo Landenfeld.

Ignition
Dipl.-Ing. Walter Gollin;
Dipl.-Ing. Werner Häming;
Dipl.-Ing. Tim Skowronek;
Dr. rer. nat. Igor Orlandini;
Dr.-Ing. Grit Vogt.

Catalytic exhaust-gas treatment
Dipl.-Ing. Klaus Winkler;
Dipl.-Ing. Detlef Heinrich.

Alternative gasoline-engine operation
LPG operation
Dipl.-Ing. Iraklis Avramopoulos,
IAV GmbH, Berlin.

Engines fueled by natural gas
Dipl.-Ing. (FH) Thorsten Allgeier.

Management for diesel engines
Management for diesel engines
Dipl.-Ing. Felix Landhäußer.

Low-pressure fuel supply
Dipl.-Ing. (FH) Stefan Kieferle;
Dr.-Ing. Thomas Kaiser.

Common-rail injection system
Dipl.-Ing. Felix Landhäußer;
Dipl.-Ing. (FH) Andreas Rettich;
Dipl.-Ing. Thilo Klam;
Dr.-Ing. Holger Rapp;
Dipl.-Ing. Anees Haider Bukhari;
Dipl.-Ing. (FH) Herbert Strahberger.

Time-controlled single-cylinder pump system
Dipl.-Ing. (BA) Jürgen Crepin,
ETAS GmbH, Stuttgart.

Diesel-distributor injection pumps
Dipl.-Ing. (BA) Jürgen Crepin,
ETAS GmbH, Stuttgart.

Start-assist systems
Dipl.-Ing. (FH) Michael Wehleit;
Dr. rer. nat. Wolfgang Dreßler;
Dipl.-Ing. Friedrich Schmid,
Daimler AG, Stuttgart.

Exhaust-gas treatment
Klaus Gottwalt.

Alternative drives
Hybrid drives
Dipl.-Ing. Thomas Huber.

Fuel cells
Dr.-Ing. Gunter Wiedemann;
Dr. rer. nat. Ulrich Gottwick;
Dipl.-Ing. (FH) Jan-Michael Grähn;
Dipl.-Ing. Dipl.-Wirt.-Ing. Nils Kaiser.

Drivetrain
Drivetrain
Dipl.-Ing. Peter Köpf,
ZF Friedrichshafen AG, Friedrichshafen;
Dipl.-Ing. (FH) Thomas Müller.

Vehicle physics
Basic terms of automotive engineering
Prof. Dr. rer. nat. Ludger Dragon,
Daimler AG, Sindelfingen.

Motor-vehicle dynamics
Dipl.-Ing. Marc Birk,
Daimler AG, Stuttgart;
Prof. Dr. rer. nat. Ludger Dragon,
Daimler AG, Sindelfingen;
Dr.-Ing. Rupert Niethammer,
Daimler AG, Stuttgart;
Dipl.-Ing. Imre Boros,
Daimler AG, Stuttgart;
Dipl.-Ing. Klaus Wüst,
Daimler AG, Stuttgart.

Aerodynamics
Dipl.-Ing. Michael Preiß,
Dr. Ing. h.c. F. Porsche AG, Weissach.

Vehicle acoustics
Dipl.-Ing. Hans-Martin Gerhard,
Dr. Ing. h.c. F. Porsche AG, Weissach.

Chassis systems
Chassis systems, suspension
Dipl.-Ing. Maciej Foltanski,
Rheinisch-Westfälische
Technische Hochschule (RWTH) Aachen.

Shock and vibration absorbers, wheel suspensions
Dipl.-Ing. Jörn Lützow,
Rheinisch-Westfälische
Technische Hochschule (RWTH) Aachen.

Wheels
Dipl.-Ing. Martin Lauer,
Daimler AG, Sindelfingen;
Dipl.-Ing. Werner Hann,
Daimler AG, Sindelfingen;
Dipl.-Hdl. Martin Bauknecht,
MAN Nutzfahrzeuge Group.

Tires
Dipl.-Ing. Dirk Vincken,
Agentur für Text&Bild, Eurasburg;
Dipl.-Ing. Reimund Müller,
Michelin Reifenwerke AG & Co.KGaA
Karlsruhe;
Thilo Baloko,
Michelin Reifenwerke AG & Co.KGaA
Karlsruhe.

Tire-pressure monitoring system
Dipl.-Ing. Norbert Polzin.

Steering
Dipl.-Ing. Peter Brenner,
ZF Lenksysteme GmbH,
Schwäbisch Gmünd.

Brake systems
Dr. rer. nat. Jürgen Bräuninger;
Werner Schneider.

Brake systems for passenger cars
Werner Schneider;
Dipl.-Ing. Bernhard Kant.

Brake systems for commercial vehicles
Werner Schneider;
Dr.-Ing. Dirk Huhn,
ZF Friedrichshafen AG, Friedrichshafen.

Wheel brakes
Werner Schneider.

Chassis control and active safety
Antilock braking system
Dipl.-Ing. (FH) Alfred Strehle;
Werner Schneider;
Dipl.-Ing. Frank Schwab,
Knorr-Bremse SfN, Schwieberdingen.

Traction control system
Werner Schneider.

Driving-dynamics control system
Dr.-Ing. Gero Nenninger;
Dipl.-Ing. (FH) Jochen Wagner;
Dr.-Ing. Falk Hecker,
Knorr-Bremse SfN, Schwieberdingen.

Integrated driving-dynamics control systems
Dr.-Ing. Michael Knoop.

Vehicle bodies
Vehicle bodies, passenger cars
Dipl.-Ing. Dieter Scheunert,
Daimler AG, Sindelfingen.

Vehicle bodies, commercial vehicles
Dipl.-Ing. Uwe Schon,
Daimler AG, Stuttgart.

Lighting equipment
Dr.-Ing. Michael Hamm,
Automotive Lighting Reutlingen GmbH;
Dipl.-Ing. Doris Boebel,
Automotive Lighting Reutlingen GmbH;
Dipl.-Ing. Tilman Spingler,
Automotive Lighting Reutlingen GmbH.

Automotive windshield and window glass
Dipl.-Kauffr. (FH) Britta Müller,
Saint-Gobain Sekurit International,
Herzogenrath.

Windshield, rear-window and headlamp cleaning systems
Dr.-Ing. Mario Hüsges;
Dipl.-Ing. Florian Hauser.

Passive safety
Occupant-protection systems
Dr. rer. nat. Alfred Kuttenberger.

Vehicle security systems
Locking systems
Dr.-Ing. Bernhard Kordowski,
Kiekert AG, Heiligenhaus.

Theft-deterrent systems
Dipl.-Ing. (FH) U. Götz.

Automotive electrics
Vehicle electrical systems
Dipl.-Ing. Clemens Schmucker;
Dipl.-Ing. Eberhard Schoch;
Dipl.-Ing. Markus Beck.

Vehicle electrical systems for hybrid drives
Dr.-Ing. Jochen Faßnacht.

Starter batteries
Dr. rer. nat. Eberhard Meißner,
Johnson Controls Autobatterie GmbH & Co.KGaA, Hannover.

Drive batteries
Dr.-Ing. Stefan Benz,
Robert Bosch Battery Systems GmbH, Stuttgart;
Dr.-Ing. Christian Pankiewitz,
Robert Bosch Battery Systems GmbH, Stuttgart;
Dr.-Ing. Holger Fink,
Robert Bosch Battery Systems GmbH, Stuttgart.

Electrical machines
Prof. Dr.-Ing. Jürgen Ulm,
Reinhold-Würth-Hochschule, Künzelsau.

Alternators
Dipl.-Ing. Reinhard Meyer.

Starting systems
Dipl.-Ing. C. Krondorfer;
Dr.-Ing. Ingo Richter.

Actuators
Dr.-Ing. Rudolf Heinz;
Dr.-Ing. Thomas Hennige.

Wiring harnesses and plug-in connections
Dipl.-Ing. (FH) Wolfgang Kircher;
Dipl.-Ing. Werner Hofmeister;
Dipl.-Ing. Andreas Simmel.

Electromagnetic compatibility (EMC)
Dr.-Ing. Wolfgang Pfaff.

Symbols and circuit diagrams
Editorial staff.

Automotive electronics
Electronic control units
Dipl.-Ing. Martin Kaiser;
Dipl.-Ing. Axel Aue.

Automotive software engineering
Dipl.-Ing. (BA) Jürgen Crepin,
ETAS GmbH, Stuttgart
Dr. rer. nat. Kai Pinnow,
ETAS GmbH, Stuttgart
Dipl.-Ing. Jörg Schäuffele,
ETAS GmbH, Stuttgart.

Automotive networking, buses in motor vehicles
Dr. rer. nat. Harald Weiler;
Dr.-Ing. Tobias Lorenz;
Dipl.-Ing. Oliver Prelle.

Architecture of electronic systems
Dr.-Ing. Wolfgang Stolz;
Dipl.-Ing. (FH) Tino Sommer.

Sensors
Dr.-Ing. Erich Zabler;
Dipl.-Ökon. Frauke Ludmann;
Dr. rer. nat Peter Spoden;
Dipl.-Ing. (FH) Cyrille Caillié;
Dr.-Ing. Uwe Konzelmann;
Dr.-Ing. Tilmann Schmidt-Sandte;
Dr.-Ing. Reinhard Neul;
Dr. Berndt Cramer;
Dipl.-Ing. Dipl.-Wirt.-Ing. Nils Kaiser;
Prof. Dr.-Ing. Peter Knoll;
Dipl.-Ing., M.S. (University of Colorado) Joachim Selinger;
Dr.-Ing. Jan Sparbert.

Mechatronics
Dipl.-Ing. Hans-Martin Heinkel;
Dr.-Ing. Klaus-Georg Bürger.

Comfort and convenience
Passenger-compartment climate control
Dipl.-Ing. Peter Kroner,
MAHLE Behr GmbH & Co. KG, Stuttgart;
Dipl.-Ing. (FH) Thomas Feith,
MAHLE Behr GmbH & Co. KG, Stuttgart;
Dipl.-Ing. Günter Eberspach,
J. Eberspächer GmbH & Co. KG, Esslingen.

Comfort and convenience systems in the door and roof areas
Dipl.-Ing. (FH) Walter Haußecker;
Dipl.-Ing. (FH) Siegfried Reichmann.

Comfort and convenience functions in the passenger compartment
Dipl.-Ing. (FH) Reiner Birkert.

User interfaces, telematics and multimedia
Display and control
Prof. Dr.-Ing. Peter Knoll.

Radio and TV reception in motor vehicles
Dr.-Ing. Jens Passoke.

Traffic telematics
Dr.-Ing. Michael Weilkes.

Driver-assistance systems
Driver-assistance systems
Dr.-Ing. Frank Niewels;
Dipl.-Math. (FH) Thomas Lich;
Dr.-Ing. Thomas Michalke;
Dr.-Ing. Thomas Maurer.

Computer vision
Dr.-Ing. Wolfgang Niehsen.

Vehicle navigation
Dipl.-Ing. Ernst Peter Neukirchner.

Night-vision systems
Prof. Dr.-Ing. Peter Knoll.

Parking and maneuvering systems
Prof. Dr.-Ing. Peter Knoll.

Adaptive Cruise Control
Dipl.-Ing. Gernot Schröder;
Prof. Dr.-Ing. Peter Knoll.

Lane departure warning and lane keeping support
Dr.rer.nat. Lutz Bürkle;
Dr.-Ing. Thomas Michalke;
Dipl.-Ing. Thomas Glaser.

Emergency-braking systems when driving in the same direction
Dr.-Ing. Thomas Gussner;
Dr.-Ing. Steffen Knoop.

Intersection assistant
Dr.-Ing. Wolfgang Branz;
Dr.-Ing. Rüdiger Jordan.

Active pedestrian protection
Dr.-Ing. Thomas Gussner;
Dr.-Ing. Steffen Knoop.

High-beam assistant
Dipl.-Ing. Doris Boebel,
Automotive Lighting Reutlingen GmbH;
Dipl.-Ing. (FH) Bernd Dreier,
Automotive Lighting Reutlingen GmbH.

Future of driver assistance
Dr.-Ing. Frank Niewels;
Dr.-Ing. Rüdiger Jordan.

Quantities and units

In order to be able to express the values of physical quantities, a system of units is required which serves as the yardstick for every measurement. The value of the physical quantity is taken as the product of numerical value and unit. Such a system of units is the SI system, which was established in 1960 by the 11th General Conference on Weights and Measures (Conférence Générale des Poids et Mesures, CGPM). It has since then been adopted by over 50 countries.

In Germany the management of units is by law under the control of the Physikalisch-Technische Bundesanstalt (PTB, based in Braunschweig), Germany's national metrology institute. International responsibility lies with the International Bureau of Weights and Measures (BIPM, Bureau International des Poids et Mesures) in Sèvres near Paris.

Further units which are to this day commonly used (e.g. liter, metric ton, hour, degree Celsius) are in Germany likewise permitted by law and are mentioned here as such.

Units permitted in other countries (e.g. inch, ounce, degree Fahrenheit) or obsolete units are discussed in a separate section.

Table 1: Base SI units

Base quantity and symbol		Base SI unit	
		Name	Symbol
Length	l	meter	m
Mass	m	kilogram	kg
Time	t	second	s
Electric current	I	ampere	A
Thermodynamic temperature	T	kelvin	K
Amount of substance	n	mole	mol
Luminous intensity	I	candela	cd

SI units

SI means "Système International d'Unités" (International System of Units). The system is laid down in ISO 80000 [1] (ISO: International Organization for Standardization) and for Germany in DIN 1301 [2] (DIN: German Institute for Standardization).

Table 1 lists the seven base SI units.

Definitions of the base SI units

The primary definitions of the base SI units are formulated in French. The translation of the original French text is shown in the following in italics (see [3]).

1. Length

The meter is defined as the distance which light travels in a vacuum in 1/299,792,458 seconds (17th CGPM, 1983).

The meter is therefore defined using the speed of light in a vacuum (c = 299,792,458 m/s) and no longer by the wavelength of the radiation emitted by the krypton nuclide ^{86}Kr. The meter was originally defined as the forty-millionth part of a terrestrial meridian (standard meter in Paris, 1875).

2. Mass

The kilogram is the unit of mass; it is equal to the mass of the international kilogram prototype (1st CGPM, 1889 and 3rd CGPM, 1901).

This prototype of a platinum-iridium alloy is kept in the BIPM. The national prototype for Germany is kept at the Physikalisch-Technische Bundesanstalt (PTB) in Braunschweig.

3. Time

The second is defined as the duration of 9,192,631,770 periods of the radiation corresponding to the transition between the two hyperfine levels of the ground state of atoms of the nuclide ^{133}Cs (13th CGPM, 1967).

This definition relates to a cesium atom at rest at a temperature of 0 K. It can be reproduced much more accurately than the earlier astronomical definitions, which were based on the length of a day and a year.

4. Electric current
The ampere is defined as the intensity of a constant electric current which, if maintained in two straight, parallel conductors of infinite length, of negligible circular cross-sections, and placed 1 meter apart in a vacuum, will produce between these conductors a force equal to $2 \cdot 10^{-7}$ newtons (9th CGPM, 1948).

Thus the magnetic field constant is established as $\mu_0 = 4\pi \cdot 10^{-7}$ H/m.

5. Temperature
The kelvin, the unit of thermodynamic temperature, is defined as the fraction 273.16 of the thermodynamic temperature of the triple point of water (10th CGPM, 1954 and 13th CGPM, 1967).

The zero point of the kelvin scale is the absolute temperature zero point. The triple point of pure water, at which all three states of aggregation (solid, liquid, gaseous) exist in equilibrium to each other, is at 273.16 K and a pressure of 611.657 Pa.

6. Amount of substance
6.1. The mole is defined as the amount of substance of a system which contains as many elementary entities as there are atoms in 0.012 kilogram of the carbon nuclide ^{12}C; its symbol is "mol".
6.2. When the mole is used, the elementary entities must be specified and may be atoms, molecules, ions, electrons, other particles, or groups of such particles of precisely specified composition (14th CGPM, 1971).

In the definition of the mole it is assumed that the atoms in question are free carbon-12 atoms which are at rest and in the ground state.

7. Luminous intensity
The candela is defined as the luminous intensity in a given direction of a source which emits monochromatic radiation of frequency $540 \cdot 10^{12}$ hertz and of which the radiant intensity in that direction is 1/683 watt per steradian (16th CGPM, 1979).

The "candela" (Latin: candle, emphasis on the 2nd syllable) superseded the "new candle", which is defined by way of the black body radiation for the melting temperature of platinum.

Decimal fractions and multiples of SI units

Decimal fractions and multiples of SI units (base SI units and derived SI units) are denoted by prefixes before the name of the unit (e.g. milligram) or by prefix symbols before the unit symbol (e.g. mg) (Table 2). The prefix symbol is placed without a gap in front of the unit symbol to form a coherent unit.

Prefixes are not used before further units of angle (degree, min, second), of time (minute, hour), and of temperature (degree Celsius).

Table 2: Prefixes for unit of measurements in acc. with DIN 1301 [2]

Prefix	Prefix symbol	Factor	Name of factor
yokto	y	10^{-24}	septillionth
zepto	z	10^{-21}	sextillionth
atto	a	10^{-18}	trillionth
femto	f	10^{-15}	thousand billionth
pico	p	10^{-12}	billionth
nano	n	10^{-9}	thousand millionth
micro	µ	10^{-6}	millionth
milli	m	10^{-3}	thousandth
centi	c	10^{-2}	hundredth
deci	d	10^{-1}	tenth
deca	da	10^{1}	ten
hecto	h	10^{2}	hundred
kilo	k	10^{3}	thousand
mega	M	10^{6}	million
giga	G	10^{9}	thousand million[1]
tera	T	10^{12}	billion[1]
peta	P	10^{15}	thousand billion
exa	E	10^{18}	trillion
zetta	Z	10^{21}	sextillion
yotta	Y	10^{24}	septillion

[1] In the USA: 10^9 = 1 billion, 10^{12} = 1 trillion.

Derived SI units

SI units are the seven base SI units and all the units derived from them, i.e. which can be represented as a product of powers of the base units. Thus, for example, the unit of force is obtained from Newton's Law $F = ma$ as

$$1\ \mathrm{kg}\,\frac{\mathrm{m}}{\mathrm{s}^2} = 1\ \mathrm{N}\ \text{(newton)}.$$

If the power product contains only the factor 1, the units are referred to as coherently derived units. There are a total of 22 coherently derived units, which like the newton have been assigned their own names (Table 3).

Legal units

The Law on Units in Metrology of 22 February 1985, amended on 29 October 2001, and the related implementing order of 13 December 1985, last amended on 12 July 2008 (as at 2013), stipulate the use of legal units in business and official transactions. Legal units are:
- the SI units,
- decimal fractions and multiples of SI units,
- other permitted units; see the overview on the following pages and [4].

The following tables provide an overview in accordance with DIN 1301 [2].

Table 3: Derived SI units with special names

Quantity	Unit	Unit symbol	Expressed in other SI units
Plane angle	radian	rad	1
Solid angle	steradian	sr	1
Frequency	hertz	Hz	1 Hz = 1/s
Force	newton	N	1 N = 1 kg m/s^2 = 1 J/m
Pressure	pascal	Pa	1 Pa = 1 N/m^2
Energy, work, quantity of heat	joule	J	1 J = 1 N m = 1 W s
Power	watt	W	1 W = 1 J/s = 1 V A
Celsius temperature	degree Celsius	°C	
Voltage	volt	V	1 V = 1 W/A
Electrical conductance	siemens	S	1 S = 1 A/V = 1/Ω
Electrical resistivity	ohm	Ω	1 Ω = 1 V/A
Electric charge	coulomb	C	1 C = 1 A s
Electrical capacitance	farad	F	1 F = 1 C/V
Inductance	henry	H	1 H = 1 Wb/A
Magnetic flux	weber	Wb	1 Wb = 1 V s
Magnetic flux density, induction	tesla	T	1 T = 1 Wb/m^2
Luminous flux	lumen	lm	1 lm = 1 cd sr
Luminous intensity	lux	lx	1 lx = 1 lm/m^2
Radioactivity	becquerel	Bq	1 Bq = 1/s
Absorbed dose	gray	Gy	1 Gy = 1 J/kg
Dose equivalent	sievert	Sv	1 Sv = 1 J/kg
Catalytic activity	katal	kat	1 kat = 1 mol/s

Table 4: Legal units

Quantity and symbol	Legal units SI	Others	Name	Relationship	Remarks and units not to be used, incl. their conversion
1. Length, area, volume					
Length l	m		meter		
Area A	m^2		square meter		
		a	are	1 a = 100 m^2	
		ha	hectare	1 ha = 100 a = 10^4 m^2	
Volume V	m^3		cubic meter		
		l, L	liter	1 l = 1 L = 1 dm^3	
2. Angle					
(Plane) angle[1] α, β etc.	rad		radian	$1\text{ rad} = \frac{1\text{ m ar}}{1\text{ m radius}}$	1^g (centesimal degree) = 1 gon 1^c (centesimal minute) = 10^{-2} gon 1^{cc} (centesimal second) = 10^{-4} gon
		°	degree	$1° = \frac{\pi}{180}\text{ rad}$	
		′	minute		
		″	second	1° = 60′ = 3,600″	
		gon	gon	$1\text{ gon} = \frac{\pi}{200}\text{ rad}$	
Solid angle[1] Ω	sr		steradian	$1\text{ sr} = \frac{1\text{ m}^2\text{ spherical surface}}{1\text{ m}^2\text{ sphere radius}^2}$	

[1] The units rad and sr can be replaced by the numeral 1 in calculations.

Quantity and symbol	Legal units SI	Others	Name	Relationship	Remarks and units not to be used, incl. their conversion

3. Mass

Quantity and symbol	SI	Others	Name	Relationship	Remarks and units not to be used, incl. their conversion
Mass m (weight)[1]	kg		kilogram		
		g	gram	$1\ \text{g} = 10^{-3}\ \text{kg}$	
		t	metric ton	$1\ \text{t} = 10^{3}\ \text{kg}$	
Density ρ	kg/m^3				Weight of unit volume γ (kp/dm^3 or p/cm^3) $\gamma = \rho \cdot g$
Moment of inertia (mass moment, 2nd order) J	$\text{kg} \cdot \text{m}^2$			$J = m \cdot r^2$ r = radius of gyration	Flywheel effect $G \cdot D^2$ in $\text{kp} \cdot \text{m}^2$ $D = 2r$, $G = m \cdot g$ $G \cdot D^2 = 4J \cdot g$

4. Time quantities

Quantity and symbol	SI	Others	Name	Relationship	Remarks and units not to be used, incl. their conversion
Time Duration Interval[2] t	s		second		In energy management one year is calculated as 8,760 hours
		min	minute	1 min = 60 s	
		h	hour	1 h = 60 min	
		d	day	1 d = 24 h	
		a	year		
Frequency f	Hz		hertz	1 Hz = 1/s	
Rotational speed (rotational frequency) n	s^{-1}			$1\ \text{s}^{-1} = 1/\text{s}$	rpm and r/min (revolutions per minute) are still permissible for expressing rotational speed, but is better replaced by min^{-1} (1 rpm = 1 r/min = 1 min^{-1})
		min^{-1} 1/min		$1\ \text{min}^{-1} = 1/\text{min} = (1/60)\ \text{s}^{-1}$	
Angular frequency ω	s^{-1}			$\omega = 2\pi f$	
Velocity v	m/s	km/h		1 km/h = (1/3.6) m/s	
Acceleration[4] a	m/s^2			Normal acceleration of free fall $g \approx 9.80665\ \text{m/s}^2$	
Angular velocity[3] ω	rad/s				
Angular acceleration[3] α	rad/s^2				

[1] The term “weight” is ambiguous in everyday usage; it is used to denote mass as well as weight (DIN 1305 [5]).
[2] Clock times: h, min, s written as superscripts. Example: $3^{h}\ 25^{m}\ 6^{s}$.
[3] The unit rad can be replaced by the numeral 1 in calculations.
[4] Acceleration is sometimes expressed in m/s^2 as a multiple of gravitational acceleration g.

Quantity and symbol	Legal units SI	Others	Name	Relationship	Remarks and units not to be used, incl. their conversion
5. Force, energy, power					
Force F Force due to weight G	N N		newton	1 N = 1 kg · m/s²	1 kp (kilopond) ≈ 9.80665 N
Impulse p	N s			1 N s = 1 kg · m/s²	
Pressure, gen. p	Pa		pascal	1 Pa = 1 N/m²	1 at (techn. atmosphere) = 1 kp/cm² ≈ 0.980665 bar p ≈ 1.01325 bar ≈ 1,013.25 hPa (standard value of air pressure)
		bar	bar	1 bar = 10^5 Pa	
Mechanical stress σ, τ	N/m²			1 N/m² = 1 Pa	1 kp/m² ≈ 9.80665 N/m²
		N/mm²		1 N/mm² = 1 MPa	
Hardness	Brinell and Vickers hardness are no longer given in kp/mm². Instead, an abbreviation of the relevant hardness is written as the unit after the numerical value used previously (including an indication of the test force, etc., where applicable).				Examples: Previously — Now HB = 350 kp/mm² — 350 HB HV30 = 720 kp/mm² — 720 HV30 HRC = 60 — 60 HRC
Energy E Work W Heat, quantity of heat[1] Q	J		joule	1 J = 1 N m = 1 W s = 1 kg · m²/s²	1 kcal (kilocalorie) = 4.1868 kJ
		W s	watt-second	1 W s = 1 J	
		kW h	kilowatt-hour	1 kW h = 3.6 MJ	
		eV	electron-volt	1 eV ≈ 1.60219 · 10^{-19} J	
Torque M	N m		newton meter		1 kp · m (kilopondmeter) ≈ 9.80665 N m
Power P Heat flow $\dot{Q}$ Radiated power Φ	W		watt	1 W = 1 J/s = 1 N m/s	1 kp · m/s ≈ 9.80665 W 1 PS (horsepower) ≈ 0.7355 kW
Apparent power P_s	VA		volt-ampere	1 VA = 1 W	
Reactive power P_q	var		var	1 var = 1 W	

[1] The quantity of heat is expressed in joules.

Quantity and symbol		Legal units SI	Others	Name	Relationship	Remarks and units not to be used, incl. their conversion
6. Viscosimetric quantities						
Dynamic viscosity	η	Pa·s		pascal-second	1 Pa·s = 1 N s/m² = 1 kg/(s·m)	1 P (poise) = 0.1 Pa·s
Kinematic viscosity	ν	m^2/s			1 m^2/s = 1 Pa·s/(kg/m^3)	1 St (stokes) = 10^{-4} m^2/s
7. Temperature and heat						
Temperature	T	K	kelvin			
	ϑ		°C	degree Celsius	$\vartheta = (T - 273.15\ \text{K}) \frac{°\text{C}}{\text{K}}$	
Temperature difference	ΔT	K		kelvin	1 K = 1 °C	In the case of composite units, express temperature differences in K
	$\Delta\vartheta$		°C	degree Celsius		
For quantity of heat and heat flow, refer to 5						
Specific heat capacity (spec. heat)	c	J/(kg·K)				1 kcal/(kg·K) = 4.1868 kJ/(kg·K)
Molar heat capacity	C	J/(mol·K)				
Thermal conductivity	λ	W/(m·K)				1 kcal/(m·h·K) = 1.163 W/(m·K)
8. Electrical quantities						
Electric current	I	A		ampere		
Electric potential (voltage)	U	V		volt	1 V = 1 W/A	
Electrical conductance: Conductance G, Susceptance B, Admittance Y		S		siemens	1 S = 1 A/V = 1/Ω	
Electrical resistance: Resistance R, Reactance X, Impedance Z		Ω		ohm	1 Ω = 1/S = 1 V/A	
Quantity of electricity, electric charge	Q	C		coulomb	1 C = 1 A s	
			A h	ampere-hour	1 A h = 3,600 C	
Electrical capacitance	C	F		farad	1 F = 1 C/V	

Quantity and symbol	Legal units			Relationship	Remarks and units not to be used, incl. their conversion
	SI	Others	Name		
Electrical flux density, displacement D	C/m^2			$1\ C/m^2 = 1\ A\,s/m^2$	
Electric field strength E	V/m			$1\ V/m = 1\ W/(A\,m)$	

9. Magnetic quantities

Quantity and symbol	SI	Others	Name	Relationship	Remarks and units not to be used, incl. their conversion
Inductance L	H		henry	1 H = 1 Wb/A	
Magnetic flux Φ	Wb		weber	1 Wb = 1 V s	1 M (maxwell) = 10^{-8} Wb
Magnetic flux density, induction B	T		tesla	1 T = 1 Wb/mm^2	1 G (gauss) = 10^{-4} T
Magnetic field strength H	A/m			1 A/m = 1 N/Wb	1 Oe (oersted) = $\frac{10^3}{(4\pi)}\frac{A}{m}$

10. Photometric quantities

Quantity and symbol	SI	Others	Name	Relationship	Remarks and units not to be used, incl. their conversion
Luminous intensity I	Cd		candela		
Luminance L	cd/m^2				1 sb (stilb) = $10^4\ cd/m^2$ 1 asb (apostilb) = $1/\pi\ cd/m^2$
Luminous flux Φ	lm		lumen	1 lm = 1 cd sr (sr = steradian)	
Illuminance E	lx		lux	1 lx = 1 lm/m^2	

11. Acoustic quantities

Quantity and symbol	SI	Others	Name	Relationship	Remarks and units not to be used, incl. their conversion
Sound pressure p	Pa		pascal		
Sound intensity I	W/m^2				
Sound pressure level L_p Sound intensity level L_I	Np		neper	1 Np = 1 (non-dimensional)	$L_p = \ln(p_1/p_2)$ Np $= L_I = 0.5 \ln(I_1/I_2)$ Np
		dB	decibel	1 dB = $\frac{1}{20}$ ln 10 Np ≈ 0.1151 Np	$L_p = 20 \lg(p_1/p_2)$ dB $= L_I = 10 \lg(I_1/I_2)$ dB At f = 1,000 Hz the (physiological) loudness level is measured in "phon": 1 phon = 1 dB and the loudness in "sone": 1 sone = 40 phon
		B	bel	1 B = 10 dB ≈ 1.151 Np	

Quantity and symbol	Legal units SI	Others	Name	Relationship	Remarks and units not to be used, incl. their conversion
Sound pressure level L_{pA} Sound power level L_{WA} A-weighted		dB (A)			Frequency-dependent weighting tuned to the human ear at 20 to 40 phon

12. Quantities used in atom physics and other fields

Quantity and symbol	SI	Others	Name	Relationship	Remarks and units not to be used, incl. their conversion
Energy W		eV	electron-volt	1 eV ≈ $1.60219 \cdot 10^{-19}$ J 1 MeV = 10^6 eV	
Activity of a radioactive substance A	Bq		becquerel	1 Bq = 1 s^{-1}	1 Ci (curie) = $3.7 \cdot 10^{10}$ Bq
Absorbed dose D	Gy		gray	1 Gy = 1 J/kg	1 rd (rad) = 10^{-2} Gy
Dose equivalent D_q	Sv		sievert	1 Sv = 1 J/kg	1 rem = 10^{-2} Sv
Absorbed dose rate D				1 Gy/s = 1 W/kg	1 rd/s = 10^{-2} Gy/s
Ion dose J	C/kg				1 R (roentgen) = $258 \cdot 10^{-6}$ C/kg
Ion dose rate $\dot{J}$	A/kg				
Amount of substance n	mol		mole		
Catalytic activity	kat		katal	1 kat = 1 mol/s	

Further units

Table 5: Conversion of units

Units of length

Name	*Conversion*
micron	1 μ = 1 μm
typographical point	1 p = 0.37607 mm
inch	1 in = 25.4 mm
foot	1 ft = 12 in = 0.3048 m
yard	1 yd = 3 ft = 0.9144 m
mile	1 mile = 1,760 yd = 1.6093 km
nautical mile (international mile)	1 NM = 1 sm = 1.852 km (≈ 1′ of degree of longitude)

Anglo-American units of length:

microinch	1 μin = 0.0254 μm
milliinch	1 mil = 0.0254 mm
link	1 link = 201.17 mm
rod	1 rod = 1 pole = 1 perch = 5.5 yd = 5.0292 m
fathom	1 fathom = 2 yd = 1.8288 m
chain	1 chain = 22 yd = 20.1168 m
furlong	1 furlong = 220 yd = 201.168 m

Units of area

Name	*Conversion*
square inch (sq in)	1 in^2 = 6.4516 cm^2
square foot (sq ft)	1 ft^2 = 144 in^2 = 0.0929 m^2
square yard (sq yd)	1 yd^2 = 9 ft^2 = 0.8361 m^2
acre	1 ac = 4,840 yd^2 = 4,046.9 m^2
square mile (sq mile)	1 mile^2 = 640 acre = 2.59 km^2
barn	1 b = 10^{-28} m^2

Units of volume

Name	*Conversion*
cubic inch (cu in)	1 in^3 = 16.3871 cm^3
cubic foot (cu ft)	1 ft^3 = 1,728 in^3 = 0.02832 m^3
cubic yard (cu yd)	1 yd^3 = 27 ft^3 = 0.76456 m^3

Further units in the United Kingdom (UK):

fluid ounce	1 fl oz = 0.028413 *l*
pint	1 pt = 0.56826 *l*
quart	1 qt = 2 pt = 1.13652 *l*
gallon	1 gal = 4 qt = 4.5461 *l*
barrel (crude oil)	1 bbl = 35 gal = 159.1 *l*
barrel (other liquids)	1 bbl = 36 gal = 163.6 *l*

Further units in the United States (US):

fluid ounce	1 fl oz = 0.029574 *l*
liquid pint	1 liq pt = 0.47318 *l*
liquid quart	1 liq qt = 2 liq pt = 0.94635 *l*
gallon	1 gal = 4 liq qt = 3.7854 *l*
liquid barrel	1 liq bbl = 31.5 gal = 119.24 *l*
barrel petroleum	1 barrel petroleum = 42 gal = 158.99 *l* (for crude oil)

Velocities

Name	*Conversion*
kilometer per hour	1 km/h = (1/3.6) m/s ≈ 0.2778 m/s
miles per hour	1 mile/h = 1.6093 km/h
knot	1 kn = 1 NM/h = 1 sm/h = 1.852 km/h = 0.5144 m/s
Mach number Ma	The Mach number Ma is the quotient of velocity and velocity of sound.

Units of mass

Name	*Conversion*
Pfund (pound)	1 Pfund = 0.5 kg
Zentner (metric hundredweight)	1 Ztr = 50 kg
Doppelzentner (metric quintal)	1 dz = 100 kg
Gamma	1 γ = 1 µg
metric carat (for precious stones only)	1 Kt = 0.2 g
atomic unit of mass	1 u = $1.6606 \cdot 10^{-27}$ kg
grain	1 gr = 64.79891 mg
pennyweight	1 dwt = 24 gr = 1.5552 g
dram	1 dr = 1.77184 g
ounce	1 oz = 16 dram = 28.3495 g
troy ounce	1 oz tr (US) = 1 oz tr (UK) = 31.1035 g
pound	1 lb = 16 oz = 453.592 g
stone (UK)	1 st = 14 lb = 6.35 kg
quarter (UK)	1 qr = 28 lb = 12.7 kg
slug (mass accelerated by 1 ft/s^2 when 1 lbf is exerted on it)	1 slug = 14.4939 kg
hundredweight (US)	1 cwt = 1 cwt sh (short cwt) = 1 quintal = 100 lb = 45.3592 kg
hundredweight (UK)	1 cwt = 1 cwt l (long cwt) = 112 lb = 50.8023 kg
ton (US)	1 ton (US) = 1 tn sh (short ton) = 0.90718 t
ton (UK)	1 ton (UK) = 1 tn l (long ton) = 1 ton dw (ton deadweight) = 1.01605 t
–	1 t dw = 1 t

Density

Name	*Conversion*
	1 lb/ft^3 = 16.018 kg/m^3
	1 lb/gal (UK) = 99.776 kg/m^3
	1 lb/gal (US) = 119.83 kg/m^3

Areometer degrees n are a measure of the density ρ of a liquid relative to the density of water at 15 °C.

$n = 144.3(\rho - 1\ \text{kg}/l)/\rho$ °Bé (degree Baumé)

$n = (141.5\ \text{kg}/l - 131.5 \cdot \rho)/\rho$ °API

(API: American Petroleum Institute)

Units of force

Name	*Conversion*
pond	1 p ≈ 9.80665 mN
kilopond	1 kp ≈ 9.80665 N
dyne	1 dyn = 10^{-5} N
sthène (French)	1 sn = 1 kN
poundforce	1 lbf = 4.44822 N
poundal (force which accelerates a mass of 1 lb by 1 ft/s^2)	1 pdl = 0.138255 N

Units of pressure and stress

Name	*Conversion*
microbar	1 µbar = 0.1 Pa
millibar	1 mbar = 1 hPa = 100 Pa
bar	1 bar = 10^5 Pa
–	1 kp/mm^2 ≈ $9.80665 \cdot 10^6$ Pa
technical atmosphere	1 at = 1 kp/cm^2 ≈ $9.80665 \cdot 10^4$ Pa
physical atmosphere	1 atm = $1.01325 \cdot 10^5$ Pa
torr	1 torr = 1 mm Hg (mercury column) = 133.322 Pa
–	1 mm water column = 1 kp/m^2 ≈ 9.80665 Pa

The following units are derived from the technical atmosphere:

p_{abs} (absolute pressure): 1 ata

p_{amb} (ambient pressure)

$p_e = p_{abs} - p_{amb}$

(pressure above atmospheric): 1 atpaa for $p_{abs} > p_{amb}$

$p_e = p_{amb} - p_{abs}$

(pressure below atmospheric): 1 atpba for $p_{abs} < p_{amb}$

Anglo-American and other units:

poundforce per square inch	1 lbf/in^2 = 1 psi = 6894.76 Pa
poundforce per square foot	1 lbf/ft^2 = 1 psf = 47.8803 Pa
tonforce per square inch (UK)	1 tonf/in^2 = 1.54443 · 10^7 Pa
poundal per square foot	1 pdl/ft^2 = 1.48816 Pa
barye (French)	1 barye = 0.1 Pa
pièce (French)	1 pz = 1 sn/m^2 (sthène/m^2) = 1,000 Pa

Units of energy

Name	*Conversion*
erg	1 erg = 10^{-7} J
calorie [1]	1 cal = 4.1868 J
	1 kp · m = 9.80665 J
	1 HP · h = 2.6478 · 10^6 J
therm	1 therm = 105.50 · 10^6 J

Anglo-American and other units:

inch ounceforce	1 in ozf = 7.062 mJ
foot poundal	1 ft pdl = 0.04214 J
inch poundforce	1 in lbf = 0.11299 J
foot poundforce	1 ft lbf = 1.35582 J
British thermal unit [2]	1 Btu = 1,055.06 J
therm	1 therm = 10^5 Btu
horsepower hour	1 hp · h = 2.685 · 10^6 J
thermie (French)	1 thermie = 1,000 frigories = 1,000 kcal = 41.868 MJ
kg coal equivalents	1 kg CE = 29.3076 MJ (for the calorific value H_u = 7,000 kcal/kg coal)
tonne of coal equivalents	1 t CE = 1,000 kg CE

Units of power

Name	*Conversion*
kilocalorie per hour	1 kcal/h = 1.163 W
calorie per second	1 cal/s = 4.1868 W
–	1 kp · m/s = 9.80665 W
horsepower, cheval vapeur (French)	1 HP = 1 ch = 735.499 W

Anglo-American units:

–	1 ft · lbf/s = 1.35582 W
horsepower	1 hp = 745.70 W
–	1 Btu/s = 1,055.06 W

Fuel consumption

Name	*Conversion*
–	1 g/(HP · h) = 1.3596 g/(kW h)
–	1 lb/(hp · h) = 608.277 g/(kW h)
–	1 liq pt/(hp · h) = 634.545 cm^3/(kW h)

$$x \text{ mile/gal (US)} \triangleq \frac{235.21}{x} \; l/100 \text{ km}$$

$$x \text{ mile/gal (UK)} \triangleq \frac{282.48}{x} \; l/100 \text{ km}$$

$$y \; l/100 \text{ km} \triangleq \frac{235.21}{y} \text{ mile/gal (US)}$$

$$y \; l/100 \text{ km} \triangleq \frac{282.48}{y} \text{ mile/gal (UK)}$$

[1] The quantity of heat which is required to heat 1 g water from 15 °C to 16 °C.
[2] The quantity of heat which is required to heat 1 lb water from 63 °F to 64 °F.

Units of temperature

Name	*Conversion*
K (kelvin)	
°C (degree Celsius)	ϑ/°C = T/K − 273.15 This scale is based on the freezing point of water at 0 °C and the boiling point at 100 °C, at normal pressure in each case.
°F (degree Fahrenheit)	T_F/°F = 1.8 · T/K − 459.67 The freezing point of water is 32 °F, human body temperature is 96 °F.
°Ra (degree Rankine)	T_{Ra}/°Ra = 1.8 · T/K This scale begins with the absolute temperature zero point at 0 K and uses a graduation like the Fahrenheit.
°Re (degree Réaumur)	T_{Re}/°Re = 0.8 · (T/K − 273.15) The freezing point of water is 0 °Re, the boiling point is 80 °Re.
Temperature differences	1 K ≙ 1 °C ≙ 1.8 °F ≙ 1.8 °Ra ≙ 0.8 °Re

Units of viscosity (kinematic)

Name	*Conversion*
	1 ft²/s = 0.092903 m²/s Measurement is carried out by viscosimeter in accordance with the standard DIN EN ISO 2431 [6].
A-seconds	Runout time from flow cup
Engler number	Relative runout time from Engler device: 1 °E ≈ 7.6 mm²/s
RI seconds	Runout time from Redwood-I viscosimeter (UK): 1 R″ ≈ 4.06 mm²/s
SU seconds	Runout time from Saybolt Universal viscosimeter (US): 1 S″ ≈ 4.63 mm²/s

Natural constants

It is currently assumed that the physical limits apply equally at any place in our world. These describe the natural constants which both contain physical quantities (e.g. speed of light in vacuum, masses of elementary particles)

Table 5: Natural constants

Avogadro constant N_A	$6.022\,141\,5(10) \cdot 10^{23}$ mol⁻¹
Faraday constant F	$F = e\,N_A \approx 96485.3365(21)\ \frac{\text{C}}{\text{mol}}$
Loschmidt number N_L	$N_L = \frac{N_A}{V_{m0}} \approx 2.6867805(24) \cdot 10^{25}\ \text{m}^{-3}$ V_{m0}: Molar volume of an ideal gas under normal conditions
Boltzmann constant k	$1.380\,650\,5(24) \cdot 10^{-23}\ \frac{\text{J}}{\text{K}}$
elementary charge e	$1.602\,176\,53(14) \cdot 10^{-19}$ C
electric field constant ε_0	$\frac{1}{\mu_0 \cdot c^2} \approx 8.854\,187\,817\,62 \cdot 10^{-12}\ \frac{\text{F}}{\text{m}}$
magnetic field constant μ_0	$4\pi \cdot 10^{-7}\ \frac{\text{Vs}}{\text{Am}} \approx 12.566\,370\,614 \cdot 10^{-7}\ \frac{\text{H}}{\text{m}}$
gravitation constant G	$6.674\,2(10) \cdot 10^{-11}\ \frac{\text{m}^3}{\text{kg} \cdot \text{s}^2}$
velocity of light c (in vacuum)	$2.997\,924\,58 \cdot 10^{8}\ \frac{\text{m}}{\text{s}}$
atomic unit of mass u	$1.660\,538\,86(28) \cdot 10^{-27}$ kg
Planck action quantum h	$6.626\,069\,3(11) \cdot 10^{-34}$ Js
rest mass of electron m_e	$9.109\,382\,6(16) \cdot 10^{-31}$ kg
rest mass of proton m_p	$1.672\,621\,71(29) \cdot 10^{-27}$ kg
Stefan-Boltzmann constant	$5.670\,400(40) \cdot 10^{-8}\ \frac{\text{W}}{\text{m}^2 \cdot \text{K}^4}$
universal gas constant R	$8.314\,472(15)\ \frac{\text{J}}{\text{mol} \cdot \text{K}}$

and represent the connections between them (e.g. field constants of gravitation, electricity and magnetism). Their values depend on the system of units and must be experimentally determined.

Table 5 lists the most important natural constants (source: [1]). The numbers in parentheses express the inaccuracy of the last two digits.

References

[1] ISO 80000: Quantities and units (2013).
[2] DIN 1301: Units – Part 1: Unit names, unit symbols (2010).
[3] 8th Edition of the SI Brochure, printed in the PTB Reports, 117th year, Volume 2, June 2007, Braunschweig and Berlin.
[4] PTB leaflet “The legal units in Germany” (2004), Braunschweig.
[5] DIN 1305: Mass, as weighed value, force, weight force, weight, load; concepts. (1988).
[6] DIN EN ISO 2431: Paints and varnishes – Determination of flow time by use of flow cups (ISO 2431:2011); German version EN ISO 2431:2011.

Mathematical signs and symbols

$+$	plus
$-$	minus
$\cdot$ or $\times$	multiplied by
$:$ or $/$	divided by
$=$	equal to
$\approx$	approximately equal to
$\neq$	not equal to
$<$	less than
$>$	greater than
$\leq$	less than or equal to
$\geq$	greater than or equal to
$\sim$ or $\propto$	proportional to
$\sum a_i$	sum over a_i
$\prod a_i$	product of a_i
$n!$	n factorial $(1 \cdot 2 \cdot 3 \ldots n)$
Δ	difference or Laplace operator
$\sqrt{\ }$	root
$\parallel$	parallel to
$\perp$	perpendicular to
$\to$	approaches
∞	infinity
d/dx	differentiation to x
$\partial/\partial x$	partial differentiation to x
$\int f(x)dx$	integral of $f(x)$

Greek alphabet

A	α	Alpha
B	β	Beta
Γ	γ	Gamma
Δ	δ	Delta
E	ε, ϵ	Epsilon
Z	ζ	Zeta
H	η	Eta
Θ	θ, ϑ	Theta
I	ι	Iota
K	$\kappa, \varkappa$	Kappa
Λ	λ	Lambda
M	μ	Mu
N	ν	Nu
Ξ	ξ	Xi
O	o	Omicron
Π	π, ϖ	Pi
P	ρ, ϱ	Rho
Σ	σ, ς	Sigma
T	τ	Tau
Y, Υ	υ	Upsilon
Φ	φ, φ	Phi
X	χ	Chi
Ψ	ψ	Psi
Ω	ω	Omega

Basic principles of mechanics

Rectilinear and rotary motion

Forces, moments and energies are required to move bodies with mass. Any number of motions can be composed of rectilinear motions (translation) and rotary motions (rotation). The most important basic equations for these motions are set out in Table 2. The quantities used are listed in Table 1.

Mass and moment of inertia

Mass m is a property of matter and the cause of the inertia which sets a resistance against a change of velocity (acceleration) in the case of a rectilinear motion. The moment of inertia J (also called mass moment of inertia or rotating mass) causes a resistance in the case of a rotary motion (Table 3).

Path, velocity and acceleration

Path s is a limited distance, velocity v is the path covered during a specific time t. For a rotary motion angular velocity ω is obtained from the angle φ covered during a specific time. A uniform motion exists when velocity v or rotational speed n (or angular velocity ω) is constant. In this case, acceleration (a or α) equals zero.

A body is accelerated when there is a change of velocity. A motion is uniformly accelerated if acceleration is constant. In the case of negative acceleration, the motion is decelerated or braked.

Force and moment

A force accelerates or deforms a body. It is referred to in classic physics as the time rate of change of linear momentum p. If a mass m is rotated at centroidal dis-

Table 1: Symbols and units

Quantity		Unit
A	Area	m^2
E	Modulus of elasticity	N/mm^2
E_k	Kinetic energy	J = N m
E_p	Potential energy	J = N m
E_{rot}	Rotational energy	J = N m
F	Force	N
F_G	Weight	N
F_m	Mean force during impulse period	N
F_R	Frictional force	N
F_U	Peripheral force	N
F_Z	Centrifugal force	N
H	Rotational impulse	$N m \cdot s = kg \cdot m^2/s$
I	Force impulse	$N s = kg \cdot m/s$
J	Moment of inertia	$kg \cdot m^2$
L	Angular momentum	$N m \cdot s$
M_t	Torque	N m
$M_{t,m}$	Mean torque during impulse period	N m
P	Power	W = N m/s
R_e	Yield point	N/mm^2
R_m	Tensile strength	N/mm^2
V	Volume	m^3
W	Work, energy	J = N m
a	Acceleration	m/s^2

Quantity		Unit
a_Z	Centrifugal acceleration	m/s^2
d	Diameter	m
e	Base of natural logarithms ($e \approx 2.781$)	–
g	Acceleration of free fall ($g \approx 9.81$)	m/s^2
h	Height	m
i	Radius of gyration	m
l	Length	m
m	Mass	kg
n	Rotational speed	1/s
p	Linear momentum	N s
p	Surface pressure	N/mm^2
r	Radius	m
s	Length of path	m
t	Time	s
v	Velocity	m/s
α	Angular acceleration	rad/s^2
β	Wrap angle	°
γ	Wedge angle	°
ε	Elongation	%
μ	Coefficient of friction	–
ν	Transverse contraction	–
ρ	Density	kg/m^3
φ	Angle of rotation	rad
ω	Angular velocity	1/s

tance r about a rotational axis at angular velocity ω, this generates a centrifugal force F_Z which acts radially outwards from the center. A force which is applied at distance r from a center of rotation generates a moment. An accelerated rotary motion with rotating mass J causes a moment in the form of an acceleration or braking torque.

Work and energy

When a force F moves a body by path s, work W is performed which is stored as energy in this body. Energy is therefore defined in physics as stored work. Conversely, an energy can perform work. A distinction is made in mechanics between kinetic and potential energy. Kinetic energy is the work that must be expended to accelerate a body with mass m or rotating mass J to velocity v or angular velocity ω. Potential energy is the work that must be expended to raise a body to a height h. When a spring is tensioned, potential energy is stored which can perform work again when the spring is released.

Table 2: Rectilinear and rotary motion

Rectilinear motion (translation)		Rotary motion (rotation)	
Mass		**Moment of inertia** (Table 3)	
$m = V\rho$		$J = m\,i^2$	
Path		**Angle**	
$s = \int v(t)\,dt$		$\varphi = \int \omega(t)\,dt$	
$s = v\,t$	[v = const.]	$\varphi = \omega\,t = 2\pi\,n$	[n = const.]
$s = \frac{1}{2}a\,t^2$	[a = const.]	$\varphi = \frac{1}{2}\alpha\,t^2$	[α = const.]
Velocity		**Angular velocity**	
$v = ds(t)/dt$		$\omega = d\varphi(t)/dt$	
$v = s/t$	[v = const.]	$\omega = \varphi/t = 2\pi\,n$	[n = const.]
$v = a\,t = \sqrt{2as}$	[a = const.]	$\omega = \alpha\,t = \sqrt{2\alpha\varphi}$	[α = const.]
		Peripheral velocity	
		$v = r\omega$	
Acceleration		**Angular acceleration**	
$a = dv(t)/dt$		$\alpha = d\omega(t)/dt$	
$a = (v_2 - v_1)/t$	[a = const.]	$\alpha = (\omega_2 - \omega_1)/t$	[α = const.]
		Centrifugal acceleration	
		$a_Z = r\,\omega^2$	
Force		**Torque**	
$F = m\,a$		$M_t = F\,r = J\alpha$	
		Centrifugal force	
		$F_Z = m\,r\,\omega^2$	
Work		**Rotational work**	
$W = F\,s$		$W = M_t\varphi$	
Translation energy		**Rotational energy**	
$E_k = \frac{1}{2}m\,v^2$		$E_{rot} = \frac{1}{2}J\omega^2$	
Potential energy			
$E_p = F_G\,h$			
Power		**Power**	
$P = dW/dt = F\,v$		$P = dW/dt = M_t\omega = M_t \cdot 2\pi\,n$	
Linear momentum		**Angular momentum**	
$p = m\,v$		$L = J\omega = J \cdot 2\pi\,n$	
Force impulse		**Rotational impulse**	
$I = \Delta p = F_m(t_2 - t_1)$		$H = M_{t,m}(t_2 - t_1)$	

Table 3: Moments of inertia

Type of solid	Moments of inertia (J_x about the x-axis [1], J_y about the y-axis [1])
Right parallelepiped, cuboid	$J_x = m\frac{b^2+c^2}{12}$, $J_y = m\frac{a^2+c^2}{12}$ $J_x = J_y = m\frac{a^2}{6}$ (cube with side length a)
Regular cylinder	$J_x = m\frac{r^2}{2}$, $J_y = m\frac{r^2+l^2}{12}$
Hollow regular cylinder	$J_x = m\frac{r_a^2+r_i^2}{2}$ $J_y = m\frac{r_a^2+r_i^2+\frac{l^2}{3}}{4}$
Circular cone	$J_x = m\frac{3r^2}{10}$ $J_x = m\frac{r^2}{2}$ Envelope of cone (excluding end base)
Truncated circular cone	$J_x = m\frac{3(R^5-r^5)}{10(R^3-r^3)}$ $J_x = m\frac{(R^2+r^2)}{2}$ Envelope of cone (excluding end faces)
Sphere and hemisphere	$J_x = m\frac{2r^2}{5}$ $J_x = m\frac{2r^2}{3}$ Surface area of sphere
Hollow sphere r_a Outer sphere radius r_i Inner sphere radius	$J_x = m\frac{2(r_a^5-r_i^5)}{5(r_a^3-r_i^3)}$
Torus	$J_x = m(R^2+\frac{3}{4}r^2)$

UAN0007-2Y

[1] The moment of inertia for an axis parallel to the x-axis or y-axis at a distance a is: $J_A = J_x + m a^2$ or $J_A = J_y + m a^2$.

Conservation of energy

The law of conservation of energy states that the total energy in a closed system is constant. Energy can be neither created nor destroyed, but instead converted between different forms of energy (e.g. kinetic energy to thermal energy) or transmitted from one body to another.

Power

Power is the work performed during a specific period of time. Because every power transfer is associated with loss, the output power is always less than the input power. The ratio of output to input power is called efficiency η and is therefore always less than 1.

$$\eta = \frac{P_{\text{out}}}{P_{\text{in}}}.$$

Linear momentum and impulse

Linear momentum p describes the motion of a body with mass, and is calculated as the product of moving mass m and velocity v. Every moving body can transmit its linear momentum during an impulse process to another body, as happens for example in a collision between two vehicles. The force acting on a body gives rise to a change of linear momentum, which is called force impulse I.

During a rotary motion angular momentum L is obtained from the product of rotating mass J and angular velocity ω. A rotational impulse H is generated for example when two disks are jerkily coupled.

Conservation of momentum

The law of conservation of momentum states that the total momentum in a closed system is constant. It follows from this that the total momentum before and after an impulse must be equal.

Statics

Statics is the science of equilibrium on a rigid body. A body is in equilibrium when it is at rest or is moving uniformly or in a straight line. Equilibrium prevails when the sum of the applied forces and moments in all directions is equal to zero.

Plane system of forces

Forces are vectors which are determined by size and direction. They are geometrically (vectorially) added:

$$\vec{F}_{\text{res}} = \vec{F}_1 + \vec{F}_2 .$$

Two forces are composed with a parallelogram of forces or a triangle of forces (Figure 1a). If there are several forces, the resultant force F_{res} is determined by means of a polygon of forces (Figure 1b). When the polygon of forces is closed, the system of forces is in equilibrium.

Figure 1: Composition of forces
a) Parallelogram of forces and triangle of forces,
b) Closed polygon of forces (system in equilibrium).
F Force,
F_{res} Resultant force.

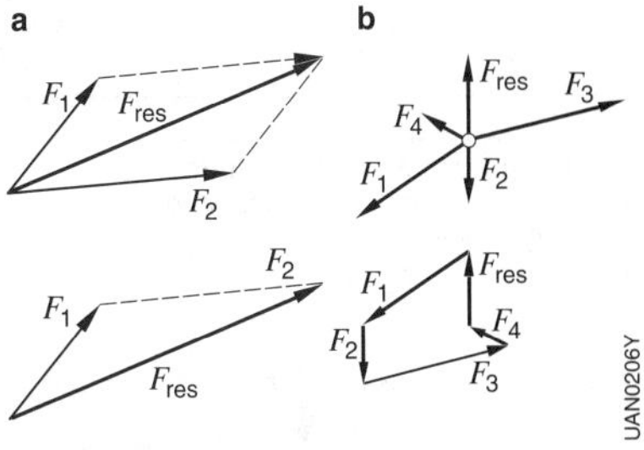

Figure 2. Force resolution
F Force,
F_x Force in x-direction,
F_y Force in y-direction.

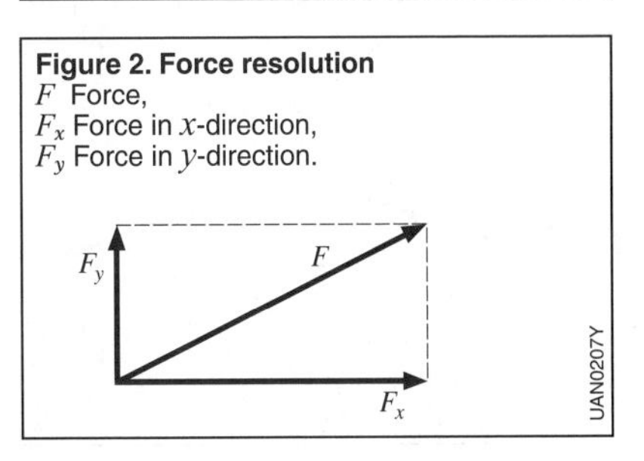

A force can also be resolved into components. A resolution into orthogonal components is sensible (Figure 2).

Transmission of force
Mechanical machines for the transmission of force can be reduced to the "lever" and "wedge" principles.

Lever
The lever principle can be derived from the equilibrium condition "sum of the moments is equal to zero". Disregarding friction, the system in Figure 3 is in equilibrium for the following condition:

$$M_{t1} = M_{t2} \quad \text{or} \quad F_1 r_1 = F_2 r_2 \,.$$

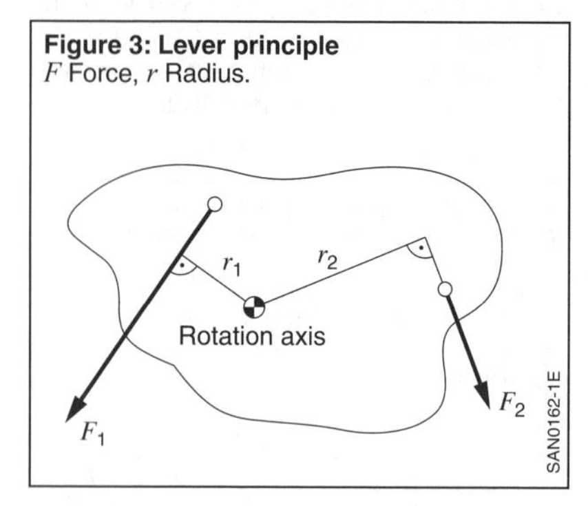

Figure 3: Lever principle
F Force, r Radius.

Figure 4: Forces on the wedge
F Insertion force, F_N Normal force,
γ Wedge angle.

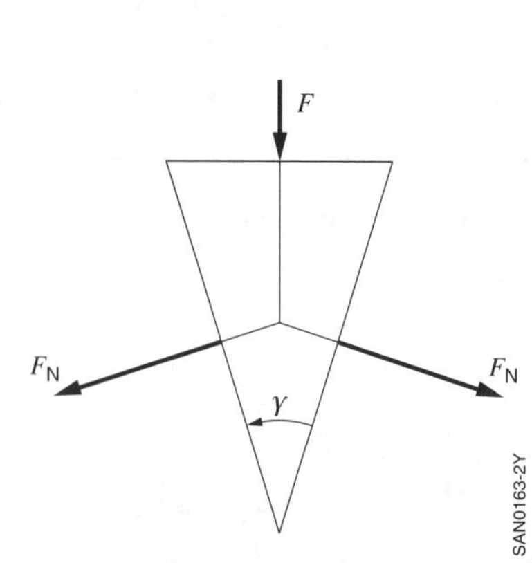

The lever principle is encountered in many applications, ranging from simple pliers, scales and wrenches through gear wheels and belt drives, right down to the connecting rods in piston engines.

Wedge
With the wedge principle, depending on the wedge angle γ, small forces (insertion force F) can be translated into large normal forces F_N (Figure 4). Without allowing for friction, the following applies:

$$F_N = \frac{F}{2 \sin \frac{\gamma}{2}} \,.$$

Very large application forces can be introduced in the smallest of spaces with the aid of wedges. Examples include wedges in shaft/hub connections and tapered connections for transmitting torques. However, both the screw and the clamping eccentric work according to the classic wedge principle.

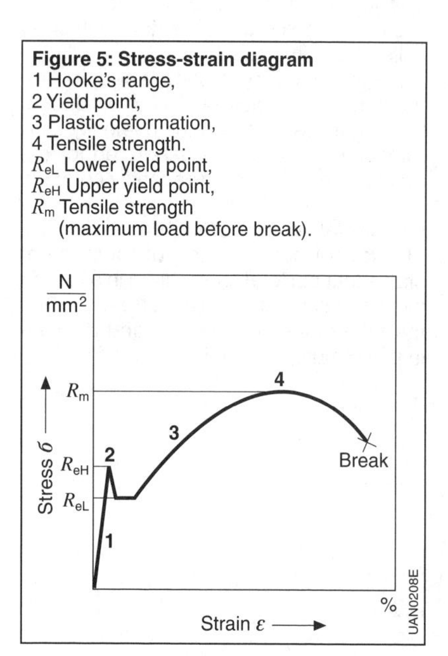

Figure 5: Stress-strain diagram
1 Hooke's range,
2 Yield point,
3 Plastic deformation,
4 Tensile strength.
R_{eL} Lower yield point,
R_{eH} Upper yield point,
R_m Tensile strength
(maximum load before break).

Calculation of strength

Hooke's law

In response to an external load a body is deformed and stresses are generated inside the material (Figure 5). Metals have linear-elastic behavior until the yield point is reached. In other words, within this Hooke's range the component assumes its original length again when the load is reduced. The yield point R_e is the limit between elastic and plastic deformation. Above the yield point the material begins to "creep" and then remains permanently deformed. The tensile strength R_m is the maximum load before the component cracks or breaks.

In the case of ductile materials with a pronounced yield point R_e is always specified as the limit for dimensioning. The tensile strength is used for dimensioning only in the case of brittle materials which do not have a pronounced yield point (e.g. gray cast iron).

The linear area is known as Hooke's straight line, in which the ratio of stress σ and strain (elongation) ε is constant. With the proportionality constant E, which is termed the modulus of elasticity, Hooke's law for the mono-axial stress state is:

$$\sigma = E\varepsilon .$$

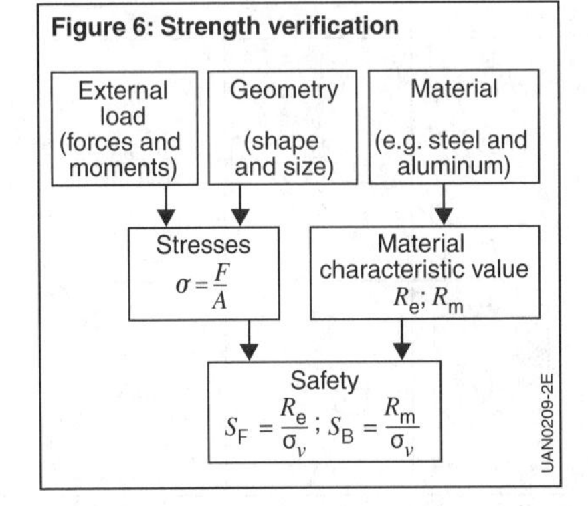

Strength verification

The aim of strength verification is to dimension components reliably and appropriately for the material involved. Strength verification in accordance with Figure 6 is conducted in four stages:

1. Determination of the external load (forces and moments),
2. Calculation of the existing stress,
3. Choice of the material characteristic value,
4. Comparison of the existing stress with the material characteristic value.

If the external forces and moments are known, the stresses in the component can be calculated in accordance with Table 4. Because tensile, compressive and bending stresses (normal stresses) lie in one plane or act in the same direction, they can be cumulatively superimposed. Even the shear and torsional stresses can be added. However, if normal and shear stresses occur simultaneously in a component, a reduced stress must be created with a strength hypothesis because the material characteristic values were calculated from mono-axial tensile and vibration fatigue tests.

There are different hypotheses to suit the material behavior (ductile or brittle). For ductile materials (most metals) the deformation energy hypothesis is used. For the mono-axial stress state (bar) the reduced stress is calculated in accordance with Table 4.

Surface pressure

Surface pressure is a compressive stress. It is generated when a force F is transmitted from one solid body to another. For effective surface areas pressure p is equal to the ratio of force F to contact surface A.

Pressure on the face of a hole

When journal and bore or shaft and friction bearing are paired, the pressure on the face of a hole is usually used for calculation. Here the load is referred to the projected area A_{proj} (Figure 7):

$$p = \frac{F}{A_{proj}} = \frac{F}{dl} \leq p_{perm} .$$

Table 4: Basic equations of strength calculation

Loading	Stress
Tensile stress	$\sigma_z = \frac{F}{A}$
Compressive stress	$\sigma_d = \frac{F}{A}$
Bending stress	$\sigma_b = \frac{M_b}{W_b}$ W_b from Table 5
Shear stress	$\tau_s = \frac{F}{A}$
Torsional stress	$\tau_t = \frac{M_t}{W_t}$ W_t from Table 5
Reduced stress	$\sigma_v = \sqrt{\sigma^2 + 3\tau^2}$

Hertzian stress

In reality, however, the maximum stress where the surfaces are curved is greater and can be calculated according to Hertz's theory. The maximum Hertzian stress is dependent on the deformation (flattening) of the touching surfaces (Figure 8). This in turn is dependent on the radii, on the modulus of elasticity E and

Figure 7: Pressure on the face of a hole
a) Projection,
b) Side view.
1 Undercut.
F Load,
d Journal diameter,
l Common contact length.

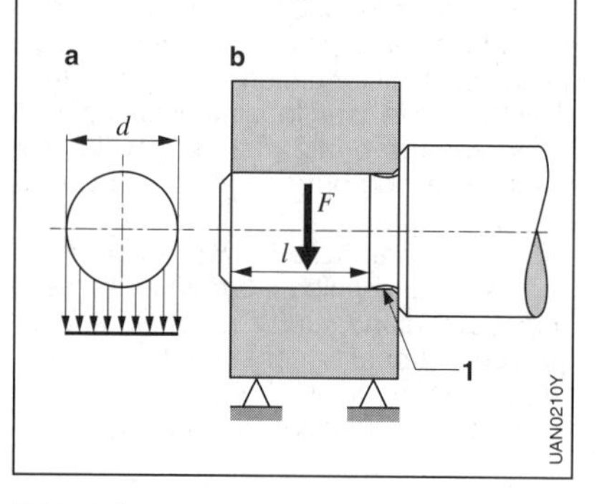

Table 5: Section moduli

Cross-section	Bending	Torsion
d	$W_b = \frac{\pi}{32} d^3$	$W_t = \frac{\pi}{16} d^3$
d, D	$W_b = \frac{\pi}{32}\left(\frac{D^4 - d^4}{D}\right)$	$W_t = \frac{\pi}{16}\left(\frac{D^4 - d^4}{D}\right)$
a	$W_b = \frac{a^3}{6}$	$W_t = 0.208\,a^3$

on the transverse contraction ν. The equations for the cases "sphere – sphere" and "cylinder – cylinder" are set out in Table 6. For the special cases "sphere – plane" and "cylinder – plane" radius $r_2 \rightarrow \infty$ or $r = r_1$.

Figure 8: Hertzian stress
F Normal force,
r Radius,
p_{max} Maximum pressure.

Table 6: Hertzian stress

Sphere – sphere (localized contact) $p_{max} = \frac{1}{\pi}\sqrt[3]{\frac{1.5FE^2}{r^2(1-\nu^2)^2}}$	Mean E modulus: $E = 2 \cdot \frac{E_1 E_2}{E_1 + E_2}$ For the radii:
Cylinder – cylinder (line contact) $p_{max} = \sqrt{\frac{FE}{2\pi r l(1-\nu^2)}}$	$\frac{1}{r} = \frac{1}{r_1} + \frac{1}{r_2} \rightarrow r = \frac{r_1 r_2}{r_1 + r_2}$ l Contact length, cylinder ν Transverse contraction

Friction

Coulomb friction

When touching bodies move relative to each other, friction acts as mechanical resistance acting in the opposite direction to the motion at velocity v (Figure 9). The force of resistance, known as frictional force F_R, is proportional to the normal force F_N. Static friction exists as long as the external force is less than the frictional force and the body remains at rest. When static friction is overcome, and the body is set in motion, frictional force is governed by Coulomb's law of sliding friction:

$$F_R = \mu F_N .$$

Coefficient of friction

The coefficient of friction μ always denotes a system property and not a material property. Coefficients of friction are, among other things, depending on material pairing (see Table 7), temperature,

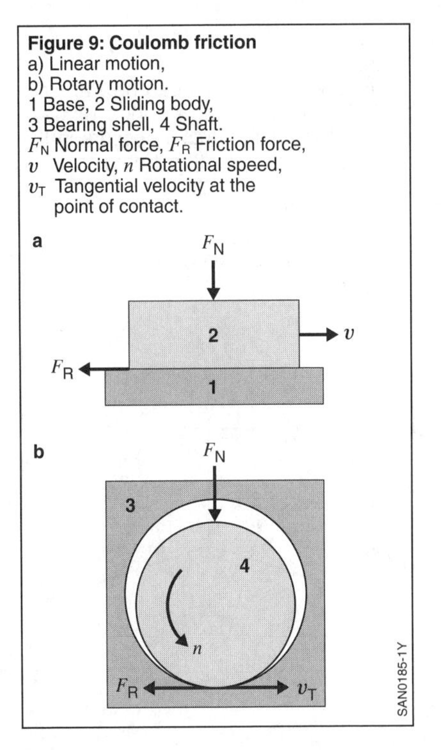

Figure 9: Coulomb friction
a) Linear motion,
b) Rotary motion.
1 Base, 2 Sliding body,
3 Bearing shell, 4 Shaft.
F_N Normal force, F_R Friction force,
v Velocity, n Rotational speed,
v_T Tangential velocity at the point of contact.

surface condition, sliding velocity, the surrounding medium (e.g. water or CO_2, which can be adsorbed by the surface), and the intermediate material (lubricant). For this reason, coefficients of friction always fluctuate between limit values and may have to be calculated experimentally. Static friction is generally greater than sliding friction. In special cases, the friction coefficient can exceed 1. For example, with very smooth surfaces where cohesion forces are predominant or with racing tires featuring an adhesion or suction effect.

Friction on the wedge

Allowing for friction, the insertion force is governed by (Figure 10):

$$F = 2\left(F_N \sin\frac{\gamma}{2} + F_R \cos\frac{\gamma}{2}\right)$$

The normal force is then

$$F_N = \frac{F}{2\left(\sin\frac{\gamma}{2} + \mu\cos\frac{\gamma}{2}\right)}.$$

Rope friction

Motions and moments can be transmitted with elastic, pliable ropes or belts (Figure 11). Sliding friction occurs in the event of relative motion between rope and pulley (e.g. belt brake or bollard with running rope). Static friction exists when rope and pulley are at rest relative to each other (e.g. belt drive, belt brake as holding brake, bollard with rope at rest). The coefficient of sliding friction μ or the static friction μ_H must be applied accordingly.

According to Euler's rope-friction formula, the force in the pulling strand (taut strand):

$$F_1 = F_2 e^{\mu\beta}.$$

The peripheral force F_U or frictional force F_R is calculated as

$$F_U = F_R = F_1 - F_2$$

and the transmittable friction torque as

$$M_R = F_R\, r.$$

Rolling friction

Rolling friction occurs when a ball, a caster or a wheel rolls on a track or roadway. Typical examples of this are the roller bearing, the pairing of wheel flange and rail for railroads or the pairing of tire and roadway for motor vehicles. Both rolling body and base are subjected to elastic deformation during rolling. This generates asymmetrical pressure (Figure 12). The

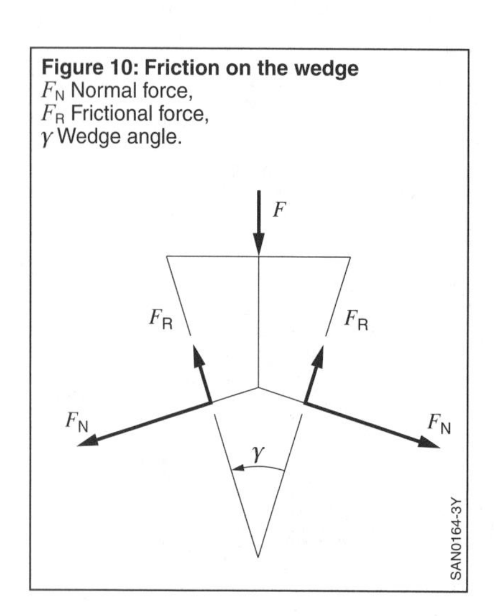

Figure 10: Friction on the wedge
F_N Normal force,
F_R Frictional force,
γ Wedge angle.

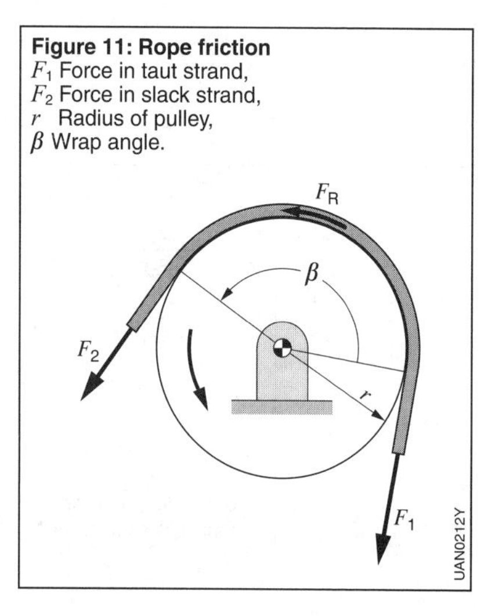

Figure 11: Rope friction
F_1 Force in taut strand,
F_2 Force in slack strand,
r Radius of pulley,
β Wrap angle.

equilibrium condition is used to calculate frictional force F_R for the dry state (i.e. without lubrication):

$$F_R = \frac{x}{R} F_N = \mu_R F_N .$$

Ratio x/R can be calculated as frictional resistance or coefficient of rolling friction μ_R. From this it can be seen that large rolling bodies roll more easily than small rolling bodies. Hard surfaces (roller bearing and railroad) give rise to small deformations and therefore result in very low coefficients of friction; soft surfaces, as in the tire/roadway pairing, on the other hand result in higher coefficients of friction. While frictional resistances of μ_R = 0.0015 can be achieved for ball bearings, the coefficients of friction of a car tire on asphalt are μ_R = 0.015.

References

[1] A. Böge: Technische Mechanik. 30th Ed., Springer-Vieweg-Verlag, 2013.
[2] R.C. Hibbeler: Technische Mechanik 1., 2. und 3., Pearson Studium.
[3] K.-H. Grote, J. Feldhusen: Dubbel – Taschenbuch für den Maschinenbau. 23rd Ed.; Springer-Verlag 2011.

Figure 12: Rolling friction
a) Asymmetrical pressure distribution,
b) Resultant forces for calculation model.
F Load, F_N Normal force, F_R Frictional force,
R Radius of rolling body,
x Lever arm for moment equilibrium.

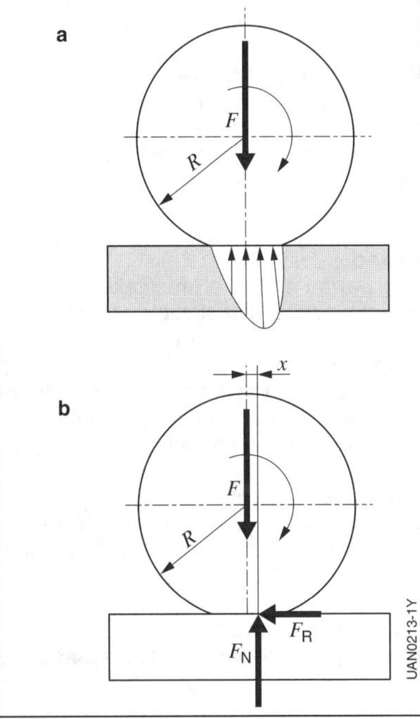

Table 7: Reference values for coefficients of static and sliding friction

Material pair	Coefficient of static friction μ_H		Coefficient of sliding friction μ	
	Dry	Lubricated	Dry	Lubricated
Steel – steel	0.15 – 0.20	0.10	0.10 – 0.15	0.05 – 0.10
Steel – cast iron	0.18 – 0.25	0.10	0.15 – 0.20	0.10
Steel – sintered bronze	0.20 – 0.40	0.08 – 0.13	0.18 – 0.30	0.06 – 0.09
Steel – brake pad	–[1]	–[1]	0.50 – 0.60	0.20 – 0.50
Steel – polyamide	0.60	0.20	0.32 – 0.45	0.10
Steel – ice	0.027	–[1]	0.014	–[1]
Cast iron – cast iron	0.18 – 0.25	0.10	0.15 – 0.20	0.10
Aluminum – aluminum	0.50 – 1.00	–[1]	0.50 – 1.00	–[1]
Wood – wood	0.40 – 0.60	0.20	0.20 – 0.40	0.10
Wood – metal	0.60 – 0.70	0.10	0.40 – 0.50	0.10
Leather – metal	0.50 – 0.60	0.20 – 0.25	0.20 – 0.25	0.12
Car tire on dry asphalt	0.50 – 0.60	–[1]	–[1]	–[1]
Car tire on wet asphalt	0.20 – 0.30	–[1]	–[1]	–[1]
Car tire on icy asphalt	0.10	–[1]	–[1]	–[1]

–[1] No practical use.

Vibrations and oscillations

Terms

(Symbols and units in Table 1, see also DIN 1311, [1], [2] and [3]).

Vibration or oscillation
A vibration or oscillation is the change in a physical quantity which repeats at more or less regular time intervals and whose direction changes with similar regularity (Figure 1).

Period
The period T is the time taken for one complete cycle of a single oscillation.

Amplitude
The amplitude $\hat{y}$ is the maximum instantaneous value (peak value) of a sinusoidally oscillating physical quantity.

Frequency
The frequency f is the number of oscillations in one second, the reciprocal value of the period of oscillation T.

Angular frequency
The angular frequency ω is 2π times the frequency f.

Particle velocity
Particle velocity v is the instantaneous value of the alternating velocity of a vibrating particle in its direction of vibration. It must not be confused with the velocity of propagation of a traveling wave (e.g. the velocity of sound).

Figure 1: Sinusoidal oscillation
(Quantities, see Table 1)

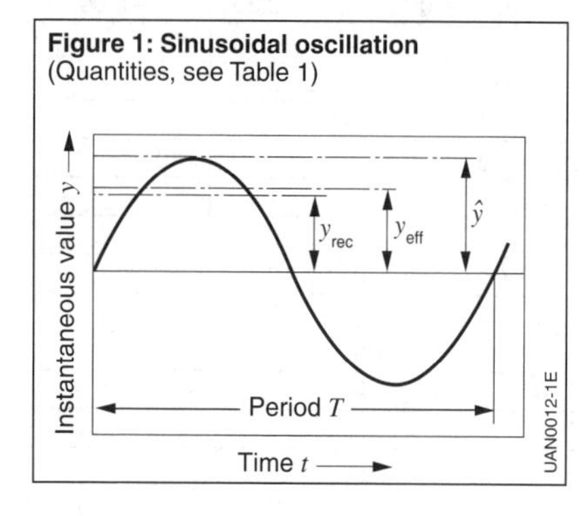

Table 1: Symbols and units

Quantity		Unit
a	Storage coefficient	
b	Damping coefficient	
c	Storage coefficient	
c	Spring constant	N/m
c_α	Torsional rigidity	N m/rad
C	Capacitance	F
f	Frequency	Hz
f_g	Resonant frequency	Hz
Δf	Half-value width	Hz
F	Force	N
F_Q	Excitation function	
I	Current	A
J	Moment of inertia	kg · m^2
L	Inductance	H
m	Mass	kg
M	Torque	N m
n	Rotational speed	1/min
Q	Charging	C
Q	Resonance sharpness	
r	Damping factor	N s/m
r_α	Rotational damping coefficient	N s · m
R	Ohmic resistance	Ω
t	Time	s
T	Period	s
U	Voltage	V
v	Particle velocity	m/s
x	Travel/displacement	m
y	Instantaneous value	
$\hat{y}$	Amplitude	
$\dot{y}$ ($\ddot{y}$)	Single (double) derivative with respect to time	
y_{rec}	Rectification value	
y_{eff}	Effective value	
α	Angle	rad
δ	Decay coefficient	1/s
Λ	Logarithmic decrement	
ω	Angular velocity	rad/s
ω	Angular frequency	1/s
Ω	Excitation-circuit frequency	1/s
D	Damping ratio	
D_{opt}	Optimum damping ratio	
	Subscripts:	
0	Undamped	
d	Damped	
T	Absorber	
U	Base support	
G	Machine	

Fourier series
Every periodic function, which is piecewise monotonic and smooth, can be expressed as the sum of sinusoidal harmonic components.

Beats
Beats occur when two sinusoidal oscillations, whose frequencies do not differ greatly, are superposed ($f_1 \approx f_2$). They are periodic. Their basic frequency is the difference between the frequencies of the superposed sinusoidal oscillations.

Natural frequency
The natural frequency is that frequency f at which an oscillating system can oscillate freely after being excited once (natural oscillation). It is dependent only on the properties of the oscillating system.

Damping
Damping is a measure of the energy losses in an oscillatory system when one form of energy is converted into another. The consequence is a decay of the oscillation (Figure 2).

Damping ratio
The damping ratio D is the measure for the degree of damping.

Logarithmic decrement
The logarithmic decrement Λ is the natural logarithm of the relationship between two extreme values of a damped natural oscillation which are separated by one period.

Forced oscillations
Forced oscillations arise under the influence of an external physical force (excitation), which does not change the properties of the oscillator. The frequency of forced oscillations is determined by the frequency of the excitation.

Transfer function
The transfer function is the quotient of amplitude of the observed state or output variable and the amplitude of excitation, plotted against the excitation frequency f or the excitation-circuit frequency ω.

Resonance
A resonance occurs when the transfer function attains its maximum value as the excitation frequency approaches the natural frequency.

Resonant frequency
The resonant frequency is the excitation frequency at which the oscillator state variable attains its maximum value. Disregarding the damping, the resonant frequency is equal to the natural frequency.

Half-value width
The half-value width is the difference between the frequencies at which the level of the observed variable has dropped to $1/\sqrt{2} \approx 0.707$ of the maximum value.

Resonance sharpness
The resonance sharpness Q (or quality factor) is the maximum value of the transfer function.

Coupling
If two oscillatory systems are coupled together – mechanically by mass or elasticity, electrically by inductance or capacitance – a periodic exchange of energy takes place between the systems.

Wave
A wave is a spatial and temporal change of state of a continuum, which can be expressed as a unidirectional transfer of location of a certain state over a period of time. The matter that may be present in the space is not necessarily also transported.

There are transversal waves (e.g. waves in rope and water) and longitudinal waves (e.g. sound waves in air).

Interference
The principle of undisturbed superposition of waves is called interference. At every point in space, the instantaneous value of the resulting wave is equal to the sum of the instantaneous values of the individual waves.

Plane wave

A plane wave is a wave in which the surfaces of the same phase (e.g. maxima or wave fronts) form a plane, i.e. the wave propagates linearly. The wave fronts are vertical to the direction of propagation.

Standing waves

Standing waves occur as a result of interference between two waves of equal frequency, wavelength, and amplitude traveling in opposite directions. In contrast to a propagating wave, the amplitude of the standing wave is constant at every point; nodes (zero amplitude) and antinodes (maximum amplitude) occur. Standing waves occur, for example, when a plane wave is reflected on a plane wall which is vertical to the direction of wave propagation.

Rectification value

The rectification value y_{rec} is the arithmetic mean value, linear in time, of the values of a periodic signal:

$$y_{rec} = \frac{1}{T}\int_0^T |y|\,dt.$$

For a sine curve:

$$y_{rec} = \frac{2\hat{y}}{\pi} \approx 0.637\,\hat{y}.$$

Effective value

The effective value y_{eff} is the time virtual value of a periodic signal. It is also known as the RMS value (root mean square):

$$y_{eff} = \sqrt{\frac{1}{T}\int_0^T y^2\,dt}.$$

For a sine curve:

$$y_{eff} = \frac{\hat{y}}{\sqrt{2}} \approx 0.707\,\hat{y}.$$

Harmonic factor

The harmonic factor is the ratio of y_{eff} to y_{rec}. For a sine curve the harmonic factor is $y_{eff}/y_{rec} \approx 1.111$.

Peak factor

For a sine curve the peak factor is $\hat{y}/y_{eff} = \sqrt{2} \approx 1.414$.

Equations

The following equations apply to simple oscillators (Table 2) if the general quantity designations in the formulas are replaced by the relevant physical quantities.

Table 2: Simple oscillatory systems

	Mechanical		Electrical
	Translational	Rotational	
Des.	Physical quantity		
y	x	α	Q
$\dot{y}$	$\dot{x} = v$	$\dot{\alpha} = \omega$	$\dot{Q} = I$
$\ddot{y}$	$\ddot{x} = \dot{v}$	$\ddot{\alpha} = \dot{\omega}$	$\ddot{Q} = \dot{I}$
F_Q	F	M	U
a	m	J	L
b	r	r_α	R
c	c	c_α	$1/C$

Differential equation

$$a\ddot{y} + b\dot{y} + cy = F_Q(t) = \hat{F}_Q \sin \Omega t,$$

Period $T = 1/f$,
Angular frequency $\omega = 2\pi f$.

Sinusoidal oscillation: $y = \hat{y} \sin \omega t$.

Figure 2: Free oscillation and damping for $0 < D < 1$

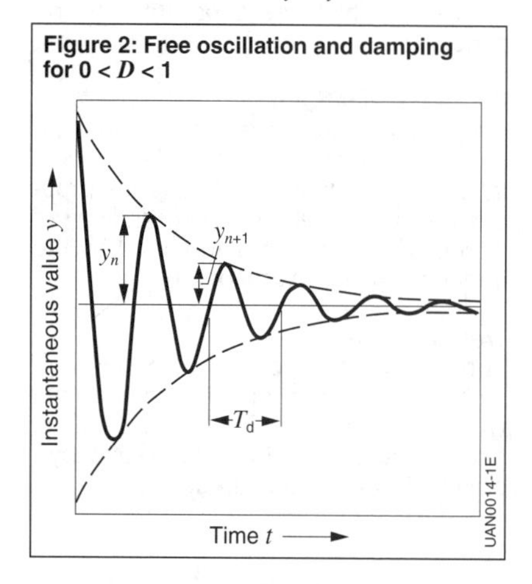

Free oscillations ($F_Q = 0$)
Logarithmic decrement (Figure 2):

$$\Lambda = \ln\left(\frac{y_n}{y_{n+1}}\right) = \frac{\pi b}{\sqrt{ca - \frac{b^2}{4}}},$$

Decay coefficient $\delta = \frac{b}{2a}$,

Damping ratio $D = \frac{\delta}{\omega_0} = \frac{b}{2\sqrt{ca}}$,

$$D = \frac{\Lambda}{\sqrt{\Lambda^2 + 4\pi^2}} \approx \frac{\Lambda}{2\pi} \text{ (low level of damping).}$$

Angular frequency of undamped oscillation ($D = 0$): $\omega_0 = \sqrt{c/a}$.

Angular frequency of damped oscillation ($0 < D < 1$): $\omega_d = \omega_0 \sqrt{1 - D^2}$.

For $D \geq 1$ no oscillations, but creepage.

Forced oscillations
Quantity of transfer function:

$$\frac{\hat{y}}{\hat{F}_Q} = \frac{1}{\sqrt{(c - a\Omega^2)^2 + (b\,\Omega)^2}}$$

$$= \frac{1}{c\sqrt{\left(1 - \left(\frac{\Omega}{\omega_0}\right)^2\right)^2 + \left(\frac{2D\Omega}{\omega_0}\right)^2}},$$

Resonant frequency $f_g = f_0\sqrt{1 - 2D^2} < f_0$,

Resonance sharpness $Q = 1/\left(2D\sqrt{1 - D^2}\right)$,

Resonant frequency $f_g \approx f_0$ (for $D \leq 0.1$),

Resonance sharpness $Q \approx \frac{1}{(2D)}$ (for $D \leq 0.1$),

Half-value width $\Delta f = 2Df_0 = \frac{f_0}{Q}$.

Figure 3: Standardized transfer function

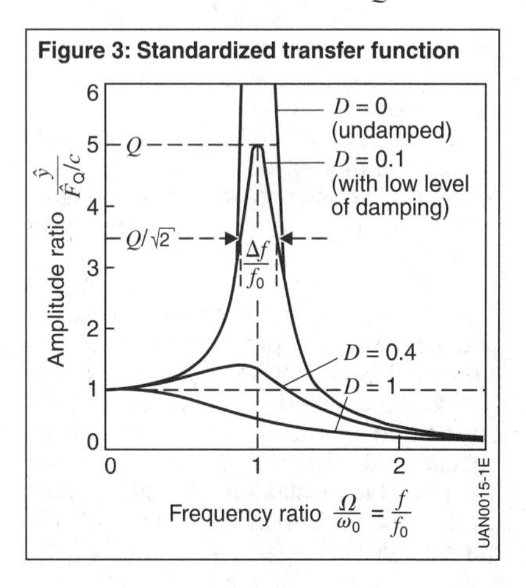

Vibration reduction

Vibration damping

If damping can only be carried out between the machine and a quiescent point, damping must be at a high level (Figure 3).

Vibration isolation

Active vibration isolation
The machines to be isolated are to be mounted so that the dynamic forces transmitted to the base support are small.

One measure to be taken: The bearing point should be set below resonance so that the natural frequency lies below the lowest excitation frequency. Damping impedes insulation. Low values can result in excessively high vibrations during run-up when the resonant range is passed through (Figure 4).

Figure 4: Vibration isolation
a) Transfer function,
b) Low setting.

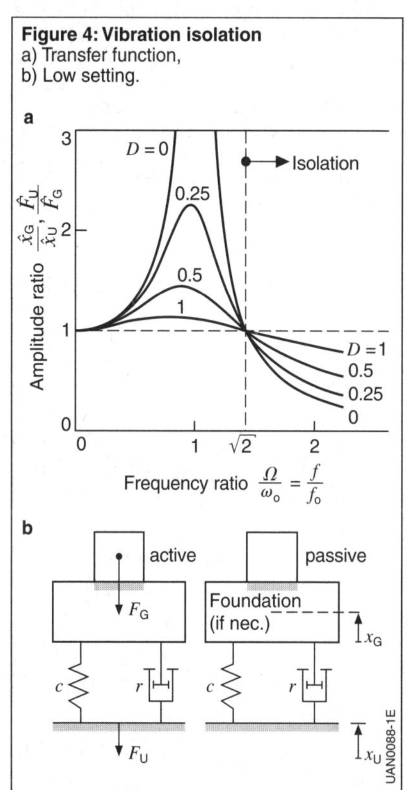

Passive vibration isolation
The machines to be isolated are to be mounted so that vibrations and shocks reaching the base support are only transmitted to the machines to a minor extent.

The measures to be taken are the same as those for active isolation. In many cases, flexible suspension or extreme damping is not practicable. To prevent the occurrence of resonance, the machine attachment should be so rigid that the natural frequency is far enough in excess of the highest excitation frequency which can occur (Figure 4).

Vibration absorption
Vibration absorber with fixed natural frequency
By tuning the natural frequency f_T of an absorption mass with a flexible, loss-free coupling to the excitation frequency, vibrations acting on the machine are completely absorbed (Figure 5). Only the absorption mass still vibrates. The effectiveness of the absorption decreases as the excitation frequency changes. Damping prevents complete absorption. However, appropriate tuning of the absorber frequency and an optimum damping ratio produce broadband vibration reduction, which remains effective when the excitation frequency changes.

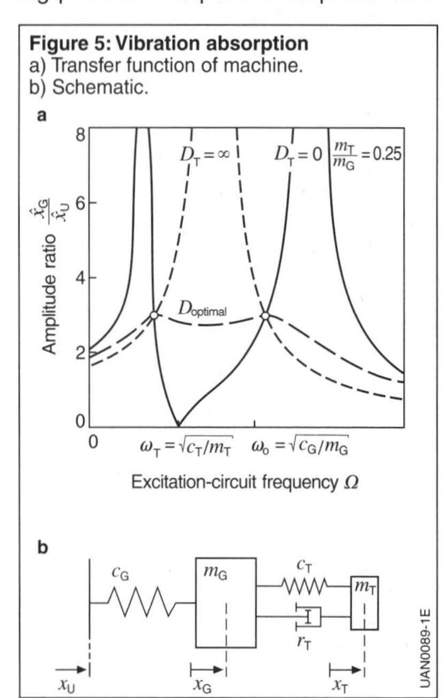

Figure 5: Vibration absorption
a) Transfer function of machine.
b) Schematic.

Vibration absorber with varying natural frequency
Rotational oscillations with excitation frequencies proportional to the rotational speed (e.g. orders of balancing in IC engines) can be absorbed by absorbers with natural frequencies proportional to the rotational speed (pendulum in the centrifugal-force field). The vibration absorption is effective at all rotational speeds. Vibration absorption is also possible for oscillators with several degrees of freedom and interrelationships, as well as by the use of several absorption masses.

Modal analysis

The dynamic behavior (natural-oscillation characteristics) of oscillating systems is determined by means of modal analysis. It is used among other things in design to optimize structures with regard to the oscillatory characteristics and to identify problem areas, and in acoustics to analyze structure-borne noise ([4]).

The oscillating structure, which as a continuum has infinitely many degrees of freedom, is replaced in a clearly defined manner by a finite number of single-mass oscillators. The modal model of the structure created in this way is described by the modal parameters

- natural-oscillation shapes (also eigenvector or mode),
- natural frequencies (also eigenvalues),
- and the associated modal damping values.

A time-invariant and linear-elastic structure is an essential precondition for model creation. Every oscillation of the structure can be represented from the eigenvectors and eigenvalues. It is, however, only observed at a limited number of points in the possible oscillation directions (degrees of freedom) and at defined frequency intervals.

The substructure-coupling process collates modal models of various structures into an overall model.

Numerical modal analysis
The geometry, material data, and marginal conditions must be known. The basis for numerical modal analysis is a multibody-system or finite-element model of the structure. Eigenvalues and eigenvectors can be calculated from this by solving an eigenvalue problem.

Numerical modal analysis manages without a prototype of the structure and can already be used at an early stage of development. However, it is often the case that precise knowledge concerning the structure's fundamental properties (damping, marginal conditions) are lacking, which means that the modal model can sometimes be inaccurate. As well as this, the error is unidentified. One remedy is to adjust the model with the results of an experimental modal analysis.

Experimental modal analysis
A prototype of the structure is required for experimental modal analysis. Analysis is based on measurements of transfer functions. To this end, either the structure is excited at one point in the frequency range of interest and the oscillation responses measured at several points, or it is excited at many points in succession, where the oscillation responses are always measured at the same point. The modal model is derived from the matrix of the transfer function (it describes the response model). An impulse hammer or an electrodynamic or hydraulic "shaker" is used as the means of excitation. The response is measured with acceleration sensors or a laser vibrometer.

Experimental modal analysis can also be used to validate numerical analysis. Simulation calculations can then be carried out on the validated numerical model. In the response calculation, the response of the structure to a defined excitation is calculated, an excitation which corresponds, for example, to test-bay conditions.

By means of structure modifications (changes in mass, damping or stiffness), the vibrational behavior can be optimized to the level required by operating conditions. When the modal models produced by both processes are compared with each other, the modal model resulting from an analytical modal analysis is more detailed than that from an experimental modal analysis, due to the greater number of degrees of freedom in the analytical process. This applies in particular to simulation calculations based on the model.

The natural-oscillation shapes resulting from a modal analysis can be shown in graphic form or animated (examples in Figure 6). The different gray stages describe the excursion vertically to the plane of projection. The partial distortion of the disks results from these excursions.

References

[1] DIN 1311: (Mechanical) vibrations, oscillation and vibration systems – Part 1: Basic concepts, survey.
[2] P. Hagedorn: Technische Schwingungslehre I, Springer-Verlag, 1987.
[3] P. Hagedorn: Technische Schwingungslehre II, Springer-Verlag, 1998.
[4] G. Natke: Einführung in Theorie und Praxis der Zeitreihen- und Modalanalyse, Vieweg-Verlag, 1992.

Figure 6: Eigenmode of round disks

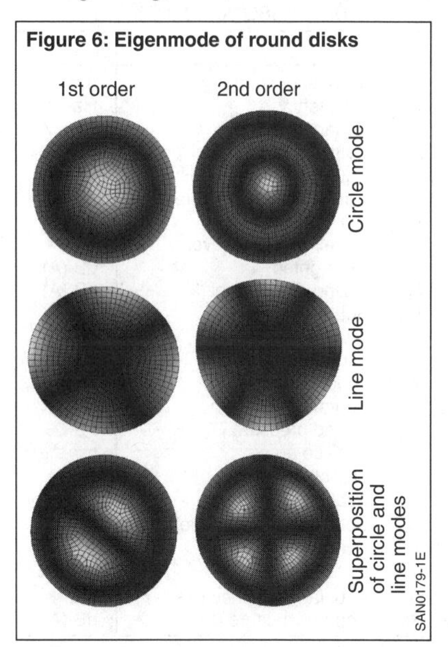

Acoustics

Acoustics (Greek "to hear") is a branch of the general science of vibration and oscillation concerned with the vibration of air particles audible to humans. This process of vibration gives rise to pressure differences in air which the human ear can detect. Vibrations are generated on an automobile by a variety of excitations which partly excite the airborne noise directly or can propagate via the vehicle structure and from there to the airborne noise.

Table 1: Quantities and units
(see also DIN EN ISO 80000-8 [2])

Quantity		SI unit
c	Velocity of sound	m/s
f	Frequency	Hz
I	Sound intensity	W/m^2
L_I	Sound intensity level	dB
L_{Aeq}	Equivalent continuous sound level, A-weighted	dB (A)
L_{den}	Noise indices	dB (A)
L_{pA}	Sound pressure level, A-weighted	dB (A)
L_r	Rating sound level	dB (A)
L_{WA}	Sound power level, A-weighted	dB (A)
L_S	Loudness level	phon
S	Loudness	sone
P	Sound power	W
p	Sound pressure	Pa
S	Surface area	m^2
v	Particle velocity	m/s
Z	Acoustic impedance	Pa · s/m
α	Sound absorption coefficient	1
λ	Wavelength	m
ρ	Density	kg/m^3
ω	Angular frequency (= $2\pi f$)	1/s
SEL	Sound exposure level	dB (A)

General terminology

(Symbols and units in Table 1, see also DIN 1320 [1])

Sound
Mechanical vibrations and waves in an elastic medium in the audible frequency band (16 to 20,000 Hz) are called sound.

Ultrasound
High-frequency vibrations above the frequency range of human hearing are called ultrasound.

Infrasound
Infrasound is the range of very low frequencies which can no longer be detected by the human hear, but can still be felt by the body.

Sound pressure
Sound pressure p is the alternating pressure generated in a medium by the vibration of sound.

Particle velocity
Particle velocity v is the alternating velocity of a vibrating particle. In a free sound field:

$$v = \frac{p}{Z}.$$

Specific acoustic impedance
Specific acoustic impedance Z denotes the wave impedance of a medium and is a measure of the ability of a medium to transmit sound waves.

In a free field the relationship between sound pressure, particle velocity and acoustic impedance is:

$$Z = \frac{p}{v} = \rho c.$$

Figuratively speaking, the formula describes the coupling of a vibrating particle to its neighboring particles. The coupling is much less pronounced in gases than in solid media; the particles can vibrate more freely.

Hearing dynamics

Hearing dynamics is the ability of the human ear to perceive noises from the threshold of audibility up to the pain threshold. It corresponds to pressure differences from the threshold of audibility, which is generally defined as 20 µPa, up to the pain threshold of approx. 200 pascal. This results in very high hearing dynamics of around 1:10,000,000.

Decibel dB

The decibel is not a unit, but the logarithmic ratio of a measured quantity to a reference value. The use of decibels makes it easier to handle the large dynamic range.

The sound pressure level is given in dB referred to the reference pressure of the threshold of audibility:

$$L_\mathrm{p} = 20 \log_{10}\left(\frac{p}{p_\mathrm{ref}}\right).$$

Frequency

Frequency f specifies the number of vibrations per second. The unit is the hertz (Hz) and is defined as follows:

1 Hz = 1/s.

Velocity of sound

Velocity of sound c is the velocity of propagation of a sound wave in a medium (Table 2).

Table 2: Sound velocities and wavelengths in different materials

Material/medium	Sound velocity c in m/s	Wave-length λ in m at 1,000 Hz
Air, 20 °C, 1,014 hPa	343	0.343
Water, 10 °C	1,440	1.44
Rubber (acc. to hardness)	60–1,500	0.06–1.5
Aluminum (rod)	5,100	5.1
Steel (rod)	5,000	5.0

Wavelength

The distance between two wave crests is called the wavelength. It is linked to the frequency:

$$\lambda = \frac{c}{f} = \frac{2\pi c}{\omega}.$$

Propagation of sound

In general, sound propagates spherically from its source. In a free sound field, the sound pressure level decreases by 6 dB each time the distance from the sound source is doubled (20 log 2). Reflecting objects influence the sound field, and the rate at which the sound level is reduced as a function of the distance from the sound source is lower. Moving vehicles have a spherical propagation of sound, but are considered to be line sources. The damping of sound each time the distance is doubled is considerably less and ranges, depending on the vehicle and the computing model, between 3 dB and 4.5 dB each time the distance is doubled.

Sound power

Sound power P is the sound energy output by a sound source per unit of time. It is not dependent on space and distance.

Sound intensity

Sound intensity I is the sound power through a plane vertical to the direction of propagation:

$$I = \frac{P}{S}.$$

In a sound field:

$$I = \frac{p^2}{\rho} c = v^2 \rho c.$$

Doppler effect

For moving sound sources: As the sound source approaches the observer, the perceived pitch is higher than the actual pitch; as distance increases, the perceived pitch falls.

Sound spectrum

Every noise is a mixture of components with different frequencies and levels. The sound-pressure level components can be broken down by frequencies with a frequency analysis. These spectra are distinguished from each other, depending on the type of frequency resolution.

Octave band spectrum

The sound pressure levels are determined and represented in terms of octave bandwidth. In the case of the octave, the fundamental frequencies behave in a ratio of 1:2. The mean frequency of the octave is:

$$f_m = \sqrt{f_1 f_2} \, .$$

Third-octave band spectrum

The sound pressure levels are determined and represented in terms of third-octave bandwidth. In the case of the third, the fundamental frequencies behave in a ratio of $1:2^{1/3}$. The bandwidth referred to the center frequency is relatively constant, as in the case of the octave band spectrum.

The frequencies are defined in DIN EN 61260 [3]. Frequency analyses in the octave and third-octave ranges produce bar charts with the frequency on the x-axis and the associated levels on the y-axis.

Narrow-band spectrum

Unlike the above spectra, it is possible to use Fourier analysis [4] to analyze frequency components with constant frequency bandwidths. This results in very many finer frequency resolutions compared with the above spectra. These are shown as line graphs, which strictly speaking are not correct, but enable different analyses to be compared.

Such narrow-band analyses are also frequently linked with a further piece of information such as, for example, the rotational speed of an engine so that a frequency analysis is available for each rotational speed for example when the engine is revved up. Such analyses are shown in three-dimensional form, either as "waterfall charts" or today frequently in color spectra, with the x-axis showing the rotational speed, the y-axis the frequency, and the z-axis in its aspect the color coding.

Sound insulation

Sound insulation is achieved by interposing a reflecting (insulating) wall (e.g. a building wall) between the sound source and the impact location. This reduces the effect of the sound source.

Sound damping, sound absorption

In the case of sound damping and sound absorption, sound energy can penetrate into the medium. Here, energy is converted into heat when reflected on peripheries, but also during propagation in the medium.

Modern noise baffles work on the principle of sound damping, i.e. sound is reflected. But parts of the sound energy are also absorbed by the baffles.

Sound absorption coefficient

The sound absorption coefficient α is the ratio of non-reflected to incident sound energy. With total reflection, $\alpha = 0$; with total absorption, $\alpha = 1$.

Low-noise design

A low-noise design is a design which is structurally optimized in accordance with acoustic considerations in order to minimize the inherent sound radiation and the propagation of sound within the structure. Simulation techniques for calculating and optimizing acoustic properties are frequently used.

Noise reduction

Noise reduction involves reducing the noise emissions of a complete system, primarily by means of low-noise designs and, secondly, through the reduction of sound propagation through the use of insulating, damping and absorbing materials.

Measured quantities for noise emissions

Sound-field quantities are usually given as effective values. Because humans perceive frequencies with varying loudness, weighting filters are used to adapt the measured levels in each frequency range to the human hearing properties. The most common weightings are A for sound as encountered, for example, in the automobile sector, and C for clearly louder noises such as, for example, in the aviation field. The corresponding weighting is appended to the symbol, such as, for example, dB(A).

Sound power level

The sound power of a sound source is described by the sound power level L_W. It is equal to ten times the logarithm to the base 10 of the ratio of the calculated sound power to the reference sound power $P_0 = 10^{-12}$ W:

$$L_W = 10 \log \left(\frac{P}{P_0} \right),$$

where log here and also in the following means the logarithm to the base 10.

Sound power cannot be measured directly. It is calculated based on quantities of the sound field which surrounds the source. The sound pressure levels L_p or the sound intensity levels L_I are usually used for the calculation.

Table 3: Overall sound level with superimposition of independent sound fields

Difference between 2 individual sound levels	Overall sound level = higher individual sound level + allowance of:
0 dB	3.0 dB
1 dB	2.5 dB
2 dB	2.1 dB
3 dB	1.8 dB
4 dB	1.5 dB
6 dB	1.0 dB
8 dB	0.6 dB
10 dB	0.4 dB

Sound pressure level

The sound pressure level L_p is ten times the logarithm to the base 10 of the ratio of the square of the effective sound-pressure value and the square of the reference sound pressure $p_0 = 20\ \mu$Pa:

$$L_p = 10 \log \left(\frac{p_{eff}^2}{p_0^2} \right) \text{ or}$$

$$L_p = 20 \log \left(\frac{p_{eff}}{p_0} \right).$$

The sound pressure level is given in decibels (dB). The frequency-dependent, A-weighted sound pressure level L_{pA} as measured at a distance of d = 1 m is frequently used to characterize sound sources.

Sound intensity level

The sound intensity level L_I is ten time the logarithm to the base 10 of the ratio of sound intensity and reference sound intensity $I_0 = 10^{-12}$ W/m²:

$$L_I = 10 \log \left(\frac{I}{I_0} \right).$$

The sound intensity can be measured directly with a probe.

Interaction of two or more sound sources

If two independent sound fields are superimposed, their sound intensities or the squares of their sound pressures must be added. The overall sound level is then determined from the individual sound levels in accordance with Table 3.

Measured quantities for noise immissions

Excessive noise

Excessive noise is the classification of a noise as an undesirable sound event and is dependent on the following factors:

- on the noise itself, measurable by physical variables (e.g. frequency, sound pressure level, or sound power),
- on the subjective attitude to the noise of the person affected,
- on the personal state of the person affected, and
- on the concrete situation in which the noise occurs.

Protection against excessive noise and in particular environmental noise is gaining increasing importance as an environmental theme. The recording and evaluation of noise immissions forms the basis for efficient noise reduction.

Rating sound level

The effect of noise on a human being is evaluated using the rating sound level L (see also DIN 45645-1 [5]; an ISO standard is currently being worked on). This is a measure of the mean noise immission over a period of time (e.g. eight working hours). With fluctuating noises it is either measured directly with integrated measuring instruments or calculated from individual sound-pressure-level measurements and the associated periods of time of the individual sound effects (see also DIN 45641, [6]). Noise-immission parameters such as pulsation (short strong deviations of the sound level from the mean level) and tonal quality (heavy dominance of one or more discrete frequencies) can be taken into account through level allowances.

Energy-equivalent continuous sound level

In the case of noises which fluctuate over time, the mean A-weighted sound pressure level resulting from the individual sound pressure levels and the individual exposure times equals the energy-equivalent continuous sound level L_{Aeq} (see also DIN 45641).

Table 4 provides guide values for the rating sound level (Germany; TI Noise [7]), measured outside the nearest residential building (0.5 m in front of an open window).

Noise indices

The EU Directive 2002/49/EC [8] defines the noise indices binding on the EU L_{den} and L_{night} as standard descriptors for offending noise over a full day (L_{den} for day, evening, night) and a night (L_{night}) along similar lines to the above-described L_{Aeq}. The assessment period here is one calendar year. Crashes for evening (5 dB) and night (10 dB) are taken into consideration for L_{den} in contrast to L_{Aeq}.

$$L_{den} = 10 \log \frac{1}{24} \cdot \left(12 \cdot 10^{\frac{L_d}{10}} + 4 \cdot 10^{\frac{L_e+5}{10}} + 8 \cdot 10^{\frac{L_n+10}{10}}\right)$$

where L_d, L_e and L_n are the values for day, for evening, and for night.

Sound exposure level

The sound exposure level SEL is used to evaluate individual excessive-noise events (e.g. an airplane starting) in sensitive conservation areas. For the purpose of calculation the sound energy of the complete event is recorded and its energy is then distributed to 1 second. It is therefore done as though the event always happens within a period of one second. The sound level is then calculated on this basis.

Table 4: Guide values for permissible noise pollution as per TI Noise [7]

	Day	Night
Purely industrial areas	70 dB (A)	70 dB (A)
Areas with predominantly commercial/industrial premises	65 dB (A)	50 dB (A)
Mixed areas	60 dB (A)	45 dB (A)
Areas with predominantly residential premises	55 dB (A)	40 dB (A)
Purely residential areas	50 dB (A)	35 dB (A)
Health resorts, hospitals, etc.	45 dB (A)	35 dB (A)

Perceived sound levels

The human ear can distinguish approximately 300 levels of acoustic intensity and 3,000 to 4,000 different frequencies (pitch levels) in rapid temporal succession and evaluate them according to complex patterns. However, there is not necessarily any direct correlation between perceived loudness levels and (energy-oriented) technically defined sound levels. Corrections (A-, B- and C-weighting) take into account the dependence on frequency of human hearing (Figure 1).

The phon unit and the definition of loudness in "sone" provide a rough approximation of subjective sound-level perception. Sound-level measurements alone do not suffice to define the nuisance and disturbance potential of noise emanating from machinery and equipment. A hardly perceptible ticking noise can thus be perceived as extremely disturbing, even in an otherwise loud environment.

Loudness level

The loudness level L_S is a comparative measure of the intensity of the subjective perception of a sound, measured in "phon". The loudness level of a sound (pure tone or noise) is the sound pressure level of a standard pure tone which, under standard listening conditions, is judged by a normal observer to be equally loud. The standard noise level is a plane sound wave at a frequency of 1,000 Hz impinging on the observer's head from the front.

Figure 1: Frequency-dependent correction values for the sound level to take into account human hearing

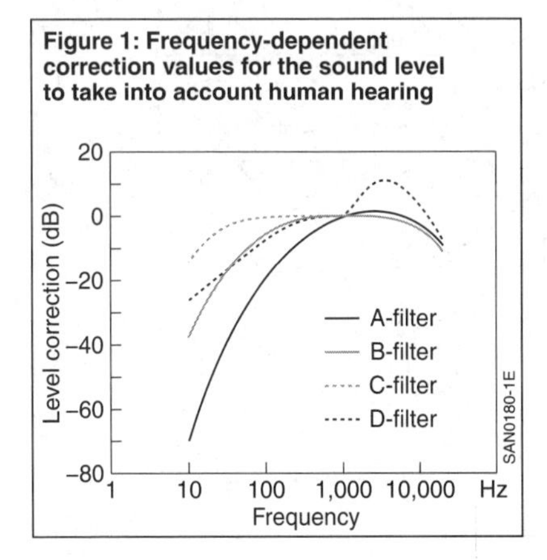

This is known internationally as the "loudness level". A difference of 8 to 10 phon is perceived as twice or half as loud.

Phon

The standard pure tone judged as being equally loud has a specific value in dB. This value is given as the loudness level of the tested sound, and has the designation "phon". As human perception of sound is frequency-dependent, the dB values of the tested sound for notes, for example, do not agree with the dB values of the standard pure tone. The connection between curves of equal sound-level perception and sound pressure in decibels is always empirically determined. These curves were published for the first time in 1933 by Fletcher and Munson. They provided the basis for the isophone curves in accordance with DIN ISO 226 [9] used today. Figure 2 shows curves based on this standard.

Loudness in "sone"

The loudness S is the measure used to define subjective noise levels. The starting point for defining the sone is: How much higher or lower is the perceived level of a particular sound relative to a specific standard. Definition: Loudness level of L_S = 40 phon corresponds to a loudness S = 1 sone. Doubling or halving the loudness is equivalent to a variation in the loudness level of approx. 10 phon.

There is a DIN loudness standard for calculating stationary sound using tertiary levels (Zwicker method [10]). This procedure takes into account both frequency weighting and the screening effects of hearing.

Pitch, sharpness

The spectrum of perceptible sound can be divided into 24 hearing-oriented frequency groups (bark). The Bark scale is a psychoacoustic scale for perceived pitch. The loudness/pitch distribution can be used to quantify other subjective aural impressions – such as the sharpness (e.g. metallic harmonic distortion) of a noise.

Articulation index

The basic prerequisite for recognizing speech is the correct transmission wherever possible of the sound waves of what is spoken from the sender (mouth) to the receiver (ear). The articulation index (AI) offers a computing model for predicting speech intelligibility in the event of interference noise. It is assumed here that the speech information is distributed to the different frequency bands of the acoustic signal. Each band then makes a contribution to speech intelligibility. If a band has a signal-to-noise ratio of more than 15 dB, the speech component in this band is evaluated as being intelligible. The sum total of all the bands with this noise ratio is multiplied by their specific weighting factor to give the articulation index AI in %.

References
[1] DIN 1320: Acoustics – Terminology.
[2] DIN EN ISO 80000-8: Quantities and units – Part 8: Acoustics .
[3] DIN EN 61260: Electroacoustics – Octave-band and fractional-octave-band filters.
[4] B. Lenze: Einführung in die Fourier-Analysis. 3rd Ed., Logos Verlag, Berlin, 1997.
[5] DIN 45645-1: Determination of rating levels from measurement data – Part 1: Noise immission in the neighbourhood.
[6] DIN 45641: Averaging of sound levels.
[7] Sixth General Administrative Regulation on the Federal Immission Protection Law (Technical Instructions on Noise Abatement – TI Noise) of 26 August 1998 (GMBl No. 26/1998 P. 503).
[8] Directive 2002/49/EC of the European Parliament and of the Council of 25 June 2002 relating to the assessment and management of environmental noise.
[9] DIN ISO 226: Acoustics – Normal equal-loudness-level contours.
[10] DIN 45631: Calculation of loudness level and loudness from the sound spectrum – Zwicker method.

Figure 2: Isophone curves in acc. with DIN ISO 226 with an allocation of noises (examples)

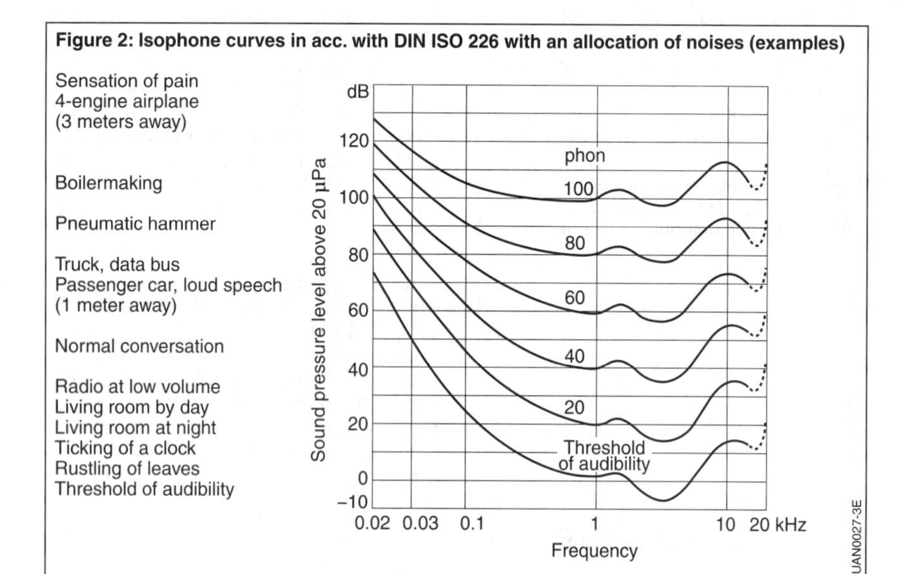

Optical technology

Electromagnetic radiation

Electromagnetic radiation denotes the wave nature of visible light, but also of radiation that is not visible to the human eye (e.g. ultraviolet light, X-radiation, radiation in the infrared range). Thus, electromagnetic radiation is subdivided by reference to its wavelength λ (Table 2).

Table 1: Quantities and units

Quantity		SI unit
A	Area A_1 Radiating area A_2 Illuminated area	m^2
E_e	Irradiance	W/m^2
I_e	Radiant intensity	W/sr
L_e	Radiance	$W/(m^2 sr)$
M_e	Radiant excitance	W/m^2
P_e	Power	W
Q_e	Radiant energy	Ws
R	Reflection coefficient	%
H_e	Irradiation	Ws/m^2
$K(\lambda)$	Absolute spectral sensitivity	lm/W
$V(\lambda)$	Spectral luminous efficiency	–
Φ_e	Radiant flux	W
r	Distance	m
t	Time	s
ε_1	Angle, incident ray (to normal of a surface)	°
ε_2	Angle, refracted ray	°
ε_3	Angle, reflected ray	°
$\varepsilon_{1,max}$	Angle of total reflection	°
η	Luminous efficiency	lm/W
Ω	Solid angle	sr
λ	Wavelength	nm
E_{phot}	Photon energy	eV
h	Planck's quantum of action	Js
c	Speed of light in a vacuum	m/s
n	Refractive index	

Geometric optics

The description of optical systems can be limited to two models – geometric optics and wave optics. Geometric optics (also called ray model) describes the propagation of light and its deflection by imaging elements with simple geometric considerations. But this is only permitted provided the imaging elements (e.g. mirrors, lenses) are much larger than the wavelength of light.

Effects such as interference, diffraction and polarization are described in physics by wave optics. This is concerned with the propagation of light in the form of a wave.

Basic principles of geometric optics

A large part of the imaging systems used in optics can be described by geometric optics. Radiation propagation is explained by "light beams" and can be described by means of simple geometric laws.

Table 2: Ranges of electromagnetic radiation

Designation	Wavelength range
Gamma radiation	0.1...10 pm
X-radiation	10 pm to 10 nm
Ultraviolet radiation	10...380 nm
Visible radiation	380...780 nm
Infrared radiation	780 nm to 1 mm
Millimeter waves (EHF)	1...10 mm
Centimeter waves (SHF)	10...100 mm
Decimeter waves (UHF)	100 mm to 1 m
Ultrashort waves (VHF)	1...10 m
High-frequency waves (HF)	10...100 m
Medium waves (MF)	100 m to 1 km
Long waves (LF)	1...10 km
Myriameter waves (VLF)	10...100 km

Basic axioms of geometric optics
Fermat's principle is the basic principle for geometric optics. The basic axioms of geometric optics can be derived from the basic statement of this principle to the effect that light always chooses the shortest path between two points [1]:
- Light beams are straight in homogenous media.
- At the boundary between two homogenous materials the beams are deflected according to the law of reflection and the law of refraction.
- Each beam path is reversible.
- Light beams can cross without influencing each other.

Law of refraction
Snell's law of refraction [2] can also be derived from Fermat's principle. This law describes the transition of a light beam from one medium (e.g. air) to another medium (e.g. glass). At the interface of the two media the light is deflected due to the change of refractive index. An incident ray is split into a refracted ray and a reflecting ray (Figure 1). The refracted ray is governed by the law of refraction:

$$n_1 \sin \varepsilon_1 = n_2 \sin \varepsilon_2 .$$

Table 3: Refractive index n of some media
(For yellow light of wavelength λ = 589.3 nm)

Medium	n
Vacuum, air	1.00
Ice (0 °C)	1.31
Water (20 °C)	1.33
Quartz glass	1.46
Polymethylmethacrylate	1.49
Standard glass for optics (BK 7)	1.51673
Window glass	1.52
Glass for headlamp lenses	1.52
Polyvinylchloride	1.54
Polycarbonate	1.58
Polystyrene	1.59
Epoxy resin	1.60
Gallium arsenide (depending on doping)	approx. 3.5

For vacuum and dielectric media (e.g. air, glass, plastics) the refractive indices n are real numbers (Table 3). Materials with a large refractive index n are referred to as optically thick, materials with a small refractive index as optically thin.

The characteristic of refractive indices of being dependent on the wavelength is called dispersion. In most cases they decreases as the wavelength increases.

Law of reflection
The following formula applies to the direction of the reflecting ray (law of reflection):

$$\varepsilon_3 = \varepsilon_1 .$$

Figure 1: Refraction of a light beam at the interface between two media with different refractive indices
a Medium 1 with refractive index n_1,
b Medium 2 with refractive index n_2.
1 Incident ray,
2 Refracted ray,
3 Reflected ray.
ε_1 Angle, incident ray,
ε_2 Angle, refracted ray,
ε_3 Angle, reflected ray.

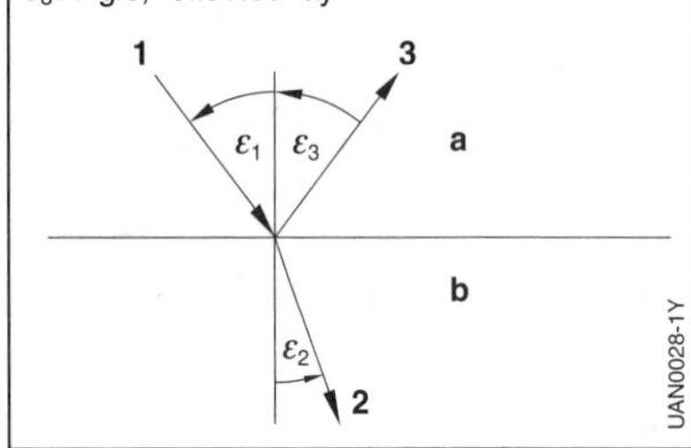

Figure 2: Dependence of the reflection coefficient on the angle of incidence
Medium 1 with refractive index n_1 = 1.00,
Medium 2 with refractive index n_2 = 1.52.

SAN0198E

Reflection coefficient R (%): 0, 20, 40, 60, 80, 100
Angle of incidence ε: 0, 20°, 40°, 60°, 80°, 100°

The angle of reflection is therefore equal to the angle of incidence. The intensity ratio of reflected to incident radiation (degree of reflection, reflection coefficient) is described by the Fresnel formula [3] and is dependent on the angle of incidence ε_1 and on the refractive indices of the adjoining media (Figure 2). In the transition from air ($n_1 = 1.00$) to glass ($n_2 = 1.52$) the upshot is that, in the case of vertical ray incidence ($\varepsilon_1 = 0$), 4.3 % of the ray energy is reflected.

Figure 3: Beam path for spherical lenses
a) Convex lens,
b) Concave lens.
A Optical axis,
F Focal point.
f Focal length,
d Thickness of lens,
r_1, r_2 Bending radii of lens.

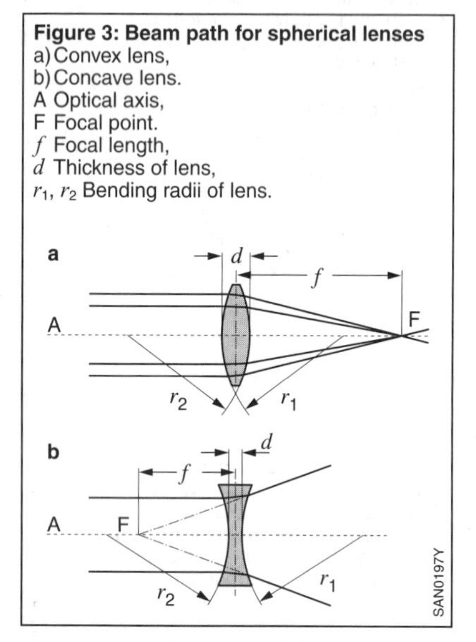

Figure 4: Beam path for a prism

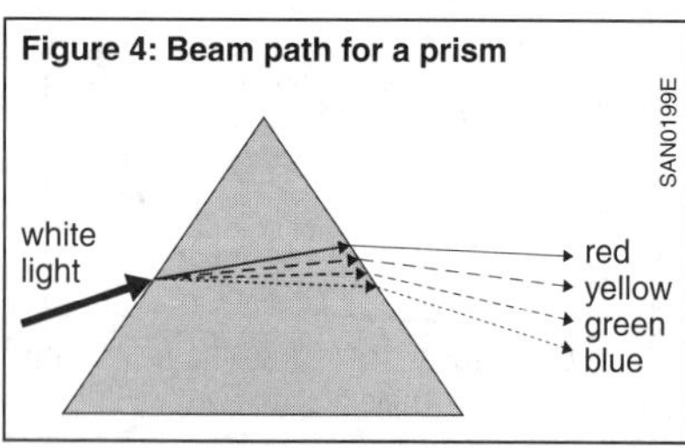

Total reflection

In the transition from an optically thicker to an optically thinner medium ($n_1 > n_2$) the angle of refraction ε_2 attains for a specific angle of incidence ε_1 its highest possible value (90°). This angle of incidence $\varepsilon_{1,max}$ is called the critical angle for total reflection. According to the law of refraction, the following is obtained for total reflection:

$$\sin \varepsilon_{1,max} = \sin 90° \frac{n_2}{n_1} = \frac{n_2}{n_1} .$$

Optical components

Spherical lenses

The optical effect and the imaging property of a lens are determined by the shape of the interfaces and by the refractive index of the lens material [4]. The imaging properties of light beams can be determined using Snell's law of refraction.

The functioning principle of a lens is divided into two subgroups based on its shape (Figure 3). Lenses that bring light to a focal point are called convergent lenses (convex lenses). In contrast, parallel beams of rays are spread out by divergent lenses (concave lenses).

Prisms

In the case of a prism, the dependence of the refractive index on the wavelength of the incident light is utilized (dispersion). Thus, blue light (λ = 420 to 490 nm) is refracted at a different angle ε_2 from red light (λ = 650 to 750 nm). When white light shines onto a prism, it is broken down into different component parts (color components) (Figure 4).

This effect also appears in rainbows. The different color components of sunlight are refracted differently on water droplets in the air.

Vehicle headlights use lenses (elements comprising cylindrical lenses and prisms) to favorably influence the radiation coming from the reflector.

Reflectors
Reflectors (mirrors) are used in optics to divert light in a particular direction (Figure 5). Reflectors are used in among other things headlights, where they specifically direct and concentrate light.

The function of reflector is to capture light from the headlight bulb, to achieve as great a range as possible, and to influence the distribution of the light on the road in such a way as to satisfy the legal requirements. Design (e.g. when installed in the bumper) places additional demands on the headlights.

Whereas previously almost exclusively paraboloids were used for reflectors, the above-mentioned, partly contradictory requirements can today sometimes only be satisfied by stepped reflectors, free-form surfaces or new headlight concepts (see Lighting equipment).

Basically speaking, the larger the lens aperture area, the greater the headlight range that can be achieved. On the other hand, the luminous efficiency increases with the size of the solid angle captured by the reflector.

Color filters
For special applications, e.g. lights and lamps on vehicles, there are precise regulations regarding the wavelength of the light used to suit the specific purpose (turn-signal lamp, stop lamp). Color filters are used here to attenuate or suppress unwanted spectral ranges.

Figure 5: Beam path for a parabolic reflector
A Optical axis,
F Focal point (focus),
P Vertex of parabola.

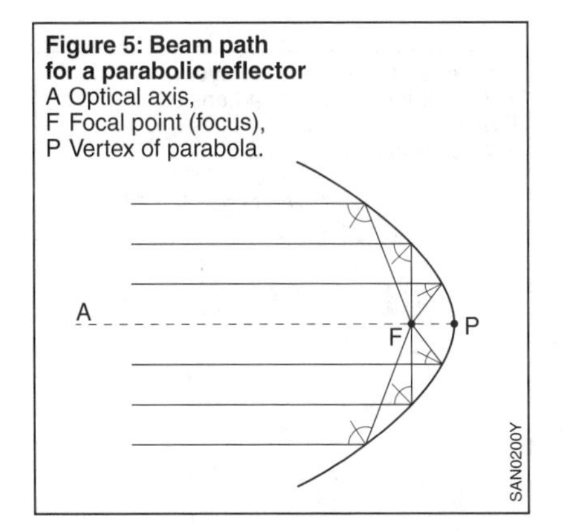

Wave optics

The previous considerations only took into account macroscopic systems, i.e. the imaging elements considered were much larger than the wavelength of the observed light beam. Wave optics is concerned with systems in which light is described as an electromagnetic wave. The color of the radiation is defined by the wavelength. Thus, monochromatic light is made up only of light waves of one wavelength, whereas white lights contains light waves of different wavelengths.

Polarization
Polarization is also taken into account when a system is considered from the wave-optics standpoint. This describes the orientation of the waves or the oscillations in relation to the plane of incidence. There are three different types of polarization: linear, circular and elliptical. In the case of linear polarization, the amplitude of the electric field varies with a constant direction of propagation; in the case of circular polarization, the direction varies while the amplitude of the electric field remains constant. Elliptical polarization describes a mixed form of linear and circular polarization.

Linear polarization is further subdivided into s-polarization (perpendicular to the plane of incidence) and p-polarization (in the plane of incidence).

As shown in Figure 6, the characteristic of the Fresnel equations calculated in Figure 2, taking into account s- and

Figure 6: Reflection coefficient
Medium 1 with refractive index $n_1 = 1.00$,
Medium 2 with refractive index $n_2 = 1.50$.

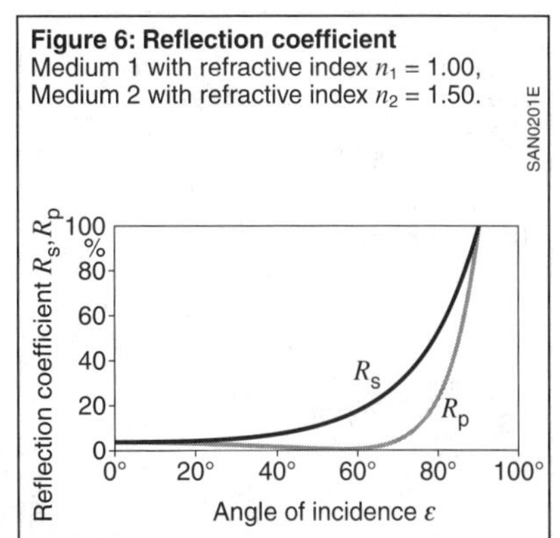

p-polarization, is divided into two different curves. The plane perpendicular to the plane of propagation is called the plane of oscillation. It follows that only waves whose electric field components oscillate in the plane of oscillation are polarizable. These are also called transversal waves (light). In contrast, longitudinal waves (e.g. sound) oscillate in the direction of propagation and are therefore not polarizable.

Interference

The superposition of two waves of equal wavelength can lead to interference. According to the superposition principle, places of constructive and places of destructive interference are formed here. An interference pattern with interference maxima and interference minima is produced. An established example of this is Young's doubles-slit experiment [4], which proved for the first time the wave nature of light.

Figure 7: Intensity curve $I(x)$ for diffraction at the single slit
L Lens, B Aperture,
D Opening diameter,
x Distance,
$I(x)$ Intensity profile.

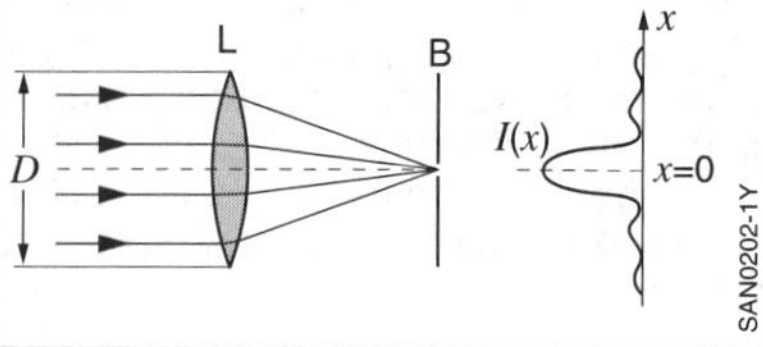

Figure 8: Resolving capacity limited by diffraction of an optical system
L Lens,
S_1, S_2 Point sources,
x_1, x_2 Imaging of point sources,
δ_{min} Minimum resolving angle
$I(x)$ Intensity profile.

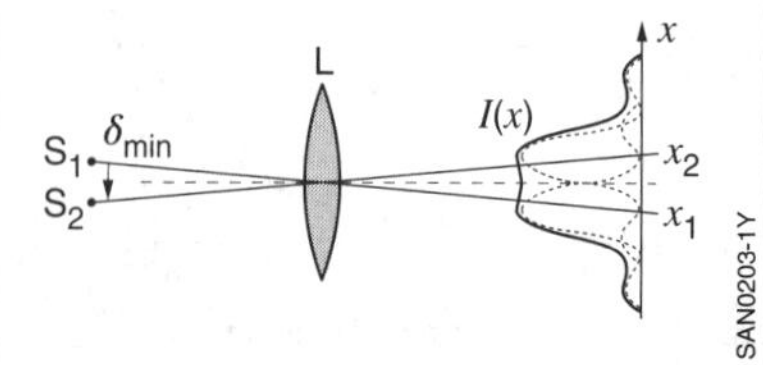

Diffraction

Diffraction phenomena of light waves can also be described using wave optics. The term diffraction is used to refer to the situation where an optical wave encounters an obstacle (e.g. aperture, single slit). The light wave is deflected at the obstacle and new elementary waves are created in accordance with Huygens'principle [4]. These waves interfere with each other and a diffraction pattern is produced in response to interference effects. Thus light penetrates into areas which should be dark according to geometric optics.

Diffraction at a single slit produces in response to interference effects the intensity curve $I(x)$ shown in Figure 7.

When diffraction is considered at an aperture plate or at a circular aperture, the resulting electric field after the aperture can be described by the Bessel functions of the first kind $J_1(2\pi\, r)$ [5]:

$$E(r) = \frac{E_0}{\pi r} J_1(2\pi\, r).$$

The resulting intensity distribution after diffraction at the aperture is obtained from

$$I(r) \sim E^2(r).$$

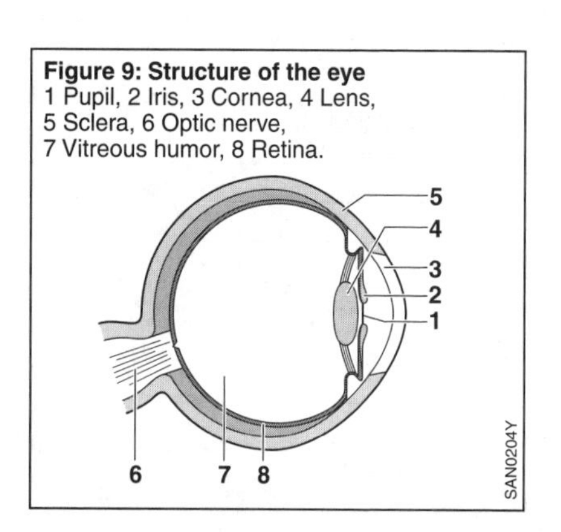

Figure 9: Structure of the eye
1 Pupil, 2 Iris, 3 Cornea, 4 Lens,
5 Sclera, 6 Optic nerve,
7 Vitreous humor, 8 Retina.

Assuming that only very weak (especially the radius) secondary maxima develop as a result of diffraction at a circular aperture, the magnitude of the principal maxima can be determined from the zero point of the Bessel function. This is obtained at $r = 0.6098$.

If diffraction is disregarded, optical systems can achieve an infinite resolution. Diffraction significantly reduces the resolving capacity of any optical system (e.g. lenses, objectives). Resolving capacity U denotes the smallest still perceptible distance between two point sources (S1 and S2).

Basically, optical elements are limited by a circular aperture. The resolving capacity of any optical system can thus be calculated from the size of the diffraction pattern of a circular aperture. The intensity curve $I(x)$ is obtained for each point source. The minimum angle δ_{min} up to which the points on the screen can still be perceived separately (Figure 8) is calculated as described in the following equation from the wavelength λ of the observed radiation and the opening diameter D:

$$\sin \delta_{min} = 1.22 \frac{\lambda}{D} .$$

Since the diameter of the aperture opening is calculated from the double radius, the factor 1.22 was determined from the zero point of the Bessel function. The resolving capacity of the system can be determined from δ_{min} at $U = 1/\delta_{min}$.

The human eye can also be considered to be an optical system. In this case the focal plane is the pupil and the shown intensity profile $I(x)$ of a point light source is imaged on the retina (Figure 9). The opening diameter D is determined by the size of the pupil. For a wavelength of $\lambda = 550$ nm and a pupil size of 1.5 mm (daylight vision) $\delta_{min} = 0.02°$ is obtained for the minimum angle. From this, at a distance of 25 cm to the eye, the minimum resolvable distance between S_1 and S_2 is calculated at 0.11 mm.

Lighting quantities

Light sources have – on the basis of their nature – radiation of different intensities and different wavelengths. The light source is classified by reference to the characteristic wavelength into the wavelength spectrum.

Physical characteristic quantities have become established for the change in intensity of the light source and to describe the effect of light on human beings. A distinction is made here between radiometric and photometric characteristic quantities. In photometry the human eye is evaluated as a detector. This is confined to the ultraviolet (UV) and the visible spectral ranges. In radiometry the power of the light source measured by a detector is determined, thereby also providing measurements in the infrared and ultraviolet ranges and with gamma rays. The lighting quantities in radiometry (radiation quantities) are identified for improved intelligibility by the indices "e" ("e" for energy).

Radiation quantities

Radiant energy Q_e

The radiant energy Q_e comprises the total energy of the light wave.

Radiant flux Φ_e

Radiant flux Φ_e denotes the amount of energy dQ_e that is transported per time unit dt by the light wave:

$$\Phi_e = \frac{dQ_e}{dt} .$$

Irradiance E_e

The proportion of the radiant flux $d\Phi_e$ which strikes a defined surface area dA is called irradiance E_e:

$$E_e = \frac{d\Phi_e}{dA} .$$

The totality of the radiated energy that strikes a bright surface element dA is determined.

Radiant intensity I_e
Radiant intensity I_e denotes the radiant flux $d\Phi_e$ emitted by a point wave in a particular direction (solid angle $d\Omega$).

$$I_e = \frac{d\Phi_e}{d\Omega}.$$

Radiance L_e
Radiance L_e denotes the radiant intensity on a vertical surface.

$$L_e = \frac{dI_e}{dA \cos \varepsilon} = \frac{d^2\Phi_e}{d\Omega\, dA \cos \varepsilon}.$$

Radiant excitance M_e
Radiant excitance M_e denotes the radiant flux $d\Phi_{e,H}$ of a light source radiated into the half-space:

$$M_e = \frac{d\Phi_{e,H}}{dA}.$$

Irradiation H_e
Irradiation H_e denotes the proportion of the radiant energy dQ that strikes per time dt a surface element dA.

$$H_e = \int E_e\, dt.$$

Lighting quantities in photometry
Spectral luminous efficiency $V(\lambda)$
Defining the lighting quantities in photometry does not take into account the fact that the eye does not have constant spectral luminous efficiency. The radiation visible to the human eye is in the wavelength range of 380 nm (blue) through 780 nm (red). The eye is at its most sensitive in daylight in the yellow-green range around 555 nm; in weaker light conditions this figure shifts to lower wavelengths. A greater radiant energy is required to achieve the impression of equal brightness for other wavelengths. Spectral luminous efficiency $V(\lambda)$ is defined as the ratio of radiant energy at 555 nm to the radiant energy for the different wavelengths.

Because not all human eyes are identical, a non-dimensional, standardized luminous-efficiency function of the human eye $V(\lambda)$ was defined in DIN 5031-3 [9] for lighting measurements and calculations. This was determined for different wavelength ranges and for daylight and nightlight conditions (Figure 10).

Luminous efficacy of radiation
The absolute spectral efficiency $K(\lambda)$, also called the luminous efficacy of radiation, is the quotient from the physiological quantity luminous flux Φ and the physical radiant flux Φ_e. For daytime vision (eye adapted to the light) the maximum value $K_{m,T}$ of $K(\lambda)$ is obtained at a wavelength of λ = 555 nm. For nighttime vision (eye adapted to the dark) the maximum $K_{m,N}$ is obtained at λ = 505 nm. The following relations are obtained for the absolute spectral efficiency $K(\lambda)$:

Figure 10: Luminous-efficiency function of the human eye
$V(\lambda)$ Day-value curve,
$V'(\lambda)$ Night-value curve.

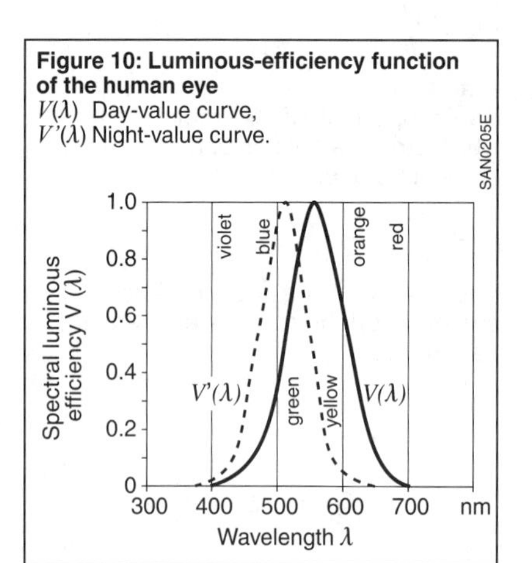

Table 4: Comparison of radiation quantities and photometric quantities

Radiation quantities		Photometric quantities	
Designation	Unit	Designation	Unit
Radiant energy Q_e	Ws	Quantity of light Q	lm s
Radiant flux Φ_e	W	Luminous flux Φ	lm
Radiant intensity I_e	W/sr	Luminous intensity I	cd = lm/sr
Radiance L_e	W/(m²sr)	Luminance L	cd/m²
Irradiance E_e	W/m²	Illuminance E	lx = lm/m²
Radiant excitance M_e	W/m²	Luminous excitance M	lx
Irradiation H_e	Ws/m²	Lumination H	lx s

Daytime vision:
$K(\lambda) = K_m\, V(\lambda)$,
$K_{m,T}$ = 683 lm/W.

Nighttime vision
$K'(\lambda) = K_m\, V'(\lambda)$,
$K_{m,N}$ = 1,699 lm/W.

$K(\lambda)$ enables the radiation quantities X_e and the photometric quantities X to be linked (Table 4). The following applies:

$$X = K_m \int X_{e\lambda}\, V(\lambda)\, d\lambda,$$

$$= \int X_{e\lambda}\, K(\lambda)\, d\lambda,$$

where $X_{e\lambda} = \dfrac{dX_e}{d\lambda}$.

Luminous flux Φ
The luminous flux Φ denotes the weighting of the radiometric radiant flux Φ_e with the wavelength-dependent efficiency curve K(λ) of the human eye. The luminous flux Φ determines the light output radiated into the totality of solid angles. All further photometric quantities are linked with the luminous flux.

Quantity of light Q
The quantity of light Q is the photometric equivalent to the radiometric radiant energy Q_e. It is calculated from the integral of the luminous flux Φ over a certain time dt:

$$Q = \int \Phi(t)\, dt.$$

Luminous intensity I
The luminous flux Φ that is radiated into a certain solid angle Ω is defined as the luminous intensity I. It offers a measure of the luminous emittance of a source radiated in a certain direction and is calculated from the radiometric radiation intensity:

$$I = \frac{\Phi}{\Omega}.$$

Illuminance E
In contrast to luminous intensity I, illuminance E denotes the luminous flux of a light source radiated to a certain surface A:

$$E = \frac{\Phi}{A}.$$

Luminance L
Luminance L defines the brightness sensation in the human eye caused by an illuminated (or also self-luminous) surface. It is obtained from the radiometric quantity for the radiant intensity:

$$L = \frac{I}{\Omega}.$$

Luminous excitance M
The proportion of the luminous flux $d\Phi$ that is generated by a defined surface element dA is known as luminous excitance. It is the photometric counterpart to radiometric radiant excitance:

$$M = \frac{d\Phi}{dA}.$$

Lumination H
The lumination H is calculated from the illuminance E_V that strikes the surface dA over a period of time dt.

$$H = \int E_V(t)\, dt.$$

Further terms
The lighting quantities are supplemented by the general terms explained in the following.

Luminous efficiency η
Luminous efficiency η is calculated from the ratio of emitted luminous flux Φ to power input P:

$$\eta = \frac{\Phi}{P}.$$

Luminous efficiency is a measure of the efficacy of converting electric power P into luminous flux Φ. It cannot exceed the maximum value of the luminous efficacy of radiation K_m = 683 lm/W for the wavelength λ = 555 nm.

Solid angle Ω
The solid angle Ω describes the ratio of the penetrated section of the spherical surface (of a sphere concentric to the radiation source) to the square of the sphere radius. The total surface of the sphere is $4\pi r^2$, thus the full solid angle is

$$\Omega = 4\pi \text{ sr} \quad (\text{sr} = \text{steradian}).$$

Steradian is a non-dimensional quantity for the solid angle.

Contrast
The contrast determines the luminance ratio between two neighboring surfaces or the maximum difference in intensity within an illuminated (or self-luminous) surface. The contrast describes the ratio of luminance or the illuminance of a bright area of an image to a dark area of the same image. According to the standard ISO IEC 21118 [10] the contrast must be determined from the values of a black-and-white checkerboard pattern (consisting of sixteen equal squares).

The small-area contrast for alternating black and white lines can also be determined. This is a measure for the specification of an optical lighting system to shown fine details on a screen. Both vertical and horizontal lines are measured for this purpose.

Table 5: Examples of some laser types

Laser type	Wavelength	Applications
He-Ne laser	632 nm	Metrology, holography
CO_2 laser	10.6 µm	Material processing
ND: YAG laser	1,064 nm	Material processing
Semiconductor laser	e.g. 670 nm e.g. 1,300 nm	Measuring technology Telecommunications
Ytterbium fiber laser	e.g. 1,070 nm	Material processing

Laser technology

Compared to other light sources, the laser (Light Amplification by Stimulated Emission of Radiation) has the following characteristic properties:
- Monochromatic radiation, i.e. a limited wavelength range
- Very high radiation density
- Low beam expansion
- Good focusability
- High time and spatial coherence of radiation

Functioning principle

A laser contains a laser-active medium that can be gaseous, solid or liquid (Table 5). The supply of energy can place the atoms or the molecules of the active medium in an excited state (Figure 11). This process is called pumping and can take place electrically (by the application of a voltage) or optically (with another light source).

After a certain period of time the excited particles of the active medium relax back into their initial state and in so doing dissipate their energy through the spontaneous emission of a photon (light particle). The energy of the photon is determined by the quantized energy states of the active medium and stipulates the wavelength λ of the laser light:

$$E_\text{P} = \frac{hc}{\lambda}.$$

Figure 11: Laser principle
1 Pumping light source,
2 Resonator mirror,
3 Laser-active material,
4 Partly transparent mirror,
5 Laser beam.

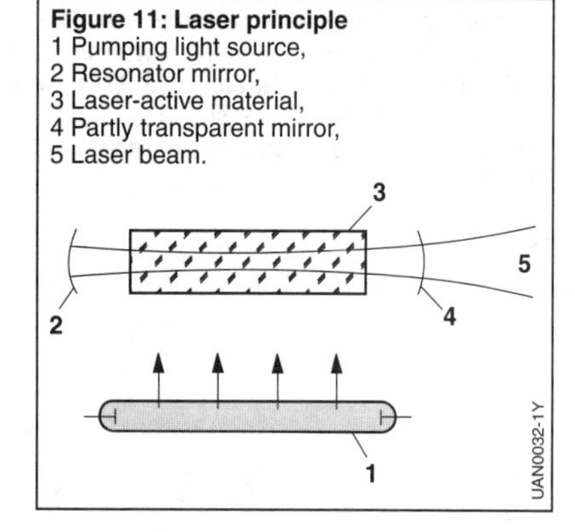

h is Planck's quantum of action and c the speed of light in a vacuum. For h and c the following values apply:
$h \approx 6.62606957 \cdot 10^{-34}$ Js
$c \approx 300,000,000$ m/s

The resonator consists of mirror surfaces at the ends of the active medium and causes the spontaneously emitted photons to be reflected back into the active medium. As they pass through the active medium again they cause new photons of equal wavelength and identical phase position to be emitted. This is called stimulated emission.

The resonator is responsible for radiation amplification and for the desired beam characteristic. The laser beam emerges at the end of the resonator via a semi-transparent mirror.

Depending on the active medium and the laser type, lasers can be operated with continuous-wave radiation or in pulsed mode with pulse lengths down to below 1 fs (10^{-15} s) [12, 13].

Areas of application

Laser measuring technology permits the noncontact, non-interacting testing of production tolerances of superfinished surfaces (e.g. fuel injectors). Resolutions in the nm range are achieved using interferometric methods.

Lasers facilitate high-precision, flexible and high-speed material processing/machining in production engineering. For example, hole diameters of 30 µm can be achieved with laser drilling.

Further laser applications in technology are holography (spatial image information), automatic character recognition (bar-code scanners), information recording (CD scanning, 3D space surveying), material processing/machining, microsurgery, and transmitters for data transmission in optical waveguides.

Specific regulations are to be observed when handling laser products. Laser products are classified according to potential hazards. For details, refer to DIN EN ISO 60825 [11].

Optical fibers/waveguides

Design

Optical fibers/waveguides transmit under controlled conditions electromagnetic waves in the ultraviolet (UV), visible, and infrared (IR) ranges of the spectrum. They are made of quartz, glass, or polymers, usually in the form of fibers or channels created in transparent materials with a core whose refractive index is usually higher than that of the cladding. Thus, light launched into the core is retained in that area by means of refraction or total reflection and routed.

Depending on the refractive index profile, a distinction is made between four types of fiber (Figure 12):

- The step-index fiber, with a sharply-defined boundary between the core and the cladding
- The graded-index fiber, with parabolic refractive index profile in the core
- The monomode fiber with a very small core diameter
- The photonic-crystal fiber with air-filled capillaries arranged periodically around the core. The arrangement corresponds to the step-index or monomode fiber.

Figure 12: Light propagation in optical fibers
a) Fibers schematically represented,
b) Refractive-index profile.
1 Step-index fiber,
2 Graded-index fiber,
3 Monomode fiber.

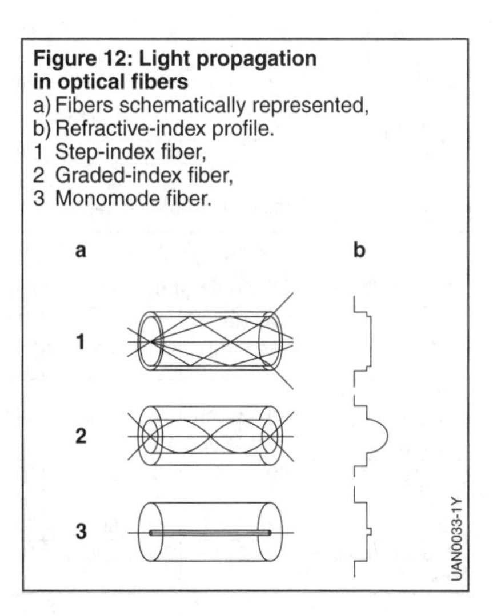

Step-index and graded-index fibers are multimode fibers, i.e. various oscillation modes of the light waves can be propagated along then at different speeds. The different oscillation modes of the light wave are dependent on the geometry and the material properties of the substrate material. Polymer fibers are always step-index fibers. Monomode fibers are, thanks to their geometry and the material used, designed in such a way that only the fundamental mode of the light wave can be propagated. Depending on the structure configuration, photonic-crystal fibers guide one or more modes.

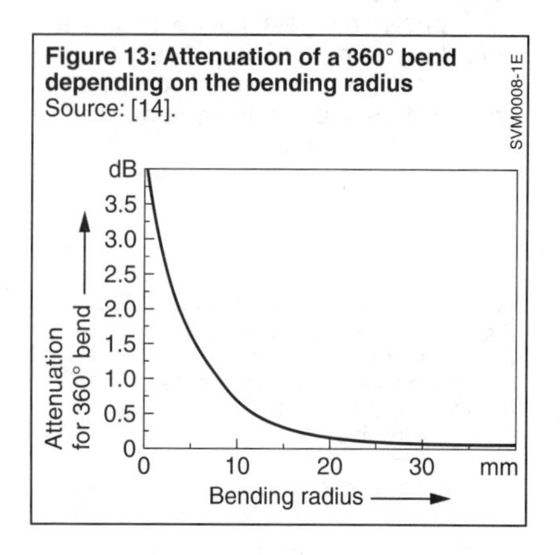

Figure 13: Attenuation of a 360° bend depending on the bending radius
Source: [14].

Properties

Glass optical fibers have a high degree of transparency in the range from ultraviolet to infrared. Losses occur primarily due to the contamination of the fiber material used in the manufacturing process. Thus, H_2O molecules absorbed from the ambient air generate absorption band in the spectrum of the optical fibers. At 950 nm, 1,240 nm and 1,380 nm the transparency of the optical fibers is reduced by the absorption of the H_2O molecules in the glass fiber. When the spectrally resolved absorption of optical fibers is considered, minima are to be seen at 850 nm, 1,310 nm and 1,550 nm, i.e. attenuation is particularly low for the wavelengths 850 nm, 1,310 nm and 1,550 nm. Synthetic fibers absorb above 850 nm and below 450 nm.

Optical fibers can only absorb light from a restricted angular range Θ. The numerical aperture $A_N = \sin(\Theta/2)$ (Table 6) serves as the measure for this.

The differences in dispersion and propagation time of the various modes cause an increasing broadening of the light pulses as the length of the fiber increases, and thus restrict the bandwidth. In photonic-crystal fibers, it is possible by means of appropriate microstructuring of the core the influence the dispersion and the efficiency of nonlinear effects to achieve desired results.

Optical fibers can be used within the temperature range of –40 to 135 °C; special versions can even be used up to 800 °C.

Table 6: Characteristic data of optical fibers/waveguides

Fiber type	Diameter Core [µm]	Cladding [µm]	Wavelength [nm]	Numerical aperture (A_N)	Attenuation [db/km]	Bandwidth [MHz · km]
Step-index fiber						
Quartz, glass	50…1,000	70…1,000	250…1,550	0.2…0.87	5…10	10
Polymer	200…>1,000	250…2,000	450…850	0.2…0.6	100…500	< 100
Graded-index fiber	50…100	100…500	450…1,550	0.2…0.3	3…5	200…10,000
Monomode fiber	3…10	100…500	850…1,550	0.12…0.21	0.3…1	2,500…10,000
Photonic-crystal fiber	1…35	250…200	300…2,000	0.1…0.8	0.2…2	≤ 160,000

Areas of application

The main area of application for optical fibers/waveguides is in data transmission. Synthetic fibers are preferred for use in the LAN (Local Area Network) field. Graded-index fibers are the most suitable for medium ranges. Only monomode fibers are used for long-distance data transmission. In fiber-optic networks, erbium-doped glass fibers serve as optical amplifiers. Here, the transported radiation can be boosted by additional optical pumping with a semiconductor source.

Optical fibers/waveguides are used in vehicles in the MOST bus. The required compliance of bending radii means that installation in motor vehicles is critical. If the bending radii are too small, the attenuation is too great (Figure 13, [14]).

Optical fibers are being increasingly used in motor-vehicle lamps and sensors. Fiber-optic sensors generate neither scatter fields nor sparks, and are themselves insensitive to that kind of disturbance. They are currently employed in potentially explosive environments, in medicine, and in high-speed trains (ICE).

Energy transport is at the forefront in the area of material processing with laser beams, in microsurgery, and in lighting engineering.

References

[1] H. Haferkorn: Optik – Physikalisch-technische Grundlagen und Anwendungen. 4th Ed., Wiley-VCH, 2002.
[2] P.A. Tipler, G. Mosca: Physik. 6th Ed.; Springer, 2009.
[3] W. Demtröder: Experimentalphysik 2 – Elektrizität und Optik. 6th Ed.; Springer, 2013.
[4] G. Litfin: Technische Optik in der Praxis. 3rd Ed.; Springer, 2005.
[5] W. Nolting: Grundkurs Theoretische Physik 5/2 – Quantenmechanik – Methoden und Anwendungen (Springer-Lehrbuch), 7th Ed., Springer-Verlag, 2012.
[6] E. Hecht: Optik. 5th Ed., Oldenbourg Wissenschaftsverlag, 2009.
[7] F. Pedrotti, L. Petrotti, W. Bausch: Optik – Eine Einführung. Reihe Prentice Hall, Markt+Technik Verlag, 1996.
[8] F. Pedrotti, L. Petrotti: Introduction to Optics. 3rd Ed., Pearson Education Limited, 2013.
[9] DIN 5031-3: Optical radiation physics and illuminating engineering; quantities, symbols and units of illuminating engineering.
[10] ISO IEC 21118: Information technology – Office equipment – Information to be included in specification sheets – Data projectors.
[11] DIN EN 60825: Safety of laser products.
Part 1: Equipment classification and requirements (2008).
Part 2: Safety of optical fibre communication systems (2011).
Part 4: Laser guards (2011).
Part 12: Safety of free space optical communication systems used for transmission of information (2004).
[12] W. Radloff: Laser in Wissenschaft und Technik. Spektrum Akademischer Verlag, 2011.
[13] F. K. Kneubühl, M. W. Sigrist: Laser. 7th Ed., Vieweg+Teubner, 2008.
[14] A. Grzemba (Editor): MOST – Das Multimedia-Bussystem für den Einsatz im Automobil. 1st Ed., Franzis-Verlag, 2007.

Hydrostatics

Density and pressure

Although fluids are compressible to a lesser extent, they can be viewed as being incompressible for most problems. In addition, since density is only slightly dependent on temperature, it can be taken as constant for many applications.

Pressure $p = dF/dA$ is non-directional in fluids which are at rest. If the pressure component produced by the difference in height (geodetic pressure) is negligible, the hydrostatic pressure is uniformly high everywhere (e.g. in a hydrostatic press).

Fluid at rest in an open vessel

In the case of open vessels, the pressure in the fluid is only dependent on the depth of the fluid (Figure 1). Closed vessels with pressure compensation, such as fuel tanks and brake-fluid reservoirs, can also be considered as open vessels.

Pressure: $p(h) = \rho g h$

Force acting on bottom: $F_\mathrm{B} = A_\mathrm{B} \rho g h$

Force acting on sides: $F_\mathrm{S} = 0.5 A_\mathrm{S} \rho g h$

Hydrostatic press

By way of example, power amplification in vehicle brakes and hydraulic power-assisted steering systems functions according to the hydrostatic-press principle (Figure 2).

Pressure: $p = \dfrac{F_1}{A_1} = \dfrac{F_2}{A_2}$.

Piston forces: $F_1 = pA_1 = F_2 \dfrac{A_1}{A_2}$,

$F_2 = pA_2 = F_1 \dfrac{A_2}{A_1}$.

Buoyancy

Buoyancy is a force acting against gravity and acts on the center of gravity of the volume of the displaced fluid. It corresponds to the weight of the fluid displaced by the submerged body:

$F_\mathrm{A} = m_\mathrm{F} g = V_\mathrm{F} \rho g$.

A body will float if $F_\mathrm{A} = F_\mathrm{G}$.

The amount of fuel available can be easily and reliably measured by analog sensors (floats) in the fuel tank with the aid of buoyancy.

Table 1: Symbols and units

Quantity		Unit
A	Cross-sectional area	m^2
A_B	Area of base	m^2
A_S	Area of side	m^2
F	Force	N
F_A	Buoyancy force	N
F_B	Force acting on bottom	N
F_G	Weight	N
F_S	Lateral force	N
V_F	Volume of displaced fluid	m^3
g	Acceleration of free fall ($g \approx 9.81\ m/s^2$)	m/s^2
h	Depth of fluid	m
m_F	Mass of displaced fluid	kg
p	Pressure	Pa = N/m^2
ρ	Density	kg/m^3

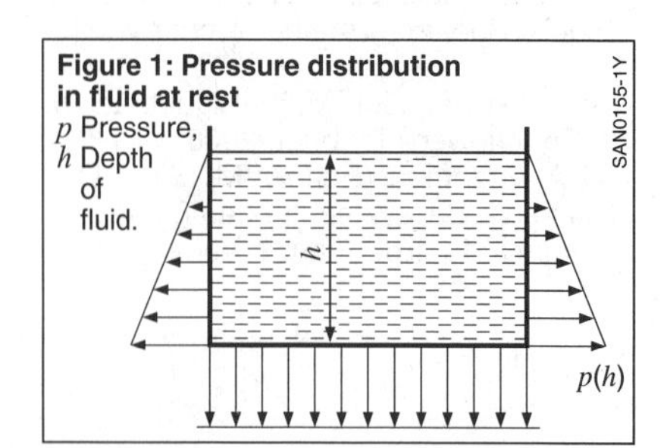

Figure 1: Pressure distribution in fluid at rest
p Pressure, h Depth of fluid.

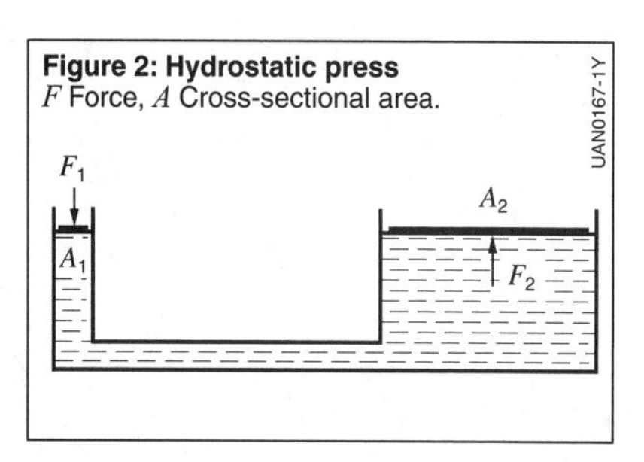

Figure 2: Hydrostatic press
F Force, A Cross-sectional area.

Fluid mechanics

Basic principles

An ideal fluid (generic term for gases and liquids) is incompressible and frictionless. This means that no shear stresses occur in the fluid, and the pressure on a fluid element is uniform in all directions. In actual fact, however, a resistance must be overcome in fluids if deformations occur that are caused by displacement of fluid elements (Figure 1). The resulting shear stress complies with Newton:

$$\tau = \frac{F}{A} = \eta\,\frac{v}{h}\,.$$

The proportionality factor η is called the dynamic viscosity and is greatly dependent on temperature. In practice, however, the kinematic viscosity

$$\nu = \frac{\eta}{\rho}$$

is often used, as it can be measured very easily with a capillary viscometer.

Flows without turbulence, in which the individual fluid layers move separately in parallel and which are predominantly determined by viscosity, are known as laminar flows. If the flow velocity exceeds a limit value, adjacent layers start to swirl, resulting in a turbulent flow. In addition to the flow velocity, the transition point between laminar and turbulent flow is also dependent on the Reynolds number

$$R_e = \frac{\rho L v}{\eta} = \frac{L v}{\nu}$$

In the case of a flow within a pipe, the pipe diameter is used for L. Flow in a pipe becomes unstable or turbulent at $R_e > 2300$.

Since, in the case of gases with low flow velocities (up to 0.5 times the velocity of sound), compression is negligible in many flow processes, they are also governed by the laws of incompressible fluids.

Table 1: Symbols and units

Quantity		Unit
A	Cross-sectional area	m^2
F	Force	N
F_A	Buoyancy force	N
F_W	Resistance force	N
L	Length in flow direction	m
Q	Volumetric flow	m^3/s
R_e	Reynolds number	–
c_W	Drag coefficient	–
d	Diameter	m
g	Acceleration of free fall ($g \approx 9.81\ m/s^2$)	m/s^2
h	Height	m
m	Mass	kg
$\dot{m}$	Mass flow	kg/s
p	Pressure	$Pa = N/m^2$
t	Thickness	m
v	Flow velocity	m/s
α	Contraction coefficient	–
η	Dynamic viscosity	$Pa \cdot s = Ns/m^2$
μ	Discharge coefficient	–
ν	Kinematic viscosity	m^2/s
ρ	Density	kg/m^3
φ	Velocity coefficient	–
τ	Shear stress	N/m^2

Figure 1: Shear stresses in fluids
τ Shear stress,
v Flow velocity,
h Height,
F Force.

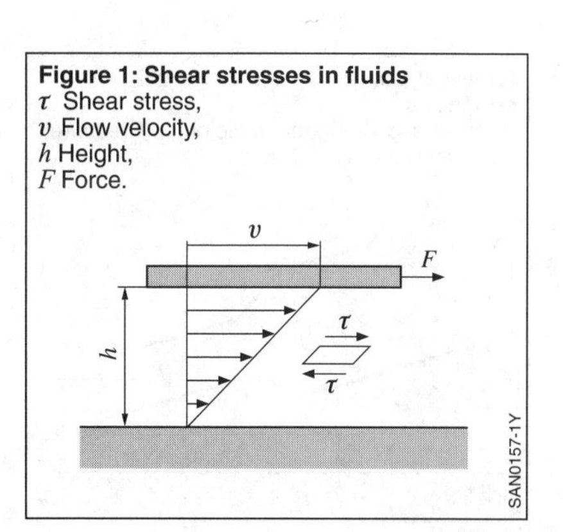

Basic equations of fluid mechanics

The most important basic equations of fluid mechanics are the continuity equation and the Bernoulli equation. They describe the conservation of mass and energy in flowing fluids.

Continuity equation

In a steady state, mass conservation requires that in a flow the mass flow rate be of equal magnitude in each cross-section (Figure 2):

$$\dot{m} = \rho A_1 v_1 = \rho A_2 v_2 = \text{const.}$$

In the case of incompressible fluids (ρ = const.), the volumetric flow must also be constant:

$$Q = A_1 v_1 = A_2 v_2 = \text{const.}$$

Bernoulli equation

From the continuity equation, it follows that an acceleration takes place between A_1 and A_2. This results in an increase in kinetic energy, which must be effected by a pressure drop, where $p_1 > p_2$ (Figure 2). According to the law of conservation of energy, the sum of the static pressure p, kinetic pressure, and geodetic pressure is constant in a flowing fluid. Ignoring friction losses, the following applies accordingly to the flowing fluid in a non-horizontal pipe:

$$p_1 + \frac{1}{2}\rho v_1^2 + \rho g h_1 = p_2 + \frac{1}{2}\rho v_2^2 + \rho g h_2 .$$

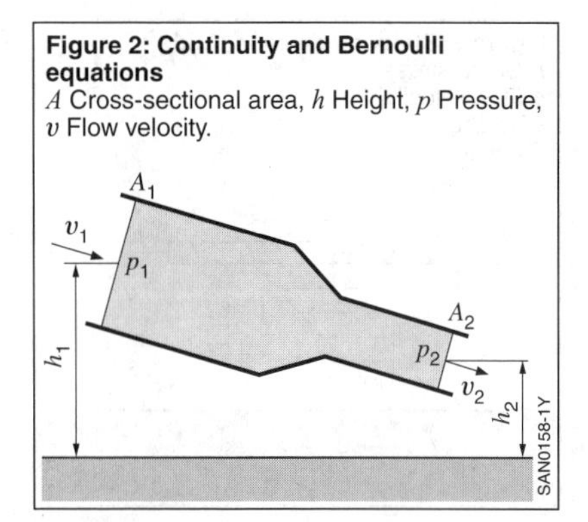

Figure 2: Continuity and Bernoulli equations
A Cross-sectional area, *h* Height, *p* Pressure, *v* Flow velocity.

Discharge from a pressure vessel

Under the precondition that the cross-sectional area of the discharge end is very much smaller than that of the vessel (Figure 3), the velocity v_1 is negligible according to the continuity equation. As derived from Bernoulli equation, the discharge velocity is governed by the following:

$$v_2 = \varphi \sqrt{\frac{2}{\rho}(p_1 - p_2) + 2gh}$$

The velocity coefficient φ takes into account the losses that occur. The jet constriction must also be taken into account for the volumetric flow or the discharge volume; this constriction is dependent on the contraction coefficient α. The following then applies to the discharge volumetric flow:

$$Q = \alpha \varphi A_2 \sqrt{\frac{2}{\rho}(p_1 - p_2) + 2gh}$$

Velocity coefficient and contraction coefficient are often expressed together as the discharge coefficient $\mu = \alpha\varphi$ (Table 2).

Table 2: Discharge openings

<table>
<tr><th>Orifice shape</th><th>Velocity coefficient φ</th><th colspan="4">Contraction coefficient α</th><th>Discharge coefficient μ</th></tr>
<tr><td>SAN0159-1Y</td><td>0.97</td><td colspan="4">0.61…0.64</td><td>0.59 … 0.62</td></tr>
<tr><td></td><td>0.97 … 0.99</td><td colspan="4">1.0</td><td>0.97 … 0.99</td></tr>
<tr><td rowspan="3"></td><td rowspan="3">0.95 … 0.97</td><td colspan="4">$(d_2/d_1)^2$</td><td rowspan="3">0.82 … 0.97</td></tr>
<tr><td>0.4</td><td>0.6</td><td>0.8</td><td>1.0</td></tr>
<tr><td>0.87</td><td>0.90</td><td>0.94</td><td>1.0</td></tr>
</table>

Resistance of bodies submerged in a fluid flow

A pressure differential occurs across a body submerged in a fluid flow (e.g. the vehicle body), resulting in a resistance force

$$F_W = \frac{1}{2} c_W A \rho v^2$$

Here, A is the cross-sectional area of the body on which the fluid flow is acting and c_d an undefined coefficient of resistance, which is dependent on the shape of the body submerged in the fluid flow.

As it is extremely complex to calculate exactly the resistance to flow even for simple bodies, resistance to flow is usually determined experimentally. If the dimensions are large, the measurements are taken on downscaled models. As well as geometrical similarity, the forms of energy that occur (kinetic energy, frictional work) in the original fluid flow and in the model flow must be proportional. This proportion is denoted by the Reynolds number R_e.

Basically: Two flows are similar in fluid-dynamic terms if their Reynolds numbers R_e are identical. Because even complex geometries are composed of simple basic bodies, streamlined surfaces can already be derived during the modeling phase as per Table 3.

Figure 3: Discharge from a pressure vessel
A Cross-sectional area, h Height, p Pressure, v Flow velocity.

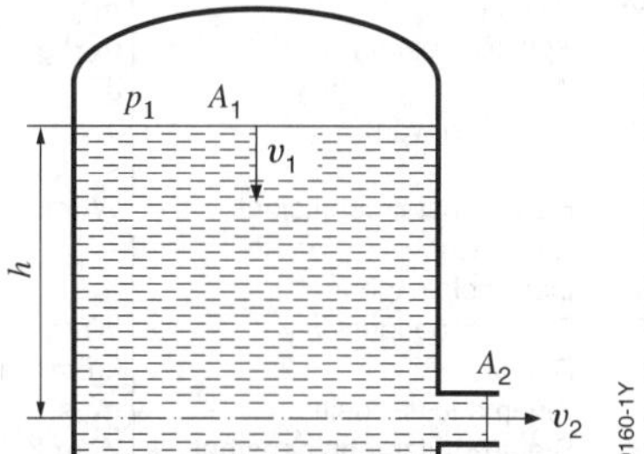

Table 3: Drag coefficients c_w

Body shape: *L Length, t Thickness,* *R_e Reynolds number.*	c_w
Circular plate 	1.11
Open dish 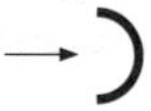	1.33
Sphere 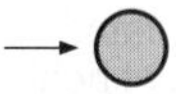 $R_e < 200{,}000$ $R_e > 250{,}000$	 0.47 0.20
Narrow rotational body $L/t = 6$ 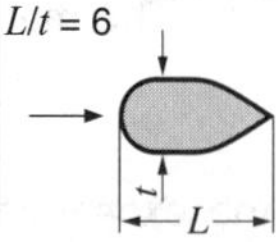	 0.05
Long cylinder 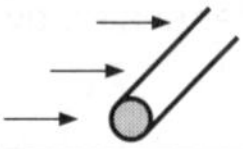 $R_e < 200{,}000$ $R_e > 450{,}000$	 1.0 0.35
Long plate $L/t = 30$ $R_e \approx 500{,}000$ $R_e \approx 200{,}000$ 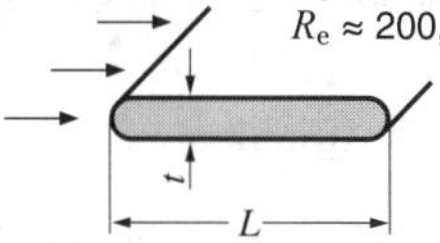	 0.78 0.66
Long wing $L/t = 18$ $R_e \approx 10^6$ $L/t = 8$ $R_e \approx 10^6$ $L/t = 5$ $R_e \approx 10^6$ $L/t = 2$ $R_e \approx 2 \cdot 10^5$ 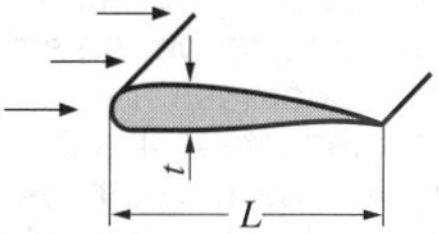	 0.2 0.1 0.08 0.2

References

[1] A. Böge: Technische Mechanik. Vieweg+Teubner Verlag, 2009.
[2] W. Bohl; W. Elmendorf: Technische Strömungslehre. Vogel-Verlag, 2008.

Thermodynamics

Basic principles

System

Thermodynamic systems are typically divided into three types. What all the systems have in common is that they are delimited from the surroundings by the system boundary within which the system is located.

Isolated system

The system boundary of an isolated system is impermeable to all types of heat Q, work W and mass m (e.g. the medium inside a perfectly insulated Thermos flask).

Closed system

A closed system allows the exchange of heat and work with the surroundings across the system boundary, but not of mass (e.g. a gas in a cylinder sealed by a moving piston).

Open system

Finally, in an open system heat, work and mass can be exchanged across the system boundary with the surroundings (e.g. an open cooking pot).

State and process variables

State variables

In thermodynamics states of systems and (changes to) processes are described by way of state variables such as for example temperature T, pressure p, volume V or mass m. The changes to these variables are called changes of state. The state of a system is described uniquely by means of the state variables. To describe the state of a system which undergoes a change of state from state 1 to state 2, it is therefore not necessary to know the path taken between initial and final states, but solely to know the relevant state variables.

Process and process variables

Generally speaking, the stringing together of changes of state is called a process. In contrast to state variables, the process variables required to describe the process depend on the path between initial and final states. Work and heat are thus process variables, the type and with it also the magnitude of the work input differ e.g. depending on process control.

Table 1: Symbols and units

Quantity		SI unit
A	Cross-sectional area	m^2
a	Thermal diffusivity	m^2/s
c	Specific heat capacity c_p Isobaric (constant pressure) c_v Isochoric (constant volume)	$J/(kg \cdot K)$
E	Energy	J
e	Specific energy	J/kg
H	Enthalpy	J
h	Specific enthalpy	J/kg
k	Heat-transmission coefficient	$W/(m^2 \cdot K)$
m	Mass	kg
n	Polytropic exponent	–
p	Pressure	$Pa = N/m^2$
Q	Heat	J
$\dot{Q}$	Heat flow dQ/dt	W
R_m	Universal gas constant $R_m \approx 8.3145$ J/(mol·K)	$J/(mol \cdot K)$
R_i	Specific gas constant $R_i = R_m/M$ (M molar mass)	$J/(kg \cdot K)$
R_λ	Thermal-conduction resistance	K/W
S	Entropy	J/K
s	Length	m
T	Thermodynamic temperature	K
ΔT	Temperature difference $\Delta T = T_1 - T_2$	K
U	Internal energy	J
u	Specific internal energy	J/kg
V	Volume	m^3
v	Specific volume	m^3/kg
W	Work	J
W_t	Technical work	J
t	Time	s
α	Heat-transfer coefficient	$W/(m^2 \cdot K)$
ε	Emissivity	–
κ	Isentropic exponent	–
λ	Thermal conductivity	$W/(m \cdot K)$
ρ	Density	kg/m^3
ν	Kinematic viscosity	m^2/s
σ	Stefan-Boltzmann constant $\sigma \approx 5.6704 \cdot 10^{-8}$ W/m^2K^4	W/m^2K^4

When a system runs through a sequence of processes and ends again in the initial state, this is called a cycle. Cycles are important for describing many technical applications (e.g. engines, power plants, power stations, air conditioners).

Intensive and extensive state variables
State variables can be further subdivided into intensive and extensive variables. When an existing system is subdivided into two parts, all the variables whose values are maintained in the two new systems are called intensive variables (e.g. temperature, pressure). Variables whose values change are called extensive variables (e.g. volume).

Often of interest are system properties which do not depend on the absolute value of a system. It can therefore be practicable to divide the extensive variables by the system mass. This results in specific state variables which retain their value when the system is subdivided (e.g. specific volume $v = V/m$ and its reciprocal value, density $\rho = 1/v$).

Forms of energy
As in classic mechanics, the state variables "kinetic energy" and "potential energy" are defined in thermodynamics. When mass m moves at constant speed c, it thus has the kinetic energy

$$E_{kin} = \frac{1}{2} m c^2.$$

A mass m that is under gravity at the acceleration of free fall g and height z has the potential energy

$$E_{pot} = m g z.$$

In addition to these forms of energy, a thermodynamic system has a further inherent type of energy, internal energy U. As well as the movement of the system, this type of energy furthermore takes into account changes within the system due for example to a change in temperature. The total energy of a system is then:

$$E_{tot} = U + E_{kin} + E_{pot} .$$

Work and heat
Work W is generally defined as the integral of the force F acting at a point of application on the system via the distance covered ds (from s_1 to s_2):

$$W_{12} = \int_{s_1}^{s_2} F\,ds .$$

The special case of volume-change work is crucially important in thermodynamics. This is defined as

$$W_{12} = -\int_{V_1}^{V_2} p\,dV .$$

Volume-change work can be shown and read as the area (integral) under the curve in a pV diagram (Figure 1).

In contrast to work, heat Q is a form of energy that occurs in an unordered way. When energy is supplied to a system for example by an electric heater, the change in the system's state of equilibrium is caused by heat, but not by work, which in this example is equal to zero.

The two variables work and heat are counted positively when they are supplied to a system. When work or heat is rejected, they are negative. This equates to a system-egotistical standpoint.

Figure 1: Volume-change work as integral in the pV diagram

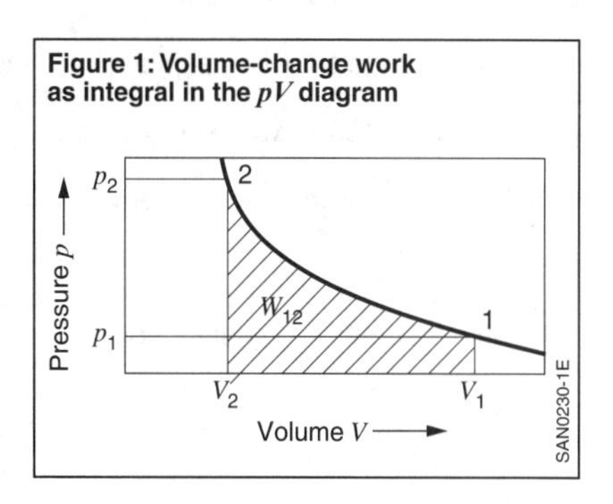

Laws of thermodynamics

Zeroth law

Two systems which are each in thermal equilibrium with a third system are also in thermal equilibrium with each other. All the systems thus have the same temperature.

First law

The first law of thermodynamics states that energy cannot be created or destroyed, but only converted from form to another, e.g. heat to work.

For an isolated system, in which there are no flows of mass or energy across the system boundaries, the sum of all the energy changes equals zero.

For a closed system, in which energy but not mass can flow across the system boundaries, the following applies:

$$Q_{12} + W_{12} = \Delta U + \Delta E_{kin} + \Delta E_{pot}.$$

The sum of the change in the internal energy ΔU and the kinetic energy ΔE_{kin}, and the potential energy ΔE_{pot} of the system is, therefore, the sum of the quantity of heat Q_{12} added and the work W_{12} done on the system (in the case of a change from state 1 to state 2).

In many technical applications a medium flows through an open system in a stationary flow process. It is therefore also necessary to consider that the intake and expulsion of the input or output mass is connected with work W_t and the input or output mass has an additional energy content. For a stationary flow process with an input and an output mass flow

$\dot{m}_1 = -\dot{m}_2 = \dot{m}$ the following applies:

$$\dot{Q}_{12} + \dot{W}_{t,12} = \dot{m}\,(\Delta h + \Delta e_{kin} + \Delta e_{pot}),$$

with the new state variable enthalpy:

$$h = u + pv.$$

e denotes the specific variable, i.e.:

$$e = \frac{E}{m}.$$

A perpetuum mobile of the first kind violates the first law of thermodynamics in that it creates "energy from nothing" or presupposes an (impossible) efficiency of more than 100%.

Second law

Once the retention of the total energy has been described by the first law, the second law of thermodynamics specifies the direction of the energy exchange and the sequences of processes. For example, heat transfer can only take place from a hot body to a cold body, never the other way round. The decisive state variable is entropy. For reversible changes of state the change dS in the entropy S of a system is defined as:

$$dS = \frac{dQ}{T}.$$

In other words, the transferred quantity of heat dQ is divided by the temperature T at the point of heat exchange. If therefore a quantity of heat dQ is isothermally added for example to a system, this results in an increase in the statistical fluctuation of the kinetic energy of the atoms (effectively the addition of heat) and also an increase in its Entropy by the value dS.

For all irreversible changes of state and thus for all real technical processes (accompanied by loss) it follows as a result of energy dissipation that entropy production is positive. In the case of a reversible process, however, no entropy is created. To summarize, it can therefore be stated that the rate of entropy production is never negative and thus:

$$dS \geq 0.$$

A perpetuum mobile of the second type violates the second law of thermodynamics, while it can perfectly comply with the first law. The second main law states that it is not possible to convert heat into work when the ambient temperature stays the same. Nor is it possible to convert a quantity of heat which exists at a temperature level above the surroundings fully into work.

Third law

The third law assigns to the entropy at absolute zero, at which the temperature $T = 0$ K, a value that is not dependent on pressure, temperature, volume, etc. This is $S_0 = 0$ J/K.

It also states that absolute zero can be approached by a series of different processes only asymptotically, but never reached.

Table 2: Specific heat capacity c_p at constant pressure for some gases
Values apply to 1 bar and 293.15 K [1].

Gas	c_p [kJ/kg·K]
Nitrogen (N_2)	1.041
Oxygen (O_2)	0.9189
Helium	5.251
Air	1.007
Carbon dioxide	0.8459
Ammonia (NH_3)	2.160

Equations of state

The thermal behavior of a gas is described by way of the correlation of pressure p, temperature T and volume V, i.e. by way of a function $F(p, V, T) = 0$. This function is called the thermal equation of state.

As well as describing the thermal behavior, a further equation, which establishes a correlation between the internal energy U and the thermal state variables, is needed to describe the caloric behavior. The caloric equation of state thus reads $U = U(V, T)$. This correlation is required when for example from the first law the resulting temperature change is to be calculated after a certain quantity of heat has been added.

Ideal gas

The thermal equation of state for an ideal gas reads:

$pV = mRT$ or specifically
$pv = RT$,

with the (specific) gas constant R ($R = R_m/M$) and the mass m. In addition, the following applies:

$u = u(T)$ and $h = h(T)$.

In other words, the internal energy u and the enthalpy h of an ideal gas only depend on the temperature T. From the differentials of the internal energy and the enthalpy it follows with $(du/dT)_v = c_v$ and $(dh/dT)_p = c_p$:

$$u(T) = \int_{T_0}^{T} c_v(T)\,dt + u_0,$$

$$h(T) = \int_{T_0}^{T} c_p(T)\,dt + h_0.$$

This correlation can be further simplified when c_v and c_p are assumed to be constant, i.e. independent of the temperature. Such a gas is called a perfect gas. Ideal gases are governed by the correlation

$$c_p - c_v = R.$$

Table 2 shows by way of example values for c_p of some gases.

Real gas
The ideal gas equation demonstrates in certain ranges, e.g. at very high pressures, significant deviations from reality. An equation which better describes this behavior is the van der Waals equation:

$$\left(p + \frac{a}{v^2}\right)(v - b) = R\,T.$$

Here a and b are substance-specific variables, taking into account the influence of internal pressure a/v^2, resulting from molecular attraction, and the specific volume b of the molecules. There are a large number of other real gas equations as well as this equation. For these and further details, refer to the further literature [2], [3], [4].

Table 3: The most important changes of state of ideal gases and their properties

Change of state	Diagram	Equations for W_{12} volume-change work and Q_{12} added heat. (For quantities used, see Table 1).
Isochoric (V = const)	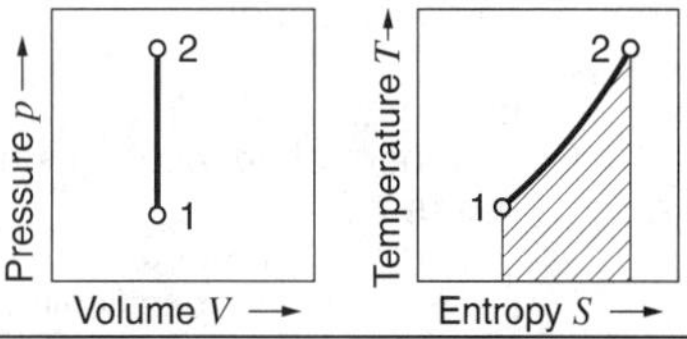 Pressure p → / Volume V →; Temperature T → / Entropy S →	$W_{12} = 0$ $Q_{12} = m\,c_v\,(T_2 - T_1)$
Isobaric (p = const)	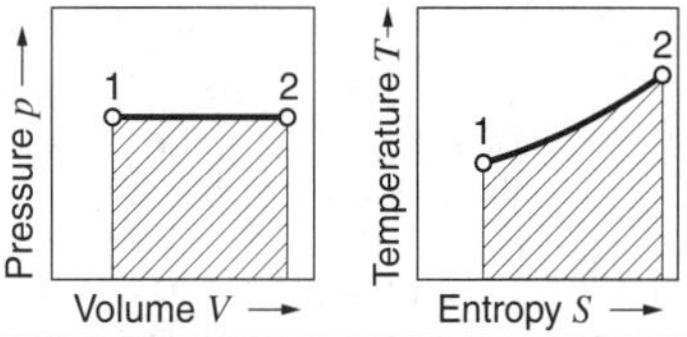 Pressure p → / Volume V →; Temperature T → / Entropy S →	$W_{12} = -p\,(V_2 - V_1)$ $Q_{12} = m\,c_p\,(T_2 - T_1)$
Isothermal (T = const)	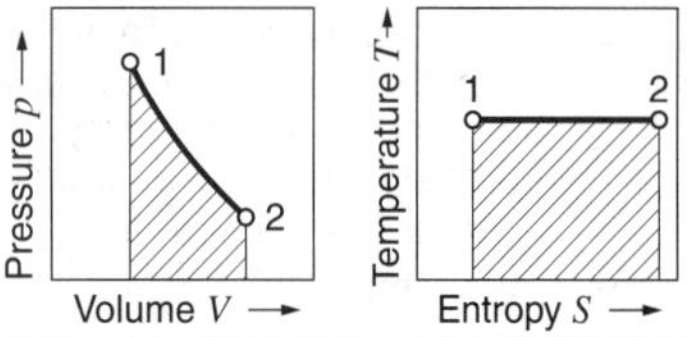 Pressure p → / Volume V →; Temperature T → / Entropy S →	$W_{12} = -p_1 V_1 \ln \frac{p_1}{p_2}$ $Q_{12} = -\,W_{12}$
Reversible adiabatic ($dS = 0$)	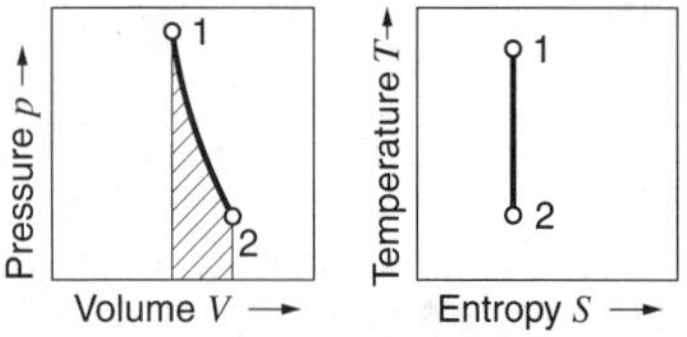 Pressure p → / Volume V →; Temperature T → / Entropy S →	$W_{12} = \frac{p_1 V_1}{\kappa - 1}\left(\left(\frac{V_1}{V_2}\right)^{\kappa - 1} - 1\right)$ $Q_{12} = 0$
Polytropic curve ($p v^n$ = const)	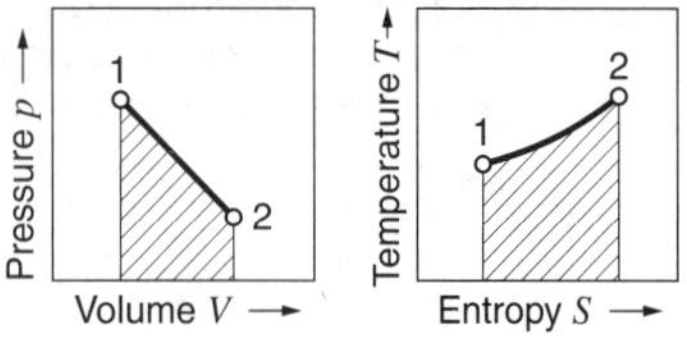 Pressure p → / Volume V →; Temperature T → / Entropy S →	$W_{12} = \frac{p_1 V_1}{n - 1}\left(\left(\frac{V_1}{V_2}\right)^{n - 1} - 1\right)$ $Q_{12} = m\,c_v \frac{n - \kappa}{n - 1}(T_2 - T_1)$

SAN0196-2E

Changes of state

Real changes of state of gases can be can be approached for many technical processes by simplifying, idealized assumptions. The technically most important changes of state include the isothermic (T = const), the isobaric (p = const), the isochoric (V = const), and the reversible adiabatic ($\dot{q} = 0$ or $dS = 0$, no heat exchange). In addition to description by these simplified idealized changes of state, which are often realized to a good degree of approximation, it is possible, assuming an ideal gas, to describe many technical processes with greater accuracy by means of a polytropic change of state of the form

$$p\,V^n = \text{const}$$

Depending on the choice of exponent n, different process paths are reproduced, thus the mentioned isothermic ($n = 1$), isobaric ($n = 0$), isochoric ($n \rightarrow \infty$), and reversible adiabatic ($n = \kappa$, with the isentropic exponent κ) changes of state. The changes of state, the associated correlations of the variables p, V and T, the associated equations for volume-change work

$$W_{12} = \int_{V_1}^{V_2} p\,dV$$

and the added or rejected heat Q_{12} are set out in Table 3. Work and heat can further be read from the illustrated pV and TS diagrams as areas.

Cycles and technical applications

Basic principles

A cycle is a sequence of different thermodynamic changes of state (processes) which when completed sees the return of the initial state, i.e. the state variables in the initial and final states are identical. Here, a cycle can take place with a constant, circulating mass flow in a closed system (e.g. in a Stirling engine or a refrigerating machine) or in an open system with media exchange (e.g. in an internal-combustion engine or a gas turbine).

A cycle is typically such that either it is deprived of work (as in a heat engine) or the heat is increased by the addition of work to a higher temperature level (as in a refrigerating machine or a heat pump). These two forms can be subdivided with regard to the successive direction of their changes of state in the pV and TS diagrams into clockwise (engines) and counterclockwise machines (machines). The

Figure 2: Clockwise cycle
a) Depiction in the pV diagram,
b) Depiction in the TS diagram.

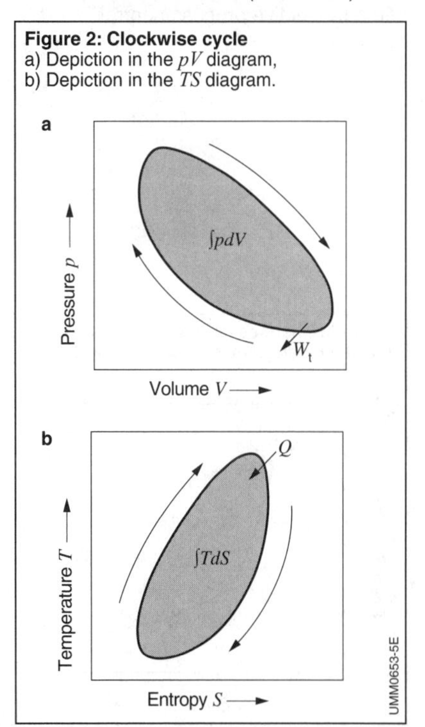

areas enclosed by the cycle represent here the input work in the pV diagram and the added heat in the TS diagram (Figure 2).

Therefore heat is added and work is output net for the situation of a clockwise cycle shown in Figure 2. The efficiency η of such a process is usually represented as the ratio of benefit (work output W_t) and expenditure (heat addition Q_{ad}), in this case therefore:

$$\eta = \frac{|W_t|}{|Q_{ad}|} .$$

For technical applications the real sequences are approached by comparison cycles. Here the entire cycle is subdivided into different reversibly accepted subcycles. The latter can be described by means of simple changes of state (e.g. isentropic, isothermal, isobaric, isochoric). In additional to simple describability, such idealized comparison cycles can be used as reference for comparing real machines as they represent the optimum in the respective process control.

Carnot cycle

The Carnot cycle constitutes the ideal comparison cycle which shows the best possible efficiency of a machine operating between two temperature levels (T_{min} and T_{max}). Heat addition and heat rejection occur in the Carnot cycle isothermally as shown in Figure 3 and thus without dissipation, work is done (compression and expansion) in reversible adiabatic fashion.

For Carnot efficiency the following equation is obtained for an ideal gas according to the above definition and by applying the first law:

$$\eta = \frac{|W_t|}{|Q_{ad}|} = 1 - \frac{T_{min}}{T_{max}} .$$

The Carnot cycle has extra significance because – although in reality it can only be approximately represented at great expense – it describes the describes the maximum amount of utilizable energy (exergy) or non-utilizable energy (anergy) and the maximum possible efficiency can be achieved with it.

Figure 3: Changes of state of the Carnot cycle
a) In the pV diagram, b) In the TS diagram.

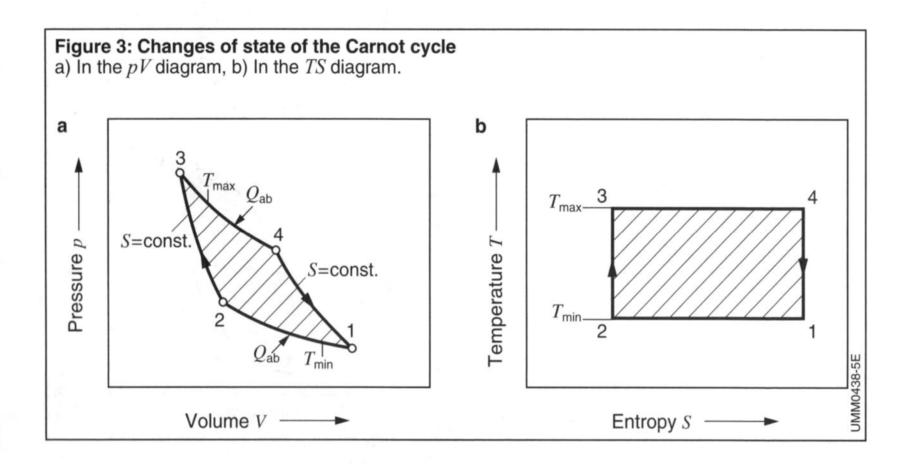

Comparison cycles of technical applications

Seiliger cycle

The Seiliger cycle describes the sequences and processes in internal-combustion engines. It also depicts the Diesel cycle (constant-pressure cycle) and the Otto cycle (constant-volume cycle) as boundary cycles. In the Seiliger cycle heat rejection takes place through combustion. This can be taken as isochoric (with a virtually stationary piston and constant volume) or isobaric (with a moving piston and constant in-cylinder pressure). The two forms and mixed forms of these two forms are depicted in the Seiliger cycle. The following five cycles steps are featured and depicted in Figure 4 in the pV and in the TS diagrams:

- reversible adiabatic compression (1→2),
- isochoric heat addition (2→3),
- isobaric heat addition (3→4),
- reversible adiabatic expansion (4→5),
- isochoric heat rejection (5→1).

Important characteristics of the Seiliger cycle are the compression ratio $\varepsilon = V_1/V_2$, the pressure-increase ratio $\psi = p_1/p_2$, and the injection ratio $\varphi = V_4/V_3$. For the thermal efficiency of the Seiliger cycle the following expression is obtained for an ideal gas:

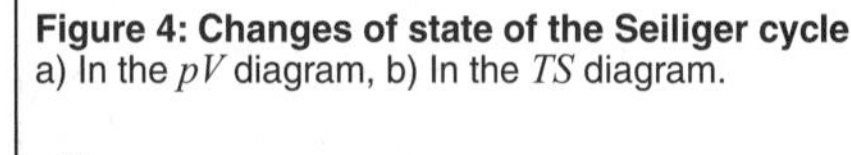

$$\eta_{th} = 1 - \varepsilon^{1-\kappa} \frac{\psi \varphi^{\kappa} - 1}{\psi - 1 + \kappa \psi(\varphi - 1)}.$$

Figure 5 shows the efficiencies that can be achieved of different process controls as a function of compression and pressure-increase ratios. It also shows the values of the boundary cases of constant-volume and constant-pressure cycles. These two

Figure 5: Efficiencies of the comparison cycles as a function of compression ratio
1 Constant-volume cycle,
2 Seiliger limit-pressure cycle, $\varphi = 1.5$, $\psi = 5.0$,
3 Seiliger limit-pressure cycle, $\varphi = 1.5$, $\psi = 1.5$,
4 Constant-pressure cycle, $\varphi = 1.5$,
5 Seiliger limit-pressure cycle, $\varphi = 2.0$, $\psi = 5.0$,
6 Seiliger limit-pressure cycle, $\varphi = 2.0$, $\psi = 1.5$,
7 Constant-pressure cycle, $\varphi = 2.0$,
ε Compression ratio,
φ Injection ratio,
ψ Pressure-increase ratio.

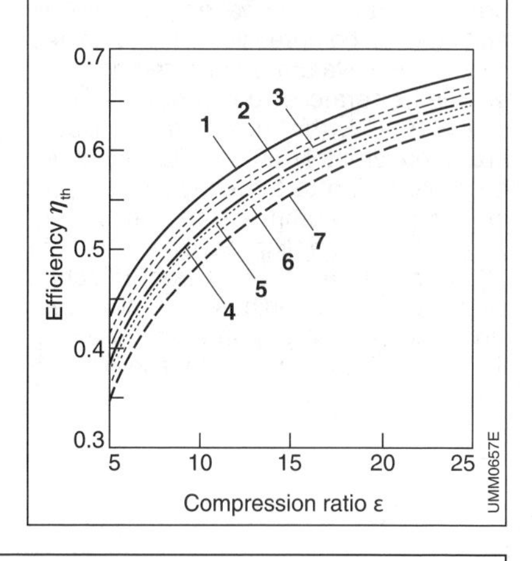

Figure 4: Changes of state of the Seiliger cycle
a) In the pV diagram, b) In the TS diagram.

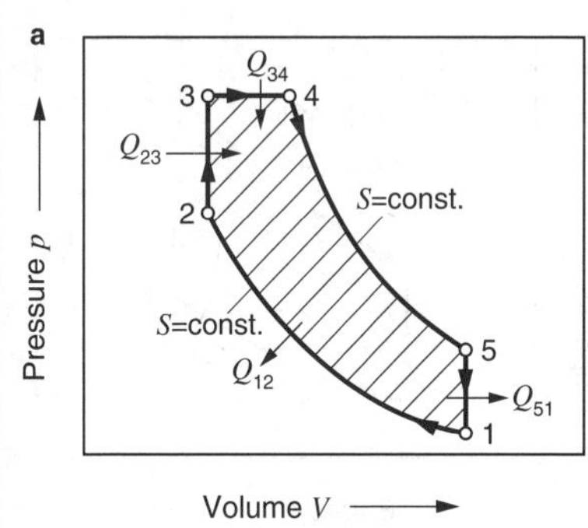

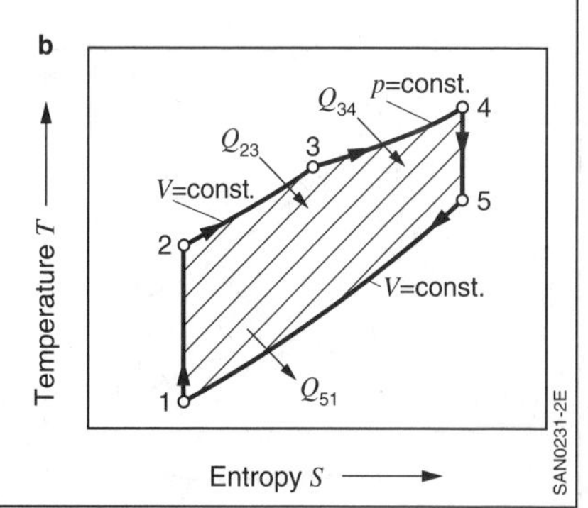

boundary cases are described in further detail in the following.

Otto cycle
The Otto cycle (constant-volume cycle) represents the boundary process of the Seiliger cycle, in which the addition of heat or combustion takes place exclusively isochorically. Compared with the Seiliger cycle, therefore, the cycle step of isobaric heat addition (3 → 4) is absent. Process control in the pV and TS diagrams is illustrated in Figure 6. The following simplified description is obtained for the efficiency of the constant-volume cycle:

$$\eta_{\text{th}} = 1 - \varepsilon^{1-\kappa}.$$

The efficiency increases as the compression ratio ε increases. In gasoline engines the compression ratio is limited by knocking. Naturally aspirated gasoline engines operate at compression ratios of $\varepsilon = 10-12$, the maximum pressures are approximately 60 bar. Turbocharged engines, owing to the knock limit, operate at lower compression ratios, e.g. at $\varepsilon = 9-10$. Peak pressures of up to 120 bar are reached. Direct-injection stratified-charge engines have compression ratios $\varepsilon > 12$; at part load the potential for variable compression ratios is even up to $\varepsilon = 14$.

Diesel cycle
The second boundary process of the Seiliger cycle is the Diesel cycle (constant-pressure cycle). Process control in the pV and TS diagrams is illustrated in Figure 7. In this cycle the addition of heat or combustion takes place isobarically, i.e. the cycle step of isochoric heat addition (2 → 3) is absent. The efficiency of the constant-pressure cycle is obtained as:

$$\eta_{\text{th}} = 1 - \frac{\varphi^{\kappa} - 1}{\varepsilon^{\kappa-1}\kappa\,(\varphi - 1)}.$$

The achieved efficiency is, for the same compression ratio, lower than in the constant-volume cycle ε. However, because diesel engines are typically operated with a higher compression ratio, their efficiency is generally better. The aim of development for diesel engines is therefore to deliver a high peak pressure.

The maximum permissible peak pressures for passenger-car engines are roughly 180 bar, and for commercial-vehicle engines over 220 bar. The compression ratios for the direct-injection processes currently used today are $\varepsilon = 16 - 19$. In the case of retarded start of injection, the final compression pressure actually corresponds to the peak pressure, and the diesel cycle resembles the constant-pressure cycle with $\varphi \approx 9$.

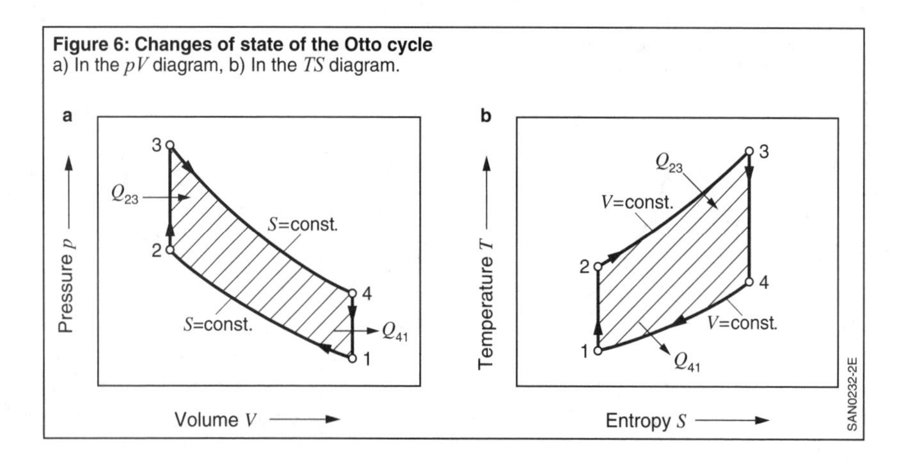

Figure 6: Changes of state of the Otto cycle
a) In the pV diagram, b) In the TS diagram.

Joule cycle
The Joule cycle (or Brayton cycle) is the comparison cycle for gas turbines. In contrast to internal-combustion engines and their comparison cycles, a gas turbine is subject to continuous flow. The Joule cycle is characterized by the following cycle steps (Figure 8):
- reversible adiabatic compression (1→2),
- isobaric heat addition (2→3),
- reversible adiabatic expansion (3→4),
- isobaric heat rejection (4→1).

An important characteristic of the Joule cycles is the pressure ratio $\pi = p_1/p_2$. The efficiency is thus:

$$\eta_{th} = 1 - \frac{T_1}{T_2} = 1 - \pi^{\frac{\kappa-1}{\kappa}}.$$

Clausius-Rankine cycle
The Clausius-Rankine cycle is usually used as the comparison cycle for steam turbines. This is a closed cycle in which the working medium undergoes among others two phase changes (evaporation and condensation). The cycle typically contains the following steps (Figure 9).
- reversible adiabatic pressure increase in the liquid phase (1→2),
- isobaric heat addition (2→3),
- isobaric and isothermal evaporation (3→4),
- isobaric overheating (4→5),
- reversible adiabatic expansion (5→6),
- isobaric and isothermal condensation (6→1).

Figure 7: Changes of state of the Diesel cycle
a) In the pV diagram, b) In the TS diagram.

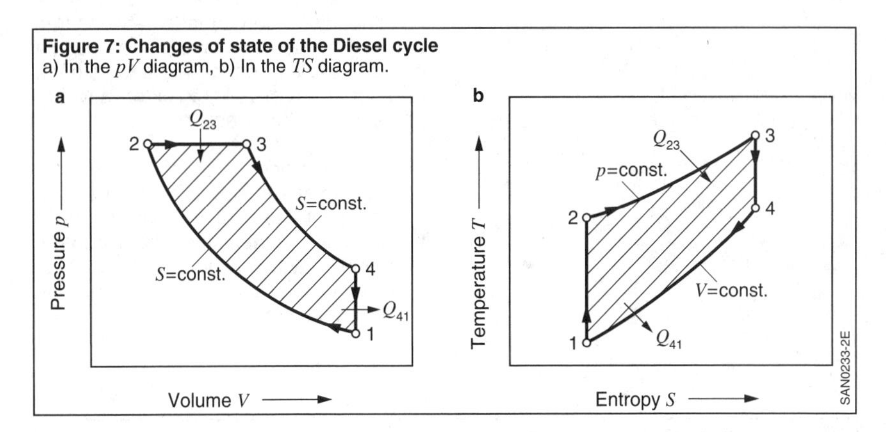

Figure 8: Changes of state of the Joule cycle
a) In the pV diagram, b) In the TS diagram.

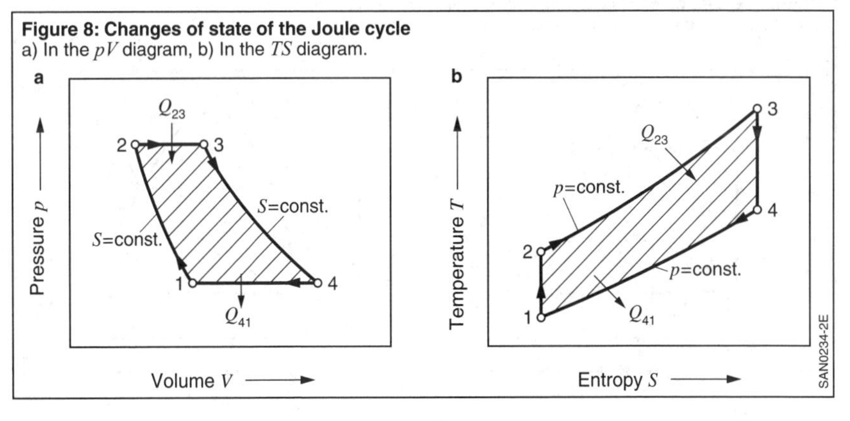

The Clausius-Rankine cycles on account of the phase changes cannot be depicted with an ideal gas; it requires a real-gas description. For the same reason, this produces a depiction which describes benefit and expenditure and thus the thermal efficiency with enthalpies. The thermal efficiency is thus [3]:

$$\eta_{th} = 1 - \frac{h_6 - h_1}{h_5 - h_2}.$$

In automotive applications the Clausius-Rankine cycle is used for example in the field of waste-heat recovery as a comparison cycle. The engine or exhaust-gas waste heat supplies the required heat here for the cycle steps 2 → 5. A turbine or a piston engine for example which makes energy available in mechanical form can be used as the engine for expansion (5 → 6).

Heat transfer

Concept formation

Essentially, heat is transferred in three different ways:

Thermal conduction

In the case of thermal conduction, heat is transferred on the molecular level. Heat is transferred here as a result of an impressed temperature gradient by a solid body or a static medium (liquid or gaseous) without macroscopic matter transport. According to the second law of thermodynamics heat is transferred here in the direction of the lower temperature.

Convective heat transfer

In the case of convective heat transfer, a flow of matter, as exists only in fluid, flowing media (liquids or gases), is required. A further distinction is made between free (caused by natural buoyancy on account of the prevailing temperature gradients) and forced (caused by externally impressed flows) convection.

Thermal radiation

In the case of thermal radiation, energy is transferred in the form of electromagnetic waves. The mechanism of heat transfer from one body to another is not matter- or substance-related and is therefore also possible in a vacuum.

Figure 9: Changes of state of the Clausius-Rankine cycle
a) In the pV diagram, b) In the TS diagram.

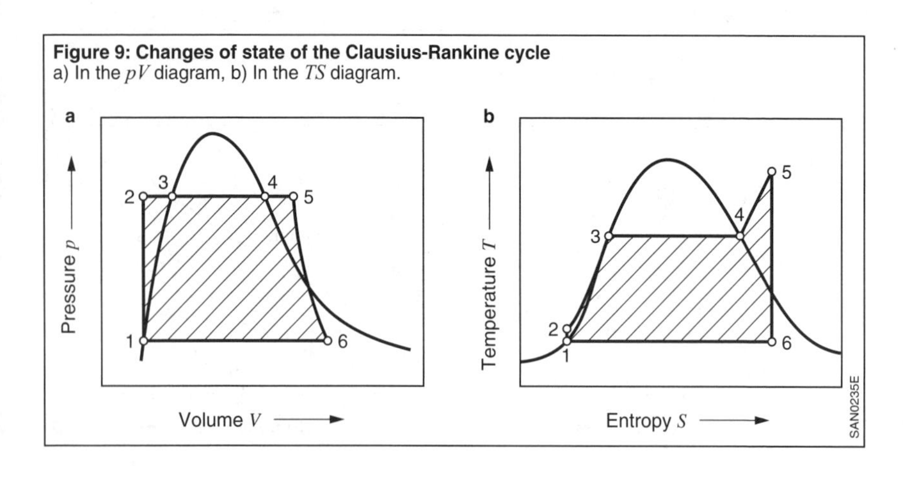

Thermal conduction

Stationary one-dimensional thermal conduction in a simple flat wall

In a homogeneous body with constant cross-sectional area A the heat flow density $\dot{q} = \dot{Q}/A$ in the one-dimensional case is given by:

$$\dot{q} = -\lambda \frac{dT}{dx}.$$

This is the Fourier thermal-conduction equation with the (substance-specific) thermal conductivity λ of the wall. For the case shown in Figure 10 (expansion in the y- and z-directions very high) it follows for a temperature difference $T_1 - T_2$ between the walls:

$$\dot{q} = \frac{\lambda}{d}(T_1 - T_2)$$

and

$$\dot{Q} = \frac{\lambda}{d}(T_1 - T_2)A.$$

By analogy to the science of electricity, the thermal-conduction resistance can be introduced:

$$R_\lambda = \frac{T_1 - T_2}{\dot{Q}} = \frac{d}{\lambda A}.$$

Table 4 shows the values for λ for some materials.

Stationary one-dimensional thermal conduction in a layered flat wall

If the wall is made not of one but of several layered materials of different thicknesses $d_1, \ldots, d_n$ and thermal conductivities $\lambda_1, \ldots, \lambda_n$ (Figure 11), this produces for a temperature difference $T_1 - T_2$ normally for the layering:

$$\dot{q} = \frac{\lambda_R}{d}(T_1 - T_2),$$

with the effective resulting thermal conductivity:

$$\lambda_R = \frac{1}{\frac{d_1}{\lambda_1} + \frac{d_2}{\lambda_2} + \ldots + \frac{d_n}{\lambda_n}}$$

and the thermal resistance:

$$R_{\lambda R} = R_{\lambda 1} + R_{\lambda 2} + \ldots R_{\lambda n}.$$

Table 4: Thermal conductivity λ of some materials at 293.15 K

Material/medium	λ [W/K·m]
Aluminum (Al)	237
Iron (Fe)	81
Copper (Cu)	399
Titanium (Ti)	22
CrNi steel (X12CrNi 18.8)	15
Glass	0.87 – 1.40
Teflon (PTFE)	0.23
Polyvinyl chloride (PVC)	0.15

Figure 10: Stationary one-dimensional thermal conduction through a flat wall

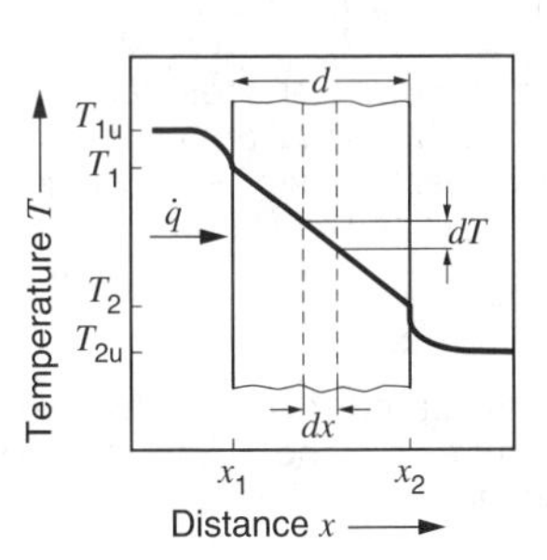

Figure 11: Stationary one-dimensional thermal conduction through a layered, flat wall

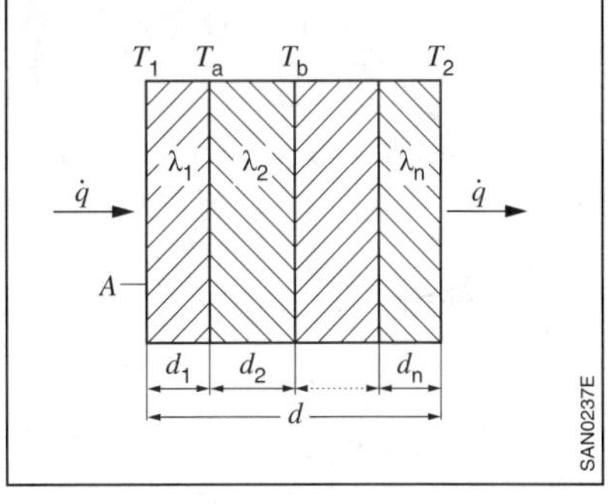

The thermal resistance therefore behaves additively (as in a series connection), with the order of the layers not playing a role.

If the temperature gradient is impressed along the plates (along a length l), it follows that:

$$\dot{q} = \frac{\lambda_p}{l}(T_1 - T_2)$$

with the effective resulting thermal conductivity

$$\lambda_p = \frac{1}{d}(\lambda_1 d_1 + \lambda_2 d_2 + \ldots + \lambda_n d_n)$$

and the thermal-conduction resistance

$$\frac{1}{R_{\lambda p}} = \frac{1}{R_{\lambda 1}} + \frac{1}{R_{\lambda 2}} + \ldots + \frac{1}{R_{\lambda n}}\ .$$

The reciprocal value of the total thermal resistance is therefore made up additively of the reciprocal values of the individual thermal resistances and behaves as in a parallel connection.

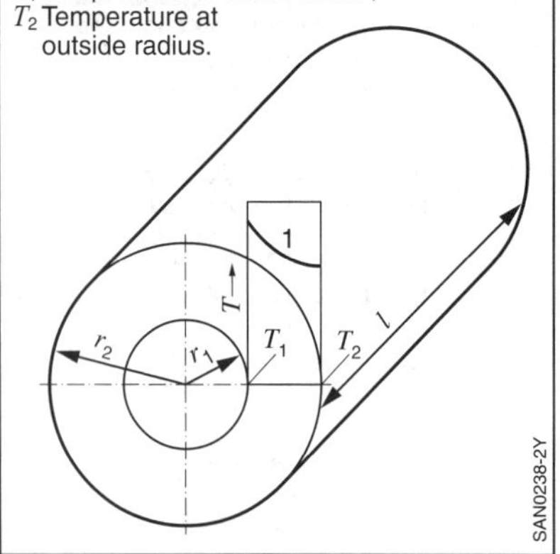

Figure 12: Stationary one-dimensional thermal conduction through a cylindrical tube wall
1 Temperature characteristic over the tube cross-section
r_1 Inside radius of tube,
r_2 Outside radius of tube,
l Length of tube,
T_1 Temperature at inside radius,
T_2 Temperature at outside radius.

Stationary one-dimensional thermal conduction in a tube wall

Similarly to thermal conduction through a wall, the following is obtained for a tube (of virtually infinite length) of length l in the one-dimensional case (thermal conduction only normally to the tube axis) as shown in Figure 12:

$$\dot{Q} = 2\pi l \lambda \frac{T_1 - T_2}{\ln\left(\frac{r_2}{r_1}\right)}\ ,$$

where T_1 is the temperature on the inner side and T_2 the temperature on the outer side.

Because of the different surface areas on the inner and outer sides of the tube the temperature assumes a logarithmic characteristic in the tube wall.

Stationary heat transmission

In many technical applications thermal conduction inside a solid body occurs in combination with a convective heat transfer (see Convective heat transfer). A (convective) heat transfer occurs as well as the thermal conduction, as shown in Figure 10; towards the wall on account of the temperature difference $(T_{1u} - T_1)$ and away from the wall on account of the temperature difference $(T_2 - T_{2u})$. The heat flows must naturally all be of equal magnitude, since heat is not lost in either convective heat transfer or in thermal conduction. The heat flows that occur are governed by:

$$\dot{Q} = \frac{\lambda}{d}(T_1 - T_2)A,$$

$$\dot{Q} = \alpha_{1u}(T_{1u} - T_1)A,$$

$$\dot{Q} = \alpha_{2u}(T_2 - T_{2u})A,$$

with the heat-transfer coefficients α_{1u} and α_{2u}. For such a situation it is possible to create the heat-transmission coefficient k, which is obtained as follows:

$$\frac{1}{k} = \frac{d}{\lambda} + \frac{1}{\alpha_{1u}} + \frac{1}{\alpha_{2u}}.$$

While taking into account the two fluid temperatures T_{1u} and T_{2u}, the following is obtained for the heat flow:

$$\dot{Q} = k(T_{1u} - T_{2u})A.$$

Convective heat transfer

The heat-transfer coefficient α has already been introduced. It is dependent on a variety of influencing variables, such as, for example, temperature, density, geometry, type of flow (laminar or turbulent), and speed. It is determined by the conservation equation (nonlinear partial differential equations) describing the (fluidic) problem for which there are only closed analytical solutions in exceptional cases. Heat-transfer coefficients are therefore often determined on the basis of numerical solutions or directly from experiments. The functional dependencies for the heat-transfer coefficients can be found from these on the basis of dimensional analysis and similarity relationships.

In view of the complexity, reference is made at this point for derivation and further details to the relevant literature (e.g. [1]). The most important dimensionless parameter for heat transfer is the Nusselt number Nu:

$$Nu = \frac{\alpha L}{\lambda}.$$

This numbers contains, in addition to the heat-transfer coefficient α, the thermal conductivity λ of the fluid, and a reference length L, which can be for example a tube diameter or a plate length. It can further be shown that the following functional correlations apply to the averaged, i.e. integrated over the body surface and thus location-independent Nusselt number Nu_∞:

- $Nu_\infty = f(Re, Pr)$ for forced convection,
- $Nu_\infty = f(Gr, Pr)$ for free convection.

The Nusselt number and thus the heat transfer can therefore be described as a function of the dimensionless parameters Reynolds number Re, Prandtl number Pr and Grashof number Gr. These must be calculated in each case as a function of the geometry and the flow (Re and Gr) or as a function of the fluid properties (Pr). The Reynolds number is defined as:

$$Re = \frac{cL}{\nu}.$$

Here c is the flow velocity, L a length characteristic of the application (e.g. tube diameter), and ν the kinematic viscosity of the fluid.

The Prandtl number is a pure substance variable and is given by:

$$Pr = \frac{\nu}{a},$$

with the thermal diffusivity a. For gases $Pr = 0.7$ can be assumed by approximation for pressures below 10 bar.

To calculate the Grashof number Gr, reference is made to the relevant literature (e.g. [1]).

The correlations for some selected cases with forced convection are set out in the following. The correlations in each case for the averaged values together with their range of application are specified here. The heat-transfer coefficients can be calculated with the Nusselt numbers determined from them and the definition of these.

Flat plate with laminar boundary layer subject to longitudinal flow

Nusselt correlation:

$$Nu_\infty = 0.664 Re^{\frac{1}{2}} Pr^{\frac{1}{3}}.$$

Characteristic length:
Plate length L.

Range of application:
$Re \leq 5 \cdot 10^5$; $0.6 \leq Pr \leq 2{,}000$.

Flat plate with turbulent boundary layer subject to longitudinal flow

Nusselt correlation:

$$Nu_\infty = 0.037 Re^{0.8} Pr^{\frac{1}{3}}.$$

Characteristic length:
Plate length L.

Range of application:
$5 \cdot 10^5 \leq Re \leq 10^7$; $0.6 \leq Pr \leq 60$.

Turbulent pipe inner flow
Nusselt correlation:

$$Nu_\infty = 0.023\,Re^{0.8}\,Pr^n,$$

with $n = 0.4$ for $T_W > T_F$
and $n = 0.3$ for $T_W < T_F$.

Here, T_W is the wall temperature and T_F the averaged fluid temperature. The Nusselt number must be evaluated for the averaged fluid temperature because this changes during an inner flow from inlet to outlet.

Characteristic length:
Tube diameter D.

Range of application:
$10^4 \le Re$; $0.7 \le Pr \le 120$.

The tube length must be at least ten times the tube diameter.

The above correlation can also be used for tubes with non-circular cross-sections; the equivalent or hydraulic diameter d_h, which is created as follows, is used here:

$$d_h = \frac{4A}{U},$$

with the through-flow cross-sectional area A and the circumference U.

Thermal radiation
Thermal radiation depends only on the type and the temperature of the radiating body. When radiation strikes a body, the following phenomena are observed: Part of the incident radiation is reflected (reflection), part is absorbed (absorption), and part is let through (transmission). Due to the conservation of energy, it follows that the sum of these three energy components is equal to the amount of energy of the incident radiation.

According to the Stefan-Boltzmann law, the heat flow radiated from a body with surface A at temperature T is:

$$\dot{Q} = \varepsilon \sigma A T^4.$$

with the Stefan-Boltzmann constant σ,

$$\sigma \approx 5.67 \cdot 10^{-8} \frac{\mathrm{W}}{\mathrm{m^2K^4}},$$

and the emissivity ε of the body. The emissivity is between 0 (total reflection) and 1 (black-body radiator) and depends among others things on the temperature and the surface condition of the body. Some emissivity values are given by way of example in Table 5.

References
[1] H.D. Baehr, K. Stephan: Wärme- und Stoffübertragung. Springer, 7th Edition, 2010.
[2] H.D. Baehr, S. Kabelac: Thermodynamik: Grundlagen und technische Anwendungen. Springer, 15th Edition, 2012.
[3] E. Hahne: Technische Thermodynamik: Einführung und Anwendung. Oldenbourg Wissenschaftsverlag, 5th Edition, 2010.
[4] B. Weigand, J. Köhler, J. von Wolfersdorf: Thermodynamik kompakt. Springer Vieweg, 3rd Edition, 2013.

Table 5: Emissivity ε of some materials (values in the range up to 300 °C)

Material	ε
Black-body radiator	1.00
Aluminum, unmachined	0.07
Aluminum, polished	0.04
Cast iron, rough, oxidized	0.94
Cast iron, turned	0.44
Copper, oxidized	0.64
Copper, polished	0.05
Brass, matt	0.22
Brass, polished	0.05
Steel, matt, oxidized	0.96
Steel, polished, oil-free	0.06
Steel, polished, oiled	0.40

Electrical engineering

Electromagnetic fields

Electrical engineering deals with electromagnetic fields and their effects. These fields are associated with electric charges (in each case an integral multiple of the electric elementary charge). Physics does not indicate whether the fields are the cause or the effect. Static charges produce an electric field, whereas moving charges give rise to a magnetic field as well. The association of electric and magnetic fields with static and moving charges is described by Maxwell's equations [1].

The presence of these fields is evidenced by the effects of their forces on other electric charges. The force on a point charge Q in an electric field is called Coulomb force. It causes a repulsion between charges of the same name. For two point charges Q_1 and Q_2 in free space at a distance a, it is:

$$F=\frac{Q_1 Q_2}{4\pi\varepsilon_0 a^2}\ .$$

$\varepsilon_0 \approx 8.854\cdot 10^{-12}$ F/m is the electric field constant, also called the dielectric constant of fee space.

The force acting on a moving charge in a magnetic field is expressed by the Lorentz force. This is responsible for two parallel conductors which carry rectified currents I_1 and I_2 being mutually attracting. Over a length l in free space at conductor distance a the force of attraction between the two conductors is:

$$F=\frac{\mu_0 I_1 I_2 l}{2\pi a}\ .$$

$\mu_0 \approx 1.257\cdot 10^{-6}$ H/m stands for the magnetic field constant, also called the permittivity of free space.

Electric field

The force acting on a static electric charge is ascribed to the effect of an electric field. An electrostatic field can be defined by the following quantities.

Electric potential φ (P) and voltage U
The electric potential φ(P) at point P is a measure of the work required per charge to move the charge Q from a reference point to point P:

$$\varphi(\mathrm{P})=\frac{W(\mathrm{P})}{Q}\ .$$

The voltage U is the potential difference (using the same reference point) between two points P_1 and P_2:

$$U=\varphi(\mathrm{P}_1)-\varphi(\mathrm{P}_2)\ .$$

Electric field strength E
The electric field strength E at point P depends on the location and its surrounding charges. It defines the maximum slope of the potential gradient at point P. The following applies to the field strength at a distance a from a positive point charge Q_1: It is directed away from the charge Q_1 and has the value

$$E=\frac{Q_1}{4\pi\varepsilon_0 a^2}\ .$$

Acting on a positive charge Q_2 at point P is a force in the direction of the electric field strength of the value

$$F=Q_2 E\ .$$

Electric field and matter
In a material which can be polarized (dielectric), an electric field generates electric dipoles (positive and negative charges $\pm Q$ at a distance a; $Q\,a$ is called the dipole moment). The dipole moment per unit volume is called the polarization M. The displacement density D indicates the density of the electric displacement flux, and is defined as follows:

$$D=\varepsilon E=\varepsilon_0 \varepsilon_r E=\varepsilon_0 E+M\ .$$

$\varepsilon = \varepsilon_0 \varepsilon_r$ is the dielectric constant of the material, ε_0 the electric field constant (dielectric constant of vacuum), ε_r the permittivity (relative dielectric constant). For air, $\varepsilon_r = 1$; see Insulating materials for further values.

The quantity

$$w_e = \frac{1}{2} E D$$

is the electrical energy density. When multiplied by the volume, it produces the electrical energy W_e.

Capacitor

Two metal bodies (electrodes) separated by a dielectric form a capacitor. When a voltage is applied to the capacitor, the two electrodes receive equal but opposite charges. The following equation holds for the received charge Q:

$$Q = C U.$$

C is the capacitance of the capacitor. It is dependent on the geometric shape of the electrodes, the distance by which they are separated, and the dielectric constant of the dielectric. Table 2 sets out the capacitances of selected arrangements.

Table 1: Quantities and units
(Additional quantities and units in the text).

Quantity		SI unit
A	Area	m^2
a	Distance	m
B	Magnetic flux density, induction	$T = Wb/m^2 = V s/m^2$
C	Electrical capacitance	F = C/V
D	Electric flux density, electric displacement	C/m^2
E	Electric field strength	V/m
F	Force	N
f	Frequency	Hz
G	Electrical conductance	S = 1/Ω
G	Antenna gain	dB
H	Magnetic H field strength	A/m
I	Electric current	A
J	Magnetic polarization	T
k	Electrochemical equivalent	kg/C (usually: g/C)
L	Inductance	H = Wb/A = V s/A
l	Length	m
M	Electric polarization	C/m^2
P	Power	W = V A
P_s	Apparent power	V A
P_q	Reactive power	var = V A
Q	Quantity of electricity, electric charge	C = A s
q	Cross-sectional area	m^2
R	Electrical resistance	Ω = V/A
T	Temperature	K
t	Time	s
r	Radius	m
r	Reflectance factor	–
S	Electromagnetic power density	W/m^2
s	Standing wave ratio	–
U	Voltage	V
V	Magnetic potential difference	A
W	Work, energy	J = W s
W_e	Electrical energy	Ws
W_m	Magnetic energy	Ws
w_e	Electrical energy density	Ws/m^3
w_m	Magnetic energy density	Ws/m^3
w	Number of turns	–
Z	Characteristic impedance	Ω
α	Geometric angle	° (degrees)
ε	Dielectric constant	F/m = C/(V m)
ε_0	Electric field constant $\approx 8.854 \cdot 10^{-12}$ F/m	
ε_r	Relative permittivity	–
λ	Wavelength	m
Θ	Current linkage	A
μ	Permeability	H/m = V s/(A m)
μ_0	Magnetic field constant $\approx 1.257 \cdot 10^{-6}$ H/m	
μ_r	Relative permeability	–
ρ	Resistivity	$\Omega m = 10^6 \Omega mm^2/m$
σ	Conductivity (= $1/\rho$)	$1/(\Omega m) = 10^{-6} m/(\Omega mm^2)$
Φ	Magnetic flux	Wb = V s
φ	Phase displacement angle	° (degrees)
φ (P)	Potential at point P	V
ω	Angular frequency (= $2 \pi f$)	Hz

Capacitance of capacitors connected in series and parallel:

$$\frac{1}{C_{\text{total}}}=\frac{1}{C_1}+\frac{1}{C_2}$$

(series connection, Figure 1a),

$$C_{\text{total}}=C_1+C_2$$

(parallel connection, Figure 1b).

In the case of a parallel-plate capacitor (which also includes a wound capacitor), the inner plates of two capacitors connected in parallel act as electrodes (Figure 2).

The energy content of a charged capacitor (charge Q, voltage U, capacitance C) is:

$$W=\tfrac{1}{2}QU=\frac{Q^2}{2C}=\tfrac{1}{2}CU^2 .$$

Figure 1: Connection of capacitors
a) Series connection,
b) Parallel connection.

Direct current and direct voltage

Moving charges give rise to a current I, which is characterized by its intensity and measured in amperes. The direction of flow and magnitude of direct current are independent of time. The electrical system in a motor vehicle is, on the other hand, a direct-voltage system; its voltage is independent of time. The currents in a vehicle electrical system are usually time-dependent. In many cases time-dependent currents such as direct current can also be handled.

Current direction and measurement

Current flowing from positive pole to negative pole outside of the current source is designated as positive (in reality, the electrons travel from the negative to the positive

Figure 2: Structure of a parallel-plate capacitor

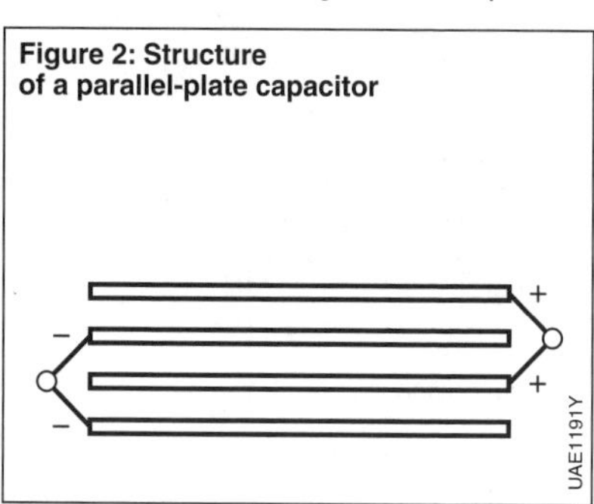

Table 2: Capacitance C of some conductor arrangements

Plate capacitor with n parallel plates	$C=(n-1)\dfrac{\varepsilon_r\,\varepsilon_0\,A}{a}$	ε_r, ε_0 n A a	Permittivity, electric field constant Number of plates Surface area of one plate Distance between plates
Parallel conductors (twin conductors)	$C=\dfrac{\pi\,\varepsilon_r\,\varepsilon_0\,l}{\ln\left(\dfrac{a+\sqrt{a^2-4r^2}}{2r}\right)}$	l a r	Length of twin conductors Distance between conductors Conductor radius
Concentric conductor (cylindrical capacitor)	$C=\dfrac{2\pi\,\varepsilon_r\,\varepsilon_0\,l}{\ln(r_2/r_1)}$	l r_2, r_1	Length of conductor Conductor radius where $r_2>r_1$
Conductor to ground	$C=\dfrac{2\pi\,\varepsilon_r\,\varepsilon_0\,l}{\ln\left(\dfrac{a+\sqrt{a^2-r^2}}{r}\right)}$	l a r	Length of conductor Distance from conductor to ground Conductor radius
Sphere with respect to distant surface	$C=4\pi\,\varepsilon_r\,\varepsilon_0\,r$	r	Sphere radius

pole). Current measurement is performed by an ammeter (A) in the current path; voltage measurement by a voltmeter (V) connected in shunt (Figure 3).

The counting direction of voltage and current identified with the arrow can in principle be selected independently of the operating state. The accordingly oriented voltages and currents are thereby assigned a sign which is dependent on the operating state. In the case of positive voltages and currents, the arrow indicates the direction from positive (+) to negative (–).

Ohm's law

Ohm's law defines the relationship between voltage U and current I in solid and liquid conductors. The following applies:

$$U = RI\,.$$

The proportionality constant R is called ohmic resistance, and is measured in ohms (Ω). The reciprocal of resistance is called conductance G, and is measured in siemens (S).

$$G = \frac{1}{R}\,.$$

Ohmic resistance

Ohmic resistance depends on the material and its dimensions. For a solid wire:

$$R = \frac{\rho l}{q} = \frac{l}{q\sigma}\,.$$

Figure 3: Current and voltage measurement
R Load,
A Ammeter in current path,
V Shunt-connected voltmeter.

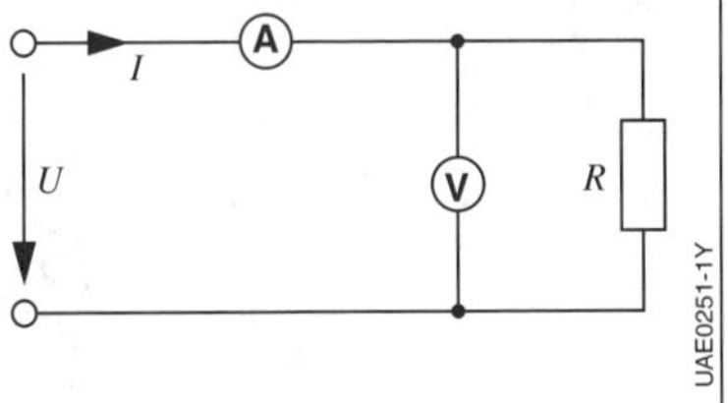

For a tube (from inside to outside):

$$R = \ln\left(\frac{r_2}{r_1}\right)\frac{1}{2\pi l\sigma}\,.$$

The equation's elements are:
ρ resistivity [$\Omega\,\text{mm}^2$/m],
$\sigma = 1/\rho$ conductivity [m/($\Omega\,\text{mm}^2$)],
l wire length, tube length [m],
q wire cross-section [mm^2],
r_2, r_1 outside and inside radii of tube [m] with $r_2 > r_1$.

In the case of metals, resistance increases with temperature. The following applies:

$$R_T = R_{20}[1 + \alpha(T - 20\,°\text{C})]$$

where
R_T resistance at T,
R_{20} resistance at 20 °C,
α temperature coefficient [1/K] (=[1/°C]),
T temperature [°C].

Work and power

In a resistor through which current passes, the following holds for the energy converted during time t into heat or another form of energy (with ohmic resistance R, voltage U and current I):

$$W = UIt = RI^2t$$

and thus for power:

$$P = UI = RI^2\,.$$

Kirchhoff's laws

First law: Current law

The sum of currents (according to their counting direction) flowing into each junction (node) is equal to the sum of currents flowing out of that junction.

Second law: Voltage law

For each closed loop of a conductor network the sum of the component voltages oriented in the direction of the loop at the individual elements (resistors and sources) is equal to the sum of the component voltages oriented against the direction of the loop.

Direct-current circuits

Circuit with load

In the circuit shown in Figure 4:

$$U=(R_a+R_l)I$$

where
R_a ohmic resistance of load,
R_l line resistance.

Battery-charging circuit

$$U-U_0=(R_v+R_i)I \quad \text{(Figure 5)}$$

Where
U line voltage,
U_0 open-circuit voltage (electromotive force) of battery,
R_v series resistance,
R_i internal resistance of battery.

Condition for charging: $U>U_0$ (charging voltage greater than battery open-circuit voltage).

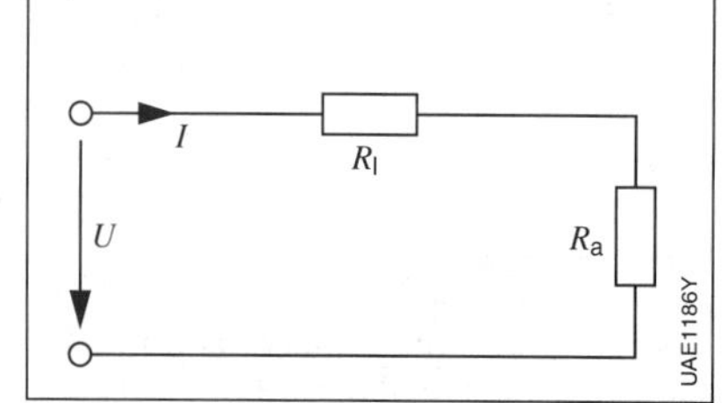

Figure 4: Circuit with load
U Voltage,
I Current,
R_a Load resistance,
R_l Line resistance.

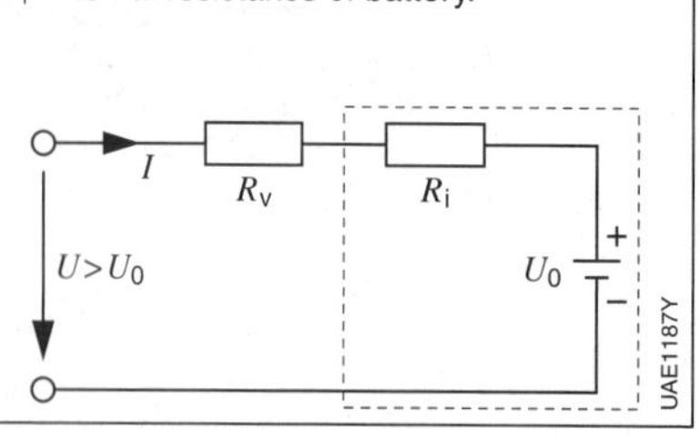

Figure 5: Battery-charging circuit
U Line voltage,
U_0 Open-circuit voltage of battery,
R_V Series resistance,
R_i Internal resistance of battery.

Series connection of resistors

$$R_{total}=R_1+R_2 \quad \text{(Figure 6)},$$

$$U=U_1+U_2\,.$$

The same current I flows in all the resistors (Kirchhoff's current law).

Parallel connection of resistors

$$\frac{1}{R_{total}}=\frac{1}{R_1}+\frac{1}{R_2} \quad \text{or} \quad G_{total}=G_1+G_2\,;$$

$$I=I_1+I_2;\ \frac{I_1}{I_2}=\frac{R_2}{R_1} \quad \text{(Figure 7)}.$$

Voltage U is the same across all the resistors (Kirchhoff's voltage law).

Measurement of a resistance

A resistance can be measured by means of current and voltage measurement, and by using direct-reading ohmmeters or measuring bridges. Measuring bridges are used for example in pressure sensors to connect strain gages.

Figure 6: Series connection of resistors
U Voltage,
I Current,
R Resistance.

R_1 R_2 U_1 U_2 U I UAE1188Y

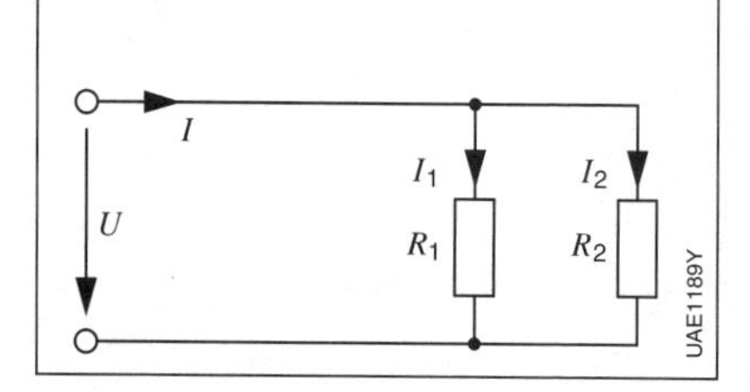

Figure 7: Parallel connection of resistors
U Voltage,
I Current,
R Resistance.

A wide used method of measurement, especially of very small resistances, is four-wire measurement (Figure 8). The measuring contacts for the test piece are configured in double form so as to prevent the resistances of the contacts from being measured as well. Only the small measured current flows through the voltage contacts to the voltmeter. The voltage drop caused by the large source current of the current source at the current contacts is no longer recorded.

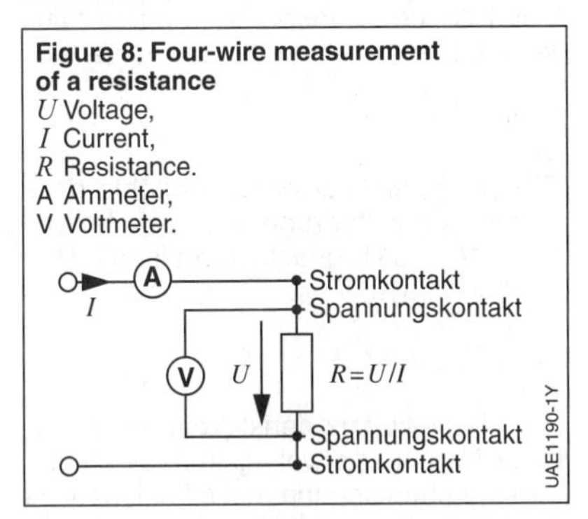

Figure 8: Four-wire measurement of a resistance
U Voltage,
I Current,
R Resistance.
A Ammeter,
V Voltmeter.

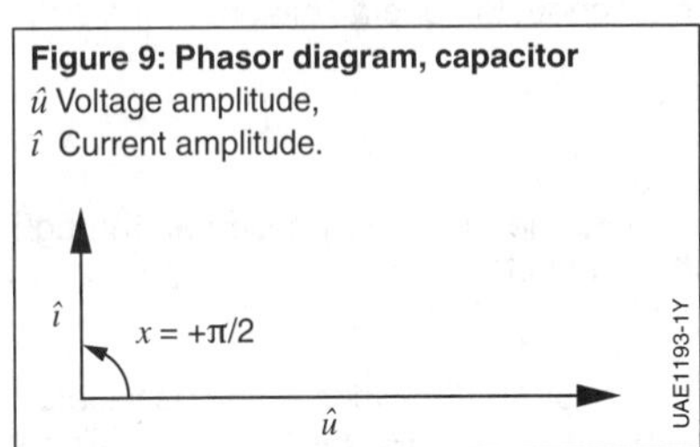

Figure 9: Phasor diagram, capacitor
$\hat{u}$ Voltage amplitude,
$\hat{\imath}$ Current amplitude.

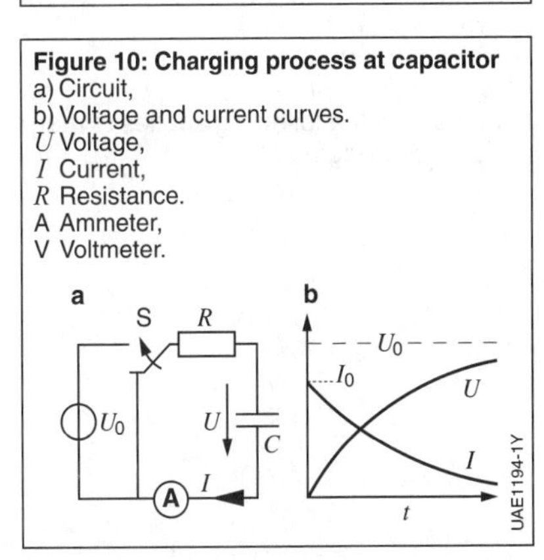

Figure 10: Charging process at capacitor
a) Circuit,
b) Voltage and current curves.
U Voltage,
I Current,
R Resistance.
A Ammeter,
V Voltmeter.

Time-dependent current

The relationships specified for direct current and direct voltage also apply in more or less modified form to time-dependent currents and voltages. In particular, Ohm's law and both of Kirchhoff's laws continue to apply unchanged.

Charging and discharging a capacitor

When a current I flows through a capacitor, its charge Q changes. The following applies:

$$I = \frac{dQ}{dt}\ .$$

In this way, there is a connection between current I and voltage U at the capacitor with capacitance C:

$$I = \frac{d(CU)}{dt}\ .$$

Here current I is supplied to that capacitor plate which has charge Q and from which voltage U is measured at the capacitor.

Examples
(C is constant in terms of time):

- Direct voltage U:
  → current $I = 0$.
- Initial voltage U_0, direct current I:
  → voltage $U = U_0 + It$.
- Voltage harmonic (sinusoidal) with $U = \hat{u}\sin(\omega t)$:
  → current cosinusoidal with
  $I = \omega C\hat{u}\cos(\omega t) = \omega C\hat{u}\sin(\omega t + \frac{\pi}{2})$.

The quantity $\omega = 2\pi f$ is called the angular frequency in relation to frequency f, $\hat{u}$ is the voltage amplitude, and $\hat{\imath} = \omega C\hat{u}$ the current amplitude. The effective values $u_{\text{eff}} = \hat{u}/\sqrt{2}$ and $i_{\text{eff}} = \hat{\imath}/\sqrt{2}$ are also used instead of the amplitudes.

In the case of harmonic excitation, the current at the capacitor is therefore phase-displaced in relation to the voltage by the angle $\varphi = +\pi/2$ (it "leads"). This property is illustrated in a phasor diagram (Figure 9).

A further special case arises when the capacitor is charged via a resistance by a direct-voltage source U_0 (Figure 10) or discharged via a resistance. The time constant $\tau = RC$ is the decisive factor in the charging and discharging of a capacitor.

Charging process

$$I=\frac{U_0}{R}e^{-\frac{t}{\tau}},$$

$$U=U_0(1-e^{-\frac{t}{\tau}}).$$

Discharging process

$$I=\frac{U_0}{R}e^{-\frac{t}{\tau}},$$

$$U=U_0\,e^{-\frac{t}{\tau}}.$$

U_0: charging voltage or voltage at start of discharge,
I charging or discharging current,
$I_0=U_0/R$ current at start of charge,
R charging or discharging resistance,
U capacitor voltage.

Charging and discharging currents have opposite directions.

Figure 11: Hysteresis loop (e.g. hard ferrite)
1 Rise path,
2, 3 Demagnetization curves.
H Magnetic field strength,
B Magnetic flux density,
J Magnetic polarization,
J_s Saturation polarization,
B_r Remanence,
H_{cB}, H_{cJ} Coercive field strength,
H_G Limiting field strength.

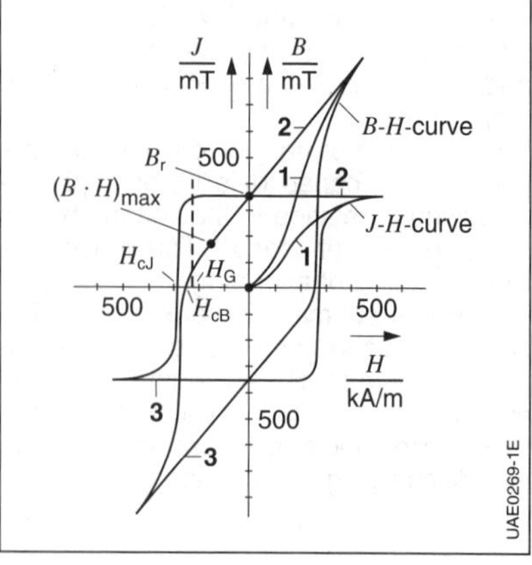

Magnetic field

Magnetic fields are produced by moving electric charges, current-carrying conductors, magnetized bodies, or by an alternating electric field. They can be detected by their effect on moving electric charges (Lorentz force) or magnetic dipoles (like poles repel, and unlike poles attract).

Magnetic fields are characterized by the vector of the magnetic flux density B (induction). A straight conductor carrying current I_1 at distance a generates a magnetic flux density directed about it of the value

$$B=B_1=\frac{\mu_0 I_1}{2\pi a}.$$

This magnetic flux density effects on a second conductor running in parallel with current I_2 equally directed over length l the force of attraction

$$F=B_1 I_2 l.$$

The magnetic flux density can be determined by means of voltage measurement, in that a changing magnetic field induces in a conductor loop a voltage:

$$U=\frac{d\Phi}{dt}$$

where
$d\Phi$ change in the magnetic flux through the conductor loop,
dt change in time.

The magnetic flux density B is associated with the other field quantities as follows (q cross-section):
Magnetic flux $\Phi=Bq$.

In free space the following applies to magnetic field strength H:

$$H=\frac{B}{\mu_0}.$$

Magnetic field and matter

In matter, induction B theoretically consists of two components. One component comes from the applied field ($\mu_0 H$), and the other from the matter (J) (see also the relationship between electric displacement density and electric field strength)

$$B=\mu_0 H+J$$

J is the magnetic polarization and describes that component of flux density contributed by the matter. In physical terms, J corresponds to one magnetic dipole moment per unit volume and is generally a function of field strength H. For many materials $J >> \mu_0 H$ and is proportional to H. Thus:

$$B=\mu_r \mu_0 H$$

with relative permeability μ_r; in free space it has the value $\mu_r=1$.

The quantity

$$w_m=\frac{1}{2} B H$$

is called magnetic energy density. When multiplied by the volume, it produces the magnetic energy W_m.

Materials are divided into three groups according to their relative permeability values:

Diamagnetic materials

μ_r is independent of magnetic field strength and smaller than 1; the values are within the range:
$(1-10^{-5})<\mu_r<(1-10^{-11})$
(e.g. Ag, Au, Cd, Cu, Hg, Pb, Zn, water, organic materials, gases).

Paramagnetic materials

μ_r is independent of magnetic field strength and greater than 1; the values are within the range:
$(1+10^{-8})<\mu_r<(1+10^{-4})$
(e.g. O_2, Al, Pt, Ti).

Ferromagnetic materials

The magnetic polarization in these materials is very high, and its change as a function of the field strength H is nonlinear; it is also dependent on hysteresis. Nevertheless, if, as is usual in electrical engineering, the relationship $B=\mu_r \mu_0 H$ is chosen, then μ_r is a function of H and exhibits hysteresis; the values for μ_r are within the range $10^2<\mu_r<5 \cdot 10^5$
(e.g. Fe, Co, Ni, ferrites).

Hysteresis loop

The hysteresis loop (Figure 11), which illustrates the relationship between B and H as well as J and H, is explained as follows: If the material is in the unmagnetized state ($B=J=0$, $H=0$) when a magnetic field H is applied, the magnetization follows the rise path (1). From a specific, material-dependent field strength, all magnetic dipoles are aligned and J reaches the value of saturation polarization J_s (material-dependent) which can no longer be increased. If H is now reduced, J decreases along section (2) of the curve and at $H=0$ intersects the B or J axis at the remanence point B_r or J_r (in which case $B_r=J_r$). The flux density and polarization drop to zero only on application of an opposing field whose field strength is H_{cB} or H_{cJ}; this field strength is called the coercive field strength.

As the field strength of the opposing field is further increased, saturation polarization is reached in the opposite direction. If the field strength is again reduced and the field reversed, curve (3), which is symmetrical to curve section (2), is traversed.

The following are usually tabulated as the most important characteristic values of a ferromagnetic material:

- Saturation polarization J_s
- Remanence B_r (residual induction for $H=0$)
- Coercive field strength H_{cB} (demagnetizing field strength where B becomes equal to 0)
- Coercive field strength H_{cJ} (demagnetizing field strength where J becomes equal to 0, of significance only for permanent magnets)
- Limiting field strength H_G (a permanent magnet remains stable up to this field strength)
- Maximum small-signal permeability μ_{max} (maximum slope of the rise path; significant only for soft magnetic materials)
- Hysteresis loss (energy converted into heat per volume during one remagnetizing cycle, corresponds to the area of the B–H hysteresis loop; significant only for soft magnetic materials).

Ferromagnetic materials

Ferromagnetic materials are divided into soft and permanent magnetic materials. What must be emphasized is the immense range of eight powers of ten covered by the coercive field strength.

Permanent-magnet materials

Permanent-magnet materials have high coercive field strengths; the values lie within the range

$$H_{cJ} > 1 \frac{kA}{m}$$

Thus high demagnetizing fields H may occur without the material losing its magnetic polarization. The magnetic state and operating range of a permanent magnet lie within the 2nd quadrant of the hysteresis loop, on the demagnetization curve.

In practice, the operating point of a permanent magnet never coincides with the remanence point, because magnetic polarization of the permanent magnet in its interior always causes a magnetic field, the demagnetizing field, which shifts the operating point into the 2nd quadrant.

The point on the demagnetization curve at which the product BH reaches its maximum value $(BH)_{max}$ is a measure of the maximum attainable air-gap energy. In addition to remanence and coercive field strength, this value is important for characterizing permanent magnets.

AlNiCo, ferrite, FeNdB (REFe), and SeCo magnets are currently the most important types of permanent magnets in terms of industrial applications; their demagnetization curves (Figure 12) exhibit characteristics typical of the individual magnet types.

Soft magnetic materials

Soft magnetic materials have a low coercive field strength

$$H_{cJ} < 1 \frac{kA}{m},$$

i.e. a narrow hysteresis loop. The flux density assumes high values (large μ_r values) already for low field strengths so that, in customary applications, $J >> \mu_0 H$, i.e. in practice no distinction need be made between $B(H)$ and $J(H)$ curves.

Due to their high induction at low field strengths, soft magnetic materials are used as conductors of magnetic flux. As they exhibit low remagnetization losses (hysteresis loss), materials with low coercive field strengths are particularly well-suited for applications in alternating magnetic fields.

The characteristics of soft magnetic materials depend essentially on their pretreatment. Machining increases the coercive field strength, i.e. the hysteresis loop becomes broader. The coercive field strength can be subsequently reduced to its initial value through material-specific annealing at high temperatures (magnetic

Figure 12: Demagnetization curves for various permanent-magnet materials
1 AlNiCo 52/6, 2 REFe 220/140, 3 AlNiCo 60/11,
4 SECo 112/100, 5 AlNiCo 30/10, 6 SECo 70/70p, 7 PICo 60/40,
8 MnAl, 9 Hard ferrite 25/25.

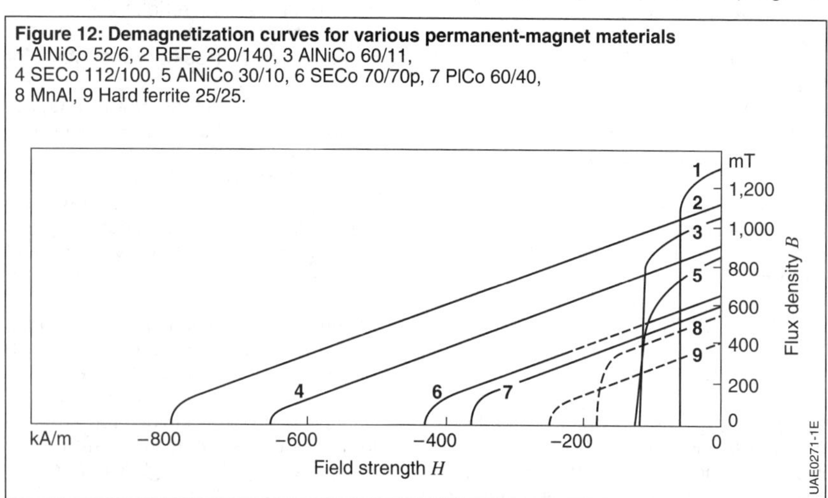

final annealing). The magnetization curves, i.e. the B-H relationships, are set out in Figure 13 for several important soft magnetic materials.

Remagnetization losses

In Table 3 P1 and P1.5 represent the remagnetization loss for inductions of 1 and 1.5 tesla respectively, in a 50 Hz field at 20 °C. These losses are composed of hysteresis losses and eddy-current losses. The eddy-current losses are caused by electric fields which are induced (law of induction) in the magnetically soft components of the magnetic circuit as a result of changes in flux during alternating-field magnetization. Eddy-current losses can be kept low by applying the following measures to reduce electric conductivity:

- Lamination of the core
- Use of alloyed materials (e.g. silicon iron)
- Use of insulated powder particles (powdered cores) in the higher frequency range
- Use of ceramic materials (ferrites).

Figure 13: Magnetization curves for soft magnetic materials
1 Pure iron, 2 78 NiFe (Permalloy), 3 36 NiFe,
4 Ni-Zn ferrite, 5 50 CoFe, 6 V360-50A (electrical sheet steel),
7 Structural steel, 8 Cast iron, 9 Powder Fe core.

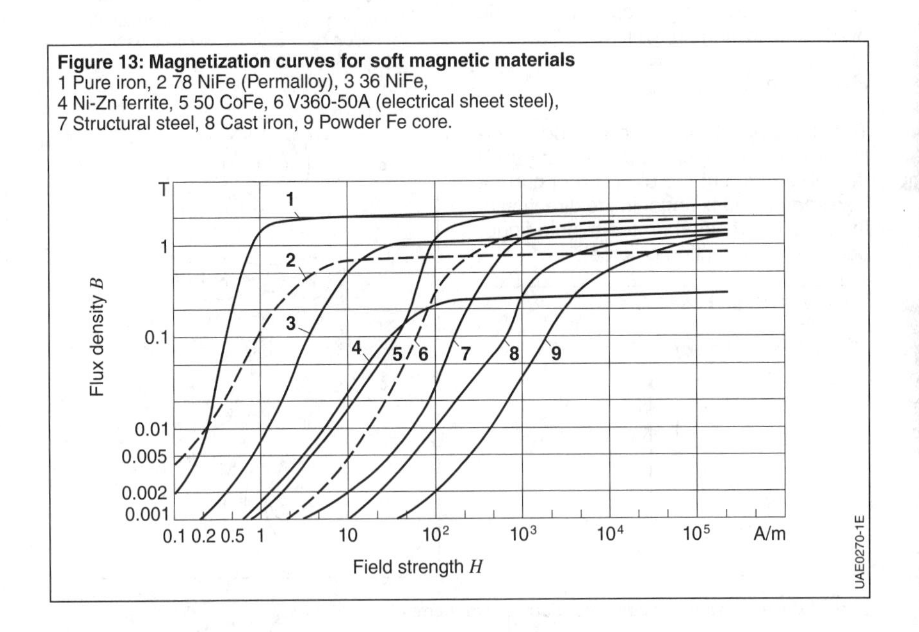

Table 3: Remagnetization losses

Sheet grade	Nominal thickness mm	Remagnetization loss W/kg P1	P 1.5	B (for H = 10 kA/m) T
M 270 – 35 A	0.35	1.1	2.7	1.70
M 330 – 35 A	0.35	1.3	33.3	1.70
M 400 – 50 A	0.5	1.7	4.0	1.71
M 530 – 50 A	0.5	2.3	5.3	1.74
M 800 – 50 A	0.5	3.6	8.1	1.77

Magnetic field and electric current

Moving charges are associated with a magnetic field, i.e. conductors through which current flows are surrounded by a magnetic field. The direction in which the current flows (⊗ current flow into the page, ⊙ current flow out of the page) in the case of positive currents and the direction of the magnetic field strength form a right-handed screw. Table 4 shows the magnetic field strength of some conductor configurations.

In a magnetic field with flux density B a force is exerted on a current-carrying conductor (current I) of length l. If the conductor and the magnetic field form an angle of α, the following applies:

$$F = B I l \sin\alpha .$$

The direction of this force can be determined using the right-hand rule (Figure 14) (thumb pointed in the direction of current flow and the index finger in the direction of the magnetic field, then the middle finger points in the direction of force).

Figure 14: Right-hand rule
I Current,
B Magnetic flux density,
F Force.

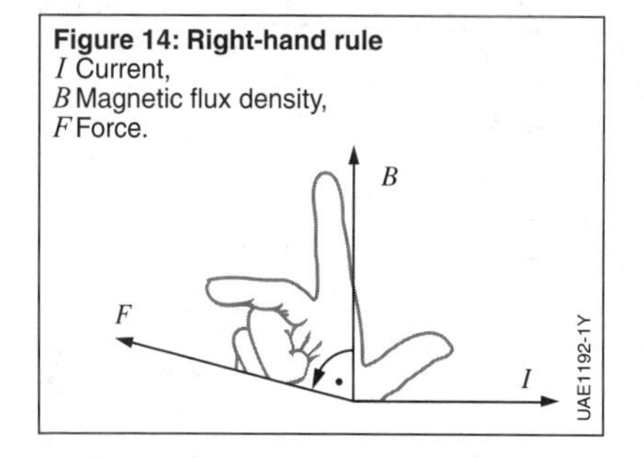

Law of induction

Any change in the magnetic flux Φ around which there is a conductor loop, caused for example by movement of the loop or changes in field strength, induces a voltage U_i at the terminals of the conductor

Figure 15: Current-carrying conductors and the associated magnetic lines of force
a) Single current-carrying conductor with magnetic field,
b) Parallel conductors, current flow in the same direction (conductors attract each other),
c) Parallel conductors, current flow in opposite directions (conductors repel each other),
d) A magnetic field of flux density B exerts a force on the current-carrying conductor (direction of force is determined using the right-hand rule).

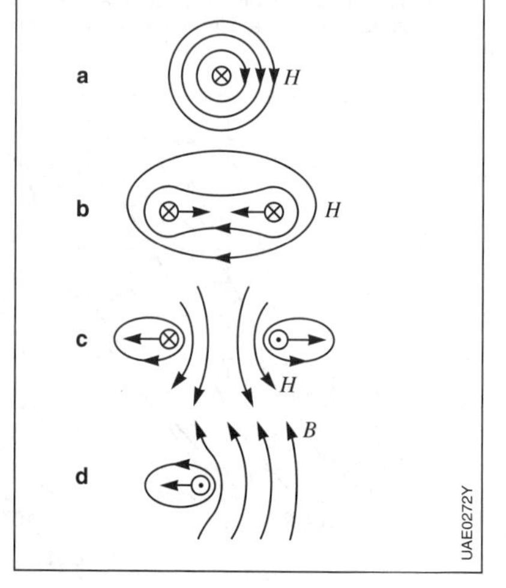

Table 4: Field strength of some conductor arrangements

Arrangement	Formula	Symbol	Meaning
Cylindrical coil, inner	$H = \frac{I w}{l}$	w I l	Number of turns Current through coil [A] Length of coil [m]
Straight wire in air, outside	$H = \frac{I}{2\pi r}$	r I	Distance from wire axis [m] Current through coil [A]
Straight wire, inside	$H = \frac{I r}{2\pi a^2}$	a r I	Wire radius [m] Distance from wire axis [m] Current through coil [A]
Circular wire, in center of circle	$H = \frac{I}{2a}$	a I	Radius of circle [m] Current through wire [A]

loop. Thus, a voltage U_i is induced also in a (de-energized) conductor moving in direction v in the magnetic field (Figure 16):

$$U_i = B v l$$

where
B Magnetic flux density,
l conductor length,
v velocity.

In a direct-current machine

$$U_i = 2 \frac{f z \Phi}{a},$$

where
U_i induced voltage in V,
Φ magnetic flux generated by the excitation (field) winding in Wb,
z number of wires on the armature surface,
a number of parallel armature-winding paths,
frequency $f = pn/60$,
p number of pole pairs,
n rotational speed in rpm.

In an alternating-current machine

$$U_i = \frac{\pi f z \Phi}{\sqrt{2}}$$

where
U_i effective value of induced voltage in V,
Φ magnetic flux generated by the excitation winding or a permanent excitation in Wb,
$f = pn/60$ frequency of alternating current in Hz,
p number of pole pairs,
n rotational speed in rpm,
z number of wires on the armature surface.

Figure 16: Motional induction
B Magnetic field,
v Direction of moving conductor,
U_i Positive induced voltage.

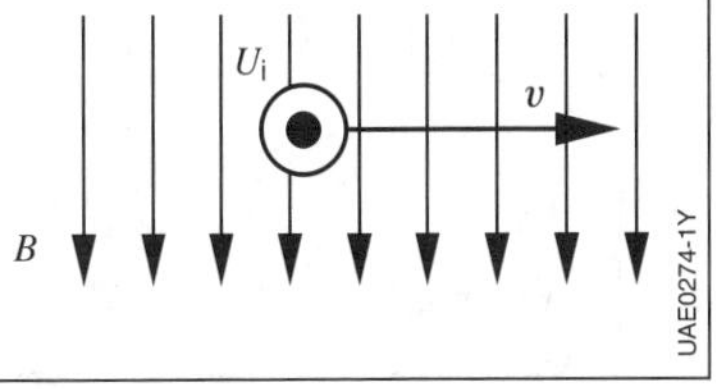

In a transformer

$$U_1 = 2\pi f w_1 \Phi,\ U_2 = 2\pi f w_2 \Phi$$

U_1, U_2 effective values of induced voltages in V,
Φ effective value of magnetic flux $\Phi(t)$ in Wb,
f frequency in Hz,
w_1, w_2 number of turns on the respective windings which surround the flux Φ.

The time-dependent flux $\Phi(t)$ is a superposition of contributions of the two time-dependent currents $i_1(t)$ and $i_2(t)$:

$$\Phi(\mathrm{t}) = A_L(w_1 i_1(t) + w_2 i_2(t)).$$

A_L is called the AL value and is dependent on the transformer design.

The terminal voltage U is smaller (alternator) or larger (motor) than U_i by the ohmic drop in the winding (about 5%).

Self-induction

The magnetic field of a current-carrying conductor or a coil changes with the conductor current. A voltage proportional to the change in current is induced in the conductor itself. The following applies:

$$U = \frac{d(LI)}{dt}.$$

The inductance L depends on the relative permeability μ_r, which for most materials is practically equal to 1 and constant. Ferromagnetic materials are an exception. In the case of iron-core coils, therefore, L is highly dependent on the operating conditions. Table 5 shows the inductance L of some conductor configurations.

At low frequencies the inductance of the conductors is further increased by the inner inductance L_i of the wires. For round wires this is

$$L_i = \frac{\mu_0 l}{8\pi}.$$

Twin conductors of two round wires have double the value ($2\,L_i$) per length l as inner inductance.

Table 5: Inductance of some conductor arrangements

Arrangement	Formula	Symbol	Meaning
Cylindrical coil	$L = w^2 \frac{\mu_r \mu_0 q}{l}$	μ_r, μ_0	Relative permeability, magnetic field constant
		w	Number of turns on coil
		q	Cross-section of coil [m²]
		l	Length of coil [m]
Parallel conductors (twin conductors) in air	$L = \frac{\mu_0 l}{\pi} \ln \frac{a+\sqrt{a^2-4r^2}}{2r}$	l	Length of twin conductors [m]
		a	Distance between conductors [m]
		r	Conductor radius [m]
Concentric line (coaxial line)	$L = \frac{\mu_r \mu_0 l}{2\pi} \ln \frac{r_2}{r_1}$	l	Length of line [m]
		r_2	Inside radius of outer tube [m]
		r_1	Radius of center tube [m]
Conductor to ground in air	$L = \frac{\mu_0 l}{2\pi} \ln \frac{a+\sqrt{a^2-r^2}}{r}$	l	Length of conductor [m]
		a	Distance from conductor to ground [m]
		r	Conductor radius [m]

Inductance of coils connected in series and parallel:

$L_{total} = L_1 + L_2$

(series connection, Figure 17a),

$\frac{1}{L_{total}} = \frac{1}{L_1} + \frac{1}{L_2}$

(parallel connection, Figure 17b).

The energy content of the coil carrying current I of inductance L is:

$W = \frac{1}{2} L I^2.$

Examples
(L is constant in terms of time):

- Direct current I:
  → voltage $U = 0$.
- Initial current I_0, direct voltage U:
  → current $I = I_0 + U t$.
- Current sinusoidal with $I = \hat{\imath} \sin(\omega t)$:
  → voltage cosinusoidal with
  $U = \omega L \hat{\imath} \cos(\omega t)$,
  $U = \omega L \hat{\imath} \sin(\omega t + \pi/2)$.

The quantity $\omega = 2\pi f$ is called angular frequency in relation to frequency f;
$\hat{\imath}$ is the current amplitude, and
$\hat{u} = \omega L \hat{\imath}$ the voltage amplitude.
Here, too, the effective values $u_{eff} = \hat{u}/\sqrt{2}$ and $i_{eff} = \hat{\imath}/\sqrt{2}$ are frequently used.

In the case of harmonic excitation, the voltage at the coil is therefore phase-displaced in relation to the current by the angle $\varphi = +\pi/2$ (it "leads"). This property is illustrated in a phasor diagram (Figure 18).

A special case arises when the coil is connected via a resistance to a direct-voltage source U_0 or discharged via a resistance. The time constant $\tau = L/R$ is the decisive factor in the switch-on and switch-off operations of the coil.

Figure 17: Connection of coils
a) Series connection,
b) Parallel connection.

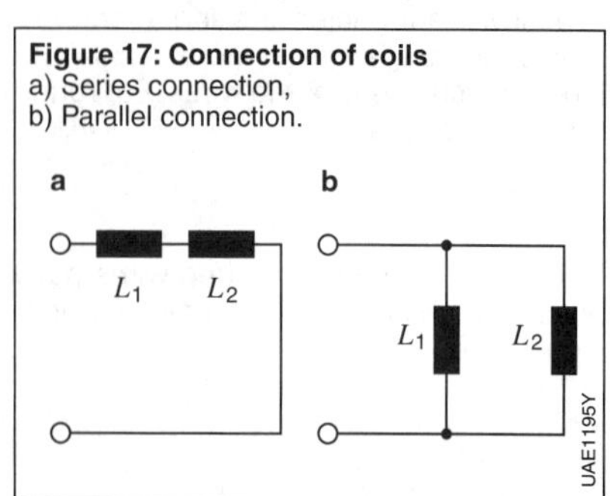

Figure 18: Phasor diagram, coil
$\hat{u}$ Voltage amplitude,
$\hat{\imath}$ Current amplitude.

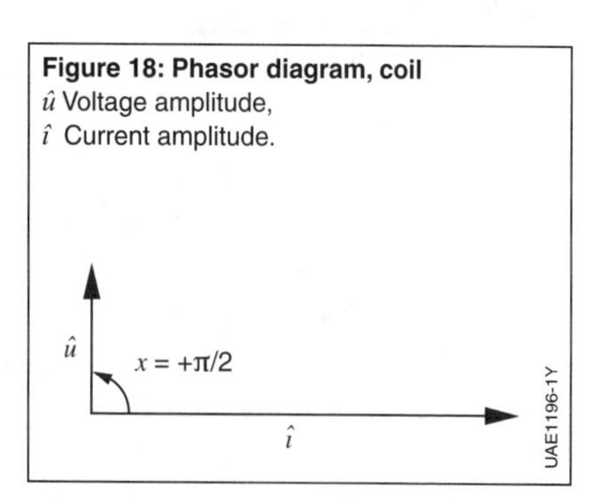

Switch-on operation (Figure 19):

$$U = U_0 \, e^{-\frac{t}{\tau}} .$$

$$I = \frac{U_0}{R} \, (1 - e^{-\frac{t}{\tau}}) .$$

Switch-off operation:

$$I = I_0 \, e^{-\frac{t}{\tau}} .$$

$$U = I_0 \, R \, e^{-\frac{t}{\tau}} .$$

Where
U_0 excitation voltage,
I coil current,
I_0 coil current at start of switch-off operation,
R resistance in series to coil,
U coil voltage.

The coil currents during the switch-on and switch-off operations have opposite directions.

The magnetic circuit
In addition to material equations, the following equations also determine the design of magnetic circuits:

1. Ampère's law
(equation of magnetic voltage)
The following equation holds true for a closed magnetic circuit:

$$\sum H_i \, l_i = \sum V_i = I w ,$$

where $I w = 0$ if no current is surrounded by the magnetic circuit.

Figure 19: Switch-on operation for coil
a) Circuit,
b) Voltage and current curves.

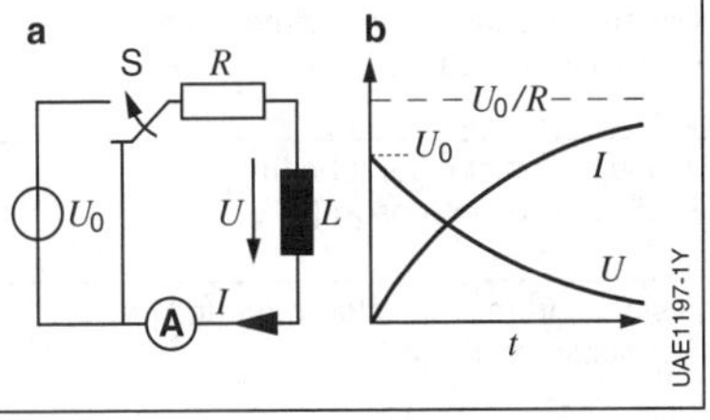

$I w = \Theta$ is the magnetomotive force (ampere turns, current surrounded in total by the magnetic circuit),
$H_i \, l_i = V_i$ magnetic potential difference ($H_i \, l_i$ is to be calculated for circuit components in which H_i is constant).

2. Law of continuity
(equation of magnetic flux).
The same magnetic total flux $\Phi = B A$ flows in the individual sections of a magnetic circuit.

Φ = const. in all sections, A is the cross-section of the respective section.

Within a section of the magnetic circuit the magnetic flux can split into partial fluxes Φ_1, Φ_2... the sum of which however is again the constant total flux Φ.

The quality of a circuit is determined by the amount of flux available in the working air gap. This flux is called useful flux. The leakage flux – the difference between the total flux and the useful flux – does not pass through the working air gap and does not add to the power of the magnetic circuit. Its ratio to total flux (flux of the permanent magnet or electromagnet) is called the leakage coefficient σ (practical values for σ are between 0.2 and 0.9).

Wave propagation

Waveguide

At higher frequencies forward and return conductors of an electric connection section are combined and called a waveguide or in short a line. Common structures are the parallel-wire line, the twisted two-wire line (twisted pair), the coaxial line, the strip line, and the microstrip line.

At higher frequencies the current is no longer distributed evenly over the line cross-section. In the individual wire the current is increasingly displaced to the edge as the frequency increases (skin effect). In lines there is an additional displacement of current to the other conductor in each case (proximity effect). In particular the ohmic loss resistance and thus the damping of the line are thereby increased.

A line has the characteristic impedance Z if it is terminated with this resistance Z and then demonstrates the same impedance Z as the input resistance. In the lines used in motor vehicles (TEM or L lines) the characteristic impedance Z is directly connected with the capacitance C per length l and the inductance L per length l:

$$Z = \sqrt{\frac{L}{C}}\,.$$

Table 6 shows characteristic impedances for some selected line types.

Even the amplitudes of the electric and magnetic fields along the line are linked via the characteristic impedance:

$$Z = \frac{\hat{E}}{\hat{H}}\,.$$

A line of characteristic impedance Z which is terminated with resistance R demonstrates at its end the reflectance factor

$$r = \frac{R - Z}{R + Z}\,.$$

The square of the absolute value of the reflectance factor indicates what fraction of the power is sent back at the end of the line.

$$P_r = |r|^2\, P_0\,.$$

Passed on to the terminating resistor, however, the power becomes

$$P_t = (1 - |r|^2)\, P_0\,.$$

When $R = Z$ applies, adaptation occurs and the reflectance factor r is zero. In this case, no power is reflected, the full power being passed on to the terminating resistor R.

At higher frequencies and with longer lines, voltage and current along the line are no longer constant. At a fixed instant both vary along the line. Here there can be points over the length of the line at which the amplitude of the voltage fluctuations is at maximum, while at other points only minimum voltage amplitudes occur. The ratio of maximum to minimum amplitude on a line is called the standing wave ratio s.

$$s = \frac{U_{max}}{U_{min}}\,.$$

If a line is supplied at one end only with a sinusoidal time function, while terminated at the other end with a resistor, it is possible to calculate the standing wave ratio from the reflectance factor.

$$s = \frac{1 + |r|}{1 - |r|}\,.$$

Table 6: Characteristic impedances for some selected line types

Line type	Formula	Symbol	Meaning
Parallel conductors (twin conductors) in air	$Z = \sqrt{\frac{\mu_0}{\varepsilon_0}}\,\frac{1}{\pi}\,\ln\frac{a+\sqrt{a^2-4r^2}}{2r}$	a r	Distance between conductors [m] Conductor radius [m]
Concentric line (coaxial line)	$Z = \sqrt{\frac{\mu_r\mu_0}{\varepsilon_r\varepsilon_0}}\,\frac{1}{2\pi}\,\ln\frac{r_2}{r_1}$	r_2, r_1	Inside radius of outer tube [m] Radius of center tube [m]
Conductor to ground in air	$Z = \sqrt{\frac{\mu_0}{\varepsilon_0}}\,\frac{1}{2\pi}\,\ln\frac{a+\sqrt{a^2-r^2}}{r}$	a r	Distance from conductor to ground [m] Conductor radius [m]

During adaptation s = 1 and at all points on the line the voltage amplitude is the same.

Couplers and matching transformers
Couplers are used in conjunction with matching transformers to forward an electromagnetic wave from one waveguide to another. Matching transformers reduce the reflectance factor during the transition from one characteristic impedance to another or couple a symmetrical waveguide (e.g. a parallel-wire line) with an asymmetrical one (e.g. a coaxial line).

Couplers and matching transformers are usually reciprocal, coupling just as well in one direction as in the other.

Wave propagation in free space
Even free space is a waveguide, with the characteristic impedance of free space (field characteristic impedance Z_0):

$$Z_0 = \sqrt{\frac{\mu_0}{\varepsilon_0}} \approx 377\ \Omega.$$

Far away from antenna structures the direction of the electric field, the direction of the magnetic field and the direction of propagation of the electromagnetic wave in free space are vertically on top of one another, and the electric field strength is linked with the magnetic field strength via the characteristic impedance Z_0. The electromagnetic power density S emitted by a transmitter is established via the amplitudes of the electric field strength $\hat{E}$ and the magnetic field strength $\hat{H}$. In free space, far away from the transmitter, the electromagnetic power density decreases with the inverse square of the distance from the transmitter:

$$S = \frac{1}{2}\hat{E}\,\hat{H}\,;$$

$$S(r) = S(r_0)\,\frac{r_0^2}{r^2}\,.$$

Electric and magnetic field strengths, on the other hand, decrease only inversely with the distance from the transmitter:

$$\hat{E}(r) = \hat{E}(r_0)\,\frac{r_0}{r}\,,$$

$$\hat{H}(r) = \hat{H}(r_0)\,\frac{r_0}{r}\,.$$

For some applications in motor vehicles the receiver is situated so near to the transmitter that near-field conditions prevail. Radio remote controls, for example, generate via a current-carrying coil magnetic fields whose field strength in the near field is inversely proportional to the cube of the distance from the transmitter.

Antennas
A coupler from a waveguide to free space is called an antenna. Antennas too are usually reciprocal; they can therefore act equally as transmitting and receiving antennas.

An electromagnetic wave in free space consists of time-variant electric and magnetic fields. These have one spatial direction – far away from transmitting antennas the field directions of these two fields are vertical to the direction of propagation of the wave. The direction of the corresponding electric field determines the polarization direction of the electromagnetic wave. Most antennas are designed in such a way that they can transmit and receive only one single direction of the electric (sometimes also the magnetic) field strength. They therefore transmit only linearly polarized waves and receive only one particular polarization of the wave.

A distinction is made between narrow-band antennas (dipole antennas, as used for example for GPS, GSM, Bluetooth, WLAN or even VHF) and wide-band antennas, which are used as transmitting and receiving antennas in EMC measurements in motor vehicles. An example of a narrow-band antenna is the half-wave dipole, a rod or wire of length $l = 0.96\ \lambda/2$ (λ: wavelength in free space). The optimum length becomes shorter for thicker rods, or if necessary also through connection with additional capacitances. In conjunction with an extended metal surface the dipole length is shortened by half to one $\lambda/4$-dipole.

An example of a wide-band antenna is the logarithmically periodic antenna. Here, several half-wave dipoles are of different lengths are arranged along a twin line. No antenna transmits in transmitting mode uniformly in all directions. Likewise, no antenna receives in receiving mode all directions equally well. This directionality is described by the directional diagram (also known as the antenna directional characteristic). Figure 20 shows the directional diagram of a half-wave dipole in a vertical section through the dipole antenna. What is depicted is the radiated power density against the angle.

Often what is of interest is not the entire directional diagram, but only its maximum value in the preferred direction. The directivity indicates by how much the radiated power per surface area S_{max} of the antenna in transmitting mode is greater in the preferred direction than for a reference antenna which does not prefer a direction:

$$D_0 = \frac{S_{max}}{S_0}.$$

The directivity in relation to the antenna gain G is given as a dB value:

$$G = 10 \log D_0 .$$

Figure 20: Directional diagram of a half-wave dipole
(Vertical section through the dipole antenna, power density against the angle).

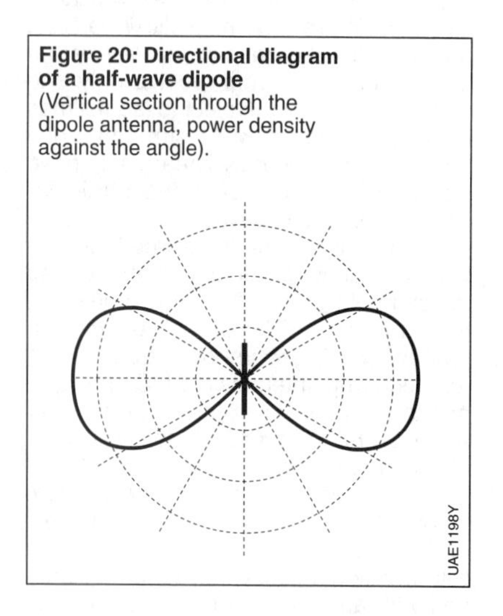

The antenna losses are usually also included in the antenna gain such that G is smaller than in the ideal case. The same directivity and the same antenna gain also apply in receiving mode. There they indicate by how much the antenna power input is greater than for the reference antenna.

In the EMC standards calculations are made not with the antenna gain but with the antenna factor AF:

$$AF = 20 \log \frac{E}{U}.$$

The antenna factor indicates what voltage amplitude U (in V) drops at the input resistance of the measuring device connected to the antenna when an electric field of amplitude E (in V/m) is received. The same antenna factor also applies to transmitting mode and links the voltage amplitude U (in V) at the input of the transmitting antenna with the amplitude E (in V/m) of the electric field of the emitted wave.

Antenna gain and antenna factor can be converted into one another:

$$AF = 10 \log \left(\frac{4\pi Z_0}{\lambda^2 R_L} \right) - G$$

where
R_L input resistance of measuring device and of antenna input (in Ω, usually 50 Ω),
Z_0 field characteristic impedance of free space (equal to 377 Ω),
λ wavelength (in m).

Electric effects in metallic conductors

Contact potential between conductors

Contact potential occurs in conductors, and is analogous to the triboelectricity or voltaic emf in insulators (e.g. glass, hard rubber). If two dissimilar metals (at the same temperature) are joined to make metal-to-metal contact with one another and are then separated, a contact potential is present between them. This is caused by the different work functions of the electrons. The magnitude of contact potential depends on the element positions in the electrode-potential series (Table 7). If more than two conductors are so joined, the resulting contact potential is the sum of the individual contact potential values.

Thermoelectricity

A potential difference, the galvanic voltage, forms at the junction of two conductors due to their dissimilar work functions. The sum of all galvanic voltages is zero in a closed conductor loop (in which the temperature is the same at all points). Measurement of these potentials is only possible by indirect means as a function of temperature (thermoelectric effect, Seebeck effect). The thermoelectric potential values are highly dependent on impurities and material pretreatment.

Table 7: Contact-potential values

Material pairing	Contact potential
Zn/Pb	0.39 V
Pb/Sn	0.06 V
Sn/Fe	0.30 V
Fe/Cu	0.14 V
Cu/Ag	0.08 V
Ag/Pt	0.12 V
Pt/C	0.13 V
Zn/Pb/Sn/Fe	0.75 V
Zn/Fe	0.75 V
Zn/Pb/Sn/Fe/Cu/Ag	0.97 V
Zn/Ag	0.97 V
Sn/Cu	0.44 V
Fe/Ag	0.30 V
Ag/Au	–0.07 V
Au/Cu	–0.09 V

The thermoelectric series (Table 8) specifies the differential thermoelectromotive forces referred to a reference metal (usually platinum, copper, or lead). At the hot junction, current flows from the conductor with the lower differential thermoelectromotive force to that with the higher force. The thermoelectromotive force η of any pair (thermocouple) equals the difference of the differential thermoelectromotive forces.

The reciprocal of the Seebeck effect is the Peltier effect, in which a temperature difference is created by electrical energy (heat pump). If current flows through an A-B-A series of conductors, one thermo-

Table 8: Thermoelectric series (referred to platinum)

Material	Thermo-electromotive force [10^{-6} V/K]
Selenium	1,003
Tellurium	500
Silicon	448
Germanium	303
Antimony	47...48.6
Nickel chromium	22
Iron	18.7...18.9
Molybdenum	11.6...13.1
Cerium	10.3
Cadmium	8.5...9.2
Steel (V2A)	7.7
Copper	7.2...7.7
Silver	6.7...7.9
Tungsten	6.5...9.0
Indium	6.5...6.8
Rhodium	6.5
Zinc	7.0...7.9
Manganin	5.7...8.2
Gold	5.6...8.0
Tin	4.1...4.6
Lead	4.0...4.4
Magnesium	4.0...4.3
Aluminum	3.7...4.1
Platinum	±0
Mercury	–0.1
Sodium	–2.1
Potassium	–9.4
Nickel	–19.4...–12.0
Cobalt	–19.9...–15.2
Constantan	–34.7...–30.4
Bismuth, perpendicular to axis	–52
Bismuth, parallel to axis	–77

junction absorbs heat while the other produces more heat than can be accounted for by the Joule effect. The amount of heat transferred between the thermojunctions is governed by the following equation:

$$\Delta Q = \pi \, I \, \Delta t \,,$$

where
π Peltier coefficient,
I current,
Δt time interval.

The relationship between the Peltier coefficient π, the temperature T and the thermo-electromotive force η is as follows:

$$\pi = \eta \, T \,.$$

Current flowing through a homogeneous conductor will also generate heat if a temperature gradient $\Delta T / l$ is maintained in the conductor (Thomson effect). Whereas the power developed by the Joule effect is proportional to I^2, the power developed by the Thomson effect is as follows:

$$P = -\sigma \, I \, \Delta T$$

where
σ Thomson coefficient,
I current,
ΔT temperature difference.

The reciprocal of the Thomson effect is the Benedicks effect, in which an electric potential is produced as a result of asymmetrical temperature distribution (particularly at points where there is a significant change in cross-sectional area).

Galvanomagnetic and thermomagnetic effects

Such effects are understood to be changes caused by a magnetic field in the flow of electricity or heat within a conductor. There are twelve different recognized effects which fall into this category, the most well-known of which are the Hall, Ettingshausen, Righi-Leduc and Nernst effects.

Of particular significance in industrial applications is the Hall effect. If a current is sent through a suitable conductor and a magnetic field is applied perpendicular to this current, a voltage is produced which is perpendicular to both the flow of current and the magnetic field. This voltage is called the Hall voltage U_H (Figure 21):

$$U_H = R_H \, I_V \, B / d \,,$$

where
R_H Hall constant,
I_V Supply current,
B Magnetic field,
d thickness of conductor.

In ferromagnetic materials, the Hall voltage is a function of magnetization (hysteresis).

The Hall effect is used in Hall-effect sensors.

References

[1] Philippow, E.: Grundlagen der Elektrotechnik. Verlag Technik Berlin, Munich, 10th Edition., 2000.
[2] Philippow, E.: Taschenbuch Elektrotechnik, Volume 1. Carl Hanser Verlag Munich, 3rd Edition, 1986.
[3] Albach, M.: Grundlagen der Elektrotechnik 1. Pearson Studium Munich, 2nd Edition., 2008.
[4] Albach, M.: Grundlagen der Elektrotechnik 2. Pearson Studium Munich, 1st Edition., 2005.

Figure 21: Hall effect
B Magnetic field,
I_H Hall current,
I_V Supply current,
U_H Hall voltage,
d thickness of conductor.

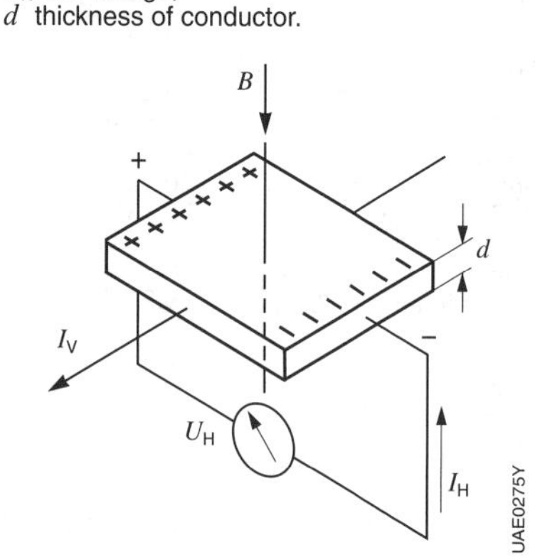

Electronics

Basic principles of semiconductor technology

Electric conductivity in solid bodies

An individual material's capacity for conducting electricity is determined by the number and mobility of the free charge carriers which it contains. The disparities in the electric conductivities displayed by various solid bodies at room temperature extend through a range defined by 10 to the 24th power. Accordingly, materials are divided into three electrical groups.

Conductors (metals)

All solid bodies contain approximately 10^{22} atoms per cubic centimeter; they are held together by electrical forces. In metals, the number of free, i.e. non-attached, charge carriers is extremely high (one to two free electrons per atom). The free carriers are characterized by moderate mobility. The electrical conductivity of metals (e.g. silver, copper, aluminum) is high. In good conductors, this amount to approx. 10^6 S/cm.

Figure 1: Doped silicon
a) n-doped silicon,
b) p-doped silicon.
o Electron,
Si Silicon,
P Phosphorus,
B Boron,
E Electric field.
The curved arrows indicate the direction of motion of the electrons.

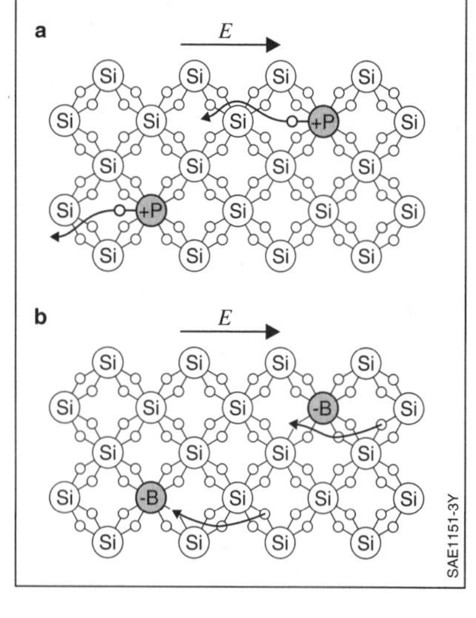

Non-conductors (insulators)

The number of free charge carriers in insulators (ex. aluminum oxide, Teflon, fused quartz) is practically zero. Accordingly, the electric conductivity is negligible. The conductivity of good insulators is approximately 10^{-18} S/cm.

Semiconductors

The amount of electrical conductivity in semiconductors (ex. germanium, silicon, gallium arsenide) is between that of metals and insulators. It is – unlike the conductivity of metals and insulators – greatly dependent on the following factors:

- the pressure affects the mobility of charge carriers,
- the temperature influences the number and mobility of charge carriers,
- the action of light also influences the number of charge carriers,
- the presence of additives also determines among others the number and type of charge carriers.

Because they are dependent on the above factors, semiconductors are also suitable for use as pressure, temperature and light sensors.

Doping of semiconductors

Doping, i.e. the controlled addition of electrically active foreign substances to the base material, makes it possible to define and localize the semiconductor's conductivity. This procedure forms the basis of semiconductor devices. The producible and also adjustable electric conductivity of silicon that can be reproduced by doping is 10^4 to 10^{-2} S/cm.

Electric conductivity of semiconductors

The following discussion focuses on the silicon-based semiconductor. In its solid state, silicon assumes the form of a crystal lattice with four equidistant contiguous atoms. Each silicon atom has four outer electrons, with two shared electrons forming the bond with the contiguous atoms. In this ideal state, silicon has no free charge carriers; thus it is not conductive. The situation changes dramatically with the addition of appropriate additives and the application of energy.

Doping will be explained here by reference to a clear model presentation. However, it is important to bear in mind that not all effects can be explained by reference to this model [2].

n Doping

As only four electrons are required for bonding in a silicon lattice (Figure 1a), the introduction of foreign atoms with five outer electrons (e.g. phosphorus) results in the presence of free electrons. Thus, each additional phosphorus atom will provide a free, negatively charged electron, where a single positively charged phosphorus nucleus remains. The silicon is transformed into an n conductor (n-type silicon), since there is an excess of negative charges (electrons). In response to an externally applied voltage, an electric field E is generated, which gives the mobile charge carriers a preferred direction for movement (Figure 1).

p Doping

The introduction of foreign atoms with three outer electrons (e.g. boron) produces electron holes which result from the fact that the boron atom has one electron less for complete bonding in the silicon lattice (Figure 1b). This hole indicates a missing electron. Holes remain in motion within the silicon; they are filled by electrons that leave behind their own hole. In an electric field, they travel in the opposite direction of the electrons. The holes exhibit the properties of a free positive charge carrier. Thus, every additional boron atom provides a free, positively charged electron hole (positive hole). The silicon is transformed into a p conductor and is called a p-type silicon.

Intrinsic conduction

Heat and light also generate free mobile charge carriers in undoped silicon; the resulting electron-hole pairs produce intrinsic conductivity in the semiconductor material. This conductivity is generally modest in comparison with that achieved by doping. Increases in temperature induce an exponential rise in the number of electron-hole pairs, ultimately obviating the electrical differences between the p and n regions produced by doping. This phenomenon defines the maximum operating temperature to which semiconductor devices may be subjected: This is for germanium 90 to 100 °C, for silicon 150 to 200 °C, and for gallium arsenide 300 to 350 °C.

A small number of opposite-polarity charge carriers is always present in both n-type and p-type semiconductors. These minority charges exert a considerable influence on the operating characteristics of virtually all semiconductor devices.

pn junction in the semiconductor

The area of transition between a p-type zone and an n-type zone within the same semiconductor crystal is referred to as the pn junction. The properties of this area exercise a major influence on the operating properties of most semiconductor devices.

pn junction without external voltage

The p-type zone has numerous holes, and the n-type zone has only very few; there is a very large number of electrons in the n-type zone, while in the p-type zone there is an extremely small number. Each type of mobile charge carrier tends to move across the concentration gradient, diffusing into the other zone (Figure 2b).

The diffusion of holes into the n-type zone gives the p-type zone in the pn-junction area a negative charge, since the negatively charged core atoms (i.e. boron atoms) remain stationary. Electron depletion results in the n-type region being positively charged, since here this is an excess of stationary, positively charged atomic residues (e.g. phosphorus). The result is an electrical potential between the p- and n-type zones (diffusion potential U_D). This potential opposes the respective migration tendencies of the charge carriers. This ultimately brings the exchange of holes and electrons to a halt. The potential current U_D created on account of diffusion cannot be directly measured from the outside and in silicon is typically just 0.6 V.

A poorly conducting zone deficient in mobile charge carriers is thus produced at the *pn* junction: This zone is called the space-charge region or depletion layer. This zone has an electric field whose strength is also dependent on the externally applied voltage.

pn junction with external voltage

Now the conditions at a diode will be explained because a *pn* junction corresponds to the structure of a diode; here the anode is situated at the *p*-doped silicon, while the cathode is situated at the *n*-doped zone.

The application of voltage U in the reverse direction (negative pole at the p-type zone and positive pole at the n-type zone) extends the space-charge region (Figure 2c). Consequently, the flow of current I is blocked except for a minimal residual current (reverse current) which stems from the minority charge carriers. The voltage U then drops within the space-charge region; this region is therefore subject to a high electric field strength.

Figure 2: *pn* junction in a diode
a) Diode circuit symbol,
b) *pn* junction without external voltage,
c) *pn* junction in reverse direction,
d) *pn* junction in forward direction.
U Applied voltage (diode voltage),
I Diode current.
⊕ Positively charged atomic residues,
⊖ Negatively charged atomic residues.

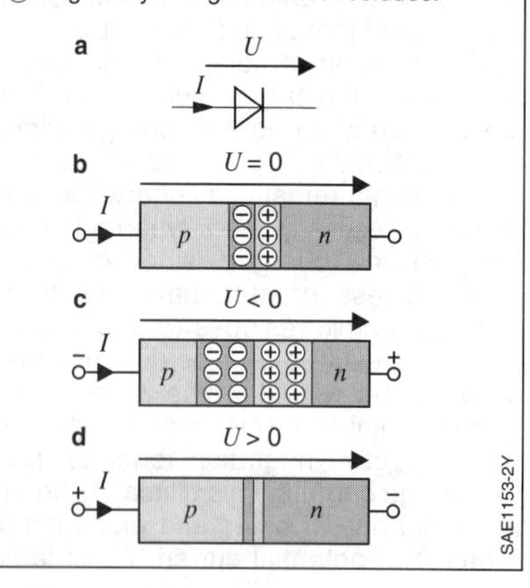

Breakdown voltage is the level of reverse-direction voltage beyond which a minimal increase in voltage will suffice to produce a sharp rise in reverse current (Figure 3). This effect can be explained as follows: Electrons which reach the space-charge region are greatly accelerated on account of the high field strength. In this way, they can for their part generate free charge carriers as a result of impacts; this is also known as collision ionization. This brings about a dramatic rise in current and results in avalanche breakdown. In addition to avalanche breakdown, there is also Zener breakdown, which is based on the tunnel effect. The breakdown of a *pn* junction can destroy it and for this reason is often not wanted. In many cases, however, breakdown is wanted. Avalanche breakdown and Zener breakdown only occur when the diode is operated in the reverse direction.

The application of a voltage U in the forward direction (positive pole at the p-type zone and negative pole at the n-type zone) reduces the space-charge region (Figure 2d). Charge carriers permeate the *pn* junction, resulting in a large flow of current in the forward direction (Figure 3), as the space-charge region no longer has a significant resistance. Only the bulk resistance, i.e. the ohmic resistance of the doped layers, is effective. The current I increases exponentially as a function of U. It is, however, important to be aware of "thermal breakdown", at which point the semiconductor can be destroyed on account of the high level of heat. This can occur, for example, if the diode is operated in the forward direction with an unacceptably high current.

Figure 3: Characteristic curve of a Si diode
U Applied voltage (diode voltage),
I Diode current.

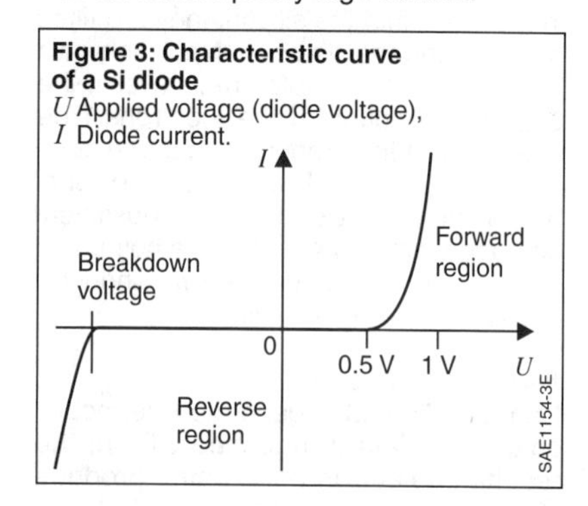

Discrete semiconductor devices

The properties of the *pn* junction and the combination of several *pn* junctions in a single semiconductor-crystal chip provide the basis for a steadily increasing array of inexpensive, reliable, rugged, compact semiconductor devices. A single *pn* junction forms a diode, two *pn* junctions are used for transistors. The planar technique makes it possible to combine numerous operating elements on a single chip to form the important component group known as integrated semiconductor circuits. Semiconductor chips measure no more than several square millimeters and are usually installed in standardized housings (metal, ceramic or plastic).

Diodes

A diode is a semiconductor device incorporating a single *pn* junction. Its specific properties are determined by the distribution pattern of the dopant in the crystal. Diodes which conduct currents in excess of 1 A in the forward direction are referred to as power diodes.

Rectifier diode

The rectifier diode acts as a form of current valve; it is, therefore, ideally suited to rectifying alternating current. The forward current can be approximately 10^7 times higher than the current in the reverse direction (reverse current, Figure 3). As the temperature rises, the power of this current increases significantly.

Rectifier diode for high reverse voltage

In the case of a rectifier, the voltage drops over the space-charge region. Because this region is generally only a few micrometers in size, the high reverse currents produce a high electric field strength and the free electrons can accelerate greatly. Accelerated electrons can cause the semiconductor to be destroyed (avalanche breakdown). To prevent this, it has proved useful to integrate an intrinsic layer between the *p*- and *n*-layers because this layer only contains few free electrons and thereby reduces the danger of a breakdown.

Switching diode

A switching diode is generally employed for rapid switching between high and low impedances. More rapid switching response can be achieved by diffusing gold into the material; this promotes the recombination of electrons and holes.

Zener diode

A Zener diode is a semiconductor diode which, once a specific initial level of reverse voltage is reached, responds to further increases of reverse voltage with a sharp rise in current flow. This phenomenon is a result of a Zener and/or avalanche breakdown. Zener diodes are designed for continuous operation in this breakdown range. These are frequently used to provide a constant voltage or reference voltage.

Variable-capacitance diode (varactor)

The space-charge region at the *pn* junction functions as a capacitor; the dielectric element is represented by the semiconductor material in which no charge carriers are present. Increasing the applied voltage extends the depletion layer and reduces the capacitance, while reducing the voltage increases the capacitance.

Schottky barrier diode (Schottky diode)

A Schottky diode contains a metal-to-semiconductor junction. As the electrons move more freely from the *n*-type silicon into the metal layer than in the opposite direction, an electron-depleted layer is created in the semiconductor material; this is the Schottky barrier layer. Charges are carried exclusively by the electrons, a factor which results in extremely rapid switching, as the minority carriers do not perform any charge storage function. The forward voltage and thus the voltage drop is at approximately 0.3 V smaller in Schottky diodes than in silicon diodes (approximately 0.6 V).

Solar cell

The photovoltaic effect is applied to convert light energy directly into electrical energy. Solar cells, consisting largely of semiconductor materials, are the basic elements of photovoltaic technology. Exposure to light can cause the formation of free charge carriers (electron-hole pairs) in the semiconductor. If the semiconductor incorporates a *pn* junction, the free charge carriers separate in its electric field before proceeding to the metal contacts on the semiconductor's surface. Depending on the semiconductor material used, a DC voltage (photovoltage) ranging from 0.5 to 1.2 V is generated between the contacts. This occurs only when the light quanta have at least the energy required to create an electron-hole pair. The theoretical efficiency of crystalline silicon solar cells is approximately 30 %.

Photodiode

The photovoltaic effect is utilized in a photodiode. The *pn* junction is operated in the reverse direction. Incident light creates additional free electrons and holes. These increase the reverse current (photovoltaic current) in direct proportion to the intensity of the light. The photodiode is thus, in principle, very similar to the solar cell.

LED (light-emitting diode)

A light-emitting diode or LED is an electroluminescent lamp which consists of a semiconductor element with *pn* junction. The charge carriers (free electrons and holes) recombine during operation in forward direction. The amount of energy released in this process is converted into electromagnetic radiation energy.

Depending on the choice of semiconductor and its doping, the LED emits in a limited spectral range. Frequently used semiconductor materials are: gallium arsenide (infrared), gallium arsenide phosphide (red to yellow), gallium phosphide (green), and indium gallium nitride (blue). In order to generate white light, either a combination of three LEDs with the primary colors Red, Green and Blue is used, or a fluorescent dye is excited with a blue or ultraviolet-emitting LED.

Bipolar transistors

Two contiguous *pn* junctions produce the transistor effect, a feature employed in the design of components used to amplify electrical signals and to assume switching duties. Bipolar transistors consist of three zones of varying conductivity, the configuration being either *pnp* or *npn*. The zones (and their terminals) are called emitter E, base B, and collector C (Figure 4).

There are different transistor classifications, depending on the fields of application: small-signal transistors (power dissipation up to 1 watt), power transistors, switching transistors, low-frequency transistors, high-frequency transistors, microwave transistors, and phototransistors. They are termed bipolar because charge carriers of both polarities (holes and electrons) are involved in the transistor effect.

Operation of a bipolar transistor

The operation of a bipolar transistor is explained here using an *npn* transistor as the example (Figure 5). The *pnp* transistor is obtained similarly by switching the *n*- and *p*-doped zones.

Figure 4: *npn* transistor
a) Diagram,
b) Layout.
E Emitter,
B Base,
C Collector.
U_{BE} Base-emitter voltage,
U_{CE} Collector-emitter voltage.
I_B Base current,
I_C Collector current,
I_E Emitter current.

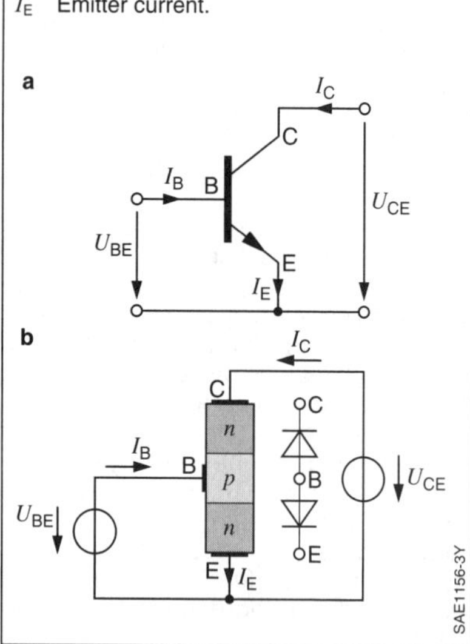

The base-emitter junction is forward-biased; this is shown in Figure 4b as a diode between base B and emitter E. In this way, with sufficient voltage U_{BE}, electrons are injected into the base zone and base current flows.

The base-collector junction is reverse-biased; this is shown in Figure 4b as a diode between base B and collector C. This creates a space-charge region at the *pn* junction between base and collector with a high electric field.

Because of the reverse-biased diode between the base and the emitter, a high current consisting of electrons flows from the emitter to the base. Here, however, only a small fraction can recombine with the (much fewer) holes and flow as base current I_B out of the base terminal; what must be borne in mind is that the technical current direction – i.e. the direction of motion of the positive charge carriers – is specified in Figure 4. The much greater amount of electrons injected into the base diffuses through the base zone to the base-collector junction and then flows as collector current I_C to the collector (Figure 5). Because the base-collector diode is operated in the reverse direction and a space-charge region is predominant, virtually all (approximately 99 %) of the electrons flowing from the emitter are "drawn off" through the strong electric field in the space-charge region from the collector. In this case, a linear connection applies approximately between the collector current I_C and the base current I_B:

$$I_C = B\,I_B$$

where B is the current gain, which is generally between 100 and 800. In the bipolar transistor, the following relation for the emitter current I_E also applies (cf. Figure 4 and Figure 5):

$$I_E = I_B + I_C\,.$$

Assuming that I_B on account of the current gain B is much smaller than I_C, it then follows:

$$I_E \approx I_C\,.$$

The very thin (and relatively low-doped) base represents a barrier that can be adjusted by means of the base-emitter voltage U_{BE} from the charge-carrier flow from the emitter to the collector. With a small change of U_{BE} and the base current I_B, it is possible to control a greater change of the collector current I_C and the collector-emitter voltage U_{CE}. Small changes in the base current I_B thus bring about large changes in the emitter-collector current I_C. The *npn* transistor is a bipolar, current-controlled semiconductor amplifier. In all, there is a power amplification.

The output characteristic curve for an *npn* transistor is shown in Figure 6. From

Figure 5: Operation of an *npn* transistor
E Emitter,
B Base,
C Collector.
U_{BE} Base-emitter voltage,
U_{CE} Collector-emitter voltage,
I_B Base current,
I_C Collector current,
I_E Emitter current.

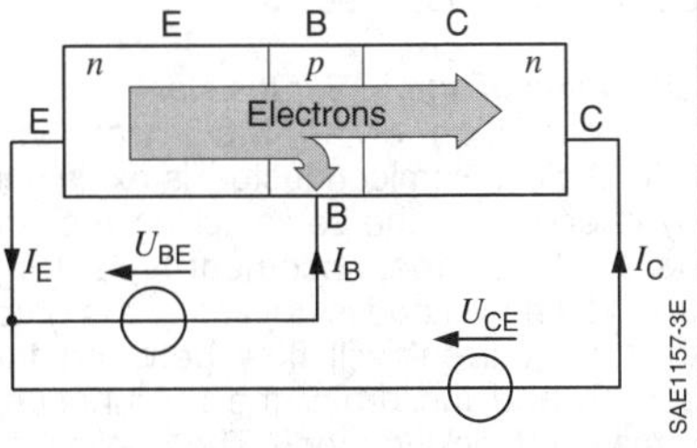

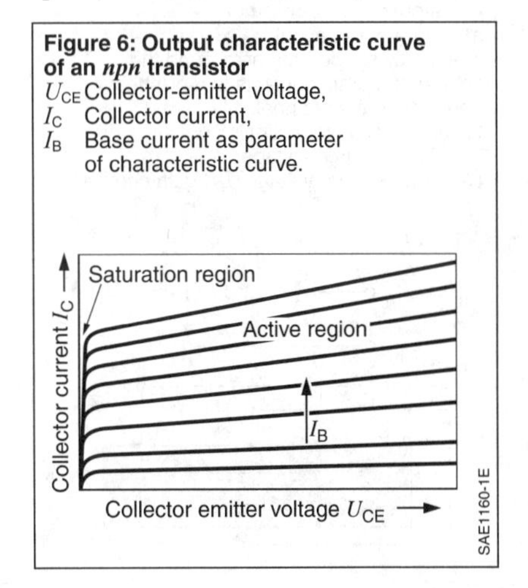

Figure 6: Output characteristic curve of an *npn* transistor
U_{CE} Collector-emitter voltage,
I_C Collector current,
I_B Base current as parameter of characteristic curve.

the saturation voltage of approximately 0.2 V for U_{CE}, the collector current I_C is virtually only dependent on the base current I_B as the parameter; this region is termed the "active region": U_{CE} here has virtually no influence on I_C and the following applies:

$$I_C = B\, I_B.$$

The region below the saturation voltage is called the "saturation region". In this region, I_C increases sharply with U_{CE}.

Field-effect transistors

In a field-effect transistor (FET), control of the current flow in a conductive path is essentially exercised by an electric field. The field, in turn, is generated by a voltage applied at a gate (Figure 7). Field-effect transistors differ from their bipolar counterparts in that they utilize only a single type of charge carrier (either electrons or holes), giving rise to the alternate designation "unipolar transistors". These are subdivided into the following categorizations: junction-gate field-effect transistors (junction FET, JFET) and insulated-gate field-effect transistors, particularly MOS field-effect transistors (MOSFET or MOS transistors).

MOS field-effect transistors are well suited for application in high-integration circuits. Power field-effect transistors represent a genuine alternative to bipolar power transistors in many applications.

The advantages of a bipolar transistor and of a field-effect transistor are utilized in power electronics in insulated gate bipolar transistors (IGBT), which exhibit a low volume resistance (small losses) and comparatively low triggering power.

Figure 7: Junction-gate field-effect transistor with *n* channel
a) Diagram, b) Layout.
The lightly shaded region around the source and drain contacts is more heavily doped than the channel.
G Gate, S Source, D Drain.
U_{DS} Drain-source voltage,
U_{GS} Gate-source voltage, I_D Drain current.

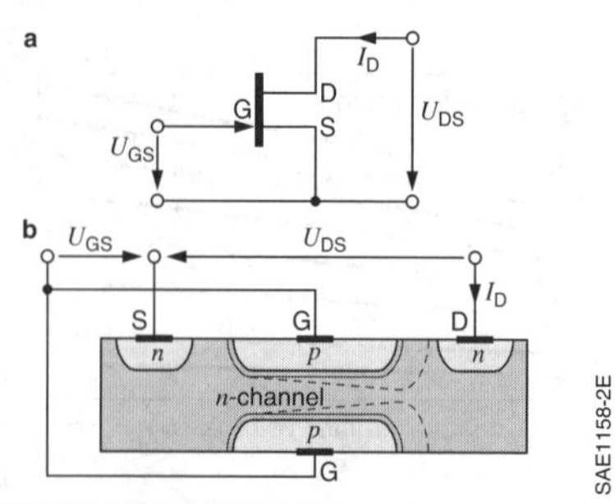

Operation of a junction FET

The operation of a junction-gate field-effect transistor is explained by reference to the n channel type (Figure 7). The terminals of a field-effect transistor are referred to as gate (G), source (S) and drain (D).

Positive direct voltage U_{DS} is applied at the ends of an n-type crystal. Electrons flow through the channel from the source to the drain. The width of the channel is defined by two laterally diffused p-type zones and by the negative gate-source voltage U_{GS} applied at them. The voltage U_{GS} between the control electrode (gate G) and the source terminal thus controls the current I_D between the source and the drain.

Only charge carriers of one polarity are required for field-effect transistor operation. The power necessary for controlling the current is virtually nil. Thus, the junction FET is a unipolar, voltage-controlled component. Increasing U_{GS} causes the space-charge regions to extend further into the channels, thereby constricting the channel and thus the current path (see dashed lines in Figure 7). If the voltage U_{GS} at the gate is zero, the channel between the two p-type zones is not constricted and the current I_D from the drain to the source is at its maximum.

The transfer characteristic curve – i.e. I_D as a function of U_{GS} – then looks exactly the same as the characteristic curve of a self-conducting n-channel MOSFET as shown in Figure 9c.

Operation of an MOS transistor

The operation of an MOS transistor (metal-oxide semiconductor) is explained by reference to the self-blocking n-channel MOSFET (enhancement type) (Figure 8). If no voltage is applied to the gate, then no current will flow between the source and the drain: the pn junctions remain in blocking mode. The application of a positive voltage at the gate causes, on account of the electrostatic induction in the p-type zone below this gate, the holes to be displaced toward the interior of the crystal and electrons – which are always present in p-type silicon as minority charge carriers – to be pulled to the surface. A narrow n-type layer, an n channel, forms below the surface. Current can now flow between the two n-type zones (source and drain). This current consists exclusively of electrons. As the gate voltage acts through an insulating oxide layer, no stationary current flows through the gate; no power is required for the control function. It is required merely to activate and deactivate electrical power in order to recharge the gate capacity. In summary, the MOS transistor is a unipolar, voltage-controlled component.

Figure 8: n-channel MOSFET, cross-section
S Source, G Gate, D Drain.
U_{DS} Drain-source voltage,
U_{GS} Gate-source voltage, I_D Drain current.

In the case of the self-conducting n-channel MOSFET (depletion type, Figure 9a), the gate-source voltage U_{GS} is between the here negative threshold voltage U_T and zero volts (Figure 9c). At $U_{GS} = 0$ V, the self-conducting n-channel MOSFET has a channel below the gates for the current flow. Figure 9c shows I_D as a function of U_{GS}, where the circuit is operated in the active region with sufficient and constant U_{DS}. The transfer characteristic curve is a parabola. In contrast, the self-blocking n-channel MOSFET (Figure 9b) conducts only from the positive threshold voltage $U_T^* > 0$ V here (see Figure 9c). The self-blocking MOSFET is much more common that the self-conducting MOSFET.

The output characteristic curve of a self-blocking n-channel MOSFET is shown in Figure 10. The region below the knee-point voltage U_K, i.e. for $U_{DS} < U_K$ is referred to, on account of the linear characteristic curve, as the linear or ohmic region; here the MOSFET behaves like an ohmic resistance. Above the knee-point voltage U_K, i.e. for $U_{DS} > U_K$, the output current I_D is virtually uninfluenced by the

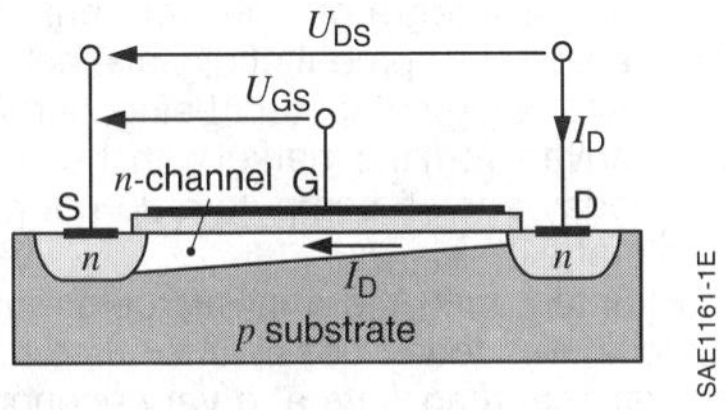

Figure 9: n-channel MOSFET
a) Diagram of self-conducting n-channel MOSFET,
b) Diagram of self-blocking n-channel MOSFET,
c) Characteristic curves.
1 Characteristic curve of self-conducting n-channel MOSFET,
2 Characteristic curve of self-blocking n-channel MOSFET.
U_{GS} Gate-source voltage, I_D Drain current,
U_T, U_T^* Threshold voltage.

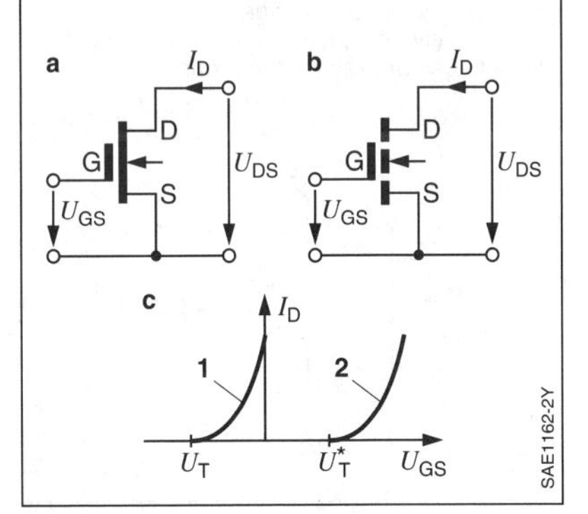

Figure 10: Output characteristic curve of a self-blocking n-channel MOSFET
U_{DS} Drain-source voltage,
U_{GS} Gate-source voltage,
I_D Drain current,
U_K Knee-point voltage.

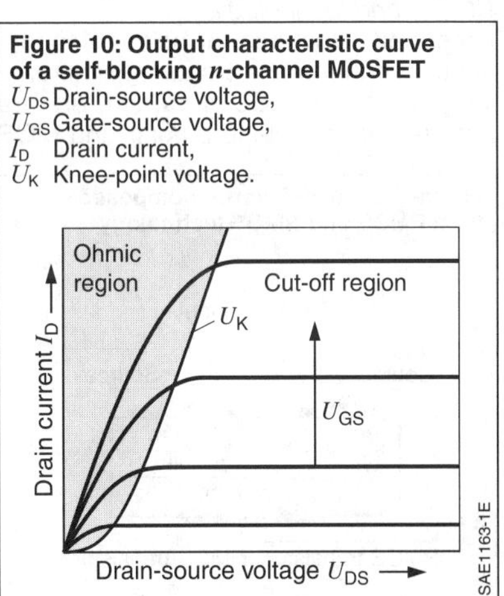

drain-source voltage U_{DS}; this region is known as the cut-off region. The amount of I_D is dependent only on the gate-source voltage U_{GS}. The formula is:

$$I_D = 0.5\,K\,(U_{GS} - U_T)^2$$

where K is the proportionality number (dependent among others on technological quantities) and U_T the threshold voltage from which the transistor conducts, i.e. a channel forms (see Figure 9c).

PMOS, NMOS, CMOS transistors
As well as the n-channel MOSFET (NMOS transistor), mixing the doping produces the PMOS transistor. As the electrons in the NMOS transistor are more mobile, it operates more rapidly than the PMOS device, although the latter was the first to become available due to the fact that it is physically easier to manufacture.

It is also possible to employ complementary MOS technology to pair PMOS and NMOS transistors in a single silicon chip; the resulting devices are called complementary MOS transistors (CMOS transistors, Figure 11). The specific advantages of the CMOS transistor are extremely low power dissipation, a high degree of immunity to interference, relative insensitivity to varying supply voltages, and suitability for analog signal processing and high-integration applications [2].

BCD hybrid technology
Integrated structures for power-electronics applications are becoming increasingly important. Such structures are realized by combining bipolar and MOS components on a single silicon chip, thereby utilizing the advantages of both technologies. BCD hybrid technology is a significant manufacturing process in automotive electronics and also facilitates the manufacture of MOS power components (DMOS). This technology is a combination of bipolar, CMOS and DMOS technologies [2].

Figure 11: CMOS inverter composed from PMOS and NMOS technology

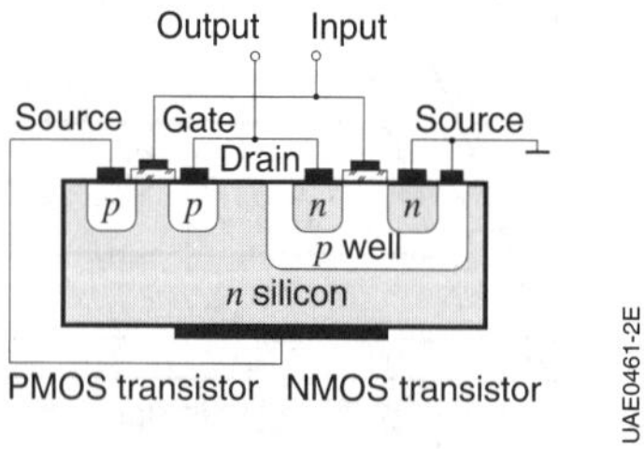

Output Input
Source Gate Source
Drain
p p n n
p well
n silicon
PMOS transistor NMOS transistor
UAE0461-2E

Operational amplifiers

Areas of application of operational amplifiers
The name "operational amplifier" (OPA) comes from analog computing technology and denotes an (almost) ideal amplifier. Because of its properties, it was used particularly in analog computers to solve nonlinear differential equations, i.e. as a summator, integrator and differentiator. The rapid development of digital electronics saw analog computers being increasingly driven from the market with the result that today analog computers do not play a role at all.

By integrating them in microelectronic circuits, it is today possible to offer such operational amplifiers at a very economical price on the market so that many amplifier applications can be put into effect. To achieve the desired properties, operational amplifiers in their integrated form

Figure 12: Basic diagram of an operational amplifier
+ Non-inverting amplifier input,
– Inverting amplifier input,
U_D Differential voltage between the two input potentials U_P and U_N where $U_D = U_P - U_N$
U_A Output voltage
U_{CC} Positive supply voltage
U_E Negative supply voltage
All voltages are referred to ground.

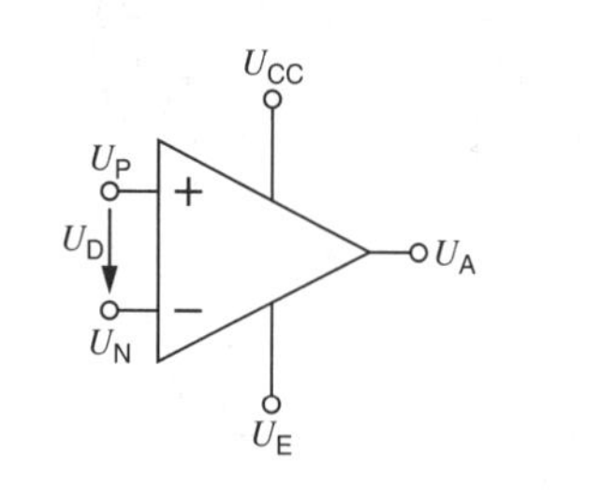

contain some (depending on the requirement 10 to 250) transistors, the number of which however is only of subordinate importance with regard to integration.

Starting out from a "normal" operational amplifier with voltage input and voltage output (VV operational amplifier), first the behavior of an ideal operational amplifier will be described and its use demonstrated. Then the real – i.e. actual – properties will be examined in more detail and its effect to be realized on the circuit analyzed.

Basic principles

The ideal standard operational amplifier is an amplifier with two inputs and (normally) one output (Figure 12). The inputs are the non-inverting input and the inverting input. The differential voltage U_D is amplified and then made available at the output as the output voltage U_A. The following formula applies:

$$U_A = A_D\, U_D.$$

A_D is the open-loop gain. The operational amplifier is connected to a positive supply voltage and to a negative supply voltage with regard to the frame potential. In the case of a unipolar supply, the negative supply voltage can be applied to the frame potential. The supply voltages are not usually indicated in many circuit diagrams. They are however very much needed in order to guarantee the energy supply to the operational amplifier.

The following variants of the operational amplifier are common (Figure 13):

- "Normal" operational amplifier (VV operational amplifier) with voltage input and voltage output
- Transconductance amplifier (VC operational amplifier) with voltage input and current output
- Transimpedance amplifier (CV operational amplifier) with current input and voltage output
- Current amplifier (CC operational amplifier) with current input and current output

Figure 13: Operational-amplifier types (diagram)
a) Normal operational amplifier (VV) with voltage input and voltage output.
b) Transconductance amplifier (VC) with voltage input and current output.
c) Transimpedance amplifier (CV) with current input and voltage output.
d) Current amplifier (CC) with current input and current output.

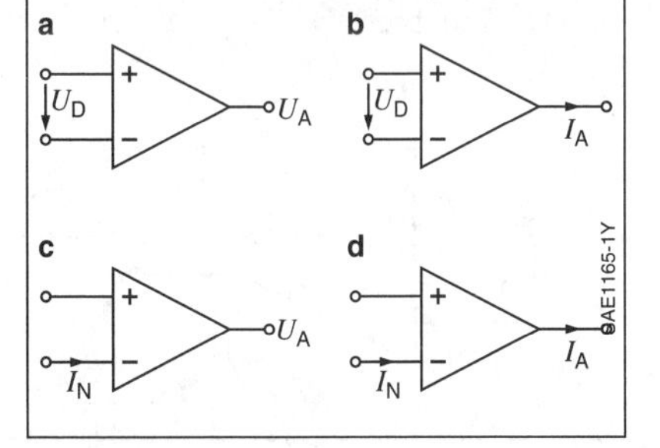

As a rule, the VV operational amplifier is used; this will now be explained in more detail in the following. Because the circuit arrangement is of crucial importance to the function of an operational amplifier, this will be addressed in more detail first. The distinction between positive feedback and negative feedback is important here. Furthermore, an ideal operational amplifier will be assumed when deriving the correlations.

Circuit arrangement: negative and positive feedback

Negative feedback counteracts the cause. In the operational amplifier, a connection is required from the output to the inverting input for this purpose (Figure 14). This connection can be implemented by a net-

Figure 14: Negative and positive feedback
U_D Differential voltage,
U_A Output voltage.

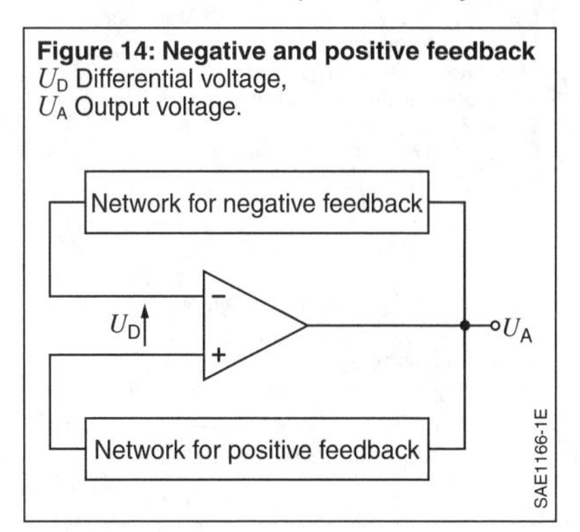

work. The cause of a change in the output voltage U_A is always a change in the differential voltage U_D at the input; negative feedback therefore always acts in such a way that the voltage U_D becomes small and ideally zero.

Positive feedback, in contrast to negative feedback, supports the cause of the change at the output. U_A is thus amplified by positive feedback, i.e. U_D increases as U_A changes and is therefore always not equal to zero. In this way, the output voltage U_A can assume only two stationary values, namely the maximum value or the minimum value.

From a control-engineering standpoint, a negative feedback results from the operational amplifier and the feedback as shown in Figure 15. Taking a high gain A_D into consideration, it follows:

$$U_A = A_D\, U_D = A_D\, (U_E - k\, U_A)$$

and for the total gain

$$A = \frac{U_A}{U_E} = \frac{A_D}{1 + k\, A_D} \approx \frac{1}{k}$$

It thus becomes clear that, in spite of a very high open-loop gain A_D on the part of the operational amplifier, with the aid of negative feedback a finite gain A can be set with the negative-feedback network. This is explained below in greater detail using examples.

Ideal and real operational amplifier

First, the properties of an ideal operational amplifier as shown in Figure 16 will be summarized. For further information, see [1].

- Common-mode input resistance between, in each case, one input and ground, where: $r_{GL\_P} = U_P/I_P$; $r_{GL\_N} = U_N/I_N$. Generally speaking, r_{GL} can be ignored.
- Differential input resistance between the two inputs; here: $r_D = (U_P - U_N)/I_P$. r_D is increased by negative feedback.
- Output resistance, differential quantity $r_A = dU_A/dI_A$. r_A is decreased by negative feedback.
- Offset voltage U_{OS}: Characteristic quantity for describing the fact that even in the event of a short circuit between the two inputs (i.e. $U_D = 0$) the output voltage U_A is not equal to zero.
- Common-mode rejection ratio (CMRR): This quantity describes the change in the output voltage U_A if the two input voltages U_P and U_N change simultaneously (in the case of periodic input signals co-phasally), i.e. U_D remains constant.
- Power-supply rejection ratio (PSRR): Change in the output voltage U_A on account of a change in the supply voltages.

The essential idealizations are:
- The open-loop gain A_D approaches infinity; in the case of negative feedback the following applies: $U_D = 0$.

Figure 15: Negative feedback
U_E Input voltage as setpoint value,
U_A Output voltage,
U_D Differential voltage is obtained at the summing point ⊕ $U_D = U_E - k\, U_A$.
Controlled system with gain A_D,
Feedback with gain k.

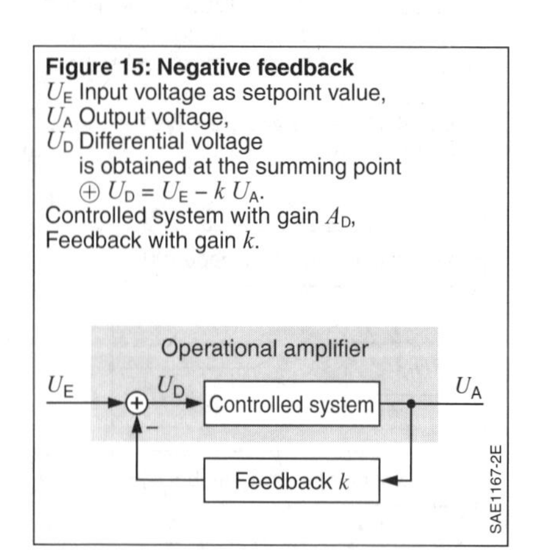

Figure 16: Ideal operational amplifier
U_D Differential voltage between the two input potentials U_P and U_N where $U_D = U_P - U_N$,
I_P, I_N Input currents,
U_A Output voltage,
I_A Output current.

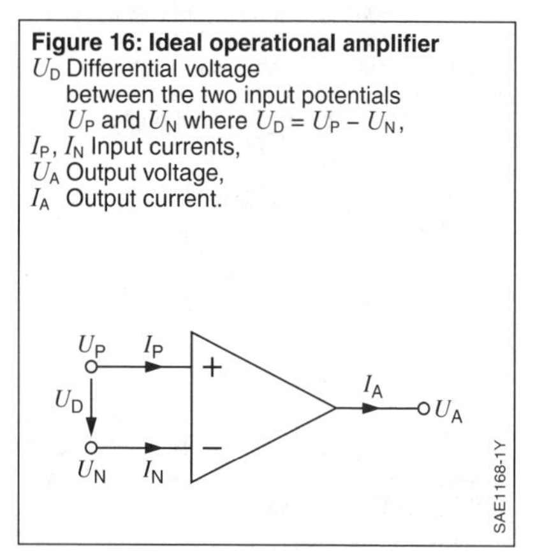

- The input currents I_N and I_P each approach zero.
- If I_N and I_P each approach zero, it follows that the common-mode and differential input resistances approach infinity.
- The offset voltage U_{OS} approaches zero.
- The output resistance R_A approaches zero.
- The common-mode rejection ratio (CMRR) approaches infinity, i.e. in the case of an equal and co-phasal change in the voltages U_P and U_N, U_A remains unchanged.
- The power-supply rejection ratio (PSRR) approaches infinity, i.e. in the case of a change in the supply voltage, U_A does not change.
- The behavior is not dependent on the frequency.

In reality, the above-mentioned idealizations do not apply entirely:

- The open-loop gain A_D is in the range of 10^4 to 10^7.
- The input currents I_N and I_P are in the range of 10 pA to 2 µA.
- The common-mode input resistance is in the range of 10^6 to 10^{12} Ω, the differential input resistance at up to 10^{12} Ω.
- The output resistance R_A is in the range of 50 Ω to 2 kΩ.
- The common-mode rejection ratio (CMRR) is in the range of 60 to 140 dB.
- The power-supply rejection ratio (PSRR) is in the range of 60 to 100 dB.
- The behavior is dependent on the frequency (low-pass behavior).

Basic circuits

The external circuit arrangement of an operational amplifier determines the behavior of the entire circuit. Here, negative feedback plays the dominant role, since it enables the gain to be set exactly through the choice of resistances. The function shall now be explained by reference to several examples.

Inverting amplifier

The basic circuit for an inverting amplifier is shown in Figure 17. The name can be attributed to the negative gain, i.e. in the case of a periodic input voltage, the output voltage U_A is always in phase opposition to the input voltage U_1. In the following, it is important that, on account of the negative feedback and the high open-loop gain A_D, the differential voltage U_D at the input be constantly zero, since the non-inverting and inverting inputs are kept at the same potential. Since, on account of the negative feedback, the differential voltage U_D is regulated to zero, this is referred to as a "virtual short circuit". "Virtual ground" is also referred to here, as the inverting input is actively kept at zero (i.e. at frame potential). In addition, the input currents are ignored, in particular $I_N = 0$ is set. The following applies:

$$I_{R1} = \frac{U_1}{R_1} \text{ and } I_{R2} = -\frac{U_A}{R_2}\,.$$

Where $I_{R1} = I_{R2}$ it then follows:

$$U_A = -\frac{R_2}{R_1}U_1\,.$$

The output voltage U_A is thus directly dependent on the input voltage U_1 and on the choice of resistances R_2 and R_1.

Non-inverting amplifier

The non-inverting amplifier can be handled along similar lines to the inverting amplifier (Figure 18). Due to negative feedback, $U_D = 0$. Where $I_{R1} = I_{R2}$, it is possible in accordance with the voltage divider – consisting of R_1 and R_2 – to calculate the voltage.

$$U_1 = \frac{R_1}{R_1 + R_2}U_A$$

Figure 17: Inverting amplifier
U_1 Input voltage,
U_D Differential voltage,
U_A Output voltage,
R_1, R_2 Network resistances,
I_{R1}, I_{R2} Currents in R_1 and R_2,
I_N Input current.

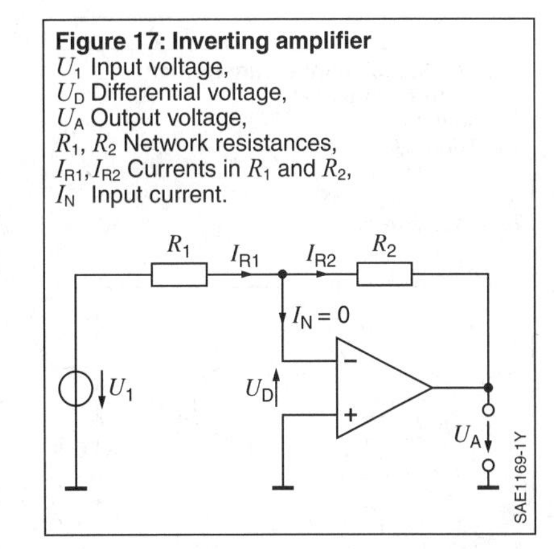

This means that

$$U_A = \frac{R_1 + R_2}{R_1} U_1 = \left(1 + \frac{R_2}{R_1}\right) U_1$$

The output voltage U_A is here also directly dependent on the input voltage U_1 and the choice of resistances R_2 and R_1; however, here the gain U_A/U_1 has at least the value one; U_A and U_1 are in phase.

A special case of the non-inverting amplifier is the isolation amplifier or impedance transformer. If R_1 assumes an infinitely high value (open loop) and R_2 is set equal to zero (short circuit) (Figure 19), then the gain is equal to one (i.e. $U_A = U_1$).

The advantage of this circuit is the property to the effect that the input-voltage source U_1 is not loaded with the internal resistance R_E, as the input current I_P is approximately equal to zero. This results in a negligible voltage drop over R_E and because $U_D = 0$, the input voltage U_1 is available at the output of the operational amplifier as U_A. This is important particularly for conditioning sensor signals, as here the sensor output voltage may not be loaded in many cases, i.e. any current flow from the sensor element can reduce the tappable voltage significantly.

Figure 18: Non-inverting amplifier
U_1 Input voltage,
U_D Differential voltage,
U_A Output voltage,
R_1, R_2 Network resistances,
I_{R1}, I_{R2} Currents in R_1 and R_2,
I_N Input current.

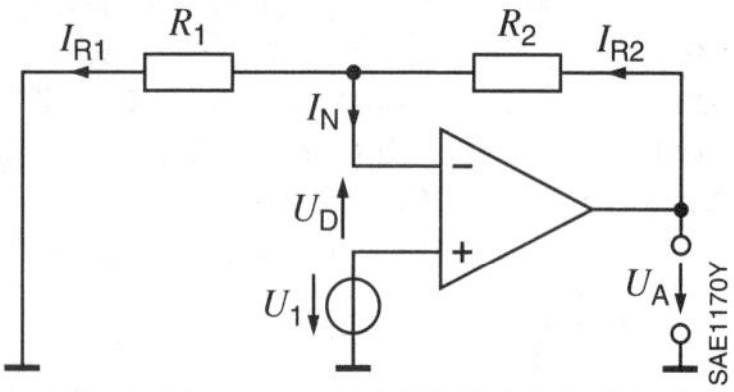

Figure 19: Impedance transformer or isolation amplifier
U_1 Input voltage,
U_D Differential voltage,
U_A Output voltage,
R_E Input resistance,
I_P Input current.

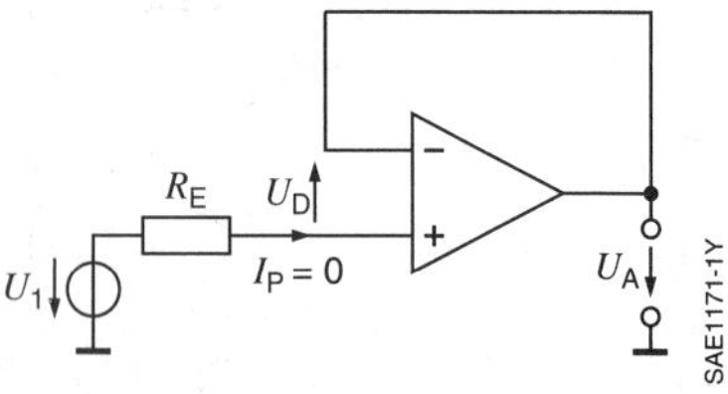

Subtracting amplifier

The subtracting amplifier (Figure 20) can be considered as a common variant of the two previously mentioned circuits. On account of the superposition theorem (superposition principle), the relationship between the output voltage U_A and the input voltages U_1 and U_2 can be derived.

$$U_A = \frac{R_2}{R_1} (U_2 - U_1) .$$

Instrument amplifier

Particularly in sensor systems, differential voltages must often be picked off at bridge circuits and amplified without the sensor voltage or the bridge voltage being subjected to an unacceptably high load. This can be implemented by means of a high-resistance voltage tap. An instrument amplifier, which outputs the difference between two potentials U_2 and U_1 as amplified output voltage U_A, can be used for this purpose. The instrument amplifier can be subdivided into two parts: the pre-amplifier and a subtracting amplifier (Figure 20) with further amplification. Figure 21 shows the basic input circuit for the pre-amplification of an instrument amplifier.

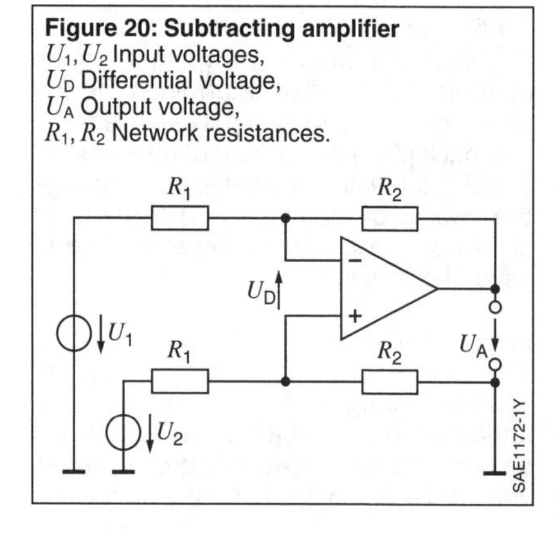

Figure 20: Subtracting amplifier
U_1, U_2 Input voltages,
U_D Differential voltage,
U_A Output voltage,
R_1, R_2 Network resistances.

According to the rule of negative feedback, the voltage difference between the inverting and non-inverting inputs is equal to zero. In each case, the current I can flow through the resistors R and R', as the input currents I_{N1} and I_{N2} can be ignored. The following applies:

$$I = \frac{U_1 - U_2}{R'} = \frac{U_{A1} - U_{A2}}{2R + R'}, \text{ thus}$$

$$U_{A1} - U_{A2} = (U_1 - U_2)\left(\frac{2R}{R'} + 1\right).$$

In this way, the amplified difference between the two voltages U_1 and U_2 is obtained as the voltage difference U_D between the two outputs of the two operational amplifiers. In order to output this voltage U_D as an output voltage referred to ground U_A, a subtracting amplifier can be connected in series (Figure 20), where U_{A1} instead of U_1 and U_{A2} instead of U_2 are supplied.

Important characteristic data

For many applications, the operational amplifiers must demonstrate properties which partly contradict each other. There are a large number of operational amplifiers, which are optimized for different fields of application. Basically, the data for specific operating points or operating ranges are specified.

Figure 21: Pre-amplification of an instrument amplifier
U_1, U_2 Input voltages
I_{N1}, I_{N2} Input currents, I Current
R, R' Resistances
U_{A1}, U_{A2} Output voltages, in relation to ground.

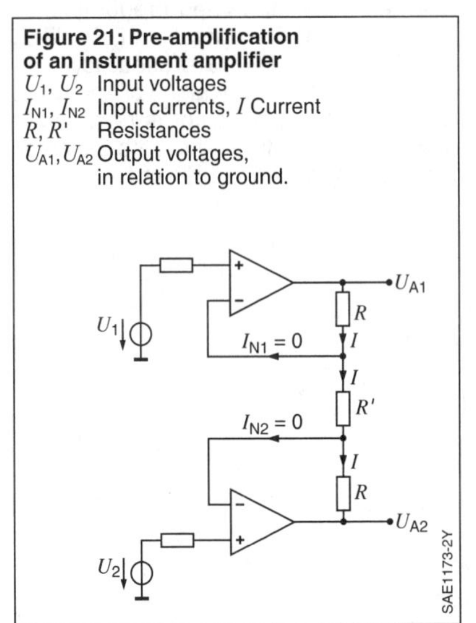

Temperature range

In the field of consumer electronics, a temperature range of between 0 °C and 70 °C is customary. For the extended industrial field, a temperature range of between –20 °C and +70 °C is often specified; this range is required above all for devices which are used outside buildings. For military applications, a temperature range of –55 °C to +125 °C is specified. These demands do not, however, cover all the requirements for use in motor vehicles; for example, even higher temperatures occur in the engine compartment or in brake systems.

Offset voltage

The characteristic quantity for describing the fact that, even in the event of a short circuit between the two inputs (i.e. for $U_D = 0$), the output voltage U_A is not equal to zero is termed the offset voltage U_{OS}. This thus acts like a voltage applied from an external source U_D and is added to this. The offset voltage U_{OS} can, for example, be determined by the fact that at the input that voltage which sets the output voltage U_A equal to zero is determined (Figure 22). The offset voltage U_{OS} stems, among other things, from asymmetries in the internal circuit arrangement of the two inputs and is typically in the range of a few V to a few mV.

However, as well as the value of the offset voltage U_{OS}, the temperature effect and long-term stability are also highly important. Some operational amplifiers offer the possibility of compensating the offset

Figure 22: Offset voltage
U_A Output voltage
U_D Differential voltage between the inputs
U_{OS} Offset voltage
$U_{A,max}$ Maximum output voltage
$U_{A,min}$ Minimum output voltage

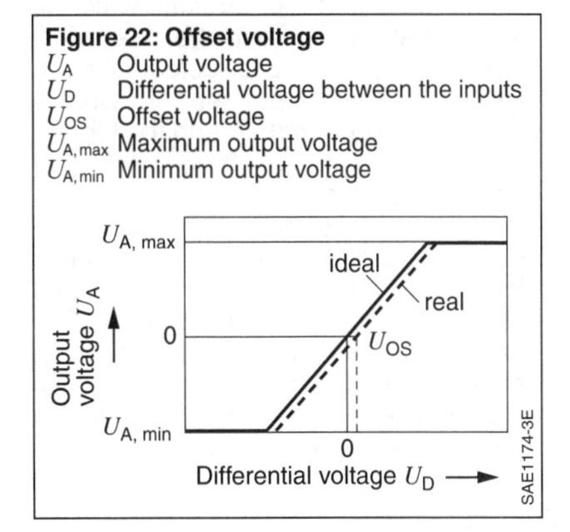

voltage by means of an external circuit arrangement – provided this is not already effected by internal circuit-engineering measures. It is important in this connection that the input voltage can also drift on account of the temperature effect; thus soldered junctions can represent thermocouples with a voltage of a magnitude of 10 to 100 µV/K.

Input resistances and currents

On account of the, as a rule, very low input currents I_N and I_P, very high input resistances, which are sometimes in the high megaohm range, are accordingly obtained. A distinction is made here between the common-mode input resistance (resistance between an input in each case and ground) and the differential input resistance between the two inputs.

The inputs of customary operational amplifiers form transistors; either bipolar transistors in which the base in each case is activated or MOS field-effect transistors in which the gate is recharged. The small input currents are explained in this way. When bipolar transistors are used, these are base currents and are in the range of A. When MOSFET are used, the corresponding gate currents are obtained which are required to recharge the associated gate capacity. The latter are proportional to the operating frequency and are usually in the range of pA.

The input bias current can cause an input-voltage error in high-resistance circuits. This can be compensated for by connecting identical impedances to the two inputs, as then in each case the same voltage drops and the differential voltage U_D remains unaffected. Like the offset voltage, the input current can also drift over the temperature and the time.

Output resistance

The output of the operational amplifier can be described by a series connection of an ideal voltage source and a resistance; the latter is then the output resistance R_A. This resistance limits the output current. In general, operational amplifiers can drive output currents of 20 mA; there are also types with an output current of up to 10 A.

Slew rate

The slew rate (SR) denotes the maximum possible change in the output voltage U_A per period of time, i.e. the maximum value for dU_A/dt. The values for the slew rate for conventional operational amplifiers are in the range of less than 1 V/µs to more than 1 V/ns.

Noise

Noise can be described by specifying the noise-voltage density or the noise-current density. Usually the noise-voltage density U_R' is given in nV/√Hz.

The effective value of the noise voltage U_R (this also applies to the noise current) is obtained from the respective key figure multiplied by the root of the bandwidth B being considered:

$$U_R = U_R' \sqrt{B}$$

For an amplifier circuit, the total effective noise-voltage density is obtained as the root of the sum of the squares of the effective values.

$$U'_{R,tot} = \sqrt{(U_{R,1})^2 + \text{to } (U_{R,m})^2} \, .$$

m here denotes the number of noise terms.

The noise is determined predominantly at the input of the operational amplifier. If JFET or MOSFET are used, low current but comparatively high voltage noise is obtained. The behavior is reversed in operational amplifiers which are based on bipolar transistors (see [1] and [4]).

Monolithic integrated circuits

Monolithic integration

In monolithic integrated circuits (IC), components are assembled on a single piece of monocrystalline silicon (substrate). Semiconductor processes are used for creating layers (e.g. epitaxy), removing layers and changing material properties (e.g. doping). This technology enables complex circuits to be housed in the smallest of spaces.

Planar technology is based on the oxidation of silicon wafers, which is a relatively simple process, and the speed at which the dopants penetrate into silicon, which is exponentially greater than the speed at which they enter the oxide. Doping only occurs at locations where openings are present in the oxide layer. The specific design requirements of an individual integrated circuit determine the precise geometric configuration, which is applied to the wafer in a photolithographic process. All processing procedures (oxidizing, etching, doping, and depositing) progress consecutively from the surface plane (planar).

Planar technology makes it possible to manufacture all circuit components (e.g. resistors, capacitors, diodes, transistors) and the associated conductor strips on a single silicon chip in a unified manufacturing process. Monolithic integrated circuits are built from semiconductor components.

This integration generally comprises a subsystem within the electronic circuit and increasingly comprises the entire system (a system on a chip).

Because of the ever-increasing component density (integration density), the third dimension, i.e. the plane vertical to the surface, is also increasingly being utilized in the design. In this way, particularly in the field of power electronics, advantages such as lower resistances, smaller losses and thus also higher current densities can be achieved.

Integration level

The integration level is a measure of the number of function elements per chip. The following technologies relate to the level of integration (and chip surface):

- SSI (Small Scale Integration) with up to several hundred function elements per chip and a mean chip surface area of 1 mm^2. But the chip surface area can be very much larger in circuits with high power outputs (e.g. smart power transistors).
- MSI (Medium Scale Integration) with a several hundred to 10,000 function elements per chip and a mean chip surface area of 8 mm^2.
- LSI (Large Scale Integration) with up to 100,000 function elements per chip and a mean chip surface area of 20 mm^2.
- VLSI (Very Large Scale Integration) with up to 1 million function elements per chip and a mean chip surface area of 30 mm^2.
- ULSI (Ultra Large Scale Integration) with over 1 million function elements per chip (flash memory today contains up to 20 billion transistors per chip), a surface area of up to 300 mm^2 and the smallest structure sizes of less than 30 nm.

Computer-aided simulation and design methods (CAE and CAD) are essential elements in the manufacture of integrated circuits. Entire function modules are used in VLSI and ULSI, otherwise the time expenditure and failure risk would make development impossible. In addition, simulation programs are used to detect any defects made.

References
[1] U. Tietze, Ch. Schenk: Halbleiter-Schaltungtechnik, 13th Edition, Springer-Verlag, 2009.
[2] A. Führer, K. Heidemann, W. Nerreter: Grundgebiete der Elektrotechnik, Bände 1–3, Carl-Hanser-Verlag.
[3] R. Ose: Elektrotechnik für Ingenieure, 4. Auflage, Carl-Hanser-Verlag, 2007.
[4] R. Müller: Rauschen, Springer-Verlag, 1989.

Chemistry

Elements

Periodic table

Structure of the periodic table

The atoms of the chemical elements are made up of positively charged protons, uncharged neutrons, and negatively charged electrons [1]. In the periodic table (Tables 1 and 2) the elements are laid out in order of increasing number of protons – i.e. increasing nuclear charge number – and in order of their atomic mass. The atomic mass is essentially determined by the total number of nuclear constituents, i.e. the sum total of protons and neutrons. The total number of protons and neutrons is also called the mass number. The number of protons corresponds to the atomic number. The neutral element atoms always have as many protons and electrons.

Groups and periods

The elements are divided in the periodic table into different groups (vertical columns) and periods (horizontal rows). The nested structure of the groups is due to the fact that the electrons always occupy the lowest energy levels. The position of these energy levels, which are also called electron orbitals and indicate the probability of electrons being present around the atomic nucleus, can be derived using quantum mechanics.

Quantum numbers

The structure of the periodic table is based on four quantum numbers (principal, secondary, magnetic, and spin quantum numbers), with which the four electron orbitals with the designations s, p, d and f can be calculated. When the electrons are arranged into these orbitals in order of increasing energy, this creates groups of elements which demonstrate a similar manner of reaction across all the periods. This manner of reaction is only minimally influenced by the electrons on the inner orbitals. The crucial factors are the energy and number of outer electrons. The outer electrons are frequently referred to as "valence electrons".

Main groups

The elements in groups Ia, IIa and IIIa... VIIIa are called main-group elements. The main-group elements include hydrogen and the alkali metals (Ia), the alkaline-earth metals (IIa), the boron (IIIa), carbon (IVa) and nitrogen groups (Va), the chalcogens (VIa), the halogens (VIIa), and the noble gases (VIIIa). The electrons of the main-group elements of the 1st period, hydrogen and helium, are found exclusively in s-orbitals. In the case of the other main-group elements in the 2nd through 7th periods, the electrons also occupy p-orbitals starting from the main group IIIa.

Secondary groups

The elements of the secondary groups Ib, IIb and IIIb...VIIIb with electrons in d-orbitals all have a metallic character. The copper group is assigned to the secondary group Ib because its electron configuration shows similarities to the main group Ia. The elements in both groups tend to form salts from monovalent ions. The same applies to the groups IIa and IIb. Divalent metal compounds are formed in the group of alkaline-earth metals and the zinc group IIb. The designations for the secondary groups IIIb...VIIb likewise have their roots in the electron configuration and provide an indication of the maximum valence of these metal ions. The elements iron, cobalt and nickel as well as the elements underneath called higher homologs are brought together as secondary group VIIIb due to their marked chemical similarity.

Table 1: Periodic table of elements

Ia	IIa	IIIb	IVb	Vb	VIb	VIIb	VIIIb	VIIIb	VIIIb	Ib	IIb	IIIa	IVa	Va	VIa	VIIa	VIIIa
1 **H** 1.008																	2 **He** 4.003
3 **Li** 6.941	4 **Be** 9.012											5 **B** 10.811	6 **C** 12.011	7 **N** 14.007	8 **O** 15.999	9 **F** 18.998	10 **Ne** 20.180
11 **Na** 22.990	12 **Mg** 24.305											13 **Al** 26.982	14 **Si** 28.086	15 **P** 30.974	16 **S** 32.066	17 **Cl** 35.453	18 **Ar** 39.948
19 **K** 39.098	20 **Ca** 40.078	21 **Sc** 44.956	22 **Ti** 47.87	23 **V** 50.942	24 **Cr** 51.996	25 **Mn** 54.938	26 **Fe** 55.845	27 **Co** 58.933	28 **Ni** 58.693	29 **Cu** 63.546	30 **Zn** 65.39	31 **Ga** 69.723	32 **Ge** 72.61	33 **As** 74.922	34 **Se** 78.96	35 **Br** 79.904	36 **Kr** 83.80
37 **Rb** 85.468	38 **Sr** 87.62	39 **Y** 88.906	40 **Zr** 91.224	41 **Nb** 92.906	42 **Mo** 95.94	43 **Tc** (98)	44 **Ru** 101.07	45 **Rh** 102.906	46 **Pd** 106.42	47 **Ag** 107.868	48 **Cd** 112.411	49 **In** 114.818	50 **Sn** 118.710	51 **Sb** 121.760	52 **Te** 127.60	53 **I** 126.904	54 **Xe** 131.29
55 **Cs** 132.905	56 **Ba** 137.327	57 **La*** 138.906	72 **Hf** 178.49	73 **Ta** 180.948	74 **W** 183.84	75 **Re** 186.207	76 **Os** 190.23	77 **Ir** 192.217	78 **Pt** 195.078	79 **Au** 196.967	80 **Hg** 200.59	81 **Tl** 204.383	82 **Pb** 207.2	83 **Bi** 208.980	84 **Po** (209)	85 **At** (210)	86 **Rn** (222)
87 **Fr** (223)	88 **Ra** (226)	89 **Ac**** (227)	104 **Rf** (267)	105 **Db** (268)	106 **Sg** (271)	107 **Bh** (267)	108 **Hs** (277)	109 **Mt** (274)	110 **Ds** (282)	111 **Rg** (280)	112 **Cn** (285)	113 **Uut** (284)	114 **Fl** (289)	115 **Uup** (291)	116 **Lv** (293)	117 **Uus** (292)	118 **Uuo** (294)
			*	58 **Ce** 140.116	59 **Pr** 140.908	60 **Nd** 144.24	61 **Pm** (145)	62 **Sm** 150.36	63 **Eu** 151.964	64 **Gd** 157.25	65 **Tb** 158.925	66 **Dy** 162.50	67 **Ho** 164.930	68 **Er** 167.26	69 **Tm** 168.934	70 **Yb** 173.04	71 **Lu** 174.967
			**	90 **Th** 232.038	91 **Pa** 231.036	92 **U** 238.029	93 **Np** (237)	94 **Pu** (244)	95 **Am** (243)	96 **Cm** (247)	97 **Bk** (247)	98 **Cf** (252)	99 **Es** (252)	100 **Fm** (257)	101 **Md** (258)	102 **No** (259)	103 **Lr** (262)

All elements are arranged sequentially according to atomic number (proton number). The horizontal rows are called periods, the vertical columns are called groups. The relative atomic masses are indicated below the element symbols.
The values given in parentheses are the mass numbers (nucleon numbers) of the stablest isotopes of radioactive elements.

Table 2: Designations of chemical elements

Element	Symbol	Atomic number
Actinium	Ac	89
Aluminum	Al	13
Americium[1]	Am	95
Antimony	Sb	51
Argon	Ar	18
Arsenic	As	33
Astatine	At	85
Barium	Ba	56
Berkelium[1]	Bk	97
Beryllium	Be	4
Bismuth	Bi	83
Bohrium[1]	Bh	107
Boron	B	5
Bromine	Br	35
Cadmium	Cd	48
Carbon	C	6
Cesium	Cs	55
Calcium	Ca	20
Californium[1]	Cf	98
Cerium	Ce	58
Chlorine	Cl	17
Chromium	Cr	24
Copernicium[1]	Cn	112
Copper	Cu	29
Cobalt	Co	27
Curium[1]	Cm	96
Darmstadtium[1]	Ds	110
Dubnium[1]	Db	105
Dysprosium	Dy	66
Einsteinium[1]	Es	99
Erbium	Er	68
Europium	Eu	63
Fermium[1]	Fm	100
Fluorine	F	9
Flerovium[1]	Fl	114
Francium	Fr	87
Gadolinium	Gd	64
Gallium	Ga	31
Germanium	Ge	32
Gold	Au	79
Hafnium	Hf	72
Hassium[1]	Hs	108
Helium	He	2
Holmium	Ho	67
Hydrogen	H	1

Element	Symbol	Atomic number
Indium	In	49
Iodine	I	53
Iridium	Ir	77
Iron	Fe	26
Krypton	Kr	36
Lanthanum	La	57
Lawrencium[1]	Lr	103
Lead	Pb	82
Lithium	Li	3
Livermorium[1]	Lv	116
Lutetium	Lu	71
Magnesium	Mg	12
Manganese	Mn	25
Meitnerium	Mt	109
Mendelevium[1]	Md	101
Mercury	Hg	80
Molybdenum	Mo	42
Neodymium	Nd	60
Neon	Ne	10
Neptunium[1]	Np	93
Nickel	Ni	28
Niobium	Nb	41
Nitrogen	N	7
Nobelium[1]	No	102
Osmium	Os	76
Oxygen	O	8
Palladium	Pd	46
Phosphorus	P	15
Platinum	Pt	78
Plutonium[1]	Pu	94
Polonium	Po	84
Potassium	K	19
Praseodymium	Pr	59
Promethium	Pm	61
Protactinium	Pa	91
Radium	Ra	88
Radon	Rn	86
Rhenium	Re	75
Rhodium	Rh	45
Roentgenium[1]	Rg	111
Rubidium	Rb	37
Ruthenium	Ru	44
Rutherfordium[1]	Rf	104

Element	Symbol	Atomic number
Samarium	Sm	62
Scandium	Sc	21
Seaborgium[1]	Sg	106
Selenium	Se	34
Silicon	Si	14
Silver	Ag	47
Sodium	Na	11
Strontium	Sr	38
Sulfur	S	16
Tantalum	Ta	73
Technetium	Tc	43
Tellurium	Te	52
Terbium	Tb	65
Thallium	Tl	81
Thorium	Th	90
Thulium	Tm	69
Tin	Sn	50
Titanium	Ti	22
Tungsten	W	74
Ununoctium[1,2]	Uuo	118
Ununpentium[1,2]	Uup	115
Ununseptium[1,2]	Uus	117
Ununtrium[1,2]	Uut	113
Uranium	U	92
Vanadium	V	23
Xenon	Xe	54
Ytterbium	Yb	70
Yttrium	Y	39
Zinc	Zn	30
Zirconium	Zr	40

[1] Artificially produced; does not occur naturally.
[2] Provisional IUPAC nomenclature (International Union of Pure and Applied Chemistry).

Periods with f-orbitals
In the 6th and 7th periods a further series of energy levels, the f-orbitals, are available in each case after the secondary group IIIb lanthanum and actinium. These are taken up by the electrons of the elements of the lanthanides (6th period) and actinides (7th period). The lanthanides are also known by the term "rare earths". All actinides are radioactive.

The arrangement of the elements in the periodic table according to the energy position of their electron orbitals becomes clear in the energy-level diagram (Figure 1). It can be seen that for higher orbital energies after the occupation of the s-orbital the p-, d- and f-orbitals subordinate to this principal quantum number are no longer automatically filled with electrons. For example, it is now understandable why the secondary-group elements scandium through zinc with the orbital energies 3d after the element calcium are placed into the 4th period of the periodic table. For higher nuclear charge numbers, however, the differences between the orbital energies are so small that the energy sequence of the orbitals shown in the energy-level diagram no longer applies to each individual element in the periodic table. In these cases the exact energy position is influenced by the partial, semi- or full occupation of the orbitals by electrons. This can be seen in the case of the lanthanides with the 4f-orbitals, which are only placed after the element lanthanum into the 6th period and not – as would be expected according to the energy-level diagram – already after the element barium.

Figure 1: Energy-level diagram of the electron orbitals

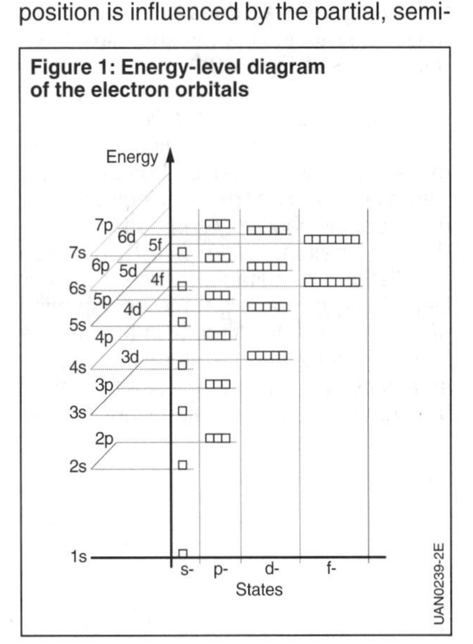

Isotopes
Isotopes are atoms of the same element with the same number of protons but with a different mass number, i.e. with a different number of neutrons. To give isotopes a unique designation, the mass number is given at top left next to the atom symbol and the number of protons is given at bottom left so that the number of neutrons can be calculated straight away by subtraction. Most naturally occurring elements are isotopic mixtures. Carbon, for example, is made up 98.89 % of $^{12}_{6}C$ and 1.11 % of $^{13}_{6}C$. The proportion of $^{14}_{6}C$ is at 10^{-10} % very low. $^{14}_{6}C$ in comparison with $^{12}_{6}C$ and $^{13}_{6}C$ not a stable isotope. The ratio of $^{14}_{6}C$ to $^{12}_{6}C$ is used in the radiocarbon method to determine the age of organic matter.

Nuclides
The term nuclide is broadly defined. All atoms that differ in the constitution of their nucleus are called nuclides. The number of nuclides therefore corresponds to the number atom types. The 8th Edition of the Karlsruhe Nuclide Chart in 2012 lists 3,847 experimentally verified nuclides and isomers. Isomers are nuclides which have the same mass number and thus the same number of protons and neutrons. Isomeric nuclides, however, show differences in the internal state of the atomic nuclei. Atomic nuclei can assume excited states as well as the normal state. Only around 270 nuclides are stable. The majority of nuclides are radioactive and are therefore called radionuclides.

Radioactive decay

α-decay

In the case of natural radioactivity [2] α-decay is observed when the atomic nucleus emits a double positively charged helium nucleus He^{2+} consisting of two protons and two neutrons. The mass number of the atomic nucleus is reduced by 4 and the nuclear charge number by 2. A different element is created (example, see Figure 2a).

Figure 2: Types of radiation for natural radioactive decay
a) α-radiation,
b) β^--radiation,
c) β^+-radiation,
d) γ-radiation,
n Neutron,
p Proton,
e^- Electron,
e^+ Positron.

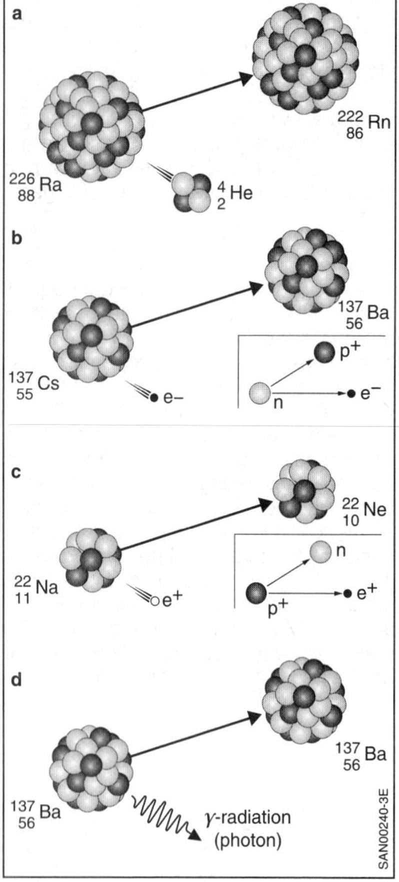

a
$^{226}_{88}Ra$ $^{4}_{2}He$ $^{222}_{86}Rn$

b
$^{137}_{55}Cs$ e− $^{137}_{56}Ba$ n p^+ e^-

c
$^{22}_{11}Na$ e^+ $^{22}_{10}Ne$ p^+ n e^+

d
$^{137}_{56}Ba$ γ-radiation (photon) $^{137}_{56}Ba$

SAN00240-3E

β-decay

An element transmutation also occurs in the case of radioactive β-decay. β^- decay is observed when a neutron from the atomic nucleus emits an electron (example, see Figure 2b). This increases the number of protons by 1 and the element with the next higher atomic number is created. When instead of an electron a positron, i.e. a positively charged particle, is emitted, β^+-decay takes place. A proton is transformed into a neutron in the case of β^+-decay. Where the mass number is the same, the nuclear charge number is therefore decreased by 1 such that the preceding element in the periodic table is created (example, see Figure 2c). β-decay also witnesses the release of antineutrinos (in the case of β^--delay) or neutrinos (in the case of β^+-decay), which cannot be discussed further at this stage.

γ-decay

In the case of γ-decay the atomic nucleus emits internal energy from excited nucleus states which for the most part arise during α- or β-decay. The γ-radiation emitted during γ-decay does not change the mass number; the element remains intact (example, see Figure 2d).

Artificial element transmutation

Elements can also be artificially transmuted into each other by bombardment with high-energy particles. Neutrons, singly or doubly positively charged hydrogen (H^+) and helium nuclei (He^{2+}), but also γ-radiation are suitable for this purpose. Depending on the type and acceleration of the particles, this results in the particle being absorbed in the atomic nucleus – sometimes also involving the emission of a proton or neutron – or in nuclear fission.

Half-life
Radioactive decay is a monomolecular reaction (first-order reaction) in which the decaying share per unit of time is directly proportional to the available quantity. Because the available quantity constantly diminishes due to the radioactive decay continuously taking place, the reaction rate decreases more and more. The half-life of the radionuclide is specified so that the rate of radioactive decay can be described independently of the quantity. The half-life is the period of time after which half the atomic nuclei have decayed. The shorter the half-life of the radioactive element, the higher its specific activity, i.e. the more nuclear decays occur per unit of mass.

Element spectroscopy
All chemical elements differ in their atomic structure and can therefore be identified by means of their electron spectra and quantified into mixtures [3]. To release electrons from their energy levels, it is necessary to supply energy to the order of their binding energy from an outside source.

Figure 3: Electron migrations in element spectroscopy

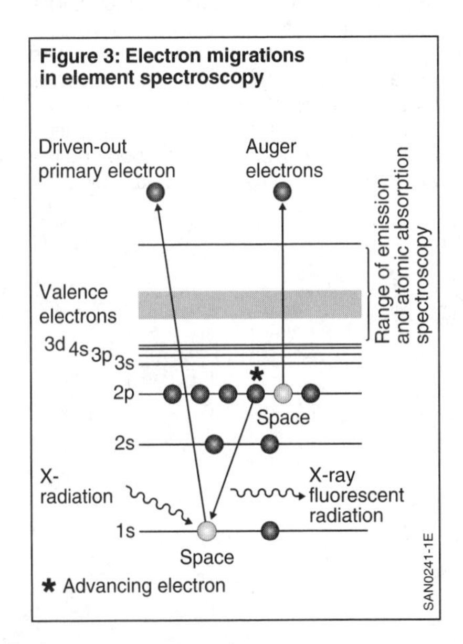

X-ray fluorescence analysis
The inner electrons more tightly bound by the attraction of the positive atomic nucleus can only be released with X-radiation or with electrons of sufficient energy. X-ray fluorescence analysis (XRF) is based on the fact that the spaces created on the more favorable inner energy levels are filled by higher-energy electrons. These advancing electrons release the excess energy in the form of fluorescent radiation. The energy of the fluorescent radiation corresponds exactly to the difference between the higher and the lower energy levels and is thus element-specific (Figure 3).

Auger electron spectroscopy
The energy released by advancing electrons can be output as fluorescent radiation, but can also be transmitted to other electrons. An electron can leave the atom with the transmitted energy. These electrons from the so-called Auger process likewise contain highly specific element information, since the following contributions determine the energy of the Auger electron: The orbital energy of the originally driven-out electron, the energy that is released by the advancing electron, and the energy of the electron emitted in the Auger process itself (Figure 3). The energy of the emitted electrons is analyzed in Auger electron spectroscopy.

Emission spectrometry
The less tightly bound outer valence electrons can already be thermally excited and thereby raised to higher, unoccupied energy levels. Because the energy from the flame of a gas torch is not sufficient to excite the valence electrons of all the elements, a plasma is often used for excitation. In the case of inductively coupled plasma optical emission spectrometry (ICP-OES), the light released during the fallback of the valence electrons that is in the visible and ultraviolet range is registered. As with the inner electrons, the energy difference between the outer occupied and unoccupied energy levels is also element-specific.

Atomic absorption spectroscopy
Instead of emitted radiation, light absorption can also be measured through the excitation of the valence electrons. To this end, in atomic absorption spectroscopy (AAS) the emitted radiation of the element to be analyzed is irradiated by a hollow-cathode lamp. When the element in the sample matches the element of the hollow cathode, the excitation of the electrons in the sample causes a weakening in the intensity of the excitation light. Because the verification of different elements accordingly requires an extensive range of materials for hollow-cathode lamps and can only be measured sequentially, atomic absorption spectroscopy has not established itself in multi-element analyses.

X-ray photoelectron spectroscopy
In X-ray photoelectron spectroscopy (XPS) electrons are driven out from all the energy levels in the atom by high-energy X-rays. The element-specific binding energy can be inferred from the kinetic energy of these electrons. Only the valence electrons are released by the ultraviolet light used instead of X-rays in ultraviolet photoelectron spectroscopy (UPS). The kinetic energy of valence electrons can be determined very precisely such that even conclusions can be made as to the type of chemical bonds by way of differences in the orbital energies.

X-ray diffraction
X-ray diffraction on solids is, as the name suggests, not a spectroscopic procedure. The X-radiation diffracted on the crystal lattice is registered with angular resolution, i.e. according to its diffraction angle [4]. The angle position, intensity and width of the diffraction peaks enable the lattice structure to be analyzed and thus crystals to be identified (crystal-structure analysis). Crystallite size, preferred orientations (textures) and lattice distortions can also be determined in mixtures. Internal mechanical stresses of components can likewise be determined.

Chemical bonds

The type of bond which tends to be adopted by the individual elements is determined by the number and arrangement of electrons around the atomic nucleus (electron configuration) [5]. The nature of these bonds can differ greatly. A distinction is made between ionic, covalent and metallic bonds. Single atoms without a chemical bond occur in nature in noble gases and in some elements in the vapor phase.

Additional, but far weaker interactions exist between molecules, i.e. particles which are composed of two or more atoms. The weak attractive forces, which have different physical causes, mean that not only ions but also molecules adopt a short-range order, which is only eliminated in the gaseous phase.

Ionic bonds
Ionic bonds tend to form in compounds between metallic and non-metallic elements. This results in the transfer of electrons. Metal atoms (electron donors) lose electrons and become positively charged cations. By gaining electrons, the non-metal atoms (electron acceptors) become negatively charged anions. The terms "cation" for a positively and "anion" for a negatively charged ion are so called because these ions move in an aqueous solution in the electric field to the oppositely charged electrodes, i.e. the cation to the negatively charged cathode and the anion to the positively charged anode.

The number of lost and gained electrons is determined by the electron configuration of the element. Ionic compounds with configurations in which the p-, d- and f-orbitals are empty, half-full and full are especially preferred.

The ionic compounds created are called salts. In a solid the cation and anion are arranged in an ionic lattice whose structure is determined by the ratio of ionic radii. When salts are melted or dissolved in water, the forces between the ions in the lattice must be overcome. Therefore heat of fusion (enthalpy of fusion) must be supplied. The energy required for dissolving (heat of solution) is determined by two opposing components. Firstly, energy to dissolve the crystal lattice must be applied (heat of dissociation); secondly, energy is released by coordination of the solvent molecules to the dissolved-out ions (heat of solvation). If the solvent is water, the heat is called heat of hydration. If the energy of dissociation is greater than the heat of solvation, the solution cools during the dissolving process. When anhydrous metal salts are dissolved in which initially water is also introduced into the ionic lattice, the heat released by hydration is often greater than the energy consumption to overcome the lattice forces, and the solution heats up.

Covalent bonds

Bonds of electron pairs which are created between two neutral atoms from the unpaired electrons in the outer orbitals (valence electrons) are called atomic bonds. The term "covalent" illustrates that two identical or similar atoms can, by forming a common electron pair, achieve an energetically more favorable electron configuration without involving an electron transfer and with it a change in the valency by the formation of ions. The configuration of the noble gas from the period in which the element is found is usually adopted as an energetically more favorable state.

In a methane molecule (CH_4) all the atoms have a noble-gas configuration. Carbon has in the p-orbitals two valence electrons and four free spaces for further electrons which are made available by the four hydrogen atoms. Carbon thus achieves the noble-gas configuration of neon. Because the two atoms share the common bonding-electron pair, the hydrogen atoms, which only have in each case one valence electron in the s-orbital, also attain the noble-gas configuration of helium through the common electron pairs with the carbon. In the case of the main-group elements of the 2nd period, the filling of empty orbitals with valence electrons of other atoms until the noble-gas configuration is attained is called the octet rule, because then a total of eight electrons, two in s- and six in p-orbitals, are present.

An atom can enter into several covalent bonds with a neighboring atom, such as for example carbon double bonds in ethylene ($H_2C{=}CH_2$) or triple bonds in acetylene ($H{-}C{\equiv}C{-}H$).

In the octet rule existing free electron pairs which have not bonding function are also taken into account. The three-dimensional structure of molecules is determined by the total number of atoms, by the type and number of bonding-electron pairs, and by the available free electron pairs. For example, methane (CH_4), ammonia (NH_3) and water (H_2O) have a tetrahedral basic structure. However, only the methane molecule represents an ideal tetrahedron with a bond angle of 109.5° (Figure 4). Free electron pairs require a greater amount of space then bonding-electron pairs. The available free electron pair in ammonia reduces the bond angle to 107.5°. The water molecule has two available free electron pairs, further reducing the bond angle to 104.5°.

Figure 4: Influence of free electron pairs on the structure of molecules
a) Methane (CH_4),
b) Ammonia (NH_3),
c) Water (H_2O).

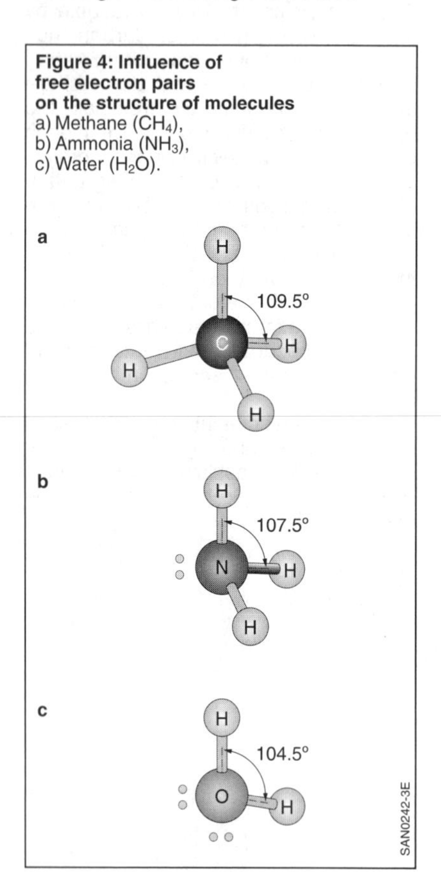

In the simplified model presented here covalent atomic bonds are put down to the formation of pairs of valence electrons from atomic orbitals. However, solely examining atomic orbitals only touches the surface of the complex conditions in a molecule. All the atomic nuclei and electrons in a molecule influence each other, and the bonding conditions are therefore better described by molecular orbitals. A quantum-mechanical examination of molecular orbitals is required to describe double and triple bonds and delocalized bonding systems in aromatic compounds; it is perfectly sufficient in this case to confine oneself to the valence electrons.

Metallic bonds

In metals the atoms are arranged in three-dimensional lattices where the valence electrons of the atoms move freely and are therefore often referred to as electron gas. Metals are good conductors of electricity and heat, which can be put down to the free valence electrons. The lattice vibrations of the atoms also control to the thermal conductivity; here the atoms vibrate freely in their positions in the metal lattice and can easily transmit heat in this way. Here too, a quantum-mechanical examination of molecular orbitals is required to better understand the greater complexities of the electric conductivity of metals. It is possible to derive from this that the electrons reside in energy bands which are formed from molecular orbitals with the smallest energy differences. Between the energy bands are zones which cannot be occupied by electrons.

Interactions between molecules

Van der Waals forces

Partial charges which are created by three-dimensional fluctuations of the positive and negative charge concentrations in the molecule induce electric dipoles (Figure 5a). Interactions that are created between the molecules polarized in this way are called van der Waals forces. The bigger the molecule, the more it can be polarized and the stronger the intermolecular forces.

Dipole-dipole interactions
In molecules which are composed of different atoms the charge concentration is continuously displaced (Figure 5b). Depending on the size and the charge of the atomic nuclei and inner electrons involved, the valence-electron pair is subjected to different degrees of attraction. The property of atoms to attract the bonding-electron pair is known as electronegativity. The more strongly attracting, more electronegative atom receives a partially negative charge. Correspondingly, the more weakly attracting atom receives a partially positive charge. The permanent dipoles brought about by a difference in the electronegativity exhibit intermolecular forces that are much stronger than van der Waals forces. These phenomena are called dipole-dipole interactions.

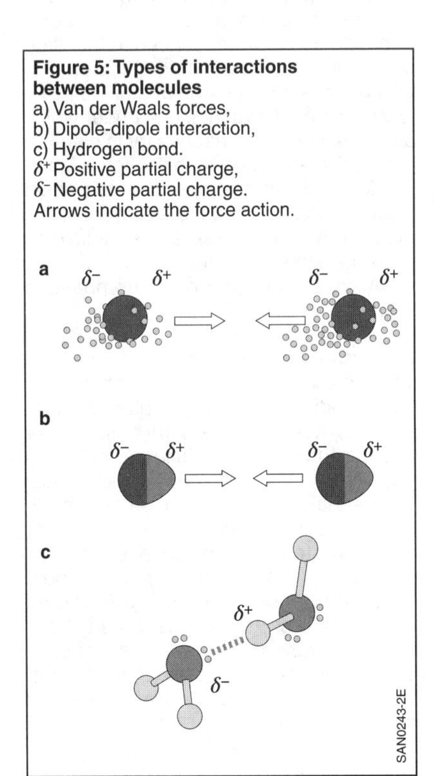

Figure 5: Types of interactions between molecules
a) Van der Waals forces,
b) Dipole-dipole interaction,
c) Hydrogen bond.
δ^+ Positive partial charge,
δ^- Negative partial charge.
Arrows indicate the force action.

The dipole-dipole interactions are also influenced by the three-dimensional structure of the molecule. Although there are differences in the electronegativity of carbon and oxygen, carbon dioxide (CO_2) for example is not a dipole on account of its linear structure. Water (H_2O) with the same atomic ratio 2:1, however, is a dipole, because the central oxygen atom has two free electron pairs, which result in an angled structure.

Hydrogen bonds
The angled structure of the water molecule facilitates additional dipole-dipole interactions (Figure 5c). The partially positively charged hydrogen atoms can interact with the free electron pairs of neighboring molecules. In liquid water the additional dipole-dipole interactions provide a particular short-range structure, which is responsible for water having the highest density at 4 °C. This enables fish to survive in winter because water with the highest density collects at the bottom of frozen lakes.

Molecular spectroscopy
Molecules, like atoms, can absorb energy and assume defined energy states. Valence electrons can be excited with ultraviolet or visible light (UV/VIS spectroscopy) and vibrations and rotations with infrared light (IR spectroscopy) respectively [6]. These procedures are used to explain structures because energy absorption allows conclusions to be drawn about molecular structure. In Raman spectroscopy the inelastic scattering of monochromatic light of molecules is analyzed. The frequency-shifted portions for example resulting from excitation of vibrations provide structural information.

Mass spectroscopy on the other hand is not based on the excitation of energy states. In this procedure molecules are ionized and the ions obtained are separated according to their mass-to-charge ratio and identified.

Substances

Substance term in chemistry

The chemical properties of a body are determined by its material and not by its size and shape. The material of a body is therefore also called substance. Even where the body has a very fine distribution the chemical properties remain the same. However, a large surface can increase the reactivity, as can be seen in nanoparticles.

Homogeneous and heterogeneous substances

Substances with a uniform structure are termed single-phase or homogeneous. A substance that is made up of two or more parts which cannot be mixed with each other is termed heterogeneous. An example of a homogeneous solid substance is elementary sulfur. Typical of a heterogeneous solid mixture is granite, which is made up of quartz, feldspar and mica.

Dispersion

Heterogeneous mixtures are always dispersions. Dispersions consist of at least two different substances which under the prevailing conditions do not or barely dissolve in each other and do not chemically react with each other. Depending on the phases, a distinction is made between suspensions (liquid and solid), emulsions (liquid and liquid) and aerosols (gas and solid or gas and liquid).

Suspension

If, for example, clay is added to pure water – a homogeneous liquid – this creates a heterogeneous mixture of a liquid and a solid which is termed a suspension.

Emulsion

Heterogeneous mixtures of two liquids, e.g. water and oil, are called emulsions.

Aerosol

A heterogeneous mixture of solid or liquid floating particles and a gas is called an aerosol. An example of an aerosol of gases and solids is exhaust gas with soot particles. Exhaust gas with white smoke, which is produced when water and sulfuric acid condense during the starting operation in the still cold exhaust gas, is on the other hand an aerosol of gases and liquids.

Colloid

The term colloid is used for particles or droplets ranging in size between 1 nm and 1 µm, irrespective of whether the heterogeneous mixtures are suspensions, emulsions or aerosols.

States of aggregation

The three classic states of aggregation (phases) are solid, liquid and gaseous – depending on whether the particles are in a fixed position in a solid body, can move in a liquid while maintaining a short-range order, or are far apart from each other in a gas. Plasma is a non-classic state of aggregation and consists of free electrons and ionized atoms.

The states of aggregation of substances are pressure- and temperature-dependent and are described in the state or phase diagrams. State diagrams, in which pressure is plotted against temperature, explain the conditions under which a solid is present, if necessary in different forms, the so-called modifications. Modifications refer to the phenomenon when a substance in the solid state occurs in different structural forms. Also following from the state diagrams are the ranges in which a liquid or a gaseous phase is present.

State diagram for carbon

The state diagram for carbon shows the solid modifications of graphite and diamond (Figure 6), where in ranges of these phases the other modification exists in parallel as a metastable form. Metastable means that the transformation of a modification is inhibited despite its higher energy content for small changes of state. Graphite is the stable modification at room temperature and atmospheric pressure; diamonds exist however as a metastable structure variant because the transformation into the more stable graphite form is strongly curbed by a high activation energy.

State diagram for water

The three areas in the state diagram for water show the existence ranges of ice, liquid water and vapor (Figure 7). Within these areas only the respective state of aggregation prevails. The curves between the areas describe the equilibrium between the liquid and gaseous phases (vaporization curve), the solid and liquid phases (fusion curve), and the solid and gaseous phases (sublimation curve). Each point on the curves corresponds to a state of equilibrium between the adjacent phases. At the point where the three curves intersect, the triple point TP (0.01 °C; 6.1 mbar), all three phases of water at equilibrium with one another. If the temperature or pressure is changed, only one state of aggregation still exists. If temperature and pressure are simultaneously changed in such a way that the new state corresponds to a point on one of the equilibrium curves, two phases exist next to each other.

An increase in temperature always results in a higher proportion of gaseous water, either through greater sublimation or through a higher vapor pressure. Conversely, a higher ambient pressure results in more condensation or resublimation of water vapor. The vapor-pressure curve demonstrates that the boiling point of water is dependent on external pressure. At low air pressure or in a vacuum water boils at temperatures below 100 °C.

What is conspicuous is the anomaly that can be observed at the phase transition from solid to liquid. An increase in both temperature and pressure causes ice to liquify. The transformation of ice into liquid water is furthered by pressure, because water assumes a smaller volume than ice.

A special thermodynamic state is the critical point CP (374 °C; 220.5 bar), at which the densities of the gaseous and liquid phases become indistinguishable. At the end of the vaporization curve the two states of aggregation – fluid and gaseous – pass into a new state, the supercritical phase. Water in the supercritical state is a liquid which has a lower density than liquid water below the critical point.

Figure 6: State diagram for carbon

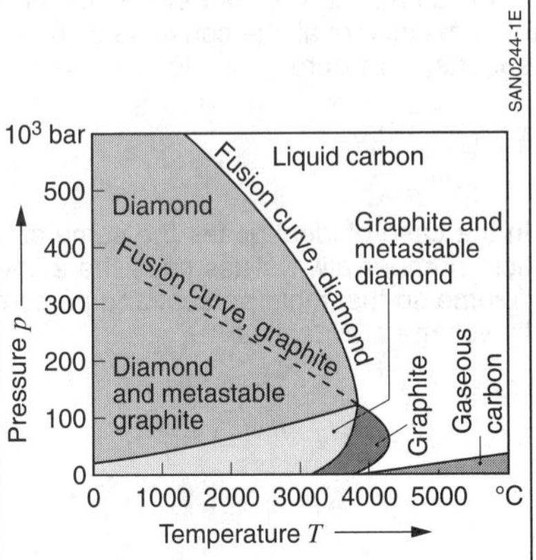

SAN0244-1E

10^3 bar

500

400

300

200

100

0

Pressure p

Diamond

Liquid carbon

Fusion curve, diamond

Fusion curve, graphite

Graphite and metastable diamond

Diamond and metastable graphite

Graphite

Gaseous carbon

0 1000 2000 3000 4000 5000 °C

Temperature T

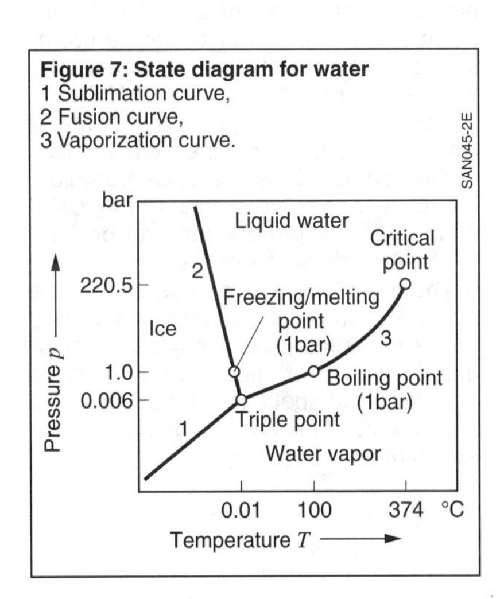

Figure 7: State diagram for water
1 Sublimation curve,
2 Fusion curve,
3 Vaporization curve.

Substance concentrations

To be able to describe the quantitative proportion of the substances involved in a chemical reaction, it is necessary to have quantitative details which refer to the number of particles involved in the conversion.

Owing to the different structure of atomic nuclei all element atoms and naturally also all the resulting molecular compounds differ in mass. It would not be very practicable to reckon with the minute masses of individual atoms or molecules directly. Because it is also helpful for chemical reactions always to consider the same number of particles, the term amount of substance was introduced with the unit "mole".

Amount of substance

An amount of substance n of 1 mole always contains the same number of particles and was therefore previously also often referred to as mole number. Irrespective of the chemical element, molecule or type of substance, 1 mole contains approximately $6 \cdot 10^{23}$ particles. This particle number can be experimentally determined and is also known as "Avogadro's number".

Molar mass

1 mole of a substance therefore always consists of around $6 \cdot 10^{23}$ particles. For element atoms the mass of these particles (molar mass M) therefore corresponds to their relative atomic mass, e.g. 1 mole of carbon has a mass of 12.011 g.

For chemical compounds the molar mass is calculated from the molar masses of the contained elements. Carbon dioxide (CO_2) has a molar mass of 44.01 g/mole (12.011 g/mole for carbon and $2 \cdot 15.9994$ g/mole for oxygen).

The molar mass of a substance – the mass of 1 mole of particles – therefore corresponds to a macroscopically manipulable amount of substance. The molar mass M is the quotient of the mass m and the amount of substance n of a substance (unit g/mol):

$$M = \frac{m}{n}.$$

In chemistry the term weight is also used instead of mass. Mass denotes the matter contained, weight the force acting as a result of the gravitational field on this matter. Since the force of gravity on Earth is approximately always the same, often no distinction is made between mass and weight. That is why the term "molar weight" is also widely used.

Molar volume

Since 1 mole of a substance always consists of the same number of particles, the volume (molar volume V_{M}) that these particles assume is always the same – provided the particles do not influence each other. This limiting case is only given for an ideal gas: 1 mole of an ideal gas assumes under normal pressure (p = 1013.25 mbar) at T = 273.15 K (0 °C) a volume of 22.414 l and at T = 298.15 K (25 °C) a volume of 24.789 l. With regard to the standard conditions, weight- and volume-specific concentration details about the molar volume can be approximately converted into each other for all gases.

Mole percent

Some concentration details are given in mole percent. A distinction is made here between the amount-of-substance- and volume-related definitions. The amount-of-substance-related concentration mole percent x with the unit % (n/n) is obtained by multiplying the mole fraction x_i – i.e. the amount of substance n_i of the constituent i referred to the sum of the amounts of substance of all the constituents of the substance mixture – by 100 %:

$$x = \frac{n_i}{\sum_{j=1}^{n} n_j} \cdot 100.$$

In the case of ideal gases the same particle number always assumes the same volume so that mole percent and percent by volume are identical.

Reactions of substances

Chemical thermodynamics

In chemical reactions starting substances are converted into reaction products with different properties. Chemical thermodynamics describes the substance conversion and the accompanying change in the internal energy ΔU, i.e. in the end whether and under what conditions reactions can take place [7]. The reaction path is determined by the amount of absorbed or output energy.

Heat of reaction

Most reactions are determine in open vessels, i.e. at a constant air pressure. At a constant pressure the change in the internal energy in a chemical reaction is made up of the two components heat of reaction Q_P and work W:

$$\Delta U = Q_P + W.$$

Mechanical work is done for example in an exothermic reaction in which a gas is produced which expands against atmospheric pressure or presses against a membrane or a moving punch.

Reactions can also be effected under a constant volume. Because then the energy content can only be changed by the heat of reaction Q_V and not by work W, the change in the internal energy ΔU corresponds to the heat of reaction Q_V. The following applies:

$$\Delta U = Q_V \, .$$

The heat of reaction heat at constant volume Q_V is therefore always greater than Q_P at constant pressure.

Enthalpy of reaction

The proportion of the heat of reaction Q_P or Q_V can also be described as the difference in heat contents of reaction and starting products and is called the enthalpy of reaction ΔH_R. Where heat is absorbed, the reaction is endothermic ($\Delta H_R > 0$). Chemical reactions in which heat of reaction heat is released ($\Delta H_R < 0$) are exothermic.

The relationship

$$\Delta U^0 = \Delta H_R^0 + W^0$$

is known as the first law of thermodynamics. As the change in the internal energy ΔU, the enthalpy ΔH_R and the work done W are dependent on the pressure and temperature, the standard conditions at 25 °C and 1,013 mbar are referenced and this is indicated at the symbols with a superscript zero (e.g. for ΔU^0).

Enthalpy of activation

Even exothermic reactions often have an initial energy demand despite the released enthalpy of reaction ΔH_R. This enthalpy of activation ΔH_A must be expended in order for example to break bonds and establish an activated complex of reacting agents before more energy is released by the creation of new bonds during the formation of the reaction product than was required to break the bonds in the starting products.

Thermodynamically and kinetically controlled reaction progression

The thermodynamically stable, i.e. energetically more favorable reaction product is reproduced in a reaction when the enthalpy of activation required for substance conversion is applied, e.g. through the supply of heat from the outside (Figure 8, path A). However, if the enthalpy of activation ΔH_{A1}^0 which is required to create the thermodynamically more stable product is not available, another chemical reaction may nevertheless under certain circumstances occur, namely the creation of the product kinetically controlled by the reaction rate. To create it an enthalpy of activation ΔH_{A2}^0 is likewise required (path B), but this is lower in terms of absolute amount. This product created by kinetic control of the reaction then has a higher internal energy. By choosing the reaction conditions, i.e. for example via the quantity of heat made available to the reaction, it is possible to influence whether the thermodynamically or kinetically controlled product tends to be created.

Catalyst decrease the enthalpy of activation of a reaction and thereby increase the reaction rate; but they cannot shift the thermodynamic equilibrium (path C).

Entropy

The enthalpy of reaction ΔH_R alone is not the only factor that determines the direction of a reaction. The distribution of the energy content also plays a role. The evaporation of a liquid takes place despite the required heat of evaporation $\Delta H_V > 0$ because the constraints for the individual molecules decrease as a result. The kinetic energy distributed to the individual molecules passes over from a more ordered, less probable distribution to a state of lesser order but greater probability. The change in the state of order is described by way of the characteristic entropy S [7]. Entropy increases in chemical reactions when more molecules are created than abreact, when the temperature rises, or when particles change to a less ordered state of aggregation (e.g. from solid to liquid or from liquid to gaseous). The increase in entropy ΔS is greater, the higher the heat input and the lower the temperature T at which the heat is transferred.

Figure 8: Enthalpy changes in thermodynamically and kinetically controlled reactions

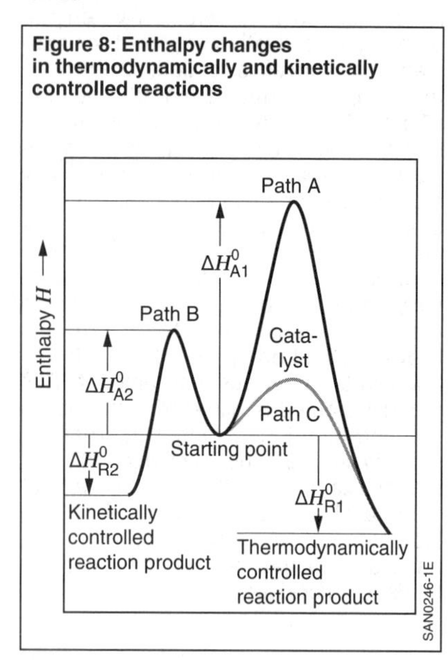

Reaction kinetics

Whereas thermodynamics describes the conversion of substance and energy, reaction kinetics is concerned with the velocity or rate of reactions [7]. The reaction rate of a substances involved in the reaction is specified as a change in concentration per time interval (Figure 9). The order of a reaction is derived from the number of starting substances whose concentration changes during the reaction.

Zero-order reaction

A zero-order reaction exists for example when the decay of a gas on a platinum surface is heterogeneously catalyzed. The concentration of the gas adsorbed on the catalyst does not change during the reaction such that the reaction rate always remains the same, regardless of the reaction time.

Figure 9: Decrease in the concentration of a starting substance in reactions of different orders (assuming equal starting rates)
$t_{1/2}$ Half-life.

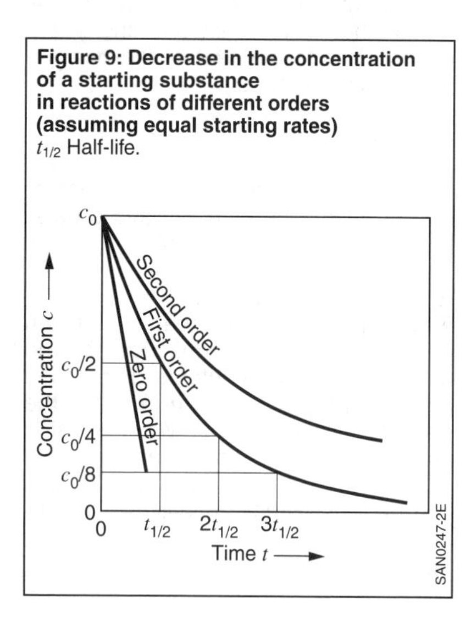

First-order reaction
In a first-order reaction the reaction rate depends only on the concentration of the or a starting product. Many decay processes such as radioactive decay for example follow a first-order reaction which is characterized by the fact that the time in which the concentration of the starting product decreases by half (half-life) is always the same. The half-life is therefore not dependent on the starting concentration.

Second-order reaction
A reaction between two molecules is called a second-order reaction when the reaction rate is determined by the concentration of the two starting substances. If one of the reactands is present in a very large surplus, as is the case for example in hydrolysis reactions in which water is simultaneously the solvent and the reacting agent, the reaction rate equates to that of a first-order reaction. This is called a pseudo-first-order reaction.

With equal starting concentration and starting rate a second-order reaction always runs more slowly than a first-order reaction. The rate of the two reaction types is dependent on the concentration of the starting substances. In a second-order reaction, however, it is necessary to factor in that for a substance transformation – unlike in a first-order reaction – a collision of molecules of the two reacting agents is required and also not every collision actually results in a transformation.

Third-order reaction
The probability of a trimolecular reaction, i.e. the simultaneous collision of three particles, is statistically low. For this reason, third-order reactions occur only very rarely.

Chemical equilibrium
Most chemical reactions are reversible, i.e. an equilibrium is obtained between the starting substances and the reaction products [8]. The position of equilibrium can naturally shift markedly from one side to the other and is influenced by a change in the concentration of a reacting agent, by removal of a reaction product, and by the choice of temperature.

In the reaction of two gaseous substances the conversion can be completed by increasing the pressure when the total number of molecules decreases due to the substance transformation. Low pressure would result in a worse conversion.

Similarly to pressure changes in gas reactions, changes in concentration also result in liquids. Dilution supports reactions which are accompanied by an increase in the number of particles. Catalysts on the other hand – as already described – cannot change the position of equilibrium, but can only influence the rate of its onset.

Law of mass action
Chemical equilibrium is described by the law of mass action [8]. The equilibrium constant K is obtained from the product of the concentrations c of substances C and D, which take part in the back reaction, divided by the product of the concentrations c of substances A and B, which are involved in the direct reaction. The law of mass action applies equally to dissolving processes, chemical reactions and changes of state:

$$\mathrm{A} + \mathrm{B} \rightleftharpoons \mathrm{C} + \mathrm{D},$$

$$K = \frac{c(\mathrm{C}) \cdot c(\mathrm{D})}{c(\mathrm{A}) \cdot c(\mathrm{B})}\,.$$

When a salt is dissolved, this creates positively and negatively charged ions, i.e. cations and anions, which are in solubility equilibrium with the undissolved sediment. In a saturated solution the number of ions which dissolve equals the number of ions which precipitate again. x describes the number of cations A and y the number of anions B in the salt A_xB_y. The dissolving process sees the creation of x dissolved cations A, whose positive charge is determined by the number y of anions B. This is accompanied by the creation of y dissolved anions B with a negative charge, which is dependent on the number x of cations A.

$$\mathrm{A_xB_y} \rightleftharpoons \mathrm{xA^{y+}} + \mathrm{yB^{x-}}\,,$$

$$K = \frac{c(\mathrm{A^{y+}})^{\mathrm{x}} \cdot c(\mathrm{B^{x-}})^{\mathrm{y}}}{c(\mathrm{A_xB_y})}\,.$$

However, the law of mass action applies in this form only to the limiting case of an ideal solution without interaction between the dissolved particles. This is approximately the case with highly diluted solutions as are created when hardly soluble salts are dissolved.

Generally speaking, the law of mass action can only be applied to solutions when the concentrations c of ions are corrected by a factor f downwards. This factor, which takes into account the mutual influence of the dissolved particles, is concentration-dependent and is known as the activity coefficient f. Thus: $f \leq 1$. The activity coefficient takes into account the tendency of ions to associate in the solution due to the difference in charge. The concentration c corrected by the activity coefficient f is called the activity a. Thus:

$$a = f \cdot c,$$

$$K = \frac{a(A^{y+})^x \cdot a(B^{x-})^y}{a(A_xB_y)} .$$

No concentration can be specified for the undissolved portion. This portion is considered to be constant and set at one. In this simplified form the law of mass action is called the solubility product K_L:

$$K_L = a(A^{y+})^x \cdot a(B^{x-})^y .$$

The unit of the solubility product depends on the number of different particle types which are created during the dissolving process. The lower the solubility product, the harder the salt is to dissolve. For improved clarity, as a rule the decimal logarithm of the numerical value of K_L multiplied by −1 is specified instead of the solubility product and denoted as pK_L. The greater the pK_L value, the lesser the solubility.

The proportion of salt that dissolves decays into cations and anions. The process of splitting into separate ions of different charge is frequently called dissociation. The dissolving of salts creates many charge carriers which help to conduct electric current. In aqueous solutions chemists refer to strong electrolytes.

In daily laboratory practice calculations are made approximately with concentrations c instead of with activities a. The term "concentration" is used instead of "activity" in the following text, even if the activity is retained as the physical variable in the formulae.

Acids in water

Even acids (HA, A stands for acid) dissociate in water with the delivery of hydrogen ions H^+ to the water molecules and creation of hydroxonium ions H_3O^+ (protolysis). Here too the law of mass action can be applied [8].

$$HA + H_2O \rightleftharpoons H_3O^+ + A^-,$$

$$K = \frac{a(H_3O^+) \cdot a(A^-)}{a(HA) \cdot a(H_2O)} ,$$

$$K_a = K \cdot a(H_2O) = \frac{a(H_3O^+) \cdot a(A^-)}{a(HA)} .$$

The concentration of water with 55.3 mol/l remains virtually constant and is incorporated into the equilibrium constant, which is then called the acidity constant or K_a value and has the unit of a concentration in mol/l. When the K_a value is high, the equilibrium moves markedly to the right. The concentration of hydroxonium ions (H_3O^+) is high. The acid therefore has a high acidic strength.

The K_a value is as a rule converted into the pK_a value, whereby the K_a value is initially divided by the standard concentration of 1 mol/l. The decimal logarithm multiplied by −1 then produces the pK_a value. Strong acids have a low or even negative pK_a value.

Acidic strength of water
Even in chemically pure water the water molecules are present in equilibrium with hydroxonium ions (H_3O^+) and hydroxide ions (OH^-).

$$2\,H_2O \rightleftharpoons H_3O^+ + OH^-.$$

The ionic product K_w of water is always 10^{-14} mol^2/l^2. Autodissociation of water is therefore very low, in which hydroxonium ions (H_3O^+) and hydroxide ions (OH^-) are created in equally high concentration of 10^{-7} Mol/l.

Disregarding the water lost through dissociation, the acidic strength K_a of water ($a(H_2O)$ = 55.3 mol/l) is obtained at:

$$K = \frac{a(H_3O^+) \cdot a(OH^-)}{a(H_2O) \cdot a(H_2O)},$$

$$K_a = K \cdot a(H_2O) = \frac{a(H_3O^+) \cdot a(OH^-)}{a(H_2O)},$$

$$-\log K_a =$$
$$-\log a(H_3O^+) - \log a(OH^-) + \log a(H_2O),$$

$$pK_a = 7 + 7 + \log 55.3 = 14 + 1.74 = 15.74.$$

Bases in water
Bases on the other hand, which dissolve in water, absorb hydrogen ions (H^+) [8]. When ammonia (NH_3) is dissolved in water, ammonia molecules adopt hydrogen ions (H^+) from the water molecules. Ammonium ions (NH_4^+) and hydroxide ions (OH^-) are created:

$$NH_3 + H_2O \rightleftharpoons NH_4^+ + OH^-.$$

According to the acidic constant K_a as the measure of acidic strength, it is also possible to describe the strength of bases by way of the base constant K_b. The basicity or base strength is derived from the law of mass action along similar lines to the acidic strength or more simply from the ionic product K_w of water via the relation

$$K_w = K_a \cdot K_b = 10^{-14}\ \text{mol}^2/l^2 \quad \text{or}$$

$$pK_w = pK_a + pK_b = 14.$$

Corresponding acid-base pairs
When an acid, e.g. hydrochloric acid (HCl), is dissociated, the hydrochloric acid represents the acid and the acid anion in equilibrium with the acid after hydrogen-ion delivery, in this example the chloride anion (Cl^-), the base. Together they are called a corresponding acid-base pair. Similarly, a base, e.g. ammonia (NH_3), in water, with its corresponding acid, the ammonium ion (NH_4^+), forms an acid-base pair.

The pK_a and pK_b values are experimentally determined for many acid-base pairs (Table 3). In the event of autodissociation of water, water acts as both an acid and a base. The property of water to act as both an acid and a base is known as amphoteric behavior. When water (pK_a = 15.74) is the acid, after hydrogen-ion delivery hydroxide ions (OH^-, pK_b = –1.74) are the corresponding base. When water (pK_b = 15.74) acts as a base, hydroxonium ions (H_3O^+, pK_a = –1.74) are the associated acid.

pH value
The pH value ("potentia hydrogenii") is defined as the negative decimal logarithm of the activity a^* of the hydroxonium ions (H_3O^+). To obtain a dimensionless quantity, the activity a of the H_3O^+ ions is divided before logarithmization by the standard concentration of 1 mol/l. The decimal logarithm then multiplied by –1 produces the pH value.

$$a^*(H_3O^+) = a(H_3O^+)\ [\text{mol}/l] \cdot \frac{1}{[\text{mol}/l]},$$

$$pH = -\log a^*(H_3O^+).$$

For simplification purposes the activity of the hydroxonium ions is often equated with the concentration.

The pH value is therefore a measure of the concentration of an acid; acid concentration (pH value) and acidic strength (pK_a value) do not necessary run parallel. A diluted hydrochloric-acid solution with a concentration of 10^{-4} mol/l can despite the high acidic strength ($pK_a = -6$) be more weakly acidic (pH = 4) than an acetic-acid solution ($pK_a = 4.75$) with a higher concentration of 10^{-1} mol/l (pH = 2.87) (Figure 10).

In the event of autodissociation of water, as already described above, hydroxonium ions (H_3O^+) and hydroxide ions (OH^-) in equal concentration of 10^{-7} mol/l are created. The pH value is therefore 7, pure water is neutral.

In the laboratory in actual fact pure water often reacts slightly acidly because CO_2 from the air dissolves with the formation of carbonic acid (H_2CO_3) in the water. Carbonic acid dissociates slightly to hydroxonium ions (H_3O^+) and hydrogen-carbonate ions (HCO_3^-), pushing the pH value slightly in the acid direction.

The ratio of hydroxonium ions (H_3O^+) to hydroxide ions (OH^-) determines whether a solution is acidic, neutral or basic (Figure 10):

Acidic $a(H_3O^+) > a(OH^-)$: pH < 7.
Neutral $a(H_3O^+) = a(OH^-)$: pH = 7.
Basic $a(H_3O^+) < a(OH^-)$: pH > 7.

Figure 10: pH scale with examples of strong and weak acids and bases

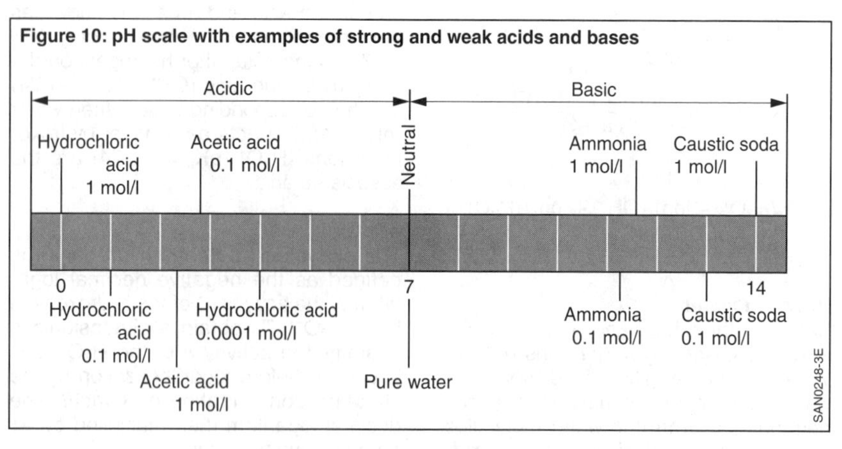

Table 3: pK_a and pK_b values of strong and weak acids

Name	Acidic strength pK_a	Corresponding acid-base pairs		Base strength pK_b
Hydrochloric acid	–6	HCl	Cl^-	20
Sulfuric acid	–3	H_2SO_4	HSO_4^-	17
Hydroxonium ion	–1.74	H_3O^+	H_2O	15.74
Nitric acid	–1.32	HNO_3	NO_3^-	15.32
Phosphoric acid	1.96	H_3PO_4	$H_2PO_4^-$	12.04
Hydrogen fluoride	3.14	HF	F^-	10.68
Acetic acid	4.75	CH_3COOH	CH_3COO^-	9.25
Carbonic acid	6.52	H_2CO_3	HCO_3^-	7.48
Ammonium ion	9.25	NH_4^+	NH_3	4.75
Hydrogen-carbonate ion	10.40	HCO_3^-	CO_3^{2-}	3.60
Hydrogen-phosphate ion	12.36	HPO_4^{2-}	PO_4^{3-}	1.64
Water	15.74	H_2O	OH^-	–1.74

pH value, strong acids
Strong acids like hydrogen chloride dissociate completely in water, during which they transfer positively charged hydrogen ions (H^+) to water molecules with the formation of hydroxonium ions (H_3O^+). 0.1 mol (3.65 g) hydrochloric acid in 1 liter of water produces 0.1 mol hydroxonium ions and 0.1 mol chloride anions (Cl^-). A 10^{-1} molar hydrochloric-acid solution therefore has a pH value of 1.

pH value, weak acids
In the case of weak acids, a certain amount of the acid remains undissociated in the water. But because the pH value of weak acids is only determined by the share of dissociated acid molecules, the calculation of the pH value is more complicated. This will be explained here using an acetic-acid solution by way of example:

In solutions of acetic acid in water less than 1 % of all the acetic-acid molecules dissolved in water (CH_3COOH) are actually dissociated to hydroxonium ions (H_3O^+) and acetate ions (CH_3COO^-). In an aqueous acetic-acid solution, in contrast to an aqueous table-salt or hydrochloric-acid solution, the electrolyte is weak. The dissociation equilibrium can again be described using the law of mass action:

$$CH_3COOH + H_2O \rightleftharpoons H_3O^+ + CH_3COO^-,$$

$$K_a = \frac{a(H_3O^+) \cdot a(CH_3COO^-)}{a(CH_3COOH) \cdot a(H_2O)}\,.$$

The concentration of water can again be set at one on account of the large surplus.

Acetic acid has an acidic strength K_a of $1.8 \cdot 10^{-5}$ mol/l or put another way a pK_a value of 4.75. During dissociation as many hydroxonium ions (H_3O^+) as acetate ions (CH_3COO^-) are created. As the dissociation is minimal, the dissociated part compared with the total concentration of acetic acid CH_3COOH can be disregarded.

With $a(H_2O) = 1$ and $a(H_3O^+) = a(CH_3COO^-)$ it follows that:

$$K_a = \frac{a(H_3O^+)^2}{a(CH_3COOH)},$$

$$a(H_3O^+)^2 = K_a \cdot a(CH_3COOH),$$

$$a(H_3O^+) = \sqrt{K_a \cdot a(CH_3COOH)},$$

$$-\log a(H_3O^+) =$$

$$-\frac{1}{2}(\log K_a + \log a(CH_3COOH)) =$$

$$\frac{1}{2}(-\log K_a - \log a(CH_3COOH)).$$

Taking into account the definitions for the pH and pK_a values, the following is obtained:

$$pH = \frac{1}{2}(pK_a - \log a(CH_3COOH)).$$

An acetic-acid solution with a concentration of 0.1 mol/l therefore has a pH value of

$$pH = \frac{1}{2}(4.75 - \log 0.1)$$
$$= \frac{1}{2}(4.75 + 1) = 2.87.$$

pH values of other weak acids and bases can be similarly calculated.

References

[1] K.-H. Lautenschläger, W. Weber: Taschenbuch der Chemie. Edition Harri Deutsch, 21st Edition, 2013.
[2] M. Borlein: Kerntechnik – Grundlagen. Vogel Buchverlag, 2nd Edition, 2011.
[3] G. Schwedt: Analytische Chemie. Wiley-VCH Verlag, 2nd Edition, 2008.
[4] L. Spieß, G. Teichert, R. Schwarzer, H. Behnken, C. Genzel: Moderne Röntgenbeugung – Röntgendiffraktometrie für Materialwissenschaftler, Physiker und Chemiker. Springer Verlag, 3rd Edition, 2013.
[5] K. Schwister: Taschenbuch der Chemie. Carl Hanser Verlag, 4th Edition, 2010.
[6] M. Hesse, H. Meier, B. Zeeh: Spektroskopische Methoden in der Organischen Chemie. Thieme Verlag Stuttgart, 8th Edition, 2011.
[7] P.W. Atkins, L. Jones: Chemie – einfach alles. Wiley-VCH Verlag, 2nd Edition, 2006.
[8] R. Pfestorf: Chemie – Ein Lehrbuch für Fachhochschulen. Edition Harri Deutsch, 9th Edition, 2013.

Electrochemistry

Electrolytic conduction and electrolysis

When a salt is dissolved in water, the constituents of the salt are present in the water in the form of ions. These ions are charged in such a way that they move when an electric field is applied and thereby generate an electric current. This phenomenon is known as electrolytic conduction. In comparison, the current in an electric conductor such as copper or iron is transported by electrons. In addition to aqueous solutions, molten salts but also certain solids (such as zirconium oxide in the catalytic converter of a passenger car) can function as electrolytes ([1], [2] and [3]).

Negatively charged ions move towards the anode and are therefore called anions, while positively charged ions (cations) migrate towards the cathode. A chemical reaction takes place at these electrodes whereby the ions either absorb electrons from the cathode or discharge electrons to the anode. These reactions can only occur when the cathode and anode have an electrically conductive connection with each other in order to facilitate the exchange of electrons between the two.

If a battery is used as the voltage source, i.e. discharged, the electrons flow from the anode via the external electric circuit to the cathode. Thus, for the user the anode is the negative pole and the cathode is the positive pole.

Electrochemical series of metals

The intensity with which this ion reaction takes place is expressed by the electrochemical series of metals, which is set out in Table 1. The normal potential E^0, applicable to an ion concentration of 1 mol/l, is specified. This "normal" concentration is indicated by the superscript index 0. The voltages specified in the table refer to the normal hydrogen electrode, to which a potential of 0 V is therefore allocated. The following therefore applies:

$$E^0(2H^+ + 2e^- \leftrightarrow H_2) = 0 \text{ V}.$$

In Table 1, a positive sign with E^0 indicates an electron absorption (reduction), and a negative sign an electron discharge (oxidation). For example, lithium by preference discharges an electron ($E^0 < 0$) and is oxidized to become the singly charged lithium ion Li^+, while fluorine absorbs electrons ($E^0 > 0$) and is thus reduced.

Oxidation: $2\,Li \rightarrow 2\,Li^+ + 2e^-$,
$E_{Li} = +3.045$ V.

Reduction: $F_2 + 2e^- \rightarrow 2\,F^-$,
$E_F = +2.87$ V.

Balance equation:
$2\,Li + F_2 \rightarrow 2\,Li^+ + 2\,F^-$,
$E = +5.915$V.

Table 1: Electrochemical series of metals ([3]) with the associated ion reactions (normal potentials at 25 °C)

Half reactions	E^0 [V]
$Li^+ + e^- \leftrightarrow Li$	−3.045
$Na^+ + e^- \leftrightarrow Na$	−2.714
$Mg^{2+} + 2e^- \leftrightarrow Mg$	−2.363
$Al^{3+} + 3e^- \leftrightarrow Al$	−1.662
$2H_2O + 2e^- \leftrightarrow H_2 + 2OH^-$	−0.828
$Zn^{2+} + 2e^- \leftrightarrow Zn$	−0.763
$Cr^{3+} + 3e^- \leftrightarrow Cr$	−0.744
$Fe^{2+} + 2e^- \leftrightarrow Fe$	−0.440
$PbSO_4 + 2e^- \leftrightarrow Pb + SO_4^{2-}$	−0.356
$Ni^{2+} + 2e^- \leftrightarrow Ni$	−0.250
$Pb^{2+} + 2e^- \leftrightarrow Pb$	−0.13
$Sn^{2+} + 2e^- \leftrightarrow Sn$	−0.136
$2H^+ + 2e^- \leftrightarrow H_2$	0
$Cu^{2+} + 2e^- \leftrightarrow Cu$	+0.337
$Cu^+ + e^- \leftrightarrow Cu$	+0.521
$Fe^{3+} + e^- \leftrightarrow Fe^{2+}$	+0.771
$Ag^+ + e^- \leftrightarrow Ag$	+0.799
$Pt^{2+} + 2e^- \leftrightarrow Pt$	+1.118
$4H^+ + O_2 + 4e^- \leftrightarrow 2H_2O$	+1.229
$Cl_2 + 2e^- \leftrightarrow 2Cl^-$	+1.360
$Au^{3+} + 3e^- \leftrightarrow Au$	+1.498
$PbO_2 + 4H^+ + 2e^- \leftrightarrow Pb^{2+} + 2H_2O$	+1.685
$F_2 + 2e^- \leftrightarrow 2F^-$	+2.87

An electrochemical reaction therefore always comprises the two partial steps of oxidation and reduction. Because the lithium discharges the electrons, the electrons appear on the right side of the reaction equation compared with the indication in the electrochemical series of metals (Table 1). The sign preceding the voltage must therefore be reversed for this oxidation equation. The total voltage E of the redox reaction is thus added up from the individual values. The balance equation thus does not feature any electrons, as these are only exchanged between the reacting agents.

An electrochemical reaction can thus only take place when the total voltage is positive. An acid (i.e. H^+ ions) therefore cannot dissolve copper, silver, platinum, and gold. These are referred to as noble metals. In contrast, base metals such as sodium, elemental iron, nickel and lead are attacked by acids and the metals are dissolved as ions.

To enable an electrochemical reaction to take place, at least two different reacting agents are required. The resulting electrochemical total voltage E is dependent on the concentration of ions.

Nernst equation

The Nernst equation reads:

$$E = E^0 + \frac{RT}{nF} \ln \frac{[\mathrm{Ox}]}{[\mathrm{Red}]}$$
$$= E^0 + \frac{0.0592\ \mathrm{V}}{n} \log_{10} \frac{[\mathrm{Ox}]}{[\mathrm{Red}]} \quad \text{(at 25 °C)}.$$

Here, [Ox] denotes the concentration of oxidized ions, [Red] the concentration of reduced ions, n the number of electrons in the reaction equation, R the gas constant, T the absolute temperature, and F the Faraday constant.

In this way, the potential can be calculated for each chemical partial reaction of the individual ion types. The total voltage of the electrolytic reactions is obtained from the sum total of all the potentials of the partial reactions. Over and above that E is dependent on the temperature.

Applications

Lead battery

A motor vehicle needs a starter battery to start an engine. This starter battery is sometimes called a storage battery in that it is repeatedly discharged and charged. Regular batteries, on the other hand, are discharged only once and cannot be recharged.

Charging and discharging processes

The following electrochemical discharging reactions take place in the lead batteries used in passenger cars:

Anode: $Pb \rightarrow Pb^{2+} + 2e^-$
Cathode: $Pb^{4+} + 2e^- \rightarrow Pb^{2+}$.

These reactions take place in sulfuric acid, where at the cathode lead oxide (PbO_2 with Pb^{4+}) is reduced to lead sulfate ($PbSO_4$ with Pb^{2+}), while at the anode elemental lead (Pb) is likewise oxidized to lead sulfate. The electrolyte is therefore depleted of SO_4 sulfate ions and the acid density decreases. During the charging process, the lead sulfate is converted back into lead and lead oxide.

The sulfuric acid H_2SO_4 was omitted from the above reaction equations in order to emphasize the central reactions. These reactions deliver a voltage of approx. 2.0 V (see Table 1). These reactions take place in the reverse direction when the battery is being charged while the car is running.

Because the electrochemical reactions described above take place very slowly, the electrodes are designed as lattices so that they have a large reaction surface. The lead and lead oxide are technically prepared as porous, sponge-like pastes and applied to these lattices.

A battery should be recharged without the voltage being allowed to drop in the process when it is fully charged. It should also be able to store a large amount of energy with minimal weight (capacity) and supply a high level of current.

Specifically the processes shown in Figure 1 occur.

Figure 1: Charging and discharging processes in a lead battery
a) Discharged cell before charging,
b) Charging process,
c) Charged cell,
d) Discharging process.

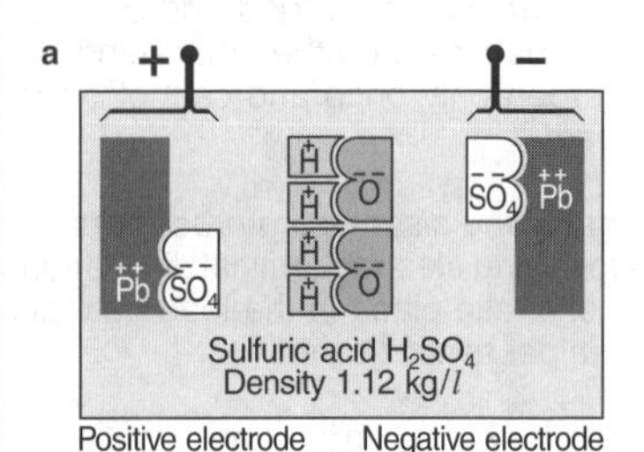

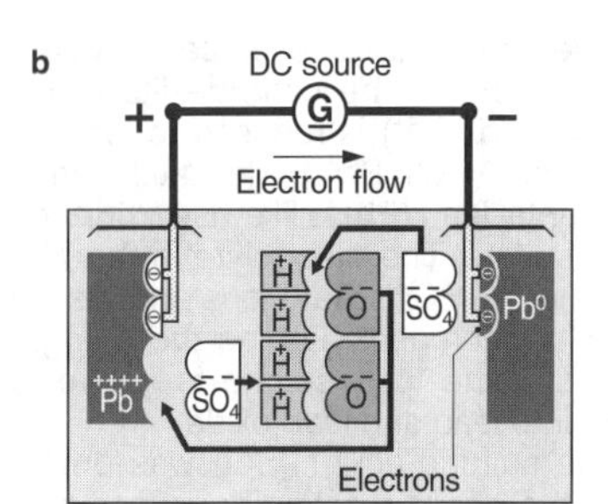

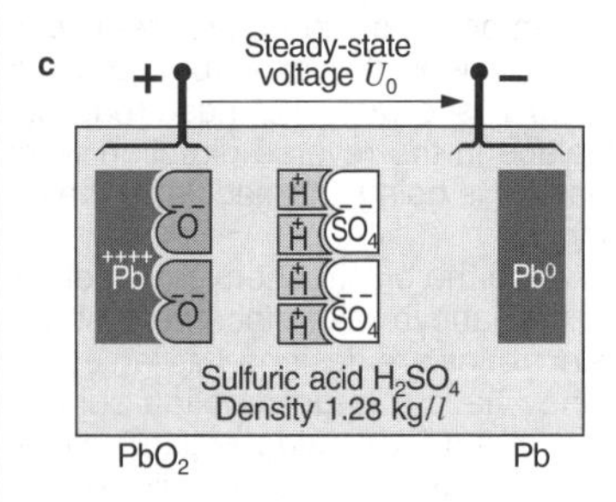

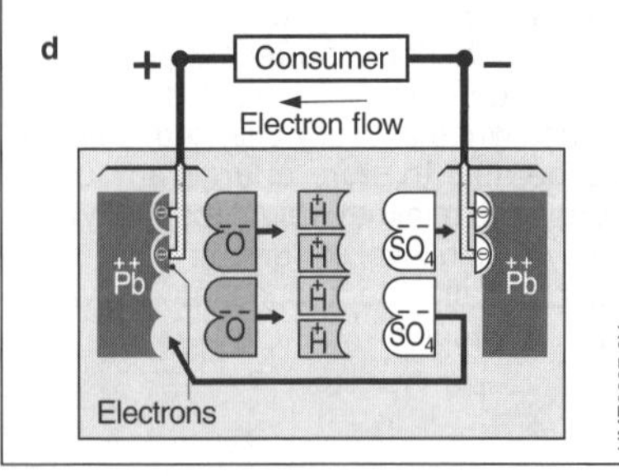

Discharged cell before charging
Located on both electrodes is $PbSO_4$, which is made up of the ions Pb^{2+} and SO_4^{2-} (Figure 1a). The electrolyte has a lower density, as it is depleted of sulfate ions by the current consumption.

Charging process
Pb^{2+} is converted at the positive electrode into Pb^{4+} due to electron donation (Figure 1b). This combines with oxygen to form PbO_2. On the other hand, elemental lead is formed at the negative electrode. Both these reactions involve the release of sulfate ions SO_4^{2-}, which with H^+ ions form sulfuric acid again and thereby increase the acid density.

The specific gravity of electrolyte can be used to indicate the state of charge of the battery (see Table 2). The accuracy of this relationship depends on battery design, electrolyte stratification, and battery wear with a certain degree of irreversible sulfating or a high degree of shedding of plate material. The electrolyte densities specified in Table 2 apply at a temperature of 20 °C; the electrolyte density drops by approximately 0.01 kg/l for every 14 K that the temperature rises and vice-versa when the temperature drops. The low value specified in Table 2 applies to high electrolyte utilization, the high value to low electrolyte utilization.

Charged cell
$PbSO_4$ on the positive electrode is converted into PbO_2 and $PbSO_4$ on the negative electrode is converted into Pb (Figure 1c). There is no further increase in electrolyte density.

If the charge voltage continues to be applied after the cell has reached a state of full charge, only the electrolytic decomposition of water occurs. This produces

Table 2: Electrolyte values of the dilute sulfuric acid in a typical auto starter battery at 20 °C

State of charge	Electrolyte density in kg/*l*	Freezing threshold in °C
Charged	1.28	−68
Semi-charged	1.61...1.20	−17...−27
Discharged	1.04...1.12	−13...−11

oxyhydrogen gas (oxygen at the positive electrode, hydrogen at the negative electrode).

Discharging process
The direction of current flow and the electrochemical processes during discharging are reversed in relation to charging, which results in the Pb^{2+} and SO_4^{2-} ions on both electrodes being combined to form the discharge product $PbSO_4$ (Figure 1d).

Nickel-metal hydride battery
In an NiMH cell (nickel-metal hydride), a metal electrode is used which can accumulate hydrogen [3]. Atomic hydrogen is created during the battery-charging process. This hydrogen is absorbed by the metal electrode (M) and a metal hydride (MH) is created. During discharging the accumulated hydrogen is oxidized on the electrode into water (Figure 2). The following reactions take place during the discharging process; these are reversed during charging:

Anode: $MH + OH^- \rightarrow M + H_2O + e^-$ (0.828 V)

Cathode: $NiOOH + H_2O + e^- \rightarrow Ni(OH)_2 + OH^-$ (0.450 V)

Redox equation:
$MH + NiOOH$ (1.278 V) $\rightarrow Ni(OH)_2 + M$

Nickel-metal hydride batteries have a high self-discharge rate of up to 1 % at room temperature, limiting their use to equipment with a short service life. They are often used in electric cars and in hybrid vehicles, since they deliver high currents and a high charging capacity while being low in weight.

Lithium-ion battery
The lithium-ion battery utilizes the reversible inclusion and withdrawal of lithium ions (Li^+) in a lattice (intercalation electrode, [3]). For example, graphite (C) in different modifications is used as anode materials and lithium metal oxides (e.g. $LiCoO_2$, $LiMn_2O_4$) are used as the cathode (Figure 3).

Figure 2: Electrochemical processes in a nickel-metal hydride battery
a) Discharging,
b) Charging.
1 Anode, 2 Cathode, 3 Electrolyte.
R Resistance, *I* Charging current.

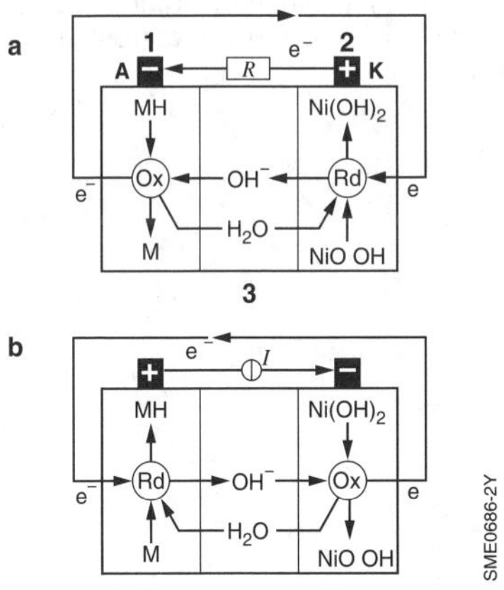

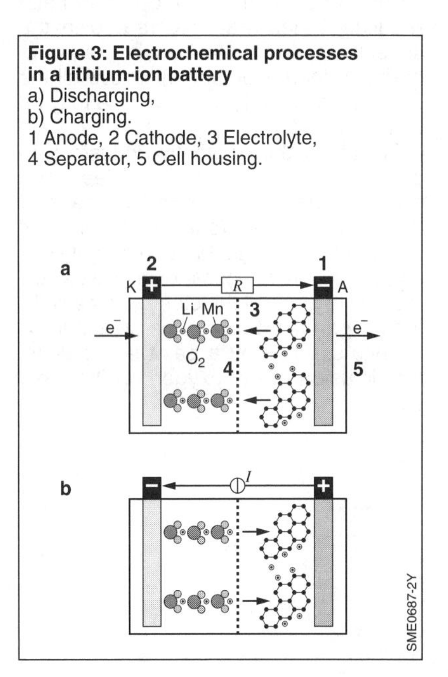

Figure 3: Electrochemical processes in a lithium-ion battery
a) Discharging,
b) Charging.
1 Anode, 2 Cathode, 3 Electrolyte,
4 Separator, 5 Cell housing.

Anode: $Li_xC_n \rightarrow n\,C + x\,Li^+ + x\,e^-$

Cathode: $Li_{1-x}Mn_2O_4 + x\,Li^+ + x\,e^-$
$\rightarrow LiMn_2O_4$

Balance equation:
$Li_{1-x}Mn_2O_4 + Li_xC_n$
$\rightarrow Li\,Mn_2O_4 + n\,C$

Depending on the anode used (Coke Carbons), which is expressed by the index n, the proportion x of the intercalated lithium changes. The electrolyte consists of waterless solvent mixtures (e.g, polyvinylidene fluoride, PVDF) and conducting salts (e.g. $LiPF_6$).

Lithium-ion batteries are characterized by high energy density, thermal stability and voltages up to 4.2 V.

Corrosion

In the case of corrosion, undesirable electrochemical reactions take place in an aqueous or gaseous environment which can damage components (e.g. the body) or even result in the component failing. In this situation, for example, iron atoms in the body dissolve as ions, as a result of which the thickness of the sheet metal reduces, holes may be created in the sheet metal and, as well as the visual impairment (corrosion spots), above all the mechanical stability of the vehicle may be compromised.

The aim of the automotive industry is to protect the vehicle's iron and steel components against corrosion by means of different measures. To this end, the metal can be coated with a baser material such as zinc, which if necessary dissolves corrosively in place of the steel. Alternatively, the oxygen reaction can be stemmed by inhibitors, whereby a paint slows down the diffusion of the oxygen on the steel surface.

Corrosive attack

Like electrochemical reactions, a corrosive attack consists of two partial reactions – an anodic oxidation and a cathodic reduction. In the case of oxidation, the metal atom discharges electrons and dissolves as an ion or forms a deposit on the surface ("tarnishing colors") with other reacting agents. This process is the central corrosion reaction. In contrast to electrochemical reactions a corrosive attack takes place by itself, i.e. it does not need a voltage source.

In the reduction process, oxygen and hydrogen ions are often the reacting agents for anodic metal dissolution. In neutral or alkaline media, oxygen is reduced to hydroxide ions:

$$O_2 + 2\,H_2O + 4\,e^- \rightarrow 4\,OH^-.$$

In acidic media, the hydrogen ions of the acids react to form hydrogen, which escapes as a gas.

$$2\,H^+ + 2\,e^- \rightarrow H_2$$

The electrochemical series of metals (Table 1) provides an indication of how great the risk of corrosion is for the different metals and which materials can protect against corrosion. This table does not provide any information about the speed of the corrosion reactions. Nor does the table take into account the fact that many metals (such as aluminum, for example) in air form an oxide layer which provides temporary protection against corrosion (see also Corrosion and corrosion protection).

Oxygen concentration sensor

The oxygen concentration sensor (λ sensor) measures the residual-oxygen content in the exhaust gas and thereby regulates the air supply in order to set an optimum air/fuel mixture for combustion in the gasoline engine ($\lambda = 1$, stoichiometric combustion). To this end, a zirconium-oxide ceramic (ZrO) is used as a solid electrolyte (Figure 4), as this conducts oxygen ions at temperatures over 300 °C. Chemically inert platinum electrodes, which are applied as a porous thick layer, are used as electrodes. The voltage generated at the electrodes U_λ can be calculated with the Nernst equation. If here [Ox] and [Red] are replaced by the oxygen partial pressures $p(O_2)$ in the reference area (ambient air) and in the exhaust-gas area, the following relationship is obtained:

$$U_\lambda = \frac{RT}{4F} \ln \frac{p_R(O_2)}{p_A(O_2)} .$$

Here $p_A(O_2)$ denotes the oxygen partial pressure in the exhaust gas, $p_R(O_2)$ the oxygen partial pressure in the reference area, R the gas constant, T the absolute temperature, and F the Faraday constant.

In the vicinity of $\lambda = 1$, the voltage change is very great, making it possible to successfully regulate to this value.

Figure 4: λ sensor (finger-type sensor) in the exhaust pipe
1 Zirconium-oxide ceramic, 2 Electrodes,
3 Exhaust pipe,
4 Porous ceramic protective layer,
5 Exhaust gas, 6 Outside air (reference air).

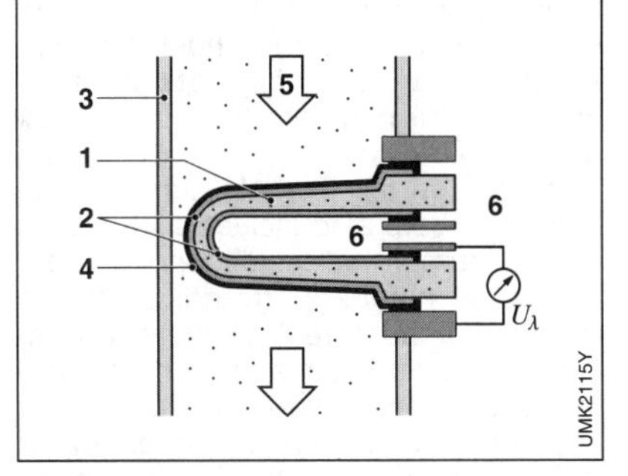

References

[1] E. Fluck, R.C. Brasted: Allgemeine und Anorganische Chemie. 6th Ed., UTB No. 53, Quelle & Meyer, Heidelberg, 1987.
[2] C.E. Mortimer, U. Müller: Chemie. 10th Ed.,Thieme, Stuttgart, 2010.
[3] C.H. Hamann, W. Vielstrich: Elektrochemie. 4th Ed., Wiley-VCH, Weinheim, 2005.

Mathematics

Numbers

Quantities

Numbers are divided into natural numbers $\mathbb{N} = \{0, 1, 2, 3, \text{etc.}\}$, whole numbers $\mathbb{Z} = \{0, \pm1, \pm2, \pm3, \text{etc.}\}$, rational numbers $\mathbb{Q}$, real numbers $\mathbb{R}$ and complex numbers $\mathbb{C}$.

In addition to all whole numbers, rational numbers also include all fractions whose numerators and denominators are whole numbers. In addition to the rational numbers, real numbers also include all (infinitely many) numbers between fractions. Two examples of real numbers are the circular constant pi ($\pi = 3.14159...$) and the Eulerian number e ($e = 2.718281...$). Complex numbers are an extension of real numbers. These are explained in detail below (for more information, refer to [1], [2] and [3]).

Number systems

Physical and technical quantities are described by their numerical value and their unit. Numbers are usually represented in the decimal system, i.e. using base 10. Other common number systems are the binary and the hexadecimal systems, which are based on the numbers 2 and 16, respectively. While the decimal system features the digits 0, 1, 2 and onwards through 9, only the digits 0 and 1 exist in the binary system. The hexadecimal system uses, in addition to the digits 0 through 9, the letters A through F to represent the numbers 10 through 15 (for conversion, see Tables 1 and 2).

The binary and hexadecimal systems are used primarily in the field of information technology (IT), since a computer can only process the two states "power off" ("0") and "power on" ("1"). These two states form the basis of the binary system. When eight binary digits are combined into one byte, it is then possible to represent the numbers 0 to 255, which corresponds to the hexadecimal digits 0 – FF.

Functions

The following section describes elementary mathematics functions. Their most important properties, such as their domain and value set, their behavior for very large and very small x values, their zero points, their derivations and the fundamental arithmetic operations with them are depicted.

These functions have been selected from a very large number of functions since they can be used to illustrate many technical processes, such as the geometric correlations between a connecting rod and moving mechanical parts (see Internal-combustion engines, Crankshaft gears), vibrations and oscillations in the vehicle (see Undercarriage, Basic principles of vibration characteristics), distance measurements of the vehicle relative to other cars and people (see Driver-assistance systems, Parking systems).

Table 1: Decimal and binary system

Decimal	Binary
0	0
1	1
2	10
3	11
4	100
8	1000
9	1001
15	1111
16	10000
32	100000
64	1000000
255	11111111

Table 2: Decimal and hexadecimal system

Decimal	Hexadecimal
0, 1–9	0, 1–9
10, 11–15	A, B–F
16	10
17	11
30	1E
31	1F
32	20
255	FF
4096	1000
65535	FFFF

Polynomial
A polynomial of the nth degree consists of $(n+1)$ summands with real (or complex) coefficients $a_0, a_1, \dots, a_n$ and the associated monomials x^i:

$$f(x)=a_0+a_1x+a_2x^2+\dots+a_nx^n,$$

where $a_i \in \mathbb{R}$ (or $a_i \in \mathbb{C}$)
and the domain range $D_f = \mathbb{R}$ (or $D_f = \mathbb{C}$) and the value set $W_f = \mathbb{R}$.

A polynomial of the nth degree may have up to n zero points and $n-1$ local extrema.

Straight line
A straight line is a polynomial of the first degree $(n=1)$:

$f(x)=a_0+a_1x$ or
$y=mx+t$ (with slope m and axis intercept t).

The zero point lies at $x_0=-\frac{t}{m}$ (for $m \neq 0$).

Parabola
A parabola is a polynomial of the second degree ($n=2$, quadratic function):

$f(x)=a_0+a_1x+a_2x^2$ or
$y=ax^2+bx+c.$

Zero points: $x_{1/2}=\frac{1}{2a}\left(-b\pm\sqrt{b^2-4ac}\right)$

Table 3: Polynomial

Definition range D_f, value set W_f
$f(x)=a_0+a_1x+a_2x^2+\dots+a_nx^n$
$D_f=\mathbb{R}$, $W_f=\mathbb{R}$ (or subset).

Figure 1: Straight line (dashed) and parabola as polynomials of the first and second degree

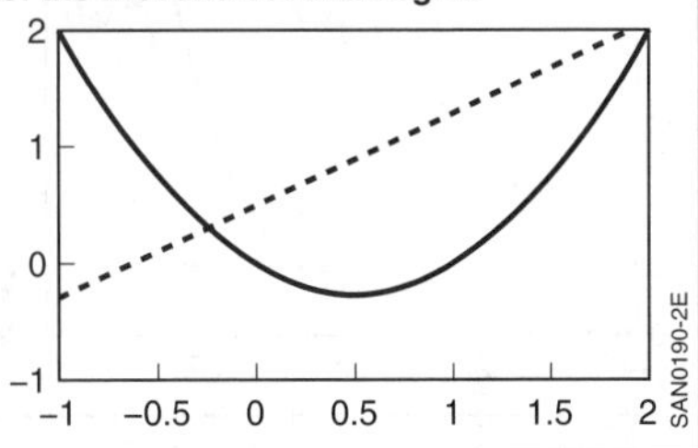

A parabola may have zero, one or two zero points.

Figure 1 shows by way of example a straight line (polynomial of the first degree) and a parabola (polynomial of the second degree).

Root function
The root function $f(x)=\sqrt{x}=x^{1/2}$ (Table 4 and Figure 2) is the inverse function of the quadratic function $f(x)=x^2$. It is needed for tasks such as solving quadratic equations (e.g. searching for the zero point of a second-degree polynomial, see Polynomial).

This also results in the root term, which is used in the calculating the resonant frequency and the damping ratio of an oscillation (see Oscillations).

Similarly, for each monomial $f(x)=x^n$, $n \in \mathbb{N}$, there is a strongly monotonically growing inverse function

$$f^{-1}(x)=\sqrt[n]{x}=x^{1/n},$$

which is only defined for $x \geq 0$.

Table 4: Root function

Definition range D_f, value set W_f and behavior.	Properties.
$f(x)=\sqrt{x}=x^{1/2}$ $D_f=\mathbb{R}_0^+$ $W_f=\mathbb{R}_0^+$ $x\to+\infty$: $f(x)\to+\infty$	$\sqrt{x}\cdot\sqrt{y}=\sqrt{xy}$ $\sqrt{x^n}=(\sqrt{x})^n$, $n\in\mathbb{N}$

Figure 2: Curve of the root function

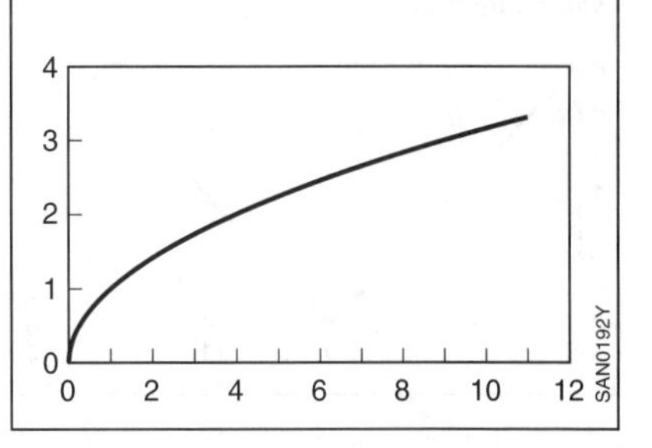

Absolute value and sign function

Each real number x can be broken down into its sign (+/–) and its absolute value $|x|$.

$$x = \text{sgn}(x) \cdot |x|.$$

For example, sgn(–3) = –1 and sgn(+4) = +1. The absolute value indicates the distance of the number x from a source of 0. The absolute value function (Table 5 and Figure 3) can be represented as

$$f(x) = |x| = \begin{cases} x & \text{for } x \geq 0 \\ -x & \text{for } x < 0 . \end{cases}$$

Exponential function

The exponential function is a very important function in mathematics, physics and technology, because it is the only function that is identical to its own derivative. This central property is used in solving linear differential equations such as the harmonics equation (see Oscillations).

Both the sine and cosine functions are related to the complex exponential function (see Complex numbers). This means the exponential function can be used to describe oscillations.

$$f(t) = e^{-\gamma t + i\omega t} = e^{-\gamma t} (\cos \omega t + i \sin \omega t).$$

Here, the real negative exponent $-\gamma t$ represents the damping of the oscillation, while the complex exponent $i\omega t$ reflects the periodic quantity, which becomes clear from the representation of the sine and cosine functions (see Complex numbers, Linear differential equations).

Furthermore, laws of growth (e.g. interest and compound interest calculations) and laws of decay (e.g. radioactive decay) can be expressed with the exponential function. The charging and discharging processes of capacitors also follow an exponential curve (see Capacitor). The final compression pressure, final compression temperature and efficiency of a reciprocating-piston engine are exponentially related to the polytropes – more specifically, the adiabatic exponents (see Reciprocating-piston engine).

Table 6 and Figure 4 show the domain and value set, as well as the properties of the exponential function.

Table 5: Absolute value function

Definition range D_f, Value set W_f and behavior.
$f(x) = \text{sgn}(x)$ $D_f = \mathbb{R}$ $W_f = \mathbb{R}_0^+$ $x \to \pm\infty : f(x) \to +\infty$

Figure 3: Curve of the absolute value function

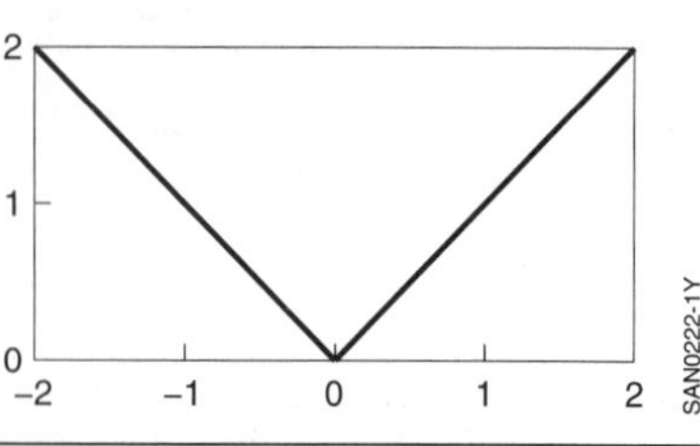

Table 6: Exponential function

Definition range D_f and behavior.	Properties.
$f(x) = e^x$; $D_f = \mathbb{R}$ $x \to -\infty : f(x) \to 0$ $x \to +\infty : f(x) \to +\infty$ $f(0) = e^0 = 1$	$e^a e^b = e^{a+b}$ $(e^a)^b = e^{ab}$ $\frac{d}{dx} e^x = e^x$
Euler's number $e = 2.71828...$	

Figure 4: Curve of the exponential function

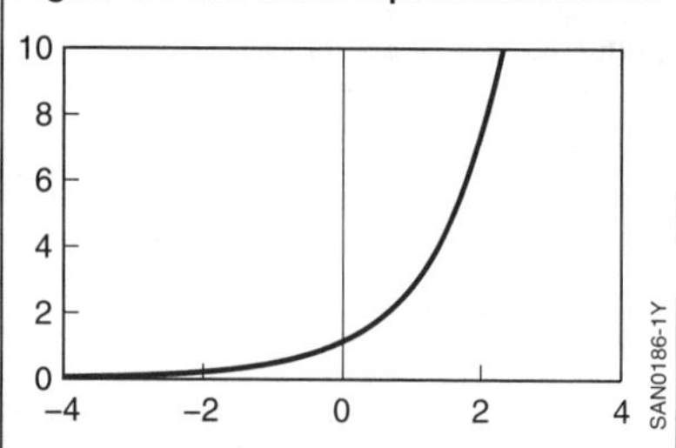

Logarithm function

The logarithm function is the inverse function of the exponential function. This is needed in order to solve equations such as the following:

$$2^x = 8 \Rightarrow x = \log_2 8 \Rightarrow x = 3.$$

Table 7 and Figure 5 show the domain and value set, as well as the properties of the logarithm function.

Application examples

The Nernst equation provides one example of application of the logarithm function. This equation calculates the voltage value from the concentration of oxygen in the surroundings and the exhaust gas, which is necessary for λ regulation (see λ sensor).

In acoustics, decibel level (dB) is defined using the logarithm of the sound pressure (see Acoustics, Decibels). Similarly, logarithms are also used to determine the sound power level and sound intensity level.

Table 7: Logarithm function

Definition range D_f and behavior.	Properties.
$f(x) = \ln x = \log_e x$; $D_f = \mathbb{R}^+$ $x \to 0: \quad f(x) \to -\infty$ $x \to +\infty: \quad f(x) \to \infty$ $f(1) = \ln 1 = 0$	$\ln a + \ln b = \ln ab$ $\ln a - \ln b = \ln \frac{a}{b}$ $c \ln a = \ln a^c$ $\frac{d}{dx} \ln x = \frac{1}{x}$

Figure 5: Curve of the logarithm function

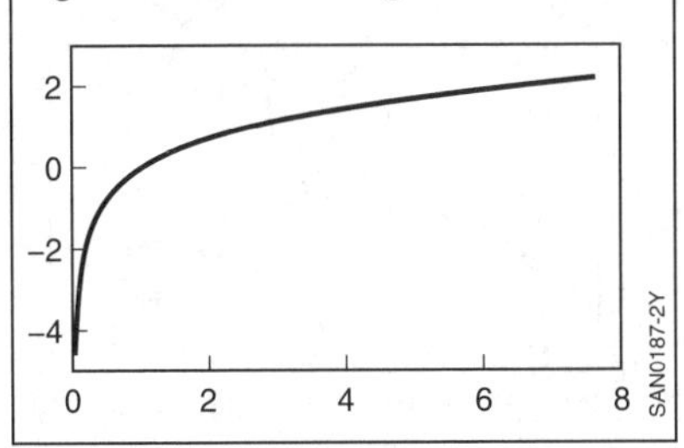

Converting different logarithms

The logarithm of the number z for the base a can be converted to the base b as follows ($a, b > 0$):

$$\log_a z = \frac{\log_b z}{\log_b a}$$

This results in $\lg z = \log_{10} z$ for the common (decimal) logarithm and $\mathrm{lb}\, z = \log_2 z$ for the binary logarithm:

$$\log_a z = \frac{\ln z}{\ln a} = \frac{\mathrm{lb}\, z}{\mathrm{lb}\, a} = \frac{\lg z}{\lg a}.$$

Trigonometric functions

Angular measurement (radians)

In mathematics, angles are usually specified as the measure of an arc in radians and rarely in degrees. So angle $\varphi = 360°$ is equivalent to a arc of $x = 2\pi$. The circumference of a circle with a radius of 1 shares this value.

This yields the following conversions between angle φ in degrees and angle x in radians:

$$\frac{\varphi}{360°} = \frac{x}{2\pi},$$

$$\Rightarrow \varphi = \frac{180°}{\pi} x,$$

$$\Rightarrow x = \frac{\pi}{180°} \varphi$$

The unit "rad" is often added to an arc measurement to make it clear that the information relates to an angle. Arc x of the angle φ is shown in Figure 6. Angle φ of the associated arc x is shown as a curved arrow.

Sine and cosine functions

In a right triangle, the sine of angle φ or arc x is equal to the ratio of the opposite side to the hypotenuse. The cosine is the ratio of the adjacent side to the hypotenuse.

In a right triangle with a hypotenuse with length $r = 1$ (unit circle), the adjacent side corresponds to the cosine and the opposite side corresponds to the sine (Figure 6).

Many periodic processes such as oscillations can be expressed through the sine or cosine function (Table 8 and Figure 7). If the variable x is replaced by

$$x = \frac{2\pi}{T}\,t \text{ and } \omega = \frac{2\pi}{T}$$

where the variable t usually stands for the time, then T is the period of the oscillation. The frequency f is the reciprocal of the period:

$$f = \frac{1}{T}$$

Figure 6: Sine and cosine in a unit circle

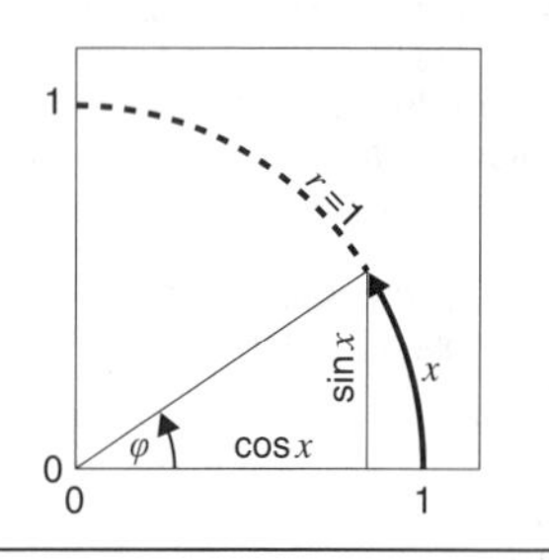

Table 8: Sine and cosine functions

Definition ranges D_f, D_g, Value sets W_f, W_g and behavior.	Properties.
$f(x)=\sin x$; $D_f=\mathbb{R}$	$\sin(x+2\pi)=\sin x$
$g(x)=\cos x$; $D_g=\mathbb{R}$	$\cos(x+2\pi)=\cos x$
$W_f=W_g=[-1;+1]$	$\sin(x\pm\frac{\pi}{2})=\pm\cos x$
Period length: 2π	$\cos(x\pm\frac{\pi}{2})=\mp\sin x$
	$\sin^2 x+\cos^2 x=1$

Figure 7: Curves of the sine and cosine functions

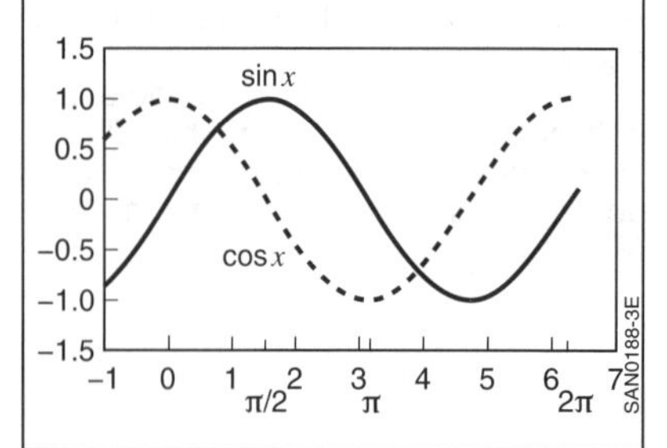

The angular frequency ω still contains the factor 2π; thus, it specifies the angle covered (in radians) per unit of time.

When a car drives on a street or a country road, it is constantly subjected to large and small bumps and shocks. Therefore, the undercarriage must be constructed so that it absorbs or compensates for the bumps in the road. In addition, the vibrational characteristics of the shock absorbers should be designed with precision (see Undercarriage, Basic principles of vibration characteristics).

Superpositions of oscillations are important in many applications, such as water waves, sound waves and the superposition of alternating electrical current (AC). In addition, the wave is usually formulated as a sine or cosine function

Table 9: Tangent and cotangent functions

Definition ranges D_f, D_g, Value sets W_f, W_g, and behavior.	Properties.
$f(x)=\tan(x)=\frac{\sin(x)}{\cos(x)}$	$\tan(x+\pi)=\tan(x)$
$g(x)=\cot(x)=\frac{\cos(x)}{\sin(x)}$	$\cot(x+\pi)=\cot(x)$
	$\tan(x+\frac{\pi}{2})=-\cot(x)$
$D_f=\mathbb{R}\setminus\{\frac{\pi}{2}+k\cdot\pi, k\in\mathbb{Z}\}$	$\cot(x+\frac{\pi}{2})=-\tan(x)$
$D_g=\mathbb{R}\setminus\{k\cdot\pi, k\in\mathbb{Z}\}$	$\cot(x)=\frac{1}{\tan(x)}$
$W_f=W_g=\mathbb{R}$	
Period length: π	

Figure 8: Curves of tangent and cotangent functions

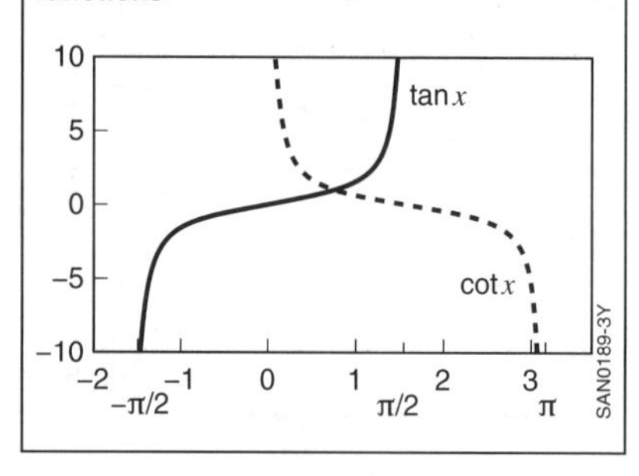

$f(t) = A \sin(\omega t + \varphi)$ or
$f(t) = A \cos(\omega t + \varphi)$

with (positive) amplitude A and phase shift φ.

If oscillations with the same angular frequency are superimposed, they can be amplified (through constructive interference), weakened or even completely eliminated (through destructive interference). This depends on the phase relation and the amplitudes of the relevant oscillations.

For example, destructive interference is utilized as "anti-noise" at airports. This involves having speakers produce sounds with the same frequency as the turbine noise from airplanes. The anti-noise and noise are out of phase with one another, canceling each other out, reducing the noise for nearby residents.

Tangent and cotangent functions
In a right triangle, the tangent results from the ratio of the opposite side to the adjacent side. Therefore, the following relationship can be derived:

$$\tan x = \frac{\sin x}{\cos x}$$

The cotangent results from the ratio of the adjacent side to the opposite side. This means that

$$\cot x = \frac{\cos x}{\sin x}$$

Table 9 and Figure 8 show the properties and the curves of both of these functions. The most important properties of trigonometric functions are summarized in Table 10.

Arc functions
The arc functions are the inverse of trigonometric functions. For example, the following equation can be solved using the arcsine:

$$\sin x = 0.5 \quad \text{where } x \in [0; \pi/2]$$

$$\Rightarrow x = \arcsin(0.5) = \pi/6.$$

Similarly, there are also an arccosine, arctangent and arccotangent.

In contrast to the sine, cosine, tangent and cotangent, which (with the exception of gaps) are defined for any $\mathbb{R}$, the arc functions may have a finite interval as a domain, in particular the value range of the original function. These properties and others are compiled in Table 11. The arc functions are not periodic.

Table 10: Properties of trigonometric functions, $k \in \mathbb{Z}$

	sin (x)	cos (x)	tan (x)	cot (x)
D_f	$\mathbb{R}$	$\mathbb{R}$	$\mathbb{R}\setminus\{x \mid x=\frac{\pi}{2}+k\pi\}$	$\mathbb{R}\setminus\{x \mid x=k\pi\}$
W_f	$[-1; +1]$	$[-1; +1]$	$\mathbb{R}$	$\mathbb{R}$
Period	2π	2π	Π	Π
Symmetry	Odd	Even	Odd	Odd
Zero points	$x_0=k\pi$	$x_0=\frac{\pi}{2}+k\pi$	$x_0=k\pi$	$x_0=\frac{\pi}{2}+k\pi$
Maxima	$x_{\text{max}}=\frac{\pi}{2}+k\cdot 2\pi$	$x_{\text{max}}=k\cdot 2\pi$	–	–
Minima	$x_{\text{min}}=\frac{3}{2}+k\cdot 2\pi$	$x_{\text{min}}=\pi+k\cdot 2\pi$	–	–
Poles	–	–	$x_{\text{Pole}}=\frac{\pi}{2}+k\pi$	$x_{\text{Pole}}=k\pi$

Table 11: Properties of arc functions

	arcsin (x)	arccos (x)	arctan (x)	arccot (x)
D_f	[–1; +1]	[–1; +1]	$\mathbb{R}$	$\mathbb{R}$
W_f	$[-\frac{\pi}{2};\frac{\pi}{2}]$	$[0;\pi]$	$]-\frac{\pi}{2};\frac{\pi}{2}[$	$]0;\pi[$
Symmetry	Odd	Point symmetry for $P(0;\frac{\pi}{2})$	Odd	Point symmetry for $P(0;\frac{\pi}{2})$
Zero points	$x_0=0$	$x_0=1$	$x_0=0$	-
Monotony	Strictly monotonically increasing	Strictly monotonically decreasing	Strictly monotonically increasing	Strictly monotonically decreasing
Asymptotes	-	-	$y=\pm\frac{\pi}{2}$	$y=0; y=\pi$

Equations in the plane triangle

Figure 9 shows a triangle with any lengths of sides a, b and c and any angles α, β and γ. The relationship between these quantities is described by the following equations.

Angular sum:
$\alpha+\beta+\gamma=180°$.

Law of sines:
$a : b : c = \sin\alpha : \sin\beta : \sin\gamma$.

Pythagorean theorem:
$c^2 = a^2 + b^2$.

Law of cosines:
$c^2 = a^2 + b^2 - 2ab\cos\gamma$.

The relationships between angles in a triangle play an important role in the design of components. One example of this would be the gear design in a reciprocating piston engine, where the piston stroke and length of the connecting rod are related to one another by the angle of the crankshaft position and the leg angle of the connecting rod (see Internal-combustion engines, Gear design).

Complex numbers

In real numbers, a root can be extracted from any positive number; this is not possible with negative numbers. In order to bypass this restriction, real numbers are changed into complex numbers. Their central element is the imaginary unit i, where:

$i^2 = -1$.

A complex number z has a real component and an imaginary component. The imaginary component is multiplied by the imaginary unit i. It is possible with the coordinates a and b to represent z in Cartesian form. Using a simple transformation, this can be converted into the polar coordinates r and φ (distance from the origin of the coordinates and the angle to the X-axis) (Figure 10).

Figure 9: Plane triangle
Vertices A, B and C.
Lengths of sides a, b and c;
Angles α, β and γ.

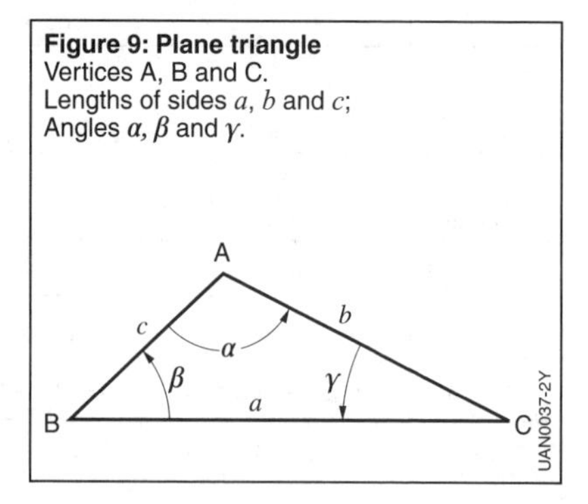

Complex number z in Cartesian form:
$z = a + ib$ where $i^2 = -1$.
$\mathrm{Re}\, z = a$, $\mathrm{Im}\, z = b$.

Complex number z in polar coordinates:
$z = re^{i\varphi}$

where $r = \sqrt{a^2 + b^2}$, $\tan \varphi = \frac{b}{a}$
$a = r\cos \varphi$, $b = r\sin \varphi$.

Complex exponential function (Euler's formula)
$e^{i\varphi} = \cos \varphi + i \sin \varphi$

Calculation rules for complex numbers
In the case of complex numbers, the same calculation rules that are used for real numbers apply, whereby addition is simpler to carry out in Cartesian notation and multiplication is simpler in polar notation.

Addition:
$z_1 + z_2 = (a_1 + a_2) + i(b_1 + b_2)$.

Multiplication:
$z_1 z_2 = (r_1 r_2) e^{i(\varphi_1 + \varphi_2)}$

In many cases, complex numbers aid in making complicated mathematical problems easier to solve. Frequently, only the real part or the imaginary part of the complex solution is necessary in finding the real solution.

For example, oscillation processes can be expressed using linear differential equations (see Differential equations), such as the oscillation and damping properties of shock absorbers or the current strength and voltage in an alternating-current circuit. A solution can be found quickly with the help of a complex exponential function. Finally, Euler's formula for exponential functions is used to convert a complex solution into a real solution.

Coordinate systems

Coordinate systems on a plane
In the previous segment, a complex number z is either represented by both of its coordinates x and y or by its absolute value $r = |z|$ and the angle φ in relation to the x-axis. This number can be graphically represented as a point on a plane (Figure 10).

Similarly, all the points on a plane can be expressed by their Cartesian coordinates (x and y) or by their polar coordinates (r and φ), and converted into one another.

Conversion of Cartesian coordinates to polar coordinates:

$$r = \sqrt{x^2 + y^2}\ ,\ x, y \in \mathbb{R},$$

$$\tan \varphi = \frac{y}{x}.$$

Conversion of polar coordinates to Cartesian coordinates:
$x = r\cos\varphi$, $r \in \mathbb{R}_0^+$, $\varphi \in [0; 2\pi]$,
$y = r\sin\varphi$.

This conversion always yields a unique value, meaning that for each x-y pair, there is an exact r-φ pair. r is only = 0 at the coordinate origin (0, 0), but the angle φ is undefined. However, in practice this does not amount to a limitation.

Cartesian coordinates can be used for straight-line movements. Naturally, circular motion can also be represented us-

Figure 10: Complex number z
Diagram in
Cartesian coordinates (a and b)
and polar coordinates (r and φ).

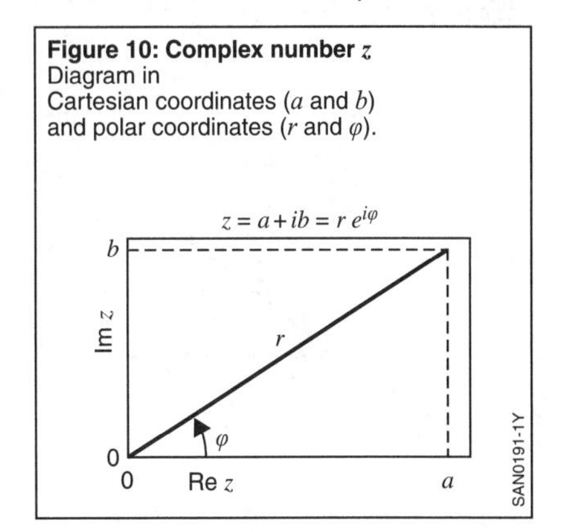

ing Cartesian coordinates. However, the corresponding equations are significantly more complicated in comparison to a description in polar coordinates (see for example Fundamentals of vehicle engineering, Translational movement).

Coordinate systems in three dimensions

Each point in a three-dimensional space has three Cartesian coordinates (x, y, z). This representation is used frequently when the system to be described is at a right or oblique angle. However, if the problem relates to a rotationally symmetrical or a spherically symmetrical system, using cylindrical or spherical coordinates is recommended.

In Figure 11, the x, y and z coordinate axes are delineated. Here, the distance r of point P to the origin and the two angles θ and φ, which can be measured from the z or the x-axis. r, θ and φ are labeled as spherical coordinates or spatial polar coordinates.

With the exception of the origin (0, 0, 0), the coordinates for each point are clearly defined in both Cartesian and polar coordinates. Conversion between both coordinate system occurs as follows.

Figure 11: Spherical coordinates of a point

Conversion of Cartesian coordinates to polar coordinates:

$$r = \sqrt{x^2 + y^2 + z^2}\,, \quad x, y, z \in \mathbb{R},$$

$$\tan \varphi = \frac{y}{x},$$

$$\cos \Theta = \frac{z}{r} = \frac{z}{\sqrt{x^2 + y^2 + z^2}}\,.$$

Conversion of polar coordinates to Cartesian coordinates:

$$x = r \cos \varphi \sin \Theta,$$
$$r \in \mathbb{R}_0^+,\ \varphi \in [0; 2\pi],\ \Theta \in [0; \pi],$$
$$y = r \sin \varphi \sin \Theta,$$
$$z = r \cos \Theta.$$

The surface of the Earth is broken down cartographically into geographic coordinates. The geographic length is the angle φ in spherical coordinates. The geographic width is determined by an angle of −90° to +90° relative to the equator. This is equivalent to the angle θ in spherical coordinates, which is measured starting from the north pole and is between 0° and 180° (0 and π in radians).

Cylindrical coordinates

If an object or a system has rotational symmetry, cylindrical coordinates are utilized (Figure 12). The x and y coordinates, just like the polar coordinates on a plane, are converted to the radius ρ; the z component of the cylindrical coordinates

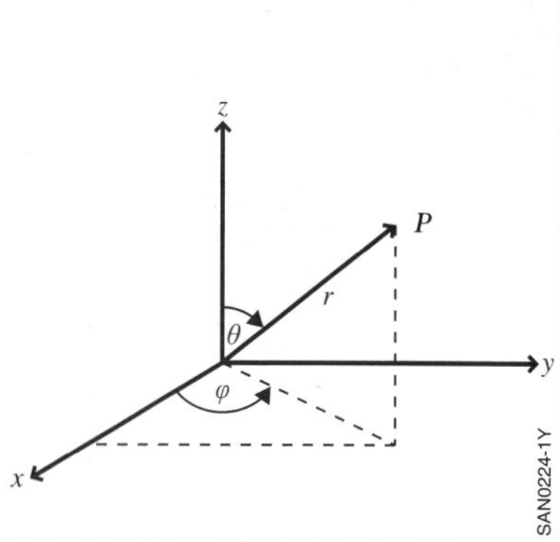

Figure 12: Cylindrical coordinates of a point

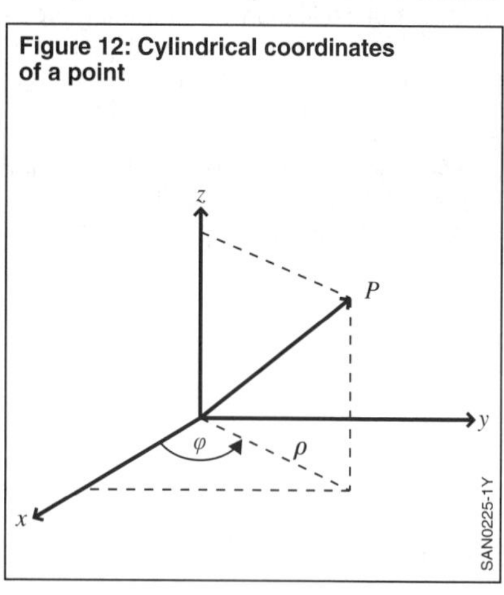

is identical to the z of the Cartesian coordinates.

Just like the spherical coordinates, any point can be clearly expressed in cylindrical coordinates, with the exception of the origin (0, 0, 0).

Conversion of Cartesian coordinates to cylindrical coordinates:

$$\rho = \sqrt{x^2 + y^2}\,,\ x, y, z \in \mathbb{R},$$

$$\tan\varphi = \frac{y}{x}.$$

$$z = z.$$

Conversion of cylindrical coordinates to Cartesian coordinates:

$$x = \rho\cos\varphi\,,\ \rho \in \mathbb{R}_0^+,\ \varphi \in [0; 2\pi],\ z \in \mathbb{R},$$
$$y = \rho\sin\varphi\,,$$
$$z = z.$$

Just like the polar coordinates, cylindrical coordinates are calculated on a plane and they are expanded by the addition of the z-axis.

Cylindrical coordinates can be used in the design of rotationally symmetrical objects, such as tubes, engine cylinders, screws, nuts, and anti-friction and friction bearings.

Vectors

A distinction is made between scalar and vector variables for physical and technical concepts. Examples of scalars include mass, temperature and pressure. By comparison, vectors have a direction, such as velocity, power and electric field. Vectors, their most important arithmetic operations and calculation rules are shown below.

Representation of vectors

A vector in a three-dimensional space has the three components a_x, a_y and a_z for the x-, y and z direction. In mathematics, this is described as follows:

$$\vec{a} = \begin{pmatrix} a_x \\ a_y \\ a_z \end{pmatrix}.$$

Calculation rules

When adding and multiplying vectors with a number, the same calculation rules for adding and multiplying numbers apply.

Addition

For adding two vectors with coordinates as follows

$$\vec{a} = \begin{pmatrix} a_x \\ a_y \\ a_z \end{pmatrix} \text{ and } \vec{c} = \begin{pmatrix} c_x \\ c_y \\ c_z \end{pmatrix}$$

results in:

$$\vec{a} + \vec{c} = \begin{pmatrix} a_x + c_x \\ a_y + c_y \\ a_z + c_z \end{pmatrix}$$

The following laws apply:

- Closure property: The sum and the difference of two vectors is also a vector.
- Commutative property: $\vec{a} + \vec{c} = \vec{c} + \vec{a}$.
- Associative property: $\vec{a} + (\vec{c} + \vec{e}) = (\vec{a} + \vec{c}) + \vec{n}$.
- For any two vectors $\vec{a}$ and $\vec{c}$ there is always one vector $\vec{z}$ so that: $\vec{a} + \vec{z} = \vec{c}$, in other words $\vec{z} = \vec{c} - \vec{a}$.

Multiplication of a vector with a scalar
When multiplying vector a by scalar $\lambda \in \mathbb{R}$, the following is true:

$$\lambda \vec{a} = \begin{pmatrix} \lambda a_x \\ \lambda a_y \\ \lambda a_z \end{pmatrix}$$

For vector $\vec{c}$, the following is true:

$$\vec{c} = \lambda \vec{a} \Rightarrow |\vec{c}| = |\lambda| \cdot |\vec{a}|.$$

For vectors $\vec{a}$ and $\vec{c} \in \mathbb{R}^3$ and scalars λ and $\mu \in \mathbb{R}$, the following laws apply:
- Closure property: The product of a vector and a scalar is also a vector.
- Associative property: $(\lambda \cdot \mu) \cdot \vec{a} = \lambda \cdot (\mu \cdot \vec{a})$.
- Distributive law:
  $(\lambda + \mu) \cdot \vec{a} = \lambda \cdot \vec{a} + \mu \cdot \vec{a}$,
  $\lambda \cdot (\vec{a} + \vec{c}) = \lambda \cdot \vec{a} + \lambda \cdot \vec{c}$.
- Multiplying by the neutral element:
  $1 \cdot \vec{a} = \vec{a}$.

Scalar product
When multiplying two vectors with one another, two different products are defined: the scalar product and the vector product.

The result of the scalar product is a number (scalar). For example, physical work is defined as a scalar product of the force vector $\vec{F}$ and the path vector $\vec{s}$ (Figure 13). Only the components of $\vec{F}$ that point in the direction of $\vec{s}$ contribute to the work. This corresponds to the offset projection depicted in Figure 13. Thus, the following is true for the absolute value of the projection

$$|\vec{F}_s| = |\vec{F}| \cos \varphi$$

Figure 13: Scalar product
The scalar product of both vectors $\vec{F}$ and $\vec{s}$ with the projection $\vec{F}_s$ of the vector $\vec{F}$ on $\vec{s}$.
φ is the angle between both vectors $\vec{F}$ and $\vec{s}$.

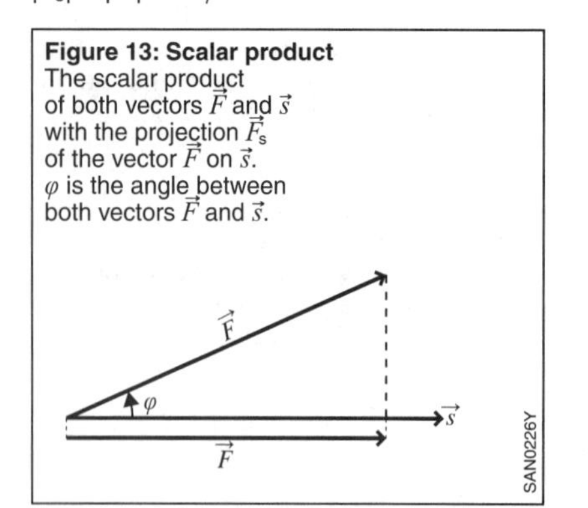

The scalar product is defined as

$$\vec{a} \cdot \vec{c} = |\vec{a}| \cdot |\vec{c}| \cdot \cos \varphi$$
$$= a_x c_x + a_y c_y + a_z c_z.$$

In order to clarify the product, the multiplication point "·" is placed between both vectors.

The calculation rules for the scalar product are the same as the calculation rules for numbers.
- Commutative property:
  $\vec{a} \cdot \vec{c} = \vec{c} \cdot \vec{a}$.
- Distributive law:
  $\vec{a} \cdot (\vec{c} + \vec{n}) = \vec{a} \cdot \vec{c} + \vec{a} \cdot \vec{n}$.
- Multiplication with a scalar, associative property:
  $\lambda(\vec{a} \cdot \vec{c}) = (\lambda \vec{a}) \cdot \vec{c} = \vec{a} \cdot (\lambda \vec{c})$.

For example, the scalar product is used to calculate the absolute value (the length) of a vector. In addition, the angles between the vectors can be determined using the same method. The scalar product in particular is used in order to find out if two vectors are perpendicular to each other. In this case, the scalar product is zero.

Cross product
A new vector results from the cross product (outer product) of two vectors; this new vector is perpendicular to both output vectors (Figure 14). Its length equals the area of a parallelogram spanning both initial vectors. The vectors $\vec{a}$, $\vec{c}$ and $\vec{m}$ form what is known as a right-handed system. According to the right-hand rule, the thumb points in the direction of $\vec{a}$, the index finger

Figure 14: Cross product
Both vectors $\vec{a}$ and $\vec{c}$ span the length of a plane E. The cross product (vector m) formed from $\vec{a}$ and $\vec{c}$ points perpendicular to both output vectors.
φ is the angle between $\vec{a}$ and $\vec{c}$.

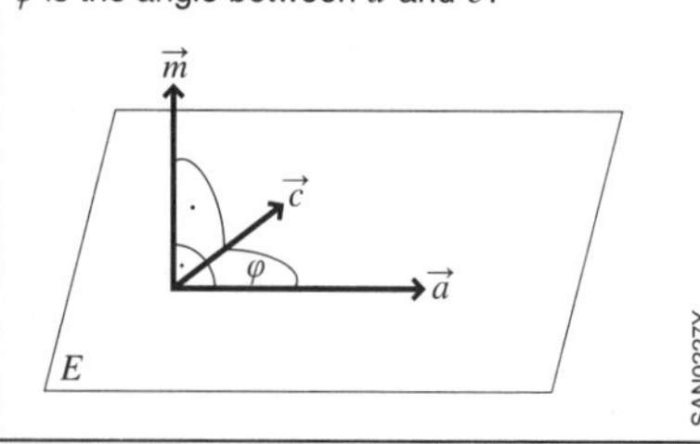

in the direction of $\vec{c}$ and the resulting vector $\vec{m}$ in the direction of the middle finger on the right hand.

In order to be able to distinguish the cross product from the scalar product, an “x” is placed between the vectors as a multiplication symbol.

The calculation rules for the cross product are summarized below:

$\vec{m} = \vec{a} \times \vec{c}$ where

$$|\vec{m}| = |\vec{a}| \cdot |\vec{c}| \cdot \sin\varphi = \begin{vmatrix} \vec{e}_x & \vec{e}_y & \vec{e}_z \\ a_x & a_y & a_z \\ c_x & c_y & c_z \end{vmatrix} = \begin{bmatrix} a_y c_z - a_z c_y \\ a_z c_x - a_x c_z \\ a_x c_y - a_y c_x \end{bmatrix}$$

$\vec{m} \perp \vec{a},\ \vec{m} \perp \vec{c}$,

$\vec{a},\ \vec{c},\ \vec{m}$ are “right-handed”

with the unit vectors $\vec{e}_x$, $\vec{e}_y$ and $\vec{e}_z$ in x, y and z direction.

When applying the calculation rules, note that the leading sign is inverted if the order of the vectors is reversed (anticommutativity):

- Anticommutativity:
  $\vec{a} \times \vec{c} = -\vec{c} \times \vec{a}$.
- Distributive law:
  $\vec{a} \times (\vec{c} + \vec{n}) = \vec{a} \times \vec{c} + \vec{a} \times \vec{n}$.
- Multiplication with a scalar, associative property:
  $\lambda(\vec{a} \times \vec{c}) = (\lambda\vec{a}) \times \vec{c} = \vec{a} \times (\lambda\vec{c})$.

The cross product becomes zero when both vectors point in the same or opposite direction (i.e. when they are collinear). This lets you use the cross product to check if two vectors span one plane. In this case, the cross product does not equal zero.

Therefore, the cross product of a vector with itself is zero.

$$\vec{a}, \vec{c} \neq \vec{0} \text{ where } \vec{a} \times \vec{c} = \vec{0} \Leftrightarrow \vec{a} \parallel \vec{c},$$

$$\vec{a} \times \vec{a} = \vec{0}.$$

Differential and integral calculus

Differentiation of functions

First derivative

The zero points, discontinuities and the behavior of a function can be determined at the margins of the domain using the function term $y = f(x)$. Finding out how steep a function is or where the function attains its maximum or minimum values is often of interest as well. This can be described using differentiation (the derivative) of a function.

The slope (steepness) of a function $y = f(x)$ is given via the tangent of this function (Figure 15). If a function attains a maximum or a minimum value, then the tangent is horizontal at that point. In order to calculate the slope of a function $f(x)$, it must be derived (differentiated). The following two notation systems have been established for the first derivative of a function $f(x)$ to accomplish this:

$$\frac{d}{dx} f(x) = f'(x).$$

The derivatives of elementary functions are summarized in Table 12.

The derivative of compound functions can be calculated using the following calculation rules:

Figure 15: Derivative of a function

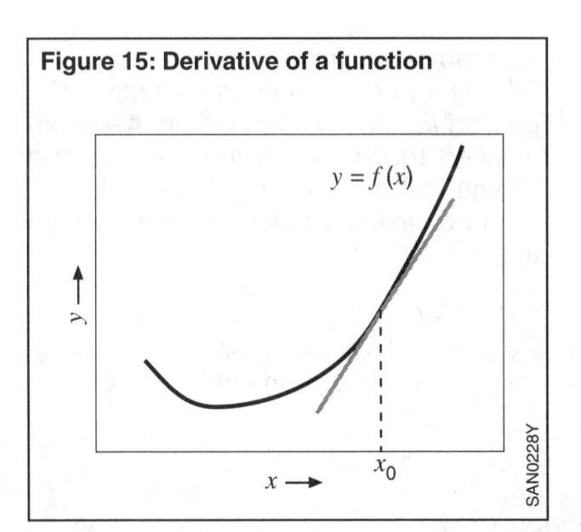

Sum rule:

$$\frac{d}{dx}(f(x) \pm g(x)) = f'(x) \pm g'(x).$$

Multiplication with a number λ:

$$\frac{d}{dx}(\lambda f(x)) = \lambda f'(x).$$

Product rule:

$$\frac{d}{dx}(f(x) \cdot g(x)) = f'(x) \cdot g(x) + f(x) \cdot g'(x).$$

Quotient rule:

$$\frac{d}{dx}\left(\frac{f(x)}{g(x)}\right) = \frac{f'(x) \cdot g(x) - f(x) \cdot g'(x)}{[g(x)]^2}$$

Chain rule:

$$\frac{d}{dx}(F(g(x))) = F'(g(x)) \cdot g'(x).$$

Higher derivatives
If the first derivative is differentiated again, then this becomes the second derivative. Higher derivatives can be formed in a similar way.

2nd derivative:

$$\frac{d}{dx}f'(x) = f''(x) \text{ or } f''(x) = \frac{d^2}{dx^2}f(x)$$

3rd derivative:

$$\frac{d}{dx}f''(x) = f'''(x) \text{ or } f'''(x) = \frac{d^3}{dx^3}f(x).$$

nth derivative($n > 3$):

$$\frac{d}{dx}f^{(n-1)}(x) = f^{(n)}(x) \text{ or } f^{(n)} = \frac{d^n}{dx^n}f(x).$$

Extrema of functions
If the first and the second derivative of a function $f(x)$ are on hand, then these can be used to determine the maxima and minima (extrema) of $f(x)$. Both of the following conditions apply to an extremum at x_0:

– $f'(x_0) = 0$,
– $f''(x_0) < 0$ for a maximum,
  $f''(x_0) > 0$ for a minimum.

Inflection point
The curvature properties of the function can be determined by using the second derivative. The point x_w at which a function switches from left-curved ($f''(x) > 0$, convex curvature) to right-curved ($f''(x) < 0$, concave curvature) and vice versa is of particular interest. These points are known as inflection points. The following is true here:

$f''(x) = 0$.

Integration of a function

A (continuous) function $f(x$ is depicted in Figure 16. Together with the X-axis in

Table 12: Elementary functions $f(x)$ and their derivatives $f'(x)$

$f(x)$	$f'(x)$
c = const	0
x^n	nx^{n-1}
$\sqrt{x} = x^{1/2}$	$\frac{1}{2}\frac{1}{\sqrt{x}} = \frac{1}{2}x^{-1/2}$
$\sin(x)$	$\cos(x)$
$\cos(x)$	$-\sin(x)$
$\tan(x)$	$\frac{1}{\cos^2(x)}$
$\cot(x)$	$-\frac{1}{\sin^2(x)}$
e^x	e^x
$a^x = e^{x \ln a}$	$(\ln a)\,a^x$
$\ln x$	$\frac{1}{x}$
$\log_a x = \frac{\ln x}{\ln a}$	$\frac{1}{\ln a}\frac{1}{x}$

Figure 16: Integral
Integration of the function $f(x)$ in the interval of a through b.

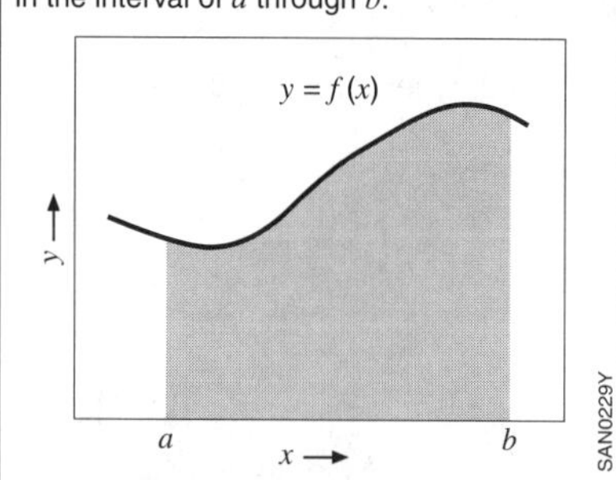

the interval [a; b], this encloses the area A marked in gray. Its dimensions can be calculated through integration.

The unknown area A is represented as an integral from a to b via the function $f(x)$ with the variable x:

$$A = \int_a^b f(x)\,dx.$$

Antiderivative

In order to find a solution to this problem, the antiderivative $F(x)$ is required for $f(x)$. Its property is expressed in the fundamental theorem of differential and integral calculus.

$$F(x) = \int_a^x f(t)\,dt$$

is an antiderivative for $f(x)$, if the following applies:

$$\frac{d}{dx}F(x) = f(x).$$

If $F(x)$ is an antiderivative for $f(x)$, then the unknown area can be calculated as

$$A = \int_a^b f(x)\,dx = F(b) - F(a).$$

Important antiderivatives are compiled in Table 13. For each function $f(x)$, there is just one antiderivative $F(x)$. This is uniquely defined with the exception of the integration constant C.

Table 13: Elementary functions and their antiderivatives (examples)

$f(x)$	$\int f(x)\,dx$
0	$c = \text{const}$
$x^n,\ c \neq -1$	$\frac{1}{n+1}x^{n+1} + C$
$\frac{1}{x}$	$\ln\lvert x\rvert + C$
$\sqrt{x} = x^{1/2}$	$\frac{2}{3}x^{3/2} + C$
$\sin x$	$-\cos x + C$
$\cos x$	$\sin x + C$
$\frac{1}{\cos^2 x}$	$\tan x + C$
$\frac{1}{\sin^2 x}$	$-\cot x + C$
e^x	$e^x + C$
$a^x = e^{x \ln a}$	$\frac{a^x}{\ln a} + C$

Calculation rules for integration

The following calculation rules apply to integration:

Sum rule:

$$\int_a^b f(x) + g(x)\ dx = \int_a^b f(x)\,dx + \int_a^b g(x)\,dx.$$

Multiplication with a number λ:

$$\int_a^b \lambda f(x)\,dx = \lambda \int_a^b f(x)\,dx.$$

Reversing the bounds of integration:

$$\int_a^b f(x)\,dx = -\int_b^a f(x)\,dx.$$

Lower parts of integration ranges in intervals ($a < c < b$):

$$\int_a^b f(x)\,dx = \int_a^c f(x)\,dx + \int_c^b f(x)\,dx.$$

Upper and lower bounds of integration are identical:

$$\int_a^a f(x)\,dx = 0.$$

Partial integration

In addition to searching for the antiderivative, there are two helpful methods for solving for integrals for partial integration and integration through substitution. Partial integration is the inverse of the product rule for the derivative.

$$\int_a^b f'(x)\cdot g(x)\,dx = f(x)\cdot g(x) - \int_a^b f(x)\cdot g'(x)\,dx.$$

The critical factor for this method is properly assigning both product terms $f'(x)$ and $g(x)$ within the integral.

Substitution

Substitution is the inverse of the chain rule. The art of substitution lies in finding the appropriate function that can be replaced (for more information, refer to [1], [2], [3]).

Linear differential equations

In many technical problems, a function is being sought that can describe the task but is not explicitly specified. Instead, information about its curvature properties (second derivative), its slope (first derivative) or a combination of the two is provided in the form of an equation.

An equation in which the function and its derivatives can be found is known as a differential equation. Equations known as common linear differential equations of the nth order with constant coefficients are covered below ([1], [2] and [3]):

$$y^{(n)}(x) + a_{n-1}y^{(n-1)}(x) + \ldots + a_0 y(x) = g(x).$$

In this equation, the nth derivative of $y(x)$ is the highest order derivative. Real numbers stand in front of individual derivatives $a_0, a_1, \ldots, a_{n-1}, a_n$, where $a_n = 1$ is selected. If the function $g(x)$ is always zero, the differential equation is homogeneous; otherwise, it is non-homogeneous. Here, x is the variable of the unknown function $y(x)$.

The differential equation above is linear, since $y(x), y'(x), \ldots y^{(n)}(x)$ are only accompanied by the coefficients and do not occur as exponents, such as $y(x)^2$, or as a nonlinear function, like $\sin(y'(x))$.

Using the formulation
$y(x) = Ae^{\lambda x}$, $A \in \mathbb{R}$, $\lambda \in \mathbb{R}$, a homogeneous differential equation is simplified to a characteristic polynomial, because the exponential term is canceled out:

$$\lambda^n + a_{n-1}\lambda^{n-1} + \ldots + a_2\lambda^2 + a_1\lambda + a_0 = 0.$$

In doing so, solving the differential equation is reduced to searching for the zero points of the characteristic polynomial. If the n zero points $\lambda_1, \lambda_2, \ldots, \lambda_n$ are determined, then the homogeneous solution $y_h(x)$ of the differential equation is yielded as

$$y_h(x) = A_1 e^{\lambda_1 x} + A_2 e^{\lambda_2 x} + \ldots + A_n e^{\lambda_n x}.$$

The numbers $A_1, A_2, \ldots A_n$ are (any) real numbers and are labeled as integration constants. If initial or ancillary conditions are given for the function, then the values $A_1, A_2, \ldots, A_n$ can be calculated from these conditions.

After finding the homogeneous solution, a special solution $y_{ih}(x)$ for the non-homogeneous differential equation needs to be found; this then provides the complete solution.

$$y(x) = y_h(x) + y_{ih}(x).$$

Linear differential equations of the second order with constant coefficients are the basis of all oscillation processes (see Oscillation system, Vibration characteristics) and also the basis for the design of safety-related sensors like acceleration and vibration sensors (see Acceleration sensors). Even regulating and control engineering technology is based on these differential equations (see Control engineering).

In addition to common linear differential equations with constant coefficients, there are also linear differential equations in which the constant coefficients are replaced by functions, such as

$$\sin(x) \cdot y'(x) + \cos(x) \cdot y(x) = 0,$$

Furthermore, the two-dimensional oscillation equation of the function $f(x,y)$ with the location coordinates x and y and time t,

$$\frac{\partial^2 f}{\partial x^2} + \frac{\partial^2 f}{\partial y^2} = \frac{1}{c^2}\frac{\partial^2 f}{\partial t^2}, c \in \mathbb{R},$$

is an example of a partial differential equation. It is significantly more complicated to solve both of these types of differential equations than it is to solve the linear differential equations described above. For more information and a more detailed approach to these differential equations, the extensive literature from the bibliography should be consulted, including [1], [2] and [3], or the specialized literature regarding special types of differential equations.

Laplace transform

There are many control loops in a vehicle (see Control engineering); in the engine (e.g. knock control, λ control), in the A/C unit or in the undercarriage (e.g. yaw-rate control). These control loops are often represented by linear differential equations. One way to solve this differential equation is to use the exponential approach (see Differential equations).

Alternatively, the Laplace transform can also be utilized if initial values like $y(0)$, $y'(0)$ etc. are given in addition to the differential equation ([1], [2], [3]). The advantage to this approach is that the Laplace transformation converts the differential equation into an algebraic equation. This can usually be resolved easily according to the Laplace transform function $Y(s)$. Then, the Laplace inverse transform must be carried out in order to find the unknown function $y(x)$.

The Laplace transform $\mathcal{L}\{y(x)\}$ or image function $Y(s)$ of a function $y(x)$ can be found as follows:

$$Y(s) = \mathcal{L}\{y(x)\} = \int_0^\infty e^{-sx}\, y(x)dx,\ s \in \mathbb{R}.$$

In order to solve for the integral, the absolute value of the function must be exponentially bound. s merely acts as a variable in the transformation [3].

Properties

Laplace transforms of the derivatives $y'(x)$, $y''(x)$ are very crucial when solving differential equations.

$$\mathcal{L}\{y'(x)\} = s\,Y(s) - y(0)$$

$$\mathcal{L}\{y''(x)\} = s^2\,Y(s) - s\,y(0) - y'(0)$$

$$\mathcal{L}\{c_1 y_1(x) + c_2 y_2(x)\} = c_1\,\mathcal{L}\{y_1(x)\} + c_2\,\mathcal{L}\{y_2(x)\} \quad \text{(linearity)}$$

$$\mathcal{L}\{y\,(ax)\} = \frac{1}{a}\,Y\left(\frac{s}{a}\right),\quad a > 0 \quad \text{(Similarity theorem)}$$

$$\mathcal{L}\{e^{-ax}\,y\,(x)\} = Y(s + a),\quad a > 0 \quad \text{(Complex shifting theorem)}$$

As you can see, the derivatives are converted into terms with the function $Y(s)$.

A few selected examples of the Laplace transforms of functions have been compiled in Table 14.

Table 14: Functions $y(x)$ and the corresponding Laplace transforms $Y(s)$

$y(x)$	$Y(s) = \mathcal{L}\{y(x)\}$
$y(x) = \begin{cases} 0 & x < 0 \\ 1 & x \geq 0 \end{cases}$	$\frac{1}{s}$
$y(x) = \begin{cases} 0 & x < 0 \\ x^{n-1} & x \geq 0 \end{cases}$	$\frac{(n-1)!}{s^n}$ $n = 1, 2, 3$
$y(x) = \begin{cases} 0 & x < 0 \\ \sin(ax) & x \geq 0 \end{cases}$	$\frac{a}{s^2 + a^2}$
$y(x) = \begin{cases} 0 & x < 0 \\ \cos(ax) & x \geq 0 \end{cases}$	$\frac{s}{s^2 + a^2}$
$y(x) = \begin{cases} 0 & x < 0 \\ e^{-ax} & x \geq 0 \end{cases}$	$\frac{1}{s + a}$

Fourier transform

During analysis, a time-dependent function is often found $f(t)$. Analysis is usually interested in the frequencies of the vibrations that underlie this function. For example, the undercarriage of a car should compensate for the bumps in the road so that the vehicle does not begin to rock back and forth. This requires a surface unevenness profile $h(x)$. If the car is driving down the road at a particular speed, it will begin to rock, which corresponds to the function $f(t)$. The characteristic frequencies must be determined from this function in order to be able to select and set the dimensions for the vehicle's shock absorbers, which will dampen the oscillations (see Undercarriage, Road excitation, Irregularity profiles).

The characteristic oscillation frequencies can be determined using the function $f(t)$, if this function $f(t)$ is subjected to a Fourier transform to form the function $F(\omega)$:

$$F(\omega) = \mathcal{F}\{f(t)\} = \int_{-\infty}^{\infty} f(t)\, e^{-i\omega t} dt,$$

where $i^2 = -1$.

This converts the time-dependent function $f(t)$ into the frequency-dependent spectral function $F(\omega)$ ([1], [2]).

This results in an inverse transformation along the lines of:

$$f(t) = \mathcal{F}^{-1}\{F(\omega)\} = \frac{1}{2\pi} \int_{-\infty}^{\infty} F(\omega) e^{i\omega t} d\omega.$$

Figure 17: Symbolic representation of a Dirac delta function

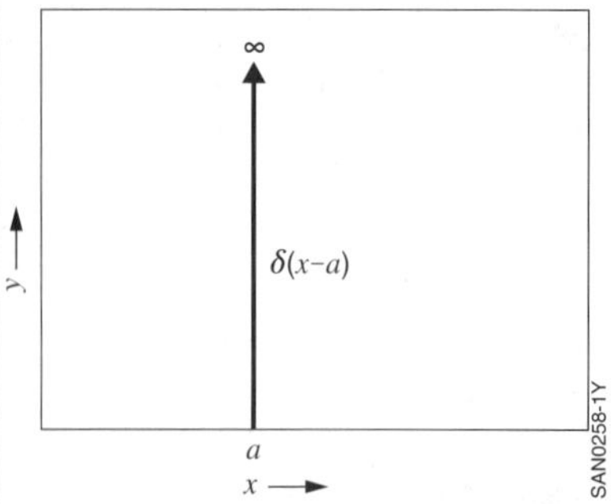

In the literature, values of 1 (like here), $1/2\pi$ or $1/\sqrt{2\pi}$ are sometimes specified as the prefactor for the Fourier transform. This also changes the factor of the inverse transform. The critical aspect here is that the product of the transform's prefactors and the inverse transform yield $1/2\pi$.

Properties

Some properties of the Fourier transform are listed below.

$$\mathcal{F}\{c_1 f_1(t) + c_2 f_2(t)\} = c_1 \cdot \mathcal{F}\{f_1(t)\} + c_2 \cdot \mathcal{F}\{f_2(t)\}$$
(Linearity)

$$\mathcal{F}\{f_1(at)\} = \frac{1}{|a|} F\left(\frac{\omega}{a}\right), \quad a \in \mathbb{R},\ a \neq 0,$$
(Similarity theorem)

$$\mathcal{F}\{f(t - t_0)\} = e^{-i\omega t_0} F(\omega), \quad t_0 \in \mathbb{R},$$
(Shift theorem in the time range)

$$\mathcal{F}\{e^{i\omega_0 t} f(t)\} = F(\omega - \omega_0), \quad \omega_0 \in \mathbb{R},$$
(Shift theorem in the frequency domain)

$$\mathcal{F}\left(\int_{-\infty}^{\infty} f(\tau) g(t-\tau)\right) d\tau = \mathcal{F}\{f(t)\} \cdot \mathcal{F}\{g(t)\}.$$

Using the notation

$$(f * g)(t) = \int_{-\infty}^{\infty} f(\tau) g(t-\tau)\, d\tau = \int_{-\infty}^{\infty} f(t-\tau) g(t)\, d\tau$$

the convolution theorem can be summarized as follows:

$$\mathcal{F}\{(f * g)\} = F(\omega)\, G(\omega).$$

The transform in the frequency domain can be explained by the Dirac delta function, which is defined as follows [5]:

$\delta(t) = 0$ for $t \in \mathbb{R}$, $t \neq 0$, and

$$\int_{-\infty}^{\infty} \delta(t - t_0) f(t)\, dt = f(t_0).$$

This can also be written as

$$\delta(t - t_0) = \frac{1}{2\pi} \int_{-\infty}^{\infty} e^{i(t-t_0)\tau}\, d\tau.$$

The Dirac delta function is always zero. However, at the point where its argument is zero, it can be said to be infinitely large. This is illustrated in Figure 17. The Fourier transform of the complex exponential function or the sine and cosine function with an angular frequency of ω_0 returns the Dirac delta function for ω_0:

$$\mathcal{F}\{e^{-i\omega_0 t}\} = 2\pi\ \delta(\omega - \omega_0)\ dx,$$

$$\mathcal{F}\{\delta(t - t_0)\} = \frac{1}{2\pi}\ e^{-i\omega t_0},$$

$$\mathcal{F}\{\sin(\omega_0\, t)\} = i\pi\ [\delta(\omega + \omega_0) - \delta(\omega - \omega_0)],$$

$$\mathcal{F}\{\cos(\omega_0\, t)\} = \pi\ [\delta(\omega + \omega_0) + \delta(\omega - \omega_0)].$$

For example, the Fourier transform is used to determine the frequencies of signals (e.g. the individual tones within the sound of a piano). In addition, the ratios of the corresponding amplitudes corresponding to the volume of sounds can be found. This allows sound waves to be broken down into individual sounds (frequencies) and their amplitude (volume) (see Undercarriage, Road excitation; Acoustics, Perceived sound levels).

Similarities and differenced between Laplace and Fourier transforms

The Laplace transform of a function $y(x)$

$$\mathcal{L}\{y(x)\} = \int_0^\infty e^{-sx}\, y(x)dx,\ s \in \mathbb{R},\ \mathbb{C}$$

and the Fourier transform of $y(x)$

$$\mathcal{F}\{y(x)\} = \int_{-\infty}^\infty e^{-i\omega x}\ y(x)\, dx, \quad \text{where } i^2 = -1$$

are similarly defined. This means that both transforms have properties that are either similar or the same. Both transforms are linear, their similarity theorems are almost identical and the complex shifting theorem of the Laplace transform is equivalent to the shift theorem of the Fourier transform.

However, both transforms can be distinguished by their bounds of integration, because the Laplace integral begins at $x = 0$ and the Fourier integral at $-\infty$. In other words, the function $y(x)$ can be defined by for the entirety of $\mathbb{R}$ in the case of the Fourier integral, while $y(x)$ can only differ from zero for $\mathbb{R}^+$. Furthermore, the exponential function of the Laplace integral usually contains an arbitrary complex exponent $-sx$, while this term $-i\omega x$ is normally completely imaginary in the case of a Fourier integral.

The inverse of the Fourier transform can usually be calculated in a manner similar to that of the Fourier transform. However, calculating the inverse Laplace transform by means of complex curvilinear integrals is often difficult [1]. Tables such as Table 14 are usually helpful when solving differential equations in order to determine the original function $y(x)$ from the Laplace transform.

The direct correlation between the Laplace and the Fourier transform is depicted as follows: For the function $y(x)$, the function $g(x)$ is defined as:

$$g(x) = \begin{cases} 0 & x<0 \\ e^{-\lambda x}\ y(x) & x \geq 0 \end{cases} \qquad \lambda \in \mathbb{C}$$

Subjecting $g(x)$ to a Fourier transform yields

$$\mathcal{F}\{g(x)\} = \int_{-\infty}^\infty e^{-i\omega x}\ g(x)\ dx$$
$$= \int_0^\infty e^{-i\omega x}\ e^{-\lambda x}\ y(x)\, dx.$$

Solving for $s = \lambda + i\omega$ then yields

$$\mathcal{F}\{g(x)\} = \int_{-\infty}^\infty e^{-i\omega x} g(x)dx = \int_{-\infty}^\infty e^{-sx} y(x)dx$$
$$= \mathcal{L}\{y(x)\}.$$

References

[1] T. Arens et al., Mathematik, 2nd Edition, Spektrum Akademischer Verlag, Heidelberg, 2011.

[2] I.N. Bronstein et al., Taschenbuch der Mathematik, 8th Edition, Verlag Deutsch (Harri), Frankfurt a.M., 2012.

[3] L. Papula, Mathematik für Ingenieure und Naturwissenschaftler, Volumes 1 to 3, Vieweg+Teubner, Wiesbaden, 2011.

[4] F. W. J. Oliver et al. (Editors), NIST Handbook of Mathematical Functions, Cambridge University Press, Cambridge 2010.

Finite-element method

Applications

Virtually all technical procedures can be simulated on a computer with the finite-element method FEM (the term having been introduced by Ray W. Clough at the start of the 1960s [1]). However, this involves breaking down any body (gaseous, liquid, or solid) into elements that are simple in shape (line, triangle, square, tetrahedron, pentahedron, or hexahedron), that are as small as possible, and that are permanently connected to each other at their corner points ("nodes"). Small elements are important because the behavior of elements formulated by approximation using linear equations is only applicable to infinitesimal elements. However, the computing time calls for finite elements. The approximation to reality is better the smaller the elements are.

The application of FEM in practice – also known as FEA (finite-element analysis) – began in the early 1960s in the aviation and aerospace industries and followed soon after in automobile manufacturing. Today the method is used in all fields of technology, including weather forecasting, medical science, and for many sectors of automobile manufacturing ranging from engine and chassis components through to body calculations and crash behavior.

There are two different types of application. Firstly, virtually fully automatic "Black Box" FEM contained in all CAD programs (computer-aided design) for rough calculations by the design engineer (e.g. in designing a bumper) and, secondly, the use reserved for specialists of special FEM programs (e.g. in body calculations, in axle development or in driving dynamics).

FEM program system

The software of an FEM system consists of a preprocessor, a postprocessor, and the actual FEM program. Network creation, i.e. breakdown into elements, is mainly performed in the preprocessor on the basis of a CAD geometry which is read directly or via neutral interfaces such as IGES (Initial Graphics Exchange Specification), VDA-FS (Verband der Automobilindustrie – Flächenschnittstelle) or STEP (Standard for the Exchange of Product Model Data). The FEM program calculates the computing model formulated in this way. The result found is then shown in graphic form in the postprocessor (e.g. stress distribution by means of isocolors, deformations as motion animation).

Basic knowledge for application

FEM is, like all numerical methods, an approximation process. In mechanics, the main area of application, the limitations caused by this are described in the following.

Small motions in one solution step

Bodies move on paths which are normally higher-order curves. With the basic principle of linearization of all processes, this motion is limited to a straight path which can then be described by linear equations. When transferred to the element corners (nodes), they also move on a straight line. Thus, the nodes are only able to realize very small motions correctly (node twists less than 3.5°). The actual motion along any path or nonlinear material behavior is thus solved linearly with many small steps.

Calculation accuracy

The linear equation system is formulated and solved with the limited computing accuracy of a computer. Usually, 8 bytes (= 64 bits) are used with a computing accuracy of 13 significant digits for the number stored, i.e. only the first 13 digits of a number can be represented exactly. The 14th digit and every further digit in this number are random numbers. As a result, this rules out the possibility of any stiffness differences of the individual components in a model. Therefore, in the deformation calculation of a body, it is necessary, as for example in the measurement of body deformation, to replace the axle springs with rigid supports.

Interpretation of results
The great danger lies in the fact that a formal computation model formulated correctly by a beginner will indeed deliver beautifully colorful images, but the results shown can center around factors far from reality.

The problems resulting from the above mentioned limitations must therefore be identified and signaled by the computing program, so that the less experienced user is also able to obtain correct results easily.

Areas of FEM application
In technology terms, physics is generally divided into five areas – mechanics with statics and kinematics (e.g. body, axle), dynamics with acoustics (e.g. vehicle noise), thermodynamics (e.g. temperature distribution in the engine), electricity with magnetism (e.g. ignition coil, sensor technology), and optics (e.g. headlamp). When it comes to FEM, a distinction is always made between
- linear and nonlinear static and dynamic problems with the deformations as an unknown for stress calculation and dynamic analysis,
- stationary (timeless) and nonstationary (time-dependent) potential problems (e.g. temperature, sound pressure, electrical or magnetic potential) with the potentials as an unknown,
- and the linking of these different fields, e.g. to calculate a temperature field and the resulting deformations, stresses and forces in linear statics when the engine is started.

Elements of FEM
The properties of the elements define the most important performance data of an FEM program. The element quality is determined by the degree of the mathematical formulation function selected. Here, a distinction is made between elements with a linear or quadratic formulation, recognizable from the midside node in the middle of the edge. The quality of a computing model is therefore dependent not just on the fineness of the used mesh, but also quite considerably on the formulation function.

A distinction is made between three different types of elements: line elements, shell elements, and volume elements.

Line elements
Line elements (Figure 1) are either straight or curved with an midside node. The cross-sections are described by specifying the numerical values for the cross-sectional area A, the reduced shear cross-sections $A_{\text{red-v-w}}$ (shear areas), the principal moments of inertia (I_v, I_w), the torsional moment of inertia (I_t) with the torsional section modulus (W_t), the sector moment of inertia for warping-force torsion, an angle α which describes the position of principal axes of inertia v and w in relation to the road plane, and the maximum four stress points (S_v, S_w) for bending-stress calculation.

Shell elements
Shell elements (Figure 2) are either triangular or quadrangular in shape – ideally an equilateral triangle or a square, usually of constant thickness. If the midside nodes are omitted, the edges are straight.

Figure 1: Line elements
With constant or linearly variable cross-section.

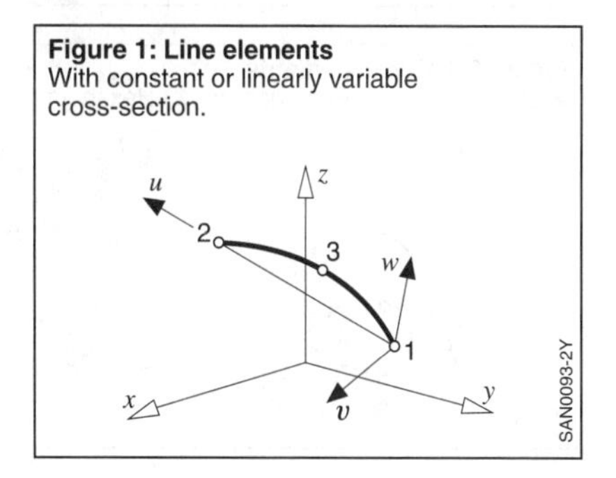

Triangles should, wherever possible, be avoided or only used in transition areas.

Volume (solid) elements
Volume elements (Figure 3) as tetrahedrons, pentahedrons and hexahedrons without midside nodes have straight edges – ideally an equilateral tetrahedron or a cube. With a sufficient number of elements in relation to the thickness (greater than or equal to three), elements with midside nodes are not necessary today. However, this does not apply to tetrahedrons, which are almost alway used in the case of a complicated geometry, such as a cylinder head for example.

Modeling and evaluation of results
The most important function in using an FEM program is the usually time-consuming task of creating the input data as a computing model with the preprocessor. The user should try to achieve this target with as few elements and nodes as possible (however, the computing model of a body today has approximately three to four million nodes). To do so, the user requires a certain level of experience, but must have an exact knowledge of the qualities of the elements used (see FEM examples). These can differ slightly in every FEM program.

The first step in modeling involves choosing the element type (line, shell or volume element) and determining the fineness of the mesh, e.g. by means of the specified middle element-edge length. The next step involves defining the properties (material data), e.g. the element thickness for unit elements, the cross-section values for line elements, and the units used (e.g. length in mm and force in N). A further step involves determining support conditions and load. The crucial factor here is considering the points where a model is fixed, and where it is loaded. In relation to loading, it is also useful to conduct a breakdown of the total load into load cases, e.g. into the mass load from gravity weight and different traffic loads.

All FEM results are available in list form or postprocessor data format, and can therefore be shown in graphic form (see FEM examples). To this end, the postprocessor offers all the conceivable forms of display.

Figure 2: Shell elements
a) Triangle, b) Quadrangular.

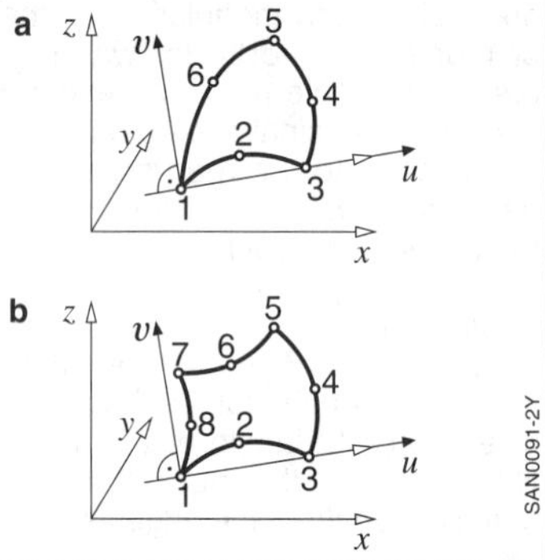

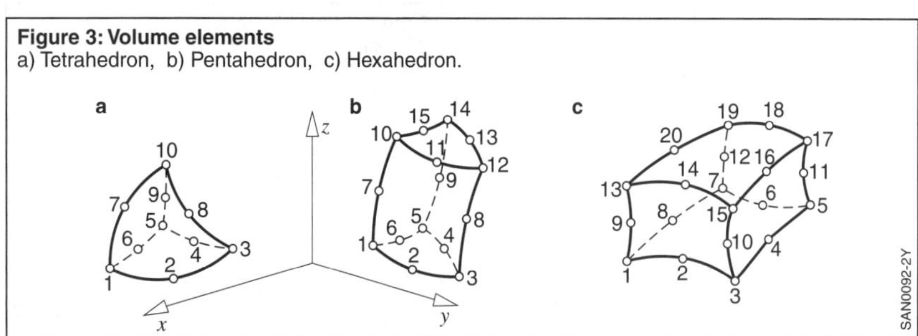

Figure 3: Volume elements
a) Tetrahedron, b) Pentahedron, c) Hexahedron.

FEM examples

For the examples, modeling is performed on the basis of CAD geometry using the FEM program TP2000. The model inputs, together with color representation of the results, can also be found on the Internet page specified in [2].

In reality, all bodies are three-dimensional. In simulation, a simplified solution is often chosen in order to save time and expense. It is much easier, for example, to realize the automatic meshing of a flat surface in shell elements than it is to realize a body in its volume elements. The frequently used tetrahedron mesh, which today every preprocessor creates for any volume geometry, does not always live up to expectations.

In automobile manufacturing, the structural components are either thick-walled and solid (e.g. engine, transmission, axles, wheels) and modeled with volume elements or they are made from thin-walled sheet metal (e.g. automobile body, truck cab) and are modeled with shell and line elements.

The first example, as a volume-element model, belongs to the first group of thick-walled, solid components. It is compared with the commonly used shell model of the second group. This also includes the line element commonly used in automobile manufacturing with the second example.

Figure 4: Cast-steel engine mount
a) CAD model with load F_x, F_y, F_z,
b) Model A1 (hexahedron),
c) Model B1 (tetrahedron),
d) Model C (shell).

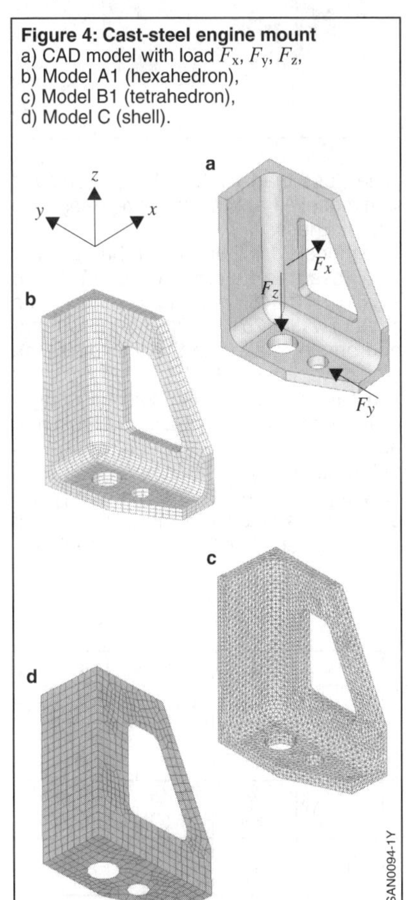

Example 1: Cast-steel engine mount as shell- and volume-element model (linear statics)

The aim is to compare shell and volume elements in linear statics on a relatively thick-walled (3.75 mm) cast-steel engine mount (50 × 25 × 57 mm^3). In the volume elements, two models A and B (each with and without midside nodes) with significantly different result qualities are compared; added to this is shell model C (Figure 4). Starting out from the CAD volume geometry, the preprocessor automatically creates volume-element networks A and B, and also shell-element model C by means of the surface model.

The material properties in the units mm, N and kg are:
Modulus of elasticity: 210,000 N/mm^2
Poisson's ratio: 0.3
Density: 0.00000785 kg/mm^3

In the case of models A and B, the element properties are defined with the element type "Solid" (volume) with three node degrees of freedom v_x, v_y and v_z. In the case of model C, the properties are defined with the element type "Plate" (shell) with six node degrees of freedom v_x, v_y, v_z, d_x, d_y, d_z and constant thickness d = 3.75 mm.

In model A, the meshing with volume elements (solid mesh) shows the preferred breakdown into hexahedrons (not possible fully automated for all geometries because here, for example, the geometry first had to be broken down into basic bodies) and in model B, the automatic tetrahedron meshing. The number of elements in relation to thickness which is crucial for accuracy is specified (at least three elements here).

Support conditions:
On the rectangular xz plane, $v_y = 0$ applies to all nodes. All the edge nodes on the right long leg are pinned by $v_x = 0$, and those of the lower short leg are pinned by $v_z = 0$.

Loading
$\sum F_x = 900$ N, $\sum F_y = 2006$ N, $\sum F_z = -550$ N. All the loads are defined as surface load ("on surface", for shell "on curve") at the recess at $F_x = 600$ N, at the small hole at $F_y = 2006$ N, and at the large hole at $F_z = -550$ N.

Note: "Isolated" individual loads are only permitted with line elements.

Figure 5: Deformation and stress distribution in model A2 as the correct model
$v = 0.064$ mm, $\sigma = 195$ MPa.

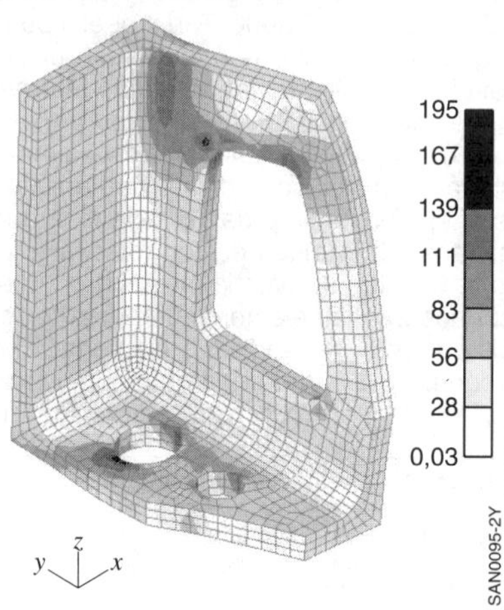

Result
The results are shown in Table 1. Figure 5 shows the result for model A with stresses as shades of gray (critical areas are shown in dark) or as isocolors in the original.

Volume elements are very sensitive to incorrectly approximated load distributions. Volume elements should therefore only ever be subjected to surface loads.

Considering Table 1 in detail: Experience shows that model A2 with midside nodes delivers the correct result ($v = 0.064$ mm, $\sigma = 195$ MPa). The deviation between averaged and maximum node stress (from all the elements at the same node) should be as small as possible (less than 15 %). This is achieved with model A1 at 11 % with three elements per thickness.

In model B1 with three tetrahedrons per thickness, the deformation is 25 % too small ($v = 0.048$ mm), and the maximum stress is also 25 %. With only two tetrahedrons in relation to the thickness, the model would even be 58 % too rigid and can barely be used with correspondingly lower stresses. However, with midside nodes (B2), it delivers virtually identical deformations and stresses to A2 with a very large number of nodes and long computing times. Caution is therefore advised with rough tetrahedron mesh without midside nodes.

Table 1: Results for engine-mount model
Values in parentheses apply to elements with midside nodes.

Model	Element type	No. of nodes	Weight in kg	Max. deformation in mm	Mean stress in MPa	Max. stress in MPa	Stress error max. in %	Computing time in sec
A1 (A2)	Solid hexahedron	4,000 (15,000)	0.119	0.061 (0.064)	145 (185)	164 (195)	15 (0)	10 (110)
B1 (B2)	Solid tetrahedron	10,000 (70,000)	0.119	0.048 (0.064)	115 (173)	146 (209)	30 (6)	60 (2,400)
C	Rectangular shell	766	0.114	0.081	197	223	14	2

Model C with shell elements without shear deformation (only intended for thin-walled structures) is clearly too soft (v = 0.081 mm, 21 % greater, as thick-walled with shear deformation even 30 % greater). Shell elements, whether thin- or thick-walled, deliver only partially usable deformations with relatively large thickness, particularly in the case of very compact bodies, as is the case here. However, at stresses of +14 %, it is for the most part on the safe side.

Example 2: Tubular body frame
The aim is to carry out an FEM analysis on a body in the form of a spaceframe without sheet paneling, including optimized weight and stiffness for the example of a mini-pickup (not a real vehicle). The material properties are:
Modulus of elasticity: 200,000 N/mm^2
Poisson's ratio: 0.3
Density: 0.00000785 kg/mm^3

For reasons of simplicity, only two shapes are used as profile sections (Figure 6). A box-type profile section (90 × 120 × 1.5 mm^3) and a tube profile section (70 x 2 mm^2), which are defined by their shape in the preprocessor in the properties directly with the element type "Bar" and its constant cross-section (different starting and end cross-sections would be "Beam"). This results in the required cross-section values as follows (in mm^2 or mm^4 and mm^3).

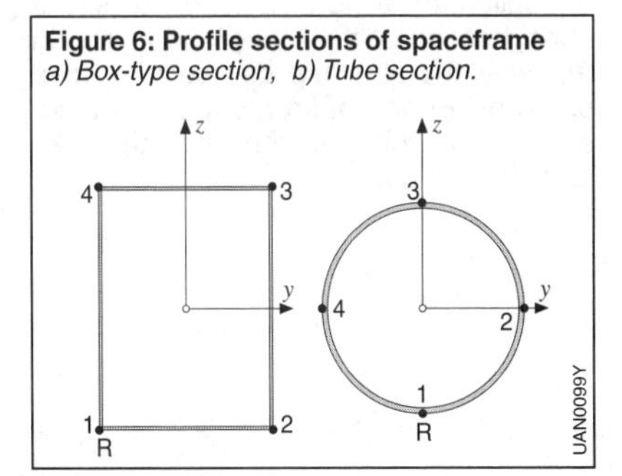

Figure 6: Profile sections of spaceframe
a) Box-type section, b) Tube section.

Box/tube (Figure 6)
- Cross-sectional area:
  A = 621/427.
- Reduced shear cross-sections (shear area):
  A_{redI} = 325/227,
  A_{redII} = 219/227.
- Principal moments of inertia:
  I_I = 1,348,306/247,168
  I_{II} = 869,596/247,168.
- Torsional moment of inertia:
  I_t = 1,606,083/494,261.
- Torsional section modulus:
  W_t = 7,334/2,177.
- Position of principal axes of inertia in relation to road:
  $\alpha = 0°$.
- Maximum four stress points:
  S_x = –45/45/45/–45 / 0/35/0/–35,
  S_y = –60/–60/60/60 / –35/0/35/0.

The main dimensions according to the side and top views (Figure 7) are (in mm):

L_1 = 4,114 (max.)
L_2 = 2,650
W_1 = 1,517 (max.)
W_2 = 1,147 (front)
W_3 = 1,374 (rear)
H_1 = 1,402 (max.)
H_2 = 1,315
H_3 = 469 (box)

Body
The tubular consists of 18 components (Figure 8), e.g.:
2 Side pillar
3 Pedal cross-pillar
7 Cockpit cross-pillar
8 Side pillar, front
10 Auxiliary cross-pillar, front
14 Roof frame, side
15 Roof frame, lateral
16 Rear subframe
17 Side pillar, box
19 Bumper, rear

Support conditions:
Linear statics, load case – bending (not shown):
A $v_y = 0$; (A – F, see Figure 7).
B, C, E, F $v_z = 0$; D $v_x = 0$, $v_y = 0$.

Figure 7: Spaceframe of a mini-pickup

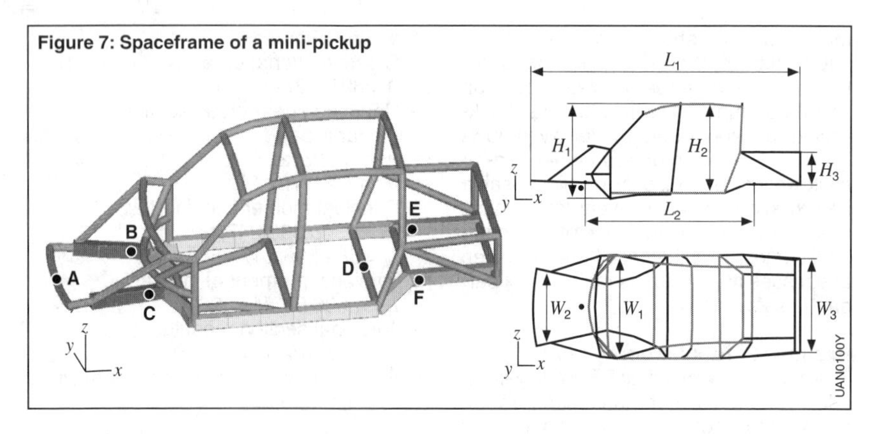

Linear statics, load case – torsion:
A, B, C = free;
E, F $v_z = 0$;
D $v_x, v_y = 0, d_y = 0, d_z = 0$.

Free/free dynamics:
No supports, body vibrates freely in the springs.

Loading
Linear statics, load case – torsion:
Torsional moment 3,000,000 Nmm as regular unit load on
Front-axle bearings B: $F = -3593.7$ N
Front-axle bearings C: $F = 3593.7$ N

Free/free dynamics:
Only masses from gravity weight with 170 kg, no supplementary masses.

Results
Free/free dynamics:
- The body can vibrate freely in its springs, the lower eigenvalues and modes of the undamped, elastic system in ascending order; for free/free beginning with several rigid-body vibration modes (here 1 through 6) with the natural frequency 0 in each case.
- The 1st eigenvalue (38 Hz) is obtained as the 1st torsional vibration, the 3rd eigenvalue (48 Hz) as the 1st flexural vibration; with in each case the normalized deformations x, y, z at all nodes (here normalized to max. 0.1 mm), the normalized stresses in all elements, and the deformation processes important for weight optimization per element and per component 1 through 18 in % as a bar chart (as shown for linear statics in Figure 8). Absolute values are only obtained in the event of an excitation calculation.

Linear statics:
- Deformations x, y, z at all nodes
- Reaction forces and moments at the support nodes, stresses at all elements, strain energy per element and per component in %

Weight and stiffness optimization
The computation formula for weight and stiffness optimization has been known since the 1960s (error max. 10 % for doubled rigidity). The stiffness change of the overall structure (in %) is obtained from the stiffness change of the component multiplied by the deformation-process ratio of the component, divided by 100.

This formula is used specifically for optimization purposes in virtually all automobile-critical "Torsion" load cases with consideration of the other load cases. In this way, the structure can be reinforced using pillars that take most of the load, and reduce weight via the pillars that take less of the load.

Application of formula
1st component (see Figure 8):
Component 15 (roof frame, lateral, G =11.85 kg with 14.57 % strain energy ratio:
Produces (14.57 x 116) / 100 = 16.9 % change (i.e. reduction) of torsion between the axles when the stiffness (planar moment of inertia) of this component is increased by a factor of 2.169 (116.9 %). The weight increase here is 3.55 kg when the tube diameter is increased from 70 to 90 mm.

2nd component (see Figure 8):
Component 6 (linear reinforcement, G=11.14 kg) with 2.64 % strain energy ratio:
Produces (2.64 x 250)/100 = 6.6 % change (i.e. only very small reduction of torsion between the axles when the stiffness (moment of inertia) of this component is reduced by a factor of 3.5 (250 %). The weight reduction is 3.34 kg when the tube diameter is reduced from 70 to 50 mm.

Result:
A minimal weight increase of 3.55 – 3.34 = 0.21 kg increases torsional stiffness by 16.9 – 6.6 = 10.3 % (checking with altered profile sections produces 9 %, this covers the low error in the computation formula).

A look at the strain energy charts of the other load cases (not pictured here) shows that this also applies to torsional vibration and is of no importance to bending. It is thus possible to significantly increase the torsional stiffness and flexural bending stiffness of this body still further and to safely reduce the total weight by reducing the cross-sections of oversized components.

References

[1] R.W. Clough: The Finite Element Method In Plane Stress Analysis; Publication, 2nd ASCE Conference on Electronic Computation,1960.
[2] www.IGFgrothTP2000.de.

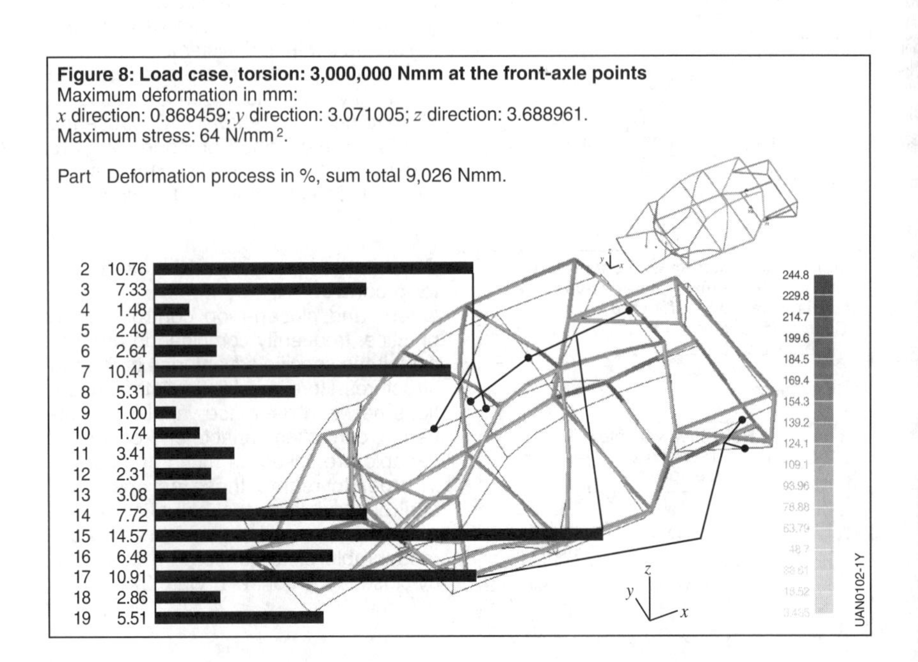

Figure 8: Load case, torsion: 3,000,000 Nmm at the front-axle points
Maximum deformation in mm:
x direction: 0.868459; y direction: 3.071005; z direction: 3.688961.
Maximum stress: 64 N/mm².

Control engineering

Terms and definitions

(in accordance with DIN 19226 [1])

Closed-loop control

In a technical process, the function of closed-loop control is to place and maintain a particular physical parameter (controlled variable y) at a specified value (e.g. the alternator voltage in the electrical system). Here, the controlled variable is continuously measured and compared with the specified value (the reference variable w, e.g. the setpoint input of the alternator voltage as a function of the state of charge of the battery) (Figure 1). In the event of a deviation, a suitable adjustment of the correcting variable u (e.g. influencing the excitation current in the alternator) on the process (controlled system) is carried out in such a way that the controlled variable y is again adapted to the specification. This process takes place in a closed control loop. Deviations can occur when disturbances z (e.g. the activation of a further electrical consumer) act on the controlled system and affect the controlled variable y in an undesirable way ([2], [3] and [4]).

Closed-loop control operations are performed at many points in a motor vehicle. Further examples include control of cooling-water temperature, A/C control, and many other control operations from the engine (knock control, λ control), transmission (clutch control) and chassis (yaw-rate control).

Figure 1: Basic structure of a control loop
y Controlled variable,
w Reference variable,
u Correcting variable,
z Disturbance.

Controlling system
Con-troller
Final controlling element
Controlled system
UAN0048-2E

Open-loop control

Open-loop control is essentially also used instead of closed-loop control. In this case, the closed-loop controlling system is replaced by an open-loop controlling system and feedback of the controlled variable is omitted. This process is only then possible if the behavior of the controlled system is known exactly and if no (non-measurable) disturbances z are acting on the controlled system.

Open-loop control is preferred here as no stability problems can arise on account of there being no feedback. Because the previously mentioned preconditions are rarely satisfied in practice, the use of closed-loop control is for the most part unavoidable.

Combination of open- and closed-loop control

Open- and closed-loop control are, in practice, frequently combined in order to exploit the specific advantages of the two structures. Here, established interconnections between reference variable, disturbance, correcting variable and controlled variable are linked as much as possible in order to realize these as open-loop control. Deviations that still arise on account of changed parameters or non-measurable disturbances are corrected by closed-loop control (Figure 2).

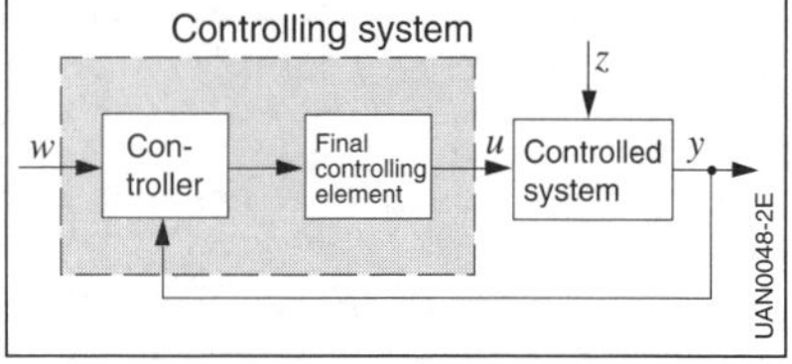

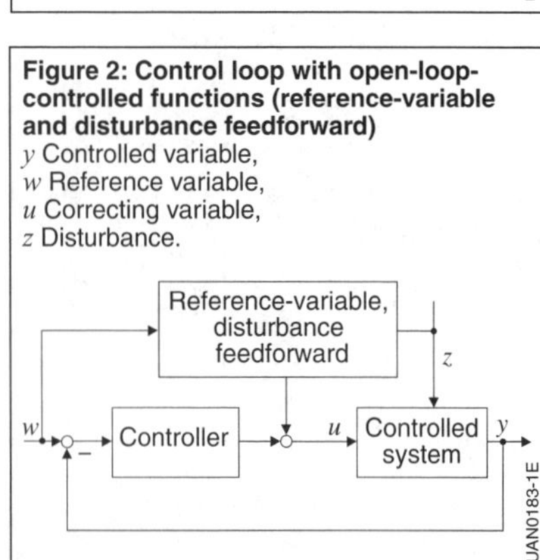

Figure 2: Control loop with open-loop-controlled functions (reference-variable and disturbance feedforward)
y Controlled variable,
w Reference variable,
u Correcting variable,
z Disturbance.

Cascade control

A structure is frequently encountered in which the controlled system is separated into two or more partial systems (e.g. into the actual process and the associated actuator). On the basis of this separation, there are one or more internal controllers and one external controller, which are designed and adjusted separately. This procedure is referred to as cascade control (Figure 3).

The controller design is simplified by this separation of the control task into several manageable subtasks. There are additional advantages in the dynamic response due to the fact that disturbances which act in the internal control loop are corrected there before they affect the external control loop. This makes the entire control operation quicker. Likewise, nonlinear characteristic curves of the internal loop can be linearized.

Cascade control is used in many control systems in motor vehicles, e.g. in current control for electrohydraulic actuators or in position control for electric-motor actuators.

Control-engineering transfer elements

A control loop must satisfy four essential requirements in terms of performance:
- The control loop must be stable
- The control loop must demonstrate a specific stationary accuracy
- The response to a step change of the reference variable must be sufficiently damped
- The control loop must be sufficiently fast.

To satisfy these partly contradictory requirements, it is first necessary to describe the static and dynamic responses

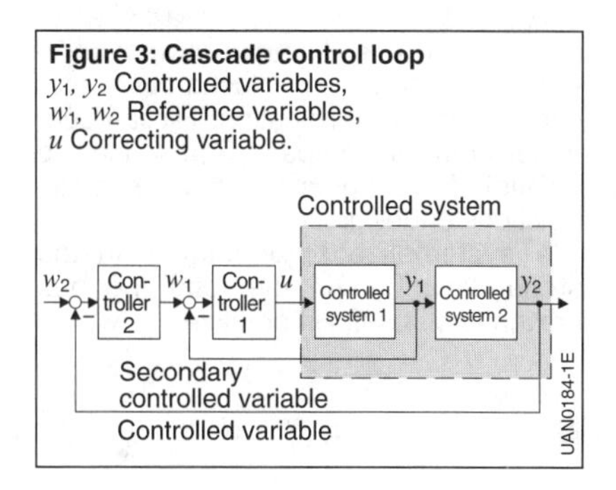

Figure 3: Cascade control loop
y_1, y_2 Controlled variables,
w_1, w_2 Reference variables,
u Correcting variable.

Table 1: Summary of important transfer elements

Transfer element	Differential equation (input u, output y)	Transfer function	Step response
P element	$y = Ku$	K	K
I element	$y = K \int_0^t u(\tau)\,d\tau$	$\frac{K}{s}$	K, 1
D element	$y = K\dot{u}$	Ks	Surface area K
Lag element	$y(t) = Ku(t - T_t)$	$Ke^{-T_t s}$	K, T_t
P-T_1 element	$T\dot{y} + y = Ku$	$\frac{K}{1 + Ts}$	K, T
P-T_2 element	$T^2\ddot{y} + 2dT\dot{y} + y = Ku$	$\frac{K}{1 + 2dTs + T^2s^2}$	K, $d<1$, $d>1$

UAN0051-2E

of the control-loop elements (controlled system and controller) with suitable methods in order to be able to analyze the response in the control loop and to design the controller according to the requirements. This description can be made in the time range (e.g. with differential equations) or in the frequency range (e.g. with a transfer function or a Bode diagram).

Many control-engineering transfer elements can be traced back to specific basic types or can be described by a linking of the same (Table 1).

The task of control-loop synthesis is to design, for a given controlled system, the matching controller (structure and parameters of the transfer element), which fulfills the above-mentioned requirements. There exists for this purpose a series of procedures (e.g. dynamic correction in the Bode diagram, root-locus curve procedures, pole specification, Riccati controller in the state space [4]), which are individually supplemented by specific functions or design steps.

A systematized procedure with the steps described in the following has proven to be useful in practice.

Designing a control task

Control task

Typically the control task is not specifically formulated as such, but must be elaborated as the result of requirements in the specific technical process. This involves defining the open- and closed-loop control tasks in the system in order to settle the questions as to what is to be achieved with the control function and with which variables the objective is described. Controlled demand-response shifting in an automatic transmission is mentioned as an example. With this function, the clutch pressure of the shifting clutch is to be aligned during the gear shift to the speed gradient in such a way that the slip time remains constant under all operating conditions, even with variable parameters (e.g. friction coefficient).

System and block diagrams

It is helpful with these considerations, to draw up a system diagram in which the essential interaction between electronics, mechanics, hydraulics and pneumatics and all the sensors, actuators and bus systems can be clearly seen. From this, the control-engineering block diagram must be derived, from which in turn the functional interaction of all the open- and closed-loop-controlled functions with the system to be controlled can be seen (Figure 4). The functions are described in headline form, but not yet formulated in detail.

Figure 4: Control-engineering block diagram, for example of controlled demand-response shifting in an automatic transmission

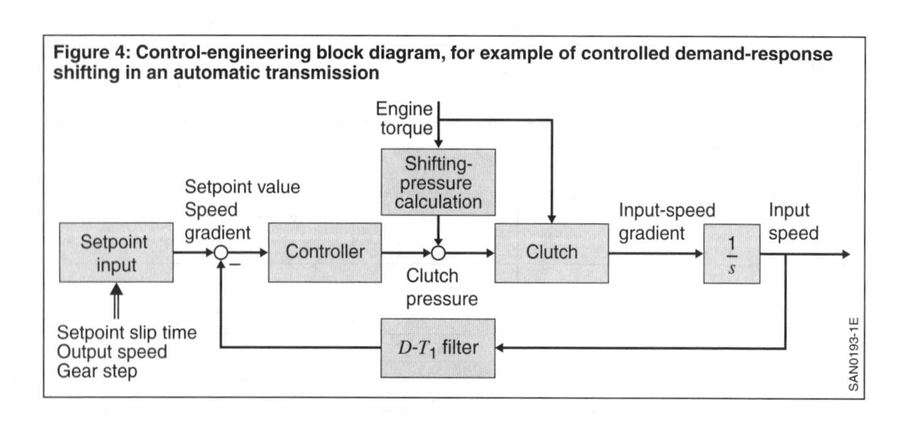

It should be possible already to obtain a fundamental system understanding of the operative connections from the system diagram and the block diagram. As long as the system (mechanics, peripherals, hardware, etc.) is still in development it should at this point still be possible to exert an influence on the structural design of the system as contemplated by an overall mechatronic approach. The filling behavior of a hydraulically actuated clutch is mentioned as an example. This should be designed in such a way that a reproducible behavior with as little dead time as possible is provided on the basis of line cross-sections, volumes and sealing behavior.

Controlled system

The controlled system is then identified. This can be done theoretically (by modeling) or practically, e.g. by measuring the step response or the frequency response. It is recommended that both methods are followed and an adjustment carried out. Identification is a very comprehensive process, depending on the task. Sometimes it is sufficient to determine only the basic type and the order of the controlled system.

Controller design

The controller is designed on the basis of the result of the identification – this is the central task of controller synthesis. It is advisable for this first to be conducted theoretically using simulations, during which the controller parameters are set at the same time. If this step is sufficiently validated, startup follows on the real controlled system on the test bench or in the vehicle. Recursion steps are usually performed once again here in order to achieve further optimizations.

Design criteria

Over and above this fundamental sequence, reference is made to the following additional criteria:

Digital (discrete-time) control

The majority of control operations in a motor vehicle are implemented on microcontrollers. In this case, it is necessary to suitably establish the sampling time, oriented towards the dynamic response of the controlled system. It is necessary to ensure here that all the function algorithms can be calculated in the time available between two samplings.

Nonlinearities

In many cases, the simple linear methods described above are not sufficient, since real controlled systems contain nonlinearities (e.g. pressure-regulator characteristics, clutch characteristics, etc.). In simple cases, where static continuous nonlinearities are involved, these can be compensated for by an additional inverse-behavior transfer element. In the case of control operations with small signal amplitudes about an operating point, the system equations can be linearized at this point, otherwise more complex procedures are called for.

Structure switchovers

Many closed-loop control operations are first initiated by open-loop-controlled signals (e.g. first fill clutch, then set shifting pressure, then start shifting control). In these cases, it is necessary when switching from the open-loop- to the closed-loop-controlled state to ensure that this takes place smoothly and that the storage devices (integrators of the I elements) are correctly initialized.

Robustness

The control is typically designed with a "nominal" controlled system. In practice, however, manufacturing tolerances dictate that controlled systems are encountered which deviate up to a defined extent from this nominal definition. Furthermore, the parameters change over the service life, e.g. due to clutch wear or depending on third variables (temperature). In none of these cases should this give rise to significant functional impairments or instability of the control loop. Different procedures from the field of "robust control" or "adaptive control" are available to master these requirements.

Adaptive controllers

Motivation

Controlled systems often do not have a constant behavior. Parameters such as time constants and gains change in many cases. Even the structure of the system can change. Adaptations match open-and closed-loop control processes to a different system behavior. Examples include:

Manufacturing tolerances
Not all the products in a batch are 100 % identical. Because an individual adjustment is complicated, the system should adapt itself automatically to different parameters (e.g. adjustment data for automatic transmissions, [5]).

Wear
Parameters change in response to wear on a reproducible (e.g. increasing clutch control travel) or random level (e.g. friction coefficient of disks). Adaptations can compensate for this different behavior (e.g. clutch-control-travel adaptation for automated clutches, [6]).

Dependence on a third variable (e.g. temperature)
The viscosity of oil is highly temperature-dependent. Because these fluctuations will also occur in the short term (e.g. repeatedly every day), they must be compensated for (e.g. interference-torque observer, lockup-clutch control, [7], [8]).

Dependence on operating point of nonlinear systems
Nonlinear systems are frequently linearized at an operating point and then controlled by a linear controller (small-signal behavior). It is possible by means of an adaptation to take into account the different behavior at the operating points (e.g. adaptation of the shifting pressure for interlaced multiple gearshifts in an automatic transmission, [9]).

The various requirements to master this problem on an open-loop control level result in the desire for adaptive systems, which are described and defined in the following.

Definition of "adaptive control"

The behavior of the control is adapted to the changing properties of the controlled system and its signals. Adaptation procedures are thus essentially divided into two steps.

- Identification: Identification of the system behavior (system parameters) from time-variable values of the system.
- Adaptation: Adaptation of the open- or closed-loop control law as a reaction to a change of the system behavior.

Open-loop-controlled adaptations

Adaptation is performed by open-loop control, a forward-pointing structure. It is assumed here that the changing properties can be recorded by means of measurable external signals z (disturbances)

Figure 5: Open-loop-control adaptive system
y Controlled variable,
w Reference variable,
u Correcting variable,
z Disturbance.

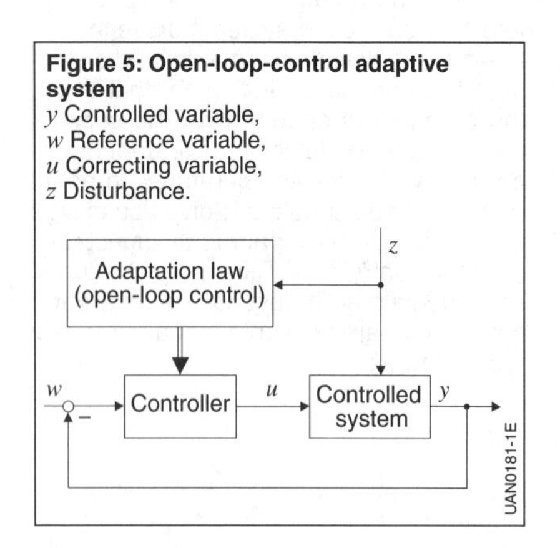

Figure 6: Adaptive system with feedback
y Controlled variable,
w Reference variable,
u Correcting variable.

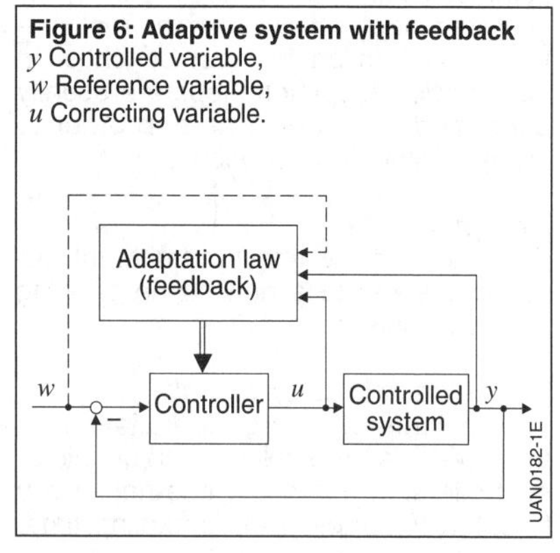

and that the dependence of the control on these signals is known (Figure 5). There is no feedback on internal signals of the control loop to the setting of the controller.

Adaptation with feedback
In the case of an adaptation with feedback, the changing properties cannot be directly recorded and must be identified from measurable signals of the control loop. The identification can be a simple measurement or even a complex algorithm to estimate dynamic process models. In addition to the basic control loop, a second feedback loop is implemented by means of the adaptation law (Figure 6).

Essentially, an open-loop-controlled adaptation is initially recommended, i.e. known and metrologically recordable connections are used in open-loop-controlled form. The advantage of this forward structure can be compared with the advantages of open-loop control over closed-loop control. A feedback loop, which can possibly cause stability problems, is omitted. For the most part, open-loop-control adaptive systems are used in practical industrial applications.

Design notes
The following questions must be clarified when it comes to designing adaptive control:
- Which parameters and features must be adapted because they cannot be covered by a robust design?
- With which signals or variables can these parameters and features be determined?
- Does the system have to be specifically excited in order to determine these parameters and features, or can the adaptation take place during ongoing control operation?
- Can the adaptation be performed by open-loop control means or does feedback have to be provided for?
- How can the stability and convergence of adaptive control be tested?

A detailed treatment of the subjects areas of "Design of adaptive control operations" and "Identification of dynamic processes" can be found in [4], [10] and [11].

References
[1] DIN 19226. Control technology, terms and designations.
[2] Winfried Oppelt: Kleines Handbuch technischer Regelvorgänge. 5th Edition 1972, Verlag Chemie GmbH, Weinheim.
[3] Otto Föllinger: Regelungstechnik. 3rd Edition 1979, AEG-Telefunken, Frankfurt.
[4] Rolf Isermann: Digitale Regelsysteme. 2nd Edition 1987, Springer-Verlag, Berlin Heidelberg New York.
[5] DE 102007040485A1, Abgleichdaten AT-Getriebe.
[6] DE 102007027702A1, Kupplungsweg-Adaption für automatisierte Kupplungen.
[7] DE 000019943334A1, Störmomentbeobachter, Wandlerkupplungsregelung.
[8] Bauer, Gunther; Schwemer, Christian: Entwurf einer Wandlerkupplungsregelung unter Berücksichtigung nichtfunktionaler Anforderungen. AUTOREG 2008, 4th Symposium Baden-Baden, 12 and 13 February 2008, VDI/VDE-Gesellschaft Mess- und Automatisierungstechnik.
[9] DE102006001899A1, Adaption Schaltdruck bei verschachtelten Mehrfachschaltungen AT- Getriebe.
[10] Rolf Isermann: Identifikation dynamischer Systeme I, II. Springer-Verlag, Berlin, 1992.
[11] Rolf Isermann: Mechatronische Systeme. 2nd Edition, Springer-Verlag Berlin, Heidelberg, New York, 2007.

Materials

Material parameters

Selecting material for a specific use is based on material parameters that describe the material properties. This process distinguishes between the physical material parameters, which result from the atomic structure, and the mechanical-technological material parameters, which are affected by manufacturing processes, for example.

Some physical material parameters are listed in tables 3 to 6 for a wide variety of materials.

Physical material parameters

Density
Density ρ is the ratio of the mass to the volume of a specific amount of substance (also see DIN 1306:1984-06 [1]). The unit of measurement is kg/m^3.

Melting temperature
The melting temperature denotes the temperature at which a material transforms from a solid to liquid state.

Boiling temperature
The boiling temperature denotes the temperature at which a material transforms from a liquid into a gas.

Heat of fusion
The specific heat of fusion of a solid (enthalpy of fusion) is the quantity of heat required to transform a material at fusion temperature from the solid to the liquid state. The unit of measurement is kJ/kg.

Heat of evaporation
The specific heat of evaporation of a liquid (enthalpy of evaporation) is the quantity of heat required to evaporate a liquid at constant boiling temperature. It is very pressure-sensitive. The unit of measurement is kJ/kg.

Thermal conductivity
Thermal conductivity is the quantity of heat that flows through a material sample with a defined surface and density due to temperature difference. In the case of liquids and gases (as opposed to solids), thermal conductivity is often highly dependent on temperature. The unit of measurement is W/(m·K).

Electric conductivity
Electric conductivity describes the physical property of a material to conduct electric current. According to the Wiedemann-Franz law, it is almost proportional to thermal conductivity. The unit of measurement is S/m (siemens per meter).

Coefficient of thermal expansion
The linear coefficient of thermal expansion α (coefficient of longitudinal expansion) indicates the relative change in length of a material during a temperature variation. For temperature change ΔT, the change in length is defined as $\Delta l = l\,\alpha\,\Delta T$. The unit of measurement is K^{-1}.

The cubic coefficient of thermal expansion (coefficient of volume expansion) is defined in a similar manner. The coefficient of volume expansion for gases is roughly 1/273 K^{-1}; For solids, it is roughly three times as large as the coefficient of linear expansion.

Heat capacity
The specific heat capacity (specific heat) is the quantity of heat required to raise the temperature of a substance. It is dependent on temperature.

In the case of gases, we distinguish between specific heat capacity at constant pressure and at constant volume (symbols c_p or c_v). This difference is usually negligible in the case of solid and liquid substances. The unit of measurement is kJ/(kg · K).

Permeability

Permeability μ describes the ratio of magnetic flux density B and magnetic field strength H (see magnetic field, relative permeability):

$$B = \mu H.$$

Permeability consists of the magnetic field constant μ_0 and the material-dependent relative permeability μ_r ($\mu_0 = 4\pi \cdot 10^{-7}$ As/Vm, $\mu_r = 1$ for vacuum). Depending on the application in which the magnetic material is used, there are roughly 15 types of permeability. These are defined according to modulation range and type of loading (direct-current or alternating-current field loading). The most important parameters are listed below.

Initial permeability μ_a

The slope of the virgin curve (figure 1, also see hysteresis loop) for $H \rightarrow 0$ is designated as the initial permeability μ_a. In most cases, however, the slope for a specific field strength is specified (in mA/cm) rather than this limit value. Notation: μ_4 is the slope of the virgin curve for $H = 4$ mA/cm.

Maximum permeability μ_{max}

The maximum slope of the virgin curve is designated as the maximum permeability μ_{max}.

Recoil permeability μ_p

Recoil permeability μ_p (or μ_{rec}) is defined as the average slope of a retrograde magnetic hysteresis loop whose lowest point usually lies on the demagnetization curve:

$$\mu_p = \frac{\Delta B}{\Delta H \, \mu_0}.$$

Figure 1: Hysteresis loop of a magnetically hard iron grade

1 Rise path,
2, 3 Demagnetization curves.
H Magnetic field strength,
B Magnetic flux density,
B_r Remanence flux density,
H_c Coercive field strength.
The arrows indicate which magnetic flux density is set if the magnetic field strength is changed.

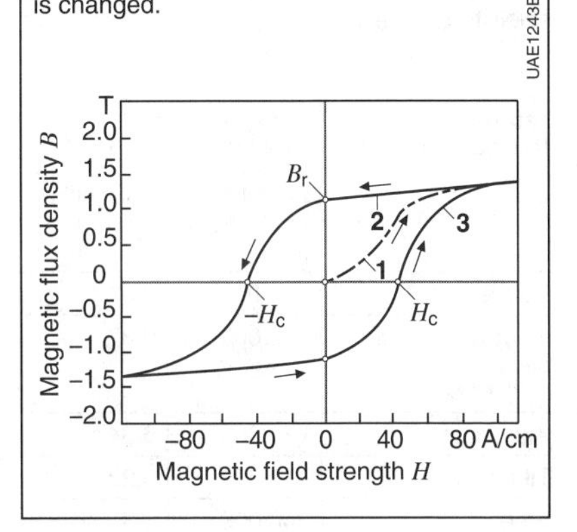

Temperature coefficient of magnetic polarization

The temperature coefficient of the magnetic polarization $TK(J_s)$ indicates the relative change in saturation polarization as temperature changes, it is given in percent per kelvin.

Temperature coefficient of coercive field strength

The temperature coefficient of the coercive field strength $TK(H_C)$ indicates the relative change in coercive field strength as temperature changes, it is given in percent per kelvin.

Curie point

The Curie point (Curie temperature T_C) is the temperature at which the magnetization of ferromagnetic and ferrimagnetic materials becomes zero and at which they behave like paramagnetic materials (see magnetic materials).

Mechanical-technological material parameters

Modulus of elasticity

The modulus of elasticity (E modulus) describes the linear relationship between stress and strain when deforming a solid body in the area of its elastic deformation (image 2). The unit of measurement is MPa. Examples of typical e-moduli are specified in table 1.

Table 1: Modulus of elasticity and Poisson's ratio for some materials

Material	Modulus of elasticity [MPa]	Poisson's ratio
Rubber	100	0.5
Fiber-reinforced plastic (e.g. PA66)	2,000	0.37
Aluminum	70,000	0.34
Titanium	110,000	0.28
Steel	200,000	0.3
Tungsten	400,000	0.28
Ceramics (e.g. Al_2O_3)	400,000	0.23

Poisson's ratio

The dimensionless Poisson's ratio is the proportionality factor between the longitudinal strain of a body under tension or compressive stress and the resulting transverse strain. Typical values are specified in table 1.

Yield point

The yield strength (yield point) is the amount of tensile stress starting at which a material experiences lasting plastic deformation. It is determined from the stress-strain curve (σ-ε curve) measured in the tensile test (see DIN EN ISO 6892-1:2009-12, [2]). The unit of measurement is MPa.

0.2% yield strength

The 0.2% yield strength ($R_{p0.2}$) is the amount of tensile stress that causes a lasting (plastic) strain of 0.2% in a material. The unit of measurement is MPa.

Tensile strength

The tensile strength R_m is the stress that is calculated in the tensile test from the maximum achieved tractive force relative to the output cross-section of a standardized material sample (figure 2 and figure 3). The unit of measurement is MPa.

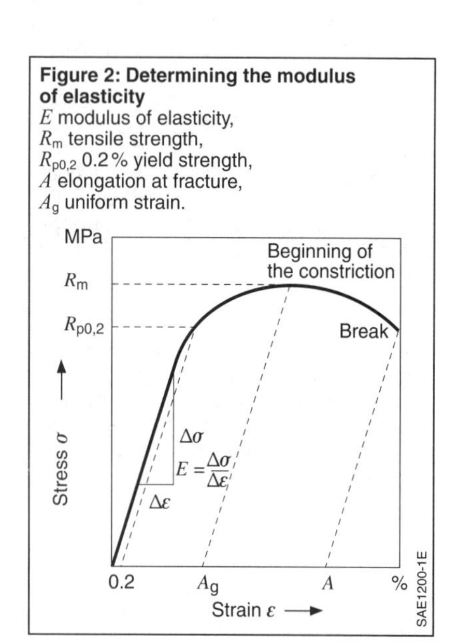

Figure 2: Determining the modulus of elasticity
E modulus of elasticity,
R_m tensile strength,
$R_{p0,2}$ 0.2% yield strength,
A elongation at fracture,
A_g uniform strain.

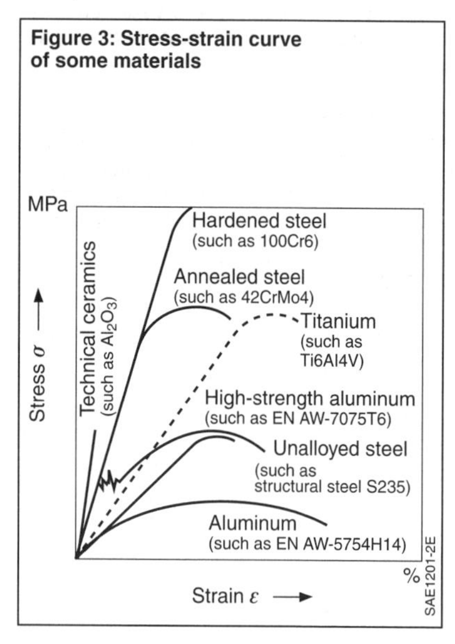

Figure 3: Stress-strain curve of some materials

Elongation at fracture

The elongation at fracture A is a dimension for the ductility (deformability) of a material. It indicates the lasting extension of a tensile specimen after the fracture relative to the initial length, specified in percent.

Fracture contraction

The fracture contraction Z, like the elongation at fracture, is a dimension for the ductility (deformability) of a material. It indicates in percent the change in cross-sectional area relative to the initial cross-section of a tensile specimen.

Uniform strain

The uniform strain A_g indicates the change in length in percent at which a tensile specimen goes through a plastic deformation without necking. For ductile materials, necking is caused on the specimen (cross-section reduction) after reaching the uniform strain (and tensile strength).

Vibration resistance

Vibration resistance is defined as the deformation and failure behavior of materials under cyclical stress. Typical vibration resistance values such as the fatigue limit have to be determined in often time-consuming Wöhler tests (fatigue tests named after August Wöhler) (image 4). These are usually carried out on electromechanical or electrohydraulic testing machines in the frequency range of 10 to 1,000 Hz.

A wide variety of test parameters influence the determined characteristic values (surface quality, test frequency, test duration, test medium, bend on tension or pressure). Here, the determined fatigue limit of a material is well below the structural strength (e.g. R_m). A simple estimate is possible using the FKM guideline (Forschungskuratorium Maschinenbau e.V., see Table 2). The tension-compression fatigue stress factor identifies the ratio of vibration resistance to tensile strength (tension-compression fatigue stress). For metals with a face-centered cubic crystal lattice, such as austenitic steel or aluminum, there is no fatigue limit since the stressability decreases when the stress duration increases.

Knowing the vibration resistance parameters is absolutely necessary for designing parts and components in mechanical engineering, since purely static stress is rare.

Figure 4: Depicting the vibration resistance with the Wöhler curve with typical values for the number of stress cycles N
R_m tensile strength,
$S_{a,D}$ Fatigue limit.

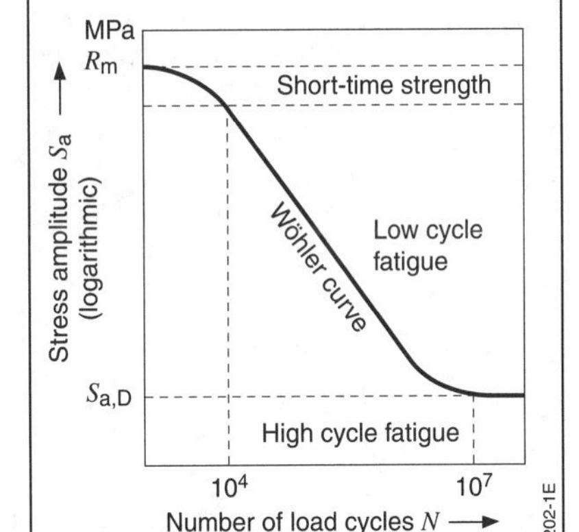

Fatigue strength under reversed bending stress

Fatigue strength under reversed bending stress identifies the stress in MPa that a material can withstand without breaking during cyclical bending stress (see vibration resistance).

Table 2: Estimation of the vibration resistance

Material group	Tension-compression fatigue stress factor for the number of stress cycles $N = 10^6$
Aluminum, wrought and casting materials	0.30
Cast iron with flake graphite (GJL)	0.30
Cast iron with nodular graphite (GJS)	0.34
Stainless steel	0.40
Structural steel, tempering steel, etc.	0.45

Hardness
The hardness of a material identifies its resistance to the penetration of a body. Various methods for determining hardness are listed in the chapter "Heat treatment of metallic materials".

Fracture toughness
The fracture toughness is the resistance of a material to the spread of cracks. The critical stress intensity factor K_{Ic} is usually used as a characteristic of this. If K_{Ic} is known, the critical fracture load can be determined from crack length, or the critical crack length can be determined from the given external stress value.

Special characteristics for sintered material
Radial crushing strength
The radial crushing strength is a strength parameter which is specified in particular for sintered friction bearings. This is determined by buckling a hollow cylinder during the pressure test.

Porosity
The porosity is a dimensionless characteristic for sintered metals, particularly friction bearings. It describes the portion of part volume that consists of cavities and is not filled with actual material.

Compressive yield point
Comparable to the yield strength in the tensile test, the compressive yield point is added in the pressure test as a material characteristic (unit of measurement MPa). This identifies the stress starting at which an irreversible plastic deformation occurs in the material.

Special characteristics for springs
Bending stress
Bending stress occurs when stress is put on torsion springs. The maximum bending stress here should not exceed $0.7\,R_m$.

Shear stress
Shear stress occurs in the material when stress is put on pressure and tension springs. The maximum shear stress (maximum stress) should not exceed $0.5\,R_m$. Here, the shear stress is dependent on the coil diameter of the spring, spring force and wire diameter. The difference between maximum stress and minimum stress is also called stress range.

References for material parameters
[1] DIN 1306:1984-06: Density; concepts, presentation of values.
[2] DIN EN ISO 6892-1:2009-12: Metallic materials – Tensile testing – Part 1: Method of test at room temperature (ISO 6892-1:2009); German version EN ISO 6892-1:2009.
[3] D. Radaj, M. Vormwald: Ermüdungsfestigkeit – Grundlagen für Ingenieure [Fatigue strength – Principles for engineers]. 3. Aufl., Springer-Verlag, 2010.
[4] G. Gottstein: Physikalische Grundlagen der Materialkunde [Physical principles of material science]. 3. Aufl., Springer-Verlag, 2007.
[5] J. Schijve: Fatigue of Structures and Materials. 2nd Ed., Springer-Verlag, 2009.
[6] S. Suresh: Fatigue of Materials. 2nd edition (Cambridge Solid State Science Series), 1998.

Table 3: Properties of solids

Material/medium		Density g/cm³	Melting point[1] °C	Boiling point[1] °C	Thermal conductivity[2] W/(m·K)	Mean specific heat capacity[3] kJ/(kg·K)	Enthalpy of fusion ΔH[4] kJ/kg	Coefficient of longitudinal expansion[3] ×10⁻⁶/K
Aluminum	Al	2.70	660	2,467	237	0.90	395	23.0
Aluminum alloys		2.60 to 2.85	480 to 655	–	70 to 240	–	–	21 to 24
Amber		1.0 to 1.1	< 300	Decomposes	–	–	–	–
Antimony	Sb	6.69	630.8	1,635	24.3	0.21	172	8.5
Arsenic	As	5.73	–	613[5]	50.0	0.34	370	4.7
Asbestos		2.1 to 2.8	< 1,300	–	–	0.81	–	–
Asphalt		1.1 to 1.4	80 to 100	< 300	0.70	0.92	–	–
Barium	Ba	3.50	729	1,637	18.4	0.28	55.8	18.1 to 21.0
Barium chloride		3.86	963	1,560	–	0.38	108	–
Basalt		2.6 to 3.3	–	–	1.67	0.86	–	–
Beef tallow		0.9 to 0.97	40 to 50	< 350	–	0.87	–	–
Beryllium	Be	1.85	1,278	2,970	200	1.88	1,087	11.5
Bismuth	Bi	9.75	271	1,551	8.1	0.13	59	12.1
Bitumen		1.05	< 90	–	0.17	1.78	–	–
Boiler scale		< 2.5	< 1,200	–	0.12 to 2.3	0.80	–	–
Borax		1.72	740	–	–	1.00	–	–
Boron	B	2.34	2,027	3,802	27.0	1.30	2,053	5
Brass CuZn37		8.4	900	1,110	113	0.38	167	18.5
Brickwork		> 1.9	–	–	1.0	0.9	–	–
Bronze CuSn 6		8.8	910	2,300	64	0.37	–	17.5
Cadmium	Cd	8.65	321.1	765	96.8	0.23	54.4	29.8
Calcium	Ca	1.54	839	1,492	200	0.62	233	22
Calcium chloride		2.15	782	> 1,600	–	0.69	–	–
Cellulose acetate		1.3	–	–	0.26	1.47	–	100 to 160
Cement, set		2 to 2.2	–	–	0.9 to 1.2	1.13	–	–
Chalk		1.8 to 2.6	Decomposes into CaO and CO_2		0.92	0.84	–	–
Chamotte (fireclay)		1.7 to 2.4	< 2,000	–	1.4	0.80	–	–
Charcoal		0.3 to 0.5	–	–	0.084	1.0	–	–
Chromium	Cr	7.19	1,875	2,482	93.7	0.45	294	6.2
Chromium oxide	Cr_2O_3	5.21	2,435	4,000	0.42[6]	0.75	–	–
Clay, dry		1.5 to 1.8	< 1,600	–	0.9 to 1.3	0.88	–	–
Cobalt	Co	8.9	1,495	2,956	69.1	0.44	268	12.4
Coke		1.6 to 1.9	–	–	0.18	0.83	–	–
Common salt		2.15	802	1,440	–	0.92	–	–
Concrete		1.8 to 2.2	–	–	< 1.0	0.88	–	–
Copper	Cu	8.96	1084.9	2,582	401	0.38	205	–
Cork		0.1 to 0.3	–	–	0.04 to 0.06	1.7 to 2.1	–	–
Corundum, fused		–	–	–	–	–	–	6.5[9]
Cotton wadding		0.01	–	–	0.04		–	–

[1] At 1.013 bar. [2] At 20 °C. ΔH of chemical elements at 27 °C (300 K). [3] At 0 to 100 °C.
[4] At melting point and 1.013 bar. [5] Sublimated. [6] Powder form. [7] At 20 to 1,000 °C.

Table 3: Properties of solids (continued)

Material/medium		Density	Melting point [1]	Boiling point [1]	Thermal conductivity [2]	Mean specific heat capacity [3]	Enthalpy of fusion ΔH [4]	Coefficient of longitudinal expansion [3]
		g/cm^3	°C	°C	$W/(m \cdot K)$	$kJ/(kg \cdot K)$	kJ/kg	$\times 10^{-6}/K$
Diamond	C	3.5	3,820	–	–	0.52	–	1.1
Duromer								
Phenolic resin without filler		1.3	–	–	0.2	1.47	–	80
Phenolic resin with asbestos fibers		1.8	–	–	0.7	1.25	–	15 to 30
Phenolic resin with wood flour		1.4	–	–	0.35	1.47	–	30 to 50
Phenolic resin with fabric threads		1.4	–	–	0.35	1.47	–	15 to 30
Melamine resin with cellulose fiber		1.5	–	–	0.35	–	–	< 60
Foam rubber		0.06 to 0.25	–	–	0.04 to 0.06	–	–	–
Germanium	Ge	5.32	937	2,830	59.9	0.31	478	5.6
Glass (fused quartz)		–	–	–	–	–	–	0.5
Glass (window glass)		2.4 to 2.7	< 700	–	0.81	0.83	–	< 8
Gold	Au	19.32	1,064	2,967	317	0.13	64.5	14.2
Granite		2.7	–	–	3.49	0.83	–	–
Graphite, pure	C	2.24	< 3,800	< 4,200	168	0.71	–	2.7
Gray cast iron		7.25	1,200	2,500	58	0.50	125	10.5
Hard coal (anthracite)		1.35	–	–	0.24	1.02	–	–
Hard metal K 20		14.8	> 2,000	< 4,000	81.4	0.80	–	5 to 7
Hard rubber		1.2 to 1.5	–	–	0.16	1.42	–	50 to 90 [5]
Heat-conductive alloy NiCr 8020		8.3	1,400	2,350	14.6	0.50 [6]	–	–
HR foam, air-filled [8]		0.015 to 0.06	–	–	0.036 to 0.06	–	–	–
HR foam, freon-filled		0.015 to 0.06	–	–	0.02 to 0.03	–	–	–
Ice (0 °C)		0.92	0	100	2.33 [6]	2.09 [6]	333	51 [7]
Indium	In	7.29	156.6	2,006	81.6	0.24	28.4	33
Iodine	I	4.95	113.5	184	0.45	0.22	120.3	–
Iridium	Ir	22.55	2,447	4,547	147	0.13	137	6.4
Iron, pure	Fe	7.87	1,535	2,887	80.2	0.45	267	12.3
Lead	Pb	11.3	327.5	1,749	35.5	0.13	24.7	29.1
Lead oxide, litharge	PbO	9.3	880	1,480	–	0.22	–	–
Leather, dry		0.86 to 1	–	–	0.14 to 0.16	< 1.5	–	
Linoleum		1.2	–	–	0.19	–	–	–
Lithium	Li	0.534	180.5	1,317	84.7	3.3	663	56
Magnesium	Mg	1.74	648.8	1,100	156	1.02	372	26.1
Magnesium alloys		< 1.8	< 630	1,500	46 to 139	–	–	24.5
Manganese	Mn	7.47	1,244	2,100	7.82	0.48	362	22
Marble	$CaCO_3$	2.6 to 2.8	Decomposes into CaO and CO_2		2.8	0.84	–	–
Mica		2.6 to 2.9	Decomposes at 700 °C		0.35	0.87	–	3
Molybdenum	Mo	10.22	2,623	5,560	138	0.28	288	5.4
Monel metal		8.8	1,240 to 1,330	–	19.7	0.43	–	
Mortar, cement		1.6 to 1.8	–	–	1.40	–	–	–
Mortar, lime		1.6 to 1.8	–	–	0.87	–	–	–

[1] At 1.013 bar. [2] At 20 °C. [3] At 0 to 100 °C. [4] At melting point and 1.013 bar.
[5] At 20 to 50 °C. [6] At −20 to 0 °C. [7] At −20 to −1 °C.
[8] Mean values for air-dried wood (humidity approximately 12 %). Radial thermal conductivity; axial is approx. twice as high.

Table 3: Properties of solids (continued)

Material/medium		Density g/cm³	Melting point[1] °C	Boiling point[1] °C	Thermal conductivity[2] W/(m·K)	Mean specific heat capacity[3] kJ/(kg·K)	Enthalpy of fusion ΔH[4] kJ/kg	Coefficient of longitudinal expansion[3] ×10⁻⁶/K
Nickel	Ni	8.90	1,455	2,782	90.7	0.46	300	13.3
Nickel silver CuNi12Zn24		8.7	1,020	–	48	0.40	–	18
Niobium	Nb	8.58	2,477	4,540	53.7	0.26	293	7.1
Osmium	Os	22.57	3,045	5,027	87.6	0.13	154	4.3 to 6.8
Palladium	Pd	12.0	1,554	2,927	71.8	0.24	162	11.2
Paper		0.7 to 1.2	–	–	0.14	1.34	–	–
Paraffin		0.9	52	300	0.26	3.27	–	–
Peat dust, air-dried		0.19	–	–	0.081	–	–	–
Phosphorus (white)	P	1.82	44.1	280.4	–	0.79	20	–
Pitch		1.25	–	–	0.13	–	–	–
Plaster		2.3	1,200	–	0.45	1.09	–	–
Platinum	Pt	21.45	1,769	3,827	71.6	0.13	101	9
Plutonium	Pu	19.8	640	3,454	6.7	0.14	11	55
Polyamide		1.1	–	–	0.31	–	–	70 to 150
Polycarbonate		1.2	–	–	0.20	1.17	–	60 to 70
Polyethylene		0.94	–	–	0.41	2.1	–	200
Polystyrene		1.05	–	–	0.17	1.3	–	70
Polyvinyl chloride		1.4	–	–	0.16	–	–	70 to 150
Porcelain		2.3 to 2.5	< 1,600	–	1.6[5]	1.2[5]	–	4 to 5
Potassium	K	0.86	63.65	754	102.4	0.74	61.4	83
Quartz		2.1 to 2.5	1,480	2,230	9.9	0.80	–	8[6]/14.6[7]
Radium	Ra	5	700	1,630	18.6	0.12	32	20.2
Red bronze CuSn5ZnPb		8.8	950	2,300	38	0.67	–	–
Red lead, minium	Pb_3O_4	8.6 to 9.1	Forms PbO		0.70	0.092	–	–
Resin bonded fabric, paper		1.3 to 1.4	–	–	0.23	1.47	–	10 to 25[8]
Resistance alloy CuNi 44		8.9	1,280	< 2,400	22.6	0.41	–	15.2
Rhenium	Re	21.02	3,160	5,762	150	0.14	178	8.4
Roofing paper		1.1	–	–	0.19	–	–	–
Rosin		1.08	100 to 130	Decomposes	0.32	1.21	–	–
Rubber, raw		0.92	125	–	0.15	–	–	
Rubidium	Rb	1.53	38.9	688	58	0.33	26	90
Sand, quartz, dry		1.5 to 1.7	< 1,500	2,230	0.58	0.80	–	–
Sandstone		2 to 2.5	< 1,500	–	2.3	0.71	–	–
Selenium	Se	4.8	217	684.9	2.0	0.34	64.6	37
Silicon	Si	2.33	1,410	2,480	148	0.68	1,410	4.2
Silicon carbide		2.4	Decomposes above 3,000 °C		9[9]	1.05[9]	–	4.0
Sillimanite		2.4	1,820	–	151	1.0	–	–
Silver	Ag	10.5	961.9	2,195	429	0.24	104.7	19.2
Slag, blast furnace		2.5 to 3	1,300 to 1,400	–	0.14	0.84	–	–
Sodium	Na	0.97	97.81	883	141	1.24	115	70.6
Soft rubber		1.08	–	–	0.14 to 0.24	–	–	–

[1] At 1.013 bar. [2] At 20 °C. ΔH of chemical elements at 27 °C (300 K). [3] At 0 to 100 °C.
[4] At melting temperature and 1.013 bar. [5] At 0 to 100 °C. [6] Parallel to crystal axis.
[7] Perpendicular to the crystal axis. [8] At 20 to 50 °C. [9] At 1,000 °C.

Table 3: Properties of solids (continued)

Material/medium		Density g/cm³	Melting point[1] °C	Boiling point[1] °C	Thermal conductivity[2] W/(m·K)	Mean specific heat capacity[3] kJ/(kg·K)	Enthalpy of fusion ΔH[4] kJ/kg	Coefficient of longitudinal expansion[3] ×10⁻⁶/K
Soot		1.7 to 1.8	–	–	0.07	0.84	–	–
Steatite		2.6 to 2.7	< 1,520	–	1.6[6]	0.83	–	8 to 9[5]
Steel, chromium steel		–	–	–	–	–	–	11
Steel, dynamo sheet		–	–	–	–	–	–	12
Steel, high-speed steel		–	–	–	–	–	–	11.5
Steel, magnet steel AlNiCo12/6		–	–	–	–	–	–	11.5
Steel, nickel steel 36 % Ni (invar)		–	–	–	–	–	–	1.5
Steel, sintered		–	–	–	–	–	–	11.5
Steel, stainless (18Cr, 8Ni)		7.9	1,450	–	14	0.51	–	16
Steel, tungsten steel (18 W)		8.7	1,450	–	26	0.42	–	–
Steel, unalloyed and low-alloy		7.9	1,460	2,500	48 to 58	0.49	205	11.5
Sulfur (α)	S	2.07	112.8	444.67	0.27	0.73	38	74
Sulfur (β)	S	1.96	119.0	–	–	–	–	–
Tantalum	Ta	16.65	2,996	5,487	57.5	0.14	174	6.6
Tellurium	Te	6.24	449.5	989.8	2.3	0.20	106	16.7
Thorium	Th	11.72	1,750	4,227	54	0.14	< 83	12.5
Tin (white)	Sn	7.28	231.97	2,270	65.7	0.23	61	21.2
Titanium	Ti	4.51	1,660	3,313	21.9	0.52	437	8.3
Tombac CuZn 20		8.65	1,000	< 1,300	159	0.38	–	–
Tungsten	W	19.25	3,422	5,727	174	0.13	191	4.6
Uranium	U	18.95	1,132.3	3,677	27.6	0.12	65	12.6
Vanadium	V	6.11	1,890	3,000	30.7	0.50	345	8.3
Vulcanized fiber		1.28	–	–	0.21	1.26	–	–
Wax		0.96	60	–	0.084	3.4	–	–
Wood[7]	Ash	0.72	–	–	0.16	2,1 to 2,9 (Ash to Walnut)	–	in fiber direction 3 to 4, transverse to fiber 22 to 43 (Ash to Walnut)
	Balsa	0.20	–	–	0.06		–	
	Beech	0.72	–	–	0.17		–	
	Birch	0.63	–	–	0.14		–	
	Maple	0.62	–	–	0.16		–	
	Oak	0.69	–	–	0.17		–	
	Pine	0.52	–	–	0.14		–	
	Poplar	0.50	–	–	0.12		–	
	Spruce, fir	0.45	–	–	0.14		–	
	Walnut	0.65	–	–	0.15		–	
Wood-wool building slabs		0.36 to 0.57	–	–	0.093	–	–	–
Zinc	Zn	7.14	419.58	907	116	0.38	102	25.0
Zirconium	Zr	6.51	1,852	4,377	22.7	0.28	252	5.8

[1] At 1.013 bar. [2] At 20 °C. ΔH of chemical elements at 27 °C (300 K). [3] At 0 to 100 °C.
[4] At melting temperature and 1.013 bar. [5] At 20 to 1,000 °C. [6] At 100 to 200 °C.
[7] HR foam of phenolic resin, polystyrene, polyethylene or the like, values dependent on cell diameter and filler gas.

Table 4: Properties of liquids

Material/medium		Density [2] g/cm³	Melting temperature [1] °C	Boiling temperature [1] °C	Thermal conductivity [2] W/(m·K)	Specific heat capacity [2] kJ/(kg·K)	Melting enthalpy ΔH [3] kJ/kg	Evaporation enthalpy [4] kJ/kg	Coefficient of volume expansion $\times 10^{-3}$/K
Acetone	$(CH_3)_2CO$	0.79	–95	56	0.16	2.21	98.0	523	–
Antifreeze-water mixture									
23 % by vol.		1.03	–12	101	0.53	3.94	–	–	–
38 % by vol.		1.04	–25	103	0.45	3.68	–	–	–
54 % by vol.		1.06	–46	105	0.40	3.43	–	–	–
Benzene	C_6H_6	0.88	+5.5 [6]	80	0.15	1.70	127	394	1.25
Common-salt solution 20 %		1.15	–18	109	0.58	3.43	–	–	–
Diesel fuel		0.81 to 0.85	–30	150 to 360	0.15	2.05	–	–	–
Ethanol	C_2H_5OH	0.79	–117	78.5	0.17	2.43	109	904	1.1
Ethanol 95 % [9]		0.81	–114	78	0.17	2.43	–	–	–
Ethyl chloride	C_2H_5Cl	0.90	–136	12	0.11 [5]	1.54 [5]	69.0	437	–
Ethyl ether	$(C_2H_5)_2O$	0.71	–116	34.5	0.13	2.28	98.1	377	1.6
Ethylene glycol	$C_2H_4(OH)_2$	1.11	–12	198	0.25	2.40	–	–	–
Fuel oil EL		< 0.83	–10	> 175	0.14	2.07	–	–	–
Gasoline/petrol		0.72 to 0.75	–50 to –30	25 to 210	0.13	2.02	–	–	1.0
Glycerin	$C_3H_5(OH)_3$	1.26	+20	290	0.29	2.37	200	828	0.5
Hydrochloric acid 10 %	HCl	1.05	–14	102	0.50	3.14	–	–	–
Kerosene		0.76 to 0.86	–70	> 150	0.13	2.16	–	–	1.0
Linseed oil		0.93	–15	316	0.17	1.88	–	–	–
Lubricating oil		0.91	–20	> 300	0.13	2.09	–	–	–
Mercury [8]	Hg	13.55	–38.84	356.6	10	0.14	11.6	295	0.18
Methanol	CH_3OH	0.79	–98	65	0.20	2.51	99.2	1,109	–
Methyl chloride	CH_3Cl	0.99 [7]	–92	–24	0.16	1.38	–	406	–
m-xylene	$C_6H_4(CH_3)_2$	0.86	–48	139	–	–	–	339	–
Nitric acid, conc.	HNO_3	1.51	–41	84	0.26	1.72	–	–	–
Paraffin oil		–	–	–	–	–	–	–	0,764
Petroleum ether		0.66	–160	> 40	0.14	1.76	–	–	–
Rapeseed oil		0.91	± 0	300	0.17	1.97	–	–	–
Silicone oil		0.76 to 0.98	–	–	0.13	1.09	–	–	–
Sulfuric acid, conc.	H_2SO_4	1.83	+10.5 [6]	338	0.47	1.42	–	–	0.55

Table 4: Properties of liquids (continued)

Material/medium		Density [2] g/cm³	Melting temperature [1] °C	Boiling temperature [1] °C	Thermal conductivity [2] W/(m·K)	Specific heat capacity [2] kJ/(kg·K)	Melting enthalpy ΔH [3] kJ/kg	Evaporation enthalpy [4] kJ/kg	Coefficient of volume expansion $\times 10^{-3}$/K
Tar		1.2	–15	300	0.19	1.56	–	–	–
Toluene	C_7H_8	0.87	–93	111	0.14	1.67	74.4	364	–
Transformer oil		0.88	–30	170	0.13	1.88	–	–	–
Trichloroethylene	C_2HCl_3	1.46	–85	87	0.12	0.93	–	265	1.19
Turpentine oil		0.86	–10	160	0.11	1.80	–	293	1.0
Water		1.00 [10]	±0	100	0.60	4.18	332	2,256	0.18 [11]

[1] At 1.013 bar. [2] At 20 °C. [3] At melting temperature and 1.013 bar. [4] At boiling temperature and 1.013 bar. [5] At 0 °C.
[6] Setting temperature 0 °C. [7] At –24 °C. [8] For conversion from Torr to Pa, use 13.5951 g/cm³ (at 0 °C).
[9] Denatured ethanol. [10] At 4 °C. [11] Volume expansion on freezing: 9 %.

Table 5: Properties of water vapor

Absolute pressure bar	Boiling point °C	Evaporation enthalpy kJ/kg
0.1233	50	2,382
0.3855	75	2,321
1.0133	100	2,256
2.3216	125	2,187
4.760	150	2,113
8.925	175	2,031
15.55	200	1,941
25.5	225	1,837
39.78	250	1,716
59.49	275	1,573
85.92	300	1,403
120.5	325	1,189
165.4	350	892
221.1	374.2	0

Table 6: Properties of gases

Material/medium		Density [1]	Melting temperature [2]	Boiling temperature [2]	Thermal conductivity [3]	Specific heat capacity [3] kJ/(kg·K)			Enthalpy of evaporation [2]
		kg/m³	°C	°C	W/(m·K)	c_p	c_v	c_p/c_v	kJ/kg
Acetylene	C_2H_2	1.17	–84	–81	0.021	1.64	1.33	1.23	751
Air		1.293	–220	–191	0.026	1.005	0.716	1.40	209
Ammonia	NH_3	0.77	–78	–33	0.024	2.06	1.56	1.32	1,369
Argon	Ar	1.78	–189	–186	0.018	0.52	0.31	1.67	163
n-butane	C_4H_{10}	2.70	–138	–0.5	0.016	1.67	1.51	1.10	–
i-butane	C_4H_{10}	2.67	–145	–10.2	0.016	–	–	1.11	–
Blast-furnace gas		1.28	–210	–170	0.024	1.05	0.75	1.40	–
Carbon disulfide	CS_2	3.41	–112	–46	0.0073	0.67	0.56	1.19	–
Carbon dioxide	CO_2	1.98	–57 [4]	–78	0.016	0.82	0.63	1.30	368
Carbon monoxide	CO	1.25	–199	–191	0.025	1.05	0.75	1.40	–
Chlorine	Cl_2	3.21	–101	–35	0.009	0.48	0.37	1.30	288
City/town gas		0.56 to 0.61	–230	–2010	0.064	2.14	1.59	1.35	–
Cyanogen (dicyan)	$(CN)_2$	2.33	–34	–21	–	1.72	1.35	1.27	–
Dichlorodifluoromethane (= Freon F 12)	CCl_2F_2	5.51	–140	–30	0.010	0.61	0.54	1.14	–
Ethane	C_2H_6	1.36	–183	–89	0.021	1.66	1.36	1.22	522
Ethanol vapor	C_2H_4	2.04	–114	+78	0.015	–	–	1.13	–
Ethylene		1.26	–169	–104	0.020	1.47	1.18	1.24	516
Fluorine	F_2	1.70	–220	–188	0.025	0.82	0.61	1.35	172
Helium	He	0.18	–270	–269	0.15	5.20	3.15	1.65	20
Hydrogen	H_2	0.09	–258	–253	0.181	14.39	10.10	1.42	228
Hydrogen chloride	HCl	1.64	–114	–85	0.014	0.81	0.57	1.42	–
Hydrogen sulfide	H_2S	1.54	–86	–61	0.013 [1]	0.96	0.72	1.34	535
Krypton	Kr	3.73	–157	–153	0.0095	0.25	0.15	1.67	108
Methane	CH_4	0.72	–183	–164	0.033	2.19	1.68	1.30	557
Methyl chloride	CH_3Cl	2.31	–92	–24	–	0.74	0.57	1.29	446
Neon	Ne	0.90	–249	–246	0.049	1.03	0.62	1.67	86
Nitrogen	N_2	1.24	–210	–196	0.026	1.04	0.74	1.40	199
Oxygen	O_2	1.43	–218	–183	0.0267	0.92	0.65	1.41	213
Ozone	O_3	2.14	–251	–112	0.019	0.81	0.63	1.29	–
Propane	C_3H_8	2.00	–182	–42	0.018	1.70	1.50	1.13	–
Propylene	C_3H_6	1.91	–185	–47	0.017	1.47	1.28	1.15	468
Sulfur dioxide	SO_2	2.93	–73	–10	0.010	0.64	0.46	1.40	402
Sulfur hexafluoride	SF_6	6.16 [3]	–50.8	–63.9	0.011	0.66	–	–	117 [1]
Water vapor, 100 °C [5]		0.60	± 0	+100	0.025	2.01	1.52	1.32	–
Xenon	Xe	5.89	–112	–108	0.0057	0.16	0.096	1.67	96

[1] At 0 °C and 1.013 bar.
[2] At 1.013 bar.
[3] At 20 °C and 1.013 bar.
[4] At 5.3 bar.
[5] At saturation and 1.013 bar, also see Table 5.

Material groups

Material requirements

Technically and economically beneficial material use requires the knowledge of the application and the stress occurring in operation to derive the material requirements. The goal is to implement the desired function in the part or component and guarantee it throughout the planned period of use. The crucial loads here are usually of mechanical nature (strength, modulus of elasticity, see Table 1), but are often supplemented by corrosion, wear or temperature loads. Furthermore, specific physical properties (e.g. magnetism, conductivity) or a combination of those mentioned can be critical for material use.

Classification of materials

Various options exist for classifying the materials now used in technical applications. One example is subdividing into the four material groups:
- Metallic materials: smelted metals, sintered metals.
- Non-metallic inorganic materials: ceramics, glass.
- Non-metallic organic materials: natural substances, plastics.
- Composite materials.

Table 1: E-modulus of some materials (example)

Material group	Material type (selection)	Abbreviation Example	E-modulus [GPa]
Cast iron and malleable cast iron	Cast iron with lamellar graphite	EN-GJL-200	80 to 140
	Cast iron with spheroid graphite	EN-GJS-400	160 to 180
	White malleable cast iron	EN-GJMW-450	175 to 195
	Black malleable cast iron	EN-GJMB-450	175 to 195
Cast steel	Cast steel for general use	GE300	≈210
Steel	Unalloyed steels	C60	≈210
	Low-alloy steels	42CrMoS4	≈210
	Austenitic steels	X5CrNi18-10	≥190
	High-alloy tool steels	HS 6-5-4	≤230
Wrought copper alloys	High-conductivity copper	Cu-ETP	110 to 130
	Brass	CuZn30	115
	Nickel silver	CuNi18Zn20	135
	Tin bronze	CuSn6	100 to 120
Casting copper alloys	Tin cast bronze	CuSn10-C	100
	Red brass	CuSn7Zn4Pb7-C	100
Aluminum alloys	Wrought aluminum alloy	EN AW-AlSi1MgMn	65 to 75
	Casting aluminum alloy	EN AC-AlSi12Cu1	65 to 80
Magnesium alloys	Wrought magnesium alloy	MgAl6Zn	40 to 45
	Cast magnesium alloy	EN-MCMgAl9Zn1	40 to 45
Titanium alloys	Wrought titanium alloy	TiAl6V4	110
Tin alloys	Cast tin alloys for composite slide bearings	SnSb12Cu6Pb	30
Zinc alloys	Die-cast zinc alloy	GD-ZnAl4Cu1	70

Metallic materials

The metals generally have a crystalline structure. Their atoms are arranged in a regular crystal lattice. The valence electrons of the atoms are not bound to a specific atom, but are able to move freely within the metal lattice – a metallic bond is present. This special metal-lattice structure explains the characteristic properties of metals: the ductility and resulting high level of formability, the high electric conductivity, the high level of thermal conductivity, the low light transmission and the great optical reflective ability (metallic gloss).

The metallic materials used most often in technical applications include ferrous material. Iron base alloys are primarily used for this. An alloy is defined as a metallic material that consists of at least two chemical elements.

The ferrous materials are divided into groups for steels and cast iron materials. The main difference between the two groups is carbon content. The carbon content of steel is generally less than 2% and over 2% for cast iron (for details, see EN metallurgy standards).

Steels

Structural compositions of steels

The majority of steels primarily consists of iron and other property-determining alloying elements such as chrome, nickel, vanadium, molybdenum and titanium. The most important alloying element is carbon, which is diffused in the iron lattice and present as iron carbide (Fe_3C), also called cementite. Various phases are present in the steel with their characteristic properties depending on the carbon content, the alloying elements and heat treatment (e.g. hardening, quenching and drawing). They can be found in the iron/carbon diagram (Figure 1). The most important phases are described below.

Ferrite

The ferritic structure exhibits a body-centered cubic atomic arrangement. It is soft, easily formable and ferromagnetic (e.g. S235). In unalloyed steels, ferrite is only present at very low carbon contents (<0.02%) as an individual phase (also called α phase).

Figure 1: Iron/carbon diagram

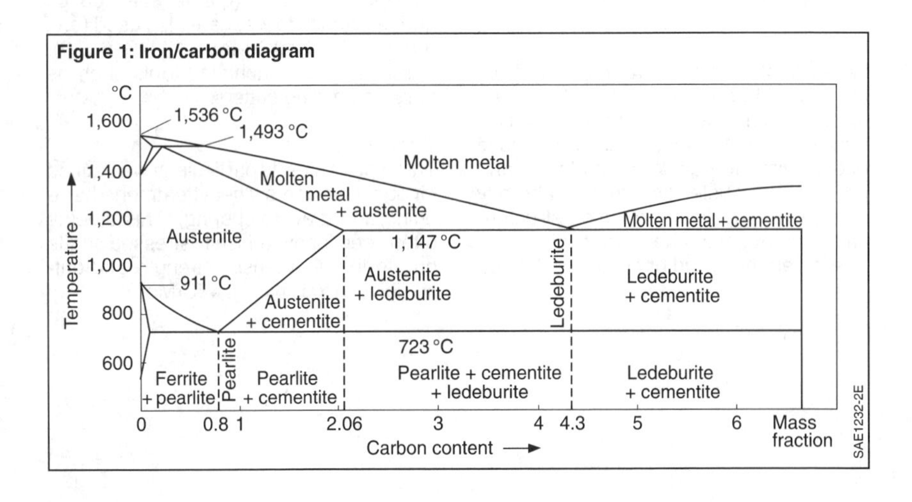

Pearlite

Pearlite has a lamellar iron-carbon structure, consisting of alternating associated areas of iron carbide (Fe_3C) and ferrite. This phase is present in steels with a carbon content higher than 0.02 %. As carbon content increase, the pearlite content present increases along with the ferrite until the structure consists entirely of pearlite at a carbon content of 0.8 % (e.g. C80). The pearlitic structure exhibits high strength.

Increasing the carbon content further results in the increased formation of iron carbide, which tends to separate out at the grain boundaries.

Austenite

If a steel is heated (between approximately 750 °C and 1,150 °C depending on the carbon content), the body-centered cubic iron lattice (α phase) transforms into a face-centered cubic iron lattice (γ phase). This austenitic structure can diffuse 100 times as much carbon as the ferritic structure. Ferrite and pearlite then form from the structure during the long cooling periods. In simple iron-carbon alloys, this lattice structure only exists at these high temperatures. Adding alloying elements such as nickel or cobalt allows the austenitic structure to be preserved stably even at room temperature. These types of austenitic alloys are very tough, easily formable, and resistant to heat and corrosion (e.g. X5CrNi1810).

Martensite

When cooling a steel very quickly from the austenite area during heat treatment (e.g. by quenching in water or oil), the carbon diffused in the lattice does not have enough time to transform into ferrite and pearlite. The austenitic structure abruptly deforms into a distorted lattice structure called martensite. The martensitic structure is very hard and brittle (e.g. 100Cr6).

Steel types

Along with many different organization and designation options, steels are often named based on their typical application, use or machining technique. Table 2 and Table 3 give an overview of properties and typical applications.

Structural steel

This is a simple steel that can usually be welded or machined easily. It is used in large quantities in general mechanical engineering and civil engineering and is reasonably priced (e.g. S235JR).

Free-cutting steel

This steel typically features high sulfur content, which causes short chips to form when machining (turning, milling). These chips can be extracted in the machinery without complications. These types of steels are usually used for volume components with low to medium-strength stresses (e.g. 11SMn30).

Tool steel

This is the umbrella term for a type of steel used for tools and molds. Depending on the application, it can feature high tensile strength, thermal stability, viscosity or corrosion resistance. This steel type can be classified as a cold-work steel (up to approx. 200 °C) or hot-work steel (up to approx. 400 °C), depending on the operating temperature. High-speed steels with high thermal stability (greater than 400 °C) and resistance to wear are also used (e.g. 90MnCrV8, X40CrMoV5-1, HS 6-5-2), particularly for machining tools such as drills and milling cutters.

Tempering steel

This type of steel is particularly well-suited for quench and draw heat treatment (hardening and draw tempering). This steel is often used for dynamically stressed shafts due to the high tensile strength and outstanding viscosity(e.g. 42CrMo4).

Table 2: Properties and typical applications for steels

Material group	Example materials	Surface hardness	Core hardness	Properties	Typical applications
Case-hardening steel (case-hardened and tempered)	C15	700 to 850 HV	200 to 450 HV	High resistance to wear	Moderately stressed gear wheels, joints
	16MnCr5	700 to 850 HV	300 to 450 HV	High resistance to wear	Piston pins, gear wheels
	18CrNiMo7-6	700 to 850 HV	400 to 550 HV	High resistance to wear, weldable	Gear wheels, drive parts
Nitriding steel (quenched, tempered and nitrided)	31CrMoV9	700 to 850 HV	250 to 400 HV	High resistance to wear	Highly-stressed piston pins, spindles
Ball and roller bearing steel (hardened and tempered)	100Cr6	60 to 64 HRC		High resistance to wear, high hardness and strength	Ball bearings, rolling elements, needle bearings
Cold-work steel (unalloyed, hardened and tempered)	C80	60 to 64 HRC		Surface hardened, tough core	Blades, chisels, striking tools
Cold-work steel (alloyed, hardened and tempered)	90MnCrV8	60 to 64 HRC		High dimensional stability, great edge retention and toughness.	Reamers, woodworking tools, thread cutters
	X210Cr12	60 to 64 HRC		Highly wear-resistant, great dimensional stability, high hardening properties	Cutting and punching tools, broaching tools, flange rollers, cold extrusion tools
Hot-work steel (hardened and tempered)	X40Cr-MoV5-1	43 to 45 HRC		High resistance to heat-related wear, great toughness and thermal conductivity	Forging dies, die-casting tools, extrusion dies
High-speed steel (hardened and tempered)	HS 6-5-2	61 to 65 HRC		High strength, great toughness, high hardness	Drills, countersinking tools
Stainless martensitic steel (hardened and tempered)	X20Cr13	≈40 HRC		Usable up to 550 °C, ferromagnetic, high resistance to wear	Pump parts, piston rods, nozzle needles, ship propellers
	X90Cr-MoV18	≥57 HRC		Good chemical resistance, resistance to wear and polishing properties	Cutting tools, ball bearings

Table 3: Properties and typical applications for steels

Material group	Example materials	Tensile strength [MPa]	Yield point [MPa]	Elongation at fracture A5 [%]	Fatigue strength under reversed bending stress [MPa]	Properties	Typical application examples
Hot-galvanized strip/sheet	DX53D	≤ 380	≤ 260	≥ 30 (A80)	≥ 190	Good deep-drawing capability, hot-galvanized, higher strength	Automotive sheet metal parts e.g. engine hood, side panel
Structural steel	S235 JR	340 to 510	≥ 225	≥ 26	≥ 170	Good weldability, no preceding or retroactive heat handling	General construction parts, profiles
Free-cutting steel	11SMn30	380 to 570	–	–	≥ 190	High sulfur content for short chips, not suitable for welding	Moderately stressed volume components from the automotive industry such as shafts
	35S20	520 to 680	–	–	≥ 260	High sulfur content for short chips, not suitable for welding	Volume parts such as moderately stressed screws
Tempering steel	C45	700 to 850	≥ 490	≥ 14	≥ 280	Tough, heat treatable, moderate to high strength	Axles, bolts, screws
	42CrMo4	1,100 to 1,300	≥ 900	≥ 10	≥ 440		Crankshafts, axle shafts
	30CrNiMo8	1,250 to 1,450	≥ 1,050	≥ 9	≥ 500		Drive parts
Stainless ferritic steel (annealed)	X6Cr17	450 to 600	≥ 270	≥ 20	≥ 200	Good corrosion resistance, deep-draw capability, polishing properties, magnetic	Household devices, medical technology, sanitary
Stainless austenitic steel (solution annealed)	X5CrNi18-10	500 to 700	≥ 190	≥ 45	–	Most frequently used corrosion resistant steel, good weldability, good deep-draw and polishing capability, non-magnetic	Chemical apparatus engineering, architecture, household objects
Cold strip (unalloyed)	DC05LC	270 to 330	≤ 180	≥ 40 (A80)	≥ 130	Good deep-draw properties	Automotive sheet metal parts e.g. engine hood, side panel

Table 4: Properties and typical applications for spring steel

Material	Diameter and thickness [mm]	Modulus of elasticity E [MPa]	Shear modulus G [MPa]	Minimum tensile strength [MPa]	Fracture contraction Z [%]	Permissible bending stress [MPa]	Permissible stress range for $N \geq 10^7$ [MPa]	Permissible maximum stress [MPa]	Properties and applications
Spring steel wire D, (patented and drawn cold-hammered)	1	206,000	81,500	2,230	40	1,590	380	1,115	Tension, compression, torsion springs, high static and low dynamic stress capacity
	3			1,840	40	1,280	360	920	
	10			1,350	30	930	320	675	
Stainless spring steel wire	1	185,000	73,500	2,000	40	1,400	–	1,000	Stainless springs
	3			1,600	40	1,130	–	800	
Heat-treated, alloyed valve-spring steel wire VD SiCr	1	200,000	79,000	2,060	50	–	430	1,030	High-strength spring material, high toughness e.g. passenger car valve springs
	3			1,920	50	–	430	960	
Spring steel strip Ck85	≤2.5	206,000	–	1,470	–	1,270	640	–	Highly stressed leaf springs
Stainless spring steel strip	≤1	185,000	–	1,370	–	1,230	590	–	Stainless leaf springs

Case-hardening steel
These steels feature a low carbon content (e.g. C15), which is usually increased up to 0.8 % by inserting the steel components into a carbonaceous furnace atmosphere (approx. 900 °C) in the edge region. The subsequent quenching causes martensite to form in the edge region and, in conjunction, high surface hardening at a simultaneously viscous core (ferritic, pearlitic). Applications include gear wheels and piston pins. Unlike nitriding, case hardening causes greater delays due to the martensitic hardening.

Nitriding steel
This steel forms a very hard and wear-resistant surface layer (approx. 10 µm) due to its chemical composition in combination with a corresponding heat treatment (nitriding at approx. 450 to 600 °C). Here, the viscous material core remains preserved, which makes this material particularly well-suited for gear wheels. There is less component deformation due to the lower temperature in comparison to case hardening (e.g. 31CrMoV9).

Spring steel
These are steels that have a particularly fine structure with high yield strength due to alloying silicon. They are especially well-suited for manufacturing springs in engine production (e.g. valve springs). Table 4 gives an overview of various spring steels.

Cast iron materials

Cast iron is defined as an iron-based alloy with a carbon content greater than approximately 2 %. The cast iron designation derives from these alloys' particularly high level of castability. One reason for this is the significantly lower melting temperature in comparison to steel (approximately 1,150 °C compared to approximately 1,500 °C).

Table 5 gives an overview of typical properties and applications for some materials.

Table 5: Properties and typical applications for cast irons

Material group	Example materials	Tensile strength [MPa]	Yield point [MPa]	Elongation at fracture A5 [%]	Fatigue strength under reversed bending stress [MPa]	Properties	Typical application examples
Cast iron with lamellar graphite (gray cast iron)	EN-GJL-200	≥ 200	–	–	90	Good castability, high vibration damping and heat capacity	Pump housing, cylinder head, machine bed, brake disc
Cast iron with spheroid graphite	EN-GJS-400-15	≥ 400	≥ 250	≥ 15	200	Higher strength and ductility than GJL	Crankshaft, crank-case, hubs
White malleable cast iron	EN-GJMW-400-5	≥ 400	≥ 220	≥ 5 (A3)	–	Good weldability, thin wall thicknesses, castable	Fittings
Black malleable cast iron	EN-GJMB-350-10	≥ 350	≥ 200	≥ 10 (A3)	–	Better machinability and hardening capability, can be manufactured with thick walls as white malleable cast iron	Fittings

Types of cast iron
The slow cooling during casting results in carbon mostly being present as graphite, typically in a ferritic-pearlitic matrix.

Gray cast iron
Depending on the formation of the graphite in layers or spheroids, the cast iron is referred to as either gray cast iron with lamellar graphite (GJL) or gray cast iron with spheroid graphite (GJS). The spheroid formation of the graphite is controlled in this process by treating the molten metal accordingly, such as with calcium. GJS features higher strength and better formability than GJL. Gray cast iron with vermicular graphite represents the middle ground between these two alloy types, both in terms of both graphite formation as well as properties.

Malleable cast iron
Along with these three types, the structure can also be affected by heat treatment (malleablizing) after casting. For malleable cast iron, the casted component made of white cast iron is subjected to an annealing process, where the iron carbide present in the matrix is converted into temper carbon. This increases the viscosity of the material. Based on the appearance of the fracture surface, these are referred to as white (annealed in an oxidizing, decarburizing atmosphere) and black malleable cast irons (annealed in a neutral atmosphere).

Cast steel
Cast steel is used for particularly highly stressed, weldable castings. In comparison to simple cast iron with lamellar graphite, castability is significantly worse while casting temperature and shrinkage are considerably higher.

Nonferrous metals
Along with the steels, some nonferrous metals (NF metals) have important technically relevant properties, such as those required in various applications. Special emphasis should be placed on copper alloys (heavy metal, density greater than 5 g/cm^3), which are used in many applications due to their high electric conductivity, and easily machinable, lightweight and nonmagnetic aluminum materials (light alloy, density less than 5 g/cm^3).

Classification of NF metals
NF metals (nonferrous metals and alloys with an iron percentage of less than 50 %) can generally be classified into wrought and cast alloys. Here, NF wrought alloys exhibit a very well malleable, ductile structure and can be brought to the desired shape by deep-drawing, bending, flanging, extruding or using similar deformation processes. Due to low ductility, shaping is limited to casting for NF casting alloys.

The properties of NF metals are frequently adapted to material requirements by adding alloying elements. Tables 6 through 9 give an overview of typical properties and applications of some materials.

Aluminum alloys
In aluminum alloys, adding alloying elements primarily aims to improve mechanical properties. For example, adding zinc or copper makes heat treatment possible, which results in high strength.

Copper alloys
Alloying zinc to copper forms brass (e.g. CuZn30), which leads to a higher strength compared to pure copper. Copper alloys with other alloying elements other than zinc are called bronzes. The most common bronzes are tin bronzes (e.g. CuSn6) with high strength and low ductility. Copper-nickel alloys (nickel-silver) are frequently used for plug contacts.

Table 6: Properties and typical applications for wrought copper alloys

Material group	Typical materials	Tensile strength [MPa]	Yield strength [MPa]	Fatigue strength under reversed bending stress (reference values) [MPa]	Properties	Application examples
High-conductivity copper	Cu-OF R220	220 to 260	≤140	70	High electric conductivity, solderable, weldable, deformable	Electronics, electrical engineering
Brass	CuZn30 R350	350 to 430	>170	110	Good strength properties, excellent cold-workability, can be soldered and brazed very well	Deep-drawn parts, fastening elements, screws, zippers
Nickel silver	CuNi18Zn20 R500	500 to 590	>410	–	Cold-workable, non-tarnishing, good spring properties	Relay springs, plugs
Tin bronze	CuSn6 R470	>470	≈ 350	190	Good cold-workability, good strength properties, good hardening properties, good wear and corrosion resistance, solderable	Springs, metal hoses, general machine and apparatus engineering

Table 7: Properties and typical applications for NF wrought alloys

Material group	Typical materials	Tensile strength [MPa]	Yield strength [MPa]	Fatigue strength under reversed bending stress (reference values) [MPa]	Properties	Application examples
Pure aluminum	EN AW-Al99,5 O	65	20	40	High conductivity, ductile, food-grade	Apparatus and tank engineering, deep-drawn parts
Self-hardening aluminum alloy	EN AW-AlMg2Mn0,8 H111	190	80	90	Seawater resistant, easily weldable	Automotive and ship engineering
Heat-treatable aluminum alloy	EN AW-AlSi1MgMn T6	310	260	90	Artificially aged, good combination of strength, corrosion resistance	Most frequently used heat-treatable aluminum alloy, profiles, vehicle frames, automobile transverse links
Heat-resistant aluminum alloy	EN AW-AlCu4MgSi (A) T4	390	245	120	Very good machinability, high temperature stability, good thermal stability	Components in hydraulic systems, aviation industry
High-strength aluminum alloy	EN AW-AlZn5,5MgCu T6	540	485	140	Good machinability, extremely high strength	Aviation industry, mechanical engineering
Universal aluminum casting alloy	EN AC-AlSi7Mg0,3 T6	290	210	80	Universal casting alloy, good mechanical properties, good corrosion resistance, very good weldability and machinability	Fittings, motor production, architecture
Eutectic casting alloy	EN AC-AlSi12Cu1 (Fe) DF	240	140	70	Excellent mold filling capability and casting characteristics, good chemical resistance	Thin-walled castings, fittings
Universal aluminum die-cast alloy	EN AC-AlSi9Cu3 (Fe) DF	240	140	70	Good processibility, low cost	Intake manifolds, transmission housings

Table 8: Properties and typical applications for NF casting alloys

Material group	Typical materials	Tensile strength [MPa]	Yield strength [MPa]	Fatigue strength under reversed bending stress (reference values) [MPa]	Properties	Application examples
High-strength casting alloy	EN AC-AlCu4Ti K T6	330	220	90	High strength and toughness	Trans-mission housings
Magnesium die-cast alloy	EN-MCMgAl9Zn1	200	140	50	Very light, low corrosion and thermal resistance, chips are com-bustible	Covers, housing, mobile telephone parts
Titanium forge alloy	TiAl6V4	890	820	–	High strength and corrosion resistance	Implants, aviation industry

Other NF metals

Along with aluminum and copper, the NF metals described below are of particular technical relevance.

Magnesium

At a density of approximately 1.75 g/cm³, magnesium is significantly lighter than aluminum. The low melting temperature makes it particularly well-suited for castings. Typical applications in automotive engineering include car wheel rims, housing units and profiles.

Titanium

Hard-to-machine titanium alloys feature great corrosion resistance and a high level of thermal stability. Due to the favorable ratio of density and strength, these alloys lend themselves well to the aviation industry (e.g. air compressor turbine blades). Another application area is medical technology (implants, e.g. hip prosthetics) due to biocompatibility.

Sintered metals

Metal powder sintering

Sintered metals are generally manufactured using near-net-shape pressing of metal powders. This allows complicated shapes to be made from sintered metals at low costs, ready-to-install immediately or only requiring little finishing. After near-net-shape pressing the metal powder, a permanent connection between the grains is established by a running diffusion process at temperatures between 60 and 80 % of the melting temperature.

Metal injection molding

During MIM (metal injection molding), a process variant of sintering, a component is formed by injection molding metal powder and plastic compounds. After removing lubricants and binders with chemicals or heat, the molded articles retain their characteristic properties just like during metal powder sintering.

In addition to its chemical composition, the degree of porosity determines the properties and application of sintered metals to a large extent (see Table 10 and Table 11).

Table 9: Other alloys and cast copper alloys

Material group	Typical materials	Tensile strength [MPa]	Yield strength [MPa]	Fatigue strength under reversed bending stress (reference values) [MPa]	Properties	Application examples
Tin alloy	SnSb12Cu6Pb	–	60	28	Low hardness, good corrosion resistance	Sliding bearings
Die-cast zinc	ZP0410	330	250	80	Excellent casting characteristics, dimensional stability, surface quality	Thin-walled, low-stressed castings
Heat-conductor alloy	NiCr8020	650	–	–	High electrical resistance, high temperature stability	Heating filament
Resistance alloy	CuNi44	420	–	–	Low temperature coefficient, high oxidation stability	Resistors, potentiometer, heating filament
Cast tin bronze	CuSn10-C-GS	250	160	90	Good corrosion, resistance to wear	Fittings, pump housings
Red brass	CuSn7Zn4Pb7-C-GZ	260	150	80	Seawater resistant, good machinability, limp-home characteristics	Friction bearings, bushings

References for metallic materials
[1] DIN 30910-3: Sintered metal materials – Sintered-material specifications – Part 3: Materials for bearings and structural parts with bearing properties.
[2] DIN 30910-4: Sintered metal materials – Sintered-material specifications – Part 4: Materials for structural parts.

Table 10: Materials for bearings and structural parts with bearing properties[1]

		Permissible ranges				Representative examples					
Material	Material code Sint-	Density ρ g/cm³	Chemical composition in % by mass Percent	Radial breaking strength K[2] N/mm²	Hardness HB	Density ρ g/cm³	chemical composition in % by mass Percent	Radial breaking strength K[2] N/mm²	Compressive yield point $\delta_{d\,0.2}$ N/mm²	Hardness HB[2]	Thermal conductivity λ W/mK
Sintered iron	A 00	5.6 to 6.0	<0.3 C; <1.0 Cu;	>150	>25	5.9	<0.2 others; rest Fe	160	130	30	37
	B 00	6.0 to 6.4	<2 others; rest Fe	>180	>30	6.3		190	160	40	43
	C 00	6.4 to 6.8		>220	>40	6.7		230	180	50	48
Sintered steel containing Cu	A 10	5.6 to 6.0	<0.3 C; 1 to 5 Cu;	>160	>35	5.9	2.0 Cu;	170	150	40	36
	B 10	6.0 to 6.4	<2 others; rest Fe	>190	>40	6.3	<0.2 others; rest Fe	200	170	50	37
	C 10	6.4 to 6.8		>230	>55	6.7		240	200	65	42
Sintered steel containing Cu and C	B 11	6.0 to 6.4	0.4 to 1.5 C; 1 to 5 Cu; <2 others; rest Fe	>270	>70	6.3	0.6 C; 2.0 Cu; <0.2 others; rest Fe	280	160	80	28
Sintered steel containing higher Cu	A 20	5.8 to 6.2	<0.3 C; 15 to 25 Cu;	>180	>30	6.0	20 Cu;	200	140	40	41
	B 20	6.2 to 6.6	<2 others; rest Fe	>200	>45	6.4	<0.2 others; rest Fe	220	160	50	47
Sintered steel containing higher Cu and C	A 22	5.5 to 6.0	0.5 to 3.0 C;	>120	>20	5.7	2.0 C[3]; 20 Cu;	125	100	25	30
	B 22	6.0 to 6.5	15 to 25 Cu; <2 others; rest Fe	>140	>25	6.1	<0.2 others; rest Fe	145	120	30	37
Sintered bronze	A 50	6.4 to 6.8	<0.2 C; 9 to 11 Sn;	>120	>25	6.6	10 Sn;	140	100	30	27
	B 50	6.8 to 7.2	<2 others; rest Cu	>170	>30	7.0	<0.2 others; rest Cu	180	130	35	32
	C 50	7.2 to 7.7		>200	>35	7.4		210	160	45	37
Sintered bronze graphitic[4]	A 51	6.0 to 6.5	0.5 to 3.0 C;	>100	>20	6.3	1.5 C[4]; 10 Sn;	120	80	20	20
	B 51	6.5 to 7.0	9 to 11 Sn;	>150	>25	6.7	<0.2 others; rest Cu	155	100	30	26
	C 51	7.0 to 7.5	<2 others; rest Cu	>170	>30	7.1		175	120	35	32

[1] According to "Material Specification Sheets for Sintered Metals": DIN 30910-3, 2004 edition [1]. [2] Measured on calibrated bearings 10/16Ø · 10.
[3] C is primarily present as free graphite. [4] C is present as free graphite.

Table 11: Sintered metals for structural parts[1]

Material	Material code	Permissible ranges			Representative examples						
		Density ρ	chemical composition in % by mass	Hardness	Density ρ	chemical composition in % by mass	Tensile strength R_m	Yield point $R_{p\,0.1}$	Elongation at break A	Hardness	E-module $E \cdot 10^3$
	Sint-	g/cm³	Percent	HB	g/cm³	Percent	N/mm²	N/mm²	Percent	HB	N/mm²
Sintered iron	C 00 D 00 E 00	6.4 to 6.8 6.8 to 7.2 >7.2	<0.3 C; <1.0 Cu; <2 others; rest Fe	>35 >45 >60	6.6 6.9 7.3	<0.5 others; rest Fe	120 170 240	60 80 120	3 8 14	40 50 60	100 130 160
Sintered steel containing C	C 01 D 01	6.4 to 6.8 6.8 to 7.2	0.3 to 0.9 C; <1.0 Cu; <2 others; rest Fe	>70 >90	6.6 6.9	0.5 C; <0.5 others; rest Fe	240 300	170 200	2 2	75 90	100 130
Sintered steel containing Cu	C 10 D 10 E 10	6.4 to 6.8 6.8 to 7.2 >7.2	<0.3 C; 1 to 5 Cu; <2 others; rest Fe	>40 >50 >80	6.6 6.9 7.3	1.5 Cu; <0.5 others; rest Fe	200 250 340	140 180 240	2 3 5	55 80 110	100 130 160
Sintered steel containing Cu and C	C 11 D 11	6.4 to 6.8 6.8 to 7.2	0.4 to 1.5 C; 1 to 5 Cu; <2 others; rest Fe	>80 >95	6.6 6.9	0.6 C; 1.5 Cu; <0.5 others; rest Fe	390 460	290 370	1 2	115 130	100 130
	C 21	6.4 to 6.8	0.4 to 1.5 C; 5 to 10 Cu; <2 others; rest Fe	>105	6.6	0.8 C; 6 Cu; <0.5 others; rest Fe	470	360	<1	140	100
Sintered steel containing Cu, Ni and Mo	C 30 D 30 E 30	6.4 to 6.8 6.8 to 7.2 >7.2	<0.3 C; 1 to 5 Cu; 1 to 5 Ni; <0.6 Mo; <2 others; rest Fe	>55 >60 >90	6.6 6.9 7.3	0.3 C; 1.5 Cu; 4.0 Ni; 0.5 Mo; <0.5 others; rest Fe	360 460 570	290 330 390	2 2 4	100 125 160	100 130 160
Sintered steel Cu, Ni and Mo	C 31 D 31 E 31	6.4 to 6.8 6.8 to 7.2 >7.2	<0.3 C; <3.0 Cu; <5.0 Ni; 0.6 to 2 Mo; <2 others; rest Fe	>50 >60 >90	6.6 6.9 7.3	0.2 C; 2.0 Ni; 1.5 Mo; <0.5 others; rest Fe	320 380 460	220 260 320	1 2 3	100 120 150	100 130 160
Sintered steel contains Mo and C	C 32 D 32	6.4 to 6.8 6.8 to 7.2	<0.3 to 0.9 C; <3.0 Cu; <5.0 Ni; 0.6 to 2 Mo; <2 others; rest Fe	>55 >60	6.6 6.9	0.6 C; 2.0 Cu; 1.5 Mo; <0.5 others; rest Fe	400 520	370 480	<1 1	140 180	100 130
Sintered steel containing P	C 35 D 35	6.4 to 6.8 6.8 to 7.2	<0.3 C; <1 Cu; 0.3 to 0.6 P; <2 others; rest Fe	>70 >80	6.6 6.9	0.45 P; <0.5 others; rest Fe	290 310	180 210	9 10	80 85	100 130
Sintered steel containing Cu and P	C 36 D 36	6.4 to 6.8 6.8 to 7.2	<0.3 C; 1 to 5 Cu; 0.3 to 0.6 P; <2 others; rest Fe	>80 >90	6.6 6.9	2.0 Cu; 0.45 P; <0.5 others; rest Fe	330 350	270 300	4 5	90 95	100 130

Table 11 (continued): Sintered metals for structural parts[1]

Material	Material code	Permissible ranges: Density ρ	Permissible ranges: chemical composition in % by mass	Permissible ranges: Hardness	Representative examples: Density ρ	Representative examples: chemical composition in % by mass	Tensile strength R_m	Yield point $R_{p\,0.1}$	Elongation at break A	Hardness	E-module $E \cdot 10^3$
	Sint-	g/cm³	Percent	HB	g/cm³	Percent	N/mm²	N/mm²	%	HB	N/mm²
Sintered steel containing Cu, Ni, Mo and C	C 39 D 39	6.4 to 6.8 6.8 to 7.2	0.3 to 0.9 C; 1 to 3 Cu; 1 to 5 Ni; <0.6 Mo; <2 others; rest Fe	>90 >120	6.6 6.9	0.5 C; 1.5 Cu; 4.0 Ni; 0.5 Mo; <0.5 others; rest Fe	480 560	350 380	1 2	140 160	100 130
Stainless sintered steel AISI 316	C 40 D 40	6.4 to 6.8 6.8 to 7.2	<0.08 C; 10 to 14 Ni; 2 to 4 Mo; 16 to 19 Cr; <2 others; rest Fe	>95 >125	6.6 6.9	0.06 C; 13 Ni; 2.5 Mo; 18 Cr; <0.5 others; rest Fe	330 400	250 320	1 2	110 135	100 130
AISI 430	C 42	6.4 to 6.8	<0.08 C; 16 to 19 Cr; <2 others; rest Fe	>140	6.6	0.06 C; 18 Cr; <0.5 others; rest Fe	420	330	1	170	100
AISI 410	C 43	6.4 to 6.8	<0.3 C; 11 to 13 Cr; <2 others; rest Fe	>165	6.6	0.2 C; 13 Cr; <0.5 others; rest Fe	510	370	1	180	100
Sintered bronze	C 50 D 50	7.2 to 7.7 7.7 to 8.1	9 to 11 Sn; <2 others; rest Cu	>35 >45	7.4 7.9	10 Sn; <0.5 others; rest Cu	150 220	90 120	4 6	40 55	50 70
Sintered aluminum AlCuMgSi	E 73	2.55 to 2.65	4 to 6 Cu; <1 Mg; <1 Si; <2 other; rest Al.	>55	2.58 [2] 2.58 [3]	4.5 Cu; 0.6 Mg; 0.7 Si; <0.5 other; rest Al.	180 285	150 [4] n.c.	1 <0.5	65 90	55 [5] 55 [5]
AlSiMgCu	F 75	2.60 to 2.66	2 to 3 Cu; <1 Mg; 13 to 16 Si; <2 other; rest Al.	>70	2.63 [2] 2.63 [3]	2.5 Cu; 0.5 Mg; 14 Si; <0.5 other; rest Al.	200 300	180 [4] n.c.	<0.5 <0.5	90 125	78 [5] 78 [5]
AlZnMgCu	F 77	2.74 to 2.78	1.5 to 2.0 Cu; 2.2 to 2.8 Mg; 5.6 to 6.4 Zn; <2 other; rest Al.	>90	2.78 [2] 2.78 [3]	1.6 Cu; 2.6 Mg; 6.0 Zn; <0.5 other; rest Al.	300 450	190 [4] 230 [4]	3 1.5	100 155	68 [5] 68 [5]

[1] According to "Material Specification Sheets for Sintered Metals": DIN 30910-4, 2010 edition [2].
[2] T1a sintered state and stored at room temperature for at least five days. [3] T6 solution annealed and artificially aged.
[4] Yield strength $R_{p0.2}$. [5] Determining the E-modulus using ultrasound.
n.c. not calculated.

EN metallurgy standards

Standardization of steels
(In accordance with DIN EN 10020, [1])
Steel is defined as an iron alloy, usually with a carbon content of ≤ 2%. Ferrous materials with a higher carbon content are usually classified as cast iron. Steels are classified in three classes, "unalloyed steels", "stainless steels" and "other alloyed steels".

Unalloyed steels
Unalloyed steels that do not reach the defined minimum alloying element content are then divided into unalloyed high-grade steels and unalloyed stainless steels. For unalloyed high-grade steels, defined requirements apply, such as those regarding toughness and malleability.

Unalloyed stainless steels feature a higher degree of purity and, therefore, improved properties such as high yield strength, hardenability, good toughness and weldability. They are usually intended for quenching and drawing or for surface hardening.

Stainless steels
Stainless steels feature a chrome content of at least 10.5 percent by mass and a carbon content of less than 1.2 percent by mass. They are subdivided further by nickel content (more or less than 2.5 percent by mass) and by the main properties, including corrosion resistance, heat resistance and temperature stability.

Other alloyed steels
These include steels with requirements in regards to aspects such as toughness, particle size or malleability. They are usually not intended for quenching and drawing or surface hardening. A distinction is made between alloyed high-grade steels and alloyed stainless steels.

Designation system for steels with material abbreviations
(In accordance with DIN EN 10027-1, [2])
The material abbreviations for steels are divided into two groups.
- Group 1: Material abbreviations that feature information on use and mechanical and physical properties.
- Group 2: Material abbreviations that contain information on chemical composition.

Abbreviated names of Group 1
These material abbreviations include notes on use and mechanical or physical properties. The prefix G indicates that it involves a cast steel material:

S, GS	For general steel construction
P, GP	For pressure-vessel construction
L	For pipeline construction
E	Engineering steels
B	Concrete reinforcing steels
Y	Prestressing steels
R	Rail steels
H	Flat products made of higher-strength steels for cold forming
D	Flat products for cold forming
T	Packaging steel sheet and strip
M	Electrical steel sheet and strip.

Examples:
S235JR

S	Generally for steel construction
235	Yield strength in MPa
JR	Notch impact work 27 J at 20 °C.

HC240LA

H	High-strength steel
C	Cold-rolling
240	Minimum yield strength in MPa
LA	Low-alloyed

Abbreviated names of Group 2
These material abbreviations contain references to the chemical composition. The prefix G indicates that this is a cast steel material:

C, GC	Unalloyed steels (Mn < 1%),
G	Unalloyed steels (Mn low-alloyed steels)
X, GX	High-alloyed steels
PM	Powder metallurgy
HS	High-speed steel

Examples:

C85S

C	Unalloyed steel (< 1% Mn)
85	0.01 times the C content, i.e. 0.85 % C
S	For springs

42CrMo4

42	Low-alloyed steel (Mn ≥ 1%) 0.01 times the C content, i.e. 0.42 % C
Cr	4 times the Cr content in percent, i.e. 1% Cr,
Mo	Non-specific percentage of alloyed constituents, Mo (< 1%).

X5CrNi18-10

X	High-alloyed steel,
5	0.05 % carbon
18	18 % Cr
10	10 % Ni

HS 7-4-2-5

HS	High-speed steel Percentage of alloyed constituents in whole percentages in the order of tungsten – molybdenum – vanadium – cobalt:
7	7 % tungsten
4	4 % molybdenum
2	2 % vanadium
5	5 % cobalt

In addition, special requirements can be defined, such as for the type of coating or notes on the treatment condition. Example information could read as follows:

+H	With hardenability
+CU	With copper coating
+Z	Hot-dip galvanized
+ZE	Electrolytically galvanized
+C	Work-hardened
+M	Thermomechanically shaped
+Q	Quenched
+U	Untreated

Designation system for steels based on the numbering system

(In accordance with DIN EN 10027-2, [3])
In addition to their material abbreviation, all steels are defined by a material number in accordance with the following structure:
Major group + steel group + counter.

Material major group number:

0	Pig iron, ferro-alloys
1	Steel
2	Nonferrous heavy metals
3	Light metals
4 to 8	Non-metallic materials
9	Unassigned for internal use

Steel group number (selection):

00	Unalloyed ordinary low-carbon steels
01 to 07	Unalloyed high-grade steels
10 to 18	Unalloyed stainless steels
40 to 49	Chemically stable alloying steels
20 to 29	Alloying tool steels

Example:
1.4301
(material abbreviation: X5CrNi18-10)

1	Steel
43	Stainless steel with more than 2.5 % Ni without any Mo, Nb or Ti
01	Count number

Standardization of cast-iron materials

The structure and form of the carbon (carbide or graphite) and the graphite structure are important to the properties. In standardization, a distinction is made between the following four groups of cast iron:

- Lamellar graphite cast iron (DIN EN 1561, [4])
- Malleable cast iron (DIN EN 1562, [5])
- Spheroidal graphite cast iron (DIN EN 1563, [6]),
- Ausferritic spheroidal graphite cast iron (DIN EN 1564, [7]).

There are two systems for designating types of cast iron: In accordance with material numbers or abbreviations.

Designation system for cast iron with abbreviations
(In accordance with DIN EN 1560, [8])
The designation is alphanumeric and consists of up to six individual positions. The first position is EN (European standard). This is followed by G (cast) and J (iron). The character for denoting the graphite structure is in the third position:

L	Lamellar
S	Spheroidal
M	Temper carbon
V	Vermicular
N	Graphite-free, ledeburitic cast chilled iron
Y	Special structure

Notes on the microstructure and macrostructure can be placed at position 4 if necessary, such as (as an excerpt):

A	Austenite
F	Ferrite
Q	Quenched

The fifth position contains information on tensile strength, impact energy with test temperature or for hardness. Alternatively, the fifth position can specify the chemical composition.

Finally, the sixth position contains additional requirements such as:

D	As-cast casting
H	Heat-treated casting

Examples:
EN-GJL-150

EN	European standard
GJ	Cast iron
L	Lamellar graphite
150	Tensile strength in MPa

EN-GJV-HV400

EN	European standard
GJ	Cast iron
V	Vermicular graphite
HV400:	Vickers hardness

EN-GJS-SiMo30-8

EN	European standard
GJ	Cast iron
S	Spheroidal graphite
Si	Silicon content 3 %
Mo	Molybdenum content 0.8 %.

Designation system for cast iron with material numbers
(In accordance with DIN EN 1560, [8])
The designation consists of a total of six positions. The first and second positions contain "5." followed by information on the graphite structure at the third position:

1	Lamellar
2	Vermicular
3	Spheroidal
4	Temper carbon

The fourth position describes the matrix structure (selection):

1	Ferrite
3	Pearlite
5	Austenite

The fourth position is followed by a double-digit count number (00-99) that identifies the individual material.

Nonferrous-metal alloys
As for ferrous materials, the EN standard also features two options for describing and identifying each nonferrous metals (NF) material and its alloys: The first designation system uses chemical symbols (abbreviations) and the second uses a numerical designation system.

Unlike steels, NF metals are each identified with their own numerical designation system. Thus, the different properties and requirements of, for example, aluminum, copper or zinc alloys are better taken into account.

Designation system for NF metals with abbreviation
The EN standard denotes nonferrous metals (NF metals) according to the following basic system:

Example:
EN AW–Al Si1MgMn T6

EN	European standard

Code letter for metal:

A	Aluminum
C	Copper
M	Magnesium

Code letter for processing:
W Wrought (wrought alloy)
C Casting (casting alloy)

Designation system with chemical symbols, e.g. AlSi1MgMn in this case:

Material condition, here, e.g.: T0

The system lists the alloy metals in order of base metal with falling percentages.

Special numerical designation system for aluminum (Al) and aluminum alloys
For wrought Al products (Al and wrought Al alloys in accordance with DIN EN 573-1, [9]), the alloy is determined by four digits (Example 1); for Al casting alloys (in accordance with DIN EN 1706, [10]), by five digits (Example 2).

Example 1:
EN AW–6082 T6
EN European standard
AW Wrought aluminum alloy
6 Indicator 6 for alloy group (Al-Mg-Si alloys)
0 Original alloy (1, 2 for modifications)
82 Designation for the alloy with about 1% Si 0.7% Mn and 0.9% Mg
T6 Material condition T6 (solution-annealed and artificially aged)

Example 2:
EN AC -45200
EN European standard
AC Casting aluminum alloy
45 AlSi5Cu alloy group
200 Number for individual alloy (here Al Si5Cu3Mn)

Code numbers for the alloy groups:
1 Pure aluminum
2 With copper
3 With manganese
4 With silicon
5 With magnesium
6 With magnesium-silicon,
7 With zinc
8 Other

Material conditions (selection):
O Soft-annealed
H Work-hardened
H14 Work-hardened, 1/2 hard (for sheet metal)
T6 Solution-annealed and artificially aged

References for EN metallurgy standards
[1] DIN EN 10020: Definition and classification of grades of steel (2007).
[2] DIN EN 10027-1: Designation systems for steels – Part 1: Steel names (2005)
[3] DIN EN 10027-2: Designation systems for steels – Part 2: Numerical system (1992).
[4] DIN EN 1561: Founding – Grey cast irons (2012).
[5] DIN EN 1562: Founding – Malleable cast irons (2012).
[6] DIN EN 1563: Founding – Spheroidal graphite cast irons (2012).
[7] DIN EN 1564: Founding – Ausferritic spheroidal graphite cast irons (2012).
[8] DIN EN 1560: Founding – Designation system for cast iron – Material symbols and material numbers (2011).
[9] DIN EN 573-1: Aluminium and aluminium alloys – Chemical composition and form of wrought products – Part 1: Numerical designation system (2005).
[10] DIN EN 1706: Aluminium and aluminium alloys – Castings – Chemical composition and mechanical properties (2013).

Magnetic materials

Materials which have ferromagnetic or ferrimagnetic properties are called magnetic materials and belong to one of two groups: metals (metals produced through melting or sintering) or nonmetallic inorganic materials. Composite materials such as soft magnetic composite materials and plastic-bonded permanent magnets are also playing an increasingly important role. They are characterized by their ability to exhibit a permanent external field or by their good magnetic flux conductivity (soft magnets).

In addition to ferromagnets and ferrimagnets, diamagnetic, paramagnetic, and antiferromagnetic materials also exist. They differ from each other in terms of their permeability μ, or the temperature dependence of their susceptibility κ. This variable gives the ratio of the magnetization of a substance to the magnetic field strength or excitation.

$$\mu_r = 1 + \kappa.$$

Classification

Ferromagnets and ferrimagnets

Both substances exhibit spontaneous magnetization which disappears at the Curie point (Curie temperature T_c). At temperatures above the Curie temperature, they behave like paramagnets. For susceptibility κ at $T > T_C$, the Curie-Weiss law applies as follows:

$$\kappa = \frac{C}{T - T_C},$$

where

C Curie constant

T Temperature in K

The saturation induction of ferromagnets is higher than for ferrimagnets because all magnetic moments are aligned in parallel. In the case of ferrimagnets, on the other hand, the magnetic moments of the two sublattices are aligned antiparallel to one another. Nevertheless, these materials are magnetic because the magnetic moments of the two sublattices have different magnitudes.

Diamagnets

For diamagnets, susceptibility κ_{Dia} is independent of the temperature.

Paramagnets

For paramagnets, susceptibility κ_{Para} drops as temperature increases. In this case, the Curie law states:

$$\kappa_{Para} = \frac{C}{T}.$$

Antiferromagnets

As in the case of ferrimagnets, adjacent magnetic moments are aligned antiparallel with respect to one another. As they are of equal magnitude, the effective magnetization of the material is zero.

At temperatures above the Néel point (Néel temperature T_N), antiferromagnets behave like paramagnets. At $T > T_N$, the susceptibility is:

$$\kappa = \frac{C}{T + \Theta},$$

where

Θ Asymptotic Curie temperature

Examples of antiferromagnets: MnO, MnS, $FeCl_2$, FeO, NiO, Cr, V_2O_3, V_2O_4.

Soft magnetic materials

The figures specified in Table 1 are from the applicable DIN standards (soft-magnetic metallic materials, DIN IEC 60404-8-6, [1]). Many material qualities defined in this standard relate to the materials in DIN 17405 (DC relays, [2]) and DIN-IEC 60740-2 (transformers and reactors, [3]).

Designation (composition)

"Code letter" "Number 1" "Number 2" – "Number 3".

The code letter specifies the alloy class: "A" pure iron, "C" silicon-iron (SiFe), "E" nickel-iron (NiFe), "F" cobalt-iron (CoFe).

Number 1 indicates the concentration of the main alloy element.

Number 2 defines the different curves: 1 indicates a round hysteresis loop, 2 indicates a rectangular hysteresis loop.

Table 1: Soft-magnetic metallic materials

Magnet type	Alloying constituents by mass	Static magnetic properties												AC test data, 50 Hz[1]		
		Coercive field strength $H_{c(max)}$ in A/m Thickness in mm		Minimum magnetic polarization in tesla (T) at field strength H in A/m									Measuring point $\hat{H}$ in A/m	Minimum amplitude permeability μ_r Sheet thickness in mm	
	%	0.4 to 1.5	>1.5	20	50	100	300	500	800	1,600	4,000	8,000		0.30 to 0.38	0.15 to 0.20
A – 240	100 Fe	240	240				1.15	1.30			1.60		Not suitable for AC applications.		
A – 120	100 Fe	120	120				1.15	1.30			1.60				
A – 60	100 Fe	60	60				1.25	1.35			1.60				
A – 12	100 Fe	12	12			1.15	1.30	1.40			1.60				
C1 – 48	0 to 5 Si (typical 2 to 4.5)	48	48			0.60	1.10	1.20			1.50				
C1 – 12	0 to 5 Si (typical 2 to 4.5)	12	12			1.20	1.30	1.35			1.50				
C21 – 09	0.4 to 5 Si (typical 2 to 4.5)												1.60	900	750
C22 – 13	0.4 to 5 Si (typical 2 to 4.5)												1.60	1,300	–
E11 – 60	72 to 83 Ni	2	4	0.50	0.65	0.70		0.73			0.75		0.40	40,000	40,000
E21	54 to 68 Ni	Not suitable for this thickness											On agreement		
E31 – 06	45 to 50 Ni	10	10	0.50	0.90	1.10		1.35			1.45		0.40	6,000	6,000
E32	45 to 50 Ni	Not suitable for this thickness											On agreement		
E41 – 03	35 to 40 Ni	24	24	0.20	0.45	0.70		1.00			1.18		1.60	2,900	2,900
F11 – 240	47 to 50 Co		240				1.40		1.70	1.90	2.06	2.15	As agreed between manufacturer and buyer.		
F11 – 60	47 to 50 Co	60					1.80		2.10	2.20	2.25	2.25			
F21	35 Co	300							1.50	1.60	2.00	2.20			
F31	23 to 27 Co	300									1.85	2.00			

[1] Data applies to laminated rings.

The significance of Number 3 following the hyphen varies according to the individual alloy. It indicates the minimum initial permeability μ_a/1,000 in nickel alloys; with other alloys, it designates the maximum coercive field strength in A/m. The properties of these materials are strongly geometry-dependent and highly application-specific. The material data quoted in extracts from the standard can therefore provide only a very general overview of the properties of these materials.

Electrical steel sheet and electrical steel strip

Applications and properties

Packages used for transformers or as stator or rotor packages for electric motors are composed of electrical steel strips, typically by packaging individual sheet metal plates. Electrical steel strips are often provided as long strips (wide strips or split strips) in a wound shape (called coils), but they are also available as panels (electrical steel sheet).

Electrical steel strips typically consist of iron-silicon alloys ([4] and [5]); other materials such as aluminum and manganese may function as alloy constituents in small quantities. Densities between 7.65 and 7.85 g/cm^3 result depending on the composition. They are characterized by low remagnetization loss, high polarization and high permeability compared to other soft magnetic ferrous materials. The static coercive field strength H_c is typically 100 to 300 A/m for A and K types (see Designations), the maximum permeability μ_{max} is at about 5,000. For S and P types, the values are in the order of magnitude of 1 A/m for H_c and about 30,000 for μ_{max}.

Grain oriented electrical steel strip

Both grain oriented and non-oriented electrical steel strips exist. Grain oriented electrical steel strips feature a crystal structure (cubic structure), meaning the grains are oriented in a preferred direction by rolling and annealing steps during manufacturing. This gives the strip material a preferred magnetic direction in the strip direction. The magnetic properties depend on the direction (anisotropic magnetic properties). Grain oriented electrical steel strip is primarily used in components where exceptionally good magnetic flux in a specific direction comes into play (e.g. transformers, choke coils and converters).

Non-oriented electrical steel strip

In non-oriented electrical steel strip, however, there is not a structure. In other words, the crystallographic orientation of the grains is almost randomly arranged. The magnetic properties do not depend on the direction (isotropic magnetic properties). For this reason, non-oriented electrical steel strip is used in components where magnetic flux is not limited to a preferred direction (e.g. electric motors and alternators) [6].

Non-oriented electrical steel strips are usually delivered in a finally annealed (fully-finished) state. In other words, final annealing for optimizing the magnetic properties takes place at the manufacturer of the electrical steel strips. For electrical steel strips that are not finally annealed (semi-finished), final annealing still has to be carried out. This is typically done to the finished laminated core.

Coating

Electrical isolation of the individual layers suppresses eddy currents for alternating-field magnetization, reducing remagnetization losses.

Identifications

The designations for electrical steel sheets and electrical steel strips (ESS) is defined in DIN EN 10027-1 [7]:

Code letter 1 Number 1 – Number 2 Code letter 2 (see below for an example).

The first code letter is "M" for all varieties. Number 1 is one hundredfold the maximum magnetic reversal loss at 50 Hz and 1.5 T (types A and K) or 1.7 T (types S and P)in W/kg. Number 2 is one hundredfold the product's nominal depth in mm.

Table 2: Properties of electrical steel strips and electrical sheet steel

Sheet grade		Nominal thickness	Maximum remagnetization loss at 50 Hz and activation of			Minimum magnetic polarization in tesla (T) in an AC field at field strength H in A/m			Applications
Abbreviated name	Material number	mm	1.0 T	1.5 T	1.7 T	2,500	5,000	10,000	
M270-35A	1.0801	0.35	1.10	2.7	–	1.49	1.60	1.70	Electric motors
M330-35A	1.0804	0.35	1.30	3.3	–	1.49	1.60	1.70	
M330-50A	1.0809	0.50	1.35	3.3	–	1.49	1.60	1.70	
M530-50A	1.0813	0.50	2.30	5.3	–	1.56	1.65	1.75	
M800-50A	1.0816	0.50	3.60	8.0	–	1.60	1.70	1.78	
M400–65A	1.0821	0.65	1.70	4.0	–	1.52	1.62	1.72	
M1000–65A	1.0829	0.65	4.40	10.0	–	1.61	1.71	1.80	
M800–100A	1.0895	1.00	3.60	8.0	–	1.56	1.66	1.75	
M1300–100A	1.0897	1.00	5.80	13.0	–	1.60	1.70	1.78	
M340-50K	1.0841	0.50	1.42	3.4	–	1.54	1.62	1.72	Low-power motors for industrial and household appliances (e.g. washing machine motors, microwave transformers, refrigerator compressors)
M560-50K	1.0844	0.50	2.42	5.6	–	1.58	1.66	1.76	
M660-50K	1.0361	0.50	2.80	6.6	–	1.62	1.70	1.79	
M1050-50K	1.0363	0.50	4.30	10.5	–	1.57	1.65	1.77	
M390-65K	1.0846	0.65	1.62	3.9	–	1.54	1.62	1.72	
M630-65K	1.0849	0.65	2.72	6.3	–	1.58	1.66	1.76	
M800-65K	1.0364	0.65	3.30	8.0	–	1.62	1.70	1.79	
M1200-65K	1.0366	0.65	5.00	12.0	–	1.57	1.65	1.77	
						At field strength $H = 800$ A/m			
M140-30S	1.0862	0.30	–	0.92	1.40	1.78			Converters, transformers, choke coils
M150-30S	1.0861	0.30	–	0.97	1.50	1.75			
M111-30P	1.0881	0.30	–	–	1.11	1.88			

Code letter 2 provides type data:

- "A" cold rolled non-oriented electrical steel sheet and strip delivered in the fully processed state (DIN EN 10106, [8]).
- "S" conventional grain-oriented electrical steel sheet or "P" grain-oriented electrical steel strip with high permeability, both types are delivered in a fully processed state (DIN EN 10107, [9]).
- "K" cold rolled electrical non-alloy and alloy steel sheet and strip delivered in the semi-processed state (DIN EN 10341, [10]).

Example: An electrical steel strip designated M330-35A is a non-oriented electrical steel strip in a finally annealed state, exhibits a remagnetization loss of 3.3 W/kg at 50 Hz and 1.5 T and has a thickness of 0.35 mm.

Materials for transformers and reactors (DIN IEC 740-2)

These materials comprise the alloy classes C21, C22, E11, E31 and E41 from the standard for soft-magnetic materials (DIN IEC 60404-8-6). The standard essentially contains the minimum values for core-sheet permeability for specified core-sheet sections (YEI, YED, YEE, YEL, YUI, and YM).

Materials for direct-current relays (DIN 17405)

Designation comprises a letter and number combination:
a) Code letter "R" (relay material).
b) Code letters for identifying alloying constituents:
Fe = unalloyed, Si = silicon steels,
Ni = nickel steels or alloys.
c) Code number for maximum coercive field strength.
d) Code letter for the stipulated delivery state: "U" = untreated, "GB" = malleable pre-annealed, "GT" = pre-annealed for deep-drawing, "GF" = final-annealed.

DIN IEC 60404-8-10 essentially contains the limit deviations for magnetic relay materials based on iron and steel. The designation code defined in this standard is as follows (example: M 80 TH):
- Code letter "M".
- Permitted maximum value for coercive field strength in A/m.
- Code letter for material composition: "F" = pure iron, "T" = steel alloy, "U" = unalloyed steel.
- Code letter for delivery state: "H" = hot-rolled, "C" = cold-rolled or cold-drawn.

Sintered metals for soft magnetic components (DIN IEC 60404-8-9)

Designation comprises a letter and number combination:
- Code letter "S" for sintered materials.
- Hyphen followed by the identifying alloy elements, i.e. Fe plus P, Si, Ni, or Co if necessary.
- The maximum permitted coercive field strength in A/m follows the second hyphen.

Soft magnetic ferrite cores (previously DIN 41280, [11])

Soft-magnetic ferrites are formed parts made of a sintered material with the general formula $MO{\cdot}Fe_2O_3$, where M is one or more of the bivalent metals Cd, Co, Ca, Mg, Mn, Ni, Zn.

Designation comprises a letter and number combination:
The various types of soft-magnetic ferrites are classified in groups according to nominal initial permeability and are designated by capital letters. Additional numbers may be used to further subdivide them into subgroups; these numbers have no bearing on material quality.

The coercive field strength H_c of soft ferrites is usually in the range of 4 to 500 A/m. Based on a field strength of 3,000 A/m, induction B is in the range of 350 to 470 mT.

Powder composite materials

Powder composite materials are not yet standardized, but are becoming increasingly more important. They consist of ferromagnetic metal powder (iron or an alloy) and an organic or inorganic grain-boundary phase as a "binder". They are manufactured in much the same way as sintered metals. The individual manufacturing stages are:
- Mixing the starting materials (metal powder and binder).
- Shaping by injection-molding, extruding or pressing.
- Heat treatment below the sinter temperature ($<600\,°C$).

Depending on the shaping process, binder type and the amount of binder used, it is possible to optimize the material to achieve high saturation polarization, higher permeability or high resistivity.

They are used primarily in fields in which all the above-mentioned characteristics are important, and where no excessively high demands are placed on mechanical strength and machinability. These fields currently consist of quick-acting actuators for diesel-injection engineering and high-speed small electric motors for motor vehicles.

Table 3: Materials for transformers and reactors

Core-sheet permeability for alloy classes C21, C22, E11, E31, and E41 for core-sheet section YEI1.

		Minimum core-sheet permeability μ_{lam} (min)											
IEC designation		C21-09 Thickness in mm				C22-13 Thickness in mm				E11-60 Thickness in mm			
		0.3 to 0.38	0.15 to 0.2			0.3 to 0.38				0.3 to 0.38	0.15 to 0.2	0.1	0.05
YEI 1	–10	630	630			1,000				14,000	18,000	20,000	20,000
	13	800	630			1,000				18,000	20,000	22,400	20,400
	14	800	630			1,000				18,000	22,400	22,400	22,400
	16	800	630			1,000				20,000	22,400	25,000	22,400
	18	800	630			1,000				22,400	25,000	25,000	22,400
	20	800	630			1,120				22,400	25,000	25,000	25,000
	22	800	630			1,120							
	25	800	630			1,120							
IEC designation		E11-100 Thickness in mm				E31-04 Thickness in mm				E31-06 Thickness in mm			
		0.3 to 0.38	0.15 to 0.2	0.1	0.05	0.3 to 0.38	0.15 to 0.2	0.1	0.05	0.3 to 0.38	0.15 to 0.2	0.1	0.05
YEI 1	–10	18,000	25,000	31,500	31,500	2,800	2,800	3,150	3,150	3,550	4,000	4,500	5,000
	13	20,000	28,000	35,500	35,500	2,800	3,150	3,150	3,550	4,000	4,500	5,000	5,000
	14	22,400	28,000	35,500	35,500	2,800	3,150	3,150	3,550	4,000	4,500	5,000	5,000
	16	25,000	31,500	35,500	35,500	2,800	3,150	3,150	3,550	4,500	4,500	5,000	5,000
	18	25,000	31,500	40,000	35,500	3,150	3,150	3,550	3,550	4,500	4,500	5,000	5,000
	20	28,000	35,500	40,000	40,000	3,150	3,150	3,550	3,550	4,500	5,000	5,000	5,000
IEC designation		E31-10 Thickness in mm				E41-02 Thickness in mm				E41-03 Thickness in mm			
		0.3 to 0.38	0.15 to 0.2	0.1	0.05	0.3 to 0.38	0.15 to 0.2	0.1	0.05	0.3 to 0.38	0.15 to 0.2	0.1	0.05
YEI 1	–10	5,600	6,300	5,600	6,300	1,600	1,800	1,800	2,000	2,000	2,240	2,500	2,240
	13	6,300	7,100	6,300	6,300	1,800	1,800	2,000	2,000	2,240	2,240	2,500	2,240
	14	6,300	7,100	6,300	7,100	1,800	1,800	2,000	2,000	2,240	2,240	2,500	2,240
	16	6,300	7,100	6,300	7,100	1,800	1,800	2,000	2,000	2,240	2,500	2,500	2,240
	18	7,100	7,100	6,300	7,100	1,800	1,800	2,000	2,000	2,240	2,500	2,500	2,240
	20	7,100	7,100	6,300	7,100	1,800	2,000	2,000	2,000	2,240	2,500	2,500	2,240

Table 4: Materials for DC relays

Material type: Abbreviated Name	Material number	Alloying constituents by mass %	Density[1] ρ g/cm³	Hardness[1] HV	Remanence[1] T (tesla)	Permeability[1] μ_{max}	Electr. resistivity[1] $\frac{\Omega \cdot mm^2}{m}$	Coercive field strength A/m max.	Magnetic polarization in T (Tesla) min. at field strength H in A/m: 20	50	100	200	300	500	1,000	4,000	Properties, application examples
Unalloyed steels																	
RFe 160	1.1011				–	–	0.15	160	–	–	–	–	1.15	1.30	–	1.60	Low coercive field strength.
RFe 80	1.1014				1.10	–	0.15	80	–	–	–	1.10	1.20	1.30	1.45	1.60	
RFe 60	1.1015	–	7.85	max. 150	1.20	–	0.12	60	–	–	–	1.15	1.25	1.35	1.45	1.60	
RFe 20	1.1017				1.20	≈ 20,000	0.10	20	–	–	1.15	1.25	1.30	1.40	1.45	1.60	
RFe 12	1.1018				1.20		0.10	12	–	–	1.15	1.25	1.30	1.40	1.45	1.60	
Silicon steels																	Direct-current relays and similar purposes.
RSi 48	1.3840	2.5	7.55	130	0.50	–	0.42	48	–	–	0.60	–	1.10	1.20	–	1.50	
RSi 24	1.3843	–	–	–	1.00	≈ 20,000	–	24	–	–	1.20	–	1.30	1.35	–	1.50	
RSi 12	1.3845	4 Si	7.75	200	1.00	≈ 10,000	0.60	12	–	–	1.20	–	1.30	1.35	–	1.50	
Nickel steels and nickel alloys																	
RNi 24	1.3911	≈ 36 Ni	8.2	130 to 180	0.45	≈ 5,000	0.75	24	0.20	0.45	0.70	–	0.90	1.0	–	1.18	
RNi 12	1.3926	≈ 50 Ni	8.3	130 to 180	0.60	≈ 30,000	0.45	12	0.50	0.90	1.10	–	1.25	1.35	–	1.45	
RNi 8	1.3927	≈ 50 Ni	8.3	130 to 180	0.60	30,000 to 100,000	0.45	8	0.50	0.90	1.10	–	1.25	1.35	–	1.45	
RNi 5	2.4596	70 to 80 Ni small quantities Cu, Cr, Mo	8.7	120 to 170	0.30	≈ 40,000	0.55	5	0.50	0.65	0.70	–	–	–	–	0.75	
RNi 2	2.4595		8.7	120 to 170	0.30	≈ 100,000	0.55	2	0.50	0.65	0.70	–	–	–	–	0.75	

[1] Standard values.

Table 5: Sintered metals for soft-magnetic components

Material Abbreviated name	Characteristic alloy elements (except Fe) % by mass	Sinter Density ρ_s	Por-osity p_s	Maximum coercive field strength $H_{c(max)}$	Magnetic polarization in tesla (T) at field strength H in A/m				Maximum permeability	Vickers hardness	Volume Electrical resistivity ρ
	%	g/cm³	%	A/m	500	5,000	15,000	80,000	$\mu_{(max)}$	HV5	$\mu\Omega$m
S-Fe-175	–	6.6	16	175	0.70	1.10	1.40	1.55	2,000	50	0.15
S-Fe-170	–	7.0	11	170	0.90	1.25	1.45	1.65	2,600	60	0.13
S-Fe-165	–	7.2	9	165	1.10	1.40	1.55	1.75	3,000	70	0.12
S-FeP-150	≈0.45 P	7.0	10	150	1.05	1.30	1.50	1.65	3,400	95	0.20
S-FeP-130	≈0.45 P	7.2	8	130	1.20	1.45	1.60	1.75	4,000	105	0.19
S-FeSi-80	≈3 Si	7.3	4	80	1.35	1.55	1.70	1.85	8,000	170	0.45
S-FeSi-50	≈3 Si	7.5	2	50	1.40	1.65	1.70	1.95	9,500	180	0.45
S-FeNi-20	≈50 Ni	7.7	7	20	1.10	1.25	1.30	1.30	20,000	70	0.50
S-FeNi-15	≈50 Ni	8.0	4	15	1.30	1.50	1.55	1.55	30,000	85	0.45
S-FeCo-100	≈50 Co	7.8	3	100	1.50	2.00	2.10	2.15	2,000	190	0.10
S-FeCo-200	≈50 Co	7.8	3	200	1.55	2.05	2.15	2.20	3,900	240	0.35

Table 6: Soft magnetic ferrites

Ferrite type	Initial permeability[1] μ_i ±25 %	Specific loss factor $\tan \delta/\mu_i$[2] 10^{-6}	MHz	Specific Power loss[3] mW/g	Amplitude permeability[4] μ_a	Curie temperature[5][6] Θ_c °C	Frequency for $0.8 \cdot \mu_i$[6] MHz	Properties, application examples
Materials in largely open magnetic circuits								Initial permeability. Compared to metallic magnetic materials high specific resistance (100 to 10^5 $\Omega \cdot$m, metals 10^{-7} to 10^{-6} $\Omega \cdot$m), therefore low eddy-current losses. Telecommunications (coils, transformers).
C 1/12	12	350	100	–	–	>500	400	
D 1/50	50	120	10	–	–	>400	90	
F 1/250	250	100	3	–	–	>250	22	
G 2/600	600	40	1	–	–	>170	6	
H 1/1,200	1,200	20	0.3	–	–	>150	2	
Materials in largely closed magnetic circuits								
E 2	60 to 160	80	10	–	–	>400	50	
G 3	400 to 1,200	25	1	–	–	>180	6	
J 4	1,600 to 2,500	5	0.1	–	–	>150	1.5	
M 1	3,000 to 5,000	5	0.03	–	–	>125	0.4	
P1	5,000 to 7,000	3	0.01	–	–	>125	0.3	
Materials for power applications								
W 1	1,000 to 3,000	–	–	45	1,200	>180	–	
W 2	1,000 to 3,000	–	–	25	1,500	>180	–	

[1] Nominal values. [2] $\tan \delta/\mu_i$ denotes the frequency-dependent material losses at a low flux density ($B < 0.1$ mT).
[3] Losses at high flux density. Measured preferably at: $f = 25$ kHz, $B = 200$ mT, $\Theta = 100$ °C.
[4] Permeability when subjected to a strong sinusoidal magnetic field. Measured at: $f \leq 25$ kHz, $B = 320$ mT, $\Theta = 100$ °C.
[5] Curie temperature Θ_c in this table is the temperature at which initial permeability μ_i drops below 10 % of its value at 25 °C.
[6] Standard values.

Permanent-magnet materials

(DIN 17410, replaced by DIN IEC 60404-8-1)
If chemical symbols are used in the abbreviated names of the materials, they refer to the primary alloying constituents of the materials. The numbers in the material abbreviation before the slash denote the $(BH)_{max}$ value in kJ/m^3 and those after the slash denote one tenth of the H_cJ value in kA/m (rounded values). Permanent magnets with binders are indicated by a final p.

Designation by material abbreviation or code number
Structure of code numbers
(DIN IEC 60404-8-1:2005-8):
Code letter (group)
+ 1st digit (material type),
+ 2nd digit of 0 (isotropic) or 1 (anisotropic),
+ 3rd digit (various quality levels).

R – Hard magnetic alloys,
such as R1: Aluminum-nickel-cobalt-iron-titanium alloys (AlNiCo).

S – Hard magnetic ceramic materials,
such as S1: Hard magnetic ferrite.

U – Combined hard magnetic materials,
such as
- U1: Combined aluminum-nickel-cobalt-iron-titanium magnets (AlNiCo).
- U2: Combined rare earth-cobalt magnets (RECo).
- U3: Combined neodymium-iron-boron magnets.
- U4: Combined hard ferrites.

Comparison of permanent and soft magnets

Figure 1 shows the magnetic characteristic value ranges of some crystalline materials in widespread use. The values of soft magnetic materials are compared to the values of hard magnets (permanent magnets).

Table 7: Permanent magnet materials
Bosch grades BTMT (not standardized)

Material Abbreviated name	Density[1)] ρ g/cm^3	$(BH)_{max}$[2)] kJ/m^3	Remanence[2)] B_r mT	Coercive field strength[2)] of flux density H_{CB} kA/m	Coercive field strength[2)] of polarization H_{CJ} kA/m
RBX HC 370	4.7 to 4.9	25	360	270	390
RBX HC 380		28	380	280	370
RBX 380 K		28	380	280	300
RBX 400		30	400	255	260
RBX 400 K		31	400	290	300
RBX HC 400		29	380	285	355
RBX 420		34	420	255	270
RBX 410 K		33	410	305	330
RBX HC 410		30	395	290	340
RBX 420 S		35	425	260	270
RBX HC 400 N		28	380	280	390

1) Standard values. 2) Minimum values.

Table 8: Permanent magnet materials

Material: Abbreviated name	Material number: DIN	Material number: IEC	Chemical composition[1)] % by weight: Al	Co	Cu	Nb	Ni	Ti	Fe	Density ρ[1)] g/cm³	$(BH)_{max}$[2)] kJ/m³	Remanence B_r[2)] mT	Coercive field strength[2)] of flux density H_{CB} kA/m	Coercive field strength[2)] of polarization H_{CJ} kA/m	Rel. permanent permeability[1)] μ_p	Curie temp.[1)] T_c K	Temp. coeff. of polar. $TK(J_s)$[1)3)] %K	Temp. coeff. of coerciv. $TK(H_c)$[1)3)] %K	Manufacture, processing, applications
Metallic magnets																			
Isotropic																			
AlNiCo 9/5	1.3728	R 1-0-3	11 to 13	0 to 5	2 to 4	–	21 to 28	0 to 1	Rest	6.8	9.0	550	44	47	4.0 to 5.0	1,030 to 1,180	– 0.02	+ 0.03 to – 0.07	Manufacture: casting or sintering. For magnets with binders: pressing or injection molding. Processing: grinding. Application: max. 400 to 500 °C.
AlNiCo 18/9	1.3756	–	6 to 8	24 to 34	3 to 6	–	13 to 19	5 to 9		7.2	18.0	600	80	86	3.0 to 4.0				
AlNiCo 7/8p	1.3715	R 1-2-3	6 to 8	24 to 34	3 to 6	–	13 to 19	5 to 9		5.5	7.0	340	72	84	2.0 to 3.0				
Anisotropic																			
AlNiCo 35/5	1.3761	–	8 to 9	23 to 26	3 to 4	0 to 1	13 to 16	–	Rest	7.2	35.0	1,120	47	48	3.0 to 4.5	1,030 to 1,180	– 0.02	+ 0.03 to – 0.07	
AlNiCo 44/5	1.3757	R 1-1-2	8 to 9	23 to 26	3 to 4	0 to 1	13 to 16	–		7.2	44.0	1,200	52	53	2.5 to 4.0				
AlNiCo 52/6	1.3759	–	8 to 9	23 to 26	3 to 4	0 to 1	13 to 16	–		7.2	52.0	1,250	55	56	1.5 to 3.0				
AlNiCo 60/11	1.3763	R 1-1-6	6 to 8	35 to 39	2 to 4	0 to 1	13 to 15	4 to 6		7.2	60.0	900	110	112	1.5 to 2.5				
AlNiCo 30/14	1.3765	–	6 to 8	38 to 42	2 to 4	0 to 1	13 to 15	7 to 9		7.2	30.0	680	136	144	1.5 to 2.5				
			Pt	Co															
PtCo 60/40	2.5210	R2-0-1	77 to 78	20 to 23						15.5	60	600	350	400	1.1	800	– 0.01 to – 0.02	– 0.35	
			V	Co	Cr	Fe													
FeCoVCr 11/2	2.4570	R 3-1-3	8 to 15	51 to 54	0 to 4	Rest				–	11.0	800	24	24	2.0 to 8.0	1,000	– 0.01	≈ 0	
FeCoVCr 4/1	2.4571	–	3 to 15	51 to 54	0 to 6					–	4.0	1,000	5	5	9.0 to 25.0				
RECo – Magnets of type $RECo_5$																			
RECo 80/80	–	R 5-1-1	Typically $MMCo_5$ (MM = Cer composition metal)							8.1	80	650	500	800	1.05	1,000	– 0.05	– 0.3	
RECo 120/96	–	R 5-1-2	Typically $SmCo_5$, typically (SmPr) Co_5							8.1	120	770	590	960	1.05	1,000	– 0.05	– 0.3	
RECo 160/80	–	R 5-1-3								8.1	160	900	640	800	1.05	1,000	– 0.05	– 0.3	
RECo – Magnets of type RE_2Co_{17}																			
RECo 165/50	–	R 5-1-11	Typically $MMCo_5$ (MM = Cer composition metal)							8.2	165	950	440	500	1.1	1,100	0.03	– 0.02	
RECo 180/90	–	R 5-1-13	Typically $SmCo_5$, typically (SmPr) Co_5							8.2	180	1,000	680	900	1.1	1,100	0.03	– 0.02	
RECo 190/70	–	R 5-1-14								8.2	190	1,050	560	700	1.1	1,100	0.03	– 0.02	
RECo 48/60p	–	R 5-1-3								5.2	48	500	360	600	1.05	1,000	– 0.05	– 0.3	

Material Abbreviated name	Material number DIN	IEC	Chemical composition[1) % by weight Al	Co	Cu	Nb	Ni	Ti	Fe	Density ρ[1) g/cm³	$(BH)_{max}$[2) kJ/m³	Remanence B_r[2) mT	Coercive field strength[2) of flux density H_{CB} kA/m	of polarization H_{CJ} kA/m	Rel. permanent permeability[1) μ_p	Curie temp.[1) T_c K	Temp. coeff. of polar. $TK(J_s)$[1)3) %K	Temp. coeff. of coerciv. $TK(H_c)$[1)3) %K	Manufacture, processing, applications
CrFeCo 12/4	–	R 6-0-1	(no data)							7.6	12	800	40	42	5.5 to 6.5	1,125	– 0.03	– 0.04	
CrFeCo 28/5	–	R 6-1-1								7.6	28	1,000	45	46	3 to 4	1,125	– 0.03	– 0.04	
REFe 165/170	–	R 7-1-1	(no data)							7.4	165	940	700	1,700	1.07	583	– 0.1	– 0.8	
REFe 220/140	–	R 7-1-6								7.4	220	1,090	800	1,400	1.05	583	– 0.1	– 0.8	
REFe 240/110	–	R 7-1-7								7.4	240	1,140	850	1,100	1.05	583	– 0.1	– 0.8	
REFe 260/80	–	R 7-1-8								7.4	260	1,180	750	800	1.05	583	– 0.1	– 0.8	

Material Abbreviated name	Material number DIN	IEC	Density[1) ρ g/cm³	$(BH)_{max}$[2) kJ/m³	Remanence[2) B_r mT	Coercive field strength[2) of flux density H_{CB} kA/m	of polarization H_{CJ} kA/m	Rel. permanent permeability[1) μ_p	Curie-temp.[1) T_c K	Temp. coeff. of polarization[1) $TK(J_s)$ %K	Temp. coeff. of coer.[1) $TK(H_c)$ %K	Manufacture, processing, applications
Ceramic magnets												
Isotropic												
Hard ferrite 7/21	1.3641	S 1-0-1	4.9	6.5	190	125	210	1.2	723	– 0.2	0.2 to 0.5	Manufacture:
Hard ferrite 3/18p	1.3614	S 1-2-2	3.9	3.2	135	85	175	1.1				sintering.
Anisotropic												Plastic-bound
Hard ferrite 20/19	1.3643	S 1-1-1	4.8	20.0	320	170	190	1.1				magnets
Hard ferrite 20/28	1.3645	S 1-1-2	4.6	20.0	320	220	280	1.1				produced by
Hard ferrite 24/23	1.3647	S 1-1-3	4.8	24.0	350	215	230	1.1	723	– 0.2	0.2 to 0.5	pressing,
Hard ferrite 25/22	1.3651	S 1-1-5	4.8	25.0	370	205	220	1.1				injection
Hard ferrite 26/26	–	S 1-1-8	4.7	26.0	370	230	260	1.1				molding,
Hard ferrite 32/17	–	S 1-1-10	4.9	32.0	410	160	165	1.1				rolling,
Hard ferrite 24/35	–	S 1-1-14	4.8	24.0	360	260	350	1.1				extruding.
Hard ferrite 9/19p	1.3616	S 1-1-3	3.4	9.0	220	145	190	1.1				Processing:
Hard ferrite 10/22p	–	S 1-2-3	3.5	10.0	230	165	225	1.1				grinding.

1) Standard values. 2) Minimum values. 3) In the range of 273 to 373 K.

References for magnetic materials

[1] DIN IEC 60404-8: Magnetic materials – Part 8: Specifications for individual materials.
8-1: Requirements for individual materials – Hard-magnetic materials (permanent magnets).
8-6: Requirements for individual materials – Soft-magnetic metallic materials.
8-9: Standard requirement for soft magnetic sintered metals.
8-10: Magnetic materials (iron and steel) for relay applications.

[2] DIN 17405: Soft magnetic materials for DC relay; technical terms of delivery.

[3] DIN IEC 60740-2: Laminations for transformers and inductors for use in telecommunication and electronic equipment – Part 2: Specification for the minimum permeabilities of laminations made of soft magnetic metallic materials.

[4] R. Boll (Bearb.): Weichmagnetische Werkstoffe – Einführung in den Magnetismus, VAC Werkstoffe und ihre Anwendungen [Soft magnetic materials – Introduction to magnetism, VAC materials and their applications]. 4. Auflage, Vacuumschmelze GmbH, 1990.

[5] L. Schneider, L. Michalowsky (ed.): Magnettechnik – Grundlagen, Werkstoffe, Anwendungen [Magnet technology – Foundations, materials, applications]. 3. Auflage, Vulkan-Verlag GmbH, 2006.

[6] Information Sheet 401, “Elektroblech und -band” [Electrical steel sheets and strips], Steel Information Center, Düsseldorf, 2005 edition.

[7] DIN EN 10027-1: Designation systems for steels – Part 1: Steel names.

[8] DIN EN 10106: Cold rolled non-oriented electrical steel sheet and strip delivered in the fully processed state.

[9] DIN EN 10107: Non-oriented electrical steel sheet and strip delivered in the fully processed state.

[10] DIN EN 10341: Cold rolled electrical non-alloy and alloy steel sheet and strip delivered in the semi-processed state.

[11] DIN 41280: Cores of soft magnetic oxides; material properties

Figure 1: Comparison of permanent and soft magnets

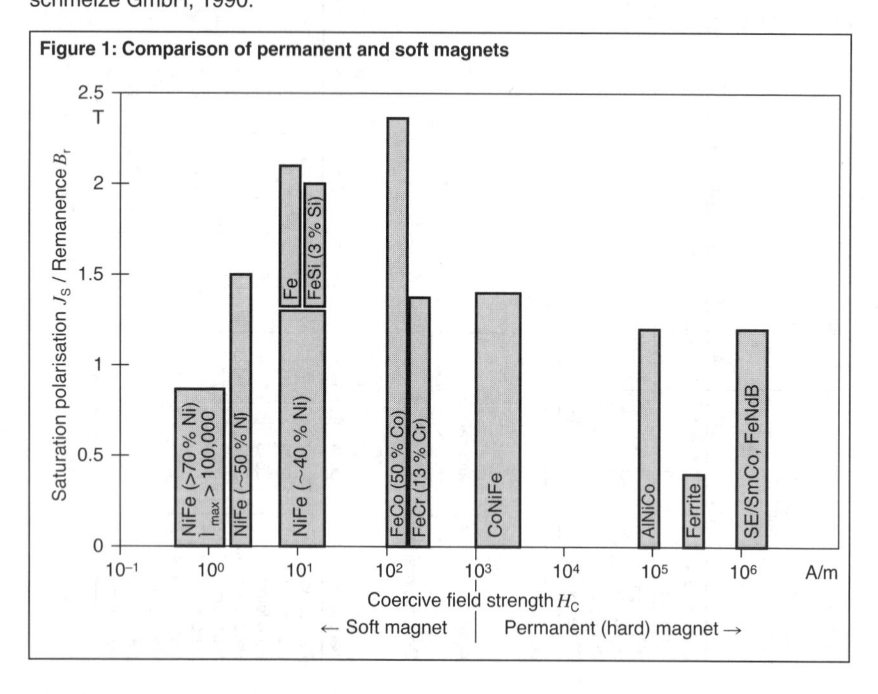

Nonmetallic inorganic materials

These materials are characterized by ionic bonds (e.g. ceramic materials), mixed (heteropolar/homopolar) bonds (e.g. glass), or homopolar bonds (e.g. carbon). These kinds of bonds, in turn, are responsible for several characteristic properties: generally poor thermal conductivity and poor electric conductivity (the latter increases with temperature), poor luminous reflectance, brittleness, and thus almost complete unsuitability for cold forming.

Ceramics

Ceramics are at least 30% crystalline in nature; most ceramics also contain amorphous components and pores. Their manufacture is similar to that of sintered metals, however nonmetallic powders or powder mixtures are used; sintering at temperatures generally higher than 1,000 °C gives ceramics their characteristic properties. Ceramic structural parts are sometimes also shaped at high temperatures or even by a melting process, with subsequent crystallization.

Table 1 gives an overview of ceramic materials and their properties.

Glass

Glass is viewed as under-cooled, frozen liquid. Its atoms are only in a short-range order. It is regarded as amorphous. Molten glass turns to solid glass at transformation temperature T_g (T_g is derived from the former designation glass formation temperature). The transformation temperature is dependent on a variety of parameters and therefore not clearly determined (better: transformation range).

Composite materials

Composite materials consist of at least two physically or chemically different components. These components must be tightly bonded together at a specific boundary layer. Under these conditions, it is possible to bond many materials together. The resulting composite material exhibits properties that neither of the original materials has.

A distinction is made between the following:
- Particle composite materials: e.g. powder-filled resins, hard metals, plastic-bonded magnets, cermets.
- Laminated composite materials: Such as thermostatic bimetals, sandwich panels.
- Fiber composite materials: Such as glass fiber or carbon fiber-reinforced plastics.

Table 1: Ceramic materials

Materials	Composition	ρ [1] g/cm³	σ_{bB} [2] MN/m²	σ_{dB} [3] MN/m²	E [4] GN/m²	α_t [5] 10^{-6}/K	λ [6] W/mK	c [7] kJ/kg · K	ρ_D [8] Ω · cm	ε_r [9]	tan ρ [10] 10^{-4}
Aluminum nitride	AlN > 97 %	3.3	250 to 350	1,100	320 to 350	5.1	100 to 220	0.8	> 10^{14}	8.5 to 9.0	3 to 10
Aluminum oxide	Al_2O_3 > 99 %	3.9 to 4.0	300 to 400	3,000 to 4,000	380 to 400	7.2 to 8.6	20 to 40	0.8 to 0.9	> 10^{11}	8 to 10	2
Aluminum titanate	$Al_2O_3 \cdot TiO_2$	3.0 to 3.7	25 to 40	450 to 550	10 to 30	0.5 to 1.5	< 2	0.7	> 10^{11}	–	–
Beryllium oxide	BeO > 99 %	2.9 to 3.0	250 to 320	1,500	300 to 340	8.5 to 9.0	240 to 280	1.0	> 10^{14}	6.5	3 to 5
Boron carbide	B_4C	2.5	300 to 500	2,800	450	5.0	30 to 60	–	10^{-1} to 10^{2}	–	–
Cordierite, e.g. C 410, 520 [11]	$2MgO \cdot 2Al_2O_3 \cdot 5SiO_2$	1.6 to 2.3	5 to 100	300	6 to 60	2.0 to 5.0	1.3 to 2.5	0.8	> 10^{11}	5.0	70
Graphite	C > 99.7 %	1.5 to 1.8	5 to 30	20 to 50	5 to 15	1.6 to 4.0	100 to 180	–	10^{-3}	–	–
Porcelain, e.g. C 110 – 120 (non-glazed)	Al_2O_3 30 to 35 % rest SiO_2 + glass phase	2.2 to 2.4	20 to 100	500 to 550	50	4.0 to 6.5	1.2 to 2.6	0.8	10^{11}	6	120
Silicon carbide pressureless-sintered SSiC	SiC > 98 %	3.1 to 3.2	400 to 600	> 1,200	400	4.0 to 4.5	90 to 120	0.8	10^{3}	–	–
Silicon carbide hot-pressed HPSiC	SiC > 99 %	3.1 to 3.2	450 to 800	> 1,500	420	4.0 to 4.5	100 to 120	0.8	10^{3}	–	–
Silicon carbide reaction-sintered SiSiC	SiC > 90 % + Si	3.0 to 3.1	300 to 400	> 2,200	380	4.2 to 4.8	100 to 160	0.8	10 to 100	–	–
Silicon nitride gas-pressure-sintered GPSN	Si_3N_4 > 90 %	3.2	800 to 1,400	> 2,500	300	3.2 to 3.5	30 to 45	0.7	10^{12}	–	–
Silicon nitride hot-pressed HPSN	Si_3N_4 > 95 %	3.2	600 to 900	> 3,000	310	3.2 to 3.5	30 to 45	0.7	10^{12}	–	–

Materials	Composition	ρ [1] g/cm³	σ_{bB} [2] MN/m²	σ_{dB} [3] MN/m²	E [4] GN/m²	α_t [5] 10^{-6}/K	λ [6] W/mK	c [7] kJ/kg · K	ρ_D [8] Ω · cm	ε_r [9]	tan ρ [10] 10^{-4}
Silicon nitride reaction-sintered RBSN	Si_3N_4 > 99 %	2.4 to 2.6	200 to 300	< 2,000	140 to 160	2.9 to 3.0	15 to 20	0.7	10^{14}	–	–
Steatite, e.g. C 220, 221	SiO_2 55 to 65 % MgO 25 to 35 % Al_2O_3 2 to 6 % Alkali oxide < 1.5 %	2.6 to 2.9	120 to 140	850 to 1,000	80 to 100	7.0 to 9.0	2.3 to 2.8	0.7 to 0.9	> 10^{11}	6	10 to 20
Titanium carbide	TiC	4.9	–	–	320	7.4	30	–	$7 \cdot 10^{-5}$	–	–
Titanium nitride	TiN	5.4	–	–	260	9.4	40	–	$3 \cdot 10^{-5}$	–	–
Titanium dioxide	TiO_2	3.5 to 3.9	90 to 120	300	–	6.0 to 8.0	3 to 4	0.7 to 0.9	–	40 to 100	8
Zirconium dioxide partially stabilized, PSZ	ZrO_2 > 90 % rest Y_2O_3	5.7 to 6.0	500 to 1.000	1,800 to 2,100	140 to 210	9.0 to 11.0	2 to 3	0.4	10^8	–	–
Standards for test procedures		DIN EN 623 Part 2	DIN EN 843 Part 1	DIN EN 993 Part 5	DIN EN 843 Part 2	DIN EN 821 Part 1	DIN EN 821 Part 2	DIN EN 821 Part 3	DIN EN 6067 Parts 2 and 3		

The characteristic values for each material can vary widely, depending on the raw material, composition, and manufacturing process. The material data relate to the information provided by various manufacturers.

The designation "KER" corresponds to DIN EN 60-672-1.

Further detailed information on ceramic materials, property tables, applications, etc., can be found on the relevant internet pages, e.g. Informationszentrum Technische Keramik (center of information for technical ceramics), www.keramikverband.de or MatWeb Material Property Data, www.matweb.com

[1] Density. [2] Flexural strength. [3] Cold compressive strength. [4] Modulus of elasticity. [5] Coefficient of thermal expansion RT to 1,000 °C.
[6] Thermal conductivity at 20 °C. [7] Specific heat. [8] Electrical resistance at 20 °C and 50 Hz. [9] Relative permittivity.
[10] Dielectric loss factor at 25 °C and 10 MHz. [11] Properties greatly dependent on porosity set specifically for the application.

Plastics

Plastics are still a relatively "young" material group, the importance of which has continuously grown since the middle of the twentieth century. Products made of plastics play a large role in the various areas of our society today. Thus plastics are used as material in areas such as supply, medical and electrical technology, for packaging, household appliances and consumer goods.

Specifically in the automotive industry, plastics have been able to gain increased use more than any other material group (Figure 1). The automotive industry is often the driver behind technological innovations for plastic.

However, the "saturation limit" for using plastics is far from having been reached. We continue to find applications in which conventional materials such as metal are being replaced by plastics.

The increasing use of plastics is mostly due to their less expensive processing options compared to other materials and to the beneficial properties of this "made-to-measure material group". In addition, new material and process developments are opening new markets and simultaneously offering enormous potential for new, innovative products made of plastic. A strong increase in the plastic content is forecast particularly due to increasing electromobilization. The greatest driver is the savings in weight compared to metallic materials.

Plastics is the general umbrella term for polymer materials. A significant characteristic is their macromolecular structure.

Figure 1: Average percentage of individual material groups in a mid-class car [1]

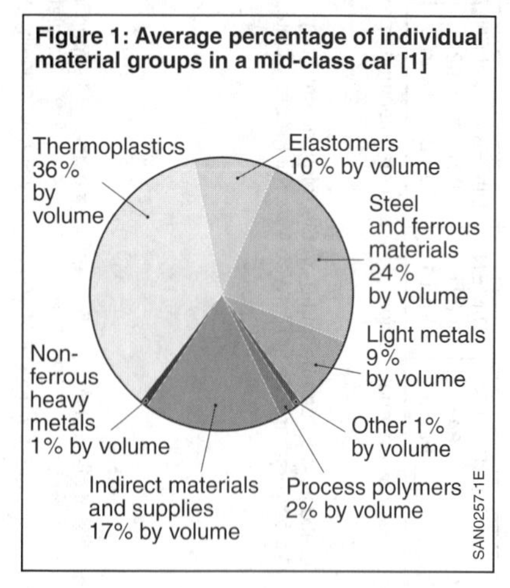

Figure 2: Classification of plastics based on [2]

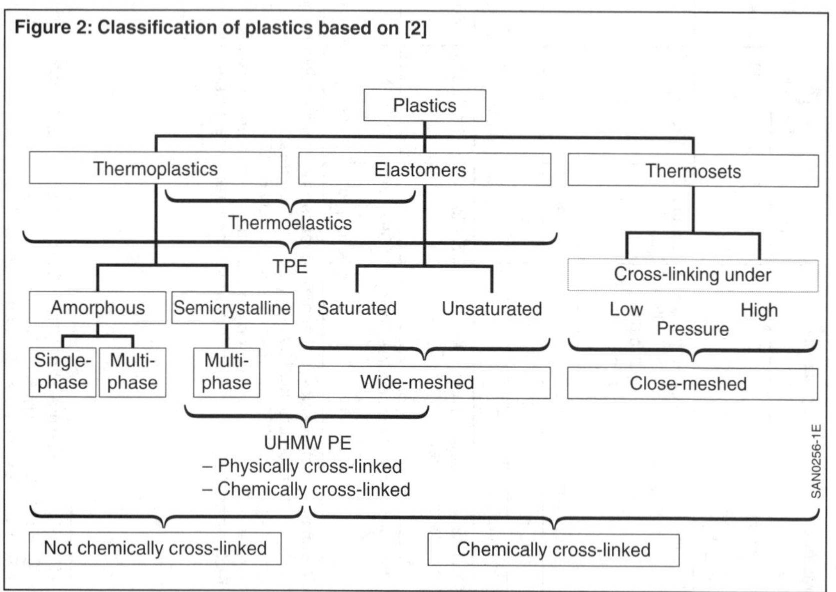

Plastics are classified into thermoplastics, thermosets, elastomers and thermoplastic elastomers (Figure 2), which are discussed in more detail in the following.

Thermoplastics
An outstanding characteristic of thermoplastics is the non-cross-linked structure between the macromolecules. This enables repeated plasticity or processability above their service temperature in the melting range. Only thermoplastics can be welded.

The material class of the thermoplastics can be further classified into amorphous thermoplastics and semicrystalline thermoplastics [3]. Semicrystalline thermoplastics are composed of amorphous and semicrystalline macromolecular structures. Unlike amorphous thermoplastics, therefore, they exist in multiple phases. Even amorphous thermoplastics can exist in multiple phases if they are modified during their synthesis into the form of a copolymer. The correlations are described in Figure 3 (based on [2]). The block copolymers existing in multiple phases form the link to the material class of thermoplastic elastomers.

The number of manufacturers and commercial types in the area of thermoplastics is very large. Most polymers come with or without different kinds and percentages of fillers and reinforcing materials. Therefore

Figure 3: Classification of thermoplastics based on [2]

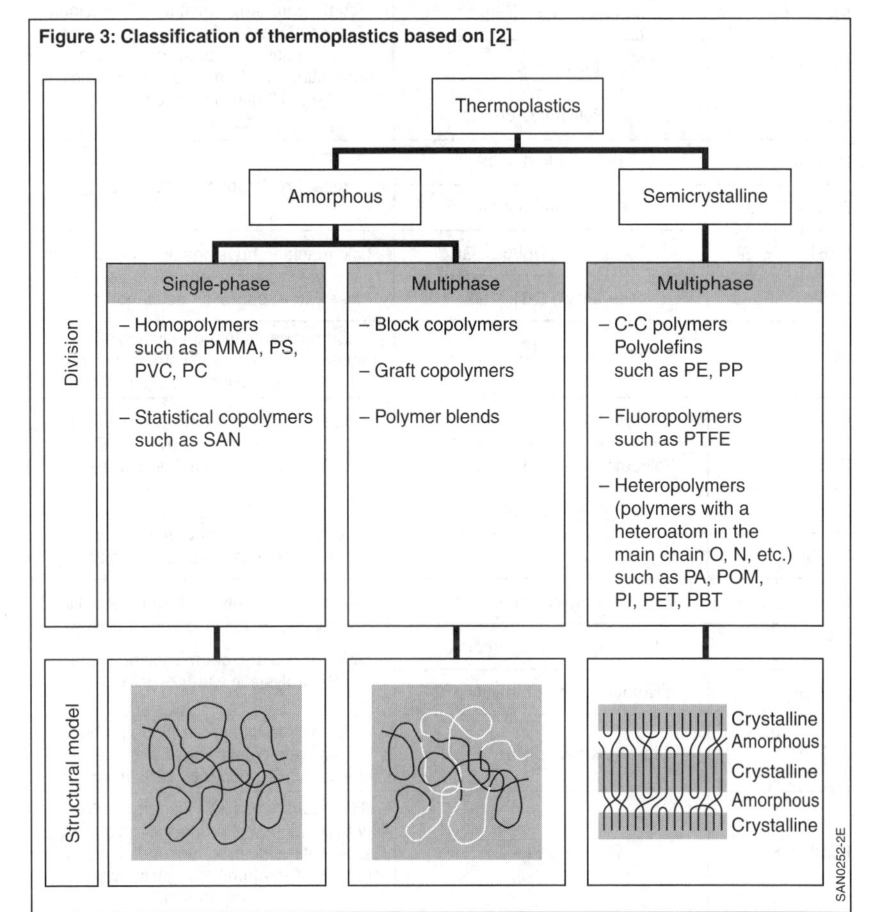

the range of properties of thermoplastics available on the market is very large. Table 1 provides an overview of names, codes and example applications of widespread thermoplastics.

Mechanical properties of thermoplastics
Compared to other construction materials, thermoplastics have exceptional viscoelastic and viscoplastic deformation behavior. Consequently, the mechanical behavior is strongly influenced by temperature, load exposure period and loading rate (Figure 4). Amorphous thermoplas-

Table 1: Chemical name and properties of thermoplastics

Code	Chemical name	Description of properties; application examples
ABS	Acrylnitrile-butadien-styrene	High gloss, some transparent types; impact-resistant housing parts
PA 11, 12	Polyamide 11, 12	Tough, hard, and resistant to abrasion, low coefficient of friction, good sound absorption, approx. 1 to 3 % water absorption required for good toughness; PA 11, 12 have much lower water absorption
PA6	Polyamide 6	
PA66	Polyamide 66	
PA6-GF	Polyamide 6 + GF	Impact-resistant machine housings
PA66-GF	Polyamide 66 + GF	
PA6T/6I/66-GF	Polyamide 6T/6I/66 +GF	Rigid machine housings and components even at high temperatures, low water absorption as standard PA
PA6/6T-GF	Polyamide 6/6T + GF	
PAI	Polyamid-imide	Components that are subjected to mechanical or electrical stress, favorable wear characteristics
PBT	Polybutylene terephthalate	Wear-resistant, chemically resistant, hydrolysis in water over 70 °C, very good electrical properties
PBT-GF	Polybutene terephthalate + GF	
PC	Polycarbonate	Tough and rigid over a broad temperature range, components have high rigidity
PC-GF	Polycarbonate + GF	
PE	Polyethylene	Acid-resistant containers and pipes, films
PET	Polyethylene terephthalate	Wear-resistant, chemically resistant, hydrolysis in water over 70 °C
PET-GF	Polyethylene terephthalate + GF	
LCP-GF	Liquid crystal polymers + GF	High dimensional stability under heat, low weld line strength, extremely thin-walled components, very anisotropic
PESU-GF	Polysulfone + GF	High continuous service temperature, low dependency of properties on temperature, not resistant to fuel and alcohol at higher temperatures, dimensionally stable components.

Table 1 (continued): Chemical name and properties of thermoplastic plastics

Code	Chemical name	Description of properties; application examples
PEEK-GF	Polyetheretherketone + GF	High-strength components for high temperatures, good antifriction properties, and low wear; very expensive.
PMMA	Polymethylmethacrylate	Transparent and in many colors, weatherproof
POM	Polyoxymethylene	Sensitive to acid-induced stress cracking; precision moldings
POM-GF	Polyoxymethylene + GF	
PPE+SB	Polyphenylene ether+SB	Resistant to hot water, flame retardant
PPS-GF	Polyphenylene sulfide + GF	High resistance to heat and media, inherently flame-retarding; underhood components
PP	Polypropylene	Household goods, battery cases, cover hoods, fan impellers (reinforced types)
PP-GF	Polypropylene + GF	
PS	Polystyrene	Transparent and in many colors
PSU-GF	Polysulfone + GF	Low dependence of properties on temperature, not resistant to fuel and alcohol.
PVC-P	Polyvinyl chloride (plasticized)	Artificial leather, flexible caps, cable insulation, tubes/hoses, seals
PVC-U	Polyvinyl chloride (unplasticized)	Weatherproof exterior parts, pipes
SAN	Styrene-acrylonitrile	Molded parts with good chemical resistance, also transparent
SB	Styrene-butadiene	Impact-resistant housing parts for many applications
SPS-GF	Syndiotactic polystyrene	Low-warpage, brittle, high mold temperatures required for processing
PI	Polyimide	High resistance to heat and radiation, can be processed only by compacting and sintering
PTFE	Polytetrafluorethylene	Strong dependence of rigidity on temperature, high resistance to heat, aging and chemicals, can be processed only by compacting and sintering

tics and semicrystalline thermoplastics behave differently. This is due to the different path of the shear modulus curve and the different transition areas associated with this. They are called glass transition temperature T_g and melting range (Figure 5, based on [2] and [3]). With thermoplastics, the combination of load and higher temperature leads to creep or relaxation due to the viscoelastic and viscoplastic material behavior. In Figure 6, the creep and relaxation behavior of thermoplastics is compared to the material behavior of metals [2].

Figure 4: Depiction of the effect of different test speeds and temperatures on the stress-strain behavior of thermoplastics
a) Dependency on the test speed
b) Dependency on the temperature

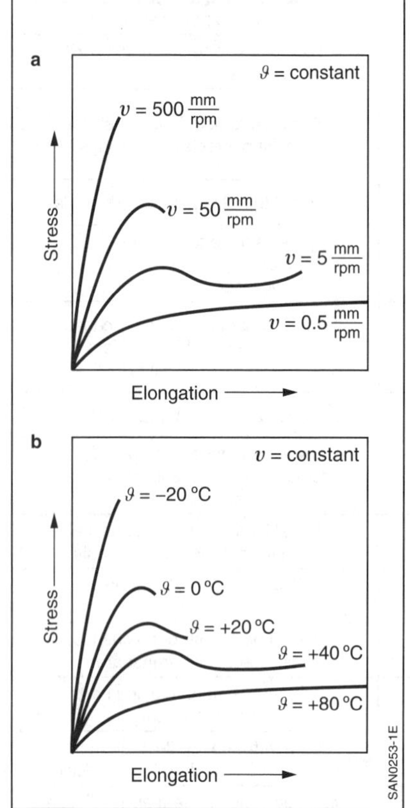

Table 2 and Table 3 show the mechanical properties of amorphous and semicrystalline materials. The values reproduce reference values from literature ([2] and [4]). The values may differ depending on the manufacturer and composition of filler or reinforcing material.

Chemical properties of thermoplastics
The chemical behavior of thermoplastics is determined by the structure of the macromolecules from which they are built. Polar plastics are attacked by polar solvents; non-polar plastics are attacked by non-polar solvents. Substances with a low molecular weight can migrate through solid thermoplastics (permeation).

Permeation of materials with a low molecular weight can trigger formation of stress cracks, much like with metals (referred to here as stress corrosion cracking). Suitable thermoplastics can be selected for a large number of application areas by selecting the right type of thermoplastic, appropriate shaping for the material and article, and optimum production parameters. Table 4 shows the chemical properties of some thermoplastics. The values are reference values from literature ([2], [4], [6]). Since the chemical behavior is difficult to estimate in some cases, consulting the plastics manufacturer or conducting your own measurements is advised.

Durability and aging of thermoplastics
Aging includes all irreversible changes to a material over time [7]. Aging processes change the properties of thermoplastics during a certain period of time. The causes of aging are distinguished between inner (such as residual stress, limited miscibility of additives) and external (energy input from heat and radiation, temperature change, mechanical loading, chemical influences). These causes lead to aging processes that manifest themselves in signs of aging and have a visible or measurable effect on thermoplastics. Examples of this are swelling, aftershrinkage, discoloration and alteration of the mechanical properties, such as brittleness.

Figure 5: Temperature dependence of the dynamic shear modulus
1 Amorphous thermoplastics
2 Semicrystalline thermoplastics
T_g Glass transition temperature
T_m Melt or flow temperature
T_s Crystalline melting temperature

Amorphous thermoplastics

Area I:
Glassy state,
Energy-elastic behavior;
Scope

Area IIa:
Entropy-elastic behavior (viscoplastic),
Hot working range

Area III:
Viscous flow behavior,
Primary forming and welding area

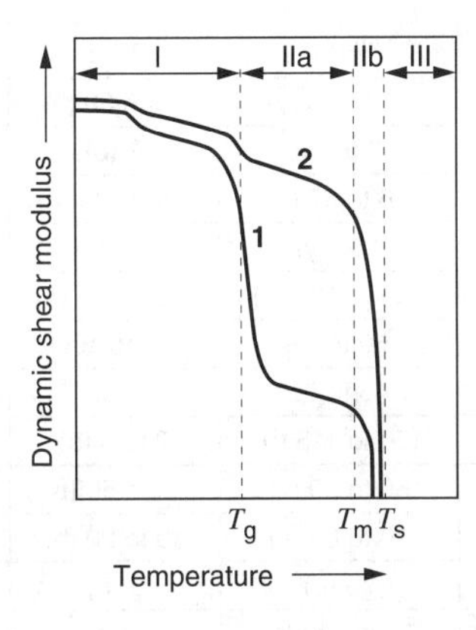

Semicrystalline thermoplastics

Area I:
Glassy state,
Energy-elastic behavior,
Amorphous areas frozen

Area IIa:
Amorphous parts Thermoelastic,
Semicrystalline parts rigid;
Scope

Area IIb:
Crystallites begin to fuse,
Hot working range

Area III:
Viscous flow behavior,
Primary forming and welding area

SAN0254-E

Figure 6: Comparison of the influence of time on the mechanical behavior of metals and thermoplastics
T_R Recrystallization temperature, T_g Glass transition temperature.

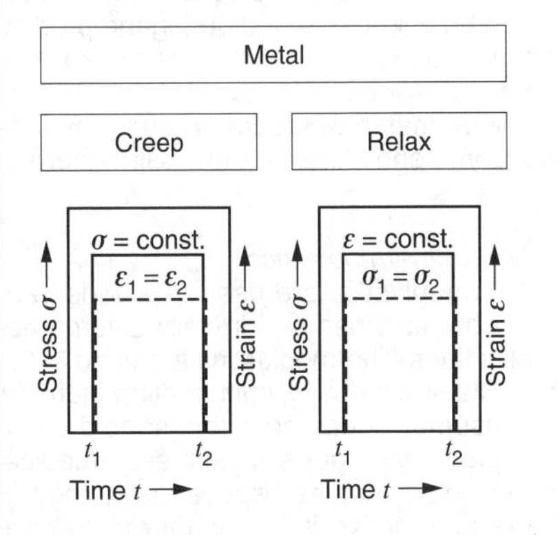

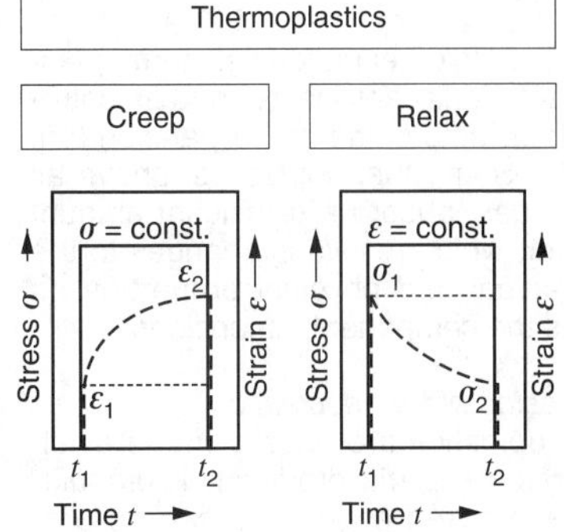

Influence of t is relatively low as long as $T < T_R$

Influence of t is to be noted particularly if $T > T_g$
The smaller the E-modulus,

SAN0255-1E

Table 2: Mechanical properties of amorphous thermoplastics

Code	E module [N/mm²]	Tensile strain at yield or tensile stress at break [N/mm²]	Tensile strain at yield or Elongation at fracture [%]	Glass transition temperature [°C]	Continuous service temperature [°C]
ABS	1,300 to 2,700	32 to 45 (y)	15 to 30 (b)	80 to 110	75 to 85
PAI-GF30	10,800	205 (b)	7 (b)	240 to 275	260
PC	2,100 to 2,400	56 to 67 (y)	100 to 130 (b)	150	130
PEI-GF30	9,000	160 (y)	3 (b)	215	170
PESU-GF20	5,700 to 7,500	105 to 130 (b)	2.5 to 3.2 (b)	220 to 225	160 to 200
PMMA	1,600 to 3,600	50 to 77	2 to 10	110	65 to 90
PS	3,200 to 3,250	45 to 65	3 to 4	95 to 100	60 to 80
PSU-GF20	6,200 to 7,000	100 to 115 (b)	2 to 3 (b)	180 to 190	160 to 180
PVC-E	2,000 to 3,000	50 to 60 (y)	10 to 50 (b)	80	65
PVC-S	2,000 to 3,000	50 to 60 (y)	10 to 20 (b)	85	65
SAN	3,600	75 (b)	5 (b)	110	85
PI	3,000* to 3,200*	75 to 100 (b)	n/s	250 to 270	260

* Flexural-E-modulus,
y Value at first maximum value during the tensile test,
b Value at which the specimen breaks,
n/s Not specified.

More physical properties of the thermoplastics

Thermoplastics have good insulation properties against electricity and heat. Compared to other materials, thermoplastics have significantly larger and anisotropic coefficients of thermal expansion ([2], [4], [5]). Some thermoplastics, above all polyamides, absorb a significant amount of water, which results in changes to the mechanical and physical properties as well as the component dimensions.

Processing of thermoplastics

Thermoplastics are usually made available by raw material producers in granular form as sacked goods and can be processed in commercially available injection molding machines and extruders. Since thermoplastics are deformable, they can also be processed using deep-draw or compression methods. Since thermoplastics are meltable and solidify again, they can be also be welded during the production process and can be recycled to a certain extent.

Raw material producers usually provide recommendations for processing thermoplastics.

Use of thermoplastics

Thermoplastics are used in a wide variety of industries and in very diverse applications. Thermoplastics are used in the packaging industry, construction industry, consumer goods area (household appliances, toys, sports equipment), medical technology and aerospace technology. In the automotive industry, thermoplastics are being used very successfully in the interior, exterior and powertrain areas.

Table 3: Mechanical properties of semicrystalline thermoplastics

Code	E module [N/mm²]	Tensile strain at yield or tensile stress at break [N/mm²]	Tensile strain at yield or Elongation at fracture [%]	Glass transition temperature [°C]	Continuous service temperature [°C]
PA 11	1,370 (k)	42 (k, y)	5 (k, y)	49 (tr)	70 to 80
PA 11-GF30	7,300 (k)	134 (k, b)	6 (k, b)	49 (tr)	70 to 80
PA 12	1,600 (tr) / 1,100 (k)	50 (tr, y) / 40 (k, y)	5 (tr, y) / 12 (k, y)	49 (tr)	70 to 80
PA 12-GF30	8,000 (tr) / 7,500 (k)	130 (tr, b) / 120 (k, b)	6 (tr, b) / 6 (k, b)	49 (tr)	70 to 80
PA 6	3,000 (tr) / 1,000 (k)	85 (tr, y) / 40 (k, y)	4.5 (tr, y) / 20 (k, y)	60 (tr)	80 to 100
PA 6-GF30	9,500 (tr) / 6,200 (k)	185 (tr, y) / 115 (k, y)	3.5 (tr, y) / 8 (k, y)	60 (tr)	100 to 130
PA 66	3,000 (tr) / 1,100 (k)	85 (tr, y) / 50 (k, y)	4.4 (tr, y) / 20 (k, y)	70 (tr)	80 to 120
PA 66-GF30	10,000 (tr) / 7,200 (k)	190 (tr, y) / 130 (k, y)	3 (tr, y) / 5 (k, y)	70 (tr)	100 to 130
PAEK-GF30	10,600	168 (b)	2.3 (b)	158	240 to 250
PBT	2,500	60 (y)	3.7 (y) / >50 (b)	60	100
PBT-GF30	10,000	135	2.5 (b)	60	150
PE-LD	200 to 500	8 to 23	300 to 1,000	–30	60 to 75
PET	2,800	80 (y)	4 (y) / 12 (b)	98	100
PET-GF35	14,000	150 (b)	1.5 (b)	98	100
PEEK-GF30	9,700	156 (b)	2 (b)	145	240
POM (H)	3,200	67 to 72 (y)	25–70 (b)	–60	90 to 110
POM (CoP)	2,800	65 to 70 (y)	25–70 (b)	–60	90 to 110
PPS-GF	14,700	195 (b)	1.9 (b)	85 to 95	200 to 240
PP	1,100 to 1,300	30 (y)	20–800 (b)	–10 to 0	100
PTFE	408	25 to 36 (y)	350–550 (b)	127	260

k Conditioned
tr Dry,
y Value at first maximum value during the tensile test
b Value at which the specimen breaks

Table 4: Chemical properties of thermoplastics

Code	Gasoline	Benzene	Diesel	Alcohol	Mineral oil	Brake fluid	Water cold/hot
ABS	+	–	+	+	+	–	+/+
PA 11	+	+	+	+	+	+	+/O
PA 12	+	+	+	+	+	n/s	+/O
PA6	+	+	+	+	+	+/O	+/O
PA6-GF30	+	+	+	+	+	+/O	+/O
PA66	+	+	+	+	+	+/O	+/O
PA66-GF30	+	+	+	+	+	+/O	+/O
PAI-GF	+	n/s	n/s	+	+	n/s	+/–
PBT	+	O	+	+	+	+	+/–
PBT-GF30	+	O	+	+	+	+	+/–
PC	+	–	O	O	+	n/s	+/O
PC-GF	+	–	O	O	+	n/s	+/O
PE	O	O	+	+	+	n/s	+/+
PET	+	+	+	+	+	n/s	+/–
PET-GF30	+	+	+	+	+	n/s	+/–
PESU-GF	+	–	+	O	+	–	+/O
PEEK-GF	+	n/s	n/s	+	n/s	n/s	+
PI	+	O	+	+	+	n/s	+/O
PMMA	+	O	+	–	+	n/s	+/+
POM (H)	+	+	+	+	+	n/s	+/+
POM-GF (H)	+	+	+	+	+	n/s	+/+
POM (CoP)	+	+	+	+	+	+	+
POM-GF (CoP)	+	+	+	+	+	+	+
PPS-GF40	+	+	+	+	+	+	+/+
PP	O	O	+	+	+	n/s	+/+
PP-GF30	O	O	+	+	+	n/s	+/+
PS	–	–	O	+	O	–	+/+
PSU	+	–	+	O	+	–	+/O
PTFE	+	+	+	+	+	n/s	+/+
PVC-P	–	–	O	–	O	n/s	+/O
PVC-U	+	–	+	+	+	n/s	+/O
SAN	–	–	O	+	+	–	+/+

+ Good resistance O Limited resistance – No resistance n/s Not specified

Thermoplastic elastomers
Thermoplastic elastomers (TPE) form their own plastic material class between thermoplastics and elastomers. They can be processed in a purely physical process that combines high shear forces, heat effects and subsequent cooling (for example, during injection molding or extrusion). Although no chemical cross-linking is required by a time-consuming and high-temperature vulcanization process such as with elastomers, the manufactured parts do in fact have rubber-elastic properties due to their special molecular structure. Renewed influence of heat and shear force causes the material to melt and deform. At the same time, however, this means that the thermoplastic elastomers can withstand far less thermal and dynamic loads than standard elastomers. The thermoplastic elastomers are not a "successor product" of conventional elastomers, but rather an interesting supplement, which combines the processing benefits of thermoplastics with the material properties of the elastomers ([2], [5], [8], [9], [10]).

Types of thermoplastic elastomers
"Thermoplastic elastomers" is the umbrella term for a whole series of different materials. They are always created through blends or block copolymers.

The blends are alloys made of a plastic matrix and a soft elastomer material. Block copolymers are molecular chains with different segments that congregate into hard and soft areas during cooling. Based on DIN EN ISO 18064 [11], the following subdivision can be made.

- TPO: Thermoplastic elastomers on an olefin basis, primarily PP/EPDM, such as Santoprene (Exxon Mobil).
- TPV: Cross-linked thermoplastic elastomers on an olefin basis, primarily PP/EPDM, such as Sarlink (DSM) and Forprene (SoFter).
- TPU: Thermoplastic elastomers on a urethane basis, such as Desmopan (Bayer) and Elastollan (BASF).
- TPC: Thermoplastic copolyester elastomers, such as Hytrel (DuPont) and Riteflex (Ticona).
- TPS: Styrene block copolymers (SBS, SEBS, SEPS, SEEPS and MBS), such as thermoplastic (Kraiburg TPE).
- TPA: Thermoplastic copolyamides, such as PEBAX (Arkema).

Properties of thermoplastic elastomers
The number of diverse commercial types with different properties on the market is very large. They, like thermoplastics, are usually made available by raw material producers in granular form as sacked goods.

Thermoplastic elastomers can be processed very well using injection molding and extrusion processes, since they pass through the plastic, molten state. They can be manufactured in all hardnesses from 5 Shore A to over 70 Shore D. The hardness and the compression set (DVR) are essential characteristics when using thermoplastic elastomers as sealants. Their thermal stability in particular is usually lower than that of elastomers. The maximum continuous service temperatures that can currently be attained for thermoplastic elastomers are approximately 150 °C.

Adhesion to nearly all technical thermoplastics can be achieved through modification. Their flowability as well as density, optics, scratch resistance and other properties can also be adjusted through compounding with a wide variety of fillers and additives.

Application of thermoplastic elastomers
Thermoplastic elastomers find a wide variety of applications in various industries while fulfilling the industry-standard requirements. Thus they are used in the automotive industry for control elements in the interior as well as for window trim in the exterior and for seals close to the engine. Furthermore, they are used in the industrial sector, for example, for tool handles or cable jackets. In the consumer sector, thermoplastic elastomers are found in toys, sports equipment, packaging, and personal care supplies such as toothbrushes and shavers. There are special compounds that satisfy the high requirements even for medical applications. They are used, for example, for drip chambers, seals and medical hoses.

Elastomers
Elastomers (or rubber compounds) are dimensionally stable, but elastically deformable plastics. The essential characteristic of these materials is that they can be stretched to at least twice their length. However, when the tensile load or compressive load is removed, they return to their initial state. This unique ability to reset is also referred to as rubber elasticity.

Elastomers are created from loose, wide-meshed and three-dimensional cross-linking of amorphous preliminary products (rubber). This loose fixation of polymer chains through chemical bonds leads to typical elastic behavior above the glass transition temperature T_g. This value usually lies significantly below 0 °C, and therefore below the service temperature of rubber compounds.

On the other hand, if the cross-linking is made solid and close-meshed, we speak of a thermoset. Cross-linked elastomers (just like thermosets) cannot be melted; that is, they decompose at high temperatures without melting.

The starting point for manufacturing elastomer materials is either natural or synthetic rubber. The rubber is mixed with various additives such as fillers, plasticizers, cross-linking chemicals, anti-aging agents and processing aids; then they are cross-linked under the influence of temperature.

During the cross-linking (vulcanization) a chemical reaction is underway, which usually takes place during the molding process under the influence of temperature (150 to 210 °C) and pressure. Not until after the vulcanization does the material have its rubber-elastic and mechanical properties, such as the desired hardness, tensile strength and elongation at break ([9], [12], [13], [14], [15]).

Classification of elastomers
Elastomers is an umbrella term for a whole material class whose individual materials can have widely differing properties. They can be classified into groups; based on the DIN ISO 1629 standard [16], the following subdivision can be made.
- R class: Rubbers with an unsaturated carbon chain, such as NR, NBR and SBR.
- M class: Rubbers with a saturated carbon chain, such as ACM, EPDM and FKM.
- O class: Rubbers with carbon and oxygen in the polymer chain, such as ECO.
- U class: Rubbers with carbon, oxygen and nitrogen in the polymer chain, such as polyurethane elastomers AU and EU.
- Q class: Rubbers with silicon and oxygen in the polymer chain, such as silicone elastomers VMQ.

Here, note that the ISO name (for example, NR for Natural Rubber) merely refers to the base rubber.

Properties of elastomers
General basic properties such as thermal and media resistance are already largely determined by the base rubber being used. These can be changed only within certain (very limited) boundaries. Yet within an elastomer class, such as NBR, there is a wide variety of different mixtures that can have significantly different specific properties such as hardness, strength behavior and elasticity. In addition, component suppliers often use their own recipes for their materials and the rubber compounds are often specially designed for the requirements of an application. Unlike thermoplastics or thermoplastic elastomers, non-cross-linked elas-

tomer mixtures are available on the free market only in a few cases and are not offered also in granular form as sacked goods or the like.

Tables 5 and 6 provide an overview of the names, application examples and some important properties of the most common kinds of elastomers. Please note: The data specified there can serve only as inexact guideline values and must be verified in each individual case of an intended application.

Use of elastomers
Elastomers have the property of being able both to absorb large deformations reversibly and to absorb mechanical energy. Ultimately, this results in diverse

Table 5: Name and application examples of elastomers

Code	Designation	Application examples
M-group		
ACM	Acrylate rubber	Oil circuit (for example, O-rings, radial shaft seal)
AEM	Ethylene acrylate rubber	Oil circuit, dampers, absorbers
EPDM	Ethylene-propylene-diene rubber	Coolant circuit, brake parts, body seals
FFKM	Perfluorinated rubber	Little use (special applications)
FKM	Fluorcarbon rubber	Standard for fuel applications (gas and diesel)
O-group		
ECO	Epichlorhydrin rubber	Fuel hoses, membranes
Q-group		
FVMQ	Fluorosilicone rubber	Membranes, fuel applications
VMQ	Silicone rubber	Turbocharger hoses, exhaust suspension, airbag coating
R-group		
CR	Chloroprene rubber	Bellows, grommets, windshield wipers, V-belts
HNBR	Hydrogenated (acrylo)nitrile-butadiene rubber	Seals in the engine compartment, hoses, drive belts
IIR	Isobutene-isoprene rubber (butyl rubber)	Gas-tight elastomer parts (inner layer of tires), dampers, membranes
NBR	(Acrylo)nitrile-butadiene rubber	Seals, dampers, absorbers, membranes, valves
NR	Natural rubber	(Truck) tires, engine suspensions, chassis mounts
SBR	Styrene butadiene rubber	Passenger car tires, brake parts
U-group		
AU / EU	Polyurethane rubber (polyester / polyether)	Gear wheels, wipers, damping elements (such as foamed material), interior

Table 6: Properties of elastomers

Code	Hardness range in Shore A	Service temperature (continuous) in °C	Resistance to					
			Weather and ozone	Oils (mineral oil, engine oil)	Gasoline	Diesel fuel	Water	Brake fluid (glycol-based)
M-group								
ACM	50 to 90	−25 to +150 [1]	1–2	1	3–4	3	4	4
AEM	50 to 90	−35 to +150	1	1–2	3–4	3	3	4
EPDM	30 to 95	−50 to +125 [2] −50 to +150 [3]	1	4	4	4	1	1
FFKM	60 to 90	−15 to +260	1	1	1	1	1	1
FKM	55 to 90	−20 to +200 [1]	1	1	1	1	2	4
O-group								
ECO	50 to 90	−40 to +120	1–2	2	2	2	3	4
Q-group								
FVMQ	30 to 80	−55 to +175	1	1	2	1–2	1–2	4
VMQ	20 to 80	−60 to +200	1	2–3	4	3	1–2	2–3
R-group								
CR	30 to 90	−40 to +110	2–3	2–3	3–4	3	2–3	3
HNBR	40 to 90	−30 to +130 [4]	2	1	2	1	1	2
IIR	40 to 85	−40 to +120	2–3	4	4	4	1	1
NBR	35 to 95	−30 to +100 [4]	3–4 [6]	1	2	1	1–2	3–4
NR	30 to 95	−55 to +80	4	4	4	4	1	1
SBR	30 to 95	−50 to +100	4	4	4	4	1	1
U-group								
AU / EU	50 to 98	−40 to +90 [5]	2	1–2	3	2–3	4	4

1 Very good resistance (no or little effect)
2 Good resistance (moderate effect)
3 Limited resistance (significant effect)
4 No resistance (unsuitable)

[1] Special materials with better cold resistance are possible
[2] Sulfur cross-linked
[3] Peroxide cross-linked
[4] Cold resistance, depending on the composition of the polymer
[5] Materials with better resistance to heat and hydrolysis are possible
[6] Materials with better ozone resistance are possible

applications for this material class. Thus rubber compounds are used to manufacture products that bridge tolerances, permit movement between various components, constitute static and dynamic seals, diminish vibrations and act as springs.

Typical technical applications include tires, seals, V-belts, hoses, flexible couplings, cable jackets, bearing elements, retaining elements, shock absorbers, windshield wipers, conveyor belts, roofing foils and shoe soles. But elastomers are also used to manufacture items for everyday use such as rubber boots, erasers, rubber bands, balloons, condoms, rubber gloves, pacifiers and wet suits.

Thermosets

Thermosets are the first plastics to have been manufactured for industrial production using a synthetic method. As early as 1910, Bakelite, a resin made of formaldehyde and phenol, came into use. In the 1920s and 1930s, molding resin masses with urea and melamine jointed it on the market. Polyester and epoxy resins were used for the first time after the end of World War II [17].

In general, thermosets refer to plastics with close-meshed molecular chains that are tightly cross-linked to each other. This results in the typical properties of thermosets:

- High strengths and rigidities with simultaneously low density
- High dimensional stability under heat and thermal resistance
- Good resistance to chemicals
- High brittleness

The varied types of thermosets find application in a wide variety of areas in the automotive industry. In addition to structural components, which will be covered in more detail below, this includes paint and adhesive systems, sealing compounds and substrates for interconnect devices in electronics.

The good thermal properties of thermosets make them particularly suitable for use in thermally stressed areas. This specifically relates to applications in the engine compartment of automobiles. Typical examples of components made of thermoset include water pump housings, belt pulleys and impeller wheels [18]. New developments in the area of tribologically modified PF molding compounds also open up further use options, such as for bearing elements, sliding elements and guide elements [19].

Due to the large number of thermosetting resin systems, the section below only covers the technically relevant materials.

Unsaturated polyester resins

Unsaturated polyester resins (UP resins) fully cure through radical polymerization under the influence of heat. They can be adjusted with almost no shrinkage, which leads to very good dimensional stability. This is why components with special requirements for precision, such as headlamp reflectors, are manufactured from these kinds of materials. Furthermore, components made of UP resins feature good electrical properties, which helped them find increased usage earlier in ignition systems of motor vehicles (ignition distributors).

Epoxy resins

Epoxy resins (EP resins) cure in a polymerization reaction and therefore, unlike the phenol-formaldehyde resins, do not split off any volatile reaction products. Compared to other thermosetting materials, EP resins have a particularly low viscosity, which simplifies the procedure for processing them. Due to the high material prices, however, use of epoxides is limited. EP resins are used, for example, for encapsulating electronic components (special low-pressure EP is used for this). Moreover, EP resin is increasingly being used in continuous-fiber-reinforced (mostly carbon fiber-reinforced) structural lightweight components as matrix materials in the automotive industry.

Table 7: Thermosets

Type	Type of resin	Filler	t_G [1] °C	σ_{bB} [2] min. N/mm²	α_n [3] min. kJ/m²	CTI [4] min. grade	Properties, application examples
Thermosets (new standards) – Pourable phenol molding compounds (PF–PMC) DIN EN ISO 14526 – Pourable melamine-formaldehyde molding compounds (MF–PMC) DIN EN ISO 14528 – Pourable melamine/phenolic molding compounds (MP–PMC) DIN EN ISO 14529 – Pourable unsaturated polyester molding compounds (UP–PMC) DIN EN ISO 14530 – Pourable epoxy-resin molding compounds (EP–PMC) DIN EN ISO 15252							
(WD30+MD20) to (WD40+MD10) (31 and 31.5) [7]	Phenol [8]	Wood flour	160/140	70	6	CTI 125	For parts subject to high electrical loads. [9]
(LF20+MD25) to (LF30+MD15) (51) [7]		Cellulose [5]	160/140	60	5	CTI 150	For parts with good insulating properties in low-voltage range. Type 74 impact-resistant. [9]
*SS40 to SS50 (74) [7]		Cotton fabric shreds [5]	160/140	60	12	CTI 150	
(LF20+MD25) to (LF40+MD05) (83) [7]		Cotton fibers [6]	160/140	60	5	CTI 150	Tougher than type 31. [9]
–		Glass fibers, short	220/180	200	12	CTI 125	High mechanical strength. Very good resistance to automotive fluids, low swelling.
–		Glass fibers, long	220/180	230	17	CTI 175	
–		Carbon fibers	220/180	250	14	–	High rigidity, low density, good wear properties. Not suitable for electrical applications (conductive).
(WD30+MD15) to (WD40+MD05) (150) [7]	Melamine	Wood flour	160/140	70	6	CTI 600	Resistant to glow heat, superior electrical properties, high shrinkage factor.

Table 7 (continued): Thermosets

Type	Type of resin	Filler	t_G [1) °C	σ_{bB} [2) min. N/mm²	α_n [3) min. kJ/m²	CTI [4) min. grade	Properties, application examples
Thermosets							
LD35 to LD45 (181) [7)	Melamine-phenol	Cellulose	160/140	80	7	CTI 250	For parts subject to electrical and mechanical loads.
(GF10 + MD60) to (GF20 + MD50) (802 and 804) [7)	Polyester	Glass fibers, inorganic fillers	220/170	55	4.5	CTI 600	Types 801, 804: low molding pressure required (large-area parts possible); types 803, 804: glow-heat resistant.
MD65 to MD75	Epoxy	Rock flour	200/170	80	5	CTI 600	Very good electrical properties. Sheathing sensors and actuators.
(GF25 + MD45) to (GF35 + MD35)		Glass fibers/ mineral	230/190	160	10	CTI 250	
–	Epoxy low-pressure molding compounds	SiO_2 (spherical)	250/200	120	6	–	Chip encapsulation (thin-bond wires).

1) Maximum service temperature, short term (100 h)/continuous (20,000 h). 2) Flexural strength. 3) Impact strength (Charpy).
4) Tracking resistance according to DIN IEC 112, Comparative Tracking Index (CTI).
5) With or without addition of other organic fillers. 6) And/or wood flour. 7) Old designation in parentheses.
8) Do not use types 13 to 83 (purely organically filled compounds) for new applications (availability no longer guaranteed).
9) Barely used any more in new applications. Supply not guaranteed in the long term.

Phenol-formaldehyde resins
Phenol-formaldehyde resins (PF resins) are manufactured from phenol and formaldehyde using polycondensation. They have high mechanical rigidities and strengths and feature high chemical resistance, thermal resistance and dimensional stability under heat. In addition, there are tribologically modified forms that, however, cost significantly more because carbon fibers are added. PF resins are inherently flame retardant, which makes them ideal for applications in engine compartments.

extenders
Filler are classified in accordance with DIN EN ISO 1043-2 [20]. A distinction is made here between material (such as carbon, glass, mineral), the form and structure (such as fibers, balls, powder) and special properties (such as flame retardant, thermal resistance) [4]. Technically relevant fillers primarily include glass fibers and glass balls as well as mineral fillers. Particularly in the aerospace industry, but also increasingly in the automotive industry, carbon fibers are also being used as reinforcement materials for thermosets.

Processing method
Thermosetting molding compounds are primarily processed using compacting or injection molding. There is also a wide variety of other processes in which the listed variants are combined or further developed.

Compacting
For compacting, pressure and temperature simultaneously act upon the molding compound. The mass is fed into the mold either without a form or in tablet form. Pressure brings the mass into the desired form, while the temperature cross-links the material. Due to the low machine-based effort, compacting is the least expensive processing method. However, this method can be used to manufacture only simple, large-format components. Above all, the manufactured components feature low level of orientation and the need for large amounts of finishing.

Injection molding
Here it is necessary to distinguish between easy-flowing granulate material and polyester bulk molding compound. Exceptionally high productivity is achieved when processing easy-flowing granulate material where the length of cycle times is largely dependent on component thickness. The barrel temperatures during injection molding are between 80 and 100 °C, while the mold temperatures range from 160 to 190 °C. While the injected mass is curing in the mold, new material for the next cycle is already being plasticized in the screw.

When processing polyester bulk molding compound (UP BMC), it is important to note that a stuffing unit is required in addition to the conventional machine equipment to convey the material into the screw. Then the material is homogenized in the injection barrel. Melting is not necessary; the barrel temperatures are usually around 25 to 35 °C.

Insulating materials
Electrical insulation plays a critical role in the proper functioning and service life of items such as alternators, engines and electrical devices, not just in motor vehicles.

Unfilled polymers have the best electrical insulation properties. Each addition of fillers reduces the dielectric strength of a polymer material by forming interfaces between the filler and polymer matrix as well as by excessive stress due to different dielectric properties.

The electrical insulation in motor vehicles not only has to reliably prevent disruptive electrical breakdowns, it also has to dissipate heat losses that arise, absorb mechanical loads and be resistant to liquids commonly used in motor vehicles. These properties typically have to be ensured in a temperature range from –40 to +180 °C. For this reason, a system of several materials is usually used as an insulating system; a single material is rarely used.

As an example here we will discuss surface insulators. These are usually used as more or less flexible combined insulation materials in electrical machines such as starters, alternators, hybrid engines and traction motors.

Plastic foils are flexible combined insulation materials that are bonded to pressboard, nonwoven materials or paper made of polymers. Flexible combined insulation materials often have a three-ply design with plastic foil as the middle layer. Depending on the combination of materials, they have different continuous service temperatures, tensile strengths and elongations at break, dielectric strengths, rigidities and impregnabilities.

The combination of plastic foils with fiber or nonwoven materials has technical advantages. Plastic foil made of unfilled polymers provides excellent electrical properties. The fiber or nonwoven materials, by contrast, have good impregnability and protect the foil against mechanical and thermal loads.

Foils made of polyester or polyimide are predominantly used as initial components as well as fiber or nonwoven materials made of organic fibers, polyester or aramide.

Flexible combined insulation materials are standardized as DIN EN 60626-1 ([21], definitions, general requirements), DIN EN 60626-2 ([21], test methods) and DIN EN 60626-3 ([21], properties of individual combinations of materials). In addition, there are standards for other surface insulators.

- Insulating foils: DIN EN 60674-1 ([22], definitions, general requirements), DIN EN 60674-2 ([22], test methods), DIN EN 60674-3-1 to -3-8 ([23], [24], properties of individual materials).
- Panel and roller pressboard: DIN EN 60641-1 ([25], definitions, general requirements), DIN EN 60641-2 ([26], test methods), DIN EN 60641-3-1, -3-2 ([26], properties of individual materials).
- Laminates: DIN EN 60893-1 ([27], definitions, general requirements), DIN EN 60893-2 ([27], test methods), DIN EN 60893-3-1 to -3-7 ([27], properties).

Sealing compounds

Sealing compounds (or cast resins) refer to reactive synthetic resins that are processed as a liquid for the finished product and solidify as this or a component part of it. While still a liquid, the resin is poured into a reusable or disposable mold. This results in either pure cast resin bodies with free form surfaces or other parts are included ([28], [29], [30], [31], [32]).

Unlike with meltable plastics (thermoplastics), the solidification occurs through a chemical cross-linking reaction and is irreversible (thermoset). They are usually poured in

- To cover and protect parts against penetration of moisture, dust, foreign matter, water, etc.
- To fix parts into place with respect to each other
- To increase the mechanical stability and the vibration and shock resistance
- To provide electrical insulation, that is, to increase the dielectric strength and contact protection
- To dissipate any heat losses that arise.

The range of applications for sealing compounds is broad and multi-faceted. As a result, the functions and requirements are also multi-faceted, ranging from processing and curing to the properties of the later application area. Selection criteria include:

- For application of liquid, not yet hardened sealing compound: viscosity, pot life, available systems engineering (manual, metering or molding system).
- The requirements that are placed on the cured molding compound in later use, such as hardness, elasticity, stretchability or pliability, thermal, mechanical, electrical and chemical loads of the molded components.

Due to their mechanical and electrical properties, sealing compounds are ideal for use in electrical engineering and electronics. Typical applications of sealing compounds include:
- Molding and manufacturing electrotechnical components (such as ignition coils, transformers, insulators, capacitors, semiconductors, assemblies)
- Molding of open contact points for cables and lines

Different material classes are used depending on the requirements profile (such as service temperatures, chemical resistance, mold geometry). The sealing compounds used most frequently in electronics potting belong to the epoxide, polyurethane and silicone material classes. Less widespread, for example, are polyester and hot melt sealing compounds (hot melts, hot melt adhesives).

Table 8: Properties of different sealing compounds (examples)

Properties	Standard	Unit	Epoxy resin, unfilled	Epoxy resin, filled (quartz powder)	Unsaturated polyester resins	Polyurethanes	Silicone
Tensile strength	ISO 527	N/mm^2	60 to 90	80 to 100	30 to 80	3 to 80	0.3 to 10
Elongation at break	ISO 527	%	3 to 8	0.8 to 1.1	2 to 4	0.5 to 80	50 to 700
Flexural strength	ISO 178	N/mm^2	60 to 140	110 to 120	60 to 140	40 to 140	n/a
Impact strength	ISO/R 179	kJ/m^2	15 to 30	10 to 12	8 to 15	40	n/a
Glass transition temperature T_G		°C	+70 to +200	+70 to +200	+70 to +150	–40 to +130	–120
E module		N/mm^2	2,000 to 4,000	5,000 to 8,000	3,500	250 to 3,000	0.005 to 5
Continuous temperature resistance	IEC 216	°C	110 to 200	110 to 200	120 to 140	90 to 140	150 to 250
Specific volume resistance		Ω cm	10^{14} to 10^{16}	10^{14} to 10^{16}	10^{13}	10^{13} to 10^{16}	10^{15} to 10^{17}
Exothermy during curing			High	Medium	Very high	Very low	Very low
Coefficient of thermal expansion		$K^{-1}\ 10^{-6}$	80 to 100	30 to 70	60 to 80	50 to 150	300
Shrinkage		%	0.5 to 2	0.1 to 1	3 to 9	0.2 to 1	0.1 to 2

\+ = Good resistance
O = Limited resistance
– = No resistance
n/a = Not applicable

There are also extensive means of using chemicals and formulas to adapt the sealing compounds to the typical requirements for an application:

- Variations in chemicals provide different resin and curing systems
- Variations in formulas provide fillers, pigments, accelerators, toughness modifiers, wetting agents, degassing agents and anti-sedimentation agents

The chemical resistance of sealing compounds depends on the components of the cast resin, the cross-link density and the degree of cross-linking. As a rule of thumb, hard sealing compounds are more resistant than soft ones.

The final properties of the cured sealing compounds largely depend on the kind and quantity of ingredients used (hardeners, diluents, fillers) and the curing conditions. Therefore generally applicable specifications about physical properties are not possible.

Epoxide systems

Epoxides have been widely used for many years. They are generally hard and capable of carrying loads, their volume shrinkage during curing is rather low.

Characteristics include their excellent mechanical properties, good temperature tolerance, good adhesive strength on a wide variety of surfaces, and likewise good chemical resistance. The cross-linking or curing process is generally slow, particularly if only small volumes react with each other. More reactive hardeners could be used, but that can cause a substantially stronger exothermic reaction, creating stress for components and the printed circuit board.

Fillers are an essential component of epoxy resin formulations. They reduce costs and improve mechanical properties. At the same time, they lead to an increase in the crack resistance and rigidity, as well as to a reduction in thermal shrinkage, lowering lower inner stresses. Quartz powder is a standard filler in electronic applications.

Epoxide systems are frequently used, for example, as a thermally conductive sealing compound for all kinds of solenoids.

Polyurethane systems

Even after curing, polyurethane sealing compounds can still expand and flex (even if only somewhat), which is particularly important when sensitive components (such as ferrite on printed circuit boards) are cast.

The curing reaction for polyurethane systems is less exothermic than for epoxide sealing compounds. The volume shrinkage after the hardening is low and a broad spectrum of hardness and/or elasticity is available. The chemical and mechanical resistances are good.

Silicone

Silicone-based sealing compounds are usually significantly more expensive than epoxide or polyurethane sealing compounds, but they are used where typical continuous service temperatures are above 180 °C. The exothermic heat buildup during the hardening is also only very low.

Polyester systems

These systems have very good resistances to a wide variety of media, but a very strong heat buildup during hardening and a high degree of shrinkage after the hardening as a result. Under certain circumstances this leads to thermomechanical damage to the cast components, even to the point of having components tear away from the substrate.

References for plastics

[1] M. Gehde, S. Englich, S. Hülder, M. Höer: Schlummerndes Potenzial für den Leichtbau [The dormant potential for lightweight construction]. In: Plastverarbeiter (10), p. 80-83, 2012

[2] P. Elsner, P. Eyerer, T. Hirth: Domininghaus – Kunststoffe: Eigenschaften und Anwendungen [Plastics: Properties and applications]. 8. Aufl., Springer, 2012.

[3] DIN 7724:1993: Polymere Werkstoffe – Gruppierung polymerer Werkstoffe aufgrund ihres mechanischen Verhaltens [Polymer materials – Grouping polymers by their mechanical characteristics].

[4] W. Hellerich, G. Harsch, S. Haenle: Werkstoff-Führer Kunststoffe – Eigenschaften, Prüfungen, Kennwerte [Plastics material guide – Properties, Testing and Characteristics]. 9. Aufl., Carl Hanser Verlag, 2004.

[5] E. Baur, S. Brinkmann, N. Rudolph, T.A. Osswald, E. Schmachtenberg: Saechtling Kunststoff Taschenbuch [Saechtling plastics pocketbook], 31. Aufl., Carl Hanser Verlag, 2013.

[6] G. Ehrenstein, S. Pongratz: Beständigkeit von Kunststoffen [The durability of plastics]. 1. Aufl., Carl Hanser Verlag, 2007.

[7] DIN 50035:2012: Begriffe auf dem Gebiet der Alterung von Materialien – Polymere Werkstoffe [Concepts in the field of material aging – Polymers].

[8] T. Dolansky, M. Gehringer, H. Neumeier: TPE-Fibel – Grundlagen, Spritzguss [TPE primer – Principles, casting]. 1. Aufl., Dr. Gupta Verlag, 2007.

[9] F. Röthemeyer, F. Sommer: Kautschuk Technologie: Werkstoffe – Verarbeitung – Produkte [Rubber technology: Materials – Processing – Products]. 3. Aufl., Carl Hanser Verlag, 2013.

[10] G. Holden, H. Kricheldorf, R. Quirk: Thermoplastic Elastomers [Thermoplastic elastomers]. 3. Aufl., Hanser Gardner Publ., 2004.

[11] DIN EN ISO 18064: Thermoplastic elastomers – Nomenclature and abbreviated terms.

[12] K. Nagdi: Gummi-Werkstoffe [Rubber materials]. 3. Aufl., Dr. Gupta Verlag, 2004.

[13] W. Hofmann, H. Gupta: Handbuch der Kautschuk-Technologie [Handbook of rubber technology]. Dr. Gupta Verlag, 2001.

[14] J. Schnetger: Lexikon Kautschuktechnik [Glossary of rubber technology]. Beuth, 2004.

[15] G. Abts: Einführung in die Kautschuktechnologie [Introduction to rubber technology]. Carl Hanser Verlag, 2007.

[16] DIN ISO 1629: Rubber and latices – Nomenclature.

[17] G.W. Becker, D. Braun (Hrsg.): Kunststoff Handbuch – Band 10: Duroplaste [Handbook of Plastics – Volume 10: Thermosets]. 2. Aufl., Hanser Verlag, 1988.

[18] E. Bittmann: Duroplaste kommen ins Rollen [Thermosets get rolling]. In: Kunststoffe 3/2003, A25-A27.

[19] E. Bittmann: Duroplaste [Thermosets]. In: Kunststoffe 10/2005, 168–172.

[20] DIN EN ISO 1043-2: Plastics – Symbols and abbreviated terms – Part 2: Fillers and reinforcing materials. [21] DIN EN 60626: Combined flexible materials for electrical insulation. Part 1: Definitions and general requirements.
Part 2: Methods of test.
Part 3: Specifications for individual materials.

[22] DIN EN 60674: Plastic films for electrical purposes. Part 1: Definitions and general requirements.
Part 2: Methods of test.

[23] DIN EN 60674-3-1: Plastic films for electrical purposes – Part 3: Specifications for individual materials.
Sheet 1: Biaxially oriented polypropylene (PP) film for capacitors.

[24] DIN EN 60674-3-2: Specification for plastic films for electrical purposes – Part 3: Specifications for individual materials.
Sheet 2: Requirements for balanced biaxially oriented polyethylene terephthalate (PET) films used for electrical insulation.
Sheet 3: Requirements for polycarbonate (PC) films used for electrical insulation.
Sheets 4 to 6: Requirements for polyimide films used for electrical insulation.
Sheet 7: Requirements for fluoroethylene-propylene (FEP) films used for electrical insulation.
Sheet 8: Balanced biaxially oriented polyethylene naphthalate (PEN) films used for electrical insulation.
[25] DIN EN 60641-1: Specification for pressboard and presspaper for electrical purposes.
Part 1: Definitions and general requirements.
[26] DIN EN 60641-2: Pressboard and presspaper for electrical purposes.
Part 2: Methods of test.
Part 3: Specifications for individual materials.
Sheet 1: Requirements for pressboard.
Sheet 2: Requirements for presspaper.
[27] DIN EN 60893-1: Insulating materials – Industrial rigid laminated sheets based on thermosetting resins for electrical purposes.
Part 1: Definitions, designations and general requirements.
Part 2: Methods of test.
Part 3-1: Specifications for individual materials – Types of industrial rigid laminated sheets.
Part 3-2: Specifications for individual materials – Requirements for rigid laminated sheets based on epoxy resins.
Part 3-3: Specifications for individual materials – Requirements for rigid laminated sheets based on melamine resins.
Part 3-4: Specifications for individual materials – Requirements for rigid laminated sheets based on phenol resins.
Part 3-5: Specifications for individual materials – Requirements for rigid laminated sheets based on polyester resins.
Part 3-6: Specifications for individual materials – Requirements for rigid laminated sheets based on silicone resins.
Part 3-7: Specifications for individual materials – Requirements for rigid laminated sheets based on polyimide resins.
[28] R. Stierli: Epoxid-Gieß- und Imprägnierharze für die Elektroindustrie [Expoxide casting and impregnating resins for the electrical industry]. In: Wilbrand Woebcken (Hrsg.): Duroplaste – Kunststoff-Handbuch [Thermosets – The plastics handbook], Band 10. 2.Aufl., Hanser Fachbuch, 1988.
[29] G. Oertel (Hrsg.): Polyurethane – Kunststoff-Handbuch [Polyurethane – The plastics handbook], Band 7. 3. Aufl., Hanser Fachbuch, 1993.
[30] Dr. Werner Hollstein, Huntsman Advanced Materials GmbH: Einführung in die Chemie der Epoxidharze und Formulierungskomponenten [Introduction to the chemistry of epoxide resins and formulation components].
[31] http://www.electrolube.com.
[32] Lackwerke Peters GmbH & Co. KG: Technische Informationen TI 15/2.

Heat treatment of metallic materials

Hardness

Hardness is a property of solid materials that defines the resistance of a materials against penetration by a harder solid body. In metallic materials the hardness is used to assess mechanical properties such as strength, machinability, malleability, and resistance to wear. DIN EN ISO 18265 [1] defines guidelines for converting hardness to tensile strength.

Hardness testing

Hardness testing is a nondestructive way of obtaining information about the mechanical properties of a material in a relatively short time.

The test data are generally derived from the size or depth of the deformation produced when a specified indentor is applied at a defined pressure. A distinction is made between static and dynamic testing. Static testing is based on measurement of the permanent impression left by the indentor. Conventional hardness tests include the Rockwell, Vickers and Brinell procedures. Figure 1 compares the fields of application of hardness testing based on these different procedures.

Dynamic testing monitors the rebound height of a test tool accelerated against the surface of the test specimen.

Another option for obtaining an index of surface hardness is to scratch the surface with a harder test tool and then measure the groove width.

Hardness-testing methods

Rockwell hardness
(DIN EN ISO 6508, [2])

This method is particularly suitable for fast, automated testing of metallic workpieces, but places specific demands on clamping the workpiece in the test equipment. It is unsuitable for workpieces which as a result of their geometry give in the test equipment (e.g. pipes).

This method involves a test tool (indentor) of defined size, shape (tapered or spherical) and material (steel, hard metal or diamond) being pressed in two stages into the test specimen. In this process, after the preliminary test force is applied, the additional test force is exerted for a defined period. Once the additional test force has been removed and with the preliminary test force retained, the Rockwell hardness HR is calculated from the remaining penetration depth h and two constant numerical values N and S according to the following equation:

$$HR = \frac{N\,h}{S}\,.$$

Figure 1: Comparison of hardness ranges for different test methods for unalloyed and low-alloy steels and cast steel
The figures at the range extremities indicate the hardness data for the respective method.
HV Vickers hardness,
HR Rockwell hardness,
HBW Brinell hardness.

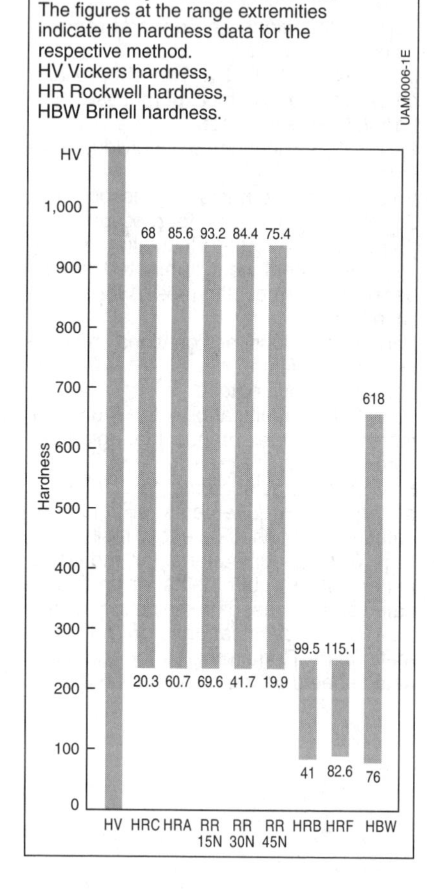

The numerical value N and the scale graduation S depend on the type of indentor and the test load.

The test specimen's surface should be smooth and as flat as possible. When testing on convex-cylindrical or spherical surfaces, the value determined must be corrected as a function of the hardness.

The abbreviation for the selected test method should be appended to the numerical value when specifying hardness (e.g.: 65 HRC, 76 HR45N). The designations provide an indication of the indentor used (diamond cone or ball), the preliminary test force and the total test force. Depending on the indentor used and the total test force applied, different hardness scales with the abbreviations HRA, HRB, HRC, HRD, HRE, HRF, HRG, HRH, HRK, HR15N, HR30N, HR45N, HR15T, HR30T and HR45T are used.

Advantages of the Rockwell test method include minimal specimen preparation and rapid measurement. This test process can also be fully automated. Any tester vibration, or shifting and movement of either the test specimen or the indentor can lead to testing errors, as can an uneven support surface or a damaged indentor.

Table 1: Application of the Brinell hardness test

Material	Brinell hardness	Load factor 0.102 F/D^2
Steel; nickel and titanium alloys		30
Cast iron (nominal diameter of ball must be 2.5, 5 or 10 mm)	< 140 ≥ 140	10 30
Copper and copper alloys	< 35	5
	35 to 200	10
	>200	30
Light metals and their alloys	< 35	2.5
	35 to 80	5 10 15
	> 80	10 15
Lead and tin		1
Sintered metals	see DIN EN ISO 4498 [5]	

Brinell hardness (DIN EN ISO 6506, [3])
This method is used for metallic materials of low to medium hardness. The test tool (indentor) is a hard-metal ball with the diameter D. It is pressed with test force F vertically into the surface of a specimen. After the test force has been removed, the Brinell hardness is calculated from the remaining indentation diameter d:

$$\text{HBW} = 0.102 \frac{2F}{\pi D^2 (1 - \sqrt{1 - d^2/D^2})},$$

with load F in N, ball diameter D in mm, and mean indentation diameter d in mm.

Test loads range from 9.81 to 29,420 N. Results obtained using balls of different diameters are only conditionally comparable, and any comparisons should be based on testing at identical force levels. Testing should always be performed using the largest possible ball, while load factors should be selected to obtain an indentation diameter of between 0.24 D and 0.6 D. Table 1 lists the recommended load factors and ball diameters for a variety of materials as laid down in DIN EN ISO 6506-1 [4].

In the Brinell-hardness designation, the numeric value is accompanied by the abbreviation for the method, the ball diameter in mm and the test force in N multiplied by a factor of 0.102 (e.g. 600 HBW 1/30).

High test loads producing deformations extending over a relatively wide surface area can be employed to gather data on materials with inconsistent structures. An advantage of the Brinell method is the relatively high degree of correlation between the Brinell hardness factor and the steel's tensile strength.

The required preparations and ensuing test procedures are more complex than those used for Rockwell testing.

Vickers hardness (DIN EN ISO 6507, [6])
This test method can be used for all metallic materials, regardless of hardness. It is especially suitable for testing minute and thin specimens, while the potential application range extends to include surface and case-hardened parts, as well as nitrided workpieces and parts carburized in nitrogen-based atmospheres.

The test tool is a square-based diamond pyramid with an apical angle of 136°. This is applied to the surface of the test specimen at an individually specified force F. The diagonals d_1 and d_2 of the indentation remaining on the test surface after test force F has been removed are measured. The Vickers hardness is obtained from the arithmetic mean value d of the two diagonals:

$$\mathrm{HV} = \frac{2F \sin \frac{136}{2}}{d^2} \approx 0.1891 \frac{F}{d^2},$$

with test force F in N and arithmetic mean value d of diagonal lengths d_1 and d_2 in mm.

In the formula for Vickers hardness data, the actual test figure is accompanied by the abbreviation HV, the force in N (multiplied by a factor of 0.102) and, following a slash, the force application period (if other than the standard 15 seconds) in seconds (e.g.: 750 HV 10/25).

The test specimen's surface should be smooth and flat. DIN EN ISO 6507 [6] stipulates that correction factors be used to compensate for any error stemming from surface curvature. Test-load levels are selected with reference to either the thickness of the workpiece to be tested or the layer to be tested.

A major advantage of this test method is that there are virtually no limitations on using it to assess thin parts or layers. It allows the use of extremely small force levels to determine the hardness of individual structural sections. The Brinell and Vickers numbers correlate well up to approximately 350 HV. However, a certain minimal degree of surface consistency is necessary to ensure accurate results.

Knoop hardness (DIN EN ISO 4545, [7])
This process closely resembles the Vickers method. The equal-sided diamond tip in the Vickers test has a rhombic shape in the Knoop test. The test tool is designed to leave an impression in the form of a thin, elongated rhombus. The long diagonal d of the indentation remaining on the test surface after test force F has been removed is measured. The hardness value HK is calculated as follows:

$$\mathrm{HK} = 1.451 \frac{F}{d^2},$$

with test force F in N and length d of the long diagonal in mm.

In the Knoop-hardness designation, the numeric value is accompanied by the abbreviation HK for the method, the test force in N multiplied by a factor of 0.102 and – if necessary separated by a slash – the application period of the test force in s (e.g. 640 HK 0.1/20).

The penetration depth is roughly 1/3 less than in the Vickers method, allowing evaluation of surface hardness in thin parts and layers. However, the test method also places heightened demands on the surface: The test must be performed on a polished, smooth and flat surface.

Knoop testing is often used on brittle materials such as, for example, ceramic or sintered materials.

Shore hardness (DIN 53505, [8])
This method is primarily used for hardness testing on elastomers and soft plastics. The test tool is a steel pin of 1.25 mm diameter; this is forced against the surface of the test specimen by a spring. The penetration depth of the steel pin into the material to be tested is a measure of Shore hardness. This is measured on a scale of 0 Shore (2.5 mm penetration depth) through 100 Shore (0 mm penetration depth).

In the Shore A method, the test tool has the shape of a truncated cone. The end face of the truncated cone has a diameter of 0.79 mm with an opening angle of 35°. In the Shore D method, measurement is performed with a needle which has an angle of 30° and a spherical tip with a radius of 0.1 mm.

Ball impression hardness
(DIN ISO 2039, [9])
This is the standard test for determining hardness levels in plastomers, and is also employed with hard rubber substances. The test tool is a hardened steel ball 5 mm in diameter. It is pressed onto the surface of the test specimen with a preliminary load of 9.8 N. Subsequent graduated rises in force provide application loads of 49, 132, 358 and 961 N. After 30 s, the penetration depth is measured. The test force F should be selected to provide a penetration depth of between 0.15 and 0.35 mm. From this, a reduced test force and finally the ball impression hardness HB in N/mm² are calculated or read off from a table.

Martens hardness
(DIN EN ISO 14577, [10])
In this method, the penetration of a pyramid-shaped test tool into materials is recorded, in the course of which both the force and the travel during plastic and elastic deformation are measured. Martens hardness is defined as the ratio of test force F to surface A_S of the indentor calculated from penetration depth h, and given in N/mm².

Martens hardness is denoted by the abbreviation HM, to which is appended the test force in N, the application time of the test force in s and the holding time of the test force in s (e.g.: HM 0.5/20/20 ≙ 8,700 N/mm²).

Scleroscope hardness
This dynamic measurement method is specially designed for heavy and large metal pieces. This process is based on the measurement of the rebound height of a steel indentor (hammer) featuring a diamond or hard metal tip. This is dropped from a stipulated height onto the surface of the test specimen. The rebound serves as the basis for determining the hardness.

The method is not standardized and there is no direct correlative relationship with any other hardness testing method.

Heat-treatment processes

Heat treatment is employed to adapt the technological material properties of metallic components and tools to the relevant requirements. Such requirements can be processing properties to suit manufacturing and usage properties to suit function.

According to DIN EN 10052 [11], heat treatment involves "subjecting a workpiece in whole or in part to time and temperature cycles in order to bring about a change in its properties of its structure. If necessary, the chemical composition of the material can be changed during the treatment".

The process modifies the microstructure to achieve the hardness, strength, ductility, wear resistance, etc. required to withstand the stresses associated with static and dynamic loads. The most significant industrial processes are summarized in Table 2 (see DIN EN 10052, [11] for terminology).

Hardening
Hardening procedures produce a martensitic microstructure of extreme hardness and strength in ferrous materials such as steel and cast iron. These procedures consist of the individual stages known as austenitizing and cooling or quenching.

Through hardening
The workpiece being treated is heated to the austenitizing, or hardening temperature (Table 3), at which it is maintained until an austenitic structure emerges, and until an adequate quantity of carbon (released in the decay of carbides, such as graphite in cast iron) is dissolved in the treated material. After austenitizing, the workpiece is cooled or quenched at a rate sufficient for hardening, which can also occur in temperature stages. A complete conversion to the martensitic microstructure should be completed as quickly as possible. The necessary cooling process is determined by the chemical composition, the austenitizing conditions, the shape and dimensions, and the desired microstructure. Reference values for the required cooling rate can be found in the time-temperature transformation chart for the steel in question.

Table 2: Overview of heat-treatment processes

Hardening	Austempering	Draw tempering	Thermo-chemical treatment	Annealing	Precipitation hardening
Through hardening Cross-section Surface Hardening Hardening of carburized parts (case-hardening)	Isothermic trans-formation in the bainite stage	Tempering and hardening of hardened parts Tempering and hardening above 540 °C Quench and draw	Carburizing Carbo-nitriding Nitriding Nitro-carburizing Boron treatment Chromating	Stress-free annealing Recrystal-lization annealing Soft annealing, spheroidi-zation Normalizing Annealing Homoge-nizing Annealing	Solution treatment and aging

The austenitizing temperature varies according to the composition of the material in question (for specific data, consult the DIN Technical Requirements for Steels). Table 3 above provides reference data. See DIN 17022 Part 1 [12] and Part 2 [13] for practical information on hardening procedures for tools and components.

Not all types of steel and cast iron are suitable for hardening. The following equation describes the hardening potential for alloyed and unalloyed steels with mass carbon contents of between 0.15 and 0.60 %, and can be applied to estimate the hardness levels achievable with a completely martensitic microstructure:

Table 3: Standard austenitizing temperatures

Type of steel	Quality specifica-tion [14, 15, 16, 17, 18]	Austenitiz-ing temperature in °C
Unalloyed and low-alloy steels	DIN EN 10083-1 10083-2 10083-3	
< 0.8 % by mass of C	10085	780 to 950
≥ 0.8 % by mass of C	–	750 to 820
Cold- and hot-working tool steels	DIN EN ISO 4957	780 to 1,150
High-speed tool steels		1,150 to 1,300

$$\text{Max. hardness} = (35 + 50x \pm 2)\,\text{HRC} \quad \text{(eq. 1)},$$

where the carbon content in % by mass is to be applied for x here. If the microstructure does not consist entirely of martensite, then the maximum hardness will not be achieved.

When the carbon content exceeds 0.6 % by mass, it may be assumed that the material's structure contains untransformed austenite (residual austenite) in addition to the martensite. Greater amounts of residual austenite can have a deleterious effect on achievable hardness and reduce wear resistance. In addition, residual austenite is metastable, i.e. there exists a potential for subsequent transformation to martensite at temperatures below room temperature or under stress. This can give rise to unwanted changes in the workpiece's dimensions and shape. Low-temperature follow-up procedures or draw-tempering operations at over 230 °C can be useful in cases where residual austenite is an unavoidable product of the hardening procedure.

Temperature differences occur between the edge and core of the workpieces during cooling and quenching. With greater cross-sections, the reduced cooling rate in the core can give rise to a decrease in hardness as the distance to the surface increases. There is a hardness progression or gradient. The hardness gradient is obtained from the material composition and the hardenability dependent on the austenitizing conditions (testing described in DIN EN ISO 642, [19]). In this case, it is necessary with regard to the required hardness to use material with sufficient hardenability. Information on choice of steel based on hardenability can be found in DIN 17021 [20].

DIN EN ISO 18265 [21] defines the method for using hardness as the basis for estimating tensile strength R_m. This method can only be applied in cases whether the surface and core hardnesses are virtually identical.

During hardening, the transformation of the microstructure to the martensitic state is combined with an increase in volume. With regard to the initial state, the volume increases by approximately 1 %. This equates to a change in length of approximately 0.3 %.

The changes in volume associated with rearrangements of the structure and thermal gradients during cooling give rise to stresses, which in turn result in distortion in the form of changes in dimension and shape. The stresses that remain in the workpiece after hardening are called internal stresses. The edge of a hardened workpiece tends to be subject to internal tensile stress, while the core tends to be subject to internal compressive stress.

Table 4: Comparison of power densities when heating with different sources

Energy source	Normal power density in W/cm^2
Laser beam	10^3 to 10^4
Electron beam	10^3 to 10^4
Induction (with medium- or high-frequency alternating current, or high-frequency pulses)	10^3 to 10^4
Flame heating	$1 \cdot 10^3$ to $6 \cdot 10^3$
Plasma beam	10^4
Molten saline solution (convection)	20
Air, gas (convection)	0.5

Surface hardening

This process is especially suited for integration within large-scale manufacturing operations, and can be adapted to fit the rhythm of the production line.

Heating and hardening are restricted to the surface, thereby minimizing alterations in shape and dimensions. Heating is generally provided by high- or medium-frequency alternating current (induction hardening) or by a gas burner (flame hardening). Friction (friction hardening) and high-energy beams (e.g. electron or laser beams) can also provide the heat required for austenitizing. Table 4 provides an overview of the specific heat energies for the individual procedures.

These methods can be used to treat both linear and flat surfaces, meaning that the parts can be heated either while stationary or in motion. The heat source itself can also be moved. Rotation is the best way of dealing with radially symmetrical parts, as it ensures concentric hardening. Either immersion or spraying arrangements can be applied for quenching. Information on performing surface hardening can be found in DIN 17022-5 [22].

Heat rise is rapid, so the temperatures must be 50 to 100 °C higher than those used in furnace heating so as to compensate for the shorter dwell period. The procedure is generally employed with low-alloy or unalloyed steels with mass carbon contents of 0.35 to 0.60 %. However, surface hardening processes can also be applied with alloyed steels, cast iron and rolling-bearing steels. The parts can be heat-treated to provide a combination of improved base strength and high surface hardness, making them suitable for high-stress applications (recessed edges, bearing surfaces, cross-sectional transitions).

Surface hardening generally results in internal compression stresses along the edge. This leads to increased fatigue resistance, especially when notched parts are exposed to inconstant vibration stress. The stress in Figure 2 corresponds to bending stress. The higher stressability results from the fact that the stress state (resulting stress) is reduced from the superposition of bending stress and internal stress.

The relationship defined in equation 1 can be employed to estimate the potential surface hardness. There is a substantial reduction in hardness between the surface and the unhardened core region. The hardening depth DS – the depth at which 80 % of the minimum Vickers surface hardness is found – can be derived from the hardness progression curve (see DIN EN 10328, [23]).

Austempering

The object of this process is to achieve a bainite microstructure. This microstructure is not as hard as martensite, but does display greater ductility, as well as smaller changes in specific volume.

After austenitizing (see hardening), the parts for austempering are first cooled to a temperature of 200 to 350 °C (depending upon the exact composition of the material) at the required rate. The parts are then held at this temperature until the microstructure's transformation into bainite has been completed. Cooling to the transformation temperature is usually performed in molten saline solutions (a typical salt is a mixture of potassium nitrate and sodium nitrite). The parts can then be cooled to room temperature (no special procedure required).

Figure 2: Cyclically alternating stress according to surface-layer hardening
$+\sigma$ Tensile, $-\sigma$ Compressive;
1 Case layer, 2 Bending stress,
3 Reduction of tensile stress,
4 Resulting stress, 5 Internal stress,
6 Increase in compressive stress.

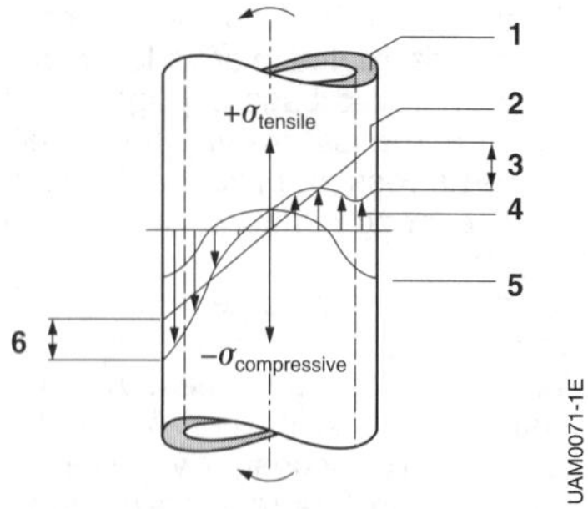

Austempering is an excellent alternative for parts whose geometrical configuration makes them sensitive to distortion or cracks, or in which high ductility is required together with substantial hardness, or which should combine hardness with a low level of residual austenite.

Application

Cylinder heads in modern diesel high-pressure pumps for common-rail systems which must be able to withstand high wear and internal compressive stresses are austempered.

Draw tempering

Draw tempering of hardened components and tools is employed to increase their deformability and to reduce their risk of cracking. According to DIN EN 10052 [11], draw tempering involves heating the part in question once or repeatedly to tempering temperature, holding the part at this temperature, and then allowing it to cool appropriately. Draw tempering is performed between room temperature and Ac_1 temperature, i.e. the temperature at which austenitic structural constituents would be created.

Tempering at temperatures as low as 180 °C is enough to reduce the hardness of unalloyed and low-alloy steels by approx. 1 to 5 HRC. The individual materials respond to higher temperatures with specific characteristic hardness loss. Figure 3 shows a characteristic tempering curve for typical types of steel. This graph illustrates the fact that the hardness of steels which are alloyed with special carbide-forming elements (Mo, V, W) – such as, for example, hot-working or high-speed tool steel – is increased by tempering in a temperature range between 400 and 600 °C to values which can be above the quenching hardness (secondary hardening).

The mutual relationships between tempering temperature on the one side, and hardness, strength, yield point, fracture contraction and elongation at fracture on the other, can be taken from the tempering diagrams for the various steels (see e.g. DIN EN 17021, [24]).

Generally speaking, draw tempering reduces hardness and strength and increases deformability. Internal stresses can also be reduced at tempering temperatures in excess of 300 °C.

The specific volume decreases when structures which are free of residual austenite are tempered. In the case of structures containing residual austenite, however, an increase in volume occurs during the transformation from residual austenite to martensite. Hardness increases, deformability is reduced, and new internal stresses can be created. The risk of cracking also increases.

It must be remembered that steels alloyed with manganese, chromium and nickel, or combinations of these elements, should not be tempered at temperatures of 350 to 550 °C, as brittleness could result. When these types of materials are cooled from tempering temperatures above 550 °C, the transition through this critical range should also be effected as rapidly as possible (see DIN 17022 Parts 1 and 2 [12, 13] for additional information). This tempering sensitivity can be avoided by adding molybdenum or tungsten by alloying.

Figure 3: Tempering response of various types of steel
1 Unalloyed tempering steel (C45),
2 Unalloyed cold-working tool steel (C80W2),
3 Low-alloy cold-working tool steel (105WCr6),
4 Alloyed cold-working tool steel (X165CrV12),
5 Hot-working tool steel (X40CrMoV51),
6 High-speed tool steel (HS6-5-2).

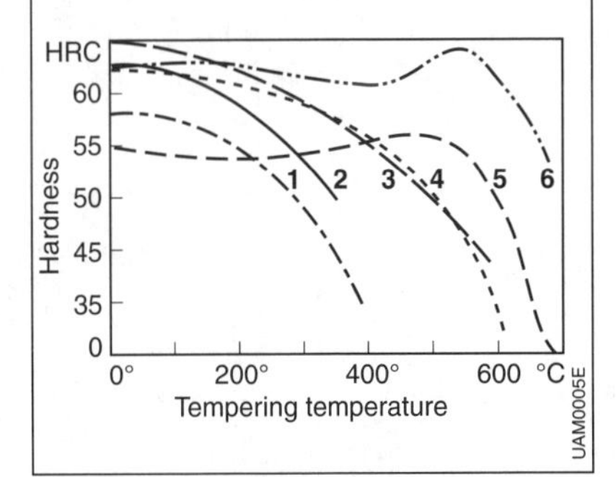

Quenching and drawing
Quenching and drawing involves a combination of hardening and tempering at a temperature which is generally between 540 °C and 680 °C. This procedure is designed to achieve an optimal relationship between strength and ductility. It is applied in cases where extreme ductility or malleability is required.

Particular care must be devoted to avoiding brittleness in the quench and draw operation (see above).

Annealing
Annealing can be applied to optimize certain operational or processing characteristics of parts. With this method, the parts are heated to a specific temperature and maintained there for an adequate period before being cooled to room temperature. The most important annealing processes in terms of technology are described in the following.

Stress-free annealing
Depending on the precise composition of the parts, stress-free annealing is carried out at temperatures ranging from 450 to 650 °C. The specific object is to achieve a reduction in internal stress in components, tools and castings.

Recrystallization annealing
Recrystallization annealing is applied with cold-form parts. The goal is to restructure the grain pattern in order to prevent increased hardening, thereby facilitating subsequent machining work.

The temperature requirement depends upon the composition of the material and the degree of deformation: it lies between 550 and 730 °C for steel.

Soft annealing and spheroidization
The purpose of soft annealing is to improve the machinability of material states which are difficult to cut or difficult to cold-form. The process involves heating the material in question to temperatures in excess of 600 °C, as briefly as possible above the Ac_1 temperature for the relevant steel, holding at this temperature, and slowly cooling to room temperature.

The temperature requirement is determined by the material's composition. It ranges from 650 to 850 °C for steel, and lower for nonferrous materials.

Spheroidization of cementite is applied when a microstructure with a granular carbide pattern is desired. The cementite loses strength and can pursue its striving for a body with as small a surface as possible (of the ball). If the initial structure is martensite or bainite, the result will be an especially homogeneous carbide distribution.

Normalizing

Normalizing is carried out by heating the parts to austenitizing temperature and then allowing them to gradually cool to room temperature. In low-alloy and unalloyed steels, the result is a structure consisting of ferrite and perlite. This process is essentially employed to reduce grain size, reduce the formation of coarse grain patterns in parts with limited reshaping, and to provide maximum homogeneity in the distribution of ferrite and perlite.

Precipitation hardening

This process combines solution treatment with aging at ambient temperature. The parts are heated and then maintained at a temperature to bring precipitated structural constituents into a solid solution, and quenched at room temperature to form a supersaturated solution. The aging process comprises one or several cycles in which the material is heated and held at above-ambient temperatures ("hot aging"). In this process, one or several phases, i.e. metallic bonds between certain base alloys, are formed and precipitated in the matrix.The precipitated particles enhance the hardness and strength of the base microstructure. The actual characteristics are determined by the temperature and duration of the aging process (option of mutual substitution). Exceeding a certain maximum will usually reduce the strength and hardness of the final product.

Precipitation hardening is mostly applied for nonferrous alloys, but some hardenable steels (maraging steels) can also be processed.

Application

Steels which can be precipitation-hardened are used for example in rail-pressure sensors in common-rail systems.

Thermochemical treatment

In thermochemical treatment, the exchange of substances with suitable agents effects a change in the chemical composition of the base material. Specific function properties can be adapted by the diffusion of specific elements into the surface layer. Of particular importance for this process are the elements carbon, nitrogen and boron.

Carburizing, carbonitriding and case hardening

Carburizing increases the carbon content in the surface layer, while carbonitriding supplements the carbon enrichment with nitrogen. This process usually takes place at temperatures ranging from 750 to 1,050 °C in gases which give off carbon or nitrogen as a result of their disintegration by heat or excited in the plasma. The actual hardening is performed subsequently, either by quenching directly from the carburizing or carbonitriding temperature (direct hardening), or by allowing the parts to cool to room temperature (single hardening), or by allowing them to cool to a suitable intermediate temperature (e.g. 620 °C) prior to reheating (hardening after isothermic conversion). This process produces a martensitic surface layer, while the degree of martensite at the core is a function of hardening temperature, hardenability and part thickness.

Specific temperatures can be selected for either surface hardening in the upper layers with higher carbon content (case refining), or for the non-carburized core (core refining) (see DIN 17022 Part 3, [25]). Carburizing and carbonitriding produce a characteristic carbon declivity, with levels dropping as the distance from the surface increases (carbon curve). The distance between the surface and the point at which the mass carbon content is still 0.35 % is normally defined as the carburization depth.

The length of the carburizing or carbonitriding process depends upon the required carburization depth, the temperature and the atmosphere's carbon-diffusion properties.

Typical surface carbon contents during case hardening range between 0.5 and 0.85 by mass. The carbon concentration essentially determines the surface hardness.

Generally, the objective is to achieve a carbon gradient with a concentration at the surface of 0.5 to 0.85 % mass carbon content in order to achieve sufficient surface hardness. Excessive concentrations of carbon can lead to residual austenite or carbide diffusion, which could have negative effects on the performance of case-hardened parts in actual use. Control of the atmosphere's carbon level is thus extremely important in process management.

The most common carburizing processes used today are gas carburizing and vacuum carburizing. In the case of gas carburizing, the carbon level of the furnace atmosphere is regulated in such a way that the surface layer of the workpiece assumes the desired carbon content. A state of equilibrium is established with the surrounding furnace atmosphere. In the case of vacuum carburizing, on the other hand, the carbon content cannot be regulated. Here, the defined surface carbon content is adjusted by means of multistage carburizing. In a first stage, carburizing is performed to a very high surface carbon content in the range of material saturation. In the subsequent stage, this high carbon content is reduced by diffusion to the desired level. In practice, vacuum-carburizing processes consist of several consecutive carburizing and diffusion stages.

The gradient defining the relationship between hardness and depth corresponds to the carbon concentration curve. The case-hardening depth CHD can be taken from this. DIN EN ISO 2639 [26] defines this as the vertical distance to the surface to the layer which has a Vickers hardness of 550 HV 1.

The case-hardened part generally exhibits compression tension at the surface, and tensile stresses at the core. As with surface-hardened materials, this distribution pattern provides enhanced resistance to vibration loads.

In carbonitriding, nitrogen is also absorbed; it serves to improve the material's tempering properties, increase its durability and enhance its wear resistance. The positive effects are especially pronounced with unalloyed steels.

For additional, more detailed information on case-hardening procedures, consult DIN 17022 Part 3 [25], and Information Sheet 452 of the Steel Information Center, Düsseldorf [27].

Application
Injection nozzles subject to wear for Bosch common-rail systems which must be able to withstand high wear and internal compressive stresses are case-hardened by vacuum carburizing.

Nitriding and nitrocarburizing
Nitriding is a thermal treatment process (temperature range: 400 to 600 °C) which can be used to enrich the surface layer of virtually any ferrous material with nitrogen. In nitrocarburizing, a certain amount of carbon is diffused into the material to form nitrogen at the same time.

Molecular nitrogen, as is present in this temperature range in gaseous nitrogen, cannot diffuse into metallic materials. It is therefore necessary to offer diffusible nitrogen via suitable donor media. In practical applications, nitriding and nitrocarburizing processes are performed in gas atmospheres containing ammonia, in plasma containing nitrogen, or even in molten saline solutions containing cyanate. While ammonia gas releases diffusible nitrogen during its thermal disintegration, nitrogen is ionized in the plasma in order to split the molecules and facilitate the diffusion of nitrogen atoms.

The enriching nitrogen in the surface layer causes the precipitation of nitrides, in response to which the surface layer hardens. In the final analysis, this results in greater resistance to wear and corrosion, and in greater endurance strength.

As the process employs relatively low temperatures, there are no volumetric changes of the kind associated with transformations in the microstructure, so that changes in dimensions and shape are minute.

The nitrided region consists of an outer layer, several millimeters in depth, and a transitional white layer, the hardness of which may be anywhere from 700 to over 1,200 HV, depending upon the composition of the material. Still deeper is a softer diffusion coating extending several tenths of a millimeter in which the nitrogen content decreases as the distance from the surface increases. The thickness of the individual layers is determined by the temperature and duration of the treatment process. The process produces a hardness gradient (similar to that which results from surface and case-hardening); this gradient furnishes the basis for determining the nitriding depth Nht. DIN 50190 Part 3 [28] defines this as the vertical distance from the surface to the point where the hardness corresponds to a defined limit. This limit hardness is usually actual core hardness + 50 HV 0.5.

The material's resistance to wear is essentially determined by the white layer, which contains up to 10 mass components of nitrogen in %. The nitriding depth and the surface hardness determine the material's resistance to alternating cyclic stress (for additional details, see DIN 17022 Part 4, [29], and Information Sheet 447 from the Steel Information Center, Düsseldorf, [30]).

The corrosion resistance of nitrided or nitrocarburized workpieces can be significantly increased by postoxidation in water vapor or other suitable gases, or in molten saline solutions at temperatures ranging between 350 and 550 °C.

Application
Injection nozzles which are used, for example, in Bosch common-rail systems and which must be able to withstand extreme operating temperatures are gas-nitrided.

Components for windshield and rear-window wiper systems are nitrocarburized and postoxidized to increase their resistance to corrosion and wear. This is how these components get their typical black color.

Boron treatment

Boron treatment involves enriching the surface layer of ferrous materials with boron. Depending upon duration and temperature (normally 850 to 1,000 °C) of the treatment, a white layer of 30 μm to 0.2 mm in depth with a hardness of 2,000 to 2,500 HV is produced. It consists of iron-boron.

Boron treatment is particularly effective as a means of protecting against abrasive wear. However, the comparatively high process temperature leads to relatively large changes in shape and dimensions, meaning that this treatment is only suitable for applications in which large tolerances can be accepted.

Application

Partly boron-treated tool holders with high wear resistance are used in Bosch hammer drills.

References

[1] DIN EN ISO 18265: Metallic materials – Conversion of hardness values.
[2] DIN EN ISO 6508: Metallic materials – Rockwell hardness test.
[3] DIN EN ISO 6506: Metallic materials – Brinell hardness test.
[4] DIN EN ISO 6506-1: Metallic materials – Brinell hardness test – Part 1: Test method.
[5] DIN EN ISO 4498: Sintered metal materials, excluding hardmetals – Determination of apparent hardness and microhardness.
[6] DIN EN ISO 6507: Metallic materials – Vickers hardness test.
[7] DIN EN ISO 4545: Metallic materials – Knoop hardness test.
[8] DIN 53505: Testing of rubber – Shore A and Shore D hardness test.
[9] DIN ISO 2039: Plastics – Determination of hardness.
[10] DIN EN ISO 14577: Metallic materials – Instrumented indentation test for hardness and materials parameters.
[11] DIN EN 10052: Vocabulary of heat treatment terms for ferrous products.
[12] DIN 17022-1: Heat treatment of ferrous materials – Methods of heat treatment – Part 1: Hardening, austempering, annealing, quenching, tempering of components.
[13] DIN 17022-2: Heat treatment of ferrous materials; heat treatment methods; hardening and tempering of tools.
[14] DIN EN 10083-1: Steels for quenching and tempering – Part 1: General technical delivery conditions.
[15] DIN EN 10083-2: Steels for quenching and tempering – Part 2: Technical delivery conditions for non-alloy steels.
[16] DIN EN 10083-3: Steels for quenching and tempering – Part 3: Technical delivery conditions for alloy steels.
[17] DIN EN 10085: Nitriding steels – Technical delivery conditions.
[18] DIN EN ISO 4957: Tool steels.
[19] DIN EN ISO 642: Steel – Hardenability test by end quenching (Jominy test).
[20] DIN 17021: Heat treatment of ferrous materials.
[21] DIN EN ISO 18265: Metallic materials – Conversion of hardness values.
[22] DIN 17022-5: Heat treatment of ferrous materials – Methods of heat treatment – Part 5: Surface hardening.
[23] DIN EN 10328: Iron and steel – Determination of the conventional depth of hardening after surface hardening.
[24] DIN 17021: Heat treatment of ferrous materials; material selection, steel selection according to hardenability.
[25] DIN 17022-3: Heat treatment of ferrous materials; heat treatment methods. case hardening.
[26] DIN EN ISO 2639: Steels – Determination and verification of the depth of carburized and hardened cases.
[27] Information Sheet 452 of the Steel Information Center, Düsseldorf: "Einsatzhärten", Edition 2008.
[28] DIN 50190-3: Hardness depth of heat-treated parts; determination of the effective depth after nitriding.
[29] DIN 17022-4: Heat treatment of ferrous materials – Methods of heat treatment – Part 4: Nitriding and nitrocarburizing.
[30] Information Sheet 447 of the Steel Information Center, Düsseldorf: "Wärmebehandlung von Stahl – Nitrieren und Nitrocarburieren", Edition 2005.

Corrosion and corrosion protection

Corrosion processes

Corrosion is the damage to a metal as a result of a reaction with substances in the environment. Corrosion processes always contain interphase reactions. An example of this type of reaction is metal scaling, i.e. oxidation in hot gases. As it proceeds, metal atoms oxidize to form nonmetallic compounds in a process emanating from the affected material's surface. In thermodynamic terms, the process can be viewed as an entropic transition from an ordered, high-energy state into a less-ordered state of lower energy, and consequently greater stability. The process takes place of its own accord.

The following deals exclusively with the corrosion that occurs at the phase boundary between the metal and aqueous phases (electrolyte), generally referred to as electrochemical corrosion.

Corrosive attack

Anodic subprocess

Two essentially distinct reactions occur in electrochemical corrosion: in the anodic subprocess, the directly visible corrosion process, the metal passes into the oxidized state as described in the reaction equation

$$Me \rightarrow Me^{n+} + n\,e^-$$

Figure 1: Free corrosion at the metal/corrosive-medium phase boundary
In an aggressive neutral or alkaline medium, oxygen is reduced at the corroding metal, and corrosion products are formed at the same time.

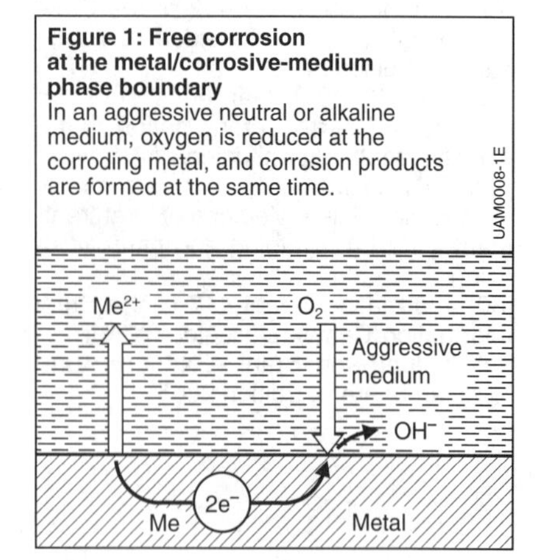

freeing an equivalent number of electrons (Figure 1). The metal ions thus formed can either be dissolved in the electrolyte, or can precipitate out on the metal as corrosion products (e.g. rust) after reacting with constituents in the attacking medium.

Cathodic counter-reactions

This anodic subprocess can continue only as long as the electrons it produces are consumed in a second process. This second subprocess is a cathodic subreaction. In neutral or alkaline media oxygen is reduced in accordance with

$$O_2 + 2\,H_2O + 4\,e^- \rightarrow 4\,OH^-$$

to hydroxyl ions, which in turn are able to react with the metal ions. In acidic media the hydrogen ions are reduced via the formation of free hydrogen, which escapes as a gas according to the following formula:

$$2\,H^+ + 2\,e^- \rightarrow H_2 .$$

Current/voltage curves

The corrosion process and the counter-reaction can be described by current/voltage curves. The potential of the metal (E_{Me}) is shifted, as a result of the counter-reaction, in the positive direction. This causes corrosion to occur. Each of the individual subreactions corresponds to a partial current/voltage curve. The total current is the sum of the two currents I_a and I_k (Figure 2):

$$I_{total} = I_a + I_k .$$

However, only the cumulative current/voltage curve can be measured (see Corrosion testing). The partial curves and the corrosion current can be derived from this.

If no voltage is supplied externally, i.e. in free corrosion, the system assumes a state in which the anodic and cathodic partial currents are precisely balanced:

$$I_a = -I_k = I_{corr} .$$

In this case, the anodic partial current is called the corrosion current I_{corr}, while the corresponding potential at which this current compensation occurs is called the "free corrosion potential" or for short the "open-circuit potential" E_{corr}.

The open-circuit potential is a mixed potential in which there is no equilibrium; matter is converted continuously, as defined in the following general equation:

$$O_2 + 2\,H_2O + \frac{4}{n}\,Me \rightarrow \frac{4}{n}\,Me^{n+} + 4\,OH^- .$$

The above is essentially applicable for contact corrosion (Figure 2). However, the interrelationships are more complicated; in addition to pairs of partial current/voltage curves for each of the two metals the resulting two cumulative current/voltage curves must be considered. Here the potential (E_{KT}) on the baser metal is shifted by the reactions on the more noble metal even further in the positive direction. Intensified corrosion (I_{KT}) takes place.

Electrochemical series of metals

Metals are often ranked in "electrochemical series of metals" corresponding to consecutively higher standard potentials (see also Electrochemistry). As the value of standard potential increases the metals are referred to as "more noble", and as the value decreases as "more base".

It should be emphasized that the table ("Standard electric potentials of metals") is limited to thermodynamic values, and does not reflect the effects of corrosion kinetics, for example as encountered in the formation of protective layers. Most corrosion-resistant structural materials, however, develop protective layers, which are crucial to corrosion resistance. Thus, for example the materials aluminum, zinc and titanium seem to be very base metals, but are highly resistant due to the formation of protective layers. The formation of protective layers is exploited in the field of electrochemical corrosion protection.

Figure 2: Partial and cumulative current/voltage curves for free corrosion (left) and contact corrosion (right) [1]

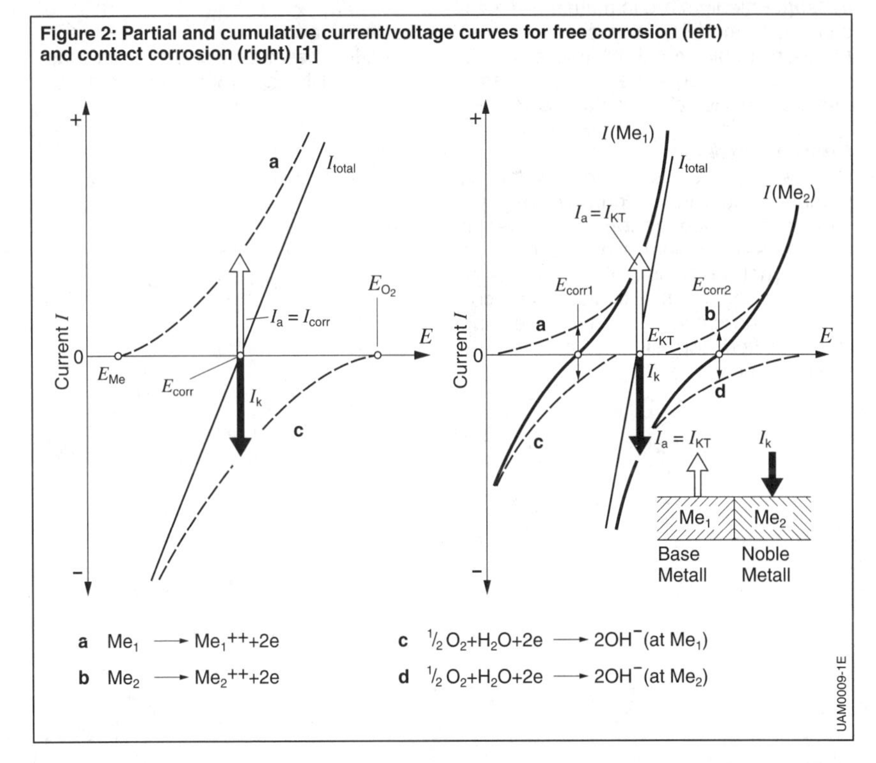

Types of corrosion

General surface corrosion
General surface corrosion is a type of corrosion with uniform attrition of material over the entire boundary surface between the material and the attacking medium. This is a very frequent type of corrosion in which the material penetration rate (removal depth) can be calculated per unit of time based on the corrosion current.

Pitting corrosion
Pitting corrosion is a limited localized attack by a corrosive medium which penetrates the material by forming holes, or pits, whose depth is almost always greater than their diameter. Practically no material is removed from the surface outside the pitted areas. Pitting corrosion is frequently caused by chloride ions (table salt).

Contact corrosion
When two different metals moistened by the same medium are in mutual electrical contact, a cathodic subprocess occurs at the more noble metal, while the anodic subprocess progresses at the baser material. This is called contact corrosion.

Crevice corrosion
Crevice corrosion is a corrosive attack primarily occurring in narrow crevices, caused by concentration differences in the corrosive medium, e.g. as a result of long oxygen diffusion paths. This type of corrosion generates potential differences between the crevice extremities, leading to intensified corrosion in more poorly ventilated areas.

Stress corrosion cracking
This is corrosion stemming from the simultaneous concerted action of a corrosive medium and mechanical tensile stress (which can also be present as internal stress in the object itself). Intergranular or transgranular fissures form, in many cases without the appearance of visible corrosion products.

Vibration corrosion cracking
Vibration corrosion cracking is corrosion caused by the simultaneous effects of a corrosive medium and mechanical fatigue stress, e.g. caused by vibrations. Transgranular cracks are formed, frequently without visible deformations.

Intergranular and transgranular corrosion
This involves types of corrosion characterized by selective formation along the grain boundaries or roughly parallel to the deformation plane in the grain interior.

Dezincification
Selective dissolution of zinc from brass, leaving behind a porous copper structure.

Corrosion testing

Electrochemical testing procedures

The primary tool for electrochemical corrosion testing is the potentiostat. Here metals or other electrically conductive materials are analyzed in different corrosive media. The typical setup with three electrodes is shown in Figure 3. This arrangement is essentially used to determine – as well as the potentials of the corroding materials – primarily the corrosion currents. In the case of uniform surface corrosion, the corrosion-current parameters are then used for the definition of the attrition mass and removal depth per unit of time. The relevant conversion factors are listed in Table 1.

The electrochemical testing procedures are a valuable supplement to the non-electrochemical methods, because they provide an understanding of corrosion mechanisms that occur. In addition to the small amount of corrosive medium required, another advantage of electrochemical procedures over non-electrochemical methods is that they provide quantitative data on attrition rates.

Current/voltage curves

The potentiostat is used to alter the potential of the material to be analyzed and measure the flowing current. The cumulative current/voltage curve is determined in the process (Figure 2). Evaluation of the curve shape enables reaction mechanisms and above all the corrosion current to be determined.

Figure 3: Schematic setup for measuring the current density/potential curves with a potentiostat
WE Working electrode,
RE Reference electrode,
CE Counter-electrode.

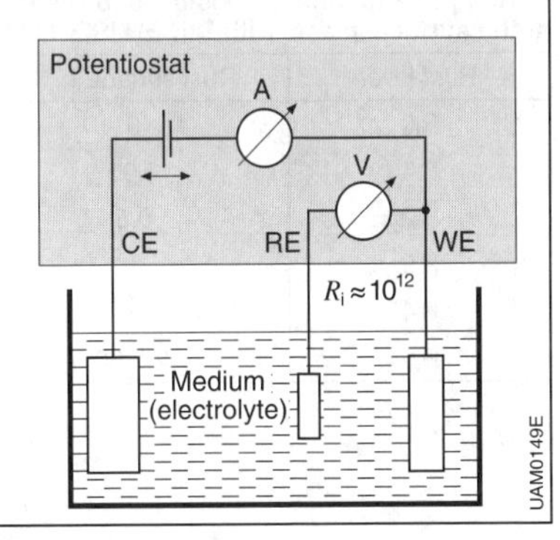

Polarization-resistance measurement

In the case of free corrosion the corrosion current is determined from the polarization resistance (slope of the cumulative current/voltage curve); testing entails subjecting the metal to minimal, alternating anodic and cathodic pulses.

Impedance spectroscopy

Electrochemical Impedance Spectroscopy (EIS) is employed to examine corrosion mechanisms. This alternating-current technique determines the AC resistance (impedance) and the phase angle of an electrochemical test object as a function of frequency. A low-amplitude sinusoidal alternating voltage is superimposed on the working electrode's potential, and the current response is measured. To interpret the measurement the system is approximated in the form of an equivalent network. By way of example, Figure 4 shows the equivalent network for the metal – coating – medium system.

The impedance elements (resistances, capacitances, inductances) are assigned physical properties. In this simple example from Figure 4 the coating is – along similar lines to the phase boundary of

Table 1: Attrition mass and removal depth due to surface corrosion of various metals with a corrosion current density of 1 μA/cm²

Metal	Relative atomic mass	Density g/cm³	Attrition mass mg/(cm²·year)	Removal depth μm/year
Fe	55.8	7.87	9.13	11.6
Cu	63.5	8.93	10.40	11.6
Cd	112.4	8.64	18.40	21.0
Ni	58.7	8.90	9.59	10.8
Zn	65.4	7.14	10.70	15.0
Al	27.0	2.70	2.94	10.9
Sn	118.7	7.28	19.40	26.6
Pb	207.2	11.30	33.90	30.0

material to corrosive medium – described as capacitor and resistor.

Direct conclusions can then be drawn from the equivalent networks and impedance-element values about various characteristics, such as the effectiveness of corrosion-protection measures, porosity, thickness, a coating's water-absorption ability, the effectiveness of inhibitors, the corrosion rate of the base metal, and so on.

Contact-corrosion current measurement
When contact corrosion is measured the two affected metals are immersed in the same corrosive medium. During measurement they are not directly contacted with each other, but instead connected via the potentiostat. Here the contact-corrosion potential (E_{KT}) and the contact-corrosion current (I_{KT}) flowing between the two affected metals are directly measured (see Figure 2) and the curve plotted against the measurement period.

Non-electrochemical corrosion-testing procedures

In non-electrochemical testing procedures test specimens essentially are first exposed to a corrosive environment (exposure tests) and then the change in the test specimens is assessed.

Figure 4: Evaluation of EIS data (electrochemical impedance spectroscopy)
1 Metal,
2 Coating,
3 Phase boundary metal to medium,
4 Corrosive medium.

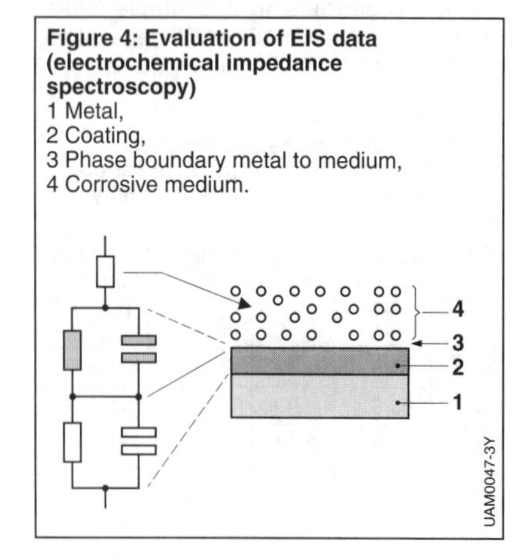

Simple standard tests (e.g. corrosion load by neutral salt spray fog testing) are conducted with the aim of testing quality or comparing corrosion-protection measures and materials. The emphasis is placed on using tests that can uncover weak points in corrosion-protection measures and describe different qualities.

After corrosion load is completed, it is possible by comparing images in accordance with DIN EN ISO 4628-3 [2] to assign rust levels (degrees of rusting) which are defined according to rust coverage or surface perforation (Table 2).

Testing procedures have evolved to reflect specific requirements, e.g. motor-vehicle testing. These are as a rule corrosion loads with cyclically alternating loading by salt spray fog, dry phase and moist phase. These tests provide reliable indices of projected service life under normal operating conditions by using short-term expose in extremely harsh conditions to simulate long-term stresses in the real world (e.g. testing in accordance with Bosch standard N42AP 226 "Climatic tests; tightened life corrosion testing").

For a further adaptation of corrosion load to practical operation products are operated under tightened conditions (e.g. vehicle tests of the different automobile manufacturers).

In addition to the corrosion loads for life testing the focus is on assessing the function of the products.

Table 2: Rust level and proportion of rust penetrations and visible sub-rusting at the surface in acc. with DIN EN ISO 4628-3

Rust level	Rust surface in %
R_i0	0
R_i1	0.05
R_i2	0.5
R_i3	1
R_i4	8
R_i5	40...50

Corrosion protection

The manifestations and mechanisms of corrosion are many and varied, so widely differing methods can be adopted to protect metals against corrosion attack. Corrosion protection means intervening in the corrosive process with the object of reducing the rate of corrosion in order to prolong the service life of the components.

Corrosion protection can be achieved by applying four basic principles:

- Measures in planning and design: the choice of suitable materials and the suitable structural design of components.
- Measures that intervene in the corrosive process by electrochemical means.
- Measures that separate the metal from the corrosive medium by protective layers or coatings.
- Measures that influence the corrosive medium, for example, the addition of inhibitors to the medium.

Corrosion protection by means of suitable structural design

Material selection

Selecting suitable materials which feature optimum resistance to corrosion under the expected conditions can be of considerable assistance in avoiding corrosion damage. When the costs that would otherwise be incurred for upkeep and repair are factored into the long-term cost of ownership equation, a more expensive material can often be the more cost-effective alternative.

Design

Design measures, too, are of major importance. A great deal of skill and expertise goes into design, particularly regarding the connections between parts that are made of the same material or different materials.

Corners and edges of sections are difficult to protect, and this is where corrosion can easily attack. A favorable installation position can avoid corrosion (Figure 5).

Beads and welts can trap dirt and moisture. Suitable surfaces and drain openings can help avoid this problem (Figure 6).

Figure 5: Good and bad installation positions for profile sections

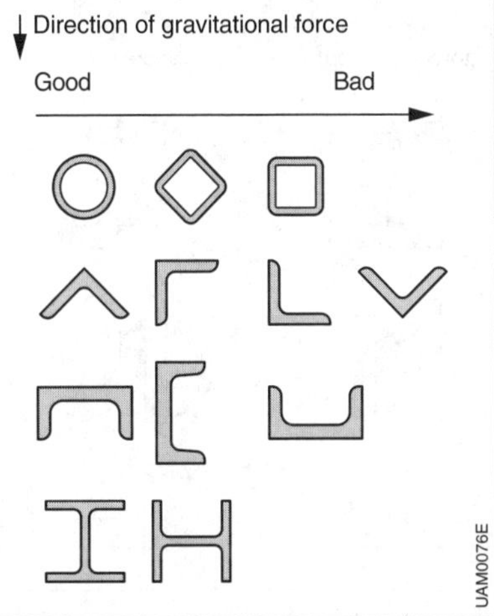

Direction of gravitational force

Good

Bad

UAM0076E

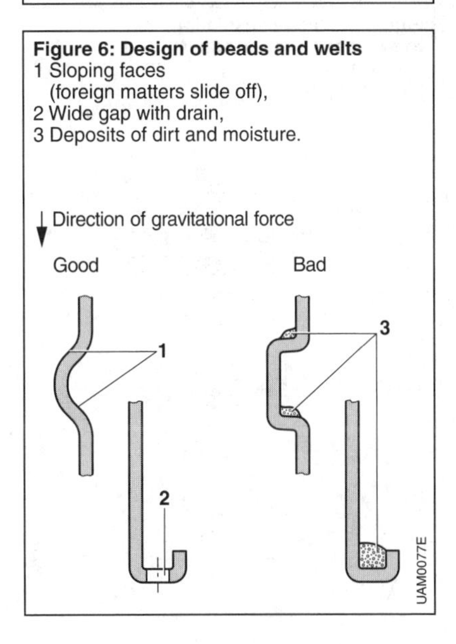

Figure 6: Design of beads and welts
1 Sloping faces (foreign matters slide off),
2 Wide gap with drain,
3 Deposits of dirt and moisture.

Welds, which generally modify the microstructure for the worse, are another weak point. In order to avoid crevice corrosion, welds have to be smooth and free of gaps (Figure 7).

Contact corrosion can be avoided by joining same or similar metals, or by installing washers, spacers, or sleeves to ensure that both metals are electrically insulated (Figure 8).

Figure 7: Design-related crevices in welds and how they can be avoided

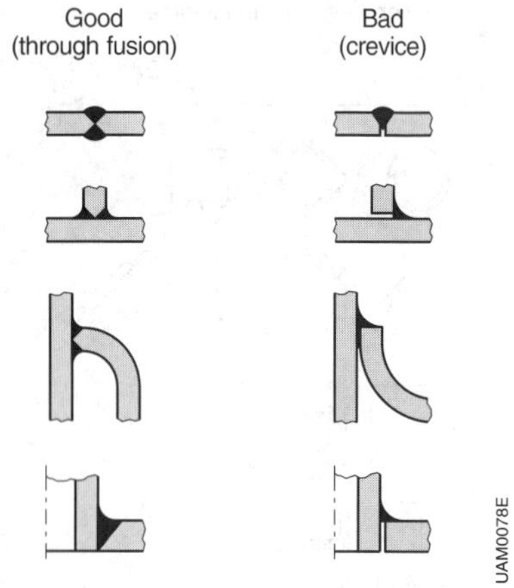

Good (through fusion) | Bad (crevice)

UAM0078E

Electrochemical processes

The schematic current/voltage curves for a metal suitable for passivation coating (e.g. by oxide layer on stainless steel) show how these processes work (Figure 9). The current-density values arranged in ascending order on the y-axis represent anodic currents corresponding to the corrosion reaction defined in the equation

$$Me \rightarrow Me^{n+} + n\,e^{-}.$$

In contrast, the descending current-density values represent cathodic currents.

The schematic indicates that external voltage can be applied to suppress corrosion. There are two basic ways of doing this.

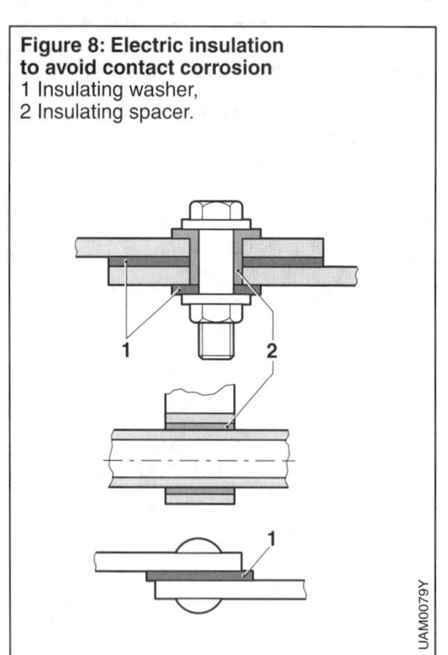

Figure 8: Electric insulation to avoid contact corrosion
1 Insulating washer,
2 Insulating spacer.

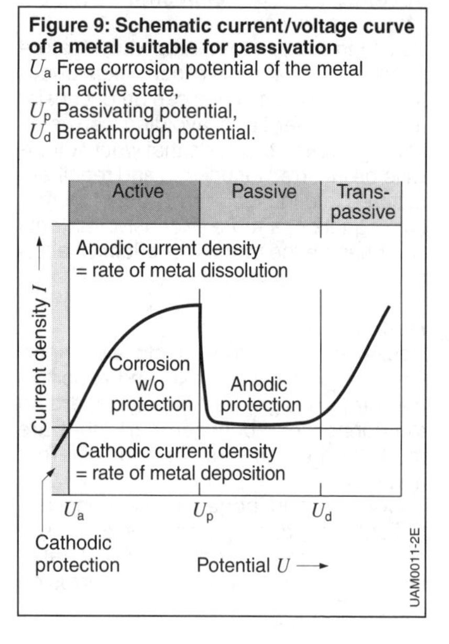

Figure 9: Schematic current/voltage curve of a metal suitable for passivation
U_a Free corrosion potential of the metal in active state,
U_p Passivating potential,
U_d Breakthrough potential.

Cathodic protection
For cathodic protection, the potential is shifted so far toward the left that no anodic currents flow, leading to $U < U_a$. As an alternative to applying voltage from an external source, it is also possible to shift tho potential by using a base metal to act as a "sacrificial" reactive anode.

Anodic protection
Another option is to shift the potential of the threatened electrode into the passive range, i.e. into the potential range between U_p and U_d. This is called anodic protection. The anodic currents, which flow in the passive range, are less than those in the active range by exponential powers of between three and six, depending on the type of metal and the corrosive medium. The result is excellent protection for the metal.

However, the potential should not exceed U_d, as oxygen would be produced in this transpassive range, potentially leading to higher rates of oxidation. Both of these effects would cause the current to increase.

Coatings
Coatings inhibit corrosion by forming protective films which are applied to the surface of the metal to be protected, where they resist attack by the corrosive medium (see also Deposits and coatings).

Inhibitors
Inhibitors are substances added to the corrosive medium in low concentrations (up to a maximum of several hundred ppm) for absorption at the surface of the protected metal. They drastically reduce the rate of corrosion by blocking either the anodic or the cathodic subprocess (frequently blocking both subprocesses simultaneously). Organic amines and amides of organic acids are the most frequent inhibitors. In automotive applications, for example, inhibitors are used in fuel additives. They are also added to antifreeze in order to inhibit corrosion damage in the coolant circuit.

Vapor-phase inhibitors
Vapor-phase inhibitors (VPI), or volatile corrosion inhibitors (VCI), are organic substances of moderate vapor pressure. They are frequently enclosed in special-purpose packing materials or as solvents in liquids or emulsions with oil. The inhibitors evaporate or sublimate over the course of time, and are adsorbed as monomolecules on the metal, where they inhibit either anodic or cathodic corrosion subreactions, or both at once. Dicyclohexylamin nitrite is a typical example.

For optimal effectiveness, the inhibitor should form a sealed coating extending over the largest possible surface area. This is why they are generally enclosed in packing materials such as special paper or polyethylene foil. An airtight edge seal is not required; the packing can be opened briefly for inspection of contents. The duration of the packing's effectiveness depends on the tightness of the seal and the temperature (normally approx. 2 years, but less in environments substantially hotter than room temperature).

Standard commercial vapor-phase inhibitors are generally a combination of numerous components capable of providing simultaneous protection for several metals or alloys; exceptions: cadmium, lead, tungsten and magnesium.

Vapor-phase inhibitors provide only temporary protection for metallic products during storage and shipping. They must be easy to apply and remove. The drawback of vapor-phase inhibitors is that they are potentially injurious to health.

References
[1] DIN 50918:1978: Corrosion of metals; electrochemical corrosion tests.
[2] DIN EN ISO 4628-3:2004: Paints and varnishes – Evaluation of degradation of coatings – Designation of quantity and size of defects, and of intensity of uniform changes in appearance – Part 3: Assessment of degree of rusting (ISO 4628-3:2003); German version EN ISO 4628-3:2003.

Deposits and coatings

Deposits and coatings are used to adapt the surface properties of components to particular requirements ("surface engineering"). It is thus possible, for example, for a component to be made from a tough, low-cost material which still has a hard and wear-resistant surface. The main uses for coatings are:
- Corrosion protection (as well as functional retention, often for decorative reasons)
- Wear protection
- Joining and bonding techniques (plug-in, welded, soldered, bonded, and crimped contacts)

In coating systems, distinctions are made between the following types of coating:
- Deposits in which one layer is applied
- Conversion coatings in which the functional coating is produced by chemical or electrochemical conversion of the base material, and
- Diffusion coatings in which the functional coating is produced by diffusion of atoms or ions into the base material.

Deposits

Electrodeposits

Electrodeposits are deposited using an external power source. The workpieces to be coated are immersed into an electrolyte (Figure 1) which contains ions of the metal to be deposited. The metal ions consumed during the deposition process are either supplied by the anode solution or added to the electrolyte (when inert anodes are used). The progression and distribution of the field lines influence the liner-thickness distribution. A uniform layer thickness can be achieved by means of an optimized configuration of anodes and screens.

Electrodeposits are widely used in corrosion protection, in wear protection and for electrical contacts. Some important deposits are described below.

Zinc and zinc alloys

Electrolytically deposited zinc coatings are widely used as corrosion-protection coatings for steel components and are a cathodically effective means of corrosion protection (sacrificial effect). To increase the corrosion-protection effect, zinc deposits are passivated in a chrome(VI)-free solution as a substitute for the customary chromalization previously used. Zinc al-

Figure 1: Electrodeposits
a) Progression and distribution of electrical field lines between anode and cathode as caused by nonuniform liner-thickness distribution,
b) Comparison of liner-thickness distribution during electroplating without auxiliary agent, with auxiliary anodes and with screens (schematic representation).
1 Anode, 2 Cathode, 3 Auxiliary anode, 4 Screen.

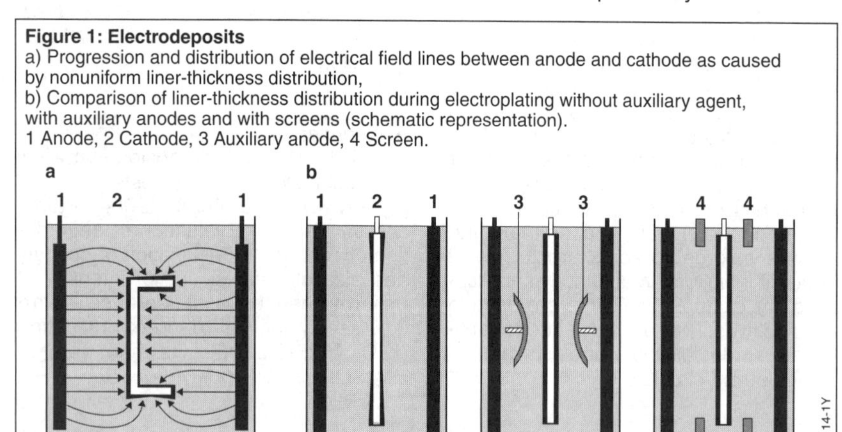

loys, such as zinc-nickel with approx. 15 % nickel offer a significantly higher degree of corrosion protection.

Nickel
Electrolytically deposited nickel coatings offer limited corrosion protection with an attractive visual appearance. One area of automotive engineering in which this is chiefly used is the nickel-plating of spark-plug shells.

Chrome
In the case of chrome deposits, a distinction is made between hard chrome and bright chrome. Bright chrome is used as a deposit approx. 0.3 µm thick with a nickel or copper/nickel intermediate deposit. In the past, bumpers and molding strips were provided with bright-chrome deposits for decorative reasons. Cu/Ni/Cr deposits had, to a large extent, declined in significance in motor-vehicle applications, but are now being used again increasingly in the premium segment.

Hard-chrome deposits are chrome deposits with a thickness greater than 2 µm. Owing to the high level of hardness of the electrolytically deposited chrome coating, it is ideally suited for use as a wear-protection coating. In the past, hard-chrome deposits were often applied in strong thicknesses and then mechanically reworked. Due to further developments in plant engineering, components today are increasingly custom-coated with coating thicknesses ranging from 5 to 10 µm, and then used without reworking (e.g. components for fuel injectors).

Tin
Electrolytically deposited tin coatings are mainly used as contact surfaces for plug-in and switch contacts, and as solder contact surfaces. A coating thickness of 2 to 3 µm is ideal for plug-in contacts. Greater coating thicknesses are required for solder applications in order to ensure solderability even after extended storage periods.

Gold
Gold deposits are normally used for contacts subject to stringent requirements. They are characterized by good conductivity, low contact resistance, and good resistance to corrosion and pollutant gases. This ensures contact stability. Hard-gold deposits (with approx. 0.5 % alloying constituents, predominantly cobalt) are harder and more abrasion-resistant than pure-gold deposits and are suitable for contacts subject to mechanical load.

Chemically deposited coatings (no external current)

In contrast to electrodeposits, chemically deposited coatings are characterized by a more uniform coating-thickness distribu-

Table 1: Areas of application for coatings

Coating system	Coating material	Main application
Electrodeposits	Zn, ZnNi Cr Sn, Ag, Au	Corrosion protection Wear protection Electrical contacts
Chemical deposits	NiP, NiP dispersions Cu, Sn, Pd, Au	Wear and corrosion protection Electronic applications
Hot-immersion deposits	Zn, Al Sn	Corrosion protection Electrical contacts
Paints (wet and powder-based paints)	Organic polymers Pigments (color particles)	(Decorative) corrosion protection Wear reduction Electrical insulation
PVD/CVD coatings (plasma technology)	TiN, TiCN, TiAlN DLC i-C(WC), a-C:H	Wear protection of tools Reducing friction and wear on components

tion on the component, since deposition does not occur under the influence of an external electric field. Owing to the slow deposition rate and the costly chemicals in the coating bath, they are more expensive than electroplated liners. The following have become widespread in industry:
- "Chemical nickel" (nickel/phosphorus) as a corrosion- and wear-protection deposit
- "Chemical copper" and "Chemical tin" in printed-board technology

Hot-immersion deposits
Hot-immersion coatings are deposited by immersing the substrates in a molten metal bath. Cathodically effective hot-immersion deposits are used to protect low-alloyed steels against corrosion. Zinc, zinc/aluminum alloys and aluminum are used. Coating start material such as sheet and strip is a cheap and widely used practice. However, free punch edges must be accepted here.

Tin and tin-alloy coatings deposited in hot-immersion priming are primarily used as surfaces for plug connectors and as soldering surfaces.

Zinc-lamella deposits
Zinc-lamella deposits are coatings based on zinc and aluminum lamellas and an inorganic binder. They are applied by immersion centrifuging or electrostatic spraying and thermal hardening. Zinc-lamella deposits are low-cost corrosion-protection coatings for mass-produced steel parts (e.g. screws/bolts).

Paints
Paints have a wide variation range because of their broad chemical bases and numerous means of application, such as brushing, spraying, and immersing, even with the use of current (CD: cathodic deposition).

Paint coatings can also fulfill a multitude of different functions. In motor vehicles, the primary function fulfilled by paint coatings is corrosion protection accompanied by a decorative effect, but other functions include wear protection, soundproofing or electrical insulation.

The body of a passenger car is protected and improved by a complex layered structure (Figure 2, Table 2, and Table 3). The assemblies in the engine compartment, on the other hand, are usually only given one or two coats of paint. Less importance is placed here on decorative properties.

Low-solvent systems, above all water-based paints, are almost always used in automotive engineering. Powder-based paints and UV-hardening paints are even entirely solvent-free.

Figure 2: Layered structure of paint coatings on car bodies

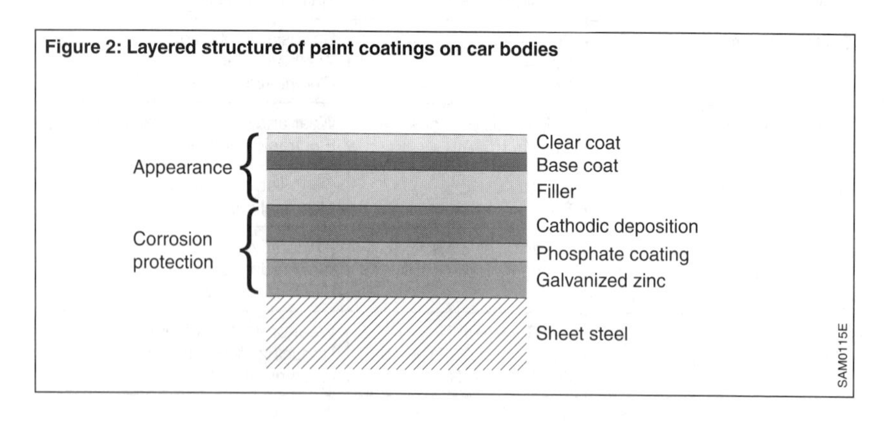

Table 2: Structure of solid-color coatings

Layer	Layer thickness in µm	Structure	Composition					Applications
			Binders	Solvents	Pigments	Extenders	Additives and SC	
1	20…25	CD	Epoxy resins	Water, small amounts of water-miscible organic solvents	Inorganic (organic)	Inorganic extenders	Surface-active substances, anti-crater agents, SC 20 %	EC
2a	approx. 35…50	Extender	Polyester, melamine, urea, and epoxy resins	Aromatic compounds, alcohols	Inorganic and organic	Inorganic solids	e.g. wetting agents, Surface-active substances SC 55…62 %	PS ESTA-HR
2b	approx. 35…50	Water extender	Water-soluble polyester, polyurethane, and melamine resins	Water, small amounts of water-miscible organic solvents			SC 43…50 %	PS ESTA-HR
3a	40…50	Solid-color top coat	Alkyd and melamine resins	Esters, aromatic compounds, alcohols		–	e.g. leveling and wetting agents	PS ESTA-HR
3b	10…35 (color-dependent)	Water-soluble solid-color base coat	Water-soluble polyester, polyurethane, polyacrylate, and melamine resins	Small amounts of water-miscible co-solvents		–	Wetting agents SC 20…40 %	PS ESTA-HR
4a	40…50	Conventional clear coat	Acrylic and melamine resins	Aromatic compounds, alcohols, esters	–	–	e.g. leveling agents and light stabilizers, SC 45 %	PS ESTA-HR
4b	40…50	2C-HS	HS acrylate resin polyisocyanates	Esters, aromatic compounds	–	–	e.g. leveling agents and light stabilizers, SC 58 %	
4c	40…50	Powder-slurry clear coat	Urethane-modified epoxy/carboxy system	–	–	–	e.g. light stabilizers, SC 38 %	
4d	40…50	Powder-clear coat	Acrylate resin	–	–	–	100 %	ESTA PS

Acronyms: TFB Thick-film build ESTA-HR Electrostatic high rotation EC Electrophoretic coating SC Solids content CD Cathodic deposition PS Pneumatic spray 2C-HS 2-Components high solid (high levels of non-volatile matter)

Table 3: Structure of metallic coatings

Layer	Layer thickness in µm	Structure	Composition					Applications
			Binders	Solvents	Pigments	Extenders	Additives and SC	
1	20…25	CD	Epoxy resins Polyurethane	Water, small amounts of water-miscible organic solvents	Inorganic (organic)	Inorganic extenders	Surface-active substances, anti-crater agents, SC 20 %	EC
2a	approx. 35…50	Extender Melamine, urea, and epoxy resins	Polyester, alcohols	Aromatic compounds, alcohols	Inorganic and organic	Inorganic extenders	e.g. wetting agents, Surface-active substances SC 55…62 %	PS ESTA-HR
2b	approx. 35…50	Water extender	Water-soluble polyester, polyurethane, and melamine resins	Water, small amounts of water-miscible organic solvents			SC 43…50 %	PS ESTA-HR
3a	10…15	Metallic base coat	CAB, polyester, melamine resins	Esters, aromatic compounds	Aluminum and mica particles	–	SC 15…30 %	PS ESTA-HR
3b	10…15	Water-soluble metallic base coat	Water-soluble polyester, polyurethane, polyacrylate, and melamine resins	Small amounts of water-miscible co-solvents	Aluminum and mica particles Organic and inorganic pigments	–	Wetting agents SC 18…25 %	PS ESTA-HR
4a	40…50	Conventional clear coat	Acrylic and melamine resins	Aromatic compounds, alcohols, esters	–	–	e.g. leveling agents and light stabilizers, SC 45 %	PS ESTA-HR
4b	40…50	2C-HS	HS acrylate resin polyisocyanates	Esters, aromatic compounds	–	–	e.g. leveling agents and light stabilizers, SC 58 %	
4c	40…50	Powder-slurry clear coat	Urethane-modified epoxy/carboxy system	–	–	–	e.g. light stabilizers SC 38 %	
4d	40…50	Powder-clear coat	Acrylate resin	–	–	–	100 %	PS ESTA

Acronyms: TFB Thick-film build ESTA-HR Electrostatic high rotation EC Electrophoretic coating SC Solids content CD Cathodic deposition PS Pneumatic spray 2C-HS 2-Components high solid (high levels of non-volatile matter)

PVD and CVD coatings

PVD and CVD coating systems are generally deposited in a vacuum supported by plasma or heat treatment on components and tools (Figure 3). There are two different procedures, depending on whether the coating-forming material is obtained from the solid (PVD, physical vapor deposition) or gaseous (CVD, chemical vapor deposition) phase. Most modern procedures combine both categories in "reactive" processes.

Hard-material coatings

Tools are coated with hard material to increase their service life and their performance. A typical example of the tool coatings widely introduced onto the market is the gold-colored titanium nitride (TiN), which is produced, for example, by the cathodic sputtering or arc evaporation of titanium in a reaction with nitrogen. More recent coating systems, such as titanium carbonitride (TiCN) or titanium aluminum nitride (TiAlN), can be used for high-speed cutting and sometimes also for material machining using low-volume lubricooling or when dry.

Figure 3: Schematic of plasma-assisted PVD/CVD coating in vacuum chamber
1 Unbalanced magnetron cathode with sputter target,
2 Gas inlet,
3 Intensive ion bombardment,
4 Substrate,
5 Substrate tension

○ Argon,
● Metal/carbon,
● Nitrogen/hydrogen.

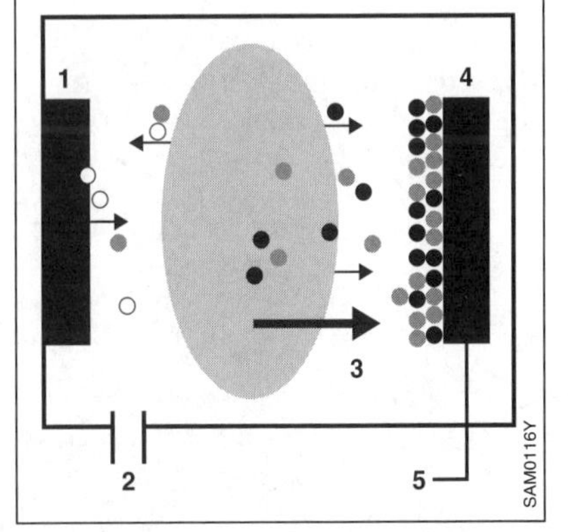

Superhard materials, such as diamond, are increasingly used to coat carbide tools.

DLC coatings

Protecting components against wear is subject to special conditions. Here, the coating should establish a low coefficient of friction for the components that come into contact with each other and minimize the wear for the entire layout. DLC (diamond-like carbon) coatings protect not only the coated component against wear, but also the uncoated opposed body. They have a very low coefficient of friction of 0.1 to 0.2, both in a dry friction pairing against steel and under media. DLC layers are media-resistant and have a corrosion-inhibiting effect. Due to these advantages, DLC coatings have become the most important in the field of coating components. However, it must be noted that coatings containing hydrogen will degrade in an oxidizing atmosphere at a local temperature of 350 °C and higher. Hard-material coatings, such as TiN, can withstand considerably higher temperatures and are also used in component coating.

Different coating compositions and processes allow materials to be adapted to different wear loads or combinations of abrasive, vibration, sliding, seizure, and adhesive wear. As there is no standardization for DLC coatings, it is necessary to carefully check the coating properties in wear tests before replacing a coating with a diamond-like coating made by a different manufacturer. Plasma-assisted PVD and CVD processes are conducted in such a way that the temperature of the component does not exceed 250 °C during coating. This avoids impairing the hardness of the steel base material.

Low-friction carbon coatings containing metal carbide i-C(WC) are electrically conductive and have a microhardness of approx. 1,800 HV at a modulus of elasticity of 150 to 200 GPa.

In comparison, metal-free carbon coatings (a-C:H) offer increased hardness of approx. 3,500 HV and significantly improved wear resistance, but this is accompanied by an increase in brittleness. They are electrically insulating.

With thicknesses of 2 to 4 µm, diamond-like carbon coatings offer very good wear protection and are especially suitable for precision components subject to high mechanical loads. There is no rework needed after coating. In high-pressure pumps for diesel and gasoline injection, the piston is sealed against the cylinder only by way of a particularly close-toleranced gap of a few micrometers. Here, DLC coatings on the piston ensure reliable operation throughout its service life.

Diffusion coatings

Surface treatment can be selectively combined with surface hardening by using the diffusion process to thermochemically carburize, carbonitride, or chromate the metal, or treat it with boron or vanadium (see Heat treatment of metallic materials, Thermochemical treatment). The metal can also be oxidized, nitrided, or sulfided without hardening.

Conversion coatings

Conversion coatings are formed not by the application of a material, but by the chemical or electrochemical conversion of the base material.

Browning coatings

Browning coatings consist of thin iron-oxide layers (predominantly Fe_3O_4) which are formed by the oxidation of the steel in an alkaline, aqueous solution containing nitrite at temperatures greater than 100 °C. With subsequent oiling, they offer temporary corrosion protection.

Phosphating coatings

Phosphating coatings are formed on steel, galvanized steel, and aluminum in solutions containing phosphoric acid by immersion or spraying. Zinc-phosphate coatings are predominantly used as a wash primer for paint coatings. Manganese phosphate serves as a wear-protection coating with anti-seizure properties, and as a wash primer for other coatings to improve component antifriction properties.

Anodized coatings

Anodized coatings are formed by the electrochemical conversion of metal into metal oxide in aqueous electrolytes. Aluminum, magnesium, and titanium can be anodized. Anodized coatings on aluminum materials are widely used for corrosion and wear protection.

Lubricants

Terms and definitions

Lubricants
Lubricants provide mutual insulation for components in a state of relative motion. The lubricant's function is to prevent direct contact between these components, thereby reducing wear and minimizing friction. Lubricants serve as coolants, friction-surface sealants, and corrosion inhibitors, and can also reduce running noise. Lubricants may be solid, consistent, liquid, or gaseous in form. Specific lubricants are selected with reference to design characteristics, materials combinations, the ambient conditions, and the stress factors encountered at the friction surface ([1], [2], [3]).

Additives
Additives are substances mixed into the lubricant in order to improve specific properties. These substances modify either the lubricant's physical characteristics (e.g. viscosity-index improvers, pour-point depressors) or its chemical properties (e.g. oxidation inhibitors, corrosion inhibitors). In addition, the properties of the friction surfaces themselves can be modified by additives which change the friction characteristics (friction modifiers), protect against wear (anti-wear agents), or provide protection against scoring and seizure (extreme-pressure additives). Great care must be exercised in order to ensure that the additives are correctly matched with each other and with the base lubricant.

ATF
ATF (automatic transmission fluids) are special-purpose lubricants specifically formulated to meet stringent requirements for operation in automatic transmissions.

Ash
Ash is the mineral residue which remains after oxide and sulfate incineration (DIN 51575, [4]).

Bleeding
Bleeding is the separation of base oil (lube oil) and the thickener in a lubricating grease (oil separation, DIN 51817 [5]). During this process the lube oil is released to the lubricating surroundings.

Bingham bodies
Bingham bodies are materials whose flow characteristics differ from those of Newtonian fluids (see Rheology, Newtonian fluids).

Fire point and flash point
The lowest temperature (referred to 1,013 hPa) at which a gaseous mineral product initially flashes is known as the flash point. The fire point is the temperature at which it continues to burn for at least 5 s (DIN ISO 2592 [6]).

Cloud point
The cloud point is the temperature at which mineral oil becomes opaque due to the formation of paraffin crystals or precipitation of other solids (DIN ISO 23015 [7]).

Detergents
These additives prevent paint- and carbon-like deposits on hot components (e.g. pistons). They serve as cleaning agents which like a tenside contain both hydrophilic and hydrophobic areas.

Dispersants
Dispersants prevent (disperse) sludging and sedimentation, particularly at low temperatures (solid substances are held finely distributed in suspension).

EP lubricants
See High-pressure lubricants (EP, Extreme Pressure).

Flow pressure
The flow pressure is the pressure required to press a consistent lubricant through a standardized test nozzle (Kesternich method, DIN 51805 [8]). The flow pressure is an index of a lubricant's starting flow characteristics, particularly at low temperatures.

Yield point

The yield point is the shear stress at which a substance begins to flow. Above the yield point, the rheological characteristics of plastic substances are the same as those of liquids (DIN 1342-3 [9]).

Friction modifiers

Friction modifiers are polar lubricant additives which reduce friction in the mixed-friction range (see Mixed friction) and increase bearing capacity after adsorption on the surface of the metal. They also inhibit stick-slip behavior.

Gel-type greases

Gel-type greases are lubricants with inorganic gelling agents (e.g. Bentonites, silica gels).

Anti-friction coatings

Anti-friction coatings are solid lubricant combinations which a binding agent holds in place on the friction surfaces (AFC, anti-friction coating).

Graphite

Graphite is a solid lubricant with layer-lattice structure. Graphite provides excellent lubrication when combined with water (e.g. high atmospheric humidity) and in carbon-dioxide atmospheres or when combined with oils. It does not inhibit friction in a vacuum.

Borderline pumping temperature

The borderline pumping temperature is the application limit for oil throughput through engines after cold starting. Below the borderline pumping temperature oil can no longer flow of its own accord to the oil pump, and thus a sufficient supply of oil can no longer be guaranteed.

High-pressure lubricants

High-pressure lubricants (EP, Extreme Pressure) contain additives to enhance load-bearing capacity, to reduce wear, and to reduce scoring (generally provide good performance in steel-to-steel and steel-to-ceramic applications).

Hydro-crack oils

Hydro-crack oils are refined mineral oils with increased viscosity index (VI 130 to 140). They are produced by hydrogenating mineral oils and are thus more thermally stable and have improved viscosity-temperature characteristics.

Induction period

The induction period is the time which elapses before substantial changes occur in a lubricant (e.g. aging of an oil containing an oxidation inhibitor).

Inhibitors

Inhibitors are active system-protecting ingredients which slow down the chemical decomposition of lubricants (e.g. due to oxidation) or protect against corrosion (corrosion inhibitors).

Low-temperature sludge

Low-temperature sludges are the products of oil degradation which form in the engine crankcase due to incomplete combustion and condensate at low engine load. Low-temperature sludge increases wear and can cause engine damage. Modern high-quality engine oils inhibit its formation.

Cold-starting reliability

Cold-starting reliability is a temperature value and based on experience is approx. 5 °C above the borderline pumping temperature.

Consistency

Consistency is a measure of the ease with which lubricating greases and pastes can be deformed (DIN ISO 2137 [10]). A standardized cone is dropped into a flattened grease sample and then the penetration depth is measured in 1/10 mm (Figure 2, see also Penetration and Table 5).

Doped lubricants

Doped lubricants contain additives for improving specific properties (e.g. aging stability, wear protection, corrosion protection, viscosity-temperature characteristics).

High-lubricity oils
High-lubricity oils are lubricating oils with multigrade properties, low cold viscosity, and special anti-friction additives. Extremely low engine friction under all operating conditions reduces fuel consumption.

Longlife engine oils
Longlife engine oils are oils for significantly extended oil-change intervals.

Multigrade oils
Multigrade oils are engine and transmission oils with good resistance to viscosity-temperature change (high viscosity index VI, Figure 1, see Viscosity). They often contain polymers which change their solid structure as a function of temperature. When the oil is cold, the molecules in the oil are "bunched together". As the temperature increases, the molecules stretch and thereby increase the friction between the particles. The viscosity of the oil increases, thereby guaranteeing a stable lubricating film even at high temperatures. Multigrade oils reduce friction and wear, and during cold starting ensure that all the engine components are quickly lubricated. These oils are formulated for year-round use in motor vehicles; their viscosity ratings extend through several SAE grades.

Metal soaps
Metal soaps are reaction products from metals or from their compounds with fatty acids. They are used as thickeners for grease and as friction modifiers.

Mineral oils
Mineral oils are distillates or raffinates produced from petroleum or coal. They consist of numerous hydrocarbons in various chemical compositions. Classification is according to the predominant component: paraffin-based oils (chain-shaped saturated hydrocarbons), naphthene-based oils (closed-chain saturated hydrocarbons, generally with five or six carbon atoms per ring) or aromatic oils (e.g. alkylbenzene). These substances are distinguished by major variations in their respective chemical and physical properties.

Molybdenum disulfide
Molybdenum disulfide (MoS_2) is a solid lubricant with layer-lattice structure. Only low cohesive forces are present between the individual layers, so their mutual displacement is characterized by relatively low shear forces. A reduction in friction is only obtained when MoS_2 is applied in suitable form to the surface of the metal (e.g. in combination with a binder such as MoS_2 anti-friction coating).

Figure 1: Viscosity/temperature curves for single- and multigrade engine oils

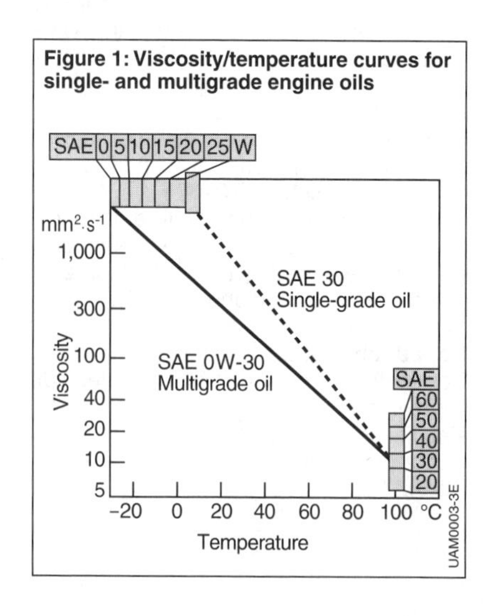

Figure 2: Determining cone penetration
1 Standardized cone,
2 Consistent lubricant at defined temperature.
x Penetration depth.

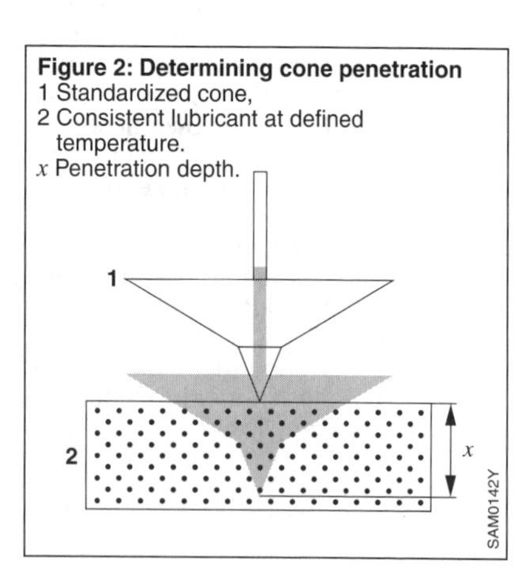

Penetration
Penetration denotes the depth (in 10^{-1} mm) to which a standardized cone penetrates into a consistent lubricant within a defined period and at a specified temperature (Figure 2). The larger the number, the softer the lubricant (DIN ISO 2137 [10]).

Polar substances
Dipolar molecules are easily adsorbed onto metal surfaces. They enhance adhesion and bearing capacity, thus reducing friction and wear. This category includes, for example, diester oils, ethers, polyglycols, and fatty acids.

Pour point
Pour point is the lowest temperature at which an oil continues to flow when cooled under defined conditions (DIN ISO 3016 [11]).

PTFE
PTFE (polytetrafluoroethylene, Teflon) is a thermoplastic with outstanding properties as a solid lubricant, particularly at very low sliding velocities (< 0.1 m/s). PTFE only becomes brittle below approx. -270 °C. The upper service temperature for use is approx. 260 °C. Above this level, it decomposes with toxic cleavage products.

Rheology
Rheology is the science dealing with the flow characteristics of materials. These are generally represented in the shape of flow curves (Figure 3). In the diagram shear stress τ is plotted against shear rate $\dot{\gamma}$.

Shear stress and shear rate
Shear stress is defined as the shear force per shear area, shear rate as the quotient of velocity v and gap height h (Figure 4).

Shear stress $\tau = F/A$ (in N/m^2 = Pa),
F shear force, A area vs. shear rate.

Shear rate $\dot{\gamma} = v/y$ (in s^{-1}),
v velocity, y thickness of lubricating film.

The flow curves are usually measured in a plate-cone measuring system. The sample to be measured is located between the plate and the cone; the torque is proportional to the shear stress and the rotational speed proportional to the shear rate.

Shear stress τ can be illustrated using the two-plate model (Figure 4). The fluid is located between the two plates. The bottom plate is fixed, the top plate moves with shear area A by means of shear force F.

Shear viscosity
Shear viscosity η is in liquids of ideal viscosity and constant temperature the ratio of shear stress τ to shear rate $\dot{\gamma}$:

$$\eta = \frac{\tau}{\dot{\gamma}} \cdot$$

The term "dynamic viscosity" is also used for η. The former unit centiPoise (cP) is equal to the unit mPa·s.

Figure 3: Flow curves
1 Rheopexic, 2 Thixotropic, 3 Newtonian, 4 Plastic, 5 Dilatant, 6 Intrinsically viscous, 7 Yield point.

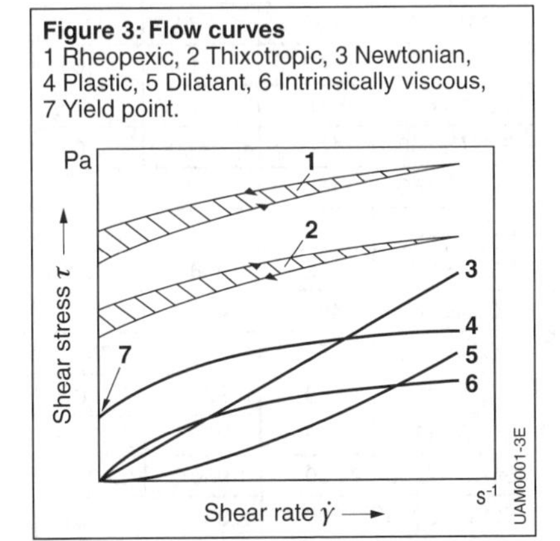

Figure 4: Velocity distribution and shear rate in the lubricating gap
1 Moving plate, 2 Fluid, 3 Fixed plate.
A Shear surface, F Shear force, h Height, v Velocity, $\dot{\gamma}$ Shear rate.

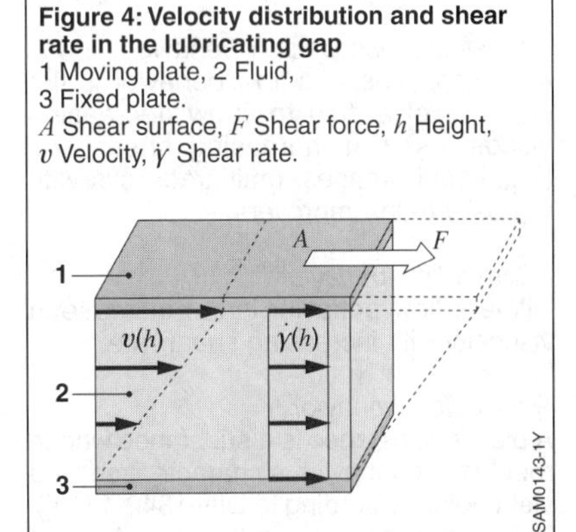

Midpoint viscosity
Each viscosity grade in accordance with DIN ISO 3448 [12] is defined by a viscosity range at a temperature of 40 °C (see Table 1). The permissible limit is ±10 %. Thus, for example, viscosity grade ISO VG 10 has the limits 9 and 11 mm²/s at a midpoint viscosity of 10 mm²/s.

HTHS viscosity
The HTHS viscosity (High Temperature High Shear) specifies the shear viscosity at a temperature of 150 °C and a shear rate of 10^6 s^{-1}.

Kinematic viscosity
The kinematic viscosity ν is usually determined in vertical, narrow glass tubes. The time the fluid needs at a defined temperature to flow through a specific section of the tube is measured. If density ρ is known, it is also possible to calculate the shear viscosity η from it.

$\nu = \frac{\eta}{\rho}$, (ν in mm²/s; ρ in kg/m³).

The former unit "centiStokes" (cSt) is equal to the unit mm²/s.

Newtonian fluids
Newtonian fluids display a linear relationship between shear stress τ and shear rate γ in the shape of a straight line through zero, with the slope increasing as a function of viscosity (Figure 3).

All materials not characterized by this kind of flow behavior are classified as non-Newtonian fluids.

Intrinsically viscous flow behavior
Intrinsically viscous flow behavior is the capacity of a fluid to show decreasing viscosities with increasing shear rate (e.g. liquid greases, multigrade oils with viscosity-index improvers).

Dilatant flow behavior
Dilatant flow behavior is the increase in viscosity with increasing shear rate.

Plastic flow behavior
A plastic substance is a substance whose rheological behavior is characterized by a yield point (according to DIN 1342-1 [13]). Plastic flow behavior is the formability of an intrinsically viscous fluid supplemented by yield point (e.g. lubricating greases).

Thixotropy
Thixotropy is a characteristic of those non-Newtonian fluids that display an decrease in viscosity proportional to shear time, and only gradually recover their original viscosity once shearing ceases.

Rheopexy
Rheopexy is a characteristic of those non-Newtonian fluids that display an increase in viscosity proportional to shear time, and only gradually recover their original viscosity once shearing ceases.

Table 1: Viscosity grades for industrial lubricating oils acc. to ISO 3448 [12]

ISO viscosity grade	Medium viscosity at 40 °C in mm²/s	Limits of kinematic viscosity at 40 °C in mm²/s	
		min.	max.
ISO VG 2	2.2	1.98	2.42
ISO VG 3	3.2	2.88	3.52
ISO VG 5	4.6	4.14	5.06
ISO VG 7	6.8	6.12	7.48
ISO VG 10	10	9.00	11.0
ISO VG 15	15	13.5	16.5
ISO VG 22	22	19.8	24.2
ISO VG 32	32	28.8	35.2
ISO VG 46	46	41.4	50.6
ISO VG 68	68	61.2	74.8
ISO VG 100	100	90.0	110
ISO VG 150	150	135	165
ISO VG 220	220	198	242
ISO VG 320	320	288	352
ISO VG 460	460	414	506
ISO VG 680	680	612	748
ISO VG 1000	1,000	900	1,100
ISO VG 1500	1,500	1,350	1,650
ISO VG 2200	2,200	1,980	2,420
ISO VG 3200	3,200	2,880	3,520

Dropping point
The dropping point is the temperature at which a lubricating grease attains a specified viscosity under specified test conditions (DIN ISO 2176 [14]).

Stribeck curve
The Stribeck curve portrays friction levels between two liquid- or grease-lubricated tribological systems separated by a narrowing gap (e.g. lubricated plain or ball bearings) as a function of sliding velocity (Figure 5). As the relative speed increases, so too does the hydrodynamic pressure in friction contact.

Solid-body friction
In the case of solid-body friction, the height of the lubricant layer is lower than that of the roughness protrusions in the material's surface. This causes wear.

Mixed friction
In the case of mixed friction, the height of the lubricant layer is approximately equal to that of the roughness protrusions. This still means increased wear because the roughness protrusions are in contact with each other.

Figure 5: Stribeck curve
R Surface roughness, F_N Normal force, d Distance between basic and opposed bodies.
Range a: Solid-body friction, high wear.
Range b: Mixed friction, moderate wear.
Range c: Hydrodynamics, no wear.

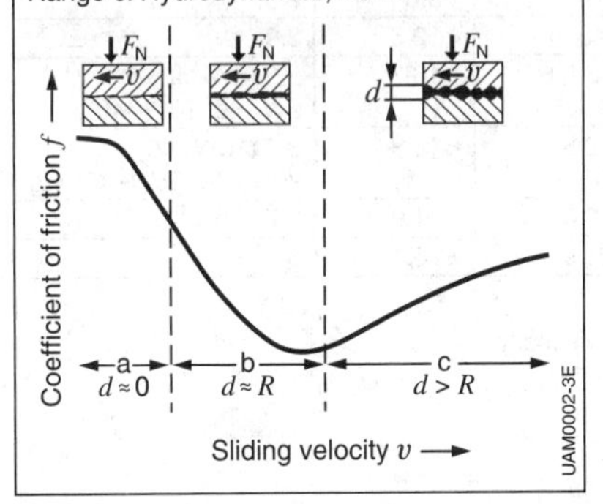

Hydrodynamics
In the case of hydrodynamics, there is complete separation between the primary body and the opposed body, e.g. when aquaplaning (virtually wear-free condition).

Synthetic oil
Synthetic oil is manufactured in a process of chemical synthesis from smaller molecules. For example, synthetic hydrocarbons in the form of poly-α-olefins, which are manufactured by the polymerization and subsequent hydrogenation of ethylene. Further synthetic oils are polyglycols, diester oils, silicone oils, and perfluoropolyether oils.

Viscosity
Viscosity is a measure of the internal friction of substances (DIN 1342 [13], DIN EN ISO 3104 [15]). It is caused by the resistance (internal friction) with which the substance's molecules oppose displacement forces (see Rheology).

As the temperature decreases, the viscosity increases and the oil becomes more viscous. At low temperatures the viscosity must not be too high so that the oil in the bearings does not exert too excessive a resistance on the rotational motion of the engine or transmission. At high temperatures the oil must still be sufficiently viscous to maintain the lubricating film.

Viscosity index
The viscosity index (VI) is a mathematically derived number expressing the change in a mineral-oil product's viscosity relative to its temperature. The greater the viscosity index, the lower the effect of temperature on the viscosity (DIN ISO 2909 [16]).

Table 2: Extract from SAE viscosity grades for transmission oils (SAE-J 306 revision June 2005)

SAE viscosity grade	Maximum temperature [°C] for dynamic viscosity at 150,000 mPa·s (ASTM D 2983 [20])	Kinematic viscosity [mm^2/s] at 100 °C (ASTM D 445 [21])	
		min.	max.
70 W	–55	4.1	–
75 W	–40	4.1	–
80 W	–26	7.0	–
85 W	–12	11.0	–
90	–	13.5	24.0
110	–	18.5	24.0
140	–	24.0	32.5
190	–	32.5	41.0
250	–	41.0	–

Viscosity grades

Oils are classified within specific viscosity ranges in viscosity grades:

- ISO viscosity grades (DIN ISO 3448 [12]): see Table 1
- SAE viscosity grades (SAE J 300 [17], SAE J 306 [18]): see Table 2 and Table 3.

Worked penetration

Worked penetration is the penetration of a grease sample after it has been processed in a grease kneader (DIN ISO 2137 [19]).

Table 3: SAE viscosity grades for engine and transmission oils (SAE J300, April 2013)

SAE viscosity grade	Viscosity (ASTM D 5293 [22]) mPa·s	Limit pumping viscosity (ASTM D 4684 [23]) with no yield point mPa·s	Kinematic viscosity (ASTM D 445 [21]) mm^2/s at 100 °C		Viscosity under high shear[1] (ASTM D 4683 [24], CEC L-36-A-90 [25], ASTM D 4741 [26] or ASTM D 5481 [27]) mPa·s at 150 °C and $10^6 s^{-1}$
	max.	max.	min.	max.	min.
0 W	6,200 at –35 °C	60,000 at –40 °C	3.8	–	–
5 W	6,600 at –30 °C	60,000 at –35 °C	3.8	–	–
10 W	7,000 at –25 °C	60,000 at –30 °C	4.1	–	–
15 W	7,000 at –20 °C	60,000 at –25 °C	5.6	–	–
20 W	9,500 at –15 °C	60,000 at –20 °C	5.6	–	–
25 W	13,000 at –10 °C	60,000 at –15 °C	9.3	–	–
16			6.1	<8.2	2.3
20	–	–	5.6	<9.3	2.6
30	–	–	9.3	<12.5	2.9
40	–	–	12.5	<16.3	2.9 (0W–40, 5W–40, 10W–40)
40	–	–	12.5	<16.3	3.7 (15W–40, 20W–40, 25W–40, 40)
50	–	–	16.3	<21.9	3.7
60	–	–	21.9	<26.1	3.7

[1] Also called HTHS (High Temperature High Shear) viscosity.

Engine oils

Engine oils are employed primarily to lubricate contiguous components in relative motion within the internal-combustion engine. The oil also removes heat generated by friction, carries abraded particles away from the friction surface, washes out contaminants, holds them in suspension, and protects metals against corrosion. The most common engine oils are mineral oils treated with additives (HD oils: Heavy Duty for high load, extreme operating conditions). Higher stress-resistance requirements combined with extended oil-change intervals have led to widespread application of fully and semi-synthetic oils (mixture of synthetic oil and mineral oil), e.g. hydro-crack oils. The quality of an engine oil is determined by its origin, the refining processes used on the mineral oil (except in the case of synthetic oils) and the additive composition.

Additives are classified according to their respective functions:
- Viscosity-index improvers
- Pour-point improvers
- Oxidation and corrosion inhibitors
- Detergent and dispersant additives
- Extreme-pressure (EP) additives
- Friction modifiers
- Anti-foaming agents.

Oil is subjected to considerable thermal and mechanical stresses in the IC engine. The data on the physical properties of engine oils provide information on their application limits (SAE viscosity grades), but are not indicative of their other performance characteristics. Therefore, there are several different test procedures for evaluating engine oils:
- ACEA (Association des Constructeurs Européens de l'Automobile) standards, replaced the CCMC standards (Comité des Constructeurs d'Automobiles du Marché Commun) at the beginning of 1996
- API classification (American Petroleum Institute)
- MIL specifications (Military)
- Company specifications (e.g. ILSAC, International Lubricants Standardization and Approval Committee).

The approval criteria include the following:
- Sulfate-ash content
- Zinc content
- Engine type (diesel or spark-ignition engines, naturally aspirated or forced-induction engines)
- Load on power-transmission components and bearings
- Wear-protection properties
- Oil operating temperature (in oil pan)
- Combustion residue and chemical stress exerted on the oil by acidic combustion products
- The oil's detergent and residue-scavenging properties
- Its suitability for use with gasket and sealing materials.

ACEA (CCMC) Specifications

Engine oils for gasoline engines
- A1: Special high-lubricity oils with reduced viscosity at high temperatures and high shear for reducing viscous friction.
- A2: Conventional and high-lubricity engine oils without any restriction on viscosity grades. Higher requirements than CCMC G4 and API SH, withdrawn.
- A3: Oils of this category meet higher requirements than A2 and CCMC G4 and G5.
- A5: Improved "fuel economy" properties compared with A3 occasioned by lower viscosity including improved additive composition. Only for use in engines which have been specifically designed for this purpose.

Engine oils for passenger-car diesel engines
- B1: Corresponding to A1 for low friction losses and, consequently, reduced fuel consumption.
- B2: Conventional and high-lubricity engine oils compliant with the current minimum requirements (higher than those of CCMC PD2), withdrawn.
- B3: Exceeds B2.
- B4: Corresponds to B2, particularly suitable for VW TDI engines.
- B5: Oils surpass B3 and B4, improved "fuel economy" properties, also satisfies VW 50600 and 50601. Only for use in engines which have been specifically designed for this purpose.

Engine oils for passenger-car diesel engines with particulate filters
- C1: Since 2004, sulfate-ash content max. 0.5 %. Lowered HTHS (Ford).
- C2: Since 2004, sulfate-ash content max. 0.8 %. With HTHS > 2.9 mPa · s (Peugeot).
- C3: Since 2004, sulfate-ash content max. 0.8 %. With HTHS > 3.5 mPa · s (MB and BMW).

Engine oils for commercial-vehicle diesel engines
- E1: Oils for naturally aspirated and turbocharged engines with normal intervals between oil changes; withdrawn.
- E2: Derived from MB Specification Sheet 228.1. Primarily for engine designs predating the Euro II standard.
- E3: For Euro II engines, derived from MB Specification Sheet 228.3. In comparison with the predecessor category CCMC D5, these oils evince a significant improvement in soot dispersion capability and a much reduced tendency to thicken; withdrawn.
- E4: Diesel engines with Euro I to Euro III standards and high requirements, particularly for extended intervals between oil changes (according to manufacturer's specification). Based to a large extent on MB Specification Sheet 228.5.
- E5: For Euro III engines, reduced ash content.
- E6: For EGR engines (exhaust-gas recirculation) with/without diesel particulate filters and SCR-NO_x engines (Selective Catalytic Reduction, i.e. reduction of NO_x to N_2). Recommendation for engines with diesel particulate filters and operation with sulfur-free fuel, sulfate ash less than 1 % by mass.
- E7: For engines without diesel particulate filters of most EGR engines and of most SCR-NO_x engines, sulfate ash max. 2 % by mass.

Example of a designation:
According to the grade specification, a numerical code is usually supplemented.

Example: An A3/B3-04 is an engine oil for spark-ignition engines (Grade A) and diesel engines (Grade B) of quality category 3, tested accord to the ACEA classification grade issued in 2004.

API classification grades
- S Grades (Service) for gasoline engines.
- C Grades (Commercial) for diesel engines.
- SF: For engines produced in the 1980s; withdrawn.
- SG: Applicable since 1988, with more stringent sludge test, improved oxidation stability and wear protection.
- SH: Since mid-1993, corresponds to API SG quality level, but with more stringent process requirements for oil grade testing.
- SJ: Since October 1996, more tests than API SH.
- SL: Since 2001, lower oil consumption compared with SJ, lower volatility, improved engine cleanness, greater resistance to aging.
- SM: Since 2004, for gasoline and light diesel engines with improved wear protection, higher aging stability, improved pumpability, and also used oil.
- SN: Since October 2010, compared with SM improvements in piston cleanliness, sludging, and exhaust-gas treatment capability. Elastomer compatibility is also defined.
- CC: Engine oils for non-turbocharged diesel engines with low stress factors; withdrawn.
- CD: Engine oils for non-turbocharged and turbocharged diesel engines, replaced by API CF in 1994.
- CD-2: API CD requirements, plus additional requirements relevant to two-stroke diesel engines.
- CF-2: Oils with special two-stroke properties (since 1994).
- CE: Oils with CD performance characteristics, with supplementary test operation in US Mack and Cummins engines.
- CF: Replaced API CD in 1994. Specially for indirect injection, even if the sulfur content of the fuel is greater than 0.5 %.
- CF-4: As API CE, but with more stringent test procedure in single-cylinder Caterpillar turbo-diesel engines.

- CG-4: For diesel engines operating under very stringent conditions. Exceeds API CD and CE. Fuel sulfur content less than 0.5 %. Required for engines compliant with post-1994 emission-control legislation.
- CH4: Modorn commercial-vehicle engine oil since 1998. Surpasses CG-4 in standards of wear, soot and viscosity. Longer oil-change intervals.
- CI-4: Since 2002 for high-speed four-stroke engines which can only still meet future exhaust-emission legislation with EGR. Suitable for sulfur content greater than 0.5 %.
- CI-4 plus: As CI-4, but with improved soot transport and higher requirements with regard to viscosity increase.
- CJ-4: Valid since 2006. For highway vehicles which must satisfy the USA 2007 emissions standards with diesel power (below 500 ppm sulfur content). Suitable for particulate-filter and NO_x reduction catalytic converters.

ILSAC

ILSAC (International Lubricants Standardization and Approval Committee) is a joint standard of General Motors, Daimler, the Japanese Automobile Manufacturers Association, and the American Engine Manufacturers Association.

ILSAC GF-3

In addition to API-SL, the standard requires a fuel-economy test.

ILSAC GF-4

Allocation to API-SM.

ILSAC GF-5

This standard has been in force since October 2010 with the aim of increased fuel reduction, elastomer compatibility, and improved protection of the exhaust-gas treatment systems (comparable with API- SN).

SAE viscosity grades

(SAE J300 [17], SAE J306 [18])

The SAE (Society of Automotive Engineers) grades are the internationally accepted standard for defining viscosity. It provides information on the temperature range in which the oils are used. The standard provides no information on the quality of the oil.

A distinction is drawn between single-grade and multigrade oils. Single-grade oils are designated for example by SAE 30. Low code numbers denote thin-bodies oils, higher code numbers denote more viscous oils.

Multigrade oils are the type in widespread use today. Two series are employed for the designation (see Table 2 and Table 3), where the letter "W" (Winter) is used to define specific cold-flow properties. These oils are also suitable for low temperatures. The viscosity grades including the letter "W" are rated according to maximum cold viscosity, maximum viscosity pumping temperature and the minimum viscosity at 100 °C.

Viscosity grades without the "W" are rated only according to viscosity at 100 °C.

Viscosity specifications at higher temperature (150 °C) and high shear stress (HTHS viscosity, see Table 3) are more practical. This value acquired particular significance with the introduction of oils with reduced high-temperature viscosity (below 3.5 mPa·s). These oils may only be used if manufacturer approvals exist.

A multigrade oil is designated for example by SAE 0W-30. This means that it has at –35 °C a viscosity of max. 6,200 mPa·s and satisfies the specification of an SAE 30 oil with a kinematic viscosity in the range of 9.3...12.5 mm^2/s at 100 °C and an HTHS viscosity of min. 2.9 mPa·s at 150 °C (Table 3, see also Figure 1).

Transmission oils

Specifications for transmission oils are defined by the type of transmission and the stresses to which it is subjected throughout the entire range of operating conditions. The requirements (high pressure resistance, high viscosity stability relative to temperature, high resistance to aging, good anti-foaming properties, compatibility with gaskets and seals) can only be satisfied by lubricants treated with special additives. In contrast to engine oils, transmission oils contain no, or only minimal, detergent additives, considerably fewer basic constituents, and mostly no viscosity-index improvers. Most of them would be sheared and thus become inactive. The use of unsuitable and qualitatively inferior oils results typically in damage to bearings and gear-tooth flanks.

The viscosity must also suit the specific application. Viscosity grades for vehicular transmissions are defined in SAE J 306 [18] and SAE J300 [17] (Table 2 and Table 3). Transmission oils with roughly the same viscosity compared with engine oils are identified by higher code numbers so that they can be clearly distinguished from engine oils.

Synthesized oils are being increasingly used to meet special requirements (e.g. poly-α-olefins). Advantages over standard mineral oils include superior temperature-viscosity properties and increased aging resistance.

API classifications of transmission oils

- GL-1...GL-3: Outdated, no longer of practical significance.
- GL-4: Transmission lubricants for moderate-stressed hypoid-gear transmissions, and for transmissions which operate at extreme speeds and impact loads, high rotational speeds and low torques, or low rotational speeds and high torques.
- GL-5: Transmission lubricants for high-stressed hypoid-gear transmissions in passenger cars and in other vehicles where they are exposed to impact loads at high rotational speeds, and at high rotational speeds and low torques, or low rotational speeds and high torques.
- MT-1: For non-synchromesh manual transmissions in US trucks.

Many truck and component manufacturers have drawn up their own specifications and no longer rely on API classification grades.

Lubricants for automatic transmissions

Automatic transmissions differ from their manual-shifted counterparts in the way they transfer torque; non-positive mechanical and hydrodynamic force transfer is supplemented by friction coupling arrangements. Thus, the friction response of automatic transmission fluids (ATF) is extremely important. Applications are basically classified according to the friction characteristics.

General Motors:

- Superseded grades: Type A, Suffix A, DEXRON, DEXRON B, DEXRON II C, DEXRON II D.
- DEXRON II E: Valid through 1994.
- DEXRON III F/G: Valid since 1994 and 1997. Features more stringent requirements for oxidation stability and consistency in frictional coefficient.
- DEXRON III H: Valid since 2005, but outdated since 2007.
- DEXRON VI: Valid since 2006.

Other manufacturers

(among others. Ford, MB, MAN, Mack, Scania, ZF): In accordance fluids and lubricants specifications.

Lubricating oils

Lubricating oils comprise the components of basic oil and additive. The additives improve the properties of the basic oils, e.g. with regard to oxidation stability, corrosion protection, protection against scoring and seizure, or viscosity-temperature characteristics. In addition, system properties such as friction and wear are optimized in the desired direction.

There are a wide range of designations in the form of letters and numbers (see among others DIN 51502 [28]) for the most varied applications. For example, hydraulic fluids:

- HL: Mineral-oil-based hydraulic fluid with additives for improving corrosion protection and resistance to aging.
- HLP: As HL, but with extra additives against scoring and seizure.
- HVLP: As HLP, with extra viscosity-index improver.

Lubricating greases

Lubricating greases are thickened lubricating oils. A great advantage that greases enjoy over oil is that they do not drain from the friction surfaces. Complicated measures to seal them in place are therefore unnecessary (e.g. application in wheel bearings pumps, alternators, windshield-wiper motors, servo motors). Table 4 provides a general overview of the components in a consistent lubricating grease as blended from three basic components – base oil, thickener, and additive.

Mineral oils are usually employed as the basic oil component, although full-synthetic oils have recently become more common as a replacement (e.g. due to more stringent requirements for aging stability, cold-flow properties, viscosity-temperature characteristics).

Thickeners are used as a binder for the base oil; metal soaps being generally employed (Figure 6). They bind the oil in a sponge-like soap structure (micelle) by means of inclusions and physical Van der Waals forces. The higher the proportion of thickener (depends on the type of

Table 4: Composition of lubricating greases
Whatever the friction pairing, the large number of lubricant components can be used to develop a high-performance lubricant.

Base oils	Thickeners	Additives
Mineral oils – Paraffinic – Naphthenic – Aromatic Poly-α-olefins Alkyl aromatics Diester oils Polyhydric alcohols Silicones Phenyletherols Perfluorpolyethers	Metal soaps of metals Li, Na, Ca, Ba, Al Normal soaps (soaps with a carbonic acid, e.g. stearic acid) Hydroxy soaps (soaps with an additional hydroxide group, e.g. 12-hydroxystearic acid Complex soaps (e.g. Ca soaps with a short- and a long-chain carbonic acid) Polyureas PTFE PE Bentonites Silica gels	Oxidation inhibitors Fe-, Cu ions, sequestering agents Corrosion inhibitors Extreme-pressure additives (EP additives) Wear-protection additives (anti-wear additives) Friction reducers (friction modifiers) Adhesion improvers Detergents, dispersants VI improvers Solid lubricants

thickener) in the grease, the less the penetration and the higher the NLGI grade (National Lubricating Grease Institute, US standard, see Table 5).

The additives serve to modify physical and chemical properties of the lubricating grease to achieve specific objectives (such as improvement of anti-oxidation properties, increased protection against scoring and seizure, and reduction of friction and wear).

Solid lubricants (e.g. MoS_2) are also added to lubricating greases (for instance for lubricating constant-velocity joints in motor vehicles).

Table 5. Consistency grades for lubricating greases (DIN 51818 [29])

NLGI grade	Worked penetration acc. to DIN ISO 2137 [19] in units of 0.1 mm
000	445...475
00	400...430
0	355...385
1	310...340
2	265...295
3	220...250
4	175...205
5	130...160
6	85...150 (unworked penetration)

Specific lubricating greases are selected with reference to their physical characteristics and their effects on the friction surface, and to minimize interaction between the grease and the contact materials. Mutually antagonistic effects with polymers are, for example: formation of stress cracks (Figure 7), consistency changes, polymer degradation, swelling, shrinkage, and brittleness.

Thus, for example, mineral-oil greases and greases based on synthetic hydrocarbons should not come in contact with elastomers used together with brake fluid (polyglycol base) (e.g. substantial swelling of EPDM elastomers).

In addition, lubricating greases with varying compositions should not be mixed (changes in physical properties, grease liquefaction due to drop-point reduction).

Thermal and mechanical stresses result in chemical and/or physical changes which may have a detrimental effect on the function of the entire tribological system (Figure 8). Oxidation, for example, results in acidification, which can trigger corrosion on metal surfaces or stress cracking on some plastics. In the event of excessive thermal load, polymerization can cause the lubricant to solidify.

Every chemical change automatically causes a change in physical properties. These include the rheological properties

Figure 6: Photo of lithium soap taken by a scanning electron microscope
The oil is retained between the twisted soap fibrils.

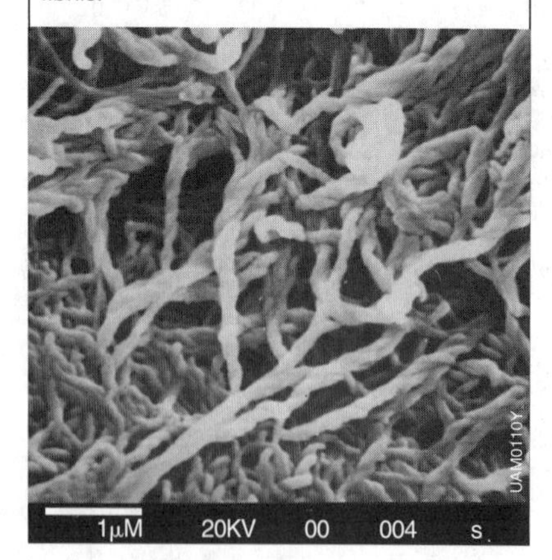

Figure 7: Stress cracks on a gearwheel made of polyoximethylene (POM), caused by poly-α-olefin (PAO)

as well as changes in the viscosity-temperature characteristics or the dropping point. A marked lowering of the dropping point would result in the lubricant flowing away from the friction surface, even at moderate heat.

It is particularly important to remember that metals such as iron or copper (or metals containing copper such as bronze or brass) catalyze the oxidation of a lubricant, i.e. oxidation occurs much more quickly than without catalyst contact. Oxidation quickly renders the lubricity of grease insufficient. Often the soap structure decomposes, the grease then becomes oily, flows away from the friction surface, or hardens due to polymerization.

By correctly matching the lubricating grease and tribological system while taking into account the load and interaction, it is possible to enhance the performance potential of products substantially with sliding-contact opposed parts (e.g. transmission, friction or roller bearings, actuator and control systems) [30].

Figure 8: Stress exerted on lubricant and the resulting effects

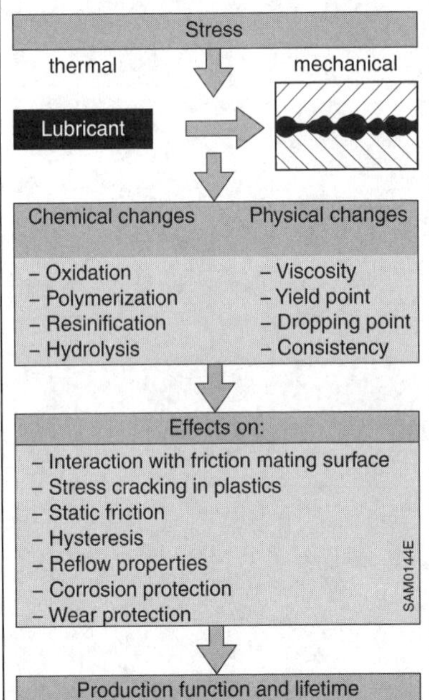

References

[1] T. Mang, W. Dresel: Lubricants and Lubrication. 2nd Edition, Wiley-VCH Verlag GmbH, 2007.

[2] Thomas Mezger: Das Rheologie-Handbuch. 3rd Edition, Vincentz-Network, 2012.

[3] Wilfried J. Bartz: Schmierfette. Expert-Verlag, 2000.

[4] DIN 51575: Testing of mineral oils – Determination of sulfated ash (publication 2011).

[5] DIN 51817: Testing of lubricants – Determination of oil separation from greases under static conditions (publication 1998).

[6] DIN ISO 2592: Petroleum products – Determination of flash and fire points – Cleveland open cup method (ISO 2592:2000); German version EN ISO 2592:2002.

[7] DIN EN 23015: Petroleum products; determination of cloud point (ISO 3015:1992); German version EN 23015:1994.

[8] DIN 51805: Testing of lubricants; determination of flow pressure of lubricating greases, Kesternich method (publication 1974).

[9] DIN 1342-3: Viscosity – Part 3: Non-newtonian liquids (publication 2003).

[10] DIN ISO 2137: Petroleum products – Lubricating grease and petrolatum – Determination of cone penetration (ISO 2137:1985) (publication 1997).

[11] DIN ISO 3016: Petroleum oils; determination of pour point (publication 1982).

[12] DIN ISO 3448: Industrial liquid lubricants – ISO viscosity classification (ISO 3448:1992) (publication 2010).

[13] DIN 1342-1: Viscosity – Part 1: Rheological concepts (publication 2003).

[14] DIN ISO 2176: Petroleum products – Lubricating grease – Determination of dropping point (ISO 2176:1995) (publication 1997).

[15] DIN EN ISO 3104: Petroleum products – Transparent and opaque liquids – Determination of kinematic viscosity and calculation of dynamic viscosity (ISO 3104:1994 + Cor. 1:1997); German version EN ISO 3104:1996 + AC:1999

[16] DIN ISO 2909: Petroleum products – Calculation of viscosity index from kinematic viscosity (ISO 2909:2002) (publication 2004).

[17] SAE J300: Engine Oil Viscosity Classification (publication 2013).
[18] SAE J306: Automotive Gear Lubricant Viscosity Classification (publication 2005).
[19] DIN ISO 2137: Petroleum products – Lubricating grease and petrolatum – Determination of cone penetration (ISO 2137:1985) (publication 1997).
[20] ASTM D 2983: Standard Test Method for Low-Temperature Viscosity of Lubricants Measured by Brookfield Viscometer (publication 2009).
[21] ASTM D 445: Standard Test Method for Kinematic Viscosity of Transparent and Opaque Liquids (and Calculation of Dynamic Viscosity) (publication 2012).
[22] ASTM D 5293: Standard Test Method for Apparent Viscosity of Engine Oils and Base Stocks Between –5 and –35 °C Using Cold-Cranking Simulator (publication 2010).
[23] ASTM D 4684: Standard Test Method for Determination of Yield Stress and Apparent Viscosity of Engine Oils at Low Temperature (publication 2012).
[24] ASTM D 4683: Standard Test Method for Measuring Viscosity of New and Used Engine Oils at High Shear Rate and High Temperature by Tapered Bearing Simulator Viscometer at 150 °C (publication 2010).
[25] CEC L-36-A-90: European standard test for HTHS viscosity, technically identical to ASTM 4741.
[26] ASTM D 4741: Standard Test Method for Measuring Viscosity at High Temperature and High Shear Rate by Tapered-Plug Viscometer (publication 2012).
[27] ASTM D 5481: Standard Test Method for Measuring Apparent Viscosity at High-Temperature and High-Shear Rate by Multicell Capillary Viscometer (publication 2010).
[28] DIN 51502: Designation of lubricants and marking of lubricant containers, equipment and lubricating points (publication 1990).
[29] DIN 51818: Lubricants; consistency classification of lubricating greases; NLGI grades (publication 1981).
[30] P. M. Lugt: Grease Lubrication in Rolling Bearings. 1st Edition, John Wiley & Sons, 2013.

Fuels

Characteristics

Gasoline and diesel fuels are the product of graduated distillation of crude oil. They consist of a multitude of individual hydrocarbons.

Boiling range and ignition temperature

The boiling range of gasoline is between 30 °C and 210 °C, and that of diesel fuel between 180 °C and 370 °C. Diesel fuel ignites on average at approximately 350 °C (lower limit 220 °C), which is very early in comparison with gasoline (on average 500 °C).

Calorific value

Normally the net calorific value H_n (former term: lower calorific value) is specified for the energy content of fuels; it corresponds to the usable heat quantity released during full combustion.

The gross calorific value H_g (former term: upper calorific value), on the other hand, specifies the total reaction heat released and therefore comprises, as well as the mechanically usable heat, the latent heat created in the water vapor. However, this component is not used in the vehicle.

At 42.9 – 43.1 MJ/kg, the net calorific value of diesel fuel is slightly higher than that of gasoline (40.1 - 41.9 MJ/kg).

Oxygenates, i.e. fuels or fuel constituents containing oxygen, such as alcohol fuels, ether, or fatty acid methyl ester, have a lower calorific value than pure hydrocarbons because the oxygen bonded in them does not contribute to the combustion process. Comparable engine power with conventional fuels therefore results in higher fuel consumption.

Calorific value of air/fuel mixture

This is the calorific value of the combustible air/fuel mixture. It is dependent on the air/fuel ratio and determines the engine's power output. With a stoichiometric air/fuel ratio, this is roughly 3.5 – 3.7 MJ/m^3 for all liquid fuels and liquified petroleum gases.

Sulfur content

In the interests of reducing sulfur-dioxide emissions (SO_2) and protecting the catalytic converters for exhaust-gas treatment, the sulfur content of gasolines and diesel fuels was limited for vehicle applications on a Europe-wide basis to 10 mg/kg as from 2009. Fuels which adhere to this limit value are known as “sulfur-free fuels”. In this way, the final stage of the desulfurization of fuels is achieved. Prior to 2009 only low-sulfur fuel (sulfur content < 50 mg/kg) – introduced at the beginning of 2005 – was permitted for use in Europe. Germany has led the way in the desulfurization of fuels and already back in 2003 had established sulfur-free fuel by way of tax measures.

In the USA, the limit value for the sulfur content of gasolines commercially available to the end user has since 2006 been set at max. 80 mg/kg, although an upper average value of 30 mg/kg for the total amount of sold and imported fuel is in place. Individual states, California for example, have laid down lower limits.

Furthermore, 2006 saw the beginning of the Introduction in the USA of sulfur-free diesel fuels for highway applications (ULSD, Ultra Low Sulfur Diesel, sulfur content max. 15 mg/kg). By the end of 2009, however, only 20 % of fuels would have a sulfur content of max. 500 mg/kg. The sulfur content of certification fuels is subject to separate regulations.

Gasolines

Fuel grades

Super fuels in Germany

In Germany two Super fuels with 95 octane, which differ in their ethanol content, are sold; the maximum permissible ethanol content is 5 % by volume for Super and 10 % by volume for Super E10. A Super Plus fuel with 98 octane is also available. Individual suppliers have replaced their Super Plus fuels with 100-octane fuels (V-Power 100, Ultimate 100, Super 100), which have had their basic quality and additives altered.

Fuels in the USA

Three different types of fuel are sold in the USA: Regular (92 octane), Premium (94 octane) and Premium Plus (98 octane). Fuels sold in the USA usually contain ethanol at 10 % by volume. The addition of components containing oxygen increases the octane number and satisfies the requirements of modern, higher-compression engines with regard to improved knock resistance.

Reformulated gasoline is the term used to describe gasoline which, through its altered composition, generates fewer evaporative and pollutant emissions than conventional gasoline. The demands placed on reformulated gasoline are laid down in the USA in federal law (Clean Air Act). This legislation prescribes, for example, lower limits for vapor pressure, benzene and overall aromatic content, and final boiling point. It also prescribes the use of additives to keep the intake system free of contamination and deposits.

Fuel standards

European Standard EN 228 [1] defines the requirements for unleaded fuel for use in spark-ignition (SI) engines. Further country-specific characteristic values are set out in the national appendices to this standard. Leaded gasoline is prohibited in Europe.

The US specifications defining fuels for spark-ignition engines are contained in ASTM D4814 (American Society for Testing and Materials) [2].

Most fuels for gasoline engines which are sold today contain components which in turn contain oxygen (oxygenates). In this respect, ethanol in particular has gained in importance, since the "EU Biofuels Directive" stipulates minimum content figures for renewable fuels (see Alternative fuels). Many countries have defined minimum quotas for biogenic components in gasolines, which are achieved for the most part with bioethanol. But also those ethers which can be produced from methanol or ethanol MTBE (methyl tertiary butyl ether) and ETBE (ethyl tertiary butyl ether) are used, of which, depending on the quality of the basic fuel, up to 22 % by volume may be added in Europe.

Adding alcohols increases the volatility of the fuel and can affect the strength of materials, e.g. the swelling of elastomers and the corrosion behavior of metals. Depending on the alcohol content and the temperature, even the admission of only small amounts of water may result in demixing and formation of an aqueous alcohol phase.

The ethers do not encounter the problem of demixing. The ethers, which have a lower vapor pressure, a higher calorific value and a higher octane number than ethanol, are chemically stable components with good material compatibility. They therefore demonstrate advantages, both from a logistical and an engine standpoint. For reasons of sustainability, because of the CO_2 savings to be attained and due to the laying down of quotas for biogenic fuels, ETBE is used in the main. Many plants originally designed for the production of MTBE have been converted to the production of ETBE.

Where the ethanol content had long been limited in the European gasoline standard EN 228 to 5 % by volume (E5), the 2013 edition contains in the first instance a specification for ethanol at 10 % by volume (E10). At present in the European market not all vehicles are yet equipped with materials to allow operation with E10. As the second quality, therefore, an exemption grade with a max. ethanol content of 5 % by volume is retained. In Germany the new edition DIN EN 228:2013 replaces the draft standard E DIN 51626-1:2010-04 [3], which was published in April 2010, to enable E10 to be introduced early to the German market. E 10 is conceived as the main grade in the gasoline market, but as yet has failed to find favor with consumers.

In the USA most gasolines contain ethanol at 10 % by volume (E 10), for which a vapor pressure higher by approximately 7 kPa is permitted according to the American standard ASTM D4814. A vapor pressure that is up to 8 kPA higher is also permitted in EN 228 (2013), depending on the ethanol content.

In Brazil, gasoline always contains ethanol at 18 to 26 % by volume.

Characteristics

Density

European standard EN 228 limits the density of gasolines to 720 – 775 kg/m^3.

Octane number

The octane number defines the gasoline's antiknock quality (resistance to pre-ignition). The higher the octane number, the greater the resistance to engine knock. Iso-octane (trimethyl pentane), which is extremely knock-resistant, is assigned the octane number 100, while n-heptane, which is extremely knock-susceptible, is assigned the number 0.

The octane number of a fuel is determined in a standardized test engine. The numerical value corresponds to the proportion (in % by volume) of iso-octane in a mixture of iso-octane and n-heptane which demonstrates the same knock resistance as the fuel to be tested.

Research and motor octane numbers

The octane number determined in testing using the Research Method ([4]) is the RON (research octane number). It serves as the essential index of acceleration knock. The octane number determined in testing using the Motor Method ([5]) is the MON (motor octane number). The MON basically provides an indication of the tendency to knock at high speeds.

The Motor Method differs from the Research Method by using preheated mixtures, higher engine speeds and variable ignition timing. This places more stringent thermal demands on the fuel under examination. MON figures are lower than those for RON.

Enhancing knock resistance

Normal (untreated) straight-run gasoline displays a low antiknock quality. Only by mixing such gasoline with different knock-resistant refinery components (reformed components, isomerisates) is it possible to produce fuels with high octane numbers suitable for modern, high-compression engines. It is possible to increase knock resistance by adding components containing oxygen such as alcohols and ethers.

Additives containing metal to increase the octane number (e.g. MMT: methylcyclopentadienyl manganese tricarbonyl) form ashes during combustion. MMT is therefore excluded in EN 228 (2013) by a limit value for manganese in the trace range.

Characteristics of fuel volatility

The volatility of gasoline has both upper and lower limits. On the one hand, they must contain an adequate proportion of highly volatile components to ensure reliable cold starting. At the same time, volatility should not be so high as to lead to starting and performance problems during operation in high-temperature environments (vapor lock). Yet another factor is environmental protection, which demands that evaporative losses be kept low.

Fuel volatility is defined by different characteristic quantities. EN 228 defines for E5 and E10 ten different volatility classes distinguished by various levels of boiling curve, vapor pressure and VLI (vapor-lock index). To meet special requirements stemming from variations in climatic conditions, European countries can incorporate specific individual classes into their own national appendices in the standard. Different values are laid down for summer and winter.

A similar provision is included in the American gasoline standard ASTM D4814. Comprehensive tables describe the scope of application of the volatility classes, depending on the calender month and state.

Boiling curve

In order to assess the fuel in vehicle operation, it is necessary to view the individual areas of the boiling curve separately. EN 228 therefore contains limit values laid down for the volume of fuel that vaporizes at 70 °C, at 100 °C and at 150 °C. The volume of fuel that vaporizes at up to 70 °C must achieve a minimum volume in order to ensure that the engine starts easily when cold (this was previously important above all for vehicles with carburetors). However, the volume of fuel that vaporizes must not be too great either, otherwise vapor bubbles may be formed when the engine is hot. While the volume of fuel that vaporizes at up to 100 °C determines the engine's warm-up characteristics, this factor's most pronounced effects are reflected in the acceleration and response provided by the engine once it warms to normal operating temperature. The volume of fuel that vaporizes at up to 150 °C should be high enough to minimize dilution of the engine oil. In particular when the engine is cold, the non-volatile gasoline components find it difficult to vaporize and can pass from the combustion chamber via the cylinder walls into the engine oil.

Vapor pressure

Fuel vapor pressure as measured at 37.8 °C (100 °F) in accordance with EN 13016-1 [6] is primarily an index of the safety with which the fuel can be pumped into and out of the vehicle's tank. Vapor pressure has upper and lower limits in all specifications. In Germany, for example, it is max. 60 kPa in summer and max. 90 kPa in winter.

In order to configure a fuel-injection system, it is also important to know the vapor pressure at higher temperatures (80...120 °C) because a rise in the vapor pressure due to the admixture of alcohol, for example, becomes apparent only at higher temperatures. If the fuel vapor pressure rises above the fuel-injection system pressure, for example during vehicle operation due to the effect of the engine temperature, this may result in malfunctions caused by vapor-bubble formation.

Table 1: DIN EN 228 (January 2013): Selected requirements of gasolines

Requirements	Unit	Characteristic	
Quality		Super E10	Super
Knock resistance: RON/MON for Super, min.	–	95/85 [1]	
Density (at 15 °C), min./max.	kg/m^3	720/775	
Sulfur, max.	mg/kg	10	
Oxygen content, max.	% by mass	3.7	2.7
Ethanol content, max.	% by vol.	10.0	5.0
Benzene, max.	% by vol.	1	
Lead, max.	mg/l	5	
Volatility			
Vapor pressure, summer, min./max.	kPa	45/60	
Vapor pressure, winter, min./max. [2]	kPa	60/90	
Evaporated volume at 70 °C, summer, min./max.	% by vol.	22/50	20/48
Evaporated volume at 70 °C, winter, min./max.	% by vol.	24/52	22/50
Evaporated volume at 100 °C, min./max.	% by vol.	46/72	46/71
Evaporated volume at 150 °C, min./max.	% by vol.	75/–	
Final boiling point, max.	°C	210	
VLI [3] transitional period [4], max. [2]		1164	1150

[1] Applicable to Super Plus: RON/MON, min. 98/88; ethanol content, max. 10.0 % by vol.
[2] National values for Germany.
[3] VLI Vapor-lock index.
[4] Spring and fall.

Vapor/liquid ratio
The vapor/liquid ratio is a measure of a fuel's tendency to form bubbles. It refers to the volume of vapor generated by a specific quantity of fuel at a defined back pressure and a defined temperature. A drop in pressure (e.g. when driving over a mountain pass), or an increase in temperature, will raise the vapor/liquid ratio and with it the possibility of operating problems. ASTM D4814 lays down, for example for each volatility class, a temperature at which a vapor/liquid ratio of 20 must not be exceeded.

Vapor-lock index
The vapor-lock index (VLI) is the mathematically calculated sum total of ten times the vapor pressure (in kPa at 37.8 °C) and seven times the volume of fuel that vaporizes at up to 70 °C. It is possible with this additional limit value to restrict the volatility of the fuel further with the result that the two maximum values for vapor pressure and boiling data cannot be achieved in the course of its production.

Additives
Additives can be added to improve fuel quality in order to counteract deteriorations in engine performance and in the exhaust-gas composition during vehicle operation. The packages generally used combine individual components with various attributes.

Extreme care and precision are required both when testing additives and in determining their optimal compositions and concentrations. Undesirable side-effects must be avoided. Basic additives are added at the refinery to protect the plants and to ensure a minimum fuel quality. To further improve quality, brand-specific multifunction additives can be added at the refinery's filling stations when the road tankers are filled (end-point dosing).

Detergents
The entire intake system (fuel injectors, intake valves) should remain free of contamination and deposits for several reasons. A clean intake tract is essential for maintaining the factory-defined air/fuel ratios, as well as for trouble-free operation and minimal exhaust emissions. To achieve this end, effective detergent additives should be added to the fuel.

Corrosion inhibitors
The ingress of water/moisture may lead to corrosion in fuel-system components. Corrosion can be effectively eliminated by the addition of corrosion inhibitors, which form a thin protective film on the metal surface.

Oxidation stabilizers
Anti-aging agents (antioxidants) are added to fuels to improve their stability during storage. They prevent rapid oxidation caused by oxygen in the air.

Diesel fuels

Fuel standards

The standard that lays down the requirements for diesel fuels in Europe is EN 590 [7]. The most important characteristics are set out in Table 2. Even the premium qualities additionally sold at some filling stations (e.g. Super diesel, Ultimate, V-Power diesel) satisfy this standard. There are differences in the basic quality, in the addition of alternative, purely paraffinic fuel components, and in the additives.

Up to 7 % by volume biodiesel (FAME) can be admixed with diesel fuel in accordance with EN 590, where the quality of the biodiesel is stipulated by standard EN 14214 [8]. The admixture of biodiesel improves lubricity, but also reduces oxidation stability. For the purpose of checking oxidation stability, EN 590 was supplemented in 2009 to include the parameter aging reserve, which is measured as an induction period at 110 °C under the test conditions defined in EN 15751 [9] and must be at least 20 hours.

The US standard for diesel fuels, ASTM D975 [10], specifies a smaller number of quality criteria and defines less stringent limits. It permits the admixture of max. 5 % by volume biodiesel, which must satisfy the requirements of standard ASTM D6751 [11].

Characteristics

Cetane number

The cetane number (CN) expresses the ignition quality of the diesel fuel [12]. The higher the cetane number, the greater the fuel's tendency to ignite. As the diesel engine dispenses with an externally supplied ignition spark, the fuel must ignite spontaneously (auto-ignition) and with minimum delay (i.e. ignition lag) when injected into the hot, compressed air in the combustion chamber. Cetane number 100 is assigned to n-hexadecane (cetane), which ignites very easily, while slow-igniting α-methyl naphthalene is allocated cetane number 0. The cetane number of a diesel fuel is determined in a standardized CFR single-cylinder test engine with a variable compressor piston (CFR, Cooperative Fuel Research). The compression ratio is measured at constant ignition lag. Comparison fuels comprising cetane and α-methyl naphthalene are operated with the determined compression ratio. The proportion of cetane in the mixture is altered until the same ignition lag is obtained. The cetane proportion in percent specifies the cetane number.

Cetane numbers over 50 are desirable for optimum operation of modern engines, particularly under cold-starting conditions. High-quality diesel fuels contain a high proportion of paraffins with high cetane numbers. Conversely, aromatics have a low ignition quality.

Table 2: DIN EN 590 (December 2013): Selected requirements of diesel fuels

Requirements		Unit	Characteristic
Cetane number, min.		–	51
Cetane index, min.		–	46
Density (at 15 °C), min./max.		kg/m³	820/845
Viscosity (at 40 °C), min./max.		mm²/s	2.0/4.5
Sulfur content, max.		mg/kg	10
Lubricity, "wear scar diameter", max.		µm	460
FAME content, max.		% by vol.	7
Oxidation stability	Insoluble [3], max.	g/m³	25
	Induction period (at 110 °C), min.	hours	20
Water content, max.		mg/kg	200
Total contamination, max.		mg/kg	24
CFPP [1] in six seasonal classes, max. [2]		°C	+5 to –20
Flash point		°C	>55

[1] Filterability limit. [2] Determined nationally, for Germany 0 to –20 °C.
[3] Sum total of insoluble substances after aging.

Cetane index
Yet another parameter of ignition quality is provided by the cetane index, which is calculated on the basis of fuel density and various points on the boiling curve. This purely mathematical parameter does not take into account the influence of cetane improvers on ignition quality. In order to limit the adjustment of the cetane number by means of cetane improvers, both the cetane number and the cetane index have been included in the list of requirements in the standard EN 590. Fuels whose cetane number has been enhanced by cetane improvers respond differently during engine combustion than fuels with the same natural cetane number.

Boiling range
The boiling range of a fuel, i.e. the temperature range at which the fuel vaporizes, depends on its composition. A low initial boiling point makes a fuel suitable for use in cold weather, but also increases the risk of cavitation damage and results in greater system wear due to poorer lubrication properties.

If, however, the end of boiling is at the high end of the temperature scale, this can result in higher soot emissions and nozzle coking. This refers to the formation of deposits as a result of the chemical decomposition of non-volatile fuel components in the spray hole and on the nozzle cone and the addition of combustion residues. When the end of boiling is higher, the ingress of fuel over the cylinder walls into the engine oil is greater. For this reason, the percentage of non-volatile fuel components should not be too high. Limiting the admixture of biodiesel to max. 7 % by volume also has its roots in the high boiling point of biodiesel (320...360 °C).

Filterability limit
Precipitation of paraffin crystals at low temperatures can result in fuel-filter blockage, ultimately leading to an interruption of the fuel flow. In worst-case scenarios, paraffin particulates can start to form at temperatures of 0 °C or even higher. The cold resistance of a fuel is assessed by means of the temperature at which the fuel filter becomes clogged under defined test conditions (CFPP, Cold Filter Plugging Point, filterability limit). European Standard EN 590 defines this limit value for various classes, and can be defined by individual member states depending on the prevailing geographical and climatic conditions.

Formerly, owners sometimes added regular gasoline to their vehicle fuel tanks to improve the cold response of diesel fuel. This practice is no longer necessary now that fuels conform to standards, and this may in any case cause damage particularly in today's widely used systems with high-pressure fuel injection.

Flash point
The flash point is the temperature at which the quantities of vapor which a combustible fluid emits to the atmosphere are sufficient to allow a spark to ignite the air/vapor mixture above the fluid. For safety reasons (e.g. for transportation and storage), diesel fuel is placed in Hazard Class A III, i.e. its flash point is over 55 °C. Less than 3 % gasoline in the diesel fuel is sufficient to lower the flash point to such an extent that ignition becomes possible at room temperature.

Density
The energy content of diesel fuel per unit of volume increases with density. Assuming constant activation of the injectors (i.e. constant injected fuel quantity), the use of fuels with widely different densities causes variations in mixture ratios (change in the air/fuel ratio λ) due to fluctuations in calorific value. When an engine runs on fuel that has a high type-dependent density, engine performance and soot emissions increase; as fuel density decreases, these parameters drop. For this reasons, the type-dependent density spread for diesel fuel is tightly limited.

Viscosity
Viscosity is a measure of a fuel's resistance to flow due to internal friction. If the viscosity of the diesel fuel is too low, this results in increased leakage losses, greater heating-up of the fuel-injection system and a heightened risk of wear and cavitation erosion. Much higher viscosity – for instance when pure biodiesel (FAME) is used – causes a higher peak

injection pressure at high temperatures in non-pressure-regulated systems (e.g. unit injector systems) compared with petroleum diesel. The fuel-injection system may not therefore be applied with petroleum diesel to the permissible peak pressure. High viscosity also changes the spray pattern due to the formation of larger droplets.

Lubricity

The hydrodynamic lubricity of diesel fuels is not as important as the lubricity in the mixed-friction range. The introduction of environment-compatible, hydrogenation-desulfurized fuels resulted in huge wear problems with distributor injection pumps in the field. Desulfurization also sees the removal of fuel components which are important to lubricity. Lubricity enhancers have to be added to many fuels to avoid these problems. The standard EN 590 prescribes a minimum lubricity, measured as wear of max. 460 μm in an oscillation wear test (High-Frequency Reciprocating Rig, HFRR [13]).

Carbon-deposit index

The carbon-deposit index describes a fuel's tendency to form carbon residue on and in the injection nozzles. The processes of carbon depositing are highly complex. Above all, components at the final boiling point of the diesel fuel contribute to coking, particularly when they come from cracking processes in the refinery.

Total contamination

Total contamination refers to the sum total of undissolved foreign particulates in the fuel, such as sand, rust, and undissolved organic components, including aging polymers of the fuel. The standard EN 590 permits a maximum total contamination of 24 mg/kg. Very hard silicates which occur in mineral dust are specially damaging to high-pressure fuel-injection systems with narrow gap widths. Even a fraction of the permissible overall contamination level of hard particulates would cause erosive and abrasive wear (e.g. at the seats of solenoid valves). Wear of this nature results in valve leakage, which lowers fuel-injection pressure and engine performance, and increases particulate emissions. Typical European diesel fuels contain about 250,000 particulates per 100 ml. Particulate sizes of 4 to 7 μm are particularly critical. High-performance fuel filters with high filtration efficiency are therefore used to prevent damage caused by particulates.

Water in diesel fuel

Diesel fuel can absorb up to approximately 100 mg/kg water at room temperature. The solubility limit is defined by the composition of the diesel fuel, its additives, and the ambient temperature. The standard EN 590 permits a maximum water content of 200 mg/kg. Although much higher water contents can occur in diesel fuel, market surveys of fuels show that water content rarely exceeds 200 mg/kg. Samples often do not detect any water, or detection is incomplete, since water is deposited on walls in a separate phase in the form of undissolved, "free" water, or it settles at the bottom. Whereas dissolved water does not damage the fuel-injection system, very small quantities of free water can cause wear or corrosion damage to fuel-injection components within a short period of time.

Additives

Additives, long a standard feature in gasolines, have attained increasing significance as quality improvers in diesel fuels. The various agents are generally combined in additive packages to achieve a variety of objectives. As the total concentration of the additives generally lies below 0.1 %, the fuel's physical characteristics – such as density, viscosity, and boiling curve – remain unchanged.

Lubricity enhancers
It is possible to improve the lubricity of diesel fuels with poor lubrication properties, caused, for example, by hydration processes during desulfurization, by adding fatty acids or glycerides. Biodiesel also contains glycerides as a byproduct. In this case, if diesel fuel already contains a proportion of biodiesel, no further lubricity enhancers are added.

Cetane improvers
Nitric-acid esters of alcohols which shorten ignition lag are frequently used as cetane improvers. They help, particularly during cold starting, to prevent an increase in combustion noise (engine noise) and extreme smoking.

Flow improvers
Flow improvers consist of polymer substances that lower the filterability limit. They are generally added in winter to ensure trouble-free operation at low temperatures. Although flow improvers cannot prevent the precipitation of paraffin crystals from diesel fuel, it can severely limit their growth. The size of the crystals produced is so small that they can still pass through the filter pores.

Detergents
Detergent additives are used to keep the nozzle holes clean. They can also inhibit the formation of deposits and reduce the buildup of carbon deposits on the injection nozzles.

Corrosion inhibitors
Corrosion inhibitors are deposited on the surfaces of metal parts and protect them against corrosion if water is entrained.

Antifoaming agents (defoamers)
Adding defoamers helps to avoid excessive foaming when the vehicle is refueled quickly.

Alternative fuels

With regard to alternative fuels, a distinction is made between fossil fuels, which are produced from coal, crude oil or natural gas, and regenerative fuels, which are created from renewable sources of energy, such as biomass, wind power or solar power.

Alternative fossil fuels include liquified petroleum gas, natural gas, synthetic liquid fuels created from natural gas (GtL, Gas-to-Liquid), and hydrogen produced from natural gas. Coal is the starting material for methanol or synthetic liquid fuels (CtL, Coal-to-Liquid) .

Regenerative fuels include methane, methanol and ethanol, provided these fuels are created from biomass. Further biomass-based regenerative fuels are biodiesel (FAME, Fatty Acid Methyl Ester) and hydrogenated vegetable oils (bioparaffins). The manufacture of synthetic liquid fuels from cellulose (BtL, Biomass-to-Liquid) is still in its infancy technologically.

Hydrogen extracted by electrolysis is then classed as regenerative if the current used comes from renewable sources (wind energy, solar energy).Biomass-based regenerative hydrogen can also be produced.

With the sole exception of hydrogen, all regenerative and all fossil fuels contain carbon and therefore release CO_2 during combustion. In the case of fuels produced from biomass, however, the CO_2 absorbed by the plants as they grow is offset against the emissions created during production and combustion in the engine.

Bioethanol

Manufacture of sugar and starch
Bioethanol can be obtained from products containing sugar and starch and is the most widely produced biofuel worldwide. Plants containing sugar (sugar cane, sugar beet) are fermented with yeast, the sugar fermenting to form ethanol in the process.

When bioethanol is obtained from starch grains such as corn, wheat or rye are pretreated with enzymes in order to partially split the long-chain starch molecules. During the subsequent saccharification there is a split into dextrose molecules with the aid of glucoamylasis. Bioethanol is created in a further process step by means of fermentation with yeast.

Manufacture of lignocellulose
Enzymes can also be used to manufacture bioethanol from lignocellulose. Lignocellulose forms the structural framework of the plant cell wall and contains the chief constituents lignin, hemicellulose and cellulose. The advantage of this process is that the entire plant can be used and not just that part containing sugar or starch. Because of this new approach, this product is also referred to as 2nd-generation bioethanol. However, the enzymes can split cellulose, but not lignin. The enzymes developed to date are still highly sensitive with regard to the usable biomass such that this process is currently still of no economic significance.

Use of bioethanol
Bioethanol is, on the basis of its properties, highly suitable for admixing in gasolines, particularly in order to increase the octane number of pure gasoline. That is why virtually all gasoline standards permit the addition of ethanol as a blend component. Even the biofuel policy of the European Union leads one to expect that the market penetration and the proportion of bioethanol in gasolines will continue to rise if it is guaranteed that bioethanol is created on a sustained basis and not in competition with foods.

Bioethanol can also be used as a pure fuel in spark-ignition engines in flexible fuel vehicles (FFVs). These vehicles can run both on gasoline and on any mixture of gasoline and ethanol. Because of the cold-start problems at low temperatures a maximum ethanol concentration of 85 % (E85) in summer and 70 - 75 % in winter has proven successful on the market. The quality of E85 is defined for Europe in the fuel specification CEN/TS 15293 [14] and in the USA in ASTM D5798 [15].

Methanol

Methanol is essentially produced not by regenerative means, but from fossil energy sources such as coal and natural gas, and therefore makes no contribution to reducing CO_2 emissions. Countries such as China, who plan to cover their high fuel demand in part from coal, will increasingly be opting for methanol in the future. Here, M 15 appears to represent an upper limit for use in conventional spark-ignition engines. In China, an M 85 analogous to E 85 is being discussed for flexible fuel vehicles.

With the same alcohol content, methanol fuels have a significantly greater corrosive effect on ferrous alloys than ethanol fuels. Even demixing occurs much more quickly when the fuel is fouled by water. Because of the negative experiences of methanol fuels during the oil crisis in 1973 and also due to its toxicity, the use of methanol as a blend component has again been abandoned in Germany. From a worldwide point of view, only occasional methanol admixtures are being conducted at present, and then for the most part with a content of well below 5 % (M 5).

Natural gas

Fossil natural gas

The main constituent of natural gas is methane (CH_4) with a content of 83 - 98 %. Further constituents are inert gases, such as carbon dioxide, nitrogen and low-chain hydrocarbons. Oxygen and hydrogen are also contained. Natural gas is available worldwide and after extraction requires relatively low expenditure for preparation. Depending on its origin, however, its composition varies, which results in fluctuations when it comes to density, calorific value and knock resistance. The properties of natural gas as a fuel are defined for Germany in the standard DIN 51624 ([16]). A European standard for natural gas which also takes into account the biomethane quality requirements is currently being drawn up.

Biomethane

Biomethane is treated biogas which is made from biogenous materials such as energy crops, liquid manure or waste containing biomass. This way of producing methane can significantly reduced overall CO_2 emissions.

Storage of natural gas

Natural gas is stored either in gas form as compressed natural gas (CNG) at a pressure of 200 bar or as a liquified gas (LNG: liquid natural gas) at –162 °C in a cold-resistant tank. LNG requires only one third of the storage volume of CNG, however the storage of LNG requires a high expenditure of energy for liquifying. Natural gas is therefore sold at natural-gas filling stations almost entirely in the form of CNG.

CO_2 emissions

The hydrogen/carbon ratio of natural gas is roughly 4:1, that of gasoline on the other hand is 2.3:1. As a result of the lower amount of carbon in natural gas, it produces less CO_2 and more H_2O than gasoline when it is burned. A spark-ignition engine converted to natural gas, without any further optimization, already creates approximately 25 % fewer CO_2 emissions than when run on gasoline (assuming a comparable power output).

Use of natural gas

Natural gas is used in vehicles with spark-ignition gas engines. Passenger cars are for the most part bivalent and can be run separately on gasoline and natural gas. Buses with spark-ignition gas engines run in monovalent mode on natural gas. In compression-ignition gas engines natural gas is burned in a pilot-injection process in mixed mode with diesel fuel.

Liquified petroleum gas

Liquified petroleum gas (LPG, also called liquid gas) is obtained during the extraction of crude oil and during different refinery processes. LPG is a mixture of the main components of propane and butane. It can be liquified at room temperature under comparatively low pressure. Because it has a lower carbon content than gasoline, approximately 10 % less CO_2 is produced during combustion. The octane number is approximately 100 - 110 RON. The demands placed on LPG for use in motor vehicles are laid down in the European standard EN 589 [17].

Hydrogen

Hydrogen can be created by chemical processes from natural gas, coal, crude oil or biomass, and by electrolysis from water. Today hydrogen is predominantly obtained on an industrial scale by steam reformation from natural gas. In this process, CO_2 is released in such a way that, overall, there is not necessarily a CO_2 advantage compared with gasoline, diesel or the direct use of natural gas in the internal-combustion engine.

A reduction in CO_2 emissions is achieved then when hydrogen is regeneratively created from biomass or by electrolysis from water provided that regeneratively generated current is used for this purpose. No CO_2 emissions occur locally when hydrogen is burned.

Storage

Hydrogen may have a very high weight-related energy density (approximately 120 MJ/kg and thus almost three times as high as that of gasoline), but its volume-related energy density is very low on account of the low specific density. When it comes to storage, this means that the hydrogen has to be compressed either under pressure (at 350 – 700 bar) or by liquifaction (cryogenic storage at –253 °C) in order to achieve an acceptable tank volume. Another possibility is for the hydrogen to be stored as a hybrid.

Use in motor vehicles

Hydrogen can be used both in fuel-cell drives and directly in internal-combustion engines. In the long term, emphasis is expected to be placed on use in fuel cells because this achieves better efficiency than in hydrogen internal-combustion engines.

Biodiesel

At present, biodiesel is the most important alternative fuel for diesel engines. The term biodiesel covers fatty acid esters which are created through transesterification of oils or greases with methanol or ethanol. This creates fatty acid methyl ester (FAME, Fatty Acid Methyl Ester) or fatty acid ethyl ester (FAEE, Fatty Acid Ethyl Ester). The molecules of biodiesel are, in terms of size and properties, far more similar to diesel fuel than to vegetable oil. Therefore, biodiesel cannot under any circumstances be equated with vegetable oil. Nevertheless, the properties of biodiesel differ significantly from time to time from those of petroleum diesel, as the fatty acid esters are polar- and chemically reactive. Conventional diesel fuel on the other hand is an inert and nonpolar mixture of paraffins and aromatic compounds.

Manufacture

Vegetable oils or animal fats can be used as the starting material for biodiesel. In Europe primarily rape oil is used, in North and South America soybean oil, in Asia palm oil, and on the Indian subcontinent to an even smaller extent the oil from the purgier nut (jatropha).Used frying oil methyl ester (UFOME) is produced worldwide. Because of the global trade in biodiesel and its raw materials, fuels containing FAME as a rule contain mixtures from different sources.

Because esterification can be technically carried out more easily with methanol than it can with ethanol, the methyl esters of these oils are manufactured almost exclusively. Methanol is generally produced from coal. Therefore, fatty acid methyl ester cannot, strictly speaking, be seen as fully biogenic. Fatty acid ethyl ester on the other hand is made up of 100 % biomass when bioethanol is used for production.

Properties

The properties of biodiesel are determined by various factors. The various vegetable oils differ in the composition of the fatty-acid blocks and demonstrate typical fatty-acid patterns. The type and quantity of unsaturated fatty acids have, for example, a decisive influence on the cold resistance and stability of biodiesel. The properties are also affected by the pretreatment of the vegetable oil and the production process of the biodiesel.

Standards

The quality of biodiesel is regulated in fuel standards. If technically tenable, limitations with regard to raw materials must be avoided. The quality requirements for biodiesel are therefore described not by way of the raw-material composition, but predominantly by way of the material properties. It is essential in particular to ensure sufficient cold resistance, good aging stability (oxidation stability) and to eliminate contamination caused by the process.

The European standard EN 14214 [8] is the most internationally comprehensive specification for biodiesel and describes the requirements relating to the use of biodiesel as a pure fuel and as a blend component for diesel fuel. Good-quality biodiesel is defined for both areas of application (Table 3).

The American biodiesel standard ASTM D6751 [11] is less quality-oriented. For example, the minimum requirement for oxidation stability in this standard is less than half the aging reserve specified in EN 14214. This increases the risk in the USA of problems arising as a result of fuel aging, in particular under limit-value application and field conditions.

Other countries such as Brazil, India and Korea have geared themselves, to a large extent, towards the European B 100 standard EN 14214.

Use in motor vehicles

Since its preferential treatment for tax purposes was discontinued in Germany, biodiesel is no longer used as a pure fuel (B 100). The introduction of diesel particulate filters, more stringent vehicle-emission requirements and their durability also rule out the use of B100 in the future for technical reasons. Originally B100 was used predominantly in commercial vehicles, because the annual high mileage ensured fast consumption, which enabled problems with insufficient oxidation stability to be avoided.

From an engine viewpoint, it is more favorable to use biodiesel as a low-percentage mixture in the blend with petroleum diesel. The high proportion of petroleum diesel ensures sufficient stability and, at the same time, the biodiesel ensures a good lubricating effect.

For practical purposes, it is important to specify not only the pure component B100, but also the diesel/biodiesel mixtures offered on the market. Here, the trend is towards admixtures up to max. 7 % biodiesel (B7 in Europe).

In closed fleets, higher proportions of biodiesel are also used (B30 in France, B20 in the USA). In the case of higher contents, however, the high boiling point of biodiesel can cause it to be heavily introduced into the engine oil after its has been injected into the combustion chamber via condensation on the cylinder walls. This affects, above all, vehicles which are fitted with diesel particulate filters, and in which regeneration occurs by way of a retarded secondary injection. Depending on the application, it is possible, particularly in slow part-load operation, to encounter an unacceptably high introduction of biodiesel, which necessitates shorter oil-replacement intervals.

Table 3: DIN EN 14214 (November 2012): Selected requirements of FAME

Requirements	Unit	Characteristic
Density (at 15 °C), min./max.	kg/m^3	860/900
Viscosity (at 40 °C), min./max.	mm^2/s	3.5/5.0
Sulfur content, max.	mg/kg	10
Content of alkali metals (Na + K), max.	mg/kg	5.0
Content of alkaline earth metals (Ca + Mg), max.	mg/kg	5.0
Content of total glycerin, max.	% by weight	0.25
Acid number, max.	mg KOH/g	0.50
Oxidation stability (induction period at 110 °C), min.	hours	8
Water, max.	mg/kg	500
Total contamination, max.	mg/kg	24
CFPP [1], in six seasonal classes in each case, max.		
Pure fuel	°C	+5...–20 [2]
Blend component	°C	+13...–10 [3]
Flash point	°C	>101

[1] Filterability limit.
[2] Determined nationally, for Germany 0 to –20 °C.
[3] Determined nationally, for Germany 0 to –10 °C.

Rape oil

Rape oil was used with great success in older diesel engines subject to minimal emission requirements. Due to the high density and viscosity of rape oil together with its high volatility, it is used in only a few applications (e.g. tractors) in modern diesel engines with high-pressure injection systems.

Paraffinic diesel fuels

Purely paraffinic fuels consist entirely of saturated hydrocarbons. Thanks to the absence of aromatics, the particulate, HC and CO emissions are significantly reduced.

Paraffinic fuels can be created in three different ways:

- Fischer-Tropsch process [18]
- Hydrogenation of vegetable oils
- COD process (Conversion of Olefins to Distillates)

Fischer-Tropsch process

The starting product required is synthesis gas, which consists of hydrogen and carbon monoxide. It can be created from natural gas, coal or biomass. By converting synthesis gas, it is possible to build up on catalysts linear, straight-chain hydrocarbons (n-paraffins). The Fischer-Tropsch catalysts function quite non-specifically such that a wealth of different components is obtained, starting with gases through short-chain gasoline components, kerosine and diesel paraffins, right down to oils and waxes of high molecular weight. For reasons of economy, splitting the production mixture is for the most part optimized to a maximum diesel yield. The fuels obtained in accordance with this process are known as synthetic diesels.

These fuels were originally also referred to as designer fuels, because the notion existed that the composition of synthetic diesel fuel could be geared exactly to the demands of diesel-engine technology. In view of the wide range of products obtained from Fischer-Tropsch synthesis, the notion of producing fuels of customized composition no longer appears justified.

The terms Gas-to-Liquid (GtL), Coal-to-Liquid (CtL) and Biomass-to-Liquid (BtL) are commonly used, depending on whether the paraffins have been created from natural gas, coal or biomass.

The production of CtL and GtL is economically significant. The production of GtL is only worthwhile with large natural-gas deposits where the natural gas cannot be attributed to any direct use. Because of the higher costs involved, the use of GtL and CtL has hitherto been limited to special markets.

CtL and GtL are based on fossil energy sources such that no reduction in CO_2 emissions is achieved. In the case of BtL, there is no CO_2 advantage. However, industrial plants that produce BtL are not yet in operation.

The approach of manufacturing fuels from whole crops through decomposition into basic chemical components and subsequent synthesis differs fundamentally from the previously customary methods, which are based on converting components present in grains such as fats, starches or sugars through chemical or enzymatic separation (transesterification, fermentation) into fuels (biodiesel or bioethanol). That is why synthetic fuels are also known as 2nd-generation fuels.

Hydrogenation of vegetable oils

Paraffinic diesel fuels can also be obtained by hydrogenating fats and oils. Unlike transesterification into biodiesel, conversion with hydrogen places fewer demands on the origin and quality of the starting materials. Hydrogenation results in a cracking of fats and oils, during which all the oxygen atoms and unsaturated bonds are also removed. Long-chain paraffins are obtained from the fatty acids, the glycerine content is converted into propane gas, and the oxygen is bound as water. Because paraffins are created from biomass in this way, they are referred to as bioparaffins. The production of bioparaffins can be implemented in separate plants or even integrated into existing refinery processes.

The hydrogenation of vegetable oils is developing increasingly into a significant alternative to the production of biodiesel.

COD process
The third way of producing paraffins is to convert olefins in accordance with the COD process (Conversion of Olefins to Distillates), frequently only as a subsequent step to previous refinery processes. Here, olefinic product fractions are converted by oligomerization and hydrogenation into paraffins.

Properties
Regardless of the type of production, paraffinic hydrocarbon mixtures with very similar chemical compositions and excellent engine properties are created. The fuels are sulfur- and aromatic-free and have high cetane numbers. But because their density is below the lower limit value defined in EN 590, the new specification CEN/TS 15940 [19] was developed (Table 4) which describes a quality of the fuel for pure use and limits use to closed fleets. In order to achieve this level of quality and especially to achieve the required cold resistance, an extra isomerization step is appended to the three production processes described.

Use in motor vehicles
Pure paraffinic fuels are used mostly in older vehicles with no diesel particulate filters and lend themselves to use above all in centers of population in order to reduce the particulate load on a local level. Paraffinic hydrocarbons are ideal for marketing as blend components in premium diesel fuels. Furthermore, diesel fuels which fail to reach the limit values established in EN 590 can be improved by the addition of paraffinic components to such an extent that they conform to the standard.

References
[1] DIN EN 228: 2013, Automotive fuels – Unleaded petrol – Requirements and test methods.
[2] ASTM D4814-13, Standard Specification for Automotive Spark-Ignition Engine Fuel.
[3] E DIN 51626-1:2010-04, Automotive fuels – Requirements and test methods – Part 1: Petrol E10 and petrol E5.
[4] DIN EN ISO 5164:2006, Petroleum products – Determination of knock characteristics of motor fuels – Research method (ISO 5164:2005).
[5] DIN EN ISO 5163:2006, Petroleum products – Determination of knock characteristics of motor and aviation fuels – Motor method (ISO 5163:2005).
[6] DIN EN 13016-1:2007, Liquid petroleum products – Vapor pressure – Part 1: Determination of air saturated vapor pressure (ASVP) and calculated dry vapor pressure equivalent (DVPE).
[7] DIN EN 590: 2013, Automotive fuels – Diesel – Requirements and test methods.

Table 4: DIN CEN/TS 15940 (December 2012): Selected requirements of paraffinic diesel fuels

Requirements		Unit	Characteristic
Cetane number, min.	Class A	–	70
	Class B	–	51
Density (at 15 °C), min./max.	Class A	kg/m^3	765/800
	Class B	kg/m^3	780/810
Viscosity (at 40 °C), min./max.		mm^2/s	2.0/4.5
Sulfur content, max.		mg/kg	5
Total aromatic content (including polyaromatics), max.		% by mass	1
Lubricity, "wear scar diameter", max.		µm	460
FAME content, max.		% by vol.	7
Water, max.		mg/kg	200
Total contamination, max.		mg/kg	24
CFPP[1] in six seasonal classes, max.[2]		°C	+5 to –20
Flash point		°C	>55

[1] Filterability limit.
[2] Determined nationally, for Germany 0 to –20 °C.

[8] DIN EN 14214:2012, Liquid petroleum products – Fatty acid methyl esters (FAME) for use in diesel engines and heating applications – Requirements and test methods.
[9] DIN EN 15751:2009, Automotive fuels – Fatty acid methyl ester (FAME) fuel and blends with diesel fuel – Determination of oxidation stability by accelerated oxidation method.
[10] ASTM D975-13, Standard Specification for Diesel Fuel Oils.
[11] ASTM D6751-12, Standard Specification for Biodiesel Fuel Blend Stock (B 100) for Middle Distillate Fuels.
[12] DIN EN ISO 5165:1999, Petroleum products – Determination of the ignition quality of diesel fuels – Cetane engine method (ISO 5165:1998).
[13] DIN EN ISO 12156-1:2008, Diesel fuel – Assessment of lubricity using the high-frequency reciprocating rig (HFRR) – Part 1: Test method (ISO 12156-1:2006).
[14] DIN CEN/TS 15293:2011, Automotive fuels – Ethanol (E85) automotive fuel – Requirements and test methods.
[15] ASTM D5798-13a, Standard Specification for Ethanol Fuel Blends for Flexible-Fuel Automotive Spark-Ignition Engines.
[16] DIN 51624: 2008, Automotive fuels – Compressed natural gas – Requirements and test methods.
[17] DIN EN 589:2012, Automotive fuels – LPG – Requirements and test methods.
[18] L. König, J. Gaube: Fischer-Tropsch-Synthese, Neuere Untersuchung und Entwicklungen. Article in Chemie – Ingenieur – Technik, Volume 55/1, 1983.
[19] DIN CEN/TS 15940:2012, Automotive fuels – Paraffinic diesel fuel from synthesis or hydrotreatment – Requirements and test methods.

Table 5: Properties of liquid fuels and hydrocarbons

Material/medium	Density kg/l	Main constituents % by weight	Boiling temperature °C	Specific heat of evaporation kJ/kg	Net calorific value MJ/kg	Ignition temperature °C	Air requirement, theoretical kg/kg	Ignition limit Lower (% by volume of gas in air)	Ignition limit Upper (% by volume of gas in air)
Gasoline,									
Regular	0.720...0.775	86 C, 14 H	25...210	380...500	41.2...41.9	≈ 300	14.8	≈ 0.6	≈ 8
Premium	0.720...0.775	86 C, 14 H	25...210	–	40.1...41.6	≈ 400	14.7	–	–
Aviation fuel	0.720	85 C, 15 H	40...180	–	43.5	≈ 500	–	≈ 0.7	≈ 8
Kerosene	0.77...0.83	87 C, 13 H	170...260	–	43	≈ 250	14.5	≈ 0.6	≈ 7.5
Diesel fuel	0.820...0.845	86 C, 14 H	180...360	≈ 250	42.9...43.1	≈ 250	14.5	≈ 0.6	≈ 7.5
Mineral oil (crude oil)	0.70...1.0	80...83 C, 10...14 H	25...360	222...352	39.8...46.1	≈ 220	–	≈ 0.6	≈ 6.5
Lignite tar oil	0.850...0.90	84 C, 11 H	200...360	–	40.2...41.9	–	13.5	–	–
Coaltar oil	1.0...1.10	89 C, 7 H	170...330	–	36.4...38.5	–	–	–	–
Pentane C_5H_{12}	0.63	83 C, 17 H	36	352	45.4	285	15.4	1.4	7.8
Hexane C_6H_{14}	0.66	84 C, 16 H	69	331	44.7	240	15.2	1.2	7.4
n-heptane C_7H_{16}	0.68	84 C, 16 H	98	310	44.4	220	15.2	1.1	6.7
iso-octane C_8H_{18}	0.69	84 C, 16 H	99	297	44.6	410	15.2	1	6
Benzene C_6H_6	0.88	92 C, 8 H	80	394	40.2	550	13.3	1.2	8
Toluene C_7H_8	0.87	91 C, 9 H	110	364	40.6	530	13.4	1.2	7
Xylene C_8H_{11}	0.88	91 C, 9 H	144	339	40.6	460	13.7	1	7.6
Ether $(C_2H_5)_2O$	0.72	64 C, 14 H, 22 O	35	377	34.3	170	7.7	1.7	36
Acetone $(CH_3)_2CO$	0.79	62 C, 10 H, 28 O	56	523	28.5	540	9.4	2.5	13
Ethanol C_2H_5OH	0.79	52 C, 13 H, 35 O	78	904	26.8	420	9	3.5	15
Methanol CH_3OH	0.79	38 C, 12 H, 50 O	65	1,110	19.7	450	6.4	5.5	26
Rape oil	0.92	78 C, 12 H, 10 O	–	–	38	≈ 300	12.4	–	–
Rapeseed methyl ester (biodiesel)	0.88	77 C, 12 H, 11 O	320...360	–	36.5	283	12.8	–	–

Viscosity at 20 °C in mm^2/s (= cSt): gasoline ≈ 0.6; ethanol ≈ 1.5; methanol ≈ 0.75.

Table 6: Properties of gaseous fuels and hydrocarbons

Material/medium		Density at 0 °C and 1,013 mbar	Main constituents	Boiling temperature at 1,013 mbar	Net calorific value		Ignition temperature	Air requirement, theoretical	Ignition limit	
					Fuel	Air-fuel mixture			Lower	Upper
		kg/m³	% by weight	°C	MJ/kg	MJ/m³	°C	kg/kg	% by volume of gas in air	
Liquified petroleum gas		2.25 ([1])	C_3H_8, C_4H_{10}	–30	46.1	3.39	≈ 400	15.5	1.5	15
Municipal gas		0.56...0.61	50 H, 8 CO, 30 CH_4	–210	≈ 30	≈ 3.25	≈ 560	10	4	40
Natural gas H (North Sea)		0.83	87 CH_4, 8 C_2H_6, 2 C_3H_8, 2 CO_2, 1 N_2	–162 (CH_4)	46.7	–	584	16.1	4.0	15.8
Natural gas H (Russia)		0.73	98 CH_4, 1 C_2H_6, 1 N_2	–162 (CH_4)	49.1	3.4	619	16.9	4.3	16.2
Natural gas L		0.83	83 CH_4, 4 C_2H_6, 1 C_3H_8, 2 CO_2, 10 N_2	–162 (CH_4)	40.3	3.3	≈ 600	14.0	4.6	16.0
Water gas		0.71	50 H, 38 CO	–	15.1	3.10	≈ 600	4.3	6	72
Blast-furnace gas		1.28	28 CO, 59 N, 12 CO_2	170	3.20	1.88	≈ 600	0.75	≈ 30	≈ 75
Sewage gas (manure gas) [2]		–	46 CH_4, 54 CO_2	–	27.2 ([2])	3.22	–	–	–	–
Hydrogen	H_2	0.090	100 H	–253	120.0	2.97	560	34	4	77
Carbon oxide	CO	1.25	100 CO	–191	10.05	3.48	605	2.5	12.5	75
Methane	CH_4	0.72	75 C, 25 H	–162	50.0	3.22	650	17.2	5	15
Acetylene	C_2H_2	1.17	93 C, 7 H	–81	48.1	4.38	305	13.25	1.5	80
Ethane	C_2H_6	1.36	80 C, 20 H	–88	47.5	–	515	17.3	3	14
Ethene	C_2H_4	1.26	86 C, 14 H	–102	14.1	–	425	14.7	2.75	34
Propane	C_3H_8	2.0 ([1])	82 C, 18 H	–43	46.3	3.35	470	15.6	1.9	9.5
Propene	C_3H_6	1.92	86 C, 14 H	–47	45.8	–	450	14.7	2	11
Butane	C_4H_{10}	2.7 ([1])	83 C, 17 H	–10; +1 ([3])	45.6	3.39	365	15.4	1.5	8.5
Butene	C_4H_8	2.5	86 C, 14 H	–5; +1 ([3])	45.2	–	–	14.8	1.7	9
Dimethyl ether	C_2H_6O	2.05 ([4])	52 C, 13 H, 35 O	–25	28.8	3.43	235	9.0	3.4	18.6

[1] Density of liquified gas 0.54 kg/*l*, density of liquid propane 0.51 kg/*l*, density of liquid butane 0.58 kg/*l*.
[2] Purified sewage gas contains 95 % CH_4 (methane) and has a calorific value of 37.7 MJ/kg.
[3] First value for isobutane, second value for n-butane and n-butene.
[4] Density of liquified dimethyl ether 0.667 kg/*l*.

Brake fluids

Brake fluid is the hydraulic medium employed to transmit actuation forces within the brake system. Compliance with stringent requirements is essential to ensure reliable brake-system operation. These requirements are defined in various standards of similar content (SAE J 1703, FMVSS 116, ISO 4925). The performance data contained in FMVSS 116 (Federal Motor Vehicle Safety Standard), mandatory in the USA, also serve as an international reference. The US Department of Transportation (DOT) has defined specific ratings for salient characteristics (Table 1).

Characteristics

Equilibrium boiling point

The equilibrium boiling point provides an index of the brake fluid's resistance to thermal stress. The heat encountered in the wheel-brake cylinders (which are subjected to the highest temperatures in the entire brake system) can be especially critical. Vapor bubbles can form at temperatures above the brake fluid's instantaneous boiling point, resulting in brake failure.

Wet boiling point

The wet boiling point is the fluid's equilibrium boiling point subsequent to moisture absorption under specified conditions (approx. 3.5 %). Hygroscopic (glycol-based) fluids respond with an especially pronounced drop in boiling point.

The wet boiling point is tested to quantify the response characteristics of used brake fluid. Brake fluid absorbs moisture, mostly by diffusion through brake-system hoses. This is the main reason why it should be replaced every one to two years. Figure 1 shows the drops in boiling point that result from moisture absorption in two different brake fluids.

Viscosity

To ensure consistent reliability throughout the braking system's extended operating range (–40 °C to +100 °C), viscosity should remain as constant as possible, with minimum sensitivity to temperature variations. Maintaining the lowest possible cold viscosity at very low temperatures is especially important in ABS, TCS and ESP systems.

Compressibility

The fluid should maintain a consistently low level of compressibility with minimum sensitivity to temperature fluctuations.

Figure 1: Brake-fluid boiling temperature as a function of water content

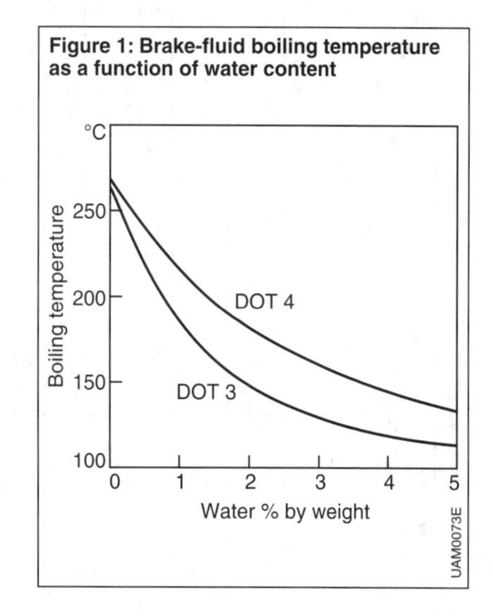

Table 1: Brake fluids

Standard	FMVSS 116			ISO 4925
Requirements/status	DOT3	DOT4	DOT5	Class 6
Dry boiling point in °C (min.)	205	230	260	250
Wet boiling point in °C (min.)	140	155	180	165
Cold viscosity at –40 °C in mm²/s	<1,500	<1,800	<900	<750

Elastomer swelling
The elastomers employed in the brake system (e.g. seals and hoses) must be able to adapt to the type of brake fluid used. Although a small amount of swelling is desirable, it should not exceed approximately 10 % under any circumstances. Otherwise, it has a negative effect on the strength of the elastomer components. Even minute levels of mineral-oil contamination (such as mineral-oil-based brake fluid, solvents) in glycol-based brake fluid can lead to destruction of elastomers (such as seals) and ultimately lead to brake-system failure.

Corrosion protection
FMVSS 116 stipulates that brake fluids shall exercise no corrosive effect on those metals generally employed in braking systems. The required corrosion protection can be achieved only by using additives.

Chemical composition
Glycol-ether fluids
Most brake fluids are based on glycol-ether compounds. These generally consist of monoethers of low polyethylene glycols. Although these components can be used to produce a brake fluid which conforms to DOT3 requirements (Table 2), their undesirable hygroscopic properties cause this fluid to absorb moisture at a relatively rapid rate, with an attendant swift reduction in the boiling point.

If the free OH (hydroxyl) groups are partially esterified with boric acid, the result is a superior DOT4 (or DOT4+, Super DOT4, DOT5.1) brake fluid capable of reacting with moisture to neutralize its effects. As the DOT4 brake fluid's boiling point drops much more slowly than that of a DOT3 fluid (Figure 1), its service life is longer.

ISO 4925 defines a further quality class, namely "Class 6". This quality is slightly better than that of DOT4, but is characterized particularly by its low cold viscosity (Table 1).

Mineral-oil fluids (ISO 7308)
The great advantage of mineral-oil-based fluids is the fact that they are not hygroscopic, so the boiling point does not drop due to moisture absorption.

The petroleum industry can also supply a range of further additives to improve other brake-fluid properties. They are of no great importance for use in the vehicle. Mineral-oil fluids should never be added to brake systems which are designed for glycol fluids (or vice versa), as this would destroy the elastomers.

Silicone-oil fluids (SAE J 1705)
As silicone oils – in the same way as mineral oils – are not hygroscopic, they formerly saw occasional use as brake fluids. The disadvantages of these products include considerably higher compressibility and inferior lubrication, both of which reduce their suitability for use as hydraulic fluid in many systems.

A critical factor with brake fluids based on silicone or mineral oils is the absorption of free water in a fluid state, as the water forms vapor bubbles when it heats up to more than 100 °C and freezes when it cools to less than 0 °C.

Table 2: Classification of brake fluids with different chemical bases

Classification / Parameter	DOT3 Glycol ether	DOT4 Glycol ether	DOT5		
			DOT5.1 Glycol ether	DOT5 SB Silicone	Mineral oil
Boiling point [°C] Wet boiling point [°C]	205 140	230 155	260 180		
Viscosity at −40 °C [mm²/s]	< 1,500	< 1,800	< 900		
Difference in color	Colorless to amber			Purple	Green

Coolants

Requirements

The cooling system must dissipate that part of the engine's combustion heat that is not converted into mechanical energy. A fluid-filled cooling circuit transfers the heat absorbed in the cylinder head to a heat exchanger (radiator) for dispersal into the air. The fluid in this circuit is exposed to extreme thermal loads; it must also be formulated to ensure that it does not attack the materials within the cooling system (corrosion).

Owing to its high specific heat and its correspondingly substantial thermal-absorption capacity, water is a very good cooling medium. Its disadvantages include its corrosive properties and limited suitability for application in cold conditions (freezing). This is why additives must be mixed with the water for satisfactory performance.

Antifreeze

It is possible to lower the coolant's freezing point by adding ethylene glycol. When glycol is added to form a mixture with water, the resulting coolant no longer freezes at a given temperature. Instead, ice crystals are precipitated in the fluid once the temperature drops to the ice flaking point. At this temperature, the fluid medium can still be pumped through the cooling circuit. Glycol also raises the coolant's boiling point (Table 1).

In car owner's manuals, automobile manufacturers usually specify various optional antifreeze mixture ratios for different levels of low-temperature frost protection.

Table 1: Ice flaking and boiling points for water-glycol mixtures

Glycol % by vol.	Ice flaking point °C	Boiling point °C
10	−4	101
20	−9	102
30	−17	104
40	−26	106
50	−39	108

Additives

Coolants must include effective additives to protect the glycol against oxidation (which forms extremely corrosive byproducts) and to protect metallic cooling-system components against corrosion. Common additives include corrosion inhibitors (nitrates, alkali salts of organic acids, benzthiazole derivates), buffers (borates), and antifoaming agents (silicones).

Many of these additives are subject to aging deterioration, leading to a gradual reduction in coolant performance. Automobile manufacturers have responded to this fact by granting official approval exclusively for coolants of proven long-term stability.

Springs

Basic principles

Functions

All elastic components to which forces are applied are spring elements. However, springs in the narrower sense mean only those elastic elements that can absorb, store, and release work over a relatively long distance. The stored energy can also be used to maintain a force. The most important applications of industrial springs are:
- Absorbing and damping shocks (shock absorbers)
- Storing potential energy (spring motors)
- Applying a force (coupling springs)
- Vibrating systems (vibrating table)
- Force measurement (spring balance)

Spring characteristic

The spring characteristic shows the behavior of a spring or spring system. This means the dependency of spring force or spring torque on deformation. Metal springs have linear characteristics (Hooke's law), rubber springs have progressive characteristics, and disc springs have degressive characteristics. The gradient of the characteristic is called the spring rate.

For translational motion: $R = \frac{dF}{ds}$.

For rotational motion: $R_t = \frac{dM_t}{d\alpha}$.

Spring duty

For frictionless springs under stress, the area under the characteristic represents the absorbed or released work (Figure 1):

$W = \int F\, ds$.

Spring damping

If friction occurs, the prevailing force when the spring is loaded is greater than when the load is removed. The area enclosed by the two characteristics represents frictional work W_R, and is thus a measure of the damping rate (Figure 1):

$\psi = \frac{W_R}{W}$.

Damping due to internal friction can be very high with rubber springs ($0.5 < \psi < 3$). With metal springs, however, it is rather low ($0 < \psi < 0.4$). This means that metal springs have a notable damping rate that is only achievable by means of

Table 1: Symbols and units

Quantity		Unit
b	Width of spring leaf	mm
d	Wire diameter	mm
D	Mean coil diameter	mm
E	Modulus of elasticity	MPa
F	Spring force	N
G	Shear modulus	MPa
h	Height of spring leaf	mm
h_0	Spring deflection (disc spring)	mm
i	Number of leaves (leaf spring)	–
i'	Number of leaves which continue to ends	–
k	Stress coefficient	–
L_c	Block length (solid length)	mm
l_f	Active length	mm
M_b	Bending moment	Nm
M_t	Torsional moment	Nm
n	Number of active coils	–
n_t	Total number of coils	–
R	Spring rate (spring constant)	N/mm
R_t	Torsional spring rate	Nm/rad
s	Spring deflection	mm
S_a	Total of minimum distances	mm
t	Thickness (disc spring)	mm
W	Spring duty	J
W_R	Frictional work	J
$\widehat{\alpha}$	Twist angle	rad
σ_A	Permissible variable stress	MPa
σ_b	Bending stress	MPa
σ_m	Mean stress	MPa
τ_t	Torsional stress	MPa
ψ	Damping	–

external friction, e.g. as occurs in layers of leaf and disc springs.

Spring combinations
A very wide variety of spring characteristics can be achieved by combining several springs (Figure 2). In principle, springs can be combined in parallel or in series. A combination of parallel and series springs is also possible.

Parallel combinations
If springs are arranged in parallel, the external load is distributed proportionally between the individual springs. However, the spring deflection (s) is equal for all springs. The spring rate of the spring system is the sum of individual spring rates:

$$R_{total} = R_1 + R_2 + R_3 + ... + R_n .$$

Accordingly, spring systems comprising parallel springs are harder than individual springs.

Series combinations
With series springs, the total external load acts on each individual spring. However, spring travel for each spring is different depending on the individual spring rates, and are added. The following applies to the resulting spring rate of the overall system:

$$\frac{1}{R_{total}} = \frac{1}{R_1} + \frac{1}{R_2} + ... + \frac{1}{R_n} .$$

Spring systems consisting of series springs are softer than the softest individual springs.

Metal springs

Normally, metal springs are classified according to their stresses (Table 2): It should be noted that the tendency of the springs to relax increases as working temperature rises. At 120 °C, relaxation is no

Figure 1: Spring characteristics and spring duty

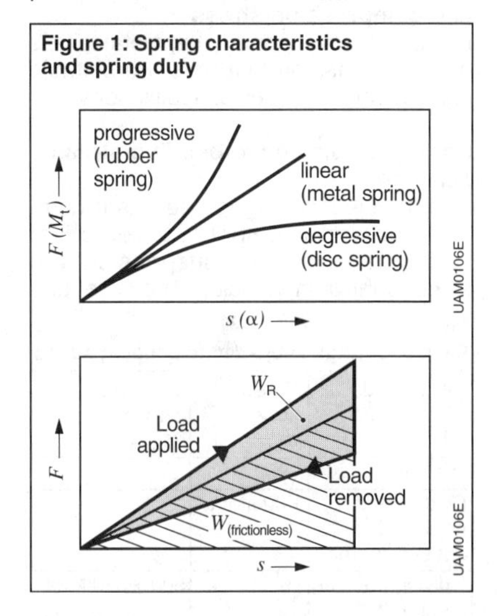

Table 2: Stress on springs

Stress	Construction
Tensile, compressive stress	Tensile test bar, ring spring
Bending stress	Leaf spring, torsion spring, spiral spring, disc spring
Torsional stress	Torsion-bar spring, helical spring

Figure 2: Spring combinations
Parallel combination of springs 2 and 3 in series with spring 1

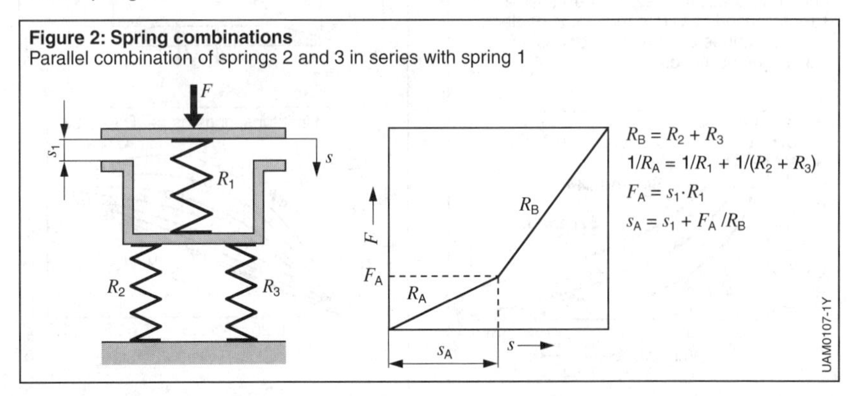

longer negligible. However, with unalloyed spring steels, settling may begin to occur at 40 °C. At higher working temperatures, springs can only be properly evaluated using relaxation-tension diagrams.

Springs subjected to tensile and compression stress

Due to their high spring rate, metal tensile and compression test bars are suitable only for very few special applications.

Springs subjected to bending stress

Leaf springs

A simple leaf spring is used as a compression or guide spring. Layered leaf springs are used for suspension and wheel control in vehicles. They are usually made of spring steel in accordance with DIN EN 10089 (hot-rolled) and DIN EN 10132 (cold-rolled strip). The draft design may assume the permissible bending stresses specified in Table 3 and Figure 3.

Table 3: Springs subjected to bending stress

Steel bands	Static stress $\sigma_{b, perm}$	Dynamic stress $\sigma_{b,perm}=\sigma_m \pm \sigma_A$
Hot-rolled	960 MPa	
Cold-rolled, hardened, and draw-tempered	1,000 MPa	
Individual leaves ground		500 ± 320 MPa
Individual leaves with rolling skin		500 ± 100 MPa
Layered leaves with rolling skin		500 ± 80 MPa

Figure 3: Permissible bending stresses
For torsion springs made of spring steel wire according to DIN EN 10270-1. With static and quasi-static stress, and if excessive stress caused by wire curvature is neglected, these σ_b values can be used for design calculation purposes.

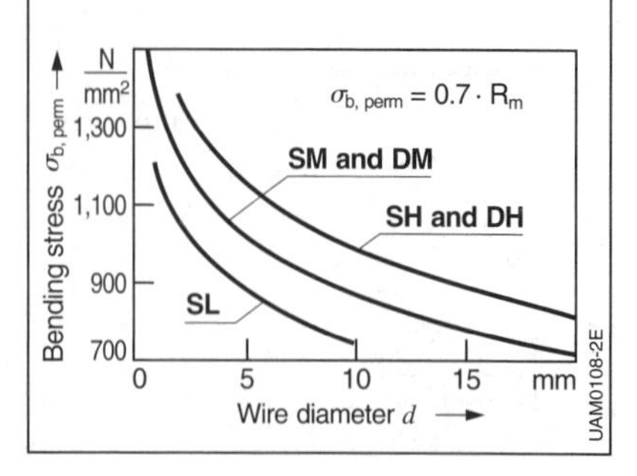

Torsion and spiral springs

In the case of deflection of torsion and spiral springs, recoil torques are generated about the axis of rotation. Due to the clamping conditions, the bending stresses in the angle are almost uniform. The same equations are used for calculating torsion and spiral springs.

Disc springs

The cone-bowl-shaped disc springs (Figure 4) are primarily subjected to bending stresses. A wide variety of applications results from the large number of possible parallel and series combinations. Disc springs are mainly used where spring force and travel must be absorbed within confined spaces. They are used among other things in regular and overload clutches and to pre-tension roller bear-

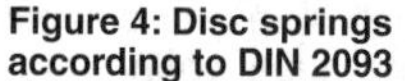

Figure 4: Disc springs according to DIN 2093
a) Without contact bearing surfaces,
b) With contact bearing surfaces,
c) Calculated spring characteristic of a disc spring acc. to DIN 2092.

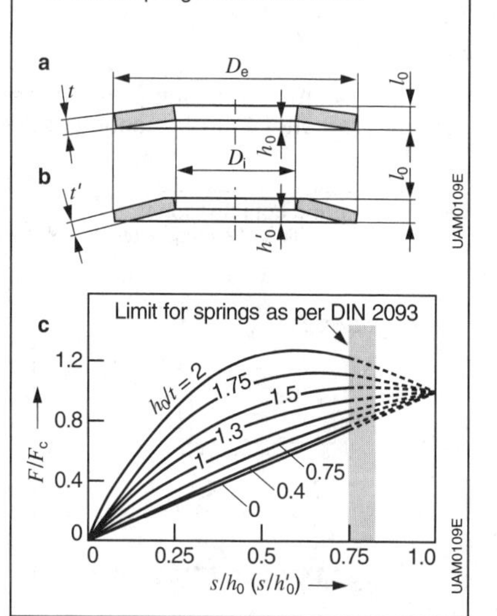

ings. Because friction occurs between the individual disc springs in spring assemblies layered in the same direction, they are also suitable for damping vibrations and shocks.

At $h_0/t > 0.4$, the nonlinearities of the springs are no longer negligible. Spring forces, spring travel, and spring rates can be calculated with sufficient precision according to DIN 2092, or they can be taken from the manufacturer's data.

For disc springs subjected to static stress ($<10^4$ stress reversals), fatigue-strength calculation is not required if the maximum spring force at $s = 0.75\ h_0$ is not exceeded.

Springs subjected to torsional stress

Torsion bars

Circular cross-sections are usually selected for torsion bars. They have a very high volume utilization factor, which means that they can absorb a lot of energy, but they occupy little space.

Helical springs

Cylindrical helical springs are manufactured as compression and extension

Table 4: Leaf spring, torsion spring, and spiral spring

Type		Spring force	Deflection
Simple straight leaf spring (constant cross-section)		$F_{max} = \frac{b \cdot h^2}{6 \cdot l} \sigma_{b,perm}$	$s = \frac{4 \cdot F \cdot l^3}{E \cdot b \cdot h^3}$
		Spring rate	Spring duty
		$R = \frac{F}{s} = \frac{E \cdot b \cdot h^3}{4 \cdot l^3}$	$W_{max} = \frac{\sigma^2_{b,perm}}{18 \cdot E} b \cdot h \cdot l$
Layered leaf spring		Spring force, spring torque	Deflection
		$F_{max} = \frac{i \cdot b \cdot h^2}{6 \cdot l} \cdot \sigma_{b,perm}$	$s = \frac{12 \cdot F \cdot l^3}{(2 \cdot i + i') \cdot E \cdot b \cdot h^3}$
		Spring rate	Spring duty
		$R = \frac{(2i + i') \cdot E \cdot h^3 \cdot b}{12 \cdot l^3}$	$W_{max} = \frac{\sigma^2_{b,perm}}{3 \cdot E} \cdot \frac{i^2}{(2i + i')} \cdot b \cdot h \cdot l$
Torsion spring (leg spring) $l_f = \pi \cdot D \cdot n$	Circular cross-section	Spring force, spring torque	Deflection
		$M_{t,max} = \frac{\pi \cdot d^3}{32} \cdot \sigma_{b,perm}$	$\widehat{\alpha} = \frac{64 \cdot M_t \cdot l}{E \cdot \pi \cdot d^4}$
		Spring rate	Spring duty
		$R = \frac{M_t}{\widehat{\alpha}} = \frac{E \cdot \pi \cdot d^4}{64 \cdot l}$	$W_{max} = \frac{\sigma^2_{b,perm}}{32 \cdot E} \cdot \pi \cdot d^2 \cdot l$
Spiral spring $l_f = 2 \cdot \pi \cdot n \cdot [r_0 + 0.5 \cdot n \cdot (h + \delta_r)]$	Rectangular section	Spring force, spring torque	Deflection
		$M_{t,max} = \frac{b \cdot h^2}{6} \cdot \sigma_{b,perm}$	$\widehat{\alpha} = \frac{12 \cdot M_t \cdot l}{E \cdot b \cdot h^3}$
		Spring rate	Spring duty
		$R = \frac{M_t}{\widehat{\alpha}} = \frac{E \cdot b \cdot h^3}{12 \cdot l}$	$W_{max} = \frac{\sigma^2_{b,perm}}{6 \cdot E} \cdot b \cdot h \cdot l$

springs. The calculation equations are identical for both types. Compression springs in conical form can optimize the use of space if the individual coils can be pushed into each other.

Force eccentricity can be minimized on compression springs by coiling the spring so that the wire end at each end of the spring touches the adjoining coil. Each end of the spring is then ground flat, perpendicular to the axis of the spring. To avoid overloading the spring, a minimum distance between the active coils must be maintained. For static stresses, the information in Table 6 applies.

For dynamic stress, S_a must be doubled. Additionally, the spring ends are arranged at 180° to each other. The total number of coils is always then a multiple of a half coil (e.g. n_t = 7.5). The effect of the wire curvature is taken into account by the stress coefficient k (Table 7).

Table 6: Helical springs

	Total number of coils	Block length (solid length)	Sum of minimum distances
Cold-formed	$n_t = n + 2$	$L_c \leq n_t \cdot d$	$S_a = (0.0015 \cdot D^2/d + 0.1 \cdot d) \cdot n$
Hot-formed	$n_t = n + 1.5$	$L_c \leq (n_t - 0.3) \cdot d$	$S_a = 0.02 \cdot (D + d) \cdot n$

Table 7: k-factor

D/d	3	4	6	8	10	14	20
k	1.55	1.38	1.24	1.17	1.13	1.10	1.07

Table 5: Torsion-bar spring and helical springs

Type	Spring force, spring torque	Deflection
Torsion-bar spring with circular cross-section (DIN 2091)	$M_{t,\,max} = \frac{\pi \cdot d^3}{16} \cdot \tau_{t,\,perm}$	$\alpha = \frac{32 \cdot M_t \cdot l_f}{G \cdot \pi \cdot d^4}$
	Spring rate	Spring duty
	$R = \frac{M_t}{\alpha} = \frac{G \cdot \pi \cdot d^4}{32 \cdot l_f}$	$W_{max} = \frac{\tau_{t,\,perm}^2}{16 \cdot G} \cdot \pi \cdot d^2 \cdot l_f$
Cylindrical helical springs with circular cross-section (DIN 2089)	Spring force, spring torque	Deflection
	$F_{max} = \frac{\pi \cdot d^3}{8 \cdot k \cdot D} \cdot \tau_{t,\,perm}$	$s = \frac{8 \cdot D^3 \cdot n}{G \cdot d^4} \cdot F$
	Spring rate	Spring duty
	$R = \frac{G \cdot d^4}{8 \cdot D^3 \cdot n}$	$W_{max} = \frac{\tau_{t,\,perm}^2}{16 \cdot G} \cdot d^2 \cdot D \cdot \pi^2 \cdot n$
Tapered helical springs with circular cross-section	Spring force, spring torque	Deflection
	$F_{max} = \frac{\pi \cdot d^3}{16 \cdot k \cdot r_2} \cdot \tau_{t,\,perm}$	$s = \frac{16 \cdot (r_1 + r_2) \cdot (r_1^2 + r_2^2) \cdot n \cdot F}{G \cdot d^4}$
	Spring rate	Spring duty
	$R = \frac{G \cdot d^4}{16 (r_1 + r_2) \cdot (r_1^2 + r_2^2) \cdot n}$	$W_{max} = \frac{\tau_{t,\,perm}^2}{32 \cdot G} \cdot \frac{d^2 (r_1 + r_2) \cdot (r_1^2 + r_2^2) \cdot \pi \cdot n}{r_2^2}$

In the case of static stress, this effect can be disregarded, e.g. $k = 1$ is then set. The following applies to the stress range prevailing in the case of dynamic stress:

$$\tau_{kh} = k \frac{8\ D}{\pi \cdot d^3} \cdot (F_2 - F_1) \leq \tau_{kH}$$

Extension springs

Extension springs are formed either with loops or with rolled-in or screwed-in end pieces. Since service life is determined primarily by the loops, it is impossible to give general fatigue limit values. Cold-formed extension springs hardened and tempered after drawing can be manufactured with an internal preload. This allows significantly higher stresses.

References

[1] DIN-Taschenbuch 29: Federn (Springs). Beuth-Verlag 2003.

[2] Haberhauer/Bodenstein: Maschinenelemente. 15th Edition, Springer-Verlag 2008.

[3] Fischer/Vondracek: Warm geformte Federn. Hoesch Hohenlimburg AG, 1987.

Figure 6: Permissible torsional stresses for helical springs with static stress

a) Cold-formed from patented drawn spring-steel wires (SL, SM, DM, SH and DH) and valve spring-steel wire (VDC) acc. to DIN EN 10270-2,

b) Hot-formed from spring steels acc. to DIN EN 10089.

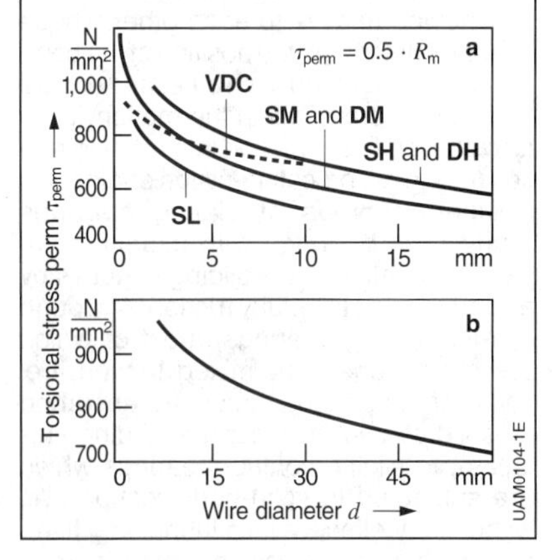

[4] Meissner/Schorcht: Metallfedern. Springer-Verlag 1997.

Figure 5: Fatigue-limit diagrams for helical compression springs

a) For cold-formed springs made from spring-steel wires SH and DH (not shot-blasted),

b) For cold-formed springs made from spring-steel wires SH and DH (shot-blasted),

c) For cold-formed springs made from valve spring-steel wire (VDC),

d) For hot-formed springs.

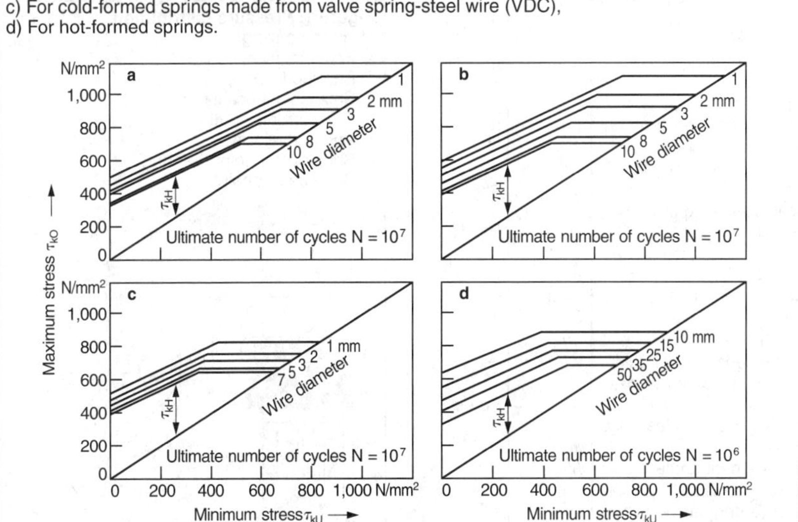

Sliding bearings

Features

Sliding bearings (also termed plain or friction bearings) serve to accommodate and transmit forces between components that are moving relative to each other. These bearings determine the position of the components in relation to each other and ensure locating accuracy within the motion. They also convert linear motion into rotating motion (e.g. in reciprocating-piston engines).

Different types of sliding bearings range from bearings with usually complete separation of the sliding surfaces by a lubrication film (fluid friction), through self-lubricating bearings, most of which are characterized by mixed friction, i.e. some of the bearing forces are absorbed by solid contact between the sliding surfaces), to sliding-contact bearings which are subjected to solid-body friction (i.e. without any effective fluid lubricating film), but which nevertheless have an adequate service life.

Table 1: Symbols and units (DIN 31652)

Description	Symbol	Unit
Axial bearing length	B	m
Inside bearing diameter (nominal diameter)	D	m
Shaft diameter (nominal diameter)	d	m
Eccentricity (displacement between shaft and bearing centers)	e	m
Bearing force (load)	F	N
Minimum lubricating film thickness	h_0	m
Local lubricating film pressure	p	Pa = N/m^2
Specific bearing load $\overline{p} = F/(B\ D)$	$\overline{p}$	Pa
Bearing clearance $s = (D - d)$	s	m
Sommerfeld number	So	–
Relative eccentricity 2 e/s	ε	–
Effective dynamic viscosity of lubricant	η_{eff}	Pa · s
Relative bearing clearance $\psi = s/D$	ψ	–
Displacement angle	β	°
Hydrodynamically effective angular velocity	ω_{eff}	s^{-1}

Hydrodynamic sliding bearings

Most of the hydrodynamic sliding-bearing types used in motor-vehicle engines are plain bearings (Figure 1) for holding the crankshaft drive (including camshafts). They are usually designed as bearing shells with special clearance (e.g. oval clearance). Thrust bearings are used as axial locators, and are predominantly not subjected to load.

A hydrodynamic sliding bearing is reliable in service if it remains sufficiently unaffected by the following:

- Wear (sufficient separation of the contact surfaces by the film)
- Mechanical stress (bearing material of sufficient strength)
- Thermal loading (observance of thermal stability of bearing material and viscosity/temperature behavior of film)

To ensure reliable operation of a hydrodynamic sliding bearing, it is necessary for there to be a load-carrying lubricating film over a wide operating range. This lubricating film is created in a radial bearing when the shaft is eccentrically positioned. This, in turn, generates a lubricant pressure.

Figure 1: Pressure distribution in plain bearing
e Eccentricity,
F Bearing force,
h_0 Smallest lubricating gap,
p Pressure distribution,
ω Angular velocity,
S Shell center point (bearing),
W Shaft center point.

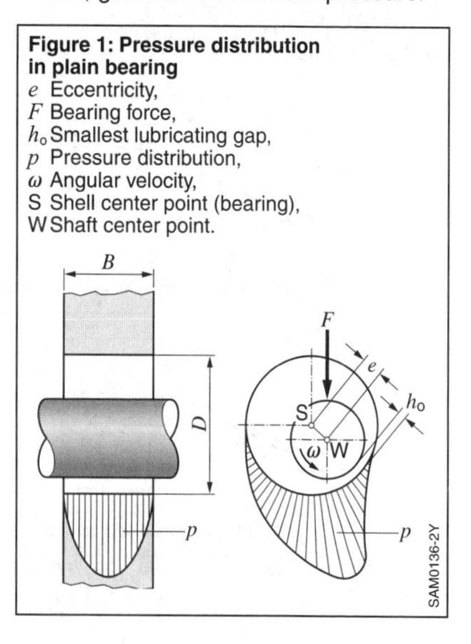

The rotating shaft delivers the lubricant into the bearing gap. The shaft eccentricity is set within the system in such a way that the integral of the lubricant pressure of the external bearing force maintains equilibrium. The hydrodynamic pressure distribution in a convergent bearing gap is determined from the solution of Reynolds' differential equation.

The load capacity of the lubricating film is determined by integrating the pressure distribution, and is expressed in the dimensionless Sommerfeld number *So*:

$$So = \frac{F\,\psi^2}{(D\,B\,\eta_{\text{eff}}\,\omega_{\text{eff}})}\ .$$

As the Sommerfeld number *So* increases, so the relative eccentricity increases and the minimum lubricating film thickness h_o decreases. The following applies:

$$h_o = \frac{(D-d)}{2} - e = 0.5\,D\,\psi\,(1-e)\ .$$

The relative eccentricity is:

$$\varepsilon = \frac{2e}{(D-d)}\ .$$

The Sommerfeld number is also used to determine the coefficient of friction in the bearing, and to calculate friction loss and thermal loading (DIN 31652, VDI Directive 2204).

The friction states in the hydrodynamic sliding bearing can be explained with the aid of the Stribeck curve (Figure 2). There are three different friction states:
- Dry friction
- Mixed friction
- Fluid friction

The coefficients of friction given in Table 2 are approximate values, and are intended solely for comparison of the different types of friction.

As hydrodynamic bearings also operate with mixed friction some of the time, must be able to withstand a certain amount of contamination without loss of function, and are also subjected to high dynamic and thermal stress (particularly in piston engines), the bearing material must meet a number of requirements, some of which are mutually exclusive.

Table 2: Orders of magnitude of the coefficients of friction for different types of friction

Type of friction	Coefficient of friction *f*
Dry friction	0.1, – >1
Mixed friction	0.01 – 0.1
Fluid friction	0.01
Sliding and rolling friction in roller bearings	0.001

Table 3: Empirical values for maximum approved specific bearing load

Bearing materials	Max. permissible spec. bearing load $\bar{p}_{\text{lim}}$
Pb and Sn alloys (babbitt metals)	5 – 15 N/mm²
Lead-base bronze	7 – 20 N/mm²
Tin-base bronze	7 – 25 N/mm²
AlSn alloys	7 – 18 N/mm²
AlZn alloys	7 – 20 N/mm²
Maximum values only apply for very low sliding velocities	

- Conformability (compensation of misalignment by plastic deformation without shortening service life)
- Wettability by fluid film
- Embeddability (ability of the bearing surface to absorb particles of dirt without increasing bearing or shaft wear)
- Wear resistance (in the case of mixed friction)

Figure 2: Stribeck curve (diagram)
1 Dry friction, 2 Mixed friction,
3 Fluid friction,
A Transition point,

f Coefficient of friction = $\frac{\text{Friction force}}{\text{Normal force}}$

v Sliding velocity.

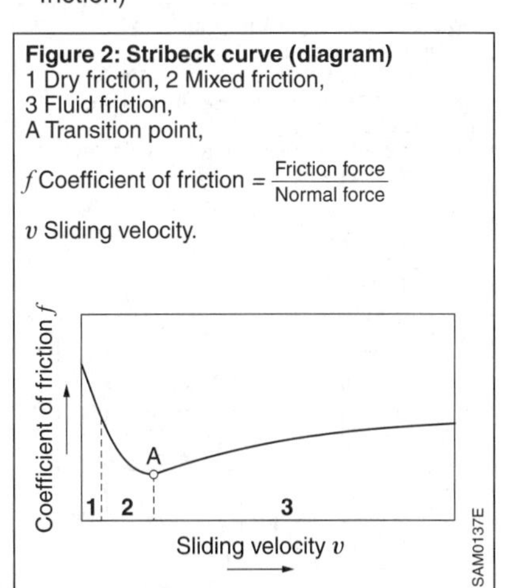

- Seizure resistance (bearing material must not weld to shaft material, even under high compressive load and high sliding velocity)
- Anti-seizure performance (resistance to wear)
- Run-in performance (a combination of conformability, resistance to wear, and embeddability)
- Mechanical loadability
- Fatigue strength (under alternating loads, particularly at high thermal load)

If a bearing (e.g. piston-pin bushing) is simultaneously subjected to high loads and low sliding velocities, high fatigue strength and wear resistance should take precedence over resistance to seizing. The bearing materials used in such cases are hard bronzes or special brass alloys.

Since they are subjected to dynamic loads with high sliding velocities, connecting rod and crankshaft bearings in internal combustion engines must fulfill a number of different requirements. In these applications, multilayer bearings (Figure 3), above all trimetal bearings, have proved themselves in practice.

The service life of sliding bearings in the crankshaft drive can be further increased through the use of special solutions such as sputter bearings or grooved sliding bearings. Sputter bearings (Figure 4) are characterized by a highly wear-resistant AlSn running layer which is applied by means of a PVD process (Physical Vapor Deposition) to the high-strength bearing material underneath.

Grooved sliding bearings (Figure 5) prove to be useful in certain applications. In these bearings, fine grooves machined into the running surface in the circumferential direction are filled with a softer running layer (galvanic layer similar to that on a trimetal bearing).

The above-mentioned solutions are used to good effect in maximum-load internal combustion engines (e.g. high pressure, supercharged diesel engines).

Figure 3: Multilayer bearing
(Design of a trimetal bearing).
1 Steel backing shell, 2 Bearing metal,
3 Diffusion barrier (e.g. 1–2 µm nickel),
4 Penetration coating
(approx. 20 µm, electroplated SnCu layer or anti-friction paint).

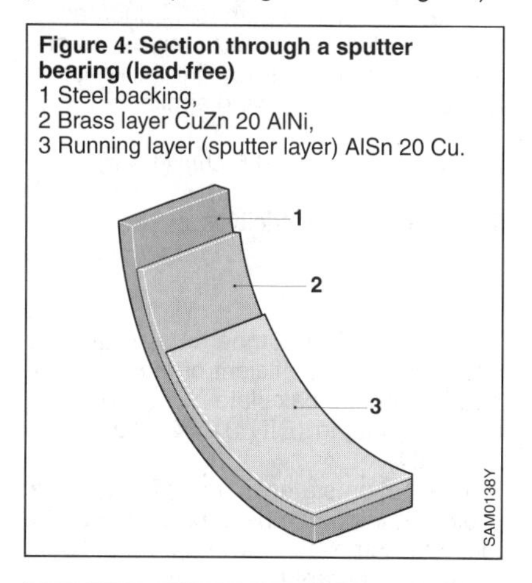

Figure 4: Section through a sputter bearing (lead-free)
1 Steel backing,
2 Brass layer CuZn 20 AlNi,
3 Running layer (sputter layer) AlSn 20 Cu.

Figure 5: Section through a grooved sliding bearing
(MIBA patent). Running surface with very fine grooves in running direction V_G.
1 Wear-resistant light alloy,
2 Soft running layer,
3 Nickel barrier.

Table 4: Selection of materials for hydrodynamic sliding bearings

Material	Alloy designation	Composition in %	HB hardness 20 °C	100 °C	Remarks Application examples
Tin-base babbitt metal	LgPbSn 80 (WM 80)	80 Sn; 12 Sb; 6 Cu; 2 Pb	27	10	Very soft, good conformance of contact surfaces to off-axis operation, very good anti-seizure performance
Lead-base babbitt metal	LgPbSn 10 (WM 10)	73 Pb; 16 Sb; 10 Sn; 1 Cu	23	9	Reinforcement necessary, e.g. as composite steel casting or with intermediate nickel layer on lead bronze.
Tin-base babbitt Lead	G-CuPb 25	74 Cu; 25 Pb; 1 Sn	50	47	Very soft, very good anti-seizure performance, less resistant to wear.
	G-CuPb 22	70 Cu; 22 Pb; 6 Sn; 3 Ni	86	79	
Lead-tin-base bronze	G-CuPb 10 Sn	80 Cu; 10 Pb; 10 Sn	75	67	Improved anti-seizure performance by alloying with lead. More resistant to off-axis operation than pure tin bronzes, therefore high-load Pb-Sn bronzes preferable for use in crankshaft drives. Composite bearings in internal-combustion engine manufacture, piston-pin bushings. $\overline{p}$ to 100 N/mm^2.
	G-CuPb 23 Sn	76 Cu; 23 Pb; 1 Sn	55	53	Composite casting for low-load bearings (70 N/mm^2). Also thick-wall bearing shells. Particularly good anti-seizure performance. Crankshaft bearings, camshafts, connecting rod bearings.
Tin-base babbitt Tin	G-CuSn 10 Zn	88 Cu; 10 Sn; 2 Zn	85		Hard material. Sliding bearing an be subjected to moderate loads at low sliding velocities. Worm gears.
	CuSn 8	92 Cu; 8 Sn	80... 220		High-grade wrought alloy. Good performance under high loads and in the absence of sufficient lubrication, steering knuckle bearings. Particularly suited for use as thin-walled sliding-bearing bushings.
	CuSn 6 Ni 6[2]	88 Cu; 6 Sn; 6 Ni	230		Good corrosion resistance, high wear resistance, can be loaded up to 200 N/mm^2, rolled small-end bushings.
Red brass	G-CuSn 7 ZnPb	83 Cu; 6 Pb; 7 Sn; 4 Zn	75	65	Tin partially replaced by zinc and lead. Can be used instead of tin bronze, but only for moderate loads (40 N/mm^2). General sliding bearings for machinery. Piston pins, bushings, crankshaft bearings, and knuckle joint bearings.

Table 4: Selection of materials for hydrodynamic sliding bearings (continued)

Material	Alloy designation	Composition in %	HB hardness 20 °C	100 °C	Remarks Application examples
Brass	CuZn 31 Si	68 Cu; 31 Zn; 1 Si	90… 200		Tin content is unfavorable at higher bearing temperatures. Can be used instead of tin bronze, low loads.
	CuZn 23 Al 5[3]	64 Cu; 23 Zn; 5 Al; 3 Mu; 3 Fe; 2 Ni	180		High static and dynamic load capacity. Perm. surface pressure up to 200 N/mm^2, thrust washers, turned piston pin bushings.
	CuZn 20 AlMnNi[4]	74 Cu; 20 Zn; 2 Al; 2 Mn; 2 Ni	120… 180		Composite steel casting, very good corrosion resistance, high wear resistance, can be loaded up to 180 N/mm^2, rolled small-end bushings. In conjunction with sputter overlay (AlSn 20) as connecting-rod bearing in turbocharged engines.
Aluminum bronze	CuAl 9 Mn	88 Cu; 9 Al; 3 Mn	110… 190		Thermal expansion comparable to that of light alloys, suitable for use as interference-fit bearings in light-alloy housings. Better wear resistance than tin bronze, but higher friction.
Aluminum alloy	AlSi 12 Cu NiMn	1 Cu; 85 Al; 12 Si; 1 Ni; 1 Mn	110	100	Piston alloy for low sliding velocities.
Rolled aluminum cladding	AlSn 6	1 Cu; 6 Sn; 90 Al; 3 Si	40	30	Liquated tin stretched by rolling, therefore high loadability, antifriction properties. Improved by galvanic layer.
	AlSn 10 Ni[5]	86 Al; 10 Sn; 2 Ni; 1 Mn; 1 Cu	50… 70		Composite rolled steel cladding, can be used as two-component bearing, high load capacity, very good corrosion resistance, crankshaft bearings, connecting-rod bearings.
	AlZn 5 Bi[6]	89 Al; 5 Zn; 3 Bi; 2 Si; 1 Cu	60… 100		Composite rolled steel cladding for two-component bearings. Can be loaded up to 90 N/mm^2, high wear resistance with good anti-seizure performance, main and connecting rod bearings.
Electro-plated layers	PbSn 10 Cu	2 Cu; 88 Pb; 10 Sn	15…20 HV		For modern trimetal bearings 10 to 30 µm thick, electroplated, very fine grain. Intermediate nickel layer on bearing metal.
	SnCu 6[7]	94 Sn; 6 Cu	15…20 HV		High load capacity in conjunction with an intermediate SnNi layer on the bearing metal, crankshaft and connecting rod bearings.

Table 4: Selection of materials for hydrodynamic sliding bearings (continued)

Material	Alloy designation	Composition in %	HB/HV hardness 20 °C 100 °C	Remarks Application examples
Anti-friction paint layers [8]		70 PAI; 20 MoS_2; 10 Graphite	25…30 HV	Good anti-seizure performance with good wear resistance. Can be loaded up to 80 N/mm², crank-shaft and connecting rod bearings.
Sputter layers	AlSn 20 Cu	79 Al; 20 Sn; 1 Cu;	85…110 HV	Deposited by means of a PVD process on the bearing metal, layer thickness 8…16 mm, very good wear resistance. Can be loaded up to 120 N/mm². For connecting rod bearings subjected to high load.

For materials, see also: DIN 1703, 1705, 1716, 17660, 17662, 17665, 1494, 1725, 1743.
ISO 4381, 4382, 4383.

Important note: Some of the alloys mentioned in Table 4 contain lead EU Directive 2000/53/EC has prohibited the use of lead in passenger cars since mid-2008 (exception provision expires). Lead is currently still permitted in commercial vehicles. This also applies to general mechanical engineering.

[1, 4, 6] KS sliding-bearing materials, [2] Material manufactured by Wieland,
[3] Material manufactured by Bögra,
[5, 7] Federal Mogul materials, [8] Anti-friction paint manufactured by Miba.

Sintered metal sliding bearings

Sintered metal sliding bearings consist of sintered metals which are porous and impregnated with liquid lubricants. For many small motors in motor automotive applications, this type of bearing is a good compromise in terms of precision, installation, freedom from maintenance, service life, and cost. They are used primarily in motors with shaft diameters from 1.5 to 12 mm. Sintered iron bearings and sintered steel bearings (inexpensive, less likely to interact with the lubricant) are preferable to sintered bronze bearings for use in motor vehicles (Table 5). The advantages of sintered bronze bearings are greater loadability, lower noise, and lower friction coefficients (this type of bearing is used in record-players, office equipment, data systems, and cameras).

The performance of sintered bearings over long periods of service is closely related to the use of optimum lubricants.

Table 5: Materials for sintered metal bearings

Material group	Designation Sint- …	Composition	Remarks
Sintered iron	B 00	Fe	Standard material which meets moderate load and noise requirements.
Sintered steel, containing Cu	B 10	<0.3 C 1…5 Cu Rest Fe	Good resistance to wear, can be subjected to higher loads than pure Fe bearing.
Sintered steel, higher Cu content	B 20	20 Cu Rest Fe	Lower-priced than sintered bronze, good noise behavior and $p \cdot v$ values.
Sintered bronze	B 50	<0.2 C 9…10 Sn Rest Cu	Standard Cu-Sn based material, good noise behavior.

Sint-B indicates 20 % P (porosity) (Sint-A: 25 % P; Sint-C: 15 % P).

Mineral oils: Inadequate cold-flow properties, moderate resistance to aging.

Synthetic oils (e.g. esters, poly-α olefins): Good cold-flow properties, high resistance to thermal stresses, low evaporation tendency.

Synthetic greases (oils which include metal soaps): Low starting friction, low wear.

Sliding-contact bearings

Solid polymer bearings made of thermoplastics

Advantages
- Inexpensive
- No danger of seizure with metals

Disadvantages
- Low thermal conductivity
- Relatively low operating temperatures
- Possible swelling due to humidity
- Low loadability
- High coefficient of thermal expansion

The most frequently used polymer materials are:
- Polyoxymethylene (POM, POM-C)
- Polyamide (PA)
- Polyethylene and polybutylene terephthalate (PET, PBT)
- Polyetheretherketone (PEEK)

The tribological and mechanical properties can be varied over a wide range by incorporating lubricants and reinforcements in the thermoplastic base material.

Lubrication additives
- Polytetrafluorethylene (PTFE)
- Graphite (C)
- Silicone oil, and other liquid lubricants, recently also enclosed in microcapsules

Reinforcement additives
- Glass fibers (GF)
- Carbon fibers (CF)

Application examples
- Windshield wiper bearings (PA and glass fiber)
- Idle actuators (PEEK + carbon fiber, PTFE and other additives)

Figure 6: Section through a self-lubricating composite bearing
1 Polymer liner,
2 Porous bronze layer,
3 Copper layer,
4 Steel backing,
5 Tin layer.

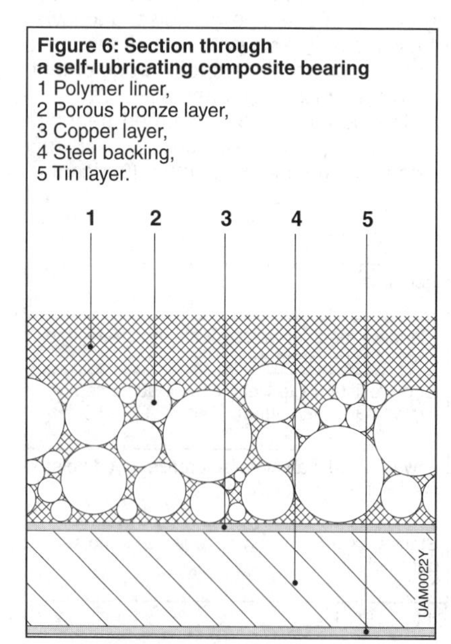

Figure 7: Section through composite bearing with fabric insert made of PTFE and fiberglass
1 PTFE fiber fabric, 2 Adhesive fibers,
3 Resin, 4 Fiberglass backing,
5 Adhesive, 6 Steel backing.

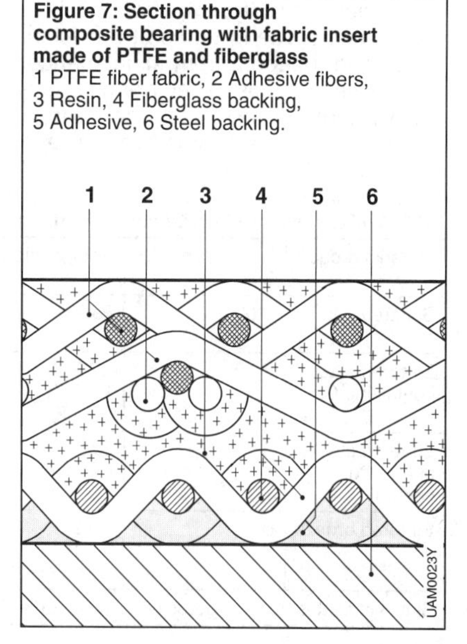

Polymer bearings made of duroplastics and elastomers

These materials, with their high levels of intrinsic friction, are seldom used as bearing materials in motor vehicles. Duroplastics include:

- Phonol resins (high friction, e.g. Resitex)
- Epoxy resins (require the addition of PTFE or C, reinforcement by fibers usually required on account of their inherent brittleness)
- Polyimides (high thermal and mechanical loadability)

Application example

- Polyimide axial stop in wiper motor.

Metal-backed composite bearings

Composite bearings are combinations of polymer materials, fibers, and metals. Depending on the bearing structure (Figures 6 and 7), they provide advantages over pure or filled polymer plain bearings in terms of loadability, bearing clearance, thermal conductivity, and installation (suitable for use with oscillating motion).

Example of bearing structure

- Tin-plated or copper-clad steel backing (several millimeters thick), onto which a layer of porous bronze is sintered 0.2 to 0.35 mm thick with a porosity of 30 to 40 %. A low-friction polymer material is rolled into this bronze layer as a liner. Liner made of:
  a) acetal resin or polyvinylide fluoride, either impregnated with oil or containing lubricating recesses, or
  b) PTFE + ZnS or MoS_2 or graphite as additive

Metal-backed composite bearings are available in a number of different shapes and compositions. Metal-backed composite bearings with woven PTFE fiber inserts have unusually high loadability and are suitable for use in ball-and-socket joints.

Application examples for motor vehicles

- Piston rod bearings for suspension struts
- Release lever bearings for clutch pressure plates
- Brake shoe bearings in drum brakes
- Ball-and-socket bearings
- Door hinge bearings
- Bearings for seatbelt winding shafts
- Steering knuckle bearings
- Gear pump bearings

Composite bearings with specially modified liners are required for heavy-duty requirements in diesel high-pressure injection pumps. The liner is made of PEEK or PPS with additives (e.g. carbon fibers, ZnS, TiO_2, and graphite). The particle size is sometimes in the nano range.

Carbon-graphite bearings

Carbon-graphite bearings are members of the ceramic bearing family due to their method of manufacture and material properties. The base materials are powdered hydrocarbons; tar or synthetic resins are used as binders.

Advantages

- Thermal stability up to 350 °C (hard-burnt carbon), up to 500 °C (electro-graphite)
- Good antifriction properties
- Good corrosion resistance
- Good thermal conductivity
- Good thermal shock resistance. They are highly brittle, however.

Application examples for carbon-graphite bearings

- Fuel pump bearings
- Bearings in drying ovens
- Adjustable blades in turbochargers

Metal-ceramic bearings

Metal-ceramic bearings consist of material manufactured by powder metallurgy processes; in addition to the metallic matrix, the bearing material also contains finely distributed solid lubricant particles.

Matrix: e.g. bronze, iron, nickel.
Lubricant: e.g. graphite, MoS_2.

These materials are suitable for use under extremely high loads, and are at the same time self-lubricating.

Application example

- Steering knuckle bearings

Table 6: Properties of maintenance-free, self-lubricating bearings

	Sintered bearings oil-impregnated		Polymer bearings		Metal-backed composite bearings Running layer		Synthetic carbons
	Sintered iron	Sintered bronze	Thermoplastic polyamide	Duroplastic polyimide	PTFE + additive	Acetal resin	
Compression strength N/mm²	80…180		70	110	250	250	100…200
Max. sliding velocity m/s	10	20	2	8	2	3	10
Typical load N/mm²	1…4 (10)		15	50 (at 50 °C) 10 (at 200 °C)	20…50	20…50	50
Perm. operating temperature °C Short-term	–60…180 (depends on oil) 200		–130 to 100 120	–100…250 300	–200…280	–40…100 130	–200…350 500
Coefficient of friction without lubrication	with lubrication 0.04…0.2		0.2…0.4 (100 °C) 0.4…0.6 (25 °C)	0.2…0.5 (unfilled) 0.1…0.4 (filled)	0.4…0.2	0.7[1]…0.2 [1] PTFE filled	0.1…0.35
Thermal conductivity W/(m · K)	20…40		0.3	0.4…1	46	2	10…65
Corrosion resistance	less good	good	very good		good	good	very good
Chemical resistance	no		very good		conditional	conditional	good
Max. $p \cdot v$ (N/mm²) · (m/s)	20		0.05	0.2	1.5 to 2		0.4…1.8
Embeddability of dirt and abraded material	less good		good	good	less good	good	less good

Roller bearings

Applications

Roller bearings are some of the most important components in machines. Great demands are made on their load capacity and operational reliability.

Roller bearings are widely used in motor vehicles, e.g. as bearings in the alternators, starters, wheel bearings, transmission, suspension struts, cardan shafts, water pumps, tensioning rollers, steering systems, windshield-wiper motors, fans and fuel-injection pumps.

Figure 1: Design of roller bearings
a) Deep-groove ball bearing,
b) Angular-contact ball bearing,
c) Needle bearing,
d) Cylindrical roller bearing,
e) Tapered-roller bearing,
f) Self-aligning roller bearing.
1 Outer race,
2 Inner race,
3 Cage,
4 Rolling element.

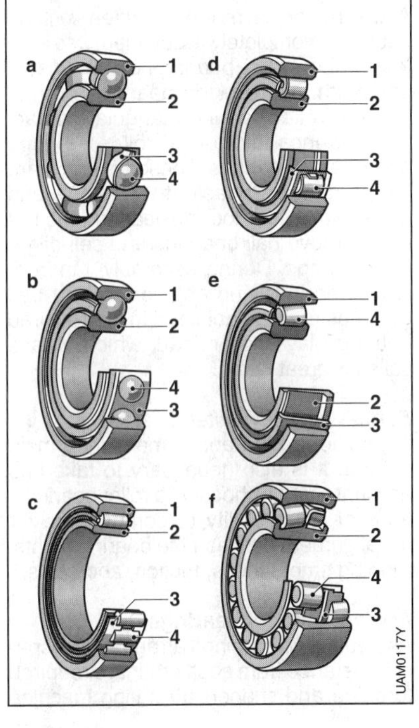

General principles

Type

Roller bearings are generally made up of two races (Figure 1), a cage and a rolling-element assembly. The rolling elements guided by the cage roll on the raceways. Balls, cylindrical rollers, needle rollers, tapered rollers, and self-aligning rollers are used as rolling elements. A roller bearing can be lubricated with grease. It is fitted with cover plates or gaskets to provide a seal against dirt.

The roller bearing transfers the outer force from one bearing race to the other via the rolling elements. A distinction is made between radial bearings and axial (thrust) bearings, depending on the main direction of load.

Structural dimensions

The roller bearing is a ready-to-install machine part. Its outer dimensions are laid down in the standards DIN 623 and DIN ISO 355.

Various ranges of outside diameters and widths are available for a hole diameter. Standardized abbreviation codes are used to identify the diameter and width ranges of a roller bearing.

Tolerances

Roller-bearing tolerances are standardized according to precision in ISO 492 and DIN 620. Roller bearings of normal precision, i.e. tolerance class P0 (also called PN), generally satisfy all the demands which are made by mechanical engineering on bearing quality. For more stringent requirements, the standard provides for more precise tolerance classes P6, P5, P4, and P2.

Tolerance tables can also be found in the catalogs of roller bearing manufacturers.

Bearing play
The bearing play of an uninstalled roller bearing refers to the distance through which the bearing races can be moved against each other. A distinction is made between radial play and axial play.

Radial play is defined in the standard DIN 620, Part 4. The normal radial play category is C0. In accordance with operating conditions, such as, for example, conversion parts and temperatures, it is possible also to use the other radial play categories C1 and C2 (<C0) or C3 and C4 (>C0).

Axial play is derived from the radial play and raceway and rolling-element geometries, and is always given as a reference parameter.

Materials
Bearing races and rolling elements are primarily composed of chrome-alloy special steel 100Cr6 (DIN 17230) or 52100 (ASTM A295) with a high degree of purity and hardness in the range of 58 – 65 HRC.

Roller bearing cages are made of metal or plastic. The metallic cage in small roller bearings is primarily composed of sheet steel.

Polyamide 66 (PA66) is used for the majority of plastic cages. This material, especially when reinforced with fiberglass, is characterized by its favorable combination of strength and elasticity. Injected PA66 cages are suitable for long-time operation at temperatures up to 120 °C.

Other thermoplastics and duroplastics are also used as cage materials for special applications subject to extremely high thermal loads.

Selection of roller bearings

It is necessary to take into account many external factors in order to choose the right bearing from the wealth of options.

Selection criteria
Load
The size and direction of the load acting on the roller bearing normally determine the type and size of the bearing. Deep-groove ball bearings are usually used in the case of low to medium loads. Roller bearings have one advantage in the case of high loads and limited installation space. With the exception of only purely radially loaded needle bearings, cylindrical-roller bearings and axial bearings, roller bearings can simultaneously accommodate radial and axial loads (combined load).

Deep-groove ball bearings transfer axial loads in both directions, while angular-contact ball bearings and tapered roller bearings can only be axially loaded in one direction.

Cylindrical roller bearings and self-aligning bearings are particularly suitable for radial loads, but less so for axial loads.

Rotational speed
Ball bearings with point contact between rolling elements and raceways have a higher speed limit than roller bearings of the same size. The permissible speed of a roller bearing is also dependent on the lubricating process. A bearing lubricated with oil has a higher speed limit than a bearing lubricated with grease.

Assembly
A distinction is made between locking bearings (completely assembled) and non-locking bearings (bearings can be disassembled). Non-locking bearings include tapered roller bearings, angular-contact ball bearings, cylindrical roller bearings and needle bearings. These bearings are for the most part easier to assemble and disassemble than locking bearings such as deep-groove ball bearings and self-aligning bearings. During assembly, tapered-roller bearings and angular-contact ball bearings must be adjusted to the required bearing play and preload, which always calls for great care.

Further selection criteria
In addition to the above-mentioned main criteria, it is also necessary to take into account – when choosing a roller bearing – angular adjustability to compensate for misalignment between the bearing points, running smoothness, friction, and costs.

Arrangement of bearings
As a rule, two bearings arranged at a specific distance from each other are required to guide and support a rotating machine

part. There are two important variations of bearing arrangement.

Locating/floating bearing arrangement
Two radial bearings are seated on the shaft and in the housing. The distance between the two bearing points is determined by manufactured conversion parts within the framework of the tolerance. In addition, the shaft, when heated to different temperatures or when made from different materials, does not expand in the same way as the housing. These differences must be compensated for in the bearing points. For this reason, one bearing is to be axially secured as a locating bearing on the shaft and in the housing and one bearing is to be movable in the axial direction as a floating bearing (Figure 2). Typical applications for locating/floating bearing arrangement: alternators, steering motors.

Single-row deep-groove ball bearings are frequently used as locating bearings. Cylindrical roller bearings, needle bearings, and deep-groove ball bearings are usually used as floating bearings.

Double-row angular-contact ball bearings and tapered-roller bearings are also used, e.g. as wheel bearings, in the case of high radial and axial loads.

Figure 2: Locating/floating bearing arrangement

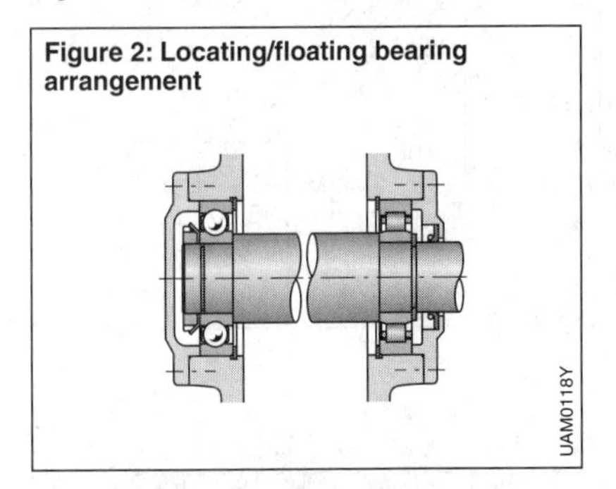

Figure 3: Preloaded bearing arrangement

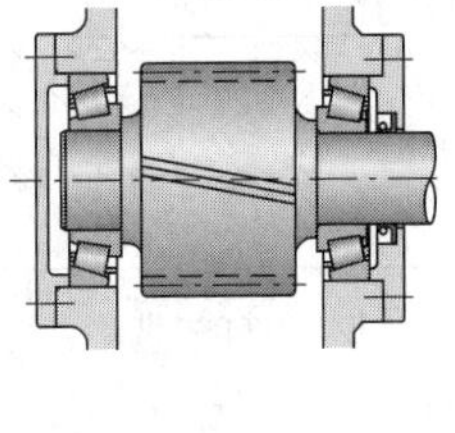

Preloaded bearing arrangement
A preloaded bearing arrangement is predominantly formed from two angular-contact ball bearings or tapered roller bearings arranged in mirror-image fashion (Figure 3). During assembly, a bearing race is displaced on its seat surface until the bearing arrangement has attained the desired or required play or preload. Because of the possibility of play regulation, a preloaded bearing arrangement is particularly suitable for applications with close guidance, e.g. bearings in transmissions.

Tolerances and fit of bearing points
Roller bearings essentially have negative tolerances for hole diameter, outside diameter and width, i.e. the nominal size is always the maximum limit of size.

Mounting the races on the bearing points (shaft and housing bore) is important for the installation of roller bearings. The roller bearings must not slip, above all tangentially, under load on the counterparts. The safest and easiest way of ensuring mounting is to make the correct choice of fit and tolerances so that the load capacity of the bearing can be fully utilized. Depending on the tolerance band of the bearing seat, the fit is referred to as clearance fit, transition fit, or interference fit (Figure 4).

The load conditions of the bearing races are of great importance to the choice of fit. A distinction is made between two loads, based on the load direction and the rotation of the bearing races.

- Rotating load: The race rotates relative to the load direction, and must be firmly seated.
- Concentrated load: The race is stationary relative to the load direction, and can have a close clearance fit to interference fit with its counterpart.

Because of the small thickness of the bearing races, the form variations of the seats are transferred to the raceways. The counterparts should therefore demonstrate as much form quality as possible, such as concentricity, cylindricity, and runout.

Calculation of load capacity

It is necessary to distinguish between static and dynamic load capacity when calculating the load capacity of a roller bearing. The basic principles are set out in ISO 76 for static and ISO 281 for dynamic calculations.

Static load capacity

If a roller bearing is subjected to load while stationary or while rotating at low speed, i.e. $n\,d_m \leq 4{,}000\,\text{mm}\cdot\text{min}^{-1}$ (n rotational speed, d_m mean value of bore and outside diameters), it is considered under static load.

If a bearing is subjected to load in both the radial and the axial directions, the static equivalent bearing load P_0 is formed from this:

$$P_0 = X_0\,F_r + Y_0\,F_a \text{ in N}$$

where

X_0 Radial factor; $X_0 = 0.6$ for single-row deep-groove ball bearing.

Y_0 Axial factor; $Y_0 = 0.5$ for single-row deep-groove ball bearing.

F_r Radial load in N.

F_a Axial load in N.

At $P_0 < F_r$, $P_0 = F_r$ must be reckoned on. For other beating types, refer to the catalogs of roller bearing manufacturers for the factors X_0 and Y_0.

As the measure of the static load capacity, a ratio

$$f_S = \frac{C_0}{P_0}$$

is formed, where C_0 is termed the static load rating. C_0 is the load at which total permanent deformation of rolling elements and raceways at the contact point subject to greatest load amounts to

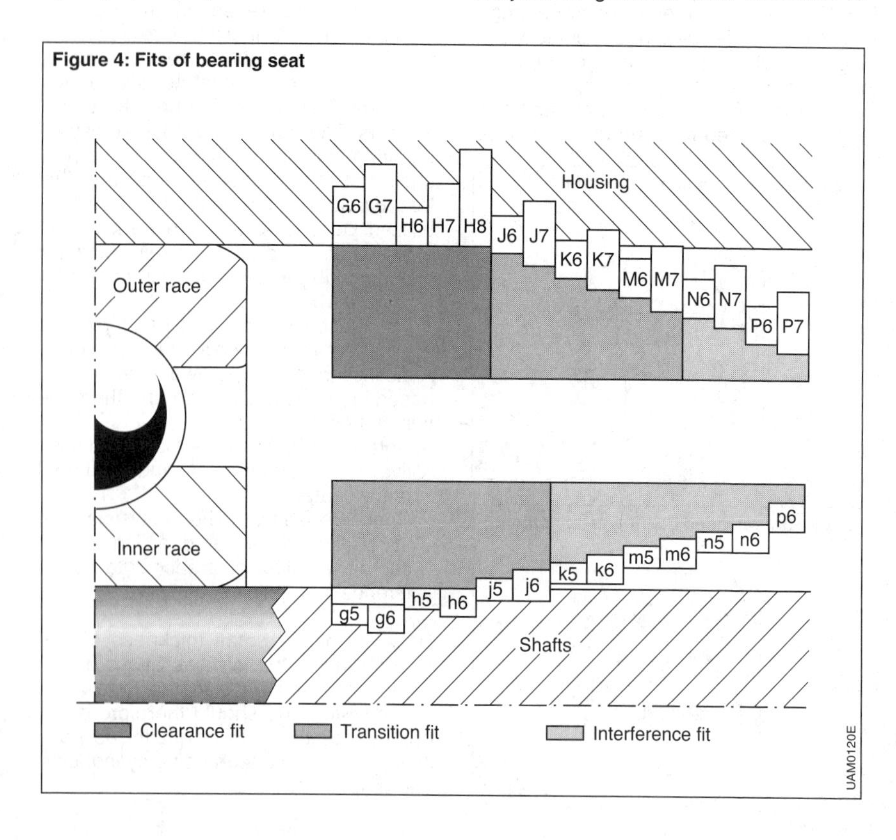

0.0001 of the rolling element diameter. Catalogs specify C_0 for all roller bearings.

In the case of normal requirements, the characteristic figure $f_s = 1$ can be approved. The requirement of lower deformation (<0.0001) corresponds to a greater characteristic figure $f_s > 1$.

Dynamic load capacity

Calculating the dynamic load capacity is based on the method of the standard ISO 281. This refers to the service life of a rotating roller bearing under load, where the running surfaces can be subject to material fatigue. The important characteristic figure of load capacity is the dynamic load rating C. This specifies the load of the roller bearing for a nominal bearing service life of one million revolutions.

For the purpose of calculating the nominal service life, the following applies in accordance with ISO 281:

$$L_{10} = 10^6 \left(\frac{C}{P}\right)^p \quad \text{in revolutions,}$$

$$L_{10h} = \frac{10^6}{60n}\left(\frac{C}{P}\right)^p \quad \text{in hours.}$$

L_{10} Nominal service life achieved or exceeded by 90% of a larger batch of identical bearings.
C Dynamic load rating in N – specified in roller bearing catalogs.
P Dynamic equivalent bearing load in N.
p Exponent, $p = 3$ for ball bearings, $p = 10/3$ for roller bearings.
n Rotational speed in min^{-1}.

The dynamic equivalent load P is defined as the imaginary load constant in size and direction which influences the service life, such as the actually acting radial load and axial load. It is calculated from:

$P = X F_r + Y F_a$ in N
F_r: Radial load
F_a: Axial load

The radial factor X and axial factor Y are dependent on bearing type, size, bearing play, and load ratio, and are specified in ISO 281 or roller-bearing catalogs.

Modified service life

In addition to tho nominal service life, a modified service life L_{na} has been introduced in the standard ISO 281, for which operating conditions can also be included in the calculation:

$$L_{na} = a_1\, a_2\, a_3\, L_{10}\,,$$

where
a_1 Factor for survival probability, e.g.:
90%: $a_1 = 1$;
95%: $a_1 = 0.62$.
a_2 Coefficient for special bearing design such as internal construction and materials.
a_3 Coefficient for operating conditions, such as bearing lubrication and operating temperature.

The coefficients a_2 and a_3 are not mutually independent, and are frequently combined as coefficient a_{23}:

$$L_{na} = a_1\, a_{23}\, L_{10}\,.$$

Numerous systematic examinations and experiences from practice make it possible to quantify the influences of bearing materials and operating conditions on the achievable service life of roller bearings. Diagrams and computing programs are provided by roller-bearing manufacturers to calculate the coefficients a_2, a_3 and a_{23}.

Gears and tooth systems

Overview of gears

Gears serve to continuously transmit the force and torque of two shafts whose layout determines the type of toothed gearing (Table 1). The law of gearing is definitive in creating the tooth shape:

"The common normal of the touching flank areas of two gears must pass in all positions of their contact through the pitch point *C*."

Thus, for a given pitch circle and defined shape of a tooth flank, it is possible to determine the structure of the associated mating flank. The number of possible profile shapes is therefore unlimited.

Involute teeth
Because it is easy to manufacture and is not sensitive to variations in the distance between centers, the most commonly used profile shape is the involute of a circle. The parameters of this involute will be discussed in greater detail here. An involute is created by unwinding a taut string from a base curve. It is also called a rolling curve. If the base curve is a circle, an involute of a circle is created (Figure 1), referred to in the following simply as involute.

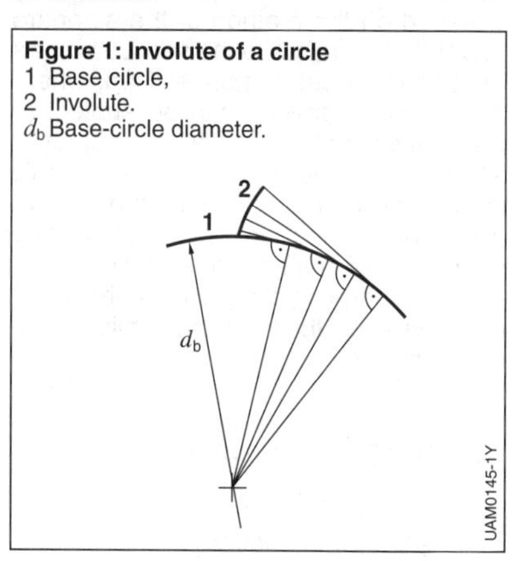

Figure 1: Involute of a circle
1 Base circle,
2 Involute.
d_b Base-circle diameter.

Table 1: Gear types

Position of shafts	Gear types	Type examples
Parallel	Spur gear	Straight- or helical-toothed external or internal gears; planetary gears
Intersecting	Bevel gear	Straight, helical- or curve-toothed differentials; axle differentials
Intersecting	Hypoid gear (offset bevel gear)	Rear-axle drives via axle-offset propshaft
	Worm gear	Windshield wiper drives
	Helical bevel gear	Seat adjustment, power-window units, power-sunroof drive units
Concentric	Involute splines (not rolling off)	Shaft-hub connections

UAM0148Y

A distinction is made between straight- and helical-tooth gearing, both in the form of external or internal teeth (Figures 2, 3 and 4). Using $\beta = 0$ in the specified calculation formulas (Table 3) for the helix angle produces the simplified formulas for calculating straight-tooth gearing.

Figure 2: Straight- and helical-tooth gearing (external teeth)
a) Straight-tooth gears,
b) Helical-tooth gears.

a b UAM0146Y

Figure 3: Internal teeth
For designations, see Table 2.

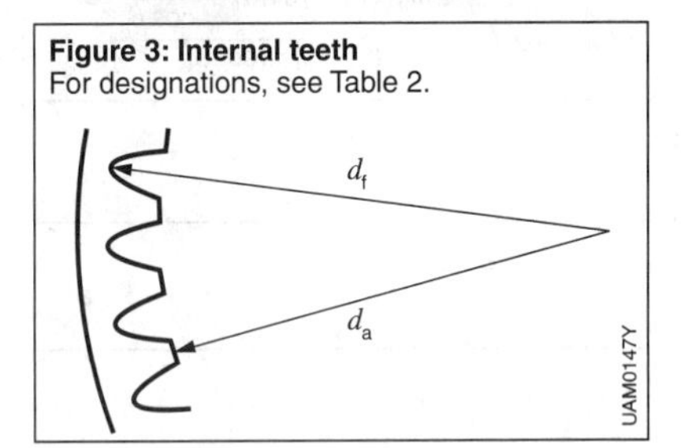

Figure 4: Parameters for spur gear
For designations, see Table 2.

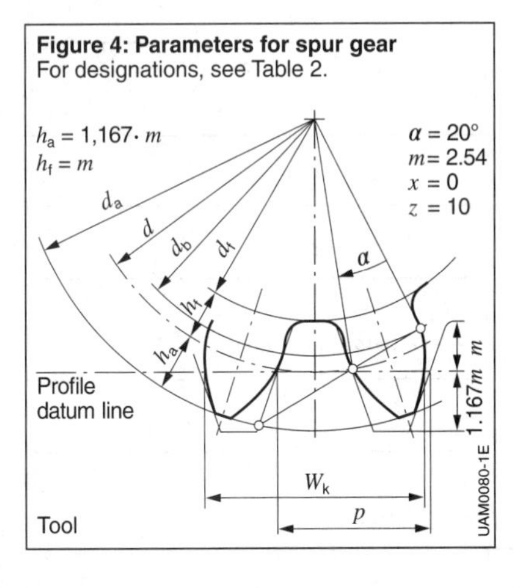

The same formulas (Table 3) used for external teeth are used to calculate internal teeth. Only the number of teeth of an internal gear is used negatively. Thus, all diameters and the distance between centers are likewise assigned negative signs; the outside diameter d_a becomes actually smaller than the root diameter d_f and the direction of positive addendum modification is reversed.

Table 2: Parameters, symbols and units

Quantity		Unit
a	Distance between centers	mm
A	Start of engagement	–
b	Face width	mm
c	Bottom clearance	mm
C	Pitch point	–
d	Pitch diameter	mm
d_a	Outside diameter	mm
d_b	Base diameter	mm
d_f	Root diameter	mm
d_w	Layout diameter	mm
E	End of engagement	–
h_a	Addendum	mm
h_{aP}	Addendum	mm
h_f	Dedendum	mm
h_{fP}	Dedendum	mm
i	Transmission ratio	–
j_n	Normal backlash	mm
m_n	Normal module	mm
m_t	Transverse module	mm
n	Rotational speed	min^{-1}
p	Pitch $p = \pi m$	mm
s	Tooth thickness	mm
W_k	Span measurement	mm
x	Addendum-modification coefficient	–
z	Number of teeth	–
α	Pressure angle	°
α_{wt}	Effective pressure angle	°
β	Helix angle	°
ρ	Fillet radius	mm
ε	Contact ratio	–
Superscripts and subscripts		
1	referred to gear 1	
2	referred to gear 2	
a	referred to tooth tip	
f	referred to root	
*	specific value referred to module m	

Table 3: Calculation formulas

Virtual number of teeth	$z_i = z \cdot \text{inv}\,\alpha_t / \text{inv}\,\alpha_v$
Transverse pressure angle	α_t from $\tan\alpha_t = \dfrac{\tan\alpha_n}{\cos\beta}$
Effective pressure angle	$\cos\alpha_{wt} = \dfrac{d_{b1}}{d_{w1}} = \dfrac{d_{b2}}{d_{w2}}$
Pitch diameter	$d = \dfrac{z \cdot m_n}{\cos\beta}$
Base diameter	$d_b = d \cdot \cos\alpha_t$
Outside diameter	$d_a = d + 2x\,m_n + 2\,h_a$
Root diameter	$d_f = d + 2x\,m_n - 2\,h_f$
Layout diameter	$d_{w1} = \dfrac{2z_1 a}{z_1 + z_2} \quad d_{w2} = \dfrac{2z_2 a}{z_1 + z_2}$
Involute	$\text{inv}\,\alpha = \tan\alpha - \text{arc}\,\alpha$
Distance between centers with backlash	$a = \dfrac{d_{b1} + d_{b2}}{2\cos\alpha_{wt}}$ where $\text{inv}\,\alpha_{wt} = \text{inv}\,\alpha_t + \dfrac{2\,m_n (x_1 + x_2)\sin\alpha_n + \dfrac{j_n}{\cos\beta_b}}{(z_1 + z_2)\,m_t \cos\alpha_t}$
Tooth thickness in reference circle (arc length)	$s_n = m_n\left(\dfrac{\pi}{2} + 2x\tan\alpha_n\right)$
Base tangent length over k teeth	$W_k = m_n\left[2x\sin\alpha_n + \cos\alpha_n \cdot [\pi(k - 0.5) + z_i\,\text{inv}\,\alpha_n]\right]$
Number of teeth spanned (rounded to the next whole number)	$k \approx \dfrac{z_i\,\alpha_{nx}}{180} + 0.5$ where $\cos\alpha_{nx} \approx \dfrac{z_i}{z_i + 2x} \cdot \cos\alpha_n$
Back-reckoning of x from W_k	$x = \dfrac{\dfrac{W_k}{m_n} - [\pi(k - 0.5) + z_i\,\text{inv}\,\alpha_n] \cdot \cos\alpha_n}{2\sin\alpha_n}$
Real pitch	$p_n = \pi\,m_n$
Transverse base pitch	$p_{te} = \dfrac{\pi\,m_n}{\cos\beta} \cdot \cos\alpha_t = p_t \cos\alpha_t$
Transverse contact ratio	$\varepsilon_\alpha = \dfrac{\sqrt{d_{a1}^2 - d_{b1}^2} + \dfrac{z_2}{\lvert z_2 \rvert}\sqrt{d_{a2}^2 - d_{b2}^2}}{2\,\pi\,m_t \cos\alpha_t} - \tan\alpha_{wt} \cdot (d_{b1} + d_{b2})$
Overlap ratio	$\varepsilon_\beta = \dfrac{g_\beta}{p_t} = \dfrac{b\sin\beta}{m_n\,\pi}$
Total contact ratio	$\varepsilon_\gamma = \varepsilon_\alpha + \varepsilon_\beta$ Apply only if: $\varepsilon_\alpha > 1$ for $d_{a\,min}$ and a_{max}

Module
The module m has been introduced to standardize (Table 4) different tooth sizes. It specifies how often the number of teeth z fits into the pitch diameter d. The following applies:

$$m = \frac{d}{z}.$$

The more teeth fit, the smaller the module.

Addendum modification
Addendum modification xm (Figure 5) is used to avoid undercut with small numbers of teeth, and thus to increase tooth-root strength or to achieve a defined distance between centers. Teeth with undercut are narrower in the tip area than in the involute profile. Addendum modification is limited in the upward direction by the tip limit (requirement of a certain tooth-tip thickness) and in the downward direction by the undercut limit of the tooth. If the profile is modified by the pitch diameter d in the direction of the tip circle d_a, the addendum modification is positive.

Table 4: Overview of selected gearing standards

Standard	Description
DIN 780	Series of modules for gears
DIN 867	Basic racks for spur gears
DIN 3960 (with supplement 1)	Definitions and parameters
DIN 3961 DIN 3962 DIN 3963	Recommended permissible deviations for spur gears
DIN 3965	Recommended permissible deviations for bevel gears
DIN 3971	Basic racks for bevel gears
DIN 3974	Recommended permissible deviation for worms and worm gears
DIN 3975	Basic racks for worms and worm gears
DIN 3990 DIN 3991	Calculations of load-bearing capacity
DIN 3993	Internal gear pairs
DIN 3999	Symbols, notation in Ge, En and Fr
DIN 5480	Involute splines
DIN 58400	Basic racks for spur gears for fine mechanics
DIN 58425	Gears with round flanks
ISO 8123	European diametral pitch starter pinions with 12° and 20°
ISO 9457-1	Metric starter pinions with 14.5°
ISO 9457-2	Metric starter pinions with 20°
NF R11-411	French starter pinions with 12° and 20°
SAE J 543	American diametral pitch starter pinions with 12° and 20°

Manufacture
Gears can be manufactured using the following processes: forming (extruding, rolling), cutting (milling, shaping, grinding), sintering, and injection molding.

All spur gears with the same module and the same pressure angle can be produced using the same milling or shaping tool, regardless of the number of teeth and addendum modification. Because tool and workpiece are in rolling engagement here, the number of teeth of the gear to be manufactured is obtained from the set pitch-circle diameter d_w. In the shaping process, the finished gear is thus created from a cylindrical blank by working out the tooth spaces.

When selecting the manufacturing quality (Table 5), it is important to ensure that all toothing errors, distance-between-centers tolerances and elongations resulting from changes in temperature are counterbalanced by backlash and bottom clearance without jamming or engagement failures occurring.

Figure 5: Effect of addendum modification
x Addendum-modification coefficient,
m Module.

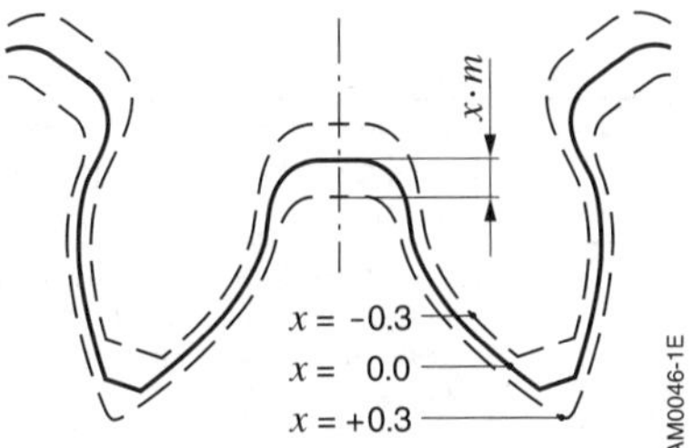

Starter-tooth designs

The "Standard distance between centers" is a system of tolerances for gears customary in mechanical engineering and is specified in DIN 3961. This system, in which the required backlash is produced by negative tooth thickness tolerances, cannot be used in starter-tooth designs. Starter-tooth designs require far more backlash than constant-mesh gears due to the starter engagement process. Such backlash is best achieved by increasing the distance between centers.

The high torque required for starting necessitates a high transmission ratio ($i = 10-20$). For this reason, the starter pinion has a small number of teeth ($z = 8-13$). The pinion generally has positive addendum modification. A selection of customary starter pinions is set out in Table 6.

The heavy impact loads during the engagement and starting sequences necessitate high-strength pinion materials. Large distance-between-centers tolerances and radial run-out of pinion and ring gear accelerate flank wear, particularly since starter-tooth designs for the most part must not be lubricated and very high flank pressures occur. However, short-time service allows starter-tooth designs to accommodate at times greater deviations than in customary transmission manufacture (Table 5).

Table 5: DIN gear qualities, manufacture and applications

Quality	Application examples	Manufacture
2	Primary standard master gears	Form grinding and selection (50–60 % scrap rate)
3	Master gears for the inspection department	Form grinding and generative grinding
4	Master gears for the workshop, measuring mechanisms	
5	Transmissions for machine tools, turbines, measuring instruments	
6	As 5 and highest gears of passenger-car and bus transmissions	
7	Motor-vehicle transmissions (highest gears), rail vehicles, machine tools, hoisting and handling equipment, turbines	Non-hardened gears (with sufficient care) by hobbing, generative shaping, and planing (subsequent shaving is desirable); additional grinding is required for hardened gears
8 and 9	Motor-vehicle transmissions (middle and lower gears), rail vehicles, and machine tools	Hobbing, generative shaping, and planing (non-ground but hardened gears)
10	Transmissions for agricultural tractors, agricultural machinery, subordinate gear units in general machine equipment, starter-tooth designs	All of the usual processes including extrusion and sintering, and injection molding for plastic gears
11 and 12	General agricultural machinery	

American gear standards

(according to SAE J 543)

Diametral pitch

The number of teeth is specified here by N. Instead of the module m, this standard uses as its basis the number of teeth on a pitch diameter of 1 inch = 1 diametral pitch (P). For converting to the module m:

$$m = \frac{25.4\ \text{mm}}{P}.$$

Addendum modification

Addendum modification is often expressed by specifying a gear ratio N_2/N_1. This means that on a gear blank with a diameter for N_2 teeth only N_1 teeth are cut; this corresponds to a positive addendum-modification coefficient of

$$x = \frac{(N_2 - N_1)}{2}.$$

As a rule $x = +0.5$; slight deviations in drawing specifications are quite common.

Tooth spacing

The tooth spacing in the pitch circle is called circular pitch (CP):

$$CP = \frac{25.4\ \text{mm}}{P}\ \pi = m\,\pi.$$

Full-depth teeth

Full-depth teeth have the addendum $h_a = m$ as in the German DIN standards; however the dedendum is frequently somewhat different.

Stub teeth

Calculation of the addendum is based on a different diametral pitch (P) from the other dimensions.

Notation (example): P 5 /7
$P = 7$ to calculate the addendum,
$P = 5$ to calculate all other dimensions.

Notation and conversions

Outside diameter
$OD = d_a$.

Pitch diameter
$PD = \frac{N}{P} = d$ (in inches).
$PD = N \cdot m = d$ (in mm).

Root diameter
$RD = d_f$

Layout diameter
$LD = \frac{(N+2x)}{P}$ (in inches).

$LD = (N+2x)\,m$ (in mm).

Table 6: Module series for customary starter pinions

Module m [mm]	Diametral pitch P [1/inch]	Pressure angle α of basic rack	American standard	European standard
2.1167	12	12°	SAE J543c	ISO 8123
2.54	10	20°	SAE J543c	ISO 8123
3.175	8	20°	SAE J543c	ISO 8123
4.233	6	20°	SAE J543c	ISO 8123
2.25	–	14.5°/20°	–	ISO 9457-1/2
2.5	–	14.5°/20°	–	ISO 9457-1/2
3	–	14.5°/20°	–	ISO 9457-1/2
3.5	–	14.5°/20°	–	ISO 9457-1/2
4	–	14.5°/20°	–	ISO 9457-1/2

Calculation of load-bearing capacity

The calculation must be made for the tooth-flank load-bearing capacity and the tooth-root load bearing capacity.

The following is suggested for rough estimations in parallel with DIN 3990. It applies to two-gear pairs in a stationary transmission unit (shafts are fixed and only rotate about themselves). The quantities and units featured in Table 7 must be used in the formulas set out in Table 8.

Table 7: Quantities and units for calculation of load-bearing capacity

	Quantity	Unit	
P	Power	kW	
M	Torque	Nm	
n	Rotational speed	min^{-1}	
F_{tw}	Peripheral force in pitch circle	N	
u	Gear ratio	–	
Φ	Life factor (flank)	–	(see Table 10)
HB	Brinell hardness	–	(see Table 13)
HRC	Rockwell hardness	–	(see Table 13)
b_n	Effective flank width	mm	
b	Face width	mm	
k	Contact pressure	N/mm^2	
L_h	Service life	h	
N_L	Number of load changes	–	(see Table 13)
S	Safety factor	–	
f_v	Velocity factor	–	(see Table 12)
Y_{Fa}	Tooth-profile factor	–	(see Fig. 6)
Y_L	Alternating load factor	–	(see Table 11)
Y_{NT}	Life factor (root)	–	(see Table 13)
σ	Root stress of tooth	N/mm^2	
d	Pitch diameter	mm	
d_w	Layout diameter	mm	
v	Peripheral velocity	m/s	
Z	Number of teeth	Z_1 for smaller gear	
β	Helix angle	°	
Pitting		Depression of tooth flank as a result of excessive contact pressure and excessive load changes	

Tooth-flank load-bearing capacity

The tooth-flank load-bearing capacity is calculated according to the formulas in Table 8. The actual contact pressure k_{act} is equal in magnitude for both gears of the pair. Different rotational speeds (numbers of teeth) can give rise to different permissible contact pressures k_{perm}. For each wheel a material as featured in Table 13 must be chosen so that the factor of resistance to pitting (S_W, see Table 8) is achieved.

The k_{perm} values in Table 9 apply when both gears are made of steel. For cast iron on steel, or bronze on steel, the values should be roughly 1.5 times higher; for cast iron on cast iron, or bronze on bronze, they should be approximately 1.8 times higher. For the gear with non-hardened surfaces, 20 % higher k_{perm} values are permissible if the other gear in the pair has hardened tooth flanks. The values in the table apply to a service life of 5,000 h ($\Phi = 1$). A different service life is allowed for by means of the life factor Φ (guidelines as set out in Table 10) in the equation for S_w (Table 8).

Tooth-root load-bearing capacity

The tooth-root load-bearing capacity is calculated according to the formulas in Table 11. Resistance to tooth fracture

Table 8: Formulas for calculation of tooth-flank load-bearing capacity

Power	$P = \frac{M \cdot n}{9{,}549} = \frac{F_{tw} d_w n}{1.9 \cdot 10^6}$
Peripheral force in pitch circle	$F_{tw} = \frac{2{,}000 \cdot M}{d_w} = \frac{19.1 \cdot 10^6 \cdot P}{d_w n}$
Gear ratio	$u = \frac{z_2}{z_1} = \frac{n_2}{n_1}$ (gear 1 is the smaller gear)
Life factor	$\Phi = \sqrt[6]{5{,}000/L_h}$
Permissible contact pressure	$k_{perm} = \frac{(HB)^2}{2{,}560 \cdot \sqrt[6]{n}} = \frac{(HRC)^2}{23.1 \cdot \sqrt[6]{n}}$
Actual contact pressure	$k_{act} = 4 \frac{F_{tw}}{b_n d_w} \cdot \frac{(u+1)\cos^2\beta}{u \sin(2\alpha_{wt})}$
Factor of resistance to pitting	$S_w = \Phi \frac{k_{perm}}{k_{act}} \geq 1$ $S_w \geq 1.2 \ldots 1.5$ for $z_1 < 20$ on account of the greater pressure at the inside single engagement point

Table 9: Permissible contact pressures k_{perm} in N/mm²
(steel against steel, L_h = 5,000 h).

Hardness of teeth		Rotational speed in rpm (with 1 load change per revolution; gear meshes only with one mating gear, planet gear makes e.g. two load changes per revolution)											
HB	HRC	10	25	50	100	250	500	750	1,000	1,500	2,500	5,000	10,000
90		2.20	1.90	1.70	1.50	1.30	1.10	1.05	1.00	0.94	0.86	0.77	0.68
100		2.70	2.30	2.00	1.80	1.60	1.40	1.30	1.20	1.15	1.06	0.94	0.84
120		3.80	3.30	2.90	2.60	2.20	2.00	1.90	1.80	1.66	1.53	1.36	1.21
140		5.20	4.50	4.00	3.60	3.90	2.70	2.50	2.40	2.26	2.08	1.85	1.65
170		7.70	6.60	5.90	5.20	4.50	4.00	3.75	3.60	3.34	3.06	2.73	2.43
200		10.70	9.10	8.10	7.30	6.20	5.60	5.20	4.90	4.60	4.24	3.78	3.37
230		14.10	12.10	10.80	9.60	8.20	7.30	6.90	6.50	6.10	5.61	5.00	4.45
260		18.00	15.40	13.80	12.20	10.50	9.40	8.80	8.40	7.80	7.17	6.39	5.69
280		20.90	17.90	16.00	14.20	12.20	10.90	10.20	9.70	9.00	8.31	7.41	6.60
300		24.00	20.60	18.30	16.30	14.00	12.50	11.70	11.10	10.40	9.54	8.50	7.60
330		29.00	24.90	22.20	19.80	17.00	15.10	14.10	13.50	12.60	11.60	10.30	9.20
400		42.60	36.60	32.60	29.00	24.90	22.20	20.70	19.80	18.50	17.00	15.10	13.50
	57	96.00	82.30	73.30	65.30	56.00	49.90	46.70	44.50	41.60	38.20	34.00	30.30
	≥ 62	112.00	96.50	86.00	76.60	65.80	58.60	54.80	52.20	48.80	44.80	39.90	35.60

Table 10: Life factor Φ

	Drives run intermittently at full load								Drives in continuous operation at full load			
Service life L_h in operating hours	10	50	150	312	625	1,200	1,500	5,000	10,000	40,000	80,000	150,000
Life factor Φ	2.82	2.15	1.79	1.59	1.41	1.27	1.12	1	0.89	0.71	0.63	0.57

Table 11: Formulas for calculation of tooth-root load-bearing capacity

Peripheral velocity $v_1 = v_2$	m/s	$v_1 = \pi d_1 n_1$
Velocity factor f_v		Take from Table 12
Root stress that can be yielded $\sigma_{F\,lim}$		Take gearing-specific material parameter from Table 13
Permissible root stress $\sigma_{F\,perm}$	N/mm²	$\sigma_{F\,perm} = \sigma_{F\,lim} Y_{NT} Y_L$ Take $\sigma_{F\,lim}$ and Y_{NT} from Table 13, estimate intermediate values. $Y_L = 1$ for pulsating loads $Y_L = 0.7$ for alternating loads
Actual root stress $\sigma_{F\,act}$	N/mm²	$\sigma_{F\,act} = \frac{F_t Y_{Fa}}{b m_n f_v \varepsilon_\alpha} \cdot \left(1 - \frac{\varepsilon_\beta \beta}{120°}\right)$; Y_{Fa} from Figure 6, $\left(1 - \frac{\varepsilon_\beta \beta}{120°}\right)$ must be ≥ 0.75.
Resistance to tooth fracture S_F	–	$S_F = \frac{\sigma_{F\,perm}}{\sigma_{F\,act}} \geq 1$

Figure 6: Tooth-profile factor Y_{Fa}

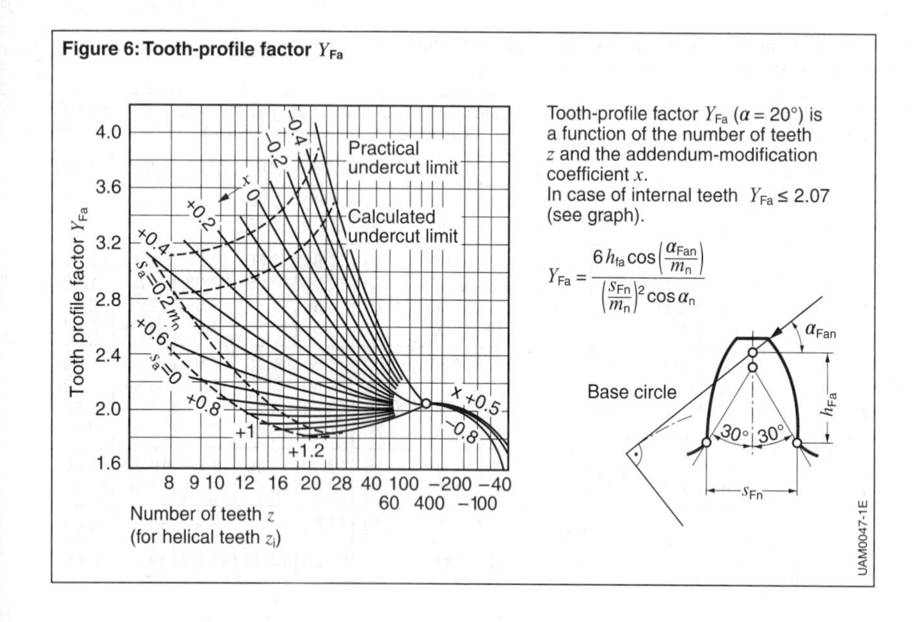

Table 12: Velocity factor f_v.

Materials	Peripheral velocity v in m/s							Base equation [1]
	0.25	0.5	1	2	3	5	10	
Steel and other metals	Velocity factor f_v 0.96	0.93	0.86	0.75	0.67	0.55	0.38	$f_v = \frac{A}{A+v}$
Fabric-based laminates and other non-metals	0.85	0.75	0.62	0.5	0.44	0.37	0.32	$f_v = \frac{0.75}{1+v} + 0.25$

exists if the equations for S_f for the smaller gear (gear 1) yield a value greater than or equal to 1. If a better material is selected for gear 1 than for gear 2, the calculation must also be made for gear 2.

Within fatigue strength of finite life, the root bending stress that can be yielded $\sigma_{F\,lim}$ is multiplied by the factor Y_{NT}, depending on the number of load changes N_L (see Table 13).

[1] Values are valid for $A = 6$ (average tooth quality). With cast gears and high-precision gears, $A = 3$ and 10 respectively.

Gear materials

Table 13: Material parameters

	Material	Condition	Tensile strength R_m in N/mm^2 min.	Fatigue strength under reversed bending stress σ_{bW} in N/mm^2 min.	Hardness HB or HRC min.	Yieldable root stress[2] $\sigma_{F\,lim}$ in N/mm^2	Life factor Y_{NT}[3] at number of load changes $N_L = L_h \cdot 60 \cdot n$: $\geq 3 \cdot 10^6$	10^6	10^5	10^4	10^3
Tempering steel	St 60, C 45	annealed	590	255	170 HB	160	1	1.25	1.75	2.5	2.5
		quenched and drawn	685	295	200 HB	185					
		surf.-hard.[4]	980	410	280 HB	245					
	St 70, C 60	annealed	685	295	200 HB	185					
		quenched and drawn	785	335	230 HB	209					
		surf.-hard.[4]	980	410	280 HB	245					
	50 Cr V 4	annealed	685	335	200 HB	185					
		quenched and drawn	1,130	550	330 HB	294					
		surf.-hard.[4]	1,370	665	400 HB	344					
	37 Mn Si 5	annealed	590	285	170 HB	160					
		quenched and drawn	785	355	230 HB	200					
		surf.-hard.[4]	1,030	490	300 HB	270					
Case-hardening steel	C 15	annealed	335	175	100 HB	110	1	1.25	1.75	2.5	2.5
		surf.-hard.[4]	590	255	57 HRC	160	1	1.2	1.5	1.9	2.5
	16 Mn Cr 5	annealed	600	–	150 HB	–	–	–	–	–	–
		surf.-hard.[4]	1,100	–	57 HRC	300	1	1.2	1.5	1.9	2.5
	20 Mn Cr 5	annealed	590	275	170 HB	172	–	–	–	–	–
		surf.-hard.[4]	1,180	590	57 HRC	330	1	1.2	1.5	1.9	2.5
	18 Cr Ni 8	annealed	640	315	190 HB	200	–	–	–	–	–
		surf.-hard.[4]	1,370	590	57 HRC	370	1	1.2	1.5	1.9	2.5
Gray iron	GG-18	–	175	–	200 HB	50	1	1.1	1.25	1.4	1.6
	GS-52.1	–	510	–	140 HB	110	1	1.25	1.75	2.5	2.5
	G-SnBz 14	–	195	–	90 HB	100	1	1.25	1.75	2.5	2.5
	GGG-60	–	600	–	230 HB	210	1	1.25	1.75	2.5	2.5
PA 66 GF30		at 60 °C	–	40	–	27	1	1.2	1.75	–	–
			140	43	–	29					
Fabric-base laminates	fine		–	–	–	75	1	1.15	1.4	1.65	2.0
	coarse	–	–	–	–	50	1	1.2	1.6	2.1	2.8

For tempering (heat-treatable) steels, two values are given in each case for the heat-treated condition. For smaller gears up to roughly module 3, the larger of the two values can be specified. With very large gears, however, only the smaller value can be achieved with certainty.

[2] For pulsating loads with a material fatigue strength ($N_L \geq 3 \cdot 10^6$). In the case of alternating loads (e.g. idler and planet gears), the alternating load factor Y_L must be taken into account.

[3] With fatigue strength of finite life, root bending stress is multiplied by the factor Y_{NT}, depending on the number of load changes $N_L = L_h \cdot 60 \cdot n$.

[4] Surface-hardened.

Belt drives

Friction belt drives

Application

Friction belt drives are used in automobiles predominantly to drive accessories (Figure 1). These drives had previously been driven primarily by V-belts. However, the significantly higher power density required in today's applications, e.g. due to space restrictions and higher power consumption of the accessories, have resulted in these having been replaced in the meantime almost exclusively by serpentine drives with ribbed V-belts (Micro-V® belts). Typical applications are the drives for the alternator, the A/C compressor, the power-steering pump, fans, mechanical chargers, or pumps for secondary air injection.

Ribbed V-belts are also used to start engines in micro-hybrid applications. Here, the conventional starter is replaced by a starter alternator, and the belt transmits the starter torque to the crankshaft to start the engine.

Forces and loads in the belt drive

The transmittable power is determined from

$$P = (F_1 - F_2) \cdot v / 1{,}000 \, .$$

Table 1: Symbols and units

Quantity	Unit
F_1 Belt-run force on load side	N
F_2 Belt-run force on slack side	N
F_U Peripheral force	N
F_{HL} Pre-tensioning force or "hubload"	N
F_C Centrifugal force of belt	N
v Belt speed	m/s
P Required power transmission	kW
α Flank angle	°
β Wrap angle	rad
μ Coefficient of friction	–
L_B Reference belt length	mm
U_B Reference circumference	mm

The Eytelwein equation describes the transition from static friction to sliding friction as:

$$R = F_1/F_2 = e^{\mu\omega\beta} \, , \text{ where } \omega = 1/\sin(\alpha/2).$$

As long as the belt-run forces are in this ratio, there will be no slip when power is transmitted. Belt run refers to that section of the belt between the belt inlet and outlet areas of two adjoining belt pulleys. A typical ratio for ribbed V-belts is $R = 4$ with a wrap angle of $\beta = 180°$.

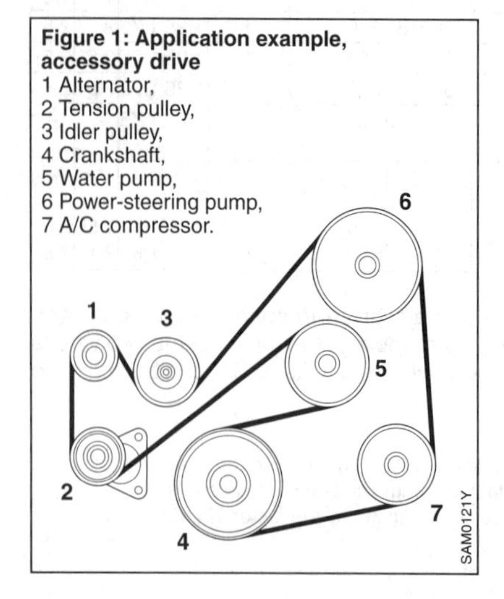

Figure 1: Application example, accessory drive
1 Alternator,
2 Tension pulley,
3 Idler pulley,
4 Crankshaft,
5 Water pump,
6 Power-steering pump,
7 A/C compressor.

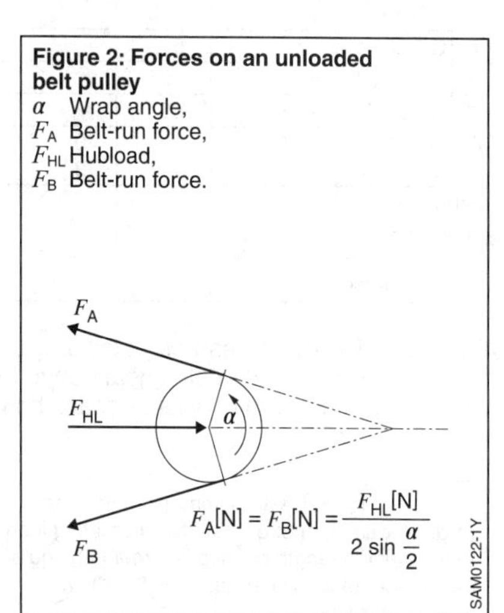

Figure 2: Forces on an unloaded belt pulley
α Wrap angle,
F_A Belt-run force,
F_{HL} Hubload,
F_B Belt-run force.

To transmit the peripheral force

$$F_U = F_1 - F_2$$

the pretensioning force F_{HL} ("hubload") is required (Figure 2). At high rotational speeds, the centrifugal force component F_c of the belt must also be taken into consideration.

Slip occurs on account of the change in tension within the belt during alternation between load and slack runs.

In spite of a constant transmitted torque, the belt tension decreases at the driving pulley and increases at the driven pulley (Figure 3). This variation in tension causes the belt to elongate over the respective pulley. In this elongation range, the belt loses its grip on the pulley and passes into a sliding state. The belt will slip if this sliding range stretches over the entire pulley wrap. Slip is below 1 % on correctly configured drives. Ribbed V-belts operate with an efficiency of more than 96 %.

Design of ribbed V-belt

The ribbed V-belt is a composite of three components (Figure 4):

- Fiber reinforced rubber compound
- Tension members
- Backing or rubber coating

The rubber compound forms the ribs and transmits the drive forces from the pulleys to the tension members. The main material used here is Ethylene Propylene Diene Monomer rubber (EPDM), and the rubber compound is fiber-filled for reinforcement.

The tension members accommodate the dynamic forces and transmit the drive power from the driving shaft (usually the crankshaft) to the accessories. They are usually made or nylon, polyester or ar-

Figure 3: Strain distribution in the belt
F_1 Force in taut strand,
F_2 Force in slack strand,
F_B Loading of belt by wrap,
F_C Loading of belt by centrifugal forces,
M Transmitted torque,
ω Angular frequency.

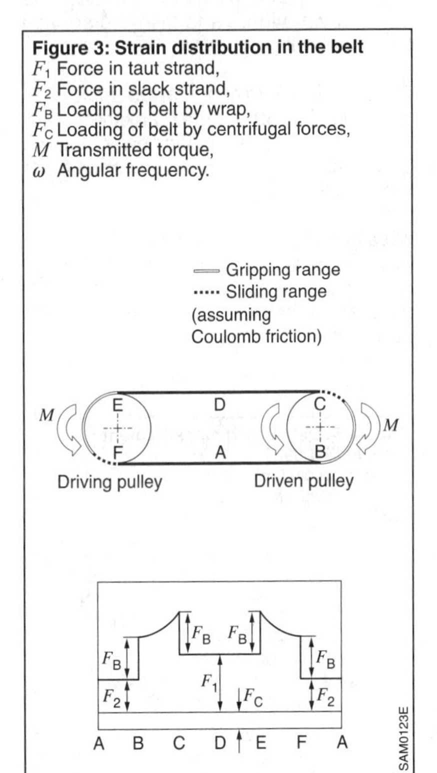

Figure 4: Structure of ribbed V-belt (cross-section)
1 Backing or rubber coating,
2 Tension members,
3 Rubber compound, fiber-filled.

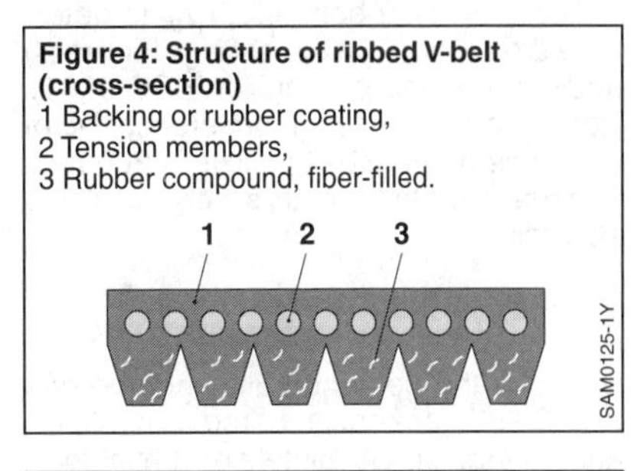

Figure 5: Moduli of elasticity of different tension-member materials
The three columns for each material represent the spread.

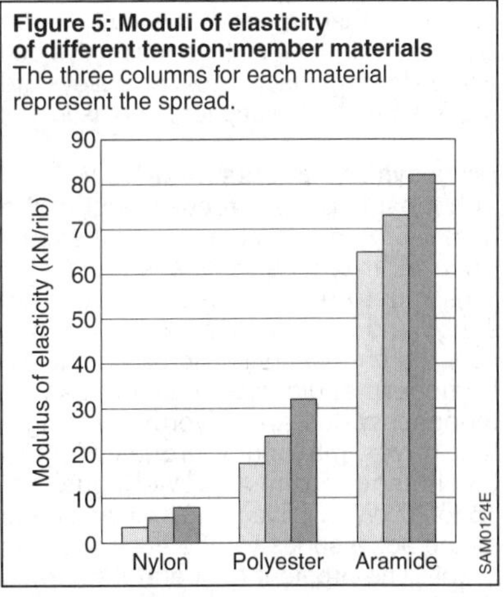

amide. These tension member materials differ primarily in their very different moduli of elasticity (Figure 5). The dynamic matching of the system can be optimized by selecting the appropriate tension member material. Tension member materials with a high modulus of elasticity (e.g. aramide) are used in highly dynamic applications to influence the resonant frequencies of the belt drive.

The back of the belt can be designed as a backing or as a rubber coating. The back of the belt forms a protective layer for the tension members. In most applications, it serves only to guide the belt quietly via the idler and tension pulleys. In some applications, however, it also serves to drive accessories which are subject to low load (e.g. water pump).

Ribbed V-belt profile and pulleys

The PK profile in accordance with ISO 9981 is usually used in automobile applications.

The reference belt length L_B is determined on a two-pulley test rig (Figure 6) under a defined pre-tension (ISO 9981). Here, the reference circumference U_B of the measurement belt pulleys is 300 mm. The reference belt length is then calculated according to

$$L_B = U_B \cdot E \; .$$

Because the belt and pulley geometries in ISO 9981 are encumbered with a tolerance band, it is absolutely essential for a detailed configuration to fall back on the characteristic values of the respective belt and pulley manufacturers. The pulleys are made from steel, aluminum or plastic.

Drive system: accessory drive

The most important requirement of the accessory drive system is that it should drive all the accessories without slip. This must be guaranteed for all load states and under all ambient conditions over the entire service life of the engine. In modern engines with full drives (i.e. all accessories are driven via a single belt drive), maximum torques of up to 30 Nm and maximum power figures of 15 to 20 kW at full load are transmitted to all the accessories via the five- or six-rib V-belt. The ambient temperatures are in the range of −40 to 100 °C. It is essential to utilize optimum system design to eliminate in particular slip noises, such as, for instance, the familiar V-belt squeaking in cold and damp weather conditions. It is also necessary to utilize design already to avoid belt noises caused by pulley misalignment.

Design criteria

Both the manufacturer's own and cross-manufacturer computing programs are used to design accessory drives. Important input parameters are the arrangement of components, i.e. the drive configuration, the torque curves, the moments of inertia of the components, the torsional vibration of the crankshaft, and the belt data. With these data, it is possible to calculate and optimize the system geometry, such as, for example, belt-run lengths and wrap angle, the system natural frequencies, the slip limit angles, the belt-run forces, the belt-run vibrations, and also the service life of the belt.

Recommended minimum wrap angle

This angle is 150° for the crankshaft, 120° for the alternator, 90° for the power-steering pump and the A/C compressor, and 60° for the tension pulley.

Misalignment and entry angle

In order to avoid unacceptable belt wear and noises, it is important that the entry angle of the belt into the grooved washers does not exceed 1°.

Figure 6: Belt-length measurement acc. to ISO 9981
Reference circumference U_B of measurement belt pulleys: 300 mm

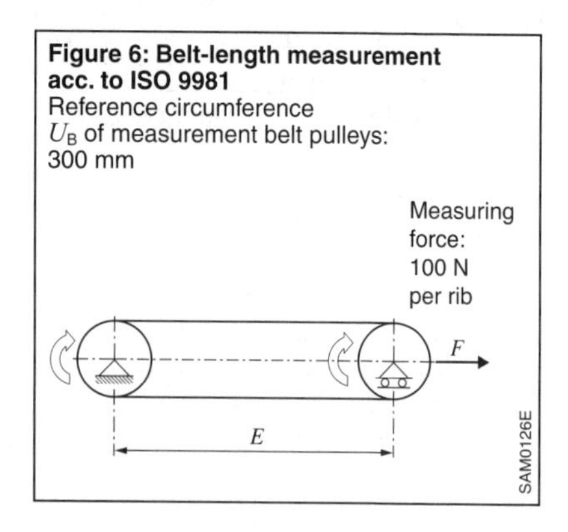

Table 2: PK belt profile and belt-pulley profile acc. to ISO 9981
Designation of a ribbed V-belt with six ribs of profile code PK and of reference length 800 mm:
Ribbed V-belt ISO 9981-6 PK 800.
Designation of corresponding ribbed V-belt pulley with reference diameter 90 mm:
Ribbed V-belt pulley ISO 9981-6 Kx 90.

Dimensions in mm	Belt	Profile grooves
Profile code	PK	K
Rib spacing s and groove spacing e (Figure 7)	3.56	3.56
Permissible deviation tolerance for e		±0.05
Sum of perm. deviation tolerance for e		±0.30
Groove angle α		40° ±0.5°
r_k at rib head and r_t at groove head	0.50	min. 0.25
r_g at rib seat and r_i at groove seat	0.30	min. 0.50
Belt height h	4 to 6	
Nominal dia. of test pin d_s		2.50
2 h_s nominal dimension		0.99
2 δ (Figure 7)		max. 1.68
f (Figure 7)		min. 2.5

Belt width $b = n\,s$ (with n = number of ribs), effective diameter for profile K
$d_w = d_b + 2\,h_b$, $h_b = 1.6$ mm

System natural frequency
The system natural frequency should not be in the engine's idle range (2nd engine order).

Minimum diameters of pulleys and idler pulleys
In practice the smallest belt pulley is often located on the alternator in order to facilitate the high rotational speeds that are required there. Typical alternator pulleys have a diameter of 50 to 56 mm. Belt fatigue increases exponentially when very small pulleys are used; this must be taken into account when the belt is designed. It is recommended to use idler pulleys with diameters of no less than 70 mm.

Figure 7: Belt profile
a) Belt cross-section,
b) Groove cross-section,
c) Determining effective diameter.
1 Position of tension member.
For quantities, see Table 2.

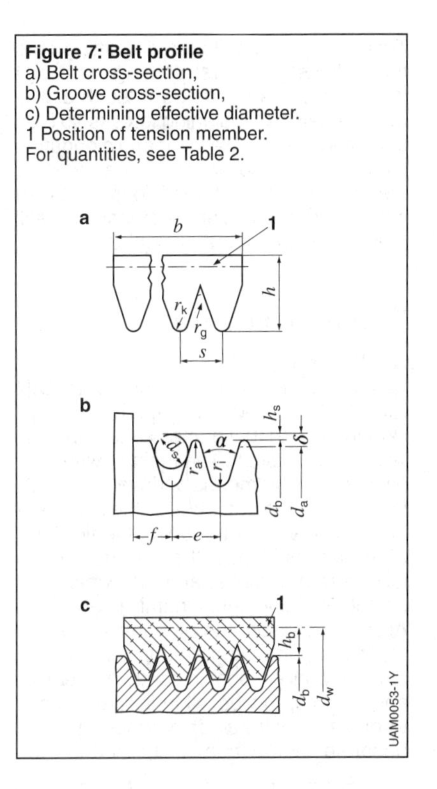

Belt-tensioning systems
Belt tension in accessory drives is today usually provided by way of automatic tension pulleys. These tension pulleys ensure that the belt is constantly pretensioned over the belt's service life by compensating belt elongation and belt wear. The design of the tension pulleys is significantly dictated by the space available. The pretensions on 6 PK belts are, depending on the system dynamics, usually in the range of 250 to 400 N.

Stretch Fit® drives
Elastic ribbed V-belts are also used occasionally in less complex belt drives. These use nylon as the tension-member material and do not require belt-tensioning systems.

This belt type is pulled during fitting onto the pulleys by overstretching. The belt length must be designed in such a way that the belt pretensioning force is so high after fitting that, while wear and elongation over the required service life as well as the relevant ambient conditions are taken into account, sufficient pretension is still maintained without retensioning.

Positive belt drives

Application

Toothed belts in accordance with DIN ISO 9010 are used in timing drives to drive camshafts and fuel pumps synchronously with the crankshaft. The main advantages over competing means of drive – such as, for example, the gear drive or the chain – lie in the simplicity of the drive, the flexibility of the belt guidance, the low friction, the low noise emissions, and the capacity to compensate for high dynamic peak loads. Accessories, such as oil pumps or water pumps, can be integrated in the drive. In modern applications, in contrast to earlier applications, it is possible in many cases to dispense with a replacement interval because of the use of innovative belt technologies and optimized design of the overall system.

Figure 8: Structure of toothed belt
a) Longitudinal section,
b) Overhead view.
1 Nylon fabric,
2 Rubber compound,
3 Tension member.

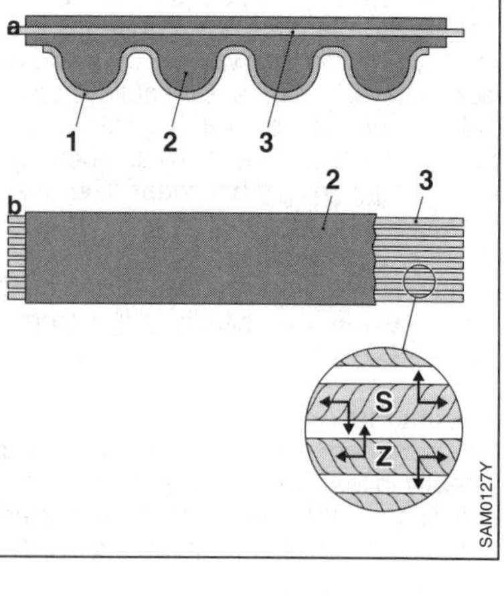

Structure of toothed belt

The toothed belt is a composite of three components (Figure 8): nylon fabric, rubber compound, and tension member.

The fabric is made of high-strength nylon and has an abrasion-proof and wear-resistant coating. It protects the rubber teeth against wear and from being shorn off.

The rubber compound consists of a high-strength polymer, and encloses the tension member on both sides. Polychloroprene (CR, chloroprene rubber) was used in the original applications. The high demands placed on resistance to temperature and aging and dynamic strength means that today only Hydrogenated Nitrile Butadiene Rubber (HNBR) is still used in today's motor vehicles. In some applications which are subject to high loads, a backing is also used to further reinforce the belt.

The tension member consists of twisted glass fibers – a material which is characterized by high tensile strength together with a high willingness to bend. As a result of the manufacturing process, the tension members are arranged in a spiral shape in the belt, and twisted in pairs in S- and Z-shapes. This ensures that the belt is subject to predominantly neutral running.

Toothed-belt profiles

The first camshaft belts were based on the classic Power Grip® trapezoidal-tooth shape (Figure 9a), as was already known in industrial applications. On account of the increasing demands with regard to load transmission, jump protection and noise, circular-arc-like profiles (e.g. Power Grip® HTD 2, High Torque Drive, Figure 9b) are almost exclusively used today. Compared with trapezoidal teeth, the forces with circular profiles are introduced more uniformly into the tooth, and stress concentrations are thereby avoided. The pitch (Figure 10) is, in most cases, 9.525 mm for diesel engines, and 8.00 mm for gasoline engines. Higher forces can be transmitted with a larger pitch, while a smaller pitch has advantages with regard to noise and space.

Double-sided toothed belts can be used for applications where the direction of rotation is reversed (e.g. balancer-shaft drives).

The toothed wheels pertaining to the toothed belts are specified in DIN ISO 9011. For the toothed wheel, the profile must be determined as a function of the diameter. The effective diameter *PD* is calculated from the number of teeth and the pitch. The outside diameter of the toothed wheel is reduced accordingly by the *PLD*.

Drive system: toothed belt

The most important requirement of the toothed-belt drive system is that it should synchronize the camshaft in relation to the crankshaft over the service life of the engine. This is an important criterion for adhering to the consumption and emission values. It is possible to keep the elongation of the toothed belt under 0.1 % of the belt length by selecting the appropriate materials for the toothed belt, and by using an automatic tensioning system and optimized system dynamics. This results in 4-cylinder engines in a timing deviation of 1 to 1.5° referred to the crankshaft.

Figure 9: Development of toothed-belt profiles
a) Power Grip® trapezoidal toothed belt,
b) Power Grip® HTD toothed belt.
1 Belt tooth,
2 Gear,
3 Load.

The service-life requirements are currently 240,000 to 300,000 km, while the temperature requirements are approx. 120 to 150 °C. Interference noise from the toothed-belt drive is not acceptable. The efficiency of a toothed-belt drive is approx. 99 %.

Design criteria

Both manufacturer's own and cross-manufacturer computing programs are used to design timing drives.

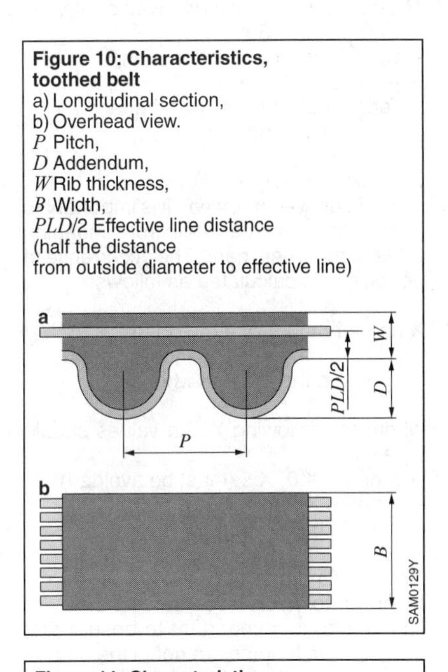

Figure 10: Characteristics, toothed belt
a) Longitudinal section,
b) Overhead view.
P Pitch,
D Addendum,
W Rib thickness,
B Width,
PLD/2 Effective line distance (half the distance from outside diameter to effective line)

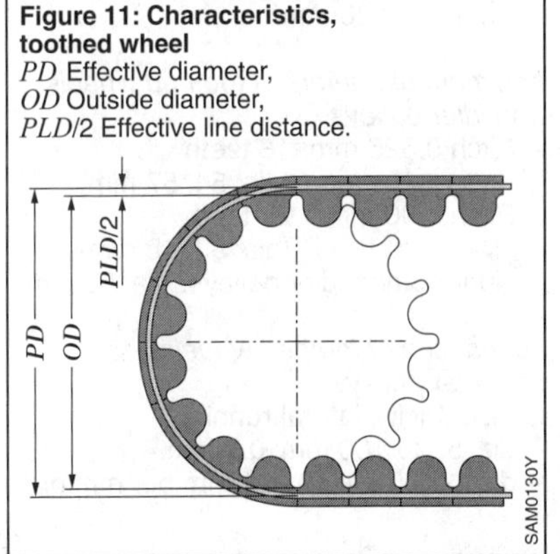

Figure 11: Characteristics, toothed wheel
PD Effective diameter,
OD Outside diameter,
PLD/2 Effective line distance.

Important input parameters are the arrangement of components, i.e. the drive configuration, the torque curves of the components, and calculated from these the dynamic peripheral forces, as well as the belt data. With these data, it is possible to calculate and optimize the system geometry, such as, for example, belt-run lengths and wrap angles, and the system dynamics.

Recommended minimum wrap angles
- Crankshaft: 150°
- Camshaft, fuel-injection pump: 100°
- Accessory wheel: 90°
- Tension pulley: min. 30°, better >70°
- Idler pulley: 30°

Periodic meshing
To avoid uneven belt wear, it is important to prevent the same belt teeth from engaging in the same wheel gaps. The occurrence of periodicity is calculated as follows:

X.nnn = number of teeth on the toothed belt/number of teeth on the toothed wheel,

where the following X.nnn values should be avoided:
- X.nnn = X.0, X.5 (must be avoided).
- X.nnn = X.25, X.333, X.666, X.75 (should be avoided).

Belt-run lengths
In order to avoid resonance noises at idle speed, it is important to ensure that free belt-run lengths are not in the area of 75 mm and 130 mm.

Minimum diameters of toothed wheels and idler pulleys
- Pitch 9.525 mm: 18 teeth (dia. 54, 57 mm).
- Pitch 8.00 mm: 21 teeth (dia. 53, 48 mm).
- Non-toothed idler pulleys: dia. 52 mm.

Tolerances of toothed wheels and idler pulleys
Concentricity, lateral running:
- dia. 50 to 100 mm: 0.1 mm.
- dia. > 100 mm: 0.001 mm per mm dia.

Conicity of outside diameter
- ≤ 0.001 mm per mm wheel thickness

Parallelism of bore to toothing
- ≤ 0.001 mm per mm wheel thickness.

Surface roughness
- $R_a \leq 1.6$ µm.

Pitch error
- dia. < 100 mm:
  ± 0.03 mm gap/gap,
  max. 0.10 mm over 90°.
- dia. 100 to 180 mm:
  ± 0.03 mm gap/gap,
  max. 0.13 mm over 90°.
- dia. > 180 mm:
  ± 0.03 mm gap/gap,
  max. 0.15 mm over 90°.

Axial guidance
A toothed belt must be guided at least at one wheel with flanged wheels to prevent it running off the belt run. Essentially, it is important to ensure that toothed wheels with flanged wheels are exactly flush with the other wheels so that the belt is not deflected from its course.

Belt-tensioning systems
The required constantly high belt tension, and compensation of the tension increase over temperature and belt elongation are usually provided in today's timing drives by tension pulleys. The design of the tension pulleys is significantly dictated by the space available. The most widely used system is the mechanical, friction-damped compact tensioner. Hydraulic tension pulleys are also used in some applications in the event of very high dynamic forces in the toothed-belt drive system. These demonstrate, on account of their asymmetrical damping, very good damping properties even with low pretensioning forces.

Chain drives

Overview

Camshafts are used to control the gas exchange valves; these camshafts open and close the valves by means of direct actuation via bucket tappets or lever assemblies. Camshafts are driven on modern overhead valve engines with the aid of a belt drive, while side-mounted camshafts are connected via a geartrain to the crankshaft pinion. Toothed belts, roller or sleeve-type chains, and tooth-type chains are used to drive overhead camshafts.

The most important criteria in deciding on the type of drive are costs, space, ease of maintenance, service life, and noise buildup. The great advantage of chain drives over toothed-belt drives is that they are completely maintenance-free over the entire service life of the engine, while toothed belts, depending on the application, have to be retensioned or changed according to the maintenance interval.

In addition to the camshaft, timing drives in modern engines frequently drive other assemblies, such as the oil pump and fuel-injection pump (Figure 1). As both camshaft and crankshaft rotate nonuniformly and are exposed to torsional vibrations, the drive is subject to highly complex dynamic stresses. Furthermore, the torque demand of the fuel-injection pump is subject to very strong periodic fluctuations, and thus generates further excitation of the timing chain drive.

Chain designs

In the case of standard chains, a distinction is made between roller and sleeve-type chains. There are also simplex and duplex chains. The tooth-type chain is a special type of chain.

Roller chain

The rotating rollers of a roller chain (Figure 2a) located over the sleeves roll with little friction on the tooth flanks of the sprocket such that a different point on the circumference is always contacted. The lubricant between the rollers and sleeves helps to muffle noise and dampen shocks.

Sleeve-type chain

In the case of a sleeve-type chain (Figure 2b), the tooth flanks of the sprocket

Figure 1: Timing-chain drive
1 Camshafts,
2 Tensioning rail,
3 Injection pump,
4 Chain-tensioner system,
5 Tensioning rail,
6 Crankshaft,
7 Guide rail,
8 Intermediate-shaft sprocket group,
9 Guide rail,
10 Crankshaft sprocket.

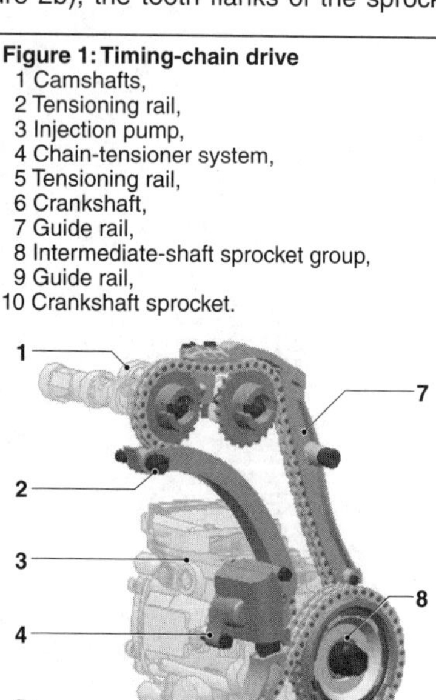

constantly touch the fixed sleeves at the same point such that the sleeves are subjected to an additional load. Perfect lubrication is therefore particularly important for these drives.

While having identical pitch and force at break, sleeve-type chains have a larger joint surface than the corresponding roller chains, since the omission of the roller means that the pin diameter can be enlarged. A larger joint surface results in a lower joint-surface pressure and thus lower wear in the joints.

Sleeve-type chains have proven useful in camshaft drives subject to high loads in diesel engines, because increased wear resistance is called for here on account of the increased entry of soot into the engine oil.

Tooth-type chain

On a tooth-type chain (Figure 2c), the links are set out in such a way that they can transmit the force between the chain and the sprocket, whereas, in the case of roller or sleeve-type chains, the connection with the sprocket is established via pin, sleeve, or roller.

Tooth-type chains can be built to virtually any desired width without the need for fundamental structural change. Guide links, which are located either in the middle or on the outside (on both sides), are fitted to prevent the sprocket from running off. A further variation of the tooth-type chain is the version toothed on both sides, which also permits negative wraps (as on roller and sleeve-type chains).

Selecting chain types

When choosing the chain types (construction and pitch), it is important, while adhering to the maximum sprocket diameter, to aim for numbers of teeth of more than 18 teeth in order to reduce dynamic influences by the polygon effect. If this cannot be implemented structurally, it is recommended to switch to a duplex chain with a smaller pitch.

Decades of experience have shown that some dimensions for roller, sleeve-type, and tooth-type chains are particularly suitable for timing drives.

Requirements of chains

Four important factors determine the usage characteristics of timing chains: ultimate tensile strength, fatigue strength, resistance to wear, and acoustic behavior.

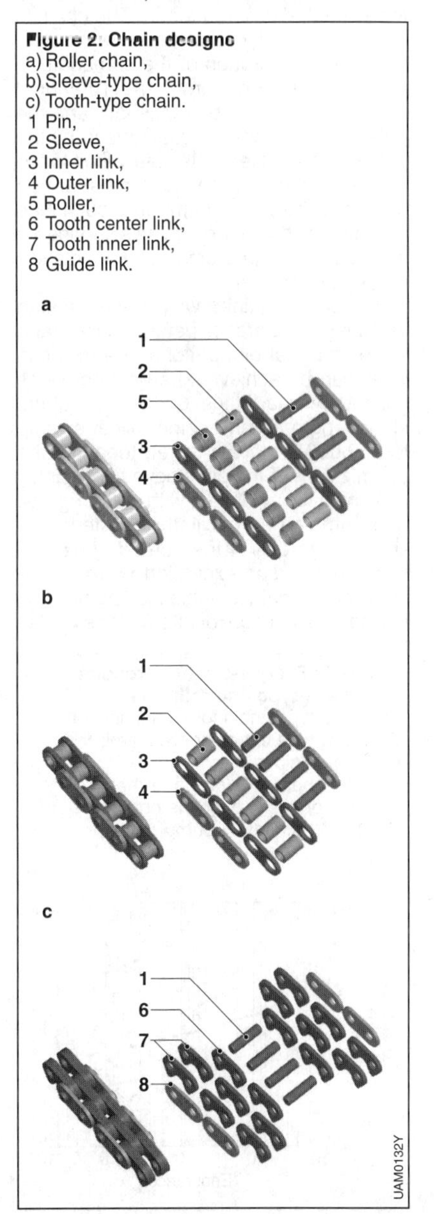

Figure 2. Chain designs
a) Roller chain,
b) Sleeve-type chain,
c) Tooth-type chain.
1 Pin,
2 Sleeve,
3 Inner link,
4 Outer link,
5 Roller,
6 Tooth center link,
7 Tooth inner link,
8 Guide link.

A break may be caused by the static or dynamic load at break being exceeded. Especially in the case of timing drives, no uniform load will be encountered such that the fatigue strength of the chain is the strength-limiting quantity. The chain is subjected to a dynamic load as a result of the pulsating torques of the camshaft(s), the fuel-injection pump, the rotational non-uniformity of the crankshaft, and the pulsating axial chain force caused by the polygon effect. Here, the safety limit to the fatigue strength of the chain must not be exceeded, since the number of such load alternations during an engine's service life is in each case greater than 10^8 load alternations (Figure 3).

In today's engines with precise timing and minimal clearance between piston and valve, minimal elongations due to chain wear can be achieved. Length increases caused by wear of just 0.2 to 0.5 % of the chain length for an engine mileage of up to 350,000 km are ensured today by the optimization of the pin/sleeve tribo-joint.

A chain timing drive is with mass, rigidity, and damping an oscillatory system with several degrees of freedom. If subjected to corresponding excitation by the camshaft, crankshaft, fuel-injection pump, etc., this can, on account of interactions, cause resonance effects which lead to extreme timing-drive load.

By employing various optimization measures (e.g. increasing chain rigidity while maintaining specific mass, utilizing special geometries on the sprocket, matching the hydraulic chain tensioner with regard to damping and rigidity), it is possible to shift the occasionally occurring points of resonance to higher engine speeds (or outside the usable speed range) (Figure 4).

Sprockets

The tooth shape of sprockets is standardized for roller chains, sleeve-type chains, and tooth-type chains:

- DIN 8187-1
  for European-type roller chains.
- DIN 8188-1
  for American-type roller chains.
- DIN 8154
  for sleeve-type chains with full pins.
- DIN 8190
  for tooth-type chains.

Optimized configuration of the tooth shape is just as important for the safe operation of a timing drive as, for example, the

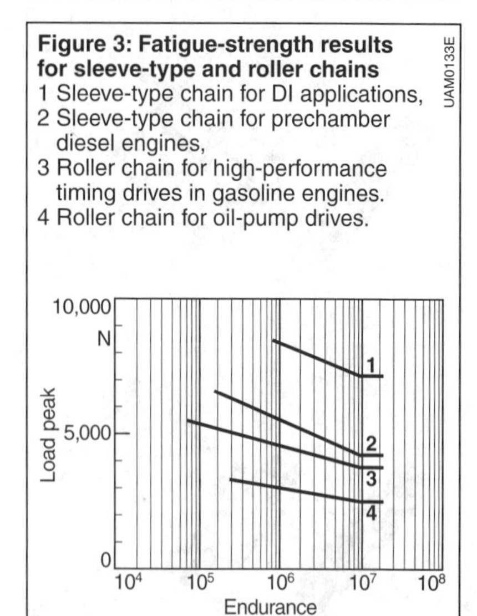

Figure 3: Fatigue-strength results for sleeve-type and roller chains
1 Sleeve-type chain for DI applications,
2 Sleeve-type chain for prechamber diesel engines,
3 Roller chain for high-performance timing drives in gasoline engines.
4 Roller chain for oil-pump drives.

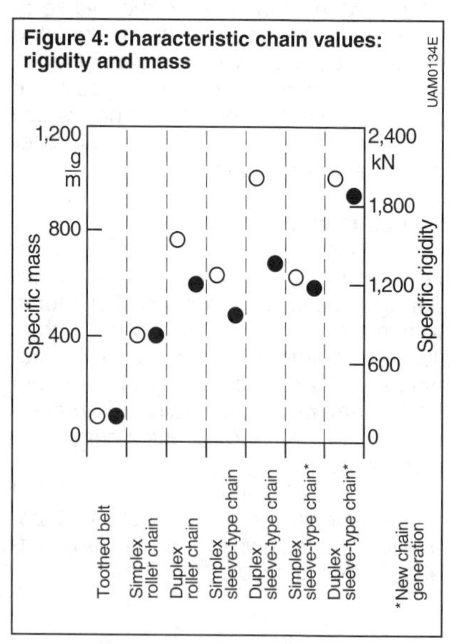

Figure 4: Characteristic chain values: rigidity and mass

resistance to wear or fatigue strength of the chain itself.

DIN permits considerable degrees of freedom in the precision configuration of the sprocket teeth. As a rule, sprockets with maximum tooth-space shape are used. This design permits, as a result of the low addendum and the larger tooth-space opening, the chain to enter and exit undisturbed even at higher chain speeds.

Depending on the available space and application, disk wheels or sprockets with one-sided or two-sided hubs and plated types are used. The choice of material is dependent on the space conditions, the operating conditions, and the power transmission of the chain drive (Figure 5).

Sprockets made of carbon steel and alloyed steels and wheels made of sintered materials are used. Precision-blanked and machine-cut wheels with the heat treatment appropriate for the material are also used. The space conditions and assembly sequence play a crucial role in the choice of material.

Figure 5: Sprockets
a) Sintered crankshaft gear,
b) Precision-blanked camshaft gear,
c) Sprocket with sealing surface,
d) Two-track sprocket.

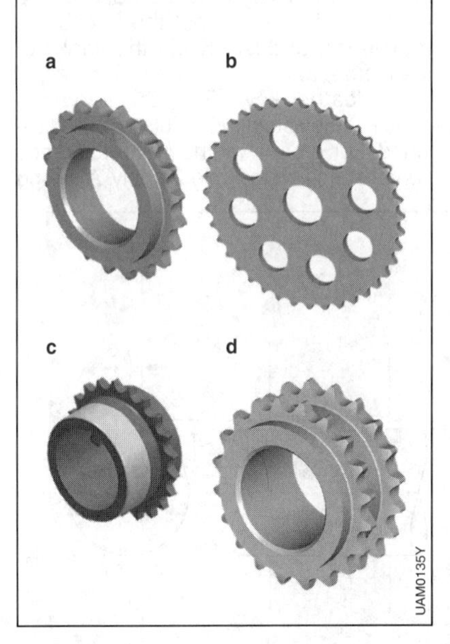

Chain-tensioning and chain-guide elements

By using permanently acting clamping and guide elements (Figure 1) which are precisely matched to the engine in question, it is possible to optimize the chain drive in such a way that its service life corresponds to that of the engine.

Chain tensioner

The (hydraulic) chain tensioner assumes a series of tasks in the timing drive. On the one hand, the timing chain is pre-tensioned in all operating conditions on the loose side under a defined load, even in the event of wear elongation that occurs during operation. A damping element – either friction or viscous damping – reduces vibrations to an acceptable level.

Mechanical chain tensioners without additional hydraulic damping are used, as a rule, in oil-pump drives which are subject to low load. In special cases, this mechanical chain tensioner can even be omitted completely.

Tensioning and guide rails

Simple rails made of plastic or metal (aluminum or sheet steel) with plastic covering and which are flat or bent – depending on the chain route – are sometimes used as tensioning and guide elements. In the case of newer types, the rails are for the most part inexpensively injection-molded from plastic and manufactured as a two-component design (base carrier and sliding lining).

Here, in the case of the tensioning rail, a friction lining of glass-fiber-free polyamide is sprayed or clipped onto a carrier of heat-resistant polyamide with 30 to 50 % glass-fiber content. The sliding rails are usually designed in plastic as one-component rails, and are used for chain guidance.

Detachable connections

Positive or form-closed joints

Operation

Positive or form-closed joints have the task of transferring forces, which themselves maintain contact via mating surfaces, by means of their geometrical shape. The forces are always transferred perpendicularly to the mating surfaces, resulting mainly in compressive and shear stress.

Form closure generally produces easily detachable connections, because as a rule there is a clearance or transition fit between shaft and bore. Depending on the choice of press fit, relative axial movements may occur during operation. If necessary, they must be prevented by suitable locking devices (e.g. locking ring according to DIN 471 [1]).

Table 1: Symbols and units

Quantity		Unit
D	Diameter	mm
F	Force	N
K_A	Application factor (service factor)	–
M_t	Torque	Nm
b	Width	mm
d	Diameter	mm
h	Height	mm
i	Number of shear surfaces	–
l	Length	mm
l_{tr}	Supporting feather-key length	mm
n	Number of drivers	–
p	Surface pressure	N/mm²
t_1	Groove depth (shaft)	mm
t_2	Groove depth (hub)	mm
σ_b	Bending stress	N/mm²
τ_s	Shear stress	N/mm²
φ	Contact-surface ratio	–

Feather-key and Woodruff-key couplings

Feather-key couplings (Figure 1a) are used for the torsion-resistant connection of belt pulleys, gears, coupling hubs, etc. to shafts. Feather keys are sometimes used to secure frictional joints or to fix a specified position in the circumferential direction.

The cheaper woodruff key (Figure 1b), the round side of which fits into the shaft, is mainly used for this purpose in automotive construction and to transfer smaller torques.

In the case of feather-key couplings, the groove faces lie against the feather-key faces. In contrast to keyed joints, there is clearance (backlash) between the rear of the feather key and the base of the groove. This means that the forces are transferred exclusively via the flanks of the feather key.

For feather-key width, the tolerance zone h 9 (key steel to DIN 6880 [2]) is provided. For groove widths b, the tolerance zones specified in Table 2 apply. A sliding

Figure 1: Positive or form-closed joints
a) Feather-key coupling,
b) Woodruff-key coupling.
Symbols, see Table 1.

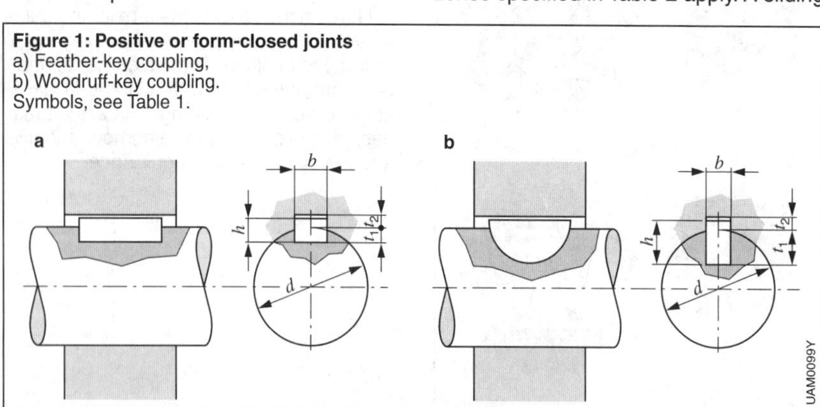

Table 2: Tolerances for groove widths

Groove fit	Fixed fit	Easy fit	Sliding fit
in hub	P 9	N 9	H 8
in shaft	P 9	J 9	D 10

seat must be used if a hub must move on the shaft in the longitudinal direction (e.g. gear in manual-shift transmission). Usually, the sliding spring is firmly bolted into the shaft groove. Round (Shape A) and angular (Shape B) feather keys (Figure 2) are manufactured. DIN 6885 [3] defines the standard in terms of their shape and dimensions, depending on the shaft diameter (Table 3).

In practice, feather keys are designed only for surface pressure. If $p \leq p_{\text{perm}}$, the required supporting feather-key length is

$$l_{\text{tr}} = \frac{2\,K_{\text{A}}\,M_{\text{t}}}{d\,(h - t_1)\,n\,\varphi\,p_{\text{perm}}}\,.$$

For round-faced feather keys (Shape A), the feather-key length is $l = l_{\text{tr}} + b$. For straight-faced ones (Shape B), it is $l = l_{\text{tr}}$. For the permissible surface pressures, the standard gives $p_{\text{perm}} = 0.9\,R_{\text{e, min}}$, where $R_{\text{e, min}}$ is the minimum yield point of the shaft, hub or feather-key material. With one feather key ($n = 1$), the contact-surface ratio is $\varphi = 1$, and with two feather keys, it is $\varphi = 0.75$.

Figure 2: Feather-key shapes
a) Shape A,
b) Shape B.
Symbols, see Table 1.

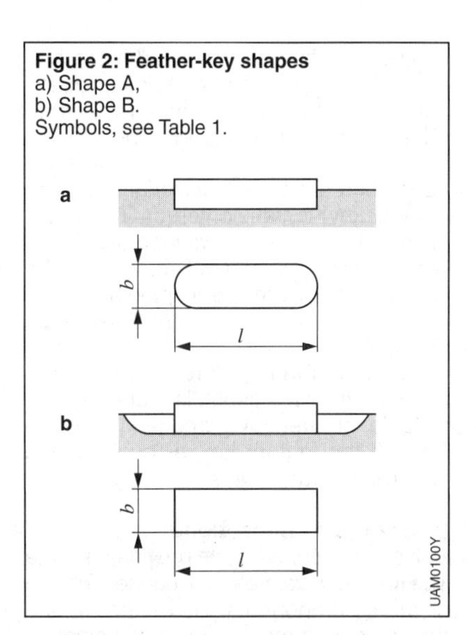

Profiled shaft-hub connections

Instead of inserting multiple feather keys into shaft grooves, the shaft cross-section can also be shaped directly in the form of a polygonal profile, and the mating hub cross-section has a corresponding shape.

Table 3: Feather-key dimensions according to DIN 6885

Shaft diameter d over mm	up to mm	Width × height $b \times h$ mm	Groove depths t_1 mm	t_2 mm	Length l mm
6	8	2 × 2	1.2	1.0	6…20
8	10	3 × 3	1.8	1.4	6…36
10	12	4 × 4	2.5	1.8	8…45
12	17	5 × 5	3.0	2.3	10…56
17	22	6 × 6	3.5	2.8	14…70
22	30	8 × 7	4.0	3.3	18…90
30	38	10 × 8	5.0	3.3	22…110
38	44	12 × 8	5.0	3.3	28…140
44	50	14 × 9	5.5	3.8	36…160
50	58	16 × 10	6.0	4.3	45…180
58	65	18 × 11	7.0	4.4	50…200
65	75	20 × 12	7.5	4.9	56…220
75	85	22 × 14	9.0	5.4	63…250
85	95	25 × 14	9.0	5.4	70…280
95	110	28 × 16	10.0	6.4	80…320
Feather-key lengths in mm:		6, 8, 10, 12, 14, 16, 18, 20, 22, 25, 28, 32, 36, 40, 45, 50, 56, 63, 70, 80, 90, 100, 110, 125, 140, 160, 180, 200, 220, 250, 280, 320			

Profiled shafts are also used if the hub must be axially movable relative to the shaft (e.g. steering column for a height-adjustable steering wheel). The profiled shaft connection has the advantage that it does not require any additional separator (feather key) to transfer torque. The hub is centered either via a cylinder jacket surface (smallest diameter of shaft) or via the flanks of the drivers. Very smooth running can be achieved with internal centering (Table 4).

Flank centering insures very low circumferential backlash. It is therefore very suitable for alternating and jerky torques. As with the feather key, a rough design is drawn up based on surface pressure.

Bolt and pin connections

Bolt and pin connections provide a simple and inexpensive means of connecting two or more components. These are among the oldest and most widely used types of connection.

Bolt connections

Bolt connections are mainly used for joining linkages (Table 5), shackles, chain links, and connecting rods, as well as axles for bearing impeller rings, rollers, levers, etc. Since relative movements occur in these connections, at least one part must be movable. The main prevailing stresses are surface pressure (Table 6) and shear. Bending stress is negligible in most cases. It only occurs to a significant extent in the case of bolt connections which are relatively long in relation to their diameter.

Pin connections

Pins are suitable for the permanent connection of hubs, levers, and set collars on shafts or axles. They also secure the precise position of two machine parts, and as guide pins to fix springs, etc. (Table 5). Since they are forced into holes as press fits with tolerances, all parts are permanent.

References

[1] DIN 471: Circlips (retaining rings) for shafts – Normal type and heavy type.
[2] DIN 6880: Bright key steel – dimensions, permissible variations, weights.
[3] DIN 6885: Drive type fastenings without taper action; parallel keys.
[4] DIN 6892: Drive type fastenings without taper action – Parallel keys – Calculation and design.
[5] Haberhauer/Bodenstein: Maschinenelemente, 15th Edition, Springer-Verlag 2008.
[6] Kollmann: Welle-Nabe-Verbindungen, Springer-Verlag 1984.

Table 4: Profiled shaft-hub connections

Designation	Standard	Illustration	Driver	Centering	Contact-surface ratio
Spline shaft	ISO 14 DIN 5464	UAM0101-1Y	Prismatic driver	Inner	$\varphi = 0.75$
				Flanks	$\varphi = 0.9$
Toothed shaft with grooved toothing	DIN 5481	UAM0101-2Y	Grooved toothing	Flanks	$\varphi = 0.5$
Toothed shaft with involute toothing	DIN 5480 DIN 5482	UAM0101-3Y	Involute teeth	Flanks	$\varphi = 0.75$

Table 5: Bolt and pin connections

Designation	Illustration	Calculations
Articulated joint	UAM0102-1aY	Surface pressure in fork: $p_G = \frac{F}{2\,b_1\,d} \leq p_{perm}$ Surface pressure in rod: $p_S = \frac{F}{b\,d} \leq p_{perm}$ Surface pressure in pin: $\tau_S = \frac{4\,F}{i\,\pi\,d^2} \leq \tau_{S,perm}$
Transverse-pin joint	UAM0102-2Y	Surface pressure in shaft: $p_{W,max} = \frac{6\,M_t}{d\,D_W^2} \leq p_{perm}$ Surface pressure in hub: $p_N = \frac{4\,M_t}{d\left(D_N^2 - D_W^2\right)} \leq p_{perm}$ Surface pressure in pin: $\tau_S = \frac{4\,M_t}{D_W\,\pi\,d^2} \leq \tau_{S,perm}$
Guide pin	UAM0102-3Y	Maximum pressure: $p_{max} = p_b + p_d = \frac{F}{d\,s}\left(1 + 6 \cdot \frac{h + s/2}{s}\right) \leq p_{perm}$ Bending stress at clamping point: $\sigma_b = \frac{32\,F\,h}{\pi\,d^3} \leq \sigma_{b,perm}$ Shear stress in clamping point: $\tau_s = \frac{4\,F}{\pi\,d^2} \leq \tau_{S,perm}$

Table 6: Permissible mean surface pressure for bolt and pin connections

Permanent fits			Sliding fits	
Material	Mean surface pressure		Material pairing	Mean surface pressure
	static	swelling		
	p_{perm} N/mm²	p_{perm} N/mm²		p_{perm} N/mm²
Gray cast iron	70	50	St / gray cast iron	5
S 235 (St 37)	85	65	St / CS	7
S 295 (St 50)	120	90	St / Bz	8
S 335 (St 60)	150	105	St hard. / Bz	10
S 369 (St 70)	180	120	St hard. / St hard.	15

Frictional joints

Operation

For frictional joints, press fits are produced in the joints (the friction surfaces being the effective areas) in which the parts to be assembled are in direct contact (Figure 1). The surface pressure p can be generated by bolt forces, keys, elastic separators, or the elasticity of the components themselves. The resulting normal force $F_N = pA$ (with friction surface A) induces a frictional force F_R, which opposes a movement caused by external forces.

Press fit

Application

For a press fit (cylindrical interference fit), the required surface pressure is produced by the elastic deformation of the shaft and hub, resulting from an interference fit. "Interference fit" is the pairing of cylindrical fitting parts which have an interference before assembly (Figure 2).

Because press fits are easy to produce and can transfer even jerky and variable torques and linear forces, they are suitable for the joints of cylindrical surfaces which do not have to be released afterwards (e.g. gear on shaft, wheel on axle, bushing in housing).

Table 1: Symbols and units

Quantity		Unit
A	Area (friction surface)	mm²
C	Taper ratio	–
D	Diameter	mm
E	Modulus of elasticity	N/mm²
F	Force	N
F_a	Axial force	N
F_N	Normal force	N
F_R	Frictional force	N
K_A	Application factor (service factor)	–
M_t	Torque	Nm
$M_{t,nom}$	Nominal load torque	Nm
Q	Diameter ratio	–
R_e	Yield point	N/mm²
R_m	Ultimate strength	N/mm²
R_z	Surface roughness	mm
S_B	Factor of safety against rupture	–
S_F	Factor of safety against yield	–
U	Tolerance	mm
Z	Allowance	mm
b	Hub width	mm
d	Diameter	mm
l	Taper or lever length	mm
n	Number of bolts	–
p	Surface pressure	N/mm²
t	Temperature	°C
α	Taper angle	°
α_A	Coefficient of linear expansion, outer part	10^{-6}/K
α_I	Coefficient of linear expansion, inner part	10^{-6}/K
μ	Coefficient of friction	–
ν	Poisson's ratio	–
ξ	Specific allowance	mm³/N
σ_{perm}	Permissible stress	N/mm²

Figure 1: Frictional joints
a) Axially stressed joint, b) Tangentially stressed joint.
1 Gripped side of joint.
F Force, F_R Frictional force, p Surface pressure, M_t Torque.

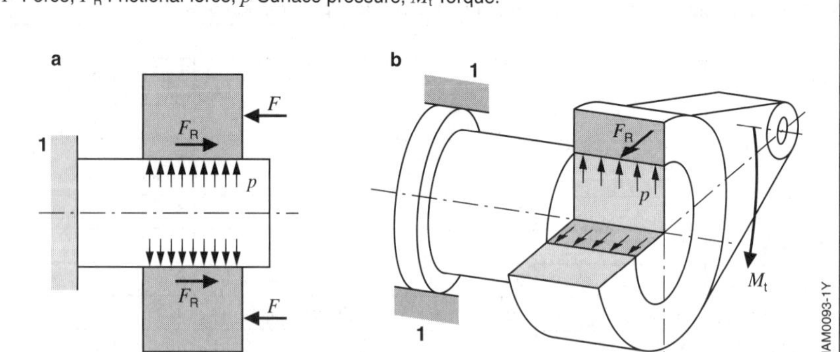

Elastic design
of cylindrical interference fits

A press fit must be designed so that at least a minimum surface pressure p_{min} is present in order to ensure the transfer of the greatest occurring stress, and so that a maximum surface pressure p_{max} is not exceeded and the components are not overstressed.

In principle, two calculations aims are possible: the required fit is defined for a given stress (Table 2), or the permissible stress is determined for a given fit (Table 3).

The diameter ratios:
$Q_A = D_F / D_{Aa}$ and
$Q_I = D_{Ii} / D_F$

and the specific allowance [1]

$$\xi = D_F \left[\frac{1}{E_I} \left(\frac{1 + Q_I^2}{1 - Q_I^2} - \nu_I \right) + \frac{1}{E_A} \left(\frac{1 + Q_A^2}{1 - Q_A^2} + \nu_A \right) \right]$$

can be used to design press fits with reference to their funotion and the required component safety.

The nominal diameter is inserted into the calculation for the joining diameter D_F. When joining an interference, loss results from plastic leveling of the roughness peaks. According to DIN 7190 [2], the smoothing of the two surfaces with 40% of the averaged peak-to-valley heights R_{zI} (shaft) and R_{zA} (bore) is taken into account in the calculation.

The maximum tensions occur at the internal diameters of hollow shaft and hub. Solid shafts are noncritical, and do not usually have to be calculated.

Table 2: Definition of fits for given stress [1]

Stress M_t or F_a given	$\sigma_{perm} = R_e / S_F$ or $\sigma_{perm} = R_m / S_B$ given
Required fit: $p_{min} = \dfrac{\sqrt{F_a^2 + \dfrac{4 M_{t,nom}^2}{D_F^2}}}{\mu \pi D_F b} K_A$	Permissible fit in hub: $p_{max} = \left(1 - Q_A^2\right) \sigma_{perm} / \sqrt{3}$ Permissible fit in hollow shaft: $p_{max} = \left(1 - Q_I^2\right) \sigma_{perm} / \sqrt{3}$
Required allowance: $Z_{min} = p_{min}\, \xi$	Permissible allowance: $Z_{max} = p_{max}\, \xi$
Required interference: $U_{min} = Z_{min} + 0.8\,(R_{zI} + R_{zA})$	Permissible interference: $U_{max} = Z_{max} + 0.8\,(R_{zI} + R_{zA})$
Select ISO fits with $U_k \geq U_{min}$ and $U_g \leq U_{max}$	

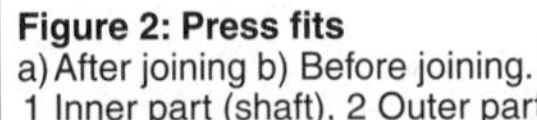

Figure 2: Press fits
a) After joining b) Before joining.
1 Inner part (shaft), 2 Outer part (hub).
F_R Frictional force, p Surface pressure, M_t Torque, b Hub width,
D_F Joining diameter, D_{Ia} Shaft outside diameter, D_{Ii} Shaft inside diameter,
D_{Aa} Hub outside diameter, D_{Ai} Hub inside diameter.

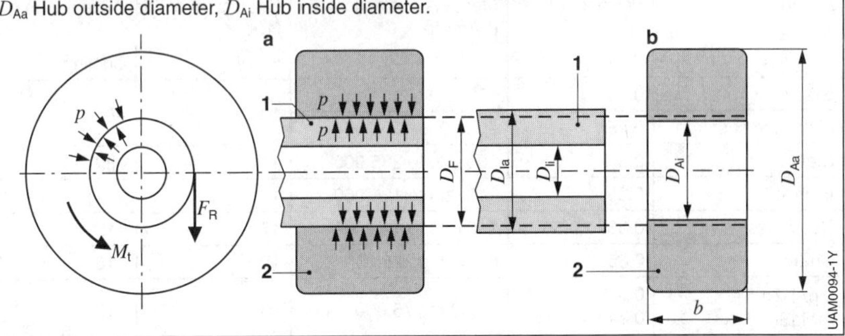

Table 3: Determination of stress for given fit

Smallest interference U_k given	Greatest interference U_g given
Smallest allowance: $Z_k = U_k - 0.8\,(R_{zI} + R_{zA})$	Greatest allowance: $Z_g = U_g - 0.8\,(R_{zI} + R_{zA})$
Smallest pressure: $P_k = \frac{Z_k}{\xi}$	Greatest pressure: $P_g = \frac{Z_g}{\xi}$
Permissible stress: $M_t = 0.5 p_k \mu \pi D_F^2 b$ (only M_t) $F_a = p_k \mu \pi D_F b$ (only F_a)	Hub safety factor: $S_F = \frac{1-Q_A^2}{\sqrt{3}\,p_g} R_e$ or $S_B = \frac{1-Q_A^2}{\sqrt{3}\,p_g} R_m$
$M_t = \frac{D_F}{2}\sqrt{(p_k \mu \pi D_F b)^2 - F_a^2}$ (F_a given) $F_a = \sqrt{p_k \mu \pi D_F b - \frac{4 M_t}{D_F^2}}$ (M_t given)	Hollow-shaft safety factor: $S_F = \frac{1-Q_I^2}{\sqrt{3}\,p_g} R_e$ or $S_B = \frac{1-Q_I^2}{\sqrt{3}\,p_g} R_m$

Assembly
There are two types of press fit depending on the assembly method: linear and transverse. Linear press fits are produced by "cold" assembly at room temperature. The large press-fit forces required are usually applied by hydrostatic presses. The press-fit velocity should not exceed 2 mm/s. For the press-fit force:

$$F_e = \frac{(U_g - 0.8\,(R_{zI} + R_{za}))\,\mu \pi D_F b}{\xi}$$

Before transverse press fits are made, either the outer part is expanded by heating, or the diameter of the inner part is reduced by supercooling, so that the parts can be assembled stress-free. If the outer part is heated, it shrinks onto the inner part as it cools (shrink fit). If the inner part is cooled so that it expands when it is heated to room temperature, this is an expansion fit. To make force-free assembly possible, a joint clearance of $\Delta D = 0.001 \cdot D_F$ must be provided (Table 4).

Assembly temperature for shrink fit:

$$t_a = t_u + \frac{U_g + \Delta D}{\alpha_i D_F}$$

Assembly temperature for expansion fit:

$$t_a = t_u - \frac{U_g + \Delta D}{|\alpha_i| \cdot D_F}.$$

Table 4: Poisson ratio, modulus of elasticity, and linear coefficient of thermal expansion for metallic materials

Material	Poisson ratio ν	Modulus of elasticity E [N/mm²]	Coefficient of linear expansion α [10^{-6}/K]	
			Heating	Cooling
Gray cast iron	0.24	100,000	10	–8
Malleable cast iron	0.25	90,000 to 100,000		
Steel	0.3	200,000 to 235,000	11	–8.5
Bronze	0.35	110,000 to 125,000	16	–14
Red brass	0.35 to 0.36	110,000 to 125,000	17	–15
CuZn	0.36	80,000 to 125,000	18	–16
$MgAl_8Zn$ AlMgSi	0.3 0.34	65,000 to 75,000	23	–18

Tapered connection

Application

A tapered connection (tapered interference fit) is suitable for transferring dynamic forces and torques. These connections are mainly used to fix parts to shaft ends (e.g. to the alternator V-belt pulley). They have the following advantages: They can be re-tensioned, are easily detachable, do not weaken the shaft, and have very good centering (i.e. no imbalance).

However, the disadvantages are the high production costs and the lack of adjustability in the axial direction. The following applies (Figure 3)

to the taper ratio $C = \frac{(d_1 - d_2)}{l}$

and to the taper angle $\tan\frac{\alpha}{2} = \frac{(d_1 - d_2)}{2l}$.

The following taper ratios are specified as guidelines (DIN 254 [3], DIN 406 [4]):

$C = 1{:}5$ Connection is easily detachable,

$C = 1{:}10$ Connection is detachable only with difficulty

$C = 1{:}20$ Tool holder for twist drill (Morse taper)

Figure 3: Tapered connection
D_a Hub outside diameter,
d_1 Large taper diameter,
d_2 Small taper diameter,
d_m Middle taper diameter,
l Common length,
p Joint pressure,
α Taper angle,
M_t Torque,
F_a Assembling force (fastener force).

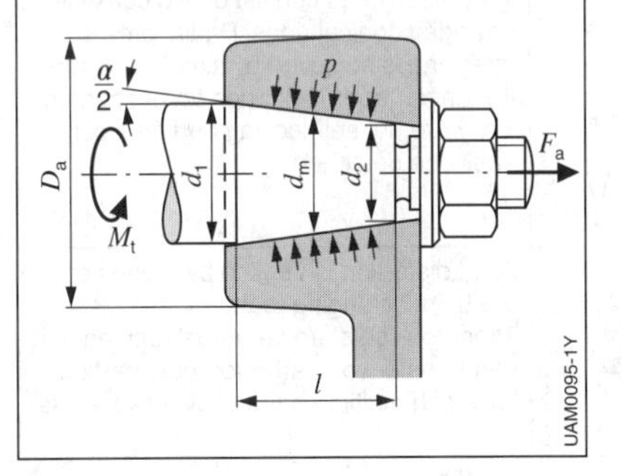

Operation

The effective area pair of a tapered connection takes the form of a truncated cone (frustum). The required surface pressure p is usually applied by an axial bolt force F_a. The relationship between the axial press-fit force F_a and the transferable torque M_t is expressed by the following equation:

$$F_a \geq \frac{2K_A M_{t,nom}}{\mu_U d_m}\left(\sin\frac{\alpha}{2} + \mu_A \cos\frac{\alpha}{2}\right).$$

This takes into account the possible difference between the coefficients of friction μ_u in the circumferential direction and μ_a in the axial direction. Self-locking exists if an expulsion force (F_a becomes negative) is required to detach the connection. This means that a torque can be transferred even if the bolt is removed after the parts are axially stressed. In contrast, with a non-self-locking tapered connection, there is no pressure between the effective areas after the axial application force is released. The condition for self-locking is:

$$\frac{\alpha}{2} \leq \arctan \mu_a$$

Component safety

The critical component is the hub. It is calculated as an open, thick-walled hollow cylinder. If $Q = d_m / D_a$, the following applies to the safety of the hub, with elastic design, depending on the modified shear-stress theory:

$$S_F = \frac{1 - Q^2}{\sqrt{3}} \frac{\left(\sin\frac{\alpha}{2} + \mu_A \cos\frac{\alpha}{2}\right) \pi d_m l}{F_{a,max}} .$$

Taper-lock joints

The required surface pressure in the effective areas can also be applied by elastic separators (see Table 5). The great advantage of these taper-lock joints is that they can be used to fix hubs, gears, couplings, etc. securely to smooth, cylindrical shafts. Unlike cylindrical interference fits, they can be freely adjusted axially and tangentially, and are above all easily detachable. They are, therefore, particularly suitable for hubs (e.g. belt pulleys), which must be adjustable and replaceable. Their disadvantages include the space required and the high costs. They are usually designed as specified by the manufacturer (see manufacturer catalogs and Table 5).

Clamp joints

In the case of clamp joints, external forces apply the required surface pressure to the joint, mostly by means of bolts. The joints with a sectional or slotted hub are preferably used for low torques which have little fluctuation. Their advantage is that the hub position is easily adjustable in the axial and tangential directions. They provide a very easy means of fixing wheels or levers to smooth shafts. However, there are also self-locking clamp joints. Here, the tilting force F_K generates edge pressures in A and B (Table 6) to prevent axial movement.

Table 5: Taper-lock joints

Designation	Illustration	Features
Clamping sleeve (Spieth)	UAM0096-1Y	Axial distortion is applied to enlarge the outside diameter of the clamping sleeve and reduce the inside diameter. The danger of loosening under dynamic loads is reduced by using long bolts.
Hydraulic hollow-jacket spring collet (Lenze)	UAM0096-2Y	Axial distortion is applied to generate a pressure in a thin-walled hollow cylinder. The spring collet is self-centering because of even pressure distribution, and thus has true-running properties. At higher temperatures, the thermal expansion of the pressure fluid must be taken into account.
Tolerance ring (Oechsle)	UAM0096-3Y	Tolerance rings are slotted and made from thin wave-shaped sheet metal. The required initial force is provided by forced deformation of the elastic connecting member. They can bridge relatively large machining tolerances, compensate for thermal expansion, and transfer torques.
Star washer (Ringspann)	UAM0096-4Y	Star washers are thin-walled, very flat conical shells with radial slots. The axial initial force is translated by forced deformation into a fivefold to tenfold radial force. Star washers do not center.
Taper-lock ring (Ringfeder)	1 Precentering 1 UAM0096-5Y	Taper-lock rings consist of two coaxially arranged conical rings. Radial prestressing is achieved by axially distorting the rings. Taper-lock rings do not center. They are not self-locking and thus can easily be detached.
Taper-lock set (Bikon)	UAM0096-6Y	Axial distortion is applied by means of the bolts belonging to the taper-lock set. Taper lock sets are very true-running and can transfer very large torques, particularly with multiple pairs of effective areas.

Keyed joints

Longitudinal keyed joints

A one-sided radial distortion is achieved by driving in a standardized key (key angle = 0.57°) between the shaft and hub. However, because of their imprecise assembly (hammer assembly) and the resulting eccentricities, keys are only of secondary importance.

Circular keyed joint

A new type of keyed joint is the 3-part circular-key sectional profile (Figure 4). Three circular keys are arranged circumferentially on the cylindrical surface of a shaft (inner part). The hub (outer part) contains an appropriate number of corresponding keys in a cylindrical hole. Twisting results in radial distortion and this can transfer large axial and tangential forces in any direction.

As opposed to press fits, circular keyed joints are detachable. They are used, for example, for shaft-hub connections, for camshafts and for hinges in vehicle construction.

References

[1] Haberhauer/Bodenstein: Maschinenelemente, 15th Ed., Springer-Verlag 2008.
[2] DIN 7190: Interference fits. Calculation and design rules.
[3] DIN 254: Geometrical product specifications (GPS) – Series of conical tapers and taper angles.
[4] DIN 406: Technical drawings; dimensioning.
[5] Kollmann: Welle-Nabe-Verbindungen, Springer-Verlag 1984.
[6] Dubbel, Taschenbuch für den Maschinenbau, 22nd Ed., Springer-Verlag 2007.

Figure 4: 3-part circular-key sectional profile

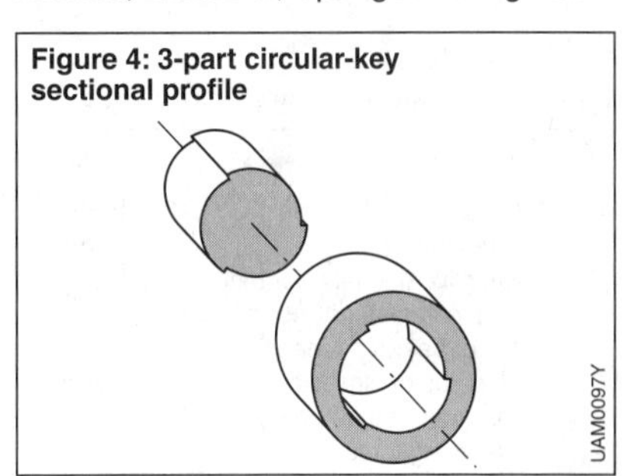

Table 6: Clamp joints

Designation	Illustration	Features
Clamp joint with separated hub.	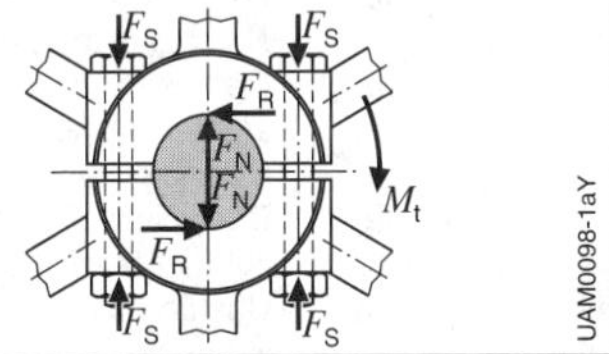	Transferable torque: $M_t = n F_S \mu \frac{\pi}{2} D_F$. n Number of bolts, F_S Fastener force, F_R Frictional force, F_N Normal force, M_t Torque.
Clamp joint with slotted hub.	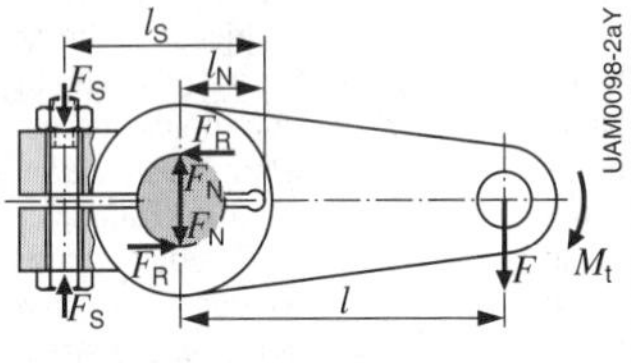	Transferable torque: $M_t = n F_S \mu \frac{\pi}{2} D_F \frac{l_s}{l_N}$. n Number of bolts, F Force, F_S Fastener force, F_R Frictional force, F_N Normal force, M_t Torque, l_S Bolt lever arm, l_N Hub-center lever arm.
Self-locking clamp joint (screw-clamp principle).	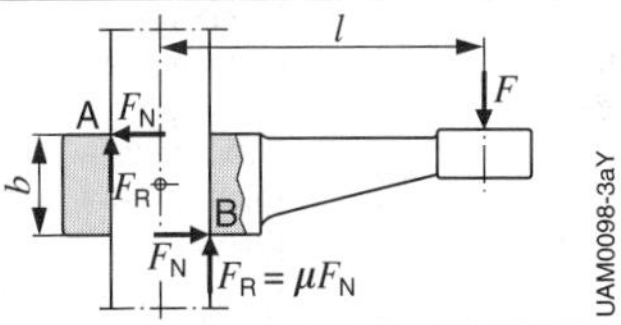	Condition for self-locking: $\frac{l}{b} \geq \frac{1}{2\mu}$. F Force, F_R Frictional force, F_N Normal force, μ Coefficient of friction, b Hub width, A, B Contact points.

Threaded fasteners

Basic principles

Operation

Threaded fasteners include “screws and bolts”. They are used to make secure joints that are detachable any number of times. The purpose of screws and bolts is to stress the mating parts so that the static or dynamic operating forces acting on the joint do not cause any relative movement between the parts.

When a fastener is tightened or loosened, a screwing motion takes place. In one full rotation of the fastener, an axial shift corresponding to the pitch P occurs. If a fastener line is uncoiled on a cylinder that has a flank diameter d_2, it will produce a straight line with a pitch angle $\varphi = \arctan(P/(\pi d_2))$.

In general, fasteners are right-threaded (the straight line rises to the right). Special applications may require left-threaded bolts.

For normal fixing bolts or screws, metric thread profiles (DIN 13, ISO 965) are used. For pipes, fittings, threaded flanges, etc., pipe threads to DIN ISO 228-1 or DIN 2999 are used.

Table 1: Symbols and units

Quantity		Unit
A	Cross-sectional area	mm^2
A_S	Stress area	mm^2
D_{Km}	Effective diameter for friction torque in fastener-head or nut bearing surface	mm
E	Modulus of elasticity	N/mm^2
F_A	Axial operating force	N
F_K	Clamping force	N
F_M	Assembly preload	N
F_N	Normal force	N
F'_N	Normal-force component in plane force polygon	N
F_{PA}	Additional plate force	N
F_Q	Transverse force, operating force applied perpendicular to fastener axis	N
F_S	Fastener force	N
F_{SA}	Additional fastener force	N
F_V	Preload	N
F_z	Loss of preload due to settling	N
M_A	Tightening torque	Nm
M_G	Effective tightening-torque component in thread	Nm
M_{KR}	Head friction torque	Nm
M_L	Release torque	Nm
P	Thread pitch	mm
R_e	Yield point	N/mm^2
$R_{p0.2}$	0.2 % yield strength	N/mm^2
R_P	Spring rate of stressed parts (plates)	N/mm
R_S	Spring rate of fastener	N/mm
R_z	Surface roughness	µm
W_t	Section modulus against torsion	mm^3
d	Nominal thread diameter	mm
d_2	Thread flank diameter	mm
d_3	Thread root diameter	mm
d_h	Hole diameter of stressed parts	mm
d_w	Outside diameter of flat fastener-head or nut bearing surface	mm
f_A	Elastic linear expansion by F_A	mm
f_{PV}	Elastic linear expansion of stressed parts by F_V	mm
f_{SV}	Elastic linear expansion of fastener by F_V	mm
f_z	Contact stress	mm
i	Friction-surface pairs	–
l	Length	mm
m	Nut height or thread reach	mm
n	Force application factor	–
n_S	Number of fasteners	–
α	Flank angle of thread	°
α_A	Tightening factor	–
μ_G	Coefficient of friction in thread	–
μ_K	Coefficient of friction in head bearing surface	–
μ_T	Coefficient of friction in parting line	–
μ'_G	Apparent coefficient of friction in thread	–
ρ'_G	Angle of friction to μ'_G	–
σ_a	Alternating stress on fastener	N/mm^2
σ_A	Permissible variable stress	N/mm^2
$\sigma_{red,B}$	Reduced stress in operating state	N/mm^2
$\sigma_{red,M}$	Reduced stress in fitted state	N/mm^2
$\sigma_{z,M}$	Max. tensile stress in fitted state	N/mm^2
σ_z	Max. tensile stress in operating state	N/mm^2
τ_t	Max. torsional stress in thread	N/mm^2
φ	Pitch angle	°
Φ	Force ratio	–
Φ_n	Force ratio at $n < 1$	–

Calculations for threaded fasteners
The prevailing basis for calculating high-stress threaded fasteners is VDI Directive 2230. It can be used as a simple, sufficiently precise calculation for a cylindrical single-threaded fastener, which can be considered as a section of a highly rigid multiple-threaded fastener. This means that, in many cases, even complex multiple-threaded fasteners can be considered as single-threaded fasteners. The precondition for this is that the fastener axes are parallel to each other and perpendicular to the parting planes. The components must also be elastic. Another important factor here is that only centrally preloaded and centrally stressed fasteners are considered. For large, eccentric stresses, which may cause the parting line to gape open, please refer to VDI 2230 (Figure 1).

Property classes
According to DIN EN 20898, the properties of a fastener are identified by two numbers separated by a decimal point. The first number equals 1/100th of the minimum tensile strength; the second is a number that is 10 times the ratio of the yield point in relation to the tensile strength. Multiplying the two numbers gives 1/10 of the minimum yield point (example: 8.8 → $R_e = R_{p0.2} = 640$ MPa).

The property class of a standard nut is identified by one number. This number corresponds to 1/100 of the minimum tensile strength of a bolt of the same property class. To optimize material exploitation, therefore, bolts and nuts of equal property classes should always be paired (e.g. bolt 10.9 with nut 10).

Tightening of threaded fasteners

Prestress
Threaded fasteners are preloaded joints in which the fastener is expanded by f_{SV} by being tightened, and the parts or plates to be tightened are pressed together by f_{PV}. The deformation depends on the dimensions (cross-section and length) and on the materials (moduli of elasticity). According to Hooke's Law, in the elastic range, these are proportional to the prevailing linear force. The ratio of force F to change in length f is the spring rate

$$R = \frac{F}{f} = \frac{E \cdot A}{l}.$$

If the rigidities of fasteners and stressed parts are known (they can be calculated according to VDI 2230), the preloaded threaded fastener can be represented by a load-extension diagram (Figure 2). After assembly, an equilibrium of forces arises so that the preload in the bolt and the stressed parts are of identical size.

Figure 1: Stresses applied to threaded fasteners
a) Centrally stressed and centrally loaded fastener,
b) Eccentrically stressed and eccentrically loaded fastener,
c) Multiple-fastener joint.
1 Area of pressure application.

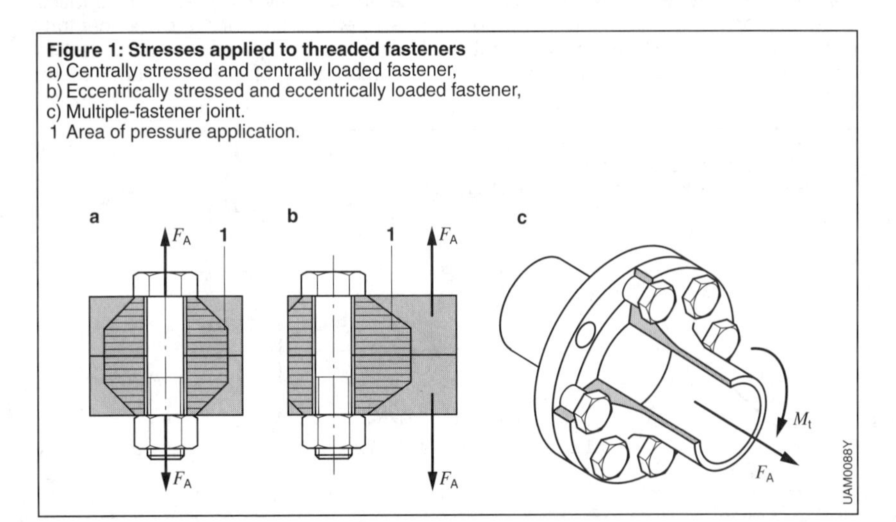

Operating forces

In the case of transversely stressed threaded fasteners (operating force F_Q perpendicular to fastener axis), the forces are transferred in the parting plane by friction. Provided that the frictional forces, which are generated by the fastener preload, are greater than the operating forces to be transferred, the assembly load-extension diagram does not change. This means that the fastener "notices" nothing about the external stress.

At a coefficient of friction μ_T in the parting line, n_S number of fasteners, i number of friction surface pairs, and S_R factor of safety against sliding, the required minimum clamping force is calculated as follows:

$$F_{K,min} = F_V \geq \frac{S_R F_Q}{\mu_T n_S i} .$$

If an external operating force F_A acts in the direction of the fastener axis according to Figure 1, the bolt is extended by f_A. At the same time the compression of the stressed parts is reduced by the same amount. The bolt is thus subjected to an additional stress of F_{SA}, whereas the stressed parts are relieved by F_{PA}. The additional fastener force F_{SA} thus depends on the rigidity and yield of the fastener.

The "softer" the fastener (tensile bolt: long and thin), the smaller the additional bolt stress F_{SA} caused by an external axial operating force F_A. This must be the objective, particularly with dynamic operating forces (e.g. cylinder-head bolts).

Force application

The rigidities of the fastener and the stressed parts also depend on force application. If the external axial operating force F_A is applied directly to the fastener head, the force-application factor $n=1$. For the application of force in the parting line $n=0$. Real force applications range between these two extreme values. In this case, only part of the stressed parts is relieved. This hardens the spring rate R_P of the stressed parts, since grip is reduced. The stressed parts of the tightened parts are hammered onto the fastener, which then becomes apparently longer and therefore softer (Figure 3). Small n values therefore result in small additional fastener forces. This has a good impact on the fastener safety factor. However, the clamping force is reduced at the same time, and this is not good for the joint function.

There is no simple method of calculating the force-application factor n, which defines the point of force application. Either n ($0 \leq n \leq 1$) is estimated, or an approximate calculation is made according to VDI 2230. The forces specified in the load-extension diagram can be calculated in accordance with Table 2.

Fastener forces and torques

Calculation model

The simplest way to represent force ratios in a threaded fastener is by concentrating the surface pressure distributed to all thread turns on a single nut element.

Figure 2: Tightening threaded fasteners
a) Load-extension diagram on assembly,
b) Load-extension diagram with axial operating force F_A.

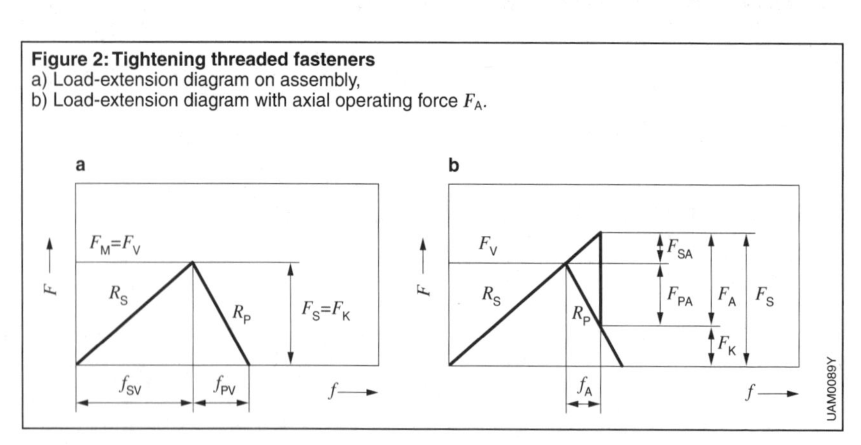

Table 2: Fastener forces (depending on force application)

Force	Force application on screw fastener head $n=1$	Force application any $0<n<1$	Force application in parting line $n=0$
Max. fastener force	$F_S=F_V+\Phi\cdot F_A$	$F_S=F_V+\Phi_n\cdot F_A$	$F_S=F_V$
Clamping force	$F_K=F_V-(1-\Phi)\cdot F_A$	$F_K=F_V-(1-\Phi_n)\cdot F_A$	$F_K=F_S-F_A$
Additional fastener force	$F_{SA}=\Phi\cdot F_A$	$F_{SA}=\Phi_n\cdot F_A$	$F_{SA}=0$
Additional plate force	$F_{PA}=(1-\Phi)\cdot F_A$	$F_{PA}=(1-\Phi_n)\cdot F_A$	$F_{PA}=F_A$

Where $\Phi=R_S/(R_P+R_S)$ and $\Phi_n=n\cdot\Phi$

During tightening and loosening, the nut element moves along the bolt thread, which, if unwound, represents a slanting plane or a wedge.

Tightening a threaded fastener
When tightened, the nut element is pushed up the wedge by the peripheral force F_U. The resulting normal force F_N causes a frictional force F_R which acts in the opposite direction and includes the angle of friction ρ. However, since all standardized thread profiles have inclined flanks, only the component $F'_N=F_N\cdot\cos\alpha/2$ appears in the plane force polygon. The frictional force is calculated as follows:

$$F_R=F_N\cdot\mu_G=F'_N\cdot\mu'_G.$$

In order to calculate the force polygon in a plane parallel to the fastener axis, an apparent coefficient of friction is introduced:

$$\mu'_G=\frac{\mu_G}{\cos\alpha/2}=\tan\rho'$$

When the peripheral force acts on the flank diameter d_2, the thread torque becomes

$$M_G=F_V\cdot\frac{d_2}{2}\cdot\tan(\varphi+\rho').$$

To tighten a bolt to preload F_V, a head friction torque M_{KR} is required to overcome the friction between the head and nut bearing surfaces, in addition to the thread torque M_G. At a coefficient of friction μ_K and a mean head friction diameter D_{Km}, the head friction torque is calculated as follows:

$$M_{KR}=F_V\cdot\mu_K\cdot\frac{D_{Km}}{2}.$$

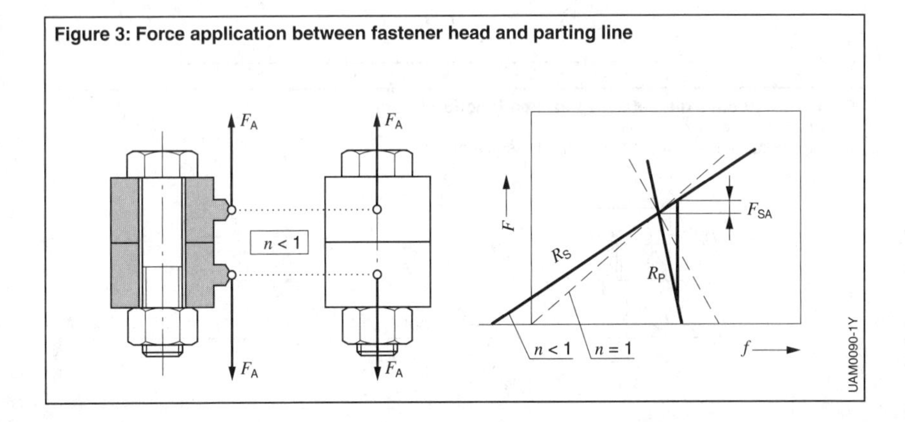

Figure 3: Force application between fastener head and parting line

The bolt tightening torque applied during assembly is then:

$$M_A = M_G + M_{KR} = F_V\left(\frac{d_2}{2}\tan(\varphi+\rho') + \mu_k\frac{D_{km}}{2}\right).$$

Loosening a threaded fastener

When a threaded fastener is loosened, the frictional force changes in the opposite direction to the tightening action. The bolt loosening torque is calculated as follows:

$$M_L = F_V\left(\frac{d_2}{2}\tan(\varphi+\rho') - \mu_k\frac{D_{km}}{2}\right).$$

In the case of self-locking threads ($\varphi < \rho'$), the loosening torque becomes negative. This means that a torque must be applied in the opposite direction to the tightening action.

Design of threaded fasteners

Overstressing

If the minimum thread reach $m = (1.0 \text{ to } 1.5) \cdot d$ is maintained, a threaded fastener fails if overstressed, not because the thread turns are stripped, but because the cylindrical screw bolt ruptures.

Assembly stress

When the fastener is tightened to preload F_V, it is stressed up to tensile stress. Due to the thread torque M_G, it is also stressed to torsion. As the friction in the thread prevents the bolt from turning back, the torsional stress also acts after tightening. According to the theory shape modification energy, the reduced stress in the bolt is

$$\sigma_{red,M} = \sqrt{\sigma_{z,M}^2 + 3\,\tau_t^2} \le v \cdot R_{p0.2}$$

at tensile stress:

$$\sigma_{z,M} = \frac{F_{V,max}}{A_S} = \frac{\alpha_A \cdot F_V}{A_S}$$

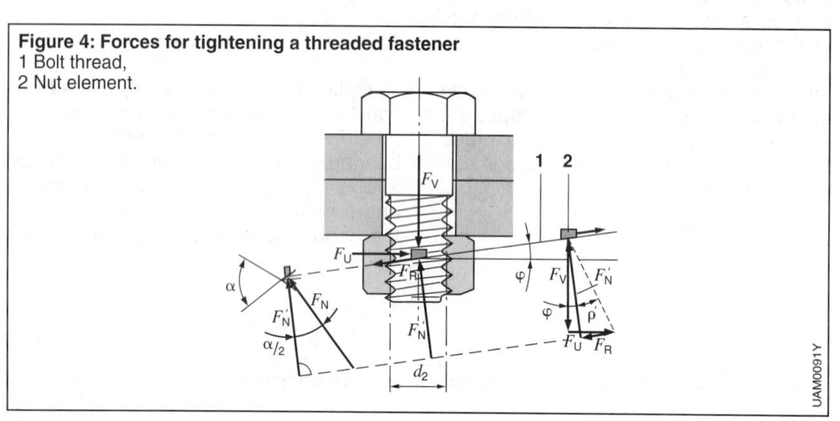

Figure 4: Forces for tightening a threaded fastener
1 Bolt thread,
2 Nut element.

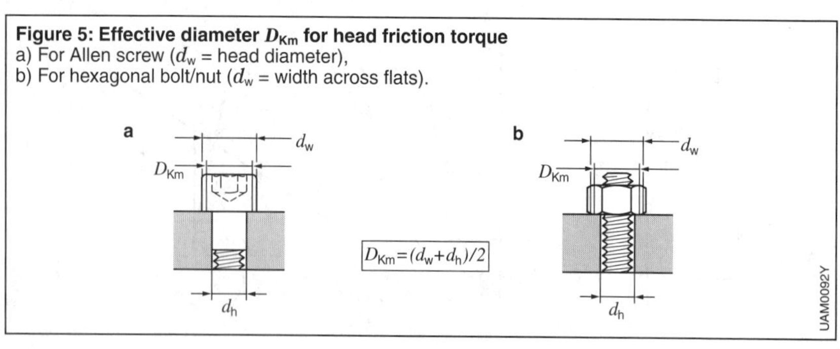

Figure 5: Effective diameter D_{Km} for head friction torque
a) For Allen screw (d_w = head diameter),
b) For hexagonal bolt/nut (d_w = width across flats).

and torsional stress:

$$\tau_t = \frac{M_{G,max}}{W_t} = \frac{16 \cdot \alpha_A \cdot F_V \cdot d_2 \cdot \tan(\varphi + \rho')}{2 \cdot \pi \cdot d_3^3}$$

The tightening factor α_A takes account of the imprecision which is unavoidable during assembly. For torque-controlled tightening (torque wrench) it is $\alpha_A = 1.4$ to 1.6; for pulse-controlled tightening (impact wrench) it is $\alpha_A = 2.5$ to 4.0.

To ensure high functional security, it is necessary to aim for the highest possible material exploitation. This is taken into account by efficiency υ. Table 3 shows the permissible preload forces and tightening torques during assembly for different coefficients of friction at 90% efficiency ($\upsilon = 0.9$) of the standardized minimum yield point.

Static stress

An axial operating force F_A increases tensile stress in the bolt. Since, in the operating state, the effect of torsional stress is less than in the fitted state, VDI 2230 applies to reduced stress:

$$\sigma_{red,B} = \sqrt{\sigma_z^2 + 3(0.5 \cdot \tau_t)^2} < R_{p0.2}$$

at tensile stress:

$$\sigma_z = \frac{F_{S,max}}{A_S} = \frac{\alpha_A \cdot F_V + F_{SA}}{A_S}.$$

Vibrational stress

At a dynamic operating force F_A, the variable stress component σ_a must not exceed the permissible variable stress σ_A.

Table 3: Permissible preloads and tightening torques for regular bolts (acc. to DIN 2230)

Thread	Assembly preload F_V for different coefficients of friction μ in thread						Tightening torque M_A with thread friction $\mu = 0.12$					
	F_V in $10^3 \cdot$ N for $\mu = 0.1$			F_V in $10^3 \cdot$ N for $\mu = 0.2$			M_A in Nm where $\mu_K = 0.1$			M_A in Nm where $\mu_K = 0.2$		
	8.8	10.9	12.9	8.8	10.9	12.9	8.8	10.9	12.9	8.8	10.9	12.9
M4	4.5	6.7	7.8	3.9	5.7	6.7	2.6	3.9	4.5	4.1	6.0	7.0
M5	7.4	10.8	12.7	6.4	9.4	11.0	5.2	7.6	8.9	8.1	11.9	14.0
M6	10.4	15.3	17.9	9.0	13.2	15.5	9.0	13.2	15.4	14.1	20.7	24.2
M8	19.1	28.0	32.8	16.5	24.3	28.4	21.6	31.8	37.2	34.3	50.3	58.9
M10	30.3	44.5	52.1	26.3	38.6	45.2	43	63	73	68	100	116
M12	44.1	64.8	75.9	38.3	56.3	65.8	73	108	126	117	172	201
M14	60.6	88.9	104.1	52.6	77.2	90.4	117	172	201	187	274	321
M16	82.9	121.7	142.4	72.2	106.1	124.1	180	264	309	291	428	501
M18	104	149	174	91.0	129	151	259	369	432	415	592	692
M20	134	190	223	116	166	194	363	517	605	588	838	980
M22	166	237	277	145	207	242	495	704	824	808	1,151	1,347
M24	192	274	320	168	239	279	625	890	1,041	1,011	1,440	1,685
M27	252	359	420	220	314	367	915	1,304	1,526	1,498	2,134	2,497
M30	307	437	511	268	382	447	1,246	1,775	2,077	2,931	2,893	3,386

The following applies:

$$\sigma_a = \frac{F_{S,max} - F_{S,min}}{2 \cdot A_S} \leq \sigma_A .$$

The permissible variable stress σ_A does not depend on the property class, but on the nominal diameter only (Table 4).

Surface pressure
between head and nut bearing surface
In the case of a large preload, the surface pressure on the head and nut bearing surfaces must be checked. Excessive surface pressures can cause plastic deformation and loss of preload. This can result in threaded fasteners coming loose.

The surface pressure p resulting from the maximum fastener force must not exceed the permissible surface pressure limit p_G (guideline values for p_G, see Table 5):

$$p = \frac{4 \cdot F_{S,max}}{\pi \cdot (d_w^2 - d_h^2)} \leq p_G .$$

Threaded-fastener locking devices
The spontaneous loosening of a threaded fastener is caused by complete or partial loss of preload, which in turn is caused by settling events (loosening) or relative movements in the parting line (unscrewing).

Loosening
Contact stress f_z caused by plastic deformation results in loss of preload

$$F_z = \frac{R_p \cdot R_S}{R_p + R_S} \cdot f_z .$$

Contact stress f_z in µm depends on the surface properties and the number of parting lines (Table 6).

Total contact stress equals the sum of individual component parts. However, contact stress determined in this way is only valid if the surface pressure limits are not exceeded. Otherwise, significantly greater settling will arise. Locking devices are intended to reduce or compensate for settling.

The following measures provide reliable protection against loosening:
- High preload
- Elastic threaded fasteners
- Low surface pressures due to large bearing surfaces and sufficient thread reach
- Low number of parting lines
- No plastic or quasi-elastic elements (e.g. seals) with stress

Table 4: Permissible variable stress σ_A

Diameter range	M6 to M8	M10 to M18	M20 to M30
Permissible variable stress σ_A in N/mm²	60	50	40

Table 5: Standard values for surface pressure limit p_G as per VDI 2230

Material	Surface pressure limit p_G in N/mm²
GD-AlSi 9 Cu 3	290
S 235 J	490
E 295	710
EN-GJL-250	850
34 CrNiMo 6	1,080

Table 6: Contact stress f_z depending on surface properties and number of parting lines

Surface	Load	Contact stress f_z in µm		
		in thread	per head or nut bearing surface	per parting line
$R_z < 10$	Tension/Compression Shear	3.0 3.0	2.5 3.0	1.5 2.0
$10 \leq R_z < 40$	Tension/Compression Shear	3.0 3.0	3.0 4.5	2.0 2.5
$40 \leq R_z < 160$	Tension/Compression Shear	3.0 3.0	4.0 6.5	3.0 3.5

Unscrewing

Dynamic stresses, particularly perpendicular to the fastener axis, may cause threaded fasteners to come loose despite sufficient preload. If transverse movements can occur, locking devices to prevent unscrewing insure that the joint retains its function. Suitable measures include:

- Avoid transverse movements by positive locking in the parting line
- Elastic threaded fasteners
- Large grip lengths
- High preload
- Suitable locking devices (locking or bonding elements, Figure 6).

References

[1] DIN-Taschenbuch 10: Mechanische Verbindungselemente – Schrauben (Mechanical threaded fasteners – Screws and bolts), Beuth-Verlag 2001.

[2] DIN-Taschenbuch 45: Gewindenormen (Thread standards), Beuth-Verlag 2000.

[3] DIN-Taschenbuch 140: Mechanische Verbindungselemente – Muttern, Zubehörteile für Schraubenverbindungen (Mechanical threaded fasteners – Nuts, accessories for threaded fasteners), Beuth-Verlag 2001.

[4] Haberhauer/Bodenstein: Maschinenelemente, 15th Edition, Springer-Verlag 2008.

[5] VDI Directive 2230: Systematische Berechnung hochbeanspruchter Schraubenverbindungen, VDI-Verlag 2003.

[6] Wiegand/Kloos/Thomala: Schraubenverbindungen, Springer-Verlag 1988.

Figure 6: Locking devices (examples)
a) Flat-headed self-locking screw,
b) Flat-headed self-locking nut,
c) Locking washer pair.

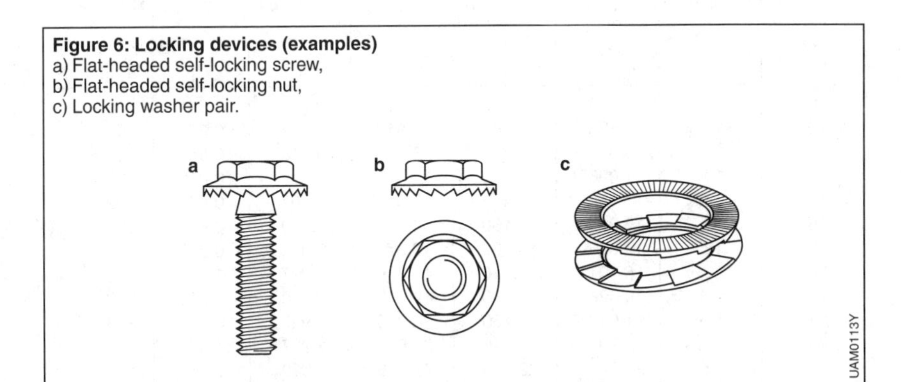

Thread selection

Figure 7: Metric ISO thread
(DIN 13, ISO 965); nominal dimensions.

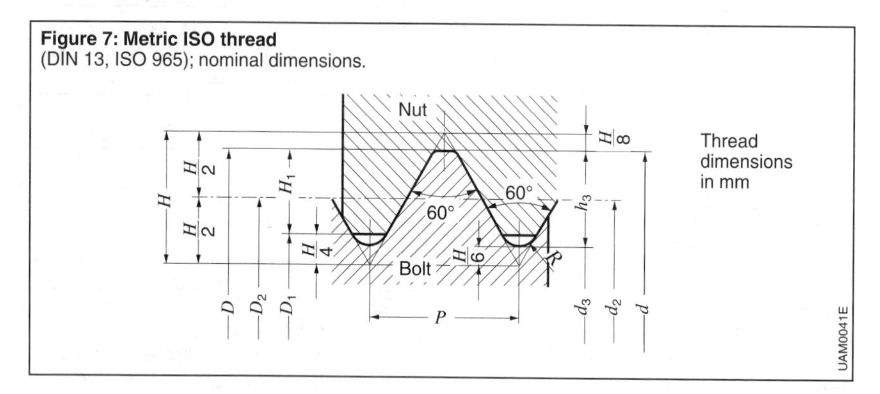

Table 7: Metric standard thread
Designation example: M8 (nominal thread diameter 8 mm).

Nominal thread dia.	Pitch	Pitch dia.	Minor dia.		Thread depth		Stress area
$d = D$	P	$d_2 = D_2$	d_3	D_1	h_3	H_1	A_s in mm²
3	0.5	2.675	2.387	2.459	0.307	0.271	5.03
4	0.7	3.545	3.141	3.242	0.429	0.379	8.78
5	0.8	4.480	4.019	4.134	0.491	0.433	14.2
6	1	5.350	4.773	4.917	0.613	0.541	20.1
8	1.25	7.188	6.466	6.647	0.767	0.677	36.6
10	1.5	9.026	8.160	8.376	0.920	0.812	58.0
12	1.75	10.863	9.853	10.106	1.074	0.947	84.3
14	2	12.701	11.546	11.835	1.227	1.083	115
16	2	14.701	13.546	13.835	1.227	1.083	157
20	2.5	18.376	16.933	17.294	1.534	1.353	245
24	3	22.051	20.319	20.752	1.840	1.624	353

Table 8: Metric fine thread
Designation example: M8 x 1 (nominal thread diameter 8 mm and pitch 1 mm).

Nominal thread dia.	Pitch	Pitch dia.	Minor dia.		Thread depth		Stress area
$d = D$	P	$d_2 = D_2$	d_3	D_1	h_3	H_1	A_s in mm²
8	1	7.350	6.773	6.917	0.613	0.541	39.2
10	1.25	9.188	8.466	8.647	0.767	0.677	61.2
10	1	9.350	8.773	8.917	0.613	0.541	64.5
12	1.5	11.026	10.160	10.376	0.920	0.812	88.1
12	1.25	11.188	10.466	10.647	0.767	0.677	92.1
16	1.5	15.026	14.160	14.376	0.920	0.812	167
18	1.5	17.026	16.160	16.376	0.920	0.812	216
20	2	18.701	17.546	17.835	1.227	1.083	258
20	1.5	19.026	18.160	18.376	0.920	0.812	272
22	1.5	21.026	20.160	20.376	0.920	0.812	333
24	2	22.701	21.546	21.835	1.227	1.083	384
24	1.5	23.026	22.160	22.376	0.920	0.812	401

Figure 8: Pipe threads for non self-sealing joints
(DIN ISO 228-1); parallel internal threads and external threads; nominal dimensions

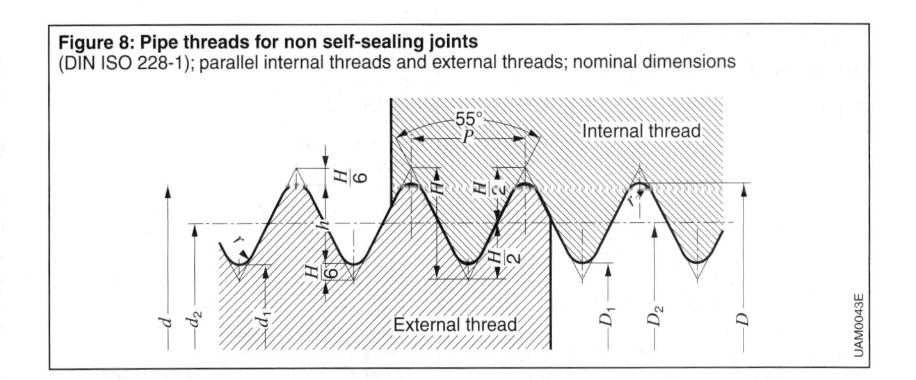

Table 9: Designation example: G1/2 (nominal thread size 1/2)

Nominal thread size	Number of threads per inch	Pitch P mm	Thread depth h mm	Major diameter $d = D$ mm	Pitch diameter $d_2 = D_2$ mm	Minor diameter $d_1 = D_1$ mm
1/4	19	1.337	0.856	13.157	12.301	11.445
3/8	19	1.337	0.856	16.662	15.806	14.950
1/2	14	1.814	1.162	20.955	19.793	18.631
3/4	14	1.814	1.162	26.441	25.279	24.117
1	11	2.309	1.479	33.249	31.770	30.291

Figure 9: Whitworth pipe threads for threaded pipes and fittings
(DIN 2999); parallel internal threads and tapered external threads; nominal dimensions (mm)

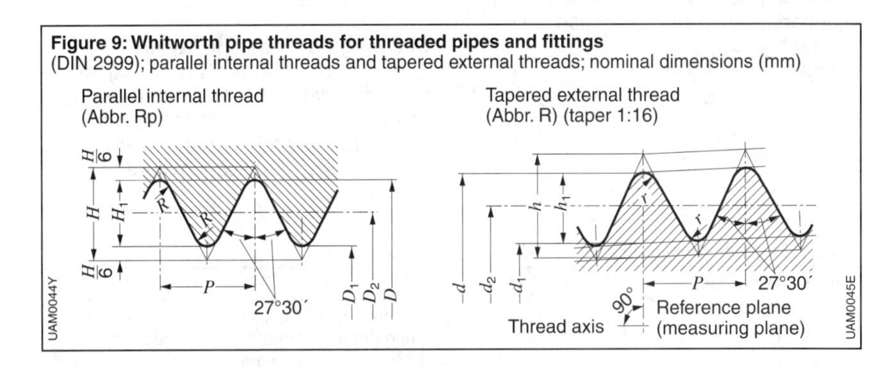

Table 10: Whitworth pipe threads for threaded pipes and fittings

Code		Major diameter	Pitch diameter	Minor diameter	Pitch	Number of threads per inch
External thread	Internal thread	$d = D$	$d_2 = D_2$	$d_1 = D_1$	P	Z
R 1/4	Rp 1/4	13.157	12.301	11.445	1.337	19
R 3/8	Rp 3/8	16.662	15.806	14.950	1.337	19
R 1/2	Rp 1/2	20.955	19.793	18.631	1.814	14
R 3/4	Rp 3/4	26.441	25.279	24.117	1.814	14
R 1	Rp 1	33.249	31.770	30.291	2.309	11

Areas of application: For joining parallel internal threads to valves and fittings, threaded flanges, etc. with tapered external threads.

Snap-on connections on plastic components

Features

Snap-on connections are a cheap and efficient way of fitting plastic components. They are used to connect housing halves, on plug connectors, and to secure mounting parts in plastic housings. They exploit the high expansibility of the plastics with relatively low rigidity.

All snap-on connections are characterized by the brief excursion of a resilient element in the joining process before it snaps into place behind a locating lug. Depending on the configuration of the joining angles at the snap-on elements, it is possible to produce nondestructively detachable and nondetachable connections (Figure 1).

The basic shapes of snap-on connections are (Table 1):

- Resilient snap-on hooks (bending springs fixed on one side)
- Resilient clips
- Ring-shaped snap-on connections, also segmented (slotted lengthways)
- Spherical snap-on connections, also segmented
- Torsion snap-on hooks

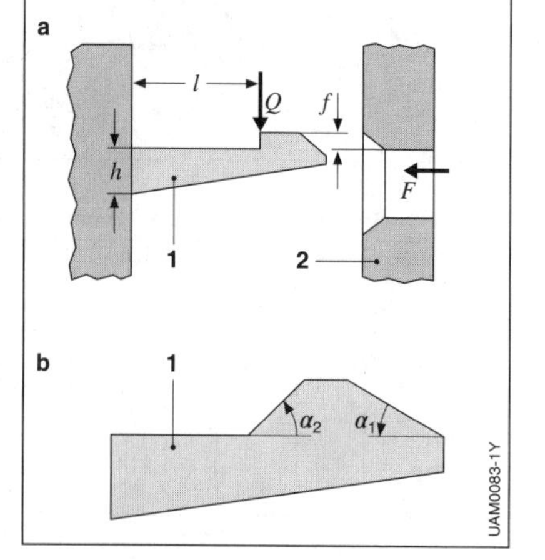

Figure 1: Snap-on connection (principle)
a) Decisive variables,
b) Joining and release angles (detachable connection: $\alpha_2 < 90°$, nondetachable connection: $\alpha_2 \geq 90°$).
1 Spring element, 2 Locating lug.
f Spring travel (rear section), l Length,
h Thickness at fixation cross-section,
F Joining force, Q Excursion force,
α_1 Joining angle, α_2 Release angle.

Design guidelines and layout

The spring elements are designed to accommodate the permitted elongation of the plastic in the joining process. The least favorable material condition must be taken into account here (e.g. dry polyamide).

Table 1: Basic shapes and types of snap-on connection

Shape	Hook shape			Ring shape		
				Annular ring, ring groove	Annular ring segmented, ring groove	Hollow-sphere section
Spring element	Bending spring	Torsion spring (+ bending spring)	Catch spring	Annular spring	Annular spring, segmented	Annular spring
Designation	(Bending) snap-on hook	Torsion snap-on hook	Resilient clip	Ring snap-on element	Ring snap-on connection	Spherical snap-on element
Type	UAM0084_1Y	UAM0084_2Y	UAM0084_3Y	UAM0084_4Y	UAM0084_5Y	UAM0084_6Y

The elongation-dependent secant modulus

$$E_s = \frac{\sigma_1}{\varepsilon_1}$$

is used as the modulus of elasticity (Figure 2). The modulus values of the different plastics can be called up, for example, from the CAMPUS database (http://www.campusplastics.com).

In order to achieve a uniform distribution of strain and optimum material utilization in the bending range of the spring elements, the thickness from the root to the free end should decrease by half. As an alternative, the width to the end of the hook can be reduced to one quarter. Radii at the point of connection of the spring element to the component can help to eliminate concentrations of strain.

When joined, the spring element must be fully returned to its initial state in order to prevent creeping under load and thus permanent deformation. Tensile stress in the snap-on element as a result of operational forces is permitted.

The permitted excursion (spring travel f) in joining is dependent on the geometry of the snap-on hook and the permitted elongation ε of the plastic (see Table 2). Formulas for different cross-sectional shapes can be taken from the relevant technical literature or are provided by special calculation programs.

Figure 2: Determination of secant modulus E_s
$E_{S1} = \sigma_1/\varepsilon_1$, E_0 Modulus of elasticity.

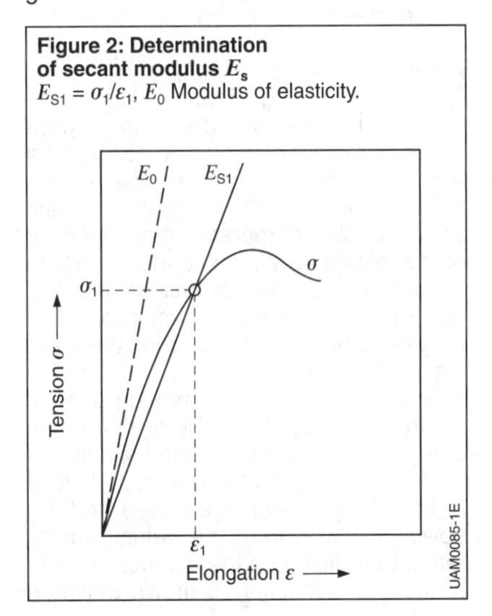

Included in the calculation of the excursion force Q are the rigidity of the plastic as the secant modulus E_s and the geometry as the bending moment/section modulus W.

The joining force F is calculated from the excursion force Q, the joining angle α_1 (usually 30°), and the coefficient of friction μ between the joining components according to the formula:

$$F = \frac{Q\,(\mu + \tan\alpha_1)}{(1 - \mu\tan\alpha_1)}\,.$$

The release force of a snap-on connection is calculated according to the same formula as the joining forced, where the release angle α_2 of the snap-on hook (usually 60°) is to be used. In the case of a nondetachable connection ($\alpha_2 \geq 90°$), the thrust capacity of the snap-on arms limits the strength.

Calculation programs

Various plastics manufacturers offer as a service to their customers easy-to-use calculation programs (e.g. "Snaps" from BASF, "FEAsnap" from Bayer, and "Fitcalc" from Ticona). Most of the material data for manufacturers' product ranges are integrated into these programs.

Table 2: Reference values for permitted elongation ε for snap-on connections
(Short-term for one-off joining, frequent activation approx. 60 % thereof).

Material	ε
Thermoplastics, semi-crystalline, unfilled	
PE	0.080
PP	0.060
PA conditioned, POM	0.060
PA dry	0.040
PBT	0.050
Thermoplastics, amorphous, unfilled	
PC	0.030
ABS/SB	0.025
CAB	0.025
PVC	0.020
PS	0.018
Thermoplastics, glass-fiber-filled	
30 % GF-PA conditioned	0.020
30 % GF-PA dry	0.015
30 % GF-PC	0.018
30 % GF-PBT	0.015
30 % GF-ABS	0.012

Permanent joints

Welding

Automotive components and subassemblies are joined using a wide and highly varied range of welding and bonding techniques [1], [2]. Resistance and fusion welding are among the most commonly applied welding methods. The overview in Figure 1 shows the most important resistance-welding procedures used in production technology (for procedure types and symbols, see DIN 1910, Part 100, [3]).

Resistance welding

Resistance spot welding

In resistance spot welding, locally applied electrical current is used to melt the contact surfaces of the parts to be joined into a soft or fluid state. The parts are then joined together under pressure (Figure 2a and 2b). The spot-welding electrodes which conduct the welding current also convey the electrode force to the parts to be joined (joining parts). The amount of heat required to create the welding spot is determined in accordance with the equation

$Q = I^2 R t$ (Joule's Law).

Figure 1: Classification of resistance-welding process in acc. with DIN 1910-100

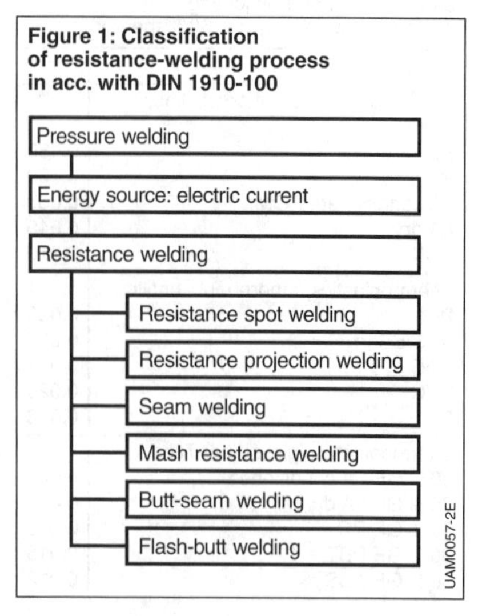

The precise amount of heat required Q is a function of welding-current intensity I, resistance R, and welding time t. A good and adequate spot diameter d_1 can be achieved by the coordination of welding-current intensity I, electrode force F, and welding time t.

According to the manner in which the current is conducted, a distinction is drawn between bilateral direct resistance spot welding (Figure 2a) and unilateral indirect resistance spot welding (Figure 2b).

The electrode for a specific spot welding operation is selected with reference to shape, outside diameter and point diameter. Since the joining parts should be as free as possible of scaling, oxides, paints, grease, and oils, they receive where necessary appropriate surface pretreatment prior to welding.

Applications:
- Joining of sheet parts up to an approximate 3 mm single sheet thickness in the lap joint or as a welding flange.
- Multiple-sheet as well as two-sheet joints of different sheet thicknesses and sheet materials.
- Spot-weld bonding in combination with adhesive.

Resistance projection welding

Projection welding (Figure 2c) is a process in which electrodes with a large surface area are employed to conduct the welding current and the electrode force to the workpiece. The projections, which are generally incorporated into the thicker of the joining parts, cause the current to concentrate at the contact points. The electrode force levels the projections partially or completely again during the welding process. A permanent, inseparable joint is produced at the welding seams, i.e. the contact points. One or more than one projection can be welded simultaneously, depending on the type of projection (round, longitudinal or annular) and the power available from the welding apparatus. Depending on the number of spots welded, a distinction is drawn between single-projection welding and multiple-projection welding.

Resistance projection welding requires high welding currents applied for short periods of time.

Applications:
- Welding parts of different thicknesses.
- Welding multiple projections in a single operation.

Seam welding
In this process, roller electrodes replace the spot-welding electrodes used in resistance spot welding (Figure 2d). Contact between the roller set and the workpiece is limited to an extremely small surface area. The roller electrodes conduct the welding current and apply the electrode force. Their rotation is coordinated with the movement of the part.

Applications:
- Production of sealed welds or seam spot welds (e.g. fuel tanks).

Flash-butt welding
In flash-butt welding, the butt ends of the workpieces are joined under light pressure (Figure 2e) while the flow of current at the contact surfaces produces localized heat and melting (high current density) (current supply via copper jaws). The metal's vapor pressure drives molten material from the contact patches (burn-off) while force is applied to form an upset butt weld.

The butt ends should be parallel to each other and at right angles to the direction in which the force is applied (or virtually so). Smooth surfaces are not required. A certain amount of extra length must be factored in to compensate for the losses incurred in the flash-butt welding process. The result is a weld with the characteristic projecting seam.

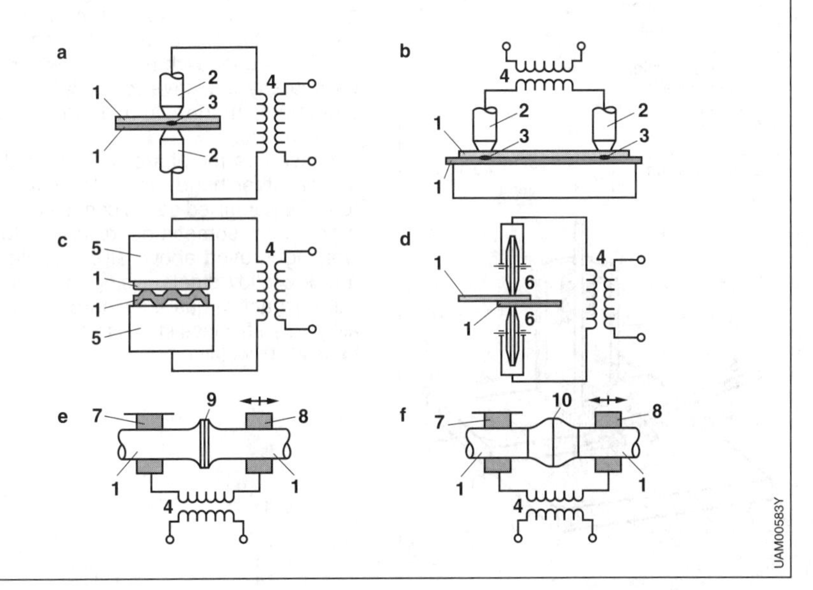

Figure 2: Resistance-welding processes
a) Bilateral resistance spot welding, b) Unilateral resistance spot welding,
c) Resistance projection welding, d) Seam welding, e) Flash-butt welding, f) Butt-seam welding.
1 Part to be joined, 2 Spot-welding electrodes, 3 Welding spot,
4 Transformer, 5 Large-surface electrodes,
6 Roller electrodes, 7 Fixed copper jaws, 8 Longitudinally movable copper jaws,
9 Weld joint with seam, 10 Weld joint with bead.

Applications:
- Joining in the butt joint, e.g. rims, link chains.
- Workshop procedures, e.g. for saw bands.

Butt-seam welding
This process (Figure 2f) employs copper jaws to conduct the welding current to the workpieces to be joined. When the welding temperature is reached, the current switches off. Constant pressure is maintained and the workpieces then weld together (requirement: properly machined butting faces). The process does not completely displace contamination that may be present at the butt ends. The result is a weld with the characteristic projecting bead.

Applications:
- Joining in the butt joint, e.g. shafts, axles.

Figure 3: Principle of tungsten inert-gas welding
1 Tungsten electrode,
2 Current contact tube,
3 Inert gas,
4 Inert-gas nozzle,
5 Welding filler,
6 Weld seam,
7 Arc,
8 Workpieces,
9 Energy source.

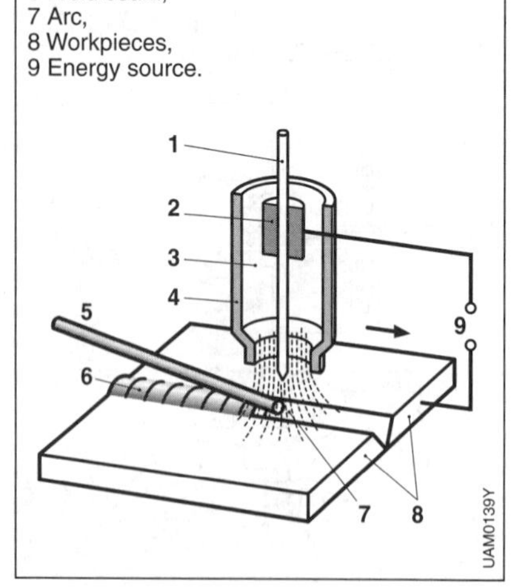

Fusion welding
The term "fusion welding" describes a process employing limited local application of heat to melt and join the parts. No pressure is applied. Shielded (inert gas) arc welding is a type of fusion welding. The electrical arc extending between the electrode and the workpiece serves as the heat source. Meanwhile, a layer of inert gas shields the arc and the melted area from the atmosphere. The type of electrode is the factor which distinguishes between various techniques:

Tungsten inert-gas welding
In this process, an arc is maintained between the workpiece and a stable, non-melting tungsten electrode. The shielding (inert) gas is argon or helium. The rod-shaped weld metal is supplied from the side (Figure 3).

Gas-shielded metal-arc welding
In gas-shielded metal-arc welding, an arc is maintained between the melting end of the wire electrode (material feed) and the workpiece. The welding current flows to the wire electrode via a current-contact nozzle in the welding torch. Inert-gas metal-arc welding (MIG welding) uses inert gases (slowly reacting and noble gases such as argon, helium, or combinations of the two) as protective gas. MIG welding is used with materials which are particularly sensitive to oxidation, e.g. aluminum, magnesium, titanium, and nickel alloys.

Active-gas metal-arc (MAG) welding, on the other hand, uses an active gas (e.g. CO_2 or mixed gases containing CO_2, argon and sometimes oxygen). MAG welding is used above all for unalloyed and low-alloy steels. Using inert gases with low active-gas admixtures for high-alloy, e.g. stainless steels is also referred to as MAG welding.

Laser-beam welding
Laser-beam welding (or laser welding for short) involves the use of light as the energy source to melt on the workpieces to be welded. The monochromatic laser radiation is generated in a laser source, where the wavelength is determined by the respective excitation medium. CO_2 lasers and NdYAG lasers (solid-state lasers) are used in industrial practice [5], [6]. Newer developments are diode lasers and fiber lasers.

Depending on the wavelength, tube guides with refraction mirrors (CO_2 lasers) or optical fibers (e.g. NdYAG lasers) are required to direct the beam from the beam generator to the weld (Figure 4). To utilize the energy of the laser beam for welding, it is necessary to implement focusing with mirror or lens systems. This allows particularly high energy flux densities to be achieved at the weld, resulting in the deep-welding effect, which provides for particularly deep, but at the same time, narrow seams. In the simplest case welding is performed without filler metal.

Feeding the workpiece relative to the focusing optics, the optics relative to the workpiece or a combination of the two is used to direct the beam along the seam. Systems with optical fibers are particularly well suited to moving the focusing optics, e.g. during robot-controlled welding of three-dimensional seam geometries.

In the case of remote welding the beam is directed over the workpiece from a relatively large distance (long focal length of the focusing optics) by moving the focusing mirror or the focusing lens inside the focusing optics.

It is possible to operate several machining stations with one beam generator by using beam switches.

Figure 4: Principle of laser-beam welding
a) Beam direction by mirror,
b) Beam direction by optical fiber.
1 Beam generator,
2 Beam direction by refraction mirror,
3 Beam direction by optical fiber,
4 Focusing optics, 5 Focused laser beam,
6 Workpieces (butt joint) with weld seam,
7 Workpieces (lap joint) with weld seam.

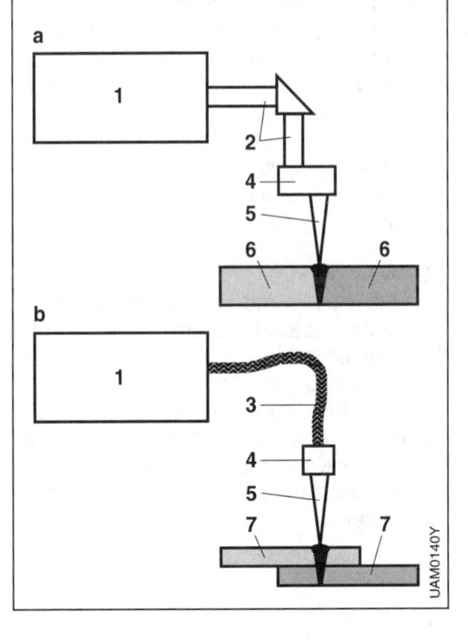

Applications:
- Joining of steel sheets using a lap joint in body manufacturing.
- Joining of unalloyed and low-alloy steels as butt joint in chassis and assembly components.
- Joining of high-alloy steels in exhaust-gas systems.
- Joints in seat systems.
- Joining of aluminum alloys (with filler metal).

Other welding processes
The following welding processes are also used in the automotive industry [5], [6]:
- Electron-beam welding
- Friction welding
- Arc pressure welding (stud welding)
- Stored-energy welding (pulsed-current arc welding)

Soldering

In soldering, a supplementary material (solder) is melted onto two or more parts of similar or varying metallic composition in order to produce a permanent connection between them. Flux or protective gas is also used [7], [8], [9].

Fluxes (non-metallic substances) are applied to remove oxide layers from the surfaces of the soldering points after cleaning and to prevent new layers from forming: This makes it possible to apply a consistent coat of solder to the joint surfaces. Information on fluxes can be found in DIN EN 29454 [10] and DIN EN 1045 [11].

The melting temperature of the solder is below that of the parts being joined. The solder is distributed along the join to produce the connection without the parts themselves being melted.

The strength of a soldered joint can be equal to that of the base material itself. To achieve the required strength, it is necessary to have narrow soldering gaps in which the adjacent, higher-strength base material prevents the solder from deforming.

Information on the temperature range, the heat source and the geometry of the joint shape (structure) is used to describe the soldered joints. In terms of the working temperature, a distinction is made between soft soldering and hard soldering. The working temperature is defined as the lowest surface temperature at the connection between the workpieces to be joined at which the solder can be melted and distributed to form a bond. Information on solders can be found in the DIN sheets DIN EN ISO 9453 [12], DIN EN ISO 12224-1 [13], DIN EN ISO 17672 [14].

In terms of the joint shapes, a distinction is basically made between close-joint soldering and V-joint brazing.

Distinction by working temperature

Soft soldering

Soft soldering is employed to form permanent solder joints at melting temperatures below 450 °C (as with soldering tin). Soft solders which melt at temperatures of 200 °C and below are also known as quick solders.

Hard soldering

Hard soldering (brazing) is used to form permanent joins at melting temperatures above 450 °C (as with copper, copper/zinc and silver alloys, e.g. silver brazing filler).

Brazing applications:
- V-joint brazing: steel sheets in body manufacturing, also of different qualities and in the case of large wall-thickness differences.
- Close-joint soldering: radiators, pipes in the assembly area.

Manufacturing methods

The method used for heating provides yet another criterion for describing soldered joints. The most important types are furnace, induction, open-flame, and iron soldering.

Sweating

Heating is performed in through-type or vacuum furnaces with defined temperature and time characteristics. Before entering the furnace, the components are secured and the solder is inserted or applied as a paste.

Induction soldering

Heating is locally limited by inductive means.

Flame soldering

Heating is performed by individual torches or in a gas-heated system. Depending on the specific soldering operation, oxyacetylene burners (familiar from gas welding), propane torches or soldering lamps are employed. The solder is usually supplied from the side in the shape of a rod. The flux is applied separately or integrated in the solder.

Iron soldering
A hand-held or mechanically guided soldering iron provides the heat. Irons can also be used to solder pretinned surfaces.

Further processes
- Salt-bath soldering
- Dip soldering
- Resistance soldering, as well as
- MIG soldering
- Plasma soldering, and
- Laser soldering

Close-joint soldering and V-joint brazing
Furthermore, with regard to the geometry of the solder and the soldering flux, a distinction is made between close-joint soldering and V-joint brazing.

Close-joint soldering
The soldering gap is filled by the capillary effect of the liquid solder. The strength of the soldered joint is determined entirely by the strength of the base materials. Typical manufacturing processes are sweating, flame soldering and induction soldering.

V-joint brazing
The soldering gap is filled by the force of gravity. The strength of the soldered joint is determined above all by the strength of the solder filler. Typical manufacturing processes are arc and laser soldering.

Adhesive technologies

Adhesives
An adhesive is a non-metallic material which can join workpieces (parts to be joined) by means of adhesion and its internal strength (cohesion) without the structure of the part to be joined being significantly altered. An adhesive enables the creation of permanent material joints between homogeneous or heterogeneous material pairings.

Organic and inorganic adhesives as they harden build up firstly adhesion with the material surfaces by way of chemico-physical interactions (adhesion or adhesive forces) and secondly the necessary structural strength (cohesion). Depending on their chemistry, these adhesives harden for example at room temperature, at increased temperatures or under UV radiation with different mechanisms.

Hardening (cross-linking) and adhesion occur, in accordance with the chemical basic structure and formulation, over a certain period of time, depending on for example temperature or air humidity. The reaction types are called polymerization, polyaddition and polycondensation. Spatially interlinked macromolecules are the result. Adhesives can, depending on the required hardening temperature, be subdivided into the category of those that harden at room temperature or at higher temperatures (100 to 200 °C).

Two different types of adhesive are used, based on their recipe and form of presentation: single-component and two-component adhesives.

Two-component adhesives
These are adhesives which are made up of two components which must be mixed together to harden. The stoichiometric mixture ratio of the two components must be maintained exactly. Component A contains the basic resin. The second component (component B) contains a hardener which initiates the cross-linking reaction and links the resin molecules from component A to each other. An accelerator may be added to the hardener. Hardening usually occurs at room tem-

perature, but can by accelerated by the additional influence of slightly increased temperatures (80 to 120 °C) and thus completed more quickly.

Single-component adhesives
Single-component adhesives contain all the constituents necessary to form a bond in one phase. These systems contain inhibitors which prevent a premature chemical reaction (hardening) between the reacting agents present in the one phase (monomers, resin and hardener; premixed two-component adhesive). Single-component adhesives do not require the additional process step of mixing the components as a two-component adhesive requires.

Hardening must, depending on the chemical formulation, be initiated by higher temperatures (in an oven, by induction or infrared radiation), ultraviolet radiation or air humidity, and completed. In this process the inhibitors are neutralized and the accelerators also contained in the adhesive are released to speed up the hardening process.

Most single-component adhesives must be stored under cooled conditions (electrical conductive adhesives down to −20 °C in an upright deep-freezer, structural adhesives and sealants at 4 to 10 °C in a cold warehouse) or in a dark environment (UV adhesives) until they are ready for use.

Structural designs of adhesive connections
Adhesive connections should be designed so that predominantly shear loads occur. Overlapping joints are particularly suitable. Butt joints subjected to tensile or sliding forces should be avoided.

Combinations of adhesive bonding with other joining methods, e.g. welding, screwing or riveting, can have beneficial effects. The joining spots secure the components during the adhesive hardening time. Stress concentrations for example at the edges of weld spots or at rivet spots can also be minimized. Constructions of this nature provide enhanced structural integrity, rigidity, and damping when subjected to dynamic loads.

Examples of adhesives with technical application potential
The most important adhesives are epoxy resins, silicones, polyurethanes, and acrylates.

Silicone and polyurethane adhesives
Silicone and polyurethane adhesives are important particularly in bonds with dynamic loads, but also where tightness against liquid media (e.g. water) is called for. Depending on the medium (polar, nonpolar), there is however a risk of flexible adhesives swelling (e.g. in fuel). This process can be reversed provided the system can be re-dried during operation and the adhesive has not suffered any damage (e.g. material embrittlement by extraction). Thanks to their flexibility (elastic properties), these adhesives can compensate the different coefficients of thermal expansion of the parts to be joined in the temperature-application range by means of their deformability (low modulus of elasticity above a crystallization temperature of −60 °C to −40 °C) such that only low mechanical stresses can be built up in the bond. Compared with polyurethanes, silicones are also highly suitable for use at high application temperatures up to approximately 220 °C.

Epoxy adhesives
Epoxy adhesives, on the other hand, are predominantly very brittle and hard adhesives which can, depending on their formulation, be used up to approximately 200 °C. Compared with flexible adhesives, they are much less susceptible to swelling in liquid media. High-strength epoxy resins lack the ability of silicones to compensate by means of deformability (elongation) mechanical stresses in the adhesive bond which are caused by different coefficients of thermal expansion of the parts to be joined in the temperature-application range. Exceptions are adhesives which, thanks tor their special formulation (e.g. impact-resistant modification), are used specifically for bonds in body manufacturing in order specifically to reduce the mechanical forces and energies that occur in a crash situation and to maintain the structural integrity of the passenger cell. Epoxy adhesives facilitate bonds with

very high adhesive strengths (high modulus of elasticity up to the glass-transition temperature; this is in the range of 80 to 200 °C, depending on the chemical recipe of the adhesive components).

Acrylate adhesives
Acrylates are capable of fast hardening reactions and therefore highly attractive for use in production processes. These adhesives can harden, depending on their chemical architecture, in response to ultraviolet radiation, mixing of components, at increased temperatures or also in response to air humidity ("superglue") within a few minutes down to a matter of seconds. On the other hand, acrylates do not demonstrate as pronounced thermomechanical performance as silicone and epoxy adhesives above approximately 120 °C or in aggressive media.

Automotive applications
Adhesive joining has become a standard technique in automotive engineering. The individual areas of application can be classified as follows:
- Electronic components: housing sealing bonds in sensors, ECUs and video systems; electrically and thermally conductive bonding of electronic component in power modules and electronic circuits.
- Electric motors: magnetic bonds in rotors and stators.
- Body shell: raised-seam and brace bonding for attached components.
- Assembly line: attachment of insulating material, appliqués, moldings, mirror-support bracket to windshield.
- Component production: bonding of brake pads, laminated safety glass (LSG), rubber-metal connections to absorb vibration.

Riveting

Classic riveting

Process
Riveting is used to produce a permanent fixed connection between two or more components made of identical or dissimilar materials. Here the components to be joined are pierced together by means of drilling or punching. The rivet is then inserted into the hole as the joining element. Depending on the method and application, riveted joints are divided into the following categories:
- Permanent rigid connections (interference fits, for example in mechanical and plant engineering)
- Permanent, sealed connections (for example in boilers and pressure vessels)
- Extremely tight seals (for example in pipes, vacuum equipment)

A distinction is drawn between cold and hot riveting, depending on the temperature used. Cold riveting is employed for rivet joints up to 10 mm in diameter in steel, copper, copper alloys, aluminum,

Figure 5: Rivet shapes
a) Mushroom-head rivet,
b) Countersunk-head rivet,
c) Oval countersunk-head rivet,
d) Flat round-head rivet,
e) Hollow rivet,
f) Tubular rivet.

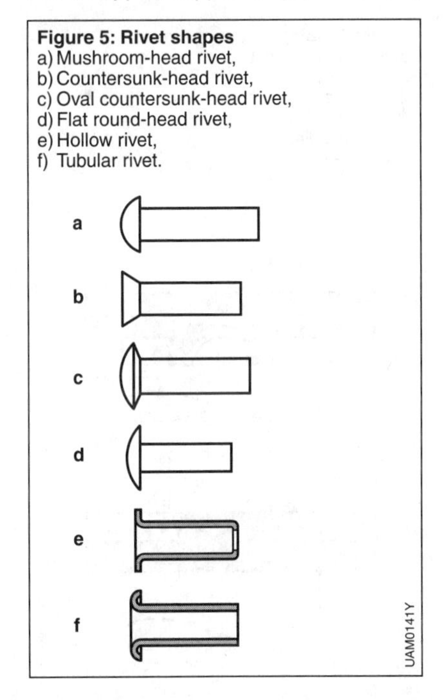

etc. Rivets with diameters in excess of 10 mm are installed hot.

The most common rivet shapes are (Figure 5, [15]...[20]) the mushroom-head rivet (DIN 660), the countersunk-head rivet (DIN 661), the oval countersunk-head rivet (DIN 662), the flat round-head rivet (DIN 674), the hollow rivet (DIN 7339), and the tubular rivet (DIN 7340).

There are also standardized rivets for specialized applications, e.g. explosive rivets or blind rivets. Blind rivets are hollow and are expanded by a drift or punch.

Furthermore, rivets are frequently designed for use as function elements. Examples are rivet nuts and clinch bolts, which serve as bolting points.

The strength properties and the chemical composition of rivet materials are laid down in numerous national and international standards. In the interests of avoiding electrochemical corrosion, it is advisable wherever possible to use materials of the same kind for rivets and component.

In mechanical engineering in general and in tank manufacture riveting has largely been displaced by welding.

Figure 6: Joint with punch rivet
a) Semi-hollow rivet,
b) Solid rivet.
1 Workpieces,
2 Punch rivet,
3 Riveting die,
4 Die plate.

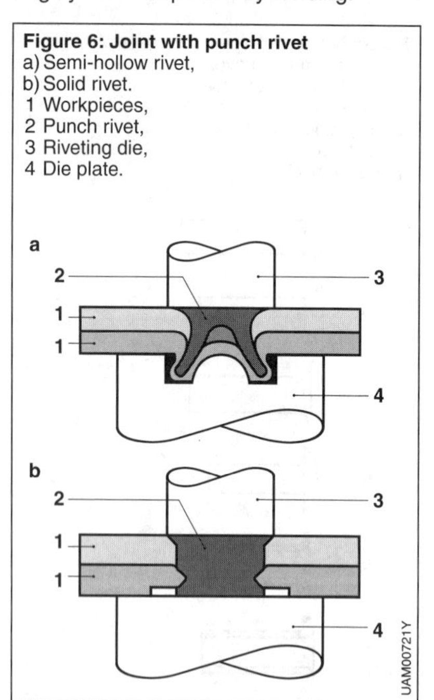

Advantages and disadvantages compared with other joining techniques
- Unlike welding, riveting exerts no influence such as hardening or structural transformation on the material.
- No distortion of the components.
- Suitable for joining dissimilar materials.
- Riveting weakens the components.
- Butt joints cannot be riveted, and
- Riveting is generally more expensive than welding when performed outside the factory.

Automotive applications
- Riveting pivot/joint pins (power-window units, hinges, windshield-wiper linkages)
- Riveting reinforcement plates (in the course of repairs)

Punch riveting

Process

The punch-riveting technique joins solid materials employing stamping and riveting elements (solid or semi-hollow rivets) without piercing in a combined cutting and joining operation. The parts to be joined do not have to be pierced or predrilled, as is the case with other riveting techniques.

Punch riveting with semi-hollow rivets

The first stage in punch riveting with a semi-hollow rivet (Figure 6a) is to position the joining point of the components to be joined on the (bottom) die plate. The riveting die descends and presses the semi-hollow rivet through the upper sheet into the lower sheet in a single stamping operation. The rivet deforms and the bottom spreads to form a securing element, usually without fully penetrating the lower sheet.

Punch riveting with solid rivets

The first stage in punch riveting with a solid rivet (Figure 6b) is to position the joining point of the components to be joined on the (bottom) die plate. The top section of the rivet unit, including the blank holder, descends and the riveting die presses the rivet through the parts to be joined in a single stamping operation.

Equipment
Hydraulic joining equipment is used to produce the joints; in equipment of this type the riveting die and the die plate are arranged in a very rigid, C-shaped frame. The rivets can be supplied to the riveting tool loose or via magazined carrier strips.

Materials
The rivets must be harder than the materials to be joined. The most common materials are steel, stainless steel, copper and aluminum with various coatings.

Features
- Possible to join similar and dissimilar materials (e.g. steel, plastic or aluminum), parts of various thicknesses and strengths, and painted sheets.
- No preliminary piercing or drilling, no heat or vacuum extraction required.
- The total material thickness that can be riveted is 6.5 mm for steel and 11 mm for aluminum.
- The joining process generates minimal heat and noise.
- Tools have a long service life (approx. 300,000 rivet applications) and deliver consistent joint quality over a long period of time.
- High process reliability through monitoring of the process parameters.
- High forces have to be applied.
- Larger tong projections of the riveting tools are only possible under limited conditions owing to the rigidity requirements.

Applications
- Punch riveting with solid rivets: joining metal sheets in automotive engineering, e.g. power-window drives in passenger cars.
- Punch riveting with semi-hollow rivets: joining materials in passenger-car body manufacturing, white goods (household appliances), joining metals to composites (heat shields).

Penetration-clinching processes

Processes
Penetration clinching ("clinching") comprises mechanical joining processes which combine penetration, cold upsetting and sometimes also cutting in a single continuous joining operation without the application of additional heat. Based on this principle, it can be assigned to the process of joining by forming (see DIN 8593-5, [21]).

Distinctions are made between processes with and without cutting and those where the joining point is round or rectangular in shape.

"Tox clinching"
Some process variations are still referred to in technical parlance by the original manufacturer's designations, e.g. "Tox clinching" for penetration clinching with a round die without cutting (Figure 7b). The tools employed for "Tox clinching" are relatively small. The diameter can be varied to suit specific applications. A force-travel curve typical of "Tox clinching" can be broken down into five characteristic phases (A-E) (Figure 8).

Penetration clinching
Penetration clinching (Figure 7a) currently allows blanks up to 3 mm thick to be joined, whereby the total thickness of the two blanks should not exceed 5 mm. The blanks being joined can be of the same material (e.g. steel to steel) or dissimilar materials (e.g. steel to nonferrous metal). Penetration clinching can be used to join coated sheets and painted parts, as well as components to which adhesives have been applied. Multiple penetration clinching can be used to produce numerous elements to be joined (up to 50) in a single process (one stroke of the press, for example).

Advantages and disadvantages of penetration-clinching processes

- No need for an acoustic enclosure.
- Tox clinching does not impair corrosion protection.
- When combined with cutting, there is a partial loss of corrosion protection.
- No distortion due to thermal stress.
- Painted, protected (oil, wax) and glued sheets can be clinched.
- Different materials can be joined (e.g. steel to plastic).
- Energy savings, since no power supply for welding and no cooling-water system are required.
- One side of the workpiece has a projection similar to that produced by a rivet head, while the other side has a corresponding depression.

Automotive applications

- Steel and aluminum bodies
- Windshield-wiper brackets
- Fastening interior door panels
- Hinges, locks
- Seat systems

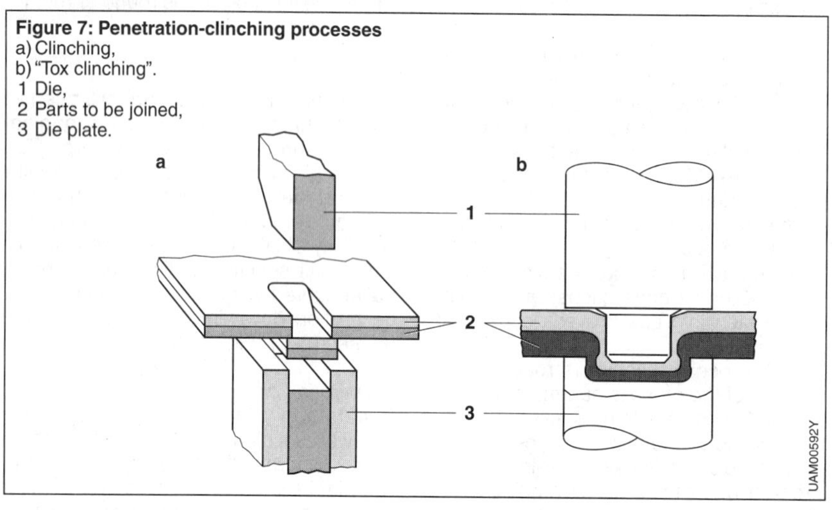

Figure 7: Penetration-clinching processes
a) Clinching,
b) "Tox clinching".
1 Die,
2 Parts to be joined,
3 Die plate.

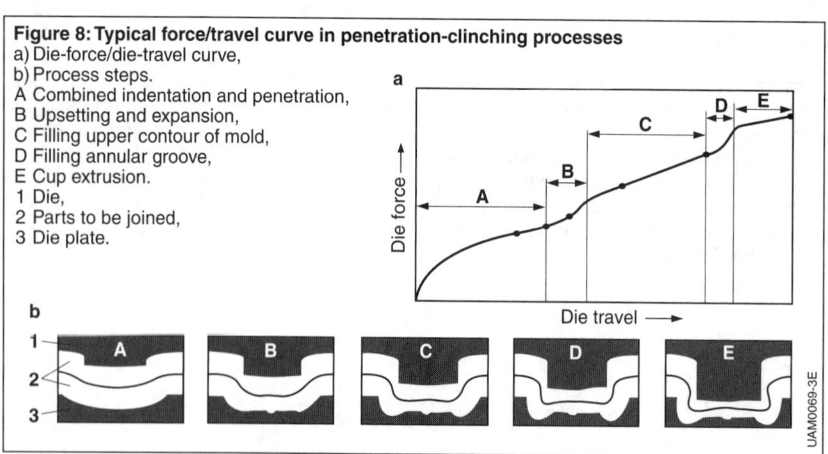

Figure 8: Typical force/travel curve in penetration-clinching processes
a) Die-force/die-travel curve,
b) Process steps.
A Combined indentation and penetration,
B Upsetting and expansion,
C Filling upper contour of mold,
D Filling annular groove,
E Cup extrusion.
1 Die,
2 Parts to be joined,
3 Die plate.

References

[1] Fügetechnik Schweißtechnik; 7th Edition, DVS Media, 2007.
[2] DIN-DVS Taschenbuch 284 – Schweißtechnik 7: Schweißtechnische Fertigung, Schweißverbindungen; 3rd Edition, DVS Media, 2009.
[3] DIN 1910-100:2008: Welding and allied processes – Vocabulary – Part 100: Metal welding processes with additions to DIN EN 14610:2005.
[4] DIN EN 14610:2005: Welding and allied processes – Definitions of metal welding processes; trilingual version EN 14610:2004.
[5] DIN-DVS Taschenbuch 283 – Schweißtechnik 6: Strahlschweißen, Bolzenschweißen, Reibschweißen; 3rd Edition. DVS Media, 2009.
[6] DIN EN 1011: Welding – Recommendations for welding of metallic materials – Part 6 (2006): Laser beam welding; German version EN 1011-6:2005. Part 7 (2004): Electron beam welding; German version EN 1011-7:2004.
[7] DIN-DVS Taschenbuch 196/1 – Schweißtechnik 5: Hartlöten; 5th Edition. DVS Media, 2008.
[8] DIN-DVS Taschenbuch 196/2 – Schweißtechnik 12: Weichlöten, gedruckte Schaltungen; 1st Edition. DVS Media, 2008.
[9] DIN ISO 857-2:2007: Welding and allied processes – Vocabulary – Part 2: Soldering and brazing processes and related terms (ISO 857-2:2005).
[10] DIN EN 29454-1:1994: Soft soldering fluxes; classification and requirements; part 1: classification, labeling and packaging (ISO 9454-1:1990); German version EN 29454-1:1993.
[11] DIN EN 1045:1997: Brazing – Fluxes for brazing – Classification and technical delivery conditions; German version EN 1045:1997.
[12] DIN EN ISO 9453:2006: Soft solder alloys – Chemical compositions and forms (ISO 9453:2006); German version EN ISO 9453:2006.
[13] DIN EN ISO 12224-1:1998: Solid wire, solid and flux cored – Specification and test methods – Part 1: Classification and requirements (ISO 12224-1:1997); German version EN ISO 12224-1:1998
[14] DIN EN ISO 17672:2010: Brazing – Filler metals (ISO 17672:2010); German version EN ISO 17672:2010.
[15] DIN 660:2012: Round head rivets – Nominal diameters 1 mm to 8 mm.
[16] DIN 661:2011: Countersunk head rivets – Nominal diameters 1 mm to 8 mm.
[17] DIN 662:2011: Mushroom head rivets – Nominal diameters 1.6 mm to 6 mm.
[18] DIN 674:2011: Flat round head rivets – Nominal diameters 1.4 mm to 6 mm.
[19] DIN 7339:2011: Hollow rivets, one piece, draw from strip.
[20] DIN 7340:2011: Tubular rivets cut from the tube.
[21] DIN 8593-5:2003: Manufacturing processes joining – Part 5: Joining by forming processes; Classification, subdivision, terms and definitions.

Internal-combustion engines

Thermal engines

Operating principles and concept

Internal-combustion engines are classified as thermal engines. The essential feature of a thermal engine is cycle direction, which is characterized by the output of work.

Contrasting with thermal engines are heat pumps, also called refrigerating machines, which are characterized by cycle direction in the opposite direction and require drive power in order to operate.

The operating principle of thermal engines is always the same. A working medium is compressed, whereupon energy is supplied while the medium is compressed with a corresponding further increase in pressure. This is followed by an expansion with power output. In open cycles the working medium which has performed the work is discharged. In closed cycles the initial state must be re-established by cooling the working medium before compression is restarted in open and closed cycles.

Table 1: Characteristics and functioning principles of important thermal engines

Thermal engine	Steam circuit	Stirling	Steam engine	Gas turbine	Jet engine	Reciprocating piston engine	Wankel rotary engine
Thermodynamic reference cycle	Rankine	Ericson	Steam cycle	Joule	Joule	Seiliger	Seiliger
Typical working medium	H_2O, ethanol	Air, helium	H_2O	Air	Air	Air, air/fuel mixture	Air, air/fuel mixture
Cycle direction	Closed		Open				
Energy input	From outside by heat transfer			From inside			
Thermodynamic energy input	Stationary	Non-stationary	Non-stationary	Stationary	Stationary	Non-stationary	Non-stationary
Typical energy sources, fuels	Coal, fuels, uranium	Any heat source	Coal, fuels	Methane, ethane, propane, butane	Kerosine	Diesel fuel, gasoline	Gasoline
Transmission of work performed	Turbine	Reciprocating piston, rotary piston (rotor)	Piston	Turbine	Linear momentum	Reciprocating piston	Rotary piston (rotor)
Typical maximum pressure	50 bar	3 bar (air)	50 bar	40 bar	40 bar	200 bar	60 bar
Typical minimum pressure	0.05 bar	1 bar	1 bar	1 bar	1 bar	1 bar	1 bar
Typical max. efficiency	40 %	30 %	~25 %	40 %	40 %	~42 %	~30 to 35 %
Operating concept	Compression, Heating, Evaporation, Superheating, Expansion, Condensing	Compression, Heating, Expansion, Cooling	Compression, Heating, Evaporation, Superheating, Expansion, Exhaust	Induction, Compression, Combustion, Expansion		Induction, Compression, Combustion, Expansion, Exhaust	

Many thermal engines are characterized by an energy input resulting from a combustion process (Table 1). During combustion the energy chemically bound in the fuel is supplied as reaction heat to the cycle. Here compounds containing carbon and hydrogen oxidize with oxygen, which is why typically ambient air with approximately 21 % volume of oxygen makes up a significant part of the working medium.

Cycle direction

The factor that is crucial to cycle direction is the energy input. Here, a distinction is initially made between stationary (or continuous) energy input and non-stationary (or cyclical) energy input. What all piston engines including the Stirling engine have in common is a non-stationary energy input which only ever occurs close to compression dead center with minimum cylinder volume.

A particular feature of all open cycles is an internal energy input which is achieved by the addition and combustion of fuel. In contrast, closed cycles require an energy input via heat exchangers. Here there is no direct contact between the working medium and combustion processes, with the exception of heat conduction. Unique in this context is the steam engine, which vaporizes the working medium by means of a heat flow from an external source and then supplies it to the piston engine.

Another particular feature of the various thermal engines is the different energy sources which are used. Three different energy sources can be used: solids, liquids and gases. The main advantage of thermal engines which operate with an open cycle and an internal energy input is that they do not require any heat exchangers for cycle direction and are therefore compact in design. This advantage can be amplified by using liquid fuels with a high energy density. Gas engines, for example when deployed in passenger cars and commercial vehicles, are also becoming more attractive (running costs, good fuel consumption). The internal-combustion engine can thus be derived from a selection of thermal engines.

Efficiency

The internal-combustion engine is characterized by open cycle direction with internal combustion. Non-stationary operation facilitates the intake and compression of the working medium at mass-averaged peak temperatures above 2,500 K and averaged peak pressures above 200 bar with a very good maximum efficiency above 40 %.

Stationary cycles, being limited by the properties of the materials, do not achieve mass-averaged peak pressures and temperatures of this magnitude, but instead achieve maximum local peak temperatures around 2,500 K. Gas turbines therefore operate at a lower level of efficiency. The closed cycle of the steam circuit achieves a higher level of efficiency than gas turbines with a likewise moderate pressure of around 50 bar. This is achieved by a significant lowering of the low-pressure level. The other thermal engines are much lower in terms of their maximum efficiency.

Reciprocating-piston engine

The internal-combustion engine, in its reciprocating-piston-engine variation, is the primary thermal engines used for mobile applications. A wide range of fuels is conceivable in principle, but diesel and gasoline are still the main energy sources used.

Real cycles

A cycle is a thermodynamic process which has identical initial and final states (see Thermodynamics). The cycle usually passes through several changes of state in such a way that with a thermal engine work is drawn from the cycle. Here the working medium of the cycle undergoes thermodynamic changes of state.

Ideal reference cycles are suitable for demonstrating basic correlations. In the case of new, as yet unknown engines, they help to provide an overview of their method of operation and efficiencies. However, a real cycle calculation is required for detailed analyses.

Real cycles are set in proportion to the ideal cycles. This means that the heat capacities are considered for example depending on temperature or pressure.

The chemically altered composition of the smoke gases is also approximated in the substance variables in order to take into consideration the changed physical properties based on combustion. In particular, what is assumed is not an adiabatic change of state, but rather at least one polytropic curve with an exponent adapted to the heat loss, or even the heat loss to the wall is used, e.g. with the Woschni approach from the Reynolds similarity theory [1].

The charge cycle is calculated with its dissipation losses (flow losses, real flow cross-sections, see Fluid mechanics) and thus the residual exhaust gas is also taken into account. Furthermore, empirical approaches are normally used for friction and the fuel calorific value is calculated depending on the air ratio. Ultimately, in particular the heat addition (combustion process, heating process) and the heat dissipation (heat transfer) are modeled in detail.

One way to quickly assess real cycles is to describe them using an efficiency chain. Here the real cycles are successively depicted by taking individual characteristic quantities into account.

Overall efficiency

The overall efficiency or effective efficiency η_{eff} sets the effectively available power P_{eff} in proportion to the energy flow $\dot{Q}_{\text{add}} = \dot{m}_{\text{B}} H_{\text{u}}$ which was added by the fuel mass flow $\dot{m}_{\text{B}}$ and its lower calorific value H_{u}:

$$\eta_{\text{eff}} = \frac{P_{\text{eff}}}{\dot{Q}_{\text{add}}} .$$

Diesel engines have at higher loads an effective efficiency of up to 45 %, large slow-running diesel engines have a much higher effective efficiency. Gasoline engines have, depending on the combustion process, an effective efficiency at their best point of over 40 %.

Mechanical efficiency

The mechanical efficiency sets the effectively measured power P_{eff} in proportion to the pressure-indicated cycle power P_{ind}. The indicated power is determined from the work W – this is the area of the real pressure characteristic against volume $\int p dV$ – and the time t per working cycle according to the following relationship:

$$P_{\text{ind}} = \frac{dW}{dt} \approx \frac{\Delta W}{\Delta t} .$$

The effective power differs from the indicated power essentially in the frictional losses (piston, bearings), the power-transmission losses from control elements (camshaft, valves), and the power of the accessories (oil and water pumps, fuel-injection pump, alternator). For the mechanical efficiency:

$$\eta_{\text{m}} = \frac{P_{\text{eff}}}{P_{\text{ind}}} .$$

Normal mechanical efficiencies are load-dependent and at full load come in at just under 90 %, while at low part load (10 % load) values around 70 % are registered.

Efficiency-of-cycle factor

The efficiency-of-cycle factor describes the efficiency with which the real cycle can be approximated by the chosen reference cycle. It therefore contains losses resulting in particular from dissipation losses. For a detailed loss analysis it is recommended to split the efficiency-of-cycle factor between the high-pressure loop and the charge-cycle loop (Table 2).

Normally, for calculation purposes, an ideal gas with temperature-dependent heat capacities is assumed and the cycle is used with geometrically identical dimensions, the same air ratio, without residual exhaust gas, complete combustion, and thermally isolated walls. An engine described in this way is also called a “perfect engine”. The efficiency-of-cycle factor at full load is in the range of approximately 80 to 90 %.

If an ideal gas with constant heat capacity is still to be used, it is possible to introduce the “efficiency of the perfect engine”, which sets the power of the “perfect engine” in proportion to the power of the “ideal cycle”.

Fuel-conversion factor

In particular, gasoline engines with rich-mixture combustion (air/fuel ratio $\lambda < 1$) are subject to high HC and CO emissions, which usually cannot be taken into account in the approach of added heat via the fuel calorific value H_{uB}. However, the exothermy of these gases H_u is considerable, which is usually also reflected in higher exhaust-gas temperatures after the oxidation-type catalytic converter. It is taken into account in the fuel-conversion factor:

$$\eta_B = \frac{(H_{uB} - H_u)}{H_{uB}} .$$

$\eta_B = 1$ is usually set for diesel engines. In gasoline engines the figure can drop to 0.95, and with very rich air ratios $\lambda < 1$ even further below that.

Efficiency chain

The entire efficiency chain can be described as follows (Table 2):

$$\eta_{eff} = \eta_i \, \eta_m = \eta_{th} \, \eta_g \, \eta_m .$$

References

[1] G. Woschni, Die Berechnung der Wandverluste und der thermischen Belastung der Bauteile von Dieselmotoren, MTZ 31 (1970).

Table 2. Graphic representations and definitions of the individual and overall efficiencies of the reciprocating-piston engine

The hatched areas indicate the newly added work component of the characteristic efficiency quantity. The efficiencies are explained in the text.

Pressure vs. volume diagram	Designation	Boundary conditions	Definition	Efficiencies		
	Theoretical reference cycle, e.g. constant-volume cycle	Ideal gas, constant specific heat, infinitely rapid heat addition and dissipation, etc.	$\eta_{th} = 1 - \varepsilon^{1-\kappa}$ Theoretical or thermal efficiency	η_{th}	η_i	η_{eff}
	Real high-pressure working cycle	Wall heat losses, real gas, finitely rapid heat addition and dissipation, variable specific heat	η_{gHD} Efficiency factor of the high-pressure cycle	η_g		
	Real charge cycle (4-stroke)	Flow losses, heating of the mixture or the air, etc.	η_{gLW} Gas-exchange efficiency			
Mechanical losses, cannot be suitably represented in the pressure vs. volume diagram	Losses due to friction, cooling, accessories	Real engine	η_m	η_m	η_m	

Mixture formation, combustion, emissions

All internal-combustion engines share the common feature of combustion occurring after induction of the fresh mixture or air and after subsequent compression. In reciprocating-piston engines with internal combustion this takes place close to top dead center (TDC). The result is a pressure increase, which is transmitted via the piston and the connecting rod to the crankshaft in the form of a crankshaft torque (Figure 1).

The sequence of compression and subsequent combustion on the one hand have a significant impact on the pressure characteristic and thus on the efficiency and the torque output. On the other hand, this sequence defines the creation of emissions inside the engine. In this respect, gasoline and diesel engines differ in their process control.

Gasoline engine

The characteristic feature of a gasoline engine is that it uses an external ignition source, normally an electrode spark plug. Ideally a suitable homogeneous air/fuel mixture is created to provide the required flammability. This is achieved by means of external mixture formation (manifold injection) or internal mixture formation (gasoline direct injection).

Mixture formation

For the most part, a homogeneous mixture preparation is effected in a gasoline engine – i.e. the intake air is fully mixed with the vaporized or atomized fuel – during the induction and compression strokes. The excellent vaporization qualities of gasoline enable it to be injected into the intake manifold. Modern stratified-charge combustion processes are, on the

Figure 1: The engine power cycle
a) Course of combustion,
b) Representation in the p-V diagram (enclosed area corresponds to indicated work),
c) Representation in the p-t or p-α diagram,
p Pressure in cylinder,
p_{max} Maximum pressure,
Q_V Combustion heat,
V_C Dead volume,
V_h Engine displacement,
t Time, α Crank angle,
TDC Top dead center,
BDC Bottom dead center,
IP Ignition point.

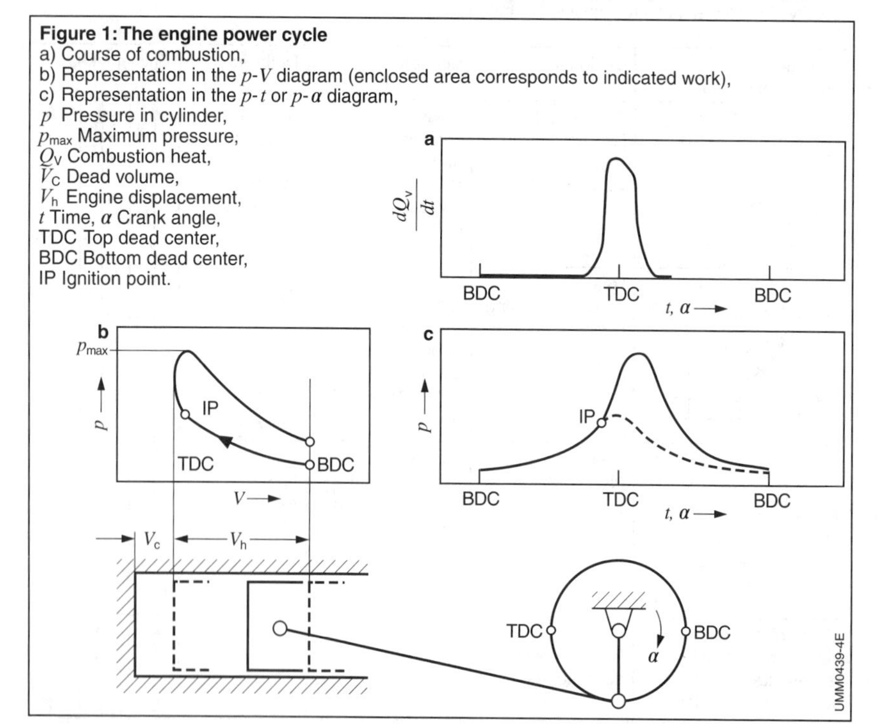

other hand, characterized by partially heterogeneous mixture preparation.

Mixture formation is decisively influenced by the vaporization conditions, the injection pressure, the cylinder-charge movement, and the time available with the aim of homogenization. Essentially, mixture formation involves the interaction of two processes: droplet vaporization caused by the temperature difference (Figure 2) and droplet disintegration by aerodynamic forces (Figure 3). Manifold injection and direct injection differ here (Table 1).

Manifold injection

In the case of manifold injection, a film of mixture is created ahead of the intake valve whose fuel mass is reduced more intensively as the air velocity increases. This air velocity changes along broadly linear lines with the engine speed. On account of the low temperature and incomplete vaporization in the intake manifold with the resulting formation of film, manifold injection occurs at a very low injection-pressure level below 10 bar.

The dynamics of the wall-applied fuel film and the mechanisms involved during vaporization are one of the main causes of inaccurate fuel metering, above all during transient engine operation. Only the smaller droplets entrained with the

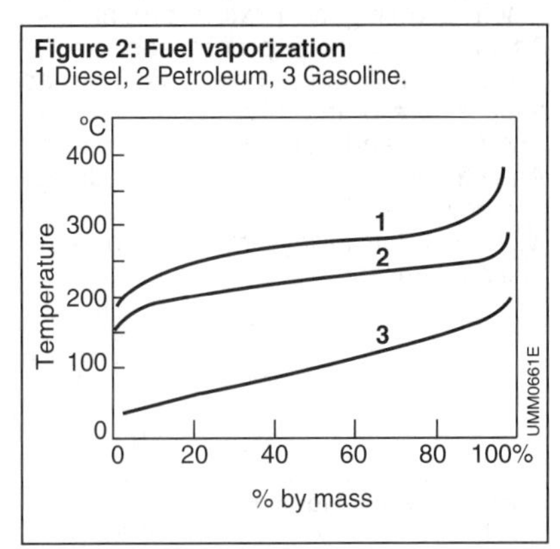

Figure 2: Fuel vaporization
1 Diesel, 2 Petroleum, 3 Gasoline.

Table 1: Operating strategies of gasoline engines

Operating strategy	Stoichiometric	Rich	Lean	Ultra-lean	Stratified-charge operation lean
Mixture formation	Homogeneous				Heterogeneous and homogeneous
Combustion-chamber composition	Fuel, Air				
Fuel injection	Manifold injection and direct injection				Direct injection
Ignition	External ignition source	External ignition source	External ignition source	Auto-ignition	External ignition source
Typical compression ratio	8...12		11...13	12...16	11...14
Load control	Quantity				Quality
Operating range	Entire program map	Full load, high speed range	Entire program map	Part load	Part load
Application, development stage	Conventional, series		Gas engines, series	Research stage	New combustion processes

intake flow reach the inside of the cylinder (Figure 3). Typically, their characteristic diameter is already less than 30 µm. Here, droplet acceleration is proportional to the ratio of relative velocity to the air and to the droplet diameter.

The very high turbulence intensity and the high flow velocities in the valve gap result in very good mixture preparation. As process control progresses the remaining small fuel droplets assume the temperature of the mixture and vaporize (Figure 4), and homogenization results. An optimum combustion-chamber design prevents intensive contact of fuel with the wall, since this always involves the risk of fuel condensation.

Figure 3: Droplet formation
a) Characteristic flow situation in valve gap,
b) Interaction between droplet diameter and relative velocity of droplets with regard to air flow.
1 0%,
2 50%,
3 Over 90%.

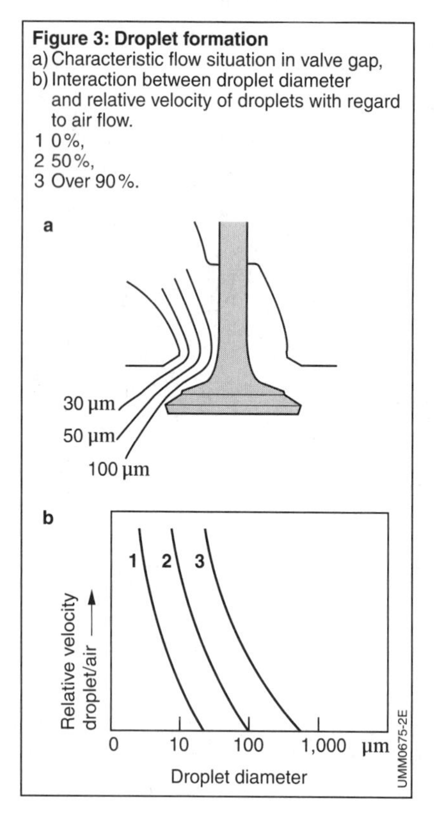

Gasoline direct injection

Gasoline direct injection does not utilize the mechanisms of mixture preparation in the valve gap. A higher injection pressure of 50 to 200 bar is therefore required. Injection is completed by no later than the gas-flow bottom dead center to allow sufficient time for homogenization.

The inducted mixture is now compressed, depending first and foremost on the throttle-valve position and the compression ratio, to a pressure level of 10 to 40 bar. This equates to a temperature level of 300 to 500 °C, depending first and foremost on the ambient temperature and the compression ratio. In heterogeneous combustion processes injection occurs only at the end of the compression phase.

The advantage of direct injection is that the fuel is metered exactly. Vaporization of the fuel in the combustion chamber also requires the cylinder charge to be adequately cooled. This enables the compression ratio to be increased by roughly one unit, resulting in greater efficiency.

In all combustion processes combustion – oxidation – occurs only at the end of the compression phase and in the early expansion phase.

Figure 4: Droplet vaporization
m Mass, D Diameter, and T Temperature of a vaporizing fuel droplet.
m and D are shown scaled to the maximum value. No scale is specified for T, because only the basic characteristic is involved.

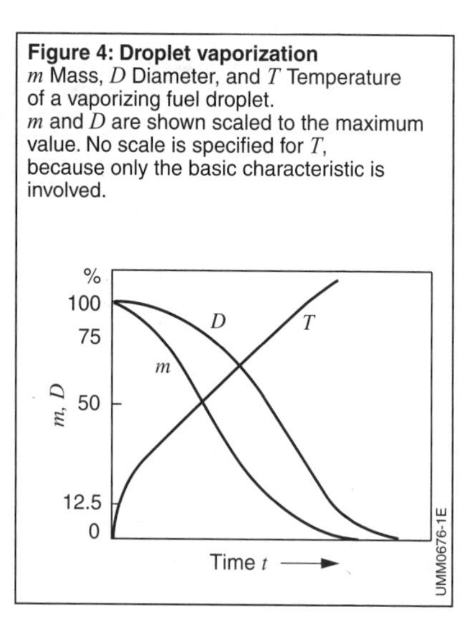

Combustion in gasoline engines

The subsequent combustion process differs depending on the nature of mixture preparation (homogeneous or heterogeneous). An entirely homogeneous mixture reacts in a premixed combustion, an entirely heterogeneous mixture in a mixture-controlled combustion. During the stratified-charge operation of modern direct-injection engines, a large proportion (>50 %) of the injected fuel is likewise homogenized up to the start of combustion.

In both homogeneous and partly heterogeneous mixture preparation, actual combustion is preceded by ignition and the ignition phase.

Ignition

Ignition is typically performed with the aid of an electrode spark plug. When a high voltage is applied, a sparkover occurs between the electrodes, depending on the mixture state (i.e. pressure, temperature, and mixture composition). Here the high voltage is typically in the two-digit kV range. First and foremost the number of molecules between the electrodes influences the ignition-voltage demand. The mixture ignited by the spark must in the course of its combustion release that amount of energy that is needed to ignite the immediately adjacent mixture. As leaning of the mixture increases, the energy content of this mixture decreases when the electrode gap remains constant. This is accompanied by an increase in the energy demand needed to ignite the adjacent – also lean – mixture. By enlarging the electrode gap, it is possible to increase the volume ignited by the spark and thereby raise the energy content. However, enlarging the electrode gap requires an increase in the ignition voltage. In this way, this increases for example in lean-combustion processes or in the event of a load increase. In the event of a load increase, the sparkover duration simultaneously decreases as the ignition-voltage demand increases (Figure 5).

On account of heat losses at the spark-plug electrodes, heat-convection losses and cyclically fluctuating mixture states, the ignition energy is above the theoretical minimum ignition energy by up to an order of magnitude (Figure 6). The sto-

Figure 5: Ignition sparks and ignition-voltage demand
a) Sparkover duration,
b) Ignition-voltage demand U_Z.
The ignition-voltage demand rises as exhaust-gas recirculation or leaning increases.

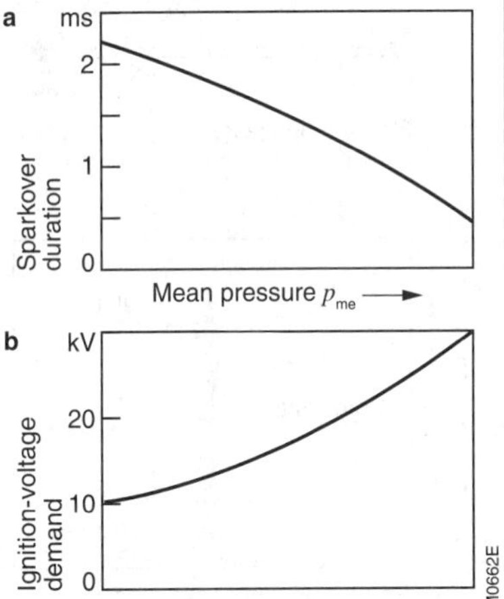

Figure 6: Minimum ignition energy for a methane/propane mixture
1 Ignition energy for a static mixture,
2 Ignition energy for a mixture with a flow velocity of 6 m/s,
3 Ignition energy for a mixture with a flow velocity of 15 m/s.
With an increase in the proportion of inert gas the ignition-voltage demand increases for all the flow states shown.

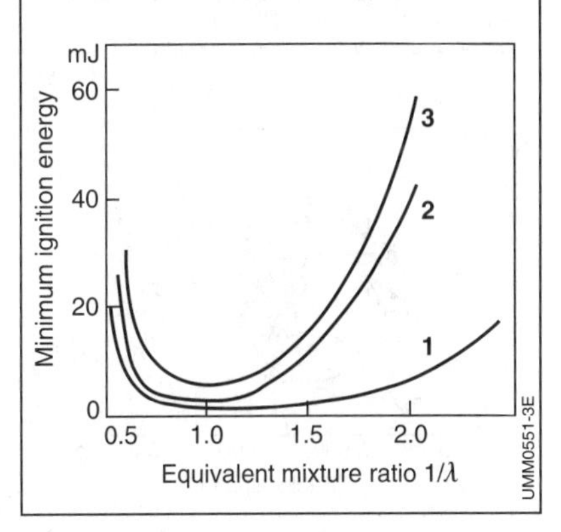

chastically fluctuating states (flow field and mixture state) between the electrodes are the main cause of large cyclic variations in a gasoline engine. This situation is improved by enlarging the electrode gap. Modern engines here already operate with a maximum value above 1 mm. Enlarging the electrode gap necessitates an increase in the ignition voltage and with it causes an increase in electrode wear.

The purpose of ignition is to ignite the air/fuel mixture and thereby initiate the actual combustion process. Depending on the velocity of the subsequent combustion and on the piston speed (and thus on the engine speed), the point of ignition must be variably adapted (Figure 7).

Combustion of homogeneous mixtures
Ignition sets in progress the actual combustion in a homogeneously operated gasoline engine. Here the flame propagates from the spark plug. It is possible to define a flame-front velocity which is composed of the flame velocity and the flame-front movement (charge movement, expansion caused by density differences). In a homogeneously operated gasoline engine a good distinction can be made between the combusted and non-combusted mixture, since the flame continuously propagates outwards (Figure 8).

It takes a few milliseconds for the flame to reach a radius of around one centimeter in order to propagate unhindered from the piston recess and cylinder head at a velocity well in excess of 10 m/s.

Figure 7: Program map of ignition point
Values in °CA before TDC for a homogeneously operated gasoline engine.

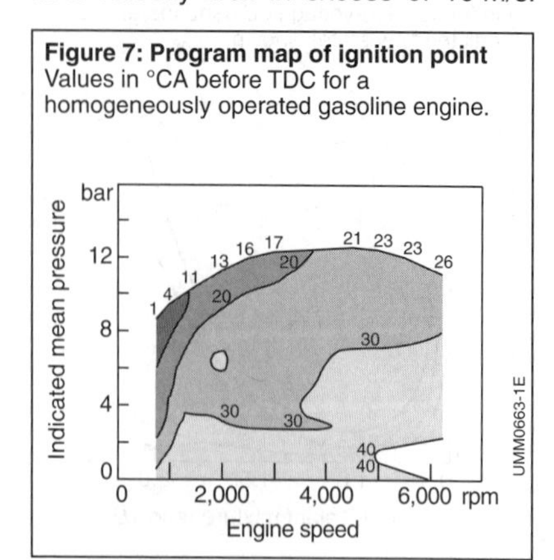

The decisive factor now is the propagation velocity of the flame, which is also referred to as the turbulent flame velocity. The faster the flame propagates, the better the combustion inside the engine. The following factors encourage a high flame velocity: a low proportion of inert gas, a temperature increase in the non-combusted mixture, a pressure increase, and a high turbulence level.

Most fuels have their maximum flame velocity in slightly rich-burn operation in the range of λ = 0.85 to 0.9. A further advantage of slightly rich-burn engine operation is the cooling effect provided by the excess fuel. Racing engines and passenger-car engines in the rated-output

Figure 8: Flame propagation and premixed combustion
a) Flame-front propagation in a gasoline engine during homogeneous operation (the figures in degrees are calculated from the ignition point),
b) Temperature and substance concentrations along the flame-propagation direction.

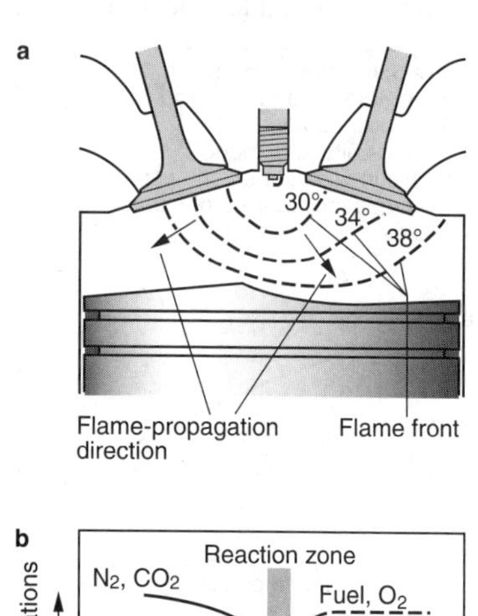

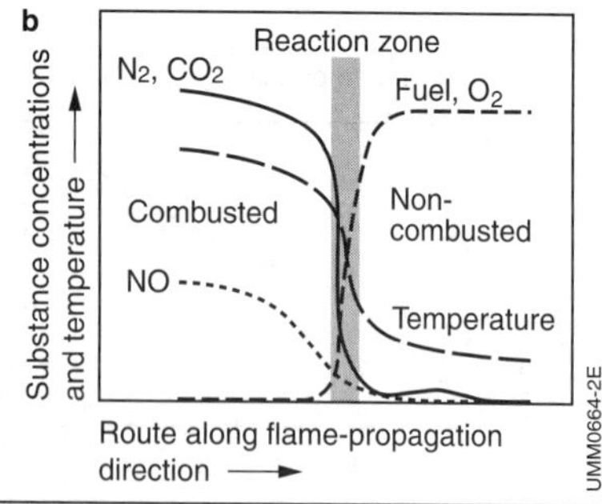

range are therefore operated with slightly rich mixtures.

Increasing the proportion of inert gas reduces the flame velocity. A practical application of inert-gas variation is exhaust-gas recirculation (EGR), in which combusted exhaust gas with the chief constituents CO_2, H_2O and N_2 is added to the air/fuel mixture. As a rule of thumb an exhaust-gas recirculation rate of 10 % already reduces the flame velocity by 20 %.

The reason why modern gasoline engines can nevertheless be operated with high internal exhaust-gas recirculation rates is the influence of temperature. Doubling the temperature increases the flame velocity by a factor of roughly 4.

Cylinder pressure has a lesser influence; an increase in pressure causes a slight acceleration of the flame velocity.

The turbulence level in the combustion chamber has the greatest influence on the combustion velocity. The flame velocity varies along roughly linear lines with the turbulence intensity. Turbulence intensity is a measure of the high-frequency change in flow velocity at a given point in the combustion chamber [1]. The turbulent kinetic energy is proportional to the square of the turbulence intensity.

Turbulence intensity is a three-dimensional quantity which is influenced above all by the flow profile of the charge movement in the combustion chamber. The velocity of the flow processes inside the engine that increases linearly to the greatest possible extent as engine speed increases is extremely important. As the flow velocity increases, so too does the turbulence intensity in the combustion chamber. This is the reason for stable engine operation over a very wide engine-speed range. The flame would otherwise no longer be able to burn cleanly at increasing engine speed and consistent flame velocity because of the shorter time available for combustion. However, the positive contribution of turbulence cannot equal completely the influence of speed, which is why combustion at high speeds nevertheless stretches over a wider crankshaft range. This is an additional reason why gasoline engines operate with less efficiency at higher speeds.

In gasoline engines the turbulence in the combustion chamber is vitally important to energy conversion. The decisive factor in causing turbulence is the charge movement in the cylinder, which is fundamentally influenced by the intake flow (depending on the configuration of the ports in the cylinder head) and the combustion-chamber shape (Figure 9).

Combustion causes a pressure increase which can also be simultaneously heard. For comfort reasons, suitable combustion-adjustment measures must be taken to minimize this pressure increase as much as possible. This poses a conflict of objectives with thermodynamic efficiency. The maximum pressure-increase gradient of gasoline engines is in the range of between 0.5 and 3 bar/°CA.

Figure 9: Turbulent and mean kinetic energy as a function of the crankshaft position, each referred to mass
1 Turbulent kinetic energy,
2 Mean kinetic energy.

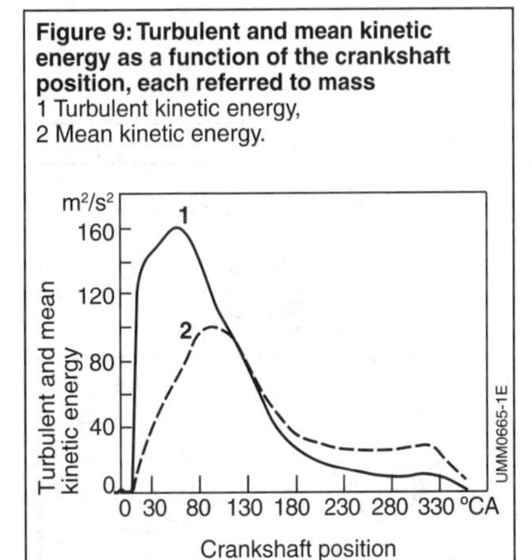

Combustion of partially homogeneous mixtures

Modern stratified-charge combustion processes facilitate operation with excess air in the part-load range. Effective mean pressures of $p_{me} < 1$ bar make even operation with a mean excess-air factor $\lambda > 5$ possible. The main benefit here is the improved charge cycle, because throttling (which has a negative impact on overall efficiency) can be largely dispensed with.

Conventional operation is not possible without throttling in the lower part-load range, because homogeneous mixtures leaned in this way burn much too slowly and therefore incompletely. The solution is to provide local fuel stratification in the area of the spark plug by means of an optimized injection strategy in the late compression phase. The difficulty here lies in optimally matching the injection strategy and ignition, since the conditions between the electrodes change (Figure 10).

Figure 10: Excess-air factor λ over time in a spherical volume (with radius r) around the electrode midpoint

1 Stratified-charge combustion process, λ within r = 2 mm (gaseous fuel),
2 Stratified-charge combustion process, λ within r = 3 mm (gaseous fuel),
3 Stratified-charge combustion process, λ within r = 5 mm (gaseous fuel),
4 Stratified-charge combustion process, λ within r = 5 mm (liquid and gaseous fuel),
5 Optimized stratified-charge combustion process,
6 Homogeneous combustion process.

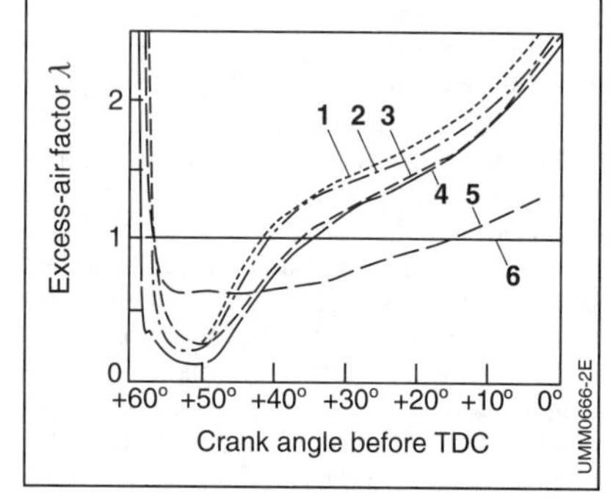

Charge movement

Large-scale, swirling and circular flows with diameters similar to the characteristic quantities of the combustion chamber provide the basis for the cylinder charge. A basic distinction is made between flows about the vertical axis (cylinder axis), which are called swirl, and flows about the two lateral axes (crankshaft axis and perpendicular to this), which are called tumble (Table 2). In reality there is an overlap of the three basic flows, which gives rise to complex three-dimensional flow fields. Tumble and swirl essentially differ in their behavior inside the engine.

Table 2: Charge cycle and flow profile

Operating strategy	Swirl	Tumble	
Flow			
Location of flow axis	Vertical axis z	Lateral axis x	Lateral axis y

UMM0660-1E

Tumble flow disintegrates up to the moment when compression dead center is reached and contributes primarily to flame propagation in the first combustion half.

Swirl flow lasts longer into the subsequent expansion phase. The disintegration of the large swirls into ever-decreasing, turbulent structures encourages the creation of turbulence. In the continuing process, however, the viscosity of the working medium results in a disintegration with a detrimental effect on the combustion velocity.

The creation of turbulence is assisted by the combustion-chamber geometry. In particular, flows in the piston-recess or squish-gap area support flame propagation (see e.g. [2]).

One of the greatest physical challenges associated with fashioning homogeneous combustion processes is the combustion behavior in the expansion phase, since typically over 10 % of the supplied fuel is still not converted 30° CA after TDC. At this point the non-combusted mixture is still only in the immediate vicinity of the wall and must still be converted after re-emerging from the top-land area (see e.g. [2]). This phenomenon results in incomplete combustion in the final phase.

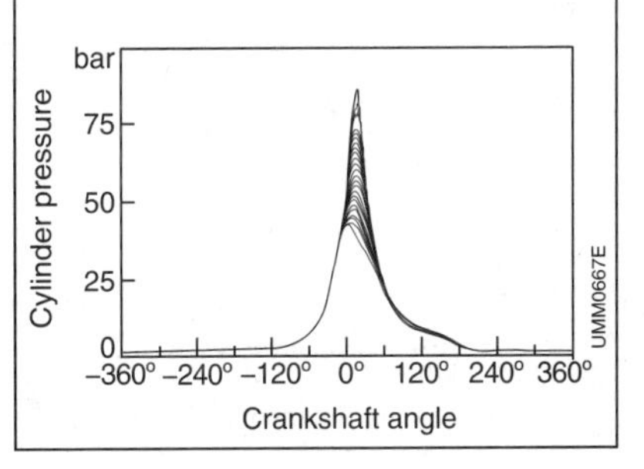

Figure 11: Cyclic fluctuations of the cylinder pressure of the individual working cycles during stoichiometric operation (λ = 1)

Uncontrolled combustion

Numerous undesirable processes make it difficult to fashion a homogeneous combustion process. In addition to cyclic variations, knock and auto-ignition characteristics have detrimental effects. Extreme forms of advanced ignitions can occur, particularly in modern, highly turbocharged engines.

Cyclic variations are a consequence of the premixed combustion of homogeneous combustion processes which reacts sensitively to numerous disturbance values (Figure 11). Such disturbance values are, for example, the mixture composition, the residual-exhaust-gas content, the thermodynamic state variables, and the flow profile. All the values vary from working cycle to working cycle and cause noticeable cyclic fluctuations in energy conversion. In particular, leaning with excess air of more than 40 % increases the cyclic variations significantly. Initiation of ignition has the greatest influence on cyclic variations.

Another challenge is posed by the undesirable phenomenon of knocking. Knocking occurs in response to increases in pressure and temperature after the onset of regular combustion until the local conditions stimulate auto-ignition. Here, the residual mixture in response to the temperature increase as a result of the progressing combustion reaches the ignition temperature and burns almost simultaneously without a further controlled flame propagation. The pressure pulsations generated in the process result in material wear in the engine bearings and cause serious engine damage in the event of prolonged operation. The temperature peaks can also result in damage to components. Typically, knocking only then occurs in a working cycle if 80 % of the fuel in the combustion chamber has not been combusted. Knocking can be observed particularly at low engine speeds with suitably sufficient time for auto-ignition and at high loads with high combustion-chamber temperatures. Fuels with high ignition temperatures such as, for example, methane or ethane reduce knock sensitivity. Knocking combustion can be reduced by shifting the energy conversion in the retardation direction (ignition retardation). High-compression and turbocharged engines are more knock-sensitive because of their higher final compression temperatures (Figure 12). Useful measures to counter knocking include effective cylinder cooling of the hot areas, for example also by the vaporization effect of gasoline direct injection, increasing the turbulence, reducing the compression ratio, and optimizing the fuel, for example with additives.

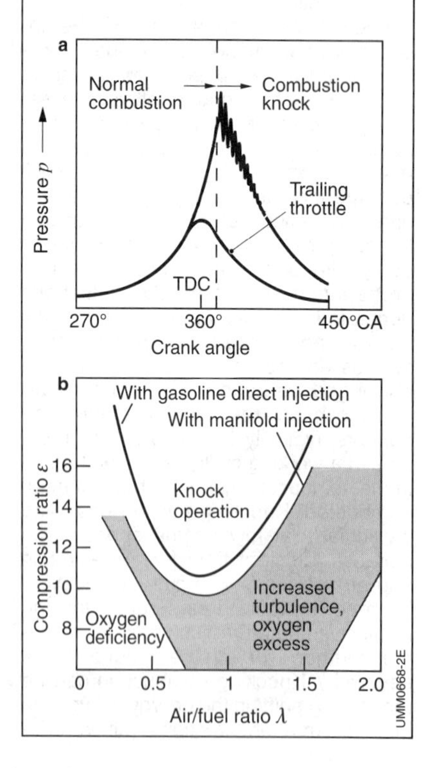

Figure 12: Compression ratio, knocking and knocking combustion
a) Example of knocking combustion,
b) Effect of compression ratio and air/fuel ratio on the operating limits (knocking and engine misfires). Direct injection instead of manifold injection and increased turbulence shift the operating limits.

In contrast to knocking, auto-ignitions can occur quite perfectly even in very retarded compression situations. Possible causes of auto-ignitions are, for example:
- Over-retarded moment of ignition with incomplete combustion, thus fuel-film formation causing auto-ignitions
- Full-load operation with high cylinder-component temperatures
- Ignition caused by abrasion and hot particulates
- Oil emissions caused by defective or worn piston rings

Extreme forms of auto-ignition can occur in turbocharged and high-compression gasoline engines; these can result in peak pressures in excess of 150 bar and serious damage. However, these extreme auto-ignitions occur only very rarely with a probability of less than 0.01 per thousand.

Pollutant formation and pollutant reduction in gasoline engines

In addition to the unavoidable combustion products of carbon dioxide (CO_2) and water (H_2O), the concentration of which is dependent on the composition of the fuel, nitrogen oxides (NO_x), non-combusted hydrocarbons (HC) and carbon monoxide (CO) are the main emissions from gasoline engines (Figure 13). Soot and sulfur-oxide compounds are of lesser importance.

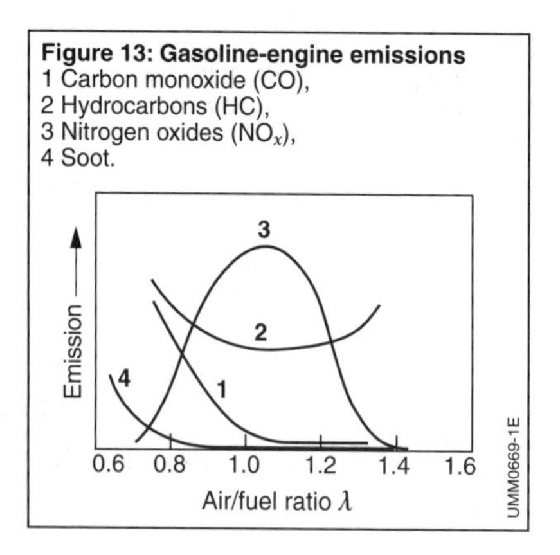

Figure 13: Gasoline-engine emissions
1 Carbon monoxide (CO),
2 Hydrocarbons (HC),
3 Nitrogen oxides (NO_x),
4 Soot.

Nitrogen oxides (NO_x) require four factors in order to be created: oxygen, nitrogen, high temperatures, and time. Because oxygen and nitrogen are determined by the mixture composition in gasoline engines and the available time is defined by way of the engine speed, nitrogen oxides in gasoline engines can only be reduced by low maximum combustion-chamber temperatures (e.g. by means of ignition retardation and exhaust-gas recirculation).

Hydrocarbons (HC) and carbon-monoxide emissions (CO) are the result of incomplete combustion. In rich-burn operation there is, on account of the lack of oxygen, an increase in both HC and CO emissions. In lean-burn operation with accordingly reduced flame temperature a more intensive extinguishing of the flame is encountered, above all in the area close to the wall, with a resulting increase in HC emissions. Because of the excess oxygen the slightly oxidizable CO emissions are nevertheless reduced.

Soot emissions are produced in homogeneous engines only in extreme rich-burn operation. Sulfur-oxide emissions are dependent on the sulfur content in the fuel.

Thanks to exhaust-gas treatment, modern homogeneous gasoline engines are, once they have reached the catalytic-converter operating temperature, virtually zero-emission machines. Three-way catalytic converters in operation at $\lambda = 1$ reduce the nitrogen-oxide emissions while simultaneously oxidizing HC and CO molecules (see Three-way catalytic converter). Alternative approaches are required in lean-burn operation. For this reason, NO_x storage catalytic converters are typically used for stratified-charge approaches. These catalysts store the nitrogen oxides. Rich-burn engine operation applied at regular intervals reduces the stored nitrogen oxides at high temperatures. Because NO_x storage catalytic converters are sensitive to sulfur contamination, desulfation cycles must additionally be completed in slight rich-burn operation at temperatures in excess of 600 °C (see NO_x storage catalytic converter).

Load control in gasoline engines

In homogeneously operated gasoline engines the load is adjusted by means of the injected fuel mass. The corresponding air mass is adapted by means of the throttle-valve position for the required operation at $\lambda = 1$. This is known as quantity control. In the part-load range this causes induction throttling, which is detrimental to the overall efficiency. This disadvantage can be partially compensated for by varying the valve timing. Advanced or retarded "intake closes", reduced valve lift or retarded "exhaust closes", in which hot exhaust gas is additionally inducted, are typical measures. Alternatively, external exhaust gas can be recirculated (exhaust-gas recirculation) in order to reduce throttling. In turbocharged gasoline engines the air flow rate and thus the fuel-mass flow rate is adjusted in the upper load range typically by means of the position of the wastegates on the turbocharger (see Exhaust-gas turbochargers).

In engines with stratified charge the load is adjusted in these map ranges by means of the injected fuel mass. This is known as quality control. The transition between stratified- and homogeneous-charge operation in the mid-load range requires control-engineering complexity and expenditure.

Power yield and efficiency

The part-load performance of a gasoline engine deteriorates as a result of charge-cycle losses (throttling), poor process control (peak pressures below 30 bar), and an increasing amount of engine friction in this map range. Because even at passenger-car driving speeds in excess of 100 km/h most vehicle engines still operate in the part-load range, measures to increase the efficiency in this operating range prove highly successful. Examples of such measures are:

- Displacement reduction (downsizing)
- Cylinder shutoff (e.g. in V8 and V12 engines)
- Dethrottling (stratified charge, exhaust-gas recirculation, valve timing)
- Increased compression ratio
- Longer transmission ratio to lower the engine-speed level

Diesel engine

A significant characteristic feature of diesel engines is the absence of external ignition. This is achieved by injecting the ignitable fuels into highly compressed and therefore hot air. High final compression temperatures and pressures to levels in excess of 600 °C and 100 bar in turbocharged engines provide for extremely stable engine operation. Fuel-spray formation, fuel vaporization, fuel mixing, and subsequent combustion can take place within a very short period of time.

Mixture formation

Mixture formation is dominated by the interaction of the injection spray with the flow field in the combustion chamber. Here, the challenge consists in quickly injecting and preparing relative large masses of fuel of up to 200 mg per liter displacement. A typical injection duration is in the range of 1 ms. The term used in relation to the fuel mass flow into the combustion chamber is the injection rate (unit: kg/s). This fuel mass flow is typically injected by multihole injection nozzles.

A combination of between four and ten injection holes with a diameter of between 120 and 150 µm is commonly used. As well as the small hole diameter, the high injection pressure up to more than 2,000 bar encourages fast fuel injection and mixture preparation.

The diameter of a characteristic spray is initially the same as that of the nozzle hole. After it has traveled a few millimeters, the spray breaks up into individual droplets which interact with the flow field. The liquid phase of the fuel spray can, especially depending on the density of the working medium, penetrate a few centimeters into the combustion chamber before the spray is completely atomized or vaporized (Figure 14).

Turbulences encourage the formation of droplets and the vaporization of the fuel. In modern diesel engines more than 80 % of the turbulence intensity in the area of the fuel spray is generated by the injection. This is further assisted by the charge movement, whereby diesel engines with flat cylinder head experience predominantly more swirl than tumble flows. An additional contribution can be made by compression-induced air flows from the outer to the inner area of the combustion chamber ("squish flows") or by the combustion-chamber design, e.g. by vaporization-inducing contact with the hot piston-recess area.

Figure 14: Spray propagation and mixture preparation in diesel engines
a) Characteristic spray penetration behavior,
b) Typical injection-spray formation (the lean edge areas with infinite λ are outside the continuous line),
c) Velocity profile of injection spray.

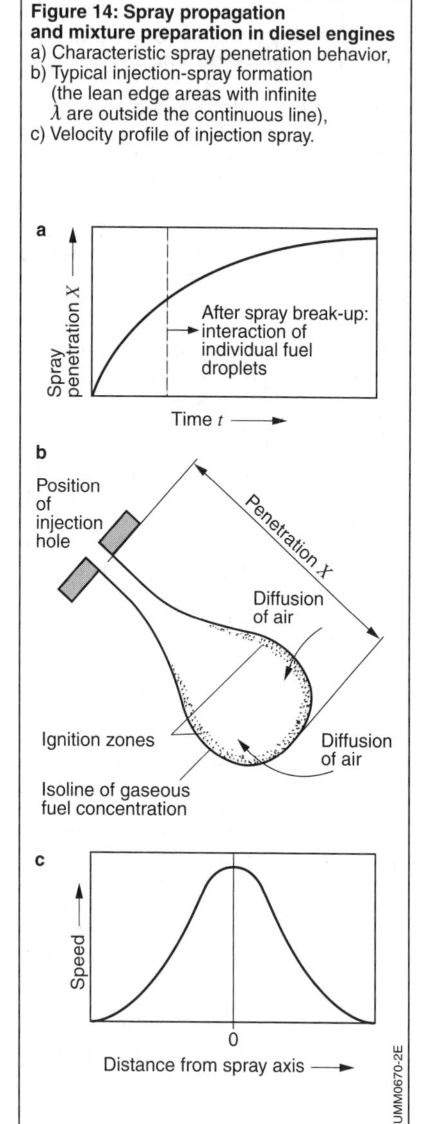

Combustion processes with direct injection have in the last few decades established themselves as the favored choice over indirect combustion processes such as whirl-chamber or prechamber systems. In indirect combustion processes the local flow in the secondary volume is essentially responsible for fuel preparation Figure 15).

Figure 15: Diesel-engine combustion processes
Combustion-chamber shape and nozzle arrangement for:
a) Direct injection,
b) Prechamber system,
c) Whirl-chamber system.

a

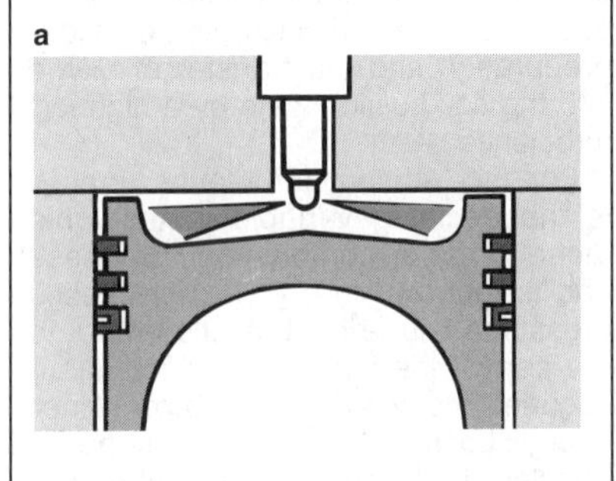

b

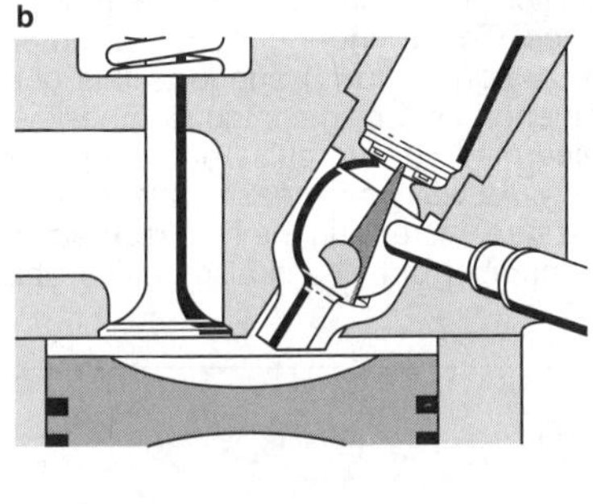

c

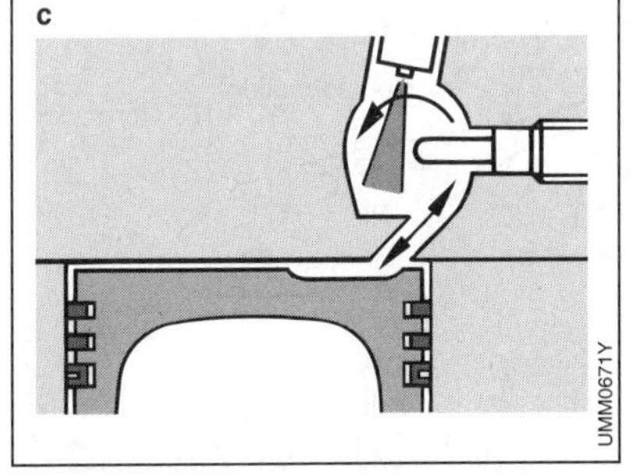

Combustion in diesel engines

Combustion in diesel engines differs from conventional gasoline-engine operation in the compression-ignition behavior. In all, diesel-engine combustion can always be described in three consecutive processes: ignition lag, premixed combustion, and mixture-controlled combustion. Depending on the operating state and the map range, the three processes have different time components (Figure 16).

Ignition lag refers to the period of time between start of injection and start of combustion. It is crucially dependent on the cylinder temperature, the cylinder pressure, and the combustibility of the fuel. Both mixture preparation and the first chemical pre-reactions of the air/fuel mixture take place in the ignition-lag phase. The ignition lag is long during cold engine operation or when a low-quality fuel with a low cetane number is used. The influence of cylinder pressure is less dominant than that of temperature. Basically, however, an increase in cylinder pressure reduces the ignition lag. The fuel injected during the ignition-lag phase does not burn yet. The ignition lag can be between 0.1 ms in the rated-output range to over 10 ms in the cold-start range.

The ignition lag defines the phase of premixed combustion by means of the injected and as yet non-combusted fuel. The longer the ignition lag, the more fuel is premixed in combusted form. This fuel mass can perfectly well be over 20 mg per liter displacement. Combustion typi-

Figure 16: Diesel-engine combustion

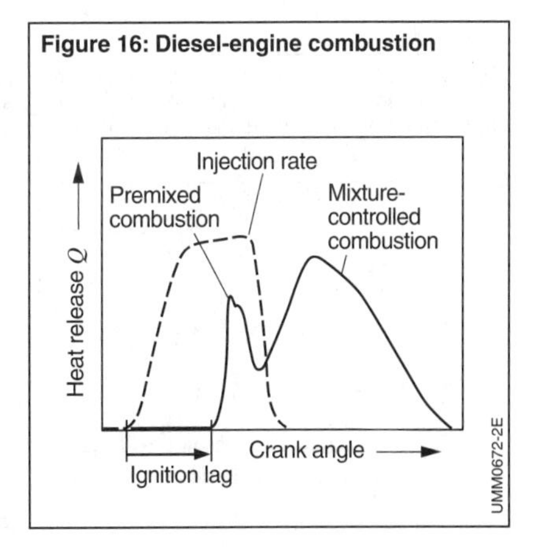

cally starts at the edge of the fuel spray, where the fuel has mixed very well with air and therefore optimum ignition conditions are present in terms of temperature and excess-air factor λ. The exothermic reaction results in a local temperature increase with temperatures above 2,300 K, which quickly converts the as yet non-combusted premixed fuel in a chain reaction. Here, the chemical reactions that take place determine the rate of combustion. The self-accelerating chain reaction results in an extremely fast combustion with large pressure-increase gradients. For this reason, the premixed converted fuel mass must be kept to a minimum in diesel engines. This is typically achieved by means of a pilot injection whose local combustion exerts an initial temperature increase with a reducing effect on the ignition lag of the subsequent main injection.

The amount of premixed converted fuel can be between less than 1 % in the upper full-load range and 100 % in the lowest part-load range. The remaining amount burns in a mixture-controlled combustion. In contrast to premixed combustion, in the case of mixture-controlled combustion – also called diffusion combustion – the transfer of oxygen into the combustion zone defines the conversion rate. It is difficult to distinguish between a combusted zone and a non-combusted zone, because there is no precisely defined flame front. Basically, the diffusion flame establishes itself at the edge of the spray and burns in a limited range at $0.8 < \lambda < 1.4$. As the boundary conditions change (e.g. further fuel vaporization, oxygen transfer, wall contact), so too the reaction zone strays to where locally stoichiometric conditions prevail (Figure 17).

Mixture-controlled combustion dominates in the high load range, in which large quantities of fuel are injected. Here, the mixture-formation and combustion processes occur in parallel. As with premixed combustion, the conversion rate can also be influenced by the injection. A lesser influence, but also of an accelerating character, is exerted by increases in both temperature and pressure and by a reduction in the proportion of inert gas. The dominant factors are mixture preparation and the transfer of oxygen into the combustion zone by a high local turbulence.

For this reason, turbulence intensity is the deciding variable in fashioning diesel-engine combustion processes and is reflected in ever higher injection pressures with a high kinetic energy on the part of the fuel spray, which is subsequently converted into turbulent kinetic energy. Local turbulence causes the vital transfer of oxygen into the local reaction zones. The cylinder-charge movement (swirl, squish flows) supports this phenomenon, the main contribution however being made by the pulse of the injection spray. As well as increasing the injection pressure, increasing the hole diameter or the number of holes is also conceivable.

Figure 17: Mixture-controlled combustion
a) Substance concentrations and temperature distribution,
b) Formation of pollutants inside the injected spray.

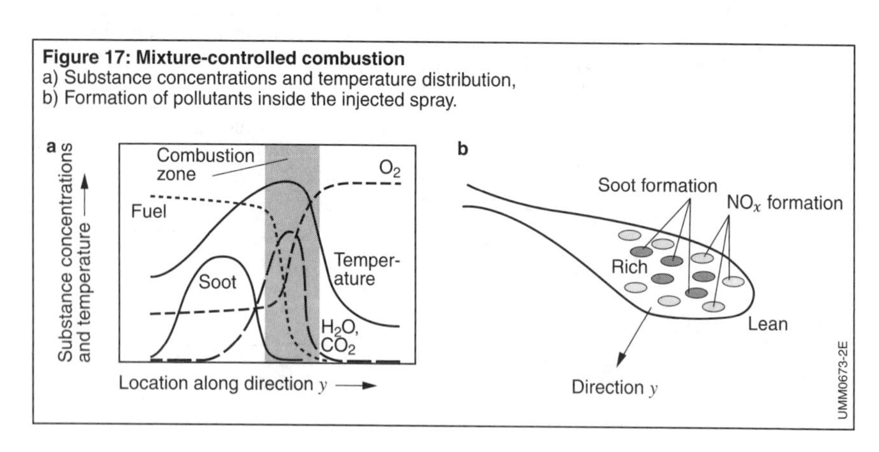

However, an increase in the injection rate results for the most part in a local over-enrichment with a negative impact on fuel conversion.

Particular characteristics of diesel-engine combustion

Cold-start processes in diesel engines pose a particular challenge, particularly at outside temperatures below −10 °C. At starter speeds down to below 100 rpm a large proportion of the cylinder charge escapes through the piston rings during the relatively slow compression phase. In addition, the cold cylinder temperatures give rise to increased wall heat losses. Results experienced are low peak pressures below 30 bar and, depending on the outside temperature, low peak temperatures below 400 °C.

Fuel vaporization at top dead center causes further cooling. The outcome is very long ignition lags. In an extreme-case scenario there may be no ignition whatsoever with the result that fuel can accumulate in the cylinder over several working cycles. Its ignition after a few working cycles can, on account of the fuel mass accumulated in the meantime, result in very high peak pressures above 150 bar.

Because the cold-start phase does not provide for enough time for an adequate hydrodynamic lubricant film to be built up at the appropriate bearing points of the crankshaft drive, the result is a negative effect on the mechanical engine systems. Typical measures to help improve the cold-start process include therefore preheating the intake air, the oil or the water. The latter, as well as influencing the combustion-chamber temperature, above all brings about lower engine friction and increased starter speeds.

A further phenomenon is encountered during operation at very hot outside temperatures or at high altitudes above 1,000 m. Because the air has a lower density, its mass in the cylinder is reduced. The effect on combustion is initially not decisive. However, the exhaust-gas temperature rises as a result of the lower amount of excess air.

This phenomenon is also encountered with turbocharged engines. A power reduction above all during operation at altitude can therefore be a necessary application measure.

A power drop in the order of 1 to 3 % is often observed in diesel engines after a break-in period. This is caused by the fuel-injection system. Deposits in the injection nozzles cause a slightly reduced nozzle-hole diameter, resulting in a smaller mass flow and thus a power loss. These deposits may be caused, for example, by copper, zinc or oil contaminants in the diesel fuel.

Pollutant formation and pollutant reduction

In contrast to gasoline engines, which are characterized with the introduction of the three-way catalytic converter during operation at $\lambda = 1$ by extremely low emissions, pollutant reduction inside the engine plays a much more important role in diesel engines. In addition to the emissions familiar from gasoline engines, i.e. CO_2, H_2O, NO_x, HC and CO, soot and particulate emissions must also be taken into consideration.

Measures which lower the combustion temperatures are useful in reducing the nitrogen oxides. This can be done effectively by reducing the oxygen concentration in the combustion zone. The combustion temperatures can also be reduced very easily by retarding injection or reducing the injection pressure.

Reducing the injection pressure or the oxygen concentration typically causes an increase in soot emissions. The formation of soot is a complex process and is dependent on both fluid-dynamic and thermodynamic boundary conditions. Initially, the locally highly rich zones ($\lambda << 1$) produce considerable amounts of soot, which are then however substantially reduced by over 70% in the subsequent combustion by oxidation processes. A high turbulence level, which encourages the soot to oxidize in the expansion phase, is crucially important. But the temperature level too is important. On account of important local interactions between the injection spray, the combustion zone, the non-combusted mixture and the recess geometry, how the combustion process is fashioned has a decisive impact on the production of emissions.

Figure 18: NO_x and soot emissions
1 Upper limit curve,
2 Lower limit curve.
A Decrease in soot and NO_x as a result of: exhaust-gas recirculation and injection-pressure increase, partial homogenization, H_2O injection,
B Increase in soot and decrease in NO_x as a result of: ignition retard, injection-pressure reduction, reduction of O_2 concentration, Miller process,
C Decrease in soot and increase in NO_x as a result of: ignition advance, injection-pressure increase, increase in O_2 concentration.

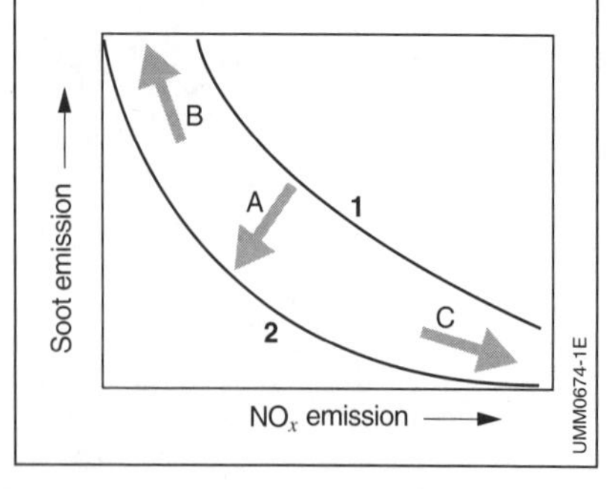

Temperature-reducing measures such as exhaust-gas recirculation, Miller processes (see e.g. [2]) or partial homogenization reduce the nitrogen-oxide levels. This overcompensates for the slight increase in soot emissions that are often observed as well (Figure 18). The level of complexity and expenditure needed to reduce both these components is high. The measure currently used involves increasingly higher exhaust-gas recirculation rates for nitrogen-oxide reduction in combination with very high injection pressures (> 2,000 bar) for soot reduction.

A distinction is made between soot and particulate emissions in this context. Soot consists of pure carbon compounds, whereas particulates also include fuel or lube-oil droplets, ash, rubbed-off metal, corrosion products, and sulfate compounds.

HC and CO compounds are typically uncritical in diesel-engine emissions. The effect of hydrocarbons on particulate emissions must be taken into consideration. HC and CO increase in concentration in particular during a highly retarded combustion with incomplete burn behavior.

Mixed forms and alternative operating strategies

The classic diesel-engine operating strategy is characterized by one or more injections in the TDC range. Established gasoline-engine combustion processes are characterized by homogeneous or partially homogeneous (with charge stratification) operation. However, alternative forms of process control, which sometimes cannot be clearly assigned to the diesel-engine or the gasoline-engine process, are currently being developed.

Homogeneous-charge compression ignition in diesel operation

With the HCCI combustion process (Homogeneous Charge Compression Ignition), which is published in a wide range of variations [3], the motivation is, through the application of very advanced injection (at least 40 to 50° CA before TDC), to achieve homogenization, significant leaning and therefore a NO_x reduction. Reliable ignition nevertheless takes place on account of the high diesel-engine compression temperature. The compression ratio must be lowered to 14 to 16 in order to maintain control of the combustion process. Exhaust-gas recirculation is typically used to enable the temperature level in the cylinder to be increased at low loads. Nevertheless, obtaining this over the entire program map, particularly in the full-load range, is very demanding, since the pressure-increase gradients become extremely high and control of transient engine operation is highly complex in view of all the engine states that arise.

Compression ignition in gasoline operation

Gasoline-engine combustion processes have been further developed along similar lines to diesel-engine HCCI operation in order to achieve in the part-load range dethrottled lean-burn operation with corresponding fuel-consumption advantages over conventional, stoichiometrically operated engines. The disadvantages of lean-burn operation with regard to conversion in the catalytic converter are compensated by extremely low untreated NO_x emissions caused by leaning. Reliable ignition of the in itself hardly combustible mixture is achieved by a high compression ratio above 13. Optimally the compression ratio is variable and can be further decreased up to full load on account of the increased combustion-chamber temperatures.

Gasoline-engine stratified charge

Direct-injection, stratified-charge gasoline-engine combustion processes share much in common in terms of combustion with classic diesel-engine operation and therefore represent a mixed form of conventional gasoline-engine and diesel-engine combustion. Combustion processes of this type are becoming increasingly established thanks to their efficiency advantages by dethrottling in part-load operation.

Multifuel engines

Multifuel engines, which are characterized by compatibility with regard to the fuel used (alternatively operation e.g. with diesel fuels or kerosene, gasolines or diesel fuels or vegetable oils), no longer play a role today, since the emission requirements for engine concepts of this type can no longer be met.

References

[1] H. Oertel; M. Böhle; U. Dohrmann: Strömungsmechanik. 5th Ed., Vieweg+Teubner, 2008.

[2] R. van Basshuysen; F. Schäfer (Editors): Handbuch Verbrennungsmotor, 4th Ed., Vieweg+Teubner, 2007.

[3] K. Boulouchos: Strategies for Future Combustion Systems – Homogeneous or Stratified Charge? SAE 2000-01-0650.

Charge cycle and supercharging

Gas exchange

In combustion engines employing open processes and internal combustion, the gas-exchange (exhaust and refill) system must serve two decisive functions:
- Replacement is employed to return the gas medium to its initial (start of cycle) condition
- The oxygen required to burn the fuel is provided in the form of fresh air.

The parameters defined in DIN 1940 [1] can be used to evaluate the gas-exchange process. For overall air flow (air expenditure $\lambda_a = m_g/m_t$) the entire charge transferred during the working cycle m_g is defined with reference to the theoretical maximum m_t for specific displacement. In contrast, the volumetric efficiency $\lambda_{a1} = m_z/m_t$ is based exclusively on the fresh charge m_z actually present or remaining in the cylinder. The difference between this and the total charge transfer m_g consists in the proportion of gas that flows directly into the exhaust tract in the overlap phase, making it unavailable for subsequent combustion.

The retention rate $\lambda_a = m_z/m_g$ is an index of the residual charge in the cylinder.

The scavenge efficiency $\lambda_S = m_z/(m_z + m_r)$ indicates the volume of the fresh charge m_z relative to the existing total charge, consisting of the fresh charge and the residual gas m_r. Here, the parameter m_r indicates the amount of residual gas from earlier working cycles remaining in the cylinder after the exhaust process.

In a 2-stroke cycle, the gas is exchanged with every rotation of the crankshaft at the end of the expansion in the area around bottom dead center. In a 4-stroke cycle, separate intake and exhaust strokes provide a supplementary gas-exchange cycle.

4-stroke cycle

Valve timing – and thus gas exchange – are regulated by a control shaft (camshaft) rotating at half the frequency of the crankshaft by which it is driven. The camshaft opens the gas-exchange valves by depressing them against the valve springs to discharge the exhaust gas and to draw in the fresh gas (exhaust and intake valves respectively) (Figure 1). Just before piston bottom dead center (BDC), the exhaust valve opens and approx. 50 % of the combustion gases leave the combustion chamber under a supercritical pressure ratio during this predischarge phase. As it moves upward during the exhaust stroke, the piston sweeps nearly all of the combustion gases from the combustion chamber.

Shortly ahead of piston top dead center (TDC) and before the exhaust valve has closed, the intake valve opens. This crankshaft top dead center position is called the gas-exchange TDC (GTDC) or overlap TDC (OTDC) because the intake and exhaust processes overlap at this point) in order to distinguish it from the ignition TDC (ITDC). Shortly after gas-

Figure 1: Representation of the 4-stroke gas-exchange process in the pV diagram
I Intake, E Exhaust,
IC Intake closes, IO Intake opens,
EC Exhaust closes, EO Exhaust opens,
V_c Compression volume, V_h Displacement,
The arrows indicate the direction of the curves.

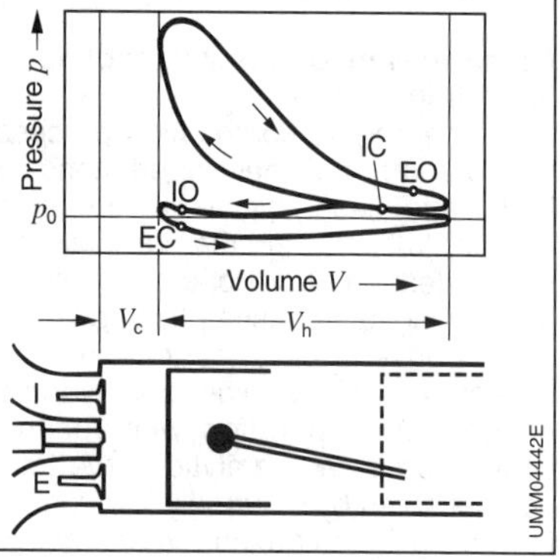

exchange TDC, the exhaust valve closes and, with the intake valve still open, the piston draws in fresh air on its downward stroke. This stroke of the gas-exchange process, the induction cycle, is completed shortly after bottom dead center (BDC) is reached. The subsequent two strokes in the 4-stroke process (Figure 2) are compression and combustion (expansion).

On throttle-controlled gasoline engines during the valve overlap period, exhaust gases flow directly from the combustion chamber into the intake passage, or from the exhaust passage back into the combustion chamber and from there into the intake passage. This internal exhaust-gas recirculation increases the combustion-chamber-charge temperature and the proportion of inert gas in the cylinder. A consequence of this is non-optimum power utilization in the upper load range. With fixed valve timing it is important to maintain a compromise, which is dependent above all in gasoline-engine applications on matching the engine operating strategy with the turbocharging unit.

Figure 2: 4-stroke gas-exchange process
E Exhaust, I Intake,
EO Exhaust opens, EC Exhaust closes,
IO Intake opens, IC Intake closes,
TDC Top dead center,
OTDC Overlap TDC, ITDC Ignition TDC,
BDC Bottom dead center,
IP Ignition point.

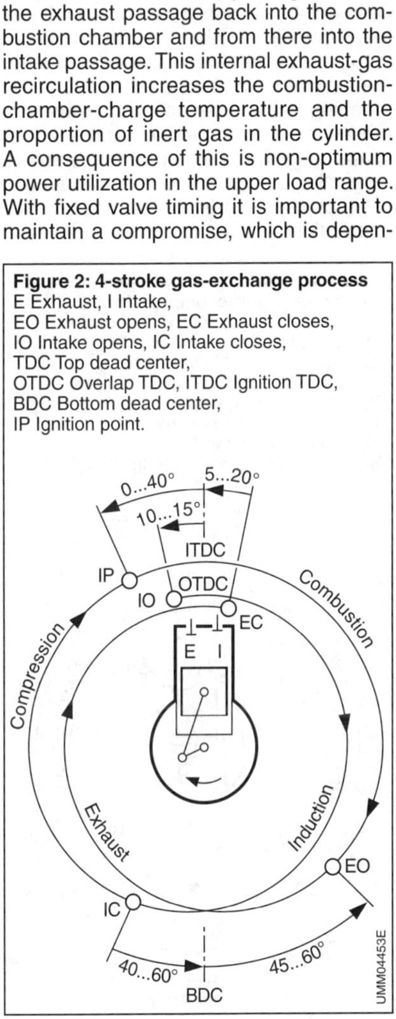

Early exhaust-valve timing allows sufficient time for the cylinder charge to escape, and thus guarantees low residual-gas compression as the piston sweeps through its upward stroke, although at the price of a reduction in the work index for the combustion gases.

The "intake valve closes" position (IC timing) exercises a decisive effect on the relationship between volumetric efficiency and engine speed. The maximum volumetric efficiency is achieved at low engine speeds when the intake valve closes early, and at high engine speeds when it closes late.

Obviously, fixed valve timing will always represent a compromise between two different design objectives: maximum brake mean effective pressure – and thus torque – at the most desirable points on the curve, and the highest possible peak output. The higher the engine speed at which maximum power occurs, and the wider the range of engine operating speeds, the less satisfactory will be the ultimate compromise. This tendency cannot be neutralized either by multi-valve cylinder heads with a larger intake-flow cross-sectional area.

At the same time, the demands for minimum exhaust emissions and maximum fuel economy mean that low idle speeds and high low-end torque (despite and along with high specific outputs for reasons of power-unit weight) are becoming increasingly important. These imperatives lead to the application of variable valve timing.

The advantages of the 4-stroke process are very good volumetric efficiency over the entire engine-speed range, low sensitivity to pressure losses in the exhaust system, and relatively good control of the speed-sensitive air expenditure through selection of appropriate valve timing and intake-system designs.

The disadvantages of the 4-stroke process are the high complexity of valve timing and the reduced power density of non-turbocharged 4-stroke engines because only every second crankshaft revolution is used for power output by combustion.

2-stroke cycle

To maintain gas exchange without an additional crankshaft rotation, the gases are exchanged in the two-stroke process at the end of expansion and at the beginning of the compression stroke. The intake and exhaust timing are usually controlled by the piston as it sweeps past the intake and exhaust ports in the cylinder housing near BDC (Figure 4). This configuration, however, requires symmetrical control times and involves the problem of directly flushing fresh mixture into the exhaust area (short-circuit scavenging). In addition, 15 to 25 % of the piston stroke cannot produce work because only charge volume V_f and not displacement volume V_h can be exploited for power generation (Figure 3).

As the two-stroke process lacks separate intake and exhaust strokes, the cylinder must be filled and scavenged using positive pressure, necessitating the use of scavenging pumps. In an especially simple and very frequently-used design, the bottom surface of the piston works in conjunction with a crankcase featuring a minimal dead volume to form a scavenging pump. Figure 4 shows a 2-stroke engine with crankcase scavenging and precompression along with the associated control processes. The processes which take place on the scavenging pump side are shown in the inner circle, while those occurring on the cylinder side are shown in the outer circle. The location of the intake and exhaust ports with the aim of maximum volumetric efficiency is also crucially dependent on the piston position. A hump-shaped increase in piston height can reduce direct throughflow from the intake area to the exhaust area.

The advantages of the 2-stroke process are high power output referred to engine weight and engine volume and a more uniform torque generation (one power stroke per revolution).

The disadvantages of the 2-stroke process are higher fuel consumption, lower mean pressures (due to poorer cylinder charge), higher thermal loads (due to the lack of a gas-exchange stroke), demanding mixture control, and higher HC emissions due to problematic cylinder scavenging. Two-stroke processes are no longer important due to the introduction of strict emission limits for most mobile applications.

Figure 3: Representation of the 2-stroke gas-exchange process in the pV diagram
p_0 Charge pressure, V_S Scavenging volume, V_c Compression volume, V_h Displacement, V_f Charge volume.

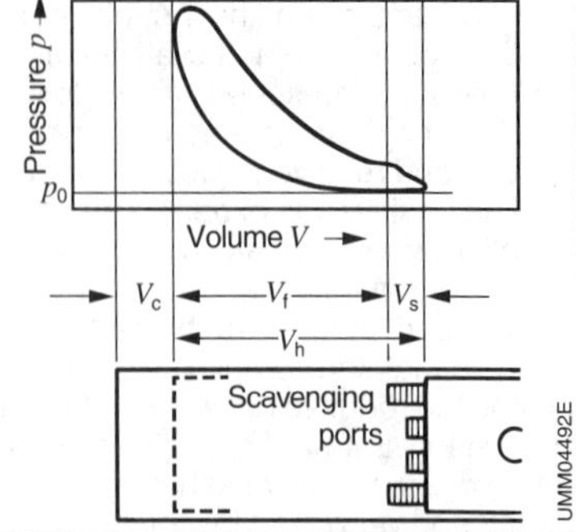

Figure 4: 2-stroke gas-exchange process with crankcase precompression
E Exhaust, I Intake, EO Exhaust opens, EC Exhaust closes, IO Intake opens, IC Intake closes, T Transfer passage, TO Transfer passage opens, TC Transfer passage closes, TDC Top dead center, BDC Bottom dead center, IP Ignition point.

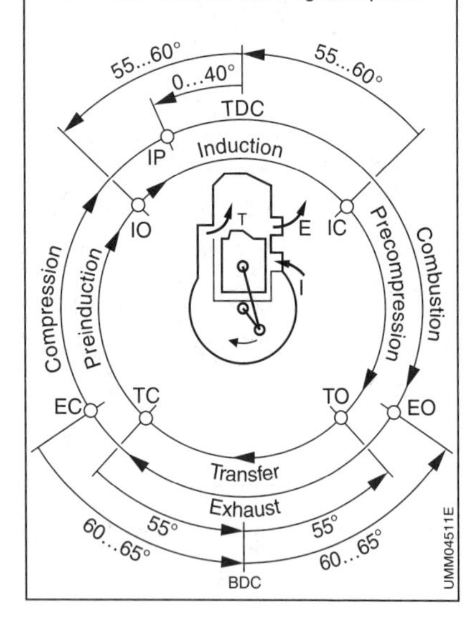

Variable valve timing

Variable valve timing in a number of design variations is gaining acceptance with different objectives. The motivations behind the use of variable valve timing are as follows: increased power and torque, dethrottling, control of the residual-exhaust-gas content, cylinder shutoff, charge movement, cold-starting and cold-running performance, control-engineering optimization (e.g. charge mass, residual-exhaust-gas content), influencing the rotational speed of the exhaust-gas turbocharger, and exhaust-gas temperature-management measures.

Today, different systems are used to provide variable valve timing. There are different motivations behind variable valve timing, depending on the combustion process and engine concept (homogeneous- or stratified-charge operation, diesel).

Camshaft adjustment

The concept of adjusting the intake and exhaust camshafts is increasingly becoming an accepted measure, offering a valuable degree of variability in the entire program map. This concept provides the option of adjusting the phase of the camshafts with respect to the crankshaft and thereby shifting the valve timing while maintaining the same form of opening. Electric or electrohydraulic activation of the camshaft adjuster is the state-of-the-art technology used.

In homogeneously operated gasoline engines the intake camshaft is subject to retarded adjustment for example in the low speed and load range while at the same time the exhaust valve is subject to very advanced closing (Figure 5). This minimal valve overlap delivers a minimal throughflow of fresh-air mixture into the exhaust area. At the same time the highly retarded opening of the intake valve also causes dethrottling by the associated retarded closing of the intake valve, because the cylinder charge is again discharged by the intake valve after bottom dead center. A rapid torque increase can be achieved for a desired load point by advancing the intake camshaft.

Exhaust-camshaft adjustment also provides for further degrees of freedom. Optimal adjustment of the intake and exhaust camshafts is dependent on numerous factors. The operating mode is crucial as well as the position in the engine map. For example, a significant factor is whether exhaust-gas turbocharging is used or whether lean-burn operation (diesel engine, gasoline-engine stratified-charge process, etc.) is applied. One example of the beneficial combination of camshaft adjustment with an exhaust-gas turbocharger is the map-dependent maximum valve overlap (very retarded closing of the exhaust valve and very advanced closing of the intake valve), at which a large proportion of mixture flows directly from the intake to the exhaust area. This can give rise to the beneficial effect of an increase in air-mass flow by the exhaust-gas turbocharger together with an advantageous increase in turbocharger speed. Rotating the intake and exhaust camshafts offers optimization opportunities for a large number of questions and is therefore being increasingly used at least in modern gasoline engines.

Systems with cam-shape control mechanism

As well as continuous phase adjustment, simpler systems are also available which alternate digitally between individual cam shapes. Typically, such systems adjust a very large cam and valve lift for ranges

Figure 5: Intake-camshaft rotation
1 Retard, 2 Standard, 3 Advance.

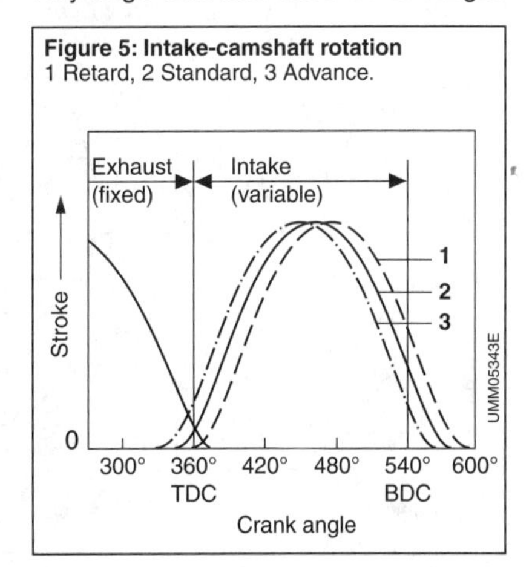

close to full load and a small cam and valve lift for dethrottling for part-load ranges. This is achieved for the most part by integrating a control mechanism on a rocker arm or bucket tappet, which thereby activates one or more (typically two) available cams on the camshaft.

Systems that can alternate between two camshaft rotation positions are also used. In contrast to fully variable systems, these systems omit an exact positioning command to the camshaft phase adjuster.

Other variable systems
Increasingly, fully variable systems for use in gasoline engines are being developed and are even already appearing in production vehicles. These systems focus on optimizing the charge-cycle phase. There are different successful approaches to suit the engine configuration and operating point, each approach optimizing charge-cycle losses, the residual-exhaust-gas content or engine power (Figure 6).

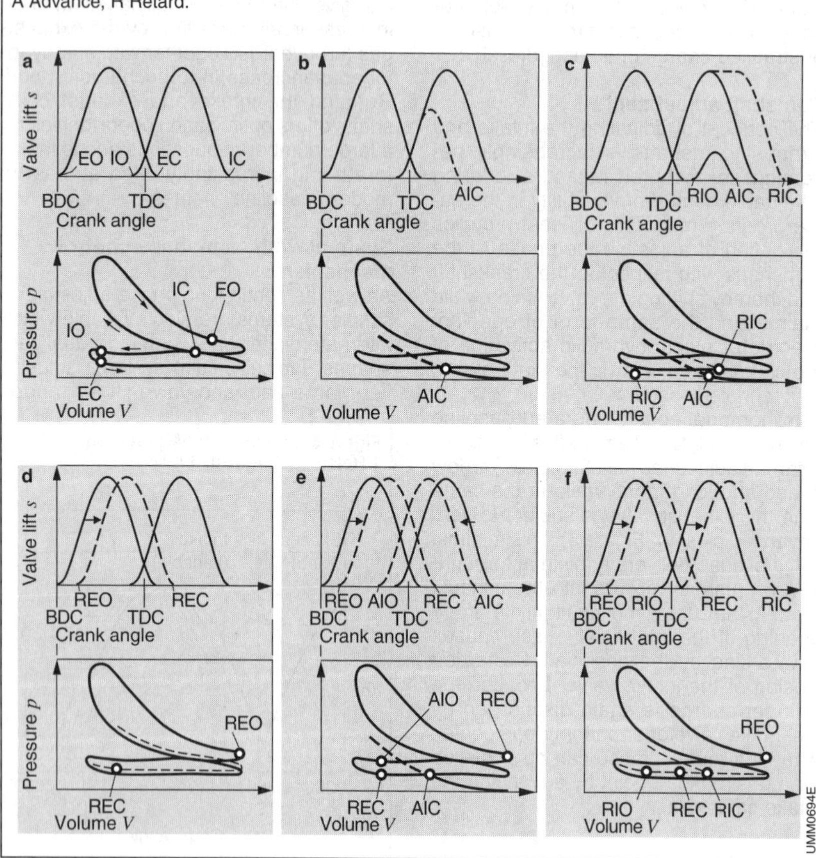

Figure 6: Influence of valve timing on the charge cycle
a) Conventional valve timing, b) Advanced "Intake closes",
c) Valve-lift reduction and retarded "Intake closes", d) Retarded "Exhaust closes",
e) Retarded "Exhaust closes" and advanced "Intake closes",
f) Retarded "Exhaust closes" and retarded "Intake closes".
BDC Bottom dead center, TDC Top dead center, IO Intake opens,
EO Exhaust opens, IC Intake closes, EC Exhaust closes,
A Advance, R Retard.

Mechanical, electromechanical, electrohydraulic or electropneumatic adjustment concepts can be used to provide a fully variable valve gear.

Mechanical systems
Mechanical, fully variable valve-gear systems typically consist of a combination of an adjustment mechanism, which permits variation of the valve lift, and a camshaft phase adjuster. The key function is the provision of variable cam lift.

A suitable design solution is provided in this case by a conventional camshaft, which does not however act directly on the valve via a bucket tappet or a rocker arm. The cam lift is adjusted by an intermediate lever whose fulcrum can be varied by means of an eccentric shaft (Figure 7 and Figure 8). Electric DC motors are typically used to drive such arrangements. BMW Valvetronic is an example of a fully variable mechanical valve gear [2]. Further advantages of a fully variable valve gear are the optimization of uniform charge distribution and the potential to increase charge movement by having the two intake valves opening at different times in the lower load range thanks to slightly differently ground eccentricities.

Figure 7: Valve-lift curves of a mechanical fully variable valve gear

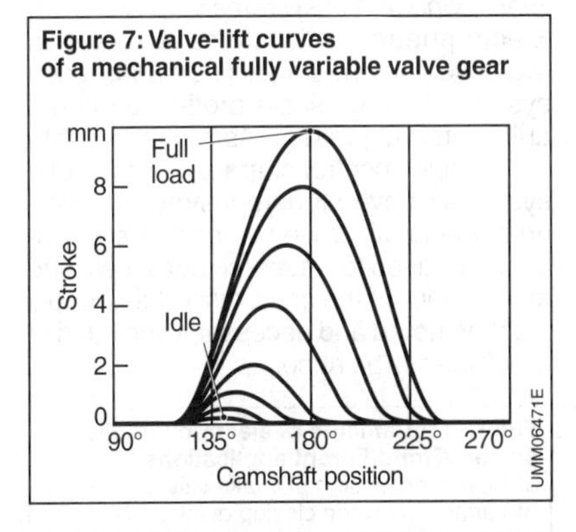

Figure 8: Basic principle of a mechanical fully variable valve gear (BMW Valvetronic)
1 Guide block,
2 Center of rotation,
3 Camshaft,
4 Hydraulic valve-clearance compensation,
5 Servo unit with spiral-toothed gear,
6 Eccentric shaft,
7 Intermediate lever,
8 Intake valve.

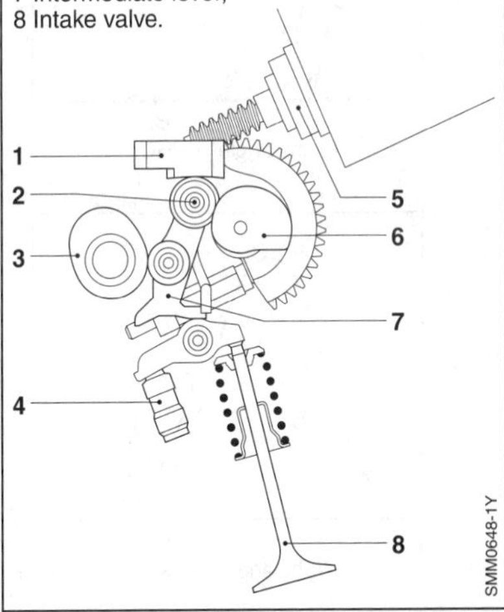

Electromechanical systems
Electromechanical systems (electromechanical valve gear) are still in the development stage and have not made it to series production yet. In an electromechanical valve gear, electrically actuated magnets are used as actuators for

Figure 9: Electromechanical systems
1 Compression spring,
2 Closing magnet,
3 Armature,
4 Opening magnet,
5 Compression spring,
6 Intake valve.

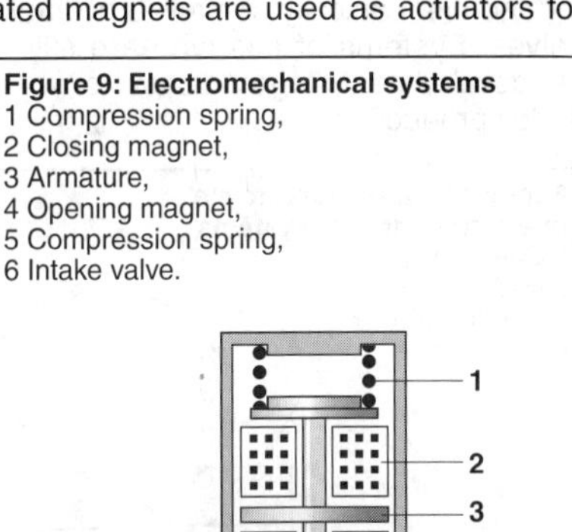

valve timing (Figure 9, see also [3]). The high power demand of the actuators must be noted in particular. The system, comprising valve, magnet and coil, is made to resonate in order to minimize this demand. Because of the high power demand and the complexity involved, electromagnetic valve gears have not yet made it to series production.

Electrohydraulic systems
Electrohydraulic systems (electrohydraulic valve timing) are an alternative to the fully variable mechanical valve gear. Different principles are applicable here.

The principle of "lost motion" is an effective approach. A motion is stipulated by a hydraulic intermediate element via a camshaft (Figure 10). An electrically actuated hydraulic valve provides the opportunity of not transferring completely the motion stipulated by the cam. The cam shape therefore stipulates an envelope (Figure 11).

An alternative is presented by a system which directly actuates the gas-exchange valves via a hydraulic pressure accumulator and electronically controlled hydraulic valves. Systems of this type are still in the development stage and are not yet in series production.

Figure 10: Working principle of electrohydraulic systems
1 Camshaft,
2 Hydraulic fluid,
3 Control valve,
4 Intake valve.

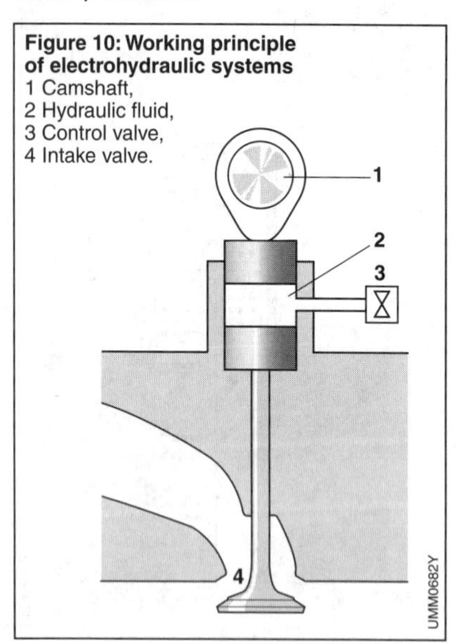

Depending on the point and duration of activation, a proportion of the hydraulic fluid escapes through the hydraulic valve. Hydraulic power loss is therefore a drawback of this system.

Electrohydraulic systems have been deployed since 2004 in Caterpillar commercial-vehicle systems and since 2010 in Fiat's MultiAir system.

Electropneumatic systems
Electropneumatic systems are still in the development stage. The use of such systems in large-scale production is not anticipated at present. As well as exhibiting complex control characteristics, these systems above all demonstrate unfavorable pneumatic drive power. Current systems are used for research purposes. The power input of the compressed-air supply must be noted and necessarily included in an efficiency balance.

Figure 11: Cam lifts of electrohydraulic systems for different applications
a) Advanced closing of intake valve,
b) Variable retarded closing of intake valve.

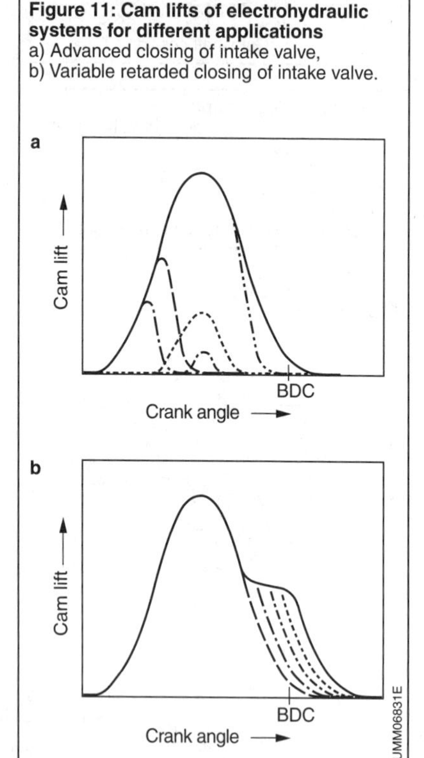

Supercharging processes

The power of an engine is proportional to its air-mass flow. As this air throughput, in turn, is linearly dependent on air density, the power of an engine, given a specific displacement and engine speed, can be increased by compressing the air before it enters the cylinders, i.e. by supercharging. The supercharging ratio indicates the density rise as compared to a naturally aspirated engine.

Thermodynamically speaking, the best results would be delivered by isothermal compression, but this cannot be technically achieved. Reversibly adiabatic compression in the supercharger unit serves as the ideal comparison process for the actual sequences; in practical terms, increased density is accompanied by losses.

The extent of the supercharging ratio is limited in gasoline engines by the occurrence of knocking combustion and in diesel engines by the maximum permissible peak pressure in the cylinder. To offset this limitation, supercharged engines usually have lower compression ratios than their naturally aspirated counterparts.

Dynamic supercharging

The gas-exchange processes are not only influenced by the valve timing, but also by the geometry of the intake and exhaust lines. Induced by the induction work of the piston, as the intake valve opens it initiates a suction wave in the intake manifold that reflects off the open end of the intake manifold and returns as a pressure wave to the intake valve. These pressure waves can be used to increase the air-mass intake (Figure 12). In addition to the geometry in the intake manifold, the impact of this supercharging effect based on gas dynamics also depends on the engine speed (see Ram-tube supercharging, Tuned-intake-tube charging, Variable-geometry intake-manifold systems).

Mechanical supercharging

In mechanical supercharging, the supercharger is driven directly by the internal-combustion engine (see Turbochargers and superchargers). The charger and the internal-combustion engine are mechanically coupled to each other. Types of mechanical supercharger unit are positive-displacement superchargers (compressors) of various designs (e.g. Roots superchargers, spiral-type superchargers) and hydrokinetic flow compressors (e.g. radial compressors).

The crankshaft and charger shaft feature a fixed transmission ratio in system designs used up to now. Mechanical or electromagnetic clutches can be used to control supercharger activation. The boost pressure is generally adjusted by a bypass device with control flap (wastegate).

Advantages of mechanical supercharging

- The supercharger is installed on the cold side of the engine.
- The engine's exhaust system is not influenced by supercharger components.
- The supercharger responds almost immediately to load changes.

Disadvantages of mechanical supercharging

- The drive power must be branched from the effective power of the engine, which results in increased fuel consumption.

Figure 12: Increase of volumetric efficiency through dynamic supercharging
The engine speed is scaled to the nominal speed.
1 System with dynamic supercharging,
2 System with standard intake manifold.

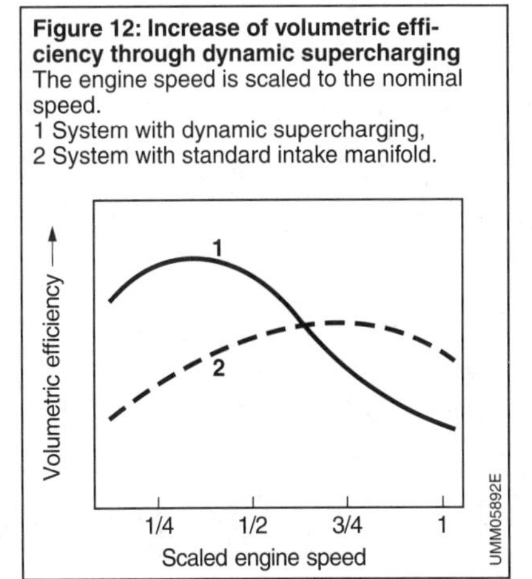

- Low-noise performance can only be achieved by additional measures.
- Comparatively high structural volume and weight.
- The supercharger unit has to be positioned at the engine's belt level.

Exhaust-gas turbocharging

In exhaust-gas turbocharging, the power for driving the charger is extracted from the engine's exhaust gas, i.e. some of the energy contained in the exhaust gas is converted into mechanical energy by an exhaust-gas turbine. In this way, the process exploits some of the enthalpy that remains unused (owing to crankshaft-assembly expansion limits) by naturally aspirated engines. Exhaust backpressure however increases in the process. Hydrokinetic flow compressors are exclusively used to compress the induction air (see Turbochargers and superchargers).

Figure 13: Comparison of power and torque curves between naturally aspirated engine and turbocharged engine
The engine speed is scaled to the nominal speed.
1 Naturally aspirated engine in stationary operation,
2 Turbocharged engine in stationary operation,
3 Torque build-up of a turbocharged engine in transient operation.
A→B Higher torque and higher power at identical speed for turbocharged engine.
C→B Identical power for turbocharged engine at lower speed.

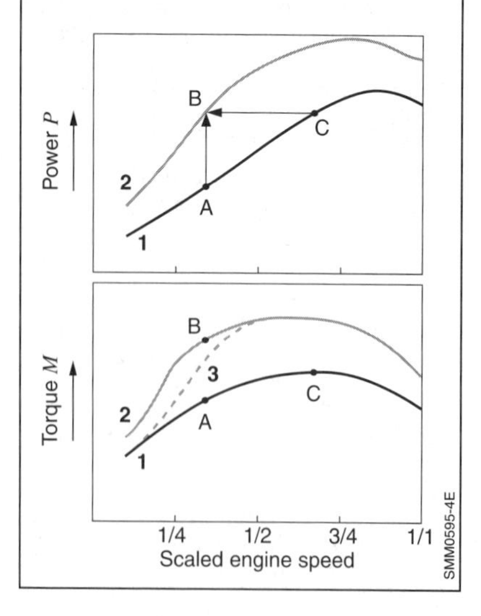

Exhaust-gas turbochargers are usually designed to generate a high boost pressure even at low engine speeds. In other words, the turbocharger turbine is generally designed for an engine mid-speed range. However, without further measures, boost pressure at the high end of the engine speed range could increase to levels that would place excessive load on the engine. The turbine is therefore equipped with a bypass valve which, from a specific operating point, diverts some of the exhaust-gas mass flow past the turbine. However, the energy of this exhaust gas remains unutilized. Much more satisfactory results can be achieved with a compromise between high boost pressure in the low engine-speed range and avoidance of engine overload at the high end of the engine-speed range by employing a turbocharger with variable turbine geometry (VTG). The adjusting blades used for this purpose adapt the flow cross-section and the angle of impact to the turbine (and thus the exhaust-gas pressure applied to the turbine) by varying the position of the guide blades (see Turbochargers and superchargers).

Advantages of exhaust-gas turbocharging
- Considerable increase in power output per liter.
- Improvement in fuel-consumption figures relative to naturally aspirated engines with the same output power.
- Improvement in exhaust-gas emission rates.
- Comparatively small structural volume.
- Can be used with low-pressure-side exhaust-gas recirculation.

Disadvantages of exhaust-gas turbocharging
- Installation of the turbine side of the turbocharger in the hot exhaust-gas system requiring materials resistant to high temperatures.
- Increased thermal inertia in the exhaust-gas system.
- Without further measures, comparatively low starting torque from low-displacement engines.

Special forms of supercharging

In electrically assisted exhaust-gas turbocharging systems, an additional electric motor serves to maintain the turbocharger at a high speed in the absence of exhaust-gas mass flow. This results in advantages above all in transient vehicle operation and at low engine speeds. These systems have not yet entered into series production in view of the very high complexity and power consumption involved. Downsizing can be successfully achieved with electrically assisted exhaust-gas turbocharging.

A special form of supercharging is pressure-wave supercharging (Comprex), which has not yet entered into series production. The working principle is based on the reflection properties of pressure waves which propagate in a rotating cell rotor (see Turbochargers and superchargers). The chief advantage is the transient response, which facilitates a very rapid torque build-up. However, the costs involved with pressure-wave superchargers are high and the required drive necessitates additional packaging measures.

Volumetric-flow map

A clear illustration of the relationship between the engine and the supercharger is provided by the diagram for pressure vs. volumetric flow (Figure 14), in which the pressure ratio π_c of the supercharger is plotted against volumetric flow $\dot{V}$.

The curves for unthrottled 4-stroke engines (diesel) are particularly descriptive because they contain sloped straight lines (engine mass-flow characteristics) which represent increasing air-throughput values as the pressure ratio $\pi_c = p_2/p_1$

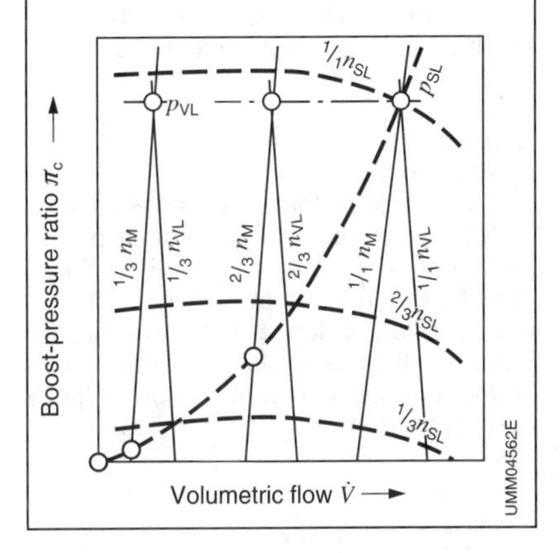

Figure 14: Pressure vs. volumetric-flow map of mechanically driven positive-displacement supercharger and hydrokinetic flow compressor
n Speed, p_L Boost pressure.
Subscripts:
VL Positive-displacement supercharger,
SL Hydrokinetic flow compressor.
M Engine.

(where p_1 denotes the ambient pressure and p_2 the boost pressure) increases at constant engine speed.

The diagram shows pressure ratios which result at corresponding constant supercharger speeds for a positive-displacement supercharger and a hydrokinetic compressor.

Only superchargers whose delivery rates vary linearly with their rotational speeds are suitable for vehicle engines. These are positive-displacement superchargers of piston or rotating-vane design or Roots blowers. Mechanically driven hydrokinetic flow compressors are not suitable.

Exhaust-gas recirculation

Operating principle

With exhaust-gas recirculation (EGR) hot exhaust gas is removed, cooled in an exhaust-gas cooler, mixed with fresh air, and returned to the combustion chamber. The exhaust-gas recirculation rate is adjusted by a valve. It is used in diesel engines and also increasingly in gasoline engines.

Exhaust-gas recirculation in diesel engines

In diesel engines the untreated emissions of NO_x are reduced significantly by exhaust-gas recirculation. Essentially, this is done by lowering the peak temperature in the cylinder, which in turn is achieved because the recirculated exhaust gas firstly reduces the combustion velocity and secondly must be heated itself.

Figure 15: High-pressure exhaust-gas recirculation
The arrows denote the flow direction.
1 EGR cooler for high-pressure EGR,
2 Engine,
3 Exhaust-gas-turbocharger turbine,
4 Exhaust-gas turbocharger compressor,
5 Intercooler,
6 EGR valve.

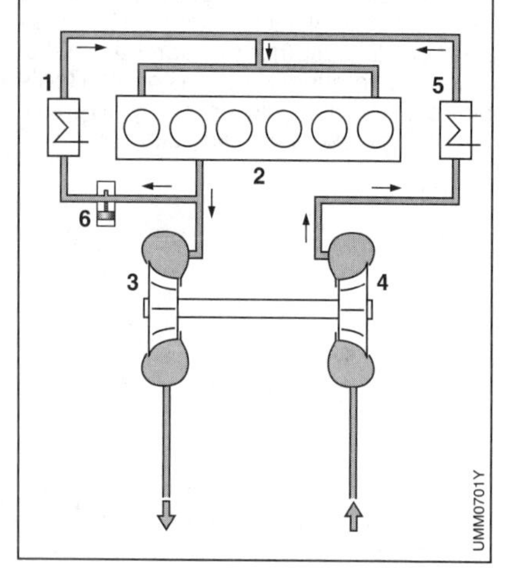

Exhaust-gas recirculation in gasoline engines

In gasoline engines the idea is not to reduce the nitrogen oxides. The aim is primarily to reduce the exhaust-gas temperature to protect components (catalytic converter and turbocharger). This is very effective particularly in engine full-load operation because the otherwise usual process of enriching the air/fuel mixture to reduce the exhaust-gas temperature can be partially or completely dispensed with. This results in a dramatic improvement in fuel consumption.

An additional consideration in gasoline engines the possibility of diluting the charge by means of exhaust-gas recirculation without compromising the function of the three-way catalytic converter. Engine dethrottling can be achieved by increasing the EGR rate particularly in engine part-load operation, which also results in improved fuel efficiency.

Exhaust-gas recirculation variants

Essentially, one of two variants – high-pressure and low-pressure recirculation – is used, depending on whether the exhaust-gas cooler is located upstream or downstream of the turbocharger. Whereas high-pressure recirculation is found primarily in commercial vehicles, both variants are used in passenger cars.

High-pressure EGR

With high-pressure recirculation (Figure 15), exhaust gas is recirculated on the high-pressure side, i.e. upstream of the turbocharger turbine. It has the advantage that the pressure difference required to transport the exhaust gas from the exhaust side to the intake side is higher and can even be altered by means of variabilities at the turbocharger (wastegate or variable turbine geometry). High EGR rates can basically be achieved because the EGR section is short and can therefore be filled and emptied quickly. Thanks to this dynamic advantage, the EGR system can respond relatively quickly to changing operating conditions.

Because the recirculated exhaust gas is not mixed with fresh air until downstream of the intercooler, the intercooler and the turbocharger compressor are not permeated and thus not fouled by the exhaust gas. The performance of these components therefore remains constant over their full service lives. By contrast, significantly greater demands are placed on the EGR cooler on the high-pressure side. This can be put down to the higher pressures and temperatures during operation and to the already mentioned component sooting.

Low-pressure EGR
This form of exhaust-gas recirculation routes the exhaust gas only downstream of the turbine through the EGR cooler and mixes it with the inducted fresh air upstream of the turbocharger compressor (Figure 16). The exhaust gas arrives at the EGR cooler only after it has passed through the diesel particulate filter (DPF, not shown in the illustration) and the soot has been removed. Thus the EGR cooler is not fouled and performs consistently over its full service life. Because of the lower temperature in this area of the exhaust system the temperature stresses in the EGR cooler are also much lower than is the case for high-pressure recirculation. The lack of fouling together with a lower inlet temperature provide for a lower cooler outlet temperature than with high-pressure recirculation. The reduction in NO_x emissions that can thereby be achieved can be sufficient in many cases to eliminate the need for further exhaust-gas treatment, such as the SCR system for example.

One drawback of low-pressure recirculation is the reduced dynamic response. The system is relatively far removed from the engine and, due to the long distances involved, cannot adapt the exhaust-gas recirculation rate quickly. It is necessary to set the EGR rates slightly lower than for high-pressure recirculation so as to prevent the engine from being "flooded" with exhaust gas.

Figure 16: Low-pressure exhaust-gas recirculation
The arrows denote the flow direction.
1 EGR cooler for low-pressure EGR,
2 Engine,
3 Exhaust-gas-turbocharger turbine,
4 Exhaust-gas turbocharger compressor,
5 Intercooler,
6 EGR valve.

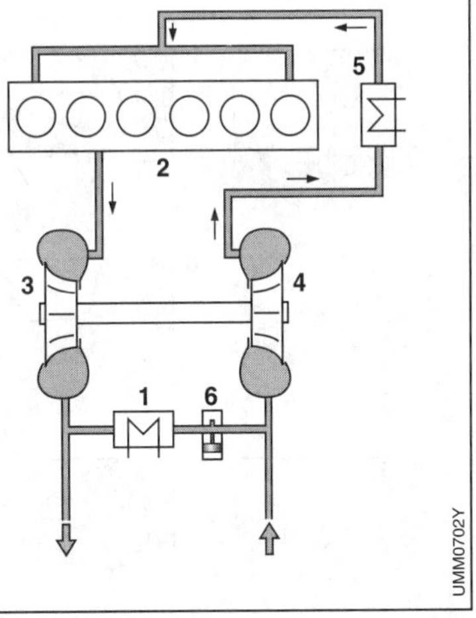

UMM0702Y

References
[1] DIN 1940: Reciprocating internal combustion engines; terms, formulae, units.
[2] R. Flierl; R. Hofmann; C. Landerl; T. Melcher; H. Steyer: Der neue BMW-Vierzylindermotor mit Valvetronic. MTZ 62 (2001), Volume 6.
[3] P. Langen; R. Cosfeld; A. Grudno; K. Reif: Der Elektromechanische Ventiltrieb als Basis zukünftiger Ottomotorkonzepte. 21st International Vienna Motor Symposium, Progress Reports VDI, Series 12, No. 420, Volume 2, 2000.
[4] R. van Basshuysen; F. Schäfer (Editors): Handbuch Verbrennungsmotor, 5th Edition, Vieweg-Verlag, 2009.

Reciprocating-piston engine

Components

In addition to the crankshaft drive, which makes a vital contribution by generating torque, the significance of the cylinder head is crucial to the efficiency of an internal-combustion engine (Table 1).

Crankshaft drive and crankcase
The crankshaft drive comprises the following components: piston, piston rings, connecting rod, and crankshaft. What all the crankshaft-drive components have in common is a translational and rotational motion. The tribological design of the crankshaft drive is already hugely significant in view of its robustness and reduced engine friction.

Piston
Because of the mechanical and thermal loads to which it is subjected, the piston is a highly complex component in an internal-combustion engine. The recess and piston-surface geometries have a crucial impact on mixture formation and combustion. In addition, the primary function of the piston is to transmit force to the connecting rod. Complex mechanical stresses occur in the process with simultaneously very high local temperatures up to and in excess of 300 °C.

The role of the piston recess is even more important in diesel engines. With the high compression ratio and a typically flat cylinder head, the entire combustion chamber must be accommodated in the piston recess. Various piston-recess designs are successfully used. Figure 1 shows examples of the design variations used in diesel engines.

The piston recess decisively defines the combustion-chamber shape and at the same time influences the strength of the piston crown, a factor which is vital to the function of force input into the piston pin. Especially when it comes to aluminum pistons, which are currently the preferred choice mainly for weight reasons, the requirements of high heat resistance, low weight and cylinder peak pressure call for technically sophisticated solutions.

Brass bushings are used to accommodate the piston pin in diesel engines subject to high loads.

Diesel engines can also utilize steel and even gray-cast-iron pistons as well as aluminum designs. Steel pistons are used in applications with peak pressures well in excess of 200 bar. An important distinguishing feature of steel is its lower heat conduction.

Modern turbocharged engines with high cylinder peak pressures and high power outputs also require a piston spray injector with an additional piston cooling duct for cooling purposes (Figure 2). Here engine oil is sprayed through a vertically arranged piston spray injector into the piston cooling duct. The inlet and outlets for the piston cooling duct are located on the underside of the piston.

Piston rings form the sealing element between the combustion chamber and the crankcase (Figure 3). As well as providing the seal, the piston rings also perform the important tasks of conducting heat to the

Figure 1: Piston shapes for diesel engines
a) Deep omega recess,
b) Flat omega recess,
c) Stepped recess,
d) Dish recess,
e) Eccentric recess design,
f) W recess.

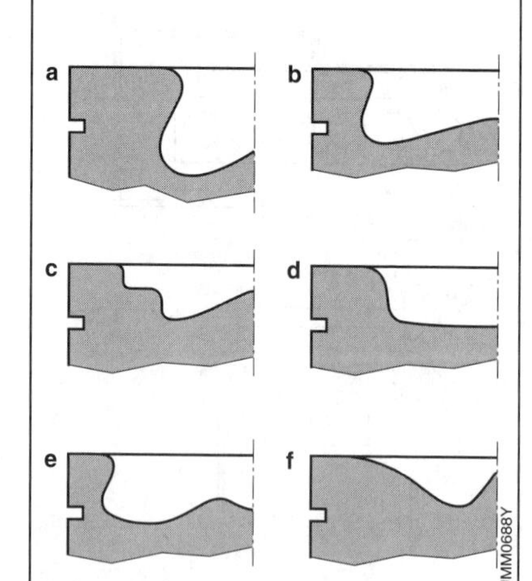

cylinder wall and controlling the entry of oil into the combustion chamber.

Three rings are typically used, the top ring acting as a compression ring and the bottom ring as an oil control ring. The middle ring frequently performs both functions. Because particulate filters become clogged with ash, which is also a product of burnt oil in the combustion chamber, oil consumption and therefore the coordination of the piston-ring set and honing are vitally important.

Table 1: Components of an internal-combustion engine

Assembly	Component	Purpose and function	Stress
Crankcase	Crankcase	– Housing for crankshaft drive – Accommodation of cylinder head and counterpart for accommodation of cylinder-head bolts – Accommodation of crankshaft bearings – Accommodation of accessories	Deformation – flow of force from cylinder-head bolts to crankshaft, bending moments, vibrations, oscillations and natural frequencies
	Bearing surface	Different approaches to structural solution of bearing surface: – Accommodation of liners with seal (wet liners) – Accommodation of cast liners (dry liners) – Piston bearing surface directly integrated in crankcase	Deformation (out-of-round), tribological counterpart to piston
Crankshaft drive	Piston	– Conversion of energy in combustion chamber (gas pressure) into a translational motion – Dissipation of heat from combustion chamber to cooling medium – Support of the normal forces occurring as a result of the connecting-rod inclination	Complex mechanical and local thermal stress above 350 °C in the recess area
	Piston rings	– Sealing of combustion chamber in relation to crankcase – Avoidance of oil entry into cylinder – Dissipation of heat to cylinder wall	High mechanical bending stress and high thermal stress, complex tribology
	Connecting rod and pin	– Absorption of the forces acting on the connecting rod and transmission to the crankshaft for torque build-up	High mechanical compressive, tensile and bending stresses, complex tribological processes
	Crankshaft	– Absorption of the forces acting on the connecting rod and conversion of oscillatory to rotational motion. Thus build-up of a torque for transmission to the drivetrain	High mechanical compressive, tensile and bending stresses (buckling), complex tribological processes
Cylinder head	Cylinder head	– Control of charge cycle – Cooling of areas close to the combustion chamber – Accommodation of injectors (in diesel engine and gasoline engine with direct injection) and of spark plug (in gasoline engine)	High stress due to thermomechanical fatigue behavior and cold-hot expansion

Table 1: Components of an internal-combustion engine (continued)

Assembly	Component	Purpose and function	Stress
Cylinder head	Camshaft (typically in cylinder head)	– Conversion of a rotational motion into a translational motion to open and close the valves	Hertzian stress
	Valve control	– Conversion of a rotational motion into a translational motion to open and close the valves	Marked valve-closing speeds cause high valve-spring forces
	Valves	– Opening of passages to intake and exhaust ports – Sealing of combustion chamber during compression phase	Valve-seat wear due to high valve-closing speeds, thermal load at exhaust valve
	Ports/ passages	– Supply and discharge of cylinder charge – Provision of the desired cylinder-charge movement	Local thermal load
	Base plate	– Accommodation of combustion-chamber pressure and sealing	– Local thermal load – Bending
	Geartrain	– Driving of camshaft by belt, chain or gear wheels	Rotational irregularities with alternating contact flanks
Other components	Oil pump	– Provision of the required oil delivery rate and provisions of the required oil pressure	Torque particularly with viscous oil
	Oil filter	– Filtering of contaminants from engine oil	–
	Water pump	– Provision of the required cooling-medium delivery rate – Provision of the required cooling-medium mass flow	Cavitation

Figure 2: Piston shapes in various engine designs
a) Commercial-vehicle aluminum diesel-engine piston with ring carrier and cooling channel,
b) Commercial-vehicle forged-steel piston,
c) Passenger-car aluminum diesel-engine piston with ring carrier and cooling channel,
d) Passenger-car aluminum diesel-engine piston with cooled ring carrier,
e) Passenger-car aluminum piston for MPI gasoline engine (Multi-Point Injection),
f) Passenger-car aluminum piston for gasoline engine with GDI (gasoline direct injection).

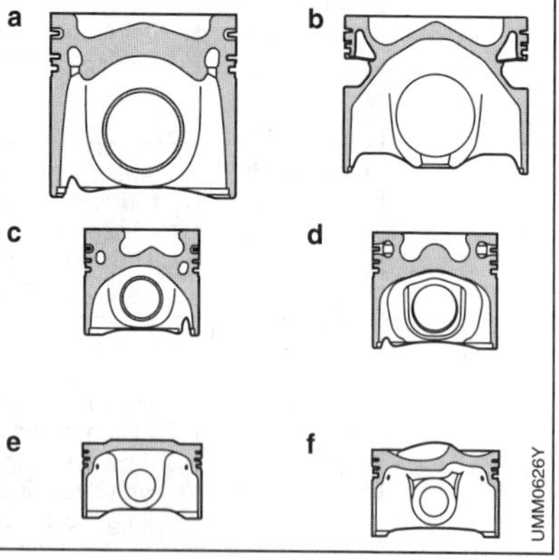

Connecting rod

The connecting rod (Figure 4) links the piston with the crankshaft. It transmits the piston's gas and inertial forces introduced into the piston pin to the crank pin. It is therefore subjected to tensile, compressive and bending stresses. It also carries the bearings for the crankshaft and usually for the piston pin. It must therefore be particularly stiff in its design, especially in the connecting-rod eyes, which accommodate the bearings. For this reason, it is usually designed with a double-T-shaped shank profile and drop-forged with tempering steels. Sintered materials can also be successfully used. Annealed cast iron can also sometimes be used in low-load applications (gasoline engines). Rarely encountered is the combination of a "standard" connecting rod and a forked connecting rod in V-engines which then permits an exactly opposed cylinder arrangement.

The length of the connecting rod is determined by the piston stroke and the counterweight radius (the geometric path of the outer contour is called the "connecting-rod violin"). In commercial-vehicle engines with a relatively large stroke/bore ratio of more than 1.1 and large bearing diameters an angled split of the big connecting-rod eye is customary. This enables the piston to be removed without the crankshaft having to be removed also. However, this angled split generates considerable shear forces even when the connecting rod is positioned vertically at top dead center. These forces are usually absorbed by toothing, groove and spring or by cracked surfaces. Cracking of the big connecting-rod eye is achieved by a "fracture separation" using extremely high (hydraulic) expansion forces and a targeted weakening of the cross-section at the big connecting-rod eye. The big-end bearing cap and the connecting rod can only be installed as a single unit.

The bolts are designed as through-bolts with or without nuts. Adapter sleeves or bolt adapter collars ensure that the bearing shells are aligned when seated.

Lateral guidance of the connecting rod can be performed in the crankshaft at the big connecting-rod eye or in the piston at the small connecting-rod eye.

The small connecting-rod eye must absorb considerable forces and is therefore often form-bored in order to avoid misalignment. For diesel engines with high peak pressures in particular and in accordance with high gas forces, a special

Figure 3: Piston-ring shapes and configurations
Diesel engine:
1 Keystone ring, crowned,
2 Taper-face compression ring with inside bevel,
3 Stepped compression ring,
4 Double-beveled ventilated oil control ring with spiral-type expander.

Gasoline engine:
5 Plain compression ring, crowned,
6 Taper-face compression ring,
7 Stepped ring,
8 Double-beveled ring,
9 Multipart steel oil ring.

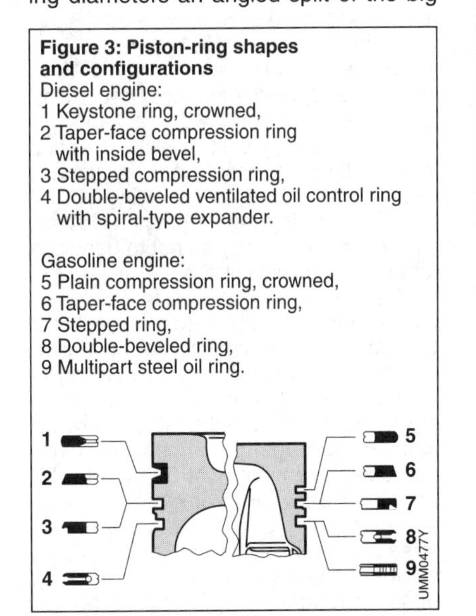

Figure 4: Connecting rod for a passenger-car engine
a) Sectional view, b) Side view.
1 Small connecting-rod eye,
2 T-shaped shank profile,
3 Straight split,
4 Big connecting-rod eye.

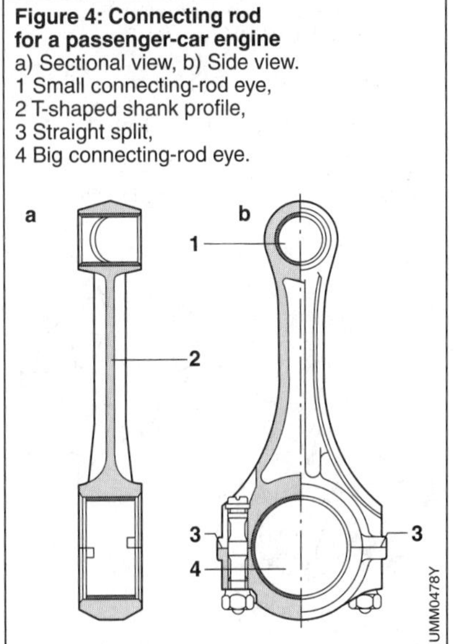

shape is chosen for the small connecting-rod eye which ensures a high contact surface on the pin without weakening the piston. This may be trapezoidal (narrow at the top, wide at the bottom) or stepped (narrow shoulder at the top, wide at the bottom).

The connecting rod is designed in line with the calculation of the gas and inertial forces (see Crankshaft-assembly design). The case of buckling is also calculated. Both buckling in the plane perpendicular to the crankshaft as a hinged beam and buckling in the crankshaft-axis plane (engine longitudinal plane) as a fixed beam must be checked. The critical cross-sections in the transition from the dimensionally stable connecting-rod eye to the double-T shank are usually recalculated with the aid of FEM simulations (finite-element method). The loads caused by the gripping of the bearing caps must not be underestimated and ignored.

Figure 5: Crankshaft for a commercial-vehicle engine
1 Front end of crankshaft,
2 Main bearing,
3 Connecting-rod bearing journal with oil bore,
4 Main web (without counterweight),
5 Main webs,
6 Intermediate web,
7 Offset connecting-rod bearing journal (split pin),
8 Counterweight (bolted).

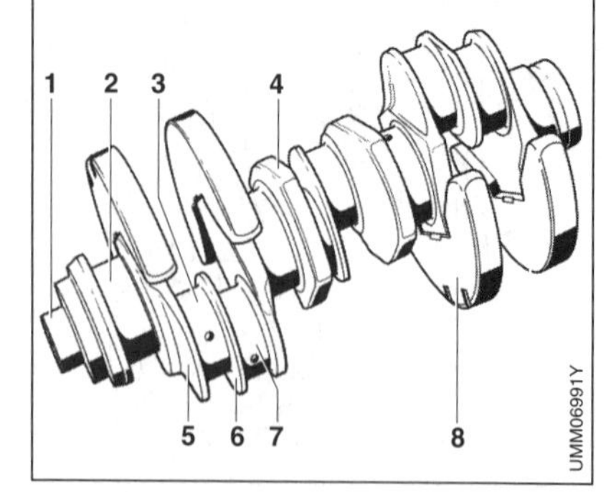

Crankshaft

The crankshaft with its throws (Figure 5) converts the oscillating motion of the piston by way of the connecting rod into a rotatory motion. The translational energy is transformed into a rotating torque, which can be picked off by a coupling.

The main bearings of the crankshaft are fixed in the crankcase, while the crank pins in the throws run in the bearings fixed to the connecting rods. The counterparts to the throws with crank pins are the counterweights, which are either cast, forged or bolted on. Depending on the engine type, the crankshaft main bearings are arranged after each cylinder – particularly in in-line engines and diesel and gasoline engines subject to high loads – or only after every second throw (gasoline engines subject to low loads). V-engines usually feature one crank pin for two adjacent connecting rods on one throw. This also applies to forked connecting rods, which permit two exactly opposing cylinder banks. In order to facilitate a uniform ignition interval in V6 engines with a 90° bank angle, "split-pin" designs which permit a 30° offset of two crank pins in a very tight space without a main bearing in between are increasingly being used. This form of design is used even in commercial-vehicle diesel engines subject to high loads.

Crankshafts are usually forged (drop forgings or open-die forgings, depending on the size) and, in lower-load applications, increasingly cast (nodular graphite iron). Built-up crankshafts are also used in large engines. Bores for supplying lube oil to the connecting-rod bearings are incorporated in such a way as not to generate further increases in stress.

The gas and inertial forces (see Crankshaft-assembly design) subject the crankshaft to bending stress in particular (Figure 6). If a main bearing is incorporated after each throw, the bending loads on the crankshaft are low on account of tangential forces. However, particular attention must be paid to the transition between crank pin and web because this is where the greatest stresses occur. Because of the large number of main bearings, the clamping of the crankshaft represents theoretically a statically overdefined system – in the simulation, however, a statically defined state which correctly describes the critical state is assumed.

Another important factor is the torsional vibrations, which, on account of the alternating forces of each individual cylinder (torsional force, see Crankshaft-assembly design) excite the overall crankshaft-assembly system and, in view of the not inconsiderable length of the crankshaft, can approach the crankshaft's natural resonance.

The oscillatory system – consisting of piston, connecting rod and crankshaft – is a complex system of constantly changing torsional forces. The further the counterweights are arranged to the outside, the higher the excitation. For this reason, heavy-metal counterweights are also bolted on in some cases. A simple simulation can be performed by assuming a smooth, inertialess and elastic shaft with attached equivalent masses. The oscillation-equation system can then be set up with the aid of the torsion-spring rate of the torque shaft. Cast shafts are ascribed a certain level of intrinsic damping. The crankshaft is thus subjected to both a locally different and a time-varying torsional load.

Figure 6: Crankshaft throw
Primary stresses and deformations due to gas pressure and inertial forces.

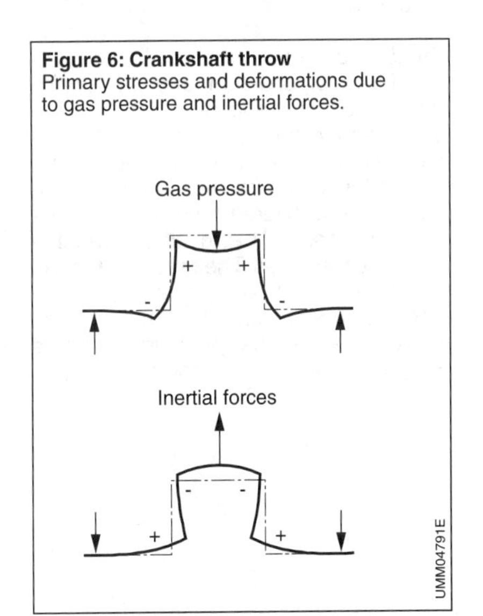

The torsional vibrations of the crankshaft, which are also called 2nd-order torsional vibrations (see Crankshaft vibration damper) must be reduced by suitable vibration dampers. This is particularly necessary to protect the crankshaft against excessively high loads.

Crankcase

The crankcase supports the force-transfer mechanism between cylinder head and crankshaft assembly. It bears the crankshaft assembly's support bearings, and incorporates (or holds) the cylinder sleeves. Also included in the block are a separate water jacket and sealed oil chambers and galleries. Furthermore, it must absorb the torque reaction forces and provide locations for mounting many add-on components.

A rigid framework-like structure (support-beam concept) is normally used to absorb the forces; this structure is designed to ensure a direct, linear and bending-moment-free flow of force. This is usually achieved by walls with reinforcing ribs and in the case of the bolt-through design by through-bolts.

The cylinder head is usually bolted as a separate component – in larger engines also as an individual cylinder head – to the crankcase. Often four bolts per head are used, with adjacent cylinders usually sharing two bolts. Six or eight bolts may also be used in engines with high peak pressures, typically above 200 bar. The purpose of this in particular is to distribute as evenly as possible the pressure force of the cylinder-head gasket.

The crankshaft-bearing caps are usually bolted from underneath to the crankcase. In the bolt-through design a long bolt performs this function from the head to the bearing cap. The advantage of this design is that it permits the use of high-tensile steels, which can transmit tensile load better than casting materials.

The crankcase is usually drawn down at the sides to deep below the crankshaft bearings in order to provide sufficient rigidity. But sometimes a separate frame ("bed plate") may also be used which is designed to reduce weight and increase rigidity. Sometimes the side walls are additionally bolted horizontally at the sides in the crankshaft-bearing area in order to achieve even greater rigidity. Bolted underneath is the oil pan, which can sometimes have a similarly bracing effect. However, this only applies to cast oil pans.

The piston bearing surface can be designed as a separate component (liner), particularly when replacement is not uncommon during the crankcase's service life, for example in commercial vehicles with high mileage ("wet liner" because coolant flows around it). However, the piston bearing surface can also be cast into the crankcase – particularly in cast-aluminum designs – or directly as a bearing surface in gray cast iron with strip hardeners.

Today's crankcases have attained a very high level of function integration and are therefore highly complex components. Frequently the oil-pump housing is also integrated in the casting, as are oil-cooler functions, the coolant-thermostat housing, the balancer-shaft locators, and all the brackets for the accessories.

Aside from the crankshaft, the crankcase is the component which contributes most to the engine weight. Many efforts are therefore made to reduce the weight by using materials other than the customary gray cast iron. These include aluminum materials, which, because of their thermally limited service life, are primarily used in smaller engines, particularly in passenger-car applications. Die casting has become an accepted cost-saving process for manufacturing aluminum, while sand casting is otherwise the regular process used. An "open-deck" construction is used to manufacture diecast crankcases efficiently, i.e. the joint face with the cylinder head features large pass-through cross-sections for the water jacket and the oil return. When cast-iron materials are used, on the other hand, the upper crankcase is closed to make the structure as rigid as possible.

Efforts are also being made to manufacture crankcases from magnesium in order to further reduce weight – but this comes at the expense of a high manufacturing cost. For this reason, a "hybrid" material concept is used where the bearing functions are performed by aluminum and the shielding functions by magnesium inserts.

Further increases in strength are provided by other cast-iron variants, compacted graphite iron (CGI) being one such variant which has been more widely used in recent years. Special heat treatment and specific cooling can be used to create in cast iron high-strength (and hard-to-machine) spheroidal graphite instead of the customary lamellar graphite. This makes it possible to produce thinner walls and higher strength with reduced weight.

The crankcase is assigned the most important function of noise emission. Because the crankcase is often joined directly to the combustion chamber, absorbs the high periodic gas and inertial forces and has a relatively large surface, it is extremely important for this surface to be rigid. The standard design methods used to provide high rigidity are ribbing and barreling, which also help to prevent a membrane effect.

Variable compression ratio

A variable compression ratio is desirable both in gasoline engines to reduce the knock tendency close to full load and in diesel engines to reduce the peak pressure at full load. However, up to now no mechanical or hydraulic system has successfully managed to provide this variability in diesel engines. The reason for this is the high loads to which diesel engines are subjected on account of the higher peak pressures and higher pressure gradients compared with gasoline engines.

The designs for variable compression ratio (VCR) can be categorized as follows:
- Solutions which are integrated into the piston (hydraulic actuation of a piston upper section that moves axially in relation to a fixed lower section)
- Solutions which alter the length of the connecting rod (sliding solution, articulated connecting rod, eccentric connecting-rod bearing in the small or big connecting-rod eye)
- Formulations for an altered crankshaft position relative to the cylinder head (swiveling or axial movement of the crankcase, eccentric crankshaft bearing)
- Coupling lever for varying the effective piston stroke (multilink, crankshaft-drive lever system, displacement alteration with ε variation)
- Additional volumes which are increased or decreased (variable additional volume in the cylinder head)

Formulations for variable compression ratios in gasoline engines cover a range of approx. $\varepsilon = 7$ to 14.

Cylinder head

The main functions of the cylinder head are to supply the cylinder with fresh mixture, to discharge the exhaust gas, and to absorb the cylinder pressure forces.

Cylinder head

The cylinder head seals off the upper end of the crankcase and the cylinder barrel. It houses the gas-exchange valves and the spark plugs in gasoline engines, the fuel injectors in direct-injection gasoline engines, and the fuel injectors and possibly the sheathed-element glow plugs in diesel engines. Together with the piston, it also provides the desired combustion-chamber shape. In the vast majority of passenger-car engines, the entire valve gear is also mounted in the cylinder head.

The cylinder head performs a large number of functions and exhibits complex load profiles. The challenge faced here, as well as precisely controlling gas exchange and charge movement, which directly influence engine emissions, efficiency and power output, is above all to cool the hot side of the cylinder head and maintain control over the complex mechanical and thermal loads involved.

Based on the gas-exchange concepts, a distinction is made between two basic design configurations. In the counterflow cylinder head intake and exhaust passages open into the same side of the cylinder head (Figure 7b). This limits the space available for the intake and exhaust-gas passages, but due to the short flow tracts, this represents a substantial advantage in supercharged applications without intercooling. This design, with the gas supply and discharge tracts on a single side, also provides practical advantages in transverse-mounted engines.

In the crossflow cylinder head intake and exhaust passages are located on opposite sides of the engine (Figure 7a), providing a diagonal flow pattern for the intake and exhaust gases. This layout's advantages include more freedom in intake and exhaust-tract design as well as less complicated sealing arrangements. Typically, vehicle-specific boundary conditions are crucial to determining whether a counterflow or crossflow cylinder head is used.

This decision also determines the valve arrangement used. Counterflow applications require a twisted valve arrangement in today's typical multi-valve designs. The main advantage of a twisted valve arrangement is that it is a simple way of delivering high cylinder-whirl values.

Figure 7: Cylinder-head identification according to intake and exhaust-tract location
a) Crossflow design,
b) Counterflow design.

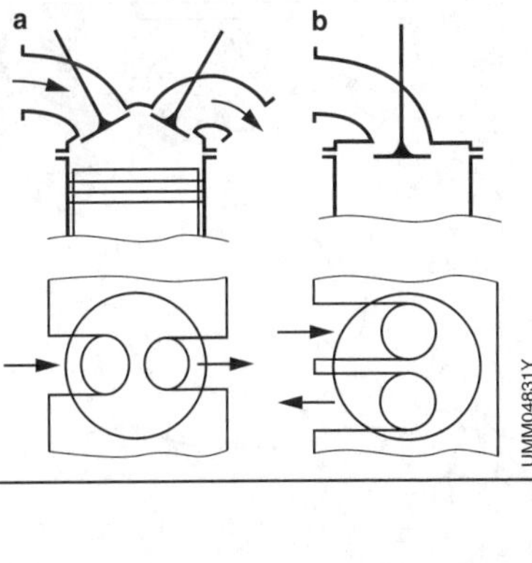

a b

UMM04831Y

A parallel valve arrangement requires greater efforts at the passage end to deliver a high cylinder whirl (Figure 8).

Today's passenger-car applications use almost exclusively a continuous single-piece cylinder head. In V-engines one cylinder head per cylinder bank is used. Individual cylinder heads are frequently still used in commercial-vehicle engines, which sometimes still feature a block-mounted camshaft with a push-rod assembly.

Figure 8: Valve arrangements
a) Twisted valve arrangements,
b) Parallel valve arrangements.
I Intake, E Exhaust.

In contrast to passenger-car applications, in which the cylinder head is often made from aluminum, which as well as weight advantages improves heat transfer from the combustion chamber, commercial-vehicle applications utilize gray-cast-iron designs in view of the fact that resistance to fatigue cannot be achieved by light alloys at hight peak pressures and with large cylinder diameters.

The cylinder head is subject to very high loads. On the one hand, frequently varying temperatures and the accompanying expansions give rise to stresses which can only be absorbed by appropriate design measures (individual cylinder head, suitable cavities in a multi-cylinder head). Thermomechanical strength is determined by the number of cold-hot load profiles.

In multi-valve designs the land between the exhaust valves is subjected to particularly high temperatures. At full load slightly critical temperatures above 350 °C are reached here. Similarly high temperatures are encountered in the recess-edge area (Figure 9). Gray-cast-iron designs

Figure 9: Piston operating temperatures in motor-vehicle engines at full load (schematic), values in °C
a) Passenger-car diesel-engine piston, 16 MPa ignition pressure, 58 kW/*l*,
b) Passenger-car gasoline-engine piston, 7.3 MPa ignition pressure, 53 kW/*l*.

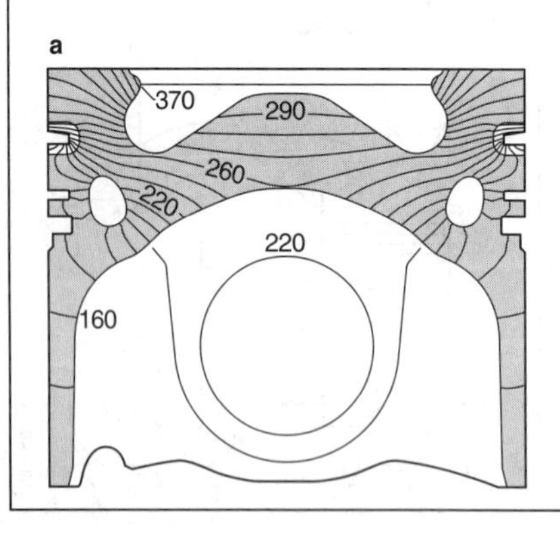

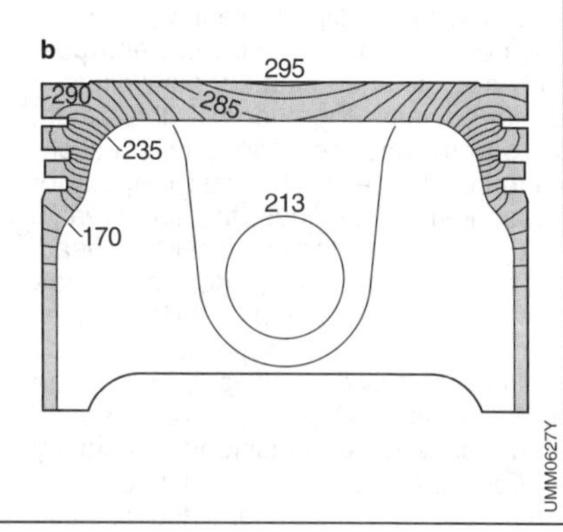

are especially prone to the risk of hot-gas corrosion, an oxidation of the surface which can culminate in component-jeopardizing cracking.

The cylinder-head gasket, which is located between the crankcase and the cylinder head, performs an important function. Firstly, it must seal the combustion chamber even in the case of high cylinder pressures and, secondly, it must ensure the flow of engine oil and cooling water between crankcase and cylinder head. Typically, metal materials are used to seal the combustion chamber; these metals adapt to the surface by means of plastic deformation and establish the seal through their elasticity. Elastomer gaskets or geometrically optimized metal-sheet steel gaskets are used.

To provide a uniform downward pressure on the cylinder-head gasket, it is possible to increase the number of cylinder-head bolts to such an extent that already commercial-vehicle engines with eight bolts per cylinder head are in use.

Engines without cylinder-head gaskets are the exception and require the crankcase deck and the underside of the cylinder head to be finished to very high levels of accuracy.

Valve gear

It is the function of the valve-gear assembly in a 4-stroke engine to permit and to control the exchange of gases in the internal-combustion engine. The valve gear includes the intake and exhaust valves, the springs which close them, the camshaft drive assembly, and the various force-transfer devices (Figure 10). In various widely-used designs, the camshaft is located in the cylinder head.

When the camshaft is installed within the block, the rocker arm is actuated not directly by the cam lobe, but by the intermediate action of a push rod and a lifter (push-rod assembly, Figure 10a).

In the case of the cam-follower or single rocker-arm assembly actuated by an overhead cam, the cam lobe's lateral and linear forces are absorbed and relayed by a cylinder-head mounted lever rocking back and forth between the cam lobe and the valve. In addition to transferring forces and absorbing lateral forces, the intermediate rocker arm can also be designed to magnify the cam pitch effect (Figure 10b).

In the case of the twin rocker-arm assembly actuated by an overhead cam, the rocker arm acts as the force-transfer element with its tilt axis located between the camshaft and the valve. Here, too, the rocker arm is usually designed as a cam-

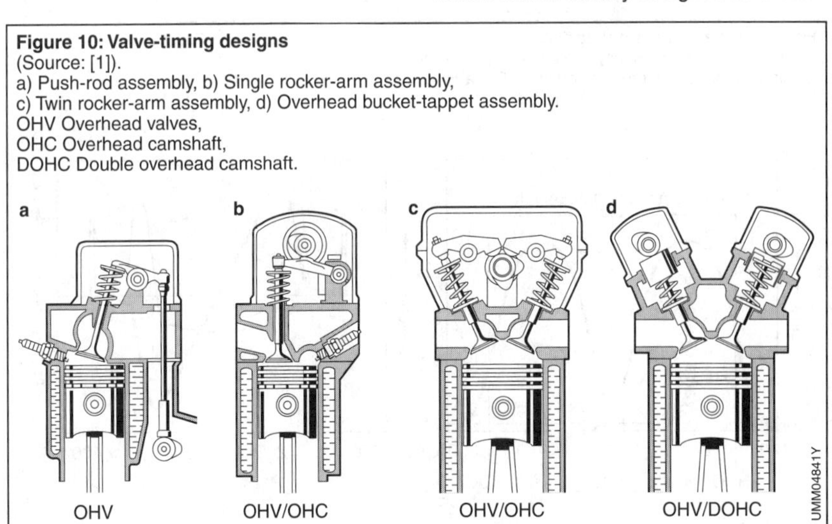

Figure 10: Valve-timing designs
(Source: [1]).
a) Push-rod assembly, b) Single rocker-arm assembly,
c) Twin rocker-arm assembly, d) Overhead bucket-tappet assembly.
OHV Overhead valves,
OHC Overhead camshaft,
DOHC Double overhead camshaft.

lift multiplier to produce the desired valve travel (Figure 10c).

In the case of the overhead bucket-tappet assembly, a "bucket" moving back and forth in the cylinder head absorbs the cam lobe's lateral force, while transferring its linear actuating pressure to the valve stem (Figure 10d).

Valve arrangements

The valve-control arrangement and the design of the combustion chamber are closely interrelated. Today, nearly all valve assemblies are overhead units mounted in the cylinder head. In diesels and simpler spark-ignition (gasoline) engines, the valves are parallel to the cylinder axis, and are usually actuated by twin rocker arms, bucket tappets or single rocker arms. With increasing frequency, current spark-ignition (SI) engines designed for higher specific outputs tend to feature intake and exhaust valves which are inclined towards each other. This configuration allows larger valve diameters for a given cylinder bore, while also providing greater freedom for optimizing intake and exhaust passage design. Twin rocker-arm assemblies actuated by overhead cams are used most often here. High-performance and racing engines are increasingly using four valves per cylinder and overhead bucket-tappet valve assemblies.

The valve-timing diagram (Figure 11) for an engine shows the opening and closing times of the valves, the valve-lift curve, the maximum valve lift, and the valve velocity and acceleration.

Typical valve-acceleration rates for passenger-car OHC (overhead camshaft) valve assemblies:

$s'' = 60...65$ mm $(b/\omega^2) \rightarrow 6{,}400$ m/s^2 at 6,000 rpm for single and twin rocker-arm assemblies.

$s'' = 70...80$ mm $(b/\omega^2) \rightarrow 7{,}900$ m/s^2 at 6,000 rpm for overhead bucket-tappet assemblies.

For heavy-duty commercial-vehicle engines with block-mounted camshafts:

$s'' = 100...120$ mm $(b/\omega^2) \rightarrow 2{,}000$ m/s^2 at 2,400 rpm.

Valve, valve guide and valve seat

The materials employed in manufacturing valves are heat and scale-resistant. The valve seat's contact surface is frequently hardened. Another proven method for improving the thermal-conductivity characteristics of exhaust valves is to fill their stems with sodium. To extend service life and improve sealing, valve-rotating systems (rotocaps) are also in common use.

The valve guides in high-performance engines must feature high thermal con-

Figure 11: Valve-timing diagram

s Valve lift, s' Valve velocity, s'' Valve acceleration.
IC Intake closes, IO Intake opens, EC Exhaust closes, EO Exhaust opens,
BDC Bottom dead center, TDC Top dead center.

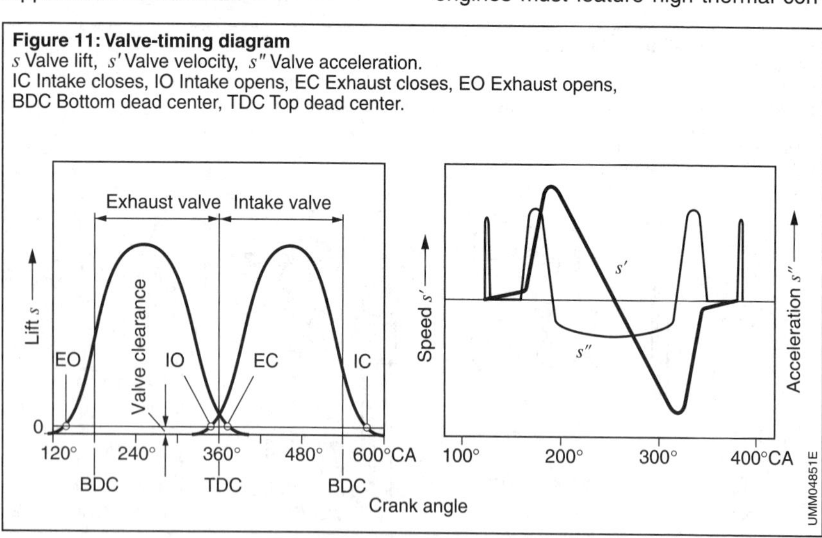

ductivity and good antifriction properties. They are usually pressed into the cylinder head and are often supplemented by valve-stem seals at their cold ends for reducing oil consumption.

Valve-seat wear is generally reduced by making the valve-seat inserts of cast or sintered materials and shrink-fitting them into the cylinder head.

Cam-lobe design and timing dynamics
The cam lobe must be able to open (and close) the valve as far, as fast and as smoothly as possible. The closing force for the valves is applied by the valve springs, which are also responsible for maintaining contact between the cam lobe and the valve. Dynamic forces impose limits on cam and valve lift.

Rarely encountered are designs which use their own mechanism (Desmotronic) to restore the valves with the advantage of increased valve velocities and thus lift curves which more closely resemble the ideal rectangular case. Conventional springs must always ensure a frictional connection between camshaft, intermediate elements and valve while adhering to the permissible surface load. A slight clearance can be encountered only when cam lift is not active. Modern engines operate here with hydraulic valve-clearance compensation. Otherwise valve clearance must be regularly checked; measurements of 0.1 to 0.2 mm are common at the intake and exhaust valves.

Camshaft drive
To drive the valve gear, the camshaft must be connected to the crankshaft in such a way that one crankshaft revolution results in a one half camshaft revolution in a 4-stroke engine.

For passenger-car applications with overhead camshaft(s) either timing-chain or toothed-belt drives are used in modern engines. Both drives require a tensioning device, which acts on the free running length, the loose side, to prevent uncontrolled vibration.

In view of the fully utilized minimum space between piston and valve in modern engines, it is very important for the camshaft drive to operate safely and reliably, because in the event of damage contact can have serious consequences for the valve gear.

For this reason, the pressure pins for the tensioning rails in a chain drive or the toothed belt are wearing parts which must be replaced on a regular basis. Modern duplex chains are highly durable and are only subject to minimal wear. The drawbacks of a timing chain are its need to be lubricated and its elongation during operation. Some major advantages, however, are that it cannot jump off the gear wheel and it is not subject to the risk of cracking.

Transmission by gear wheels in a geartrain is typically used in commercial-vehicle engines or engines with block-mounted camshafts. In engines with overhead camshafts in particular, this is a very expensive solution, but one which provides for extending engine running times on account of the reliable transmission.

Expensive vertical shafts, whose axes run at right angles to the crankshaft and camshaft, are rarely used.

Oil supply

As well as lubricating the tribologically critical pairings of the crankshaft drive, the cylinder head and other components, the oil supply serves to remove local contaminants, combustion residues and wear particles which are filtered out in the oil-filter unit. Additional functions include the dissipation of heat in areas subject to thermal load, such as the friction bearings in the crankshaft drive or for oil-cooled pistons, and damping of vibrations in bearings.

With the system of forced-feed lubrication conventionally used (see also “Engine lubrication”), the oil pump, a gear pump that typically operates as a positive-displacement pump, delivers a defined volumetric flow from the oil sump through an oil-filter unit (Figure 12a). For safety reasons, the oil-filter unit is often equipped with a bypass valve and a pressure valve in a full-flow design. The oil cooler provided in engines subject to higher loads is cooled by either air or coolant. The engine oil flows though oil ducts and by the force of gravity back into the oil sump and into the oil pan typically located underneath the crankcase. That is why forced-feed lubrication is also known as wet-sump lubrication. In addition to the direct supply of oil by the oil pump, the rotational motion of the crankshaft effects a fine conditioning of the oil mist in the crankcase.

Figure 12: Oil supply
a) Forced-feed lubrication (wet-sump lubrication),
b) Dry-sump lubrication.
1 Oil reservoir, 2 Oil pump, 3 Oil cooler,
4 Filter unit, 5 Bypass valve,
6 Engine with lubrication points,
7 Second oil pump.

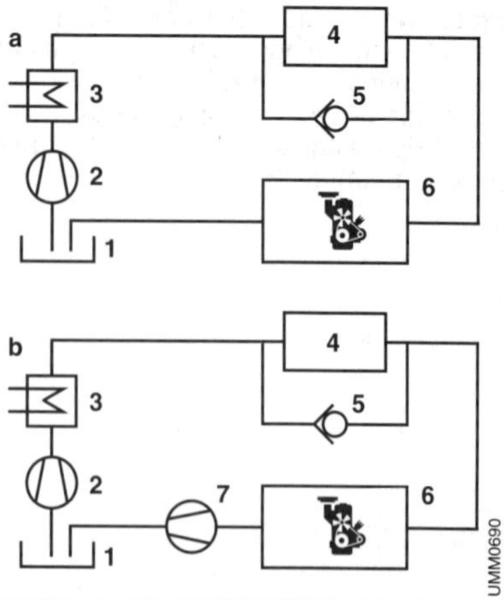

The opposite system to conventional wet-sump lubrication is the more expensive dry-sump lubrication, because it requires a second oil pump to remove the oil from the engine compartment (Figure 12b). The advantages are a constantly guaranteed lube-oil supply under conditions of high lateral acceleration or inclination and freedom in deciding on where to locate the oil-supply system. These give rise to lower engine heights and a possible increase in the engine-oil quantity, which is beneficial to engine cooling.

The mixture and total-loss lubrication frequently used in 2-stroke and Wankel engines is no longer relevant to present-day automobile applications.

Cooling

In order to avoid thermal overload, combustion of the lubricating oil on the piston's sliding surface, and uncontrolled combustion due to excessive component temperatures, the components surrounding the hot combustion chamber (cylinder liner, cylinder head, valves and in some cases the pistons themselves) must be intensively cooled (see also “Engine cooling”).

As water has a high specific heat capacity and provides efficient thermal transition between the materials, most contemporary vehicle engines are water-cooled. Moreover, it is possible, under very hot local boundary conditions when the cooling medium is locally evaporated, to achieve through the heat of evaporation an additional cooling effect which together with the condensation behavior in colder adjacent areas reduces unwanted high local temperature gradients.

Air/water recirculation cooling is the most prevalent system (Figure 13). It comprises a closed circuit allowing the use of anticorrosion and antifreeze additives. The coolant is pumped through the engine and through an air/water radiator. The cooling air flows through the radiator in response to vehicle movement and/or is forced through it by a fan. The coolant temperature is regulated by a thermostatic valve which bypasses the radiator as required.

Air-cooled concepts still play a subordinate role today. Modern emission regulations call for effective engine cooling with simultaneously high specific power output. This can practically no longer be achieved with air cooling, a system which is indeed robust and low-maintenance, but is also not as effective. A crucial drawback of air-cooled concepts is their acoustic performance. The cooling fins act like a resonator and therefore result in a marked increase in sound emissions.

Crankshaft vibration damper

The crankshaft is subjected to high-frequency vibrations (2nd-order torsional vibrations), particularly in the region of resonant frequencies. Vibration dampers are therefore used to moderate dangerous crankshaft resonance phenomena.

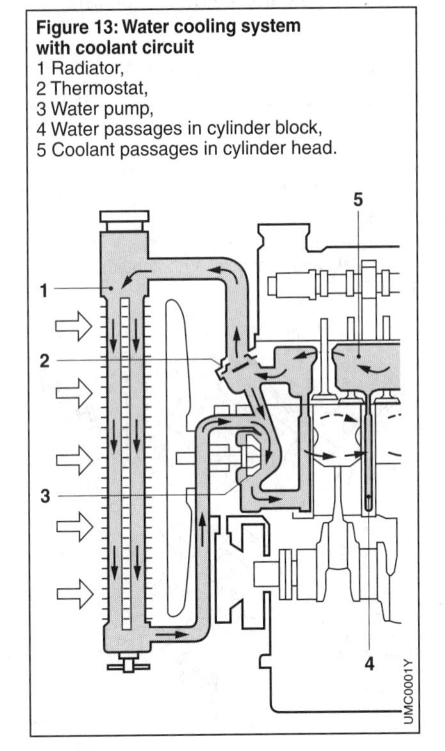

Figure 13: Water cooling system with coolant circuit
1 Radiator,
2 Thermostat,
3 Water pump,
4 Water passages in cylinder block,
5 Coolant passages in cylinder head.

The choice of vibration-damper design ranges from simple flywheel masses, which through elastically damping layers – e.g. rubber – provide a countervibration system, through more sophisticated oil-damped flywheel masses, in which the viscosity of the oil and friction surfaces provide the damping effect, up to complex pendulum adsorbers. The vibration damper is generally mounted at the free crankshaft end (front).

1st-order torsional vibrations, i.e. vibrations resulting from ignition and inertial forces, which can result in excitation of transmission and input-shaft vibrations, are reduced by dual-mass flywheels or other damping technologies. Phenomena such as clutch grabbing, transmission chatter and starting jerks can be optimized by such vibration absorbers, since all the following components – clutch, transmission and input shafts – likewise represent an oscillatory system. In a dual-mass flywheel the two flywheel masses are coupled to each other by means of springs. Torsional-vibration dampers are also commonly featured in clutches – both components must naturally be tuned together.

Design and tuning are difficult to simulate and must in the final analysis be conducted in tests, both in the vehicle with the entire drivetrain (1st order: low-frequency drivetrain resonance phenomena) and on the test bench (2nd order: high-frequency crankshaft resonances).

Reciprocating-piston engine types

The internal-combustion engine is characterized by its great variability and application-specific adaptability. In particular, the arrangement of the individual cylinders provides for numerous variants.

Arrangements

In principle, a huge number of possible cylinder arrangements is conceivable. Few of these arrangements have proven to be particularly effective for automotive applications (Figure 14).

Radial engines are not suitable for automotive applications on account of their height and are therefore not used.

It is important to note the difference between the opposed-cylinder (boxer) engine and the V-engine with 180° included angle between cylinder banks. In the boxer engine the pistons always move in opposite directions, which is why the inertial forces cancel each other out. In the V-engine, on the other hand, the pistons of two opposed cylinders always move in the same direction.

It is also important to note the design of V- and W-engines, in which the connecting-rod/crankshaft link (large connecting-rod eye) of associated cylinders is effected at slightly offset crank pins.

In multi-piston units compression is typically generated by several working pistons. In the U-engine the pistons undertake an approximately parallel motion. In the opposed-piston engine the pistons undertake a typically opposed motion. Engine variants with more than one piston per combustion chamber have not gained acceptance. Weight and engine size are not suitable for automotive applications.

Definitions

The following definitions in accordance with DIN 73021 [2] as viewed looking at the end opposite the power-output end apply only to motor-vehicle engines. For internal-combustion engines for general and marine use, the reverse direction – i.e. as viewed looking at the power-output end – is standardized (ISO 1204 [3]).

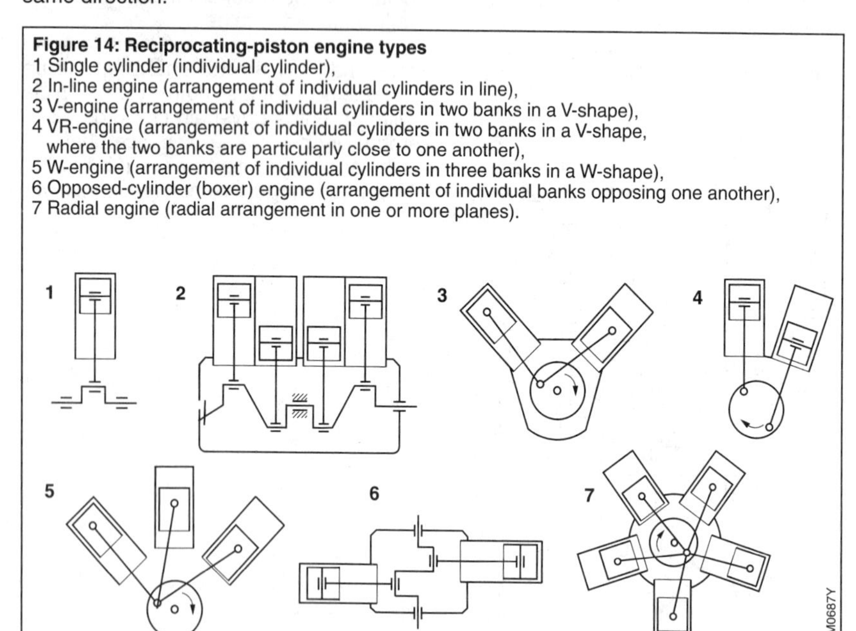

Figure 14: Reciprocating-piston engine types
1 Single cylinder (individual cylinder),
2 In-line engine (arrangement of individual cylinders in line),
3 V-engine (arrangement of individual cylinders in two banks in a V-shape),
4 VR-engine (arrangement of individual cylinders in two banks in a V-shape, where the two banks are particularly close to one another),
5 W-engine (arrangement of individual cylinders in three banks in a W-shape),
6 Opposed-cylinder (boxer) engine (arrangement of individual banks opposing one another),
7 Radial engine (radial arrangement in one or more planes).

Clockwise rotation
The direction of rotation is clockwise, as viewed looking at the end opposite the power-output end.

Counterclockwise rotation
The direction of rotation is counterclockwise, as viewed looking at the end opposite the power-output end.

Numbering the cylinders
The cylinders are numbered consecutively 1, 2, 3, etc. in the order in which they would be intersected by an imaginary reference plane, as viewed looking at the end of the engine opposite the power-output end. This plane is located horizontally to the left when numbering begins; the numerical assignments then proceed clockwise about the longitudinal axis of the engine. If there is more than one cylinder in a reference plane, the cylinder nearest the observer is assigned the number 1, with consecutive numbers being assigned to the following cylinders. Cylinder 1 is to be identified by the number 1.

Firing sequence
The firing sequence is the sequence in which combustion is initiated in the cylinders. Engine design configuration, uniformity in ignition intervals, ease of crankshaft manufacture, optimal crankshaft load patterns, etc. all play a role in defining the firing sequence.

Figure 15: Crankshaft assembly for reciprocating-piston engine (principle)
1 Valve gear,
2 Piston,
3 Connecting rod,
4 Crankshaft.

1 2 3 4 UMM0463Y

Crankshaft-assembly design

Crankshaft-drive kinematics

The kinematics of the single-cylinder crankshaft assembly can be determined from the geometrical assignments of piston and piston-pin axis, connecting rod and crankshaft (crankshaft radius equal to one half stroke) (Figure 15). If piston travel x at top dead center is assumed to be zero, the following is obtained with crank radius r and connecting-rod length l (Figure 16):

$$x = r\,(1 - \cos\alpha) + l\,(1 - \cos\beta)\ .$$

Where
$r \sin\alpha = l \sin\beta$
and
$\lambda = r/l$
the following is obtained:

$$x = r\left(1 - \cos\alpha + \frac{1}{\lambda}\left(1 - \sqrt{1 - \lambda^2 \sin^2\alpha}\right)\right)$$

An offset of the piston pin is performed by some manufacturers. Noise and friction advantages can be expected on account of the changing piston location and depending on the position of the connecting rod. The offset can be effected either in the piston by shifting the piston pin from the center position or by offsetting the crankshaft.

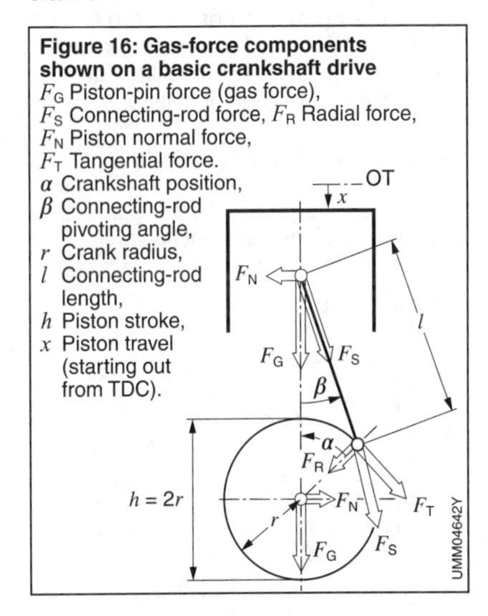

Figure 16: Gas-force components shown on a basic crankshaft drive
F_G Piston-pin force (gas force),
F_S Connecting-rod force, F_R Radial force,
F_N Piston normal force,
F_T Tangential force.
α Crankshaft position,
β Connecting-rod pivoting angle,
r Crank radius,
l Connecting-rod length,
h Piston stroke,
x Piston travel (starting out from TDC).

If the offset for positive crank angles is assumed to be positive and the quantity

$$\delta = \frac{\text{Offset}}{\text{Connecting-rod length}}$$

is introduced, this leads to the following function of the piston travel:

$$x = r\left(1 - \cos\alpha + \frac{1}{\lambda}\left(1 - \sqrt{1 - (\lambda\sin\alpha - \delta)^2}\right)\right)$$

Figure 17 shows by way of example the influence of the stroke/connecting-rod ratio and the offset. However, the differences with normal offset values in the millimeter range ($\delta < 0.04$) are clearly smaller.

By developing the root function into a Taylor series (around $x = 0$: Maclaurin series) and replacing the powers of the trigonometric functions with the multi-harmonic functions, the following expression can be obtained [4]:

$$x = r\,[1 + \tfrac{1}{4}\lambda + \tfrac{3}{64}\lambda^3 + \ldots - \cos\alpha$$
$$- (\tfrac{1}{4}\lambda + \tfrac{3}{64}\lambda^3 + \ldots)\cos 2\alpha$$
$$+ (\tfrac{3}{64}\lambda^3 + \ldots)\cos 4\alpha + \ldots].$$

This expression shows that there are higher harmonics on account of the crankshaft-drive kinematics, and these are also called engine orders (multiples of the engine speed).

Because normal values for λ are around 0.3, the higher-level λ terms are readily ignored and the following simplified function is assumed for further calculations:

$$x = r\,[1 + \tfrac{1}{4}\lambda - \cos\alpha - \tfrac{1}{4}\lambda\cos 2\alpha\,].$$

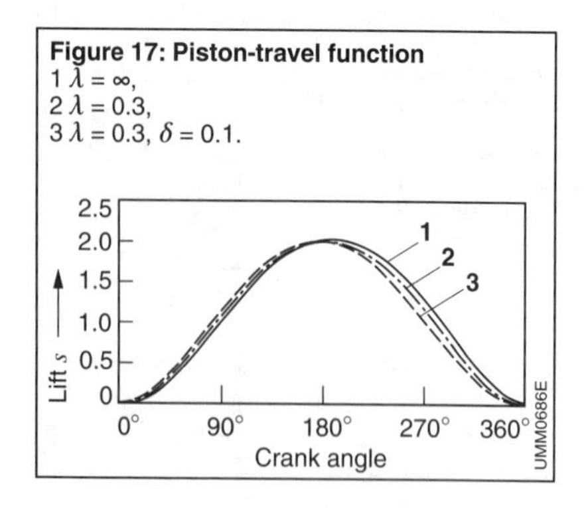

Figure 17: Piston-travel function
1 $\lambda = \infty$,
2 $\lambda = 0.3$,
3 $\lambda = 0.3$, $\delta = 0.1$.

This simplification may not be used if detailed vibration analyses and resonance phenomena are to be examined.

From the simplified equation the following piston speeds v and accelerations a are obtained, where the angular velocity $d\alpha/dt = \omega = 2\pi n$ (n speed) was introduced:

$$v = r\omega\,(\sin\alpha + \frac{\lambda}{2}\sin 2\alpha),$$

$$a = r\omega^2\,(\cos\alpha + \lambda\cos 2\alpha).$$

Here too there are higher harmonics (orders), which must not be disregarded when resonances are examined.

Crankshaft-assembly kinetics

The forces acting on the crankshaft assembly and the resulting moments can initially be derived without inertial forces as follows (Figure 16).

The piston-pin force results as gas force from the combustion-chamber pressure and the piston area. The following applies:

$$F_G = (p - p_{KGH})\,A_{piston}\,.$$

The connecting-rod force is obtained from the vectorial analysis of the piston-pin force in the direction of the connecting rod. The following applies:

$$F_S = \frac{F_G}{\cos\beta} = \frac{F_G}{\sqrt{1 - \lambda^2\sin^2\alpha}}$$

The piston normal force F_N is the vectorial component of the piston-pin force normal to the cylinder wall and for balancing the connecting-rod force:

$$F_N = F_G\tan\beta = \frac{F_G\,\lambda\sin\alpha}{\sqrt{1 - \lambda^2\sin^2\alpha}}\,.$$

This force contributes significantly to the friction of the piston on the cylinder barrel. The side which is touched by the piston after top dead center on account of the combustion pressure is termed the major thrust face, the opposite side the minor thrust face. The greatest friction is therefore created shortly after TDC on the major thrust face.

The tangential force at the crankshaft crank pin contributes to an acceleration of the crankshaft and thus to the build-up of torque at the crankshaft. It is created from the vectorial analysis of the connecting-rod force.

$$F_T = \frac{F_G \sin(\alpha+\beta)}{\cos\beta}$$
$$= F_G\left(\sin\alpha + \frac{\lambda}{2}\sin 2\alpha/\sqrt{1-\lambda^2\sin^2\alpha}\right).$$

Here too the radicand can be simplified in a series expansion as

$$F_T \approx F_G\,(\sin\alpha + \tfrac{\lambda}{2}\sin 2\alpha).$$

The radial force F_R at the crankshaft crank pin is:

$$F_R = F_G\,\frac{\cos(\alpha+\beta)}{\cos\beta}$$
$$= F_G\left(\cos\alpha + \lambda\frac{\sin^2\alpha}{\sqrt{1-\lambda^2\sin^2\alpha}}\right)$$

or approximately

$$F_R \approx F_G\,(\cos\alpha - \tfrac{\lambda}{2} + \tfrac{\lambda}{2}\cos 2\alpha).$$

The inertial forces can be split into oscillating and rotating components. The piston, piston-ring and piston-pin weights m_K belong to the oscillating component and can be punctually assigned to the piston pin.

The crankshaft web with crank pin belong to the rotating component. Here the mass is usually reduced to the crank radius and punctually assigned to the crank-pin center point. The following applies:

$$m_W = \Sigma\,\frac{m_i\,r_{si}}{r},$$

where m_i is the respective mass component of web, pin, etc. and r_{si} is the corresponding center-of-mass radius.

Because of the connecting rod's swinging motion, it is useful to split the connecting-rod mass into an oscillating component and a rotating component. This is done exactly where the connecting-rod-mass center of gravity and mass moment of inertia are known by assuming two dynamically equal individual masses at the small and big connecting-rod eyes and by calculating the equilibrium formulation of force, moment and mass inertia. Usually one third of the connecting-rod mass m_{Pl} can be assumed to be oscillating and two thirds to be rotating. Then, with $m_o = m_K + 1/3\, m_{Pl}$ as the oscillating mass and the piston acceleration (see below), an oscillating inertial force is obtained at:

$$F_o \approx m_o\, r\, \omega^2\,(\cos\alpha + \lambda\cos 2\alpha).$$

The oscillating inertial force therefore increases quadratically with the engine speed ($\omega = 2\pi n$) and has a first-order component and a lower second-order component.

The rotating inertial force is obtained as centrifugal force from the reduced mass $m_r = m_W + 2/3\, m_{Pl}$ and the rotation speed at:

$$F_r = m_r\, r\, \omega^2.$$

The rotating inertial force likewise increases with the square of the speed, but has not higher orders. The rotating inertial force can therefore be easily balanced by counterweights which rotate at engine speed. The rotational irregularities of the crankshaft are so small compared with these forces that they can be ignored in the balancing of masses.

As shown in the crankshaft-assembly kinematics, higher harmonics – higher engine orders – occur on account of the crankshaft-drive geometry. Aside from the 1st and 2nd engine orders, however, the amount of the 4th order and the higher components decreases rapidly, and is generally ignored for a balancing of masses.

Balancing of masses in a single-cylinder engine

The component of rotating masses in a single-cylinder engine can be fully balanced by appropriate counterweights on the crank pin. Weights are normally provided for on both sides and the weights must be balanced merely with the center-of-mass radius. The oscillating forces can be represented with force vectors (Figure 18) when they are modeled rotating in the opposite direction in each case with half the amount.

Therefore two contra-rotating shafts with weights can be used to balance the oscillating inertial forces. The horizontal component then cancels out and at least the first order of the oscillating inertial-force component can be compensated.

A virtually full balancing of masses requires further balancer shafts, which must rotate at twice the engine speed in order to balance completely the second-order component of the oscillating mass component.

A compromise is often reached due to the fact that the expenditure associated with contra-rotating shafts is high and at least with the first order already not inconsiderable weights would have to be realized in the shafts. For example, half the oscillating mass can be integrated in the counterweights. The free inertial forces acting outwards are now reduced in the cylinder longitudinal direction by one half; however, additional transverse forces are now created on account of the excessive rotating components (see Table 2). A half compensation of this kind is termed a 50 % balancing rate. Usual figures are 100 % rotatory and 50 % oscillatory balancing of masses.

Balancing of masses in a multiple-cylinder engine

The inertial forces of a multiple-cylinder engine are composed of the inertial forces of each individual cylinder which are superimposed according to the crankshaft throw. In addition, free moments of inertia are created on account of the cylinder spacing. Table 3 shows a summary of the possible transverse and longitudi-

Figure 18: Complete balancing of 1st- and 2nd-order masses in a single-cylinder assembly

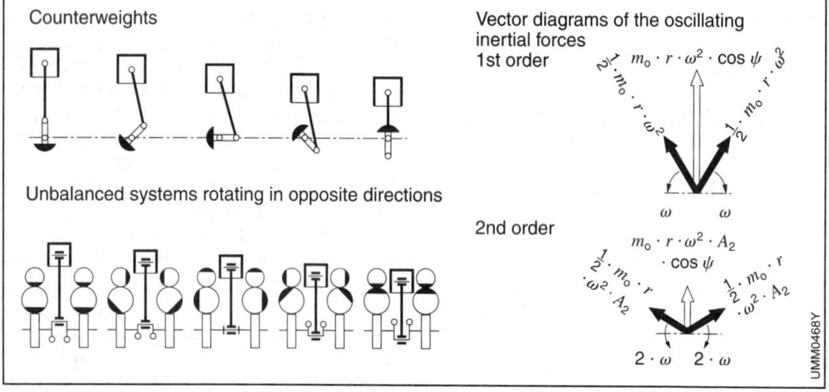

Table 2: Balancing of masses on a single-cylinder engine, dependent on the balancing rate

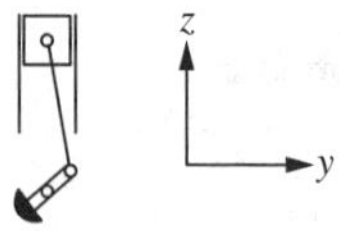 UMM0470Y		Balancing rate		
		0 %	50 %	100 %
Size of counterweight	$m_G \triangleq$	m_r	$m_r + 0.5 m_0$	$m_r + m_0$
Residual inertial force (z) 1st order	$F_{1z} =$	$m_0 \cdot r \cdot \omega^2$	$0.5 \cdot m_0 \cdot r \cdot \omega^2$	0
Residual inertial force (y) 1st order	$F_{1y} =$	0	$0.5 \cdot m_0 \cdot r \cdot \omega^2$	$m_0 \cdot r \cdot \omega^2$

nal tilting moments that occur and the free inertial forces.

The mutual balancing of the inertial forces is one of the essential factors determining the selection of the crankshaft's configuration, and with it the design of the engine itself. The inertial forces are balanced if the common center of gravity for all moving crankshaft-assembly components lies at the crankshaft's midpoint, i.e. if the crankshaft is symmetrical (as viewed from the front). This is represented with the 1st-order and 2nd-order star diagrams (Table 4).

The 2nd-order star diagram for the four-cylinder in-line engine is asymmetrical, meaning that this order is characterized by free inertial forces. These forces can be balanced using two countershafts rotating in opposite directions at double the rate of the crankshaft (Lanchester system).

Table 5 shows a summary of free forces and moments based on balancing of masses for different numbers of cylinders and throw variants.

Torsional force

Moving masses are exposed to constantly changing acceleration, thereby generating inertial forces. The cyclically occurring pressure forces in the cylinders are termed gas forces. Both form in total internal and external forces and moments. The internal forces and moments must be

Table 3: Transverse and longitudinal tilting moments, free inertial forces in multiple-cylinder engines

Forces and moments at the engine				UMM0469Y
Designation	Oscillating torque, transverse tilting moment, reaction torque	Free inertial force	Free inertial moment, longitudinal tilting moment about the y-axis (transverse axis) ("pitching" moment) about the z-axis (vertical axis) ("rolling" moment)	Internal bending moment
Cause	Tangential gas forces as well as tangential inertial forces for the ordinals 1, 2, 3 and 4	Unbalanced oscillating inertial forces 1st order in 1- and 2-cylinders; 2nd order in 1-, 2- ,4-cylinders	Unbalanced oscillating inertial forces as a composite of 1st- and 2nd-order forces	Rotating and oscillating inertial forces
Design factors	Number of cylinders, ignition intervals, displacement, p_i, ε, p_z, m_0, r, ω, λ	Number of cylinders, crank configuration m_0, r, ω, λ	Number of cylinders, crank configuration, cylinder spacing, counterweight size influences inertial-torque components about the y- and z-axes m_0, r, ω, λ, a	Number of throws, crank configuration, engine length, engine-block rigidity
Remedy	Can only be compensated for in exceptional cases	Free mass effects can be eliminated by rotating balancing systems, however, this process is complex and therefore rare; crank sequences with limited or no free mass effects are preferable		Counterweights, rigid engine block
	Shielding of the environment through flexible engine mounts (in particular for orders ≥ 2)			

Table 4: Star diagram for in-line engines

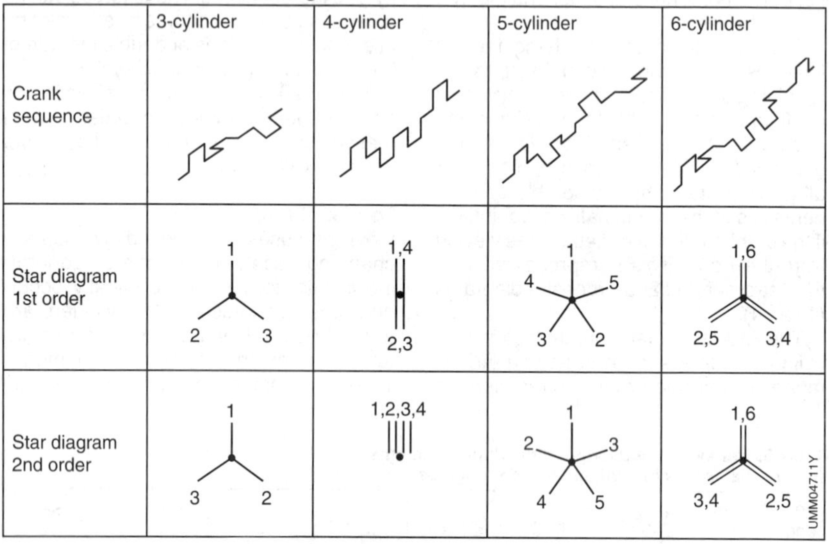

	3-cylinder	4-cylinder	5-cylinder	6-cylinder
Crank sequence				
Star diagram 1st order	1 2 3	1,4 2,3	1 4 5 3 2	1,6 2,5 3,4
Star diagram 2nd order	1 3 2	1,2,3,4	1 2 3 4 5	1,6 3,4 2,5

UMM04711Y

absorbed by the components, especially the crankshaft and crankcase, while the external effects act via the engine bearings on the support structure and impart oscillations to the chassis or engine foundation.

If the periodic gas force acting on the piston and the periodic mass inertial forces acting on the piston, connecting rod and crankshaft assembly are grouped together, they generate a sum of tangential force components at the crankshaft journal. When multiplied by the crank radius, this produces a periodically variable torque value.

With multiple-cylinder engines, the tangential-pressure curves for the individual cylinders are superimposed with a phase shift dependent on the number of cylinders, crankshaft-journal configuration, crankshaft design, and the firing sequence. The resulting composite curve is characteristic for the engine design, and covers a full working cycle (i.e. two crankshaft rotations for 4-stroke engines, Figure 18). It can be depicted in a torsional-force diagram. This variable torsional force and the resulting torque generate, depending on the available moment of inertia J, a variable rotation speed ω:

$$\frac{d\omega}{dt} = \frac{M(t)}{J}$$

with all the superimposed and newly created engine orders (there are also half orders). This deviation from the speed constant is called the coefficient of cyclic variation and is defined as follows:

$$\delta_S = \frac{\omega_{max} - \omega_{min}}{\omega_{min}}.$$

This coefficient of cyclic variation is reduced to a level suitable for the application by energy-storage mechanisms, such as a flywheel mass for example. The torsional vibrations which can be traced back to the torsional force depicted are also called 1st-order torsional vibrations. These are not to be confused with the (high-frequency) torsional vibrations resulting from elastic deformations and natural resonances of the crankshaft (see below), which are referred to as 2nd order.

Table 5: Free forces and moments of the 1st and 2nd order, and ignition intervals of the most common engine designs

$F_r = m_r\, r\, \omega^2 \quad F_1 = m_0\, r\, \omega^2 \cos\alpha \quad F_2 = m_0\, r\, \omega^2\, \lambda \cos 2\alpha$

Cylinder arrangement	Free forces of 1st order[1]	Free forces of 2nd order	Free moments of 1st order[1]	Free moments of 2nd order	Ignition intervals
3-cylinder					
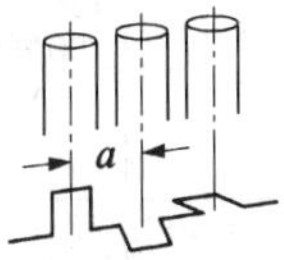 In-line, 3 throws (VM384-01)	0	0	$\sqrt{3} \cdot F_1 \cdot a$	$\sqrt{3} \cdot F_2 \cdot a$	240°/240°
4-cylinder					
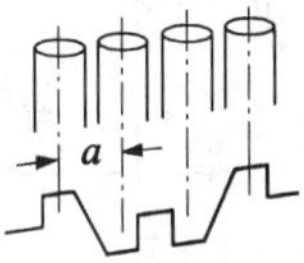 In-line, 4 throws (VM384-02)	0	$4 \cdot F_2$	0	0	180°/180°
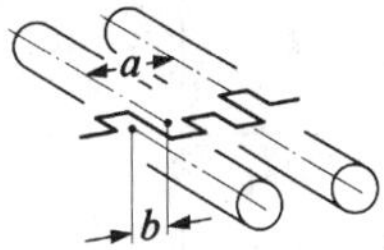 Opposed-cylinder (boxer), 4 throws (VM384-03)	0	0	0	$2 \cdot F_2 \cdot b$	180°/180°
5-cylinder					
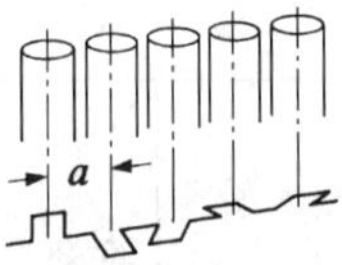 In-line, 5 throws (VM384-05)	0	0	$0.449 \cdot F_1 \cdot a$	$4.98 \cdot F_2 \cdot a$	144°/144°
6-cylinder					
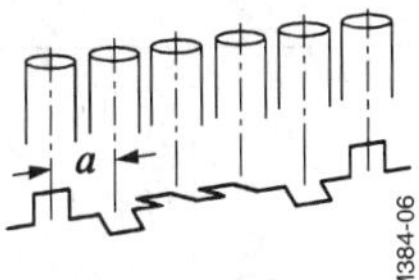 In-line, 6 throws (VM384-06)	0	0	0	0	120°/120°

[1] Without counterweights.

Table 5 (continued): Free forces and moments of the 1st and 2nd order, and ignition intervals of the most common engine designs

Cylinder arrangement	Free forces of 1st order[1]	Free forces of 2nd order	Free moments of 1st order[1]	Free moments of 2nd order	Ignition intervals
6-cylinder (continued)					
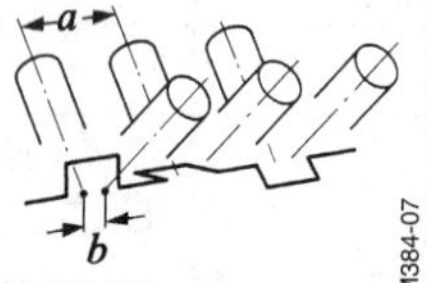 V 90°, 3 throws (VM384-07)	0	0	$\sqrt{3} \cdot F_1 \cdot a$ [2]	$\sqrt{6} \cdot F_2 \cdot a$	150°/90° 150°/90°
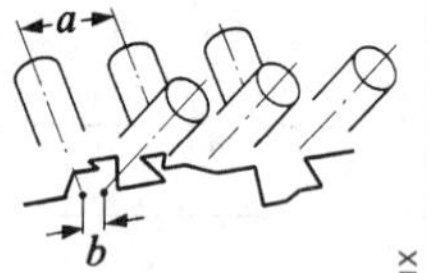 Normal balance V 90°, 3 throws, 30° crank offset (UMM0531X)	0	0	$0.4483 \cdot F_1 \cdot a$	$(0.966 \pm 0.256) \cdot \sqrt{3} \cdot F_2 \cdot a$	120°/120°
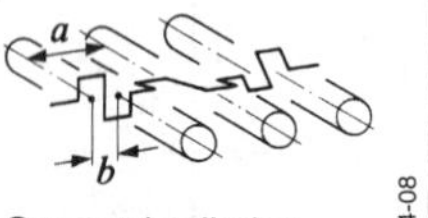 Opposed-cylinder (boxer), 6 throws (VM384-08)	0	0	0	0	120°/120°
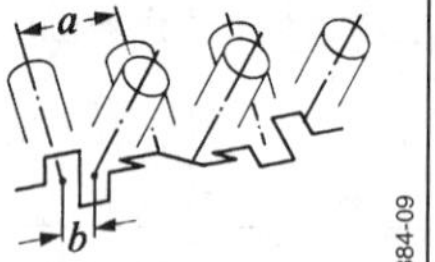 V 60°, 6 throws (VM384-09)	0	0	$3 \cdot F_1 \cdot a/2$	$3 \cdot F_2 \cdot a/2$	120°/120°
8-cylinder					
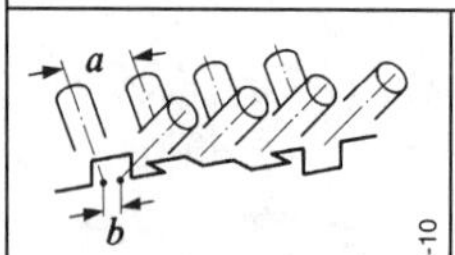 V 90°, 4 throws in two planes (VM384-10)	0	0	$\sqrt{10} \cdot F_1 \cdot a$ [2]	0	90°/90°
12-cylinder					
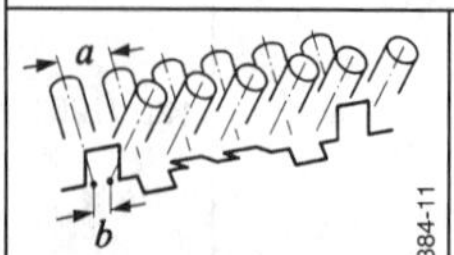 V 60°, 6 throws (VM384-11)	0	0	0	0	60°/60°

[1] Without counterweights, [2] Can be completely balanced by using counterweights.

Tribology and friction

The piston with piston rings and bearing surface is a self-contained and highly complex tribological system. This also applies to the piston pin and the crankshaft friction bearings. Detailed simulation formulations are necessary in view of the periodically changing application of force both in terms of amount and direction.

To be able to ensure wear resistance, it is necessary to build up a hydrodynamic lubricating film which is higher than the surface roughness of the two contact surfaces (see also Stribeck curve). The oil film either is already present in the lubricating gap (friction bearing with compressed-oil supply) or must be built up dynamically as an oil wedge (cambered piston ring on liner). However, the oil wedge always collapses when the relative speed of the contact surfaces assumes a value of zero. This is the case for piston rings at the piston reversal points (top and bottom dead centers), and for the piston pin between these positions. It is therefore also important to ensure that a sufficient oil-retaining volume can be maintained in the contact surfaces in order to retain by forces of adhesion in the short time of standstill as much oil as to avoid seizing or minimize increased wear. In the tribological system for the piston this is bearing-surface honing, and for the piston pin shaping of the piston-pin bearing.

Traced back to Newton is the formulation to the effect that the friction between two adjacent oil elements is virtually independent of the prevailing pressure and only proportional to the change in speed from one element to the next.

$$\tau = \frac{\eta\, dv}{dz}.$$

$\tau = F/A$ Shear stress,
F Shear force,
A Contact surface,
v Speed,
z Coordinate direction normal to speed direction v.

Figure 19: Pressure forces in a one-dimensional stress state

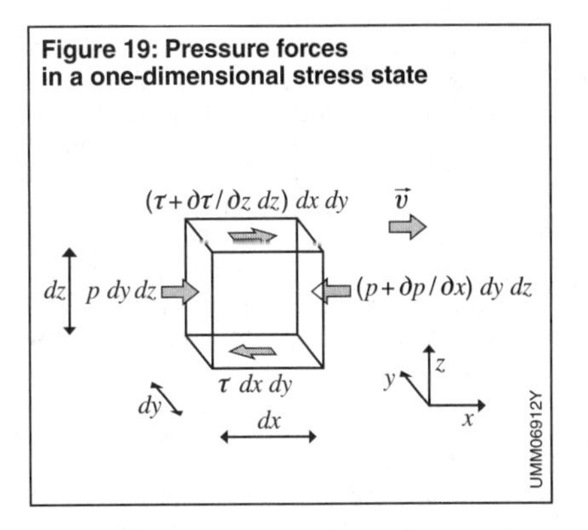

Sometimes the kinematic viscosity ν is also used; this corresponds to the dynamic viscosity referred to density and is defined by

$$\nu = \frac{\eta}{\rho}\ [\mathrm{m^2/s}].$$

The dynamic viscosity η [$\mathrm{N\,s/m^2}$ or $\mathrm{Pa\cdot s}$] decreases sharply with temperature. It is important to see to it that the oil temperature ensures a viscosity η which facilitates sufficient shear stress to build up a supporting oil film. It is also necessary to provide for a sufficiently high relative speed v on the part of the contact surfaces. It can be recognized from the formula that hydrodynamic shear stress collapses when the speed assumes a value of zero (reversal point of the piston, pivoting-angle end of the small connecting-rod eye). If can further be recognized that the lubricating film Δy must not become too large, as otherwise the shear stress will become too small and the lubricating film will break.

The equilibrium of forces for the simplified case of a one-dimensional stress state in the x direction (Figure 19) shows that $\partial p/\partial x = \partial\tau/\partial z$ must apply and thus $\partial p/\partial x = \eta\, \partial^2 v/\partial z^2$. It is thus demonstrated on the one hand that the pressure loss of a flow is dependent on the viscosity and the speed. In addition, the speed profile can thus be derived.

On account of the adhesion condition which states that the speed of the oil on the wall is identical to the wall speed, a speed distribution (linear for laminar flow) and a pressure distribution build up in the lubricating gap (see Friction bearings).

It is attempted with the aid of (thermo-) elastohydrodynamic simulation (EHD, elastohydrodynamics) to determine the lubricating-film thickness by solving the mathematical equations in the multi-dimensional case for mass balance and momentum equilibrium.

- Mass balance: supplied oil mass and oil quantity escaping through the bearing gap in equilibrium.
- Momentum equilibrium: equilibrium of forces from direct stress (pressure force from oil pressure), shear stress (viscosity force from oil viscosity) including apparent viscosity stresses on account of turbulent flow and the inertial force (acceleration component).

The equation system is established as a Reynolds differential equation or a Navier-Stokes equation to solve the flow problem. By using a finite-element simulation, the bearing deformations are added and the highly complex equations (because they are simultaneous) are numerically solved.

An adequate supply of oil must be guaranteed in the bearings, since generally the oil pressure in the bearings (> 100 bar) far exceeds the static oil pressure provided by the oil pump (< 10 bar). This is achieved by means of pressure-oil bores and special grooves in friction bearings or by means of initial forces at the piston rings.

Attempts have also been made in the meantime to simulate the case of mixed friction. The influence of surface fine geometry (consideration of roughness on microhydrodynamics by “flow tensors”) is taken into account in an extended Reynolds differential equation. By using contact-pressure models of rough surfaces, it is possible to determine, as well as the hydrodynamic contact-area ratios, the solid contact-area ratio and to take this into account accordingly. Intensive work is being conducted on models to determine the running-in behavior of contact surfaces. This will finally enable the wear behavior to be assessed.

Because the shear stresses and thus also the frictional forces are proportional to the speed, the share of friction of the piston assembly at the dead centers and that of the piston-pin bearing is small. Instead, the wear parameters are decisive here.

Sliding bearings

The bearings most commonly encountered in an internal-combustion engine are friction (sliding) bearings. Crankshaft main bearings, big-end bearings, piston-pin bearings, and camshaft bearings are generally oil-lubricated friction bearings (see Friction bearings).

It is important to ensure that the seating pressure is precisely configured in both split bearings (crankshaft, connecting rod) and non-split bearings (piston pin, camshaft) so as to prevent them from jamming and rotating also at higher temperatures or under emergency-lubrication conditions. However, material overloading must also be reliably prevented.

The lubricating gap of the crankshaft main bearings changes in response to the constantly varying bearing forces (see also Crankshaft-assembly design). This generates a kind of pumping effect and the oil in the bearing is constantly replaced.

The swinging motion of the connecting rod effects an increased relative speed when the piston travels up the cylinder and a reduced relative speed when the piston travels down the cylinder in the lubricating gap of the big-end bearing or crank pin. This alone causes oil to be pumped into the oil gap. Furthermore, naturally periodic forces acting on the bearing cause the connecting rod to change position constantly and thereby assist the build-up of lubricating-film pressure.

The small-end bearing is merely subject to a swinging motion. A supply of oil is therefore particularly important, but difficult to achieve. In practical terms, mixed friction cannot be avoided. If the arrangement involves a floating pin bearing (bearing in the piston and connecting rod), the pin performs a slight rotary motion in the

Figure 20: Roughness values and oil-film theory (roughness definitions, see DIN 4760 [5])
1 Piston, 2 Oil film, 3 Liner.

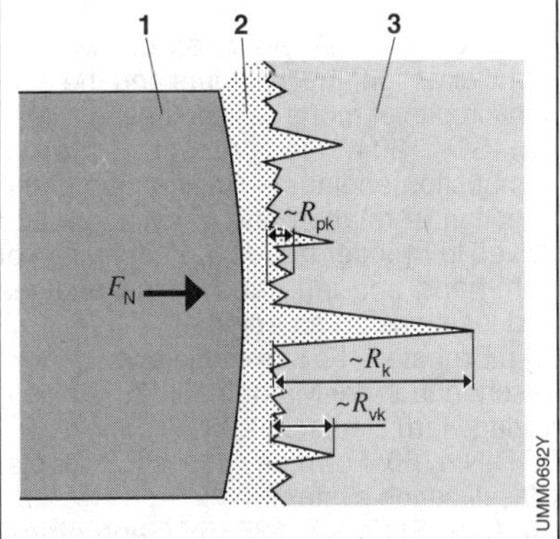

piston bearing of just a few degrees – here too a good supply of oil to the bearings must be ensured.
Oil is generally supplied as splash oil, and partly also via a specific locating bore or via a pressure-oil supply from the crankshaft through grooves in the crankshaft-journal bearing and a vertical bore in the connecting rod.

Rolling pairs

Ball, roller or needle bearings are used in 2-stroke engines or sometimes in camshafts (see Roller bearings).

A roller follower or the roller of an injection-pump plunger-and-barrel assembly is a special case in point. The system of rotating roller and moving cam is a rolling-element pair with linear contact. A characteristic load variable is the Stribeck rolling-contact pressure:

$$K = \frac{F}{D_l \, l_{eff}},$$

F Force load,
$\varphi = D_l/D$ Osculation coefficient,
D_l Substitute diameter,
D Roller diameter,
l_{eff} Effective roller width.

The effective substitute diameter D_l is obtained from the instantaneous bend radius of the cam R_{cam} and the roller radius R_{roller}:

$$D_l = 2\,\frac{R_{cam}\,R_{roller}}{R_{cam} + R_{roller}}.$$

It takes into account the osculation of the two rolling bodies.

The Hertzian stress p_{Hertz} is also relevant:

$$p_{Hertz} = \sqrt{\frac{KE}{2.86}}.$$

with modulus of elasticity E:

$$E = 2\,\frac{E_{cam}\,E_{roller}}{E_{cam} + E_{roller}}.$$

By slightly angling the roller in relation to the cam or by offsetting the stroke function, it is possible to ensure that an oil wedge can be built up and the roller rotates continuously. Pitting is a typical indicator of overloading in rolling pairs that are subject to high loads. It results from molecular deformations of the metal and possible embedding of oil, which gives rise to extremely high local pressures and plastic deformations or material erosion. Harder materials and decreasing the load reduce the tendency to pitting.

Piston ring and bearing surface

The piston rings together with the bearing surface constitute a complex tribological system. On the one hand, the pressure force of the piston ring is very much dependent on the cylinder pressure and thus on the crank angle and the power stroke. On the other hand, the piston rings make axial secondary movements in their grooves which also have a significant impact on the lubricating-film thickness.

The oil is generally applied as splash oil to the bearing surface (crankshaft churning or splash oil for piston cooling and pin lubrication) and allowed back through the oil control ring as a thin film. The oil-retaining volume of the liner is dependent on the honing. This includes the surface roughness (furrow depth) and the honing pattern (cross-furrow angle).

Honing the bearing surface serves two functions. Firstly, it is intended to ensure with the aid of honing furrows of sufficient depth that oil can be retained by adhesion forces for an adequate length of time (Figure 20). Secondly, surface roughness must be so low as to prevent contact with the piston ring. Various roughness values are used to characterize honing; these values are standardized on a generic level in DIN 4760 [5] and in detail in DIN EN ISO 13565 [6]. For initial wear and initial friction the R_{pk} value ("reduced peak height", see DIN 4760 and DIN EN ISO 11562 [7] or DIN EN ISO 13565 for an explanation) is decisive – but the value is swiftly and drastically minimized after run-in (e.g. from a few micrometers to 0.2 to 0.8 µm). The R_{vk} value (1.0 to 3.5 µm) is the decisive parameter for the possible oil-retaining volume after run-in. It is to be noted that the honing usually takes the form of a cross-shaped furrow system. In this form, the furrows constitute a communicating system of oilways and the oil pressure is partly forced out of the gap, but this also permits a uniform distribution of the oil film over the surface. Flat honing angles (referring to the angle formed by the cross-shaped furrows) reduce the amount of oil forced out, but there is a tendency to seizing on account of the unsatisfactory axial oil distribution. Steep honing angles result in increased oil consumption. Honing angles between 30° and 90° are customary.

Mixed friction occurs at any rate at the piston reversal points. The least that is required in this area is micro-pockets of oil, which can be produced for example by laser honing or by bringing out amounts of alloy during the honing procedure.

Usually, such high temperatures are generated during the combustion stroke of a piston engine that much of the bearing-surface oil vaporizes or in diesel engines even burns off. This vaporized oil constitutes the majority of the oil consumption by modern-day internal-combustion engines.

The special shape of the piston rings is designed to facilitate the build-up of an oil wedge. A critical factor is, on the one hand, the piston speed, which must not be allowed to become too high, as otherwise the shear forces can no longer be absorbed by the oil and the oil film will break. On the other hand, the reversal points, particularly at top dead center, must be classed as critical with regard to wear.

References

[1] H. Hütten: Motoren – Technik, Praxis, Geschichte. Motorbuch-Verlag, Stuttgart, 1997.

[2] DIN 73021: Designation of the rotational direction, the cylinders and the ignition circuits of motorcar engines.

[3] ISO 1204: Reciprocating internal combustion engines – Designation of the direction of rotation and of cylinders and valves in cylinder heads, and definition of right-hand and left-hand in-line engines and locations on an engine.

[4] I. Bronstein; K. Semendjajew: Taschenbuch der Mathematik. 7th Edition, Verlag Harri Deutsch, 2008.

[5] DIN 4760: Form deviations; Concepts, Classification system.

[6] DIN EN ISO 13565: Geometrical Product Specifications (GPS) – Surface texture: Profile method – Surfaces having stratified functional properties.

[7] DIN EN ISO 11562: Geometrical Product Specifications (GPS) – Surface texture: Profile method – Metrological characteristics of phase correct filters.

Empirical values and data for calculation

Comparisons

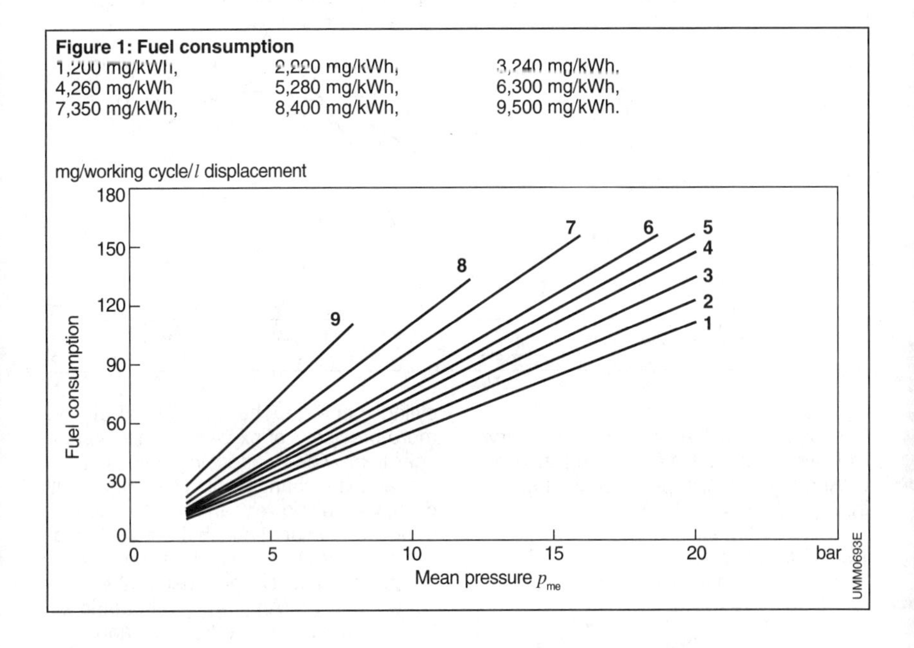

Figure 1: Fuel consumption
1,200 mg/kWh, 2,220 mg/kWh, 3,240 mg/kWh,
4,260 mg/kWh 5,280 mg/kWh, 6,300 mg/kWh,
7,350 mg/kWh, 8,400 mg/kWh, 9,500 mg/kWh.

Table 1: Comparative data

Engine type/ application			Engine speed n_{nom} rpm	Compression ratio ε	Max. mean pressure p_e bar	Power output per liter kW/l	Weight-to-power ratio kg/kW	Specific fuel consumption g/kWh	Torque increase %
Gasoline engines	Motorcycles	4-stroke	5,000–13,000	9–12	9–13	50–150	2.5–0.5	230–280	10–15
	Pass. cars	NAE	5,000–8,000	9–13	11–14	40–80	2.0–0.8	220–270	15–20
		IC/ SCE	5,000–7,500	9–12	15–22	60–110	1.5–0.5	220–250	20–40
Diesel engines	Pass. cars/ Light comm. veh.	NAE	3,500–4,500	19–24	7–9	20–35	4.0–2.0	220–260	5–10
		SCE/ IC	3,500–4,500	18–22	12–20	35–55	3.0–1.3	200–220	20–40
	Comm. veh.	SCE/ IC	1,800–2,600	15–18	18–24	25–40	4.0–2.5	180–210	20–40

NAE = Naturally aspirated engine; SCE = Supercharged engine; IC = Intercooling

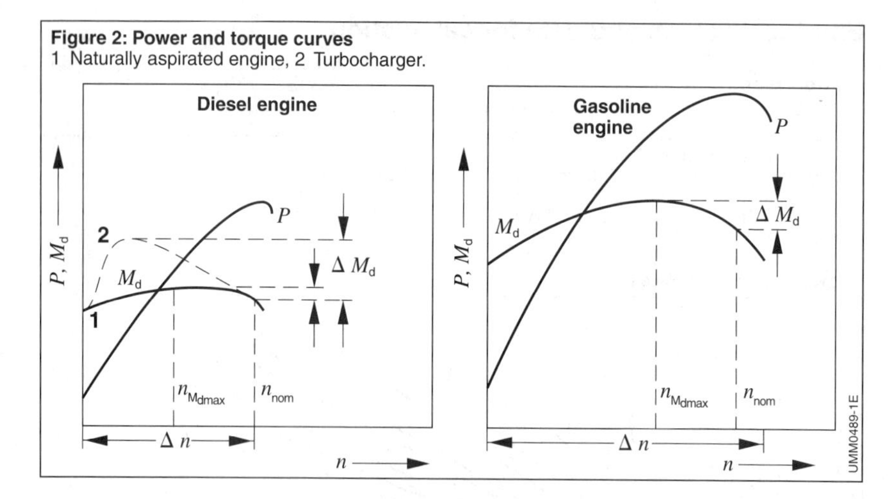

Figure 2: Power and torque curves
1 Naturally aspirated engine, 2 Turbocharger.

Torque position
The position of the engine-speed curve (relative to rpm for max. output) at which maximum torque is developed, specified in % ($n_{Mdmax}/n_{nom} \cdot 100$).

Useful speed range
(minimum full-load speed/nominal speed)

Engine type		Useful speed range Δn_N	Torque position %
Diesel engine	Pass. cars	3.5 to 5	15 to 40
	Comm. veh.	1.8 to 3.2	10 to 60
Gasoline engine		4 to 7	25 to 35

Torque increase

Engine type		Torque increase M_d in %
Diesel engine pass. cars	Nat. asp. engine	15 to 20
	SC [1] + IC [2]	25 to 35
Diesel engine comm. veh.	SC [1] + IC [2]	25 to 40
Gasoline engine	Nat. asp. engine	25 to 30
	SC [1] + IC [2]	30 to 35

[1] with supercharging, [2] with intercooling.

Engine output, atmospheric conditions
The torque and thus the power output of an internal-combustion engine are essentially determined by the heat content of the cylinder charge. The flow rate of air (or, more precisely, of oxygen) in the cylinder charge provides a direct indication of heat content. The change which the engine will display at full power can be calculated as a function of variations in the condition of the ambient air (temperature, barometric pressure, humidity), provided that engine speed, air/fuel (A/F) ratio, volumetric efficiency, combustion efficiency, and total engine power loss remain constant. The A/F mixture responds to lower atmospheric density by becoming richer. The volumetric efficiency (pressure in cylinder at BDC relative to pressure in ambient air) only remains constant for all atmospheric conditions at maximum throttle-valve aperture (WOT). Combustion efficiency drops in cold thin air as vaporization rate, turbulence, and combustion speed all fall. Engine power loss (friction losses + gas-exchange work + boost power drain) reduces the indicated power.

Effect of atmospheric conditions
The quantity of air which an engine draws in, or is inputted to the engine by supercharging, depends upon ambient-air density; colder, heavier, denser air increases engine output. Rule of thumb: Engine power drops by approximately 1 % for each 100 m rise in elevation. Depending upon engine design, the cold intake air is normally heated to some degree while traversing the intake passages, thereby reducing its den-

sity and thus the engine's ultimate output. Humid air contains less oxygen than dry air and therefore produces lower engine power outputs. The decrease is generally modest to the point of insignificance. The warm humidity of air in tropical regions can result in a noticeable engine power loss.

Definitions of power

The effective power is the engine's power as measured at the crankshaft or ancillary mechanism (such as the transmission) at the specified engine speed. When measurements are made downstream from the transmission, the transmission losses must be factored into the equation. Rated power is the maximum effective power of the engine at full throttle. Net power corresponds to effective power.

Conversion formulas are used to convert the results of dynamometer testing to reflect standard conditions, thereby negating the influences of such factors as time of day and year while simultaneously allowing the various manufacturers to provide mutually comparable data. The procedure converts air density – and thus the effective volume of air in the engine – to defined "standard conditions" for air mass.

The cross-references in the following table show the most important standards used in power correction.

Table 2: Power correction standards (comparison)

Standard (date of publication)	EEC 80/1269 (4/81)	ISO 1585 (5/82)	JIS D 1001 (10/82)	SAE J 1349 (5/85)	DIN 70 020 (11/76)
Barometric pressure during testing (* vapor pressure subtracted!)					
Dry p_{PT}* kPa	99	99	99	99	–
Absolute p_{PF} kPa	–	–	–	–	101.3
Temperature during testing					
Absolute T_p K	298	298	298	298	293
Gasoline engines, naturally aspirated and supercharged					
Correction factor α_a	$\alpha_a = A^{1.2} \cdot B^{0.5}$ $A = 99/p_{PT}$ $B = T_p/298$				$\alpha_a = A \cdot B^{0.5}$ $A = 101.3/p_{PF}$ $B = T/293$
Corrected power: $P_0 = \alpha_a \cdot P$ (kW) (P measured power)					
Diesel engines, naturally aspirated and supercharged					
Atmospheric engine correction factor f_a	$f_a = A \cdot B^{0.7}$ ($A = 99/p_{PT}$; $B = T_p/298$) (naturally aspirated and mechanically supercharged engines) $f_a = A^{0.7} \cdot B^{1.5}$ ($A = 99/p_{PT}$; $B = T_p/298$) (turbocharged engines with/without intercooling)				As α_a for gasoline engines
Engine correction factor f_m	$40 \leq q/r \leq 65$: $f_m = 0.036 \cdot (q/r) - 1.14$ $q/r < 40$: $f_m = 0.3$ $q/r > 65$: $f_m = 1.2$				$f_m = 1$
$r = p_L/p_E$ Boost pressure response at p_L absolute boost pressure, p_E Absolute pressure upstream of air compressor, q Spec. fuel consumption (SAE J 1349). 4-stroke engines: q = 120,000 F/DN, 2-stroke engines: q = 60,000 F/DN, with F Fuel flow (mg/s), D Displacement (l); N Engine speed (rpm).					
Corrected power: $P_0 = P \cdot f_a^{fm}$ (kW) (P measured power).					
Mandatory accessories					
Fan	Yes, with electric/viscous-drive fan at max. slip				Not defined
Emission control	Yes				Not defined
Alternator	Yes, loaded with engine-current draw				Yes
Servo-pumps	No				No
Air conditioner	No				No

Calculation

Quantity		Unit
a_K	Piston acceleration	m/s²
B	Fuel consumption	kg/h; dm³/h
b_e	Spec. fuel consumption	g/kWh
D	Cylinder diameter $2 \cdot r$	mm
d_v	Valve diameter	mm
F	Force	N
F_G	Gas force in the cylinder	N
F_N	Piston side thrust	N
F_o	Oscillating inertial force	N
F_r	Rotating inertial force	N
F_s	Connecting-rod force	N
F_T	Tangential force	N
M	Torque	Nm
M_o	Oscillating moments	Nm
M_r	Rotating moments	Nm
M_d	Engine torque	Nm
m_p	Power-to-weight ratio	kg/kW
n	Engine speed	rpm
n_p	Fuel-injection pump speed	rpm
P	Power	kW
P_{eff}	Effective power[1]	kW
P_H	Power output per liter	kW/dm³
p	Pressure	bar
p_c	Final compression pressure	bar
p_e	Mean piston pressure (mean pressure, mean working pressure)	bar
p_L	Boost pressure	bar
p_{max}	Peak injection pressure in the cylinder	bar
r	Crankshaft radius	mm
s_d	Injection cross-section of the nozzle	mm²
S, s	Stroke, general	mm
s	Piston stroke	mm
s_f	Induction stroke of a cylinder (2-stroke)	mm
s_F	Induction stroke, 2-stroke engine	mm
S_k	Piston clearance from TDC	mm
S_s	Slot height, 2-stroke engine	mm
T	Temperature	°C, K
T_c	Final compression temperature	K
T_L	Charge-air temperature	K
T_{max}	Peak temperature in combustion chamber	K
t	Time	s
V	Volume	m³

[1] Effective power P_{eff} is the effective horsepower delivered by the internal-combustion engine, with it driving the auxiliary equipment necessary for operation (e.g. ignition equipment, fuel-injection pump, scavenging-air and cooling-air fan, water pump and fan, supercharger) (DIN 1940). This power is called net engine power in DIN 70 020.

Quantity		Unit
V_c	Compression volume of a cylinder	dm³
V_E	Injected fuel quantity per pump stroke	mm³
V_f	Charge volume of a cylinder (2-stroke)	dm³
V_F	Charge volume of a 2-stroke engine	dm³
V_h	Displacement of a cylinder	dm³
V_H	Displacement of the engine	dm³
v	Velocity	m/s
v_d	Mean velocity of injection spray	m/s
v_g	Gas velocity	m/s
v_m	Mean piston speed	m/s
v_{max}	Max. piston speed	m/s
z	Number of cylinders	–
α_d	Injection time (in °CA at fuel-injection pump)	°
β	Pivoting angle of connecting rod	°
ε	Compression ratio	–
η	Efficiency	–
η_e	Net efficiency	–
η_{th}	Thermal efficiency	
ν, n	Polytropic exponent of real gases	–
ρ	Density	kg/m³
φ, α	Crank angle (φ_o = top dead center)	°
ω	Angular velocity	rad/s
λ	$= r/l$ Stroke/connecting-rod ratio	–
λ	Excess-air factor	–
κ	$= c_p/c_v$ Adiabatic exponent of ideal gases	–

Superscripts and subscripts

0, 1, 2, 3, 4, 5	Cycle values/main values
o	Oscillating
r	Rotating
1st, 2nd	1st, 2nd order
A	Constant
′, ″	Subdivision of main values, derivations

Conversion of units

1 g/hp · h	= 1.36 g/kWh
1 g/kWh	= 0.735 g/hp · h
1 kpm	= 9.81 Nm ≈ 10 Nm
1 Nm	= 0.102 kpm ≈ 0.1 kpm
1 hp	= 0.735 kW
1 kW	= 1.36 hp
1 at	= 0.981 bar ≈ 1 bar
1 bar	= 1.02 at ≈ 1 at

Table 3: Calculation equations

<table>
<tr><th>Mathematical relationship between quantities</th><th>Numerical relationship between quantities</th></tr>
<tr><td>

Piston displacement

Displacement of a cylinder

$V_h = \frac{\pi \cdot d^2 \cdot s}{4}$; $V_f = \frac{\pi \cdot d^2 \cdot s_f}{4}$ (2-stroke)

Displacement of an engine

$V_H = V_h \cdot z$; $V_F = V_f = z$ (2-stroke)

</td><td>

$V_H = 0.785 \cdot 10^{-6}\, d^2 \cdot s$

V_h in dm³, d in mm, s in mm

$V_H = 0.785 \cdot 10^{-6}\, d^2 \cdot s \cdot z$

V_h in dm³, d in mm, s in mm

</td></tr>
<tr><td>

Compression

Compression ratio

$\varepsilon = \frac{V_h + V_c}{V_c}$

Final compression pressure

$p_c = p_o \cdot \varepsilon^{\nu}$

Final compression temperature

$T_c = T_o \cdot \varepsilon^{\nu - 1}$

</td><td>

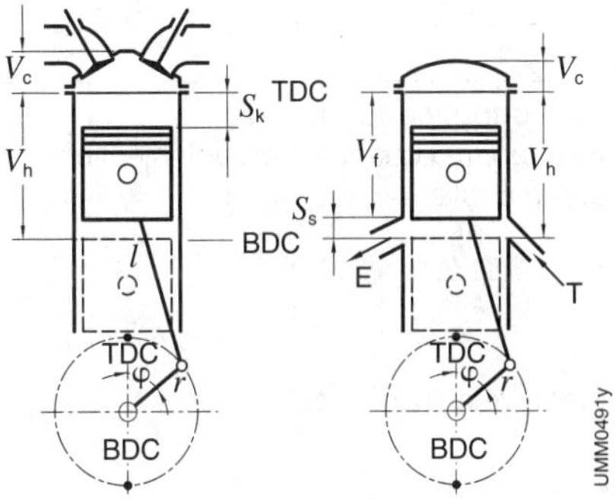

4-stroke engine **2-stroke engine**

</td></tr>
<tr><td>

Piston stroke

Piston clearance from top dead center

$S_k = r\left[1 + \frac{l}{r} - \cos\varphi - \sqrt{\left(\frac{l}{r}\right)^2 - \sin^2\varphi}\right]$

Crank angle

$\varphi = 2 \cdot \pi \cdot n \cdot t$ (φ in rad)

Piston speed (approximation)

$v \approx 2 \cdot \pi \cdot n \cdot r\left(\sin\varphi + \frac{r}{2l}\sin 2\varphi\right)$

Mean piston speed

$v_m = 2 \cdot n \cdot s$

Maximum piston speed (approximate, if connecting rod is on a tangent with the big-end trajectory; $a_k = 0$)

<table>
<tr><td>l/r</td><td>3.5</td><td>4</td><td>4.5</td></tr>
<tr><td>v_{max}</td><td>$1.63 \cdot v_m$</td><td>$1.62 \cdot v_m$</td><td>$1.61 \cdot v_m$</td></tr>
</table>

</td><td>

$\varphi = 6 \cdot n \cdot t$

φ in °, n in rpm, t in s

$v \approx \frac{n \cdot s}{19{,}100}\left(\sin\varphi + \frac{r}{2l}\sin 2\varphi\right)$

v in m/s, n in rpm, l, r and s in mm

$v_m = \frac{n \cdot s}{30{,}000}$

v_m in m/s, n in rpm, s in mm

</td></tr>
<tr><td>

Piston acceleration (approximation)

$a_k \approx 2 \cdot \pi^2 \cdot n^2 \cdot s\left(\cos\varphi + \frac{r}{l}\cos 2\varphi\right)$

</td><td>

$a \approx \frac{n^2 \cdot s}{182{,}400}\left(\cos\varphi + \frac{r}{l}\cos 2\varphi\right)$

a_k in m/s², n in rpm, l, r and s in mm

</td></tr>
</table>

[1] The variables can be entered in the equations in any units. The unit of the quantity to be calculated is obtained from the units chosen for the terms of the equation. The numerical equations apply only with the units of measure specified under the equation.

Table 3: Calculation equations (continued)

Mathematical relationship between quantities	Numerical relationship between quantities
Gas velocity Mean gas velocity in the valve section $v_g = \frac{d^2}{d_v^2} \cdot v_m$	$v_g = \frac{d^2}{d_v^2} \cdot \frac{n \cdot s}{30{,}000}$ v_g in m/s, d, d_v, and s in mm, n in rpm
Fuel supply and delivery Injected fuel quantity per fuel-injection pump stroke $V_E = \frac{P_{eff} \cdot b_e}{\rho \cdot n_p \cdot z}$	$V_E = \frac{1{,}000 \cdot P_{eff} \cdot b_e}{60 \cdot \rho \cdot n_p \cdot z}$ V_E in mm³, P_{eff} in kW, b_e in g/kW · h (or also P_{eff} in hp, b_e in g/hp · h), n_p in rpm, ρ in kg/dm³ (for fuels $\rho \approx 0.85$ kg/dm³)
Mean velocity of injection spray $v_d = \frac{2 \cdot \pi \cdot n_p \cdot V_E}{S_d \cdot \alpha_d}$ (α_d in rad)	$v_d = \frac{6 n_p \cdot V_E}{1{,}000 \cdot S_d \cdot \alpha_d}$ v_d = in m/s, n_p in rpm, V_E in mm³, S_d in mm², α_d in °
Engine power $P = M \cdot \omega = 2 \cdot \pi \cdot M \cdot n$	$P = M \cdot n / 9{,}549$ P in kW, M in $N \cdot m$ (= W · s),
$P_{eff} = V_H \cdot p_e \cdot n / K$ $K = 1$ for 2-stroke engine $K = 2$ for 4-stroke engine	$P_{eff} = \frac{V_H \cdot p_e \cdot n}{K \cdot 600} = \frac{M_d \cdot n}{9{,}549}$ P_{eff} in kW, p_e in bar, n in rpm. M_d in N · m
Power per unit displacement (power output per liter) $P_H = \frac{P_{eff}}{V_H}$	$P = M \cdot n / 716.2$ P in hp, M in kp · m, n in rpm
Power-to-weight ratio $m_p = \frac{m}{P_{eff}}$	

Table 3: Calculation equations (continued)

Mathematical relationship between quantities		**Numerical relationship between quantities**	
Mean piston pressure (mean pressure, mean working pressure)			
4-stroke engine	2-stroke engine	4-stroke engine	2-stroke engine
$p=\frac{2\cdot P}{V_H\cdot n}$	$p=\frac{P}{V_H\cdot n}$	$p=1{,}200\frac{P}{V_H\cdot n}$	$p=600\frac{P}{V_H\cdot n}$
		p in bar, P in kW, V_H in dm³, n in rpm	
		$p=833\frac{P}{V_H\cdot n}$	$p=441\frac{P}{V_H\cdot n}$
		p in bar, P in hp, V_H in dm³, n in rpm	
$p=\frac{4\cdot\pi\cdot M}{V_H}$	$p=\frac{2\cdot\pi\cdot M}{V_H}$	$p=0.1257\frac{M}{V_H}$	$p=0.628\frac{M}{V_H}$
		p in bar, M in N · m, V_H in dm³	
Engine torque		$M_d=\frac{V_H\cdot P_e}{0.12566}$	$M_d=\frac{V_H\cdot P_e}{0.06284}$
$M_d=\frac{V_H\cdot P_e}{4\pi}$	$M_d=\frac{V_H\cdot P_e}{4\pi}$	M_d in N·m, V_H in dm³, p_e in bar	
		$M_d = 9{,}549\cdot P_{eff}/n$	
		M_d in N·m, P_{eff} in kW, n in rpm	
Fuel consumption B = Measured values in kg/h $b_e = B/P_{eff}$ $b_e = 1/(H_u\cdot\eta_e)$		B in dm³/h or kg/h V_B = Measured volume on test dynamometer t_B = Elapsed time for measured volume consumption $b_e=\frac{V_b\cdot\rho_B\cdot 3{,}600}{t_b\cdot P_{eff}}$ ρ_B = Fuel density in g/cm³, t_B in s, V_B in cm³, P_{eff} in kW.	
Efficiency $\eta_{th} = 1-\varepsilon^{1-\kappa}$ $\eta_e = P_{eff}/(B\cdot H_u)$		$\eta_e = x/b_e$ where $x = 82$ for $H_u = 44$; $x = 86$ for $H_u = 42$; $x = 90$ for $H_u = 40$; $x = 120$ for $H_u = 30$ H_u Specific calorific value in MJ/kg b_e Specific fuel consumption in g/kWh	

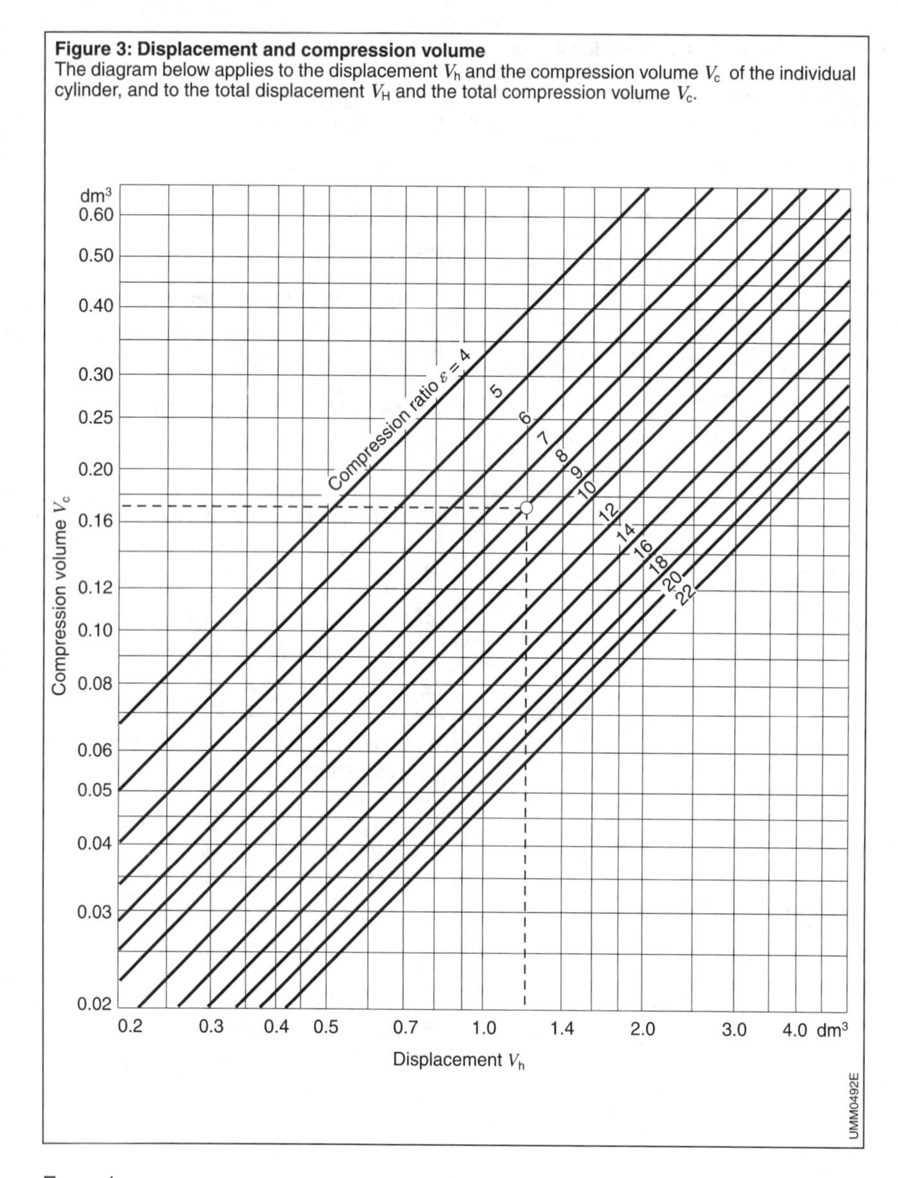

Figure 3: Displacement and compression volume
The diagram below applies to the displacement V_h and the compression volume V_c of the individual cylinder, and to the total displacement V_H and the total compression volume V_c.

Example:
An engine with a displacement of 1.2 dm³ and a compression ratio $\varepsilon = 8$ has a compression volume of 0.17 dm³.

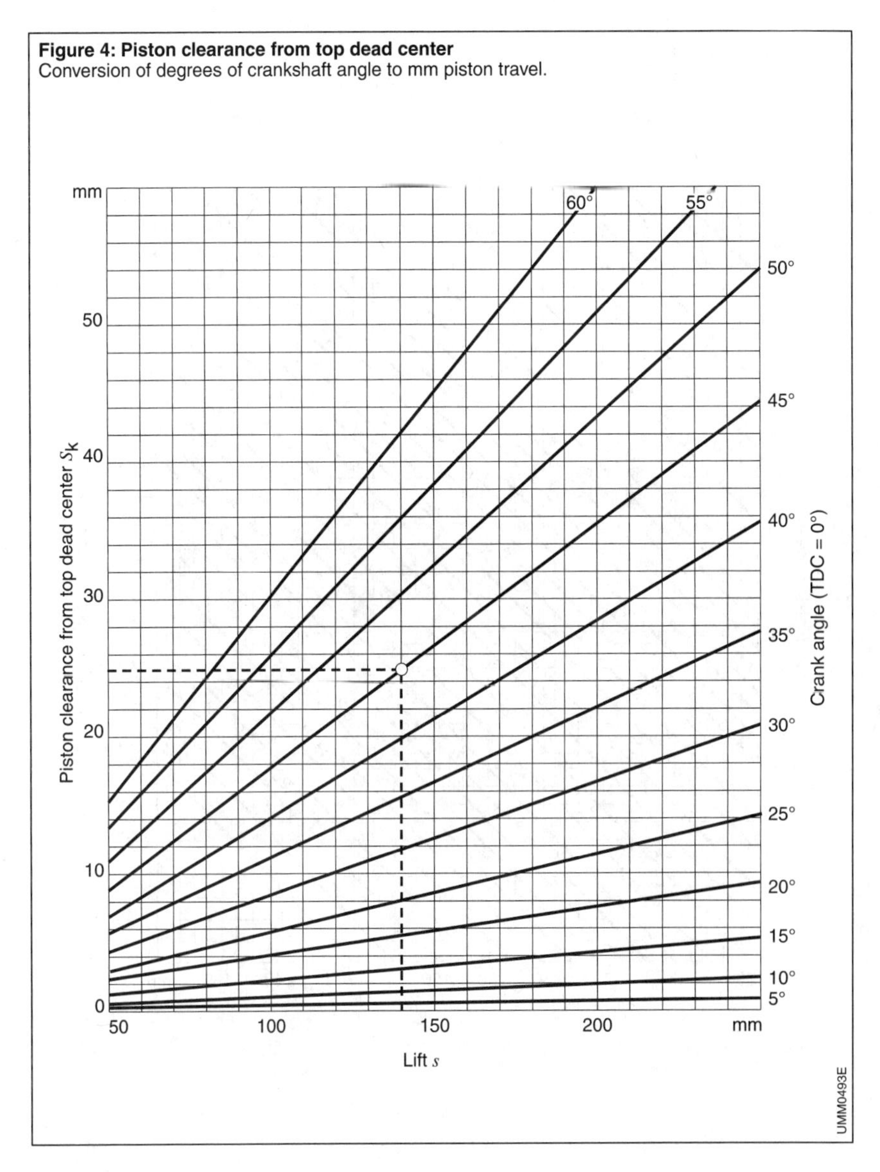

Figure 4: Piston clearance from top dead center
Conversion of degrees of crankshaft angle to mm piston travel.

Example:
The piston clearance from top dead center is 25 mm for a stroke of 140 mm at 45° CA.

The diagram is based on a crank ratio $l/r = 4$
(l connecting-rod length, r one half of the stroke length).
However, it also applies with good approximation (error less than 2 %) to all ratios l/r between 3.5 and 4.5.

Figure 5: Piston speed

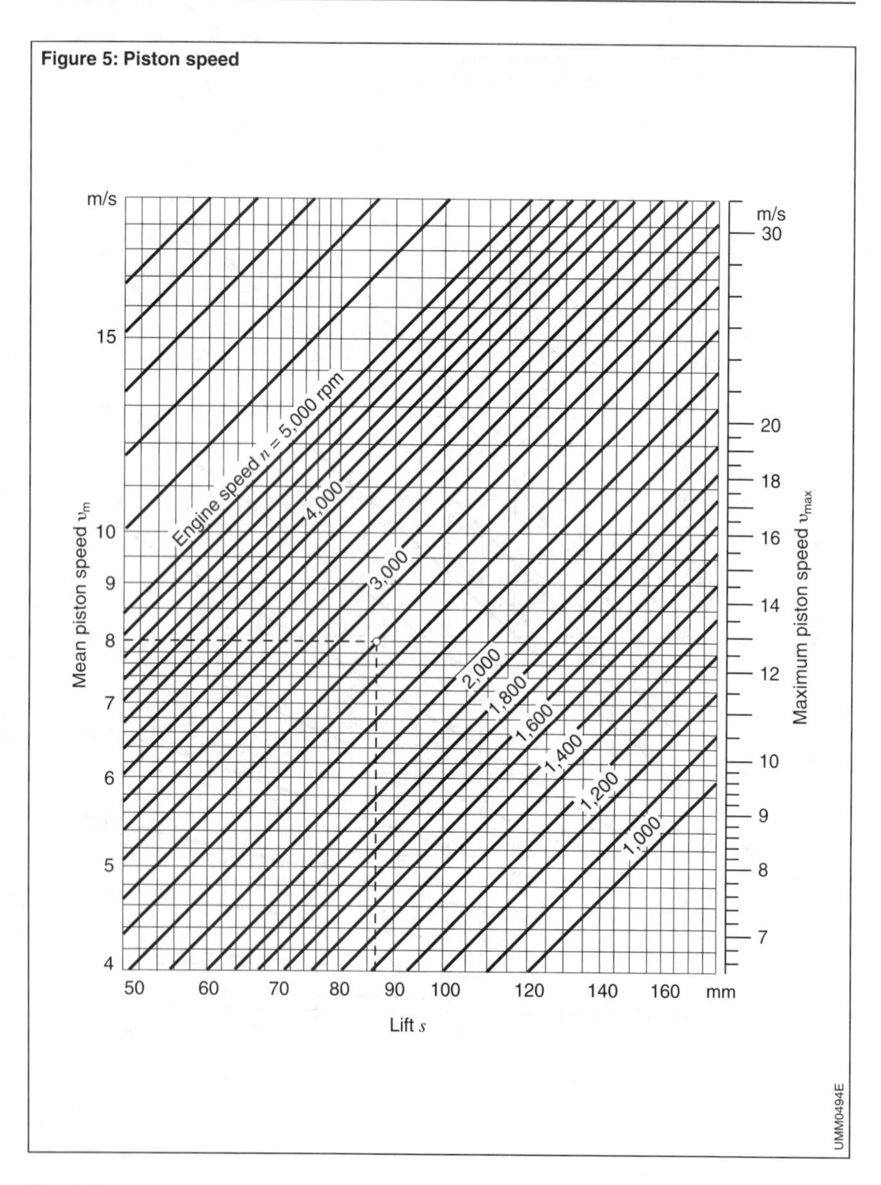

Example:
For stroke s = 86 mm and at an engine speed of n = 2,800 rpm, mean piston speed v_m = 8 m/s and maximum piston speed v_{max} = 13 m/s.

The diagram is based on $v_{max} = 1.62\ v_m$.

Figure 6: Density increase of combustion air in cylinder on turbo/supercharging

Increase in density on supercharging as a function of the pressure ratio in the compressor, of the compressor efficiency and of the intercooling rate for intercooling (IC).

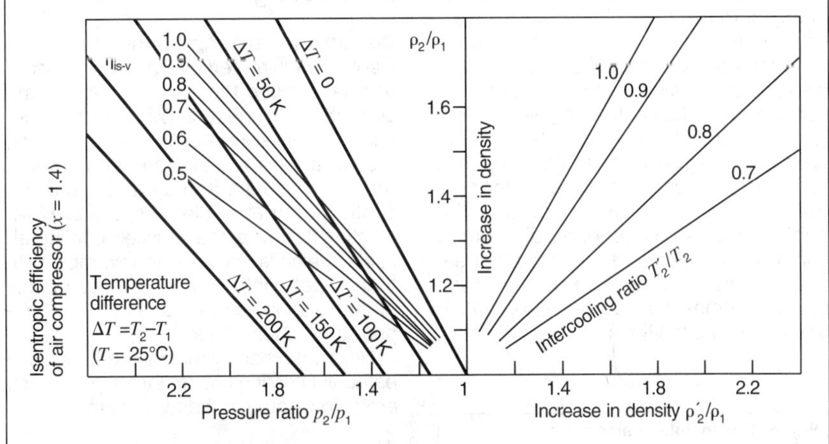

$p_2/p_1 = \pi_c$ = Pressure ratio during precompression
ρ_2/ρ_1 = Increase in density, ρ_1 = Density upstream of compressor,
ρ_2 = Density downstream of compressor in kg/m³
T_2'/T_2 = Intercooling rate, T_2 = Temperature before IC, T_2' = Temperature after IC in K
$\eta_{is\text{-}v}$ = Isentropic compressor efficiency

UMM0495-1E

Figure 7: Final compression pressure and temperature

Final compression temperature as a function of the compression ratio and the intake temperature.

$t_c = T_c - 273.15\text{K},\ T_c = T_A \cdot \varepsilon^{n-1},\ n = 1.35$

Final compression pressure as a function of the compression ratio and boost pressure.

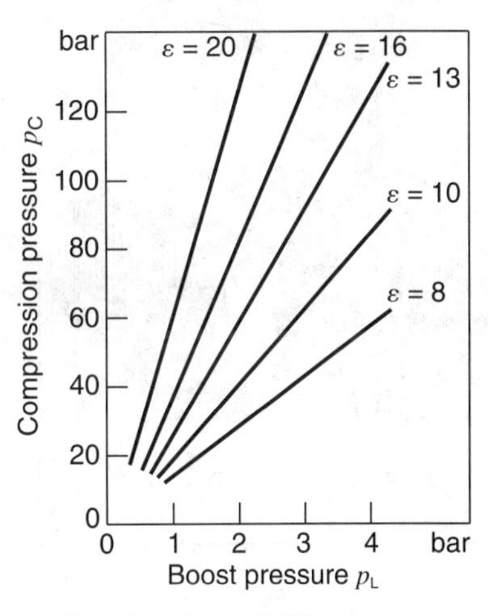

$p_c = p_L \cdot \varepsilon^n,\ n = 1.35$

UMM0496E

Engine cooling

The process of converting any type of energy into mechanical energy to drive a vehicle creates heat as a waste product, whether it be by friction of different engine components against each other or by the combustion of fuel in the engine. In order to optimally utilize the conversion of the chemical energy of fuel into kinetic energy, this process must take place under controlled conditions and, to protect the engine and its components, also under controlled temperatures. The waste heat that is generated must be reliably dissipated to atmosphere under all operating and ambient conditions.

Figure 1: Principle of air cooling
1 Oil cooler,
2 Cooling-air fan,
3 Cylinder with cooling fins,
4 Oil pan,
5 Pressure-relief valve,
6 Oil pump (gear pump).
7 Lube-oil filter,
8 Thermostat.

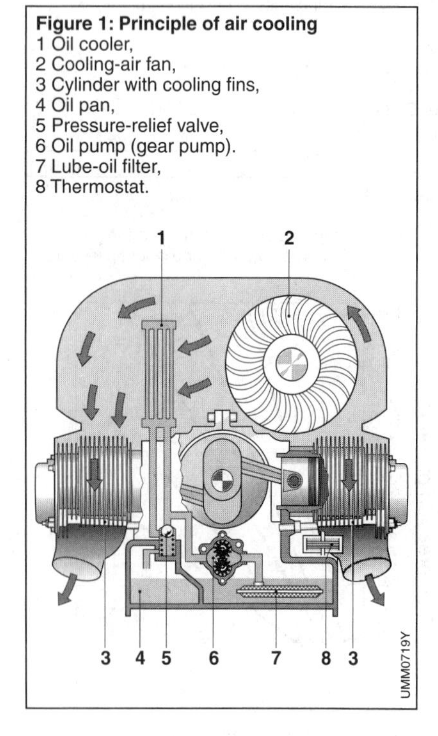

Air cooling

Design and operating principle

Cooling air is routed by dynamic pressure or a fan around the external walls of the cylinder casing (Figure 1). These walls are finned in order to achieve better cooling results thanks to the increased surface area. The cooling-air quantity can be controlled for example by specifically restricting its flow at the vehicle's cooling-air inlets, which is achieved for example with thermostat-controlled flaps or by means of the rotational speed of a load- or temperature-dependent fan.

Heat absorbed by the engine oil is dissipated via cooling fins on the oil pan which are also located in the air stream.

Advantages and disadvantages of air cooling

The advantages of air cooling are simple and cost-effective construction, reliable operation, and lower weight when compared with water-cooling systems.

The disadvantages are higher noise generation, lower temperature consistency of all engine parts, and poorer thermal behavior at high specific power outputs.

Air-cooled engines struggle to comply with the currently required exhaust-emission limits.

Applications

Today, air cooling is mainly used for motorcycle and aircraft engines and in special applications. Extra-small engines in model-making applications also use air cooling.

The expected operating conditions for a vehicle can only dictate which cooling principle is chosen. In Arctic regions, where temperatures down to –50 °C are experienced, water-cooled engines cannot be used owing to the risk of the water freezing; for this reason air cooling must be used in these regions.

Water cooling

Design and operating principle

Cooling circuit

In contrast to air cooling, with water cooling the waste heat from the engine is not dissipated directly to the ambient air. Instead, the waste heat is absorbed by the coolant and transported via this medium to the main coolant radiator (Figure 2). Here, heat is exchanged between the coolant and the ambient air. The cooling-air flow rate is controlled by a fan, and the coolant is circulated in the circuit by a coolant pump.

Coolant cooling, also called water cooling, has become the standard in both passenger cars and commercial vehicles.

Coolant

Coolants are a mixture of water, antifreeze (mostly ethylene glycol), and corrosion inhibitors. The inhibitors prevent the components through which coolant flows from corroding (see Coolants). Adding antifreeze in concentrations of 30 to 50 % firstly stops the coolant from freezing even at temperatures down to – 25 °C and secondly raises the coolant boiling temperature to allow coolant temperatures of up to 120 °C at a pressure of 1.4 bar in passenger cars.

Radiator designs

The cores of the coolant radiators in modern passenger cars are almost exclusively made of aluminum. Aluminum radiators are also being used to an ever-increasing extent in a wide range of commercial vehicles and trucks throughout the world. There are two assembly variants: brazed radiators and mechanically joined or assembled radiators.

Flat-tube/corrugated-fin systems

For cooling high-performance engines, or when space is limited, the best solution is a brazed flat-tube and corrugated-fin radiator layout with minimized aerodynamic resistance on the air-intake side.

Construction

The radiator core consists of series of flat tubes with corrugated fins located in between (Figure 3). The flat tubes lead into the openings in the radiator base. During the manufacturing process the tips of the corrugated fins are soldered to the

Figure 2: Passenger-car cooling system
1 Engine,
2 Fan,
3 Coolant line (arrow indicates direction of flow),
4 Coolant pump,
5 Main coolant radiator (flat-tube/corrugated-fin system),
6 Bypass line (secondary cooling circuit),
7 Expansion tank with pressure-relief valve (integrated in cap),
8 Thermostat.

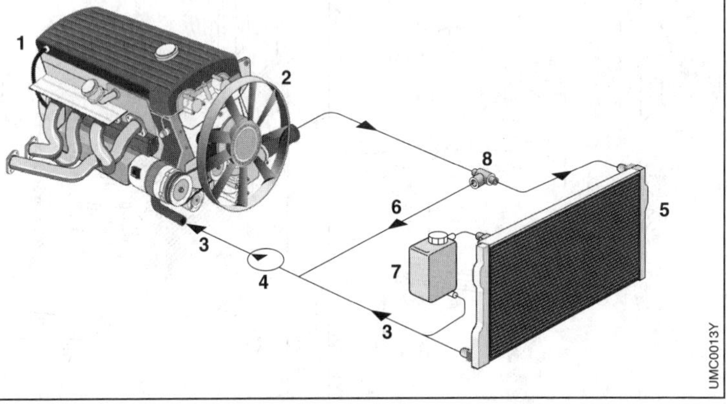

UMC0013Y

flat tubes and the flat tubes to the radiator base in an oven. The solder material is applied as a layer to the materials.

The radiator tank ensures that the coolant is distributed throughout the core. These tanks are made of fiberglass-reinforced polyamides, and are injection-molded with all connections and mountings in a single unit. They are flange-mounted to the radiator core and sealed by an integrated elastomer sealing element.

Operating principle
The coolant is routed through the radiator tank to the radiator core. It flows through the flat tubes before emerging again from the opposite radiator tank. The hot coolant in the flat tubes dissipates heat to the corrugated fins, while the cooling air flowing through the radiator core cools the corrugated fins. The soldered connections guarantee a good heat transfer from the tubes to the fins. Turbulators or gills in the fins, which are combined into so-called gill zones, induce additional cooling-air turbulence. This results in improved heat dissipation.

Radiator tanks are designed in such a way as to distribute the coolant as uniformly as possible to all the flat tubes. The tubes have minimal depth (approx. 2 mm with wall thicknesses of 0.2 to 0.3 mm) so that the full coolant flows past the tube walls as closely as possible, thereby ensuring a good heat transfer to the fins.

Tube/fin systems
The less expensive, mechanically assembled tube/fin system is generally employed for applications with less powerful engines or when more space is available.

Construction
When the radiator is assembled mechanically, the cooling grid is formed by mounting stamped cooling fins around round, oval or flat-oval tubes (Figure 4). The distance between the cooling fins is maintained by trapezoidal cutouts. In the manufacturing process the tubes are

Figure 3: Flat-tube/corrugated-fin system
1 Radiator tank,
2 Radiator base,
3 Gasket,
4 Flat tube,
5 Corrugated fins,
6 Gill zones.

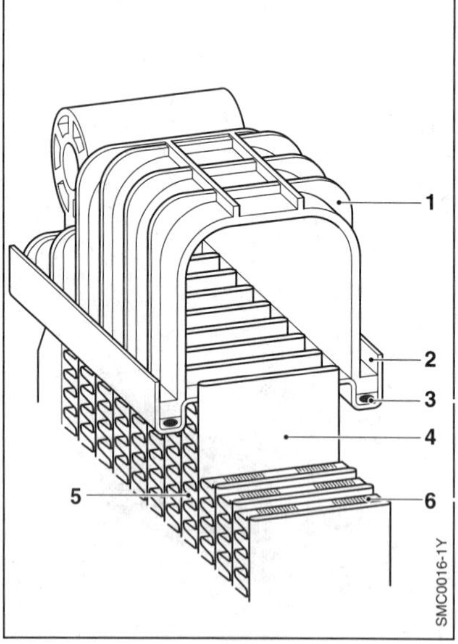

Figure 4: Tube/fin system
1 Radiator tank,
2 Radiator base,
3 Gasket,
4 Round tube,
5 Cooling fin,
6 Gill zones,
7 Turbulators.

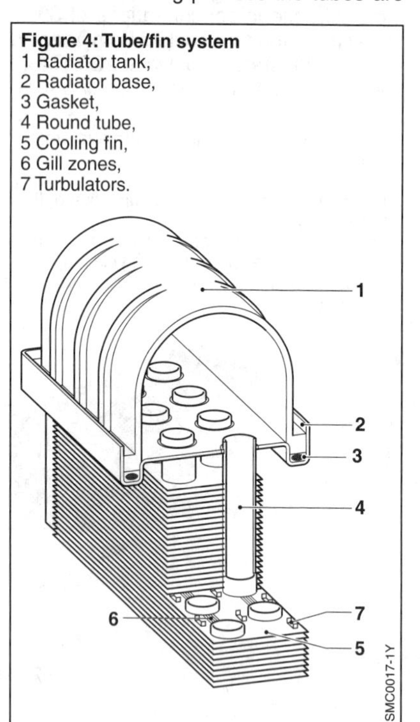

initially inserted loosely into the cooling fins and flared with a tool. This creates an internal connection between the tubes and the fins. The connection is established exclusively by the mechanical flaring of the tube. However, the heat transfer between the tubes and the fins is much worse than in the soldered systems.

Operating principle
The cooling air flows through the radiator core between the cooling fins. These are corrugated or slotted at right angles to the direction of air flow. They are – like the corrugated fins in soldered radiators – also fitted with turbulators and gill zones, inducing air turbulence and improving the cooling effect.

Options for increasing radiator performance
Gills and a larger number of corrugations are integrated into the cooling fins on the cooling-air side to enhance thermal transfer. The gills induce turbulence in the cooling air flowing through, and the larger number of corrugations produces a larger surface area. These measures increase the quantity of heat that can be dissipated.

Further measures to increase cooling efficiency include the use of tubes with the lowest possible width and wall thickness, and turbulators on the coolant side, provided the attendant pressure losses remain within acceptable limits. The negative effect of these performance-enhancing measures is the higher pressure drop on both the cooling-air and coolant sides, which must be compensated by a higher-capacity fan and on the coolant side by a more powerful coolant pump.

Coolant expansion tank
Mounted at the highest point of the cooling circuit is the airtight expansion tank; this tank has a filler neck through which coolant is poured into the cooling system. When the engine is running the expansion tank takes up excess coolant that expands due to heating. Where necessary, it returns it to the cooling circuit. As the coolant heats up, so too the system pressure in the cooling system increases. This causes the coolant boiling point to rise.

The expansion tank's air volume must be large enough to absorb the coolant's thermal expansion during rapid pressure buildup, and prevent the coolant from boiling over under the permitted operating conditions. At excessively high operating temperatures and thus also excessively high operating pressures the expansion tank protects the cooling circuit by means of a pressure-relief valve.

The expansion tank also provides a reliable escape channel for pressurized gases, preventing cavitation of the kind that tends to occur on the intake side of the coolant pump.

Expansion tanks are injection-molded in plastic (generally polypropylene), although simple designs can also be blow-molded to shape. A system of hoses is normally used to connect the expansion tank to the cooling system. The expansion tank is mounted in the engine compartment at the location representing the highest point in the cooling system to ensure air is expelled effectively. In some cases, the expansion tank can form a single unit with the radiator tank, or the two can be joined by means of a flange or plug connection.

The position and shape of the filler neck can be used to limit capacity, thus preventing overfilling. An electronic level sensor can be fitted to monitor the level of the coolant. The coolant level can also be monitored by manufacturing the expansion tank completely or partially in natural-color, transparent plastic with molded level markers. However, colorless polypropylene is sensitive to ultraviolet rays. For this reason, the transparent part of the expansion tank should not be exposed to direct sunlight.

Cooling-air fan
Design
Since motor vehicles also require substantial cooling capacity at low speeds, force-air ventilation is required for the radiator. Injection-molded plastic fans with drive-power ratings extending up to 30 kW are now used as standard in commercial vehicles. A fan of this type is driven by a mechanical coupling with the internal-combustion engine, for example via a belt drive. The fan can also be mounted directly on the crankshaft.

Single-piece injection-molded plastic fans are generally used in passengers cars; these are usually driven by DC brush motors or brushless DC motors. These motors are installed in the fan hub. The electrical drive power is around 400 W in the small-car segment, and extends up to 1 kW in the luxury-class segment and in offroad vehicles. Although blade design and arrangement can be selected to provide relatively quiet running, these fans have considerable noise levels at high rotation speeds.

In some passenger-car applications, in particular offroad vehicles with very high engine power outputs in combination with equipment options for operation in hot climates and equipment variants for diesel engines and air-conditioning systems, the capacity of the electric drive is no longer sufficient to provide the volume of air necessary to ensure effective cooling. Fan-output figures in excess of 1 kW can only be achieved when the fan is driven mechanically by the engine. However, this is only possible with a longitudinally installed engine.

Figure 5: Bimetall-controlled viscous clutch
1 Cover with cooling fins,
2 Primary disk,
3 Valve bore,
4 Valve lever,
5 Thermal bimetallic strip,
6 Operating pin,
7 Seal,
8 Supply chamber,
9 Working chamber,
10 Return bore,
11 Fan,
12 Mounting bolt,
13 Basic body,
14 Ball bearing,
15 Flanged shaft

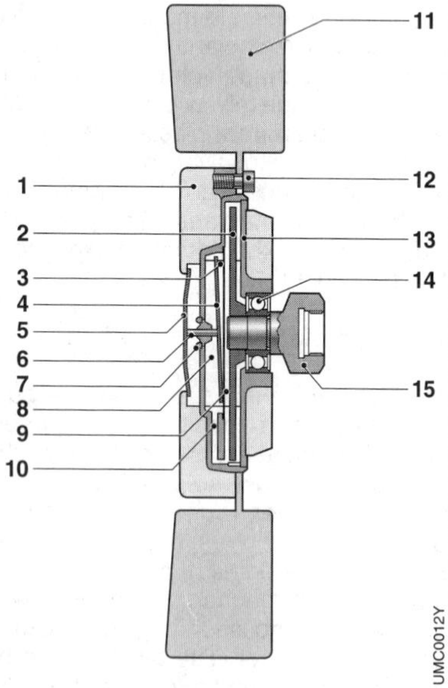

Control of electrically driven fans

Depending on vehicle and operating conditions, the unassisted air stream can provide sufficient cooling up to 95 % of the time. It is thus possible to economize on energy that would otherwise have to provide the power to drive the fan. For this purpose, electric fans use a multistage or continuous control system that specifically adapts the operating periods and speed of the fan to the required cooling capacity. A multistage control system may consist of relays and series resistors, while continuous variable control necessitates the use of power electronics. Electric thermostatic switches or the engine control unit supply the input signals for the control system.

Drive of mechanical fans

The fluid-friction clutch (viscous clutch) is a mechanical-drive arrangement of proven effectiveness for use in commercial vehicles. It essentially consists of the following assemblies (Figure 5):

- Input or primary section (flanged shaft and primary disk)
- Output or secondary section (cover, basic body)
- Control facility for controlling and regulating the oil filling (working fluid)

The secondary section of the viscous clutch is divided into a working chamber and a supply chamber. When the viscous clutch is not cut in, only a small amount of working fluid is in the working chamber, such that only a very small amount of torque can be transmitted due to the high degree of slip between primary disk and secondary section. The viscous working fluid is a silicone oil.

As the temperature increases, the bulging of the bimetallic strip causes the operating pin to open the valve. Silicone oil flows from the supply chamber into the working chamber. This reduces the slip between primary disk and secondary section – in turn increasing the fan speed. The fan speed and thus the cooling capacity are infinitely increased.

As the temperature decreases, the bimetallic strip cools and causes the operating pin to close the valve slowly. The silicone oil flows through the pump body back into the supply chamber. The fan speed decreases. The amount of silicone oil located in the working chamber determines the output power transmitted by the clutch and therefore the speed of the fan.

A distinction is made between types of viscous clutch, based on how the valve is actuated: Firstly, as already described, the temperature-dependent, self-governing clutch which varies its speed infinitely by means of a bimetallic element, an operating pin, and a valve lever. The controlled variable is the temperature of the air leaving the radiator, and thus indirectly the temperature of the coolant. The electrically activated clutch is now used in 95% of heavy-duty commercial vehicles. In this clutch type the valve lever is electromagnetically actuated and thus the oil quantity in the working chamber is regulated. Instead of just one controlled variable, a wide range of input variables is used for control purposes. These are usually the temperature limits of the various cooling media.

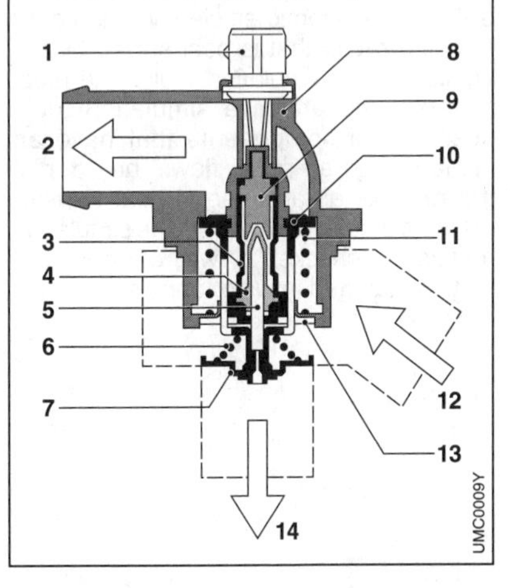

Figure 6: Map-controlled thermostat
1 Connector, 2 Connection to radiator,
3 Housing of working element,
4 Elastomer insert,
5 Plunger,
6 Bypass spring,
7 Bypass valve,
8 Housing,
9 Heating resistor,
10 Double-acting disk valve (main valve),
11 Main disc spring,
12 Connection from engine,
13 Tie bar,
14 Connection to engine (bypass).

Regulation of coolant temperature

A motor vehicle's engine operates in a very wide range of climatic conditions and with major fluctuations in engine load. The consequences of this are significant variations in coolant and engine temperatures, which in turn result in increased engine wear, unfavorable exhaust-gas composition, higher fuel consumption, and unreliable vehicle heating. The temperature of the coolant must be regulated in order to counteract these unwanted side effects and to keep the coolant and engine temperature as constant as possible.

Expansion-element-regulated thermostat

A temperature-sensitive thermostat incorporating an expansion element proves to be a robust regulator operating independently of the changing pressure conditions in the cooling system. It is very similar in design to the map-controlled thermostat (Figure 6), but does not have a heating resistor. The expansion element used in this type of thermostat actuates a double-acting disk valve (main valve), which – until the operating temperature is reached – closes the connection to the radiator while simultaneously releasing coolant flow from the engine outlet to the bypass line (see Figure 2). The coolant flows uncooled back into the engine ("secondary cooling circuit").

Both sides of the double-acting disk valve are partially opened within the control range of the thermostat. This allows a mixture of cooled and uncooled coolant to flow to the engine at such a rate as to maintain a constant operating temperature. At wide-open throttle (full load) the opening to the radiator is fully opened and the bypass line closed off ("primary cooling circuit").

Electronic map-controlled thermostat
Further possibilities to regulate the coolant temperature are permitted when a map-controlled thermostat is used. An electronically controlled thermostat differs from purely expansion-element-regulated thermostats in that the opening temperature can be controlled. A map-controlled thermostat features a heating resistor that is used to additionally heat the expansion element (Figure 6). This has the purpose of enlarge the opening of the double-acting disk valve to the radiator, thereby reducing the coolant temperature. The heating resistor is controlled by the engine management system in order to ensure that the engine operating temperature is optimally adapted to the operating conditions. The information necessary for this purpose is stored in the form of program maps in the engine-management system (Figure 7).

Raising the operating temperature in the part-load range and reducing the operating temperature at wide-open throttle provide the following benefits:

Figure 7: Electronic control of coolant temperature

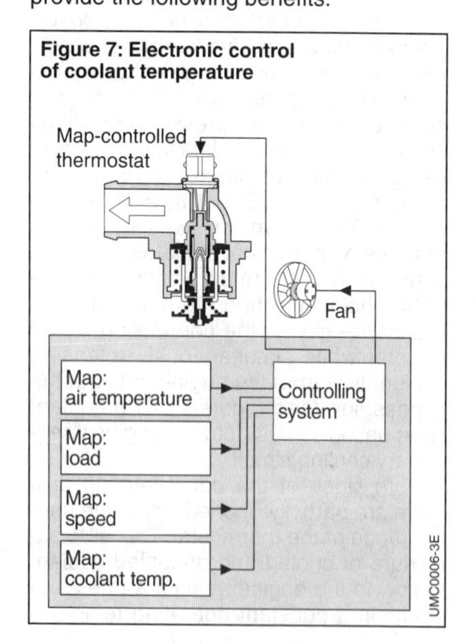

- Lower fuel consumption
- Low-pollutant exhaust-gas composition
- Reduce engine wear
- Improved heating efficiency of the vehicle interior

Radiator design
The size and therefore the cooling capacity of a specific radiator can be determined by calculation, based on correlation equations derived from tests relating to thermal transfer and flow-pressure loss. The mass of air that flows through the radiator in the vehicle is a decisive factor and depends on the driving speed, the resistance to flow in the engine compartment, the resistance to flow of the radiator, and the efficiency of the fan.

The primary objective of radiator design is to maintain the coolant temperature at the engine outlet below a maximum permissible value under given operating conditions. Because at low driving speeds the dynamic pressure of the cooling air and thus the cooling-air flow rate in the radiator are very low, the cooling-air flow required for adequate cooling must be ensured either by a high-output fan or by a radiator with low flow resistance, while a high air-mass flow rate facilitates the use of smaller radiators with high flow resistance. However, the latter causes high energy consumption if the air-mass flow is generated by a powerful fan.

The task of determining the most favorable solution in terms of technical feasibility and economic efficiency is a matter of optimization that is best solved by the application of simulation tools. The most suitable and effective simulation tools describe all components that have an influence on air-mass flow. They depict the radiator as an integral heat-transfer medium or heat exchanger. The simulation results are checked by way of on-vehicle tests conducted in wind tunnels.

Intercooling (charge-air cooling)

Trends in engine development show a constant increase in specific engine output power. This development goes hand in hand with the current transition from naturally aspirated engines to supercharged engines and ultimately to supercharged, intercooled engines. The need for intercooling (charge-air cooling) is attributed to the higher air density levels associated with supercharging systems and therefore the amount of oxygen available in the combustion air. Intercooling also reduces exhaust-gas emissions from supercharged diesel engines. If there was no intercooling (charge-air cooling) employed on supercharged spark-ignition (SI) engines, appropriate steps would need to be taken to prevent engine knocking attributed to mixture enrichment or retarded ignition timing. Consequently, intercooling indirectly serves to reduce fuel consumption and exhaust emissions.

Design variations

Basically speaking, both the ambient air and the engine coolant can be employed to cool the charge air. With only few exceptions, air-cooled intercoolers are now used both on passenger cars and commercial vehicles.

Air-cooled intercoolers

Air-cooled intercoolers can be mounted in front of or next to the engine radiator, or even completely separate from the adiator at a different location in the engine compartment. A separately positioned intercooler can utilize either the unassisted vehicle air stream or its own fan. With the intercooler located in front of the engine radiator, the cooling-air fan ensures sufficient air flow even at low vehicle speeds. However, a drawback of this arrangement is that the cooling air is itself heated in the process. To compensate for this effect, the higher temperature of the inflowing cooling air must be taken into account in the design of the engine radiator.

The system of corrugated aluminum fins and tubes employed for the intercooler core is similar to the design of the engine coolant radiator. Wide tubes with internal fins provide superior performance and structural integrity in actual practice. The fin density on the cooling-air side is relatively low and corresponds approximately to the density of the inner fins in order to achieve a good distribution of thermal-transfer resistance.

Wherever possible, the plenum chamber is injection-molded in fiberglass-reinforced polyamide as a single casting incorporating all connections and mounts. The plenum chambers which are subject

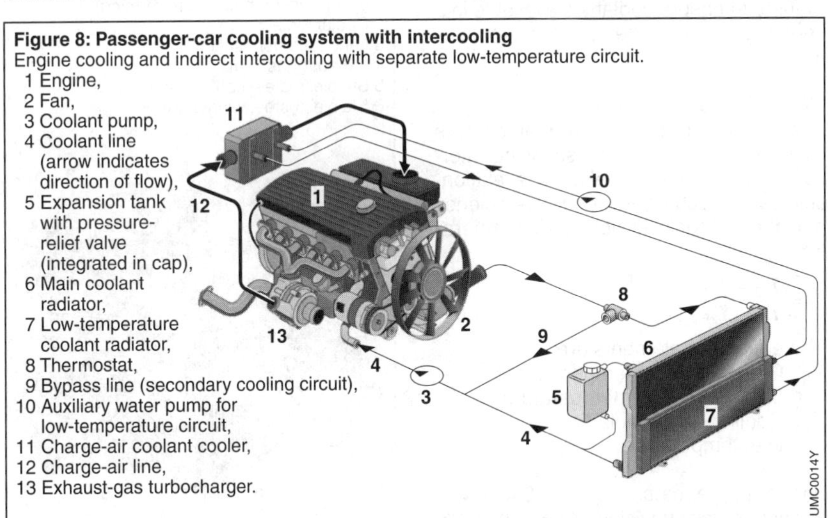

Figure 8: Passenger-car cooling system with intercooling
Engine cooling and indirect intercooling with separate low-temperature circuit.
1 Engine,
2 Fan,
3 Coolant pump,
4 Coolant line (arrow indicates direction of flow),
5 Expansion tank with pressure-relief valve (integrated in cap),
6 Main coolant radiator,
7 Low-temperature coolant radiator,
8 Thermostat,
9 Bypass line (secondary cooling circuit),
10 Auxiliary water pump for low-temperature circuit,
11 Charge-air coolant cooler,
12 Charge-air line,
13 Exhaust-gas turbocharger.

to increased stresses on the charge-air inlet side are injection-molded from highly heat-resistant PPA (polyphthalamide) or PPS (polyphenylene sulfide). They are flange-mounted to the radiator core and sealed by an integrated elastomer sealing element. Plenum chambers which feature undercut shapes or are intended for high-temperature applications are cast in aluminum and welded to the core.

Coolant-cooled intercoolers
Coolant-cooled intercoolers can be installed in virtually any location in the engine compartment, as there are no technical difficulties to supplying the system with coolant. In addition, thanks to its modest dimensions, this type of intercooler requires substantially less space than the air-cooled intercooler. Coolant-cooled intercoolers have a high power density. However, coolant at a very low temperature must be available in order to effectively cool the charge air. This requirement is of particular significance in commercial vehicles and heavy-duty trucks as, in this case, it is necessary to heat the charge air to a level of 15 K above the ambient temperature. Because this stipulation cannot be implemented with the temperature level of the normal cooling circuit at approx. 100 °C, it is necessary, in order to cool the charge air, to install a low-temperature radiator to ensure coolant is available in a separate circuit at the required temperature level.

Diffusion rate, intercooler
The diffusion rate Φ of an intercooler is particularly important in assessing intercooler performance. It defines the relationship between charge-air cooling efficiency and the charge-air/cooling-air temperature differential:

$$\Phi = \frac{T_{1E} - T_{1A}}{T_{1E} - T_{2E}}.$$

The equation's elements are:
T_{1E} Charge-air inlet temperature
T_{1A} Charge-air outlet temperature
T_{2E} Cooling-air or coolant inlet temperature

For passenger cars: $\Phi = 0.4...0.7$
For commercial vehicles: $\Phi = 0.9 - 0.95$

Exhaust-gas cooling

Exhaust-gas recirculation is used in diesel engines to reduce raw nitrogen-oxide emissions (NO_x). In gasoline engines it serves both to protect components and to dethrottle in the part-load range by charge dilution (see Exhaust-gas recirculation). Exhaust-gas recirculation delivers the best results when the recirculated exhaust gas is cooled in an EGR cooler.

Design of an EGR cooler
A central aspect in the design of an EGR cooler is, as well as its output, its durability. The former is achieved either by specially shaped tubes (winglets) which generate a high level of turbulence in the tube or by internal fins in the tubes which increase the surface area for heat exchange. The essential characteristic here is the tendency of an EGR cooler towards fouling. This can be significantly influenced by a suitable design of the winglets or tube fins. Figure 9 shows such an EGR cooler in longitudinal section. The winglet tubes direct the hot exhaust gas and are themselves surrounded by coolant.

It must be borne in mind with regard to durability that the hot exhaust gas may

Figure 9: EGR cooler
1 Stainless-steel housing (manufactured by internal high-pressure forming),
2 Coolant inlet,
3 Length compensation,
4 Exhaust-gas inlet,
5 Stainless-steel cooler base.
6 Stainless-steel winglet tubes.

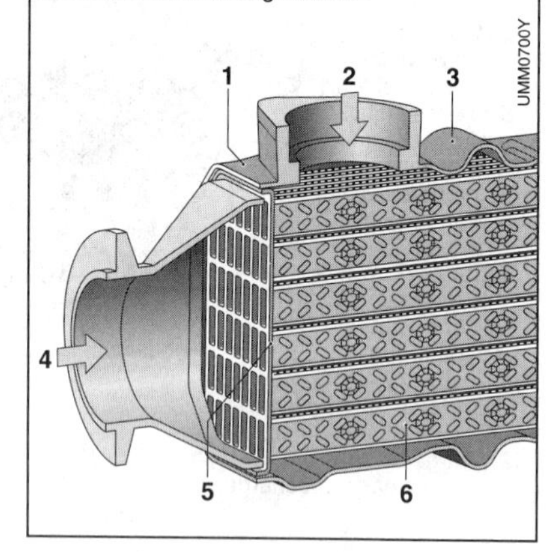

condense in the cooler. The condensate produced has a lower pH value and attacks the material like a strong acid. For this reason, the gas-directing parts of EGR coolers are frequently made of stainless steel.

When the EGR cooler is in operation the tubes get hotter than the housing. Suitable design measures must be used to compensate for the linear difference in the cooler.

The issue of vibration also plays an important role in view of the fact that the cooler is usually bolted directly on the engine. The EGR cooler must therefore be as rigid as possible in its design. The first natural frequency of an EGR cooler should at least be above the first natural frequency of the engine block. For this reason, the brackets of such EGR coolers are specifically very rigid and sometimes multiply ribbed in design.

The cooler housing is often made of stainless steel, which can be strengthened further by cold-working. The exhaust pipes connected to the cooler should be isolated at the engine end. This isolation is achieved by bellows or by a comparable structure.

Cooling circuit

The coolant comes primarily from the engine cooling circuit, with the EGR cooler being cooled by coolant flowing out of the engine. A variant of this arrangement is "two-stage exhaust-gas cooling". Here the exhaust gas is cooled in a first stage as described above. There is also a second cooling stage, which is supplied by a cooling circuit independent from the engine (low-temperature circuit). Because the temperatures in this second cooling stage are much lower (approx. 10 to 20 K above ambient temperature), the exhaust gas can also be cooled significantly below the otherwise normal values, increasing the effect of reducing nitrogen oxide. The two cooling stages are either integrated together in a single housing or designed separately as two independent components.

Oil and fuel cooling

Oil coolers are often needed in motor vehicles to cool both the engine oil and transmission oil. They are used when the heat losses from the engine or transmission can no longer be dissipated via the surface of the oil pan or the transmission with the result that the permitted oil temperatures are exceeded.

Air- or coolant-cooled oil coolers are used to suit their application profile.

Oil-to-air coolers

Oil-to-air coolers are predominantly made of aluminum. In most cases, soldered flat-tube/corrugated-fin systems (Figure 10, similar to engine radiators) with a high power density are used. Mechanically assembled systems with round tubes and fins are less commonly used. Turbulence inserts are brazed in the flat-tube system to increase cooling capacity and strength (to resist high internal pressures).

Figure 10: Oil-to-air cooler (flat-tube/corrugated-fin system)
1 Collector tank with oil connections,
2 Base,
3 Partition,
4 Flat tube with turbulence insert,
5 Corrugated fin,
6 Side section,
7 Bracket,

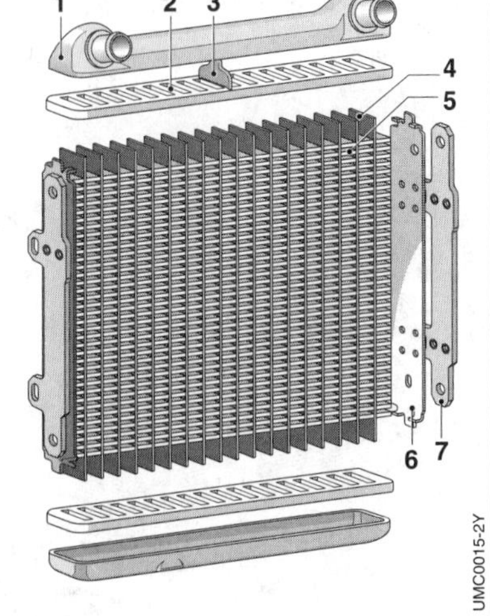

The principle of oil-to-air cooling is ideal for use in commercial vehicles and powerful passenger cars to cool the transmission oil. To ensure good ventilation, they are ideally mounted upstream of the coolant radiator, but can also be mounted elsewhere in the engine compartment. Oil-to-air coolers which are not located upstream of the main coolant radiator and are thus outside the range of action of the main fan must be supplied with cooling air accordingly, for example by exposure to dynamic air pressure or by a separate electric fan.

Oil-to-coolant coolers

Aluminum stack designs employed in oil-to-coolant coolers have largely replaced stainless-steel disk coolers, double-tube oil coolers, and aluminum forked-tube coolers.

Disk-stack oil coolers

Disk-stack oil coolers are made of individual disks with turbulence inserts inserted between the disks (Figure 11). The upright edges of the disks fit together in a casing. Passages connect the channels formed by the disks in such a way that coolant and oil flow through alternate channels.

Figure 10: Oil-to-coolant cooler (disk-stack oil cooler)
1 Oil connections,
2 Coolant connections,
3 Cover,
4 Disk stack (oil passage),
5 Disk stacks (coolant passage),
6 Reinforcement plate,
7 Base plate.

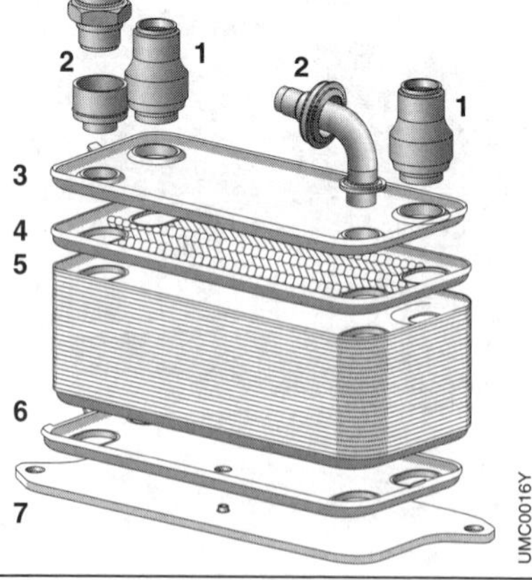

Double-tube oil coolers and flat-tube oil coolers

These cooler types are mounted directly in the radiator tank on the outlet side of the main radiator (coolant radiator). The radiator tank thus assumes the function of the coolant-side housing for the oil coolers. The double tube of the oil cooler is formed by an outer tube and an inner tube inside it with a turbulence insert in between. The two tubes are soldered to each other at the ends. The transmission oil passes through the space between the inner and outer tubes, around or through which already cooled coolant flows.

Double-tube coolers are used as transmission-oil coolers in passenger cars and commercial vehicle in the lower power segment up to approx. 2.5 kW. As power requirements increase up to 4 kW the double tube is replaced by numerous flat tubes connected to each other by turbulence plates on the coolant side. The flat tubes are interconnected by openings at their ends. The flat tubes are also provided on the oil side with turbulence plates, which are soldered to increase both cooling capacity and strength.

Disk oil coolers

Disk oil coolers are mounted between the engine block and the oil filter. They have a separate casing and a central channel for the oil to pass through. The oil flowing back from the oil filter is routed through a labyrinth of perforated disks separated by turbulence inserts. This labyrinth is cooled by coolant flowing from the main circuit through the casing.

Forked-tube coolers

Forked-tube coolers are made of finned forked tubes through which the coolant flows. On the oil side, they have no casing and must therefore be integrated into the oil-filter housing or the oil pan.

Engine-oil coolers for commercial-vehicle applications

The engine oil in commercial vehicles is generally cooled by stainless-steel disk-stack coolers or aluminum flat-tube coolers without a casing on the coolant side. They are accommodated in an extended coolant duct in the engine block.

Fuel coolers

Fuel coolers are installed in modern diesel engines in order to cool excess diesel fuel down to permissible levels. This excess diesel fuel heats up during the injection process as the result of compression in the high-pressure pump, before it is routed back via the return line to the fuel tank. The excess diesel fuel that is returned without cooling to the tank has a temperature of well over 70 °C and heats the diesel fuel in the tank. If now the tank were almost empty, the temperature of the diesel fuel in the tank would therefore exceed the maximum permissible temperature. The excess fuel must therefore be cooled before being introduced into the tank.

The fuel can be cooled by means of an air-cooling or coolant-cooling system. Coolers operating along similar lines to oil-to-air coolers or disk-stack oil coolers are used to cool fuel.

Figure 12: Cooling module
1 Transmission-oil-to-coolant cooler,
2 Coolant radiator,
3 Intercooler,
4 Condenser,
5 Air-to-power-steering-fluid cooler,
6 Module frame,
7 Module bearing,
8 Electric fan,
9 Double fan cowl,
10 Transmission-oil lines.

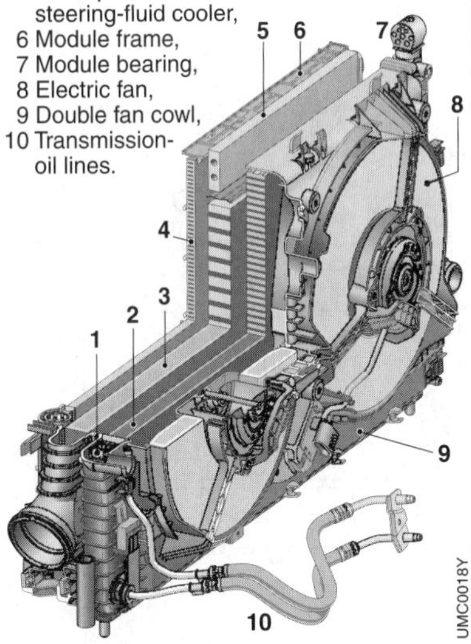

Modularization

Cooling module

A cooling module is a structural unit which can consist of various components for cooling (e.g. main radiator, intercooler, oil-to-air cooler) and a condenser for air-conditioning a passenger car, and can include a fan unit with drive (e.g. electric motor) (Figure 12).

The design of a cooling module must take into consideration the interaction of the individual components, the dimensioning of the components with respect to the package space in the vehicle, and the procedure for dealing with interfaces. Important factors here are the mounting technology, the cooling-air ducts and the seals on the cooling-air side, the component connections on the fluid side, and the electrical plug-in connections.

The standard modular construction used in both passenger cars and commercial vehicles essentially offers a whole range of technical and economic benefits:
- Simplified logistics by combining components to form one structural unit
- Reduced number of interfaces
- Simplified mounting and assembly
- Optimum component design by using matched components
- Modular systems, encompassing various engine and equipment variants
- Improved quality with regard to assembly

Simulation and test methods are employed for the purpose of achieving optimized component design and layout in the cooling module. Assuming exact knowledge of the fan characteristics, the fan drive and the heat exchangers, simulation programs are created to replicate both the cooling-air side and the fluid side. By integrating the individual components into the simulation models, it is possible to examine the interactions of the individual components under various operating conditions. Ever greater significance is being attached to this type of virtual analysis, which is characterized by the use of computer-aided development tools. In line with this development, all geometric data are entered and processed in a CAD system (Computer-Aided Design). CFD analyses (Computational Fluid Dynamics) are conducted to examine the flow of cooling air in the engine compartment, while FEM analyses (Finite-Element Method) provide statements concerning the strength and stability of the design layout. The design analysis phase concludes with verification tests that may also be performed in a wind tunnel and on vibration test rigs.

Cooling-system technology

While the cooling module comprises a structural unit of components with defined functions, the cooling system encompasses all components that are associated with the functions of the cooling system, even if they do not form complete structural units (Figure 13). This includes for example, beyond the cooling module, lines, pumps, control components, and the expansion tank, if it is not already a component part of the cooling module.

Cooling-system technology, in which all the components are matched to each other, offers a whole range of technical and economic benefits:

- Reduced parasitic losses through appropriate hydraulic design matching
- Consideration of control systems and dynamics
- Consideration of the passenger-compartment heating system
- Larger scope of intervention options for the purpose of optimizing the design
- Standardized assembly concept for all cooling-system components
- Reduced development expenditure by cutting the number of development interfaces

Figure 13: Controlled cooling system
Example architecture with heat exchangers and actuators
1 Condenser, 2 Coolant radiator, 3 Shutter, 4 Fan,
5 Thermostat, 6 ECU, 7 Electric motor (fan drive),
8 Electronically controlled thermostat,
9 Electric motor with pump, 10 Internal-combustion engine,
11 Electric motor with pump, 12 Heater core,
13 Stepper motor, 14 Low-temperature circuit,
15 Low-temperature regulator, 16 Oil cooler,
17 Transmission, 18 Transmission-oil cooler.

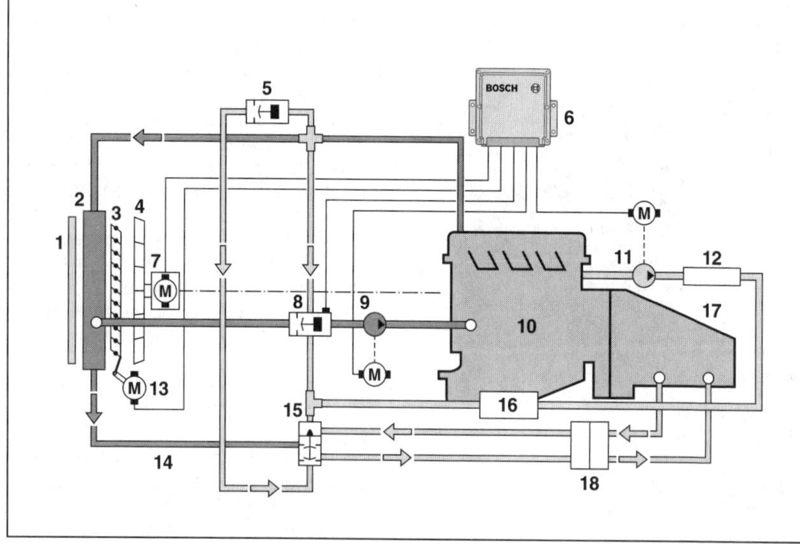

Intelligent thermal management

Function

Future trends are heading toward the system optimized regulation of various heat and substance flows. Thermal management goes beyond cooling-system technology in that it takes into consideration all material and heat flow systems in the vehicle, i.e. in addition to the flow structures of the cooling system, it also deals with those of the air-conditioning system. The objectives of optimization consist in reducing fuel consumption and exhaust emissions, increasing air-conditioning comfort, increasing the service life of components, and improving cooling capacity in part-load states.

Optimization objectives

One of the basic principles of thermal management concerns itself with the fact that auxiliary energy employed to operate the cooling system always represents a loss for the vehicle energy balance, and component efficiency cannot be increased arbitrarily at a constant supply of auxiliary energy. To achieve the optimization objectives, the cooling system is therefore equipped with "intelligence", installed in familiar and new types of actuators, as well as in microprocessor-controlled control systems that operate these actuators. One example is demand-triggered regulation of cooling-air flow by radiator shutters and controllable fan drives so that cooling-air throughout is kept to a minimum under all operating conditions. In addition to improving the vehicle's drag coefficient c_d), this measure also ensures that all media achieve their operating temperature more effectively during the warm-up phase after a cold start, and that the passenger compartment is heated more efficiently. Cutting back the use of auxiliary energy in this way means that auxiliary energy can be diverted for use in operating states that are critical to the cooling output, while still achieving the optimization objectives.

Another important basic principle is to maintain a constant temperature in the components to be cooled as far as possible, irrespective of the operating state and the ambient conditions. An example of this temperature control principle is to use coolant to regulate the transmission oil temperature. Heating the transmission oil during the warm-up phase and employing an efficient cooling system to prevent the transmission oil from overheating reduce friction losses in the transmission, increase the service life of the transmission, and extend the service intervals for the transmission oil.

Ultimately, considering the cooling and air-conditioning systems in their entirety opens up the option of utilizing "thermal integration". Heat flow from one of the systems can be utilized or dissipated by another system without the need for any major additional input in auxiliary energy. An example of this is the utilization of waste heat from the exhaust-gas cooling system to heat up the vehicle interior.

Applications

In the field of engine cooling thermal management is used for the following:
- Transmission-oil temperature equalization
- Map-controlled thermostat
- Electrically controlled viscous clutch
- Controllable electric coolant pump
- Cooling-air control, e.g. radiator shutter
- Exhaust-gas cooling
- Coolant-cooled charge-air cooling (intercooling)

Advantages

The fuel-saving potential based on the sum of all measures is in the range of approx. 5 % (for passenger vehicles). On top of this, there is a range of further advantages corresponding to the above-mentioned optimization objectives. The extent to which engine management utilizes the cooling system control options is of decisive significance in leveraging this potential.

In the meantime, individual measures have been implemented to achieve system-optimized temperature equalization in motor vehicles. Nevertheless, thermal management as an all-encompassing optimization principle has to date not been fully implemented and remains the reserve of future vehicle generations.

Engine lubrication

Force-feed lubrication system

The force-feed lubrication system (Figure 1) in combination with splash and oil mist lubrication is the most commonly used system for lubricating motor-vehicle engines. An oil pump (usually a gear pump) conveys pressurized oil to all bearing surfaces in the engine. while the sliding parts are lubricated by splash lubrication systems and oil mist.

After flowing through the bearing surfaces and sliding parts, the oil collects below the piston, connecting rod and crankshaft assembly in the oil pan. The oil pan is a reservoir where the oil cools, and the foam dissipates and settles. Engines subject to high stresses are fitted additionally with an oil cooler.

Engine service life can be prolonged drastically by keeping the oil clean.

Components

Oil filters

Function

Oil filters remove and reduce particulate (combustion residues, metal abrasion, dust, etc.) from the engine oil which would otherwise cause damage or wear in the lubrication circuit. Since the engine oil is constantly circulating in the lubricating oil system, inadequate filtration could cause particulate to accumulate and this would accelerate the rate of wear. The oil filter does not filter out liquid or soluble constituents such as water, additives or decay products attributed to the effects of oil aging.

In terms of wear, the significance of particulate in the oil circuit depends on the quantity and size of the particles. The typical size of particles in engine oil

Figure 1: Forced-feed lubrication system
1 Pressure-relief valve, 2 Oil filter, 3 Gear pump, 4 From main bearing to connecting-rod bearing, 5 Intake bell housing with strainer, 6 Oil pan, 7 Main oil feed line to crankshaft bearings, 8 Return flow from timing case to crankcase, 9 To camshaft bearings.

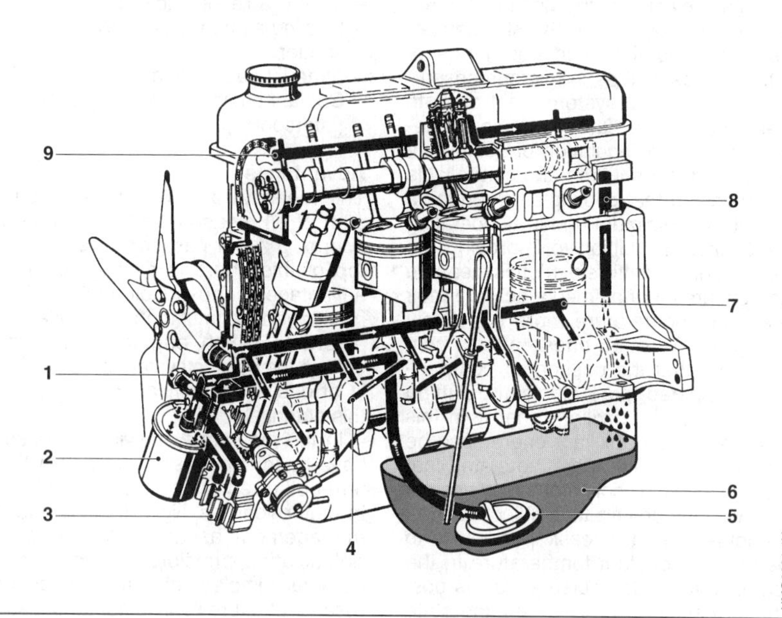

ranges from 0.5 to 500 µm. The fineness of the oil filter is therefore specifically adapted to the requirements of a particular engine.

Different types and designs
In principle, oil filters are based on two specific designs: easy-change filters and housing filters. In easy-change filters (Figure 3), the filter element is located in a housing that cannot be opened and which is secured by means of a threaded stud to the engine block. The complete easy-change filter unit is replaced as part of the oil service.

The housing filter (Figure 2) is made up of a housing that is permanently connected to the engine block and can be opened to access the replaceable filter element. During the oil service, only the filter element is replaced; the housing is a permanent component. The filter element used on recent engines is based on a metal-free design. This means that the filter element can be completely incinerated.

Besides the filter element, the two filter designs normally feature a filter bypass valve, which opens at high differential pressures to ensure effective lubrication at the necessary points in the engine. Typical opening pressures range from 0.8 to 2.5 bar. Elevated differential pressures can occur in connection with high oil viscosities or when the filter element is heavily contaminated.

Depending on the specific engine requirements, the two filter designs may also feature a non-return or backflow check valve on the filtered or unfiltered oil side (contaminated oil side). These valves prevent the oil filter housing from draining empty after the engine is turned off.

Currently, the oil and oil-filter change intervals for passenger-car engines are between 15,000 and 50,000 km, and for commercial vehicles are between 60,000 and 120,000 km.

Filter media
Various types of deep-bed filter media are employed in oil filtration. They mainly consist of fiber-pile structures that are arranged in various configurations. The filter material most commonly used is flat media which, in the majority of cases, is pleated, but in some applications they are also wound or used in the form of fiber packings, especially in bypass filters. The material generally used for the fibers is cellulose. Additional quantities of plastic or glass fiber in virtually any proportions may be added to the cellulose. These filter media are impregnated with resin to en-

Figure 2: Housing oil filter
1 Oil-mist separator, 2 Control valve,
3 Filter element, 4 Plastic cover,
5 Oil-to-water heat exchanger, 6 Housing,
7 Bypass valve.

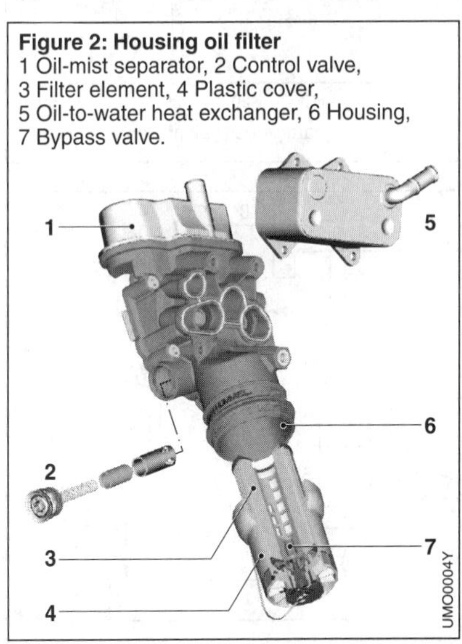

Figure 3: Easy-change oil filter
1 Threaded cover, 2 Filter element,
3 Bypass valve, 4 Spring,
5 Seal, 6 Non-return diaphragm,
7 Housing, 8 Center tube.

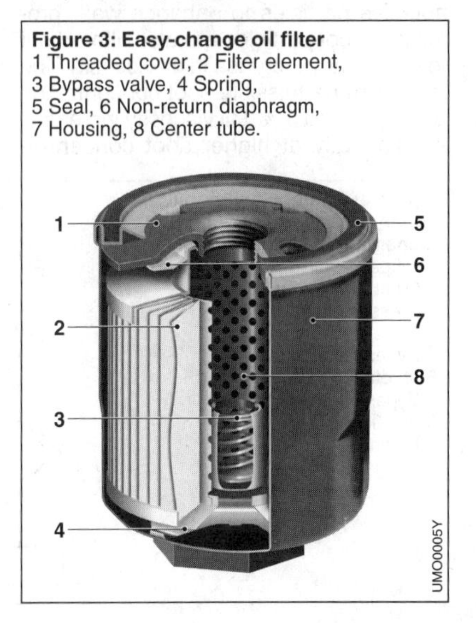

hance their resistance to oil and their operational stability. However, filter media made of purely synthetic fibers are being used even more, as they have a much greater resistance to chemicals that allows for longer service intervals. They also provide better options for structuring the three-dimensional fiber matrix to optimize filtration and increase particulate retention efficiency.

Full-flow filters
All recent motor vehicles are equipped with full-flow filters. Based on this filtration principle, the entire volumetric flow of oil that is pumped to the lubricating points in the engine is routed through the filter (Figure 4). Consequently, all particles which could cause damage and wear due to their size are trapped the first time they pass through the filter.

The decisive factors governing the filter area are oil volumetric flow and particulate retention capacity.

Bypass filters
Bypass filters, designed as deep-bed filters or as centrifuges, are used for superfine or microfiltration of the engine oil. These filters remove much finer particles from the oil than is possible using full-flow filters (Figure 5). They can remove minute abrasive particles to enhance wear protection. Soot particles are also filtered out to reduce any increase in oil viscosity. The maximum permissible soot concentration is roughly 3 to 5 %. Oil viscosity increases substantially at higher soot concentrations, resulting in a drop in the operational effectiveness of the oil.

For this reason, bypass filters are used mainly on diesel engines. Only part of the oil flow (8 to 10 %) from the engine is routed via the bypass filter.

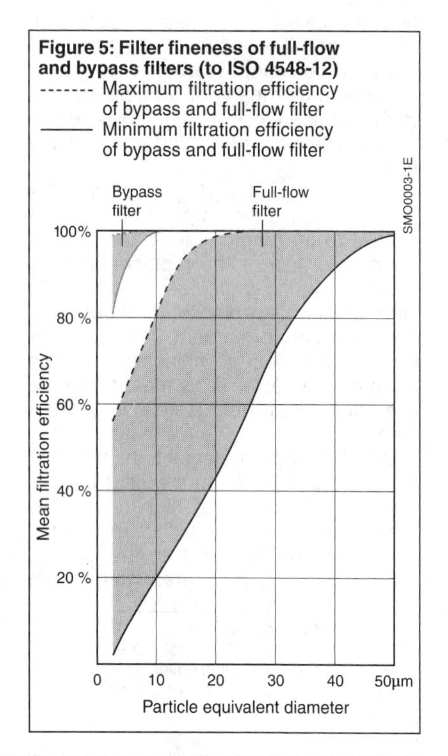

Figure 5: Filter fineness of full-flow and bypass filters (to ISO 4548-12)

Figure 4: Oil circuit with full-flow and bypass filters (schematic diagram)
1 Oil pan,
2 Oil pump,
3 Pressure-control valve,
4 Oil cooler,
5 Bypass filter,
6 Throttle,
7 Bypass valve,
8 Full-flow filter,
9 Engine.

Air-intake systems and intake manifolds

Overview

Functions
The air-intake system and the intake manifold are designed to supply the engine with air that is as cold and particle-free as possible, and to distribute this air to the individual cylinders. The main functions of the air-intake system are to direct the air from the front of the vehicle to the engine, to separate particulates contained in the intake air from the surrounding area, and to muffle the noises emitted by the engine.

Components
Basic functions
The air-intake system (Figure 1) consists of an air duct on the raw-air side, the air filter as the central components, and the lines to which the throttle valve and the intake manifold are connected. The intake manifold distributes the intake air to the individual cylinders and ensures uniform and effective combustion. The air filter prevents mineral dust and particles from being drawn into the engine and contaminating the engine oil. This reduces wear in the bearings, piston rings, cylinder walls, etc. In this way the individual system components can have a crucial positive effect on the performance and service life of the engine.

Supplementary functions
The air-intake system often incorporates further function components such as, for example, sensors (air-mass meter, pressure sensor, and temperature sensor), HC adsorber, water separator, and even anti-snow systems.

Figure 1: Design of an air-intake system with exhaust-gas turbocharging
1 Raw-air line,
2 Air-filter housing,
3 Clean-air line 1 (low-pressure side) with hot-film air-mass meter (hidden),
4 Clean-air line 2 (low-pressure side), with introduction of blow-by gas,
5 Clean-air line 3 (low-pressure side),
6 Exhaust-gas turbocharger,
7 Charge-air line (hot high-pressure side)
8 Intercooler,
9 Charge-air line cold high-pressure side)
10 Throttle valve,
11 Attachment points.

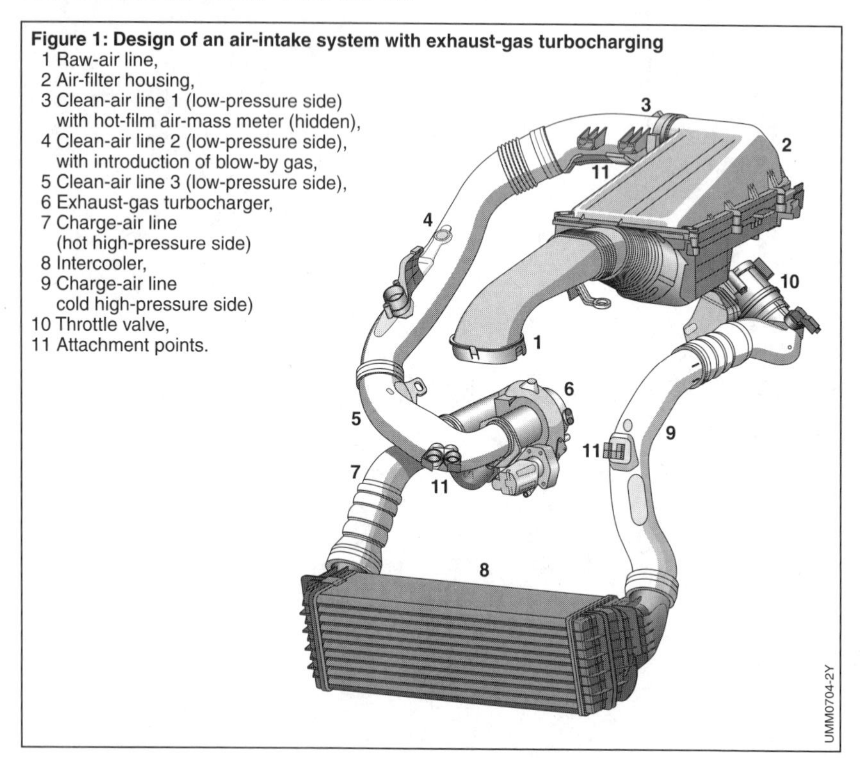

Requirements

Functional requirements

The development of the air-intake system and its components is based on numerous specifications which are shaped on the one hand by space considerations and on the other hand by functional requirements. Specifications define for example temperatures in the engine compartment which influence the choice of plastic materials and elastomers for the individual components, permissible leakage rates and pressure losses for the system or the individual components, the acoustic performance of the intake tract, and the separation performance with regard to filtration as service life and as filtration-efficiency requirement.

With regard to the intake manifold, further emphasis is placed on pressure-pulsation resistance and the level of uniform distribution of intake air. The air flow into the combustion chamber can if necessary be positively influenced by different types of flap.

Visual appearance

The visual appearance and the design of surfaces in the engine compartment gain in importance as soon as surfaces are located in the visible area of the engine compartment (e.g. design air filters or engine-mounted air filters).

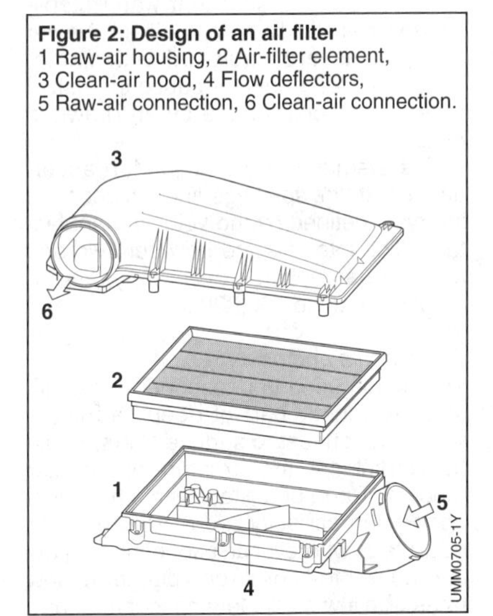

Figure 2: Design of an air filter
1 Raw-air housing, 2 Air-filter element,
3 Clean-air hood, 4 Flow deflectors,
5 Raw-air connection, 6 Clean-air connection.

Passenger-car air-intake system

Air duct

The intake point of the raw-air line is usually located at the front of the vehicle. It is particularly important in that it has a crucial influence on the ingress of particles and the unwanted ingress of water droplets when it is raining and of snow crystals when it is snowing. However, this position is generally stipulated by the automobile manufacturers.

Classic internal-combustion engines are divided into non-supercharged naturally aspirated engines and supercharged engines, in which an exhaust-gas turbocharger for example is deployed. The air-intake system (Figure 1) in both variants consists of a raw-air line, which feeds into an air-filter housing with integrated filter element, and a clean-air line on the low-pressure side, which in a naturally aspirated engine serves as the interface to the throttle valve. In a turbocharged engine the clean-air line on the low-pressure side feeds into the turbocharger, while the charge-air lines on the hot and cold sides lead to the intercooler and to the throttle valve respectively. Low pressure losses can be achieved by large line cross-sections (this influences space and acoustics) and by optimized flow routing.

The lines are usually made of plastic and elastomer materials, while the air-filter housings are generally injected-molded plastic parts. Depending on the application, the air filter can be mounted to the body and to the engine in such a way that the lines compensate not only the tolerances between the components and their attachment points, but also above all the movement of the engine relative to the body. The components are joined with vibration isolation to the vehicle by way of rubber buffers.

Air filtration

Engine air-intake filters (Figure 2) reduce the particles contained in the intake air which can be generated for example by abrasion and incomplete combustion or condensation processes or are even of natural origin in the form of organic and mineral dust.

Composition of air pollutants
Typical air pollutants includes oil mist, aerosols, diesel soot, industrial waste gases, pollen, and dust. These particles vary greatly in size. The dust particles drawn in together with the intake air have a diameter of between 0.01 µm (mostly soot particles) and 2 mm (sand grains). Approximately 75 % of the particles (referred to the mass (weight)) are between 5 µm and 100 µm.

The amount of dust contained in the intake air depends heavily on the environment in which the vehicle is used (e.g. freeway/interstate or dirt track). In extreme cases, over a period of ten years, the mass concentration accumulated in a passenger car can range between a few grams through to several kilograms of dust.

Figure 3: Filter media
a) Flow through a folded/pleated filter medium,
b) Folded/pleated filter medium.

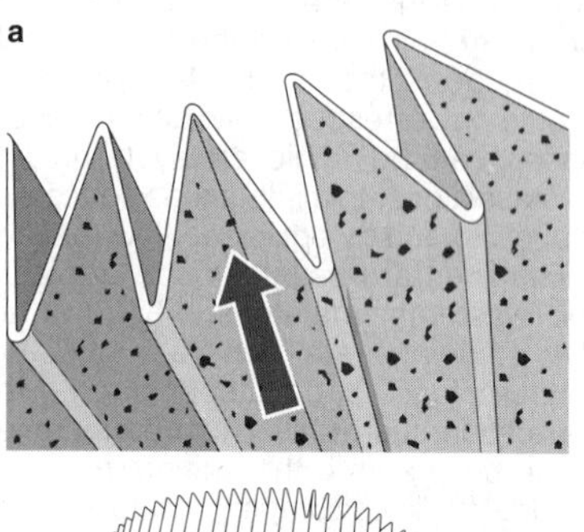

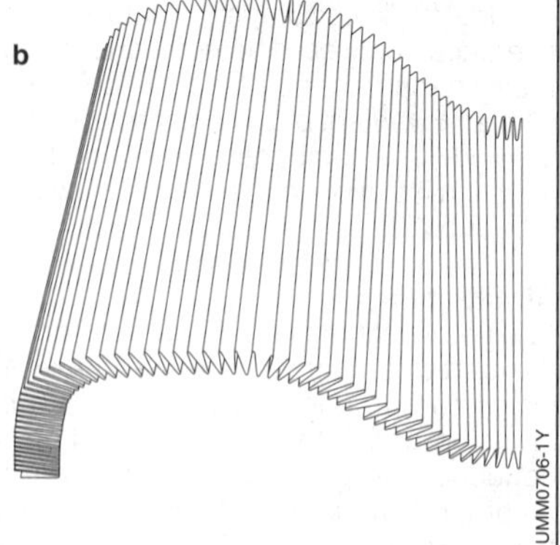

Filter elements
Filter elements which incorporate the latest technology achieve total mass filtration rates of up to 99.8 %. It is necessary to maintain these values under all prevailing conditions – also under dynamic conditions as found in the engine's intake tract. Poor-quality filters will exhibit an increased rate of dust breakdown or rupture.

Design
Filter elements are designed to meet the requirements of each individual type of engine. This ensures that pressure losses remain minimal and that high levels of filtration efficiency are not dependent on air throughput. The filter medium that makes up the filter elements in flat or cylindrical filters is installed in folded or pleated layers in order to achieve a maximum filter surface area in the smallest possible space (Figure 3). Conical, oval, stepped, as well as trapezoidal geometries complement the standard structures with the aim of optimizing the use of package space that is becoming very scarce and confined in the engine compartment. These media, mostly consisting of cellulose fibers, go through a special embossing and impregnation process to achieve the necessary mechanical and thermal strength, sufficient water stability, and resistance to chemicals. Newer applications are equipped with flame-inhibiting filter media to reduce the risk of fire caused for example by an drawn-in cigarette.

The elements are designed in accordance with the spacings and replacement intervals defined by the vehicle manufacturer. The intervals are between two and four, and sometimes even six years, or 30,000 km up to 100,000 km.

Filter media and design
The media used in passenger-car air filters are mostly part of deep-bed filters which, in contrast to surface filters, retain the particles in the structure of the filter medium. Deep-bed filters with a high dust absorption capacity are always used to advantage wherever an economic solution is required to filter large volumetric flow rates with low particulate concentrations.

The demands for compact, high-performance filter elements (with reduced package space), as well as longer service intervals, are the driving force behind the development of new, innovative air filter media. To further increase the dust retention capacity of deep-bed filters, design engineers are increasingly using media which have a gradient structure with increasing fiber density to the clean-air side.

Fully synthetic filter media
New air-filter media consisting of synthetic fibers with much improved performance data have already been phased into series production. Figure 4 shows a photograph of a synthetic high-performance filter medium (felt) with continuously increasing density and decreasing fiber diameter across the filter section from the input side to the output side.

Semi-synthetic filter media
Better results than with purely cellulose-based media can be achieved using composite materials. Here, for example, a cellulose-based paper layer is combined with a synthetic layer of melt-blown fibers. The melt-blown fibers constitute a layer of fibers formed from a polymer melt in the air flow which is either deposited directly on the paper layer or laminated on in a separate process step as a separate layer.

Figure 4: Photographic image of an air-filter medium consisting of synthetic fibers, taken with a scanning electron microscope

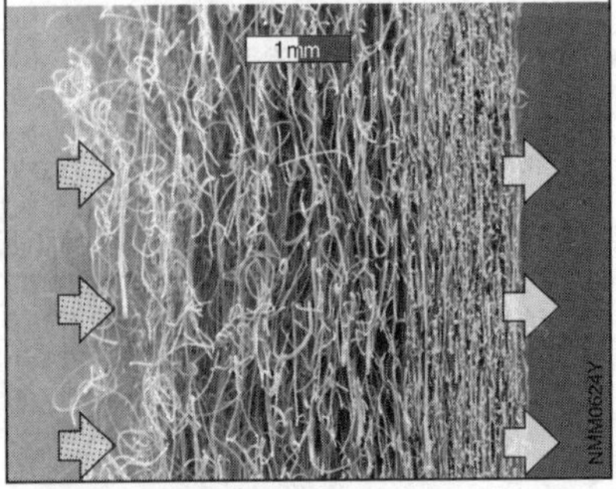

1 mm

NMM0624Y

Acoustics

The air-intake system contributes in a similar way to the exhaust system to the overall noise of the vehicle. It is therefore necessary to introduce noise-reduction measures already in the early development stages. In the past these measures

Figure 5: Acoustic muffling measures with different resonator shapes

a) Resonator-type muffler (relatively wide-band, suitable for low and medium frequencies),
b) Absorption-type muffler (wide-band, suitable for medium and higher frequencies),
c) Resonance muffler (narrow-band, suitable for low and medium frequencies),
d) Whistle muffler (narrow-band, high muffling, suitable for medium frequencies above the Helmholtz resonance),
e) Branch whistle muffler (λ/4-pipe, narrow-band, suitable for medium frequencies),
f) Wide-band muffler for turbocharged engines (wide-band, suitable for higher frequencies).

1 Air duct, upstream,
2 Air duct, downstream,
3 Acoustically absorbent material,
4 Inserted pipe.

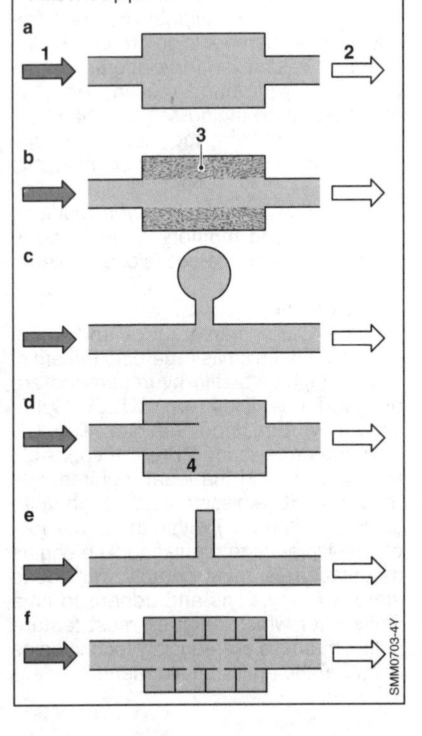

were confined to designing a sufficiently sized air-filter housing as an intake muffler. In the meantime the excitation spectrum has shifted from low frequencies (f < 200 Hz) to medium to high frequencies with the result that separate measures in the form of Helmholtz resonators and $\lambda/4$ pipes are used. Computer simulation programs that determine the optimum number of resonators and their frequency are used to design these measures.

Wide-band mufflers for turbocharged engines

The increasingly widespread use of turbocharged engines has led design engineers to focus their attention on new noise sources. The whistling and whining noises are subjectively perceived to be disruptive and annoying and now measures are often integrated in the form of wide-band mufflers in the clean-air line of the air-filter system. Wide-band mufflers comprise several individual chambers through which a perforated central pipe passes (Figure 5). The individual resonant frequency of a chamber is determined by the chamber volume and the degree of perforation; a wide-band muffling effect is achieved by arranging several chambers in succession. Here too modern simulation programs are used to optimize the number.

In addition to resonator-type mufflers, absorption-type mufflers as featured in exhaust systems are occasionally used.

HC adsorption

For the Californian market and some other US states limits regarding the emission of highly volatile hydrocarbons are defined for vehicles (e.g. SULEV, Super Ultra Low Emission Vehicle). The US environment agency EPA is responsible for implementing the legal requirements regarding HC emission limits (High Volatile Hydro Carbons) within the framework of a multiple-stage regulation. The engine manufacturers must comply with these statutory provisions and adhere to time limits after which engines must feature pollutant-emission-reducing technologies to satisfy the emission standards.

Some of the unburned hydrocarbons (primarily fuel components) escape after the engine is turned off from the engine through the intake manifold and from the crankcase ventilation into the air-intake system, travel against the normal direction of air flow, and exit the system to atmosphere. This can mean that it is necessary to integrate an HC adsorber on the clean-air side of the system so that the hydrocarbons are separated by adsorption on activated-carbon media. An equivalent, the "Butane Working Capacity" (BWC), is defined in specifications in order to define the absorption capacity comparably. The adsorber elements can be used in the "full flow" (main path, full throughflow) or in a bypass configuration.

Unlike air-filter elements, adsorber elements according to the requirements are lifetime components which may not be removed from the system. The elements are charged while the engine is stopped with the air-borne hydrocarbons of minimal droplet size, and desorption takes place the next time the engine is operated, i.e. the hydrocarbons are returned to the combustion process in this cycle and the HC adsorber is regenerated for the next charging cycle. The adsorptive medium used is activated carbon, which is located as a single or double layer between the carrier media. These are for the most part elements which are manufactured in a process as plastic coating.

Water separation

When driving in the rain, not inconsiderable amounts of water may be drawn in, depending on the position and layout of the intake fitting at the front of the vehicle. Because the filter medium only has a limited water absorption capacity, the drawn-in liquid may reach the clean-air side of the system after a certain time. Signal deviations or even damage to the hot-film air-mass meter may occur if droplets touch the hot sensor surface and result in a local, unwanted cooling of this surface. The signal deviation suggests an incorrect air-mass flow and gives rise to an incorrect setting of the mixture composition, to power losses, and ultimately to increased consumption (and increased

CO_2 emissions). The time until water penetration at the filter element can be significantly extended or even avoided entirely by way of downstream measures.

Finned separators, baffle plates or measures based on the cyclone principle such as the "peeling collar" (Figure 6) integrated in the intake air are used to separate water droplets. The shorter the raw-air line – i.e. the distance from the air inlet to the filter element – the more difficult the task. Essentially, the aim is to achieve additional flow pressure losses that are as low as possible. The measures must be combined with suitable water-discharge devices which divert the separated water from the system to atmosphere. However, separating principles are also used which carry the liquid in combination with measures on the housing side and with additional water-separating fleeces attached to the filter element from the system. An optimized design of the raw-air line and the air-filter housing clearly enhance the above-mentioned measures such that outstanding separation results are achieved even with comparatively short raw-air lines.

Figure 6: Principle of water separation
1 Raw-air intake,
2 Raw-air line,
3 Water droplets,
4 Swirl generator (optional),
5 Wall-applied water film,
6 Peeling collar,
7 Water discharge.

UMM0709Y

Anti-snow system

In winter when it is snowing snow crystals may be drawn into the system which charge the filter element on the raw-air side. If this situation persists, the initial upshot will be an increase in pressure loss. In due course the filter element may become blocked on the raw-air side such that the engine stops receiving sufficient intake air for the combustion process. This will ultimately cause the engine to stall.

To prevent this, anti-snow systems (ASS) – essentially a second intake – are used (Figure 7). If the filter element becomes blocked, hot, dry and snow-free air is drawn in from the engine compartment, whereby this reduced air flow maintains the engine function. The supplementary intake can be opened by temperature- or pressure-controlled control mechanisms. In the case of thermal control a flap is moved by an expanding-wax element. Pressure control is based on a spring-loaded valve integrated in the

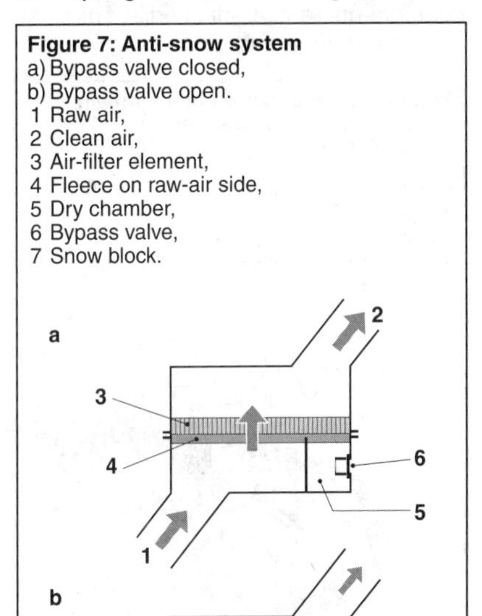

Figure 7: Anti-snow system
a) Bypass valve closed,
b) Bypass valve open.
1 Raw air,
2 Clean air,
3 Air-filter element,
4 Fleece on raw-air side,
5 Dry chamber,
6 Bypass valve,
7 Snow block.

air filter on the raw-air side. The filter element is basically fitted with a preliminary fleece which assumes the sealing function and protects the "dry" chamber of the second intake against snow ingress.

Sensor technology
Sensors which measure the pressure, the temperature and even the air-mass flow are often integrated on the clean-air side of the air filter. The exhaust-gas-relevant air-mass flow is measured by a hot-film air-mass meter. As the emission-control legislation is tightened the requirements with regarding to signal quality are becoming increasingly stringent. Signal deviations for an air-filter system in the range of ±2.5% are difficult to achieve, but not unusual. This requires an increased simulation outlay to identify disruptive turbulent areas and instabilities in the flow close to the sensor and to minimize them by means of suitable product design and additional measures or components (e.g. air-duct grills, fins or air deflectors) (Figure 8).

Figure 8: Flow optimization upstream of the hot-film air-mass meter
1 Raw-air connection,
2 Air-filter element,
3 Deflector geometry,
4 Rectifying grill,
5 Hot-film air-mass meter,
6 Clean-air connection,
7 Flow direction.

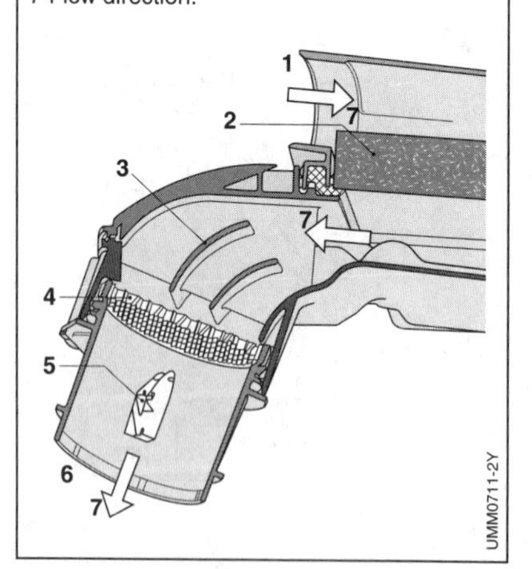

Visual design of surfaces
The structuring of plastic surfaces in the visible area called for by customers for visual reasons can be achieved in a variety of ways, such as for example by erosion, blasting or photoetching of the injection-molding die. Engine-mounted design air filters are a special example here.

Interfaces
The type of connection point between the air-filter housing and the clean-air line is often geared towards customer requirements. The production plant of the automobile manufacturer often specifies the shape of the interface. If an elastomer component is called for, lines based on the material PP-EPDM with integrated bellows for isolation are manufactured by injection molding. These can be connected to a line section made using blown-part technology.

Crankcase ventilation
The clean-air line on the low-pressure side can incorporate further interfaces such as for example the blow-by introduction point from the crankcase ventilation or even the extraction point for the PCV (Positive Crankcase Ventilation). The blow-by introduction point is frequently fitted with a heating tube which is intended to prevent icing.

Low-pressure exhaust-gas recirculation
A measure specially suited to smaller diesel engines to meet the strict Euro 6 limits for nitrogen oxides (NO_x) is low-pressure exhaust-gas recirculation (see Exhaust-gas recirculation, EGR). But even larger-sized engines utilize low-pressure exhaust-gas recirculation, often in conjunction with SCR systems (Selective Catalytic Reduction). With low-pressure exhaust-gas recirculation the exhaust gas is diverted after the diesel particulate filter, cooled in an EGR cooler, and introduced into the clean-air line upstream of the turbocharger compressor. To obtain high EGR rates in wide ranges of the engine map, it is very important for the EGR system to have a low-pressure-loss

design including mixing of air with exhaust gas. For special operating points in the part-load range an additional pressure drop us generated in the EGR system by an exhaust-gas restrictor (downstream of the EGR extraction point) or by an air-intake restrictor in the clean-air line (upstream of the EGR introduction point). Configuring the mixer geometry appropriately prevents local temperature gradients from damaging the compressor and compressor efficiency from being reduced. After the EGR cooler the exhaust gas reaches maximum temperatures of 150 to 200 °C, which enable the mixing and control functions to be executed cost-effectively and space-efficiently with the aid of plastic components. However, it is important when choosing the material to ensure that it is resistant to acidic condensate (pH value of 2 to 3) which can be produced at exhaust-gas temperatures below 85 °C.

The use of low-pressure exhaust-gas recirculation is being analyzed even in gasoline engines; however, the objectives here are to reduce CO_2 and the knock tendency rather than to reduce NO_x emissions. The first applications in series production are expected in the next few years.

Validation

Full product validation, consisting of simulation, acoustics and component testing, is an essential part of the development process for air-filter systems in order to achieve a high degree of product maturity as early as possible. Already in the early concept phase – within the framework of virtual product validation – simulations are conducted to optimize the functions and properties of air-filter systems. Here the acoustic properties such as intake-opening noise and surface sound radiation are calculated and optimized.

As well as acoustic behavior the fluid properties such as pressure loss and approaching flow to the measuring-sensor system are analyzed and specifically improved. In addition, the vibration characteristics are tested under roadway or engine excitation using finite-element simulations to fulfill the later life test without any problems. The filtration properties among others are checked as soon as the first sample components are available. The scope of testing within the framework of component testing includes tests on individual components and on the complete system. This is rounded off by functional tests where the ambient conditions prevailing in the vehicle (such as temperature, moisture, and media pressurization) are superimposed.

The component life is tested in comprehensive vibration and pulsation tests which simulate the real load in the vehicle. In view of the strict emission-control legislation, the flow behavior in the air filter must also be very closely analyzed with a dust charge also taken into consideration. Influencing factors such as fold/pleat geometry of the air-filter element and spread of the filter medium are taken into consideration here. Component validation is rounded off by simulations of the injection-molding process to optimize component quality and the manufacturing process.

Trends – further development of air-intake systems

Three main trends can currently be observed in the market: Downsizing of engines, platform strategies to utilize synergy effects, and stricter requirements with regard to emission-control legislation.

Downsizing necessitates a higher proportion of supercharged engines with the following effects: Specifically higher air-mass flows, smaller and more complex air-filter housings, increased conflict between acoustics and pressure loss, higher temperatures in the engine compartment, and higher charge-air pressures. This necessitates a higher simulation outlay for example to optimize flow and acoustics, new air-filter media, innovative filter-element designs, and higher-quality plastic materials.

Platform strategies of the automobile manufacturers necessitate more complex package spaces and a high-grade identical-part strategy. Stricter requirements with regard to emission-control legislation give rise to heightened specifications with regard to sensor technology, particularly the quality of the signal stability of the hot-film air-mass meter. This likewise necessitates more complex simulation tools, new test procedures, and more complex product designs and constructions.

The weights of components can be reduced with new production technologies. Air filters produced by injection molding on the basis of foamed plastics help to achieve this. A further possibility is to reduce wall thicknesses, where the acoustic behavior in particular of the air-filter housing must be observed. Structure-borne noise radiation can be counter-compensated by cambering (curved geometry) surfaces and integrating ribs or beads (groove-shaped depressions), even when smaller wall thicknesses are used. New sealing concepts for the interface between filter element and housing, e.g. designed as a radial concept (Figure 9), render metal compounds superfluous, and result on the one hand in reduced costs and on the other hand in increased service friendliness when replacing the filter element. The two housing sections can be connected to each other by injected snap-on hooks.

Figure 9: Radial sealing principle
1 Air-filter bellows,
2 Radially acting seal,
3 Raw-air housing,
4 Clean-air hood.

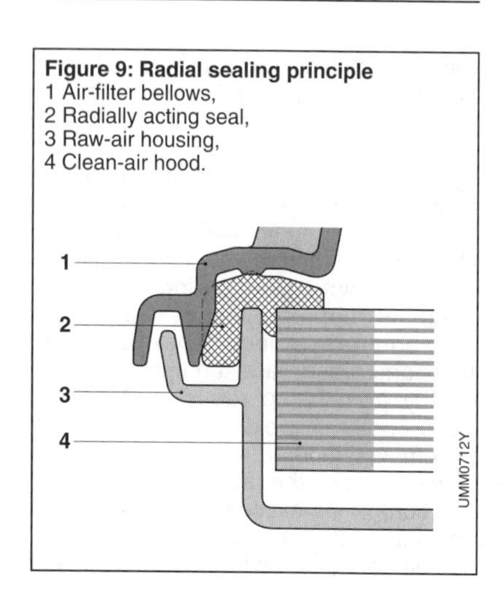

Hybrid vehicles and active noise measures

In the years to come hybridization in the automotive industry will present new challenges. Switches in operation between the internal-combustion engine and the electric motor, but also purely electric operation generate unusual acoustic states. Active systems (with ANC, Active Noise Control) constitute an interesting remedial measure in that the can not only reduce noise and thus replace passive components, but also create new degrees of freedom with regard to noise generation, e.g. for warning pedestrians in purely electric operation. Some countries have already initiated legal initiatives prescribing a warning function for purely electric operation. There are also opportunities to support a specific marque sound or for noise control in order to acoustically mask unwanted phenomena. This approach makes active systems interesting for future engine concepts.

Passenger-car intake manifolds

The main function of the intake manifold is to distribute the air homogeneously to the individual cylinders in order to ensure combustion that is as uniform and effective as possible in the individual cylinders. The quality of uniform distribution has a decisive effect on the engine's performance and emission characteristics.

In the case of gasoline engines a distinction is made between naturally aspirated engines and supercharged engines; in the case of diesel engines, on the other hand, only supercharged engines are still relevant. All three categories feature both passive and active intake manifolds. Active intake manifolds are usually called variable induction systems.

Figure 10: Ram-tube supercharging
1 Cylinder,
2 Individual ram tube,
3 Plenum chamber,
4 Throttle valve.

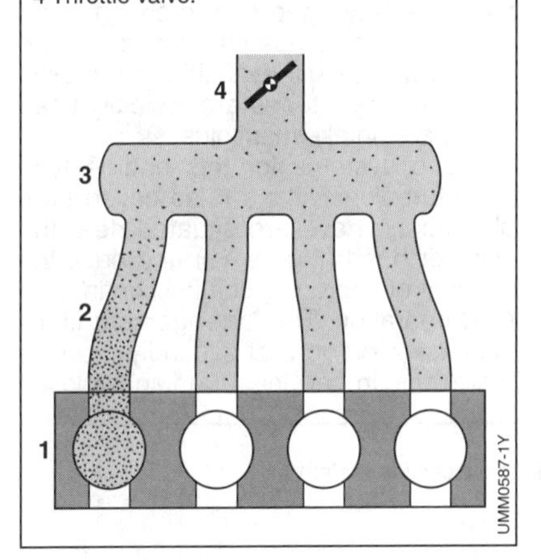

Intake manifolds for naturally aspirated engines

In these types of intake manifolds gas-dynamic supercharging effects are used to increase the air mass in the cylinder and thus the engine output or torque. There are two different supercharging effects here.

Ram-tube supercharging

Ram-tube supercharging is based on the principle that each cylinder has its own ram tube that feeds into a common plenum chamber, also called the intake plenum (Figure 10).

The intake pulse in the induction stroke triggers a vacuum wave. This propagates in the ram tube, is reflected at the opening to the intake plenum, and returns as a pressure wave to the combustion chamber. Because pressure waves always propagate at the speed of sound, the length of the ram tubes can be used to determine the engine-speed range in which the supercharging effect is wanted.

To be able to utilize this supercharging effect in numerous engine-speed ranges, there is the option of realizing active intake manifolds of this type (variable induction system, Figure 11). The optimum solution

Figure 11: Variable induction system
a) Intake-manifold geometry with changeover flap closed (torque setting),
b) Intake-manifold geometry with changeover flap open (power setting).
1 Changeover flap,
2 Plenum chamber,
3 Long, thin oscillatory intake passage with changeover flap closed,
4 Short, wide oscillatory intake passage with changeover flap open.

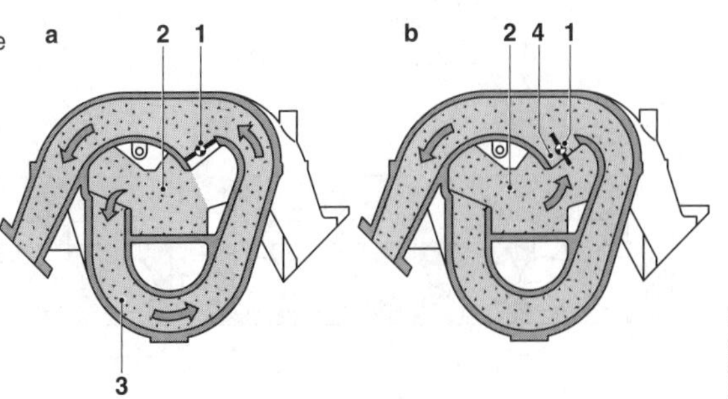

would be an infinitely variable induction system in which an optimum ram-tube length is assigned to each engine-speed range. But this is barely feasible, both technically and economically. For this reason mostly two- or in rare cases also three-stage induction systems are on the market.

Tuned-intake-tube charging

Tuned-intake-tube charging is based on the principles of a Helmholtz resonator. Two groups of cylinders with the same ignition interval are established. The reason for this is that the intake pulses within the groups must not overlap. These groups are connected by short intake manifolds to the two resonance chambers (resonance plenums) (Figure 12). The two resonance plenums are connected by resonance (tuned) tubes to the common main intake plenum. The periodic intake pulses induce the gas oscillations to resonate at a specific engine speed and thereby generate the supercharging effect. The engine-speed range in which the supercharging effect acts is determined by the length and the diameter of the tuned intake tubes. This principle is preferably used in 3-, 6- and 12-cylinder engines because it is relatively easy, because of the firing order, to create independently operating cylinder groups (each consisting of three cylinders with 240° ignition interval).

Incorporating a resonance flap between the two resonance chambers creates an active system that provides further advantages. The intake manifold with closed resonance flap generates a torque increase in the lower engine-speed range. In the upper speed range this valve is opened to create an intake manifold with short ram tubes which increases power output in this speed range.

Intake manifolds for supercharged gasoline engines

Intake manifolds for supercharged gasoline engines are in most cases passive intake manifolds. Power and torque are primarily determined via the charge-air pressure generated by the charger and thus by the air mass introduced into the cylinders. The still available, but mostly very short ram tubes are typically used for passive intake manifolds.

Active systems for this engine type have turbulence flaps (also called tumble flaps). These are situated near the cylinder head (Figure 13) and have the function of improving combustion in part-load operation. This arrangement firstly increases torque and secondly reduces emissions. In part load the flap is closed

Figure 12: Tuned-intake-tube charging
a) Resonance induction system, b) Resonance variable induction system.
1 Cylinder, 2 Short intake manifold, 3 Resonance chamber,
4 Resonance tube, 5 Plenum chamber (main intake plenum),
6 Throttle valve, 7 Resonance flap.
A, B Cylinder groups.

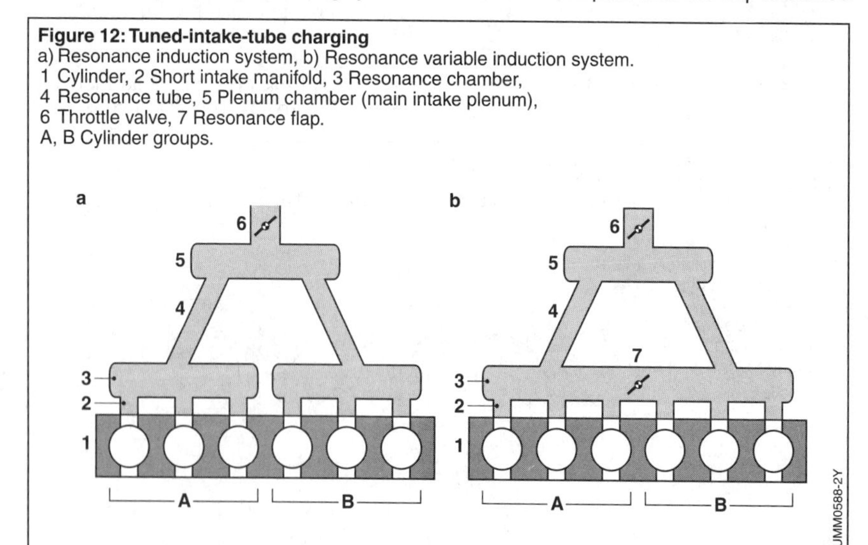

and generates a cylinder flow about the horizontal axis (tumble). In the power setting the flap is opened and the engine receives through the opening cross-section as much air mass as possible into the cylinder.

A special form here is the trough flap, which in the open setting disappears completely into the duct wall and thus ensures unrestricted flow into the cylinder.

Intake manifolds for supercharged diesel engines

In the case of intake manifolds for supercharged diesel engines there is no such clear distinction between passive and active variants. These intake manifolds practically do not have manifolds any more. All that remain are short inlet tulips that connect the intake plenum to the cylinder head.

In the case of active intake manifolds the cylinder head always has two intake ducts per cylinder, where one is a swirl duct and the other a filling duct (Figure 14). The filling duct is fitted in the intake manifold with a duct shutoff flap. The function of this flap is to fill the cylinder via the swirl

Figure 13: Tumble arrangement
a) Intake manifold in tumble setting,
b) Intake manifold in power setting.
1 Intake manifold,
2 Turbulence flap (tumble flap).

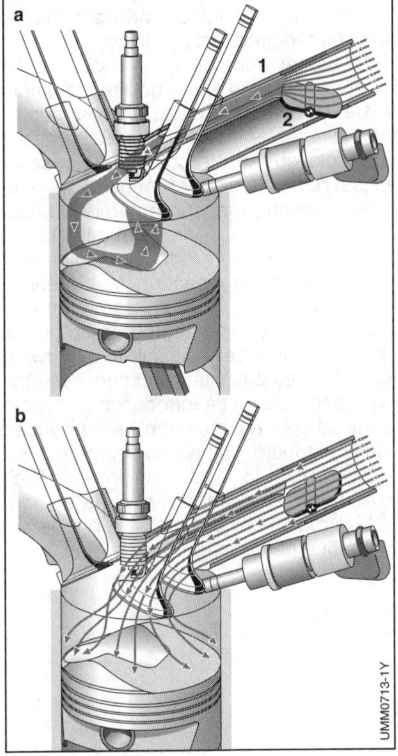

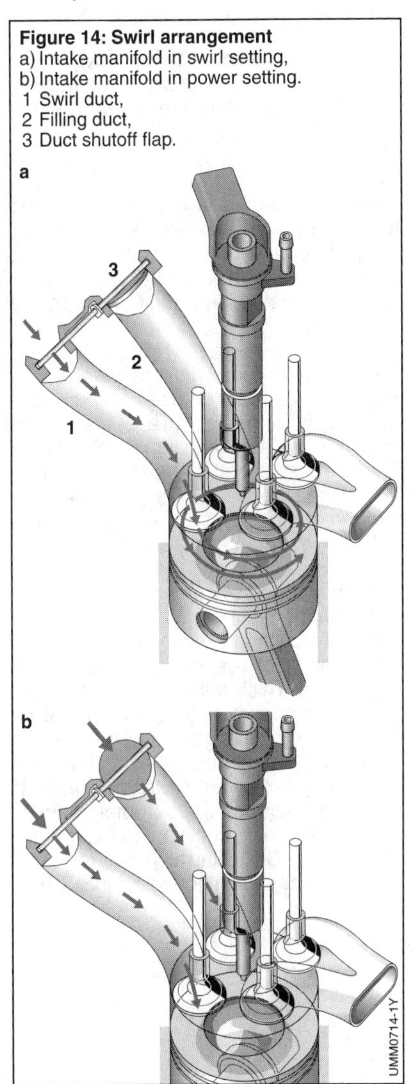

Figure 14: Swirl arrangement
a) Intake manifold in swirl setting,
b) Intake manifold in power setting.
1 Swirl duct,
2 Filling duct,
3 Duct shutoff flap.

duct in part-load operation and thus create a pronounced swirl about the vertical axis. This swirl ensures that air and fuel are thoroughly mixed and thereby helps to reduce emissions. In the transition to full-load operation the flap is then opened incrementally and the swirl is nullified by the filling duct. This results in as much air mass as possible being delivered to the cylinder and thus in high torque and high power output.

Actuator and sensor technology

Primarily electrical actuators or vacuum units are used to drive the different switching elements.

Electric actuators

Electric actuators are preferably used to activate swirl flaps in diesel engines. They enable any intermediate settings to be accurately and rapidly approached. Control is provided in most cases by the engine ECU.

Vacuum units (pneumatic actuators)

A vacuum unit can be used for pure two-point switchings, as are sufficient for length, resonance, and tumble switchings. These are switched by means of an electric changeover valve. The prerequisite is that the engine or a vacuum pump makes sufficient vacuum available.

Sensor technology

The sensor technology that features in variable induction systems has increased significantly in response to the OBD II legislation, which states that all exhaust-gas-relevant switching organs must be monitored by the engine ECU and that the driver must be alerted if they malfunction. Exhaust-gas-relevant switching elements are the turbulence flaps (tumble flaps in gasoline engines, duct shutoff flaps in diesel engines). Switching elements for length or resonance switching are not included.

For electric actuators the internal angle-of-rotation sensors of the actuators can be used in most cases. For vacuum units a separate sensor is required, but this can also be integrated in the vacuum unit.

For active intake manifolds, in addition to the sensor technology for angle-of-rotation monitoring, pressure, temperature, or combined pressure/temperature sensors are mounted on the intake manifolds. These are necessary to supply the engine ECU with the information from the intake manifold required for combustion.

Introduction of gases

A further function of the intake manifold is to distribute introduced gases uniformly into the respective cylinders. In gasoline engines these are primarily crankcase-ventilation and tank-ventilation gases. Fittings are attached to the intake manifold for these gases and then connected by the engine manufacturer to the relevant hoses. When crankcase blow-by gases are introduced, an additional heating tube may be required to avoid icing.

In diesel engines it is important for high-pressure exhaust-gas recirculation to be correctly introduced. Often uncooled exhaust gas is introduced into the intake manifold during the cold-start phase or diesel particulate-filter regeneration. Suitable shielding measures must be taken here where plastic intake manifolds are used. A central introduction point with adequate mixing section is beneficial to the extremely important uniform distribution here. The uniform-distribution requirement is very difficult to achieve in cylinder-selective introduction systems where the gas to be introduced (e.g. from exhaust-gas recirculation) is introduced via a distributor line with a separate introduction point for each cylinder (i.e. each cylinder is supplied with gas via a separate introduction point) at the very different operating points in the engine map.

Gaskets

A trend towards higher-quality materials can be discerned in gaskets. This is mainly because low-pressure exhaust-gas recirculation is more frequently used in diesel engines and this results in an aggressive condensate in the induction system. Peroxide-cured fluorocarbon rubber, for example, must be used in such applications. Due to the fact that the operating temperatures can extend both up and down, sometimes special low-temperature materials are required.

Validation

Full product validation, consisting of simulation, acoustics and component testing, is an essential part of the development process for air-filter systems in order to achieve a high degree of product maturity as early as possible. Already in the early concept phase – within the framework of virtual product validation – simulations are conducted to optimize the functions and properties of air-filter systems. Here the acoustic properties such as intake-opening noise and surface sound radiation are calculated and optimized.

As well as acoustic behavior the fluid properties such as pressure loss and approaching flow to the measuring-sensor system are analyzed and specifically improved. In addition, the vibration characteristics are tested under roadway or engine excitation using finite-element simulations to fulfill the later life test without any problems. The filtration properties among others are checked as soon as the first sample components are available. The scope of testing within the framework of component testing includes tests on individual components and on the complete system. This is rounded off by functional tests where the ambient conditions prevailing in the vehicle (such as temperature, moisture, and media pressurization) are superimposed.

The component life is tested in comprehensive vibration and pulsation tests which simulate the real load in the vehicle. In view of the strict emission-control legislation, the flow behavior in the air filter must also be very closely analyzed with a dust charge also taken into consideration. Influencing factors such as fold/pleat geometry of the air-filter element and spread of the filter medium are taken into consideration here. Component validation is rounded off by simulations of the injection-molding process to optimize component quality and the manufacturing process.

Development trends for intake manifolds

Due to the switch from direct to indirect intercooling and the resulting benefits with regard to CO_2 emissions, the main trend in the intake-manifold market is towards the integration of intercoolers.

This trend has above all an effect on the intake manifolds in supercharged gasoline engines. Virtually all manufacturers pursue solutions involving full integration of the intercooler in the induction system. The cuboid-shaped intercooler must be integrated in the plastic intake manifold and, despite the large flat surfaces and sometimes for plastic extremely high temperatures (upstream of the intercooler), satisfy the requirements with regard to pressure-pulsation resistance. Materials that meet these requirements are currently in development.

The diesel-engine market is split two ways: firstly the intercooler directly integrated in the intake manifold, and secondly the arrangement of the intercooler as a “single box” upstream of the throttle valve.

Commercial-vehicle air-intake system

Air duct

The design of the air-intake system for commercial vehicles is essentially comparable to that of passenger-car air-intake systems (see Figure 1). What must be taken into account here is the much higher service life of commercial vehicles, the higher rated volumetric flow, and the broad range of uses in local, long-distance and construction-site applications. The different volumes and types of dust must be taken into account in filter design.

Raw-air duct, air filters and clean-air pipes up to the turbocharger are made of thermoplastics. After the turbocharger for the most part come (still) combinations of aluminum and rubber parts. However, more plastic parts will be used in future for charge-air pipes and intake manifolds (boost manifolds).

Raw-air intake

The positioning of the intake point is extremely important. A roof or side intake is usually chosen on heavy commercial vehicles. Both are generally mounted with rubber mounts for acoustic isolation to the cab rear wall. The dust and water concentrations in the air are lowest with the roof intake. There are however many raw-air duct variants to cater for the different driver's-cab heights and widths. A good compromise in terms of costs is the side intake (Figure 15): Here air is still inducted from height at the side rear of the driver's cab, irrespective of the different roof heights.

A low-height front intake or side intake, along similar lines to that used in passenger cars, is usually chosen for light utility vehicles. To keep the volumes of inducted dust a water in check, air is inducted through baffles on the vehicle sides. An air duct that delivers the best flow possible must be provided in all cases. Hot

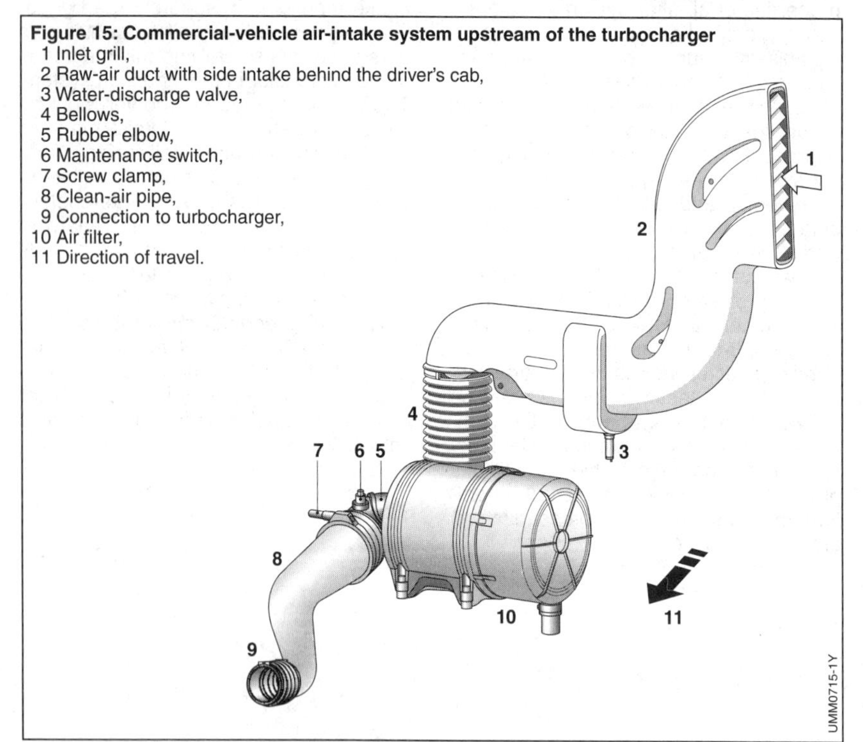

Figure 15: Commercial-vehicle air-intake system upstream of the turbocharger
1 Inlet grill,
2 Raw-air duct with side intake behind the driver's cab,
3 Water-discharge valve,
4 Bellows,
5 Rubber elbow,
6 Maintenance switch,
7 Screw clamp,
8 Clean-air pipe,
9 Connection to turbocharger,
10 Air filter,
11 Direction of travel.

engine waste air must not be inducted as this mean a performance drop and higher fuel consumption.

Because most truck cabs are tilted forwards to access the engine for servicing, the raw-air duct is connected via bellows to the rest of the system. The function of the bellows is to seal the interface when the cab is tilted back and to compensate the vibrating motion of the sprung cab.

Water separation

Measures for water separation are integrated in the raw-air system. Water separation prevents dust from being washed through the filter medium, which can increase engine wear, but also cause sensors to malfunction. Because as a rule cellulose-based filter media are used, at worst the filter element may be destroyed. Measures for water separation include for example using a large intake cross-section to reduce the intake velocity so that basically as little water as possible is inducted. Baffles in the system cause the water droplets in the intake air to be thrown against the inner manifold walls, where they form a film. The water droplets are then separated via a peeling collar or a change in volume and discharged via a water-discharge valve.

Figure 16: Dust preseparator (cyclone)
1 Outlet guide vanes,
2 Housing,
3 Inlet guide vanes,
4 Dust outlet.

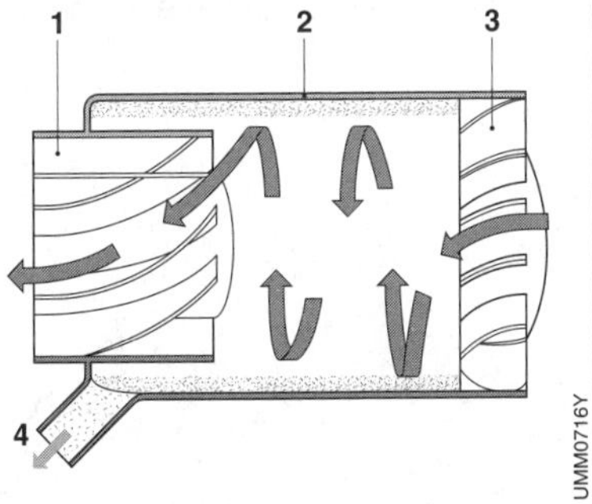

UMM0716Y

Dust preseparation

In vehicle applications with a higher dust concentration in the air (e.g. construction sites or operation in dusty countries) a dust preseparator is installed in the raw-air supply, either as an individual cyclone (Figure 16) or as a cell cyclone. Cell cyclones are small cyclones which are integrated in a housing to save space. These separate out a high proportion of the coarse dust contained in the intake air. The air is induced to rotate at the inlet by guide vanes; the heavy dust particles are thrown against the inner wall of the housing and discharged through the dust outlet. The precleaned air is directed by a further guide-vane assembly (optional), which rectifies the air flow again, to the air filter.

Air filters

The most widely used air filters in commercial vehicles at present are single-stage round air filters with spiral vee-shaped filter elements. They are characterized by low pressure loss, high dust capacity, and thus long service intervals. Depending on

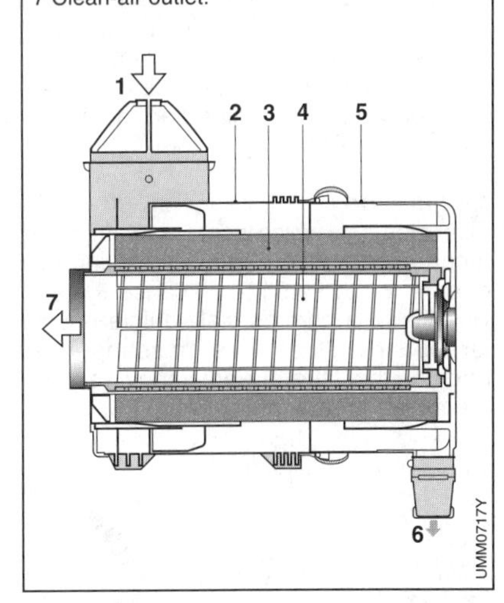

Figure 17: Commercial-vehicle air filter
1 Raw-air inlet,
2 Housing,
3 Filter element,
4 Safety element,
5 Cover,
6 Dust-discharge valve,
7 Clean-air outlet.

how they are used, they can extend a typical oil-change interval by a factor of two. A further advantage of round filters is that they can also be fashioned as two-stage filters. In these variations dust separation takes place inside the air filter itself (Figure 17). The tangential air inlet generates a swirl in the housing which separates out the coarse particles.

Depending on the application, different filter media are available today which are optimized for example for maximum filtration efficiency of fine dust and soot.

A safety element is optionally fitted inside the main filter element which prevents dust from reaching the clean-air side during servicing. It also protects the engine even if the main filter element has been damaged during servicing. Basically, however, air-filter elements should be replaced and not cleaned.

To optimize vehicle acoustics, the air filter is also fitted where necessary with resonators to minimize the opening or engine-braking noise.

Clean-air line as connection to the turbocharger

The clean-air line is as a rule a combination of metal, plastic, and rubber parts. Sensors for air mass, vacuum and air moisture can be optionally integrated. It is important for all the interfaces on the clean-air side to be permanently tight. The vacuum switch indicates to the driver that the servicing limit for the filter element has been reached. Flexible rubber connections compensate the relative movement between the engine and the frame-mounted air filter.

Air filtration

Filter elements which incorporate the latest technology achieve total mass filtration rates of up to 99.95 %. Because the service life of a commercial-vehicle engine far outstrips that of a passenger-car engine, the filtration-efficiency requirements are greater accordingly.

The filter elements have to be replaced at the intervals specified by the vehicle manufacturer. Usual replacement intervals are, depending on the application, 40,000 to 320,000 km or two years. In most cases a service-interval indicator monitors when the service interval is reached.

The media used in commercial-vehicle air filters are comparable to filter media in passenger cars, but are characterized by higher filtration efficiency and are usually designed as surface-type filters. The separated particles thus precipitate mainly on the surface of the cellulose-based medium.

Better values than with pure cellulose-based media are obtained with special nanofiber-based filter media, in which ultrathin, synthetic fibers with diameters of just 30 to 40 nm are applied to a relatively coarse backing layer of cellulose. The filtration-efficiency figures improved in this way enable such systems to satisfy the constantly increasing requirements of the commercial-vehicle manufacturers regarding the protection of sensors, turbochargers and other components, and by protecting these components help to comply with the ever more stringent exhaust-emissions standards.

Further development

The space allotted to the air-intake system is becoming increasingly limited, particularly due to the comprehensive exhaust-gas treatment system. For this reason, there will in future also be air filters that are flatter than round filters. This opens up the possibility of new installation positions in the vehicle, for example in front of the front wheel or underneath the driver's cab.

Commercial-vehicle intake manifolds

Supercharged diesel engines require one intake manifold, usually referred to simply as a boost manifold. The air passes through the air filter to the turbocharger, and from there through the intercooler to the boost manifold. This distributes the air to the cylinders. For the most part the valve for EGR introduction (exhaust-gas recirculation) and possibly an electric auxiliary heater are flange-mounted at the inlet. The auxiliary heater helps to reach the required exhaust-gas values as quickly as possible during engine starting.

The introduced exhaust gas is mixed with the fresh air in the boost manifold and distributed as uniformly as possible to all the cylinders. The exhaust gas cooled by the EGR cooler contains, depending on the operating state and EGR rate, sulfurous acid and soot. The aluminum used for the boost manifold must be protected against the acid by for example painting. If the boost manifold is made of plastic, the plastic must be chosen so that temperature, charge-air pressure and acid do not damage the components over their service lives. The same requirements also apply to the flange gaskets.

Figure 18: Boost manifold for commercial vehicles
1 Flange for EGR valve or auxiliary heater,
2 Flange for cylinder head.

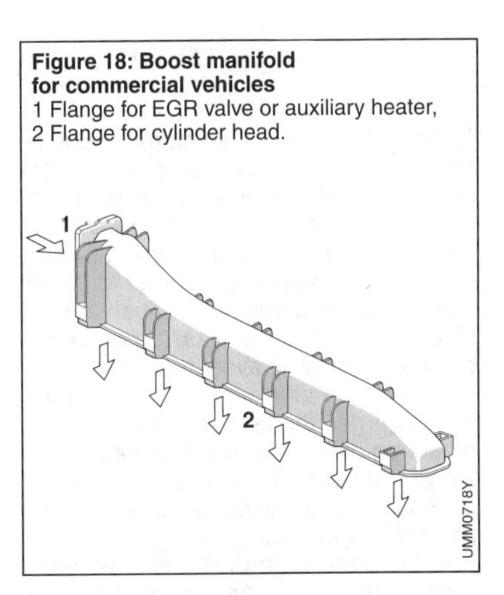

There are both simple boost manifolds (Figure 18) which can be manufactured as a single plastic injection-molded part and also parts which due to their more complex geometry are manufactured in two parts and welded. Charge-air pressure, size of the part and tolerances must be taken into account when configuring the welding. The use of plastic also contributes to the lightweight construction of the engine.

Turbochargers and superchargers

Turbochargers and superchargers increase the air-mass flow for a given engine displacement and a given engine speed, thereby increasing the power density. These "compressors" are generally systematically divided into three different types: the mechanically driven supercharger, the exhaust-gas turbocharger, and the pressure-wave supercharger.

In the case of the mechanically driven supercharger, the required drive power is branched directly from the crankshaft via a belt or a gear drive from the engine, i.e. the supercharger and the engine are mechanically coupled.

In the case of the exhaust-gas turbocharger, the drive power is generated after banking up the exhaust gas through expansion of the exhaust gas – i.e. by using part of the exhaust-gas energy. The turbocharger and the engine are thus purely thermodynamically coupled. In this respect, an exhaust-gas turbocharger with an electric, mechanical or hydraulic supplementary drive is a special case.

In the case of the pressure-wave supercharger, the compression power is derived from the exhaust gas, but an additional drive is required to synchronize the supercharger, meaning that the supercharger is simultaneously coupled mechanically and thermodynamically to the engine.

Figure 1: Mechanical centrifugal supercharger (schematic)
1 Variator primary pulley,
2 Variator secondary pulley, 3 Solenoid clutch,
4 Step-up planetary-gear set,
5 Compressor, 6 Air inlet, 7 Air outlet.

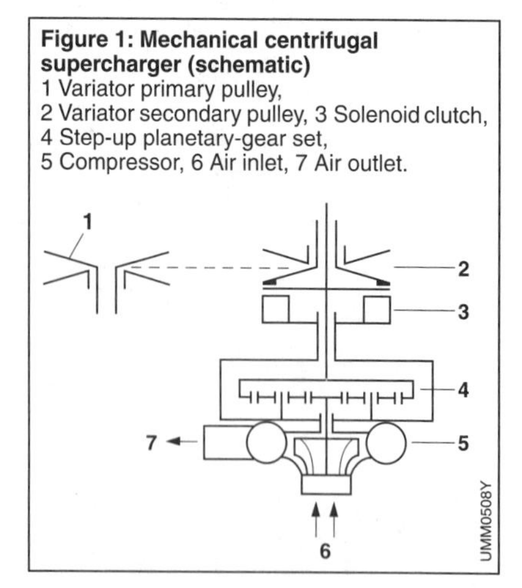

Superchargers (mechanically driven)

Mechanically driven superchargers [1] fall into two categories: Those which operate according to the positive-displacement principle and those which compress air according to the flow principle and the principle of momentum. Superchargers on IC engines are usually belt-driven (toothed or V-belt). They are driven either directly (continuous-running operation) or via a clutch. The gear ratio has been constant over the engine-speed range in applications up to now. A variable-speed drive would deliver technical advantages. However, previous attempts to achieve this have for various reasons translated into series production only in special cases.

Mechanical centrifugal supercharger

The mechanical centrifugal supercharger essentially consists of a compressor impeller which is mounted on a shaft with a belt pulley. In order to adapt supercharger operation to the requirements of effective engine operation, latest developments have seen the use of a speed variator and a switchable clutch ("Turmat" system, Figure 1). The primary pulley of the speed variator is driven directly by the engine crankshaft. The spring forces of the variator pulleys and the pre-tension of the drive belt are matched to each other in such a way that at low engine speed the secondary pulley on the compressor rotates faster than the primary pulley. This creates a dependence between the engine speed and the secondary-pulley speed such that the gear ratio decreases continuously as engine speed increases.

The compressor of the mechanical centrifugal supercharger operates according to the flow principle. This type of device is very efficient, providing the best ratio between unit dimensions and volumetric flow when compared with other mechanically driven superchargers. The pressure ratio that can be achieved is dependent on the peripheral velocity of the compressor impeller. The air throughputs typical of passenger-car engines give rise to small compressor-impeller dimensions, which result in high necessary compressor speeds with regard to the pressure ratios required for effective engine supercharging.

As the secondary pulley (conversion ratio relative to the primary pulley is 2:1) does not rotate fast enough to drive a mechanical centrifugal supercharger, a gear unit with a speed-increasing ratio of 15:1 or more is employed to achieve the required compressor peripheral velocities. In addition, this gear unit must be equipped with a highly efficient speed variator in order to provide a reasonably constant pressure ratio of a wide engine-speed range and to make high boost pressures available already at low engine speeds.

The range of application of the mechanical centrifugal supercharger is limited technically by the high rotational speeds required and the transmittable drive power, and economically by the comparatively high costs. It is used in small numbers in performance-oriented engine concepts.

Positive-displacement supercharger

A comparatively large number of designs has been proposed for units operating according to the positive-displacement principle, but only a few have succeeded in being employed in current series applications. Positive-displacement superchargers can operate with or without internal compression. Internal-compression superchargers include the reciprocating-piston, screw-type and rotary-piston compressors. The Roots supercharger is an example of a unit without internal compression. All of these positive-displacement superchargers share certain characteristics as shown here in the program map of a Roots supercharger (Figure 2):

- The speed curves n_{COM} = const in the p_2/p_1-$\dot{V}$ program map are very steep, i.e. Wthe volumetric flow $\dot{V}$ decreases only slightly as the pressure ratio p_2/p_1 increases. The drop in volumetric flow is basically determined by the efficiency of the gap seal (backflow losses). It is a function of the pressure ratio p_2/p_1 and of time, and is not influenced by rotational speed.
- The pressure ratio p_2/p_1 does not depend on the rotational speed. In other words, high pressure ratios can also be generated at low volumetric flow rates.
- The volumetric flow $\dot{V}$ is independent of the pressure ratio and, roughly formulated, directly proportional to rotational speed
- The unit retains stability throughout its operating range. The positive-displacement compressor operates at all points of the p_2/p_1-$\dot{V}$ program map as determined by supercharger dimensions.

Mechanical positive-displacement superchargers generally must be substantially larger than mechanical centrifugal superchargers in order to produce a given volumetric flow.

Roots supercharger

The Roots supercharger is a rotary-piston machine with contra-rotating twin- or multi-bladed rotary pistons mounted in roller bearings and synchronized by a gear set. The rotary pistons mesh at identical speed without directly contacting each other in the housing (Figure 3). The gaps between these components essentially determine the machine's efficiency.

Roots superchargers operate without internal compression. Mufflers are generally fitted on the suction and pressure sides to limit their acoustic impact. Their system and design limits the pressure ratios that can be achieved to levels below 2. Coating the functional components has helped to improve efficiency. Current development

Figure 2: Program map of a Roots supercharger
p_1 Total pressure before compressor,
p_2 Total pressure after compressor,
n_L Supercharger speed.

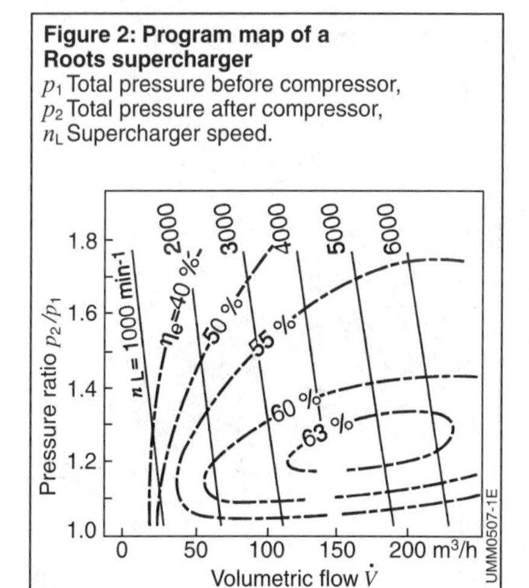

work is focused on among other things variable-speed gearboxes on the drive side.

Screw-type supercharger
The screw-type supercharger (Figure 4) is very similar in design to the Roots supercharger, i.e. it is a twin-shaft contra-rotating rotary-piston machine. However, it differs from the Roots supercharger in that it generally operates with internal compression. It can achieve higher pressure ratios than a Roots supercharger. On the suction side (inlet) the rotation of the pistons opens a profile-gap space which fills with inducted air. As the rotors continue to contra-rotate, the profile-gap space continuously decreases in size until it reaches the outlet control edges. At this point internal compression is completed and the compressed volume is forced into the pressure connection (outlet). In order to minimize internal leakage losses, the tolerances between the rotors and the walls must be very closely adhered to. Measures along similar lines to the Roots supercharger are considered for improving the function and efficiency of this type of supercharger.

Figure 3: Cross-section through a Roots supercharger
1 Housing,
2 Rotary piston.

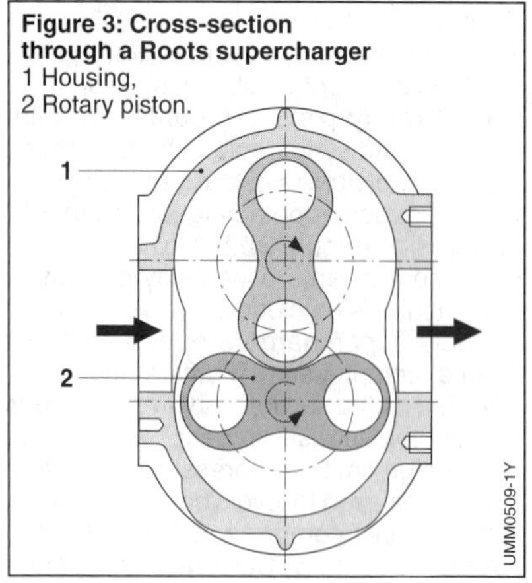

Figure 4: Screw-type supercharger
1 Inlet,
2 Rotors,
3 Outlet.

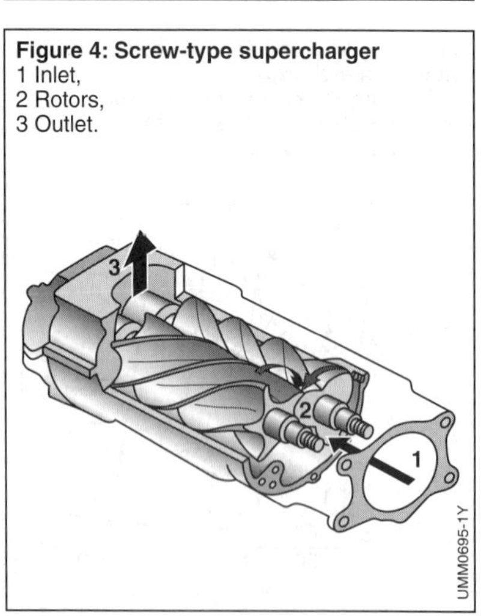

Spiral-type supercharger
The spiral-type supercharger (Figure 5) is a compressor in which a rotor with spiral lands eccentrically describes a circular path in a housing also with spiral lands. The spiral-type supercharger operates according to the principle of a circulatory displacer element. Working chambers in sequence open for charging, close for transport, and open once again for discharge at the hub.

The G-supercharger (G-Lader) is an example of this type. The rotor of the G-supercharger is eccentrically driven by a cam on the central main shaft. An auxiliary shaft driven by a V-belt ensures a uniform eccentric rotation on the part of the rotor. The drive shaft is lubricated by the engine's oil circuit. Maintaining very tight tolerances and the seal has a function-determining effect. Radial sealing is provided by the narrowest gaps possible, axial sealing by

Figure 5: Cross-section through a spiral-type supercharger
1 Air inlet into second working chamber,
2 Drive shaft,
3 Displacer element,
4 Air inlet into primary working chamber,
5 Housing,
6 Displacer element.

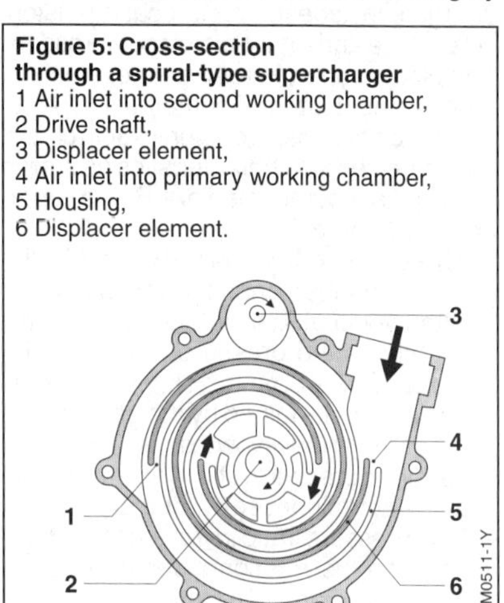

touching sealing strips inserted on the end faces. These sealing strips are subject to wear and are if necessary replaced during regular vehicle servicing. Internal compression can be achieved by an appropriate configuration of spirals.

More recent developments are heading towards a simplification of the design (omission of the auxiliary shaft) and the integration of a switchable clutch.

Wankel supercharger

The Wankel supercharger (Figure 6) is a rotary-piston machine operating on an internal axis. The driven inner rotor (rotary piston) turns through an eccentric pattern in the cylindrical outer rotor. The rotor ratio for the designs mentioned is either 2:3 or 3:4. The rotors turn in opposing directions about fixed axes without contacting each other or the housing. The eccentric motion makes it possible for the unit to ingest the maximum possible volume (chamber I) for compression and discharge (chamber III). The internal compression is determined by the position of the outlet edge A.

A ring and pinion gear with sealed grease lubrication synchronizes the motion of the inner and outer rotors. Permanent lubrication is also employed for the roller bearings. Inner and outer rotors use gap seals, and usually have some form of coating. Piston rings provide the seal between working chamber and gear case.

Figure 6: Cross-section through a Wankel supercharger
1 Housing, 2 Outer rotor,
3 Inner rotor, 4 Outlet edge A,
5 Chamber III, 6 Chamber II,
7 Chamber I.

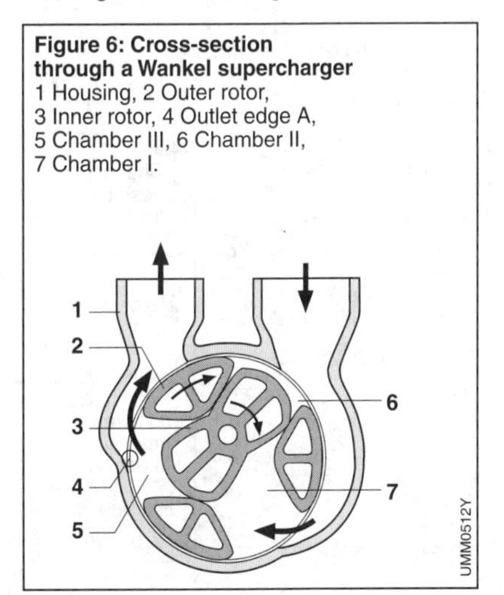

Pressure-wave superchargers

Design and operating principle

The pressure-wave supercharger (Figure 7) is a gas-dynamic machine whose core component is a rotor with open channels arranged coaxially around its circumference ("cell wheel" or "rotor") [2, 3]. The inlet and outlet openings for fresh air and exhaust gas pressurize the channels via the rotor end faces. The fresh air is compressed in the rotor channels by way of gas-dynamic processes. Fresh gas and exhaust gas briefly come into direct contact with each other in the process. Essential to the function is the physical fact that the gas-dynamic compression process takes place in a much shorter period of time than the mixing of the two gas flows.

The operating principle of the pressure-wave supercharger is based on the fact that a pressure wave is reflected at an open end of a line as a vacuum wave and at a closed end as a pressure wave; this also applies accordingly to the reflection of a vacuum wave. To control and maintain the pressure-wave process, the channel openings pass over "open ends" and "closed ends"; the cell rotor must rotate for this purpose. The

Figure 7: Pressure-wave supercharger
1 Engine, 2 Cell rotor, 3 Belt drive,
4 High-pressure exhaust gas,
5 High-pressure air,
6 Low-pressure air inlet,
7 Low-pressure gas outlet.

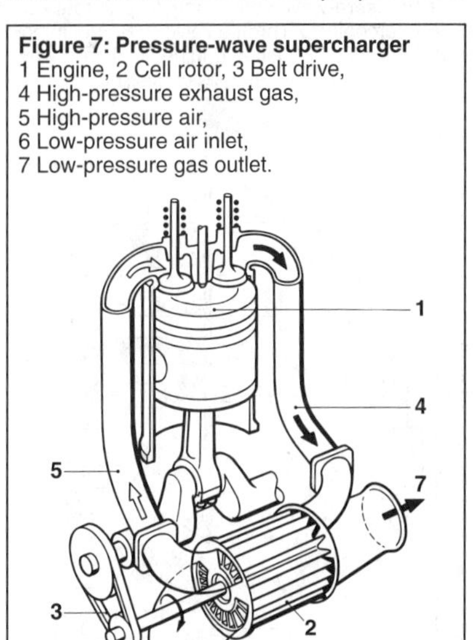

drive power serves merely to cover the rotor bearing and ventilation losses and to accelerate the rotor in the event of a sudden load variation. By configuring the routing of gas accordingly in the housing it is possible to set a sufficiently homogeneous temperature distribution in the rotor so that sufficiently small gaps can be maintained. The acoustics can be improved by configuring the cells accordingly.

The gas-flow and state diagrams (Figure 8) illustrate the pressure-wave process in a basic pressure-wave supercharger at WOT and moderate engine speed. Energy is exchanged in the channels at the velocity of sound, and, owing to the principles involved, the pressure-wave supercharger responds very quickly to changes in engine demand, with the actual response times of the supercharger being determined by the charging processes in the air and exhaust pipes. The velocity of sound is, as well as the physical characteristics, a function of temperature, meaning that essentially it is dependent on engine torque, but not engine speed.

Comprex supercharger

If a constant gear ratio is employed between engine and cell rotor, as is achieved with the belt-driven Comprex supercharger, the pressure-wave process is, on account of this boundary condition, optimally tailored to only a single operating point. To get around this disadvantage, appropriately designed "pockets" can be incorporated in the gas routing of the end housings. These achieve high efficiency levels extending through a relatively wide range of engine operating conditions and provide a good overall boost curve.

The Comprex supercharger's rotor is over-mounted and is provided with permanent grease lubrication, with the bearing located on the unit's air side. The air housing is made of aluminum, whereas the gas housing is made of NiResist materials. The rotor with its axial cells is cast using the lost-wax method. An integral bypass valve in the supercharger regulates boost pressure according to demand.

Figure 8: Gas-flow and state diagrams for a pressure-wave supercharger
a) Gas-flow diagram, b) State diagram.
1 Pressure-wave supercharger, 2 Throttle valve 1, 3 Air filter, 4 Rotor drive, 5 Housing offset, 6 Intercooler, 7 Throttle valve 2, 8 Fuel injector, 9 Engine ECU, 10 Internal-combustion engine, 11 Three-way catalytic converter, 12 Gas-pocket valve, 13 Oxidation-type catalytic converter.

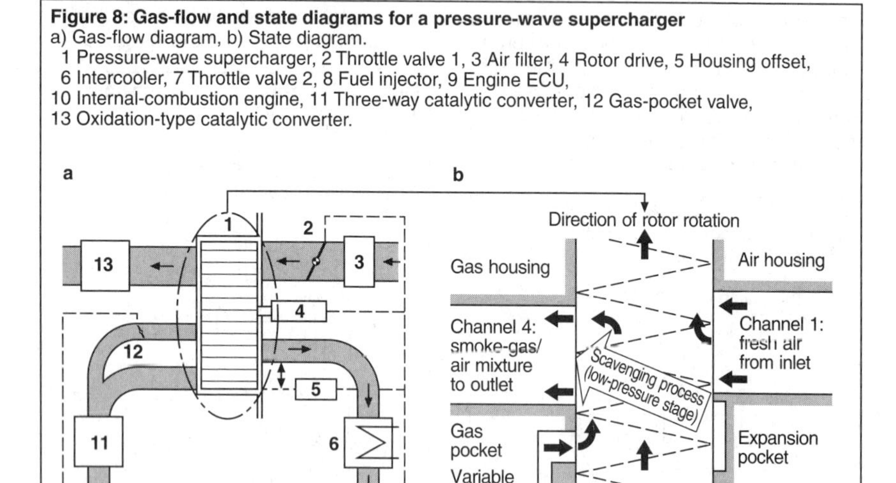

Hyprex supercharger
A further development of the Comprex is the Hyprex, which provides an effective supercharging process particularly for small-capacity gasoline engines. However, its high exhaust-backpressure sensitivity has hitherto come out against series application.

The Hyprex supercharger's rotor is driven independently of engine speed by an electric motor, thanks to which the rotor speed can be better adapted to the engine's operating state. In addition to other modifications which improve cold starting, asymmetrical arrangement of the cells has further improved acoustics. A variable gas-pocket inflow provides for improved efficiency in the lower engine speed range with a resulting increase in boost pressure. Comprehensive application of a Hyprex supercharger requires the use of modern electronic engine management.

Exhaust-gas turbochargers

Of all the supercharger units in use, the exhaust-gas turbocharger is by far the most widespread. While nowadays virtually 100% of passenger-car and commercial-vehicle diesel engines are turbocharged, the proportion of supercharged gasoline engines will increase dramatically in the next few years. Here, too, the exhaust-gas turbocharger will be the dominant supercharger.

Design and operating principle
The exhaust-gas turbocharger consists of two turbo elements: a turbine and a compressor, where the impellers for the two elements are mounted on a common shaft (Figure 9 and Figure 10). The operating concept of a turbo element is based physically on the principle of momentum. The turbine converts some of the enthalpy contained in the exhaust gas into mechanical energy to drive the compressor. The compressor draws in fresh air through the air filter and compresses it.

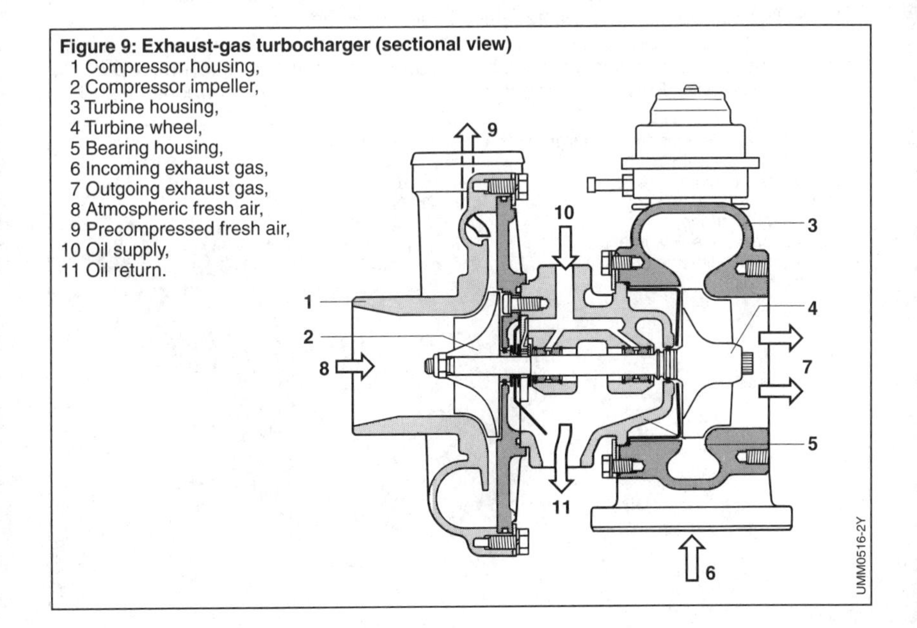

Figure 9: Exhaust-gas turbocharger (sectional view)
1 Compressor housing,
2 Compressor impeller,
3 Turbine housing,
4 Turbine wheel,
5 Bearing housing,
6 Incoming exhaust gas,
7 Outgoing exhaust gas,
8 Atmospheric fresh air,
9 Precompressed fresh air,
10 Oil supply,
11 Oil return.

The exhaust-gas turbocharger is coupled only thermodynamically, not mechanically, to the engine. Turbocharger speed does not depend on engine speed, but is rather a function of the balance of drive energy between the turbine and the compressor. The turbine generates the compressor output and the output which is essentially used up in the bearings and ultimately dissipated as heat (mechanical losses).

Applications

The exhaust-gas turbocharger in its present form can trace its roots back to the work of Alfred Büchi (1905) [2], who had already recognized the potential of combining supercharging and valve overlap to scavenge residual exhaust gas (1915). Exhaust-gas turbochargers have traditionally been used for the purpose of supercharging large diesel engines, originally being used primarily for truck, marine and locomotive power plants as well as for agricultural- and construction-machinery applications.

Use in diesel vehicles

The mid-1970s saw the advent of the first production diesel engines supercharged by exhaust-gas turbochargers in passenger cars. The introduction of a "wastegate" for controlling boost pressure finally facilitated a torque-oriented design and thereby improved driveability. Further increases in the performance of passenger-car diesel engines have been achieved through the introduction of direct injection (1987) and by exhaust-gas turbocharging with variable turbine geometry (1996) or two-stage exhaust-gas turbocharging (2004). These developments have been reflected in a marked increased in the market share for diesel engines in Europe. Today in Europe all passenger-car and commercial-vehicle diesel engines are equipped with exhaust-gas turbocharging and intercooling (charge-air cooling).

Use in passenger cars with gasoline engines

Exhaust-gas turbocharging of gasoline engines was originally the reserve of only high-output sports engines for increased performance and, because of inadequate driveability ("turbo lag"), was rarely encountered on the market. In the meantime, gasoline-engine supercharging has become an integral part of engine development, predominantly in connection with small to medium-sized engines. In addition to the improvement in efficiency, one of the main objectives is to avoid the increases in the number of engine cylinders, thus positively influencing package space and fuel consumption.

In contrast to diesel engines, mechanical supercharging was also employed and today to a lesser extent continues to be

Figure 10: Exhaust-gas turbocharger with wastegate (view)
1 Compressed-air outlet,
2 Compressor impeller,
3 Shaft,
4 Exhaust-gas turbine,
5 Exhaust-gas mass-flow intake,
6 Wastegate with linkage and flap (in turbine inlet).

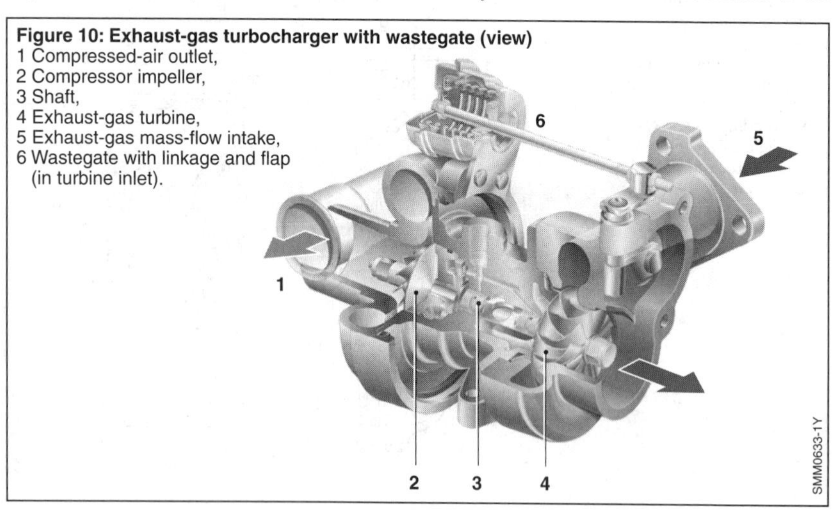

employed on gasoline engines for marketing reasons and on account of the excellent transient build-up of boost pressure. In the meantime, turbocharged direct-injection gasoline engines have practically reached the levels of equivalent mechanically supercharged gasoline engines in terms of the transient build-up of boost pressure.

In addition, a combination of mechanical supercharging and exhaust-gas turbocharging (combination supercharging) is currently employed to provide small-capacity gasoline engines with high rated output and high torque at a comparatively low specific speed.

While the exhaust-gas turbocharger with variable turbine geometry is today the standard supercharger used in diesel engines, the high exhaust-gas temperature and the costs associated with this technology means that up to now it has only been used on gasoline engines in niche applications.

With regard to the statutory exhaust-emission and consumption guidelines and the existing performance figures required by customers, the importance of exhaust-gas turbocharging to the reduced-capacity and reduced-cylinder engines currently in development ("downsizing") will increase further. Today, the number of turbocharged gasoline engines is rising dramatically and in the next few years a further dramatic growth of the market segment is expected.

Exhaust-gas turbocharger design

The exhaust-gas turbocharger essentially consists of three assemblies: the central housing, the compressor, and the turbine. Depending on the design, there may be a fourth assembly: a boost-pressure control facility.

Bearing housing

The bearing housing accommodates the bearings and the elements for sealing the shaft. State-of-the-art designs usually feature a specially developed plain bearing both in the radial as well as the axial (thrust) bearing assembly. The radial bearings are designed either as rotating double plain bushings or as stationary plain-bearing bushings. The requirements relating to stability, power loss and noise-emission characteristics govern what type of bearing system is used.

The bearing bushings are back-flushed by a film of oil ("floating-bushing bearing") to improve bearing damping. The oil also performs a cooling function, particularly for the shaft. The thrust bearing in conventional designs is made up of a multiple-spline surface bushing that is subject to load from both sides and which is lubricated either centrally or individually for each spline surface. The lubricating oil is supplied by connecting the turbocharger to the engine's oil circuit. The oil outlet is connected directly to the oil pan in the crankcase. Today, this type of bearing assembly is used to control rotational speeds of more than 300,000 rpm reliably.

Roller-bearing exhaust-gas turbocharger rotors are expected to deliver consumption-related efficiency advantages which provide for a better transient build-up of boost pressure and higher boost pressures in the part-load range. Until now roller-bearing exhaust-gas turbocharger rotors have been used primarily in motor-racing applications. To enable these rotors to be used in standard vehicles, it will be necessary to resolve issues relating to service life and long-term acoustic behavior. With the aims of achieving further efficiency improvements or an extension of the permissible installation location and of completely avoiding any oil overflow in order to reduce overall emission levels, intensive development is currently being conducted on creating alternative bearing assemblies such as air or magnet bearings.

To seal off the oil chamber to the exterior and to minimize the entry of charge air ("blow-by") and exhaust gas to the charger interior, the shaft is equipped at the casing openings with piston rings which are tensioned in the bearing housing and with the shaft groove form a simple labyrinth. In some special applications, the sealing effect can be enhanced by implementing additional measures such as providing an additional piston ring on the compressor and turbine sides, or an air seal or slide ring seals (up to now only possible on the cold compressor side). The state-of-the-art sealing technology used today limits the permissible installation location for the exhaust-gas turbocharger to a comparatively small range of inclination. Touching seals would be useful in this respect, but have not been able to be produced up to now at reasonable cost owing to the high relative speeds between shaft and housing.

No additional cooling measures are necessary to maintain efficient operation of the bearing assemblies under suitable ambient conditions in the engine compartment, i.e. at exhaust-gas temperatures up to approximately 820 °C. The relevant temperatures can generally be maintained below critical levels using devices such as a heat shield, and by thermally isolating the hot turbine housing, supplemented by incorporating suitable design elements in the bearing housing itself. Water-cooled bearing housings are employed for higher temperatures, for example, on gasoline engines operating at up to 1,050 °C exhaust-gas temperature or in certain diesel-engine applications.

Compressor

The compressor assembly in passenger car and commercial-vehicle applications consists of an axial-inflow and radial-outflow impeller (centrifugal design) and a generally guide-vane-free compressor housing made of cast aluminum. The flow is further slowed down in the housing, in the course of which some of the pressure build-up occurs. Having the smallest possible contour gap between the impeller and the housing is crucially important to efficiency. The impeller is mass-produced in a special casting process from an aluminum alloy. Today, impellers milled from an aluminum forging alloy are increasingly being used. In the event of more rigorous requirements, particularly for reasons of service life in the case of critical application profiles, impellers milled from a titanium alloy are also used in the commercial-vehicle sector – a comparatively cost-intensive measure. For certain applications – for instance on engines which are equipped with low-pressure-side exhaust-gas recirculation such that the compressor has to deliver medium containing exhaust gas – the compressor impeller is provided with a coating.

Although the compressor impeller contributes much less than the turbine wheel to the rotor's mass moment of inertia, a value which is as low as possible is also important on the cold side. Because of the heightened requirements with regard to performance and service life, attempts to mass-produce plastic compressor impellers have hitherto failed.

The characteristic of a compressor is described by a program map (Figure 11). In contrast to a positive-displacement supercharger, the program map of a flow compressor features ranges in which stable operation is not physically possible. By implementing appropriate measures, the effective range as well as the speed and efficiency characteristics of the compressor can be adapted to the required boost-pressure curve. The effective range of the compressor is determined on the "left" side of the program map (i.e. maximum achievable pressure ratio for a given throughput) by the surge limit and on the "right" side (maximum possible throughput, limited by the attainment of the velocity of sound in the narrowest cross-section) by the choke limit. The surge limit is defined as the transition from the stable to the unstable operating range. Unstable means that, triggered by a severance in flow – generally at the compressor-impeller inlet – the air-mass flow is periodically interrupted and re-established to produce a pumping effect. Among other factors, the extent of the surge limit is also governed by the

Figure 11: Compression graph with typical engine operation curves valid for all displacements
1 u = 150 m/s, 2 u = 300 m/s, 3 u = 450 m/s.
u Peripheral velocity,
p_1 Total pressure at compressor inlet,
p_2 Total pressure at compressor outlet,
η_{isv} Isentropic compressor efficiency.
Volumetric-flow factor is a dimensionless volumetric flow.

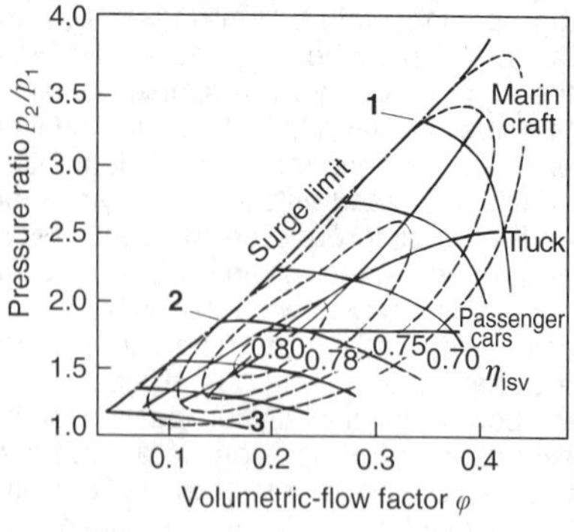

design of the intake line. The surge limit is therefore not a compressor property, but rather a system property. The choke limit, identifiable by the range of steeply descending speed curves in the program map, is determined based on the free inlet cross-section of the compressor impeller and therefore across the impeller diameter.

In view of the proportionality of the engine speed and volumetric air flow of the compressor, it is easy to deduce that compressors for turbocharging gasoline engines must have a substantially larger effective map width than, for instance, compressors which supply large diesel engines with air. The dimensionless compressor program map in Figure 11 shows the characteristic air requirements along the WOT for passenger cars, commercial vehicles, and large heavy-duty engines.

Precise aerodynamic adaptation of the compressor to the respective requirements is effected by modifying the component geometry, where impeller and housing must be viewed as a single component and strength criteria and natural-oscillation responses must be taken into account in the design of the impeller. A diffuser with vanes may indeed increase efficiency at the design point, but with an otherwise identical geometry gives rise, apart from a blocking effect, to a downturn in efficiency when the design point is deviated from. This is why only compressors without guide vanes are used for passenger-car turbocharging in view of the comparatively large mass-flow spread here.

Recirculation channels integrated into the housing ("blow-off" by returning some of the delivered mass flow to the compressor inlet) or geometrically matched configuration of the compressor inlet channel (preliminary volumes) are also used to extend and stabilize the program map.

The program map can be positively influenced in the surge-limit range by variabilities – for example by installed components which generate a pre-swirl or by diffusers which are equipped with pivoted adjustable vanes in order to remove the swirl after the impeller from the flow as a function of the operating point. In other words, thanks to these devices the compressor can be run at an operating point at which without this measure the surge limit would already be exceeded. Variable compressors have up to now not been series-produced – apart from a simple preliminary diffuser whose vane angles twist into a specific position as the mass flow increases (flow forces) and are released in the course of the resetting of the twisted structure as the mass flow decreases.

Compressors for turbocharging gasoline engines have until now been equipped with a blow-off valve, whose task it is to prevent compressor surge when load is quickly removed (i.e. throttle valve closes). The blow-off valve – originally pneumatically actuated, but now mainly electrically actuated – for this purpose creates a bypass between the compressor outlet and inlet so that it pumps briefly as a blower in the circuit. Today it is possible, above all by introducing electrical actuation of the wastegate and through further measures, to prevent surge without the use of a blow-off valve and guarantee reliable operation.

Turbine

The turbine consists of a diffuser and a wheel. The diffuser of the turbocharger turbine is integrated in the flow housing (spiral housing) and for passenger-car applications, if the turbocharger is one without variable turbine geometry, is designed without vanes. Fixed-geometry diffusers with vanes are frequently used in exhaust-gas turbochargers for large engines for better fine graduation of the turbine throughput. The flow is accelerated in the diffuser and distributed as evenly as possible to the turbine wheel.

The turbine wheel for standard applications is designed as a centripetal turbine, i.e. radial inflow and axial outflow. Semi-axial wheels are used on account of their favorable mass moment of inertia. These have diagonal inflow and axial outflow. Axial wheels with axial inflow and axial outflow are only used in heavy-duty turbine applications. The type of construction impacts on pressure ratios that can be implemented and achieved, efficiency, and other characteristics. Generally speaking: The greater the mass flow, the more advantages the axial machine offers and vice versa. In other words, the smaller the mass flow, the more optimal the radial inflow.

The arrangement of the exhaust-gas pipe plays an important role in supercharging. Classically, a distinction is made between pulse supercharging and constant-pressure supercharging. In the case of pulse supercharging, the exhaust-gas pipes are routed separately to the turbine housing. Those cylinders whose exhaust pulses influence each other and the scavenging process of one cylinder minimally are brought together on the exhaust-gas side. The turbine housing is designed in such a way that the separation of channels is maintained as far as possible up to the turbine-wheel inlet. The channel cross-section is the size of the exhaust cross-section. The exhaust pulse is essentially transmitted as a pressure wave to the turbine such that the exhaust-gas energy is applied at the turbine inlet in highly pulsating form. This effect can be utilized particularly well at low engine speeds where pressure surges can develop effectively due to the long time interval from one exhaust pulse to the other.

In the case of constant-pressure supercharging, the exhaust discharges are directed in a common, comparatively large-volume collector to the turbine housing. This process smooths out the individual exhaust pulses to a large extent.

With constant-pressure supercharging the more uniform enthalpy over time can be processed with greater efficiency. Pulse supercharging offers advantages for the engine's part-load and accelerating performance.

The design of the turbine housing differs significantly, depending on whether constant-pressure or pulse supercharging is used. Pulse supercharging for the most part is used in commercial-vehicle applications. The turbine housing is designed as double-flow housing in which the two flows join only just before the wheel inlet.

Constant-pressure supercharging is used on high-speed engines such as the diesel engines in passenger vehicles. The turbine housings are single-flow ("single-scroll") in design and partially integrated with the exhaust manifold as a single component. This provides for a particularly compact and streamlined component geometry.

Gasoline engines utilize both constant-pressure and pulse supercharging. The exhaust systems are configured accordingly and the housings are single-flow or twin-flow ("twin-scroll") in design. Partially manifold-integrated turbine housings are also used in exhaust-gas turbochargers for gasoline engines. Even welded constructions (single-walled or air-gap-insulated, i.e. double-walled sheet-metal manifolds with welded-on cast turbine housings) have been series-produced.

To withstand the high temperatures and high loads involved, the turbine wheels are made of a material containing high levels of nickel using the lost-wax process and joined to the steel rotor shafts by means of friction, laser or electron-beam welding. Because of its outstanding high-temperature properties, the material titanium aluminide, an intermetallic phase, is being discussed as a potential material for the turbine wheel. Thanks to its low density, it reduces significantly the mass moment of inertia and allows the wheel to be fashioned with a raised wheel back without negatively impacting on the turbocharger's run-up performance. On account of its brittleness at room temperature and problematic castability, development and production issues must still be resolved in this regard.

The turbine housings made of high-alloy spheroidal or spherulitic cast iron, depending on the specific application temperature, are produced in an open sand-casting process. Alloyed cast steels, normally cast in thin walls using the lost-wax process, are used in high-temperature applications (up to 1,050 °C). Modern low-pressure casting processes are also used to manufacture thin-walled components (weight, costs, thermal inertia). Twin-scroll housings, on account of the filigree geometry of the partition wall, require cast-steel materials. Turbine housings built up from sheet-steel stampings or from internal high-pressure forming parts and welded together in single-walled form or as AGI (air-gap-insulated) components have also been produced. The AGI versions reduce wall heat losses and have a low thermal inertia.

Built-up turbine housings have hitherto not been introduced into series production; they could however make a contribution to complying with future emission legislation. To ensure reliable application, it will be necessary, in addition to mass-production capability, to resolve above all the issues of durability and guaranteeing the smallest possible turbine contour gap under all operating conditions.

Boost-pressure control

In view of the large rotational speed spread of passenger-car engines, a boost-pressure control facility is indispensable to maintain the maximum permissible boost pressure if the design torque is to remain acceptable. Standard practice presently favors turbine output control on the exhaust-gas side.

Bypass control

A simple and widely used method is that of bypass control, in which generally some of the exhaust-gas flow is routed through a flap valve ("wastegate") around the turbine (Figure 10 and Figure 12). The wastegate is actuated by a pneumatic unit or an electric actuator. The pneumatic unit can take the form of an overpressure unit (supplied by the boost pressure itself) or a vacuum unit (vacuum supply from the vehicle system). If the boost pressure is used, the wastegate cannot however be actuated independently of the engine operating state. In most cases, the control pressures are regulated by using clocked-pulse valves. Because of the thermal load, the actuators are generally mounted on the compressor side and connected via a control rack to the wastegate lever.

More recent developments are clearly heading towards the use of electric actuators that can adjust the control characteristics faster and with greater precision, and towards actuating the wastegate independently of the engine operating state. This is advantageous when it comes to compliance with emission and consumption legislation. In addition, with electric actuators it is possible to create locking forces in a tenable space which ensure that the wastegate is firmly seated in all operating states, which in turn provides for a rapid buildup of boost pressure.

Variable turbine geometry

Compared to bypass control, variable turbine geometry (VTG, Figure 13) offers by far the most efficient options for adapting the turbocharger within the overall program map. The entire mass exhaust-gas flow is routed here via the turbine, providing specific benefits in exploiting the available energy. By varying the flow cross-sections (turbine cross-sections), the turbine's resistance to flow is adjusted corresponding to the required boost-pressure level.

Figure 12: Boost-pressure control via exhaust-side boost-pressure control valve (wastegate)
1 Engine,
2 Exhaust-gas turbocharger,
3 Boost-pressure control valve (wastegate).

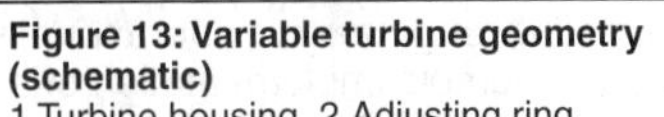

Figure 13: Variable turbine geometry (schematic)
1 Turbine housing, 2 Adjusting ring,
3 Control cam, 4 Adjusting blade,
5 Adjusting blade with control lever,
6 Air inlet.

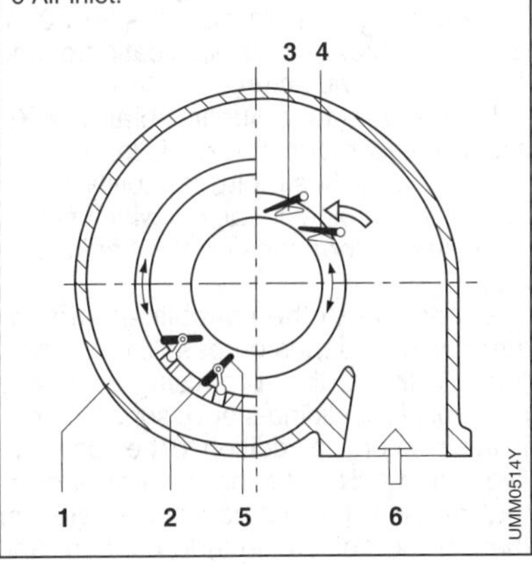

Of all the potential designs, adjustable blades have achieved general acceptance, as they combine a wide control range with high efficiency levels. The blade angle can be easily adjusted by means of a swiveling movement. The blades, in turn, are set to the required position using adjusting cams, or directly via control levers attached to the individual blades. All blades engage in an adjusting ring which can be turned by means of a lever. The lever is actuated by pneumatic or electric actuators. Where a pneumatic unit is fitted, today a position sensor is increasingly integrated in the unit, and the signal from this sensor ensures that the VTG blade position is clearly identified in the engine management.

Where originally symmetrical blades with straight median lines were installed, today profiled blades with pronounced pressure and suction sides and curved median lines are used. This helps to improve efficiency above all in the closed position, i.e. when accelerating and in the part-load range. Maintaining small gaps in the guide-vane area is of vital importance to engine performance.

Turbochargers with variable turbine geometry are state-of-the-art technology in diesel-engine applications. Using a turbocharger with variable turbine geometry is also of interest in gasoline engines. Owing to the high exhaust-gas temperatures of a gasoline engine, the operationally reliable and durable implementation of an adjusting-blade exhaust-gas turbocharger involves very intensive thermomechanical matching and the use of highly developed materials, and up to now (as at 2010) has only featured in an application of one sports-car manufacturer.

In addition to adjusting-blade VTG, a Japanese manufacturer is employing a comparatively simple variability in the field of the turbine spiral, but which fails to attain the performance levels of adjusting-blade VTG.

In addition to the variabilities already introduced in series on gasoline engines, there is the further promising operating concept of the sliding-sleeve supercharger. In this concept, in addition to the comparatively high efficiency, the number of parts that move in the hot-gas flow is smaller than in the adjusting-blade concept and an internal bypass can be comparatively easily opened to bypass the turbine with just one actuator in addition to actuating the variability.

Exhaust-gas turbochargers with supply of additional energy

Modern exhaust-gas turbochargers are characterized by high efficiency levels and a low mass moment of inertia on the part of the rotor such that a very good transient build-up of boost pressure is achieved, particularly in conjunction with measures on the engine, such as for instance "scavenging" [4]. Both in the past and today various solutions have been proposed, when minimal exhaust-gas energy is available, for supplying additional mechanical or electrical power to the shaft for the purpose of accelerating the rotor more quickly.

For example, the incorporation of a gear on the turbocharger shaft allows mechanical energy to be supplied to the turbocharger shaft via a gearbox and a switchable mechanical connection to the crankshaft. With regard to the differences in speed between this shaft and the engine, the gearbox and clutch pose a challenge to engineers, particularly with regard to the high efficiency levels required and durability.

A further approach provides for installing on the rotor between the bearings a Pelton turbine wheel which is pressurized with high-pressure oil (approx. 100 bar) from the engine circuit or separately with hydraulic fluid. In this approach functions must be solved associated with the abrasive effect of the soot particles finely distributed in the engine oil which strike the walls of the necessarily filigree nozzle and blade structures at high speed. A separate hydraulic circuit would mean a further space in the bearing housing which must be effectively sealed against the other media.

In a further approach it is proposed to install on the turbocharger shaft, for instance between the bearings, an electric motor which is supplied via power electronics by the vehicle electrical system ("electrically assisted exhaust-gas turbocharger"). Limits are set to this approach in connection with the possibilities of a 12-V vehicle electrical system and with regard to the required power to be supplied to ensure effective operation. The electrically assisted turbocharger is able in certain operating phases to function as a generator and thereby return electrical energy to the vehicle electrical system, but a wastegate cannot be dispensed with.

What all the approaches have in common is the fact that in operating phases in which no additional energy is supplied the components integrated on the turbocharger shaft must be carried along, thereby causing losses and being subject to considerable thermomechanical load. The potential of these approaches must further be considered against the background that additional energy, partially isolated from the engine's operating state, can indeed be supplied, but the program-map limits of the compressor used cannot thereby be extended or shifted. In addition, all the solutions described involve a significant increase in expenditure on equipment and thereby in the costs. None of these approaches has been introduced into series production.

Complex supercharging systems

With a parallel or multistage connection of supercharger units – partly with complex valves for control, for operating-point-dependent cut-in or defined distribution of the exhaust-gas mass flow – the power limits can be significantly extended compared with supercharging with just one single-stage supercharger unit. The objective here is to improve the air supply on both a stationary and a non-stationary basis and at the same time to improve the specific consumption of the engine.

Bi-turbo supercharging

In the case of bi-turbo supercharging, the air supply, instead of using one large exhaust-gas turbocharger, is distributed to two or more exhaust-gas turbochargers connected in parallel. These are generally permanently assigned to specific cylinders on the exhaust-gas side, for example the bank of a V-engine. Each turbocharger pressurizes the engine with air over the full rotational-speed range. The advantage of this arrangement lies primarily in the rapid transient build-up of boost pressure and in the comparatively compact routing of pipework. The turbochargers operating in parallel must be identical in terms of their control engineering.

Sequential supercharging

Sequential supercharging is predominantly used in ship propulsion systems or generator drives. However, because of the high levels of power it can deliver, this arrangement variant is also used in passenger cars. In the case of passenger-car sequential supercharging, as engine load and speed increase, an additional turbocharger is cut in to the basic supercharging process (Figure 14). Thus, in comparison with mono-supercharging (i.e. a single charger which is geared to the rated output), two supercharging optima are achieved.

The sequential-supercharging system contains, in addition to the turbochargers, valves and sensors which facilitate a smooth switchover from one- to two-charger operation and back from two- to one-charger operation. When the engine is in the lower speed range, only the basic turbocharger is active, supplying all the cylinders with air. When a certain speed is reached, the second turbocharger is run up and cut in. Both turbochargers are now operating in parallel to pressurize all the cylinders. Sharing the air supply between two turbochargers helps to deliver a highly responsive torque curve and high rated output. Furthermore, the significantly lower mass moment of inertia of the smaller turbocharger shafts gives rise to a favorable transient response.

Even an integral design for the two turbochargers including the switchover valves is possible, but complex. Controlling the switchover process (by means of control engineering) is comparatively complex, but technically well within the realms of feasibility with the methods of modern electronic engine management.

Sequential supercharging is used in both gasoline and diesel engines [5].

Figure 14: Sequential supercharging
a) Run-up in one-charger operation,
b) Controlled one-charger operation,
c) Two-charger operation.
1 Turbocharger 1,
2 Turbocharger 2,
3 Compressor cut-in valve,
4 Turbine cut-in valve,
5 Transverse pipe,
6 Bypass valve.

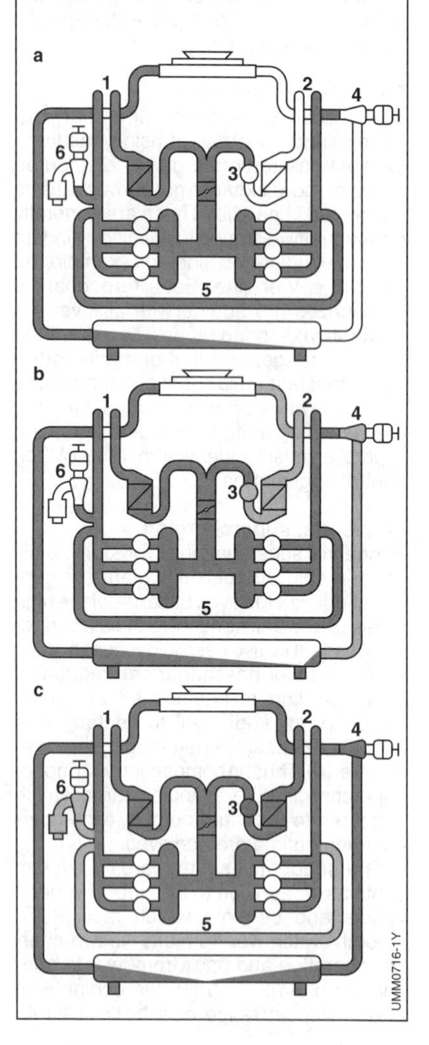

Dual-stage controlled supercharging

In contrast to sequential supercharging, with dual-stage controlled supercharging the turbochargers are connected in such a way as to effect serial operation in certain operating phases (Figure 15). The advantage of this supercharging system over the single-stage method lies in the increase in rated output with a simultaneous improvement in the stationary torque at low speeds and in the engine's accelerating performance thanks to the rapid build-up of boost pressure. This extremely powerful system features the use of two turbochargers of significantly different sizes. There is no dip in torque when this system is appropriately tuned and controlled.

The entire inducted fresh-air mass flow is precompressed by the low-pressure stage. The charge air is further compressed in the high-pressure stage. As a result of precompression, the relatively small high-pressure compressor works at a higher pressure level so it is able to generate the required air-mass flow. Performance can be further improved by cooling the charge air after the first compression stage, but as yet this has not been done in passenger-car applications for space and cost reasons.

Particular importance is attached to the sealing and the control quality of the turbine bypass valve. By way of this valve, depending on the operating point, either the entire exhaust-gas mass flow is expanded in two stages in the high-pressure turbine and then in the low-pressure turbine, or some of the exhaust-gas mass flow is diverted past the high-pressure turbine and delivered directly to the low-pressure turbine. The thermomechanical and vibrational demands placed on the valve system are very high. Routing flow between the compressors and the turbines with minimal pressure losses is also crucially important.

The bypass valve is closed in the engine's low speed range, i.e. with small exhaust-gas mass flow rates. The exhaust-gas mass flow expands through the small high-pressure turbine. This results in a very good transient build-up of boost pressure. As engine speed increases, the turbine bypass valve routes an increasing exhaust-gas mass flow directly to the low-pressure turbine until finally the low-pressure turbine too is relieved by a further wastegate.

In this operating mode dual-stage controlled supercharging enables infinitely variable adaptation of the turbine side to the requirements of engine operation. When a specific engine speed is reached, the high-pressure compressor in spite of precompression is no longer able to throughput the air-mass flow and would in this case act as a throttle. In this event, an (uncontrolled) valve opens a channel directly from the low-pressure compressor to the intercooler and bypasses the high-pressure compressor.

This supercharging system was introduced as standard in passenger-car diesel engines in 2004. It is technically possible to implement this approach in gasoline engines too.

Combination supercharging

Combination supercharging combines a mechanically driven supercharger unit with an exhaust-gas turbocharger (Figure 16). This powerful supercharging system was used for the first time in a passenger car in 1985 (Lancia Delta). It is used today in comparatively small-capacity gasoline engines with high rated output and high torque already at comparatively low speeds and high dynamic response.

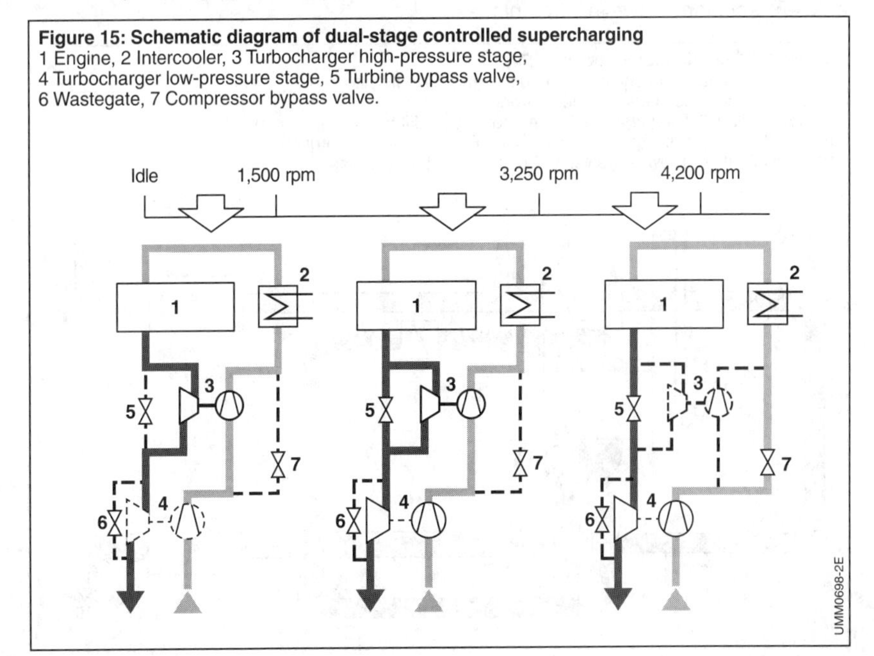

Figure 15: Schematic diagram of dual-stage controlled supercharging
1 Engine, 2 Intercooler, 3 Turbocharger high-pressure stage,
4 Turbocharger low-pressure stage, 5 Turbine bypass valve,
6 Wastegate, 7 Compressor bypass valve.

In the lower engine-speed range in which little exhaust-gas energy is available to drive the turbocharger turbine, the air supply is practically managed by the mechanically driven supercharger on its own. Because the exhaust-gas turbocharger is designed for the mean-speed and rated-output ranges, this results in a comparatively high-flow-rate charger with correspondingly high levels of efficiency. In a certain operating range the system operates as a series connection of two supercharger units, i.e. the total pressure ratio is the product of the individual pressure ratios.

The mechanically driven supercharger can be positioned in the flow direction before or after the turbocharger compressor. When the engine is in the upper speed range and there is therefore an adequate supply of exhaust-gas energy, supercharging is performed by the turbocharger and the mechanical supercharger is bypassed. Because friction losses are branched off as drive power from the crankshaft even when the mechanical supercharger is idling, disengagement of the mechanical supercharger is the most efficient solution. However, this places

Figure 17: Electrically driven compressor in series with an exhaust-gas turbocharger
1 Internal-combustion engine, 2 Intercooler, 3 Bypass valve for turbocharger, 4 Turbocharger turbine, 5 Turbocharger compressor, 6 Electric motor, 7 Triggering electronics for electric motor, 8 Bypass valve for eBooster, 9 eBooster compressor, 10 Exhaust gas, 11 Intake air.

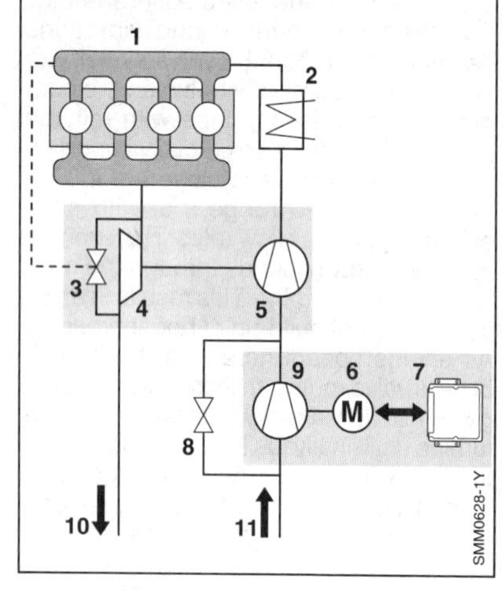

Figure 16: Combination supercharging
1 Internal-combustion engine, 2 Turbocharger, 3 Wastegate, 4 Compressor (mechanical supercharger), 5 Control flap, 6 Throttle valve, 7 Intercooler, 8 Air filter, 9 Catalytic converter, 10 Belt drive for accessories with clutch, 11 Belt drive for compressor, 12 Crankshaft, 13 Blow-off valve, 14 Fresh air, 15 Exhaust gas.

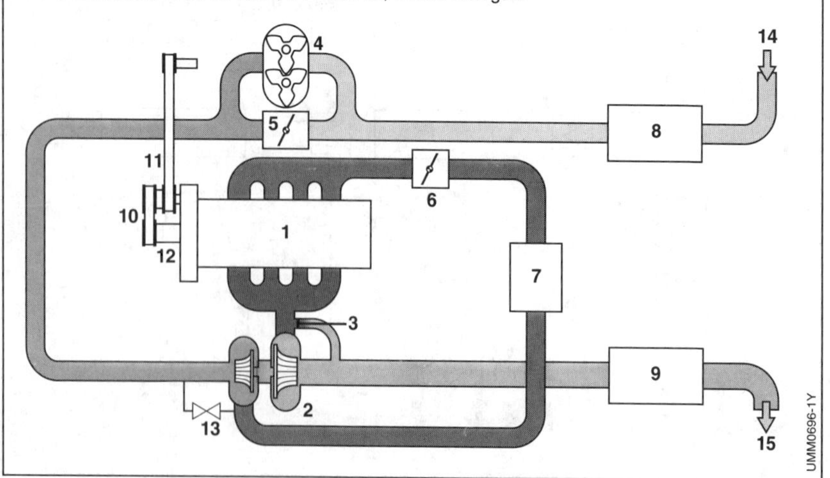

great demands on the clutch, since high speed gradients must be consistently accommodated on account of the high transmission ratios required.

Electrically driven compressor in series with an exhaust-gas turbocharger

One possible solution for obtaining the benefits of exhaust-gas turbocharging in quasi-stationary operation and in the process improving the response of a turbocharger designed for high rated output is the series connection of an exhaust-gas turbocharger with an electrically driven flow compressor ("booster", Figure 17). The advantages of this solution over the electrically assisted exhaust-gas turbocharger are that the effective program-map range is extended by the series connection of two flow compressors and the booster can be placed in suitable locations (thermo-mechanical load) in the engine compartment. The booster can be positioned in the flow direction before or after the turbocharger. A bypass allows the booster to be bypassed when it is not in operation. Finally, the electric booster is mechanically coupled to the engine via the alternator and its belt drive; direct coupling is bypassed for instance by means of a capacitor battery.

One potential mode of operation for this system, especially with regard to the electrical energy available in conventional vehicles, is operation of the electrically driven stage exclusively in transient operating phases in the engine's lower speed range.

For the system to operate effectively, the booster must reach a pressure ratio of 1.3 within roughly 0.3 s; when the acceleration of the shaft is taken into consideration, this produces a brief booster power peak – depending on other technical boundary conditions – of between 2 and 3 kW. Especially in view of the prevailing vehicle system voltage of 12 V, this entails considerable, cost-intensive additional equipment in the vehicle electrical system. The quality of the boost process and thus the effectiveness and the performance of this supercharging system are crucially dependent on the provision of electrical energy from the vehicle electrical system. This supercharging system has been intensively researched and analyzed, but has not yet been developed beyond the prototype stage.

References

[1] H. Hiereth, P. Prenninger: Aufladung der Verbrennungskraftmaschine. Springer-Verlag, 2003.
[2] K. Zinner: Aufladung von Verbrennungsmotoren: Grundlagen, Berechnungen, Ausführungen. 3rd Edition, Springer-Verlag, Berlin, 1985.
[3] L. Flückiger, S. Tafel, P. Spring: Hochaufladung mit Druckwellenlader für Ottomotoren. MTZ 12/2006, No. 67.
[4] R. v. Basshuysen (Editor): Ottomotor mit Direkteinspritzung. ATZ/MTZ Verlag, 2007.
[5] J. Portalier, JC. Blanc, F. Garnier, N. Schorn, H. Kindl: Conception of PSA Twin Turbo Boosting System for 2.2 l Euro IV Diesel Engine. 11th Supercharging Conference 2006.
[6] M. Mayer: Abgasturbolader – sinnvolle Nutzung der Abgasenergie. 5th Edition, Verlag Moderne Industrie, 2003.
[7] G. Hack, G.-I. Langkabel: Turbo- und Kompressormotoren – Entwicklung und Technik. 3rd Edition, Motorbuch Verlag, 2003.

Exhaust-gas system

Purpose and design

In compliance with legal requirements, the exhaust-gas system reduces the pollutants in the exhaust gas that are generated by an internal-combustion engine. The exhaust-gas system also helps to muffle exhaust-gas noise and to discharge the exhaust gas at a convenient point on the vehicle. Engine power should be reduced as little as possible during the process.

Components

An exhaust-gas system consists of the exhaust manifold, the components for exhaust-gas treatment and for sound absorption, and the connections between these components.

The design and the layout of these components are very different for passenger cars and commercial vehicles. In the case of passenger-car exhaust-gas systems the components are usually individually connected by pipes and routed under the vehicle chassis (Figure 1). Depending on engine displacement and the type of muffler used, a passenger-car exhaust-gas system weighs between 8 and 40 kg. The components are primarily made of high-alloy steels, as they are exposed to corrosion attacks from the inside by hot gas and condensate, and from the outside by moisture and splash and salt water.

Since the introduction of Euro IV, commercial vehicles have also been required to carry components for exhaust-gas treatment. These are, however, usually integrated into a large system and secured to the frame (see Commercial-vehicle exhaust-gas systems). The individual components are explained in the following using a passenger-car exhaust-gas system as the example

Exhaust-gas treatment

The components of the exhaust-gas treatment include:
- the catalytic converter to break down the gaseous pollutants in the exhaust gas, and
- the particulate filter (or soot filter) to filter out the fine, solid particulates in the exhaust gas (especially in diesel engines).

Catalytic converters are installed in the exhaust-gas system as close as possible to the engine so that they can quickly reach their operating temperature and therefore be effective even at low operating temperatures (e.g. in urban driving). The decisive factor here is the light-off temperature of the catalytic converter, i.e. the temperature at which the catalytic converter begins to break down pollutants (approx. 250 °C for three-way convert-

Figure 1: Exhaust-gas system of a passenger car (example with three mufflers)
1 Exhaust manifold, 2 Near-engine catalytic converter, 3 Front pipe, 4 Front muffler, 5 Intermediate pipe, 6 Center muffler, 7 Rear muffler, 8 Exhaust-gas flap, 9 Tailpipe.

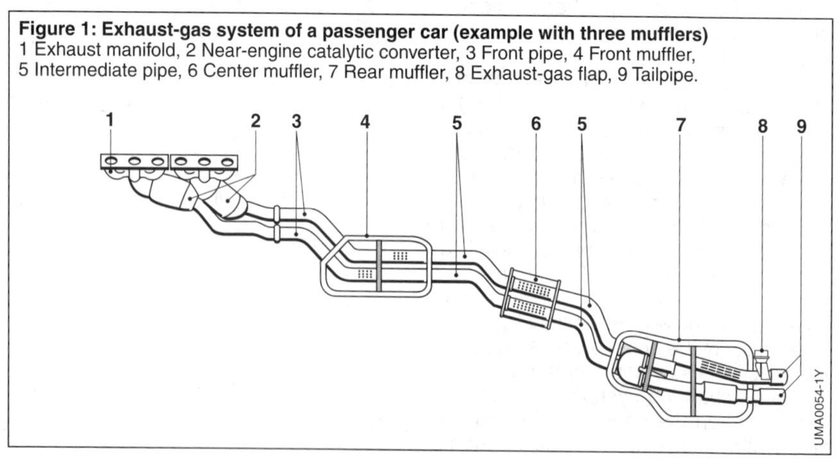

ers). Some catalytic-converter coatings, such as those for NO_x storage converters for example, are highly sensitive to temperature. These catalytic converters are therefore sooner installed in the cooler underbody area.

Diesel particulate filters are also installed in the front area of the exhaust-gas system to ensure that the soot particles they have retained are burnt off more effectively at the higher exhaust-gas temperatures.

Essentially three-way catalytic converters are used in spark-ignition engines. During engine operation with a stoichiometric air-fuel mixture ($\lambda = 1$) they convert the hydrocarbons (HC), nitrogen oxides (NO_x) and carbon monoxide (CO) present in the untreated exhaust gas by more than 99%. An additional NO_x storage catalytic converter is only installed in lean-burn-operated gasoline direct-injection engines. This reduces the nitrogen oxides (NO_x) increasingly produced in this operating state.

Diesel engines require an oxidation-type catalytic converter to oxidize the hydrocarbons (HC) and carbon monoxide (CO) and a particulate filter to hold back the solid exhaust-gas components. For low NO_x emissions a further catalytic converter for NO_x reduction is required – an NO_x storage catalytic converter or an SCR catalytic converter (Selective Catalytic Reduction).

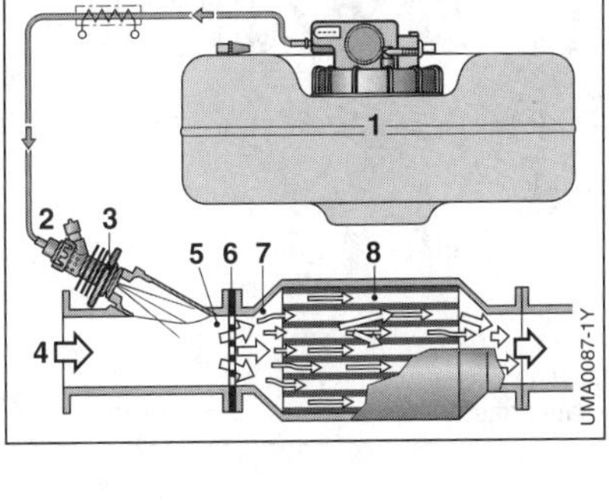

In order to ensure that the emissions are converted as much as possible in the catalytic converter or are well filtered in the particulate filter, it is essential to optimize the flow against these components in the exhaust-gas system. This is generally achieved by the shaping of the inflow funnel. Additional components such as swirl or mixing elements are required to optimize uniform distribution further. Especially SCR systems in which a reducing agent (Adblue) is injected in liquid form into the exhaust-gas system require a mixer in order to ensure vaporization or atomization of the reducing agent to NH_3. The aim is to ensure uniform and gaseous distribution of NH_3 ahead of the catalytic converter (Figure 2).

Figure 2: Atomization and vaporization of reducing agent in SCR system
1 Reservoir for reducing agent,
2 Connection, injector for reducing agent,
3 Injector for reducing agent,
4 Gas flow, 5 Exhaust pipe, 6 Mixer,
7 Inflow funnel, 8 SCR catalytic converter.

Sound absorption

The primary cause of exhaust-gas noise is the gas pulsations of the internal-combustion engine, i.e. the gas vibrations which are generated by the combustion process and by the exhaust gas being forced out through the exhaust valves during each engine power cycle. This pulsation noise is in fact damped slightly by the catalytic converters and particulate filter. But this is not sufficient to keep the noise below the pass-by limit value prescribed by law (see Vehicle acoustics). For this reason, mufflers are installed in the middle or rear section of the exhaust-gas system.

Depending on the number of cylinders and engine output, generally one, two or three mufflers are used in an exhaust-gas system. In V-engines, the left and right cylinder banks are often run separately, each being fitted with its own catalytic converters and mufflers.

The noise-emission limit for the complete vehicle is defined by legislation. The noise produced by the exhaust-gas system represents a substantial source of noise emission in a vehicle. This fact makes it necessary to devote particular attention and resources to the development of mufflers. Although the aim is to reduce noise in compliance with the legislation, they can also create the sound specific to the type of vehicle (sound design).

Exhaust manifold

An important component in the exhaust-gas system is the exhaust manifold (Figure 3). It routes the exhaust gas out of the cylinder outlet ports into the exhaust-gas system. The geometric design of the manifold (i.e. length and cross-section of the individual pipes) has an impact on the performance characteristics, the acoustic behavior of the exhaust-gas system, and the exhaust-gas temperature. In some cases, the exhaust manifold is air-gap-insulated (i.e. double-walled) in order to achieve high exhaust-gas temperatures quickly so as to enable the catalytic converter to work at an early stage during cold start. The exhaust manifold must consistently endure very high temperatures (up to 1,050 °C in gasoline engines). It is therefore produced from very high-quality material (high-alloy casting or high-alloy stainless steel).

Figure 3: Exhaust manifold with near-engine catalytic converter
1 Exhaust manifold, 2 λ sensor,
3 Metal monolith, 4 Insulating shell,
5 OBD sensor.

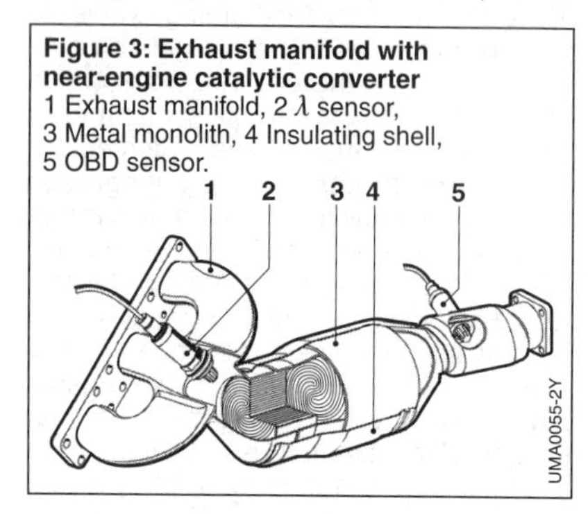

Catalytic converter

A catalytic converter is made up of an inflow funnel, an outflow funnel and a monolith (Figure 4). The monolith contains a large number of very fine, parallel channels covered with an active catalytic coating. The number of channels ranges from 400 to 1,200 cpsi (cells per square inch). The functional principle of the active catalytic-converter layer is described elsewhere (see Catalytic exhaust-gas treatment). For operational reasons, several differently coated monoliths are often used in one catalytic converter. Particular attention must be paid to the shape of the inflow funnel to ensure that the exhaust gas flows uniformly through the monolith. The external shape of the monolith depends on the installation space available in the vehicle, and may be triangular, oval or round.

The monolith can be made of metal or ceramic material.

Metal monolith

The metal monolith is made of finely corrugated, 0.05 mm thick metal foil, wound and brazed in a high-temperature process. Due to the very thin walls between the channels, the metal monolith offers an extremely low resistance to the exhaust gas. It is therefore frequently used in high-performance vehicles. The metal monolith can be welded directly to the funnels.

Figure 4: Catalytic converter with ceramic monolith
1 λ sensor,
2 Inflow funnel,
3 Ceramic monolith,
4 Mounting mat,
5 Metal housing,
6 Outflow funnel.

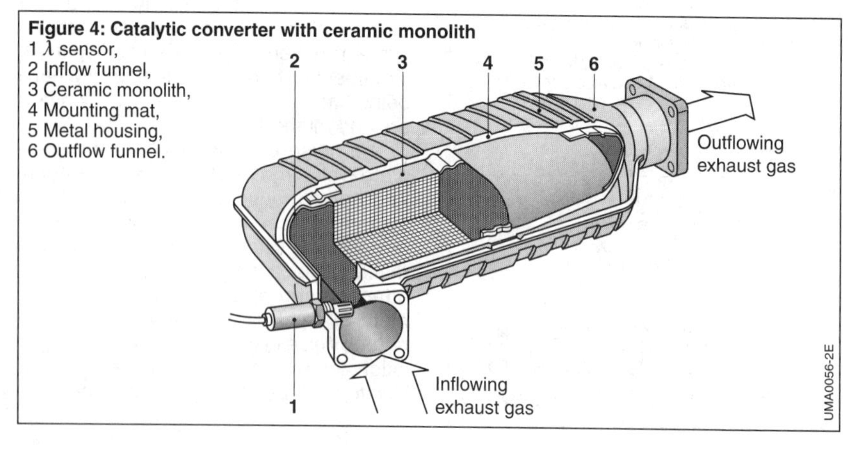

Ceramic monolith

The ceramic monolith is made of cordierite. Depending on the cell density, the wall thickness between the channels ranges from 0.05 mm (at 1,200 cpsi) to 0.16 mm (at 400 cpsi).

Ceramic monoliths have an extremely high stability to temperature and thermal shock. However, they cannot be installed directly in the metal housing and require a special mounting. This mounting is necessary in order to compensate for the difference between the thermal-expansion coefficients of steel and ceramics, and to protect the sensitive monolith against shocks. Extreme care and attention is required in the mounting and production process, particularly for thin-walled monoliths (< 0.08 mm). The monolith is mounted on a mat located between the metal housing and the ceramic monolith. The mounting mat is made up of ceramic fibers. It is extremely flexible to ensure that the pressure load exerted on the monoliths is at a minimum. It also serves as a heat insulator.

Figure 5: Ceramic particulate filter
1 Exhaust-gas inlet,
2 Ceramic plug,
3 Cell partition,
4 Exhaust-gas outlet.

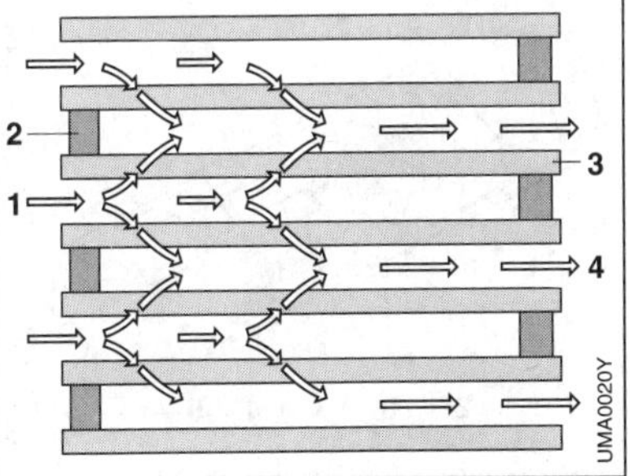

Particulate filter

As with catalytic-converter monoliths, there are metallic and ceramic particulate-filter systems. The method of installing and mounting this filter in the metal housing is the same as the process used for the catalytic converter.

Design

In the same way as the ceramic monolith for the catalytic converter, the ceramic particulate filter is made up of a large number of parallel channels. However, these channels are alternately closed (Figure 5). Consequently, the exhaust gas is forced to flow through the porous walls of the honeycomb structure. The solid particles are deposited in the pores. Depending on the porosity of the ceramic body, the filtration efficiency of these filters can attain up to 97%.

Regeneration

The soot deposits in the particulate filter induce a steady rise in flow resistance. For this reason, the particulate filter must be regenerated at certain intervals in two different processes. Details on these processes are described elsewhere (see Exhaust-gas treatment for diesel engines).

Passive process

In the passive process, the soot is burnt off by a catalytic reaction. For this purpose, an additive in the diesel fuel reduces the flammability of the soot particles to normal exhaust-gas temperatures.

Other passive regeneration options include catalytically coated particulate filters or the CRT process (see Continuous Regeneration Trap).

Active process

In the active process, external measures are implemented to heat the filter to the temperature necessary for burning off the soot. This rise in temperature can be achieved by a burner mounted upstream of the filter or by secondary injection initiated by the engine management and the use of a preliminary catalytic converter.

Mufflers

Mufflers (or silencers) are intended to smooth out exhaust-gas pulsations and make them as inaudible as possible. There are essentially two physical principles involved: reflection and absorption. Mufflers also differ according to these principles. However, they mostly comprise a combination of reflection and absorption.

In a reflection muffler (Figure 6a) the sound waves travel through the pipes and are reflected at the pipe outlet in the chambers. In an absorption muffler (Figure 6b) the sound waves spread over the perforation in the absorption material, where they are then muffled. Figure 6c shows a combination of reflection muffler and absorption muffler.

As mufflers and the exhaust-gas system pipes together form an oscillating system with its own natural resonance, the position of the mufflers is highly significant for the quality of sound-damping. The objective is to tune the exhaust-gas systems as low as possible, so that their natural frequencies do not excite bodywork resonances. To avoid structure-borne noise and to provide heat insulation for the vehicle underbody, mufflers often have double walls and an insulating layer.

Figure 6: Muffler principles
a) Reflection muffler,
b) Absorption muffler,
c) Combination of reflection and absorption mufflers.
1 Inlet pipe,
2 Perforated pipe,
3 Chamber,
4 Venturi nozzle,
5 Absorption material,
6 Outlet pipe.

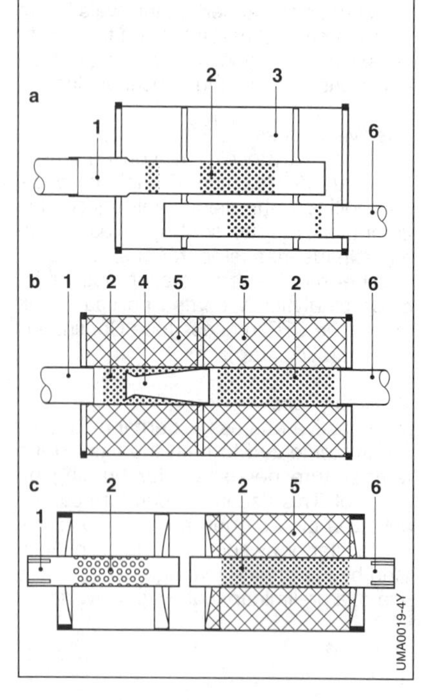

Reflection mufflers

Reflection (resonator-type) mufflers consist of chambers of varying lengths which are interconnected by pipes (Figure 6a and Figure 7). The pipes and partitions are perforated, thereby allowing exhaust gases to pass through. The differences in the cross-sections between the pipes and the chambers, the diversion of the exhaust gases, and the resonators formed by the connecting pipes and the chambers generate superpositions of the sound waves moving back and forth, causing the sound waves to be partially extinguished. In this way, effective muffling is achieved, particularly in the middle and low frequency ranges. The more such chambers are used, the more efficient is the muffler.

Figure 7: Muffler with integrated catalytic converter
1 Inlet pipe,
2 Mounting mat,
3 Ceramic monolith,
4 Reflection muffler,
5 Tailpipe.

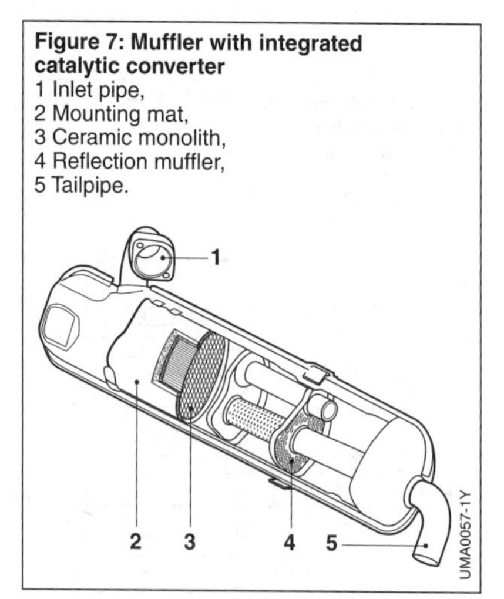

Absorption mufflers
Absorption mufflers are designed with one chamber, through which a perforated pipe passes (Figure 6b). The chamber is fillod with absorption material (mineral wool or glass fibers). The sound enters the absorption material through the perforated pipe and is converted into heat by friction.

The absorption material usually consists of long-fiber mineral wool or glass fibers with a bulk density of 100 to 150 g/*l*. The level of muffling depends on the bulk density, the sound-absorption grade of the material, and on the length and coating thickness of the chamber. Damping takes place across a very broad frequency band, but only begins at higher frequencies.

The shape of the perforations, and the fact that the pipe passes through the wool, ensures that the material is not blown out by exhaust-gas pulses. Sometimes the mineral wool is protected by a layer of stainless-steel wool around the perforated pipe.

Because the exhaust gas in absorption mufflers essentially passes through a straight pipe, the pressure loss here is very low compared with the pressure loss at the differences in the cross-sections in reflection mufflers.

Muffler design
Depending on the space available under the vehicle, mufflers are produced either as spiral-wound casing or from half-shells.

To produce the jacket for a spiral-wound muffler, one or several metal sheet blanks are shaped over a round mandril and joined together either by the longitudinal folds or by laser welding. The completely assembled and welded core is then installed in the jacket casing. It consists of internal tubes, baffles, and intermediate layers. The outer layers are then connected to the jacket in a folding or laser-welding process.

It is often not possible to effectively accommodate a spiral-wound muffler in view of the complicated space conditions in the floor assembly. In such cases, a shell-type muffler made of deep-drawn half-shells is used as it can assume virtually any required shape.

The total volume of the mufflers in a passenger-car exhaust-gas system corresponds to approximately eight to twelve times the engine displacement.

Connecting elements

Pipes are used to connect the catalytic converters and mufflers together. Arrangements where the catalytic converter and muffler are integrated in a single housing may also be used on very small engines and vehicles (Figure 7).

The pipes, the catalytic converter and muffler are connected to form an integrated system by means of plug-in connections and flanges. Many original-equipment systems are fully welded for faster mounting.

The entire exhaust-gas system is connected to the vehicle underbody via elastic mounting elements (Figure 8). The fixing points must be carefully selected, otherwise vibration may be transmitted to the bodywork and generate noise in the passenger compartment. The wrong fixing points may also create strength and therefore durability problems. In some cases, these problems are counteracted by the use of vibration absorbers. These components oscillate at the critical frequency in precisely the opposite direction of the exhaust-gas system, thereby absorbing system vibration.

The exhaust-gas system noise at the exhaust-emission point (tailpipe), as well as sound radiation from the mufflers, can also cause bodywork resonance. Depending on the intensity of the engine vibrations, decoupling elements (Figure 9) are used to isolate the exhaust-gas system from the engine block and to relieve the stress load on the exhaust-gas system. A decoupling element comprises the liner, i.e. an inner pipe with pipe segments that move in each other. The corrugated pipe, also known as the boot, is mounted on top. The liner serves to protect the boot and limit linear expansion. The corrugated pipe exercises the decoupling/isolating effect by way of its soft structure. The corrugated pipe is protected against external influences by a wire-fabric jacket.

Finally, the mounting arrangement of an exhaust-gas system is tuned so that it is rigid enough to withstand vibrations reliably on the one hand, and it exhibits sufficient flexibility and damping properties to reduce the transfer of forces to the bodywork effectively on the other.

Figure 8: Mounting element
1 Rubber mount,
2 Metal bracket,
3 Muffler shell.

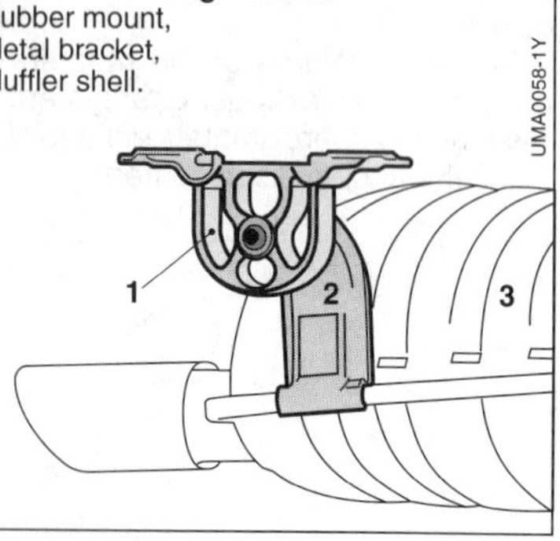

UMA0058-1Y

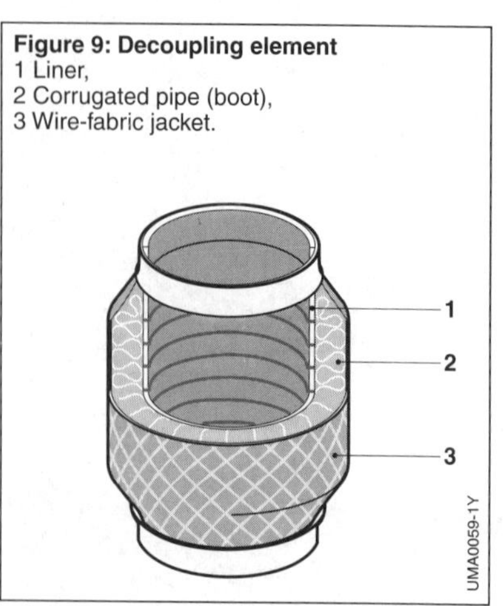

Figure 9: Decoupling element
1 Liner,
2 Corrugated pipe (boot),
3 Wire-fabric jacket.

Acoustic tuning devices

A number of different components can be used to eliminate disturbing frequencies in the noise emitted from the tailpipe. These components can be used specifically to damp individual frequency ranges very effectively.

Helmholtz resonator
The Helmholtz resonator consists of a pipe arranged along the side of the exhaust-gas train and a defined volume connected to it (Figure 10). The gas volume acts as a spring, while the gas in the pipe section acts as a mass. At its resonant frequency, this spring-mass system provides a very high degree of sound absorption but in a narrow frequency band. The resonant frequency f is dependent on the size of the volume V as well as the length L and the cross-sectional area A of the pipe:

$$f = \frac{c}{2\pi} \sqrt{\frac{A}{L \cdot V}}$$

The value c is the speed of sound.

$\lambda/4$ resonators
$\lambda/4$ resonators consist of a pipe, sealed at the end, branching off from the exhaust-gas system. The resonant frequency f of these resonators is derived from the length L of the pipe branch. It is expressed as:

$$f = \frac{c}{4L}\ .$$

Those resonators also feature a very narrowband damping range about their resonant frequency.

Exhaust-gas flaps
Exhaust-gas flaps are most commonly found in rear mufflers. Depending on the engine speed or exhaust-gas throughput, they close off a bypass pipe in the muffler or a second tailpipe (Figure 11). As a result, exhaust-gas noise can be substantially damped at lower engine speeds without the need to trade off power losses at high engine speeds.

Exhaust-gas flaps can be either self-controlling based on pressure and flow, or they can be controlled externally. An interface to the engine management system must be provided for externally controlled flaps. This makes them more complex than self-controlled flaps. However, their application range is also more flexible.

Figure 10: Helmholtz resonator
1 Helmholtz volume,
2 Gas-carrying exhaust pipe,
3 Helmholtz pipe,
4 Muffler shell.

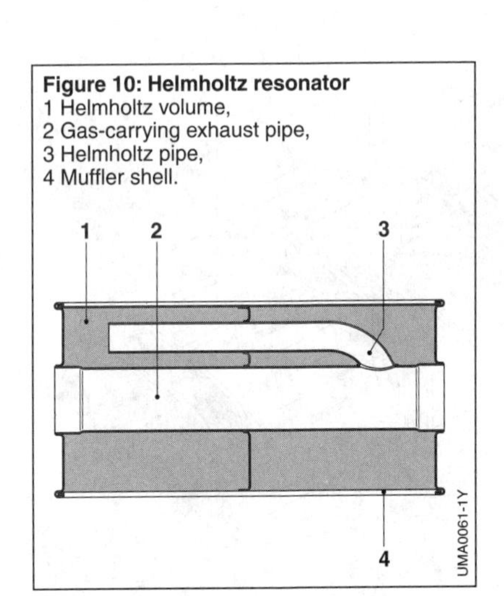

Figure 11: Exhaust-gas flap, controlled by vacuum
1 Vacuum unit,
2 Butterfly valve,
3 Tailpipes.

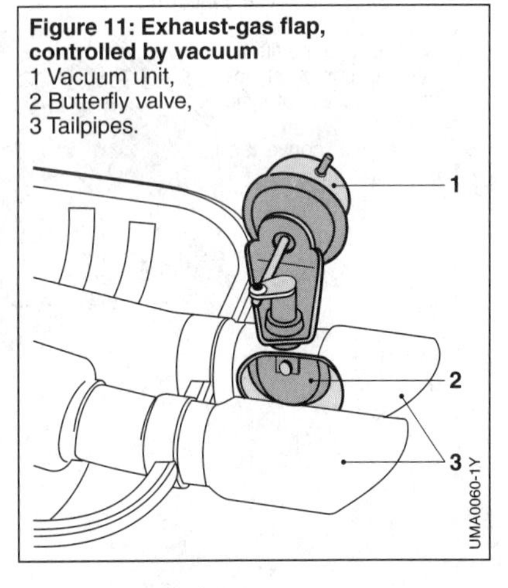

Commercial-vehicle exhaust-gas systems

In commercial-vehicle exhaust-gas systems most of the previously described components are integrated in a housing which is mounted to the vehicle frame. The number of catalytic converters and particulate filters depends on which emissions-control legislation the exhaust-gas system is designed for.

Figure 12: Commercial-vehicle exhaust-gas system for Euro IV
1 SCR catalytic converters, 2 Inlet pipe, 3 Outlet pipe, 4 Tailpipe, 5 Exhaust-gas inlet.

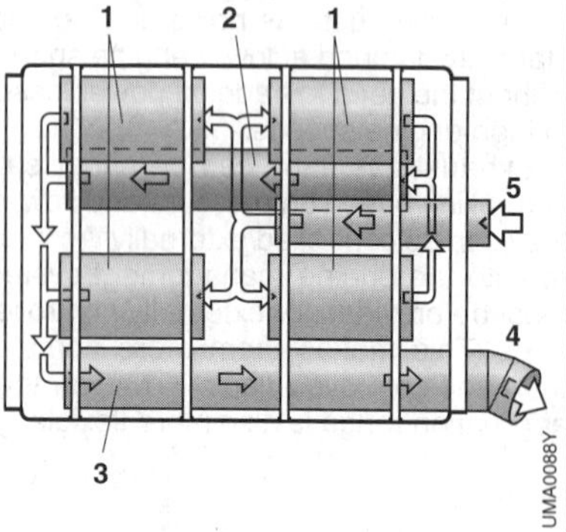

Exhaust-gas systems for Euro IV and Euro V

Generally speaking, a particulate filter is not required for Euro IV and Euro V. Only oxidation and SCR catalytic converters are used (see Exhaust-gas treatment for diesel engines). Alternatively, the diesel engine can also be tuned in such a way that the untreated NO_x emissions comply with the limit values for Euro IV and Euro V. But then a particulate filter is required in the exhaust-gas system.

Figure 12 shows a Euro IV exhaust-gas system with SCR catalytic converters. In contrast to passenger-car exhaust-gas systems, several catalytic converters are frequently arranged in parallel in order to accommodate the required catalytic surface in the space available. Bypass pipes and holes in the bases serve to route the gases and muffle the sound. Depending on the engine size, these systems attain a volume of 150 to 200 l and a weight of 150 kg.

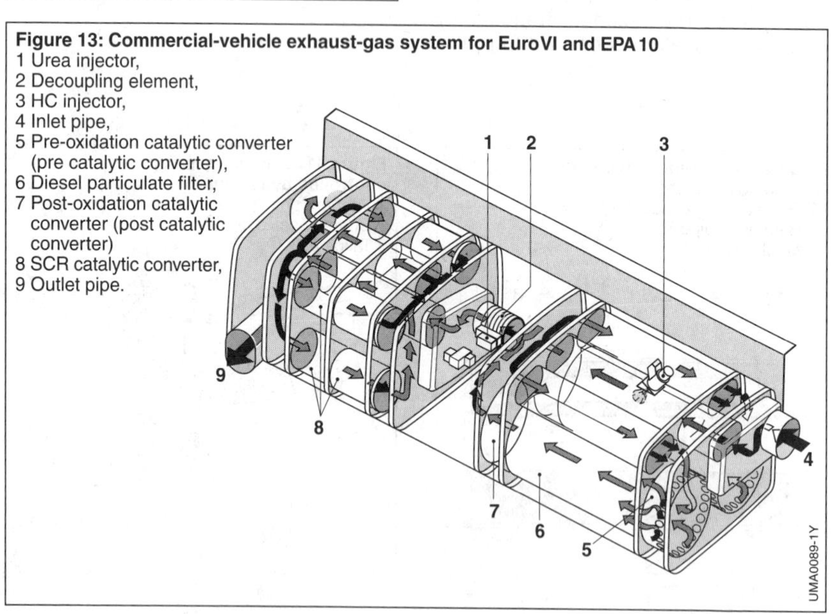

Figure 13: Commercial-vehicle exhaust-gas system for Euro VI and EPA 10
1 Urea injector,
2 Decoupling element,
3 HC injector,
4 Inlet pipe,
5 Pre-oxidation catalytic converter (pre catalytic converter),
6 Diesel particulate filter,
7 Post-oxidation catalytic converter (post catalytic converter)
8 SCR catalytic converter,
9 Outlet pipe.

Exhaust-gas systems for Euro VI and EPA 10

Exhaust-gas systems for the latest legislation (Euro IV in Europe and EPA 10 in the USA) require all the components, i.e. oxidation catalytic converters, particulate filters and SCR catalytic converters (Figure 13). They are thus even bigger in terms of volume and weight.

Two concepts are currently used. Either all the components are accommodated in a single housing or the SCR catalytic converters and the particulate filters are divided between two housings. The following components are additionally required to ensure correct functioning of exhaust-gas treatment: The SCR function requires a urea metering system (see SCR system), whose nozzle (injector) is located at a suitable location in the exhaust-gas system. Furthermore, to ensure reliable regeneration of the particulate filter, an HC metering unit is frequently needed to inject fuel (see HCI system). In both cases a position must be chosen which ensures that the liquid urea solution and the fuel are adequately vaporized and distributed. If necessary, mixers are used here to prepare the liquids better (see section on Exhaust-gas treatment).

A variety of sensors must also be accommodated in the housings. In addition to pressure sensors for monitoring the filter load, temperature sensors and NO_x sensors for monitoring NO_x conversion are required. The position of the sensors in the exhaust-gas system must, depending on the system design, be optimized to guarantee an adequate signal quality in all operating states.

References

[1] C. Hagelüken: Autoabgaskatalysatoren – Grundlagen, Herstellung, Entwicklung, Recycling, Ökologie; 3rd Edition, Expert-Verlag, 2012.

Emission-control legislation

Overview

California introduced the first emission-control legislation for gasoline engines in the mid-1960s. These regulations became progressively more stringent in the ensuing years. In the meantime, all industrialized countries have introduced emission-control laws which define limits for gasoline and diesel engines, as well as the test procedures employed to confirm compliance. In some countries, regulations governing exhaust emissions are supplemented by limits on evaporative losses from the fuel systems of vehicles with gasoline engines.

The most important legal restrictions on exhaust emissions are listed below. Figure 1 provides an overview of the regions where the various types of legislation apply:

- CARB legislation (California Air Resources Board)
- EPA legislation (Environmental Protection Agency), USA
- EU legislation (European Union) and the corresponding UN/ECE regulations (United Nations / Economic Commission for Europe)
- Japanese legislation

Classification

Countries with legal limits on motor-vehicle emissions divide vehicles into various classes:

- Passenger cars: Emission testing is conducted on a chassis dynamometer
- Light commercial vehicles: Depending on national legislation, the top limit for gross weight rating is 3.5 to 6.35 t. Emission testing is performed on a vehicle chassis dynamometer (as for passenger cars).

Figure 1: Overview of the regions where the various types of emission-control legislation apply
The map shows the regulations for passenger cars.

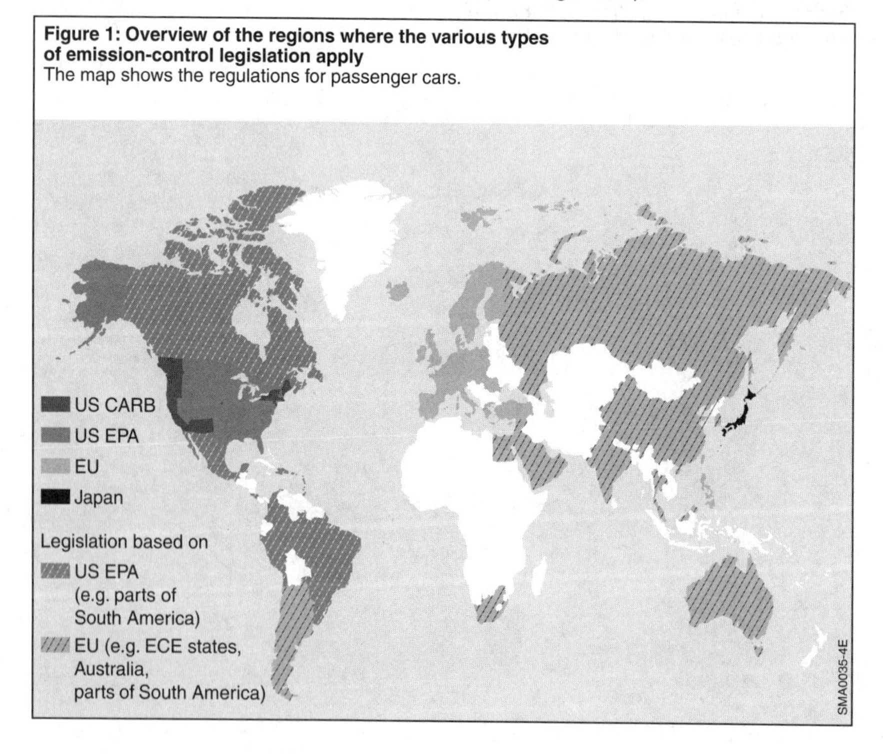

- Heavy commercial vehicles: Gross weight rating over 3.5 to 6.35 t (depending on national legislation). Emission testing is performed on an engine test bench.
- Non-road (e.g. construction, agricultural, and forestry vehicles): Emission testing is performed on an engine test bench, as for heavy commercial vehicles.

Test procedures

Japan and the European Union have followed the lead of the United States by defining test procedures for certifying compliance with emission limits. These procedures have been adopted in modified or unrevised form by other countries.

Legal requirements prescribe any of three different tests, depending on vehicle class and test objective:

- Type approval (TA) to obtain General Certification
- Random testing of vehicles from serial production conducted by the approval authorities (COP, Conformity of Production)
- In-field monitoring for checking the emission-reduction systems of privately owned production vehicles under real driving conditions

Type approval

Type approvals are a precondition for granting General Certification for an engine or vehicle type. For this purpose, test cycles must be driven under specific operating conditions and emission limits must be complied with. The test cycles and emission limits are specified individually by each nation.

Dynamic test cycles are specified for passenger cars and light commercial vehicles. The country-specific differences between the two procedures are rooted in their respective origins (see Test cycles for passenger cars and light-duty trucks):

- Test cycles designed to mirror conditions recorded in actual highway operation, e.g. Federal Test Procedure (FTP) test cycle in the USA
- Synthetically generated test cycles consisting of phases at constant cruising speed and acceleration rates, e.g. Modified New European Driving Cycle (MNEDC) in Europe

The mass of toxic emissions from each vehicle is determined by operating it in conformity with speed cycles precisely defined for the test cycle. During the test cycle, the exhaust gases are collected for subsequent analysis to determine the pollutant mass emitted during the driving cycle (see Exhaust-gas measuring techniques).

For heavy-duty trucks and non-road applications, stationary (e.g. 13-stage test) and dynamic test cycles (e.g. US HDDTC or ETC) are carried out on the engine test bench (see Test cycles for heavy commercial vehicles).

Serial testing

Vehicle manufacturers normally perform serial testing themselves during production as part of quality control. The same test procedures and the same limits are generally applied as for type approval. The approval agency may request rechecks as often as it deems necessary. EU regulations and ECE directives take account of production tolerances by carrying out random testing on a minimum of 3 to a maximum of 32 vehicles per vehicle type. The strictest requirements are applied in the USA where, particularly in California, very nearly 100 % quality monitoring is required.

In-field monitoring

Privately owned production vehicles are selected at random for emission testing under real driving conditions. The mileage and age of the vehicle must be within defined limits. The emission-test procedure is partly simplified compared with type approval.

Emission-control legislation for passenger cars and light commercial vehicles

CARB legislation

The emission limits of CARB (California Air Resources Board) for passenger cars and light commercial vehicles (light-duty trucks; LDT) are specified in the emission-control standards LEV I and LEV II (LEV, Low Emission Vehicle, i.e. vehicles with low exhaust and evaporative emissions). LEV III legislation is scheduled to be introduced between 2015 and 2025.

Since model year 2004 the LEV II standard has applied to all new vehicles up to a gross weight rating of 14,000 lbs (lb: pound; 1 lb = 0.454 kg, 14,000 lb = 6.35 t).

Originally the CARB legislation only applied in the US state of California, but has now been adopted in some other states.

Vehicle classes

Figure 2 shows an overview of the categorization of vehicles into vehicle classes.

Emission limits

The CARB legislation specifies emission limits for carbon monoxide (CO), nitrogen oxides (NO_x), non-methane organic gases (NMOG), formaldehyde (HCHO), and particulate matter (diesel: LEV I and LEV II; gasoline: only LEV II) (Figure 3). On account of the extremely low emission limits, the LEV III legislation provides for an aggregate value of NMOG and NO_x.

Exhaust emissions are measured in the FTP 75 driving schedule (Federal Test Procedure). The emission limits are correlated with the route driven during the test and are expressed in grams per mile.

Within the period 2001 through 2004 the SFTP (Supplement Federal Test Procedure) standard was introduced together with two other test cycles. There are also further limits that require compliance in addition to FTP emission limits.

Exhaust-gas categories

Automotive manufacturers are at liberty to deploy a variety of vehicle concepts within the permitted limits, providing they maintain a fleet average (see the section entitled "Fleet average"). The concepts are allocated to the following exhaust-gas categories, depending on their emission values for NMOG, CO, NO_x, and particulate emissions:

- LEV (Low-Emission Vehicle)
- ULEV (Ultra-Low-Emission Vehicle)
- SULEV (Super Ultra-Low-Emission Vehicle)

Since 2004 new vehicle registrations have been governed by the LEV II exhaust-emission standard. SULEV with significantly lower limits has been added. The LEV and ULEV categories will remain. The CO and NMOG limits from LEV I

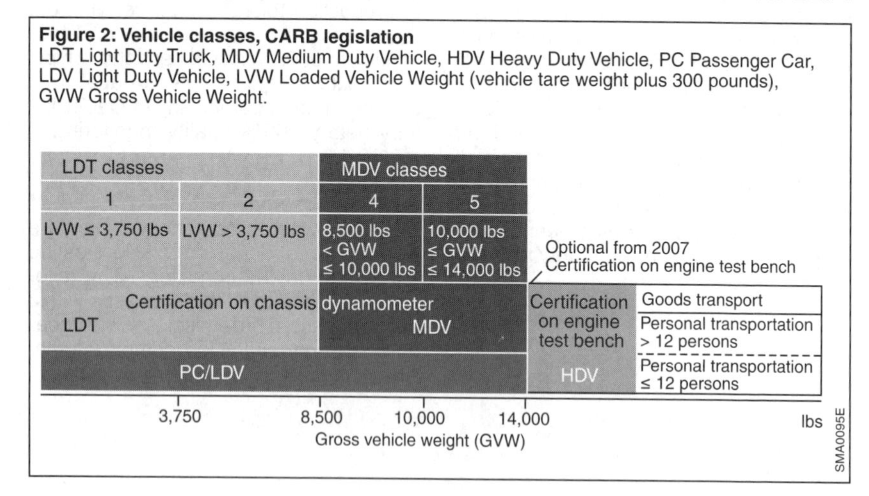

Figure 2: Vehicle classes, CARB legislation
LDT Light Duty Truck, MDV Medium Duty Vehicle, HDV Heavy Duty Vehicle, PC Passenger Car, LDV Light Duty Vehicle, LVW Loaded Vehicle Weight (vehicle tare weight plus 300 pounds), GVW Gross Vehicle Weight.

remain unchanged, but the NO_x limit is substantially lower for LEV II.

LEV III provides for a total of 6 vehicle categories (Figure 3), one of which is below SULEV. In addition to the categories of LEV I and LEV II, three categories of zero-emission or partial zero-emission vehicles are defined in the ZEV legislation:

- ZEV (Zero-Emission Vehicle, i.e. vehicles without exhaust and evaporative emissions)
- PZEV (Partial ZEV) which is basically SULEV, but with more stringent limits on evaporative emissions (zero evap) and stricter long-term performance criteria
- AT-PZEV (Advanced Technology PZEV), e.g. PZEV gas-powered vehicles or PZEV hybrid vehicles

Phase-in

Following introduction of the LEV II standard, at least 25 % of new vehicle registrations had be certified to this standard. The phase-in rule stipulated that an additional 25 % of new vehicle registrations then had to conform to the LEV II standard in each consecutive year. All new vehicles have been required to meet LEV II standard since 2007.

Durability

In order to gain approval for vehicle types, vehicle manufacturers must certify that the emissions of limited pollutants do not exceed the respective limits over 50,000 miles or 5 years ("half useful life") and 100,000 miles (LEV I) or 120,000 miles (LEV II) or 10 years ("full useful life"). Manufacturers also have the option of certifying vehicles for 150,000 miles using the same limits that apply to 120,000 miles. The manufacturer then receives a bonus when the NMOG fleet average is defined (see the section entitled "Fleet average").

The relevant figures for the ZEV, AT-PZEV and PZEV emission-limit categories are 150,000 miles or 15 years ("full useful life").

LEV III will basically extend durability to 150,000 miles.

Durability test

The vehicle manufacturer must supply two vehicle fleets from its production line for the durability test: One fleet in which each vehicle has driven 4,000 miles before the test, and one fleet for the endurance test with which the deterioration factors of individual pollutant components are measured.

Figure 3: Exhaust-gas categories and emission limits for NO_x and NMOG of CARB legislation for passenger cars and light commercial vehicles

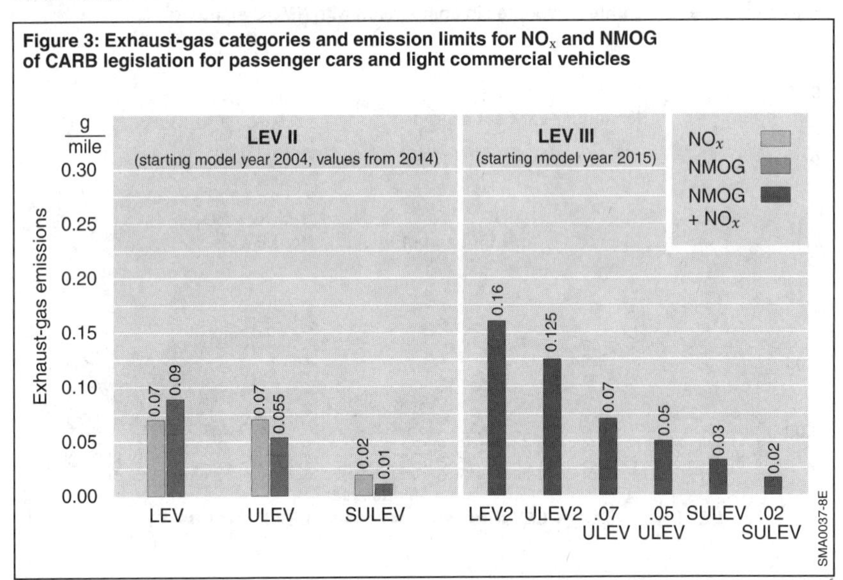

Endurance testing entails subjecting the vehicles to specific driving cycles over distances of 50,000 and 100,000 or 120,000 miles. Exhaust emissions are tested at intervals of 5,000 miles. Servicing and maintenance tasks may only be performed at the specified intervals.

Alternatively, the vehicle manufacturers may also use deterioration factors defined by legislation. To this end the vehicle manufacturer runs endurance tests and determines the percentage by which the emissions have deteriorated for each pollutant component. Then factors are defined by which the emissions from the new vehicle must be better than the legal limit so that the limits can still be maintained after the endurance run.

Fleet average (NMOG)

Each vehicle manufacturer must ensure that its vehicles do not exceed an average specific emission limit for exhaust-gas emissions (Figure 4). The NMOG emissions are used as the criteria in this regard (with LEV III NMOG + NO_x). The fleet average is calculated for LEV I and LEV II from the average of the NMOG limit for "half useful life" produced by all of the manufacturer's vehicles are sold within one year. The limits for "full useful life" apply to LEV III. The emission limits for the fleet average are different for passenger cars and light commercial vehicles.

The emission limit for the NMOG fleet average is reduced every year. To meet the lower fleet limit, manufacturers must produce progressively cleaner vehicles in the more stringent emissions categories in each consecutive year.

Fleet consumption (fuel consumption)

US lawmakers specify mandatory requirements on vehicle manufacturers with regard to the mean fuel consumption of their vehicle fleets, or the number of miles driven per gallon (Federal Law, the relevant authority is the National Highway Traffic Safety Administration NHTSA). Here it is the metric "miles per gallon of fuel (mpg)" and not as otherwise usual the "amount of fuel used per unit distance" that is used. This representation of "fuel economy" corresponds to the reciprocal value of consumption per unit of distance. The prescribed CAFE value (Corporate Average Fuel Economy) is until 2010 27.5 mpg for passenger cars. This equates to a fuel consumption of 8.55 *l* of gasoline per 100 km.

Figure 4: Passenger-car fleet averages in comparison with NMOG standard

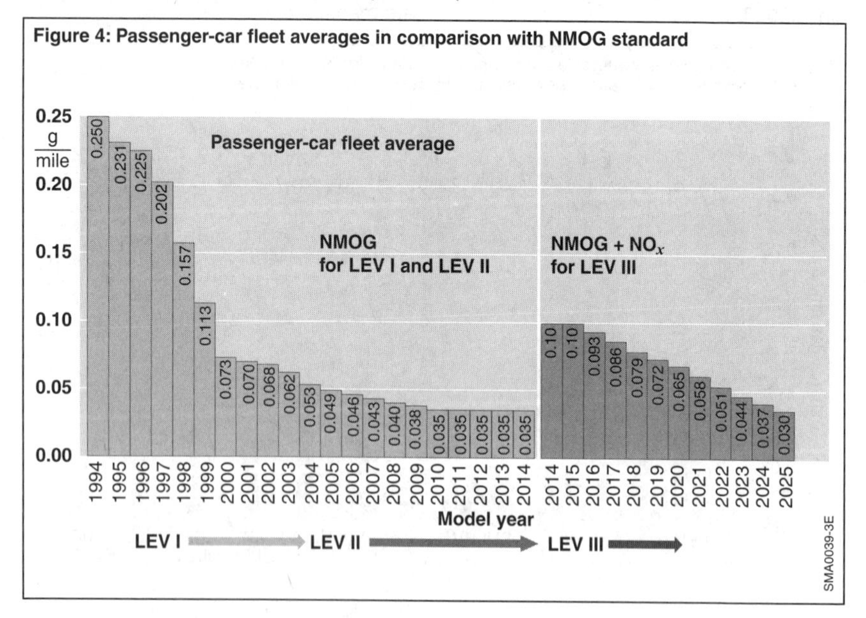

Up to 2004 the value for light commercial vehicles was 20.7 mpg or 11.36 *l* of gasoline per 100 km. Between 2005 and 2010 fuel economy was raised each year by 23.5 mpg.

For model year 2011 the CAFE system was restructured for passenger cars and light commercial vehicles (among others the definitions of passenger cars and light trucks) and ambitious target values were set: 33.3 mpg or 22.8 mpg for 2011, 40.1 mpg or 25.4 mpg for 2014, 43.4 mpg or 26.8 mpg for 2016, 46.8 mpg or 33.3 mpg for 2021, and 56.2 mpg or 40.3 mpg for 2025.

At the end of each year the average "fuel economy" for each vehicle manufacturer is calculated based on the numbers of vehicles sold. The manufacturer must remit a penalty fee of $5.50 per vehicle for each 0.1 mpg its fleet falls short of the target. Buyers will also have to pay a gas-guzzler tax on vehicles with especially high fuel consumption. Here, the limit is 22.5 mpg (10.45 *l* per 100 km). These penalties are intended to spur development of vehicles offering greater fuel economy.

The FTP 75 test cycle and the highway cycle are applied to measure CAFE fuel economy (see the section entitled "USA test cycles").

A Fuel Economy Label provides vehicle buyers with information on fuel consumption. Model year 2008 sees the introduction of "5 cycle fuel economy" (also called "5 cycle method"), which is intended to better reflect real driving conditions. This takes into account measurements in the SFTP schedules and in the FTP at −7 °C, which include among others aggressive acceleration, high final speed, and also operation with air conditioning.

Emission-free vehicles

With effect from 2005 in California, 10 % of new vehicle registrations will have to meet the requirements of the ZEV (Zero-Emission Vehicle) legislation. Genuine ZEV vehicles may not release any emissions when they are in operation. These vehicles are electric cars which are operated by a battery or a fuel cell.

A 10 % percentage may partly be covered by vehicles of the PZEV (Partial Zero-Emission Vehicles) exhaust-gas category. These vehicles are not zero-emission, but they emit very few pollutants. They are weighted using a factor of more than 0.2, depending on the emission-limit standard. The following requirements must be satisfied for the minimum factor of 0.2:

- SULEV certification for a durability of 150,000 miles or 15 years
- Warranty coverage extending over 150,000 miles or 15 years on all emission-related components
- No evaporative emissions from the fuel system (0 EVAP, zero evaporation). This is achieved by extensive encapsulation of the tank and fuel system. This results in greatly reduced evaporative emissions by the overall vehicle

Special provisions apply to hybrid vehicles with a gasoline or diesel engine and an electric motor and to gas-powered vehicles (compressed natural gas, hydrogen). These vehicles may as AT-PZEV (Advanced Technology PZEV) also contribute to the 10 % quota.

In-field monitoring

Non-routine inspection

Random emission testing is conducted on in-use vehicles using the FTP 75 test cycle and – for gasoline vehicles – an evaporative-emission test. Depending on the relevant exhaust-gas category, vehicles with mileage readings below 90,000 or 112,500 miles are tested.

Vehicle monitoring by the manufacturer
Official reporting of claims or damage to specific emissions-related components and systems has been mandatory for vehicle manufacturers since model year 1990. The reporting obligation remains in force for a maximum period of 15 years, or 150,000 miles, depending on the warranty period applying to the component or assembly.

The reporting method is split into three reporting levels, each with an increasing requirement to supply detailed information: Emissions Warranty Information Report (EWIR), Field Information Report (FIR), and Emission Information Report (EIR). The California Air Resources Board is notified of information regarding complaints, fault quotas, fault analysis, and impacts on emissions. The agency uses the Field Information Report as the basis to decide whether to enforce a recall action on the vehicle manufacturer.

EPA legislation
EPA (Environmental Protection Agency) legislation applies to all of the states where the more stringent CARB stipulations from California are not in force. CARB regulations were already adopted in 2004 by some northeastern states, such as Maine, Massachusetts, and New York. As of 2014 this has extended to a total of 16 states (including California). EPA regulations in force since 2004 conform to the Tier 2 standard.

Vehicle classes
With the transition to Tier 2, an additional vehicle class has been introduced in the form of the MDPV (Medium-Duty Passenger Vehicle) (Figure 5). Thus, all vehicles up to a permissible total weight of 10,000 lbs (4.54 t) which are designed to carry up to 12 persons, are certified on a chassis dynamometer.

Light commercial vehicles are divided into two groups: LLDT (Light-Light-Duty Truck) with a permissible total weight up to 6,000 lbs (2.72 t) and heavier HLDT (Heavy-Light-Duty Truck) with a permissible total weight up to 8,500 lbs (3.86 t).

Since 2007 certification on a chassis dynamometer has also been optionally possible for vehicles weighing up to 14,000 lbs (6.35 t).

Emission limits
EPA legislation specifies limits for the pollutants carbon monoxide (CO), nitrogen oxides (NO_x), non-methane organic gases (NMOG), formaldehyde (HCHO), and particulate matter (PM). Exhaust emissions are measured in the FTP 75 driving schedule. The emission limits are correlated with the route driven during the test and are expressed in grams per mile.

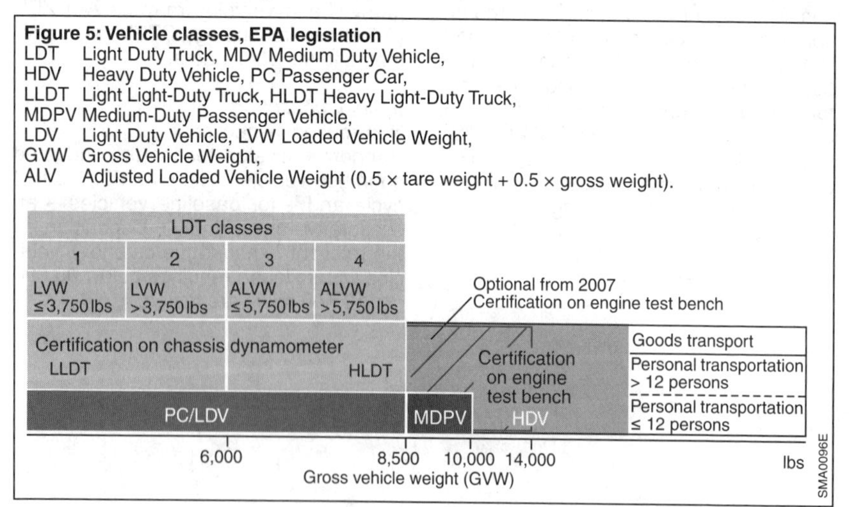

Figure 5: Vehicle classes, EPA legislation
LDT Light Duty Truck, MDV Medium Duty Vehicle,
HDV Heavy Duty Vehicle, PC Passenger Car,
LLDT Light Light-Duty Truck, HLDT Heavy Light-Duty Truck,
MDPV Medium-Duty Passenger Vehicle,
LDV Light Duty Vehicle, LVW Loaded Vehicle Weight,
GVW Gross Vehicle Weight,
ALV Adjusted Loaded Vehicle Weight (0.5 × tare weight + 0.5 × gross weight).

The SFTP (Supplemental Federal Test Procedure) standard, comprising two further test cycles, has been in force since 2000. Prevailing emission limits must be met in addition to FTP emission limits.

Since the introduction of Tier 2 standards in 2004, vehicles with diesel and gasoline engines have been subject to identical exhaust-emission limits.

Exhaust-gas categories
For Tier 2, the limits are divided into 10 (for passenger cars and LLDT) and into 11 (for HLDT and MDPV) emission standards (Bins) (Figure 6). For passenger cars and LLDT, Bin 9 and Bin 10 ceased to apply in 2007, and, for HLDT and MDPV, Bin 9 through Bin 11 ceased to apply in 2009.

The transition to Tier 2 has produced the following changes:
- Introduction of fleet averages for NO_x
- Formaldehyde (HCHO) is subject to a separate pollutant category
- Passenger cars and LLDT are treated identically to the greatest possible extent with regard to FTP limits
- "Full useful life" is increased, depending on the emission standard (Bin), to 120,000 or 150,000 miles

Phase-in
At least 25% of all new passenger-car and LLDT registrations were required to conform to Tier 2 standards which took effect in 2004. The phase-in rule stipulated that an additional 25% of vehicles would be required to conform to the Tier 2 standards in each consecutive year. All vehicles have been required to conform to Tier 2 standards since 2007. The phase-in period for HLDT and MDPV terminated in 2009.

Durability
The same criteria as for CARB apply to durability.

Fleet average
NO_x emissions are used to determine fleet averages for individual manufacturers under EPA legislation. Up to 2008 the value was 0.2 g/mile, since 2008 it has been 0.07 g/mile. The CARB provisions, however, are based on the NMOG emissions (LEV II) or NMOG + NO_x (with planned LEV III).

Figure 6: Tier 2 emission limits for EPA legislation compared with CARB emission limits for LEV II

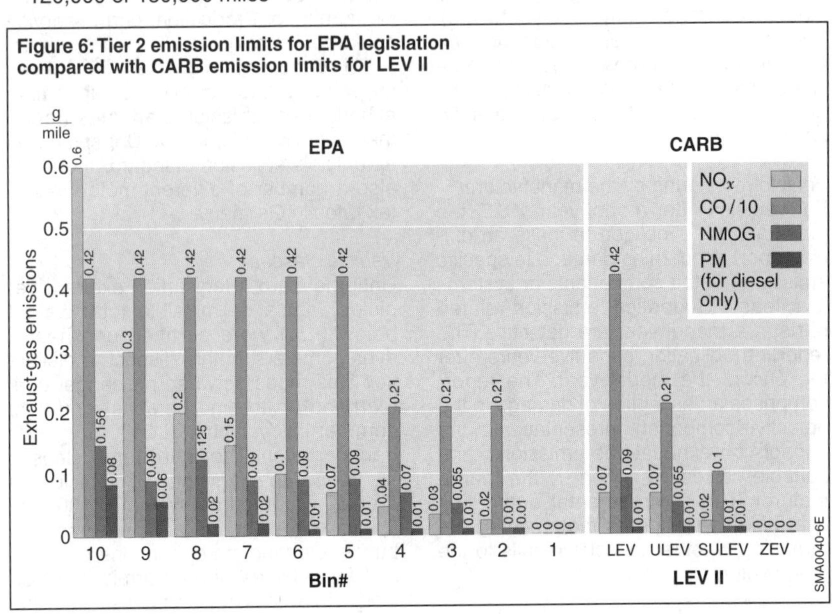

Fleet consumption (fuel consumption)
New vehicles registered in the area of responsibility are governed by the same regulations for determining fleet fuel consumption as for CARB (see CARB legislation, section "Fleet consumption").

For model years 2012 through 2016, in parallel with the CAFE legislation, EPA greenhouse-gas legislation will be introduced, with target values of 34.1 mpg (CAFE) or 250 g CO_2 equivalent per mile (EPA) for 2016. In the period from 2017 through 2025 the target values will be further increased in stages, for EPA up to 54.5 mpg (163 g CO_2 per mile) and for CAFE up to 49.7 mpg (179 g CO_2 per mile).

In-field monitoring
Non-routine inspection
EPA legislation, just like CARB legislation, provides for an exhaust-gas emission test in accordance with the FTP 75 test method to be carried out on in-use vehicles on a random-test basis. Testing is conducted on low-mileage vehicles (10,000 miles, roughly one year old) and higher-mileage vehicles (50,000 miles, however at least one vehicle per test group with 90,000 or 105,000 miles, depending on the emission standard; roughly four years old). The number of vehicles is dependent on the number sold. For vehicles with gasoline engines, at least one vehicle per test group is also tested for evaporative emissions.

Vehicle monitoring by the manufacturer
For vehicles after model year 1972, the manufacturer is obliged to make an official report concerning damage to specific emission-related components or systems if at least 25 identical emission-related parts in a model year are defective. The reporting obligation ends five years after the end of the model year. The report comprises a description of damage to the defective component, presentation of the impacts on exhaust-gas emissions, and suitable corrective action by the manufacturer. The environmental authorities use this information as the basis for determining whether to issue recalls to the manufacturer.

EU legislation
The Directives of European emission-control legislation are proposed by the EU Commission and ratified by the Council of Environment Ministers and the EU Parliament. The basis of emission-control legislation for passenger cars and light commercial vehicles is Directive 70/220/EEC [1] from 1970. For the first time it defined exhaust-emission limits, and the provisions have been updated ever since.

The emission limits for passenger cars and light commercial vehicles LCV (light-duty trucks LDT) are contained in the emission-control standards Euro 1 (from 07/1992), Euro 2 (01/1996), Euro 3 (01/2000), Euro 4 (01/2005), Euro 5 (09/2009), and Euro 6 (09/2014).

Instead of a "phase-in" over several years as in the USA, a new emission-control standard is introduced in two stages. In the first stage newly certified vehicle types must adhere to the newly defined emission limits. In the second stage – generally one year later – every newly registered vehicle (i.e. all types) must adhere to the new limits. Legislators can inspect production vehicles for compliance with emission limits (COP, Conformity of Production, and in-service conformity check).

EU directives allow tax incentives for vehicles that comply with upcoming exhaust-gas emission standards before they actually become law. Depending on a vehicle's emission standard, there are also a number of different motor-vehicle tax rates in Germany.

Vehicle classes
Until the Euro 4 legislation expired, vehicles with a permissible total weight below 3.5 t were certified on chassis dynamometers; in this respect, a distinction was made between passenger cars (carrying up to nine persons) and light commercial vehicles (LCV) for goods transportation. There are three classes for LCV, depending on the vehicle reference weight (tare weight + 100 kg). Engine certifications are performed for buses (carrying more than nine persons) and for vehicles with a permissible total weight in excess of 3.5 t. LCV engines can

optionally also be certified on an engine test bench.

When the Euro 5 and Euro 6 legislation comes into force, the vehicle reference weight (tare weight + 100 kg) will be the distinguishing criterion with regard to the certification procedure. Vehicles with a reference weight of up to 2.61 t will be certified on a chassis dynamometer. Vehicles with a reference weight in excess of 2.61 t will be certified on an engine test bench. Flexible variations will be possible.

Emission limits

The EU standards specify limits for carbon monoxide (CO), hydrocarbons (HC), nitrogen oxides (NO_x), and particulate matter (PM, for gasoline engines as from Euro 5) (Figures 8 and 9).

The limits for hydrocarbons and nitrogen oxides for the Euro 1 and Euro 2 stages are combined into an aggregate value ($HC+NO_x$). Since Euro 3 a special NO_x limit in addition to the aggregate value has applied to diesel vehicles; in the case of gasoline vehicles, the aggregate has been replaced by separate HC and NO_x limits. The Euro 5 stage was introduced in two steps as Euro 5a and Euro 5b. With Euro 5a (from September 2009) an NMHC limit (Non-methane Hydrocarbons) was also included for gasoline engines, and with Euro 5b (from September 2011) a particulate limit was included for diesel which from Euro 6b (September 2014) and Euro 6c (September 2017) also applies to vehicles with gasoline engines (a higher interim value is possible on request for Euro 6b).

The limits are defined based on mileage and indicated in grams per kilometer (g/km). Since EU 3, emissions are measured on a chassis dynamometer using the MNEDC (Modified New European Driving Cycle).

The limits are different for vehicles with diesel and gasoline engines, but with Euro 6 will be brought further into line.

The limits for LCV class 1 are the same as for cars. Passenger cars with a permissible total in excess of 2.5 t were for Euro 3 and Euro 4 treated as LCVs and were therefore likewise categorized into one of the three LCV classes. This option ceases to apply from Euro 5.

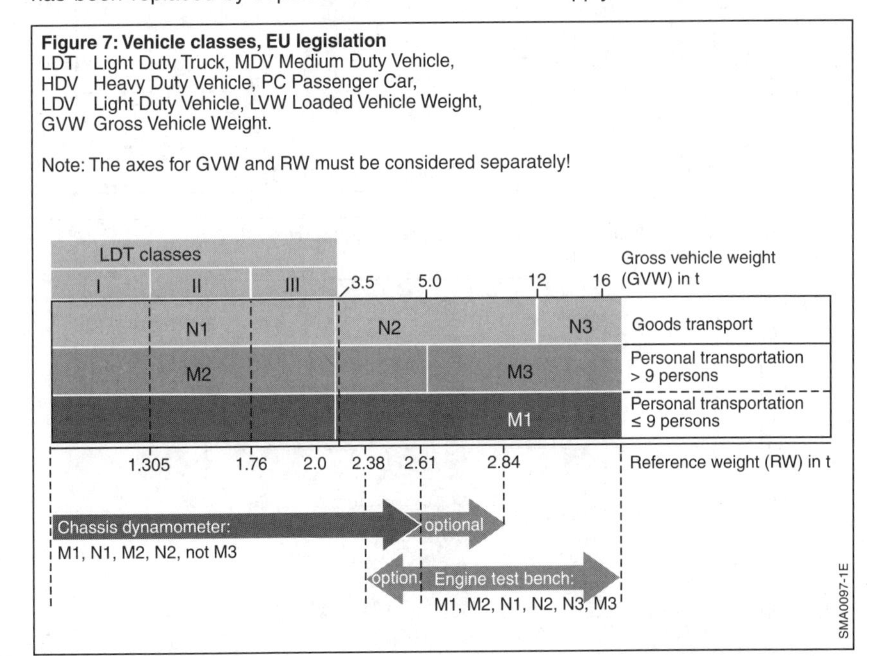

Figure 7: Vehicle classes, EU legislation
LDT Light Duty Truck, MDV Medium Duty Vehicle,
HDV Heavy Duty Vehicle, PC Passenger Car,
LDV Light Duty Vehicle, LVW Loaded Vehicle Weight,
GVW Gross Vehicle Weight.

Note: The axes for GVW and RW must be considered separately!

Type approval

While type approval testing basically corresponds to the US procedures, deviations are encountered in the following areas: measurements of the pollutants HC, CO, NO_x are supplemented by particulate and exhaust-gas opacity measurements on diesel vehicles.

Test vehicles absolve an initial run-in period of 3,000 kilometers before testing. The limit values in the Type I test are "full useful life" limit values, i.e. they must be complied with even when the durability distance is reached. In order to take into account the aging of components up to the durability distance, deterioration factors are applied to the values measured in the type approval. These are defined in the legislation for every pollutant component; manufacturers are also allowed to present documentation confirming lower factors obtained during specified endurance testing (Type V test) over 80,000 km. From Euro 5 the endurance distance has been increased from 80,000 km to 160,000 km, where further alternative test procedures are possible.

Type tests

There are four essential type tests for type approval. Type I, Type IV, Type V and Type VI tests are used for vehicles with gasoline engines; only Type I and Type V tests are used for diesel vehicles.

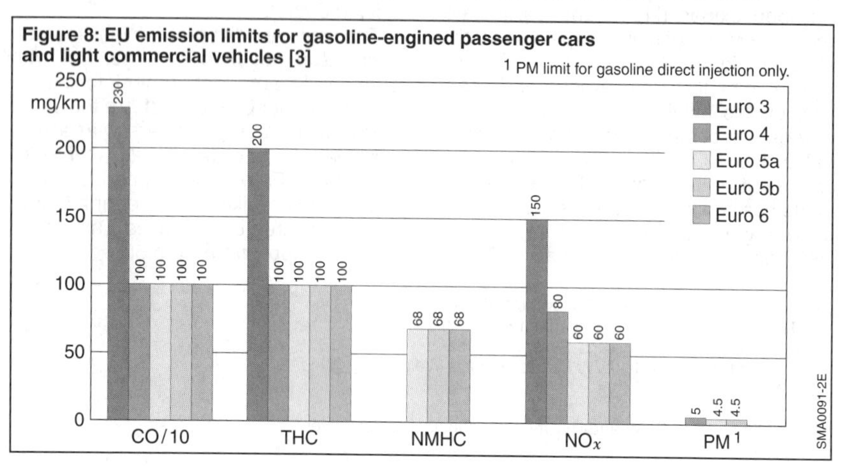

Figure 8: EU emission limits for gasoline-engined passenger cars and light commercial vehicles [3]

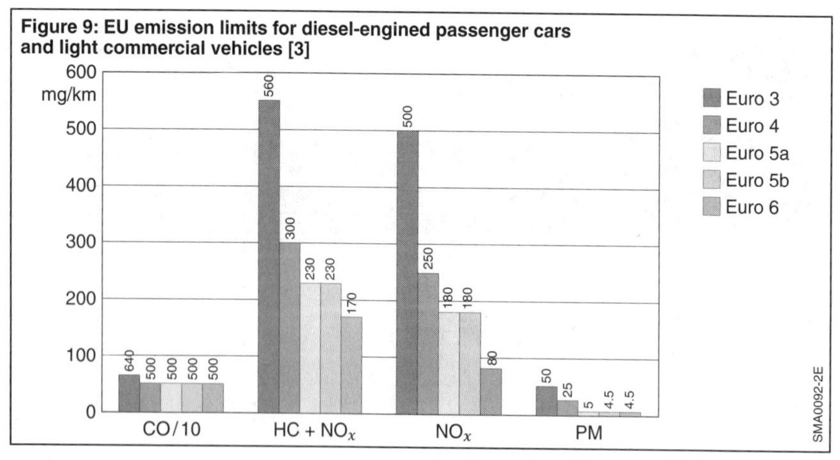

Figure 9: EU emission limits for diesel-engined passenger cars and light commercial vehicles [3]

The Type I test, the primary exhaust-gas test, evaluates exhaust emissions after cold starting in the MNEDC (Modified New European Driving Cycle). In addition, the opacity of the exhaust gas is recorded for diesel-engined vehicles.

Type IV testing measures evaporative emissions from parked vehicles. This primarily concerns fuel vapor that evaporates from the fuel-carrying system (fuel tank, pipes, etc.).

Type VI testing embraces hydrocarbon and carbon monoxide emissions immediately following cold starts at −7 °C. Only the first section (urban section) of the MNEDC is driven for this test. This test has been binding since 2002.

The Type V test assesses the long-term durability of the emissions-reducing equipment. In addition to the specified endurance test, alternative test procedures are possible from Euro 5 (e.g. test-bench aging).

CO_2 emissions

Until 2011 there is no legislation that lays down limits for CO_2 emissions, although vehicle manufacturers in Europe (ACEA, Association des Constructeurs Européens d'Automobiles), Japan (JAMA, Japan Automobile Manufacturers Association), and Korea (KAMA, Korea Automobile Manufacturers Association) have imposed voluntary limits. The target for 2008 (KAMA 2009) was a CO_2 emission of max. 140 g/km for passenger cars – equating to a fuel consumption of 5.8 l/100 km (gasoline) or 5.3 l/100 km (diesel). Because the manufacturers failed to reach their target, a fleet target value has been laid down by law for passenger cars. Within an introductory phase from 2012 through 2015 a fleet value of 130 g/km must be achieved (equating to 5.3 l/100 km gasoline or 4.9 l/100 km diesel). For light commercial vehicles a similar regulation stipulates a target value of 175 g/km for 2017. A further reduction for passenger cars to 95 g/km and for light commercial vehicles to 147 g/km has been decided on for 2020. Similarly to the US CAFE regulations, fines must be paid if the target value is not met.

In-field monitoring

EU legislation also calls for conformity-verification testing on in-use vehicles as part of the Type I test cycle. The minimum number of vehicles of a vehicle type under test is three, while the maximum number varies according to the test procedure.

Vehicles under test must meet the following criteria:

- Mileages vary from 15,000 km to 80,000 km, and vehicle age from 6 months to 5 years (from Euro 3)
- Regular service inspections were carried out as specified by the manufacturer
- The vehicle must show no indications of non-standard use (e.g. tampering, major repairs, etc.)

If emissions from an individual vehicle fail substantially to comply with the standards, the source of the high emissions must be determined. If several vehicles display excessive emissions in random testing for the same reason, the results of the random test must be classified as negative. If there are various reasons, the test schedule may be extended, providing the maximum sample size is not reached.

If the type-approval authorities detect that a vehicle type fails to meet the criteria, the vehicle manufacturer must devise suitable action to eliminate the defect. The action catalog must be applicable to all vehicles with the same defect. If necessary, a recall action must be started.

Periodic emission testing

In Germany all passenger cars and light commercial vehicles and vans are required to undergo emissions inspections three years after their initial registration, and then at subsequent intervals of two years. For gasoline-engined vehicles, the main focus is on CO levels and λ closed-loop control, while for diesel vehicles, the opacity test is the main criterion. Data from the diagnostic system are taken into consideration for vehicles with On-Board Diagnostic systems (OBD).

Comparable tests are also available in other countries; in Europe, for example, in Austria, France, Spain, and Switzerland, and in many parts of the USA in the form of "Inspection and Maintenance".

Japanese legislation

The permitted emission values are also subject to gradual stages of severity in Japan. The limit values have since September 2007 been subject to further tightening within the framework of the "New Long Term Standards" (Table 1).

Diesel vehicles have since September 2010 been subject to further tightening of the limit values in the form of the "Post New Long Term Regulation" (Table 2). In the case of vehicles with gasoline engines, the previous synthetic test cycles are being replaced in two stages (2008 and 2011) by the more realistic JC08 cycle.

Table 1: Emission limits laid down by Japanese legislators for passenger cars (New Long Term Standards)

	CO g/km	NO_x g/km	NMHC g/km	Particulates g/km
Gasoline	1.15	0.05	0.05	–
Diesel	0.63	0.14	0.024	0.013

Table 2: Emission limits laid down by Japanese legislators for passenger cars (Post New Long Term Standards, planned targets)

	CO g/km	NO_x g/km	NMHC g/km	Particulates g/km
Gasoline	–	–	–	–
Diesel	0.63	0.08	0.024	0.005

Vehicle classes

Vehicles with a permissible total weight up to 3.5 t are essentially divided into three classes (Figure 10): passenger cars (up to ten seats), LDVs (Light-Duty Vehicles) up to 1.7 t, and MDVs (Medium-Duty Vehicles) up to 3.5 t. MDVs have higher limits for CO and NO_x emissions (for vehicles with gasoline engines) than the other two vehicle classes. For diesel engines, the vehicle categories are distinguished by NO_x and particulate limits.

Emission limits

The Japanese legislation specifies limits for carbon monoxide (CO), nitrogen oxides (NO_x), non-methane hydrocarbons (NMHC), particulate matter (for diesel vehicles, from 2009 also for gasoline direct injection with leaner NO_x reducing technology), and smoke opacity (diesel vehicles only) (Tables 1 and 2).

Exhaust emissions are determined with a combination of 11-mode and 10•15-mode test cycles (see section entitled "Japanese test cycle"). Cold-start emissions are thus also taken into account. A new test cycle was introduced in 2008 (JC08). This is designed to replace initially the 11-mode and as from 2011 also the 10•15-mode, so that only the JC08 will be used as the cold- and hot-start test.

Evaporative emissions

The exhaust-gas regulations in Japan include limits on evaporative emissions in vehicles with gasoline engines, which are measured using the SHED method (see chapter entitled "Exhaust-gas measuring techniques").

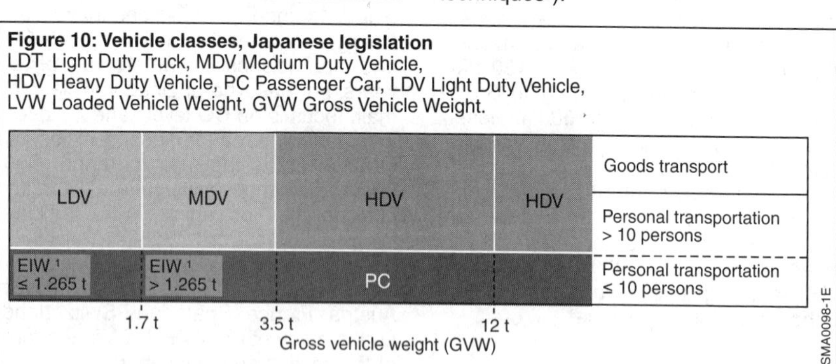

Figure 10: Vehicle classes, Japanese legislation
LDT Light Duty Truck, MDV Medium Duty Vehicle, HDV Heavy Duty Vehicle, PC Passenger Car, LDV Light Duty Vehicle, LVW Loaded Vehicle Weight, GVW Gross Vehicle Weight.

Durability
Manufacturers must demonstrate a durability of 45,000 km (New Long Term Standards) or 80,000 km (Post New Long Term Standards) for diesel vehicles. 80,000 km apply to all stages for gasoline vehicles.

Fleet consumption
In Japan targets apply to the fleet consumption of a manufacturer for 2010 and 2015, based on target values for vehicle weight classes. For tax-incentive purposes (green tax program) there are two stages rewarding fuel consumption which is improved by 15 % and 25 % respectively.

Test cycles for passenger cars and light commercial vehicles

US test cycles

FTP 75 test cycle
The FTP 75 test cycle (Federal Test Procedure, Figure 11a) consists of speed cycles that were actually recorded in commuter traffic in Los Angeles. This test is also in force in some countries of South America and in Korea besides the USA (including California).

Figure 11: US test cycles for passenger cars and light commercial vehicles

	a	b	c	d
Test cycle	FTP 75	SC03	US06	Highway
Cycle distance:	17.87 km	5.76 km	12.87 km	16.44 km
Cycle duration:	1,877 s + 600 s pause	594 s	600 s	765 s
Mean cycle speed:	34.1 km/h	34.9 km/h	77.3 km/h	77.4 km/h
Maximum cycle speed:	91.2 km/h	88.2 km/h	129.2 km/h	96.4 km/h

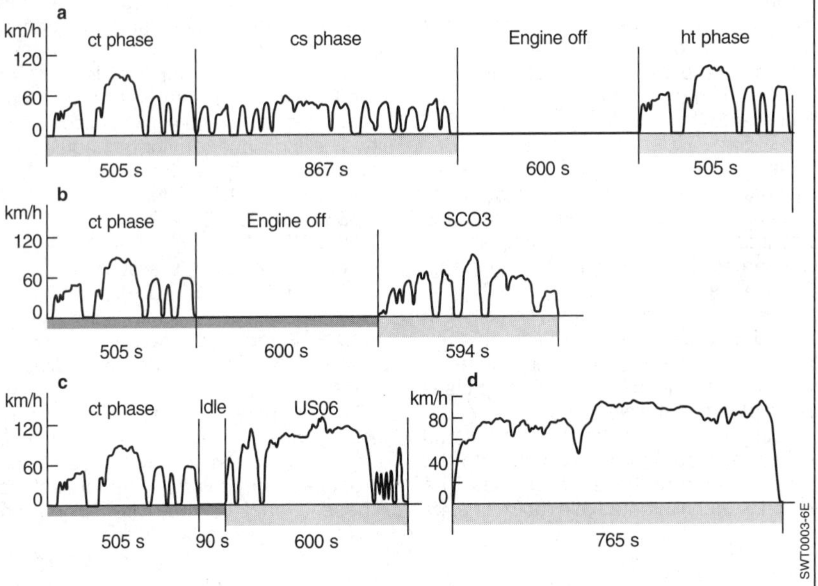

Conditioning
The vehicle is subjected to an ambient temperature of 20 to 30 °C for a period of 6 to 36 hours.

Collection of pollutants
The vehicle is started and driven on the specific speed cycle. The emitted pollutants are collected in separate bags during defined phases (see Exhaust-gas measuring techniques).

Phase ct (cold transient)
The exhaust gas is collected during the cold test phase (seconds 0 to 505 s).

Phase cs (cold stabilized)
The stabilized phase begins 506 seconds after start. The exhaust gas is collected without interrupting the driving cycle. Upon termination of phase cs, after a total of 1,372 seconds, the engine is switched off for a period of 600 seconds (hot soak).

Phase ht (hot transient)
The engine is restarted for the hot test. The speed cycle is identical to the cold transient phase (Phase ct).

Phase hs (hot stabilized)
For hybrid vehicles, a further phase hs is driven. It corresponds to the progression of phase cs. For other vehicles, it is assumed that the emission values are identical to the cs phase.

Assessment
The bag samples from the first two phases are analyzed during the pause before the hot test. This is because samples may not remain in the bags for longer than 20 minutes.

The sample exhaust gases contained in the third bag are also analyzed on completion of the driving cycle. The total result includes emissions from the three phases rated at different weightings.

The pollutant masses of phases ct and cs are aggregated and assigned to the total distance of these two phases. The result is then weighted at a factor of 0.43. The same process is applied to the aggregated pollutant masses from phases ht and cs, related to the total distance of these two phases, and weighted at a factor of 0.57. The test result for the individual pollutants (HC, CO, and NO_x) is obtained from the sum of the two previous results.

The emissions are specified as the pollutant emission per mile.

SFTP schedules
Tests according to the SFTP standard (Supplemental Federal Test Procedure) were phased in from 2001 to 2004. These are composed of two driving cycles, the SC03 cycle (Figure 11b) and the US06 cycle (Figure 11c). The extended tests are intended to examine the following additional driving conditions:
- Aggressive driving
- Radical changes in vehicle speed
- Engine start and acceleration from a standing start
- Operation with frequent minor variations in speed
- Periods with vehicle parked
- Operation with air conditioner on

For preconditioning, the SC03 and US06 cycles proceed through the ct phase from FTP 75 without exhaust-gas collection. However, other conditioning procedures are also possible.

The SC03 cycle (for vehicles with air conditioning only) is carried out at a temperature of 35 °C and 40 % relative humidity. The individual driving schedules are weighted as follows:
- Vehicle with A/C systems:
  35 % FTP 75 + 37 % SC03 +
  28 % US06.
- Vehicles without A/C systems:
  72 % FTP 75 + 28 % US06.

The SFTP and FTP 75 test cycles must be successfully completed on an individual basis.

Cold-start enrichment, which is necessary when a vehicle with a gasoline engine is started at low temperatures, produces particularly high emissions. These cannot be measured in current emissions testing, which is conducted at ambient temperatures of 20 to 30 °C. An additional exhaust-gas test is performed at –7 °C on vehicles with gasoline engines in order to limit these pollutants. However, this test only prescribes a limit for carbon monoxide;a fleet limit was introduced for NMHC emissions in 2013.

Test cycles for determining fleet consumption

Each vehicle manufacturer is required to provide data on corporate average fuel economy. Manufacturers that fail to comply with the target values are required to pay penalties.

Fuel consumption is determined from the exhaust-gas emissions produced during two test cycles – the FTP 75 test cycle (weighted at 55 %) and the highway test cycle (weighted at 45 %). An unmeasured highway test cycle (Figure 11d) is conducted once after preconditioning (vehicle allowed to stand with engine off for 12 hours at 20 to 30 °C). The exhaust emissions from a second test run are then collected. The CO_2 emissions are used to calculate fuel consumption.

Further test cycles

FTP 72 test

The FTP 72 test routine – also known as the UDDS (Urban Dynamometer Driving Schedule) – corresponds to the FTP 75 test, but does not include the ht test component (hot test). This cycle is driven during the running-loss test for vehicles with a gasoline engine.

New York City Cycle (NYCC)

This cycle is also an element in the running-loss test (for vehicles with a gasoline engine). It simulates low speeds in urban traffic with frequent stops.

Hybrid cycle

For hybrid vehicles the phase hs (progression corresponding to phase cs) is attached to the FTP 75 cycle. This driving cycle thus corresponds twice to the UDDS cycle, which is why it is called 2UDDS.

European test cycle

MNEDC

The “Modified New European Driving Cycle” (MNEDC, Figure 12) has been in force since Euro 3. Contrary to the “New European Driving Cycle” (Euro 2), in which measurement of emissions only began 40 seconds after the vehicle was started, the MNEDC also includes a cold-start phase (including engine starting).

Conditioning

The vehicle is subjected to an ambient temperature of 20 to 30 °C for a minimum period of 6 hours. Since 2002 the starting temperature has been lowered to –7 °C for the Type VI test (only for vehicles with gasoline engines).

Collection of pollutants

The exhaust gas is collected in bags in two phases: the Urban Driving Cycle (UDC) at a max. speed of 50 km/h and the Extra Urban Driving Cycle (EUDC) at a max. speed of 120 km/h.

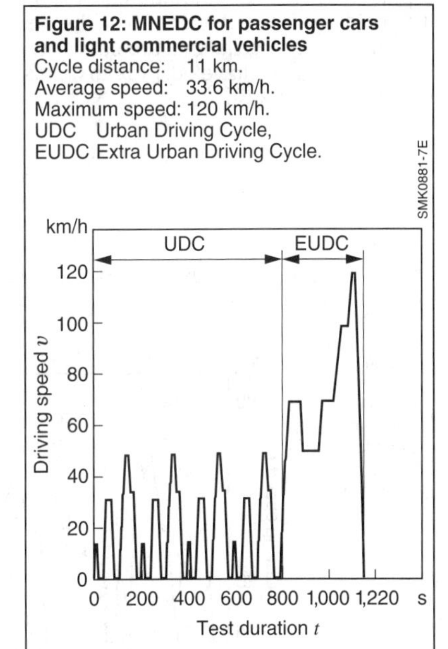

Figure 12: MNEDC for passenger cars and light commercial vehicles
Cycle distance: 11 km.
Average speed: 33.6 km/h.
Maximum speed: 120 km/h.
UDC Urban Driving Cycle,
EUDC Extra Urban Driving Cycle.

Assessment
The pollutant mass measured by analyzing the bag contents is referred to the distance covered (see Exhaust-gas measuring techniques).

WLTP
In the medium term the MNEDC is to be replaced by a new, more realistic test cycle (WLTC, World Harmonized Light Duty Test Cycle) and the associated test procedure (WLTP, World Harmonized Light Duty Test Procedure). The EU, China, India, Japan, Korea, and the USA, among others, are working towards this goal on a UNECE level (United Nations Economic Commission for Europe).

Japanese test cycle

JC08 test cycle
In 2008 a new exhaust-gas test was introduced in the form of the JC08 (Figure 13), which initially replaced the 11-mode test as the cold test. Since 2011 only the JC08 has continued to be used (both as the cold-start and the hot-start tests). The cold test is weighted at 25%, the hot test at 75%. The pollutants are converted based on distance traveled, i.e. into grams per kilometer (g/km).

Emission-control legislation for heavy commercial vehicles

US legislation

Vehicle classes
Heavy commercial vehicles are defined in EPA legislation as vehicles with a gross weight rating over 8,500 lbs or 10,000 lbs (equivalent to 3.9 t or 4.6 t), depending on vehicle type (see Figure 5).

In California, all vehicles over 14,000 lbs (6.4 t) are classified as heavy commercial vehicles (see Figure 2). To a great extent, Californian legislation is identical to parts of EPA legislation. However, there is an additional program for city buses.

Emission limits
The US standards specify limits for diesel engines for hydrocarbons (HC), carbon monoxide (CO), nitrogen oxides (NO_x), particulate matter (PM), exhaust-gas opacity, and in parts for non-methane hydrocarbons (NMHC).

The permissible limits are related to engine power output and specified in g/kW (Figure 14). The emissions are measured on the engine test bench during the dynamic test cycle with cold-starting sequence (HDDTC, Heavy-Duty Diesel Transient Cycle); the exhaust-gas opacity is measured using the Federal Smoke Test (FST).

Figure 13 : Japanese JC08 test cycle for passenger cars and light commercial vehicles

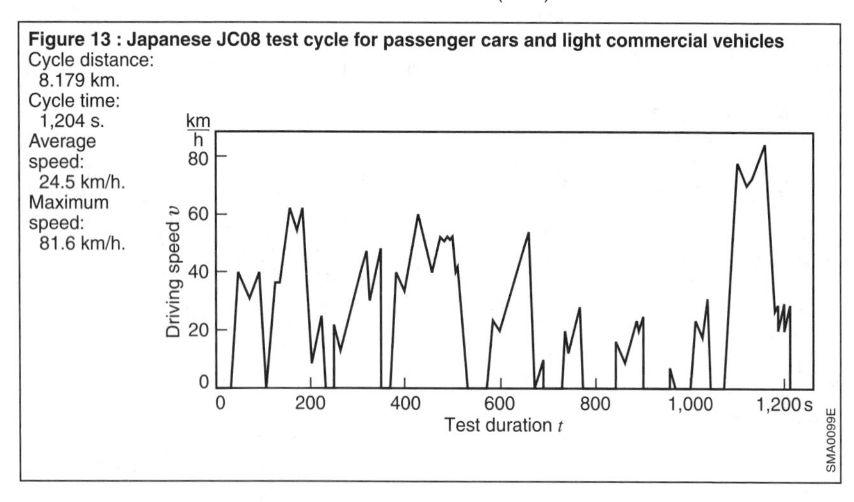

New, more stringent regulations apply to vehicles starting model year 2004, with significantly reduced NO_x emission limits. Non-methane hydrocarbons and nitrogen oxides are grouped together in one aggregate (NMHC + NO_x). CO and particulate emission limits remain at the same level as model year 1998.

Another very drastic tightening of emission restrictions came into force in model year 2007. The NO_x and particulate emissions are separately limited and are now a tenth of the previous values. This is not achievable without the use of emission-control systems (e.g. DeNOx catalytic converters or particulate filters).

A gradual phase-in will take place for NO_x and NMHC emission limits between model years 2007 and 2010.

To help compliance with severe particulate limits, the maximum permitted sulfur content in diesel fuel was reduced to 15 ppm from mid-2006.

For heavy commercial vehicles – in contrast with cars and light commercial vehicles – there are no limits specified for average fleet emissions and fleet consumption.

Consent decree

In 1998 a legal agreement was reached between EPA, CARB, and a number of engine manufacturers. It provides for sanctions against manufacturers if they make illegal modifications to engines to achieve optimized consumption in the highway cycle, resulting in higher NO_x emissions. The "Consent Decree" specifies that the applicable emission limits must also undercut the steady-state European 13-stage test in addition to the dynamic test cycle. Furthermore, emissions are not allowed to exceed the limits for model year 2004 by more than 25%, regardless of driving mode within a specified engine-speed/torque range (Not-to-Exceed Zone).

These additional tests are mandatory for all diesel commercial vehicles starting with model year 2007. However, emissions in the not-to-exceed zone may be up to 50% above the emission limits.

Durability

Compliance with emission limits must be demonstrated over a defined mileage or a specific time period in which a distinction is drawn between three classes, each with increasing durability requirements:

Figure 14: Comparison of NO_x and PM emission limits for diesel commercial vehicles: EU, USA, Japan [4]

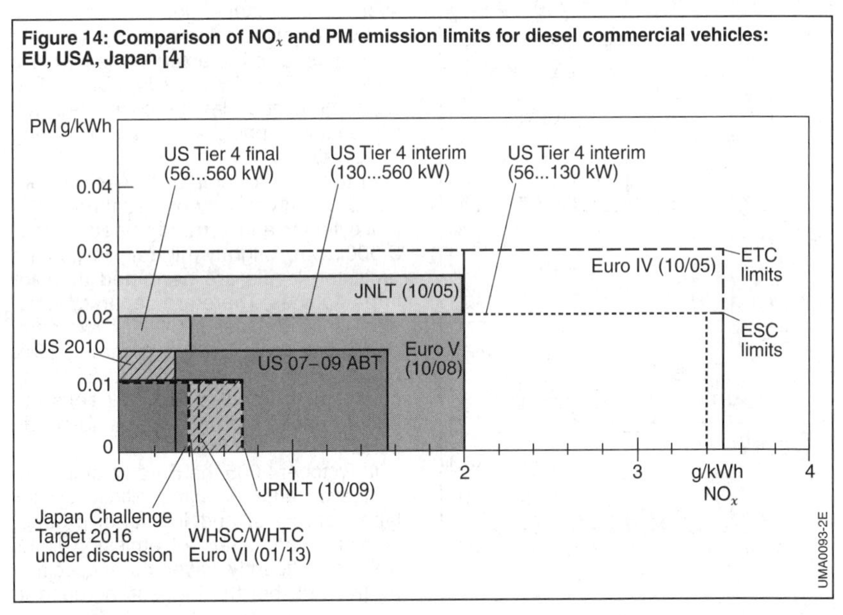

- Light commercial vehicles from 8,500 lbs (EPA) or 14,000 lbs (CARB) to 19,500 lbs: 10 years or 110,000 miles
- Medium-heavy commercial vehicles from 19,500 lbs to 33,000 lbs: 10 years or 185,000 miles
- Heavy commercial vehicles in excess of 33,000 lbs: 10 years or 290,000 miles

Fuel-consumption requirement
In the USA consumption regulations for heavy commercial vehicles are currently in the planning stage and are expected to come into force in 2017.

EU legislation

Vehicle classes
In Europe, all vehicles with a permissible gross weight of over 3.5 t, or capable of transporting more than nine persons, are classified as heavy commercial vehicles (see Figure 7). The emission-limit regulations are set down in Directive 88/77/EEC [2], which is subject to continuous updating.

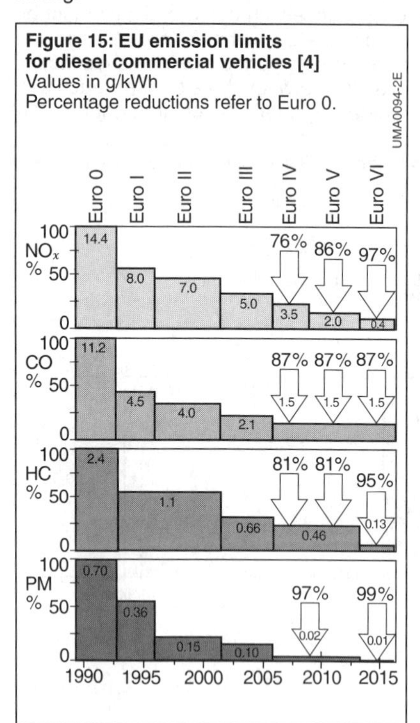

Figure 15: EU emission limits for diesel commercial vehicles [4]
Values in g/kWh
Percentage reductions refer to Euro 0.

Emission limits
As for passenger cars and light commercial vehicles, new emission-limit levels for heavy commercial vehicles are introduced in two stages. New engine designs must meet the new emission limits during type approval. One year later compliance with the new emission limits is a prerequisite for registering a new vehicle. The legislator can inspect conformity of production (COP) by taking engines out of serial production and testing them for compliance with the new emission limits.

For commercial-vehicle diesel engines, the Euro standards define emission limits for hydrocarbons (HC and NMHC), carbon monoxide (CO), nitrogen oxides (NO_x), particulates, and exhaust-gas opacity. The permissible limits are related to engine power output and specified in g/kW (Figure 15).

The Euro III emission-limit level has applied to all new engine type approvals since October 2000 and also to all production vehicles since October 2001. Emissions are measured during the 13-stage European Steady-State Cycle (ESC), and exhaust-gas opacity in the supplementary European Load Response (ELR) test. Diesel engines that are fitted with systems for exhaust-gas treatment (e.g. DeNOx catalytic converter or particulate filter) must also be tested in the dynamic European Transient Cycle (ETC). The European test cycles are conducted with the engine running at normal operating temperature.

In the case of small engines, i.e. engines with a capacity of less than 0.75 l per cylinder and a rated speed of over 3,000 rpm, slightly higher particulate emission levels are permitted than for large engines. There are separate emission limits for the ETC – for example, particulate limits are approximately 50 % higher than specified for the ESC because of the soot-emission peaks expected under dynamic operating conditions.

In October 2005 the Euro IV emission-limit level came into force initially for new type approvals, and for serial production one year later. All emission limits were significantly lower than specified by Euro III, but the biggest increase in severity applied to particulates, for which the limits were reduced by approximately 80%. The following changes also applied after introduction of Euro IV:

- The dynamic exhaust-gas emission test (ETC) is obligatory – in addition to ESC and ELR – for all diesel engines
- The continued functioning of emissions-related components must be documented for the entire service life of the vehicle

The Euro V emission-limit level was introduced in October 2008 for all new engine approvals, and one year later for all new serial-production vehicles. Only the NO_x emission limits were more severe compared to Euro IV.

In January 2013 the Euro VI emission-limit level for new engine types came into force (year later for all newly produced engines). Compared with Euro V, the nitrogen-oxide emissions are again reduced by 80% and the particulate emissions by more than 60% (referred to ETC limits for Euro V). New harmonized engine tests are to be introduced with Euro VI. Here, too, there will be a stationary test (WHSC, World Harmonized Stationary Cycle) and a dynamic test (WHTC, World Harmonized Transient Cycle). Unlike the previous Euro V regulations, from Euro VI no specific particulate limits will be given for the transient test; the limits given will be identical to those for the stationary test.

Durability
Compliance with emission limits must be demonstrated over a defined mileage or a specific time period in which a distinction is drawn between three classes, each with increasing durability requirements:

- Light commercial vehicles up to 3.5 t permissible gross weight: 6 years or 100,000 km (Euro IV and Euro V) or 160,000 km (Euro VI)
- Medium-heavy commercial vehicles under 16 t permissible gross weight: 6 years or 200,000 km (Euro IV and Euro V) or 300,000 km (Euro VI)
- Heavy commercial vehicles over 16 t permissible gross weight: 7 years or 500,000 km (Euro IV and Euro V) or 700,000 km (Euro VI)

Fuel consumption
Regulations for CO_2 emissions are currently in the planning stage; measurements are expected to be made with transportable measuring equipment on the vehicle while driving.

Very low-emission vehicles
The EU Directives allow for tax incentives for early compliance with the limits specified by a particular phase of the EU standards, and for EEVs (Enhanced Environmentally Friendly Vehicles). Voluntary emission limits are defined for the EEV category for the ESC, ETC, and ELR emission-limit tests. The NO_x and particulate limits are equivalent to those specified by Euro for the ESC. The standards for HC, NMHC, CO, and exhaust-gas opacity are stricter than required by Euro V.

Japanese legislation
Vehicle classes
In Japan, vehicles with a permissible gross weight of over 3.5 t, or capable of transporting more than ten persons are classified as heavy commercial vehicles (see Figure 10).

Emission limits
The “New Long-Term Regulation” came into force in October 2005. It stipulated emission limits for hydrocarbons (HC), nitrogen oxides (NO_x), carbon monoxide (CO), particulates, and exhaust-gas opacity. Emission levels are measured in the newly introduced transient JE05 test cycle (hot test), and exhaust-gas opacity in the Japanese smoke test.

The “Post New Long-Term Regulation” introduced in the meantime has been in force since September 2009. The particulate and NO_x limits were reduced by almost two-thirds compared with 2005 levels.

For 2016 the “Challenge Target” of a further NO_x reduction of around two-thirds is being discussed.

Durability
Compliance with the emission limits must be demonstrated over a defined mileage in which a distinction is drawn between three classes, each with increasing durability requirements:
- Commercial vehicle under 8 t permissible gross weight: 250,000 km
- Medium-heavy commercial vehicles under 12 t permissible gross weight: 450,000 km
- Heavy commercial vehicles over 12 t permissible gross weight: 650,000 km

Fuel-consumption requirement
Fuel-consumption limits are prescribed for trucks and buses with a permissible gross weight of over 3.5 t. Two driving cycles (urban and extra-urban) are used.

Fuel consumption is determined on the engine test bench. Because consumption is very much dependent on the individual vehicle engine and equipment specification (e.g. drivetrain, rolling resistance, vehicle weight), the calculation is made using a conversion program. The requirements are for commercial vehicles with a permissible gross weight of under 10 t 6.5 km/*l* (or 7.4 km/*l*), for semitrailers with a permissible gross weight of under 20 t 2.7 km/*l* (2.9 km/*l*), and for buses with a permissible gross weight of under 14 t 4.5 km/*l* (5.0 km/*l*). The values in parentheses apply to 2015.

Regional programs
In addition to the nationwide regulations for new vehicles, there are also regional requirements for the overall vehicle population aimed at reducing existing emission levels by replacing or upgrading old diesel vehicles.

The "Vehicle NO_x Law" has been in force since 2003 within, among other places, the greater urban area of Tokyo to vehicles with a permissible gross weight of over 3,500 kg. It states that 8 to 12 years after a vehicle is first registered, the NO_x and particulate limits of the relevant preceding phase of emission limits must be adhered to (e.g. the 1998 limits from 2003). The same principle also applies to particulate emissions. Here, the regulation will already apply seven years after first vehicle registration.

Test cycles for heavy commercial vehicles

For heavy commercial vehicles, all test cycles are run on the engine test bench. In the transient test cycles, the emissions are collected and evaluated according to the CVS principle. The untreated emissions are measured in the stationary test cycles. Emissions are specified in g/kWh.

Europe
All European test cycles start with a hot engine.

European Steady-State Cycle
For vehicles over 3.5 t permitted vehicle weight and more than 9 seats, the 13-stage test ESC (European Steady-State Cycle, Figure 16), has been in force in Europe since the introduction of Euro III (October 2000). The test procedure specifies measurements in 13 steady-state operating states calculated from the engine full-load curve. The emissions measured at each operating point are weighted according to certain factors. This also applies to power output. The test results are obtained for each pollutant by calculating the total of the weighted emissions divided by the total of the weighted power output.

An additional three NO_x tests may be performed in the test range when certification is performed. The NO_x emissions may not vary by a significant degree from the levels measured at the adjacent operating points. The additional measuring has the goal of preventing engine modifications performed specially for the test.

European Transient Cycle
As well as Euro III, the ETC (European Transient Cycle, Figure 17) was also introduced to determine gaseous emissions and particulate, and the ELR (European Load Response) test to measure exhaust-gas opacity. Under the Euro III standards, the ETC applies only to commercial vehicles with exhaust-gas treatment (particulate filters, DeNOx catalytic converter); starting with Euro IV (October 2005), it will be obligatory for all vehicles.

The test cycle is derived from realistic road-driving patterns and is subdivided into three sections: an urban section, an extra-urban section, and an expressway section. The length of the test is 30 minutes, and the periods of time for which engine speeds and torque levels must be maintained are specified in seconds.

World harmonized cycles
From 2013 world harmonized engine test cycles are to be applied with the introduction of the Euro VI emission-limit level. The prescribed limits must be equally met both in the WHSC (World Harmonized Stationary Cycle) and in the WHTC (World Harmonized Transient Cycle). A new feature is a WNTE zone (World Harmonized Not To Exceed Zone), as was previously customary only in the USA. The NTE test is conducted in any driving mode within a specified engine-speed/torque range, where an approx. 25 % higher emission limit is permissible. Compared with the current European tests, the harmonized engine tests are conceived in the direction of lower loads (fewer full-load operating points and during the transient test clearly more overrun phases). The associated lower exhaust-gas temperatures pose a challenge to active exhaust-gas treatment systems, which must be regularly regenerated.

Figure 16: Steady-state 13-stage test ESC (Europe)

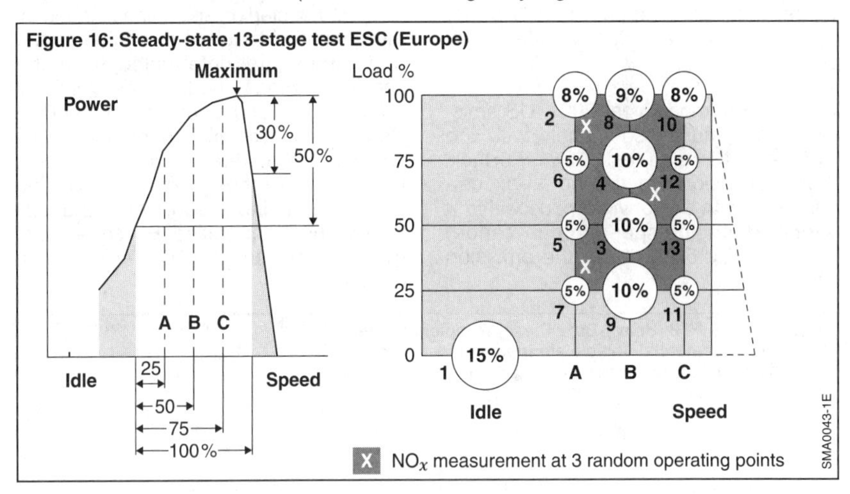

Figure 17: Transient driving cycle ETC (Europe)

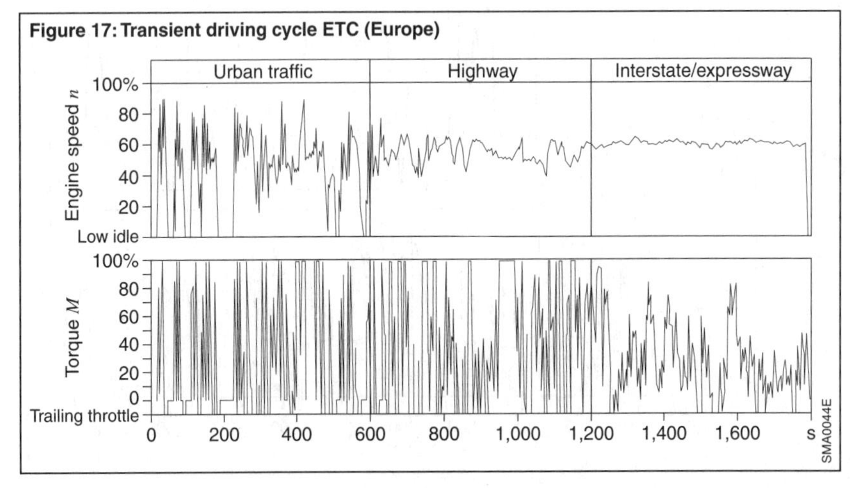

USA

Transient FTP dynamometer test cycle
Since 1987 engines for heavy commercial vehicles have been tested from a cold start on an engine test bench in a transient driving cycle (US HDDTC, Heavy Duty Diesel Transient Cycle). The test cycle is basically equivalent to operating an engine under realistic road-traffic conditions (Figure 18). It includes significantly more idle sections than the European ETC.

Federal Smoke Cycle
An additional test, the Federal Smoke Cycle, tests exhaust-gas opacity under dynamic and quasi steady-state conditions.

Starting with model year 2007, US emission limits must also comply with the European 13-stage test (ESC). Furthermore, emissions in the not-to-exceed zone (i.e. with any driving mode within a specified engine-speed/torque range) may be max. 50 % above the emission limits.

Japan

JE05 test cycle
Since October 2005, exhaust emissions have been determined in the transient JE05 test cycle. Similarly to the European transient test for passenger cars, the JE05 test cycle for commercial vehicles comprises an extra-urban part, an urban part, and an expressway part. The test lasts 1,830 seconds and is started when the engine is hot.

Unlike the European and US commercial-vehicle tests, the JE05 test specifies the driving speed instead of the engine speed and engine torque. Because the test is conducted on an engine test bench, the quantities required of engine speed and torque are determined from the specified speeds and from the individual vehicle data using a conversion program. Required quantities include, among others, vehicle weight, tire rolling resistance, transmission ratios, torque curve, and maximum engine speed. For future limit levels, the harmonized European test cycles WHSC and WHTC are under discussion.

Figure 18: US FTP (Heavy-Duty Diesel Transient Cycle, HDDTC) for heavy commercial-vehicle engines
Both the nominal engine speed n^* and the nominal torque M^* are taken from tables specified by legislation

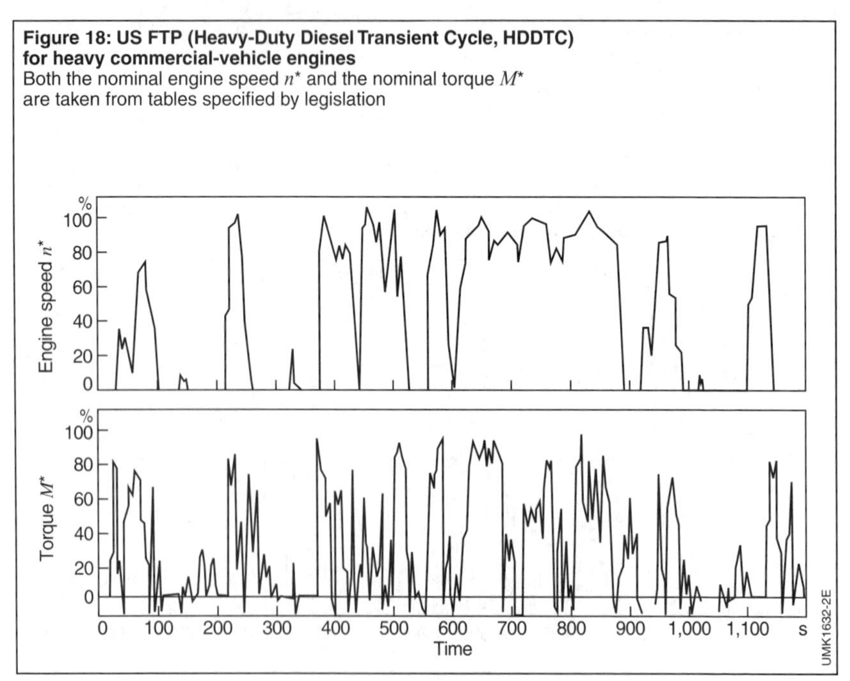

References

[1] 70/220/EEC: Council Directive (of the European Communities) of 20 March 1970 on the approximation of laws of the Member States relating to measures to be taken against air pollution by gases from positive-ignition engines of motor vehicles.

[2] 88/77/EEC: Council Directive of 3 December 1987 on the approximation of the laws of the Member States relating to measures to be taken against the emission of gaseous pollutants from diesel engines for use in vehicles.

[3] Regulation (EC) No. 715/2007 of the European Parliament and of the Council of 20 June 2007 on type approval of motor vehicles with respect to emissions from light passenger and commercial vehicles (Euro 5 and Euro 6) and on access to vehicle repair and maintenance information.

[4] Regulation (EC) No. 595/2009 of the European Parliament and of the Council of 18 June 2009 on type approval of motor vehicles and engines with respect to emissions from heavy duty vehicles (Euro VI) and on access to vehicle repair and maintenance information, and amending Regulation (EC) No. 715/2007 and Directive 2007/46/EC and repealing Directives 80/1269/EEC, 2005/55/EC and 2005/78/EC.

[5] Commission Directive 2010/26/EU of 31 March 2010 amending Directive 97/68/EC of the European Parliament and of the Council on the approximation of the laws of the Member States relating to measures against the emission of gaseous and particulate pollutants from internal combustion engines to be installed in non-road mobile machinery.

[6] 40 CFR Part 1039: Control of emissions from new and in-use nonroad compression-ignition engines.

[7] 40 CFR Part 86: Control of emissions from new and in-use Highway Vehicles and engines.

[8] K. Reif (Editor): Ottomotor-Management – Bosch Fachinformation Automobil. 4th Edition, Springer Vieweg, 2014.

Exhaust-gas measuring techniques

Emissions testing

Requirements

Exhaust-gas tests on chassis dynamometers are used for the type approval to attain General Certification, as well as to develop engine or other components. They differ from exhaust-gas tests that are conducted in the course of general and partial inspections using, for example, workshop measuring devices. In addition, exhaust-gas tests are carried out on engine test benches, for instance for the type approval of heavy commercial vehicles.

The exhaust-gas test on chassis dynamometers is performed on vehicles. The methods used are defined to simulate actual vehicle operation on the road as far as possible. Measurement on a chassis dynamometer offers the following advantages here:

- Highly reproducible results, as environmental conditions can be kept constant.
- Good comparability of tests, as a defined speed-time profile can be driven independently of traffic flow.
- Stationary setup of the measuring techniques required.

Figure 1: Exhaust-gas test on the chassis dynamometer
1 Roller with dynamometer, 2 Primary catalytic converter, 3 Main catalytic converter, 4 Filter, 5 Particulate filter, 6 Dilution tunnel, 7 Mix-T, 8 Valve, 9 Dilution-air conditioner, 10 Dilution air, 11 Exhaust-gas/air mixture, 12 Blower, 13 CVS system (Constant Volume Sampling), 14 Dilution-air sample bag, 15 Exhaust-gas sample bag (from Mix-T), 16 Exhaust-gas sample bag (from dilution tunnel), 17 Particulate counter.
① Path for exhaust-gas measurement via Mix-T (without determination of particulate emission),
② Path for exhaust-gas measurement via dilution tunnel (with determination of particulate emission).

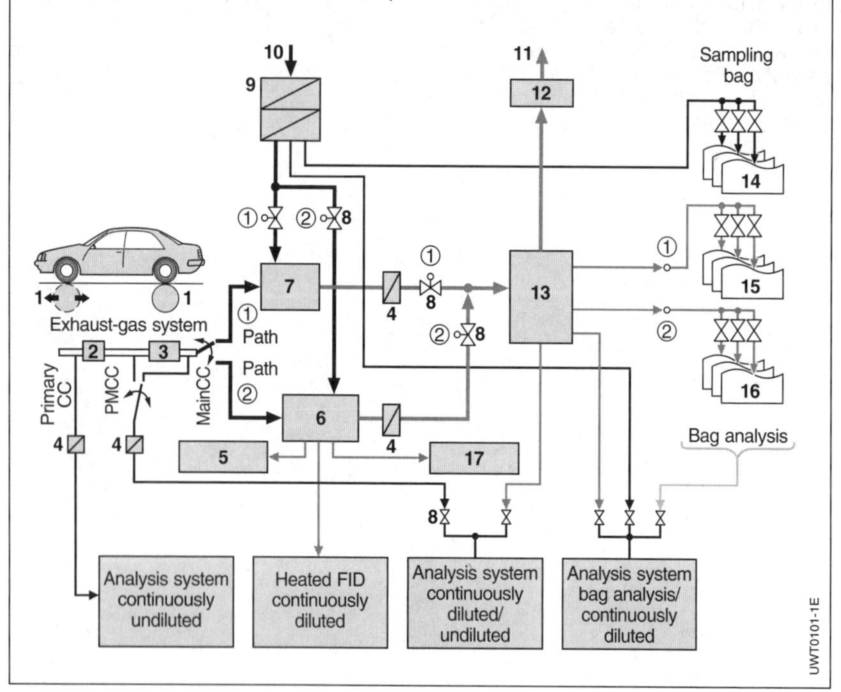

Test setup

General setup

The test vehicle is parked on a chassis dynamometer with its drive wheels on the rollers (Figure 1). This means that the forces acting on the vehicle, i.e. the vehicle's moments of inertia, rolling resistance and aerodynamic drag, must be simulated so that the trip on the test bench reproduces emissions comparable to those obtained during an on-road trip. For this purpose, asynchronous machines, direct-current machines, or even electrodynamic retarders on older test benches, generate a suitable speed-dependent load that acts on the rollers for the vehicle to overcome. More modern machines use electric flywheel simulation to reproduce this inertia. Older test benches use real flywheels of different sizes attached by rapid couplings to the rollers to simulate the vehicle mass. A blower mounted in front of the vehicle provides the necessary engine cooling.

The test-vehicle exhaust pipe is generally a gas-tight attachment to the exhaust-gas collection system – the dilution system is described below. A proportion of the exhaust gas is collected there. At the end of the driving test, the gas is analyzed for gaseous emission-limit components (hydrocarbons, nitrogen oxides and carbon monoxide) and carbon dioxide (to determine fuel consumption).

Following the introduction of the emission-control legislation, particulate emissions were initially limited in diesel-engined vehicles only. In the last few years, legislators have also begun to limit these emissions in vehicles with gasoline engines. To determine particulate emissions, a "dilution tunnel" is used with a high internal flow turbulence (Reynolds number > 40,000). Particulate filters are also used to calculate particulate emission based on load.

In addition, and for development purposes, part of the exhaust gas flow can be extracted continuously from sampling points in the vehicle's exhaust-gas system or dilution system to analyze the pollutant concentrations.

The test cycle is repeated by a driver in the vehicle. The required and current driving speeds are displayed on a driver control-station monitor. In some cases, an automated driving system replaces the driver to increase the reproducibility of test results.

Test setup for diesel-engined vehicles

To determine the exhaust emissions from diesel vehicles, it is necessary to make some changes to the test-bench setup and to the measuring techniques used. The complete sample-taking system, including the exhaust-gas measuring device for hydrocarbons, must be heated to 190 °C. This is to prevent condensation of hydrocarbons which have high boiling points, and to evaporate the hydrocarbons that have already condensed in the diesel exhaust gas.

Dilution system

Objectives of the CVS method

The most commonly used method of collecting the exhaust gases emitted from an engine is the CVS dilution procedure (Constant Volume Sampling). It was introduced for the first time in the USA in 1972 for passenger cars and light-duty commercial vehicles. In the meantime, it has been improved in several stages. The CVS method is used in other countries, such as Japan. It has also been in use in Europe since 1982. It is therefore an exhaust-gas collection method that is recognized throughout the world.

In the CVS method, the exhaust gas is only analyzed at the end of the test. Therefore the condensation of water vapor and the resulting nitrogen-oxide losses as well as secondary reactions in the collected exhaust gas must be prevented from occurring.

Principle of the CVS method

The exhaust gases emitted by the test vehicle are diluted in the Mix-T or in the dilution tunnel with ambient air at a mean ratio of 1:5...1:10, and extracted using a special system of pumps in such a way that the total volumetric flow composed of exhaust gas and dilution air is constant. The admixture of dilution air is therefore dependent on the momentary exhaust-gas volumetric flow. A representative sample is continuously extracted from the diluted exhaust-gas flow and is collected in one or more exhaust-gas sample bags. The sampling volumetric flow is constant during the bag-filling phase. Therefore, the pollutant concentration in a sample bag at the end of the filling process is identical to the mean value of the concentration in the diluted exhaust gas during the bag-filling process.

While the exhaust-gas sample bags are being filled, a sample of the dilution air is taken and collected in one or more air sample bags in order to measure the pollutant concentration in the dilution air.

Filling the sample bags generally corresponds to the phases in which the test cycles are divided (e.g. the ht phase in the FTP 75 test cycle).

The pollutant mass emitted during the test is calculated from the total volume of the diluted exhaust gas and the pollutant concentrations in the exhaust-gas and air-sample bags.

Dilution systems

There are two alternative methods to achieve constant volumetric flow in the diluted exhaust gas:

- PDP method (Positive Displacement Pump): A rotary-piston blower (Roots blower) is used.
- CFV method (Critical Flow Venturi): A venturi tube and a standard blower are used in the critical state.

Advances in the CVS method

Diluting the exhaust gas causes a reduction in pollutant concentrations as a factor of the dilution. The concentrations of some pollutants (especially hydrocarbon compounds) in the diluted exhaust gas are comparable to the concentrations in the dilution air (or lower) in certain test phases, since exhaust emissions have been significantly reduced in recent years as emission limits have become more stringent. This poses a problem from the measuring-process aspect as the difference in the two values is crucial for the exhaust emissions. A further challenge is presented by the precision of the measuring devices used to analyze the pollutants.

The following measures have been generally implemented in order to counter the problems described above:

- Lowering the dilution; this requires precautions to prevent water from condensing, e.g. by heating sections of the dilution systems, or drying or warming dilution air on vehicles with a gasoline engine.
- Reducing and stabilizing pollutant concentrations in the dilution air, e.g. by using activated charcoal filters.
- Optimizing the measuring devices (including dilution systems), e.g. by appropriately selecting or preconditioning the materials used and system setups or by using modified electronic components.
- Optimizing processes, e.g. by applying special purge procedures.

Bag Mini Diluter

As an alternative to the improvements in CVS technology described above, a new type of dilution system was developed in the USA: the Bag Mini Diluter (BMD). Here, part of the exhaust gas flow is diluted at a constant ratio with dried, heated, pollutant-free zero gas (e.g. cleaned air). During the test, part of this diluted exhaust-gas flow that is proportional to the exhaust-gas volumetric flow is filled in exhaust-gas sample bags and analyzed at the end of the driving test.

In this procedure, dilution is performed with a pollutant-free zero gas free of pollutants and not with air containing pollutants. This has the purpose of avoiding the air-sample bag analysis and the subsequent differential formation of exhaust-gas and air-sample bag concentrations. However, a more complex procedure is required than that for the CVS method, e.g. one requirement is to determine the (undiluted) exhaust-gas volumetric flow and the proportional sample-bag filling.

Exhaust-gas measuring devices

In vehicles with gasoline engines, the emissions of limited gaseous pollutants are calculated from the concentrations in the exhaust-gas and air sample bags (CVS method). Emission-control legislation defines globally standard test procedures for this purpose (Table 1).

Essentially, the same devices are used to measure the concentrations of gaseous pollutants in the exhaust gas of gasoline-engined vehicles as for diesel-engined vehicles. However, there is a difference when it comes to measuring hydrocarbon emissions (HC): It is not performed in the exhaust-gas sample bag but by continuous analysis of part of the diluted exhaust-gas flow. The concentration measured throughout the driving test is then added. The reason for this is that the hydrocarbons (which have a high boiling point) condense in the (non-heated) exhaust-gas sample bag.

Table 1. Discontinuous test procedures

Components	Procedure
CO, CO_2	Non-Dispersive Infrared Analyzer (NDIR)
Nitrogen oxides (NO_x)	Chemiluminescence Detector (CLD).
Total hydrocarbon (THC)	Flame Ionization Detector (FID)
CH_4	Combined design of gas chromatographic procedure and flame ionization detector (GC FID)
CH_3OH, CH_2O	Combined design of impinger or cartridge process and chromatographic analysis techniques; mandatory in the USA when certain fuels are used
Particulates	1.) Gravimetric process: weighing of particulate filters before and after the test drive 2.) Particulate counting

For development purposes, many test benches also include the continuous measurement of pollutant concentrations in the vehicle exhaust-gas system or the dilution system. The reason is to capture data for the components under control, as well as for other components not subject to legislation. Other test procedures than those listed in Table 1 are required for this, e.g.:
- Paramagnetic method (to measure O_2 concentration).
- Cutter FID: a combination of flame-ionization detector and absorber for non-methane hydrocarbons (to measure the CH_4 concentration).
- Mass spectroscopy (multi-component analyzer).
- FTIR spectroscopy (Fourier Transform Infrared, multi-component analyzer).
- IR laser spectroscopy (multi-component analyzer).

A description of the main measuring devices is given below.

NDIR analyzer

The NDIR (Non-Dispersive Infrared) analyzer utilizes the property of certain gases to absorb infrared radiation within a narrow wavelength range. Absorbed radiation is converted to vibration or rotation energy by the absorbing molecules. In turn, this energy can be measured as heat. The phenomenon described occurs in molecules that are formed from atoms of at least two different elements, e.g. CO, CO_2, C_6H_{14} or SO_2.

There are a number of variants of NDIR analyzers; the main component parts are a source of infrared light (Figure 2), an absorption cell (cuvette) through which the test gas is routed, a reference cell generally positioned in parallel (filled with inert gas, e.g. N_2), a rotating chopper, and a detector. The detector comprises two chambers connected by a membrane and containing samples of the gas components under analysis. Radiation from the reference cell is absorbed in one chamber and radiation from the cuvette in the other. Radiation from the cuvette may have already been reduced by absorption in the test gas. The difference in radiant energy causes a flow movement that is measured by a flow sensor or a pressure sensor. The rotating chopper interrupts the infrared radiation in cycles, causing the flow movement to change direction and therefore a modulation of the sensor signal.

NDIR analyzers possess strong cross sensitivity to water vapor in the test gas since H_2O molecules absorb a wide range of infrared radiation wavelengths. This is the reason why NDIR analyzers are positioned downstream of a test-gas treatment system (e.g. a gas cooler) to dry the exhaust gas when they are used to make measurements on undiluted exhaust gas.

Chemiluminescence Detector (CLD)

In a reaction chamber, the test gas is mixed with ozone that is produced from oxygen in a high-voltage discharge (Figure 3). The nitrogen monoxide content in the test gas oxidizes to nitrogen dioxide in this environment; some of the molecules produced are in a state of excitation. When these molecules return to their basic state, energy is released in the form of light (chemiluminescence). A detector (e.g. photomultiplier) measures the emitted luminous energy; under specific conditions, it is proportional to the nitrogen-monoxide concentration (NO) in the test gas.

It is a requirement to measure the NO and NO_2 molecules as the legislation regularizes the emission of the total nitrogen oxides. However, since the test principle of the chemiluminescence detector is limited to measuring the NO concentration, the test gas is channeled through a converter that reduces the nitrogen dioxide to nitrogen monoxide.

Flame Ionization Detector (FID)

The test gas is burned in a hydrogen flame (Figure 4), where carbon radicals are formed and some of these radicals are temporarily ionized. The radicals are discharged at a collector electrode. The current produced is measured and is proportional to the number of carbon atoms in the test gas.

GC FID and Cutter FID

There are two generally common methods to measure the methane concentration (CH_4) in the test gas. Each method consists of the combination of a CH_4-separating element and a flame ionization detector. In these methods, either a gas-chromatography column (GC FID) or a heated catalytic converter oxidizing the non-CH_4 hydrocarbons (cutter FID) are used to separate the methane.

Unlike the cutter FID, the GC FID can only determine the CH_4 concentration dis-

Figure 2: Measuring chamber for the NDIR method
1 Gas outlet, 2 Absorption cell,
3 Test-gas inlet, 4 Optical filter,
5 Infrared light source,
6 Infrared radiation, 7 Reference cell,
8 Rotating chopper, 9 Detector.

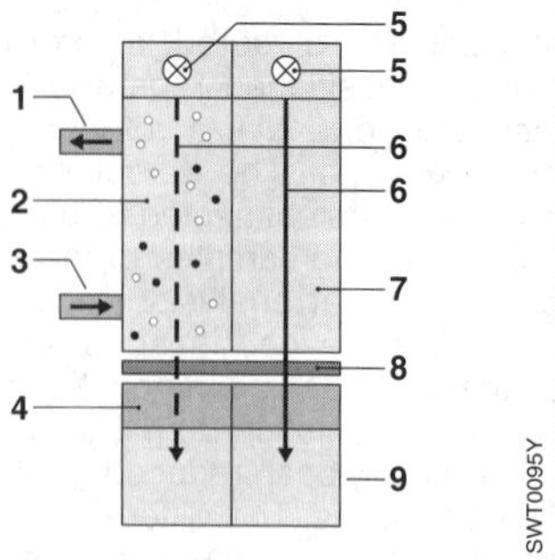

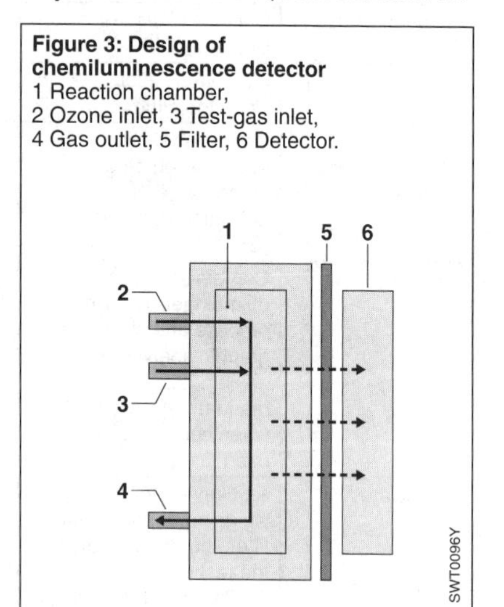

Figure 3: Design of chemiluminescence detector
1 Reaction chamber,
2 Ozone inlet, 3 Test-gas inlet,
4 Gas outlet, 5 Filter, 6 Detector.

continuously (typical interval between two measurements: 30 to 45 seconds).

Paramagnetic detector (PMD)
There are different constructions of paramagnetic detectors (dependent on the manufacturer). The constructions are based on the phenomenon that forces with paramagnetic properties (such as oxygen) act on molecules in inhomogeneous magnetic fields. These forces cause the molecules to move. The movement is sensed by a special detector and is proportional to the concentration of molecules in the test gas.

Measuring particulate emission
In addition to gaseous pollutants, solid particulates are also measured, as they are also pollutants subject to legislation. Currently, the gravimetric process is the process specified by law to measure particulate emissions.

Gravimetric process (particulate-filter process)
Part of the diluted exhaust gas is sampled from the dilution tunnel (CVS method) during the driving test and then channeled through particulate filters. The particulate loading is calculated from the weight of the particulate filters before and after the test. The particulate emission during the driving test is then calculated from the load, the total volume of the diluted exhaust gas, and the partial volume channeled through the particulate filters.

The gravimetric process has the following disadvantages:

- Relatively high detection limit, only reducible to a limited extent by using complex instrumentation (e.g. to optimize the tunnel geometry).
- It is not possible to measure particulate emissions continuously.
- The process is complex as particulate filters have to be conditioned in order to minimize environmental influences.
- Selection with regard to the chemical composition of particulates or particulate size is not possible.

Particulate counting
Because of the disadvantages mentioned and the progressive reduction of the limit values, the number of emitted particulates will in future be limited as well as the particulate emission (particulate mass per distance traveled).

The "Condensation Particulate Counter" (CPC) has been earmarked as the measuring device for calculating the number of particulates in compliance with legislation (particulate counting). In this counter, a small partial flow of the diluted exhaust gas (aerosol) is mixed with saturated butanol vapor. The process of the butanol condensing on the solid particulates causes the particulates to increase dramatically in size so that it is possible to calculate the number of particulates in the aerosol with the aid of scattered-light measurement.

The number of particulates in the diluted exhaust gas is continuously calculated; integration of the measured values produces the number of particulates over the driving test.

Figure 4: Design of flame ionization detector
1 Gas outlet, 2 Collector electrode,
3 Amplifier outlet, 4 Combustion air,
5 Test-gas inlet, 6 Combustion gas (H_2/He),
7 Burner.

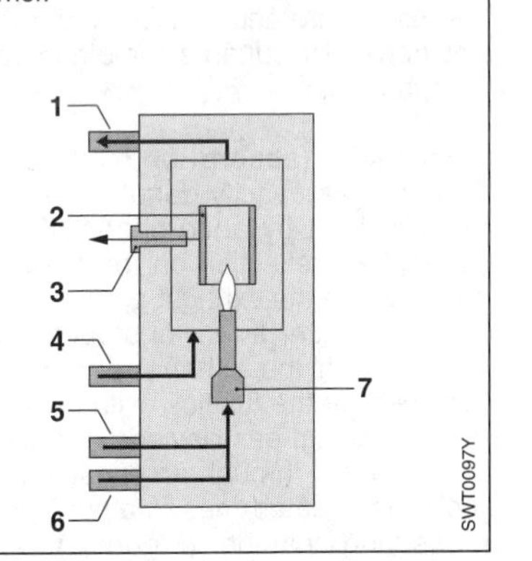

Determination of particulate size distribution
There is increasing interest in acquiring knowledge of the size distribution of particulates in the exhaust gas of a vehicle. Examples of devices that supply this information are the Scanning Mobility Particle Sizer (SMPS), the Electrical Low Pressure Impactor (ELPI), and the Differential Mobility Spectrometer (DMS).

Testing commercial vehicles
The transient test method has been prescribed in the USA since model year 1987 for testing emissions from diesel engines in heavy-duty commercial vehicles with a maximum permissible mass of over 8,500 lbs (EPA) or 14,000 lbs (CARB). It is performed on dynamic engine test benches. In Europe this test method was introduced with Euro III for engines with active exhaust-gas treatment (particulate filter or Selective Catalytic Reduction, SCR) and with Euro IV for all engines in heavy-duty commercial vehicles with a maximum permissible weight of over 3.5 t. Since the introduction of Euro V testing on the engine test bench is no longer dependent on the maximum permissible weight, but instead on the reference weight (empty weight plus 100 kg).

The CVS procedure is also used in the transient test method. However, the size of the engines demands a test setup with a substantially higher throughput in order to keep to the same dilution ratios as for cars and light-duty commercial vehicles. Double dilution (through a secondary tunnel) approved by legislators helps to limit the increased complexity of instrumentation. The diluted exhaust-gas volumetric flow can be adjusted with a Roots blower or critical-flow venturi nozzles.

Another possibility would be to calculate the particulate emissions with a partial-flow dilution system, provided that the remaining pollutants in the undiluted exhaust gas are measured.

With Euro VI (2013) limit values for the number of particulates are binding on all new model types. The value is 8 to 10^{11} particles in stationary and 6 to 10^{11} particles per kWh in transient operation.

Diesel smoke-emission test

Processes
Separate legislation for testing the smoke emissions of diesel-engined vehicles came into force long before the introduction of legislation for testing gaseous pollutants. All existing smoke tests are coupled closely with the measuring devices used. One measure of smoke emission (soot, particulates) is the smoke number.

Two methods are essentially customary for measuring this value. In the absorption method (opacity measurement), the opacity of the exhaust gas is indicated by the degree to which it blocks the passage of a beam of light shining through it (Figure 5). In the filter method (measurement of reflected light), a specified quantity of exhaust gas is routed through a filter element. The degree of filter discoloration provides an indication of the amount of soot contained in the exhaust gas (smoke tester, Figure 6).

Diesel smoke-emission measurement is relevant only if the engine is under load, since it is only when the engine is operated under load that emission of significant levels of particulates occur. Two different test procedures are also commonly used here. Firstly, measurements under WOT, e.g. on a chassis dynamometer, or on a defined test circuit. Secondly, measurements under unrestricted acceleration using a defined throttle burst under the load of the engine flywheel (Figure 7).

As the results of testing for diesel smoke emissions vary according to both test procedure and type of load, they are not generally suitable for direct mutual comparisons.

Opacimeter (absorption method)
Exhaust-gas opacity denotes the attenuation of light by absorption, diffraction, scatter and reflection on the particulates contained in the exhaust gas.

A transmitter and a photodetector are mounted on the exhaust pipe for measurement of the full flow with an opacimeter. In partial-gas devices, the exhaust gas is routed through an exhaust-sample probe and heated lines by a pump into the measuring chamber (Figure 5). Using a longer measuring chamber improves the device's detection limit.

During unhindered acceleration some of the exhaust gas is routed to the measuring chamber. A beam of light passes through the measuring chamber, which is now filled with exhaust gas. The attenuation of the light is measured photoelectrically and is indicated as opacity T in % or as absorption coefficient k in m^{-1} (Figure 7). A precisely defined measuring-chamber length and keeping the optical window clear of soot (by air curtains, i.e. tangential air flows) are basic requirements for high levels of accuracy and reproducibility of the measurements.

During testing under load, measurement and display are a continuous process. The opacimeter automatically determines the maximum value and calculates the mean from several gas pulses.

Smoke tester (filter method)

In soot measurement according to Bosch, a defined exhaust-gas volume (330 cm^3) is extracted via a hand pump through a strip of white filter paper. The blackening of the paper is determined by means of a reflective photometer. The degree to which the paper is blackened is subdivided into

Figure 6: Smoke tester (filter method)
1 Filter paper, 2 Gas passage,
3 Heater, 4 Reflective photometer,
5 Paper transport,
6 Volume measurement,
7 Changeover valve for purge air,
8 Pump.

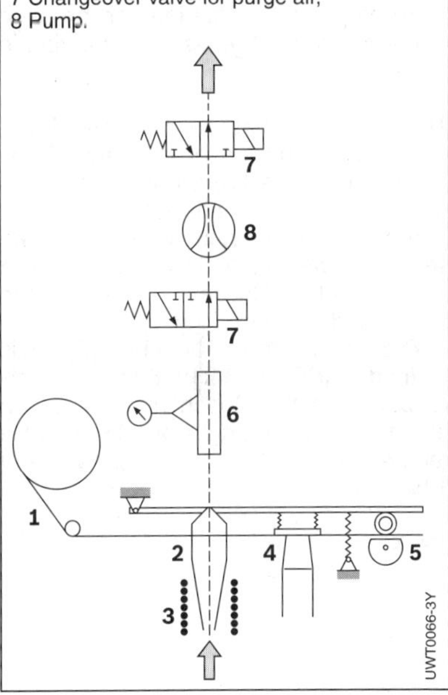

Figure 5: Opacimeter (absorption method)
1 Sample probe, 2 Light source,
3 Purge air for calibration,
4 Calibrating valve, 5 Heater,
6 Receiver, 7 Measuring chamber,
8 Electronic analyzer and display,
9 Optical window,
10 Measuring-chamber outlet to pump.
➧ Exhaust-gas path,
⇨ Purge air flowing past to keep the optical window clear.

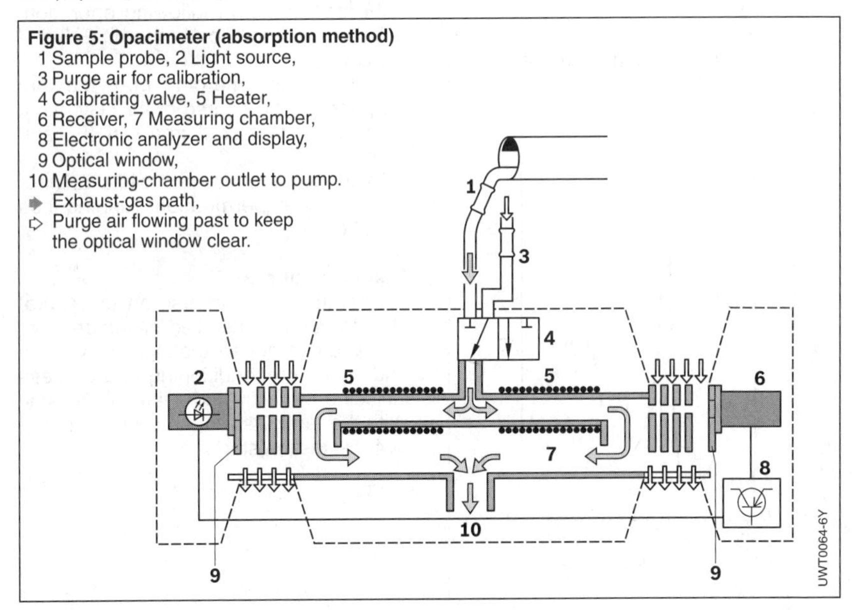

numbers between 0 and 10 (smoke number according to Bosch, Bosch number), where 0 stands for white, unused papers and 10 stands for completely blackened measuring points.

An empirical correlation can be used to convert to soot-mass concentration in mg/m^3.

The procedure has been further developed over a period of decades. Thus, for example, the hand pump has been replaced by a continuously operating pump, the dead volume between the sample probe and the filter paper is taken into account, and the testers are heated to avoid condensation (Figure 6).

The Filter Smoke Number (FSN) is defined in ISO 10054:1998 [1]. This corresponds to the Bosch number for an effective suction length of 405 mm at a temperature of 298 K and a pressure of 1 bar.

Figure 7: Throttle-burst measurement with an opacimeter

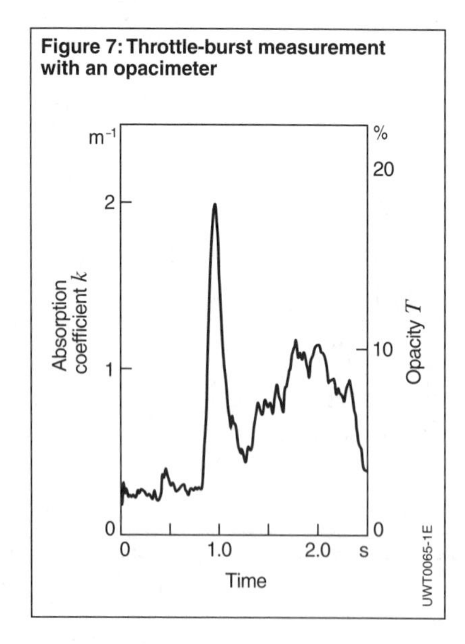

Evaporative-emissions test

Independently of the combustion pollutants produced in the engine, a gasoline-engined motor vehicle emits additional quantities of hydrocarbons (HC) through evaporation of fuel from the fuel tank and fuel system. The quantities evaporated are dependent on the design of the vehicle and on the fuel temperature. Some countries (e.g. the USA and Europe) have regulations which limit these evaporative losses.

Test methods

These evaporative emissions are usually quantified with the aid of a hermetically sealed SHED (Sealed Housing for Evaporative Determination) chamber. For the test, HC concentrations are measured at the beginning and the end of the test with a flame ionization detector (FID), with the difference representing the evaporative losses.

The evaporative losses must – depending on the country – be measured in some or all of the operating states below and satisfy emission limits:

- Evaporation that emerges from the fuel system when the vehicle is parked with the engine warm following operation: "hot parking test" or "hot soak" (EU and USA).
- Evaporation emerging from the fuel system in the course of the day: "tank ventilation test" or "diurnal test" (EU and USA).
- Evaporative emissions while driving, e.g. due to permeation: running-loss test (USA).

Test procedures

Evaporations are measured in several phases during a detailed prescribed test procedure. Prior to testing, the vehicle undergoes preconditioning in a process including the activated charcoal canister. With the tank filled to the stipulated level of 40 %, testing starts.

1st test: Hot-soak losses (hot soak)
To measure the evaporative emissions in this test phase, the vehicle is heated up before the test by running through the test cycle valid for the country concerned. The engine is then turned off when the vehicle is in the SHED chamber. The increase in HC concentration is measured for a period of one hour as the vehicle cools.

The vehicle's windows and trunk lid must remain open during the test. This allows evaporative losses from the vehicle interior to be included in measurements.

2nd test: Tank-breathing losses
For this test, a typical temperature profile for a warm summer day (maximum temperature for EU: 35 °C, for EPA: 35.5 °C, for CARB: 40.6 °C) is simulated within the hermetically sealed climate chamber. The hydrocarbons emitted by the vehicle under these conditions are then collected.

In the USA, both a 2-day diurnal test (48 hours; in-use test) and a 3-day diurnal test (72 hours; certification) must be carried out (the highest value over the course of the day is used in each case). EU legislation provides for a 24-hour test.

3rd test for running losses
The running-loss test is conducted prior to the hot-soak test. It is used to assess the hydrocarbon emissions generated during vehicle operation in the prescribed test cycles (1 FTP-72 cycle, 2 NYCC cycles, 1 FTP-72 cycle; refer to section on "US test cycles").

Limits
EU legislation
The total of the measurement results from the first and second tests produces the evaporative losses. This total must lie below the current limit value of 2 g of evaporated hydrocarbons for all the measurements.

USA
In the USA (legislation in accordance with CARB LEV II and EPA Tier 2), the evaporative losses monitored in the running-loss test must remain below 0.05 g per mile. The limits for the hot-soak losses and the tank-breathing losses are defined as follows:
- 2-day diurnal + hot soak: 1.2 g (EPA) / 0.65 g (CARB),
- 3-day diurnal + hot soak: 0.95 g (EPA) / 0.50 g (CARB).

These limits must be adhered to over 120,000 miles (EPA) or 150,000 miles (CARB). They have been introduced in stages since model year 2004 and apply in full as from model year 2007 (EPA) or 2006 (CARB). With model year 2009, EPA alternatively permits certification in accordance with CARB limits and CARB regulations (harmonization).

Refueling emissions
Refueling test
In the refueling test, HC emissions are measured during refueling in order to monitor the evaporation of hydrocarbon vapors that are dispelled during refueling (emission limit: 0.053 g HC per liter of fuel filled).

In the USA, this test applies to both CARB and EPA.

Spitback test
In the spitback test, the quantity of fuel splashed out during each refueling operation is measured. The tank must be refueled to at least 85 % of its total volume.

This test is conducted only in response to failure of successfully passing the refueling test (limit: 1 g HC per test).

References
[1] ISO 10054: Internal combustion compression-ignition engines – Measurement apparatus for smoke from engines operating under steady-state conditions – Filter-type smokemeter (1998).

Diagnostics

The rise in the sheer amount of electronics in the vehicle, the use of software to control the vehicle, and the increased complexity of modern electronic systems place high demands on the diagnostic concept, monitoring during vehicle operation (on-board diagnostics), and workshop diagnostics.

As emission-control legislation becomes more and more stringent and continuous monitoring in driving mode is now called for, lawmakers have now acknowledged on-board diagnostics as an aid to monitoring exhaust-gas emissions, and have produced manufacturer-independent standardization. This additional system is termed the OBD system (On-Board Diagnostic system). Diagnosis of engine management systems is thus particularly important.

Monitoring in driving mode

The diagnosis integrated in the control unit is a basic feature of electronic engine management systems. Besides a self-test of the control unit, input and output signals, and control-unit intercommunication are monitored.

Monitoring algorithms check input and output signals during vehicle operation, and check the entire system and all the relevant functions for malfunctions and disturbances. Any errors or faults detected are stored in the control-unit fault memory. When the vehicle is serviced in the dealer's workshop, the stored information is exported over a serial interface. This allows troubleshooting and repairs to be carried out quickly and reliably (Figure 1).

Figure 1: Diagnostic system consisting of a diagnostic tester for reading out OBD data and an external offboard tester for further selective troubleshooting and fault localization

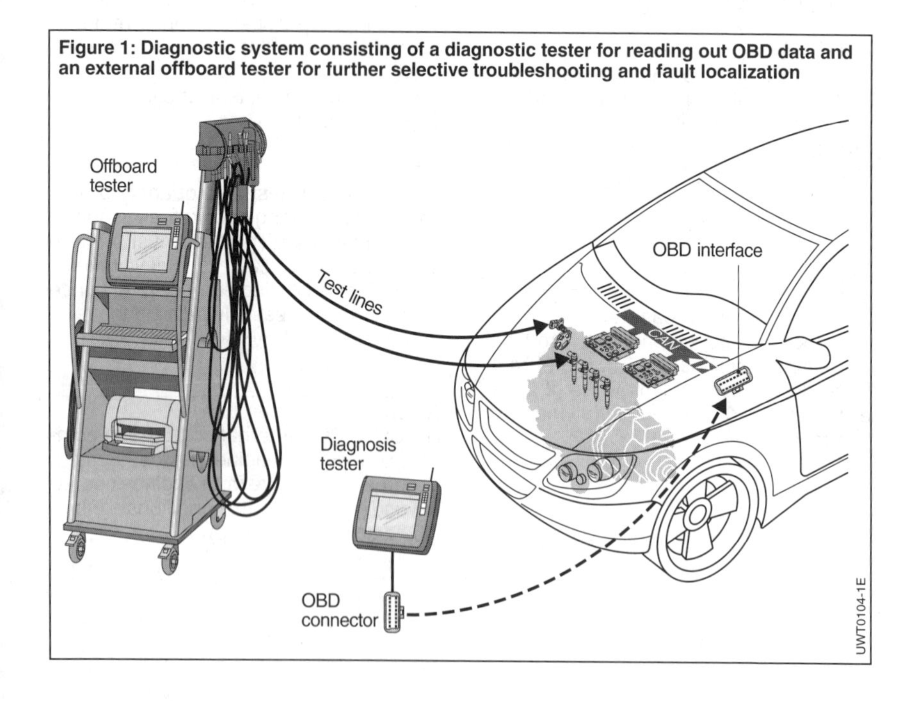

Monitoring input signals
The sensors, connectors and connecting lines (in the signal path) to the control unit are monitored by using the evaluated input signals. With this monitoring strategy, it is possible to detect sensor faults, short circuits to battery voltage U_B and to ground, and open circuits. The following procedures are used for this purpose:
- Monitoring sensor supply voltage (if applicable)
- Examining measures recorded for permissible value ranges (e.g. 0.5 to 4.5 V)
- Plausibility check of different physical signals (e.g. comparison of crankshaft and camshaft speeds)
- Plausibility check of a physical variable which is detected redundantly with different sensors (e.g. pedal-travel sensor).

Monitoring output signals
Actuators triggered by a control unit via output stages are monitored. The monitoring functions detect open circuits and short circuits in addition to actuator faults. The following procedures are used for this purpose: On the one hand, the electric circuit of an output signal is monitored by the output stage; the circuit is monitored for short-circuits to battery voltage U_B, to ground, and for open circuit. On the other hand, impacts on the system by the actuator are detected directly or indirectly by a function or plausibility monitor. System actuators, e.g. the exhaust-gas recirculation valve or the throttle valve, are monitored indirectly via closed-control loops (e.g. for continuous control variance), and also partly by means of position sensors (e.g. position of throttle valve).

Monitoring of internal ECU functions
Monitoring functions are implemented in control-unit hardware (e.g. "intelligent" output-stage modules) and software to ensure that the control unit functions correctly at all times. The monitoring functions check each of the control-unit components (e.g. microcontroller, flash EPROM, RAM). Many tests are performed immediately after switch-on. Further monitoring functions are repeated at regular intervals during normal operation so that component failure can also be detected during operation. Test procedures that require a high amount of computing capacity, or which cannot be performed during vehicle operation for other reasons, are carried out in the after-run after "engine off". This avoids impacting the other functions. An example of such a function is the checksum check of the flash EPROM.

Monitoring ECU communication
Communication with the other control units normally takes place over the CAN bus. Control mechanisms for error detection are incorporated in the CAN protocol so that transmission errors can be detected in the CAN chip already. The control unit also runs a variety of other tests. Since the majority of CAN messages are sent at regular intervals by the individual control units, the failure of a CAN controller in a control unit is detectable by checking these regular intervals. If redundant information is stored in the control unit, all input signals are checked against this information.

Error/fault handling

Error/fault detection

A signal path (e.g. sensor with plug connector and connecting line) is categorized as totally defective if a fault occurs over a specific length of time. The system will continue to use the last valid value until the defect is categorized. When the defect is categorized, a standby function is triggered (e.g. engine-temperature substitute value $T = 90$ °C).

A healing or "restored-signal recognition" feature is available during driving mode for most errors. The signal path must be detected as intact for a specific period of time for this purpose.

Error/fault storage

All faults are stored as a fault code in the non-volatile area of the data memory. The fault code also describes the fault type (e.g. short circuit, open circuit, plausibility, value range exceeded). Each fault-code input is accompanied by additional information, e.g. the operating conditions (freeze frame) at the time of fault occurrence (e.g. engine speed, engine temperature).

Limp-home functions

If a fault is detected, limp-home strategies can be triggered in addition to substitute values (limp home, e.g. engine output power or speed limited). These measures are used to maintain driving safety, prevent consequential damage (e.g. catalytic converter overheating), and minimize exhaust-gas emissions.

On-board diagnostics

The engine system and components must be continuously monitored in driving mode so that compliance with the emission limits required by law can be achieved in everyday use. Therefore, starting in California, regulations were adopted to monitor exhaust-gas-related systems and components. This has standardized and expanded manufacturer-specific On-Board Diagnostics (OBD) with respect to the monitoring of emission-related components and systems.

OBD I (CARB)

In 1988 the first stage of CARB legislation (California Air Resources Board) came into force in California with OBD I. This first OBD stage requires the monitoring of emission-related electrical components (short-circuits, line breaks) and storage of the faults in the control-unit fault memory as well as a malfunction indicator lamp (MIL) that alerts the driver to detected faults. On-board means (e.g. flashing code on a diagnosis lamp) must also be in place to provide a readout of which component has malfunctioned.

OBD II (CARB)

In 1994 the second stage of diagnosis legislation was introduced in California with OBD II. OBD II became mandatory for diesel-engine cars with effect from 1996. In addition to the scope of OBD I, system functionality was now monitored (e.g. plausibility check of sensor signals).

OBD II stipulates that all emission-related systems and components must be monitored if they cause an increase in toxic exhaust-gas emissions (and thus the OBD limits to be exceeded) in the event of a malfunction. In addition, all components used to monitor emission-related components or which affect the diagnosis result must be monitored.

Normally, the diagnostic functions for all components and systems under surveillance must run at least once during the exhaust-gas test cycle (e.g. FTP 75, Federal Test Procedure).

The OBD II legislation furthermore prescribes standardization of the fault-memory information and access to the information (connector, communication) compliant with ISO 15031 [1] and the corresponding SAE standards (Society of Automotive Engineers), e.g. SAE J1979 [2] and SAE J1939 [3]. This permits a fault-memory readout over standardized, commercially available testers (scan tools).

OBD II expansions

From model year 2004

The law has been revised several times since OBD II was introduced. The legal requirements are generally revised by the authorities every two years ("biennial review"). Since model year 2004 it has been necessary, in addition to meeting tighter and additional functional requirements, to check the diagnostic frequency (from model year 2005) in everyday operation (In-use Monitor Performance Ratio, IUMPR).

From model year 2007 through 2013 for gasoline passenger cars

The last revision applies from model year 2007. New requirements for gasoline engines are essentially the diagnosis of cylinder-individual mixture trimming (air/fuel imbalance), extended requirements with regard to diagnosis of the cold-start strategy, and permanent error/fault storage, which also applies to diesel systems.

From model year 2007 through 2013 for diesel vehicles

For diesel passenger cars and light commercial vehicles the tightened OBD emission limits are classified into three stages (up to model year 2009, model years 2010 through 2012, from model year 2013). Significantly extended functions are additionally required for the fuel-injection system, the air system, and the exhaust-gas treatment system. Thus, for example, the fuel-injection system is required to monitor the injected fuel quantity and the injection timing. The air system is required, for example, to monitor boost-pressure control and additionally the dynamic response of exhaust-gas recirculation control and boost-pressure control. In the exhaust-gas treatment system, new monitoring functions are required of the oxidation-type catalytic converter, the particulate filter, the NO_x storage catalytic converter, and the SCR dosing system (Selective Catalytic Reduction) with SCR catalytic converter. Thus, to be monitored are, for example, the regeneration frequency in the particulate filter and the dosing quantity of the NO_x reducing agent in the SCR dosing system.

One of the new requirements for diesel systems since 2009 is that, as well as the regulators, controlled functions – where they are relevant to exhaust gas – are to be monitored as well. Likewise, the monitoring of cold-start functions is subject to extended requirements.

From model year 2014/2015

Enhanced requirements have already been formulated for model year 2015 for individual components in diesel passenger cars, light commercial vehicles, and general commercial vehicles. These relate to the monitoring of the oxidation catalytic converter for "feedgas" (ratio between NO and NO_2 to operate the SCR catalytic converter), to the monitoring of the coated particulate filter for NMHC conversion (non-methane hydrocarbons), and to the monitoring of the injection system for quantity-encoded injectors.

Furthermore, the OBD II law is currently being revised within the framework of the LEV III emission legislation. The OBD II limit for NO_x and NMHC will be defined as from LEV III as a quantity (NO_x+NMHC). The limit for particulate matter will also be tightened. Likewise, in the course of the revision of the LEV III emission legislation some requirements pertaining to hybrid vehicles were defined more precisely in 2012 with an impact on the IUMPR calculation.

Concrete considerations are being given to extending the OBD requirements to CO_2 monitoring. More precise requirements with regard to hybrid vehicles are also expected. The definition of adapted OBD II limits with regard to the LEV III emission limits is expected in 2013 and will come into force successively as from model year 2014/2015.

Scope of application

The previously presented OBD regulations for the CARB and apply to all pas-

Table 1: OBD emission limits

	Gasoline passenger cars	Diesel passenger cars	Diesel commercial vehicles
CARB	– Relative limits – Mostly 1.5 × the limit of the respective exhaust-gas category	– Relative limits – Mostly 1.5 or 1.75 × the limit of the respective exhaust-gas category, but from 2007 through 2013 introduction of more stringent limits in three stages, e.g. for particulate filter: 2007–2009: 5 × limit 2010–2012: 4 × limit From 2013: 1.75 × limit	2010–2012: CO: 2.5 × limit NMHC: 2.5 × limit NO_x: +0.4/0.6 g/bhp-hr [3] PM: +0.06/0.07 g/bhp-hr From 2013: CO: 2.0 × limit NMHC: 2.0 × limit NO_x: +0.2/0.4 g/bhp-hr [3] PM: +0.02/0.03 g/bhp-hr Transition phase for some monitors up to 2016.
EPA (US Federal)	– Relative limits – Mostly 1.5 × the limit of the respective exhaust-gas category	– Relative limits – Mostly 1.5 × the the limit of the respective exhaust-gas category CARB certificates with corresponding limits are recognized by EPA	2010–2012: CO: 2.5 × limit NMHC: 2.5 × limit NO_x: +0.6/0.8 g/bhp-hr [3] PM: +0.04/0.05 g/bhp-hr From 2013: CO: 2.0 × limit NMHC: 2.0 × limit NO_x: +0.3/0.5 g/bhp-hr [3] PM: +0.04/0.05 g/bhp-hr
EOBD	Euro 5 (09/2009): CO: 1,900 mg/km NMHC: 250 mg/km NO_x: 300 mg/km PM: 50 mg/km [1] Euro 6-1 (09/2014): CO: 1,900 mg/km NMHC: 170 mg/km NO_x: 150 mg/km PM: 25 mg/km [1] Euro 6-2 (09/2017): [2] CO: 1,900 mg/km NMHC: 170 mg/km NO_x: 90 mg/km PM: 12 mg/km [1]	Euro 5 (09/2009): CO: 1,900 mg/km NMHC: 320 mg/km NO_x: 540 mg/km PM: 50 mg/km Euro 6 interim (09/2009): CO: 1,900 mg/km NMHC: 320 mg/km NO_x: 240 mg/km PM: 50 mg/km Euro 6-1 (09/2014): CO: 1,750 mg/km NMHC: 290 mg/km NO_x: 180 mg/km PM: 25 mg/km Euro 6-2 (09/2017): [2] CO: 1,750 mg/km NMHC: 290 mg/km NO_x: 140 mg/km PM: 12 mg/km PN [4]: under discussion, still no limit established	Euro IV (10/2005)/ Euro V (10/2008): NO_x: 7.0 g/kWh PM: 0.1 g/kWh NO_x control-system monitoring (since 11/2006): NO_x emission limit + 1.5 g/kWh Euro IV: (3.5+1.5) g/kWh Euro V: (2.0+1.5) g/kWh Euro VI-A (2013): NO_x: 1.5 g/kWh PM: 0.0254 g/kWh Functional alternative for DPF monitor NO_x control system: SCR reagent NO_x: 0.9 g/kWh Euro VI-B (09/2014): As Euro VI-A, but with no alternative for DPF monitor Euro VI-C (2016): NO_x: 1.2 g/kWh PM: 0.025 g/kWh NO_x control system: SCR reagent NO_x: 0.46 g/kWh

[1] For gasoline direct injection.
[2] EU Commission proposal, final definition by 09/2014.
[3] g/bhp-hr: grams per "break horse power" times "hour".
[4] Number of particulates.

senger vehicles with up to 12 seats as well as small commercial vehicles up to 14,000 lbs (6.35 t). Heavy commercial vehicles are governed as from model year 2007 by EMD (Engine Manufacturer Diagnostics). The requirements for EMD are less comprehensive than those for OBD. As from model year 2010 heavy commercial vehicles will be subject to similar regulations to passenger cars (see OBD requirements for heavy commercial vehicles).

The current CARB OBD II legislation for California is at present also in force in some other US states. Furthermore, other US states are planning to adopt this legislation in future.

EPA OBD

Laws enforced by the EPA (Environmental Protection Agency) have been in force since 1994 in those US states which have not adopted the CARB legislation. The requirements of this diagnosis are essentially equivalent to the CARB legislation (OBD II). It is planned as from model year 2017, within the framework of revising the Tier 3 emission legislation, to adapt the EPA-OBD requirements to the CARB OBD requirements. A CARB certificate is already recognized now by the EPA.

EOBD (European OBD)

On-board diagnostics attuned to European conditions is termed EOBD. EOBD has been valid since January 2000 for passenger cars and light commercial vehicles equipped with gasoline engines and weighing up to 3.5 t with up to 9 seats. The regulation has been valid for passenger cars and light commercial vehicles equipped with diesel engines since 2003, and for heavy commercial vehicles since 2005 (see OBD requirements for heavy commercial vehicles).

In 2007 and 2008 new EOBD requirements were adopted for gasoline and diesel passenger cars within the framework of the Euro 5 and Euro 6 emission and OBD legislation (Euro 5 emission level from September 2009; Euro 6 from September 2014).

A general new requirement for gasoline and diesel passenger cars is for checking the diagnostic frequency in everyday operation (In-Use Performance Ratio, IUPR) in accordance with the CARB OBD legislation (In-use Monitor Performance Ratio, IUMPR) as from Euro 5+ (September 2011).

EOBD EU5 and EU5+ requirements for diesel and gasoline engines

For gasoline engines, the introduction of Euro 5 as from September 2009 involved primarily a reduction in the OBD limits. In addition to a particulate-matter OBD limit (for direct-injection engines only), an NHMC OBD limit was introduced (non-methane hydrocarbons, instead of the previous HC). Direct functional OBD requirements result in the monitoring of the three-way catalytic converter for NMHC. Since September 2011 the Euro 5+ level has applied with unchanged OBD limits compared with Euro 5. A significant functional requirement with regard to EOBD is the additional monitoring of the three-way catalytic converter for NO_x.

For diesel passenger-car engines, Euro 5 involved a reduction of the OBD limits for particulate matter, CO and NO_x. In addition, there are extended requirements with regard to the monitoring of the exhaust-gas recirculation system (cooler) and above all with regard to the exhaust-gas treatment components. Here, monitoring of the SCR DeNOx system (dosing system and catalytic converter) is subject to very stringent requirements. Functional monitoring of the particulate filter is mandatory, irrespective of the untreated emissions.

EOBD EU6 requirements for diesel and gasoline vehicles

With Euro 6-1 from September 2014 and Euro 6-2 from September 2017 a further two-stage reduction of some OBD limits has been decided on (see Table 1), where for Euro 6-2 the limits can still be revised up until September 2014. Furthermore, diesel systems are governed by tighter regulations for monitoring the oxidation catalytic converter and the NO_x exhaust-gas treatment system (NO_x storage catalytic converter or SCR catalytic converter with dosing system).

Other countries

Some other countries have already adopted the EU or US OBD legislation or are planning to introduce them (e.g. China, Russia, South Korea, India, Brazil, Australia).

OBD system requirements

The engine control unit must use suitable measures to monitor all on-board systems and components whose malfunction may cause a deterioration in exhaust-gas test specifications stipulated by law. A malfunction must be displayed to the driver by means of the malfunction indicator lamp (MIL) if a fault results in an excess in OBD emission limits.

Limits

US OBD II (CARB and EPA) prescribes thresholds that are defined based on emission limits. Accordingly, there are different permissible OBD limits for the various exhaust-gas categories that are applied during vehicle certification (e.g. LEV, ULEV, SULEV). The European EOBD regulations are based on absolute limits (Table 1).

Functional requirements

All exhaust-gas-related systems and components must, within the framework of On-Board Diagnostics (OBD) required by law, be monitored for malfunctions and for exceeded emission limits.

Legislation demands the monitoring of electrical functions (short-circuit, line breaks), a plausibility check for sensors, and a function monitoring for actuators.

The pollutant concentration expected as the result of a component failure (can be measured in the exhaust-gas cycle) and the monitoring mode partly required by law determine the type of diagnostics. A simple functional test (black/white test) only checks system or component operability (e.g. swirl control valve opens or closes). The extensive functional test provides more detailed information about system operability and also identifies if necessary the quantitative influence of defective components on emissions. As a result, the limits of adaptation must be monitored when monitoring adaptive fuel-injection functions (e.g. zero delivery calibration for a diesel engine, λ adaptation for a gasoline engine).

Diagnostic complexity has constantly increased as emission-control legislation has evolved.

Malfunction indicator lamp

The malfunction indicator lamp (MIL), also called the warning lamp, informs the driver that a component has malfunctioned. When a malfunction is detected, the CARB and the EPA stipulate that it must light up no later than after one driving cycle of its occurrence. In the area where EOBD applies, it must light up no later than the third driving schedule after the fault was detected.

If the malfunction disappears (e.g. a loose contact), the malfunction remains entered in the fault memory for 40 trips ("warm-up cycles"). The malfunction indicator lamp goes out after three fault-free driving schedules. The malfunction indicator lamp flashes for faults in the gasoline system if such a fault could cause damage to the catalytic converter (e.g. combustion misses).

Communication with the scan tool

The OBD legislation prescribes standardization of the fault-memory information and access to the information (connector, communication interface) compliant with the standard ISO 15031 and the corresponding SAE standards (e.g. SAE J1979, [2]). This permits a fault-memory readout over standardized, commercially available testers (scan tools).

The protocols for passenger cars used for the diagnostics interface are also permissible today for OBD diagnostics. However, diagnostics will only be permitted over CAN (ISO 15765 [4]) as from 2008 for the CARB and as from 2014 for the EU.

Vehicle repair

Any workshop can use a scan tool to read out emission-related fault information from the control unit. This permits even non-franchised workshops to carry out repairs.

Manufacturers are obliged to provide the required tools and information (repair manuals on the internet), for a suitable fee, to ensure that repairs can be carried out with the correct degree of expertise.

Switch-on conditions
The diagnostic functions are only executed if the physical switch-on conditions are fulfilled. These include, for example, torque thresholds, engine-temperature thresholds, and engine-speed thresholds or limits.

Inhibit conditions
Diagnostic functions and engine functions cannot always operate simultaneously. There are inhibit conditions that prohibit the performance of certain functions. For instance, tank ventilation (with evaporative-emissions control system) in the gasoline engine cannot function while catalytic-converter diagnosis is in operation. In the diesel engine the hot-film air-mass meter can only be monitored satisfactorily if the exhaust-gas recirculation valve is closed.

Temporary interruption of diagnostic functions
Diagnostic capabilities may only be disabled under certain conditions in order to prevent false diagnosis. Examples of such conditions are high elevation (low air pressure), during engine starting low ambient temperature, or low battery voltage.

Readiness codes
When the fault memory is checked, it is important to know that the diagnostic functions have run at least once. This can be checked by reading out the readiness codes over the diagnostics interface. These readiness codes are set for the most important monitored components on completion of the relevant diagnoses required by law.

Diagnostic System Management
The diagnostic capabilities for all components and systems checked must regularly run in driving mode, but also at least once during the exhaust-gas test cycle (e.g. FTP 75, NEDC). The Diagnostic System Management (DSM) can dynamically change the sequence for running the diagnostic functions depending on the driving condition. The objective here is to run the diagnostic functions frequently in everyday vehicle operation. Diagnostic System Management consists of the following components:
- Diagnostic Fault Path Management for storing fault states and associated ambient conditions (freeze frames)
- Diagnostic Function Scheduler for coordinating the engine and diagnostic functions
- Diagnostic Validator for deciding centrally when faults are detected whether they are the cause or a consequential fault. As well as central validation there are also systems with decentralized validation, i.e. validation is performed in the diagnostic function.

Vehicle recall
If vehicles fail to comply with OBD requirements by law, the authorities may demand the vehicle manufacturer to start a recall at their own cost.

OBD functions

Overview

Whereas EOBD only contains detailed monitoring specifications for individual components, the specific requirements in CARB OBD II are much more detailed. The list below shows the current state of the CARB requirements (from model year 2010) for gasoline-engined and diesel-engined vehicles. The requirements that are also described in detail in the EOBD legislation are marked by (E):

Gasoline and diesel system:
- Exhaust-gas recirculation system (E)
- Cold-starting emission-control system
- Crankcase ventilation
- Combustion misses/misfiring (E, for gasoline system only)
- Fuel system
- Variable valve timing
- Exhaust-gas sensors (λ oxygen sensors (E), NO_x sensors (E), particulate sensor)
- Engine cooling system
- Other emission-related components and systems (E)
- In-use Monitor Performance Ratio (IUMPR) for checking the frequency of diagnostic functions in everyday operation (E)

Gasoline system only:
- Secondary-air injection
- Three-waycatalyticconverter(E),heated catalytic converter
- Tank-leak diagnosis, with (E) at least electrical testing of the canister-purge valve
- Air-conditioning system (in the event of influence on emissions or on OBD)
- Direct ozone-reduction system

Diesel system only:
- Oxidation-type catalytic converter (E)
- SCR DeNOx system (E)
- NO_x storage catalytic converter (E)
- Particulate filter (E)
- Fuel-injection system (injected fuel quantity and injection timing)
- Cooler for exhaust-gas recirculation (E)
- Boost-pressure control
- Intercooler

Other emission-related components and systems refer to components and systems not mentioned in this list that, if they malfunction, may cause the exhaust-gas emissions to be increased (CARB-OBD II), the OBD limits to be exceeded (CARB OBD II and EOBD), or the diagnostic system to be negatively influenced (e.g. by inhibiting other diagnostic functions). Minimum values must be maintained with regard to the frequency of diagnostic functions.

Examples of OBD functions

Catalytic-converter diagnosis

Gasoline system

This diagnostic function monitors the conversion efficiency of the three-way catalytic converter. This is measured by the catalytic converter's oxygen retention capability. Monitoring is performed by observing the signals from the λ oxygen sensors in reaction to a specific alteration of the setpoint value of the λ closed-loop control.

Additionally, the NO_x storage capacity (catalytic-converter quality factor) must be assessed for the NO_x storage catalytic converter. For this purpose, the actual NO_x accumulator content resulting from consumption of the reduction agent during regeneration of the catalytic converter is compared with an expected value.

Diesel system

In the diesel system, carbon monoxide (CO) and unburned hydrocarbons (HC) are oxidized in the oxidation-type catalytic converter (pollutant minimization). Diagnostic functions are used to monitor the operation of the oxidation-type catalytic converter on the basis of the temperature differential before and after the catalytic converter (exothermy).

The storage and regeneration capacity of the NO_x storage catalytic converter is monitored. The monitoring functions run based on loading and regeneration models, and the measured regeneration duration. This requires the use of λ or NO_x sensors.

The SCR DeNOx catalytic converter is monitored using efficiency diagnostics. NO_x sensors located before and after the catalytic converter are required for this purpose. The components of the dosing system and the quantity and dosing of the reducing agent are monitored separately.

Tank-leak diagnosis
Gasoline system
Tank-leak diagnosis detects evaporation from the fuel system that may cause an increase in HC values, in particular. EOBD is limited to simply testing the electrical control circuit of the tank-pressure sensor and the canister-purge valve (evaporative-emissions control system). In the USA, on the other hand, it must be possible to detect leaks in the fuel system. There are two different methods of doing this.

The low-pressure method observes the tank pressure and first tests its operability by deliberately actuating the tank ventilation and carbon-canister check valves. A conclusion can then be drawn on the leak size using the time curve of the tank pressure – again by deliberately actuating the valves.

The overpressure method uses a diagnosis module with an integrated electrically powered vane pump that can be used to pump up the tank system. The flow from the pump is high when the tank is hermetically sealed. A conclusion can then be drawn on leak size by evaluating the flow from the pump.

Particulate-filter diagnosis
Diesel system
The diesel particulate filter is currently monitored for the most part for filter breakage, removal, or blockage. A differential-pressure sensor is used to measure the pressure differential (exhaust-gas back-pressure downstream and upstream of the filter) at a specific volumetric flow. The measured value can be used to verify whether the filter is defective.

An extended function uses load models to monitor the efficiency of the particulate filter.

Since model year 2010 the regeneration frequency has also had to be monitored.

As from model year 2013 a particulate sensor will be used to monitor the particulate filter in response to tightened OBD requirements in the USA. The particulate sensor (from Bosch) operates according to the "collection principle", i.e. the soot collected over a specific distance driven is evaluated using a model for a limit filter. If the collected soot matter, as a function of different parameters, exceeds a certain threshold, the particulate filter is detected as faulty. Combined particulate-filter faults (e.g. broken and melted filters) can also be detected with the particulate sensor.

It is therefore expected that in Europe too a particulate sensor will have to be used to monitor the particulate filter no later than from the introduction of Euro 6-2.

Exhaust-gas recirculation system diagnosis
Diesel system
In the exhaust-gas recirculation system (EGR) the regulator, the exhaust-gas recirculation valve, the exhaust-gas cooler, and other individual components are monitored.

Functional system monitoring is performed by air-mass regulators and position controllers. They check for permanent control variances. An excessively high or low EGR throughflow must be detected. System response ("slow response") must also be monitored.

The exhaust-gas recirculation valve itself is monitored for its electrical and functional operability.

The EGR cooler is monitored by means of additional temperature measurement after the cooler and with model values. This allows the cooler efficiency to be calculated.

Comprehensive components
On-board diagnostics requires that all sensors (e.g. air-mass sensor, speed sensor, temperature sensors) and actuators (e.g. throttle valve, high-pressure pump, glow plugs) having either an impact on emissions or are being used to monitor other components or systems (and consequently may if necessary disable other diagnoses) have to be monitored.

Sensors monitor the following faults (Figure 1):

- Electrical faults, i.e. short-circuits and line breaks ("Signal Range Check").
- Range faults ("Out of Range Check"), i.e. undercutting or exceeding voltage limits set by the sensors' physical measurement range.
- Plausibility faults ("Rationality Check"); these are faults that are inherent in the components themselves (e.g. drift), or which may be caused by shunts, for instance. Monitoring is carried out by a plausibility check on the sensor signals, either by using a model or directly by other sensors.

Actuators must be monitored for electrical faults and – if technically possible – also for function. Functional monitoring means that, when a control command (setpoint value) is given, it is monitored by observing or measuring (e.g. by a position sensor) the system reaction (actual value) in a suitable way by using information from the system.

Figure 1: Sensor monitoring
1 Sensor curve,
2 Upper threshold for "Signal Range Check",
3 Upper threshold for "Out of Range Check",
4 Lower threshold for "Out of Range Check",
5 Lower threshold for "Signal Range Check",
6 Plausibility check "Rationality Check".

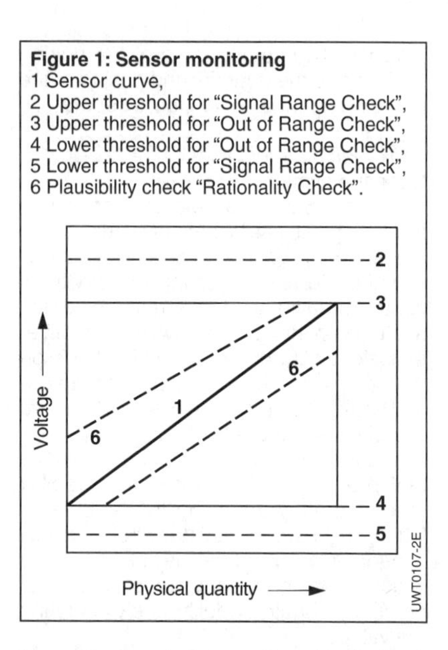

The actuators to be monitored include all output stages, the throttle valve, the exhaust-gas recirculation valve, the variable turbine geometry of the exhaust-gas turbocharger, the swirl flap, the injectors, the glow plugs (for diesel systems), the tank-ventilation system (for gasoline systems), and the active-charcoal check valve (for gasoline systems).

OBD requirements for heavy commercial vehicles

Europe

For commercial vehicles, the first stage of On-Board Diagnostics was introduced in the EU (EOBD) together with Euro IV (10/2005), the second stage together with Euro V (10/2008). A new OBD regulation came into force together with Euro VI in 2013.

Monitoring requirements in Stage 1
- Fuel-injection system: monitoring for electrical faults and for total failure.
- Engine components: monitoring of emission-related components for compliance with the OBD limit.
- Exhaust-gas treatment systems: monitoring for serious faults.

Additional requirements in Stage 2
- Exhaust-gas treatment systems: monitoring for compliance with the OBD limit.

Additional requirements

Since November 2006 it has been required to monitor the NO_x control systems for correct operation. The systems are monitored for their own emission limits, which are more stringent than the OBD limits.

SCR system

The aim is to ensure that the systems are supplied with the correct reagent (urea/water solution, the customary brand name is AdBlue). The availability of the reagent must be monitored by way of the tank fill level. To check the correct quality, it is necessary to monitor the NO_x emissions either with an exhaust-gas sensor or alternatively via a quality sensor. In the latter case, it is also necessary to monitor for correct reagent consumption.

Exhaust-gas recirculation system

The exhaust-gas recirculation system is monitored for correctly recirculated exhaust-gas mass flow and for deactivation of exhaust-gas recirculation.

NO_x storage catalytic converters

Exhaust-gas sensors are used to monitor NO_x emission.

Monitoring of NO_x control systems

Faults in NO_x control systems must be permanently stored (i.e. non-erasable) for 400 days (9,600 hours). Engine power must be throttled if the NO_x OBD limit is exceeded or if the urea tank is empty.

Euro VI

The OBD part of the Euro VI regulation is based on the Global Technical Regulation (GTR) "World Wide Harmonized OBD" (WWH OBD). In terms of structure this WWH OBD GTR corresponds to the Californian OBD legislation (passenger cars and commercial vehicles).

WWH OBD leaves opens which monitoring operations are actually selected in a national regulation (here Euro VI) which implements WWH OBD.

In addition, emission limits and OBD limits as well as the choice of test cycles are established by way of the national regulations.

Special aspects of WWH OBD involve the introduction of new fault storage and new scan-tool communication (ISO 27145 [5]).

Faults must be classified according to their severity with regard to emission deterioration. Emission-related faults can be distinguished by means of malfunction-indicator-lamp behavior and scan-tool communication. The following different categories are applied:
- A: Emissions above the OBD limit.
- B1: Emissions above or below the OBD limit.
- B2: Emissions below the OBD limit, but above the emission limit.
- C: Influence on emissions below the emission limit.

According to this principle all emission-related faults are output, even those which exert a very small influence.

Data for Euro VI:
- Drastic reduction of the emission limits and OBD limits for NO_x and particulate matter compared with Euro V.
- Emission limits to be introduced for NH_3 and the number of particulates.
- Use of the harmonized WHSC and WHTC test cycles.
- The OBD demonstration is performed with a double WHTC hot start part.
- Introduction of a facility for checking system conformity with regard to emissions in the field by means of random measurements with portable emission measurement systems (PEMS).
- Checking of the diagnostic frequency of OBD monitoring in everyday operation (In-Use Monitoring, IUMPR).

Euro VI A:
- For new type approvals mandatory since 31 December 2012.
- Valid up to 31 August 2015.
- Strict OBD limits for NO_x and particulate matter. For particulate filter monitoring a functional, non-emission-correlated diagnosis is possible as an alternative to OBD limit diagnosis.

Euro VI B:
- For new type approval mandatory as from 1 September 2014.
- Valid up to 31 December 2016.
- Change in respect of Euro VI A: Particulate filter monitoring must be effected with emission correlation to the OBD limit.

Euro VI C:
- For new type approval mandatory as from 31 December 2015.
- Change in respect of Euro VI B: Tightened NO_x OBD limit and tightening of the NO_x control-system requirements for SCR reagent quality and consumption monitoring. Monitoring is performed with regard to the long-term drift behavior of the fuel injectors. Monitoring of the OBD diagnostic-frequency rate is mandatory.

Required diagnoses from WWH OBD

These diagnoses are compulsory for particulate filter, SCR catalytic converter, NO_x storage catalytic converter, oxidation catalytic converter, exhaust-gas recirculation, fuel-injection system, charge-air pressure system, variable valve control, cooling system, exhaust-gas sensors, idle control system, and components.

Required diagnoses outside the WWH OBD scope

For the particulate filter, the exhaust-gas recirculation system and the charge-air pressure control system no exceptions are permitted in monitoring for specific diagnoses. The relevant faults must not be defined as Class C.

Furthermore, as from Euro VI C the monitoring of possible component-damaging effects of a long-term drift of fuel injectors is required.

Fundamentally the definition of the non-emission-correlated "performance monitors" in WWH OBD has been changed. In Euro VI these diagnoses must be demonstrated with emission correlation for the first certification of an engine from an engine family.

Gas engines

Gas engines are covered by specific monitoring requirements with regard to compliance with the λ setpoint value, to NO_x and CO conversion of the three-way catalytic converter, and to the λ sensor. Furthermore, catalytic-converter-damaging combustion-miss detection is required.

Requirements with regard to NO_x control systems

For SCR systems, monitoring of the reagent tank fill level, of the reagent quality, of reagent consumption, and of dosage interruption is required.

For exhaust-gas recirculation systems, monitoring of the exhaust-gas recirculation valve is required.

Furthermore, all the NO_x control systems must be monitored for system deactivation by tampering.

Detected faults in the NO_x control system result in a reduction in stages of vehicle drivability. Torque limitation as the first stage is followed by vehicle-speed limitation to crawling speed as the second stage.

USA

CARB, model year 2007

In California "Engine Manufacturer Diagnostics" (EMD) has been called on for heavy commercial vehicles since model year 2007. This can be viewed as a precursor to an OBD regulation. EMD calls for the monitoring of all components and of exhaust-gas recirculation.

The requirements are not reflected in a separate emission limit. Furthermore, no standardized scan-tool communication is required.

Model year 2010 and following

With model year 2010 an OBD system as for passenger-car OBD II was introduced. The technical requirements are at the same level as the respective requirements for passenger cars.

Differences ensue due to the fact that commercial vehicles are subject to engine certification. All the emission and OBD limits apply here to engine cycles. The absolutely applicable values are scaled with the work performed in the cycle.

In contrast to passenger-car LEV III, a new emission regulation (introducing an NO_x and NMHC aggregate limit value) is not planned for commercial vehicles. The OBD limits for NO_x and NMHC thus remain unchanged, i.e. separate, for commercial vehicles.

Time frame for introducing OBD requirements

Model year 2010

A performance variant of a manufacturer's top-selling engine family must be equipped with an OBD system. The other performance variants of this engine family are covered by a simplified certification procedure.

Model year 2013

An engine family of a manufacturer must be equipped in all the performance variants with an OBD system. Furthermore, an OBS system is required in each case for a performance variant in every engine family. The other performance variants of these engine families are covered by a simplified certification procedure.

Model year 2016

All the engine families of a manufacturer must have an OBD system in all the performance variants.

Model year 2018

Engines powered by alternative fuels (e.g. gas) are subject to the OBD requirements.

Japan

Japan has had its own OBD regulation for commercial vehicles in force since 2004. The requirements are comparable in terms of content with EMD in California for model year 2007.

Other countries

Further countries have in the meantime introduced OBD for commercial vehicles. These include China, India, Korea, Australia, Brazil, and Russia. These countries have adopted EU regulations in this regard (Euro IV or Euro V).

References

[1] ISO 15031: Road vehicles - Communication between vehicle and external equipment for emissions-related diagnostics (2011).

[2] SAE J 1979: E/E Diagnostic Test Modes (2012).

[3] SAE J 1979: E/E Diagnostic Test Modes (2012).

[4] ISO 15765: Road vehicles - Diagnostics over Controller Area Network (2011).

[5] ISO 27145: Road vehicles – Implementation of World-Wide Harmonized On-Board Diagnostics (WWH-OBD) communication requirements (2012).

Workshop diagnostics

Function

The function of diagnostics in the workshop is to locate quickly and reliably the smallest replaceable defective unit. In modern engines it is essential to use a generally PC-based diagnostic tester. Diagnostics in the workshop utilizes the results of diagnostics conducting in driving mode (fault-memory entries), employing special workshop diagnostic modules in the vehicle ECU or diagnostic tester and additional test and measuring equipment. These diagnostic options are integrated in the diagnostic tester in prompted troubleshooting.

Prompted troubleshooting

The main element is the prompted troubleshooting procedure. The workshop employee is guided, starting from the symptom or from the fault-memory entry, through fault diagnosis with the aid of an event-driven sequence. Prompted troubleshooting links all the diagnostic options with a specifically target troubleshooting sequence. These include symptom analysis (fault-memory entry or vehicle symptom), workshop diagnostic modules in the ECU, workshop diagnostic modules in the diagnostic tester, test equipment, and supplementary sensors.

All the workshop diagnostic modules can be used only when a diagnostic tester is connected and generally only when the vehicle is stationary. The operating conditions are monitored in the ECU.

Symptom analysis

Faulty vehicle behavior can be either perceived directly by the driver or documented by way of a fault-memory entry. At the beginning of fault diagnosis the workshop employee must identify the existing symptom as the starting point for prompted troubleshooting.

Read-out and deletion of fault-memory entries

All the faults that occur during driving are stored in the fault memory together with defined ambient conditions prevailing at the time of their occurrence, and can be read out via an interface protocol generally specific to the vehicle manufacturer. This protocol is based on one of the established standards and is generally extended to include manufacturer-specific components. The fault memory can also be deleted with the diagnostic tester.

Additional test equipment and sensors

The diagnostic options in the workshop are expanded by using additional sensors (e.g. clamp-on current pickup, clamp-on pressure pickup) or test equipment (e.g. Bosch vehicle-system analyzer). In the event of a fault detected in the workshop, the equipment is adapted to the vehicle. The measurement results are generally evaluated by the diagnostic tester.

ECU-based workshop diagnostic modules

These diagnostic modules integrated in the ECU, after being started by the diagnostic tester, run completely autarkically in the ECU and on completion signal the result back to the diagnostic tester. ECU-based workshop diagnostic modules differ from simple actuator tests with acoustic feedback in that they can place the vehicle to be diagnosed in the workshop into predetermined no-load operating points, apply actuator excitations, and independently evaluate the result via sensor values with an evaluator logic circuit. Examples of such modules are the high-pressure test as system test for the diesel injection system (Figure 1) and the run-up test as component test for the diesel injectors (Figure 2).

A facility for parameterizing the diagnostic modules using the diagnostic tester is provided and offers the opportunity for diagnostic adaptation even after a vehicle has been launched onto the market.

Diagnostic-tester-based workshop diagnostic modules

The functional sequence and evaluation are executed in the case of tester-based diagnostic modules in the diagnostic tester, where the measurement data used for evaluation are determined with the aid of the ECU by sensors in the vehicle or by additional test sensors. An example of such a module is the EGR diagnostic test for the exhaust-gas recirculation valve.

Maximum flexibility is offered by dynamic test modules, whose structure can be variably specified by way of input values by the diagnostic tester. Such test modules permit, for example, the "ramping" of definable actuators, deactivation of corrective functions, deactivation of injections, and buffer storage and read-out of definable sensor values (Figure 3). The test modules mentioned are stored in the ECU. Coordination and thus time resolution of individual modules are performed in the ECU by the "test coordinator". The measurement results from the different sensors can be transmitted either in real time or buffered to the diagnostic tester and evaluated there.

Dynamic test modules make it possible for vehicle manufacturers to adapt diagnostic tests flexibly to altered boundary conditions even after a vehicle has been launched onto the market, because new diagnostic tests can be made available to workshops within the framework of customary software-update cycles.

Figure 1: High-pressure test for the diesel injection system
Detection of leaks in the high-pressure system and efficiency of high-pressure generation.
Steps:
Start test – Increase setpoint pressure – Measure pressure build-up time – Reduce setpoint pressure – Measure pressure reduction time – Variation of engine speed and pressure – Measure pressure reduction time with engine stopped at end of test – Diagnostic results.

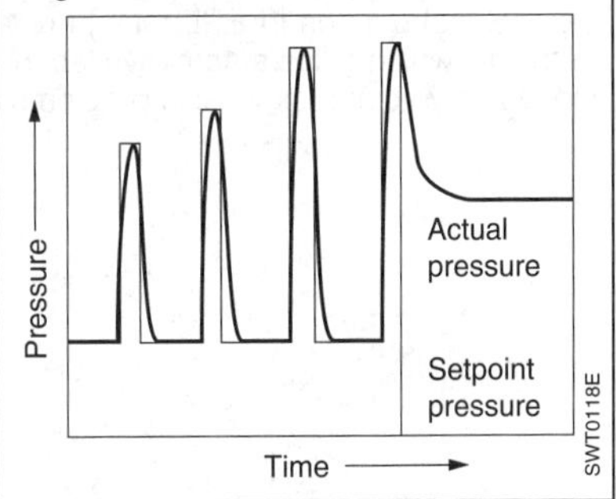

Figure 2: Run-up test for diesel injectors
Detection of deviations of injected fuel quantity of individual injectors.
Steps:
Start test – Cutout of individual cylinder – Quantity jump – Measure maximum speed – Repeat with further cylinders – Diagnostic result.

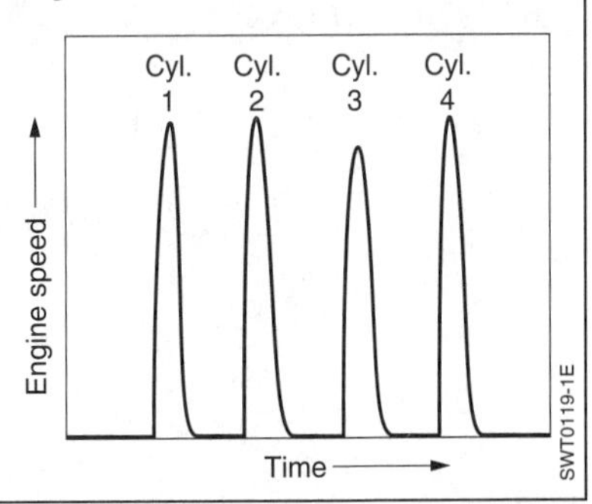

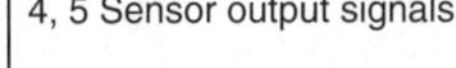

Figure 3: Functioning principle of dynamic test modules
Flexible configuration of function sequences.
Steps:
Select and parameterize test modules – Store constellation on tester – Start test, transmit parameters to ECU –
Test sequence runs in ECU – Transmission of results to tester – Evaluation in tester.
1, 2, 3 Ramp activation of actuators,
4, 5 Sensor output signals.

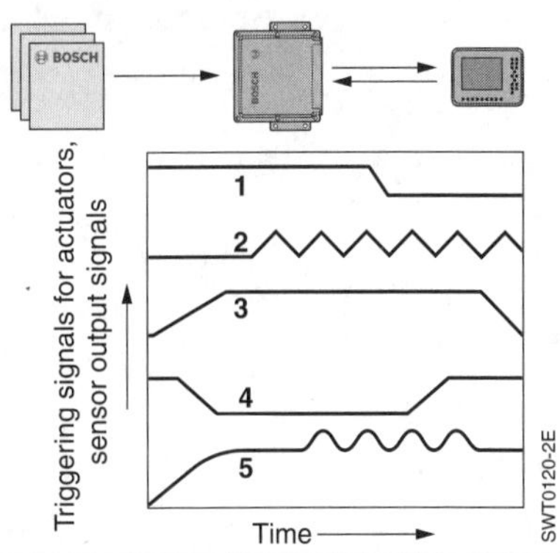

ECU diagnostics and Service Information System

Fault diagnosis and repair

Bosch KTS Series diagnostic testers (Diagnostics Hardware small testers) and the ESI[tronic] software enable comprehensive troubleshooting to be carried out on vehicles (diagnostics). The Service Information System (SIS) is an important part of ESI[tronic] and supports fast and accurate diagnostics.

ESI[tronic] offers different SIS information types. Fault-finding instructions are the type most used. These instructions provide automotive technicians with the most important information needed to diagnose complex problems in different electronic vehicle systems and to eliminate faults.

The test is accessed either via the vehicle with a code number (different designations in different countries, in Germany the KBA number) or via the exact vehicle type. All the SIS fault-finding instructions available for this vehicle are displayed. The relevant instructions, e.g. for Motronic MED 9.5.10 engine management, can then be selected.

If the customer has described a fault symptom, the automotive technician can gain access via prompted troubleshooting and now see displayed which components he has to check in order to find the cause of the fault.

Or he gains access by reading out the fault memory with the diagnostic tester. If one or more faults are stored, he gains access via these fault codes to the component check of the repair instructions.

Now he can use the instructions to check the function of the components. All the components are listed with setpoint values in the system check (subchapter of the instructions). The automotive technician now checks as described all the components to be considered. At the end of the check he ascertains whether a component needs to be replaced.

The fault-finding instructions also include function descriptions, electrical terminal diagrams, installation positions for components, read-outs of actual values, and actuator diagnostics.

The SIS instructions in ESI[tronic] are a useful tool which helps automotive technicians to solve complex vehicle problems.

Management for spark-ignition engines

Description of the engine management system

The engine management system ensures that the driver command is implemented. The driver can request acceleration, deceleration or driving at a constant speed – the engine management system ensures that the spark-ignition engine's drive output required to achieve this is set. The system controls all engine functions in such a way that the engine delivers the required level of torque, but fuel consumption and exhaust emissions are kept low.

The power output from a spark-ignition (SI) engine is determined by the available clutch torque and the engine speed. The clutch torque is the torque developed by the combustion process less friction torque (friction losses in the engine), pumping losses, and the torque needed to drive the auxiliary equipment (Figure 1). The drive torque is available at the wheels. This is developed from the clutch torque reduced by the losses in the clutch and in the transmission. This resulting torque is offset by the tractive resistances such as rolling resistance and aerodynamic drag. Depending on the driver command, a state of equilibrium or imbalance exists between these resistances and the drive torque. When the resistances and the drive torque are equal, a constant driving speed is adopted, otherwise the vehicle is subject to acceleration or deceleration.

Combustion torque is generated in the engine's power cycle and is mainly determined by the following variables:

- The air mass which is available for combustion when the intake valves close
- The fuel mass available in the cylinder
- The point at which the combustion takes place

Furthermore, there are also minor influences, for example, due to mixture composition (amount of residual exhaust gas) or combustion processes.

The primary function of engine management is to coordinate the various subsystems (air, fuel and ignition systems) in order to adjust the torque requested by the engine and at the same time meet the exacting standards placed on exhaust-gas emissions, fuel consumption, power output, comfort, and safety. The engine-management system also performs diagnoses on the subsystems.

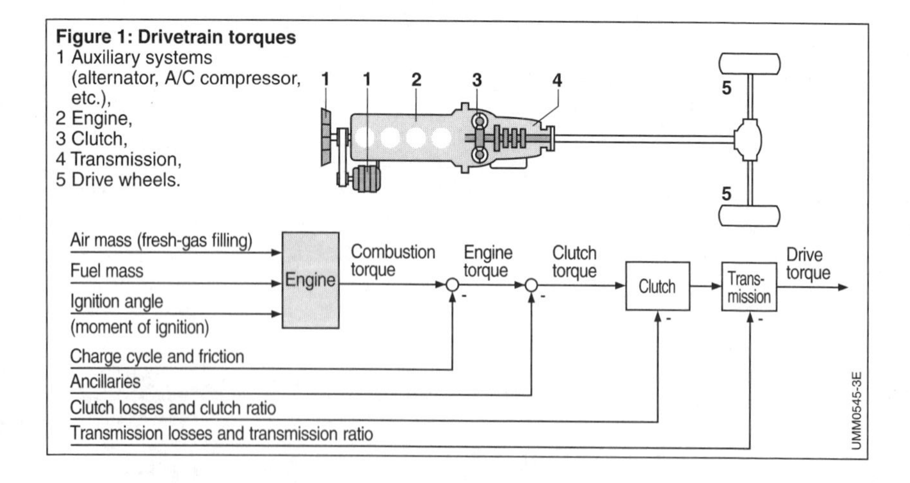

Figure 1: Drivetrain torques
1 Auxiliary systems (alternator, A/C compressor, etc.),
2 Engine,
3 Clutch,
4 Transmission,
5 Drive wheels.

System overview

Electrical system overview

Motronic is the name given to Bosch systems for the open- and closed-loop control of spark-ignition (SI) engines. The Motronic system (Figure 2) comprises all the sensors for detecting ongoing operating data from the engine and vehicle, and all the actuators that perform the required adjustments to the SI engine. The control unit uses data from the sensors to scan the status of the vehicle and engine at very short intervals (every few milliseconds to meet the system's real-time requirements). Input circuits suppress sensor-signal interference and convert the signals to a single uniform voltage scale. An analog/digital converter then transforms the conditioned analog signals into digital values. Other signals are received via digital interfaces (e.g. CAN bus, FlexRay) or via pulse-width-modulated (PWM) interfaces.

The nerve center of the engine control unit is a microcontroller with a program memory (e.g. flash EPROM), which stores all the process control algorithms – i.e. arithmetical processes performed according to a specific pattern – and data (parameters, characteristics, program maps). The input variables derived from the sensor signals influence the calculations in the algorithms, and thus the triggering signals for the actuators. From these input signals, the microcontroller detects the sort of vehicle response the driver wants and calculates, for example, the required torque, the resulting cylinder charge with the associated injected fuel quantity, the correct ignition timing, and the triggering signals for the actuators (e.g. the evaporative-emissions control system, the exhaust-gas turbocharger, and the secondary-air system).

The low-level signal data from the microcontroller outputs are adapted by the driver stages to the levels required by the various actuators.

Another important function of Motronic is the monitoring of the operational ability of the entire system, using On-Board Diagnostics (OBD). Legal requirements (diagnosis legislation) have placed extra demands on Motronic, with the result that about half of Motronic system capacity (in terms of computing power and memory requirements) is dedicated to diagnostics-related tasks.

Figure 2: Components used for electronic control of a spark-ignition engine

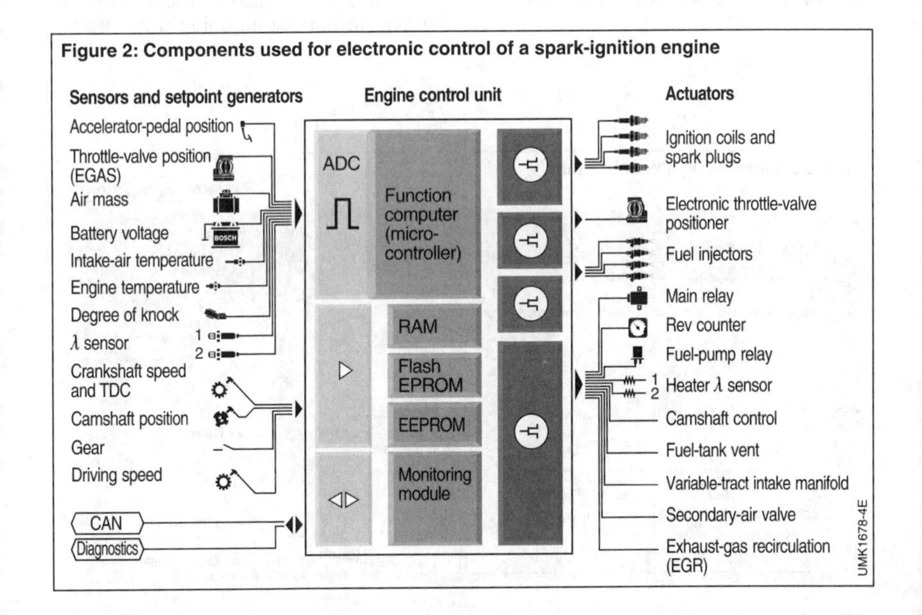

Functional system overview
An engine management system has many other secondary functions in addition to its primary functions of cylinder-charge control, fuel supply, mixture formation, and ignition. The entire system is broken down into subsystems so that they can be described more clearly and in greater detail. The full description is shown in the system structure (Figure 3).

Torque Demand (TD) subsystem
The driver issues a concrete driver command through the accelerator-pedal position. The accelerator-pedal position is converted into a setpoint value for the drive torque.

As well as a direct torque input, the driver can also input an indirect driver command via the cruise-control system. A setpoint drive torque is calculated, depending on the current driving condition.

When the accelerator pedal is not pressed, that amount of engine torque which is needed to maintain idle speed is calculated.

Drivability filters, the surge-damping function, the electrical system (starter, alternator, battery) and other electrical consumers such as the air-conditioning system issue further torque demands.

Torque Structure (TS) subsystem
The varied torque demands from the Torque Demand subsystem, demands from the transmission system, driving-dynamics demands, and other engine-specific torque demands (e.g. catalytic-converter heating) are coordinated in the Torque Structure subsystem. The result is the generation of a torque demand made of the internal-combustion engine.

Setpoint values for charge, injection and ignition are generated from the resulting torque demand of the engine.

The charge is input as a relative air mass. The relative air mass (standardization for all engine-performance classes) is the ratio between the real air mass in the cylinder and the maximum possible air mass in the cylinder at the current engine speed.

The setpoint value for ignition is described by an ignition angle.

Torque reductions can be represented with injection blank-outs (e.g. traction control system demand). The number of blank-outs is determined for this purpose.

In systems with gasoline direct injection, it is possible to set lean-burn operating modes (e.g. stratified fresh charge in the combustion chamber). In these operating modes, it is also possible to set the engine torque by inputting a λ setpoint value.

Figure 3: Motronic system structure

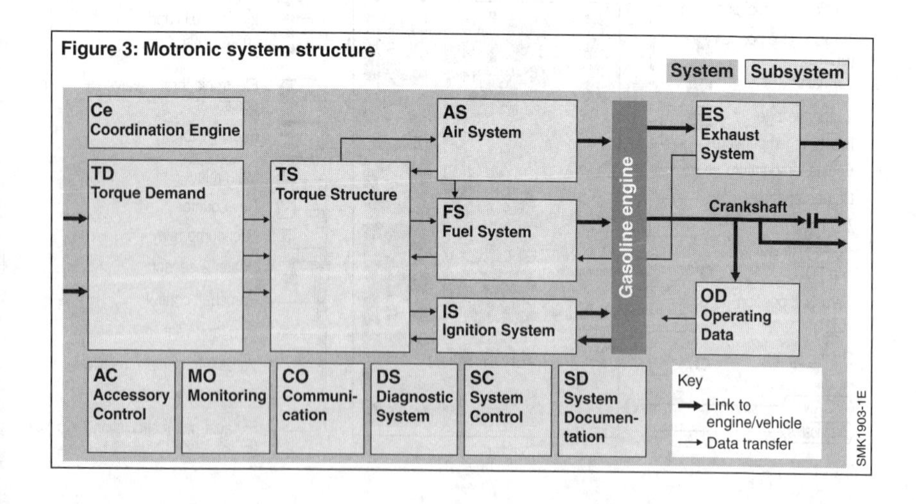

Physical models calculate the actual engine torque at the clutch from various sensor signals. The actual torque is used to monitor the Motronic system and is required by other systems such as transmission control.

Air System (AS) subsystem
The input of the relative setpoint air mass from the Torque Structure subsystem is converted into concrete quantities for the actuators which are used for cylinder-charge control.

The throttle valve is the main actuator for charge. Models are used to calculate the throttle-valve opening angle from the setpoint air mass and from this opening angle, a pulse-width-modulated activation of the actuator.

There are systems in which the main setting path is represented by the activation of intake and exhaust valves. The throttle valve usually remains constantly open in such systems. Only in special situations (e.g. in limp-home mode) is the throttle valve used in these systems as the setting path for charge.

In turbocharged engines, wastegate activation for the exhaust-gas turbocharger or control of mechanical superchargers are also taken into consideration.

The camshaft adjustment systems and exhaust-gas recirculation valves are further setting devices.

In addition, the current actual charge of the internal-combustion engine is determined. Sensor signals such as pressure and temperature in the intake manifold are used as basic variables for this purpose.

Fuel System (FS) subsystem
The function of the fuel system is to supply the fuel rail with fuel from the fuel tank in the requested quantity and at the prespecified pressure.

By using the current actual charge, the fuel pressure in the rail and the pressure in the intake manifold, the opening duration of the fuel injectors is calculated from the λ setpoint value.

The fuel injectors are activated synchronously with the crank angle in order to optimize the air/fuel mixture.

Longer-term adaptations of the actual value for λ ensure greater pilot-control accuracy of fuel metering.

Ignition System (IS) subsystem
The resulting ignition angle is calculated from the setpoint input for ignition, from the engine operating conditions and on the basis of interventions (e.g. knock control), and an ignition spark is generated at the spark plug at the desired moment of ignition.

The ignition angle is set in such a way that the engine runs with optimal consumption. The subsystem only deviates from this in a few special situations (e.g. catalytic-converter heating or fast torque reduction when changing gear).

The knock-control system permanently monitors combustion in all the cylinders. It ensures that the engine runs with optimal consumption close to the knock limit. At the same time, damage caused by combustion knock is safely avoided. Failures in the knock-detection path are subject to continuous monitoring so that ignition can be effected in the event of a fault at sufficient distance from the knock limit.

Exhaust System (ES) subsystem
The open- and closed-loop control interventions for optimum operation of the three-way catalytic converter are calculated in this subsystem. The combustion mixture must be regulated within narrow limits around the stoichiometric mixture ratio.

The operational capability of the catalytic converter is also monitored. The signals from exhaust-gas sensors (e.g. λ oxygen sensor) serve as the basis for this monitoring.

Component protection functions ensure that the exhaust system is not subject to thermal overloads. The actual temperatures in the exhaust system required for this purpose are usually modeled.

In lean-burn operating modes with stratified fresh charge (in the case of gasoline direct injection), the combustion mixture is also controlled to ensure optimum operation of the NO_x storage catalytic converter.

Coordination Engine (CE) subsystem
In the case of gasoline direct injection, the operating modes (e.g. operation with homogeneous or stratified mixture distribution in the combustion chamber) are coordinated and switched over. In order to determine the required operating mode, the demands of various functionalities must be coordinated on the basis of defined priorities.

Operating Data (OD) subsystem
The Operating Data subsystem evaluates engine operating-condition variables (e.g. engine speed, temperatures), ensures digital conditioning and plausibilization, and makes the result available to the other subsystems.

Adaptation of tolerances in speed sensing helps to control injection and ignition more accurately.

Engine misfires – as a precondition for catalytic-converter protective functions – are detected.

Accessory Control (AC) subsystem
Additional functionalities such as A/C-compressor control, fan control or engine-temperature regulation are frequently integrated into the engine management system. They are coordinated in the Accessory Control subsystem.

Communication (CO) subsystem
The vehicle network contains many other systems (e.g. transmission control, Electronic Stability Program), as well as the Motronic system. The systems exchange information via standardized interfaces (e.g. CAN communication).

In addition, workshop testers can read out signals from the engine management system and perform defined actuator adjustments (actuator diagnosis).

Diagnostic System (DS) subsystem
The operational capability of the Motronic system is subject to continuous monitoring by diagnostic functions. These include both electrical and plausibility checks though comparisons of sensor signals with models. Faults are stored and managed (e.g. faults are assigned "time stamps"). The faults can be called up at a later stage on workshop testers.

Some diagnostic functions are only operational under certain boundary conditions (e.g. in specific temperature or load ranges). There are also diagnostic function which must occur in a prespecified sequence. Coordination of this sequence control is performed by the Diagnostic System.

Monitoring (MO) subsystem
"Drive by wire" systems are relevant to monitoring. The core function is torque comparison. This compares the permissible torque calculated on the basis of driver command with the actual torque calculated from the engine data.

On further levels, the computer core and its peripherals are monitored.

System Control (SC) subsystem
The run-up of the Motronic system is regulated. Computing frameworks must be provided before individual functionalities are calculated. Different computing frameworks (e.g. frameworks with angle or time synchronism) are required for the resource optimization of the computing time.

Defined functionalities (e.g. function diagnosis of output stages) occur before the engine starts. Sequence control also handles resets and ECU run-on.

System Document (SD) subsystem
In addition to the extensive open- and closed-loop control functions of the Motronic system, numerous documents are required in order to describe a concrete project in detail. These documents include system descriptions of the ECU hardware and software, the wiring harness, the engine data, components, and connection-pin assignments.

Versions of Motronic

Originally, Motronic's function essentially combined electronic fuel injection and electronic ignition in a single control unit. Gradually, more and more functions were added in response to legislative requirements to reduce exhaust-gas emissions, reduce fuel consumption, and meet increased demands in terms of performance, driving comfort, and driving safety. Examples of these additional functions are:

- Idle-speed control
- λ control
- Control of the evaporative-emissions control system
- Exhaust-gas recirculation to reduce NO_x emissions and fuel consumption
- Control of the secondary-air system to reduce HC emissions in the starting and warm-up phases
- Control of the exhaust-gas turbocharger and variable-tract intake manifold to increase engine performance
- Camshaft control to reduce exhaust-gas emissions and fuel consumption and to increase performance
- Component protection (e.g. knock control, engine-speed limitation, exhaust-gas temperature control)

The Motronic system has undergone substantial development since its introduction in 1979. In addition to systems with electronic multipoint fuel injection, the following simpler and, at that time, cost-effective systems were also developed to allow Motronic to be used in mid-range and compact vehicles:

- KE-Motronic, based on KE-Jetronic continuous gasoline injection
- Mono-Motronic, based on Mono-Jetronic intermittent single point injection

Now, only multipoint fuel-injection systems are used for new vehicles. They include:

- M-Motronic for controlling ignition and fuel injection on intake-manifold injection systems with conventional throttle valves. However, this Motronic system is becoming an increasingly less popular choice.
- ME-Motronic with electronic throttle control (ETC) for controlling injection, ignition, and fresh-air charge on intake-manifold injection systems (Figure 4).
- DI-Motronic (Direct Injection), with additional open- and closed-loop control functions for the high-pressure fuel circuit on gasoline direct-injection systems and to implement the various operating modes of this engine type (Figure 5 for homogeneous operation, Figure 6 for lean-burn concept).
- Bifuel-Motronic for controlling components for running the spark-ignition engine on natural gas or gasoline (see Engines fueled by natural gas).

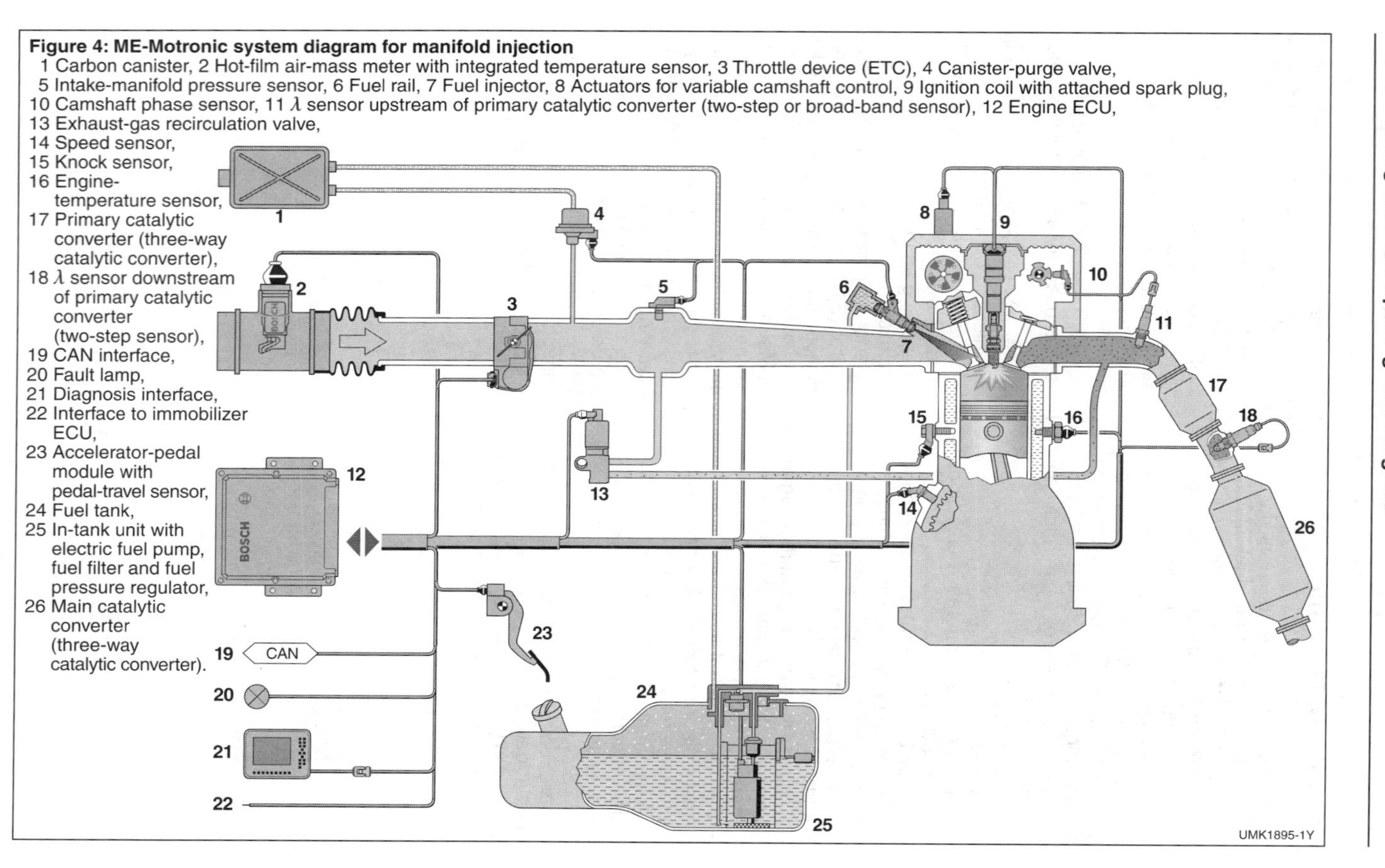

Figure 4: ME-Motronic system diagram for manifold injection
1 Carbon canister, 2 Hot-film air-mass meter with integrated temperature sensor, 3 Throttle device (ETC), 4 Canister-purge valve, 5 Intake-manifold pressure sensor, 6 Fuel rail, 7 Fuel injector, 8 Actuators for variable camshaft control, 9 Ignition coil with attached spark plug, 10 Camshaft phase sensor, 11 λ sensor upstream of primary catalytic converter (two-step or broad-band sensor), 12 Engine ECU, 13 Exhaust-gas recirculation valve, 14 Speed sensor, 15 Knock sensor, 16 Engine-temperature sensor, 17 Primary catalytic converter (three-way catalytic converter), 18 λ sensor downstream of primary catalytic converter (two-step sensor), 19 CAN interface, 20 Fault lamp, 21 Diagnosis interface, 22 Interface to immobilizer ECU, 23 Accelerator-pedal module with pedal-travel sensor, 24 Fuel tank, 25 In-tank unit with electric fuel pump, fuel filter and fuel pressure regulator, 26 Main catalytic converter (three-way catalytic converter).

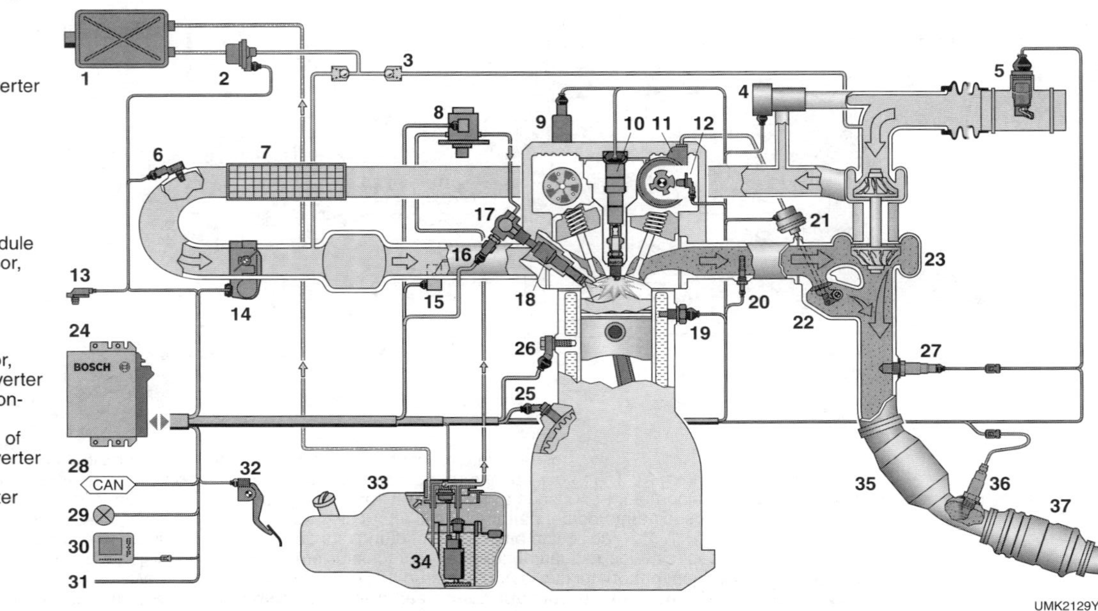

Figure 5: DI-Motronic system diagram for gasoline direct injection (homogeneous operation) with exhaust-gas turbocharging
1 Carbon canister, 2 Canister-purge valve, 3 Non-return valves, 4 Blow-off valve,
5 Hot-film air-mass meter with integrated temperature sensor, 6 Charge-air pressure sensor (optionally in combination with charge-air temperature sensor),
7 Intercooler, 8 High-pressure pump with integrated fuel-supply control valve, 9 Actuators for variable camshaft control,
10 Ignition coil with attached spark plug, 11 Vacuum pump for wastegate activation, 12 Camshaft phase sensor,
13 Ambient-pressure sensor, 14 Throttle device (ETC),
15 Swirl control valve, 16 Fuel-pressure sensor, 17 Fuel rail (high-pressure rail),
18 High-pressure fuel injector, 19 Engine-temperature sensor, 20 Exhaust-gas temperature sensor, 21 Boost-pressure control valve,
22 Wastegate (bypass valve),
23 Exhaust-gas turbocharger,
24 Engine ECU,
25 Speed sensor,
26 Knock sensor,
27 λ sensor upstream of primary catalytic converter (two-step or broad-band sensor),
28 CAN interface,
29 Fault lamp,
30 Diagnosis interface,
31 Interface to immobilizer ECU,
32 Accelerator-pedal module with pedal-travel sensor,
33 Fuel tank,
34 In-tank unit with electric fuel pump, fuel filter and fuel pressure regulator,
35 Primary catalytic converter (three-way catalytic converter),
36 λ sensor downstream of primary catalytic converter (two-point sensor),
37 Main catalytic converter (three-way catalytic converter).

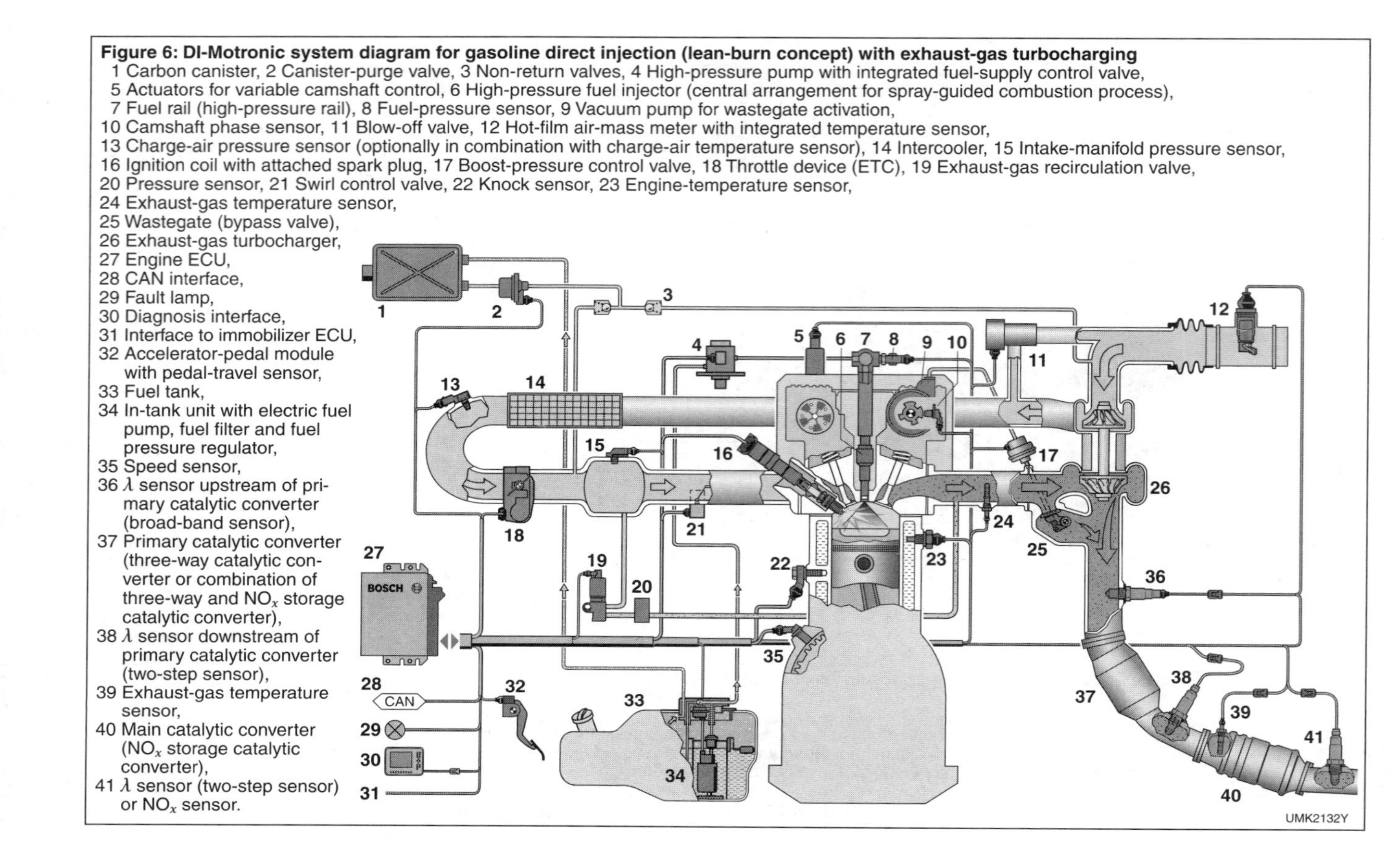

Figure 6: DI-Motronic system diagram for gasoline direct injection (lean-burn concept) with exhaust-gas turbocharging
1 Carbon canister, 2 Canister-purge valve, 3 Non-return valves, 4 High-pressure pump with integrated fuel-supply control valve,
5 Actuators for variable camshaft control, 6 High-pressure fuel injector (central arrangement for spray-guided combustion process),
7 Fuel rail (high-pressure rail), 8 Fuel-pressure sensor, 9 Vacuum pump for wastegate activation,
10 Camshaft phase sensor, 11 Blow-off valve, 12 Hot-film air-mass meter with integrated temperature sensor,
13 Charge-air pressure sensor (optionally in combination with charge-air temperature sensor), 14 Intercooler, 15 Intake-manifold pressure sensor,
16 Ignition coil with attached spark plug, 17 Boost-pressure control valve, 18 Throttle device (ETC), 19 Exhaust-gas recirculation valve,
20 Pressure sensor, 21 Swirl control valve, 22 Knock sensor, 23 Engine-temperature sensor,
24 Exhaust-gas temperature sensor,
25 Wastegate (bypass valve),
26 Exhaust-gas turbocharger,
27 Engine ECU,
28 CAN interface,
29 Fault lamp,
30 Diagnosis interface,
31 Interface to immobilizer ECU,
32 Accelerator-pedal module with pedal-travel sensor,
33 Fuel tank,
34 In-tank unit with electric fuel pump, fuel filter and fuel pressure regulator,
35 Speed sensor,
36 λ sensor upstream of primary catalytic converter (broad-band sensor),
37 Primary catalytic converter (three-way catalytic converter or combination of three-way and NO_x storage catalytic converter),
38 λ sensor downstream of primary catalytic converter (two-step sensor),
39 Exhaust-gas temperature sensor,
40 Main catalytic converter (NO_x storage catalytic converter),
41 λ sensor (two-step sensor) or NO_x sensor.

Cylinder charge

Component parts

The gas mixture located in the cylinder after the intake valves have closed is termed the cylinder charge. It consists of the supplied fresh A/F mixture and the residual exhaust gas.

Fresh A/F mixture

The component parts of the fresh A/F mixture drawn in are fresh air and – on systems with external mixture formation – the fuel suspended in it (Figure 1). Most of the fresh air flows through the throttle valve; additional fresh A/F mixture can be drawn in through the evaporative-emissions control system. The air mass present in the cylinder after the intake valves have closed is the decisive factor in the work performed above the piston during combustion, and ultimately in the torque delivered by the engine. Therefore, measures for increasing maximum torque and maximum engine power almost always require an increase in the maximum possible cylinder charge. The theoretical maximum charge is predetermined by the piston displacement and, in the case of turbocharged engines, by the achievable charge-air pressure, as well.

Figure 1: Cylinder charge in a spark-ignition engine
1 Air and fuel vapor,
2 Canister-purge valve,
3 Connection to evaporative-emissions control system,
4 Exhaust gas,
5 Exhaust-gas recirculation valve (EGR valve),
6 Air-mass flow (ambient pressure),
7 Air-mass flow (intake-manifold pressure),
8 Fresh-mixture charge (combustion-chamber pressure),
9 Residual-gas charge (combustion-chamber pressure),
10 Exhaust gas (exhaust-gas back pressure),
11 Intake valve,
12 Exhaust valve,
13 Throttle valve.
α Throttle-valve angle.

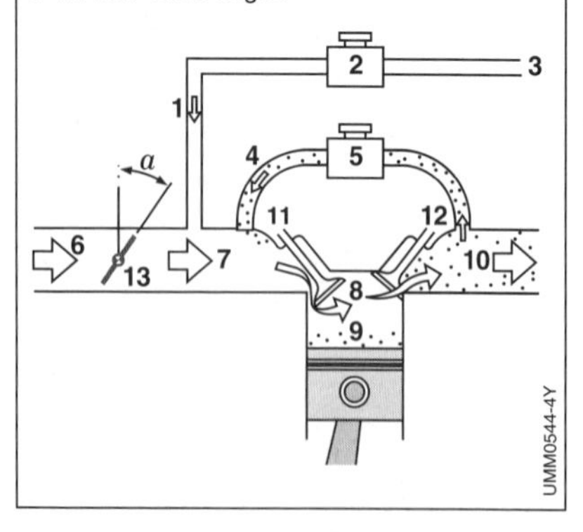

Residual exhaust gas

The residual exhaust gas in the charge is formed by:
- The mass of exhaust gas that remains in the cylinder and is not displaced during the open time of the exhaust valve
- In the case of systems with exhaust gas recirculation by the mass of the returned exhaust gas

The amount of residual exhaust gas is determined by the charge cycle. It does not contribute directly to the combustion process, but does influence ignition and the combustion process as a whole. At wide-open throttle, the amount of residual exhaust gas generally needs to be as small as possible in order to maximize the fresh-air mass and also the power output of the engine.

At part load, however, the amount of residual exhaust gas is desirable in order to reduce fuel consumption. This is achieved by a more favorable cycle, resulting from a change in mixture composition, and also by a reduction in pump losses during the charge cycle, as a higher intake-manifold pressure is required for the same air charge. A specifically introduced amount of residual exhaust gas can likewise reduce the emission of nitrogen oxides (NO_x) and unburned hydrocarbons (HC).

Controlling the air charge

On spark-ignition (SI) engines with external mixture formation (manifold injection) and on systems with internal mixture formation (gasoline direct injection) and homogeneous cylinder charge, the torque delivered by the engine is determined by the air charge. In contrast to this, in the case of internal mixture formation with an air surplus, the engine torque can also be controlled directly by varying the injected fuel mass (stratified-charge operation).

Throttle valve

The critical adjustment mechanism for controlling air-mass flow is the throttle valve. When the throttle valve is not fully open, the air flow drawn in by the engine is throttled, thus reducing maximum engine torque. This throttling effect is dependent on the position, and thus the opening cross-section of the throttle valve, as well as on engine speed (Figure 2). Maximum torque is developed with the throttle valve wide open.

On systems with electronic throttle control (ETC), the required air charge is calculated from the desired engine torque (accelerator-pedal position), and the throttle valve is activated accordingly.

On mechanical systems, the driver directly controls the opening of the throttle valve via a Bowden cable by pressing the accelerator pedal.

Charge cycle

The charge cycle of fresh A/F mixture and residual exhaust gas is controlled by the opening and closing of the intake and exhaust valves. The opening and closing times of the valves (control times) and the valve lift curve are critical factors.

Valve timing

Valve overlap, i.e. the overlap of the opened times of the intake and exhaust valves, has a decisive influence on the mass of residual exhaust gas remaining in the cylinder. The quantities of fresh A/F mixture and residual exhaust gas in the cylinder can be controlled by changing the valve-lift time curves.

In the overlap phase exhaust gas flows through the intake valve into the intake system due to the pressure differential between the exhaust system and the intake manifold. In the induction stroke it then returns to the combustion chamber (internal exhaust-gas recirculation). The longer the valve overlap lasts (early opening of the intake valve), the more mass of exhaust gas remains in the cylinder. This results in shortened combustion temperatures and thus a reduction in NO_x emissions.

However, since recirculated exhaust gas displaces fresh A/F mixture, opening the intake valve early also leads to a reduction in the maximum torque. In addition, excessively high exhaust-gas recirculation, particularly at idle, can lead to combustion misses which, in turn, cause an increase in HC emissions. With variable valve timing, varying the timing dependent on the operating point can optimize emissions.

Figure 2: Throttle map of a spark-ignition engine

--- Intermediate position of the throttle valve

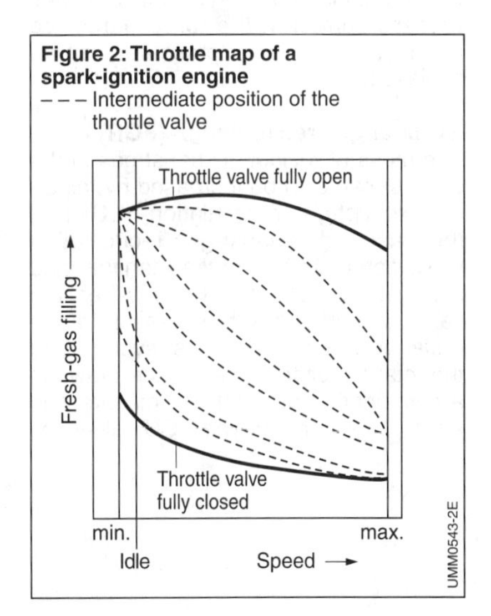

By varying the valve timing accordingly (e.g. continuous adjustment of the phase position and valve lift), it is possible to control the air-mass flow, and thus engine performance, without having to use the throttle valve. The amount of residual exhaust gas can also be adjusted by means of valve timing.

On today's systems, the valves are actuated mechanically via the camshaft. Additional measures can be used to vary this activation to a certain degree (e.g. camshaft adjustment or camshaft shifting). However, these mechanical systems cannot dispense with the throttle valve completely.

Scavenging

In modern turbocharged engines valve-gear variabilities are additionally used to increase significantly the torque in the low-rpm range. This timing strategy is called scavenging. For this purpose the phase position of the intake valves and if necessary the exhaust valves are adjusted in the full-load range in such a way as to produce a valve overlap. During the valve overlap a partial flow of air passes directly again through the exhaust valve to the exhaust side, improving the scavenging of residual gas and with it enhancing anti-knock performance. At the same time the additional scavenging air improves process control at the turbocharger so that higher charge-air pressures can be achieved.

Exhaust-gas recirculation (EGR)

The mass of residual exhaust gas in the cylinder can also be increased by "external exhaust-gas recirculation" (EGR). In this case, an exhaust-gas recirculation valve connects the intake manifold and exhaust manifold (Figure 1). When the valve is open and with a pressure differential between the exhaust system and the intake manifold, the engine draws in a combination of fresh A/F mixture and exhaust gas. The engine ECU calculates the level of exhaust-gas recirculation for a specific operating state, and activates the exhaust-gas recirculation valve accordingly.

Fuel reduction

Exhaust gas recirculation increases intake-manifold pressure. The higher intake-manifold pressure leads to a reduction in the charge cycle work and this lowers fuel consumption.

NO_x reduction

Exhaust-gas recirculation is used on engines with gasoline direct injection which operate in lean-burn mode (stratified-charge operation) to reduce NO_x emissions. Exhaust-gas recirculation is the main way to minimize untreated NO_x emissions, and thus prolong the duration of lean-burn operation, which is limited by the NO_x storage catalytic converter.

Exhaust gas can be conducted back to the combustion chamber to reduce peak combustion temperatures. The reduction in temperature is based on the fact that the returned exhaust gas does not participate in the combustion and thus delivers no combustion energy. It is also an additional thermal mass, which means that the combustion energy distributes itself to a higher overall mass.

Higher combustion temperatures induce an over-proportional increase in NO_x formation, and as exhaust-gas recirculation (EGR) reduces combustion temperatures, it represents a particularly effective means of controlling NO_x emissions.

Supercharging

The torque obtainable is proportional to the charge of fresh A/F mixture. This enables an increase in the maximum torque in that the air in the cylinder is compressed by means of dynamic charging, mechanical charging or exhaust turbocharging (see Supercharging).

Electronic throttle control (ETC) components

The major components that influence the air charge are shown in Figure 1. The throttle valve is the most important component on modern systems.

The electronic throttle control system (Figure 3) consists of the accelerator-pedal module, the engine control unit, and the throttle device. The throttle device mainly consists of the throttle valve, the electric throttle-valve drive element, and the throttle-valve position sensor. The drive element is a DC motor, which acts on the throttle-valve shaft via a gear unit. The throttle-valve position sensor, designed as a redundant unit, detects the position of the throttle valve.

The driver command is detected by the redundant sensor system in the accelerator-pedal module, and the signal is sent to the engine control unit. This calculates the required cylinder charge, based on the current engine operating point, and regulates the opening angle of the throttle valve by means of the throttle-valve drive element and the throttle-valve position sensor.

The redundancies incorporated in the accelerator-pedal module and the throttle device are part of the ETC monitoring concept for avoiding malfunctions.

Figure 3: ETC system

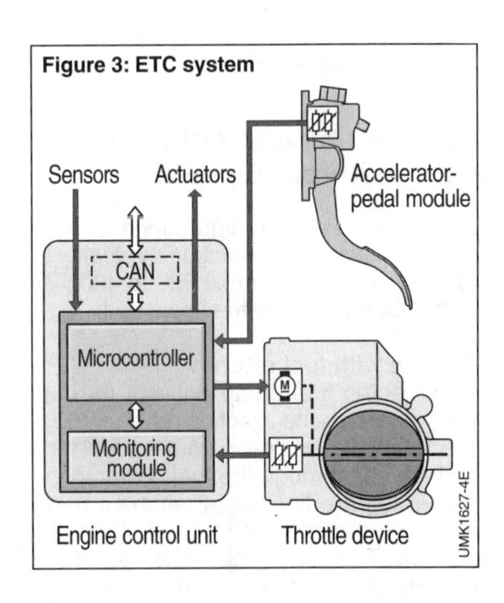

Fuel supply

Fuel supply and delivery with manifold injection

Different system configurations operating with manifold injection typically at fuel pressures of approx. 300 to 400 kPa (3 to 4 bar) are used to supply fuel.

System with fuel return

The electric fuel pump delivers the fuel and generates the injection pressure (Figure 1). The fuel is drawn from the fuel tank and passes through the fuel filter into the high-pressure line, from where it flows to the engine-mounted fuel rail (fuel distributor). The rail supplies the fuel to the fuel injectors. The mechanical pressure regulator mounted on the rail keeps the differential pressure between the fuel injectors and the intake manifold constant, regardless of the absolute intake-manifold pressure, i.e. the engine load.

The fuel not needed by the engine flows through the rail via the return line connected to the pressure regulator back to the fuel tank. This excess fuel heated in the engine compartment causes the fuel temperature in the tank to rise. Fuel vapor is generated as a function of fuel temperature. Ensuring adherence to environmental-protection regulations, the vapors are routed through a tank-ventilation system for intermediate storage in a carbon canister until they can be returned through the intake manifold for combustion in the engine (see Evaporative-emissions control system).

Figure 1: Fuel supply and delivery with manifold injection (system with fuel return)
1 Fuel tank, 2 Electric fuel pump,
3 Fuel filter, 4 Pressure line,
5 Fuel-pressure regulator, 6 Fuel injectors,
7 Rail (fuel flowing through), 8 Return line,
9, 10 Flow direction of fuel.

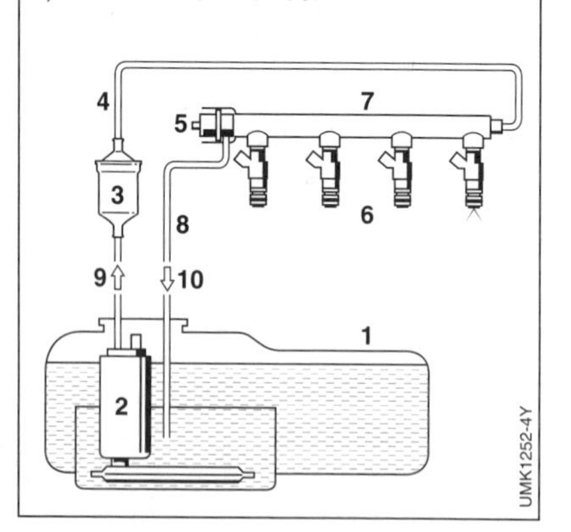

Returnless system

In a returnless fuel-supply system, the pressure regulator is located in the fuel tank or in its immediate vicinity, which means that the return line from the engine to the fuel tank is no longer required. Since the pressure regulator on account of its installation location has no reference to the intake-manifold pressure, the relative injection pressure is not dependent on the engine load. This is taken into account in the calculation of the injection duration in the engine control unit.

Only the amount of fuel which is to be injected is delivered to the rail. The excess flow volume delivered by the electric fuel pump returns directly to the fuel tank without taking the circuitous route through the engine compartment. In this way, fuel heating in the fuel tank and thus also evaporative emissions are significantly lower than in systems with fuel return. Because of these advantages, it is returnless systems which are predominantly used today.

Demand-regulated returnless system

In a demand-regulated system, the fuel-supply pump delivers only that amount of fuel that is currently used by the engine and that is required to set up the desired pressure. Pressure control is effected by means of a closed control loop in the engine control unit, whereby the current fuel pressure is recorded by a pressure sensor (Figure 2). This dispenses with the mechanical pressure regulator. To vary the delivery volume of the fuel-supply pump, its operating voltage is altered by means of a clock module that is triggered by the engine control unit.

The system is equipped with a pressure-limiting valve to prevent the buildup of excessive pressure even during overrun fuel cutoff or after the engine has been switched off.

As a result of demand control, no excess fuel is compressed and thus the capacity of the electric fuel pump is minimized. This results in reduced fuel consumption compared with system with non-controlled electric fuel pumps. The in-tank fuel temperature can also be further reduced.

Further advantages of a demand-regulated system are derived from the variably adjustable fuel pressure. On the one hand, the pressure can be increased during hot starting to prevent the formation of vapor bubbles. On the other hand, it is possible above all in turbocharging applications to meter better the very large injected quantities required at high loads and the very small quantities at idle (injected-fuel-quantity spread) by effecting a pressure increase at full load and a pressure decrease at very low loads. This resolves the problem of metering small quantities in the fuel-injector configuration.

Furthermore, the measured fuel pressure provides for improved diagnostic options for the fuel system compared with previous systems. More precise fuel metering is achieved by taking into account the current fuel pressure when calculating the injection time.

Fuel supply and delivery with gasoline direct injection

Compared with injecting fuel into the intake manifold, there is only a limited time window available for injecting fuel directly into the combustion chamber. Mixture formation is therefore more important and fuel must be injected in the case of direct injection at a pressure approximately 50 times greater than is the case with manifold injection.

The fuel system is divided into a low-pressure circuit and a high-pressure circuit.

Low-pressure system

With gasoline direct injection the low-pressure system serves to supply the high-pressure system, involving the use of the same fuel systems and components as manifold injection. Because of the high temperatures at the high-pressure pump in hot-start and hot-running conditions, a higher predelivery pressure (admission pressure) is required to prevent the formation of vapor bubbles. It is therefore an advantage to use systems with variable low pressure. Demand-regulated low-pressure systems are therefore particularly suitable here, because they allow the optimum admission pressure to be set for every engine operating state; this admission pressure usually varies in a range of 300 to 600 kPa (3 to 6 bar) relative to the ambient pressure.

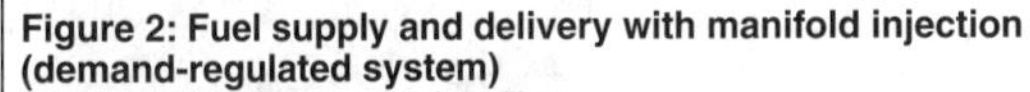

Figure 2: Fuel supply and delivery with manifold injection (demand-regulated system)
1 Electric fuel pump with fuel filter (fuel filter may also be mounted outside of the tank),
2 Pressure-limiting valve and pressure sensor (pressure sensor alternatively in the rail),
3 Clock module for controlling the electric fuel pump,
4 Pressure line, 5 Rail (returnless),
6 Fuel injectors, 7 Fuel tank,
8 Suction jet pump,
9, 10 Flow direction of fuel.

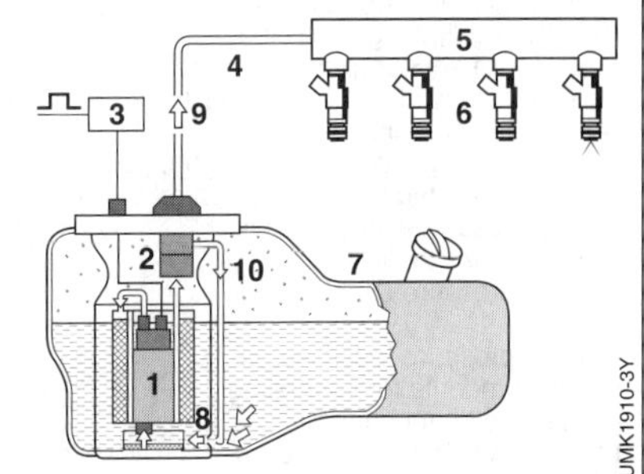

Demand-controlled systems are increasingly being used as a further variant. In contrast to demand-regulated systems, in which the pressure is adjusted via a closed control loop by the pressure sensor, the pressure sensor is omitted here. The electric fuel pump is activated 100 % by pilot control as a function of the engine operating point.

High-pressure system

High-pressure fuel circuits use predominantly demand-regulated high-pressure pumps as well as continuous-delivery high-pressure pumps. The system comprises the fuel distributor (high-pressure rail) with high-pressure fuel injectors and a high-pressure sensor (Figures 3 and 4). The continuous-delivery system also requires a separate pressure-control valve.

Depending on the engine's operating point, the pressure varies typically in a continuous-supply system in a range of 5 to max. 11 MPa (50 to 110 bar), and in a demand-regulated system up to max. 20 MPa (200 bar). The high-pressure sensor provides the information on the currently prevailing pressure.

The signal from the high-pressure sensor is also used for injection calculation and for fuel-system diagnostics.

Continuous-delivery system

The high-pressure pump driven by the engine camshaft, normally a three-barrel radial-piston pump (see High-pressure pumps for gasoline direction injection), forces fuel into the rail against the system pressure (Figure 3). The pump delivery quantity cannot be adjusted. The excess fuel not required for fuel injection and to

Figure 3: Fuel supply and delivery with gasoline direct injection (continuous-delivery system)

1 Suction jet pump,
2 Electric fuel pump with fuel filter,
3 Pressure regulator,
4 Low-pressure line,
5 Continuous-delivery high-pressure pump,
6 High-pressure line,
7 High-pressure sensor,
8 Rail,
9 Pressure-control valve,
10 High-pressure fuel injectors,
11, 12 Flow direction of fuel.

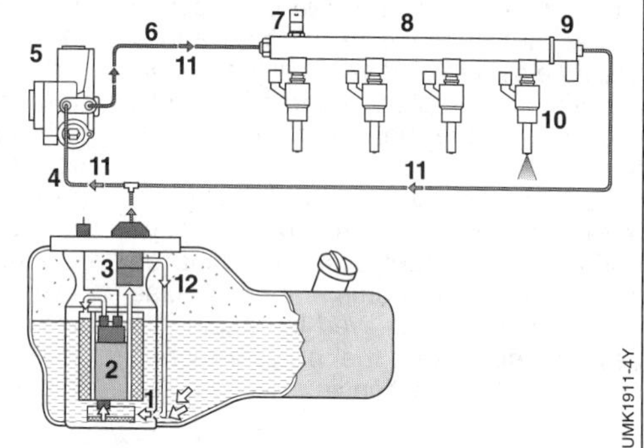

Figure 4: Fuel supply and delivery with gasoline direct injection (demand-regulated system)

1 Suction jet pump,
2 Electric fuel pump with fuel filter,
3 Pressure-limiting valve and pressure sensor (pressure sensor alternatively in the low-pressure line),
4 Low-pressure line,
5 Demand-regulated high-pressure pump with integrated fuel-supply control valve and pressure-limiting valve,
6 High-pressure line,
7 High-pressure sensor,
8 Rail,
9 High-pressure fuel injectors,
10 Clock module for controlling the electric fuel pump,
11, 12 Flow direction of fuel.

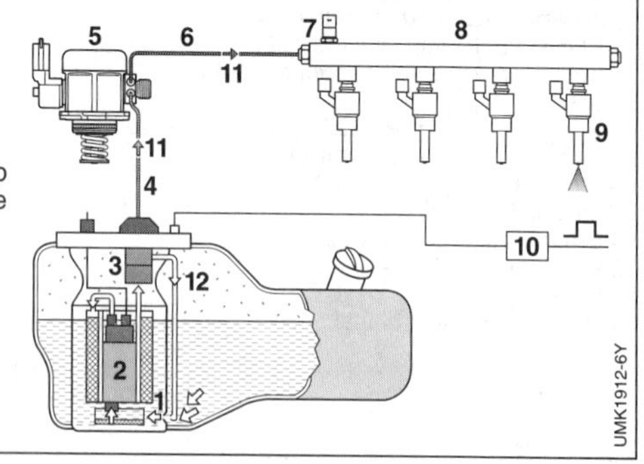

maintain the pressure is depressurized by the pressure-control valve and returned to the low-pressure circuit. For this purpose, the pressure-control valve is actuated by the engine control unit in such a way as to obtain the injection pressure required at a given operating point. The pressure-control valve also serves as a mechanical pressure-limiting valve.

In continuous-delivery systems, most of the operating points cause significantly more fuel to be compressed to high system pressure than is needed by the engine. This results in unnecessary energy consumption and thus higher fuel consumption than demand-regulated systems. The excess fuel released via the pressure-control valve also contributes to increasing the temperature in the fuel system. For these reasons, only demand-regulated high-pressure systems are used in modern direct-injection engines.

Demand-regulated system
In a demand-regulated system (Figure 4), the high-pressure pump – usually a single-barrel radial-piston pump (see High-pressure pumps for gasoline direct injection) – delivers to the fuel rail only the precise amount of fuel which is actually needed for injection and to set the desired pressure. The pump is usually driven by the engine camshaft, single-barrel pumps being driven via special cams which drive the pump plunger. The delivery quantity is varied by a fuel-supply control valve which is integrated in the high-pressure pump. The engine control unit actuates this valve with each pump lift with such precision as to provide the required delivery quantity in order to set the system pressure needed in the rail for the given operating point.

For safety reasons the high-pressure circuit incorporates a mechanical pressure-limiting valve, which is usually integrated directly in the high-pressure pump. Should the pressure exceed the permissible level, fuel is returned via the pressure-limiting valve to the low-pressure circuit.

Evaporative-emissions control system

Function

An evaporative-emissions control system (tank ventilation) is required for vehicles equipped with a spark-ignition (SI) engine. Its purpose is to intercept and collect fuel vapors from the fuel tank and to comply with the legal stipulations governing emission limits for evaporative loss. More fuel vapor escapes from the fuel tank when the fuel is heated. This may be caused by the higher ambient temperatures, the power loss of the fuel pump integrated in the fuel tank or – depending on the fuel supply system – the return of fuel heated in the engine and no longer required for the combustion process. Fuel vapor also escapes when the ambient pressure drops, for example due to weather influences or when driving uphill.

Figure 5: Evaporative-emissions control system
1 Engine ECU,
2 Fuel-tank vent line,
3 Fuel tank, 4 Throttle valve,
5 Purge valve, 6 Carbon canister,
7 Line to intake manifold,
8 Fresh air, 9 Intake manifold.

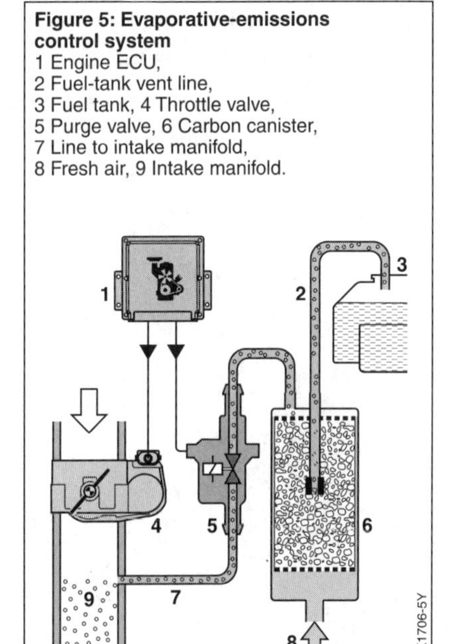

Design and operating principle

The evaporative-emissions control system consists of a carbon canister, into which the vent line from the fuel tank projects, as well as a regeneration valve (canister-purge valve) that is connected both to the carbon canister as well as to the intake manifold (Figure 5). The activated carbon absorbs the fuel vapor. Due to the vacuum prevailing in the intake manifold, fresh air is drawn through the activated carbon when the purge valve frees up the line between the carbon canister and the intake manifold when driving. The fresh air picks up the absorbed fuel and feeds it to the combustion process. This is know as purging the carbon canister.

The engine control unit controls the purge-gas volume depending on the engine operating point. In order to ensure that the carbon canister is always able to absorb fuel vapor, the activated carbon must be regenerated at regular intervals. In systems with gasoline direct injection which also operate in stratified-charge mode, it is necessary, because of the small difference between intake-manifold and ambient pressures in this operating mode, to switch to homogeneous mode for purging.

Fuel filter

Function

The function of the fuel filter is to filter the fuel that flows into the fuel system. Contaminants in the fuel must be filtered out in order to protect the system and in particular the fuel injectors.

Body

Fuel filters for spark-ignition engines (gasoline filters) are located on the pressure side after the fuel-supply pump. In-tank filters are the preferred choice in newer vehicles, i.e. the filter is integrated in the fuel-delivery module in the fuel tank. In this case, it must always be designed as a lifetime filter, which does not need to be changed over the full service life of the vehicle. Furthermore, in-line filters, which are installed in the fuel line, continue to be used. These can be designed as replacement parts or lifetime parts.

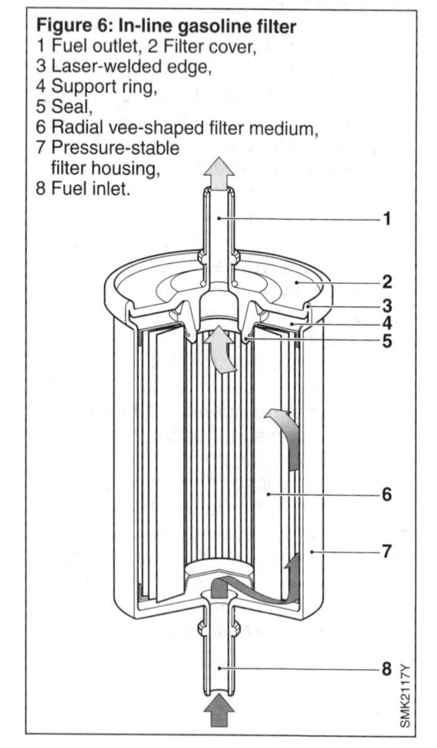

Figure 6: In-line gasoline filter
1 Fuel outlet, 2 Filter cover,
3 Laser-welded edge,
4 Support ring,
5 Seal,
6 Radial vee-shaped filter medium,
7 Pressure-stable filter housing,
8 Fuel inlet.

The filter housing is manufactured from steel, aluminum or plastic. It is connected to the fuel feed line by a thread, tube or quick-action connection. The housing contains the filter element, which filters the dirt particles out of the fuel (Figure 6). The filter element is integrated in the fuel circuit in such a way that fuel passes through the entire surface of the filter medium as much as possible at the same flow velocity.

Filter medium

Special resin-impregnated cellulose-fiber papers which are also bonded for heavier-duty applications to a synthetic-fiber (meltblown) layer are used as the filter medium. This bond must ensure high mechanical, thermal and chemical stability. The paper porosity and the pore distribution of the filter paper determine the filtration efficiency and throughflow resistance of the filter.

Fuel filters for gasoline engines are either spiral vee-form or radial vee-form in design. In a spiral vee-form filter, an embossed filter paper is wrapped round a support tube. The fuel flows through the filter in the longitudinal direction.

In a radial vee-form filter, the filter paper is folded and inserted into the housing in the shape of a star (Figure 6). Plastic, resin or metal end rings and, if necessary, an inner protective jacket provide stability. The fuel flows through the filter from the outside inwards, during which the dirt particles are separated from the filter medium.

Requirements

The required filter fineness is dependent on the fuel-injection system. For systems with manifold injection, the filter element has a mean pore width of approx. 10 µm. Finer filtering is required for gasoline direct injection. The mean pore width here is in the range of 5 µm. Particles larger than 5 µm must be separated off by up to 85 %. In addition, a filter for gasoline direct injection, when new, must satisfy the following residual-dirt requirement: Metal, mineral and plastic particles and glass fibers with diameters of more than 400 µm must not be flushed by the fuel out of the filter.

Filter efficiency depends on the through-flow direction. When replacing in-line filters, it is imperative that the flow direction specified by the arrow be observed.

The interval for changing conventional in-line filters is, depending on filter volume and fuel contamination, normally between 30,000 km and 120,000 km. In-tank filters can generally be used for at least 250,000 km before they have to be replaced. This corresponds to today's design lifetime for a spark-ignition engine. There are in-tank and in-line filters available for use with gasoline direct-injection systems which feature service lives in excess of 250,000 km. If the in-tank filter has to be exchanged, this is only possible by completely changing the fuel-delivery module.

Electric fuel pump

The electric fuel pump must deliver sufficient quantities of fuel to the engine and maintain the pressure required for efficient fuel injection in all operating states. Essential requirements include:

- Delivery quantity between 40 and 300 *l*/h at nominal voltage
- Pressure in the fuel system between 300 and 650 kPa (3.0 to 6.5 bar)
- Buildup of system pressure from 50 to 60 % of nominal voltage; the decisive factor here is operation during cold starting.

Apart from this, the electric fuel pump is increasingly being used as the pre-supply pump for the modern direct-injection systems used on diesel and gasoline engines. In the case of gasoline direct injection, sometimes pressures of up to 650 kPa must be provided during hot-delivery operation.

Figure 7: Electric fuel-pump design using a flow-type pump as an example
A End cover,
B Electric motor,
C Plunger-and-barrel assembly.
1 Electrical connection,
2 Hydraulic connection (fuel outlet),
3 Non-return valve,
4 Pressure-limiting valve,
5 Commutator with carbon brushes,
6 Motor armature with windings which are energized by the commutator,
7 Magnet,
8 Impeller ring of flow-type pump,
9 Hydraulic connection (fuel inlet).

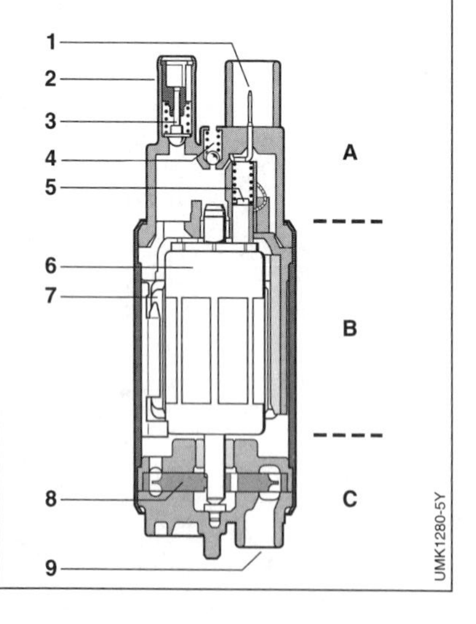

Electric motor

The plunger-and-barrel assembly (pump element) of the electric fuel pump (Figure 7) is driven by an electric motor. An armature with copper or carbon commutator is used as standard in this engine. Electronic commutating systems without commutators and carbon brushes are increasingly being used in new vehicles that appear on the market. The design of the electric motor depends on the desired delivery quantity for a given system pressure. It is continuously surrounded by flowing fuel and thereby cooled.

End cover

The end cover contains the electrical connection and the pressure-side hydraulic connection. The non-return valve prevents the fuel lines from draining after the electric fuel pump is switched off; this ensures that the system pressure is maintained for a certain period of time after switch-off.

A pressure-limiting valve is integrated as required. Most end covers also include the carbon brushes for the commutation system and interference-suppression elements (inductance coils, with capacitors in some applications).

Plunger-and-barrel assembly

The plunger-and-barrel assembly is designed as a positive-displacement or flow-type pump.

Positive-displacement pump

In a positive-displacement pump, volumes of liquid are basically drawn in and transported in a closed chamber (apart from leaks) by rotation of the pump element to the high-pressure side. A roller-cell pump (Figure 8a), an internal-gear pump (Figure 8b) or a screw-spindle pump may be used for the electric fuel pump.

Roller-cell pump

The slotted rotor arranged eccentrically in the pump housing has metal rollers around its circumferences which are loosely guided in the slot-shaped openings. The centrifugal force generated by the rotation of the slotted rotor and the fuel pressure press the rollers against the pump housing and the driving flanks of the slots. The rollers operate here as circulating seals. The fuel is delivered in the chamber created between in each case two rollers of the slotted rotor and the pump housing. The pumping effect is created by the fact that, after the inlet opening is sealed, the chamber volume is continuously reduced until the fuel leaves the pump through the outlet opening.

Figure 8: Electric fuel-pump principles
a) Roller-cell pump,
b) Internal-gear pump,
c) Flow-type pump.

A Fuel inlet (suction opening),
B Outlet (pressure side).
1 Slotted rotor (eccentric), 2 Roller,
3 Inner drive wheel, 4 Rotor (eccentric),
5 Impeller ring, 6 Impeller blades,
7 Passage (peripheral),
8 Gas-discharge orifice.

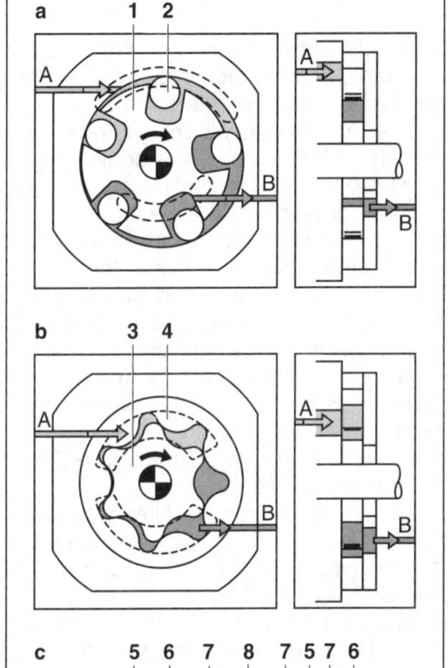

Internal-gear pump
The internal-gear pump consists of an internal drive wheel which engages an eccentrically arranged outer rotor. This outer rotor has one tooth more than the drive wheel. The tooth flanks that are sealed against each other create during the rotation process in their spaces variable chambers which provide the pumping effect.

Positive-displacement pumps are advantageous when used with high-viscosity media, e.g. cold diesel fuel. They have a good low-voltage characteristic, i.e. they have a relatively “flat” delivery-rate characteristic over the operating voltage. Efficiency can be as high as 35 %. The unavoidable pressure pulses may cause noise; the extent of this problem varies according to the pump's design configuration and mounting location.

Whereas in electronic gasoline-injection systems the positive-displacement pump has to a large extent been superseded by the flow-type pump for the classical electric-fuel-pump requirements, it has gained a new field of application as the pre-supply pump on diesel common-rail systems with their significantly greater pressure requirements and viscosity range.

Flow-type pump
The flow-type pump has become the accepted solution for gasoline applications. An impeller ring equipped with numerous peripheral vanes rotates within a chamber consisting of two fixed housing sections (Figure 8c). Each of these sections features a passage along the path of the impeller blades, with openings at one end of the passage starting at the suction openings. From there, they extend to the point where the fuel exits the plunger-and-barrel assembly at system pressure. A small gas-discharge orifice,

located at a specified angular distance from the suction opening, improves performance when pumping hot fuel; this orifice facilitates the discharge of any gas bubbles that may have formed (with minimal leakage).

Pressure builds up along the passage as a result of the exchange of pulses between the impeller blades and the liquid particles. This leads to spiral-shaped rotation of the liquid volume trapped in the impeller and in the passages.

Flow-type pumps feature a low noise level since pressure buildup takes place continuously and is practically pulsation-free. Pump design is also much less complex than the positive-displacement unit. System pressures of up to 650 kPa can be achieved. The efficiency of these pumps can be as high as 26 %.

Fuel-supply module

Whereas in the early stages of electronic gasoline injection the electric fuel pump was mounted exclusively outside the fuel tank (in-line), it is common practice today to install the pump inside the tank itself. The electric fuel pump forms an integral part of a fuel-delivery module (in-tank unit, Figure 9) which may comprise further elements:

- a swirl pot,
- a fuel-level sensor (see Sensors),
- a fuel-pressure regulator for returnless fuel supply systems,
- a prefilter for protecting the pump,
- a pressure-side fine-mesh fuel filter, which will not need replacing for the entire life of the vehicle,
- and electrical and hydraulic connections.
- Tank pressure sensors (for diagnosing tank leaks), fuel-pressure sensors (for demand-regulated systems) and valves for tank ventilation may also be incorporated.

Figure 9: Fuel-delivery module
1 Swirl pot, 2 Fuel filter,
3 Electric fuel pump,
4 Suction jet pump,
5 Fuel-pressure regulator,
6 Fuel-level sensor, 7 Prefilter.

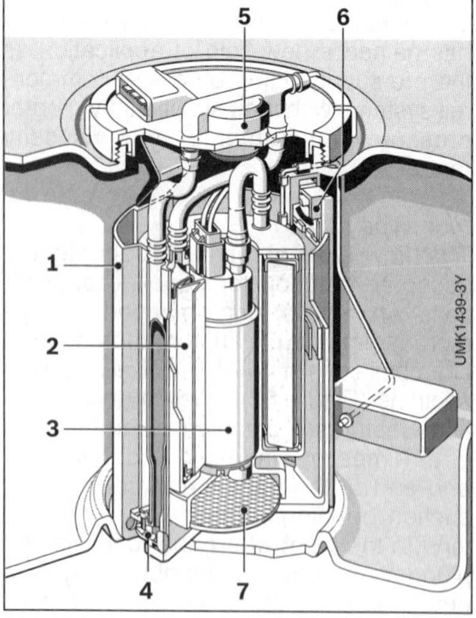

The swirl pot is situated centrally in the fuel tank. The suction side of the electric fuel pump is inserted in this beaker-shaped container. When the tank level is high, fuel flows over the beaker edge, when the level is low, fuel passes through a small opening in the beaker base or through the return line of the electric fuel pump into the swirl pot. When the vehicle is cornering, accelerating and braking, the fuel in the fuel tank flows to one side, which, when the tank level is low, could cause the fuel supply to be interrupted. The fuel can only escape slowly from the swirl pot through the small opening such that the fuel supply is safeguarded for a certain period of time.

The swirl pot can be filled passively, for example by a flap system or a change-over valve, or actively by a suction jet pump. This pump is designed as a venturi nozzle and does not contain any moving parts. The returning fuel flows as a driving medium into this pump and is accelerated after being discharged through the nozzle. In accordance with Bernoulli's law this produces a pressure drop. In this way, fuel from the fuel tank is drawn in through a second inlet and delivered under pressure with the returning fuel to the pump outlet.

High-pressure pumps for gasoline direct injection

The function of the high-pressure pump is to compress the fuel delivered in sufficient quantity by the electric fuel pump at an rpm- and temperature-dependent admission pressure (program-map value) to the level required for high-pressure injection.

Demand-regulated high-pressure pump

Design and operating principle

The Bosch demand-regulated high-pressure pump is used in systems for gasoline direction injection for injection pressures up to 20 MPa (2nd-generation gasoline direct injection). It is a cam-driven single-barrel pump running in oil (Figure 10) with an integrated fuel-supply control valve (also called the metering unit), an integrated pressure-limiting valve on the high-pressure side and an integrated fuel-pressure attenuator on the low-pressure side. To comply with future stringent fuel standards and in response to emission-control legislation, the pump is made of stainless steel and is welded at all the emission-relevant connection points.

The high-pressure pump is mounted as a plug-in pump on the cylinder head. The interface between the engine camshaft and the delivery barrel is a bucket tappet for a double cam (Figure 11a), and

Figure 11: High-pressure pump drives
a) Drive via bucket tappet,
b) Drive via roller tappet.
1 Pump plunger,
2 Bucket tappet,
3 Roller tappet,
4 Double cam,
5 Triple cam,
6 Drive shaft (engine camshaft).

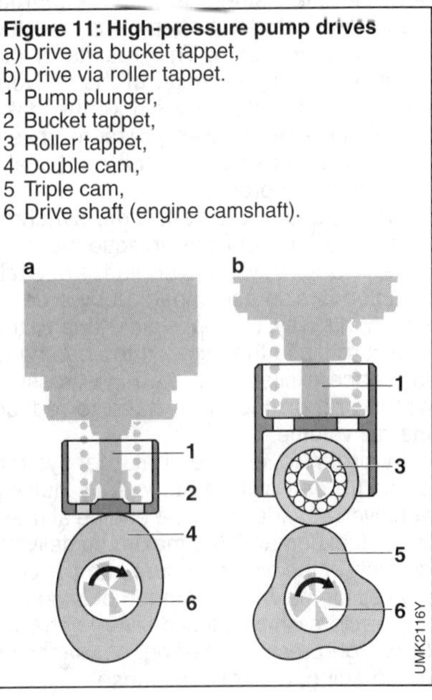

Figure 10: Demand-regulated single-barrel high-pressure pump for 2nd-generation gasoline direct-injection systems
a) View with high-pressure port,
b) Detailed view with low-pressure port (offset at angle on the same plane to the high-pressure port).
1 Variable pressure attenuator,
2 Pressure-limiting valve,
3 High-pressure port,
4 Low-pressure port,
5 Outlet valve,
6 Fuel-supply control valve,
7 Inlet valve,
8 Mounting flange,
9 O-ring,
10 Passage to delivery plunger (pressure-attenuation function),
11 Delivery plunger,
12 Plunger seal,
13 Plunger spring.

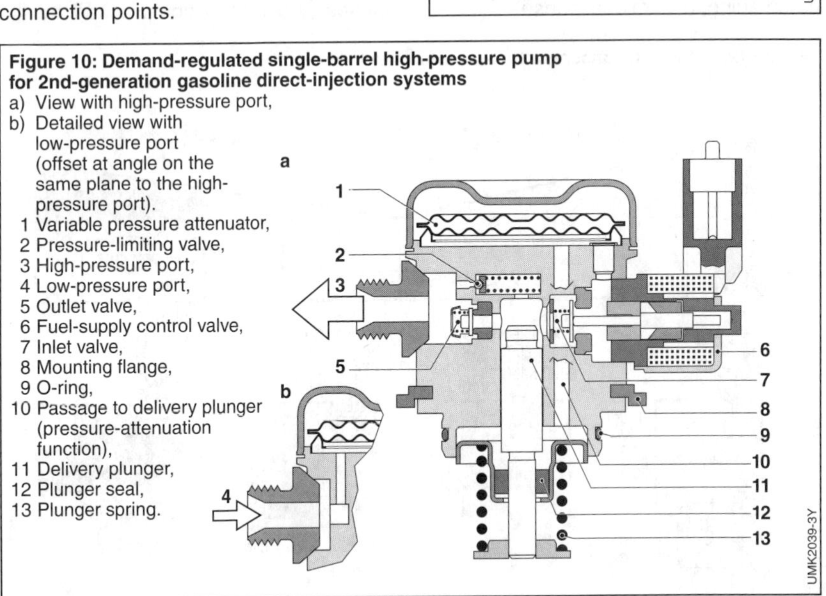

a roller tappet for triple and quadruple cams (Figure 11b). This ensures that the cam lift curve is transferred to the delivery plunger of the high-pressure pump with the requirements with regard to lubrication, Hertzian stress and mass inertia. During the cam lift the tappet follows the contour of the cam. This results in the vertical motion, or stroke, of the delivery plunger. In the delivery stroke the push rod absorbs the applied pressure, mass, spring and contact forces. It is rotationally secured in the process.

With a quadruple cam, time synchronization of delivery and injection is possible in a four-cylinder engine, i.e. each injection is also accompanied by a delivery. In this way, it is possible – due to the pressure dip in the rail – on the one hand to reduce excitation of the high-pressure circuit and on the other hand to reduce the rail volume.

In order to ensure that the system pressure can still be sufficiently quickly achieved or varied with the engine at maximum fuel demand, the maximum delivery quantity of the high-pressure pump is configured for maximum demand plus necessary factors which affect delivery behavior (e.g. high-pressure starting, hot gasoline, pump aging, dynamic response).

The volumetric efficiency is derived from the ratio of actually delivered fuel quantity to theoretically possible quantity. This is dependent on the delivery-plunger diameter and stroke. The volumetric efficiency is not constant across all engine speeds. It is dependent on the following factors:

- In the lower speed range: plunger and other leakages
- In the upper speed range: inertia and opening pressure of the inlet valve
- In the total speed range: dead volume of the delivery chamber and temperature dependence of fuel compressibility.

Fuel-supply control valve

Demand control of the high-pressure pump is effected with the fuel-supply control valve (Figure 12). The fuel delivered by the electric fuel pump is drawn into the delivery chamber via the inlet valve of the open fuel-supply control valve. In the subsequent delivery stroke, the fuel-supply control valve remains open after bottom dead center so that fuel that is not needed at the relevant load point is returned at admission pressure to the low-pressure circuit. After the fuel-supply control valve is actuated, the inlet valve closes, the fuel is compressed by the

Figure 12: Activation concept of fuel-supply control valve

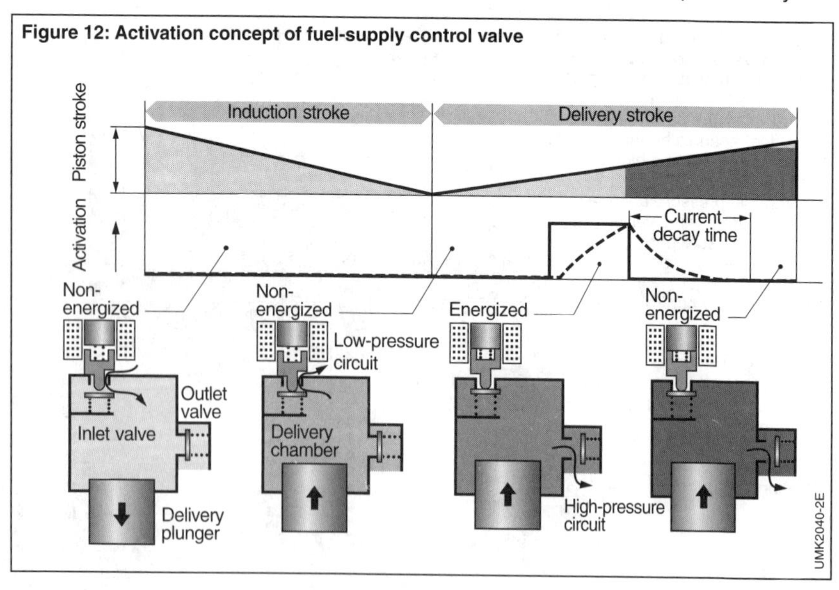

pump plunger and delivered to the high-pressure circuit. The intake valve then remains closed during the delivery stroke on account of the pressure in the delivery chamber even when the fuel-supply control valve is non-energized. The engine-management system calculates the time from which the fuel-supply control valve is actuated as a function of delivery quantity and rail pressure. The start of delivery is varied here for demand control.

Fuel-pressure attenuator
The variable fuel-pressure attenuator integrated in the high-pressure pump, together with the stepped plunger (back-pumping in the induction stroke due to the smaller diameter in the lower area causes a repumping effect), attenuate the pressure pulsations stimulated by the high-pressure pump in the low-pressure circuit and at high rotational speeds ensure a good delivery-chamber charge. The pressure attenuator takes up the fuel quantity diverted at the relevant operating point via the deformation of its gas-filled metal diaphragms and releases it again in the induction stroke to fill the delivery chamber. Operation with variable admission pressure – i.e. the use of demand-regulated low-pressure systems – is possible here.

Continuous-delivery high-pressure pump

Design and operating principle
The Bosch continuous-delivery high-pressure pump is used in systems for gasoline direction injection for injection pressures up to 12 MPa (1st-generation gasoline direct injection). It is a radial-piston pump with three delivery barrels situated at circumferential offsets of 120° (triple-barrel radial-piston pump, Figure 13).

The drive shaft is driven via a clutch by the engine camshaft and rotates with the eccentric cam. The eccentric cam converts the rotational motion via the cam ring and the slipper in a vertical motion of the pump plungers. The drive runs in gasoline for cooling and lubrication purposes.

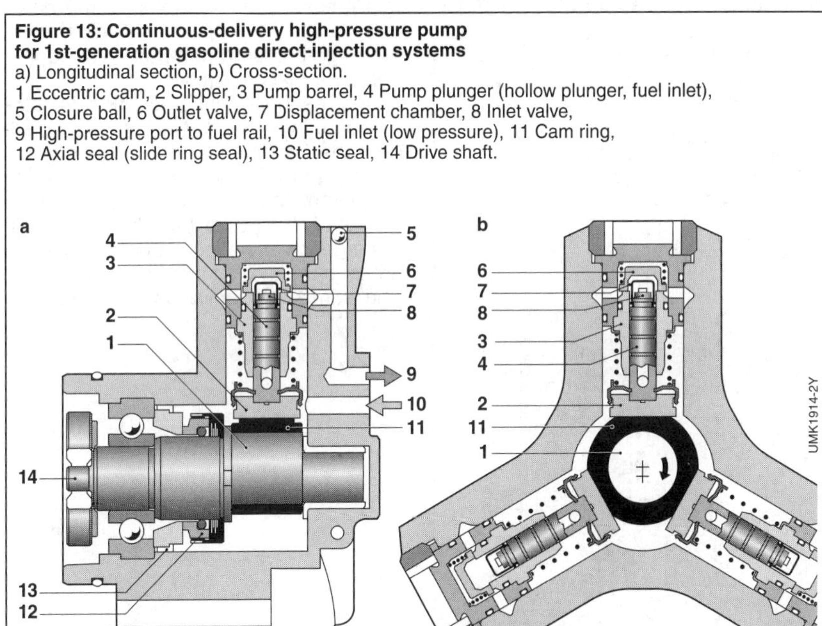

Figure 13: Continuous-delivery high-pressure pump for 1st-generation gasoline direct-injection systems
a) Longitudinal section, b) Cross-section.
1 Eccentric cam, 2 Slipper, 3 Pump barrel, 4 Pump plunger (hollow plunger, fuel inlet), 5 Closure ball, 6 Outlet valve, 7 Displacement chamber, 8 Inlet valve, 9 High-pressure port to fuel rail, 10 Fuel inlet (low pressure), 11 Cam ring, 12 Axial seal (slide ring seal), 13 Static seal, 14 Drive shaft.

The fuel delivered by the electric fuel pump enters the high-pressure pump through the fuel inlet. The pump plungers contain transverse and longitudinal ports, through which the fuel enters the displacement chambers of the three delivery barrels. As the pump plunger travels from top to bottom dead center, the fuel is drawn in through the inlet valve. In the delivery stroke, the drawn-in fuel is compressed as the pump plunger travels from bottom to top dead center and delivered through the outlet valve into the high-pressure area.

The delivery quantity of the continuous-delivery high-pressure pump is proportional to the rotational speed. The three barrels deliver fuel at circumferential offsets of 120° in order to ensure overlapping and therefore continuous delivery. This gives rise to minimal pressure pulsations only. This means that, when compared with demand-regulated systems with single-plunger pumps, less demands have to be placed on the pump connections and piping. Furthermore, there is no need for a low-pressure attenuator. On the downside, continuous delivery at high pressure means that it has a higher power loss than a demand-regulated pump.

When the pump operates at constant rail pressure or at part load, the pressure of the excess fuel is reduced to admission pressure level by the pressure-control valve mounted on the rail and returned to the intake side of the high-pressure pump. The pressure level in the high-pressure circuit is regulated and adjusted by means of the engine control unit, which specifically actuates the pressure-control valve.

Figure 14: Pressure-control valve
1 Electrical connection,
2 Valve spring,
3 Solenoid coil,
4 Solenoid armature,
5 Valve needle,
6 Sealing rings (O-rings),
7 Drain hole,
8 Valve ball,
9 Valve seat,
10 Inlet with inlet strainer.

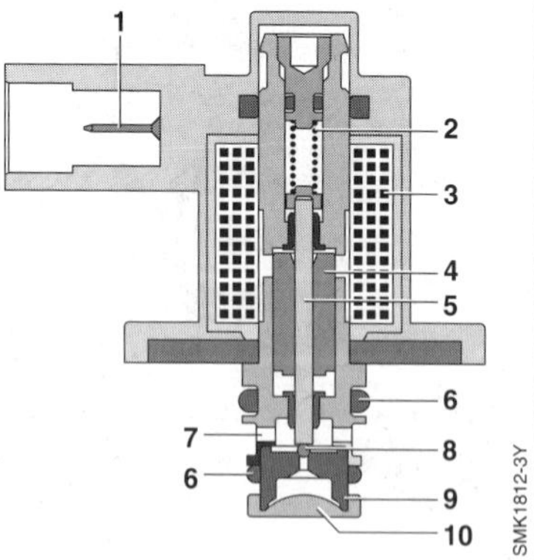

SMK1812-3Y

Pressure-control valve

The pressure-control valve (Figure 14) is a proportional control valve which is closed at zero current and actuated by means of a pulse-width-modulated signal. During operation, the energizing of the solenoid coil sets a magnetic force which relieves the load on the spring, lifts the valve ball off the valve seat, and thereby alters the flow cross-section. The pressure-control valve sets the desired rail pressure as a function of the pulse duty factor. The excess fuel delivered by the high-pressure pump is diverted into the low-pressure circuit.

A pressure-limiting function is integrated by way of the valve spring to protect the components against unacceptably high rail pressures, for example, in the event of actuation failure. If one or more of the pump barrels should fail, emergency operation is possible with the intact barrels or by means of the electric fuel pump with admission pressure.

Fuel rail

Intake-manifold injection

The function of the fuel rail is to store the fuel required for injection, to attenuate pulsations and to ensure uniform distribution to all the fuel injectors. The fuel injectors are mounted directly on the rail. Next to the fuel injectors it is possible in systems with fuel return to integrate a fuel-pressure regulator and if necessary a fuel-pressure attenuator in the rail.

Local pressure fluctuations caused by resonance when the fuel injectors open and close is prevented by careful selection of the fuel-rail dimensions. As a result, irregularities in injected fuel quantity which can arise as a function of load and engine speed are avoided.

Gasoline direct injection

The fuel rail has the task of storing and distributing the volume of fuel required for each operating point. The fuel is stored by way of its volume and compressibility. In this way, the volume is engine-power-dependent and must be adapted to the relevant engine demand (injected fuel quantity) and pressure range. The volume ensures attenuation in the high-pressure range, i.e. pressure fluctuations in the rail are compensated.

The add-on components for gasoline direct injection are mounted on the rail: the high-pressure fuel injectors, the pressure sensor for regulating the high pressure and – for 1st-generation fuel-injection systems – the pressure-control or pressure-limiting valve.

The rail for 1st-generation fuel-injection systems is designed for a pressure range up to 12 MPa (plus 0.5 MPa pressure-limiting valve opening pressure). For the 2nd generation, the pressure range stretches up to 20 MPa (plus 5.0 MPa pressure-limiting valve opening pressure). The burst pressure is accordingly higher.

Fuel-pressure regulator

Function

With manifold injection the amount of fuel injected by the fuel injector depends on the injection period and the pressure differential between the fuel pressure in the fuel rail and the back pressure in the manifold. On fuel systems with fuel return, the influence of pressure is compensated for by a fuel-pressure regulator which maintains the differential between fuel pressure and manifold pressure at a constant level. This pressure regulator permits just enough fuel to return to the fuel tank so that the pressure drop across the fuel injectors remains constant. In order to ensure that the fuel rail is efficiently flushed, the fuel-pressure regulator is normally located at the end of the rail.

On returnless fuel systems, the fuel-pressure regulator is part of the fuel-delivery module in the fuel tank. The fuel-rail pressure is maintained at a constant level with reference to the ambient pressure. This means that the pressure differential between fuel-rail pressure and manifold pressure is not constant and must be taken into account when the injection duration is calculated.

Mounting on the rail

The fuel-pressure regulator (Figure 15) is of the diaphragm-controlled overflow type. A rubber-fabric diaphragm divides the pressure regulator into a fuel chamber and a spring chamber. Through a valve holder integrated in the diaphragm, the spring

Figure 15: Fuel-pressure regulator for mounting on the rail

1 Intake-manifold port (omitted for application in fuel-delivery module),
2 Compression spring,
3 Compression spring,
4 Valve holder,
5 Diaphragm,
6 Valve,
7 Nozzle,
8 Fuel supply,
9 Fuel return.

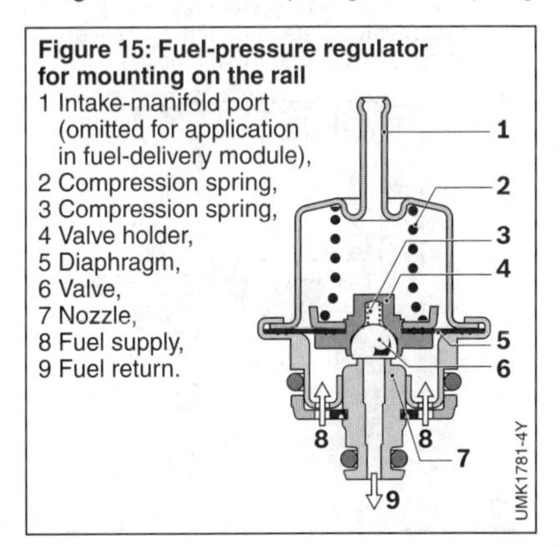

forces a movable valve plate against the valve seat so that the valve closes. As soon as the pressure applied to the diaphragm by the fuel exceeds the spring force, the valve opens again and permits just enough fuel to flow back to the fuel tank that equilibrium of forces is achieved again at the diaphragm.

The spring chamber is pneumatically connected with the intake manifold behind the throttle valve. As a result, the intake manifold vacuum also has an effect in the spring chamber. There is therefore the same pressure ratio at the diaphragm as at the fuel injectors. This means that the pressure drop across the fuel injectors is solely a function of spring force and diaphragm surface area, and therefore remains constant.

Installation in the fuel-delivery module

For applications in the fuel-delivery module the pressure regulator can have a simpler design that the pressure regulator mounted on the rail. These simple pressure regulators (Figure 16) have therefore become the accepted choice. The fuel flows through a coarse prefilter (protecting the pressure regulator against fouling) against the valve plate. An annular gap opens when the fuel pressure exceeds the spring force that presses the valve plate against the seal seat. The fuel flows through this annular gap and returns through the pressure regulator housing directly to the swirl pot.

Figure 16: Fuel-pressure regulator for applications in the fuel-delivery module
1 Adjusting screw (for system pressure), 2 Spring, 3 Valve plate, 4 O-ring, 5 Prefilter, 6 Housing, 7 Discharge to swirl pot.

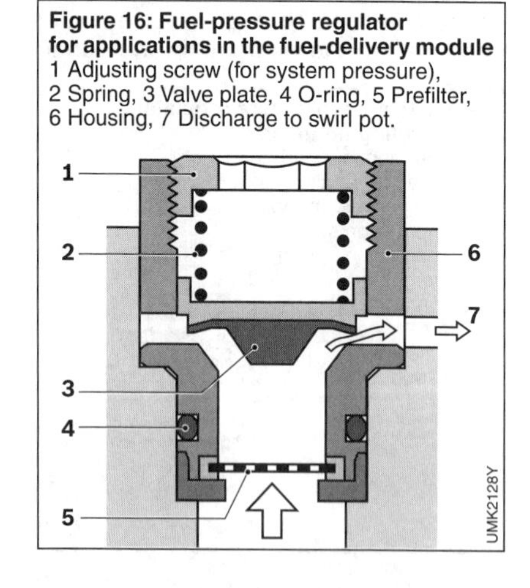

Fuel-pressure attenuator

Function, design and operating principle

The repeated opening and closing of the fuel injectors, high-pressure pump pulsations (the principal cause today) and the periodic supply of fuel when electric positive-displacement fuel pumps are used lead to fuel-pressure oscillations. These can cause pressure resonances which adversely affect fuel-metering accuracy. It is even possible that under certain circumstances noise can be caused by these vibrations being transferred to the fuel tank and the vehicle bodywork through the mounting elements of the electric fuel pump, fuel lines, and fuel rail. These problems are alleviated by the use of special-design mounting elements and fuel-pressure attenuators.

The fuel-pressure attenuator has a similar design to the fuel-pressure regulator. Here too, a spring-loaded diaphragm separates the fuel chamber from the air chamber. The spring force is set in such a way that the membrane rises from its seat as soon as the fuel delivery pressure reaches its operating range. This means that the fuel chamber is variable and not only absorbs fuel when pressure peaks occur, but also releases fuel when the pressure drops. In order to always operate in the most favorable range when the absolute fuel pressure fluctuates due to conditions at the manifold, the spring chamber can be provided with an intake-manifold connection.

Similar to the fuel-pressure regulator, the fuel-pressure attenuator can also be attached to the rail or installed in the fuel line. In the case of gasoline direct injection, it can also be attached to the high-pressure pump.

Mixture formation

Basic principles

A/F (Air/fuel) mixture

To be able to operate, a spark-ignition engine requires an air-fuel mixture. The mixture for ideal, theoretically complete combustion requires a mass ratio of 14.7:1 (stoichiometric ratio), i.e. 14.7 kg of air are required to burn 1 kg of fuel. Or: 1 *l* of fuel burns completely in roughly 9,500 *l* of air.

The specific fuel consumption is essentially dependent on the air-fuel ratio. Excess air is required in order to ensure genuine complete combustion, and thus as low a fuel consumption as possible. However, limits are imposed due to the flammability of the mixture and the available combustion time.

The A/F mixture also has a decisive impact on the efficiency of exhaust-gas treatment systems. State-of-the-art technology is represented by the three-way catalytic converter. However, it needs a specific stoichiometric A/F ratio in order to operate at maximum efficiency. This can reduce damaging exhaust gas components by higher than 99 %. The homogeneous engines available today are therefore operated with a stoichiometric mixture as soon as their operating status allows this. By way of departure, individual direct-injection engines are also operated with a lean mixture (stratified-charge operation) in the interest of fuel economy.

Certain engine operating statuses require mixture adaptation. Selective modifications to the mixture composition are required e.g. in the case of a cold engine.

Excess-air factor λ

The excess-air factor or air ratio λ (lambda) has been chosen to designate the extent to which the actual air/fuel mixture differs from the theoretically necessary mass ratio (stoichiometric ratio 14.7:1). λ indicates the ratio of inducted air mass to air requirement with stoichiometric combustion.

Stoichiometric ratio

$\lambda = 1$: The inducted air mass corresponds to the theoretically required air mass.

Air deficiency

$\lambda < 1$: This indicates air deficiency and therefore a rich A/F mixture. The maximum engine output results at $\lambda = 0.85$ to 0.95.

Enriching the air-fuel mixture may be necessary to protect components against thermal overload (component protection). This reduces the exhaust-gas temperature, since the fuel is converted not completely to CO_2, but to CO. The enthalpy for the reaction to CO is lower than for complete conversion to CO_2. A lower thermal load e.g. for exhaust-gas turbocharger, λ sensor and catalytic converter results from the lower exhaust-gas temperature.

Figure 1: Effects of excess-air factor λ on specific fuel consumption b_e and uneven running at constant power output

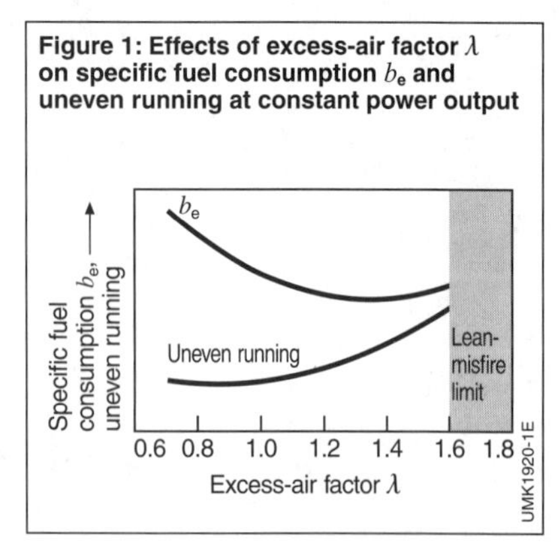

Figure 2: Effect of excess-air factor λ on pollutant composition

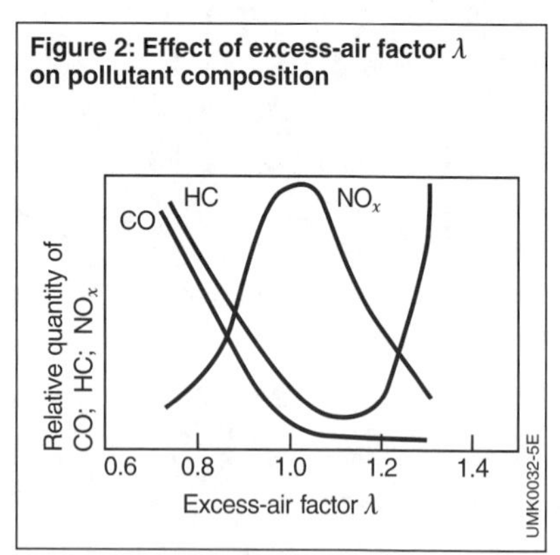

However, this also entails increased consumption. Enrichment is therefore only performed for high-load points for which an exhaust-gas temperature of over 900 °C would be reached.

Air surplus
$\lambda > 1$: In this range, there is an air surplus and thus a lean mixture.

This excess-air factor is characterized by reduced fuel consumption (Figure 1), but also by reduced power output. The maximum value for λ – the "lean-misfire limit" – that can be achieved is very strongly dependent on the construction of the engine and the mixture-formation system used. The mixture is no longer ignitable at the lean-misfire limit because of the lower number of HC molecules. Combustion misses occur, and this is accompanied by a marked increase in uneven running.

Influence of λ on fuel consumption and exhaust-gas composition
Spark-ignition engines achieve the lowest fuel consumption at constant engine output dependent on the engine at 20 to 50 % air surplus (λ = 1.2 to 1.5). This potential can however only be achieved with lean-burn concepts (homogeneous-lean or stratified-charge operation)

For a typical engine with intake-manifold injection, Figures 1 and 2 show the dependency of the specific fuel consumption and uneven running, as well as development of pollutants, on the excess-air factor at constant engine output. It can be deduced from these graphs that there is no ideal excess-air factor at which all the factors assume the most favorable value. In order to implement "optimal" consumption at "optimal" power output, excess-air factors of λ = 0.95 to 1.05 have proven conducive to the achievement of objectives for engines with intake-manifold injection and for homogeneous direct-injection engines – above all in view of exhaust-gas treatment with the three-way catalytic converter, which can optimally convert all three pollutant components only in a narrow range around $\lambda = 1$.

To achieve this, the inducted air mass is recorded precisely and an exactly metered fuel mass is apportioned by means of λ regulation in the engine-management system.

Intake-manifold injection
For optimum combustion in engines with intake-manifold injection that are common today, not only is a precise injected fuel quantity necessary, but also a homogeneous A/F mixture. This requires efficient fuel atomization. If this precondition is not satisfied, large fuel droplets will precipitate on the intake manifold or on the combustion-chamber walls. These large droplets cannot fully combust and will result in increased hydrocarbon emissions (HC).

Gasoline direct injection
Direct-injection engines with charge stratification involve different combustion conditions so that the lean-misfire limit occurs at significantly higher global λ values. In the part-load range, these engines can therefore be operated with considerably higher excess-air factors (up to λ = 4). Here too the area around the spark plug requires locally a stoichiometric mixture for good flammability.

Unfortunately, in this case of a globally lean mixture, only two of the three ways of the three-way catalytic converter function. CO emissions and unburnt hydrocarbons can still be oxidized since sufficient oxygen is available. However, the third chief reaction, reducing the NO_x emissions, cannot be carried out sufficiently in lean-operation mode, since insufficient reduction partners are available. For this reason, in addition to catalytic exhaust-gas treatment with a three-way catalytic converter, a further exhaust-gas treatment facility must be installed for NO_x reduction (e.g. an NO_x storage catalytic converter).

Mixture-formation systems

It is the job of fuel-injection systems, or carburetors, to provide an A/F mixture which is adapted as well as possible to the relevant engine operating status. Without fuel-injection systems, particularly electronic systems, it would not be possible to maintain the increasingly narrowly defined limits for mixture composition and thus to adhere to the stipulations of the emission-control legislation. These requirements and the significant improvements in fuel metering and mixture formation together with the benefits in terms of fuel consumption, driving performance and power output have meant that, in the automotive sector, gasoline injection has completely superseded the carburetor in moder engines.

Until the beginning of this century, the automotive industry almost exclusively used systems in which A/F mixture formation takes place outside the combustion chamber (intake-manifold injection, Figure 3a). Systems with internal A/F mixture formation, i.e. where the fuel is injected directly into the combustion chamber (gasoline direct injection, Figure 3b), are designed to reduce fuel consumption even further and to increase power output, and have therefore become increasingly more important.

Figure 3: Schematic representation of fuel-injection systems
a) Intake-manifold injection,
b) Gasoline direct injection.
1 Fuel,
2 Air,
3 Throttle device,
4 Intake manifold,
5 Fuel injectors,
6 Engine,
7 High-pressure fuel injectors.

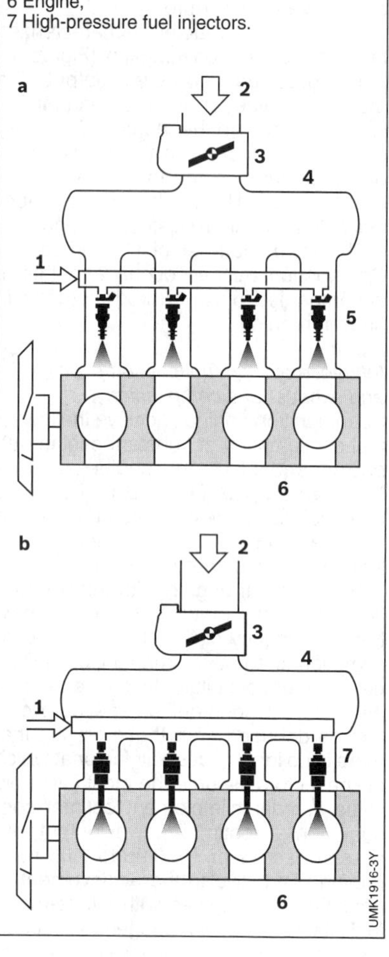

Intake-manifold injection

Gasoline injection systems for external mixture formation are characterized by the fact that the A/F mixture is created outside of the combustion chamber (in the intake manifold). The state of the art is represented by electronic intake-manifold injection systems where the fuel is injected cyclically for each individual cylinder directly ahead of the intake valves (Figure 4).

Systems that are based on mechanical-continuous fuel injection (K-Jetronic) or a central fuel injection arranged upstream of the throttle valve (Mono-Jetronic) are no longer of any significance for new developments.

Requirements

The high requirements for engine smooth-running and exhaust emissions make high demands on the A/F mixture composition in each power cycle. Precise injection timing is vital, as well as precise metering of the injected fuel mass as a factor of the engine intake air. In electronic multipoint fuel-injection systems, therefore, not only is each engine cylinder assigned an electromagnetic fuel injector, but this fuel injector is also activated individually for each cylinder. The engine control unit has the task of calculating both the required fuel mass for each cylinder and the correct start of injection for the fuel mass drawn in and the current engine operating status. The injection time required to inject the calculated fuel mass is a function of the opening cross-section of the fuel injector and the pressure differential between the intake manifold and the fuel supply system.

Fuel system

In intake-manifold injection systems, fuel is sent via the electric fuel pump (see Fuel supply), the fuel supply lines, and the fuel filter at typical system pressures of 3 to 7 bar to the fuel rail, which ensures that the fuel is evenly distributed to the fuel injectors. How the fuel is prepared by the fuel injectors is extremely important for the quality of the A/F mixture. It is essential that the fuel is atomized into very fine droplets. The spray shape and spray dispersal angle of the fuel injectors are adapted to the geometric shape of the intake manifold and cylinder head (see Intake-manifold fuel injector).

Figure 4: Principle of intake-manifold injection
1 Cylinder with piston, 2 Exhaust valves, 3 Ignition coil with spark plug, 4 Inlet valves, 5 Fuel injector, 6 Intake manifold.

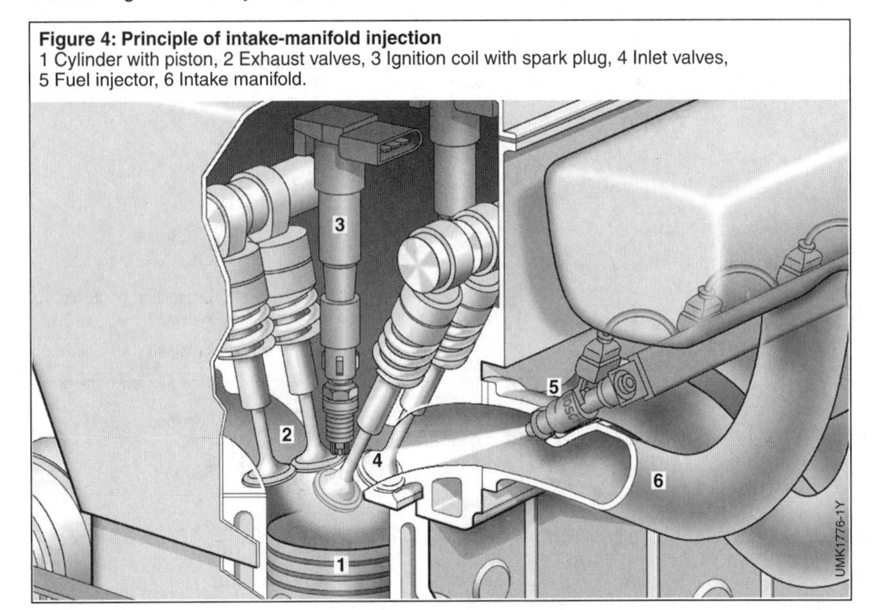

Fuel injection

The exactly metered fuel mass is injected directly upstream of the cylinder intake valve(s). As the air enters through the throttle valve there is a pulse exchange with the finely atomized fuel, and the droplets on the surface mix increasingly with the air and therefore evaporate to a large extent (Figure 5). The required air-fuel mixture can therefore be formed at the right time by varying the point of injection.

The amount of time available for mixture formation can be increased by injecting the fuel into the intake valves that are still closed. How much air-fuel mixture is created directly after injection depends on the quality of the fuel injector (primary droplet size). In this case, the primary evaporation occurs only when the intake valves open in response to the high flow velocities with an initially very small valve gap.

A proportion of the fuel precipitates on the intake duct (part of the intake manifold leading directly to the combustion chamber) and on the lower edge of the intake valves as wall film; the thickness of this film essentially depends on the pressure in the intake manifold, and thus on the engine load condition. In the case of non-stationary (dynamic) engine operation, this precipitation can lead to temporary deviations in the desired λ value ($\lambda = 1$), which means that the fuel mass stored in the wall film is to be kept as low as possible.

Wall-coating effects in the intake duct, particularly in cold-start conditions, cannot be neglected either. As fuel does not evaporate sufficiently, more fuel is required initially in the starting phase in order to create an ignitable A/F mixture. When the intake-manifold pressure then drops, parts of the previously formed wall film will vaporize. This may result in increased HC emissions if the catalytic converter is not running at operating temperature.

Irregular fuel injection may also result in the formation of wall films in the combustion chamber, and may in turn become critical emission sources. Defining the geometric alignment of fuel sprays ("spray targeting") will allow the selection of suit-

Figure 5: Mechanisms and factors influencing mixture formation in intake-manifold injection

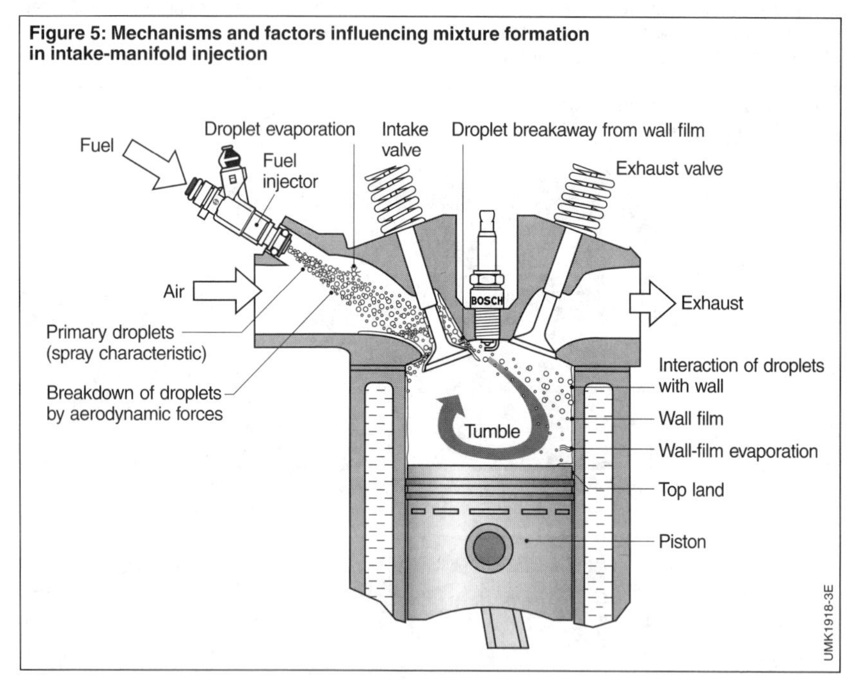

able fuel injectors which will control and minimize manifold-wall fuel condensation in the area of the intake duct and the intake valves.

The intake manifolds used can be optimally adapted to the combustion air flow and the dynamic gas requirements of the engine.

Gasoline direct injection

In gasoline direct injection (GDI), unlike intake-manifold injection, pure air flows through the intake valves into the combustion chamber. Only then is the fuel injected into the air via an injector (see high-pressure fuel injector) located directly in the cylinder head (internal mixture formation, Figure 6). There are basically two main operating modes. Fuel injection in the intake stroke is called homogeneous operation, while fuel injection during compression is called stratified-charge operation. There are also various special modes, which are either a mixture of the two main operating modes or a slight variation of them.

In stratified-charge operation, the air is not restricted; the A/F mixture is lean. The excess air in the exhaust gas prevents conversion of the nitrogen oxides by means of a three-way catalytic converter. These direct-injection systems therefore require exhaust-gas treatment with an additional NO_x storage catalytic converter. For this reason, mostly direct-

Figure 6: Principle of gasoline direct injection
1 Cylinder with piston, 2 Intake valves, 3 Ignition coil with spark plug, 4 Exhaust valves, 5 High-pressure fuel injector, 6 Rail.

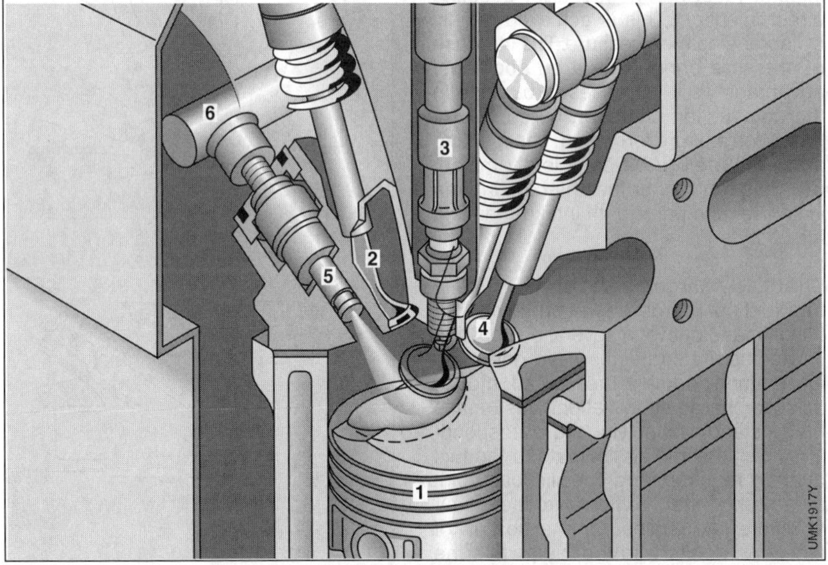

UMK1917Y

injection systems that work exclusively in the homogeneous mode are currently being placed on the market.

Homogeneous operation

In homogeneous operation, mixture formation is similar to intake-manifold injection. The mixture is formed in a stoichiometric ratio (λ = 1). However, from a mixture formation point of view, there are some differences. For instance, there is no flow process around the intake valve to promote mixture formation, and there is much less time available for the mixture formation process itself. Whereas injection can take place across the overall 720° crankshaft (stored and synchronous with induction) of the four operating cycles in the case of intake-manifold injection, only an injection window of 180° crankshaft remains in the case of gasoline direct injection.

Fuel injection is only permitted in the induction stroke. This is because, prior to this, the exhaust valves are open and unburned fuel would otherwise escape into the exhaust-gas train. This would cause high HC emissions and catalytic converter problems. In order to deliver a sufficient quantity of fuel in this shortened period, the fuel flow through the injector must be increased for gasoline direct injection. This is achieved mainly by increasing fuel pressure. The increase in pressure brings with it an additional advantage as it increases turbulence in the combustion chamber, which in turn promotes mixture formation. The fuel and air can therefore be completely mixed, even though the air/fuel interaction time is shorter compared with intake-manifold injection.

Stratified-charge operation

In stratified-charge operation, a distinction is made between several combustion strategies. All strategies have one thing in common, namely they all attempt to achieve charge stratification. This means that, instead of supplying the corresponding stoichiometric air flow rate to the fuel quantity required for a specific load point by adjusting the throttle valve, the full air flow rate is supplied. Only a portion of the air interacts with the fuel, while the rest of the fresh air surrounds this stratified-charge cloud. This dethrottling action facilitates, together with the cooling effects of the directly injected fuel (evap-

Figure 7: Combustion processes for gasoline direct injection
a) Wall-directed combustion,
b) Air-directed combustion,
c) Spray-directed combustion.
1 High-pressure fuel injector,
2 Spark plug.

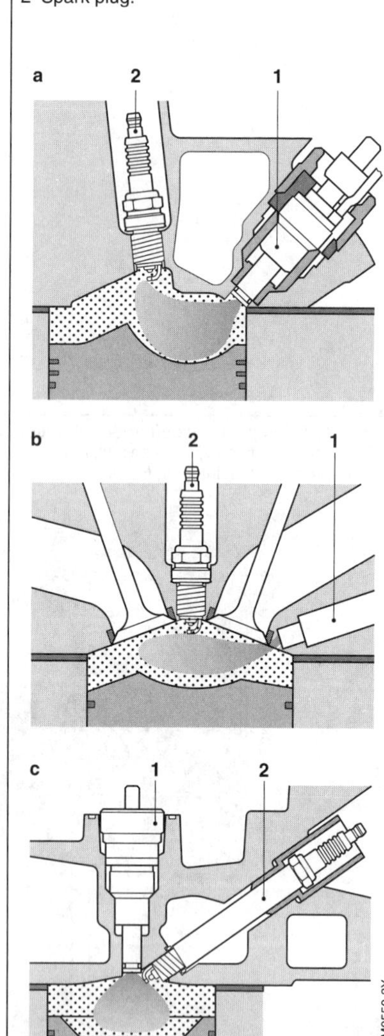

orative emission), a possible increase in compression and thus considerable fuel-saving potential.

Wall-directed combustion process
In a wall-directed combustion process, fuel is injected into the combustion chamber from the side (Figure 7a). A recess in the piston crown deflects the fuel spray in the direction of the spark plug. Mixture formation takes place on the path from the injector tip to the spark plug. As the mixture-formation time is very short during injection in the compression stroke (stratified-charge operation), the fuel pressure for this system must often be even higher than for homogeneous operation. The increased fuel pressure shortens the injection time and increases interaction with the air because pulse reflection is greater.

The disadvantage of the wall-directed combustion process is that fuel condenses on the piston wall, which increases HC emissions. As mixture-formation time is short, the stratified-charge cloud usually contains rich mixture zones at higher loads, and this increases the risk of soot production. At low loads, the pulse of the fuel droplets, which is used as a means of transporting the stratified-charge cloud to the spark plug, is low due to the small fuel mass. As a result, the flow must usually be restricted here so that the fuel meets with a lower density of air and thus a lower air resistance.

Air-directed combustion process
In principle, an air-directed combustion process works in exactly the same way as a wall-directed process. The main difference is that the stratified-charge cloud does not interact directly with the piston recess. Instead, it moves on a cushion of air (Figure 7b). The cushion of air is generated by the air present in the cylinder. In contrast to the wall-directed combustion process, with air-directed combustion the fuel-spray injection angle is much flatter such that the fuel spray cannot penetrate the full amount of air up to the piston. This solves the problem of fuel condensing on the piston recess.

Compared with wall-directed combustion processes, the air flows in air-directed combustion processes cannot be fully reproduced. This gives rise to large variations from injection to injection and thus frequently to poor combustion stability right through to individual misfires.

Often, the actual combustion processes are a mixture of wall-directed and air-directed processes, depending on the operating point in each case.

Spray-directed combustion process
The spray-directed combustion process is visually different from the other two processes in that the injector is installed at a different location. It is located top center and injects vertically down into the combustion chamber (Figure 7c). The fuel spray is not deflected. As a result of the high injection pressure (200 bar), an air-fuel mixture forms around the fuel spray already during the injection process. Located directly next to the injector is the spark plug, which ignites the air-fuel mixture.

As a result, with the spray-directed combustion process the mixture-formation time is very short. This requires an even higher fuel pressure for the spray-directed combustion process. This combustion process can eliminate the disadvantages of fuel condensing on the manifold walls, air-flow dependency, and flow restriction at low loads. It therefore has the greatest potential for fuel saving. Nevertheless, the short mixture formation time is a huge challenge for the fuel-injection and ignition systems.

Other operating modes
In addition to homogeneous and stratified-charge operation, there are also special operating modes. They include "changeover between operating modes" (homogeneous-stratified mode), "catalytic-converter heating", and "knock protection" (homogeneous-split mode), and "homogeneous-lean mode". These operating modes are sometimes very complex and only ever represent short, temporary states. Therefore they are frequently not operating modes for longer operation.

Fuel injectors

Essentially, the mixture-formation components must ensure that the air-fuel mixture is formed in the most optimal way possible to facilitate combustion. In intake-manifold injection, this is mainly the task of the fuel injector, while in gasoline direct injection, the high-pressure fuel injector can be assisted by a turbulence flap.

Intake-manifold fuel injector

Design and function

Essentially, electromagnetic fuel injectors are comprised of the following components: the valve housing with electrical connection and hydraulic port (Figure 8), the solenoid coil, the moving valve needle with solenoid armature and valve ball, the valve seat with injection-orifice plate, and the valve spring.

A filter strainer in the fuel inlet protects the fuel injector against contamination. The sealing ring (O-ring) on the hydraulic port seals off the injector at the fuel rail. The lower sealing ring provides the seal between the injector and the intake manifold.

When the solenoid coil is de-energized, the valve needle and valve ball are pressed against the cone-shaped valve seat by the valve spring and the force exerted by the fuel pressure. The fuel-supply system is thus sealed off from the intake manifold.

When the solenoid coil is energized, this generates a magnetic field which attracts valve-needle solenoid armature. The valve ball lifts off the valve seat (Figure 10) and the fuel is injected. When the excitation current is switched off, the valve needle closes again due to spring force.

The fuel is atomized by an injection-orifice plate. The orifices are stamped out of the plate and ensure that the injected fuel quantity remains highly reproducible. The spray pattern and the atomization quality of the fuel leaving the injector is produced by the number of orifices and their configuration (maximum twelve).

The injected fuel quantity per unit of time is determined mainly by the system pressure in the fuel-supply system (typically 3 to 4 bar, corresponding to 300 to 400 kPa), the back pressure in the intake manifold, and the geometry of the fuel-exit area.

Spray formation and spray targeting

An injector's spray formation, i.e. its spray shape, spray angle, and fuel-droplet size, influences the formation of the air/fuel mixture. Individual geometries of intake manifold and cylinder head make it necessary to have different types of spray formation. Different spray-formation variants

Figure 8: Fuel injector (Bosch) for intake-manifold injection

1 Hydraulic port,
2 Sealing ring (O-ring),
3 Valve housing,
4 Electrical connection,
5 Plastic clip with inserted pins,
6 Filter strainer,
7 Internal pole (stop for solenoid armature),
8 Valve spring,
9 Solenoid coil,
10 Valve needle with solenoid armature,
11 Valve ball,
12 Valve seat,
13 Injection-orifice plate.

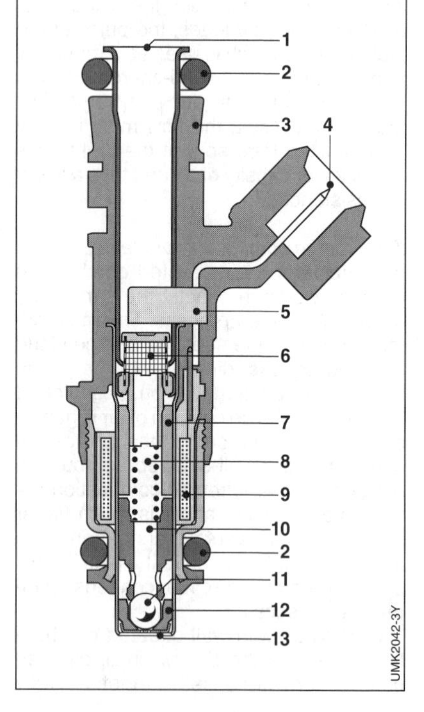

are available in order to satisfy with these requirements.

Tapered spray
Individual fuel sprays emerge through the openings in the injection-orifice plate (Figure 9a). These fuel sprays combine to form a spray taper.

Tapered-spray injectors are typically used in engines which feature only one intake valve per cylinder. However, the tapered spray is also suitable for two intake valves.

Dual spray
Dual-spray formation (Figure 9b) is often used in engines which feature two intake valves per cylinder. The holes in the injection-orifice plate are arranged in such a way that two fuel sprays – which are each formed from a number of individual sprays that produce a homogeneous spray shortly after leaving the injection orifices – impact ahead of the intake valves or against the web between the intake valves.

Gamma angle
Compared to the injector's principle axis, the fuel spray in this case (single spray and dual spray) is at an angle: the offset spray angle (Figure 9c).

Injectors with this spray shape are mostly used when installation conditions are difficult, for example when the injectors are installed so steeply in the intake manifold that the required spray target-

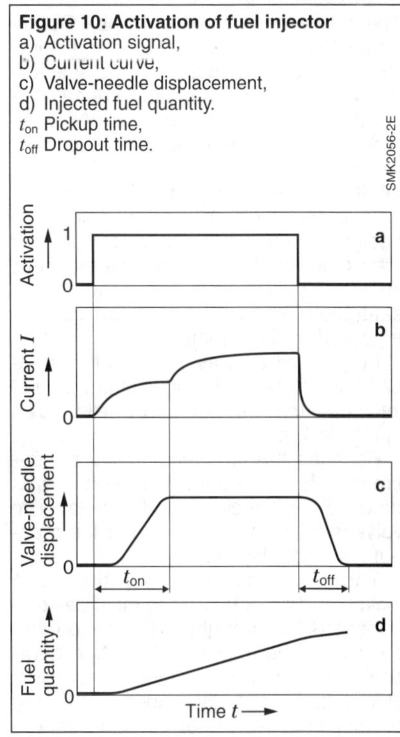

Figure 10: Activation of fuel injector
a) Activation signal,
b) Current curve,
c) Valve-needle displacement,
d) Injected fuel quantity.
t_{on} Pickup time,
t_{off} Dropout time.

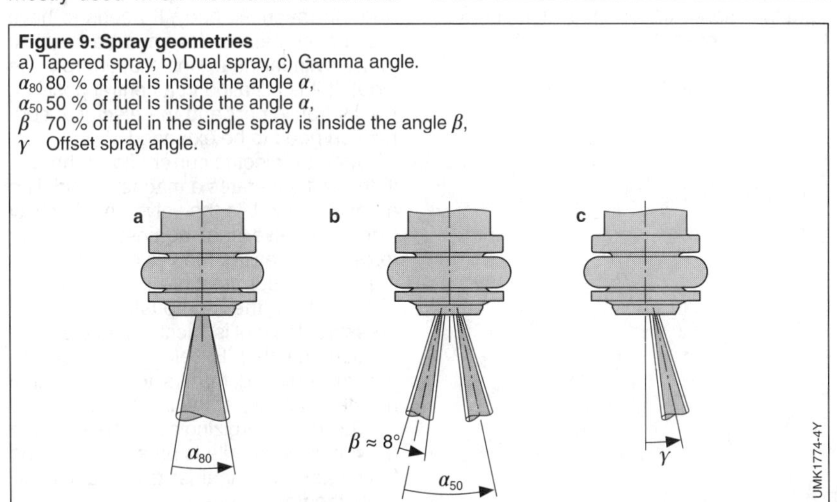

Figure 9: Spray geometries
a) Tapered spray, b) Dual spray, c) Gamma angle.
α_{80} 80 % of fuel is inside the angle α,
α_{50} 50 % of fuel is inside the angle α,
β 70 % of fuel in the single spray is inside the angle β,
γ Offset spray angle.

ing can only be achieved by the use of a gamma-angle spray.

Electrical activation

The output module in the engine ECU actuates the injector with a switching signal (Figure 10a). The current in the solenoid coil rises (Figure 10b) and causes the valve needle to lift (Figure 10c). After the pickup time has elapsed, the maximum valve-needle displacement is achieved at the solenoid-armature stop at the internal pole. Depending on the solenoid-valve design, the current continues to rise after the armature stop because the magnetic circuit at this point is not yet at 100 % saturation. The kink in the current curve is triggered by the armature stop.

Fuel is sprayed as soon as the valve ball lifts off its seat. The quantity of fuel injected during an injection pulse is shown in Figure 10d.

Because the magnetic field is not decreased abruptly after activation is shut down, the valve closes with a delay. The valve is fully closed again after the dropout time has elapsed.

The non-linearity during the valve pickup and dropout phases must be compensated for throughout the period that the injector is activated (injection duration). The speed at which the valve needle lifts off its seat is also dependent on the battery voltage. Battery-voltage-dependent injection-duration extension corrects these influences.

Electromagnetic high-pressure fuel injector

The higher fuel pressure required for gasoline direct injection means that the fuel injector has to meet additional requirements. As a result, special high-pressure fuel injectors have been developed for gasoline direct injection.

Function

The function of the high-pressure fuel injector is to meter and atomize the fuel. Atomization ensures that the fuel is specifically mixed with the air in a specific area in the combustion chamber. Depending on the desired operating mode, the fuel is either concentrated in the vicinity of the spark plug (stratified charge) or evenly distributed throughout the combustion chamber (homogeneous distribution).

To ensure optimum mixture preparation, the fuel is injected at a pressure of up to 200 bar (20 MPa). The required quantity of fuel per combustion process is partly split into several single shot injections.

Design and function

The electromagnetic high-pressure fuel injector (Figure 11) comprises the housing, the valve seat, the valve needle with solenoid armature, the compression spring, and the magnetic circuit. This consists of the solenoid coil, the internal pole, and the magnetic body. It needs to have a compact, narrow design to be installed in the cylinder head (installation from the side). If it is installed in a central position, the hydraulic port and the electrical connection need to be extended.

When an electric current flows through it, the coil generates a magnetic field. The magnetic field lifts the valve needle away from the valve seat against the spring force and releases the valve outlet opening. As the fuel pressure is considerably higher than the combustion-chamber pressure, the fuel is injected into the combustion chamber. Excellent fuel atomization is achieved thanks to the suitable nozzle geometry at the injector tip.

When the energizing current is switched off, the valve needle is pressed by spring force back down against its seat and interrupts the flow of fuel.

With needle valves that open inward, the pressure in the rail helps the closing process. On opening, therefore, it acts in the opposite direction to the opening direction. As a result, due to the higher rail pressure, a stronger magnetic field is required than with conventional intake-manifold fuel injectors.

Spray formation

Reproducible fuel quantities can be metered by defining the opening and keeping the opening cross-section constant with the valve needle fully raised. Here, the fuel quantity is dependent on the pressure in the fuel rail, the backpressure reaction in the combustion chamber, and the valve opening time. The fuel-atomization process is supported by an appropriate valve-seat geometry and flow guidance.

In the second generation of high-pressure fuel injectors, the swirl-generating single-hole nozzles, which create a hollow-tapered spray, are replaced by multi-hole nozzles, which provide a clearly more flexible "spray layout" (spray pattern, droplet size, mass distribution of the individual sprays, etc.). These nozzles use alignment of the individual sprays adapted to the engine geometry (spray targeting) to increase "engine performance" (performance, engine behavior, emissions, smooth running, etc.).

Electrical activation

To ensure a defined and reproducible injection process, the high-pressure fuel injector has to be activated by a complex current curve. The microcontroller in the engine ECU delivers a digital signal for this purpose (Figure 12a). An output module uses this signal to generate the actuating signal (Figure 12b) for the injector.

A DC/DC converter in the engine ECU generates the booster voltage of 65 V. This voltage is required in order to bring the current up as quickly as possible in the booster phase to a high current value and thus to accelerate the injector needle as quickly as possible.

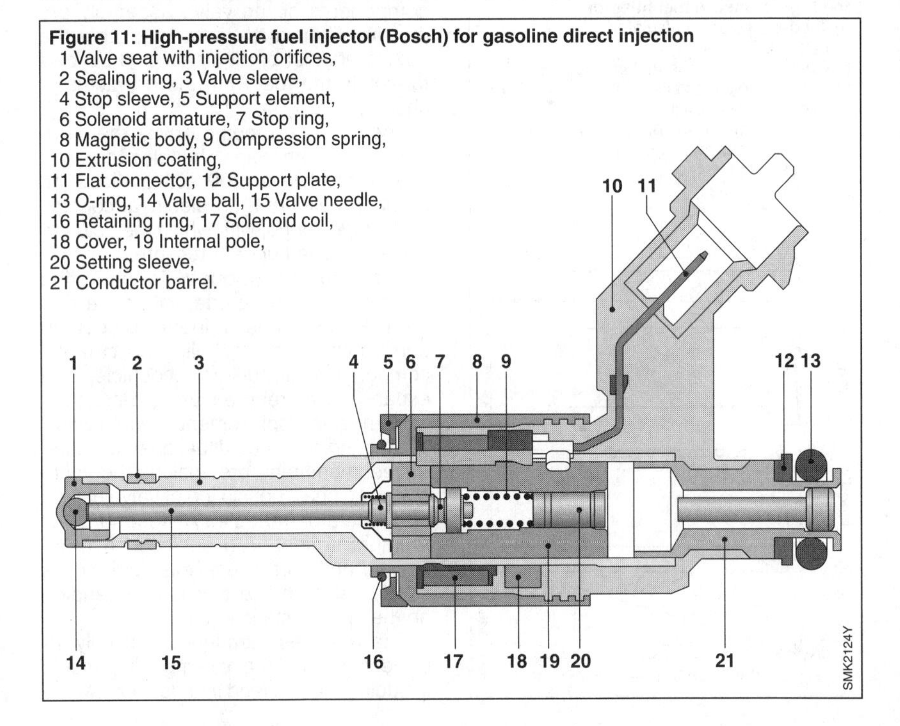

Figure 11: High-pressure fuel injector (Bosch) for gasoline direct injection
1 Valve seat with injection orifices,
2 Sealing ring, 3 Valve sleeve,
4 Stop sleeve, 5 Support element,
6 Solenoid armature, 7 Stop ring,
8 Magnetic body, 9 Compression spring,
10 Extrusion coating,
11 Flat connector, 12 Support plate,
13 O-ring, 14 Valve ball, 15 Valve needle,
16 Retaining ring, 17 Solenoid coil,
18 Cover, 19 Internal pole,
20 Setting sleeve,
21 Conductor barrel.

In the pickup phase, the valve needle then achieves the maximum opening lift (Figure 12c). When the fuel injector is open (maximum valve-needle lift), a lower control current (holding current) is sufficient to keep the valve open. With a constant injector-needle lift, the injected fuel quantity is proportional to the injection duration (Figure 12d).

Figure 12: Signal curves for activating the high-pressure fuel injector
a) Triggering signal calculated by the ECU,
b) Current curve in the fuel injector,
c) Valve-needle displacement,
d) Injected fuel quantity.
I_{max} Maximum current in the booster phase,
I_h Holding current,
t_{on} Pickup time,
t_{off} Dropout time.

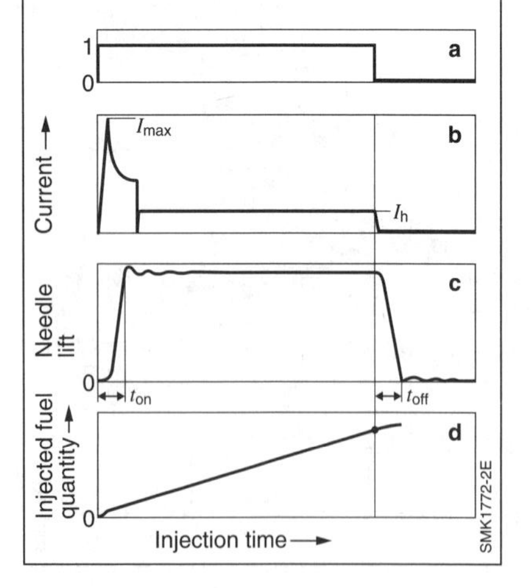

Piezo fuel injector

Design and function

In addition to displacement reduction (downsizing), the highly-effect spray-directed lean-burn process is used in gasoline direct injection as a means of reducing CO_2 emissions. The only mass-produced injector that does this currently is the piezo fuel injector.

The piezo fuel injector, which is operated similarly to the solenoid valve at 200 bar (20 MPa) fuel pressure, has a modular design. It contains the following function elements: valve assembly, actuator module, compensation element (coupler), and housing and connection components (Figure 13).

The valve assembly is responsible for forming the fuel as a defined spray and for proportioning the fuel. It essentially consists of the outward-opening, moving valve needle which is pretensioned by a closing spring against a housing body. The nominal valve-needle displacement is 33 µm, which is reached after an opening period of 180 µs. Further important components of the valve assembly are the fine filter in the inlet to protect the nozzle against particulates and bellows to isolate the fuel from the dry actuator chamber.

The actuator module drives the valve needle. It comprises the piezoceramic actuator with over 400 active slices, an insulating element, an electrical contact, and a tube spring, with which the actuator is pretensioned under pressure.

The coupler, through its operating principle, effects under all occurring thermal conditions a linear compensation between the metallic and ceramic components with different coefficients of expansion, thereby ensuring a constant valve-needle displacement (lift) and with it a reproducible injected fuel quantity. Here, a steel membrane pretensions the entire switching chain consisting of valve needle and actuator module. A second steel membrane keeps an enclosed hydraulic fluid under pressure; this fluid moves slowly between two cavities, depending on the applied static force.

The modules mentioned are fully installed by way of the housing components including the fuel feed in a defined switching-force chain. The direct mechanical coupling between actuator module and valve needle makes possible short switching times of up to 80 µs, which also permit partial-lift injections with exact fuel metering with only minimal lift losses.

A coding process integrated into the production sequences determines the electrical charge requirement of each piezo fuel injector, the value of which is stored by way of a data matrix code for later programming in the engine ECU. This compensates for the manufacturing tolerances and improves significantly the accuracy of fuel-quantity metering during operation.

Spray formation and spray targeting

The spray issued by an outward-opening nozzle differs fundamentally from that by a multi-hole nozzle. The free flow cross-section created when the valve needle opens ensures that fuel emerges in a taper shape at very high speed and a positionally stable 85° hollow cone with a spatially assigned boundary vortex is formed (Figure 14). The spray shape is to a large extent not dependent on the back pressure in the combustion chamber and provides for a stable ignition of the mixture in the case of injection during the compression stroke shortly before ignition and partly in the form of several, closely consecutive partial quantities.

Figure 14: Tapered spray pattern of piezo fuel injector
Fuel pressure 200 bar,
chamber pressure 6 bar (lab measurement).
a) Realistic representation,
b) Schematic representation.
1 Inner boundary vortex,
2 Outer boundary vortex.

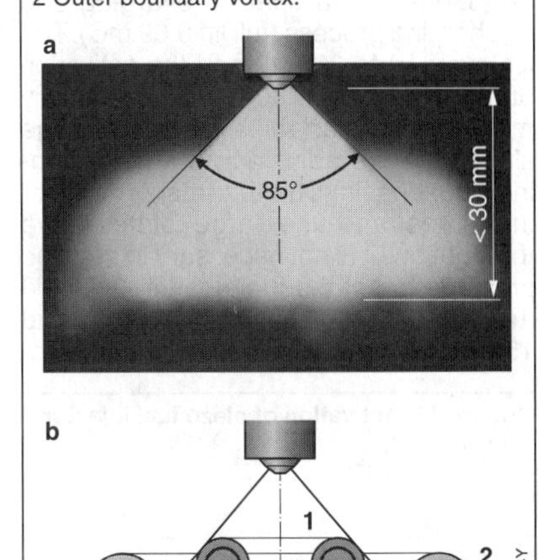

Figure 13: Function elements of piezo fuel injector (Bosch)
A Valve assembly, B Piezo actuator module, C Coupler.
1 Fine filter, 2 Closing spring, 3 Piezoceramic actuator,
4 Outward-opening nozzle, 5 Valve needle, 6 Bellows,
7 Tube spring (around the actuator),
8 Housing body, 9 Coupler,
10 Electrical connection,
11 Hydraulic port.

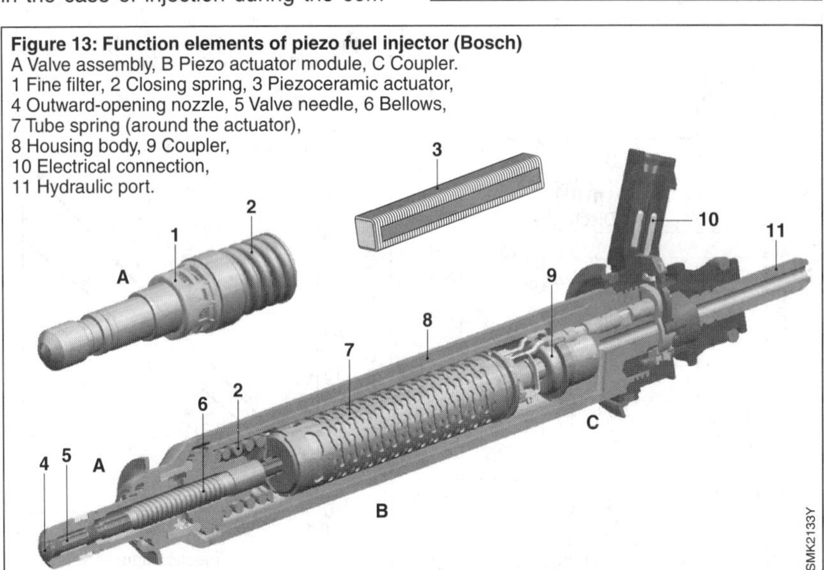

The droplets range in size from 10 to 15 μm. The outward-opening nozzle is on principle very robust against deposits.

Electrical activation

The piezo fuel injector is operated under charge-control conditions by a special output stage integrated in the ECU including a central DC/DC converter with up to 200 V (Figure 15). The valve is opened and closed by a specific charging and discharging process (full lift 0.69 mC). The lift level (10 to 33 μm) and the activation duration (80 to 5,000 μs) as primary manipulated variables and the edge slopes can be adjusted on an injection-individual basis. Due to the short switching times and the large static valve throughflow, it is not necessary to vary the pressure to set the fuel quantity. Injected fuel quantities of between 0.7 mg and 150 mg (Figure 16) and up to five injections per working cycle are achieved. Characteristic linearity is provided up to the shortest actuation times of 80 μs (0.7 mg at 200 bar). The minimum rest periods are only 50 μs; they much shorter than for a solenoid drive.

Several adaptation functions are used during engine operation to ensure that the smallest quantities are metered. On the one hand, physically model-based compensations for example of the temperature-dependent lift behavior take effect. On the other hand, the smallest quantity drifts and thus moment differences can be corrected individually for each cylinder and as a function of operating time. These functions act on the basis of the causative principle either on the valve-needle displacement or on the activation duration.

Figure 15: Activation of piezo fuel injector
a) For full lift,
b) For partial lift.

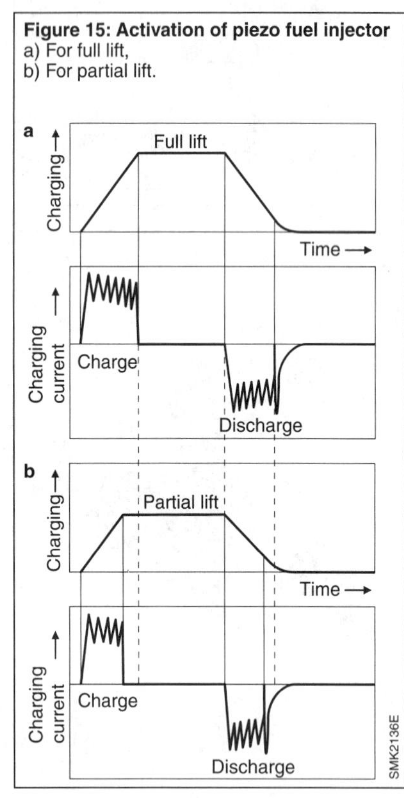

Figure 16: Injected fuel quantity as a function of the injection time
1 Full lift,
2 Partial lift.

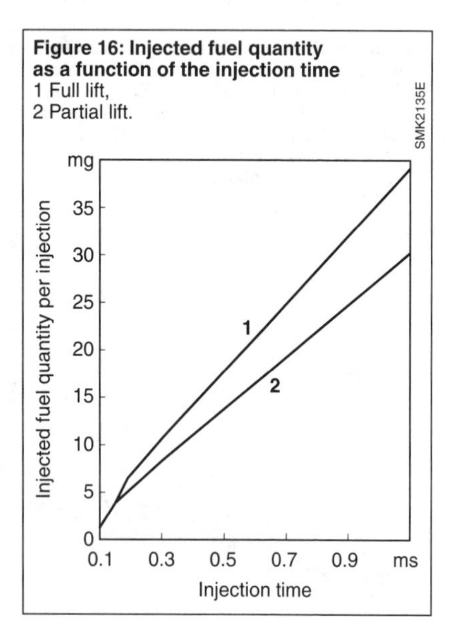

Ignition

Basic principles

Function

On a spark-ignition (SI) engine, the combustion process is initiated by an externally supplied ignition. The ignition is responsible for igniting the compressed A/F mixture at the right time. This is done by producing an electric spark between the electrodes of a spark plug in the combustion chamber.

Consistent, reliable ignition under all conditions is essential to ensure fault-free engine operation. Misfiring leads to combustion misfires and damage or destruction of the catalytic converter, poor exhaust emission figures, higher consumption and lower engine output.

Ignition spark

An electric spark can only occur at the spark plug if the necessary ignition voltage is exceeded (Figure 1). The ignition voltage is dependent on the spark-plug electrode gap and the density of the A/F mixture at the moment of ignition. It can be up to several 10 kV. After flashover, the voltage at the spark plug drops to the firing voltage. The firing voltage depends on the length of the spark plasma (electrode gap and excursion by the A/F mixture flow) and can rise from several 100 V to a few kV.

During the ignition-spark combustion time (spark duration), the ignition-system energy is converted into the ignition spark. After the spark has broken away, the voltage is damped and drops to zero.

Figure 1: Spark-plug voltage characteristic with static or semi-static A/F mixture
1 Ignition voltage,
2 Firing voltage,
t Spark duration.

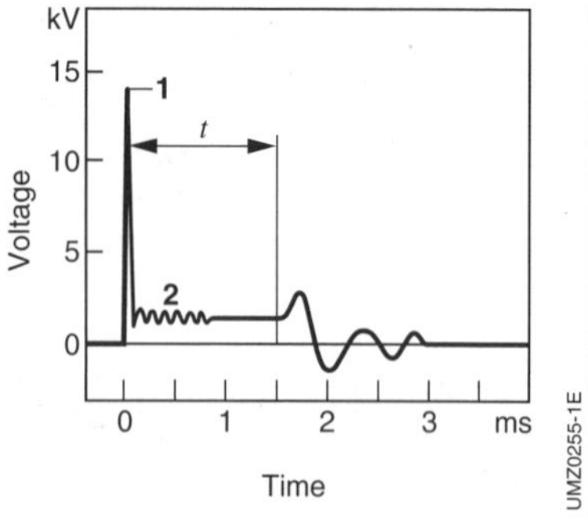

Mixture ignition and ignition energy

The electrical spark between the spark-plug electrodes generates a high-temperature plasma. If mixture conditions at the spark plug are suitable and sufficient energy has been supplied by the ignition system, the resulting arc develops into a flame front that propagates independently.

The ignition must guarantee this process under all engine operating conditions. Under ideal conditions (e.g. in a simulation of a combustion chamber, a "combustion bomb"), provided that the A/F mixture is stationary, homogeneous and stoichiometric, an energy of approx. 0.2 mJ is required to ignite the mixture by means of electric spark for each individual ignition process. In real engine operation, however, much higher energy levels are required. Some of the spark energy is converted at flashover, and the rest in the spark combustion phase.

Larger electrode gaps generate a larger arc, but require higher ignition voltages. Lean A/F mixtures or turbocharged engines have a higher ignition-voltage demand. At a given level of energy, the spark duration shortens as the ignition voltage increases. A longer spark duration generally stabilizes the combustion; lack of mixture homogeneity at the moment of ignition in the area of the spark plug can be compensated for by means of a longer spark duration. A/F-mixture turbulences such as those which occur in the stratified-charge mode with gasoline direct injection can divert the ignition spark to such an extent that it extinguishes. Follow-up sparks are then required to ignite the A/F mixture again.

The need for higher ignition voltages, longer spark durations, and the provision of follow-up sparks have resulted in the design of ignition systems with higher ignition energy. If not enough ignition energy is produced, ignition will not occur, resulting in engine misfire. The system must therefore deliver enough ignition energy to ensure reliable ignition of the A/F mixture under all operating conditions.

Efficient fuel atomization and good access of the A/F mixture to the ignition sparks enhance ignitability, extend spark duration and spark length, and lengthen the electrode gap. The dimensions of the spark plug determine the position (see Spark position, spark plug) and length of the spark; spark duration depends on the type and design of the ignition system, as well as on the instantaneous ignition conditions in the combustion chamber. Dependent on the engine requirements (intake-manifold injection, gasoline direct injection or exhaust-gas turbocharging), the spark energy of ignition systems lies in a range of approx. 30 to 100 mJ.

Figure 2: Combustion-chamber pressure characteristic for various moments of ignition
1 Ignition (Z_a) at the correct point in time,
2 Ignition (Z_b) too advanced,
3 Ignition (Z_c) too retarded.

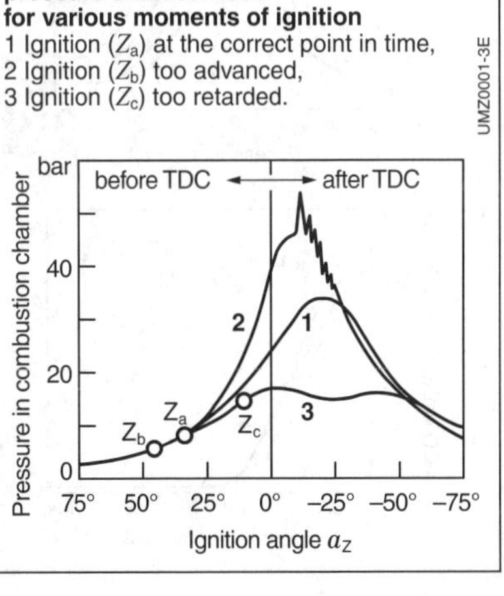

Moment of ignition

The start of combustion in a spark-ignition (SI) engine can be controlled by selecting the moment of ignition. The moment of ignition always refers to the top dead center of the spark-ignition (SI) engine power cycle. The earliest possible moment of ignition is determined by the knock limit, and the latest possible moment of ignition by the combustion limit or the maximum permissible exhaust-gas temperature. The moment of ignition affects the torque output, exhaust-gas emissions and fuel consumption.

Basic ignition point

Maximum combustion, and thus maximum combustion pressure, should occur shortly after top dead center (Figure 2) to deliver maximum engine torque. Since a certain amount of time passes from the moment of ignition to complete combustion of the A/F mixture, ignition must occur before top dead center. The moment of ignition must be advanced as rotational speed increases and charge decreases.

Likewise, the moment of ignition must be advanced in the case of lean A/F mixtures ($\lambda > 1$) because the flame front propagates more slowly. Ignition timing adjustment, therefore, essentially depends on rotational speed, charge, and the air/fuel

Figure 3: Ignition map as a function of engine speed and relative air charge

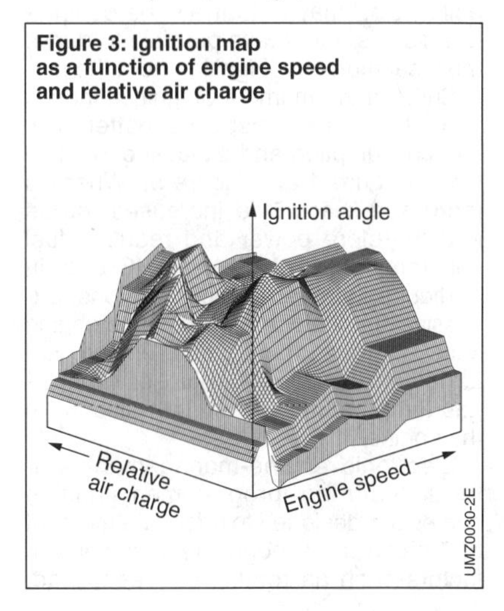

ratio (excess-air factor λ). The moments of ignition are determined on the engine test bench and, in the case of electronic engine management systems, are stored in program maps (Figure 3).

Ignition timing corrections and operating-point-dependent moments of ignition

Electronic engine-management systems can take other effects on the moment of ignition into consideration in addition to rotational speed and charge. The basic moment of ignition can either be modified by means of additive corrections or replaced for certain operating points or ranges by special ignition angles or ignition maps. Examples of ignition timing corrections are knock control, the correction angle for the gasoline direct injection homogeneous-lean operation, and warm-up. Examples of special ignition angles or ignition maps are gasoline direct injection stratified-charge operation, and starting operation. Final implementation depends on the prevailing application.

Exhaust gas and fuel consumption

The moment of ignition has a considerable impact on exhaust gas because it can be used to control the various untreated exhaust-gas constituents directly. However, the various optimization criteria, such as exhaust gas, fuel economy, drivability, etc., may not always be compatible, so it is not always possible to derive the ideal moment of ignition from them.

Shifts in the moment of ignition induce mutually inverse response patterns in fuel consumption and exhaust-gas emissions (Figure 4 and Figure 5). Whereas more spark advance increases torque and therefore power and reduces fuel consumption, it also raises HC and, in particular, nitrogen-oxide emissions. Excessive spark advance can cause engine knock that may damage the engine. Retarded ignition results in higher exhaust-gas temperatures, which can also harm the engine.

Electronic engine-management systems featuring programmed ignition curves are designed to adapt the moment of ignition in response to variations in factors such as rotational speed, load, temperature, etc. They can thus be employed to achieve the optimum compromise between these mutually antagonistic objectives.

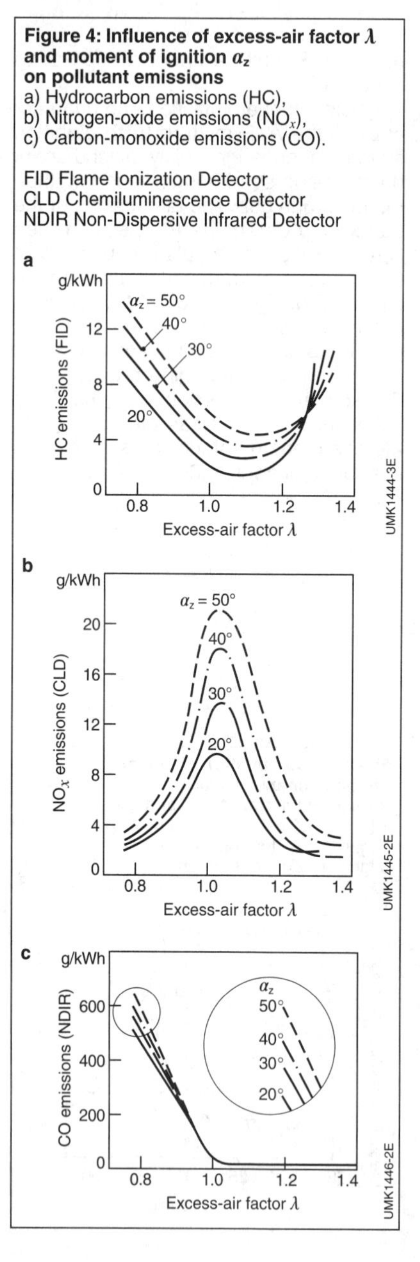

Figure 4: Influence of excess-air factor λ and moment of ignition α_z on pollutant emissions
a) Hydrocarbon emissions (HC),
b) Nitrogen-oxide emissions (NO_x),
c) Carbon-monoxide emissions (CO).

FID Flame Ionization Detector
CLD Chemiluminescence Detector
NDIR Non-Dispersive Infrared Detector

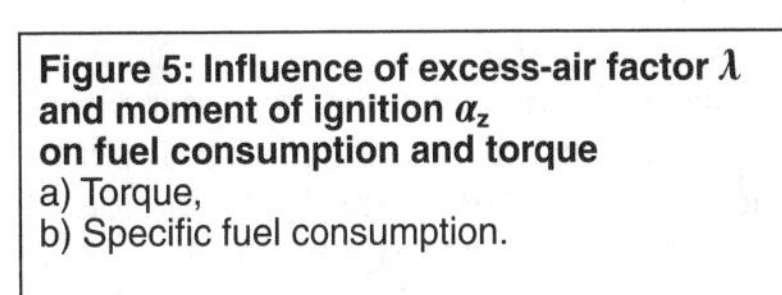

Figure 5: Influence of excess-air factor λ and moment of ignition α_z on fuel consumption and torque
a) Torque,
b) Specific fuel consumption.

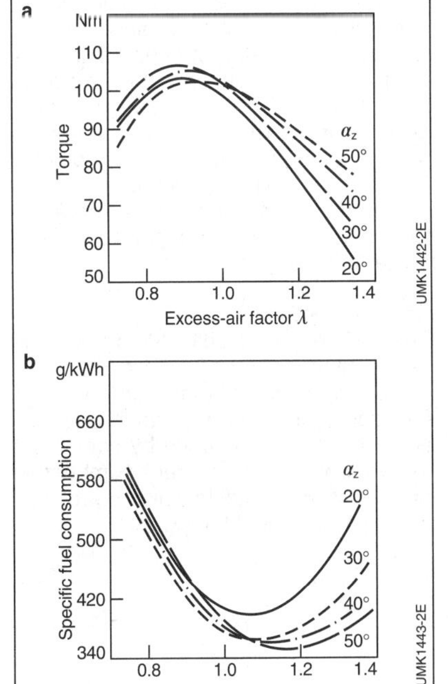

Figure 6: Schematic of a knock-control system

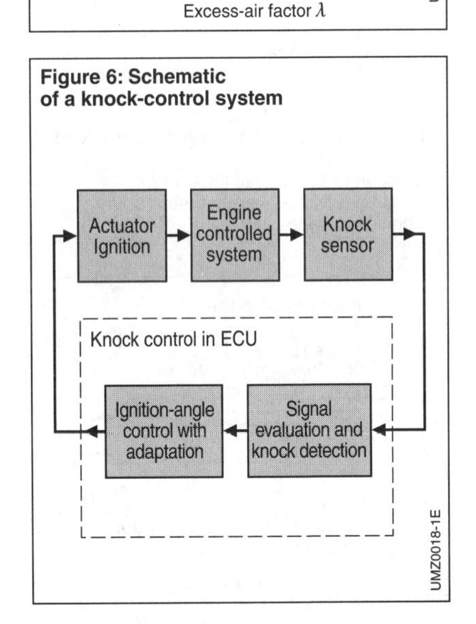

Knock control

Basic principles

Electronic control of the moment of ignition offers the possibility of accurate control of the ignition angle as a function of rotational speed, load, temperature, etc. Nevertheless, if there is no knock control, there must still be some means to define a clear safety margin to the knock limit.

This margin is necessary to ensure that, even in the most knock-sensitive case with regard to engine tolerances, engine aging, environmental conditions, and fuel quality, no cylinder can reach or exceed the knock limit. The resulting engine design leads to lower compression, retarded moments of ignition, and thus worsening fuel consumption and torque.

These disadvantages can be avoided through the use of knock control. Experience shows that knock control increases engine compression. This results in lower fuel consumption and higher torque. Now, however, the pilot control ignition angle no longer has to be determined for the conditions most sensitive to knocking but rather for the conditions least sensitive to knocking (e.g. compression of the engine at lowest tolerance limit, best possible fuel quality, cylinder least sensitive to knocking). Each individual engine cylinder can now be operated throughout its service life in virtually all operating ranges at its knock limit, and thus at optimum efficiency. For this type of ignition angle adjustment, a reliable method of knock detection is essential. It should detect knock for each cylinder throughout the engine's operating range starting from a specified knock intensity.

Knock-control system

A knock-control system consists of knock sensor, signal evaluation, knock detection, and ignition-angle control system with adaptation facility (Figure 6).

Knock sensor

A typical symptom of combustion knock is high-frequency vibrations which are superimposed on the high-pressure curve in the combustion chamber (Figure 2). These vibrations are best detected directly in the combustion chamber by means of pressure sensors. As fitting these pressure sensors in the cylinder head for each cylinder is still relatively costly, these vibrations are usually picked up using knock sensors fitted to the exterior of the engine. These piezo-electric acceleration sensors (Figure 7) pick up the characteristic vibrations of knocking combustion and convert them into electrical signals.

There are two types of knock sensor. A wide-band sensor, with a typical frequency band of 5 to 20 kHz, and a resonance sensor, which preferably transmits only one knock-signal resonant frequency. When combined with the flexible signal-evaluation system in the control unit, it is possible to evaluate different or several resonant frequencies from one wide-band knock sensor. This improves knock detection performance, which is why the wide-band knock sensor is increasingly replacing the resonance sensor.

To ensure sufficient knock detection in all cylinders and across all operating ranges, the number and location of the required knock sensors must be carefully determined for each engine type. Four-cylinder in-line engines are usually fitted with one or two knock sensors, while 5- and 6-cylinder engines are fitted with two, and 8- and 12-cylinder engines with four knock sensors.

Signal evaluation
For the duration of a timing range in which knock can occur, a special signal evaluation circuit in the control unit evaluates from the wide-band signal the frequency band(s) with the best knock information and generates a representative variable for each combustion process. This very flexible signal evaluation of the wide-band sensor enables high detection quality. When using a resonance knock sensor that transmits just one resonant frequency for analysis of all cylinders assigned to it across the entire engine map, knock detection is normally no longer possible at higher engine speeds.

Knock detection
The variable produced by the signal-evaluation circuit is classified in a knock-detection algorithm as "knock" or "no knock" for each cylinder and for each combustion process. This is done by comparing the variable for the current combustion process with a variable which represents combustion without knock.

Figure 7: Knock sensor
1 Seismic mass, 2 Sealing compound, 3 Piezoceramic, 4 Connection, 5 Electric connection.

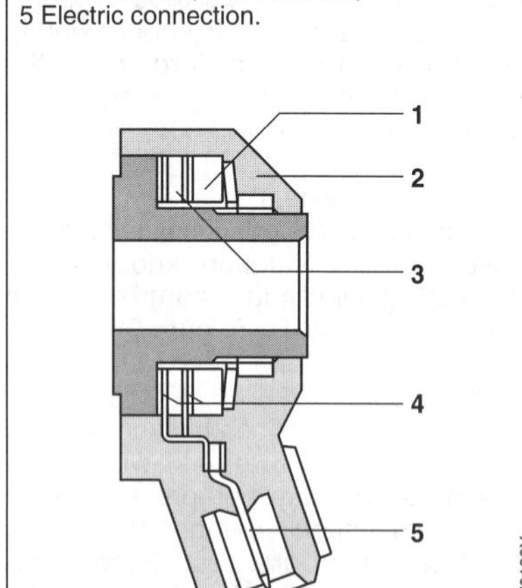

Figure 8: Knock control
Control algorithm for ignition adjustment on a 4-cylinder engine.
$K_{1 \text{ to } 3}$ Occurrence of knocking at cylinders 1 to 3, at cylinder 4 no knocking,
a Map ignition angle,
b Ignition retard step at knocking,
c Ignition advance waiting time,
d Ignition advance step.

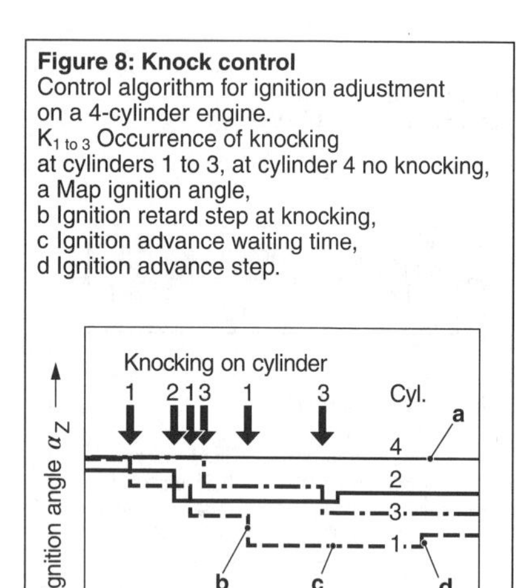

Ignition-angle control system with adaptation facility

If combustion knock is detected in a cylinder, the moment of ignition for that cylinder is retarded (Figure 8). When knock stops, the moment of ignition is advanced again in stages up to the precontrol value. The knock-detection and knock-control algorithms are matched in such a way as to eliminate any knock that is audible and damaging to the engine, even though each cylinder is operated at knock limit within the optimum efficiency range.

Real engine operation produces different knock limits, and thus different moments of ignition for individual cylinders. In order to adapt precontrol values for the moment of ignition to a particular knock limit, the ignition retard values are stored for each cylinder dependent on the operating point. They are stored in non-volatile program maps in the permanently powered RAM for load and engine speed. In this way, the engine can be operated at optimum efficiency for each operating point and without audible combustion knocks, even if there are rapid load and engine-speed changes.

This adaptation even enables the use of fuels with lower antiknock properties (e.g. regular instead of premium grade petrol).

Phenomenon of pre-ignition

Basic principles

There is currently a clear trend in the development of modern gasoline engines towards downsizing (displacement reduction at same engine performance) and downspeeding (speed reduction by means of longer transmission ratio) in combination with direct injection and supercharging. Supercharging allows for a reduction in displacement without lowering performance level. The engine can therefore be operated in part load at higher loads with higher part-load efficiency, and fuel consumption can be reduced. The charge-air pressure increase to improve efficiency is however limited by the phenomenon of pre-ignition.

A few years ago the terms “extreme knocker” or “superknocker” were used, but they described the symptom and not the cause of the phenomenon of pre-ignition. However, the term “pre-ignition” has become the established term for some time now. In the name there is therefore also a clear distinction from normal combustion knock.

Pre-ignition

Pre-ignitions are uncontrolled auto-ignitions of the air-fuel mixture, which occur before the ignition that is triggered by the spark plug. Compared to normal combustion, this premature initiation of combustion leads to a considerable increase in pressure and temperature, which then leads to severe combustion knock in the subsequent combustion process. In doing so, individual events can pre-damage the engine.

Detection of pre-ignitions

Reliably detecting pre-ignitions is absolutely necessary in order to prevent engine damage. Knocking events as a result of pre-ignitions can be clearly evaluated and detected based on knock sensors by means of position and frequency ranges. This process allows for a high level of detection quality with a clear distinction from normal combustion knock.

Measures to prevent pre-ignitions
If pre-ignitions are detected, countermeasures are necessary in order to prevent further pre-ignitions. However, since pre-ignition is an uncontrolled auto-ignition, there is no direct manipulated variable – e.g. the ignition angle at combustion knock – that can be used to reliably prevent further pre-ignitions. For this reason, the pre-ignition functionality involves initiating multiple countermeasures after detected pre-ignitions in order to quickly reduce the temperature level in the combustion chamber and thus prevent further pre-ignitions. The various measures, such as mixture enrichment or charge lowering, can be activated in the combination that is ideal for the respective engine type.

Pre-ignitions are heavily dependent on fuel and on oil emissions. This functionality is therefore absolutely necessary to protect the engine, especially for turbocharged gasoline engines. This enables vehicle manufacturers to develop efficiency-optimized, supercharged gasoline engines and to market them worldwide despite different fuel and oil grades.

Risk of accidents
All electronic ignition systems are high-voltage systems. To avoid any risks, always switch off the ignition, or disconnect the power supply when working on the ignition system. Such work includes, for example:
- Replacing parts such as spark plugs, ignition coils, or ignition transformers, ignition distributors, high-tension ignition cables, etc.
- Connecting engine analyzers, such as stroboscopic lamps, dwell angle/rotational speed testers, ignition oscilloscopes, etc.

When checking the ignition system, remember that dangerously high levels of voltage are present within the system whenever the ignition is on. All tests and inspections should therefore only be carried out by qualified professional personnel.

Ignition systems

On modern vehicles, the ignition systems are almost always incorporated as subsystems of the engine-management system. Autonomous ignition systems are now only used for special applications (e.g. small engines). In the case of ignition systems, coil ignition (inductive ignition) with a separate ignition circuit per cylinder (static high-voltage distribution with single-spark coils, Figure 9) has come to the fore.

Alongside this, but to a much lesser extent, high-voltage capacitor ignitions (capacitive ignition) or other special designs such as magnetos are used for small engines. The next section will focus on coil ignition alone.

Coil ignition (inductive ignition)

Principle of coil ignition
The ignition circuit of a coil-ignition system (Figure 10) consists of:
- An ignition coil with a primary and a secondary winding
- An ignition driver stage to control the current by means of the primary winding, predominantly integrated as an IGBT (Insulated Gate Bipolar Transistor) in the engine control unit or in the ignition coil
- A spark plug connected to the high-voltage connection point of the secondary winding

Figure 9: Ignition system with single spark coils
1 Ignition lock, 2 Ignition coil, 3 Spark plug, 4 Control unit, 5 Battery.

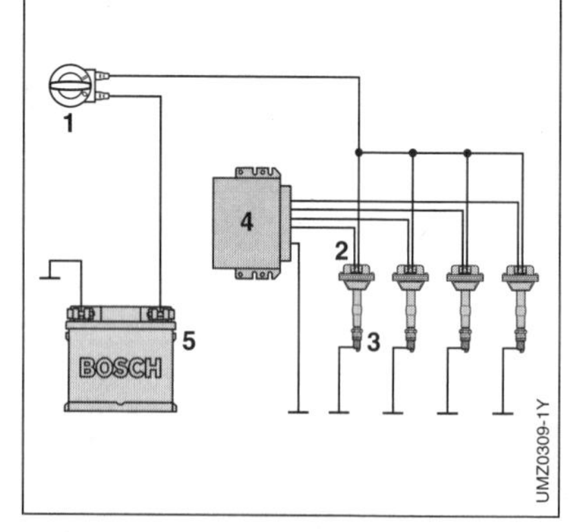

Before the desired moment of ignition, the ignition driver stage switches a current from the vehicle electrical system through to the primary winding of the ignition coil. While the primary current circuit is closed (dwell period), a magnetic field builds up in the primary winding.

At the moment of ignition, the current through the primary winding is interrupted again, and the magnetic-field energy is discharged, mainly via the magnetic-coupled secondary winding (induction). In the process, a high voltage is produced in the secondary winding. Flashover occurs if the ignition voltage supply of the ignition system exceeds the ignition-voltage demand of the spark plug. After flashover, the remaining energy is converted at the spark plug while the spark is present.

Functions of an ignition system with coil ignition

Determining the moment of ignition

The current moment of ignition in each case is determined from program maps dependent upon the operating point and output.

Determining the dwell period

The required ignition energy is made available at the moment of ignition. The amount of ignition energy is dependent upon the amount of primary current at the moment of ignition (cutoff current) and the inductance of the primary winding. The amount of cutoff current is mainly dependent upon the operating time (dwell period) and upon the battery voltage at the ignition coil. The dwell periods required to achieve the desired cutoff current are contained in characteristic curves or program maps as a function of the steady-state voltage. The change in dwell period with temperature can also be compensated for.

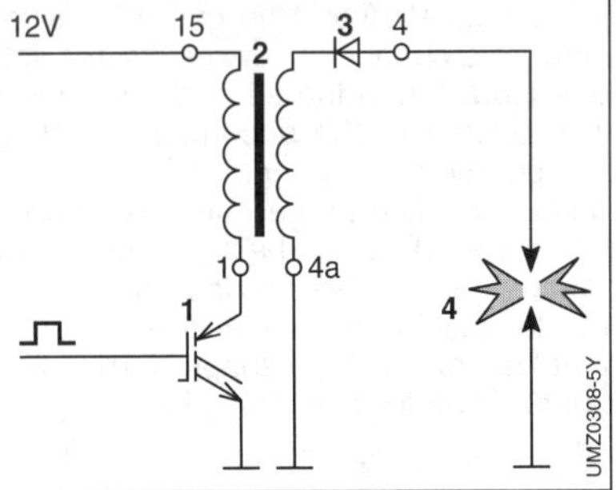

Ignition release

The ignition release ensures that the ignition spark occurs at the right cylinder at the right time and with the required level of ignition energy. On electronic-controlled systems, a trigger wheel with a fixed-angle reference mark (typically 60−2 teeth, two teeth missing at the reference mark) on the crankshaft is usually scanned (sensor system with hall or inductive sensor). From this, the control unit can calculate the crankshaft angle and the momentary rotational speed. The ignition coil can be switched on and off at any required crankshaft angle. An additional phase signal from the camshaft is required for the unambiguous identification of the cylinder.

For each combustion, the control unit uses the desired moment of ignition, the required dwell period and the current engine speed to calculate the switch-on time and switches on the driver stage. The moment of ignition, or the switchoff point for the driver stage, can be triggered either when the dwell period expires or when the desired angle is reached.

Figure 10: Structure of an ignition circuit with single spark coils
1 Ignition driver stage,
2 Ignition coil,
3 Activation arc diode (suppression of activation spark),
4 Spark plug.
15, 1, 4, 4a Terminal designations,
⎍ Triggering signal.

Ignition coil

Function

The ignition coil is principally an energy-charged high-voltage source similar in structure to a transformer. Energy is supplied by the vehicle electrical system during the dwell period or charging time. At the moment of ignition, which at the same time is the end of the charging time, the energy is then transferred with the required high voltage and sparking energy to the spark plug (inductive ignition system).

Figure 1: Design of the compact ignition coil
1 Printed-circuit board (optional),
2 Ignition output stage (optional),
3 Activation arc diode (optional),
4 Secondary-winding housing,
5 Secondary winding, 6 Contact plate,
7 High-voltage pin
(connecting element to contact spring),
8 Connector, 9 Primary winding,
10 I-core, 11 Permanent magnet, 12 O-core,
13 Contact spring (spark-plug contact),
14 Silicone jacket (high-voltage insulation).

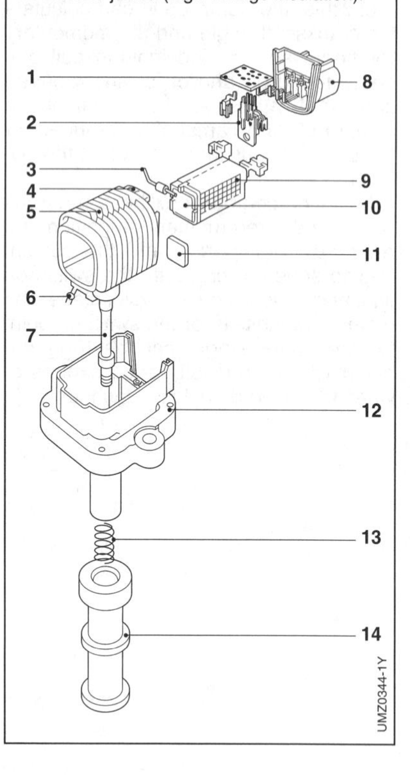

Body

The ignition coil (Figure 1) comprises two windings that are magnetically linked by an iron core (I-core and O-core). This iron core may contain a permanent magnet for energy optimization. Compared with the secondary winding, the primary winding has significantly fewer coils. The turns ratio $\ddot{u}$ is in the range of $\ddot{u}$ = 80 to 150.

The windings must have good electrical insulation to prevent electrical discharge and flashovers either to the inside or to the outside. For this purpose, the windings are usually cast in epoxy resin in the ignition coil housing.

As a rule, the iron core consists of stacked, ferromagnetic steel-plate fins in order to minimize above all eddy-current losses.

Alternatively, the ignition output stage can be incorporated into the ignition coil instead of the engine control unit. Interference-suppression elements may also be incorporated into the ignition coil together with the activation arc diode (activation spark suppression). An interference-suppression resistor is commonly used on the high-voltage output to the spark plug.

Operation

The ignition output stage switches the primary current in the ignition coil. The current rises with a delay, in accordance with the inductance. Energy is stored in the magnetic field created in the process. The operating time (charging time) is calculated in such a way that a specific cutoff current, and thus a specific level of stored energy, will be reached by the end of the operating time.

The current is cut off by the ignition output stage, which causes a swift change in the magnetic flux in the ignition coil's iron core. This change in flux results in a voltage induction in the secondary winding. The design of the secondary winding, the geometric arrangement of the secondary winding to the iron core and the primary winding and the materials used in the insulation system and the iron core give rise to inductive and capacitive properties that lead to a voltage supply in excess of 30,000 V on the secondary side.

When the ignition-coil voltage supply equals the ignition-voltage demand at the spark-plug electrodes, the voltage drops to a spark voltage of about 1,000 V. A spark current then flows and decreases as spark duration increases, until the spark is finally extinguished.

As the current changes over time when the primary current is cut off, an induction voltage occurs at the ignition coil output, similar to the start of charging time. However, this voltage is much smaller than the voltage at the moment of ignition and has an inverse polarity to that voltage. To prevent this "transient voltage" causing unwanted ignition, it is usually suppressed by means of a high-voltage diode in the secondary circuit (activation arc diode, activation spark suppression).

The design of an ignition coil can determine its electrical characteristics. In this respect, the requirements in terms of installation space (e.g. geometry of cylinder head and cylinder-head cover, position of fuel injector and intake manifold) and the two specified interfaces, the ECU with output stage (e.g. cutoff current) and the spark plug (e.g. ignition voltage, spark data), are decisive.

Design variations

There are several types of ignition coil that are distinguishable by their various features.

Individual coils and modules

In addition to the individual coils, which normally sit directly on the spark plug, several ignition coils can be grouped together in a single housing as a module. They are then mounted directly on the spark plugs, or they may be located a short distance away, in which case the high voltage must be supplied via appropriate cables.

Single-spark and dual-spark ignition coils

In addition to ignition coils with just one high-voltage output (single-spark ignition coil), there are also coils that use both ends of the secondary winding as an output (dual-spark ignition coil). The electric circuit must always be closed by both spark gaps to prevent discharge on the secondary side. One feasible application for this design is dual ignition, i.e. two spark plugs per cylinder supplied by one ignition coil. Another application involves splitting up the two high-voltage outputs between two spark plugs on different cylinders. In this case, one of the two spark plugs will always be in the ignition stroke. As a result, the voltage and energy requirements of the "passive spark" (backup spark) are reduced substantially. Above all, this variant offers cost advantages; however, it must be matched to the overall system to prevent damage from unwanted ignition events caused by backup sparks.

Compact coil and pencil coil

Ignition coils are also distinguishable by their basic design. For example, there is the conventional compact coil, that features an equal-sided coil body and an O-I core or C-I core magnetic circuit. The coil body sits in the engine above the spark-plug well.

Another type is the pencil coil whose coil body projects into the spark-plug well. Here, too, the windings are located on an I or pencil core, with a plate (yoke plate) arranged concentrically around the windings serving as a magnetic yoke.

Ignition coils with ignition output stage

Ignition coils are available with or without the ignition output stage. The reason for integrating the output stage is to relieve the strain on the engine ECU. Additionally, further electronic circuitry is used in the ignition coil in order to meet the existing requirements (Figure 2).

Requirements

The principal requirements of modern ignition systems are derived indirectly from necessary emission and fuel reductions. Requirements of ignition coils are derived from relevant engine solutions, such as high-pressure supercharging and lean-burn and stratified-charge operation (spray-guided direct injection) in combination with increased exhaust-gas recirculation rates (EGR).

It is necessary to represent an increased ignition-voltage and energy demand with increased temperature requirements. This is effected among others by means of

- High-energy coils with high voltage supply (in excess of 40,000 V)
- Multispark ignition (MSI), i.e. spark-sequence control with integrated electronics (ASIC), controlled by way of primary- and secondary-current evaluation
- Diagnostic functions (e.g. charge-time monitoring, ionic-current measurement for combustion diagnosis)
- Protective functions (e.g. overtemperature cutoff, cutoff-current regulation)

Furthermore, motor vehicles are subject to increased demands with regard to electromagnetic compatibility (EMC). Precisely because of the increased ignition-voltage demand, the increased ignition frequencies necessitated by multispark ignition and the increased cutoff currents, it is now necessary to reduce the interference emitted by the ignition system so as not to disrupt the functioning of other automotive components (ECUs, microcontrollers, sensors, actuators).

Figure 2: Installation of ignition output stage in a pencil ignition-coil housing
1 Connector,
2 SMD components (Surface Mounted Device),
3 Electronic circuitry for ignition functions,
4 Primary-winding contacts,
5 Yoke plate of the pencil ignition-coil transformer,
6 Attachment lug,
7 Ignition output stage.

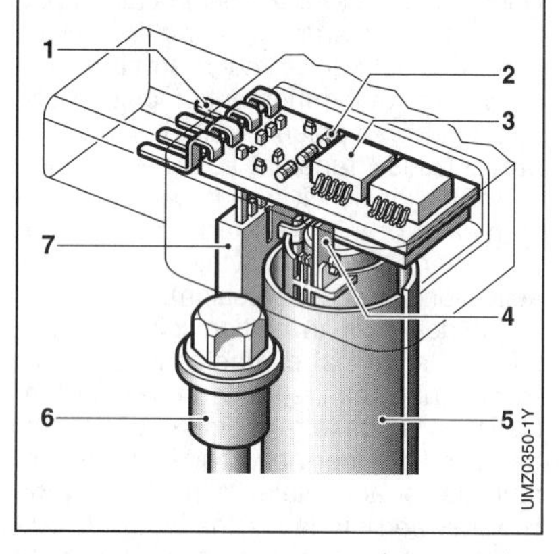

The integration of increasingly more electronic circuitry in ignition coils gives rise to more stringent demands with regard to the interference immunity of these components. In the interests of avoiding impaired functioning and malfunctions, the electronic circuity must be immune to the emissions of the ignition system itself and to the interference emitted by other vehicle components.

Spark plug

Function

The spark plug introduces the ignition energy generated by the ignition coil into the combustion chamber. The high voltage applied creates an electric spark between the spark-plug electrodes, which ignites the compressed A/F mixture. As this function must also be guaranteed under extreme conditions (e.g. cold start, wide-open throttle), the spark plug plays a decisive role in the performance and reliable operation of a spark-ignition engine. These requirements remain the same over the entire service life of the spark plug.

Requirements

The spark plugs must satisfy a variety of extreme performance demands. It is exposed to the varying periodic processes within the combustion chamber, as well as external climatic conditions.

When spark plugs are used with electronic ignition systems, ignition voltages of up to 42 kV can occur; these high voltages must not cause ceramic or insulator-head flashovers. This insulation capability must be maintained for the entire service life and must be guaranteed even at high temperatures (up to approx. 1,000 °C).

Mechanically, the spark plug is subjected to the pressures (up to 150 bar) occurring periodically in the combustion chamber; however, gas tightness may not be impaired. In addition, the spark-plug electrode materials exhibit extreme resistance to thermal loads and continuous vibratory stress. The shell must be able to absorb the forces occurring during assembly without any lasting deformation.

At the same time, the section of the spark plug that protrudes into the combustion chamber is exposed to high-temperature chemical processes, making resistance to aggressive combustion deposits essential. As it is subjected to rapid variations between the heat of combustion gases and the cool A/F mixture, the spark-plug insulator must feature high resistance to thermal stresses (thermal shock). The electrodes and the insulator at the cylinder head must have good heat dissipation properties – essential for reliable spark-plug performance.

Design

In a special high-grade ceramic insulator (Al_2O_3) an electrically conductive glass seal forms the connection between the center electrode and the terminal stud (Figure 1). This conductive glass seal acts as a mechanical support for the components, while providing a gas seal against the high combustion pressure. It can also incorporate resistor elements for interference suppression and burn-off.

At the connection end, the insulator has a lead-free glaze to repel moisture and dirt. This largely prevents insulation flashovers. The connection between the insulator and the nickel-plated steel shell must also be gas-tight.

The ground electrode(s), like the center electrode, is (are) usually manufactured from nickel-based alloys to cope with the

Figure 1: Spark-plug design
1 Terminal stud (here with SAE thread),
2 Insulator head,
3 Nickel-plated steel shell,
4 Heat-shrinkage zone,
5 Electrically conductive glass seal,
6 Captive gasket (seal seat),
7 Thread,
8 Compound center electrode (Ni/Cu),
9 Insulator nose,
10 Ground electrode (here Ni/Cu compound electrode).

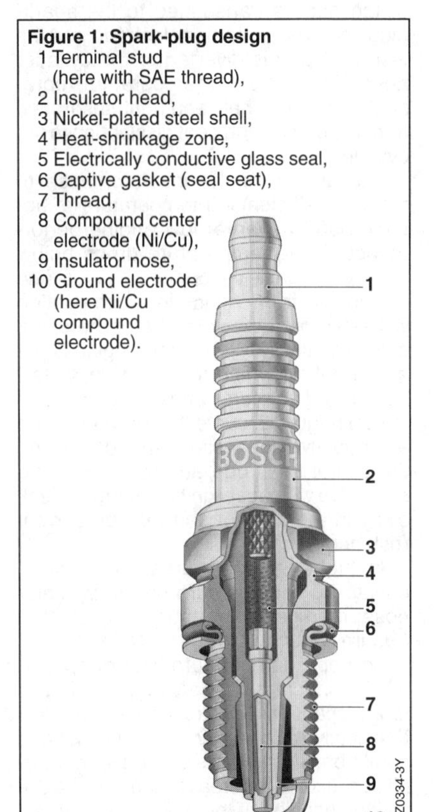

high thermal stresses, and welded to the shell. Compound electrodes with a jacket material made of a nickel alloy and a copper core are used to improve heat dissipation for both center and ground electrodes. Silver, platinum or platinum alloys are employed as electrode material for special applications.

The spark plugs have either an M4 or a standard SAE thread, depending on the type of high-voltage connection. Spark plugs with metal shields are available for watertight systems and for maximum interference suppression.

Heat range

When the engine is operating, the spark plug is heated by combustion heat. Some of the heat absorbed by the spark plug is diverted to the fresh A/F mixture. Most of the heat is transmitted to the spark-plug shell via the center electrode and the insulator, and is diverted to the cylinder head. The operating temperature represents a balance between heat absorption from the engine and heat dissipation to the cylinder head.

The aim is for the insulator nose to reach a self-cleaning temperature of approx. 500 °C even at low engine performance. If the temperature drops below this level, there is the danger that soot and oil residue from incomplete combustion will settle on the cold areas of the spark plugs (particularly when the engine is not at normal operating temperature, at low outside temperatures and during repeated starts) (Figure 2, curve 3). This can create a conductive connection (shunt) between the center electrode and the spark-plug shell. This will cause ignition energy to leak away in the form of short-circuit current (risk of misfiring).

At higher temperatures, the residue containing carbons burn on the insulator nose; the spark plug thus "cleans" itself (Figure 2, curve 2).

An upper temperature limit of approx. 900 °C should be observed since, in this range, spark-plug electrode wear increases drastically (due to oxidation and hot-gas corrosion). If this limit is exceeded by a significant extent, it increases the risk of auto-ignition (ignition of the A/F mixture on hot surfaces) (Figure 2, curve 1). Auto-ignition subjects the engine to extreme loads, and may result in engine destruction within a short period of time. The spark plug must therefore be adapted to the engine type in terms of its heat-absorbing property.

The identifying feature of a spark plug's thermal loading capacity is its heat range, which is defined by a code number and determined in comparison measurements with a reference standard source.

The Bosch ionic-current measurement procedure uses combustion characteristics to determine the engine's heat range requirements. The ionizing effect of flames is used to assess how combustion develops over time; this is done by measuring conductivity in the spark gap. Character-

Figure 2: Spark-plug temperature response
1 Spark plug with excessively high heat-range code number (hot plug),
2 Spark plug with suitable heat-range code number,
3 Spark plug with excessively low heat-range code number (cold plug).

The working range temperature should be from 500 °C to 900 °C for different engine performance ratings.

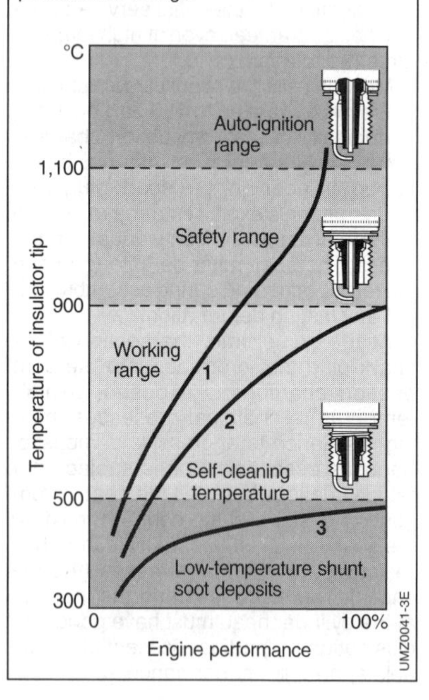

istic changes in the combustion process due to increased thermal loading of the spark plugs can be detected using ionic current and used in the assessment of the auto-ignition process. The spark plug must be adapted in such a way as to prevent thermal ignition (pre-ignition) before the actual moment of ignition.

The use of materials with a higher thermal conductivity (silver or nickel alloys with copper core) for center electrodes makes it possible to increase the insulator nose length substantially without changing the heat-range code number. This extends the spark-plug operating range down to a lower thermal-load range and reduces the probability of soot formation. These advantages are inherent in all Bosch Super spark plugs.

Reducing the likelihood of combustion misses and misfiring – with their attendant massive increases in hydrocarbon emissions – is beneficial in exhaust emissions and fuel consumption in part-throttle operation at low load factors.

Electrode gap and ignition voltage

The electrode gap is the shortest gap between center and ground electrodes, and determines, amongst other things, the ignition-spark length. The electrode gap should be, on the one hand, as large as possible so that the ignition spark activates a large volume element, and thus results in reliable ignition of the A/F mixture by developing a stable flame core. On the other hand, the ignition voltage to produce a spark is decreased with a smaller electrode gap. If the electrode gap is too small, however, only a small flame core will be created around the electrode. Since energy will be drawn from the flame core via the contact areas with the electrodes (quenching), the flame core will only be able to propagate very slowly. In extreme cases, so much energy may be drawn off that ignition misses will occur.

As electrode gaps increase (e.g. due to electrode wear), the conditions for ignition are improved, but the ignition voltage requirement rises. As the ignition coil's voltage supply is a fixed amount, the voltage reserves are reduced and the risk of misfiring increases.

The ignition-voltage demand is influenced not only by the size of the electrode gap, but also by electrode shape and temperature, and the electrode material. Parameters specific to the combustion chamber, such as mixture composition (λ value), flow velocity, turbulence, and density of the ignitable gas, also play an important role.

On modern high-compression engines, which frequently feature high charge turbulence, electrode gaps must be carefully defined in order to guarantee reliable ignition, and thus misfire-free operation throughout the required service life.

Spark position

The position of the spark gap relative to the combustion chamber wall defines the spark position. On modern engines (and particularly with gasoline direct-injection engines), the spark position has a considerable influence on combustion. A perceptible improvement in ignition response is observed when the spark position projects deeper into the combustion chamber. Combustion can be characterized by smooth-running engine performance, which can be derived directly from engine speed fluctuations.

However, because the ground electrodes are longer, higher temperatures are achieved. This, in turn, has an effect on electrode wear and electrode durability. It is possible to achieve the required service life by implementing design measures (extending the spark-plug shell beyond the combustion chamber wall), or by using compound electrodes, or high temperature-resistant materials.

Spark-plug concepts

One or more ground electrodes can be advantageous, depending on the demands placed on the spark plug (wear, ignition response, etc.). The spark-plug type is determined by the relative location of the electrodes to each other and the position of the ground electrodes with respect to the insulator.

Spark-air-gap concepts

In spark-air-gap concepts (Figure 3a), the ground electrode is positioned with respect to the center electrode in such a way that the ignition spark jumps directly between the electrodes, igniting the A/F mixture between the electrodes.

Surface-gap concepts

As a result of the defined position of the ground electrode relative to the ceramic, the spark initially travels from the center electrode across the surface of the insulator nose before jumping across a gas-filled gap to the ground electrode (Figure 3b). As less ignition voltage is required for discharging across the surface than for discharging across an air gap of the same size, the surface-gap spark can bridge bigger electrode gaps than the air gap spark given the same ignition voltage. The resulting larger flame core improves ignition properties considerably.

These spark-plug concepts also have much better repeat cold-starting performance because the surface-gap spark cleans the insulator end-face, and prevents soot from settling there.

Surface-air-gap concepts

In these spark-plug concepts, the ground electrodes are positioned at a specific distance from the center electrode and the end face of the ceramic insulator (Figure 3c). Two alternative spark gaps are created as a result, thus allowing both forms of discharge with different ignition-voltage requirements. Depending on operating conditions, the spark behaves either as an air-gap spark or a surface-gap spark (e.g. Bosch Super 4).

Designation codes for Bosch spark plugs

Individual spark-plug specifications are contained in the designation code. This code includes all the spark-plug characteristics (Figure 4). The electrode gap is also indicated on the packaging. The spark plug which is suitable for a given engine is specified or recommended by the engine manufacturer and by Bosch. For detailed information, see [1].

Spark-plug operating performance

Electrode wear

As spark plugs operate within an aggressive atmosphere, sometimes at extremely high temperatures, electrodes are subject to wear. This material erosion causes the electrode gap to increase noticeably the longer the spark plug is in service, therefore

Figure 3: Spark-plug concepts
a) Air-gap spark concept,
b) Surface-gap spark concept,
c) Surface-air-gap spark concept.

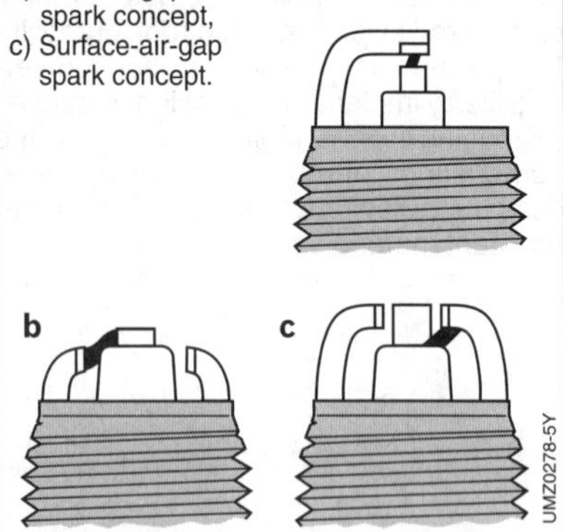

b
c
UMZ0278-5Y

Figure 4: Designation code
Each position in the designation code describes the following characteristics:

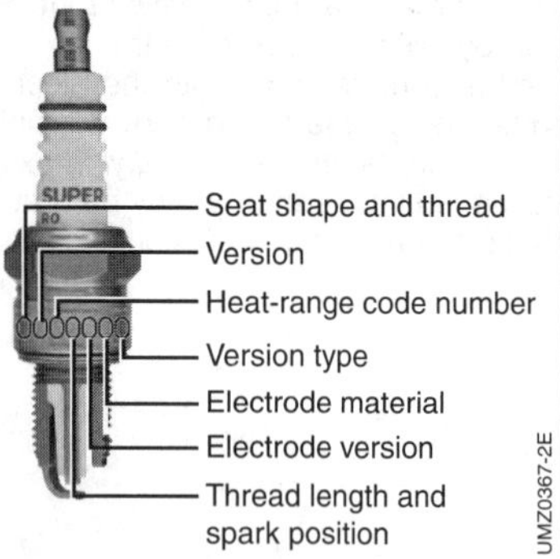

SUPER
Seat shape and thread
Version
Heat-range code number
Version type
Electrode material
Electrode version
Thread length and spark position
UMZ0367-2E

causing the ignition-voltage requirement to increase. When this requirement can no longer be met by the supply from the ignition coil, misfiring will occur.

There are essentially two mechanisms that are responsible for electrode wear: spark erosion and corrosion in the combustion chamber. Materials with a high thermal resistance (special-steel alloys of platinum or iridium) are used to minimize electrode wear. Material wear can also be reduced for the same period of use, by appropriate selection of electrode geometry and spark-plug concept. In a surface-air-gap spark plug with four ground electrodes, there are eight possible spark gaps. In this way, the wear is evenly distributed across all four electrodes.

The resistor in the conductive glass seal reduces burn-off, and thus helps to reduce wear.

Changes in operation

Dirt and changes in the engine caused by aging (e.g. higher oil consumption) can also affect operation of the spark plug. Deposits on the spark plug can result in shunts, and thus in misfiring. This, in turn, may cause a considerable rise in pollutant emissions, and even damage the catalytic converter.

For these reasons, spark plugs have a defined service life after which the spark plugs must be replaced.

Spark plugs for direct-injection gasoline engines

Direct-injection engines place especially high requirements on the spark plugs, which is why they have to be specifically adapted to the needs of the respective engine (e.g. performance, mean pressure). This results in different requirements of the spark-plug concept for combustion processes with stratified or homogeneous operation.

Through the combination of direct injection with turbocharging, the emphasis in spark-plug development is on expanding ignition-voltage demand, electrode temperature and service life (wear reduction). For this reason, spark plugs with precious metal electrodes and M12 thread are normally used, and they are also specifically aligned in the combustion chamber when used for spray-guided combustion processes with stratified operation.

Abnormal operating conditions

Incorrectly set ignition systems, the use of spark plugs whose heat range is unsuitable for the engine, or the use of unsuitable fuels can cause abnormal operating conditions to arise (auto-ignition, combustion knock, etc.) and damage the engine and the spark plugs.

Auto-ignition

Auto-ignition is an uncontrolled ignition process. It occurs because the temperature in one spot (e.g. at the tip of the insulator nose, at the spark plug, at the exhaust valve, or at the cylinder-head gaskets) can rise to such an extent that the air/fuel mixture ignites uncontrollably in that spot. Severe damage can be caused to the engine and spark plug as a result of auto-ignition.

Combustion knock

Knocking is characteristic of an uncontrolled combustion process with very sharp rises in pressure (see Knocking). The combustion process is considerably faster than normal combustion. Due to high pressure gradients, the components (cylinder head, valves, pistons, and spark plugs) are subjected to high temperature loads. This may result in damage to one or several of the components. Knocking can be prevented by retarded ignition timing (see Knock control).

References

[1] www.bosch-zuendkerze.de.

Catalytic exhaust-gas treatment

Catalytic converter

Emission-control legislation defines limits governing pollutant emissions from motor vehicles. Engine-design measures alone are not sufficient to comply with these limits. In addition to reducing untreated engine emissions, the emphasis in spark-ignition engines is on catalytic aftertreatment of the exhaust gas in order to convert the pollutants. Catalytic converters convert the pollutants produced during combustion into harmless components.

Three-way catalytic converter

Function

The state-of-the-art technology for engines operating with a stoichiometric A/F mixture is the three-way catalytic converter. Its task is to convert pollutant components HC (hydrocarbons), CO (carbon monoxide), and NO_x (nitrogen oxides), which arise during the combustion process, into harmless components. The end products are H_2O (water vapor), CO_2 (carbon dioxide), and N_2 (nitrogen).

Design and operating principle

The catalytic converter comprises a steel-plate container as housing, the substrate and the substrate coating (washcoat) made of aluminum oxide (Al_2O_3) on which the precious metal is finely spread. The substrate is usually a ceramic monolith, though metallic monoliths are used for special applications (see Ceramic monolith and Metal monolith). On the monolith is a substrate coating, which enlarges the effective area of the catalytic converter by a factor of 7,000. The catalytic layer on top of this contains precious metals such as platinum or palladium, and rhodium. Platinum and palladium accelerate the oxidation of HC and CO, while rhodium is responsible for reducing NO_x. The amount of precious metal contained in a catalytic converter is approx. 1 to 10 g depending on the engine displacement and the exhaust emissions standard to be met.

For example, oxidation of CO and HC takes place according to the following equations:

$$2\,CO + O_2 \rightarrow 2\,CO_2\,,$$
$$2\,C_2H_6 + 7\,O_2 \rightarrow 4\,CO_2 + 6\,H_2O\,.$$

Reduction of nitrogen oxides takes place, for example, according to the following equations:

$$2\,NO + 2\,CO \rightarrow N_2 + 2\,CO_2\,,$$
$$2\,NO_2 + 2\,CO \rightarrow N_2 + 2\,CO_2 + O_2\,.$$

The oxygen required for the oxidation process is either present in the exhaust gas as exhaust-gas oxygen (resulting from incomplete combustion), or it is taken from the NO_x, which is reduced at the same time.

The concentrations of pollutants in the untreated exhaust gas (upstream of the catalytic converter) depend on the excess-air factor λ (Figure 1a). In order for the conversion of the three-way cata-

Figure 1: Catalytic-converter efficiency as a function of the excess-air factor λ
a) Exhaust-gas emissions upstream of three-way catalytic converter,
b) Exhaust-gas emissions downstream of three-way catalytic converter,
c) Electrical signal of two-step λ sensor.
U_λ Sensor voltage.

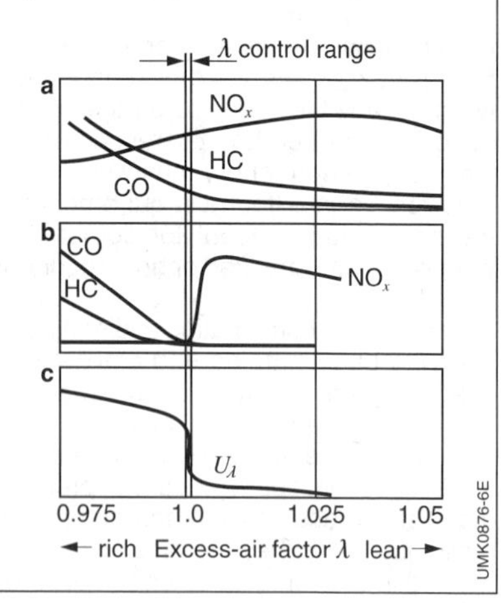

lytic converter to be as high as possible for all three pollutant components, a mixture composition in the stoichiometric ratio $\lambda = 1$ is required (Figure 1b). At $\lambda = 1$, there is a state of equilibrium between oxidation and reduction reactions which facilitates a full oxidation of HC and CO and, at the same time, reduces the NO_x. In this way, HC and CO act as reducing agents for NO_x. The "window" (λ control range) in which the mean time value of λ must lie is very small. Mixture formation must therefore be corrected in a λ control loop with the aid of the signal of a λ oxygen sensor (Figure 1c) (see λ control).

Oxygen storage

The λ accuracy in the dynamic range is typically 5 %, i.e. fluctuations of $\lambda = 1$ in this scale are inevitable. The catalytic converter can compensate for minor mixture fluctuations itself. It has the ability to store excess oxygen in the lean phase in the catalytic converter in order to release it again in the subsequent rich phase. Its substrate coating contains ceroxide, which can make oxygen available via the following balance reaction:

$$2\,Ce_2O_3 + O_2 \leftrightarrow 4\,CeO_2\,.$$

The task of the engine-management system is, therefore, clear: The mean time value of the resulting λ upstream of the catalytic converter must be very precise at one (a deviation of only a few thousandths is permitted). The mean value deviations, converted to oxygen input and output, must not overtax the available oxygen retained in the catalytic converter. Typical oxygen-storage values are in the range of 100 mg to 1 g; these values decrease as the catalytic converter ages. All conventional methods of diagnosing catalytic converters are based on the direct or indirect determination of this oxygen storage capacity (osc).

NO_x storage catalytic converter

Function, design and operating principle

During lean-burn operation, it is impossible for the three-way catalytic converter to convert the nitrogen oxides (NO_x) which have been generated during combustion. CO and HCs are oxidized by the high residual-oxygen content in the exhaust gas and are therefore no longer available as reducing agents for the nitrogen oxides.

The catalytic layer of the NO_x storage catalytic converter also contains substances which can store NO_x (e.g. barium oxide). All conventional NO_x storage coatings also have the properties of a three-way catalytic converter, with the result that the NO_x storage catalytic converter operates like a three-way catalytic converter at $\lambda = 1$.

In stratified lean-operation mode (see Gasoline direct injection), NO_x is not converted continuously, but instead in three stages.

NO_x storage:

During storage, NO_x is initially oxidized to NO_2, which then reacts with the special oxides of the catalytic-converter surface and oxygen (O_2) to form nitrates (e.g. barium nitrate):

$$2\,NO + O_2 \rightarrow 2\,NO_2\,,$$
$$2\,BaO + 4\,NO_2 + O_2 \rightarrow 2\,Ba(NO_3)_2\,.$$

Regeneration:

As the quantity of stored NO_x (load) increases, the ability to continue binding NO_x decreases. At a predefined laden state, the NO_x accumulator must be regenerated, i.e. the nitrogen oxides stored in it must be released again and converted. For this purpose, the engine switches briefly to rich homogeneous operation ($\lambda < 0.8$) to reduce NO to N_2, without emitting CO and HC in the process. In a second stage, the rhodium coating reduces the nitrogen oxides with CO:

$$Ba(NO_3)_2 + 3\,CO \rightarrow 3\,CO_2 + BaO + 2NO\,.$$
$$2\,NO + 2\,CO \rightarrow N_2 + 2\,CO_2\,.$$

The end of the storage and release phase is either calculated using a model-based method or measured with a λ oxygen sensor after the catalytic converter.

Desulfating

The sulfur contained in the fuel also reacts with the accumulator material in the catalytic layer. The result is that, over time, the amount of accumulator material available for NO_x accumulation diminishes. This creates sulfates (e.g. barium sulfate) which are very temperature-resistant and are not reduced during NO_x regeneration. For desulfating, special measures must be used to heat the catalytic converter to 600 to 650 °C before exposing it alternately to rich (λ = 0.95) and lean exhaust gas (λ = 1.05) for a few minutes. This process reduces the sulfates.

The various methods to heat up the NO_x storage catalytic converter located in the underfloor regions must be careful not to overheat the primary catalytic converter.

Catalytic-converter operating temperature

Catalytic converters do not start significant conversion until they have reached a specific operating temperature (light-off temperature). On a three-way catalytic converter, this is approximately 300 °C. Ideal conditions for a high conversion rate are achieved at 400 to 800 °C.

In the case of the NO_x storage catalytic converter, the favorable temperature range is lower; it reaches maximum storage capacity at 300 to 400 °C. The reason for the lower temperature range is that the maximum storage capacity is reduced at higher temperatures. At temperatures higher than 500 to 550 °C the barium compound is no longer stable, which means that nitrogen oxides can no longer be stored.

Operating temperatures of 800 °C to 1,000 °C result in intensified thermal aging of the catalytic converter. This is caused by sintering of the precious metals and the substrate coating, which reduces the active surface. At temperatures above 1,000 °C, thermal aging increases considerably until the catalytic converter has no effect at all.

Catalytic-converter configurations

The required operating temperature of the three-way catalytic converter limits the installation options. Engine-proximate catalytic converters quickly reach operating temperature, but can be exposed to very high thermal loads.

A widely used configuration of the three-way catalytic converter is the split

Figure 2: Catalytic-converter configurations

a) Deployment of an engine-proximate primary catalytic converter and a main catalytic converter.
b) 4-in-2 exhaust manifold for power-optimized engine design: The positioning of the main catalytic converter only after the second junction is unfavorable for the heating-up characteristics, which means that preferably two engine-compartment primary catalytic converters are deployed.
c) Engine with more than one cylinder bank (V engine): The exhaust system. Runs completely as twin-branch with a primary and main catalytic converter for each.
d) Engine with more than one cylinder bank (V engine): Y-shaped junction in the underfloor area to form an overall exhaust-gas train with a joint main catalytic converter for both banks.

1 Primary catalytic converter,
2 Main catalytic converter.

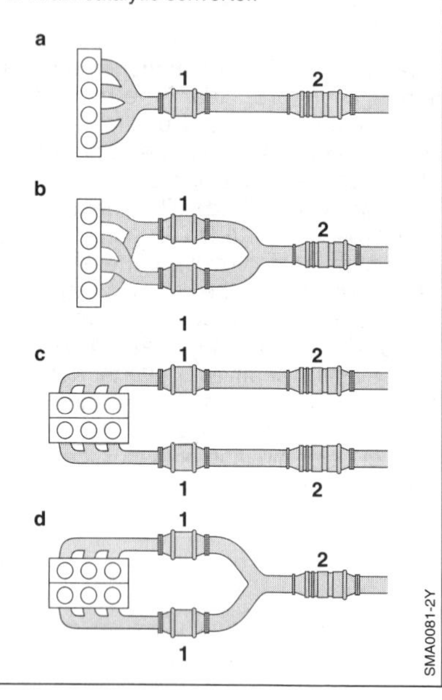

arrangement with an upstream primary catalytic converter and an underfloor catalytic converter (main catalytic converter). The primary catalytic converter is optimized with regard to high-temperature stability, the underfloor catalytic converter with regard to "low light off" (low activation temperature). Figure 2 shows different possible arrangements of the primary and underfloor catalytic converters. Owing to their lower maximum permissible operating temperature, NO_x storage catalytic converters are always installed in the underfloor area.

Catalytic-converter heating

Whereas a catalytic converter at operating temperature reaches very high conversion rates of almost 100%, considerably greater amounts of pollutants are emitted in the cold start phase and in the warm-up phase. The HC and CO emissions are particularly high when the engine is cold because fuel condenses onto the cold cylinder walls, leaves the combustion chamber unburned and cannot be converted by the cold catalytic converter.

It is therefore important to minimize untreated emissions during the warm-up phase before the catalytic converter lights off (operating temperature reached). Measures are also required to bring the catalytic converter up to operating temperature quickly. The required heat is provided by raising the exhaust-gas temperature and increasing the exhaust-gas mass flow. This is made possible by the following measures.

Ignition-timing adjustment

The main measure for increasing the exhaust-gas heat flow is ignition-timing retardation. Combustion is initiated as late as possible and takes place during the expansion phase. The exhaust gas has a relatively high temperature at the end of the expansion phase. Retarded combustion has an unfavorable effect on engine efficiency.

Idle speed increase

A supporting measure is to raise the idle speed and thereby increase the exhaust-gas mass flow. The increased engine speed permits a greater ignition-angle retardation. However, in order to ensure reliable ignition, the ignition angles are limited to roughly 10° to 15° after TDC. The heat output limited in this way is not always enough to achieve the current emission limits.

Exhaust-camshaft adjustment

A further contribution to increasing the heat flow can be achieved, if necessary, with exhaust-camshaft adjustment. The process of the exhaust valves opening as early as possible interrupts the retarded combustion early and the mechanical work generated is thus reduced further. The quantity of energy not converted into mechanical work is available as a quantity of heat in the exhaust gas.

Homogeneous split

With gasoline direct injection, there is, in principle, the possibility of multiple injection. This enables the catalytic converter to be quickly heated up to operating temperature without the need for additional components. The "homogeneous-split" measure involves initially creating a homogeneous, lean basic mixture by means of injection during the induction stroke. Subsequent injection during the compression stroke generates a stratified charge cloud. This facilitates retarded moments of ignition and results in high exhaust-gas heat flows. The exhaust-gas heat flows that can be achieved are comparable with those of a secondary-air injection.

Secondary-air injection

Thermal afterburning of unburnt fuel constituents increases the temperature in the exhaust system. A rich ($\lambda = 0.9$) to very rich ($\lambda = 0.6$) air/fuel basic mixture is adjusted for this purpose. A secondary-air pump supplies oxygen to the exhaust system (Figure 3) to produce a leaner exhaust-gas composition. Where the basic mixture is very rich ($\lambda = 0.6$), the unburnt fuel constituents oxidize above a specific temperature threshold exothermically before entering the catalytic converter. To achieve this temperature, it is necessary on the one hand to raise the temperature level with retarded ignition angles and on the other hand to introduce the secondary air as closely as possible to the exhaust valves. The exothermic reaction in the exhaust system increases the heat flow to the catalytic converter and therefore shortens its heating period. The HC and CO emissions are for the most part reduced before entering the catalytic converter.

Where the basic mixture is less rich ($\lambda = 0.9$), there is no significant reaction ahead of the catalytic converter. The unburnt fuel constituents oxidize in the catalytic converter and heat it up from the inside. For this purpose, however, the end face of the catalytic converter must first be brought up to a level above the light-off temperature by means of conventional measures (e.g. ignition-timing retardation).

Figure 3: Secondary-air system
1 Secondary-air pump,
2 Inducted air,
3 Relay,
4 Engine ECU,
5 Secondary-air valve,
6 Control valve,
7 Battery,
8 Inflow point in exhaust pipe,
9 Exhaust valve,
10 To intake-manifold connection.

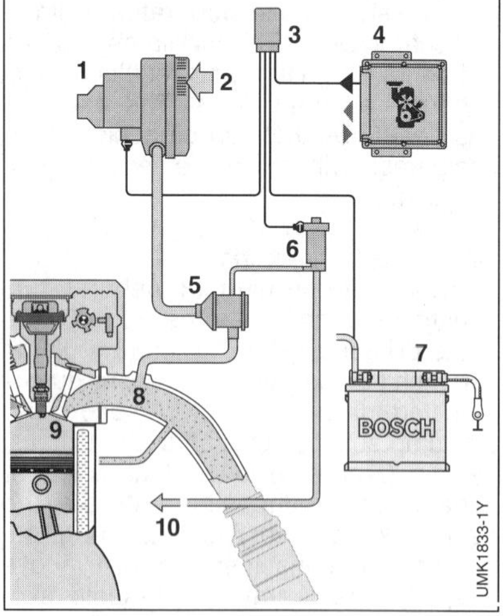

As a rule, a less rich basic mixture is adjusted, because, if the basic mixture is very rich, the exothermic reaction ahead of the catalytic converter occurs reliably only under stable boundary conditions.

Secondary-air injection is performed with an electric secondary-air pump, which is switched by means of a relay on account of the high power demand. Since the secondary-air valve prevents backflow of exhaust gases into the pump, it must remain closed when the pump is deactivated. Either it is a passive non-return valve or it is actuated by purely electrical or (as shown in Figure 3) pneumatic means with an electrically actuated control valve. When the control valve is actuated, the secondary-air valve opens in response to the intake-manifold vacuum. The secondary-air system is coordinated by the engine ECU.

Alternative concepts for active heating
As a supplement in special cases, an electrically heated catalytic converter is used for quick heating up of the catalytic converter. This has been used previously in individual small-series projects.

λ control

To insure that the conversion rates of the three-way catalytic converter are as high as possible for the pollutant components HC, CO and NO_x, the reaction components must be present in the stoichiometric ratio. This requires a mixture composition of $\lambda = 1.0$, i.e. the stoichiometric air/fuel ratio must be adhered to very precisely.

Mixture formation must be followed up in a control loop, because sufficient accuracy cannot be achieved solely by controlling the metering of the fuel. When the λ control loop is used, deviations from a specific air/fuel ratio can be detected and corrected through the quantity of fuel injected. The residual-oxygen content in the exhaust gas, which is measured with λ oxygen sensors (see Two-step λ oxygen sensor and Broad-band λ oxygen sensor), serves as the measure for the composition of the air/fuel mixture.

Two-step λ control

Two-step λ control adjusts the mixture to $\lambda = 1$. A manipulated variable, composed of the voltage jump and the ramp, changes its direction of control for each voltage jump of the two-step λ sensor. This indicates a change from rich to lean or from lean to rich (Figure 4). The typical amplitude of this manipulated variable has been set in the range of 2 to 3%. The result is a limited controller dynamic, which is predominantly determined by the sum of the response times (pre-storage of fuel in the intake manifold, four-stroke principle of the spark-ignition (SI) engine, and gas travel time).

The typical shift of the oxygen zero transition point (theoretically at $\lambda = 1.0$) and therefore the jump of the λ sensor, caused by the variation in exhaust-gas composition, can be control-compensated by shaping the manipulated variable's characteristic curve asymmetrically (rich or lean shift). Here, the preferred method is to hold the ramp value at the voltage jump for a controlled dwell time t_V after the sensor is subjected to a voltage jump. A rich or lean shift is performed depending on the operating point. When a shift is made towards "rich", the manipulated variable holds at the rich position for a dwell time t_V although the sensor signal has already jumped towards "rich" (Figure 4a). Jump and ramp of the manipulated variable do not also move towards "lean" until the dwell time has elapsed. If the sensor signal then jumps towards "lean", the manipulated variable directly counteracts this (with jump and ramp) without dwelling on the lean position. The dwell time t_V is the result of a pre-control coefficient, which is corrected by a proportion from the two-sensor λ control system (post cat control).

The behavior is reversed in a shift to "lean". If the sensor signal indicates a lean mixture, the manipulated variable holds at the lean position for the dwell time t_V (Figure 4b) and then adjusts towards

Figure 4: Manipulated-variable curve of a two-step control with a λ sensor upstream of the catalytic converter and λ shift (delay time t_v) due to pre-control proportion and post cat control
a) Shift to "rich", b) Shift to "lean".
t_v Dwell time after sensor jump.

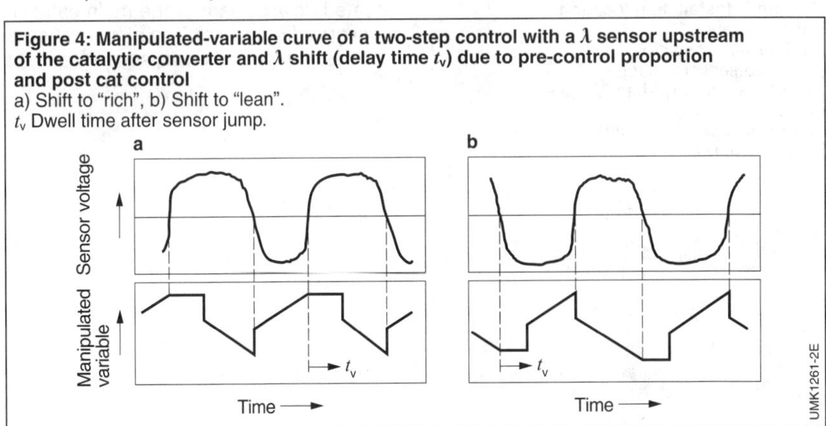

"rich". A sensor signal jump from "lean" to "rich" is, however, immediately counteracted.

Continuous λ control

The defined dynamic response of a two-step control can only be improved if the deviation from $\lambda = 1$ can actually be measured. The broad-band λ sensor can be used to achieve continuous-action control at $\lambda = 1$ with a stationary, very low amplitude in conjunction with high dynamic response. The control parameters are calculated and adapted as a function of the engine's operating points. Above all, with this type of λ control, compensation for the unavoidable offset of the stationary and non-stationary pilot control is far quicker.

The broad-band λ oxygen sensor also enables adjustment to mixture compositions that deviate from $\lambda = 1$. Thus, controlled enrichment ($\lambda < 1$), e.g. for component protection, can be effected. The broad-band λ sensor also enables controlled leaning ($\lambda > 1$), e.g. for a leaner warm-up during catalytic-converter heating.

Two-sensor λ control

When it is situated upstream of the catalytic converter, the λ sensor is heavily stressed by high temperatures and untreated exhaust gas, and this leads to limitations in λ sensor accuracy. The voltage-jump point of a two-step sensor or the characteristic curve of a broad-band sensor can shift, for example, as a result of changed exhaust-gas compositions. A λ sensor downstream of the catalytic converter is subjected to these influences to a much lesser extent. However, λ control on its own with the sensor downstream of the catalytic converter demonstrates disadvantages in dynamic response on account of the gas travel times, and responds to mixture changes more slowly.

Greater accuracy is achieved with two-sensor control. Here, a slower correction control loop is superimposed on the two-step or continuous-action λ control described by means of an additional two-step λ sensor (Figure 5a). The voltage of the two-step sensor downstream of the catalytic converter is compared with a setpoint value (e.g. 600 mV) for this purpose. On this basis the control evaluates the deviations from the setpoint value and additionally alters over a dwell time t_v additively the controlled rich or lean shift of the first control loop of a two-step control or the setpoint value of a continuous-action control.

Figure 5: Installation locations of λ sensors
a) Two-sensor control,
b) Three-sensor control.
1 Two-step or broad-band λ sensor,
2 Two-step λ sensor,
3 Primary catalytic converter,
4 Main catalytic converter.

SMA0090Y

a
1 3 2 4

b
1 3 2 4 2

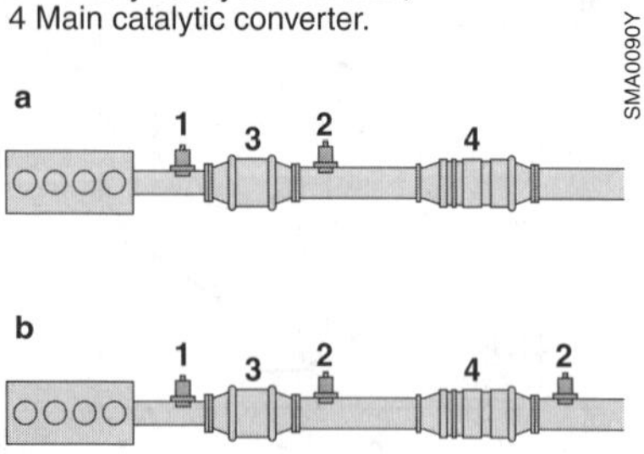

Three-sensor λ control

There are SULEV concepts (Super Ultra-Low-Emission Vehicle) with a third sensor downstream of the main catalytic converter. The two-sensor control system (single cascade) has been supplemented here by an extremely slow closed-loop control system using a third sensor located downstream of the main catalytic converter (Figure 5b). The aim of development is, however, the two-sensor concept.

LPG operation

Applications

Liquified petroleum gas (LPG) is obtained during the extraction of crude oil and natural gas and during different refinery processes. It is made up primarily of propane and butane (see “Alternative fuels”).

As the conventional fuels gasoline and diesel have become more expensive in the last few years, the cheaper liquified petroleum gas – not least due to regional tax reductions (e.g. for Germany, Italy, Poland) – is becoming increasingly more widespread. Liquified petroleum gas is widely used in Europe and in South Korea, Australia, and China. At present (as at 2014),

Figure 1: Bivalent gasoline-engine operation with gaseous LPG injection
1 Activated charcoal canister, 2 Canister-purge valve, 3 Gasoline rail with gasoline injectors,
4 Camshaft adjuster, 5 Ignition coil with spark plug, 6 Camshaft sensor,
7 λ sensor upstream of primary catalytic converter, 8 Hot-film air-mass meter,
9 Electronically actuated throttle valve (ETC), 10 Intake-manifold pressure sensor,
11 LPG ECU, 12 Engine ECU, 13 Accelerator-pedal module,
14 LPG shutoff valve, 15 Evaporator with pressure regulator,
16 Combined pressure and temperature sensor, 17 LPG rail with gas injectors,
18 LPG filter, 19 Engine-temperature sensor, 20 Primary catalytic converter,
21 Crankshaft speed sensor, 22 Knock sensor,
23 Gasoline tank with integrated electric fuel pump, 24 Gasoline filler neck,
25 LPG tank (steel tank), 26 Electromagnetic shutoff valve,
27 LPG filler neck, 28 80 % filler stop valve, 29 LPG level indicator,
30 Pressure-relief valve, 31 λ sensor downstream of primary catalytic converter,
32 Main catalytic converter.

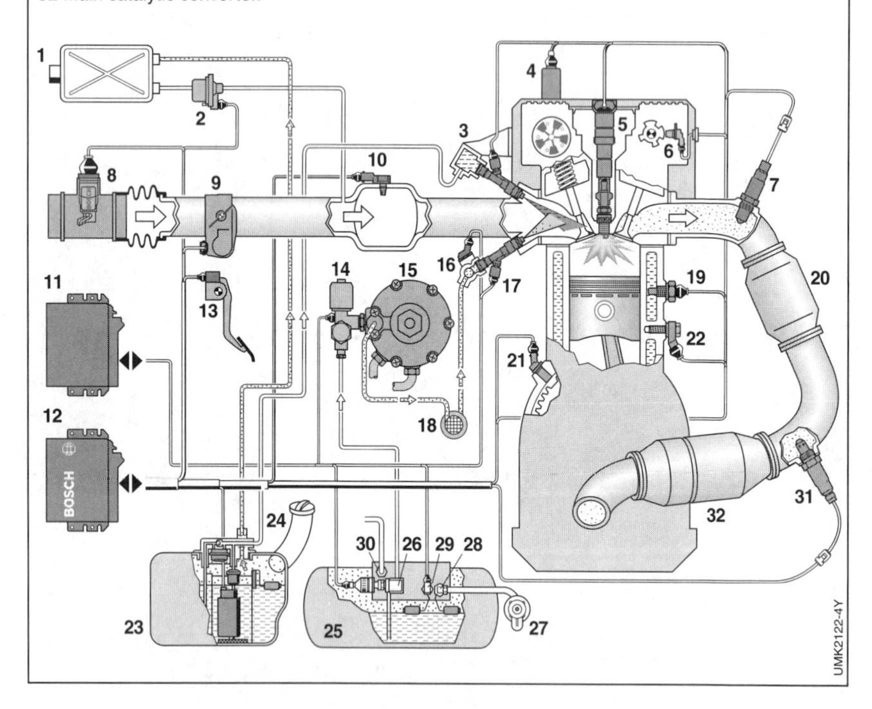

there is estimated to be a population of more than 21 million LPG-fueled vehicles worldwide [1]. Up-to-date information on the number of LPG-fueled vehicles and the LPG filling-station network in Germany can be found on the Internet (see [2] and [3]).

The additional volumetric consumption of LPG-fueled vehicles is, depending on engine concept, engine management and gas system 20–30 %, compared with gasoline-fueled vehicles. In spite of this additional costs, the running costs of LPG vehicles are cheaper than those of gasoline and diesel vehicles.

LPG systems are distinguished by their mixture formation with gaseous injection or liquid LPG injection and by the ECU concepts with one common or two separate engine ECUs (master/slave concept). Systems with venturi mixers (operating according to the carburetor principle) can only still be used in older vehicles or in markets with laxer exhaust-emission regulations.

Vehicles with spark-ignition engines are virtually exclusively converted to bifuel LPG/gasoline operation. Most LPG systems are installed as a retrofit. In the meantime, however, vehicle manufacturers are also offering models with bifuel LPG/gasoline operation off the assembly line. In these vehicles, the LPG tank – as in the retrofit concepts – is installed for the most concepts in the spare-wheel compartment. The luggage-compartment load volume remains unchanged with this form of installation. The spare wheel is simply omitted and replaced by a blow-out repair kit (Tire-Fit).

Design

Storage of LPG

LPG can liquify at ambient temperatures of –20 to +40 °C under pressure (3–15 bar), depending on the propane-to-butane mixture ratio. It is stored in a liquid state in steel tanks. The filling system of the LPG tanks incorporates a fill-volume restrictor which ensures that the fill volume does not exceed 80 % of the liquid phase of the maximum tank volume. This ensures that the internal tank pressure does not rise to unacceptable levels when the tank is subjected to heat. The tanks are subject (also in some non-European countries) to European safety regulations governing their properties, equipment specification and installation – currently ECE-R67-01 [4].

The safety devices of an LPG tank comprise a pressure-relief valve, an electromagnetic shutoff valve with integrated flow limiter (pipe-fracture protection), and optionally a thermal safety valve which allows the gas to escape under controlled conditions in the event of a fire.

Fuel delivery

The LPG is delivered from the tank to the engine in systems with gaseous injection by the internal tank pressure (LPG vapor pressure). Because this pressure drops as ambient temperature decreases, the operation of such systems is partly limited with regard to their suitability for winter operation.

In systems with liquid injection, the LPG is delivered by an LPG fuel-supply pump installed in the LPG tank which generates a delivery pressure in addition to the internal tank pressure. In terms of their suitability for winter operation, these systems are comparable with gasoline injection systems.

Components

The following components are integrated into LPG vehicles (Figures 1 and 2):

- LPG filler neck with integrated non-return valve
- Steel tank (cylindrical or toroidal)
- Pressure-relief valve (27 bar)
- LPG level sensor
- Valve with integrated 80 % tank filler stop
- Electromagnetic shutoff valve (in the tank)
- LPG shutoff valve (in the engine compartment)
- Evaporator with pressure regulator (only for systems with gaseous LPG injection)
- Pressure regulator (optional in systems with liquid LPG injection)
- LPG fuel rail with gas injectors
- Pressure and temperature sensors (on the rail)
- LPG filter
- If necessary, a slave ECU
- Demand-controlled LPG supply pump (only in systems with liquid LPG injection)

Figure 2: Bivalent gasoline-engine operation with liquid LPG fuel injection into the intake manifold

1 Activated charcoal canister, 2 Canister-purge valve, 3 Gasoline rail with gasoline injectors, 4 Camshaft adjuster, 5 Ignition coil with spark plug, 6 Camshaft sensor, 7 λ sensor upstream of primary catalytic converter, 8 Hot-film air-mass meter, 9 Electronically actuated throttle valve (ETC), 10 Intake-manifold pressure sensor, 11 LPG ECU, 12 Engine ECU, 13 Accelerator-pedal module, 14 LPG shutoff valves for supply and return, 15 Combined pressure and temperature sensor, 16 LPG fuel rail with gas injectors, 17 Engine-temperature sensor, 18 Primary catalytic converter, 19 Crankshaft-speed sensor, 20 Knock sensor, 21 Gasoline tank with integrated electric fuel pump, 22 Gasoline filler neck, 23 LPG tank with integrated electric LPG pump, 24 Electromagnetic shutoff valve, 25 LPG filler neck, 26 80 % filler stop valve with integrated LPG level sensor, 27 Pressure-relief valve, 28 Non-return valve, 29 λ sensor downstream of primary catalytic converter, 30 Main catalytic converter.

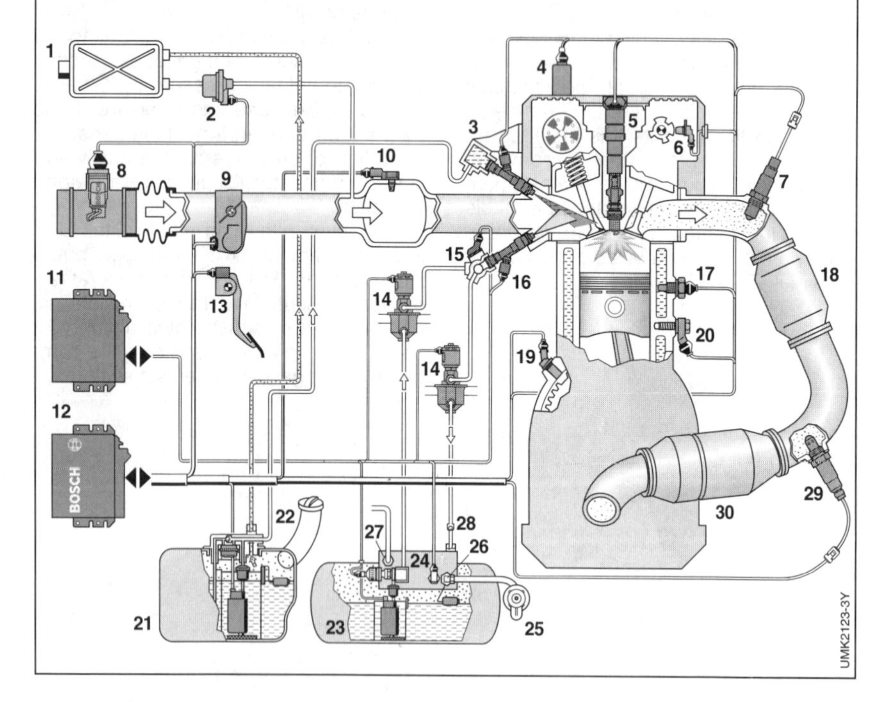

LPG systems

Systems with gaseous LPG injection

Operating principle

The LPG is delivered in liquid state from the tank via the electromagnetic shutoff valve to the engine compartment (Figure 1), where an additional shutoff valve is installed. After this, the LPG is converted into the gaseous state with pressures of 0.5 to 3.5 bar in the evaporator and the pressure regulator. The evaporator is heated by the engine coolant in order to compensate for the cooling encountered during the phase transition. The gaseous LPG is directed from the evaporator through a pressure-resistant flexible hose to the LPG rail and from here injected by the gas injectors into the intake manifold.

The air-intake section of the basic engine remains unchanged except for the connection of the LPG gas injectors to the intake manifold.

Mixture formation

Mixture formation is effected in the same manner as conventional multipoint gasoline injection. The LPG rail supplies the individual gas injectors with gaseous LPG. Engine management determines the required injected fuel quantity, which is corrected by means of λ control. The injected fuel quantity is additionally corrected by means of the pressure and the temperature of the fuel in the LPG rail.

LPG injection displaces some of the inducted air volume in the intake manifold, which results in charge losses in non-turbocharged engines. Engine performance reduced by a factor of 2–4% compared with gasoline operation can thus be expected for LPG operation.

Systems with LPG liquid injection into the intake manifold

Operating principle

In systems with LPG liquid injection into the intake manifold (Figure 2) the LPG ECU regulates the system pressure via a fuel pump installed in the LPG tank.

Some systems may also need a pressure regulator to regulate the injection pressure. Because of the tendency for vapor bubbles to form in the fuel system, systems with LPG liquid injection must be fitted with a tank return line in order if necessary to expel vapor bubbles that occur.

Mixture formation

As with conventional multipoint gasoline injection, the LPG is injected into the intake manifold in front of the engine intake valves. The evaporation that occurs as the LPG is injected (according to the Joule-Thomson effect) lowers the temperature of the inducted air/gas mixture, which has a positive impact on the engine's volumetric efficiency and performance figures.

Advantages of LPG liquid injection

LPG liquid injection offers the following advantages:

- Gas operation is also possible at extremely low temperatures.
- Monofuel gas operation is possible.
- Engine start in LPG mode is also possible in bifuel systems.
- A better cylinder charge is obtained thanks to the cooling effect in the intake manifold and combustion chamber, thus producing identical engine-performance figures to gasoline without further measures. Engines with LPG liquid injection can even deliver slightly higher performance figures when compared with gasoline operation.

Systems with direct LPG liquid injection into the combustion chamber

LPG behaves very similar to gasoline at higher system pressures. Therefore LPG direct injection (LPG-DI) into the combustion chamber is also possible. LPG direct injection offers some significant advantages:

- A better exhaust-gas composition is obtained (virtually no particulate emissions) with comparable engine-performance figures.
- Very good over all thermodynamic behavior.

LPG direct liquid injection into the combustion chamber is possible in all modern gasoline engines with direct injection. Such systems are currently still in the development phase (see e.g. [5]).

Exhaust emissions

With regard to the limited pollutant components HC, CO and NO_x, the use of LPG as fuel in modern spark-ignition engines offers some advantages over gasoline operation. The situation when it comes to CO_2 emissions is even better. Here, LPG vehicles can achieve a reduction of approximately 10%. This reduction is founded on the lower carbon content in LPG. Furthermore, when compared with diesel engines, there are virtually no particulates. LPG drives also have advantages when it comes to other exhaust-gas components which are not limited by law but which are harmful to health. Thus, for instance, LPG drives emit significantly lower amounts of aromatic hydrocarbons (e.g. benzene).

Unlike engines fueled by CNG (Compressed Natural Gas), engines fueled by LPG are not subject to any particular demands on the exhaust-gas treatment system of the existing gasoline concept. The essential exhaust-gas components of LPG operation are comparable with those of gasoline operation.

Engine management for LPG systems

Both double-ECU concepts (one ECU for gasoline operation and one for LPG operation) and single-ECU concepts (gasoline and gas functions integrated into a single ECU) are used. Because the vehicles are bifuel, the driver can select between gasoline and LPG operation using a switch.

The level sensor installed in the LPG tank signals the current LPG tank level to the engine-management system. A combined low-pressure and temperature sensor mounted on the gas rail enables the engine-management system to correct the injection timing in such a way that a stoichiometric mixture can be pilot-controlled in the intake manifold even if the gas density fluctuates. The engine-management systems uses the values from the λ sensor and different adaptation algorithms to adapt the mixture formation for different gas qualities. Even functions for switching between gas and gasoline operation are integrated.

The other sensors and actuators of the engine-management system are for the most part identical to those in a gasoline engine.

Systems with LPG liquid injection into the combustion chamber require, in addition to the functions for mixture formation and gas-quality adaptation, functions for controlling the LPG fuel pump.

Figure 3: Electromagnetic liquid LPG injector
1 Electrical connection,
2 Field winding,
3 O-rings,
4 Valve housing,
5 Filter strainer,
6 LPG supply,
7 Armature,
8 O-rings,
9 Valve body,
10 Valve needle,
11 Injection nozzle.

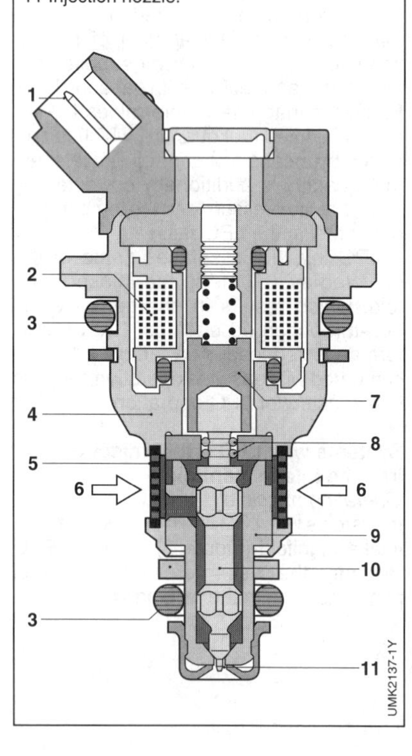

Components

LPG gaseous injector

Function and requirements

When LPG is injected in the gaseous state of aggregation, the given energy density and the low gas pressures resulting from the state of aggregation of approx. 0.5 to 3.5 bar dictate that the following parameters must be taken into account in the design of LPG injection valves:

- Configuration of the injection valves for large gas volumes
- Adaptation of the sealing mechanisms and materials to the gaseous medium
- Consideration of potentially aggressive fuel contaminants in the choice of material

Design and operating principle

Essentially, the design and function of a gas injector configured for operation with gaseous LPG are very similar to those of a CNG injector (see Engines fueled by natural gas).

LPG liquid injector

Function and requirements

In systems where the LPG is injected in the liquid state of aggregation the LPG must be kept in the liquid state in the entire pipe and rail system by means of pressure build-up by a pump installed in the LPG tank. Only in this state can it be injected.

LPG injectors must satisfy the following requirements:

- Operation at variable pressures of 2 to 25 bar.
- Absolute tightness in the case of systemic pressure variability (e.g. on shutdown of gas operation and spontaneously occurring pressure peaks due to the LPG evaporating in the rail system).
- Possibility of a flushing process to eliminate vapor bubbles that occur (e.g. during hot start or switching from gasoline to LPG operation with the engine at operating temperature).

Design and operating principle

LPG injectors (Figure 3) are, in response to the above-mentioned requirements, designed according to the bottom-feed principle. The injector is designed in such a way that the fitting for the fuel inlet into the injector housing is as close as possible to the injector outlet. This layout enables the valve to be easily flushed with fresh fuel in order to quickly remove possible vapor bubbles.

The fuel injectors normally used in gasoline engines, on the other hand, operate for the most part according to the top-feed principle. Because of the supply of fresh fuel located at the top end of the injector housing, the flush process for the purpose of removing vapor bubbles in the injector would be very diffecult or not possible at all. Under such conditions it would not be possible to provide a correct metering of the fuel quantity and thus optimum combustion, which is why the top-feed principle is not suitable for use in liquid LPG injection.

Installation position

The position of the LPG injectors in the intake manifold is subject to special design requirements according to the generation of cold by the evaporation of the LPG during the injection process (as a result of the Joule-Thompson effect). The liquid LPG evaporates abruptly on entry into the intake manifold and extracts the required energy in the form of heat out of the fresh air in the intake manifold. This specific-point cold in the intake manifold can cause unwelcome operating phenomena (e.g. misfires), which can be put down among other things to the formation of ice crystals.

LPG filler neck

Neck variants

The filler neck is not internationally standardized. For this reason, different fueling adapters are required from country to country. Four different types of LPG fuel nozzle are used in filling stations. Different filler necks are therefore needed depending on the field of application. The following necks are widely used:

- ACME neck (named for the American trapezoidal-thread standardization) is used in Germany, Belgium, Ireland, England, Luxembourg, Scotland, Austria, China, Canada, USA, Australia, and Switzerland.
- DISH neck (named for its dish shape) is used in Denmark, France, Poland, Greece, Hungary, Italy, Austria, Portugal, the Czech Republic, Turkey, China, Korea, and Switzerland.
- Bayonet neck is used in the Netherlands, Belgium and the UK.
- Euronozzle is currently used in Spain only.

The Euronozzle is supposed to replace all the other necks on a Europewide basis in future. The EU Regulation for this standardized neck has not yet been adopted by the majority of EU member states.

Design

Each filler neck in the vehicle is designed in such a way that the LPG vehicle can be refueled at all LPG filling stations by using a suitable adapter.

Figure 4: Evaporator with pressure regulator
1 Solenoid coil of electromagnetic shutoff valve,
2 Seal,
3 Inlet for liquid LPG and filter,
4 Inlet for engine coolant,
5 Outlet for engine coolant,
6 Temperature sensor,
7 Outlet for gaseous LPG at constant pressure,
8 Safety pressure-relief valve,
9 Evaporation chamber,
10 Area for gaseous LPG,
11 Connection for intake-manifold vacuum,
12 Pressure diaphragm with mounting,
13 Control rod,
14 Adjusting screw,
15 Spring,
16 Control plunger and throttling orifice.

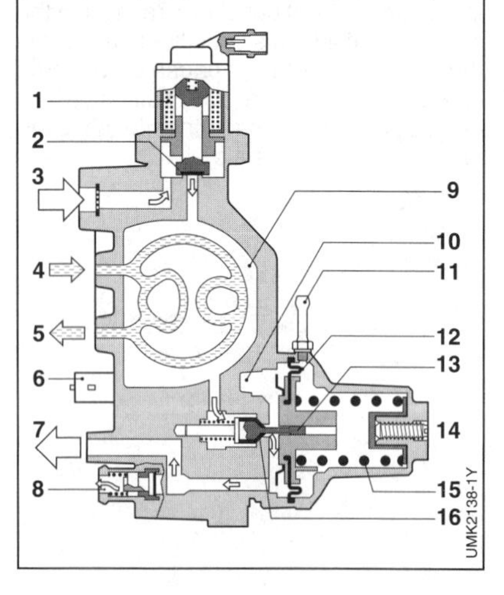

Evaporator for systems with gaseous LPG injection

Function

Evaporators are used in systems with gaseous LPG injection. The LPG is delivered to the evaporator by the vapor pressure in the gas tank via an electromagnetic shutoff valve.

The evaporator is one of the main components of every LPG injection system and performs two functions: Firstly, the liquid gas is converted from the liquid to the gaseous state in the first part of the evaporator. Secondly, the pressure of the gaseous LPG is then regulated to approx. 1 bar differential pressure in relation to the intake manifold. This pressure, with which the injection valves are supplied, must be kept constant through all the operating states and the varying vapor pressures of the different LPG qualities within minimum tolerances. The most widely used evaporators are those with single-stage pressure regulation. Figure 4 shows an evaporator with single-stage pressure regulation.

Operating principle
The LPG enters through the inlet connection into the evaporation area. This consists of a labyrinth that is heated by the vehicle's heating circuit to convert the incoming liquid gas into the gaseous state. Pressure of 4 to 15 bar are achieved here. The pressure varies depending on the temperature of the engine coolant and the composition of the LPG (mixture ratio of propane and butane).

The now gaseous medium passes through a transfer passage into the pressure-regulation area, which is closed by a diaphragm. A pre-tensioned spring is seated on the back of the diaphragm. The pre-tension can be adjusted with an adjusting screw in order to effect deviations from the factory output-pressure pre-setting.

The diaphragm is connected via a lever mechanism to a valve on the transfer-passage connection. When the pressure increases, the movement of the diaphragm actuates a control rod, which in turn alters the opening cross-section at the outlet of the pressure-regulation area. Thus the supply is reduced or closed off and no more gas can flow. When the gas pressure decreases in response to increased fuel reduction by the engine, the diaphragm is moved in the opposite direction. The valve is opened via the lever mechanism to allow gas to flow until again an equilibrium of forces acts on the diaphragm and the lever mechanism can close the transfer passage again.

An important characteristic of the quality of the evaporator is the lowest possible fluctuation of the operating pressure in different engine operating states together with the highest possible control precision over the operating time.

In certain countries, owing to contaminants in the LPG, the manufacturer recommends that the diaphragm be replaced after a certain operating time.

Intake-manifold vacuum support
In addition, the response of evaporator pressure regulation is optimized by connecting the intake-manifold pressure to the area above the diaphragm.

Operating values
The inlet pressure is 4 to 15 bar, depending on ambient temperature and gas quality. The output pressure is 0.5 to 3.5 bar, depending on the setting.

Electromagnetic shutoff valve and temperature sensor
The electromagnetic shutoff valve interrupts the gas supply when the engine is not in gas mode. The changeover from gasoline to LPG operation is initiated by the engine ECU as a function of the engine coolant temperature. The minimum coolant-temperature value for changeover is selected to ensure reliable evaporation of the LPG in the evaporator. A temperature sensor is integrated in the evaporator coolant circuit for this purpose. A cold start in gas mode is not possible with these systems.

References
[1] World LPG Gas Association, www.worldlpgas.com/autogas.
[2] www.gastankstellen.de.
[3] www.autogastanken.de.
[4] ECE-R67: Standard conditions on the: I. Approval of specific equipment of motor vehicles using LPG in their propulsion system. II. Approval of a vehicle fitted with specific equipment for the use of LPG in its propulsion system with regard to the installation of such equipment.
[5] K. Reif: Ottomotor-Management, 4th Ed., Springer-Vieweg, 2014.

Engines fueled by natural gas

Applications

In view of the worldwide efforts to reduce CO_2 emissions (carbon dioxide), natural gas is becoming increasingly important as an alternative fuel. Compressed natural gas (CNG), not to be confused with liquified petroleum gas (LPG), consists primarily of methane. Liquified petroleum gas, on the other hand, consists of propane and butane (see Alternative fuels).

Emissions

Compared to gasoline, the combustion of CNG produces approximately 25% less CO_2. CNG thus generates the lowest local CO_2 emissions of the fossil fuels. Because of the lower CO_2 emissions, the petroleum tax on CNG is reduced in many countries.

Use of biogas

Using biogas (biomethane) can achieve a regenerative fuel supply that further reduces global greenhouse-gas emissions. Biomethane can be produced from plant remains, organic waste and liquid manure, or alternatively from electricity generated from renewable sources. Here, hydrogen is obtained by electrolysis from water and then reacts in a chemical reaction with CO_2 to produce methane. This methane gas obtained from renewable sources is called solar gas, wind gas or simply RE gas (renewable-energy gas).

Natural-gas vehicles

Various vehicle manufacturers have in the meantime started to offer natural-gas versions of automobile models from the factory. This ensures that the larger CNG tanks are accommodated more efficiently and conveniently, thereby preventing – as is practically inevitable when these tanks are installed subsequently – the loss of luggage-compartment volume.

The latest information on natural-gas vehicles and the filling-station network in Germany and Europe can be found at the Internet addresses [1] and [2]. Natural-gas vehicles are, for the most part, bifuel vehicles which allow the driver to switch between natural-gas and gasoline operation. There are variants in the form of monovalent vehicles and the concepts referred to as "Monovalent plus", in which the engine is optimized for natural-gas operation so as to be able to utilize to optimum effect all the favorable properties of natural gas (higher knock resistance, lower CO_2 and pollutant emissions). "Monovalent plus" vehicles still feature a small gasoline tank (less than 15 *l*) so that they can continue to be driven in gasoline mode in the event of there being no natural-gas filling station nearby.

Storage of natural gas

Natural gas can be stored in compressed form at a pressure of 200 bar as CNG (compressed natural gas) or in liquid form at −162 °C as LNG (liquified natural gas). In view of the great expense involved in storing the gas in liquid form, storing the gas in compressed form at 200 bar has become the standard method. In spite of being stored at high-pressure, natural gas has a lower energy-storage density than gasoline such that a tank with four times the volume is required for the same energy content.

Figure 1: Spark-ignition engine with natural-gas or gasoline operation
1 Carbon canister, 2 Canister-purge valve, 3 Exhaust-gas recirculation valve,
4 Camshaft adjuster, 5 Hot-film air-mass meter,
6 Electronically actuated throttle valve (ETC),
7 Intake-manifold pressure sensor, 8 Gasoline rail with gasoline injectors,
9 Ignition coil with spark plug, 10 Camshaft sensor,
11 λ sensor before catalytic converter,
12 Bifuel Motronic ECU, 13 Accelerator-pedal module,
14 Natural-gas pressure-control module with integrated gas shutoff valve and high-pressure sensor,
15 Gas rail with natural-gas pressure and temperature sensor, 16 Natural-gas injector,
17 Crankshaft speed sensor, 18 Engine-temperature sensor, 19 Knock sensor,
20 Primary catalytic converter, 21 CAN interface,
22 Engine warning light, 23 Diagnosis interface, 24 Interface to immobilizer ECU,
25 Gasoline tank with integrated electric fuel pump, 26 Gasoline filler neck,
27 Natural-gas filler neck, 28 High-pressure shutoff valve on natural-gas tank,
29 Natural-gas tank, 30 Main catalytic converter,
31 λ sensor after primary catalytic converter.

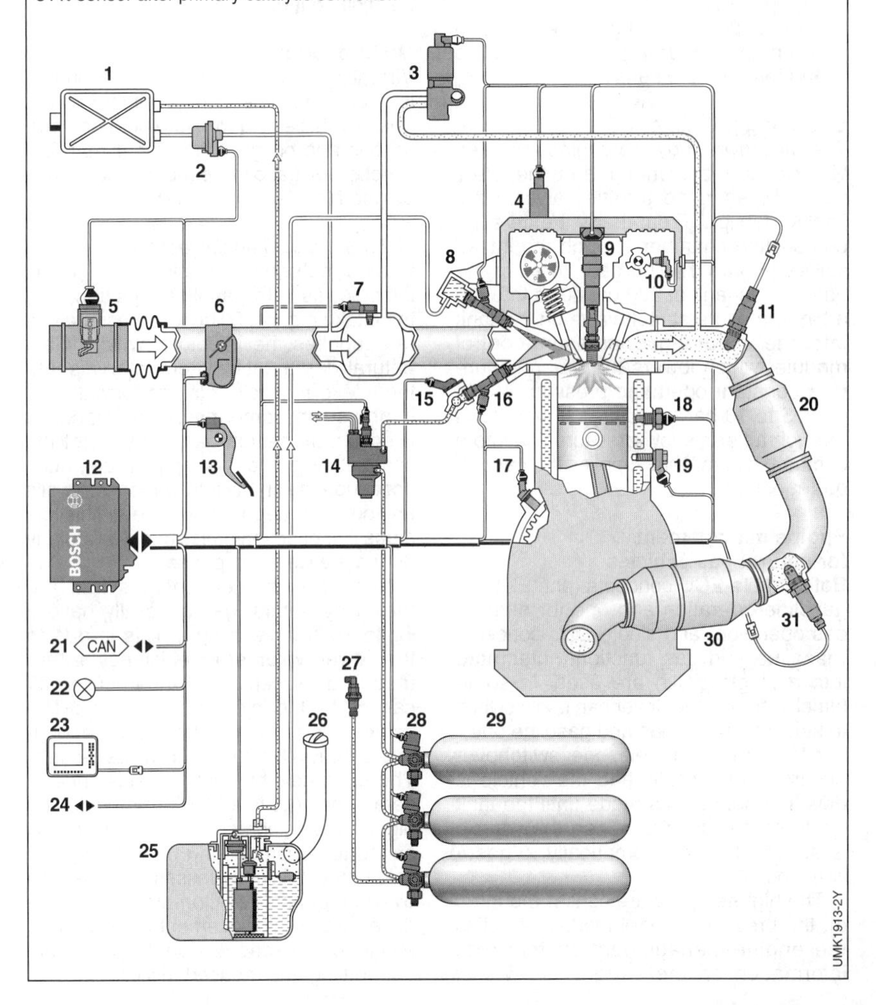

Design

Components

Natural-gas automobiles are equipped almost exclusively with spark-ignition engines. They are extended to include the following natural-gas components (Figure 1, [3]):

- Natural-gas filler neck
- Natural-gas tanks
- High-pressure shutoff valves on the natural-gas tanks
- Natural-gas pressure-control module with integrated gas shutoff valve and high-pressure sensor
- Gas rail with natural-gas injectors and combined natural-gas low-pressure and temperature sensor

Operating principle

The air drawn in by the engine is routed via the air-mass meter and the electronically actuated throttle valve to the intake manifold. From there, it enters the combustion chamber through the intake valves (Figure 1). The gas stored in the natural-gas tank at 200 bar flows through a high-pressure shutoff valve on the tank into the natural-gas pressure-control module, which lowers the gas pressure to a constant operating pressure of approx. 5 to 10 bar (absolute). It then flows through a flexible low-pressure line into a common gas rail that supplies the natural-gas injectors.

Engine management for natural-gas vehicles

Both double-ECU concepts (one ECU for gasoline operation and one for natural-gas operation) and single-ECU concepts (gasoline and gas functions integrated into a single ECU) are used. In some bifuel vehicles, the driver can use a switch to select between gas and gasoline operation. In most models, the switchover occurs automatically and the vehicle is driven in natural-gas mode until the tank contents are used up. Then the vehicle is switched over automatically to gasoline mode.

The high-temperature sensor mounted on the pressure-control module supplies the engine-management system with information on the current fill level of the natural-gas tank and also facilitates leakage diagnostics. The combined natural-gas low-pressure and temperature sensor fitted on the gas rail enables the engine-management system to correct the valve injection timing in such a way that a stoichiometric mixture can be pilot-controlled in the intake manifold even if the gas density fluctuates. The engine-management system uses an adaptation algorithm to adapt to changing gas qualities.

The other sensors and actuators of the engine-management system are mostly identical to those in a gasoline engine.

Mixture formation

Virtually exclusively bifuel and "Monovalent plus" vehicles are available today. These vehicles can operate both on natural gas and on gasoline. Either manifold injection or gasoline direct injection may be used to inject the gasoline.

Natural-gas manifold injection

In most natural-gas engines, the gas is injected, as with gasoline injection, into the intake manifold. The low-pressure gas rail supplies the injectors, which inject natural gas at intermittent intervals into the intake manifold. Mixture formation is improved in comparison with gasoline injection, as natural gas, because of the completely gaseous supply of fuel, does not condense on the intake-manifold walls and does not deposit a film there. This has a positive effect on emissions, particularly during the warm-up phase.

Non-supercharged engines that are fueled by natural gas generally have a 10 to 15% lower output. This is due to the lower volumetric efficiency attributed to displacement of the inducted air caused by the injected natural gas. The vehicle engines can however be optimized specially for natural gas as the fuel. The extremely high knock resistance of natural gas (up to 130 RON) provides for higher engine compression and is ideally suited to supercharging. In this way, the volumetric-efficiency losses can even be overcompensated. Combined with reduced piston displacement (downsizing), efficiency increases due to additional dethrottling and reduced friction.

Natural-gas direct injection
Natural gas can also be injected directly into the combustion chamber. In this way, the volumetric-efficiency losses can be completely avoided and in turbocharged engines the supercharging efficiency, especially at low speeds, can be improved by scavenging. Scavenging involves engine operation with extended overlapping of the intake- and exhaust-valve opening times. The drawbacks of natural-gas direct injection are the greater complexity of the injectors (tightness, temperature stability) and the fact that, due to the necessary higher natural-gas injection pressure dictated by the concept, the tank cannot be drained as far as in engines with natural-gas manifold injection.

Exhaust emissions
Compared with gasoline engines, natural-gas vehicles are characterized by approximately 25% lower CO_2 emissions. The reason for this is the better hydrogen/carbon ratio (H/C ratio) of almost 4:1 (gasoline: approximately 2:1). This results in the creation of more water and less CO_2 during combustion.

Apart from the virtually particulate-free combustion, in conjunction with a three-way closed-loop catalytic converter, only very low levels of the pollutants NO_x, CO and HC are produced. As a rule, the catalytic converter has a higher precious-metal load than is the case in gasoline engines, enabling it to better convert the HC emissions in the exhaust gas that mainly consist of chemically stable methane and to counteract the higher "light-off temperature" (minimum temperature of the catalytic converter for conversion) for natural gas. Methane is, however, classified as nontoxic.

Natural-gas vehicles comply with the current exhaust-emission limits, while trucks and natural-gas buses also comply with the EEV limits (Enhanced Environmentally friendly Vehicle, [4]). The natural-gas engine offers distinct advantages over gasoline and diesel engines, also in particular with regard to the non-limited pollutant emissions that are in part carcinogenic and contribute to smog and acid formation.

Components

Natural-gas injector for manifold injection

Function
In order to supply the internal-combustion engine with gaseous fuel, it is necessary to meter through the natural-gas injectors a much greater volume of gas than gasoline in a conventional gasoline engine. This requirement places specific demands on the design of the natural-gas injector, which must be adapted to the greater gas volume in its cross-sections. Even the high flow velocities that are encountered call for a special form of flow routing in order to reduce pressure losses in the injector.

In highly supercharged engines, the intake-manifold pressure can rise up to 2.5 bar (absolute). In order to suppress the influence of the intake-manifold pres-

Figure 2: Natural-gas injector (Bosch)
1 Pneumatic port, 2 Sealing ring,
3 Valve housing, 4 Filter strainer,
5 Electrical connection, 6 Sleeve,
7 Solenoid coil, 8 Valve spring,
9 Solenoid armature with elastomer seal,
10 Valve seat.

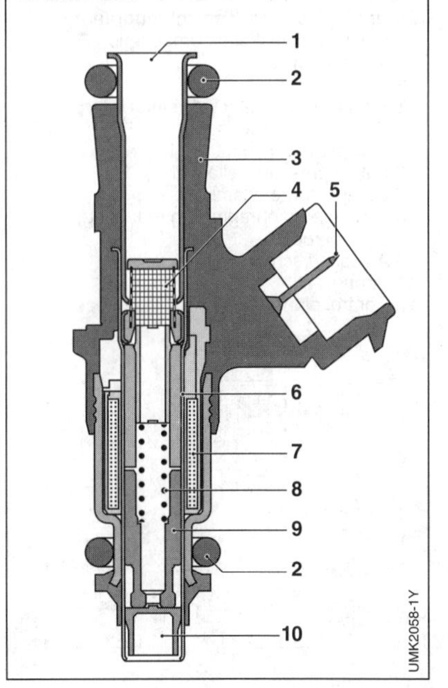

sure on the mass flow, it is necessary for the pressure ahead of the nozzle at the narrowest cross-section (throttling point) accepted as the nozzle above critical to be at least twice as high as the maximum intake-manifold pressure (pressure after the nozzle). The gas then flows at the speed of sound, irrespective of the absolute pressure after the nozzle. The mass flow is therefore not influenced by the variable intake-manifold pressure. While possible pressure losses ahead of the throttling point are taken into account, a minimum operating pressure of 7 bar (absolute) has proven to be advantageous.

Design and operating principle
The solenoid armature (Figure 2) is guided in a sleeve. The armature has fuel flowing through it on the inside and has an elastomer seal at the discharge end. This seal closes on the valve seat and thereby seals off the fuel supply from the intake manifold. When energized, the solenoid coil effects the necessary force to lift the solenoid armature and open the metering cross-section (throttling point in the valve seat). When the coil is at zero current, the valve spring holds the injector closed.

Figure 3: Pressure-control module
1 Solenoid coil of electromagnetic gas shutoff valve,
2 Seal,
3 High-pressure inlet with sinter filter,
4 High-pressure sensor,
5 Low-pressure outlet,
6 Safety pressure-relief valve,
7 Low-pressure chamber,
8 Pressure diaphragm with mounting,
9 Control rod,
10 Adjusting screw,
11 Spring,
12 Control plunger and throttling orifice.

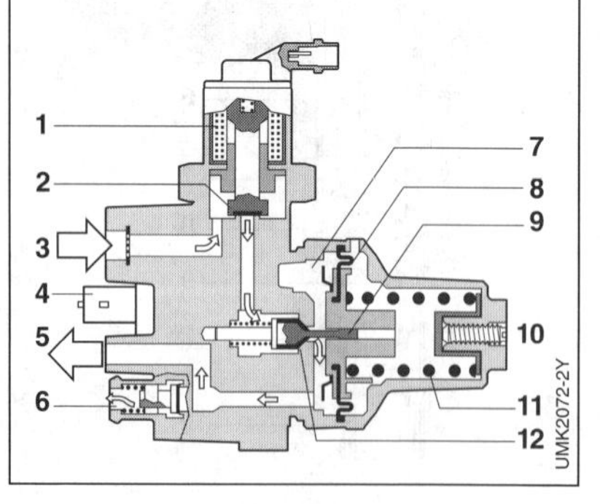

Flow-optimized geometry
Thanks to design measures in the routing of the flow, the pressure loss is minimized ahead of the throttling point in order to ensure the greatest possible mass flow. Furthermore, the narrowest cross-section and thus the throttling point is located at the discharge end after the seal. The speed of sound prevails here such that the valve represents approximately an ideal nozzle in physical terms.

Sealing geometry
The natural-gas injector is fitted with an elastomer seal and is similar in terms of its seal-seat geometry to shutoff valves for pneumatic applications. The elastomer increases the tightness of the seal against metallic needle valves.

Damping in the elastomer also prevents "rebounding", i.e. a repeated, unwanted opening of the solenoid armature during the closing operation, and thus increases metering precision.

Pressure-control module

Function
The function of the pressure-control module is to reduce the pressure of the natural gas from tank pressure (up to 200 bar) to the nominal operating pressure (typically 5 to 10 bar). At the same time, the operating pressure should be kept constant within specific tolerances through all operating states or, depending on the engine's operating point, regulated to a specified pressure. The typical operating pressure of present-day system is usually about 5 to 10 bar (absolute). There are also systems which operate at pressures starting from 2 bar ranging up to 11 bar.

Design
Mechanical version:
Today, mainly mechanical diaphragm- or plunger-type pressure regulators are used. Pressure reduction is effected by means of throttle action and can occur either in one single stage or in several stages.

Figure 3 shows the sectional view of a single-stage diaphragm-type pressure regulator. A sinter filter (pore size approx. 40 μm), a gas shutoff valve, and a high-pressure sensor are integrated into the pressure-control module on the high-pressure side. The sinter filter is designed to retain solid particulates in the gas flow. The gas shutoff valve enables the gas flow to be interrupted, for example when the vehicle is stopped. The high-pressure sensor is used to determine the natural-gas supply in the tank and for diagnostics.

A pressure-relief valve is mounted on the pressure regulator on the low-pressure side. In the event of a fault in the pressure regulator, this valve prevents damage to components in the low-pressure system.

In normal control operation, the gas cools down very dramatically as it expands as a result of the Joule-Thompson effect. In order to prevent freezing, the pressure-control module has an integrated heater, which is connected to the vehicle's heating circuit (not shown in the figure).

The controlled operating pressure is dependent on the diaphragm size and the initial spring tension. To adjust the pressure level, the initial spring tension is set and sealed by means of an adjusting screw in the factory.

Electromechanical version:
As well as purely mechanical pressure-control modules there are also electromechanical versions. These typically consist of an initial mechanical pressure-control stage, which reduces the tank pressure to a mean pressure of approximately 20 bar. In a second stage the pressure is further reduced by an electromagnetically actuated control valve to the operated pressure specified by the engine-management system, and electronically regulated. The advantages of electromechanical pressure regulators are the increased pressure-control precision and the variability of the control pressure. In this way, the injection pressure can be decreased and thus the metering precision of the natural-gas injectors increased in engine low-load phases, e.g. at idle.

Operating principle of mechanical pressure-control module
The gas flows from the high-pressure side through a variable throttling orifice into the low-pressure chamber, where the diaphragm is situated. The diaphragm controls the opening cross-section of the throttle via a control rod. When the pressure in the low-pressure chamber is low, the diaphragm is forced by the spring in the direction of the throttling orifice, which opens to allow the pressure to increase on the low-pressure side. In the event of excessive pressure in the low-pressure chamber, the spring is compressed more sharply and the throttling orifice closes. The decreasing cross-section of the throttling orifice reduces the pressure on the low-pressure side. In stationary operation, the throttling cross-section required for the operating pressure is set, while the pressure in the low-pressure chamber remains extensively constant.

Minimal operating-pressure fluctuations in the event of load changes testify to the quality of the pressure regulator.

References
[1] www.erdgasfahrzeuge.de.
[2] www.gibgas.de.
[3] T. Allgeier, J. Förster: 2nd International CTI Forum, Stuttgart, March 2007; Einspritzsysteme – Motormanagementsystem und Komponenten für Erdgasfahrzeuge.
[4] www.umweltbundesamtumwelt-deutschland.de.

Management for diesel engines

Description of the engine management system

In the diesel combustion process, the fuel is injected directly into the combustion chamber at a nozzle pressure ranging from 200 to more than 2,200 bar. Depending on the combustion process, fuel is injected into a prechamber in indirect-injection engines (at a relatively low pressure of less than 350 bar). In the meantime, direct injection has become the most common process, in which fuel is injected (at high pressure up to more than 2,200 bar) into the non-divided combustion chamber.

Diesel-engine management

Requirements

The power output P from a diesel engine is determined by the available clutch torque and the engine speed. The clutch torque is produced from the torque generated by the combustion process, reduced by the friction torque and the charge-cycle losses, and the torque required for operating the auxiliary systems driven directly by the engine. The combustion torque is generated in the power cycle and is determined by the following variables if the excess air is sufficient: the supplied fuel mass, the start of combustion determined by the start of injection, and the injection and combustion processes.

In addition, the maximum speed-dependent torque is limited by the smoke emissions, the cylinder pressure, the temperature load of different components, and the mechanical load of the complete drivetrain.

Primary function of engine management

The primary function of engine management is to adjust the torque generated by the engine or, in some applications, to adjust a specific engine speed within the permitted operating range (e.g. idling).

In a diesel engine, exhaust-gas treatment and noise suppression are performed to a great extent inside the engine, i.e. by controlling the combustion process. This, in turn, is performed by engine management by changing the following variables:

- cylinder charge,
- tempering of the cylinder charge during the induction stroke,
- composition of the cylinder charge (exhaust-gas recirculation),
- charge motion (intake swirl),
- start of injection,
- injection pressure, and
- rate-of-discharge curve control (e.g. pilot injection, divided fuel injection).

Until the 1980s fuel injection in vehicle engines, i.e. injected fuel quantity and start of ignition, was governed exclusively by mechanical means. In an in-line fuel-injection pump, for example, the injected fuel quantity is varied as a function of load and speed by rotating the pump plunger, which has an angled helix. In the case of mechanical control, the start of injection and start of delivery are adjusted by flyweight governors (speed-dependent), or hydraulically by pressure control as a function of speed and load.

Emission-control legislation calls for highly precise control of the injected fuel quantity and start of injection as a function of such variables as engine temperature, engine speed, load, and height. This can only be provided effectively by electronic control. Now electronic control has fully replaced mechanical control. This is the only form of control that permits continuous monitoring of the emission-related function of the fuel-injection system. Legislation also requires on-board diagnosis in some applications.

Control of injected fuel quantity and start of injection is performed in EDC systems (Electronic Diesel Control) by means of low- or high-pressure solenoid valves or other electrical actuators. Rate-of-discharge curve control within an injection sequence, i.e. division of the fuel quantity into several partial injections, is performed for example indirectly by means of a servo valve and needle-lift control.

Electronic Diesel Control

Electronic control of a diesel engine (Electronic Diesel Control, EDC) enables precise and differentiated modulation of fuel-injection parameters. This is the only means by which a modern diesel engine is able to satisfy the many demands placed upon it.

System overview

Requirements

The lowering of fuel consumption and pollutant emissions (NO_x, CO, HC, particulates) combined with simultaneous improvement of engine power output and torque are the guiding principles of current development work on diesel-engine design. In recent years this has led to an increase in the popularity of the direct-injection (DI) diesel engine, which uses much higher fuel-injection pressures than indirect-injection (IDI) engines with whirl or prechamber systems. In addition, diesel-engine development has been influenced by the high levels of comfort and convenience demanded in modern cars. Noise levels, too, are subject to more and more stringent requirements. As a result, the performance demanded of fuel-injection and engine-management systems has also increased, specifically with regard to

- high injection pressures,
- rate shaping,
- pre-injection and, if necessary, post-injection
- variation of injected fuel quantity, charge-air pressure, and start of injection to suit operating conditions,
- temperature-dependent excess-fuel quantity for starting,
- load-independent idle-speed control,
- controlled exhaust-gas recirculation,
- cruise control, and
- tight tolerances for start of injection and injected fuel quantity and maintenance of high precision over the service life of the system (long-term performance).

Conventional mechanical governing of engine speed uses a number of adjusting mechanisms to adapt to different engine operating conditions. Nevertheless, it is restricted to a simple engine-based control loop and there are a number of important influencing variables that it cannot take account of or cannot respond quickly enough to. As demands have increased, what was originally a straightforward system using electric actuators has developed into present-day Electronic Diesel Control (e.g. distributor-type injection pump), a complex electronic engine-control system capable of processing large amounts of data in real time. And as a result of the increasing integration of electronic components, the control-system circuitry can be accommodated in a very small space.

Operating principle

Electronic Diesel Control is capable of meeting the requirements listed above as a result of microcontroller performance that has risen considerably in the last few years.

In contrast to diesel-engine vehicles with conventional in-line or distributor injection pumps, the driver of an EDC-controlled vehicle has no direct influence, for instance through the accelerator pedal and Bowden cable, upon the injected fuel quantity. Instead, the injected fuel quantity is determined by a number of influencing variables. These are, for example, driver command (accelerator-pedal position), operating status, engine temperature, intervention by other systems (e.g. traction control), and effects on exhaust emissions.

Start of injection can also be varied. This demands a comprehensive monitoring concept that detects inconsistencies and initiates appropriate actions in accordance with the effects (e.g. torque limitation or limp-home mode in the idle-speed range). Electronic Diesel Control therefore incorporates a number of control loops.

Electronic Diesel Control allows data communication with other electronic systems, such as the Traction Control System (TCS), Electronic Transmission Control (ETC), or Electronic Stability Program (ESP). As a result, the engine-management system can be integrated in the vehicle's overall control system network, thereby enabling functions such as reduction of engine torque when the automatic transmission changes gear, regulation of engine torque to compensate for wheel spin.

The EDC system is fully integrated in the vehicle's diagnosis system. It meets all OBD (On Board Diagnosis) and EOBD (European OBD) requirements.

System blocks
Electronic Diesel Control is divided into three system blocks (Figure 1):

Sensors and setpoint generators detect operating conditions (e.g. engine speed) and setpoint values (e.g. switch position). They convert physical variables into electrical signals.

The engine ECU processes the information from the sensors and setpoint generators in its function computer using open- and closed-loop control algorithms. It controls the actuators by means of electrical output signals. In addition, the ECU acts as an interface to other systems and to the vehicle diagnosis system.

Actuators convert electrical output signals from the control unit into mechanical parameters (e.g. the solenoid valve for the fuel-injection system).

Data processing
The main function of Electronic Diesel Control is to control the injected fuel quantity and the injection timing. The common-rail fuel-injection system also controls injection pressure.

The ECU analyzes the incoming signals from the sensors and limits them to the permissible voltage level. Some input signals are also checked for plausibility. Using these input data together with stored program maps, the function computer calculates injection timing and its duration. This information is then converted to a signal characteristic which is aligned to the engine's piston strokes. This calculation program is termed the "ECU software". For all components to operate efficiently, the EDC functions must be precisely matched to every vehicle and every engine. This is the only way to optimize component interaction (Figure 2).

The required degree of accuracy together with the diesel engine's outstanding dynamic response requires high-level computing power. The output signals are applied to output stages which provide adequate power for the actuators

Figure 1: Components for Electronic Diesel Control (EDC) on the example of a common-rail system

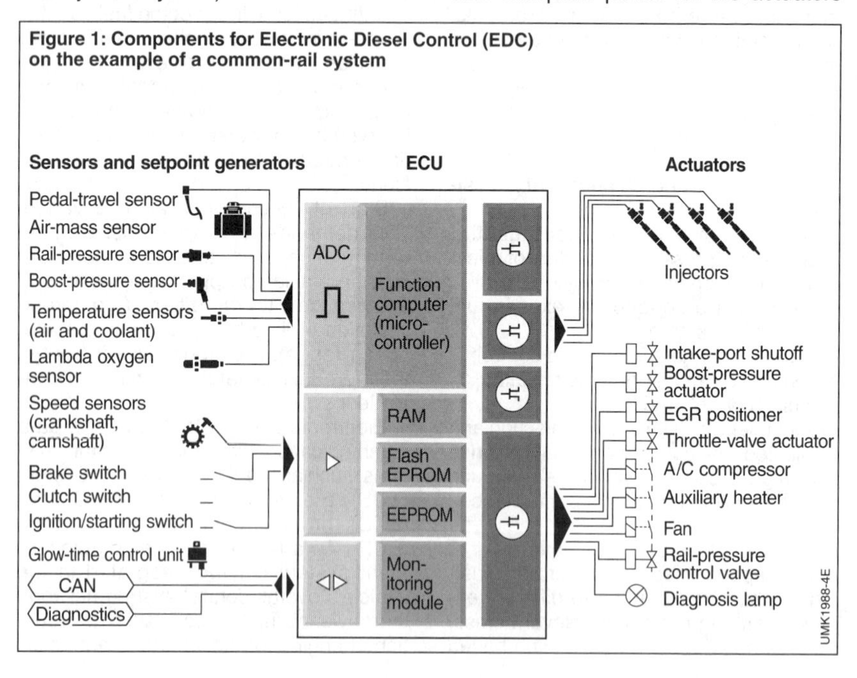

(for instance, the high-pressure solenoid valves for fuel injection, EGR positioner, or boost-pressure actuator). Apart from this, a number of other auxiliary-function components (e.g. glow relay and air-conditioning system) are triggered.

Faulty signal characteristics are detected by output-stage diagnostic functions for the solenoid valves. Furthermore, signals are exchanged with other systems in the vehicle via the interfaces. The engine ECU monitors the complete fuel-injection system as part of a safety strategy.

Figure 2: Basic sequence of Electronic Diesel Control

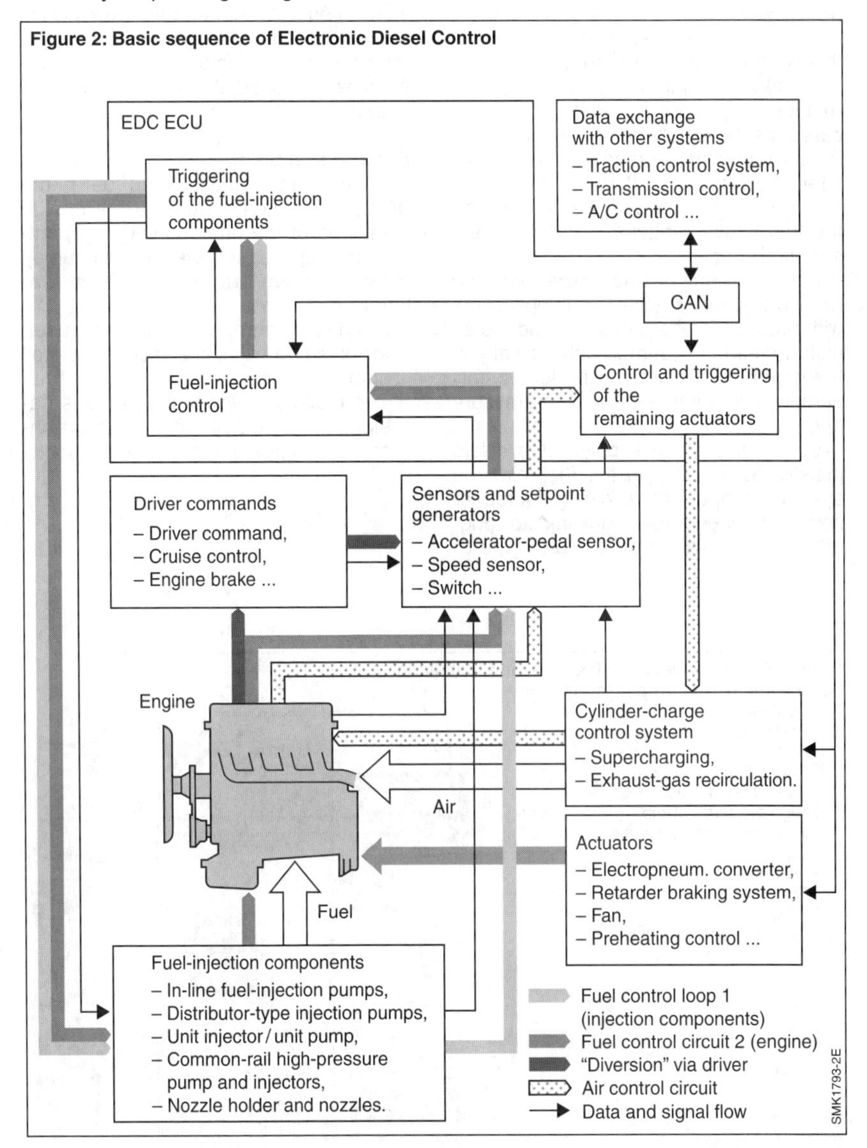

Low-pressure fuel supply

The function of the fuel-supply system is to store and filter the required fuel and to make fuel available to the fuel-injection installation at a specific supply pressure under all operating conditions. The fuel return flow is cooled in some applications. The fuel supply system comprises the following essential components (Figs. 1 to 3): the fuel tank, the preliminary filter (not for unit-injector system for passenger cars), the presupply pump (optional, also in-tank pump for passenger cars), the fuel filter, the fuel pump (low pressure), the pressure-control valve (overflow valve), the fuel cooler (optional), and low-pressure fuel lines.

Individual components can be combined to form assemblies (e.g. fuel-supply pump with pressure limiter). In axial- and radial-piston distributor pumps and partially in the common-rail system, the fuel-supply pump is integrated into the high-pressure pump.

Essentially, the fuel-supply system differs greatly, depending on the fuel-injection system used, as the following figures for common rail (Figure 1), unit injector (Figure 2) and radial-piston pump (Figure 3) show.

Fuel supply and delivery

The function of the fuel pump in the low-pressure stage is to supply the high-pressure components with sufficient fuel in every operating state, with a low level of noise, at the required pressure, and over the entire service life of the vehicle. Different types are used, depending on the field of application.

Electric fuel pump

The structure of the electric fuel pump (Figure 4) is comparable to the design variations of the pumps used on spark-ignition engines (see Electric fuel pump). However, there are some different features:

- Roller-cell pumps are usually used for diesel applications (see Roller-cell pump),
- The electric motor of the pump is designed with carbon commutators instead of copper commutators.

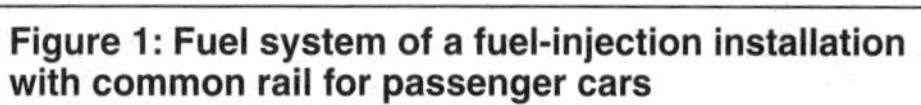

Figure 1: Fuel system of a fuel-injection installation with common rail for passenger cars

1 Fuel tank,
2 Preliminary filter,
3 Fuel pump,
4 Fuel filter,
5 Low-pressure fuel lines,
6 High-pressure pump,
7 High-pressure fuel lines,
8 Rail,
9 Injector,
10 Fuel return line,
11 ECU,
12 Pressure-control valve.

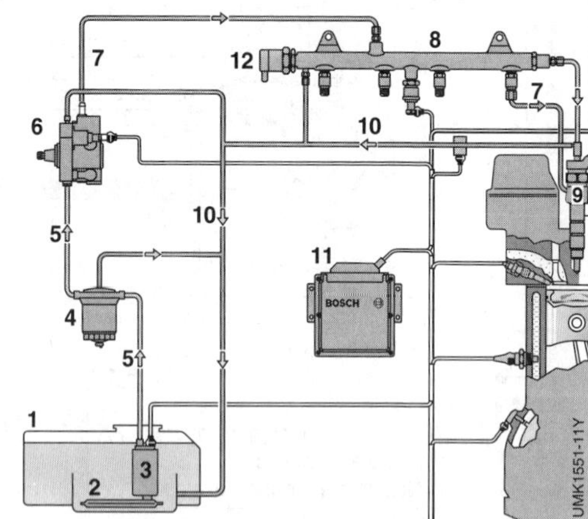

Use of electric fuel pumps:
- Optional for distributor injection pumps (only in event of long fuel lines or great difference in height between fuel tank and fuel-injection pump)
- Unit-injector system (passenger cars)
- Common-rail system (passenger cars)

Gear pump

The gear pump is mounted directly onto the engine or, in the case of a common rail, it is integrated into the high-pressure pump. It is driven mechanically by way of a clutch, a gear wheel, or a toothed belt.

The primary components are two intermeshing, counter-rotating gear wheels (Figure 5) which deliver the fuel into the tooth spaces from the suction side to the

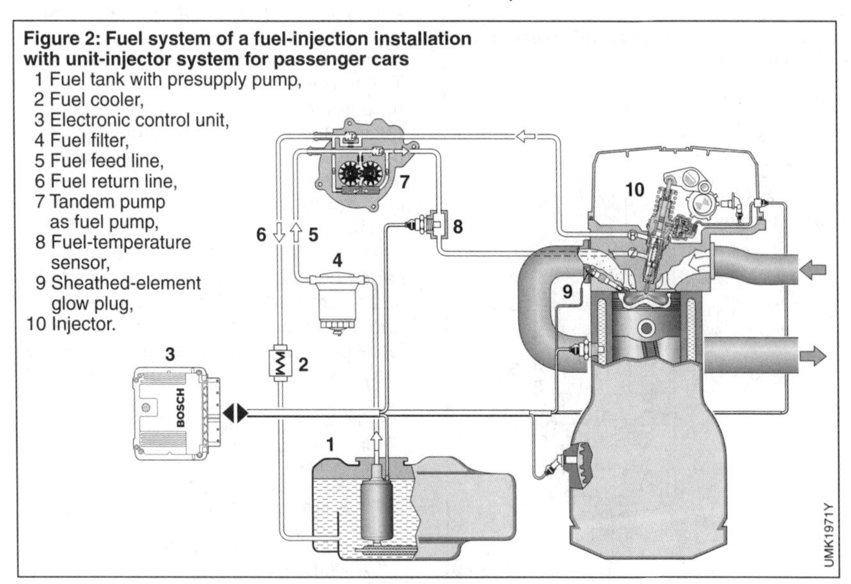

Figure 2: Fuel system of a fuel-injection installation with unit-injector system for passenger cars
1 Fuel tank with presupply pump,
2 Fuel cooler,
3 Electronic control unit,
4 Fuel filter,
5 Fuel feed line,
6 Fuel return line,
7 Tandem pump as fuel pump,
8 Fuel-temperature sensor,
9 Sheathed-element glow plug,
10 Injector.

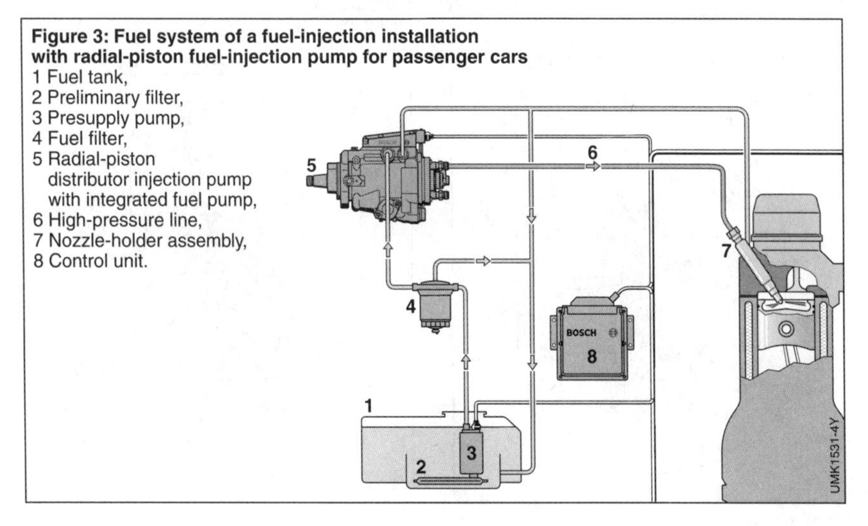

Figure 3: Fuel system of a fuel-injection installation with radial-piston fuel-injection pump for passenger cars
1 Fuel tank,
2 Preliminary filter,
3 Presupply pump,
4 Fuel filter,
5 Radial-piston distributor injection pump with integrated fuel pump,
6 High-pressure line,
7 Nozzle-holder assembly,
8 Control unit.

pressure side. The line of contact of the gear wheels provides the seal between the suction and pressure sides and prevents fuel from flowing back.

The delivery quantity is roughly proportional to the engine speed. At high rotational speeds, therefore, the quantity is limited by throttle action on the intake side or by an overflow valve on the pressure side, which discharges the feed volume to the inlet inside the pump. This prevents unnecessarily high flows in the vehicle's conduit system.

Figure 4: Single-stage electric fuel pump
A Pump element,
B Electric motor,
C Connecting cover.
1 Pressure side,
2 Motor armature,
3 Pump element,
4 Pressure limiter,
5 Suction side,
6 Non-return valve.

Figure 5: Gear fuel pump (diagram)
1 Suction side,
2 Drive gear,
3 Pressure side.

Applications:
- For single-cylinder pump systems for commercial vehicles (unit-injector and unit-pump system, individual injection pump (PF))
- for common-rail systems (commercial vehicles, in part for passenger cars).

Vane-type supply pump

The vane-type supply pump (Figure 6) is mounted on the drive shaft in the distributor injection pump. The impeller is centered on the drive shaft and supported by a Woodruff key. An eccentric ring mounted in bearings in the housing surrounds the impeller.

The fuel passes through the inlet passage and a kidney-shaped recess into the space created by the impeller, the vane and the eccentric ring. The centrifugal force generated by the motion of rotation presses the four vanes of the impeller

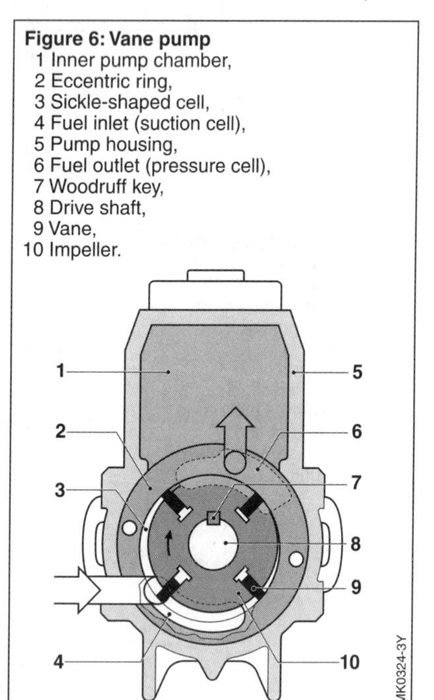

Figure 6: Vane pump
1 Inner pump chamber,
2 Eccentric ring,
3 Sickle-shaped cell,
4 Fuel inlet (suction cell),
5 Pump housing,
6 Fuel outlet (pressure cell),
7 Woodruff key,
8 Drive shaft,
9 Vane,
10 Impeller.

outwards against the eccentric ring. The fuel between the underside of the vanes and the impeller supports this outward movement of the vanes. The rotary motion forces the fuel located between the vanes into the upper kidney-shaped recess and through a bore to the outlet.

Application: Integrated presupply pump in the distributor injection pump.

Locking-vane pump

In a locking-vane pump (Figure 7), springs press two locking vanes against a rotor. As the rotor turns, the volume is increased on the suction side and fuel is drawn into two chambers. The volume is reduced on the pressure side and fuel is delivered from the two chambers. The locking-vane pump delivers fuels at very low speeds.

Application: Unit-injector system for passenger cars.

Tandem fuel pump

The tandem fuel pump (Figure 8) is an assembly consisting of a fuel pump and a vacuum pump, for example, for the brake booster. From a functional perspective, these are two separate pumps, which, however, are driven by a common drive shaft. It is integrated into the engine cylinder head and driven by the engine camshaft.

Figure 7: Locking-vane pump (diagram)
1 Rotor,
2 Suction side (inlet),
3 Spring,
4 Locking vane,
5 Pressure side.

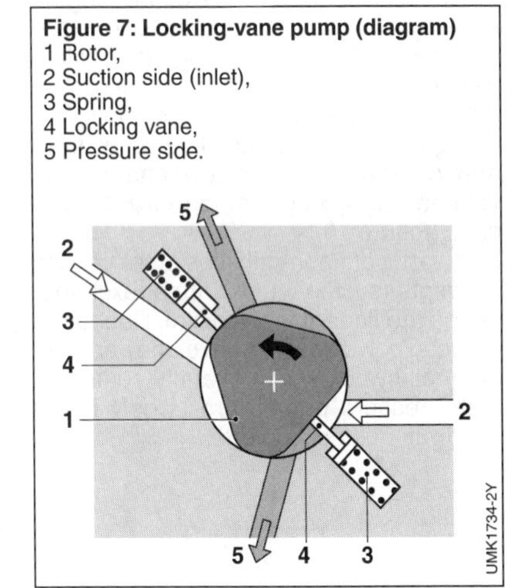

The fuel pump itself is either a locking-vane pump or a gear pump, and even at low speeds (cranking speeds) delivers enough fuel to ensure that the engine starts reliably. The pump's delivery quantity is proportional to the rotational speed.

The pump contains a variety of valves and throttling orifices. The pump's maximum delivery quantity is limited by the suction throttling orifice so that not too much fuel is delivered. The pressure-relief valve limits the maximum pressure in the high-pressure stage. Vapor bubbles in the fuel feed are eliminated in the fuel-return throttling bore. If there is air in the fuel system (e.g. if the vehicle has been driven until the fuel tank is empty), the low-pressure control valve remains closed. The air is forced out of the fuel system through the bypass by the pressure of the pumped fuel.

Application: Unit-injector system for passenger cars.

Figure 8: Fuel pump in a tandem pump
1 Return to tank,
2 Supply from tank,
3 Pump element (gear wheel),
4 Throttling bore,
5 Filter,
6 Suction throttle,
7 Pressure-relief valve,
8 Connection for pressure measurement,
9 Supply, injector,
10 Return, injector,
11 Non-return valve,
12 Bypass.

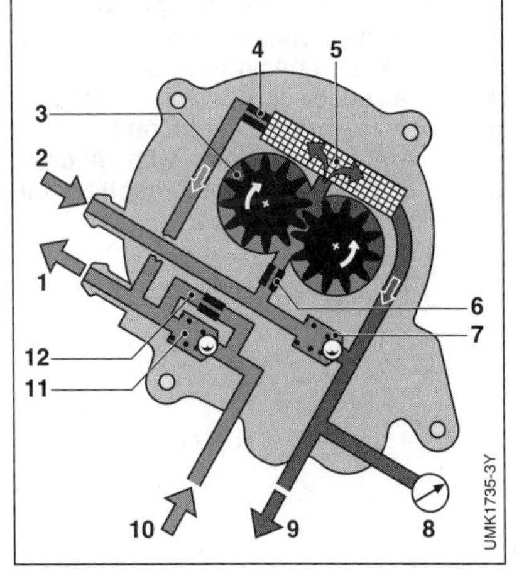

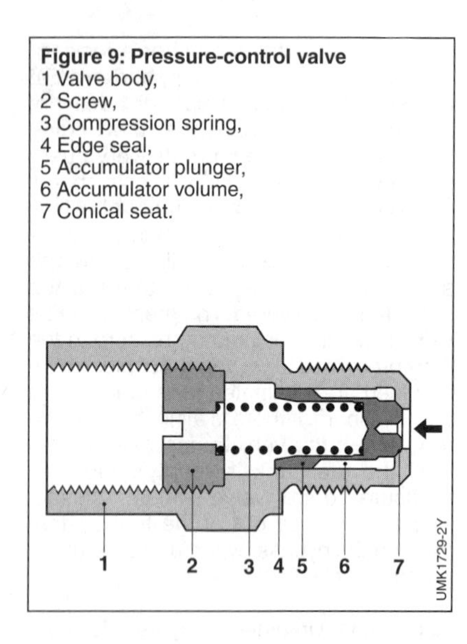

Figure 9: Pressure-control valve
1 Valve body,
2 Screw,
3 Compression spring,
4 Edge seal,
5 Accumulator plunger,
6 Accumulator volume,
7 Conical seat.

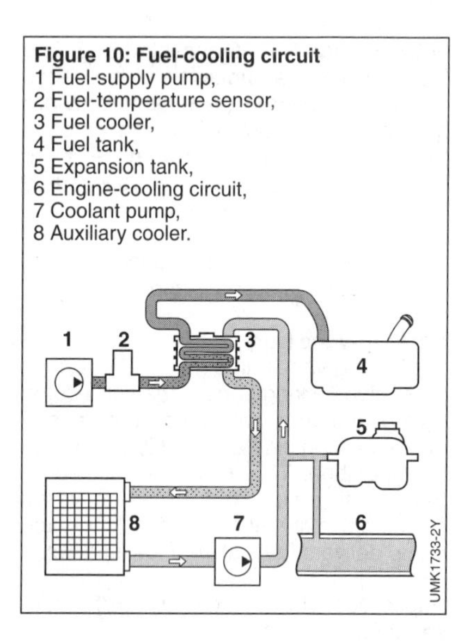

Figure 10: Fuel-cooling circuit
1 Fuel-supply pump,
2 Fuel-temperature sensor,
3 Fuel cooler,
4 Fuel tank,
5 Expansion tank,
6 Engine-cooling circuit,
7 Coolant pump,
8 Auxiliary cooler.

Low-pressure control valve

The pressure-control valve (Fig. 9, also called overflow restriction) is built into the fuel return, for common-rail usually into the high-pressure pump. It provides an adequate operating pressure in the low-pressure stage of the injection systems under all operating conditions and ensures that the pumps are filled uniformly.

For example, the accumulator plunger of the valve for the unit injector and unit pump opens at the initial opening pressure of 300 to 350 kPa (3 to 3.5 bar). A compression spring compensates for minor pressure variations in accumulator volume. At an opening pressure of 4 to 4.5 bar, an edge seal opens and produces a noticeable increase in flow rate.

There are two screws with different spring-force settings for presetting the opening pressure.

Fuel cooler

The high pressure in the injector of the unit-injector system for passenger cars and some common-rail systems heats up the fuel to such an extent that it has to be cooled before it flows back. The fuel returning from the injector flows through the fuel cooler (heat exchanger) and dissipates thermal energy to the coolant in the fuel-cooling circuit. This is separate from the engine-cooling circuit because the coolant temperature of an engine running at operating temperature is too high to cool the fuel. The fuel-cooling circuit is connected to the engine-cooling circuit at the compensating reservoir. This helps to keep the fuel-cooling circuit filled and compensates for any changes in volume caused by temperature fluctuations (Figure 10).

In part, common-rail uses other cooling concepts (such as fuel-air heat exchangers on the floor of the vehicle). Due to the reduced power loss, however, newer common-rail systems with quantity-controlled high-pressure pumps are largely being used without fuel coolers.

Fuel filter

Function

As in spark-ignition (SI) engines, it is essential in a diesel engine to ensure that the fuel system is protected against contamination. Contaminants are introduced into the system during refueling or can get into the tank via the tank ventilation and thus into the fuel itself. The function of the fuel filter is to reduce contamination from particulates so as to protect the fuel-injection system.

Compared with gasoline-injection systems, diesel fuel-injection systems require increased wear protection and finer filters to cater for the much higher injection pressures. Diesel fuel is also more heavily contaminated than gasoline.

Design

Diesel filters are designed as easy-change filters (Figure 11). Spin-on filters, in-line filters and nonmetal filter elements are widely used as replacement parts in aluminum, plastic or sheet-steel filter cases (to comply with increased requirements in crashes). Spiral vee-shape filter elements are the preferred choice. The diesel fuel filter is installed in the low-pressure circuit: in suction systems upstream of the fuel-supply pump and in pressure systems downstream of the electric fuel pump. The trend is now towards pressure systems.

Requirements

In the last few years, the requirements for filter fineness have become more stringent owing to the introduction of common-rail systems with higher injection pressures and advanced unit-injector systems for passenger cars and commercial vehicles. Depending on the application (fuel contamination, periods of engine standstill), the new systems require a filtration efficiency of between 85 % and 98.6 % (particulate interval 3...5 µm, ISO/TS 13353 [1] and ISO 19438 [2]). The fuel filters fitted to the latest generation of automobiles must be capable of storing larger volumes of particulates due to the longer service intervals, and of separating superfine particulates efficiently. This can only be achieved by using special filter media, e.g. consisting of multiple layers made of synthetic microfiber. These filter media exploit the effects of a fine prefilter, and guarantee maximum particulate-retention capability by separating particulates inside each filter layer.

Typical replacement intervals today are between 60,000 and 90,000 km. These intervals are significantly shortened in markets where the diesel quality is poor, such as Eastern Europe, China, India, and the USA. Likewise, when biodiesel is used, halving the replacement interval is recommended.

Figure 11: Diesel line filter
1 Water drain plug,
2 Fuel inlet,
3 Fuel outlet,
4 Filter cover with galvanized coating,
5 Radial vee-form filter element with two-ply filter medium,
6 Compression-proof housing made from Galfan (hot-dip-modified sheet with a zinc-aluminum alloy coating on both sides),
7 Tube to drain,
8 Water accumulation chamber,
9 Water-level and temperature sensors.

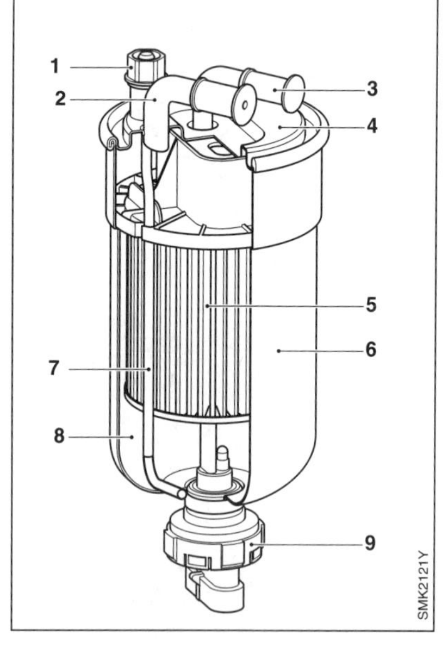

Water separation
A second essential function of the diesel filter is to separate emulsified and free water in order to prevent corrosion damage. An effective water separation of more than 93% at rated flow (testing as per ISO 4020 [3]) is particularly important for distributor injection pumps and common-rail systems. Water separation takes place by coalescence on the filter medium (formation of droplets due to the different surface tensions of water and fuel). Separated water collects in the chamber at the bottom of the filter housing (Figure 11). Conductivity sensors are used in some cases to monitor the water level. The water is drained manually by opening a water drain plug or pressing a pushbutton switch.

For extra heavy-duty requirements, it is beneficial to fit an additional preliminary filter with water separator on the suction or pressure side. Preliminary filters of this type are mostly used for commercial vehicles in countries where diesel quality is poor.

Modular additional functions
New-generation diesel fuel filters integrate modular additional functions such as, for example, fuel preheating to prevent paraffin clogging during winter operation. Fuel preheating can be effected electrically or by utilizing the hot fuel return flow from the engine. In the first instance, PTC heaters (Positive Temperature Coefficient) are installed in the filter. The second instance requires the fitting of a bimetal valve or a wax expansion element which opens at low temperatures and allows fuel to flow back to the filter.

Further additional functions are maintenance indication by way of differential-pressure measurement and filling and venting devices.

References

[1] ISO/TS 13353:2002: Diesel fuel and petrol filters for internal combustion engines – Initial efficiency by particle counting.
[2] ISO 19438.2003: Diesel fuel and petrol filters for internal combustion engines – Filtration efficiency using particle counting and contaminant retention capacity.
[3] ISO 4020:2001: Road vehicles – Fuel filters for diesel engines – Test methods.

Common-rail injection system

System overview

Requirements

The demands placed on diesel-engine fuel-injection systems are continually increasing. Higher injection pressures, faster injector switching times, and a variable rate-of-discharge curve modified to the engine operating state have made the diesel engine economical, clean, and powerful.

The main advantage of the common-rail system (CR) over conventional fuel-injection systems is its ability to vary injection pressure and timing over a broad scale. This was achieved by separating pressure generation (in the high-pressure pump) from the fuel-injection system (injectors). The rail here acts as a pressure accumulator (Figure 1).

The common-rail system is a highly flexible system for adapting fuel injection to the engine. This is achieved by an injection pressure adapted to the operating status of 200 to 2,200 bar, a variable start of injection, and the possibility of using pre-injection and secondary-injection events (even highly retarded secondary-injection events are possible). In this way, the common-rail system helps to raise specific power output, lower fuel consumption, reduce noise emission, and decrease pollutant emission in diesel engines.

Common rail is today the most commonly used fuel-injection system for modern passenger-car and commercial-vehicle diesel engines.

Design

The common-rail system consists of the following main component groups (Figure 1): the low-pressure system comprising the components of the fuel-supply system, the high-pressure system comprising the high-pressure pump, the rail, the injectors and the high-pressure fuel lines, and Electronic Diesel Control (EDC).

The key components of the common-rail system are the injectors connected to the rail. They are fitted with a rapid-action valve (solenoid valve or piezo actuator) which opens and closes the nozzle. This permits control of the injection process for each cylinder.

Figure 1: System areas of engine management with the common-rail fuel-injection system
1 High-pressure pump, 2 Rail, 3 Injectors.

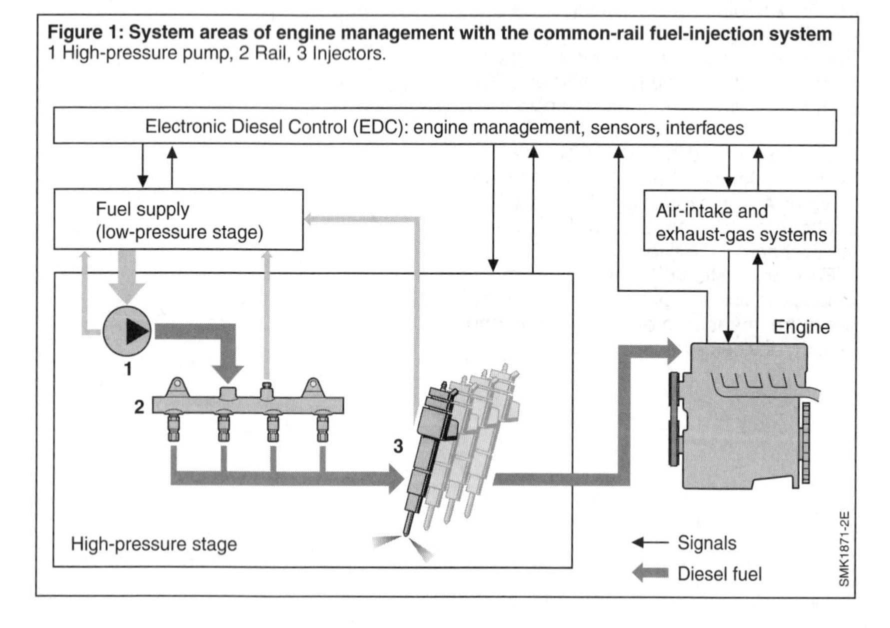

Fuel supply

In common-rail systems for passenger cars and light-duty commercial vehicles, electric fuel pumps or gear pumps are used to deliver fuel to the high-pressure pump. It is exclusively gear pumps that are used for heavy-duty commercial vehicles (see Low-pressure fuel supply).

Systems with electric fuel pump

Systems with an electric fuel pump are increasingly gaining acceptance in the passenger-car market. There are several reasons for this:

- Using a controlled electric fuel pump results in a lower power demand (CO_2 advantage).
- As well as supplying the fuel-injection system, an electric fuel pump permits in parallel the operation of suction jet pumps in the tank; these pumps fill the swirl pot and thus ensure the presence of fuel in the intake area (see Fuel-delivery module).
- The fact that the pressure build-up in the low-pressure system is not dependent on engine speed improves the starting response.

The electric fuel pump is usually part of the in-tank unit (in the fuel tank) or occasionally is fitted in the fuel line (in-line). It intakes fuel via a pre-filter and delivers it to the high-pressure pump at a pressure of 4 to 6 bar (Figures 2a and 2c). The maximum delivery rate is, depending on engine performance, 150 to 240 *l*/h.

Systems with gear pump

The gear pump is flanged to the high-pressure pump and is driven by its input shaft (Figure 2b). In this way, the gear pump starts delivery only after the engine has started. Delivery rate is dependent on the engine speed and reaches rates up to 400 *l*/h at pressures up to 7 bar.

Combination systems

There are also applications where the two pump types are used. The electric fuel pump improves starting response, in particular for hot starts, since the delivery rate of the gear pump is lower when the fuel is hot, and therefore, thinner, and at low pump speeds. In addition, the electric

Figure 2: Common-rail systems for passenger cars (examples)

a) Continuous-delivery radial-piston pump: pressure control on the high-pressure side with pressure-control valve.
b) Demand-controlled radial-piston pump: pressure control on the suction side with metering unit flanged to the high-pressure pump.
c) Demand-controlled radial-piston pump: two-actuator system with pressure control on the suction side via the metering unit and pressure control on the high-pressure side via the pressure-control valve.

1 Fuel tank, 2 Fuel filter,
3 Preliminary filter, 4 Electric fuel pump,
5 Rail, 6 Rail-pressure sensor,
7 Solenoid-valve injector,
8 Pressure-control valve,
9 Continuous-delivery high-pressure pump,
10 Demand-controlled high-pressure pump with mounted gear presupply pump and metering unit,
11 Demand-controlled high-pressure pump with metering unit,
12 Piezo injector,
13 Pressure limiter.

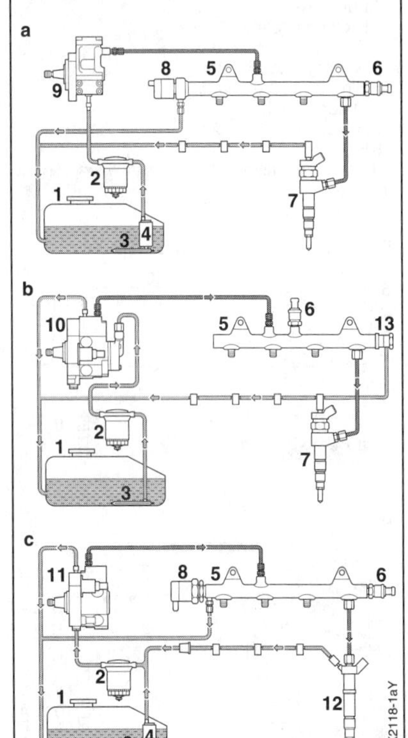

fuel pump ensures the operation of suction jet pumps in the tank.

Combination systems are however rarely used, due to the increased costs and the increased power demand for two pumps.

Fuel filtering

In passenger-car systems with an electric fuel pump the fuel filter is located on the pressure side between the electric fuel pump and the high-pressure pump. In systems with a gear pump the fuel filter is fitted between the fuel tank and the gear pump. In contrast, the fuel filter (fine filter) is fitted on the pressure side in commercial-vehicle systems. For this reason, an exterior fuel inlet is required, in particular when the gear pump is flanged to the high-pressure pump (Figure 3).

Figure 3: Common-rail systems for commercial vehicles (examples)

a) Demand-controlled radial-piston pump with pressure control on the suction side via the metering unit,
b) Demand-controlled two-plunger in-line pump with pressure control on the suction side via the metering unit.

1 Fuel tank,
2 Preliminary filter,
3 Fuel filter,
4 Gear presupply pump,
5 High-pressure pump,
6 Metering unit,
7 Rail-pressure sensor,
8 Rail,
9 Pressure limiter,
10 Injector.

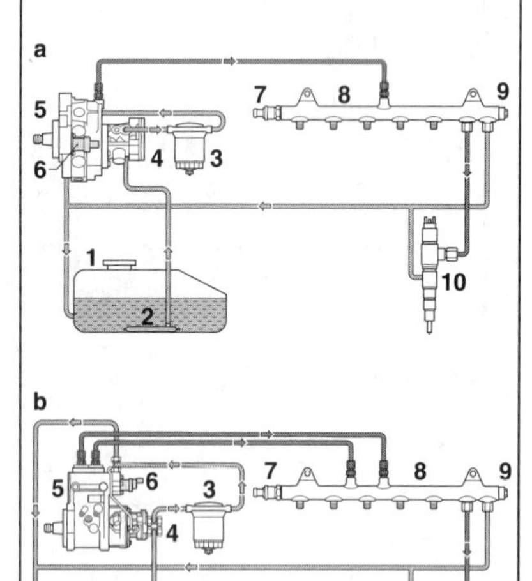

Pressure generation

In the common-rail pressure-accumulator fuel-injection system, the functions of pressure generation and fuel injection are separate. The injection pressure is generated independent of the engine speed and the injected fuel quantity. Electronic Diesel Control controls each of the components.

The functions of pressure generation and fuel injection are separated by the rail's accumulator volume. The pressurized fuel is held in the accumulator volume ready for injection.

A continuously operating high-pressure pump driven by the engine produces the desired injection pressure. Pressure in the fuel rail is maintained irrespective of engine speed or injected fuel quantity.

The high-pressure pump is a radial-piston pump. On commercial vehicles, in contrast to passenger cars, an in-line pump is sometimes fitted.

Pressure control

Different procedures are used for pressure control.

Control on the high-pressure side

The desired rail pressure is regulated on the high-pressure side by a pressure-control valve (Figure 2a). Fuel not required for injection flows back to the low-pressure circuit via the pressure-control valve. This type of control loop allows rail pressure to react rapidly to changes in operating point (e.g. in the event of load changes).

Fuel-delivery control on the suction side

Another way of controlling rail pressure is to control fuel delivery on the suction side (Figures 2b and 3). The metering unit flanged on the high-pressure pump makes sure that the pump delivers exactly the right quantity of fuel to the fuel rail in order to maintain the injection pressure required by the system. In a fault situation, the pressure-relief valve prevents rail pressure from exceeding a maximum.

Fuel-delivery control on the suction side reduces the quantity of fuel under high pressure and lowers the power input of the pump. This has a positive impact on fuel consumption. At the same time, the temperature of the fuel flowing back to the fuel tank is reduced in contrast to the control method on the high-pressure side.

Two-actuator system
The two-actuator system (Figure 2c) combines pressure control on the suction side via the metering unit and control on the high-pressure side via the pressure-control valve, thus marrying the advantages of suction-side fuel-delivery control with the dynamic response of control on the high-pressure side. Another advantage compared with control on the low-pressure side only is that the high-pressure side is also controllable when the engine is cold. The high-pressure pump then delivers more fuel than is injected and pressure is controlled by the pressure-control valve. Compression heats the fuel, thus eliminating the need for an additional fuel heater.

Fuel injection
The injectors spray fuel directly into the engine's combustion chambers. They are supplied by short high-pressure fuel lines connected to the fuel rail. The engine control unit controls the switching valve integrated into the injector to open and close the injector nozzle.

The injector opening times and system pressure determine the quantity of fuel delivered. At a constant pressure, the fuel quantity delivered is proportional to the switching time of the solenoid valve. This is, therefore, independent of engine and pump speeds (time-based fuel injection).

Hydraulic performance potential
Separating the functions of pressure generation and fuel injection opens up further degrees of freedom in the combustion process compared with conventional fuel-injection systems; the injection pressure is more or less freely selectable within the program map.

The common-rail system allows a further reduction in exhaust-gas emissions by introducing pre-injection events and multiple injection events and also attenuating combustion noise significantly. Multiple injection events of up to seven per injection cycle can be generated by triggering the rapid-action switching valve several times. Hydraulic pressure is used to augment nozzle-needle closing, ensuring rapid termination of the injection process.

Control and regulation
Operating principle
The engine control unit detects the accelerator-pedal position and the current operating states of the engine and vehicle by means of sensors (see "Electronic Diesel Control"). The data collected includes:
the crankshaft speed and the crankshaft angle, the rail pressure, the charge-air pressure, the intake-air, coolant and fuel temperature, the inducted air mass, and the wheel speed (for calculating the driving speed). The electronic control unit evaluates the input signals. In sync with combustion, it calculates the triggering signals for the pressure-control valve or the metering unit, the injectors, and the other actuators (e.g. the EGR valve, exhaust-gas turbocharger actuators, etc.).

Basic functions
The basic functions involve the precise control of diesel-fuel injection timing and fuel quantity at the reference pressure. In this way, they ensure that the diesel engine has low consumption and smooth running characteristics.

Correction functions for calculating injection time
A number of correction functions are available to compensate for tolerances between the fuel-injection system and the engine.

Fuel-balancing control
These tolerances result in different torque build-up of individual cylinders. The consequences of these torque differences are irregular engine running and increased exhaust-gas emissions. Injected-fuel-quantity corrections are calculated from the resulting speed fluctuations. Smooth running is increased by specifically adapting the injection time for each cylinder (smooth-running control).

Injector delivery compensation
Injector delivery compensation enables the fuel-quantity deviations of brand-new injectors to be corrected. To this end a wide range of measured data is recorded during injector production for each injector and then applied to the injector in the form of a data matrix code. Information about the lift behavior is also added for a piezo in-line injector. These test data are transmitted during vehicle production to the engine ECU. These values are used during engine operation to compensate deviations in metering and switching performance.

Zero delivery calibration
To achieve both noise reduction and emission targets it is particularly important to be safely in control of small pre-injections over the service life of the vehicle. Injector fuel-quantity drifts must therefore be compensated. To this end a small quantity of fuel is specifically injected into a cylinder in overrun conditions. The speed sensor detects the resulting torque increase as a small dynamic change in speed. This torque increase which is not felt by the driver is clearly linked to the injected fuel quantity. The process is repeated in succession for all the cylinders and for different operating points. A learning algorithm establishes the smallest changes in the pre-injection quantity and corrects the activation duration for the injectors accordingly for all the pre-injections.

Fuel-quantity mean-value adaptation
The deviation of the actually injected fuel quantity from the setpoint value is required to correctly adapt exhaust-gas recirculation and charge-air pressure. Fuel-quantity mean-value adaptation determines the value for the fuel quantity averaged over all the cylinders from the signals from the λ sensor and the air-mass meter. Correction values for the injected fuel quantity are calculated from the comparison of setpoint and actual values.

Supplementary functions
Additional open- and closed-loop control functions perform the tasks of reducing exhaust-gas emissions and fuel consumption, or providing added safety and convenience. Examples of such functions are control of exhaust-gas recirculation, boost-pressure control, cruise control, and the electronic vehicle immobilizer.

The integration of Electronic Diesel Control in a vehicle's overall system network facilitates the exchange of data with, for example, transmission control or air conditioning. A diagnosis interface facilitates the analysis of stored system data when the vehicle is serviced.

Injectors

Solenoid-valve injectors

Design and operating principle

The solenoid-valve injector can be subdivided into a number of function modules: the injection nozzle, the hydraulic servo valve to actuate the valve plunger, the nozzle needle, and the solenoid valve.

The fuel is routed from the high-pressure port (Figure 4a) via an inlet passage to the injection nozzle and via the inlet restrictor into the valve control chamber. The valve control chamber is connected by the outlet restrictor, which can be opened by the solenoid valve, to the fuel return.

The function of the injector can be subdivided into four operating states when the engine and the high-pressure pump are operating. The operating states are caused by the balance of forces acting on the injector components. The nozzle spring closes the nozzle when the engine is not running and there is no pressure in the rail.

All the solenoid-valve injectors have a hydraulic port on the solenoid group for the return flow to the low-pressure system. The return flow is made up of the control quantity (only while the valve is open) and the leakage of the guides at the nozzle, the valve plunger and the solenoid valve (injectors with a maximum pressure greater than 1,600 bar). To increase hydraulic efficiency, injectors with a maximum injection pressure greater than 2,000 bar have no leakage at the nozzle guide and valve plunger.

Nozzle closed (rest position)

In its rest position, the solenoid valve is not triggered (Figure 4a). The solenoid valve, in response to the force of the solenoid-valve spring, closes the valve seat, thereby stopping the flow through the outlet restrictor. Inside of the valve control chamber, the pressure rises to the pressure in the fuel rail. The same pressure is also present in the nozzle chamber. The forces acting on the end faces of the control plunger resulting from the fuel-rail pressure and the force of the nozzle spring hold the nozzle needle in the closed position against the opening force acting on the pressure shoulder of the needle.

Nozzle opens (start of injection)

During the opening phase (Figure 6) of triggering the magnetic force of the electromagnet exceeds the spring force of the solenoid-valve spring. The armature

Figure 4: Functioning principle of solenoid-valve injector (schematic representation)
a) Nozzle closed (rest position),
b) Nozzle opens (start of injection),
c) Nozzle closes (end of injection).
1 Fuel return,
2 Solenoid coil (solenoid valve),
3 Overstroke spring,
4 Solenoid armature,
5 Valve seat,
6 Valve control chamber,
7 Nozzle spring,
8 Pressure shoulder of nozzle needle,
9 Chamber volume of injection nozzle,
10 Nozzle-body seat with injection orifices,
11 Solenoid-valve spring,
12 Outlet restrictor,
13 High-pressure port,
14 Inlet restrictor,
15 Valve plunger (control plunger),
16 Injection-nozzle needle.

a 1 2 3 4 5 6 7 8 9 10 11 12 13 14 15 16

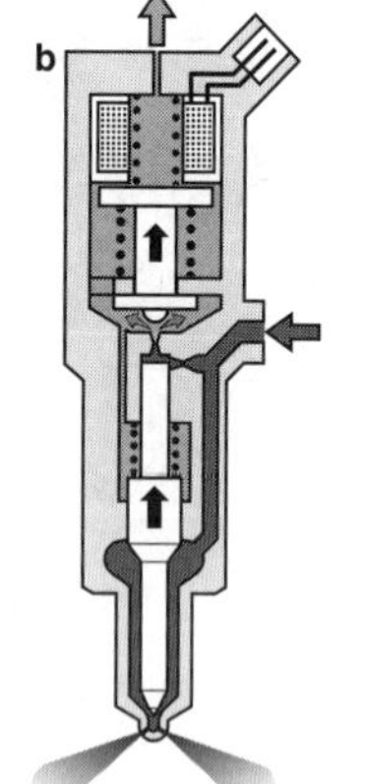

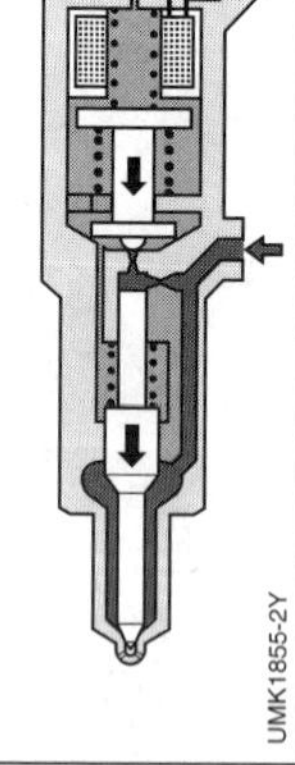

opens the valve completely in the ensuing pickup-current phase. This releases the flow through the outlet restrictor (Figure 4b).

The rapid opening of the solenoid valve and the required rapid switching times are achieved by controlling solenoid-valve triggering in the ECU at high voltages and currents (Figure 6). After a short time, the increased pickup current is reduced to a lower holding current.

When the outlet restrictor opens, fuel flows from the valve control chamber to the armature chamber and then via the fuel-return line to the fuel tank. The pressure in the valve control chamber drops. Fuel flows continuously through the inlet restrictor into the control chamber. This prevents the pressure in the control chamber from dropping completely. The flows of the outlet and inlet restrictors are adapted to the dynamic response (switching times) of the solenoid valve. The pressure at the nozzle needle remains at the level of that in the rail. The reduction in pressure in the valve-control chamber reduces the force acting on the control plunger and opens the nozzle needle. Fuel injection commences.

Nozzle open

The rate of movement of the nozzle needle is determined by the difference in the flow rates through the inlet and outlet restrictors. Fuel is injected into the combustion chamber at a pressure approaching that in the fuel rail.

The balance of forces in the injector is similar to that during the opening phase. At a given rail pressure, the fuel quantity injected is proportional to the length of time that the solenoid valve is open. This is entirely independent of the engine or pump speed (time-based injection system).

Nozzle closes (end of injection)

When the solenoid valve is no longer triggered, the solenoid-valve spring presses the armature down, closes the valve seat, and thus stops the flow through the outlet restrictor (Figure 4c). When the outlet restrictor closes, pressure in the control chamber rises again to that in the fuel rail via the inlet restrictor. The higher pressure exerts a greater force on the control plunger. The force on the valve-control chamber and the nozzle-spring force then exceed the force acting on the nozzle needle, and the nozzle needle moves in the direction of the nozzle-body seat. The flow from the inlet restrictor determines the speed with which the nozzle needle closes. The fuel-injection cycle comes to an end when the nozzle needle is resting against its seat, thus closing off the injection orifices.

This indirect method is used to trigger the nozzle needle by means of a hydraulic servo system because the forces required to open the nozzle needle rapidly cannot be generated directly by the solenoid valve. The control quantity required in addition to the injected fuel quantity passes through the valve control chamber restrictors into the fuel return.

Multiple injections of up to eight injection pulses are possible (pilot injection, main injection, secondary injection) within one injection cycle. The minimum possible time interval is approximately 150 μs.

Injector variants

A distinction is made between two different valve concepts with solenoid-valve injectors:

- Injectors with a pressure-loaded ball valve (valve forces act against the applied rail pressure)
- Injectors with a pressure-compensated valve (valve forces are virtually independent of the rail pressure)

In the case of a pressure-loaded ball valve, the pressure of the compressed fuel acts on the area resulting from valve-seat angle and ball diameter (Figure 5a). This pressure generates an opening force. The spring force must be at least of sufficient magnitude that valve remains closed in the non-active state. In practice, the spring force is approximately 15 % greater than the hydraulic force at maximum injection pressure in order on the one hand to achieve a sufficient dynamic response when the valve closes, and on the other hand to ensure sufficient leak-tightness when closed.

In the case of a pressure-compensated valve, there is no area on which the pressure acts and a force can be generated in the opening direction (Figure 5b).

Bosch offers solenoid-valve injectors for passenger-car diesel engines from a maximum rail pressure of more than 1,600 bar with a pressure-compensated valve. The demands placed on the injectors can be met with these valves for modern diesel engines, since a maximum valve opening cross-section can also be attained at pressures in excess of 1,600 bar. This guarantees the hydraulic stability of the servo valve with which the needle lift is controlled. In addition, the valve lift is also significantly reduced at high rail pressures in order also to achieve the required dynamic response for injection pressures in excess of 1,600 bar and to reduce the sensitivity of the valve dynamic response to external influences. The required minimum time intervals between two injections are achieved on account of the high dynamic response. The power demand for triggering the solenoid valves can also be reduced.

These injector variants with a pressure-compensated valve have a potential for rail pressures in excess of 2,000 bar. There is also the possibility of damping pressure pulsations in the injector with a mini-volume and thereby increasing metering precision in the case of multiple injections.

Triggering the solenoid-valve injector

Triggering of the solenoid valve is divided into five phases (Figure 6; the following details apply to injectors with a pressure-loaded ball valve; small values are sufficient for pressure-compensated valves).

Initially, in order to ensure tight tolerances and high levels of reproducibility for the injected fuel quantity, the current for opening the solenoid valve features a steep, precisely defined flank and increases rapidly up to approx. 20 A (opening phase). This is achieved by means of a booster voltage of up to 50 V which is generated in the ECU and stored in a capacitor (boost-voltage store). When this voltage is applied across the solenoid valve, the current increases several times

Figure 5: Valves of injector variants
a) Pressure-loaded ball valve,
b) Pressure-compensated valve.
1 Valve ball with diameter d,
2 Valve-seat diameter D,
$D = d \sin(90° - \alpha/2)$,
3 Valve section with valve seal seat,
4 Armature with valve seat,
5 Armature guide (on valve section),
6 Seat diameter (guide diameter of armature).
α Valve-seat angle,
p Rail pressure,
F_p Hydraulic force ($F_p = \pi/4 \cdot D^2 p$).

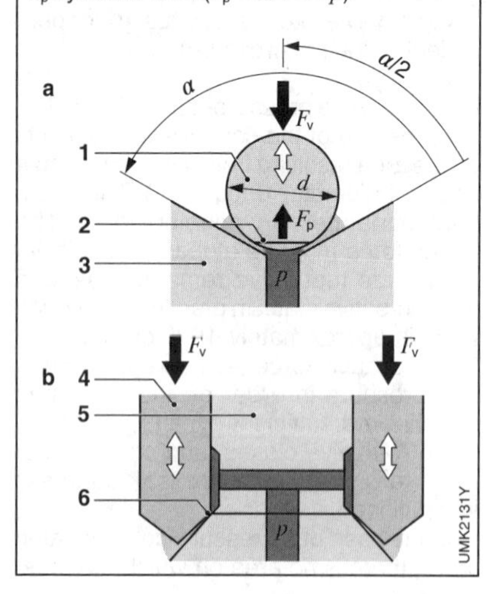

Figure 6: Triggering sequence of a high-pressure solenoid valve for a single injection event
a Opening phase,
b Pickup-current phase,
c Transition to holding-current phase,
d Holding-current phase,
e Switch-off.

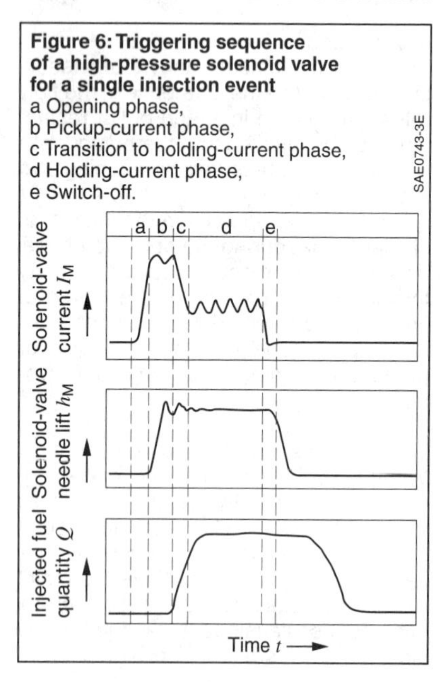

faster than it does when only battery voltage is used.

During the pickup-current phase, battery voltage is applied to the solenoid valve. Current control limits pickup current to approx. 20 A.

In order to reduce power loss in the ECU and injector, the current is dropped to approx. 13 A in the holding-current phase. The energy which becomes available when pickup current is reduced to holding current is routed to the booster-voltage capacitor.

The solenoid valve is closed when the current is switched off, and energy is also released in the process. This energy is also routed to the booster-voltage capacitor. The difference between the energy drawn from the booster-voltage capacitor and the energy returned to it is routed to the booster-voltage capacitor via a step-up chopper integrated in the ECU from the vehicle electrical system. Recharging from the vehicle electrical system is performed until the original voltage level required to open the solenoid valve is reached.

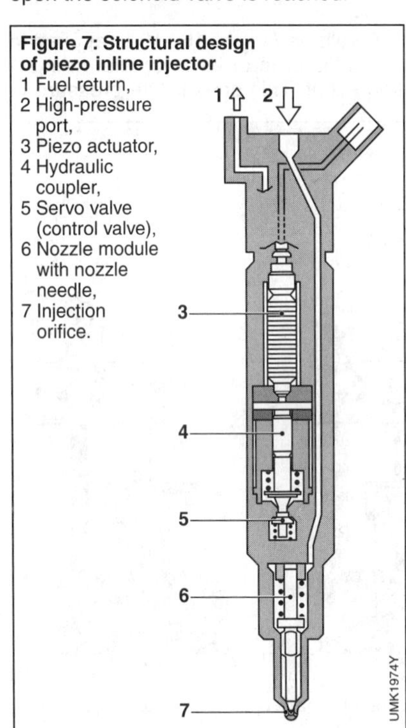

Figure 7: Structural design of piezo inline injector
1 Fuel return,
2 High-pressure port,
3 Piezo actuator,
4 Hydraulic coupler,
5 Servo valve (control valve),
6 Nozzle module with nozzle needle,
7 Injection orifice.

Piezo inline injector

Design and requirements

The design of the piezo inline injector is divided into its main modules in the schematic (Figure 7):

- actuator module (piezo actuator and encapsulation, contact, components for support and force transmission of the actuator),
- hydraulic coupler,
- servo valve (control valve), and
- nozzle module.

A direct response of the nozzle needle to piezo-actuator operation is achieved by coupling the servo valve closely to the nozzle needle. The delay between the electric start of triggering and hydraulic response of the nozzle needle is about 150 µs. At the same time, a servo-valve flight time (time between end stops) of just 50 µs is achieved on account of the high switching force of the piezo actuator. This meets the contradictory requirements with regard to high nozzle-needle speeds and extremely small reproducible injected fuel quantities.

Similarly to the solenoid-valve injector, a control quantity is terminated via the servo valve to activate an injection. Necessitated by the design, however, the piezo inline injector otherwise contains no leakage points between the high-pressure and low-pressure circuits. The result is an increase in the hydraulic efficiency of the overall system.

Up to eight injection pulses can be set per injection cycle. In this way, the injection can be adapted to the requirements of the respective engine operating point. The maximum possible number of injection pulses decreases in the upper rpm range.

Operating principle of the servo valve

The nozzle needle on a piezo inline injector is controlled indirectly by the servo valve. The desired injected fuel quantity is set, taking into account the prevailing rail pressure, via the activation duration of the valve. In its non-triggered state, the injector is in the starting position and the servo valve is closed (Figure 8a), i.e. the high-pressure section is separated from the low-pressure section. The nozzle is kept closed by the rail pressure exerted in the control chamber.

When the piezo actuator is triggered, it increases in length and this increase is transmitted via the hydraulic coupler to the servo valve. This process opens the servo valve and closes the bypass passage (Figure 8b). The flow-rate ratio between the outlet and inlet restrictors lowers pressure in the control chamber and the nozzle opens. The control quantity flows via the servo valve to the low-pressure circuit of the overall system and from there back to the tank.

To start the closing process, the actuator is discharged, and the servo valve closes to release the bypass passage at the same time (Figure 8c). The control chamber is then refilled by reversing the inlet and outlet restrictors, and the nozzle needle is closed again. The injection process is completed as soon as the nozzle needle reaches the nozzle seat again.

Operating principle of the hydraulic coupler

Another vital component in the piezo inline injector is the hydraulic coupler (Figure 9). It provides for compensation of length tolerances of the steel and ceramic components (e.g. caused by the varying thermal expansion of ceramic and steel or by clamping forces on the holder body). On the other hand, it adjusts the translation of actuator stroke and actuator force to the level required on the servo-valve side. The translation ratio is obtained from the diameters of the coupler plunger and valve plunger.

The actuator module and the hydraulic coupler are immersed in the diesel fuel flow, which through the system low-pressure circuit at the injector return is at a pressure of about 10 bar. When the actuator is not triggered, pressure in the hydraulic coupler is in equilibrium with its surroundings, and the coupler does not exert any force on the valve pin. Changes in length caused by temperature or clamping forces acting on the holder body are compensated by small leakages that flow via the guide clearances of the coupler and valve plungers between the coupler gap and the coupler surrounding area. This maintains a coupling of forces between the piezo actuator and the servo valve at all times.

A voltage (110 – 160 V) is applied to the actuator in order to create an injection event. This causes the pressure in

Figure 8: Operating principle of servo valve
a) Start position,
b) Nozzle needle opens (bypass closed),
c) Nozzle needle closes (bypass open)
1 Valve pin, 2 Outlet restrictor,
3 Control chamber, 4 Inlet restrictor,
5 Nozzle needle, 6 Bypass.

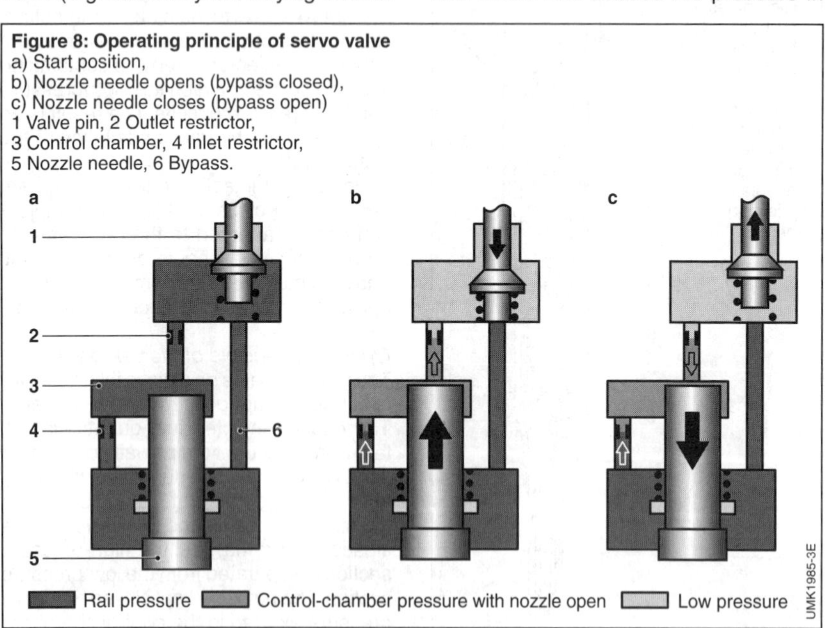

Figure 9: Operating principle of hydraulic coupler
1 Low-pressure rail with pressure-holding valve (fuel return), 2 Piezo actuator, 3 Coupler plunger, 4 Hydraulic coupler, 5 Valve plunger (lower coupler plunger), 6 Valve-plunger spring, 7 Valve pin.

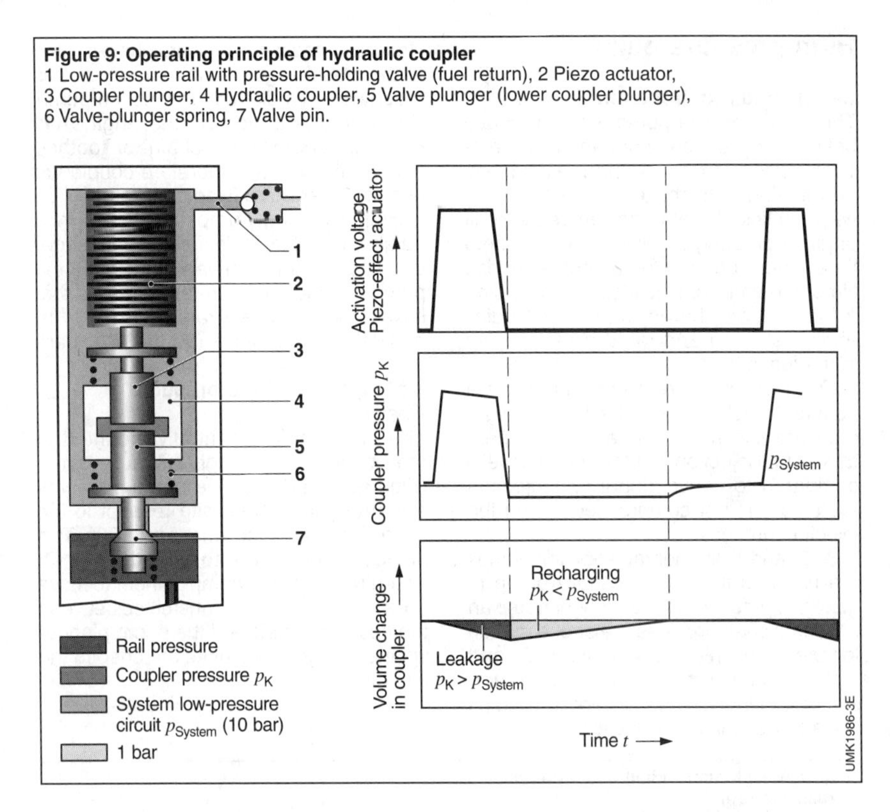

the coupler to increase and a switching force to be exerted on the valve pin. The servo valve opens if this switching force exceeds the closing force on the valve pin caused by the rail pressure. On account of the higher pressure in the coupler than in the surroundings, a small leakage volume flows through the plunger guide clearances out of the coupler into the low-pressure circuit (10 bar) of the injector. Even when the coupler is repeatedly actuated in quick succession during an engine combustion cycle, there is no effect on the injector function when the coupler is drained.

The quantity missing in the hydraulic coupler is refilled in the pauses between the triggering pulses of the piezo actuator. This occurs in the reverse direction via the plunger guide clearance, whereby the valve-plunger spring generates a vacuum pressure inside the coupler against the surroundings. The guide clearances and the low-pressure level are matched to refill the hydraulic coupler entirely before the next engine combustion cycle starts.

High-pressure pumps

Design and requirements

The high-pressure pump is the interface between the low-pressure and high-pressure stages of the common-rail system. Its function is to make sure there is always sufficient fuel under pressure in all engine operating conditions. At the same time, it must operate for the entire service life of the vehicle. This includes providing a fuel reserve that is required for quick engine starting and rapid pressure rise in the fuel rail.

The high-pressure pump generates a constant system pressure for the high-pressure accumulator (fuel rail) independent of fuel injection. For this reason, fuel – compared to conventional fuel-injection systems – is not compressed during the injection process.

3-, 2- and 1-plunger radial-piston pumps are used as the high-pressure pump to generate pressure. In a 3-plunger pump an eccentric shaft provides the stroke motion for the pump plungers, while in 2- and 1-plunger pumps this is provided by a camshaft. 2-plunger in-line fuel-injection pumps are also used in commercial vehicles.

Preferably, the high-pressure pump is fitted to the diesel engine at the same point as a conventional distributor injection pump. The pump is driven by the engine via coupling, gearwheels, chain, or toothed belt. Pump speed is therefore coupled to engine speed via a fixed gear ratio.

High-pressure pumps are used in a number of different designs in passenger cars and commercial vehicles. Within the pump generations there are versions that have different delivery rates (50 – 550 l/h) and delivery pressures (900 – 2,500 bar).

3-plunger radial-piston pump

Design

The drive shaft is centrally mounted in the housing of the high-pressure pump (Figures 9 and 10). The pump elements are arranged radially with respect to the central bearing and offset by 120°. The polygon ring fitted to the drive-shaft eccentric forces the pump plunger to move up and down. Force is transmitted between the eccentric shaft and the pump plunger via the plunger base plate attached to the plunger base.

Figure 9: 3-plunger radial-piston pump (cross-section)
1 Drive shaft with eccentric,
2 Polygon ring,
3 Pump plunger,
4 Suction valve (inlet valve),
5 Fuel supply,
6 High-pressure outlet,
7 Outlet valve,
8 Element chamber,
9 Valve spring,
10 Plunger base plate.

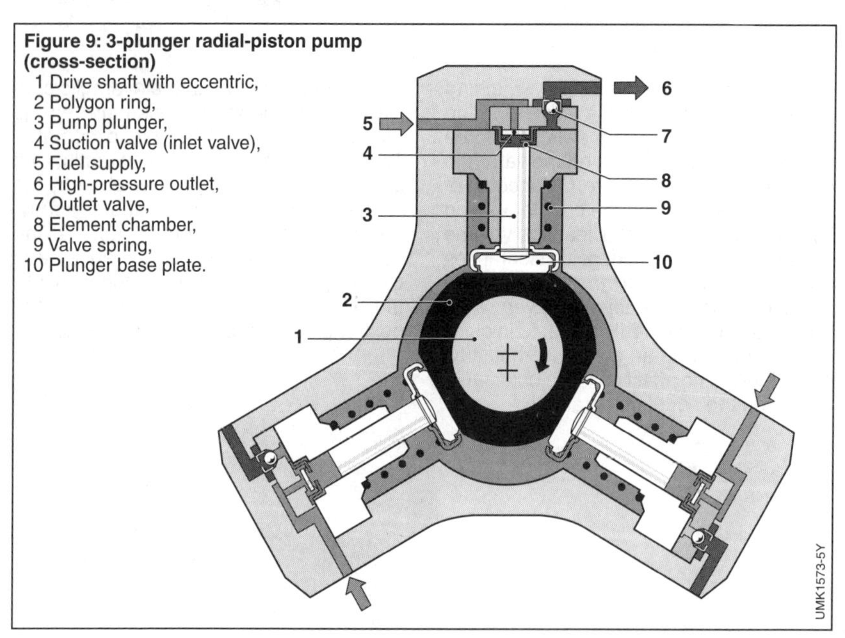

Fuel delivery and compression
The presupply pump – an electric fuel pump or a mechanically driven gear pump – delivers fuel via a filter and water separator to the inlet of the high-pressure pump. The inlet is located inside the pump on passenger-car systems with a gear pump flanged to the high pressure pump. An overflow valve is fitted behind the inlet. If the delivery pressure of the presupply pump exceeds the opening pressure (0.5 – 1.5 bar) of the overflow valve, the fuel is pressed through the restriction bore of the overflow valve into the lubrication and cooling circuit of the high-pressure pump. With its eccentric, the drive shaft moves the pump plunger up and down to mimic the eccentric lift. Fuel passes through the suction valve into the element chamber and the pump plunger moves downward (induction stroke).

When the bottom-dead center of the pump plunger is exceeded, the suction valve closes, and the fuel in the element chamber can no longer escape. It can then be pressurized beyond the delivery pressure of the presupply pump. When the rising pressure exceeds the back pressure from the rail, the outlet valve opens and the compressed fuel passes into the high-pressure circuit. The high-pressure ports of the three pump elements are combined inside the pump housing so that only one high-pressure line runs to the rail.

The pump plunger continues to deliver fuel until it reaches its top-dead center position (delivery stroke). The pressure then drops so that the outlet valve closes. The pump plunger moves downward in response to the force of the valve spring and the remaining fuel in the dead volume is depressurized.

When the pressure in the element chamber drops below the difference between the pre-delivery pressure and the opening pressure of the suction valve, the suction valve reopens and the process starts over.

Transmission ratio
The delivery quantity of a high-pressure pump is proportional to its rotational speed. In turn, the pump speed is dependent on the engine speed. The transmission ratio between the engine and the pump is determined in the process of adapting the fuel-injection system to the engine so as to limit the volume of excess fuel delivered. At the same time, it makes sure that the engine's fuel demand at WOT is covered to the full extent. Possible transmission ratios relative to the crankshaft are between 1:2 and 5:6, i.e. the high-pressure pump is geared up. Pump speed is thus higher than engine speed. Higher transmission ratios are required for commercial vehicles on account of the low engine speeds.

Delivery rate
As the high-pressure pump is designed for high delivery quantities, there is a surplus of pressurized fuel when the engine is idling or running in part-load range. This excess

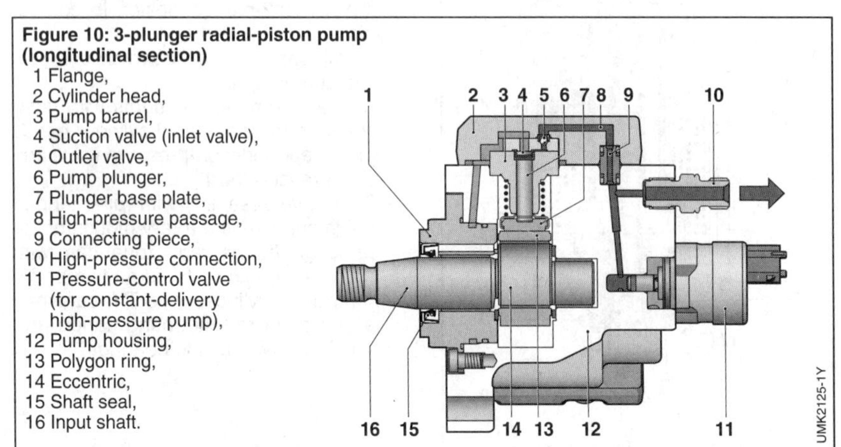

Figure 10: 3-plunger radial-piston pump (longitudinal section)
1 Flange,
2 Cylinder head,
3 Pump barrel,
4 Suction valve (inlet valve),
5 Outlet valve,
6 Pump plunger,
7 Plunger base plate,
8 High-pressure passage,
9 Connecting piece,
10 High-pressure connection,
11 Pressure-control valve (for constant-delivery high-pressure pump),
12 Pump housing,
13 Polygon ring,
14 Eccentric,
15 Shaft seal,
16 Input shaft.

fuel is returned in first-generation systems to the fuel tank via the pressure-control valve seated on the rail or flanged to the pump. As the compressed fuel expands, the energy imparted by compression is lost; overall efficiency drops. Compressing and then expanding the fuel also heats the fuel.

Demand control

An improvement in energy efficiency is possible by means of fuel-delivery control of the high-pressure pump on the fuel-delivery side (suction side). Fuel flowing into the pump elements is metered by an infinitely variable solenoid valve (metering unit) mounted on the high-pressure pump (Figure 11). This valve adapts the fuel quantity delivered to the rail to system demand via the metering slots in the plunger. The plunger operated by the solenoid valve frees up a metering orifice depending on its position via the metering slots. The solenoid valve is activated by means of a PWM signal (pulse width modulation).

This fuel-delivery control not only drops the performance demand of the high-pressure pump, it also reduces the maximum fuel temperature.

Figure 11: Design of metering unit
1 Plug with electrical interface,
2 Solenoid housing,
3 Bearing,
4 Armature with tappet,
5 Solenoid-valve winding with coil body,
6 Bowl,
7 Residual air-gap washer,
8 Magnetic core,
9 O-ring,
10 Plunger with metering slots,
11 Spring,
12 Fuel supply,
13 Fuel outlet.

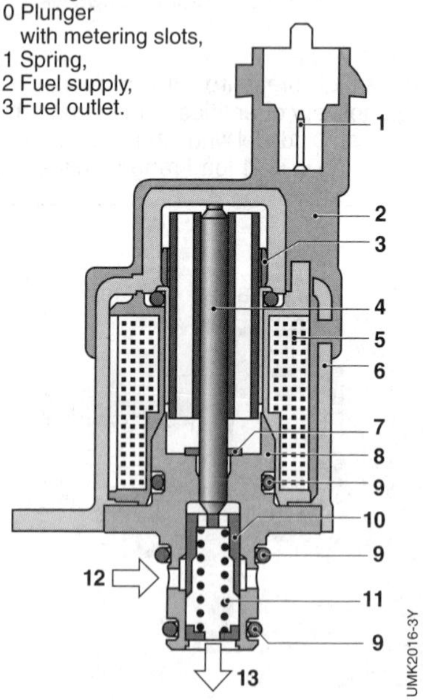

1- and 2-plunger radial-piston pumps

Requirements

The delivery stroke of the pump elements causes pressure pulsations in the rail which give rise to fluctuations in injected fuel quantity in the 3-plunger pumps. In order to comply with the ever more stringent emissions limits, precision of injection coupled with minimal fluctuations in injected fuel quantity is increasingly gaining in importance. 1- and 2-plunger radial-piston pumps facilitate injection-synchronous delivery, i.e. the delivery stroke of the pump elements is synchronous with the induction stroke of the engine cylinders. In this way, the pump delivery for each engine cylinder is always at the same crankshaft angle.

With one or two pump elements, all engines with three to eight cylinders can be injection-synchronously operated by adapting the transmission ratio of 1:2 to 1:1 between engine and pump speeds.

Design

This high-pressure pump is a radial-piston pump of 1- or 2-plunger design. Its components are (Figure 12):

- Aluminum housing, to which only low pressure is applied
- One or two pump elements with high-pressure-resistant steel cylinder heads with integrated high-pressure valve and high-pressure port
- Cam drive assembly with roller tappet, which converts the rotary motion of the camshaft via the cams (dual cam with 180° offset) into a stroke motion of the pump in the cylinder head. The camshaft is guided in the mounting flange and housing in two friction bearings.

The high pressure is generated in the pump element. 1- or 2-plunger pumps are used, depending on the engine's displacement and number of cylinders. Two pump elements are required to cover the fuel demand of larger engines. In the 2-plunger version, the pump elements are arranged in a vee configuration at 90° to each other.

The great overlap length between the cylinder wall and the pump plunger results in low leakage losses when the fuel is compressed. On the other hand, short leakage times result – on account of the high delivery frequency (two strokes per rotation per plunger) and the small dead volume in the cylinder head – in further optimization of efficiency and thus reduced fuel consumption.

Figure 12: 1-plunger radial-piston pump
1 Metering unit,
2 Pump element,
3 Pump housing,
4 Mounting flange,
5 Friction bearing,
6 Input shaft (camshaft),
7 Shaft seal,
8 Cylinder head,
9 Suction valve (inlet valve),
10 High-pressure valve (non-return valve) in high-pressure port (fuel inlet not visible in this illustration),
11 Pump plunger,
12 Roller tappet,
13 Roller support,
14 Drive roller,
15 Dual cam.

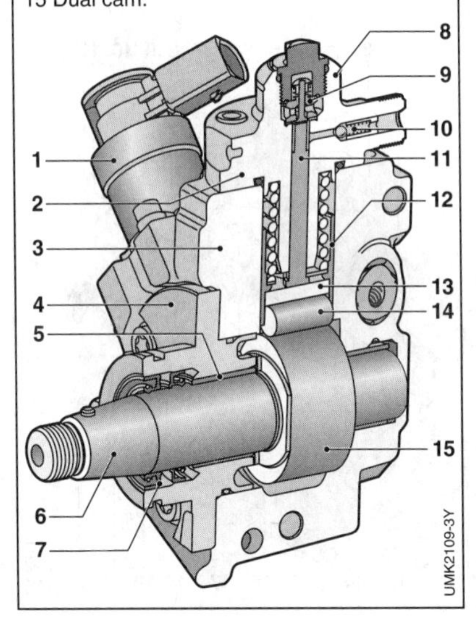

The 90° vee configuration of the cylinder heads in the 2-plunger pump means that there is no overlapping of the induction strokes. The charge of the two pump elements is therefore identical (uniformity of delivery).

The high-pressure port is connected to the rail by way of one (in the 1-plunger pump) or two (in the 2-plunger pump) high-pressure lines. The high pressure is not condensed in the housing, but instead routed directly outwards from the cylinder head. The housing therefore does not require any high-pressure- and strength-increasing measures.

Low-pressure circuit

The total fuel delivered by the presupply pump (electric fuel pump or gear pump flanged to the high-pressure pump) is routed through the pump interior to the overflow valve and to the metering unit. The volume of fuel used for lubrication and cooling is thus greater than in previous pumps. The overflow valve regulates the pump internal pressure and thereby protects the housing against overpressure.

The entire low-pressure path is dethrottled on account of the large cross-sections in such a way as to ensure that the pump elements are filled even at high engine speeds. Volume metering is performed on the low-pressure side by the metering unit. The concept of the metering unit used here corresponds to that of the metering unit used for the 3-plunger radial-piston pump (Figure 11); however, they differ in their design.

High-pressure circuit

The fuel pilot-controlled by the metering unit passes in the induction phase through the suction valve into the element chamber, and is compressed to high pressure during the ensuing delivery phase and delivered by the high-pressure valve and the high-pressure line to the rail.

2-plunger in-line piston pump

Design

This demand-controlled high-pressure pump for rail pressures up to 2,500 bar is only used in commercial vehicles. This is a 2-plunger pump with an in-line design, i.e. the two pump elements are arranged one after the other in relation to the camshaft axial direction (Figure 13). This high-pressure pump comes in both oil-lubricated and fuel-lubricated variants.

A spring seat forms the positive link between the roller tappet and the pump plunger. The rotary motion of the camshaft is converted via the cams into a stroke motion of the pump plungers. The plunger spring ensures pump-plunger return. The combined inlet and outlet valve is located at the top of the pump element.

A gear presupply pump with a high gear ratio is located on the camshaft extension. Its function is to draw fuel via the fuel inlet from the fuel tank and route it through the fuel outlet to the fuel fine filter. From there, the fuel passes through another line to the metering unit located on the upper section of the high-pressure pump.

Lube oil is supplied either directly via the mounting flange of the pump or via a side-mounted inlet. The lube oil is returned to the engine oil pan.

Operating principle

When the pump plunger moves from top dead center towards bottom dead center, the suction valve opens on account of the fuel pressure (pre-delivery pressure). The fuel is drawn into the element chamber in response to the downward movement of the pump plunger. The outlet valve is closed by the valve spring.

As the pump plunger moves upwards, the suction valve closes and the enclosed fuel is compressed. When the rail pressure is exceeded, the outlet valve opens and the fuel is delivered through the high-pressure port to the rail. This increases the pressure in the rail. The rail-pressure sensor measures the pressure, from which the engine ECU calculates the triggering signals (PWM) for the metering unit. The metering unit regulates the fuel quantity provided for compression according to the current demand.

Figure 13: 2-plunger in-line piston pump
1 Speed sensor (pump speed),
2 Metering unit,
3 Fuel supply for metering unit (from fuel filter),
4 Fuel return line to fuel tank,
5 High-pressure port,
6 Valve body,
7 Valve holder,
8 Outlet valve with valve spring,
9 Suction valve (inlet valve) with valve spring,
10 Fuel supply to pump element,
11 Plunger spring,
12 Fuel supply from fuel tank,
13 Fuel outlet to fuel filter,
14 Gear presupply pump,
15 Overflow valve,
16 Concave cam,
17 Camshaft,
18 Roller bolt with roller,
19 Roller tappet,
20 Pump plunger,
21 Mounting flange.

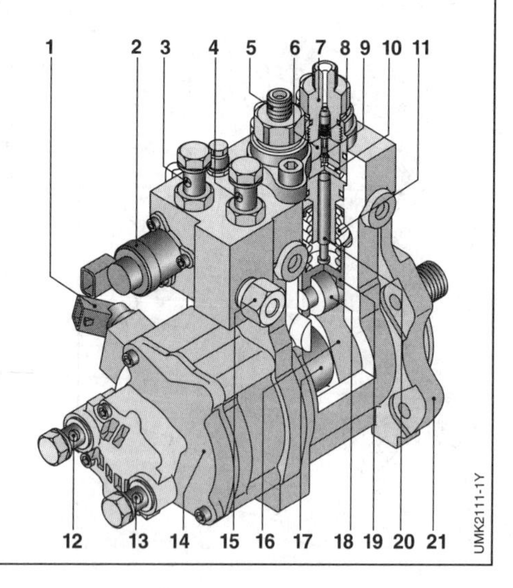

Rail

Function

The function of the rail is to maintain the fuel at high pressure. In so doing, the accumulator volume dampens pressure fluctuations caused by fuel pulses delivered by the pump and the fuel-injection cycles. This ensures that, when an injector opens, the injection pressure remains constant. On the one hand, the accumulator volume must be large enough to meet this requirement. On the other hand, it must be small enough to ensure a fast enough pressure rise on engine start.

Besides acting as a fuel accumulator, the fuel rail also distributes fuel to the injectors.

Application

The tubular rail can vary depending on the various constraints which determine the way it is fitted to the engine. It has connections for fitting the rail-pressure sensor and the pressure limiter or the pressure-control valve (Figure 13).

The pressurized fuel delivered by the high-pressure pump passes via one or two high-pressure fuel lines to the rail inlet. From there, it is distributed via high-pressure lines to the individual injectors. The cavity inside the fuel rail is permanently filled with pressurized fuel during engine operation.

Figure 13: Design of a rail with pressure limiter
1 Fuel return (low pressure),
2 High-pressure ports injectors,
3 High-pressure port to high-pressure pump (one or two ports),
4 Rail-pressure sensor.
5 Pressure limiter,
6 Rail body,
7 Restrictor (pressed-in, optional),
8 Mounting lug (engine mounting).

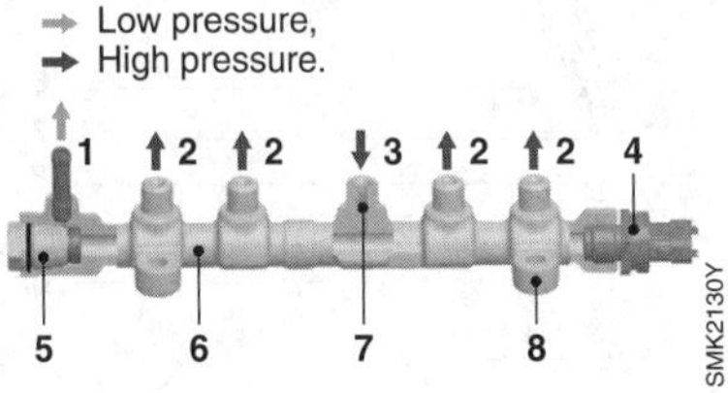

→ Low pressure,
→ High pressure.

The fuel pressure is controlled by means of Electronic Diesel Control (EDC), whereby the fuel pressure is measured by the rail-pressure sensor and – depending on the system – regulated by demand control or via the pressure-control valve to the desired value. The pressure limiter is used as an alternative to the pressure-control valve – depending on system requirements – and its function is to limit fuel pressure in the fuel rail to the maximum permissible pressure in the event of a fault.

Some rail types make use of restrictors in the inlet and outlet which additionally dampen pressure pulsations caused by pump delivery and injections. If fuel is now drawn from the rail for an injection, the pressure is the rail remains virtually constant.

Rail types

Two different rail types are used: the forged rail (hot-forged rail) and the welded rail (laser-welded rail). Both types are used by Bosch, the hot-forged rail being the preferred variant.

In the case of a hot-forged rail, the blank for machining is produced by forging from bar stock. The internal geometry and the rail interfaces of the rail body are produced by deep-hole boring, drilling and milling. A corrosion-resistant surface is then applied. Finally, the add-on components are mounted and functionally tested.

The forging process affords more possibilities in the shaping of the external geometry than laser welding. One advantage is the possibility of shaping the external geometry with regard to weight optimization.

The hot-forged rail is used in series production up to 2,200 bar, and a further pressure increase is planned for the next generations.

Time-controlled single-cylinder pump systems

Unit-injector system for passenger cars

System requirements

The electronic controlled unit injector is a single-cylinder pump injection system with integrated high-pressure pump and injection nozzle (Figure 1). The high-pressure line required in other injection systems between the injection pump and the injection nozzle is omitted. This means that the system has particularly good hydraulic performance.

The unit injector is installed in the cylinder head between the valves, with the nozzle protruding into the combustion chamber. It is operated by rocker arms driven by the overhead engine camshaft. Each engine cylinder is allocated a separate unit injector. The start of injection and injection duration are calculated by an electronic control unit and controlled by the high-pressure solenoid valve mounted externally on the pump body.

The unit-injector system (UIS) is designed to meet the demands of modern direct-injection diesel engines with high levels of power density. It is characterized by its compact design, high injection pressures of up to 2,200 bar at full load, and mechanical-hydraulic pilot injection throughout the entire program-map range to substantially reduce combustion noise. The compact design enables a very low high pressure volume with correspondingly high hydraulic efficiency.

Unit-injector systems are no longer used for new developments.

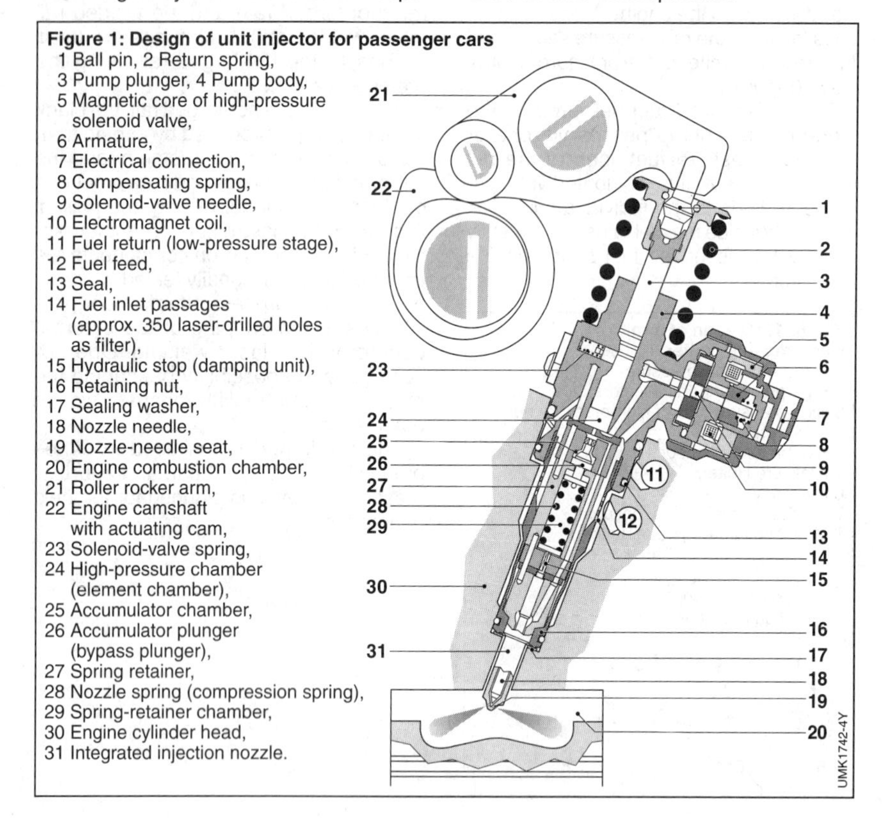

Figure 1: Design of unit injector for passenger cars
1 Ball pin, 2 Return spring,
3 Pump plunger, 4 Pump body,
5 Magnetic core of high-pressure solenoid valve,
6 Armature,
7 Electrical connection,
8 Compensating spring,
9 Solenoid-valve needle,
10 Electromagnet coil,
11 Fuel return (low-pressure stage),
12 Fuel feed,
13 Seal,
14 Fuel inlet passages (approx. 350 laser-drilled holes as filter),
15 Hydraulic stop (damping unit),
16 Retaining nut,
17 Sealing washer,
18 Nozzle needle,
19 Nozzle-needle seat,
20 Engine combustion chamber,
21 Roller rocker arm,
22 Engine camshaft with actuating cam,
23 Solenoid-valve spring,
24 High-pressure chamber (element chamber),
25 Accumulator chamber,
26 Accumulator plunger (bypass plunger),
27 Spring retainer,
28 Nozzle spring (compression spring),
29 Spring-retainer chamber,
30 Engine cylinder head,
31 Integrated injection nozzle.

Operating principle

Intake stroke

The unit injector is filled during the intake stroke of the upward-moving pump plunger. Fuel flows from the low-pressure stage of the fuel supply through the inlet holes into the unit injector. The fuel enters the high-pressure chamber when the solenoid-valve seat is opened.

Prestroke

The rotation of the actuating cam causes the pump plunger to move downwards. When the solenoid valve is still open, the fuel is forced back by the pump plunger from the high-pressure chamber into the low-pressure stage of the fuel supply. Heat is also dissipated from the unit injector (cooling) with the fuel as it flows back.

Delivery stroke and pilot injection

The electronic control unit energizes the coil of the solenoid valve at a particular point in time so that the solenoid-valve needle is pressed onto the solenoid-valve seat and the connection between high-pressure chamber and low-pressure stage is closed (beginning of injection period, BIP). The downward movement of the pump plunger increases the fuel pressure in the high-pressure chamber. The nozzle-opening pressure for pilot injection is approximately 180 bar. When this pressure is reached, the nozzle needle is raised and pilot injection begins. In this phase the nozzle-needle lift is hydraulically limited by a damping unit located between nozzle needle and nozzle spring so that the low injected fuel quantity required can be accurately metered. The accumulator plunger remains on its seat for the time being because the nozzle needle opens first due to its greater hydraulically effective area on which the pressure acts.

As the pressure increases, the accumulator plunger (bypass plunger) is forced downwards and now lifts off its seat. A connection is established between high-pressure chamber and accumulator chamber. The resulting pressure drop in the high-pressure chamber, the increased pressure in the accumulator chamber and the simultaneous increase in the pre-tension of the nozzle spring cause the nozzle needle to close. Pilot injection is terminated. However, the accumulator plunger does not return to its starting position because in its open state it offers the fuel pressure a greater working area than the nozzle needle.

Main injection

The continuing movement of the pump plunger causes the pressure in the high-pressure chamber to increase further. The nozzle-opening pressure for main injection is at approximately 300 bar higher than pilot injection. There are two reasons for this. Firstly, the excursion of the accumulator plunger increases the pre-tension of the nozzle spring. Secondly, due to the bypassing of the accumulator plunger fuel must be forced from the spring-retainer chamber via a restrictor into the low-pressure stage of the fuel supply so that the fuel in the spring-retainer chamber is compressed to a greater extent (pressure-backing). The pressure-backing level is derived from the size of the restrictor in the spring retainer and can be varied. It is therefore possible to achieve a sensible compromise between a low opening pressure for pilot injection (for noise reasons) and as high an opening pressure as possible for main injection, especially at part load (reducing emissions). The stroke and shaft diameter of the bypass plunger determine the length of the interval, the so-called injection interval, between the end of pilot injection and the start of main injection.

When the nozzle-opening pressure is reached, the nozzle needle is raised and fuel is injected into the combustion chamber (actual start of injection). The pressure during the entire injection process continues to rise due to the high delivery rate of the pump plunger.

The current flow through the solenoid-valve coil is deactivated to terminate main injection. The solenoid valve opens and releases the connection between high-pressure chamber and low-pressure stage. The pressure collapses. When the nozzle-closing pressure is undershot, the injection nozzle closes and the injection process ends. At this point the accumulator plunger also returns to its starting position. The remaining fuel is returned to the low-pressure stage during the continuing downward movement of the pump plunger (residual stroke). Heat is also dissipated from the unit injector in the process.

Electronic control allows the values for the start of injection and the injected fuel quantity to be selected as desired from those stored in the program map. This feature, together with the high injection pressures, makes it possible to achieve high power densities combined with low emission levels and low fuel consumption.

Figure 2: Design of unit injector for commercial vehicles
1 Sliding disk, 2 Return spring,
3 Pump plunger, 4 Pump body,
5 Electrical connection,
6 High-pressure chamber (element chamber),
7 Engine cylinder head,
8 Fuel return (low-pressure stage),
9 Fuel feed, 10 Spring retainer,
11 Pressure pin,
12 Intermediate ring,
13 Integrated injection nozzle,
14 Retaining nut,
15 Armature, 16 Electromagnet coil,
17 Solenoid-valve needle,
18 Solenoid-valve spring,
19 Nozzle needle.

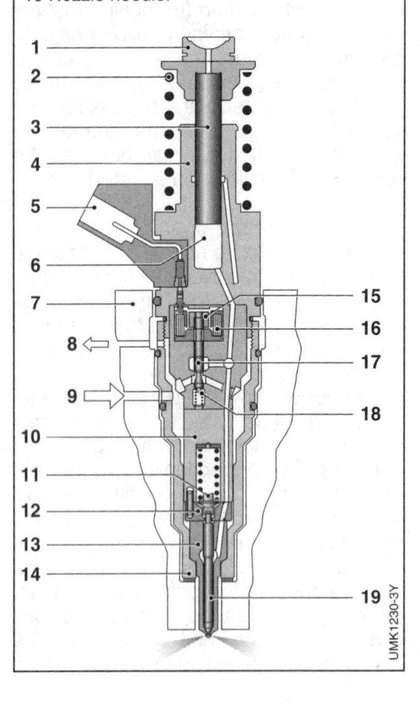

Unit-injector system for commercial vehicles

The unit-injector system for commercial vehicles is largely the same as the system for passenger cars. Due to the larger dimensions for the commercial-vehicle system, the solenoid valve can be integrated in the unit-injector here (Figure 2).

With regard to main injection, the commercial-vehicle system has the same operating principle as the passenger-car system. The systems differ with regard to pilot injection. In the unit-injector System for commercial vehicles this can be electronically controlled in the lower speed and load range. This significantly reduces combustion noise and improves cold-starting performance.

Figure 3: Design of unit pump
1 Hydraulic high-pressure port,
2 Solenoid-valve needle-travel stop,
3 Engine block, 4 Pump body,
5 High-pressure chamber, 6 Pump plunger,
7 Tappet spring, 8 Roller-tappet shell,
9 Roller tappet, 10 Roller-tappet pin,
11 Solenoid-valve spring, 12 Armature plate,
13 Solenoid-valve housing,
14 Solenoid-valve needle, 15 Filter,
16 Fuel feed, 17 Fuel return,
18 Pump-plunger retainer, 19 Spring seat,
20 Locating groove, 21 Tappet roller.

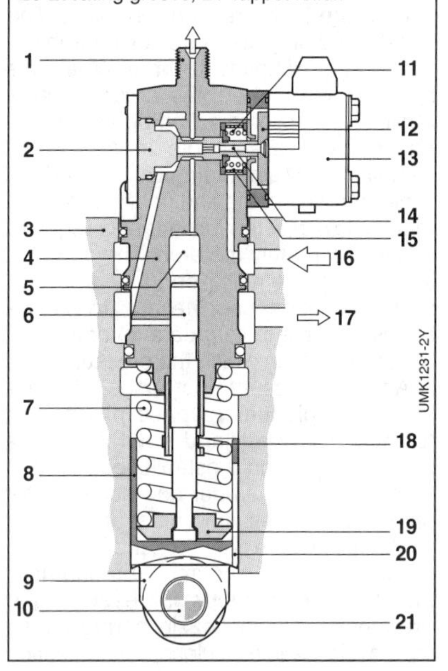

Unit-pump system for commercial vehicles

The unit-pump system (UPS) is also a modular, time-controlled single-cylinder pump injection system, and is closely related to the unit-injector system. It is used in commercial-vehicle engines and in large engines. Each of the engine's cylinders is supplied by a separate module consisting of a high-pressure plug-in pump with integral, high-speed solenoid valve, a short high-pressure line, and an injection nozzle (Figures 3 and 4). This system is therefore also called Pump-Line-Nozzle (PLN). It permits injection pressures of up to 2,100 bar.

The separation of high-pressure generation and injection permits a simpler mounting on the engine. The hydraulic performance is nevertheless very good due to the shortest line possible. The unit pump is mounted on the side of the engine block. It is operated directly by an injection cam on the engine's camshaft via a roller tappet. The injection nozzle is installed with a nozzle holder in the cylinder head. The lines are made of high-strength, seamless steel tubes. They must be of equal length for the individual pumps of an engine.

The method of solenoid-valve actuation is the same as that of the unit-injector system. When the solenoid valve is open, the fuel can be drawn into the pump barrel during the pump plunger's intake stroke, and return during the delivery stroke. Only when the solenoid valve is energized, and thus closed, can pressure build up in the high-pressure system between the pump plunger and the nozzle during the pump-plunger delivery stroke. Fuel is injected into the combustion chamber of the engine once the nozzle-opening pressure is exceeded.

Electronic control

The solenoid valves are actuated by an electronic control unit. The ECU analyzes all of the relevant status parameters in the system relative to the engine and its environment, and defines the start of injection and injected fuel quantity for the operating state of the engine at any given time. The start of injection is also controlled by a BIP signal (beginning of injection period) in order to balance out the tolerances in the overall system. Start of injection is synchronized with engine piston position by analysis of the signals from an incremental trigger wheel.

In addition to the basic fuel-injection functions, there are further functions for improving driving smoothness (e.g. surge dampers, idle-speed governors, adaptive cylinder equalization). Diagnosis of the fuel-injection system and the engine is also included in the range of functions. The ECU communicates with other electronic vehicle components (e.g. antilock braking system (ABS), the traction control system (TCS) or transmission-shift control system) via a CAN data bus.

Figure 4: Unit-pump system
1 Engine, 2 Nozzle holder, 3 Injection nozzle, 4 High-pressure line, 5 Solenoid valve, 6 Fuel feed, 7 High-pressure pump (unit pump), 8 Camshaft.

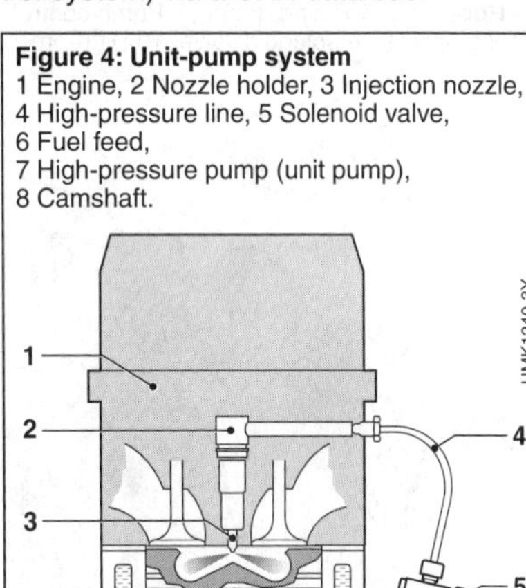

Diesel distributor injection pumps

Diesel distributor injection pumps were used in large numbers from 1962 until after 2000. These pumps are used in 3-, 4-, 5- and 6-cylinder diesel engines in passenger cars, tractors and light and medium commercial vehicles. They generate up to 50 kW per cylinder, depending on engine speed and combustion system. Distributor injection pumps for direct-injection (DI) engines achieve a peak injection pressure of up to 1,950 bar in the nozzle at engine speeds up to 4,500 rpm.

A distinction is made between distributor injection pumps with mechanical control, and those with an electronic governor available in port-controlled versions with a rotary-magnet actuator and versions with solenoid-valve open-loop control.

In passenger cars and commercial vehicles, distributor injection pumps have been superseded by common-rail systems.

Axial-piston distributor pumps

Body

Fuel-supply pump

If no presupply pump is fitted in the fuel-injection system, this integral vane-type supply pump draws fuel from the tank and, together with a pressure-control valve, generates an internal pump pressure which increases with engine speed.

High-pressure pump

The axial-piston distributor pump (VE pump) incorporates only one pump element for all cylinders. The distributor plunger displaces the fuel during its axial stroke and rotates at the same time to distribute the fuel through passages to the individual high-pressure connections (Figure 1).

Figure 1: Solenoid-valve-controlled axial-piston distributor pump
1 Angle-of-rotation sensor, 2 Drive shaft, 3 Support ring of vane-type supply pump, 4 Roller ring, 5 Timing device, 6 Pump control unit, 7 Cam plate, 8 Distributor plunger, 9 High-pressure solenoid valve, 10 High-pressure connection, 11 Distributor bore.

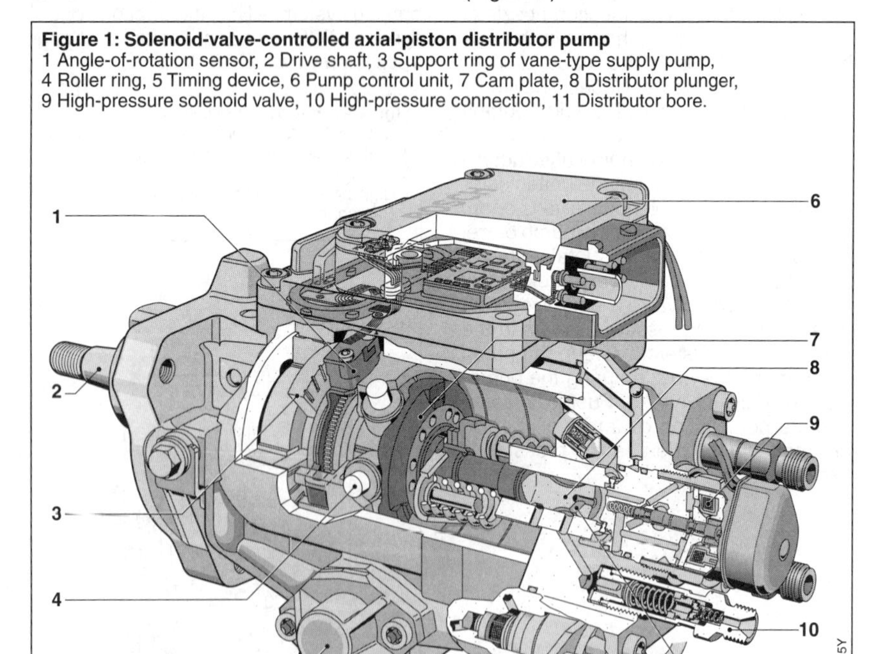

A clutch unit transfers the rotation of the drive shaft to the cam plate and the distributor plunger fixed to it. When doing so, the claws of the drive shaft and the cam plate grip into the yoke positioned in between. The cam lifts on the underside of the cam plate turn against the rollers of the roller ring. As a result, the cam plate and distributor plunger perform a stroke movement in addition to the rotation. During each rotation of the drive shaft, the distributor plunger completes a number of strokes equal to the number of engine cylinders to be supplied.

The pump delivers fuel for as long as the spill port in the distributor plunger remains closed off during the working stroke. Delivery ends when the spill port is uncovered by the control collar (Figure 2).

Electronic control for distributor injection pumps with rotary-magnet actuator

In contrast to the mechanically controlled distributor injection pump, the pump with rotary-magnet actuator has an electronic governor and an electronically controlled timing device (Figure 2).

Electronic governor

An eccentrically mounted ball head provides the connection between the distributor injection pump control collar and the rotary-magnet actuator. The actuator's rotary setting determines the position of the control collar, and with it the effective stroke of the pump. A non-contacting position sensor is connected to the rotary-magnet actuator.

The control unit receives various signals from the sensors: accelerator-pedal position; engine speed; air, coolant and fuel temperature; charge-air pressure; atmospheric pressure; etc. It uses these input variables to determine the correct injected fuel quantity, which is then converted to a specific control-collar position with the aid of program maps stored in the unit's memory. The control unit varies the excitation current to the rotary-magnet actuator until it receives a signal indicating convergence between the setpoint and actual values for control-collar position.

Electronically controlled timing device

The hydraulic timing device with the timing-device solenoid valve rotates the roller ring, depending on the load state and engine speed in such a way that the start of delivery – in relation to the engine piston position – is advanced or retarded.

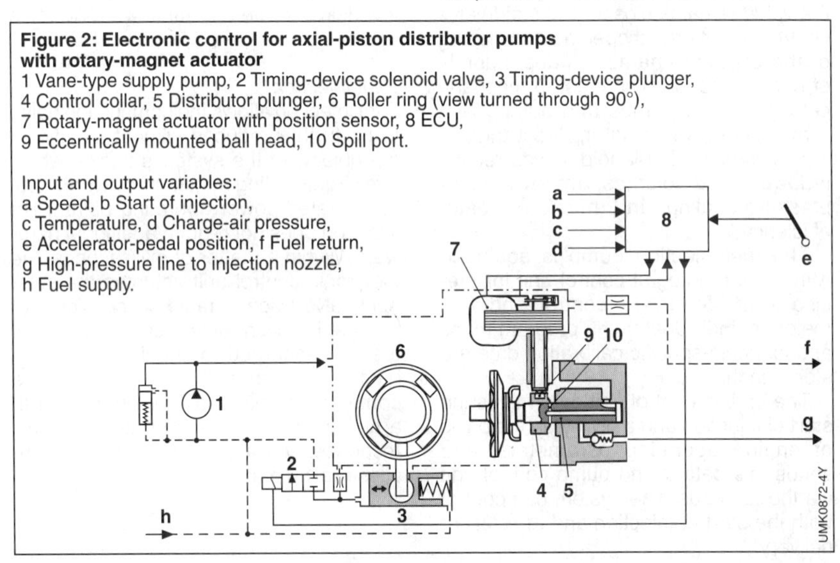

Figure 2: Electronic control for axial-piston distributor pumps with rotary-magnet actuator
1 Vane-type supply pump, 2 Timing-device solenoid valve, 3 Timing-device plunger, 4 Control collar, 5 Distributor plunger, 6 Roller ring (view turned through 90°), 7 Rotary-magnet actuator with position sensor, 8 ECU, 9 Eccentrically mounted ball head, 10 Spill port.

Input and output variables:
a Speed, b Start of injection, c Temperature, d Charge-air pressure, e Accelerator-pedal position, f Fuel return, g High-pressure line to injection nozzle, h Fuel supply.

The signal from a sensor in the nozzle-holder assembly, which indicates when the nozzle begins to open, is compared with a programmed setpoint value. The timing-device solenoid valve connected to the working chamber of the plunger in the timing device varies the pressure above the timing-device plunger and thus the position of the timing device. The actuation clock ratio of the timing-device solenoid valve is varied until the setpoint and actual values agree.

Electronic control for solenoid-valve-controlled distributor injection pumps

In the case of solenoid-valve-controlled distributor injection pumps (Figure 1), the fuel is metered by a high-pressure solenoid valve which directly closes off the pump's element chamber. This permits even greater flexibility in fuel metering and for varying the start of injection. The main assemblies are the high-pressure solenoid valve, the ECU, and the incremental angle/time system for angle/time control of the solenoid valve using an angle-of-rotation sensor integrated in the pump.

The solenoid valve closes to define the start of delivery, which then continues until the valve opens. The injected fuel quantity is determined by the length of time the valve remains closed. Solenoid-valve control permits rapid opening and closing of the element chamber independently of engine speed. In contrast to mechanically governed pumps and pumps with a rotary-magnet actuator, direct triggering by means of solenoid valves results in lower dead volumes, improved high-pressure sealing, and therefore greater efficiency.

The fuel-injection pump is equipped with its own, integral control unit for precise start-of-delivery control and fuel metering. Individual pump program maps and example-specific calibration data are stored in this ECU.

The engine control unit determines the start of injection and delivery on the basis of engine operating parameters, and sends this data to the pump control unit via the data bus. The system can control both the start of injection and the start of delivery.

The pump control unit also receives the injected fuel quantity signal via the data bus. This signal is generated by the engine control unit according to the accelerator-pedal signal and other parameters for torque demand. In the pump control unit, the injected fuel quantity signal and the pump speed for a given start of delivery are taken as the input variables for the pump map on which the corresponding actuation period is stored as degrees of cam rotation.

And finally, the actuation of the high-pressure solenoid valve and the desired period of actuation are determined on the basis of the angle-of-rotation sensor integrated in the distributor injection pump. This sensor is used for angle/time control. It consists of a magnetoresistive sensor and a reluctor ring divided into 3° increments interrupted by a reference mark for each cylinder. The sensor determines the precise angle of cam rotation at which the solenoid valve opens and closes. This requires the pump control unit to convert timing data to angular position data and vice versa.

The low fuel-delivery rates at the start of injection, which result from the design of the distributor injection pump, are further reduced by the use of a two-spring nozzle holder. With a hot engine, these low delivery rates permit low basic noise levels.

Pilot injection

Pilot injection allows the combustion noise to be further reduced without sacrificing the objects of the system's design which aim at generating maximum power output at the rated-power operating point. Pilot injection does not require additional hardware. Within a matter of milliseconds, the electronic control unit actuates the solenoid valve twice in rapid succession. For the first injection, only a small amount of fuel is discharged to condition the combustion chamber. The solenoid valve controls the injected fuel quantity with a high degree of precision and dynamic response. Typical pilot-injection fuel quantities are 1.5 mm^3.

Radial-piston distributor pumps

Body

High-pressure pump

The radial-piston high-pressure pump (VR pump, Figure 3) is driven directly by the distributor-pump drive shaft. It comprises the cam ring, the roller supports and rollers, the delivery plunger, the drive plate, and the front section (head) of the distributor shaft.

The drive shaft drives the drive plate by means of radially positioned guide slots. The guide slots simultaneously act as the locating slots for the roller supports. The roller supports and the rollers held by them run around the inner cam profile of the cam ring that surrounds the drive shaft. The number of cams corresponds to the number of cylinders in the engine.

The drive plate drives the distributor shaft. The head of the distributor shaft holds the delivery plungers which are aligned radially to the drive-shaft axis (hence the name "radial-piston distributor pump").

The delivery plungers rest against the roller supports. As the roller supports are forced outwards by centrifugal forces, the deliver plungers follow the profile of the cam ring and describe a cyclical-reciprocating motion.

Figure 3: Solenoid-valve-controlled radial-piston distributor pump
1 Sensor (angle/time system), 2 Pump control unit, 3 Distributor shaft, 4 Solenoid-valve needle, 5 Valve body, 6 Timing device, 7 Radial-piston pump, 7.1 Cam ring, 7.2 Roller, 7.3 Distributor shaft, 7.4 Delivery plunger, 7.5 Roller support, 8 Timing-device solenoid valve, 9 Delivery valve, 10 High-pressure solenoid valve, 11 Drive shaft.

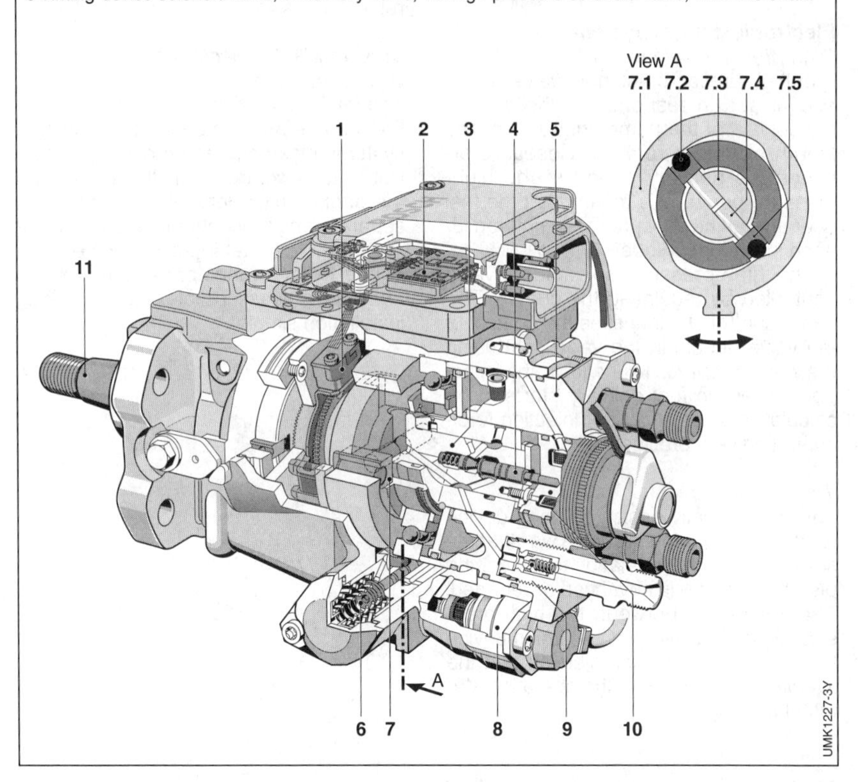

When the delivery plungers are pushed inwards by the cams, the volume in the central plunger chamber between the delivery plungers is reduced. This compresses and pumps the fuel when the solenoid valve is closed. The fuel is directed through passages in the distributor shaft at defined times to the appropriate outlet delivery valves.

As the cam-drive design employs a direct, positive link, flexibility and compliance remain minimal, so the performance potential is greater. Fuel delivery is shared between at least two radial plungers. The small inertial forces involved mean that steep (fast) cam profiles are possible. The fuel-delivery rate can be further increased by increasing the number of delivery plungers.

Radial-piston distributor pumps for direct-injection engines achieve element-chamber pressures up to 1,100 bar and pressures in the nozzle up to 1,950 bar.

Electronic control system

High-pressure solenoid valve

The high-pressure solenoid valve opens and closes in response to the triggering signals of the pump control unit. The length of time it remains closed determines the delivery period of the high-pressure pump. This means that the fuel quantity can be very precisely metered for each individual cylinder.

The high-pressure solenoid valve is controlled by regulating the current. The pump control unit identifies the contact of the valve needle in the valve seat using the current curve. This allows the exact point at which fuel delivery starts to be calculated and the start of injection to be controlled very precisely.

Timing devices

The hydraulically assisted timing device rotates the cam ring in such a way that the start of delivery – in relation to the engine piston position – is advanced or retarded. The interaction between the high-pressure solenoid valve and the timing device thus varies the start of injection and the injection pattern to suit the operating status of the engine.

The cam ring engages in a cross-slot in the timing-device plunger by means of an adjuster lug so that the axial movement of the timing-device plunger causes the cam ring to rotate. In the center of the timing-device plunger is a control collar which opens and closes the control ports in the timing-device plunger. In axial alignment there is a spring-loaded hydraulic control plunger which defines the required position for the control collar. Under the control of the pump control unit, the timing-device solenoid valve modulates the pressure acting on the control plunger.

This hydraulically assisted timing device can apply higher displacement forces than the hydraulic timing device of the axial-piston distributor pump.

The timing-device solenoid valve acts as a variable throttle. It can vary continuously the control pressure so that the control plunger can assume any position between the fully advanced and fully retarded positions.

Version with full electronic circuitry on injection pump

The latest generation of distributor injection pumps are compact, self-contained systems incorporating an electronic control unit to control both the pump and the engine-management functions. As a separate engine control unit is no longer required, the fuel-injection system requires fewer connectors and the wiring harness is less complex, thus making installation simpler.

Fuel-injection system

The injection pump is a component part of the fuel-injection system (Figure 4). A diesel fuel-injection system comprises the fuel-supply system (low-pressure stage), the high-pressure component, the injection components and the control system. The fuel-supply system accumulates and filters the fuel. If necessary, an additional fuel pump is present. The injection pump with the lines is the high-pressure component. It generates the high pressure and distributes it at the right time to the corresponding engine cylinders.

In distributor-pump injection systems, the injection component comprises the injection nozzles and the nozzle holders. These nozzle-holder assemblies come in a wide range of types. One nozzle-holder assembly is used on each cylinder. They are mounted with brackets or with hollow screws in the cylinder head. The functions of the injection nozzles are to inject metered quantities of fuel and to prepare the fuel, to shape the injection pattern, and to provide a seal against the combustion chamber. Each injection nozzle consists of the nozzle body with several injection orifices (up to 0.12 mm) and the nozzle needle. The needle is guided in the nozzle-body guide bore to ensure correct alignment between the injection orifices (which are at different angles in the nozzle body) and the engine combustion chamber.

The mechanical or electronic control system for the distributor injection pump is mounted on the pump itself. Some systems have a separate engine control unit. Versions with electronic control are equipped with numerous sensors and setpoint generators.

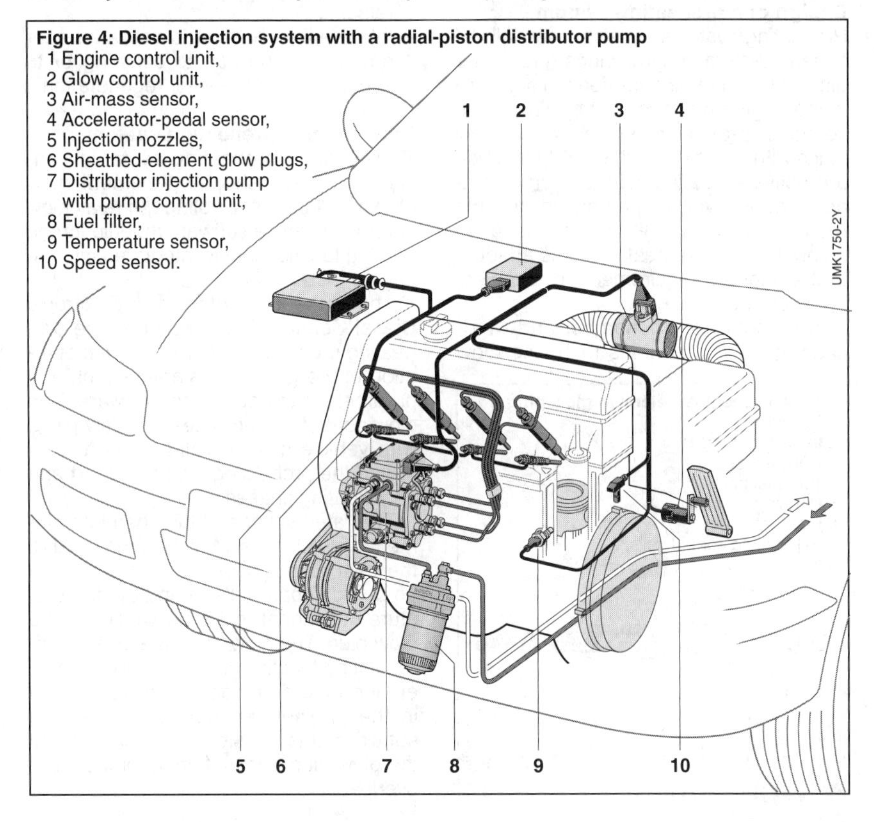

Figure 4: Diesel injection system with a radial-piston distributor pump
1 Engine control unit,
2 Glow control unit,
3 Air-mass sensor,
4 Accelerator-pedal sensor,
5 Injection nozzles,
6 Sheathed-element glow plugs,
7 Distributor injection pump with pump control unit,
8 Fuel filter,
9 Temperature sensor,
10 Speed sensor.

Start-assist systems for diesel engines

Preheating systems for passenger cars and light utility vehicles

Warm precombustion-chamber and whirl-chamber diesel engines and direct-injection (DI) engines will start spontaneously at low outside temperatures above 0 °C. Here, the spontaneous-ignition temperature for diesel fuel of 250 °C is achieved with the engine turning at starting speed. Cold precombustion-chamber and whirl-chamber diesel engines require starting assistance at ambient temperatures below 40 °C or 20 °C, DI engines only below 0 °C. Preheating systems are used for passenger cars and light commercial vehicles.

Design of a preheating system

Preheating systems consist essentially of sheathed-element glow plugs (GLP), the glow control unit and preheating software in the engine-management system. Conventional preheating systems use glow plugs with a nominal voltage of 11 V which are activated by the vehicle system voltage. New low-voltage preheating systems require glow plugs with nominal voltages below 11 V whose heat output is adapted to the engine's requirements by an electronic glow control unit.

In precombustion-chamber and whirl-chamber diesel engines (IDI), the glow plug extends into the secondary combustion chamber, while in DI engines, it extends into the main combustion chamber of the engine cylinder. The air/fuel mixture is directed past the hot tip of the glow plug and heated. The ignition temperature is reached in combination with the heating of the intake air during the compression cycle.

Figure 1: Sheathed-element glow plug
1 Terminal,
2 Insulator shim,
3 Double seal,
4 Terminal pin,
5 Housing,
6 Element seal,
7 Heater and control coils,
8 Glow tube,
9 Filling powder.

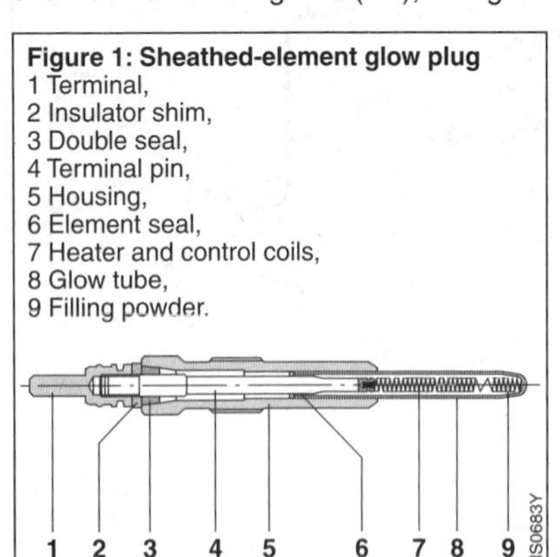

Preheating phases

- Preheating: The glow plug is heated to operating temperature.
- Standby heating: For the start, the preheating system maintains a required temperature for a defined period.
- Glow-plug start assist is applied during warming up of the engine.
- Post-heating phase begins after starter release.
- Intermediate heating is activated after engine cooling due to coasting or to support particulate-filter regeneration.

Conventional preheating system

Conventional preheating systems consist of a metal sheathed-element glow plug with 11 V rated voltage, a relay glow control unit and a software module for the heating function that is integrated into the engine ECU.

The preheating software in Electronic Diesel Control starts and ends the preheating process in accordance with operation of the glow-plug starter switch and parameters stored in the software. The glow control unit activates the glow plugs with vehicle system voltage via a relay during the preheating, standby, start and post-heating phases.

The rated voltage of the sheathed-element glow plugs is 11 V. This means that the heat output depends on the current on-board supply voltage and the temperature-dependent resistance (PTC) of the glow plug. The glow plug thus has a self-regulating function. In conjunction with an engine-load-dependent cutout function in the engine-management preheating software, it is possible to safely prevent the glow plug from suffering temperature overload.

Design and characteristics

The main component in the sheathed-element glow plug is the tubular heating element (Figure 1). The tubular heating element consists of a hot-gas and corrosion-resistant element sheath which encloses a filament surrounded by compressed magnesium oxide powder. That filament is made up of two resistors connected in series – the heating filament located in the tip of the sheath, and the control filament.

Whereas the heating filament has an electrical impedance that is independent of temperature, the control filament has a positive temperature coefficient (PTC). Its impedance increases as the temperature rises. The sheathed-element glow plug is thus faster at reaching the temperature required for igniting the diesel fuel and also has a lower steady-state temperature (Figure 2). This means that the temperature is kept below the critical level for the glow plug. Consequently, it can remain in operation for up to three minutes after the engine has started. This post-heating function results in a more effective engine cold-idle phase with substantially lower noise and emission output.

Figure 2: Comparison of preheating curves, from $t = 0$ s with a flow velocity of 11 m/s
1 Low-voltage ceramic sheathed-element glow plug (7 V nominal voltage),
2 Low-voltage metal sheathed-element glow plug (5 V nominal voltage),
3 Metal sheathed-element glow plug (11 V nominal voltage).

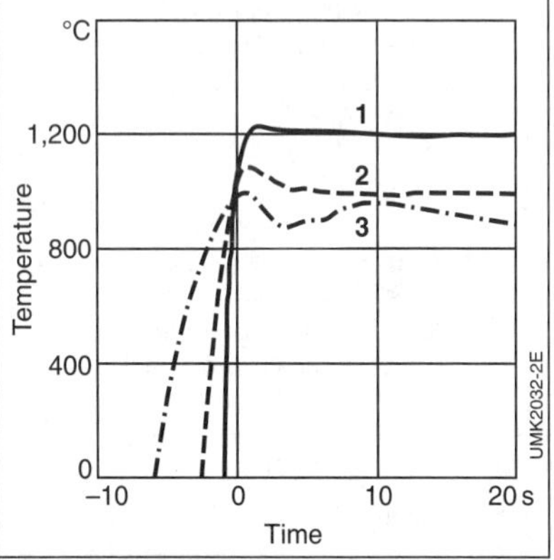

The heating filament is welded into the cap of the element sheath for grounding. The control filament is contacted at the terminal stud, which establishes the connection to the vehicle electrical system.

Operation

When voltage is applied to the glow plug, most of the electrical energy of the heating filament is initially converted into heat; the temperature at the tip of the glow plug increases sharply. The temperature of the control filament – and with it also the impedance – increase with a time delay. The current draw and thus the total heating output of the glow plug decrease and the temperature approaches a steady-state condition.

Low-voltage preheating system

The low-voltage preheating system contains:

- ceramic sheathed-element glow plugs of low-voltage configuration for operation below 11 V,
- an electronic glow control unit,
- a software module for the heating function integrated into the engine ECU.

In order to achieve as quickly as possible the heating temperature required for engine starting during preheating, the glow plugs are briefly operated in this phase with push voltage, which is above the nominal voltage. The activating voltage is then reduced to the nominal voltage during start-standby heating.

During glow-plug start assist, the activation voltage is increased again in order to compensate for the cooling of the glow plug by the cold intake air. This is also possible in the post- and intermediate-heating phases. The required voltage is taken from a characteristic map that is adapted to each engine. The characteristic map contains the parameters engine speed, injected fuel quantity, time after starter release, and coolant temperature.

Map-controlled activation reliably prevents thermal overloading of the glow plug in all engine operating states. The heating function implemented in Electronic Diesel Control contains an overheating-protection facility in the event of repeat heating.

These preheating systems facilitate a rapid start or an immediate start similarly to a gasoline engine up to –28 °C.

Low-voltage metal sheathed-element glow plug

The basic design and operation are the same as those of a conventional sheathed-element glow plug. The heating and control filaments are designed here for a lower nominal voltage and a high preheating rate.

The slender shape is designed to suit the restricted space in four-valve engines. The glow element is tapered at the front in order to accommodate the heating filament closer to the element sheath. This allows preheating rates of up to 1,000 °C within 3 s with the push mode. The maximum heating temperature is in excess of 1,000 °C. The temperature during start-standby heating and in post-heating mode is approx. 980 °C. These functional properties are adapted to the requirements of diesel engines with a compression ratio of over 18.

Low-voltage ceramic sheathed-element glow plug

Low-voltage ceramic sheathed-element glow plugs have glow plugs made of a ceramic material that is highly resistant to temperatures. Due to their very high resistance to oxidation and thermal shocks, they enable an immediate start, as well as minute-long post-heating and intermediate heating at 1,300 °C. They are designed for a nominal voltage of 7 V.

Reduced emissions in diesel engines with a low compression ratio

By lowering the compression ratio in modern diesel engines from $\varepsilon = 18$ to $\varepsilon = 16$, it is possible to reduce the NO_x and soot emissions while simultaneously increasing the specific power. However, the cold-starting and cold-running performance is problematic in these engines. In order to obtain minimal exhaust-gas opacity values and heightened smooth running during cold starting and cold running, temperatures at the glow plug of over 1,150 °C are required – 850 °C is sufficient for conventional engines. During the cold-starting phase, these low emission values – blue-smoke and soot emissions – can only be maintained by post-heating lasting several minutes. Compared with standard preheating systems, the ceramic preheating system from Bosch reduces the exhaust-gas opacity values by up to 60 % (Figure 3).

Figure 3: Influence of the sheathed-element glow-plug surface temperature on the exhaust-gas opacity with optimal alignment. Cold idling of a four-cylinder engine with common-rail system and a compression of 16:1 at –20 °C

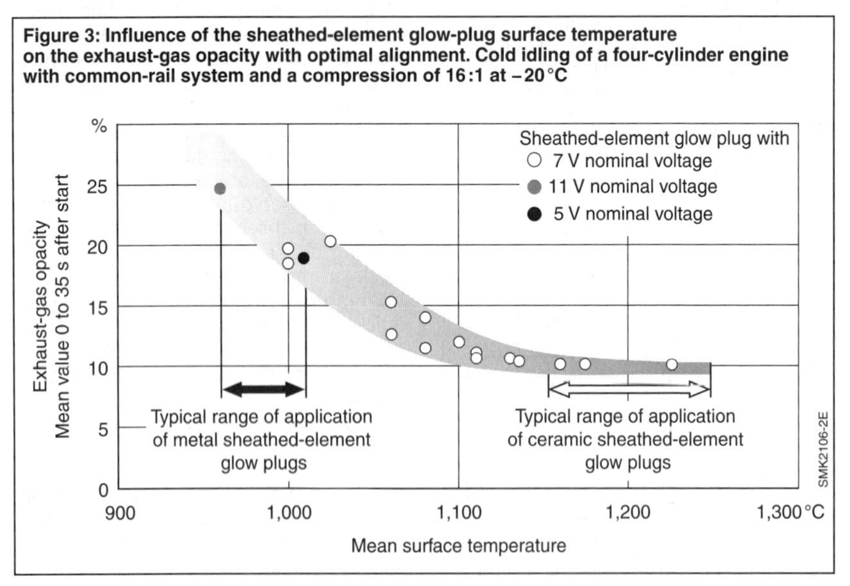

Flame start systems for commercial-vehicle diesel engines

Requirements

Commercial-vehicle diesel engines can start down to temperatures of around −20 °C with start assistance. The start durations are dependent on the compression ratio ε and the fuel grade used (cetane number CN, Figure 4). To prevent misfiring and emissions of unburnt hydrocarbons (HC), it is however possible, depending on the compression ratio and the fuel grade used, to utilize already from a coolant temperature of 10 °C air preheating as start assistance (flame start system or grid heater).

Air preheating provides for safe, quick starting down to low outside temperatures. The battery and the starter are protected and fuel consumption is reduced in the starting and warm-up phases. At higher temperatures air preheating would not be necessary for safe starting, but does prevent the formation of white smoke (HC emission, Figure 5).

Figure 4: Starting times without start assistance as a function of compression ratio ε and cetane number CN
1 Starting-time curve for ε = 16.5 and CN = 41,
2 Starting-time curve for ε = 16.5 and CN = 45,
3 Starting-time curve for ε = 16.5 and CN = 51,
4 Starting-time curve for ε = 17.0 and CN = 51,
5 Starting-time curve for ε = 17.75 and CN = 51.

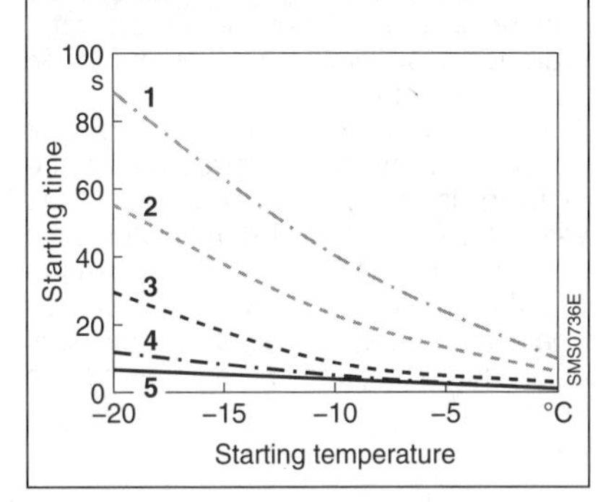

Figure 5: Use of start assistance for starting and to reduce HC emissions

Diesel fuel (CN < 45)
Start assistance
White-smoke reduction
Diesel fuel (CN > 50)
Start assistance
White-smoke reduction
−30
−20
−10
0
°C
Starting temperature
SMS0737E

Design

There are two different types of flame start system: the flame start system with burner chamber and the flame glow plug.

Flame start system with burner chamber
The flame start system with burner chamber (Figure 6) consists of a nozzle holder on which a solenoid valve, a nozzle and a rod-type glow plug are mounted. The nozzle holder is attached together with gaskets and the burner chamber by three screws to the charge-air housing.

The fuel is drawn after the fuel-system fuel filter and routed to the solenoid valve via a fuel line. The solenoid valve controls the fuel flow while the flame start system is in operation. The glow plug is supplied with electrical power by the flame-start control unit.

Figure 6: Flame start system with burner chamber
1 Nozzle, 2 Fuel line,
3 Nozzle holder, 4 Solenoid valve,
5 Rod-type glow plug,
6 Burner chamber,
7 Charge-air housing.

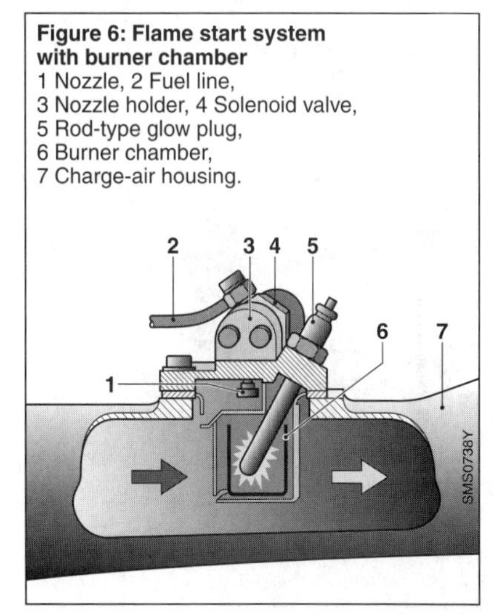

Flame glow plug

In contrast to the flame start system with burner chamber, the solenoid valve in the case of the flame glow plug is not mounted directly on the flame start system, but instead is connected by a fuel line to the flame glow plug. The flame glow plug also has a much more compact design (Figure 7).

The glow plug integrated in the flame glow plug is enclosed in a vaporizer tube in the upper area and extends up to the protective tube at the end of the flame glow plug (Figure 8). The fuel is drawn after the fuel filter, routed through the separately mounted solenoid valve, and supplied to the flame glow plug via the fuel line. The fuel enters the flame glow plug via the metering device – to control the fuel quantity. The flame-start control unit assumes the job of controlling the flame start system (current supply to the glow plug and switching on the solenoid valve).

Use of flame start systems

The flame start system with burner chamber is used primarily in commercial-vehicle engines with a displacement $V_H > 10\ l$, while the flame glow plug is used in engines with a displacement V_H of 4...10 l. The flame start system is installed in the engine charge-air tube in such a way that the heated intake air is uniformly distributed to all the cylinders.

Figure 7: Installed flame glow plug
1 Fuel line,
2 Throttle line,
3 Flame glow plug,
4 Ground connection,
5 Charge-air tube.

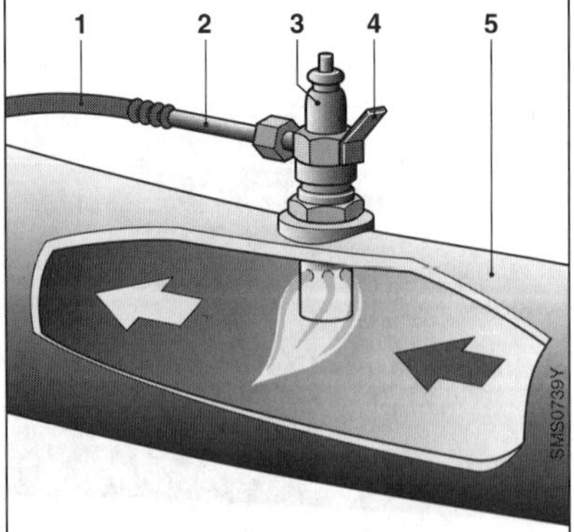

Operating principle

The flame start system is activated below the switch-on temperature defined in the control unit. When the ignition key is turned to the Drive position, the "Flame start system" indicator lamp lights up and the glow plug is energized and preheated. The indicator lamp goes out after a preheating time dependent on the vehicle system voltage (approx. 25 s). The flame start system is operational and the engine can be started. At the outset of the starting operation the solenoid valve opens and the fuel flows to the hot glow plug, where it ignites. The flame that is now produced heats the intake air flowing past. The flame start system is operated after the engine start in the post-flaming time until a stable burn is guaranteed and shuts down according to the parameters defined in the control unit. The parameters for post-flaming are dependent on the coolant temperature.

The flame start system is also shut down if the engine is started before the indicator lamp goes out or is not started within 30 seconds of the indicator lamp going out. In this case the driver can put the flame start system into operation again by turning the ignition key back and then turning into to the drive position again.

Figure 8: Flame glow plug (Beru system)
1 Fuel-supply connection,
2 Metering device,
3 Glow plug,
4 Vaporizer tube,
5 Protective tube.

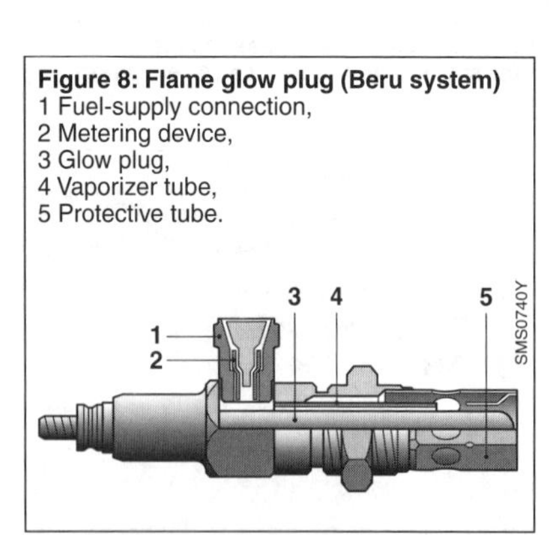

Control of the glow plug

Glow plug for the flame start system with burner chamber

The glow plug is preheated depending on the vehicle system voltage until a temperature of approximately 1,050 °C is reached. This is followed by the beginning of the period of start readiness (duration 30 s), during which the glow plug is cyclically energized. The cycle ratio is set depending on the current vehicle system voltage in such a way that the glow-plug temperature remains between 1,020 °C and 1,080 °C. With the glow plug in this temperature range the fuel is safely ignited and the glow plug itself is not exposed to thermal overload.

As soon as the engine is started, the solenoid valve opens, the fuel is directed through the nozzle and sprayed onto the spark-plug grooves, where it ignites. In this phase the cycle ratio of the glow plug is increased by the control unit so that the glow plug does not cool in response to the heat dissipated from the vaporizing fuel. This higher cycle ratio is maintained after starting for a further period of around 30 s until all the components are heated and the flame is safely burning. The glow plug is now additionally heated by the flame. The control unit decreases the cycle ratio again to protect against overheating. After post-flaming the glow plug is deactivated and monitored until it has cooled down entirely. Start and cold-running support by the flame start system is concluded.

Flame glow plug

This glow plug is preheated in exactly the same way before starting, depending on the applied vehicle power supply. The preheating times are slightly longer than for the glow plug with burner chamber, since the vaporizer tube is additionally heated during preheating (see Figure 8). When the engine is started, the solenoid valve is opened, fuel flows via the fuel-supply connection and the metering device into the flame glow plug, strikes the hot glow plug, vaporizes between the glow plug and the vaporizer tube, and mixes with the air at the evaporator-tube outlet. The air/fuel mixture ignites at the hot glow plug. The flame produced emerges from the protective tube and heats the intake air flowing past.

After preheating the flame glow plug is likewise cyclically energized. The cycle ratio is decreased in the start-readiness period (30 s) and increased again when the engine is started. In this phase the cycle ratio for the flame glow plug is higher than for the flame start system with burner chamber, since the fuel is vaporized via the supplied electrical energy and not via the burning flame.

Cooling-down phase

The glow plug of the flame start system is monitored in the cooling-down phase. If preheating is performed again in this phase, the control unit calculates the preheating time from a program map depending on the cooling-down time and the voltage in such a way that the glow plug reached the temperature range described above.

Because the glow plugs have heating and control filaments (see Preheating systems for passenger-car diesel engines), the optimum preheating time is set as a function of the ambient conditions.

Grid heater for commercial-vehicle diesel engines

The grid heater is used as a further cold-start aid, particularly for engines which are started under aggravated starting conditions such as constant load when starting (e.g. for hydraulic output) or at altitudes above 2,000 m (e.g. snow groomers).

Design

The grid heater consists of an electric heater element which is installed in the charge-air tube and powered by a relay. A separate ECU or the engine ECU switches on the grid heater on the basis of the stored parameters.

There are different versions which are used, depending on the engine displacement V_H and the vehicle system voltage (12 V or 24 V). Grid heaters with a power input of 1.8 kW are used for engines with a displacement V_H of 4...10 l, while grid heaters with a power input of 2.7 kW are used for engines with $V_H > 10\ l$.

Figure 9: Grid heater
1 Housing,
2 Mounting flange,
3 Cover plate,
4 Cover,
5 Mounting bore,
6 Lock nut,
7 Connecting-thread bolt,
8 Retaining frame,
9 Heating filament,
10 Insulating plate.

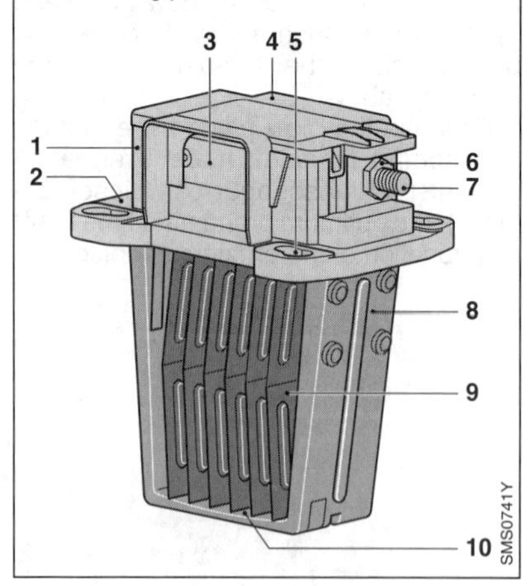

Operating principle

Depending on the coolant and charge-air temperatures, the grid heater is preheated for periods varying in duration (preheating time t_v = 2...28 s) and remains active in the post-heating phase until stable engine running is ensured. The switch-on temperature is 5 °C for engines with V_H = 4...10 l and –4° for engines with $V_H > 10\ l$.

Once the preheating time has elapsed, the indicator lamp goes out and the engine can be started. On starting the grid heater is switched off for approximately 2 s so as to make full battery capacity available to the starter. For start and idle support the grid heater is then switched back on and remains in operation until it is switched off after the applied post-heating time.

The grid heater remains in start readiness for only around 5 s (as opposed to 30 s for the flame start system) in order to conserve battery capacity.

Advantages and disadvantages of the grid heater

The grid heater offers the following advantages over the flame start system:

- The intake air is heated before starting
- There are shorter starting times thanks to good ignition conditions from the first revolution
- No burning of oxygen is required to heat the intake air

On the down side, the grid heater requires greater battery capacity.

Exhaust-gas treatment

Catalytic converters

In diesel engines operating with excess air, the three-way catalytic converter can be used not only for nitrogen-oxide reduction NO_x. This is because, in the lean diesel exhaust gas, hydrocarbon and carbon-monoxide emissions (HC and CO) in the catalytic converter prefer not to react with nitrogen oxide but with exhaust-gas oxygen.

HC and CO emissions can be removed relatively easily from diesel exhaust by means of an oxidation catalytic converter. The removal of nitrogen oxides in the presence of oxygen is more complicated; as a general principle, it is possible to remove nitrogen oxides with an NO_x storage catalytic converter or an SCR (Selective Catalytic Reduction) catalytic converter.

Diesel oxidation catalytic converter

Diesel oxidation catalytic converters (DOC) consist primarily of a carrier structure made of ceramic, an oxide mixture (washcoat) composed of aluminum oxide (Al_2O_3), cerium (IV) oxide (CeO_2), zirconium oxide (ZrO_2), and active catalytic noble metals, such as platinum (Pt), palladium (Pd) and rhodium (Rh).

The oxidation catalytic converter performs a number of functions:

- CO and HC are oxidized at the catalytic converter to form CO_2 and H_2O. Oxidation is almost complete, starting from a specific limit temperature, i.e. the light-off temperature (170–200 °C).
- The emitted particles consist in part of hydrocarbons, which desorb from the particle core with increasing temperature. Oxidation of these hydrocarbons in the oxidation catalytic converter reduces the particle mass.
- Oxidation of NO to NO_2; a high NO_2 proportion in the NO_x is important for several downstream components (particulate filter, NO_x storage catalytic converter, SCR catalytic converter).
- The oxidation catalytic converter can be used as a catalytic burner ("cat burner") to raise the exhaust-gas temperature (e.g. in the case of particulate-filter regeneration).

NO_x storage catalytic converter

As the NO_x storage catalytic converter (NSC: NO_x storage catalyst) is only able to store NO_2, but not NO, the NO proportions are initially oxidized in a downstream or integrated oxidation catalytic converter into NO_2 (Figure 1).

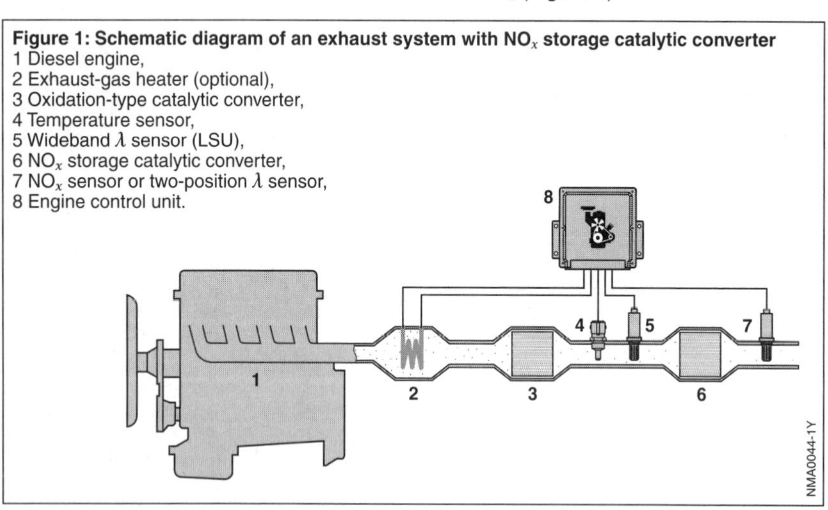

Figure 1: Schematic diagram of an exhaust system with NO_x storage catalytic converter
1 Diesel engine,
2 Exhaust-gas heater (optional),
3 Oxidation-type catalytic converter,
4 Temperature sensor,
5 Wideband λ sensor (LSU),
6 NO_x storage catalytic converter,
7 NO_x sensor or two-position λ sensor,
8 Engine control unit.

NO_x accumulation (storage)
The NO_2 is stored in that it reacts with the compounds of the catalytic converter surface (e.g. barium carbonate $BaCO_3$ as storage material) and oxygen (O_2) from the lean diesel exhaust gas to form nitrates.

The storage (load) is only optimal in a material-dependent temperature interval of the exhaust gas from 250–450 °C; below this, the oxidation from NO to NO_2 is very slow, and above it the NO_2 is unstable.

The charging phase takes 30–300 seconds depending on the operating point.

NO_x removal and conversion
At the end of the storage phase, the catalytic converter must be regenerated. To achieve this, rich conditions ($\lambda < 1$) must be set in the exhaust gas. There are so many reducing agents in the exhaust gas (CO, H_2 and various hydrocarbons) that the nitrate compound is abruptly dissolved and the released NO_2 is reduced directly to N_2 in the noble-metal catalytic converter.

Regeneration takes place in 2–10 seconds.

Desulfating
One problem faced by the NO_x storage catalytic converter is its sensitivity to sulfur. SO_2 is removed from the exhaust gas even more effectively than NO_x and stored in the catalytic converter. The sulfate compound that is formed is not dissolved during normal regeneration, which means that the amount of stored SO_2 rises continuously. This reduces the number of storage places for NO_x storage and NO_x conversion decreases. In order to achieve an adequate NO_x storage capacity, desulfating (sulfur regeneration) must be carried out regularly.

During the desulfating process, the catalytic converter is heated to a temperature of over 650 °C for a period of more than five minutes, and it is purged with rich exhaust gas ($\lambda < 1$). The combustion control aims to achieve complete removal of O_2 from the exhaust gas. Under these conditions, the sulfur compound is dissolved again.

The choice of a suitable desulfating process control (e.g. oscillating λ about $\lambda = 1$) must ensure that the SO_2 removed is not reduced to hydrogen sulfide (H_2S) by a continuous deficiency of exhaust-gas oxygen.

Selective catalytic reduction of nitrogen oxides
Selective catalytic reduction (SCR) is based on the principle that selected reducing agents selectively reduce nitrogen oxides (NO_x) in the presence of oxygen. Here, “selective” means that the reducing agent prefers to oxidize selectively with the oxygen contained in the nitrogen oxides instead of with the molecular oxygen present in much greater quantities in the exhaust gas.Ammonia (NH_3) has proven to be a highly selective reducing agent in this case.

As ammonia is toxic, the actual reducing agent for motor-vehicle applications is obtained from the non-toxic catalyst carrier urea $(NH_2)_2CO$. Urea is highly soluble in water, and can therefore be added to the exhaust gas as an easy-to-meter urea/water solution (Figure 2). The urea/water solution is marketed under the brand name AdBlue and DEF in the USA (Diesel Exhaust Fluid, SAE name).

At a mass concentration of 32.5 % urea in water, the freezing point has a localized minimum at –11 °C: A eutectic solution forms, but does not separate when frozen.

Chemical reactions
Urea first forms ammonia before the actual SCR reaction starts. This takes place in two reaction steps, which together are referred to as a hydrolysis reaction. Firstly, NH_3 and isocyanic acid are formed in a thermolysis reaction:

$$(NH_2)_2CO \rightarrow NH_3 + HNCO \text{ (thermolysis).}$$

Then isocyanic acid is converted with water in a hydrolysis reaction to form ammonia and carbon dioxide:

$$HNCO + H_2O \rightarrow NH_3 + CO_2 \text{ (hydrolysis).}$$

To prevent the precipitation of solids, the second reaction must take place rapidly by selecting suitable catalysts and temperatures that are sufficiently high (starting at 250 °C). Modern SCR catalytic converters handle catalytic hydrolysis at the same time.

Ammonia produced by thermohydrolysis reacts in the SCR catalytic converter according to the following equations:

$$4\ NO + 4\ NH_3 + O_2 \rightarrow 4\ N_2 + 6\ H_2O \quad \text{(Eq. 1)},$$
$$NO + NO_2 + 2\ NH_3 \rightarrow 2\ N_2 + 3\ H_2O \quad \text{(Eq. 2)},$$
$$6\ NO_2 + 8\ NH_3 \rightarrow 7\ N_2 + 12\ H_2O \quad \text{(Eq. 3)}.$$

At low temperatures (below 300 °C), conversion mainly takes place using the reaction set out in equation 2. For good low-temperature conversion, it is therefore required to set an NO_2/NO ratio of approx 1:1. Under these circumstances, this reaction can take place at temperatures starting at 170 to 200 °C.

Oxidizing NO to form NO_x occurs in an upstream oxidation catalytic converter, and this is necessary to achieve optimized efficiency.

If more reducing agent is dispensed than is converted during the reduction with NO_x, unwanted NH_3 slip may result. NH_3 is removable by placing an additional oxidation catalytic converter downstream of the SCR catalytic converter. This blocking catalytic converter oxidizes any ammonia that may occur to form N_2 and H_2O. In addition, a careful application of metered AdBlue is essential.

One key parameter for the application is the feed ratio α, which is defined as the molar ratio of metered NH_3 as a factor of NO_x present in the exhaust gas. Under ideal operating conditions (no NH_3 slip, no secondary reactions, no NH_3 oxidation), α is directly proportional to the NO_x reduction rate: At $\alpha = 1$, NO_x reduction of 100 % is achievable in theory. In

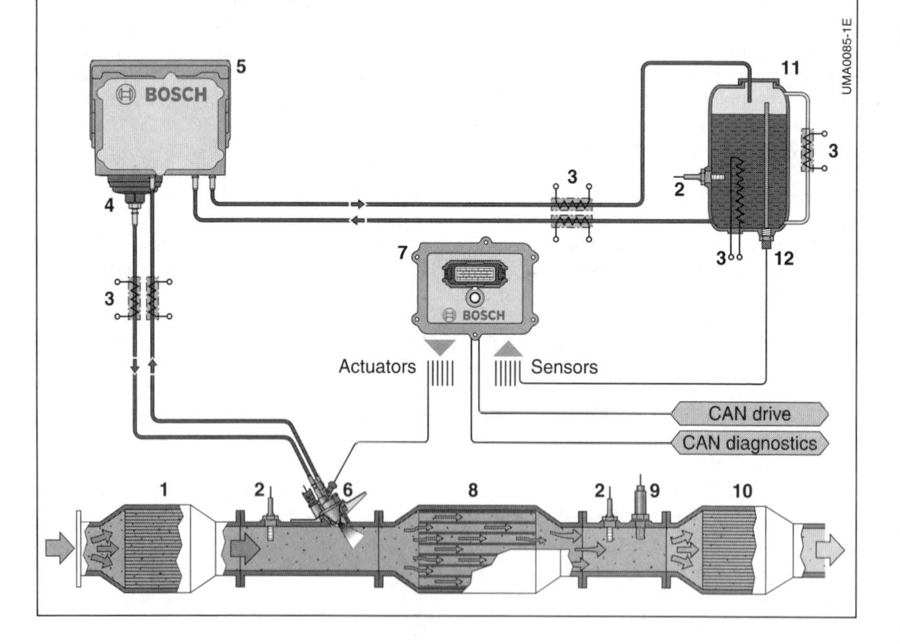

Figure 2: Exhaust system with catalytic reduction of nitrogen oxides (SCR)
1 Diesel oxidation catalytic converter, 2 Temperature sensor,
3 Heater, 4 Filter, 5 Delivery module,
6 AdBlue dosing module, 7 Dosing control unit, 8 SCR catalytic converter,
9 NO_x sensor, 10 Slip catalytic converter,
11 AdBlue tank (urea/water solution), 12 AdBlue level sensor

practice, however, an NO_x reduction of 90 % in fixed and mobile operation is achievable at an NH_3 slip of less than 20 ppm. The quantity of AdBlue required for this is approximately equivalent to 5 % of the quantity of diesel fuel used.

By arranging the hydrolysis reaction upstream, modern SCR catalytic converters achieve an NO_x conversion rate of greater than 50 % only at temperatures above approx. 250 °C. Optimized conversion rates are attained within a temperature window of 250 to 450 °C.

SCR system
The modular SCR system (Figure 2) meters the reducing agent. The delivery module brings the urea/water solution to the required pressure by means of a diaphragm pump and delivers it to the dosing module. The dosing module meters the precise volume of urea/water solution and ensures that it is atomized and distributed in the exhaust pipe. The primary function of the control unit (functionality featured in a separate dosing control unit, optionally integrated into the engine ECU) is the model-based calculation of the required dosing quantity in accordance with a prescribed dosing strategy.

Dosing strategy
The dosing quantity for the reducing agent is stored as a function of injected fuel quantity and engine speed in program map A, measured either on the test bench, or calculated by “a priori” assumptions. The system feeds in correction parameters, such as engine temperature (influence on NO_x production) and the number of system operating hours (to consider aging).

A correction factor for dosing on switching between two stationary operating points is determined from the difference between the stationary catalytic converter temperature and the measured exhaust-gas temperature after the catalytic converter.

In particular in the case of catalytic converters with high NH_3 storage capacity, it is recommended to model the transient operations and the quantity of actually stored NH_3, as the NH_3 storage capacity of SCR catalytic converters falls with increasing temperature.

Particulate filter

Soot particles emitted from a diesel engine can be efficiently removed from the exhaust gas by diesel particulate filters (DPF).

Closed particulate filters
Ceramic particulate filters consist essentially of a honeycomb structure made of silicon carbide or cordierite, which has a large number of parallel channels. The thickness of the channel walls is typically 300–400 µm. Channel size is specified by their cell density (channels per square inch, cpsi) (typical value: 100–300 cpsi).

Adjacent channels are closed off at each end by ceramic plugs to force the exhaust gas to penetrate through the porous ceramic walls (Figure 3). As soot particles pass through the walls, they are transported into the pore walls by diffusion (inside of the ceramic walls) where they adhere (deep-bed filtration). As the filter becomes increasingly saturated, a layer of soot forms on the surface of the channel walls (on the side opposite to the inlet channels). This provides highly efficient surface filtration for the following operating phase.

As opposed to deep-bed filters, wall-flow filters store the particles on the surface of the ceramic walls (surface filtration).

Besides filters with a symmetrical arrangement of square inlet and outlet channels, ceramic “octosquare substrates” are also available (Figure 4). They have larger octagonal inlet channels and smaller square outlet channels. The large inlet channels considerably increase the storability of the particulate filter for ash, non-combustible residue from burned engine oil, and additive ash.

Figure 3: Diesel particulate filters
1 Housing,
2 Extruded honeycomb ceramic,
3 Ceramic plug.

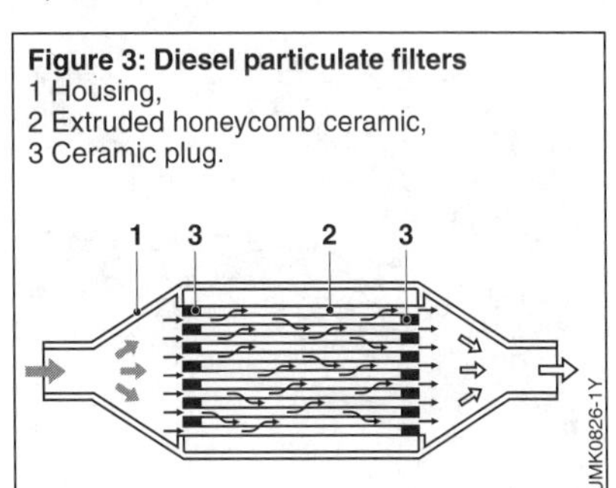

Ceramic filters achieve a retention efficiency of more than 95 % for particulates across the entire spectrum range in question (10 nm – 1 µm). In these closed particulate filters, the entire exhaust gas passes through the pore walls.

Open particulate filters
In the case of open particulate filters, only part of the exhaust gas is directed through a filter wall, while the remainder flows past unfiltered. Depending on the application, open filters achieve a filtration efficiency of 30–80 %.

As the particulate load increases, so too does the amount of exhaust gas which passes the filter unfiltered, thereby preventing it from blocking the filter. This, however, reduces the filtration efficiency. Open filters are mainly used as retrofit filters, since regulated filter cleaning is not required. Open filters are cleaned by the CRT effect.

Regeneration
The growing amount of soot deposited in the filter gradually increases the exhaust-gas backpressure. The particulate filter must therefore be regenerated regularly.

The filter is regenerated by burning off the soot that has collected in the filter. The particle carbon component can be oxidized (burned) using the oxygen constantly present in the exhaust gas above a temperature of approx. 600 °C to form nontoxic CO_2.

Figure 4: Designs of ceramic particulate filter
a) Square channel cross-section,
b) Octosquare design.

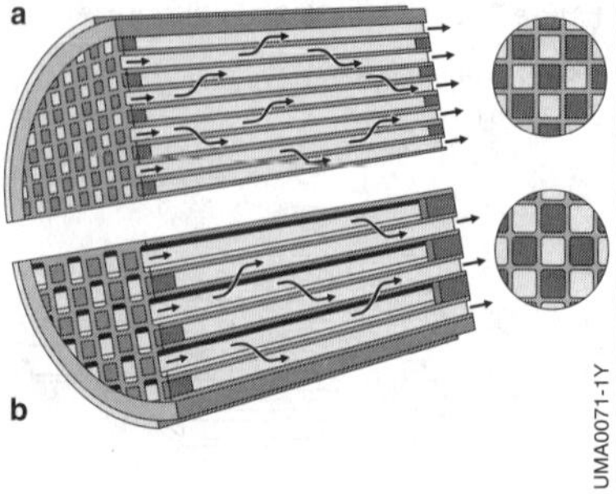

a

b

UMA0071-1Y

Such high temperatures occur only when the engine is operating at rated output. It is highly rare in normal vehicle operation. For this reason, measures must be taken to lower the soot burnoff temperature or raise the exhaust-gas temperature.

Measures inside of the engine to raise the exhaust-gas temperature
The main measures taken inside of the engine ("engine burner") to increase exhaust-gas temperature are advanced "burnoff" or "additive" secondary injection, retarded main injection and intake-air throttling. Depending on the engine operating point, one or several of these measures are triggered during regeneration. In some operating points, these measures are supplemented by retarding secondary injection. This leads to a further increase in exhaust-gas temperature due to oxidation of fuel in the diesel oxidation catalytic converter ("cat burner") no longer converted in the combustion chamber.

Additive system
The soot oxidation temperature of 600 °C can be lowered to approx. 450–500 °C by using an additive – usually cerium or iron compounds – in diesel fuel (Figure 5). But even this temperature is not always reached in the exhaust-gas system when the vehicle is operating. The result is that the soot is not burned off continuously. Thus, active regeneration is triggered above a specific level of soot saturation in the particulate filter. To achieve this, the combustion control of the engine is changed, e.g. by means of retarded fuel injection, in such a way that the exhaust-gas temperature rises up to soot burnoff temperature.

The additive in the fuel is retained in the filter as a residue (ash) after regeneration. This ash, as well as ash from engine-oil and fuel residue, gradually clogs the filter, thus raising the exhaust-gas backpressure. To reduce the pressure rise, the ash storability of ceramic octosquare filters is increased by making the cross-sections of the inlet channels as large as possible. This provides the filters with sufficient capacity to accommodate all the ash residue occurring on burnoff during the normal service life of the vehicle.

With conventional ceramic filters, it is assumed that the filter has to be removed and mechanically cleaned approximately every 120,000 km when additive-based regeneration is used.

Catalyzed diesel particulate filter
Soot-particle burnoff can also be improved by coating the particulate filter with noble metals (mainly platinum). However, the effect is less than when using an additive.

The catalyzed diesel particulate filter (CDPF) requires further measures for regeneration to raise the exhaust-gas temperature, similar to the measures taken with the additive system. Compared with the additive system, however, the catalyzed coating has the advantage that no additive ash occurs in the filter.

The catalyzed coating fulfills several functions: The oxidation of CO and HC, the oxidation of NO to NO_2 and the oxidation of CO to CO_2.

CRT system
Truck engines run close to maximum torque more frequently than car engines, i.e. causing comparatively high NO_x emissions. On trucks, therefore, it is possible to perform continuous regeneration of the particulate filter based on the CRT principle (Continuously Regenerating Trap).

According to this principle, soot combusts with NO_2 at temperatures as low as 300 to 450 °C. The process is reliable at these temperatures if the mass ratio between NO_2 and soot is greater than 8 : 1. To use this process, an oxidation catalytic converter, which oxidizes NO into NO_2, is located upstream of the particulate filter. In most cases, this provides ideal conditions for regeneration using the CRT system on trucks at normal operation. The method is also termed "passive regeneration", since soot is burned continuously without the need for active measures.

Figure 5: Particulate filter with additive system
1 Additive control unit, 2 Engine control unit,
3 Additive pump, 4 Fill level sensor, 5 Additive tank,
6 Additive injection nozzle, 7 Fuel tank,
8 Diesel engine, 9 Oxidation catalytic converter,
10 Particulate filter, 11 Temperature sensor,
12 Differential pressure sensor.

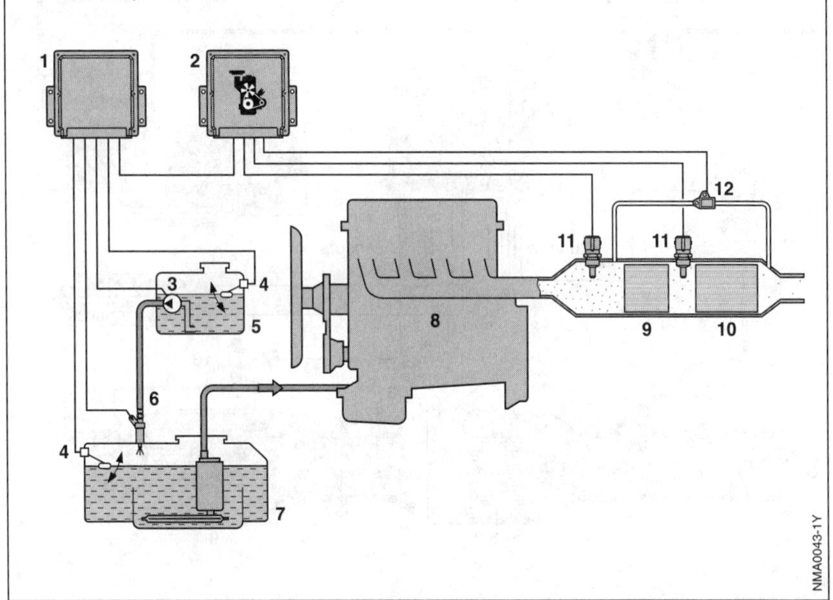

HCI system

To actively regenerate particulate filters, it is necessary to increase the temperature in the filter to over 600 °C. This can be achieved by means of internal engine settings. In the case of unfavorable application – e.g. if the distance between particulate filter and engine is too large – the internal engine measures become very complicated. In this case, an HCI system (hydrocarbon injection) is used, whereby diesel fuel is injected and vaporized upstream of a catalytic converter (Figure 6), and then catalytically combusted in the latter. The heat generated during combustion is used to regenerate the downstream particulate filter.

The control algorithms for the HCI system are integrated into the separate dosing control unit or optionally into the engine ECU. An important input variable is the saturation of the particulate filter.

Saturation detection

Two processes are used in parallel for particulate-filter saturation detection. Flow resistance in the filter is calculated from the pressure drop across the filter and volumetric flow. This is a measure of filter permeability, and thus soot mass.

In addition, a model is used to calculate the soot mass stored in the particulate filter. The soot-mass flow of the engine is integrated into this model. Corrections for system dynamics, such as the residual oxygen part in the exhaust gas, are taken into consideration, as well as continuous particulate oxidation by NO_2. During thermal regeneration, soot burnoff is calculated in the control unit as a factor of particulate-filter temperature and oxygen mass flow.

Soot mass is calculated by a coordinator using the soot masses determined in both processes, and this becomes the key factor in the regeneration strategy.

Figure 6: HCI system (hydrocarbon injection)
1 Fuel pump, 2 Fuel tank, 3 Temperature sensor, 4 HC dosing module,
5 HC metering unit, 6 Fuel filter, 7 Engine ECU,
8 Diesel oxidation catalytic converter, 9 Diesel particulate filter,
10 Differential-pressure sensor.

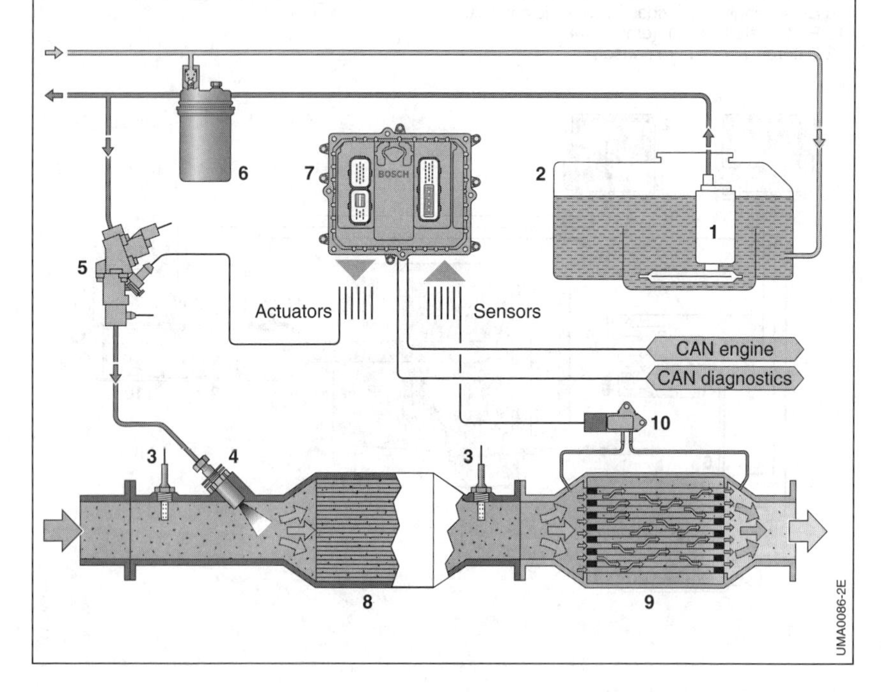

Hybrid drives

Features

A hybrid electric vehicle (HEV) uses both an internal-combustion engine and at least one electric machine for its means of propulsion. There are in this respect a multitude of drive structures which partly pursue different optimization objectives and which utilize to differing extents electrical energy to drive the vehicle. There are essentially three objectives being pursued by hybrid electric drives: reduced fuel consumption (CO_2 emissions), reduced exhaust emissions, and increased torque and power to improve driving dynamics.

Hybrid vehicles require an electric energy accumulator to supply the electric drive. Current solutions use a nickel-metal-hydride or lithium-ion traction battery at a comparatively high voltage level in the range of 200 to 400 V.

The electric drive consists of an electric machine and a pulse-controlled inverter. The electric machines used are generally permanent-field synchronous machines with high power density. The electric drive offers constantly high torques at low rotational speeds. In this way, it ideally supplements the internal-combustion engine, whose torque only starts to increase at mid-range rotational speeds. The electric drive and internal-combustion engine together are thus able to deliver a high dynamic response from every driving situation (Figure 1).

Figure 1: Torque curve of different vehicle drives
1 Hybrid drive, consisting of 3 and 4,
2 Standard IC engine with 1.6 *l* displacement,
3 Turbocharged IC engine with 1.2 *l* displacement,
4 Electric machine, 15 kW.

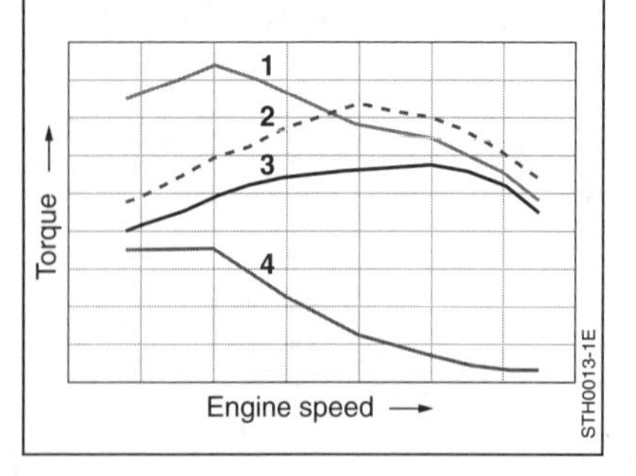

The combination of electric and combustion-engine drives has the following advantages over a conventional drivetrain:
- The assistance provided by the electric drive makes it possible to operate the internal-combustion engine predominantly in the range of its best efficiency or in ranges in which only low pollutant emissions occur (operating-point optimization occurs).
- The combination with an electric drive facilitates the use of a smaller internal-combustion engine while retaining the same overall power output (power-neutral downsizing).
- Furthermore, a higher-geared transmission can be used while retaining the same levels of driving performance. This shifts the combustion engine's operating points into ranges with improved efficiency (downspeeding).
- Through generator operation of the electric machine, it is possible when braking to convert part of the vehicle's kinetic energy into electrical energy. The electrical energy is stored in the energy accumulator and can be used at a later stage to drive the vehicle.
- In certain drive structures the electric drive can be use for purely electric driving. Here the internal-combustion engine is switched off and the vehicle is locally operated with zero emissions.

Functions

Depending on the operating state and required drive power, the internal-combustion engine and the electric drive contribute to propelling the vehicle to different extents. The hybrid control system determines the distribution of power between the two drives. The way in which the internal-combustion engine, the electric drive and the energy accumulator (battery) interact determines the different functions.

Start/stop function

The start/stop function switches the internal-combustion engine off temporarily without the driver having to turn the ignition key. The engine is typically switched off when the vehicle is stationary. It is then restarted automatically as soon as the driver is ready to resume driving.

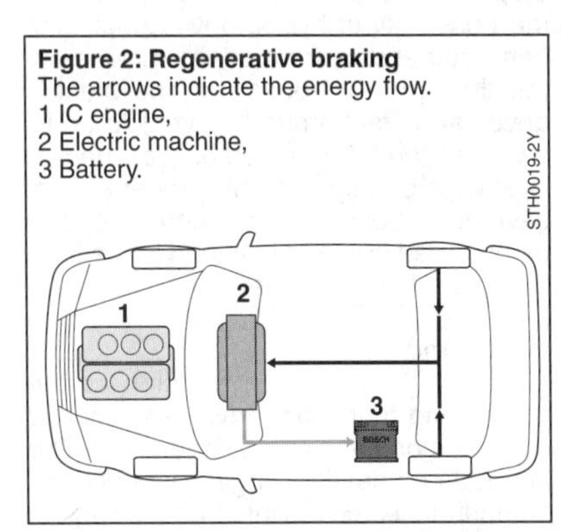

Figure 2: Regenerative braking
The arrows indicate the energy flow.
1 IC engine,
2 Electric machine,
3 Battery.

Regenerative braking

During regenerative braking the vehicle is not – or not only – braked by the service brake's friction torque, but instead by a generator braking torque of the electric machine. Here it converts the vehicle's kinetic energy into electrical energy, which is stored in the energy accumulator (Figure 2). Regenerative braking is also known as recuperative braking, recuperation or regeneration.

Hybrid driving

Hybrid driving refers to those states in which the internal-combustion engine and the electric machine exert the drive torque together. Hybrid driving can be further subdivided into generator and motor modes of the electric machine. The electric energy accumulator is charged in generator mode (Figure 3). For this purpose the internal-combustion engine is operated in such a way as to deliver a greater amount of power than is needed for the desired propulsion of the vehicle. The excess amount of power is fed to the electric machine and converted into electrical energy, which is stored in the energy accumulator.

In motor mode (Figure 4) the electric energy accumulator is discharged. The electric drive assists the internal-combustion engine in providing the desired propulsive power.

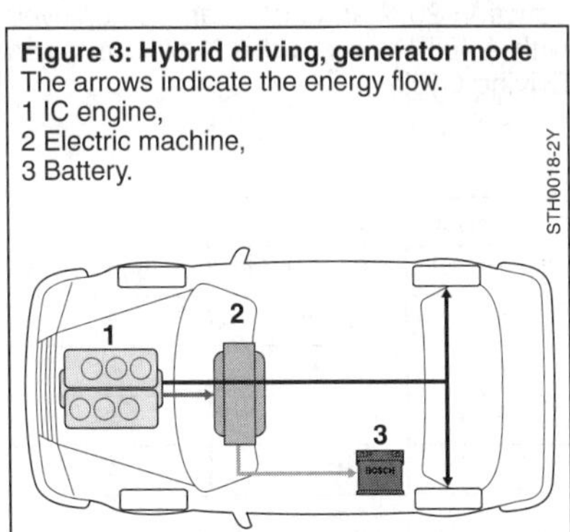

Figure 3: Hybrid driving, generator mode
The arrows indicate the energy flow.
1 IC engine,
2 Electric machine,
3 Battery.

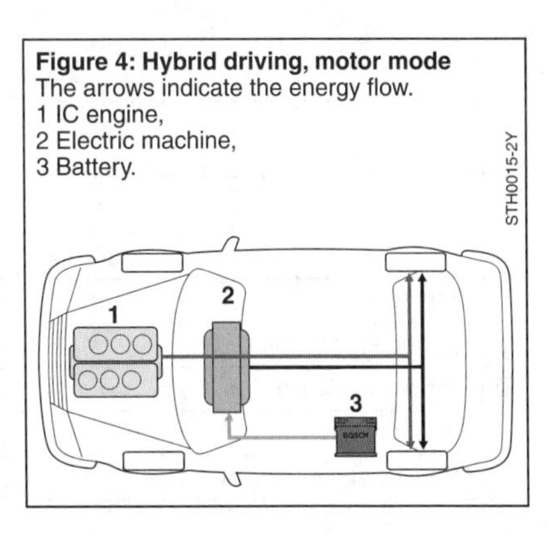

Figure 4: Hybrid driving, motor mode
The arrows indicate the energy flow.
1 IC engine,
2 Electric machine,
3 Battery.

Purely electric driving

In the case of purely electric driving, the vehicle is driven solely by the electric drive. To this end the internal-combustion engine is decoupled from the vehicle drive and switched off (Figure 5). In this operating mode the vehicle can run virtually noiselessly and locally without emissions.

Recharging at the power socket

In the case of recharging at the power socket, the vehicle can be connected via a charger to the mains power supply and thereby the electric energy accumulator recharged (plug-in hybrid).

Figure 5: Purely electric driving
The arrows indicate the energy flow.
1 IC engine,
2 Electric machine,
3 Battery.

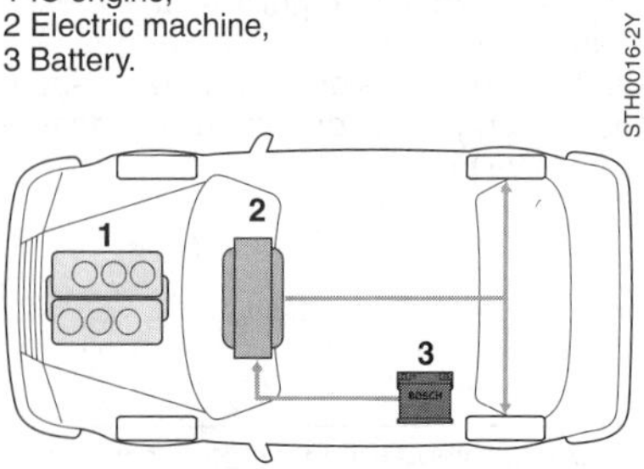

STH0016-2Y

Functional classification

Hybrid vehicles can be divided into different categories based on the effected functions (Table 1).

Start/stop system

A start/stop system effects the "start/stop" and regenerative-braking functions. Alternator control in the conventional vehicle is adapted for this purpose. In normal driving mode the alternator operates with low output. In overrun phases the alternator output is increased in order to utilize a greater proportion of the vehicle deceleration to "generate power". Fuel savings of between 4 % and 5 % can be achieved with a start/stop system in the New European Driving Cycle (NEDC).

Mild hybrid

The mild hybrid offers, as well as the start/stop function and regenerative braking, the possibility of hybrid driving including generator and motor modes. Purely electric driving is not possible. The electric drive can in fact propel the vehicle on its own, but the internal-combustion engine is always engaged in the process. Fuel savings of between 10 % and 15 % can be achieved with a mild hybrid in the New European Driving Cycle.

Full hybrid

The full hybrid can, in addition to the mild-hybrid functions, drive over shorter distances on the electric drive alone. The internal-combustion engine is switched off during electric driving. Fuel savings of between 20 % and 30 % can be achieved with a full hybrid in the New European Driving Cycle.

Table 1: Functions and hybrid systems

		Hybrid system			
		Start/stop	Mild hybrid	Full hybrid	Plug-in hybrid
Function	Start/stop function	●	●	●	●
	Regenerative braking	●	●	●	●
	Electric assistance		●	●	●
	Electric driving			●	●
	Charging at power socket				●

Plug-in hybrid

Full hybrids can alternatively also be designed as plug-in hybrids. These offer the possibility of charging the traction battery from the power socket via a corresponding charger. It is useful here to use a battery with a higher energy content so that longer distances can be covered by purely electric driving. As a rule the power output of the electric drive is increased to such an extent as to facilitate normal driving on the electric drive alone. Fuel savings of between 50 % and 70 % can be achieved with a plug-in hybrid in the New European Driving Cycle. The figures are of this magnitude because some of the energy for propelling the vehicle comes from the mains power supply and is not directly attributed to the fuel consumption.

Drive structures

There are different ways of configuring the internal-combustion engine, transmission and electric machines in hybrid vehicles. The different drive structures can be divided using possible energy flows into the following three categories: parallel, series, and power-split hybrid drives.

Parallel hybrid drive

In parallel hybrid drives the internal-combustion engine and an electric machine contribute to driving the vehicle independently of one another. The two energy flows from the internal-combustion engine and the battery therefore run parallel to each other, the two individual components of power adding up to a total drive power. Parallel hybrid drives come in a mild-hybrid variant (with start/stop function, regenerative braking and hybrid driving) or in a full-hybrid variant (additionally with electric driving).

A fundamental advantage of the parallel hybrid is the possibility of maintaining the conventional drivetrain in wide ranges. The extent of development and installation work for parallel drive structures is lower when compared with series and power-split structures, in that for the most part only one electric machine with lower electric power is required and the adaptations necessary in converting a conventional drivetrain are fewer.

Figure 6: Parallel hybrid with one clutch
a) Drive structure, b) Energy flow.
1 IC engine, 2 Tank, 3 Electric machine, 4 Clutch, 5 Transmission,
6 Final-drive unit, 7 Pulse-controlled inverter, 8 Battery.

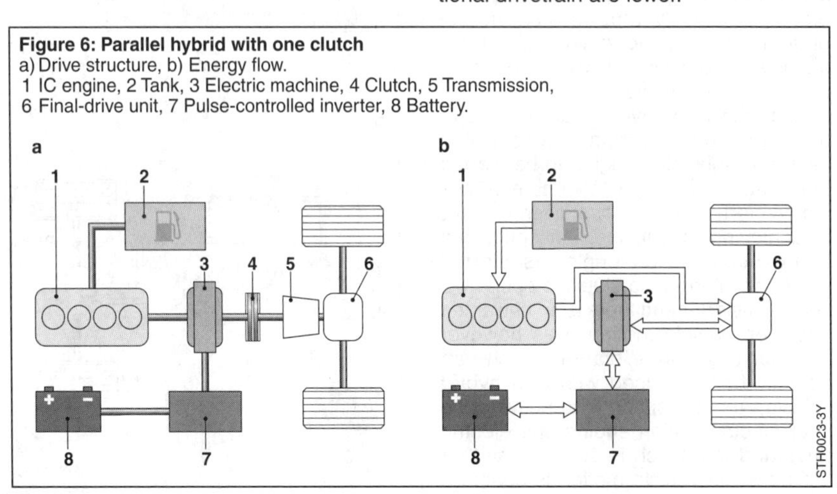

Parallel hybrid with one clutch
In the variant shown in Figure 6 the electric machine is connected directly to the internal-combustion engine. In contrast to series and power-split drive structures, the speed of the internal-combustion engine cannot be adjusted independently of the speed of the electric machine. In vehicle deceleration phases the internal-combustion engine cannot be decoupled from the electric machine and is therefore always engaged. The associated drag torque of the engine reduces the potential for regenerative braking.

Purely electric driving is not possible with this drive structure. The electric drive can in fact be used as the sole drive source, but the internal-combustion engine is always engaged when driving. The electric drive can be used to assist the internal-combustion engine and thereby improve dynamic performance.

Parallel hybrid with two clutches
A parallel full hybrid can be structured in several ways. The obvious structure is the following extension (Figure 7): integrated between the internal-combustion engine and the electric machine is an additional clutch which allows the engine to be cut in and out as required. Purely electric driving is therefore possible. The internal-combustion engine can also be decoupled in deceleration phases. Firstly, it increases the potential for regenerative braking. Secondly, it permits sailing operation, whereby the vehicle rolls freely and is slowed down only by aerodynamic drag and rolling friction.

To make this drive structure acceptable, it is very important to enable the internal-combustion engine to be started from electric driving without compromising comfort. There are two different ways of achieving this. In the first option the internal-combustion engine is started when the interrupting clutch is open by a separate starter and there is no unwanted effect on vehicle motion. This however calls for a separate starter, which can actually be dispensed with in a hybrid vehicle. Another option is to activate the internal-combustion engine, the electric drive and the clutch in such a way that the effect on vehicle motion is compensated for during engine starting. To this end an intelligent control system requires access to measured values from the internal-combustion engine, the electric drive and the clutch. The clutch must be able to adapt automatically to the changing conditions in ongoing operation and to follow the control-system inputs.

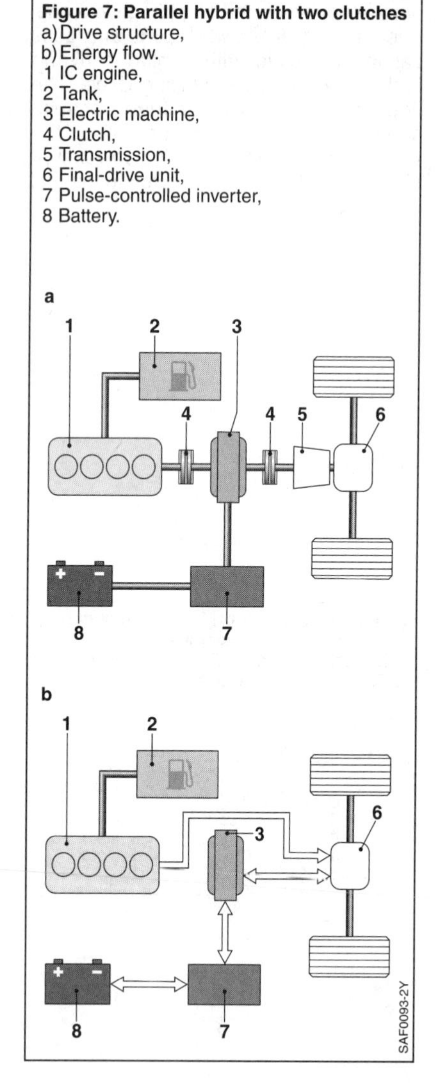

Figure 7: Parallel hybrid with two clutches
a) Drive structure,
b) Energy flow.
1 IC engine,
2 Tank,
3 Electric machine,
4 Clutch,
5 Transmission,
6 Final-drive unit,
7 Pulse-controlled inverter,
8 Battery.

Parallel hybrid with double-clutch transmission
Installing the additional clutch between the internal-combustion engine and the electric machine results in the drivetrain being increased in length. In some vehicles the installation space required for this drive configuration is not available. This problem can be remedied by integrating the electric machine in a double-clutch transmission (Figure 8). The electric machine is no longer connected to the crankshaft of the internal-combustion engine, but is instead connected to a subunit of the double-clutch transmission. In this configuration there is no additional clutch between the engine and the electric machine. Purely electric driving with the engine stopped is possible by opening the double clutch of the transmission. Therefore this configuration is also a parallel full hybrid. Depending on the gear selected in the transmission subunit, with the electric machine a different transmission ratio is possible between the engine and the electric machine. This produces an additional degree of freedom for the hybrid control, which can be used to further reduce fuel consumption.

Figure 8: Parallel hybrid with double-clutch transmission
a) Drive structure,
b) Energy flow.
1 IC engine,
2 Tank,
3 Clutch,
4 Double-clutch transmission,
5 Final-drive unit,
6 Electric machine,
7 Pulse-controlled inverter,
8 Battery.

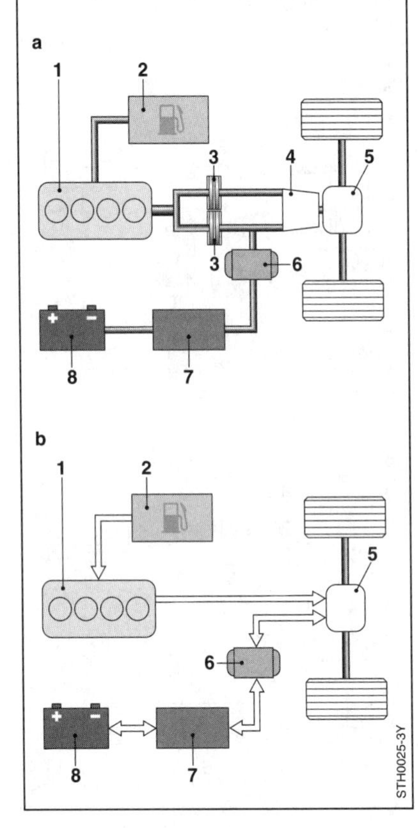

Axle-split parallel hybrid
Another parallel drive structure is produced by electrifying a separate axle (Figure 9). Here, a conventional drivetrain comprising internal-combustion engine and transmission on a powered axle is combined with an electrically powered axle. The drive configuration becomes a full hybrid as soon as the internal-combustion engine can be switched off and decoupled while the electric drive propels the vehicle. A semi-automatic transmission and a start/stop system for the

Figure 9: Electrification of a separate axle (axle-split parallel hybrid)
1 IC engine,
2 Tank,
3 Electric machine,
4 Pulse-controlled inverter,
5 Battery.

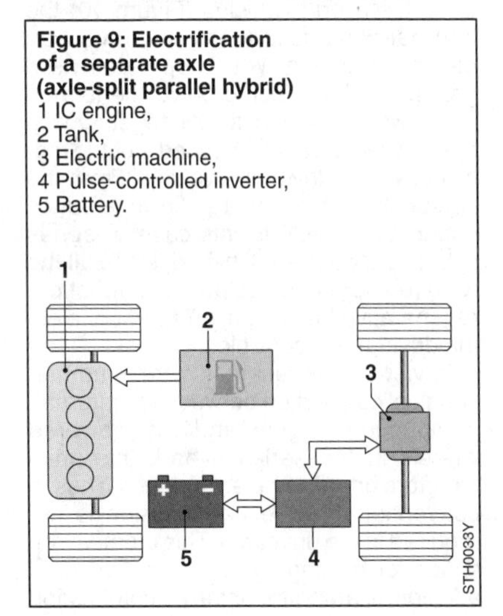

internal-combustion engine are required for this purpose. This drive structure is classed as a parallel hybrid drive because the individual components of power from the engine and the electric drive add up. In contrast to the drive structures described above, the addition point is not within the drivetrain, but on the plane of the powered wheels.

In this case the traction battery is recharged by regenerative braking. The traction battery cannot be recharged when the vehicle is stationary. The interaction of the internal-combustion engine and the electric drive can deliver an all-wheel drive to the vehicle. Distribution of the drive torques can be adjusted by specific activation of the electric drive within wide limits. Permanent all-wheel drive can however only be realized when not only the original electric drive is supplied by the battery, but also a second electric machine can deliver the required electrical energy.

A second electric machine directly connected to the internal-combustion engine (either on the crankshaft or in the belt drive) means not only that permanent all-wheel drive is realized but also that the battery is recharged even when the vehicle is stationary.

Series hybrid drive

In series hybrid vehicles (Figure 10) the internal-combustion engine drives an electric machine, which operates as a generator. The electric power generated in this way is made available together with the battery power to a second electric machine, which drives the vehicle. From the perspective of the energy flows, a series connection exists in this case. A series hybrid is always a full hybrid, since all the required functions (start/stop function, regenerative braking, hybrid driving, electric driving) are possible.

In view of the fact that there is no mechanical connection between the internal-combustion engine and the powered wheels in the series hybrid, this drive structure offers some advantages. Thus, a conventional range transmission is not required in the drivetrain. This creates free space for packaging the overall drive. In addition, starting the internal-combustion engine from electric driving does not cause any unwanted effect on vehicle motion. The chief advantage when driving is the ability to freely select the operating point of the engine. This supports fuel-saving and low-emission operational management of the vehicle. The engine can also be optimized to a limited operating range.

A downside of the series hybrid is that electrical energy has to be converted twice. The losses incurred through double energy conversion are higher than in the case of a purely mechanical transfer by

Figure 10: Series hybrid drive
a) Drive structure,
b) Energy flow.
1 IC engine,
2 Tank,
3 Electric machine,
4 Final-drive unit,
5 Pulse-controlled inverter,
6 Battery.

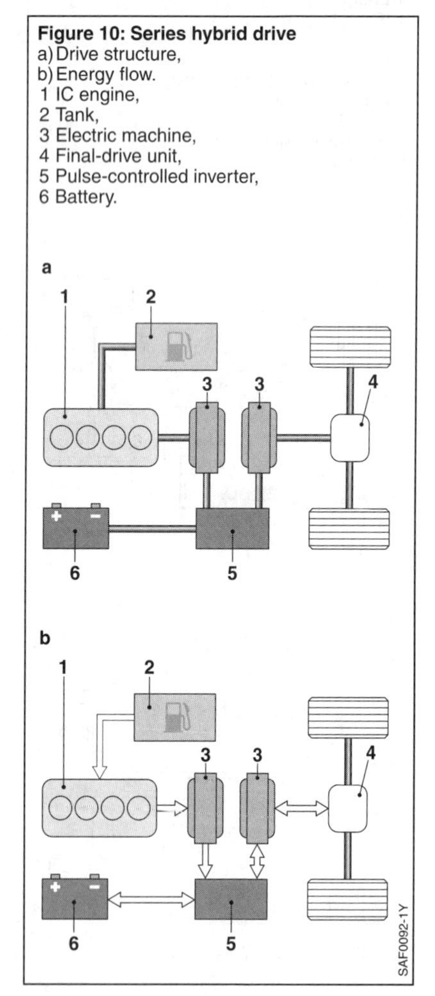

a transmission. Furthermore, two electric machines of the same power category as the internal-combustion engine are needed to transmit the engine's power.

At low speeds a series hybrid, despite the higher losses, offers a fuel-consumption advantage because here the positives afforded by the ability to freely select the engine's operating point outweigh the negatives. The higher losses are predominant at medium and higher speeds.

Series hybrids are currently used predominantly in diesel-electric locomotives and city buses. In the passenger-car sector series drive structures are being encountered more and more frequently in electric vehicles whose ranges are extended where necessary by an internal-combustion engine as a "range extender".

Series-parallel hybrid

A series hybrid is extended to become a series-parallel hybrid (Figure 11) by establishing a mechanical connection between the two electric machines which is connected or separated by a clutch. The series-parallel hybrid can utilize the advantages of the series hybrid at low speeds and circumvent the disadvantages at higher speeds by closing the clutch. When the clutch is closed, the series-parallel hybrid behaves like a parallel hybrid. Since the double energy conversion is limited to the range of lower speeds and power outputs, smaller electric machines are sufficient for the series-parallel hybrid than are required for the series hybrid. Compared with the series hybrid, the advantage in packaging is lost on account of the mechanical connection between the internal-combustion engine and the powered wheels. Compared with the parallel hybrid, two electric machines are required for the same task.

Power-split hybrid drive

Power-split hybrid vehicles combine features of parallel and series hybrid vehicles with those of a power split. Some of the engine power is converted by a first electric machine into electric power, the remainder – together with a second electric machine – driving the vehicle. A power-split hybrid is always a full hybrid, since all the required functions (start/stop function, regenerative braking, hybrid driving, electric driving) are possible.

Figure 11: Series-parallel hybrid drive
a) Drive structure,
b) Energy flow.
1 IC engine,
2 Tank,
3 Electric machine,
4 Clutch,
5 Final-drive unit,
6 Pulse-controlled inverter,
7 Battery.

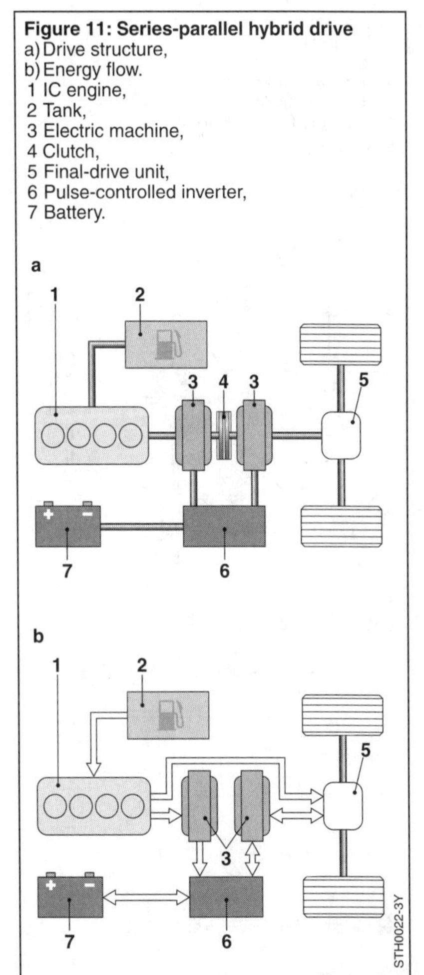

The design is shown in Figure 12. The central element is a planetary-gear set, which is connected via its three shafts to the internal-combustion engine and two electric machines. Because of the kinematic boundary conditions at the planetary-gear set, the engine speed can be adjusted within certain limits independently of the vehicle speed. Following a continuously variable transmission (CVT), the terminology used here is an electric continuously variable transmission (ECVT).

Figure 12: Power-split hybrid drive
a) Drive structure,
b) Energy flow.
1 IC engine,
2 Tank,
3 Planetary-gear set,
4 Electric machine,
5 Pulse-controlled inverter,
6 Battery.

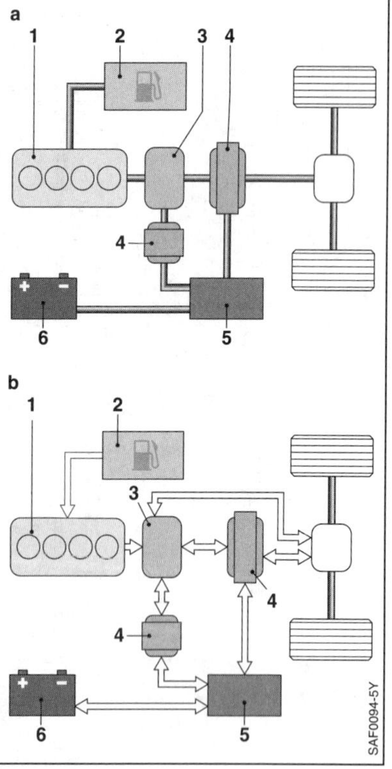

Some of the engine power is transmitted by the planetary-gear set via a mechanical path to the powered wheels. The remainder of the power is transmitted via an electrical path with double energy conversion to the powered wheels. Similarly to the series hybrid, the electric transmission path can be used in the case of low power demands. For higher power demands, the mechanical transmission path is also available. However, it is not possible to switch arbitrarily between the mechanical and electrical transmission paths. Depending on the configuration of the planetary-gear set, the electric machine and the internal-combustion engine, without additional transmissions only certain combinations between the mechanical and electrical transmission paths are ever possible. In this way, the power-split hybrid can achieve substantial fuel savings at low and medium speeds. No additional fuel savings can be achieved at high speeds.

Similar to the series hybrid, the power-split hybrid requires electric machines with relatively high power outputs in the range of the installed engine power.

The power-split hybrid can be expand by mechanical, fixed gears by using a second planetary-gear set. The mechanical complexity increases, while the electrical complexity decreases. Smaller electric machines are then sufficient for a comparable concept. Furthermore, fuel consumption at medium and higher speeds can be improved.

Control of hybrid vehicles

The efficiency which can be achieved with the relevant hybrid drive is crucially dependent on the higher-level hybrid control. Figure 13 uses the example of a vehicle with a parallel hybrid drive to show the function and software structure and the networking of the individual components and ECUs in the drivetrain. The higher-level hybrid control coordinates the entire system, the subsystems of which have their own control functions. These are battery management, engine management, management of the electric drive, transmission management, and management of the brake system. In addition to pure control of the subsystems, the hybrid control also includes an operating strategy which optimizes the way in which the drivetrain is operated. The operating strategy brings influence to bear on the consumption- and emission-reducing functions of the hybrid vehicle, i.e. on start-stop operation of the engine, regenerative braking, and hybrid and electric driving.

Operating strategies for hybrid vehicles

The operating strategy determines how the drive power is shared between the engine and the electric drive. It decides the extent to which the potentials for fuel saving and for reducing the emissions of a vehicle are utilized. The operating strategy must also implement the different hybrid functions such as regenerative braking and hybrid and electric driving.

Selection of and switching between the individual states is performed in accordance with numerous conditions relating to, for example, the accelerator-pedal position, the state of charge of the battery, and the vehicle's current speed. The components in the hybrid vehicle behave

Figure 13: Networking of control systems in the drivetrain
a) Function and software structure,
b) Drivetrain components and associated ECUs.

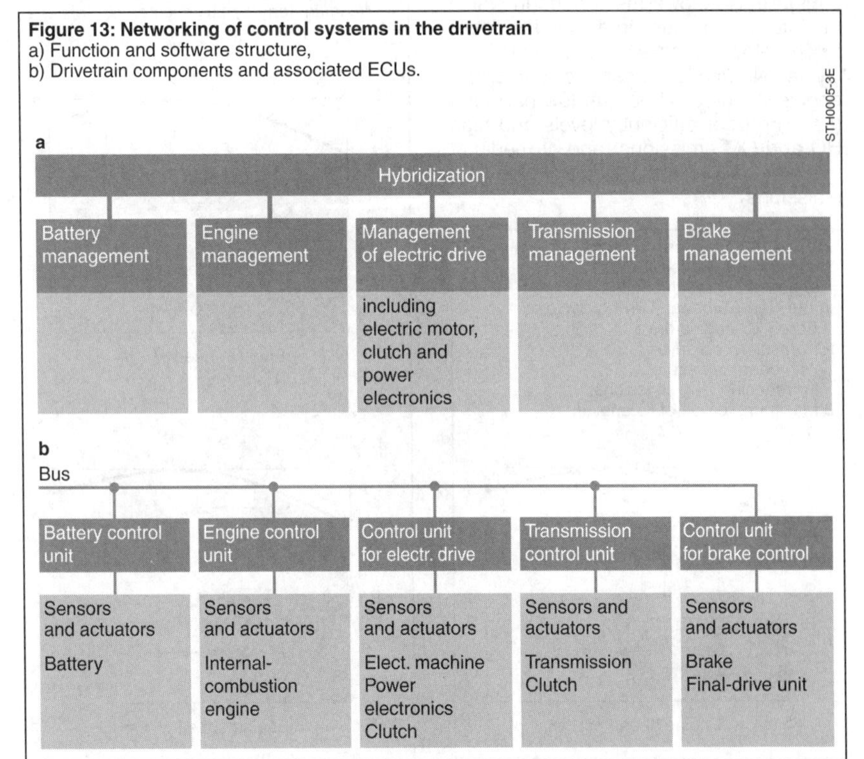

differently, depending on the optimization objective (e.g. fuel saving or emission reduction).

Operating strategy for reducing NO_x

Vehicles with lean-running internal-combustion engines already achieve relatively low consumption values in part-load operation. However, in part-load operation the influence of friction loss is huge, so much so that even the specific fuel consumption is high. In addition, low combustion temperatures and a local oxygen deficiency in the low part-load range result in high carbon-monoxide and hydrocarbon emissions. Already an electric drive with relatively low power can replace the internal-combustion engine as drive in the low load range. If the required electrical energy can be obtained through regenerative braking, this simple strategy can deliver a huge benefit with regard to fuel consumption and emissions.

Figure 14 shows the ranges in which the internal-combustion engine is primarily operated in the New European Driving Cycle (NEDC). A passenger-car diesel engine is operated both at low part load (i.e. with poor efficiency levels and high HC and CO emissions) and at medium/higher load (i.e. in the range of high NO_x emissions).

Figure 14: Operating strategy for reducing NO_x emissions
Ranges of operating points in the New European Driving Cycle.
1 Purely IC-engine drive,
2 Parallel hybrid drive with operating strategy for reducing NO_x emissions,
3 Maximum torque of IC engine.

Torque
Engine speed
STH0006-5E

Figure 14 shows furthermore the range of operating points for a parallel hybrid which bypasses low internal-combustion engine loads through purely electric driving or through load-point increase. This on the one hand reduces the fuel consumption, but on the other hand reduces the CO, HC and NO_x emissions – which are high in this range. To achieve a further lowering of NO_x emissions, it is possible to lower the load points in the medium load range by simultaneously operating the electric drive and the engine.

Figure 15: Operating strategy for reducing CO_2 emissions
Ranges of operating points in the New European Driving Cycle.
a) Comparison of a purely IC-engine drive with parallel hybrid drive,
b) Comparison of a purely IC-engine drive with a power-split drive.
1 Purely IC-engine drive,
2 Parallel hybrid drive,
3 Power-split hybrid drive,
4 Maximum torque of IC engine.

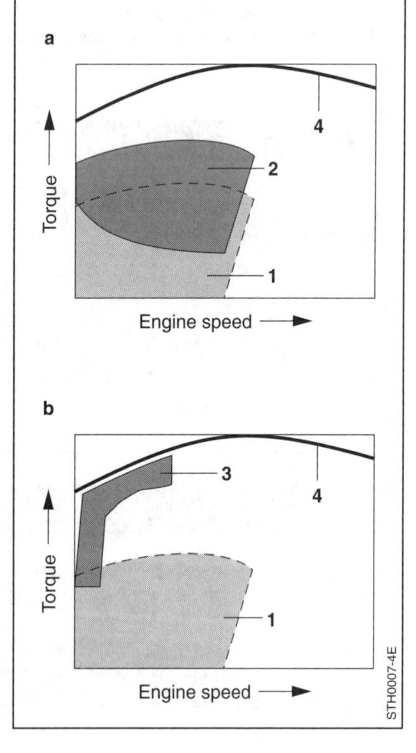

Operating strategy for reducing CO_2

With vehicles with stoichiometrically running gasoline engines, extremely low emission values can be realized on account of the three-way catalytic converter used. In these vehicles the focus is on reducing the fuel consumption and thus also the CO_2 emissions. Figure 15 shows for different drive structures a possible optimization of the engine's operating range with regard to minimal CO_2 emissions.

In the New European Driving Cycle (NEDC) internal-combustion engines in conventional vehicles are operated at low part load and thus with poor efficiency. In vehicles with parallel hybrid drives low engine loads can be avoided by means of purely electric driving (Figure 15a). Since the required electrical energy as a rule cannot be obtained exclusively through recuperation, the electric machine is then operated as a generator. This results, when compared with a conventional vehicle, in a shift in engine operation to higher loads and thus in better levels of efficiency.

In the case of the power-split hybrid vehicle (Figure 15b), the operating range of the internal-combustion engine is subject to greater limitations than the parallel hybrid vehicle. As a rule it is operated as a function of engine speed at the load at which the entire drivetrain operates under optimum energy conditions.

Regenerative braking system

In the case of regenerative braking (also known as recuperation), the vehicle's kinetic energy during deceleration is converted by the electric machine, which is operated as a generator for this purpose, into electrical energy. In this way, some of the energy which is normally lost as frictional heat during braking is fed in the form of electrical energy to the battery and then utilized.

Drag-torque simulation

A simply way of realizing regenerative braking is drag-torque simulation. Here the electric machine is operated as a generator as soon as the driver releases the accelerator. The driver does not need to press the brake pedal for this purpose. In the case of a full hybrid, the internal-combustion engine is decoupled and the electric machine exerts a generator torque of the magnitude of the engine's drag torque. If the engine cannot be decoupled (as in the case of a mild hybrid), alternatively a lower generator torque can be exerted on the drivetrain in addition to the engine drag torque (drag-torque increase). The vehicle behavior changes only slightly compared with a non-hybrid vehicle.

Regenerative braking system

During braking the electric machine can exert an additional generator torque as well as drag-torque simulation and increase. The vehicle decelerates with the brake pedal in the same position faster than a comparable conventional vehicle. The available generator torque depends on the driving speed, the selected gear and the state of battery charge. Therefore the vehicle can be subject to different levels of braking response, even with the brake pedal in the same position. This difference in braking response is perceived by the driver to be all the more disruptive, the greater the generator torque is part of the vehicle deceleration. For this reason, only small power levels can be recuperated with this simple regenerative braking system.

Cooperatively regenerative braking system

The service-brake system must be modified in the case of higher decelerations in order to further exploit the kinetic energy. To this end all or some of the service-brake friction torque must be replaced with a regenerative braking torque without the vehicle deceleration changing when the brake-pedal position and force are kept constant. This is realized in a cooperatively regenerative braking system, where the vehicle control and the braking system interact in such a way that constantly as much friction braking torque is recovered as generator braking torque can be replaced by the electric machine.

Fuel cells for the vehicle drive

Fuel cells are electrochemical converters which convert the chemical energy contained in the fuel directly into electrical energy. In a hydrogen/oxygen fuel cell, the hydrogen reacts with oxygen to create water, producing electrical energy in the process.

Fuel-cell drives are very appealing within the field of electromobility because they permit the accustomed convenience of conventional vehicles such as short refueling time and long range with expanded H_2 infrastructure. They allow for zero local emissions and cause little to no CO_2 emissions in the overall chain when using hydrogen from renewable energy sources.

Figure 1: Functioning principle of the PEM fuel cell
a) Hydrogen oxidation,
b) Oxygen reduction,
c) Water production.
1 Anode,
2 Membrane,
3 Cathode.

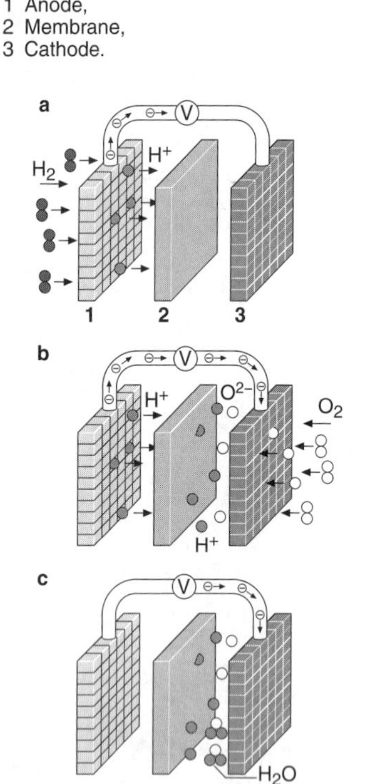

Functioning principle

A fuel cell consists of two electrodes (anode and cathode) which are separated from each other by an electrolyte. The electrolyte is not permeable for ions. The electrodes are connected to each other by way of an external electric circuit.

Polymer electrolyte membrane fuel cells (PEM-FC) are mostly used for mobile applications (Figure 1 and Figure 2). The functioning principle of the fuel cell is described in the following using this type.

Functioning principle of the PEM fuel cell

In the PEM fuel cell, hydrogen is directed to the anode, where it is oxidized. H^+ ions and electrons are created (Figure 1a).

Anode: $2\,H_2 \rightarrow 4\,H^+ + 4\,e^-$.

The electrolyte is fashioned as a proton-conducting polymer membrane. It is permeable for protons, but not for electrons. The H^+ ions (protons), which are formed on the anode, pass through the membrane and reach the cathode. The polymer membrane must be sufficiently moist-

Figure 2: Structure of the PEM fuel cell
1 Membrane (electrolyte layer),
2 Electrodes (catalyst layer),
3 Gas-diffusion layers,
4 Seals,
5 Bipolar plates.

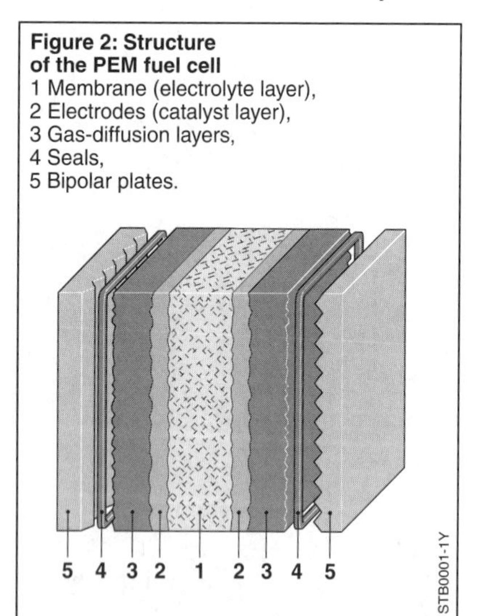

ened for this proton conduction. Air is directed onto the cathode – in automotive applications, to simplify matters, air is used with a oxygen content of around 21 %. Oxygen (O_2) is reduced at the cathode (Figure 1b). Reduction occurs using the electrons, which pass from the anode via the external circuit to the cathode.

Cathode: $O_2 + 4\,e^- \rightarrow 2\,O^{2-}$.

In a further reaction stage, the O^{2-} ions and the protons react to form water.

Cathode: $4\,H^+ + 2\,O^{2-} \rightarrow 2\,H_2O$.

The overall reaction of the fuel cell results in the transformation of hydrogen and oxygen into water (Figure 1c). Unlike an electrolytic-gas reaction, in which hydrogen and oxygen react with each other explosively, the transformation here takes place in a controlled reaction, since the reaction steps occur in separate spaces on the anode and cathode.

Overall reaction: $2\,H_2 + O_2 \rightarrow 2\,H_2O$.

The reactions described occur on catalysts in the fuel-cell electrodes. Platinum is frequently used for this.

The voltage of an individual hydrogen/oxygen fuel cell is theoretically 1.229 V at a temperature of 25 °C. This value is obtained from the standard electrode potentials (see Electrochemistry). This open-circuit voltage is however not reached during operation. Voltage losses can be put down, for example, to internal resistance or to gas-diffusion limitations (Figure 3). Essentially, the voltage is dependent on the temperature, the stoichiometric ratios of hydrogen and oxygen to the quantity of electricity produced, the hydrogen and oxygen partial pressures, and the current density.

In order to achieve the high voltages required for a technical application, individual cells are electrically connected in series to form a stack (Figure 4). These fuel cell stacks are made up of approximately 40 to 450 cells so that the maximum operating voltages are between 40 V and 450 V.

Stacks in the power range of 5 to 120 kW are typically used for automotive applications. High currents are achieved by appropriately sized membrane surfaces. Currents up to 500 A flow in automotive applications.

Figure 3: Electrical characteristic quantities of the fuel cell
1 Voltage,
2 Power density.

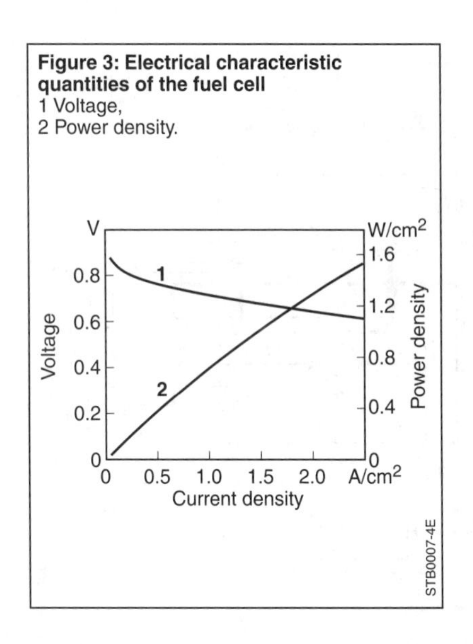

Figure 4: Structure of a fuel-cell stack
1 End plate
2 Membrane electrode unit with membrane, catalyst material and seals
3 Bipolar plates
4 Gas channels
5 Repeating units
6 End plate

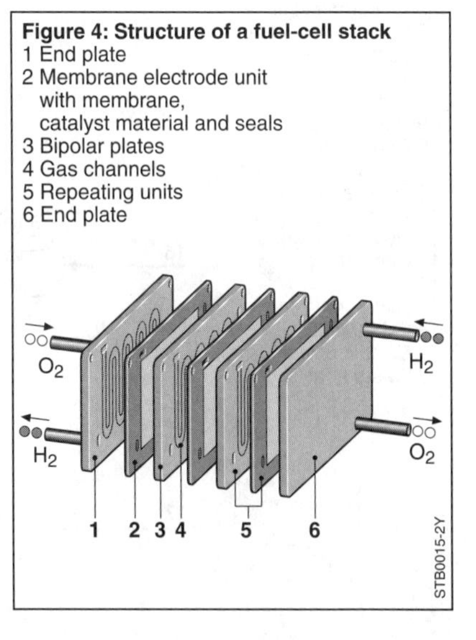

Functioning principle of the fuel-cell system

Subsystems for oxygen and hydrogen supply and for thermal conditioning are required to operate a fuel-cell stack (Figure 5). There are, in principle, many ways of realizing these subsystems. The description given here corresponds to the state of the art of all major vehicle manufacturers.

Hydrogen supply

The hydrogen is stored in a 700 bar high-pressure tank. It is expanded to approximately 10 bar via a pressure reducer and fed to the anode via the hydrogen gas injector.

The hydrogen gas injector is an electrically actuated valve used to set the pressure of the hydrogen on the anode side. In contrast to fuel injectors in internal-combustion engines, the hydrogen gas injector must set continuous mass flows. A typical flow value for 100 kW is 2.1 g/s hydrogen. The pressure to be adjusted is max. 2.5 bar.

The anode side of the membrane has to be supplied with hydrogen everywhere equally. This is ensured by recirculating the hydrogen.

Disruptive foreign gases on the anode side are regularly removed via an electrically actuated valve, the drain valve. This prevents an enrichment of foreign gases from the fuel tank or of diffusion gases from the cathode side (e.g. nitrogen, water vapor). The valve is mounted on the anode side on the stack outlet. The valve open at zero current is also used to discharge unwanted quantities of water in the anode path.

The hydrogen necessarily discharged during draining is either heavily diluted with air or catalytically converted to water. Since these are only small amounts, the overall efficiency is only slightly reduced.

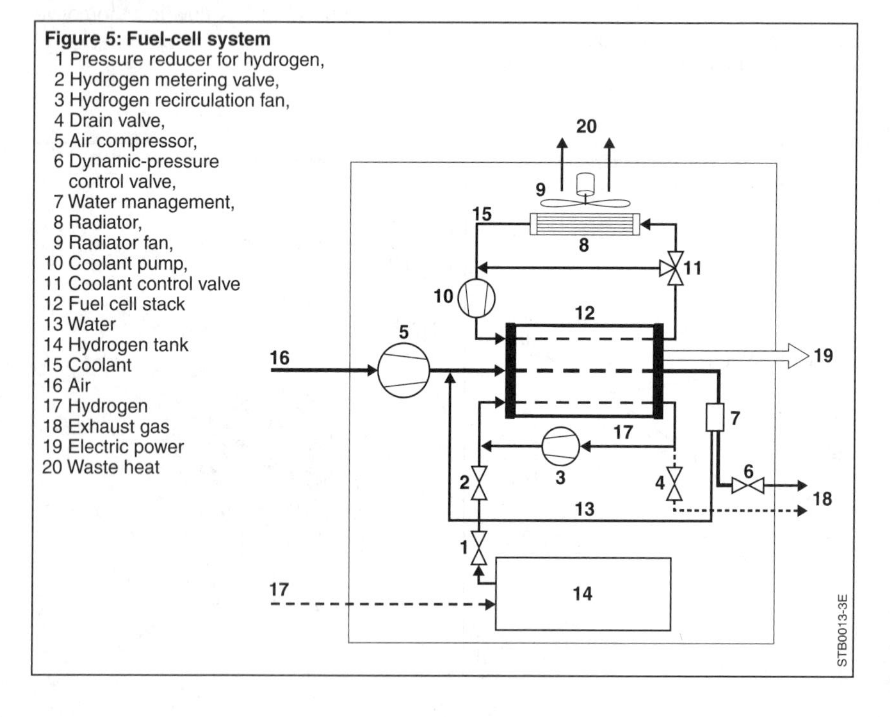

Figure 5: Fuel-cell system
1 Pressure reducer for hydrogen,
2 Hydrogen metering valve,
3 Hydrogen recirculation fan,
4 Drain valve,
5 Air compressor,
6 Dynamic-pressure control valve,
7 Water management,
8 Radiator,
9 Radiator fan,
10 Coolant pump,
11 Coolant control valve
12 Fuel cell stack
13 Water
14 Hydrogen tank
15 Coolant
16 Air
17 Hydrogen
18 Exhaust gas
19 Electric power
20 Waste heat

Oxygen supply
The oxygen required for the electrochemical reaction is taken from the ambient air. The required air mass of up to 100 g/s is inducted by a compressor in accordance with a desired fuel-cell power of 100 kW, compressed to max. 2.5 bar and fed to the cathode side of the fuel cell. The pressure in the fuel cell is adjusted by a dynamic-pressure control valve in the exhaust-gas path downstream of the fuel cell.

In order to ensure sufficient moisture for the polymer membrane, the supplied air is humidified using water management, either from the moist exhaust gases using an exchange membrane or by injecting condensed water.

Thermal management
Fuel cells have an electrical efficiency of approximately 60%. This means that a large proportion of heat is released during the conversion of chemical energy. The heat must be dissipated. PEM fuel cells operate with an operating temperature of approximately 85 °C at a lower temperature level than internal-combustion engines. The radiator and radiator fan must therefore be designed to be bigger in fuel-cell systems in the vehicle drive in spite of the higher efficiency.

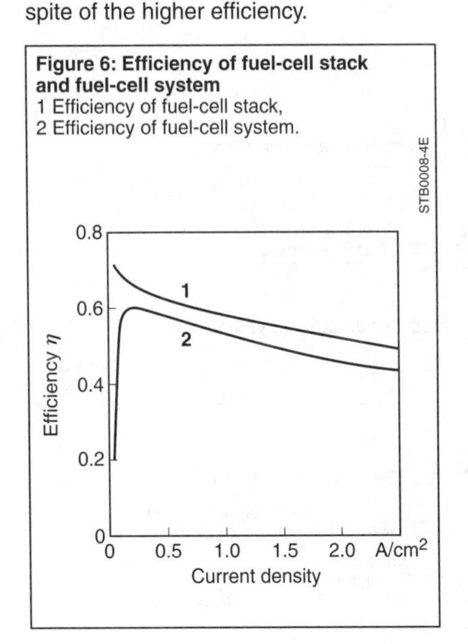

Figure 6: Efficiency of fuel-cell stack and fuel-cell system
1 Efficiency of fuel-cell stack,
2 Efficiency of fuel-cell system.

Because the coolant used is in direct contact with the fuel cells, it must be electrically non-conducting (deionized). The electric coolant pump directs a flow of coolant through all the incorporated components. The flow rate is up to 12,000 *l*/h. A coolant control valve splits the flow of coolant between the radiator and the radiator bypass.

Coolant based on a deionized water-glycol mixture is used. The coolant has to be deionized in the vehicle. The coolant flows through an ion exchanger filled with mixed-bed resin (not shown in Figure 5) and is cleaned by ion removal. The setpoint value for coolant conductivity is less than 5 µS/cm.

Efficiency of the fuel-cell system
In addition to quickly providing the requested power to the stack under the best possible operating conditions, it is important to design the system for primarily operating at high efficiency.

In Figure 6, the efficiency of a fuel-cell stack is compared with that of a fuel-cell system. The secondary loads (e.g. compressor) take up some of the electric power and thereby reduce the overall efficiency of the system. Fuel-cell systems demonstrate substantially higher efficiency than an internal-combustion engine, particularly in the frequently used part-load range.

Safety
Hydrogen is colorless and odorless. It can be ignited starting at a concentration of 4 % by volume in the air. Therefore, hydrogen concentration sensors are placed at central positions in the vehicle (e.g. cab, tank, engine compartment). These sensors are intended to help implement appropriate measures starting at 1 % by volume.

Functioning principle of the drivetrain

Expansion of the fuel-cell system with a battery

Fuel-cell vehicles are electric vehicles in which the energy for the electric drive is made available by the fuel-cell system. For several reasons, it is sensible to expand the drivetrain to include a traction battery:

- The energy gained through recuperative braking can be temporarily stored
- The dynamic response of the drivetrain can be further increased
- The efficiency of the drivetrain can be further increased by shifting the load point between the fuel-cell system and the battery

The topology with a fuel-cell system and traction battery corresponds to a series hybrid.

The ratio of battery power and total power varies for different applications. Typically, fuel-cell systems are used as the primary energy source in the drivetrain. Such vehicles are known as Fuel Cell Hybrid Vehicles (FCHV). Typically, fuel-cell systems have a power rating of 50 to 120 kW. Batteries have a power rating of up to 30 kW with an energy content of 1 to 2 kWh.

Alternatively, the battery can have a much greater power rating and energy content, and be recharged by the fuel-cell system where necessary. Fuel-cell power ratings of between 10 and 30 kW are sufficient for this. This drive concept is called Fuel Cell Range Extender (FC-REX).

Electrical-system typologies

One or more DC/DC converters spread the power between the fuel-cell system, the traction battery and the electric drive. Figure 7 shows various converter configurations. Depending on the configuration, the drive voltage is identical to one of the two voltage sources (Figures 7a and 7b) or isolated from the battery and stack voltages (Figure 7c).

Components of the drivetrain

Electric motor

The electric motor consists of power electronics (a converter) and a synchronous or

Figure 7: Converter configurations of fuel-cell drives
a) DC/DC converter on the traction battery,
b) DC/DC converter on the fuel-cell stack,
c) Configuration with two DC/DC converters.

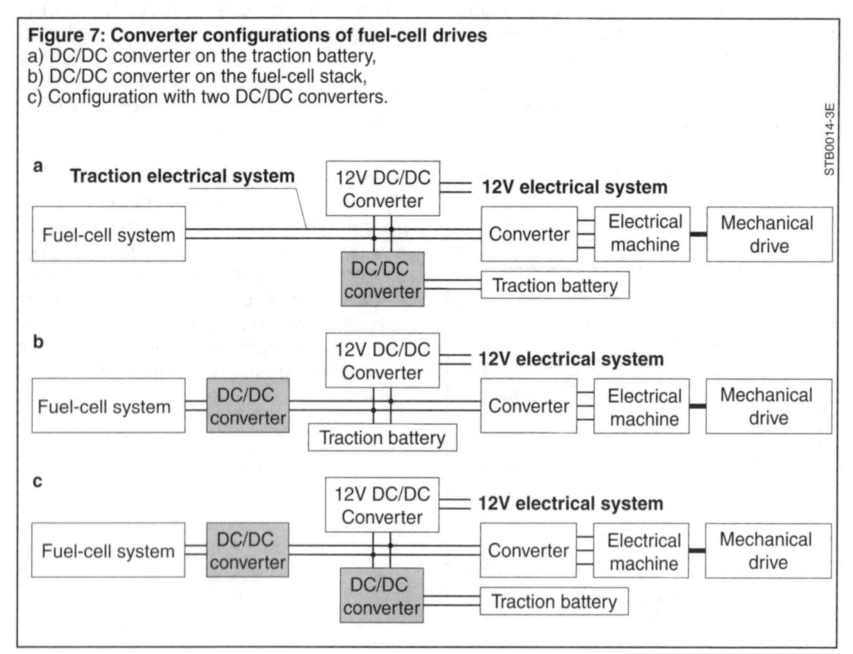

asynchronous machine that is energized by the converter to generate the required engine torque. Because the electric motor has a high power rating (approximately 150 kW), it is operated at voltages of up to 450 V; a vehicle manufacturer also selects a voltage up to 700 V. For safety reasons, the traction electrical system is isolated from the vehicle ground.

When the vehicle brakes, the electric motor switches to generator mode and generates electric current stored that is in the traction battery.

The direct voltage of the traction electrical system is converted in the converter into a multi-phase alternating current where the amplitude is regulated as a function of the desired drive torque. Cyclically actuated IGBT output stages (Insulated Gate Bipolar Transistor) are the preferred choice for use as actuators.

Traction battery
Depending on the application, high-capacity or high-energy batteries are used with voltages of between 150 and 400 V. Nickel-metal-hydride or lithium-ion batteries are used in high-capacity applications, while only lithium-ion batteries are used in high-energy applications. A battery-management system monitors the state of charge and the capacity of the battery.

Traction-battery DC/DC converter
A DC/DC converter regulates the traction battery's charge and discharge currents where maximum currents of up to 300 A flow through the battery. This converter may be omitted in certain system configurations.

Fuel-cell DC/DC converter
Another DC/DC converter regulates the current from the fuel-cell stack, where maximum currents of up to 500 A flow through the stack. This converter too may be omitted in certain system configurations.

12 V DC/DC converter
There is also – as in conventional vehicles – a 12 V electrical system for the low-power electric loads/consumers. The 12 V electrical system is supplied from the traction electric system. A DC/DC converter is used between the two systems for this purpose. This converter is electrically isolated for safety reasons. It operates unidirectionally or bidirectionally with a power rating of up to 3 kW.

Outlook
Drives with fuel cells have already proven that they are suitable for everyday use. However, for commercial use as vehicle drives, they must be improved further with regard to economy and readiness for series production.

System simplifications deliver both cost and reliability advantages. One approach is to develop new polymer membranes for fuel cells where it is not necessary to humidify the reaction gases and which at the same time allows the operating temperature to be increased.

The cost of all components must be significantly reduced. For example, great potential savings can be made by reducing the platinum content in the catalytic layer in the fuel cell. Many large automakers are planning a market launch for small series runs in the period from 2015 to 2017.

References
[1] Pukrushpan, Stefanopoulou, Peng: Control of Fuel Cell Power Systems: Principles, Modeling, Analysis and Feedback Design. Springer-Verlag, London, 2004.

Drivetrain

Table 1: Quantities and units

Quantity		Unit
a	Acceleration	m/s^2
c_d	Drag coefficient	–
e	Rational inertia coefficient	–
f	Coefficient of rolling resistance	–
g	Gravitational acceleration	m/s^2
i	Conversion ratio	–
m	Vehicle mass	kg
n	Revolutions per minute	rpm
r	Dynamic tire radius	m
s	Wheel slip	–
v	Driving speed	m/s
A	Frontal area	m^2
D	System diameter	m
I	Overall conversion range	–
J	Mass moment of inertia	$kg \cdot m^2$
M	Torque	N m
P	Power	kW
α	Gradient angle	°
φ	Overdrive factor	–
η	Efficiency	–
λ	Performance index	–
μ	Conversion	–
ρ	Density	kg/m^3
ω	Angular velocity	rad/s
ν	Speed ratio	–

Subscripts:

eff	Effective	0	Associated with maximum output
tot	Total		
hydr	Hydraulic	A	Drivetrain
max	Maximum	G	Gearbox
min	Minimum	P	Pump
h	Final drive	R	Roadwheel
m	Engine	T	Turbine

Overview

Function

The function of the automotive drivetrain is to provide the thrust and tractive forces required to induce motion. Chemical energy (from the fuel) or electrical energy is converted into mechanical energy in the powerplant, with gasoline and diesel engines representing the powerplants of choice. Powerplants operate within a specific speed range as defined by two extremities: idle speed and maximum engine speed. Torque and power are not delivered at uniform rates throughout the operating range; the respective maxima are available only within specific bands. The drivetrain's conversion ratios adapt the available torque to the momentary requirement for tractive force.

Design

The dynamic condition of a motor vehicle is described by the running-resistance equation (see Formula 1). It equates the forces generated by the drivetrain with the forces required at the driving wheels (running resistance).

From the running-resistance equation, it is possible to calculate the acceleration, the top speed, the climbing ability, and also the overall conversion range I of the transmission:

$$I = \frac{(i/r)_{max}}{(i/r)_{min}} = \frac{\tan \alpha_{max} \cdot v_0}{(P/m\,g)_{eff}\ \varphi}$$

Formula 1: Equilibrium relation between drive forces and running resistance

The equation defining the equilibrium relation between drive forces and resistance factors is applied to determine various quantities, such as acceleration, top speed, climbing ability, etc.

Available power = Running resistance at drive wheels (power requirement)

$$M_m \frac{i_{tot}}{r} \eta_{tot} = m g f \cos\alpha + m g \sin\alpha + e m a + c_d A \frac{\rho}{2} v^2$$

$M_m \frac{i_{tot}}{r} \eta_{tot}$	$m g f \cos\alpha$	$m g \sin\alpha$	$e m a$	$c_d A \frac{\rho}{2} v^2$
Driving force applied to tire footprint	Rolling resistance	Climbing resistance	Acceleration resistance	Aerodynamic drag

Where rotational inertia coefficient $e = 1 + \frac{J}{m\,r^2}$ and mass moment of inertia $J = J_R + i_h^2 J_A + i_h^2 i_G^2 J_m$

The overdrive factor φ is defined as:

$$\varphi = \frac{(i/r)_{\text{min}}}{\omega_0/v_0}.$$

Calculations of effective specific output should always be based on the power P which is actually available for tractive application (net power minus driven ancillaries, power losses, altitude loss). The weight mg must also take account of special cases such as trailers on passenger cars; $\varphi = 1$ applies when in top gear the running-resistance curve is just moving through the point of maximum output (Figure 1). The factor φ determines the relative positions of the curves for running resistance and engine output in top gear. It also defines the efficiency level at which the engine operates.

By transferring the running-resistance curve in top gear to the engine map (full-load curve, engine characteristic curves, power hyperbolas), it is possible to assess the interaction between engine and vehicle in this gear.

$\varphi > 1$ displaces engine operation to an inefficient performance range, but also enhances acceleration reserves and hill-climbing ability in top gear. In contrast, selecting $\varphi < 1$ will increase fuel economy, but only at the price of much slower acceleration and lower climbing reserves. Minimum fuel consumption is achieved along the operating curve η_{opt}. $\varphi > 1$ reduces, $\varphi < 1$ increases the required transmission conversion range I.

In the tractive-force/speed diagram (Figure 2) the engine full-load curve is entered for each gear in the hyperbola field for drive power. It is possible to identify how well an engine/transmission combination (the diagram depicts a five-speed manual transmission) completes the driving map and what hill-climbing ability can be achieved in the individual gears.

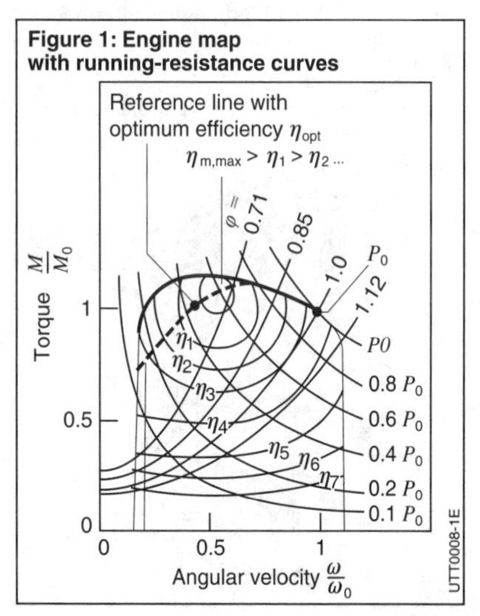

Figure 1: Engine map with running-resistance curves

Drivetrain configurations

The layout of the automotive drivetrain varies according to the position of the engine and the powered axle (Table 2).

Table 2: Drivetrain configurations

Type of drive	Engine position	Powered axle
Standard drive	Front	Rear axle
Front-wheel drive	Front, longitudinal or traverse	Front axle
All-wheel drive	Front, rarely rear or center	Front and rear axle
Rear-wheel drive	Rear	Rear axle

Figure 2: Tractive force/speed diagram
The physical quantities are referred in each case to the value corresponding to maximum power P_0.

Drivetrain elements

The elements of the drivetrain must perform the following functions:

- Keep the vehicle stationary even with the engine running
- Achieve the transition from stationary to mobile state
- Convert torque and rotational speed
- Provide for forward and reverse motion
- Compensate for drive-wheel speed variations when cornering
- Ensure that the power unit remains within the program map that permits minimum fuel consumption and exhaust emissions.

Stationary idle, starting off and interruptions in the power flow are all made possible by the clutch. The clutch slips to compensate for differences in the rotational speeds of engine and the drivetrain when the vehicle is starting off from standstill. When different conditions demand a change of gear, the clutch disengages the engine from the transmission while the gearshift operation takes place. On automatic transmissions, the hydrodynamic torque converter assumes the start-off procedure. The transmission modifies the engine torque and engine speed and adapts them to the vehicle's momentary tractive requirements.

The overall conversion of the drivetrain is the product of the constant transmission ratio of the axle differential and the variable transmission ratio of the transmission – assuming there are no other transmission stages involved. Transmissions are almost always multiple fixed-ratio gearboxes, though some have continuously variable ratios.

Transmissions generally fall into one of two categories: manually shifted and double-clutch transmissions with spur gears in a countershaft arrangement, and load-actuated converter-type automatic transmissions with planetary-gear sets. The transmission also allows the selection of different rotational directions for forward and reverse operation.

The differential allows laterally opposed axles and wheels to rotate at varying rates when cornering to provide uniform distribution of the driving forces. Limited-slip final drives respond to slip at one of the wheels by limiting the differential effect. This shifts additional power to the wheel at which traction is available. Newly developed torque-vectoring differentials allow the drive torque to be distributed specifically as required to the drive wheels.

Torsion dampers, hydrodynamic transmission elements, controlled-slip friction clutches or mass-suspension systems dissipate high vibration amplitudes. They protect against overload and provide added ride comfort.

Figure 3: Function of dry-plate friction clutch
a) Disengaged clutch,
b) Engaged clutch.
1 Engine flywheel, 2 Clutch disk,
3 Pressure plate, 4 Drive connector plate,
5 Throwout bearing, 6 Torsion damper.

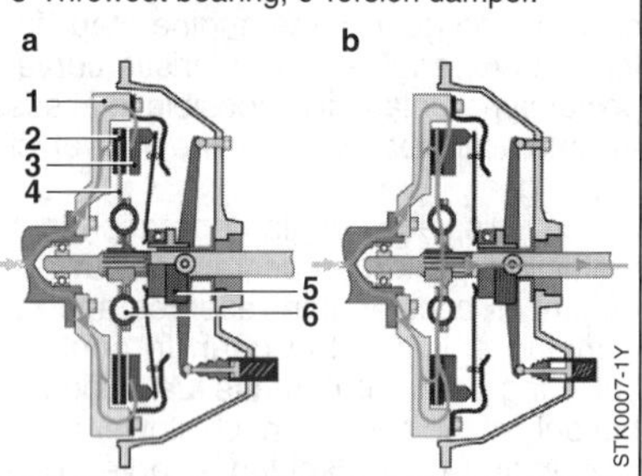

Figure 4: Clutch with dual-mass flywheel
1 Dual-mass flywheel,
2 Flexible element,
3 Pressure plate, 4 Spring plate,
5 Drive connector plate, 6 Throwout bearing.

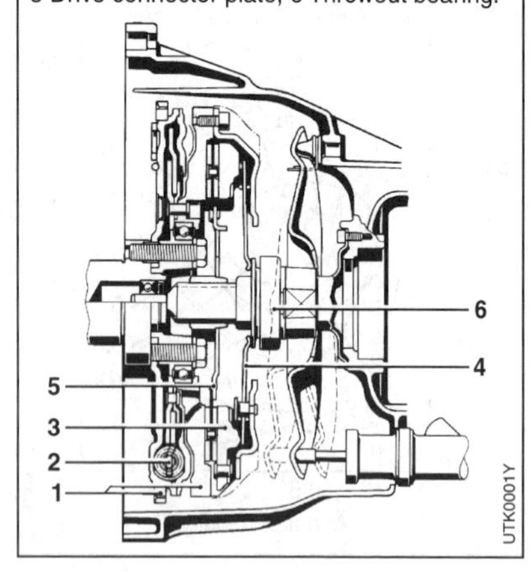

Power take-up elements

Dry-plate friction clutch

The friction clutch consists of a pressure plate, a clutch disk – featuring bonded or riveted friction linings – and a mating friction surface represented by the engine-mounted flywheel. The flywheel and pressure plate provide the thermal absorption required for friction operation of the clutch; flywheel and pressure plate are connected directly to the engine, while the clutch disk is mounted on the transmission's input shaft.

A spring arrangement, frequently in the form of a central spring plate, applies the force which joins the flywheel, pressure plate and clutch disk for common rotation; in this state, the clutch is engaged for positive torque transfer. To disengage the clutch (e.g. when gearshifting), a mechanically or hydraulically actuated throwout bearing applies force to the center of the pressure plate and releases pressure at the periphery (Figure 3). The clutch is controlled either by a clutch pedal or an electrohydraulic, electropneumatic or electromechanical final-control element. A single- or multi-stage torsion damper, with or without predamper, may be integrated in the clutch disk to absorb vibration.

A two-section (dual-mass) flywheel featuring a flexible intermediate element can be installed forward of the clutch for maximum insulation against vibrations (Figure 4). The resonant frequency of this spring/mass system is below the excitation frequency (ignition frequency) of the engine at idle speed, and is a therefore outside the operating speed range. It acts as a vibration insulating element between the engine and the other drivetrain components (similar to low-pass filter).

Wet-plate friction clutch

The wet-plate friction clutch has the advantage over the dry-plate version that its thermal load capacity is better as it can be flooded with oil to enhance heat dissipation. On the down side, its drag losses when disengaged are higher than those of a dry clutch. The wet-plate friction clutch is used as standard in combination with double-clutch transmissions, some continuously variable transmissions, and occasionally in modified converter-type transmissions in place of the hydrodynamic torque converter.

Hydrodynamic torque converter

The hydrodynamic torque converter consists of an impeller which is the drive element, a turbine which is the driven component, and a stator which assists the torque-converter function (Figures 5 and 6). The torque converter is filled with oil and transmits engine torque by means of the flowing forces of the oil. It compensates for speed differences between the engine and the other drivetrain components, and is therefore ideally suited for the start-off function. An impeller converts kinetic energy into fluid energy. This is converted back to kinetic energy in a second transformation that takes place in the turbine impeller.

The impeller's input torque M_P and the power input P_P are calculated as follows:

$$M_P = \lambda \, \rho \, D^5 \, \omega_P{}^2 \,, \quad P_P = \lambda \, \rho \, D^5 \, \omega_P{}^3 \,.$$

The equation's elements are:

λ Performance index
ρ Density of medium
(≈ 870 kg/m³ for hydraulic fluid)
D Circuit diameter in m
ω_p Angular velocity of impeller

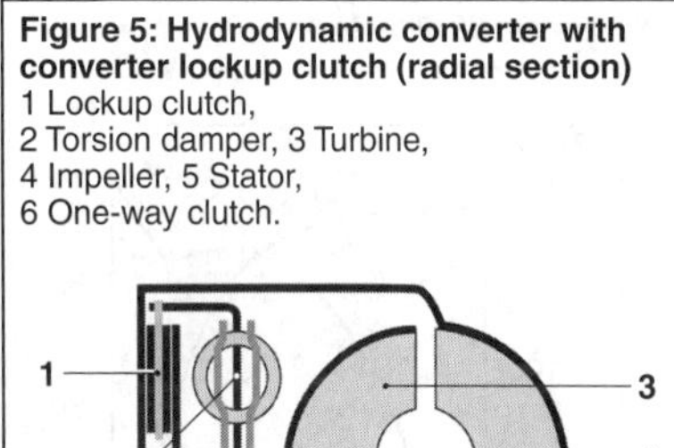

Figure 5: Hydrodynamic converter with converter lockup clutch (radial section)
1 Lockup clutch,
2 Torsion damper, 3 Turbine,
4 Impeller, 5 Stator,
6 One-way clutch.

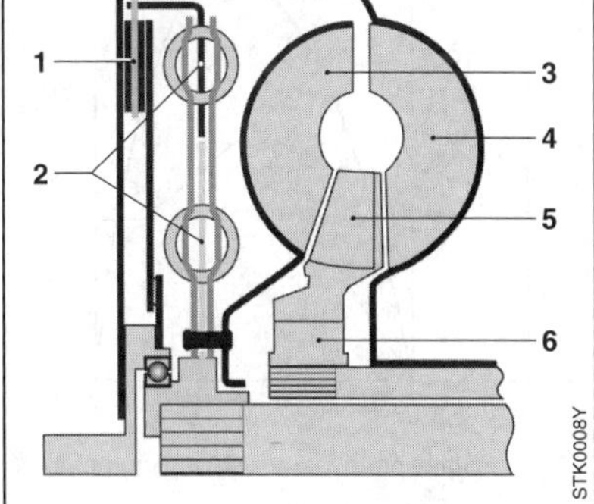

A stator located between impeller and turbine diverts the hydraulic oil back to the input side of the impeller. This means the torque output is higher than the impeller torque taken up by the engine. Torque multiplication or torque conversion is then:

$$\mu = -\frac{M_T}{M_P}.$$

The performance index λ and the torque multiplication μ are dependent on the speed ratio ν between turbine and impeller:

$$\nu = \frac{\omega_T}{\omega_P}.$$

The slip factor $s = 1 - \nu$ and the torque conversion together determine the hydraulic efficiency:

$$\eta_{hydr} = \mu\,(1 - s) = \mu\,\nu.$$

Maximum torque multiplication is achieved at $v = 0$, i.e. with the turbine at stall speed. Further increases in turbine speed are accompanied by a virtually linear drop in multiplication until a torque ratio of 1:1 is reached at the coupling point. Above this point, the stator, which is housing-mounted with a one-way clutch, freewheels in the flow.

In motor-vehicle applications, the two-phase Föttinger torque converter with centripetal flow through the turbine, i.e. the "Trilok converter", has become the established design. The geometrical configuration of the unit's blades is selected to provide torque multiplication in the range of 1.7 to 2.5 at stall speed ($\nu = 0$) (Figure 7). The curve defining the hydraulic efficiency factor $\eta_{hyd} = \nu\mu$ in the conversion range is roughly parabolic. Above the coupling point, which is at 10 to 15 % wheel slip, the efficiency is equal to the speed ratio ν and reaches levels of around 97 % at high speeds.

The hydrodynamic torque converter is a fully automatic, infinitely variable transmission with virtually zero-wear characteristics; it eliminates vibration peaks and absorbs vibration with a high degree of efficiency.

However, its conversion range and efficiency, particularly at high levels of slip, are not sufficient for motor-vehicle applications. As a result, the torque converter can only be usefully employed in combination with multi-speed or continuously variable transmissions.

Figure 6: Sectional view of a hydrodynamic converter with lockup and torsion damper
1 Torsion damper,
2 Lockup clutch,
3 Turbine,
4 Impeller,
5 Stator,
6 One-way clutch.

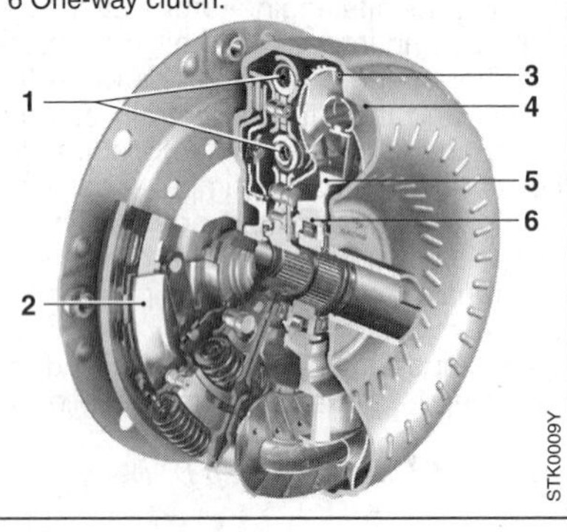

Figure 7: Trilok converter
(typical passenger-car program map)
M_{P2000} Input torque M_P at n = 2,000 rpm.
μ Torque conversion,
η_{hydr} Hydraulic efficiency,
ν Speed ratio.
1 Efficiency improvement with regard to clutch.

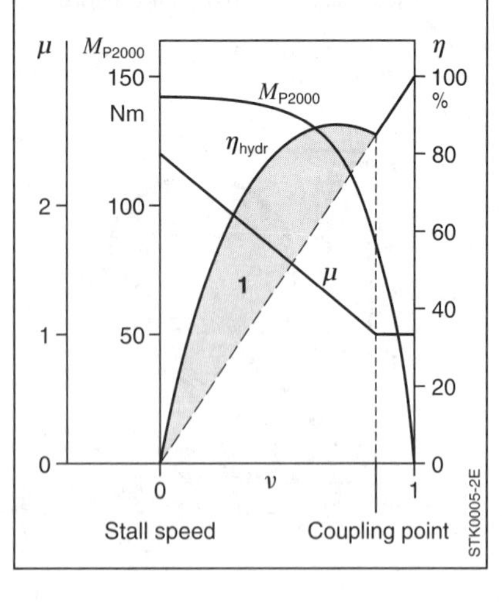

Converter lockup clutch

In order to improve efficiency, the impeller and turbine can be locked together by a converter lockup clutch after startup has ended. The converter lockup clutch consists of a plunger with a friction lining that is connected to the turbine hub. The transmission-shift control regulates the direction in which the fluid flows through the converter to regulate clutch lockup.

The converter lockup clutch normally requires additional means of vibration absorption such as

- a torsion damper,
- controlled-slip operation of the converter lockup clutch at critical vibration levels, or
- a combination of both of the above.

Multi-speed gearbox

Multi-speed gearboxes have become the established means of power transmission in motor vehicles. The main reasons for its success include excellent efficiency characteristics dependent on the number of gears and engine torque characteristics, medium to good adaptation to the traction hyperbola, and easily mastered technology.

Gearshifting on multi-speed gearboxes is performed using either disengagement of power transmission (positively interlocking mechanism) or under load by a friction mechanism. The first group includes manually shifted and semi-automatic transmissions, while the second group encompasses automatic transmissions.

The manually shifted transmissions installed in passenger cars and in most heavy vehicles are dual-shaft units with main shaft and countershaft. Transmissions in heavy commercial vehicles sometimes incorporate two or even three countershafts. In such cases, special design features are required in order to ensure

Figure 8: Planetary-gear set with various conversion ratios
A Sun gear, B Internal ring gear, C Planet gears with carrier,
Z Number of teeth, i Conversion ratio.

Basic equation of the basic planetary-gear transmission: $n_A + (Z_B/Z_A)\, n_B - (1 + Z_B/Z_A)\, n_C = 0$.

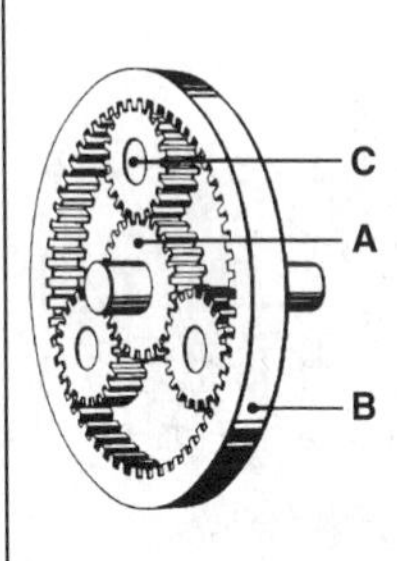

Input	Output	Fixed	Conversion ratio	Remarks
A	C	B	$i = 1 + Z_B/Z_A$	$2.5 \leq i \leq 5$
B	C	A	$i = 1 + Z_A/Z_B$	$1.25 \leq i \leq 1.67$
C	A	B	$i = \frac{1}{1 + Z_B/Z_A}$	$0.2 \leq i \leq 0.4$ overdrive
C	B	A	$i = \frac{1}{1 + Z_A/Z_B}$	$0.6 \leq i \leq 0.8$ overdrive
A	B	C	$i = -Z_B/Z_A$	Stationary transmission with reversal of direction $-4 \leq i \leq -1.5$
B	A	C	$i = -Z_A/Z_B$	Stationary transmission with reversal of direction $-0.67 \leq i \leq -0.25$

UTA0003Y

that power is evenly distributed to all countershafts.

Automatic transmissions for cars and commercial vehicles are, in the majority of cases, planetary-gear transmissions, and only in rare cases are countershaft arrangements used. The planetary gears generally take the form of a planetary-gear link mechanism. They frequently involve the use of Ravigneaux or Simpson planetary gears.

Planetary gears

The basic planetary-gear set consists of the sun gear, internal ring gear and the planet gears with carrier. Each element can act as input or output gear, or may be held stationary. The coaxial layout of the three elements makes this type of unit ideal for use with friction clutches and brake bands, which are employed for selective engagement or fixing of individual elements (Figure 8). The engagement pattern can be changed – and a different conversion ratio selected – without interrupting torque flow; this capability is of particular significance in automatic transmissions.

As several gear wheels mesh under load simultaneously, planetary-gear transmissions are very compact. They have no free bearing forces, permit high torque levels, and feature very good efficiency levels.

Manually shifted transmissions

The basic elements of the manually shifted transmission are:

- Single or multi-plate dry clutch for interrupting and engaging the power flow; actuation may be power-assisted to deal with high operating forces
- Variable-ratio gear transmission unit featuring permanent-mesh gears in one or several individual assemblies
- Shift mechanism with shift lever

The force required for gear selection is transmitted via gearshift linkage or a cable, while dog clutches or synchronizer assemblies lock the active gears to the shafts. Before a shift can take place, it is necessary to synchronize the rotating speeds of the transmission elements to be joined. When the transmission incorporates dog clutches (of the type still sometimes used in transmissions for heavy commercial vehicles), the driver performs this task by double-clutching when upshifting and briefly touching the accelerator when downshifting.

Virtually all passenger-car transmissions and the majority of those in commercial vehicles employ locking synchronizer assemblies, which use friction between the synchronizer ring and the clutch body for initial equalization of rotating speed (Figure 9). The lockout toothing on the outside of the synchronizer ring permits further sliding of the gearshift sleeve for positive

Figure 9: Gearshift mechanism with locking synchromesh
a) Single cone, b) Dual cone, c) Triple cone.
1 Synchronizer ring, 2 Pressure pin with compression spring, 3 Gearshift sleeve, 4 Clutch body, 5 Synchronizer body.

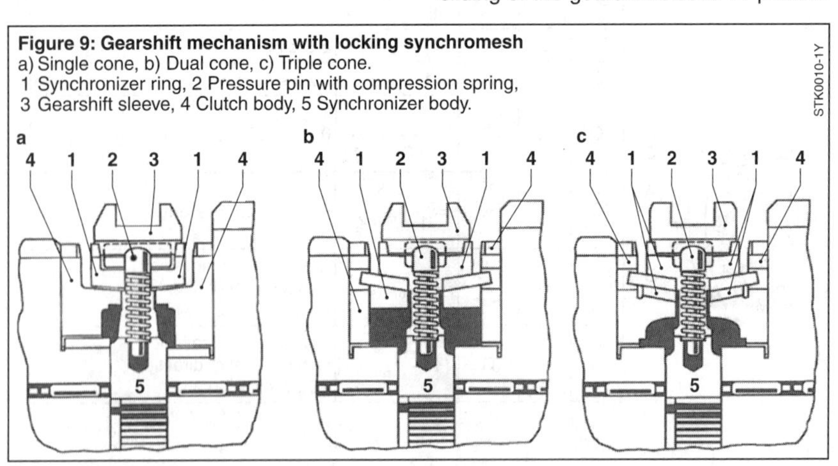

gear engagement only on completion of the synchronization process. By far the majority use single-cone synchromesh clutches. In cases where there are particularly high demands for performance and/or reduction of gearshifting force, double-cone or even triple-cone synchromesh clutches or multi-plate synchromesh clutches are used.

Most transmissions in passenger cars have 5 and increasingly 6 forward gears (Figure 10). The transmission-ratio range (depending on the number of gears and closeness of the ratios) is approximately between 4 and 6.3 while the transmission efficiency can be as high as 99 %. The transmission layout depends on the vehicle's drive configuration (standard rear-wheel drive, front-wheel drive with inline or transverse engine, or four-wheel drive). Accordingly, the input and output shafts may share a single axis, or they may be mutually offset; the final-drive and differential assembly may also be included in the unit (Figure 11).

Transmissions in commercial vehicles can have between 5 and 16 gears, depending on the type of vehicle and the specific application. For up to 6 gears, the transmission consists of a single gearbox. The transmission-ratio range is between 4 and 9. Transmissions with up to 9 gears are two-case transmissions in which the range-selector case is pneumatically operated. The transmission-ratio range extends to 13.

For still higher numbers of ratios – up to 16 – three transmission elements are employed (Figure 12). These are a multi-speed basic gearbox as the main unit, a two-stage splitter group for transmission ratios between the stages of the main unit, and a two-stage range-selector group for increasing the transmission-ratio range of

Figure 11: Manual-transmission designs
a) With axial offset,
b) Coaxial arrangement of input and output shafts.
The numbers denote the gears.

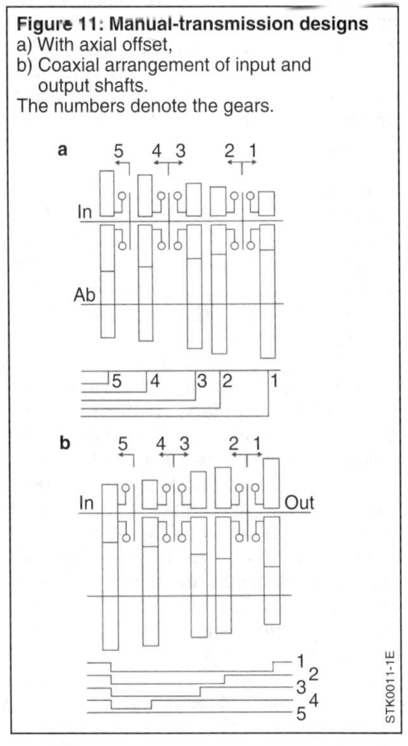

Figure 10: 6-speed manual transmission for passenger cars with standard drive layout (ZF S6-37)
1 Input shaft,
2 Gearshift rail,
3 Countershaft,
4 Output shaft.

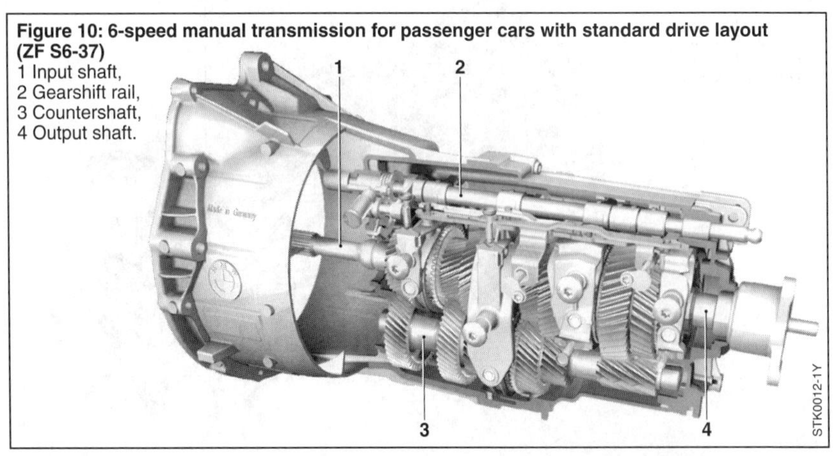

the main units. The splitter and range-selector groups are pneumatically actuated. The transmission-ratio range is as high as 16.

Power take-offs

Commercial-vehicle transmissions are fitted with a variety of power take-off connections for driving ancillary equipment. A basic distinction is made between clutch and engine-driven PTOs. The individual selection depends upon the specific application.

Retarder

Hydrodynamic or electrodynamic retarders are non-wearing auxiliary brakes for reducing the thermal load on the road-wheel brakes under continuous braking. They can be fitted on both the drive input side (primary retarders) or the drive output side (secondary retarders), either as a separate unit or integrated in the transmission. The advantages of the integrated designs are compact dimensions, low weight and fluid shared with the transmission in a single circuit. Primary retarders have specific advantages when braking at low speeds and are therefore widely used on public-transport buses. Secondary retarders have particular advantages on long-distance trucks for adjustment braking at higher speeds or when traveling downhill.

Automatic transmissions

There are two types of automatic transmission depending on their effect on vehicle handling dynamics. Semi-automatic transmissions are manually shifted transmissions on which all operations normally performed by the driver when changing gear are carried out by electronically controlled actuator systems. In terms of vehicle dynamics, this means that a gear change always involves disengaging the clutch and therefore interrupting the drive to the driving wheels.

Fully automatic transmissions, usually referred to simply as automatic transmissions, change gear under load, i.e. power continues to be transmitted to the driving wheels during a gearshift operation.

That difference in vehicle handling dynamics is the essential factor which determines the types of application for these two transmission types. Fully automatic transmissions are used in situations where disengagement of power transmission would be associated with a significant reduction in comfort (above all on cars with powerful acceleration), or where it is unacceptable for reasons of handling dynamics (mainly on off-road vehicles). Semi-automatic transmissions are equipped on long-distance trucks, tour buses and more recently on small

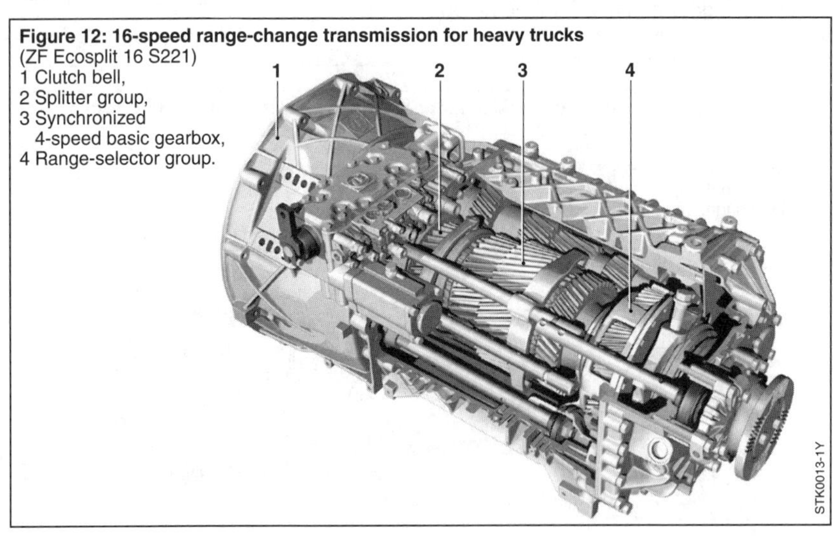

Figure 12: 16-speed range-change transmission for heavy trucks
(ZF Ecosplit 16 S221)
1 Clutch bell,
2 Splitter group,
3 Synchronized 4-speed basic gearbox,
4 Range-selector group.

cars, racing cars and very sporty road vehicles.

Semi-automatic transmissions

Partially or fully automated gearshifting systems make a substantial contribution to simplifying control of the gears and increasing fuel economy. Particularly when used on trucks, the disadvantages inherent in the interruption of power transmission are compensated by a number of decisive advantages:

- Narrower spacing of ratios, with up to 16 gears
- Enhanced efficiency power transfer
- Reduced costs
- Same basic transmission unit for manual and semi-automatic designs

Operating concept

An electric, hydraulic or pneumatic positioner module on the transmission shifts the individual gears and activates the clutch. The electronic transmission control generates the control signals for the gearshift operation (Figure 13).

Design variations

The simplest system merely replaces the mechanical linkage with remote shifting. The shift lever then merely sends out electrical signals. Start-off and clutching procedures are identical to a standard manually shifted transmission. More complex versions combine these systems with a recommended shift-point function. Advantages are:

- Reduced shifting effort
- Simplified installation (no shift linkage)
- Prevention of incorrect operation (overrevving of engine)

On fully automated gearshifting systems, both the transmission and the power take-up element are automated. The driver's control element consists of either a lever or pushbuttons, with an override provision in the shape of a driver-selected manual mode or ± buttons.

Complex shift programs are required to control a multi-ratio transmission. A system which engages the gears according to a fixed pattern will not be adequate. Current running resistance (as determined by payload and road conditions) must be factored in to achieve the optimum balance between drivability and fuel

Figure 13: Semi-automatic transmission for commercial vehicles (system diagram)
1 Engine management (EDC, Electronic Diesel Control), 2 Electronic transmission control, 3 Transmission actuator, 4 Diesel engine, 5 Dry-plate interrupting clutch, 6 Clutch positioner, 7 Intarder electronics, 8 Display, 9 Gear selector lever, 10 ABS/TCS, 11 Transmission, 12 Air supply.
—— Electrical system, – – – – Pneumatic system, —— CAN communications.

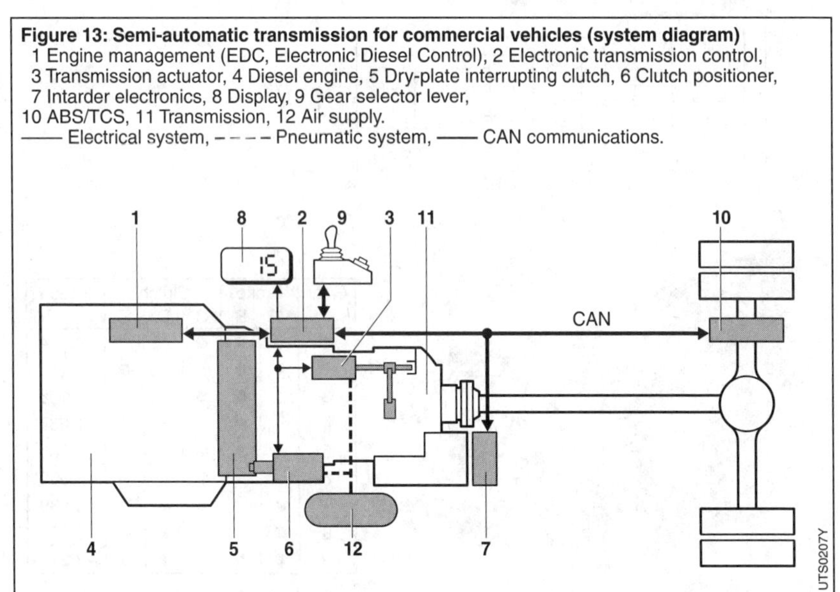

economy. This task and adaptation of the speed during gearshifting (synchronization) are performed by the transmission control system. Engine speed is adjusted by an electronic throttle-control system (ETC) to the speed requested by the transmission control system via the data communication bus. As a result, mechanical synchromesh systems can be partly or entirely dispensed with in the gearbox. Advantages are:

- Optimum fuel economy through automatic, computer-controlled shifting
- Reduced driver stress
- Lower weight and smaller dimensions
- Enhanced safety for both driver and vehicle

Fully automatic transmissions

Fully automatic transmissions perform start-off and ratio selection (shifting) operations with no additional driver input. In the majority of cases, the power take-up unit is a hydrodynamic torque converter that generally features a mechanical converter lockup clutch. Alternatively, wet multi-plate clutches are used as the power engagement facility.

The power-transmission efficiency of such fully automatic transmissions is slightly lower than that of manual and semi-automatic transmissions due to its operating principle. However, this is compensated by shift programs designed to keep engine operation inside the maximum fuel economy range.

Figure 14: 8-speed automatic transmission for passenger cars with standard drive layout (ZF 8 HP70)
A, B Multi-plate brakes,
C, D, E Multi-plate clutches,
P1...P4 Planetary gear sets,
GWK Controlled converter lockup clutch.

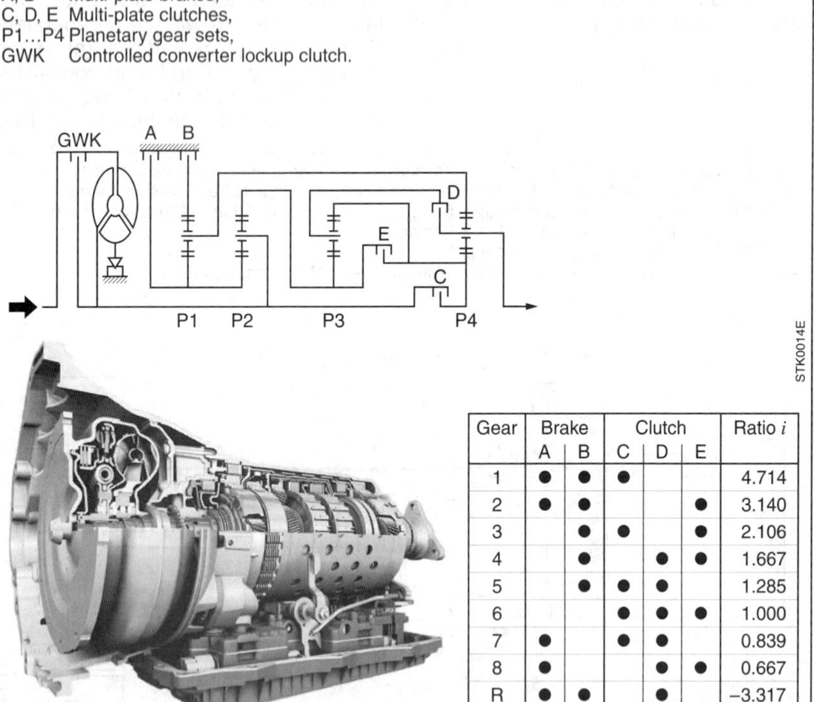

Gear	Brake		Clutch			Ratio i
	A	B	C	D	E	
1	●	●	●			4.714
2	●	●			●	3.140
3		●	●		●	2.106
4		●		●	●	1.667
5		●	●	●		1.285
6			●	●	●	1.000
7	●		●	●		0.839
8	●			●	●	0.667
R	●	●		●		–3.317

Common fully automatic transmission components (Figure 14) include:

- An engine-driven hydraulic-fluid pump provides hydraulic pressure for the shift elements and the transmission-shift control system as well as supply fluid to the power take-up unit. It also supports lubrication and cooling in the transmission.
- Hydraulically actuated multi-plate clutches, plate or band brakes to execute shifts without interrupting the flow of power.
- A transmission-shift control system to define gear selections and shift points and to regulate demand-response shifting, as dictated by the driver-selected shift program (selector lever, tap shift), accelerator-pedal position, engine operating conditions and vehicle speed. The transmission-shift control system works electronically and hydraulically.

Converter-type automatic transmissions with hydrodynamic torque converter
Converter-type automatic transmissions feature a hydrodynamic torque converter (always used on passenger-car transmissions, while commercial vehicles generally use the Trilok design); for starting off, torque multiplication and absorbing harmonic vibrations. Several planetary-gear sets are arranged downstream of the hydrodynamic torque converter. The number and arrangement of the planetary-gear sets depend on the number of gears and transmission ratios.

Passenger-car converter-type automatic transmissions have up to 8 gears. The mechanical transmission-ratio range goes from about 3.5 for 4-speed transmissions, through 5 for 5-speed transmissions, to 6 for 6- and 7-speed transmissions, and 7 for 8-speed transmissions. The start-off conversion of the hydrodynamic torque converter ranges from 1.7 to 2.5. Converter-type automatic transmissions in commercial vehicles have from 3 to 6 gears. The mechanical transmission-ratio range extends from 2 to 8. These transmissions frequently feature integrated hydrodynamic retarders on the primary or secondary side that can use the peripheral systems, i.e. oil pump, oil pan and oil cooler, that are already in place in the transmission system.

Double-clutch transmissions
Double-clutch transmissions are increasingly being used as an alternative to converter-type transmissions in front-wheel-drive vehicles and in sporty vehicles (Figure 16). Double-clutch transmissions feature two multi-plate clutches arranged on the drive side to regulate demand-response shifting and power take-off (start-off). Each clutch is connected to a synchronized reduction gear mechanism with one subunit responsible for shifting the even gears and the other for shifting the odd gears (Figure 15). Both subunits of the reduction gear are based on a mutually interlaced design. To facilitate the gearshift operation, the next gear is preselected in the transmission unit currently not transmitting power, with subsequent load delivery from the other transmission part.

Double-clutch transmissions have the advantage over their converter-type counterparts in being suitable for maximum engine speeds in excess of around 6,500 to 7,000 rpm. Furthermore, thanks to their countershaft arrangement, double-clutch transmissions can have advantages based on the drive configuration (weight, space).

6- and 7-speed transmissions have been introduced in series production. Both clutches are usually of the wet type. A version with a dry clutch and electromechanical actuation has also been intro-

Figure 15: Schematic diagram of double-clutch transmission

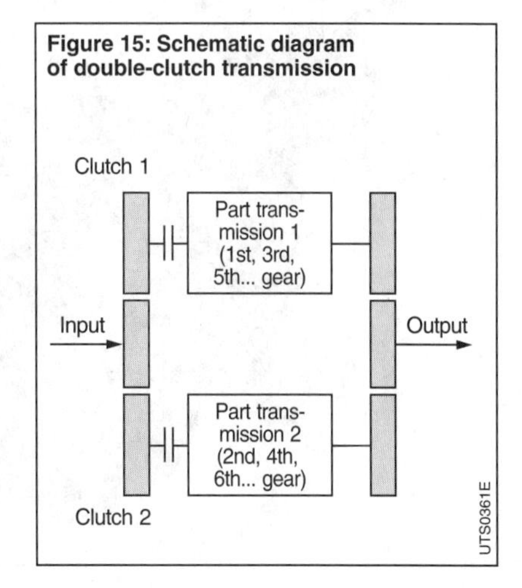

duced in series production to reduce the drag losses of each clutch when disengaged.

Electronic transmission control
Automatic transmissions are almost exclusively controlled by electronically operated hydraulic systems. Hydraulic actuation is retained for the clutches, while the electronics assume responsibility for gear selection, for gearshifting, and for adapting the pressures in accordance with the torque flow. The advantages are:
- A number of gearshift programs including adaptive drive functions
- Improved, constant gearshift comfort over the full service life
- Adaptability to different types of vehicle
- Simplified hydraulic control

Sensors detect the transmission output- and input-shaft speeds, the selector-lever position, and the positions of the program selector and the kickdown switches. The information about the engine (speed, driver command, and engine torque) is transmitted via the CAN bus. The control unit processes this information according to a predefined program and uses the results in determining the control variables to be transmitted to the gearbox. Electrohydraulic converter elements form the link between the electronic and hydraulic circuits, while standard solenoid valves activate and disengage the clutches. Pressure regulators ensure precise control of pressure levels at the clutches.

Shift-point control
In selecting the gear to be engaged, the system refers to the rotating speed of the transmission output shaft and the driver command before triggering the appropriate solenoid valves. Either the driver may select from among different drive programs (e.g. for optimum fuel

Figure 16: 7-speed double-clutch transmission (ZF 7DT50)
1 Wet double clutch,
2 Gearshift rails for synchronization,
3 7-speed countershaft transmission,
4 Oil-pump drive,
5 Electrohydraulic transmission-shift control.

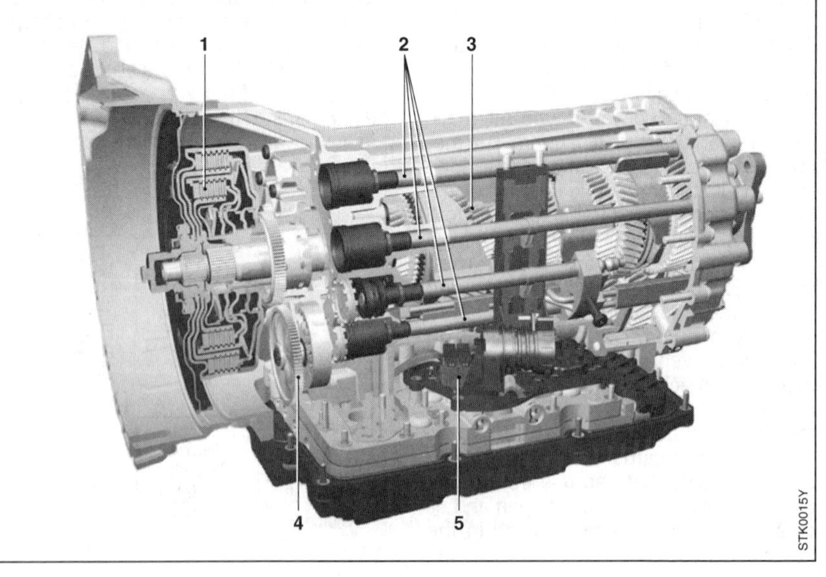

economy or optimum performance) or an adaptive drive program is implemented which selects the optimum gear from the driving situation and driver type. In addition, the selector lever allows manual input from the driver. Driving-situation detection includes variables such as linear and lateral acceleration, and results in the detection of hills or corners. The speed with which the accelerator pedal is pressed and the number of kickdown operations in a defined period of time or the frequency of brake-pedal operation serve to identify the type of driver. A complex control program selects the appropriate gear for the current operating conditions and driving style. In other words, it suppresses trailing-throttle upshifts on the approach to and during corners, or automatically responds to slow throttle openings by activating a drive program for low engine-speed upshifts.

Concepts which combine the high level of convenience of such "intelligent" gearshift programs with facilities for active adaptation to individual driver preferences have become very widespread. In addition to the normal positions for neutral, drive and reverse, the selector levers for such systems have a second parallel channel in which one touch of the lever produces an immediate gear change (provided no engine-speed limits would be exceeded). Alternative solutions come in the form of paddles or ± buttons on the steering wheel.

Converter lockup (for multi-speed and continuously variable transmissions)

A mechanical lockup clutch can be employed to improve the efficiency of the transmission unit by eliminating torque-converter slip. The variables employed to determine when conditions are suitable for activation of the converter lockup mechanism are engine load, transmission output-shaft speed, and state of the transmission.

Control of shift quality

The accuracy with which the pressure at the friction elements of the clutches is adjusted to the level of torque being transmitted (determined with reference to engine load and speed) has a decisive influence on shift quality; this pressure is regulated by a pressure regulator. Shifting comfort can be further enhanced by briefly reducing engine torque for the duration of the shift (e.g. in a gasoline engine by retarding the ignition timing); this also reduces friction loss at the clutches and extends component service life. Crucial parameters such as clutch fill time and slip time are permanently monitored to ensure a constant gearshift quality throughout the transmission's service life. In the event of deviations from specified limits, pressure corrections are made in the system in order to maintain an optimum time characteristic during the next gearshift.

Safety strategy

Special monitoring circuits and functions prevent transmission damage stemming from operator error, while the system responds to malfunctions in the electrical system by reverting to a backup mode.

Final-control elements

Electrohydraulic converter elements such as solenoid valves and pressure regulators form the interface between electronic and hydraulic circuits. Electric motors are used as final-control elements in semi-automatic manual transmission with electromechanical control.

Continuously variable transmissions

The continuously variable transmission (CVT) can convert every point on the engine's operating curve to an operating curve of its own, and every engine operating curve into an operating range within the field of potential driving conditions. Its advantage over conventional fixed-ratio transmissions lies in the potential for enhancing performance and fuel economy while reducing exhaust emissions (e.g. by maintaining the engine in the performance range for maximum fuel economy).

The continuously variable transmission can operate by mechanical, hydraulic or electrical means. To date, only mechanical CVT systems have been realized for passenger vehicles, mainly in the form of a belt-wrap drive and, to a limited extent, a toroidal rolling transmission.

Belt-wrap transmissions have V-pulley halves that can be axially shifted by fluid pressure. Either a linked thrust belt or a band chain running under fluid pressure is used as the belt-wrap element for power transmission. Axial shifting of the V-pulleys alters the running radii of the belt-wrap element and with it the transmission ratio (Figure 17).

This transmission type is particularly suitable for installation in front-wheel-drive vehicles with transversally or longitudinally mounted engines. The transmission ratio is between 5.5 and 6. Today, transmissions for engine torque ranges between 300 Nm and 350 Nm are produced as standard.

As well as the V-pulley set and the belt-wrap element, the following assemblies are essentially to a continuously variable transmission:

- A wet engagement clutch or a hydrodynamic converter as power take-up mechanism
- An engine-driven oil pump
- Electronic-hydraulic transmission-shift control
- A reversing mode for forward-reverse shift
- Final-drive unit with differential

Figure 17: Continuously variable passenger-car transmission
(ZF CFT23)
a) Schematic diagram, b) Design.
1 Converter, 2 Pump, 3 Constant transmission ratio, 4 Differential, 5 Shift elements,
6 Reversing set, 7 V-pulley set, 8 Electrohydraulic transmission-shift control.
WK Converter lockup clutch, R Reverse, V Forward.

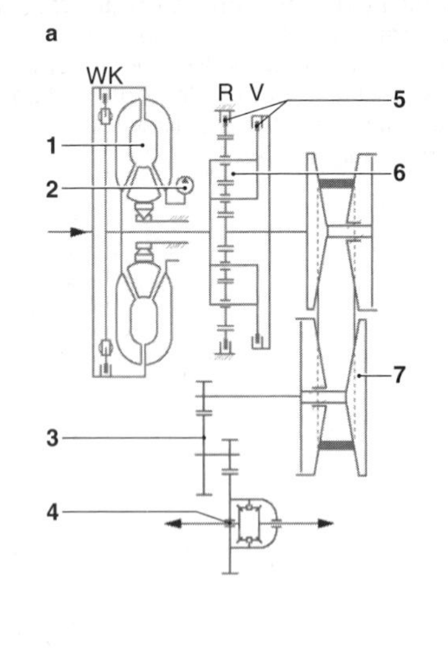

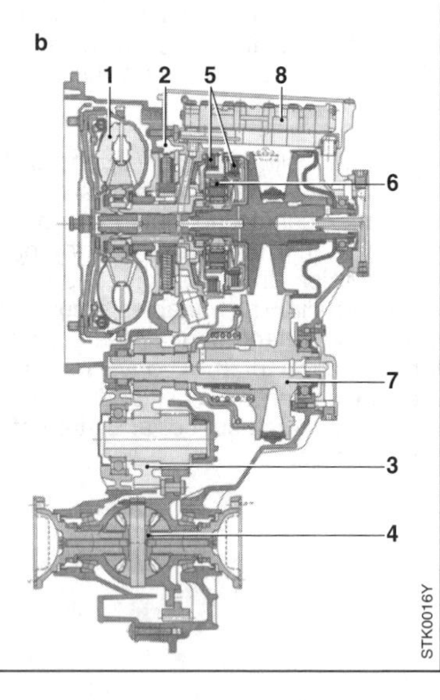

Electric power transmission in vehicles is gaining in importance in conjunction with hybrid drives. Passenger cars with hybrid drives feature the first series-production electric-mechanical power-split transmissions. City buses also feature solutions with electric power transmission. They either operate by diesel-electric means or are alternatively powered by batteries, an overhead power line, and in future possibly also by fuel cells. The advantages of electric drive technology in buses are nonintegrated (distributed) design of the overall drivetrain, the possibility of single-wheel drive and, in particular, simpler realization of low-floor vehicle designs.

Hydrostatic-mechanical power-split transmissions are in use on production agricultural tractors. Use on road vehicles is unlikely due to the high noise levels produced.

Figure 18: Passenger-car rear-axle final-drive unit
1 Elastomer bearing,
2 Differential,
3 Differential bearing,
4 Half-shaft flange,
5 Housing (two-part),
6 Drive flange,
7 Half-shaft flange,
8 Bevel-gear set,
9 Bevel-pinion bearing.

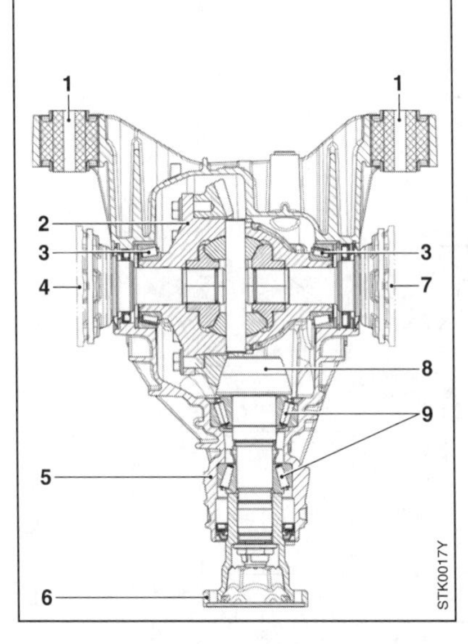

Final-drive units

Design and elements

The overall conversion ratio between engine and drive wheels is produced by several elements operating in conjunction: a transmission with variable ratios (manual, automatic, CVT), an intermediate transmission in some applications (AWD transfer case), and the final-drive unit.

Longer distances between transmission and final drive are bridged by the propshaft (in one piece or in several sections with intermediate bearings). Angular offsets resulting from nonaligned connecting drive shafts are compensated for by means of universal joints, constant-velocity joints and flexible-disk joints.

The central element of a car final drive (Figure 18) is either a hypoid-gear crown wheel and pinion (inline engine) or a spur-gear crown wheel and pinion (transverse engine). These components are integrated in the transmission on front-wheel-drive vehicles. On standard-drive vehicles they are located separately on the powered axle.

The chief components of a final-drive differential are the crown wheel and pinion, the planetary gears, bearings, drive shaft and half-shaft flanges and differential housing. The final-drive transmission ratio is usually between 2.3 and 4.0.

The crown wheel is normally bolted to the differential case which holds the planetary gears, and the pinion shaft and differential case run in taper roller bearings. To reduce the transmission of structure-borne noise to the bodyshell, the final drive unit is attached to the vehicle frame by flexible (rubber) mountings.

In addition to its torque transmission capabilities, mechanical efficiency and weight, the noise-producing characteristics of a final drive unit have become a decisive criterion in modern automobile development. In this regard, the crown wheel and pinion are of primary significance in terms of noise generation. The quietness of the mechanism is essentially dependent on the way in which the gears are manufactured. The best results are obtained by grinding the gear teeth after heat treatment (case hardening), as in this way reproducible tooth-flank topographies can be achieved particularly well.

On commercial vehicles, direct-drive axles with hypoid bevel gears are most commonly used. The final-drive transmission ratio ranges from 3 and 6 to 1. In cases where smooth-running characteristics are particularly important, e.g. on buses, the gears are ground.

On public transport buses, which nowadays are almost always low-floor designs, hub reduction axles (Figure 19) are used as they allow very low floor levels. In addition to the helical bevel-gear differential, there is also an indirect-drive spur-gear reducer stage. This indirect-drive arrangement allows the required high torque levels to be transmitted with the limited reducer offset.

If there are special ground clearance requirements (e.g. construction-site vehicles), planetary axles are used. Propeller shaft and half-shafts can be made smaller by splitting the power transmission, and the available space is increased.

Differential

The differential unit compensates for differences in the rotation rates of the drive wheels between inside and outside wheels during cornering, and between different drive axles on AWD vehicles (Figure 20).

Apart from a few special cases, bevel gears are the preferred design for differentials. The differential gears act as a balance arm to equalize the distribution of torque to the left and right wheels. When lateral variations in the road surface produce different coefficients of friction at the respective wheels, this balance effect limits the effective drive torque to a level defined as twice the tractive force available at the wheel (tire) with the lower coefficient of friction. If torque then exceeds frictional resistance, the wheel will spin.

This undesirable effect can be eliminated by locking the differential either by a positively interlocking or friction mechanism. Positively interlocking differential locks are switched-in by the driver. Their disadvantage lies in the stress applied to the drivetrain which occurs during corner-

Figure 19: Hub reduction axle for buses
(ZF AV132)
1 Wheel hub, 2 Disk brake, 3 Portal offset, 4 Power-split spur stage,
5 Differential, 6 Bevel-gear set,
7 Drive flange.

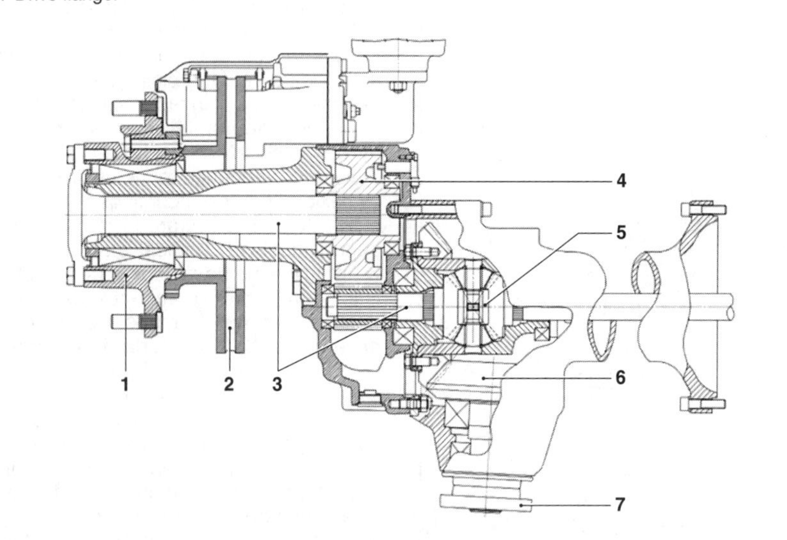

ing. Friction-type differential locks operate automatically using friction plates, cones or a combination of worm and spur-gear drives and thus have a variable locking action depending on torque. The friction-type locking action can also be achieved with a viscous coupling and is then dependent on the differential speed.

Other systems employ electronically controlled multi-plate clutches to produce the friction-type locking action variable up to full lock.

Figure 20: Schematic diagram of differential
1 Drive pinion,
2 Crown wheel,
3 Differential cage,
4 Differential gears.

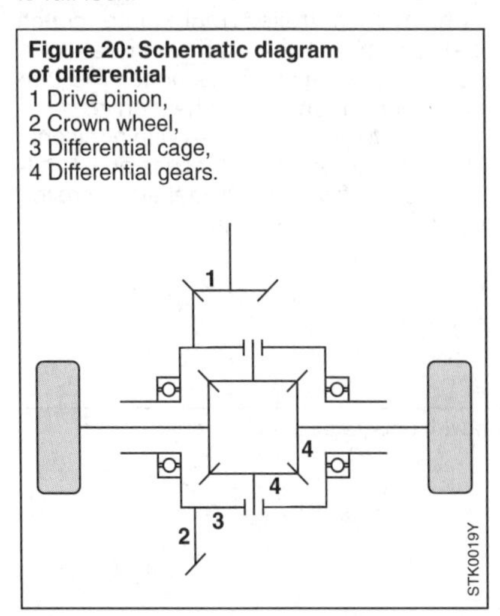

Today, self-locking differentials are in competition with electronic systems which slow down a spinning wheel by applying the brake and thereby transfer power transmission to the wheel with more grip (e.g. Traction Control System (TCS)).

To improve vehicle handling, the differential is sometimes combined in some high-end and sporty vehicles with torque-vectoring units (Figure 21). In this way, the differential's torque equilibrium can be specifically altered by electronic control to suit the driving situation. When the vehicle is cornering, the inside wheel is relieved of load and more torque is applied to the outside wheel. The yaw moment thereby created supports cornering maneuvers and increases agility.

To achieve this, the differential is assigned a torque-vectoring unit for each wheel side. These are designed, for example, as planetary gear sets. A braking torque can be applied to each planetary-gear carrier by a multi-plate brake; this torque then influences the differential's torque equilibrium to the desired extent.

Figure 21: Extended differential with torque-vectoring units
a) Schematic diagram, b) Design.
A Torque-vectoring unit, B Differential, C Torque-vectoring unit.
1 Multi-plate brake, 2 Planetary drive, 3 Input.

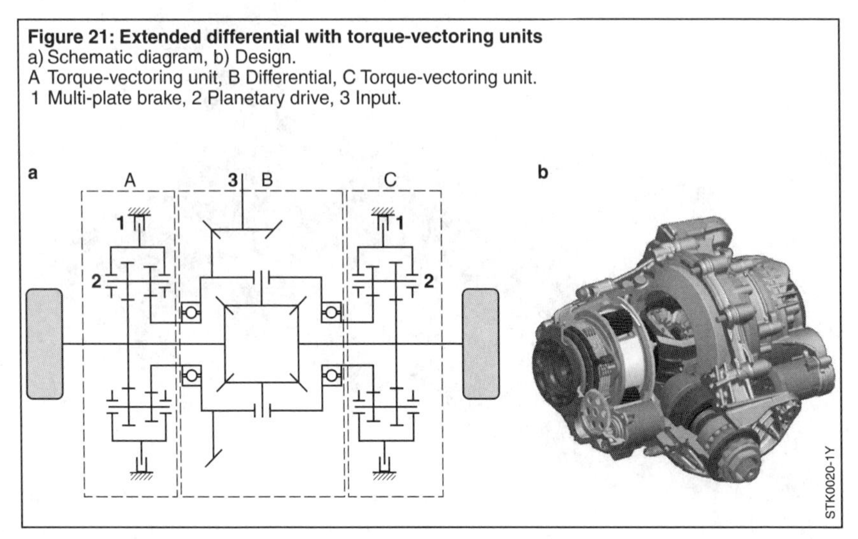

All-wheel drive and transfer case

All-wheel drive (AWD) improves traction on cars, off-road vehicles and commercial vehicles on wet and slippery road surfaces and rough terrain. Two different types of all-wheel drive, disengageable and permanent, may be used.

Disengageable all-wheel drive operates either with direct transmission to front and rear axles, or with a transfer case. The transfer case and final-drive differentials can also have a disengageable lock. Transfer cases for off-road vehicles incorporate an additional driver-controlled conversion range for steep gradients and low speeds.

With permanent all-wheel drive, all wheels are driven at all times. The central transfer case is either non-locked or locked by a torque-dependent frictional force, a Torsen lock or a viscous coupling. Torque distribution between the front and rear axles is either 50:50 or is higher to the rear axle. Additional crawler-gear ratios are also possible.

The all-wheel-drive transfer case can be mounted as a separate assembly on the vehicle transmission and can then be universally used with different transmission types. Integrating the transfer case in the vehicle transmission (Figure 22) reduces installation space, costs and weight. However, the integrated solution cannot be universally applied.

Designs with viscous coupling or electronically controlled multi-plate clutch instead of the central all-wheel-drive transfer case also come under the umbrella of permanent all-wheel drive.

Some vehicles dispense with additional locks on the AWD transfer case or axle differentials in favor of intelligently controlled intervention of the brakes.

Figure 22: Automatic passenger-car transmission with integrated all-wheel drive (ZZ 8 HP)
1 Torsen differential,
2 Primary spur gear,
3 Beveloid gears for side-shaft drive,
4 Side shaft,
5 Central transmission,
6 Front-axle drive unit with differential.

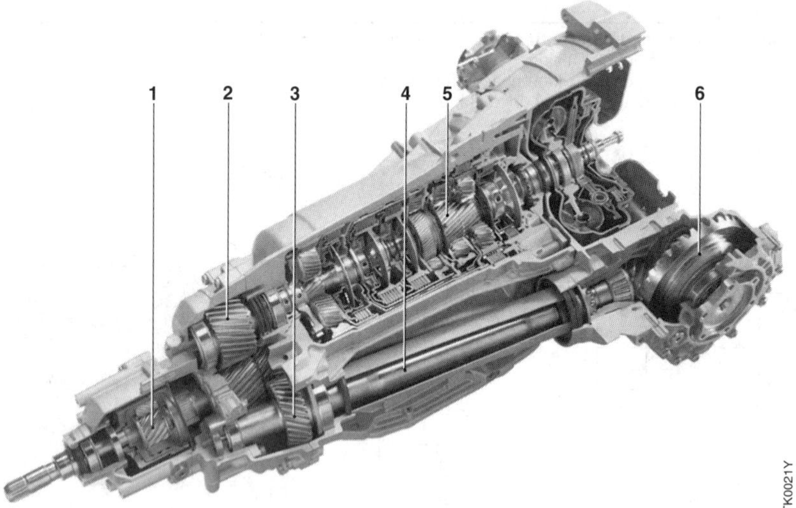

Basic terms of automotive engineering

Basic terms of vehicle handling

Basic principles and systematics

Vehicle handling covers the dynamics of lateral, linear and vertical motion of passenger cars and commercial vehicles. Essentially, the dynamics of lateral motion deals with the steering behavior of the vehicle, the dynamics of linear motion with the acceleration behavior and braking response, and the dynamics of vertical motion with handling caused by excitations of road irregularities. The basic terms explained in the following (for quantities and units, see Table 1) describe vehicle handling in the important frequency range up to approx. 8 Hz. Many of these quantities are established in the German standard DIN 70000 [1] and the international standard ISO 8855 [2]. The quantities used can be subdivided into three groups.

Body

The motion of the body can generally be described as a rigid body. Even in the bodies of convertibles, deviations from rigid-body motion only occur above the frequency range of 10 to 15 Hz.

Wheel suspension

The front wheels have two degrees of freedom, i.e. the degree of freedom for compression and rebound and that for steering. The rear wheels generally only have the one degree of freedom for compression and rebound. In the case of vehicles with four-wheel steering, the rear wheels also have the degree of freedom for steering. These degrees of freedom are defined by the kinematics and elastokinematics of the axle. Kinematics of the axle is the pure rigid-body motion of the individual suspension arms, while elastokinematics describes the behavior of the axle in response to the effect of forces and moments.

Tires

The tire is the most important contact between the vehicle and its environment, generating forces for propelling, braking and steering the vehicle in interaction with the road.

Table 1: Quantities and units

Quantity		Unit
α	Slip angle	degree
β	Float angle	degree
δ_H	Steering-wheel angle	degree
δ_R	Toe angle of right wheel	degree
δ_L	Toe angle of left wheel	degree
δ_A	Axle steering angle (steering angle)	degree
ε_V	Longitudinal inclination at wheel center	degree
ε_{BV}	Anti-dive angle	degree
ε_{AV}	Anti-lift angle	degree
ψ	Yaw angle	degree
φ	Roll angle	degree
θ	Pitch angle	degree
γ	Camber angle	degree
σ	Kingpin angle	degree
τ	Caster angle	degree
λ	Tire slip	–
λ_B	Braking-force distribution	–
ω_R	Wheel angular speed	s^{-1}
a_x	Longitudinal acceleration	m/s^2
a_y	Lateral acceleration	m/s^2
a_z	Vertical acceleration	m/s^2
a_t	Tangential acceleration	m/s^2
a_c	Centripetal acceleration	m/s^2
F_S	Lateral force	N
F_U	Longitudinal force	N
F_Z	Axle load	N
h	Height of center of gravity	m
h_W	Height of roll pole	m
i_s	Steering ratio	–
l	Wheelbase	m
M_H	Steering-wheel torque	Nm
M_R	Tire aligning torque	Nm
n_τ	Caster offset	m
n_V	Caster offset in wheel center	m
n_R	Tire caster offset	m
r_σ	Kingpin offset at wheel center	m
r_{st}	Deflection-force lever arm	m
r_l	Scrub radius	m
r_{dyn}	Dynamic rolling radius	m
s	Track width	m
v_x	Longitudinal velocity	m/s
v_y	Lateral velocity	m/s
v_z	Vertical velocity	m/s
v_{RAP}	Speed at wheel contact point	m/s
v_{RMP}	Speed at wheel center	m/s
X_A	Starting-torque compensation	–
X_B	Braking-torque compensation	–

Basic terms for the body

Translational (uniform) motion

The body is described by three translational and three rotational degrees of freedom (Figure 1). Generally, the system of coordinates (rectangular right-handed system) has its origin at the vehicle's center of gravity. The x-axis points forwards, i.e. in the direction of travel. It is situated in the plane that is perpendicular to the road surface. This plane is called the vehicle center plane. Situated perpendicular to the vehicle center plane is the y-axis, which points, when viewed in the direction of travel, to the left. The z-axis points upwards. The translational motions are designated as follows:

- In x-direction: linear motion.
- In y-direction: lateral motion.
- In z-direction: lift.

Translational velocities and accelerations

Single time derivation of the translational motion variables determines the linear, lateral and vertical velocities (v_x, v_y, v_z). Renewed time derivation produces the linear acceleration a_x, lateral acceleration a_y and vertical acceleration a_z.

Float angle

In the dynamics of lateral motion the center of gravity of the body does not always move along the x-axis. The angle that forms between the vehicle center plane and the trajectory is called the float angle β (Figure 2). It is counted from the vehicle center plane to the trajectory. The float angle is calculated from the linear velocity v_x and the lateral velocity v_y:

$$\beta = \arctan \frac{v_y}{v_x}.$$

Because linear velocity during forward driving is by definition positive, the sign preceding the lateral velocity in this case also determines the sign preceding the float angle.

Centripetal acceleration and tangential acceleration

Acceleration is proportioned in the horizontal plane along the instantaneous vehicle trajectory into centripetal acceleration a_c and tangential acceleration a_t. The centripetal acceleration is that part which is perpendicular to the trajectory, while the tangential acceleration is that part in the tangential direction of the trajectory.

Yaw, pitch and roll

The following terms have been established to cover rotational motion:

- pure rotation about the z-axis is called yaw and is described by ψ,
- pure rotation about the y-axis is called pitch and is described by θ,
- pure rotation about the x-axis is called roll and is described by φ.

The signs preceding these rotations correspond to those of a rectangular right-handed system. The counting directions

Figure 1: Translational and rotational degrees of freedom of the body
ψ Yaw angle,
φ Roll angle,
θ Pitch angle,
S Center of gravity.

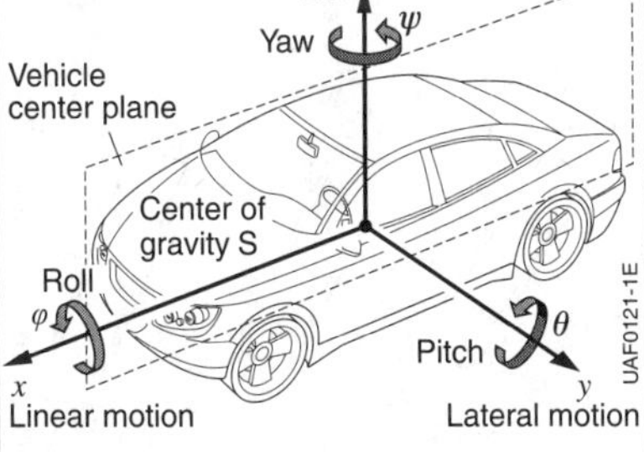

Lift z, Yaw ψ, Vehicle center plane, Center of gravity S, Roll φ, x Linear motion, Pitch θ, y Lateral motion, UAF0121-1E

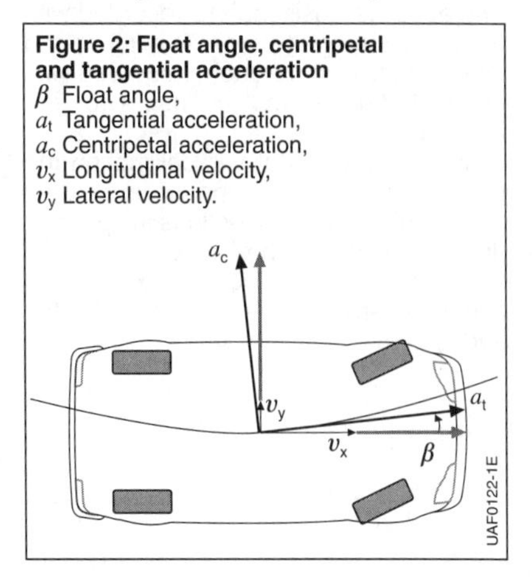

Figure 2: Float angle, centripetal and tangential acceleration
β Float angle,
a_t Tangential acceleration,
a_c Centripetal acceleration,
v_x Longitudinal velocity,
v_y Lateral velocity.

are shown in Figure 1. In many important driving maneuvers the body simultaneously moves about more than one axis. Because, when viewed mathematically, the three rotations are not commutative, the following order for the rotations has been defined in DIN 70000:

1. Yaw
2. Pitch
3. Roll

Rotational velocities and accelerations
Single time derivation of the rotational motion variables produces the yaw, pitch and roll velocities. Renewed time derivation produces the yaw, pitch and roll accelerations.

Yaw moment, pitch moment, rolling moment
External forces acting on the vehicle can generate a moment about the center of gravity. This moment is broken down into three components. The portion of moment along the z-axis is called yaw moment, that along the y-axis pitch moment, and along x-axis rolling moment.

Measured variables
In the interests of vehicle development it is sensible to use a special measurement to enable vehicle handling to be measured with greater accuracy. In the dynamics of lateral motion the translational accelerations and the attitude angles are often measured by gyro-stabilized platforms. The absolute positions are recorded with GPS measuring systems. The linear and lateral velocities are each measured with proximity-type velocity sensors for the float angle.

In the dynamics of vertical motion the translational accelerations are measured in each of the three directions in space at different points of the body. From these the most important body accelerations can be determined, i.e. those for lift, pitch and roll.

Basic terms for the wheel suspension

Steering-wheel angle and steering-wheel torque
The steering-wheel angle δ_H is the turning angle of the steering wheel measured from the straight-ahead position. The angle is positive for a left-hand curve.

The driver must apply a torque when adjusting the steering-wheel angle. This torque is called the steering-wheel torque M_H and is also positive for a left-hand curve.

Toe angle
Moving the steering wheel causes the two front wheels to turn in the same direction. This alters the toe angle of the right and left wheels (δ_R and δ_L) respectively. The toe angle is the angle between the vehicle center plane and the wheel center plane when projected onto the road surface (Figure 3). In the case of a positive turn, i.e. in the counterclockwise direction about the wheel's z-axis, the toe angle is also positive.

On vehicles with four-wheel steering the rear wheels also move.

Axle steering angle, steering angle
When the steering wheel is moved, different toe angles are obtained at the two front wheels. The difference between the toe angles can amount to several degrees with the steering-wheel angle at maximum. When viewed geometrically, the toe angle of the inside wheel is greater than the toe of the outside wheel. The mean

Figure 3: Toe angle and axle steering angle
1 Vehicle center plane,
2 Wheel center plane.
δ_L Toe angle of left wheel,
δ_R Toe angle of right wheel,
δ_A Axle steering angle (steering angle), mean toe angle.

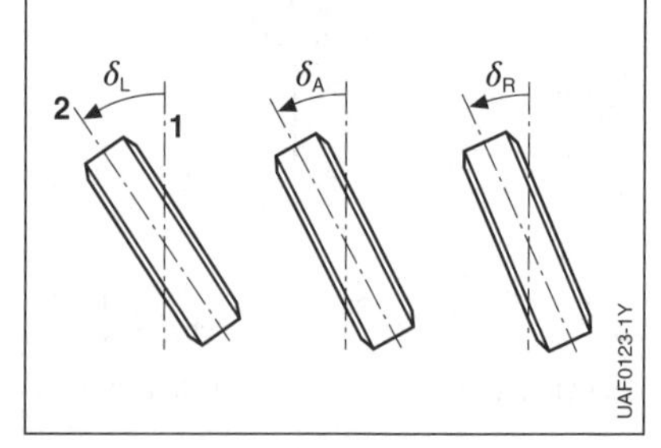

toe angle is called the axle steering angle δ_A or simply the steering angle (Figure 3).

Steering ratio

Essentially, on account of the steering gear but also on account of the kinematics of the front axle, the toe angle obtained at the wheels is significantly smaller than the steering-wheel angle. In such a case where there are no acting forces or torques and where the vehicle is barely loaded, the steering ratio i_s is governed by the following equation:

$$i_s = \frac{2\,\delta_H}{(\delta_L + \delta_R)}\ .$$

Toe-in and toe-out

When the steering wheel is in the straight-ahead position, the toe angles of the front wheels are in the dynamically advantageous range of 0.1 to 0.3°. When the distance of the rim flanges ahead of the wheel centers is smaller than that of the rim flanges after the wheel centers, this is referred to as wheel toe-in (Figure 4). When the situation is reversed, this is referred to as wheel toe-out. Both toe-in and toe-out are given in degrees (°).

The terms toe-in and toe-out are also used to refer to a single wheel. In this case, toe-in means that the wheel has a toe angle in the direction of the vehicle center plane and toe-out that it has a toe angle against the direction of the vehicle center plane.

Figure 4: Toe-in
1 Wheel center.
d_v Distance between rim flanges, front,
d_h Distance between rim flanges, rear,
δ_L Toe angle of left wheel,
δ_R Toe angle of right wheel,
⇨ Direction of travel.

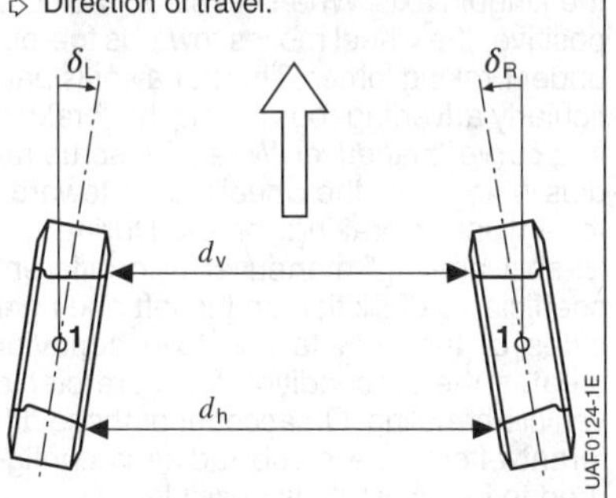

Camber angle

The camber angle γ is the angle between the vehicle center plane and the wheel center plane when projected onto the z-y plane. The camber angle is positive when the wheels are further away from the vehicle center plane at the top than at the bottom (Figure 5a). On account of the axle kinematics the camber angle relative to the body is dependent on the spring travel.

Further to this definition, the camber angle relative to the road is likewise important to vehicle handling. The camber angle relative to the road is the angle between the wheel center plane and the normal line to the road surface (Figure 5b). The preceding sign is determined according to the rectangular right-handed system. If the vehicle center plane is perpendicular to the road surface, both definitions of the camber angle are equal in terms of amount. Otherwise, it is important to observe the precise definition.

Steering axis, kingpin axis

During steering the wheels move not about their z-axis, but rather about the steering axis, also known as the kingpin axis. The position of the kingpin axis is

Figure 5: Camber
a) Relative to the body,
b) Relative to the road.
1 Vehicle center plane,
2 Wheel center plane,
3 Normal line to the road surface.
γ Camber angle.

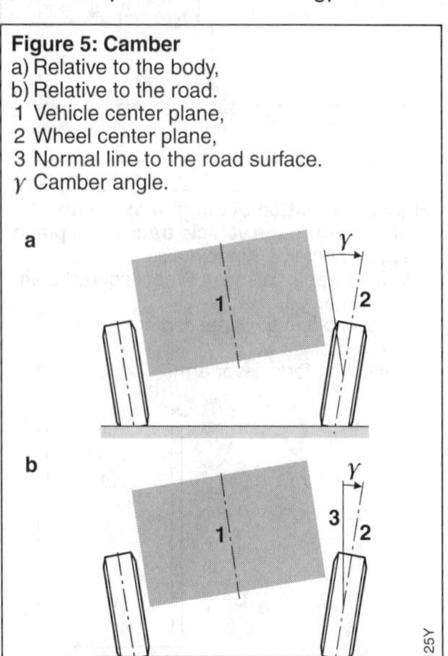

essentially determined by the kinematics of the axle (Figure 6 and Figure 7).

Caster angle, caster offset in wheel center and caster offset
When the wheel and the kingpin axis are projected onto the vehicle center plane, the kingpin axis is inclined by the caster angle τ. It is positive when the top end of the kingpin axis is inclined towards the rear (Figure 6). In this projection the kingpin axis generally does not pass through the wheel center; instead, it is offset by the caster offset in wheel center n_v to the rear. The distance between the wheel contact point and the point where the kingpin axis intersects the road surface is called the caster offset n_τ. If this point of intersection is located in front of the wheel center (Figure 6), the caster offset is positive. Caster offsets are typically in the range of 15 to 30 mm.

Figure 6: Position of kingpin axis when projected onto the vehicle center plane
1 Wheel center, 2 Kingpin axis.
τ Caster angle,
n_v Caster offset in wheel center,
n_τ Caster offset.
◁ Direction of travel.

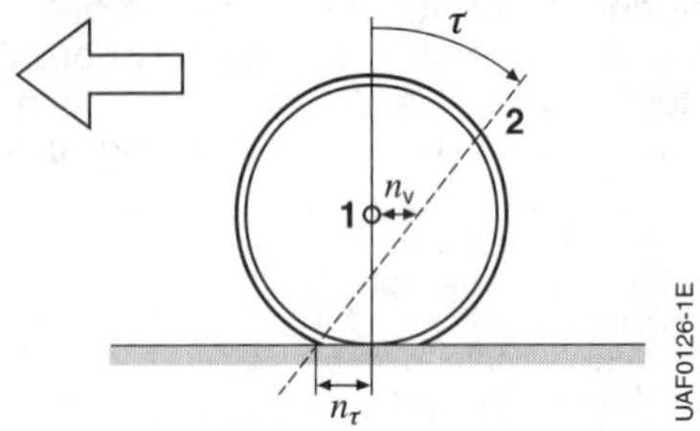

Figure 7: Position of kingpin axis when projected onto the vehicle transverse plane
1 Wheel center, 2 Kingpin axis,
3 Vehicle center plane, 4 Wheel contact point.
σ Kingpin angle,
r_σ Kingpin offset at wheel center,
r_l Scrub radius,
r_{st} Deflection-force lever arm.

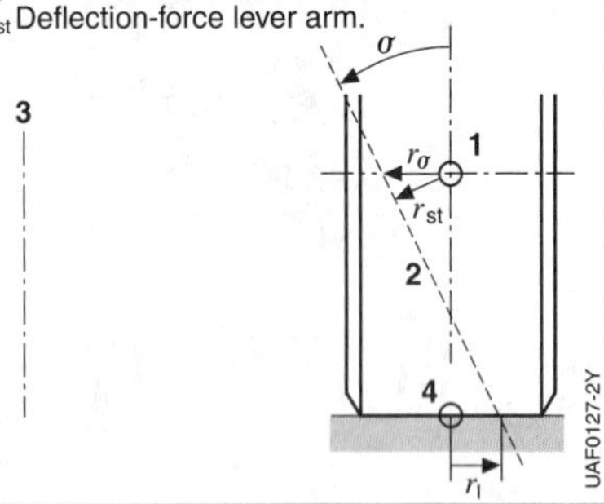

Kingpin angle, kingpin offset at wheel center, deflection-force lever arm and scrub radius
When the wheel and the kingpin axis are projected onto the vehicle transverse plane, the kingpin axis is inclined by the kingpin angle σ (Figure 7). The kingpin angle (also known as the kingpin inclination) is positive when the kingpin axis is inclined to the vehicle center. Kingpin angles are generally positive.

The distance between the wheel center and the kingpin axis parallel to the road is called the kingpin offset at wheel center r_σ. The kingpin offset at wheel center is positive when the wheel center is further from the vehicle center plane than the kingpin axis (Figure 7). The shortest connection between the wheel center and the kingpin axis is called the deflection-force lever arm r_{st}. The deflection-force lever arm is positive when the wheel center is further from the vehicle center plane than the kingpin axis.

The axes are arranged in such a way that the kingpin offset at wheel center and the deflection-force lever arm are as small as possible. This arrangement prevents deflection forces from being generated in the steering.

The distance between the wheel contact point and the point where the kingpin axis intersects the road surface is called the scrub radius r_l (Figure 7). The scrub radius is positive when the wheel contact point is further from the center plane than the kingpin axis. When the scrub radius is positive, the wheel moves towards toe-out under braking forces. This behavior is particularly advantageous during the "braking in a curve" maneuver. When the scrub radius is negative, the wheel moves towards toe-in under braking forces. During the "μ-split braking" maneuver with different coefficients of friction on the left and right sides of the vehicle this axle behavior creates the precondition for more stable vehicle handling. On account of these different effects the scrub radius is configured to be as small as possible.

Transverse pole, roll pole, roll axis
When the axle is subjected to compression and rebound the position of the tires is determined primarily by the kinematics and elastokinematics. The tire moves transversally to the direction of travel about the transverse pole (Figure 8). The speeds for example at the wheel contact point (v_{RAP}) and at the wheel center (v_{RMP}) are during compression and rebound perpendicular on the connecting line to the transverse pole. The position of the transverse pole changes during compression and rebound.

The vehicle body moves in the case of low lateral accelerations about the roll pole of the respective axle (Figure 8). The roll pole is on the connecting line for wheel contact point and transverse pole in the vehicle center plane, i.e. at half the track width ($s/2$). The height h_W of the roll pole can thus be easily calculated:

$$h_W = \frac{v_{RAP,y}}{v_{RAP,x}} \frac{s}{2} .$$

The height of the roll pole is typically below 120 mm. In order in the case of high lateral accelerations to avoid the support effect, i.e. the jacking effect, the height of the roll pole decreases with compression. The roll pole is also known as the roll center or instantaneous center.

Figure 8: Transverse pole and roll pole
1 Wheel center,
2 Tire center plane,
3 Vehicle center plane.
Q Transverse pole, W Roll pole.
s Track width, h_W Height of roll pole.
v_{RAP} Speed at wheel contact point,
v_{RMP} Speed at wheel center.

The connection between the roll pole of the front axle and the roll pole of the rear axle is called the roll axis (Figure 9). The center of gravity of the body is usually above the roll axis. Center-of-gravity heights for sedans range between 550 and 650 mm. This roll axis applies for lower lateral accelerations. In the case of higher lateral accelerations, both the suspension adjustment and the axle behavior must be taken into consideration. The roll axis is then not inevitably in the vehicle center plane.

Longitudinal pole, pitch pole, pitch axis, braking-torque compensation, and starting-torque compensation
When the spring movement of the axle is projected onto the vehicle center plane, the tire moves about the longitudinal pole L (Figure 10). During compression the wheel center moves upwards at the longitudinal inclination at wheel center ε_V. The speeds at the wheel contact point (v_{RAP}) and the wheel center (v_{RMP}) are during compression and rebound perpendicular on the connecting line to the longitudinal pole. The position of the longitudinal pole can change during compression and rebound. The angle between the connection of the wheel contact point to the longitudinal pole and the road is called

Figure 9: Roll axis
1 Roll axis.
S Center of gravity,
W_V Roll pole, front axle,
W_H Roll pole, rear axle,
s Track width,
l Wheelbase,
h_{WV} Roll-pole height, front axle,
h_{WH} Roll-pole height, rear axle.
⇦ Direction of travel.

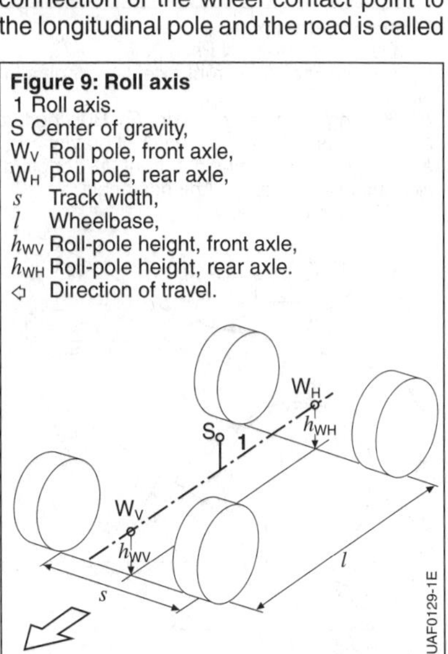

the anti-dive angle ε_{BV}. The angle between the connection of the wheel center to the longitudinal pole and the parallel line to the road is called the anti-lift angle ε_{AV}.

The longitudinal pole of the front axle is behind the front wheels and the longitudinal pole of the rear axle is in front of the rear wheels (Figure 11).

Figure 10: Longitudinal pole, anti-lift angle and anti-dive angle
L Longitudinal pole,
h_V Longitudinal-pole height, front,
ε_V Longitudinal inclination at wheel center,
ε_{BV} Anti-dive angle,
ε_{AV} Anti-lift angle,
v_{RAP} Speed at wheel contact point,
v_{RMP} Speed at wheel center.
◁ Direction of travel.

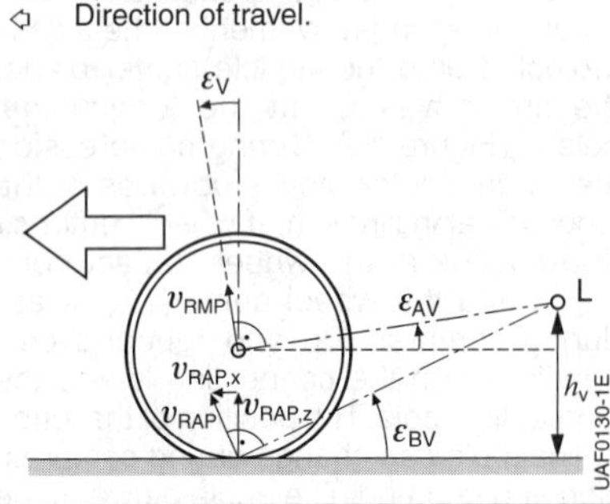

When the vehicle is braked, the front axle load is increased by ΔF_z and the rear axle load decreased by ΔF_z. At the vehicle center of gravity the force $F = m a$ is applied. With braking-force distribution λ_B the braking forces $F_{xV} = \lambda_B F_x$ and $F_{xH} = (1 - \lambda_B) F_x$ are applied to the front and rear axles respectively. In the optimum case where the resulting force from F_{xV} and ΔF_z on the front axle passes exactly through the longitudinal pole and similarly on the rear axle, no spring movement occurs in the vehicle's body springs. For the optimum anti-dive and anti-rise angles:

$$\tan(\varepsilon_{BV,opt}) = \frac{1}{\lambda_B}\ \frac{h}{l}\,,$$

$$\tan(\varepsilon_{BH,opt}) = \frac{1}{(1-\lambda_B)}\ \frac{h}{l}\,,$$

with wheelbase l and height of center of gravity h.

The two connecting lines between the wheel contact points and the longitudinal pole intersect at the pitch pole N (Figure 11). The pitch pole N is always below the height of the center of gravity so that an expected vehicle pitching motion towards the rear can always be effected when

Figure 11: Pitch pole
S Center of gravity, N Pitch pole,
L_V Longitudinal pole, front axle, L_H Longitudinal pole, rear axle,
m Mass, a Acceleration,
F_{xV} Braking force on front axle, F_{xH} Braking force on rear axle, ΔF_z Axle-load change,
l Wheelbase, h Height of center of gravity,
h_V Longitudinal-pole height, front, h_H Longitudinal-pole height, rear,
ε_{BV} Anti-dive angle, ε_{BH} Anti-rise angle.
◁ Direction of travel.

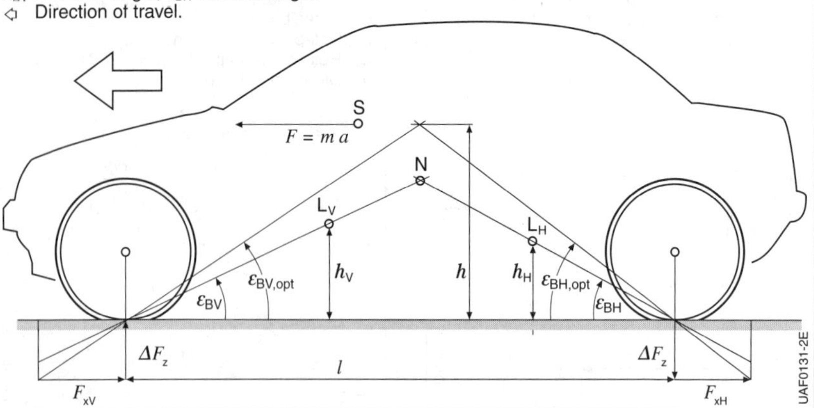

braking. The pitch axis passes through the pitch pole and is perpendicular to the vehicle center plane. Braking-torque compensation X_{BV} and X_{BH} is a measure of how effectively optimum braking compensation has been implemented. The following applies:

$$X_{BV} = \frac{\tan(\varepsilon_{BV})}{\tan(\varepsilon_{BV,opt})} \, 100\,\% \,,$$

$$X_{BH} = \frac{\tan(\varepsilon_{BH})}{\tan(\varepsilon_{BH,opt})} \, 100\,\% \,.$$

The same applies to acceleration. It is necessary to take into account firstly the type of drive, i.e. four-, front- or rear-wheel drive, and secondly the fact that the tractive force is applied at the wheel center. In the general case of four-wheel drive the following applies to starting-torque compensation X_{AV} and X_{AH}:

$$X_{AV} = \frac{\tan(\varepsilon_{AV})}{\tan(\varepsilon_{AV,opt})} \, 100\,\% \,,$$

$$X_{AH} = \frac{\tan(\varepsilon_{AH})}{\tan(\varepsilon_{AH,opt})} \, 100\,\% \,.$$

Measured variables
Steering-wheel angle and steering-wheel torque are measured using special measuring steering wheels. If the measuring accuracy is sufficient, the steering-wheel-angle sensor fitted as standard in many vehicles can also be used.

Special measuring devices can be used in both mobile and stationary applications to measure toe and camber angles. These angles are typically measured on special test benches.

Special test benches are also used to measure the axle steering angle and the steering ratio.

The variables for the position of the kingpin axis are generally not measured directly. The axle pivot points are often recorded by means of geometric measurements and the following variables are calculated from them: caster angle, caster offset in wheel center, caster offset, kingpin angle, kingpin offset at wheel center, deflection-force lever arm, and scrub radius. The roll pole can be determined by measuring the track change during the reciprocal compression and rebound of an axle. The roll axis is obtained from the roll poles of the front and rear axles.

The pitch pole and starting- and braking-torque compensation are usually not measured directly, but rather determined from the measured kinematic points of the axle. The same applies to the longitudinal inclination at wheel center, and the anti-lift/anti-squat and anti-dive/anti-rise angles.

Basic terms for tires
The most important external forces and moments which act on a vehicle occur during the transmission of force between tire and road surface. Added to this are the wind forces which act on the vehicle special situations.

Tire contact area
Force is transmitted between tire and road surface by friction in the contact area. The two most important types of friction are adhesive friction (intermolecular adhesive force) and hysteretic friction (gearing force).

Lateral force, longitudinal force
Forces can be generated in the road plane by friction. Lateral force F_S is the force perpendicular to the wheel center plane; longitudinal force F_U runs in the direction of the wheel center plane (Figure 12). The forces are generally not applied exactly at the wheel contact point such that moments are likewise generated in relation to the wheel contact point (see Tire aligning torque).

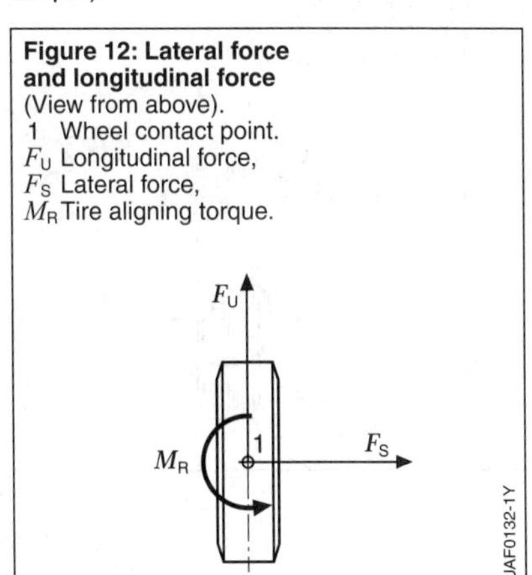

Figure 12: Lateral force and longitudinal force
(View from above).
1 Wheel contact point.
F_U Longitudinal force,
F_S Lateral force,
M_R Tire aligning torque.

Wheel load and slip angle

When lateral and longitudinal forces occur simultaneously, the forces may be mutually influenced. The situation where lateral and longitudinal forces occur but not combined is considered in the following. The generated lateral force F_S is dependent on the wheel load and the slip angle α. A dependence of lateral force F_S on speed can generally be ignored. The wheel load is the force with which the wheel center is pressed towards the road-surface plane. The slip angle α is the angle between the direction of motion of the wheel contact point and the wheel center plane (Figure 13).

When the wheel load is kept constant and the slip angle α is increased, the lateral force initially increases linearly. The lateral force reaches its maximum level at a slip angle of approx. 5° and then decreases slightly (Figure 14).

Tire aligning torque and tire caster offset

With small slip angles the lateral force is applied after the wheel contact point. When the slip angle is increased, the lateral force moves increasingly towards the wheel contact point and can also be located ahead of the wheel contact point. The distance between the lateral application point and the wheel contact point is called the tire caster offset n_R. The lateral force therefore generates a moment about the tire's vertical axis, the tire aligning torque M_R:

$$M_R = F_S \; n_R \,.$$

This produces, at constant wheel load, a curve as shown in Figure 15 for the tire aligning torque.

If the tire aligning torque is positive, it helps to make the slip angle smaller in terms of amount. This behavior helps to make the turned wheels return to the straight-ahead position when the steering wheel is released.

Slip and rolling radius

Similarly to the slip angle α for lateral force F_S, slip λ is the variable which at constant wheel load determines longitudinal force F_u. Slip occurs when the speed v_{xR} at which the wheel center moves in the longitudinal direction differs from the speed v_U at which the circumference rolls. The circumferential speed is calculated from the angular speed ω_R of the wheel and the dynamic rolling radius r_{dyn}:

$$v_U = \omega_R \; r_{dyn} \,.$$

A distinction is made between static and dynamic rolling radius. The static rolling

Figure 13: Slip angle
(View from above).
F_S Lateral force,
M_R Tire aligning torque,
α Slip angle.
R Wheel contact point.

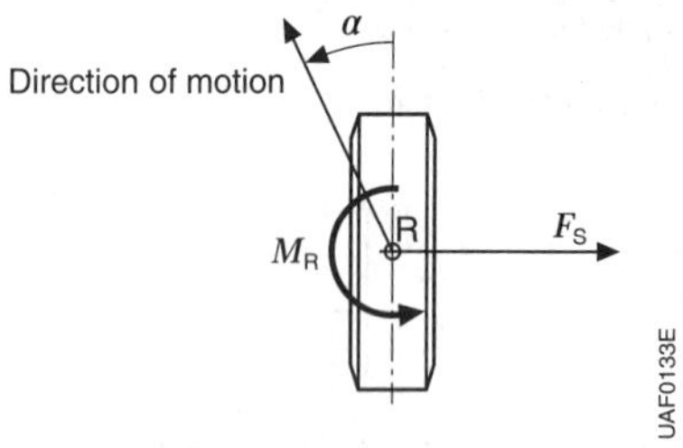

Direction of motion
α
M_R
R
F_S
UAF0133E

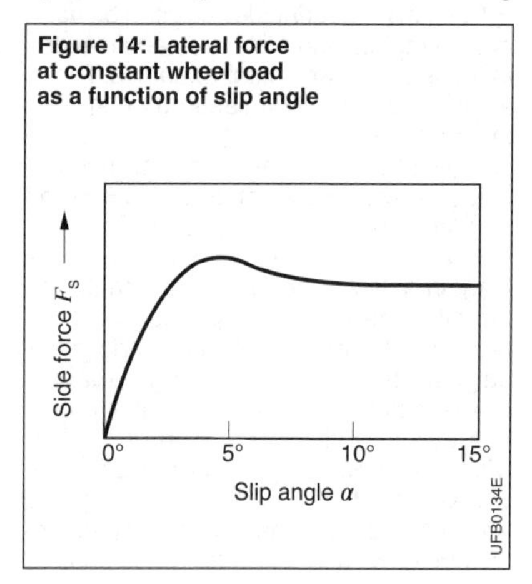

Figure 14: Lateral force at constant wheel load as a function of slip angle

radius is the shortest distance between wheel center and tire contact area. The dynamic rolling radius r_{dyn} is calculated by means of the circumference U:

$$r_{dyn} = \frac{U}{2\pi}.$$

Slip λ_A for tractive forces is defined as:

$$\lambda_A = \frac{\omega_R\, r_{dyn} - v_{xR}}{\omega_R\, r_{dyn}}.$$

Similarly, slip λ_B for braking forces is defined as:

$$\lambda_B = \frac{\omega_R\, r_{dyn} - v_{xR}}{v_{xR}}.$$

According to this definition, drive slip is always positive and brake slip always negative. These two slip definitions ensure that when a wheel is locked ($\omega_R = 0$, $\lambda_B = -1$) a slip of –100% is obtained and when a wheel is spinning ($v_{xR} = 0$, $\lambda_A = 1$) a slip of 100% is obtained.

If the drive slip is increased at constant wheel load, the tractive force (longitudinal force) increases linearly. At approx. 10% drive slip the longitudinal force reaches its maximum level and then falls again (Figure 16). The same applies to brake slip. Here the maximum braking force is created at approx. –10%.

Measured variables

The slip angle α is measured, along similar lines to the float angle, with two proximity-type speed sensors. The wheel speed and the longitudinal velocity are measured for slip λ. The dynamic rolling radius r_{dyn} is determined on test benches.

The tire aligning torque M_R, the lateral force F_S, and the longitudinal force F_U can be recorded in mobile operation by multi-component measuring wheels. Because this is very costly, tire forces and tire moments are generally measured on stationary test benches or determined using special vehicles directly on the road. Up to now, precise measurement of tire forces and tire moments has been accompanied by many systematic failures.

References

[1] DIN 70000 (earlier standard) Road vehicles – Vehicle dynamics and road-holding ability – Vocabulary.
[2] ISO 8855: Road vehicles – Vehicle dynamics and road-holding ability – Vocabulary.
[3] B. Heissing, M. Ersoy (Editors): Fahrwerkhandbuch. Vieweg+Teubner Verlag, 2008.

Figure 15: Tire aligning torque at constant wheel load as a function of slip angle

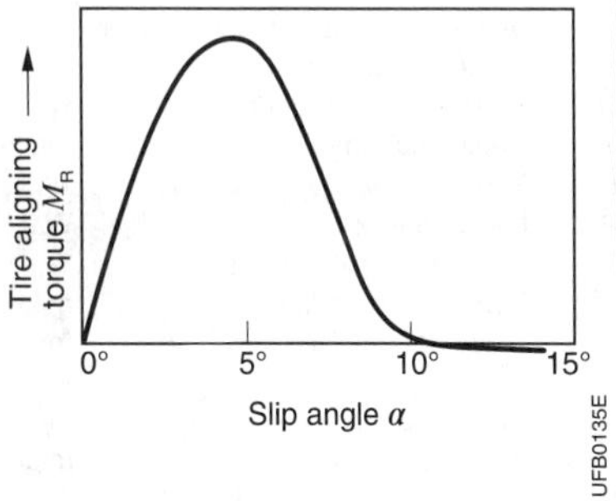

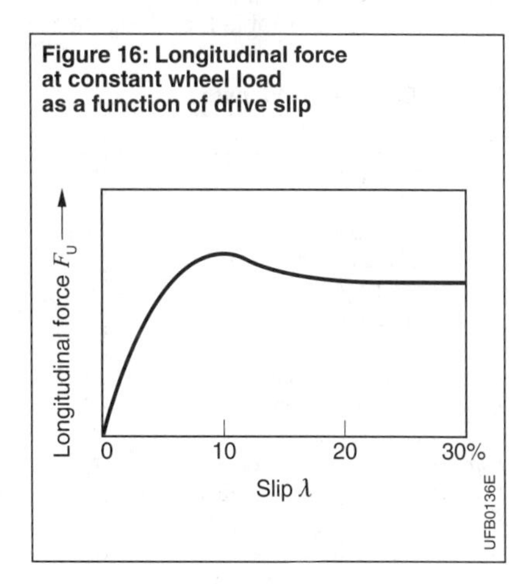

Figure 16: Longitudinal force at constant wheel load as a function of drive slip

Motor-vehicle dynamics

Dynamics of linear motion

Total running resistance

The running resistance is calculated as (Figure 1):

$$F_W = F_{Ro} + F_L + F_{St}.$$

The power which must be transmitted through the drive wheels to overcome running resistance (running-resistance power) is:

$P_W = F_W v$ or

$$P_W = \frac{F_W \cdot v}{3{,}600}$$

with P_W in kW, F_W in N, v in km/h.

Rolling resistance

The rolling resistance F_{Ro} is the product of deformation processes which occur at the contact patch between tire and road surface. The following applies:

$$F_{Ro} = f\,G \cos\alpha = f\,m\,g \cos\alpha.$$

An approximate calculation of the rolling resistance can be made using the coefficients provided in Table 2 and Figure 2.

The increase in the coefficient of rolling resistance f is directly proportional to the level of deformation, and inversely proportional to the radius of the tire. The coefficient will thus increase in response to greater loads, higher speeds and lower tire pressure.

In turns, the rolling resistance is augmented by the cornering resistance.

$$F_K = f_K\ G.$$

The coefficient of cornering resistance f_K is a function of vehicle speed, curve

Figure 1: Running resistance
C Center of gravity, α Gradient angle,
F_L Aerodynamic drag, F_{Ro} Rolling resistance,
F_{St} Climbing resistance,
G Vehicle weight.

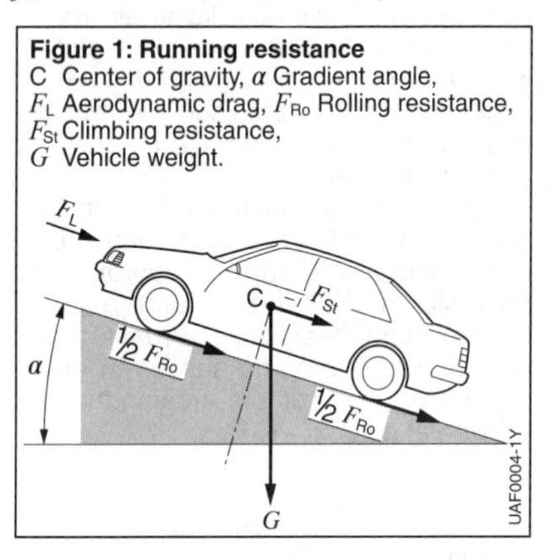

Table 1: Quantities and units

Quantity		Unit
A	Largest cross-section of vehicle	m^2
a	Acceleration, braking (deceleration)	m/s^2
c_w	Drag coefficient	–
F	Motive force	N
F_{cf}	Centrifugal force	N
F_L	Aerodynamic drag	N
F_{Ro}	Rolling resistance	N
F_{St}	Climbing resistance	N
F_W	Running resistance	N
f	Coefficient of rolling resistance	–
G	Weight = $m\,g$	N
G_B	Sum of wheel forces on driven or braked wheels	N
g	Gravitational acceleration = 9.81 $m/s^2 \approx$ 10 m/s^2	m/s^2
i	Gear or transmission ratio between engine and drive wheels	–
M	Engine torque	N m
m	Vehicle mass	kg
n	Engine speed	min^{-1}
P	Power	W
P_W	Motive power	W
P	Gradient (= $100 \cdot \tan\alpha$)	%
r	Dynamic tire radius	m
s	Distance traveled	m
t	Time	s
v	Driving speed	m/s
v_0	Headwind speed	m/s
W	Work	J
α	Incline angle, Gradient angle	°
μ_r	Coefficient of static friction	–

Additional symbols and units in text.

radius, suspension geometry, tires, tire pressure, and the vehicle's response under lateral acceleration.

Aerodynamic drag
Aerodynamic drag is calculated as:

$F_L = 0.5 \rho c_w A (v + v_0)^2$ or

$F_L = 0.0386 \rho c_w A (v + v_0)^2$

with v in km/h, F_L in N, ρ in kg/m³, A in m², air density ρ (ρ = 1.202 kg/m³ at 200 m altitude).

The aerodynamic drag is

$P_L = F_L v = 0.5 \rho c_w A v (v+v_0)^2$

or

$P_L = 12.9 \cdot 10^{-6} c_w \cdot A v (v+v_0)^2$

with P_L in kW, F_L in N, v and v_0 in km/h, A in m², ρ = 1.202 kg/m³. The largest cross section of an automobile is:

$A \approx 0.9 \times$ lane width $\times$ height.

Empirical determination of coefficients for aerodynamic drag and rolling resistance
The vehicle is allowed to coast down in neutral under windless conditions on a

Table 2: Coefficients of rolling resistance

Road surface	Coefficient of rolling resistance f
Pneumatic car tires on	
Large set pavement	0.013
Small set pavement	0.013
Concrete, asphalt	0.011
Rolled gravel	0.02
Tarmacadam	0.025
Unpaved road	0.05
Field	0.1–0.35
Pneumatic truck tires	
on concrete, asphalt	0.006–0.01
Strake wheels in field	0.14–0.24
Track-type tractor in field	0.07–0.12
Wheel on rail	0.001–0.002

Figure 2: Rolling resistance coefficient for adial car tires on smooth, level road surfaces under normal load and at correct tire pressure

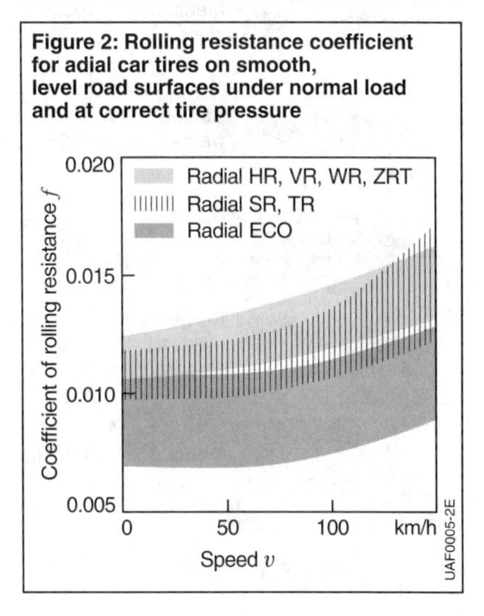

Table 3: Empirical determination of coefficients for aerodynamic drag and rolling resistance

	1st trial (high speed)	2nd trial (low speed)
Initial velocity	v_{a1} = 60 km/h	v_{a2} = 15 km/h
Terminal velocity	v_{b1} = 55 km/h	v_{b2} = 10 km/h
Interval between v_a and v_b	t_1 = 7.8 s	t_2 = 12.2 s
Mean velocity	$v_1 = \frac{v_{a1}+v_{b1}}{2} = 57.5$ km/h	$v_2 = \frac{v_{a2}+v_{b2}}{2} = 12.5$ km/h
Mean deceleration	$a_1 = \frac{v_{a1}-v_{b1}}{t_1} = 0.64 \frac{\text{km/h}}{\text{s}}$	$a_2 = \frac{v_{a2}-v_{b2}}{t_2} = 0.41 \frac{\text{km/h}}{\text{s}}$
Drag coefficient	$c_W = \frac{6m\,(a_1-a_2)}{A\,(v_1^2-v_2^2)} = 0.29$ (with m = 1,450 kg, A = 2.2 m²)	
Coefficient of rolling resistance	$f = \frac{28.2(a_2\,v_1^2 - a_1\,v_2^2)}{10^3 \cdot (v_1^2-v_2^2)} = 0.011$	

Table 4: Drag coefficients and aerodynamic drag for various body configurations

	Coefficient of aero-dynamic drag c_W	Frontal area in m^2	Drag area in m^2	Aerodynamic drag in kW at a driving speed of	
				40 km/h	120 km/h
Subcompact	0.30 to 0.37	2.04 to 2.15	0.65 to 0.72	0.5 to 0.6	14.4 to 16.0
Compact class	0.27 to 0.32	2.21 to 2.26	0.62 to 0.72	0.5 to 0.6	13.8 to 16.0
Medium class	0.22 to 0.35	2.22 to 2.38	0.51 to 0.66	0.4 to 0.5	11.3 to 14.7
Station wagon	0.28 to 0.35	2.27 to 2.35	0.64 to 0.82	0.5 to 0.7	14.2 to 18.2
Van	0.30 to 0.35	2.60 to 3.25	0.80 to 1.14	0.7 to 0.9	17.8 to 25.3
Convertible – closed	0.28 to 0.37	1.91 to 2.20	0.57 to 0.81	0.5 to 0.7	12.7 to 18.0
Convertible – open	0.35 to 0.49	1.84 to 2.04	0.71 to 0.98	0.6 to 0.8	15.8 to 21.8
Off-road vehicle	0.32 to 0.55	2.34 to 3.15	0.77 to 1.38	0.6 to 1.1	17.1 to 30.7
Sports cars	0.27 to 0.40	1.61 to 2.25	0.56 to 0.79	0.5 to 0.7	12.4 to 17.6
Luxury class	0.24 to 0.38	2.33 to 2.81	0.62 to 1.07	0.5 to 0.9	13.8 to 23.8

UAF0006-2E

Table information:
The midsize class market segment features the lowest c_W values, followed by the luxury class, compact class, sports cars, closed convertibles, station wagons, vans, small cars up to the highest values for open-top convertibles and off-road vehicles.
However, the frontal area of a vehicle also has a significant influence on the power requirement for overcoming the drag. This changes the order. Midsize vehicles also require the lowest power here, followed by the compact class, sports cars and closed-top convertibles, which are nearly the same. This is followed by small vehicles and station wagons. Due to the size, luxury vehicles have fallen behind significantly and then open-top convertibles due to their high c_W value. Vans and off-road vehicles are found at the end.
This also shows that the aerodynamic drag at 40 km/h hardly differs between the classes.
At 120 km/h, however, there are substantial differences because the power requirement increases with the speed cubed.

level road surface. The time that elapses while the vehicle coasts down by a specific increment of speed is measured from two initial velocities, v_1 (high speed) and v_2 (low speed). This information is used to calculate the mean deceleration rates a_1 and a_2. The formulae and example from Table 3 are based on a vehicle weighing m = 1,450 kg with a cross-section A = 2.2 m².

The method is suitable for application at vehicle speeds of less than 100 km/h.

Table 5: Gradient angle and climbing resistance

Gradient angle α	Pitch: p in %	Pitch: *Gradient*	Climbing resistance at m = 1,000 kg in N
45°	100	1 in 1	
	90		6,500
40°	80		6,000
35°	70	1 in 1.5	5,500
30°	60		5,000
			4,500
25°	50	1 in 2	4,000
	40	1 in 2.5	3,500
20°		1 in 3	3,000
15°	30	1 in 4	2,500
10°	20	1 in 5	2,000
			1,500
5°	10	1 in 10	1,000
		1 in 20	500
		1 in 50	
0°	0	1 in 100	0

UAF0007E

Climbing resistance and downgrade force

Climbing resistance (F_{St} with positive sign) and downgrade force (F_{St} with negative sign) are calculated as:

$$F_{St} = G \sin\alpha = m\, g \sin\alpha$$

or, for a working approximation:

$$F_{St} \approx 0.01\, m\, g\, p\,.$$

These equations apply to gradients up to $p \leq 20\%$, because at small angles the following applies:

$\sin\alpha \approx \tan\alpha$ (less than 2% error).

Climbing power is calculated as:

$P_{St} = F_{St}\, v$ or

with P_{St} in kW, F_{St} in N and v in km/h:

$$P_{St} = \frac{F_{St}\, v}{3{,}600} = \frac{m\, g\, v \sin\alpha}{3{,}600}$$

or, for a working approximation:

$$P_{St} = \frac{m\, g\, p\, v}{360{,}000}$$

Table 6: Climbing resistance and climbing power

Values at m = 1,000 kg

Climbing resistance F_{St} in N	Climbing power P_{St} in kW at different speeds: 20 km/h	30 km/h	40 km/h	50 km/h	60 km/h
6,500	36	54	72	–	–
6,000	33	50	67	–	–
5,500	31	46	61	–	–
5,000	28	42	56	69	–
4,500	25	37	50	62	–
4,000	22	33	44	56	67
3,500	19	29	39	49	58
3,000	17	25	33	42	50
2,500	14	21	28	35	42
2,000	11	17	22	28	33
1,500	8.3	12	17	21	25
1,000	5.6	8.3	11	14	17
500	2.3	4.2	5.6	6.9	8.3
0	0	0	0	0	0

The gradient is:

$p = (h/l) \cdot 100\,\%$ or
$p = (\tan \alpha) \cdot 100\,\%$,

with h as the height of the projected distance l.

In English-speaking countries, the *gradient* is calculated as follows: The following relationship exists between gradient in % and *gradient*:

Gradient 1 in $100/p$.

Example: 1 in 2.

Example of calculating motive force and climbing power
To climb a hill with a gradient of $p = 21\,\%$, a vehicle weighing 1,500 kg will require approximately $1.5 \cdot 2{,}000$ N = 3,000 N of motive force at the wheels (value from Table 5) and at $v = 40$ km/h roughly $1.5 \cdot 22$ kW = 33 kW of climbing power (value from Table 6).

Motive force
The higher the engine torque M and overall transmission ratio i between the engine and drive wheels, and the lower the power-transmission losses, the higher the motive force F available at the driven wheels.

$$F = \frac{M\ i}{r} \cdot \eta \quad \text{or} \quad F = \frac{P\ \eta}{v}.$$

η indicates the drivetrain efficiency. For a longitudinal engine $\eta \approx 0.88\text{–}0.92$, for a transverse engine $\eta \approx 0.91\text{–}0.95$.

The motive force F is partially consumed in overcoming the running resistance F_W. Numerically higher transmission ratios are applied to deal with the substantially increased running resistance encountered on gradients (variable speed transmission).

Figure 3: Determining the rotational inertia coefficient k_m
V_H Engine displacement in liters,
m Vehicle weight,
i Overall ratio between engine and drive wheels
r Wheel radius
1 m/V_H = 500 kg/l,
2 m/V_H = 750 kg/l,
3 m/V_H = 1,000 kg/l.

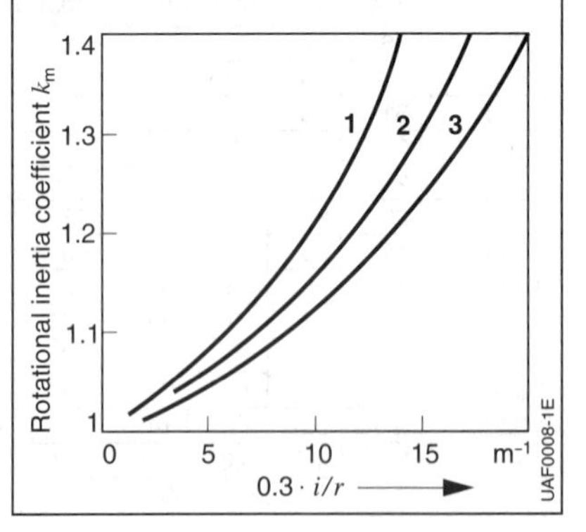

Vehicle and engine speeds
The following applies to engine speed:

$$n = \frac{60\ v\ i}{2\ \pi\ r},$$

or with v in km/h:

$$n = \frac{1{,}000\ v\ i}{2\ \pi\ 60\ r}.$$

Acceleration
The surplus force $F - F_W$ accelerates the vehicle. Or decelerates it when F_W exceeds F.

$$a = \frac{F - F_W}{k_m\ m} \quad \text{or} \quad a = \frac{P\ \eta - P_W}{v\ k_m\ m}.$$

The rotational inertia coefficient k_m (Figure 3) compensates for the apparent increase in vehicle mass due to the rotating masses (wheels, flywheel, crankshaft, etc.).

Motive force and road speed on vehicles with automatic transmissions
When the formula for motive force is applied to automatic transmissions with hydrodynamic torque converters or hydrodynamic clutches, the engine torque M is replaced by the torque at the converter turbine, while the rotational speed of the converter turbine is used in the formula for engine speed. The relationship between $M_{Turb} = f(n_{Turb})$ and the engine characteristic $M_{Mot} = f(n_{Mot})$ is determined using the characteristics of the hydrodynamic converter.

The tractive force and the motive force at the wheels for the individual gears as a function of the driving speed can be read from the tractive-force/running diagram (Figure 4). The kink typical of a hydrodynamic converter resulting from the end of torque multiplication can be seen. The top speed in each case as a function of gear and gradient can be determined from the points of intersection of the tractive-force lines with the running-resistance lines.

Figure 4: Tractive-force running diagram for a car with automatic transmission and hydrodynamic Trilok converter at full throttle
1 Running-resistance lines for different gradient,
2 Motive-force lines (tractive-force lines) for different gears.

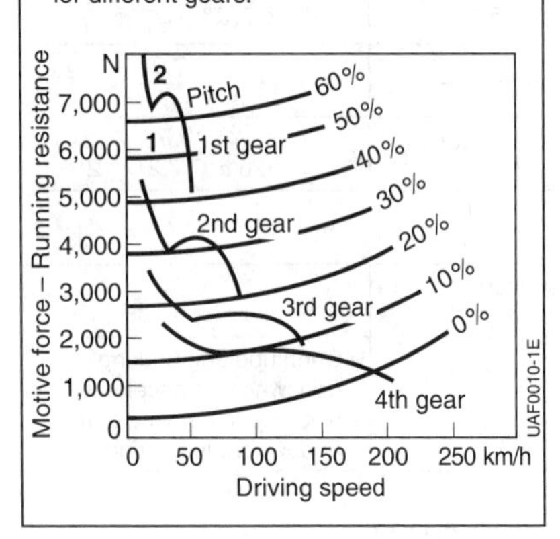

Adhesion to road surface

Coefficient of static friction

The coefficient of static friction μ_r (between the tires and the road surface) is determined by the vehicle's speed, the condition of the tires and the state of the road surface (Table 7). The figures cited apply for concrete and tarmacadam road surfaces in good condition. The coefficients of sliding friction (with wheel locked) are usually lower than the coefficients of static friction.

Special rubber compounds providing friction coefficients of up to μ_r = 1.8 are employed in racing tires.

Aquaplaning

Aquaplaning has a particularly dramatic influence on the contact between tire and road surface. It describes the state in

Figure 5: Aquaplaning
1 Tire, 2 Wedge of water, 3 Road surface

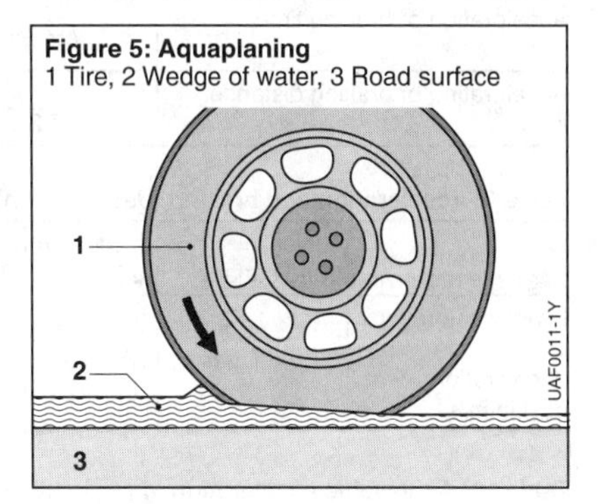

Table 7: Coefficients of static friction for pneumatic tires on various road surfaces

Vehicle speed in km/h	Condition of the tires	State of road surface Dry	Wet Water level approx. 0.2 mm	Heavy rain Water level approx. 1 mm	Puddles Water level approx. 2 mm	Iced (black ice)
		Coefficient of static friction μ_r				
50	new	0.85	0.65	0.55	0.5	0.1 and less
	worn[1]	1	0.5	0.4	0.25	
90	new	0.8	0.6	0.3	0.05	–
	worn[1]	0.95	0.2	0.1	0.05	
130	new	0.75	0.55	0.2	0	–
	worn[1]	0.9	0.2	0.1	0	

[1] Worn to ≥ 1.6 mm tread depth
(legal minimum in Germany as per § 36.2 StVZO (German Road Traffic Licensing Regulations)).

which a layer of water separates the tire and the (wet) road surface (Figure 5). The phenomenon occurs when a wedge of water forces its way underneath the tire's contact patch and lifts it from the road. The tendency to aquaplane is dependent upon such factors as the depth of the water on the road surface, the vehicle's speed, the tread pattern, the tread wear, and the load pressing the tire against the road surface. Wide tires are particularly susceptible to aquaplaning. An aquaplaning vehicle cannot transmit control and braking forces to the road surface, and may as result go into a skid.

Accelerating and braking

The vehicle is regarded as accelerating or braking (decelerating) at a constant rate when a remains constant. The equations set out in Table 8 apply to an initial or final speed of $v = 0$.

Maxima for acceleration and braking (deceleration)

When the motive or braking forces exerted at the vehicle's wheels reach such a magnitude that the tires are just still within their limit of adhesion (maximum adhesion is still present), the relationships shown in

Table 8: Accelerating and braking

	Equations for v in m/s	Equations for v in km/h
Acceleration or braking (deceleration) in m/s²	$a=\frac{v^2}{2s}=\frac{v}{t}=\frac{2s}{t^2}$	$a=\frac{v^2}{26s}=\frac{v}{3.6t}=\frac{2s}{t^2}$
Acceleration or braking time in s	$t=\frac{v}{a}=\frac{2s}{v}=\sqrt{\frac{2s}{a}}$	$t=\frac{v}{3.6a}=\frac{7.2s}{v}=\sqrt{\frac{2s}{a}}$
Accelerating or braking distance in m	$s=\frac{v^2}{2a}=\frac{vt}{2}=\frac{at^2}{2}$	$s=\frac{v^2}{26a}=\frac{vt}{7.2}=\frac{at^2}{2}$

Table 9: Acceleration and braking (deceleration)

	Level road surface	Inclined road surface α [°]; $p = 100 \tan \alpha$ [%]	
Limit acceleration or deceleration a_{max} in m/s²	$a_{max} = k g \mu_r$	$a_{max} = g(k\mu_r \cos\alpha \pm \sin\alpha)$ approximation[1]: $a_{max} \approx g(k\mu_r \pm 0.01p)$	+ when upgrade braking or downgrade acceleration – when upgrade acceleration or downgrade braking

Table 10: Achievable acceleration a_e (P_a in kW, v in km/h, m in kg)

Level road surface	Inclined road surface	
$a_e = \frac{3{,}600\,P_a}{kmv}$	$a_e = \frac{3{,}600\,P_a}{kmv} \pm g \sin\alpha$	+ Downgrade acceleration – Upgrade acceleration for $g \sin\alpha$ the approximation[1] $g\,p/100$

Table 11: Work and power

	Level road surface	Inclined road surface α [°]; $p = 100 \tan \alpha$ [%]	
Acceleration or Braking work W in J[2]	$W = kmas$	$W = ms(ka \pm g \sin\alpha)$ approximation[1]: $W = ms(ka \pm g\,p/100)$	+ Downgrade braking or upgrade acceleration – Downgrade acceleration or upgrade braking
Acceleration or braking power in W at velocity v	$P_a = kmasv$	$P_a = mv(ka \pm g \sin\alpha)$ approximation[1]: $P_a = mv(ka \pm g\,p/100)$	v in m/s. For v in km/h, use $v/3.6$.

[1] Valid to approx. $p = 20\,\%$ (under 2 % error), [2] 1 J = 1 N m = 1 W s.

Tables 9 and 10 exist between the gradient angle α, coefficient of static friction μ_r and maximum acceleration or deceleration. The real-world figures are always somewhat lower, as all the vehicle's tires do not simultaneously exploit their maximum adhesion during each acceleration or deceleration. Electronic traction control and braking systems (TCS, ABS, ESP) maintain the traction level in the vicinity of the coefficient of static friction.

k indicates the ratio between the load on driven or braked wheels and the total weight. If all wheels are driven or braked, $k = 1$. At 50 % load distribution $k = 0.5$.

Example:

$k = 0.5$; $g = 10\ \text{m/s}^2$;
$\mu_r = 0.6$; $p = 15\,\%$;
$a_{max} = 10 \cdot (0.5 \cdot 0.6 \pm 0.15)\ \text{m/s}^2$
upgrade braking (+): $a_{max} = 4.5\ \text{m/s}^2$,
downgrade braking (–): $a_{max} = 1.5\ \text{m/s}^2$.

Work and power
The power required to maintain a consistent rate of acceleration (deceleration) varies according to vehicle speed (Table 11). The power available for acceleration is:

$P_a = P\,\eta - P_W$, where

P Engine output
η Efficiency
P_W Motive power

Actions: Reaction, braking and stopping

(in accordance with ÖNORM V 5050 [1] and [2])

Hazard recognition time
The hazard recognition time, also known as the danger reaction time, is the period of time that elapses between perceiving a visible obstacle or its movement and the time required to recognize it as a hazard (Figure 6). If, as part of this danger recognition and response process, it is necessary for the driver to turn his eyes towards the hazardous situation, the hazard recognition and danger reaction time will extend by roughly 0.4 s.

Prebraking time
The prebraking time (t_{VZ}) is the period of time that elapses between the moment the hazard is recognized and the start of braking, defined by way of calculation. Based on the following formula, the prebraking time is in the range from approx. 0.8 to 1.0 s:

$$t_{VZ} = t_R + t_U + t_A + t_S/2\ .$$

The reaction time (t_R) is the period of time that elapses between the moment a defined incitement to action occurs and the start of the first specifically targeted action. Instinctive hazard recognition triggers an inherent, automatic reaction (spontaneous reaction), enabling the vehicle driver to determine both the point of the reaction as well as the position of the reason for the reaction, delayed by the distance covered during the prebraking time. Human beings require about 0.2 s for the spontaneous reaction; however, the reaction time will be at least 0.3 s if the driver needs to make a decision to perform a preventive or evasive action in response to conscious hazard recognition (choice reaction).

The transfer time (t_U) is the period of time the driver requires to transfer the foot from the accelerator pedal to the brake pedal. The transfer time is in the range of about 0.2 s.

The response time (t_A) is the period of time it takes to transmit the pressure applied at the brake pedal via the brake system through to the point when the braking action becomes effective (complete build-up of the application force and incipient increase in vehicle deceleration).

The pressure build-up time (t_S) is the period of time that elapses between the braking action taking effect and reaching fully effective braking deceleration. Alternatively, half the pressure build-up time ($t_S/2$) may be assumed as the start of braking, determined by way of calculation.

In accordance with the EU Council of Ministers Directive EEC 71/320 Addendum 3/2.4 [3], the sum of the response time and pressure build-up time may not exceed 0.6 s. A poorly maintained braking system will lengthen the response and pressure build-up time.

Braking time

The braking time (t_B) is the period of time that elapses between the mathematically calculated start of braking and the moment the vehicle comes to a complete stop. This period of time comprises half the pressure build-up time ($t_S/2$) (calculation assumption: only half the pressure build-up time but complete braking deceleration), as well as the full braking time (t_V), during which maximum braking deceleration is actually effective.

$$t_B = t_S/2 + t_V.$$

Stopping time and stopping distance

The stopping time (t_{AH}) is the sum of the prebraking time (t_{VZ}) and braking time (t_B).

$$t_{AH} = t_{VZ} + t_B.$$

The stopping distance (s_{AH}) can be calculated by way of integration (Tables 12 and 13).

Figure 6: Actions: Reaction, braking and stopping
a) Deceleration while braking,
b) Distance covered while braking.

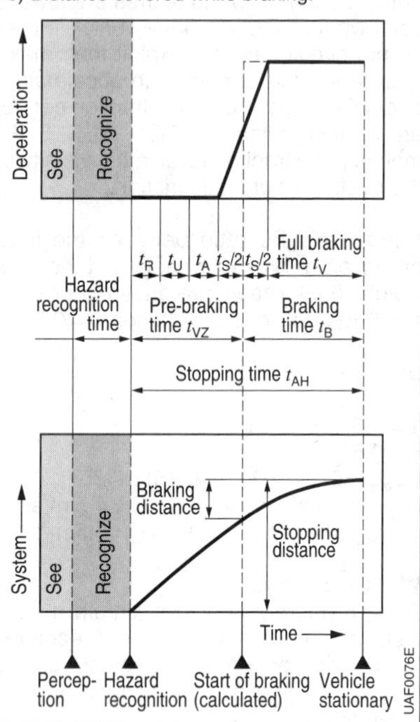

Table 12: Stopping time and stopping distance

	Equations for v in m/s	Equations for v in km/h
Stopping time t_{AH} in s	$t_{AH} = t_{VZ} + \frac{v}{a}$	$t_{AH} = t_{VZ} + \frac{v}{3.6\,a}$
Stopping distance s_{AH} in m	$s_{AH} = v\,t_{VZ} + \frac{v^2}{2\,a}$	$s_{AH} = \frac{v}{3.6}\,t_{VZ} + \frac{v^2}{25.92\,a}$

Table 13: Stopping distance as a function of driving speed and deceleration

Deceleration a in m/s²	Vehicle speed prior to braking in km/h												
	10	30	50	60	70	80	90	100	120	140	160	180	200
	Distance during prebraking time (delay) of 1 s in m												
	2.8	8.3	14	17	19	22	25	28	33	39	44	50	56
	Stopping distance in m												
4.4	3.7	16	36	48	62	78	96	115	160	210	270	335	405
5	3.5	15	33	44	57	71	87	105	145	190	240	300	365
5.8	3.4	14	30	40	52	65	79	94	130	170	215	265	320
7	3.3	13	28	36	46	57	70	83	110	145	185	230	275
8	3.3	13	26	34	43	53	64	76	105	135	170	205	250
9	3.2	12	25	32	40	50	60	71	95	125	155	190	225

Passing (overtaking)

The complete passing maneuver involves pulling out of the lane, overtaking the other vehicle, and returning to the original lane (Figure 7). Passing can take place undor a widc variety of very different cir-cumstances and conditions, so precise calculations are difficult. For this reason, the following calculations, graphs, and illustrations will confine themselves to an examination of two extreme conditions: passing at a constant velocity and passing at a constant rate of acceleration. We can simplify graphic representation by treating the passing distance s_u as the sum of two (straight-ahead) components, while disregarding the extra travel involved by pulling out of the lane and back in again.

Passing distance

This passing distance is

$s_u = s_H + s_L$ (variables in Table 14).

The distance s_H which the more rapid vehicle must cover compared to the slower vehicle (considered as being stationary) is the sum of the vehicle lengths l_1 and l_2 and the safety margins s_1 and s_2.

$$s_H = s_1 + s_2 + l_1 + l_2 .$$

During the passing time t_u, the slower vehicle covers the distance s_L, this is the distance that the overtaking vehicle must also travel in order to maintain the safety margin.

$$s_L = \frac{t_u\, v_L}{3.6} \quad \text{(for } v \text{ in km/h)}.$$

Table 14: Quantities and units

Quantity		Unit
a	Acceleration	m/s²
l_1, l_2	Vehicle length	m
s_1, s_2	Safety margin	m
s_H	Relative distance traveled by overtaking vehicle	m
s_L	Distance traveled by vehicle being passed	m
s_U	Passing distance	m
t_U	Passing time	s
v_L	Speed of slower vehicle	km/h
v_H	Speed of faster vehicle	km/h

Safety margin

The minimum safety margin corresponds to the distance covered during the pre-braking time t_{VZ}. The figure for a prebraking time of t_{VZ} = 1.08 s for velocities in km/h would be (0.3 v) meters. However, a minimum of 0.5 v is advisable outside of built-up areas.

Passing at constant speed

On highways with more than two lanes, the overtaking vehicle will frequently be traveling at a speed adequate for passing before the actual process begins. The passing time (from initial lane change until return to the original lane has been completed) is then:

$$t_U = \frac{3.6\, s_H}{v_H - v_L}$$

with t in s, s in m and v in km/h.

The passing distance is

$$s_U = \frac{t_U\, v_H}{3.6} \approx \frac{s_H\, v_H}{v_H - v_L} .$$

Passing with constant acceleration

On narrow roads, the vehicle will usually have to slow down to the speed of the preceding car or truck before accelerating to pass. The attainable acceleration figures depend upon engine output, vehicle weight, speed, and running resistance. These generally lie within the range of 0.4–0.8 m/s², with up to 1.4 m/s² available in lower gears for further reductions in passing time. The distance required to complete the passing maneuver should never exceed half the visible stretch of road.

Figure 7: Passing distance
Quantities, see Table 14.
1 Overtaken vehicle,
2 Overtaking vehicle.

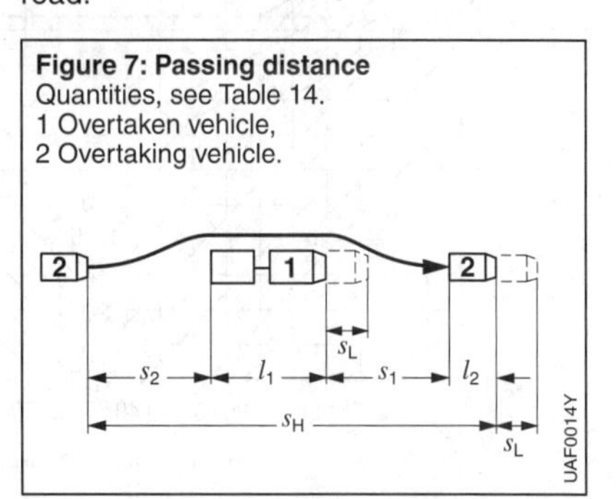

Operating on the assumption that a constant rate of acceleration can be maintained for the duration of the passing maneuver, the passing time will be:

$t_U = \sqrt{2\, s_H / a}$.

The distance which the slower vehicle covers within this period is defined as $s_L = t_U v_L / 3.6$. This gives a passing distance of:

$$s_U = s_H + \frac{t_U v_L}{3.6} .$$

with t in s, s in m and v in km/h.

The left side of Figure 8 shows the relative distances s_H for speed differentials $v_H - v_L$ and acceleration rates a, while the right side shows the distances s_L covered by the vehicle being passed at various speeds v_L. The passing distance s_U is the sum of s_H and s_L.

First, determine the distance s_H to be traveled by the passing vehicle. Enter this distance on the left side of the graph between the Y axis and the applicable ($v_H - v_L$) line or acceleration line. Then extrapolate the line to the right, over to the speed line v_L.

Example (represented by dash-dot lines in the graph):
$v_L = v_H = 50$ km/h,
$a = 0.4$ m/s²,
$l_1 = 10$ m, $l_2 = 5$ m,
$s_1 = s_2 = 0.3\, v_L = 0.3\, v_H = 15$ m.

Solution:
Enter intersection of a = 0.4 m/s² and
s_H = 15 + 15 + 10 + 5 = 45 m
in the left side of the graph.
Indication: t_U = 5 s, s_L = 210 m.
The following applies: $s_U = s_H + s_L$ = 255 m.

Visual range

For safe passing on narrow roads, the visibility must be at least the sum of the passing distance plus the distance which would be traveled by an oncoming vehicle while the passing maneuver is in progress. This distance is approximately 400 m if the vehicles approaching each other are traveling at speeds of 90 km/h, and the vehicle being overtaken at 60 km/h.

Figure 8: Graph for determining passing distance

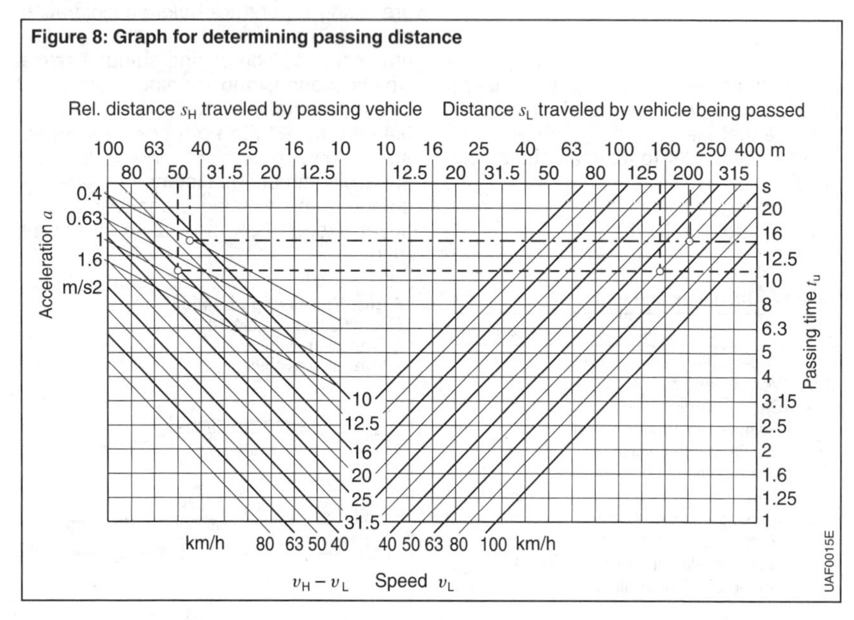

Fuel consumption

Determining fuel consumption

The official data for standard fuel consumption are determined on the basis of dynamometer tests conducted on exhaust-emission test benches in which statutory test cycles (e.g. MNEDC for Europe, FTP 75 and Highway for the USA and JC08 for Japan) are run. The exhaust gas is collected in sample bags and its constituents HC, CO and CO_2 are subsequently analyzed for the purpose of determining consumption (see Exhaust-gas measuring techniques). The CO_2 content of the exhaust gas is proportional to the fuel consumption.

The following reference values apply to Europe:
Diesel: 1 *l*/100km ≈ 26.5 g CO_2/km.
Gasoline (Euro 4):
1 *l*/100km ≈ 24.0 g CO_2/km.
Gasoline (Euro 5):
1 *l*/100km ≈ 23.4 g CO_2/km.

The change from Euro 4 to Euro 5 is based on a 5% admixture of ethanol in gasoline.

Vehicle weights are simulated on the test bench as an alternative by test weights which, depending on the country and the curb weight of the vehicle, is graded in increments of 55–120 kg. When the vehicle weight is allocated to the corresponding inertia weight class, the ready-for-operation weight of the vehicle (including all the fillers, tool kit and the fuel

Table 15: Quantities and units

Quantity		Unit
B_e	Distance consumption	g/m
b_e	Specific fuel consumption	g/kWh
m	Vehicle mass	kg
A	Frontal area of vehicle	m^2
f	Coefficient of rolling resistance	–
c_d	Drag coefficient	–
g	Gravitational acceleration	m/s^2
t	Time	s
v	Driving speed	m/s
a	Acceleration	m/s^2
B_r	Braking resistance	N
$\eta_Ü$	Transmission efficiency of drivetrain	–
ρ	Density of the air	kg/m^3
α	Gradient angle	°

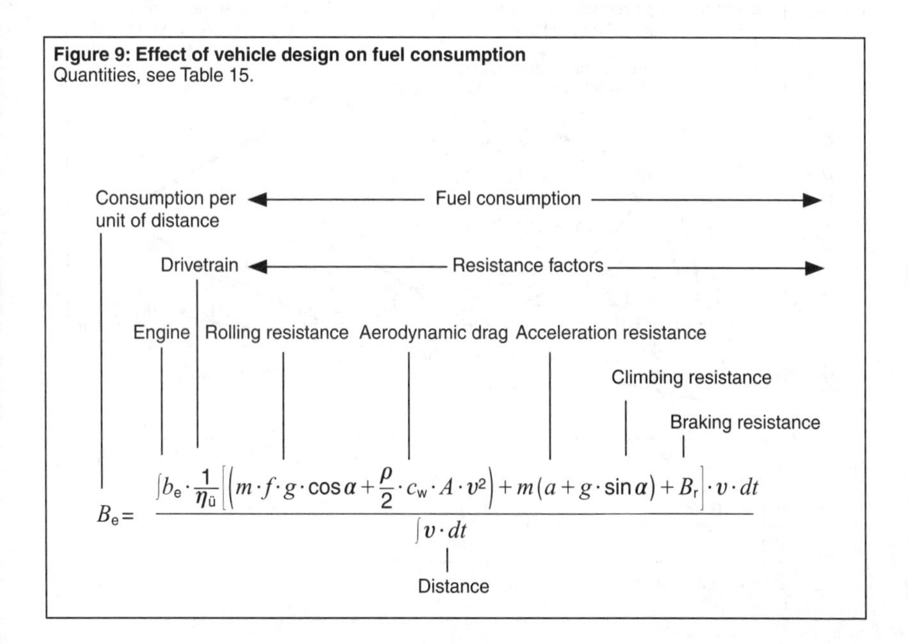

$$B_e = \frac{\int b_e \cdot \frac{1}{\eta_Ü}\left[\left(m \cdot f \cdot g \cdot \cos\alpha + \frac{\rho}{2} \cdot c_w \cdot A \cdot v^2\right) + m(a + g \cdot \sin\alpha) + B_r\right] \cdot v \cdot dt}{\int v \cdot dt}$$

Figure 9: Effect of vehicle design on fuel consumption
Quantities, see Table 15.

tank filled to 90%) plus 100 kg as substitute for a driver and luggage is applied. The difference in consumption in the jump to an adjacent inertia weight class is – depending on the vehicle – between 0.15 and 0.25 l/100km.

Units of fuel consumption
Standard fuel consumption is given in different units, depending on the country and the test cycle. In Europe, the figure is given in g CO_2/km or l/100 km, in the USA in mpg (miles per gallon), and in Japan in km/l.

Conversion examples:
30 mpg → 235.215/30 → 7.8 l/100 km,
22.2 km/l → 100/22.2 → 4.5 l/100 km.

Running resistances
If one disregards the influence of the driver on fuel consumption, which can run to 30%, there are, based on the consumption formula (Figure 9), three distinct groups of influencing factors:

- The engine (including belt drive and ancillaries)
- The internal running resistances of the drivetrain (e.g. transmission, differentials)
- The external running resistances.

External resistance factors
The external running resistances determine a vehicle's minimum energy demand in a specified driving profile. They can be reduced by measures such as reduced vehicle weight, lower tire rolling resistance, and improved aerodynamics. On an average production vehicle, 10% reductions in weight, drag and rolling resistance result in fuel-consumption reductions of roughly 6%, 3%, and 2%, respectively.

The formula in Figure 9 distinguishes between acceleration resistance and braking resistance. This illustrates clearly that consumption per unit of distance increases above all when the brakes are subsequently applied and not the overrun fuel cutoff of the engine or even a hybrid

Figure 10: Engine fuel-consumption map

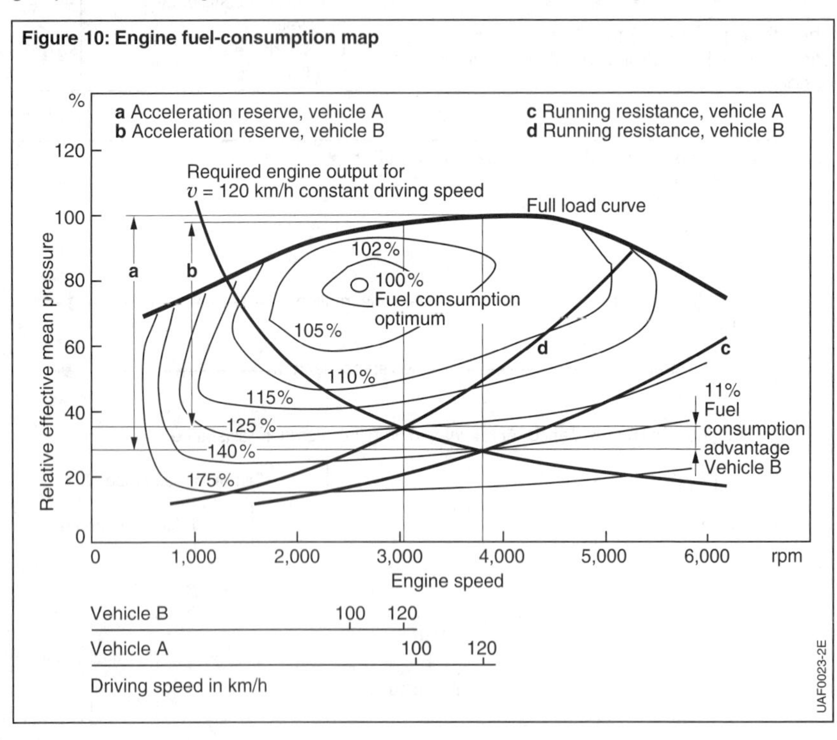

system is used to recover some of the brake energy.

Internal running resistances
The internal running resistances are made up of the losses in the drivetrain from the crankshaft to the wheels. In Figure 10, the sum total of the internal and external running resistances are shown by curves c and d, where vehicle A shows significantly lower resistances than vehicle B.

As well as the transmission losses in the drivetrain, the overall ratio also affects fuel consumption. This is calculated from the product of the transmission and differential ratios. The choice of overall ratio results in different operating points in the engine fuel-consumption map for a specific driving speed. A "longer" ratio, i.e. smaller overall ratio, will generally move the operating points to map areas corresponding to lower fuel consumption. At the same time, it must be noted that the accelerating performance is reduced and the NVH behavior worsens ("Noise Vibration Harshness", impacting on driving comfort). A useful choice of ratio is therefore only possible within limits.

The most common way of presenting engine fuel-consumption maps is the "graph diagram", in which the effective mean pressure and, as family parameters, lines of constant specific consumption (fuel throughput referred to power output in g/kWh) are plotted against the engine speed (Figure 10). This makes it possible to compare the efficiency of engines of different sizes and types.

Another way of presentation is, as a family parameter, the fuel throughput or mass flow (e.g. in kg/h). This form of presentation is particularly suitable as an input variable for CAE programs (computer-aided engineering), which can be used to simulate fuel consumption (corresponding to CO_2 emissions).

Entered in both presentations as additional information are the full-load torque curve as the upper limiting line, the idle and breakaway speeds as limiting speeds, and often also lines of constant power (power hyperbolas on account of $P \sim p_{\text{me}} \cdot n$).

Dynamics of lateral motion

Ranges of lateral acceleration

Today passenger vehicles can reach lateral acceleration levels of up to 10 m/s^2. Lateral acceleration is subdivided into the following ranges (Figure 11):

The range from 0 to 0.5 m/s^2 is known as the small-signal range. The phenomenon to be considered in this range is the straight-running behavior, caused by crosswind and irregularities in the road, such as ruts. Crosswind excitation is induced by blustery, gusty wind and by driving into and out of wind shadow areas.

The range from 0.5 to 4 m/s^2 is known as the linear range, as the vehicle behavior that occurs in this range can be described with the aid of the linear single-track model. Typical maneuvers involving dynamics of lateral motion include sudden steering input, changing driving lanes as well as combinations of maneuvers involving dynamics of both lateral and longitudinal motion, such as load change reactions in turns.

Table 16: Quantities and units

Quantity		Unit
δ	Axle steering angle	rad
δ_H	Steering-wheel angle	rad
α_v	Slip angle of front axle	rad
α_h	Slip angle of rear axle	rad
β	Float angle	rad
ψ	Yaw angle	rad
ω_e	Undamped natural frequency	s^{-1}
l	Wheelbase	m
l_v	Distance between front axle and center of gravity	m
l_h	Distance between rear axle and center of gravity	m
v	Longitudinal velocity	m/s
v_r	Resulting wind impact velocity	m/s
C_v	Rear cornering stiffness Front axle	N/rad
C_h	Rear cornering stiffness Rear axle	N/rad
D	Damping factor	1/rad
m	Total mass (weight)	kg
i_l	Steering ratio	–
F_{SV}	Lateral force on front axle	N
F_{SH}	Lateral force on rear axle	N
a_y	Lateral acceleration	m/s^2
θ	Yaw moment of inertia	Nms^2
ρ	Air density	kg/m^3
A	Frontal area	m^2
τ	Angle of impact	rad
F_S	Crosswind force	N
M_Z	Crosswind yaw moment	Nm

In the lateral acceleration range from 4 to 6 m/s^2, depending on their design features, passenger vehicles are categorized as either still linear or already nonlinear. This range is therefore considered to be a transition range. In this range, vehicles with maximum lateral acceleration of 6 to 7 m/s^2 (e.g. true offroad vehicles) already feature nonlinear characteristics, while vehicles that achieve higher levels of lateral acceleration (e.g. sports cars) still behave in line with linear characteristics.

The lateral acceleration range above 6 m/s^2 is reached only in extreme situations and is therefore referred to as the limit range. In this range, the vehicle characteristics are predominantly nonlinear, with the main emphasis shifted to vehicle stability. This range is reached on racing circuits or in situations leading up to accidents in normal road traffic.

The average driver generally drives in the range up to 4 m/s^2. This means both the small signal range as well as the linear range are relevant to the car driver when subjectively assessing the situation (Figure 11). For the average car driver, the probability of lateral acceleration occurring decreases exponentially with lateral acceleration.

Figure 11: Lateral-acceleration ranges
A Small-signal range,
B Linear range (relevant to normal car drivers),
C Transition range,
D Limit range (emphasis on stability, relevant to press and experts).

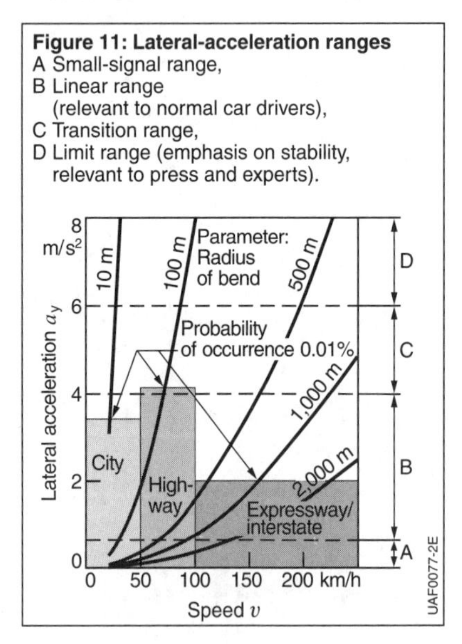

Linear single-track model

Important deductions relating to the dynamic characteristics of lateral motion can be gained from the linear single-track model. The single-track model combines the lateral dynamic properties of one axle and its wheels to form one effective wheel. In the simplest version, as illustrated here, the characteristics under consideration are positioned in the linear range, which explains why this type of model is referred to as a linear single-track model. The most important model assumptions are:

- Kinematics and elastokinematics of the axle are only modeled linearly
- The lateral force buildup of the tire is linear and the aligning torque is ignored
- The center of gravity is assumed to be at road level. The only rotational degree of freedom of the vehicle is therefore the yaw motion. Rolling, pitching and bounce are not taken into consideration

Self-steering effect

Figure 12 represents the single-track model in connection with the fast and slow skidpad. This representation results in the following interrelationships describing the kinematics of slip angles [4, 5, 6]:

$$\alpha_{\mathrm{v}} = \delta - \beta - \frac{\dot{\psi}\ l_{\mathrm{v}}}{v}\ ;\ \alpha_{\mathrm{h}} = -\beta - \frac{\dot{\psi}\ l_{\mathrm{h}}}{v}\,.$$

Together with the torque balance, it is possible to calculate the change in the steering-wheel angle in connection with increasing lateral acceleration for the skidpad maneuver at a constant radius. This results in the definition of the self-steering gradient EG:

$$EG = \frac{d\delta}{da_{\mathrm{y}}} = \frac{m}{l}\left(\frac{l_{\mathrm{h}}}{C_{\mathrm{v}}} - \frac{l_{\mathrm{v}}}{C_{\mathrm{h}}}\right).$$

All passenger vehicles are designed so that they understeer in the linear lateral acceleration range. The EG value for passenger vehicles is in the range of about 0.25 degrees · s²/m.

In terms of the dynamics of lateral motion, the self-steering gradient characterizes the stability and damping of the vehicle. In addition, the significance of the self-steering gradient for the average car driver becomes apparent in that the steering angle requirement increases the faster the cornering

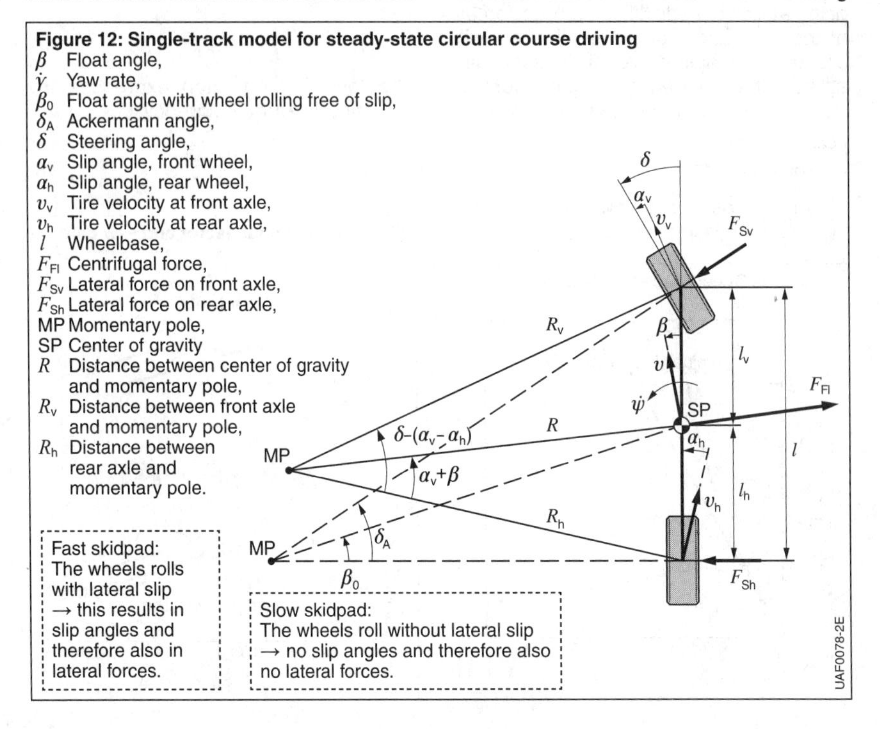

Figure 12: Single-track model for steady-state circular course driving

β Float angle,
$\dot{\gamma}$ Yaw rate,
β_0 Float angle with wheel rolling free of slip,
δ_A Ackermann angle,
δ Steering angle,
α_v Slip angle, front wheel,
α_h Slip angle, rear wheel,
v_v Tire velocity at front axle,
v_h Tire velocity at rear axle,
l Wheelbase,
F_{Fl} Centrifugal force,
F_{Sv} Lateral force on front axle,
F_{Sh} Lateral force on rear axle,
MP Momentary pole,
SP Center of gravity
R Distance between center of gravity and momentary pole,
R_v Distance between front axle and momentary pole,
R_h Distance between rear axle and momentary pole.

speed. This draws the driver's attention to the increasing lateral acceleration.

The float-angle gradient (SG) can be calculated from Figure 12. The float angle gradient should be as low as possible in order to increase the stability of the vehicle [4, 5, 6].

$$SG = \frac{d\beta}{da_y} = \frac{m\ l_v}{C_h\ l}.$$

Yaw gain

The yaw gain defines the degree of yaw response that a vehicle executes in response to a steering angle in the quasi-steady-state range. The yaw gain factor can be determined by conducting the follow test procedure: When driving at a constant speed, the steering wheel is turned with sinusoidal motion at a frequency of less than 0.2 Hz. The steering-angle amplitude is selected in order to achieve a maximum lateral acceleration of about 3 m/s^2. Starting at a speed of 20 km/h, the maneuver is repeated at a speed that is increased by 10 km/h each time. Provided no aerodynamic influences occur at high speeds (lift or upthrust forces at the front and rear axles), the test will produce yaw gain curves that essentially agree with the following equation, derived from the linear single-track model [4, 5, 6]:

$$\left(\frac{\dot{\psi}}{\delta}\right)_{stat} = \frac{v}{l + EG\ v^2}.$$

Figure 13 shows the yaw gain for a vehicle that tends to oversteer ($EG < 0$), that has neutral steering ($EG = 0$), and that understeers ($EG > 0$). Only the vehicle that understeers is acceptable at high speeds and therefore has the right vehicle dynamics even when driving in a straight line. The speed at which a vehicle that tends to understeer has the greatest yaw response is known as the characteristic speed v_{char}. In the linear single-track model, this speed is expressed as:

$$v_{char} = \sqrt{\frac{l}{EG}}$$

Figure 13: Speed-dependent yaw gain

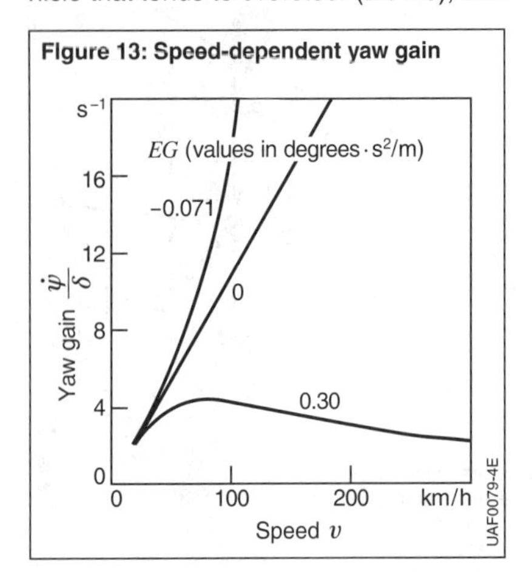

Damping factor

The following equilibrium of forces in the lateral direction is derived for the linear single-track model:

$$m\ a_y = F_{sv} \cos\delta + F_{sh}.$$

For the torque balance:

$$\theta\ \ddot{\psi} = F_{sv}\ l_v \cos\delta + F_{sh}\ l_h.$$

The damping factor D for excitation in terms of the dynamics of linear motion can be derived from the two equations:

$$D = \frac{1}{\omega_e}\left(\frac{C_v + C_h}{m\ v} + \frac{C_v\ l_v^2 + C_h\ l_h^2}{\theta\ v}\right).$$

The following equation expresses the undamped natural frequency:

$$\omega_e = \sqrt{\left(\frac{C_h\ l_h - C_v\ l_v}{\theta} + \frac{C_v\ C_h\ l^2}{\theta\ m\ v^2}\right)}.$$

Figure 14: Damping factor and yaw gain

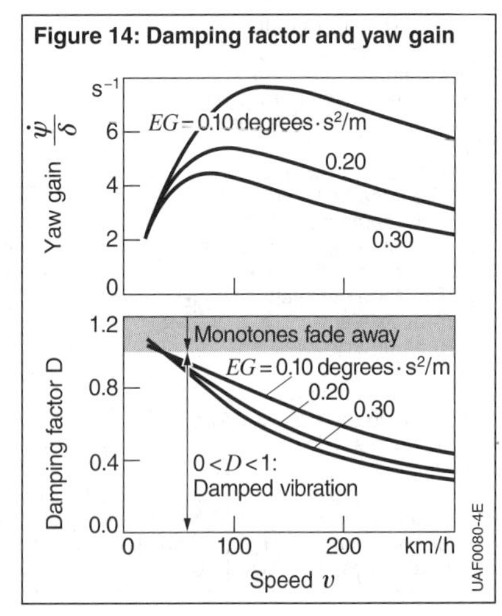

The damping factor of a vehicle can be identified, for example, from the yaw response of sudden steering or step input. The vehicle is designed to ensure that damping is as high as possible.

Figure 14 shows the damping factor and yaw gain for various self-steering gradients. This results in the following conflict of objectives:

- A high self-steering gradient is required if the vehicle is to have good straight-running characteristics
- The self-steering gradient must be as low as possible to facilitate a high damping factor, particularly at high speeds

Lateral-agility diagram

A further important variable governing vehicle balance is the total steering ratio i_l. The steering-wheel angle, together with the total steering ratio i_l, is calculated from the axle steering angle:

$$\delta_H = i_l\ \delta.$$

This results in the following equation for maximum yaw gain:

$$\left(\frac{\dot{\psi}}{\delta_H}\right)_{max} = \frac{1}{2\, i_l\, \sqrt{l\ EG}}.$$

This maximum is plotted in the lateral-agility diagram (Figure 15) as a function of the steering ratio. The diagram additionally shows the EG isolines. The self-steering gradient is constant along these curves. The desired target ranges for the yaw gain and steering ratio can be plotted in this diagram for the purpose of defining the necessary self-steering gradients.

If only the steering ratio is changed in a vehicle, the maximum yaw gain can be determined in the lateral-agility diagram by shifting the baseline along the EG isolines. The shift will be along the vertical axis if the axis characteristics are varied.

Dynamics of lateral motion caused by crosswind

Wind can induce lateral dynamics in motor vehicles. The vehicle responds to this external influence by drifting off course, lateral acceleration, and a change in yaw angle and roll angle. The driver then attempts to take corrective action to counteract this discrepancy. Consequently, the response capabilities of the driver as well as the correctability of the vehicle are taken into consideration in a second stage. According to current findings, the direct response of the vehicle to crosswind is the principal variable for subjectively assessing overall vehicle stability in crosswinds. This offers the advantage that the interaction between crosswind and vehicle response can be effectively observed by analysis.

Characteristically, the average car driver perceives two states induced by wind excitation:

- Natural crosswind, which can vary in terms of direction and wind velocity while driving
- Driving into and out of wind shadow areas where forces of greatly varying strength can act on the vehicle

The automotive industry strives to minimize the effects of excitations triggered by wind forces by taking the following vehicle factors into consideration:

- "Cornering stiffness" of the tires, i.e. to what extent the lateral force changes as the slip angle increases. The wheel load of the tire remains constant in this consideration
- Total weight rating of the vehicle
- Position of center of gravity
- Axle characteristics
- Equilateral and reciprocal suspension
- Damping
- Kinematics and elastokinematics of the axles
- Aerodynamic shape and frontal area of the vehicle

Figure 15: Lateral-agility diagram
EG Self-steering gradient.

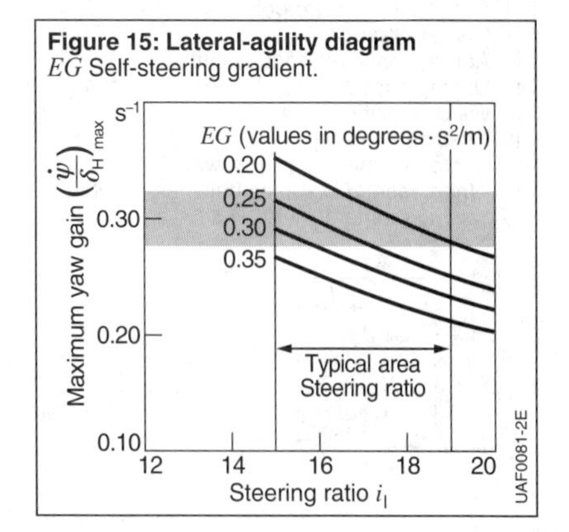

Aerodynamic forces and moments

When a vehicle moves at velocity v in a wind at a speed v_w, the vehicle will be subject to wind impact applied at the resulting velocity v_r. In connection with natural crosswind, the angle of impact τ generally differs from 0 degrees and therefore generates a lateral force F_S and a yawing moment M_Z that acts on the vehicle.

In aerodynamics, it is standard practice to specify dimensionless coefficients instead of forces and moments. Therefore:

$$F_s = c_S \frac{\rho}{2} v_r^2 A \;;\; M_z = c_M \frac{\rho}{2} v_r^2 A\, l.$$

The moment M_Z and the lateral force F_S, which is defined at the mid-point of the wheelbase, can be represented by a single lateral force F_S when the point of impact is located at the center of the pressure point D (Figure 16). The distance d between the aerodynamic reference point B and the center of the pressure point D is calculated as follows:

$$d = \frac{M_z}{F_s} = \frac{c_M\, l}{c_S}.$$

To keep aerodynamic influences as low as possible, appropriate steps should be taken to ensure that the center of the pressure point D is as close to the vehicle's center of gravity S as possible. This consequently reduces the effective impact of the moment.

Figure 17 represents the aerodynamic coefficients of the two most typical structural shapes of vehicles, i.e. the station wagon and sedan, as a function of the angle of impact τ. The resulting distance d is considerably lower for stations wagons than for sedans (Figure 16). For vehicles with the center of gravity located at the mid-point of the wheelbase, the station wagon structure is therefore less sensitive to crosswind than the sedan structure.

Figure 16: Vehicle exposed to crosswind
D Center of pressure point,
S Center of gravity
B Aerodynamic reference point
v Driving speed
v_w Wind velocity
v_r Resulting velocity
τ Angle of impact, F_S Crosswind force
l Wheelbase, d Distance between B and D
M_z Yawing moment

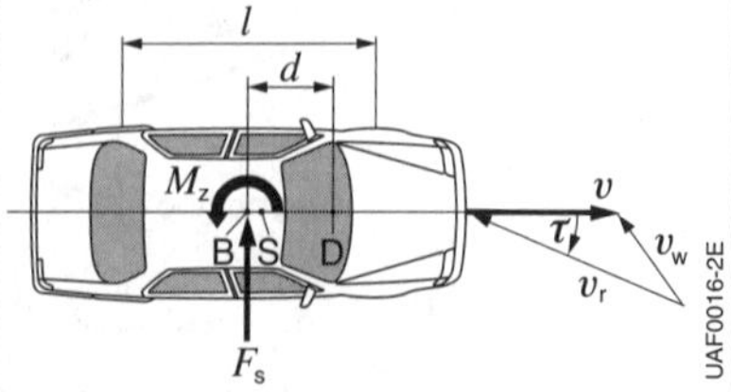

Cornering behavior

Centrifugal force in turns

$F_{cf} = \frac{m\, v^2}{r_k}$ (see Figure 18).

Body roll in turns

In turns, the centrifugal force which concentrates around the center of gravity causes the vehicle to tilt away from the path of travel. The magnitude of this rolling motion depends upon the rates of the springs and their response to alternating compression, and upon the lever arm of the centrifugal force (distance between the roll axis and the center of gravity). The roll axis is the body's instantaneous axis of rotation relative to the road surface.

Figure 17: Lateral-force coefficient and pressure point
1 Station wagon,
2 Sedan.
d Distance between aerodynamic reference point B and center of pressure point D.

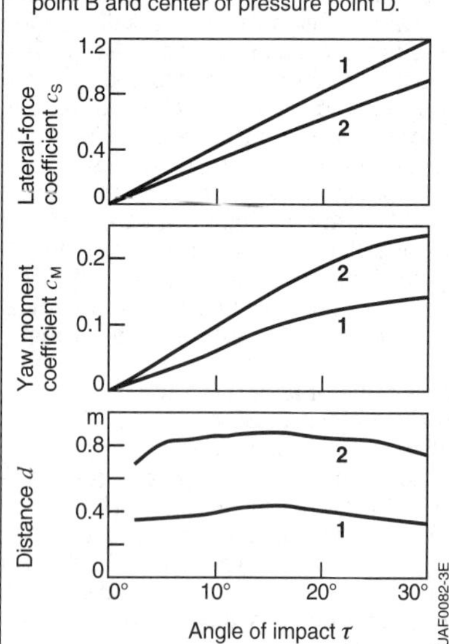

Like all rigid bodies, the vehicle body consistently executes a screwing or rotating motion; this motion is supplemented by a lateral displacement along the instantaneous axis.

The higher the roll axis, i.e. the closer it is to a parallel axis through the center of gravity, the greater will be the transverse stability and the less the roll in turns. However, this generally implies a corresponding upward displacement of the instantaneous axes of the wheels, resulting in a change in track width (with negative effects on driving safety). For this reason, designs seek to combine a high instantaneous roll center with minimal track change. The goal therefore is to place the instantaneous axes of the wheels as high as possible relative to the body, while simultaneously keeping them as far from the body as possible.

A frequently applied means for finding the approximate roll axis is based on determining the centers of rotation of an equivalent body motion. This body motion takes place in those two planes through the front and rear axles which are vertical relative to the road. The centers of rotation are those (hypothetical) points in the body which remain stationary during the rotation. The roll axis, in turn, is the line which connects these centers (instantaneous roll centers). Graphic portrayals of instantaneous roll centers are based on a rule according to which the instantaneous poles of rotation of three systems in a state of relative motion lie along a common pole line.

The complexity of the operations required for a more precise definition of the spatial relationships involved in wheel motion make it advisable to employ general, three-dimensional simulation models [4, 5, 6].

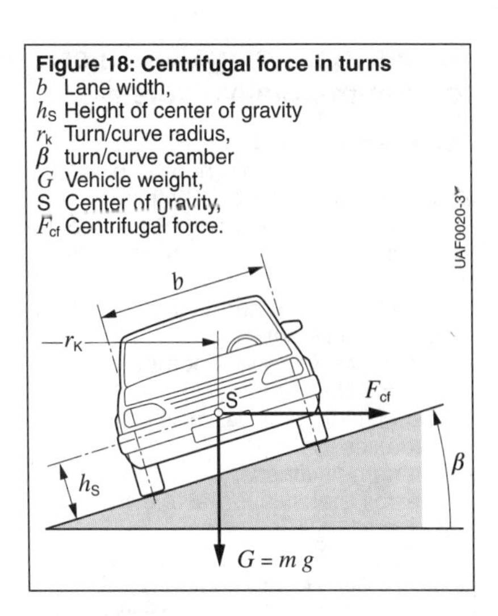

Figure 18: Centrifugal force in turns
b Lane width,
h_S Height of center of gravity
r_K Turn/curve radius,
β turn/curve camber
G Vehicle weight,
S Center of gravity,
F_{cf} Centrifugal force.

References:

[1] ÖNORM V 5050: Straßenverkehrsunfall und Fahrzeugschaden; Terminologie.

[2] Fritz Sacher in Fucik, Hartl, Schlosser, Wielke (Editors): Handbuch des Verkehrsunfalls, Part 2, Manz-Verlag Vienna, 2008.

[3] Council Directive of 26 July 1971 on the approximation of the laws of the Member States relating to the braking devices of certain categories of motor vehicles and their trailers.

[4] B. Heissing, M. Ersoy: Fahrwerkhandbuch, 2nd Edition, Vieweg + Teubner Verlag, 2008.

[5] H.-P. Willumeit: Modelle und Modellierungsverfahren in der Fahrzeugdynamik, Teubner-Verlag, 1998.

[6] M. Mitschke, H. Wallentowitz: Dynamik der Kraftfahrzeuge, 4th Edition, Springer-Verlag 2004.

Table 17: Critical speeds in turns/curves (numerical-value equations)

	Flat curve	Banked curve
Speed at which the vehicle exceeds the limit of adhesion (skid)	$v \leq 11.28\sqrt{\mu_r\, r_K}$ km/h	$v \leq 11.28\sqrt{\dfrac{(\mu_r + \tan\beta)\; r_K}{1 - \mu_r\, \tan\beta}}$ km/h
Speed at which the vehicle tips	$v \geq 11.28\sqrt{\dfrac{b\; r_K}{2\; h_s}}$ km/h	$v \geq 11.28\sqrt{\dfrac{\left(\dfrac{b}{2\; h_s} + \tan\beta\right) r_K}{1 - \dfrac{b}{2\; h_s}\, \tan\beta}}$ km/h

h_S Height of center of gravity (in m), μ_r Max. coefficient of friction, b Lane width (in m), r_K Turn/curve radius (in m), β Turn/curve camber

Special operating dynamics for commercial vehicles

Self-steering effect

The aim is to achieve non-problematic, understeering handling. Stationary vehicle handling is defined by the self-steering effect. The self-steering effect is determined by:

- The mechanical and hydraulic parameters of the steering gear
- The stiffness and geometry of the steering linkage
- The elastokinematics of the front and rear axles
- The frame stiffness
- The roll stabilization at the front and rear axles

Figure 1: Self-steering angle of an 18-ton truck cornering at 3 m/s² of lateral acceleration

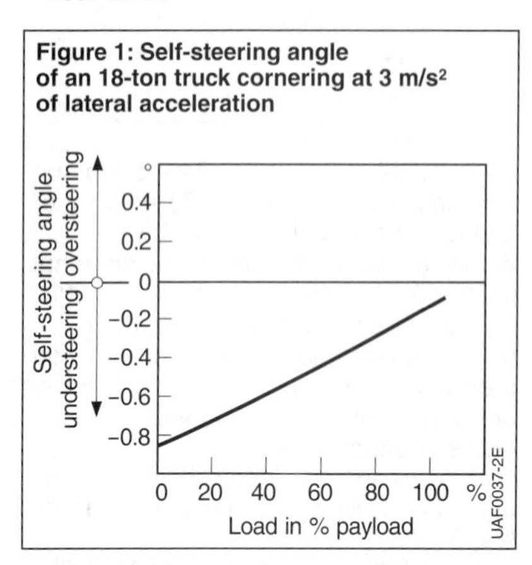

External interference factors such as road-surface irregularities or crosswind, for example, must not give rise to any great disruptions of vEehicle motion. The elastokinematic design of the axles serves to minimize these interference factors. The aim is to minimize steering movements as a result of parallel and alternating compression and rebound and as a result of braking. The kinematic points of the steering or the connecting points of the leaf springs, for example, act as the setting levers in the configuration of the elastokinematics. Geometrically nonlinear finite-element programs are used to configure the elastokinematics. The configuration is monitored on elastokinematic test benches.

Heavy-duty trucks with air suspension are normally equipped with solid or rigid axles controlled with suspension arms and links. Such axles are generally designed to ensure that the proportion of self-steering effects in axle control is virtually constant for all laden states since there is no difference in level between the "unladen" and "laden" states. To date, axle control concepts involving independent wheel control have only been implemented on light utility vans and buses.

Wheel loads at the truck's rear axle vary dramatically, depending on whether the truck is "unladen" or "laden". This leads to the vehicle responding to reductions in load with more pronounced understeer (Figure 1).

Figure 2: Multibody-system model of an all-wheel drive vehicle for determining self-steering effects and simulating operating dynamics.

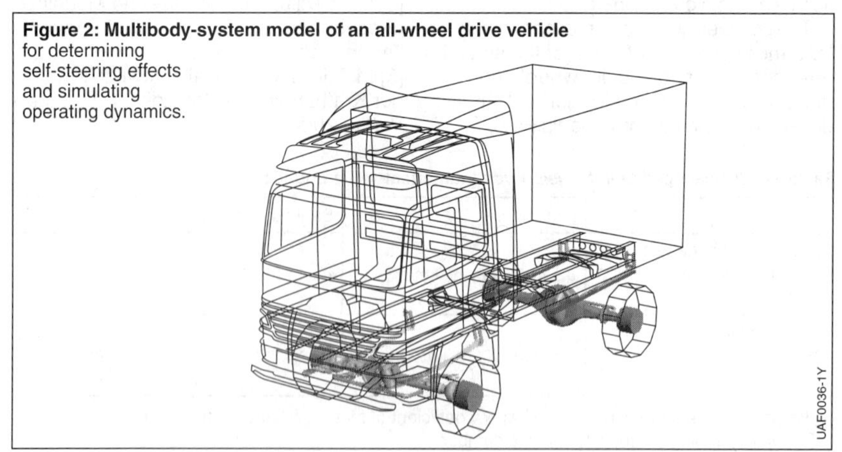

In vehicles with multiple non-steered rear axles, such as 6×4 or 6×2 (six wheels, four or two of which are driven), a constraining torque is generated around the vehicle's vertical axis on account of the different slip angle at the first and second rear axles. The lateral force on the axles that results from this is relevant for the maneuvering system and can be determined as follows (Figure 3).

Cornering forces resulting from constraint for small angles α(For sizes, see Figure 3):
$F_{S1} = F_{S2} - F_{S3}$, where
$F_{S2} = c_{p2} n_2 \alpha_2$,
$F_{S3} = c_{p3} n_3 \alpha_3$.

Slip angle:

$$\alpha_2 = \frac{1}{r} \cdot \frac{c_{p3} n_3 b (a+b)}{c_{p3} n_3 (a+b) + c_{p2} n_2 a},$$

$$\alpha_3 = \frac{b}{r} - \alpha_2.$$

α_i is in the radiant, c_{p2} and c_{p3} are the slip stiffnesses from the tire maps, n_2 and n_3 the number of tires per axle.

Figure 3: Cornering forces F_S and slip angle α on a three-axle vehicle with non-steered tandem axle
a Distance, front axle to first rear axle
b Distance, first rear axle to second rear axle
c Distance, first rear axle to instantaneous center
d Distance, second rear axle to instantaneous center
F_{Si} Cornering forces, i = 1, 2, 3, L,
α_i Slip angle, i = 1, 2, 3,
δ Steering angle,
r Radius of curve,
M Instantaneous center.

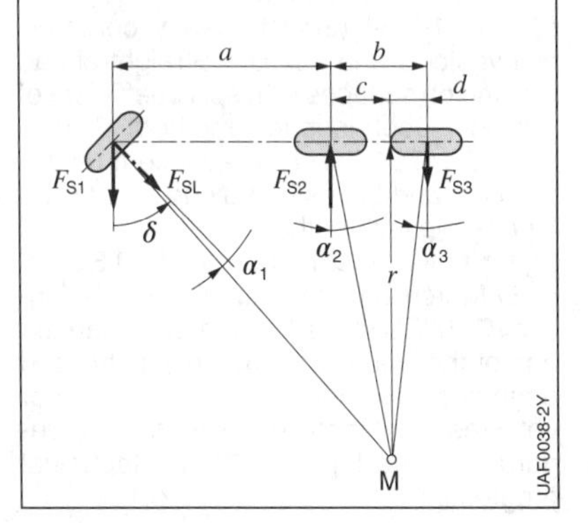

Tipping resistance
As the vehicle's total height increases, the vehicle will also have an increasing tendency to tip to the side in a turn before it starts to slide. In the same way as the overall dynamic behavior, the tipping limits of vehicles are determined by means of multibody-system simulations (multibody system, Figure 2). The simulations investigate various stationary and non-stationary maneuvers, e.g. steady-state circular-course driving ([1]) and double lane change [2]. The achievable lateral acceleration b at the tipping limit during steady-state circular-course driving is b = 6 to 8 m/s² for light utility vans, b = 4 to 6 m/s² for trucks and $b \approx 3$ m/s² for double-decker buses.

The increasing use of ESP systems (Electronic Stability Program) also on commercial vehicles can largely reduce the risk of tipping by estimating or determining the load status and load distribution.

Width requirement
The width requirement of motor vehicles and semi-trailer combinations is greater during cornering than when the vehicle moves in a straight line. With respect to selected driving maneuvers, it is necessary to determine the radius described by the vehicle's outer extremities during cornering, both in order to ascertain its suitability for certain applications (e.g. narrow transit routes through constricted areas) and to confirm compliance with legal regulations (Figure 4). This is determined using multibody-system simulations.

Figure 4: Turning-circle regulation [8] on example of a semi-trailer unit
Representation of the permissible steady-state cornering radii.

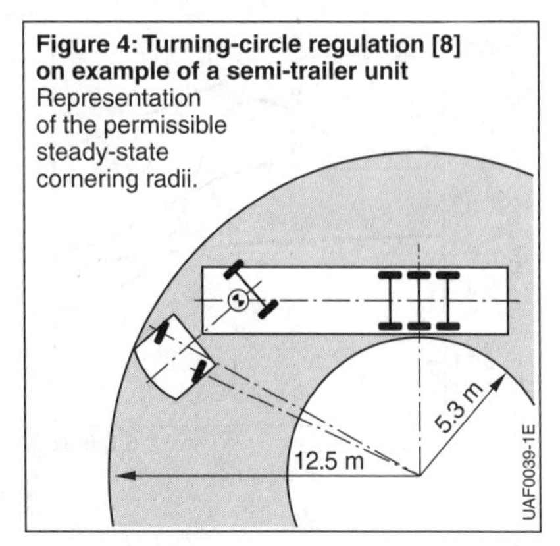

Handling characteristics

Objective analyses of vehicle handling are based on various driving maneuvers such as steady-state circular-course driving [1], single and double lane change [2], step input [3], weave test [4], sinusoidal steering input and frequency response [5] and braking in a straight line on μ split [6] and braking during a turn/curve [7].

The dynamic lateral response of truck-trailer combinations (e.g. semitrailer, truck-trailer) generally differs from that of rigid vehicles. Particularly significant are the distribution of loads between truck and trailer or semitrailer, and the design and geometry of the mechanical coupling device within a given combination.

Impairments of straight-running stability due to vehicle yaw are induced by rapid steering corrections associated with evasive maneuvers, gusts of crosswind, uneven road surfaces, obstacles on one side, ruts in the road surface, and cross-fall cambers. The yaw oscillations caused by these impairments must decay rapidly if vehicle stability is to be maintained. These yaw oscillations can be assessed on the basis of the yaw-velocity frequency response (Figure 5). The yaw-velocity frequency responses of different truck-trailer combinations demonstrate excessive increase in resonance in the least favorable case (tractor-trailer unladen, center-axle trailer laden, curve 2 in Figure 5). This type of combination demands a high degree of driver skill and a defensive driving style.

Figure 5: Yaw-velocity frequency responses
1 Semi-trailer unit (laden)
2 Truck-trailer unit (unladen with laden center-axle trailer)
3 Truck-trailer unit (laden),
4 Truck (laden)

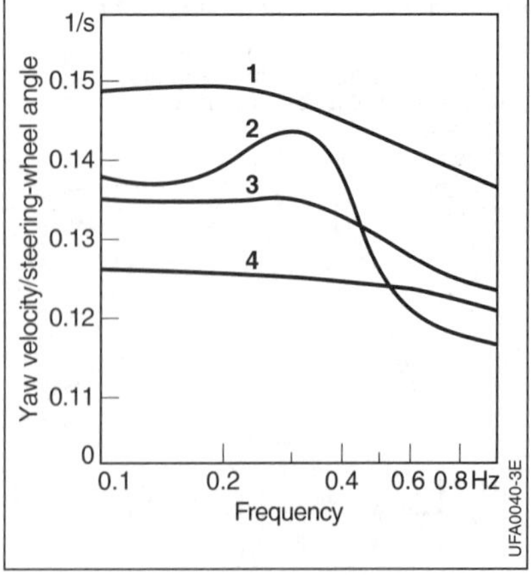

Yaw velocity/steering-wheel angle (1/s): 0, 0.11, 0.12, 0.13, 0.14, 0.15
Frequency: 0.1, 0.2, 0.4, 0.6, 0.8 Hz
1, 2, 3, 4
UFA0040-3E

With semitrailer units, braking maneuvers undertaken under extreme conditions can induce jackknifing. This process is initiated when, on a slippery road surface (μ-low), loss of lateral force is induced by excess braking force applied to the tractor's rear axle or due to excess yaw moment under μ-split conditions. Installation of an ESP system (Electronic Stability Program) represents the most effective means of preventing jackknifing.

References

[1] ISO 14792 (2011): Road vehicles – Heavy commercial vehicles and buses – Steady-state circular tests.

[2] ISO 3888-1 (1999), ISO 3888-2 (2011): Passenger cars – Test track for a severe lane-change maneuver –
Part 1: Double-lane change
Part 2: Obstacle avoidance

[3] ISO 14793 (2011): Road vehicles – Heavy commercial vehicles and buses – Lateral transient response test methods.

[4] ISO 11012 (2009): Heavy commercial vehicles and buses – Open-loop test methods for the quantification of on-center handling – Weave-test and transition test.

[5] ISO 7401 (2011): Road vehicles – Lateral transient response test methods – Open-loop test methods.

[6] ISO 16234 (2006): Heavy commercial vehicles and buses – Straight-ahead braking on surfaces with split coefficient of friction – Open-loop test method.

[7] ISO 14794 (2011): Heavy commercial vehicles and buses – Braking in a turn – Open-loop test methods.

[8] Commission Directive 2003/19/EC of 21 March 2003 for changing the Guideline 97/27/EC of the European Parliament and of the Council about the weights and dimensions of certain classes of motor vehicles and motor-vehicle trailers regarding the adaptation to the technical progress.

Operating-dynamics test procedures as per ISO

Overview

The science devoted to studying the dynamics of vehicle handling generally defines its subject as the overal l behavior of the entire system represented by "driver – vehicle – environment" (Figure 2). The driver assesses the vehicle's handling qualities based on the sum total of his/her subjective impressions. Internationally standardized test procedures have been developed since the end of the 1970s. On the one hand, these test procedures are intended to describe vehicle handling as objectively and consistently as possible. On the other hand, they serve to determine characteristic variables which correlate well with the subjective driving impressions. Furthermore, it is possible to use the standardized test procedures to compare test results with simulation results under identical boundary conditions.

Most of the internationally standardized procedures are carried out in an open loop, i.e. with a defined steering input without controlling intervention by the driver. In this way, vehicle handling is determined without the influence of different steering strategies of individual drivers. Only a few procedures are carried out in a closed loop with the input of a specific course which can be taken by the driver in an individual way.

Figure 2: Overall system "Driver – Vehicle – Environment" as a closed-loop control system

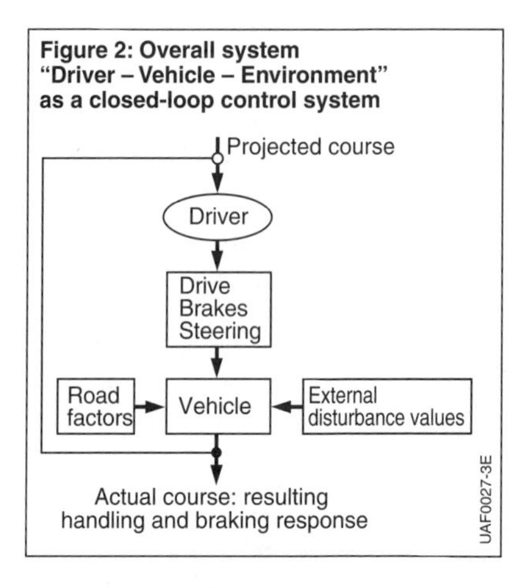

There are currently a total of roughly 20 ISO standards for operating-dynamics procedures, of which the following maneuvers used most frequently in vehicle development will be described here:
- Steady-state circular-course driving [1], [2],
- Test procedures for transient response [3], [4],
- Weave test and transition test [5], [6],
- Braking in a turn/curve [7], [8].

Further test procedures not explicitly described here are, for example:
- Double lane change (closed loop, [11]),

Figure 1: Assessment quantities for operating dynamics
C Center of gravity,
x, y, z Directions in space.

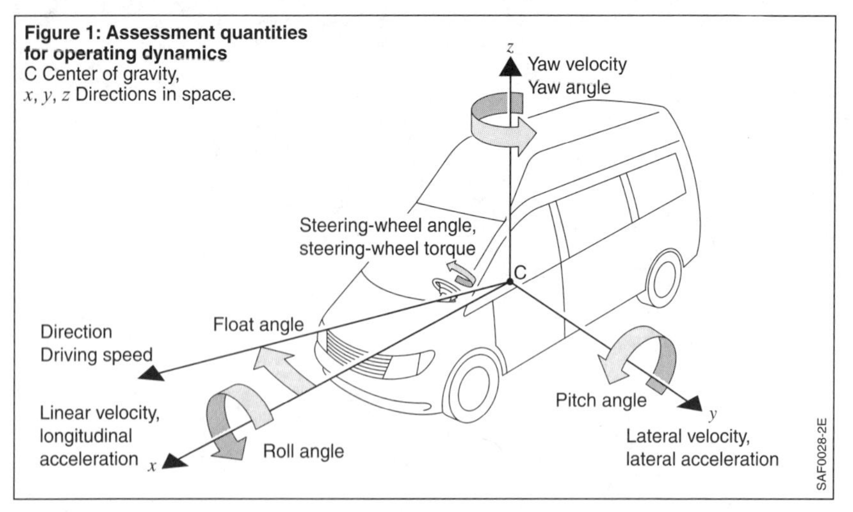

– Lateral stability of vehicle combinations [12], [13],
– Crosswind stability [14],
– Load changes when cornering [15].

These test procedures were originally created for automobiles ([9], [10]). Test methods based on these standards were later developed for heavy-duty commercial vehicles in order to cater for the unique handling characteristics of heavy-duty commercial vehicles with their large mass and inertia figures.

General boundary conditions which must be equally observed for all operating-dynamics test procedures, such as, for example, the condition of the road surface, environmental conditions and tires, are defined in a separate ISO standard [16], which additionally defines what is required of the measuring techniques for operating-dynamics measurements.

Assessment quantities

The following measurable quantities are primarily used to assess vehicle handling (Figure 1):

– Steering-wheel angle and steering-wheel torque
– Lateral acceleration
– Yaw velocity
– Roll angle
– Float angle

Depending on the test procedure defined, characteristic values are determined from these and other quantities for the purpose of describing and assessing vehicle handling. These measured values and characteristic quantities are defined in a separate ISO standard [17].

Steady-state circular-course driving

Performance

The "Steady-state circular-course driving" test procedure ([1], [2]) is generally car-

Figure 3: Steady-state circular-course driving
a) Passenger car, 40 m circle radius,
b) 40-ton semitrailer unit, loaded, 80 m circle radius.

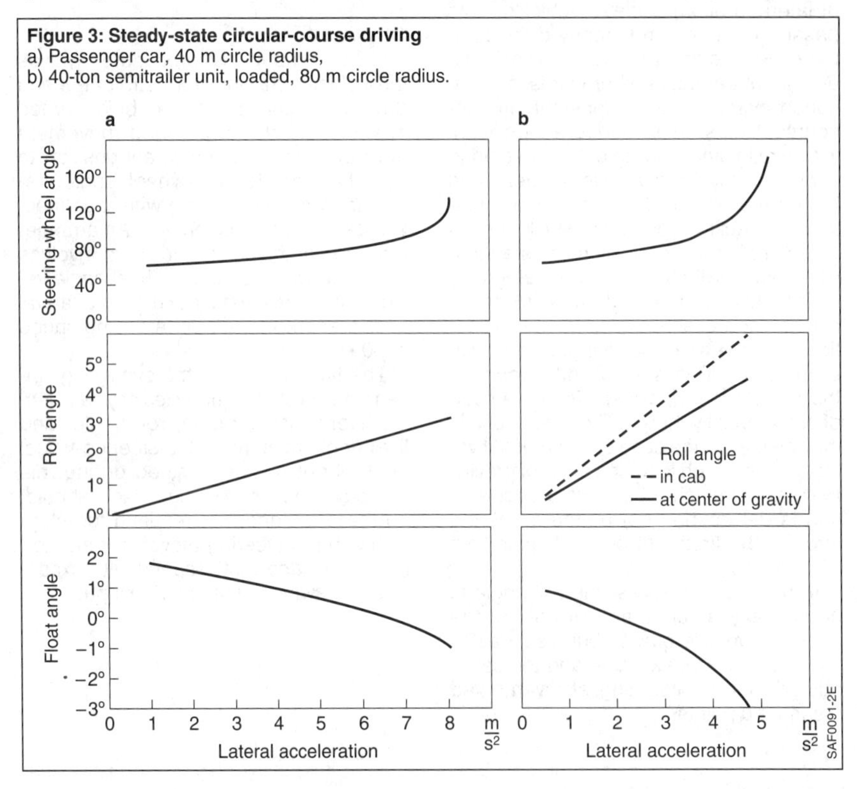

ried out in such a way that the vehicle is accelerated on a circular course with a constant radius (a radius of 40 m is usually selected for passenger cars, 80 m for trucks) starting from a very slow starting speed up to the maximum attainable vehicle lateral acceleration. The longitudinal acceleration here should not exceed approximately 1 m/s^2 so that the driving state can be viewed as sufficiently steady-state.

Assessment

The assessment criteria primarily used are the curves of steering-wheel angle, roll angle and float angle plotted against lateral acceleration, which are shown in Figure 3 for a passenger car and for a truck with semitrailer.

From the curve of steering-wheel angle against lateral acceleration it is possible to determine the vehicle's self-steering effect using a single-lane model (see Dynamics of lateral motion). Today's vehicles, both passenger cars and heavy-duty commercial vehicles, are basically designed as understeering. In other words, as their speed increases on a constant circular-course radius, they require a significant increase in steering angle. The limit range of an understeering vehicle is determined by the maximum transferable cornering force of the front wheels, which results in a sharply progressively rising steering-wheel angle with high lateral acceleration.

The curve of roll angle against lateral acceleration describes the lateral inclination of the vehicle that can be perceived by the driver. This is greatly dependent on the vehicle load, particularly in the case of commercial vehicles. The upshot of this may be that, in the case of commercial vehicles with a high loading center of gravity, the maximum achievable lateral acceleration is determined not by the cornering force of the tires, but by the tipping limit of the vehicle.

In the case of trucks, the roll angle is additionally determined not only at the center of vehicle gravity, but also, due to the torsionally weak frame and the separate driver's cab mounting, at several fixed measuring points.

The curve of float angle against lateral acceleration is an indicator of the lateral stability of the vehicle that can be perceived by the driver and is determined above all by the properties of the tires.

In addition to the tire properties, the following vehicle parameters have a significant influence on stationary vehicle handling:

- Load and axle-load distribution
- Stiffness of springs and stabilizers
- Kinematic and elastokinematics movements of the wheel suspension

Transient response

The test procedures for transient response ([3], [4]) serve to determine the vehicle response to rapid, dynamic steering stimulations, e.g. during rapid evasion maneuvers. Procedures frequently used in tests and simulation are "Step input" and "Sinusoidal steering input and frequency response".

Step input

The "Step input" test procedure involves the vehicle, starting from straight-ahead driving at a constant speed, being turned in with a very fast steering movement from the zero steering-wheel position to a stationary steering-wheel angle, resulting in circular driving with a defined, constant lateral acceleration. An arrangement usual for passenger cars involves turning in with a steering-wheel angle velocity of approximately 360°/s to a lateral acceleration of 4 m/s^2 at a driving speed of 80 km/h.

The time delay and the overswing values with which the quantities of yaw velocity, lateral acceleration, roll angle, and float angle respond to the steering-wheel angle input are determined during this maneuver (Figure 4). The vehicle should, on the one hand, not respond too sluggishly to the steering movement and, on the other hand, not respond with excessive overswing to the steering input.

Sinusoidal steering input and frequency response

The "Sinusoidal steering input and frequency response" test procedure involves the vehicle being excited at a constant driving speed with a sinusoidal steering-wheel angle signal whose amplitude is kept constant and whose frequency is increased starting at slow steering movements of 0.2 Hz to fast steering movements up to 2.0 Hz. The steering amplitude is as a rule selected in such a way that the vehicle remains in the linear driving range, for example in the case of passenger cars usually in such a way that a lateral acceleration of max. 4 m/s² is obtained at the slowest steering frequency and a driving speed of 80 km/h. This makes it possible to assess the dynamic vehicle handling in the entire steering-frequency range to be excited by the driver.

For assessment, the amplitudes referred to the input variable of steering-wheel angle and phase angles of the quantities of steering-wheel torque, yaw velocity, lateral acceleration, roll angle, and float angle are determined and plotted against steering frequency. The position of the natural frequencies, the banking that occurs and the size of the phase angles can be used as assessment criteria for vehicle agility and stability in the case of dynamic steering excitations. Figure 5 shows such an assessment for an 18-ton truck using the example of lateral acceleration, yaw velocity, and roll angle.

As well as the vehicle parameters which influence the stationary handling, above all the damping properties and the moments of inertia of the vehicle as well as the dynamic properties of tires, steering system and wheel suspension are crucial to vehicle handling in the case of dynamic driving maneuvers.

Figure 4: Step input
(Unladen, driving speed 60 km/h).
δ_H Steering-wheel angle,
φ Roll angle,
a_{y0} Lateral acceleration,
$\dot{\psi}$ Yaw velocity.

100°
δ_H
0°
18 t truck
Passenger cars
2.5°
φ
0°
4.5
a_{y0} [m/s²]
0
15
$\dot{\psi}$
10
[°/s]
5
0
0 0.5 1 1.5 s
Elapsed time after disturbance input t_n
UAF0032-3E

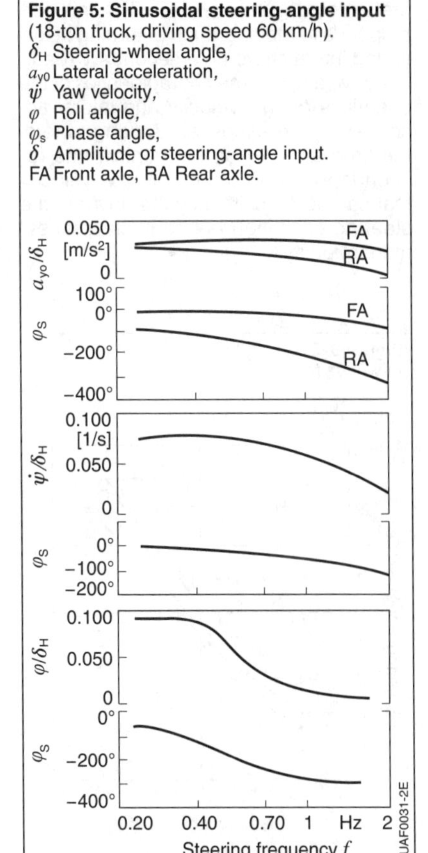

Figure 5: Sinusoidal steering-angle input
(18-ton truck, driving speed 60 km/h).
δ_H Steering-wheel angle,
a_{y0} Lateral acceleration,
$\dot{\psi}$ Yaw velocity,
φ Roll angle,
φ_s Phase angle,
δ Amplitude of steering-angle input.
FA Front axle, RA Rear axle.

Weave test and transition test

The "Weave test" and "Transition test" assessment procedures ([5], [6]) were developed to develop the vehicle's response to small, slow steering movements about the steering zero position. The behavior of the vehicle and steering system during the maneuvers correlates well with the driver's subjective impression of the monitorability of the vehicle in everyday driving.

Transition test

In the transition test the vehicle is turned in, starting from straight-ahead driving at a constant driving speed, with a slow steering movement (e.g. at 80 km/h with a steering-wheel angle velocity of 5°/s) to a circular course with a low lateral acceleration of 1 to 2 m/s². In practice this corresponds to the operation of turning at low speed into a curve on a secondary road.

Above all, the time characteristics of the quantities of yaw velocity and lateral acceleration are assessed. Among others, the time after which the vehicle has responded with a quantity of yaw velocity that can be perceived by the driver to the steering excitation is a suitable assessment quantity.

Weave test

In the weave test the vehicle is excited at a constant driving speed (e.g. 80 km/h) with a sinusoidal steering excitation and a slow steering frequency of 0.1 to 0.2 Hz. The maximum lateral acceleration here ranges between 2 m/s² (usual for trucks) and 4 m/s² (usual for passenger cars), i.e. the vehicle is constantly in the linear driving range.

To assess the vehicle handling and steering behavior, the curves for the quantities of steering-wheel torque, yaw velocity, lateral acceleration, and float angle as a function of the input steering-wheel angle are used. The non-linearities (among others friction) in the steering system, tires and wheel suspension give rise to hysteresis loops, like the hysteresis loop of steering-wheel torque vs. steering-wheel angle shown by way of example in Figure 6. The ranges of the hystereses and the rise gradients of the curves are suitable assessment criteria. For example, the steering behavior of a vehicle which demonstrates on overlapping hysteresis loop, like vehicle B in Figure 6, is perceived by the driver to be "indifferent about the center position" and "poor in straight-running stability".

Figure 6: Steering hysteresis in weave test

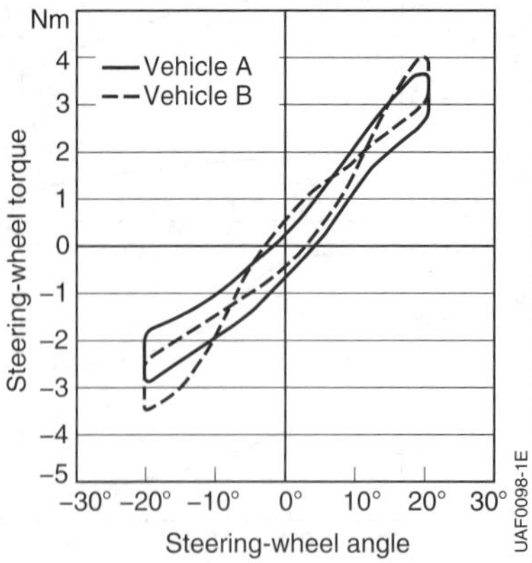

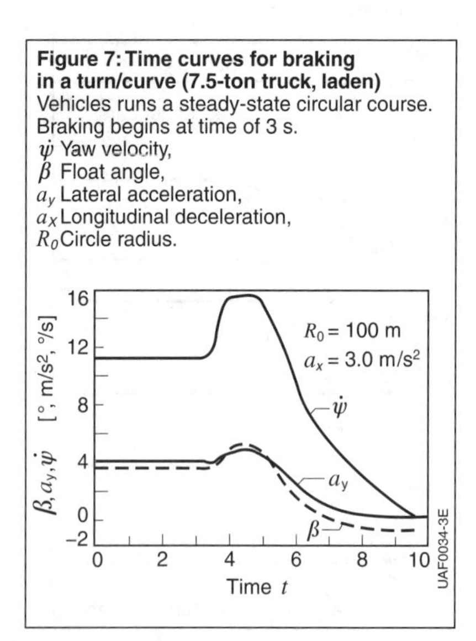

Figure 7: Time curves for braking in a turn/curve (7.5-ton truck, laden)
Vehicles runs a steady-state circular course. Braking begins at time of 3 s.
$\dot{\psi}$ Yaw velocity,
β Float angle,
a_y Lateral acceleration,
a_x Longitudinal deceleration,
R_0 Circle radius.

Braking in a turn/curve

The "Braking in a turn/curve" test procedure ([7], [8]) is likewise geared towards a situation that occurs frequently in real driving operation. Starting out from stationary cornering on a circle with a constant radius with a lateral acceleration of between 3 m/s² (usual value for trucks) and 5 m/s² (usual value for passenger cars), the vehicle is braked with a defined deceleration, set by the brake-pedal travel required for this purpose. Several tests are carried out in which the brake-pedal setting and thus the braking deceleration are varied from light braking through to full braking.

Figure 7 shows the time characteristics of yaw velocity, lateral acceleration and float angle of a 7.5-ton truck during a braking operation with a deceleration of 3 m/s². Once the braking has started (time 3.0 s), a clear banking of the operating-dynamics quantities can be made out. It shows that the vehicle turns in during the braking operation to the inside of the curve when the steering-wheel angle is kept constant at the value which was set before the start of braking during stationary cornering.

The extent of the banking of the operating-dynamics quantities, which are usually plotted against the varied longitudinal deceleration, are used as an assessment criterion for vehicle stability [18].

As well as the vehicle parameters which influence vehicle handling in the driving maneuvers described up to now, the braking-force distribution between front and rear axles and the design of the ABS and ESP brake-control systems have a significant influence on vehicle handling during this maneuver. It is therefore used among other things to tune the brake system and brake-control systems.

References

[1] ISO 4138 (2012): Passenger Cars – Steady-state circular driving behaviour – Open-Loop test methods.
[2] ISO 14792 (2011): Road vehicles – Heavy commercial vehicles and buses – Steady-state circular tests.
[3] ISO 7401 (2011): Road vehicles – Lateral transient response test methods – Open loop test methods.
[4] ISO 14793 (2011): Road vehicles – Heavy commercial vehicles and buses – Lateral transient response test methods.
[5] ISO 13674: Road vehicles – Test method for the quantification of on-centre handling – Part 1: Weave test (2010), Part 2: Transition test (2006).
[6] ISO 11012 (2009): Heavy commercial vehicles and buses – Open-loop test methods for the quantification of on-centre handling – Weave test and transition test.
[7] ISO 7975 (2006): Passenger cars – Braking in a turn – Open-loop test method.
[8] ISO 14794 (2011): Heavy commercial vehicles and buses – Braking in a turn – Open-loop test methods.
[9] A. Zomotor: Fahrwerktechnik – Fahrverhalten. 2nd Ed., Vogel-Verlag, 1991.
[10] M. Mitschke, H. Wallentowitz: Dynamik der Kraftfahrzeuge. 4th Ed., Springer-Verlag 2004.
[11] ISO 3888: Passenger cars – Test track for a severe lane-change manoeuvre – Part 1: Double lane-change (1999), Part 2: Obstacle avoidance (2011).
[12] ISO 9815 (2010): Road vehicles – passenger-car and trailer combinations – Lateral stability test.
[13] ISO 14791 (2000): Road vehicles – Heavy commercial vehicle combinations and articulated buses – Lateral stability test methods.
[14] ISO 12021 (2010): Road vehicles – Sensitivity to lateral wind – Open-loop test method using wind generator input.
[15] ISO 9816 (2006): Passenger cars – Power-off reaction of a vehicle in a turn – Open-loop test method.
[16] ISO 15037: Road vehicles – Vehicle dynamics test methods – Part 1: General conditions for passenger cars (2006), Part 2: General conditions for heavy commercial vehicles and buses (2012).
[17] ISO 8855 (2011): Road vehicles – Vehicle dynamics and road-holding ability – Vocabulary.
[18] E.-C. von Glasner: Einbeziehung von Prüfstandsergebnissen in die Simulation des Fahrverhaltens von Nutzfahrzeugen. Postdoctoral thesis, University of Stuttgart, 1987.

Vehicle aerodynamics

Aerodynamic forces

Vehicle aerodynamics concerns all phenomena that arise as a surrounding medium (air) flows around or through moving vehicles and their components. The drag, forces and moments (see Table 1) are evaluated and usually optimized by changing the shape of the vehicle.

Aerodynamic drag

Aerodynamic drag W_L is given a lot of attention in developing motor vehicles, because it has a direct impact on the vehicle performance and fuel consumption. A critical variable here is the drag coefficient c_W, which describes the aerodynamic quality of the shape of a body in a flow. The aerodynamic drag is also influenced by the frontal area A_{fx} and the dynamic pressure q of the flow, which depends on the air density ρ and the airflow velocity v.

Dynamic pressure is described by this equation:

$$q = 0.5\,\rho\, v^2.$$

Aerodynamic drag is calculated as (also refer to fluid mechanics):

$$W_L = \frac{\rho}{2}\, v^2 c_W A_{fx}.$$

Aerodynamic drag is composed of the skin friction drag and form drag:

$$W_L = W_R + W_D.$$

Skin friction drag W_R
Skin friction drag arises due to shear stress τ_w on the walls as the air flows by. Figure 1 shows an example of this on a wing. Skin friction drag is calculated as an integral of the shear stress over the surface:

$$W_R = \int \tau_w \cos\varphi\, dF.$$

Table 1: Aerodynamic forces and moments
q Dynamic pressure, ρ Air density, v Airflow velocity, A_{fx} Frontal area,
c_A Lift coefficient, l Wheelbase.
$c_{AV} = 0.5\, c_A + c_M$ (relative to the front axle),
$c_{AH} = 0.5\, c_A - c_M$ (referred to rear axle).

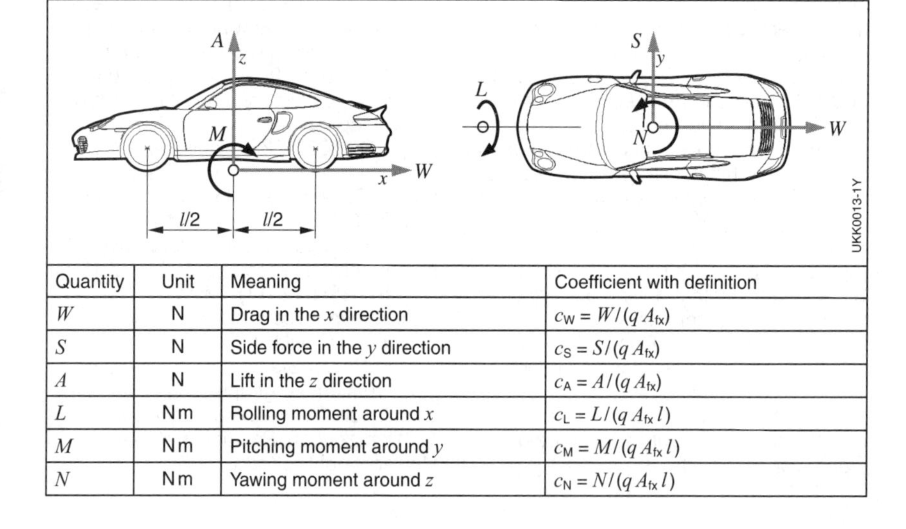

Quantity	Unit	Meaning	Coefficient with definition
W	N	Drag in the x direction	$c_W = W/(q\,A_{fx})$
S	N	Side force in the y direction	$c_S = S/(q\,A_{fx})$
A	N	Lift in the z direction	$c_A = A/(q\,A_{fx})$
L	Nm	Rolling moment around x	$c_L = L/(q\,A_{fx}\,l)$
M	Nm	Pitching moment around y	$c_M = M/(q\,A_{fx}\,l)$
N	Nm	Yawing moment around z	$c_N = N/(q\,A_{fx}\,l)$

Form drag W_D

Form drag arises through pressure changes on blunt bodies caused by shedding. It is calculated as an integral of the pressure over the surface:

$$W_D = \int p \sin\varphi \, dF.$$

Figure 2 shows the flow pattern for flow around a wing and a hemisphere. Shedding results in vortices, and therefore pressure differences. Form drag is predominant for blunt bodies. For streamlined bodies, the friction component dominates; there is no turbulent shedding from the body's surface.

For the c_W values for different bodies and car body shapes, refer to the various tables (see Drag coefficients, Fluid mechanics and Drag coefficients, Motor-vehicle dynamics).

Figure 1: Formation of skin friction drag from airflow
v_∞ Airflow velocity
φ Angle of impact to the surface element,
τ_W Shear stress,
p Pressure,
dF Surface element.

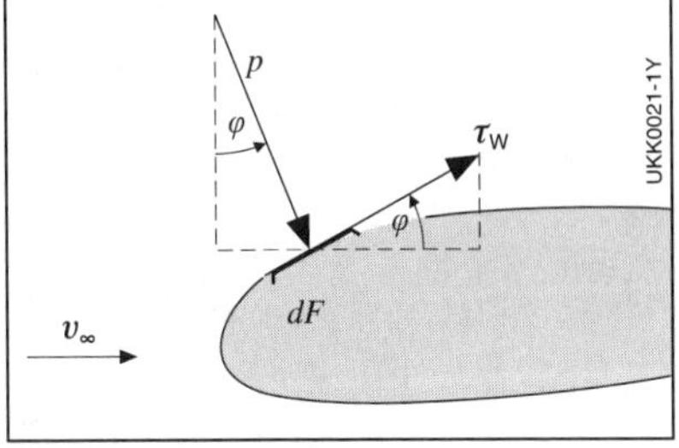

Figure 2: Flow conditions for various bodies
a) Streamlined body (wing),
b) Blunt body (hemisphere).
1 Stagnation point,
2 Separated flow, vortices.

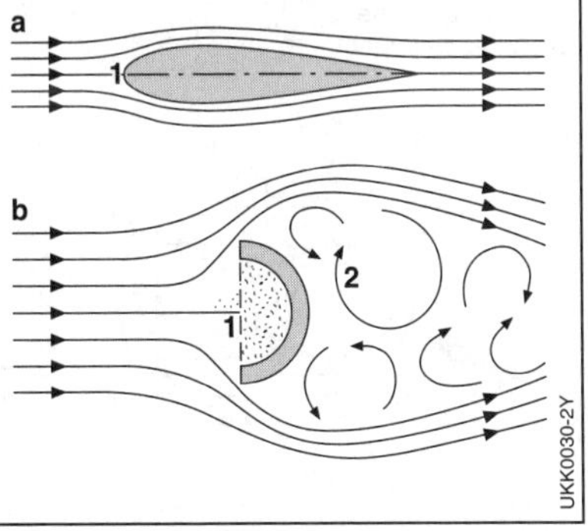

Induced drag

The difference in pressure above and below a vehicle in flow lead to superimposition of the main horizontal flow around the body. A vertical flow component arises through pressure equalization via the body sides, and wake turbulence develops, which causes the induced drag.

Internal drag

Internal drag arises from pressure loss as air flows through the engine compartment.

Interference drag

Interference drag is produced by the interaction of attachments (such as the wheel suspension, wheels, exterior mirrors, antennas, windshield wipers, spoilers, vanes). The total drag is not the sum of the drag forces of the individual parts, because the mutual interference has to be taken into account. Thus attachments can also cause the c_W value to drop.

The impact of an exterior rearview mirror on the car body is shown here as an example (Fig. 3). It causes the stagnation point to move toward the attachment; this results in asymmetrical flow around the car body. Consequently, there is a greater risk of shedding vortices, and that leads to an increasing c_W value.

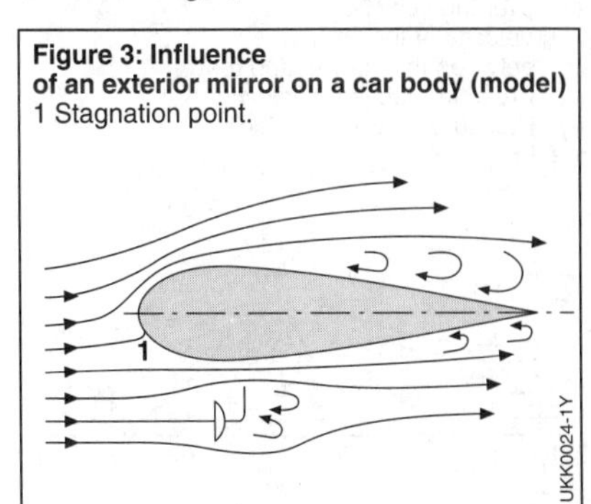

Figure 3: Influence of an exterior mirror on a car body (model)
1 Stagnation point.

Buoyancy force

The curvature of the car's top surface causes the air to flow faster above the vehicle than under it, which generates an undesirable lift force (Fig. 4). This reduces the tire contact forces and thereby has a negative impact on the directional stability.

The lift coefficient c_{A} is the sum of the lift coefficients on the front axle c_{AV} and the rear axle c_{AH}. The difference between the lift on the front and rear axles is referred to as "lift balance" and is a variable influencing directional stability.

The pitching moment M, which acts around the y-axis, is often used as a design variable instead. A positive pitching moment requires understeer, and a negative pitching moment requires oversteer. The rolling moment L around the x-axis is disregarded with respect to aerodynamics.

Lateral force

A car has an almost symmetrical shape when viewed from the front, which means that side forces generated by air flows are small. As soon as the approaching air flow deviates from the x-axis (i.e. in a crosswind), the airflow generates side forces that can affect vehicle behavior considerably.

The yawing moment N, which acts around the z-axis, is also used as an indicator for crosswind sensitivity. This value is used to derive the yaw-angle velocity and yaw angular acceleration, which provide information about the negative impact of crosswind.

Figure 4: Lift resulting from the airflow
v_{o} Velocity of the airflow over the vehicle,
v_{u} Velocity of the airflow between the vehicle and road,
p_{o} Pressure above the vehicle,
p_{u} Pressure below the vehicle,
A Lift.

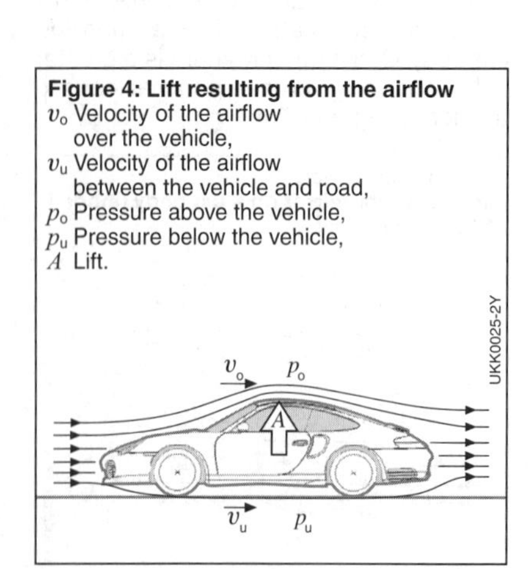

Figure 5: Side forces
S Side force,
W Drag in the longitudinal direction,
N Yawing moment,
β Angle of impact.

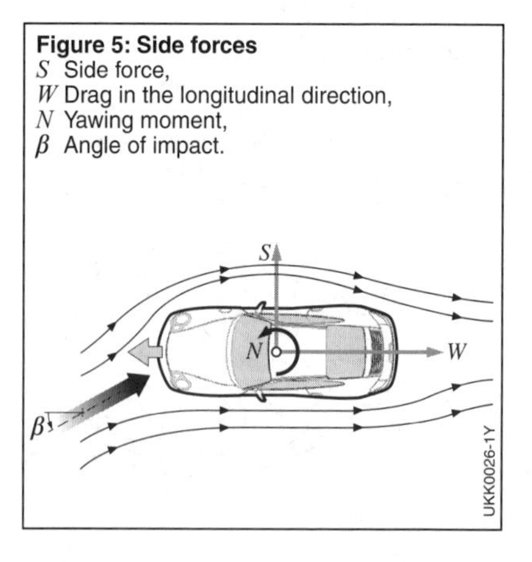

Tasks of vehicle aerodynamics

Aerodynamics has other tasks besides reducing the c_W value (Fig. 6).

Motive power

Impact of velocity on tractive resistance
In order to set the vehicle in motion and keep it moving, the engine has to produce a tractive force Z. This is composed of the aerodynamic drag W_L, rolling resistance W_R, climbing resistance W_S, drag during acceleration W_B and losses in the powertrain W_A. This results in the tractive resistance equation:

$$Z = W_L + W_R + W_S + W_B + W_A.$$

Climbing resistance and acceleration resistance are zero during constant motion on a level stretch. The mileage is substantially determined by the aerodynamic drag, because it increases quadratically with the velocity and surpasses the other tractive resistances as of 80 km/h (Fig. 7).

The lower the product of the c_W value and frontal area A_{fx} ($A_{fx} \cdot c_W$ is often called the drag area), the faster a vehicle can move with a suitable transmission ratio. At a constant velocity, a smaller drag area leads to lower fuel consumption. Consequently, the vehicle produces fewer emissions.

As a function of the low driving speeds in the New European Driving Cycle (average speed of 33.4 km/h, see European test cycle), however, the influence of the aerodynamic drag is greatly underestimated in the manufacturer/consumption data, which can lead to higher fuel consumption in everyday operation for the average driver.

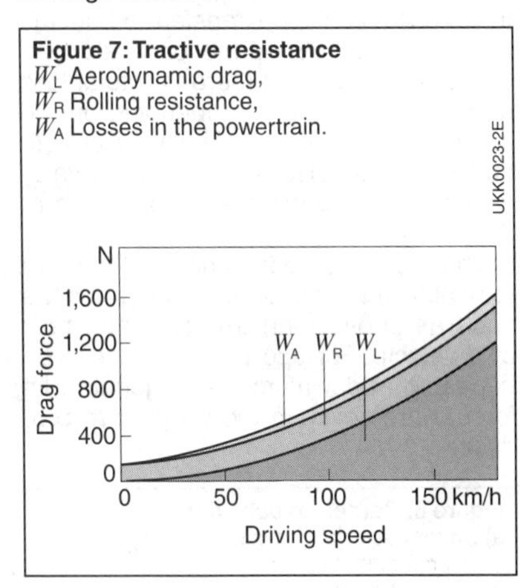

Figure 7: Tractive resistance
W_L Aerodynamic drag,
W_R Rolling resistance,
W_A Losses in the powertrain.

Figure 6: Vehicle aerodynamics: requirements and influences

Vehicle performance
Consumption, emissions, top speed, acceleration, driving dynamics

Design
Design that is characteristic of the brand, visible or invisible aerodynamic measures (retractable rear spoiler)

Component forces
Doors and cover, windows and sliding sunroof, vibration of mirrors, rear window fluttering, puffing up behavior of convertible top

Convenience
Draft prevention (mobile roofs), wind noise, prevention of dirt buildup

Cooling and ventilation
Ventilation of brakes, engine compartment and components, engine cooling, cooling of units (such as gearbox), intercooling, air conditioning, removing condensation from headlights

Directional stability
Lift forces, lift balance, straight-running stability, lane-change behavior, crosswind stability, handling

UKK0027-3E

Directional stability

Directional stability, sensitivity of the steering behavior, lane-change behavior, handling and crosswind sensitivity are critically influenced by the lift and side forces. The forces that can be transferred from the wheels to the road in the longitudinal and transverse directions can be no larger than the vertical force (normal force).

The lift acts against the vertical forces resulting from the weight. Consequently, the forces that can be transferred from the tires, above all during motion in a curve, can be reduced until there is a total loss of directional stability, depending on the driving speed and lift. The transferable braking forces also depend on the vertical wheel load, and therefore on the aerodynamic lift.

The task of aerodynamics is therefore to minimize the lift forces through means such as spoilers on the front and rear of a vehicle. For sports cars it can even happen that downforces (i.e. negative lift forces) greater than the weight are produced.

Figure 8: Cornering behavior
a) Understeering behavior,
b) Oversteering behavior.
1 Correct trajectory,
2 Trajectory of an understeered vehicle,
3 Trajectory of an oversteered vehicle.

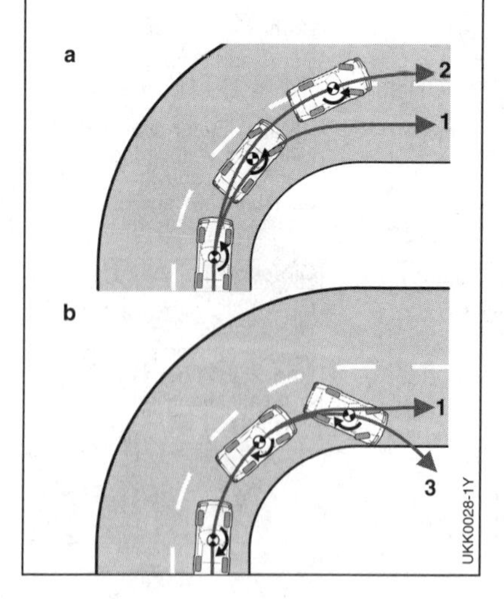

Lift balance

If the lift balance has a lift coefficient c_{AV} on the front axle greater than c_{AH} on the rear axle, there is a positive pitching moment and the design produces understeering driving behavior. Then if the speed in a curve is too high it will be noticed immediately when the steering angle is too high; the vehicle should be stabilized again by backing off the accelerator (Fig. 8a).

If c_{AV} is less than c_{AH}, a negative pitching moment is produced and there is an oversteering design (Fig. 8b). Then if the speed in a curve is too high it will be perceived not through the steering wheel, but through the seat. Backing off the accelerator makes the vehicle unstable; it can be kept in the lane only with targeted countersteering and depressing the accelerator.

Aerodynamic efficiency

A significant reduction of the lift forces usually comes at the expense of a favorable c_W value. The smaller this impact, the greater the aerodynamic efficiency E. It describes the relationship between the lift and drag coefficients:

$$E = \frac{c_A}{c_W}.$$

Figure 9 shows a comparison of the values for the aerodynamic efficiency of different vehicles. Due to the freedom of body design and the very fast racetrack, current Le Mans prototypes have the best

Figure 9: Aerodynamic efficiency for various vehicle types
1 Passenger car,
2 Sports car,
3 Formula 3 racecar,
4 Indy car series,
5 Formula 1 racecar,
6 Le Mans racecar.

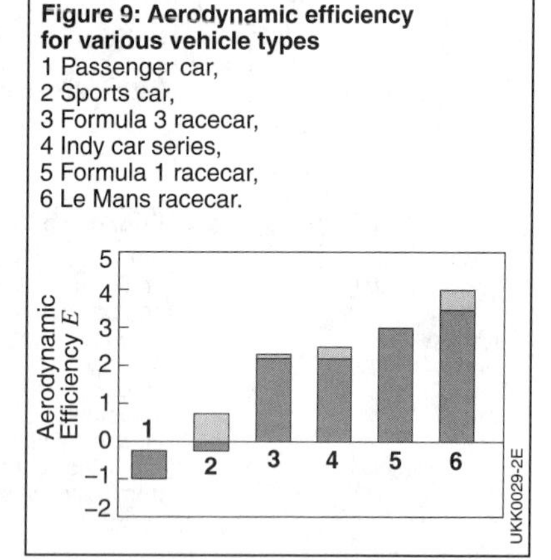

efficiency (E = 3.5 to 4). Despite a high downforce, a low c_W value is achieved.

Due to the fuel consumption advantages, passenger cars usually have very low c_W values, but positive lift, and therefore a negative efficiency down to $E = -1$.

Cooling and ventilation

The responsibility of aerodynamics consists in requirements-based air intake, so that the heat generated by operating the vehicle is dissipated to the environment (see Engine cooling).

An intensive flow of cooling air can lead to a high internal drag, and therefore increase the aerodynamic drag. Therefore a great deal of attention is given to achieving an optimum balance between cooling requirements and aerodynamics (Fig. 10). To optimize the air intake, air inlets and outlets are located on the vehicle in zones with the greatest possible pressure difference for propulsion. The flow of cooling air can be controlled and adapted by changing the size of the respective openings. The cooling air is conducted through channels to the radiators, which should be designed to be as free of pressure loss and leaks as possible.

The same criteria also apply to an exhaust air duct up to the air outlet. The air outlet should be positioned downstream with the least possible pressure loss. If it can be placed in a zone where there are already aerodynamic losses, these can be favorably reduced by the arising interference. For example, blowing out the cooling air ahead of the front wheel can form an air cushion and thereby guide the flow around the wheel with fewer losses, which nearly equalizes the cooling air drag (Fig. 11).

The aerodynamic pressure losses can also be controlled by choosing suitable radiator cores and their dimensions, as well as fans and their frames. To entirely prevent cooling air drag with a low power requirement, electrically closable radiator shutters were introduced in series production back in 1987 (for the Porsche 928). Today this measure is widely used in the premium class.

Figure 11: Interference from blowing out the cooling air ahead of the front wheel
1 Airflow approaching the radiator,
2 Airflow approaching the wheel,
3 Radiator,
4 Exhaust flow after the radiator and air cushion ahead of the wheel.

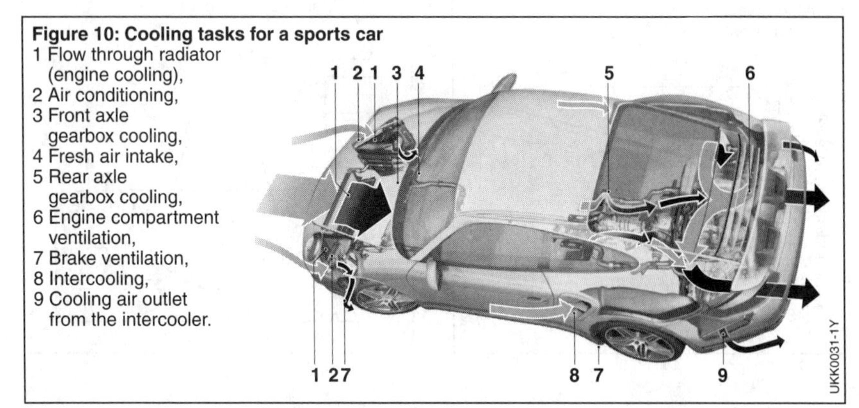

Figure 10: Cooling tasks for a sports car
1 Flow through radiator (engine cooling),
2 Air conditioning,
3 Front axle gearbox cooling,
4 Fresh air intake,
5 Rear axle gearbox cooling,
6 Engine compartment ventilation,
7 Brake ventilation,
8 Intercooling,
9 Cooling air outlet from the intercooler.

Comfort and convenience

Aerodynamics helps provide the passengers with comfort primarily by reducing wind noise (aeroacoustics), dirt and drafts.

Aeroacoustics

Wind noise in the passenger compartment of a car is the primary noise source for speeds over 120 km/h; above 200 km/h it is approximately 6 dB(A) louder than all other acoustic sources (Fig. 12). This corresponds to four times the sound power.

The term "aeroacoustics" refers to the production of noise from

- Flow around the body (flow separation, broadband noise from unsteady pressure fluctuations, noise with discrete frequencies such as from antennas and exterior mirrors),
- Component excitation with low-frequency noise excitation due to large surfaces in a flow (such as convertible top, roof, door and cover panels),
- Overflow with noise created by a Helmholtz resonance (such as with open windows, sliding sunroof),
- Leakage due to pressure differences between the passenger compartment and flow around the body as the result of aerodynamic and mechanical component loads.

Improvement of the aeroacoustics

By giving the basic body an aerodynamic shape you can reduce the local flow velocities, influence the flow direction, and prevent or limit turbulent shedding. Generally speaking, whatever reduces the c_W value also leads to lower wind noise.

Freestanding attachments (such as mirrors, wipers, antennas, roof transport system) disrupt the flow around the basic body by creating additional flow separation. Therefore they should be positioned where they will produce the least possible disruption and their shape should be optimized. With component excitation, aerodynamics analyses the necessity and placement of reinforcement plates or damping measures.

In the case of flow over cavities (such as an open sliding sunroof), low-frequency pulsing of the air volume in the passenger compartment can occur (humming). That is the resonance case of periodic oscillation of the shear layer between the air overflow and the static air in the passenger compartment (Fig. 13) with the natural frequency of the passenger compartment (Helmholtz resonator principle). As an aerodynamic measure, the shear layer buildup is upset by interference flows, for example, using a wind deflector on the sliding sunroof.

Figure 12: Wind noise components in total noise
Measurement of the sound pressure level with an artificial head in the driver's seat, left ear.
1 Total noise,
2 Wind noise,
3 Other noise.

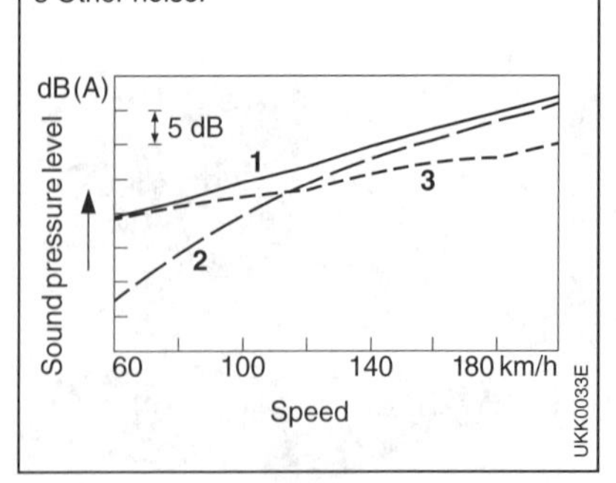

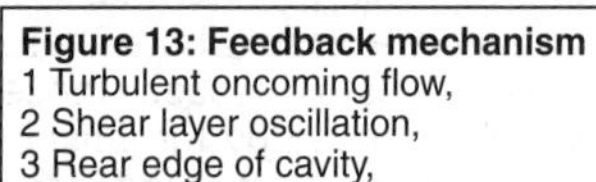

Figure 13: Feedback mechanism
1 Turbulent oncoming flow,
2 Shear layer oscillation,
3 Rear edge of cavity,
4 Vortex,
5 Pressure wave.

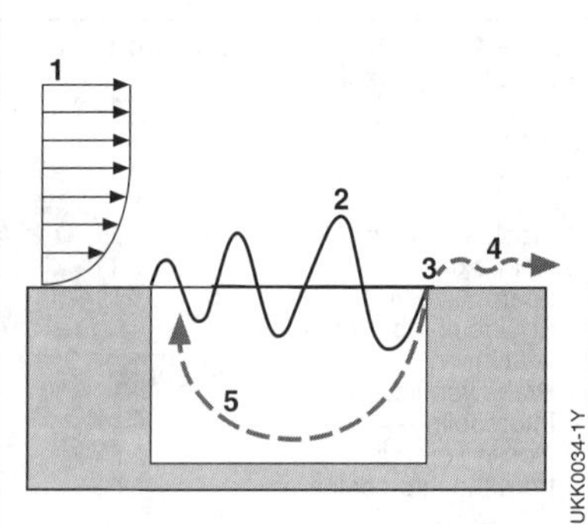

The aeroacoustic quality is evaluated not only by the spectral analysis, but also by taking into account psychoacoustic parameters. These parameters measure higher-frequency noise components; consequently, subjective perception (loudness, sharpness) and speech intelligibility are taken into account better. The sharpness is evaluated independently from loudness and roughness. The articulation index is a measure of speech intelligibility.

Dirt
The wheels of your own vehicle kick up dirt particles, which settle on the surface of your vehicle (contamination of your vehicle) or they mix with the turbulent wake and settle on following vehicles (contamination of other vehicles). The magnitude and intensity of the contaminated surfaces are caused by local wake turbulence of the surrounding flow. The task of vehicle aerodynamics is to identify such turbulence and reduce it or transpose it to unproblematic zones. Great care is given to avoiding impaired visibility by keeping the windows and exterior mirrors free of dirt buildup.

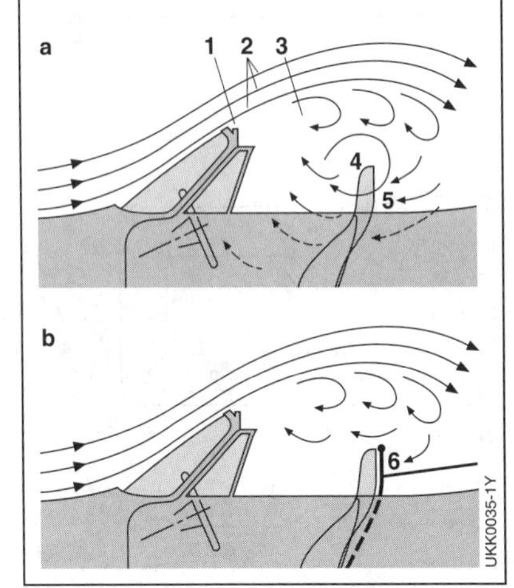

Figure 14: Reduction of drafts for a convertible by using a wind deflector
a) Vehicle without wind deflector,
b) Vehicle with wind deflector.
1 Flow separation at the top edge of the windshield,
2 Flow around vehicle,
3 Unsteady vortex sheet,
4 Vortex direction of rotation in passenger compartment,
5 Airflow between the seats and between the seats and side panels,
6 Wind deflector for reducing airflow between the seats.

Draft prevention
Usually there are bothersome drafts when driving with the windows and roof open. This can be clearly observed in a convertible when the top is open, because there is shedding of the flow over a large area behind the windshield frame. A low-pressure region behind the windshield is filled by a backflow. Backflows and turbulence cause the bothersome drafts (Fig. 14).

The vortices can be damped and significantly reduced by the mesh fabric of a wind deflector. The task of vehicle aerodynamics is to determine the size and shape of this component as well as its mesh density.

Component forces
Component forces arise from the pressure forces produced by the flow around the body (measured quantity is the c_p value), which vehicle aerodynamics has to determine empirically and increasingly also through calculations. The positive or negative pressure forces acting perpendicular to the outer skin increase along with the airflow velocity and act as a load on the components located there.

In areas where there is shedding (such as A-pillars, exterior mirrors, rear window, rear spoiler), the turbulence results in stochastically varying loads, which have to be taken into account for determining the function and durability of the components.

Vehicle wind tunnels

Application

Aerodynamic measurements on roads are difficult, because the environmental conditions are neither uniform nor steady. The direction and steadiness of the conditions are constantly changing due to factors such as natural wind, buildings, and traffic. A vehicle wind tunnel is used to represent the air flow affecting a vehicle traveling on the road in an environment that is as realistic and reproducible as possible.

The advantages of using a vehicle wind tunnel as an experimental development tool compared with road measurements are the reproducibility of test conditions, the uncomplicated, reliable, and fast measuring technology, and the ability to disassociate effects (such as driving noise) that occur only in conjunction with road travel. Design prototypes that are not yet roadworthy can be optimized aerodynamically in a wind tunnel with guaranteed secrecy.

Wind tunnel designs

Types of wind tunnel

Aerodynamic parameters are calculated in wind tunnels. They vary in the type of air guidance, test section, size, and road-surface simulation (Table 2). Closed-circuit (closed return) wind tunnels are referred to as "Göttingen type", while open-circuit (no return) wind tunnels are referred to as "Eiffel type" (Figure 15). For vehicle aerodynamics, Göttingen wind tunnels with an open test section or with slotted walls are used predominantly.

Standard wind tunnel equipment

The test section design can be open, have slotted walls, or be closed. It is characterized by the outlet cross-section of the nozzle, the cross-section of the collector, and the length of the test section (Fig. 19, Table 2).

Obstruction $\Phi_N = A_{fx}/A_N$ is also an important parameter. This is the ratio between the frontal area of the vehicle A_{fx} and the nozzle cross-section A_N. On the road, this ratio is $\Phi_N = 0$, therefore it should also be as small as possible in the wind tunnel. Taking the construction and operating costs of a wind tunnel into account, $\Phi_N = 0.1$ is a common figure in practice. This corresponds to a diffuser cross-section of approx. 20 m^2.

The contraction and contours of a diffuser determine the velocity and steadiness of the wind-tunnel air flow. A large contraction ratio κ of the cross-sectional areas of the prechamber to the nozzle outlet ($\kappa = A_V/A_D$) leads to uniform velocity distribution and low turbulence, as well as to fast acceleration of the airflow.

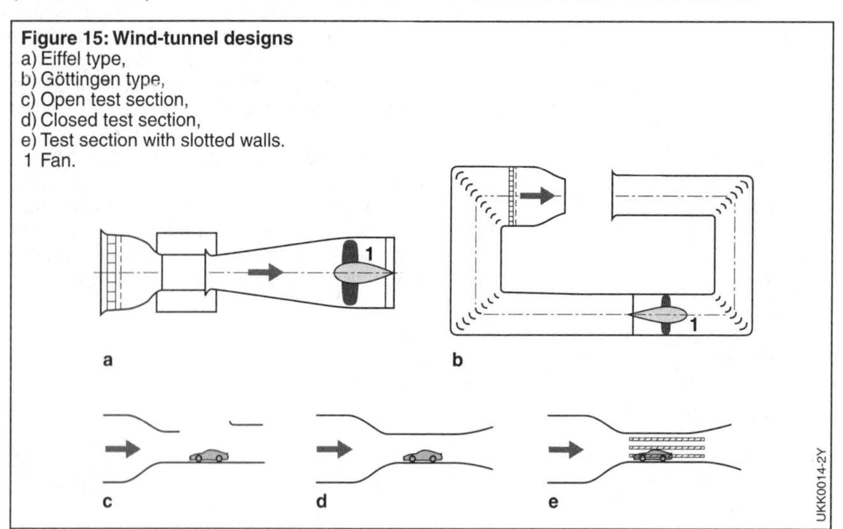

Figure 15: Wind-tunnel designs
a) Eiffel type,
b) Göttingen type,
c) Open test section,
d) Closed test section,
e) Test section with slotted walls.
1 Fan.

The diffuser contour can influence the steadiness of the velocity profile at the diffuser outlet in the test section, and parallelism in relation to the geometrical tunnel axis.

The settling chamber or prechamber is positioned upstream of the nozzle and in the largest cross-sectional area of the wind tunnel. The prechamber contains flow rectifiers, filters, and heat exchangers that improve the quality of the air flow in terms of steadiness and direction, and keep the channel temperature constant.

The fan in most wind tunnels can generate wind speeds of well over 200 km/h. Speeds of this size are only rarely used, for example, for testing the functional safety and stability of body components under wind loads. This is because such tests require the full fan capacity of up to 5,000 kW. Normal measurements are taken at 140 km/h. At this speed, the aerodynamic coefficients can be determined reliably and at low operating cost. The wind speed is regulated either by changing the fan speed, or by adjusting the fan blades at a constant speed (Tablo 3).

A wind tunnel balance records the aerodynamic forces at work on the test vehicle and the moments of all four tire contact surfaces; it breaks these forces into x, y, and z components to calculate the aerodynamic parameters.

A wind tunnel balance is normally positioned below a turning platform, which is used to turn the vehicle in relation to the wind direction, and thus simulate the effect of side winds.

In contrast to a real situation, a vehicle in a wind tunnel is stationary and has air blown toward it. The influence of the relative movement between the vehicle and

Table 2: Vehicle wind tunnels in Germany (examples)

Wind tunnel operator	Nozzle cross-section	Collector cross-section	Test section length	Test section design	Simulation of the moving road
Audi	11.0 m²	37.4 m²	9.5 to 9.93 m	Open	with 5 moving belts
BMW	18.0 to 25.0 m²	41.4 m²	17.9 m	Open	with 5 moving belts
BMW Aerolab	14 m²	25.9 m²	15.7 m	Open	1 conveyor belt, width greater than vehicle width
Daimler	28.0 m²	61 m²	19 m	Open	with 5 moving belts
IVK Stuttgart	22.5 m²	26.5 m²	9.9 m	Open	with 5 moving belts
Ford	20.0 m²	28.2 m²	9.7 m	Open	–
Porsche	22.3 m²	37.7 m²	13.5 m	Slotted walls, open	–
Volkswagen	37.5 m²	44.8 m²	10.0 m	Open	–

Table 3: Wind-tunnel fans

Wind tunnel	Drive power	Fan diameter	Wind velocity	Control
Audi	2,600 kW	5.0 m	300 km/h	Revolutions per minute
BMW	4,140 kW	8.0 m	300 km/h	Revolutions per minute
BMW Aerolab	3,440 kW	6.3 m	300 km/h	Revolutions per minute
Daimler	5,200 kW	9.0 m	265 km/h	Revolutions per minute
IVK Stuttgart	3,300 kW	7.1 m	260 km/h	Revolutions per minute
Ford	1,950 kW	6.3 m	185 km/h	Revolutions per minute
Porsche	2,600 kW	7.4 m	220 km/h	Revolutions per minute
Volkswagen	2,600 kW	9.0 m	180 km/h	Blade adjustment

the road cannot be taken into account. Therefore, in addition to wind tunnels with a stationary floor, there is now increased use of wind tunnels with moving belts incorporated into the floor to imitate motion of the car over the road and rotating wheels (Figure 16). The quality of the air flow between the vehicle and the road can thus be improved and made significantly more realistic.

Wind-tunnel auxiliary systems

Frontal-area measuring system

The frontal-area measuring system (laser or CCD system) measures the frontal area of the vehicle by optical means. The results are then used to calculate the aerodynamic coefficients from the forces measured in the wind tunnel.

The ideal pressure transducer system of a wind tunnel can simultaneously record the pressure curve at up to 100 pressure measuring points, for example, using the flat sensors affixed to the car body surface for the pressure distribution. The miniature pressure sensors used for each measuring point have no wearing parts

Figure 16: Turning platform on test section floor with flat belts incorporated

1 Moving belt between the wheels,
2 Balance,
3 Turning platform,
4 Wheel drive unit with small moving belt.

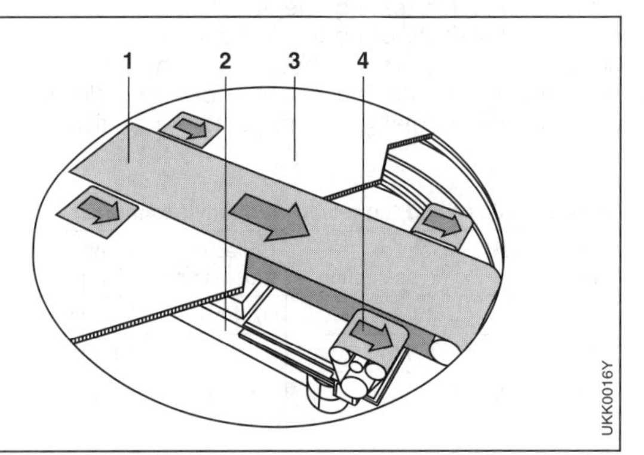

Figure 17: Measurement of pressure distribution (application example)

a) Pressure distribution on midsection of vehicle surface (y = 0), measured using 63 flat sensors,
b) Flat sensor.

The lines show the static pressure acting perpendicular to the measurement point.
The arrow length indicates the magnitude of the pressure.
The line under $c_p = 1$ shows the scale.
c_p is a dimensionless pressure coefficient;
it describes the pressure p_∞ at any point on the vehicle at the velocity v_∞ with respect to the dynamic pressure p prevalent at this velocity.

$$c_p = \frac{p - p_\infty}{0.5 \cdot \rho \cdot v_\infty^2}.$$

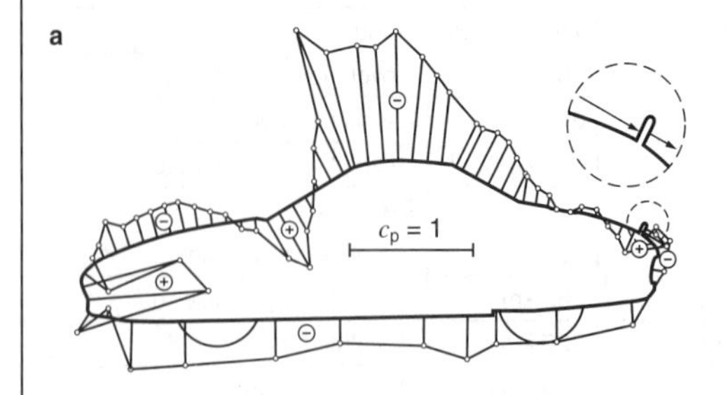

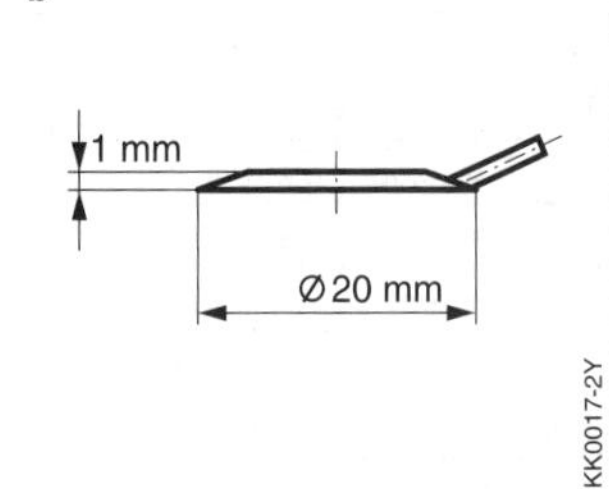

and can therefore electronically scan at a high frequency (Figure 17).

Traversing cradle
The traversing cradle allows the measurement of the complete vehicle airflow field. Each position in the test section can be located by coordinates and reproduced. According to the sensors present at each point, the pressure, velocity, or noise sources at each location can then be determined.

Smoke tubes
Smoke tubes are used to visualize the airflow, which would otherwise be invisible. The smoke pattern enables detection of flow separation, the vortices of which worsen the coefficients (Fig. 18). The non-toxic "smoke" is usually created by heating a glycol mixture in an oil-vapor generator. Other methods for visualizing the air flow include:
- Tufts on the vehicle surface
- Tuft sensors
- Photographs of flow patterns with quick-drying liquid mixture of kerosene or talc
- Helium bubble generator
- Laser sheet

Pollutant dispersion system
The pollutant dispersion system can be used to spray the vehicle in the wind tunnel with water ranging from light mist to heavy rain. The flow patterns can be visualized and documented by mixing chalk or a fluorescent agent to the pollutant.

Hot-water unit
The hot-water unit provides heated water at a constant rate for measuring radiator cooling capacity in prototypes that are not yet roadworthy.

Wind tunnel variants
A vehicle wind tunnel is a major investment. This investment, combined with high operating costs, make it an expensive test environment with a high hourly rate. Only very frequent utilization of vehicle wind tunnels for aerodynamic, aeroacoustic, and thermal experiments can justify the construction of several wind tunnels specialized for different tasks.

Model wind tunnel
A model wind tunnel can considerably reduce operating costs, due to the lower construction requirements and technical complexity. Depending on the scale (1:5 to 1:2), it is easier, faster, and more cost-effective to change the shape of vehicle models.

Model tests are primarily used in the early development phase for optimizing the basic aerodynamic shape. With the support of designers, plasticine models are used in a wind tunnel to optimize shapes or develop complete alternative shape variants and establish their aerodynamic potential.

Once there is a data set, new production methods (such as rapid prototyping) can be used to quickly produce models to any scale with accurate details. This allows for important investigations of the models in order to optimize details, even after the shaping phase.

Figure 18: Use of a smoke rake consisting of smoke tubes to make the airflow visible
1 Rake,
2 Smoke stream.

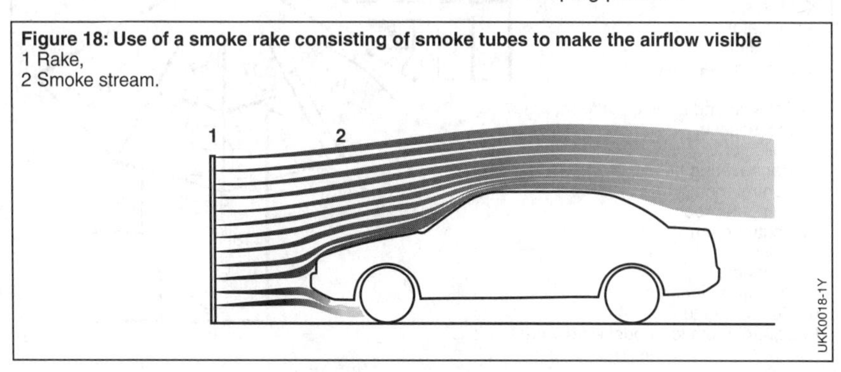

Acoustic wind tunnel
In an acoustic wind tunnel, extensive sound insulation measures mean that the sound pressure level is approximately 30 dB (A) lower than in a standard wind tunnel. This provides a sufficiently large signal-to-noise ratio of more than 10 dB (A) from the useful signal of the vehicle to allow identification and evaluation of wind noises generated by the air circulation and throughflow.

Climate-controlled wind tunnels
Climate-controlled wind tunnels are used for thermal analysis and protection of vehicles in defined temperature ranges at different load conditions.

Large-scale heat exchangers are used to set a temperature range, e.g. from –40 °C to +70 °C at a low control tolerance limit of approx. ± 1 K.

The vehicle is positioned on dynamometer rollers and driven at the required load condition or load cycle. Wind speed and roller speed must be exactly synchronized, even at low speeds. If necessary, uphill or downhill gradients can be simulated to introduce realistic road factors to the vehicle operating cycle.

There is also the option of regulating the air humidity or simulating solar radiation (using a bank of lamps).

The extensive aerodynamic challenges in vehicle design are not all covered by enhanced test-bench simulations as described above. In addition to experimental testing, manufacturers are therefore making increasing use of CFD models (Computational Fluid Dynamics). They allow preliminary decisions to be made to reduce the test workload.

Figure 19: Vehicle wind tunnel (example of the Göttingen type from Dr. Ing. h.c. F. Porsche AG)

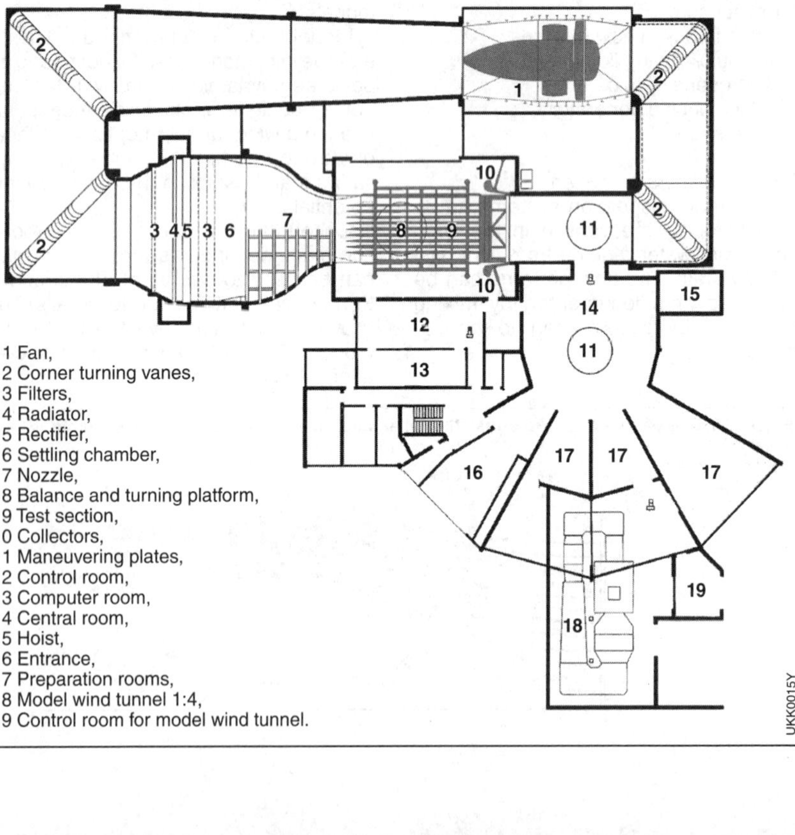

1 Fan,
2 Corner turning vanes,
3 Filters,
4 Radiator,
5 Rectifier,
6 Settling chamber,
7 Nozzle,
8 Balance and turning platform,
9 Test section,
10 Collectors,
11 Maneuvering plates,
12 Control room,
13 Computer room,
14 Central room,
15 Hoist,
16 Entrance,
17 Preparation rooms,
18 Model wind tunnel 1:4,
19 Control room for model wind tunnel.

UKK0015Y

Vehicle acoustics

Vehicle acoustics deals with all vibrational excitations and their transmission in the vehicle and in airborne noise. The primary sources here are the engine, excitation of the tires and chassis due to the texture and irregularities of the road, as well as excitation due to the airflow around the vehicle with increasing driving speed.

Table 1: Limits in dB(A) for noise emission from motor vehicles as per [1]

Vehicle category	since October of 1995 dB(A)
Passenger cars	
With spark-ignition or diesel engine	74
– with direct-injection diesel engine	75
Trucks and buses	
Permissible total weight below 2 t	76
– with direct-injection diesel engine	77
Buses	
Permissible total weight 2 t to 3.5 t	76
– with direct-injection diesel engine	77
Permissible total weight above 3.5 t:	
– engine power output up to 150 kW	78
– engine power output above 150 kW	80
Trucks	
Permissible total weight 2 t to 3.5 t	76
– with direct-injection diesel engine	77
Permissible total weight above 3.5 t (German Road Traffic Licensing Regulations: above 2.8 t):	
– engine power output up to 75 kW	77
– engine power output up to 150 kW	78
– engine power output above 150 kW	80
Off-road and four-wheel drive vehicles These vehicles are governed by higher additional limit values for engine brake and compressed-air noises.	

Legal requirements

Noise emission from vehicles and its testing

Test procedures used to monitor compliance with legal requirements are concerned exclusively with exterior-noise levels. The EC Directive 70/157/EEC enacted in 1970, together with its last revision 2007/34/EC [1], defines measurement procedures for noise emissions from stationary and moving vehicles, as well as limit values for noise emissions from moving vehicles (Table 1) for various vehicle categories. Due to the inadequate effectiveness of the statutory provisions in real traffic situations, EU legislators are currently working on revising the measurement procedure with the aim of better reproducing urban traffic situations in the test procedure. New legislation is expected to be introduced in 2014, with the first obligatory application of it in 2016.

In addition, tread noise requirements were defined in the regulation 661/2009/EC [2]. Depending on the tires, the test area is 70 or 80 km/h.

Measuring noise emissions from moving passenger cars and trucks up to a permissible total weight of 3.5 t

The vehicle approaches line AA, which is located up to 10 m from the microphone plane, at a constant velocity of 50 km/h (Figure 1). After the vehicle reaches line AA, it continues under full acceleration as far as line BB (placed 10 m behind the microphone plane), which serves as the end of the test section. Passenger cars with manual transmissions and a maximum of four forward gears are tested in second gear. Consecutive readings in second and third gear are employed for vehicles with more than four forward gears, while sports cars are tested in third gear in accordance with the definition in the Directive. Vehicles with automatic transmissions are tested in the D position.

The noise-emissions level is defined as the maximum occurring sound level which is measured on the left and right sides of the vehicle at a distance of 7.5 m from the center of the lane. In a test in two gears, the noise-emissions level is the arithmetic mean from the measurement of both gears.

Measuring noise emissions from moving trucks starting from a permissible total weight of 3.5 t

The vehicle approaches line AA, which is located up to 10 m from the microphone plane, at a constant velocity (Figure 1). After the truck reaches line AA, it continues at full acceleration as far as line BB (also placed 10 m from the microphone plane), which serves as the end of the test section. The driving speed adopted is dependent on the gear tested and the rated engine speed. The choice of gears selected is based on the fact that at least the rated speed must be reached within the test section, however, the vehicle must pass through the test section without reaching the engine-speed limitation. The noise-emissions level is defined as the maximum occurring sound level from all the measurements.

Measuring noise emissions from stationary vehicles

Following noise approval of a vehicle, measurement of noise emissions from stationary vehicles makes it possible to determine a reference noise value for the noise emissions of the vehicle in traffic when in "proper" condition. This makes it easier for the vehicle's noise emissions to be checked, for example, by the police or regulatory official. Noise emission from stationary vehicles is measured at a distance of 50 cm from the exhaust outlet at a horizontal angle of 45° ± 10° relative to the direction of exhaust flow. During the measurement, the engine has to be brought up to a speed dependent on its nominal speed. Upon reaching this test speed, it is to be maintained for 3 sec. and then the throttle response switch is to be quickly moved to the idle position. The maximum A-weighted sound pressure level determined during this measurement is entered in the vehicle documentation with the suffix P´ (to distinguish it from data of earlier test procedures).

A maximum deviation of 5 dB from the entered reference value is permitted for tests in traffic.

Figure 1: Test layout for driving-noise measurement
1 Road surface as specified by ISO 10844 [3],
2 Left microphone,
3 Right microphone.

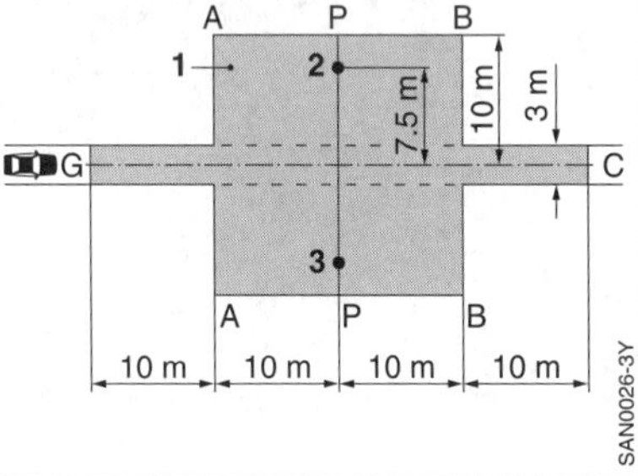

Minimum noise from vehicles

When quiet electric motors began being widely used as a drive source for vehicles, it became a problem for pedestrians: They have great difficulty recognizing these vehicles in traffic in a timely manner. Investigations have shown that this is particularly the case in the low-speed range up to about 20 to 30 km/h. To reduce the danger of collisions between vehicles and pedestrians, work is currently being done to specify an international standard for minimum noise emissions. It is anticipated that these requirements will become obligatory in the EU as of 2019.

Development work on vehicle acoustics

Measuring equipment for acoustics

The measuring techniques listed in the following are used to measure and evaluate not only the exterior noise level, but also the interior noise level and in particular the various vehicle vibrations [4]:

- Sound-pressure recording is performed with capacitor microphones, e.g. using sound-level meters, in dB(A).
- The dummy-head measurement technique enables noise to be recorded by microphones built into the ears of an artificial head. Artificial- or dummy-head technology provides the opportunity of making realistic recordings, so that they can be compared with other similar recordings at a different place and time.
- Measuring rooms for standard sound measurements are generally equipped with highly sound-absorbent walls. Depending on the objective, these rooms can be equipped with chassis dynamometers, so that a vehicle can be evaluated even while in driving mode.
- For vibrations and structure-borne noise acceleration, sensors (mass partly under 1 g) are used which operate, for example, according to the piezoelectric principle. Laser vibrometers are used for rapid, non-contact measurements based on the Doppler principle.
- Position measuring devices are used, for example, to record movement of the shock absorbers and chassis relative to the body.
- Vibration test benches provide the option of measuring the various cases of vibrational excitation in the overall vehicle, the body, or individual components. Hydropulsers permit excitation of low frequencies smaller than 150 Hz with large vibration paths for simulating road excitation. Shakers, on the other hand, permit high frequencies up to 2 kHz for simulating engine excitation.
- Ride comfort segments simulate various kinds of road excitation, ranging from different kinds of asphalt to manhole covers or concrete transverse joints.
- Reverberation rooms enable the sound insulation of individual materials or components to be determined, for example, for doors or whole bulkheads in the engine compartment.

Calculating methods in acoustics

Vibrations and oscillations

Natural-vibration calculations are performed using the finite-element method (FEM). Model adjustment with experimental modal analysis or modeling of forces acting during operation enables the calculation of real operational vibration shapes. In this way, structures can be optimized in early design phases with regard to their vibration response and sound radiation.

Airborne noise and fluid-borne noise

Sound-field calculations, for example of cabinet radiation or in cavities, are carried out using the finite-element method (FEM) or boundary-element method (BEM) ([5]).

Ride comfort simulation

Effects of vibration from the road on the chassis and body can be simulated by using rigid body models in combination with the road excitation. What is interesting here is that it is also possible to connect the drive unit via the engine mount and transmission mount. Such models enable variable coordination of the chassis mount and power train mount, typically in the frequency range up to 50 Hz.

Sources of noise

Engine and transmission acoustics

Engine noises are caused primarily by the mechanical movement of components. Even the combustion process in the cylinders causes excitations of the engine structure. In all cases, the vibrations are, on the one hand, directed through the structure and, on the other hand, radiated as airborne noise. In the process, noises can also occur dominantly at a clear distance from the excitation location. This is determined by the sensitivity of the membrane surfaces and the natural-vibration response of the local structure.

Important sources of noise on the engine are the crankshaft drive, the camshaft drive, the crankcase, the oil pan, the cylinder-head cover, the auxiliary power take-offs (e.g. alternator and water pump), and the belt drives. All of these elements together shape the basic acoustic character of the unit (engine/transmission combination).

The objective of the acoustician is to achieve harmonization of the vibration and noise behavior in the power train by means of suitable technical modifications. Some of the important influencing factors to take into account include the following:

- Tolerances, clearances and structures
- Thermal expansion in clearances
- Avoiding resonances
- Choice of materials
- Vibration shapes of the engine/transmission unit
- Muffler design

Here, most of these are in a conflict of interests with other development disciplines, particularly space design (package), energy management and driving dynamics. Thus, in a four-cylinder engine, the often dominant second engine order, which is frequently perceived as humming, can be significantly reduced with balancer shafts and very rigid units. But this comes at the expense of additional weight, more complicated structures and a considerable dip in performance.

Engine and transmission acoustics also has the task of largely reducing the transmission of rigid-body sound vibrations to the body and the airborne noise. Starting points for this are the design of the power train mount with respect to insulation, low-vibration designs, short, compact connections of auxiliary units, and prevention of large, smooth vibration surfaces by means of ribbing or additional damping materials.

The engine mount deserves special attention. Stiff assembly bearings improve driving dynamics, but convey to the driver a very raw engine noise and increase the vibration input into the body. If the bearings are too soft, it means the engine can move large distances, which can lead to unpleasant bucking motion of the whole vehicle. Highly insulating, hydraulically damped elements can be used to bring about a certain softness with good vibration comfort for the whole vehicle.

Gas exchange noise (exhaust and intake system)

The energy transmitted by the combustion processes to the inducted air is output in time with the ignition processes and the valve movements via the exhaust-gas system to the surroundings again. The design of the exhaust manifolds, the exhaust pipes and the mufflers decisively affects which frequency components are radiated with which levels. Here, the sound output, not only at the orifice but also at the surface of all the components, is important. The noise dynamics of this acoustic source are enormous, easily reaching 15 dB and higher when the level difference is considered between idle and full load. On the basis of these dynamics, this source is ideally suitable for sound designs.

Chassis acoustics
Excitation from the road introduces vibrations into the chassis. These are strongly attenuated at first by the damping of the tires, but vibrations can still be transmitted in the chassis and into the body. Moreover, the road can incite natural resonance in the chassis, which can reach the passenger compartment in the form of unpleasant droning if the body has a certain sensitivity. Chassis acoustics has the task of finding an optimum acoustical configuration of the chassis without impairing the dynamic handling characteristics of the vehicle.

Body acoustics
The body is the interface between the chassis, power train and mounts of many auxiliary drives. These include, for example, fans, servomotors and even loudspeakers. The body, however, is also a point of attack for airflow forces. It surrounds the driver and passengers and protects them from external influences. But it also transmits vibrations and introduces them to the passenger compartment, where the occupants can both feel and hear them.

Body acoustics deals specifically with minimizing the transmission of sound and preventing airborne noise excitations, which, along with resonance in the passenger compartment, can lead to unpleasant droning noise. It also deals with optimizing the natural-oscillation characteristics of the body, because the body's eigenmodes – such as the torsion of the basic body or the first bending mode of the front end – can lead to shaking of the vehicle, which feels like when the vehicle is very fluttery on the road.

High rigidity of the attachment points for the chassis and engine provide for low excitation, sealing the byways reduces the conduction of sound within the body, and damping materials provide for sufficient insulation.

Body acoustics also includes aeroacoustics, in which optimized transitions at vehicle edges and flows around components should prevent turbulent flows from developing. These can make themselves apparent as unpleasant interior and exterior noise at high speeds.

Ride comfort
Ride comfort concerns all of the ride-related components that the driver and passengers touch. In particular these include the steering wheel, shift lever, pedals, and of course the seat. Excitations can come from the road, body, or engine and are in the frequency range under 50 Hz. The objective is to minimize all vibrations.

Dynamic strength
All vehicle components are subject to vibrational excitation, and therefore to dynamic loads. To ensure that all components have a sufficient dynamic load capacity over a long time period, they have to be investigated with respect to their dynamic strength. To do so, individual components as well as entire assemblies are subjected to continuous loading on test benches by electrodynamic shakers or hydraulic punches. The excitation profiles are gained from load measurements from real driving profiles. If necessary, the components can also be tested with climatic adaptations, such as low or extremely high temperatures. Exhaust systems can, for example, have hot gases pumped through them to simulate the engine exhaust gases. These tests provide valuable indications for optimizing details, as a result of which materials, flanges or ribs are often optimized again.

Sound design

Definition of sound design

Sound design is the active and specific matching of a product to a desired noise behavior. The aim is to lend the product a typical "sound" expected by the customer and corresponding to the brand identity and to arouse quite specific emotional associations with the product.

The most well-known field is the automobile sector, where both customers and the press show great interest in the result of the sound development of products. However, intensive work is also being carried out on the sound of products in many other fields, such as, for example, household appliances (vacuum cleaners and washing machines) and even foods (the "crunch" of potato chips).

Terms other than sound design are also commonly encountered. Sound engineering is, as a rule, the concrete technical realization on the product, i.e. the specific technical implementation of a target sound elaborated in sound design.Sound cleaning is the suppression of unwanted noise to achieve an optimally neutral noise behavior. This is often carried out for luxury vehicles as a starting point for developing a discreet sound.

Sound at the vehicle

The steps for implementing sound at a product, are generally speaking, always the same. For clarification purposes, this is explained by reference to the automobile sector.

Sound is the audible feedback of the vehicle at a specific operating state arrived at by the driver using the accelerator pedal and choosing a gear. Sound is accordingly described by the engine's operating point with regard to engine revs and load, as well as their time variance – i.e. acceleration, steady-state driving or deceleration. Starting and idle noises are added.

Depending on the design, the subjective perception of the throttle response is "spontaneous" or "sluggish", regardless of the real physically measurable acceleration.

A distinction is made between a vehicle's interior and exterior noises. Exterior noise is shaped by the airborne noise radiation of all the components, primarily by the engine and the exhaust-gas and air-intake systems and their orifices. In addition to these, there is disruptive noise from rolling and the wind. The acoustic conditions in the passenger compartment are additionally influenced by the structure-borne noise and the acoustic transfer paths through the body.

Sound quality

Sound quality refers to discrete components that can be actively moved or changed, such as switches, levers, servomotors, or flaps. Here, the actuating noise is intended to underline the quality and the soundness of the product. Opening and closing doors and hoods also has to sound good and must not be jarring. Moreover, sound quality also includes prevention of rattling, grinding and squeaking noises produced by components touching and rubbing against each other.

Assessment methods

Psychoacoustics

Sound cannot be assessed with the standard assessment method based on loudness, roughness or tonality. When the sound of a vehicle is mentioned, for the most part terms such as "sporty", "sonorous", "aggressive" or "inspiring" are used. These must be translated into a technical language so that the development engineers know at a later stage what spectral content, what dynamics and what time variances are responsible for them. For this reason, these subjective factors are being increasingly brought into play and structured in order to establish assessment criteria from them. In the practical course of a psychoacoustic assessment (Figure 2), sound samples are assessed according to antagonistic value pairs (such as good – bad, loud – quiet, weak – strong) so as to arrive at a characterization of the sound.

Virtual prototyping
Analyses and measurements conducted on components can be used to make predictions about sound behavior and the leeway for making modifications. Vehicle measurements are also made using artificial-head technology. Variations can then be simulated in a narrow range with modern computer technology, always with a view toward whether a "virtually created sound" can also be technically realized.

Now the virtual sounds processed using computer technology can be evaluated by test subjects with respect to previously developed evaluation criteria. From this, it is possible to derive assessor preferences, which are now, however, technically stored and provide indications as to which frequency elements are necessary or offending and which components come into consideration.

Figure 2: Psychoacoustic assessment

Dummy head
Test person
Assessment
exciting
pleasant
loud
sporty
interesting
hard
dynamic
rough
UAN0154E

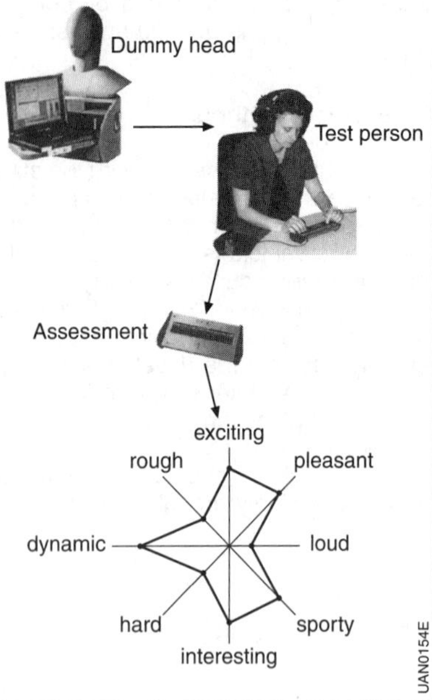

Active sound
Of course, missing sound elements can also be provided with active speaker systems inside and outside the vehicle. This technology is currently gaining ever more importance, because electric vehicles and hybrid vehicles in electric driving mode are frequently evaluated as being too quiet. Significant improvements can be achieved here by adding electronic sound. For the aforementioned vehicles, such systems will even be prescribed by law as of 2019 (see Minimum noise from vehicles).

References
[1] Commission Directive 2007/34/EC of 14 June 2007 amending, for the purposes of its adaptation to technical progress, Council Directive 70/157/EEC concerning the permissible sound level and the exhaust system of motor vehicles.
[2] Regulation (EC) No 661/2009 of the European Parliament and of the Council of 13 July 2009 concerning type-approval requirements for the general safety of motor vehicles, their trailers and systems, components and separate technical units intended therefor.
[3] ISO 10844:2011: Acoustics – Specification of test tracks for measuring noise emitted by road vehicles and their tyres.
[4] Klaus Genuit: Sound-Engineering im Automobilbereich. 1st ed., Springer, 2010.
[5] L.C. Wrobel, M.H. Aliabadi:
The Boundary Element Method. 1st ed., Wiley, 2002.

Chassis systems

Basic principles

The roads normally used by motor vehicles feature irregularities that cover a frequency range of up to around 30 Hz and are the most intensive source of excitation in the vehicle. The resulting excitation leads to vertical movements (vertical accelerations) of the vehicle and its occupants. The interface between the road surface and the vehicle is the chassis system, the main task of which is power transmission between the environment and the vehicle body. This means that both the dynamic characteristics of the vehicle and driving safety, as well as driving comfort, are heavily influenced by the choice of chassis system.

The dynamics of the vehicle response, however, are not exclusively determined by the chassis system components, but rather are much more a consequence of the combination of different overall vehicle parameters. As a rule, influencing the vehicle dynamics with measures on the chassis system involves high complexity, in particular because the effects of parameter variations have to prevail among the conflicting priorities of driving safety and driving comfort.

Driving safety depends decisively on the contact relationships between the tires and the road surface, and thus on the longitudinal and lateral forces that can be transferred. A fundamental aim in chassis system configuration with regard to driving safety is therefore always to minimize the dynamic wheel load fluctuations that cause a reduction in the level of force that can be transferred.

The driving comfort, on the other hand, depends on the movements and accelerations that affect the occupants (above all in the vertical direction). Depending on the area of application, comfort is of great relevance and should in no way be regarded merely as a concomitant of system development. In the case of professional drivers in particular, adequately high driving comfort is to be ensured in order to prevent long-term damage to health. The effective value of the body acceleration has

Figure 1: Requirements for a passenger car chassis system

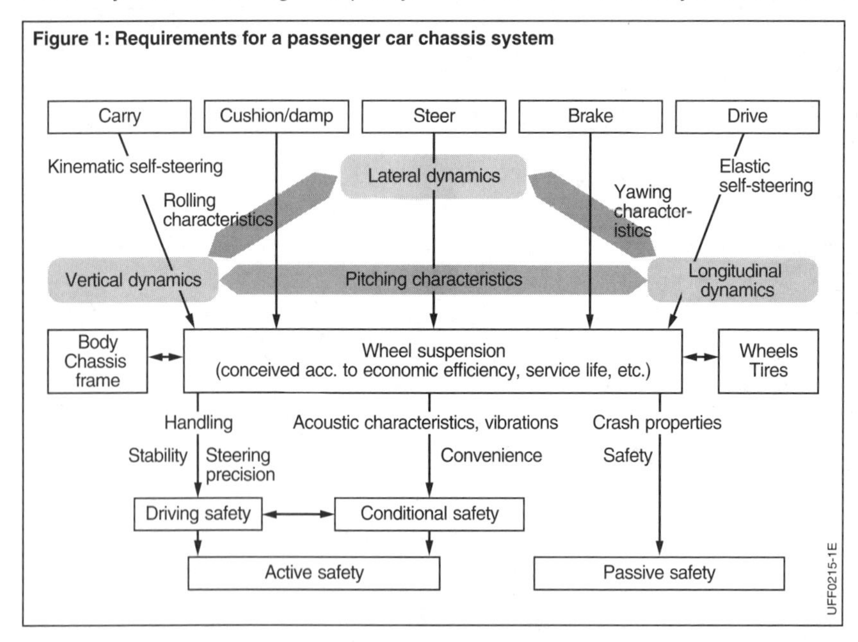

proven to be a good evaluation parameter in this context.

These core parameters, however, initially only describe the potentials of a chassis system to meet the corresponding requirements. The actual dynamics of the vehicle, tho driving safety, and the comfort also depend decisively on the choice of road surface parameters (environment) and vehicle-internal manipulated variables (for example, steering angle and accelerator pedal position) set by the driver. The fundamental requirements for a chassis system are shown in Figure 1 [1]. The chassis as a system (without electronics) is traditionally divided into the following subsystems:
- Suspension
- Shock absorbers
- Wheel suspension and suspension linkage
- Tires
- Steering
- Brake system

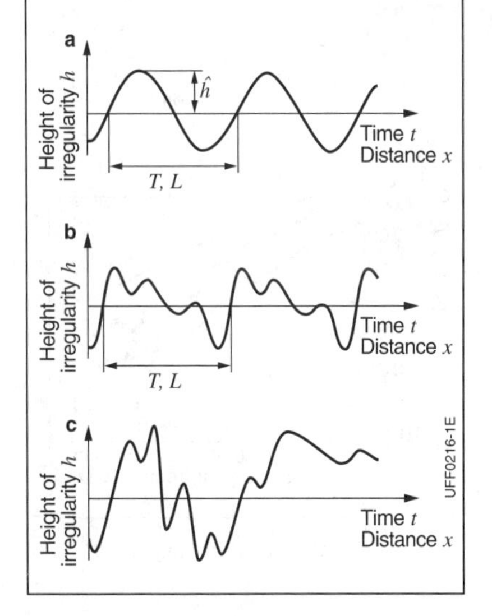

Figure 2: Types of courses of irregularity
a) Sinusoidal course of irregularity,
b) Periodical course of irregularity,
c) Stochastic course of irregularity.
T Period duration,
L Period length,
$\hat{h}$ Amplitude of irregularity height.

Road excitation

Before the basic principles of individual components of the "chassis" as a system are presented below, first of all the quantification of the road excitation will be discussed and then the effect of the excitation on the vibrating overall "vehicle" system. Alongside the adaptation of the oscillatory system to the external road excitations, vehicles also require the influence of internal sources of excitation on the vibration characteristics (internal-combustion engine, wheel, and tires) to be minimized [2].

Knowledge and a description in the form of objective variables of the road excitation that causes vibrations are required in order to examine the vibration characteristics and to configure the suspension/shock absorber system of a chassis. Whereas minor irregularities can already be compensated for by the suspension characteristics of the tire, an element between the wheels and body that changes its length is required to reduce greater body movements. Steel springs are used most frequently here, delivering a return force that depends on the change in length. The result is – taking account of the wheel and body masses – a system that can vibrate and requires other elements for damping.

Courses of irregularity

When describing courses of irregularity in a road depending on the path or time, a distinction is made between harmonious (sinusoidal), periodical, and stochastic courses of irregularity (Figure 2).

Harmonious course of irregularity

In this case (Figure 2a), the irregularity height h following distance driven x or point in time t results in:

$$h(x) = \hat{h} \sin(\Omega x + \varepsilon) \quad \text{(Equation 1)},$$

$$h(t) = \hat{h} \sin(\omega t + \varepsilon) \quad \text{(Equation 2)},$$

where:
$\hat{h}$ Amplitude of irregularity height,
L Period length,
v Speed,
ε Phase shift,
$\Omega = 2\pi/L$ Breakdown angular frequency,
$\omega = v\,\Omega = 2\pi\, v/L$ Time angular frequency.

Periodical course of irregularity
In the next step, it is no longer only purely sinusoidal but also periodical courses of irregularity (Figure 2b) that are assumed, which means that expressing the relationship as a Fourier series (periodical functions can be described as an infinite series of sinusoidal oscillations) results in:

$$h(x) = h_0 + \sum_{k=1}^{\infty} \hat{h}_k \sin(\Omega x + \varepsilon_k) \quad \text{(Equation 3)},$$

$$h(t) = h_0 + \sum_{k=1}^{\infty} \hat{h}_k \sin(\omega t + \varepsilon_k) \quad \text{(Equation 4)},$$

where:
h_0 Base amplitude,
$\hat{h}_k$ Amplitude,
ε_k Phase shift,
$\Omega = 2\pi/L$ Breakdown angular frequency,
L Period length.

Stochastic course of irregularity
For real road surfaces, i.e. not only for periodical but also for random (stochastic) courses of irregularity (Figure 2c), in the transition into the complex syntax of the summation formula the integral is arrived at, thus moving from the discrete to the continuous spectrum [2]:

$$h(x) = \int_{-\infty}^{\infty} \underline{\hat{h}}(\Omega)\, e^{j\Omega x} d\Omega \quad \text{(Equation 5)}$$

with the continuous amplitude spectrum:

$$\underline{\hat{h}}(\Omega) = \frac{1}{2\pi} \int_{-\infty}^{\infty} h(x)\, e^{-j\Omega x} dx \quad \text{(Equation 6)},$$

or in the time range:

$$h(t) = \int_{-\infty}^{\infty} \underline{\hat{h}}(\omega)\, e^{j\omega t} d\omega \quad \text{(Equation 7)},$$

where

$$\underline{\hat{h}}(\omega) = \frac{1}{v}\, \underline{\hat{h}}(\Omega) \quad \text{(Equation 8)}.$$

Power spectral density of the road irregularities
As a rule, however, the course of irregularity as a function of path or time is less suitable for theoretical examinations of the road excitation. What is considerably more interesting is knowing the statistical mean of the excitations that occur when driving on a certain road surface, particularly in comparison with different road surfaces. This is why the power spectral density $\Phi_h(\Omega)$ (PSD) depending on the breakdown angular frequency and power spectral density depending on the time angular frequency $\Phi_h(\omega)$ have prevailed as the measure for evaluation of road excitations. A mathematical definition and detailed discussion of these variables are not provided here. Refer to [1, 2]. For the following examination, the descriptive significance of the power spectral density that exists in the "energy in an infinitesimal frequency band" (with regard to path or time frequency) or the "power output in the natural frequency band" is sufficient.

Taking account of $L = vT$, $\Phi_h(\Omega)$ and $\Phi_h(\omega)$ can be converted into one another. The following equation applies:

$$\Phi_h(\omega) = \frac{1}{v}\, \Phi_h(\Omega) \quad \text{(Equation 9)}.$$

The power spectral densities specify the distribution of the power output of the excitation spectrum within the entire irregularity spectrum. This is usually represented in the double-logarithmic scale (Figure 3). An approximation of the courses of power spectral density on real road surfaces is possible here with straight lines, as described by the following equation:

Figure 3: Irregularity spectrum depending on the breakdown angular frequency
1 Highway,
2 Freeway,
3 Upper range limit for poor roads,
4 Lower range limit for good roads.

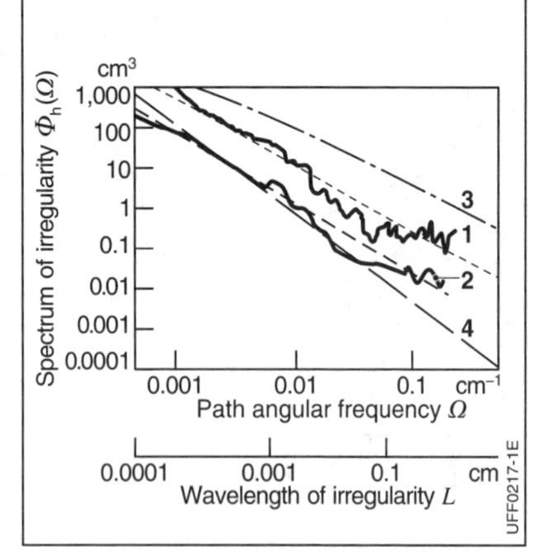

$$\Phi_h(\Omega) = \Phi_h(\Omega_0) \left(\frac{\Omega}{\Omega_0}\right)^{-w} \quad \text{(Equation 10)}$$

where:
$\Phi_h(\Omega_0)$ Power spectral density with a reference angular frequency Ω_0 (as a rule $\Omega_0 = 1\ m^{-1}$, which corresponds to a reference wavelength of $L_U = 2\pi/\Omega_0 = 6.28\ m$),
w Surface undulation of the road surface (from 1.7 to 3.3; for standard roads $w = 2$).

If the degree of irregularity and the surface undulation of a road surface are known, all the parameters are set in the above linear equation and it can then be used to characterize different types of road surface. The degree of irregularity is defined as the power spectral density at a reference angular frequency, i.e $\Phi_h(\Omega_0)$.

Basic principles of the vibration characteristics of vehicles

The road excitations take effect on the vehicle body via the tires, wheel suspension, and the suspension / shock absorber system. For theoretical analyses, vibration models of different complexity can be applied. With rising model complexity, there is an increase in the degrees of freedom and linked differential equations. To improve clarity, the fundamental relationships within vibrating vehicle systems will be explained on the basis of the dual-mass oscillator (Figure 4). More detailed information can be found in [1], [2], and [3].

On specification of the masses, the rigidities, and damping factors, all of the parameters have been set in the dual-mass model for technical analyses of the vibrations and two differential equations can be set up with the variables designated in Figure 4:

$$m_A \ddot{z}_A + k_{A,R}(\dot{z}_A - \dot{z}_R) + c_{A,R}(z_A - z_R) = 0 \quad \text{(Equation 11a)},$$

$$m_R \ddot{z}_R - k_{A,R}(\dot{z}_A - \dot{z}_R) - c_{A,R}(z_A - z_R) + c_R z_R + k_R \dot{z}_R = c_R h + k_R \dot{h} \quad \text{(Equation 11b)}.$$

The division of Equation 11a and Equation 11b by the mass in each case leads to the normal form of a differential equation of the 2nd order and delivers the damped natural frequency ω_g and the undamped natural angular frequency ω_u as well as the damping factors for the wheel D_R and the body D_A.

The following applies to the uncoupled and undamped natural angular frequency at the wheel:

$$\omega_R = \sqrt{\frac{c_R + c_{A,R}}{m_R}} \approx \sqrt{\frac{c_R}{m_R}} \quad (\text{as } c_R \approx 10\, c_A) \quad \text{(Equation 12)}.$$

The following applies accordingly to the body:

$$\omega_A = \sqrt{\frac{c_{A,R}}{m_A}} \quad \text{(Equation 13)}.$$

In general, the damped natural angular frequency ω_g is calculated according to:

$$\omega_g = \omega_u \sqrt{1 - D^2} \quad \text{(Equation 14)},$$

whereby the following is also assumed as an approximation:

$$\omega_g \approx 0.9\, \omega_u .$$

Figure 4: Dual-mass oscillator as quarter-vehicle model
m_R Wheel mass, including components that also vibrate (brake, proportional axle mass etc.),
m_A The body mass supported by a wheel,
c_R Rigidity of the tire springs,
$c_{A,R}$ Wheel-related rigidity of the body springs,
k_R Dampening rate of the tire (usually negligible as $k_R << k_{A,R}$),
$k_{A,R}$ Wheel-related dampening rate of the body damper,
h Vertical excitation,
z_A Body coordinates in vertical direction,
z_R Wheel coordinates in vertical direction.

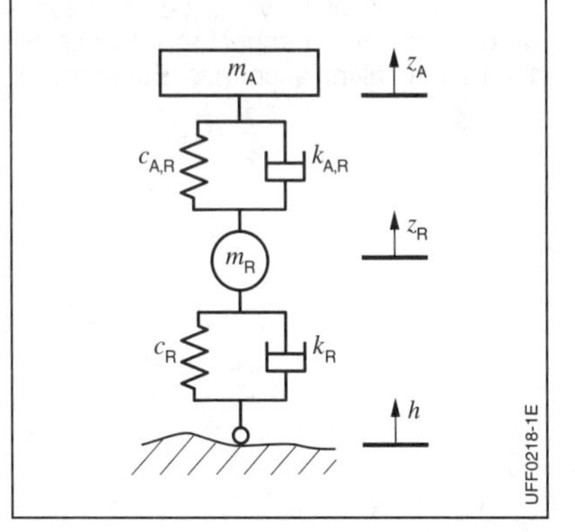

The following applies to the damping factor D_R at the wheel:

$$D_R = \frac{k_{A,R}}{2 m_R \omega_R} = \frac{k_{A,R}}{2\sqrt{(c_R + c_{A,R})\, m_R}} = \frac{m_A \omega_A}{m_R \omega_R} D_A \quad \text{(Equation 15)},$$

whereby experience has shown that $D_R \approx 0.4$ is to be aimed for. The same applies to the body:

$$D_A = \frac{k_{A,R}}{2 m_A \omega_A} = \frac{k_{A,R}}{2\sqrt{(c_{A,R} m_R)}} \quad \text{(Equation 16)}.$$

In this case, $D_A \approx 0.3$ has proven to be effective. The dynamic wheel load fluctuations ΔG result in:

$$\Delta G = m_R \ddot{z}_R + m_A \ddot{z}_A = c_R (h - z_R) + k_R (\dot{h} - \dot{z}_R) \quad \text{(Equation 17)}.$$

These variables together form a basis for the rough configuration of the suspension/shock absorber system of a motor vehicle.

If a body natural frequency (usually $f_A \approx 1$ Hz) is specified and the body mass (or the proportion of the body mass on a wheel) is known, the wheel-related body spring rigidity can be determined:

$$c_{A,R} = \omega_A^2\, m_A \quad \text{(Equation 18)}.$$

Figure 5: Spring ratio
ΔF_F Actual spring force,
ΔF_R Wheel force,
Δz_R Spring compression,
d_2 Distance from wheel center plane to the pivot,
$d_2 - d_1$ Spring force lever arm.

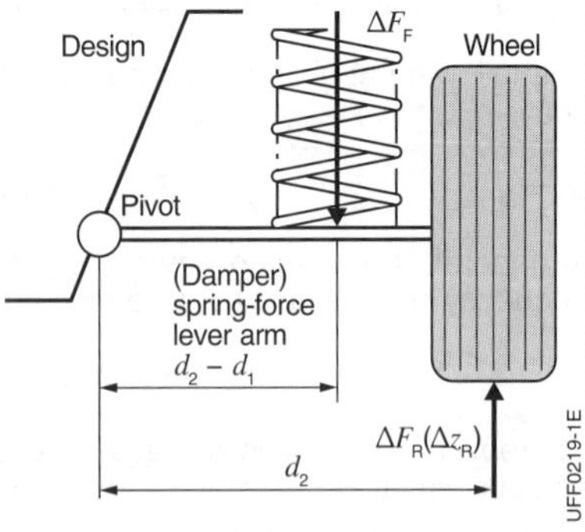

The conversion to the actual rigidity of the body springs takes place taking account of the ratio i between the wheel and spring movement in accordance with Figure 5. First of all, the actual spring force

$$\Delta F_F = c_A \Delta z_F \quad \text{(Equation 19)}$$

and the wheel force ΔF_R are formulated with a spring compression Δz_R. The following applies to ΔF_R:

$$\Delta F_R = c_{A,R} \Delta z_R \quad \text{(Equation 20)}.$$

The torque equilibrium around the pivot in Figure 5 leads to:

$$c_{A,R} \Delta z_R d_2 = c_A \Delta z_F (d_2 - d_1) \quad \text{(Equation 21)}.$$

This can be used to convert the actual spring rigidity c_A in accordance with the geometric relationships to the wheel-related rigidity $c_{A,R}$:

$$c_{A,R} = c_A i^2 \quad \text{(Equation 22)}$$

with the spring ratio i

$$i = \frac{(d_2 - d_1)}{d_2} = \frac{\Delta z_F}{\Delta z_R} \quad \text{(Equation 23)}.$$

The same holds true for the vibration damper. For calculation of the body dampening rate (its effect on the wheel), the following applies in relation to the wheel in accordance with Equation 16:

$$k_{A,R} = 2 D_A \sqrt{c_{A,R} m_R} \quad \text{(Equation 24)}.$$

With $D_A = 0.3$ (see above) and m_A as a known variable of the vehicle under examination, the body dampening rate can be determined taking account of Equation 23.

Table 1: Effects of changes to the suspension/shock absorber system on the vertical vehicle vibration characteristics

	Design parameters	Effects in the low-frequency range (body natural frequency)	Effects in the middle frequency range	Effects in the higher-frequency range (axle natural frequency)
Body data	Spring constant	Body natural frequency and body acceleration drop radically when the spring rigidity is reduced	Body acceleration drops slightly on reduction of the spring rigidity	Body acceleration remains virtually constant on reduction of the spring rigidity
Body data	Dampening rate	Body acceleration rises radically on reduction of the dampening rate	Body acceleration drops radically on reduction of the dampening rate	Dynamic wheel load fluctuation rises radically on reduction of the dampening rate
Tire data	Spring constant	Natural frequency and amplitude remain approximately constant		Natural frequency and amplitude of the body acceleration and wheel load fluctuations drop approximately proportionally to the reduction in vertical wheel spring rigidity
Tire data	Dampening rate	Natural frequency and amplitude remain approximately constant		Amplitude of the body acceleration and wheel load fluctuation drops slightly with increasing damping at consistent wheel natural frequency

Using Equation 15, the estimate of optimal relationships between wheel and body mass with the help of the damping factors $D_R = 0.4$ and $D_A = 0.3$ to be aimed for delivers

$$D_R = \frac{m_A \omega_A}{m_R \omega_R} D_A$$

the following relationship:

$$\frac{m_A}{m_R} = \frac{0.4\,\omega_R}{0.3\,\omega_A} = \frac{0.4 f_R}{0.3 f_A} \qquad \text{(Equation 25)},$$

whereby $f_R = \omega_R/2\pi$ and $f_A = \omega_A/2\pi$ were set. With $f_R = 12$ Hz and $f_A = 1$ Hz, the result is

$$m_R = \frac{1}{16}\, m_A\,.$$

The influence of different suspension/shock absorber parameters and their effects on different frequency ranges is shown in Table 1.

Transfer function and power spectral density of body acceleration and wheel load fluctuations

Equation 11a and Equation 11b can be sued to calculate the transfer functions (enlargement functions). The amount of the transfer function is the quotient from the stationary proportion of the amplitude of the output variables (body acceleration or wheel load fluctuation) and the amplitude of the input variable (vertical excitation, see Figure 4). The principle of the progression is shown in Figures 6c and 6d.

These transfer functions are now linked to the road irregularities. To do so, the distribution of the road irregularities is assumed to be a function of the breakdown angular frequency (Figure 7a) and the breakdown angular frequency is converted using the driving speed v (Figure 7b) into a time-dependent exciter angular frequency (Figure 7c). It can be seen from

Figure 7c that the exciter amplitudes at low frequencies are significantly greater than is the case at high frequencies. Transfer functions for the body acceleration (Figure 6c) and the wheel load fluctuation (Figure 6d) are squared and multiplied by this excitation spectrum. The results are the power spectral densities for the body acceleration (Figure 6a) and for the wheel load fluctuation (Figure 6b) [2].

The higher excitation amplitudes of the road surface at low frequencies raise the vibration amplitudes in the natural frequency range of the body (1 to 2 Hz). In the natural frequency range of the wheel (10 to 14 Hz), however, the low irregularity amplitudes lead to drops in the power spectral densities compared to the transfer functions, which means the body movements become more dominant than the axle movements.

Figure 6: Transfer function and power spectral density for body acceleration and wheel load fluctuations
(According to [2])
a) Power spectral density, body acceleration,
b) Power spectral density, wheel load fluctuation,
c) Amount of transfer function, body acceleration,
d) Amount of transfer function, wheel load fluctuation.

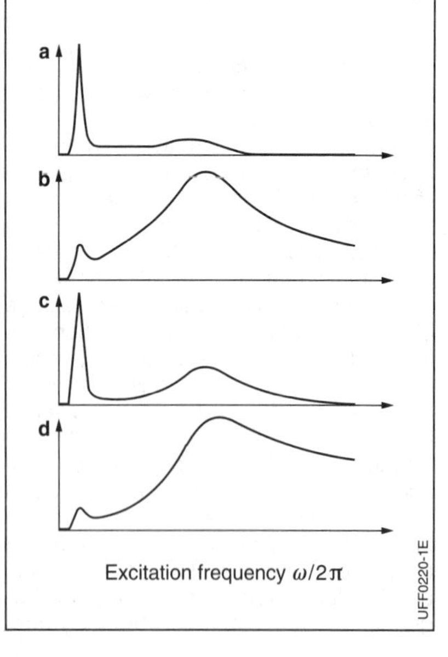

Quantification of vehicle vibrations

Only a short time after the introduction of the automobile, the first approaches to quantifying driving comfort in vehicles were developed at the start of the 20th century. The influence of comfort – particularly in the case of professional drivers – on fitness (i.e. ensuring good physical and mental condition) as well as on the health of the driver and thus on overall road safety was recognized very quickly. Accordingly, attempts were made to shape meaningful variables to describe comfort and to compare these in the form of objective comfort factors for different variants of vehicles and roads. Since then, a number of different assessment methods have been defined, based almost exclusively on acceleration signals at and around the driver's seat. Here, the frequency-dependent sensitivity of people to vibrations and oscillations initiated at different positions as well as different types and directions (translatory or rotatory; x, y, z direction) is taken into account with the help of the corresponding frequency weighting filters.

Two widely used approaches are the VDI guideline 2057 [4] and the standard ISO 2631 [9]. Despite the relatively low

Figure 7: Relationship between spectral densities dependent on time and breakdown angular frequencies

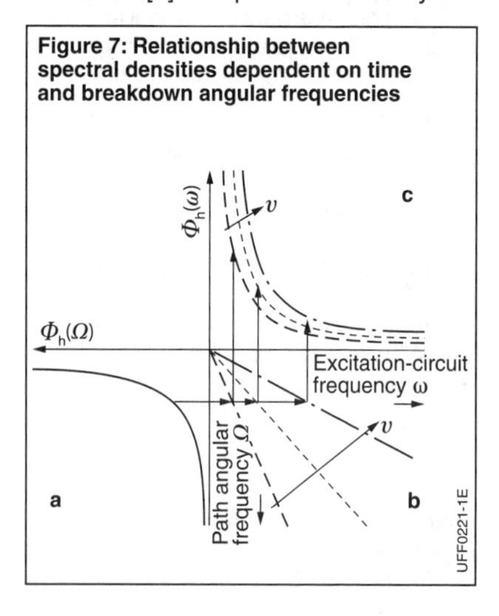

overhead involved in these analyses compared to other methods, these have high informative value. At the same time, these form the basis for many approaches to comfort assessment that were developed in the 1990s and later (see [4], [5], [6], [7]) In the methods used to date, it is normally the case that the accelerations caused by the excitation at different measuring points (initiation positions) are used to determine an overall comfort factor (possibly from a number of partial comfort factors at different measuring points) that is to represent a measure of the load on the occupants. The accelerations are first of all frequency-weighted (i.e. amplitudes of different frequencies are linked to frequency-dependent factors) and subsequently reduced to a value, for example with the RMS method (Root Mean Square, see [4]) in accordance with:

$$\overline{a}_{\mathrm{RMS}} = \sqrt{\frac{1}{T}\int_0^T a_{\mathrm{w}}^2(t)\,dt} \qquad \text{(Equation 26)}$$

or with the RMQ method (Root Mean Quad, see [6]) in accordance with:

$$\overline{a}_{\mathrm{RMQ}} = \sqrt[4]{\frac{1}{T}\int_0^T a_{\mathrm{w}}^4(t)\,dt} \qquad \text{(Equation 27).}$$

Depending on the number of evaluated measuring points, an overall factor has to be formed in a subsequent operation. For the VDI guideline 2057, only the evaluation of the measuring point with the highest load is relevant. All translatory accelerations of this initiation position (hand, seat or foot) are taken into account equally in the calculation of the factor. However, no uniform specifications for weighting individual initiation positions have yet been developed. The reason for this is that many of the general conditions of the subject tests used to determine the totaling rule for the overall comfort factor formed from the partial comfort factors deviated greatly from one another.

In order to ensure comparability of the factors of individual calculation methods, the overall comfort factor is frequently standardized to the standard VDI scale in a subsequent operation. At the same time, the objective factors are assigned verbalized levels of load intensity (as an aid to evaluation for the test subjects).

The differences between the individual comfort assessment methods (see [4], [5], [6], [7]) lie mainly in the number of measuring points to be evaluated, in the calculation specification for formation of the partial comfort factor from the temporal acceleration signal (RMS and RMQ method), and finally in the scattered weighting of different acceleration signals on merging to form an overall comfort factor. Over and above this, the question regarding the frequency weighting filter to be used depending on the measuring point has not been clarified conclusively. The frequency weighting functions, most of which have been determined empirically, deviate strongly from one another, particularly in low frequency ranges (below 1 Hz).

Experts also criticize the fact that the grouping of different acceleration types and directions to form a single comfort factor involves an information loss. This means that, under certain circumstances, effects such as a partial factor rise with a simultaneous drop in other partial factors at the same measuring point (for example of another direction) and the resulting compensatory influence on the overall comfort value remain undetected. This can be avoided by evaluating the temporal progressions of the frequency-evaluated accelerations immediately before formation of the partial comfort value.

Measures for vibration minimization

Road excitations lead to axle and body vibrations in vehicles. These are to be avoided in the course of minimizing dynamic wheel load fluctuations (safety) and body accelerations (comfort). They are reduced by means of the corresponding components within the wheel suspension, whereby different systems for suspension, damping, and suspension linkage have proven effective depending on the area of application.

Body springs and body dampers are traditionally used for these tasks. Coupled with kinematics and elastokinematics in the wheel suspension that are configured in line with requirements, these are intended to ensure optimal power transmission between the tires and road surfaces and, simultaneously, high levels of comfort.

However, with the increasing use of actuators in the area of chassis systems (for example superimposed steering or torque vectoring), the application overhead for optimization of the subsystems is becoming increasingly important. One of the most important reasons for this is the increase in the product added value due to low-cost function integration on the software side (i.e. the complete implementation of various functions of different subsystems in the software environment without being forced to increase the technological overhead). In addition, the technical and economic synergy potentials in overall vehicle networking can be increased significantly and the differentiation potential of various derivatives can be enhanced using the same electronic and software modules. This means that electronics and software are becoming increasingly important in the area of chassis systems, requiring interdisciplinary cooperation between hardware, electronics, and software developers at very early phases of development to enable the efficient and low-cost development of mechatronic systems (see [8]), particularly where safety is critical.

References

[1] B. Heissing, M. Ersoy (Editors): Fahrwerkhandbuch. 1st Edition, Vieweg Verlag, 2007.
[2] M. Mitschke, H. Wallentowitz: Dynamik der Kraftfahrzeuge. 4th Edition, Springer-Verlag 2004.
[3] J. Reimpell, J.W. Betzler: Fahrwerktechnik – Grundlagen. 5th Edition, Vogel Verlag, 2005.
[4] VDI guideline 2057: Human exposure to mechanical vibration, whole-body vibration. VDI-Gesellschaft Entwicklung Konstruktion Vertrieb, Düsseldorf 2002.
[5] S. Cucuz: Auswirkung von stochastischen Unebenheiten und Einzelhindernissen der realen Fahrbahn. Dissertation, University of Braunschweig, 1992.
[6] D. Hennecke: Zur Bewertung des Schwingungskomforts von Pkw bei instationären Anregungen. VDI Progress Reports, Düsseldorf, 1995.
[7] M. Mitschke, S. Cucuz, D. Hennecke: Bewertung und Summationsmechanismen von ungleichmäßig regellosen Schwingungen. ATZ 97/11 (1995).
[8] ISO 26262: 2011: Road vehicles – Functional safety.
[9] ISO 2631 AMD1 (2010): Mechanical vibration and shock – Evaluation of human exposure to whole-body vibration.

Suspension

Basic principles

The suspension system of a vehicle has a decisive influence on the vibration characteristics and therefore on both comfort and driving safety. Depending on the vehicle category and use case, different solutions have prevailed in the meantime. An overview of the different suspension design elements is shown in Figure 1, using a quarter vehicle as an example.

As a general principle, suspension design elements include all parts of the wheel suspension of a motor vehicle that deliver return forces in the case of elastic deformation. The media that perform the suspension work on the different suspension systems are either steel (spring steel), polymer materials (rubber), or a gas (air).

Tires

As the connecting element between the road surface and vehicle, the tire is the first suspension design element in the transfer chain from excitation to the occupants that has a decisive influence on both the comfort (acoustics, rolling characteristics) and driving safety (longitudinal and lateral force potential). It influences both suspension and damping characteristics, whereby these are not sufficient to eliminate the need for other vibration-absorbing elements in modern vehicles. An exception here is for example construction machines, where changed comfort requirements mean that in almost all cases the suspension and damping are achieved by means of the tires.

Figure 1: Suspension design elements within the wheel suspension (McPherson split axle as an example)
1 Dome mount (rubber mount),
2 Body spring,
3 Tires,
4 Rubber mounts,
5 Stabilizer.

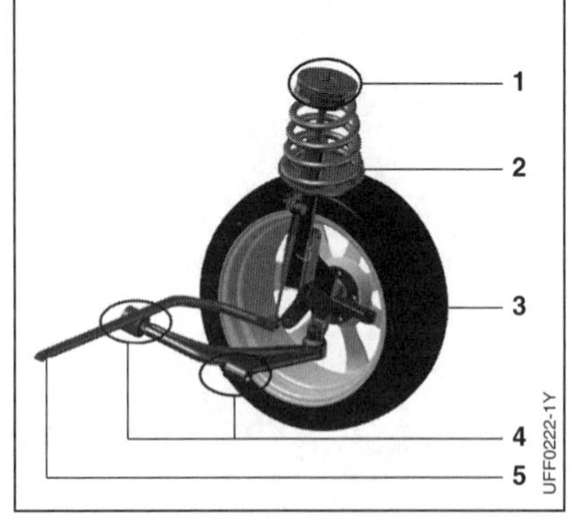

Elastomer mounts

Elastomer mounts are rubber elements with different functions and properties that interconnect individual components of a chassis system or secure them to the body.

The rubber mounts are used to provide insulation against vibrations and thus enhance comfort, particularly in the case of higher-frequency excitation (acoustics). At the same time, a suitable configuration (geometry, rigidity of the rubber mounts etc.) can decisively influence the driving dynamics.

In contrast to series-production vehicles, uniball joints (not rubber-mounted connections between wheel suspension and body) are deployed in motor racing; this improves driving dynamics to the detriment of comfort.

In order to reconcile the conflicting objectives of soft mounts for high comfort and rigid mounts for high driving dynamics, in the meantime increasing use is made of adaptive or active suspension mounts that are able to adapt their properties to each driving situation.

Body springs

Body springs are parts of the chassis system that provide most of the vertical return forces between the wheel and body. Depending on the use case, various types of spring with very varied properties are used. A summary of the characteristics of suspension design elements used in vehicle construction can be found in Table 1.

Table 1: Suspension design elements in vehicle construction

Suspension design element	Load influence on body natural frequency	Advantage	Disadvantage	Properties
Steel springs				
Leaf spring	– Natural frequency drops with increasing load – Characteristic curves are generally linear	– Good transfer of forces to chassis (trucks) – Low cost	– Maintenance requirement – Friction damping usually inadequate – Acoustic influences	– Assumption of the suspension linkage function is possible – Single-layer or multilayer version – Depending on type, subject to friction (in the passenger car, an intermediate plastic layer reduces friction → positive influence on acoustics)
Coil spring		– Great leeway for configuration – Low cost – No intrinsic damping – Lower space requirement – Low weight – Maintenance-free	– Separate elements required for suspension linkage – Spring characteristic curve is non-variable	– Arrangement of the shock absorbers within the springs is possible – Progressive characteristic curve due to corresponding geometry of the springs can be implemented (variable upward gradient or conical wire)
Torsion-bar spring		– Wear-free and maintenance-free – Depending on design, ride height adjustment also possible	– Long springs – Wheel-related spring rigidity depends on suspension arm arrangement	– Made of round steel (for low weight) or flat steel (with increased stress)
Stabilizer	– No influence with suspension on the same side – Half of stabilizer rigidity effective with one-sided suspension – Entire stabilizer rigidity effective with alternating suspension	– Simple possibility to influence driving dynamics of a vehicle – Reduction in the roll angle – With deployment of active systems improved comfort and enhanced driving dynamics	– Additional weight – Costs	– Influence on cornering characteristics (oversteer or understeer) – U-shaped, bent fully round or tube material is usual – Stub often flat-rolled due to bending stress – Stabilizer attachment points located far out on the axle to achieve small diameter – Axes of rotation of suspension arms configured so that stabilizer load is only torsion (not flexion)

Table 1: Suspension design elements in vehicle construction (continued)

Suspension design element	Load influence on body natural frequency	Advantage	Disadvantage	Properties
Air springs and hydropneumatic springs				
U-type bellows gas springs and air springs with bellows	– Natural frequency remains constant with increasing load – Characteristic curves depend on gas properties, form of rolling piston shape, and cord angle in the bellows	– Comfort characteristics independent of payload	– Separate elements required for suspension linkage – Low pressure (< 10 bar) requires high volumes	– Implementation of soft, vertical spring rigidity – As spring strut or individual spring, can be found above all in commercial vehicles and buses – Increasingly, deployment in passenger cars for ride height control of the rear axle and all-round suspension
Hydro-pneumatic spring	– Natural frequency rises with increasing load due to nonlinear spring rigidity	– Hydraulic damping and ride height control easy to integrate	– Maintenance requirement for rubber membrane due to diffusion tendency	– Gas volume in the spring accumulator determines the suspension characteristics – Power flux by means of gas and oil – Integration of the damper valves in the shock absorber and in the connection between the suspension strut and accumulator
Rubber springs				
Rubber spring	– Natural frequency is influenced by increased load due to the nonlinear spring rigidity	– Design is very freely definable – Low cost	– Limited temperature range – Aging	– Vulcanized rubber shear spring between metal parts – Used as assembly mountings (engine and transmission), suspension-arm mountings, additional springs etc. – Increasingly with integrated hydraulic damping

Figure 2: Example of a leaf spring with suspension linkage function
1 Stabilizer, 2 Damper, 3 Leaf springs, 4 Panhard rod, 5 Rigid axle.

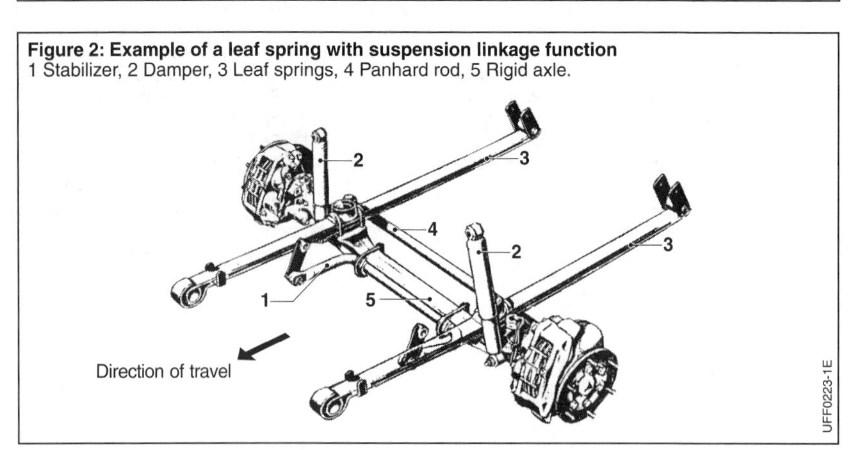

Types of spring

Leaf springs

The oldest types of spring used in vehicle construction are leaf springs, which were even used on horse-drawn coaches (Figure 2). Alongside the suspension functionality, major advantage of this type of spring is their possible use as a suspension linkage engineering design element to connect the body and axle. Multilayer leaf springs also have damping characteristics which, however, can lead to negative responses (acoustic influences). The damping forces that can be achieved are not sufficient to completely eliminate the need for conventional shock absorbers.

The influences on comfort and their weight mean that leaf springs in the meantime no longer meet market requirements for personal transportation and they are therefore only used now in a few passenger cars (light utility vans, off-road vehicles). It is still usual to deploy this type of spring in the area of commercial vehicles on account of the low costs and high reliability.

Helical springs

The great leeway for configuration with simultaneously low costs means that helical springs are the most frequently deployed type of body springs in the field of passenger cars. On this type of spring, the return forces are generated by the elastic torsion of individual coils during the change in length.

As helical springs are mainly able to absorb forces in the direction of the longitudinal axis of the spring, when they are used as body springs the other force components are braced by the suspension linkage.

A geometric configuration of the springs that is in line with requirements (wire thickness, coil diameter and spacing, Figure 3) means that not only different design envelopes but also different spring rigidity progressions over the compression path can be achieved. This can be used in turn to influence the load-dependent body natural frequency and thus the driving comfort.

Torsion-bar springs

This type of spring is encountered mainly in passenger cars and light utility vans. They are bars made of spring steel that are subjected to torsion. The fixed clamping of one end of the bar and the rotatable mounting of the other end means that the shaft is elastically torsioned when subjected to a load in the form of torque applied in the direction of its axis. In the motor vehicle, the elastic twisting of the torsion bar is achieved with the help of a crank secured to the rotatable end of the rod (Figure 4). As a rule, the crank arms are suspension arms of the axle or wheel suspension. The torsion-bar springs are usually arranged in the bearing center of the suspension arms on the body side on the opposite end of which the vertical wheel force F_R takes effect as the external load.

Figure 3: Design examples of different types of helical springs
a) Changeable coil diameter,
b) Changeable wire diameter,
c) Changeable coil spacing,
d) Mini-block spring (combination of a, b, and c).

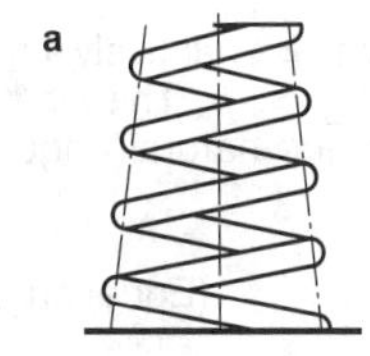

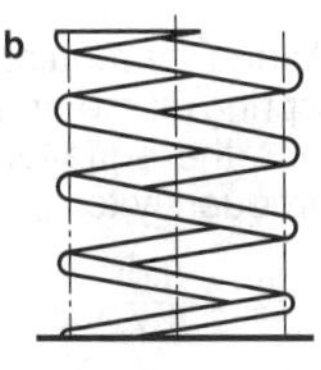

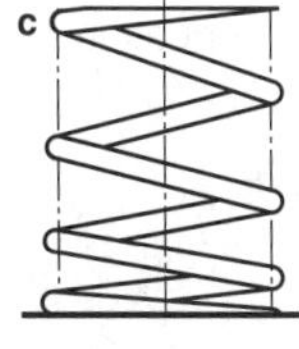

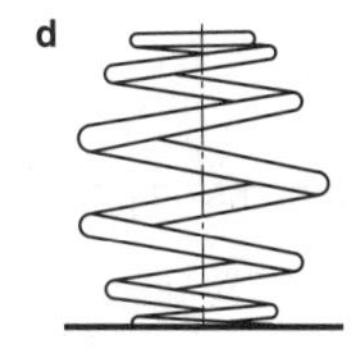

Figure 4: Design of a torsion-bar spring
1 Connection with the body,
2 Wheel connection.
l Length of torsion bar,
r Length of the crank arm,
F_R Wheel force,
z_R Compression path,
ψ Torsion angle.

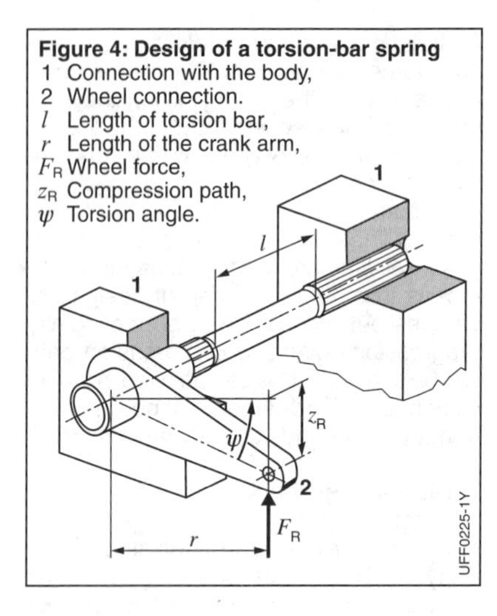

Figure 5: Structure of a gas spring
1 Structure,
2 Gas spring,
3 Wheel.
h_{th} Theoretical spring length,
m_A Body mass,
m_R Wheel mass,
p_i Pressure in the gas spring (internal pressure),
p_a Ambient pressure,
V Working volume of the gas spring,
A Area to which gas pressure is applied

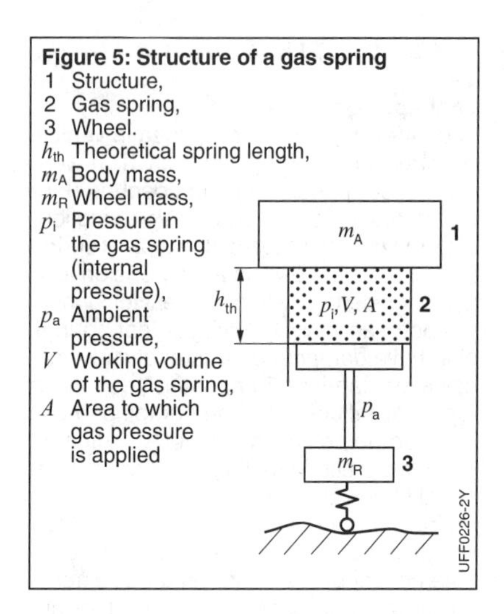

Gas springs

The body springs presented so far are fixed, springing media, whereby the work was performed by the change in shape of the steel springs. In contrast to this, the spring work in the case of gas springs is provided by a change in volume of the gas. The body of the vehicle is decoupled by an effective gas volume (possibly also by an additional fluid, see hydropneumatic springs) from the excitation and it vibrates on the gas cushion within the gas spring (Figure 5). This results in a favorable possibility to integrate a ride height control function that can be implemented by pumping the intermediate medium (gas or fluid) in or out.

A characteristic parameter of the gas spring is the "theoretical spring length" h_{th}, which results as a quotient from the compression-dependent working volume $V(z)$ (including any additional volume) and the effective surface area A to which gas pressure is applied:

$$h_{th} = \frac{V(z)}{A} \qquad \text{(Equation 1)}$$

where:
z Compression path.

The equation for the spring force F:

$$F = (p_i - p_a)\,A \qquad \text{(Equation 2)}$$

Where
p_a Ambient pressure,
p_i Internal pressure,

generally leads to the following for the spring rigidity of a gas spring [1]:

$$c(z) = A\,n\,p(z)\,\frac{1}{h_{th}} \qquad \text{(Equation 3).}$$

The polytropic exponent for isothermic and slow spring movements is $n = 1$ – for adiabatic and faster spring movements it is $n = 1.4$. The natural angular frequencies are calculated in the same way as for steel springs:

$$\omega_{Gas} = \sqrt{\frac{c}{m}} = \sqrt{\frac{c(z)\,g}{(p - p_a)\,A}} = \sqrt{\frac{g\,n\,p(z)}{(p - p_a)\,h_{th}}} \qquad \text{(Equation 4).}$$

When the requirement for a relatively small spring diameter is met $p_i >> p_a$. This simplifies the equation for the natural angular frequency to:

$$\omega_{Gas} = \sqrt{\frac{g\,n}{h_{th}}} \qquad \text{(Equation 5).}$$

However, the above-mentioned theoretical piston cylinder gas spring is only used in vehicles in a modified form, whereby in principle a distinction is made between two types of gas spring, the bellows air spring and the hydropneumatic spring. The fundamental difference with regard to the vertical dynamics lies in the influence of the load on driving smoothness and in the different effects on the spring rigidity in the level balancing of both systems. Whereas in the case of the hydropneumatic type level balancing is achieved by pumping in fluid (with constant mass of the gas in the spring), level balancing on air springs with bellows takes place by pumping a gas (air) into the spring, thus restoring the original suspension volume. The change in spring rigidity of the hydropneumatic spring type that this causes also leads to load dependence of the body natural frequency. In contrast to this, the air spring with bellows has a virtually constant body natural frequency in the entire load range.

To conclude, Figure 6 shows the influence of different suspension systems on the natural frequency, and thus also indirectly on the comfort with rising loads. The reason for the influence of the body natural frequency on comfort lies in the different resonance ranges of different organs in the human body and the consequence that an excitation of human body parts with their natural frequency impairs well-being. This is why a body natural frequency, below the resonant frequencies of the human body, that is as independent as possible of the load is to be ensured.

However, Figure 6 also clearly indicates that an approximately constant natural frequency with rising load is only present in the case of air springs. With steel springs, the natural frequency drops due to the constant spring rigidity; with the hydropneumatic system, on the other hand, it increases.

Figure 6: Comparison of different suspension systems depending on payload
1 Steel spring,
2 Air spring,
3 Hydropneumatic spring.

Air springs with bellows

Gas springs with bellows with pneumatic ride height control are suspension systems with constant gas volume (see above), whereby these are in turn divided into two categories. These are firstly gas springs with bellows and, secondly, U-type bellows gas springs (Figure 7) which, in a similar way to pneumatic tires, consist of rubber material reinforced by woven textiles. On

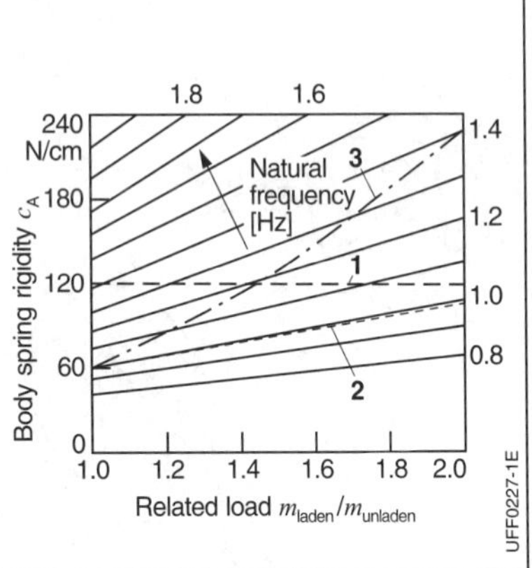

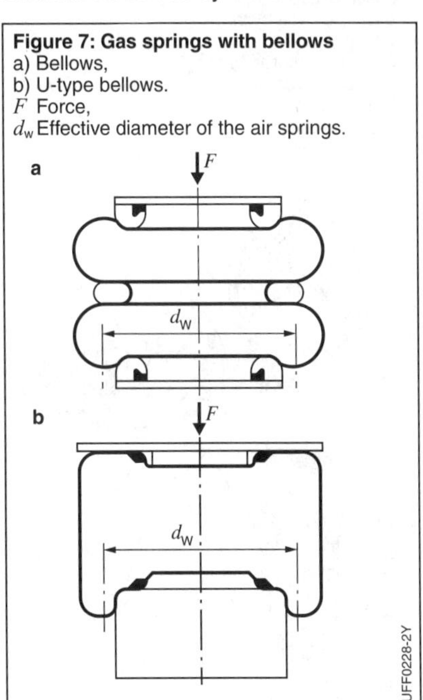

Figure 7: Gas springs with bellows
a) Bellows,
b) U-type bellows.
F Force,
d_w Effective diameter of the air springs.

these systems, the ride height control is implemented by pumping gas into or releasing gas from the spring with the suspension volume remaining constant as a general principle. The effective surface area of the air springs (and thus the gradient of the return force) that is affected by the overpressure is not usually constant, rather it changes throughout the stroke. This enables specific influence on the load-bearing capacity by designing the contours of the rolling piston of the U-type bellows gas springs (and thus a change in the effective surface area A in Equation 3 via the range of the spring) accordingly. The effective surface area A of the air spring can be determined via the effective diameter. With integration of an additional volume (increase of h_{th}, see Figure 5), a less progressive characteristic curve can be achieved.

Figure 8: Hydropneumatic spring
1 Suspension ball,
2 Membrane,
3 Connection to ride height controller,
4 Piston,
5 Screw cap,
6 Suspension cylinder,
7 Cup seal,
8 Shock absorber valve,
9 Rebound stage,
10 Bypass,
11 Compression stage.

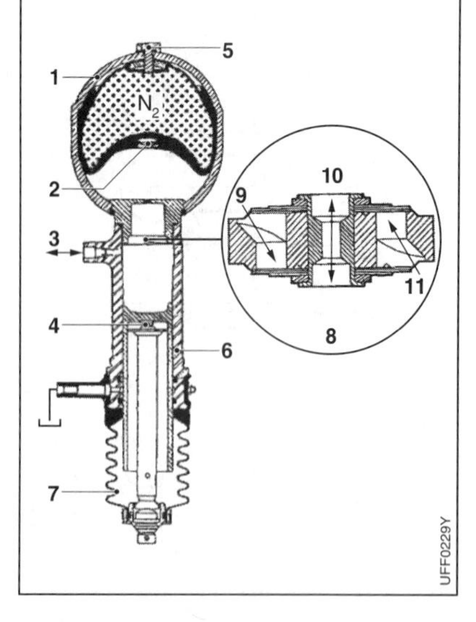

Hydropneumatic springs
In accordance with the considerations above, hydropneumatic springs (Figure 8) with integrated ride height control are gas springs with constant gas weight (see above), whereby the power flux is routed not only through a gas but additionally through a fluid. Here, the fluid and gas are separated by an impermeable rubber membrane. It is only when the fluid is placed in between that a wear-resistant and low-friction seal is achieved between the piston and cylinder.

Another advantage of this system lies in the possibility to integrate hydraulic damping in the suspension design element. A disadvantage, on the other hand, is the dependency of the natural frequency on the load (influence on comfort). The reason for this is the pumping in and out of the fluids at a constant gas weight that is required for ride height control. The load-sensitive volume change of the gas leads to a shift in the spring rigidity in such a way that with rising load there is a fundamental increase in natural frequency (see Figure 6).

Stabilizers

The suspension systems described above are deployed primarily for the vertical suspension of a vehicle. For the roll suspension, on the other hand, additional passive or active stabilizer springs (under certain circumstances with additional roll damping)

Figure 9: Function principle of the stabilizer spring

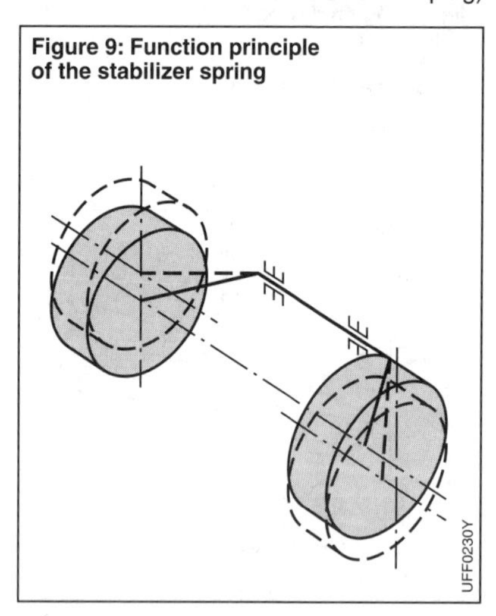

are used alongside classical body springs. A diagram of the principle is provided in Figure 9. In the event of a rolling movement of the body, i.e. spring compression of the wheels of one axle in opposite directions, the stabilizer is torsioned and delivers an aligning torque around the roll axis. In the case of purely vertical motions of one axle, on the other hand, this has no effect. If the proportions of the rolling moment braced by the stabilizers on the front and rear axles have a different ratio to the proportions braced by the body springs, not only is the roll angle reduced but the breakdown of the differences in wheel load of an axle on cornering is also influenced.

On a vehicle with a corresponding stabilizer configuration, this can shift the driving characteristics towards understeer (increase in the roll rigidity on the front axle or reduction in the roll rigidity on the rear axle) or oversteer with all other parameters unchanged. In the case of active stabilizers, the stabilizer force can also be actively influenced and adapted to suit the driving status. This enables a reduction in for example copy effects on one axle (due to decoupling between the right-hand and left-hand sides) when the vehicle is driven straight ahead but it also increases the driving dynamics on cornering by minimizing the inclination of the body. Here, the stabilizers do not influence the vertical vibration characteristics of the vehicle.

Figure 10: Function principle of the compensating spring

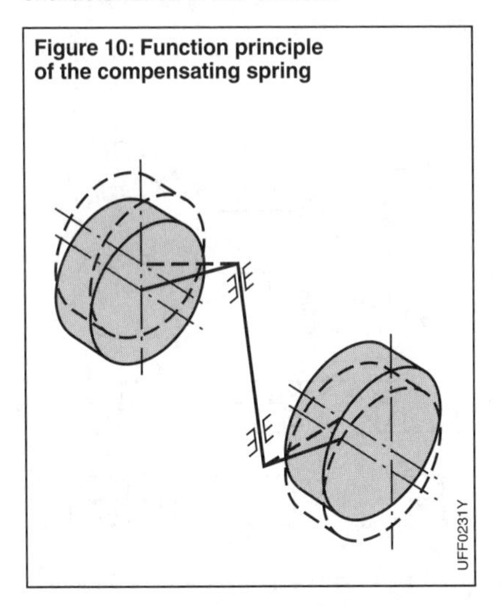

Compensating springs

Compensating springs have the opposite effect to that of the stabilizer (Figure 10). As a pure stroke element, they have no effect during rolling motions of the body.

The compensating spring was used in the past on axle designs where the wheel suspension kinematics required wheel load differences that were as low as possible to suppress the "resting effect" (stronger rebound of the wheel on the inside of the curve compared to the spring compression of the wheel on the outside of the curve when cornering). The rigidity of the body springs and this braced proportion of the rolling moment on the axle under examination could then be reduced accordingly. Compensating springs are no longer in modern passenger car wheel suspensions.

Suspension systems

Increasing customer requirements (comfort and driving dynamics) with regard to passenger cars and the strongly fluctuating load states of commercial vehicles mean that exclusive deployment of conventional steel springs is often insufficient. In such cases, either partially loaded or fully loaded suspension systems are used.

The integration of the additional functions of partially loaded or fully loaded systems enables increases in both comfort and driving dynamics (for example transverse locking of the springs of one axle to enhance stability on cornering).

Partially loaded systems

These systems are characterized by the fact that the forces to be braced by the suspension system are divided between steel and air springs according to a specified ratio.

In the case of soft body springs (to enhance driving comfort), wide spring ranges occur for example when a vehicle is loaded. In order to prevent the vehicle body from being lowered too much, ride height control systems with air springs or hydropneumatic springs are used. Here, sensors determine the ride height and provide this information to the control system. By pumping in or releasing air (air springs with bellows) or oil (hydropneumatic springs), the ride height can be then adapted in line with requirements. Depending on the vehicle segment (passenger cars or commercial vehicles), ride height control systems offer different additional functions.

In the passenger car, for example, a speed-dependent ride height control of the body in order to save fuel is possible. The adjustable ride height can also be used on poor-quality road surfaces to enhance the vehicle's capability to handle rough terrain.

In the case of commercial vehicles, on the other hand, ride height control enables variable adaptation of the loading area to different loading ramps. Other functions can also be implemented by networking with other systems. These include, for example, an automatic ride height increase if the lifting axle is raised, lowering on exceeding the maximum axle load, or brief raising of the lifting axle to increase the wheel load on the driven axle.

Figure 11: System architectures of ride height control, fully loaded systems
a) Open system, b) Closed system.
1 Filter, 2 Compressor, 3 Drier, 4 2/2-directional-control valve,
5 Bellows, 6 Non-return valve,
7 Pressure container, 8 Pressure switch, 9 3/2-directional-control valve.

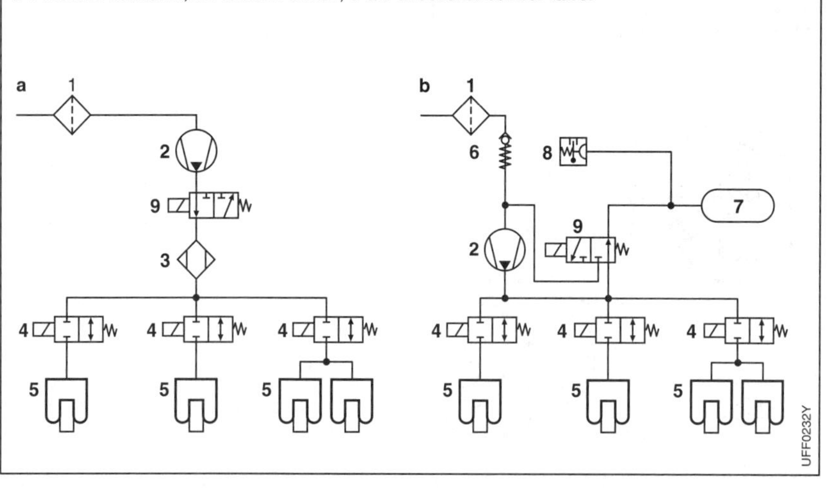

Fully loaded suspension systems

In the case of fully loaded systems, as opposed to partially loaded solutions, the task of suspension is assumed solely by the gas springs and the helical springs are completely eliminated. Depending on the available hardware and control strategy, the ride height control can be on selected axles or on all axles. Embedding the control system architecture in the global vehicle control architecture ensures that a negative mutual influence of the axles is excluded, thus preventing, for example, the vehicle body from slanting.

As a general principle, the ride height control for fully loaded systems can be designed in the form of an open or closed system.

In the case of the open system, a compressor draws air from the atmosphere and provides it in compressed form to the air springs if required. The increase in pressure raises the vehicle body. To lower the ride height, air is expelled into the environment, thus lowering the pressure in the springs again. Although this system involves relatively low overhead in construction and has a simple control system, it has the decisive disadvantage that high compressor output has to be provided for the short periods of control operation. Moreover, an air drier is required and an acoustic load can be expected when air is drawn in and expelled.

The closed system draws air from a pressure accumulator of the suspension system and feeds this directly to the air springs. When the ride height is lowered, the compressed air is returned to the pressure tank. Although the compressor output requirements for this system are lower and the air drier can be eliminated because the working medium is already dry, other components (accumulator, pressure switch, non-return valve, return line etc.) are required, fundamentally increasing the overhead in construction compared to an open system.

The system architectures of the open and closed ride height control systems are compared in Figure 11. This clearly illustrates the greater complexity of the closed system.

Chassis systems

Compared to passive systems, active vehicle chassis systems enable optimal adaptation of spring and damper forces to all driving states and road irregularities. With the aid of external sources of energy, forces are generated that stabilize both the axles and the vehicle body. In the meantime, a number of different system architectures have been developed. These differ above all with regard to the overhead involved (design envelope and costs), the energy requirement, and the quality of control operations. A few of these systems are presented briefly below.

Systems with hydraulic cylinders

With this version, the body motion and body position are regulated with quickly adjustable hydraulic cylinders (Figure 12), whereby different items of sensor information (wheel load, range of spring, acceleration etc.) are used as input variables for the control operation.

The control system achieves virtually constant wheel loads while maintaining a constant mean ride height. The static wheel load in this case is carried by steel springs or hydropneumatic springs.

Figure 12: Active chassis system with hydraulic cylinder
1 Vehicle body,
2 Wheel-load sensor,
3 Travel sensor,
4 Accumulator,
5 Pump circuit,
6 Servo valve,
7 Positioning cylinder,
8 Acceleration sensor.

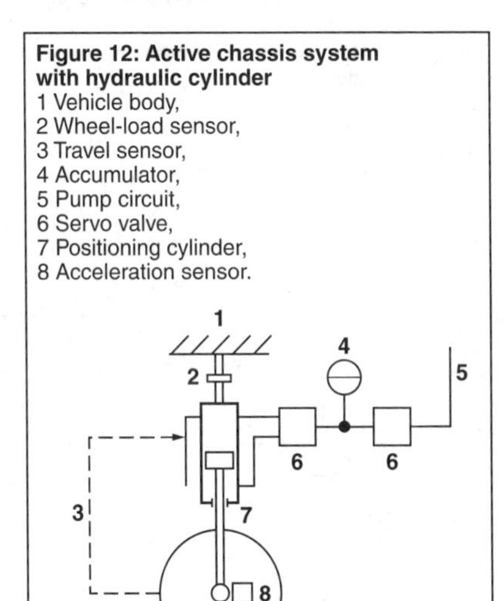

Systems with hydropneumatic suspension system
To stabilize the vehicle, a hydropneumatic suspension system uses specific oil regulation. This is done by pumping hydraulic oil into the suspension struts or draining it off the suspension struts (Figure 13). To limit the energy absorption, the control strategy of this system consists of control operations to eliminate longwave irregularities (low-frequency excitations). In the case of higher-frequency proportions, a gas volume takes effect in the proximity of the suspension strut. Here, the shock absorber is essentially geared to the wheel movements.

Version with spring-mounted point adjustment
In this system, the vehicle body is held horizontally in the low-frequency range in that a conventional helical spring is designed as variable at its mounting point (either against the vehicle body or in relation to the axle, Figure 14). The base is raised for spring compression and lowered for spring rebound. Control operations are continuous by means of a fluid pump and proportioning valves. The helical spring, however, must be longer compared to its original design.

In this type of system, the shock absorbers can be fitted with constant adjustment parameters and, above all, they can be matched to the wheel dampers.

Version with an air suspension system
For high dynamics of an active air suspension system, rapid pressure changes (buildup and reduction) are required in the wheel-mounted air-spring bellows. This can be implemented by means of coupled shift elements (compressor output alone is not sufficient here), which, depending on the driving state, takes air out of the air spring on the inside of the curve and feeds it into the bellows on the outside of the curve (Figure 15) to keep the body horizontal. The drive of the shift unit can be either electric, hydraulic or a combination of both. Links to other vehicle systems are also possible.

Electromagnetic system
In the case of an electromagnetic active chassis system, linear electromagnetic motors are fitted on each wheel. These are able to actively compensate for the road irregularities. The linear motors are supplied with electrical energy by power amplifiers, whereby as a general principle force control or path control can be imple-

Figure 13: Hydropneumatic system
1 Vehicle body,
2 Travel sensor,
3 Accumulator,
4 Pump circuit,
5 Acceleration sensor,
6 Throttle,
7 Proportioning valve,
8 Shock-absorber piston fitted with valves.

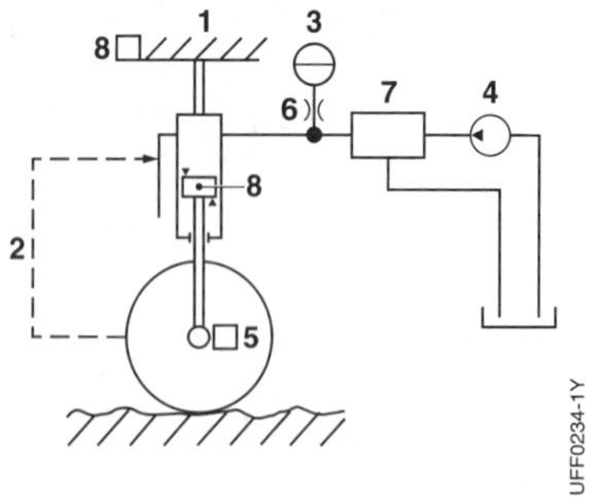

UFF0234-1Y

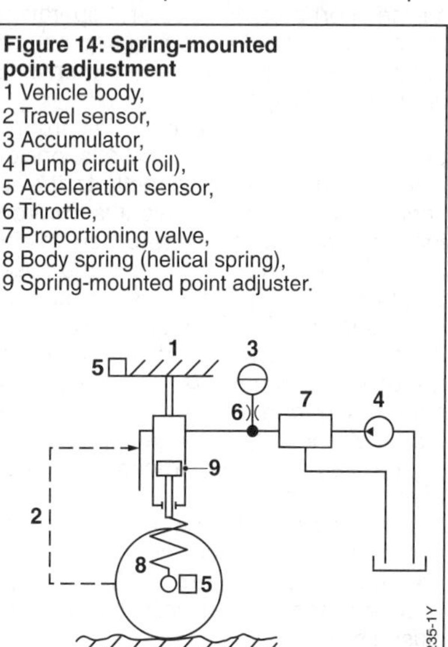

Figure 14: Spring-mounted point adjustment
1 Vehicle body,
2 Travel sensor,
3 Accumulator,
4 Pump circuit (oil),
5 Acceleration sensor,
6 Throttle,
7 Proportioning valve,
8 Body spring (helical spring),
9 Spring-mounted point adjuster.

mented. Cross-axle system networking also enables compensation of rolling and pitching vibrations. In the solution that is currently implemented, the static wheel loads are absorbed via torsion springs at the wheels to limit the electrical energy requirement. A passive shock absorber is also used.

The advantages of the electromagnetic solution lie above all in the high adjustment speed. Despite the low outer dimensions, the electric motors have sufficient power output to ensure driving safety in all driving situations. In comparison with conventional systems, the disadvantages lie in the greater weight and in the increased costs, as an additional, non-electronic damping system is required for safety reasons. As a general principle, however, the linear electromagnetic motor can also be operated as an alternator, which means that energy can be recuperated. This enables a lowering of the power requirement of the overall system, which is specified for standard road surfaces as less than 1 kW.

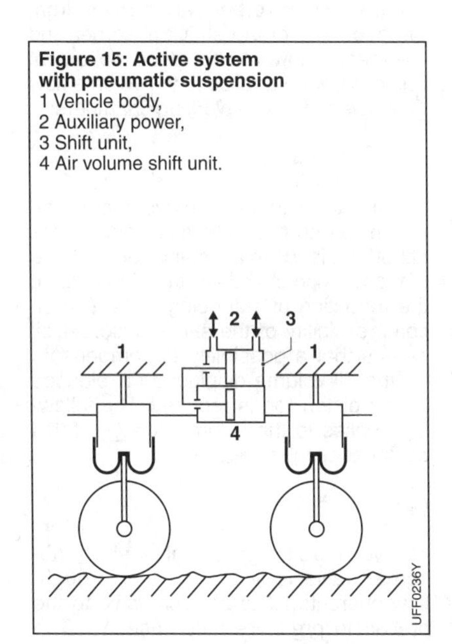

Figure 15: Active system with pneumatic suspension
1 Vehicle body,
2 Auxiliary power,
3 Shift unit,
4 Air volume shift unit.

References

[1] B. Heissing, M. Ersoy (Editors): Fahrwerkhandbuch. 1st Edition, Vieweg Verlag, 2007.

[2] M. Mitschke, H. Wallentowitz: Dynamik der Kraftfahrzeuge, 4th Edition, Springer-Verlag 2004.

[3] J. Reimpell, J.W. Betzler: Fahrwerktechnik – Grundlagen. 5th Edition, Vogel Verlag, 2005.

[4] L. Eckstein: Vertikal- und Querdynamik von Kraftfahrzeugen. ika/fka 2010.

Shock absorbers and vibration absorbers

Shock absorbers

The masses of the vehicle body and wheels connected by the body springs form a system that can vibrate and is excited by irregularities on the roadway and the dynamic movements of the vehicle. Shock absorbers are required to dampen the vibrating system. Nowadays, it is almost exclusively hydraulic telescopic shock absorbers that are used in motor vehicles as body shock absorbers; these convert the kinetic energy of the body and wheel vibrations into heat. The shock absorbers are configured taking account of the partially conflicting requirements for comfort (minimization of body accelerations) and driving safety (minimization of wheel load fluctuations).

Basic principles of hydraulic telescopic shock absorbers

The damper effect of hydraulic telescopic shock absorbers (Figure 1) is based on the movement subject to flow resistance of a shock absorber piston equipped with throttle elements (damper valves) within an oil-filled working cylinder. In the process, mechanical work is converted into heat that is released into the environment via the shock absorber surface. The pressure difference Δp between the two working chambers and the effective surfaces caused by the throttle elements on both sides of the shock absorber piston generates the resulting shock-absorber force F_D when the shock absorber moves in or out. The area on which the pressure prevailing in each working chamber is applied corresponds to an annular surface A_{KR} for the working chamber through which the piston rod of the shock absorber runs (see working chamber 1 in Figure 1). The outer diameter of the annular surface corresponds to the diameter D of the shock absorber piston; the inner diameter corresponds to the diameter d of the piston rod.

The following applies:

$$A_{KR} = \frac{\pi}{4}(D^2 - d^2)\ .$$

In the other working chamber (see working chamber 2 in Figure 1), the effective area corresponds to the piston surface A_K which results from the diameter D of the shock absorber piston.

$$A_K = \frac{\pi}{4}D^2\ .$$

When there is a movement of the shock absorber piston (i.e. an inward or outward movement of the shock absorber), the changes to the volumes in the two working chambers lead to a flowing movement of the incompressible damping fluid between the working chambers of the shock absorber (in the case of twin-tube shock absorbers, additionally between a working chamber and the compensating chamber). The oil volume flows in each case must pass the corresponding shock absorber valves.

The oil volume flows through the relevant shock absorber valves result from the geometry of the shock absorber and the speed of inward or outward movement $\dot{z}$. The following applies to the volume flow $\dot{Q}_1$ between the two working chambers:

$$\dot{Q}_1 = \frac{\pi}{4}(D^2 - d^2)\,\dot{z}\ .$$

The inward or outward movement of the piston rod on hydraulic telescopic shock absorbers leads to a variable total volume of the working chambers that depends on the retracting or extending state. The incompressibility of the shock absorber oil means that a possibility to compensate for the oil volume displaced or released by the piston rod is required. The following applies to the volume flow $\dot{Q}_2$ of this compensating volume:

$$\dot{Q}_2 = \frac{\pi}{4}\,d^2\,\dot{z}\ .$$

The volume flow $\dot{Q}$ through a shock absorber valve is linked via the through-flow characteristics of each valve to the prevailing pressure difference Δp. The throughflow characteristics of a valve result from the joint effect of the throttle geometry (for example bore diameter of the flow-through channel) and any spring load (i.e. pressure-dependent variation of the discharge opening, Figure 1). The throughflow characteristics can be adapted to the needs of each situation by configuring and coordinating these parameters. The characteristics of the valves are to be designed in such a way that no cavitation whatsoever occurs (formation and implosion of gas bubbles in the working medium due to static pressure fluctuations in the range of the vapor pressure of the working medium) inside the shock absorber. Cavitation leads to acoustic problems and also to damage – ultimately to failure of a shock absorber.

Types of hydraulic telescopic shock absorbers

Single-tube shock absorbers

To balance out the retracting or extending piston rod volume, single-tube shock absorbers have an enclosed gas volume which is separated from the working chambers filled with shock absorber oil with the help of a moving dividing piston (Figure 1a). In the compression phase (moving in) of the shock absorber, the gas volume is compressed according to the volume flow $\dot{Q}_2$; in the rebound phase (moving out), it is relaxed according to the volume flow $\dot{Q}_2$. As a rule, the pressure of the gas volume is between 25 and 35 bar, which means that the maximum occurring retraction forces can be absorbed. The working volume flow $\dot{Q}_1$ flows through each of the corresponding shock absorber valves in the shock absorber piston. On moving in, it flows though the compression stage valve and on moving out through the rebound stage valve.

The high gas pressure means that the tendency for cavitation to occur is low in the case of the single-tube shock absorber. The heat that is generated can be released into the environment directly via the outer surface of the working cylinder. The advantages of lean design, the low weight and discretionary installation position of the single-tube shock absorber are offset by the great length, increased friction, and high requirements with regard to sealing the piston rod and gas volume.

Twin-tube shock absorbers

Twin-tube shock absorbers have a compensating chamber resulting from the arrangement of an outer tube around the

Figure 1: Structure of single-tube and twin-tube shock absorbers
a) Single-tube shock absorber,
b) Twin-tube shock absorber.
1 Piston rod, 2 Working cylinder,
3 Shock absorber piston, 4 Piston seal,
5 Working chamber 1,
6 Working chamber 2,
7 Shock absorber valve (rebound stage valve),
8 Shock absorber valve (compression stage valve),
9 Dividing piston, 10 Gas volume,
11 Compensating chamber,
12 Reserve volume,
13 Outer tube,
14 Bottom valve (compression stage valve),
15 Bottom valve (rebound stage valve).
D Inner diameter of working cylinder and diameter of shock absorber piston,
d Diameter of piston rod.

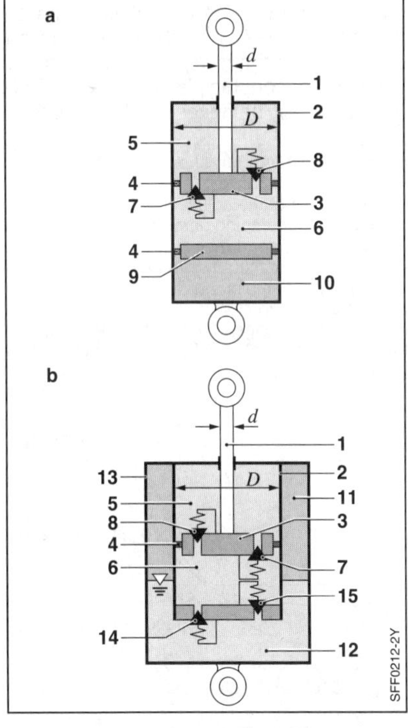

working cylinder (Figure 1b). The compensating chamber balances out the retracting or extending piston rod volume. To achieve this, it is connected via bottom valves to the lower working chamber of the shock absorber. The compensating chamber is partly filled with shock absorber oil and partly with a gas (as a rule, air). The gas volume is usually at atmospheric pressure or slight overpressure (6 to 8 bar). The piston and bottom valves must be coordinated is such a way that no cavitation occurs. In the compression stage, i.e. when the shock absorber is moving in, the damping work is therefore done at the corresponding bottom valve through which the volume flow $\dot{Q}_2$ flows. On the other hand, the oil volume flow $\dot{Q}_1$ from the lower working chamber into the upper working chamber can only flow through the compression stage valve in the shock absorber piston with low flow resistance. This prevents a radical drop in pressure in the upper working chamber. In contrast, in the rebound stage, i.e. when the shock absorber is moving out, the damping work is essentially done by the volume flow $\dot{Q}_1$ from the upper working chamber into the lower working chamber at the corresponding piston valve. The bottom valve only balances out the extending piston rod volume in that shock absorber oil flows virtually without resistance from the compensating chamber into the lower working chamber (volume flow $\dot{Q}_2$).

The compensating chamber means that twin-tube shock absorbers have poorer heat dissipation in comparison with single-tube shock absorbers. Furthermore, the installation position of twin-tube shock absorbers is restricted, as it must be ensured at all times that there is compensating fluid at the bottom valves. Advantages compared to the single-tube shock absorber are the lower shock absorber length, the softer responsiveness, as well as the lower requirements with regard to seals. In the area of passenger cars, the twin-tube shock absorber has prevailed as the standard shock absorber – also due to its lower costs.

Figure 2: Adjustable shock absorber characteristic curves
1 Upper limit of the adjustment range (maximum damping hardness, for example sport mode),
2 Lower limit of the adjustment range (minimum damping hardness, for example comfort mode),
3 Adjustment range of the shock absorber,
4 Characteristic curve of a passive shock absorber.

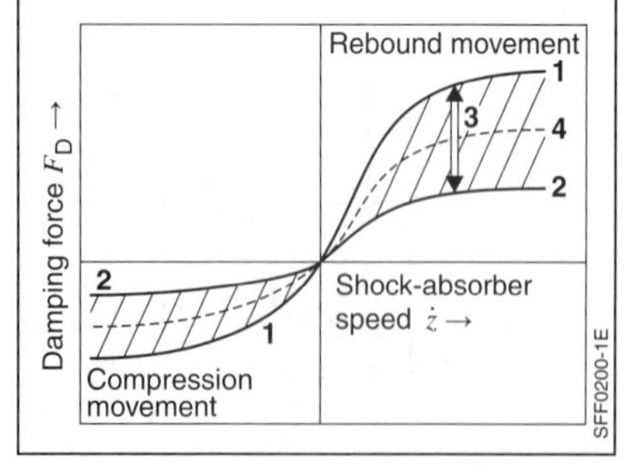

Adjustable shock absorbers

The conflict of objectives with regard to the coordination of body shock absorbers between driving comfort and driving safety can be mitigated using adaptive or semi-active shock absorbers. In comparison with passive shock absorbers with fixed shock absorber characteristics (i.e. defined force-speed characteristics, cf. section "Damping characteristics"), adaptive shock absorbers provide the possibility of discrete to infinitely variable adjustment of the damping characteristics (Figure 2).

Alongside manual adjustment of the shock absorbers (for example soft damping in the comfort mode or hard damping in the sport mode), adjustable shock absorbers can also be activated automatically depending on the driving state in each case (see section "Damping control").

Adaptive hydraulic absorbers

In the case of adaptive or semi-active shock absorbers of conventional design, the adjustability of the damping characteristics is implemented by means of adjustable shock absorber valves, bypass boreholes that can be activated (located on the outside or inside), or with the help of double pistons [1]. As a rule, activation is electronic. Major system features are the adjustment times that can be achieved, the spread of the adjustment range, as well as the number of definable shock absorber characteristics. Whereas systems of the first generation only permitted adjustment between a few characteristic curves, today's adaptive shock absorbers can usually be set to a large number of characteristics [2]. Newer systems usually even have infinitely variable adjustment between a minimum and maximum shock-absorber force characteristic (Figure 2).

Rheological shock-absorber systems

The adjustability of the damping characteristics in the case of rheological shock-absorber systems is based on the change in the flow properties of the working medium that is used. Magneto-rheological or electro-rheological fluids that change their viscosity under the influence of a magnetic or electrical field are used here instead of the usual mineral oils. The viscosity of the working medium has a direct influence on the flow resistance through the shock absorber valves. If, for example, creating a magnetic field increases the viscosity of a magneto-rheological working medium, the flow resistance through the shock absorber valves increases. Rheological shock-absorber systems not only provide the possibility for infinitely variable adjustment of the damping characteristics but also the implementation of very short adjustment times [2].

Damping characteristics

Damping force is a function of the speed of the inward or outward movement of the shock absorber, whereby the direction of force is opposed to the direction of speed at all times. It is generally applicable that the damping force F_D and the speed $\dot{z}$ are linked via the damping constant k_D and the damping exponent n. The following applies:

$$F_D = -\text{sign}(\dot{z}) \cdot k_D \cdot |\dot{z}|^n .$$

The damping constant and damping exponent are essentially dependent on the design of the shock absorber (valve characteristics, geometry). With the corresponding configuration of the individual parameters, progressive to degressive damping characteristic curves can be created. Modern body shock absorbers have mainly degressive characteristics. This achieves a high damping effect at low excitation speeds as well as a limitation of the maximum shock-absorber forces.

Damping characteristic curves are usually determined with the help of mechanical or servo-hydraulic testing units.

Figure 3: Damping characteristics

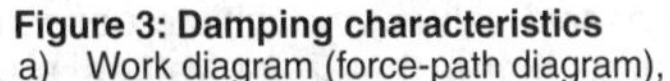

a) Work diagram (force-path diagram),
b) Damping characteristic curve (force-speed diagram).
f Variable excitation frequency,
f_1 Excitation frequency 1,
f_2 Excitation frequency 2,
$\dot{z}(f_1)$ Maximum shock-absorber speed with f_1,
$\dot{z}(f_2)$ Maximum shock-absorber speed with f_2,
A Constant vibration amplitude.

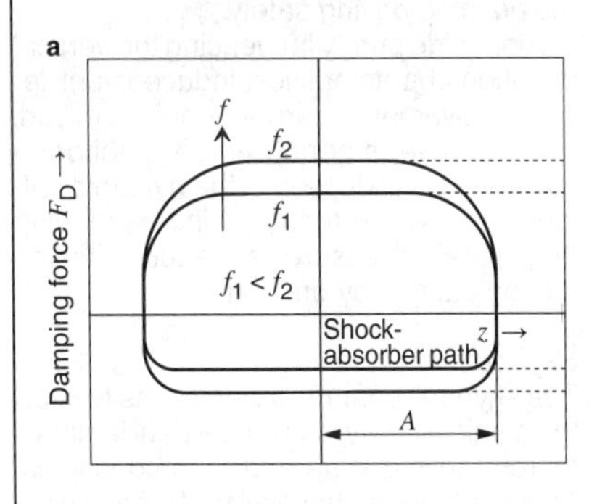

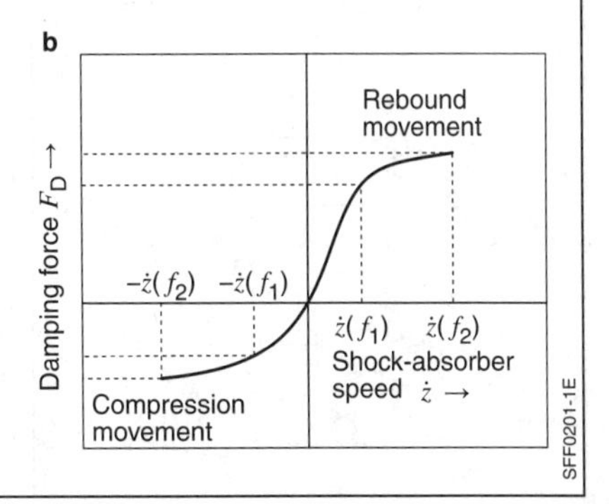

A sinusoidal path excitation of constant amplitude and variable frequency or constant frequency and variable amplitude results in various maximum speeds of the inward or outward movement. The recorded path and force signals can be applied in a force-path diagram (work diagram) (Figure 3a). The force-speed characteristic curve of the shock absorber (damping characteristic curve) can be derived from the work diagram by transferring the maximum force and speed values (Figure 3b).

It is mainly for reasons related to comfort that the configurations of the rebound and compression stage differ. The shock-absorber forces created in the rebound stage are usually more than double the correspondingly created forces on compression (i.e. in the compression stage) of the shock absorber (Figure 3). This limits the impact forces on the vehicle body during the compression phase (comfort) and simultaneously ensures the system is strongly dampened (system relaxation) in the rebound phase.

Figure 4: Theoretical principles of the skyhook approach
k_S Damping constant of the sky shock absorber,
m_A Vehicle-body mass,
z_A Vertical vehicle-body motion,
c_A Spring rigidity of the vehicle-body springs,
k_A Damping constant of the vehicle-body shock absorber,
m_R Wheel mass (unsprung mass),
z_R Vertical wheel movement,
c_R Vertical tire spring rigidity,
k_R Damping constant of the tire,
h Vertical road excitation.

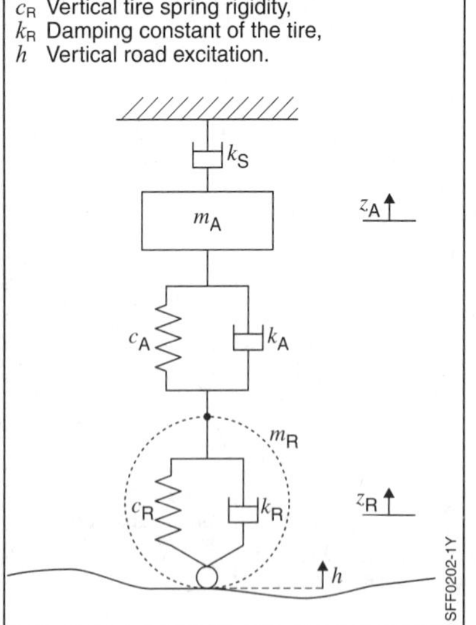

Damping control
In conjunction with electronically adjustable shock absorbers, damping control systems are being used to an increasing degree nowadays. The major component parts of such damping control systems are the adaptive shock absorbers and sensors (for example acceleration sensors on the wheel and body mass) and intelligent algorithms and control strategies. With the help of the sensors and algorithms, the current driving state is continuously determined and evaluated. In accordance with the stored control strategies, this enables the control system to adapt the shock-absorber characteristics to each driving state by activating the shock absorbers, thus, for example, influencing and optimizing driving comfort or driving safety.

Control strategies
Threshold-value strategy
Threshold-value controllers compare one or a number of relevant driving state variables (for example body acceleration, steering angle) with the corresponding threshold values and initiate defined measures if the threshold values are exceeded or not reached. The shock-absorber forces are usually influenced axle by axle simultaneously in the rebound and compression directions. The main focus of threshold-value controllers is on increasing comfort while simultaneously maintaining driving safety.

Alongside purely influencing the vertical vibration characteristics, induced vehicle-body movements can also be optimized. For example, steering angle monitoring can lead to a reduction in the dynamic rolling or hardening the damping depending on the brake pressure can reduce pitching motion caused by braking.

Skyhook
The skyhook control strategy aims to keep the vehicle body calm independently of current driving state and road conditions. This is intended above all to increase driv-

ing comfort. In contrast to the threshold-value strategy, the skyhook control strategy regulates the damping characteristics at each individual wheel. The basic principle is to decouple the movement of the vehicle body from the road excitation. To achieve this, it is imagined that the vehicle body is connected by means of a shock absorber to the sky (Figure 4). The shock-absorber force F_{DS} of the skyhook shock absorber results from the link of the body speed $\dot{z}_A$ and the damping constant k_S of the imaginary sky shock absorber:

$$F_{DS} = k_S \dot{z}_A .$$

For the conventional vibration system, on the other hand, the shock-absorber force F_D would result from the link of the damping constant k_A of the vehicle-body shock absorber and the difference between the vertical body speed $\dot{z}_A$ and the vertical wheel speed $\dot{z}_R$:

$$F_D = k_A (\dot{z}_A - \dot{z}_R) .$$

In order to brace the vehicle body against the sky, in the real implementation the additional portion of force F_{DS} of the sky shock absorber must be applied by the body shock absorber. The proportional damping factor k_{AS} this requires is calculated to:

$$k_{AS} = \frac{k_S \dot{z}_A}{\dot{z}_A - \dot{z}_R} .$$

As an adaptive (semi-active) shock absorber is only able to extract energy from the system in the form of heat, but is unable to supply heat to the system, a case distinction is required [1], [2]. The following applies:

$$F_{Dtot} = \left(k_S \frac{\dot{z}_A}{\dot{z}_A - \dot{z}_R} + k_A \right) \cdot (\dot{z}_A - \dot{z}_R)$$

where $\dot{z}_A (\dot{z}_A - \dot{z}_R) \geq 0$, and

$$F_{Dtot} = k_A (\dot{z}_A - \dot{z}_R)$$

where $\dot{z}_A (\dot{z}_A - \dot{z}_R) < 0$.

Depending on the amount and direction of the vehicle-body speed and the shock absorber movement (rebound or compression movement), the control strategy for the skyhook approach shown in Figure 5 is pursued for comfortable damping of the vehicle body. However, specific damping of rolling, pitching, and wheel vibrations is not taken into account by this approach. These movements are also highly relevant with regard to driving comfort and

Figure 5: Control strategy of skyhook control (case distinction)

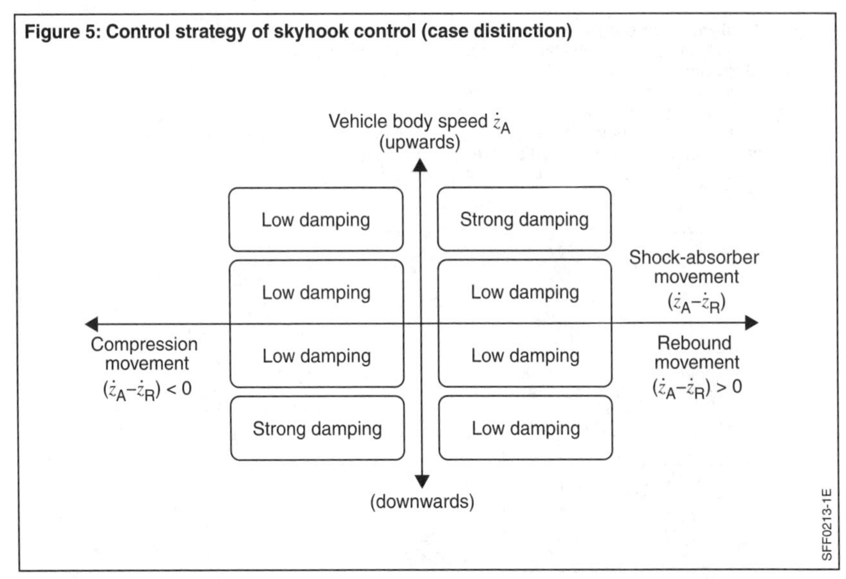

Figure 6: Theoretical principles of the groundhook approach
m_A Vehicle-body mass,
z_A Vertical vehicle-body motion,
c_A Spring rigidity of the vehicle-body springs,
k_A Damping constant of the vehicle-body shock absorber,
m_R Wheel mass (unsprung mass),
z_R Vertical wheel movement,
c_R Vertical tire spring rigidity,
k_R Damping constant of the tires,
h Vertical road excitation,
k_G Damping constant of the groundhook shock absorber.

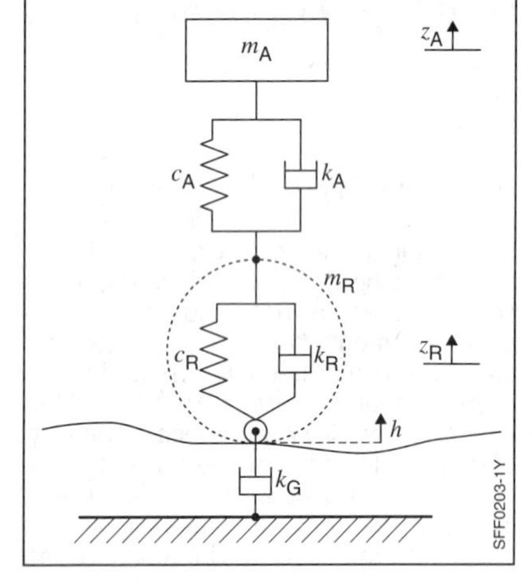

driving safety, which means that the skyhook controller is usually overridden by other controllers.

Groundhook
A groundhook controller aims to improve driving safety by reducing the wheel load fluctuations. In much the same way as the considerations for the skyhook strategy, it is imagined that the wheel is connected by a shock absorber to the roadway (Figure 6) and a proportional damping factor k_{AG} is derived. The following applies in the same way as the derivations for the skyhook controller:

$$k_{AG} = k_G \frac{\dot{z}_R - \dot{h}}{\dot{z}_R - \dot{z}_A}$$

where:
$\dot{z}_A$ Vertical body speed,
$\dot{z}_R$ Wheel speed
$\dot{h}$ Vertical excitation speed,
k_GDamping constant.

Also for the groundhook controller, a case distinction is made depending on the directions of movement of the wheel and vehicle body. This case distinction takes place on the basis of the term:

$$(\dot{z}_R - \dot{h})\ (\dot{z}_R - \dot{z}_A)\ .$$

Figure 7: Vibration absorber
a) Vibration absorber in the chassis (diagram), b) Substitute system.
k_A Damping constant of the vehicle-body shock absorber,
c_A Spring rigidity of the vehicle-body springs,
k_T Damping constant of the vibration absorber, c_T Spring rigidity of the absorber springs,
m_T Absorber mass, m_A Vehicle-body mass, z_A Vertical vehicle-body motion, m_R Wheel mass,
z_R Vertical wheel movement, c_R Vertical tire spring rigidity,
k_R Damping constant of the tire, h Vertical road excitation.

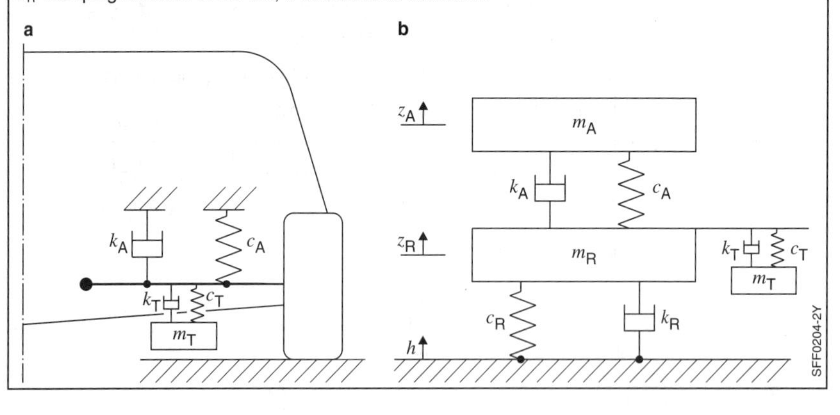

Vibration absorber

To specifically influence the vibration properties of the vibration system consisting of the wheel and vehicle-body mass, vibration absorbers (see chapter "Vibrations and oscillations") are deployed in some cases in the area of the chassis.

Depending on the configuration and arrangement of the vibration absorber, the comfort, acoustics, or driving safety can be influenced. A distinction is made between passive and active vibration absorbers. A passive vibration absorber is a mass attached to the chassis by sprung and damped mountings (Figure 7). The absorbing effect is created by the corresponding mass forces and in the case of passive absorbers is restricted to a certain frequency range. The effective range can be enlarged using an active absorber with an actuator that can be activated.

On excitation of the vibration system, the vibrations and oscillations of the main system are taken over by the appropriately coordinated vibration absorber, i.e. the main system only vibrates very slightly whereas the absorber absorbs a large portion of the energy. Figure 8 shows an example of the progression of the vibration amplitude of a wheel movement with and without vibration absorbers. With the deployment of a vibration absorber geared to the frequency range of the natural wheel frequency, a significant drop in the vibration amplitude can be observed in the corresponding frequency range.

Figure 8: Vibration amplitude of a wheel movement as a function of the excitation angular frequency with and without vibration absorbers
1 Progression without vibration absorber,
2 Progression with vibration absorber.
ω_R Natural angular frequency of wheel.

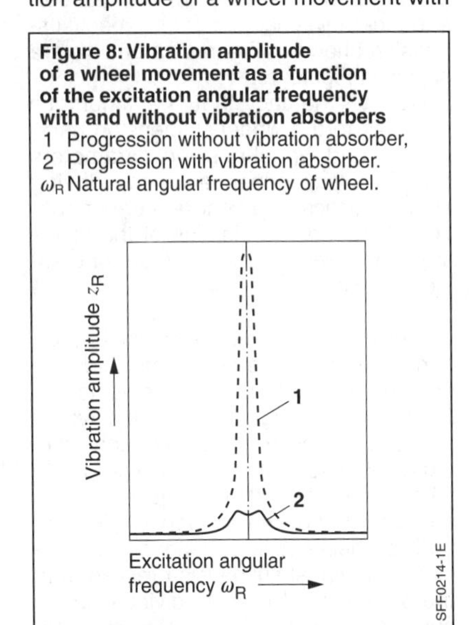

References

[1] B. Heissing, M. Ersoy (Editors): Fahrwerkhandbuch. 1st Edition, Vieweg Verlag, 2007.
[2] L. Eckstein: Aktive Fahrzeugsicherheit. ika/fka 2010.

Wheel suspensions

Basic principles

Vehicle wheels and the vehicle body are connected via wheel suspensions. On the one hand, a wheel suspension has the function of guiding the relevant wheel in relation to the vehicle body in such a way that a movement that is essentially directed vertically relative to the vehicle body remains possible and, on the other hand, that the tire forces exerted in the wheel contact point in the horizontal plane and the torques generated by these forces can be transferred to the body. On the front axle and also on the rear axle of vehicles with rear-axle steering, additional steerability of the wheels is to be ensured.

Alongside the tires, the suspension and shock absorber system, the vehicle body mass, and the individual wheel masses, the wheel suspensions have a major influence on the driving characteristics of a vehicle, as they influence the parameters of the vehicle axle concerned that are relevant to driving dynamics. These are, for example:

- The track width
- The toe-in or toe-out angle
- The camber angle
- The caster angle
- The caster offset in wheel center
- The spread angle
- The spread offset
- The kingpin offset
- The disturbing force lever arm
- The position of the rolling pole of the axle and thus the orientation of the roll axis
- The location of the pitching pole
- The braking and anti-squat control
- The longitudinal and transversal springs

For the definitions of the individual wheel suspension or vehicle parameters, refer to the chapter “Basic terms of automotive engineering: ”.

Kinematics and elastokinematics

During operation of the vehicle, the geometry and kinematics of a wheel suspension lead to changes in the characteristic parameters of the wheel suspension (for example, caster angle, location of the rolling pole) and wheel-position parameters (for example, camber angle and toe angle) of the corresponding wheel. This is the case, for example, as a result of a movement of the wheel within the framework of the vertical degree of freedom permitted by the wheel suspension (i.e. compression or rebound movement of the wheel) or with steerable wheel suspensions as a result of a steering movement. Figure 1 shows examples of the camber- and toe-angle changes of a wheel suspension due to kinematics during compression by the compression path Δz in comparison with the design position. As a rule, the kinematic wheel-position changes are shown with the help of diagrams of the compression and rebound path of the wheel. The so-called “wheel paths” for the camber- and toe-angle changes of the wheel suspension shown in Figures 1a and 1b are shown in Figure 1c.

Due to the fact that the kinematic changes to the wheel-suspension parameters and wheel-position parameters have a great influence on the driving characteristics of a vehicle, correct configuration and coordination of the steering and wheel lift kinematics are of great significance.

Alongside the changes to the wheel-position parameters due to kinematics during compression or rebound movements, the forces and torques affecting the wheel suspension (for example drive and braking forces, lateral and vertical forces in the wheel contact point) in conjunction with the elasticity of the suspension lead to further wheel-position changes. The elasticity of a wheel suspension results from the deformability of the individual wheel suspension components (for example links) and the bearings used when forces and torques are applied. For reasons related to driving comfort and acoustics, the mounts used in modern wheel suspensions are usually elastic mounts (for example rubber mounts). Figure 2 shows an example of a wheel suspension with two rubber mounts fitted on the vehicle body side; their elasticity when longitudinal force occurs at the wheel contact point leads to an elastokinematic change in the toe angle of the wheel.

Alongside the kinematic wheel-position changes, the elastokinematic effects also influence the driving characteristics of a vehicle. When coordinating the kinematics and elastokinematics of a wheel suspension, this is why it is usually the aim that the kinematic and elastokinematic effects supplement one another when influenced by forces and springs.

For example, elastokinematic steering is used on a number of modern rear-axle suspensions to reduce load change reactions (for example by increasing the toe-in when braking force affects the rear wheel on the outside of the bend) [1]. It is possible to influence the elastokinematic properties of a wheel suspension by, for example, coordinating the individual mount elasticities or adjusting individual mounting points.

Figure 1: Kinematic wheel-position changes in the case of compression movement
a) Kinematic camber change (view from rear),
b) Kinematic toe-angle change (view from above),
c) Wheel path.
Δz Compression path from the design position,
$\Delta\gamma_{kin}$ Kinematic camber-angle change,
$\Delta\delta_{kin}$ Kinematic toe-angle change.

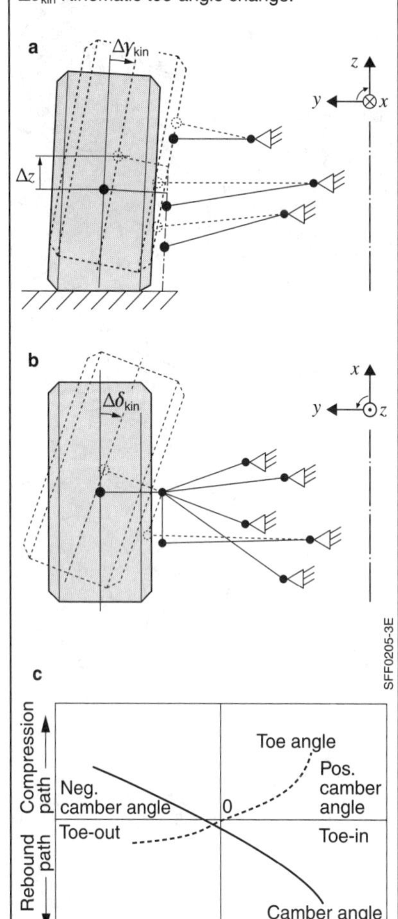

Figure 2: Elastokinematic toe-angle change due to effect of longitudinal force
F_x Longitudinal force (braking force),
$\Delta\delta_{ekin}$ Elastokinematic toe-angle change,
1 Elastic shift of pivot point 1 (rubber mounts),
2 Elastic shift of pivot point 2 (rubber mounts).

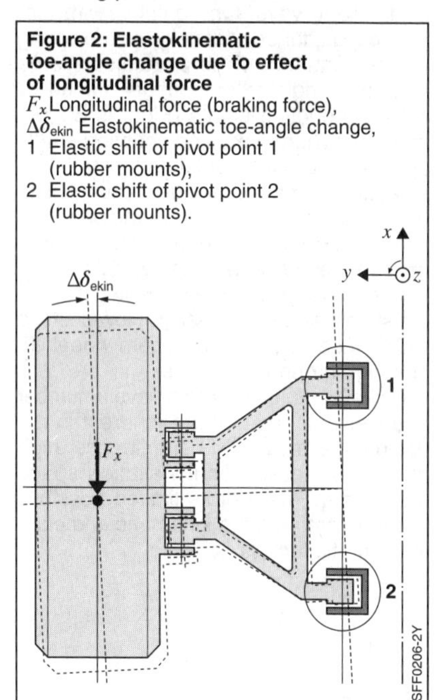

Basic categories of wheel suspensions

There are a large number of different wheel suspensions. They are classified primarily according to the type of suspension concept. An initial distinction is made between rigid axles (dependent wheel control), semi-rigid axles, and independent wheel suspensions (independent wheel control).

Rigid axles

In the case of a rigid axle, the wheels of an axle are firmly interconnected by a rigid axle body, which leads to mutual influences on the wheels. Rigid axles are used as both driven and non-driven rear axles on heavy vehicles (for example off-road vehicles, light utility vans, trucks). Occasionally, however, their sturdy construction and high ground clearance mean that steerable variants are also used as front axles (for example on off-road vehicles or off-road trucks).

Guidance of a rigid axle in relation to the vehicle body can be implemented in different ways. On vehicles with leaf springs, guidance is usually via the spring leaves (Figure 3a). There are also a large number of rigid axle concepts guided by links or coupling shafts (Figures 3b, 3c, and 3d). Where links and coupling shafts are used, statically undefined mounts are selected to make linking at the vehicle body easier and to reduce the required space [3]. Refer to [2] for detailed explanations of the individual axle variants.

The major advantages of rigid axles are the simple and sturdy design, low costs, a high roll center, high maximum wheel lift, and high ground clearance.

However, rigid axles also have a number of disadvantages that are inherent in the design: the mutual wheel influence, high unsprung mass, high installation space requirement, as well as limited possibilities to coordinate the kinematic and elastokinematic factors.

Figure 3: Design examples for rigid axles
a) Longitudinal leaf-spring suspension,
b) Trailing and coupling-shaft links,
c) Coupling shaft with watt linkage,
d) Trailing link with panhard rod.
1 Rigid axle, 2 Leaf spring,
3 Trailing link, 4 Coupling-shaft link,
5 Coupling shaft, 6 Watt linkage,
7 Panhard rod.

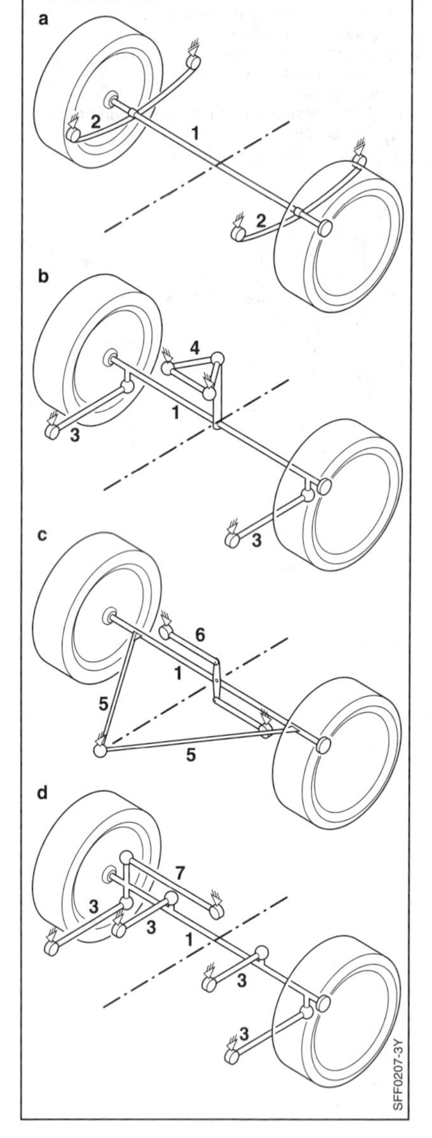

Semi-rigid axles
Semi-rigid axles also involve a mechanical coupling of the wheels. In contrast to rigid axles, however, this coupling is not rigid. The elasticity of the coupling profile that is used enables relative movements between the wheels. The coupling profile forms a cross-connection between two trailing links to which it is firmly connected. Longitudinal forces are absorbed via the trailing links. The bracing of lateral forces is supported by the stiffening effect of the coupling profile. In order to guarantee a relative movement between the two wheels of the axle, the coupling profile has to be designed as weak. Depending on the arrangement of the coupling profile, distinctions are made between torsional-link axles, twist-beam axles and semi-independent axles (Figure 4).

Due to their simple and low-cost design, semi-rigid axles are widely used as rear axles on vehicles with front-wheel drive.

The advantages of this axle concept include the low installation space requirement, the low unsprung masses, easy assembly and removal, the stabilizing effect of the coupling profile, the low track width and toe-angle changes, as well as the good anti-dive properties.

These advantages are offset by a number of disadvantages that are inherent in the principle: the mutual wheel influence, the low suitability for driven axles, the high tension peaks at the transition points between the trailing link and coupling profile, the increase in the tendency to oversteer in the event of influencing lateral forces (lateral force oversteer) due to link deformations, as well as limited kinematic and elastokinematic optimization potential.

Torsional-link axles
In the case of torsional-link axles (Figure 4a), the two wheel carriers are connected by means of a coupling profile arranged close to the wheel center. As a rule, the lateral guidance of the axle is supported by an additional guide element (for example a panhard rod) [2]. There are great similarities to a rigid axle with regard to both the structure and properties.

Twist-beam axles
In contrast, twist-beam axles (Figure 4b) have kinematic properties similar to those of trailing-link wheel suspensions. The coupling profile is arranged at the height of the pivot points of the trailing link. The deployment and arrangement of the cou-

Figure 4: Design examples for semi-rigid axles
a) Torsional-link axle with panhard rod,
b) Twist-beam axle,
c) Semi-independent axle.
1 Trailing link,
2 Coupling profile,
3 Panhard rod.

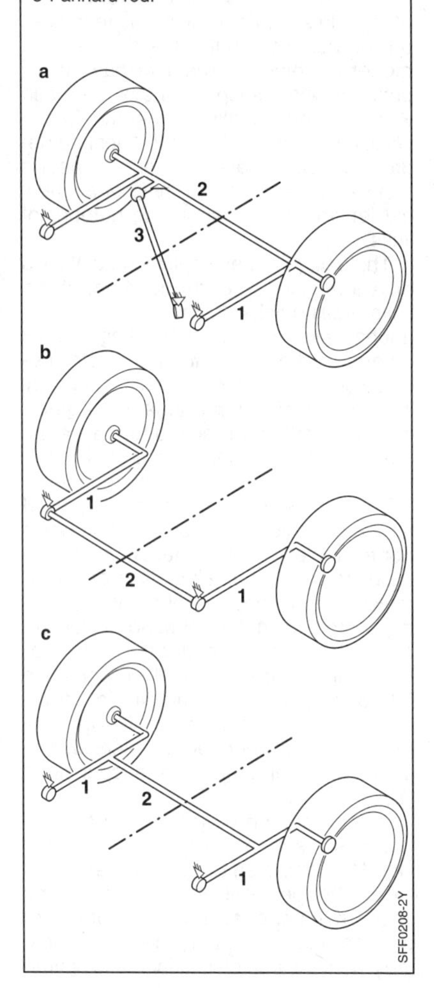

pling profile greatly simplifies the mounting of the trailing links compared to a trailing-link independent wheel suspension.

Semi-independent axles
In comparison with twist beam axles, the coupling profile on a semi-independent axle (Figure 4c) is not at the height of the link pivot points rather is offset towards the rear. This improves above all lateral force bracing in comparison with the twist-beam axle.

Independent wheel suspensions
Alongside semi-rigid axles – as rear axles on vehicles with front-wheel drive – most modern vehicles nowadays have independent wheel suspensions where each wheel is individually connected to the vehicle body according to the desired degrees of freedom of movement. A wheel is connected here with the help of a wheel carrier and a corresponding number of links.

The design (for example two-point links or A-arm links) and arrangement of the links (trailing links, transverse links or diagonal links) and the connecting mounts determine the kinematic and elastokinematic properties of the wheel suspension. The design of the individual links determines the number required to reduce the freedom of movement of a wheel to the desired number of degrees of freedom.

The number of links is frequently used as the classification of suspension types (for example, five-link independent wheel suspension). The resulting type of spatial movement (kinematics) of the wheel during compression and rebound movement is also frequently used for the classification of independent wheel suspensions [1], [3]. Depending on the type of movement of the wheel carrier, distinctions are made here between level, spherical, and spatial independent wheel suspensions [1], [3].

The proportion of independent wheel suspensions in modern vehicles is rising steadily. Compared to rigid and semi-rigid axles, independent wheel suspensions offer a number of advantages. For example, there is no mutual wheel influence, the kinematic and elastokinematic optimization potential is high, and the space requirement and unsprung mass are low in some cases.

However, independent wheel suspensions also have some disadvantages. In some cases, they lead to a complex design, the costs are high, the maximum wheel lift is low, and the configuration and coordination process is more complex in some cases.

There are large numbers of different designs of independent wheel suspensions. The basic principles of selected designs are to be explained briefly in the following section and their structures shown in diagrams. For detailed explanations of the individual designs and of other independent wheel suspensions and specific design examples, refer to [2].

Trailing-link independent wheel suspension
On a trailing-link independent wheel suspension, a wheel is connected to the vehicle body by means of a single link arranged in longitudinal direction (Figure 5a). The trailing link transfers both the longitudinal and lateral forces, which means that high mount forces occur and the mounts have to be designed accordingly.

The axis of rotation of the links runs parallel to the vehicle transverse axis. The advantages of this suspension form are usually low installation space requirement as well as low costs. The disadvantages are the limited kinematic optimization possibilities, the instantaneous center located at road height that causes high roll torque on cornering, as well as the high stresses placed on the links and their mounts.

Diagonal-link independent wheel suspension
In the same way as on the trailing-link wheel suspension, on the diagonal-link wheel suspension the wheel is also connected to the vehicle body by means of a single link. However, to achieve better bracing of the longitudinal and above all lateral forces, the link is arranged diagonally (Figure 5b) and there is more space between the mounting points. To achieve more favorable kinematic properties, on modern suspensions the axis of rotation of the link is arranged diagonally both in

Figure 5: Trailing- and diagonal-link independent wheel suspension
a) Trailing-link independent wheel suspension,
b) Diagonal-link independent wheel suspension.
1 Trailing link,
2 Diagonal link.

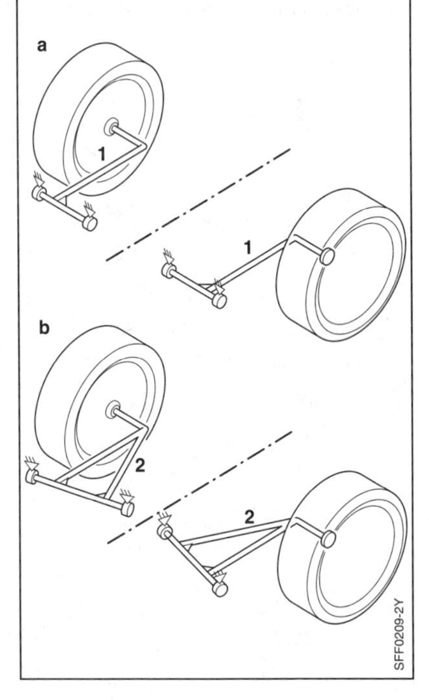

Figure 6: Double-wishbone and spring-strut independent wheel suspension
a) Double-wishbone independent wheel suspension,
b) Spring-strut independent wheel suspension.
1 Wheel carrier, 2 Upper A-arm link,
3 Lower A-arm link,
4 Spring shock absorber strut,
5 Tie rod (steering).

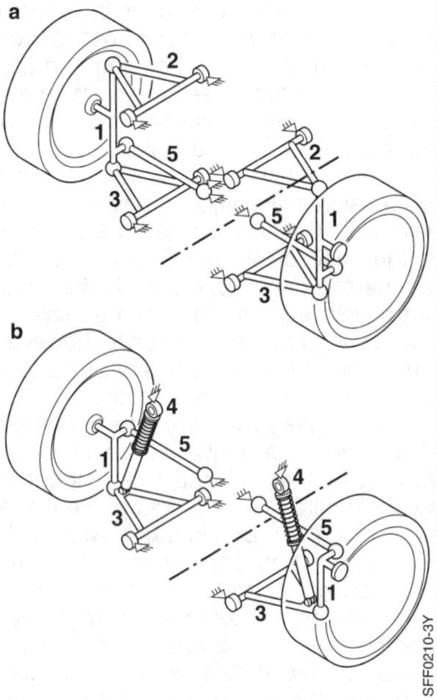

the projection to the vehicle lateral plane (roof angle) and in the projection to the road (V-shaped angle) [1].

Double-wishbone independent wheel suspension
A double-wishbone independent suspension is when a wheel is connected to the vehicle body via two A-arm links. One link is arranged below the wheel center and the other is arranged above it (Figure 6a), which enables the suspension to brace all of the forces and torques that occur at the wheel. As a rule, the high articulation forces mean that the transverse links are not directly connected to the vehicle body structure but rather are secured to a "chassis subframe" that interconnects both wheel suspensions and thus relieves the load of inner forces on the vehicle body.

With adaptation of the mounts and design of the links, double-wishbone independent suspensions provide very high kinematic optimization potential [2]. Depending on the location of the axes of rotation of the links, level, spherical or spatial wheel suspension kinematics are achieved [3]. The disadvantages of double-wishbone independent suspensions are the higher costs as well as the greater installation space requirement.

Spring-strut independent wheel suspension

The kinematics of a spring strut wheel suspension correspond to those of a double-wishbone independent suspension in which the upper transverse link is replaced by a sliding guide (Figure 6b). This sliding guide corresponds to the spring strut on other suspensions (in the case of a combined spring-shock absorber unit) or the shock absorber strut (in the case of separate spring and shock absorber arrangement) where the housing is rigidly connected to the wheel carrier. With this design, the shock absorber rod also assumes wheel guidance tasks.

The lower link plane of a spring-strut suspension is usually formed by two two-point links (radius links) or an A-arm link. In the case of a suspension using the McPherson principle, originally the lower A-arm link was formed from a transverse link and a stabilizer. Nowadays, however, other spring-strut suspensions are also referred to as McPherson axles.

Advantages of the spring-strut wheel suspension are above all the low construction overhead and low amount of space required at the height of the wheel axes which can be used in particular in the case of passenger cars with front-wheel drive with laterally installed engine-transmission units. Other advantages are the cost-saving and weight-saving design, simple installation and assembly, as well as the high degree of integration. On specifying the kinematics, however, spring-strut suspensions offer slightly less leeway for configuration in comparison with double-wishbone independent suspensions.

Multilink independent wheel suspension

Wheel suspensions with four or five single links are generally referred to as multilink axles. Multilink axles result, for example, from breaking down an A-arm link into two individual two-point links in the case of a double wishbone axle (Figure 7a). Breaking down three-point links and the use of independent two-point links usually result in greater leeway for configuration of the kinematic and elastokinematic properties of the axle. On the one hand, this increases the optimization of the

Fig. 7: Design examples for multilink independent wheel suspensions
a) Multilink independent wheel suspension,
b) Trapezoidal-link independent wheel suspension,
c) Control-blade independent wheel suspension,
d) Five-link independent wheel suspension.
1 Wheel carrier, 2 Two-point link,
3 A-arm link, 4 Tie rod (steering),
5 Trapezoidal link, 6 Control blade.

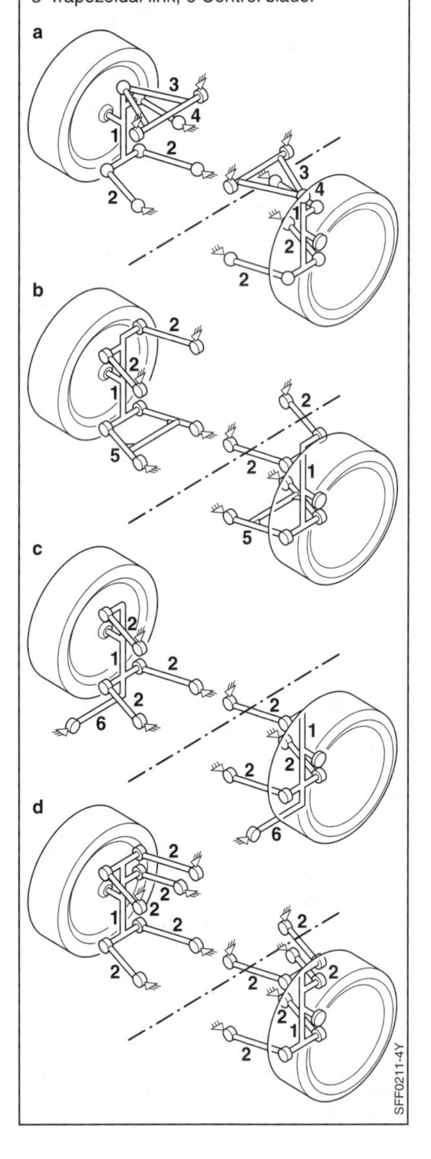

axle with regard to comfort and driving safety requirements. On the other hand, the somewhat complex design increases the overhead involved in the configuration and coordination processes for the wheel suspension.

Trapezoidal-link independent wheel suspension

Trapezoidal-link wheel suspensions (Figure 7b) are a special form of the multilink independent wheel suspension that is mainly used as rear axles. The lower plane is formed by a trapezoidal link that has two connection points on the wheel side and an axis of rotation on the vehicle body side. This means the lower link sets a total of three degrees of freedom of the wheel. Another two degrees of freedom are eliminated by two two-point links arranged accordingly so that only the desired compression degree of freedom of the wheel remains.

Control-blade independent wheel suspension

Another design of multilink independent wheel suspensions is the control blade axle on which the wheel is guided by one trailing link and three transverse links (Figure 7c). The trailing link (control-blade link) is connected to the vehicle body in such a way that it can be rotated and is firmly connected to the wheel carrier. As a rule, it has elastic properties in order to enable kinematic track and camber changes. The lateral forces are braced via three two-point transverse links which are usually arranged in two planes (one above and one below the center of the wheel). The arrangement and orientation of the transverse links determines the kinematic properties of the wheel suspension.

Five-link independent wheel suspension

A wheel suspension with completely detached links requires five individual two-point links to reduce the movement of a wheel to the desired vertical degree of freedom (Figure 7d). Five-link rear axles are generally referred to as multilink suspensions, whereby on the front axle this is referred to as a four-link axle plus tie rod [2].

References

[1] L. Eckstein: Vertikal- und Querdynamik von Kraftfahrzeugen. ika/fka 2010.
[2] B. Heissing, M. Ersoy (Editors): Fahrwerkhandbuch. Vieweg+Teubner Verlag, 2008.
[3] M. Matschinsky: Radführungen der Straßenfahrzeuge. Springer-Verlag, 2007.

Wheels

Function and requirements

All of the vehicle-specific or axle-specific tasks are performed via the wheel, e.g. transfer of dynamic forces between the vehicle and road surface. These include taking up the vehicle load and the impact forces of the road surface, transferring the rotary motion of the axles to the tires, and taking up and transferring braking and acceleration forces as well as lateral forces when cornering. Wheel size is mainly determined by the space required by the braking system, the axle components, and the size of the tires used.

Wheels have primarily a technical function. But the increasingly booming light-alloy wheel market calls for visually attractive designs.

Figure 1: Structure of a disk wheel
1 Rim inner flange,
2 Rim base,
3 Rim outer flange,
4 Ventilation hole,
5 Wheel disk,
6 Pitch-circle diameter,
7 Center hole,
8 Dish.

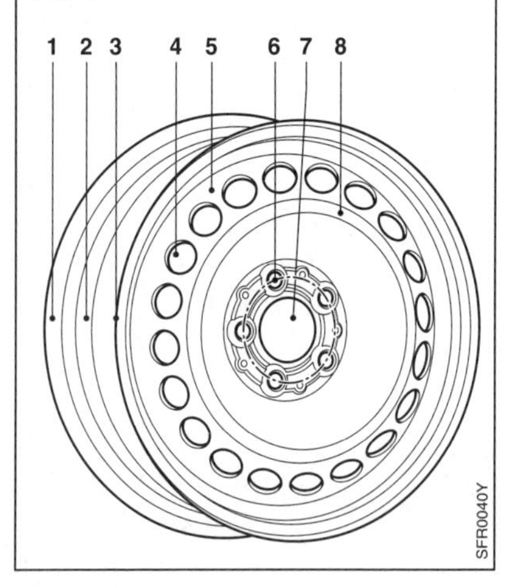

Structure

The wheel is a load-bearing, rotating part between the tire and the axle. It usually consists of two main components – the rim and the wheel disk. These two components can be made from a single part, and can also be permanently or non-permanently attached to each other. A permanent connection of a rim with a wheel disk is called a disk wheel.

Figure 1 shows the basic structure of a steel wheel. Here the rim mounts the tire and the wheel disk connects the wheel to the axle.

In daily usage the terms rim and wheel are often interchanged. The term rim is often used when actually the complete wheel is meant. The "wheel" in general usage often also refers to the tire. However, as a technical term in automotive engineering "wheel" generally means the wheel without the tire.

Wheel disk

The wheel disk (wheel nave) is the part that connects the rim to the axle hub. In the case of a steel wheel, the wheel disk consists of a formed steel-sheet blank. This has holes for ventilating the brake system and is usually curved (dish, see Figure 1). The center of the wheel disk contains the center hole and the wheel-bolt or stud holes. The wheel is secured to the axle through these holes. The center hole is provided with a locating bore by means of which the wheel is radially centered on the axle. This bore determines together with the rim bead seat (for the bead ledge) the wheel's true-running quality (via the radial run-out). The attachment face is, together with the rim flanges, responsible for lateral wheel run-out.

In the case of an aluminum wheel, the wheel disk can, due to its design, have different shapes such that the term wheel disk is no longer applicable.

Rim

The term rim strictly speaking describes only the radially outermost part of the wheel which holds the tire. The rim is therefore the fundamental connecting element between the wheel disk and the tire. In a tubeless tire it provides the air seal and is geometrically matched to the tire. The most commonly used shape of rim is divided into four zones (Figure 2):
- rim flange (inner and outer),
- hump (inner and outer),
- rim bead seat (inner and outer),
- rim base and drop center.

There are also different shapes for passenger cars (Figure 3).

Rim flange

The rim is limited on the inner and outer sides by a rim flange (rim inner flange and rim outer flange). It acts as the side stop for the tire bead (see Tires) and absorbs the forces resulting from the tire pressure and the axial tire load. The rim flange is specified in the guidelines of the ETRTO (European Tyre and Rim Technical Organisation) by, for example, J, K, JK or B. In this way, the geometry of the rim flange and the ratio to the drop center are dimensionally described. This is based on how the wheel is used.

The most common rim-flange shape for passenger cars is the J flange shape. The lower B flange can be found on smaller vehicles and in inflatable spare tire systems. K and JK flanges are rarely used any more; they used to be the domain of heavy, comfortable vehicles of the upper price segment.

Figure 2: Rim zones
1 Rim flange,
2 Rim bead seat,
3 Hump,
4 Rim base,
5 Drop center.

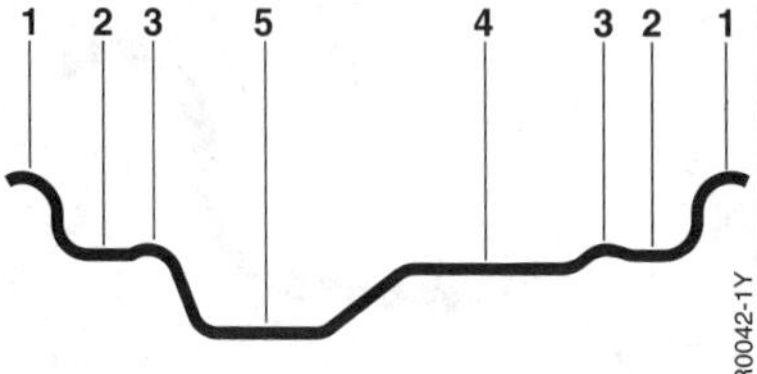

1 2 3 5 4 3 2 1

SFR0042-1Y

Rim bead seat

The rim bead seat describes the contact zone of the tire with the rim. It centers the tire radially. In this zone the tire is given the correct position for radial and lateral true running. All the dynamic driving forces are transferred here. In the tubeless tires mainly used today in passenger cars the wheel/tire system is sealed at the rim bead seat.

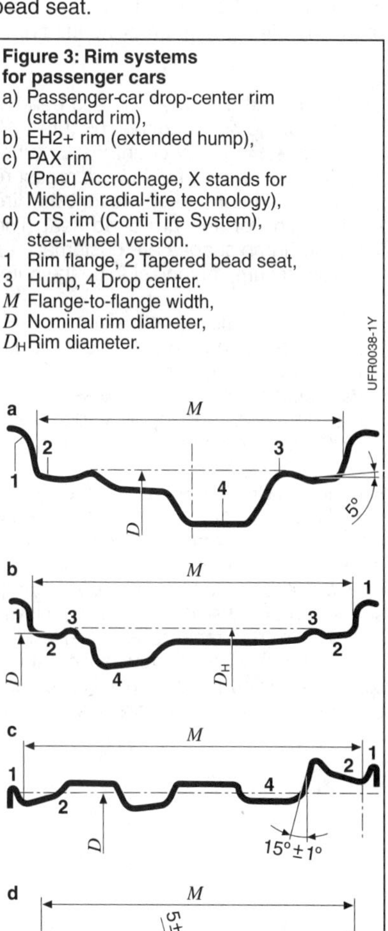

Figure 3: Rim systems for passenger cars
a) Passenger-car drop-center rim (standard rim),
b) EH2+ rim (extended hump),
c) PAX rim (Pneu Accrochage, X stands for Michelin radial-tire technology),
d) CTS rim (Conti Tire System), steel-wheel version.
1 Rim flange, 2 Tapered bead seat,
3 Hump, 4 Drop center.
M Flange-to-flange width,
D Nominal rim diameter,
D_H Rim diameter.

Rim base
The rim base connects the inner and outer rim bead seats. The drop-center rim is primarily used in passenger cars. Drop-center rims have a clearly defined shape with a drop-center rim base (drop center). When the tire is mounted on the wheel, the tire is initially positioned with one side of the tire bead in the drop center so that it can be pulled over the rim flange on the opposite side. The drop center is the required shaping of the rim base to mount the tire base (tire bead, tire inner ring) when fitting and removing the tire.

Hump
The hump is the all-round raised bead in the area of the rim bead seat (Figure 4). It is prescribed in many countries for tubeless tires. In the event of low tire pressure the hump is intended to prevent the tire from coming off the rim bead seat. The following hump shapes are customary:
- H: round hump on one side on the outer bead seat,
- H2: round hump on both sides,
- FH: flat hump on the outer bead seat,
- CH: combination hump, flat hump on the outer bead seat, round hump on the inner bead seat,
- EH2: extended hump on both sides.

Primarily H2 rims are used in passenger cars. H rims (formerly also called H1) are also used in commercial vehicles and older vehicles as well. The major differences are between the standard hump shape (H) and the flat hump (FH). Newly included in the standard is the extended hump (EH2) with slightly larger hump diameter, which is used in some cases, particularly with "run flat" tire systems.

Rim and wheel dimensions

Terms
The most important terms for the function and design of a wheel are (Figure 5):
- rim diameter (nominal diameter, dimension from rim bead seat to rim bead seat),
- rim circumference (measured value, determined with bead tape around the rim bead seat),
- flange-to-flange width (rim width, inside dimension between the rim flanges),
- center-hole diameter (centering diameter, as size of fit),
- rim offset ET (dimension in mm from the rim center to the attachment face of the disk wheel),
- pitch-circle diameter (diameter of the circle on which the center points of the bolt holes are situated),
- flange height (measured from the rim nominal diameter to the saddle point of the flange radius).

The rim offset determines the position of the wheel in the vehicle. The wheel is generally secured to the brake-disk chamber or to the brake drum. With a rim offset $ET = 0$ wheel attachment is exactly in the middle of the rim width. Changing the rim offset changes the position of the wheel and with it the track width of the vehicle. A smaller rim offset equates to a wider

Figure 4: Hump shapes
a) Hump,
b) Flat hump.
D Nominal rim diameter,
D_H Rim diameter.

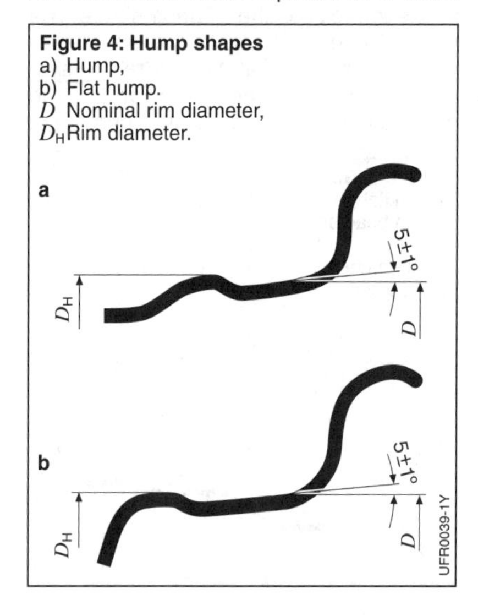

track width, a larger rim offset equates to a narrower track width.

Rim size
The basic dimensions of a rim size using a commercial-vehicle disk wheel with 15° tapered bead seat rim by way of example is 22.5 × 8.25 inches. The first value gives the rim diameter in inches, the second value the flange-to-flange width in inches.

Rim designs
Depending on the intended purpose and the tire design, there are a range of different cross-sectional rim shapes available (Figure 6): Rims can be made from a single part or from multiple parts. According to EU standardization single-part rims must be identified with a "×" (e.g. 6J × 15H2) and multiple-part rims with a "-".

Drop-center rims for passenger cars and light utility vehicles
The rim base is recessed for tire mounting. The drop-center rim can be made

Figure 5: Rim and wheel dimensions
(Picture source: MAN Nutzfahrzeuge Group),
1 Bolt holes,
2 Rim center,
D Rim diameter,
ET Rim offset,
L Pitch-circle diameter,
M Flange-to-flange width,
N Center-hole diameter,
S Tire width.

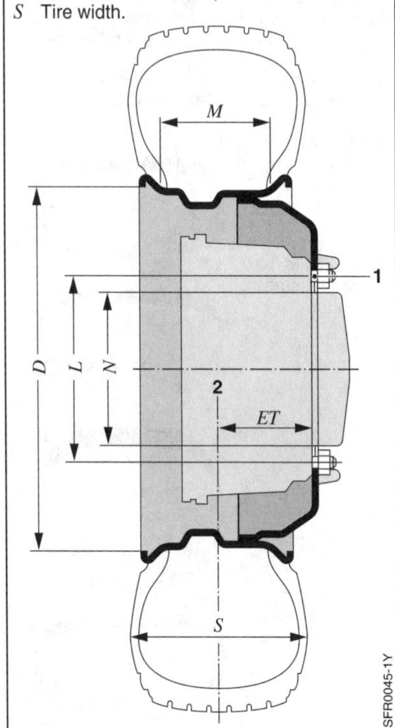

Figure 6: Rim design types
a) Drop-center rim,
b) 15° tapered bead seat rim,
c) Wide 15° tapered bead seat rim,
d) 5° tapered bead seat rim,
e) Two-part passenger-car rim.
1 Hump,
2 Rim base,
3 Flange-to-flange width,
4 Rim flange,
5 15° tapered bead seat,
6 5° tapered bead seat,
7 Sealant,
8 Wheel disk,
9 Bolt.

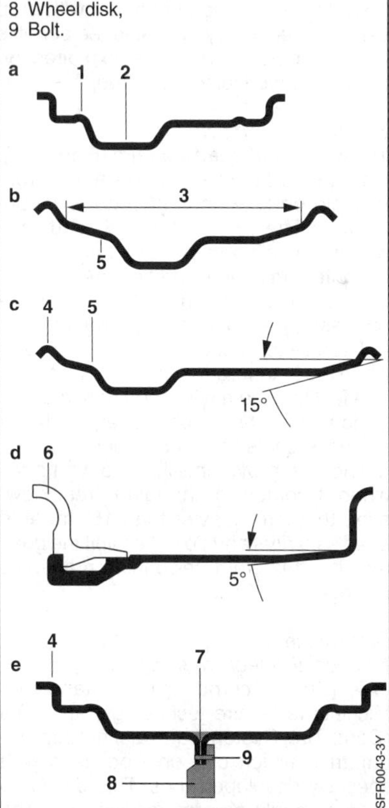

from one part (Figure 6a) or two parts (Figure 6e). The two-part rim is divided into a front half and a rear half which are bolted to each other in the drop center through an all-round pitch circle and to the wheel disk. A sealing ring or a sealing compound is used to provide the seal. The two-part variant traces its origins back to motor sport and offered the advantage of being able to replace one rim half in the event of damage. The tire is mounted in exactly the same way as on the single-part rim.

15° tapered bead seat rim
for commercial vehicles

The 15° tapered bead seat rim (Figures 6b and 6c) is made from a single part. The rim base is for tire-mounting purposes provided with a drop center to which the 15° tapered bead seats are connected. The 15° tapered bead seat rim with drop center is necessary to enable the benefits of the tubeless tire to be exploited on heavy commercial vehicles too.

5° tapered bead seat rim

The 5° tapered bead seat rim (Figure 6d), also called flat-base rim, is made from multiple parts. This is needed to mount the tires. The outer 5° tapered bead seat, which is non-permanently connected to the outer rim flange, can be removed. The outer 5° tapered bead seat is held on the rim base by an all-round sealing ring.

The tire is pushed onto the rim base when the 5° tapered bead seat is removed. The fact that the rim is made from multiple parts means that a tube is required. The two bead seats have a 5° taper.

These rim systems have advantages when it comes to changing tires. However, they are heavier than 15° tapered bead seat rims and do not exhibit the good radial and lateral true-running properties of single-part rims.

Special trims
for special wheel/tire systems

PAX (Pneu Accrochage, X stands for Michelin radial tire technology) and CTS (Conti Tire System) are distinct rim geometries which can only be used with specially developed tires. This design is used primarily on armored vehicles. The complete system is designed to permit continued driving with a punctured tire without the tire coming off the rim. The two systems also prevent the tire from being destroyed by thermal load during driving when punctured. With a conventional rim base the sidewall folds up when the tire is punctured and friction occurs.

In the PAX system (Figures 3c and 7b) from Michelin an additional support ring is mounted on the rim base; the punctured tire rests on this support ring.

The CTS system from Continental does not have an additional support ring. The tire reaches round the rim base and can rest when punctured on the rim base (Figures 3d and 7c). Both run-flat systems are required by law to have a tire-pressure control system.

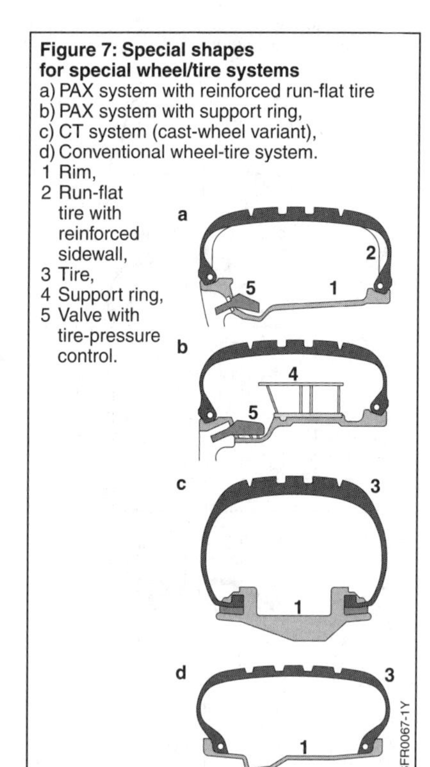

Figure 7: Special shapes for special wheel/tire systems
a) PAX system with reinforced run-flat tire
b) PAX system with support ring,
c) CT system (cast-wheel variant),
d) Conventional wheel-tire system.
1 Rim,
2 Run-flat tire with reinforced sidewall,
3 Tire,
4 Support ring,
5 Valve with tire-pressure control.

Design criteria

Passenger-car wheels

Design criteria for passenger-car wheels include:
- High durability
- Good support for brake cooling
- Reliable wheel mounting
- Low radial and lateral run-out
- Small space requirement
- Good corrosion protection
- Low weight
- Low costs
- Problem-free tire fitting
- Good tire seating
- Good balance-weight seating (see Wheel with tire)
- Appealing design (for aluminum wheels)
- In part requirements to improve vehicle aerodynamics (c_d value).

Special requirements for commercial-vehicle wheels

Wheels for commercial vehicles are technologically highly sophisticated. Whereas racing-car wheels are geared towards maximum speed, commercial-vehicle wheels carry high tonnages at simultaneously – compared with excavators for example – high speed. In Europe, for example, long-distance commercial vehicles travel with a total weight of 40 tons at a speed of 80 km/h. It is essential when establishing the minimum tire size for a commercial-vehicle axle always to start out from the permissible axle load and the top speed dictated by the vehicle design.

The larger the load to be transported and the rougher the terrain to be negotiated, the more important it is to take into account the – sometimes competing – rim-design requirements. High load capacity is to be achieved by designing an appropriate wheel shape and using optimum materials. High fatigue strength is essential to increasing road safety. Low wheel weight optimizes payload and is required because the wheel, as an unsprung, rotating mass, influences the overall vehicle oscillatory system.

Designations for passenger-car wheels

A typical wheel designation for passenger-car wheels is, for example:
6 ½ J x 16 H2 ET30.

6 ½	Dimension of rim width in inches
J	Rim-flange geometry
×	Single-part rim base
16	Dimension of rim diameter in inches
H2	Rim hump at the inner and outer rim bead seats
ET	Rim offset
30	Dimension of rim offset in mm

The designation and the associated dimensions with permitted tolerances have been prescribed as binding and standardized by worldwide recognized standards organizations such as ETRTO (European Tyre and Rim Technical Organization) or ISO (International Standards Organization) to harmonize rim and tire dimensions.

Materials for wheels

There are basically steel and light-alloy wheels. The material must always be specified in connection with the manufacturing technology. The following overview is intended to simplify systematization.

Classification of wheels

Steel wheel

A steel wheel consists of two parts, the rim and the wheel disk. Steel wheels are made from hot-rolled sheet steel in a forming process under rolling and bending conditions and joined by welding.

Light-alloy wheel

Light-alloy wheels are usually made of aluminum or magnesium alloys. They are produced using a variety of technologies. The aluminum wheel is produced as a cast wheel, forged wheel, sheet-metal wheel, or hybrid wheel. The magnesium wheel is only produced as a cast wheel.

The advantages of lightweight wheels are improved vibrational behavior, responsive suspension, reduced fuel consumption, and higher payloads. Light-alloy wheels are used in commercial-vehicle applications in particular in weight-sensitive transportation jobs. These jobs include the transportation of tanks and silos, where not exceeding the maximum transportation weight is of primary importance. In these situations the higher-cost light-alloy wheels usually pay for themselves within the first year of use.

Plastic wheel

Plastic wheels are made in an injection-molding process from mineral-fiber-reinforced polyamide and with metal inserts.

Materials

Sheet steel

In all the cheapest variant of passenger-car wheels is manufactured from hot-rolled and pickled sheet steel strip unrolled from the coil. The very good mechanical properties of this material provide for thin-walled wheel designed which are manufactured in highly automated, high-precision forming processes under bending conditions to finishing dimensions with close tolerances.

The continuous trend towards lightweight construction particularly since the CO_2 discussion has accelerated the use of high-strength, fine-grained structural steel. Its high tensile strength (600...750 N/mm^2) and very good formability and weldability enable lightweight, and cost-effective wheels to be manufactured efficiently.

Further potential for weight savings can be opened up by the use of "tailored blanks" for rim production. Here the sheet thickness of the starting material is adapted to the stresses in the wheel, in the course of which material strips of sheets of different thicknesses are joined by laser welding to create a blank.

Aluminum sheet

Aluminum sheet as an alternative to sheet steel can be more easily formed and is also – albeit thanks to more expensive methods (MIG welding) – more readily weldable. The manufacturing overhead is greater relative to steel wheels and the material costs are comparatively high, which prevents a wider range of applications. The use of high-strength sheet steels has greatly reduced the original weight advantage of aluminum sheet with the result that a cost/benefit analysis comes out in favor of steel wheels.

Light-metal alloys

Light-metal alloys are based in the main on aluminum alloys and in rare cases (e.g. in motor sport) on magnesium alloys. In the case of aluminum wheels, a distinction is made between cast and forge alloys depending on the manufacturing method used.

Cast alloys

Aluminum cast wheels are manufactured in low-pressure diecasting from aluminum alloys. The cast blank is formed in a steel mold, which is filled with liquid molten mass and cooled under controlled conditions to solidify. Aluminum alloys with a silicon content of between 7 and 11 % are used, depending on whether good castability or high strength is to be achieved.

Two alloys have proven successful. GK-AlSi11 is used for small wheels (up to 16 inches) with low wheel loads. The outstanding castability thanks to the high silicon content provides for highly efficient production with low reject rates due to casting defects. This alloy cannot be hardened by heat treatment. The wheels are therefore designed with greater wall thicknesses, which is reflected in slightly higher weights.

GK-AlSi7Mg is used for large wheels with a higher wheel load and for weight-optimized wheels. Adding 0.2...0.5 % magnesium to the aluminum alloy increases the strength of the cast wheel as a result of subsequent heat treatment (solution heat treatment and aging at warm temperature). This advantage is exploited to meet high load requirements during vehicle operation with a minimum of material usage.

To ensure that the high requirements with regard to strength, tightness and ductility placed on these safety-related vehicle components are met, only pure primary aluminum is used as the starting material. Contamination of the alloy with iron would cause acicular structures to form, which in turn would weaken the mechanical properties (elongation at fracture and tensile strength). Contamination with copper would reduce the chemical stability.

Forge alloy

Aluminum forged wheels are used in passenger cars when light, weight-optimized wheels are required and the target weight cannot be achieved with cast wheels. The increased hardness of the aluminum wrought alloy caused by the forging process (increase in mechanical strength due to plastic deformation) enables the wheel to be designed with a thinner wall, which means that less material needs to be used and thus the wheel is lighter.

The starting material takes the form of round continuously cast bars of AlSi1Mg which are sawn into precisely "portioned" disks. These are subjected to a three- to four-stage forging process to create the visible side (design side, disk, spider) and a flow-forming process to create the rim. As well as being hardened by plastic deformation the material is modified by a heat-treatment process.

Magnesium alloys

Whereas magnesium alloys have been unable to establish themselves in volume production – due to higher production costs in view of special safety precautions (risk of fire during cutting) – they are used in individual cases for special-purpose vehicles and racing cars.

Plastics

The use of plastic as a material for wheels is still in the developmental stage due in particular to insufficient high-temperature strength and difficult wheel mounting and manufacture. In particular the insufficient impact strength and thermal load capability as well as the incalculable long-term properties mean that plastic currently appears to make little sense as a material for a safety component such as the disk wheel in automobile construction.

Manufacturing processes

Steel wheels

When the text refers to passenger-car and commercial-vehicle steel wheels, exclusively sheet-steel wheels are what are meant. Other manufacturing processes such as casting and forging are not used for this material to manufacture wheels.

Passenger-car steel wheels consists of two parts: the wheel disk and the rim. They are welded to each other at the end of the production process. Both the wheel disk and the rim are manufactured in volume production on fully linked, highly automated production lines by means of forming processes under bending conditions. Smaller quantities can be produced more economically on individual presses.

Manufacturing the wheel disk

To manufacture the wheel disk the material required is unrolled directly from the coil, straightened flat and fed to a large transfer press with a force of pressure of approx. 40,000 kN. The press is equipped with a nine- to eleven-stage follow-on composite tool, in which the sheet blank is automatically routed with each press stroke from station to station.

A square blank with rounded corners is punched in the first machining step. Then in three to four stages the wheel disk and the center-hole area are formed by deep-drawing and stamping operations. In the next two or three stations the ventilation holes are punched with wedge driving tools and then stamped on the tool outlet side. The stamping replaces deburring of the pointed cutting edges and reduces the wheel's susceptibility to cracking under load.

Finally, the wheel designations are stamped on the reverse side and the calotte- or cone-shaped wheel-stud attachment faces are formed. In the last stage the wheel disk is calibrated to the finishing dimension for joining with the rim.

Manufacturing the rim

The rim material is likewise unrolled from the sheet coil, straightened and cut to length. The strips are stacked and forwarded to the automated rim production line. Here the sheet strip is shaped in a bending machine between three cylinder rolls into a ring and joined at the junction point in a butt-seam welding machine. The upsetting flash created in the process is then planed away on the inside and outside and the weld seam is smoothed by rollers. The unavoidable upsetting flash also on the side edges is then deburred and rounded.

The final rim contour is created on roller-burnishing machines by three consecutive forming steps, each with three profile rollers. The contour of the tool rollers is transferred to the rim here.

Where necessary, flow-forming is used to adapt the wall thickness and material distribution in the rim profile to the loads involved, thus saving more weight.

In the next operation the valve bore is produced on a rotary table. In the first station a flat attachment face is struck, from which the bore is punched in the second cycle. The cutting edges are then rounded on both sides by stamping.

Finally, the rim is drawn in a press onto a gage with the exact finishing dimension in order to create the tight concentricity and lateral-running tolerances (calibration).

Joining the wheel disk and rim

The process of joining the wheel disk to the rim creates the steel wheel. The two wheel parts – wheel disk and rim – are routed to a fully automated welding installation. On a small press they are aligned to each other and then joined to each other in a precisely specified position.

In the next station the assembly is connected by shielded arc welding with four to eight weld seams. These are deslagged and cleaned by rotating brushes adapted to the wheel. Then the wheel is calibrated to the finishing dimensions and made available for surface treatment.

The manufacturing process for steel wheels is so precise that the blank wheel is able to move on to the painting plant without any mechanical reworking. There, as a general principle, it runs through cathodic dip coating and, if stipulated, is given an additional visually appealing finishing coat.

Sheet-steel wheels are characterized in particular by their robustness and, thanks to their low manufacturing costs, are offered in the entry-level equipment specs of vehicle models.

Aluminum wheels

Aluminum-sheet wheels

For the most part, the basic manufacturing process is identical to that of the sheet-steel wheel. In view of the lower strength, the wall thickness must be greater compared with the sheet-steel wheel. Despite engineers having full mastery of the technology, aluminum-sheet wheels have not gained acceptance because steel wheels as a whole are the more economical wheel variant and the design options are very limited.

Aluminum forged wheels

As the name suggests, a forged wheel is created by the hot forming of an aluminum round blank between two forming tools. The forming process takes place in two stages – the forging of the wheel front and rear sides between two defined tools and the flow-forming of the rim contour.

The starting material for aluminum forged wheels takes the form of 6 m long continuously cast bars of AlSi1Mg with a diameter of 200...300 mm, depending on the planned wheel size. After an ultrasonic test the cavity-free bar sections are sawn off to a predefined length. The saw support – a cylinder approx. 250 mm in diameter and 150 mm high – is fed to the automated forging line. This consists of a heating furnace and up to four consecutive forging presses with forces of pressure of 8,000...40,000 kN. Parts are handled between the presses by robots. The result of this first forming process is blank with the finished design of the wheel disk, a punched-out center hole and a ring-shaped material reservoir positioned around the circumference with the material scheduled for the rim.

In flow-forming operation involving three rollers the ring-shaped disk is split open (hence the term "split wheel") and rolled out on a bell-shaped tool to make the rim.

Before being machined, the forged blank is subjected to heat treatment to improve the mechanical properties. The complete wheel contour is turned on two consecutive lathes (i.e. machined to finishing dimension). Then the bolt holes and the valve hole are drilled and milled on a machining center. The concluding polishing process gives the wheel its reflecting luster.

This high-precision machining ensures that every wheel runs absolutely true. There is no radial or lateral run-out.

If small batch sizes do not justify the manufacture of a forging tool, a special form of forged-wheel production is used. This involves the manufacture of a forged blank in the shape of a thick-walled cylinder whose base is a disk representing the rotation contour of the design and of the inner side of the wheel. With high machining overhead, the wheel design is usually 100% milled and the rim base first flow-formed and then turned.

Aluminum cast wheels

The most common method used is low-pressure mold casting. In the casting machine the aluminum molten mass tempered under controlled conditions is located in a crucible underneath the mold. The mold and crucible are connected by means of a feed tube. Once the mold is closed, the pressure in the crucible is increased to approx. 1 bar, causing the molten mass in the feed tube to rise and fill the mold.

Fusion heat is drawn off during the solidification process through precisely defined cooling channels in the mold. The specific cooling and heat removal during the solidification process and the casting parameters in their entirety (pressure, temperature and time) are decisive factors for the casting quality.

The production process is completely automated from the casting stage. The cast blanks removed by robot arms pass via linked conveyors through the following machining stages until they are automatically stacked and packed as wheels in the dispatch area:

- casting,
- removal of riser bore,
- X-ray test,
- heat treatment,
- machining,
- brushing and deburring,
- leak test,
- painting,
- dispatch.

In the X-ray test all the blanks are tested for casting defects according to specifications stipulated by the customer. The defective parts with casting defects that are not visible from the outside, e.g. porosity or shrink holes (cavities, material breaks) and inclusions or contaminants, are separated out and returned to the melting furnace.

The automated run is interrupted only before the painting stage so as to form production batches which are painted the same color.

Flow-forming process

Where necessary, it is possible to use a modified, slightly more sophisticated process to manufacture weight-optimized cast wheels, so-called "flow-forming wheels". This produces a weight saving of around 0.9 kg for a 19″ wheel. For the flow-forming process, the cast blank is made in a similar way to that used in forging. A ring is provided around the design surface instead of the formed rim contour as a material deposit for the rim. This ring is processed in a specially designed machining cell as follows:

- preturning for rolling,
- heating,
- rolling out of the rim (flow-forming).

The blank formed in this way is returned to the "normal" manufacturing process prior to heat treatment.

Another way of manufacturing weight-optimized wheels is to insert lost cores in zones of the wheel subject to less stress. Thus, aluminum is replaced by cavities, for example in the spoke, but also much more rarely in the hump. Accordingly there are hollow-spoke wheels and hollow-hump wheels.

Squeeze-cast process

The squeeze cast process attempts to exploit the advantages of diecasting for aluminum wheels. An exactly portioned amount of molten aluminum is pressed under high pressure into a diecasting mold under exactly defined casting parameters. The great advantage lies in the high solidification speed with positive effects on the material structure. Other advantages are the significantly lower machining overhead – and thus less material usage – and the relatively high output and longer mold service life. This casting process is used in individual cases, but it requires special, relatively complex casting machines and molds. This process has yet to gain acceptance.

Wheel design variations

Single- or multiple-part design variation

Sheet-steel and aluminum-sheet wheels consist of two parts. In this design wheel disk and rim are welded to each other. In the case of forged light alloy wheels and cast wheels, the one-piece version dominates.

Multiple-part versions, even those which are made of different materials (e.g. magnesium wheel disk and aluminum rim), are available mainly for tuning applications and for sports vehicles. Multiple-part wheels trace their origin back to motor sport. Mechanics exploited the advantage of being able to replace the damaged parts. In tuning applications, for example, the possibility of standardized rim rings and disks is used to create a large number of different wheel dimensions. For the most part, however, multiple parts no longer have a technical background and are now only used for visual reasons. Multiple-part wheels are subdivided into two- and three-part wheels.

Wheel spider

The wheel spider on cast wheels refers to the area of the spokes, which is represented in a steel wheel as the wheel disk and is for the most part provided with openings. Ventilation holes, slits and openings in the wheel disk or the wheel spider serve on the one hand to reduce weight, and on the other hand to ventilate the brake system and to enhance the visual design of the wheel. Set against this is the present-day requirement also to optimize the effects of the wheel with regard to the entire aerodynamics of the vehicle. Depending on the dynamic behavior of the body, a positive effect can be achieved here by keeping the area of the openings small and fashioning the geometry of the spokes as flat as possible. This measure has a more negative impact on the weight of the wheel such that it is now more common to find plastic parts on an aluminum wheel which are intended to aerodynamically optimize the wheel.

Rim variants

Rims used for passenger cars, vans and light utility vehicles are almost always drop-center rims with H2 double humps (rarely with FH or FH2 flat humps), tapered bead seats, and J section rim flange. Less common on smaller vehicles is the lower flange shape B, which today is used primarily in compact spare wheels. The higher flange shapes, JK and K, are rare on modern vehicles, and then only on heavy vehicles.

Lightweight-construction technologies

Hollow-spoke technology with sand casting cores or ceramic cores remaining (lost) in the wheel also provides good possibilities to reduce weight, but it requires a suitable design and special manufacturing equipment. This process is also associated with higher costs.

More widely used today is the flow-forming process for cast wheels, in which the rim base is only partly precast and then rolled out by machine to the corresponding rim width. The compressed material provides for thinner wall thicknesses with reduced weight in the rim base.

"Structural wheels" are used, among other things, as spare wheels or as road wheels with plastic wheel covers. Unrestricted by design conditions, the aim here is to use the minimum quantity of material possible to guarantee operating and functional safety, as well as streamline production costs for these wheels.

Wheel mounting

The design of the wheel and the mounting elements must meet safety requirements in all vehicle operating conditions. The wheel forces resulting from motive force, brakes, wheel load and wheel location must be supported by the overall mounting system (wheel bolt, wheel hub, brake-disk chamber, wheel-bolt holes, possibly coatings of parts) without impairing fatigue limits or the function of the wheel and axle components. Careful coordination of the friction parameters and geometry at the wheel bolts or the wheel nuts and contact zone of the wheel (bolt head to wheel-bolt hole) is essential when specifying the tightening torques in engineering design and in practice.

The geometric configuration of the wheel mounts in pitch diameter, number and dimensioning of the mounting elements are subject to the needs and requirements of each vehicle manufacturer. The wheel on a passenger car is secured to the axle hub by three to five wheel bolts or wheel nuts inserted through the mounting holes. Off-road vehicles and light utility vehicles often have six wheel bolts or wheel nuts. Commercial vehicles as a rule have ten wheel nuts, but this number can sometimes be even higher (e.g. tractors and excavators). The design of the contact surface of the nuts varies, depending on the vehicle manufacturer (e.g. calotte, cone, flat head). The longitudinal bolt forces that are decisive for the durability of the bolted connection must be reached and adhered to, both when new and used, in all dynamic operating states.

The high degree of true-running is achieved by means of a central wheel mount at the wheel hub with a precise alignment shoulder.

At present, wheels mounted with a central nut and interlocking driving pins are used almost exclusively on racing cars.

Wheel trims

Wheel trims (wheel caps) are mainly used for visual reasons on steel wheels and are affixed to the wheels using elastic retaining-spring elements which are easily detachable. But today aluminum cast wheels also feature wheel trims, which are often intended to improve aerodynamics. The design of aluminum wheels is for the most part kept simple and the weight kept low. Bolted solutions are also used in rare cases. The material used most often for wheel caps is heat-resistant plastic, e.g. polyamide 6. However, in some cases, aluminum and stainless pressed steel are also used.

Special wheel/tire systems

TRX rim

More recent rim developments, which have produced in limited series, are the TR rim (in metric dimensions). They were developed by MICHELIN for use with matching TRX tires and provide more room for the brakes.

Rims made by DUNLOP with a Denloc groove also require special tires; at low tire pressure and also in the case of pressure loss, the system is supposed to prevent the tire from coming off the rim, enhancing safety and mobility.

The TD system (TRX-Denloc) brings together both wheel/tire systems. As opposed to common practice, the two above designs are have rim and tire matched to each other and are unable to combine rim and tire with other tire versions, or only to a very limited extent.

CTS and PAX systems
The CTS/CWS and PAX systems were able to dispense with the need for a spare wheel. The original idea behind using these two systems to save on the spare wheel failed to catch on in the market and today these are used primarily on armored vehicles (see Rim designs).

Compact spare wheels
For space-saving reasons a compact spare wheel (mini spare) is often used as the spare wheel. This can be stowed in combination with a collapsible spare tire in an even smaller space (e.g. in roadsters, convertibles). All compact spare wheel systems are equipped with a specially designed tire where the driving properties are only suitable for emergency operation and limited top speed (approx. 80 km/h). Its benefits are the subject of debate, but it is becoming increasingly popular compared with a full-size spare wheel.

In many countries it is no longer required by law to carry a spare wheel. Instead, vehicles are equipped with a puncture kit (Tire-Fit) to repair any tire damage. The puncture kit consists of an electrically driven compressor and a sealant, which is pumped into the tire through the valve.

Stress and testing of wheels

The extremely varied and complex stress conditions in the wheel as a component in conjunction with a wide variety of operating conditions in the vehicle require specific endurance tests in order to be able to confirm the durability of a wheel with acceptable overhead. In general, the dynamic tests are run in test laboratories on standardized testing units, whereby a simulation of road operation that is close to reality is simulated and a good correlation of the test results to pure road operation is achieved. Country-specific legal requirements make special tests necessary, e.g. in the case of light-alloy wheels the simulation of a side curb impact (impact test).

Testing of sheet-steel wheels

The critical zones on a sheet-steel wheel are in particular the zones around the weld seams, mounting boreholes, dish (curvature of the wheel disk, see Figure 1), and ventilation holes. The operating conditions in each case, for example straight-ahead driving and cornering, generate different damage patterns in the area of the welding seam on the drop center well and in the wheel disk. Tests of the material quality and of the welded joints as well as surface tests are backed up by the endurance tests and indicate the need to optimize wheel manufacturing.

Testing of light-alloy wheels

Light-alloy wheels run through a similar testing process whereby, unlike sheet-steel wheels, the more varied influencing parameters of material, manufacturing, and design mean that the test requirements are at a significantly higher level. This ensures that fluctuations in material and manufacturing are unable to lead to premature failure. The maximum stresses occur mainly on the back of the wheel in the supporting structure of the ribs and spokes, in rare cases on the visible side.

The material quality and processing have a great influence on the durability of aluminum cast wheels. Inadequate physical values such as elasticity (during expansion) and tensile strength can be caused by poor heat application during casting or during heat treatment. This leads to porosity and shrink holes and deficient structural formation. The burrs that occur in high-stress zones during machine-cutting represent preliminary damage similar to notches and are often the starting point for incipient cracks. Careful machine deburring of these zones or specific constructive countermeasures, e.g. generously molded radii, are essential.

Testing of wheel with tire

Concentricity and lateral running

To assess the concentricity (true running) quality of a wheel on a vehicle, it is necessary to assess the wheel with the tire fitted, i.e. as a wheel with tire. During the manufacture of a wheel for concentricity the hub centering is in proportion with the two areas for the inner and outer tire seats. Likewise the contact face on the wheel hub and the inner areas of the rim flanges are responsible for the lateral running of the wheel. As a result of production these areas are endowed with tolerances (0.3 mm is usually given for concentricity and lateral running for passenger-car wheels), with which the tolerances of the tire now overlap. This can positively or negatively influence the concentricity of the wheel with tire. “Matching” is used to facilitate optimum concentricity of a wheel with tire. In this process the wheel and the tire are positioned during fitting in relation to each other in such a way that the “concentricity high point” of the wheel matches up with the “low point” of the tire.

On the wheel the high point is determined from the concentricity measurements of the two tire-seat areas. For each area an individual high point is obtained with different angular positions on the circumference of the wheel. These two values produce through vector addition a common value with a resulting angular position. This position is marked on the wheel with a colored dot or an adhesive dot.

On the tire the low point corresponds to the position where it reaches the lowest force variation while rolling. It is also marked with a colored dot. From a technical viewpoint the tire can also be compared with a spring that exhibits a radial stiffness. Based on production the tire can never be manufactured to such precision that it exhibits a uniform stiffness over its entire circumference. Wheels with tires with poor concentricity make themselves felt on the vehicle not only through a radial movement of the body (i.e. in the z direction). In the direction of travel a minimal variable force from acceleration and braking is also experienced with each wheel rotation.

Good wheel centering is extremely important for commercial vehicles that travel at higher speeds, but also when the wheels in question are large and heavy. On commercial vehicles that travel at higher speeds in particular, the lowest possible radial and lateral run-out on both rim bead seats and flanges is essential to ensuring smooth running. This increases safety and fuel economy.

Imbalance

Just as important as concentricity and lateral running to a smooth-rolling wheel with tire is compensation of the differently distributed masses on the wheel and tire. To this end it is necessary to minimize the influences of the masses on the rotating wheel with balance weights by balancing. Normally passenger-car wheels are dynamically balanced due to the rim width, i.e. measurements are taken on two planes (inner and outer tire seats) and the required compensation mass is determined. This is then applied with balance weights at the point indicated by the balancing machine. Adhesive, clip-on or drive-on balance weights are used for this purpose. The ideal position for the balance weights on the wheel for dynamic balancing is the maximum distance to the rim center at as large a diameter as possible.

An imbalance, referred to as "residual imbalance", of 5 g per balance plane depending on the vehicle type and suspension is not discernible in the majority of vehicles. For each balance plane a balance weight should be attached at one position only. If a relatively high balance mass (over 80 g) is required in a balance plane, it is recommended to turn the tire on the wheel and repeat the balancing process. The lower the balance-weight mass on the wheel, the lower the residual-imbalance potential also.

Narrow wheels for two-wheeled vehicles are balanced on one plane only (with a balance weight in the rim center). This method is called static balancing.

Wheels for commercial vehicles and compact spare wheels with a restricted top speed are not balanced.

Figure 8: Balance-weight positions

a) Clip-on or drive-on weight on the inside and outside (visible),

b) Clip-on or drive-on weight inside in combination with an adhesive weight under the drop center (concealed),

b) Clip-on or drive-on weight inside in combination with an adhesive weight under the tire seat (concealed),

d) Two clip-on drive-on weights (concealed).

1 Balance weight with retaining spring clipped onto the rim flange,

2 Balance weight glued to the inside of the wheel drop center,

3 Balance weight glued to the inside of the rim bead seat.

a Distance of balance weight to wheel center,

D Distance of balance weight to axis of rotation.

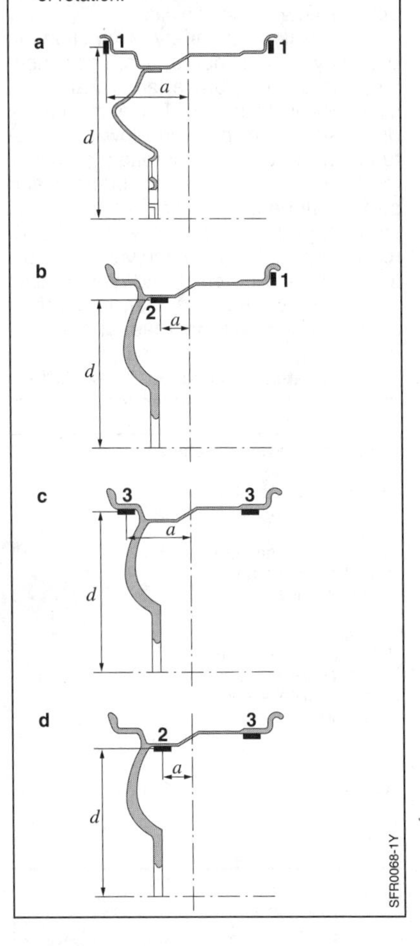

Tires

Function and requirements

The tire is the only component of a vehicle that comes into contact with the road. It thus assumes a key driving-dynamics position. Downstream driving-dynamics control systems such as the Antilock Braking System (ABS), the Traction Control System (TCS) and the Electronic Stability Program (ESP) are only ever as effective as the tire allows within the framework of its instantaneous power-transmission potential. When it comes to the tire, ultimately the crucial factor is the active safety of the vehicle.

Tires perform a variety of functions in everyday driving applications: They cushion, damp, steer, brake, accelerate, and simultaneously transmit forces in all three dimensions – at high and low temperatures, in the wet, on dry roads, on snow, mud and ice, on asphalt, concrete and pebble stones. They are meant to roll straight, permit precise steering, absorb road irregularities, bring the vehicle safely to a stop, and be quiet and comfortable. They are also meant to last, retain their characteristics with increasing age and decreasing tread depth, and produce as little rolling resistance as possible. Furthermore, an inflated tire performs supporting, vibration-damping and comfort-giving functions, and is therefore an active and fully integrated suspension element. The requirements resulting from these functions to be met by tires can be summarized as follows:

- high-speed resistance,
- durability,
- abrasion resistance (mileage),
- low rolling resistance,
- good properties in the wet (aquaplaning, wet braking, wet handling),
- good tire comfort, quiet rolling noise,
- good running characteristics in the limit range,
- resistance to aging,
- precise steering properties (handling),
- short braking distances,
- simple installation and assembly,
- true running and uniformity,
- economy,
- resistance to damage,
- resistance to chemicals.

The fundamental and visible material of a tire is rubber, an elastic to viscous material to which a tire owes the majority of its typical properties that are so important to the vehicle.

Figure 1: Tire construction
1 Nylon binding,
2 Steel-belt assembly,
3 Radial textile cord plies (casing),
4 Tire tread (tread rib),
5 Tread (tread groove),
6 Tire shoulder,
7 Sidewall,
8 Bead apex,
9 Bead with bead core (steel core, numerous thin steel cables twisted to other).

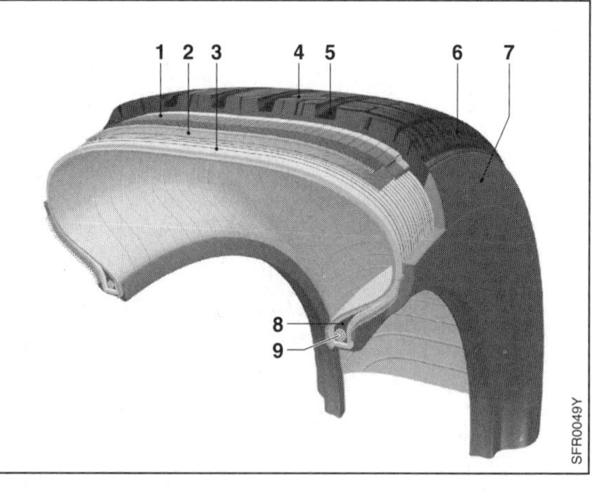

Tire construction

Design and components

Tires are a complex construction of different, mutually influencing raw materials, components and chemicals. A standard passenger-car tire consists of up to 25 different components and up to 12 different rubber compounds.

Today, only the tubeless, radial-ply steel-belted tire constructed in two stages satisfies the stringent demands imposed by the automotive industry and consumers.

Ingredients

The ingredients of a radial tire are:

- natural and synthetic rubber (approx. 40 %),
- fillers such as e.g. soot, silica, silane, carbon, and chalk (approx. 30 %),
- strength members such as e.g. steel, aramide, polyester, rayon, and nylon (approx. 15 %),
- softeners, e.g. oils and resins (approx. 6 %),
- vulcanization accelerators, e.g. sulfur, zinc oxide, stearin (approx. 6 %),
- anti-aging agents, e.g. UV and ozone blockers (approx. 2 %).

Since 2010 the toxicologically serious softeners and paraffined oils have been subject to particularly stringent limits values in the EU. For this reason manufacturers are increasingly using uncritical natural oils (e.g. sunflower oil).

Casing

The casing is stretched over a thin inner liner of airtight butyl rubber (Figure 1). Around 1,400 rubberized cords of rayon, nylon or polyester are combined in one or more casing plies to form the decisive strength member, the elastic "shell" of the tire. The cords run radially, i.e. at right angles to the tire plane from bead to bead – hence the designation radial tire. Cross-ply tires, in which casing cords are placed diagonally to the tire plane, in practical terms no longer play a role in modern-day applications.

Bead

The bead performs the important function of ensuring that the tire is securely and tightly seated on the rim. Driving and braking torques are transmitted via this crucial connecting point from the rim to the tire tread and thus to the road surface. Seated in the bead core is a cable of numerous steel wires, each of which can bear a load of up to 1,800 kg [1].

Sidewall

A thin and highly flexible rubber flank forms the sidewall and thus the flexible zone of the tire. The sidewall (tire flank) is however also the area of the tire that is most sensitive to damage.

Trouble-proof run-flat tires on the other hand have much thicker sidewalls than conventional designs (Figure 2). In the event of a blow-out the rim does not slump onto the tire casing and thus cannot damage it. In addition, fully deflated run-flat tires ensure a certain degree of steerability and directional stability for a further 80 km at speeds of up to 80 km/h.

Figure 2: Comparison of run-flat tire with standard tire
a) Standard tire,
b) Run-flat tire.
1 Tire with normal inflation pressure,
2 Tire without inflation pressure,
3 Reinforced sidewall.

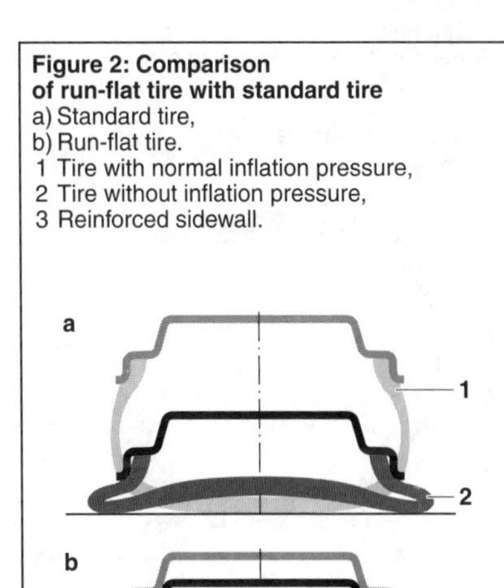

Cambering to blank

The finished, cylindrical combination of casing, inner liner, bead and sidewalls is pushed over a drum, the outside diameter of which corresponds to the inside diameter of this preliminary tire stage and to that of the subsequent tire. On this drum the cylindrical combination is cambered into the "real" tire shape (inflated and fixed) and then built further.

Because the casing cords run radially, i.e. transversely, to the rolling direction, the casing on its own would not be able sufficiently to transmit lateral forces when cornering and peripheral forces when accelerating and braking. It therefore needs support. This job is performed by the steel-belt assembly placed on top. Two or more plies of twisted and brass- and rubber-coated steel wires (steel cord) run not in the peripheral direction, but alternately at acute angles of between 16° and 30° to each other. High-speed-resistant tires are additionally stabilized by a nylon or aramide binding which suppresses a peripheral increase caused by centrifugal force. The tread surrounds the casing.

Figure 3: Design of rim with tire
1 Hump, 2 Rim bead seat,
3 Rim flange,
4 Casing (cord carcass),
5 Airtight rubber liner,
6 Steel belt,
7 Tread,
8 Sidewall,
9 Bead (with bead foot, bead core and bead apex),
10 Bead apex,
11 Bead core with steel core,
12 Valve.

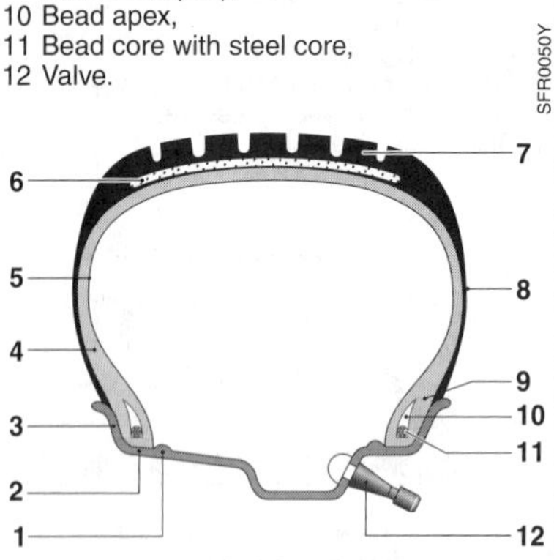

From blank to finished tire

The tires in this penultimate stage is called a blank and is now placed in a heating press. Inside this press is an exchangeable recess – an exactly shaped negative mold of the later finished tire. In this heating mold the tire blank is "baked" under steam pressure (approx. 15 bar) and heat (to 180 °C) for up to 30 minutes and acquires its final, typical appearance. The tread rubber creeps on heating exactly and without cavities into the tire negative mold of the heating press, thus creating the tread pattern and the sidewall markings. As a result of sulfur added in a previous process the hitherto plastic rubber vulcanizes into elastic rubber and acquires its desired operating characteristics.

The tread pattern ensures low rolling resistance, water expulsion, good grip, and high mileage.

A finished tire of the standard size for the medium-size car class 205/55 R 16 91 H weighs around 8.5 kg. A commercial-vehicle tire of the standard size 385/65 R22.5 weighs around 75 kg.

Tire with wheel

Together with the rim, the tire valve and the balance weights the tire forms the car's operational wheel (Figure 3). The rubber-elastic tire is inflated with compressed air, at which point it is able to absorb and transmit forces. Tire inflation pressure is usually 2...3.5 bar for a passenger car and 5...9 bar for a commercial vehicle. It is not the tire itself but rather the inflation air that supports the weight of the vehicle.

Differences between commercial-vehicle and passenger-car tires

Commercial-vehicle tires and generally similar in design to passenger-car tires, but are bigger, wider and heavier. Tire inflation pressure at 5...9 bar is much higher than approx. 2...3.5 bar for passenger-car tires.

The primary development objective is, as with passenger-car tires, to have all the parameters in proper proportion and above all the mileage. Commercial-vehicle tires therefore have comparatively hard, low-wear treads which are also regroovable and retreadable. Retreading of a bald tire identified as such is possible if the tire casing is undamaged.

Although the rolling resistance of a truck tire is lower than that of a passenger-car tire, its influence on the truck fuel consumption is greater due to the higher vehicle weight and the number of axles. In addition to high load capacity (up to 3...4 t per tire), other important tire characteristics include good straight-running stability, good cornering stability and traction. The trend in truck tires is towards increasingly smaller tire dimensions. This increases the useful load height and thus the transport volume.

Tire inflation pressure

Motor-vehicle manufacturers specify two values for tire inflation pressure for every vehicle – the part-load air pressure for a partly laden vehicle and the full-load air pressure for a fully laden vehicle or high driving speeds. The values are based first and foremost on the vehicle weight, on its top speed, on the tire design, and on the tire size. As a rule, the front and rear axles are subject to different desired pressures. The tire inflation pressure may only be measured and adjusted on a tire that is cold, i.e. that has not been heated up by vehicle operation.

A correct tire inflation pressure is important for

- optimum tire contact patch and optimum ground contact,
- shortest possible braking distance,
- optimum wet grip,
- balanced cornering stability,
- low rolling noises,
- low rolling resistance,
- low flexing work and heat production.

An excessively low tire inflation pressure will result in each case in the opposite end of the above parameters and in

- reduced service life,
- increased and sometimes uneven abrasion,
- progressive structural damage,
- danger of sudden tire blow-out,
- increased risk of accident,
- increased fuel consumption.

An excessively high tire inflation pressure (much less critical compared with low pressure)

- results in impaired tire comfort,
- reduces the ground contact patch (tire "stands") and thus diminishes cornering-stability and braking-force potential,
- causes increased central abrasion,
- but only reduces the rolling resistance slightly.

Tire tread

A tire is provided over its entire circumference with geometrically shaped tread grooves, ribs and channels as well as additional notches forming gripping edges (sipes) (Figure 4).

The most important function of the integrated tread (not cut, but heated) is to adequately absorb and disperse water on the road surface (also snow and mud in the case of winter tires), since wet and even surface moisture alone have a negative effect on grip characteristics. The braking distance is therefore dependent not only on the effects of interlocking and adhesion produced by the two friction partners tread rubber and road surface (see Tire grip). The braking distance on wet roads increases as tire wear increases.

Minimum tread depth

Summer tires

The minimum tread depth prescribed by law (EU Directive 89/459 of 1989 [2]) is established in most European countries at 1.6 mm (for passenger cars).

Figure 4: Tire tread
a) Typical tread of a summer tire,
b) Typical tread of a winter tire.
1 Wear marker.

Winter tires

The minimum tread depth for winter tires varies greatly from country to country. In Austria, for example, winter tires for passenger cars are required to have a minimum tread depth of 4.0 mm.

Wear-detection aid

The main tread-groove base features several rubber bumps exactly 1.6 mm high spread across the entire tread for the purpose of checking the minimum tread depth prescribed by law of 1.6 mm. If such a wear marker (TWI, tread wear indicator) on the tire makes contact with the road surface, the tire may no longer be used in road traffic.

Truck tires bearing the wording "Regroovable" on their sidewalls may be regrooved by the regrooving depth approved by the tire manufacturer (depending on the tire version 2...4 mm), ideally when the remaining tread depth is still 2 to 4 mm. Essentially, passenger-car tires are not permitted to be regrooved.

Aquaplaning

At higher speeds or when there is an enclosed layer of water on the road the tread the tread is no longer able to absorb sufficient water in itself and disperse it to the sides and to the rear. A wedge of water forces its way between the tires and the road, the tires lose contact with the road, and the vehicle loses its controllability – aquaplaning occurs (Figure 5).

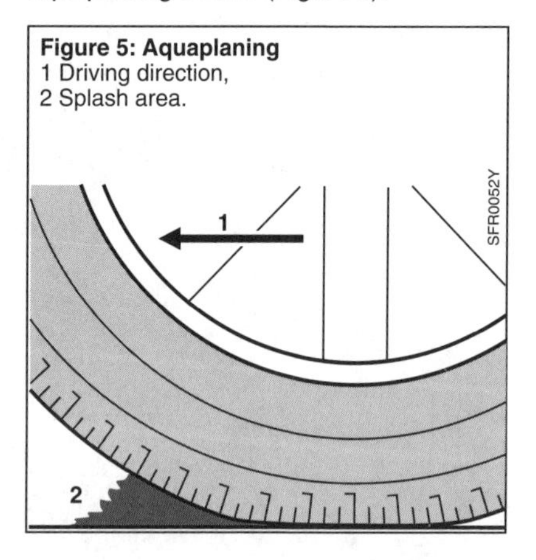

Figure 5: Aquaplaning
1 Driving direction,
2 Splash area.

The tire loses contact at precisely the stage when the pressure of the wedge-shaped splash water in front of the tire exceeds the pressure of the tire on the road. This pressure is squared as the driving speed increases. Because the critical pressure at which the tire aquaplanes is approximately equal to the internal tire pressure, passenger-car tires with an inflation pressure of approx. 2.3 bar aquaplane at a much lower speed than truck tires with 8 bar. Driving with a lower inflation pressure than prescribed decreases the already low speed from which aquaplaning occurs, once again significantly in passenger-car tires.

Tire contour, tread design and tread depth can defer the speed from which the risk of aquaplaning arises. Narrow tires, due to the higher pressure on the road (surface pressure, weight per contact patch), essentially aquaplane at higher speeds than wide tires, and they also heave to channel a much smaller volume of water. Drainage channels and rounded contact surfaces on wide tires reduce their risk of aquaplaning to an acceptable level. To compare: A tire with a nominal width of 220 mm must at 80 km/h and a rainwater height of 3 mm disperse roughly 15 liters per second so as not to aquaplane. This figure is 10 liters for a 140 mm "narrow" tire.

Figure 6: Braking distance from 80 km/h to a step on a wet road with new tires and with bald tires [3]
A Braking distance with 8 mm tread depth: 42.3 m.
B Braking distance with 3 mm tread depth: 51.8 m.
C Braking distance with 1.6 mm tread depth: 60.9 m.
v_R Residual speed.

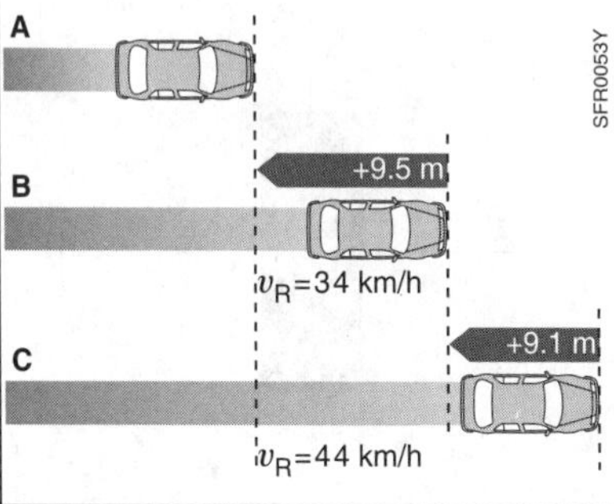

Wet braking

Figure 6 shows the further point at which the tread depth is of crucial importance to road safety: The braking distance on a road wet with rain is roughly 50 % longer with an almost bald tire (tread depth 1.6 mm) than with a new tire of identical size (tread depth 8 mm).

The residual speed v_R denotes the speed of the worse braking vehicle at the moment when the vehicle with better tires comes to a stop. It is calculated as

$$v_R = \sqrt{v_0^2 \cdot \left(1-\frac{s_1}{s_2}\right)} \text{ in m/s,}$$

where
v_0 Driving speed at start of braking in m/s,
s_1 Braking distance with vehicle 1 (with better tires) in m,
s_2 Braking distance with vehicle 2 (with bald tires) in m.

The residual speed is a measure of the theoretical accident severity to be expected in the event of a collision.

These calculated differences however can in practice only be effected by highly experienced drivers with extremely fast reactions. Many car drivers are not accustomed to ABS full braking; the total braking distance is lengthened significantly. Potentially shorter-braking tires only get a look-in if the vehicle is fitted with a brake assistant.

Force transmission

The four tire contact patches (footprint) are the direct and sole interface between the road surface and the vehicle.

Slip angle and slip

Only the tires rolling under an angle to the wheel rolling plane (slip angle α, Figure 7) and in the process deforming and constantly more or less slipping (see slip) transmit simultaneously and within physical limits the forces requested by the driver through steering, braking and accelerating. Conversely: A tire that does not roll or slip at an angle does not transmit any forces.

The tire transmits ever higher forces as the slip angle and slip increase. This relationship however is not linear. The effect is reversed after a relevant maximum value is reached (Figure 8). For passenger-car tires this reversal point for the slip angle is roughly 4...7°, corresponding to a pronounced steering-wheel angle. The reversal point for slip is 10...15% (on snow up to 30%). Excessive slip is a consequence of excessive acceleration or excessively heavy braking. If the steering angle or brake pressure is increased further, the wheels lock if the vehicle is not fitted with an antilock braking system. The slip is then 100%.

Figure 7: Slip angle
1 Wheel rolling plane,
2 Tangent to driving direction,
3 Driving direction.
α Slip angle.

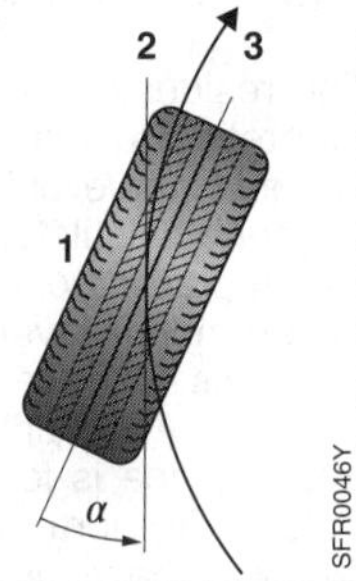

1 2 3 α SFR0046Y

Longitudinal and lateral forces

If a force F_x in the peripheral direction and a side force F_y occur simultaneously (e.g. when braking while cornering), the resulting transmitted horizontal force

$$F_{\text{h}} = \sqrt{F_x^2 + F_y^2}$$

cannot exceed the value $\mu_{\text{h}} F_z$. This situation can be explained by reference to the Kamm (friction) circle (Figure 9). The radius of the Kamm circle is equal to the maximum horizontal force $\mu_{\text{h}} F_z$ that can be transmitted via the tire. The maximum side force F_y is therefore smaller is at the same time a force F_x occurs in the peripheral direction. With the forces F_x and F_y marked in Figure 9 the wheel is exactly at the limit of the maximum horizontal force that can be transmitted.

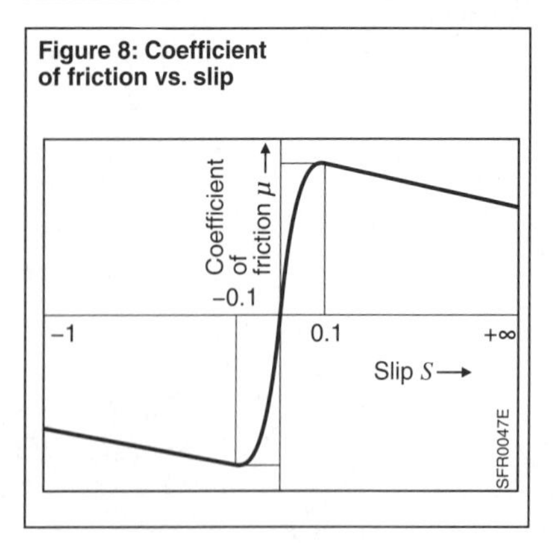

Figure 8: Coefficient of friction vs. slip

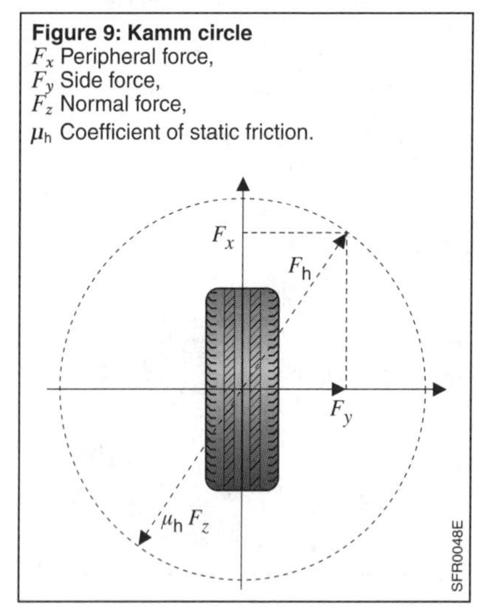

Figure 9: Kamm circle
F_x Peripheral force,
F_y Side force,
F_z Normal force,
μ_{h} Coefficient of static friction.

Tire grip

Generation of grip

Tires must transmit all the dynamic forces to only four roughly postcard-sized areas. The grip required for this purpose between the tire contact patches and the road surface is generated by several simultaneously occurring phenomena. These are essentially positive locking (also referred to in this context as the interlocking effect) and adhesion by molecular forces of attraction.

When a vehicle driving past at a constant speed is considered from the outside, as the tire rolls the continuously changing ground contact patch remains apparently fixed in relation to the vehicle (Figure 10) – while each individual rubber block of the tire tread runs into this forcibly flattening contact patch, deforms, and is "ejected" again at the other end. Relative movements and thus slip are generated in this contact patch: Each individual rubber block more or less slips during the dwell time in the contact patch.

Figure 10: Flattening in the contact patch between tire and road surface

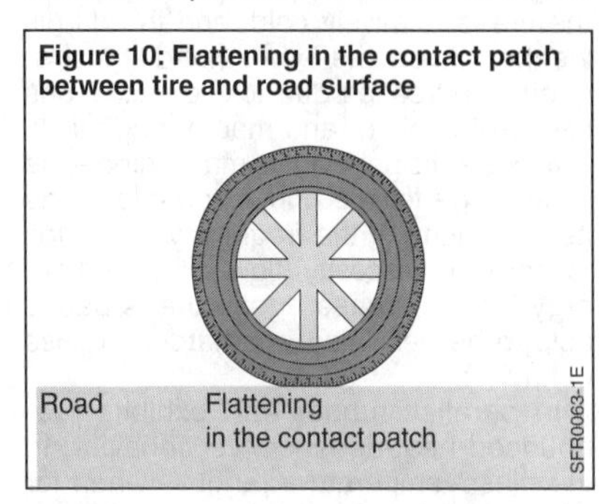

Figure 11: Micro- and macro-roughness of a dry road surface

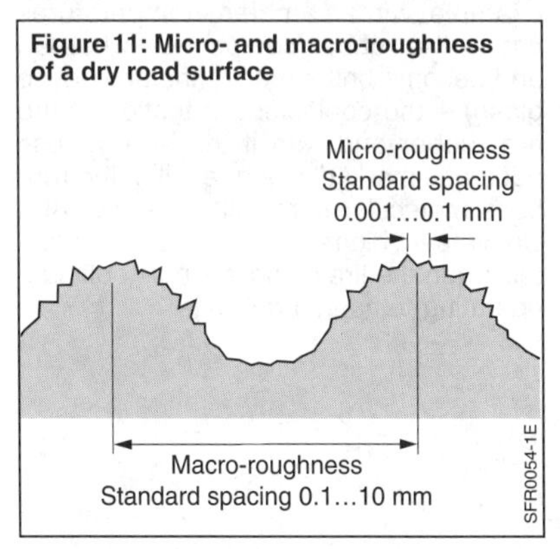

Viscoelasticity

Viscoelasticity describes the time-, temperature- and frequency-dependent elasticity and the viscosity of polymer and elastomer substances (e.g. of plastics, rubber). Internal damping, molecular interlocking and creeping processes prevent the two extreme states "fully elastic" (e.g. like an elastic spring) and "highly viscous" (like a solid body). Deformation and the force that causes it as well as the mechanical strain pass off at different times.

Interlocking effect

The interlocking effect is created by the direct and intensive contact of the tire with the road, depending on the micro- and macro-roughness of the road surface (Figure 11). In the passage of the contact patch the tread block under consideration comes up against a bump in the asphalt, is upset, and slips off again on the other side of the bump at accelerated speed. Only when it produces slip in the process can it build up in the tangential direction a counterforce contrary to the rolling direction which counteracts the sliding and thus permits the transmission of steering, drive or braking forces.

Figure 12: Load application and load removal of a viscoelastic substance

1 Strain, force per unit of area,
2 Deformation, elongation or upsetting relative to output variable.
δ Phase lag.

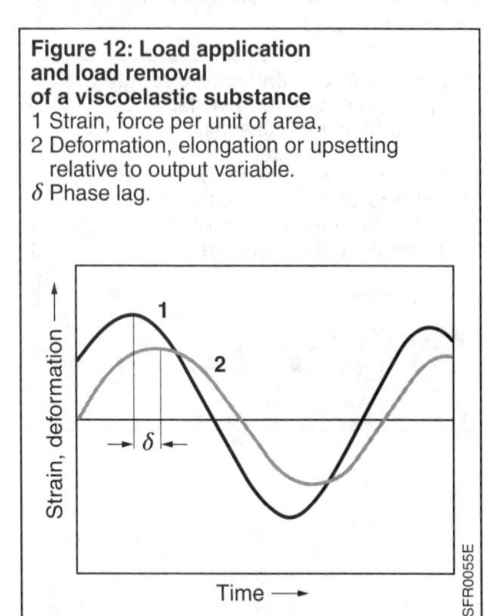

On account of its viscoelastic properties a rubber block does not revert immediately to its original shape after deformation; the strain follows on from the deformation that causes it (Figure 12). This effect typical of rubber of hysteresis results, on account of the cyclic deformation of the viscoelastic rubber, in a loss of energy in the form of non-utilizable heat and thus in a contribution to friction (hysteresis friction). The component of the friction force parallel to the road surface thus facilitates the transmission of drive or braking forces.

The principle of positive-locking grip also functions on damp and wet micro- and macro-rough road surfaces, but with limited effectiveness. Figure 13 shows that microfine asphalt points can penetrate the layer of moisture and thus the interlocking effect is maintained. However, an enclosed layer of water forms over the more rounded local bumps.

Essential to the interlocking effect is the presence of micro- and macroscopically small road irregularities. On a completely smooth surface (the coefficient of friction μ approaches zero) there would be no interlocking effect at all.

The interlocking effect and inner-molecular friction cause the tire to heat up. The resulting loss of energy is jointly responsible for the tire's rolling resistance, which makes up around 20...25 % of a vehicle's fuel consumption.

Figure 13: Micro- and macro-roughness on a damp or wet road surface
1 Macro-roughness channels and stores the water, but cannot penetrate the water layer.
2 Micro-roughness generates local pressure peaks and can thus penetrate the residual layer.

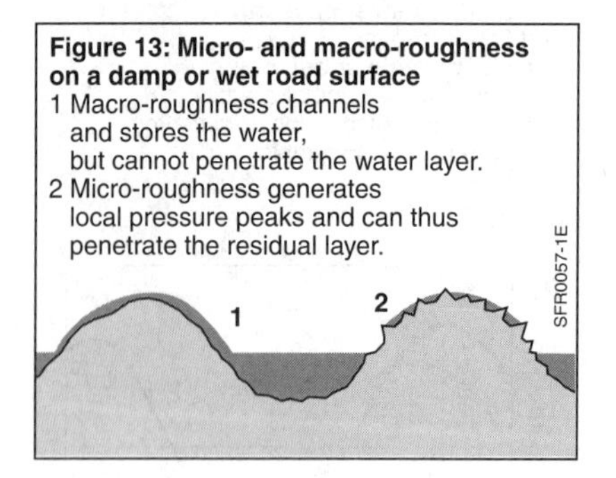

Adhesion
Molecular adhesion is created by the interaction and intensive contact between tire and dry road. The creation and subsequent breaking up of adhesive connections at the contact points result in a contribution to the coefficient of friction (adhesion friction). On a wet road molecular adhesion fails while the interlocking effect remains effective.

The frequency range of molecular adhesion, excited by the micro-rough road surface during drastic braking and extreme cornering, encompasses the spectrum of $10^6...10^9$ Hz.

Load frequency and temperature
Two further important influencing factors on the quality of grip must be mentioned: If the tire when rolling is only excited with a low frequency (exposure frequency), the rubber behaves elastically (low energy loss, Figure 14a). In this low frequency range the rolling resistance is very low, the tire is relatively cold, and the grip is weaker. If, on the other hand, the frequency increases due to the excitations caused by micro- and macro-roughness, a viscoelastic behavior is manifested – the ideal range for tire grip (energy loss has its maximum). If the frequency continues to increase, viscosity (flowability) and energy loss decrease again, the is barely able to be deformed and hardens (glass behavior).

In parallel, rubber part exhibits pronounced heat-dependence behavior: In the glass-temperature range – when for example winter ambient temperatures cause the rubber compound to harden and become brittle (hence the analogy to glass) – the coefficient of friction of the tire rubber and with it the energy loss decrease markedly as a result of the rubber's molecular immobility (Figure 14b). Conversely, higher forces can be transmitted when the tire is moved in its optimum operating-temperature range.

Figure 14: Energy loss in the tire as a function of frequency and temperature
a) Behavior as a function of frequency,
b) Behavior as a function of temperature.

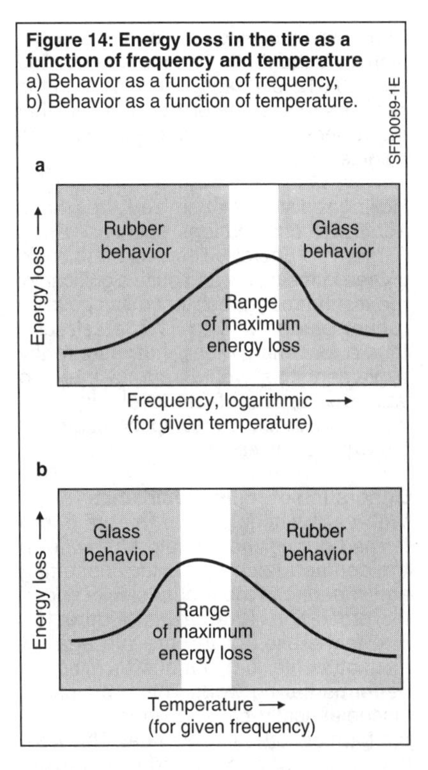

For rubber, therefore, an inversely proportional dependence between temperature increase and exposure frequency can be ascertained. Thus the glass temperature of an elastomer rises from –20 °C at only 10 Hz to +10 °C at 10^5 Hz (Figure 15). Compound developers are able to design rubber compounds with a glass temperature of between –60 °C and 0 °C at a frequency of 10 Hz.

Figure 15: Rubber behavior as a function of frequency and temperature

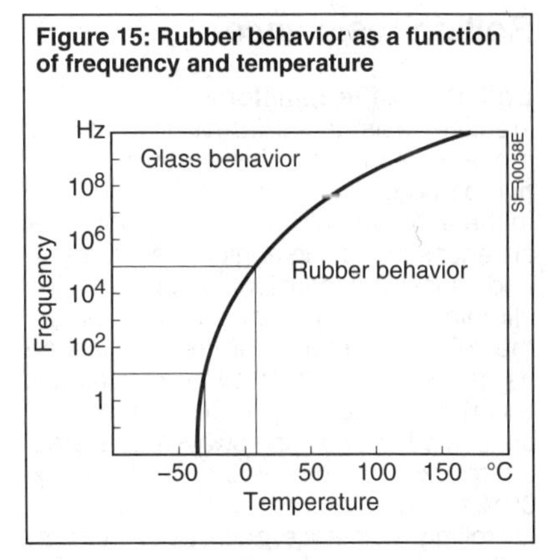

Special case: Rubber friction
Coulomb's classic law of friction ($F_R = \mu m g$) does not apply to vulcanized tire rubber. Its coefficient of friction μ (also called friction value or friction factor) is not constant, but
- increases with decreasing surface pressure (force per contact patch, given in N/mm^2),
- decreases with increasing surface pressure,
- depends on the sliding velocity,
- depends on the temperature and exposure frequency.

The phenomenon of increased grip with decreasing surface pressure explains the importance of very wide (and thus low-surface-pressure) tires in motor sport.

Rolling resistance

Definition of terminology

Tractive resistances inhibit the forward movement of the vehicle and must be overcome by motive means. In addition to the aerodynamic drag, the frictional resistances in moving engine, transmission and chassis/suspension components, the climbing resistance, and inertia forces, the rolling resistance of tires is classed as one of the main tractive resistances. It contributes to approx. 20 % of the fuel consumption on expressways/interstates, approx. 25 % on orbital roads and approx. 30 % on urban and ordinary roads [4].

Rolling-resistance decreases thus result directly in consumption and emission reductions. Rolling resistance (RR) corresponds to the energy loss per unit of distance and is given like every force in N (newtons). The dimensionless rolling-resistance coefficient c_{RR} denotes the ratio of rolling-resistance force to vehicle weight.

Example: Assuming a rolling-resistance force F_{RR} = 120 N and a vehicle weight G = 10,000 N ($G = m\,g$; vehicle mass m in kg, gravitational acceleration $g \approx 9.81$ m/s^2) c_{RR} amounts to 0.012 = 1,2 %. Standard values for c_{RR} for passenger-car tires on asphalt amount to 0.006 to 0.012, the (lower) value for truck tires to 0.004 to 0.008.

Figure 16: Shearing, upsetting and flexion in the contact lead-in

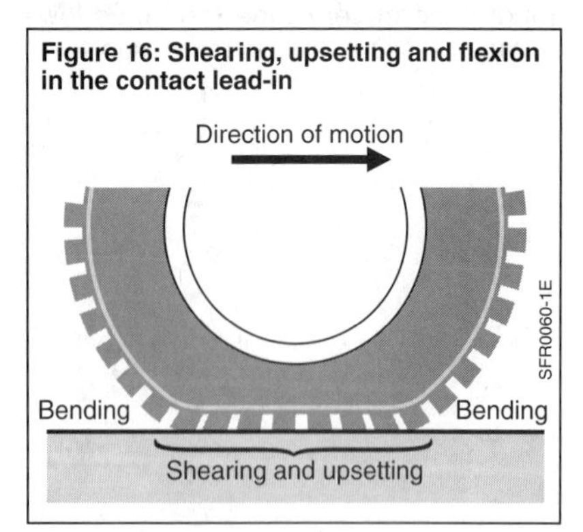

The "unit" kg/t (kilogram per ton) is occasionally used. In the example above the result 0.012 is 12 kg/t. This means in the case of a wheel load of 1 t = 1,000 kg that the rolling-resistance force F_{RR} assumes a value of 120 N.

Rolling-resistance-optimized tires have additional designations such as "Eco", "Green" or "Energy". Tire designers could indeed immediately and quite significantly reduce the rolling resistance by choosing rubber grades with low hysteresis and thus a low energy loss, but, as already explained, this would reduce the grip values to unacceptable levels. In other words, lower fuel consumption comes at the expense of less grip.

Generation of rolling resistance

With each wheel rotation the tire is deformed during the forcible flattening in the contact patch by flexion, upsetting and shearing of the rubber blocks and of the proportional tire casing (Figure 16). The fabric plies of the tire rub against each other (flexion), during which the tire performs flexing work. This produces a viscoelastically caused energy loss in the form of non-utilizable heat. This heat loss makes up 90 % of the rolling resistance.

Narrow tires and higher tire inflation pressure do increase the rolling resistance because contact patch and flexing work are reduced. However, designers are subject to very tight limits in terms of their scope of action in that the catalog of requirements with regard to handling performance, grip level and comfort are in direct conflict. Nevertheless, the technical specifications of the tire industry for future tire generations already feature tire dimensions such as 115/65 R 15 for subcompact-size cars and 205/50 R 21 for medium-size cars. Further drastic rolling-resistance reductions are not achievable with the requirements still standard today with regard to vehicle size, weight, maximum speed, sportiness, and comfort.

Even increasing the tire inflation pressure by 1 bar above the recommended value will only deliver a rolling-resistance reduction of 15 %. In a vehicle with an assumed fuel consumption of 10 *l*/100 km this would only produce a saving of 1.6 %. But because drivers still neglect to check tire pressure, the rolling-resistance reductions achieved in the most recent development cycles are not always implemented in real driving conditions. In addition, there is a direct conflict of aims between rolling-resistance optimization and the wet grip essential to road safety.

Conflict of aims between rolling resistance and grip

Frequency ranges

The deformations of the tread and rubber blocks caused by micro- and macro-roughness in the contact area between road surface and tire surface which generate the grip potential due to viscoelasticity occur in the very high frequency range of $10^3...10^{10}$ Hz. The energy loss caused by hysteresis is high here, resulting in the creation of high grip values.

However, the frequency spectrum which is important to rolling resistance is much lower at 1...100 Hz – precisely the range in which the inner tire structure is excited with each wheel rotation. At a driving speed of 100 km/h a passenger-car tire is deformed roughly 15 times per second, equating to a load frequency of 15 Hz. Tire designers nonetheless refer here to low-frequency excitation of the tire structure, particularly the casing.

Optimization incompatibility

These greatly differing frequency ranges explain the basic optimization incompatibility of simultaneously high values for grip and low rolling resistance. In conventional tires using industrial soot as the primary filler (up until the mid-1990s) a rubber compound with high hysteresis in the high-frequency grip range automatically results in a high energy loss in the tire components subjected to low-frequency load and thus in high rolling resistance.

Silica to resolve the conflict of aims

The solution in the late 1990s involved the introduction of silica (trade name for refined silicate) as a gray-powdered filler, which increasingly replaced the standard industry soot used up that point. Together with auxiliary binding materials, or silanes, the conflict of interests between rolling resistance, grip and abrasion resistance can be elevated to a high level of compromise.

Silica-based rubber compounds exhibit low energy losses in the low-frequency range relevant to rolling resistance, but high energy losses in the high-frequency rubber-grip range (Figure 17). The curve for energy absorption has this rise steeply and thus advance in frequencies of $10^2...10^4$ Hz. This results in tires which have low rolling resistance but nevertheless very good grip.

Figure 17: Frequency dependence of energy loss

1 Rubber compound with distinct hysteresis (high tire-grip values),
2 Latest-generation rubber compound (combines low rolling resistance, good grip and high abrasion resistance),
3 Rubber compound with weak hysteresis (low rolling resistance, poor grip).

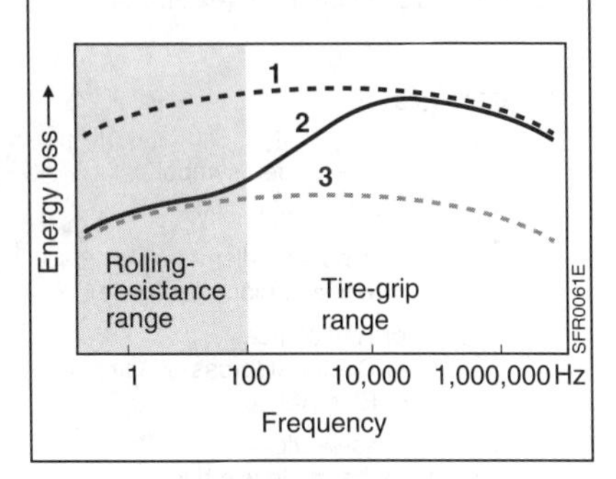

Rolling resistance and air pressure
Reduced tire pressure means increased flexing work and thus higher rolling resistance and impaired steering precision and braking stability.

Safety checks carried out in 2009 by the tire industry (Goodyear, Dunlop, Fulda) on 52,400 vehicles in 15 EU countries established that 81 % of all car drivers drive with excessively low tire pressure. Of these, 26.5 % were driving with clearly (down to 0.3 bar) and 7.5 % with significantly reduced pressure (0.75 bar and higher). The upshot of excessively low tire pressure is that fuel is wasted.

Tire designation

Definition of terminology
According to EU Directives ECE 30 (for passenger cars, [5]), ECE 54 (for trucks, [6]) and ECE 75 (for motorcycles, [7]) tires must be provided with internationally agreed, standardized tire designations. This applies in particular to the tire sidewall (Figure 18). Tires tested in accordance with ECE (mandatory since October 1998) carry a burned-in circle with a capital "E" or a lower-case "e" and the code of the approving authority, e.g. E4, and followed by a release number.

The lettering, codes and symbols on the sidewall indicate, as well as the tire manufacturer and the type designation, the origin, the production date, the dimension, the load capacity, the maximum permissible speed, the tire design, and the ratio of tire width to tire height (tire cross-section).

Reading the information is made more difficult by the fact that units of measurement of the metric system standard in Central Europe (mm, bar) and of the British imperial system (1 inch = 25.4 mm) are jointly used.

The service description must feature near the size specification. This consists of the load index (LI, Table 1) and the speed symbol (SSY; also used is SI, speed index, Table 2) and provides information about the maximum load capacity of the tire in question for the top speed corresponding to the speed symbol. This

Figure 18: Tire designation (example)

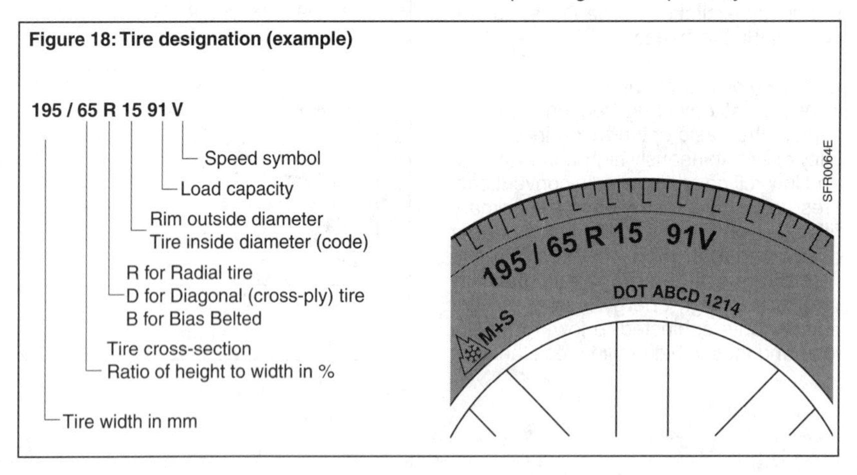

regulation is binding in all EU member states and in Switzerland.

On tires with directional treads (often on winter and summer tires with V-shaped treads) an arrow on the sidewall indicates the prescribed rolling direction. If the tires are already mounted on the rim, they can only be mounted on one side of the vehicle and may not for example be exchanged cross-wise (diagonally) together with the rim.

Example: A 205/55 R 16 91 H size tire has a nominal width of 205 mm, the height of the sidewall is 55% of the nominal width, which in this case in roughly 112 mm. The diameter of the matching rim to be mounted is 16 inches, i.e. 406 mm. The load capacity corresponds with a load index LI = 91 to the table value 615 kg – the vehicle axle load therefore according to the certificate of registration must not exceed 2 x 615 kg, i.e. 1,230 kg.

With the speed index SI = H the vehicle may be driven with this tire at a speed not exceeding 210 km/h, even if the vehicle is designed to reach an higher final speed.

Table 2: Speed Index SI (speed symbol)

A1	up to 5 km/h	L	up to 120 km/h
A2	up to 10 km/h	M	up to 130 km/h
A3	up to 15 km/h	N	up to 140 km/h
A4	up to 20 km/h	P	up to 150 km/h
A5	up to 25 km/h	Q	up to 160 km/h
A6	up to 30 km/h	R	up to 170 km/h
A7	up to 35 km/h	S	up to 180 km/h
A8	up to 40 km/h	T	up to 190 km/h
B	up to 50 km/h	U	up to 200 km/h
C	up to 60 km/h	H	up to 210 km/h
D	up to 65 km/h	V	up to 240 km/h
F	up to 80 km/h	W	up to 270 km/h
G	up to 90 km/h	ZR	over 240 km/h
J	up to 100 km/h	Y	up to 300 km/h
K	up to 110 km/h	(Y)	over 300 km/h

Table 1: Load index LI (values up to 3,350 kg, table open in upward direction)

LI	kg	LI	kg	LI	kg	LI	kg	LI	kg	LI	kg
1	46.2	26	95	51	195	76	400	101	825	126	1,700
2	47.5	27	97.5	52	200	77	412	102	850	127	1,750
3	48.7	28	100	53	206	78	425	103	875	128	1,800
4	50	29	103	54	212	79	437	104	900	129	1,850
5	51.5	30	106	55	218	80	450	105	925	130	1,900
6	53	31	109	56	224	81	462	106	950	131	1,950
7	54.5	32	112	57	230	82	475	107	1,000	132	2,000
8	56	33	115	58	236	83	487	108	1,030	133	2,060
9	58	34	118	59	243	84	500	109	1,060	134	2,120
10	60	35	121	60	250	85	515	110	1,090	135	2,180
11	61.5	36	125	61	257	86	530	111	1,120	136	2,240
12	63	37	128	62	265	87	545	112	1,150	137	2,300
13	65	38	132	63	272	88	560	113	1,180	138	2,360
14	67	39	136	64	280	89	580	114	1,215	139	2,430
15	69	40	140	65	290	90	600	115	1,250	140	2,500
16	71	41	145	66	300	91	615	116	1,285	141	2,575
17	73	42	150	67	307	92	630	117	1,320	142	2,650
18	75	43	155	68	315	93	650	118	1,360	143	2,725
19	77.5	44	160	69	325	94	670	119	1,400	144	2,800
20	80	45	165	70	335	95	690	120	1,450	145	2,900
21	82.5	46	170	71	345	96	710	121	1,500	146	3,000
22	85	47	175	72	355	97	730	122	1,550	147	3,075
23	87.5	48	180	73	365	98	750	123	1,600	148	3,150
24	90	49	185	74	375	99	775	124	1,650	149	3,250
25	92.5	50	190	75	387	100	800	125	1,700	150	3,350

Load-capacity reductions must be taken into account from speed index V and upwards.

Production date

There is usually a pressed-in four-digit number in an oval field on at least one of the two sidewalls next to the acronym DOT (US Department of Transportation) and a sequence of letters (code for manufacturing plant). This number denotes the production date. The first two digits indicate the calendar week, the last two digits the final digits of the production year (Figure 18).

Example: 1214 means the 12th week of the year 2014. Prior to the year 2000 the designation of the production date had only three digits.

Special case: Winter-tire designation

M+S tires

Winter tires must carry the M+S symbol (Mud and Snow) (see Figure 18). EU Regulation No. 661/2009 [8] denotes as an M+S tire a tire whose tread pattern, tread compound or construction is designed first and foremost "to achieve compared with a summer tire better values for winter handling performance and traction on snow" – an extremely woolly definition.

To compare: The previously applicable EU Regulation from 1992 [9] stated that "M+S tires" are such tires "on which the pattern of the tread and the structure are designed in such a way that they guarantee above all in mud and fresh or melting snow better handling performance than normal tires. The pattern of the tread of M+S tires is generally characterized by larger tread grooves and lugs which are separated from each other by larger spaces than is the case on normal tires".

"M+S" is to date not a protected or precisely defined designation and for this reason may also be used on tires not suitable for winter driving conditions (i.e. also on summer tires). The M+S symbol no longer has any significance with regard to suitability for winter driving conditions, a clear definition of the term for "winter tire" does not exist to date.

Snow-flake symbol

A revision of the M+S designation was called for in response to the widespread traffic chaos in the USA in 1995. A winter tire should satisfy certain criteria with regard to its suitability for winter driving conditions and verify this by way of appropriate tests. This gave rise to the designation 3PMSF (Three Peak Mountain Snow Flake), which today is firmly established in North American legislation.

This designation has also be used voluntarily for a few years in Europe. This is intended to verify to the consumer suitability for winter driving conditions substantiated by tests. The test criteria are defined by the European Union in UN-ECE R 117 [11]. Currently the designation with the snow-flake symbol for use in winter conditions is not legally binding, but the idea of incorporating it into European legislation is being discussed.

Figure 19: Snow-flake symbol (3PMSF) in acc. with UN-ECE R117

SFR0065Y

Sound identification

The noise classification so designated has been binding since 1 October 2011. It is positioned next to the ECE approval mark and can be recognized by a lowercase "e" followed by an "-s" after the homologation number. Tires with this identification comply with ECE 2001/43/EC [10], which lays down the maximum values for rolling noise (will be leveled again in 2016).

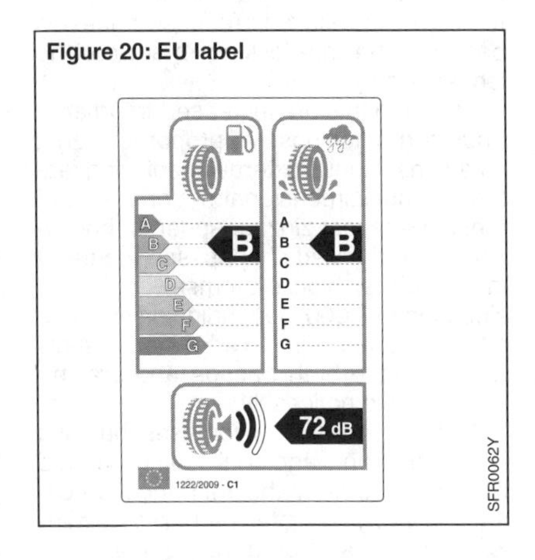

Figure 20: EU label

EU tire label

Definition of terminology

Since November 2012 new tires sold in the EU which were produced from July 2012 onwards must be provided with a standardized tire label (7.5 cm × 11 cm) (Figure 20). This label is meant to provide the buyer with quick and unmistakable information on the three tire properties of rolling resistance, wet grip (restricted to wet braking distance) and pass-by noise (not passenger compartment), thereby helping them to make a more informed decision to purchase. The winter properties of winter tires are currently not recorded by the EU tire label.

The tire label records tires in the categories C1 (passenger cars), C2 (light utility vehicles) and C3 (heavy commercial vehicles) as set out in Table 3. Excluded from the regulation are retreaded tires, professional off-road tires, racing tires, spikes, compact spare-wheel tires, vintage-car and modern classic-car tires (for vehicles first registered before 1 October 1990), tires for top speeds below 80 km/h, tires with inside diameters of less than 254 mm or more than 635 mm, and motorcycle tires.

Naturally, the tire label cannot show all the tire criteria (there are up to fifty), but the criteria selected do represent a certain combination of many other linked properties.

Table 3: Energy-efficiency classes of the different tire categories

Class C1 tires (passenger cars)		Class C2 tires (light utility vehicles)		Class C3 tires (heavy commercial vehicles)	
c_{RR} in kg/t	Energy-efficiency class	c_{RR} in kg/t	Energy-efficiency class	c_{RR} in kg/t	Energy-efficiency class
$c_{RR} \leq 6.5$	A	$c_{RR} \leq 5.5$	A	$c_{RR} \leq 4.0$	A
$6.6 \leq c_{RR} \leq 7.7$	B	$5.6 \leq c_{RR} \leq 6.7$	B	$4.1 \leq c_{RR} \leq 5.0$	B
$7.8 \leq c_{RR} \leq 9.0$	C	$6.8 \leq c_{RR} \leq 8.0$	C	$5.1 \leq c_{RR} \leq 6.0$	C
not assigned	D	not assigned	D	$6.1 \leq c_{RR} \leq 7.0$	D
$9.1 \leq c_{RR} \leq 10.5$	E	$8.1 \leq c_{RR} \leq 9.2$	E	$7.1 \leq c_{RR} \leq 8.0$	E
$10.6 \leq c_{RR} \leq 12.0$	F	$9.3 \leq c_{RR} \leq 10.5$	F	$c_{RR} \geq 8.1$	F
$c_{RR} \geq 12.1$	G	$c_{RR} \geq 10.6$	G		
Example: $c_{RR} = 10.5$ kg/t (label class E) corresponds to $c_{RR} = 0.0105$ or $F_{RR} = 105$ N.					

Fuel consumption
Letters from A (highest efficiency) through G (lowest efficiency) and the traffic-light colors green, yellow and red in the EU label denote the efficiency of the tire with regard to rolling resistance and thus fuel consumption. The range for rolling resistance from class A through G represents a difference in fuel consumption of up to 7.5% [12]. The potential for savings is even higher, depending on the driving situation. The difference between the individual stages of the tire classes with regard to rolling resistance is clearly defined: for instance 0.11 *l*/100 km for a vehicle with an average consumption of approx. 6.6 *l*/100 km. The difference between an A-rated tire and a G-rated tire adds up to approx. 0.5 *l*/100 km (in each case provided a consistent driving style and identical tire inflation pressure).

Wet grip
Letters A (shortest braking distance) through G (longest braking distance) provide information on the tire's wet grip when braking. The difference in wet grip between a particularly good tire and a poor tire when applying full brakes from 80 km/h to zero produces an 18-meter shorter braking distance.

Tire noise
Pass-by noise is depicted in the EU label by a pictograph together with an indication of the dB value. The symbol for external tire noise is based on the binding noise-emission limit values. The more sound waves in the EU label are shown in black, the louder and thus the more environmentally harmful the tire in acoustic terms.

Winter tires

Technical characteristics
Passenger-car winter tires exhibit no or only slight differences in structural design from summer tires. They are characterized by good force transmission (traction) on snow and mud, satisfactory grip on ice, good adhesion on wet and dry road surfaces, safe handling, comfortable rolling, and low noise.

The feature that is unique to them is primarily the softer and elastic-when-cold rubber compound with high natural-rubber content. Unlike the summer rubber compound, it does not become brittle and harden ("vitrify") at minus temperatures (which results in reduced adhesion because the interlocking effect can no longer occur).

Added to this are increased tread area (thus a greater positive proportion) and, as a conspicuous external identifying feature, a multidimensional superfine tread (design shapes: zigzag, spheres, honeycombs, and mixed shapes, see Figure 4) in the rubber blocks themselves. These sipes (up to 2,000, depending on tire size) offer additional gripping edges in the snow and increase both traction and braking performance noticeably.

As the tread depth decreases, but also with increasing age (hardening caused by among others the influence of UV and ozone) the winter properties diminish: Traction and cornering-stability potential deteriorate, the braking distance is extended, and the risk of aquaplaning increases. Winter tires identified for example in Austria under a remaining tread depth of 4 mm are always classed as summer tires.

Standard winter tires customary in Europe are offered in speed ratings up to speed index W (maximum speed 270 km/h).

Development aims

The tire is always a compromise product: If a particular property (e.g. rolling resistance) is predominantly developed, this is inevitably at the expense of other properties (e.g. with the upshot of reduced wet grip) and thus of balance. This is only desirable or permissible in special cases (motor sport, special tires, industrial requirements). There is a conflict of aims when on principle contrasting properties are to be simultaneously optimized. Modern rubber compounds with a high silica filler content and optimized ground contact patches raise this balancing act to a higher level, but this does not eliminate the conflict.

The conflicting development aims which can negatively influence each other include grip potential and rolling resistance, dry braking and wet braking, aquaplaning and dry handling, and grip and abrasion.

Tire tests

Tire manufacturers conduct around fifty objective laboratory tests and subjective road tests on in-house test and racing circuits. The most important tests are:

Handling on a dry road
Criteria are directional stability, steering precision, straight-running stability, noise, and tire comfort.

Properties in the wet
Criteria are handling on a wet course with bends, braking, aquaplaning, longitudinal and lateral, and circular-course driving.

Machine tests
Criteria are top speed, continuous running, and abrasion.

Winter properties
Criteria are snow handling, driving on passes, traction measurement, accelerating performance, and braking.

References

[1] Source: Goodyear Dunlop, 2012.

[2] Council Directive 89/459/EEC of 18 July 1989 on the approximation of the laws of the Member States relating to the tread depth of tyres of certain categories of motor vehicles and their trailers.

[3] Source: Continental AG, 2011. The specified brake differences were determined with a Mercedes C-Class car on 205/55 R 16 V size tyres in over 1,000 brake tests.

[4] Source: Michelin tire plants, 2010.

[5] ECE 30: Regulation No. 30 of the United Nations Economic Commission for Europe (UN/ECE) – Uniform provisions concerning the approval of pneumatic tyres for motor vehicles and their trailers.

[6] ECE 54: Regulation No. 54 – Uniform provisions concerning the approval of pneumatic tyres for commercial vehicles and their trailers.

[7] ECE 75: Regulation No. 75 – Uniform provisions concerning the approval of pneumatic tyres for motor cycles and mopeds.

[8] Regulation (EC) No. 661/2009 of the European Parliament and of the Council of 13 July 2009 concerning type-approval requirements for the general safety of motor vehicles, their trailers and systems, components and separate technical units intended therefore.

[9] Council Directive 92/23/EEC of 31 March 1992 relating to tyres for motor vehicles and their trailers and to their fitting installation.

[10] Directive 2001/43/EC of the European Parliament and of the Council of 27 June 2001 amending Council Directive 92/23/EEC relating to tyres for motor vehicles and their trailers and to their fitting.

[11] Regulation No. 117 of the United Nations Economic Commission for Europe (UN/ECE) – Uniform provisions concerning the approval of tyres with regard to rolling sound emissions and/or to adhesion on wet surfaces and/or to rolling resistance.

[12] Source: Tire manufacturers.

Tire-pressure monitoring systems

Application

Tire-Pressure Monitoring Systems (TPMS) are used to monitor the tire pressure on vehicles to prevent tire defects due to insufficient tire pressure, thus reducing the number of accidents resulting from defective tires.

If a vehicle is operated with insufficient tire pressure, this leads to increased flexing energy on the tire sidewalls and thus to increased wear of the tire. When operated at full load or at high speed, the greater flexing energy results in increased thermal load, which can even cause tire bursts. Following a spate of severe fatal accidents in the USA due to tire bursts caused by insufficient inflation pressure, legislation was passed (NHTSA Tread Act) to regulate the nationwide introduction of tire-pressure monitoring systems in the USA in order to warn drivers about low tire inflation pressure at an early stage in the future. Since September 2007, all new cars have been required to be fitted with tire-pressure monitoring systems which detect both tire damage and slow pressure losses through the tire rubber caused by gas diffusion.

The tire inflation pressure, however, is not only an important variable for traffic safety. Ride comfort, tire service life, and fuel consumption are also significantly influenced by the inflation pressure. Inflation pressure reduced by 0.6 bar can increase fuel consumption by up to 4 % in urban traffic and shorten the service life of the tire by up to 50 %. In the European Union (EU), the decision has been taken to prescribe the fitting of tire-pressure monitoring systems in all new cars as from October 2012 in order to help reduce CO_2 emissions.

Already today the rising proportion of tires with run-flat properties necessitates the deployment of tire-pressure monitoring systems, as the car driver is no longer able to detect a tire with a considerable pressure deficiency ("flat") on the basis of the drivability. In order to prevent the driver from inadvertently exceeding the speed and range limits that apply in this case, run-flat tires may only be used in conjunction with tire-pressure monitoring systems.

As a general principle, two different types of tire-pressure monitoring system are used: directly measuring and indirectly measuring systems.

Directly measuring systems

In directly measuring systems, a sensor module with a pressure sensor is installed in each tire of the vehicle. This transfers data such as the tire pressure and tire temperature from inside the tire across a coded high-frequency transmission link to a control unit. The control unit evaluates these data in order to detect not only pressure losses in individual tires ("puncture detection") but also slow pressure losses in all the tires ("diffusion detection"). If the tire pressure falls below a specified threshold or if the pressure gradient exceeds a certain value, the driver is warned by a visual or acoustic signal.

The sensor modules are usually integrated into the tire valve. As a rule, they are supplied by a battery. In comparison with other applications, this results in additional requirements with regard to power consumption, media resistance, and sensitivity to acceleration. Micromechanical absolute-pressure sensors are used as sensor elements.

The data measured with the pressure and temperature sensor in the tire are processed in the sensor module, modulated on a HF carrier signal (433 MHz in Europe, 315 MHz in the USA), and emitted via an antenna. This signal is either detected via individual antennas on the wheel arches or in a central receiver (e.g. in the control unit of existing Remote Keyless Entry systems).

Directly measuring systems do not need a reset function if they have a fixed, constant pressure-loss warning threshold. For such a system to work, the vehicle must only have one prescribed inflation pressure, regardless of vehicle load and tire size. As soon as different inflation pressures have to be set in the vehicle, a directly measuring system also needs a reset function to be able to adapt the warning threshold accordingly.

The advantages of directly measuring systems are that they provide precise, real measurement of the tire pressure and temperature, and their functioning is not dependent on specific tire types, vehicle conditions and road conditions. The disadvantages of direct systems compared with indirect systems are the much higher system costs, the additional logistical costs in the field of maintaining the availability of all the design variants, the follow-up costs for each new rim, and their battery-dependent, limited service life.

Indirectly measuring systems

In indirectly measuring systems, pressure loss in the tires is not determined directly, but rather by means of a derived variable. To achieve this, these systems perform a mathematical-statistical evaluation of the speed differences of all wheels for "puncture detection", and if necessary also an evaluation of a wheel natural-frequency shift for "diffusion detection". In vehicles with antilock braking or driving-dynamics control systems, the wheel speed required for this is determined by sensors that are already present and transferred to the control unit. Speed differences occur when pressure loss reduces the rolling circumference of the corresponding tire, thus increasing its speed relative to the other three wheels. The subtraction, which can be implemented using a low-cost extension of the software algorithms in the antilock braking or driving-dynamics control system, enables detection of high pressure losses on up to three tires. The wheel natural-frequency spectrum of the individual wheels is evaluated to enable a simultaneous pressure loss at all four wheels to be detected. Typically, the maximum wheel natural frequency shifts at a pressure loss of 20% from 40 Hz to approximately 38 Hz.

Indirectly measuring systems must necessarily be calibrated to the nominal pressure. A calibration is initiated by operating the reset button. When the reset function is activated, the system stores the current learning values on the next few kilometers as new reference values, based on the current rolling circumferences and wheel natural-frequency characteristics. The warning capability takes effect after approx. ten minutes of driving time.

The driver is required to activate the reset function to recalibrate the system when one or more tires are changed, the tire positions are changed (e.g. switching the front and rear wheels), the tire pressure is altered (e.g. when the vehicle is fully laden), or work has been carried out on the wheel suspension (e.g. adjustment work, shock-absorber replacement).

Advantages of indirectly measuring systems are their lower system costs and their robustness over the service life of vehicles, since no additional components are required. Because the system is linked to the vehicle and not to the wheels, no further costs are incurred in the field for logistics and spare parts. A disadvantage is the system's dependence on the specific tire, resulting in a wider variation of the detection times and in higher costs for adapting the system's function to the tire dimensions permitted for a particular vehicle. The system's dependence on mileage and road surface also influences the detection times.

Fulfillment of statutory requirements

As things stand, both systems satisfy the statutory requirements in North America and Europe with regard to "puncture detection" and "diffusion detection". Statutory provisions for tire-pressure monitoring systems are also being drafted in China and Korea.

Steering

Definitions for motor-vehicle steering systems

Motor-vehicle steering systems can be classified as follows:

Muscular-energy steering system

The required steering forces are generated exclusively by the muscular energy of the driver. These steering systems are currently used in the smallest vehicle categories.

Power-assisted steering system

The steering forces are generated by the muscular energy of the driver and by an additional auxiliary force hydraulically and increasingly electrically. This steering system is currently the type typically used in passenger cars and commercial vehicles.

Power-steering system

The steering forces are generated exclusively by non-muscular (external) energy (e.g. in machinery).

Friction steering system

The steering forces are generated by forces which act on the tire contact patch. The trailing axles in trucks are an example of this type. The transmissions for the steering and auxiliary forces are effected by mechanical, hydraulic or electrical means or by combinations of the three.

Steering-system requirements

General requirements

The steering system converts the driver's rotational input at the steering wheel into a change in the steering angle of the vehicle's steered road wheels. The design and layout of the system is intended to facilitate comfortable and safe steering of the vehicle in all situations and at all driving speeds. The entire steering system, from the steering wheel to the steered road wheels, must demonstrate the following properties for this purpose.

Transmission without play of the driver-initiated steering movement at the steering wheel is particularly important in the straight-ahead driving range. This ensures that the vehicle is handled safely and without the driver incurring fatigue, above all at medium and high driving speeds.

The steering train must therefore be highly rigid. This is necessary to ensure precise vehicle handling and to negotiate a bend at an identical steering-wheel angle with varying return forces, for example caused by changing lateral acceleration.

Low friction in the entire steering train ensures that the driver receives via the reaction and disturbance forces at the road wheels haptic feedback on the coef-

Table 1: Regulations for steering operating force

Vehicle category	Intact system			Faulty system		
	Maximum operating force in daN [1]	Time in s	Turning circle in m	Maximum operating force in daN [1]	Time in s	Turning circle in m
M1	15	4	12	30	4	20
M2	15	4	12	30	4	20
M3	20	4	12 [2]	45	6	20
N1	20	4	12	30	4	20
N2	25	4	12	40	4	20
N3	20	4	12 [2]	45 [3]	6	20

[1] 1 daN = 10 N.
[2] Or steering lock in case this value is not reached.
[3] 50 daN for non-articulated vehicles, with two or more steered axles, excluding friction-steered axle.

ficient of friction between the road and the tires. Low friction also assists the road wheels in aligning themselves to the straight-ahead driving position. In muscular-energy steering systems, it provides for low operating forces; in power-assisted steering systems it increases the efficiency.

Steering kinematics and axle design must be such that, although the driver receives feedback on the adhesion between wheels and road surface, the steering wheel is not subjected to any forces from the spring motion of the wheels or from motive forces (front-wheel drive).

Steering-behavior requirements

The requirements in terms of steering behavior can be summarized as follows:

Light, safe steering of the vehicle must be facilitated. This includes, for example, the tendency for the steering to align into straight-ahead driving and travel and time synchronization during operation.

Jolts from irregularities in the road surface should be damped as much as possible during transmission to the steering wheel. But, in the process, the required haptic feedback from the road to the driver must not be lost.

To ensure that the road wheels roll cleanly and are thereby not subjected to excessive tire wear, the entire steering kinematics must satisfy the Ackermann condition. This means that the extensions of the wheel axes of the steered wheels intersect at the same point on the extension of the wheel axis of the non-steered wheels (Figure 1).

An appropriately rigid layout of all the components in the steering train means that the smallest steering movements by the driver are converted into changes in direction at the steered wheel, thereby facilitating safe and precise handling of the vehicle.

The steering-angle requirement for turning the steering wheel from lock to lock should, for comfort reasons, be as small as possible when parking and driving at low speed. However, the direct steering ratio should not make the vehicle susceptible to instability at medium and high speeds.

Statutory requirements

The statutory requirements imposed on steering systems in motor vehicles are described in the international regulation ECE-R79 [1]. These requirements include, as well as the basic functional requirements, the maximum permissible control forces for an intact steering system and for a faulty steering system. These requirements govern, above all, the behavior of the vehicle and the steering system when driving into and out of a circle. For vehicles of all categories: After the steering wheel is released when the vehicle is driven on a circular course at half lock and at a speed of 10 km/h, the driven radius of the vehicle must become larger or at least remain the same.

For M1 category vehicles (passenger cars with up to eight seats): When the vehicle is driven tangentially out of a circle with a radius of 50 m at a speed of 50 km/h, no unusual vibrations may occur in the steering system. In M2, M3, N1, N2 and N3 category vehicles, this behavior must be substantiated at a speed of 40 km/h or, if this value is not reached, at top speed.

This behavior is also prescribed in the event of a fault in vehicles with power-assisted steering systems. For M1 category vehicles, it must be possible in the event of a steering-servo failure to drive at a speed of 10 km/h within four seconds into a circle with a radius of 20 m. The control force at the steering wheel must not exceed 30 daN in the process (Table 1).

Figure 1: Ackermann condition
1 Front axle,
2 Rear axle.
S Point of intersection of extension of wheel axes.

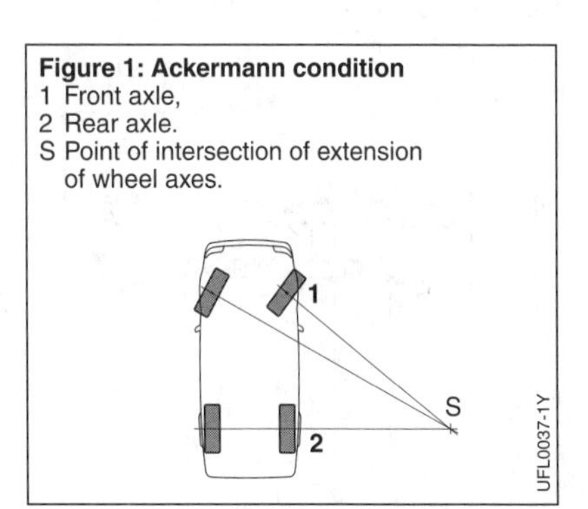

Types of steering box

The specified steering-system requirements have given rise above all to two fundamental types of steering box. Both types can be utilized in pure muscular-energy steering systems or, in combination with appropriate servo systems, as power-assisted steering systems.

Rack-and-pinion steering

Basically, as the name implies, the rack-and-pinion steering consists of a steering pinion and a rack (Figure 2). The steering ratio is defined by the ratio of pinion revolutions (steering-wheel revolutions) to rack travel.

As an alternative to a constant reduction ratio on the rack, suitable toothing of the rack allows the ratio to be varied as a function of travel. In this way, the straight-running stability of the vehicle can be improved by a suitably indirect ratio around the center of the steering. At the same time, it is possible with a direct ratio arrangement in the range of medium and large steering angles (e.g. when parking) to reduce the necessary steering-angle requirement when turning from lock to lock.

Recirculating-ball steering

The forces generated between steering worm and steering nut are transmitted via a low-friction row of recirculating balls (Figure 3). The steering nut acts on the steering shaft via gear teeth. A variable ratio is also possible with this type of steering box.

The increasing performance of rack-and-pinion steering has meant that recirculating-ball steering is practically no longer used in passenger cars.

Figure 3: Recirculating-ball steering
1 Steering worm, 2 Recirculating balls,
3 Steering nut,
4 Steering shaft with toothing segment.

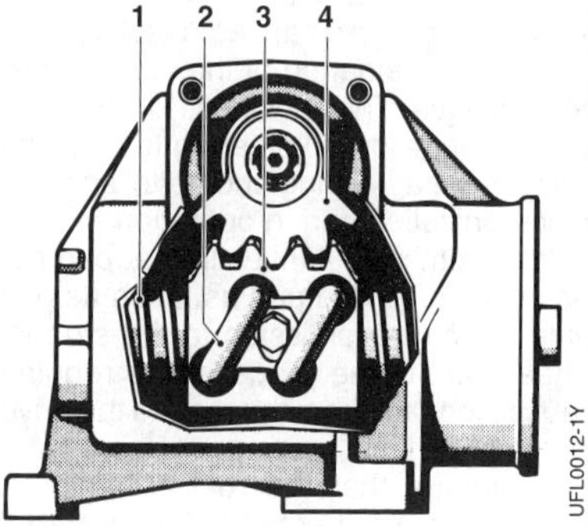

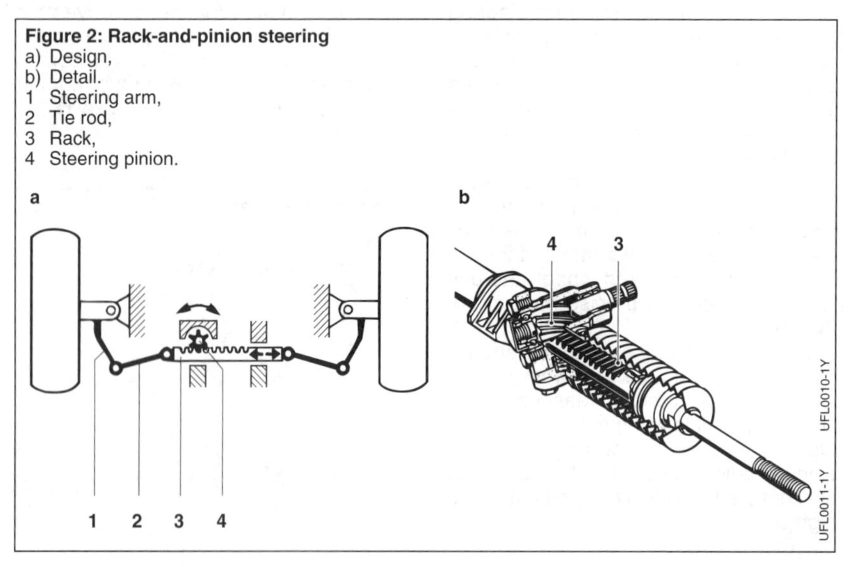

Figure 2: Rack-and-pinion steering
a) Design,
b) Detail.
1 Steering arm,
2 Tie rod,
3 Rack,
4 Steering pinion.

Power-assisted steering systems for passenger cars

The upshot of the increasing size and weight of vehicles and the heightened comfort and safety requirements involved is that in the last few years power-assisted steering has gained acceptance in all vehicle categories down to compact cars. These steering systems are, but for a few exceptions, already installed as standard. The steering forces exerted by the drive are boosted by a hydraulic or electric servo system. This servo system must be such that the driver receives good feedback on the adhesion conditions between the tires and the road surface at all times, and yet negative influences caused by road-surface jolts are effectively damped.

Hydraulic power-assisted steering

Combining the mechanical type of steering box with a hydraulic servo system produces rack-and-pinion power steering (Figure 4) and ball-and-nut power steering.

Control valve

The control valve provides the steering cylinder with an oil pressure that corresponds to the rotary force of the steering wheel (Figure 5). A flexible torque sensor, usually a torsion bar (Figure 4), translates

Figure 5: Functioning principle of control valve of hydraulic power-assisted steering
a) Control valve in neutral position,
b) Control valve in working position.
1 Power-steering pump, 2 Control bushing, 3 Rotary slide, 4 Left cylinder chamber, 5 Right cylinder chamber, 6 Oil reservoir.

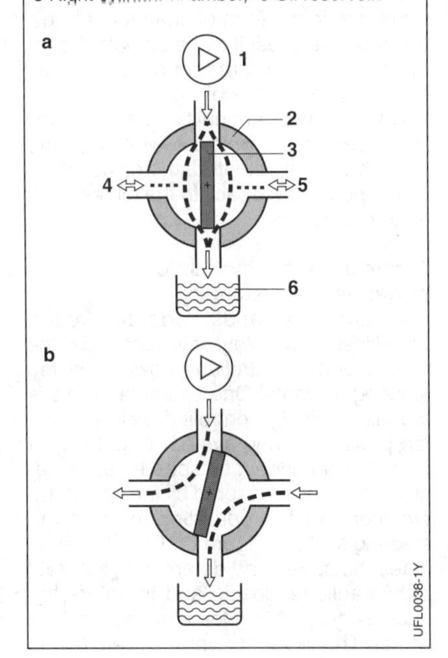

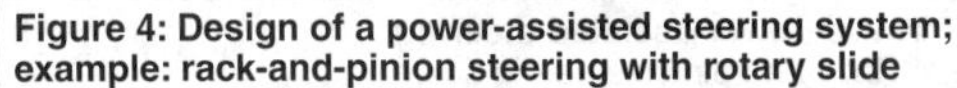

Figure 4: Design of a power-assisted steering system; example: rack-and-pinion steering with rotary slide
a) Steering assembly, b) Rotary distributor cross-section (enlarged), c) Oil supply (energy source).
1 Working cylinder, 2 Drive pinion, 3 Rack, 4 Torsion bar, 5 Lower steering shaft, 6 Control groove, 7 Rotary slide, 8 Control bushing, 9 Return line, 10 Oil reservoir, 11 Pressure- and oil-flow-limiting valve, 12 Pressure line, 13 Vane pump.

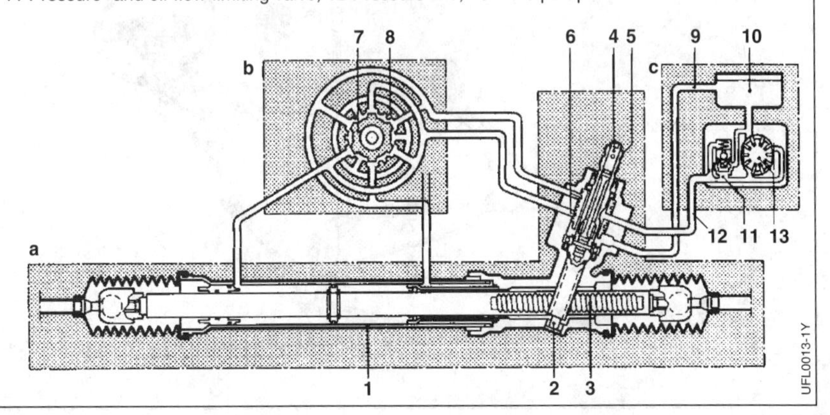

the applied torque, proportionally in most cases, precisely and without any degree of play, into as small an actuator travel as possible. The actuator travel is the movement made by the rotary slide in relation to the control bushing. The control edges, which are in the form of chamfers or bevels, move as a result of the actuator travel, and vary the corresponding opening cross-section for the oil flow.

Control valves are built according to the "open center" principle, i.e. when the control valve is not actuated, the oil delivered by the pump flows back to the oil reservoir at zero pressure.

Parameterizable hydraulic power-assisted steering

Increasing demands regarding user-friendliness and safety have resulted in the introduction of controllable power-assisted steering systems. One example of this is the electronically controlled rack-and-pinion power-steering system (Figure 6). It operates dependent on speed, i.e. the vehicle speed as measured by the speedometer controls the operating force of the steering system (Figure 7). The ECU evaluates the speed and determines the level of hydraulic feedback, and therefore the necessary operating force on the steering wheel. This level of hydraulic reaction is transmitted to the steering-system control valve via an electrohydraulic converter. This modifies the hydraulic reaction in relation to vehicle speed.

The special design of the steering characteristic means that, when parking and when moving the steering wheel at standstill, only minimal forces need to be applied by the driver to the steering wheel. The level of power assistance is reduced as speed increases. In this way, precise and accurate steering is possible at high speeds. With this system, it is important that oil pressure and volumetric

Figure 7: Characteristic curves of parameterizable power-assisted steering
The characteristic curve can be adjusted to the vehicle data.

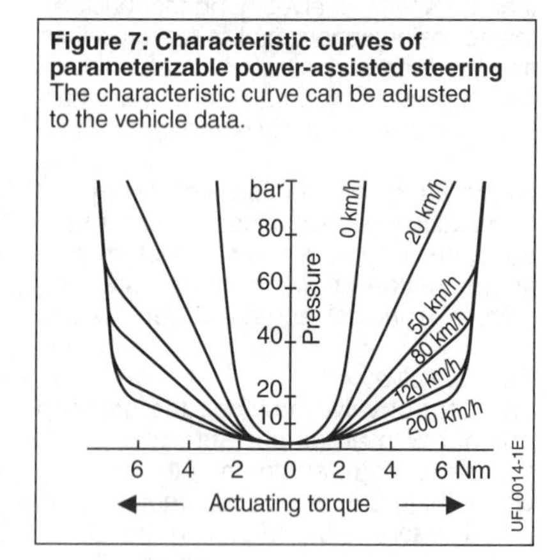

Figure 6: Design of parameterizable hydraulic power-assisted steering
1 Driving speed, 2 ECU, 3 Steering valve, 4 Rack-and-pinion steering box, 5 Power-steering pump, 6 Oil reservoir, 7 Hoses, 8 Steering column.

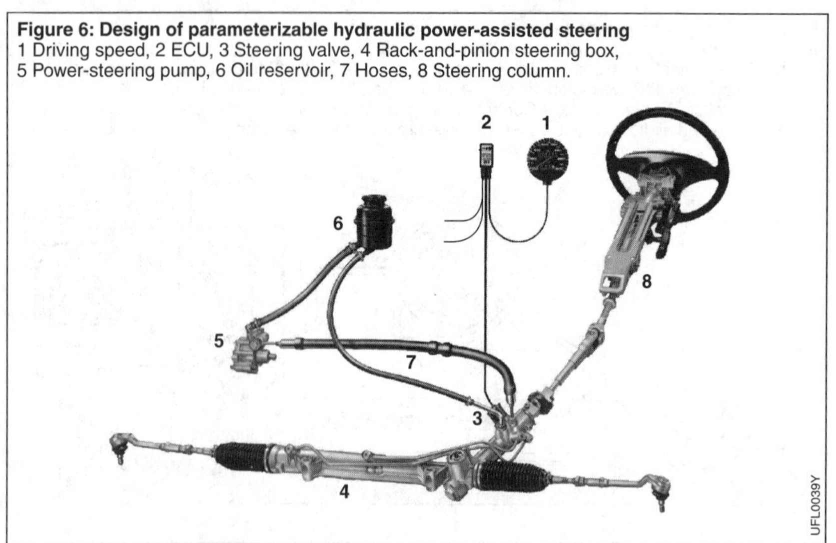

flow are at no time reduced, so that they are immediately available in emergency situations.

Working cylinder
The double-action steering cylinder converts the applied oil pressure into an assisting force which acts on the rack and intensifies the steering force exerted by the driver. The steering cylinder is normally integrated into the steering box. As the steering cylinder has to have extremely low friction, particularly high demands are made on the piston and rod seals.

Energy source
The energy source consists of a vane-type supply pump (generally driven by the internal-combustion engine) with an integrated oil-flow regulator, an oil reservoir, and connecting hoses and pipes. The pump must be dimensioned so that it makes available the required oil pressure and the required oil quantity for the parking maneuver, even when the engine is just idling.

A pressure-limiting valve is required in the steering system to protect against overload. This valve is usually integrated into the pump. The pump and system components must be designed so that the operating temperature of the hydraulic fluid does not rise to an excessive level, so that no noise is generated, and the oil does not foam.

Alternatively, the pump for the energy supply of the steering can also be driven by an electric motor. Here, the pump is usually designed as a gear or roller-cell pump. Because of the limited power available from the vehicle's electrical system, these systems are used primarily in vehicles ranging from the compact to the lower mid-size categories. Elimination of the belt drive to the internal-combustion engine means that the pump can be variably arranged, thereby favoring modular vehicle construction. The control electronics and the evaluation of signals, for example vehicle speed or steering speed, enable adaptation of the pump speed to the current energy requirement of the steering and the driving situation in order to save energy.

Electric power-assisted steering
Electromechanical power-assisted steering systems are also used in mid-size category passenger cars and smaller vehicles. Such systems feature an electric motor powered by the vehicle electrical system. The mechanical coupling of the motor to the steering box can be set up as a steering-column, steering-pinion or rack drive. The system consists of the following components (Figure 8):

- Steering column that connects the steering pinion with the steering wheel inside the vehicle.
- Steering pinion that converts the rotating steering movement into the linear movement of the rack.
- Rack connected to the wheels via the tie rods and links.
- Sensors to record the information required to calculate the necessary supporting steering torque.
- Servo unit, consisting of an ECU and a servo motor (electric motor), that generates the supporting steering torque.

If the driver moves the steering wheel, a sensor registers the steering torque exerted and sends this information as an electrical signal (analog or digital) to the ECU. This calculates the supporting steering torque and activates the servomotor on the basis of the calculated result. Currently, commutator or brushless DC motors or three-phase asynchronous motors are used as servomotors. Depending on the required performance capability of the steering, the torque generated by these motors is 3 to 6 Nm.

The direction of rotation of the motor depends on the direction of motion of the steering wheel. The return movement of the steering can also be supported. This takes place when the driver comes out of a curve. In this situation, the servomotor generates torque that supports return rotation of the steering back to the straight-ahead position.

The servomotor transfers the supporting steering torque via a worm or recirculating-ball steering gear. Depending on the steering variant, it is transferred to the steering column, pinion, dual-pinion, or the rack of the mechanical rack-and-pinion steering.

The control electronics take into account different signals and parameters, e.g. driving speed, steering angle, steering torque, and steering speed. With the help of other sensors located in the vehicle and due to the networking of the steering ECU with other ECUs in the vehicle framework, this steering system can be used to implement assistance functions to enhance comfort and safety.

Demand-oriented control of the electric motor leads to considerable fuel savings of 0.3 *l*/100 km on average compared to hydraulic power steering with a pump driven by the vehicle engine. In urban driving, the fuel savings increase to 0.7 *l*/100 km.

In the event of a failure of the energy supply or steering support, the driver can continue to steer mechanically, but with higher manual steering torque.

Superimposed steering

With a superimposed-steering system, an additional steering angle can be added to or subtracted from the steering-wheel angle set by the driver at the steering wheel. This system is usually combined with a parameterizable hydraulic or electric power-assisted steering system. Superimposed does not facilitate autonomous driving, but it does provide for a steering characteristic optimally adapted to the driving situation, and thus maximum comfort and directional stability. When networked with driving-dynamics control systems, it can further increase safety in critical driving situations by means of driver-independent steering adjustments. Such steering systems are already in series production, known as Active Steering at BMW and Dynamic Steering at Audi.

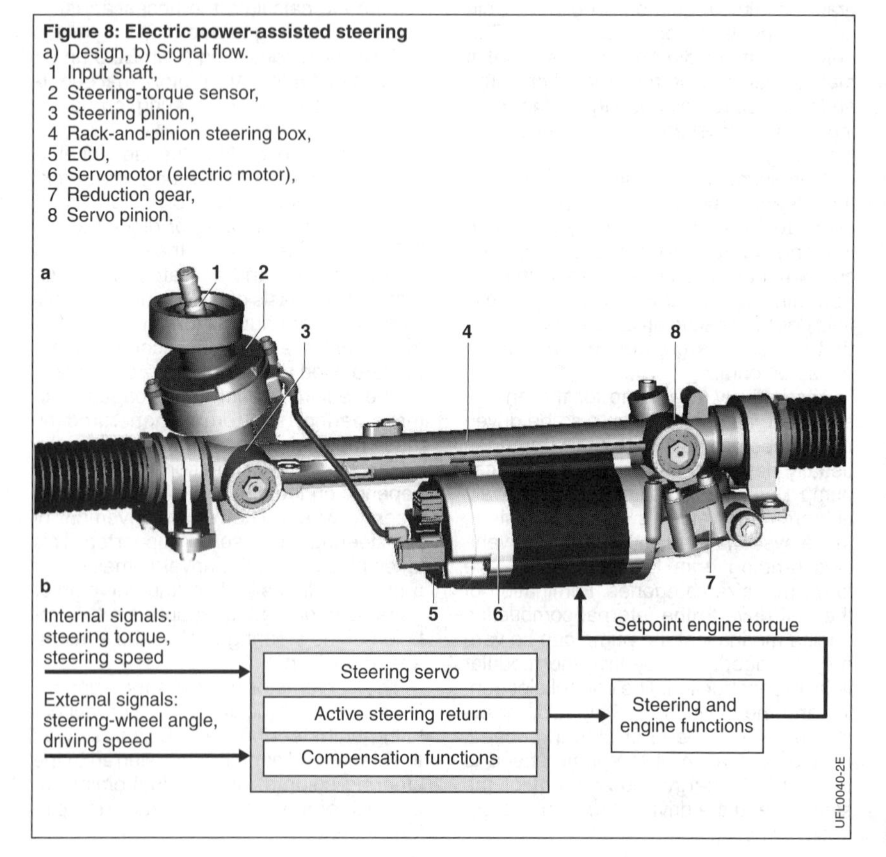

Figure 8: Electric power-assisted steering
a) Design, b) Signal flow.
1 Input shaft,
2 Steering-torque sensor,
3 Steering pinion,
4 Rack-and-pinion steering box,
5 ECU,
6 Servomotor (electric motor),
7 Reduction gear,
8 Servo pinion.

The angle superimposition independent of the driver's steering angle is currently effected by two different technical solutions.

Figure 9: Planetary-gear set of superimposed steering
1 Valve,
2 Electromagnetic lock,
3 Worm,
4 Electric motor,
5 Rack,
6 Planetary gear,
7 Worm gear.

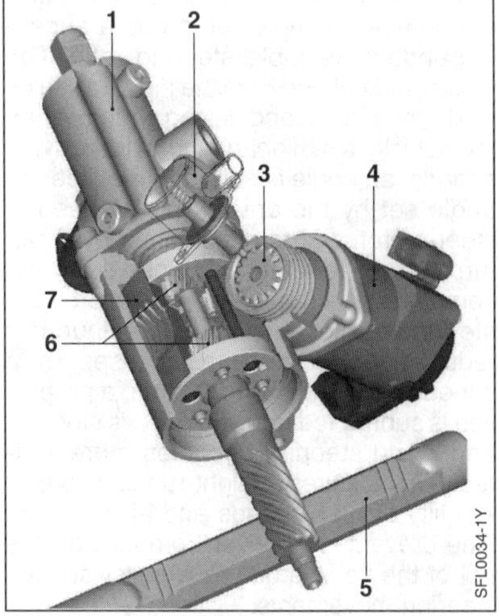

Planetary gears
A twin planetary-gear set with different gear ratios of the gear stages is integrated into a common planetary-gear carrier in the steering train (Figure 9). This means there is always a mechanical link between the steering wheel and the steered wheels. The different gear ratios of the gear stages mean that turning the planetary-gear carrier sets the additional steering angle. The angle setting is made by a electric motor that drives the worm gear of the planetary-gear carrier.

Harmonic drive
The steering-angle superimposition unit (Figure 10) consists in this case of a harmonic drive and an electric motor with a hollow shaft (Figure 11). The highly compact design enables it to be integrated into the steering column without compromising the requirements with regard to installation space and crash behavior. The shaft at the steering-wheel end is positively connected to a flex spline. The rotary movement of the steering wheel is transmitted via the toothing to an internal gear (circular spline) for the output shaft. An elliptical inner rotor (shaft generator) located in the flex spline, which is driven by the electric motor, generates the superimposed steering angle via the different number of teeth between the flex spline and the circular spline. Here, too, there is

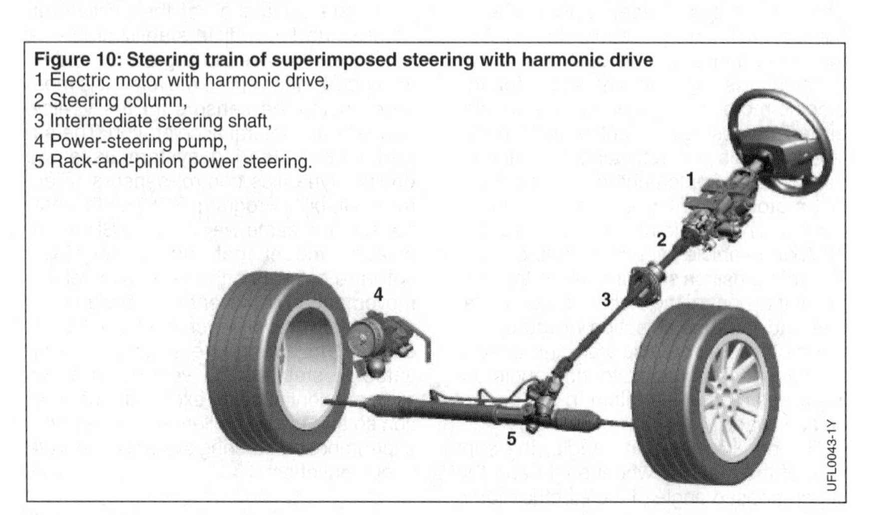

Figure 10: Steering train of superimposed steering with harmonic drive
1 Electric motor with harmonic drive,
2 Steering column,
3 Intermediate steering shaft,
4 Power-steering pump,
5 Rack-and-pinion power steering.

Figure 11: Actuator of superimposed steering with harmonic drive
1 Input shaft,
2 Output shaft,
3 Electric motor,
4 Rotor-position sensor,
5 Elliptical inner rotor (shaft generator),
6 Internal gear (circular spline),
7 Flex spline.

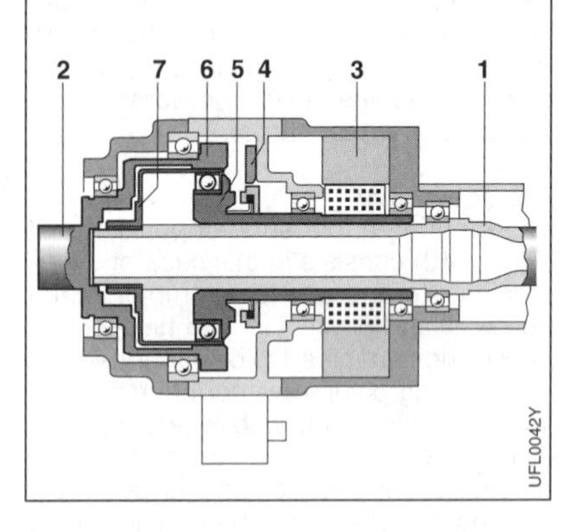

always a mechanical link between the steering wheel and the steered wheels via the toothing of the harmonic drive.

In the passive state, the electric motor is blocked by an electromechanical lock, thereby ensuring direct mechanical through-drive for the steering movement.

Activation concept
The superimposed-steering ECU checks the plausibility of the required sensor information and evaluates this information. It calculates the setpoint angle for the electric motor and generates via an integrated driver stage the pulse width-modulated signals for activating the electric motor. This is implemented as a brushless DC motor with integrated rotor position sensor. The maximum motor current is 40 A at a vehicle system voltage of 12 V. The rotor position sensor enables the control unit to control the electronic commutation and thus the direction of rotation of the motor. It also calculates and checks the total set additional steering angle using a summation algorithm in the control unit software.

The effective steering angle, the sum total of the steering-wheel angle and the superimposed angle of the electric motor, is calculated by the ECU, and made available on the vehicle communication bus to the partner ECUs.

Setpoint value
The setpoint value for the effective steering angle formed in the superimposed-steering ECU is made up of the partial setpoint value for steering comfort and the partial setpoint value for vehicle stabilization. The signals required to calculate these variables are read in by the control unit via the CAN bus.

The partial setpoint value for the steering comfort is implemented as a speed-dependent, variable steering ratio. The value is calculated from the input variables of steering angle and driving speed. When the vehicle is stationary and at low driving speeds, an angle is added to the steering angle set by the driver. This makes the steering ratio more direct. The driver can turn the wheels fully with less than one complete steering wheel revolution. This steering-angle addition is continuously reduced as driving speed increases. From speeds of roughly 80–90 km/h, a proportion is subtracted from the driver's steering angle and steering becomes more indirect. This ensures straight-running vehicle stability at high speeds and at the same time prevents the driver from losing control of the vehicle due to excessively fast steering movements.

For the calculation of the partial setpoint value for vehicle stabilization – in addition to the steering angle and the driving speed – the vehicle movement is measured using the sensors for the vehicle yaw rate and lateral acceleration. The superimposed-steering system uses the driving-dynamics control sensors (Electronic Stability Program, ESP) for this purpose. In the same way as the ESP, a calculation model that runs in the ECU software calculates the reference vehicle movement. In the event of a deviation of the actual vehicle movement from the reference movement, the steering is activated to stabilize the vehicle. The two systems continuously exchange information so that the controllers of ESP and the superimposed-steering system cooperate to optimal effect.

Safety concept
All of the internal and external signals that are used are continuously monitored in the control unit and their plausibility checked. If a sensor signal is no longer plausible, the additional steering function on which it is based is first disabled. For example, if the yaw-rate sensor that measures the rotation of the vehicle around its vertical axis (yaw rate) fails, the yaw-rate control of the superimposed-steering system is disabled. The variable steering ratio remains active.

If safe activation of the electric motor is no longer possible due to a fault, the system is completely shut down and direct steering-wheel through-drive is ensured by self-inhibiting of the gear stage and by the locking pin of the electromechanical lock. This fallback level is also active when the internal-combustion engine is shut down or there is no electrical supply voltage, thus permitting the vehicle to be towed, for example.

Power-assisted steering systems for commercial vehicles

Power-assisted steering with all-hydraulic transmission

Hydrostatic steering systems are hydraulic power-assisted steering systems. The steering force of the driver is hydraulically boosted and transmitted exclusively by hydraulic means to the steered wheels. Because there is no mechanical connection, the maximum permissible speed is limited by national regulations. In Germany, this is 25 km/h. Depending on the system configuration and emergency steering properties, approval up to a speed of 62 km/h is possible. Use of these systems is therefore confined to machinery and special vehicles.

Single-circuit power-assisted steering system for commercial vehicles

Commercial vehicles are usually equipped with ball-and-nut power steering (Figure 12). The control valve is integrated into the steering box and together with the steering worm forms a single unit. The rotary movement of the steering wheel is transmitted via an endless ball chain to the ball-and-nut. Short toothing on the ball-and-nut meshes with the toothing on the segment shaft. The generated rotary movement of the segment shaft is transmitted via a steering arm to the steering linkage and the steered wheels.

The servo force is applied, as with rack-and-pinion power steering, by a rotary-slide valve. The working cylinder is formed by a sealing surface between the ball-and-nut housing and the steering box. Because no additional lines are required outside the housing, a robust and compact steering box with high power output is created.

Dual-circuit power-assisted steering system for heavy-duty commercial vehicles

Dual-circuit steering systems (Figure 13) are required when the operating forces needed at the steering wheel exceed the statutory requirements in ECE-R79 [1] if the power-assistance system fails. These steering systems feature hydraulic redundancy. Both steering circuits in these systems are functionally tested by means of flow indicators and a fault is indicated to the driver. The pumps for supplying the independent steering circuits must be driven in various ways (e.g. engine-dependent, vehicle-speed-dependent, or electrically). If a circuit fails, caused for example by a fault in the steering system or engine failure, the vehicle can be steered with the still operational redundant circuit in accordance with the statutory requirements.

Dual-circuit systems usually take the form of ball-and-nut power steering with an integrated second steering valve. This second valve controls an additionally installed working cylinder and thereby provides the redundancy in relation to the existing servo system in the ball-and-nut steering.

References

[1] ECE-R79: Uniform provisions concerning the approval of vehicles with regard to steering equipment.

Figure 13: Dual-circuit power-assisted steering
1 Steering pump 1, 2 Steering pump 2,
3 Oil reservoir 1, 4 Oil reservoir 2,
5 Working cylinder, 6 Left cylinder chamber,
7 Right cylinder chamber,
8 Dual-circuit steering valve.

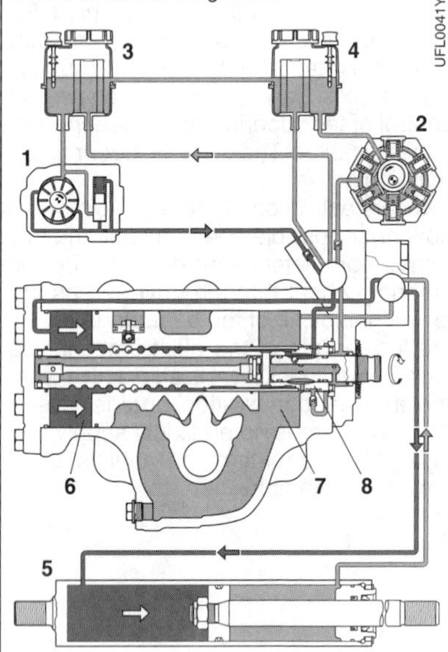

Figure 12: Ball-and-nut power steering
a) Steering assembly, b) Rotary distributor cross-section (enlarged), c) Oil supply (energy source).
1 Housing, 2 Piston, 3 Torsion bar, 4 Rotary slide (steering shaft), 5 Control bushing (worm),
6 Segment shaft, 7 Pressure-limiting valve, 8 Replenishing valve, 9 Inlet slot,
10 Return slot, 11 Axial groove, 12 Return groove,
13 Vane pump, 14 Pressure- and oil-flow-limiting valve,
15 Oil reservoir.

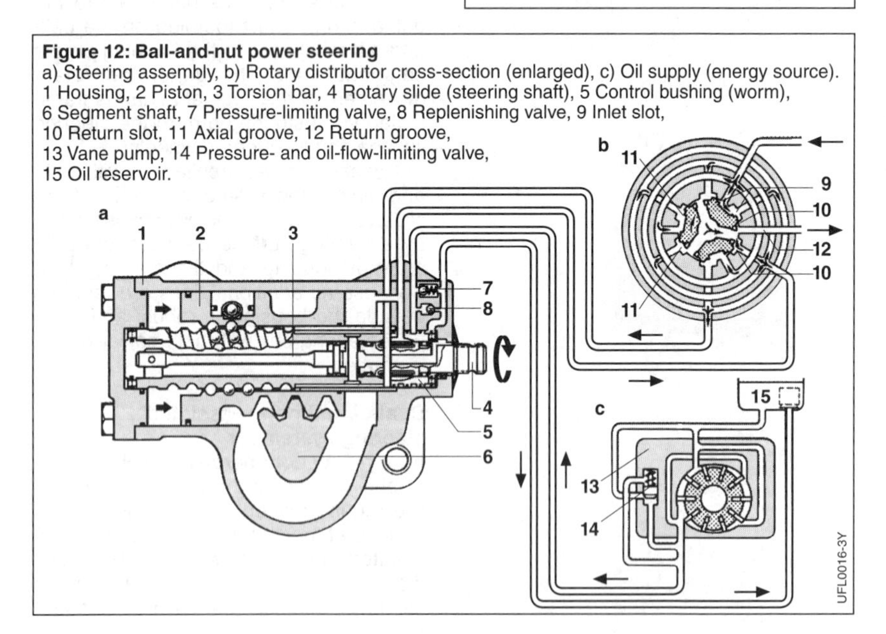

Brake systems

Definitions and principles

(based on ISO 611 [1] and DIN 70024 [2])

Brake equipment
All the vehicle brake systems whose functions are to reduce vehicle speed or bring the vehicle to a halt, or to hold the vehicle stationary if already halted.

Brake systems
Service brake system
All the elements, the action of which may be regulated, allowing the driver to reduce, directly or indirectly, the speed of a vehicle during normal driving or to bring the vehicle to a halt.

Secondary brake system
All the elements, the action of which may be regulated, allowing the driver to reduce, directly or indirectly, the speed of a vehicle or to bring the vehicle to a halt in case of failure of the service brake system.

Parking brake system
All the elements allowing the vehicle to be held stationary mechanically even on an inclined surface, and particularly in the absence of the driver.

Continuous-operation brake system
System of components which allows the driver to reduce the vehicle's speed or descend a long downhill gradient at a virtually constant speed with practically no wear to the friction brakes. A continuous-operation brake system may incorporate one or more retarders.

Automatic brake system
All the elements which automatically brake the trailer as a result of intended or accidental separation from the tractor vehicle.

Electronic brake system (ELB, EHB)
Brake system controlled by an electrical signal generated and processed by the control transmission system. An electrical output signal controls components which generate the application force.

Component parts
Energy-supplying device
Parts of a brake system which supply, regulate and, if necessary, condition the energy required for braking. It terminates at the point where the transmission device starts, where the various circuits of the brake systems, including the circuits of accessories if fitted, are isolated either from the energy supplying device or from each other.

The energy source is that part of an energy-supplying device which generates the energy. It may be located remotely from the vehicle (e.g. in the case of a compressed-air braking system for a trailer) or may be the muscular force of the driver.

Control device
Parts of a brake system which initiate the operation and control the effect of this brake system. The control signal can be conveyed within the control device by, for example, mechanical, pneumatic, hydraulic or electrical means, including the use of auxiliary energy or non-muscular force.

The control device is defined as starting at the component to which the control force is directly applied. It can be operated:
- By direct application of force by the driver by hand or foot
- By indirect action of the driver or without any action (only in the case of trailers)
- By varying the pressure in a connecting line, or the electric current in a cable, between the tractor vehicle and the trailer at the time when one of the brake systems on the tractor vehicle is operated, or if it fails
- By the inertia of the vehicle or by its weight or of one of its main component parts

The control device is defined as ending at the point at which the braking energy is distributed, or where part of the energy is diverted to control braking energy.

Transmission device
Parts of a brake system which transmit the energy distributed by the control device. It starts either at the point where the control device terminates or at the point where the energy supplying device terminates. It terminates at those parts of the brake system in which the forces opposing the vehicle's movement, or its tendency towards movement, are generated. It can, for example, be mechanical, hydraulic, pneumatic (pressure above or below atmospheric), electric, or combined (for example hydromechanical, hydropneumatic).

Brake
Parts of a brake system in which the forces opposing the vehicle's movement, or its tendency towards movement, are developed, such as friction brakes (disk or drum) or retarders (hydrodynamic or electrodynamic retarders, exhaust brakes).

Auxiliary device of the tractor vehicle for a trailer
Parts of a brake system on a tractor vehicle which are intended to supply energy to, and control, the brake systems on the trailer. It comprises the components between the energy supplying device of the tractor vehicle and the supply-line coupling head (inclusive), and between the transmission device(s) of the tractor vehicle and the control-line coupling head (inclusive).

Brake-system types relating to the energy supplying device
Muscular-energy brake system
Brake system in which the energy necessary to produce the braking force is supplied solely by the physical effort of the driver.

Energy-assisted brake system
Brake system in which the energy necessary to produce the braking force is supplied by the physical effort of the driver and one or more energy supplying devices.

Non-muscular-energy brake system
Brake system in which the energy necessary to produce the braking force is supplied by one or more energy-supplying devices excluding the physical effort of the driver. This is used only to control the system.

Note: A brake system in which the driver can increase the braking force, in a state of totally failed energy, by muscular effort acting on the system, is not included in the above definition.

Inertia brake system
Brake system in which the energy necessary to produce the braking force arises from the approach of the trailer to its tractor vehicle.

Gravity brake system
Brake system in which the energy necessary to produce the braking force is supplied by the lowering of a component part of the trailer (e.g. trailer drawbar) due to gravity.

Definitions of brake systems relating to the arrangement of the transmission device
Single-circuit brake system
Brake system having a transmission device embodying a single circuit. The transmission device comprises a single circuit if, in the event of a failure in the transmission device, no energy for the production of the application force can be transmitted by this transmission device.

Multi-circuit brake system
Brake system having a transmission device embodying several circuits. The transmission device comprises several circuits if, in the event of a failure in the transmission device, energy for the production of the application force can still be transmitted, wholly or partly, by this transmission device.

Definitions of brake systems relating to vehicle combinations

Single-line brake system
Assembly in which the brake systems of the individual vehicles act in such a way that the single line is used both for the energy supply to, and for the control of, the brake system of the trailer.

Dual- or multi-line brake systems
Assembly in which the brake systems of the individual vehicles act in such a way that several lines are used separately and simultaneously for the energy supply to, and for the control of, the brake system of the trailer.

Continuous brake system
Combination of brake systems for vehicles forming a road train. Characteristics:
- From the driving seat, the driver can operate a directly operated control device on the tractor vehicle and an indirectly operated control device on the trailer by a single operation and with a variable degree of force.
- The energy used for the braking of each of the vehicles forming the combination is supplied by the same energy source (which may be the muscular effort of the driver).
- Simultaneous or suitably phased braking of the individual units of a road train.

Semi-continuous brake system
Combination of brake systems for vehicles forming a road train. Characteristics:
- The driver, from his driving seat, can gradually operate a directly operated control device on the tractor vehicle and an indirectly operated control device on the trailer by a single operation.
- The energy used for the braking of each of the vehicles forming the road train is supplied by at least two different energy sources (one of which may be the muscular effort of the driver).
- Simultaneous or suitably phased braking of the individual units of a road train.

Non-continuous brake system
Combinations of the brake systems of the vehicles forming a road train which is neither continuous nor semi-continuous.

Brake-system control lines
Wiring and conductors: These are employed to conduct electrical energy.

Tubular lines: Rigid, semi-rigid or flexible tubes or pipes used to transfer hydraulic or pneumatic energy.

Lines connecting the brake equipment of vehicles in a road train
Supply line: A supply line is a special feed line transmitting energy from the tractor vehicle to the energy accumulator of the trailer.

Brake line: A control line is a special control line by which the energy essential for control is transmitted from the tractor vehicle to the trailer.

Common brake and supply line:
Line serving equally as brake line and as supply line (single-line brake system).

Secondary-brake line: Special actuating line transmitting the energy from the tractor vehicle to the trailer essential for secondary braking of the trailer.

Braking mechanics
Mechanical phenomena occurring between the start of actuation of the control device and the end of the braking action.

Gradual braking
Braking which, within the normal range of operation of the control device, permits the driver, at any moment, to increase or reduce, to a sufficiently fine degree, the braking force by operating the control device. When an increase in braking force is obtained by the increased action of the control device, an inverse action must lead to a reduction in that force.

Brake-system hysteresis: Difference in control forces between application and release at the same braking torque.

Brake hysteresis: Difference in application force between application and release at the same braking torque.

Forces and torques

Control force F_c: Force exerted on the control device.

Application force F_s: On friction brakes, the total force applied to a brake lining and which causes the braking force by the effect of friction.

Braking torque: Product of frictional forces resulting from the application force and the distance between the points of application of these forces and the axis of rotation of the wheels.

Total braking force F_f: Sum of the braking forces at the tire contact patches of all the wheels and the ground, produced by the effect of the brake system, and which oppose the movement or the tendency of the vehicle to move.

Braking-force distribution: Specification of braking force according to axle, given in % of the total braking force F_f. Example: front axle 60%, rear axle 40%.

Figure 1: Times and deceleration during braking to a stop

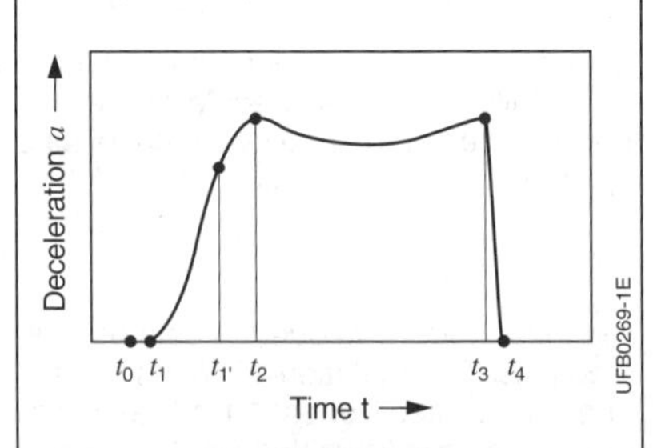

before t_0: Reaction time,

t_0: Initial application of force on control device,
t_1: Start of deceleration,
t_1': End of pressure build-up time,
t_2: Fully developed deceleration,
t_3: End of maximum retardation,
t_4: End of braking operation (vehicle stationary),

$t_1 - t_0$: Initial response time,
$t_1' - t_1$: Pressure build-up time,
$t_3 - t_2$: "Mean maximum retardation" range,
$t_4 - t_1$: Active braking time,
$t_4 - t_0$: Total braking time.

Brake coefficient C^*: Defines the relationship between the total peripheral force of a given brake and the brake's application force.

$$C^* = F_u / F_s$$

F_u Total peripheral force,
F_s Application force.
The mean is employed when there are variations in application forces at individual brake shoes (i number of brake shoes):

$$F_s = \Sigma F_{si} / i$$

Time periods

Reaction time (see Figure 1): The time that elapses between perception of the state or object which induces the response, and the point at which the control device is actuated (t_0).

Actuating time of the control device: Elapsed time between the moment when the component of the control device (t_0) on which the control force acts starts to move, and the moment when it reaches its final position corresponding to the applied control force (or its travel). (This is equally true for application and release of the brakes).

Initial response time $t_1 - t_0$: Elapsed time between the moment when the component of the control device on which the control force acts starts to move and the moment when the braking force takes effect.

Pressure build-up time $t_1' - t_1$: Period that elapses between the point at which the braking force starts to take effect and the point at which a certain level is reached (75% of asymptotic pressure in the wheel-brake cylinder as per EU Directive 71/320/ EEC [3], Annex III/2.4).

Initial response and pressure build-up time: The sum of the initial response and pressure build-up times is used to assess how the brake system behaves over time until the moment at which full braking effect is reached.

Active braking time $t_4 - t_1$: Elapsed time between the moment when the braking force starts to take effect and the moment when the braking force ceases. If the

vehicle stops before the braking force ceases, the time when motion ceases is the end of the active braking time.
Release time: Elapsed time between the moment when the control device starts to release and the moment when the braking force ceases.

Total braking time $t_4 - t_0$: Elapsed time between the moment when the control device on which the control force acts starts to move and the moment when the braking force ceases. If the vehicle stops before the braking force ceases, the time when motion ceases is the end of the active braking time.

Braking distance s
Distance traveled by the vehicle during the total braking time. If the time when motion ceases constitutes the end of the total braking time, this distance is called the "stopping distance".

Braking work W
Integral of the product of the instantaneous braking force F_f and the elementary movement ds over the braking distance s:

$$W = \int_0^s F_f \, ds$$

Instantaneous braking power P
Product of the instantaneous total braking force F_f and the vehicle's road speed v:

$$P = F_f \, v \, .$$

Braking deceleration
Reduction of speed obtained by the brake system within the considered time t. A distinction is made between the following:

Instantaneous braking deceleration

$$a = dv/dt \, .$$

Mean braking deceleration over a period of time:
The mean braking deceleration between two points in time t_B and t_E is

$$a_{mt} = \frac{1}{t_E - t_B} \int_{t_B}^{t_E} a(t) \, dt \, ;$$

This means that:

$$a_{mt} = \frac{v_E - v_B}{t_E - t_B} \, ,$$

where v_B and v_E are the vehicle speeds at the times t_B and t_E.

Mean braking deceleration over a specific distance:
The mean braking deceleration over the distance between two points s_B and s_E is:

$$a_{ms} = \frac{1}{s_E - s_B} \int_{s_B}^{s_E} a(s) \, ds \, ;$$

This means that:

$$a_{ms} = \frac{v_E^2 - v_B^2}{2\,(s_E - s_B)}$$

where v_B and v_E are the vehicle speeds up to the points s_B and s_E.

Mean braking deceleration over the total braking distance:
The mean braking deceleration is calculated according to the equation:

$$a_{ms0} = \frac{-v_0^2}{2\,s_0},$$

where v_0 relates to the time t_0 (special instance of a_{ms} where $s_E = s_0$).

Mean fully developed deceleration d_m:
Mean fully developed deceleration over the distance determined by the conditions $v_B = 0.8\,v_0$ and $v_E = 0.1\,v_0$ thus:

$$d_m = \frac{v_B^2 - v_E^2}{2\,(s_E - s_B)}$$

The mean fully developed deceleration is used in ECE Regulation 13 [6] as a measure of the effectiveness of a brake system. Since positive values for d_m are used here, the mathematic sign has been reversed in this case. (In order to establish a relationship between braking distance and braking deceleration, braking deceleration must be expressed as a function of the distance traveled.)

Braking factor z
Ratio between the total braking force, F_f, and the permissible total static weight, G_s, exerted on the axle or axles of the vehicle:

$$z = \frac{F_f}{G_s} \, .$$

Legal regulations

General Certification for a vehicle with regard to its brake system may only be granted when the brake system complies with the following regulations:

- §41 StVZO [4] (German Road Vehicle Registration Regulation) in conjunction with §72 StVZO [5] and the associated directives.
- Council Directive of the European Community (RREG) 71/320/EEC [3], associated Amending Directives and Annexes.
- ECE Regulations R13 [6], R13H [7], R78 [8].

In §41 StVZO, the requirements placed on the brake system differ depending on the type, gross weight rating, application, date of registration, and type-determined top speed. In the EC Directives, the requirements are allocated to individual vehicle categories. The vehicle categories are as follows (see also Road-vehicle systematics):

- M1, M2, M3: Passenger cars with at least four wheels
- N1, N2, N3: Commercial vehicles with at least four wheels
- O1, O2, O3, O4: Trailers and semitrailers
- L1, L2, L3, L4: Motorcycles, threewheelers

The values stipulated in §41 StVZO with regard to mean fully developed deceleration do not apply, for example, to the recurring inspections required in Germany for registered vehicles on the road (general inspection, safety inspection). In these inspections, the requirements of §29 StVZO [9], s. 1, Annex VIII in conjunction with Annex VIIIa, Guideline for performing the general inspection and Guideline for performing the safety inspection, apply.

The requirements in §41 StVZO and ECE-R13H with regard to brake equipment are essentially identical. However, ECE Regulations R13, R13H and R78 have been further updated and also contain, for example, regulations covering electronically controlled brake systems. The prescribed braking effects must be determined in accordance with Directive 71/320/EEC, s. 1.1.2, Annex II, amended by Directive 98/12/EC [10], or in accordance with §41 StVZO, s. 12.

Requirements placed on brake systems

(as per §41 StVZO, EC Directive 71/320/EEC, ECE-R13, as at 2009)

Category M and N vehicles must comply with the provisions which pertain to the service, secondary and parking brake systems. The brake systems may have common components. Such vehicles must have at least two mutually independent control devices for the brake systems, one of which must be lockable. The control devices must be fitted with separate transmission devices, each must be able to continue operating when the other fails. The distribution of braking force between the individual axles is prescribed and must be sensible. If a malfunction occurs, it must be possible to achieve the prescribed secondary braking effect with the remaining operational part of the brake system or with the vehicle's other brake system without the vehicle departing from its lane.

Vehicles in categories M2 and N2 and above must be fitted with automatic antilock braking facilities. Regulation EC 661/2009 [11] stipulates that from 1 November 2011 all new vehicle models and from 2014 all new category M1 and N1 vehicles to be brought onto the road must be equipped with an electronic driving-dynamics control system (ESP, Electronic Stability Program). This regulation also applies, with the exception of off-road vehicles as per Directive 2007/46/EC [12], Annex II, Part A, to vehicles in the following categories:

- M2 and M3, except vehicles with more than three axles, articulated buses and buses of category 1 or A
- N2 and N3, except vehicles with more than three axles, tractor units with a total weight of between 3.5 and 7.5 t and special-purpose vehicles as per Directive 2007/46/EC, Annex II, Part A
- O3 and O4, with pneumatic suspension, except vehicles with more than three axles, trailers for heavy transports and trailers with areas for standing passengers

The implementation of this regulation for the vehicle categories, except categories M1 and N1, is defined in Regulation EC 661/2009 [11], Annex V, and set out in Tables 1 and 2.

Continuous-operation brake systems
Continuous-operation brake systems are additionally used to relieve the strain on the service brake on long downhill gradients. Category M3 vehicles for local and long-distance duty (buses weighing more than 5.5 t (except city buses)) and other category N2 and N3 vehicles with a gross weight rating of more than 9 t (Directive 71/320/EEC, §41 StVZO, s. 15) must be equipped with a retarder. Exhaust brakes or similar facilities are classed as retarders. The retarder must be designed to hold the vehicle when fully laden when driving on a downhill gradient of 7% and a distance of 6 km at a speed of 30 km/h.

Category O trailers
Category O1 trailers are not required to have their own brake system; a securing connection to the tractor vehicle is sufficient. Category O2 trailers and above must be fitted with service and parking brake systems, which may have common components. The distribution of braking force between the individual axles is prescribed in Directive 71/320/EEC. It must be sensibly distributed between the axles.

Category O3 trailers and above (ECE), as well as trailers and semitrailers with a gross weight rating of more than 3.5 t and a type-determined top speed of more than 60 km/h (§41b StVZO, s. 2), must be equipped with an antilock braking facility. Semitrailers only need to be equipped with ABS if the gross weight rating reduced by the fifth-wheel load exceeds 3.5 t.

Category O3 trailers (existing types) which are registered for use on public roads after 11.07.2014 or 01.11.2014 must be equipped with an electronic driving-dynamics control system (Electronic Stability Program, ESP). For new types, this regulation will already apply from 01.11.2011 or 11.07.2012 (Regulation EC 661/2009 [11], Annex V, see Tables 1 and 2).

Category O2 trailers and below must be equipped with inertia brake systems. The trailer must brake automatically if it becomes decoupled from the tractor vehicle while moving, or (for trailers weighing less

Table 1: Implementation data for new types as per EC 661/2009, Annex V [11]

Vehicle category		Date of implementation
M2		11. 07. 2013
M3	Category III	1. 11. 2011
M3	< 16 tons, pneumatic transmission	1. 11. 2011
M3	Category II and B, hydraulic transmission	11. 07. 2013
M3	Category III, hydraulic transmission	11. 07. 2013
M3	Category III, pneumatic signal and hydraulic energy transmission	11. 07. 2014
M3	Category II, pneumatic signal and hydraulic energy transmission	11. 07. 2014
M3	(miscellaneous)	1. 11. 2011
N2	Hydraulic transmission	11. 07. 2013
N2	Pneumatic signal and hydraulic energy transmission	11. 07. 2014
N2	(miscellaneous)	11. 07. 2012
N3	Tractor units with 2 axles	1. 11. 2011
N3	Tractor units with 2 axles and pneumatic signal transmission (ABS)	1. 11. 2011
N3	3 axles with electronic signal transmission (EBS)	1. 11. 2011
N3	2 and 3 axles with pneumatic signal transmission (ABS)	11. 07. 2012
N3	(miscellaneous)	1. 11. 2011
O3	Combined axle load between 3.5 and 7.5 t	11. 07. 2012
O3	(miscellaneous)	1. 11. 2011
O4		1. 11. 2011

than 1.5 t) it must be equipped with a securing connection to the tractor vehicle.

Category L vehicles
Motorized two- and three-wheeled vehicles must be equipped with 2 mutually independent brake systems. In the case of duty category L5 three-wheel vehicles, the two brake systems must both act on all the wheels. A parking brake system must be fitted.

Tractor vehicles and trailers with compressed-air brake systems
The compressed-air connections between the individual vehicles must be of the dual- or multi-line design. This ensures that the compressed-air brake system in the trailer can also be refilled during the braking operation. When the service brake system on the tractor unit is operated, the service brake system on the trailer must also be operated with a variable degree of force. If a fault occurs in the service brake system of the tractor unit, that part of the system not affected by the fault must be capable of braking (controlling) the trailer with a variable degree of force. If one of the connecting lines between tractor unit and trailer is interrupted or develops a leak, it must still be possible to brake the trailer, or it must brake automatically.

The energy accumulators of the service brake systems must be designed such that, after eight full operations of the service brake, at least the requested secondary braking effect is furnished on the ninth braking. The energy accumulators must not be replenished during this test. The braking effect of the individual vehicles is prescribed as a function of the pressure at the "Brake" coupling head in Directive 71/320/EEC.

Vehicles with antilock braking facilities
Antilock braking facilities must comply with Directives 71/320/EEC, Annex X and ECE-R13, Annex 13. An antilock braking facility is part of a service brake system which automatically controls the slip in the direction of wheel rotation at one or more of the vehicle's wheels during the braking operation. The requirements placed on the antilock braking facility differ, depending on the category, for vehicles ABS categories 1, 2 and 3, for trailers ABS categories A and B.

Table 2: Implementation data for new vehicles as per EC 661/2009, Annex V [11]

Vehicle category		Date of implementation
M2		11. 07. 2015
M3	Category III	1. 11. 2014
M3	< 16 tons, pneumatic transmission	1. 11. 2014
M3	Category II and B, hydraulic transmission	11. 07. 2015
M3	Category III, hydraulic transmission	11. 07. 2015
M3	Category III, pneumatic signal and hydraulic energy transmission	11. 07. 2016
M3	Category II, pneumatic signal and hydraulic energy transmission	11. 07. 2016
M3	(miscellaneous)	1. 11. 2014
N2	Hydraulic transmission	11. 07. 2015
N2	Pneumatic signal and hydraulic energy transmission	11. 07. 2016
N2	(miscellaneous)	11. 07. 2014
N3	Tractor units with 2 axles	1. 11. 2014
N3	Tractor units with 2 axles and pneumatic signal transmission (ABS)	1. 11. 2014
N3	3 axles with electronic signal transmission (EBS)	1. 11. 2014
N3	2 and 3 axles with pneumatic signal transmission (ABS)	11. 07. 2014
N3	(miscellaneous)	1. 11. 2014
O3	Combined axle load between 3.5 and 7.5 t	11. 07. 2014
O3	(miscellaneous)	1. 11. 2014
O4		1. 11. 2014

Essential requirements placed on the anti-lock braking facility (Category 1) are:

- Locking of the directly controlled wheels under braking must be prevented at speeds of over 15 km/h on all road surfaces.
- Directional stability and maneuverability must be maintained. Under a μ-split condition (extremely different coefficients of friction between the left and right wheels), steering corrections of 120° during the first two seconds and 240° in total are permissible.
- There must be a special visual warning system (yellow warning signal) to indicate electrical faults.
- Motor vehicles (except Categories M1 and N1) with ABS which are equipped to tow a trailer with ABS must be fitted with a separate visual warning system (yellow warning signal) for the trailer. Transfer must be effected via pin 5 of the electrical plug-in connection as per ISO 7638 [13].
- The energy accumulators of the service brake system in vehicles with ABS must be designed such that the prescribed secondary braking effect is still achieved even after a controlled braking operation of longer duration ($t = v_{max}/7$, at least 15 seconds) and then four uncontrolled full-braking operations without energy replenishment.

Requirements and test conditions

The required values and test conditions must be applied during the test in accordance with §19 [14] and §20 StVZO [15]. Departures from the test method described in Directive 71/320/EEC, Annex II, s. 1.1.2, last amended by Directive 98/12/EC and the test method specified in §41 StVZO, s. 12, are permitted – especially during verification checks as per §29 StVZO – if the condition and the effect can be ascertained by other means (§41 StVZO, s. 12). When vehicles to be newly registered are tested, a higher braking deceleration corresponding to the usual easing of the braking effect must be achieved; furthermore, sufficient state-of-the-art continuous brake duty must be guaranteed for longer stretches of downhill driving (Tables 3 to 6).

Minimum retardation and max. permissible control forces during the general inspection as per §29 StVZO

Tables 5, 6 and 7 provide an overview of the requested retardation values and the maximum control forces at the control device during the safety and general inspections as per §29 StVZO [9], Annex VIII and Annex VIIIa. The required retardation values are determined on test benches and only in exceptional cases in road tests. The required values are maximum values, because the time response which is needed to determine the mean fully developed deceleration is not measured in these recurring inspections.

Table 5 applies to vehicles which have been tested and approved after 1 January 1991 (entry into force of §41 StVZO, s. 18 in conjunction with §72 StVZO, s. 2). The values quoted in parentheses apply to vehicles which were tested and registered for road use before 1 January 1991. If no value in parentheses is quoted for these vehicles, the values in Table 6 apply. Table 6 applies to vehicles which were tested and approved before 1 January 1991 and for which no values (values in parentheses) are quoted in Table 5. The values quoted in Table 7 apply to the safety inspection.

Table 3: Required values for mean fully developed deceleration as per §41 StVZO [4] (excerpt: as at 2009)

Vehicle type and description	Service brake system	Parking brake system
Motor vehicles, except motorcycles, type-determined top speed > 25 km/h, which were registered for the first time for road use after 01.01.2001.[1,7]	5.0 m/s^2	1.5 m/s^2
Motor vehicles, except motorcycles, type-determined top speed < 25 km/h, which were registered for the first time for road use before 01.01.2001.[1]	3.5 m/s^2	1.5 m/s^2
Motor vehicles, except motorcycles, type-determined top speed > 25 km/h, which were registered for the first time for road use before 01.01.2001.[1]	2.5 m/s^2	1.5 m/s^2
Motor vehicles, except motorcycles, type-determined top speed < 25 km/h, which were registered for the first time for road use before 01.01.2001.[1]	1.5 m/s^2	1.5 m/s^2
Two- or multi-axle trailers, except two-axle trailers with a wheelbase of less than 1.0 m, which were registered for the first time for road use after 01.01.2001.[2,6]	5.0 m/s^2	[5]
Two- or multi-axle trailers, except two-axle trailers with a wheelbase of less than 1.0 m, which were registered for the first time for road use after 01.01.2001. For trailers after motor vehicles with a speed < 25 km/h (operating instructions) if the trailer is identified for a top speed of not more than 25 km/h.[2,4]	3.5 m/s^2	[5]
Semitrailers which were registered for the first time for road use after 01.01.2001.[3,6]	4.5 m/s^2	[5]
Semitrailers which were registered for the first time for road use before 01.01.2001.[3]	2.5 m/s^2	[5]

[1] For all other motor vehicles – except motorcycles – §41 StVZO, s. 4 applies in the version valid before 01. 04. 2000. See also §72 StVZO, s. 2.
[2] For all other trailers – except two-axle trailers – with a wheelbase of less than 1.0 m – §41 StVZO, s. 9 applies in the version valid before 01.04.2000. See also §72 StVZO, s. 2.
[3] For all other semitrailers §41 StVZO, s. 9 applies in the version valid before 01.04.2000. See also §72 StVZO, s. 2.
[4] Identification must be completed in accordance with the requirements in §58 StVZO.
[5] The brake must be lockable. The locked brake must use solely mechanical means to secure the fully laden trailer against rolling on uphill and downhill gradients of 18% on dry roads (§41 StVZO, s. 9).
[6] The trailer's brake system must automatically bring the trailer to a stop on uphill and downhill gradients of 18% when the trailer becomes decoupled from the towing vehicle (§41 StVZO, s. 9).
[7] On motor vehicles, except motor vehicles as per §30a StVZO, s. 3, which were registered for the first time for road use after 01.01.2001, if a part of the brake system malfunctions, it must be possible to achieve with the remaining operational part of the brake system at least 44% of the braking effect prescribed in §41 StVZO, s. 4 for the service brake without the vehicle departing from its lane (§41 StVZO, s. 4a and §72 StVZO,. 2).

Table 4: Requirements and test conditions for brake systems as per Directive 71/320/EEC [3], ECE-R13 [6], ECE-R13H [7]

Vehicle category		Passenger cars and buses			Commercial vehicles			Trailers			
		M_1	M_2	M_3	N_1	N_2	N_3	O_1	O_2	O_3	O_4
Service brake system		Acting on all wheels, prescribed braking-force distribution to the axles						No braking system or as O_2	Inertia braking system or as O_3		
ABS as per EU Dir. or ECE[1] ($v_{max} \geq 25$ km/h)		–	+	+	–	+	+	–	–	+	+
Type 0 test (drive disengaged)											
Test speed	km/h	80.0	60.0	60.0	80.0	60.0	60.0	–	60	60	60
Braking distance	≤ m	50.7	36.7	36.7	61.2	36.7	36.7		$z \geq 0.50$, semitrailer: $z \geq 0.45$		
Braking-distance formula		$0.1v + \frac{v^2}{150}$	$0.15v + \frac{v^2}{130}$								
Mean fully developed deceleration	≥ m/s²	5.8	5.0								
Actuating force	≤ N	500	700						at ≤ 6.5 bar		
Type 0 test (drive engaged)		Behavior of vehicle under braking from 30 % – 80 % v_{max} and braking efficiency									
Test speed $v = 80\,\%\ v_{max}$, but	≤ km/h	160	100	90	120	100	90	–	–	–	–
Braking distance	≤ m	212.9	111.6	91.8	157.1	111.6	91.8				
Braking-distance formula		$0.1v + \frac{v^2}{130}$	$0.15v + \frac{v^2}{103.5}$								
Mean fully developed deceleration	≥ m/s²	5.0	4.0								
Actuating force	≤ N	500.0	700								
Type I test		Repeated braking at 3 m/s² fully laden, drive engaged							Continuous braking, fully laden, 40 km/h, 7 % gradient, 1.7 km, $z \geq 0.36$ and $z \geq 60\,\%$ of the value measured in the Type 0 test at 40 km/h		
$v_1 = 80\,\%\ v_{max}$, but	≤ km/h	120	100	60	120	60	60	–			–
$v_2 = 1/2\ v_1$											
Number of braking cycles	n	15	15	20	15	20	20				
Braking cycle duration	s	45	55	60	55	60	60				
Hot-brake efficiency at end of Type I test		≥ 80 % of the braking effect prescribed for the Type 0 test (declutched) and ≥ 60 % of the braking effect achieved in the Type 0 test (declutched)									
Type II test on long descents; Hot-brake efficiency at end of Type II test		Energy corresponding to 30 km/h, 6 % downhill gradient and 6 km, fully laden, drive engaged, continuous-operation brake system operated. Measured as for Type 0 test (drive disengaged)									at 40 km/h
Braking-distance formula		$M_3{:}\ 0.15v + \frac{1.33\,v^2}{130}$			$N_3{:}\ 0.15v + \frac{1.33v^2}{115}$			–	–	–	
Braking distance	≤ m			45.8			50.6				
Mean fully developed deceleration	≥ m/s²	–	–	3.75	–	–	3.3				$z \geq 0.33$

[1] Exceptions may still be allowed for older models.

Table 4 (contd.): Requirements and test conditions for brake systems

Vehicle category		Passenger cars and buses			Commercial vehicles			Trailers			
		M_1	M_2	M_3	N_1	N_2	N_3	O_1	O_2	O_3	O_4
Type IIa test For additional retarding braking systems		Energy corresponding to 30 km/h, 7 % downhill gradient, 6 km, fully laden, only continuous-action brake system in operation. Only permitted with M_3 [2] and for towing O_4, N_3.									
Type III test		–						–	–	–	[3]
Residual braking effect After transmission system/brake circuit failure, drive disengaged								The brakes of the trailer must be fully or partially operable with graduated effect.			
Test speed	km/h	80.0	60.0	60.0	70.0	50.0	40.0				
Braking distance, laden	≤ m	150.2	101.3	101.3	152.5	80.0	52.4				
Braking distance, unladen	≤ m	178.7	119.8	101.3	180.9	94.5	52.4				
Mean fully developed deceleration, laden	≥ m/s^2	1.7	1.5	1.5	1.3	1.3	1.3				
Unladen	≥ m/s^2	1.5	1.3	1.5	1.1	1.1	1.3				
Actuating force	≤ N	700	700	700	700	700	700				
Secondary brake system (tested as for Type 0 test, drive disengaged)								The brakes of the trailer must be operable with graduated effect.			
Test speed	km/h	80.0	60.0	60.0	70.0	50.0	40.0				
Braking distance	≤ m	93.3	64.4	64.4	95.7	54.0	38.3				
Braking-distance formula		$0.1v + \frac{2v^2}{150}$	$0.15v + \frac{2v^2}{130}$		$0.15v + \frac{2v^2}{115}$						
Mean fully developed deceleration	≥ m/s^2	2.9	2.5		2.2						
Actuating force, by hand	≤ N	400	600		600						
with foot	≤ N	500	700		700						
Parking brake system (test fully laden) Holding stationary on incline (downgrade or upgrade)	≥ %	18			18			–	18		
Together with unbraked vehicles of Category O	≥ %	12			12			–	–		
Actuating force, by hand	≤ N	400	600		600			–	600		
with foot	≤ N	500	700		700			–	–		
Type 0 test [4] (drive disengaged, laden) Test speed	km/h	80	60	60	70	50	40	–			
Mean maximum retardation and deceleration prior to standstill	≥ m/s^2	1.5			1.5			–			
Automatic braking system With compressed-air systems automatic trailer braking in case of pressure loss in supply line Test speed	km/h							–	40.0		
Braking factor	≥ %							–	13.5		

[2] Except city buses, [3] Repeated braking as for Type I test for N_3. Afterwards, braking effect ≥ 40 and ≥ 60 % of level achieved in Type 0 test, [4] With parking brake system or via auxiliary control device for service brake system.

Table 5: Minimum retardation and max. permissible control forces during the general inspection as per §29 StVZO [9]

Applies to vehicles which have been tested and approved after 1 January 1991 (entry into force of §41 StVZO, s. 18 in conjunction with §72 StVZO, s. 2). The values quoted in parentheses apply to vehicles which were tested and approved before 1 January 1991. If no value in parentheses is quoted for these vehicles, the values in Table 6 apply.

Vehicle categories		Service brake system					Parking brake system		
		Braking factor in %			Manual force in N	Pedal force in N	Retardation in %	Manual force in N	Pedal force in N
M1	Pass. cars, motorhomes	50			–	500	16 (15)	400	500
M2, M3	Motor buses	50 (48)[1]			–	700	16 (15)	600	700
L5	Vehicles w/ 3 wheels up to 1,000 kg gross weight rating	45			200	500	15	200	500
N1	Trucks/tractors up to 3.5 t gross weight rating	50 (45)[2]			–	700	16 (15)	600	700
N2, N3	Trucks/tractors over 3.5 t gross weight rating	45 (43)[2]			–	700	16 (15)	600	700
O3	Trailers and semitrailers up to 25 km/h	25			–	–	15	600	–
	over 25 km/h	43 (40)			$p_m \le$ 6.5 bar	$p_m \le$ 6.5 bar	16 (15)	600	–
Lof	Tractor units $v_{max} \le 30$ km/h	35			400	600	15	400	600
	Motorcycles (when tested w/out passenger)	f	r	f&r					
L2	3 wheels / ≤ 50 km/h	15[4]	15[4]	35/40[5]	200	400	–	–	–
L3	2 wheels / > 50 km/h	35	30	50	200	500	–	–	–
L4	3 wheels / asym. / > 50 km/h	–	–	45	200	500	–	–	–
	Other motor vehicles ≤ 25 km/h	25			–	700	15	600	700
	> 25 km/h	40			–	700	16 (15)	600	700
		Continuous-action braking effect (cf. Directive 71/320/EEC Annex II)							
M_3	$G \le 10$ t	as per Type II							
M_3	$G > 10$ t	as per Type II a							
N_3		as per Type II							

G Gross weight rating, f front, r rear, f&r front and rear, p_m Pressure at the "Brake" coupling head.

1) Values for vehicles which were tested and registered for road used before 1 January 1991.

2) For vehicles with a wheelbase-related center-of-gravity height of $h/E \ge 0.5$ a retardation of 40% is sufficient

3) For category O1 trailers only if a brake system is provided,

4) Only if independent brake systems for front and rear wheels are provided,

5) 35% for asymmetrically arranged and 40% for symmetrically arranged wheels,

6) ≤ 2.5 t gross weight rating, 7) > 2.5 t gross weight rating.

Table 6: Minimum retardation and max. permissible control forces during the general inspection as per § 29 StVZO [9]
Applies to vehicles which were tested and approved before 1 January 1991 and for which no values (values in parentheses) are quoted in Table 5.

Vehicle categories	Service brake system			Secondary brake system			Parking brake system		
	Retardation in %	Manual force in N	Pedal force in N	Retardation in %	Manual force in N	Pedal force in N	Retardation in %	Manual force in N	Pedal force in N
Two-wheeled vehicles	30	250	500	–	–	–	–	–	–
Other motor vehicles ≤ 25 km/h > 25 km/h	 25 40	 – –	 80 80	 20 20	 40 [1] 60 [2]	 80 80	 20 20	 40 [1] 60 [2]	 80 80
Trailers ≤ 25 km/h > 25 km/h	 25 40	 40 40	 – –	 – –	 – –	 – –	 20 20	 60 60	 – –

Table 7: Minimum retardation and max. permissible control forces during the safety inspection as per § 29 StVZO [9]

Vehicle categories		Service brake system			Parking brake system		
		Retardation in %	Manual force in N	Pedal force in N	Retardation in %	Manual force in N	Pedal force in N
M		48	–	700	15	60	70
N		43	–	700	15	60	70
O	Trailers ≤ 25 km/h > 25 km/h	 25 40	 – $p_m \leq$ 6.5 bar	 – $p_m \leq$ 6.5 bar	 15 15	 – 60	 – –
	Other motor vehicles ≤ 25 km/h > 25 km/h	 24 40	 – –	 70 70	 15 15	 60 60 [2]	 70 70

[1] ≤ 2.5 t gross weight rating,
[2] > 2.5 t gross weight rating.

Structure and organization of brake systems

Essential requirements
The brake systems in motor vehicles must comply with the requirements of Directive 71/320/EEC, ECE-R13 Part 1, ECE-R13 Part 2 and ECE-R13 H, and other country-specific regulations. Motor vehicles must be equipped with two separate brake systems, one of which must be lockable. The brake systems must have separate control devices. In the event of a fault in the service brake system, it must still be possible for at least two wheels (not on the same side) to be braked.

Types of brake system
The brake systems comprise the service brake and parking brake systems and (in commercial vehicles and motor buses) the continuous-operation (retarder) brake system. The requested secondary brake system normally comes into play when a fault occurs in the service brake system. Special-purpose vehicles with special requirements may also have special braking functions such as a hill-climbing brake or an anti-jackknifing brake.

Type of force generation
When it comes to how the force is generated, there are three different types of system: muscular-energy, energy-assisted, and non-muscular-energy brake systems. In muscular-energy systems, the muscular force of the driver alone is effective, in energy-assisted systems this is boosted by booster systems (brake booster), and in non-muscular-energy systems, the driver's control force acts only as a control variable. The maximum required control forces are prescribed for each type of vehicle.

Transmission device
Force is transmitted from the control device to the wheel brakes by mechanical, hydraulic, pneumatic or electric means. Mechanical force transmission is only customary and prescribed for parking brake systems (§41 StVZO, s. 5).

Force transmission for the service brake system is performed via two separate brake circuits by hydraulic or pneumatic means so that at least one brake circuit remains in operation in the event of a fault.

Electrical brake operation has up to now only been used in electrically acting parking brake systems (see Electromechanical parking brake systems).

Brake-circuit configuration
Configuration of the brake circuits is governed by DIN 74000 [16]. In category M1 vehicles (passenger cars), the brake-circuit configuration is often diagonal (Figure 2b). But this is only possible in conjunction with suitable front-axle geometry (steering offset negative or neutral). In all other vehicle categories, the II configuration is used (Figure 2a). Here the front axle forms one of the brake circuits and the rear axle forms the other. All other brake-circuit configurations as per DIN 74000 are rarely used today and are therefore no longer described in these pages. The direct demand for a dual-circuit design of the transmission device is prescribed in §41 StVZO, s. 16 only for motor buses.

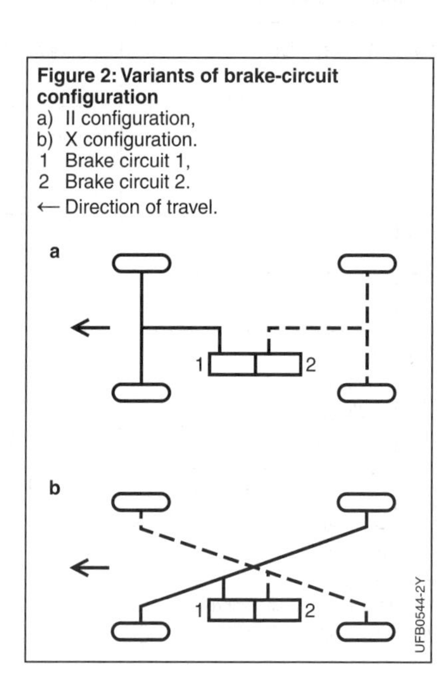

Figure 2: Variants of brake-circuit configuration
a) II configuration,
b) X configuration.
1 Brake circuit 1,
2 Brake circuit 2.
← Direction of travel.

Braking-force distribution

Directives 71/320/EEC, ECE-R13 and ECE-R13H also place requirements on braking-force distribution between the individual axles. This must be sensibly distributed in all load states between the axles. Braking-force distribution can be effected on the one hand through an assembly-related configuration of the wheel brakes and on the other hand through a vehicle-related configuration. Among other things, the center-of-gravity height, the wheelbase and the empty-empty ratio of the vehicle are taken into consideration.

In commercial vehicles, according to the diagrams in Directive 71/320/EEC, braking-force distribution is also dependent on the pressure at the "Brake" coupling head. Vehicle-related configuration of braking-force distribution is effected by the integration of a braking-force limiter or an automatically acting braking-force metering device (automatic load-sensitive braking-force metering).

In modern vehicles braking-force distribution is integrated as an additional function into the electronic wheel-slip control system (antilock braking facility, driving-dynamics control).

Assemblies

Brake systems in motor vehicles consist of the following assemblies, which differ in design depending on whether the system is hydraulic or pneumatic: energy supply, control devices, transmission devices, control facilities, wheel brakes, and auxiliary devices.

References

[1] ISO 611: Road vehicles – Braking of automotive vehicles and their trailers – Vocabulary.
[2] DIN 70024: Vocabulary for components of motor vehicles and their trailers.
[3] EC Directive 71/320/EEC: Council Directive of 26 July 1971 on the approximation of the laws of the Member States relating to the braking devices of certain categories of motor vehicles and their trailers.
[4] §41 StVZO: Brakes and wheel chocks.
[5] §72 StVZO: Entry into force and transitional provisions.
[6] ECE-R13: Standard conditions for approval of category M, N and O vehicles with regard to brakes.
[7] ECE-R13H: Standard conditions for approval of passenger cars with regard to brakes. Day of entry into force: 11 May, 1998.
[8] ECE-R78: Standard conditions for approval of category L1, L2, L3, L4 and L5 vehicles with regard to brakes.
[9] §29 StVZO: Inspection of motor vehicles and trailers.
[10] EC Directive 98/12/EC of the Commission of 27 January 1998 adapting to technical progress Council Directive 71/320/EEC on the approximation of the laws of the Member States relating to the braking devices of certain categories of motor vehicles and their trailers.
[11] Regulation (EC) No. 661/2009 of the European Parliament and of the Council of 13 July 2009 concerning type-approval requirements for the general safety of motor vehicles, their trailers and systems, components and separate technical units intended therefore.
Published in the Official Journal of the European Union L200 of 31. July 2009.
[12] Directive 2007/46/EC of the European Parliament and of the Council of 5 September 2007 establishing a framework for the approval of motor vehicles and their trailers, and of systems, components and separate technical units for such vehicles (Framework Directive).
[13] ISO 7638: Road vehicles – Connectors for the electrical connection of towing and towed vehicles.
[14] §19 StVZO: Granting and effectiveness of design certification.
[15] §20 StVZO: General Certification for types.
[16] DIN 74000: Hydraulic braking systems; dual circuit brake systems; symbols for brake circuits diagrams.

Brake systems for passenger cars and light utility vehicles

Subdivision of passenger-car brake systems

Brake systems for passenger cars and light utility vehicles must comply with the requirements of various directives and statutory provisions, e.g. 71/320/EEC [1], ECE R13 [2], ECE R13-H [3] and in Germany §41 StVZO [4] (see Brake systems, Legal regulations). The requirements with regard to functioning, effect and test methods are set out in these regulations.

The entire system is subdivided into the service-brake system, the parking-brake system and the secondary brake system.

Service-brake system

The service-brake system allows the driver to reduce with graduable effect the speed of a vehicle during normal driving or to bring the vehicle to a halt. In passenger cars and light utility vehicles it is normally designed as an energy-assisted brake system.

The driver meters the braking effect steplessly by pressing on the brake pedal. Force is transmitted to the wheel brakes via the tandem brake master cylinder to two mutually independent hydraulic transmission devices (Figure 1). The service-brake system acts on all four wheels.

Parking-brake system

The parking-brake system ("handbrake") is an independent brake system which holds the vehicle stationary, even on a gradient and especially when the driver is not in the vehicle. The holding-stationary mechanism is integrated in the wheel brake. Legal requirements stipulate that the parking brake must have a consistently mechanical connection between the control device and the wheel brake, e.g. by means of a linkage or a control cable.

The parking brake is usually actuated by a handbrake lever next to the driver's seat, and in some cases also by a pedal. In the case of electrically actuated parking-brake systems, the parking brake is locked or released by means of an

Figure 1: Hydraulic dual-circuit brake system with ABS
1 Wheel-speed sensors,
2 Wheel brakes (disk brakes, drum brakes also possible on the rear axle),
3 Hydraulic modulator (for antilock braking system or driving-dynamics control system),
4 Control device with brake booster, tandem brake master cylinder and expansion reservoir,
5 ECU (can be directly mounted on the hydraulic modulator),
6 ABS warning lamp.

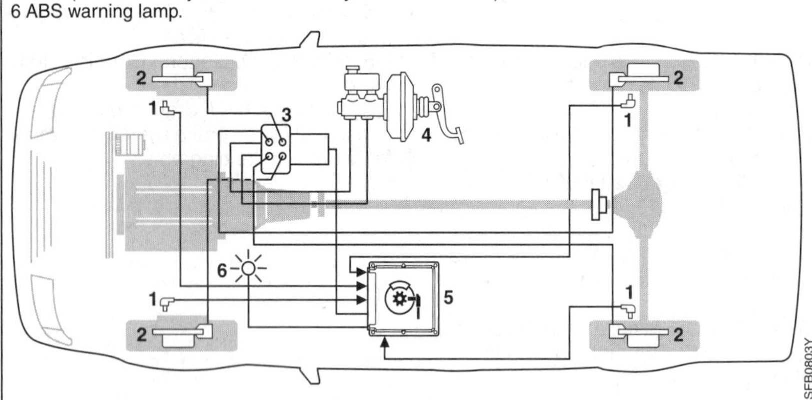

electrical operating device (switch). The service- and parking-brake systems are thus equipped with separate, individual control and transmission devices. The parking-brake system may be graduable in design, acting on the wheels of only one axle.

The holding-stationary effect is calculated according to ECE R13-H on a downhill gradient with fully laden vehicles. The downhill gradient for single vehicles is 18%. If the vehicle is equipped to tow a trailer, the holding-stationary effect must also be achieved with an unbraked trailer on a downhill gradient of 12%.

Secondary brake system

In the event of a fault, e.g. leak or fractured pipe, it must still be possible with the operational part of the brake system to achieve at least the secondary braking effect – with the identical control force at the control device. It must be possible to meter the secondary braking effect, which in turn must be at least 50% (ECE R13-H) or 44% (§41 s. 4a). The vehicle must not leave its lane when the secondary brake is applied.

The secondary brake system does not need to be an independent third brake system (in addition to the service- and parking-brake systems) with a special control device. Either the intact brake circuit of a dual-circuit service-brake system or a graduable parking-brake system can be used as the secondary brake system.

Components of the passenger-car brake system

Control device

The control device comprises those parts of the brake system which initiate the effect of this brake system. When the service brake is applied, the force of the driver's foot acts on the brake pedal. The lever-transmitted pedal force is boosted in the brake booster, depending on its design, by a factor of 4 to 10 and acts on the piston in the brake master cylinder (Figure 1). The control force is converted into hydraulic pressure. Under full braking, this pressure ranges between 120 and 180 bar, depending on the system configuration.

Brake booster

Function

The brake booster reduces the applied control force required for the braking operation, but must not impair the sensitive graduation of the braking force and the feeling of the measure of braking.

Design

Brake boosters function as vacuum-operated boosters or also hydraulically. Hydraulic brake boosters are supplied

Figure 2: Vacuum brake booster
1 Push rod,
2 Vacuum chamber with vacuum connection,
3 Diaphragm,
4 Working piston,
5 Valve unit,
6 Air filter,
7 Piston rod,
8 Working chamber,
9 Backing plate.

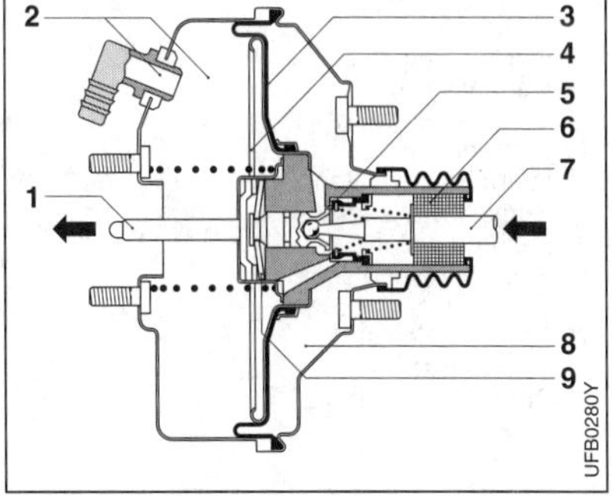

by the power steering or by a separate hydraulic pump and pressure-accumulator devices.

Passenger-car brake systems are usually equipped with vacuum brake boosters. These vacuum boosters utilize the negative pressure generated in the intake manifold during the induction stroke on gasoline engines or the vacuum (0.5 – 0.9 bar) produced by a vacuum pump on diesel engines and on electric vehicles or hybrid vehicles to amplify the force applied by the driver's foot. A diaphragm separates the vacuum chamber with vacuum connection from the working chamber (Figure 2). The piston rod transmits the applied foot pressure to the working piston and the amplified force is passed to the brake master cylinder via the push rod.

Operating principle
If the brake is not operated, the vacuum chamber and working chamber are connected via the valve unit. Given that the vacuum pipe connection is connected to a vacuum source, this means that there is a vacuum in both chambers.

As soon as a braking operation is initiated, the piston rod moves forwards in the direction of the arrow. After a short stroke the connection between the working chamber and the vacuum chamber is blocked. As the piston rod continues to move, the inlet valve in the valve unit is opened and atmospheric air flows into the working chamber. The pressure in the working chamber is then greater than in the vacuum chamber. The atmospheric pressure acts via the diaphragm on the diaphragm disk with which it is in contact. Because the diaphragm disk is attached to the valve unit, the latter moves when the disk moves, thereby assisting the foot pressure transmitted by the connecting rod. Maximum boost is dependent on the effective diaphragm or piston area, on the atmospheric pressure and on the effective vacuum.

When the braking operation has ended, the inlet valve is closed and the vacuum and working chambers are connected via the valve unit. In this way, the pressure (vacuum) is identical in both chambers.

Vacuum non-return valve
In all brake systems which have a vacuum brake booster a non-return valve is incorporated in the vacuum line between vacuum source and brake booster. While there is a vacuum present, the non-return valve remains open. It closes when the vacuum source ceases to produce a vacuum (engine is switched off) so that the vacuum inside the brake booster is retained. Thus, brake boost is effective even when the engine is switched off for several brake actuations.

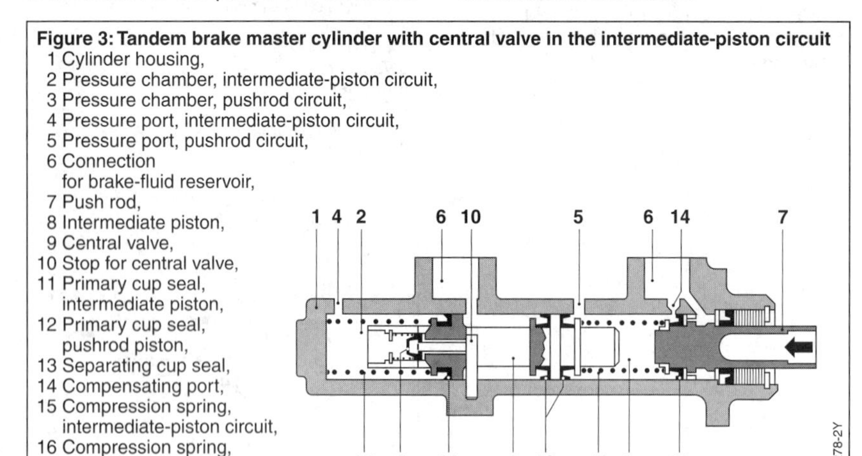

Figure 3: Tandem brake master cylinder with central valve in the intermediate-piston circuit
1 Cylinder housing,
2 Pressure chamber, intermediate-piston circuit,
3 Pressure chamber, pushrod circuit,
4 Pressure port, intermediate-piston circuit,
5 Pressure port, pushrod circuit,
6 Connection for brake-fluid reservoir,
7 Push rod,
8 Intermediate piston,
9 Central valve,
10 Stop for central valve,
11 Primary cup seal, intermediate piston,
12 Primary cup seal, pushrod piston,
13 Separating cup seal,
14 Compensating port,
15 Compression spring, intermediate-piston circuit,
16 Compression spring, pushrod circuit.

Brake master cylinder

The brake master cylinder converts the foot force applied by the driver and boosted by the brake booster into hydraulic brake pressure.

Brake master cylindor with central valve

Design

In order to comply with statutory safety requirements, service-brake systems are equipped with two separate service-brake circuits. If a leak occurs (circuit failure), the other circuit remains intact (secondary braking effect). This can be achieved by a tandem brake master cylinder (Figure 3). The compression spring of the intermediate-piston circuit in the rest state holds the intermediate piston and the pushrod piston against the rear stop. The compensating port and the central valve are opened. Both hydraulic service-brake circuits are depressurized (drive position).

Operating principle

The force applied at the brake pedal and boosted by the brake booster acts directly on the pushrod piston and pushes it to the left. After short piston travel the compensating port is sealed and a pressure build-up can take place in the pushrod circuit. This also causes the intermediate piston to be pushed to the left.

Brake master cylinder with captive piston spring

Design

The "captive" piston spring – a compression spring – in the rest state holds the pushrod piston and the intermediate piston always at the same distance. This prevents the piston spring in the rest state from pushing the intermediate piston and the latter from overrunning the compensating port with the primary cup seal. In this situation pressure compensation via the compensating port would no longer be possible in the secondary circuit, and in the event of a residual pressure the brake shoes would not lift off the brake drums when the brake releases.

Operating principle

When the brake is actuated, the pushrod piston and the intermediate piston move in the direction of the arrow to the left, overrun the compensating ports and force brake fluid through the pressure ports into the brake circuits. As pressure increases the intermediate piston is no longer moved by the captive piston spring, but instead by the pressure of the brake fluid.

Expansion reservoir

The expansion reservoir, also called the brake-fluid reservoir, is mounted directly on the brake master cylinder and connected to it via two ports. It is both

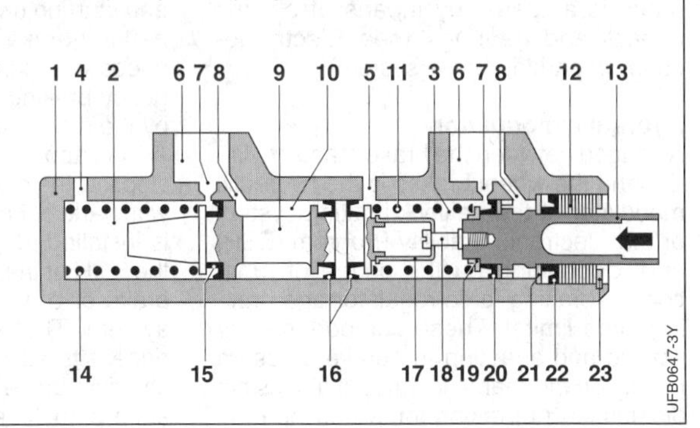

Figure 4: Tandem brake master cylinder with captive piston spring
1 Cylinder housing, 2 Pressure chamber, intermediate-piston circuit,
3 Pressure chamber, pushrod circuit, 4 Pressure port, intermediate-piston circuit,
5 Pressure port, pushrod circuit, 6 Connection for brake-fluid reservoir,
7 Compensating port, 8 Replenishing port,
9 Intermediate piston, 10 Space, 11 Captive piston spring, 12 Plastic bush,
13 Pushrod piston,
14 Compression spring, intermediate-piston circuit,
15 Primary cup seal, intermediate piston,
16 Separating cup seal,
17 Stop sleeve,
18 Stop screw,
19 Support ring,
20 Primary cup seal, pushrod piston,
21 Stop disk,
22 Secondary cup seal,
23 Snap ring.

the reservoir for the brake fluid and the expansion reservoir. It compensates volume fluctuations in the brake circuits which occur after the brake is released, in response to wear of the brake linings and to temperature differences in the brake system, and during intervention by the antilock braking system (ABS) or the Electronic Stability Program (ESP).

Transmission device
The hydraulic pressure is transmitted by the brake fluid via brake pipes as per DIN 74234 [5] and brake hoses as per SAE J 1401 [6] to the wheel-brake cylinders. Brake fluids must comply with the requirements laid down in SAE J 1703 [7] or FMVSS 116 [8] (see Brake fluids).

Wheel brakes
Floating-caliper disk brakes are usually used on the front wheels, but fixed-caliper disk brakes may also be used. Both floating-caliper disk brakes with integrated locking mechanism and Simplex drum brakes are used on the rear wheels (see Wheel brakes). Combinations of disk brakes and drum brakes (drum-in-head systems) can also be used on the rear wheels. In this case, the drum brake accommodated in the brake-disk chamber is used exclusively for the parking-brake system.

The parking-brake control device can be mechanically designed as a handbrake lever or a footbrake pedal with locking mechanism. Force is generally transmitted via cables or linkage to the wheel brakes on the rear axle. In the case of electromechanical parking brakes, the brake is actuated by means of electric motors and gearings (see Electromechanical parking-brake system).

Hydraulic modulator
Arranged between the brake master cylinder and the wheel brakes is the hydraulic modulator of the antilock braking system or the Electronic Stability Program (ESP) and, depending on the scope of functions, a braking-force regulator or a braking-force limiter. These components, by limiting and adapting the brake pressure mostly on the rear axle, ensure a sensible distribution of braking force between the front and rear axles. This function can, especially in vehicles with markedly different load states, be executed on a load-sensitive basis (automatic load-sensitive braking-force metering).

The hydraulic modulator modifies the brake pressure during the braking process in such a way that the wheels are prevented from locking. Depending on the control variation, this job is performed by several solenoid valves and an electrically driven supply pump. In passenger-car brake systems, the front axle is individually controlled, i.e. each wheel is braked according to the respective grip. The rear wheels are controlled according to the select-low principle so that both rear wheels are braked together according to the wheel which has the lower grip (see also Antilock braking system and Driving-dynamics control).

Electromechanical parking brake
System overview
Conventional parking-brake systems are muscular-energy brake systems and are operated by purely mechanical means via lockable hand levers or foot pedals or via a crankgear. In electromechanical parking-brake systems, also referred to simply as electromechanical parking brakes or automatic parking brakes, the control (operating) force is generated by an electric drive.

Operation and control are effected electrically by means of a switch. The electromechanical parking brake can only be operated when the vehicle is stationary or at low speeds of less than 10 km/h. This must also be possible when the ignition and starting switch is turned off. If electric parking-brake systems are operated at speeds in excess of 10 km/h, an emergency-braking operation is first executed by the driving-dynamics control system.

The application force depends on the slope of the gradient on which the vehicle was parked. For this purpose, a tilt sensor is installed, depending on the system, in the ECU of the electromechanical parking brake or of the driving-dynamics control system. The retensioning of the brake necessitated by the cooling of the mechanical brake components is performed according to a calculated temperature model or after vehicle movement has been detected.

A safety concept must be provided to prevent the system from being unintentionally activated by an electrical fault, and from being improperly or accidentally released by children. Furthermore, intentional activation (emergency braking, only necessary if the control device of the service brake system is broken) of the electromechanical parking brake must not give rise to critical driving situations. If the operating unit of the electromechanical parking brake is deliberately actuated on a permanent basis, the driving-dynamics control system takes over the task of braking the vehicle at a speed in excess of 10 km/h. This ensures an optimally safe braking operation even in critical road situations. The electromechanical parking brake is activated only after the vehicle speed has dropped below a specific threshold. The systems communicate with each other through a CAN data link.

Figure 5: Electric parking-brake system
a) System with servomotor on brake caliper,
b) System with cables.

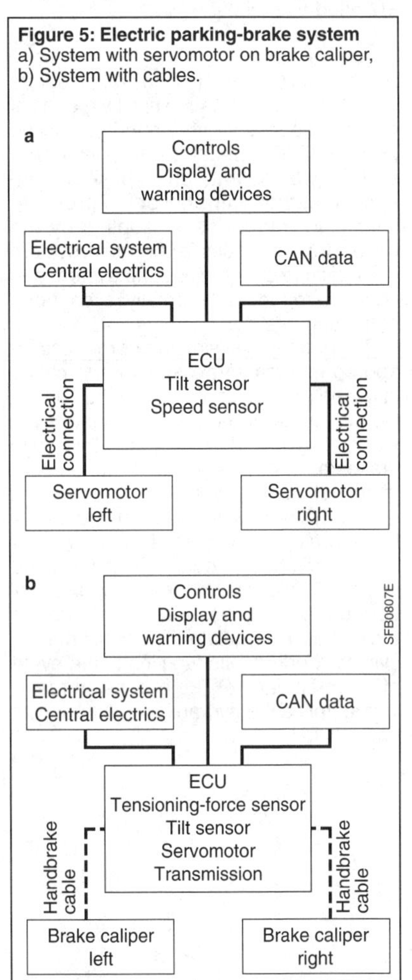

Electric parking-brake systems may also include additional functions such as automatic braking (for example, when a door is opened) or automatic release when driving off.

Electric parking-brake systems are energy-assisted systems and are equipped with an emergency release device. Electric operation must be designed in such a way as to make it possible to prevent unintentional braking while driving. It must also be possible for the system to be activated even when the ignition and starting switch is turned off, and the system may only be released when the ignition and starting switch is turned on and the brake pedal is simultaneously being pressed.

Self-diagnosis detects malfunctions and faults, and indicates them via a warning lamp. A text message can also appear in a driver information display. The fault memory can be read out with a diagnosis tester and cleared after the fault has been corrected.

Diagnosis testers and relevant software may be required for servicing work, e.g. when replacing the brake pads.

Electromechanical parking brake with servomotor on the brake caliper

The electromechanical parking brake with servomotor comprises the following components (Figure 5a): operating unit, ECU, display and warning devices, tilt sensor (can be installed in the driving-dynamics control system), floating caliper with electric motor and multiple-stage gearing.

In the case of a brake caliper with an electric servomotor, force is transmitted for the parking brake effect via a multiple-stage gearing and a threaded spindle. It is activated by way of an electrical switch (operating unit), which forwards the control commands to the ECU redundantly and in accordance with the safety concept. The ECU, taking into account further boundary conditions (e.g. road gradient), activates the electric servomotors via separate driver stages and electrical connecting leads.

A very high gear ratio means that very high application forces can be generated. These forces are approximately 15 to 20 kN and correspond to the application force when the hydraulic nominal pressure is applied in the hydraulic section of the brake.

Electromechanical parking brake with cables

In the case of an electromechanical parking brake with cables, the following components are combined in a centrally arranged assembly – above the rear axle, in the passenger compartment or in the fender (Figure 5b): electric drive motor with gearing, required sensors (depending on the scope of functions, e.g. force, tilt, temperature and position sensors), ECU, and cable mechanism (if necessary with emergency release device).

This system too is activated by way of an electrical switch, which forwards the control commands to an ECU. The ECU activates the electric servomotor(s) via a driver stage. The application force can vary, depending on the road gradient. The system is automatically retensioned when the vehicle is stopped either after a cooling phase corresponding to a temperature model or after vehicle movement has been detected.

Electrohydraulic brake

Function

The electrohydraulic brake (EHB, also known by the term "Sensotronic Brake Control", SBC) is an electronic brake-control system with hydraulic actuator engineering. Like a conventional hydraulic brake, its function is to reduce vehicle speed, bring the vehicle to a halt, or keep the vehicle stationary. As an active braking system, it takes control of brake operation, braking-force boosting and braking-force control. Hydraulic standard wheel brakes are used as brakes.

Actuation unit

Mechanical operation of the brake pedal is detected by the actuation unit by means of electronic sensors with redundant backup (Figure 6). The pedal-travel sensor consists of two separate angle-position sensors. Together with the brake-pressure sensor for the pressure applied by the driver, this produces a threefold system for detecting driver input. The system can continue to function normally, even if one of the sensors fails.

The pedal-travel simulator produces an appropriate force/travel curve and calculates the amount of brake-pedal damping. Consequently, the driver experiences the same "brake feel" with electrohydraulic brakes as with a very well designed conventional braking system.

The brake booster used in a conventional brake is not necessary here. Only the driver's brake request is determined in the actuation unit during normal operation; the brake pressure is generated in the hydraulic modulator. The brake master cylinder performs its function in the event of a system failure. The expansion reservoir supplies the hydraulic modulator with brake fluid.

Electronic open- and closed-loop control

The driver's brake request is determined in the remote-mounted ECU from the sensor signals of the actuation unit. Braking characteristics can be adapted to the driving conditions (e.g. sharper response at high speeds or with more dynamic driving styles). A "duller" pedal response can be used to alert the driver to a reduction in braking effect, when the brakes reach the limits of their effectiveness before overheating induces brake fading.

The functions for the antilock braking system, the traction control system and the Electronic Stability Program (ESP) are also integrated in this ECU. Also featured are comfort and convenience functions such as, for example, hill-start assist (prevents the vehicle from rolling back when starting on a hill), enhanced brake assist (the brake pads are gently applied by metered brake-pressure build-up when the accelerator is abruptly released), the chauffeur brake (soft stop, stopping without jolting by automatic brake-pressure reduction shortly before standstill), and dry-brake function (water film is regularly removed from brake disks in wet conditions, which can reduce the stopping distance).

Thanks to complete electronic pressure control, the electrohydraulic brake can be easily networked with vehicle guidance systems (e.g. Adaptive Cruise Control, ACC).

An intelligent interface with the CAN bus provides the link between the remote-mounted ECU and the add-on ECU.

Hydraulic modulator

Operating principle in normal operation

Figure 6 shows the electrohydraulic-brake components as a block diagram. An electric motor drives a hydraulic pump. This charges a high-pressure accumulator to a pressure of between approx. 90 and 130 bar, monitored by an accumulator pressure sensor. The four separate wheel-pressure modulators are supplied by the accumulator and set the required pressure at the wheel-brake cylinders separately for each wheel. The pressure modulators themselves each consist of two valves with proportional-control characteristics and a pressure sensor. Brake-pressure modulation and active braking are silent and generate no brake-pedal feedback.

In normal mode, the isolating valves isolate the brakes from the actuation unit. The system is in "brake-by-wire" mode. It

Figure 6: Interaction of electrohydraulic-brake function modules

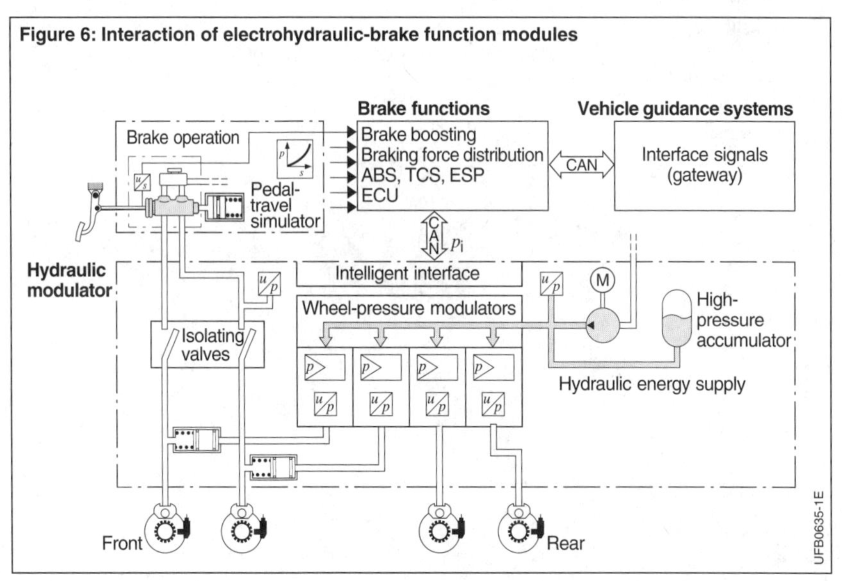

electronically detects the driver's braking request and transmits it "by wire" to the wheel pressure modulators. The interaction between electric motor, valves and pressure sensors is controlled by hybrid-technology electronic circuits in the form of an add-on ECU. This has two microcontrollers which monitor each another. The essential feature of these electronics is their extensive self-diagnosis, which monitors the plausibility of every system state at all times. It means that any faults can be displayed to the driver before a critical condition arises. If components fail, the system automatically provides the optimal remaining partial function to the driver.

Braking in the event of system failure
For safety reasons, the electrohydraulic braking system is designed so that in the event of serious faults (e.g. power-supply failure), it switches to a state in which the driver can brake the vehicle without using the active brake-booster function. When de-energized, the isolating valves establish a direct connection to the actuation unit and thus allow a direct hydraulic connection from the actuation unit to the wheel-brake cylinders (hydraulic fallback level).

References

[1] 71/320/EEC: Council Directive of 26 July 1971 on the approximation of the laws of the Member States relating to the braking devices of certain categories of motor vehicles and their trailers.
[2] ECE R13: Regulation No. 13 of the United Nations Economic Commission for Europe (UN/ECE) – Uniform provisions concerning the approval of vehicles of categories M, N and O with regard to braking.
[3] ECE R13-H: Regulation No. 13-H of the United Nations Economic Commission for Europe (UN/ECE) – Uniform provisions concerning the approval of passenger cars with regard to braking.
[4] §41 StVZO (road traffic licensing regulations, Germany) – Brakes and wheel chocks.
[5] DIN 74234: Hydraulic braking systems; brake pipes, flares.
[6] SAE J 1401: Road Vehicle Hydraulic Brake Hose Assemblies for Use with Nonpetroleum-Base Hydraulic Fluids.
[7] SAE J 1703: Motor Vehicle Brake Fluid.
[8] FMVSS 116: Federal Motor Vehicle Standard No. 116: Motor Vehicle Brake Fluids.
[9] B. Breuer, K. Bill (Editors): Bremsenhandbuch. 3rd Edition, Vieweg+Teubner, 2006.

Brake systems for commercial vehicles

System overview

Brake systems for commercial vehicles and trailers must satisfy the requirements of various regulations such as, for example, RREG 71/320 EEC and ECE R13. Essential functions, effects and test methods are set out in these regulations. The entire system is subdivided into the service-brake, parking-brake, secondary-brake systems, and continuous-action brake systems.

Service-brake system

Service-brake system, tractor vehicle
The service-brake system, designed as an energy-assisted brake system in commercial vehicles (Figures 1 and 2), can operate with compressed air or also with a combination of compressed air and hydraulics.

In the event of a fault, e.g. brake-circuit failure, it must still be possible with the operational part of the system to achieve at least the secondary braking effect – with the identical control force at the normal control device. It must be possible to meter the effect, and the trailer must not be affected by this malfunction, i.e. the trailer control (trailer control valve) must have a dual-circuit design. The secondary braking effect must achieve at least 50 % of the braking effect of the service-brake system. It is therefore customary to split the system into two brake circuits already separate on the supply side, even though this configuration is only legally required in motor buses.

The energy supply to the trailer must be guaranteed even during the braking operation. The dual-line system became mandatory when RREG 71/320 came into force, but had already been available prior to this and was known by the name "Nato" brake.

Figure 1: Structure of a compressed-air brake system with trailer control
1 Air compressor, driven by engine, 2 Pressure regulator, 3 Four-circuit protection valve,
4.1 Air reservoir V1 for circuit 1, 4.2 Air reservoir V2 for circuit 2,
4.3 Air reservoir V3 for circuit 3 (trailer, pneumatic suspension),
5 Overflow valve with limited return flow, 6 Trailer control valve with throttle valve,
7 "Supply" coupling head (red), 8 "Brake" coupling head (yellow),
9 Parking-brake valve with test position, 10 Relay valve,
11.1 Combination brake cylinder, rear right, 11.2 Combination brake cylinder, rear left,
12 Load-dependent braking-force regulator (ALB), 13 Service-brake valve,
14.1 Brake cylinder, front right, 14.2 Brake cylinder, front left,
15 Secondary loads/ancillaries (e.g. pneumatic suspension, door-closing system).

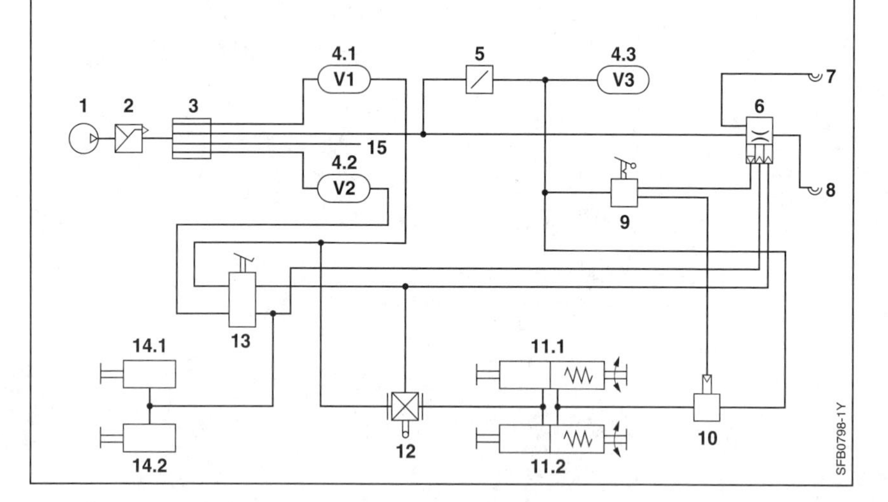

The trailer is continually supplied with a defined pressure via the supply line. This pressure must be between 6.5 and 8.0 bar on an intact tractor vehicle, irrespective of the tractor vehicle's operating pressure established by the manufacturer. The trailer must be exchangeable. The trailer's service-brake system is controlled by a second line, the brake line. This line is also governed by the regulations pertaining to trailer exchangeability. Thus the pressure in the brake line must be 0 bar in driving mode, and 6.0 to 7.5 bar in fully-braked mode.

Service-brake system, trailer
The trailer has an independent service-brake system, which is only partly subject to the demand for a secondary braking effect. According to the requirements in RREG 71/320, the braking effects of the service-brake systems in the tractor vehicle and in the trailer must be within closely set tolerances as a function of the control pressure in the brake line to the trailer, i.e. they must be approximately the same (design tolerance band RREG 71/320 and ECE R13).

If the supply line or the brake line is fractured, the trailer must be able to be fully or partly braked or it must initiate an automatic braking. Commercial vehicles with electronically controlled brake systems have – in addition to the pneumatic brake line – an electrical signal-transmission feature for electrically controlling the service-brake system in the trailer. This is

Figure 3: Compatibility diagram
Tractor vehicle and trailer RREG 71/320, ECE R13.
T_R Sum total of braking forces at circumference of all trailer wheels,
P_R Total static normal force of trailer,
T_M Sum total of braking forces at circumference of all tractor-vehicle wheels,
P_M Total static normal force of tractor vehicle,
P_m Pressure at "Brake" coupling head.

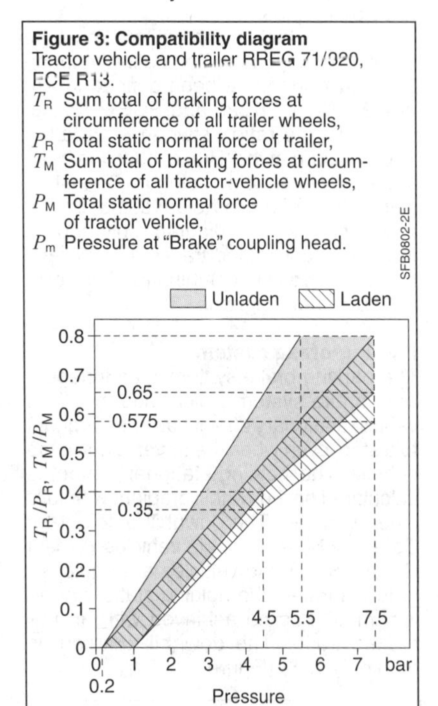

Figure 2: Compressed-air brake system of a 2-axle semitrailer with ABS (simplified representation)
1 "Supply" coupling head (red), 2 Double release valve, 3 1-channel ABS pressure-control valve, 4 Combination brake cylinder, 5 Shuttle valve, 6 Air reservoir, 7 Load-sensitive braking-force regulator, 8 Drain valve, 9 Trailer brake valve, 10 "Brake" coupling head (yellow), 11 Test connection, 12 Line filter.

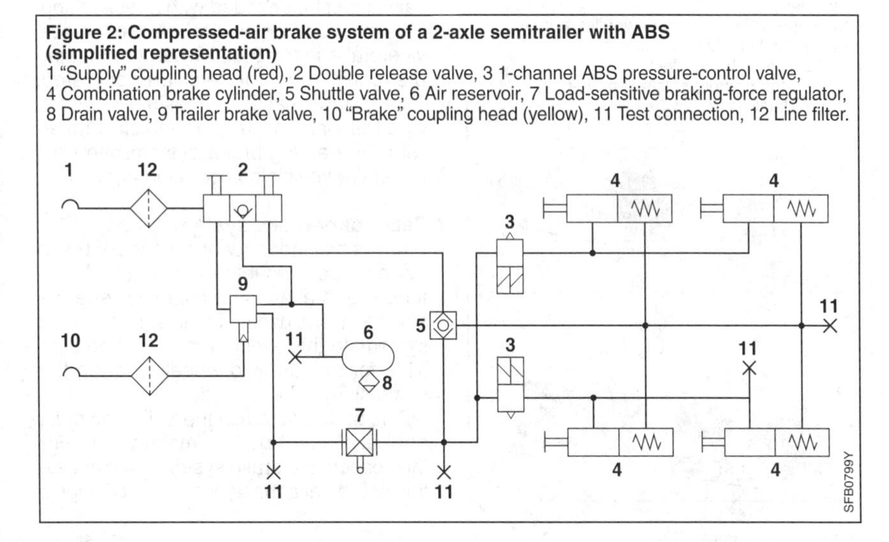

performed by a standardized electrical plug-in connection in accordance with ISO 7638; this plug-in connection may have 5 or 7 pins.

Tractor vehicles and trailers must be freely exchangeable. Compatibility conditions have therefore been defined in Annex 2 RREG 71/320 and ECE R13. Accordingly, the ratio between retardation and pressure at the "Brake" coupling head in the range depicted in Figure 3 must be in the range of 0.2 to 7.5 bar at the "Brake" coupling head. The diagram only applies to tractor vehicle and trailer. All other vehicles and vehicle combinations are covered by other diagrams.

Parking-brake system

The parking-brake system is an independent brake system which must hold the vehicle stationary after it has been brought to a stop, even when the driver is not in the vehicle. The holding-stationary effect is calculated on a downhill gradient with fully laden vehicles. The downhill gradient for Category M, N, O single vehicles (except O1) is 18%. If the vehicles are equipped to tow a trailer, the holding-stationary effect must also be achieved with an unbraked trailer. The downhill gradient is then only 12% (Figure 4).

Figure 4: Test condition for parking-brake system
a) Single vehicle, 18% downhill gradient.
b) Tractor vehicle and trailer, 12% downhill gradient; only the tractor vehicle is braked.
γ Gradient angle.

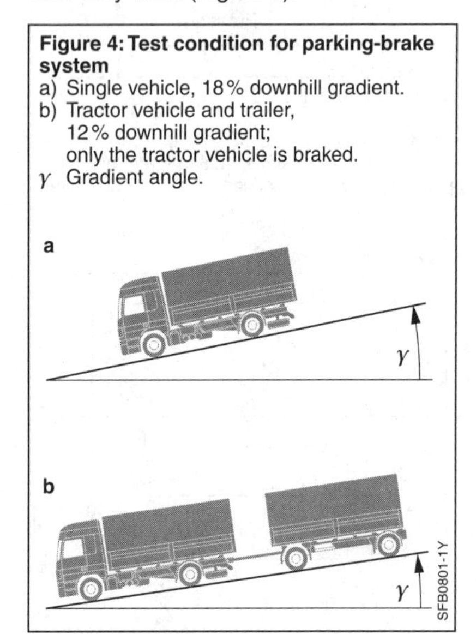

The parking-brake system in commercial vehicles and motor buses is usually designed as a spring-loaded brake system. The springs in the spring-type brake cylinders, when the wheel brakes are adjusted in accordance with the regulations, generate the same force as the pneumatic brake cylinders in the service-brake system when the nominal pressure (brake-system calculated pressure) acts on their nominal effective area. If certain malfunctions occur, e.g. brake-circuit failure or energy-source failure, the spring-loaded brakes may not brake automatically and must therefore be protected and designed accordingly.

Energy-assisted parking-brake systems (spring-loaded brake systems) must be equipped with at least one emergency release device. This may be a mechanical, pneumatic or hydraulic device. The parking-brake system must then only be designed for graduated (metered) operation if it has to be engaged to achieve the prescribed secondary braking effect.

In the trailer, the parking-brake system frequently operates as a muscular-energy brake system. If the trailer control system in the tractor vehicle is configured in such a way that the service brake in the trailer also responds when the parking brake is operated in the tractor vehicle (trailer control valve with port 4.3), the parking-brake valve must be provided with a test setting. This makes it possible to release the service brake in the trailer when the parking brake in the tractor vehicle is operated. This in turn makes it possible to check whether only the tractor vehicle braked with the parking-brake system can hold the entire vehicle combination.

Secondary-brake system

There is no independent secondary-brake system. It comes into effect if a malfunction, e.g. brake-circuit failure or energy-source failure, occurs in the service-brake system. In this event, it must still be possible for at least two wheels (not on the same side) to be braked.

The brake system in the trailer too must not be affected by this malfunction. For this reason, the brake systems and activation of the trailer have dual-circuit designs.

The supply volume must be designed such that, in the event of an energy-source failure, after eight full operations of the service brake there is still enough pressure available to achieve the secondary braking effect on the ninth full operation. In the event of a failure of a brake circuit on the supply side, it is essential to ensure that, when the energy source is intact, the pressure in the intact brake circuits does not drop permanently below the nominal pressure. This is achieved by using special protective devices, e.g. a four-circuit protection valve or an electronic unit.

Continuous-operation brake system
The wheel brakes used are not designed for continuous operation. Prolonged braking (e.g. on hill descents) can result in the brakes being thermally overloaded. This causes a reduction in braking effect ("fading") or, in extreme cases, in complete brake-system failure.

A wear-free brake system is referred to as a continuous-operation (retarder) brake system. In Germany, they are required in accordance with StVZO §41 s. 15 for motor buses with a gross weight rating of more than 5.5 t and for other motor vehicles with a gross weight rating of more than 9 t. A retarder must hold a fully laden vehicle over a distance of 6 km and a downhill gradient of 7% at a speed of 30 km/h.

The service brake must be designed accordingly for trailers. Operation of the retarder in the tractor vehicle must not cause the service brake in the trailer to be operated (see also StVZO §72 and Federal Law Gazette 1990 II p. 885, 1102).

Components of commercial-vehicle brake systems

Air supply and air processing
Air supply and processing comprises the energy source, pressure control, air processing, and compressed-air distribution.

Compressor
A compressor is the source of energy. It takes in air and compresses it to compressed air, the working medium for the brake systems and the ancillaries (e.g. pneumatic suspension, door-closing system).

The compressor is a plunger pump where the crankshaft is driven directly by the vehicle engine (Figure 5). It is fitted to the vehicle engine by means of a flange. Its component parts are:

Figure 5: Compressor
a) Induction,
b) Compression and discharge,
c) Compression into the additional clearance.
1 Cylinder head, 2 Intermediate plate (with inlet and outlet valves), 3 Cylinder, 4 Piston, 5 Connecting rod, 6 Crankcase, 7 Drive 8 Crankshaft, 9 ESS valve, 10 Additional clearance.

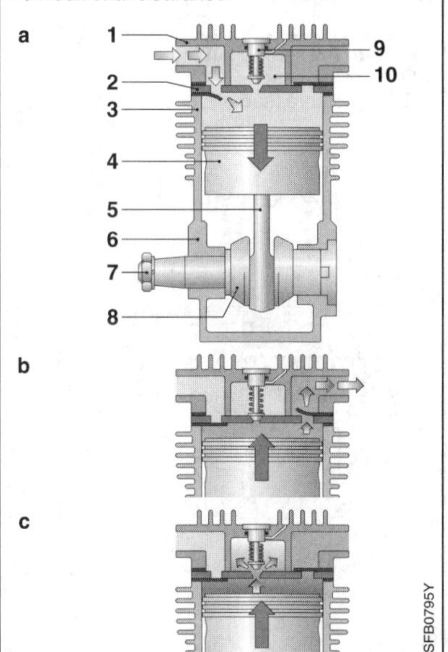

- The crankcase, which forms a monobloc unit with the cylinder. It contains the crankshaft with connecting rod and piston.
- The cylinder head with intake and pressure connections, as well as connections for water cooling.
- The intermediate plate with inlet and outlet valves.

To reduce losses in the idle mode (opening and flow resistances in valves and lines), an energy-saving system (ESS) is used; this activates a clearance and thus reduces compression work. This reduces fuel consumption.

During its return stroke, the piston draws in air after the inlet valve opens automatically due to vacuum. The inlet valve closes at the start of the piston's return stroke. In the fore stroke, the piston compresses the air. When a certain pressure is reached, the outlet valve opens and compressed air is supplied to the brake system.

Nowadays, compressors achieve a stroke displacement of up to 720 cm^3, a pressure level of up to 12.5 bar, and a maximum speed of 3,000 rpm. Their features include high efficiency, low oil consumption, and a long service life.

Figure 6: Pressure regulator
1 From compressor,
2 To the compressed-air reservoirs, 3 Venting

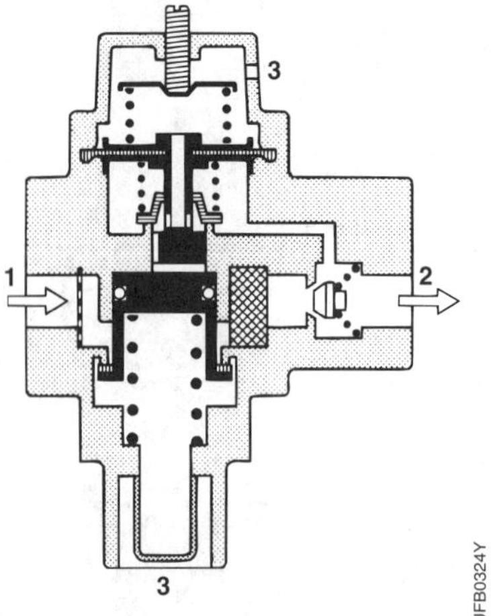

Pressure regulator

The pressure regulator controls the compressed air supplied by the compressor in such a way that the operating pressure lies within the activation and cutoff pressure (Figure 6).

As long as the pressure in the compressed-air reservoirs lies below the cutoff pressure, connections 1 and 2 are connected and compressed air passes the pressure regulator. Once the cutoff pressure is reached, the pressure regulator switches to idle mode. Here, the venting piston is activated and connection 1 is connected to the atmosphere (venting).

Air drier

The air drier cleans the compressed air and dries it to prevent corrosion and freezing in the brake system during winter operation.

Basically, an air drier consists of a desiccant box and a housing. The housing incorporates the air passage, a bleeder valve, and a control element for granulate regeneration (Figure 7). The granulate is regenerated by activating a regeneration-air tank.

When the bleeder valve is closed, compressed air from the compressor flows through the desiccant box and from there to the supply-air reservoirs. At the same time, a regeneration-air tank is filled with dry compressed air. As compressed air flows through the desiccant box, water is removed by means of condensation and adsorption.

The granulate in the desiccant box has a limited water absorption capacity and must therefore be regenerated at regular intervals. In the reverse process, dry compressed air from the regeneration-air reservoir is reduced to atmospheric pressure via the regeneration throttle upstream of the air drier, flows back through the moist granulate from which it draws off the moisture, and flows as moist air via the open bleeder valve to the atmosphere.

The pressure regulator and air drier can be combined into one unit.

Four-circuit protection valve

The four-circuit protection valve distributes the compressed air to the various brake and ancillary circuits, isolates the circuits from one another, and ensures the water supply for the remaining circuits in the event of failure of a circuit (Figure 8).

The function of the four-circuit protection valve is provided by overflow valves specially developed for this application. In contrast to a normal overflow valve, the design of this type of overflow valve features two different effective areas on the intake-flow side. The incoming pressure from the pressure regulator acts on one effective area, while the pressure available in the pneumatic circuit acts on the other. The opening pressure of the overflow valves is thus dependent on the (residual) pressure of the assigned pneumatic circuit.

Figure 7: Air drier with integrated pressure regulator
1 Desiccant box, 2 Compression spring,
3 Desiccant, 4 Cup (control valve),
5 Compression spring, 6 Pin,
7 Diaphragm, 8 Compression spring,
9 Heating element, 10 Bleeder valve,
11 Drain connection, 12 Throttle,
13 Non-return valve, 14 Preliminary filter,
15 Secondary filter.

Ports:
1 From compressor,
21 To air reservoir,
22 To regeneration-air tank,
3 Vent.

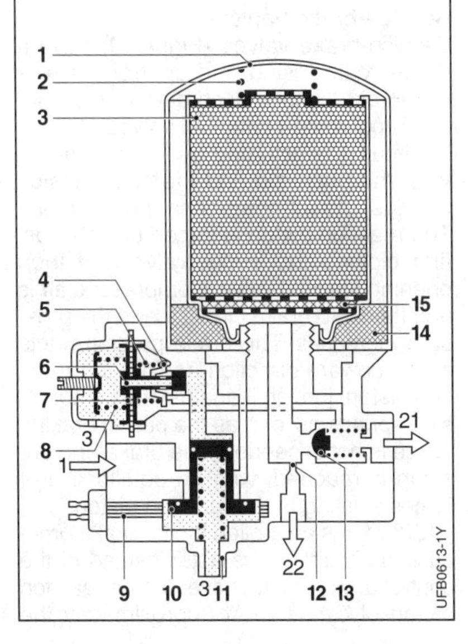

The overflow valves can be arranged differently. Often service-brake circuits 1 and 2 and ancillary circuits 3 and 4 are connected in pairs in succession. This ensures that at least one of the two service-brake circuits is filled as a matter of priority. The ancillary circuits for this type of valve are additionally protected by two non-return valves. These non-return valves can be omitted from four-circuit protection

Figure 8: Four-circuit protection valve
a) Filling a compressed-air reservoir,
b) Filling all compressed-air reservoirs.
1 Housing,
2 Compression spring,
3 Diaphragm piston,
4 Valve seat,
5 Non-return valve,
6 Fixed throttle.

I – IV overflow valves

Ports:
1 Energy input,
21 – 24 Energy output to circuits 1 – 4.

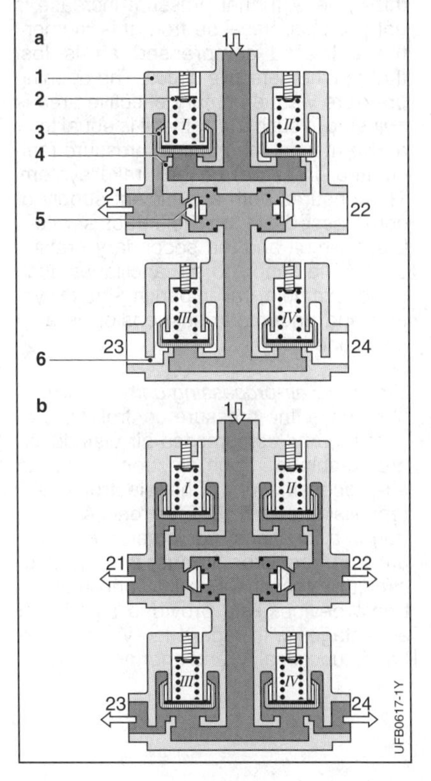

valves with central intake flow. These overflow valves may also be provided with variable flow restrictors. These enable an empty system to be filled with small quantities of air.

If a malfunction occurs for example in circuit 1 (circuit failure due to a leak), the pressure drops initially only in circuit 1 to 0 bar and in circuit 2 to the closing pressure. The pressure in circuits 3 and 4 is initially maintained by the effect of the non-return valves, but will also drop through consumption to the closing pressure. The intact circuits continue to be supplied under subsequent delivery by the compressor, because the residual pressure in circuits 2, 3 and 4 acts on the secondary effective area of the corresponding overflow valves. The intact circuits are filled again until the opening pressure of the defective circuit (circuit 1) acts on the primary effective area of the corresponding overflow valve, opening this valve. A further pressure increase is not possible, because from this moment the delivered compressed air is lost through the defective circuit. The opening pressure via the primary effective area is adjusted in such a way that it is equal to or above at least the nominal pressure (calculated pressure) of the brake system. This ensures both a sufficient supply of compressed air for the intact service-brake circuit and the secondary braking effect. The supply to the ancillaries such as, for example, trailer, parking-brake system and pneumatic suspension is also maintained.

Electronic air-processing unit

Nowadays, the pressure control, air processing, and compressed-air distribution are combined in one electronic unit, the air-processing unit. The electronic air-processing unit (EAC, Electronic Air Control) is a functional agglomeration of the pressure regulator, air drier and multiple-circuit protection valve into one mechatronic device. This provides significant advantages with regard to system overhead, functionality, and energy saving.

Energy storage

The energy required for the braking operation and for the function of the ancillaries is provided and stored in sufficient quantities in compressed-air reservoirs approved for use in road vehicles. The volume must be designed such that, without subsequent delivery, after eight full brakings the secondary braking effect prescribed for this vehicle is still achieved at least by the ninth full braking.

Despite the use of an air drier, the compressed-air reservoirs are equipped with manual or automatically acting drain devices. Compressed-air reservoirs are subject to the requirements of §41a s.8 in conjunction with §72 StVZO, and must be approved for use and permanently identified.

The supply systems for the brake systems must be fitted with warning devices. The following requirements apply:

- Red warning light
- Visible to the driver at all times
- Comes on no later than on brake application or if the pressure in the supply system for the service brake has dropped to 65% nominal pressure. 80% nominal pressure applies to the supply system for the parking-brake system (spring-loaded brake).

Service-brake valve

Service-brake valves (Figure 9) have a dual-circuit design and control the service-brake circuits according to the control force (force-controlled valves).

Circuit 1 is actuated by the control device, the push rod and the compression springs (travel-compensating springs). The reaction piston is forced downwards, first closing the outlet valve and then opening the inlet valve. Compressed air is admitted into brake circuit 1 and the pressure increases. The brake pressure acts in the upward direction against the reaction piston, forcing it against the compression springs as long as the partial braking range is not exceeded. The brake end position is reached, with an equilibrium of forces existing at the reaction piston.

Circuit 2 is controlled by the brake pressure in circuit 1. This acts, instead of the control device from above, on the reaction piston of circuit 2. At approximately the same time, the brake end position is reached in circuit 2 as well. In the full-braking position or in the event of a failure of circuit 1, both reaction pistons are mechanically moved to their full extent by means of the control device. The outlet valves are closed, and the inlet valves remain open. Circuits 1 and 2 are pneumatically fully and safely isolated from each other.

Special designs facilitate different controlled brake pressures for circuits 1 and 2. These are required if a dual-circuit booster cylinder is actuated by the service-brake valve, or if circuit 2 is subject to load-sensitive control. This is made possible by installing an appropriate spring assembly or a reaction piston with several effective areas.

Figure 9: Service-brake valve
1 Push rod,
2 and 3 Compression springs,
4 Reaction piston,
5 and 9 Inlet-valve seat,
6 and 8 Outlet-valve seat,
7 and 10 Valve plates,
11 Valve springs,
12 Return spring,
13 Control plunger,
14 Spring seat,
15 Connecting rod.

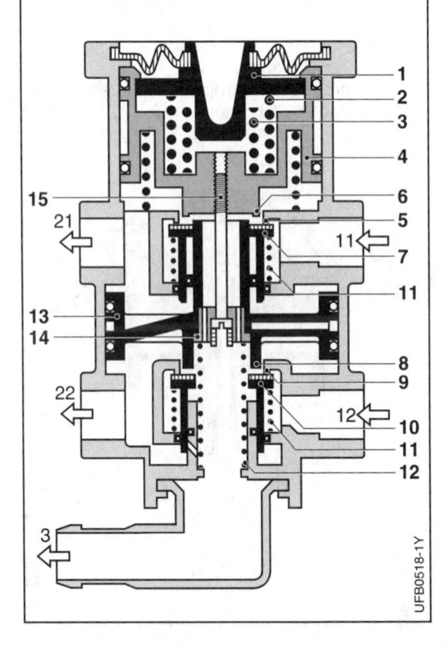

Parking-brake valve

Parking-brake valves (Figure 10) control the pressure in the spring-type brake cylinders as a function of the lever travel (travel-controlled valves). The lever must be permanently and securely lockable in the brake-applied position. Parking-brake valves must only operate under graduated (metered) application when the effect of the parking brake is required to achieve the secondary braking effect. Parking-brake valves must be provided with a test setting when the service-brake system is activated in the trailer on actuation of the parking brake.

Different variants of parking-brake valve exist, depending on their application: non-graduated, graduated or graduated with steep-droop characteristic curve. The latter variant provides for a highly sensitive graduable effect, because the operating range of spring-type brake cylinders, considered over the lever angle

Figure 10: Parking-brake valve (driving mode)
1 Actuating lever, 2 Detent element,
3 Eccentric element,
4 Return spring, 5 Outlet valve seat,
6 Inlet-valve seat, 7 Valve plate,
8 Reaction piston, 9 Reaction spring,
10 Compression spring.

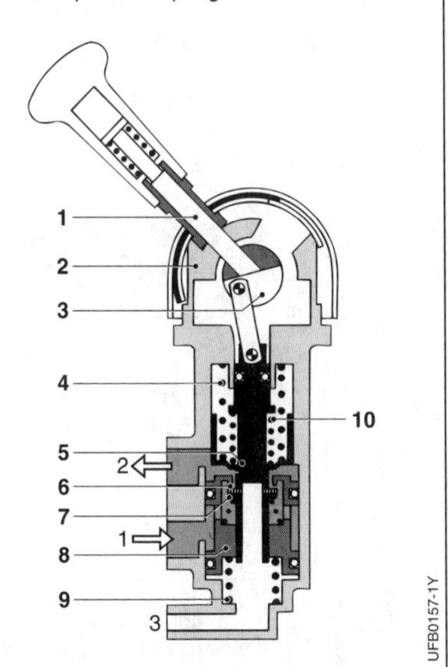

of the parking-brake valve of approximately 80°, is optimally used. The operating range of spring-type brake cylinders is between approximately 5 bar (start of braking) and approximately 2 bar (end of braking, see diagrams in Figure 11).

In pneumatic high-pressure brake systems (operating pressure greater than 10 bar), the parking-brake valve can be fitted with a pressure limiter so that standard spring-type brake cylinders can be used. The facility in parking-brake valves for attaining the capability of metering the controlled pressure is similar to the facility in service-brake valves, but operates in the opposite direction, because the spring-type brake cylinders are ventilated in driving mode and the brake-applied mode is achieved by bleeding.

Figure 11: Working range of spring-type brake cylinder
a) Normally metered parking-brake valve,
b) Metered parking-brake valve with steep-droop characteristic curve.
1 Pressure characteristic.
a Free travel (valve lift), b Actuation range,
c Start of braking, d End of braking.

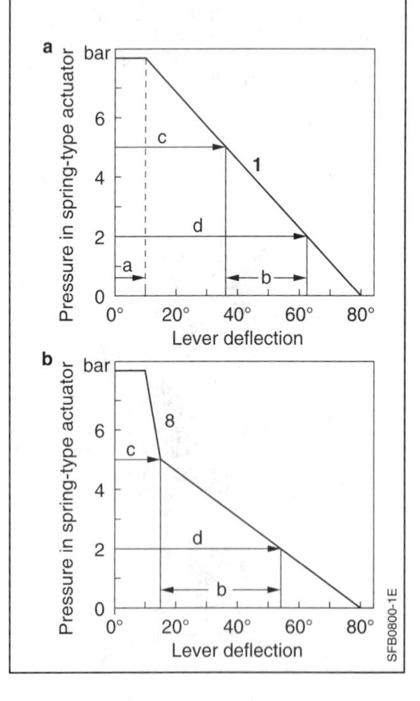

Parking-brake valves can have a dual-circuit design. The system is supplied in this case from circuit 3 and the pneumatic auxiliary release device of the spring-type actuators from circuit 4. An additionally required rotary-knob, shuttle or check valve can be omitted.

In the version with steep-droop characteristic curve (Figure 11), the start of braking is attained earlier and the actuation range is significantly greater. This is particularly advantageous when the parking brake is used as a secondary brake.

Automatic load-sensitive braking-force regulator

The automatic load-sensitive braking-force control system (ALB) is a vital element in the transmission device of a service brake system on a commercial vehicle. Valves that handle the braking-force distribution enable adjustment of the braking forces to the low axle loads in the partially laden and unladen state and thus a correction of the braking-force distribution on the axles of an individual vehicle or a certain braking level in road trains or semitrailers.

Figure 12: Braking-force regulator with relay valve
1 Vent, 2 Rake, 3 Transfer diaphragm,
4 Energy input from air reservoir,
5 Vent, 6 Uncontrolled pressure from service-brake valve, 7 Control valve,
8 Relay piston, 9 Controlled brake pressure to brake cylinders, 10 Rotary cam.

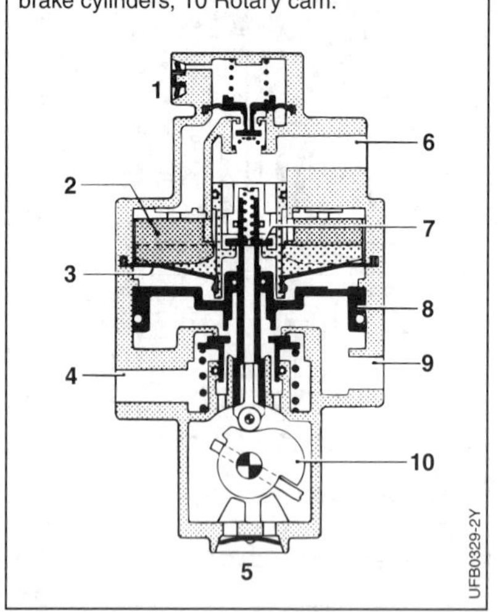

The braking-force regulator (Figure 12) is connected between service-brake valve and brake cylinder. Depending on vehicle payload, it regulates the applied braking pressure. The device has a transfer diaphragm with variable effective area. The diaphragm is held in two radially arranged, interlocking rakes. Depending on the vertical position of the control-valve seat, there is a large reaction area (valve position at bottom) or a small reaction area (valve position at top). Consequently, the brake cylinders are supplied via an integrated relay valve with a reduced pressure which is lower than (unladen), or which is the same as (fully laden) the pressure coming from the service-brake valve. The control-valve can be moved into the load-sensitive position by means of an eccentric element that is connected via linkage to the vehicle axle or by means of a wedge (in the case of vehicles with pneumatic suspension).

Figure 13: Combination brake cylinder for disk brake (driving mode)
1 Pressure pin, 2 Piston rod,
3 Bellows with seal to disk brake,
4 Compression spring (diaphragm cylinder),
5 Piston (diaphragm cylinder),
6 Housing with fastening bolts,
7 Diaphragm, 8 Intermediate flange,
9 Cylinder housing (spring-type brake actuator),
10 Piston (spring-type brake actuator),
11 Bleeder valve (spring-type brake actuator chamber),
12 Compression spring (spring-type brake actuator),
13 Release device (spring-type brake cylinder).
Air ports: 11 Service brake, 12 Parking brake.

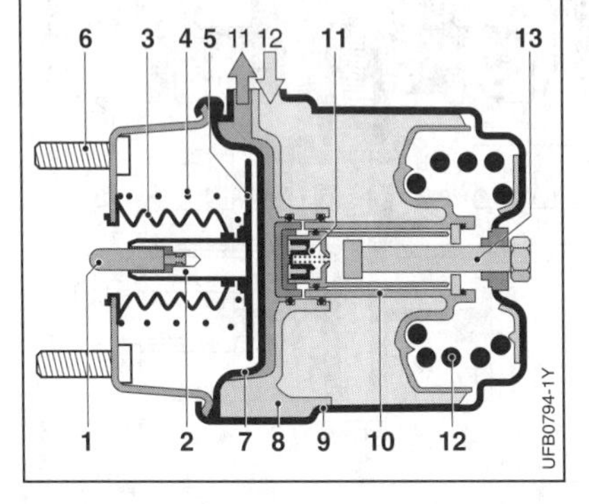

The pressure limiter which is integrated into the device at the top allows a small partial pressure (approx. 0.5 bar) to flow in to the top of the diaphragm. Thus, up to this pressure there is no reduction in brake-cylinder pressure. This results in the synchronous application of the brakes on all vehicle axles.

Combination brake cylinder

The combined cylinder in the commercial vehicle consists of a diaphragm cylinder part for the service brake and a spring-type actuator part for the parking brake (Figure 13). They are arranged one behind the other and exert force on a joint push rod. A distinction can be made between combined cylinders for S-cam brakes, wedge-actuated brakes, and disk brakes based on the type of wheel brake.

The two cylinders can be actuated independently of one another. Simultaneous actuation results in the addition of their forces. This can be prevented by installing a special relay valve in order to automatically prevent mechanical overloading of other downstream components (e.g. brake drums).

A central release screw allows for a tensioning of the spring of the spring-type brake cylinder without compressed air having to be applied (mechanical emergency release device). This is necessary to assist fitting or, in the event of failure of the compressed air, to be able to maneuver the vehicle.

When the service brakes are operated, compressed air flows into the diaphragm cylinder and presses the plunger disk and the push rod against the lever in the disk brake. A drop in air pressure releases the brake.

When compressed air flows into the spring-type actuator part, the piston presses the springs together and the brake is released. If the chamber is vented, the spring-type brake cylinder exerts a force via the piston rod on the diaphragm part and presses the push rod into the mechanism of the disk brake via the piston disk.

Trailer control valve

The trailer control valve installed in the tractor controls the trailer's service brake. This multi-circuit relay valve is triggered by both service-brake circuits and by the parking brake (Figure 14). In the driving mode, supply chamber III and chamber IV of the parking-brake circuit are under the same pressure. The brake line to the trailer is connected to the atmosphere via the central venting. A pressure increase in chamber I of brake circuit 1 and in chamber V of brake circuit 2 leads to the corresponding pressure increase in chamber II for the brake line to the trailer. A pressure drop in both brake circuits also leads to the same pressure drop in the brake line. Operation of the parking-brake system leads to venting of the parking-brake circuit (chamber IV). This increases the pressure in chamber II for the brake line to the trailer. When air enters chamber IV, the brake line is vented again.

If the brake line to the trailer is pulled off, it is prescribed that the pressure in the supply line to the trailer must have fallen to a pressure of 1.5 bar in less than two seconds (RREG 71/320). To achieve this, the compressed-air supply to the supply line is throttled by means of an integrated valve.

Electronically controlled brake system

Requirements and functions

With the further development of the dual-line compressed-air brake system, an electronic (or electronically controlled) brake system (EBS) was created in the mid-1990s. On account of the modular design, it is possible to cover different vehicle types with just a few components. Vehicle-specific differences and factors can be covered to a large extent by programming the central ECU accordingly. The control arrangement is determined by the number of axles and their arrangement and the required scope of functions, and ranges from 4S/4M to 8S/6M (S wheel-speed sensor, M pressure-control module).

Design and operating principle

The electronic brake system (Figure 15), like a conventional compressed-air brake with antilock braking system (ABS), comprises a compressed-air supply system, but the pressure-regulator, air-drier and multiple-circuit protection-valve functions can be combined in an electronic unit (EAC, Electronic Air Control). It is thus possible to adapt certain functions – such as

Figure 14: Trailer control valve with decoupling function (driving mode)
1 and 2 Compression spring, 3 Control plunger, 4 Spring assembly, 5 Outlet-valve seat, 6 Disk, 7 Inlet-valve seat, 8 Compression spring, 9 Throttle pin, 10 Housing, 11 and 12 Control plunger, 13 Adjusting screw, 14 Compression spring, 15 Valve disk, 16 Reaction piston, 17 Collar, 18 Control plunger.

I – VIII Chambers.

Ports:
1.1 Energy input from circuit 3,
2.1 Energy output to "Supply" coupling head (red),
2.2 Energy output to "Brake" coupling head (yellow),
4.1 Control port uncontrolled pressure circuit 1,
4.2 Control port uncontrolled pressure circuit 2,
4.3 Control port parking brake,
3 Central vent.

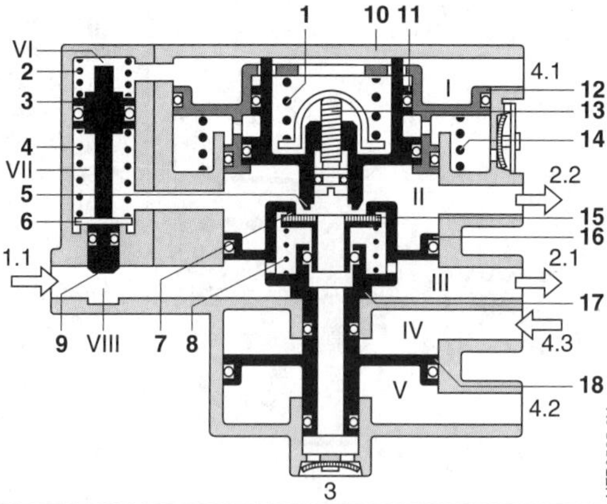

UFB0525-2Y

e.g. fill sequences or regeneration – better to the required conditions and to guarantee even greater functional reliability.

In the electronic brake system too, energy is stored in compressed-air reservoirs and from there made available to the pressure-control modules and to the service-brake valve. The service-brake valve consists of an electric pedal-travel sensor and a pneumatic section which is functionally identical to the previous design. The pedal-travel sensor consists of two redundantly arranged potentiometers which are deflected by means of the control device and supplies two items of opposing voltage information to the central ECU. This in turn activates the pressure-control modules on the front and rear axles so that the required brake pressure is directed into the brake cylinders downstream of the pressure-control modules. The directed brake pressure is monitored by integrated pressure sensors in the pressure-control modules. At the same time, there is a feedback of brake pressure to the pneumatic section of the service-brake valve in order to convey the intensity of the braking.

The brake-pressure modules are available as one-channel or two-channel designs. If the vehicle is set up to tow a trailer, a trailer control module is also provided as a substitute for the trailer control valve. This trailer control module is likewise activated in the braking operation by the central ECU and makes an adapted control pressure available at the "Brake" coupling head (yellow). This makes it possible to carry a conventionally braked trailer. If a trailer is carried with an independent electronic brake system, this is controlled by an electrical connection via the plug-in connection as per ISO 7638 (ABS connector). The trailer must nevertheless also be pneumatically coupled, because this is the only way that the trailer can be supplied with pressure and in the event of a system failure pneumatically controlled. By controlling the electronic brake system in the trailer, it is possible to provide optimum matching with regard to braking performance between the tractor vehicle and the trailer. Simultaneous and matched braking performance facilitate optimized coupling-forcing matching.

Figure 15: Service-brake system of an electronically controlled braking system
a) Tractor vehicle, b) Trailer.
1 Four-circuit protection valve, 2 Air reservoir, 3 Service-brake valve with braking-level sensor,
4 Single-channel pressure-control module, 5 Brake cylinder, 6 Wheel-speed sensor,
7 Brake-lining wear sensor, 8 EBS ECU in tractor vehicle,
9 Two-channel pressure-control module, 10 Pressure sensor, 11 Air-spring bellows,
12 Trailer control valve, 13 "Supply" coupling head (red),
14 "Brake" coupling head (yellow), 15 ISO 7638 plug-in connection (7-pin), 16 Line filter,
17 Trailer brake valve with release device, 18 EBS ECU in trailer.

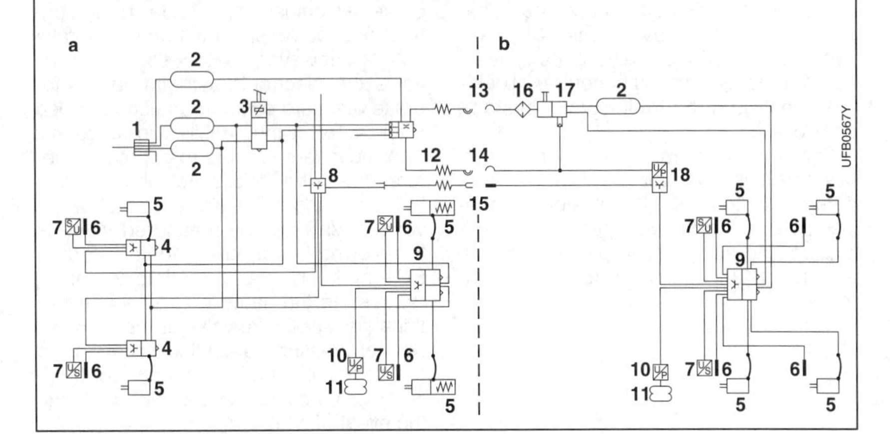

Further functions, such as antilock braking system (ABS), traction control system (TCS) and driving-dynamics control system (Electronic Stability Program, ESP), are integrated into the electronic brake system's scope of functions. The turning behavior of the wheels is monitored by the wheel-speed sensors and the antilock braking system. Depending on the design, the information is made available to the central ECU or to the pressure-control module, where it is processed. In the event of incipient wheel locking, depending on the system arrangement and design, a control intervention is effected via the pressure-control modules or by downstream pressure-control valves in accordance with the control variants known to the ABS system (individual control, modified individual control, or select-low control). Intervention by the traction control system when the wheels spin takes the form of an engine and brake intervention. Further sensors are needed for the functions of the driving-dynamics control system. The steering-wheel angle is recorded by a steering-angle sensor. A yaw-velocity sensor, also known simply as a yaw sensor, records the rotational speed about the vehicle vertical axis. A lateral-acceleration sensor also records the lateral acceleration. When the data have been evaluated, swerving or jackknifing is detected and stabilized by the specific introduction of brake pressure into the relevant brake cylinders and intervention in other systems.

If an electrical fault occurs, the vehicle can be braked by means of one or two redundant pneumatic circuits with at least the demanded secondary braking effect, and the trailer brake system controlled.

Optimum cooperation between all the systems can be achieved through data communication with other systems in the vehicle and the trailer. Optimized deceleration and acceleration processes and additional functions can be realized in this way.

Advantages of electronic brake systems in a commercial vehicle are:

- Fast and simultaneous brake-pressure buildup in all the brake cylinders
- Good metering capability, thus optimum braking comfort
- Optimum matching between tractor and trailer through control of coupling forces
- Exact braking-force distribution
- Uniform brake-lining wear
- ABS, TCS and ESP functions are integrated (brake and engine interventions), traction control for offroad applications can be easily realized
- Driving-dynamics control via engine and brake interventions when oversteering or understeering is detected, in response to the risk of swerving (articulated road train, articulated bus), and intervention in response to the risk of overturning
- Ease of servicing thanks to extensive diagnosis functions.

Components of the electronic brake system

Electronic control unit (ECU)

The control center of an electronic brake system consists of one or more ECUs. At present, there are both systems with a centralized structure (i.e. all software functions are run on a single ECU) and those with a decentralized configuration involving several ECUs.

Service-brake valve

A service-brake valve for the electronic brake system is similar in design to conventional, purely pneumatic service-brake valves. However, in the service-brake valve the electronic setpoint values for brake-pressure control are also recorded (Figure 16). It thus fulfills two functions: Two redundant sensors (e.g. potentiometers) detect the driver's braking command by measuring the actuation travel of the valve push rod. The measured value is transferred redundantly to the central ECU and converted there into a braking request. In the same way as a conventional service-brake valve, the pneumatic control pressure is applied according to the actuation travel. These control pressures are required for "backup" control in the event of a fault.

Pressure-control modules
The pressure-control modules (Figure 17) are the interface between the electronic brake system and the pneumatically actuated wheel brakes. They convert the required braking pressures transmitted via the CAN bus into pneumatic pressures. Conversion is carried out by "proportional solenoids" or an inlet/outlet solenoid combination. A pressure sensor measures the braking pressure delivered. Thus, braking pressure can be controlled in a closed control loop. The electrically activated "backup" valve shuts off the pneumatic control pressures of the service-brake valve in order to permit interference-free electrical pressure control.

Mounting the pressure-control modules close to the wheels means that the electrical wires for connecting the wheel-speed sensors and the brake-lining wear sensors can be kept short. The information is prepared in the pressure-control module and transmitted via the CAN bus to the central ECU. This reduces the outlay on cabling in the vehicle.

Figure 16: Service-brake valve with two pneumatic control circuits
1 Braking-level sensor,
2 Service-brake valve,
3 Supply connection,
4 Potentiometer connection,
5 Ground connection.

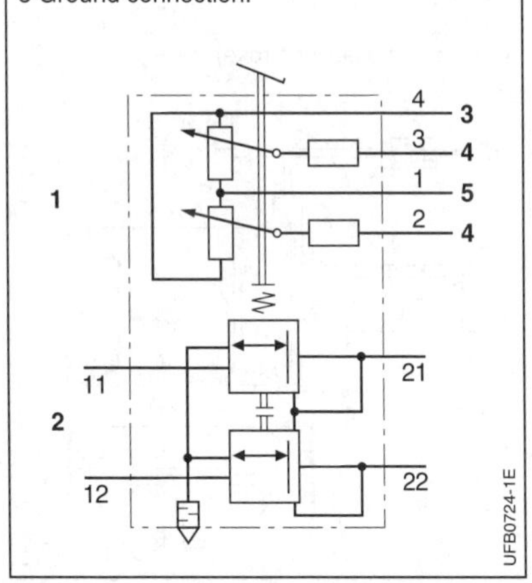

Trailer control module
The electronic trailer control module enables modulation of the trailer control pressure according to the functional requirements of the electronic brake system. The limits of the electrical control ranges are defined by legal requirements. The electronically specified setpoint value is converted into a physical braking pressure by means of a solenoid arrangement similar to that in the pressure-control module. The "backup" pressure is shut off either by a "backup" solenoid or by pneumatic retention, depending on the type of design adopted.

Under all normal conditions, the trailer control module must be activated by two independent control signals. This may be two pneumatic signals from two control circuits, or one pneumatic and one electrical control signal.

Figure 17: Single-channel pressure-control module
1 ECU,
2 Speed sensor,
3 Brake-lining sensor,
4 CAN,
5 "Backup" valve,
6 Inlet valve,
7 Outlet valve,
8 Pressure sensor,
9 Filter,
10 Relay valve,
11 Muffler,
12 "Backup" circuit,
13 Supply,
14 Brake cylinder.

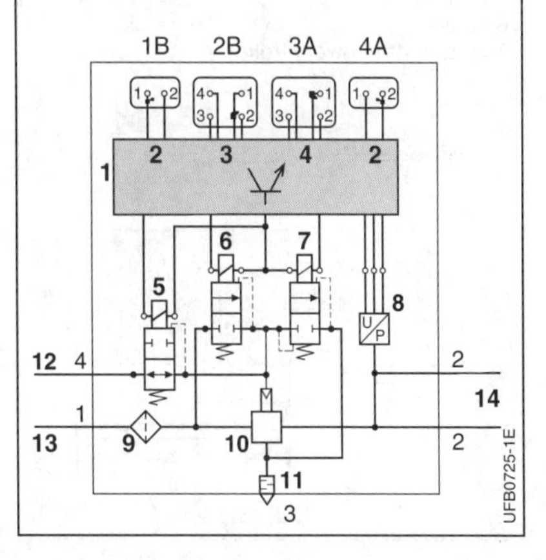

Continuous-operation brake systems

Commercial vehicles essentially use two types of continuous-operation brake systems, deployed separately or in combination: the exhaust-brake system and the retarder.

Exhaust-brake systems

The resistance which an engine brings to the speed imposed from the outside without a fuel supply is termed the engine or exhaust brake, or drag power. The drag power of standard engines is 5 to 7 kW per liter displacement. The requirements as laid down in §41, s.15 cannot be observed with a pure exhaust brake. Further measures can be used to increase the effect of the exhaust brake.

Exhaust-brake system with exhaust flap

In the exhaust brake with exhaust flap, a valve with a flap closes the exhaust train. The fuel supply is interrupted at the same time. As a result, back pressure is generated in the exhaust-gas system and must be overcome by each piston during its exhaust stroke (Figure 18). The braking power can be regulated by means of a pressure-control valve in the exhaust train. Additionally, this valve ensures that at high revs an excessively high pressure does not result in valve or valve-gear damage.

The exhaust brake is the most common variant used in trucks and buses, delivering a braking power of 14 to 20 kW per liter displacement.

Exhaust-brake system with constant throttle

The exhaust brake with constant throttle is also known as a decompression brake. In this system, the work performed by the engine in the compression phase is not utilized. The exhaust valves or an additional valve (constant throttle, Figure 19) is/are specifically opened at the end of the compression cycle, thereby relieving the pressure built up in the compression phase. Thus, no further work can be delivered to the crankshaft in the expansion phase.

Engine brake system with exhaust flap and constant throttle

Braking power can be further improved by a combination of exhaust flap and constant throttle (Figure 19). This combination can deliver braking power of 30 to 40 kW per liter displacement.

Figure 18: Exhaust brake with exhaust flap and additional pressure-control valve
1 Exhaust-flap actuation (compressed air),
2 Exhaust flap,
3 Bypass,
4 Pressure-control valve,
5 Exhaust,
6 Intake,
7 Piston (4th power stroke).

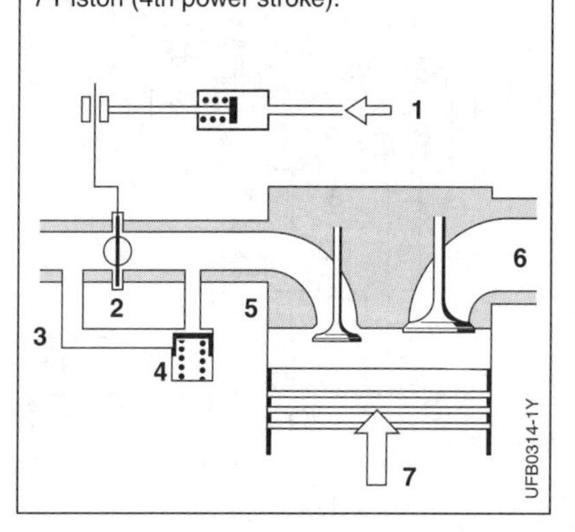

Figure 19: Exhaust brake with exhaust flap and constant throttle
1 Compressed air,
2 Exhaust flap,
3 Exhaust,
4 Constant throttle,
5 Intake,
6 Piston (2nd power stroke).

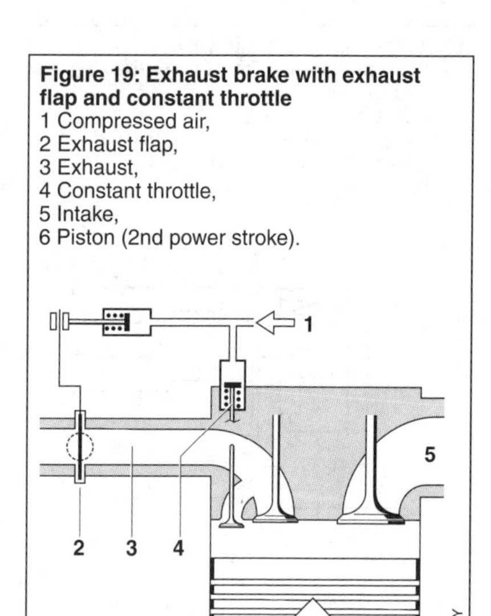

Retarder
Retarders are wear-free continuous-operation brakes. There are two types that differ in how they operate: hydrodynamic and electrodynamic retarders. Both systems, and the exhaust-brake systems, relieve the load on the service brake system and thereby increase the economic efficiency of the vehicle. The use of a hydrodynamic retarder can increase the service life of the service brake by a factor of 4 to 5.

In modern vehicles, retarders are incorporated into the brake-management system. The exhaust brake and the retarder are often combined as a continuous-operation brake in a vehicle. The brakes must then be activated by means of the electronic brake-management system.

Hydrodynamic retarder
Hydrodynamic retarders, also known simply as hydraulic retarders, can be subdivided into the categories of primary retarders and secondary retarders.

The primary retarder is located between the engine and the transmission, the secondary retarder between the transmission and the powered axle. Both primary and secondary retarders operate in the same way. When the retarder is activated, oil is pumped into the working area. The driven rotor accelerates this oil and transfers it at the outside diameter to the stator (Figure 20). There the oil strikes the static stator blades and is decelerated. The oil flows at the inside diameter to the rotor. The rotor's rotary motion is inhibited, and the vehicle is decelerated.

The kinetic energy is primarily converted into heat. For this reason, some of the oil must be permanently cooled by a heat exchanger.

The braking torque can be input using a hand lever or the brake pedal (in the case of a retarder integrated into the EBS electronically controlled brake system). A retarder's braking torque is dependent on the degree of fill in the working area between the rotor and the stator. The degree of fill is regulated by an ECU via a control pressure which is adjusted by proportioning valves.

A retarder can be activated hydraulically or pneumatically, whereby the braking torque can be achieved in discrete braking stages and also steplessly. Oil is primarily used as the working medium in a retarder. Current hydraulic retarders can deliver braking power up to 600 kW for brief periods. However, the continuous braking power of a retarder is dependent on the cooling capacity of the vehicle's cooling system. Modern vehicles can dissipate a continuous braking power from a retarder of 300 to 350 kW via the cooling system. Sensors are used to record overheating of the retarder or the cooling system, and if necessary the braking power is reduced under controlled conditions until the braking power equals the dissipatable quantity of heat.

Primary retarder
In the case of a primary retarder located in the drivetrain between the engine and transmission after the converter, force is

Figure 20: Functioning principle of a retarder as illustrated by a ZF intarder by way of example
1 High-speed stage,
2 Output flange,
3 Stator, 4 Rotor.

transmitted through the powered axles and the transmission in such a way that the total overrun torque is directed through the transmission. The braking effect of the primary retarder is dependent on the engine speed and the selected gear, but is not dependent on the vehicle's output speed and driving speed. This lack of dependence on the output speed is one of the major advantages of primary retarders. These are highly effective at speeds below 25 to 30 km/h (Figure 21). This is the reason why primary retarders are primarily used in vehicles which are driven at lower average speeds such as, for example, city buses and municipal vehicles. Their compact design is another advantage. A disadvantage of primary retarders is that the braking force is interrupted during a gear change. The braking force must be reduced during gear changing.

Figure 21: Operating ranges of primary and secondary retarders
1 Primary retarder, 2 Secondary retarder.

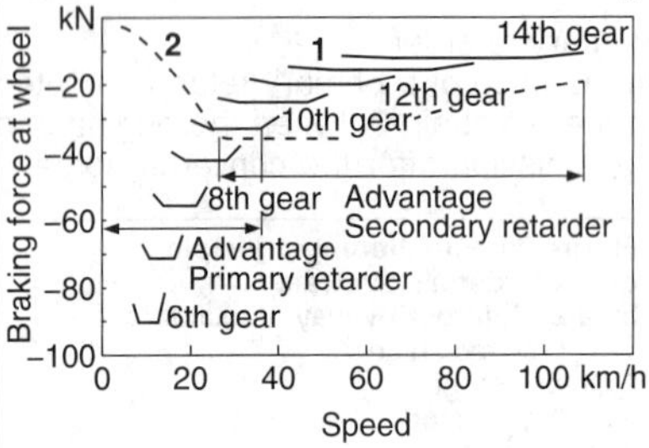

SFB0805-3E

Secondary retarder
In the case of a secondary retarder (Figure 22), which is located after the engine, clutch and transmission, force is transmitted directly via the output shaft. Unlike a primary retarder, there is no interruption of the braking force during a gear change with a secondary retarder. The braking effect is dependent on the ratio of the output shaft and on the driving speed. It is not dependent on the selected gear. The braking torque of a secondary retarder is very much dependent on the rotor speed. For this reason, the rotor speed is often increased by means of a high-speed stage.

The secondary retarder demonstrates great efficiency at speeds over 40 km/h (Figure 21); at speeds below 30 km/h the braking torque falls off dramatically. Because of its design, the secondary retarder can also be subsequently adapted to a transmission. The extra weight of a secondary retarder with accompanying heat exchanger and oil fill is often cited as a disadvantage, since the additional mass reduces the vehicle payload.

Secondary retarders are primarily used in long-distance vehicles which are driven at high average speeds, such as trucks and tour buses.

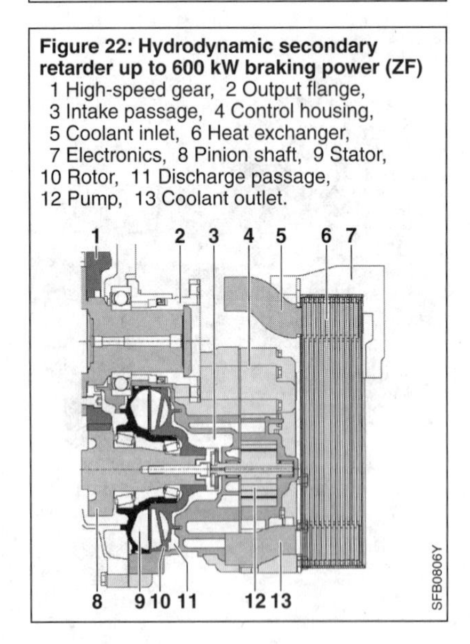

Figure 22: Hydrodynamic secondary retarder up to 600 kW braking power (ZF)
1 High-speed gear, 2 Output flange, 3 Intake passage, 4 Control housing, 5 Coolant inlet, 6 Heat exchanger, 7 Electronics, 8 Pinion shaft, 9 Stator, 10 Rotor, 11 Discharge passage, 12 Pump, 13 Coolant outlet.

Electrodynamic retarder (eddy-current brake)

The electrodynamic retarder (Figure 23) contains two non-magnetizable steel disks (rotors) which are connected with torsional strength to the input and output shafts (here: propshaft), and a stator which is equipped with 8 or 16 coils and fastened by means of a star-shaped bracket to the vehicle frame. As soon as electric current (from the alternator or battery) flows through the coils, magnetic fields are generated which are closed by the rotors. These magnetic fields induce eddy currents in the rotors as they turn. These eddy currents in turn generate magnetic fields in the rotors which counteract the exciting magnetic fields and thus build up a braking effect. The braking torque is determined by the strength of the excitation field, the rotational speed, and the air gap between the stator and the rotors. The braking torque decreases as the air gap increases; this air gap can be adjusted by means of spacers. Shift stages with different braking torques (Figure 24) are obtained by interconnecting the field coils in different configurations. The heat generated is dissipated by convection and radiation to atmosphere via the internally ventilated rotor disks.

As the rotors are increasingly heated, the braking power of the electrodynamic retarder decreases significantly (Figure 25). The retarder's braking power is reduced by thermal protection in order to prevent the retarder from being destroyed by excessive temperature during braking operation.

Figure 23: Electrodynamic retarder
1 Star-shaped bracket,
2 Rotor, transmission side,
3 Spacers (for adjusting air gap),
4 Stator with coils, 5 Intermediate flange,
6 Rotor, rear-axle side, 7 Transmission cover,
8 Transmission output shafts, 9 Clearance gap.

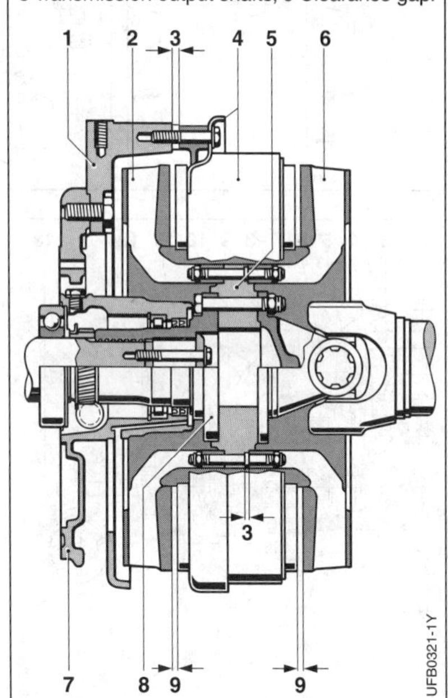

Like the primary retarder, the electrodynamic retarder is distinguished by high braking power at low engine speeds and relative design simplicity. On the downside, however, it can weigh up to 350 kg, depending on its size.

References

[1] E. Hoepke, S. Breuer (Editors): Nutzfahrzeugtechnik, 4th Edition, Vieweg Verlag, 2006.

Figure 24: Braking-torque characteristic of an electrodynamic retarder
4a Braking power when the cooling power limit has been reached (switching stage 4)

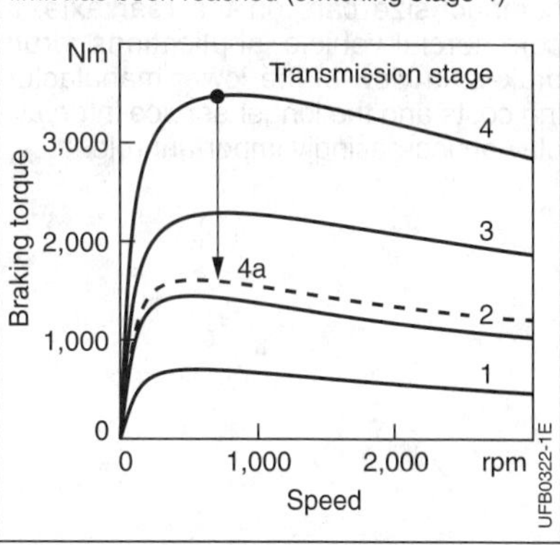

Figure 25: Influence of transmission ratio and rotor temperature on the performance of electrodynamic retarders
17 t commercial vehicle, laden.

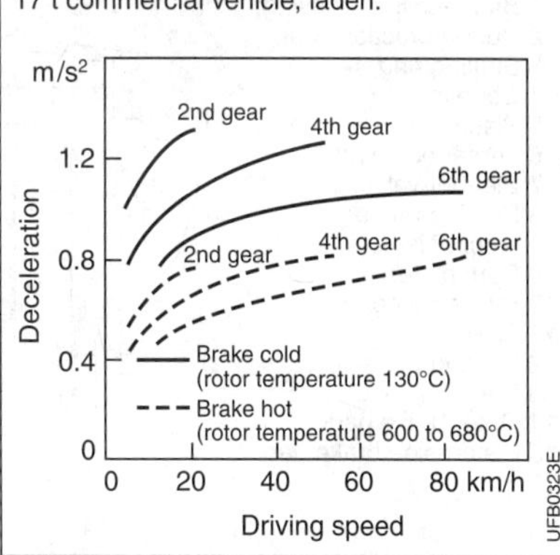

Wheel brakes

Wheel brakes are friction brakes that convert kinetic energy into heat energy during braking. Disk and drum brakes are used as wheel brakes. Hydraulic pressure (for passenger cars) and pneumatic pressure and spring force (spring-loaded brake, for commercial vehicles) are converted into an application force to press the brake pads and linings against the brake disks and drums respectively.

In passenger-car applications the thermal demands placed on wheel brakes can, in view of ever-increasing vehicle weights and higher attainable driving speeds, only be satisfied by disk brakes. Drum brakes are now only used in subcompact-size cars on the rear axle. In commercial-vehicle applications drum brakes, in view of the lower manufacturing costs and the longer service intervals, play an increasingly important role.

Disk brakes

Functioning principle

Disk brakes generate the braking forces on the surface of a brake disk that rotates with the wheel (Figure 2). The U-shaped brake caliper with the brake pads is mounted to non-rotating vehicle components (wheel carrier). The floating-caliper design with or without parking-brake mechanism has proven successful.

Figure 2: Disk brakes (diagram)
a) Fixed-caliper brake,
b) Floating-caliper brake.
1 Brake pads,
2 Piston,
3 Brake disk,
4 Brake-caliper housing,
5 Brake anchor plate.

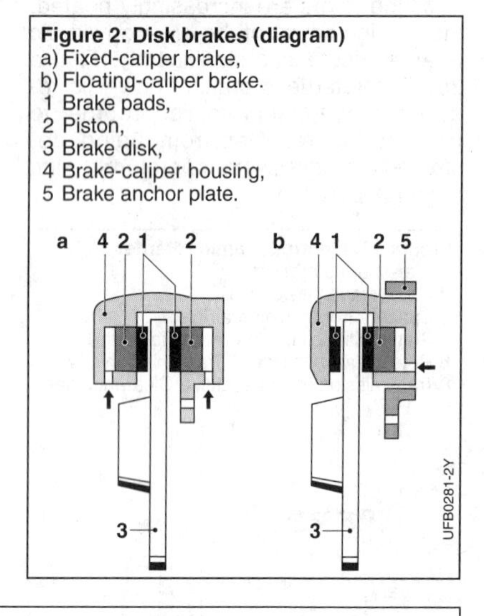

Figure 1: Floating-caliper brake with parking-brake mechanism
1 Brake-caliper housing,
2 Dust-protection seal,
3 Sealing ring,
4 Coupling,
5 Piston,
6 Threaded spindle,
7 Bleeder valve,
8 Parking-brake mechanism,
9 Cam plate,
10 Hydraulic port,
11 Shaft,
12 Parking-brake lever,
13 Spring,
14 Outer brake pad,
15 Piston-side brake pad.

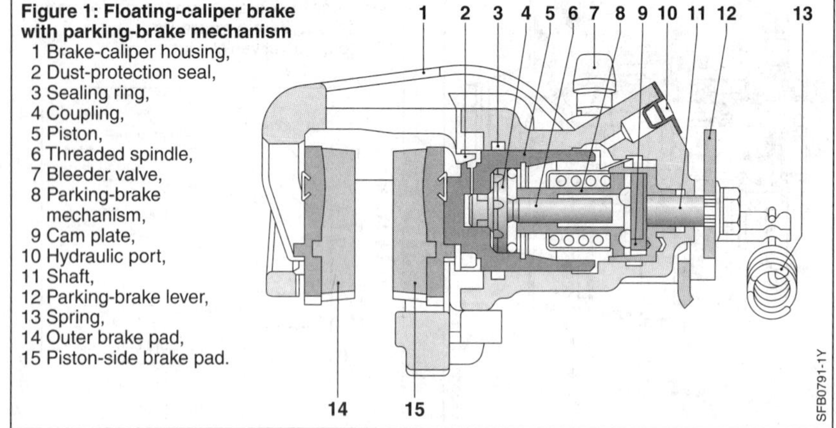

Principle of the fixed-caliper brake
In a fixed-caliper brake, both halves of the housing (flange and cover parts) are joined by the housing connecting bolts. Each half of the housing contains a piston to press the brake pad against the brake disk (Figure 2a). Ports in the housing halves connect the two pistons hydraulically.

Principle of the floating-caliper brake
In a floating-caliper brake, a piston presses the piston-side (inner) brake pad against the brake disk (Figure 2b). The generated reaction force moves the brake-caliper housing and thereby presses the outer brake pad indirectly against the brake disk. In this brake caliper the piston is therefore only seated on the inner side.

Floating-caliper brake for passenger cars
The brake caliper can be moved axially in the brake anchor plate and is guided by two sealed guide pins in the brake anchor plate (Figure 1).

Braking with the service brake
The hydraulic pressure generated by the brake master cylinder enters the cylinder chamber behind the piston via the hydraulic connection. The piston is shifted forwards and the brake pad on the piston side is applied to the brake disk. The reaction force that arises shifts the brake-caliper housing mounted in a bolt guide against the direction of piston movement. This also means that the outer brake pad is applied to the brake disk. The path of the brake pads and of the piston covered up to that point is referred to as clearance. Another increase in pressure increases the downforce of the brake pads.

Releasing the service brake
When the piston moves through the clearance, the sealing ring, which is rectangle in its initial position, is deformed. The deformed sealing ring pulls the piston back by the clearance when the hydraulic pressure drops (roll-back effect).

Braking with the parking brake
When the parking brake is operated, the force is transferred via the handbrake cable to the parking-brake lever. This is then twisted and the rotary motion is transferred via the shaft to the cam plate. As the balls run onto the cam plate, the piston is shifted via the pressure sleeve in the parking-brake mechanism; the threaded spindle bolted in this mechanism is shifted towards the brake pad. After crossing the clearance, first the brake pad on the piston side and then the outer brake pad are pressed against the brake disk.

Releasing the parking brake
After releasing the handbrake lever, the parking-brake lever, the shaft and the cam plate turn back to their initial positions. The pressure sleeve, the threaded spindle and the piston are pressed back into their initial position by the springs in the parking-brake mechanism. The final clearance is reached as the sealing ring reassumes its shape.

Automatic self-adjusting mechanism
Wear on the brake pads and brake disks increases the clearance and thus has to be balanced out. This automatic clearance compensation takes place during braking. The inside diameter of the rectangular piston sealing ring is slightly smaller than the piston diameter. The sealing ring thus surrounds the piston with a pre-tension. During braking the piston moves towards the brake disk and tensions the sealing ring, which as a result of its static friction can then slip on the piston only when the piston travel between brake pad and brake disk has in response to abrasion on the brake pads has become greater than the envisaged clearance. When the brake is released the piston is pulled back only by the envisaged clearance. In this way, stepless readjustment to a constant clearance is possible.

The clearance compensation of the parking-brake mechanism also takes place on application of the service brake.

The clearance of a brake caliper is approx. 0.15 mm and is thus in the range of the maximum permissible static disk run-out (axial movement per brake-disk rotation on account of manufacturing tolerances or bearing clearances).

Brake disks

The energy converted during braking into heat is mainly absorbed by the brake disk and then dissipated to the ambient air. Perforated or grooved solid disks improve the cooling effect and also reduce water susceptibility. Internally ventilated brake disks with radial cooling ducts are used – primarily on the front wheels – to further improve heat dissipation (Figure 3).

Brake disks are usually manufactured from gray cast iron. Alloys with chrome and molybdenum increase the resistance to wear, carbon improves the heat-absorption capability. More thermally resistant brake disks made of ceramic containing silicon carbide reinforced with carbon fiber are also used on sports cars.

Brake pads

Brake pads are essentially made from four raw-material groups (Table 1 shows an example of main constituents), where the proportions of these raw-material groups differ depending on the area of application and the required coefficient of sliding friction (coefficient of friction for parking brakes). Thus, for example, the brake pads of a luxury-class car have a different make-up from the drum-brake linings of a subcompact-size car. The friction-pad recipes are closely guarded secrets of the pad manufacturers.

Wear sensors in the disk-brake pads complete a circuit when a minimum pad thickness is reached through contact with the brake disk; this causes a warning lamp to light up, alerting the driver that a pad change is necessary.

Brake pads and brake disks must be replaced in pairs on each axle so as to ensure an identical braking effect at both wheels on an axle. Only parts approved by the vehicle manufacturer may be used.

Table 1: Recipe for a disk-brake pad (example)

Raw-material group	Raw materials	% by vol.
Metals	Steel wool, copper wool	14
Fillers	Aluminum oxide, mica powder, heavy spar, iron oxide	23
Lubricants	Antimony sulfide, graphite, coke powder	35
Organic constituents	Aramide fiber, resin filler powder, binding resin	28

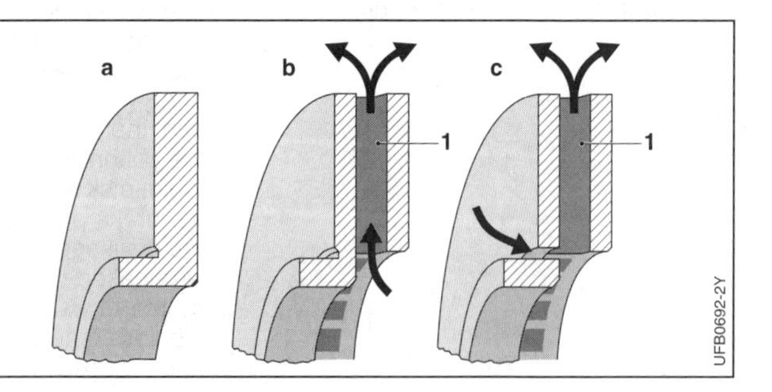

Figure 3: Brake disks
a) Solid brake disk,
b) Internally ventilated brake disk,
c) Externally ventilated brake disk.
1 Cooling duct.

Disk brakes for commercial vehicles
Special disk brakes have been developed for commercial vehicles. These disk brakes are actuated with compressed air. Because the pressure here is much smaller than that in a hydraulic brake, the brake cylinders cannot be integrated into the brake calipers. They must be flanged into position (Figure 4).

Operating concept of the service brake
When air enters the service brake cylinder, the eccentrically mounted brake lever is actuated. The brake cylinder force is boosted by the lever ratio and transferred via the bridge and plungers to the inner brake pad. The reaction force that arises at the brake caliper is transferred by shifting the brake caliper to the outer brake pad.

Operating concept of the parking brake
When the spring-type brake cylinder is vented, the force of the pre-tensioning springs is released. Via the spring-type brake piston, this moves the piston and the push rod of the service brake cylinder to operate the brake. In the case of the parking brake, the pressure in the spring-type brake actuator is completely eliminated, thus releasing the force of the pre-tensioning springs to achieve maximum braking effect.

Automatic self-adjusting mechanism
Disk brakes pneumatically or mechanically actuated by the spring-type actuator are fitted with automatic clearance compensation.

Wear monitoring
Continuous wear monitoring may also be provided. This is required in the case of electronic brake systems for wear adaptation and for service information systems.

Brake disks
Solid brake disks are rarely used on commercial vehicles because they can only dissipate the heat slowly. Internally ventilated brake disks have a larger surface through which heat exchange can take place. In this design two friction rings are connected via bridges. The rotation of the brake disk creates on the inside a radial ventilation effect in the outward direction.

Brake disks for commercial vehicles are usually manufactured from gray cast iron. The carbon content increased up to the saturation limit provides good thermal conductivity.

Figure 4: Disk brake with combination brake cylinder
1 Brake caliper, 2a Inner brake pad, 2b Outer brake pad,
3 Brake disk, 4 Plunger, 5 Bridge,
6 Eccentrically mounted brake lever,
7 Service-brake cylinder, 8 Spring-type actuator.

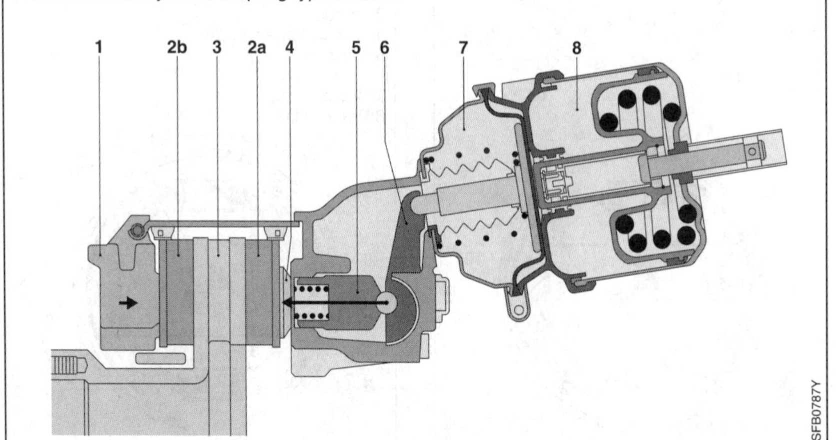

Drum brakes

Drum brakes are radial brakes with two brake shoes. They generate their braking force on the inner friction surface of a brake drum.

Drum-brake designs

There are two different designs of drum brake, based on how the brake shoes are guided:
- brake shoes with fixed pivot (Figures 5a and 5b),
- brake shoes as sliding shoes (Figure 6).

Figure 5: Principle of simplex brake
a) Brake shoe with two single pivots,
b) Brake shoe with one double pivot.
1 Direction of rotation of brake drum with vehicle moving forwards,
2 Self-augmentation, 3 Self-inhibition,
4 Torque,
5 Double-acting wheel-brake cylinder,
6 Leading brake shoe (primary shoe),
7 Trailing brake shoe (secondary shoe),
8 Fulcrum (pivot),
9 Brake drum, 10 Brake lining.

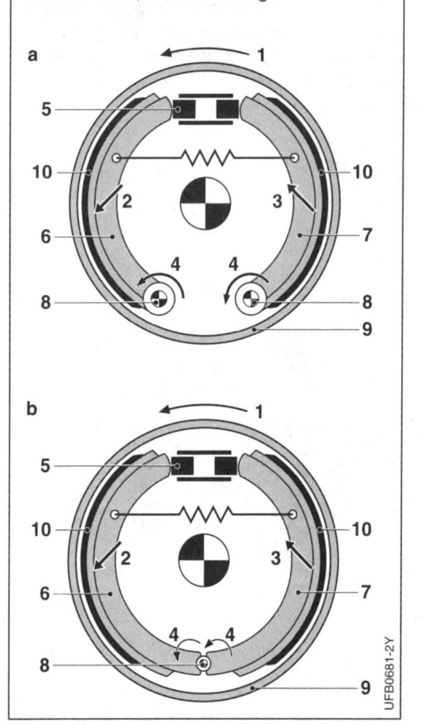

At the brake shoe that rotates in the direction of the brake drum (primary shoe, leading brake shoe, Figure 5), the friction force during the braking operation creates a turning force around the brake-shoe fulcrum which in addition to the application force presses the shoe against the drum. This generates a self-augmenting effect. In a simplex brake a turning force is created around the fulcrum of the trailing brake shoe (secondary shoe) which diminishes the applied application force. This therefore creates a self-inhibiting effect.

Sliding-shoe guides are used in simplex, duplex, duo-duplex, servo and duo-servo brakes. Brake shoes with a fixed pivot are subject to unequal levels of wear in that they cannot center themselves like sliding shoes.

Principle of the simplex brake

A double-acting wheel-brake cylinder actuates the brake shoes (Figures 5a and 5b). The fulcrums of the brake shoes are pivots (two single pivots or one double pivot). When the vehicle is moving forwards, self-augmentation affects the leading brake shoe and self-inhibition affects the trailing brake shoe; the pattern is the same when the vehicle is backing up.

Figure 6: Principle of duplex brake
1 Direction of rotation of brake drum with vehicle moving forwards,
2 Self-augmentation,
3 Torque,
4 Wheel-brake cylinder,
5 Fulcrums,
6 Brake shoes,
7 Brake drum,
8 Brake lining.

Principle of the duplex brake
Each brake shoe is actuated by a single-acting wheel-brake cylinder (Figure 6). The brake shoes designed as sliding shoes are supported on the back of the opposing wheel-brake cylinder. The duplex brake is single-acting, i.e. it has two leading self-augmenting brake shoes when the vehicle is moving forwards. There is no self-augmentation when the vehicle is backing up.

Simplex drum brake
Functioning principle of passenger-car brake
The principle of a drum brake is explained using a hydraulically operated simplex drum brake with integrated parking-brake system and automatic self-adjusting mechanism as an example (Figure 7). Other drum brake designs (e.g. duplex brake, duo-duplex brake) are rarely used today.

When driving extension springs pull the two brake shoes away from the brake drum so that a clearance is created between the drum friction surface and the brake linings. In the case of simplex brakes, a two-sided hydraulic wheel-brake cylinder generates the application force for the brake shoes during braking by

Figure 7: Simplex drum brake with integrated parking brake
1 Wheel-brake cylinder, 2 Brake lining, 3 Extension spring (for brake shoes),
4 Extension spring (for adjuster), 5 Trailing brake shoe,
6 Brake drum, 7 Parking-brake lever, 8 Brake cable,
9 Direction of drum rotation, 10 Thermocouple (adjuster),
11 Adjuster wheel (with elbow lever), 12 Leading brake shoe,
13 Brake anchor plate, 14 Extension spring (for brake shoes),
15 Brake-shoe pin bushing.

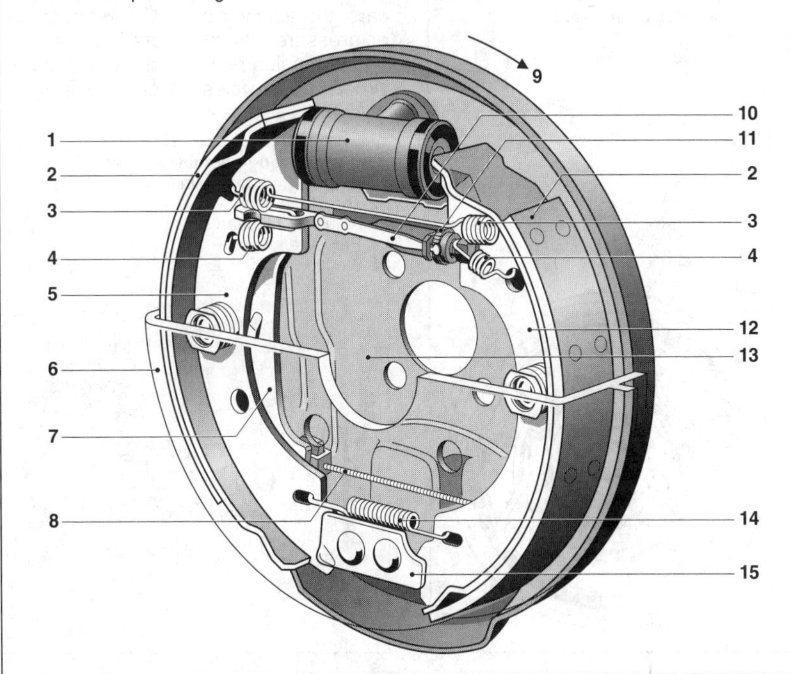

converting the hydraulic pressure into mechanical force. Here, the leading and the trailing brake shoes with the brake pads press against the brake drums. The other ends of the brake shoes on the opposite side to the wheel-brake cylinder are braced by a support bearing that is attached to the brake anchor plate.

The leading brake shoe (primary shoe) generates a higher proportion of braking torque than the trailing brake shoe (secondary shoe). Wear is therefore greater on the primary lining. This lining is thicker or longer in design to compensate.

Functioning principle of simplex brake with S-cam for commercial vehicles
In commercial vehicles with compressed-air brake systems, the application force is frequently generated by a rotatable S-cam. S-cam rotation is effected by the brake cylinder, the brake lever (slack adjuster), and the brake-cam shaft (Figure 8).

Figure 8: Simplex drum brake with S-cam
1 Diaphragm-type cylinder,
2 S-cam,
3 Brake shoes,
4 Return spring,
5 Brake drum.

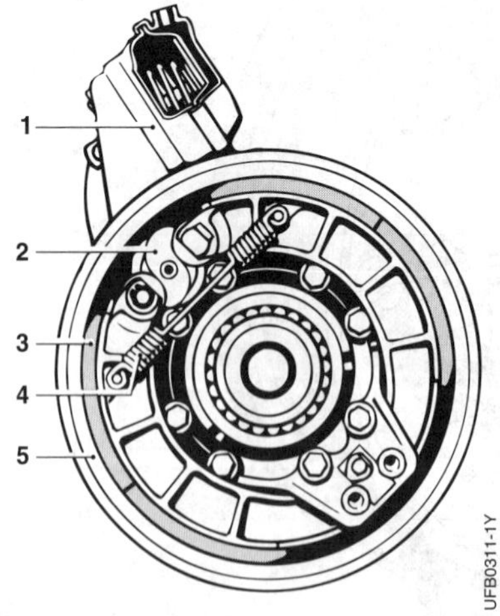

Functioning principle of wedge-actuated brake
Wedge-actuated brakes are also used in commercial vehicles. Here, the application force for the brake shoes is generated by a wedge actuated by the brake cylinder (Figure 9).

During braking the diaphragm brake cylinder is exposed to compressed air. This displaces the wedge to the right. The wedge slides between the pressure rollers. These roll on the wedge and the thrust members. The generated application force is transferred via the thrust members to the brake shoes. The excessive clearance created by brake-lining wear is compensated for by the adjusting mechanism.

Automatic adjusters
Wheel brakes must be fitted with adjusters to compensate for the increased clearance caused by lining wear. The brakes must be easily adjustable or have an automatic adjuster (§ 41 s. 1 StVZO, ECE R13-H [1]).

On simplex drum brakes for passenger cars, the adjuster is part of the push rod or pressure sleeve situated under initial spring tension between the brake shoes. When the permissible clearance is ex-

Figure 9: Wedge-actuated brake
1 Diaphragm brake cylinder,
2 Thrust member,
3 Adjusting mechanism,
4 Pressure rollers,
5 Wedge.

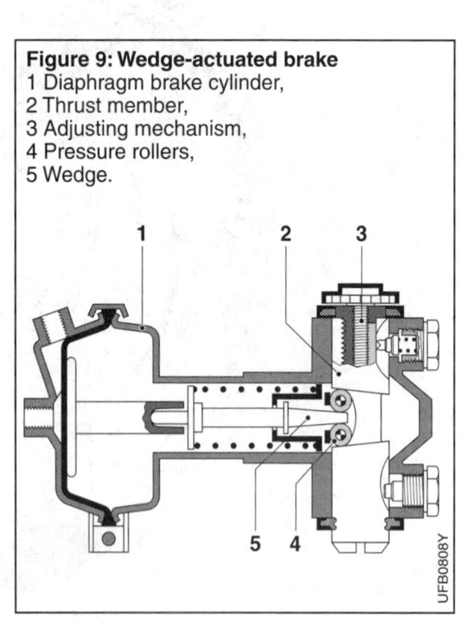

ceeded, the adjuster automatically lengthens the push rod or pressure sleeve (to different extents depending on the adjuster design) and thereby adjusts the clearance between brake shoe and brake drum. Automatic adjusters mostly operate in conjunction with a thermocouple on a temperature-sensitive basis in order to prevent adjustment when the brake drum is hot (expanded).

In commercial vehicles with S-cams, the adjuster is part of the brake lever (manual adjustment). Adjustment is automatic when automatic slack adjusters are fitted.

In wedge-actuated brakes, an automatically acting adjuster is integrated into the wedge mechanism.

Parking brake
The drum brake can also be operated as a parking brake by means of the handbrake lever and handbrake cable. The handbrake lever is mounted at the top of the trailing brake shoe. When the parking-brake system is operated, the brake cable pulls the handbrake lever downwards to the right, causing the handbrake lever to press the brake shoes via the push rod against the brake drum.

References
[1] ECE R13-H: Regulation No. 13-H of the United Nations Economic Commission for Europe (UN/ECE) – Uniform provisions concerning the approval of passenger cars with regard to braking.
[2] B. Breuer, K. Bill (Editors): Bremsenhandbuch: Grundlagen, Komponenten, Systeme, Fahrdynamik. 4th Ed., Vieweg+Teubner, 2012.

Antilock braking system

Function and requirements

Antilock braking systems (ABS) are braking-system closed-loop control devices which prevent wheel lock when braking and, as a result, retain the vehicle's steerability and stability. In general, they also shorten braking distances compared with braking scenarios when the wheels lock completely. This is particularly the case on wet roads. The reduction in braking distance may be 10% or several times this figure, depending on how wet the conditions are and the road/tire friction coefficient. Under certain, very specific road-surface conditions, braking distances may be longer, but the vehicle still retains vehicle stability and steerability.

The requirements placed on an ABS system are described in the regulations ECE-R13 [1]. This regulation defines ABS as a component of a service-brake system (Figure 1) which automatically controls wheel slip in the direction of wheel rotation on one or more wheels when braked.

ECE-R13 Annex 13 defines three categories. The present generation of ABS meets the highest level of requirements (Category 1).

Operating principle

Pressure modulation

A 2/2-way solenoid valve (inlet valve) with two hydraulic connections and two switching positions is fitted between the brake master cylinder and the wheel-brake cylinder of a conventional braking system (Figure 2). When the valve is open (normal setting for standard braking action), braking pressure can be generated in the wheel-brake cylinder. The outlet valve, also a 2/2-way solenoid valve, is closed at this point.

If the wheel-speed sensor detects an abrupt deceleration of the wheel (risk of wheel lock), the system prevents any further increase of braking pressure at the wheel concerned. The inlet and outlet valves are closed, and the braking pressure remains constant.

If the wheel deceleration rate continues to increase, the outlet valve has to be opened. The pressure in the wheel-brake cylinder then drops and the wheel is braked less heavily. The brake fluid escaping into the intermediate reservoir is then pumped back to the brake master cylinder by the return pump.

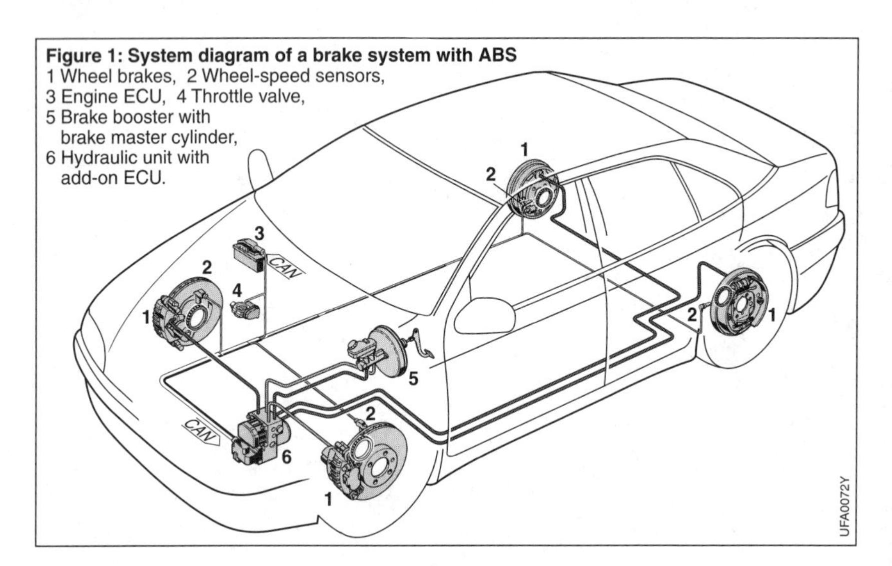

Figure 1: System diagram of a brake system with ABS
1 Wheel brakes, 2 Wheel-speed sensors,
3 Engine ECU, 4 Throttle valve,
5 Brake booster with brake master cylinder,
6 Hydraulic unit with add-on ECU.

Figure 2: Design of ABS
1 Brake master cylinder,
2 Wheel-brake cylinder,
3 Hydraulic unit,
4 Pulsation-damping features,
5 Flow restrictor,
6 Return pump,
7 Inlet valve,
8 Outlet valve,
9 Intermediate accumulator.

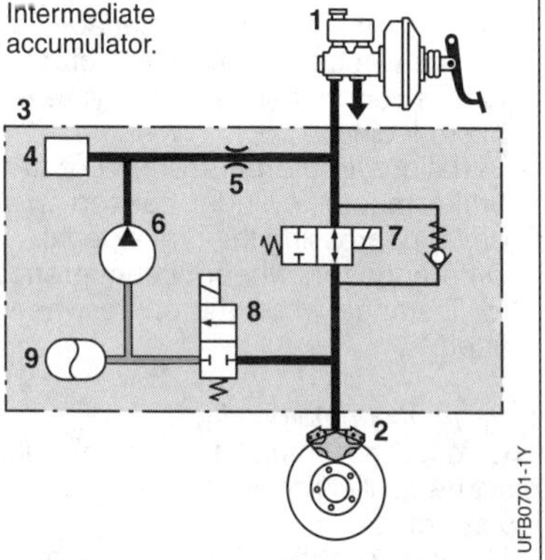

Slip

Slip occurs when the speed v_R at which the wheel center moves in the longitudinal direction (vehicle speed) differs from the speed v_U at which circumference rolls. Brake slip λ is calculated as follows:

$$\lambda = \frac{(v_U - v_R)}{v_R} \cdot 100\,\% .$$

In the case of a locked wheel, brake slip $\lambda = -100\,\%$ according to this definition.

On initial braking, braking pressure increases; brake slip λ rises in terms of amount and at the maximum point on the adhesion/slip curve (Figure 3), it reaches the limit between the stable and unstable ranges. From this point on, any further increase in braking pressure or braking torque does not cause any further increase in braking force F_B (Figure 4). In the stable range, brake slip is largely skidding, it increasingly tends to slipping in the unstable range.

There is a more or less sharp drop in the coefficient of friction μ_{HF}, depending on the shape of the slip curve. Without ABS, the resulting excess torque causes the wheel to lock very quickly when braked.

Basic closed-loop control process

Control processes

The wheel-speed sensor senses the state of motion of the wheel (Figure 5). If one of the wheels shows signs of incipient lock, there is a sharp rise in peripheral wheel deceleration and in wheel slip. If these exceed defined critical levels, the ABS controller sends commands to the solenoid-valve unit (hydraulic unit) to stop increasing or to reduce wheel brake pressure until the danger of wheel lock is averted. The braking pressure must then rise again to ensure that the wheel is not underbraked. During automatic brake control, the stability or instability of wheel motion must be detected constantly, and kept within the slip range at maximum

Figure 3: Adhesion/slip curve
Curve shape for dry road surface, $\mu_{HF} \approx 0.8$.
1 Braking or tractive force, 2 Lateral force.

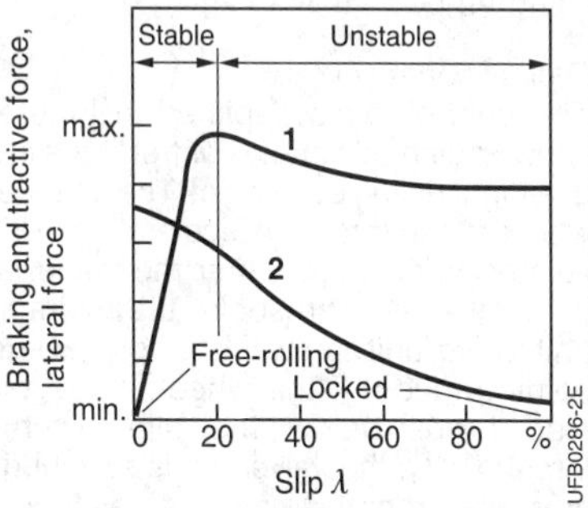

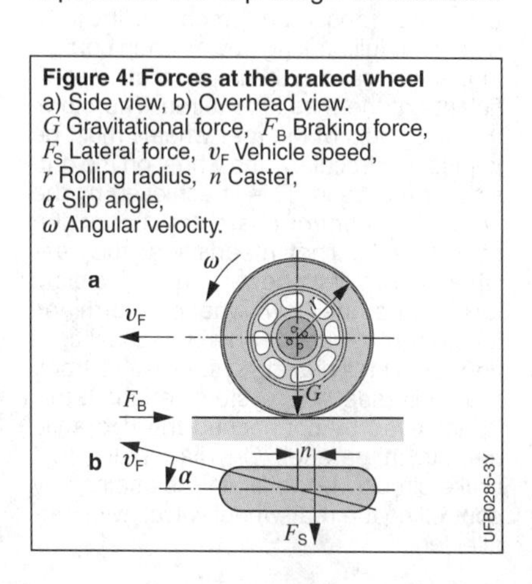

Figure 4: Forces at the braked wheel
a) Side view, b) Overhead view.
G Gravitational force, F_B Braking force,
F_S Lateral force, v_F Vehicle speed,
r Rolling radius, n Caster,
α Slip angle,
ω Angular velocity.

Figure 5: ABS control loop
1 Brake pedal,
2 Brake booster,
3 Brake master cylinder with fluid reservoir,
4 Wheel-brake cylinder,
5 Wheel-speed sensor,
6 Indicator lamp.

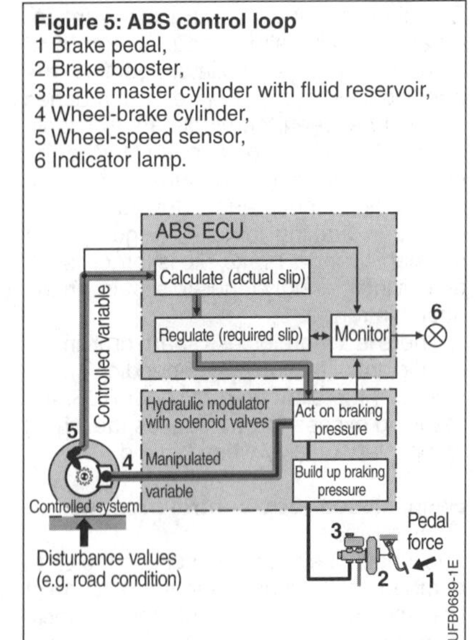

braking force by a sequence of pressure-rise, pressure-retention and pressure-drop phases.

With reference to the front wheels, this control sequence is performed individually, i.e. separately and independently for each wheel. For reasons of handling stability, a different control strategy is required for the rear wheels. In order to be able to maintain lateral acceleration, and therefore transverse forces, on the rear wheels at full braking power when cornering, an increase in the lateral friction coefficients of the tires is required. Therefore, the slip levels of the rear wheels must be kept low, particularly the wheel on the outside of the bend. This is achieved by the select-low control characteristic for the rear wheels. That means that the rear wheel which first shows signs of incipient locking, i.e. the "low" wheel, determines the control sequence. In a 3-channel configuration for a braking system with a front/rear split (see ABS system variants), this is achieved by connecting the hydraulic circuits in parallel. On diagonally split brake circuits, however, this is attained by controlling the rear-wheel valves with parallel logic.

Disturbances in the closed control loop
The ABS system must take the following disturbances into account:
- Changes in the adhesion between the tires and the road surface caused by different types of road surface and changes in the wheel loadings, e.g. when cornering.
- Irregularities in the road surface causing the wheels and axles to vibrate.
- Out-of-roundness, brake hysteresis, brake fading.
- Variations in the pressure input to the brake master cylinder caused by the driver depressing the brake pedal.
- Differences in wheel circumferences, for instance when the spare wheel is fitted.

Criteria of control quality
The following criteria for control quality must be fulfilled by efficient antilock braking systems:
- Maintain directional stability by providing sufficient lateral forces at the rear wheels.
- Maintain steerability by providing sufficient lateral forces at the front wheels.
- Reduce the stopping distance as opposed to braking with locked wheels by optimizing adhesion between tires and the road surface.
- Rapid adjustment of braking force to different friction coefficients, for instance when driving through puddles or over patches of ice or compacted snow.
- Insure low braking-torque control amplitudes to prevent vibrations in the suspension.
- Achieve a high level of comfort by using silent actuators and slight feedback through the brake pedal.

Typical control cycle
The control cycle depicted in Figure 6 shows automatic brake control in the case of a high friction coefficient. The change in wheel speed (braking deceleration) is calculated in the ECU. After the value falls below the $(-a)$ threshold, the hydraulic-unit valve unit is switched to pressure-holding mode. If the wheel speed then also drops below the slip-switching threshold λ_1, the valve unit is switched to pressure drop; this continues as long as

Figure 6: ABS control cycle for high friction coefficients
v_{Ref} Reference speed,
v_R Wheel speed,
v_F Vehicle speed,
a,A Wheel-deceleration thresholds.

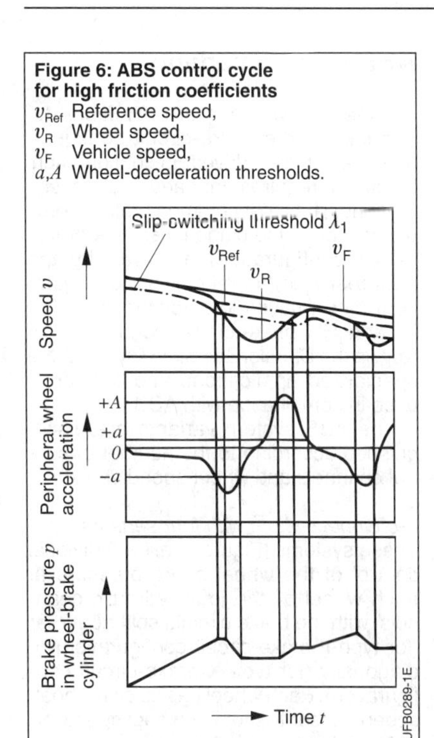

the (–a) signal is applied. During the following pressure-holding phase, peripheral wheel acceleration increases until the (+a) threshold is exceeded; the braking pressure is then kept at a constant level.

After the relatively high (+A) threshold has been exceeded, the braking pressure is increased, so that the wheel is not accelerating excessively as it enters the stable range of the adhesion/slip curve. After the (+a) signal has dropped off, the braking pressure is slowly raised until, when the wheel acceleration again falls below the (–a) threshold, the second control cycle is initiated, this time with an immediate pressure drop.

In the first control cycle, a short pressure-holding phase was initially necessary to filter out any faults. In the case of high wheel moments of inertia, low friction coefficient and slow pressure rise in the wheel-brake cylinder (cautious initial braking, e.g. on black ice), the wheel might lock without any response from the deceleration switching threshold. In this case, therefore, the wheel slip is also included as a parameter in the brake-control system.

Under certain road-surface conditions, passenger cars with all-wheel drive and with differential locks engaged pose problems when the ABS system is in operation; this calls for special measures to support the reference speed during the control process, lower the wheel-deceleration thresholds, and reduce the engine-drag torque.

Control cycle with yaw-moment build-up delay

When the brakes are applied on a road surface with uneven grip (for instance, μ split: left-hand wheels on dry asphalt, right-hand wheels on ice), vastly different braking forces at the front wheels result and induce a turning force (yaw moment) about the vehicle's vertical axis (Figure 7).

On smaller cars, ABS must be supplemented by an additional yaw-moment build-up delay device to ensure that control is maintained during panic braking on asymmetrical road surfaces. Yaw-moment build-up delay holds back the pressure rise in the wheel-brake cylinder on the front wheel with the higher coefficient of friction at the road surface ("high" wheel).

The yaw-moment build-up delay concept is demonstrated in Figure 8: Curve 1 represents the brake-master-cylinder

Figure 7: Yaw-moment build-up induced by large differences in friction coefficients
M_{yaw} Yaw moment,
F_B Braking force,
μ_{HF} Coefficient of friction.
1 "High" wheel,
2 "Low" wheel.

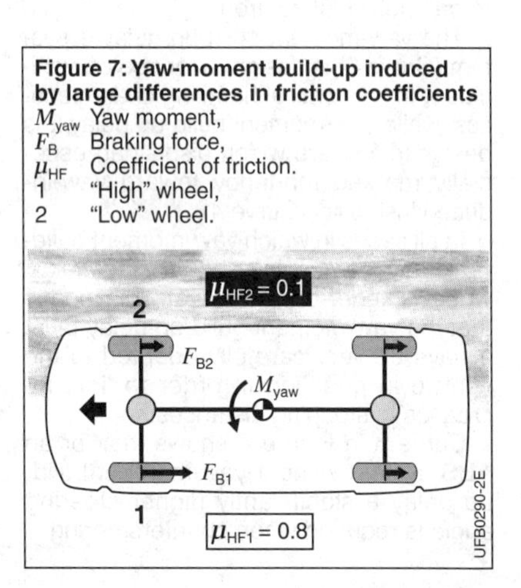

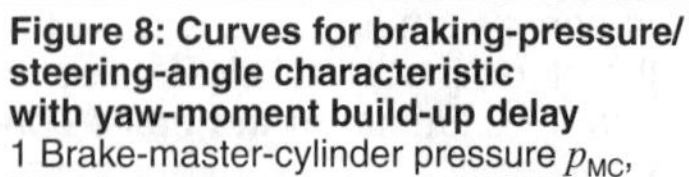

Figure 8: Curves for braking-pressure/ steering-angle characteristic with yaw-moment build-up delay
1 Brake-master-cylinder pressure p_{MC},
2 Brake pressure p_{high} w/o YMBD,
3 Brake pressure p_{high} w/ YMBD 1,
4 Brake pressure p_{high} w/ YMBD 2,
5 Brake pressure p_{low} at "low" wheel,
6 Steering angle α w/o YMBD,
7 Steering angle α w/ YMBD.

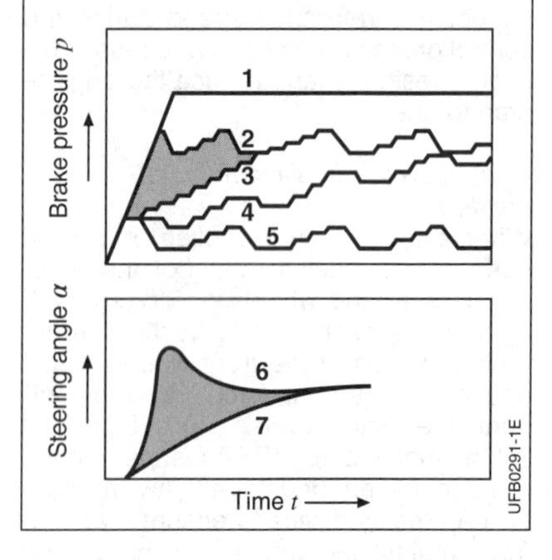

pressure p_{MC}. Without yaw-moment build-up delay, the braking pressure at the wheel running on asphalt quickly reaches p_{high} (Curve 2), while the braking pressure at the wheel running on ice rises only to p_{low} (Curve 5); each wheel brakes with the maximum transferable braking force (individual control).

The yaw-moment build-up delay 1 system (Curve 3) is designed for use on vehicles with less critical handling characteristics, while yaw-moment build-up delay 2 is designed for cars which display an especially marked tendency toward yaw-induced instability (Curve 4).

In all cases in which yaw-moment build-up delay comes into effect, the "high" wheel is under-braked at first. This means that the yaw-moment build-up delay must always be very carefully adapted to the vehicle in question in order to limit increases in stopping distances.

Curve 6 in Figure 8 shows that for an ABS system without yaw-moment build-up delay a significantly higher steering angle is required when countersteering.

ABS system variants

A variety of versions are available depending on the brake-circuit configuration, the vehicle's drivetrain configuration, functional requirements and cost considerations. The most common braking-force distribution is the diagonal split (X brake-circuit configuration), followed by the front-rear split (II brake-circuit configuration) (see Braking-force distribution). The HI and HH brake-circuit configurations (e.g. in the DaimlerChrysler Maybach) are specialized applications and are rarely used in combination with ABS.

The ABS system variants are distinguished according to the number of control channels and wheel-speed sensors.

4-channel system with four sensors
These systems (Figure 9) allow individual control of the wheel brake pressure at each wheel by the four hydraulic channels, with the brake circuits split front/rear (for type-II brake-circuit configuration) or diagonally (for type-X brake-circuit configuration). Each wheel has its own wheel-speed sensor to monitor wheel speed.

A much simplified ABS variant was developed for a particular section of the Japanese car market, the MIDGET segment (small cars with engines smaller than 660 cm^3). It did away with damping chambers and return pumps/pump motors. The smaller number of components compared with conventional systems offers considerable cost savings but also involves significant functional tradeoffs. These types of system have been phased out.

3-channel system with three sensors
Instead of the familiar arrangement with a separate speed sensor on each wheel, the rear wheels with this variant share a single sensor which is fitted in the differential. Due to the characteristics of the differential, it allows the measurement of wheel-speed differences with certain restrictions. Due to the select-low control characteristic for the rear wheels, i.e. parallel connection of the two rear-wheel brakes, a single hydraulic channel is sufficient for (parallel) control of the rear braking pressures.

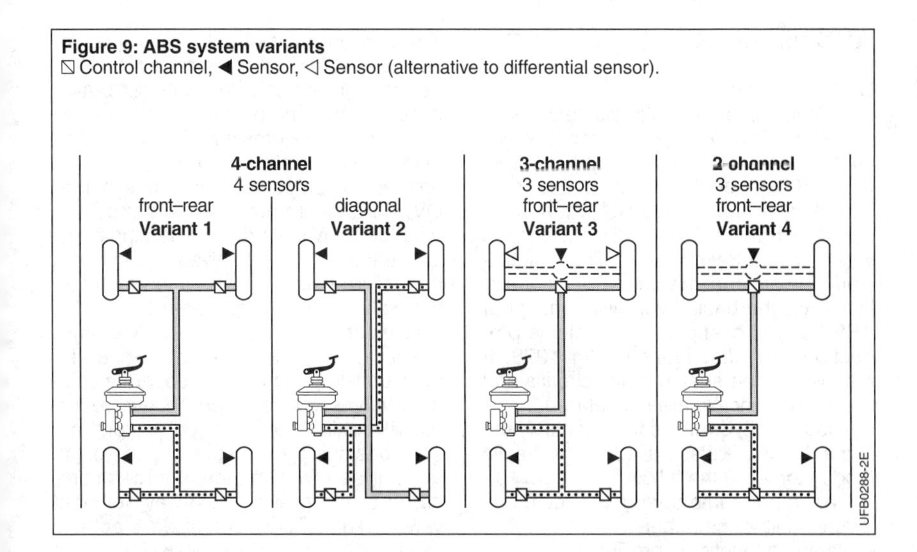

Figure 9: ABS system variants
⧅ Control channel, ◀ Sensor, ◁ Sensor (alternative to differential sensor).

Hydraulic 3-channel systems require a type-II brake-circuit configuration (front and rear split).

3-sensor systems can only be used on vehicles with rear-wheel drive, i.e. primarily small commercial vehicles and light trucks. The number of vehicles fitted with such systems is generally dropping off.

2-channel system with one or two sensors

Two-channel systems were produced because of the smaller number of components required and the resulting potential for cost savings. Their popularity was limited as their functionality does not match that of "full-fledged" systems. These systems are now hardly ever used in cars.

Some light trucks sold on the American market with front/rear-split brake circuits are (still) fitted with RWAL systems (Rear wheel anti-lock brake system) – special simplified versions of the 2-channel system consisting of a sensor on the rear-axle differential and a single control channel (with no return pump) which prevents the rear wheels from locking. If sufficiently high braking pressure is applied, the front wheels may still lock, i.e. loss of steerability under certain conditions is an accepted risk.

This arrangement does not meet the functional requirements expected of a Category 1 ABS system, as mentioned at the start of this section.

Use of ABS on motorcycles

It has been possible to reduce the size and weight of car ABS systems substantially in recent years. As a result, volume-production ABS systems are now a very attractive option for motorcycles. Consequently, this class of vehicle will also be able to benefit from the advantages of ABS as a safety system.

The car system is modified for use on motorcycles. Instead of the usual eight 2/2-way valves in the hydraulic unit for cars (with X-configuration brake circuits), motorcycles normally only require four valves. The control algorithm also differs fundamentally from that of a car ABS system.

Other system variants have arisen from the demand for combined brake systems (CBS), i.e. systems in which both the front and rear brakes can be operated either by a foot pedal or a hand-operated lever, possibly in combination with a separate means of actuating the front brakes. This type of special case requires a 3-channel hydraulic unit. However, the CBS variant is very model-specific in design.

ABS versions

Hydraulic unit

The development of solenoid valves with two hydraulic switching positions (2/2-way valves, used in Bosch ABS systems from the ABS 5 generation onwards) allowed a complete redesign of ABS when compared with the ABS-2S/ABS-2E version which used 3/3-way valves. This rationalized design and manufacture radically. However, the basic hydraulic concept of ABS has not changed since volume production was first launched in 1978. It means that the sealed brake circuits and the return principle are unchanged.

The main hydraulic components of the hydraulic unit – also called the hydraulic modulator – are the following (Figure 10):

- One return pump per brake circuit
- Accumulator chamber
- Damping functions, previously performed by an accumulator chamber and a flow restrictor, are now performed both hydraulically and by control systems, i.e. software
- 2/2-way solenoid valves with two hydraulic positions and two hydraulic connections

There is one pair of solenoid valves for each wheel (except in the case of 3-channel configurations with front/rear brake-circuit split) – one of which is open when de-energized for pressure rise (inlet valve, IV) and one which is closed when de-energized for pressure drop (outlet valve, OV). In order to achieve rapid pressure relief of the wheel brakes when the pedal is released, the inlet valves each have a non-return valve which is integrated into the valve body (e.g. non-return valve sleeves or unsprung non-return valves).

The assignment of pressure-rise and pressure-drop functions to separate solenoid valves with only one active (energized) setting has resulted in compact valve designs, i.e. smaller size and weight, as well as lower magnetic forces compared to the previous 3/3-way solenoid valves. This allows optimized electrical control with low electrical power loss in the solenoid coils and the control unit. In addition, the valve block (Figure 11) can be made smaller. This results in quite significant savings in weight and size.

The 2/2-way solenoid valves are available in a variety of designs and specifications, and, because of their compact di-

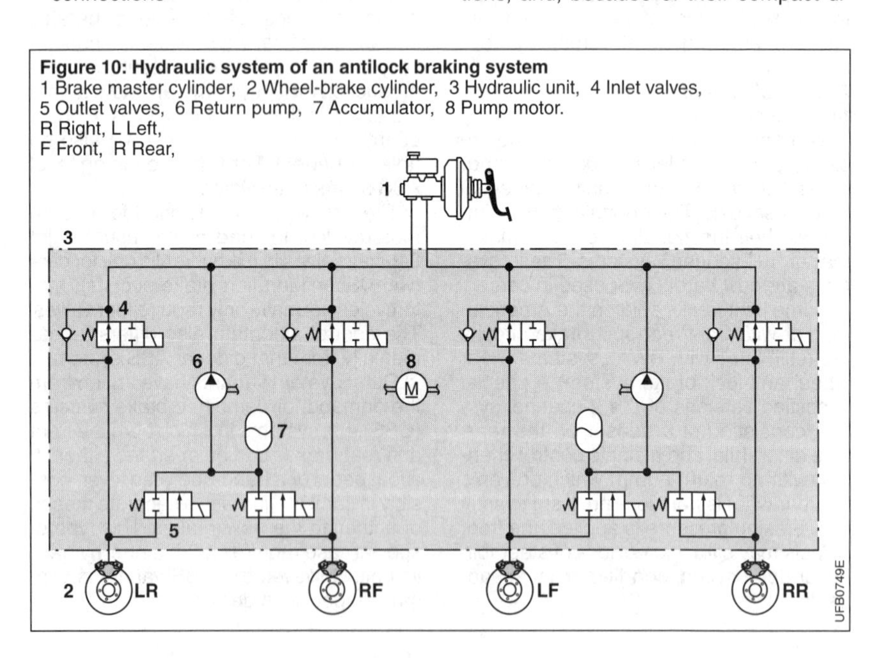

Figure 10: Hydraulic system of an antilock braking system
1 Brake master cylinder, 2 Wheel-brake cylinder, 3 Hydraulic unit, 4 Inlet valves,
5 Outlet valves, 6 Return pump, 7 Accumulator, 8 Pump motor.
R Right, L Left,
F Front, R Rear,

Figure 11: Design of ABS 8 hydraulic unit
1 Control unit, 2 Coil pack, 3 Hydraulic unit, 4 Pump motor.

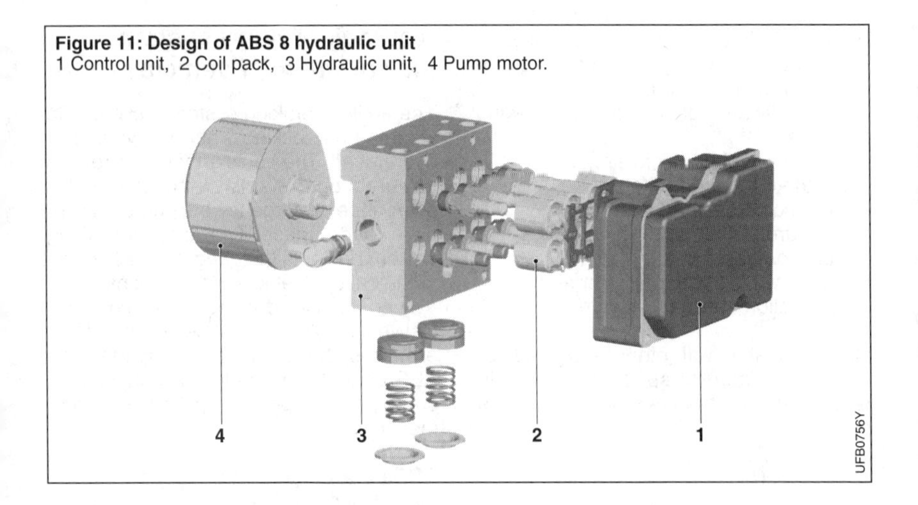

mensions and excellent dynamics, they allow fast electrical switching times sufficient for pulse-width-modulated cyclic operation. In other words, they have “proportional-valve characteristics”.

The ABS 8 benefits from current-signal-modulated valve control which substantially improves function (e.g. adaptation to changes in coefficient of friction) and ease of control (e.g. smaller deceleration fluctuations with the aid of pressure stages and analog pressure control). This mechatronic optimization has positive effects not only on function but also on user-friendliness, i.e. noise and pedal feedback.

ABS 8 is capable of specific adaptation to individual vehicle-class requirements by varying the components (e.g. using motors of different power ratings, varying accumulator chamber size, etc.). The power of the return motor can vary within a range of approx. 90 to 200 watts. The accumulator-chamber size is also variable.

Electronic control unit (ECU)

The progress achieved in the continuing development of ABS is primarily a factor of the enormous advances in the field of electronics. The times when an ABS ECU was made up of more than 1,000 components (ABS 1 generation of 1970, analog design) are long gone. The integration of functions into LSI circuits, the use of high-performance microcomputers and the use of hybrid ECU technology allows a high packing density and therefore further miniaturization. At the same time, this results in a significant increase in system performance and functionality. The use of microcontrollers has lead to substantial optimization of control algorithms incorporating customization to vehicle-manufacturer and model requirements.

The control unit is designed as an add-on ECU and is mounted directly on the hydraulic unit. The advantage of this arrangement is that external wiring can be minimized. The wiring harness contains fewer wires. This results in a reduction in space requirements and less complicated installation. The design layout requires only a single plug-in connection between the ECU and the hydraulic unit and for connecting up the return-pump motor.

The ECU illustrated schematically (Figure 12) represents a 4-channel version with 4 sensors. The two microcontrollers process the control program. In ABS ECUs, they have a frequency rating of approximately 20 MHz and a ROM of around 128 KB. Memory capacities up to about 256 KB are sufficient for ABS versions with special functions.

In highly complex systems such as the Electronic Stability Program (ESP), the ROM may be as large as 1 MB. Depending on the required processing capacity, microcontrollers with higher timing frequencies may also be used.

The software consists of the following modules:

- Hardware-related software, i.e. the operating system
- Self-monitoring and diagnostic software
- Function-specific software
- Purely vehicle-manufacturer and application-specific software

Communication with other ECUs and vehicle-manufacturer-specific diagnosis takes place via the CAN or FlexRay bus system.

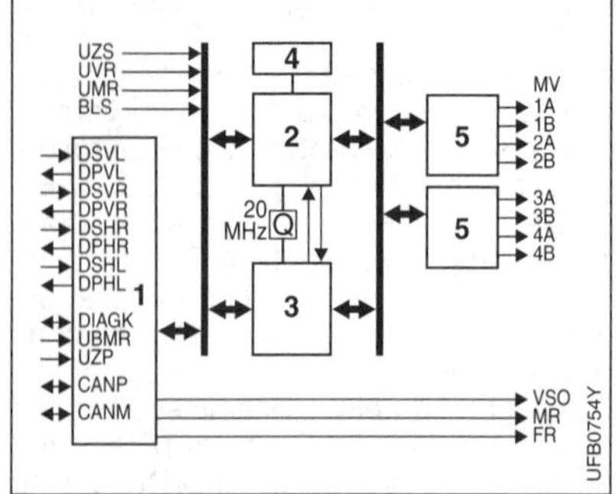

Figure 12: Block diagram of ECU
1 Multifunction circuit with input circuit, speed-signal conditioning circuit, diagnostic circuit, voltage regulator, CAN, relay control circuit, etc.
2 Microcontroller 1,
3 Microcontroller 2,
4 EEPROM,
5 Output-stage module.

DS Wheel-speed sensors
DP Speed-signal outputs (e.g. for instrument cluster)
F Front, R Rear, L Left, R Right,
MV Solenoid valve

Antilock braking system for commercial vehicles

The antilock braking system prevents the wheels from locking when the vehicle is overbraked. The vehicle therefore retains its directional stability and steerability even under emergency braking on a slippery road surface. The antilock braking system prevents the danger of jackknifing in the case of vehicle combinations.

In contrast to passenger cars, commercial vehicles have air-brake systems. Nevertheless, the functional description of an ABS control process for passenger cars onwards also applies in principle to commercial vehicles.

ABS control methods

Individual control (IR)

This process, which sets and controls the optimum braking pressure individually for each wheel, produces the shortest braking distances. Under μ-split conditions (different adhesion on the drive wheels, e.g. asphalt on one side, icy road surface on the other side), braking produces a high yaw moment about the vertical axis of the vehicle, thus making short-wheelbase vehicles difficult to control. In addition, this is coupled with high steering moments as a result of the positive steering roll radii in commercial vehicles. Individual control is normally used on commercial vehicles on the rear axle.

Select-low control (SL)

This process reduces the yaw and steering moments to zero. It is achieved by applying the same braking pressure to both wheels on the same axle. Only one pressure-control valve is required for each axle. The pressure applied is dependent on the wheel with less grip (select low), the wheel with more grip is braked less heavily compared to the individual control (IR) method. Under μ-split conditions, braking distances are longer, but steerability and directional stability of the vehicle are enhanced. If grip conditions (friction coefficients) are the same on both sides of the vehicle, then braking distances, steerability and directional stability are virtually identical with individual control systems (IR).

Individual control modified (IRM)
This process requires a pressure-control valve at each wheel of the axle. It reduces yaw and steering moments only as far as necessary, and limits braking-pressure difference between the left and right sides to permissible levels. As a result, the wheel running at a high friction coefficient is braked slightly less. This compromise results in a braking distance which is only a little longer than that for individual control (IR), but it does ensure that vehicles with critical handling characteristics remain controllable.

ABS equipment for commercial vehicles
The current state of the art is that ABS ECUs for tractor units (commercial vehicles, tractor vehicles, buses) can be used on two and three-axle vehicles (Figure 13). During a learning operation on initial commissioning, the control unit adjusts to the corresponding vehicle depending on the connected components. This involves detecting the number of axles, the ABS control method, and any additional functions that may be required, such as TCS. A similar situation applies to ABS ECUs for trailers/semitrailers. The same ECU can be used on trailers and semitrailers with one, two or three axles, and adapts itself to the level of equipment present.

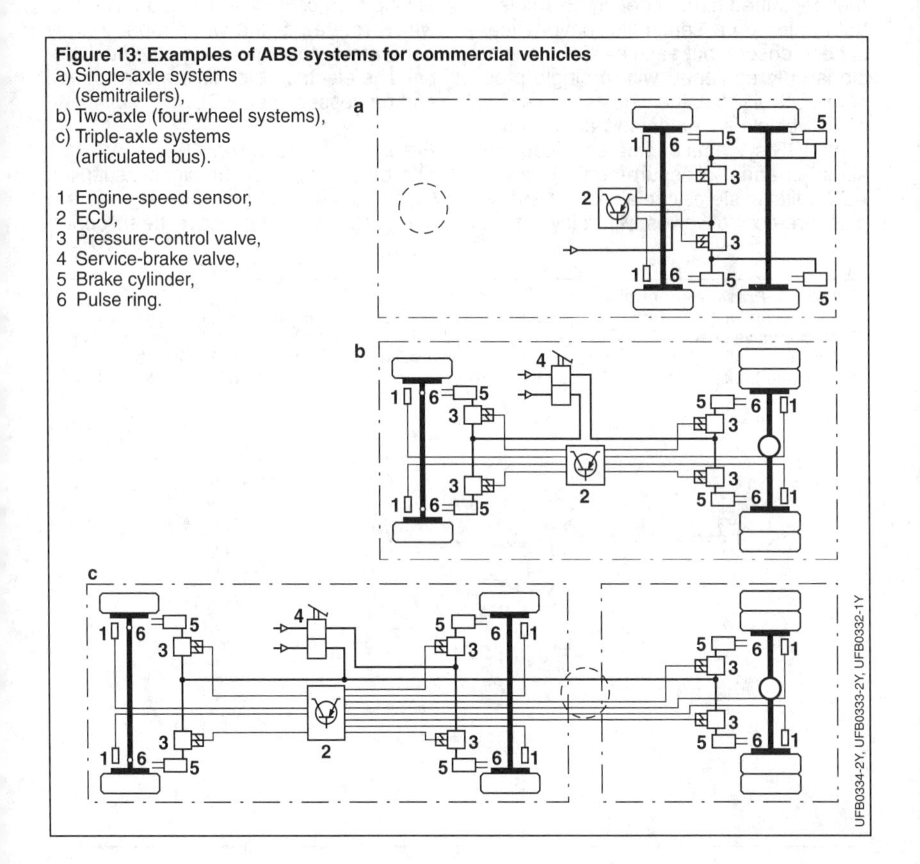

Figure 13: Examples of ABS systems for commercial vehicles
a) Single-axle systems (semitrailers),
b) Two-axle (four-wheel systems),
c) Triple-axle systems (articulated bus).

1 Engine-speed sensor,
2 ECU,
3 Pressure-control valve,
4 Service-brake valve,
5 Brake cylinder,
6 Pulse ring.

If one axle is a lifting axle, it is automatically excluded from the ABS control process when lifted.

When two axles are close together, often only one of them is fitted with wheel-speed sensors. The brake pressure of two neighboring wheels is regulated jointly by a single pressure-control valve. On multi-axle vehicles with axles that are further apart, e.g. articulated buses, triple-axle control is preferred.

The IRM (individual control modified) control process is most commonly used on steering axles; the select-low control method is sometimes adopted, though very rarely. On tractor-unit rear axles, the individual-control method is normally selected.

The range of available control equipment permits further control combinations (not described here in detail). Example: If both axles on a semitrailer have wheel-speed sensors, but each side of the vehicle is only equipped with a single pressure-control valve, the wheels of one side of the vehicle are select-low-controlled.

All ABS systems can be equipped with single-channel pressure-control valves. ABS trailer systems can also be fitted with pressure-control valves with relay action.

In light commercial vehicles with pneumatic-hydraulic converters, ABS intervenes in the pneumatic brake circuit via single-channel pressure-control valves and defines the hydraulic braking pressure.

When the vehicle is running on a low-friction-coefficient road surface, the operation of an additional retarding brake (exhaust brake or retarder) can lead to excessive slip at the driven wheels. This would impair vehicle stability. ABS therefore monitors brake slip and controls it to permissible levels by switching the retarder on and off.

ABS components

Wheel-speed sensor

The rotational behavior of the wheels is monitored by speed sensors which operate inductively or according to the Hall principle. In conjunction with a pulse ring which rotates at the wheel speed, they generate a corresponding electrical signal. The electrical signals are conditioned and processed in an ECU.

Electronic control unit (ECU)

The ECU processes the signals supplied by the speed sensors. The signals are compared after they have been condi-

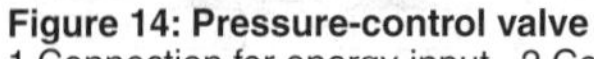

Figure 14: Pressure-control valve
1 Connection for energy input, 2 Connection for energy output, 3 Venting, 4 Diaphragm, 5 Inlet, 6 Valve seat, 7 Solenoid valve for pressure-holding valve, 8 Valve seat, 9 Valve seat, 10 Diaphragm, 11 Outlet, 12 Solenoid valve for outlet valve, 13 Valve seat, 14 Service-brake valve, 15 ABS/TCS ECU, 16 Pressure-control valve, 17 Wheel brake.

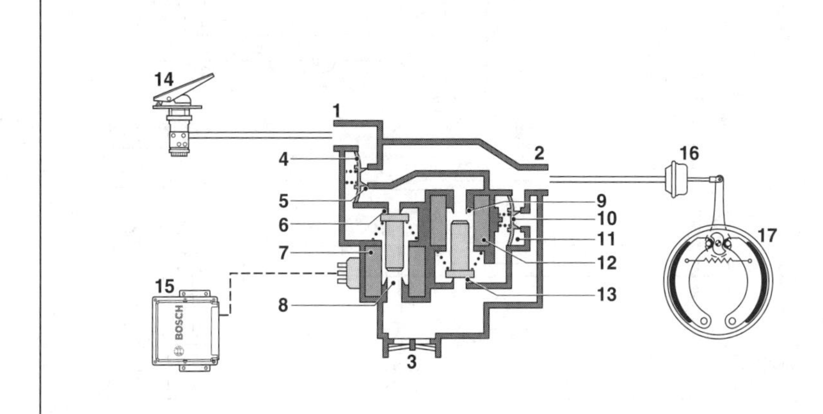

tioned. What are always compared are a driven wheel and a non-driven wheel, a wheel on the inside of the bend and a wheel on the outside of the bend, or a dynamically loaded wheel and a dynamically unloaded wheel. The slip of the individual wheel is calculated from these and the corresponding pressure-control valves are activated.

Further functions, such as automatic deactivation of the retarder, can be executed in the control process. ABS ECUs are equipped with a safety circuit which continuously monitors the complete system. Fault detection results in partial or complete shutdown of the system. The faults are stored in a fault memory which can be interrogated by a diagnosis tester and erased after the faults have been cleared.

Some ECUs contain not only the ABS function, but are also equipped with other functions, e.g. traction control system (TCS) or engine drag-torque control (MSR).

Pressure-control valve

The pressure-control valves are located between the service-brake valve and the brake cylinders, and control the brake pressure of one or more wheels (Figure 14). They consists of a combination of solenoid valves and pneumatic valves. They usually contain one outlet valve and one pressure-holding valve (single-channel pressure-control valve), but a combination of one outlet valve and two pressure-holding valves may also be used (double-channel pressure-control valve). The electronics control the solenoid valves in the appropriate combination so that the required function is performed ("pressure holding" and "pressure drop"). If no pilot-valve actuation takes place, "pressure rise" is the result.

When braking normally (that is, without ABS response = no incipient wheel locking), air flows through the pressure-control valves unhindered in both directions when pressure is applied to or vented from the brake cylinders. This ensures fault-free functioning of the service braking system.

References

[1] ECE-R13: Uniform provisions concerning the approval of category M, N and O vehicles with regard to brakes.

Traction control system

Function and requirements

When starting off, accelerating and braking, the efficiency required to transfer forces to the road depends on the traction available between the tires and the road surface. The adhesion/slip curves for acceleration and braking have the same basic patterns (Figure 1).

The vast majority of acceleration and braking operations involve only limited amounts of slip, allowing response to remain within the stable range in the adhesion/slip curves. Up to a certain point, any rise in slip is accompanied by a corresponding increase in useful adhesion. Beyond this point, any further increases in slip take the curves through the maxima and into the unstable range where any further increase in slip generally results in a reduction in adhesion. When braking, this results in wheel lock within a few tenths of a second. When accelerating, one or both of the driven wheels start to spin more and more as the drive torque exceeds the adhesion by an ever increasing amount.

The antilock braking system responds in the first case (braking) by inhibiting wheel lock. The traction control system reacts to the second scenario by holding drive slip within acceptable levels to prevent wheel spin. The traction control system actually performs two functions:

- Increasing traction (electronic locking-differential function)
- Maintaining vehicle stability (directional stability)

The functions create the requirements demanded of the traction control system. It must reliably prevent wheel spin, even under variable μ conditions, i.e.:

- It must prevent the driving wheels from spinning under μ-split conditions and on slippery road surfaces.
- It must prevent wheel spin when the vehicle pulls out of icy parking spaces and lay-bys.
- It must prevent wheel spin when the vehicle accelerates in a corner.
- It must prevent wheel spin when the vehicle pulls away on a hill.
- It must improve cornering stability.

TCS control loops

The traction control system is a component part of the electronic wheel-slip control systems. Components which are already in place and are required anyway for ABS control, such as wheel-speed sensors for example, can thus also be used for TCS control.

Traction control is performed essentially by means of two different control interventions adapted to the individual circumstances – brake intervention and engine intervention.

Brake intervention

Brake intervention is normally performed only in the lower speed range when one of the two drive wheels spins due to the lack of adhesion. During brake intervention, a braking pressure adapted to the given situation is applied to the wheel brake of the spinning wheel, and thereby the drive torque is transferred via the differential to the other, non-spinning wheel. A kind of differential lock is created.

Brake intervention on commercial vehicles

On commercial vehicles with compressed-air braking systems a "TCS solenoid valve" and a pneumatic shuttle valve are required for brake intervention (Figure 2).

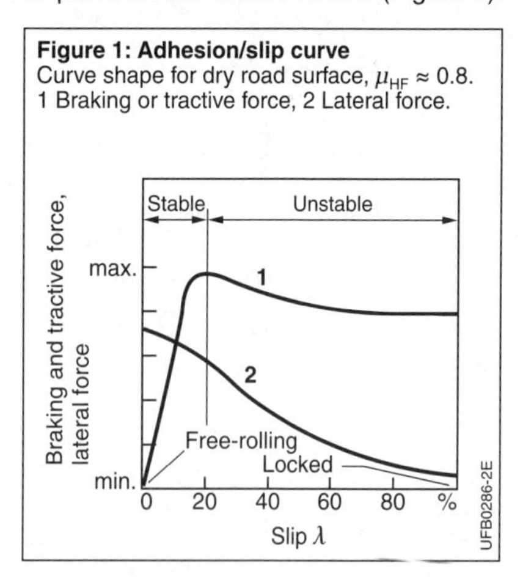

Figure 1: Adhesion/slip curve
Curve shape for dry road surface, $\mu_{HF} \approx 0.8$.
1 Braking or tractive force, 2 Lateral force.

During a necessary brake intervention, the electrically actuated TCS solenoid valve delivers supply pressure via the shuttle valve to the ABS pressure-control valves. The shuttle valve simultaneously blocks the connection to the service-brake valve. At the same time, the solenoid valve is electrically actuated for the pressure-holding valve in the ABS pressure-control valve which is assigned to the non-spinning wheel. This rules out the possibility of a pressure build-up in the wheel-brake cylinder.

The solenoid valves in the ABS pressure-control valve of the spinning wheel are initially not actuated. Braking pressure is built up in the corresponding wheel-brake cylinder, as a result of which the wheel is braked and prevented from spinning. Braking pressure is built up, as dictated by the situation and adapted by continuous monitoring of the control process, by alternating and electrically clocked actuation of the corresponding solenoid valves in the pressure-control valves.

TCS control shuts down brake intervention when homogeneous slip conditions are attained. The TCS valve and the solenoid valves in the ABS pressure-control valves are no longer actuated. The braking pressure in the corresponding wheel-brake cylinder is reduced via the bleeder valve in the pressure-control valve to atmosphere.

Using the brake-control function described above, the driving wheels can also be synchronized so that a mechanical differential lock, if fitted, can be engaged automatically, e.g. with the aid of a pneumatic cylinder. The ABS/TCS ECU calculates the correct point and conditions for releasing the differential lock.

In contrast to mechanical differential locks, the tires do not scrub on tight corners. A fundamental observation about this type of system (when it assumes an electronic brake-control function) is that it is not intended for continuous use on difficult offroad terrain. Since the brake-control function is achieved by braking the relevant wheel, brake heating is an inevitable consequence.

Brake intervention on passenger cars

On passenger cars with hydraulic braking systems, an expanded ABS hydraulic unit is required for TCS brake intervention. Depending on the variant, the expansion can

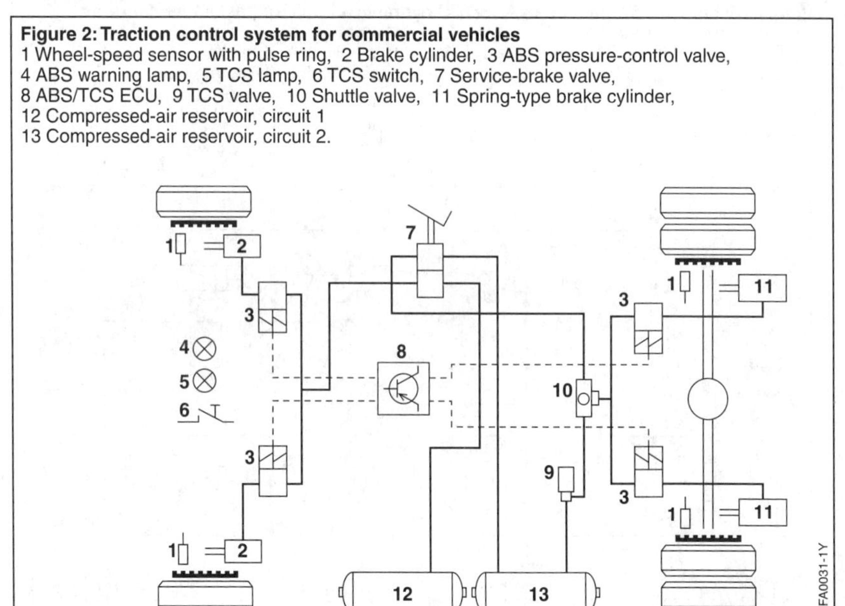

Figure 2: Traction control system for commercial vehicles
1 Wheel-speed sensor with pulse ring, 2 Brake cylinder, 3 ABS pressure-control valve, 4 ABS warning lamp, 5 TCS lamp, 6 TCS switch, 7 Service-brake valve, 8 ABS/TCS ECU, 9 TCS valve, 10 Shuttle valve, 11 Spring-type brake cylinder, 12 Compressed-air reservoir, circuit 1
13 Compressed-air reservoir, circuit 2.

comprise an intake valve and a changeover valve (Figure 3). An additional hydraulic presupply pump and a pressure accumulator may be required. During a necessary brake intervention, the intake and changeover valves assigned to the spinning wheel and the ABS return pump are electrically actuated. The return pump can draw brake fluid from the brake master cylinder through the intake valve. The changeover valve blocks the return flow to the brake master cylinder. The pressure generated by the return pump passes through the inlet valve to the wheel-brake cylinder of the spinning wheel, as a result of which the wheel is braked and prevented from spinning. Braking pressure is built up, as dictated by the situation and adapted by continuous monitoring of the control process, by alternating and electrically clocked actuation of the inlet and outlet valves in the hydraulic unit.

On completion of the control phase, electrical actuation is terminated and the braking pressure applied for TCS control, as following a normal braking operation, is reduced via the intake and changeover valves and the brake master cylinder.

Engine intervention

Engine intervention is performed when both drive wheels spin. This is the case when the drive torque is higher than the transmittable torque at the wheels. The drive torque is reduced accordingly by engine intervention.

On commercial vehicles and passenger cars with diesel engines, engine intervention is performed, depending on the variant, by means of electronic diesel control or the ETC (electronic throttle control) system (reduced delivery).

On passenger cars with spark-ignition engines, torque reduction is usually performed by means of a combination of several functions. In this way, it is possible to reduce the engine torque according to the requirements by specifically suppressing injection pulses, retarding the ignition timing or closing the throttle device (ETC).

The engine-management systems receive the TCS request via signal or CAN data lines from the TCS control system.

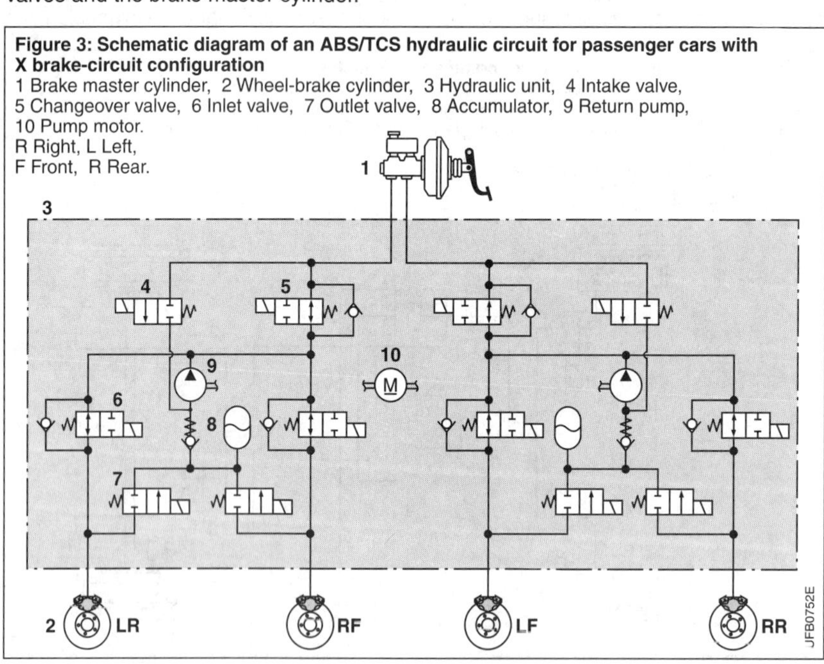

Figure 3: Schematic diagram of an ABS/TCS hydraulic circuit for passenger cars with X brake-circuit configuration
1 Brake master cylinder, 2 Wheel-brake cylinder, 3 Hydraulic unit, 4 Intake valve, 5 Changeover valve, 6 Inlet valve, 7 Outlet valve, 8 Accumulator, 9 Return pump, 10 Pump motor.
R Right, L Left,
F Front, R Rear.

Driving-dynamics control system

Function

Human error is the cause of a large portion of road accidents. Even under normal driving conditions, a driver and his vehicle can reach their physical operating limits on account of, for example, an unexpected bend in the road, a suddenly appearing obstacle or an unanticipated change in the condition of the road surface. Increased speed can also result in the driver not being able to control his vehicle safely, since the lateral-acceleration forces acting on the vehicle in such a situation reach levels which make excessive demands on him.

If the tires' coefficients of friction are exceeded, the vehicle will suddenly behave differently from what the driver with his driving experience expects. In such situations of operating limits, the driver is often no longer able to stabilize the vehicle himself; as a rule, he will intensify the instability through reactions arising from fear and panic. As a result, a significant discrepancy is built up between the longitudinal motion of the vehicle and the longitudinal axis of the vehicle (float angle β). Even by steering in the opposite direction, a normal driver will barely be able to restabilize his vehicle on his own at float angles in excess of 8°.

The Electronic Stability Program (ESP) – as Bosch's driving-dynamics control system is known – makes a significant contribution to defusing such situations by helping the driver to keep his vehicle under control within the physical operating limits. Sensors constantly record the behavior of both the driver and the vehicle. By comparing the actual state with a target state appropriate for the relevant situation, the system, in the event of significant discrepancies, makes interventions in the braking system and in the drivetrain to stabilize the motion of the vehicle (Figure 1).

The integrated functionality of the antilock braking system (ABS) prevents the wheels from locking when the brakes are applied, while the similarly integrated traction control system (TCS) inhibits excessive wheel spin during acceleration. ESP as an overall system, however, embraces capabilities extending far beyond those of either ABS or ABS and TCS combined. The system ensures that the vehicle does not swerve outwards with its rear end (oversteering) or does not push excessively outwards with its front end (understeering), but instead follows the driver's steering input as far as is physically possible.

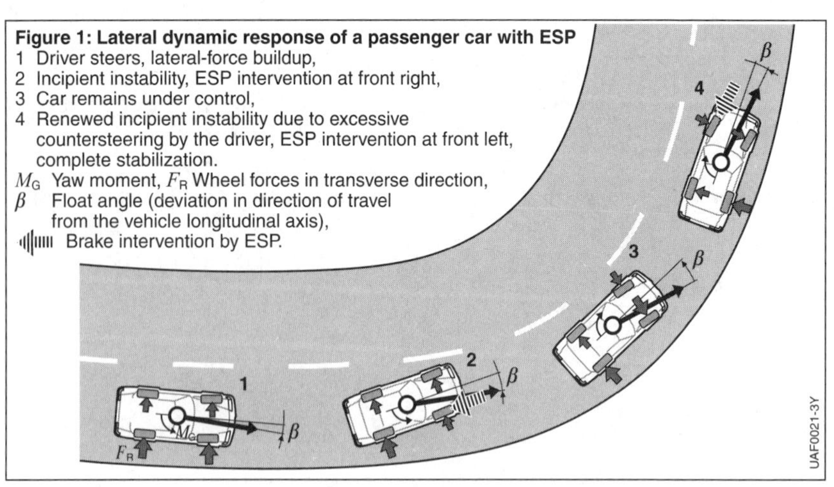

Figure 1: Lateral dynamic response of a passenger car with ESP
1 Driver steers, lateral-force buildup,
2 Incipient instability, ESP intervention at front right,
3 Car remains under control,
4 Renewed incipient instability due to excessive countersteering by the driver, ESP intervention at front left, complete stabilization.
M_G Yaw moment, F_R Wheel forces in transverse direction,
β Float angle (deviation in direction of travel from the vehicle longitudinal axis),
Brake intervention by ESP.

ESP relies on tried and proven ABS and TCS components. In this way, the individual wheels can be actively braked with high dynamic response. The engine torque and thus the traction-slip values at the wheels can be influenced by means of the engine-management system. The systems communicate via, for example, the CAN bus.

Table 1: Terms and quantities

a_y	Measured vehicle lateral acceleration
F_x	Tire force in longitudinal direction
F_y	Tire force in lateral direction (lateral force)
F_N	Tire force in normal direction (normal force)
L	Distance between front and rear axles
M_{BrNom}	Nominal braking torque
M_{DifNom}	Nominal differential torque
M_{EngNom}	Nominal engine torque
$M_{MWhlNom}$	Nominal sum torque
ΔM_{RedNom}	Setpoint change of engine-torque reduction
ΔM_Z	Stabilizing yaw moment
p_{Whl}	Wheel-cylinder pressure
p_{Adm}	Admission pressure
r	Radius of bend
v_{ch}	Characteristic vehicle speed
v_{Dif}	Wheel-speed differential of drive wheels (on one axle)
v_{DifNom}	Nominal wheel-speed differential of drive wheels (on one axle)
v_{MWhl}	Mean wheel speed of driven axle
$v_{MWhlNom}$	Setpoint value of mean wheel speed
v_{Whl}	Measured wheel speed
v_x	Vehicle linear velocity
v_y	Vehicle lateral velocity
α	Tire slip angle
β	Float angle
δ	Steering-wheel angle
λ	Tire slip
λ^i_{Nom}	Setpoint value of tire slip at wheel i
$\Delta\lambda_{DifTolNom}$	Setpoint change of permissible slip differential of driven axle(s)
$\Delta\lambda_{Nom}$	Setpoint slip change
μ	Coefficient of friction
$\dot{\psi}$	Yaw velocity
$\dot{\psi}_{Nom}$	Nominal yaw velocity

Requirements

The Electronic Stability Program (ESP) helps to increase driving safety. It improves vehicle behavior up to the physical operating limits. The vehicle's reaction remains foreseeable to the driver and can thus be better controlled even in critical driving situations.

At the vehicle's physical driving limits, vehicle and directional stability are enhanced in all operating states, such as full braking, partial braking, coasting, accelerating, overrunning and load changes, and also for example in the case of extreme steering maneuvers (fear and panic reactions). The risk of skidding is drastically reduced.

In a variety of different situations, further improvements are obtained in the exploitation of traction potential when ABS and TCS come into action, and when engine drag-torque control is active (automatic increase in engine speed to inhibit excessive engine braking torque). This leads to shorter braking distances and greater traction with enhanced stability and higher levels of steering response.

Incorrect system interventions could have an impact on safety. A comprehensive safety concept ensures that all faults that are not essentially avoidable are detected in time and the ESP system is shut down fully or partially depending on the type of fault.

Numerous studies (e.g. [1] and [2]) have demonstrated that ESP drastically reduces the number of accidents caused by skidding and the number of associated fatalities. The upshot of this is that ESP will become mandatory in vehicles in North America by September 2011. In the European Union (EU), all new passenger cars and light commercial vehicles have had to be fitted with a driving-dynamics control system since November 2011 (integral part of ECE-R 13-H [12]). Other new cars will be subject to a transition period which must be complied with by the end of 2014. Other regions such as Japan and Australia, for example, will also be introducing such a provision.

Operating principle

The Electronic Stability Program (ESP) is a system which uses a vehicle's brake system and drivetrain to deliberately influence the vehicle's longitudinal and lateral motion in critical situations. When the stability-control function assumes operation, it shifts the priorities that govern the brake system. The basic function of the wheel brakes – to decelerate and/or stop the vehicle – assumes secondary importance as ESP intervenes to keep the vehicle stable and on course. ESP can also accelerate the drive wheels by means of engine interventions to contribute to the vehicle's stability.

Both mechanisms act on the vehicle's intrinsic motion. During steady-state circular-course driving, there is a defined connection between the driver's steering input and the resulting vehicle lateral acceleration and thus the tire forces in the lateral direction (self-steering effect). The forces acting on a tire in the longitudinal and lateral directions are dependent on the tire slip. It follows from this that the vehicle's intrinsic motion can be influenced by the tire slip. The specific braking of individual wheels, e.g. of the rear wheel on the inside of the bend in the case of understeering or of the front wheel on the outside of the bend in the case of oversteering, helps the vehicle to remain on the course determined by the steering angle as precisely as possible.

Typical driving maneuver

To compare how a vehicle handles at its operating limits with and without ESP, the following example is given. The driving maneuver reflects actual operating conditions, and is based on simulation programs designed using data from vehicle testing. The results have been confirmed in subsequent road tests.

Rapid steering and countersteering
Figure 2 demonstrates the handling response of a vehicle without ESP and of a vehicle with ESP negotiating a series of S-bends with rapid steering and countersteering inputs on a high-grip road surface (coefficient of friction $\mu = 1$), without the driver braking and at an initial speed of 144 km/h. Figure 3 shows the curves for dynamic-response parameters. Initially, as they approach the S-bend, the conditions for both vehicles, and their reactions, are identical. Then come the first steering inputs from the drivers (phase 1).

Vehicle without ESP
As can be seen, in the period following the initial, abrupt steering input the vehicle without ESP is already threatening to become unstable (Figure 2a, Phase 2). Whereas the steering input has quickly generated substantial lateral forces at the front wheels, there is a delay before the rear wheels start to generate similar forces. The vehicle reacts with a clockwise

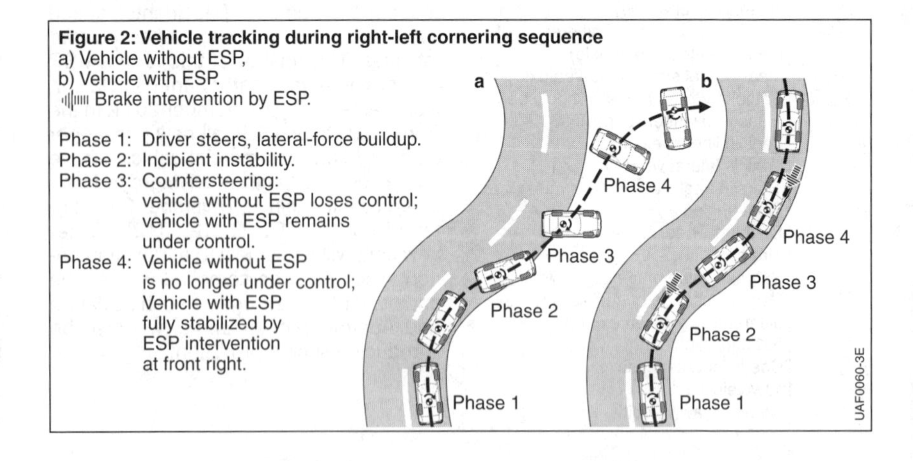

Figure 2: Vehicle tracking during right-left cornering sequence
a) Vehicle without ESP,
b) Vehicle with ESP.
Brake intervention by ESP.

Phase 1: Driver steers, lateral-force buildup.
Phase 2: Incipient instability.
Phase 3: Countersteering: vehicle without ESP loses control; vehicle with ESP remains under control.
Phase 4: Vehicle without ESP is no longer under control; Vehicle with ESP fully stabilized by ESP intervention at front right.

movement around its vertical axis (inward yaw). The vehicle barely responds to the driver's attempt to countersteer (second steering input, Phase 3), because it is no longer under control. The yaw velocity and the side-slip angle rise radically, and the vehicle breaks into a skid (phase 4).

Vehicle with ESP
The vehicle with ESP is stabilized after the initial steering input by active braking of the front left wheel to counter the threat of instability (Figure 2b, Phase 2): This occurs without any intervention on the driver's part. This action limits the inward yaw with the result that the yaw velocity is reduced and the float angle is not subject to an uncontrolled increase. Following the change of steering direction, first the yaw moment and then the yaw velocity reverse their directions (between Phases 3 and 4). In Phase 4, a second brief brake application – this time at the right front wheel – restores complete stability. The vehicle remains on the course defined by the steering-wheel angle.

Figure 3: Dynamic-response curves during a right-left cornering sequence
1 Vehicle without ESP, 2 Vehicle with ESP.
Phases 1-4 see Figure 2.

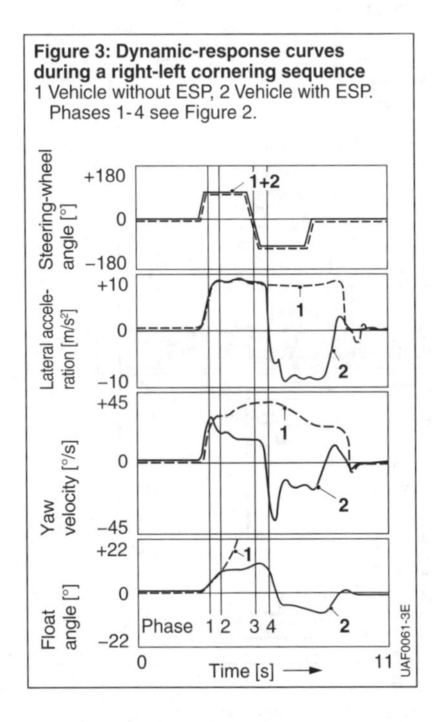

Structure of the overall system

Objective of driving-dynamics control
The control of the handling characteristics at the vehicle's physical driving limits is intended to keep the vehicle's three degrees of freedom in the plane of the road – linear velocity v_x, lateral velocity v_y and yaw velocity $\psi^{\cdot}$ about the vertical axis – within the controllable limits. Assuming appropriate operator inputs, driver demand is translated into dynamic vehicular response that is adapted to the characteristics of the road in an optimization process designed to ensure maximum safety.

System and control structure
The ESP system comprises the vehicle as a controlled system, the sensors for determining the controller input variables, the actuators for influencing the braking, motive and lateral forces, as well as the hierarchically structured controller, comprising a higher-level transverse-dynamics controller and lower-level wheel controllers (Figure 4). The higher-level controller determines the setpoint values for the lower-level controller in the form of moments or slip, or their changes. Internal system variables that are not directly measured, such as the float angle β for example, are determined in the driving-condition estimation ("observer").

In order to determine the nominal behavior, the signals defining driver command are evaluated. These comprise the signals from the steering-wheel-angle sensor (driver's steering input), the brake-pressure sensor (desired deceleration input, obtained from the brake pressure measured in the hydraulic unit) and the accelerator-pedal position (desired drive torque). The calculation of the nominal behavior also takes into account the utilized friction-coefficient potential and the vehicle speed. These are calculated in the observer from the signals sent by the wheel-speed sensors, the lateral-acceleration sensor, the yaw sensor, and the brake-pressure sensor. Depending on the control deviation, the yaw moment, which is necessary to make the actual-state variables approach the desired-state variables, is then calculated.

In order to generate the required yaw moment, it is necessary for the changes in desired braking torque and slip at the wheels to be determined by the transverse-dynamics controller. These are then set by means of the lower-level brake-slip and traction controllers together with the brake-hydraulics actuator and the engine-management actuator.

Driving-condition estimation

To determine the stabilization interventions, not only knowledge of the signals from the sensors for wheel speeds v_{Whl}, admission pressure p_{Adm}, yaw rate (yaw velocity) $\dot{\psi}$, lateral acceleration a_y, steering-wheel angle δ and engine torque is important, but also knowledge of a series of further internal system variables which can be measured indirectly with appropriate effort. These include, for example, the tire forces in the longitudinal, lateral and normal directions (F_x, F_y and F_N), the vehicle linear velocity v_x, the tire slip values λ_i, the slip angle α on one axle, the float angle β, the vehicle lateral velocity v_y, and the coefficient of friction μ. They are estimated on a model-supported basis from the sensor signals in the observer.

The vehicle linear velocity v_x is of crucial importance to all the wheel-slip-based controllers and must therefore be calculated with great accuracy. This is done on the basis of a vehicle model using the measured wheel speeds. Numerous influences must be taken into account here. The vehicle speed v_x differs for example already in normal driving situations on account of brake or drive slip from the wheel speeds v_{Whl}. For all-wheel-drive vehicles, special linking of the wheels must be taken into account in particular. During cornering, the

Figure 4: ESP overall control system
1 Wheel-speed sensors,
2 Brake-pressure sensor (integrated in hydraulic unit),
3 Steering-wheel-angle sensor,
4 Yaw sensor (yaw-rate sensor) with integrated lateral-acceleration sensor,
5 ESP hydraulic unit (hydraulic modulator) with mounted ECU,
6 Wheel brakes, 7 Engine ECU.

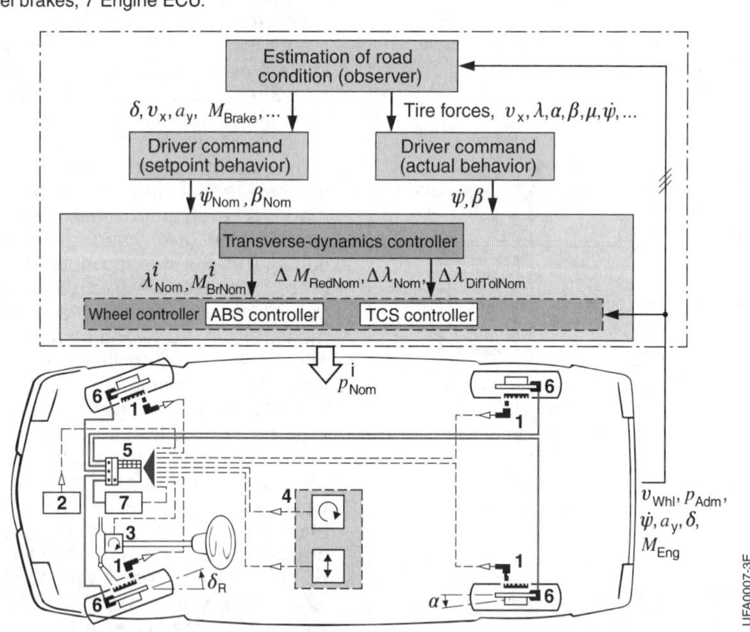

wheels on the inside of the bend follow a different course from the wheels on the outside of the bend, and consequently rotate at a different speed.

Vehicle handling changes during normal use in response to varying load, altered tractive resistance (e.g. road gradient or surface, wind) or wear (e.g. of the brake pads).

Under all these boundary conditions, the vehicle linear velocity must be estimated with a deviation of a few % in order to ensure the enabling and intensity of stabilization interventions to the necessary extent.

Basic transverse-dynamics controller

The function of the transverse-dynamics controller is to calculate the actual behavior of the vehicle from, for example, the yaw-velocity signal and the float angle estimated in the observer, and to bring the driving behavior in the driving-dynamic limit range into line with behavior in the normal range as closely as possible (nominal or setpoint behavior).

The connection that exists during steady-state circular-course driving between the yaw velocity and the steering-wheel angle δ, the vehicle linear velocity v_x and characteristic vehicle variables are used to determine the nominal behavior. The single-track vehicle model (see e.g. [3]) is used to produce

$$\dot{\psi} = \frac{v_x}{l}\;\delta\;\frac{1}{1+\left(\frac{v_x}{v_{\text{ch}}}\right)^2}$$

as the basis for calculating the vehicle nominal motion. In this formula, l denotes the distance between the front and rear axles. Geometric and physical parameters of the vehicle model are summarized in the "characteristic vehicle speed" v_{ch}.

The variable $\dot{\psi}$ is then limited according to the current friction-coefficient conditions and to the special properties of the vehicle dynamics and the driving situation (e.g. braking or acceleration by the driver) and to the particular conditions such as a sloping road surface of different friction coefficients under the vehicle (μ split). The driver command is thus known as the nominal yaw velocity $\dot{\psi}_{\text{Nom}}$.

The transverse-dynamics controller compares the measured yaw velocity with the associated setpoint value and in the event of significant deviations calculates the yaw moment that is required to match the actual state variable to its setpoint state. At a higher level, the float angle β is monitored and, as the values rise, increasingly taken into consideration in the calculation of the stabilizing yaw moment ΔM_z. This controller output variable is applied by means of braking-torque and slip inputs to the individual wheels which must be adjusted by the lower-level wheel controllers.

Stabilization interventions are performed at the wheels, the braking of which generates a yaw moment in the required direction of rotation and at which the limit of the transmittable forces has not yet been reached. For an oversteering vehicle, the physical limit is first exceeded on the rear axle. Stabilization interventions are therefore performed via the front axle. For an understeering vehicle, the situation is reversed (see e.g. [6]).

The nominal slip values requested by the transverse-dynamics controller λ^{i}_{Nom} at individual wheels are set with the aid of lower-level wheel controllers (see Figure 4). A distinction is made between the following three application cases.

Wheel control in the coasting case

In order to exert as accurately as possible the yaw moments required to stabilize the vehicle, the wheel forces must be altered under defined conditions by controlling the wheel slip. The nominal slip requested by the transverse-dynamics controller at a wheel is adjusted in the unbraked case by the lower-level brake-slip controller by way of an active pressure build-up. The current slip at the wheel must be known as precisely as possible for this purpose. This is calculated from the measured wheel-speed signal and the vehicle linear velocity determined in the observer v_x. The nominal braking torque at the wheel is formed from the deviation of the actual wheel slip from its setpoint value using a *PID* control law.

It is not only in the event of an active pressure build-up by transverse-dynamics control that a wheel can be subject to brake slip. Following downshifts and when the accelerator is suddenly released, inertia in the engine's moving parts exerts a degree of braking force at the drive wheels. Once this force and the corresponding reactive torque rise beyond a certain level, the tires will lose their ability to transfer the resulting loads to the road and will tend to lock (e.g. because the road is suddenly slippery). The brake slip for the driven wheels can be limited in the coasting case by engine drag-torque control. This acts like "gentle acceleration" by the driver.

Wheel control in the braked case

In the braked case, different actions overlap at individual wheels, depending on the driving situation:

- Driver input via the brake pedal and the steering wheel
- Effect of the ABS controller, which prevents individual wheels from locking
- Interventions of the transverse-dynamics controller which ensures vehicle stability by specifically braking individual wheels if necessary.

These three requirements must be coordinated in such a way that the driver's deceleration and steering inputs are implemented as much as possible. If wheel control is performed primarily with the objective of maximum vehicle deceleration, it can be performed on the basis of wheel acceleration which can be robustly determined with minimal sensor information (instability control). In order to specifically adjust the longitudinal and lateral tire forces to stabilize the vehicle, the principle of slip control [4] must be applied because it also permits wheel control in the unstable range of the friction-coefficient/slip characteristic. From the available sensor information, however, it must be possible to determine the absolute wheel slip to a few %, depending on the vehicle speed.

The function of the ABS controller is to ensure vehicle stability and steerability in all road conditions and in so doing exploit as much as possible the friction between wheels and road. It does this also in its capacity as the lower-level controller to the transverse-dynamics controller by modulating the brake pressure at the wheel in such a way that the maximum possible longitudinal force can be exerted while maintaining sufficient lateral stability. However, more variables are measured in ESP than in a pure ABS configuration, which only contains the wheel-speed sensors. Thus, individual vehicle-motion information, such as for example yaw rate or lateral acceleration, is available through direct measurement of greater accuracy than is the case with model-supported estimation of the basis of few measured values.

In certain situations, it is possible to increase performance by adapting ABS control by means of inputs from the transverse-dynamics controller. When a vehicle decelerates on unequal road surfaces (μ split), very different braking forces occur at the wheels on the left and right sides of the vehicle. This generates a yaw moment about the vehicle vertical axis, to which the driver must react by countersteering in order to stabilize the vehicle. How quickly this yaw moment is built up – and how fast the driver must consequently react – depends on the vehicle's moment of inertia about the vertical axis. ABS features yaw-moment build-up delay to hold back the pressure rise at the front wheel with the higher coefficient of friction at the road surface ("High" wheel). This ABS facility can also use information from the higher-level transverse-dynamics controller (on the driver's reaction and the vehicle behavior) and thereby react even better to the actual vehicle motion.

If, when braking in a bend, the vehicle starts to turn under certain conditions, the oversteer tendency can be counteracted by means of electronic braking-force distribution through pressure reduction in individual wheels. If this is not sufficient on its own, the transverse-dynamics controller helps by actively building up pressure at the front wheel on the outside of the bend (reduction of lateral force). If, on the other hand, the vehicle understeers, the braking torque is increased at the rear on the inside (provided the wheel is not subject to ABS control) and decreased slightly at the front on the outside.

If the vehicle starts to oversteer during a fully or partially braked lane change, the pressure at the rear wheel on the inside of the bend is specifically decreased (increase in lateral force) and the pressure at the front wheel on the outside of the bend increased (decrease in lateral force). If the vehicle understeers while braking in the bend, the braking torque is increased at the rear on the inside (provided the wheel is not yet in the ABS control range) and decreased slightly at the front on the outside.

Wheel control in the drive case
The lower-level traction controller (TCS) is activated as soon as the drive wheels start to spin in the drive case. The measured wheel speed and the respective drive slip can be influenced by changing the torque balance at each drive wheel. The TCS controller limits the drive torque at each drive wheel to the drive torque that can be transferred there to the road. In this way, the driver command is implemented after acceleration as well as is physically possible and, at the same time, fundamental directional stability is ensured, since the lateral forces at the wheel are not too greatly reduced.

In a vehicle with a powered axle, the mean wheel speed of the driven axle

$$v_{\mathrm{MWhl}} = \frac{1}{2}\left(v^{\mathrm{L}}_{\mathrm{Whl}} + v^{\mathrm{R}}_{\mathrm{Whl}}\right)$$

and the wheel-speed differential

$$v_{\mathrm{Dif}} = v^{\mathrm{L}}_{\mathrm{Whl}} - v^{\mathrm{R}}_{\mathrm{Whl}}$$

between the measured wheel speed of the left wheel $v^{\mathrm{L}}_{\mathrm{Whl}}$ and the right wheel $v^{\mathrm{R}}_{\mathrm{Whl}}$ are used as controlled variables.

The structure of the entire TCS controller is depicted in Figure 5. The reference variables of the transverse-dynamics controller are included in the setpoint-value calculation for the mean wheel speed and the wheel-speed differential, as well as the setpoint slip values and the coasting wheel speeds. In the calculation of the setpoint values v_{DifNom} (setpoint wheel-speed differential of the drive wheels on one axle) and v_{MWhlNom} (setpoint value of the mean wheel speed), the inputs for changing the setpoint slip $\Delta\lambda_{\mathrm{Nom}}$ and the permissible slip differential $\Delta\lambda_{\mathrm{DifTolNom}}$ of the drive axle(s) act in the form of an offset on the basic values calculated in the TCS. In addition, an understeer or oversteer tendency identified by the transverse-dynamics controller directly influences, through the setpoint change of the en-

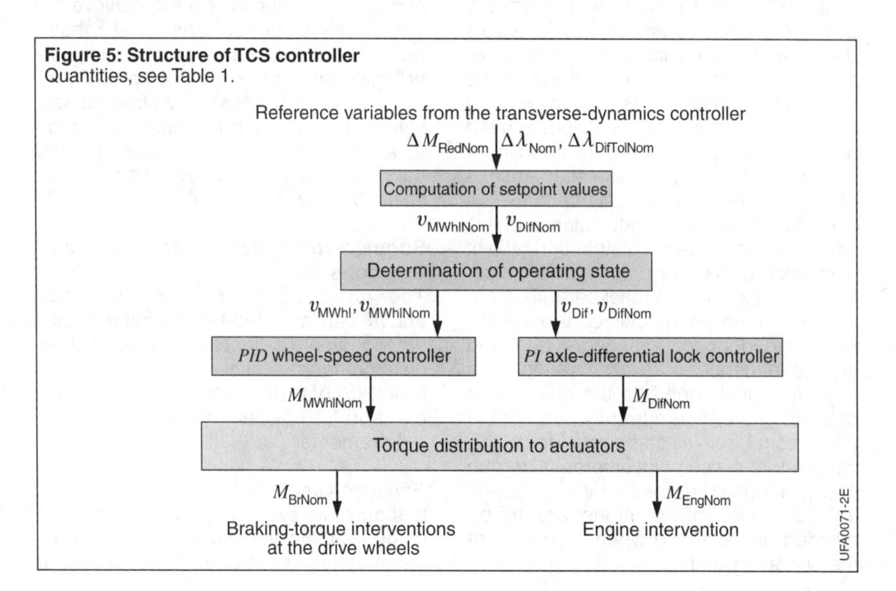

Figure 5: Structure of TCS controller
Quantities, see Table 1.

gine-torque reduction ΔM_{RedNom}, the determination of the maximum permissible drive torque.

The dynamic response of the drivetrain depends on the highly differing operating states. It is therefore necessary to determine the current operating status (e.g. selected gear, clutch actuation) in order to be able to adapt the controller parameters to the controlled system's dynamic response and to nonlinearities.

Because the mean wheel speed is affected by variable inertial forces originating from the drivetrain as a whole (engine, transmission, drive wheels, and the propshaft itself), a relatively large time constant is employed to describe its correspondingly leisurely rate of dynamic response. The mean wheel speed is controlled by means of a nonlinear *PID* controller, whereby in particular the gain of the *I*-component (dependent on the operating status) can vary over a wide range. In the stationary case, the *I*-component is a measure for the torque which can be transferred to the road surface. The output variables of this controller is the setpoint sum torque $M_{MWhlNom}$.

In contrast, the time constant for the wheel-speed differential is relatively small, reflecting the fact that the wheels' own inertial forces are virtually the sole determining factor for their dynamic response. Furthermore, in contrast to the mean wheel speed, it is influenced only indirectly by the engine. The wheel-speed differential v_{Dif} is controlled by a nonlinear *PI* controller. Because brake interventions at a drive wheel initially only become noticeable through the torque balance of this wheel, they change the distribution ratio of axle differential and thereby emulate a differential lock. The controller parameters of this axle-differential-lock controller are only dependent on the engaged gear and engine influences to a minimal extent. If the differential speed on the driven axle deviates more than currently permissible (dead zone) from its setpoint value v_{DifNom}, calculation of a setpoint differential torque M_{DifNom} starts. The dead zone is widened if TCS brake interventions are to be avoided, for example when cornering at the operating limits.

Setpoint sum and setpoint differential torques are the basis for distributing the positioning forces to the actuators. The setpoint differential torque M_{DifNom} is set by the braking-torque difference between the left and right drive wheels by means of corresponding valve actuation in the hydraulic unit (asymmetrical brake intervention). The setpoint sum torque $M_{MWhlNom}$ is adjusted by both the engine interventions and a symmetrical brake intervention.

With a gasoline engine, adjustments undertaken through the throttle valve are relatively slow to take effect (lag and the engine's transition response). Retarding the ignition timing and, as a further option, selective suppression of injection pulses are employed for rapid engine-based intervention. In diesel-engine vehicles, the electronic diesel control system (EDC) reduces the engine torque by modifying the quantity of fuel injected. Symmetrical brake intervention can be applied for brief transitional support of engine-torque reduction.

Traction plays a special role in off-road applications. Normally, in vehicles with off-road requirements, traction control is automatically adapted by way of special situation identification in order to achieve the best levels of performance and robustness possible. Other vehicle manufacturers give the driver the opportunity to choose different adjustments from deactivation of engine-torque limitation through to adaptations tailored to special road conditions (e.g. ice, snow, grass, sand, slush, and rocky ground).

Supplementary transverse-dynamics functions

The basic ESP functions described above can also feature supplementary driving-dynamics functions for special vehicle categories, such as sport utility vehicles (SUVs) and small vans for example, and for special vehicle-stabilization requirements.

Enhanced understeering control

It is possible even in normal driving conditions for the vehicle to fail to comply adequately with the driver's steering input (it understeers) if, for example, the road surface in a bend is suddenly wet or contaminated. ESP can therefore increase the yaw rate by exerting an additional yaw moment. This enables the vehicle to negotiate a bend at the maximum speed physically possible. The expected frequency of interventions and the comfort requirements of the vehicle manufacturer differ from vehicle type to vehicle type, and there are accordingly different expansion stages for executing such brake interventions which influence the vehicle's understeering behavior.

If the driver requests a smaller radius of bend than is physically possible, then only the reduction of the vehicle speed remains. This can be read from the connection applicable during steady-state cornering between the radius of bend r, the vehicle linear velocity v_x and the yaw rate $\dot{\psi}$:

$$r = \frac{v_x}{\dot{\psi}} .$$

In order to ensure a desired track course, the vehicle is then – without a yaw moment being applied – braked as far as necessary by specific braking of all the wheels (Enhanced Understeering Control, EUC).

Rollover prevention

In particular, light commercial vehicles and other vehicles with a high center of gravity, such as sport utility vehicles (SUVs), can overturn when high lateral forces are generated by a spontaneous steering reaction by the driver in the course of an evasive maneuver on a dry road for example (highly dynamic driving situations) or when the lateral acceleration of a vehicle slowly increases into the critical range as it negotiates a freeway exit with a decreasing radius of bend at excessive speed (quasi-stationary driving situations).

There are special functions (Rollover Mitigation Functions, RMF) which identify these critical driving situations by using the normal ESP sensors and stabilize the vehicle by intervening in brake and engine control. In order to ensure intervention on time, in addition to the driver's steering input and the measured reaction of the vehicle (yaw rate and lateral acceleration), a predictive process is used to estimate the vehicle's behavior in the near future. The two wheels on the outside of the bend, in particular, are braked if an imminent danger of overturning is identified. This action reduces the lateral forces on the wheels and thereby reduces the critical lateral acceleration. Particularly in the event of highly dynamic evasive maneuvers, wheel control must be effected with such high levels of sensitivity that, in spite of the wildly fluctuating vertical forces F_N, vehicle steerability is not diminished by the tendency of individual wheels to lock. The reduction of wheel speed by individual wheel braking also ensures that the driver is able to keep the vehicle in lane. In quasi-stationary driving situations, punctual reduction of the engine torque also prevents the driver from provoking a critical situation.

The moment of intervention and the intensity of the stabilizing interventions must be adapted as accurately as possible to the current vehicle behavior. This behavior can change significantly with the load, for example in the case of vans or sport utility vehicles fitted with roof racks. Such vehicles therefore make use of additional estimation algorithms which calculate the vehicle mass and the change in the center of mass caused by load distribution, if this is required to adapt the ESP functions (Load Adaptive Control, LAC).

Trailer sway mitigation

Depending on the vehicle speed, combinations of towing vehicle and trailer are prone to swaying about their vertical axis. If the vehicle is traveling at a slower speed than the "critical speed" (normally between 90 km/h and 130 km/h), these swaying motions are adequately damped and are quick to die down. If, however, the combination is traveling at a higher speed, small steering movements, crosswinds or driving over a pothole can suddenly induce such swaying motions, which then quickly intensify and can ultimately cause an accident due to the combination jackknifing.

Clear periodic oversteering triggers normal ESP stabilization interventions, but these normally arrive too late and on their own are not sufficient to stabilize the combination. The Trailer Sway Mitigation (TSM) function identifies swaying motions in good time on the basis of the customary ESP sensors; it does this by model-based analysis of the towing vehicle's yaw rate while taking into account the driver's steering movements. When these swaying motions reach a critical level, the combination is automatically braked in order to reduce the speed to such an extent that not even the smallest subsequent excitation will cause an immediate critical oscillation again. In order to damp the oscillation as effectively as possible in a critical situation, in addition to symmetrical deceleration through all the towing vehicle's wheels, individual wheel interventions are performed which swiftly damp the swaying motion of the combination. Limitation of the engine torque prevents dangerous acceleration by the driver during the stabilization process.

Activation of further driving-dynamics actuators

In addition to utilization of the hydraulic wheel brakes, other actuators are provided by means of which the driving-dynamic properties of a vehicle can be specifically influenced. When active steering and chassis systems are linked with ESP to form the composite system known as Vehicle Dynamics Management (VDM), they can, in their entirety, support the driver even better and thereby improve safety and driving dynamics even further.

While the combination of the steering or roll-stabilization system with the brake system has been introduced in the last few years [5], systems for activating differential locks in the drivetrain have been established on the market for some time now. The large number of such systems means that linking with ESP is possible in many cases. The supplementary actuator can basically be activated either directly from the extended ESP function (cooperation approach) or via a separate ECU which exchanges information with the ESP ECU (coexistence approach).

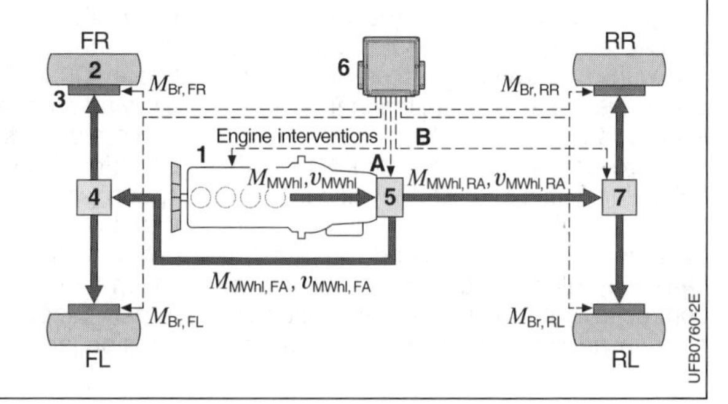

Figure 6: Drive concept of an all-wheel-drive vehicle with ESP

1 Engine with transmission,
2 Wheel,
3 Wheel brake,
4 Axle differential,
5 Central differential,
6 ECU with enhanced ESP function,
7 Axle differential.

Engine, transmission, gear ratios of differentials and losses are combined into one unit.

A Lock interventions with active central differential,
B Torque-vectoring interventions.

v Wheel speed,
v_{MWhl} Mean wheel speed,
M_{MWhl} Driving sum torque,
M_{Br} Braking torque,
FA Front axle
RA Rear axle

R Right, L Left,
F Front, R Rear,

In all-wheel-drive vehicles, the drive torque is distributed via a central element to both powered axles (Figure 6). When the engine acts first and foremost on one axle and the second axle is linked via the central element, this is known as a hang-on system. If this central element is an open differential (without a locking action), drive torque is limited when one axle demonstrates increased slip. In the most unfavorable case, propulsion cannot be achieved if a wheel spins. In combination with ESP, symmetrical brake interventions by the all-wheel TCS controller can limit the differential speed between the axles and thereby achieve a longitudinal locking action.

The traction control of ESP can also be matched to the special operating concept of other types of central elements such as Torsen and viscous couplings. Basically, all the controllable drivetrain actuators must demonstrate a defined locking moment and dynamic response when opening and closing in order to specifically adapt the vehicle's self-steering properties with them.

If the drivetrain of a vehicle can be manually switched over between different configurations, ESP can be automatically adjusted to the operating mode selected by the driver. Because ESP is based on individual wheel control, cooperation with mechanical differential locks for specific off-road conditions is only possible if the differential lock can be automatically opened during interventions by the transverse-dynamics controller. The system must otherwise be switched to an ABS fallback level when the lock is engaged, because driving-dynamics interventions at one wheel would also affect other wheels if the axles were rigidly linked.

In addition to simple links between the two axles, there are controllable central locks in which an electric or hydraulic actuator activates a coupling and thereby adapts the locking moment (Figure 6, A). In this way, it is possible with the ESP driving-dynamics information (e.g. wheel speeds, vehicle speed, yaw rate, lateral acceleration, and engine torque) and by also taking into account actuator-specific variables (such as e.g. the mechanical load on the component) to optimally adapt the linking of the two axles to the current driving situation (Dynamic Coupling Torque at Center, DCT-C).

The example in Figure 7 shows how variable drive-torque distribution influences the vehicle behavior. If, in the event of a risk of oversteering in a bend, provisionally more drive torque can be shifted to the front axle, it is necessary only much later to avoid instability, to lower the engine torque or even stabilize the vehicle with brake interventions (the maximum possible drive-torque shift is shown). If a vehicle tends to understeer, this tendency can be lessened by shifting drive torque to the rear axle. In both cases, vehicle behavior with improved response and better stability is achieved. The limits within which shifting the drive torque is actually possible are dependent on the concrete drivetrain configuration.

Figure 7: Influence of drive-torque distribution on vehicle behavior
a) Oversteer: Stability limit is first exceeded on the rear axle,
b) Understeer: Stability limit is first exceeded on the front axle.
1 Standard distribution during stable driving,
2 Incipient instability, drive torque is shifted to the axle which still has stability potential,
3 Maximum shift of drive torque,
4 Withdrawal of shift,
5 Standard distribution is re-established after instability has been reduced.

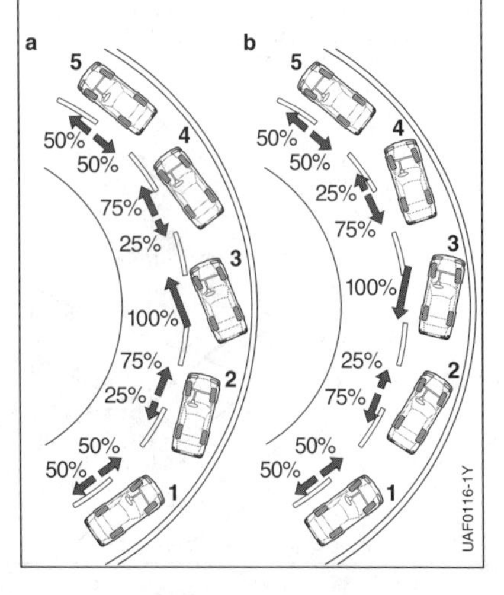

A controllable element on one axle can be activated by ESP along similar lines to the flexible linking of the two axles described. In terms of basic operation, the Dynamic Wheel Torque Distribution (DWT) function barely differs from the axle-differential lock effected by TCS via the hydraulic wheel brakes. However, such a supplementary actuator also actively distributes the drive torque between the wheels on one axle in normal situations. This is done with minimal losses and which much greater sensitivity and comfort than can be achieved by traction control in combination with braking-torque control and engine-torque reduction while taking into account the wear of the ESP hydraulic unit.

System components

The hydraulic unit, the ECU directly connected to it (add-on ECU) and the speed sensors are suitable for the rugged ambient conditions that are encountered in the engine compartment or the wheel arches. The yaw sensor and the lateral-acceleration sensor are either integrated into the ECU or like the steering-angle sensor installed in the passenger cell. Figure 8 shows by way of example the installation locations of the components in the vehicle with the electrical and mechanical connections.

Electronic control unit (ECU)

The ECU of printed-circuit board design comprises, as well as a dual-core computer, all the drivers and semiconductor relays for valve and pump activation, as well as interface circuits for sensor-signal conditioning and corresponding switch inputs for supplementary signals (e.g. brake-light switch). There are also interfaces (CAN, FlexRay) for communicating with other systems, such as engine and transmission management for example.

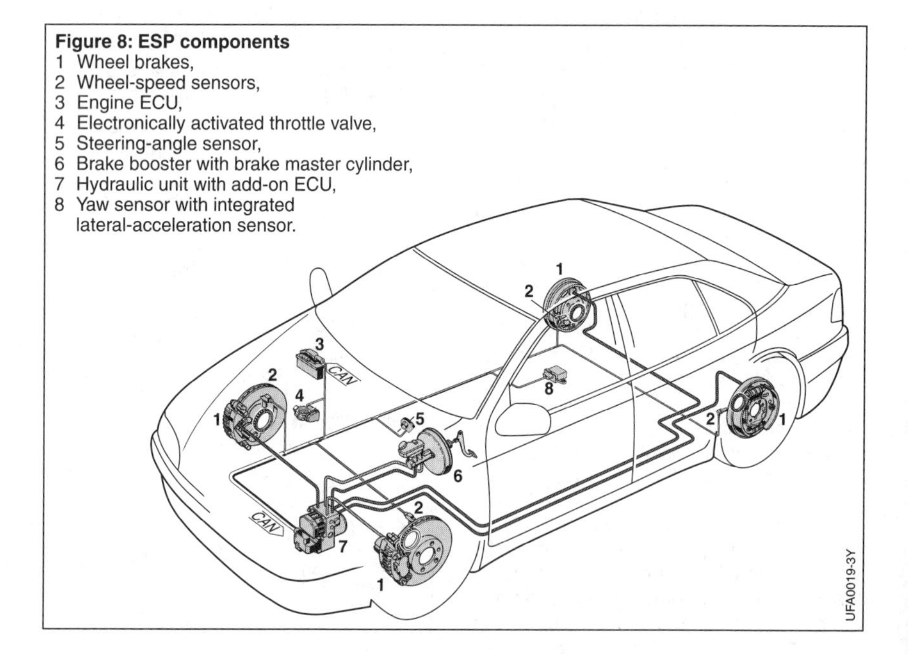

Figure 8: ESP components
1 Wheel brakes,
2 Wheel-speed sensors,
3 Engine ECU,
4 Electronically activated throttle valve,
5 Steering-angle sensor,
6 Brake booster with brake master cylinder,
7 Hydraulic unit with add-on ECU,
8 Yaw sensor with integrated lateral-acceleration sensor.

Hydraulic unit
The hydraulic unit (also called the hydraulic modulator), as in ABS or ABS/TCS systems, forms the hydraulic connection between the brake master cylinder and the wheel-brake cylinders. It converts the control commands of the ECU and uses solenoid valves to control the pressures in the wheel brakes. The hydraulic circuit is completed by bores in an aluminum block. This block is also used to accommodate the necessary hydraulic function elements (solenoid valves, plunger pumps and reservoir chambers).

ESP systems require twelve valves irrespective of the brake-circuit configuration (Figure 9). In addition, a pressure sensor is usually integrated which measures the driver's deceleration command by way of the brake pressure in the brake master cylinder. This increases the performance of vehicle stabilization in partially active maneuvers. Pressure is modulated during ABS control (passive control) using ESP hydraulics in the same way as described for the ABS system.

But because ESP systems must also actively build up pressure (active control) or increase a brake pressure input by the driver (partially active control), the return pump used for ABS is replaced by a self-priming pump for each circuit. The wheel-brake cylinders and the brake master cylinder are connected via a changeover valve open at zero current and a high-pressure switching valve.

An additional non-return valve with a specific closing pressure prevents the pump from drawing unwanted brake fluid from the wheels. These pumps are driven by a DC motor based on demand. The motor drives an eccentric bearing located on the shaft of the motor.

Three examples of pressure modulation are shown in Figure 10. In order to build up pressure independently of the driver (Figure 10c), the switchover valves are closed and the high-pressure switching valves opened. The self-priming pump now pumps brake fluid to the relevant wheel or wheels in order to build up pressure. The inlet valves of the other wheels

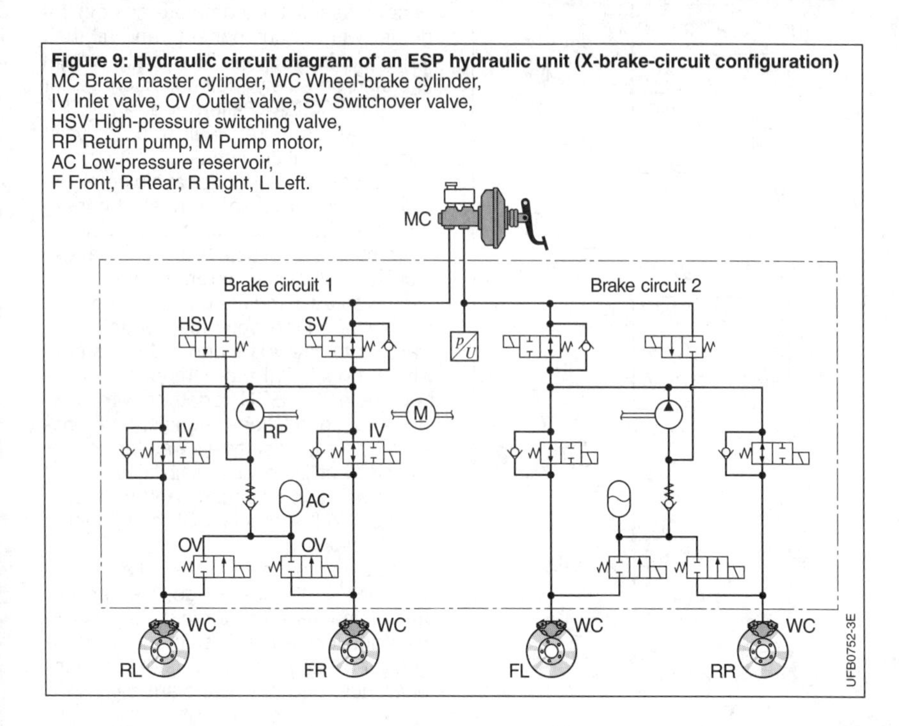

Figure 9: Hydraulic circuit diagram of an ESP hydraulic unit (X-brake-circuit configuration)
MC Brake master cylinder, WC Wheel-brake cylinder,
IV Inlet valve, OV Outlet valve, SV Switchover valve,
HSV High-pressure switching valve,
RP Return pump, M Pump motor,
AC Low-pressure reservoir,
F Front, R Rear, R Right, L Left.

remain closed. To reduce the pressure, the outlet valves are opened and the high-pressure switching valves and switchover valves return to their original position (Figure 10b). The brake fluid escapes from the wheels into the low-pressure reservoirs, which are run empty by the pumps. Demand-based control of the pump motor reduces noise emission during pressure generation and regulation.

Figure 10: Pressure modulation in the ESP hydraulic unit
a) Pressure build-up when braking,
b) Pressure reduction with ABS control,
c) Pressure build-up via self-priming pump due to TCS or ESP intervention.
IV Inlet valve,
OV Outlet valve,
SV Switchover valve,
HSV High-pressure switching valve,
RP Return pump, M Pump motor,
AC Low-pressure reservoir,
F Front, R Rear, R Right, L Left.

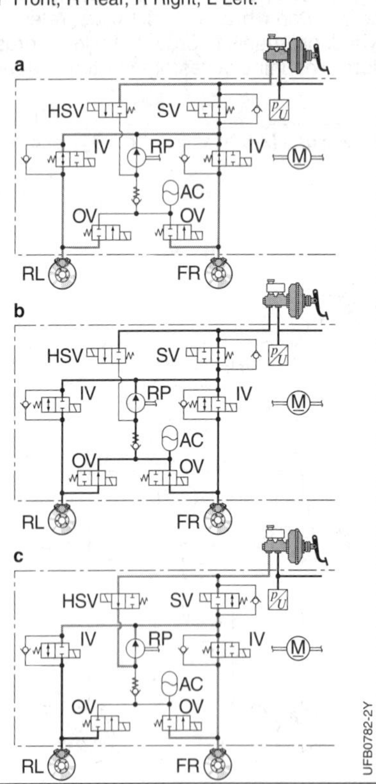

For partially active control (Figure 10a), the high-pressure switching valve must be able to open the suction path of the pump against higher differential pressures (> 0.1 MPa). The first stage of the valve is opened via the magnetic force of the energized coil; the second stage via the hydraulic area difference. If the ESP controller detects an unstable vehicle state, the switchover valves (open at zero current) are closed and the high-pressure switching valve (closed at zero current) is opened. The two pumps then generate additional pressure in order to stabilize the vehicle. When the intervention is finished, the outlet valve is opened and the pressure in the controlled wheel discharged to the reservoir. As soon as the driver releases the brake pedal, the fluid is pumped from the reservoir back to the brake-fluid reservoir.

Monitoring system

A comprehensive safety-monitoring system is of fundamental importance for reliable ESP functioning. The system used encompasses the complete system together with all components and all their functional interactions. The safety system is based on safety methods such as, for example, FMEA (Failure Mode and Effects Analysis), FTA (Fault Tree Analysis) and error-simulation studies. From these, measures are derived for avoiding errors which could have safety-related consequences. Extensive monitoring programs guarantee the reliable and punctual detection of all sensor errors which cannot be prevented completely. These programs are based on the well-proven safety software from the ABS and TCS systems which monitor all the components connected to the ECU together with their electrical connections, signals, and functions. The safety software was further improved by utilizing the possibilities offered by the additional sensors, and by adapting them to the special ESP components and functions.

The sensors are monitored at a number of stages. In the first stage, the sensors are continuously monitored during vehicle operation for line break, signal implausibility ("out-of-range" check), detection of interference, and physical plausibility. In a second stage, the most important sensors are tested individually. The yaw-rate sensor is tested by intentionally detuning the sensor element and then evaluating the signal response. Even the acceleration sensor has internal background monitoring. When activated, the pressure-sensor signal must show a predefined characteristic, and the offset and amplification are compensated for internally. The steering-angle sensor has its own integrated monitoring functions and directly delivers a message to the ECU in the event of error. In addition, the digital signal transmission to the ECU is permanently monitored. In a third stage, analytical redundancy is applied to monitor the sensors during the steady-state operation of the vehicle. Here, a vehicle model is used to check that the relationships between the sensor signals, as determined by vehicle motion, are plausible. These models are also frequently applied to calculate and compensate for sensor offsets as long as they stay within the sensor specifications.

In case of error, the system is switched off either partially or completely depending on the type of error concerned. The system's response to errors also depends on whether the control is activated or not.

Special Electronic Stability Program for commercial vehicles

Function

Heavy commercial vehicles essentially differ from passenger cars in their much greater mass combined with higher centers of gravity and in additional degrees of freedom resulting from trailer operation [7]. They can thus assume unstable states that extend far beyond the skidding familiar to passenger cars. Such states include, as well as the jackknifing of multiple-stage vehicle combinations, caused for example by trailer sliding, overturning caused by high lateral acceleration. A vehicle-dynamics control system for commercial vehicles must therefore, in addition to providing the stabilization functions familiar to passenger cars, also prevent jackknifing and overturning.

Requirements

The following requirements, in addition to those for passenger cars, can be derived from the extended functions of driving-dynamics control for commercial vehicles:

- Improvement in directional stability and response of a vehicle combination (e.g. articulated road train or articulated-train combination) at physical driving limits in all operating and laden states. This includes preventing jackknifing on vehicle combinations.
- Reduced risk of overturning for a vehicle or vehicle combination in both quasi-stationary and dynamic vehicle maneuvers.

These requirements implemented in commercial-vehicle ESP lead, as is the case with passenger cars, to a significant improvement in driving safety. For this reason, European law will from 2011 require the gradual introduction of a driving-dynamics control system for heavy commercial vehicles (from 7.5 t) (integral part of ECE-R 13 [11]).

Application

In the meantime, commercial-vehicle ESP has become available for virtually all vehicle configurations (except all-wheel-drive vehicles):

- Vehicles with the wheel formulas 4×2, 6×2, 6×4 and 8×4
- Combinations of tractor unit and semitrailer (articulated vehicle or simply semitrailer unit)
- Combinations of rig and drawbar trailer (articulated road train)
- Multiple trailer combinations (Eurokombi), e.g. combinations of rig, dolly and semitrailer or semitrailer unit with additional center-axle trailer or tractor unit with B-link and semitrailer.

Operating principle

Driving-dynamics control for commercial vehicles can be divided according to requirements into the two function groups described in the following.

Stabilizing the vehicle in the event of imminent skidding or jackknifing

Directional stabilization of a commercial vehicle is initially performed according to the same principles as for a passenger car. The controller compares current vehicle motion with vehicle motion desired by the driver, taking into account physical driving limits. The physical model of motion in the horizontal plane – for a single vehicle characterized by three degrees of freedom (longitudinal, lateral and yaw motion) – is however extended for an articulated vehicle to include the articulation angle between the rig and the trailer (an additional degree of freedom). There are further degrees of freedom involved for combinations with fifthwheel trailers.

To calculate the vehicle motion desired by the driver, the ECU uses simplified mathematical-physical models (single-track vehicle model, [8]) to determine the nominal yaw velocity of the rig. The parameters that are encountered in these models (characteristic vehicle speed v_{ch}, wheelbase l and steering ratio i_L) are either parameterized at the end of the vehicle assembly line or adapted to the vehicle's behavior during vehicle operation with the aid of special adaption algorithms (e.g. Kalman filters or recursive least-squares estimators, [9]). "Online" adaptation of parameters is particularly important in commercial vehicles, because the variety of variants and loads is much greater than for passenger cars.

Parallel to this, ESP calculates current vehicle motion from the measured variables available for yaw rate and lateral acceleration plus the wheel speeds. A significant deviation between the current vehicle motion and the motion expected by the driver leads to a control fault that is transformed by the actual controller into a corrective nominal yaw moment.

The level of the nominal yaw moment for a commercial vehicle depends on the control fault, the current vehicle configuration (wheelbase, number of axles, operation with or without trailer etc.) and the laden state (mass, center of gravity in the linear direction, etc.). As these parameters are variable, they are continuously determined by ESP. This is achieved, for example, in the laden state with the aid of an estimation algorithm that uses the signals from the engine management (engine speed and engine torque) and the vehicle linear motion (wheel speeds) to permanently identify the current vehicle mass.

On the basis of the current driving situation, the nominal yaw moment by braking individual or several wheels and of the trailer is transformed in a suitable manner. This is depicted by way of example in Figure 11a and in Figure 11b for a clearly defined oversteer and understeer situations respectively.

In addition to these clearly defined situations, there are other critical dynamic situations in which other wheels or wheel combinations are braked depending on the desired stabilization effect. Thus, for example in the case of sharper understeer, the entire vehicle is braked along similar lines to Enhanced Understeering Control (EUC) in passenger cars.

Because of the high center of gravity of commercial vehicles, skidding and jackknifing by such vehicles primarily occur at low-to-medium coefficients of friction at which the tire static-friction limit is already exceeded at an early stage. At high coefficients of friction, laden commercial vehicles, on account of their high center of gravity, normally start to overturn before the static-friction limit of the tires is reached.

Reducing the risk of overturning
The overturning limit (lateral-acceleration limit) of a vehicle depends not only on the height of the center of gravity but also on the chassis systems (axle suspension, stabilizers, springs, etc.) and the type of payload (fixed or moving) [10].

The situation which causes a commercial vehicle to overturn is, aside from a relatively low overturning limit, an excessive cornering speed. ESP makes use of this scenario to reduce the probability of the vehicle overturning. As soon as the vehicle approaches the overturning limit, it is slowed down by reducing engine torque and, if necessary, also applying the brakes. The overturning limit is determined here depending on the load of the vehicle and the load distribution, whereby the laden state of the vehicle is estimated "online".

Depending on each driving situation, the overturning limit that is determined is modified. Thus, the overturning limit in high-speed dynamic situations (e.g. obstacle-avoidance maneuvers) is reduced in order to permit early intervention. In very slow maneuvers, on the other hand (e.g. negotiating tight hairpin bends on uphill stretches), it is increased in order to prevent unnecessary and disruptive ESP intervention.

Determining the overturning limit is based on various assumptions regarding the height of the center of gravity and the dynamic response of the vehicle combination with a known axle-load distribution. This covers the largest portion of the usual vehicle combinations.

In order to ensure stabilization even in the case of strong deviations from these assumptions (e.g. extremely high centers of gravity), ESP additionally detects the lift of the wheels on the inside of a bend. This is achieved by monitoring the wheels

Figure 11: Principle of ESP braking intervention on semitrailers
a) Vehicle oversteering,
b) Vehicle understeering.

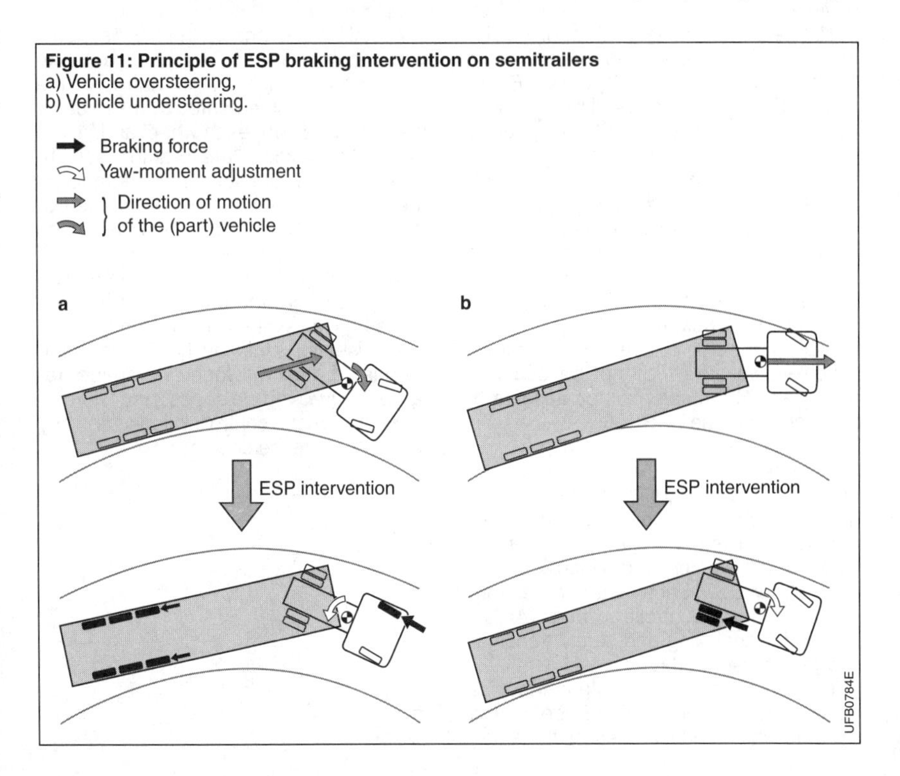

for implausible rotation speed. If necessary, the entire vehicle combination is then heavily decelerated by brake intervention.

A trailer wheel lifting on the inside of a bend is indicated by the trailer's electronically controlled braking system (ELB) via the CAN communication line (ISO 11992 [13]) by activating the ABS controller. For combinations with trailers equipped with ABS only, wheel-lift detection on the inside of a bend is limited to the tractor unit (rig).

System design

On the European market, the electronically controlled braking system ELB has come to the fore as the standard for brake control in heavy commercial vehicles. ESP is based on this system, extending it to include regulation of the driving dynamics. To do so, ESP uses the ELB capability of generating varying braking forces for each individual wheel independently of driver action.

The very different general conditions for commercial vehicle brake systems in North America mean that purely ABS or ABS/TCS systems are used as standard. An ESP based on ABS/TCS is therefore used for these and similar markets. Here, ESP uses the method already applied with TCS on the drive axle to generate braking force individually for each wheel independently of the driver by means of a TCS valve and the downstream ABS valves. In addition, for ABS-based ESP, the driver brake command must be measured by means of pressure sensors, which would otherwise not be possible during an ESP intervention on account of the function of the TCS valve.

Sensor systems

Like passengers cars, commercial vehicles use a combined yaw-rate and lateral-acceleration sensor and a steering-wheel-angle sensor as driving-dynamics sensors for ESP. Each of these sensors contains a microcontroller with a CAN interface for analyzing and safely transmitting the measured data.

The steering-wheel-angle sensor is usually mounted immediately below the steering wheel and it measures the angle of rotation of the steering wheel. This is then converted in the ECU into a wheel steering angle.

In order to pick up the lateral acceleration as close as possible to the center of gravity of the rig, the combined yaw-rate and lateral-acceleration sensor is usually mounted in the vicinity of the center of gravity.

Even though commercial vehicles essentially use the same sensors as passenger cars, the yaw-rate and lateral-acceleration sensor must have a much more robust design to cope with the rougher ambient conditions, particularly on the commercial-vehicle frame.

Electronic control unit (ECU)

The ESP algorithms are run together with the other algorithms for brake control (e.g. ABS and TCS) in the brake control unit. This control unit is constructed using conventional circuit-board technology with correspondingly powerful microcontrollers.

A CAN bus connects the ESP sensors with the control unit. The nominal brake pressures and wheel slip values of the ESP are then implemented by the relevant braking system for each wheel and for the trailer. In addition, the braking system transmits the requested engine torque via the vehicle CAN bus (usually standardized as per SAE J 1939 [14]) to the engine ECU for implementation.

Moreover, relevant information is also transferred from the engine and retarder to the braking system via the vehicle CAN bus. Essentially, this involves current and requested engine torque and speed, retarder torque, vehicle speed, and information from various control switches and any trailer that may be coupled.

Safety and monitoring functions

The extensive possibilities for ESP intervention in the handling characteristics of the vehicle and vehicle combination require a comprehensive safety system to ensure proper system functioning. This extends not only to the basic ELB or ABS/TCS system respectively, but also to the additional ESP components, including all sensors, ECUs and interfaces.

The monitoring functions used for ESP are essentially based on the functions used in passenger cars and are adapted to the characteristics of commercial vehicles.

There is also mutual monitoring of the microcontrollers distributed in the overall system. This means that the brake control unit contains a main computer and a monitoring computer that mainly performs plausibility checks alongside minor functional tasks. Furthermore, the corresponding algorithms permanently check memory and other internal computer hardware components to detect any defects that occur at an early stage.

The occurrence of faults results, depending on the nature and significance of the fault, in the shutdown of individual function groups through to a complete switching to "backup mode", in which the brakes are controlled by purely pneumatic means (fail-silent response). This ensures that incorrect sensor signals cannot cause implausible and possibly dangerous operating states in braking operations.

The occurrence of a fault is indicated to the driver by suitable means (e.g. warning lamp or display) so that suitable action can be taken.

Furthermore, any faults that occur are assigned a time stamp in the control unit and stored in the fault memory. The workshop can evaluate these with the help of a suitable diagnosis system.

References

[1] E.K. Liebemann, K. Meder, J. Schuh, G. Nenninger: Safety and Performance Enhancement: The Bosch Electronic Stability Control. SAE Paper Number 2004-21-0060.

[2] National Highway Traffic Safety Administration (NHTSA) FMVSS 126: Federal Motor Vehicle Safety Standards; Electronic Stability Control Systems; Controls and Display. Vol. 72, No. 66, April 6, 2007.

[3] M. Mitschke, H. Wallentowitz: Dynamik der Kraftfahrzeuge. 4th Edition, Springer-Verlag, 2004.

[4] A. van Zanten, R. Erhardt, G. Pfaff: FDR – Die Fahrdynamikregelung von Bosch. ATZ Automobiltechnische Zeitschrift 96 (1994), Volume 11.

[5] A. Trächtler: Integrierte Fahrdynamikregelung mit ESP, aktiver Lenkung und aktivem Fahrwerk. at – Automatisierungstechnik 53 (1/2005).

[6] K. Reif: Automobilelektronik, 3rd Edition, Vieweg+Teubner, 2009.

[7] E. Hoepke, S. Breuer (Editors): Nutzfahrzeugtechnik – Grundlagen, Systeme, Komponenten. 4th Edition; Vieweg Verlag, 2006.

[8] C.B. Winkler: Simplified Analysis of the Steady State Turning of Complex Vehicles. International Journal of Vehicle Mechanics and Mobility, 1996.

[9] Ali H. Sayed: Adaptive Filters. John Wiley & Sons, 2008.

[10] D. Odenthal: Ein robustes Fahrdynamik-Regelungskonzept für die Kippvermeidung von Kraftfahrzeugen. Dissertation TU München, 2002.

[11] ECE-R 13: Uniform provisions concerning the approval of category M, N and O vehicles with regard to brakes.

[12] ECE-R 13-H: Standard conditions for approval of passenger cars with regard to brakes.

[13] ISO 11992: Road vehicles – Interchange of digital information on electrical connections between towing and towed vehicles.

[14] SAE J 1939: Serial Control and Communications Heavy Duty Vehicle Network Top Level Document.

Supplementary functions (automatic brake-system operations)

The supplementary functions described here utilize the infrastructure of the ESP system, excluding the pneumatic-mechanical brake assistant. This comprises the sensors, the actuators and the ECU.

Brake-assistant operations

Investigations into driver braking behavior in the 1990s showed that car drivers differ in the way that they respond to braking situations. The majority – the "average drivers" – do not brake hard enough when faced with an emergency situation, in other words, they require an unnecessarily long braking distance (Figure 1). This can be remedied by a system first launched in 1995, the brake assistant. Its main purposes are as follows:

- It interprets a certain rate of pedal movement (rapid application of the brakes) that fails to apply maximum braking force as an intention by the driver to carry out full braking. In this case, it generates the braking pressure required to achieve full braking effect.
- It allows the driver to "cancel" full braking operation at any time.

Figure 1: Comparison between braking with and without brake-assistant operations

A longer braking distance is required without a brake assistant than with a brake assistant.

1 "Average driver",
2 Experienced driver,
3 "Average driver" with brake assistant.
t_v Delay time for braking results in longer braking distance.

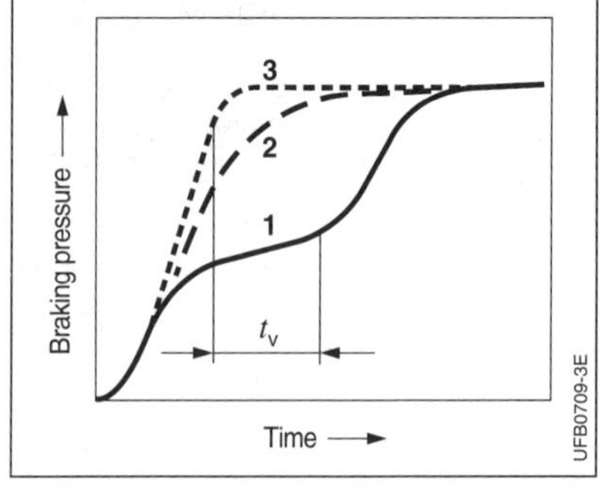

- The behavior of the brake booster and, therefore, the pedal feedback, is not altered under normal braking conditions.
- The basic braking-system function is not diminished if the brake assistant fails.
- The system is designed to prevent accidental activation.

Pneumatic brake assistant

This system requires a modified brake booster which increases the level of amplification according to the rate of pedal movement and pedal force. This results in faster and greater buildup of pressure in the wheel brakes.

In an alternative version, the brake booster is extended to include an electronically activated valve. This enables an ECU to influence the pressure difference between the brake-booster chambers and thus the boosting of the braking force. This provides better opportunities for optimizing the trigger threshold and response characteristics.

Hydraulic brake assistant

The hydraulic brake-assistant function makes use of the ESP system's hardware. A pressure sensor detects the driver's braking intention; the ECU analyzes the signal on the basis of the defined triggering criteria, and initiates an appropriate brake-pressure build-up in the hydraulic system. The upstream brake booster is a standard unit and does not require any modification.

As a general observation, it should be stated that it is an absolute requirement that all brake-assistant system variants referred to are used in conjunction with an ABS, TCS or ESP system due to the actively generated rapid brake-pressure rise beyond the wheel-lock limit.

Automatic brake-pressure increase at the rear wheels

This is a function which provides the driver with additional brake servo assistance for the rear wheels if the front wheels are controlled by the ABS system. This function was introduced because many drivers do not increase the pedal force at the start of ABS control even though the situation would require this. Once ABS control has been initiated on the front wheels, the wheel pressures at the wheels on the rear axle are increased by way of the return pump of the hydraulic modulator until these too have reached the lockup-pressure level and ABS control is initiated (Figure 2). The brake application is therefore at the physical optimum. The pressure in the rear axle wheel-brake cylinders can then exceed the pressure in the brake master cylinder, also during ABS control.

The cutoff condition is fulfilled when the wheels on the front axle are no longer under ABS control or when the pressure in the brake master cylinder falls below the cutoff threshold.

Figure 2: Automatic brake-pressure increase
1 Brake-pressure increase by brake booster,
2 Further brake-force increase by pedal force,
3 Brake-pressure increase by ESP hydraulic system,
4 ABS control range.

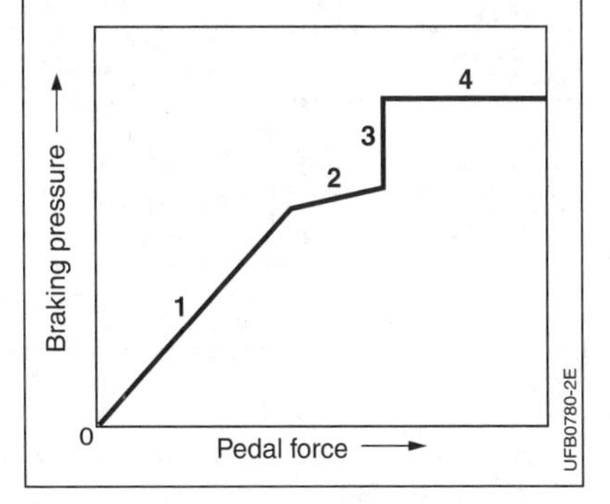

Automatic brake-pressure increase with forceful pressing of brake pedal

This function provides the driver with additional brake servo assistance. It is activated if the maximum possible vehicle decoloration is not achieved even if the driver forcefully presses the brake pedal to the point that would normally cause the lockup pressure to be reached (primary pressure over approx. 80 bar). This is the case, for example, at high brake-disk temperatures or if the brake pads have a considerably reduced coefficient of friction.

When this function is activated, the wheel pressures are increased until all wheels have reached the lockup pressure level and ABS control is initiated (Figure 2). The brake application is therefore at the physical optimum. The pressure in the wheel-brake cylinders can then exceed the pressure in the brake master cylinder, also during ABS control.

If the driver reduces the desired level of braking to a value below a particular threshold value, the vehicle deceleration is reduced in accordance with the force applied to the brake pedal. The driver can therefore precisely modulate the vehicle deceleration when the braking situation has passed. The cutoff condition is fulfilled if the primary pressure or vehicle speed falls below the respective cutoff threshold.

Brake-disk wiper

This function ensures that, in the event of rain or a wet road, the splash water is cyclically removed from the brake disks. This is achieved by automatically setting a low brake pressure at the wheel brakes. In this way, the function helps to ensure minimum brake-response times when driving in wet conditions. Evaluated signals from the windshield wipers or rain sensor are used to detect wet conditions.

The pressure level is adjusted so that the vehicle deceleration cannot be perceived by the driver. Actuation is repeated at a defined interval for as long as the system detects rain or a wet road. If required, just the disks at the front axle can be wiped. The wiping procedure is terminated as soon as the driver applies the brakes.

Automatic prefill

This function reduces the total braking distance in emergency situations whereby the driver operates the brake pedal immediately after the accelerator pedal is released. This is achieved by prefilling and thereby preloading the brake system after release of the accelerator pedal, which means that in the subsequent brake operation the pressure build-up is considerably more dynamic. Accordingly, high vehicle deceleration sets in earlier.

Prefilling of the brake system is adjusted by the return pump of the ESP hydraulic modulator. The brake shoes are then firmly applied to the brake disks. If there is no operation of the brake pedal directly after a rapid release of the accelerator, the pressure in the brake system is reduced again. This does not impair driveability.

Electromechanical parking brake

The electromechanical parking brake (EMP) generates the force for applying the parking brake by electromechanical means. The function of the hand- or foot-operated parking-brake lever is performed by a control knob with an electric-motor-and-transmission combination. When the driver operates the control knob, the electric motor (actuator) is activated when the system detects that the vehicle is at a standstill. When the vehicle is parked on a level surface, the holding forces are set lower than when the vehicle is fully laden and parked on a gradient. Active wheel sensors are used to detect that the vehicle is parked (i.e. stationary). As an option, it is also possible for the road gradient to be detected by a tilt sensor.

The parking brake is released by means of the same control knob. However, various safety regulations and requirements have to be met, e.g. to prevent inappropriate or inadvertent release of the parking brake by children or animals (see also Brake systems for passenger cars).

Controlled braking with ESP hydraulic system

If the electromechanical parking brake is activated while driving, the vehicle must be safely decelerated to a stop. The brake pressure required for this is built up by the return pump of the ESP hydraulic modulator. The ABS and ESP systems ensure safe braking even on slippery or wet road surfaces. When the vehicle has come to a stop, the electromechanical parking brake assumes the function of keeping the vehicle stationary.

The driver must press the activation knob of the automatic parking brake continuously during the deceleration phase.

Hill hold control

This system simplifies hill starts. It prevents the vehicle from rolling back after the driver has released the brake pedal. This is particularly helpful on heavily laden vehicles with manual transmission and vehicles that are towing trailers. There is no need to operate the parking brake. The function also works when performing a hill start in the reverse direction.

The system detects the driver's intention to pull away (Figure 3). After the brake pedal is released, there is approximately two seconds to start pulling away. The

Figure 3: Hill hold control

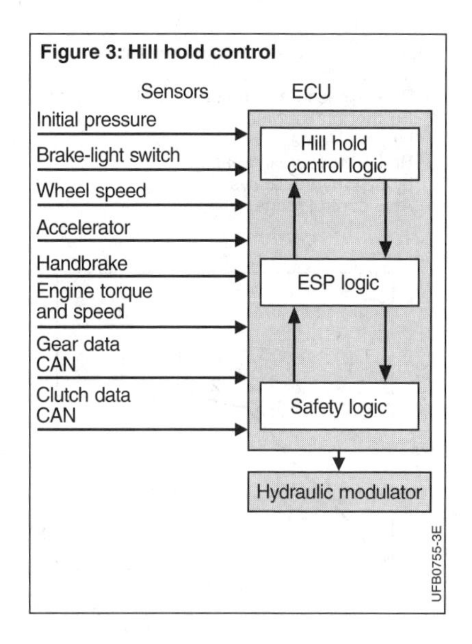

brake is released automatically here if the drive torque is greater than the torque resulting from downgrade force.

The HHC system is based on the ESP system hardware with additional sensors – a tilt sensor detects the road gradient, a gear switch detects whether the driver has shifted to reverse gear, and a clutch switch recognizes whether the driver has depressed the clutch pedal.

Automatic braking on hill descent – hill descent control

This system is a convenience function which assists the driver on offroad descents with gradients of approximately 8 to 50 % by automatically operating the brakes. The driver can then concentrate fully on steering the vehicle and is not distracted by the need to operate the brakes at the same time. The brake pedal does not have to be operated.

When this system is activated, e.g, by pressing a button or switch, a preset speed is maintained over the extent of the specified brake pressure. If required, the driver can vary the predetermined speed by pressing the brake and accelerator pedal or using the control buttons of a speed control system.

The system remains active until it is switched off by pressing the button or switch again, i.e. it is not automatically deactivated.

Automatic brake application for driverassistance systems

This function is an additional function for active brake application with adaptive cruise control (ACC), i.e. for automatic vehicle-to-vehicle ranging. The brakes are applied automatically without the driver pressing the brake pedal as soon as the distance to the vehicle in front falls below a predetermined distance (Figure 4). This is based on a hydraulic brake system and an ESP system.

The function receives a request to decelerate the vehicle by a desired amount (input). This is calculated by the upstream ACC system. Automatic brake application maintains vehicle deceleration by means of appropriate brake pressures which are adjusted with the aid of the ESP hydraulic modulator.

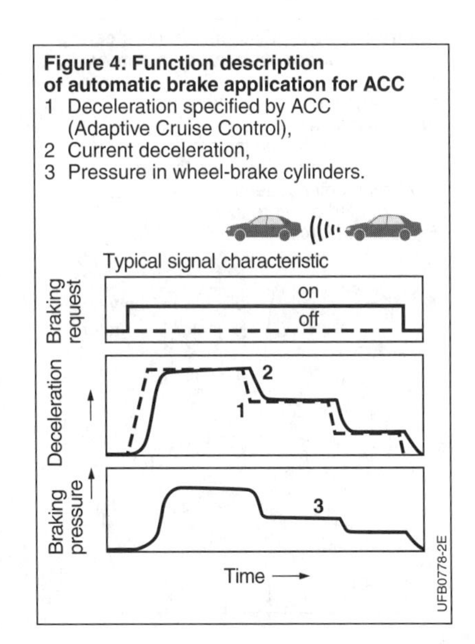

Figure 4: Function description of automatic brake application for ACC
1 Deceleration specified by ACC (Adaptive Cruise Control),
2 Current deceleration,
3 Pressure in wheel-brake cylinders.

Integrated driving-dynamics control systems

Overview

In addition to the brakes and engine, as used by the driving-dynamics control system (Electronic Stability Program, ESP), there are other actuators in the chassis and in the drivetrain which are used to purposefully influence driving dynamics. The functional combination of actuators relevant to driving dynamics is referred to by different names by the various automobile manufacturers and suppliers: Vehicle Dynamics Management (VDM), Integrated Chassis Control (ICC), Integrated Chassis Management (ICM), and Global Chassis Control (GCC).

Primary uses

The functions are used primarily in the fields of directional stability, agility and reducing "driver workload", i.e. the efforts exerted by the driver to operate and control the vehicle.

Improving directional stability

In critical driving situations, these functions take on stabilization tasks which are normally performed by the driver himself. Typical examples are the oversteer situation or braking on roads with different grip factors on the left and right sides of the vehicle (μ split).

Increasing agility

Some of these functions improve the acceleration response of the lateral-dynamic response of the vehicle to steering inputs by the driver. These make the vehicle more agile, i.e. it responds more spontaneously and dynamically to driver inputs.

Reducing the effort to control/operate the vehicle

An improved response by the vehicle to movements of the steering wheel and automatic stabilization interventions relieve the driver's workload, the steering effort in particular being reduced.

Functions

Driving-dynamic driver steering recommendation

Function

This function utilizes electrical power steering as the actuator. An additional, driving-dynamically motivated steering torque is superimposed on the steering-servo torque of the power steering in order to provide the drive with a steering recommendation.

Limitation of supplementary steering torque

The driving-dynamic supplementary steering torque is limited, depending on the vehicle, to values of approximately 3 Nm so that the driver can oversteer the steering recommendation at any time.

Driving-dynamics applications

The steering recommendation is activated in different driving situations. In a oversteer situation, a steering torque is introduced in the countersteer direction (Figure 1a). In the event of understeer, the function motivates the driver not to increase the steer

Figure 1: Driving-dynamic driver steering recommendation with supplementary steering torque

a) Oversteer,
b) Braking on road with different grip factors.
M_L Supplementary steering torque,
M_Z Yaw moment based on intended driver reaction,
F_B Braking force.

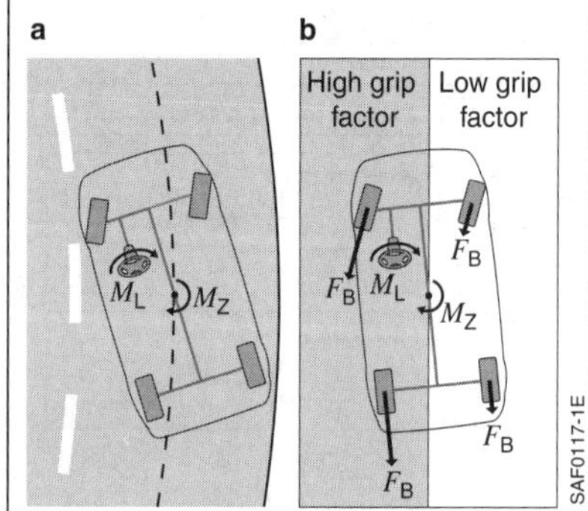

angle further, since the side-force potential on the front wheels is already exhausted and further steering may even reduce the side force.

When braking or accelerating on roads with different grip factors on the left and right sides of the vehicle (μ split), the driver is assisted in compensating the vehicle yaw motion by countersteering (Figure 1b).

Active steering stabilization on the front axle

Function

This function intervenes via the angle-superimposing steering actuator of the override steering directly in the vehicle movement. An additional steering angle adjusted by an electric motor is added to the driver's steering angle in an override gearbox.

Driving-dynamics applications

The function operates, for example, in an oversteering driving situation. Yaw-rate control with automatic steering-angle intervention returns the excessively high yaw-rate value to its setpoint value.

During braking on roads with different grip factors, the function provides for compensation of the yaw motion. The driver's workload is significantly relieved when stabilizing the vehicle. Because yaw compensation is automatically initiated and happens much more quickly than the steering operation of a typical driver, the measures in the Electronic Stability Program (ESP) for yaw-moment build-up delay can be reduced. The braking distance is also reduced.

A further benefit can be achieved in the event of understeer and starting on roads with different grip factors.

Demands on the sensor system

Because of the high positioning rate of an override steering system, the sensor signals must be monitored with low fault latencies, i.e. a fault must be signaled quickly. This means that the inertial sensor system must have a redundant design.

Active steering stabilization on the rear axle

Function

In vehicles with rear-axle steering, the primary function controls the rear-axle steering angle as a function of the steering-wheel angle and the driving speed. At low speeds, the rear wheels are steered in the opposite direction to the front wheels, which improves vehicle maneuverability.

At high speeds, the rear wheels are steered in the same direction as the front wheels. In this way, the yaw motion is less intensely excited during obstacle-avoidance maneuvers. This increases directional stability considerably.

Driving-dynamics applications

The associated functions act in situations similar to active front-axle steering interventions. The effectiveness of oversteer interventions is crucially dependent on the current utilization of the side-force potential on the rear wheels.

When braking on roads with different grip factors, it is necessary on account of the reduction of load on the rear axle to make interventions with larger angles in order to suppress the yaw motion. On the other hand, braking can take place without a considerable attitude angle.

Demands on the sensor system

The functions for rear-axle steering likewise require a redundant inertial sensor system, since the positioning rate of a steering actuator on the rear axle is comparable with the override steering.

Torque distribution in the longitudinal direction for four-wheel drives

Function

In vehicles with four-wheel drive, interaxle differential locks or transfer cases within the framework of their basic function improve the traction and, with it, the accelerating performance of the vehicle, whereby the wheels of both axles transfer tractive forces.

Driving-dynamics applications
In situations where the grip factor is not fully utilized by traction, the self-steering properties can be altered by shifting the tractive forces between the front and rear axles. The more tractive force an axle transfer, the more the side-force potential on the wheels of this axle is weakened. Shifting the tractive force to the front axle increases the understeer tendency, while shifting it to the rear axle reduces the understeer tendency (see Drivetorque distribution). Through dynamic control of drive-torque distribution, lateraldynamic agility can be increased without the vehicle tending towards oversteer.

Wheel-torque distribution in the lateral direction

Function
Functions based on a "torque-vectoring actuator" make a significant contribution to increasing agility. The actuator permits an extensively free shift of the wheel torque between the left and right wheels on an axle. In this way, the total turning force excluding friction losses is not reduced in the actuator meaning that this intervention is virtually neutral with regard to the driving speed.

Figure 2: Wheel-torque distribution in the lateral direction
a) Understeer,
b) Oversteer.
F_U Wheel peripheral force (driving, braking),
M_Z Yaw moment based on intervention.

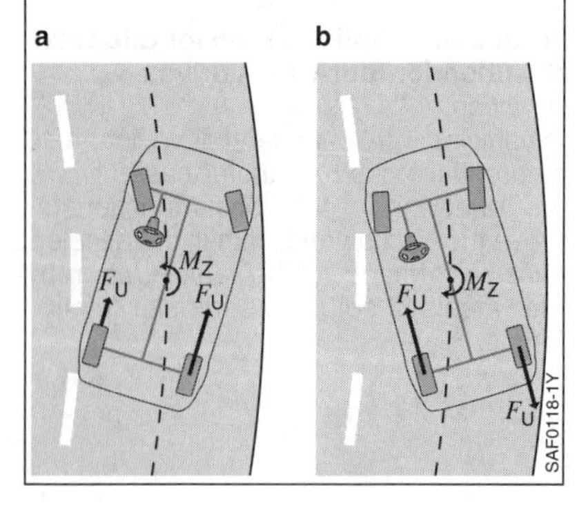

Driving-dynamics applications
In an understeering driving situation, the drive torque of the wheel on the outside of the curve is increased and the drive torque of the wheel on the inside of the curve is decreased. In this way, an additional yaw moment turning into the curve acts on the vehicle (Figure 2a). The understeer tendency decreases; the vehicle's lateral-dynamic agility increases.

When accelerating on roads with different grip factors, the drive torque is specifically directed to the wheel with a high grip factor. Brake interventions by the traction control system on the wheel with a low grip factor are dispensed with to a large extent. This increases the mean acceleration on roads with different grip factors.

Even in an oversteering situation, the brake interventions are partly replaced by a wheel-torque shift (Figure 2b). In this way, the speed loss can be reduced by brake interventions of the driving-dynamics control in the event of mild oversteer. In critical oversteering situations, the brakes still intervene, since in this case the speed loss is desired to alleviate the driving situation.

Replacement of wheel-torque distribution by brake and engine-control interventions

The understeer interventions by means of a torquevectoring actuator can be emulated by brake interventions on the wheels on the inside of the curve. The engine torque is increased to compensate the deceleration by the one-sided brake intervention. This produces an effect similar to understeer intervention with a torque-vectoring actuator. This only requires a hydraulic modulator with a long service life and low noise generation instead of an additional torque-vectoring actuator.

Influencing self-steering properties through roll stabilization

Function
In line with their comfort-oriented basic function, systems for roll stabilization serve to compensate the rolling motion of the vehicle body when cornering.

Driving-dynamics applications
Where the roll-stabilization system has a two-channel design with separate activation of the actuators on the front and rear axles, there is an additional option for influencing the self-steering properties. The roll-compensating torques can – within the framework of the performance of the individual channels – be actively distributed between the front and rear axles. This results in a change of the wheel downforces.

The effect on the self-steering properties is based on the degressive increase in the lateral forces with the normal forces (Figure 3). When the rolling moment is extensively supported over the rear axle, the normal force at the rear on the wheel on the outside of the curve increases dramatically; the associated side force, however, increases only underproportionally. The side force on the rear axle is weakened, and the understeer tendency decreases. As a countermove, the understeer tendency increases when the rolling moment is extensively supported over the front axle.

Influencing self-steering properties through variable dampers

Function
The use of variable dampers for adaptive damping of the vertical movement of the vehicle body has become significantly widespread. Pitching and rolling motions which are caused by driver inputs such as braking and steering or by road-surface defects, as well as vertical movements caused by road-surface defects, can be noticeably reduced.

Driving-dynamics applications
It is also possible with limited effect to influence the self-steering properties. The action mechanism corresponds to influencing the self-steering properties through roll stabilization. However, the function only has a transient effect during a rolling motion of the vehicle body, since a damper movement is a prerequisite for damping forces. The damping forces can be modulated by adjusting the damping hardness so that the wheel downforces can be semi-actively distributed as described above.

System architecture

Allocation of functions to ECUs

The actuators in the chassis and in the drivetrain are generally activated not only by functions of the integrated driving-dynamics control systems, but also first and foremost by basic functions which are very closely connected with the actuator. Examples of such basic functions include servo assist in an electrical power-steering system or variable steering ratio in an override steering system.

The basic functions are characterized by moderate networking with other vehicle systems. Because of their close connection to the actuator, they are usually integrated into the associated ECU, which also activates and monitors the actuator.

On the other hand, the functions of the integrated drivingdynamics control systems are highly networked, in particular with driving-dynamics control (ESP) and other chassis control systems. They calculate a setpoint driving-dynamic value for the actuator, which is transmitted via a data bus to the ECU. An arbitration with the setpoint values of the basic functions is performed in the ECU.

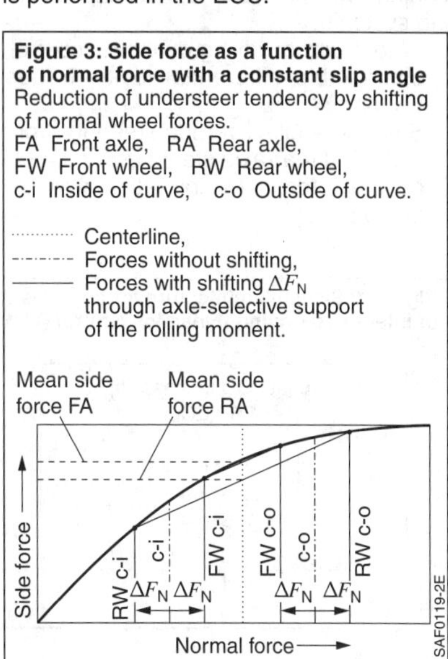

Figure 3: Side force as a function of normal force with a constant slip angle
Reduction of understeer tendency by shifting of normal wheel forces.
FA Front axle, RA Rear axle,
FW Front wheel, RW Rear wheel,
c-i Inside of curve, c-o Outside of curve.

The ECU for driving-dynamics control (ESP) or a central ECU of the "Chassis" functional area is suitable for use as the integration platform for the functions of the integrated driving-dynamics control systems (Figure 4).

Interaction of various functions

Various functions and actuators are increasingly being installed in a single vehicle. This calls into question a function structure which guarantees good interaction of the individual functions and which above all prevents mutual interference.

The original trend observed of implementing a separate controller for each actuator is reaching its limits. However, fully bringing together all the algorithms in a central controller limits flexibility in the distributed development and in the use of ECU resources.

A more promising compromise is bringing together all the controllers which utilize a related positioning principle, since interventions with a related positioning principle require particularly intensive coordination.

The following are mentioned as positioning principles and allocated positioning systems:

- Wheel torque: driving-dynamics control, controllable differential locks, torque vectoring
- Steering angle: override steering (front axle) and rear-axle steering
- Normal force: roll stabilization, variable dampers.

The driving-dynamic steering-torque recommendation plays a special role. It must be matched in terms of its effect on the driver with functions for override steering.

References

[1] Isermann (Editor): Fahrdynamikregelung – Modellbildung, Fahrerassistenzsysteme, Mechatronik. Vieweg Verlag, Wiesbaden, 2006.

[2] Deiss, Knoop, Krimmel, Liebemann, Schröder: Zusammenwirken aktiver Fahrwerk und Triebstrangsysteme zur Verbesserung der Fahrdynamik. 15th Aachen Colloquium Automotive and Engine Technology, Aachen 2006, PP. 1671...1682.

[3] Klier, Kieren, Schröder: Integrated Safety Concept and Design of a Vehicle Dynamics Management System. SAE Paper 07AC-55, 2007.

[4] Erban, Knoop, Flehmig: Dynamic Wheel Torque Control – DWT. Agility enhancement by networking of ESP and torque vectoring. 8th European All-wheel Congress, Graz 2007, PP. 17.1...17.10.

[5] Flehmig, Hauler, Knoop, Münkel: Improvement of Vehicle Dynamics by Networking of ESP with Active Steering and Torque Vectoring. 8th Stuttgart International Symposium Automotive and Engine Technology, Report Vol. No. 2, PP. 277...291, Vieweg Verlag, Wiesbaden 2008.

Figure 4: System architecture for implementing functions of integrated driving-dynamics control (VDM)

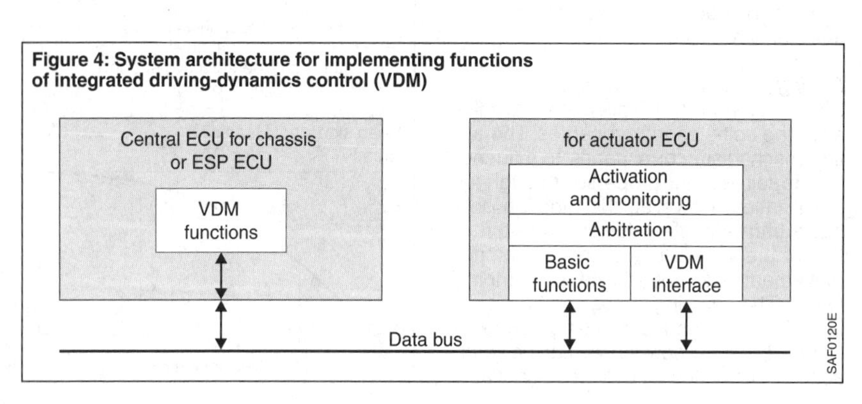

Vehicle bodies, passenger cars

Main dimensions

Interior dimensions

The dimensional layout depends on body shape, type of drive, scope of aggregate equipment, desired interior size, trunk volume, and other considerations such as driving comfort, driving safety, and operating safety (Figure 2). The seating positions are designed according to ergonomic findings and with the aid of templates or 3D CAD dummy models (DIN, SAE, RAMSIS): body template as per DIN 33408, for men (5th, 50th, and 95th percentile) and women (1st, 5th and 95th percentile). For example, the 5th percentile template represents "small" body size, i.e. only 5% of the population has smaller bodies, while 95% has larger body dimensions.

SAE H-point template in accordance with SAE J826 (May 1987): 10th, 50th, and 95th percentile thigh segments and lower leg segments. For legal reasons, motor-vehicle manufacturers in the ECE, the USA and Canada must use the SAE H-point template for determining the seating reference point. Most motor-vehicle manufacturers around the world use the 3D CAD dummy model RAMSIS (computer-aided anthropological-mathematical system for passenger simulation) in development.

The hip point (H-point) is the pivot center of torso and thigh, and roughly corresponds to the hip-joint location. The seating reference point (SRP, ISO 6549 and US legislation) or R point (ISO 6549 and EEC Directives/ECE regulations) is the position of the H-point in the seat adjustment field on variable seats, again taking account of the heel point (Figure 2). In determining the design H-point position, many vehicle manufacturers use the 95th percentile adult-male position or, if this position is not reached, the position with the seat adjusted to its rearmost setting. In order to check the position of the measured H-point relative to the vehicle, a three-dimensional, adjustable SAE H-point machine weighing 75 kg is used. The seating reference point, heel point, vertical, and horizontal distance between these two points, and body angles specified by the vehicle manufacturer form the basis for determining the dimensions of the driver's seating position.

The seating reference point is used:
- To define the positions of the eye ellipse (SAE J 941) and the eye points (RREG 77/649) as a basis for determining the driver's direct field of view
- To define hand reach envelopes in order to correctly position controls and actuators
- To determine the accelerator heel point (AHP) as a reference point for positioning the pedals.

The space required by the rear axle as well as location and shape of the fuel tank primarily determine the rear-seating arrangement (height of the seating reference point, rear seating room, headroom) and thus the shape of the roof rear portion. Depending on the type of vehicle under development, the projected main dimensions and the required passenger sizes, there are different body angles for the 2D templates or body postures (RAMSIS) and different distances between the seating reference points of the driver's and rear seats. The longitudinal dimensions are greatly influenced by the height of the seating reference point above the heel point. Lower seats require a more

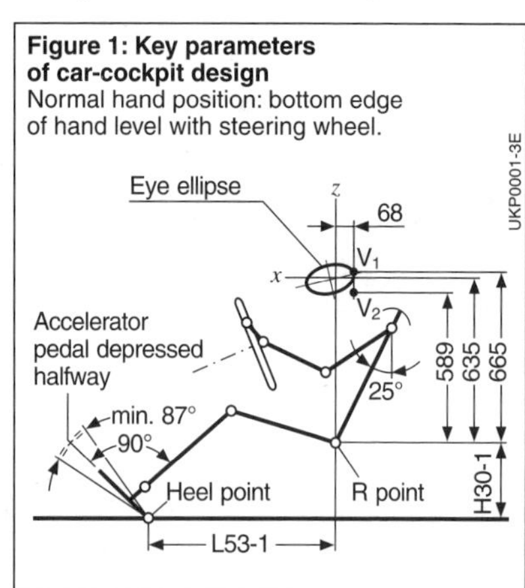

Figure 1: Key parameters of car-cockpit design
Normal hand position: bottom edge of hand level with steering wheel.

Figure 2: Typical internal and external dimensions (as per DIN 70020, Part 1)

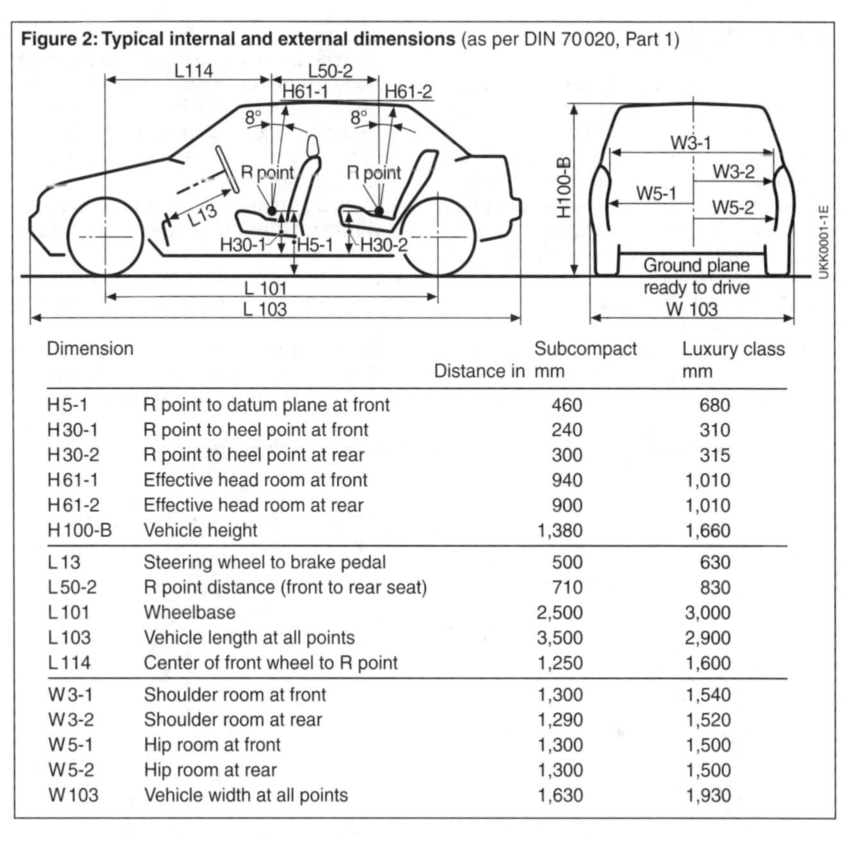

Dimension	Distance in	Subcompact mm	Luxury class mm
H 5-1	R point to datum plane at front	460	680
H 30-1	R point to heel point at front	240	310
H 30-2	R point to heel point at rear	300	315
H 61-1	Effective head room at front	940	1,010
H 61-2	Effective head room at rear	900	1,010
H 100-B	Vehicle height	1,380	1,660
L 13	Steering wheel to brake pedal	500	630
L 50-2	R point distance (front to rear seat)	710	830
L 101	Wheelbase	2,500	3,000
L 103	Vehicle length at all points	3,500	2,900
L 114	Center of front wheel to R point	1,250	1,600
W 3-1	Shoulder room at front	1,300	1,540
W 3-2	Shoulder room at rear	1,290	1,520
W 5-1	Hip room at front	1,300	1,500
W 5-2	Hip room at rear	1,300	1,500
W 103	Vehicle width at all points	1,630	1,930

stretched passenger seating position, and thus greater interior length.

The passenger-cell width, and thus the shoulder room, elbow room and hip room, are dependent on the projected external width, the shape of the sides (curvature), and the space required for door mechanisms, passive restraint systems, and various assemblies (propshaft tunnel, exhaust-gas system, etc.).

Trunk dimensions

The size and shape of the trunk are dependent on the design of the rear end of the vehicle, the position of the fuel tank and its volume, the entire rear package, the positions of the axle and the resulting wheel arches, and the location of the main muffler.

The trunk capacity is determined as per DIN ISO 3832 or, more commonly, using the VDA (Association of German Engineers) method with the VDA module (cuboid $200 \times 100 \times 50$ mm^3 – corresponds to 1 dm^3 volume).

External dimensions

The following factors must be taken into consideration:

- Seating arrangement and trunk
- Engine, transmission, radiator
- Auxiliary systems and special equipment
- Space requirement for compressed or pivoted wheels (allowance for snow chains)
- Type and size of powered axle
- Position and volume of fuel tank
- Front and rear fenders
- Aerodynamic considerations
- Ground clearance (approx. 100 to 180 mm)
- Visibility legislation.

Body design

The following technical requirements must be met in interior and exterior body design:
- Mechanical functions (lowering of side windows, opening of hood, trunk lid, and sliding sunroof, positions of lamps)
- Manufacturability and ease of repair (gap widths, bodywork assembly, window shape, protective molding rails, paint feature lines)
- Safety (position and shape of fenders, no sharp edges or points)
- Aerodynamics (air forces and moments if they affect performance, fuel consumption, and emissions as well as vehicle dynamics/directional stability, wind noise, adhesion of dirt to the outer body panels, enjoyment of open-topped driving, cabin ventilation, windshield wiper function, cooling of fluids and components; see vehicle aerodynamics, also see Figure 3)
- Optics (visual distortion caused by window type and slope, glare due to reflection)
- Legal requirements (position and size of lamps, rearview mirror, license plates)
- Design and layout of controls (positions, shapes, and surface contours)
- Visibility of vehicle extremities (parking).

Figure 3: Aerodynamic effects on vehicle operations and vehicle characteristics

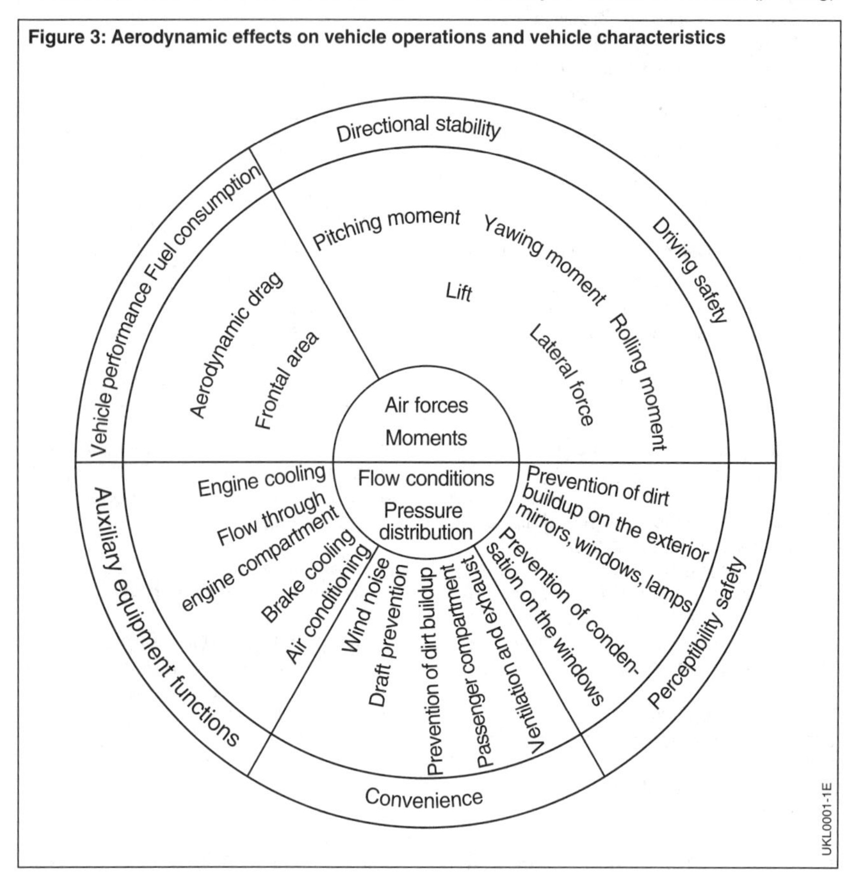

Body

Body structure

Unitized body (standard design)
The unitized body consists of sheet-metal panels, hollow tubular members, and body panels that are joined together by multi-spot welding machines or welding robots. Individual components may also be bonded, riveted, or laser-welded.

Depending on vehicle type, roughly 5,000 spot welds must be made along a total flange length of 120 to 200 m. The flange widths are 10 to 18 mm. Other parts (front fenders, doors, hood, and trunk lid) are bolted to the body supporting structure. Other types of body construction include frame and sandwich designs.

Increasing use is being made of hybrid bodyshell construction, where the individual structural components of the bodyshell are made of different materials according to function and required load capacity. The bodyshell illustrated in Figure 4 (MBW 211), for example, consists of 47% high-strength steel, 42% standard steel, 10% aluminum, and 1% plastic.

The general requirements placed on the body structure are as follows:

Rigidity
Torsional and bending rigidity should be as high as possible in order to minimize elastic deformation of the apertures for the doors, hood, and trunk lid. The effect of body rigidity on the vibrational characteristics of the vehicle must be taken into consideration.

Vibrational characteristics
Body vibrations as well as vibrations of individual structural components as a result of excitation by the wheels, wheel suspension, engine, and drivetrain can severely impair driving comfort if resonance occurs.

The natural frequency of the body, and of those body components which are capable of vibration, must be detuned by means of creasing (impressions in the component to stiffen flat structures), and changing the wall thicknesses and cross sections to minimize resonance and its consequences.

Operational integrity
The alternating stresses that occur during vehicle operation place a strain on the body over time. The layout of the body (focal points are mounting points of the

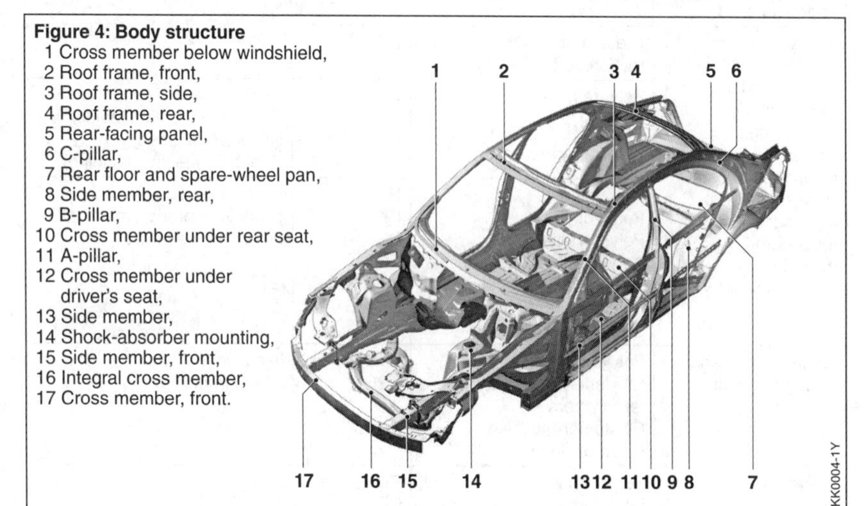

Figure 4: Body structure
1 Cross member below windshield,
2 Roof frame, front,
3 Roof frame, side,
4 Roof frame, rear,
5 Rear-facing panel,
6 C-pillar,
7 Rear floor and spare-wheel pan,
8 Side member, rear,
9 B-pillar,
10 Cross member under rear seat,
11 A-pillar,
12 Cross member under driver's seat,
13 Side member,
14 Shock-absorber mounting,
15 Side member, front,
16 Integral cross member,
17 Cross member, front.

chassis, steering and drive units) must take account of the alternating stresses so that the body also remains undamaged during rough vehicle operation.

Body stresses due to accidents
In the event of a collision, the body must be capable of transforming as much kinetic energy as possible in the deformation process, while minimizing deformation of the passenger cell.

Ease of repair
The components which are most susceptible to damage as a result of minor accidents ("fender-benders") must be easily replaceable or repairable (access to exterior body panels from the inside, access to bolts, favorable location of joints, feature lines for repainting individual components).

Body materials

Sheet steel
Steel plates with different qualities are normally used for the bodyshell. The bodyshell plate thicknesses are 0.6 to 3.0 mm; the highest proportion are steel sheets of 0.75 to 1.0 mm in thickness. Due to the mechanical properties of steel with regard to rigidity, strength, economy, and ductility, alternative materials for the vehicle body structure are not yet fully available.

High Strength Low Alloy (HSLA) sheet steel is used for high-stress structural components. The resulting high strength of these components allows their thickness to be reduced.

Aluminum
In order to reduce weight, aluminum can be used for separate body components, such as the hood, trunk lid, etc.

Since 1994 an aluminum body has been in use on one of the German luxury

Table 1: Examples of alternative materials

Typical applications	Material	Abbreviation	Processing method
Supporting parts, e.g. fender spring support	Glass-mat reinforced thermoplastic	PP-GMT	Injection molding
Trim panels and finishers, e.g. front apron, spoiler, front section, radiator grill, wheel-housing cladding, hubcaps. Bodyshell parts, e.g. hood, mudguards, trunk lid, sliding sunroof	Glass-mat reinforced thermoplastic	PP-GMT	
	Polyurethane	PUR	RIM (Reaction-Injection Molding) RRIM (Reinforced-Reaction-Injection Molding)
	Polyamide Polypropylene Polyethylene Acrylonitrile-butadiene-styrene (ABS) copolymer Polycarbonate (with poly-buteneterephthalate)	PA PP PE ABS PC-PBT	Injection molding, glass-fiber proportion determines elasticity
Elastic fender and front guard strips	Polyvinylchloride Ethylene-propylene terpolymers Elastomer-modified Polypropylene	PVC EPDM PP-EPDM	Injection molding and extruding
Energy-absorbing foam	Polyurethane Polypropylene	PUR PP	Reaction foams
Fenders	Duromers with reinforcing fibers (Sheet Molding Compound)	SMC	Pressing

classes. The vehicle's frame is constructed from aluminum extruded sections, and the panel parts are integrated as self-supporting parts (ASF Audi Space Frame). The implementation of this principle required the use of suitable aluminum alloys, as well as new production processes and special repair facilities. According to the manufacturers, the rigidity and deformation characteristics are identical to those of steel, or are even superior.

Plastics
Plastics replace steel in a limited number of cases for separate body components to reduce weight (see Table 1).

Body surface
Corrosion protection
Allowance must be made for corrosion protection as early as the body-design phase. The "Anti Corrosion Code", Canada, is an agreement which was developed by Scandinavian car manufacturers and consumer-protection organizations and establishes that all vehicles after January 1983 may not exhibit either surface rust for a period of three years or rust penetration or solid structural weakening for a period of six years.

Corrosion-protection measures:
- Minimize the number of flanged joints, sharp edges and corners
- Avoid areas where dirt and humidity can accumulate
- Provide holes for pretreatment and electrophoretic enameling
- Provide good accessibility for the application of corrosion inhibitor
- Allow for ventilation of hollow spaces/cavities
- Prevent the penetration of dirt and water to the greatest extent possible; provide water drain openings
- Minimize the area of the body exposed to stone-chip impact
- Prevent contact corrosion.

Precoated sheet steel (inorganic zinc, electrolytically galvanized, hot-dip galvanized) is often used for those components at particularly high risk, such as doors and load-bearing members at the front of the vehicle. Inaccessible structural areas are coated with spot-welding paste (PVC or epoxy adhesive, approx. 10 to 15 m seam length per vehicle) prior to assembly.

Painting
Measures subsequent to electrophoretic enameling:
- Covering the spot-welded seams (up to 90 to 110 m), welts and joints with PVC sealing compound
- Coating the underbody against stone-chip impact damage with PVC underbody protection (0.3 to 1.4 mm thick, 10 to 18 kg per vehicle); alternatively, using paneling sections made of plastic
- Filling cavities with penetrating, non-aging water-based wax
- Using corrosion-resistant, attached plastic components in high-risk areas, such as the front fenders (PVC coating not used in those places)
- Sealing of underbody and engine compartment after final assembly.

Body finishing components
Fenders
The front and rear of the vehicle should be protected so that low-speed collisions will only damage the vehicle slightly, or not at all. Prescribed fender evaluation tests (US Part 581, Canada CMVSS 215, and ECE-R42) specify minimum requirements in terms of energy absorption and installed fender height. Compliance with fender tests in the USA to US Part 581 (4 km/h barrier impact, 4 km/h pendulum tests) and in Canada (8 km/h) requires a fender system with energy absorbers that automatically regenerate. The requirements of the ECE standard are satisfied by plastic-deformable retaining elements located between the bendable bar (fender) and the vehicle body structure. In addition to sheet steel, many bendable bars are

Table 2: Paint/coating thicknesses

Paint application, total thickness	≈ 120 μm
Zinc-phosphate coat Electrophoretic enameling (cathodic) Filler Finishing paint	≈ 2 μm 13 to 18 μm ≈ 40 μm 35 to 45 μm
Clear coat finish (only for metallic and water-based paint)	40 to 45 μm

made using fiber-reinforced plastics and aluminum sections.

Exterior trim panels, protective molding rails
Plastics have become the preferred materials for external protective molding rails, trim panels, skirts, spoilers, and particularly for components whose purpose is to improve the aerodynamic characteristics of the vehicle. Criteria used in the selection of the proper material are flexibility, high-temperature shape retention, coefficient of linear expansion, notched-bar toughness, scratch resistance, chemical resistance, surface quality, and paintability.

Glazing
The windshield and rear window are retained in rubber strips, and sealed or bonded in place. The weight of glazing per vehicle is 25 to 35 kg. The substitution of glass with plastics (PC, PMMA) to reduce weight has not succeeded due to various disadvantages. Laminated safety glass is sometimes used in door windows to provide better sound and heat insulation. Sliding sunroofs are also increasingly made of glass (usually single-pane toughened safety glass).

Door locks
Door locks are of immense importance to passive accident safety. Various manufacturers have produced a variety of solutions to the issues of ease of operation, theft deterrence, and child safety. The legal requirements are as follows:

ECE (ECE-R11):
Every lock must have a latched position and a fully closed position.
- Longitudinal force: capable of withstanding 4,440 N in latched position, 11,110 N in fully closed position.
- Lateral force: capable of withstanding 4,440 N in latched position, 8,890 N in fully closed position.
- Inertial force: The lock should not be forced out of the fully closed position by linear or lateral acceleration of 30 g acting on the lock in both directions, and on the striker and the operating device, when the locking mechanism is not engaged.

USA (FMVSS 206):
Every lock must have a fully closed position. Doors which are attached by hinges must have a latched position.
- Longitudinal force: capable of withstanding 4,450 N in latched position, 11,000 N in fully closed position.
- Lateral force: capable of withstanding 4,450 N in latched position, 8,900 N in fully closed position.
- Inertial force: The lock should not be forced open from the fully closed position by linear or lateral acceleration of 30 g acting on the door lock system in both directions (lock and operating device).

Trunk locks
(Excerpt from FMVSS 401, in force since 9.1.2002)

These safety regulations for passenger cars with trunks contain the requirements for a release mechanism for the trunk. Such a mechanism should allow a person locked inside the trunk of a car to free himself/herself from the trunk. According to these regulations, a manually operated release mechanism must be provided with a function (e.g. lighting or phosphorescence) by which the release mechanism can easily be seen inside the closed trunk.

Seats
The strength requirements which must be met by seats in a collision pertain to the seat frame (seat cushions, backrest), the head restraints, the seat adjustment mechanism, and the seat anchors (regulations: FMVSS 207, 202; ECE-R 17, 25; RREG 74/408, 78/932, and others).

One component of active safety is seating comfort. Seats must be designed so that vehicle occupants with different body dimensions are able to sit for long periods of time without becoming tired. Parameters:
- Support of individual body areas (pressure distribution)
- Lateral support when cornering
- Seat climatic conditions
- Freedom of movement so that an occupant can change his/her sitting position without having to readjust the seat
- Vibrational and damping characteristics (matching the natural frequency within the excitation frequency band)
- Adjustability of seat cushion, backrest and head restraint.

The above parameters are affected by the following:
- Dimensions and shapes of the upholstery of seat cushions and backrests
- Distribution of the spring rates of individual padded zones
- Overall spring rate and damping capacity of the seat cushions in particular
- Thermal conductivity and moisture-absorption capacity of the covering material and upholstery
- Operation and range of the seat adjustment mechanisms.

Interior trim
A section of trim consists of a dimensionally stable core (sheet steel, sheet aluminum, or plastic) with fastening elements, and energy-absorbing padding made of foam material (e.g. PUR), and a flexible surface layer. One-piece plastic trim sections made of injection-molded thermoplastic material are also used.

The headliner is made either as a stretched liner or finished liner. The materials used must be flame-retardant and slow burning (FMVSS 302).

Safety

Active safety

The purpose of active safety is to prevent accidents. Driving safety is the result of a harmonious chassis and suspension design with regard to wheel suspension, springing, steering, and braking, and is reflected in optimum dynamic vehicle behavior.

Conditional safety results from keeping the physiological stress to which vehicle occupants are subjected by vibration, noise, and climatic conditions down to as low a level as possible. It is a significant factor in reducing the probability of misactions in traffic.

Vibrations within a frequency range of 1 to 25 Hz (stuttering, shaking, etc.) induced by wheels and drivetrain assemblies reach the occupants of the vehicle via the car body, seats, and steering wheel. The effect of these vibrations is more or less pronounced, depending on their direction, amplitude, and duration.

Noises produced as acoustical disturbances in and around the vehicle can come from internal sources (engine, transmission, propshafts, axles) or external sources (tires on roads, wind noises), and are transmitted through the air or by structures. The sound pressure level is measured in dB (A).

Noise reduction measures are concerned, on the one hand, with the development of quiet-running components and the insulation of noise sources (e.g. engine encapsulation), and on the other hand, with noise damping by means of insulation or anti-noise materials.

The major climatic influences are air temperature, air humidity, air flow rate, and air pressure.

Perceptibility safety
Measures which increase perceptibility safety are concentrated on:
- Lighting equipment
- Acoustic warning equipment
- Direct and indirect visibility (driver's view: the angle of binocular obscuration – i.e. for both of the driver's eyes – caused by the A-pillars must not be more than 6 degrees).

Operational safety
Low driver stress, and thus a high degree of driving safety, requires optimum design of the driver's surroundings with regard to ease of operation of the vehicle controls.

Passive safety
The purpose of passive safety is to mitigate the consequence of accidents.

Exterior safety
The term "exterior safety" covers all vehicle-related measures which are designed to minimize the severity of injury to pedestrians, and bicycle and motorcycle riders struck by the vehicle in an accident. The determining factors are the vehicle-body deformation behavior and the exterior vehicle-body shape.

The primary objective is to design the vehicle so that its exterior design minimizes the consequences of a primary collision (a collision involving persons outside the vehicle and the vehicle itself).

The most severe injuries sustained by pedestrians result from impact with the front of the vehicle or the road, in addition to which the precise accident sequence greatly depends on body size. The consequences of collisions involving two-wheeled vehicles and passenger cars can only be slightly improved by passenger-car design due to the two-wheeled vehicle's often considerable inherent energy component, its high seat position, and the wide dispersion of contact points. The design features that can be incorporated in a passenger car are, for example:
– Movable front lamps
– Recessed windshield wipers
– Recessed drip rails
– Recessed door handles
– Deformable vehicle front, including hood.

See also ECE-R26, RREG 74/483, 2003/102/EC, 2004/90/EC.

Interior safety
The term "interior safety" covers vehicle measures whose purpose is to minimize the acceleration and forces acting on vehicle occupants in the event of an accident, provide sufficient survival space, and ensure the operability of vehicle components critical to the extrication of occupants from the vehicle after the accident has occurred. The determining factors for passenger safety are:
– Deformation behavior of vehicle body
– Passenger-cell strength, size of the survival space during and after impact
– Restraint systems

Figure 5: Safety in traffic
Terms and influencing factors

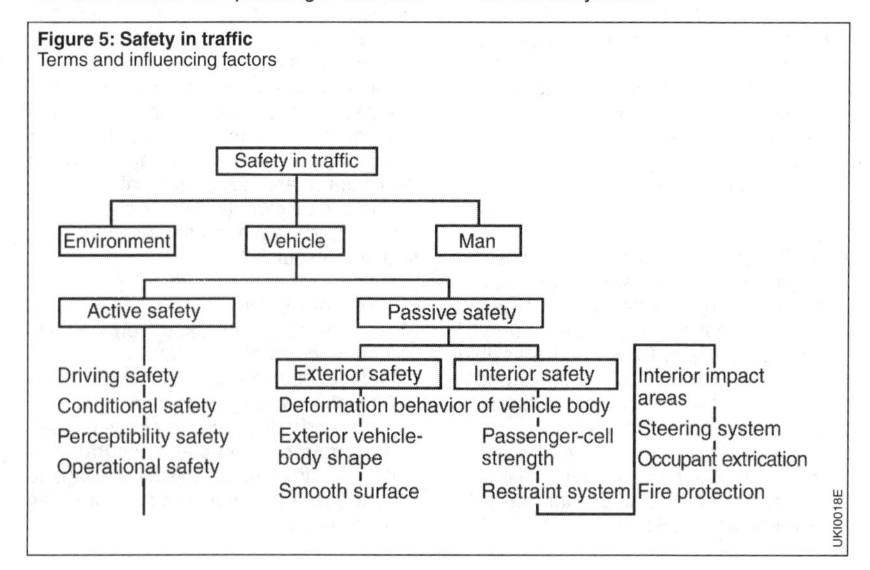

- Impact area in the interior (FMVSS 201)
- Steering system
- Occupant extrication
- Fire protection.

Laws governing inner safety (frontal and side impacts):
- Protection of vehicle occupants in the event of an accident, in particular restraint systems (FMVSS 208 amended version, FMVSS 214, ECE R94, ECE R95, injury criteria)
- Windshield mounting (FMVSS 212)
- Penetration of the windshield by vehicle body components (FMVSS 219)
- Storage compartment lids (FMVSS 201)
- Fuel leakage prevention (FMVSS 301).

Deformation behavior of vehicle body
Due to the frequency of frontal collisions, an important role is played by the legal regulations for frontal impact tests in which a vehicle is driven at a speed of 30 mph (48.3 km/h) into a rigid barrier which is either perpendicular or inclined at an angle of up to 30° relative to the car's longitudinal axis. Figure 6 shows the distribution of collision types for accidents resulting in injuries to vehicle occupants. Source: GIDAS, German In-Depth Accident Study (research project by BASt and FAT).

As virtually 50 % of frontal collisions mainly affect only one half of the front of the vehicle, offset frontal impact with a coverage of 30 to 50 % of the vehicle width is carried out worldwide.

Excerpt from ECE-R 94: "The barrier must be configured in such a way that the vehicle contacts it on the driver's side first. If the test can be conducted with a vehicle with either right- or left-hand steering, it must be conducted with the least favorable type of steering which is established by the technical agency responsible for the test".

In a frontal collision, kinetic energy is absorbed through deformation of the fender, the front of the vehicle, and in severe cases, the forward section of the passenger cell (engine compartment bulkhead). Axles, wheels (rims) and the engine limit the deformable length. Adequate deformation lengths and displaceable drivetrain assemblies are necessary, however, in order to minimize passenger-cell acceleration. Depending on vehicle design (body shape, type of drive, and engine position), vehicle mass and size, a frontal impact with a barrier at approx. 50 km/h results in permanent deformation in the

Figure 6: Risk to pedestrians in collisions with passenger cars
Frequency of involvement of injury-causing contact areas according to GIDAS (2006); 100 % equals 2,338 injuries

No. in figure	Vehicle area	Proportion
1	Front fender	15 %
2	Radiator grill, headlamps, and wings	3 %
3	Hood edge	3 %
4	Hood	11 %
5	Windshield with frame	18 %
6	Ground in front of vehicle (secondary impact)	37 %
–	Other	11 %
–	Unknown	2 %

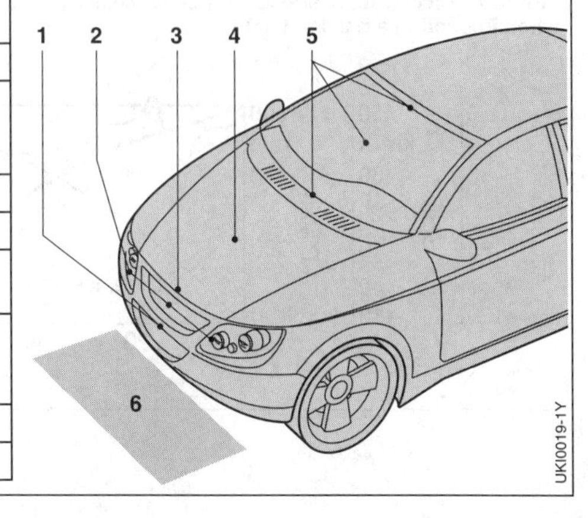

forward area of 0.4 to 0.7 m. Damage to the passenger cell should be minimized. This concerns primarily:
- The engine-bulkhead area (displacement of steering system, instrument panel, pedals, constriction of footwell)
- The underbody (lowering or tilting of seats)
- The side structure (ability to open the doors after an accident).

Acceleration measurements and evaluations of high-speed films allow a precise analysis of deformation behavior. Dummies of various sizes are used to simulate vehicle occupants and provide measured data for forces acting on the head, neck, chest, and legs.

The side impact, as the next most frequent type of accident, places a high risk of injury on vehicle occupants due to the limited energy-absorption capacity of trim and structural components, and the resulting high degree of vehicle interior deformation.

The risk of injury is largely influenced by the structural strength of the side of the vehicle (pillar/door joints, top/bottom pillar points), load-carrying capacity of floorpan cross-members and seats, and the design of inside door panels (FMVSS 214 and 301, ECE R95, Euro NCAP, and US SINCAP). Additional airbags in the doors or seats and the headliner have considerable potential for reducing the risk of injury.

In the rear impact test, deformation of the passenger cell must be minor at most. It should still be possible to open the doors, the edge of the trunk lid should not penetrate the rear window or enter the vehicle interior, and fuel-system integrity must be preserved (FMVSS 301).

Roof structures are investigated by means of rollover tests and quasi-static car-roof crush tests (FMVSS 216).

In addition, some manufacturers subject their vehicles to the inverted vehicle drop test in order to test the fatigue strength of the roof structure (survival space) under extreme conditions (the vehicle falls from a height of 0.5 m onto the left front corner of its roof).

Integral safety

The overlap between active and passive safety is increasing on account of the sensor technology used by both disciplines. That is why more and more systems are being developed which precondition the occupants better for a possible accident (e.g. Pre-Safe).

Figure 7: Acceleration, speed, and distance traveled of a passenger cell when impacting a barrier at 50 km/h

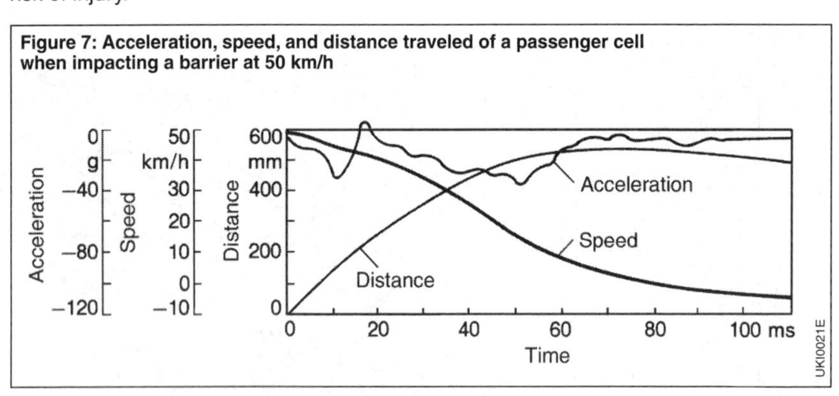

Vehicle bodies, commercial vehicles

Classification of commercial vehicles

Commercial vehicles are used for the safe and efficient transportation of persons and freight. In this respect, the degree of economic efficiency is determined by the ratio of usable space to overall vehicle volume, and of payload to gross vehicle weight. Dimensions and weights are limited by legal regulations.

A wide variety of vehicle types meet the demands of local and long-distance transportation, as well as the demands encountered on building sites and in special applications (examples in Figure 1).

Due to the wide diversity of vehicle types, the process of calculating the dimensions of the body structures (unitized body, cab, chassis, etc.) takes on a major importance right from the earliest stages of design. Building on experience with comparable vehicles, benchmark designs (volume-sales units, worst-case configurations) are defined by simulation and calculation with the aid of gradually refined complete-vehicle models using FEM (Finite Element Method) or MKS (multiple-body simulation). In this way, the rigidity, operational integrity, and vibration, acoustic and crash characteristics, etc. of relevant body-structure variants can be obtained to a substantial degree by computational means, even before testing starts. Structural calculations also take account of the requirements of (international) statutory safety standards.

Figure 1: Overview of commercial vehicles (Examples).
a) Light utility van, b) Truck,
c) Articulated road train,
d) High-capacity road train,
e) Semitrailer (Europe),
f) Semitrailer (NAFTA), g) Bus.

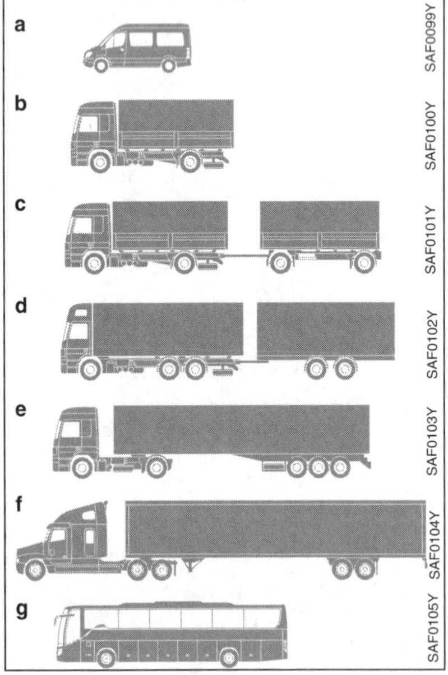

a SAF0099Y
b SAF0100Y
c SAF0101Y
d SAF0102Y
e SAF0103Y
f SAF0104Y
g SAF0105Y

Figure 2: Light utility van, load-bearing unit

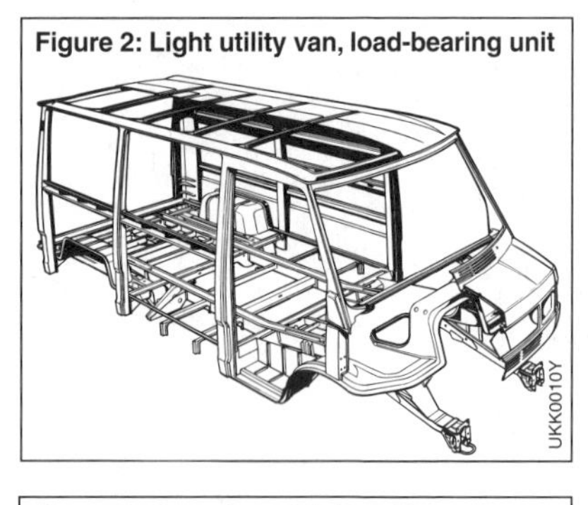

Figure 3: Overview of light utility vans (Examples).
a) Panel van, b) Flatbed van,
c) Twin cab, d) Chassis.

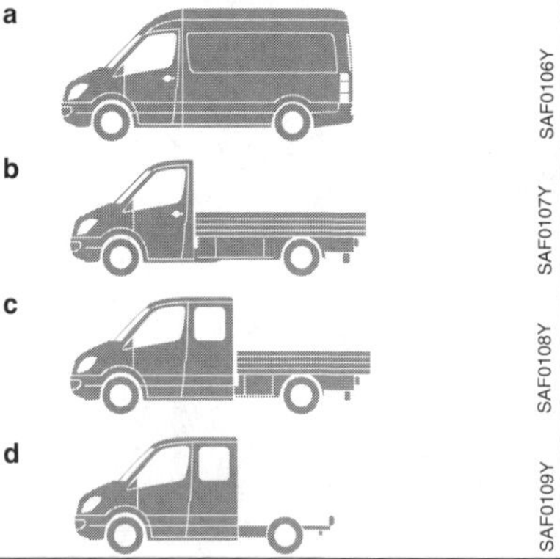

a SAF0106Y
b SAF0107Y
c SAF0108Y
d SAF0109Y

Light utility vans

Areas of application

These are light utility vans (2 to 7 t) used in the transportation of persons and in local freight distribution. Light utility vans with more powerful engines are also increasingly deployed in pan-European long-distance transport duties involving high mileages (express delivery services, overnight courier services). In both cases, stringent demands are made on the vehicle in terms of agility, performance, user-friendliness, and safety.

Body-structure variants of light utility vans

The design concepts are based on front-mounted engine, front or rear-wheel drive, independent suspension, or rigid axle and, with gross vehicle weights of 3.5 t and higher, twin tires on the rear axle.

The product range includes enclosed-body multipurpose panel vans and platform-body vehicles, as well as low-bed and high-bed platform trucks with special superstructures and crew cabs (examples in Figure 3).

In light utility vans (up to approx. 6 t gross vehicle weight), the bodies form an integral load-structure unit together with the chassis (Figure 2). The body and chassis frameworks consist of sheet-metal pressed elements and flanged profiles similar to passenger cars.

Light utility vans with platform bodies have a ladder-type frame with open or closed side members and cross-members as the primary load-bearing structure (similar to trucks, next section, Figure 6).

Large light utility vans usually have a separate chassis similar to trucks (see next section). They have a separate body, while, for reasons of comfort and noise reduction, the cab is generally attached by flexible mountings and is thus partially isolated from chassis vibration.

Medium- and heavy-duty trucks and tractor units

Body

The vehicles in this sector have either a load-bearing chassis or partially load-bearing body. The maximum dimensions and maximum permitted vehicle weights are subject to statutory provisions, which sometimes differ significantly from country to country. The maximum dimensions with regard to overall and semitrailer lengths are laid down, for example, in Europe by the Regulation 96/53/EU. In the NAFTA zone, however, only the semitrailer length, but not the overall length, is subject to restriction.

In most cases, the engine is at the front. It is seldom fitted as an underfloor engine between the axles. The vehicle is driven via one or more twin-tire axles. In individual cases, the rear axle is fitted with

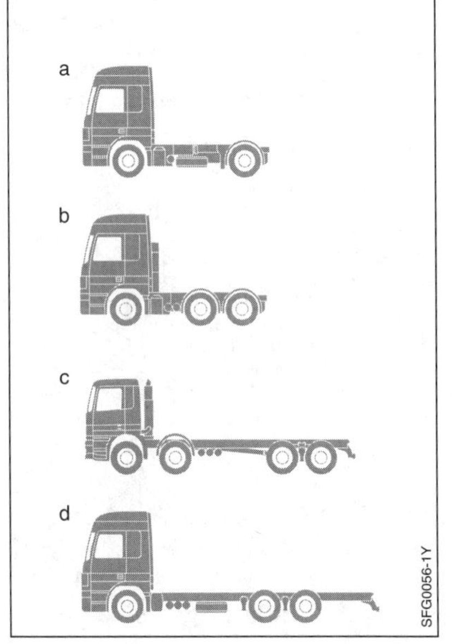

Figure 4: Types of truck undercarriage
(Examples).
a) 4×2 (four wheels, two of which driven),
b) 6×2/4 (six wheels, two of which driven, four steered),
c) 8×6/4 (eight wheels, six of which driven, four steered),
d) 6×2 (six wheels, two of which driven).

single tires. For building-site (offroad) applications with high traction requirements, all-wheel drive with inter-axle and cross-axle differential lock technology is used.

Undercarriage types

The type of truck undercarriage (Figure 4) is given its designation in accordance with the convention N×Z/L, where N denotes the number of wheels, Z the number of driven wheels, and L the number of steered wheels (twin wheels count as one wheel). If L is not specified (e.g. 4×2), the vehicle has two steered wheels.

Chassis

Normal chassis have leaf- or air-sprung rigid front and rear axles. Pneumatic suspension reduces body acceleration forces (better driving comfort, less strain on the cargo, better ride) and also allows easy swapping of interchangeable bodies (e.g. containers) and decoupling of semitrailers.

Three-axle vehicles (6×2) are fitted with either a leading or a trailing axle (nonpowered axles in front of or behind the driven axle) to increase the payload capacity. High-traction 6×4 vehicles for construction-site use often have two axles combined to form a steel-sprung tandem-axle assembly. The axle load is compensated mechanically on these two axles by a pivoted (center) bearing point between the axles. In the case of non-combined pneumatically spring single axles, the axle load is normally compensated pneumatically by altering the spring stiffness of the pneumatic suspension of the single axles.

Figure 5: Truck assemblies
1 Cab, 2 Engine, 3 Transmission, 4 Axle, 5 Chassis frame, 6 Body.

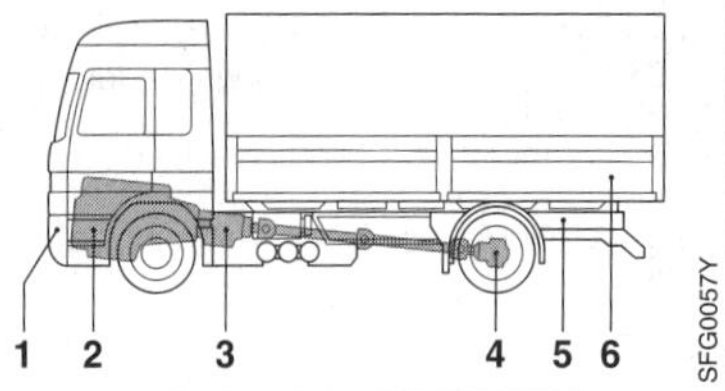

Chassis frames

The chassis frame is the commercial vehicle's actual load-bearing element, on which the body is bolted and the driver's cab located (Figure 5). It is designed as a ladder-type frame, consisting of side members and cross-members (Figure 6). The dimensions of the members are chosen to suit the required severity of application and load capacity (light- and heavy-duty commercial vehicles), but also in line with cost and weight consid-

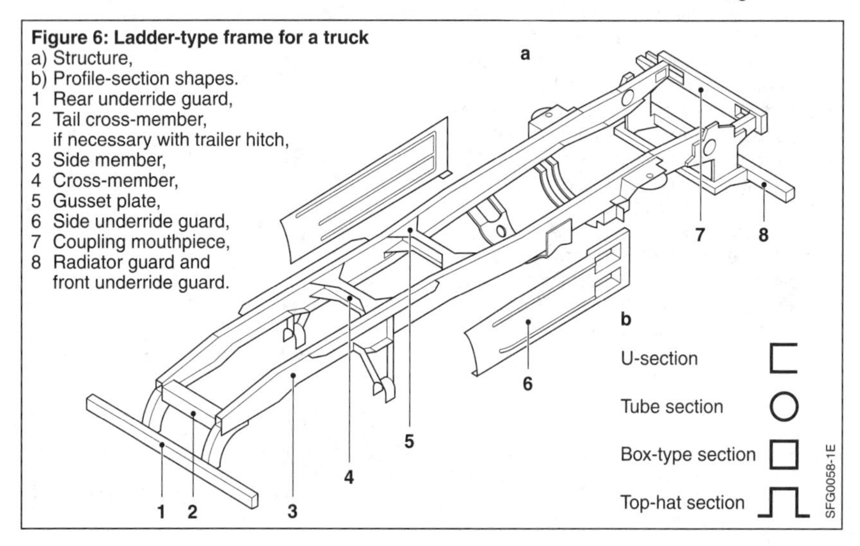

Figure 6: Ladder-type frame for a truck
a) Structure,
b) Profile-section shapes.
1 Rear underride guard,
2 Tail cross-member, if necessary with trailer hitch,
3 Side member,
4 Cross-member,
5 Gusset plate,
6 Side underride guard,
7 Coupling mouthpiece,
8 Radiator guard and front underride guard.

erations. The choice of profile sections (number and thickness) determines the level of torsional stiffness. Torsionally flexible frames are preferred in medium- and heavy-duty trucks because they allow the suspension to cope better with uneven terrain. Torsionally stiff frames are more suitable for light-duty delivery vehicles.

Apart from the force introduction points, critical points in the chassis-frame design are the side- and cross-member junctions (Figure 7). Special gusset plates or pressed cross-member sections form a broad connection basis. The junctions are riveted, bolted, and welded. U- or L-shaped side-member inserts provide increased frame flexural strength and reinforcement at specific points.

The chassis frame also serves to accommodate a wide range of add-on components such as tanks, battery device holders, air bottles, exhaust system, or spare wheel. The configuration varies here, depending on the application profile required. Special mounted installations such as, for example, loading cranes or platform lifts are likewise mounted on the chassis frame for corresponding applications.

Driver's cab

There are a variety of cab designs available depending on the vehicle concept. In delivery vehicles and for public-service use, low, convenient entrances are an advantage, whereas in long-distance transport applications, space and comfort are more important. Modular design concepts allow for short, medium and long cab versions while retaining the same front, rear and doors.

The cab is connected to the chassis frame by the cab mounting. A distinction is made here between comfort and standard mountings with different spring-and-damper combinations or transverse leaf-spring links with natural frequencies of 1 to 6 Hz.

From the design-concept viewpoint, a distinction must be made between cab-over-engine (COE) and cab-behind-engine (CBE) vehicles (Figure 8). In the case of cab-over-engine (COE) vehicles, the bulkhead and steering system are positioned right at the front of the vehicle. The engine sits under the high-level cab (special long-distance cab with flat floor) or under an engine tunnel between the driver and co-driver. The entrance is positioned in front of or above the front axle. A mechanical (by pretensioned torsion bars) or hydraulic cab-tipping mechanism provides access to the engine.

In cab-behind-engine (CBE) vehicles, the engine/transmission assembly is mounted ahead of the cab firewall under a steel or plastic hood (which is usually tiltable for reasons of accessibility). The driver enters the cab behind the front axle.

The standards required of the driver's cab with regard to aerodynamics, choice of materials, corrosion or equipment must be viewed as the same as the standards required of an automobile body.

Figure 7: Junctions
a) Cap cross-member, b) U-cross-member.
1 Side member, 2 Cross-member,
3 Gusset plate.

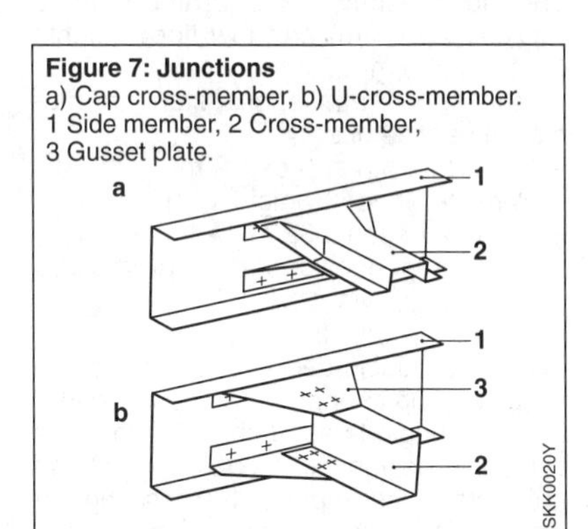

Figure 8: Driver's cab
a) Cab over engine,
b) Cab behind engine.

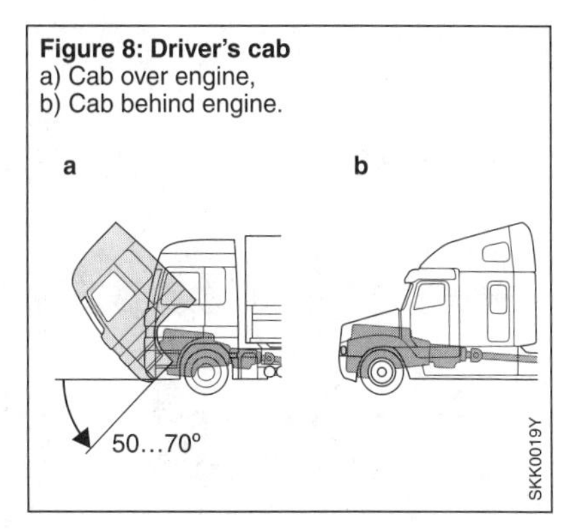

Body structures

Specific body structures such as flatbeds, panel vans, box vans, dump-truck deep-beds, tankers, and concrete mixers permit the economical and efficient transportation of a wide variety of freight and materials. The body and load-bearing chassis frame are joined sometimes by means of auxiliary frames with non-positive or positive attachments. Special design features (e.g. sprung mountings in the forward body area) are required to connect the chassis frame (which usually has low torsional strength) to a rigid body (e.g. box-type).

Articulated road trains and semitrailer rigs are used in long-distance transport (Figure 1). As the size of the transportation unit increases, the costs relative to the freight volume decrease. Load volume is increased by reducing the empty spaces between the cab, body, and trailer (high-capacity road train, Figure 1d). Advantages of semitrailer rigs lie in the greater uninterrupted loading length of the cargo area and the shorter inoperative times of the tractor units. Measures to improve aerodynamics, such as front and side panels on the vehicle and specially adapted air deflectors from the cab to the body, are applied to minimize fuel consumption.

Figure 9: Overview of buses
(Examples).
a) Microbus,
b) Mini- and midibus,
c) City bus,
d) Tour bus.

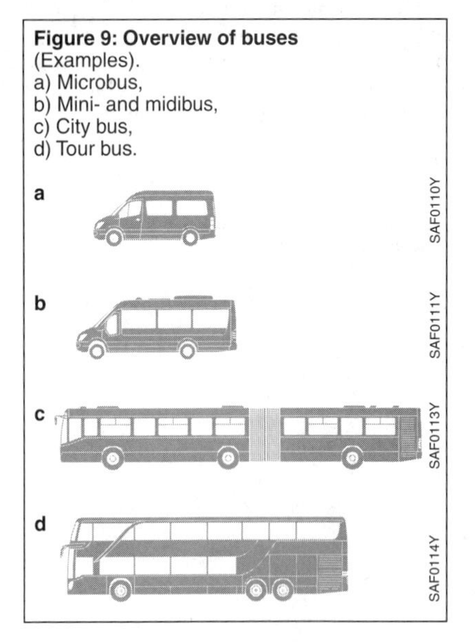

Buses

The bus market offers a specific vehicle for practically every application. This has resulted in a wide range of bus types, which differ in their overall dimensions (length, height, width) and appointments (depending on the application) (Figure 9).

Bus types

Microbuses

Minibuses carry up to approx. 20 passengers. These vehicles have been developed from light utility vans weighing up to approx. 4.5 t.

Mini- and midibuses

Buses which can carry up to approximately 25 persons are called mini- or midibuses. The transition from minibus to midibus is fluid. These vehicles have for the most part been developed from light utility vans weighing up to approx. 7.5 t. They are occasionally built on ladder-type chassis of light-duty trucks, or a unitized integral structure is used. A modified suspension design and special measures carried out on the body (e.g. flexible mountings) result in optimum ride comfort and low noise levels.

City buses

City buses are equipped with seating and standing rooms for scheduled routes. The short intervals between stops in suburban passenger transportation operations require rapid passenger turnover. This is achieved by wide doors that open and close swiftly, low boarding heights (approx. 320 mm) and low floor heights (approx. 370 mm).

Main specifications for a standard municipal-service bus

- Vehicle length approx. 12 m
- Permissible total weight 18.0 t
- Number of seats 32 to 44
- Total passenger capacity approx. 105 persons.

The use of double-decker buses (length 12 m carrying up to approx. 130 passengers), three-axle rigid buses (length up to 15 m carrying up to approx. 135 passengers) and articulated buses (approx. 160 passengers) provides increased transport capacity.

Intercity/overland buses
Depending on the application (standing passengers are not permitted at speeds over 60 km/h), low-floor designs featuring low boarding and floor heights (as in city buses) or higher floors and small luggage compartments (already very much like tour buses) are used. Intercity buses come in lengths of 11 to 15 m as rigid vehicles or in lengths of 18 m as articulated vehicles.

Tour buses (coaches)
Tour buses are designed to provide comfortable travel over medium and long distances. They range from the low, two-axle standard bus through to the double-decker luxury bus measuring approx. 10 to 15 m in length.

Body
Lightweight design based on a unitized body (integral structure): The body and base frames, which are firmly welded together, consist of pressed grid-type support elements and rectangular tubes (Figure 10).

In the case of the chassis structure, the body is positioned on a supporting ladder-type frame (similar to trucks). Except for mini- and midibuses, this design is not commonly encountered in Europe.

Chassis systems
The vertically or horizontally mounted engine drives the rear axle. Pneumatic suspension on all axles permits ride-level stabilization and a high degree of ride comfort. Intercity buses and tour buses are mainly equipped with independent suspension on their front axles. Disk brakes, frequently supported by retarders, are used on all axles.

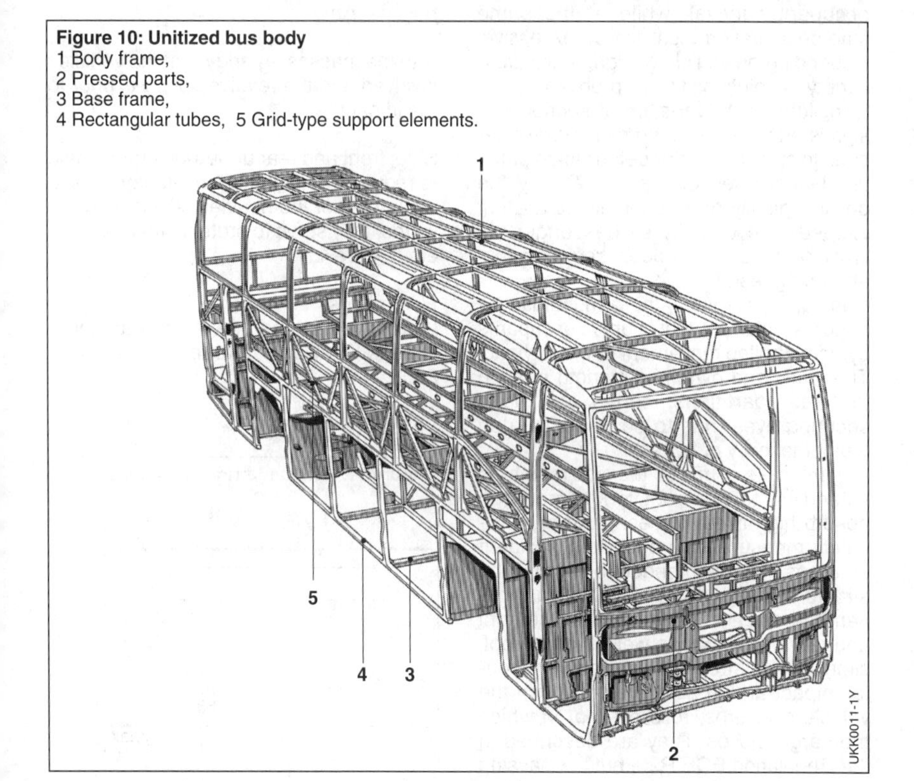

Figure 10: Unitized bus body
1 Body frame,
2 Pressed parts,
3 Base frame,
4 Rectangular tubes, 5 Grid-type support elements.

Passive safety in commercial vehicles

Passive safety is intended to limit the consequences of accidents and to protect other road users. Systematic recording of accidents, accident tests with complete commercial vehicles, and intensive computer optimization help to devise safety measures.

Requirements

Generally, the effectiveness and strength of occupant restraint systems has to be demonstrated. Therefore, the dimensioning of commercial-vehicle body structures must take account of aspects, such as the strength and rigidity of seat-belt anchor points on the seats and of the related body structures (seat rails, floor, frame, etc.).

In the event of a collision, the driver's cab and the passenger cabin must maintain the amount of room necessary for occupant survival, while at the same time deceleration must not be excessive. Depending on vehicle design, there are a variety of solutions to this problem.

In light utility vans, front-section design is energy-absorbing as in passenger cars. In spite of shorter deformation paths and higher levels of released energy, the physiologically permissible limits are not exceeded in virtually all passenger-car crash-test standards (legal requirements and rating tests).

In light utility vans, there must also be features which prevent injury to occupants by uncontrolled movement of the payload. The static and dynamic strength of such features (partitions, cages and nets, securing eyes) has to be demonstrated mathematically or by testing.

In the case of trucks, the side members extend up to the front fender and can absorb high linear forces. Such passive-safety measures are based on accident analyses and are intended to improve the structural design of the cab. Static and dynamic stress and impact tests at the front and rear of the cab, as well as on its roof, simulate the stresses involved in a frontal impact and in accidents in which the vehicle overturns, rolls over, or in which the cargo moves. They are described in the Regulation ECE R29, but the passing of these tests as a condition of type approval is only required in a few European countries.

Statistical analyses have proved that the bus is the safest means of passenger transportation. Static roof-load tests and dynamic overturning tests provide evidence of body strength. The use of flame-retardant and self-extinguishing materials for the interior equipment of the vehicle minimizes the risk of fire.

As road traffic involves many different kinds of vehicles, collisions between light and heavy vehicles are unavoidable. As a result of the differences in vehicle weight, and incompatibility in terms of vehicle geometry and structural stiffness, the risk of injury in the lighter vehicle is greater.

The change is speed Δv during a central plastic impact for front-end or rear-end collisions between two vehicles (vehicle 1 and vehicle 2) is defined as

$$\mu = \frac{m_2}{m_1}:\ \Delta v_1 = \frac{\mu v_r}{1+\mu},\ \Delta v_2 = \frac{v_r}{1+\mu},$$

with the masses m_1 and m_2 of the vehicles involved and the relative speed v_r prior to the impact.

Side, front and rear underride guards help to reduce the danger of the lighter vehicle driving under the heavier vehicle. In other words, they serve to protect other road users (Figure 11).

References

[1] E. Hoepke, and others: Nutzfahrzeugtechnik. Vieweg+Teubner, 5th Edition, 2008.

Figure 11: Rear underride guard for trucks

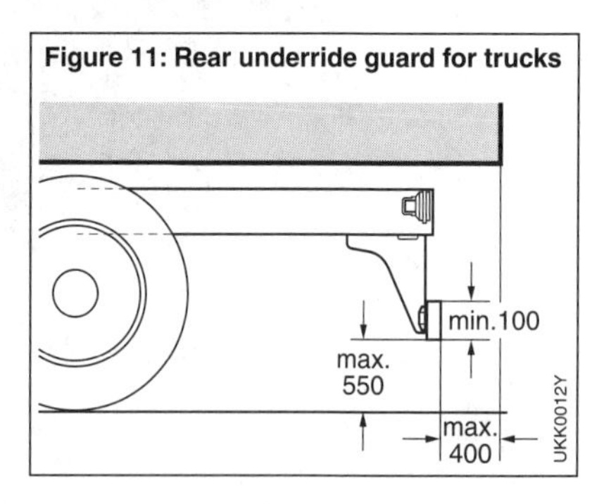

Lighting equipment

Functions

Ever since the invention of the automobile, it has been accompanied by vehicle lighting. Initially candles, then petroleum and carbide lamps were used to provide lighting; today, these lamps would be classed as position or marker lamps. Only with the introduction of automotive electrics in the form of the Bosch alternator in the Adler vehicle (1913) were manufacturers able to introduce systems which generated good ranges and lived up to being called "headlamps". Further significant milestones have been
- the introduction of the asymmetrical low-beam pattern, characterized (RHD traffic) by an extended visual range along the right side of the road (1957),
- the introduction of new headlamp systems featuring complex geometrical configurations (PES, Poly-Ellipsoid System, free-form surfaces, facetted reflectors) offering efficiency-level improvements of up to 50% (1985),
- the "Litronic" headlamp system with gas-discharge lamps (xenon lamps with luminous arc) which supplied more than twice the light generated by comparable halogen units (1990),
- adaptive front-lighting systems (AFS) with moving, dynamic PES modules (Poly-Ellipsoid System) or static, activated reflectors for turning (2003).

Lighting at the vehicle front end

The primary function of the headlamps at the vehicle front end is to illuminate the roadway so that the driver can detect traffic conditions and recognize any obstacles and hazards in good time. They also serve to identify and mark out the vehicle to oncoming traffic. The turn-signal lamps serve to show the driver's intention to change direction or to indicate a hazardous situation.

The headlamps and lights at the front end include the following:
- Low-beam headlamps
- High-beam headlamps
- Fog lamps
- Auxiliary driving lamps
- Turn-signal (direction-indicator) lamps
- Parking lamps
- Position/clearance lamps (for wide vehicles)
- Daytime running lamps (if required by law in individual countries).

Lighting at the vehicle rear end

Lights are turned on at the vehicle's rear end in accordance with the lighting and weather conditions and indicate the vehicle's position. They also indicate how the vehicle is moving and in which direction, e.g. whether the brakes are applied or the driver is intending to change direction, or whether a hazardous situation exists. The reversing lamps illuminate the roadway while the vehicle is reversing.

The lamps and lights at the rear end include the following:
- Stop lamps
- Tail lamps
- Rear fog warning lamps
- Turn-signal (direction-indicator) lamps
- Parking lamps
- Clearance lamps (for wide vehicles)
- Reversing (backup) lamps
- License-plate lamps.

Lighting in the vehicle interior

In the vehicle interior, priority over all other functions is given to the ease and reliability with which the switch elements can be reached and operated, and to provide the driver with sufficient information on the vehicle's operating states (while distracting him/her from driving as little as possible). These priorities dictate effectively illuminated instrument panels and discrete lighting for various control clusters (such as the sound and navigation systems), where they satisfy a prime requirement for relaxed and safe vehicle operation. Optical and acoustic signals must be prioritized according to their urgency and then relayed to the driver.

Regulations and equipment

Approval codes and symbols, Europe/ECE

Automotive lighting equipment is governed by national and international design and operating regulations, according to which the equipment in question must be manufactured and tested. For every type of lighting equipment, there is a special approval code and symbol which has to be legibly displayed on the equipment concerned. The preferred locations for approval codes and symbols are places that are directly visible when the hood is open, such as the lenses of headlamps and other lights, and the headlamp unit components. This also applies to approved replacement headlamps and lights.

If an item of equipment carries such an approval code/symbol, it has been tested by a technical inspectorate (e.g. in Germany the Lighting Technology Institute of Karlsruhe University) and approved by a licensing authority (in Germany the Federal Road Transport Office). All volume-production units which carry the approval code/symbol must conform in all respects to the type-approved unit. Examples of approval symbols:

(E1) ECE approval mark,

[e1] EU approval mark.

The number 1 following each letter indicates the type-approval test carried out and the award of approval according to ECE (Economic Commission for Europe) Regulations for Germany with Europe-wide recognition. In Europe, installation of all automotive lighting and visual signaling equipment is governed not only by national guidelines but also by the higher European directives (ECE: whole of Europe, EU, New Zealand, Australia, South Africa and Japan). In the course of the ongoing union of Europe, the implementation regulations are being increasingly simplified by the harmonization of directives and legislation.

Right-hand-drive or left-hand-drive traffic
The ECE Regulations apply by analogy to driving on the right or left. The technical requirements for lighting are mirrored around the central perpendicular of the test screen (see Figure 3). According to the Vienna Global Treaty of 1968, when traveling in countries which drive on the other side of the road, all road users are obliged to adopt measures to prevent increased glare of oncoming traffic at night due to the asymmetrical light pattern. This can be achieved either by self-adhesive strips obtainable from the vehicle manufacturer or two-way switches in the headlamps (in the case of PES).

Regulations for the USA

In the USA, lighting equipment is governed by regulations that are very different from those in Europe. The principle of self-certification compels each manufacturer as an importer of lighting equipment to ensure, and in an emergency to furnish, proof that its products conform 100 % with the regulations of FMVSS 108 [27] (Federal Motor Vehicle Safety Standard) laid down in the Federal Register. There is therefore no type approval in the USA. The regulations of FMVSS 108 are partly based on the SAE industry standard (Society of Automotive Engineers).

Upgrading and conversion

Vehicles imported to Europe from other regions must be modified to comply with European directives. This applies in particular to the lighting equipment. Identical components available for the European market can be used as direct replacements. Other solutions, such as retail products or, in certain cases, retention of the original equipment, require an engineer's report. In Germany, Article 22a of the StVZO (Road Traffic Licensing Regulations) [1] requires "Approximation Certificates" for lighting equipment. Such certificates are issued by the Lighting Technology Institute of Karlsruhe University.

Subsequent alterations to type-approved headlamps and sockets invalidate the type approval and consequently the general operating license of the vehicle.

Light sources

With regard to automotive light sources, a distinction is made between thermoluminescent radiators (thermal radiators) and electroluminescent radiators. The outer electron shells of atoms of certain materials can absorb varying levels of energy by excitation (energy input). The transition from higher to lower levels may lead to the emission of electromagnetic radiation.

Thermal radiators

In the case of this type of light source, the energy level of the crystal system is increased by adding heat energy. Emission is continuous across a broad wavelength range. The total radiated power is proportional to the power of 4 of the absolute temperature (Stefan-Boltzmann law). The distribution-curve maximum is displaced to shorter wavelengths as temperature increases (Wien's displacement law) [39].

Filament lamps

Filament lamps, with tungsten filament (fusion temperature 3,660 K), are also thermal radiators. The evaporation of the tungsten and the resulting blackening of the bulb restrict the service life of this type of lamp.

Halogen lamps

A halogen filling (iodine or bromine) in the lamp allows the filament temperature to rise to close to the melting point of the tungsten. Close to the hot bulb wall, the evaporated tungsten combines with the filler gas to form tungsten halide. This is gaseous, light-transmitting and stable within a temperature range of 500 K to 1,700 K. It reaches the filament by means of convection, decomposes as a result of the high filament temperature, and forms an even tungsten deposit on the filament. In order to maintain this cycle, an external bulb temperature of approx. 300 °C is necessary. To achieve this, the bulb, made of fused silica (quartz), must surround the filament closely. A further advantage of this measure is that a higher filling pressure can be used, thereby providing additional resistance to tungsten evaporation.

Gas-discharge lamps

Gas-discharge lamps are electroluminescent radiators and distinguished by their higher luminous efficiency. A gas discharge is maintained in an enclosed, gas-filled bulb by applying a voltage between two electrodes. The atoms of the emitted gas are excited by collisions between electrons and gas atoms. The atoms excited in the process give off their energy in the form of luminous radiation.

Examples of gas-discharge lamps are sodium-vapor lamps (street lighting), fluorescent lamps (interior illumination) and D lamps for motor-vehicle applications (Litronic).

Light-emitting diodes

A light-emitting diode or LED is an electroluminescent lamp. Thanks to their robustness, high energy efficiency, fast reaction time, and compact construction, LEDs are used as illuminators or displays in a wide range of applications. They are used in motor vehicles in the passenger compartment to provide illumination, as displays or for display backlighting. In the exterior areas LEDs are fitted in particular in auxiliary stop lamps and tail lamps and, with further increases in their luminous efficiency, can increasingly be used in lamps and main functions at the front end of the vehicle.

Motor-vehicle bulbs

Replaceable filament bulbs for motor-vehicle lighting must be type-approved in accordance with ECE R37 [10], replaceable gas-discharge light sources in accordance with ECE R99 [19]. Other light sources which do not comply with these regulations (LEDs, neon tubes, special bulbs) are permitted, but can only be installed as a fixed component part of a lamp or as a "light-source module".

Bulbs complying with ECE R37 are generally available in 12 V versions, some bulbs also in 6 V and 24 V versions (Table 1). To help avoid mix-ups, different bulb types are identified by different base shapes. Bulbs of differing operating voltages are labeled with this voltage in order to avoid mix-ups in the case of identical bases. The bulb type suitable in each case must be indicated on the equipment.

A voltage increase in halogen lamps of 10 % results in a 70 % reduction in service life and a 30 % increase in luminous flux (Figure 1, [40]).

The luminous efficiency (lumens per watt) represents the bulb's photometrical efficiency relative to its power input. The luminous efficiency of vacuum bulbs is 10 to 18 lm/W. The higher luminous efficiency of halogen lamps (22 to 26 lm/W) is primarily a consequence of increasing the filament temperature. Gas-discharge bulbs provide a luminous efficiency in the order of 85 lm/W for substantial improvements in low-beam performance.

LEDs today achieve a luminous efficiency in the range of 50 lm/W (LEDs with high power consumption) or 100 lm/W (LEDs with low power consumption). An increase in luminous efficiency of up to 25 % is expected in the next few years.

Figure 1: Influence of operating voltage on some data of halogen lamps
(Source: [40])
L Service life (the dispersion width during operation with undervoltage is necessitated by the presence of halogen),
U Operating voltage,
I Lamp current,
P Lamp power output,
Φ Luminous flux.

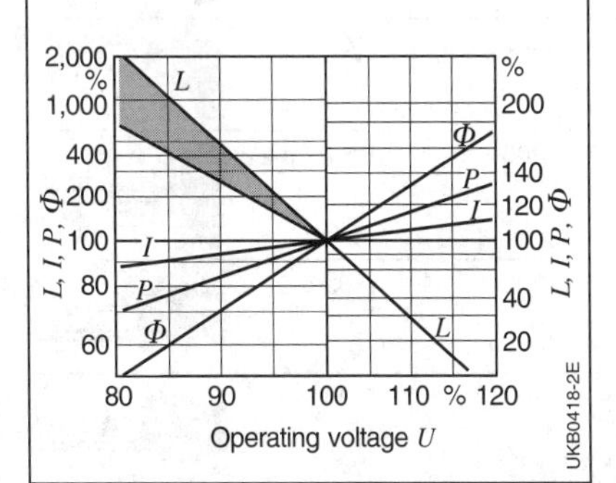

Main-light functions

Low beam (dipped beam)

The main light for driving at night is provided by low-beam headlamps. The creation of the characteristic light-dark boundary was one of the technological milestones in lighting technology.

The "dark above/bright below" distribution pattern resulting from the light-dark boundary furnishes acceptable visual ranges under all driving conditions. This configuration reduces glare, to which approaching traffic is exposed, within reasonable limits, and at the same time it supplies relatively high luminous intensity in the area below the light-dark boundary.

The light distribution pattern must combine maximum visual ranges with minimum glare effect. These demands are supplemented by other requirements affecting the area directly in front of the vehicle. For instance, the headlamps must provide assistance when cornering, i.e. the light distribution pattern must extend beyond the left and right-side extremities of the road surface.

High beam

High-beam headlamps illuminate the road to the maximum range. This creates a high luminous intensity, depending on the distance, acting on all objects in the space reserved for traffic. High-beam headlamps are therefore only permitted when they do not dazzle oncoming traffic (glare).

The high traffic density on modern roads severely restricts the use of high-beam headlamps.

Installation and regulations

Designs

Global regulations mandate two headlamps for low beam and at least two (or the option of four) high-beam units for all dual-track vehicles. The light color is white.

Dual-headlamp system

The dual-headlamp system (Figure 2a) uses lamps with two light sources (halogen double-filament bulbs (H4), US sealed beam) for high and low beams using shared reflectors. In headlamps with gas-

Table 1: Specifications for the main motor-vehicle bulbs (not including motorcycle bulbs)

Application	Category	Voltage Nominal values V	Power Nominal values W	Luminous flux Reference values Lumen	Base type IEC	Illustration
Fog lamps, high beam, low beam in 4 HL	H1	6 12 24	55 55 70	1,350[2]) 1,550 1,900	P 14.5 e	
Fog lights, high beam	H3	6 12 24	55 55 70	1,050[2]) 1,450 1,750	PK 22s	
High beam, low beam	H4	12 24	60/55 75/70	1,650/ 1,000[1]), [2]) 1,900/1,200	P 43 t - 38	
High beam, low beam (dipped beam) in 4 HL, fog lights	H7	12 24	55 70	1,500[2]) 1,750	PX 26 d	
Fog lights, static cornering headlamps	H8	12	35	800	PGJ 19-1	
High beam	H9	12	65	2,100	PGJ 19-5	
Low beam, fog lights	H11	12 24	55 70	1,350 1,600	PGJ 19-2	
Fog lights	H10	12	42	850	PY 20 d	
High beam, daytime running lamps	H15	12 24	55/15 60/20	260/1,350 300/1,500	PGJ 23t-1	
Low beam (dipped beam) in 4-HL systems	HB4	12	51	1,095	P 22 d	
High beam in 4-HL systems	HB3	12	60	1,860	P 20 d	
Low beam, high beam	D1S	85 12[5])	35 approx. 40[5])	3,200	PK 32 d-2	
Low beam, high beam	D2S	85 12[5])	35 approx. 40[5])	3,200	P 32 d-2	
Low beam, high beam	D2R	85 12[5])	35 approx. 40[5])	2,800	P 32 d-3	

Table 1 (contd.): Specifications for the main motor-vehicle bulbs (not including motorcycle bulbs)

Application	Category	Voltage Nominal values V	Power Nominal values W	Luminous flux Reference values Lumen	Base type IEC	Illustration
Stop lamps, turn-signal lamps, rear fog light, backup lamps	P 21 W PY 21 W[6])	6, 12, 24	21	460[3])	BA 15 s	
Brake lamp/ tail lamp	P 21/5 W	6 12 24	21/5[4]) 21/5 21/5	440/35[3]), [4]) 440/35[3]), [4]) 440/40[3])	BAY 15d	
Position lamp, tail lamp	R 5 W	6 12 24	5	50[3])	BA 15 s	
Tail lamp	R 10 W	6 12 24	10	125[3])	BA 15 s	
Daytime running lamps	P 13 W	12	13	250[3])	PG 18.5 d	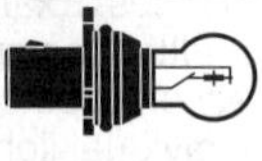
Brake lamp, turn signal	P 19 W PY 19 W	12 12	19 19	350[3]) 215[3])	PGU 20/1 PGU 20/2	
Rear fog light, backup lamp, front turn signal	P 24 W PY 24 W	12 12	24 24	500[3]) 300[3])	PGU 20/3 PGU 20/4	
Stop, turn-signal, rear fog, backup lamps	P 27 W	12	27	475[3])	W 2.5 x 16 d	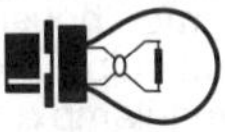
Brake lamp/ tail lamp	P 27/7 W	12	27/7	475/36[3])	W 2.5 x 16 q	
License-plate lighting, tail lamp	C 5 W	6 12 24	5	45[3])	SV 8.5	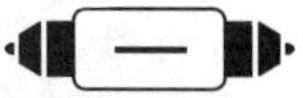
Position lamp	H 6 W	12	6	125	BAX 9 s	
Position lamp, license-plate lighting	W 5 W	6 12 24	5	50[3])	W 2.1 x 9.5 d	
Position lamp, license-plate lighting	W 3 W	6 12 24	3	22[3])	W 2.1 x 9.5 d	

[1]) High/low beam. [2]) Specifications at test voltage 6.3; 13.2 or 28.0 V.
[3]) Specifications at test voltage of 6.75; 13.5 or 28.0 V. [4]) Main/secondary filament.
[5]) With ballast unit. [6]) Yellow-light version.

discharge bulbs the dual function is achieved by focusing or defocusing the xenon burner in a shared reflector. In bi-xenon projection systems a screen is moved out of or into the beam path.

Quad-headlamp system
Two of the headlamps in a quad-headlamp system provide both high and low beam, while the second pair only provides high-beam illumination (Figure 2b). The light functions of projection and reflection systems can be used in any combination. The low-beam headlamps can additionally be combined with the fog lamps (Figure 2c).

Important definitions of device design
Grouped design
A single housing, but with different lenses and bulbs. Example:
- Multiple-compartment rear-lamp assemblies containing different individual light units.

Combined design
A single housing and bulb assembly with more than one lens. Example:
- Combined tail lamp and license-plate lamp.

Nested design
Common housing and lens, but with individual bulbs. Example:
- Headlamp assembly with nested position lamp.

Main-light functions for Europe

Regulations and directives for Europe
The most important regulations and directives are laid down in ECE R112 [20], ECE R113 [21], ECE R48 [12], 76/756/EEC [24], ECE R98 [18], and ECE R123 [23].
- ECE R112: Headlamps for asymmetrical low beam and/or high beam that are fitted with filament bulbs or LED modules (cars, buses, trucks).
- ECE R113: Headlamps for symmetrical low beam and/or high beam that are fitted with filament bulbs or LED modules (motorbicycles, motorcycles).
- ECE R48 and 76/756/EEC: for attachment and application.
- ECE R98: Headlamps with gas-discharge lamps as per ECE R99.
- ECE R123: Adaptive front-lighting systems (AFS) for motor vehicles.

The installation regulations described in the following refer to passenger cars.

Figure 2: Headlamp systems
a) Dual-headlamp system,
b) Quad-headlamp system,
c) Quad-headlamp system with additional fog lamps.

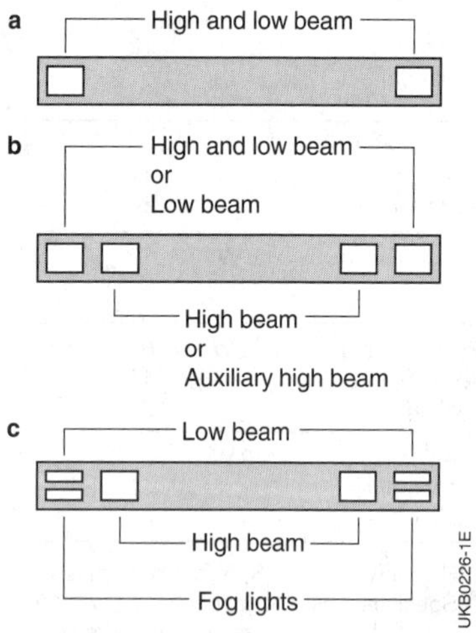

a High and low beam
b High and low beam or Low beam
High beam or Auxiliary high beam
c Low beam
High beam
Fog lights
UKB0226-1E

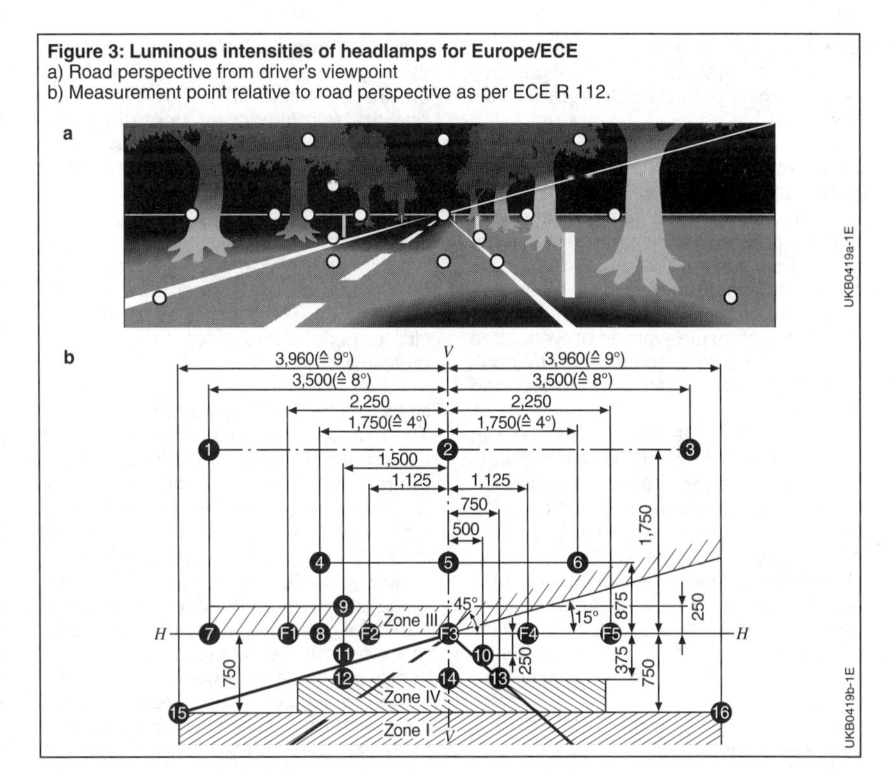

Figure 3: Luminous intensities of headlamps for Europe/ECE
a) Road perspective from driver's viewpoint
b) Measurement point relative to road perspective as per ECE R 112.

Table 2: Measurement points and luminous intensities for headlamps

Low beam (dipped beam)				**High beam**		
Measurement points on graphic			Luminous intensity	Measurement points		Luminous intensity
Figure no.	RHD traffic	LHD traffic	Class B (lx)	Figure no.	Point	Class B (lx)
01	8L/4U		≤ 0.7		E_{max}	48 < E
02	V/4U		≤ 0.7			< 240
03	8R/4U		≤ 0.7	F1	$E_{H-5.15°}$	> 6
04	4L/2U		≤ 0.7	F2	$E_{H-2.55°}$	> 24
05	V/2U		≤ 0.7	F3	$E_{HV^{9}}$	≥ 0.8
06	4R/2U		≤ 0.7			E_{max}
07	8L/H	8R/H	≥ 0.1; ≤ 0.7	F4	$E_{H+2.55°}$	> 24
08	4L/H	4R/H	≥ 0.2; ≤ 0.7	F5	$E_{H+5.15°}$	> 6
09	B50L	B50R	≤ 0.4			
10	75R	75L	≥ 12			
11	75L	75R	≤ 12	For low beam:		
12	50L	50R	≤ 15	Total 1 + 2 + 3 ≥ 0.3 lx		
13	50R	50L	≥ 12	Total 4 + 5 + 6 ≥ 0.6 lx		
14	50V	50V	≥ 6			
15	25L	25R	≥ 2			
16	25R	25L	≥ 2			
Any point in zone III			≤ 0.7	[1] E is the current measured value in point 50R or 50L.		
Any point in zone IV			≥ 3			
Any point in zone I			≤ 2E[1]			

Low beam, installation

Regulations prescribe 2 white-light low-beam headlamps for multiple-track vehicles (Figure 4).

Low beam, lighting technology

Automotive headlamp performance is subject to technical assessment and verification before they are put into volume production. Among the requirements are minimum luminous intensity, to ensure adequate road-surface visibility, and maximum intensity levels, to prevent glare (see measurement points and luminous intensities for headlamps, Figure 3 and Table 2).

Homologation testing is carried out under laboratory conditions using test lamps manufactured to more precise tolerances than those installed in production vehicles. The lamps are operated at the specified test luminous flux for each lamp category. The laboratory conditions apply across the board to all headlamps, but only take limited account of the specifics of individual vehicles, such as headlamp fitted height, vehicle power supply, and adjustment.

Low beam, switching

All high-beam headlamps must extinguish simultaneously when the low beams are switched on. Dimming (gradual deactivation) is permitted, with a maximum dimming period of 5 seconds. A 2-second response delay is required to prevent the dimming feature from activating when the headlamp flashers are used. When the high-beam headlamps are switched on, the low-beam units may continue to operate (simultaneous operation). H4 bulbs are generally suitable for short periods of use with both filaments in operation.

High beam, installation

A minimum of two and a maximum of four headlamps are prescribed for the high-beam mode. The prescribed instrument-cluster high-beam indicator lamp is blue or yellow in color.

High beams, lighting technology

The high beam is usually generated by a light source located at the focal point of the reflector (Figure 5). This causes light to be reflected outward in the direction of the reflector axis. The maximum luminous intensity achievable by the high beam is largely a function of the reflector's illuminated area. In quad-headlamp systems, in particular, roughly paraboloid high-beam reflectors can be replaced by units with complex geometrical configurations designed to supply a "superimposed" high-beam pattern. The calculations employed to design these units seek to achieve a high-beam distribution that harmonizes with the low-beam pattern (simultaneous activation). The pure high beam is "superimposed" as it were on the low-beam projection. The annoying overlap area close to the front of the vehicle is done away with in this case.

Figure 4: European headlamp system (low beam)

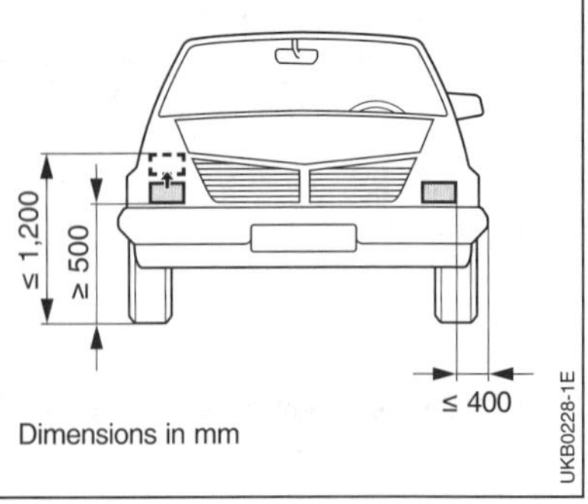

Dimensions in mm

Figure 5: Parabolic reflector
F Focal point,
S Vertex of the parabola
f Focal length

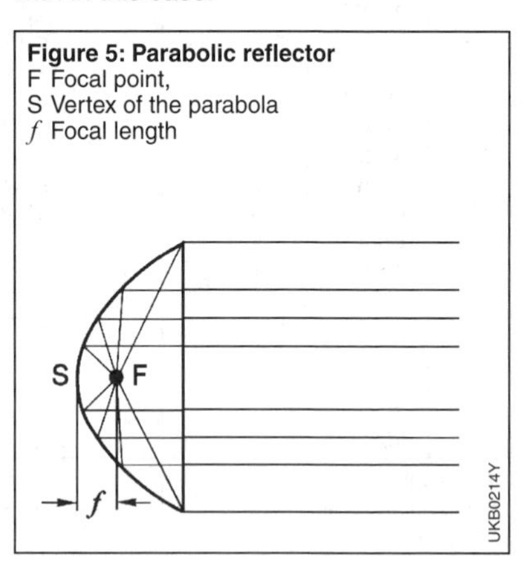

High-beam light distribution pattern is defined in the regulations and guidelines together with stipulations governing the low beams.

The maximum approved luminous intensity, a composite of the intensity ratings of all high-beam headlamps installed on the vehicle, is 430,000 cd. This value is indicated by identification codes located adjacent to the homologation code on each headlamp. 430,000 cd corresponds to the figure 100. The luminous intensity of the high beam is indicated by the number, e.g. 25, stamped next to the round ECE approval mark. If these are the only headlamps on the vehicle (no auxiliary driving lamps), then the composite luminous intensity is in the range of 50/100 of 430,000 cd, i.e. 215,000 cd.

Auxiliary driving lamps

Auxiliary driving lamps are used to complement the effectiveness of the high beam in standard high-beam headlamps.

Auxiliary driving lamps are mounted and aimed in the same way as standard headlamps, and the underlying lighting technology is the same. Auxiliary driving lamps are also subject to the regulations governing maximum luminous intensity in vehicle lighting systems, according to which the sum of the reference numbers of all headlamps fitted to a vehicle must not exceed 100. For older headlamps without approval number, the number 10 is used for general assessment purposes.

Main-light functions for the USA

Regulations and directives

The national standard is the Federal Motor Vehicle Safety Standard (FMVSS) No. 108 [27] and the SAE Ground Vehicle Lighting Standards Manual (Standards and Recommended Practices) to which it refers.

The regulations governing installation and control circuits for headlamps are comparable in parts to those in Europe. Since 5/1/1997, however, headlamps with light-dark boundaries have also been authorized in the USA, but these require manual adjustment. It is now possible to develop headlamps which conform to legal requirements in both Europe and the USA.

As in Europe, dual- and quad-headlamp systems are used in the USA. The fitting and use of fog lamps and additional high-beam headlamps, however, are subject to a variety of, in some cases, widely diverging, local laws passed by the 50 individual states.

Up to 1983 the allowable sealed-beam headlamp sizes (Figure 6) in the USA were as follows:

Dual-headlamp systems:
- 178 mm diameter (round)
- 200 × 142 mm (rectangular)

Figure 6: American sealed-beam headlamps
a) Low beam,
b) High beam.
1 Low-beam filament,
2 Focal point,
3 High-beam filament (at focal point).

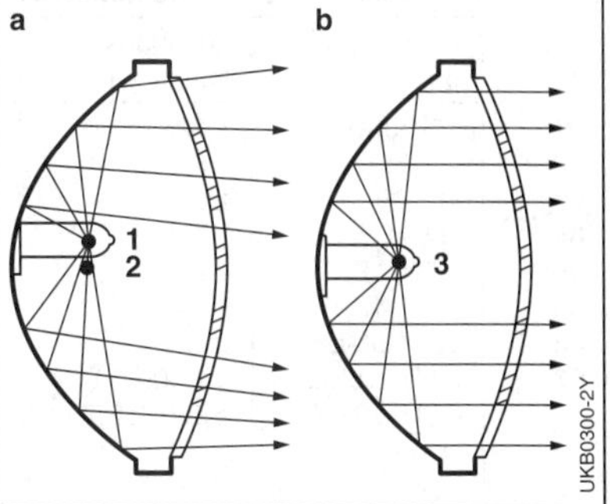

Quad-headlamp systems:
- 146 mm diameter (round)
- 165 × 100 mm (rectangular).

Low beam, lighting technology
The light-pattern requirements in America differ to a greater or lesser degree from the European system depending on the design type. In particular, the minimum glare levels are higher in the USA, and the maximum low-beam width is closer to the vehicle. The basic setting is generally higher (see measurement points in Figure 7 and Table 3).

High beam
The designs for high-beam headlamps are the same as in Europe. Differences exist in the required dispersion width of the light pattern, and there is a lower maximum figure on the axis of the high-beam headlamp.

Figure 7: Luminous intensities of headlamps for the USA
a) Road perspective from driver's viewpoint
b) Measurement point relative to road perspective as per FMVSS 108 (excerpt).

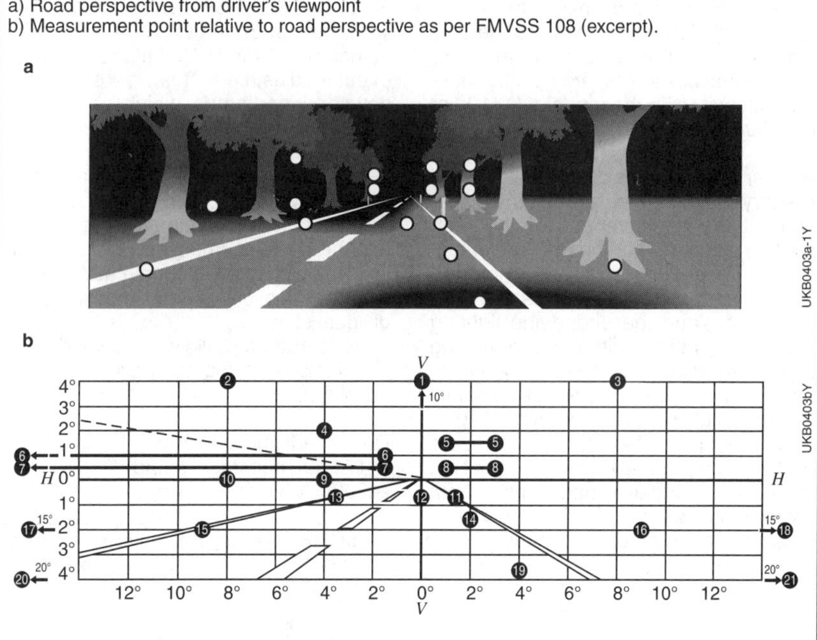

Table 3: Measurement points and luminous intensities for headlamps, low beam

Figure no.	Measurement points	Luminous intensity (cd)
01	10U-90U	≤ 125
02	4U, 8L	≥ 64
03	4U, 8R	≥ 64
04	2U, 4L	≥ 135
05	1.5U, 1R-3R	≥ 200
05	1.5U, 1R-R	≤ 1,400
06	1U, 1.5L-L	≤ 700
07	0.5U, 1.5L-L	≤ 1,000
08	0.5U, 1R-3R	≥ 500; ≤ 2,700
09	H, 4L	≥ 135
10	H, 8L	≥ 64
11	0.6D, 1.3R	≥ 10,000
12	0.86D, V	≥ 4,500
13	0.86D, 3.5L	≥ 1,800; ≤ 12,000
14	1.5D, 2R	≥ 15,000
15	2D, 9L	≥ 1,250
16	2D, 9R	≥ 1,250
17	2D, 15L	≥ 1,000
18	2D, 15R	≥ 1,000
19	4D, 4R	≥ 12,500
20	4D, 20L	≥ 300
21	4D, 20R	≥ 300

Designs

Sealed-beam design

In this design, which is no longer used (Figure 6), the aluminized glass reflector and the lens must be sealed gas-tight on account of the light sources that are not encapsulated. The whole unit is sealed and filled with an inert gas. If a filament burns through, the entire light source must be replaced. Units with halogen lamps are also available.

The limited range of available sealed-beam headlamps on the market severely restricted the freedom of design for vehicle front ends.

Vehicle headlamp aiming device (VHAD)

This design involves replaceable-bulb headlamps which are adjusted vertically with the aid of a spirit levels integrated in each headlamp, and horizontally by means of a system comprising a needle and dial. In fact, this is equivalent to "on-board aiming".

Headlamps for visual aim (VOL / VOR)

These systems have been in use since 1997. These are replaceable-bulb headlamps whose low beam has a light-dark boundary line which allows visual aiming of the headlamps (as is standard in Europe).

Either the left horizontal light-dark boundary (VOL, visual optical aim left; VOL mark on the headlamp) or, as is more often the case in the USA, the right horizontal light-dark boundary (VOR, visual optical aim right; VOR mark on the headlamp) is used. What is notable about the US systems is the position of the light-dark boundary, which is much closer to the horizon (inclination depending on type 0.4 to 0 %). This increases the potential dazzling risk of such systems.

There is no horizontal aiming with this type of headlamp.

Definitions and terms

Photometrical terms and definitions

Headlamp range

This is the distance at which the light beam continues to supply a specified luminous intensity – mostly the 1 lux line at the right side of the road (LHD traffic).

Geometric range of a headlamp

This is the distance to the horizontal portion of the light-dark boundary on the road surface (see Table 4). A low-beam inclination of 1 %, or 10 cm/10 m, results in a geometric range equal to 100 times the headlamp's fitted height (as measured between the center of the reflector and the road surface).

Visual range

The visual range is the distance at which an object (vehicle, object, etc.) within the luminance distribution of the human visual field is still visible.

The visual range is influenced by the shape, size, and reflectance of objects, the road-surface type, headlamp design and cleanliness, and the physiological condition of the driver's eyes. Due to the large number of influencing factors, it is not possible to quantify this range using precise numerical definitions. Under extremely unfavorable conditions (with RHD traffic, on the left side of a wet road surface) the visual range can fall to below 20 m. Under optimum conditions, it can extend outward to more than 100 m (with RHD traffic, on the right side of the road).

Disability glare

This is the quantifiable reduction in visual performance that occurs in response to light sources emitting glare. An example would be the reduction in visual range that occurs as two vehicles approach one another.

Discomfort glare

This condition occurs when a glare source induces discomfort without, however, causing an actual reduction in visual performance. Discomfort glare is assessed according to a scale defining different levels of comfort and discomfort.

Headlamp technology

Reflector focal length

Conventional reflectors for headlamps and other automotive lamps are usually parabolic in shape (Figure 5). The focal length f (distance between the vertex of the parabola and the focal point) is 15 to 40 mm.

Free-form reflectors

The geometrical configurations of free-form reflectors are generated using complex mathematical calculations (HNS, Homogeneous Numerically Calculated Surface). Here, the low mean focal length f is defined relative to the distance between the reflector vertex and the center of the filament. Typical values range from 15 to 25 mm.

In the case of reflectors partitioned with steps or facets, each partition can be created with its own mean focal length f.

Reflector illuminated area

This is the parallel projection of the entire reflector opening on a transverse plane. The standard reference plane is perpendicular to the vehicle's direction of travel.

Effective luminous flux, efficiency of a headlamp

The first of the above is the portion of the light source's luminous flux that is capable of supplying effective illumination via its reflective or refractive components (for instance, as projected on the road surface via the headlamp reflector). A reflector with a short focal length makes efficient use of the filament bulb and has high efficiency because the reflector extends outward to encompass the bulb, allowing it to convert a large proportion of the luminous flux into a useful light beam.

Angles of geometric visibility

These are angles that are defined relative to the axis of the lighting device, at which the illuminated area must be visible.

Technical design variations of headlamps

Components

Reflector

Reflectors direct the light from the light source either directly onto the road (reflection system) or into an intermediate plane which is further projected by a lens (projection system). The reflectors are made of plastic, die-cast metal or sheet steel.

Plastic reflectors are manufactured by injection molding (thermosetting plastics), which offers a considerably better precision of geometry reproduction than the deep-drawing process. The geometric tolerances that can be achieved lie in the range of 0.01 mm. Furthermore, reflectors in stages and any desired facet distributions can also be implemented. The base material requires no corrosion-proofing treatment.

The die-cast metal used is usually aluminum, or occasionally magnesium. The advantages are high thermal resistance and the ability to produce shapes with a high degree of complexity (shaped bulb holders, screw holes and bosses).

The surfaces of thermosetting plastic and die-cast metal reflectors are given a smooth finish by spray painting or powder-based paint before a layer of aluminum 50 to 150 nm thick is applied. An even thinner transparent protective coat prevents the aluminum from oxidizing.

Sheet-steel reflectors are manufactured using deep-drawing and punching dies. A powder-based paint is then applied. This process hermetically seals the sheet steel and gives it excellent surface smoothness. The base coat created in this way is, like other reflectors, coated with aluminum.

Lens

A large proportion of contoured lenses is manufactured using high-purity glass (free of bubbles and streaks). During the lens molding process, high priority is given to surface quality in order to prevent undesirable upward light deflection in the final product – this would tend to create glare for oncoming traffic. The type and configuration of the lens prisms depend on the reflector and the desired light distribution pattern.

The clear lenses used on modern headlamps are usually made of plastic. Besides reducing weight, plastic lenses provide other advantages for automotive applications, including greater freedom in headlamp and vehicle design. Since around 2007 multicolor plastic lenses (2-component lenses) have also been used in which the edge area is sprayed in a different color, usually black or gray. The advantage here is that the spray tools are designed in such a way that no inner slides are needed and therefore no separating lines are created on the visible surface. Light dispersion from the edge areas is also avoided.

There are several reasons why plastic lenses should not be cleaned with a dry cloth:

- Despite the scratch-proof coating, rubbing with a dry cloth can damage the surface of the lens
- Rubbing with a dry cloth can produce an electrostatic charge in the lens, which can then allow dust to build up on the inside of the lens

Figure 8: Low beam (beam-path vertical section, H4 lamp)
1 Low-beam filament,
2 Bulb cover.

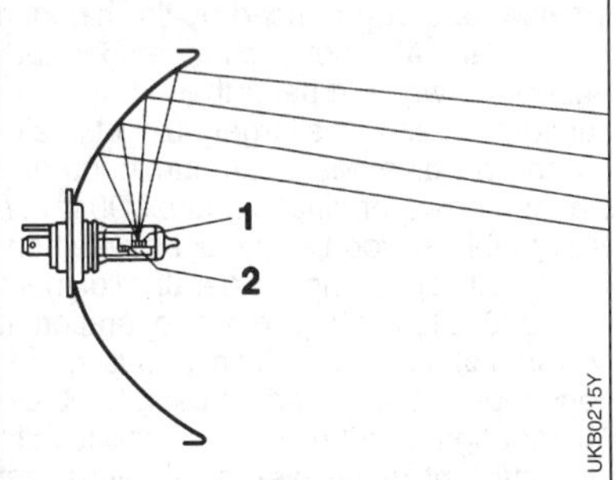

Headlamp versions

Reflection headlamps

For conventional headlamp systems with virtually parabolic reflectors (Figure 5 and Figure 8), the quality of the low beam increases in direct proportion to the size of the reflector. At the same time the geometric range increases as a function of installation height.

These factors must be balanced against the aerodynamic constraints according to which the vehicle's front-end profile must be kept as low as possible. Under these circumstances, increasing the size of the reflector results in wider headlamps.

Reflectors of a given size, but with different focal lengths, also perform differently. Shorter focal lengths are more efficient and produce wider light beams with better close-range and side illumination. This is of particular advantage during cornering.

Specially developed lighting programs (CAL, computer-aided lighting) enable the implementation of infinitely variable reflector shapes with non-parabolic sections and of faceted reflectors.

Headlamps with facet-type reflectors

In the case of facets, the reflector surface is partitioned and each individual segment is individually optimized. The im-

Figure 9: Free-form or facet-type reflector
Filament pattern reflection with mirror optics.
1 Filament,
2 Mirror optics.

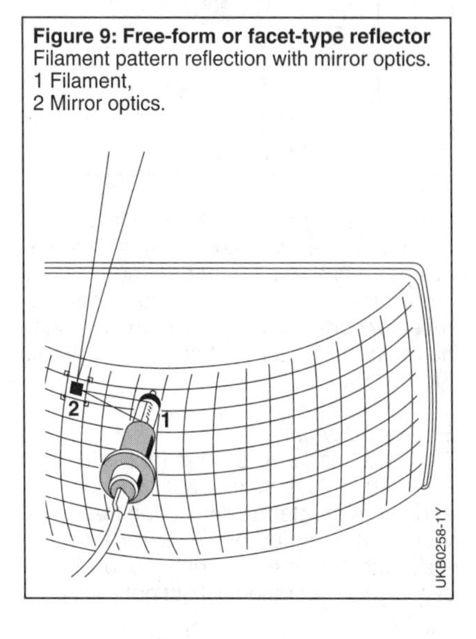

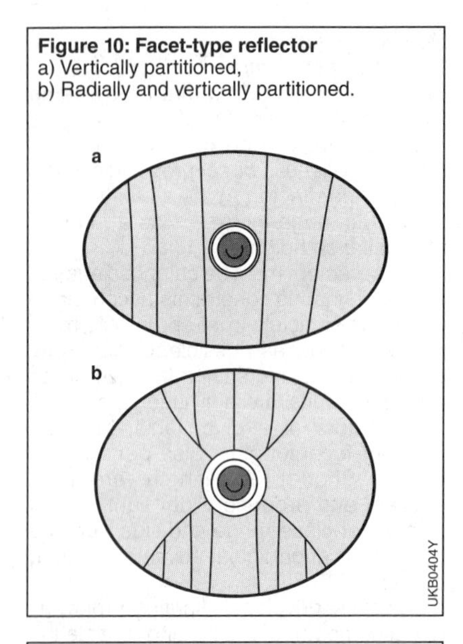

Figure 10: Facet-type reflector
a) Vertically partitioned,
b) Radially and vertically partitioned.

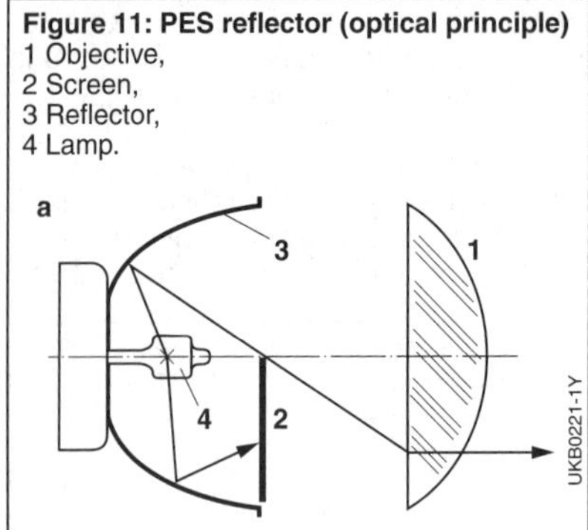

Figure 11: PES reflector (optical principle)
1 Objective,
2 Screen,
3 Reflector,
4 Lamp.

portant feature of surfaces with facet-type reflectors is that discontinuity and steps are permitted at all boundary surfaces of the partition. This results in freely shaped reflector surfaces with maximum homogeneity and side illumination (Figure 9 and Figure 10).

PES headlamps

The PES headlamp system (poly-ellipsoid system) employs imaging optics (Figure 11) and offers greater design scope than conventional headlamps. A lens aperture area with a diameter of only 40 to 70 mm allows the generation of light patterns previously only achievable with large-area headlamps. This result is obtained using an elliptical (CAL designed) reflector in combination with optical projection technology. A screen reflected with the objective projects precisely defined light-dark boundaries. Depending on specific individual requirements, these transitions can be defined as sudden or gradual intensity shifts, making it possible to obtain any geometry required.

PES headlamps can be combined with conventional high beams, position lamps and PES fog lamps to form lighting-strip units in which the entire headlamp is no higher than approximately 80 mm.

In PES headlamps, the beam path can be configured in such a way that the surroundings of the objective are also used in the signal image. This enlargement of the signal image is used above all with small objective diameters to reduce the psychological glare for the oncoming traffic. The additional surface can also be made into a lens, a partially vapor-deposited screen or a design feature with illuminated round or rectangular gaps, or illuminated three-dimensional objects.

Xenon headlamps

The headlamp system with a xenon gas-discharge lamp as its central component (Figure 12) generates high illumination-intensity levels with minimal frontal-area requirements, making it ideal for vehicles with aerodynamic styling with exceptional c_W values. In contrast with the conventional filament bulb, light is generated by a plasma discharge inside a burner the size of a cherry stone (Figure 13).

The arc of the 35 W gas-discharge lamp generates a luminous flux twice as intense as that produced by the halogen H1 bulb and at a higher color temperature (4,200 K), which means that – similar to sunlight – it contains larger proportions of green and blue. Maximum luminous efficiency, corresponding to approx. 90 lm/W, is available as soon as the quartz element reaches its operating temperature of more than 900 °C. Brief high-power operation at currents of up to 2.6 A (continuous operation: approx. 0.4 A) can be used to obtain “instant light”. 2,000 hours of service life are sufficient for the average required total duration of operation in passenger cars. As no sudden failure occurs as in the case of a filament, diagnosis and replacement in good time are possible.

At present gas-discharge lamps with the type designations D1 and D2 are used and from 2012 only the type designations D3 and D4 will be used. In the D3 and D4 families dosing of the heavy metal mercury (approx. 1 mg) can be dispensed with. These lamps are distinguished by lower lamp voltage, a different plasma composition, and different arc geometries. The electronic control units for the individual lamp types are generally developed for a specific design type and are not universally interchangeable.

The D2 and D4-series automotive gas-discharge lamps feature high-voltage-proof bases and UV glass shielding elements. On the D1 and D3-series models, the high-voltage electronics necessary for operation are also integrated in the lamp base. All systems feature two subcategories: S-lamps for projection-system headlamps and R-lamps for reflection headlamps with an integrated light shield for producing the light-dark boundary comparable with the bulb cover used for halogen H4 low beams. At present the D1S and D3S lamps are the type most widely used.

An integral part of the Litronic headlamp (contraction of Light and Electronic) is the electronic ballast unit responsible for activating and monitoring the lamp (Figure 14). Its functions include ignition of the gas discharge (voltage 10 to 20 kV), controlled power supply during the warm-up phase when the lamp is cold, and demand-oriented supply during stationary operation.

Figure 13: Gas-discharge lamp D2S/DS4
1 UV glass shielding bulb,
2 Lead-in insulator,
3 Discharge chamber (burner),
4 Electrodes,
5 Bulb base.

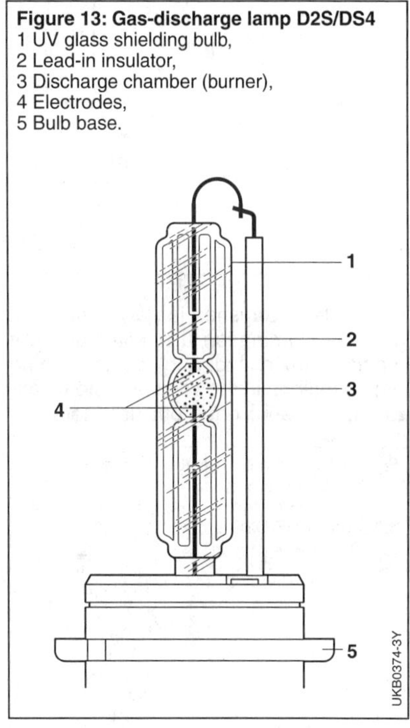

Figure 12: System components for PES-design Litronic headlamp
1 ECU,
2 To vehicle electrical system,
3 Shielded cable,
4 Ignition unit of D1S/D3S lamp,
5 Projection module,
5a Burner of D1S/D3S lamp,
5b Lens.

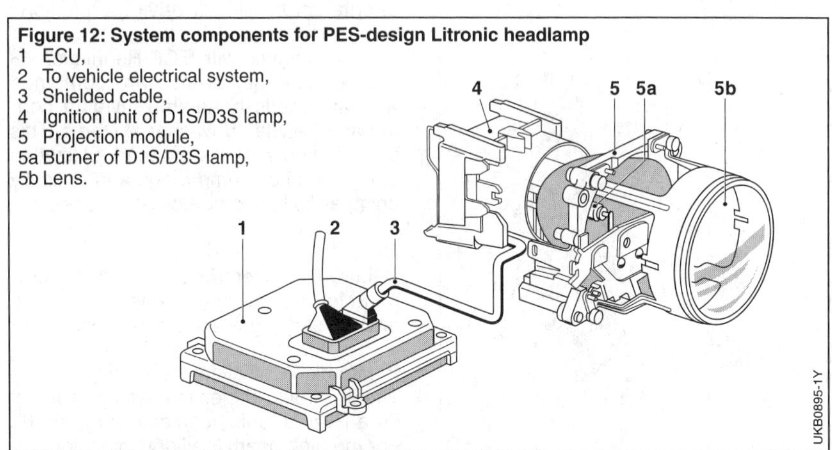

Figure 14: Electronic ballast unit for 400-Hz AC supply and pulse triggering of the bulb
1 ECU, 1a DC/DC converter, 1b Shunt, 1c DC/AC converter, 1d Microprocessor,
2 Ignition unit, 3 Lamp socket, 4 D2S lamp, U_B Battery voltage.

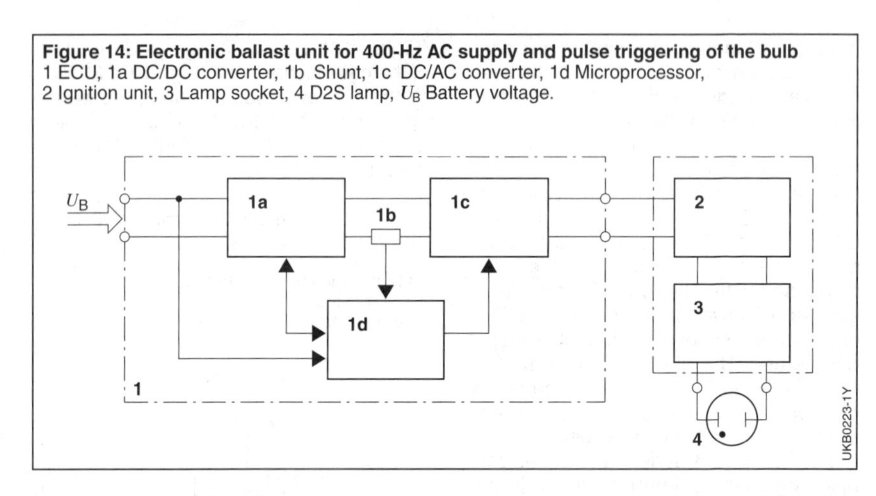

The system furnishes largely consistent levels of illumination (i.e. elimination of luminous-flux changes) by compensating for fluctuations in the vehicle system voltage. If the lamp goes out (for instance, due to a momentary voltage drop in the vehicle electrical system), re-ignition is spontaneous and automatic.

Figure 15: Light distribution pattern on the road (comparison)
a) Halogen H4 lamp,
b) Litronic PES D2S lamp.

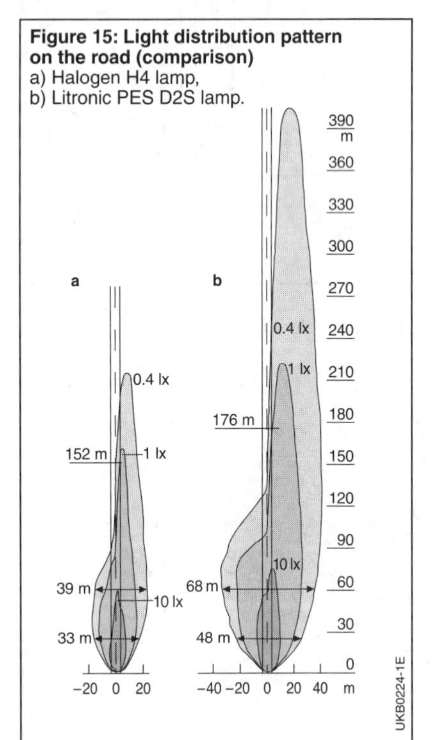

The electronic ballast unit responds to defects (such as a damaged lamp) by interrupting the power supply to help avoid injury in the event of contact.

The xenon light emitted by Litronic headlamps produces a broad carpet of light in front of the vehicle combined with a long range (Figure 15). This has made it possible to achieve substantially wider road-illumination patterns for lighting the edges of bends and wide roads as effectively as a halogen unit illuminates straight stretches of road. The driver enjoys substantial improvements in both visibility and orientation in difficult driving conditions and bad weather.

In accordance with ECE Regulation 48 [12], Litronic headlamps are combined with automatic headlamp leveling control and headlamp washer systems. This combination ensures optimum utilization of the long headlamp range and optically unimpaired light emission at all times.

Bi-Litronic

Bi-Litronic systems permit both the low and high beams to be generated by the arc of a single gas-discharge lamp.

Bi-Litronic "Projection" *(bi-xenon)*:

The Bi-Litronic "Projection" system is based on a PES Litronic headlamp (Figure 16). For the high beam the light shield for creating the light-dark boundary (for the low beam) is moved out of the light beam. With lens diameters of 70 mm, Bi-Litronic "Projection" allows the present highly compact headlamp design with combined high and low beam and at the same time outstanding luminous efficiency (Figure 18).

The primary advantage of Bi-Litronic "Projection" is the xenon light for high-beam operation.

Bi-Litronic "Reflection":
Both mono- and bi-xenon systems use a single DxR lamp for both headlamp functions.

In the bi-xenon version, when the high-/low-beam switch is operated, an electromechanical actuator moves the gas-discharge lamp in the reflector into the appropriate position for producing the high or low beam as required (Figure 17).

Figure 16: Bi-Litronic "Projection"
1 Low beam,
2 High beam.

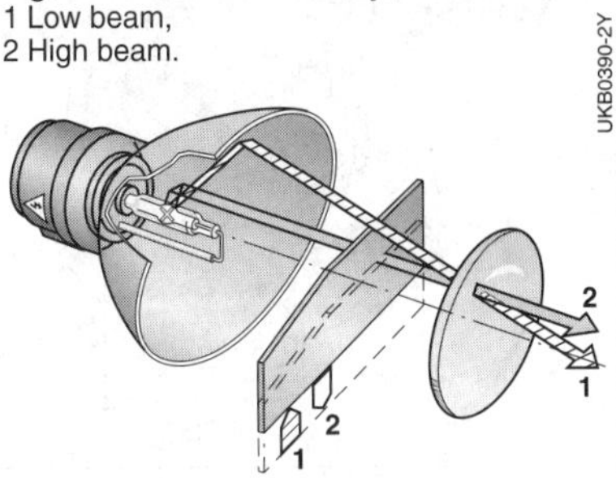

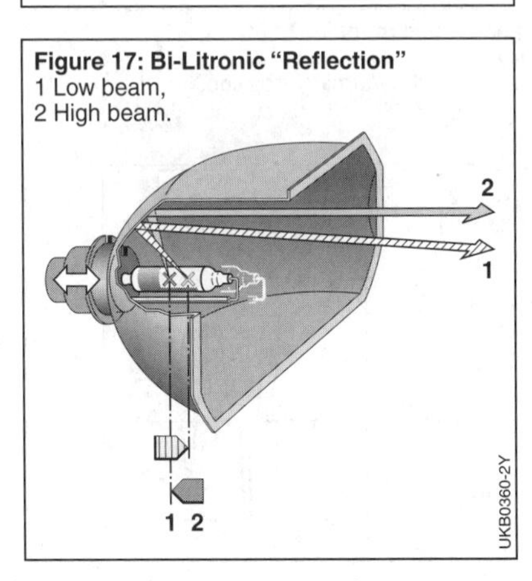

Figure 17: Bi-Litronic "Reflection"
1 Low beam,
2 High beam.

Cornering headlamps
Cornering headlamps have been approved for public use since the beginning of 2003. Whereas previously only high-beam headlamps were allowed to turn in response to changes in steering angle (1960s Citroën DS), swiveling low-beam headlamps are now also permitted (dynamic cornering headlamps or adaptive headlamps) or a supplementary light source (static cornering headlamps). This provides for a larger visual range on winding roads.

Static cornering headlamps
Static cornering headlamps are used mainly to illuminate areas close to the side of the vehicle (switchbacks, turning maneuvers). For this purpose, activating additional reflector elements is generally the most effective way.

Figure 18: Light distribution patterns of Bi-Litronic
1 Low beam,
2 High beam.

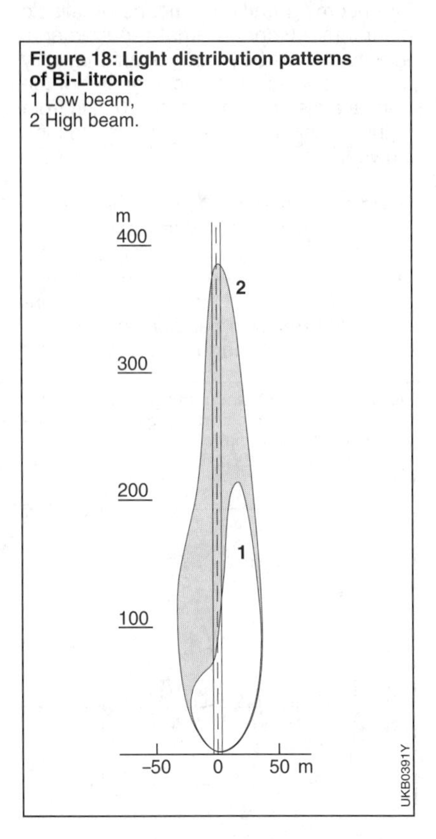

Dynamic cornering headlamps
Dynamic cornering headlamps are used to illuminate the changing course of the road, e.g. on winding overland highways (Figure 19).

In contrast with the directly linked swiveling action of the cornering headlamps of the 1960s, modern high-end systems electronically control the rate of swivel and the swivel angle in response to the vehicle's speed. This optimizes "harmonization" between the headlamps and the vehicle attitude, and eliminates "jerky" headlamp movements. Headlamp positioning is performed by a positioner unit (stepping motor) which moves the basic or low-beam module or the reflector elements in response to changes in steering-wheel angle or the steering angle of the front wheels (Figure 20). Sensors detect those movements to prevent glare for oncoming traffic by means of failsafe algorithms. General legal requirements specify that the headlamp beam may only be turned as far as the center line of the road at a distance of approx. 70 m in front of the vehicle to prevent glare for oncoming vehicles.

Road safety and driving convenience
The introduction of dynamic cornering headlamps is a significant improvement in the safety and convenience of driving at night (Figure 21). Compared with conventional low-beam headlamps, improvements in visual range of approximately 70% are achieved, representing an extra 1.6 seconds of travel time. With cornering headlamps, a motorist can assess hazards better and start braking sooner. As a result, the severity of an accident can be significantly reduced. Static cornering headlamps double the visual range for turning maneuvers.

Figure 20: Cornering-headlamp modules
1 Supporting frame,
2 Mounting frame,
3 Drive motor for horizontal rotation,
4 Bi-Litronic PES.

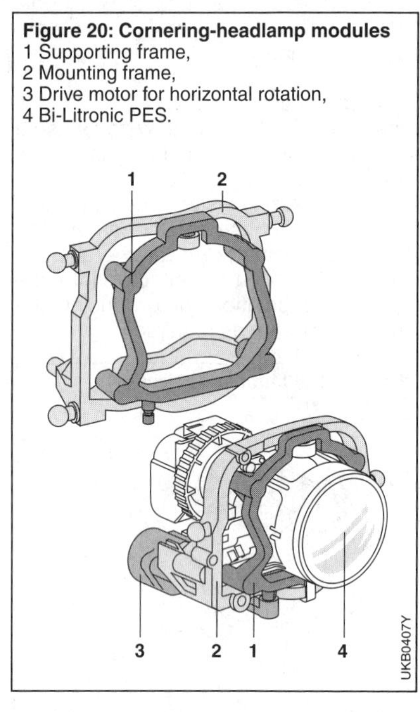

Figure 19: Switch and aiming strategy of turning and basic modules of a static/dynamic cornering headlamp (left side)
a) "Highway/cornering" position, b) "Expressway" position, c) "City/turning" position.
1 Turning module, 2 Basic module.

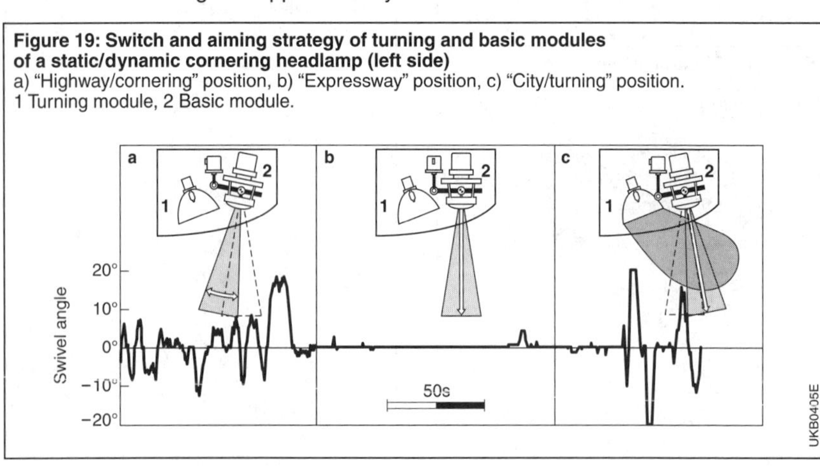

AFS functions

Expressway beam

For special driving situations, altered light patterns (AFS, adaptive front-lighting system) have been developed to enable a better view for the driver for each driving state. In the development of the expressway beam (Figure 22), special attention was paid to achieving a better range for the driver without dazzling oncoming traffic to an excessive degree. The increase in the detection distance to up to 150 m enables an extension of driving time up to the detected object of approx. 2 seconds (comparison at 100 km/h with halogen headlamps). This enables the driver to better assess a critical situation and, possibly, to initiate braking much earlier.

Poor-weather light

In the case of the poor-weather light beam, the special focus is on improving the optical guidance on the road. Especially the zones around the sides of the road are better illuminated.

Most variants of the poor-weather light feature movement of the left cornering-headlamp module by 8° to the side and simultaneously a slight lowering or activation of the static cornering lights. This provides for very wide illumination of the road and the edge of the road. In future, component elements, for example the elements responsible for widened side illumination, will be activated sequentially. Control parameters are, for example, steering-angle information and direction-indicator actuation. The individual segments are then "quasidynamically" activated.

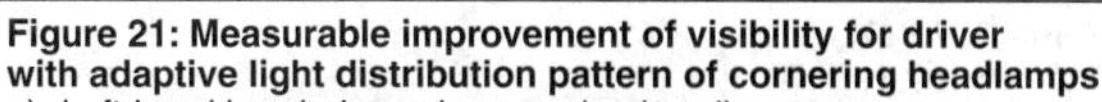

Figure 21: Measurable improvement of visibility for driver with adaptive light distribution pattern of cornering headlamps

a) Left-hand bend, dynamic cornering headlamps,
b) Right turn, static cornering headlamps.
1 Halogen headlamps, 2 Xenon headlamps,
3a Adaptive light distribution pattern with dynamic cornering headlamps,
3b Adaptive light distribution pattern with static cornering headlamps.

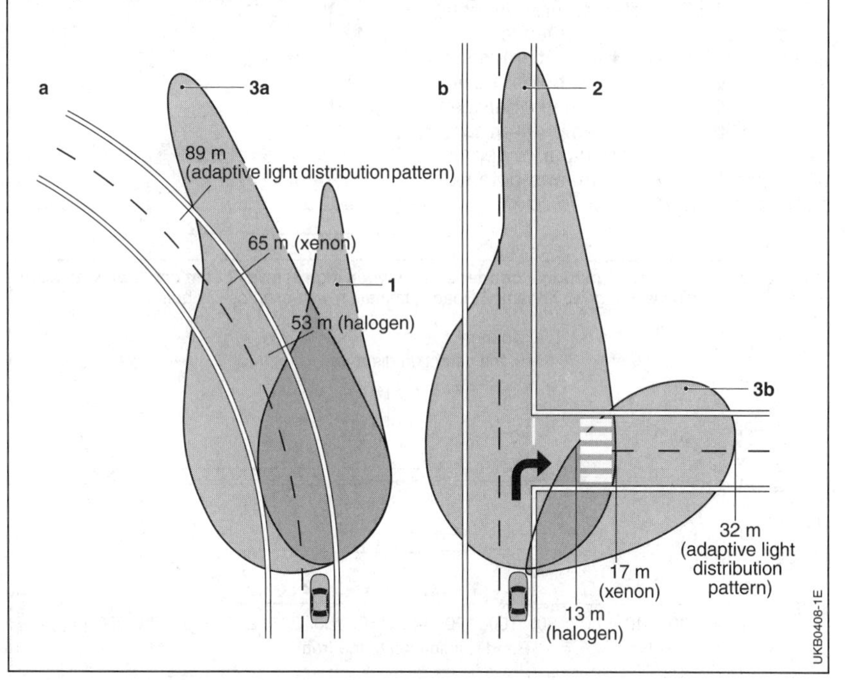

Light functions and driver-assistance systems
The introduction of video technology in automotive engineering makes it possible to implement camera-based headlamp functions as well. When the position of an oncoming vehicle is identified by the camera, the headlamp or the AFS system can adapt the range of the driving lights in such a way that it is increased for large distances and reduced for smaller distances (dynamic range function). This ensures optimum illumination without dazzling oncoming traffic.

LED headlamps

Potential for reducing energy consumption
LEDs are increasingly being used as economical alternatives in avoiding CO_2 emissions and fuel consumption. The energy consumption of the main-light functions will play a significant role in future low-energy vehicles. Xenon and LED alternatives offer the levels of optimized consumption and improved road safety called for by the EU.

Today's LED systems already consume, depending on performance (luminous flux, range, side illumination), much less energy than halogen bulbs. Power consumption currently fluctuates between 28 W and 50 W per headlamp. When compared with bulb power output of approx. 65 W (for 13.2 V), this means potential savings of 30 to 70 W per vehicle.

Figure 23: Basic design of an LED low-beam headlamp
Interaction of projection and reflection systems.
a) Projection system for the spot,
b) Reflection system for the basic light,
c) Total light distribution.
1 LEDs,
2 Light shield for light-dark boundary,
3 Projection system,
4 Reflector,
5 Spot,
6 Basic light.

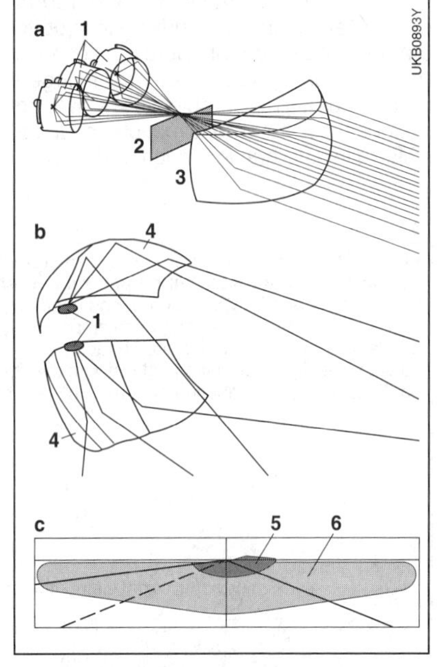

Figure 22: Metrological-functional connection between expressway beam and highway beam
1 Road illumination with highway beam, 2 Road illumination with expressway beam.

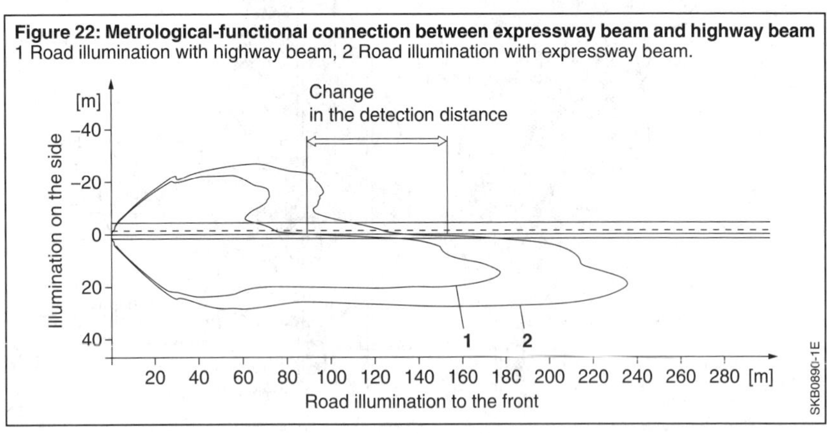

One projection system and two reflection elements are used for the low-beam spot. The light of three multichip LEDs each with two LEDs is concentrated by three primary optical elements and imaged by a projection lens. The optical system incorporates a screen to guarantee the quality of the light-dark boundary. A reflector is positioned both above and below the lens (Figure 23).

The optical efficiency of the LED low-beam headlamp is roughly 45 %. Compared with this, the efficiency of a bi-xenon system is around 33 %. This can be explained by the nature of LEDs, which give off their light only to the half-space and do not, like conventional light sources, illuminate the entire space. Because of the higher efficiency of an LED system, less luminous flux is needed in the LEDs to deliver the same luminous flux to the road with one LED low-beam headlamp.

The better LEDs will become in future, the less power will be output by the control unit, the light performance always being kept at a constantly high level.

The first LED headlamps on the market also show the increasing role of design elements in headlamps. In 2008, for example, a headlamp received an internationally renowned design award for the first time.

Around 75 % of total worldwide mileage is completed during daytime hours. Great importance is therefore also attached to the energy consumption of the daytime driving lights function. A typical LED daytime driving light has an energy consumption of 14 W (0.36 g CO_2/km) per vehicle. Use of vehicle lights during the day consumes up to 300 W (7.86 g CO_2/km) (low-beam headlamps, tail lamps, position lamps, number-plate lamps, switch and instrument lighting).

Fog lamps

Fog lamps (white light) are intended to improve orientation in fog, snow, heavy rain, and dust. A beam of light with a particularly high side dispersion is generated for this purpose. This ensures that the side of the road which is close to the vehicle is particularly well illuminated. The brightness levels achieved on close objects are significantly higher. Unlike the usually dark road surface far ahead of the vehicle, these high brightness levels help drivers to find their bearings better despite the poor weather conditions.

Designs

DIY and dealer-installed fog lamps are designed as individual projection units in their own housings. These are installed either upright on the fender, or suspended from it (Figure 24). Stylistic and aerodynamic considerations have led to increased use of integrated fog lamps, designed either for installation within body openings or included as a component within a larger light assembly (with adjustable reflectors when the fog lamps are combined with the main headlamps).

Present fog lamps produce white light. There is no substantive evidence that yellow lamps provide any physiological benefits. A fog lamp's effectiveness depends

Figure 24: Fog lamp (upright mounting)
1 Bulb cover,
2 Lens,
3 Reflector,
4 Vertical adjustment spindle.

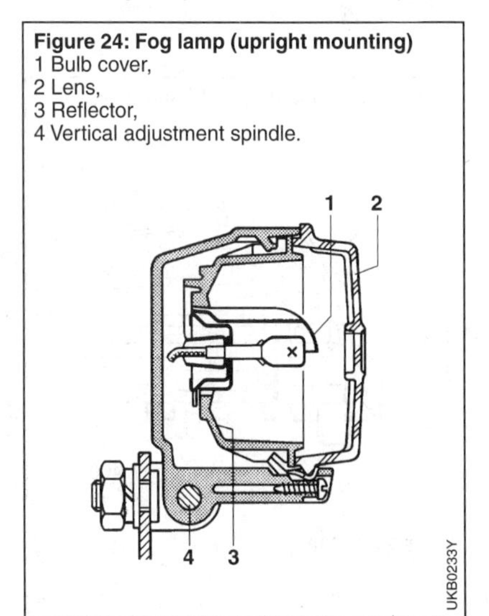

on the size of the illuminated area and the focal length of the reflector. Assuming the same illuminated area and focal length, the differences between round and rectangular fog lamps from the technical viewpoint are negligible.

Regulations
Design is governed by ECE R19 [8], installation by ECE R48 [12] (and StVZO (Road Traffic Licensing Regulations) Article 52 [3] in Germany); two white or yellow fog lamps are permitted. The control circuit for switching the fog lamps must be independent of the high- and low-beam circuits. In Germany, the StVZO allows fog-lamp installation in positions more than 400 mm from the widest point of the vehicle's width, provided they are wired so that they can only be switched on when the low-beam headlamps are on (Figure 25).

Paraboloid with lens
A parabolic reflector featuring a light source located at the focal point reflects light along a parallel axis (as with the high-beam headlamp), and the lens extends this beam to form a horizontal band (Figure 24). A special bulb cover prevents the beam from being projected upward.

Free-form technology
Calculation methods, such as CAL (computer-aided lighting), can be used to design reflector shapes in such a way that they scatter light directly (i.e. without optical lens contouring) and also generate (without separate shading) a sharp light-dark boundary. The fact that the lamp features pronounced envelopment of the bulb leads to an extremely high volume of light combined with maximum dispersion width (Figure 26).

PES fog lamps
This technology minimizes reflected glare in fog. The screen, the image of which is projected onto the road surface by the lens, provides a light-dark boundary with minimum upward light dispersion.

Innovations
The technical-functional importance of the fog lamp has been pushed into the background slightly by the introduction of powerful halogen, xenon and AFS systems. The emphasis is still on the demand for improved lighting in traffic situations with adverse ambient conditions. This can be solved by separate fog lamps of the type used up to now (as per ECE R19 [8]), by the poor-weather light function of AFS systems (as per ECE R123 [23]), or by combinations of the two functions.

Figure 25: Fog lamp (positioning)

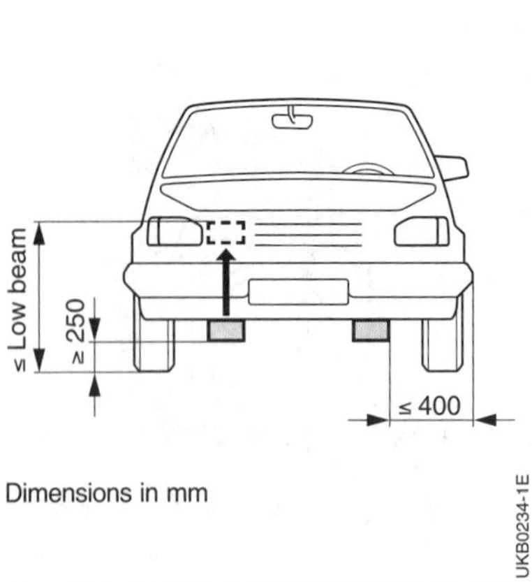

≤ Low beam
≥ 250
≤ 400

Dimensions in mm

UKB0234-1E

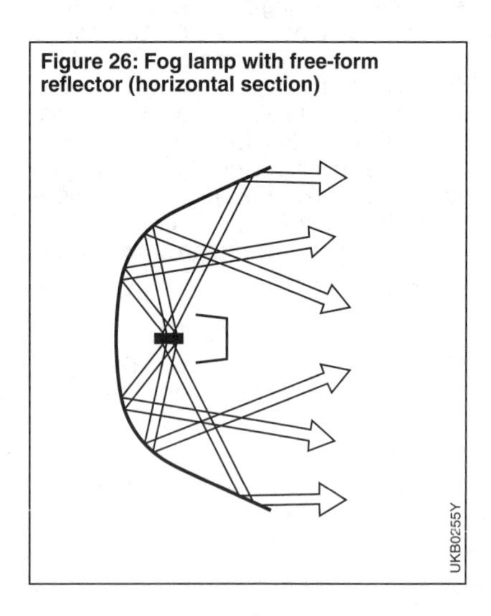

Figure 26: Fog lamp with free-form reflector (horizontal section)

Installation and regulations for signal lamps

These types of lamp are intended to facilitate recognition of the vehicle, and to alert other road users to any intended or present changes in direction or motion. Uniform, distinctive colors in the red, yellow or white color range are prescribed for these lamps to denote their application. White, yellow and red lights are used to mark the positions of the vehicle at the front, the sides and the rear respectively. Stop lamps and rear fog warning lamps are also red. Yellow lights are used in most applications for turn-signal lamps. Only in the USA are red turn-signal lamps also permissible at the rear.

As projected along the reference axis, minimum and maximum luminous intensities for all lamps must remain within a range calculated to guarantee signal recognition without, however, causing glare nuisance for other road users.

Turn-signal lamps and hazard-warning flashers

(as per ECE R6 [6])
ECE R48 and 76/756/EEC specify Group 1 (front), Group 2 (rear), and Group 5 (side) turn-signal lamps for vehicles with three or more wheels. For motorcycles, Group 2 turn-signal lamps are sufficient.

The lamps are electrically monitored. A function indicator is required inside the vehicle. The dashboard-mounted monitoring lamp may be in any color desired.

The flashing frequency is 90 ± 30 cycles per minute with a relative illumination period of 30 to 80 %. Light must be emitted within 1.5 s of lamp switch-on. All the turn-signal lamps on one side of the vehicle must flash synchronously. If one lamp fails, the remaining lamps must continue to generate visible light.

All the turn-signal lamps flash synchronously in hazard-warning mode, also operational when the vehicle is stopped. An operation indicator is mandatory.

Two lamps each (color: yellow) are prescribed for the front, the rear and the side turn-signal lamps. In the USA the color red or yellow is permitted for the rear and side turn-signal lamps (SAE J588, Nov. 1984, [33]).

Figure 27: Positioning of front turn-signal lamps as per ECE
Dimensions in mm
[1]) Less then 2,100 mm if vehicle body type prohibits compliance with regulations on maximum height.

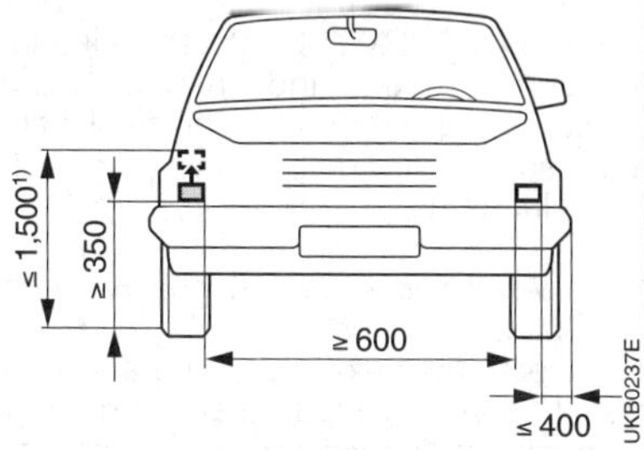

Figure 28: Positioning of rear turn-signal lamps as per ECE
Dimensions in mm Height and width same as front turn-signal lamp.
1 Tail lamp.

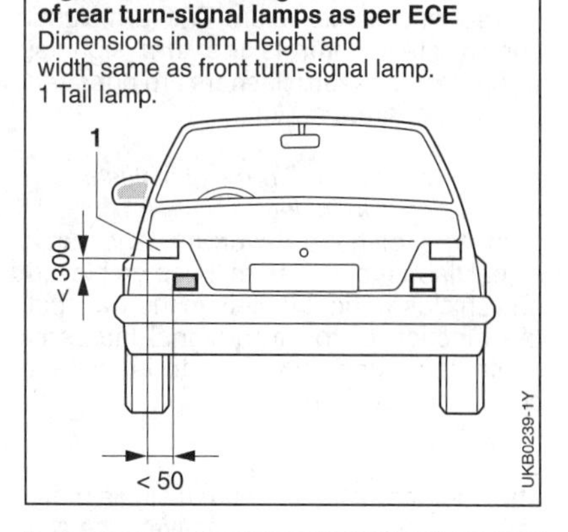

Figure 29: Positioning of side turn-signal lamps as per ECE
Dimensions in mm
If the type of vehicle body does not permit adherence to the maximum dimensions:
[1]) or 2,500 mm.
[2]) or 2,300 mm.

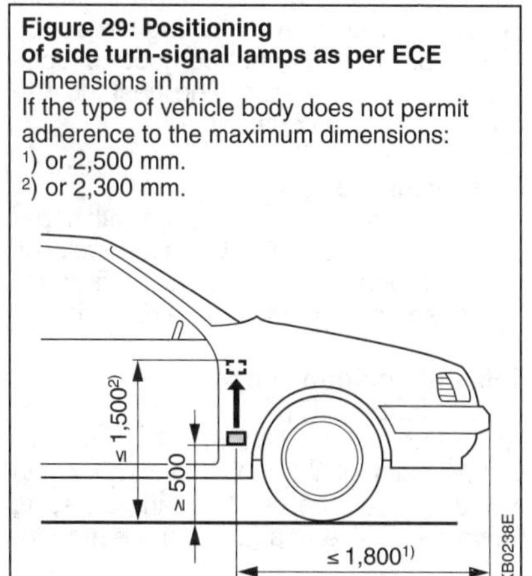

Design regulations
The design requirements for Europe are set out in the ECE Regulations R6, R7, R23, R38 and R87 [6, 7, 9, 11, 16], and the installation requirements in ECE R48 [12] (Figures 27, 28 and 29).

In the USA, FMVSS 108 specifies the number, location, and color of signal lamps. The design and technical lighting requirements are defined in the relevant SAE standards.

Hazard-warning and turn-signal flashers for vehicles without trailer
The electronic hazard-warning and turn-signal flashers include a pulse generator designed to switch on the lamps via a relay and a current-controlled monitoring circuit to modify the flashing frequency in response to bulb failure. The turn-signal control stalk controls the turn signals, whereas the hazard flashers are operated using a separate switch.

Hazard-warning and turn-signal flashers for vehicles with trailer
This type of hazard-warning and turn-signal flasher differs from those employed on vehicles without trailers in the way that the function of the turn-signal lamps is controlled when they flash to indicate a change in direction.

Single-circuit monitoring
The tractor and trailer share a single monitoring circuit designed to activate the two indicator lamps at the flashing frequency. This type of unit cannot be used to localize lamp malfunctions. The flashing frequency remains constant.

Dual-circuit monitoring
Tractor and trailer are equipped with separate monitoring circuits. The malfunction is localized by means of the indicator lamp. The flashing frequency remains constant.

Tail and position lamps
(as per ECE R7 [7])
According to ECE R48 and 76/756/EEC, vehicle and trailer combinations wider than 1,600 mm require position lamps (facing forwards). Tail lamps (at rear) are mandatory equipment on vehicles of all widths. Vehicles wider than 2,100 mm (e.g. trucks) must also be equipped with clearance lamps visible from the front and rear.

Position lamps
Two white-light position lamps are stipulated. The regulations set out in SAE J222, Dec. 1970, [29] apply in the USA.

Tail lamps
Two red tail lamps are stipulated. When the tail and stop lamps are combined in a nested design, the luminous-intensity ratio for the individual functions must be at least 1:5. Tail lamps must operate together with the position lamps.

The regulations set out in SAE J585e, Sept. 77, [30] apply in the USA.

Clearance lamps
(as per ECE R7 [7])
Vehicles wider than 2,100 mm require two white lamps facing forward, and two red lamps facing to the rear. They must be positioned as far outward and as high as possible.

The regulations set out in SAE J592e, [34] apply in the USA.

Side-marker lamps
(as per ECE R91 [17])
According to ECE R48, vehicles of any length exceeding 6 m must have yellow side-marker lamps (SML) except on vehicles with cab and chassis only.

Type SM1 side-marker lamps may be used on vehicles of all categories; type SM2 side-marker lamps, on the other hand, may only be used on cars.

The regulations set out in SAE J592e] apply in the USA.

Rear reflectors
(as per ECE R3 [4])
According to ECE R48, two red, non-triangular, rear reflectors are required on motor vehicles (one on motorcycles and motorbicycles).

Additional reflective items (red reflective tape) are permitted if they do not impair the function of the legally required lighting and signaling equipment.

Two colorless, non-triangular front reflectors are required on trailers and on vehicles on which all forward-facing lamps with reflectors are concealed (e.g. retractable headlamps). They are permitted on all other types of vehicle.

Yellow, non-triangular, side reflectors are required on all vehicles with a length exceeding 6 m, and on all trailers. These are permitted on vehicles shorter than 6 m.

Two red triangular rear reflectors are required on trailers, but are banned on motor vehicles.

There may be no light fitted inside the triangle.

The regulations set out in SAE J594f, [36] apply in the USA.

Parking lamps
(as per ECE R77 [15])
ECE R48 permits either two parking lamps at the front and rear, or one parking lamp on each side. The prescribed colors are white facing forwards and red facing rearwards. Yellow may also be used at the rear if the parking lamps have been designed as single units together with the side turn-signal lamps.

The parking lamps must be designed to operate even when no other vehicle lights (headlamps) are on. The parking-lamp function is usually assumed by the tail and position lamps.

The regulations set out in SAE J222, Dec. 1970, [29] apply in the USA.

License-plate lamps
(as per ECE R4 [5])
According to ECE R48, the rear license plate must be illuminated so as to be legible at a distance of 25 m at night.

Across the complete license-plate, luminance must be at least 2.5 cd/m^2. The luminance gradient of $2 \times B_{min}$/cm should not be exceeded between any of the test points distributed across the surface of the license plate. B_{min} is defined as the smallest luminance measured at the test point.

The regulations set out in SAE J587, Oct. 1981, [32] apply in the USA.

As an alternative to the license-plate lamp, self-illuminating license plates will also be permitted in future.

Stop lamps
(as per ECE R7 [7])
According to ECE R48, all cars must be fitted with two type S1 or S2 stop lamps, and one type S3 brake lamp, colored red in each case (Figure 30).

When a nested design with stop and tail lamps is used, the luminous-intensity ratio between individual functions must be at least 5 : 1.

The Category S3 brake lamp (central high-level brake lamp) must not be a combined unit incorporating any other lamp.

The regulations set out in SAE J586, Feb. 1984, [31] and SAE J186a, Sept. 1977, [28] apply in the USA.

Figure 30: Positioning of stop lamps as per ECE
Dimensions in mm
1 Central high-level brake lamp (Category S3),
2 Two stop lamps (categories S1 and S2).
1) ≥ 400 mm if width < 1,300 mm,
2) ≤ 2,100 mm if compliance with max. height not possible, or
3) ≤ 150 mm below bottom edge of rear window,
4) However, the lower edge of the central high-level brake lamp must be higher than the upper edge of the main stop lamps.

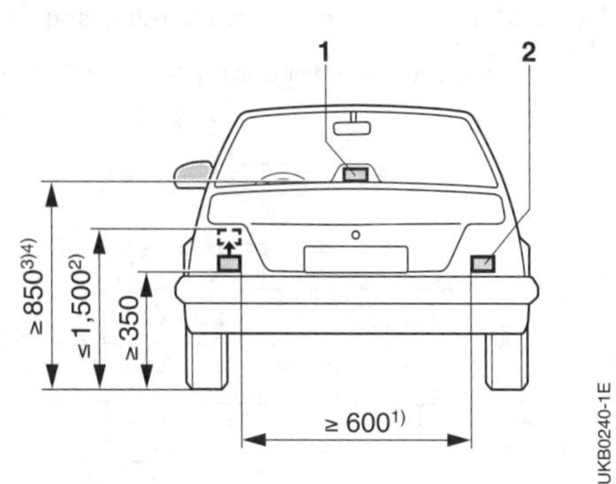

Rear fog warning lamps

(as per ECE R38 [11])
For EU/ECE countries, ECE R48 prescribes one or two red rear fog warning lamps for all new vehicles. They must be distanced at least 100 mm from the brake lamp (Figure 31).

The visible illuminated area along the reference axis may not exceed 140 cm^2. The circuit must be designed to ensure that the rear fog warning lamp operates only in conjunction with the low beam, high beam and/or front fog lamp. It must also be possible to switch off the fog warning lamps independently of the front fog lamps.

Rear fog warning lamps may only be used if the visual range due to fog < 50 m, as rear fog warning lamps have a high luminous intensity that can severely dazzle drivers at the rear when there is a clear view. The required indicator lamp is yellow in color.

Reversing (backup) lamps

(as per ECE R23 [9])
According to ECE R48, one or two white reversing lamps are permitted (Figure 32). The switching circuit must be designed to ensure that the reversing lamps operate only when reverse gear is engaged and the ignition on.

The regulations set out in SAE J593c, Feb. 1968, [35] apply in the USA.

Daytime running lamps and daytime driving lights

(as per ECE R87 [16])
ECE R87 authorizes the installation of daytime running lamps in Europe (Figure 33). Their use or the use of low-beam headlamps for daytime driving is currently being regulated in the individual states (as at 2009: Prescribed in Denmark, Norway, Finland, Sweden, Latvia, Estonia, Lithuania, Austria, Hungary, Poland, Slovakia, Slovenia, Czech Republic, Spain (outside of urban areas), and Italy (outside of urban areas)).

As of February 2011, a daytime running lamp as per ECE R87 will be required for vehicle approval of passenger cars and about a year later also for the other vehicle classes (Figure 34). This will replace the national solutions.

The regulations set out in SAE J2087, Mar. 2006, [38] apply in the USA.

Figure 32: Positioning of reversing lamps as per ECE
Dimensions in mm
Quantity: 1 or 2.

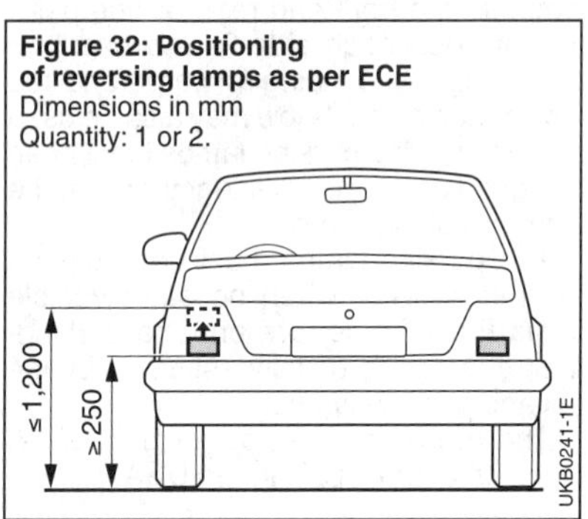

Figure 31: Positioning of rear fog warning lamps as per ECE
Dimensions in mm
1 Brake lamp, 2 Two rear fog warning lamps, 3 One rear fog warning lamp for driving on right.

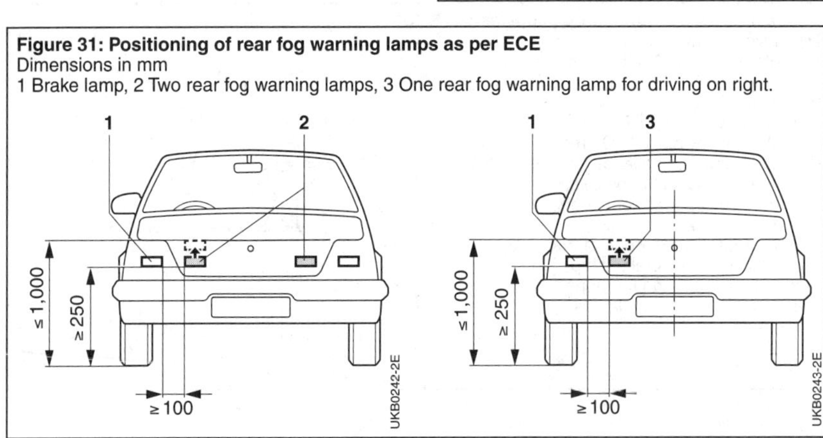

Front cornering lamps

(as per ECE R119 [22])
Two cornering lamps are permitted on the front of the vehicle which must radiate light at least 60° to the outside of the vehicle. They are used in cornering maneuvers up to a maximum speed of 40 km/h to illuminate better the destination road (e.g. side roads or garage entrances) which is usually inadequately lit by the regular headlamps. The front cornering lamps are activated via the turn-signal control stalk or when the steering wheel is turned.

Front cornering lamps typically have the same light distribution as static cornering headlamps. The switching conditions, however, are different. For the most part the same reflector is used for both the static cornering headlamps and the front cornering lamps.

The regulations set out in SAE J852, Apr. 2001, [37] apply in the USA.

Identification lamps

According to ECE R65 [14], identification lamps must be visible from any direction, and create the impression of flashing. The flashing frequency is 2 to 5 Hz. Blue identification lamps are intended for installation on official vehicles. Yellow identification lamps are designed to warn of dangers or the transport of dangerous freight.

Figure 33: Positioning of daytime running lamps as per ECE
Dimensions in mm

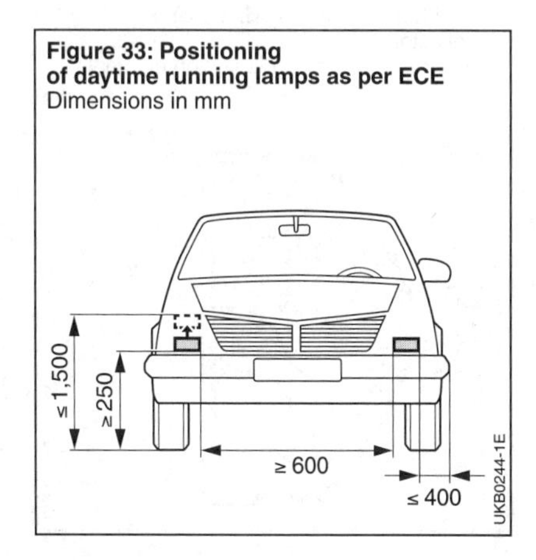

Technical design variations for lamps

Light color

Depending on their application (e.g. stop lamps, turn-signal lamps, or rear fog warning lamps), motor-vehicle lamps must display uniform, distinctive colors in the red or yellow color range. These colors are defined in specific ranges of a standardized color scale (color location).

Since white light is composed of various colors, filters can be used to weaken or filter out the emission of undesirable spectral ranges (colors) completely. The color-filter functions may be performed by either the tinted lenses of the lamp, or the colored glass bulb of the lamp (e.g. yellow bulb in turn-signal lamps with clear lens).

Filter technology can also be used to design lamp lenses in such a way that, when the lamp is switched off, the color is matched to the vehicle's paintwork. Nevertheless, existing homologation regulations are complied with when the lamp is on. Color locations have been laid down in the EU/ECE region. They correspond to a wavelength of approximately 592 nm for "yellow/orange" turn-signal lamps, for example, and to a wavelength of approximately 625 nm for "red" brake and tail lamps.

Figure 34: Headlamps and lights at the vehicle front end
A Low-beam headlamp,
F High-beam headlamp,
T Daytime running lamp,
P Position/parking lamp,
B Turn-signal lamp,
N Fog lamp (optional).
T and P are often combined.

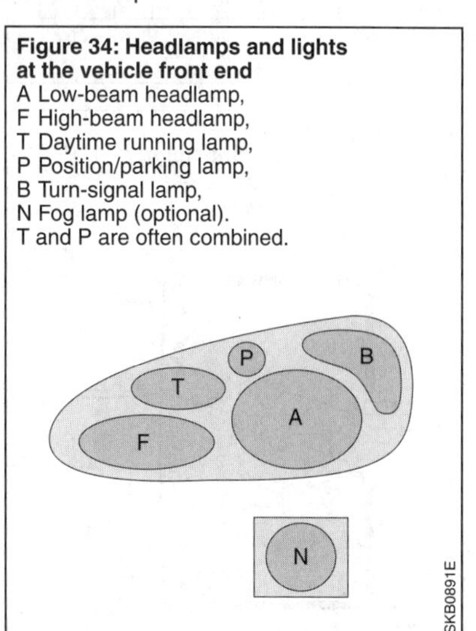

Luminous intensities

As projected along different axes, minimum and maximum luminous intensities for all lamps must remain within a range calculated to guarantee signal recognition without, however, causing glare nuisance for other road users. The same percentile basis ("unified spatial light distribution pattern") is used for most lamps. With regard to the value along the reference axis, luminous-intensity levels to the sides and above and below may be lower (Figure 35). However, these values may, depending on the installation height or for special lamps (e.g. daytime running lamps), differ from this basis.

Figure 35: ECE "lamp" test screen
Schematic diagram of
spatial light distribution pattern (values in %)

Lamps with Fresnel optics

The light from the bulb is projected directly (without diversion) on the lens, which uses Fresnel optics to shape the beam in the desired manner (Figure 36). This concept is a highly cost-effective solution in that no vapor-deposited surface is needed. The drawbacks are low efficiency and the limited vehicle-design options.

Lamps with reflector optics

Lamps with approximately parabolic reflectors or stepped reflectors direct the light from the bulb in an axial beam, and shape the beam by means of the optical dispersion elements in the lens (Figure 37).

Lamps with free-form reflectors achieve the required beam spread or light pattern, completely or partially, by directing the light by means of a reflector. The outer lens can thus be designed as a clear lens, or supplemented by cylindrical lenses in the horizontal or vertical direction.

Designs combining both the above principles have also been successfully employed. The principle of the free-form lamp with Fresnel cap combines excellent luminous efficiency with a variety of possible stylistic implementations. It is essentially the reflector that shapes the light pattern. The Fresnel cap improves luminous efficiency by diverting a proportion of the light, which would not otherwise contribute to the function of

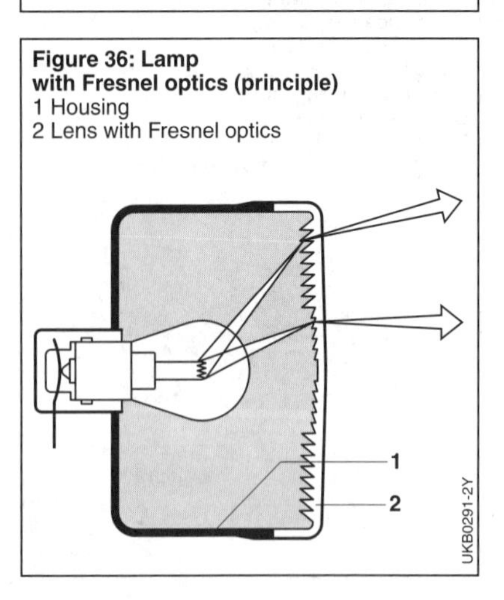

Figure 36: Lamp with Fresnel optics (principle)
1 Housing
2 Lens with Fresnel optics

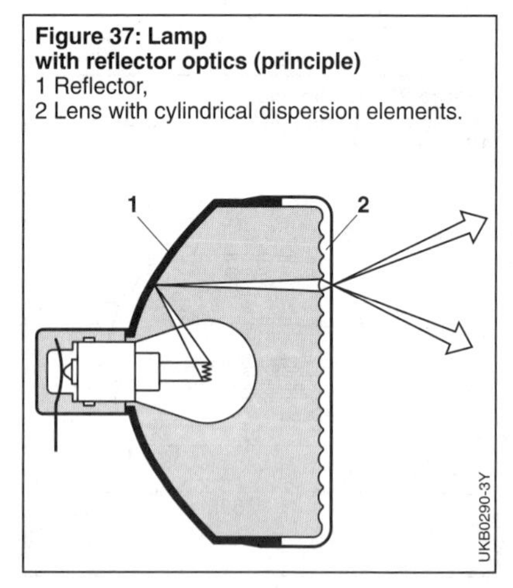

Figure 37: Lamp with reflector optics (principle)
1 Reflector,
2 Lens with cylindrical dispersion elements.

the lamp, in the desired direction (Figure 38).

Mainly lamps with free-form reflectors are used in modern vehicle designs. These lamps offer the best opportunities of adaptation to the body shape and therefore to the available structural space, and at the same time of meeting stylistic requirements.

Figure 38: Free-form lamp with Fresnel cap (principle)
1 Reflector,
2 Fresnel cap,
3 Clear lens.

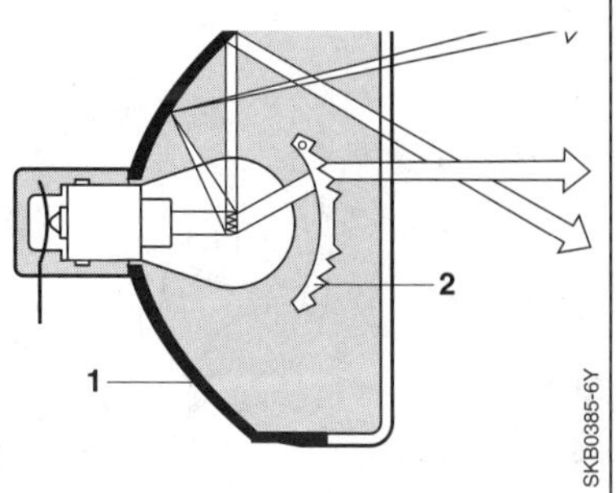

Figure 39: LED lamp with Fresnel optics (principle)
a) Vertical section,
b) Overhead view.
1 Fresnel optics, 2 LED.

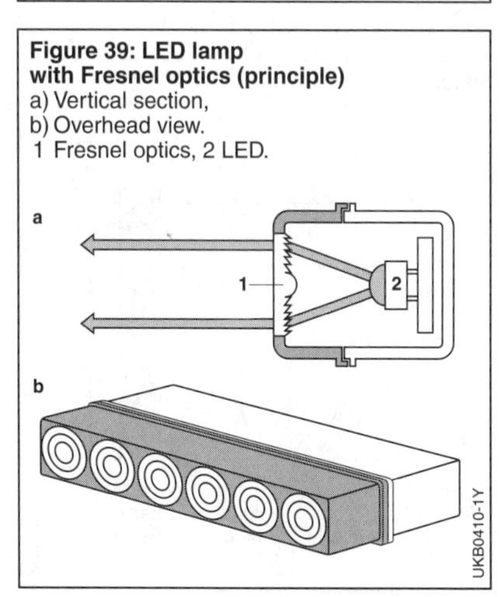

Lamps with light-emitting diodes

Light-emitting diodes (LEDs) are increasingly used as the light source for lamps. LEDs have been used for years as standard in auxiliary stop lamps. The crucial factor here is the design freedom to deliver a slimline design with a number of light sources (Figure 39 and Figure 40). In addition, LEDs in stop lamps provide an added safety aspect. An LED reaches maximum light output in less than 1 ms, whereas filament bulbs require about 200 ms to reach their nominal luminous flux. This means that LEDs emit the brake signal sooner, and this in turn reduces reaction time.

With their improved efficiency with regard to luminous flux, luminance, thermal performance and mechanical design, LEDs can also be used for functions with higher lighting requirements.

Figure 40: LED lamp
a) With attachment optics,
b) With reflector,
c) Example of a daytime running lamp.
1 Heat sink,
2 LED,
3 Optics,
4 ECU,
5 Flexboard,
6 Holder,
7 Cover.

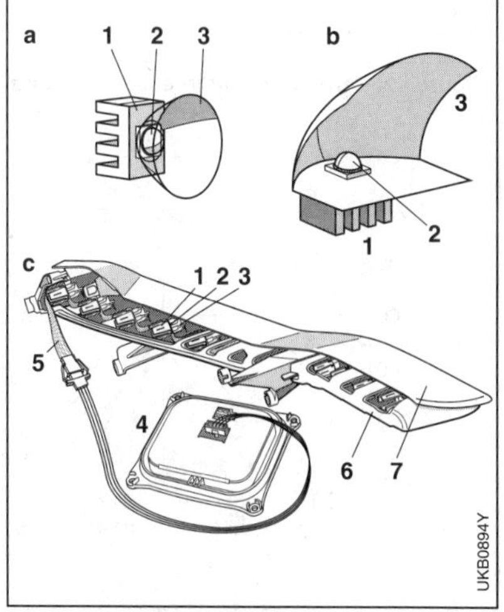

When it comes to being used as light sources in vehicle lights, LEDs differ crucially from filament and halogen lamps in that they cannot be operated directly from the vehicle electrical system. They require both a defined voltage, which ranges depending on the semiconductor material between 2.2 V and 3.6 V, and a defined current, by means of which the luminous intensity is adjusted. Resistor solutions can be used for activating functions with very low photometric requirements. Most applications require linear controllers or DC/DC converters. The electronics can be either integrated as individual electronics in the lamp or in a self-contained ECU, which is either mounted to the lamp or in the vehicle.

Lamps with fiber-optic technology

By using optical waveguides, the light source can be separated from the light emission point. In order to obtain the desired light pattern, special beam launchers in the optical waveguides, or optically active elements on or in front of the optical waveguide are required. Filament lamps can be used as light sources, but the high infrared levels are a drawback of these lamps. This necessitates the use of heat-resistant materials such as glass or heat shields. When using LEDs as "cold" light sources, it is possible to launch directly into transparent plastics such as PC (polycarbonate) or PMMA (polymethyl methacrylate). Lamps with fiber-optic technology are used primarily to produce

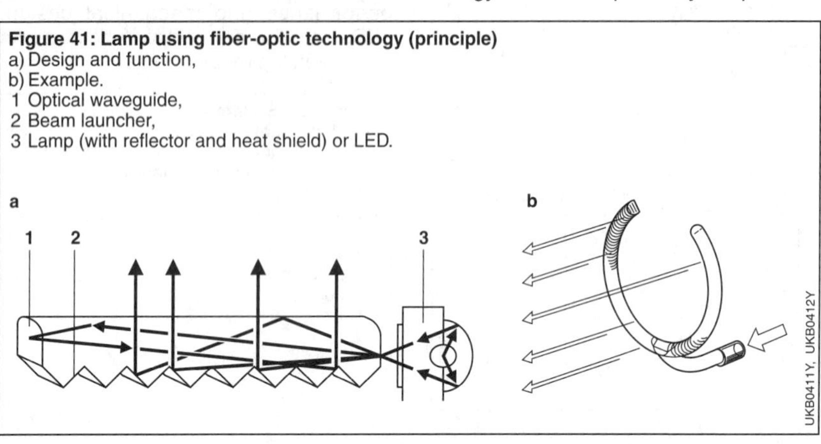

Figure 41: Lamp using fiber-optic technology (principle)
a) Design and function,
b) Example.
1 Optical waveguide,
2 Beam launcher,
3 Lamp (with reflector and heat shield) or LED.

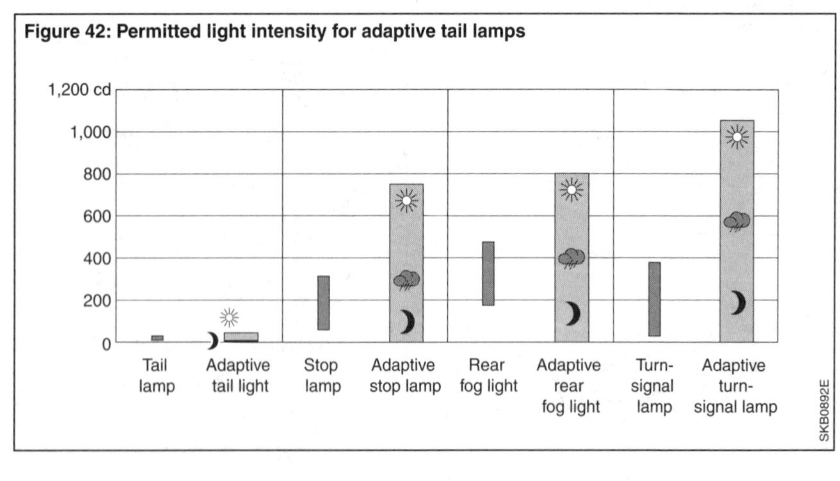

Figure 42: Permitted light intensity for adaptive tail lamps

stylistic elements such as narrow strips or thin rings (Figure 41).

Adaptive rear-lighting systems

Until now, rear-lighting functions were produced with a one-level circuit. Depending on the version and design, this resulted in a fixed luminous intensity that had to lie within the legal limit values to guarantee minimum visibility.

A large number of sensors (brightness, dirt, visual range, wetness, etc.) mean that the vehicle is nowadays able to ascertain the ambient parameters and lighting conditions more exactly. So that the optimal visibility (e.g. sufficient luminous intensity without excessive glare) can be achieved, rear light functions will be able to vary the luminous intensity they emit depending on the ambient conditions detected around the vehicle. For example, a brake lamp can be operated with high luminous intensity during strong sunlight and with lower values at night in order to ensure optimal detection and attribution of the driver's actions (Figure 42).

Convenience lamps

Lights for use when the vehicle is stationary are increasingly being fitted. Typical applications are the illumination of areas immediately outside the doors when the doors are open or closed, lights that identify the vehicle outline in the underbody area, or lights which identify the door handles.

Combinations of such lights with position lamps or fog lamps are referred to as "coming home" functions, which are activated when the doors are unlocked, for instance.

Headlamp leveling control

Headlamp adjustment for low and high beams

The correct adjustment of vehicle headlamps is an essential factor in nighttime road safety for both the driver of the vehicle concerned and for oncoming vehicles. If the beam is set only a fraction too low, there will be a substantial reduction in headlamp geometric range (Table 4). If the beam is set only a fraction too high, oncoming traffic will suffer much greater glare.

Installation and regulations

Headlamp leveling, Europe

Since 1/1/1998, an automatic or manual headlamp leveling control (beam-height adjustment) device has been mandatory in Europe for all first-time vehicle registrations, except in cases where other equipment (e.g. hydraulic suspension leveling) guarantees that light-beam inclination will remain within the prescribed tolerances. Although this equipment is not mandatory in other countries, its use is permitted.

Automatic headlamp leveling control must be designed to compensate for vehicle laden states by lowering or raising the low beam by between 5 cm/10 m (0.5%) and 25 cm/10 m (2.5%).

A manual headlamp leveling control device is operated from the driver's seat and must incorporate a detent position in the base setting; beam adjustment is also performed at this position. Units with infinitely variable and graduated control must both feature visible markings in the vicinity of the hand switch for vehicle load conditions that require vertical aiming.

Table 4. Geometric range of the horizontal component of the low beam's light-dark boundary
Headlamp installed at height of 65 cm

Inclination of the light-dark boundary (1% = 10 cm/10 m)	%	1.0	1.5	2.0	2.5	3.0
Adjustment dimension e	cm	10	15	20	25	30
Geometric range for the horizontal portion of the light-dark boundary	m	65.0	43.3	32.5	26.0	21.7

Technical design variations

All design variations employ an adjustment mechanism to provide vertical adjustment of the headlamp reflector (housing design) (Figure 43). Manually operated units employ a switch near the driver's seat to control the setting (Figure 44), while automatic units rely on level sensors on the vehicle axles to monitor suspension spring compression, and relay the proportional signals to the adjustment mechanisms.

Hydromechanical systems

Hydromechanical systems operate by transmitting a fluid through the connecting hoses between the hand switch (or level sensor) and the adjustment mechanisms. The degree of adjustment corresponds to the quantity of pumped fluid.

Vacuum systems

With vacuum systems, the hand switch (or level sensor) regulates vacuum from the intake manifold, and transmits it to the adjustment mechanisms to achieve varying degrees of adjustment.

Electrical systems

Electric systems employ electric gear motors as the adjustment mechanisms. They are switched either by switches in the vehicle or by axle sensors.

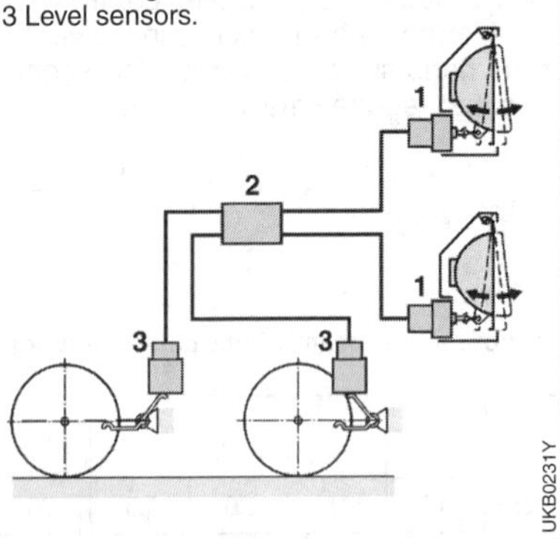

Headlamp adjustment

Headlamp adjustment, Europe

Regulations and procedure

Correct adjustment of motor-vehicle headlamps should ensure the best possible illumination of the roadway at the same time as minimizing glare for other traffic. The EU Directives (and therefore the requirements of Article 50 of the StVZO [2] (Road Traffic Licensing Regulations) in Germany) specify the required horizontal and vertical alignment of the headlamp beams. Low-beam glare is regarded as eliminated if the luminous intensity is not more than 1 lux at a distance of 25m in front of each headlamp when the beam is projected on a surface vertical to the road surface at the height of the center of the headlamp and above. If, however, the vehicle is subject to extreme variations in attitude due to changes in payload, the headlamps must be adjusted so that the desired aim is achieved.

EECE R48 [12] and EEC 76/756 [24] define the basic setting and the adjustment dimension required on the vehicle. For vehicle categories not covered by these directives, the regulations contained in ECE R48 or ECE R53 [13] shall apply.

Figure 43: Automatic headlamp leveling control (principle)
1 Adjustment mechanism,
2 Processing unit,
3 Level sensors.

Figure 44: Manual headlamp leveling (principle)
1 Adjustment mechanism,
2 Hand switch.

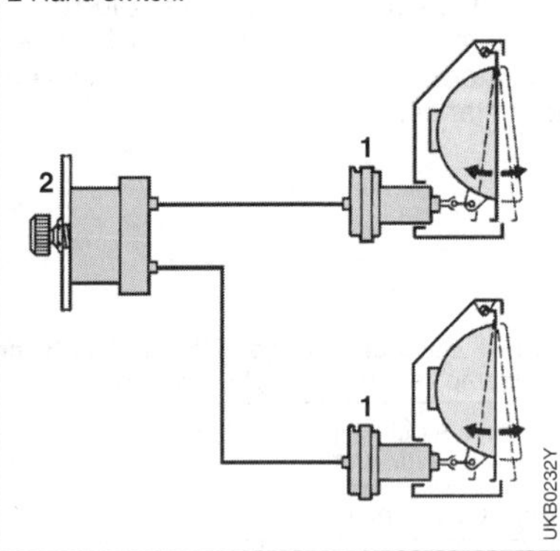

Preparations for adjustment

Laden state of vehicle

- Motor vehicles excluding motorcycles: unladen, 75 kg (one person in driver's seat)
- Motorcycles: for EC (as per 93/92/ EEC [25]) unladen, without a person on the driver's seat; for ECE and StVZO (Road Traffic Licensing Regulations) unladen, 75 kg (one person on the driving seat).

Suspension

- Vehicles without self-leveling suspension are rolled for few meters or rocked so that the suspension settles to a level position
- Vehicles with self-leveling suspension are set and operated to the correct level according to the operating instructions.

Tire pressure

The tire pressure must be adjusted as per the vehicle manufacturer's instructions in accordance with the laden state.

Compulsory function checks on headlamp-range adjustment systems

- Systems that operate automatically should be set or operated according to the manufacturer's instructions.
- On vehicles registered before 1/1/1990, a positively locating adjuster position is not required.
- Manually operated systems with two positively locating adjuster positions: On vehicles on which the headlamp beam is raised as the vehicle payload is increased, the adjuster must be set to the position at which the beam is at its highest (smallest degree of dip). On vehicles on which the headlamp beam is lowered as the vehicle payload is increased, the adjuster must be set to the position at which the beam is at its lowest (smallest degree of dip).

Surface and testing environment

- The vehicle and headlamp tester must be standing on a flat surface (based on ISO 10604 [26]).
- Adjustments and tests should be carried out in an enclosed space where the lighting conditions are not too bright.

Adjusting and testing with a headlamp aiming device

- The headlamp aiming device should be aligned at the specified distance in front of the headlamp under test (except in the case of automatic equipment).
- In the case of headlamp aiming devices which do not run on tracks, vertical alignment relative to the vehicle's longitudinal center axis should be carried out separately for each headlamp. The test should then be carried out without moving the unit sideways again.
- In the case of headlamp aiming devices which run on tracks (or the like), alignment with the vehicle's longitudinal center axis need only be carried out once at the most favorable position (e.g. in a central position in front of the vehicle).
- The specified adjustment dimension for the headlamp concerned should then be set on the headlamp aiming device, and the headlamp adjustment checked and the headlamp adjusted to the correct setting.

Adjusting and testing with a test surface

- The test surface must be vertical to the surface on which the vehicle is standing and at right angles to the vehicle's longitudinal center axis.
- The test surface should be finished in a light color, adjustable vertically and horizontally, and marked with the markings shown in Figure 45.
- The test surface should be positioned at a distance of 10 m in front of the vehicle so that the center mark is aligned with the center of the headlamp under test or adjustment (Figure 46). In the case of lamps with very low beams (e.g. fog lamps), a shorter distance can be chosen by calculating the adjustment dimension accordingly.
- Each headlamp must be set individually. Therefore, the other headlamp(s) must be covered over.
- The vertical position of the test surface must be set so that the top cutoff (parallel to the ground) is at the height $h = H - e$. If the surface is not 10 m from the vehicle, the setting e must be converted accordingly.

Notes on adjustment

In the case of headlamps with asymmetrical lower beams and fog lamps, the highest position of the light-dark boundary must be on the top cutoff and run as horizontally as possible across the minimum width of the test surface. The lateral adjustment of the headlamps must be such that the light pattern is positioned as symmetrically as possible about the perpendicular line running through the center mark.

In the case of headlamps for asymmetrical lower beam, the light-dark boundary must touch the top cutoff to the left of the center. The intersection between the left section (which should be as horizontal as possible) and the sloping section on the right of the light-dark boundary must be located at the perpendicular line passing through the center mark.

Figure 45: Test surface for headlamp beam
1 Top cutoff, 2 Center mark, 3 Test surface, 4 Break point.
H Height of headlamp center above standing surface in cm.
h Height of test-surface top cutoff above standing surface in cm.
$e = H - h$ Adjustment dimension.

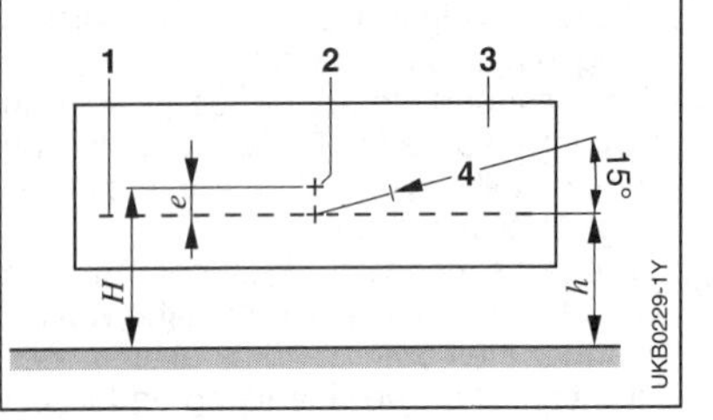

Figure 46: Relative positions of test surface and vehicle longitudinal axis
1 Center mark,
2 Test surface.
A Distance between headlamp centers.

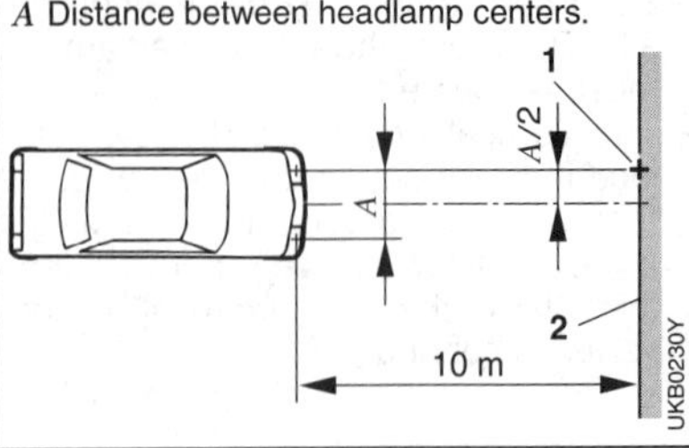

The center of the high beam must be on the center mark.

For headlamps which feature a common adjustment facility for low beam and fog lamps or for high beam, low beam and fog lamps, it is always necessary to use the low-beam headlamp as the basis for adjustment (adjustment dimension e, see Table 5).

Headlamp aiming devices

Function

Correct adjustment of motor-vehicle headlamps should ensure the best possible illumination of the roadway by the low beam while minimizing glare for oncoming traffic at the same time. For this purpose, the inclination of the headlamp beams with respect to a level base surface, and their direction to the vertical longitudinal center plane of the vehicle, must satisfy official directives.

Equipment design

Headlamp aiming devices are portable imaging chambers (Figure 47). They comprise a single lens and an aiming screen located in the focal plane of the lens, and are rigidly connected to it. The aiming screen has markings to facilitate correct headlamp adjustment, and can be viewed by the equipment operator using suitable devices such as windows and adjustable

Table 5: Headlamp adjustment (excerpt from StVZO (Road Traffic Licensing Regulations))

Vehicle type: Motor veh., > 2 whs. Headlamp position: Height above road surface	Adj. dimension "e"	
	Dipped beam	Fog beam
Approved to 76/756/EEC or ECE R48 and StVZO, first registered on 1/1/1990 or later, < 1,200 mm	Setting on vehicle e.g. ≣D1.0%	–2.0 %
First registered 12/31/1989 or earlier, ≤ 1,400 mm, and first registered after 12/31/1989, > 1,200 mm, but ≤ 1,400 mm	–1.2 %	–2.0 %

refraction mirrors. The prescribed headlamp adjustment dimension e, i.e. the inclination relative to the centerline of the headlamp in cm at a fixed distance of 10 m, is set by turning a knob to move the aiming screen (Tables 4 and 5).

The aiming device is aligned with the vehicle axis using a sighting device, such as a mirror with an orientation line. It is turned and aligned so that the orientation line uniformly touches two external vehicle-reference marks. The imaging chamber can be moved vertically and clamped at the level of the vehicle headlamp.

Headlamp testing

The headlamp can be tested after the equipment has been correctly positioned at the front of the lens. An image of the light pattern emitted by the headlamp appears on the aiming screen. Some test devices are also equipped with photodiodes and a display to measure the luminous intensity.

On headlamps with asymmetrical lower-beam patterns, the light-dark boundary should touch the horizontal top cutoff; the intersection between the horizontal and sloping sections must be located on the perpendicular line running through the center mark (Figure 48). After adjusting the lower-beam light-dark boundary in accordance with the regulations, the center of the upper beam (assuming that high and low beam are adjusted together) should be located within the rectangular border about the center mark.

Headlamp adjustment, USA

For headlamps compliant with US Federal legislation, visual (vertical only) adjustment as permitted since 5/1/1997 has become increasingly widespread in the US since mid-1997. There is no horizontal aiming here.

Before then, the use of mechanical aiming devices was the most common method of headlamp adjustment in the USA. The headlamp units were equipped with three pads on the lenses – one for each of the three adjustment planes. A calibrating unit is placed against these pads. Aiming is checked using spirit levels. With the VHAD (Vehicle Headlamp Aiming Device) method permitted since 1993, the headlamps are adjusted relative to a fixed vehicle reference axis. This procedure is carried out using a spirit level firmly attached to the headlamp. As a result, the three lens pads were no longer required.

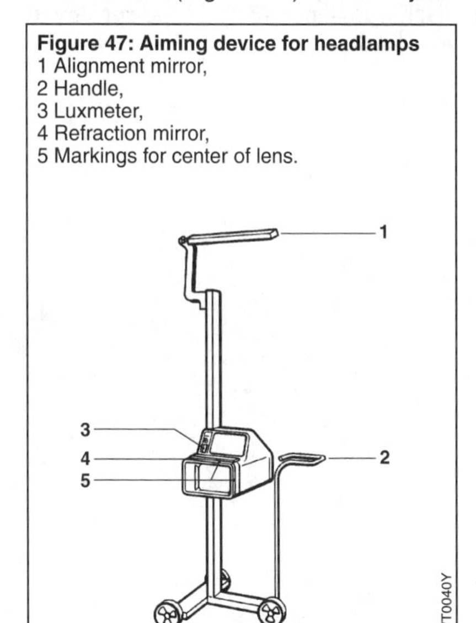

Figure 47: Aiming device for headlamps
1 Alignment mirror,
2 Handle,
3 Luxmeter,
4 Refraction mirror,
5 Markings for center of lens.

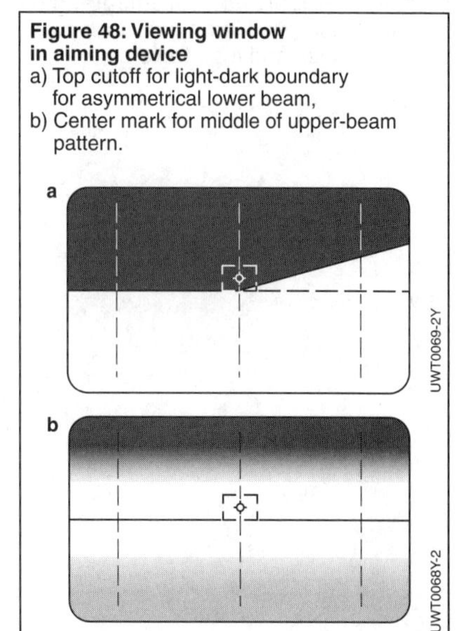

Figure 48: Viewing window in aiming device
a) Top cutoff for light-dark boundary for asymmetrical lower beam,
b) Center mark for middle of upper-beam pattern.

References

[1] StVZO §22a: Design certification for vehicle parts.
[2] StVZO §50: Headlight for high and low beams.
[3] StVZO §52: Zusätzliche Scheinwerfer und Leuchten.
[4] ECE R3: Uniform provisions concerning the approval of retro-reflecting devices for power-driven vehicles and their trailers.
[5] ECE R4: Uniform provisions concerning the approval of devices for the illumination of rear registration plates of power-driven vehicles and their trailers.
[6] ECE R6: Uniform provisions concerning the approval of direction indicators for power-driven vehicles and their trailers.
[7] ECE R7: Uniform provisions concerning the approval of front and rear position (side) lamps, stop-lamps and end-outline marker lamps for motor vehicles (except motorcycles) and their trailers.
[8] ECE R19: Uniform provisions concerning the approval of motor vehicle front fog lamps.
[9] ECE R23: Uniform provision concerning the approval of reversing lamps for power-driven vehicles and their trailers.
[10] ECE R37: Uniform provisions concerning the approval of filament lamps for use in approved lamp units on power-driven vehicles and their trailers.
[11] ECE R38: Uniform provisions concerning the approval of rear fog lamps for power-driven vehicles and their trailers.
[12] ECE R48: Uniform provisions concerning the approval of vehicles with regard to the installation of lighting and light-signalling devices.
[13] ECE R53: Uniform provisions concerning the approval of category L3 vehicles with regard to the installation of lighting and light-signaling devices.
[14] ECE R65: Uniform provisions concerning the approval of special warning lamps for motor vehicles.
[15] ECE R77: Uniform provisions concerning the approval of parking lamps for power-driven vehicles.
[16] ECE R87: Uniform provisions concerning the approval of daytime running lamps for power-driven vehicles.
[17] ECE R91: Uniform provisions concerning the approval of side-marker lamps for motor vehicles and their trailers.
[18] ECE R98: Uniform provisions concerning the approval of motor vehicle headlamps with gas-discharge light sources.
[19] ECE R99: Uniform provisions concerning the approval of gas-discharge light sources for use in approved gas-discharge lamp units of power-driven vehicles.

[20] ECE R112: Uniform provisions concerning the approval of motor vehicle headlamps emitting an asymmetrical passing beam or a driving beam or both and equipped with filament lamps and/or light-emitting diode (LED) modules.
[21] ECE R113: Uniform provisions concerning the approval of motor vehicle headlamps emitting a symmetrical passing beam or a driving beam or both and equipped with filament lamps.
[22] ECE R119: Uniform provisions concerning the approval of cornering lamps for power-driven vehicles.
[23] ECE R123: Uniform provisions concerning the approval of adaptive front-lighting systems (AFS) for motor vehicles.
[24] 76/756/EEC: Council Directive of 27 July 1976 on the approximation of the laws of the Member States relating to the installation of lighting and light-signaling devices for motor vehicles and their trailers.
[25] 93/92/EEC: Council Directive of 29 October 1993 on the installation of lighting and light-signalling devices on two or three-wheeled motor vehicles.
[26] ISO 10604: Road vehicles; measurement equipment for orientation of headlamp luminous beams.
[27] FMVSS 108: Lamps, reflective devices, and associated equipment.
[28] SAE J186: Supplemental High Mounted Stop and Rear Turn Signal Lamps for Use on Vehicles Less than 2032 mm in Overall Width.
[29] SAE J222: Parking Lamps (Front Position Lamps).
[30] SAE J585: Tail Lamps (Rear Position Light) for Use on Motor Vehicles Less than 2032 mm in Overall Width.
[31] SAE J586: Stop Lamps for Use on Motor Vehicles Less Than 2032 mm in Overall Width.
[32] SAE J587: License Plate Illumination Devices (Rear Registration Plate Illumination Devices).
[33] SAE J588: Turn Signal Lamps for Use on Motor Vehicles Less Than 2032 mm in Overall Width.
[34] SAE J592e: Clearance, Side Marker and Identification Lamps.
[35] SAE J593c: Back-up Lamps.
[36] SAE J594: Reflex Reflectors.
[37] SAE J852: Front Cornering Lamps for Use on Motor Vehicles.
[38] SAE J2087: Daytime Running Light.
[39] D. Meschede: Gerthsen Physik. 24th Edition, Springer-Verlag, 2010.
[40] R. Baer: Beleuchtungstechnik – Grundlagen. 3rd Edition, Huss-Medien-GmbH, Verlag Technik, 2006.

Automotive glazing

The material properties of glass

Basic components

Window panes for automotive use are made of silica glass. The basic chemical constituents and their proportions are as follows:

- 70 to 72% silicic acid (SiO_2) as the basic component of glass
- Approx. 14% sodium oxide (Na_2O) as flux
- Approx. 10% calcium oxide (CaO) as stabilizer.

These substances are mixed in the form of quartz sand, soda ash, and limestone. Other oxides such as magnesium and aluminum oxide are added to the mixture in proportions of up to 5%. These additives improve the physical and chemical properties of glass.

Manufacturing flat glass

The glass panes are made out of the basic product, flat glass. Flat glass that is cast using the float-glass process is used. This process involves melting the mixture at a temperature of 1,560 °C. The melt then passes through a refining zone at 1,500 to 1,100 °C, and is then floated on a float bath of molten tin. The molten glass is heated from above (smoothing of the surface by fire-finishing). The flat surface of the molten tin creates flat glass with flat parallel surfaces of a very high quality (tin bath underneath, fire-finishing on top). The glass is cooled to 600 °C before it is lifted out of the float bath into the cooling section. After a further period of slow, non-stress cooling, the glass is cut into sheets measuring 6.10 × 3.20 m^2.

Tin is suitable for the float-glass process because it is the only metal that does not produce any vapor pressure at 1,000 °C and is liquid at 600 °C.

Table 1: Material properties and physical data of glass and finished windshields and windows

Property	Dimension	TSG	LSG
Density	kg/m³	2,500	2,500
Hardness	Mohs	5 to 6	5 to 6
Resistance to pressure	MN/m²	700 to 900	700 to 900
E module	MN/m²	68,000	70,000
Flexural strength			
Before pre-tension	MN/m²	30[2]	30[1]
After pre-tension	MN/m²	50[2]	
Specific heat	kJ/kg · K	0.75 to 0.84	0.75 to 0.84
Thermal conduction coefficient	W/m · K	0.70 to 0.87	0.70 to 0.87
Coefficient of thermal expansion	K^{-1}	$9.0 \cdot 10^{-6}$	$9.0 \cdot 10^{-6}$
Relative permittivity		7 to 8	7 to 8
Light transmittance (DIN 52306) clear[3]	%	≈ 90	≈ 90[1]
Refractive index[3]		1.52	1.52[1]
Deviation angle of wedge[3]	Arc minute	< 1.0 flat < 1.5 curved	≤ 1.0 flat[1] ≤ 1.5 curved[1]
Dioptric divergence DIN 52305[3]	Diopters	< 0.03	≤ 0.031[1]
Thermal stability	°C	200	90[1] (max. 30 min)
Resistance to temperature shocks	K	200	

[1] Properties of finished laminated safety glass (LSG).
In calculating the permissible bending stress, the coupling effect of PVB film is to be disregarded.

[2] Calculated values; these values already contain the necessary safety margins.

[3] Figures for optical properties depend very heavily on the type of window.

Automotive windshield and window glass

Glass used for automotive glazing is of two types:
- Single-pane toughened safety glass (TSG), which is chiefly used for side-window, rear-window and sunroof glazing.
- Laminated safety glass (LSG), which is used primarily for windshields and rear windows, but also for sunroofs. LSG is also increasingly being fitted in vehicle side and rear windows.

The materials from which TSG and LSG panes are made are the basic glass types:
- Transparent float glass: This glass offers the best possible light transmittance.
- Tinted float glass: This glass has a homogeneous green or gray tint within the material; the tint blocks heat from the sun.
- Coated float glass: The glass is coated on one side with noble-metal and metal oxides; the coating reduces heat and UV radiation entering the vehicle, thus providing thermal insulation.

TSG panes

TSG panes differ from LSG panes as they have greater mechanical and thermal strength, and their breaking and shattering behavior is different. They pass through a toughening process which greatly prestresses the surface of the glass. In case of breakage, these panes shatter into many small blunt-edged pieces (to eliminate the risk of injury).

Post-processing (by grinding or drilling) of TSG panes is not possible. The standard thicknesses are 3, 4 and 5 mm.

LSG panes

An LSG pane is made of a crack-proof, flexible plastic intermediate layer of polyvinyl butyral (PVB) bonded between two sheets of glass. When subjected to impact or shock, the glass splits into web-like crack patterns. The plastic intermediate layer holds the broken pieces of glass together (to eliminate the risk of injury). The laminate retains its integrity and transparency when the glass is shattered.

The standard thicknesses for LSG panes are 4.5 to 5.6 mm.

Optical properties

The requirements for the optical quality of automotive glazing are as follows.
- Unimpeded vision
- Flawless vision
- Undistorted vision.

Achievement of optimum optical quality has to be balanced against structural requirements and the vehicle-body design, taking account of such factors as
- glazing with large surface areas,
- glazing that is fitted laying flat,
- cylindrical or spherical panes,
- panes with a high degree of curvature.

Possible quality impairments arise from:
- Optical deflection
- Optical distortion
- Double imaging.

Optical deflection increases with:
- Increasing obliqueness of the angle of incidence, i.e. increasing slope of the window
- Increasing pane thickness
- Reducing radius of curvature (increasing degree of bend)
- Increasing divergence from perfect surface parallelism of the original glass sheet.

Green- or gray-tinted glass is used as heat absorption glass because it blocks the transmission of infrared rays (heat radiation) more strongly than shorter wavelengths. On the other hand, it also reduces transmittance within the visual spectrum. The PVB film in LSG absorbs light in the ultraviolet range.

The optical properties of TSG and LSG panes are roughly the same because the optical properties of the intermediate plastic layer in LSG are very similar to those of glass in the visible spectrum.

Functional design glazing

The demands placed on glazing are continually increasing. The flat panes used in the past simply served to protect occupants from wind and weather. Now, automotive glazing performs a wide variety of functions.

Tinted glazing

This is made from glass which is tinted within the material, and reduces the direct penetration of solar radiation into the vehicle interior.

The reduction in the transmittance of solar energy occurs chiefly in the long-wavelength spectrum (infrared, Figure 1), and this mainly lowers the transmittance of energy, thus reducing the heat transmitted to the vehicle interior. The degree to which light transmittance within the visible spectrum is affected depends on the degree of color tinting and the thickness of the glass.

For windshields, light transmittance must be at least 75 %. Strongly tinted glass with a light transmittance of less than 70 % can be used in windows from the B-pillar to the rear, if the vehicle has two exterior mirrors. For sunroof glazing, tinted glass with a significantly lower light transmittance and a UV transmittance of ≤ 2 % is used.

Coated glazing

This type of glazing uses glass with a metal or metal-oxide coating. Depending on the production process, the coating may be carried out before or after the sheet of glass is bent and prestressed. The coating is applied to the inner glass surface of an LSG pane.

This type of coated glazing has a light transmittance of less than 70 %. Consequently, it can be fitted from the B-pillar to the rear, if the vehicle has two exterior mirrors.

Coated glass can also be used for sunroof glazing. Another type of sunroof glazing uses pyrolytically coated windows which are post-processed after the coating has been applied.

Coated glazing reduces direct solar radiation into the vehicle interior and absorbs solar energy in the infrared and ultraviolet ranges.

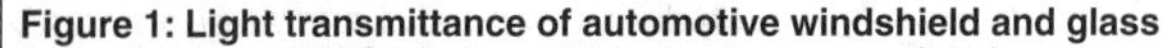

Figure 1: Light transmittance of automotive windshield and glass
1 Float glass and TSG windows, thickness 4 mm, non-tinted,
2 LSG windows, overall thickness 5.5 mm, non-tinted,
3 LSG windows, overall thickness 5.5 mm, green.

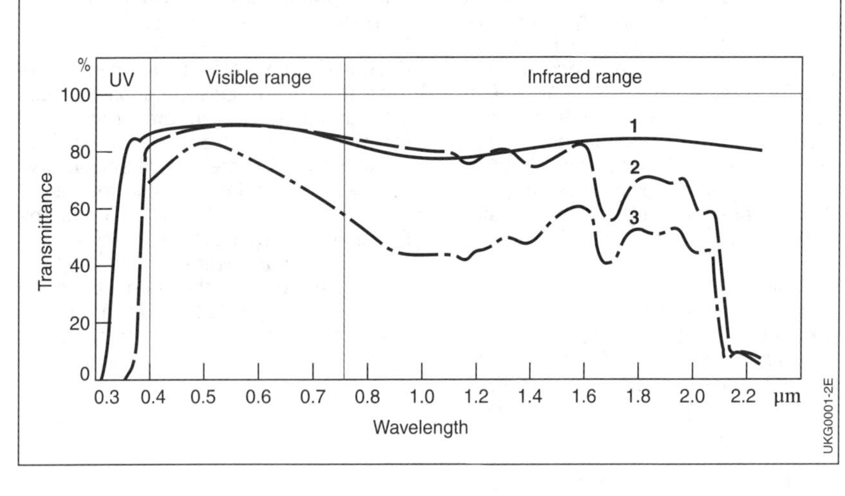

Windshields with sunshield coating

A coating is applied to the inner surface of the outer or inner layer of glass in a laminated glass pane. The coating is a multilayer interference system with silver as its base layer. As the coating is on the inside of the laminate, it is permanently protected against corrosion and scratching.

The purpose of the coating is to reduce the transmittance of solar energy by more than 50%. thus reducing the heat transmitted to the vehicle interior. Transmittance is reduced primarily in the infrared range so that visible light transmittance is altered negligibly. The reduction is achieved largely by reflection, so that secondary reflection into the interior is low. The light transmittance of this glass in the UV range is very low, i.e. less than 1%.

Sunroof glazing made from laminated safety glass

The bent laminated safety glass consists of two tinted-glass layers which are thermally partially prestressed in order to increase mechanical strength. They are bonded to either side of a highly crack-resistant and specially tinted film. The overall thickness depends on the surface area of the glass, and the overall design of the sunroof.

Absorption that takes place mainly in the infrared range guarantees minimum heat penetration. The coating also provides low light transmittance and complete filtering of UV light.

Automotive insulation glass

Automotive insulation glass is made of two flat or curved sheets of single-pane toughened safety glass (3 mm) separated by an air gap (3 mm). This type of insulation glass reduces heating of the vehicle interior, particularly in combination with a tinted coating. Transmittance is reduced primarily in the infrared range so that visible light transmittance is altered negligibly.

Insulation glass also provides better thermal insulation in the winter and improves sound insulation.

Insulation glass is now only used on commercial vehicles such as buses, trains and airplanes.

Heated laminated safety glass

Heated laminated safety glass can be used for windshields or rear windows. They prevent icing and misting of the glass pane, even in extreme winter temperatures, and ensure clear visibility.

Heated laminated safety glass is made up of two more sheets of glass with a PVB film bonded between them. Inside the PVB film, there are heater filaments which may be less than 20 μm thick, depending on the heat output required. The heater filaments may be laid in a waveform pattern or in straight lines. They may also run vertically or horizontally. The heated area may cover the whole of the pane or may be divided into separate zones with varying heat outputs. In this way, the windshield wipers can be prevented from freezing to the windshield in subzero temperatures.

Heated windows can also be created by layering.

Automotive antenna glass
This type of glass has antenna wire embedded in it. In the case of windows made of single-pane toughened safety glass (roof and side windows), the antenna is printed on the glass and is located – almost invisibly – on the inner surface of the window. With laminated safety glass windows (windshields), the antenna wiring system is embedded or printed onto the inner film.

Acoustic glazing
An acoustic windshield is composed, like every windshield, of two individual sheets of glass – an outer and an inner sheet which are firmly connected to each other by way of a sturdy plastic film of polyvinyl butyral (PVB). In the case of acoustic glazing, however, the conventional PVB film is replaced by acoustic PVB which has a highly damping acoustic core located in the middle between two layers of standard PVB. The two outer layers of film of conventional plastic thus safeguard the mechanical properties, while the core layer of damping material absorbs vibrations.

Acoustic windshields reduce the transmission of low-frequency engine vibrations into the vehicle interior. Engine-vibration noises are absorbed by the windshield, these booming noises thus being reduced at low frequencies by up to 5 dB.

Water-repellent glazing
A coating on the glass causes tiny pearls of water to form on the glass surface when it rains. These are blown off the glass by the air stream.

All glass surfaces can be fitted with this function. After several years of use, the effectiveness of water-repellent glazing can be fully restored with a regenerating kit.

Electrochromic glazing
The standard version of electrochromic glass consists of two 2.1 mm thick glass panes and a thin intermediate layer with an overall thickness of 5.8 mm and a weight of 16 kg/m^2. Depending on the size of the glass, the light transmittance can be changed entirely in 30 to 60 seconds. This reaction time remains constant over a temperature range of −25 to +90 °C. The glass remains transparent in every tint, i.e. it never becomes opaque. A ±1.5 V DC power supply is required to activate the glass. Power consumption is extremely low at less than 0.1 Wh/cycle/m^2.

Preferred areas of application for electrochromic glazing are glass roofs, rear windows and triangular windows from the C-pillar.

Panorama roofs
There is a wide selection of panorama roofs available, ranging from the closed, large-surface glass roof, through the twin- or multi-panel roof, to the lamellar sunroof.

The following glass technologies can be used for panorama roofs:
- Single-pane safety glass, 5 mm or 4 mm thick
- Conventional laminated safety glass, 6 mm and 5 mm thick
- Partially prestressed laminated safety glass, 5 mm thick.

Windshield, rear-window and headlamp cleaning systems

The function of systems for cleaning the windshield and rear window is to provide the driver with sufficient visibility in the motor vehicle at all times. The following types of system are used:
- Windshield wiper systems
- Rear-window wiper systems
- Washer systems in combination with wiper systems
- Headlamp washer systems.

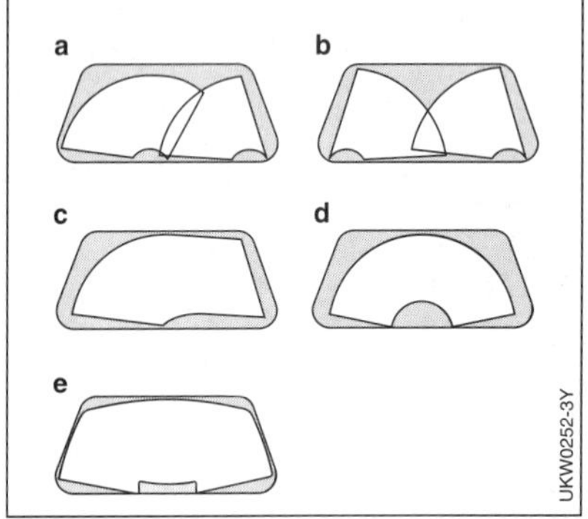

Figure 1: Windshield cleaning systems
a) Tandem system,
b) Opposed-pattern system,
c) Wiper-blade-controlled single-arm system,
d) Non-controlled single-arm system (rear wiper),
e) Stroke-controlled single-arm system.

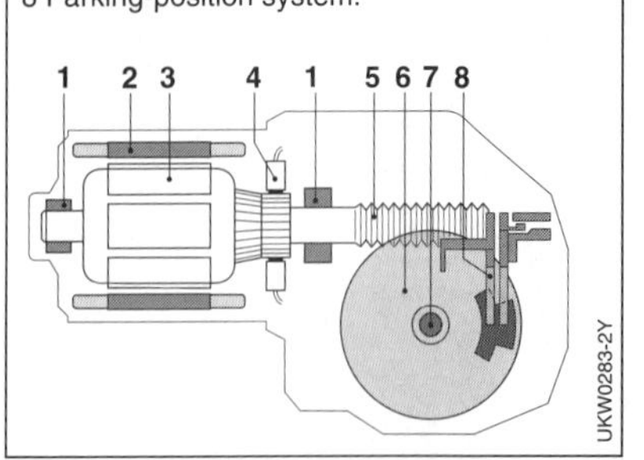

Figure 2: Wiper drive with helical-gear mechanism
1 Bearing, 2 Magnet, 3 Armature,
4 Commutator with carbon brushes,
5 Worm, 6 Helical gear, 7 Output shaft,
8 Parking-position system.

Windshield wiper systems

Function and requirements

The function of wiper systems is to remove water, snow and dirt (mineral, organic or biological) from the windshield and rear window. Boundary conditions are:
- Operational at high temperature (+80 °C) and low temperature (–40 °C)
- Corrosion resistance against acids, bases, salts, ozone
- Endurance strength under load (e.g. due to high snow load)
- Pedestrian protection
- Cleaning effect at high driving speeds
- Low operating noise.

Legislative bodies have translated the demand for adequate visibility into standardized areas of vision on the windshield (e.g. EEC for Europe [1] and FMVSS for the USA [2]). These are subdivided into several areas and must be cleaned by the wiper system to fixed percentage levels. The most important cleaning systems for passenger-car windshields which satisfy these requirements are shown in Figure 1.

Wiper systems for commercial vehicles are similar to those for passenger cars, but are subject to different requirements especially with regard to driving speed and windshield shape.

Drive

Windshield wiper systems consist of an electric motor with a helical-gear mechanism (drive), a linkage, the wiper bearings, the wiper arms, and the wiper blades.

Motor design

Permanent-magnet DC motors with integrated helical-gear mechanisms (Figure 2) are used to drive wiper systems. The motors are for cost, weight and space reasons high-speed in design. Adaptation to the speed and torque requirements of the wiper system is performed via the helical-gear mechanism.

The primary function of wiper drives is to guarantee visibility under anticipated conditions, i.e. sufficiently frequent windshield wiping. For windshield wiper systems, this generally means roughly 40 times per minute and under more extreme conditions approximately 60 times. In the standard SAE J903 [3], this is laid down with the requirement of a speed setting of 45 and a speed jump of at least 15 wiping cycles per minute.

The classification by power output often used for electrical machines is not suitable for drives for wiper systems. Wiper systems are normally operated on a wet windshield; in this case, minimal torque is demanded of the motor at high rotational speeds. On the other hand, in the case of high-load states such as when the wiper blade is frozen, a high torque is required at low rotational speed. In both cases, the power output (product of torque and rotational speed) is low. Because the drive variable is dependent on the required torque at the high load point, this torque is used as the classification feature.

Conventional drives

Conventional drives (rotary drives) are characterized by the fact that the output shaft permanently rotates in the same direction. The actual wiping movement on the windshield is determined by the wiper linkage and its kinematic layout (Figure 3).

The different speed settings are achieved in such systems due to the fact that there is a third carbon brush connected to the battery voltage in addition to the carbon brushes connected to ground and to the battery voltage. This third carbon brush is at a defined angle to the other brushes and results through its commutation angle in a pull-out of the motor characteristic (Figure 4a).

The wiper drives have a sensor system to ensure that the wiper blades adopt the correct parking position. Because the position of the output shaft of the conventional drive has a 1:1 correlation with the wiping angle, this function can be guaranteed

Figure 3: Conventional (rotary) drive
a) Parallel drive, b) Series drive.
1 Motor crank, 2 Articulated rod,
3 Oscillating-crank link as ball pivot,
4 Oscillating crank,
5 Wiper bearing,
6 Wiper arm.
α Angle of motor rotation,
β, γ Wiping angle (β and γ can be different).
The arrows indicate the direction of motion.

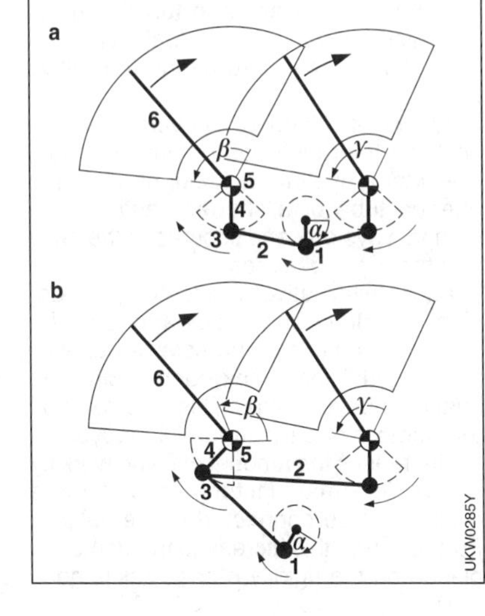

Figure 4: Drive characteristics of wiper drives
a) Conventional wiper drive,
b) Reversing wiper drive.
1 Speed setting 1,
2 Current setting 1,
3 Speed setting 2,
4 Current setting 2,
5 Current, reversing mode.

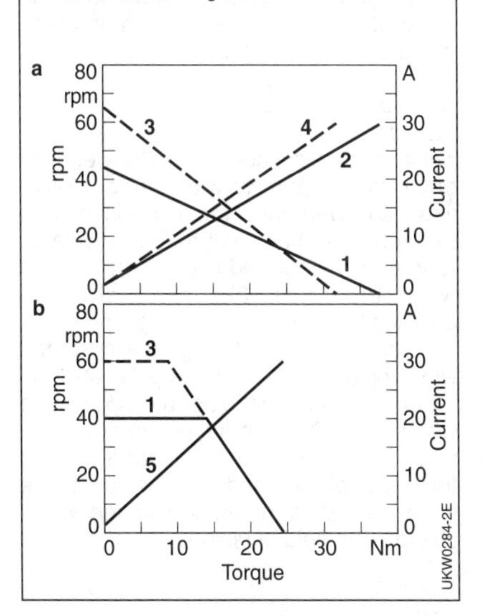

with a reference position at the output. Microswitches with cam activation, sliders with contact disks or Hall-effect sensors with a sensor magnet are used as the parking-position system.

Wiper drives are designed with high-load and anti-blocking protection. This protection can be provided conventionally by a thermostatic switch or by blocking or heavy-load detection in the triggering electronics.

Reversing drives
In the case of reversing drives, the output shaft oscillates at a defined angle, usually less than 180°. Output to the wiper shafts is provided, as is also the case with conventional systems, via a wiper linkage.

Reversing drives, unlike conventional rotary drives, have only two carbon brushes. The motor voltages required to effect the different wiping frequencies (Figure 4b) are provided via the integrated control electronics.

The position and speed of the output is relevant to controlling the drive. These variables are recorded by sensors (e.g. Hall-effect sensors with sensor magnet) and processed in the control electronics. The signals from this sensor system are usually used to detect the parking position and spare the parking-position system of conventional drives. High-load and blocking protection is likewise guaranteed by the integrated electronics.

Linkage

Function
The linkage is the connection between the drive and one or more wiper arms. It transfers the movement of the drive via the motor crank with ball pivot to the articulated rods and via these to the ball pivot of the wiper arm (Figure 3). The decisive variable for transfer is the wiping angle, i.e. the angle between the upper stop of the wiper blade close to the A-pillar and the lower stop close to the bottom edge of the windshield. This produces - in combination with the length of the wiper arms and the wiper blades and the position of the wiper bearings in relation to the windshield – the field of vision.

System concepts
With regard to the linkages, a distinction is made between purely mechanical wiper systems (rotary drives) and electronically controlled wiper systems based on reversing technology. The following list sets out the four most important concepts for current wiper systems.

- Rotary drive: The drive runs in a circle and the linkage transfers the circular motion into the oscillating motion of the wiper arms (Figure 3).
- Reversing technology: The electronically controlled drive rotates less than a half rotation. The linkage transfers the force from the drive to the wiper arms. The advantage of such systems over the rotary drive lies in the fact that the space required for the linkage kinematics is roughly halved.
- Two-motor wiper system: Each wiper arm is moved by its own drive with linkage.
- Direct wiper drive: The wiper arm is connected directly to the drive shaft; the linkage is omitted.

Attachment to the body
There are, depending on the wiper-system concept, different ways of attaching the system to the body. The wiping angle is influenced by the position of the drive in relation to the wiper arms and the position of the wiper bearings in relation to the vehicle. This angle in turn is one of the deciding variables for satisfying the field of vision required by the legislative bodies.

To simplify fitting on the vehicle and to limit the tolerances of the wiping angle, the wiper system is designed with a shaped tube or a shaped casting as a compact system, and secured to the drive and the wiper bearings.

As an alternative to this compact design, the drive and the bearings of the wiper arms can also be screwed directly to the body (loose-link connection); the shaped tube is omitted. This cuts down on the number of parts in the wiper system, but increases the amount of fitting work by the vehicle manufacturer in that more parts must be connected to the vehicle. Furthermore, this increases the demands placed on the rigidity of the vehicle body and necessitates increased precision when installing the components in order to ensure wiping-angle precision.

Depending on the position of the drive in relation to the wiper bearings, either parallel coupling (i.e. both wiper arms are driven directly by the drive, Figure 3a) or series coupling (i.e. the drive moves only one wiper arm, while the second wiper arm is connected to the first arm, Figure 3b) is used.

In the case of two-motor wiper systems and the direct wiper drive, the drives are connected directly to the body.

It is important to optimize the linkage to ensure that the wiper system works harmoniously, i.e. the wiper blade moves uniformly over the windshield. Smooth operation is achieved and the reversing noise of the wiper blade on the windshield is reduced by matching the maximum values of angular acceleration and the force-transmission angles near to wiper-arm reversal points at the outer and lower edges of the windshield.

Large wiping angles or difficult transfer ratios give rise under certain circumstances to greatly changing wiper speed on the windshield in the edge areas, caused by the changing lever ratios. For this reason, crosslays are used in some vehicles (Figure 5).

The current trend towards weight reduction and associated CO_2 reduction is taken into account with electronically controlled wiper systems. Electronic control of the drives significantly reduces the space needed by the articulated rods of the linkage for movement. The installation space can be further reduced by the use of a two-motor system with two considerably smaller drives.

With the direct wiper drive, it is also possible to dispense with the linkage of the two-motor system.

Additional functions, such as, for example, an extended parking position, an overload-protection function (e.g. for snow), an infinitely variable wiping speed, and a consistently large wipe pattern over varying operating conditions (e.g. driving speed), can be implemented with electronically controlled wiper systems.

Wiper arms

The wiper arm is the connecting link between the wiper linkage and the wiper blade. It is screwed to the tapered wiper bearing shaft by its mounting end, which is usually made of diecast aluminum or steel sheet. The other end is generally a steel strip and holds the wiper blade (Figure 6). A distinction is made here between hook-type, side and front fastenings (Figure 7).

The wiper bearing is screwed to the body. The position of the wiper arm in relation to the windshield is established by way of the position of the wiper bearing relative to the windshield. The field of vision is established in combination with the wiper-blade length and the wiping angle.

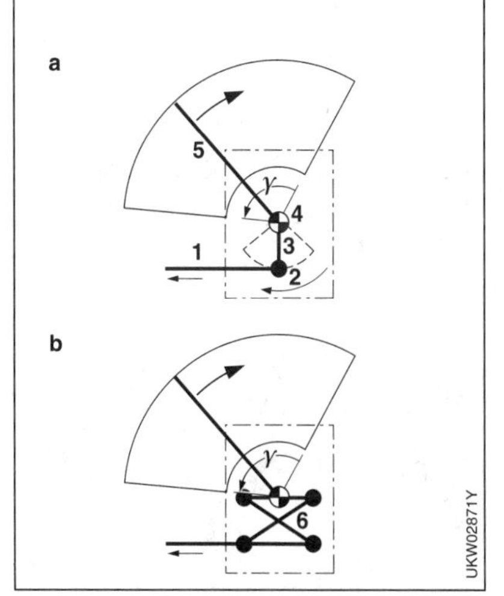

Figure 5: Wiper-system linkage mechanism
a) Conventional linkage,
b) Crosslay linkage.
1 Articulated rod,
2 Oscillating-crank link as ball pivot,
3 Oscillating crank,
4 Wiper bearing,
5 Wiper arm,
6 Crosslay.
γ Wiping angle.
The arrows indicate the direction of motion.

In addition to the standard wiper-arm design, there are a number of design variations. These are also special designs of wiper arms, which perform, for example, the following additional functions:

Wiper arm with quadruple-link reciprocating action
The wiper arm with quadruple-link reciprocating action specifically shifts the wipe pattern on the windshield, usually on the passenger side. This reduces the size of the unwiped area in the top corner of the windshield.

Wiper arm with wiper-blade control
The wiper arm with wiper-blade control provides an additional rotation of the wiper blade relative to the wiper arm, in order, for example, to wipe parallel to the A-pillar on wiper systems with only one wiper blade (Figure 1c).

Parallelogram wiper arm
The parallelogram wiper arm is a special type with wiper-blade control. It holds the wiper blade in a fixed position over the entire wipe movement (e.g. in the vertical position on local-transport buses).

Wiper-blade positioning
A second optimization step involves the work position of the wiper-element lip relative to the surface of the windshield or rear window (see Wiper-blade element). The position of the wiper blades is defined by the angular position of the wiper bearings to the windshield and by applying additional torsion to the wiper arms. The aim is to incline the wiper arms laterally at their reversal points towards the wipe-angle bisector. This helps the wiper-blade elements to swivel into their new working position. This, in turn, reduces wiper-blade wear and the reversing noise.

Figure 6: Wiper arm (side and top views)
1 Steel strip with hook-type fastening,
2 Joint section,
3 Tension spring,
4 Attachment part with taper for mounting on wiper bearing shaft.

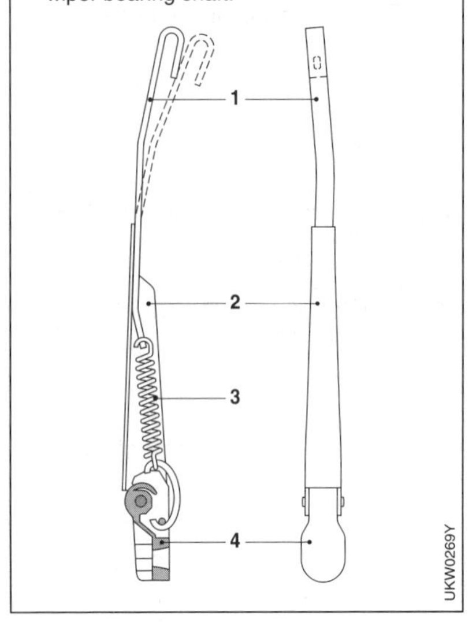

Figure 7: Attachment of wiper blade to wiper arm
a) Hook-type fastening,
b) Side fastening,
c) Front fastening.

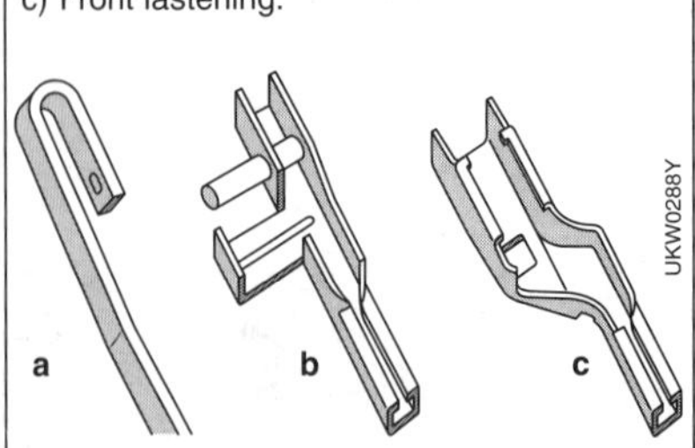

Figure 8: Wiper-blade element in working position
1 Claw,
2 Spring strip,
3 Wiper-element lip,
4 Windshield,
5 Double microedge.

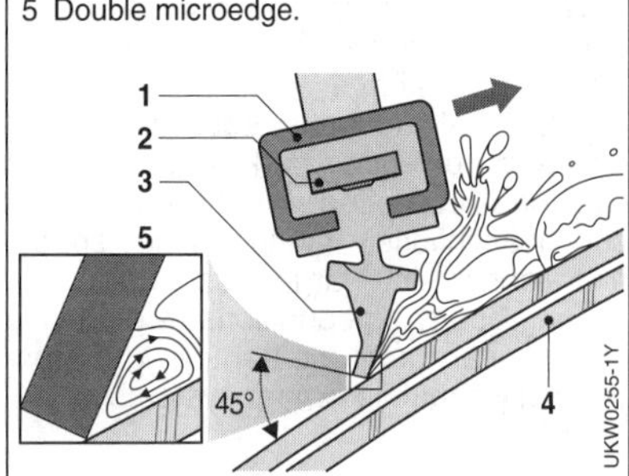

Wiper blades

Wiper-blade element

The most important component of the wiper system is the rubber wiper-blade element. It is retained by the claws of the bracket system (Figure 9) or supported by spring strips. The edge of the wiper-element lip touches the windshield over a width of 0.01 to 0.015 mm. When moving across the windshield, the wiper-blade element must overcome coefficients of dry friction of 0.8 to 2.5 (depending on air humidity), and coefficients of wet friction of 0.6 to 0.1 (depending on frictional velocity). The correct combination of wiper-element profile and rubber properties must be chosen so that the wiper lip can cover the complete wipe pattern on the windshield surface at an angle of approx. 45° (Figure 8).

The twin wiper, with a two-component wiper-blade element made of synthetic rubber, consists of a specially hardened, abrasion-resistant wiper-element lip which merges into an extra-soft spine. The soft spine ensures that the wiper element has optimum reversing characteristics and wipes smoothly.

Conventional wiper blade

The wiper blade (Figure 9) holds the wiper-blade element and directs its movement across the windshield. Wiper blades are used in lengths of 260 to 1,000 mm. Their dimensions for mounting (e.g. hook-type or snap-in fastening) are standardized. Low-wear operation is achieved by compensating the play in their mountings and linkages. The tops of the center brackets are perforated to prevent blade liftoff at high speeds. In special cases, wind deflectors are integrated into the wiper arms or blades to press the blades against the windshield.

Figure 9: Conventional wiper blade
a) Wiper blade under load,
b) At zero load, c) Cross-section.
1 Wiper-blade element, 2 Claw bracket,
3 Joint, 4 Center bracket, 5 Adapter,
6 Spring strip.

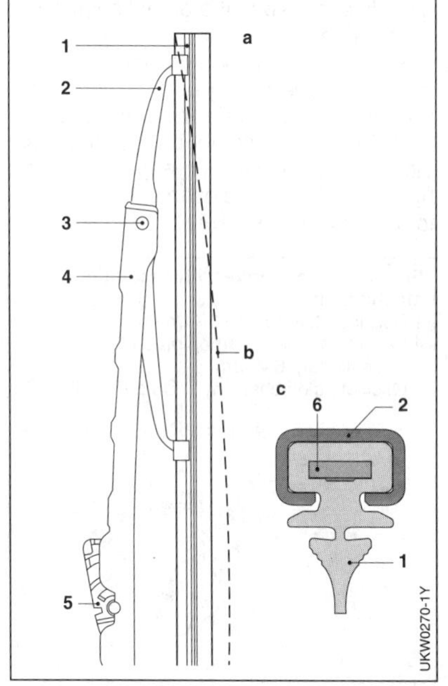

Figure 10: Flat wiper blade (aero-wiper blade)
a) Wiper blade under load, b) At zero load,
c) Cross-section.
1 End clip, 2 Spring strip,
3 Wiper-blade element,
4 Spoiler, 5 Wiper arm.

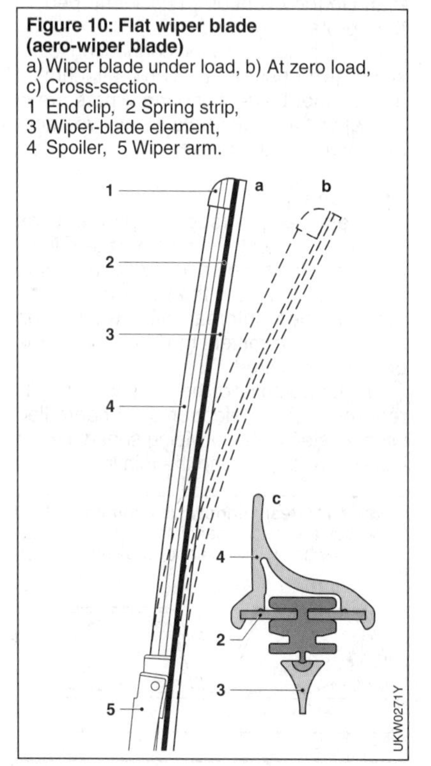

Flat wiper blade (aero-wiper blade)
The flat wiper blade (aero-wiper blade) is the present trend in wiper-blade design (Figure 10). The contact pressure over the wiper-blade element is no longer distributed by the claws of the wiper bracket, but by two preshaped spring strips (sprung rails) specially adapted to the shape of the windshield. They press the wiper-element lip more evenly against the windshield. This reduces wear on the wiper-element lip and increases wipe quality. In addition, the elimination of the bracket system means that there is no linkage wear, the overall height of the wiper system is substantially reduced, the weight is lower, and the wiper is quieter (there is also less wind noise).

The top edge of the wiper blade is in the shape of a spoiler (air deflector) and allows the blade to be used even at very high speeds without any modification. The flexible material of the spoiler is also much less likely to cause injury to pedestrians in the event of an accident (pedestrian protection).

An adapted, simplified connection to the wiper arm ensures a reliable attachment of the wiper blade when the wipers are in operation, and also allow easy replacement when required.

Rain sensor

The rain sensor (see chapter entitled "Sensors") detects how heavy the rain is, and converts this data into an appropriate signal that is sent to the wiper motor. The motor is then automatically switched on and set to intermittent mode or speed one or two as required.

The full potential of the rain sensor is fully realized with an electronically controlled wiper system, whose wiping speed can be infinitely re-adjusted to the rain intensity.

Figure 11: Rear-window wipe patterns
The gray areas represent impaired visibility when vehicles are overtaking.

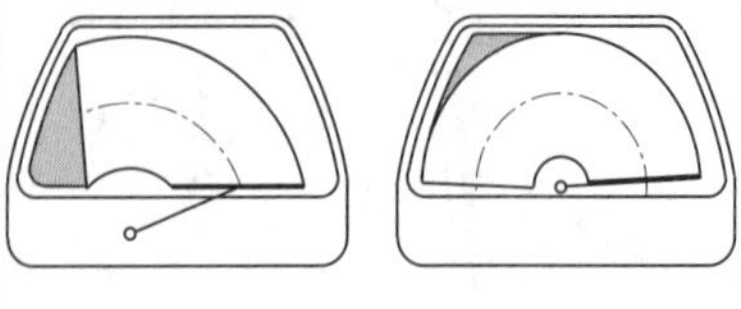

UKW0256-1Y

Rear-window wiper systems

Function and requirements

Rear-window wiper systems are used when the rear window, on account of its sloping angle or the body shape, is prone to getting very dirty and rear visibility is compromised. The principle of rear-window cleaning is generally the same as that for windshield cleaning systems.

A rear-window cleaning system is subject to much lower demands than that of a windshield cleaning system. A rear-window cleaning system is therefore frequently operated in intermittent mode and its field of vision is not subject to any statutory requirements. The wiping angle typically ranges between 60° and 180° (Figure 11).

Rear drives

Rear drives are essentially the same as front drives. A rear-window wiper system is usually driven by an electric motor with integrated oscillating mechanism, which assumes the function of the linkage in windshield cleaning systems and performs the oscillating movement of the output shaft (Figure 12). The wiper arm is directly attached to the output shaft of the wiper drive.

To increase the wiping angle while at the same time taking up minimal space, solutions have been implemented whereby the classic four-bar mechanism increases the oscillation angle to a value up to 180° by means of an additional rotatory relative movement.

Figure 12: Rear drive with oscillating mechanism
1 Bearing, 2 Magnet, 3 Armature,
4 Commutator with carbon brushes,
5 Helical gear, 6 Worm,
7 Oscillating mechanism, 8 Output shaft.

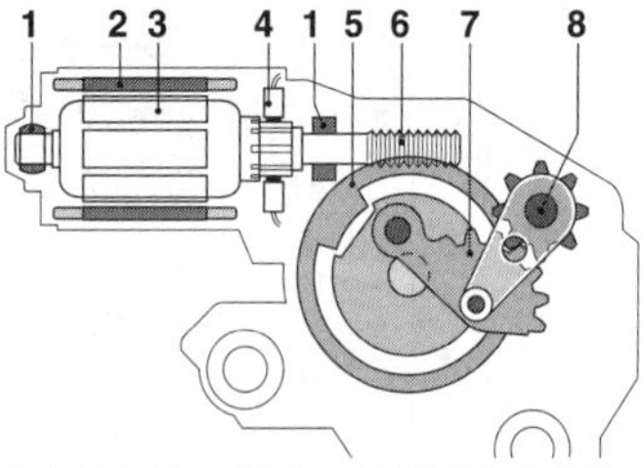

1 2 3 4 1 5 6 7 8

UKW0289-1Y

Windshield and rear-window washer systems

Body

To ensure good visibility in the wipe pattern, it is imperative that the wiper system is backed by a washer system. Electrically driven centrifugal pumps are used to direct water mixed with detergent additive from the washer-fluid reservoir through 2 – 4 nozzles in a pointed spray or through sprinkler nozzles as water mist onto the windshield (Figure 13). The capacity of the washer-fluid reservoir is usually 1.5 to 2 l. If the same washer-fluid reservoir is also used to supply the headlamp washers, a capacity of up to 7 l may be required. A separate reservoir may be provided for the rear-window cleaning system.

Electronic control

The washer system is often linked to the cleaning system by means of an electronic control system so that water is sprayed onto the rear window or windshield for as long as a pushbutton remains pressed. The wiper system then continues to operate for several additional cycles after the pushbutton is released.

Figure 13: Windshield washer system and high-pressure washer system for headlamps
1 Washer-fluid reservoir, 2 Pump,
3 Non-return valve, 4 T-joint,
5 Nozzle holder (horn) for headlamp cleaning,
6 Tube,
7 Nozzles of headlamp washer system.

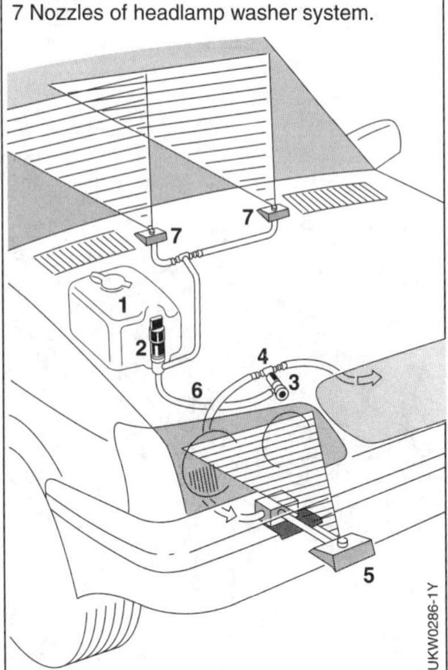

Headlamp washer systems

Pure washer systems have become established for cleaning headlamps. The advantages of the headlamp washer system over the previously used wipe/wash systems lie in its simple design and in the fact that it is easier to adapt to the vehicle styling concept.

A headlamp washer system is prescribed by law in Germany for xenon headlamps to prevent oncoming traffic from being dazzled by light diffusion.

Body

High-pressure washer systems (Figure 13) consist of the washer-fluid reservoir (the washer fluid required is taken from the reservoir for the windshield washer system), the pump, tubes with a non-return valve, and the nozzle holders (horn) with one or more nozzles. As well as fixed nozzle holders on the bumpers, there are also telescopically extending nozzle holders. The telescope improves the cleaning effect because it can assume an optimum spray position. In addition, when not in use, the nozzle holder can be concealed, e.g. inside the bumper.

Cleaning effect

The cleaning effect is chiefly determined by the cleaning pulse of the water droplets. The decisive factors here are the distance between the nozzles and the lens, the size, the contact angle, the contact velocity of the washer-fluid droplets, and the amount of washer fluid. It is important that the nozzles be positioned correctly so that water jets properly cover the headlamps at all driving speeds.

References
[1] Council Directive 78/318/EEC: Wiper and washer systems of motor vehicles.
[2] FMVSS Part 571, Standard No. 104: Windshield Wiping and Washing Systems.
[3] SAE J903: Passenger Car Windshield Wiper Systems.

Occupant-protection systems

In the event of an accident, passive safety systems are intended to keep the accelerations and forces that act on the passengers low and thereby lessen the consequences of the accident. Vital contributions in this context are made by the following occupant-protection systems (Figure 1):
- Seat belts with belt pretensioners and belt-force limiters
- Various airbags
- Rollover protection systems.

Seat belts with belt pretensioners provide the majority of the protective effect, absorbing 50 to 60 % of the kinetic energy of the occupants. With front airbags, the energy absorption is about 70 % if deployment timing is properly synchronized.

In order to achieve optimum protection, the response of all components of the complete occupant-protection system must be adapted to one another. This is made possible by suitable sensors and with high-speed signal processing. The control algorithms for the belt pretensioners, the airbags and if necessary the rollover protection systems are stored in a combined ECU.

Seat belts and seat-belt pretensioners

Function

The function of seat belts is to restrain the occupants of a vehicle in their seats when the vehicle impacts against an obstacle. In this way, the occupants are already involved at an early stage in the deceleration of the vehicle during an impact (Figure 2). The standard equipment is the three-point belt with inertia-reel device, which is increasingly also used on the middle seat in the rear passenger compartment. With adjustable seat systems, the seat-belt buckle is attached directly to the seat, while the inertia-reel device is attached to the B- or C-pillar (Figure 1).

A loosely fitting belt (for example, when thick winter clothing is worn) results in belt slack. The softness of the clothing prevents the occupant from being involved at an early stage in the deceleration of the vehicle in the event of a collision. The occupant initially continues moving without being restrained, which reduces the protective effect of the belt. In addition, the decelerating effect of the inertia-reel device

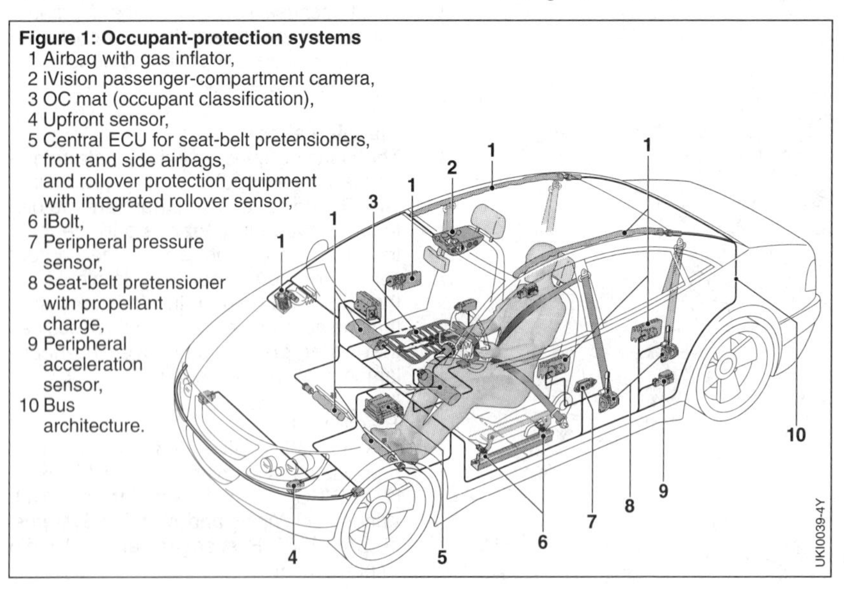

Figure 1: Occupant-protection systems
1 Airbag with gas inflator,
2 iVision passenger-compartment camera,
3 OC mat (occupant classification),
4 Upfront sensor,
5 Central ECU for seat-belt pretensioners, front and side airbags, and rollover protection equipment with integrated rollover sensor,
6 iBolt,
7 Peripheral pressure sensor,
8 Seat-belt pretensioner with propellant charge,
9 Peripheral acceleration sensor,
10 Bus architecture.

("film-reel effect") and the elongation of the belt webbing contributes to the belt slack.

Because of belt slack three-point inertia-reel belts only afford limited protection in the event of a frontal impact at speeds in excess of 40 km/h against solid obstacles, because they cannot safely prevent the head and body from striking the steering wheel and the dashboard. In the event of a frontal impact, seat-belt pretensioners pull the seat belts tighter against the body and hold the upper body as closely as possible against the seat backrest. This prevents excessive free forward displacement of the occupants caused by inertia. Seat-belt pretensioners improve the restraining characteristics of a three-point inertia-reel belt and increase protection against injury.

A prerequisite for optimum protection is that the occupants' forward movement away from their seats must be minimal as they decelerate with the vehicle. Activation of the seat-belt pretensioners takes care of this virtually from the moment of impact, and ensures restraint of occupants as early as possible. The maximum forward movement with pretensioned belts is approx. 2 cm; the mechanical pretensioning takes 5 to 10 ms.

Seat-belt pretensioners are activated particularly in the event of frontal impacts, but are increasingly also being activated in the event of side impacts. This is because, in the event of accidents of this type too, the occupants are better protected by the more tightly fitting belts.

Design and operating principle

Shoulder-belt tightener

In an impact, the shoulder-belt tightener eliminates the seat belt slack and the "film-reel effect" by rolling up and tightening the belt webbing. This initiates the protective effect of the seat belt at an earlier stage. At an impact speed of 50 km/h, this system achieves its full effect within the first 20 ms of impact; this supports the protective effect of the airbag, which needs approx. 40 ms to inflate completely.

On activation, the system electrically fires a pyrotechnic propellant charge (Figure 3). The gas charge released in this process acts on a plunger, which turns the belt reel via a steel cable in such a way that it is held tightly against the occupant's body. The belt webbing is therefore tightened already before the occupant starts their forward movement. With these belt tighteners the belt webbing can be pulled back within a period of 10 ms by up to 12 cm.

Activation of the belt tightener is triggered by acceleration sensors which are integrated in the ECU. Further sensors

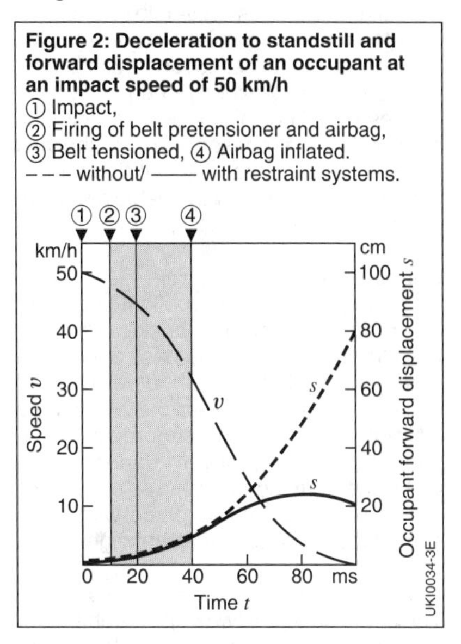

Figure 2: Deceleration to standstill and forward displacement of an occupant at an impact speed of 50 km/h
① Impact,
② Firing of belt pretensioner and airbag,
③ Belt tensioned, ④ Airbag inflated.
--- without/ —— with restraint systems.

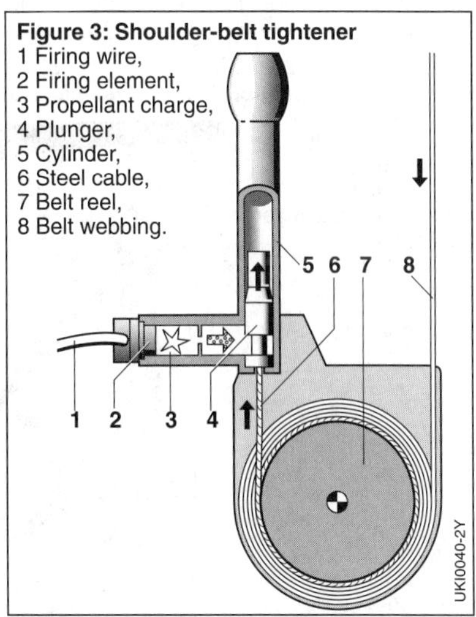

Figure 3: Shoulder-belt tightener
1 Firing wire,
2 Firing element,
3 Propellant charge,
4 Plunger,
5 Cylinder,
6 Steel cable,
7 Belt reel,
8 Belt webbing.

are installed at the front end of the vehicle to ensure that accidents are detected quickly and safely. Micromechanical acceleration sensors are used in the main. The analysis algorithms in the ECU read in the data from the sensors permanently and compute whether an accident is occurring.

Because the firing of the pyrotechnic propellant charge is irreversible, the decision to effect firing must be covered and distinguished from situations such as, for example, driving over the edge of curbs, as these must not initiate deployment. Signals must be processed in the shortest possible time so that the belt tightener is deployed on time after the vehicle has impacted against an obstacle.

Buckle tightener

When triggered by a propellant charge or by spring systems, the buckle tightener pulls the seat-belt buckle back and simultaneously tightens the shoulder and lap belts. It further improves the restraining effect and the protection to prevent occupants from sliding forward under the lap belt ("submarining effect").

The tightening takes place within the same time as for shoulder-belt tighteners.

Combination of two belt tighteners

The combination of two belt tighteners per belt, consisting of a shoulder-belt tightener and a belt-buckle tightener, provides a greater tightener path to achieve a better retention effect. The buckle tightener is activated either when a specific crash severity level is reached or in response to a specific time delay after the shoulder-belt tightener has deployed.

Belt-force limiter

In this case, the seat belt pretensioners initially tighten fully (using the maximum force of approx. 4 kN, for example) and restrain the occupants. If a certain belt tension is exceeded, the belt gives and allows a greater degree of forward movement by the occupant. The occupant's kinetic energy is converted by deformation elements into deformation energy. Examples of deformation elements used in this situation are a torsion bar in the inertia-reel device shaft or a deployment control seam in the belt.

A further variant is an electronically controlled single-stage belt-force limiter, which reduces the belt tension to 1 to 2 kN by firing a detonator (e.g. by way of deployment control seams in the belt) a specific period after deployment of the second front airbag stage – and thus with the airbag fully inflated – and after a specific extent of forward movement is reached.

The belt-force limiter prevents the occurrence of acceleration peaks and thereby prevents the risk of broken collarbones and ribs with the resulting internal injuries.

Figure 4: Occupant-protection systems with seat-belt pretensioner and front airbags

1 Inertia-reel device with seat-belt pretensioner,
2 Front airbag for front passenger,
3 Front airbag for driver,
4 ECU.

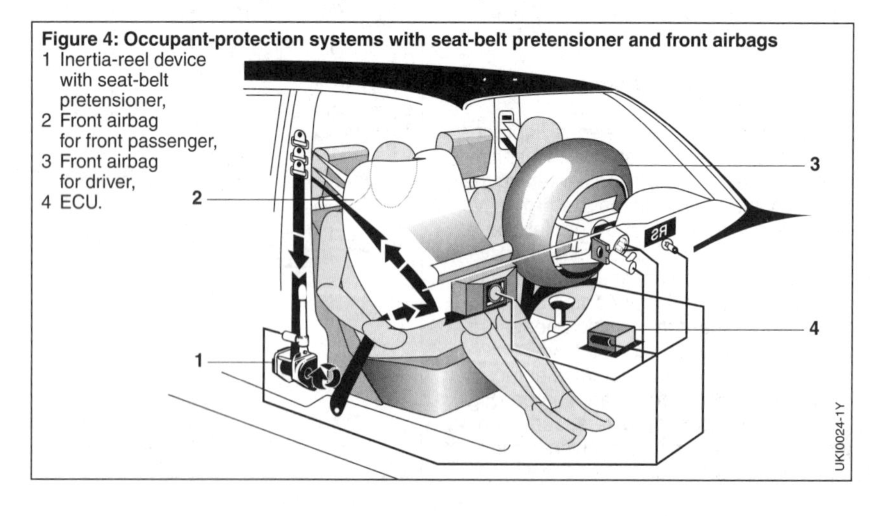

Airbag

Front airbag

Function

Front airbags have the task of protecting the driver and front-seat passenger against injuries to the head and upper body in the event of a vehicle colliding with obstacles (i.e. front impact) (Figure 4). In a serious accident, a seat belt with belt pretensioner cannot keep the head from striking the steering wheel or the dashboard. To perform this task, airbags have different fill volumes and shapes adapted to the vehicle conditions, depending on the installation point, vehicle type and structure deformation properties (vehicles are deformed in different ways in the event of a crash). Occupants are afforded the most effective protection when the belt and front-airbag systems operate in optimum coordination.

Operating principle

Pyrotechnic gas inflators inflate the driver and passenger airbags using dynamic pyrotechnics after a vehicle impact detected by the acceleration sensors (Figure 5). In order to achieve maximum protection, the airbag must be fully inflated before the occupant comes into contact with it. When the occupant comes into contact with the airbag, the airbag is partially deflated via deflation vents. The impact energy acting on the occupant is "softly" absorbed with noncritical (in terms of injury) surface pressures and deceleration forces.

The inflation speed and the hardness of the inflated airbag can be influenced in the case of two-stage gas inflators by a delayed firing of the second stage.

The maximum permitted forward motion of the driver until the airbag on the driver's side has inflated is approx. 12.5 cm. This corresponds to a time of approx. 40 ms after the start of impact (in the case of an impact at 50 km/h on a hard obstacle). It takes 10 ms for the electronics to sense the impact and trigger the detonation; it takes 30 ms to inflate the airbag. The airbag deflates after another 80 to 100 ms through the deflation vents. The entire operation thus takes only slightly more than a tenth of a second.

Figure 5: Highly dynamic deployment of the driver airbag

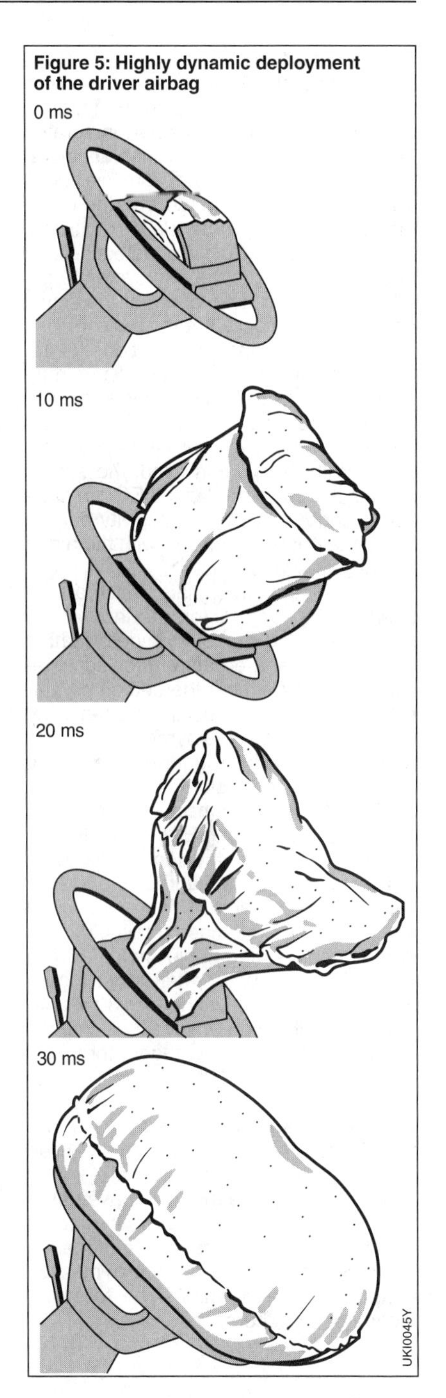

Impact detection
One or two acceleration sensors which measure along the vehicle longitudinal axis and are for the most part integrated in the ECU register the braking deceleration that arises on impact. From this deceleration the speed change and the occupant's forward movement are calculated. In order to be able to better detect oblique and offset impacts, the deployment algorithm can also take account of the signal from the lateral acceleration sensor.

In addition to impact detection, the impact must also be evaluated. The airbag should not trigger from a hammer blow in the workshop, gentle impacts, bottoming out, or driving over curbstones or potholes. With this goal in mind, the sensor signals are processed in the ECU in analysis algorithms whose sensitivity parameters have been optimized with the aid of crash-data simulations.

The acceleration signals, which are influenced by such factors as the vehicle equipment and the body's deformation characteristics, are different for each vehicle. They determine the setting parameters which are of crucial importance for sensitivity in the deployment algorithm and, ultimately, for triggering the airbag and seat belt pretensioners.

At least one further acceleration sensor is located in the airbag ECU to prevent an incorrect airbag deployment in the event of a defective acceleration sensor. In the event of an accident, this sensor must also exceed a predefined threshold in order to enable airbag deployment.

The first seat-belt pretensioner trigger threshold is reached within 8 to 30 ms depending on the type or severity of impact, and the first front airbag trigger threshold after approx. 10 to 50 ms.

Adapted airbag deployment
In order to prevent injuries being caused by airbags to occupants who are "out of position" (e.g. are leaning far forwards) or to small children in reboard child seats (pointing towards the rear), the triggering of the front airbag and its inflation must be adapted to the situation. The following measures exist here:

- Deactivation switch: A manually actuated deactivation switch can be used to disable the passenger airbag
- There is an increasing availability of standardized anchoring systems (ISOFIX child seats). Switches integrated in the anchoring locks initiate an automatic passenger-airbag shutoff, which must be indicated in the instrument cluster.

Depowered airbags
In the USA attempts are being made to reduce the active force of the inflation by introducing "depowered airbags". These are airbags whose gas-inflator power has been reduced by 20 to 30 %, which itself reduces inflation speed, inflation severity (hardness) and the risk of injury to "out-of-position" occupants. It is easier for large and heavy occupants to push through these airbags, i.e. they have reduced energy absorption properties.

In the USA the "low-risk" deployment method is currently preferred. This means that, in "out-of-position" situations, only the first front-airbag stage is triggered. In heavy impacts, the full gas-inflator output can then be brought into effect by triggering both inflator stages.

Another way of implementing "low-risk" deployment with single-stage inflators is to keep the deflation valves constantly open.

Intelligent airbag systems
The introduction of improved sensing functions and control options for the airbag inflation process, with the accompanying improvement in protective effect, is intended to result in a gradual reduction in the risk of injury. Such functional improvements are:

- Impact-severity detection through optimization of the deployment algorithm or using one or two upfront sensors. The latter are acceleration sensors installed in the vehicle's crumple zone (e.g. on the radiator cross-member) which facilitate early detection and distinction of the different impact types such as ODB crashes (offset deformable barrier crash, offset crash into soft barriers),

pole or underride impacts under a truck. They also allow an assessment of the impact energy. In accidents which are less serious, where the protective effect afforded by the seat-belt pretensioner is sufficient, the airbag does not have to be triggered (reduced repair costs)
- Seat-belt usage detection
- Occupant-presence, position and weight detection (occupant classification)
- Use of seat-belt pretensioners with occupant-weight-dependent belt-force limiters
- Seat-position and backrest-inclination detection
- Use of front airbags with multi-stage gas inflators or with a single-stage gas inflator and pyrotechnically activated gas-discharge valve. The inflation speed and the hardness of the airbags can be adapted with several trigger thresholds to the severity and nature of the accident
- The data interchange with other systems, e.g. ESP (Electronic Stability Program) and environment sensors means that information from the phase shortly before impact can be used to optimize triggering of the retention systems. Example: By using ESP data it is possible to activate the side window bag earlier in certain vehicle rollover accidents.

Knee airbag

In a few vehicle types, front airbags also operate in conjunction with "inflatable knee pads", which safeguard the "ride down benefit", i.e. the speed decrease of the occupants together with the speed decrease of the passenger cell. This ensures that the upper body and head describe the rotational forward motion which is needed for the airbag to provide optimum protection. Furthermore, the knee airbag prevents contact with the dashboard support and thereby reduces the risk of injury in this area.

Side airbag

Function

Side airbags which inflate along the length of the roof lining for head protection (e.g. inflatable tubular system, window bag, inflatable curtain) or for upper-body protection from the door or seat backrest (thorax bag) are designed to cushion the occupants and protect them from injury in the event of a side impact.

Operating principle

Due to the lack of a crumple zone, and the minimum distance between the occupants and the vehicle's side structural components, it is particularly difficult for side airbags to inflate in time. The time for crash sensing and activation of the side airbag must therefore be approx. 5 to 10 ms in the case of severe side impacts. The inflation of the approx. 12 *l* thorax bags may take a maximum of 10 ms.

These requirements can be fulfilled through the evaluation of peripheral, laterally measuring acceleration and pressure sensors. These sensors are installed at appropriate points on the bodywork, e.g. B-pillar or door.

The peripheral acceleration sensors (PAS) transmit acceleration data to the central ECU via a digital interface. The ECU triggers the side airbags provided the lateral sensor has confirmed a side impact by means of a plausibility check.

Alternatively, the pressure changes caused by door deformation (air pressure in the door cavity) are measured by a peripheral pressure sensor (PPS). This will result in rapid detection of door impacts. Confirmation of plausibility is now provided by acceleration sensors mounted on supporting peripheral structural components. This is now unquestionably faster than the central lateral acceleration sensors.

Rollover protection systems

Function

In the event of an accident where the vehicle rolls over, open-top vehicles such as convertibles lack the protecting and supporting roof structure of closed-top vehicles. Initially, therefore, rollover sensing and protection systems were only installed in convertibles and roadsters without fixed rollover bars.

Now rollover sensing is being used in closed-top passenger cars. If a car turns over, there is the danger that non-belted occupants may be thrown through the side windows, or that parts of the bodies of belted occupants (e.g. arms) may protrude from the vehicle and incur serious injury. To provide protection in such cases, already existing restraint systems such as seat-belt pretensioners and side and head airbags are activated. In convertibles, the extendable rollover bars or the extendable head restraints are triggered.

Operating principle

Current sensing concepts trigger the system at a threshold that conforms to the situation and only in the event of the most frequently occurring type of vehicle rollover (rollover about the longitudinal axis). In the Bosch concept sensing is performed by a surface-micromechanical yaw-rate sensor and high-resolution acceleration sensors in vehicle transverse and vertical directions (y- and z-axes). The yaw-rate sensor is the main sensor, while the y and z acceleration sensors are used both to check plausibility and to identify the type of rollover (embankment, curb-trip or soil-trip rollover). On Bosch systems, these sensors are incorporated in the airbag triggering unit.

Deployment of occupant-protection systems is adapted to the situation according to the type of turnover, the yaw rate and the lateral acceleration, i.e. systems are triggered after 30 to 3,000 ms by automatic selection and use of the algorithm module appropriate to the type of rollover.

Components

Airbag control unit

Optimum occupant protection against the effects of frontal, offset, oblique or pole impact is obtained through the precisely coordinated interaction of electronically detonated pyrotechnical front airbags and seat-belt pretensioners. To maximize the effect of both protective devices, they are activated with optimized time response by a common ECU (airbag control unit) installed in the passenger cell. The control for the side airbags and the rollover protection facility are also integrated. The following functions are currently incorporated in this central ECU, also referred to as the trigger unit:

- Crash sensing by acceleration sensor and safety switch (mechanical acceleration switch with older ECUs) or by two acceleration sensors without safety switch (redundant, fully electronic sensing)
- Prompt activation of front airbags and seat-belt pretensioners in response to different types of impact in the vehicle longitudinal direction (e.g. frontal, oblique, offset, pole, rear-end)
- Rollover detection through yaw-rate and acceleration sensors that pick up the y and z acceleration (lateral acceleration and acceleration in the direction of the vertical axis) in the low g range (up to approx. 5 g)
- Activation of rollover protection equipment
- To activate the side airbags the ECU operates in conjunction with a central lateral-acceleration sensor, two or four peripheral acceleration sensors, and a peripheral pressure sensor (PPS) installed in each door cavity
- Voltage transformer and energy accumulator in case the supply of power from the vehicle battery should fail
- Selective triggering of the seat-belt pretensioners depending on the seat-belt buckle queries: The airbag is only detonated when the seat-belt buckle has been fastened (detection by belt-buckle switch)
- Setting of multiple triggering thresholds for two-stage seat-belt pretensioners and two-stage front airbags depending on the status of belt use and seat occupation

- Reading in the signals of occupant classification (iBolt force sensor) and corresponding triggering of the retention systems
- To place an emergency call after an impact and to activate secondary safety systems (hazard warning lights, opening the central locking system, deactivating the fuel-supply pump, disconnecting the battery, etc.), the airbag ECU sends the signal regarding a detected impact via e.g. the CAN bus.

Gas inflators

Body

The pyrotechnic propellant charges of the gas inflators to generate the airbag gas and operate the seat belt pretensioners are activated by an electrical firing element. The gas inflator inflates the airbag with gas.

The firing pellet (Figure 6) contains a reservoir holding the propellant charge and a priming wire. The firing pellet is connected to the airbag ECU via the connector pins and a two-wire circuit. To trigger the airbag, the ECU uses firing output stages to generate a current, which flows inside the firing pellet through the priming wire. This wire glows and activates the propellant charge.

The driver airbag fitted in the steering-wheel hub (volume approx. 60 *l*) and the passenger airbag fitted in the glovebox space (approx. 120 *l*) are inflated approx. 30 ms after detonation.

Passenger-compartment sensing

Occupant classification

A method of measuring the absolute weight, the iBolt ("intelligent" bolt), is available for occupant classification. These iBolts (Figure 1) measure the forces involved and secure the seat frame (suspended seat) on the sliding base, replacing the four mounting screws that are otherwise used. They measure the weight-dependent change in the gap between the bolt sleeve and the internal bolt with integral Hall-element IC connected to the sliding base.

Out-of-position detection

The following optical methods are conceivable for out-of-position detection:

- "Time of Flight" principle (TOF): The system transmits infrared light pulses and measures the time until their reflection returns depending on the distance of the occupants. The time intervals being measured are of the order of picoseconds
- "Photonic Mixer Device" method (PMD): An imaging sensor of this nature transmits light pulses and enables spatial vision and triangulation
- "iVision" passenger-compartment stereo-video camera using CMOS technology: This detects occupant position, size and restraint method and can also control convenience functions (seat, mirror and radio settings) to suit the individual occupant.

No unified standard for passenger-compartment sensing has yet been able to establish itself. Occupant-classification (OC) mats combined with ultrasonic sensors may also be used.

Figure 6: Firing pellet
1 Propellant charge,
2 Priming charge,
3 Cap,
4 Charge holder,
5 Priming wire,
6 Firing head,
7 Housing,
8 Connector pins.

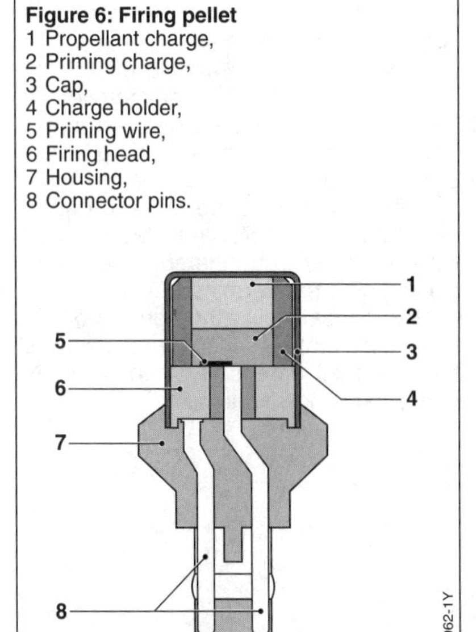

Further developments

The following additional improvements in the field of occupant protection are being developed.

Airbags with active ventilation system
These airbags have a controllable deflation valve to maintain the internal pressure of the airbag constant even if an occupant falls against it and to minimize occupant trauma. A simpler version is an airbag with "intelligent vents". These vents remain closed (so that the airbag does not deflate) until the pressure increase resulting from the impact of the occupant causes them to open and allow the airbag to deflate. As a result, the airbag's energy absorption capacity is fully maintained until the point at which its motion-damping function comes into effect.

Networking of passive and active safety features
Examples of the synergetic use of the sensors of various safety systems (here: Electronic Stability Program, ESP) are the functions Advanced Rollover Sensing, Early Pole Crash Detection, and Secondary Collision Mitigation.

Advanced Rollover Sensing
Advanced Rollover Sensing utilizes the ESP signals via the CAN for improved detection of soil-trip rollover situations. These data are used by the airbag ECU to calculate the speed vector and the lateral velocity. From these the deviation of the vehicle-motion vector from the vehicle's longitudinal axis and thereby lateral vehicle motion are determined.

The ESP can utilize the signals from the low g acceleration sensors (y and z axes) for improved detection of unstable dynamic handling situations.

Early Pole Crash Detection
Early Pole Crash Detection also utilizes the ESP signals for better detection of a side-on pole impact. The system utilizes the fact that in the case of a side-on pole impact the vehicle skids before the impact, which is detected with the ESP signals. The information regarding the lateral movement of the vehicle is then used in the triggering algorithm to accelerate triggering of the side airbags.

Secondary Collision Mitigation
In the event of accidents the initial collision may be followed by further collisions by the accident vehicle, for example as a result of the driver losing control of the vehicle. This phenomenon puts both the vehicle occupants and other road users at risk. The Secondary Collision Mitigation function provides assistance in the event of such accidents. If an impact occurs, the airbag ECU transmits this information to the ESP control unit. The ESP uses specifically targeted brake interventions to slow down the vehicle or where necessary to shut down the engine. Consequential accidents can thus be avoided or their severity reduced.

By networking ESP and airbag ECUs, it is possible also to utilize the detection of unstable or critical states to initiate more specifically targeted safety measures. When an unstable driving state is detected, safety measures can be introduced in stages; these measures can include closing the windows and the sliding sunroof, and tightening a reusable (reversible) motor-driven seat-belt pretensioner. In a critical driving situation, this reduces the belt slack and thus also the uncontrolled lateral movement of the occupant, enabling the side airbag to provide optimized protection in a subsequent side impact.

The trend is towards a standardized safety system in which the functions of active and passive safety are combined in a single unit.

Pre-crash sensing

For further improvement of the deployment function and better advance detection of the type of impact (pre-crash detection), microwave radar, ultrasound or lidar sensors (optical system using laser light) are used to detect relative speed, distance and angle of impact for frontal impacts.

In connection with pre-crash sensing, reversible seat-belt pretensioners are deployed. These are electromechanically actuated. The fact that they are reversible means that they can already be tightened prior to a possible collision. In this way, belt slack is already eliminated at the start of the crash, enabling the occupants to participate in the vehicle's deceleration from the outset.

Further airbag variants

A further improvement in the restraining effect will be provided by airbags integrated in the thorax section of the seat belt ("air belts", "inflatable tubular torso restraints" or "bag-in-belt" systems), which reduce the risk of broken ribs.

The same path for improving protective functions is being pursued by engineers developing "inflatable headrests" (adaptive head restraints for preventing whiplash trauma and cervical injuries), "inflatable carpets" (prevention of foot and ankle injuries), two-stage seat-belt pretensioners and "active seats". Here, an airbag is inflated in the front section of the seat surface to increase the tilt angle and diminish the submarining effect (sliding forwards) of the occupant.

Locking systems

Function

The locking system comprises the latches in the vehicle doors, trunk lid, engine hood, filler cap and glovebox, and the associated electronic control units. The system is complemented on the electronic side by the radio remote control or solutions to keyless unlocking with transponders. The system is rounded off in the premium segment by drive motors for automatic movement and stepless arresting of the side door.

The following typical areas of application must be covered by the locking system:

- Access control, theft deterrence and occupant protection in respect of third parties
- Actuation of opening handles with higher operating comfort (haptics and acoustics)
- Lifting, braking and locking of the vehicle door during the closing operation
- Immobilization of the door during driving (creaking-noise problems).

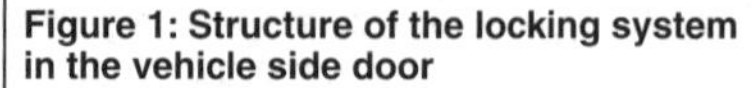

Figure 1: Structure of the locking system in the vehicle side door
1 Outer locking (to lock barrel),
2 Lock case,
3 Lock holder,
4 Cable with outside door handle,
5 Electrical plug-in connection,
6 Cable to inside opener.

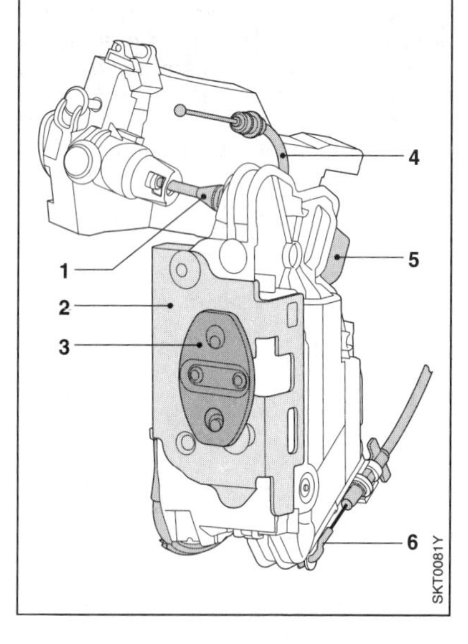

The wealth of access scenarios encountered in everyday vehicle use results in a series of different logic functions. For emergency situations all the electronic function logic circuits are made available with full mechanical redundancy. The amount of precision-mechanical assemblies required is accordingly high.

Beyond the applications mentioned above, the components in the side door make a passive contribution to occupant protection, since in the event of a crash a large proportion of the structural forces in the vehicle door is dissipated by the door lock and the lock holder to the vehicle body. The extreme accelerations during these processes must neither disable the locking mechanism nor alter the logic state of the locking chain. For example, a childproof lock, once it has been engaged, must prevent the door from being opened from the inside even after a crash. To achieve this, the mechanical assemblies are mass-balanced or secured by means of high spring loads against self-actuation.

In everyday vehicle use the locking system in combination with the door seal plays an important role in conveying an impression of the vehicle's high-end qualities. Good actuation haptics and an agreeable closing noise are much sought-after qualities in modern vehicles.

Structure (side door as example)

The central component in the locking system is the electromechanical door lock. The system is extended at the mechanical interfaces by a lock holder, inner and outer actuators, and Bowden cables or connecting linkage (Figure 1). The lock is electrically activated externally, for example by means of the door control unit.

The main functions of the door lock are closing, locking, double-locking, and opening. On the basis of this functional categorization, the door lock consists of the following assemblies: locking mechanism, lever mechanism, and double-locking actuator system.

Closing function

In the locking mechanism the combination of rotary latch, locking pawl and lock housing acts on the work arm of the lock-holder striker (Figure 2). When the door shuts, the lock holder moves into the rotary latch. Its rotary motion forces the vehicle door into the correct vertical end position (centering function). This lifting effect by the rotary latch is crucial to immobilizing the door when the vehicle is moving. In the opposite direction the sprung design of the lock housing (catcher) acts on the lock holder. On rough roads excessive relative movement between the door lock and the lock holder would encourage a disruptive creaking noise.

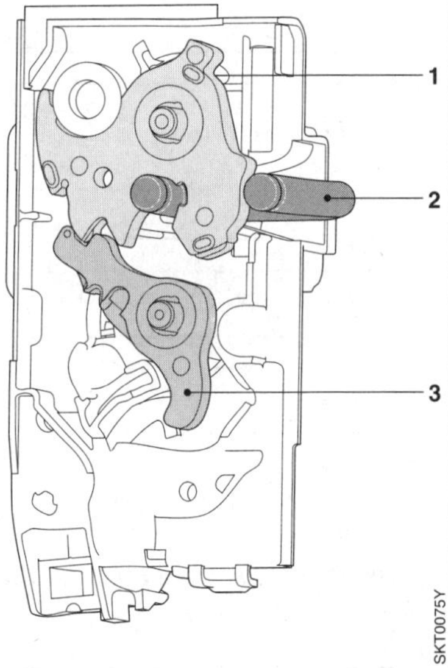

The door is braked in the horizontal direction during the closing process. The high kinetic energy (at a typical speed of 1.2 m/s) is dissipated to a few millimeters of lock-holder travel. The door-sealing system and the locking mechanism are very closely matched to each other in order to achieve an excellent closing noise. Pulse-minimizing contact geometries have proven effective inside the locking mechanism. The contact geometries are optimized in such a way that the kinetic energy can be reduced uniformly and over a longer engagement distance. Sound-deadening elements reduce the residual energy.

When the lock holder has fully retracted, the locking pawl drops into the primary detent of the rotary latch. The locking mechanism is now locked. Structural forces between the vehicle door and the body are transmitted reliably and without disruptive vibrations or relative movements.

For safety reasons the law stipulates a redundant preliminary detent. If the system fails to reach the primary detent (e.g. due to insufficient closing energy or a trapped seat belt), the door does not fall open entirely, but is instead reliably retained in the preliminary-detent position.

Figure 2: Locking mechanism with lock holder
1 Rotary latch,
2 Lock holder,
3 Locking pawl.

1
2
3
SKT0075Y

Opening function

To open the vehicle door, mechanical work is performed on the outside door handle or on the inside opener, transmitted by an intermediate lever mechanism, and finally directed to the locking pawl. The positive locking of the locking pawl and the rotary latch must then be overcome. Here the applied door-seal pressure (usually 300 N, in the premium segment up to 700 N) is abruptly released. Here too it is possible to deliver acoustically flawless solutions with a special detent and by carefully avoiding alternating-load cycles at the locking pawl.

A harmonic force characteristic and low maximum force are the quality criteria for vehicle users. The lever ratio of the lock holder and locking pawl at the rotary latch has the greatest influence on this opening comfort. Today's conventional designs produce lever ratios of 1:2 to 1:2.5 (Figure 3). Figure 4 depicts the influence of the locking-element ratio on the total opening force. The effect of the return springs in the lever mechanism, however, cannot be ignored.

Outlook: i-comfort two-pawl locking mechanism

The conventional pawl mechanism is self-locking in order to ensure reliable locking even during rough vehicle operation or in the event of a crash. Engagement is positive-locking, the locking functions being performed by just one pawl.

In a variation on this basic principle, the future two-pawl locking mechanism will feature an intermediate pawl. This measure facilitates considerably better lever ratios

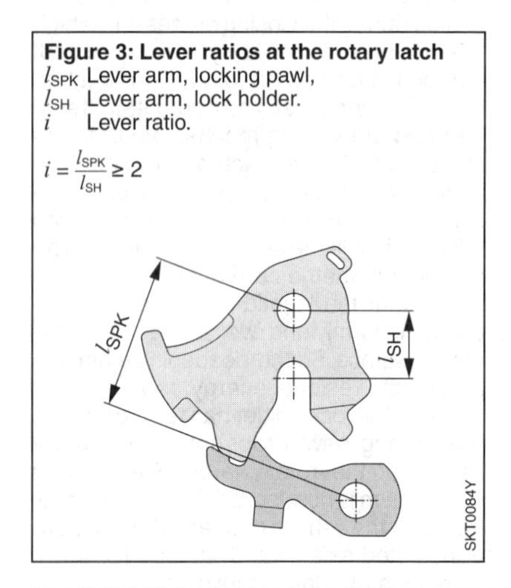

Figure 3: Lever ratios at the rotary latch
l_{SPK} Lever arm, locking pawl,
l_{SH} Lever arm, lock holder.
i Lever ratio.

$$i = \frac{l_{SPK}}{l_{SH}} \geq 2$$

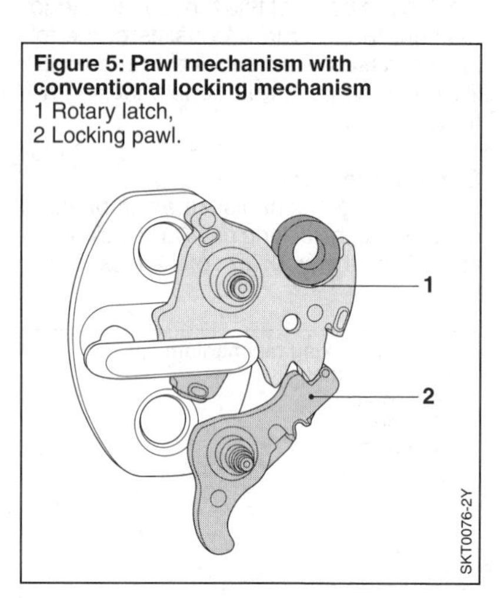

Figure 5: Pawl mechanism with conventional locking mechanism
1 Rotary latch,
2 Locking pawl.

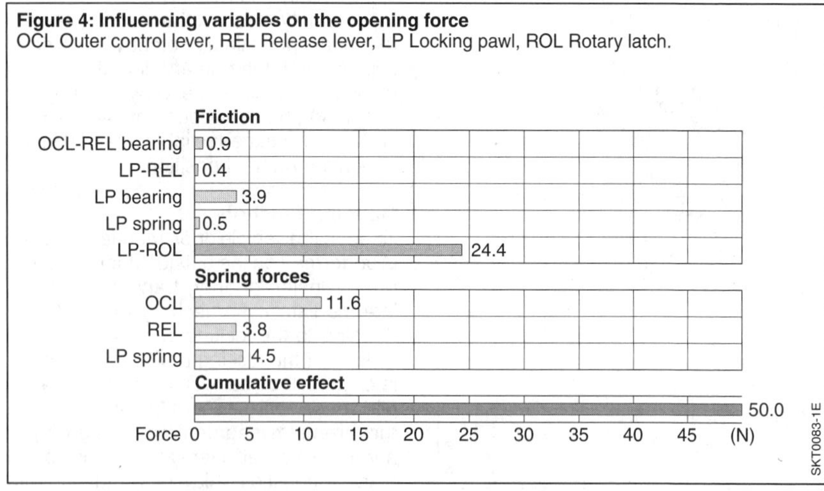

Figure 4: Influencing variables on the opening force
OCL Outer control lever, REL Release lever, LP Locking pawl, ROL Rotary latch.

(direction of force) at the rotary latch. The required opening forces are drastically reduced.

Even more importantly, the force characteristic can be specified with the intermediate-pawl contour over the opening travel (Figure 6).

Figure 6: Opening force against release travel (haptics)
1 Conventional locking mechanism,
2 Two-pawl locking mechanism.
A Inclusion of outer control lever,
B Inclusion of release lever,
C Pre-tension of locking pawl,
D Movement of locking pawl,
E Opening of rotary latch,
F Locking pawl free.

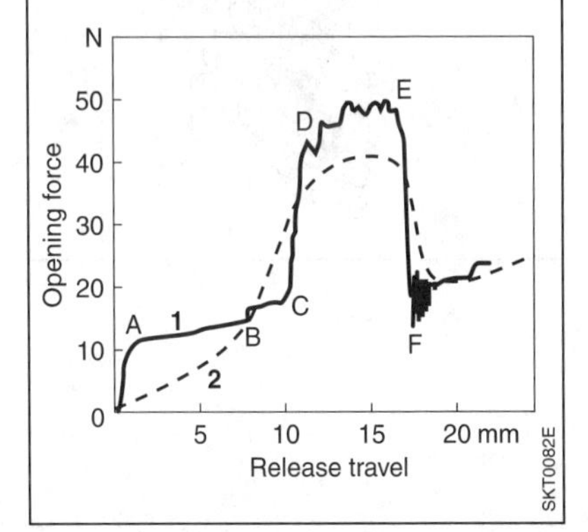

Double-locking functions

Double-locking means that the outer opening mechanism is deactivated under defined conditions. Because the locking system must continue to function reliably during and after a power failure, the entire double-locking actuator system is mechanically redundant. In each case the double-locking state is stored mechanically. Positive-locking coupling mechanisms featuring a bistable spring and a locking lever have proven successful here. Actuation is generally by electrical means (operating comfort). Only in exceptional situations is the mechanical redundancy engaged: Double-locking can then be performed at an actuating socket in the end panel while the door is open, unlocking is performed necessarily each time the inside door opener is operated while the occupant alights from the vehicle. Only the driver's door still features a – concealed – lock barrel to facilitate access from the outside with an emergency key.

For cost reasons, but also to eliminate disruptive mechanical joints on the inside of the door, the classic locking knob in the door capping is on the decline. In future an LED will indicate the double-locking status.

Further functions

Theft deterrence

In addition to the double-locking function, the inside door openers and locking knobs are deactivated. The vehicle can now only be unlocked by the remote control or with a mechanical emergency key.

Childproof lock

Only the inside-opening chain is deactivated. Occupants are however able to alter the double-locking status from the inside (rescue and protection scenarios). Only the opening function from the inside is disabled.

Electrical opening

The lock in the detent mechanism is overridden by an electric motor (locking-pawl drive).

Override
To prevent the user from being locked out, a double-locking mechanism that may be engaged is overridden from the inside during opening.

Lock-out protection
When the rotary latch is opened, the double-locking function at the driver's door is not possible.

Double actuation
Two independent actuations are required at the rear door when alighting from the vehicle (US requirement).

i-close power-assisted closing
This function originates from the premium segment, where it was used to deliver the much higher door-seal pressures. When the preliminary detent has been reached, the door is electromechanically pulled against the building seal pressure into the primary detent (rotary-latch drive). The closing operation can be mechanically interrupted at any time (Figure 7).

The annoying noise created when the door is closed is eliminated. Users no longer need to slam doors shut, but can guide them to their starting positions with negligible effort.

Another advantage is that, after power-assisted closing, vehicle doors are reliably closed in the primary detent in all conceivable situations (taxi passengers, children, carrying luggage in both hands, etc.).

Figure 7: i-close power-assisted closing
1 Motor with gear set,
2 Drive lever for rotary latch,
3 Mechanical release.

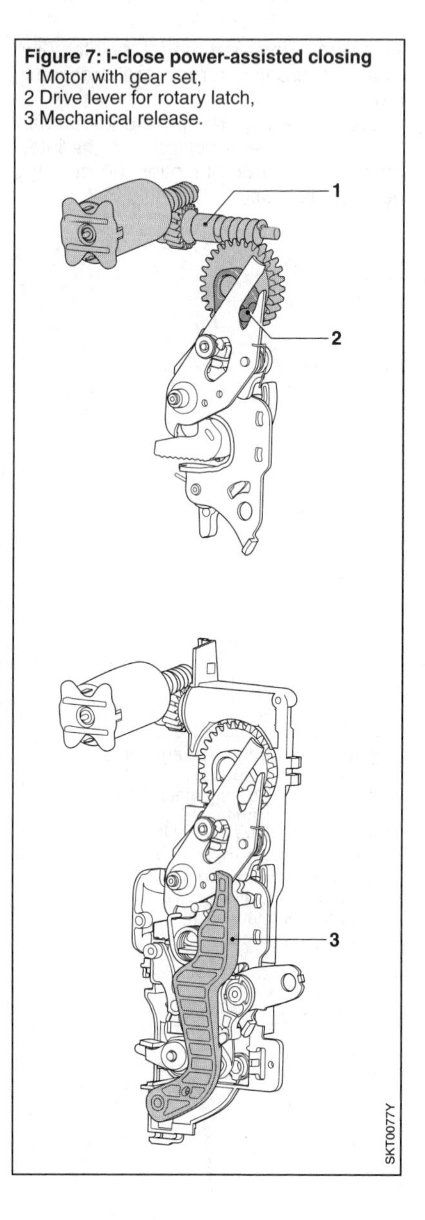

Lock acoustics

A crucial quality characteristic of present-day door locks is their acoustic properties. Suitable design measures can be specifically used to achieve a rich closing sound or an agreeable and at the same time clearly perceptible locking sound. These measures include a minimized mass moment of inertia and low spring force, the avoidance of load-change reactions, pulse-minimized contact geometry, sound-deadening elements, and the isolation of resonant surfaces (Figure 8). The development work and outlay involved is considerable.

Matching the friction parameters of the rotary latch and the lock holder with the locking-mechanism stiffness helps to effectively eliminate annoying clicking and creaking noises. The critical displacements between the rotary latch and the lock holder are then moved into ranges which do not arise under normal driving conditions.

Flow inside the vehicle door

Diverting the flow around the vehicle while it is moving creates a pressure differential between the passenger cell and the door outer skin. Differential pressures of up to 80 MPa are recorded. The upshot of this phenomenon is a secondary flow through the interior of the vehicle door. The flow usually contains dust and its direction depends on the relative position of the sealing line to the entry opening of the door lock.

The door-lock design dictates that it cannot be completely sealed (entry opening for the lock holder). It is therefore necessary to take specific measures to divert the dust flow away from the precision-mechanical assemblies. In practice the prototype is introduced into a flow system for dust flow. The flow optimum is then elaborated using methods of visualization, differential-pressure measurement, etc.

Figure 8: Example of acoustic optimization of a rotary latch
a) Potential noise sources at rotary latch,
b) Time characteristic of closing pulse,
c) Changes to rotary latch.
1 Original contour of rotary latch,
2 Optimized contour.

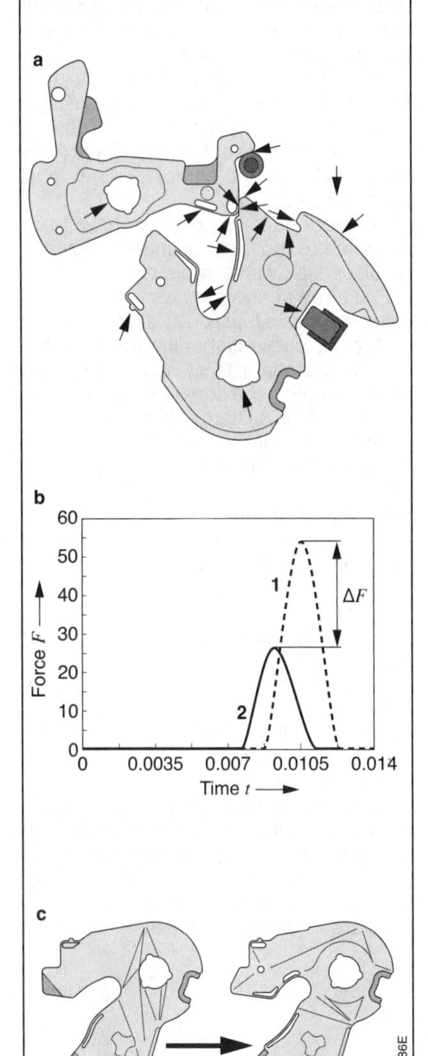

Electrical locking system

With the current trend towards electromobility, the electric lock is gaining in importance again. This type of lock consists of only the locking mechanism and the opening actuator. The door handles and other operating elements are equipped with sensors. Electrical wires replace the mechanical connections to the lock.

The advantages of an electrical locking system are considerable: Reduced size and weight of the lock, symmetrical design, and only one lock variant per vehicle (variant encoding at the end of the line) must be mentioned here.

The door handles no longer have to be designed to move, or ultimately they can be dispensed with completely.

Fault-tree analysis shows that electrical locking systems function at least as reliably as today's conventional locking systems. What is essential in this respect is a reliable power supply, i.e. a separate battery powers the locking system, while the vehicle's main battery (active redundancy) is only engaged in exceptional situations.

i-move door comfort systems

Future expansion of the locking system will focus on automatic movement of the door (i-move). This will make it much easier for the user to get into the vehicle.

These solutions involve the door being electrically driven. The door initially moves under operator control to the preliminary detent. Once it has reached this point, the preliminary-detent signal automatically initiates system transfer to power-assisted closing (i-close).

To deliver smooth operation but also as a comfort feature in its own right (e.g. in tight parking spaces), the door stop will in future be steplessly arrestable (i-stay). Hydrodynamic and viscomechanical solutions are currently being developed.

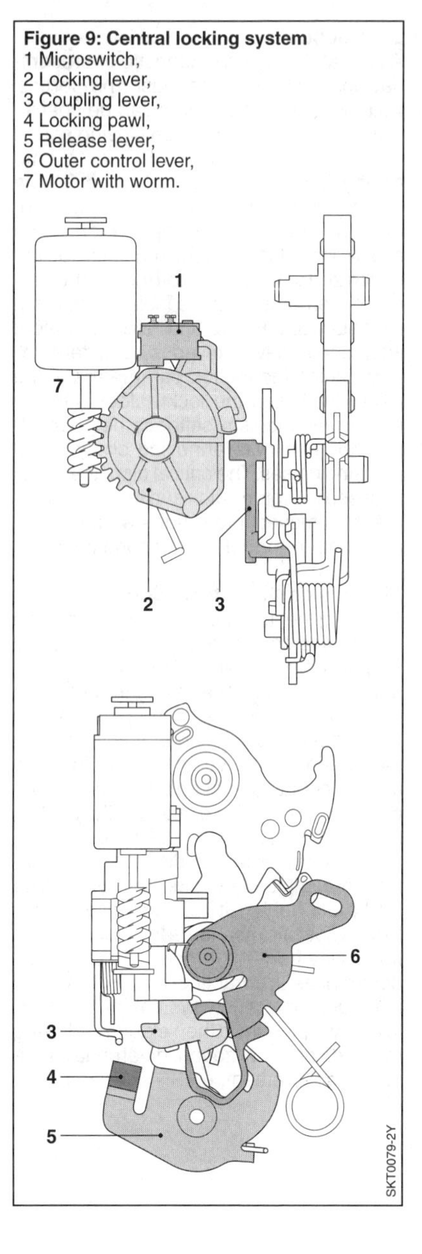

Figure 9: Central locking system
1 Microswitch,
2 Locking lever,
3 Coupling lever,
4 Locking pawl,
5 Release lever,
6 Outer control lever,
7 Motor with worm.

Technology

Lock components are made of steel materials (precision stampings). Elaborate heat and surface treatment methods (finishing, quenching and drawing, coating) are required to reduce the potential for corrosion, friction and wear.

The rotary latch and the locking pawl are sheathed in plastic materials to provide effective sound-deadening in order to satisfy requirements to minimize closing noise. Higher-grade door locks normally feature extensive sound-deadening measures incorporated in the lever mechanisms and actuators.

Cost-intensive electroplating is on the decline. Lock cases, for instance, are today coated in painting processes (zinc-lamella systems).

The metallic components in the lever mechanism have to a large extent been superseded by plastic components made of reinforced PBT (polybutylene terephthalate). The main reasons behind this are the absence of surface treatment, shortened process times, and reduced weight.

Large plastic components are made of microcellular foam. The components feature the desired bonelike porous structure on the inside; the zones close to the surface are undisrupted (Figure 10). Especially when it comes to the bending stress to which the components are for the most part subjected, the reduced strength resulting from this porous structure is technologically negligible. The 10 % reduction in weight and the 35 % reduction in process time more than make up for the increased complexity and expenditure of the associated mechanical engineering and process monitoring.

In the past, operation of automotive locks was purely mechanical. Central locking was then introduced in the form of flange-mounted locking actuators (electrically and pneumatically operated control elements). These actuators are today integrated in the housings of higher-grade locks. The pneumatic solutions have been replaced to the greatest possible extent by electric actuators. Miniature motors carry out the electrical commands from the access authorization facility in the form of mechanical actuating movements. The high-speed drives are converted by worm gears (Figure 9). The crucial benefits of this type of gearing are a high conversion ratio (space requirement) and inner sound deadening (positioning noise).

Sensors in the lock effect electrical evaluation of the lock status. Hall sensors and microswitches are used here.

The signals are transmitted between the door lock and the door control unit by conventional means. Digital conversion first takes place in the door ECU unit so as to facilitate communication with the other doors and the central unit in the overall system. Data transfer via a LIN bus represents state-of-the-art technology.

Figure 10: Microcellular-foam components
a) Maximum cell size 50 µm,
b) Maximum cell size 300 µm,
c) Maximum cell size 1,000 µm.

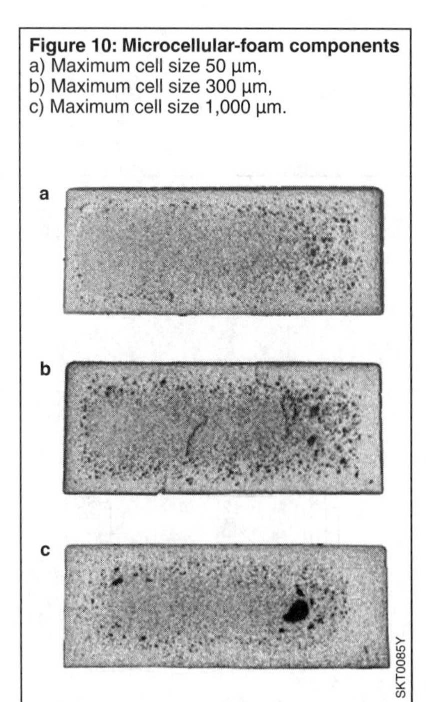

Theft-deterrent systems

Regulations

Theft-deterrent (alarm) systems must meet the requirements of ECE regulations R18 [1] and R116 [2]. An armed alarm system must be capable of issuing acoustic and visual alarm signals in the event of unauthorized intervention in the vehicle. Extended requirements result from insurance regulations in various countries.

Permissible alarm signals
- Intermittent sound signals (25 to 30 s duration, 1,800 to 3,550 Hz; min. 105 dB(A), max. 118 dB(A) at 2 m distance
- Visual flashing signals (max. 5 minutes).

System design

The alarm system comprises a central control unit, sensors that detect intervention, and an alarm siren (Figure 1).

Alarm detectors
- Door and hood contacts
- Interior monitoring
- Tilt sensor
- Self-monitoring of the alarm siren.

Alarm system control unit
Control of the alarm system is assumed by a control unit of the comfort electrical systems in the vehicle. The alarm system is armed or disarmed via the mechanical locks or via the radio remote control. When the command "Arm" is received, monitoring of the alarm detectors is activated. If required, selection buttons can be used to disable individual sensors. This deactivation applies once to the next "Arm" command.

Alongside the sensors, the alarm siren is also armed. Once the siren has assumed the "Armed" status, the control unit starts cyclical communication to monitor the lines. If an alarm is triggered, deactiva-

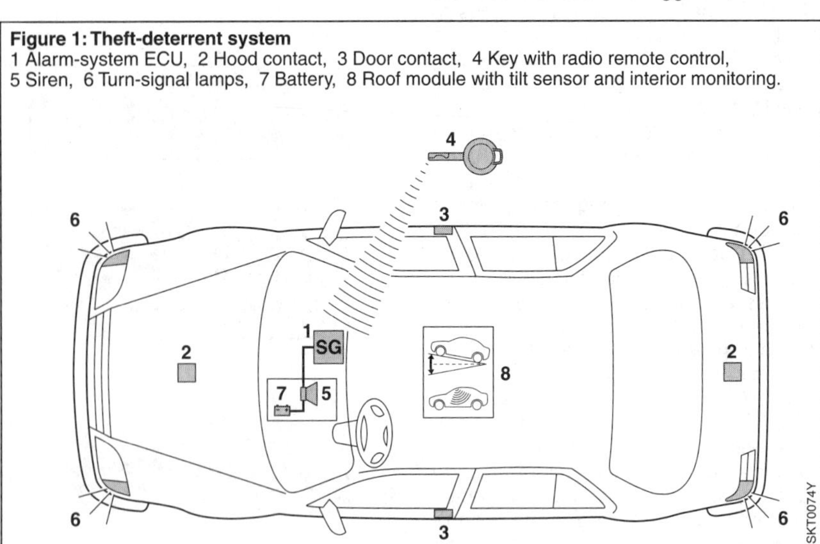

Figure 1: Theft-deterrent system
1 Alarm-system ECU, 2 Hood contact, 3 Door contact, 4 Key with radio remote control, 5 Siren, 6 Turn-signal lamps, 7 Battery, 8 Roof module with tilt sensor and interior monitoring.

tion of the alarm (e.g. via the radio remote control) must lead to immediate termination of the alarm.

The control unit must ensure that the max. permitted number of alarms per alarm detector is not exceeded. Moreover, overlapping alarms must not lead to prolonged alarm output.

Alarm siren

The alarm siren comprises control-system electronics, a diaphragm as sound source, and a rechargeable battery for redundant power supply. The alarm tone is generated by, for example, a piezo-loudspeaker. Communication with the control unit of the alarm system takes place across a serial single-wire bus (e.g. LIN). On the basis of a command from the control unit, the alarm siren can trigger the alarm or autonomously detect a manipulation. Here the following signals are monitored: cyclical communication, line contact to terminal 30 and terminal 31, voltage gradient of the vehicle battery, and overvoltage e.g. due to external supply.

The aim is to recognize manipulations reliably and to prevent false alarms.

Figure 2: Electronic wheel-theft and tow-away protection

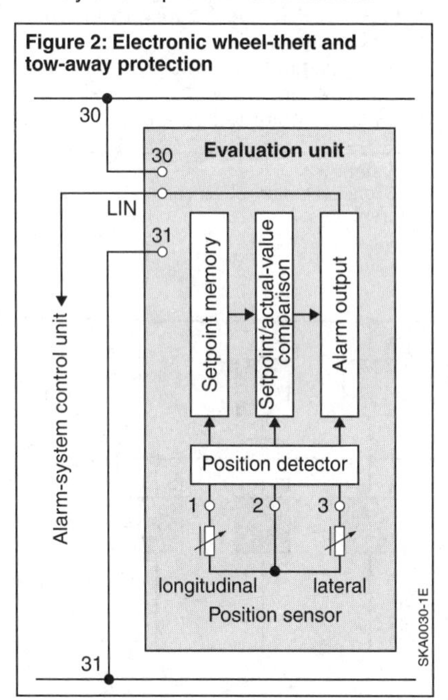

Tilt sensor

The task of the tilt sensor is to detect changes in the vehicle position caused by jacking up or towing the vehicle. A very sensitive acceleration sensor is deployed here. Its signal depends on the angle in relation to gravity. In order to monitor both the longitudinal and transverse axes of the vehicle, twin-axis micro-mechanical sensors are deployed (Figure 2).

Particularly important is the reliable prevention of false alarms, as the vehicle rolling, e.g. due to wind, may not lead to an alarm.

Interior monitoring

The reflections of an ultrasound or a microwave field are analyzed (Figure 3). If the modification exceeds a comparative value, the alarm is triggered. Here, too, reliable prevention of false alarms is an important criterion.

References

[1] ECE-R18: Uniform provisions concerning the approval of motor vehicles with regard to their protection against unauthorized use.
[2] ECE-R116: Uniform technical regulations on the protection of motor vehicles against unauthorized use.

Figure 3: Ultrasonic passenger-compartment protection
1 Ultrasound detector,
2 Ultrasound field.

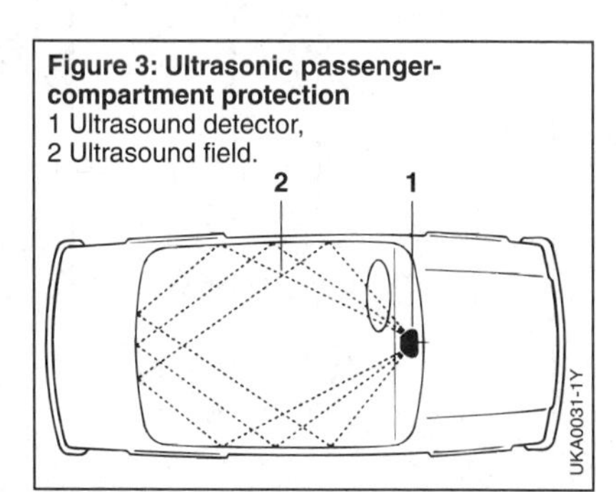

Conventional vehicle electrical systems

Electrical power supply

Function and requirements

The vehicle electrical system of a motor vehicle comprises the alternator as the energy converter, one or more batteries as the energy accumulators, and the electrical equipment as consumers (Figure 1). The energy from the battery is supplied to the starter, which then starts the vehicle engine. During vehicle operation, the ignition and fuel-injection system, the control units, the safety and comfort and convenience electronics, the lighting, and other equipment have to be supplied with power. The alternator supplies the required electrical power for these components and also to charge the battery.

Heightened comfort and safety requirements result in a significant increase in the power demand in the vehicle electrical system. Added to this is the continuing trend towards the electrification of ever more vehicle components (e.g. seat adjustment, electric parking-brake system, electric steering servo). The nominal output of alternators ranges from approx. 1 kW in subcompact-size cars to over 3 kW in the upper class. This is less than the consumers need in total. In other words, the battery must also supply power at times during vehicle operation.

An equalized battery charge balance must always be guaranteed through the choice and dimensioning of battery, alternator, starter and the other consumers in the electrical system so that the internal-combustion engine can always be started and, when the engine has been switched off, specific electrical consumers can still be operated for an appropriate period of time.

The alternator supplies current (I_G, Figure 1) when the engine is running. To be able to charge the battery, the alternator must increase the vehicle-system voltage above the battery open-circuit voltage. However, the alternator is only able to do this when the cut-in consumers do not demand more current than it can supply. If the equipment current draw I_V in the vehicle electrical system is greater than the alternator current I_G (e.g. when the engine is idling), the battery is discharged. The vehicle system voltage falls to the voltage level of the battery from which current is drawn.

The maximum alternator current is very much dependent on the rotational speed and the alternator temperature. When the

Figure 1: Schematic diagram of a vehicle electrical system
1 Battery, 2 Alternator, 3 Alternator regulator, 4 Starter, 5 Ignition switch, 6 Consumers.
I_B Battery current, I_G Alternator current, I_V Equipment current draw.
B Battery, G Alternator, S Starter.

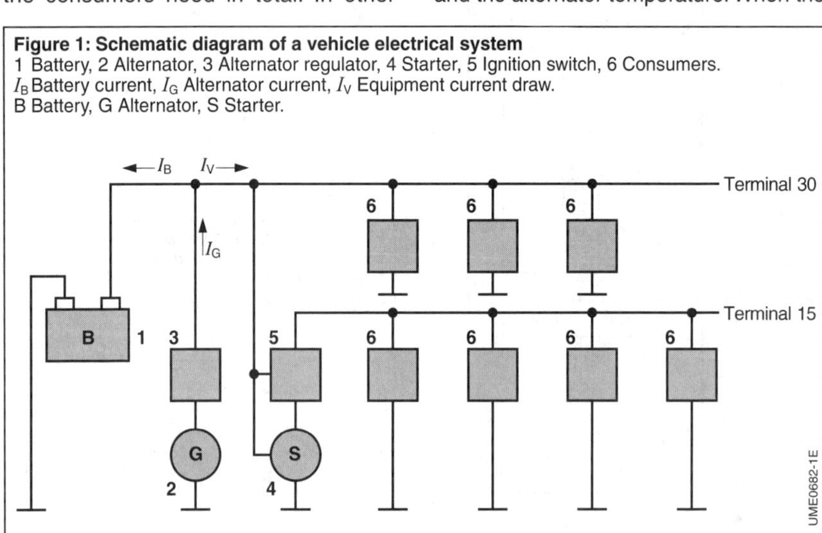

engine is idling, the alternator can only supply 55 to 65 % of its nominal output. However, immediately after a cold start at low outside temperatures, the alternator is able from medium engine speed to supply the vehicle electrical system with up to 120 % of its nominal output. When the engine is hot, the engine compartment heats up, depending on the outside temperature and engine load, to 60 to 120 °C. High engine-compartment temperatures cause high winding resistances, which reduce the maximum alternator output.

Layout and operating principle of the 14 V electrical system

Schematic diagram

The electric system in a motor vehicle can be represented as the interaction of the energy converter (alternator), the energy accumulator (battery) and the consumers (Figure 1).

The alternator is driven via the V-belt by the crankshaft and converts mechanical energy into electrical energy. The alternator regulator limits the output power such that the desired voltage set in the regulator is not exceeded (14.0 to 14.5 V).

When the ignition key is removed, only a few consumers are supplied with voltage (e.g. anti-theft alarm system, car radio, auxiliary heating). The terminal via which these consumers are supplied is called "terminal 30" (continuous plus).

The other consumers are connected to "terminal 15". When the ignition switch is in the "ignition on" position, the battery voltage is applied to this terminal so that now all the consumers are connected to the power supply.

Battery installation locations

In most cars the battery is located in the engine compartment. However, a large battery (e.g. 100 Ah) takes up a lot of space and sometimes, when the amount of space available in the engine compartment is restricted, may not be able to be installed there. Another argument against installing the battery in the engine compartment is the potentially high ambient temperatures. As an alternative, the battery may be installed in the luggage compartment or in the passenger compartment (e.g. under the front passenger seat).

Effect of installation location on charge voltage

The line between a battery installed in the engine compartment and the alternator is shorter than for a battery installed in the luggage compartment. This affects the line resistance and thus directly the voltage drop on the cable. Voltage drops can be minimized by suitable conductor cross-sections and good connections with low contact resistance, even after a long service life.

Figure 2a shows the conditions for installation in the engine compartment. A battery installed in the luggage compartment requires a longer supply line with the additional line resistance R_{L2} (Figure 2b). Because of the higher voltage drop, the charge voltage for a battery installed in the luggage compartment is

Figure 2: Battery installation locations
a) Installation in engine compartment,
b) Installation in luggage compartment.
G Alternator, B Battery, S Starter,
R_L Line resistances,
R_V Equipment resistances,
I_G Alternator current,
I_V Equipment current draw,
I_B Battery charging current.

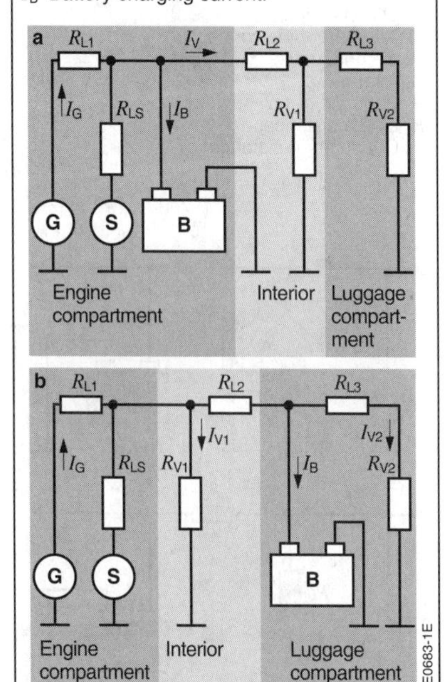

lower. The additional voltage difference caused by R_{L2} can be balanced by increasing the setpoint alternator-voltage value. This increases the alternator output.

Effect of installation location on startability

Startability is dependent on the voltage applied at the starter. The higher this voltage, the higher the starter speed during the starting sequence. Because of the high starter current, the resistance of the supply line has a crucial effect on this voltage. In the case of the variant with the battery installed in the luggage compartment, the line between the battery and the starter is longer than for the battery installed in the engine compartment; accordingly, the resistance and with it the voltage drop are higher. Better startability is assured when the battery is installed in the engine compartment with short lines to the starter.

Figure 3: Connection options for consumers
a) Battery-side connection of consumers,
b) Alternator-side connection of consumers.
G Alternator,
B Battery,
R_L Line resistance,
R_V Consumer resistance.

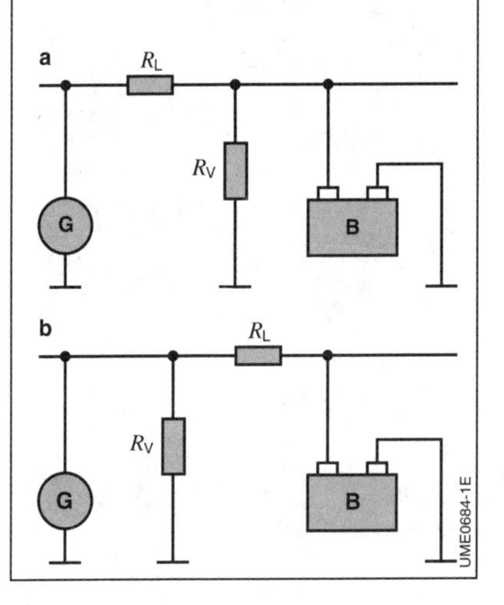

Effect of ambient temperature

High temperatures, as can be encountered in the engine compartment, intensify aging effects in the battery (e.g. corrosion of the battery and water loss due to gassing), which have a negative impact on battery service life. High temperatures in the battery can be reduced by shielding.

The charge acceptance of the batteries is restricted at low battery temperatures, i.e. the battery takes in less charge, which can lead to a low state of charge. Low states of charge in turn facilitate aging effects such as sulfating, which lead to reduced usable capacity and therefore shorter service life.

Effect of installation location on voltage stability

Because the battery can only be charged with direct current, the alternating current generated in the alternator must be rectified. This job is performed by a diode rectifier integrated in the alternator. Rectification of the alternating voltage creates a pulsating direct voltage. Furthermore, the switching of the diodes – when the current is commutated from one diode to the next – generates high-frequency voltage oscillations, which are smoothed to the greatest possible extent by the interference-suppression capacitor installed in the alternator.

Electronic consumers (e.g. ECUs) can be disrupted or even damaged by the voltage peaks or the voltage ripple. The battery can use its large capacity to smooth the voltage fluctuations. However, because of the line resistance R_L between the alternator and the battery (Figure 3), they are not completely suppressed at the alternator. When the consumers are connected on the battery side (Figure 3a) or after the battery (e.g. R_{V1} and R_{V2} in Figure 2a), they are supplied with the extensively smoothed system voltage. When the consumers are connected on the alternator side, i.e. directly to the alternator (Figure 3b), they are exposed to a greater voltage ripple and greater voltage peaks.

Electrical devices (consumers) which feature a high power input and relative insensitivity to overvoltage should therefore be connected close to the alternator, whereas voltage-sensitive loads with a low power input should be connected close to the battery.

Power output of the consumers

Consumer classification

The electrical consumers have a variety of switch-on durations. Here the operative distinction is between:

- Continuous loads, which are always switched on (e.g. electric fuel pump, engine management)
- Long-term loads, which are switched on where necessary and then remain on for an extended time (e.g. low-beam headlamps, car radio, electric radiator fan)
- Short-term loads, which are only switched on for brief periods (e.g. turn-signal lamps, stop lamps, electrical seat adjustment, power-window units).

Running-time-dependent electrical load requirements

The electrical load requirements encountered during vehicle operation are not constant. The first minutes after startup are generally characterized by high demand (e.g. heated rear window, seat heating, door-mirror heating), followed by a sharp drop in electrical load requirements.

These consumers are switched off after a few minutes. The electrical load requirements are then determined predominantly by the continuous loads and the long-term loads.

No-load-current consumers

Different ECUs and consumers require a power supply even when the vehicle is parked. The no-load current is composed of the sum total of all these switched-on consumers. Most of these consumers switch off shortly after the engine is turned off (e.g. interior lights). Some, however, are always active (e.g. anti-theft alarm system).

An unexpectedly high no-load current can also occur when, for example, shutdown of the ECU interconnection does not work correctly or recurring "Wake-Ups" cause an ECU to frequently reactivate the interconnection when the vehicle is parked. The control unit networking must be configured and validated for this.

The no-load current must be supplied by the battery. The maximum value for the no-load current is defined by the vehicle manufacturers. Battery dimensioning is based among other things on this value. Typical values for the no load current in a passenger car are approx. 3 to 30 mA.

Current output of the alternator

Important component parts of the alternator are the stator (Figure 5) and the rotor, which is driven via the V-belt by the crankshaft. An alternating voltage is induced in the three stator windings (see Three-phase alternator) when a current flows in the rotor coil (excitation current) and a magnetic field is thereby built up. The excitation current is tapped from the generated alternator current (self-excitation). The induced voltage is dependent on the rotor speed and the level of the excitation current. The generated alternating voltage is rectified by diodes.

The voltage induced in the alternator is dependent on the alternator speed and thus on the engine speed. Therefore the voltage is low at low speeds. For an engine idling speed of n_L, the alternator can only supply some of its rated current if it has a conventional turns ratio (crankshaft speed to alternator speed) ranging from 1:2.5 to 1:3 (values for passenger cars;

Figure 4: Alternator current output I_G as a function of alternator speed
I_V Equipment current draw,
I_G Alternator current,
n_L Engine idle speed.

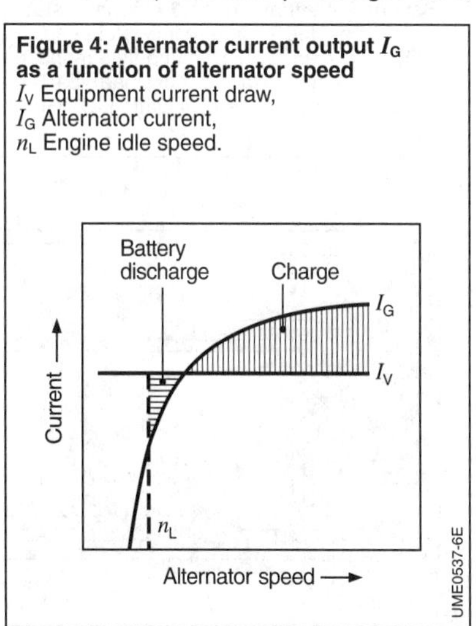

turns ratio is much higher for commercial vehicles) (Figure 4). The rated current is achieved at full load at an alternator speed of 6,000 rpm. The mean speed attained during vehicle operation must be sufficiently high in order to achieve the nominal alternator output. Driving cycles involving extensive levels of idling are particularly critical, because the available alternator output is so low that the battery is discharged when the electrical load requirements are high.

If the alternator voltage is higher than the battery voltage, a battery charging current flows into and charges the battery. The voltage is limited by the alternator regulator so as to maintain a system voltage of approx. 14 V.

Power generation by the alternator also affects fuel consumption. The increased consumption for 100 W output is in the order of 0.17 *l* per 100 km driven and is dependent on the efficiency of both the alternator and the internal-combustion engine.

Voltage regulation in the vehicle electrical system

Generating the excitation field when starting

A magnetic field is required in the rotor to enable a voltage to be induced in the stator windings of the alternator. Self-excitation is not possible at low speeds after starting. In this phase the starter battery delivers the excitation current (external excitation).

The torque of an alternator running under load would hinder the starting sequence and idle stabilization of the internal-combustion engine. Today's regulators therefore correct the excitation current during the starting phase to a low level (controlled pre-excitation). Current generation is delayed until the engine has run up to speed (Load-Response Start, LRS). Until this point the consumers are supplied by the battery.

Voltage regulation during vehicle operation

The excitation current is tapped from the generated alternator current (self-excitation). The alternator regulator adjusts the excitation field by way of a pulse-width-modulated (PWM) current in the rotor winding in such a way that the voltage at

Figure 5: Interaction of alternator, alternator regulator and battery
1 Battery, 2 Alternator rotor, 3 Alternator stator, 4 Alternator regulator,
5 Rectifier diodes, 6 Charge indicator lamp, 7 Ignition switch.
DF Dynamo field, DFM Dynamo-field monitor, L Lamp connection, 15 Terminal 15,
S Sense (battery voltage supply lead), B+ Battery plus, B– Battery minus,
V Phase signal for speed evaluation, D– Ground connection.

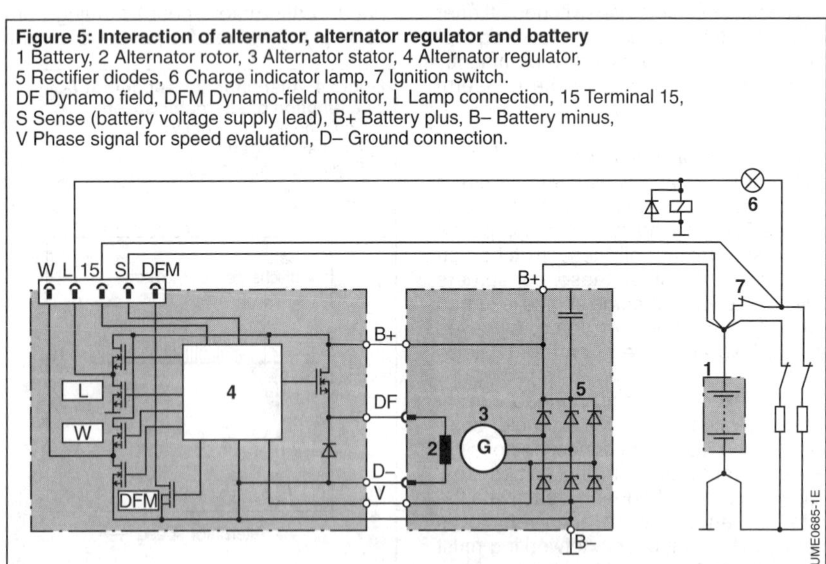

B+ corresponds to the specified setpoint value. The frequency of the PWM signal is 40 to 200 Hz; the pulse duty factor is dependent on how much power is demanded by the consumers. In the case of a load variation the vehicle system voltage changes, whereupon the regulator adjusts the excitation field by adapting the PWM signal in such a way that the voltage is compensated.

The connection of the excitation winding is known as DF (dynamo field). The alternator regulator outputs the PWM signal as a DFM (DF monitor) to notify other ECUs of the alternator capacity utilization.

The regulator requires the battery-voltage value in order to regulate. It obtains this value via terminal B+. In the case of a long supply line and high currents on the line between alternator and battery, the voltage drop between the battery and the regulator can be high to the extent that power generation by the alternator is too low and the battery is possibly insufficiently charged. This problem can be remedied by the S-terminal, which supplies the regulator with battery voltage via a cable which is separately connected to the battery positive terminal.

The regulator bus connection (e.g. LIN bus) allows the system to vary the setpoint value to which the voltage is to be regulated. This makes possible functions such as, for example, recuperation. The Load-Response Driving function effects ramp regulation of the alternator voltage back to the setpoint value during vehicle operation after the connection of a high load and the associated sudden drop in the alternator voltage. This function prevents the alternator from abruptly subjecting the engine to load.

Charge indicator lamp

The charge indicator lamp is activated by the alternator regulator. It comes on with "ignition on" and goes out when the alternator is supplying current. The regulator activates the charge indicator lamp as soon as it detects a fault (e.g. alternator failure due to V-belt break, open circuit or short circuit in the excitation circuit, open circuit in the charging cable between the alternator and the battery).

Charging the battery

Due to the chemical processes that take place in the battery, the ideal battery charge voltage has to be higher at low temperatures and lower at high temperatures. The gassing voltage curve shows the maximum permissible voltage at which the battery does not produce gas significantly. The alternator regulator limits the voltage if the alternator current I_G is greater than the sum of the consumer equipment current draw I_V and the temperature-dependent, maximum permissible battery charging current I_B.

Regulators are normally mounted on the alternator. If there are significant deviations between the temperatures of the voltage regulator and the battery electrolyte, it is better to monitor the voltage-regulation temperature directly at the battery. This is possible in vehicles with a battery sensor. The temperature value is conveyed by means of a communication interface (e.g. LIN bus) (see Battery sensor).

The arrangement of the alternator, battery and consumers influences the voltage drop on the charging cable and thus the charge voltage. The total current $I_G = I_B + I_V$ flows through the charging cable if all electrical equipment is connected to the battery. The charging voltage is accordingly less due to the high voltage drop. If all consumers are connected on the alternator side, the voltage drop is lower and the charging voltage is higher. The regulator can take into account the voltage drop by measuring the actual voltage value directly at the battery.

Electrical-system structures

One-battery vehicle electrical system

Figure 1 depicts a one-battery vehicle electrical system, as featured predominantly in passenger-car applications. Working as an energy accumulator is a battery which both supplies the current for the starting sequence and delivers the power supply to the consumers in the event of no alternator output (engine switched off) or insufficient alternator output (idling phases). This is the most widely used concept today in that it is the most cost-effective solution for supplying power in motor vehicles.

In the design of a vehicle battery for the one-battery electrical system which supplies both the starter and the other consumers in the vehicle electrical system, a compromise has to be found between different requirements. During the engine starting sequence, the battery is subjected to high current loads (300 to 500 A). The associated voltage drop has an adverse effect on certain electrical equipment (e.g. undervoltage reset on units with microcontrollers) and should be as low as possible.

By contrast, only comparatively low currents flow during standard vehicle operation. Battery capacity is the salient factor in providing a reliable current supply. Neither properties – output nor capacity – can be optimized simultaneously.

Two-battery vehicle electrical system

In vehicle electrical systems with two batteries – starter battery and general-purpose battery – the "high power for starting" and "general-purpose electrical supply" functions are separated by the vehicle electrical system control unit (Figure 6) to make it possible to avoid the voltage drop during the starting process, while ensuring reliable cold starts, even when the charge level of the general-purpose battery is low.

Starter battery

The starter battery must supply a high amount of current for only a limited period of time (during starting). It is therefore designed for a high power density (high power for low weight). Compact dimensions allow installation in the immediate vicinity of the starter motor with short connecting cables (low voltage drop on the line). The capacity is reduced.

General-purpose battery

This battery only supplies the vehicle electrical system (excluding the starter). It supplies currents to the electrical-system consumers (e.g. approx. 20 A for the engine-management system) and must therefore be able to store and provide large amounts of energy. It also has a high cycle stability. This means it can be charged and discharged very often before the performance criteria of the battery are

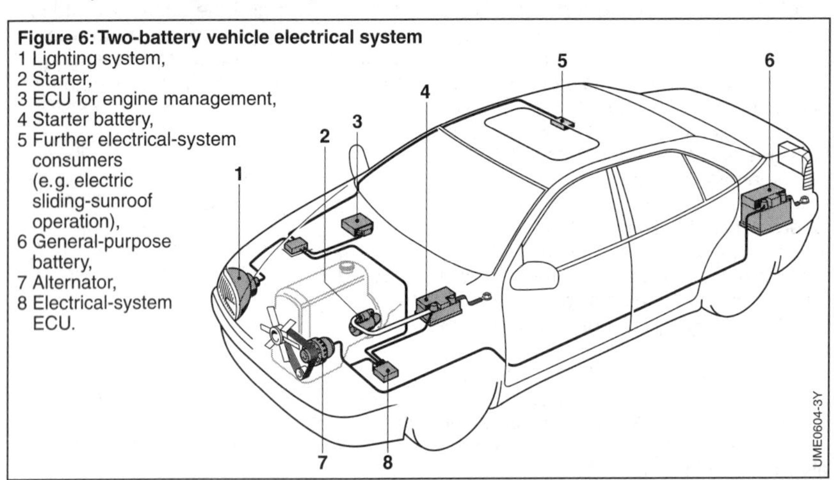

Figure 6: Two-battery vehicle electrical system
1 Lighting system,
2 Starter,
3 ECU for engine management,
4 Starter battery,
5 Further electrical-system consumers (e.g. electric sliding-sunroof operation),
6 General-purpose battery,
7 Alternator,
8 Electrical-system ECU.

no longer met. Dimensioning is based essentially on the capacity reserves required for activated consumers, the consumers that operate with the engine switched off (no-load-current consumers, e.g. receiver for central-locking radio remote control, anti theft alarm system), and the minimum permissible charge level.

Vehicle power supply control unit
The vehicle power supply control unit in a two-battery vehicle electrical system separates the starter battery and the starter from the rest of the vehicle electrical system provided this can be supplied with sufficient power by the general-purpose battery. It therefore prevents the voltage drop that occurs during starting, affecting the performance of the vehicle electrical system. When the vehicle is parked, this prevents the starter battery from becoming discharged by electrical equipment that draws current when switched on and by no-load-current consumers.

The 12 V battery can be charged at higher voltage than the usual alternator voltage in the vehicle. The voltage level is limited by consumers, e.g. the bulbs, whose service life severely deteriorates with the voltage. By separation of the starter battery from the remainder of the vehicle electrical system, there are theoretically no limits for the charge voltage level within the starter battery. It can therefore be raised to the "ideal" value by means of a DC/DC converter (e.g. 15 V). This is dependent on the temperature and on the state of charge. Charging duration is minimized with the increased charge voltage.

If there is no charge in the general-purpose battery, the control unit is capable of provisionally connecting both vehicle electrical systems. This means that the vehicle electrical system can be sustained using the fully-charged starter battery. In another possible configuration, the control unit for the starting operation would connect only the start-related consumers to whichever battery was fully charged.

Electrical-system parameters

State of charge

The state of charge (SOC) of the battery is one of the most important parameters in the vehicle electrical system. It can be defined as the ratio of the amount of charge still stored in the battery (actual state of charge) to the maximum amount of charge which the fully charged battery when new can store:

$$SOC = \frac{Q_{act}}{Q_{max}} .$$

The value Q_{max} is obtained when a fully charged battery is discharged at discharge current I_{20} – corresponding to a twentieth of the nominal capacity in amperes (5 A for a 100 Ah battery) – until the cutoff voltage of 10.5 V is reached. The amount of charge which was drawn during this discharge process corresponds to Q_{max}.

Since in this way Q_{max} is only accessible by means of one measurement (integral of the current over the time, i.e. current and time measurement), the definition can also frequently be provided by the nominal capacity of the battery, which can be found on the label, where: $Q_{max} = K_{20}$ (nominal).

The currently stored amount of charge Q_{act} is obtained from the difference between Q_{max} and the amount of charge drawn during the discharge of the fully charged battery.

The state of charge of the battery correlates directly with the electrolyte density, where furthermore the steady-state voltage of the battery is proportional to the electrolyte density. The voltage value which is obtained when a stable voltage final value is assumed after the battery charging or discharging process is referred to as steady-state voltage. This can, owing to the slow diffusion and polarization processes in the battery, take several days, especially after long charging phases. The steady-state voltage is measured at the terminals.

The state of charge is defined as follows:

$$SOC = \frac{(U_{current} - U_{min})}{(U_{max} - U_{min})}$$

where

$U_{current}$: Current steady-state voltage.
U_{max}: Steady-state voltage of the fully charged battery (SOC = 100 %).
U_{min}: Steady-state voltage of the battery at SOC = 0 %. Because the dependence of the steady-state voltage on the state of charge is nonlinear for low states of charge (approx. less than 20 %), the value linearly extrapolated to SOC = 0 % must be used here.

In this way, it is possible to infer the state of charge from the measured steady-state voltage.

Functional and performance capability

The capability of the battery to meet a specified functional or performance requirement in its current state – e.g. providing the starting power for the internal-combustion engine – is called the SOF (State of function). The SOF is application-specific and can therefore not be defined generally. For example, the SOF can be used not only as a measure of startability, but also for evaluating the current performance capability of the battery to supply other electrical heavy current consumers, e.g. electrical power steering.

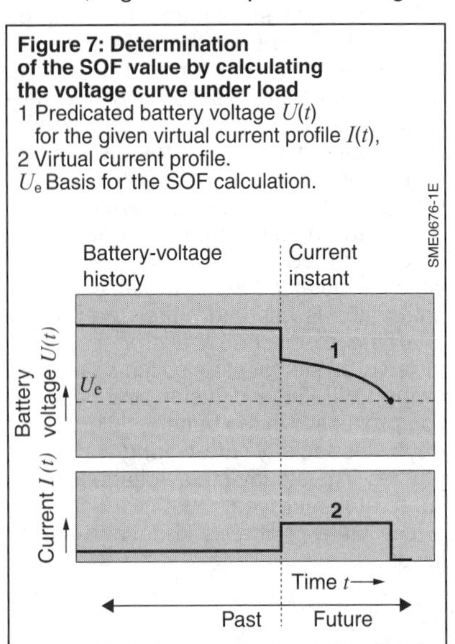

Figure 7: Determination of the SOF value by calculating the voltage curve under load
1 Predicated battery voltage $U(t)$ for the given virtual current profile $I(t)$,
2 Virtual current profile.
U_e Basis for the SOF calculation.

The current performance capability of the battery is evaluated using the SOF by precalculating the voltage drop at battery load with a predefined current profile. If the precalculated battery voltage U_e drops below a specified threshold, this means that the battery is no longer delivering the required performance (e.g. for an engine start) (Figure 7).

Moreover, it is also possible to query the current remaining charge or the discharge duration which states how long the battery can continue to deliver the required performance. An application example of this is how long a vehicle can be parked at a known no-load current consumption without losing startability. In this case, the SOF does not provide the voltage drop at high-current load, rather the remaining charge reserve or the discharge duration until the startability threshold is no longer reached (Figure 8).

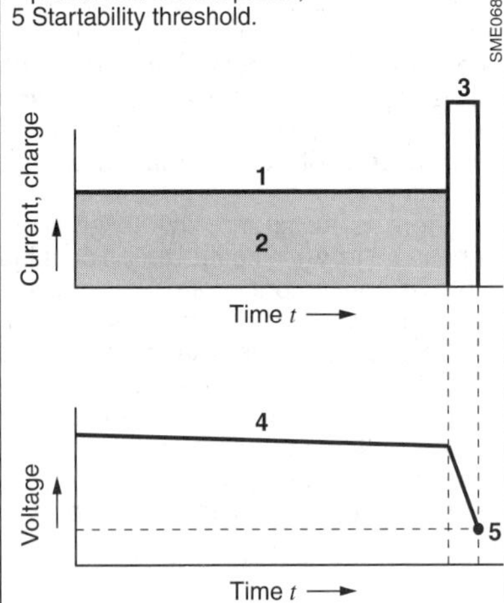

Figure 8: Calculation of the SOF value based on the available charge reserve
This calculation is to determine the charge reserve or the time duration after which under no-load current the startability threshold will be reached.
1 No-load current, 2 Charge reserve,
3 Starting current,
4 Voltage curve under load with predefined current profile,
5 Startability threshold.

As already mentioned, the SOF is determined on the basis of the current battery state, i.e. it depends on the current state of charge, the temperature and the state of health of the battery. These variables have to be determined by the battery-status recognition (see Battery sensor) for SOF calculation.

State of health

Depending on the respective operating and ambient conditions, different aging processes take place in a lead storage battery, which can, for example, lead to an increase in internal resistance (caused by corrosion) or to capacity losses (caused by active mass loss or sulfating) and therefore deteriorate the general performance and storage capability of the battery, i.e. its state of health (SOH) over the service life.

The state of health of the battery with regard to its storage capability can, for example, be expressed by the ratio of current capacity K_{20} to nominal capacity of the battery when new K_{20new}:

$$SOH = K_{20} / K_{20new} \cdot 100\ \%.$$

Further (application-dependent) SOH quantities can be derived from the corresponding SOF variables by calculating them not for the current battery state, rather with defined values for state of charge (e.g. SOC = 100 %) and battery temperature (e.g. 25 °C) to achieve the desired aim of only factoring the dependence on the state of health into the SOH quantity.

Whereas the SOF states how well a performance requirement can be met in the current state of the battery, the corresponding SOH variable states how well the battery will meet this performance requirement generally in the current state of health. For example, in this way an SOH variable which calculates the battery's remaining minimum performance capability for the start/stop function can be used to permanently deactivate this function when the battery is severely aged, in order to prevent the risk of breakdowns.

Alternator capacity utilization

The current flowing in the alternator excitation winding determines the voltage induced in the stator windings. The alternator regulator regulates the required excitation current by means of a pulse duty factor (PWM signal). DF (dynamo field) is the terminal through which the excitation current is supplied. The pulse duty factor of the PWM signal indicates the alternator capacity utilization, i.e. whether the alternator still has reserves and can supply even more current for additionally connected loads.

The alternator regulator additionally outputs this signal as a DFM signal (DF monitor). Regulators with a bus interface apply this pulse duty factor to the bus. The excitation current is also output in amperes. Different ECUs evaluate the DFM signal in order, for example, to switch off the seat heating or the windshield heater when the alternator is subject to high capacity utilization.

Electrical energy management

Motivation

Reduced fuel consumption

One of the primary goals of motor-vehicle manufacturers is to reduce fuel consumption and greenhouse gases, in particular CO_2. This is achieved by optimizing the energy flows in motor vehicles. Measures to achieve this goal include:

- Avoiding idling losses by using the stop-start function (automatic engine switch-off and restarting for example when waiting at traffic signals on red).
- Increasing the efficiency of electrical power generation by optimizing the alternator and applying intelligent alternator activation (recuperation).
- Electrically driven accessories to facilitate activation to suit demand by means of isolation from the internal-combustion engine.

Electrical power demand

Additional comfort and convenience functions and electrified accessories give rise to increasing electrical power requirements and at the same time to a reduced range of speed for electrical power generation (e.g. due to stop-start operation). New comfort, convenience and safety functions (e.g. electrical power steering, electric water pump, PTC auxiliary heater, electric climate control in vehicles with stop-start function) require additional electrical power, so much so that it is useful to integrate an electrical energy management system (EEM).

Purpose of electrical energy management

Electrical energy management controls the energy flows and at the same time safeguards the electrical power supply in order to maintain vehicle startability and to reduce breakdowns caused by flat (discharged) batteries. Electrical energy management also stabilizes the battery voltage and optimizes the availability of comfort and convenience systems – even when the engine is stopped. This can be achieved by safeguarding a positive or at least equalized charge balance during vehicle operation and by monitoring the power demand when the engine is stopped. Furthermore, peak loads can be reduced by the coordinated switching of electrical consumers. This is coordinated in the electrical energy management system (Figure 9).

The effects of the measures taken sometimes conflict with one another. For example, switching off comfort and conve-

Figure 9: Electrical energy management (EEM)

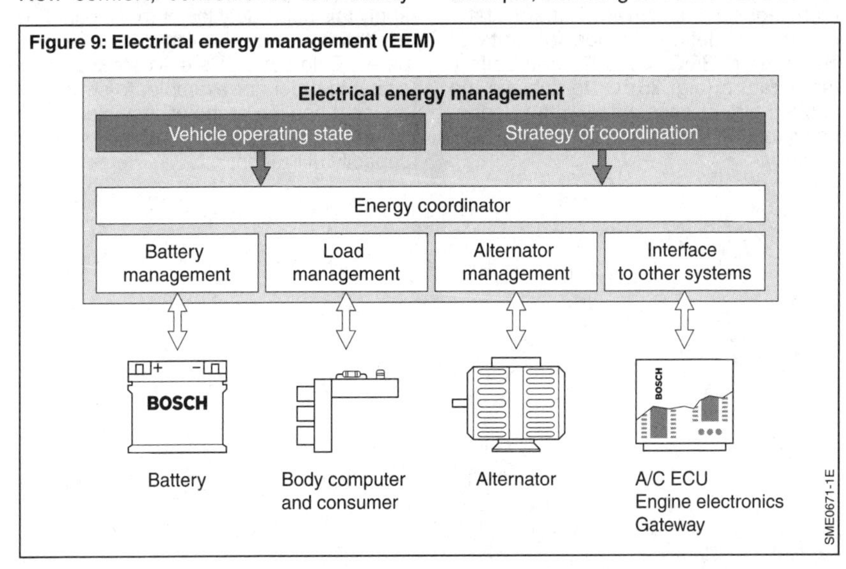

nience consumers results in loss of comfort and convenience, while disabling the stop-start function results in increased fuel consumption. Different vehicle manufacturers prefer different measures, and the possible measures for safeguarding the charge balance are prioritized accordingly.

Functions of electrical energy management

Load management in no-load mode (no-load-current management)

The no-load-current management implemented in the battery sensor by means of software regularly monitors the battery status and thus startability when the engine is stopped. With the aid of precise battery-status recognition it is possible to optimize the availability of consumers by means of no-load-current management, i.e. the operating time of the comfort and convenience consumers can be maximized. In the event of an imminent loss of startability, electrical energy management can, for example, send a message to the display module in order to notify the user. In addition, when the startability limit is approached, load management will reduce the power consumption (e.g. by reducing the power consumption of the A/C fan) right down to shutting down individual consumers in order to maintain startability for as long as possible. Examples of such comfort and convenience consumers are the auxiliary heater, infotainment, navigation system, radio, and telephone.

Load management during vehicle operation

The function of electrical energy management with the alternator active is, in addition to load management, principally alternator management including a recuperation function and the energy-management interface to other systems such as, for example, engine management.

Switching of consumers

Load management coordinates the activation and deactivation of consumers in order to reduce power peaks. Load management is also involved in controlling the high-performance heating systems (windshield heater and PTC auxiliary heater).

In vehicle operation, too, safeguarding restartability is the primary function of electrical energy management. In critical battery states, load management reduces the electrical power demand in order to recharge battery as quickly as possible. Reservoir-type comfort and convenience consumers (heating systems) are switched back by preference, because intelligent activation can be used to delay for as long as possible perceptible deviations from the nominal performance.

There are limits to deactivating comfort and convenience functions in that this will only be accepted by the user in rare exceptional circumstances. The vehicle electrical system must therefore be configured in such a way that these situations only occur rarely. Noticeable effects must be indicated to the user in order to explain the performance that is deviating from normal operation.

Increasing the alternator output

As an alternative or complement to reducing the electrical power demand, it is possible by increasing the engine speed to increase the electrical power generation by the alternator (e.g. idle-speed increase or disabling of engine stop during stop-start operation). For example, to increase the idle speed, electrical energy management issues a request via the data bus to the engine management. These measures have a direct influence on fuel consumption and noise generation, and must therefore be optimally matched to the individual vehicle.

During a recuperation the vehicle's kinetic energy is converted at least partially into electrical energy and delivered to the battery for storage. This function requires an alternator that can be controlled via an interface to input the desired operating voltage and a battery sensor for recognizing the battery status. The function itself can be partitioned in the engine electronics, a gateway or a body computer.

During overrun fuel cutoff an increased desired voltage is input for the alternator in order to charge the battery intensively. Electrical power is generated at this operating point without fuel consumption. In driving states with poor efficiency in terms of electrical power generation the alternator voltage is decreased and the battery slowly discharged again in order to minimize the fuel demand for electrical power generation.

A fully charged battery cannot absorb charge. For this reason, recuperation is only possible with a partially charged battery (partial state of charge, PSOC). This is a departure from the conventional charging strategy, the objective of which is to achieve as fully charged a battery as possible. A minimum battery status required for startability must not under any circumstances be undershot, i.e. the current battery status must be known to electrical energy management.

The recuperation function results in an increased cyclical variation of the battery, the influence of which on battery aging must be tested on an application-specific basis. The use of AGM batteries (absorbent glass mat) to increase the possible energy throughput (throughput in Ah over the service life, the throughput critical to service life increases by a factor of 3) is therefore recommended.

The recuperation algorithm must take into account the influence of voltage changes on the consumers, since these can be observable (e.g. change in the speed of the A/F fan or light flickers).

Recuperation provides for fuel savings in the range of 1 to 4 %, depending on the cycle and configuration of the function.

Figure 10: Interaction of battery sensor, battery-status recognition and electrical energy management
EEM Electrical energy management, BSR Battery-status recognition,
EBS Electronic battery sensor
SOC State of charge, SOH State of health, SOF State of function,
U Battery voltage, I Battery current, T Battery temperature.

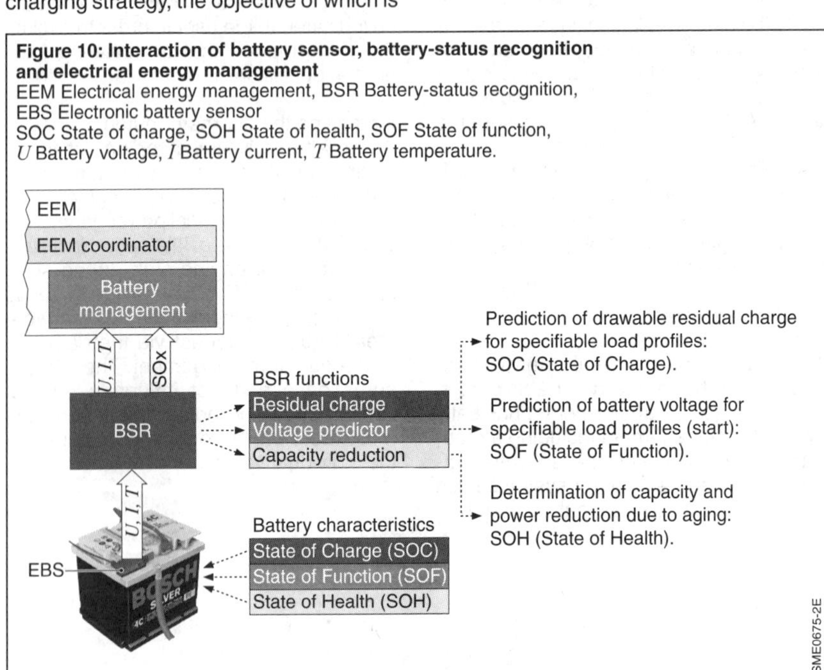

Battery-status recognition and battery management

Function

A crucial element to sound electrical energy management is a battery-status recognition facility (BSR) which reliably calculates the capability of the battery. Algorithms for battery-status recognition normally use as input variables the measured variables of battery current, voltage and temperature. On the basis of these variables the state of charge (SOC), the state of function (SOF) and the state of health (SOH) of the battery are determined and made available to electrical energy management as input variables (Figure 10).

A battery sensor which measures the battery current and voltage directly is used to measure the battery variables. The battery temperature is determined by way of a temperature measurement in the vicinity of the battery, since direct measurement of the battery electrolyte temperature in the vehicle would require an intervention in the battery, which is currently not possible.

Example

An example of a battery-status-recognition function is determining startability on the basis of the SOF. With the SOF the future behavior of the battery when subjected to starting current is predicted. In other words, battery-status recognition determines the battery-voltage dip for a given starting-current profile (Figure 7). Because the minimum voltage level for a successful start is known, the predicted voltage dip provides a measure of current startability. Depending on the interval between the predicted voltage dip and the startability limit, electrical energy management defines measures for maintaining or improving startability.

Battery sensor

The battery measured variables of current, voltage and temperature must be sensed very precisely, dynamically and synchronously. Particularly the measurement of currents in the range of a few mA through to starting currents of more than 1,000 A makes great demands on the sensor system. An electronic battery sensor (EBS) is mounted directly on and combined with the battery terminal. Because the terminal compartment is standardized in accordance with DIN 50342-2 [2], application on different batteries is not necessary.

Current is measured using a special manganin shunt. The core of the battery sensor's electrical wiring is an ASIC which contains among others a powerful microprocessor for recording and processing measured values. The battery-status-recognition algorithms are also processed on this microprocessor. Communication with higher-level ECUs is via the LIN bus.

As well as calculating the battery status for electrical energy management, the battery sensor can also be used for other functions. For example, precise sensing of current and voltage can also be used for prompted troubleshooting in production and workshops (e.g. searching for faulty no-load-current consumers).

References

[1] K. Reif (Editor): Bosch Autoelektrik und Autoelektronik, 6th Edition, Vieweg+ Teubner, 2011.

[2] DIN EN 50342-2 (2008): Dimensions of batteries and marking of terminals.

Vehicle electrical systems for hybrid and electric vehicles

The electrical system of a vehicle with start/stop system is very similar to a conventional vehicle electrical system (see Conventional vehicle electrical systems). In contrast, vehicle electrical systems for mild, full and plug-in hybrid vehicles and those for electric vehicles usually have a high-voltage level and thus differ significantly from the electrical system of a conventional vehicle.

The electrical system of a hybrid or electric vehicle has the following chief functions:

- Storing excess electric energy from the powertrain
- Outputting electric energy to the powertrain
- Reliable supply of the electrical consumers
- For plug-in hybrid and electric vehicles, storing energy from the public power supply and supplying the vehicle with it during operation.

Vehicle electrical systems for mild and full hybrid vehicles

The function of a mild or full hybrid vehicle (see Hybrid drives) requires a large amount of electric power, 8 to 60 kW. This cannot be provided on the 14 V voltage level in a sensible way. Therefore, an additional high-voltage vehicle electrical system with voltage in the range from 42 to 750 V is required. However, the 14 V standard electrical system is still required to supply the 14 V consumers in the vehicle. The individual consumers are supplied from the corresponding vehicle electrical system according to their power requirements (Figure 1). Generally, for cost reasons, automakers attempt to make do with standard 14 V components, as these are available cheaply and in large quantities.

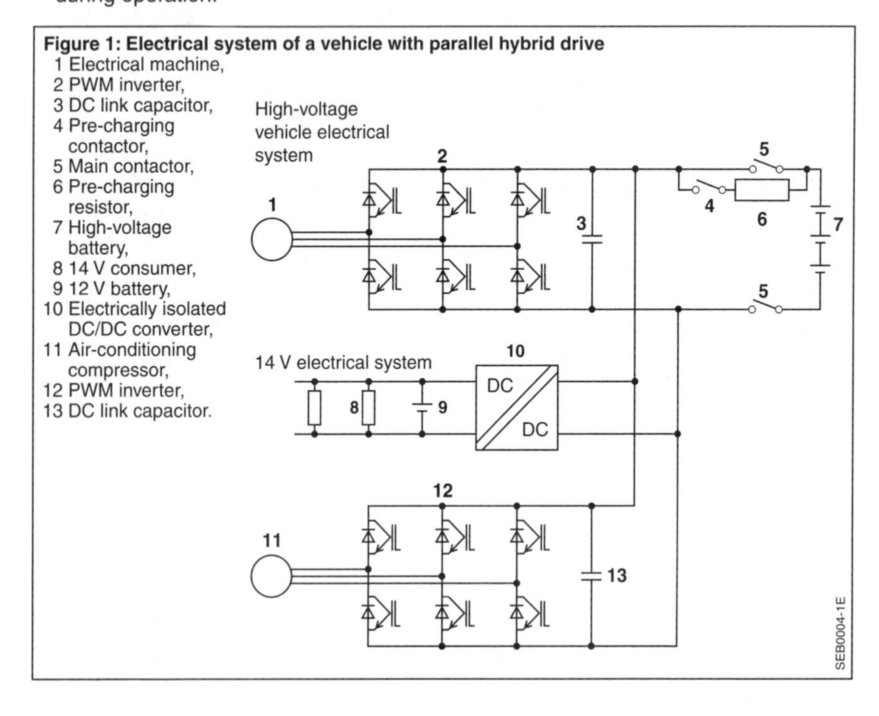

Figure 1: Electrical system of a vehicle with parallel hybrid drive
1 Electrical machine,
2 PWM inverter,
3 DC link capacitor,
4 Pre-charging contactor,
5 Main contactor,
6 Pre-charging resistor,
7 High-voltage battery,
8 14 V consumer,
9 12 V battery,
10 Electrically isolated DC/DC converter,
11 Air-conditioning compressor,
12 PWM inverter,
13 DC link capacitor.

High-voltage vehicle electrical system

Layout

The high-voltage vehicle electrical system consists of a high-voltage battery (HV battery), at least one PWM inverter to power the electric motor (electrical machine, e-motor), other high output or high-voltage consumers (e.g. electric air-conditioning compressor) and an electrically isolated DC/DC converter to supply the 14 V vehicle electrical system (Figure 1).

Operating principle

The battery cells are energized via the contactors integrated into the high-voltage battery. They provide buffer storage during operation. When the vehicle is switched off or if there is an accident, the high-voltage vehicle electrical system is de-energized via the contactors, limiting the hazardous high voltage to the battery block.

The PWM inverter in the power class from 10 to 200 kVA generates a three-phase system from the high-voltage direct voltage with a variable current intensity and rotating field frequency for the electrical machine.

Power is supplied to the high-voltage vehicle electrical system when the electrical machine is in generator mode.

The electrically isolated DC/DC converter transmits electrical energy from the high-voltage vehicle electrical system to the 14 V vehicle electrical system. Power is thus supplied to this low-voltage vehicle electrical system from the high-voltage vehicle electrical system.

An example of an additional component of the high-voltage vehicle electrical system can be an electric air-conditioning compressor, which requires a maximum electric power of 3 to 6 kW depending on the vehicle.

HV electrical system for parallel or quasi-parallel hybrid drives

The topologies of the vehicle electrical systems for mild and full hybrids with only one electric motor each are very similar. The vehicle electrical system of a mild hybrid (compared to the full hybrid) makes do with a lower energy accumulation and lower output capacity, as the vehicle can "crawl" electrically for a very short time at most.

The primary flow of energy of a vehicle with parallel hybrid drive is from the electrical machine to the high-voltage battery and vice versa. There is also a small flow of energy via the DC/DC converter to the 14 V vehicle electrical system or to other high-voltage consumers. The structure in Figure 1 is useful for implementing this in the most efficient and cost-effective manner possible, i.e. with few converters.

HV electrical system for power-split or semi-serial hybrid drives

Vehicles with two electrical machines that are run mostly serially (e.g. power-split hybrids) require a different electrical system topology.

In power-split hybrid vehicles or serial or parallel-serial hybrid vehicles, for the majority of the operating time, the electrical machines run in serial or partially serial operation. This means that one of the electric motors runs primarily in generative mode and the other in motor mode. Due to the large amounts of energy that are transmitted via the two electrical machines and the PWM inverters, these components should be operated in their optimal working range.

The battery voltage is set at the defined output of the battery. The intermediate circuit – consisting of the DC capacitor (DC link capacitor) and the associated conductors and, optionally, an electrically conductive DC/DC converter (high-output DC/DC converter) – distributes the energy on the high-voltage direct current level in the vehicle. It uses DC link capacitors to stabilize the direct voltage for sudden load variations (e.g. at the PWM inverter). For this concept, however, there are two principal DC links that are separated by the high-output DC/DC converter (Figure 2). A high-output DC/DC converter between the battery and DC link is useful for decoupling the DC link voltage for the electric motor from the battery voltage and enabling higher machine speed at a given battery voltage (Figure 2). Thus the DC link voltage can be adjusted as necessary between the level of the battery voltage and a significantly higher voltage value (2 to 2.5 times the battery voltage) [1]. The maximum voltage is determined via the required maximum output and the design of the electrical machines. The currently configured DC link voltage can be selected such that it is just above the maximum value of the rectified induced voltages of the electrical machines. This makes it possible to minimize the switching frequency of the inverter switches and thus the electrical inverter losses.

Low-voltage vehicle electrical system

The low-voltage vehicle electrical system has a similar design for all hybrid vehicles that have both a high-voltage and a low-voltage vehicle electrical system. It is very similar to the 14 V electrical system of a conventionally powered vehicle except for the fact that it usually does not have a starter and the power is supplied not by an alternator, but by an electrically isolated DC/DC converter from the high-voltage vehicle electrical system.

If the vehicle can drive or crawl electrically, all assisting functions – such as power steering, cooler pumps or braking system – must be operated electrically so that they are also available when the internal-combustion engine is stopped.

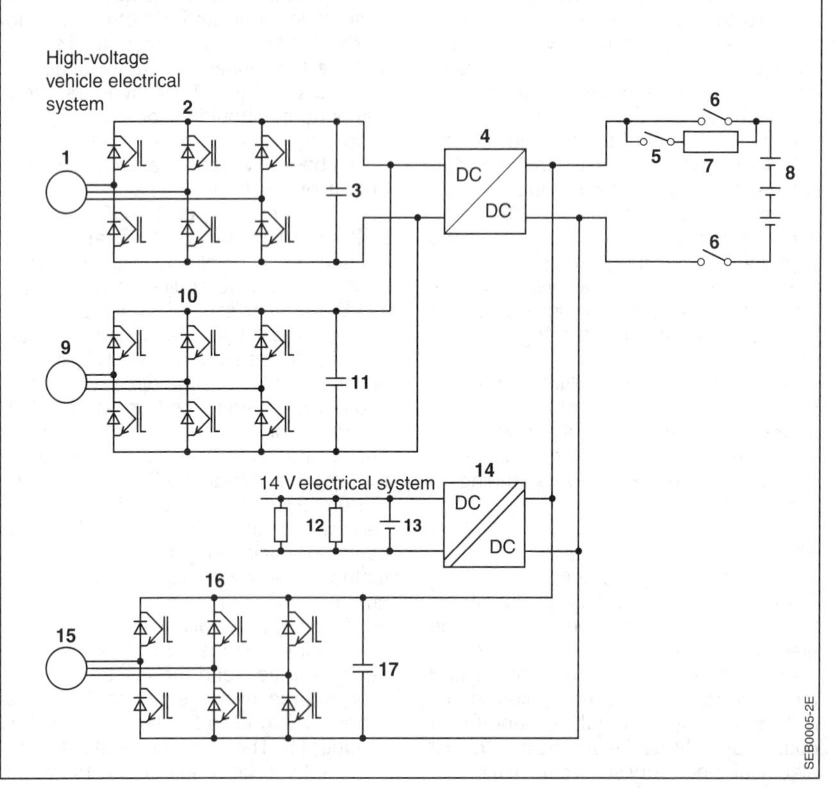

Figure 2: Electrical system of a power-split or serial vehicle
1 Electrical machine, 2 PWM inverter, 3 DC link capacitor,
4 High-output DC/DC converter, 5 Pre-charging contactor,
6 Main contactor, 7 Pre-charging resistor, 8 High-voltage battery,
9 Electrical machine, 10 PWM inverter,
11 DC link capacitor, 12 14 V consumer, 13 12 V battery,
14 Electrically isolated DC/DC converter,
15 Air-conditioning compressor, 16 PWM inverter,
17 DC link capacitor.

Vehicle electrical systems for plug-in hybrid and electric vehicles

Topology

The vehicle electrical system of a plug-in hybrid vehicle (see Plug-In hybrid) mostly corresponds to that of the comparable hybrid vehicle, other than having a larger battery with a greater energy storage capacity and a charging device.

For a purely electric vehicle, the topology of the parallel hybrid vehicle with a larger battery and a charging device is usually used (Figure 3). A charging device can be used that draws electrical energy from a public three-phase or alternating current power supply and uses it to charge the battery in a controlled manner. The battery management system tells the charging device the charging output it needs. For safety and to protect against overvoltage, this charging device usually has an electrically isolated design.

Alternatively, the battery of an electric vehicle can also be charged with direct current from an externally controlled charging station. In this case, all that is required is a plug connection to the high-voltage battery and a communication interface between the battery management and charging station to specify the charging output.

Efficiency

For electric vehicles, the efficiency of the vehicle electrical system is critical, as it has a direct effect not only on consumption, but also the range and thus the battery size needed. As a result, more expensive and more efficient components can reduce the total costs of the vehicle, as they make the same range possible with a smaller battery. Therefore, for many components of the vehicle electrical system, a review should be carried out to determine whether a less-expensive standard component can be used or it is worthwhile to develop a new, more efficient component.

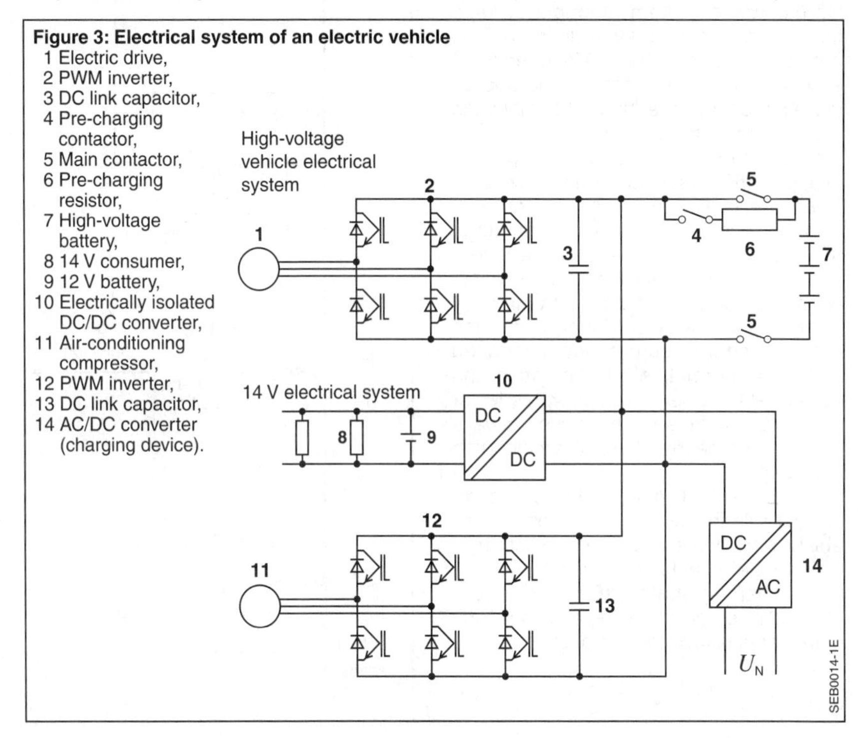

Figure 3: Electrical system of an electric vehicle
1 Electric drive,
2 PWM inverter,
3 DC link capacitor,
4 Pre-charging contactor,
5 Main contactor,
6 Pre-charging resistor,
7 High-voltage battery,
8 14 V consumer,
9 12 V battery,
10 Electrically isolated DC/DC converter,
11 Air-conditioning compressor,
12 PWM inverter,
13 DC link capacitor,
14 AC/DC converter (charging device).

Charging strategy

Cyclical variation

Generally, cyclical variation (cyclical charging and discharging) damages the battery. The higher the cycle strokes, the greater the damage will be. However, cyclical variation is required to increase the efficiency of the powertrain, for example with electric driving followed by recuperation. Thus the charging strategy and the selected size of the battery are a compromise between battery life, battery costs and weight on the one hand and good efficiency of the hybrid powertrain or longer range for electric vehicles on the other.

Usually, the thermal aging processes of the battery cells are accelerated significantly at a high state of charge (SOC). Therefore, the charging strategy should avoid combining a high state of charge and high temperature of the cells.

Operating strategy for hybrid vehicles

Normally, the system attempts to keep the battery of a hybrid vehicle in an SOC window of approx. 50 to 70 %. If the SOC exceeds this window, there is no operating point shift of the internal-combustion engine and no more recuperation.

When the lower SOC limit ($SOC_{min\,charge}$) of approx. 50 % is reached, it is necessary to ensure that an adequate battery discharge power is still possible. Therefore, when the lower SOC limit is reached, battery recharging increases if no boost energy is required at the moment. Only when a much lower SOC limit is reached is the discharge output slowly reduced to zero. In normal vehicle operation, the vehicle almost never reaches this lower discharge limit, and the driver always experiences nearly identical acceleration behavior.

The lower SOC limit also plays an important role for reliably maintaining startability and preventing deep discharge, which is detrimental to service life.

The charging strategy (Figure 4) is usually implemented in the engine control unit or in a special hybrid control unit.

Operating strategy for electric vehicles

In a pure electric vehicle, there is only one source of energy – the battery. Therefore, the operating strategy must fulfill the driver's command from this source, regardless of the state of charge. Therefore, strong cyclical variation is inevitable to attain the necessary range of the vehicle at a reasonable battery size. Here, the SOC limits are placed such that there are no effects that are very detrimental to service life, such as those due to electrolyte decomposition in case of overcharge.

Figure 4: Charging strategy of a hybrid vehicle in driving mode
SOC (State of Charge).

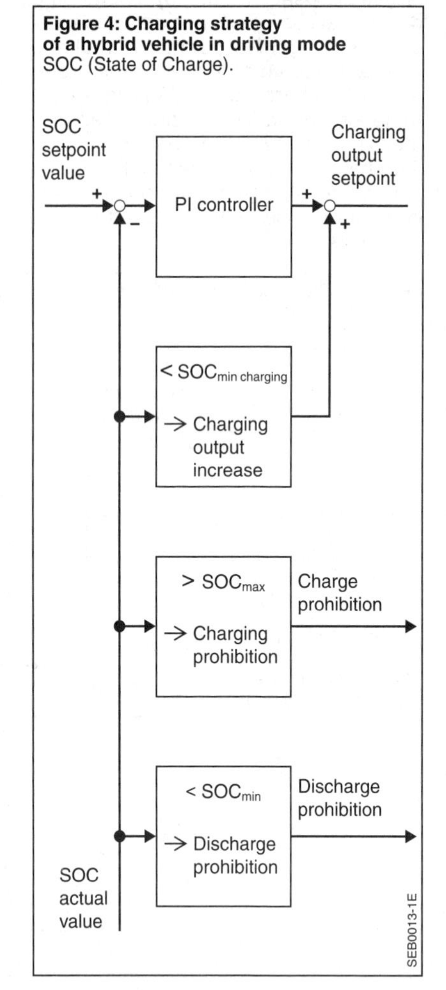

Depending on the battery system, the usable SOC range is between approx. 10 % and 90 %.

Operating strategy for electric vehicles with range extender

For an electric vehicle with range extender – a small internal-combustion engine with an alternator that supplies electrical energy to the high-voltage vehicle electrical system – the range extender is activated during a longer journey or if a lower SOC limit value is reached. As the output of the range extender is usually low, it is operated constantly at full load and recharges the battery where applicable.

Operating strategy for plug-in hybrid

A plug-in hybrid can still drive with the combustion engine. Therefore, its charging strategy is a mix of that of an electric vehicle and that of a hybrid vehicle. To optimize fuel consumption, the strategy attempts to maximize the amount of electric driving. Only if this is not possible, such as during a long journey, does the system switch to hybrid operation with a charging strategy that is similar to that of a hybrid vehicle, usually with lower SOC limits.

When charging the battery from the power grid, the battery charger or the public power grid is usually the limiting element. Therefore, in many cases, the battery is charged using the currently available charging power or charging current until a voltage limit is reached; then, the charging power is then successively withdrawn until the end-of-charge voltage is reached.

References

[1] S. Sasaki, E. Sato, M. Okamura: The Motor Control Technologies for the Hybrid Electric Vehicle; Toyota Motor Corporation.

Starter batteries

Requirements

The starter battery is used to start the internal-combustion engine and to provide electrical energy for lighting and other components (SLI, Starting, Lighting, Ignition) if the generator does not provide any energy or enough energy.

The battery in the vehicle electrical system

The performance requirements for starter batteries in modern motor vehicles are very high. Diesel engines and large-volume gasoline engines have high cold-cranking power requirements with high starting currents, particularly at low temperatures. When the engine is running, the electric components are usually supplied directly from the generator. When the engine speed is low or at a standstill, on the other hand, it is supplied partially or completely from the battery. If a vehicle is parked for days or weeks, a peak coil current of typically 5 to 20 milliamperes must be supplied. This also includes the much higher temporary flow current consumption for ventilators, pumps and electronic components immediately after switching off the internal-combustion engine, as well as for comfort and convenience consumers operated while the vehicle is parked, such as entertainment and communicat ion electronics and for the auxiliary heater, if present. The battery must also be able to start the vehicle, even after it has been parked for extended periods.

In addition to the uniform energy supply, the battery in the vehicle electrical system covers high dynamic current pulses which cannot be delivered by the alternator as quickly (for transient processes, such as switch-on processes in the electrical power steering). Furthermore, due to the very high inherent natural capacitance of the double-layer capacitor of a few farad, the battery smooths ripples in the auto-

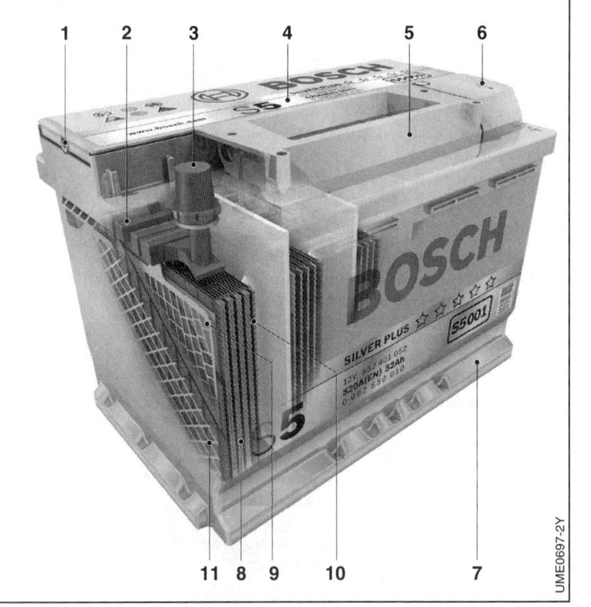

Figure 1: Structure of a starter battery
1 Gas outlet opening,
2 Electrode connector,
3 Battery terminal,
4 Double cover with labyrinth structure,
5 Carrying handle,
6 Terminal-post cover,
7 Bottom rail,
8 Electrode block (consisting of a set with positive electrodes and a set with negative electrodes),
9 Positive electrode, with stamped grid and positive active materials,
10 Envelope separator,
11 Negative electrode, with expanded grid and negative active materials.

motive electrical system current, which helps minimize EMC problems (electromagnetic compatibility).

The rated voltage of the vehicle electrical system is 12 V for passenger cars and 24 V for commercial vehicles; two identically sized 12 V batteries are connected in series for this.

The battery in innovative vehicle electrical systems

Operating strategies of new vehicles use the battery in an increasingly active manner for reducing fuel consumption and emissions (also see conventional vehicle electrical system). In vehicles with active alternator control, the battery is operated in a partially charged state. The alternator output power is reduced in a targeted manner in operating phases (if necessary, all the load is taken off the alternator is completely discharged), in which the engine is working at poor efficiency or engine output is particularly necessary (e.g. during vehicle acceleration, "Passive Boost"). During these phases, the battery supplies the vehicle with electricity. In phases where the motor is highly efficient and when braking, the battery is then recharged in a targeted manner by raising the charge voltage (see Recuperation).

A start/stop system is used to switch off the engine if the vehicle is not moving (e.g. at a red light). The starter battery then covers the vehicle's entire power consumption of approximately 25 to 70 A. The engine is restarted automatically when the driver wants to pull away.

Depending on the operating conditions, using the battery's storage capacity in this way significantly reduces fuel consumption. The increased charging capacity represents an additional load on the battery, which is accommodated by special designs such as AGM and EFB (see Battery designs). Here, charging the battery well and sufficiently is always a requirement for a level charge balance in order to fulfill these tasks in the long term. A battery status detector is used to monitor the battery status. It generally consists of a sensor attached to the negative battery terminal (EBS, electronic battery sensor for measuring current, voltage and temperature) and software that uses the determined measurements to identify the status and capacity of the starter battery. Based on these results, the system can make the necessary intervention into the engine management system based on an operating strategy and, for example, decide whether the engine is to be shut off when the vehicle is stationary, whether it is to be restarted automatically when the vehicle is parked for long periods, or how the alternator voltage must be controlled (see Conventional vehicle electrical systems).

Numerous vehicles are also equipped with two batteries, for example, to guarantee startability even at a high standby current consumption or to prevent inconveniences (e.g. resetting the infotainment systems) due to the short-term voltage drop in vehicles with start-stop function when restarting automatically (see Two-battery vehicle electrical system).

Hybrid and electric vehicles also have a low-voltage vehicle electrical system of 14 V, on which the majority of the electric components work. This is supported by a 12 V battery, in many cases a high-performance starter battery. In some hybrid vehicles, cold cranking is carried out by a starter motor supplied from this battery (see Vehicle electrical systems for hybrid and electric vehicles).

The battery as component

The requirements mentioned define the electrical properties of the accumulator such as starting power, capacities and charge current absorption. Furthermore, depending on operating conditions, thermal requirements (e.g. due to the installation point in the vehicle and the climate zone) and mechanical requirements (e.g. regarding attachments and vibration resistance) must be observed. The battery should also be maintenance-free, safe in use and environmentally friendly in production. The lead battery features excellent recyclability and achieves the highest recycling rates of all consumer durables at well over 95 %, where lead and plastic from old batteries are re-used for battery manufacturing.

The battery in operation

The design of the vehicle's components and the electrical operating concept, as well as the usage profile and driver affect the proper function of the battery and, therefore, the vehicle electrical system. Despite the excellent charge acceptance properties of modern starter batteries, a positive battery charge balance sometimes cannot be achieved during regular, short trips through the city in winter (involving high power consumption and low engine speed) and the state of charge decreases. At low states of charge, not only does the amount of energy still available decrease, but the capability of producing currents sufficient to start the engine decreases as well. Low battery charge levels over longer time periods worsen the cold-cranking characteristics and shorten the service life of the battery.

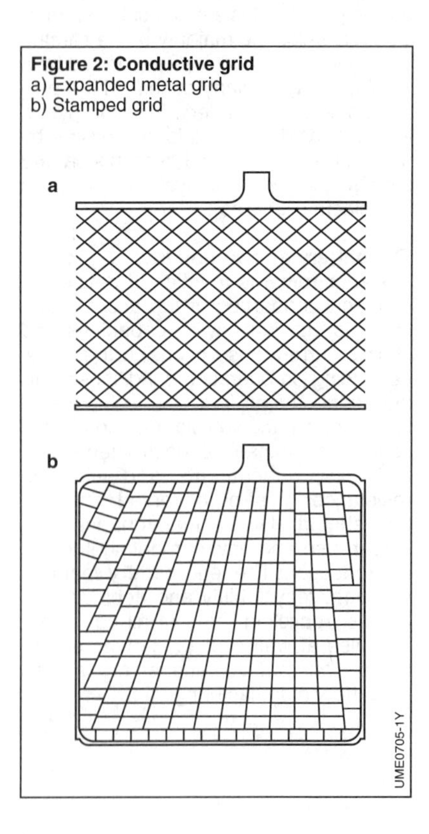

Figure 2: Conductive grid
a) Expanded metal grid
b) Stamped grid

Battery design

Components

A 12 V starter battery contains six series-connected individually partitioned cells in a polypropylene case (Figure 1). A cell consists of a stack of positive and negative electrodes arranged in alternation, each designed from a lead grid as a carrier and electrical arrestor, filled with porous active materials. Microporous separators isolate the electrodes from each other, where they usually enclose the electrodes of one polarity in a pocket-size format. The separators are made from porous polyethylene filled with silica or from a glass-fiber fleece (AGM, Absorbent Glass Mat).

The electrolyte is in the form of diluted sulfuric acid which fills the pores in the electrodes and separators, and the voids in the cells. The electrodes of one polarity are connected in parallel by plate straps and, using cell connectors, are connected and sealed from openings in the partitions to the neighboring cells or the battery terminals. Battery terminals, cell and plate straps are made of lead alloys.

A lead battery develops an ever-increasing amount of oxygen at the positive electrode and hydrogen at the negative electrode, and this is increased when charging. During this electrolysis process, water is consumed from the electrolytes. The charging gases must be released. The one-piece cover welded to the battery case provides the battery's upper seal and contains various ventilation openings depending on design. On conventional batteries, each cell has a plug, which is used for initially filling the battery with electrolyte, maintenance and the release of charging gases. Maintenance-free batteries usually no longer have a plug. Degassing occurs via an advanced labyrinth system, which prevents liquid from escaping, even if the battery is tilted. However, the ventilation openings, often arranged on the side of the cover, must not be sealed on both sides. Porous sintered bodies permeable to gas (called frits) are installed in front of the ventilation opening inside the cover. These sintered bodies prevent potential flames or sparks outside from backfiring into the inside of the battery.

Conductive grid

Active materials

In the manufacturing process, the active materials are coated (pasted) into the conducted grid. The positive active materials contain porous lead dioxide (PbO_2, dark brown). The negative active materials contain porous metallic lead sponge (Pb, metallic gray-green).

Manufacturing and geometry

The lead grid is used for the mechanical mount and electric contact of the active materials. Optimizing the grid structure in terms of electrical conductivity improves the utilization of the active materials. The standard grid manufacturing methods include casting liquid lead into a mold or continuous processes like drawing out and punching from a strip (Figure 2).

Alloy materials

Alloying elements are added to the lead used for the grid to optimize the production process and increase the mechanical strength and corrosion resistance. The strong oxidizing electrochemical potential exposes the positive grid to a continuous corrosion attack. As a result, the cross-sectional area of the webs detaches throughout the duration of use, the electrical resistance is increased and alloying elements are in released the electrolytes. Negative grids, on the other hand, do not have corrosive potential and, therefore, are not affected by corrosion (see Corrosion).

Antimony

If, for example, antimony assumes the function of the hardener for the positive grid, it is gradually released by corrosion and travels diagonally through the electrolyte and the separator to the negative electrode. It "poisons" the negative active material, which greatly increases the spontaneous development of hydrogen there. This increases the self-discharge of the negative electrodes and water consumption if overcharged. Overall, this causes the capacity to be constantly reduced throughout the duration of use. The battery does not achieve the required state of charge, and the electrolyte has to be checked at frequent intervals.

This is why, for grids, lead antimony alloys (PbSb) with a low percentage of antimony by weight are generally still used only for battery types for which regular maintenance is accepted.

Calcium, tin and silver

Today, calcium, tin and silver are usually standard alloying elements for maintenance-free batteries, because they do not noticeably affect hydrogen development, even when released due to corrosion. To ensure the required strength, terminal posts are made from lead alloys with up to around 10 percent of antimony by weight.

Lead-calcium alloys (PbCa) with around 0.1 percent calcium by weight are usually used for negative grids during any manufacturing process. Calcium is electrochemically inactive in lead storage battery and, therefore, the self-discharge and water consumption do not increase.

Lead-calcium-tin alloys (PbCaSn) are used for positive grids that are manufactured in a continuous stretching process or rolling and punching process. Due to the increased percentage of tin (0.5 to 2%), these alloys feature very high corrosion resistance, which allow lower grid weights.

Casted positive grids are often made from a lead-calcium-silver alloy (PbCaAg). In addition to 0.06 percent calcium by weight and tin, this alloy also contains a proportion of silver (Ag). It has a finer grid structure and has proved to be extremely durable even at high temperatures – which have the effect of accelerating corrosion.

Charging and discharging

Chemical reactions

The active materials of the lead storage battery include the lead dioxide (PbO_2) of the positive electrode, the sponge-like, highly porous lead (Pb) of the negative electrode and the electrolyte, and diluted sulfuric acid (H_2SO_4). The electrolyte fills the pores of the active materials and the separator and is simultaneously an ionic conductor for charging and discharging. Compared with the electrolyte, PbO_2 and Pb adopt typical electrical voltages (individual potentials). Their difference is equal to the cell voltage of approximately 2 V (at rest) measurable from the outside (Figure 3).

When the cell discharges, PbO_2 and Pb combine with H_2SO_4 to form $PbSO_4$ (lead sulfate). This conversion causes the electrolyte to lose SO_4^{2-} (sulfation) ions, and its specific gravity decreases. During the charging process, the active materials PbO_2 and Pb are reconstituted from $PbSO_4$, H_2SO_4 is released and the electrolyte density increases again (see Electrochemistry, lead battery).

When a discharge current is applied to the battery, this reduces the battery voltage depending on the magnitude of current and the duration of discharge (Figure 4). Therefore, the charge quantity that can be drawn from the battery decreases when the current level increases.

Behavior at low temperatures

The chemical reactions in a battery always take place more slowly at lower temperatures and when internal resistance is increasing. The voltage at the specified discharge current and, therefore, the starting power of a fully charged battery decreases as the temperature drops. The further the discharge progresses, the more the concentration of acid drops, up to the freezing point of the electrolyte. A discharged battery can only supply a low current which is not sufficient to start the vehicle.

Figure 3: Electrical parameters inside and on the battery
I_E Discharge current
R_i Internal resistance
R_V Load resistance
U_0 Steady-state voltage
U_K Terminal voltage
U_i Voltage drop across internal resistance

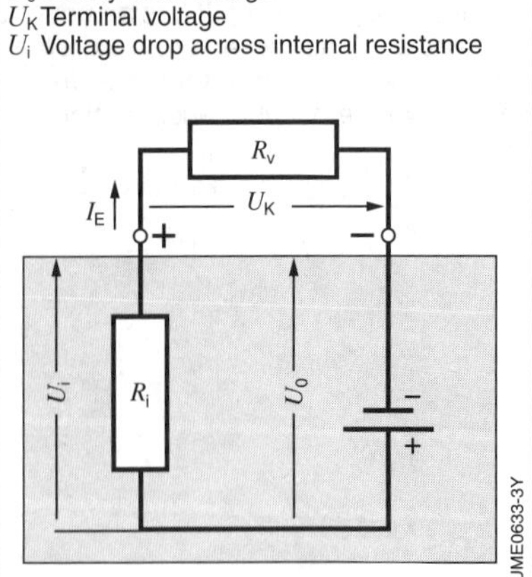

Figure 4: Battery-voltage curves as a function of the discharge duration for various discharge currents
1 I=0.05 A per Ah,
2 I=0.2 A per Ah,
3 I=1.0 A per Ah,
4 I=4.0 A per Ah.

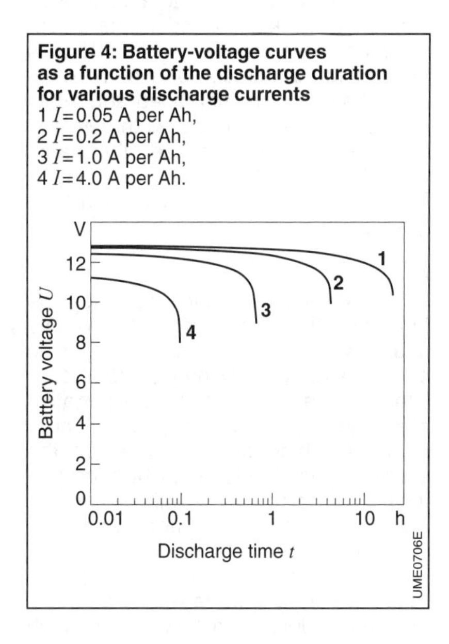

Battery characteristics

In addition to mechanical parameters, such as physical dimensions, attachment, and terminal design, characteristic electrical ratings are defined in testing standards (e.g. DIN EN 50342, [1], [2], [3]). General terminology on batteries is described in DIN 40729 [4].

For the cross-manufacturer marking of starter batteries, a nine-digit ETN (European Type Number) was described in the appendix of the revised EN 50342-1:2001 to compile the most important properties of a starter battery (including voltage, nominal capacity, cold cranking performance, terminal details, etc.). However, this representation cannot map expanded differences regarding construction characteristics, design and areas of application, which are described additionally in requirements outside of EN 50342-1. Therefore, starting with the version from 2006, the ETN nomenclature was withdrawn as a standard in coordination with battery manufacturers and users. Since then, the ETN only continues to be used occasionally.

Capacity

Capacitance is the charge quantity in ampere hours (A h) which can be drawn from the battery under specified conditions. It decreases as discharge current increases and temperature decreases.

Nominal capacity

EN 50342-1 defines the nominal capacity K_{20} as the charge the battery is able to deliver within 20 h up to a cutoff voltage of 10.5 V (1.75 V per cell) at a specified, constant discharge current $I_{20} = K_{20}/20$ at 25 °C. Nominal capacity is calculated from the quantities of active material used (positive material, negative material, electrolyte), and is only marginally affected by the number of electrodes.

Low-temperature test current

The low-temperature test current I_{CC} indicates the battery's current output capability at low temperatures. According to EN 50342-1, the battery terminal voltage with I_{CC} at −18 °C must be at least 7.5 V (1.25 V per cell) up to 10 seconds after discharging has started. Further details relating to the discharge period are specified in EN 50342. The short-term response of the battery when discharged at I_{CC} is largely determined by the number, geometric surface area of electrodes, the gap between electrodes, and the separator thickness and material.

The battery internal resistance R_i and other resistances in the starter circuit determine the cranking speed of the engine and also identify the starting response. For a charged battery (with 12 V), R_i at −18 °C is in the order of magnitude of $R_i \leq 4{,}000/I_{CC}$ (in mΩ), where I_{CC} must be applied in amps.

Water consumption

Lead battery loses water from the electrolytes through electrolysis, particularly when charging. EN 50342-1 defines the test conditions for quantification. A limit value of water consumption of 1 g/A h, for example, means that, under test conditions, a battery with a 50 A h nominal capacity may lose up to 50 g of water. By comparison, the electrolyte of this kind of battery in new condition, depending on type, contains approximately 1.8 to 2.7 kg of water.

Cycle stability

Lead batteries undergo a certain amount of aging during repeated charging and discharging. EN 50342-1 defines the test conditions for quantification.

Battery types

Maintenance-free battery in accordance with EN

Whether batteries require maintenance by refilling with water depends primarily on the grid alloy used. Batteries with lead antimony alloy (conventional and low-maintenance batteries) are very durable in cyclical charging and discharging operations, but require frequent maintenance because of the high degree of water loss and, therefore, are only used in some commercial vehicles.

Longer maintenance intervals result from low-maintenance batteries with negative grids made from a lead-calcium alloy (PbCa) and positive grids from an antimony alloy (PbSb). However, due to the antimony content in the positive grid, these hybrid batteries also seldom meet the high demand for very low water consumption (in accordance with EN 50342-1, less than 1 g/A h). Batteries with a water consumption under 4 g/A h in accordance with EN 50342-1 are considered maintenance-free in accordance with the EN standard.

Completely maintenance-free battery

Both grids in completely maintenance-free batteries are made from the lead-calcium alloy. When the vehicle electrical system is operating normally, this water decomposition is reduced to such an extent (in accordance with EN 50342-1, less than 1 g/A h) that the electrolyte reserves will last for the entire service life of the battery. Therefore, completely maintenance-free batteries do not require electrolyte level monitoring and do not usually provide a facility for doing this. They have a safety labyrinth cover that prevents electrolyte from escaping even when the battery is at a strong slope and are tightly sealed with the exception of two ventilation openings.

Completely maintenance-free batteries are already filled with electrolyte in the manufacturer's plant and can be stored in a fully charged state for up to 18 months after delivery due to the very low self-discharge.

A "dry" storage without electrolyte occurs only for very few designs (particularly for motorcycles). The electrolyte is first poured in from a delivered acid pack when commissioning in workshops or at dealers' locations.

AGM battery

AGM (Absorbent Glass Mat) batteries are often used for applications with high charging capacities. In these batteries, the electrolyte is bound in a microporous glass-fiber fleece, which, instead of the conventional separators, is located between the positive and negative electrodes (Figure 5).

"Fixing" the electrolyte prevents, for example, the sulfuric acid of high concentration and density released during the charge from sinking and accumulating in the lower area of the cell. This "acid layer" with excess H_2SO_4 in the lower cell area and a shortage in the upper cell area occurs gradually during repetitive charging and discharging in conventional lead storage batteries with freely moving electrolyte. This impairs the charging behavior, accelerates the coarsening of the $PbSO_4$ discharge product's crystalline structure (sulfating), reduces the storage capacity and accelerates the overall aging process of the battery.

In AGM batteries, these effects are reliably prevented. The elastic fleece also slightly pressurizes the electrode pair. This reduces the shedding and separation effect of the active material considerably. Overall, this enables charging capacities more than three times higher than in comparable starter batteries with free electrolyte by charging and discharging throughout the entire service life. This makes it particularly well-suited for vehicles with a start/stop system that require the consumer to be supplied when the engine is stopped and then restarted reliably. Other typical cases of application for AGM batteries include vehicles with many electrical consumers that are also used when the vehicle is stationary (e.g. auxiliary heater, consumer electronics), taxis, special vehicles and logistics vehicles.

In the AGM battery, the oxygen that occurs at the positive electrode in an internal circuit is used up again, the amount of hydrogen that is produced there is suppressed and the amount of water loss is therefore minimized further. This circuit is enabled by small gas channels in the separator fleece, through which the oxygen is transported. The individual cells are separated from the environment by valves that open only if the operation causes the internal pressure to exceed a value of approximately 100 to 200 mbar. Under normal operating conditions, the valves are closed, which reduces water consumption even further. Lead batteries with sealed valves, such as the AGM battery, are called VRLA batteries (Valve Regulated Lead Acid Battery) and are completely maintenance-free.

The inner resistance is particularly low and high cold start currents are achieved because of the high porosity of the glass-fiber fleece. Therefore, AGM batteries are also often used in diesel vehicles.

Even if the battery housing is destroyed (e.g. in case of an accident), the electrolyte bound in the glass-fiber fleece (diluted sulfuric acid) does not generally leak. This and the improved gassing characteristics make the AGM battery particularly well-suited for installation points in the passenger cell.

EFB battery

The EFB battery (Enhanced Flooded Battery) is a battery with free electrolyte optimized for use in vehicles with a start/stop system. Although an acid layer is also formed when cyclic charging batteries without fixed electrolyte, such as the AGM battery, special supporting separators and other design characteristics of these batteries make them more robust than standard batteries in this operating mode.

EFB batteries are generally used in vehicles with a start/stop system and low consumer loads.

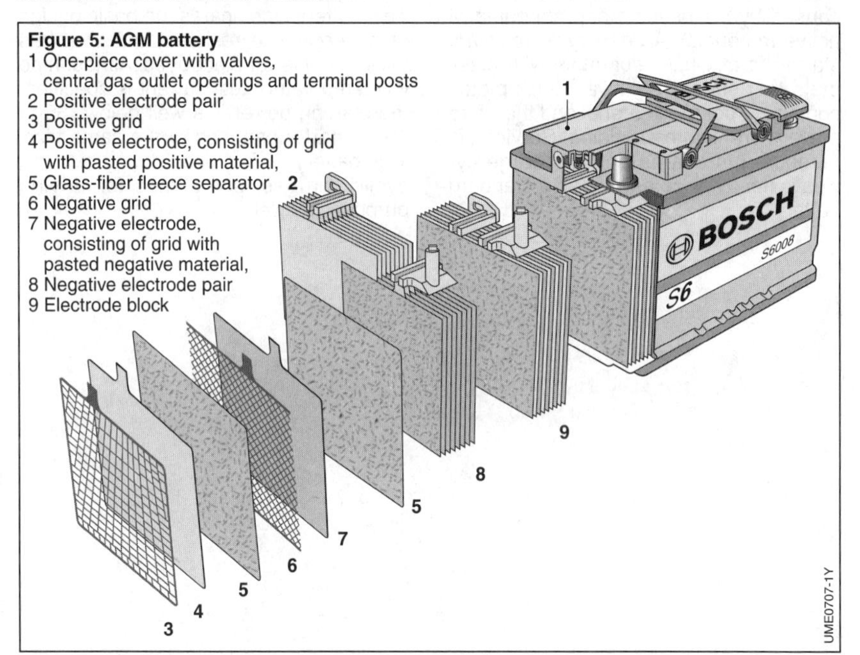

Figure 5: AGM battery
1 One-piece cover with valves, central gas outlet openings and terminal posts
2 Positive electrode pair
3 Positive grid
4 Positive electrode, consisting of grid with pasted positive material,
5 Glass-fiber fleece separator
6 Negative grid
7 Negative electrode, consisting of grid with pasted negative material,
8 Negative electrode pair
9 Electrode block

Commercial vehicle batteries
Bosch also provides completely maintenance-free batteries for commercial vehicles. Specifically in commercial vehicle applications, these batteries offer a substantial cost advantage. These battery types have a special labyrinth cover, which guarantees leakproof performance and features central venting. Porous frits integrated into the outlet opening prevent flames or sparks outside from backfiring into the inside of the battery. There are special design variations for applications with extreme cycle or vibration stress.

The housing dimensions of commercial vehicle batteries are described in the standard EN 50342-4.

Deep-cycle resistant battery
Due to their design (relatively thin electrodes, lightweight separator material), starter batteries are less suitable for applications which involve repeated exhaustive discharging, as this results in heavy wear on the positive electrodes (mainly caused by a loosening and shedding of active materials). A deep-cycle resistant starter battery has separators with glass mats that support relatively thick plates containing positive material and therefore prevent premature shedding. Service life measured in charge and discharge cycles is nearly twice as long as a standard battery.

Vibration-proof battery
In a vibration-proof battery, the plate block is fixed to the battery case by means of cast resin or plastic to prevent any relative movement between the two components. In accordance with DIN EN 50342-1, this battery type must pass a 20-hour sinusoidal vibration test (at a frequency of 30 Hz) and be capable of withstanding a maximum acceleration of 6 g. Requirements are therefore about 10 times greater than for standard batteries.

Vibration-proof batteries are mainly used, for example, in commercial vehicles, construction machines or tractor vehicles.

Heavy-duty battery
A heavy-duty battery combines the attributes of the deep-cycle resistant and vibration-proof types. It is used in commercial vehicles which are subjected to extreme vibrations, and where cyclic discharge patterns are commonplace.

Battery for extended current output
This battery type shares the basic design of a deep-cycle resistant unit, but has thicker and fewer electrodes. Although no low-temperature test current is specified, the starting power lies well below (35 to 40%) starter batteries of the same size. This battery type is used during extreme cyclic charges, sometimes for driving purposes as well.

Using the battery

Charging

In vehicle electric systems, the alternator regulator specifies the voltage. In terms of the battery, this corresponds to the IU-charging method, where the battery charge current I is initially limited by the output of the alternator and then reduces automatically if the battery voltage reached the control value (Figure 6). Since the current consumption capacity of the battery decreases at low temperatures, the charge voltage in the vehicle is usually regulated depending on the temperature (Figure 7). At battery temperatures well under 0 °C, current is consumed only very slowly.

The IU charging method prevents damage due to overcharging, and makes sure that the battery has a long service life. Modern battery chargers operate according to similarly defined characteristic curves.

Older devices, on the other hand, still operate using constant current or a W-characteristic curve. In both cases, once the full state of charge has been reached, charging continues with only a slightly reduced current, or possibly a constant current. This leads to high water consumption and to subsequent corrosion of the positive grid. In particular, maintenance-free batteries can be damaged by this in the long term.

Discharging

Shortly after discharging at a continuous current begins, the voltage in the battery drops to a value which only changes relatively slowly if discharging continues. Only shortly before the end of the discharging process will the voltage collapse sharply due to exhaustion (d.h. complete electrochemical implementation) of one or more of the active components (positive or negative material, electrolyte).

Self-discharge

Batteries discharge over a period of time if no electrical equipment is connected. Causes include the spontaneous development of oxygen at the positive electrode and hydrogen at the negative one, as well as the continuously expiring slow corrosion of the positive grid. Completely maintenance-free batteries with lead calcium grids have a self-discharge of approximately 3 % per month at 25 °C, which remains roughly constant over the entire service life and nearly doubles for each temperature increase of 10 degrees.

Modern batteries using the lead-antimony alloy lose approximately 4 to 8 % of their charge every month when new. As battery ages, this value can increase by up to 1 % or more each day due to the migration of antimony to the negative electrode until a point is reached when the battery finally stops functioning.

Figure 6: Charging according to the IU characteristic
1 Charge voltage
2 Charging current

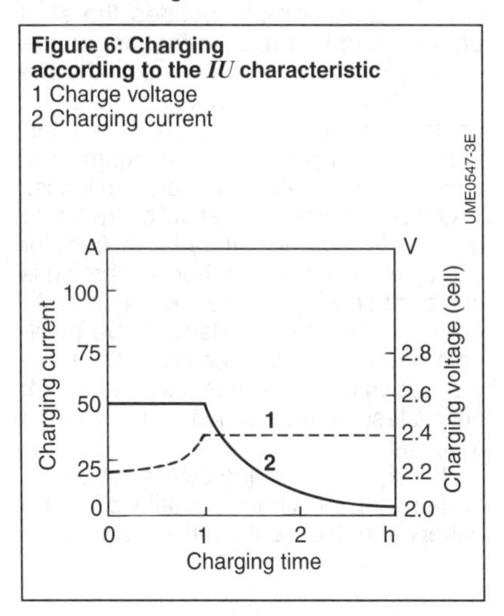

Figure 7: Temperature-dependent charging characteristic
1 Low-maintenance standard battery
2 AGM battery

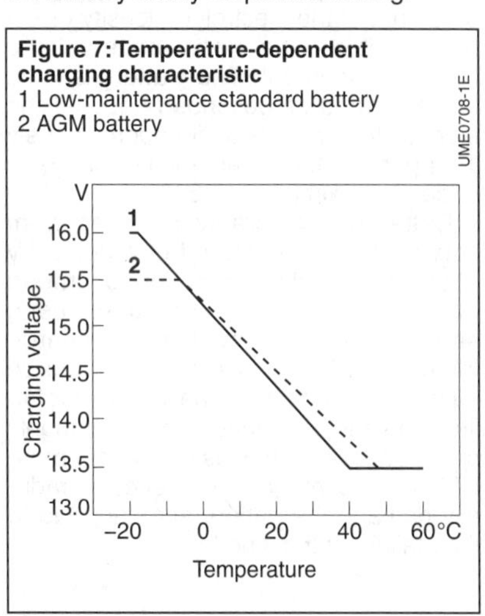

Battery maintenance

On low-maintenance batteries, the electrolyte level should be inspected in accordance with the manufacturer's operating instructions; when required, it should be replenished to the MAX mark with distilled or demineralized water. This is generally not required for maintenance-free batteries in accordance with EN standard. This step is omitted entirely for completely maintenance-free batteries.

All batteries should be kept clean and dry to minimize self-discharge. Terminal posts, terminal clamps, and fixings should be coated with acid-protection grease.

Before the onset of winter, it is recommend to check the battery status with a modern battery tester. If this tester recommends recharging, the battery should be recharged at a maximum voltage of approximately 14.4 to 14.8 V with a suitable charger with a regulated charging characteristic curve (IU charging characteristic curve or similar to prevent overcharging). Here, note the information in the user manual for the vehicle, battery and charging device. The charging area should be well ventilated (risk of explosion due to oxyhydrogen gas, risk of explosion, no naked flames or sparks). The electrolyte is corrosive. Therefore, gloves and protective eyewear must be worn when handling.

If no battery tester is available, a measurement of the electrolyte density or – if not possible – the steady-state voltage as a replacement. The battery should be recharged as described above if the electrolyte density is under approximately 1.24 g/ml or the steady-state voltage is under approximately 12.5 V.

Batteries temporarily removed from service should be stored in a cool, dry place. If the above-mentioned criteria exist, recharging is also required. Long service lives at a low state of charge increase grid corrosion and reinforce sulfating. During this, the finely crystalline lead sulphate transforms into roughly crystalline, which makes the battery more difficult to recharge. These effects gradually damage the battery and can lead to their failure in the long term.

Battery malfunctions

Functional disturbances due to damage inside the battery cannot be remedied through repair. The battery has to be replaced.

A battery failure usually does not happen abruptly; instead, steadily decreasing starting power can provide an indication of the problem. Causes usually include a combination of consumed or sulfated active materials, grid corrosion and short-circuits due to separator wear. Modern battery testers can be used to evaluate whether a weak battery needs to be replaced or can simply be recharged.

If no battery defects are determined and the battery's charge is still always low, there may be a fault in the vehicle electrical system (defective alternator, electrical consumers remain switched on when the motor is not running, etc.). The battery can also be insufficiently charged as a result of using the vehicle primarily for short-distance driving. This can be balanced by occasionally driving longer distances or, alternatively, by occasionally recharging with an external charging device.

Battery testers

The starter battery is a wearing part. The service life is heavily dependent on the type and use of the vehicle and the climatic conditions. Aging becomes noticeable due to an increase in the internal resistance R_i (which impedes the start performance) and a reduction of the storage capacity by loosening of the active materials or sulfating. A low state of charge, on the other hand, is not a characteristic of aging, but a consequence of a usage with insufficient charge balance.

Often, a quick evaluation of the state of an older starter battery is desired, for example, to decide whether recharging is sufficient or a replacement is required. To do so, battery testing devices are available that can be connected to the powered-down battery and make a statement after a test duration of generally less than a minute.

Usually, these testing devices imprint a certain current charge profile onto the battery and deduce the battery state from the voltage response. To do so, they use partially simplified electrochemical impedance spectroscopy approaches (EIS, see electrochemical impedance spectroscopy). In general, the battery's correlation to the steady-state voltage is used to estimate the state of charge. Therefore, the battery must not be charged with other currents during the test.

Either a very rough evaluation (e.g. "ok", "charge", "replace", "test again") or a quantifying specification (e.g."state of charge 82 %", "startability 87 %") is given, depending on the design of the tester. Different manufacturers use a wide variety of methods.

Often, nominal values of the battery (capacities, cold cranking performance) must be specified and relative evaluations are made. The battery design (free electrolyte or AGM) and the battery temperature are also often requested or measured directly to also take into account this important influencing factor.

Battery testers cannot replace a complete test according to the applicable standards (e.g. EN 50342-1), but instead quickly give guideline for additional actions for batteries that have been in use for an extended amount of time. This test is not valid for new or like-new batteries because the structure of the active materials is often not yet balanced immediately after manufacturing.

Replacing the battery

Specifications in the vehicle's operating instructions must be taken into account when replacing batteries. These operating instructions often describe the specified or permissible sizes and design. The replacement battery should at least provide the specified capacity and cold cranking values to ensure functional and operational reliability. If a special battery design, such as AGM or EFB, is specified, such as in vehicles with a start/stop system, a replacement battery of the same design must be used.

References

[1] EN 50342-1:2006 + A1:2011: Lead-acid starter batteries – Part 1: General requirements and methods of test.

[2] EN 50342-2:2007: Lead-acid starter batteries – Part 2: Dimensions of batteries and marking of terminals.

[3] EN 50342-4: 2009: Lead-acid starter batteries – Part 4: Dimensions of batteries for commercial vehicles.

[4] DIN 40729:1985: Accumulators; Galvanic Secondary Cells; General Terms And Definitions.

Drive batteries

Requirements

The drive battery in modern electric vehicles often accounts for more than two-thirds of the weight, size and cost of all components in the electric powertrain. Therefore, the main challenge for the future is to improve and optimize the battery in terms of these three criteria.

A drive battery, also called the battery pack, consists of many battery cells. The (sometimes conflicting) requirements described below must be taken into account when developing and designing the cells and other components in the battery pack.

Safety

The safety of vehicle occupants and the environment must be guaranteed in all situations. This applies to hazards that result from the electrochemical properties of the battery, for example, preventing fire. It must also take electrical safety into account, particularly in batteries with voltages larger than 60 V.

Performance

For use in vehicles, low weight and high performance are required simultaneously. Achieving this requires batteries with high energy and power density. There are different requirements for this depending on the vehicle. Figure 1 shows the typical values for the respective characteristics and the ratio of performance P to energy E. This results in various requirements for the battery cells used. Various cell sizes and designs are required to be able to cover all vehicle segments.

Operation in the automotive sector requires a service life usually lasting more than ten years or a running performance of typically 250,000 km. In purely electric vehicles, approximately 3,000 complete charging cycles occur throughout the life of the vehicle, while hybrid applications often have over a million partial charging cycles. The largest challenge by far is to reduce the costs with the goal of reaching total costs comparable to those in systems with internal-combustion engines.

Figure 1: Requirements for energy content and capacity of the battery for various vehicle concepts
HEV Hybrid Electric Vehicle,
PHEV Plug-in Hybrid Electric Vehicle,
EV Electric Vehicle.

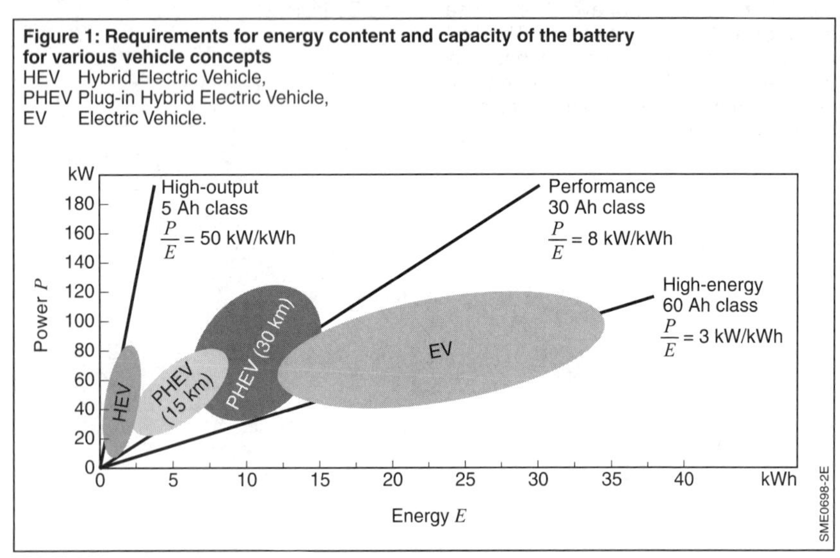

Storage technologies

Today, lithium-ion (Li-Ion) batteries are usually used as electrical energy accumulators in electric vehicles, hybrid vehicles and plug-in hybrid vehicles. Nickel-metal hydride batteries (NiMH) are still used in isolated cases. However, it is to be expected that these will be gradually replaced by lithium-based battery systems.

Figure 2 shows a comparison of specific output and specific energy of various technologies.

Nickel-metal hydride technology

Nickel-metal hydride batteries have already been in use as drive batteries for a few years, in particular for hybrid vehicles. The primary disadvantage of cells in nickel-metal hydride technology, however, is the lower energy density at a specific power that is also usually lower compared to lithium-ion batteries. Nickel-metal hydride cells achieve a specific energy of up to 80 Wh/kg at a voltage of typically 1.25 V per cell. Nickel-metal hydride cells also have higher self-discharge than lithium-ion cells.

Lithium-ion technology

Lithium-ion systems feature a rated voltage of approximately 3.7 V per cell, while the voltage limit in use is between 2.8 V and 4.2 V.

The positive electrode (cathode) of a lithium-ion cell is based on a metal oxide structure that can store lithium ions (see Electrochemistry). These metal oxides – the active material – currently usually consist of compounds from nickel, manganese or cobalt. The lithium ions move around in a reversible mechanism during the discharging process towards the negative electrode (anode) and during the charge operation towards the positive electrode (cathode). Graphite is frequently used as the active material in the anode.

New materials are in development in order to improve safety characteristics and increase the energy and power density. This includes, for instance, using lithium iron phosphate (Li_xFePO_4) in the cathode or lithium-titanate oxide ($Li_4Ti_5O_{12}$) in the anode instead of graphite.

Figure 2: Attainable specific energies and capacities for various storage technologies (Ragone chart)
1 Development goal for the year 2020.

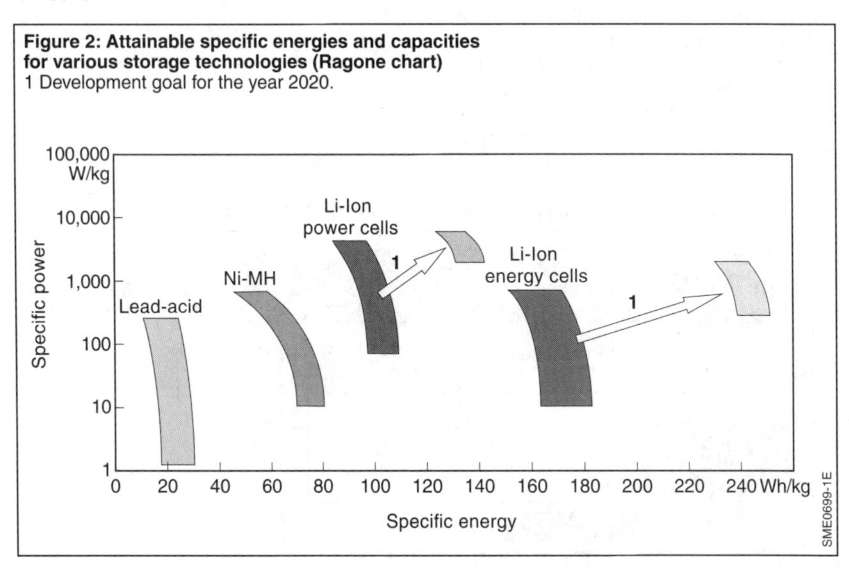

The simultaneous optimization to a high output and energy density is a technical challenge. A high power density requires large electrode surfaces with low-resistance conductors, while a high energy density requires the opposite: compact electrodes with a large amount of active material.

Typical characteristic values of today's automotive cells include 5,000 W/kg for the specific performance of power cells and 180 Wh/kg for the specific energy of energy cells. Lithium-ion systems with a high specific energy of 250 Wh/kg are in development (see Figure 2).

Lithium-ion systems, compared to other technologies, require more elaborate system monitoring and control systems using a battery management system. A lithium-ion battery must be kept within strictly defined charge, current and temperature limits for service life and safety reasons. In particular, lithium-ion cells are not capable of tolerating overcharging. Monitoring electronics are used to prevent this.

It is to be expected that cells based on innovative electrochemical approaches, such as lithium-sulfur or lithium-air, are able to provide significantly higher energy and power densities at equal or even lower costs. However, systems based on this are still being researched. The capability for mass production for use in the automotive industry has yet to be proven.

Basic structure of a battery system

Drive batteries also differ in their internal structure due to each of the different technical requirements and individual installation situations in the vehicle. However, most follow a similar basic concept, which is depicted below.

A battery system usually consists of multiple modules, each of which, in turn, contains multiple cells. The modules are connected to each other via a high-voltage wiring harness and are integrated together with the battery management system and cooling system in a battery housing. Interfaces for the vehicle include the high-voltage terminals of the battery, a data interface and usually an additional connection for the cooling system (Figure 3).

Figure 3: Schematic representation of the architecture of a battery system

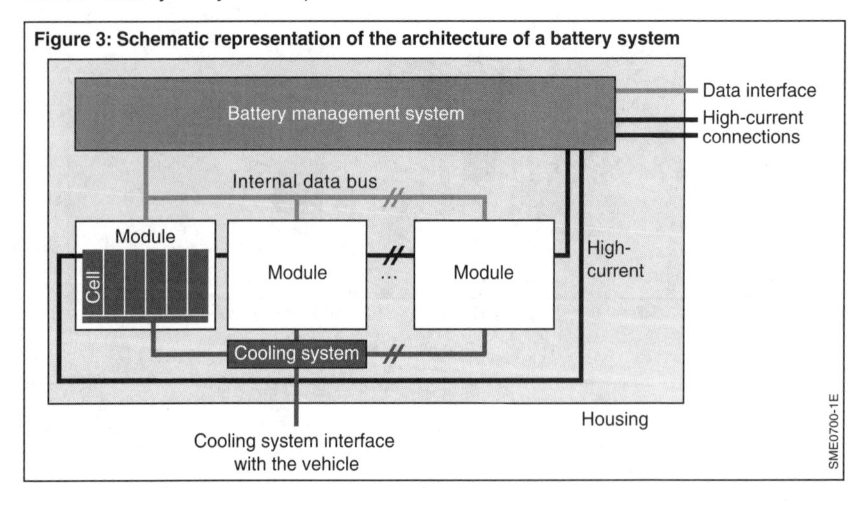

Components of a lithium-ion battery system

Lithium-ion battery cells

Depending on the application, a drive battery consists of battery cells, ranging from ten to one hundred and, in individual cases, even thousands. Here, battery cells are used in three different designs (shown in Figure 4) for use in automotive industry.

Designs

Battery cells with cylindrical metal housing have major cost advantages in manufacturing when compared to the other designs.

Prismatic cells with a housing made of laminated metal foil (pouch cells) have a particularly homogeneous temperature distribution and permit highly compact battery systems of flexible sizes.

Prismatic hardcase cells also permit a very compact construction and are also pressure-sealed and stable in comparison.

It is still uncertain whether one of the three cell types will prevail for automotive applications. Currently, there is an identifiable trend toward pouch cells and prismatic hardcase cells.

There are global efforts to standardize the types of cell housing for automotive applications as well.

Layout of a lithium battery cell

Figure 5 shows the cross-section of a prismatic hardcase cell. The active materials required for the chemical process in the anode (for example graphite) and the cathode (lithium compound) are present as coating on a copper film (anode) and an aluminum film (cathode). The metal films are each used here as shunts. The films, coated on both sides, are separated by a separator film that prevents a direct electrical connection between the anode

Figure 4: Lithium-ion battery cell designs
a) Battery cells with a cylindrical metal case.
b) Prismatic cells with case made of laminated metal foil.
c) Prismatic hardcase cell.

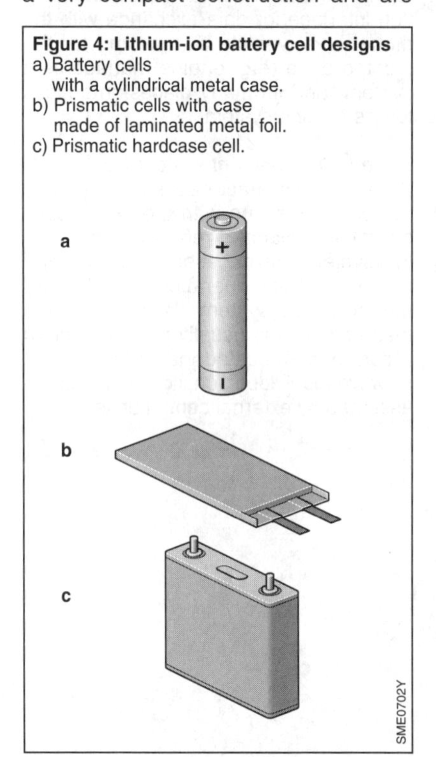

Figure 5: Cross-section of a prismatic hardcase cell
1 Positive terminal (pole),
2 Safety valve,
3 Negative terminal (pole),
4, 6 Shunts,
5 Cell case,
7 Cell winding with anode, cathode and separator.

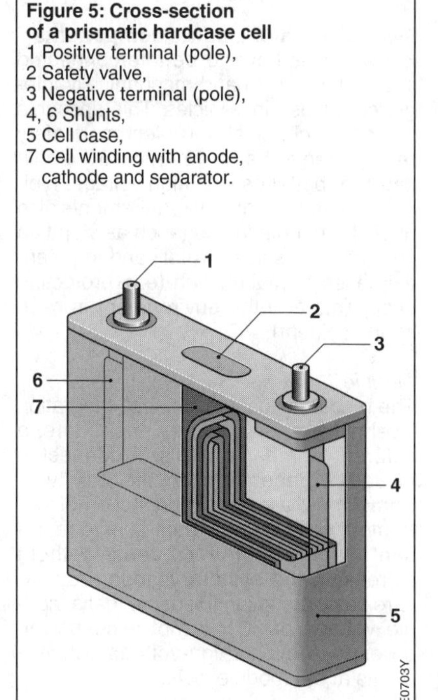

and cathode, but is conductive for ions. The ion-conductive electrolyte is also located between the anode and cathode. This sequence of anode, cathode and separator is wound into a flat coil of cells, the electric shunts at the anode and cathode lead to the negative/positive electric terminal of the cell.

An additional safety valve is also located on the cell. If the pressure in the cell rises in the case of a malfunction, it opens to prevent the cell from bursting.

Safety requirements

When a technology-dependent temperature threshold in the cell (typically around 140 °C) is exceeded, a self-reinforcing, irreversible exothermic process called "Thermal Runaway" starts and can no longer be stopped by external influences. This can cause the cell to start on fire or burst. Therefore, corresponding measures must be taken to ensure that this process is not triggered by improper operation.

Mechanical and electric components

Battery cells have to be interconnected in a suitable form and mechanically integrated for use in vehicles. This accounts for the mechanical and electrical safety requirements that are particularly applicable for batteries with high voltage. Typical automotive industry requirements also apply here, both in use (such as vibration protection or splash water) and in potential cases of malfunction (e.g. protecting occupants and the environment in case of an accident).

Module

The module is usually the first integration level of cells for a battery pack. Here, a different number of cells is connected in series depending on the cell type. Sometimes, a parallel connection of two or more cells is used to be able to represent larger energy contents without increasing the system voltage. Usually, care has to be taken to ensure that a module voltage of 60 V is not exceeded in order to prevent high-voltage requirements at the module level.

The cells are mechanically integrated into a network and their terminals are connected by a high-voltage cable harness. Additionally, sensors and evaluation electronics are frequently integrated into the module. The focus here is to ensure the most compact design engineering possible that allows flexible, cost-effective and high-quality battery production.

While small battery systems can only consist of one module, battery packs for electric vehicles often use over ten modules. The modules are connected to each other electrically via a high-voltage cable harness.

Housing

The battery housing, along with holding together the cells and protecting them from mechanical loads, is also used for contact protection against high voltages. On the outside of the battery housing, there are high-voltage connections for connecting to the electric powertrain, a communication interface for data exchange with the higher-level vehicle control unit for the electric drive (e.g. engine management system) and, in many places, interfaces for the cooling system.

Battery management system

Determining characteristics of the battery system, complying with the permissible operating limits and regulating the cooling system are necessary for safe and reliable battery operation. The battery management system (BMS) assumes these tasks. It is usually implemented in a control unit located inside the battery. However, individual functions can also be assumed by external control units.

Operation

The battery management system measures the battery current, the temperature of the battery cells and the voltage of individual cells, the voltage of individual modules and the voltage of the overall battery. This is used to determine the battery cells' state of charge (SOC) and state of health (SOH). The currently available battery rating and permissible battery current can be determined based on this. The battery management system can also carry out "Balancing" to match the cells' state of charge. This information is transmitted to the central vehicle control unit in the electric powertrain of the vehicle. The battery management system cannot directly affect the discharge current, charging current and state of charge during this process. The vehicle control unit controls the power electronics in the electric powertrain or the charging device based on the information sent by the battery management system to configure a corresponding current.

The battery management system protects the battery cells by preventing the current flow in cases when the battery is overcharged, the safe operating window is exited due to over- or undervoltage, or if the battery is overheated. This occurs by instructing consumers to switch off the vehicle control unit in case of emergency by disconnecting the battery on its own.

Even if the battery itself has to be assigned to the electrochemical domain, the functional safety standard (ISO 26262 [1]) for electrical and electronic components must be taken into account when developing the battery's safety concept for the battery management system. Safety-related components of the battery management system, particularly the sensor technology, are, therefore, redundant.

Architecture

In Figure 6, the distributed architecture of a battery management system for plug-in hybrid electric vehicles (PHEV) and electric vehicles (EV) is shown as an example.

Figure 6: Overview of the battery system
1 Main contactor,
2 Charging contactor,
3 Pre-charging resistor.
CSC Cell Supervisory Circuits,
BCU Battery Control Unit.

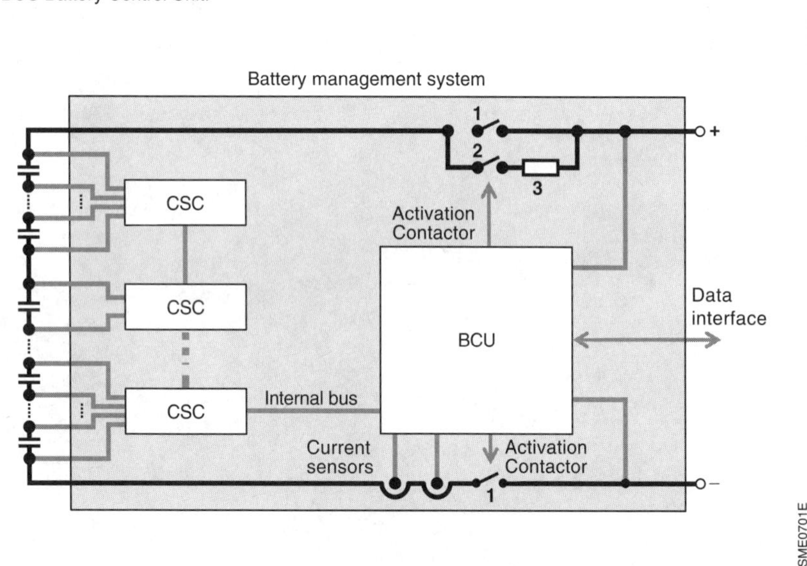

Other architecture variants are also used, particularly for hybrid electric vehicles (HEV).

The battery management system consists of a central battery control unit, cell supervisory circuits, current sensors and a contactor circuit. The cell supervisory circuits monitor the battery cells in terms of their voltage and temperature.

Manner in which current is conducted

The contactor current allows the vehicle electrical system battery of the electric powertrain to be electrically isolated. If an accident is detected or another potential hazardous state is identified, a control logic opens the contactor that is located at each positive or negative battery terminal. To prevent a safety-critical condition, this logic can open the contactor even if the battery system is overloaded.

The battery is connected to the vehicle electrical system of the electric powertrain with its buffer capacitor via a charging contactor and a pre-charging resistor. This prevents an excessively large initial charge current when the buffer capacitor is discharged. During normal operation, the battery is connected to the vehicle electrical system with low impedance via the main contactor.

Balancing

Due to manufacturing tolerances and side reactions that depend on cell parameters and the cell temperature, the individual cells eventually have a different state of charge. This is problematic insofar as the cell with the lowest state of charge sets the discharge limit, and the cell with the highest state of charge sets the charge limit. For this reason, the state of charge of a battery's cells has to be balanced from time to time. This is done by a balancing circuit, which is usually integrated in the cell supervisory circuits. In automotive applications, balancing is typically a passive process; that is, it occurs by discharging individual cells via a resistor connected in parallel.

Thermal management

To ensure safe and reliable operation of the battery even when the load is fluctuating and to maximize the service life of the cells, the temperature of the cells should be kept in a range typically from 20 °C to 40 °C. The service life of the battery cells is largely reduced as the temperature increases, because many aging processes are temperature-dependent. This is particularly relevant for use in hot climate zones. At low temperatures, the capacity of the cells noticeably decreases.

In vehicles, there are various systems used to control the temperature of cells.

Liquid-cooling system
Liquid-cooling systems are usually used in systems with high requirements for regulating temperature. The battery is equipped with a separate cooling circuit, in which circulates a special coolant, such as a water-glycol mixture. The coolant temperature is controlled by a heat exchanger, which is connected to the vehicle's air-conditioning system. Frequently a heater element is also integrated, so that the battery can be heated at low temperatures.

It is connected to the battery cells via cooling plates. Temperature sensors monitor the temperature of the cells. The central thermal management system, often integrated in the battery management system, controls the flow rate and coolant temperature depending on the situation and temperature.

Batteries with a liquid-cooling system can be built very compactly, but require additional effort due to the cooling plates, the piping and the connection to the vehicle's air-conditioning system.

Refrigerant cooling system
An alternative to a liquid-cooling system is the refrigerant cooling system, in which the cooling components of the battery are designed as evaporators and directly integrated into the refrigerant circuit of the vehicle's air-conditioning system. Systems of this kind have been used only rarely to date.

Air cooling
Air-cooled systems are the most cost-effective designs. However, they require more space due to the need for space between the cells. Moreover, the air interfaces mean that hermetically sealed battery cases are not possible, or they can be implemented only with great effort.

Air-cooled systems are often used for batteries with low cooling requirements, for example, in systems with low specific power requirements, such as batteries for electric vehicles and for use in moderate climate zones.

In air-cooled systems, fans controlled by a thermal management system take in air that is usually precooled. This is blown through the battery system; the fan speed can be controlled depending on the cell temperature determined by the temperature sensors.

Some systems for low power requirements have no fans; in that case, the battery cells are simply cooled by natural convection.

References
[1] ISO 26262: Road vehicles – Functional safety.

Electrical machines

Systematics of rotating electrical machines

Electrical machines [7, 8] are electro-magneto-mechanical energy converters. With the energy W_m stored in the magnetic field, discharge according to the angle of rotation γ generates the magnetic force F_t:

$$F_t = \frac{\partial W_m}{\partial \gamma}.$$

It serves as the tangential force to generate the torque calculated with the rotor radius r (Figure 1). The following applies:

$$M = F_t\, r.$$

Electrical machines can be systematically categorized with regard to their guidance (Table 1). Part of the systematic approach is from DIN 42027 [1].

Table 1: Systematic approach to electrical machines

Self-commutated machines (commutator machines)	
Alternating current machine (universal machine)	Direct-current machine – Series-wound – Shunt
Externally-commutated machines (frequency-commutated machines)	
Asynchronous machines	Synchronous machines

Figure 1: Operating principle of rotating electrical machines
1 Stator, 2 Rotor.
r Rotor radius, F_t Tangential force,
γ Angle of rotation.

UAE1119-1Y

Direct-current machines

Direct-current machines are frequently preferred for motor operation. They are used, for example, as drives for electric fuel pumps, fan motors, starter motors, windshield wiper motors, and power-window motors.

The direct-current machine (Figure 2) consists of the stator (frame), which bears the excitation, compensation, and commutating winding as well as the rotor (armature, inductor) with the rotor winding. The rotor is supplied with current via the commutator brushes and commutator lamina. The windings can be combined in parallel or in series.

The direct-current machines are classified according to their different characteristics (series or shunt). The calculations apply to self-commutated motors. The connection designations of direct-current machines comply with DIN EN 60034, Part 8 [2].

Figure 2: Structure of the two-pole direct-current machine (cross-section)
1 Stator, 2 Exciter pole (pole shoe),
3 Exciter winding,
4 Compensation winding (preferably on machines with higher power output),
5 Commutating pole,
6 Commutating winding,
7 Rotor, 8 Rotor winding,
9 Commutator lamina,
10 Commutator brush (position in the neutral zone).

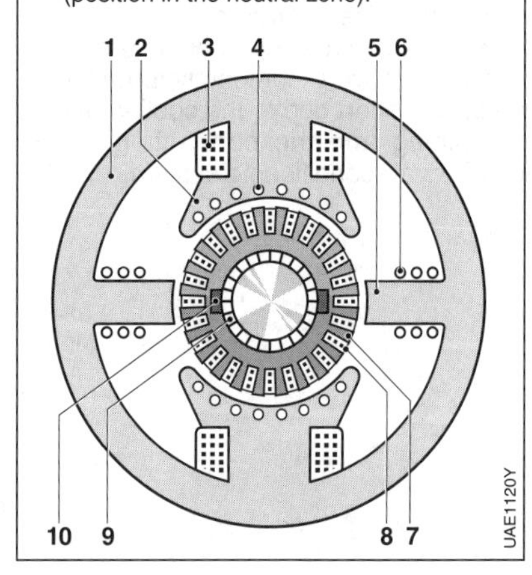

Commutation

For the function of the direct-current machine, it is required that the direction of the current in the rotor remains constant in relation to the stator poles (Figure 3). The process of changing the direction of the current in the rotor takes place in the neutral zone and is called commutation. The commutator is supplied with current I_R via the commutator brushes. This is divided into the branch currents I_{ZW}. The following applies in general to the voltage induced in coils:

$$u = -L\ \frac{di}{dt}\,.$$

The tangential speed v_t of the commutator surface is:

$$v_t = \omega\ \frac{d_c}{2}\,.$$

If the angle speed ω is used to determine the cycle duration T_c and K commutator lamina are taken into account, the following applies:

$$T_c = \frac{\pi d_c}{v_t K}\,.$$

The current change in the commutating coil takes place in the time T_c. If it is taken into account that only the current I_{ZW} takes effect, the following applies for the induced voltage:

$$u = -L\ \frac{I_{ZW} K v_t}{\pi d_c}\,.$$

The designations are taken from DIN 1304, Part 7 [3].

Figure 3: Principle of current commutation
(Diagram of the rotor)
1 Stator pole, 2 Rotor, 3 Rotor winding,
4 Commutator brushes,
5 Commutator lamina.
I_R Rotor current,
I_{ZW} branch current ($I_{ZW} = I_R/2$),
d_c Commutator diameter,
Φ Magnetic stator flux,
N North pole, S South pole.

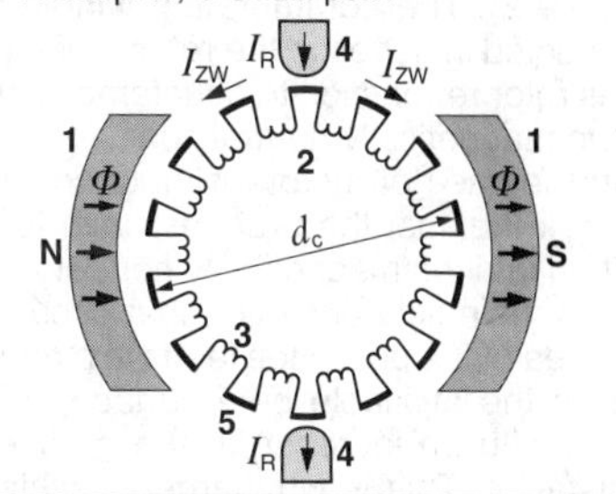

Figure 4: Superposition of fields
a) Main field: exciting current switched on, rotor current switched off,
b) Armature quadrative-axis field: exciting current switched off, rotor current switched on,
c) Complete field: superposition of main and armature quadrative-axis field; magnetic neutral zone is deflected by angle β.
1 Exciter pole (stator pole, main), 2 Rotor, 3 Rotor winding.
Φ_S Magnetic stator flux (magnetic circuit via the motor housing),
Φ_R Magnetic rotor flux (magnetic circuit via the exciter pole or motor housing),
Φ_{RS} Magnetic flux of the complete field.

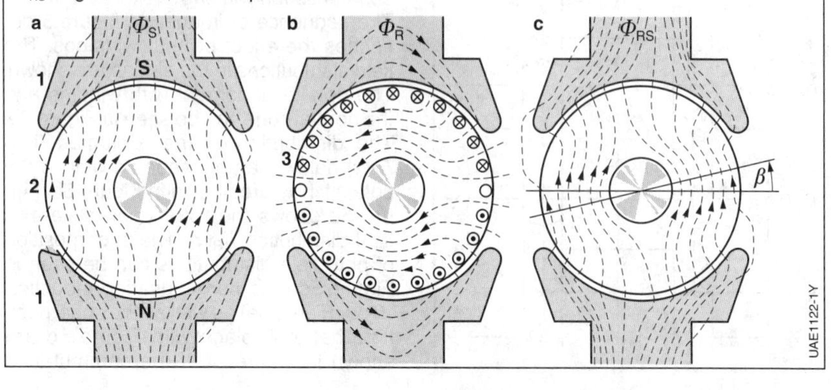

Commutating and compensation winding

Field distribution

The main field that penetrates the rotor without hindrance when no current is present has a symmetrical distribution (Figure 4a). Likewise, there is a symmetrical flux division when current is applied to only the rotor (Figure 4b). If both fields are superpositioned, the neutral zone is deflected by the angle β (Figure 4c). This magnetically neutral zone thus no longer corresponds to the geometrically neutral zone (position of the commutator brushes).

Figure 5: Effect of the compensation and commutating winding
a) Pole arrangement,
b) Distribution of the excitation field $B_E(x)$,
c) Distribution of the rotor quadrative-axis field $B_R(x)$,
d) Superposition of $B_E(x)$ and $B_R(x)$,
e) Compensation induction $B_K(x)$,
f) Superposition of $B_E(x)$, $B_A(x)$ and $B_K(x)$,
g) Commutating induction $B_W(x)$,
h) Superposition of all field distributions.
1 Pole shoe, 2 Neutral zone.
Φ Stator flux.

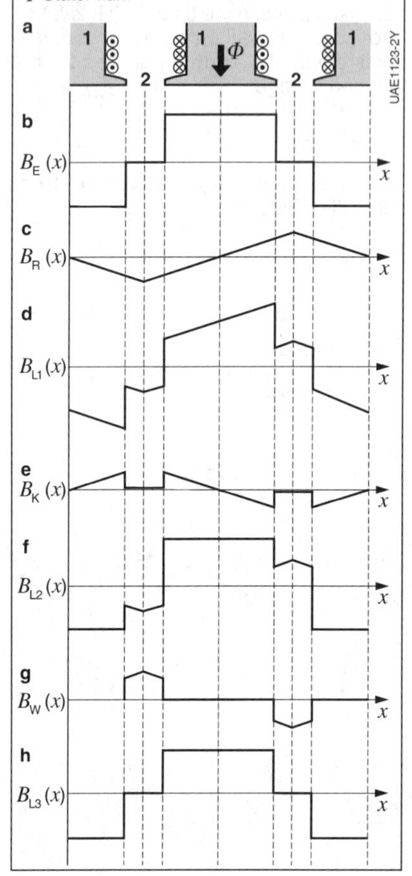

In the geometrically neutral zone, this creates a magnetic field through which a voltage is induced during the commutation process in the coil to be commutated, creating brush sparking (break sparks) between the brush and the running commutator lamina. To prevent this, another voltage is induced during commutation in the coil concerned where the amplitude and direction cancel out the effect of the originally induced voltage. This is achieved by means of the commutating winding (Figure 2). The commutating winding is connected in series to the rotor winding. It uses rotor retroaction to counteract a shift in the magnetically neutral zone.

In the case of motors without commutating windings, the brushes have to be shifted into the magnetically neutral zone.

The main field distortion that occurs in the area of the pole shoe leads to a reduction in the available pole surface, combined with an increase in the magnetic resistance. This is why larger machines are given a compensation winding that is integrated in the pole shoes (Figure 2). The compensation winding is connected in series with the rotor winding and its dimensions are such that it compensates for the rotor quadrative-axis field.

Effect of the commutating and compensation winding

The sequence of images in Figure 5 describes the effect of both windings. The field distributions in the air gap are shown. The pole arrangement with winding and the neutral zone can be seen in Figure 5a. The distribution of the excitation field $B_E(x)$ below the pole shoe as well as the pole pitch τ_P are shown in Figure 5b. Figure 5c shows the rotor quadrative-axis field distribution $B_R(x)$. The superposition of both field distributions can be seen in Figure 5d. The compensation induction $B_K(x)$ (Figure 5e) as well as the superposition that took place with Figure 5d are shown in Figure 5f. If the commutating induction $B_W(x)$ in Figure 5g is superpositioned with the field distribution of Figure 5f, this results in the desired field distribution according to Figure 5h.

Tangential-force calculation

A tangential force engaging at the rotor is required to generate torque. The object of examination is a stator and a rotor with groove (Figure 6) in which there is a current-carrying conductor (winding strand). The rotor moves from position 1 into position 2. The stator flux Φ_S creates the induction B_S in the air gap while the conductor in the rotor through which current passed causes the induction B_R. On the left of the groove, there is a destructive superposition and on the right of the groove there is a constructive superposition of both inductions. The energy stored in the air gap on the left of the groove amounts to:

$$dW_2 = \frac{\delta l_L}{2\mu_0} \cdot (B_S^2 - B_R^2) dx . \qquad \text{(Equation 1)}$$

On the right of the groove, it amounts to:

$$dW_1 = \frac{\delta l_L}{2\mu_0} \cdot (B_S^2 + B_R^2) dx . \qquad \text{(Equation 2)}$$

The work dW done on that section results from the difference between the energies dW_1 and dW_2:

Figure 6: Stator-rotor arrangement
1 Stator, 2 Rotor, 3 Groove,
4 Conductor through which current passed.
Φ_S Stator flux, Φ_R Rotor flux, δ Air gap,
B_S Induction of the stator flux,
B_R Induction of the rotor flux,
F_t Tangential force, F_n Normal force,
F_{res} Resulting force,
P1 Position 1, P2 Position 2.

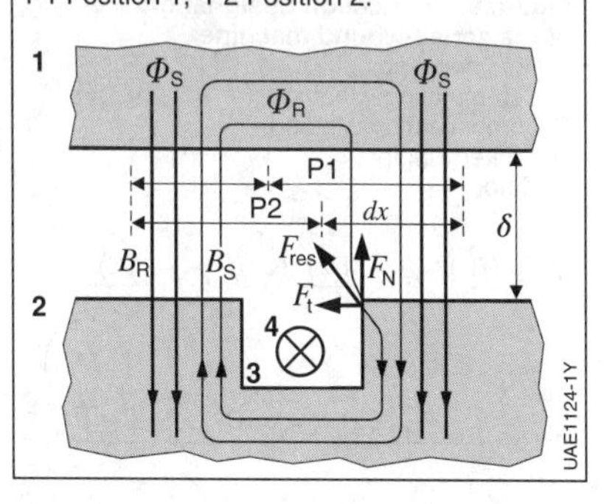

$$dW = dW_2 - dW_1 = F_t dx . \qquad \text{(Equation 3)}$$

Equations 1, 2, and 3 can be used to calculate the force F_t. The following applies:

$$F_t dx = \frac{l_R \delta}{2\mu_0} [(B_S^2 - B_R^2) - (B_S^2 + B_R^2)] dx .$$

This means that:

$$F_t = \frac{l_R \delta}{\mu_0} B_R^2 .$$

The magnetically effective rotor length is l_R. The force F_t is a quadratic function of the induction of the wire through which current passed. The influences of stator induction cancel each other out. As forces on boundary layers always apply in the direction of low permeability, forces affect the torque due to fluxes that emerge sideways from the groove.

Shunt-wound machine

A feature of the shunt-wound machine is that the rotor winding is connected parallel to the exciter winding. The following connection conditions apply to the shunt-wound machine (Figure 7):

- A indicates the rotor winding
- B indicates the commutating winding
- C indicates the compensation winding
- E indicates the exciter winding

The calculation of the shunt-wound machine includes the rotor resistance R_A and the voltage induced by the rotor winding. The terminal voltage U_{KL} is composed of the rotor voltage $U_A = I_A R_A$ and the induced voltage $U_{ind} = c_1 n \Phi_S$:

Figure 7: Connection designation of the shunt-wound machine
A Rotor winding,
B Commutating winding,
C Compensation winding,
E Exciter winding,
M Motor.

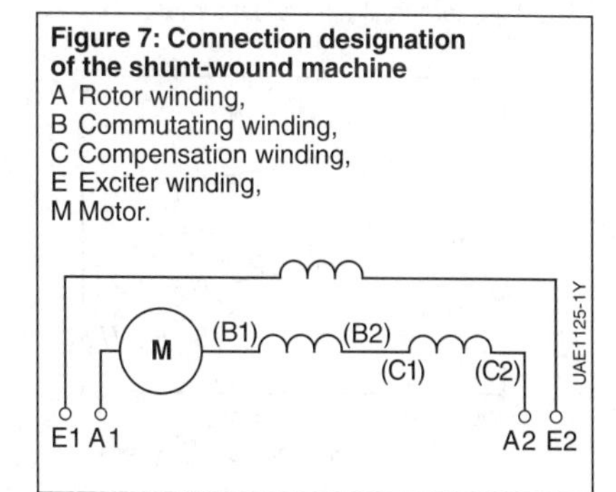

$$U_{KL} = I_A R_A + c_1 n \Phi_S \qquad \text{(Equation 4).}$$

Here, c_1 designates the constructively specified motor constant (no unit) that depends on the number of coils in the rotor winding. n is the rotational speed, Φ_S the stator flux, and I_A is the rotor current.

Solving Equation 4 for n yields the rotational speed (rotational speed-rotor current equation):

$$n = -\frac{R_A}{c_1 \Phi_s} I_A + \frac{U_{KL}}{c_1 \Phi_s} \qquad \text{(Equation 5).}$$

If the motor torque

$$M_M = c_2 \Phi_S I_A$$

is inserted in the rotational speed-rotor current equation (Equation 5), the rotational speed-torque equation follows with

$$n = -\frac{R_A}{c_1 c_2 \Phi_S^2} M_M + \frac{U_{KL}}{c_1 \Phi_S} \qquad \text{(Equation 6).}$$

c_2 is also a constructively specified motor constant (no unit) depending on the number of coils of the rotor.

Figure 8 shows the operating characteristics of the shunt-wound machine. In order to overcome the friction, the motor must apply the frictional torque M_R. At the point in time of switching on, the rotational speed equals zero. This enables calculation of the starting torque

$$M_A = \frac{U_{KL} c_2 \Phi_S}{R_A}$$

Figure 8: Rotational speed – torque characteristic curves of the shunt-wound machine
n Motor speed, n_n Rated speed,
U_{Kl} Terminal voltage, U_N Rated voltage,
M_M Motor torque, M_R frictional torque,
M_N Rated torque.

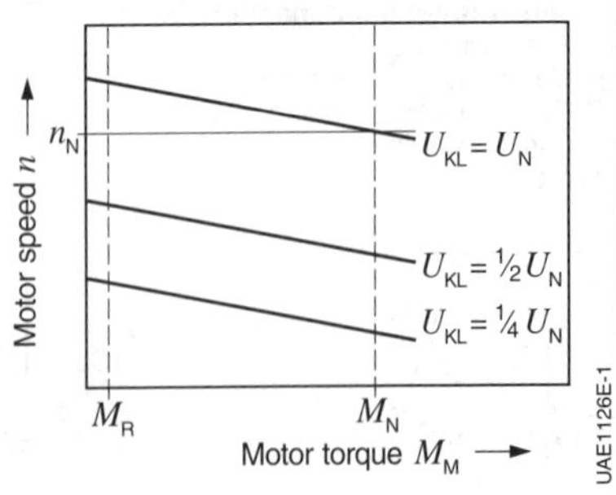

in Equation 6. At the theoretically highest rotational speed of a characteristic curve (idle speed), the motor does not provide any torque. The idle speed n_0 is

$$n_0 = \frac{U_{KL}}{c_1 \Phi}.$$

Series-wound machine

In the case of the series-wound machine, the commutating, compensation, exciter, and rotor winding are connected in series (Figure 9). D designates the exciter winding of the series-wound machine. To determine the operating characteristics, the ohmic resistances of the windings are added up to form the resistance R_A.

In the same way as with the shunt-wound machine, the terminal voltage U_{KL} is composed of the rotor voltage and the induced voltage:

$$U_{KL} = I_A R_A + c_1 n \Phi_S \qquad \text{(Equation 7).}$$

This results in the rotational speed-current equation:

$$n = -\frac{R_A}{c_1 \cdot \Phi} I_A + \frac{U_{KL}}{c_1 \cdot \Phi} \qquad \text{(Equation 8).}$$

The magnetic flux can be calculated with the other motor constant c_3:

$$\Phi = c_3 I_A \qquad \text{(Equation 9).}$$

The motor constant c_3 has the unit of an inductivity and is therefore dependent on the geometry, the number of coils, and the permeability. The motor torque M_M is calculated in accordance with

$$M_M = c_2 c_3 I_A^2 \qquad \text{(Equation 10).}$$

Figure 9: Connection designation of the series-wound machine
A Rotor winding,
B Commutating winding,
C Compensation winding,
D Exciter winding,
M Motor.

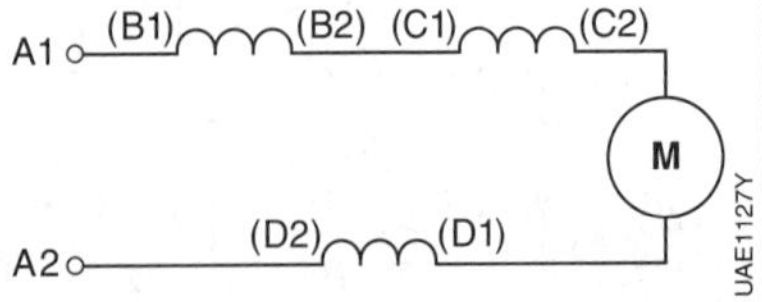

A1 (B1) (B2) (C1) (C2)
M
A2 (D2) (D1)
UAE1127Y

If Equation 10 is adjusted according to I_A, the expression takes effect in Equation 9 and this in turn in the rotational speed-current equation (Equation 8), and the result is:

$$n = -\frac{R_A}{c_1\, c_3} + \frac{U_{KL}\sqrt{c_2\, c_3}}{c_1\, c_3} \cdot \frac{1}{\sqrt{M_M}}$$

(Equation 11).

In contrast to the shunt-wound machine, the rotational speed is proportional to the reciprocal value of the root of the torque (Figure 10). The motor is characterized by a large drop in rotational speed at low load. If the external load on the motor is set to equal zero, the rotational speed theoretically goes towards infinite.

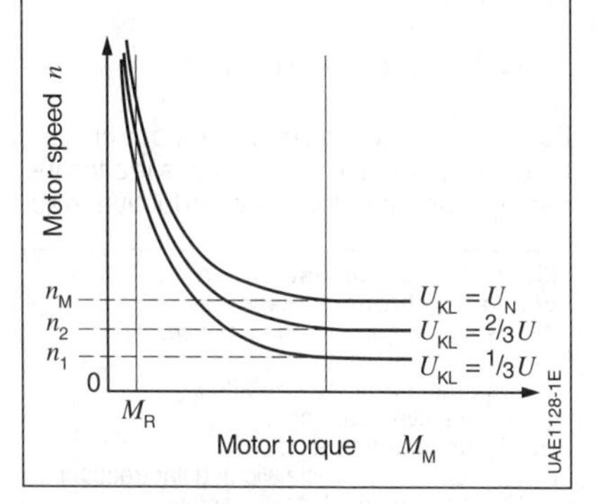

Figure 10: Rotational speed-torque characteristic curves of the series-wound machine
n Motor speed, n_n Rated speed,
U_{Kl} Terminal voltage, U_N Rated voltage,
M_M Motor torque, M_R frictional torque,
M_N Rated torque.

Figure 11: Principle of an asynchronous machine
1 Stator winding, 2 Rotor.
Axis of rotation vertical to drawing plane.

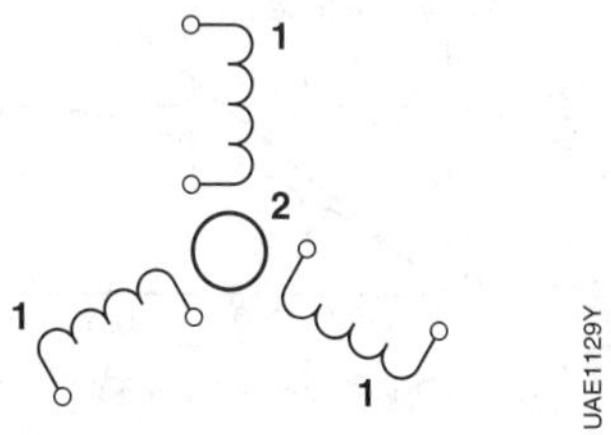

Asynchronous machine

The asynchronous machine is the main drive used in industry. In the automotive sector, for example, it is used in electrical power-assisted steering and in hybrid vehicles. The following section presents the operating concept of the asynchronous machine as an induction machine. There is also an energetic evaluation of the asynchronous machine because of its strong prevalence in drive technology.

General setup

A distinction is made between external rotor and internal rotor machines. In the case of the external rotor machine, the rotor encloses the stator; on the internal rotor machine, the stator encloses the rotor. The diagram of the principle (Figure 11) indicates the fundamental structure of an internal rotor asynchronous machine.

The rotor consists of a short-circuit cage with lamination stack (Figure 12, as an example with four short-circuit rods). The lamination stack fills the space within the short-circuit cage completely (not shown in the illustration). It consists of individual steel sheets insulated against one another to keep the eddy current losses to a minimum.

Figure 12: Short-circuit cage of the asynchronous machine
1 Short-circuit ring,
2 Short-circuit rod.

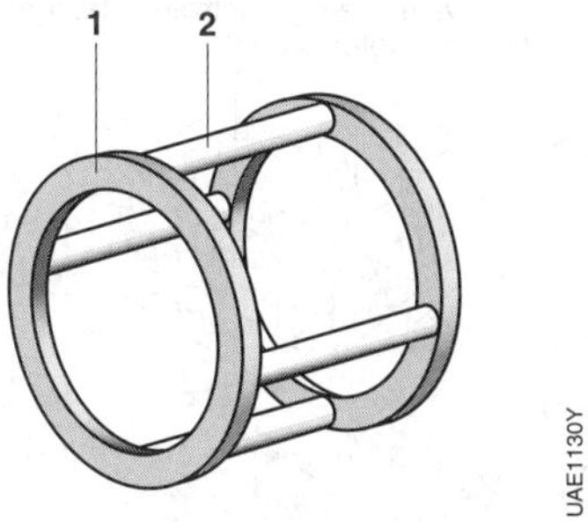

Operating characteristics

The stator winding generates a rotating field with a three-phase alternating current. There is a speed difference between the rotating field speed and the rotor speed that enables the induction of a magnetically effective current in the rotor, which in turn contributes to the generation of torque.

The physical operating concept is based on the law of magnetic induction. Figure 13 shows the rotor as a simplified conductor loop on rotating mounts. The relative movement between the stator and rotor is described by the slip angular frequency ω_S. The magnetic field B_E with the slip angular frequency that surrounds the rotor loop induces a voltage in the short-circuit rotor in accordance with the second Maxwell equation [4, 5, 6]:

$$\oint \vec{E} d\vec{s} = -\frac{d}{dt}\iint B_E dA \qquad \text{(Equation 12).}$$

This equation and the constructive variables of the motor (cf. Figure 13) result in the relationship

$$2\vec{E}(l+2r)\vec{n} = 2lr\hat{B}_E\omega\sin(\omega t) \qquad \text{(Equation 13).}$$

Where $E = I \cdot A_{nom}/\kappa$ (κ specific electrical conductivity), the voltage drives the magnetically effective power

$$i = \frac{\kappa A_{nom} l r \hat{B}_E \omega_S}{(l+r)}\sin(\omega_S t) \qquad \text{(Equation 14)}$$

Figure 13: Two-bar rotor as short-circuit rotor
1 Short-circuit rod, 2 Short-circuit bridge.
B_E Flux density of the stator rotating field,
i Induced current, l Length of short-circuit rod,
r Rotor radius, A_{nom} Rod cross-sectional area,
ω_S Slip angular frequency (results from the relative movement between stator rotating field and rotor speed).

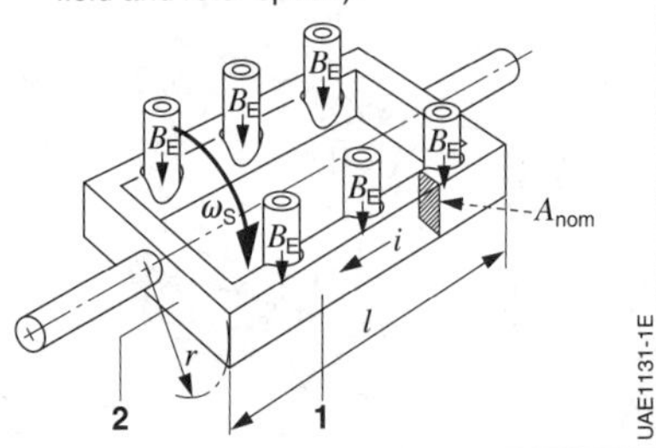

in the conductor loop, whose magnetic field

$$H_{ind} = \frac{iN}{l_{Fe}}$$

and its induced flux density

$$B_{ind} = \mu H_{ind}$$

weakens the original rotating field B_E (inductive reactance). The following applies to the resulting flux density:

$$B_R = B_E - B_{ind}\ .$$

In the Maxwell equation (Equation 13), instead of B_E, the resulting magnetic flux density B_R remains. At the conductor loop, the tangential force F_t

$$F_t = il\hat{B}_R\sin(\omega_S t) \qquad \text{(Equation 15)}$$

is applied (Lorentz force). This is used to calculate the torque. The following applies with Equations 14 and 15:

$$M = 2F_t r$$

$$= \frac{2\kappa A_{nom}\omega_S}{(l+2r)}[lr\hat{B}_R\sin(\omega_S t)]^2\ .$$

Figure 14 shows two characteristic curves. One characteristic curve shows the torque path under the influence of inductive reac-

Figure 14: Torque distribution of an asynchronous machine
1 Distribution under the influence of inductive reactance,
2 Distribution without the effect of inductive reactance.
ω_S Slip angular frequency,
ω_{Smax} Maximum possible slip angular frequency,
ω_K Breakdown angular frequency,
M Motor torque, M_K Breakdown torque.

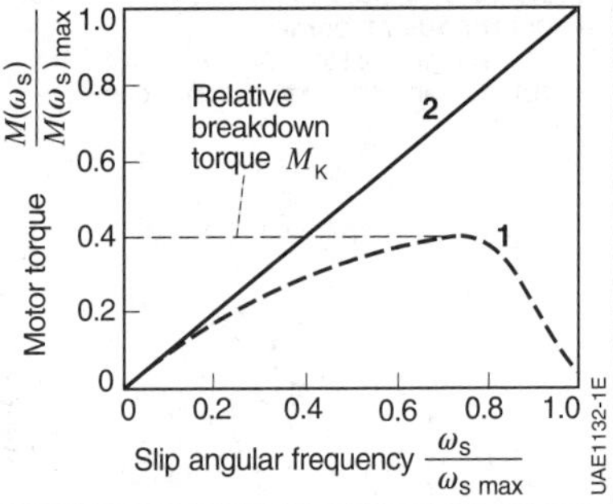

Motor torque $\frac{M(\omega_S)}{M(\omega_S)_{max}}$; Slip angular frequency $\frac{\omega_S}{\omega_{S\,max}}$; Relative breakdown torque M_K; 1; 2; UAE1132-1E

tance. It can be influenced by rotor rod geometries and the choice of materials. The other characteristic curve shows the torque path without the effect of inductive reactance. This represents the limiting case for technical implementation.

If there is an increase in the slip angular frequency, inductive reactance initially leads to a rise in torque until the breakdown torque is reached. This is the maximum possible motor torque. Subsequently, the motor torque drops due to the increasing influence of the inductive reactance. The power dissipation P_V that arises in the conductor loop is calculated from the conductor loop resistance R_S and the current induced in the loop i_S:

$$P_V = R_S \cdot i_S^2 \, .$$

The power dissipation thus increases squarely with the induced current.

Efficiency classes

The CEMEP (European Committee of Manufacturers of Electrical Machines and Power Electronics) has introduced an efficiency classification based on three classes (EFF1, EFF2, and EFF3). The efficiency classes apply to three-phase current asynchronous motors with two and four poles as well as a power output between 1.1 and 90 kW (Figure 15).

Figure 15: Efficiency-power output diagram
Required efficiency for
1 EFF1 (slight differences between two-pole and four-pole motors, not discernible in the scaling shown),
2 EFF2 (two-pole and four-pole motors),
3 EFF3 (two-pole and four-pole motors).
Nonlinear application of the rated power, higher spreading in the lower power output range.

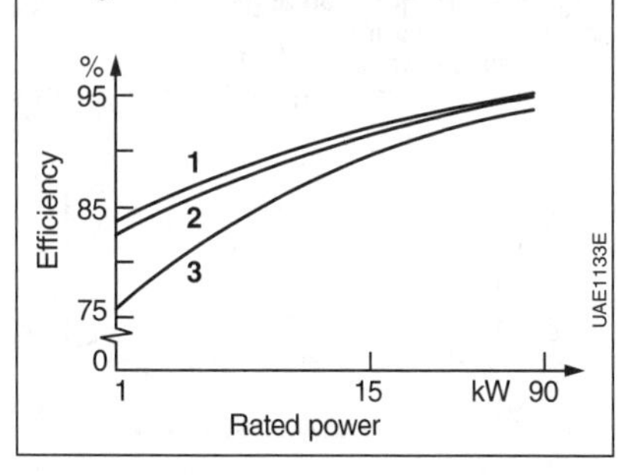

Synchronous machine

Synchronous machines are used preferably as claw pole generators. In motor operation, they are used for example in electrical power-assisted steering, to drive hybrid vehicles, and in electrically driven turbochargers.

General setup

In contrast to the asynchronous machine, in the synchronous machine the rotor rotates synchronously with the excitation field at the angle speed $\omega_{\Phi S}$. The flux Φ_R generated by the rotor winding and the stator flux Φ_S superposition themselves in relation to the resulting flux Φ_{RS} (Figure 16):

$$\Phi_{RS} = \Phi_R + \Phi_S \, .$$

As the rotor and stator material are operated far below the magnetic saturation ($\mu_r \rightarrow \infty$), the air gap d between the rotor and stator as well as the angle α determine the magnetic circuit resistance R_m:

Figure 16: Basic design of the synchronous machine
1 Stator,
2 Rotor,
3 Rotor winding with N coils.
Φ_S Stator flux,
Φ_R Rotor flux,
Φ_{RS} Superpositioned flux,
$\omega_{\Phi S}$ Angle speed of the excitation field,
I_{er} Exciting current in the rotor,
A_S Magnetically effective stator surface,
A_R Magnetically effective rotor surface,
r Radius of the rotor,
d Distance between rotor and stator,
δ Air gap length,
α Deflection angle.

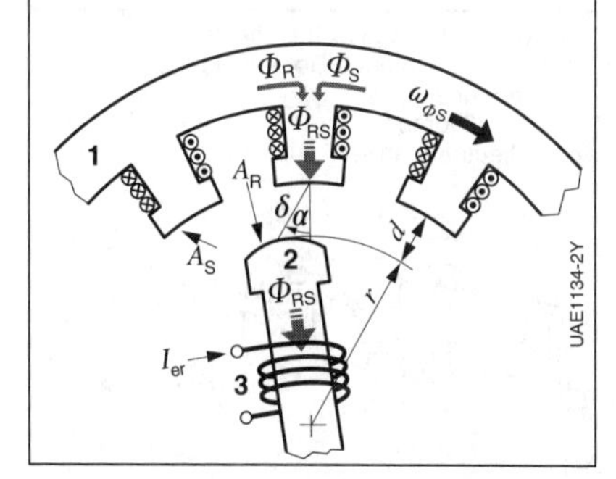

$$R_m = 2\frac{\delta}{\mu_0 A_R} = 2\frac{d}{\mu_0 A_R \cos\alpha} \quad \text{(Equation 16).}$$

The factor 2 results from the fact that there are two air gaps between the rotor and stator. If the motor provides torque, the rotor rotates with the angle α from its idle position (Figure 17).

The resulting flux Φ_{RS} is calculated as:

$$\Phi_{RS} = \frac{\Theta_{er}}{R_m} + \Phi_S .$$

With R_m from Equation 16, this results in

$$\Phi_{RS} = \frac{\Theta_{er}\mu_0 A_R \cos\alpha + 2d\Phi_S}{2d} .$$

With $\Theta_{er} = NI_{er}$, the result is

$$\Phi_{RS} = \frac{NI_{er}\mu_0 A_R \cos\alpha + 2d\Phi_S}{2d} \quad \text{(Equation 17).}$$

Θ_{er} is the magnetic rotor flooding and I_{er} is the exciting current fed to the rotor via the slip rings. The tangential force F_t affecting the torque is calculated using the Maxwell pole-force formula

$$F_t = -\frac{\Phi_{RS}{}^2}{\mu_0 \; A_R}\sin\alpha \quad \text{(Equation 18)}$$

[4, 5, 6]. The tangential force is used to calculate the motor torque M_M:

$$M_M = 2F_t r \quad \text{(Equation 19).}$$

Equation 17 inserted in Equation 18 and the result in Equation 19 yields the following relationship:

$$M_M = -\frac{r\sin\alpha}{2\mu_0 \; A_R \; d^2} \cdot [(N\,I_{er}\,\mu_0 \; A_R \cos\alpha)^2$$
$$+4N \; I_{er} \; \mu_0 \; A_R d \; \Phi_S \cos\alpha$$
$$+4d^2 \; \Phi_S{}^2)] .$$

The first term is only dependent on the exciting current I_{er} and corresponds to the cogging torque. The second term generates the motor torque to a decisive degree. The linear dependency of the rotor flooding $\Theta = I_{er}N$ and stator flux Φ_S can be seen here. The third term also generates a torque and is only dependent on the stator flux.

An increase in the outer load torque leads to an increase in the load angle α and thus to a change in motor torque M_M to be delivered (Figure 18). The maximum motor torque to be delivered is referred to as the breakdown torque M_K at the position α_K. If α_K is exceeded, the machine slips out.

Operating characteristics

A drawing of the synchronous machine can be made in a single-phase electrical equivalent circuit diagram in that the voltage induced by the rotor in the stator (pole wheel voltage U_P) is assumed as the voltage source and remaining reactances (inductive resistances) are summarized to form the synchronous reactance X_S (Figure 19). The voltage above the synchronous reactance is designated U_S and the terminal voltage U_0. The direction of the current is specified in accordance with the consumer unit counting arrow system. Whereas current flows into the consumer units during motor operation, current flows out of the generator during generator operation. Setting up the mesh equation yields the current I

$$I = \frac{U_0 - U_P}{X_S} \quad \text{(Equation 20).}$$

Figure 17: Forces at the rotor
1 Stator, 2 Rotor, 3 Rotor winding.
F_t Tangential force,
F_n Normal force,
α Deflection angle.

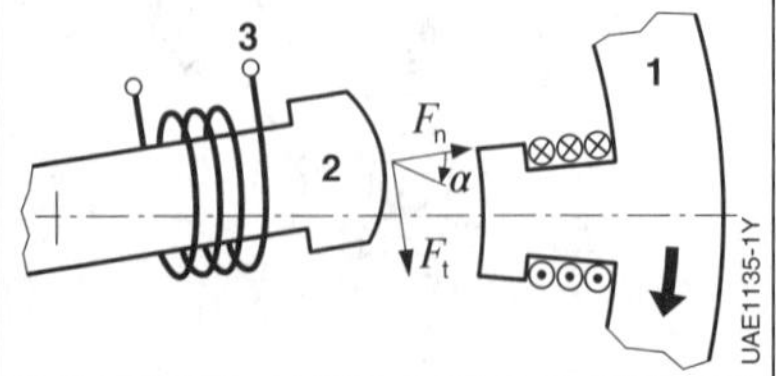

Figure 18: Torque-load angle characteristic curve
M_K Breakdown torque,
α_K Deflection on reaching the breakdown torque.

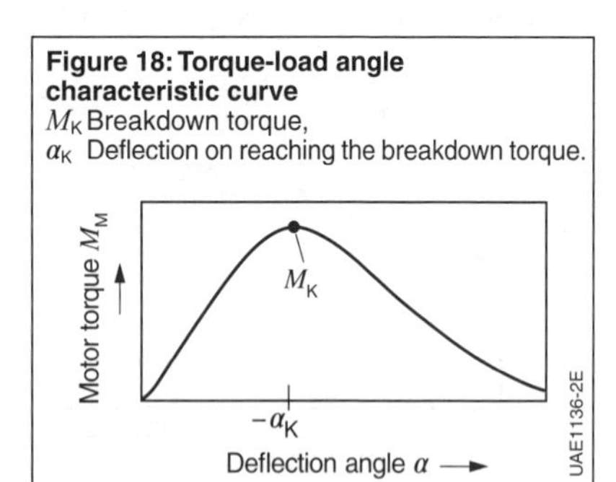

The amount of the pole wheel voltage is influenced by the exciting current. The relationships are to be derived below. The following applies:

$$U_P = -\frac{d\Phi_R}{dt} .$$

With the cosinoidal flux Φ_R and the relationship

$$\Phi_R = B A_S ,$$

including its temporal derivation, this leads to

$$\begin{aligned} u_P &= \Phi_R \omega_{\Phi S} \sin(\omega_{\Phi S} t) \\ &= B_R A_S \omega_{\Phi S} \sin(\omega_{\Phi S} t) \\ &= \mu H_R A_S \omega_{\Phi S} \sin(\omega_{\Phi S} t) . \end{aligned}$$

The magnetic field intensity created in the rotor is described by Ampère's Law. The pole wheel voltage

$$\begin{aligned} u_P &= \mu \frac{\Theta_R}{2\delta} A_S \omega_{\Phi S} \sin(\omega_{\Phi S} t) \\ &= I_{er} \frac{\mu N}{2\delta} A_S \omega_{\Phi S} \sin(\omega_{\Phi S} t) \\ &= \hat{u}_P \sin(\omega_{\Phi S} t) \end{aligned}$$

is then linear-dependent on the exciting current I_{er}. The temporally changeable pole wheel voltage is converted into its effective value using

$$U_P = \frac{\hat{u}_P}{\sqrt{2}} .$$

Figure 19: Single-phase equivalent circuit diagram of the synchronous machine
U_P Pole wheel voltage,
U_S Reactance voltage,
U_0 Terminal voltage, I Current.

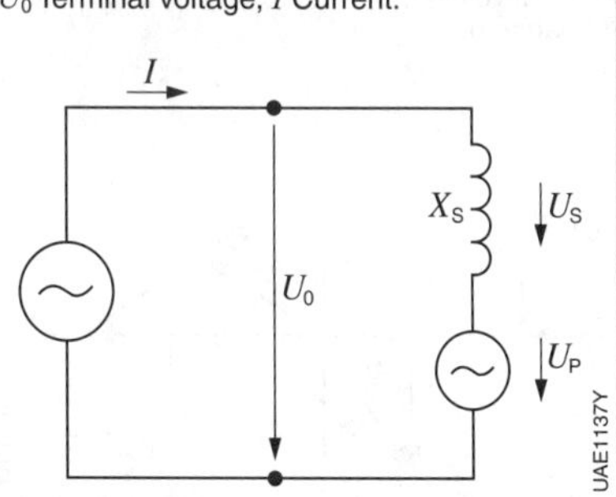

On the basis of the mesh equation (Equation 20), three operating states of the synchronous machine can be derived depending on the pole wheel voltage (Figure 20):

Case 1: $U_P < U_0$, underexcitation, inductive

Case 2: $U_P = U_0$, idling

Case 3: $U_P > U_0$, overexcitation, capacitive.

The first case occurs as long as U_P is $< U_0$. If $I_{er} = 0$, only the self-induction voltage is effective as induced voltage. If current is applied to the rotor, the mutual induction caused by the rotor also takes effect. The first case is referred to as underexcitation. The current lags behind the voltage by 90° ($\varphi(I,U) < 0$). The synchronous machine shows inductive characteristics.

Another increase in the exciting current leads to $U_P = U_0$. This produces the second operating case (idling). The current I_1 becomes zero if no more voltage is supplied via the synchronous reactance.

Another increase in the rotor current leads with $U_P > U_0$ to the third operating case (overexcitation).

Figure 20: Operating states of the synchronous machine
a) Underexcitation (inductive),
b) Idling,
c) Overexcitation (capacitive).
U_P Pole wheel voltage,
U_S Reactance voltage,
U_0 Terminal voltage,
I_1 Stator current.

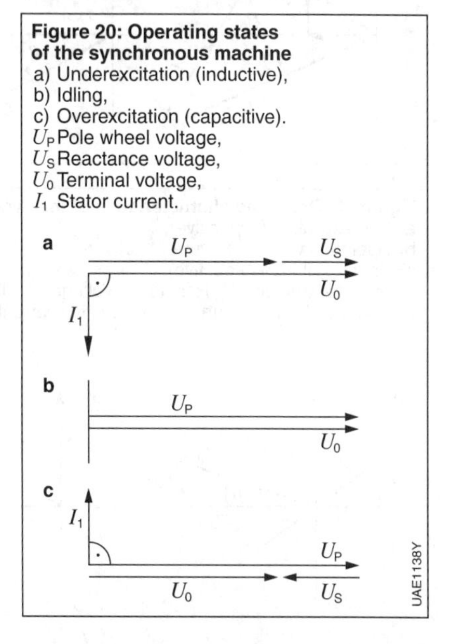

All three cases apply to motor and generator operation. For the single-phase equivalent circuit diagram, the pointers are applied to the voltages and currents. Furthermore, the load angle β that sets in between the voltages U_0 and U_S is defined. For motor operation, the load angle β is < 0 (Figure 21a). The voltage triangle is closed by the voltage U_S.

The synchronous reactance means that the current I_1 leading by 90° in relation to the voltage U_S flows. This is broken down into its components: the active current I_W and reactive current I_B (Figure 21a).

If the pole wheel voltage is reduced so that the pointer of the reactance voltage is vertically on the pointer of the terminal voltage U_0, the motor only consumes active current (Figure 21b).

A further reduction in the pole wheel voltage leads to underexcitation. The current I_1 lags behind the voltage U_S by 90°, which is the equivalent of inductive characteristics of the motor (Figure 21c).

If torque is applied to the motor, it switches to generator operation. Generator operation features the positive load angle β (Figure 22). The sign of the current becomes negative. Current flows out of the machine. In the case of overexcitation, the machine behaves like a capacitor. It delivers reactive power (Figure 22a).

If the pole wheel voltage is reduced so that the pointer of the reactance voltage U_S is vertically on the pointer of the terminal voltage, the generator only delivers active current (Figure 22b).

Figure 21: Operating characteristics of the synchronous machine with motor operation
a) Overexcitation,
b) Motor operation with consumption of active current,
c) Underexcitation.
U_0 Terminal voltage, U_S Reactance voltage, U_P Pole wheel voltage,
I_1 Current, I_W Active current, I_B Reactive current, β Load angle.

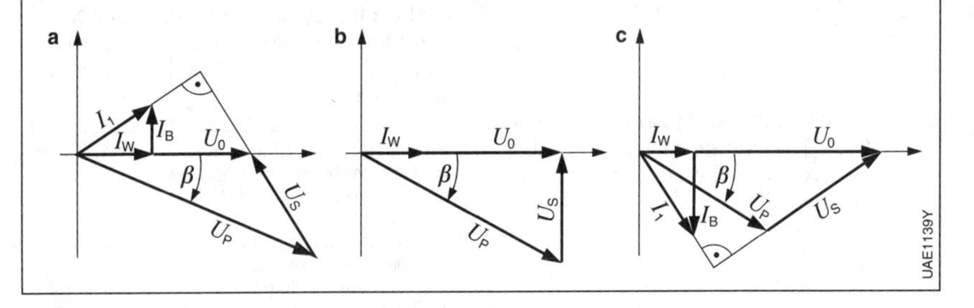

Figure 22: Operating characteristics of the synchronous machine during generator operation
a) Overexcitation (capacitive),
b) Operation with active current delivery,
c) Underexcitation (inductive).
U_0 Terminal voltage, U_S Reactance voltage, U_P Pole wheel voltage,
I_1 Current, I_W Active current, I_B Reactive current, β Load angle.

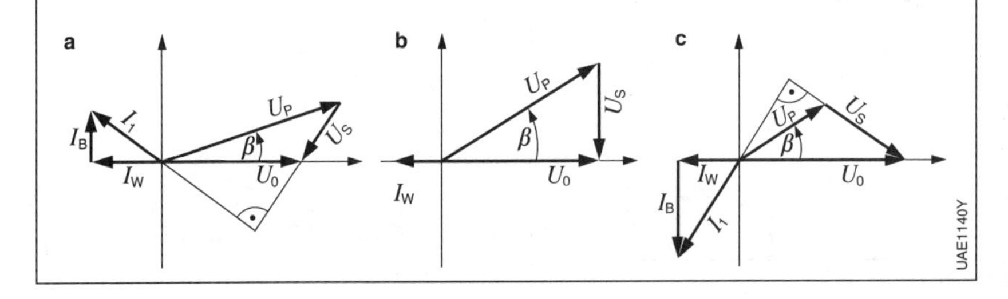

A further reduction in the pole wheel voltage leads to a case of underexcitation. The machine behaves inductively. It consumes reactive power (Figure 22c).

Figure 23: Variations of rotors for the electronic motor
a) Rotor with surface magnets,
b) Rotor with embedded magnets (buried magnets).
Φ Rotor flux.

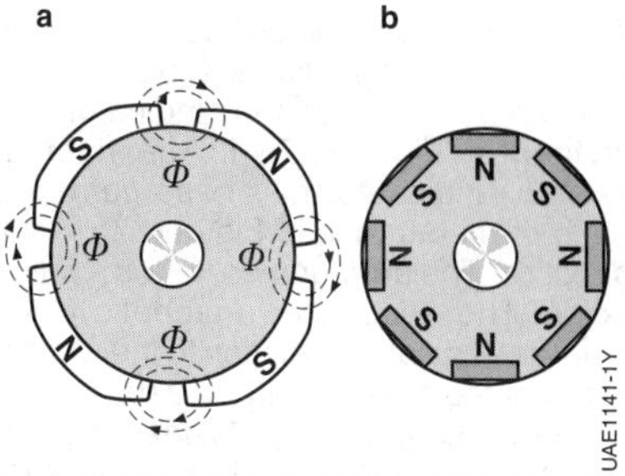

a b

UAE1141-1Y

Electronic motors

In the case of electronically commutating motors (electronic motor, EC motor), the rotor excitation winding, including the electrical contacting with slip rings, is not required. Electronic motors are brushless synchronous motors where the rotors are fitted with permanent magnets. The permanent magnets can be arranged, for example, on the rotor surface or in the rotor (Figure 23). The commutation of the current takes place generally in the fixed stator winding by means of an electronic assembly (Figure 24).

The rotational speed of the electronic motor is set by the frequency of the surrounding stator field. Sensors are required to pick up the rotor position. Widespread use is made of Hall sensors fitted in the working air gap to enable cyclic relaying between the winding strands with the help of activation electronics.

Figure 24: Activation electronics of the electronic motor

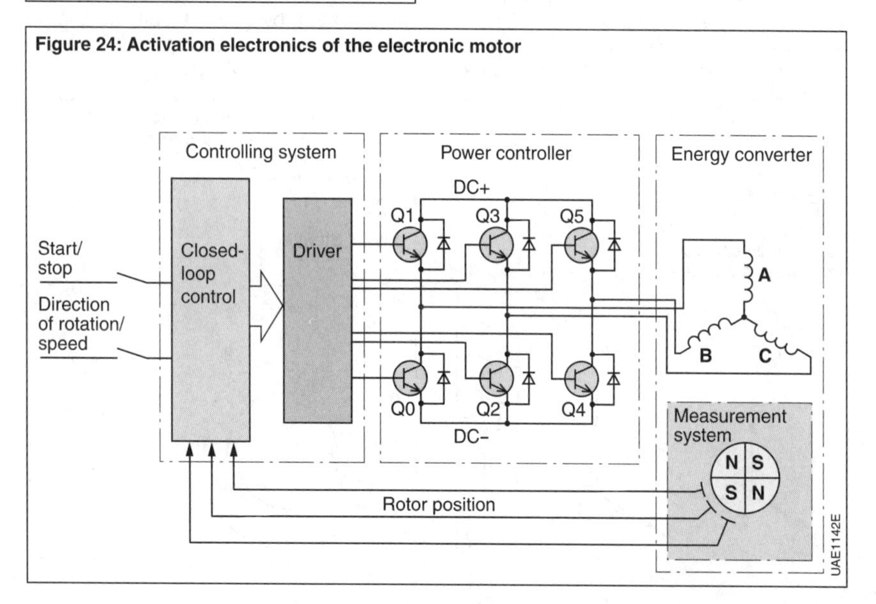

Three-phase current system

Technically relevant is the application of the three-phase alternating current system as a three-phase current system, the major feature of which is that the total of all voltages and currents is zero at all times.

Definitions

The electric circuits are referred to as phases m. The total of the electric circuits in which voltages of the same frequency have an effect and are phase-shifted are referred to as multiphase systems. A multiphase system consists of winding strands. On a three-phase system, $n = 3$ symmetrical systems are possible (Figure 25). On all symmetrical systems – with the exception of the zero system – the total of all pointers is zero. m phases result in n different symmetrical systems depending on the phase difference α:

$$\alpha = \frac{2\pi\, n}{m}\,.$$

The task of windings is to generate a rotating field. Asynchronous and synchronous machines have the same stator structure. In the air gap, a magnetic field with constant amplitude is to be created, rotating at constant angle speed. For this rotating field to be generated, the temporal phase positions of the currents must match the spatial position of the corresponding strands.

Figure 25: Symmetrical systems
a) Co-system, $n = 1$, $\alpha = 2\pi/3$ (120°),
b) Counter system, $n = 2$, $\alpha = 4\pi/3$ (240°),
c) Zero system, $n = 1$, $\alpha = 0$.
n Symmetrical systems,
α Phase difference.

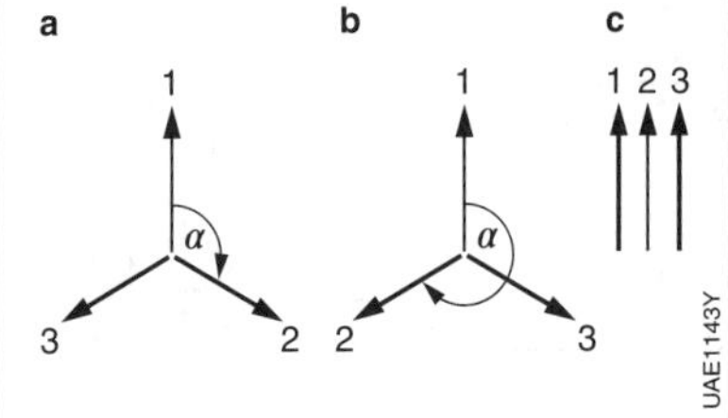

For a simple symmetrical system ($n = 1$) with $m = 3$, the three strands (designated with U, V and W) and therefore the windings must be evenly distributed over the circumference. Figure 26 shows the arrangement of a three-strand winding with one coil per pair of poles and strand. The connection designations of the phases comply with DIN EN 60034, Part 8 [2].

Rotating field generation

In order to generate a rotating field in the case of a simple symmetrical system ($n = 1$) with the number of strands $m = 3$, the strands must be geometrically offset by the electrically effective angle

$$\alpha_{el} = 360° \cdot \frac{1}{3} = 120°\,.$$

With one coil per pair of poles and strand, the resulting magnetic field rotates counterclockwise in that the "indicator bar" shifting in Figure 27a to the right (at $\alpha = 90°$) indicates the phase current in each of the strands in Figure 27b in flux direction. The arrangement forms a pair of poles. The associated magnetic fluxes emerge vertically in relation to the winding plane of the strands (Figure 27b).

The flux Φ_{Res} (Figure 27c) resulting from the three strands as well as its direction are achieved by geometrically adding up the three individual fluxes Φ_U, Φ_V and Φ_W.

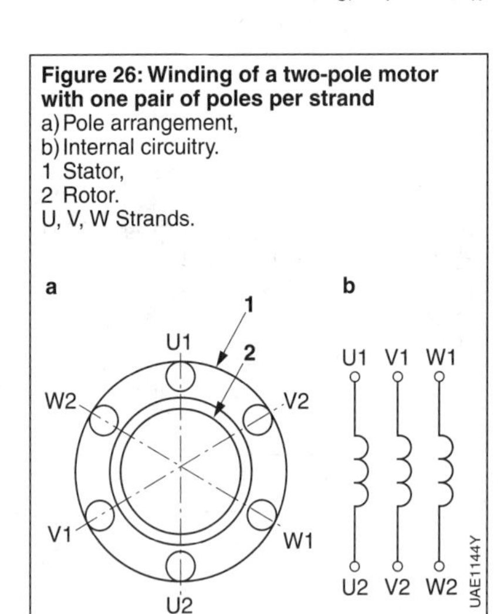

Figure 26: Winding of a two-pole motor with one pair of poles per strand
a) Pole arrangement,
b) Internal circuitry.
1 Stator,
2 Rotor.
U, V, W Strands.

A progression of the indicator bar up to $\alpha = 180°$ leads to a current direction reversal in strand W and therefore to a further turn to the right of the resulting field Φ_{Res} (Figure 28).

Figure 27: Generating a rotating field with one coil per strand
a) Strand currents,
b) Strand currents with angle $\alpha = \alpha_1$,
c) Direction of the rotating field (spatial directions).
1 Stator, 2 Rotor.
U, V, W Strands.

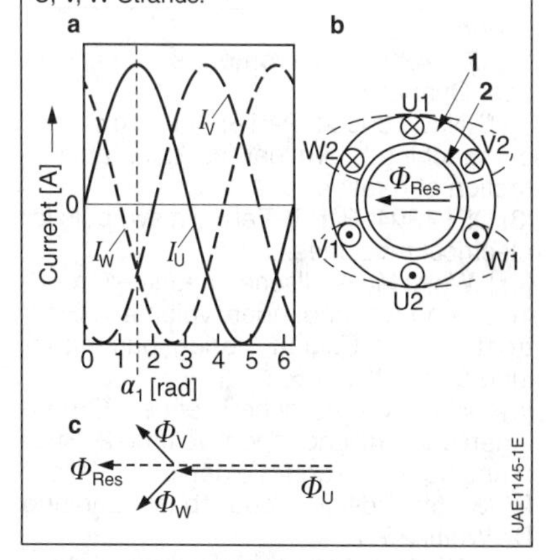

Figure 28: Generating a rotating field with one coil per strand
a) Strand currents,
b) Strand currents with $\alpha = \alpha_2$,
c) Direction of the rotating field (spatial directions).
1 Stator, 2 Rotor.
U, V, W Strands.

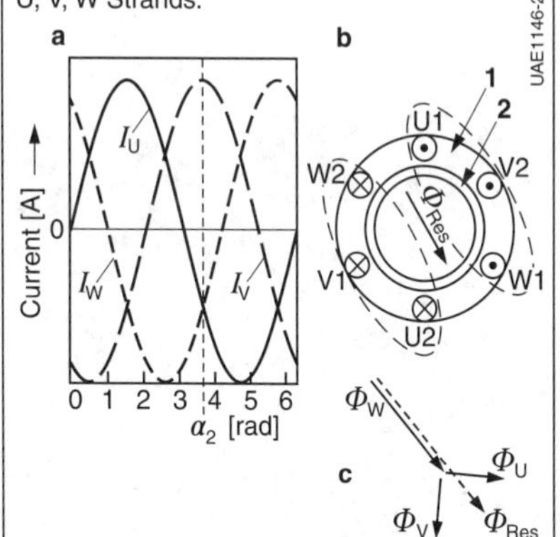

When two coils per strand are used, the conductor arrangement is "doubled". If the winding is to form two pairs of poles ($p = 2$), the windings have to be divided into groups (Figure 29). With this, the mechanically effective angle

$$\alpha_m = 360° \cdot \frac{1}{m\,p} = 60°$$

sets in. The electrically effective angle remains unchanged. In the case of both the two-pole and four-pole arrangement, the field rotates counterclockwise (Figure 30). The rotating field speed

$$n_d = \frac{f_n}{p}$$

can be calculated with the mains frequency f_n and the number of pole pairs p. For $p = 1$, the rotating field speed is equal to the mains frequency (Table 2).

Table 2: Rotating-field speeds

Pair of poles p	n_0 [rpm] at f = 50 Hz
1	3,000
2	1,500
3	1,000

Figure 29: Winding with two pairs of poles per strand
a) Pole arrangement,
b) Internal interconnections (example). The interconnections in parentheses are internal and thus inaccessible.
1 Stator, 2 Rotor.
U, V, W Strands.

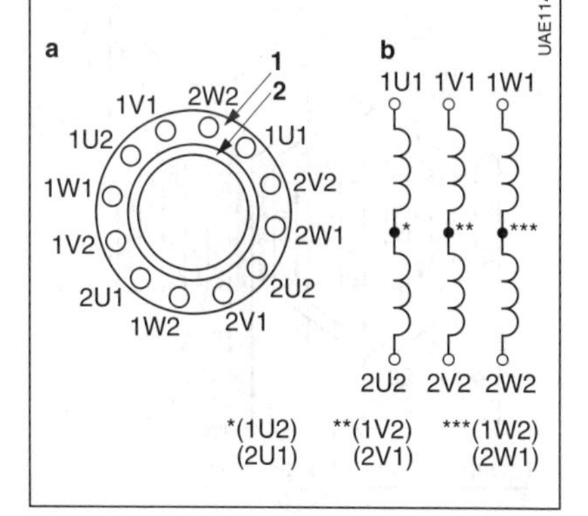

Together with the number of pole pairs, the pole pitch

$$\tau_p = \frac{d_{si}}{2\pi}$$

can be calculated as a stator circumference proportion, where d_{si} is the inner diameter of the stator. It corresponds to the length of a sinusoidal half wave that corresponds to the induction distribution of the rotor field. In the case of a two-pole machine ($p=1$), the pole pitch is $\alpha_{el}=180°$ (electrical angle) at all times and matches the mechanical angle α_m. The relationship of the two angles is indicated by $\alpha_{el} = p \cdot \alpha_m$. So that voltages of equal size are induced in the windings, the winding strands have to be offset in relation to one another by $\alpha_{el} = 120°$ or $2\tau_p/3$ and the coil structure and number of coils must be the same. One third of the pole pitch is apportioned to each strand.

Figure 30: Rotating field generation with two coils per strand
a) Strand currents with angle $\alpha=\alpha_1$,
b) Resulting magnetic field with $\alpha=\alpha_1$.
1 Stator, 2 Rotor.
U, V, W Strands.

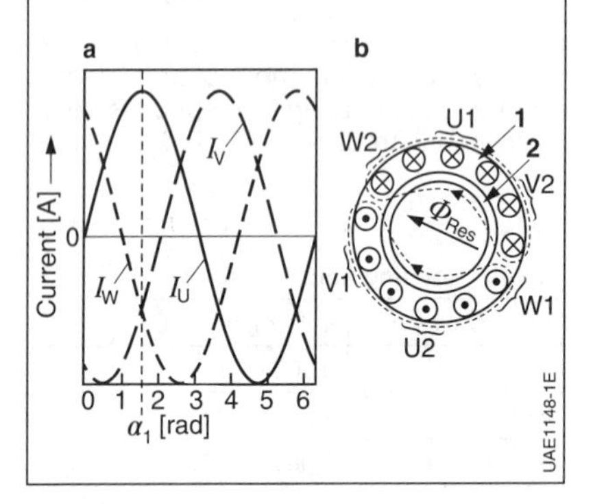

Figure 31: Structure of a stator three-phase current winding
1 Stator tooth,
2 Stator groove with winding strand.

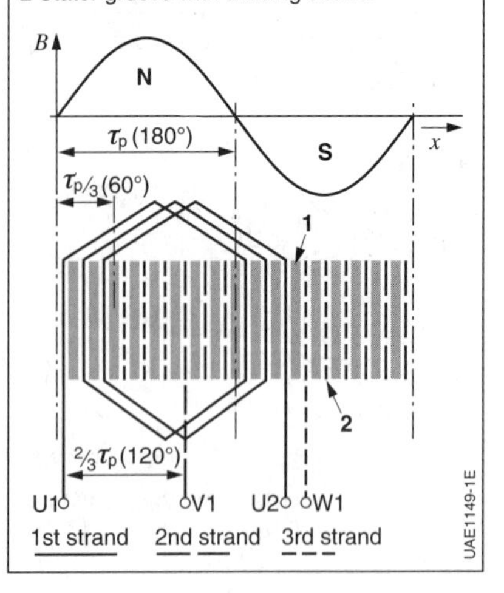

References

[1] DIN 42027: Servomotors; Classification, Overview.
[2] DIN EN 60034, Part 8: Rotating electrical machines; terminal markings and direction of rotation.
[3] DIN 1304, Part 7: Formula symbols for electrical machines.
[4] I. Wolff: Maxwellsche Theorie – Grundlagen und Anwendungen; Volume 1, Elektrostatik, 5th Edition, Verlagsbuchhandlung Dr. Wolff, 2005.
[5] I. Wolff: Maxwellsche Theorie – Grundlagen und Anwendungen; Volume 2, Strömungsfelder, Magnetfelder und Wellenfelder, 5th Edition, Verlagsbuchhandlung Dr. Wolff, 2007.
[6] G. Wunsch: Elektromagnetische Felder, 2nd Edition, Verlag Technik, 1996.
[7] R. Fischer: Elektrische Maschinen, 13th Edition, Carl Hanser Verlag, 2006.
[8] K. Fuest, P. Döring: Elektrische Maschinen und Antriebe, 6th Edition, Vieweg-Verlag, 2008.

Alternators

Electric power generation

Motor vehicles need an alternator to charge the battery and to supply the electrical equipment such as ignition and fuel-injection systems, ECUs, lights, etc. with power. To charge the battery, it is necessary for the alternator to generate more current than is required by the electrical equipment when switched on. The alternator output, the battery capacity and the power demand of the electrical equipment must be matched to each other to ensure that under all operating conditions enough current is supplied to the vehicle electrical system and the battery is always sufficiently charged. This results is a consistent charge balance.

Because the battery and many items of electrical equipment have to be supplied with direct current, the alternating current generated by the alternator is rectified. The alternator is equipped with a voltage regulator so that it can supply the battery and the electrical equipment with a constant voltage. Alternators in passenger cars are designed for charge voltages of 14 V and those in many commercial vehicles for 28 V (28 V electrical system).

Figure 1: Generation of an induced voltage in a coil permeated by a magnetic field
U_{ind} Induced voltage.

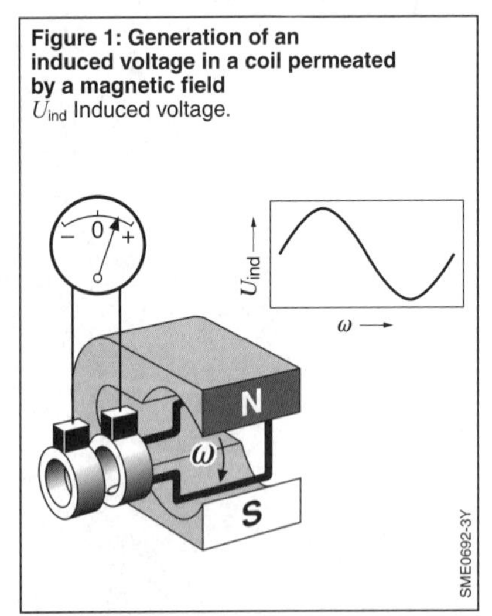

Requirements

The essential alternator requirements are:
- Maintenance of a direct-voltage supply to all electrical equipment in the system
- Charging reserves for (re)charging the battery, even at a constant load from electrical devices in continuous operation
- Maintenance of a constant alternator voltage throughout the entire ranges of engine speed and load conditions
- High level of efficiency
- Low operating noise
- Robust design capable of withstanding external stresses (such as vibration, high ambient temperatures, temperature cycles, dirt, moisture)
- Long service life, comparable with that of the vehicle itself (for passenger cars)
- Low weight
- Compact dimensions.

Principle of electromagnetic induction

When the magnetic field in a coil changes, a voltage is induced in that coil (Figure 1). According to Faraday's law of induction, the induced voltage U_{ind} increases as the speed v of the motion vertical to the field lines increases and as the magnetic flux Φ passing through the conductor cross-section increases. The following applies:

$$U_{ind} \sim \frac{d\Phi}{dt} .$$

The magnetic field for generating the induced voltage (excitation field) can be generated by permanent magnets. The advantage of these is that, thanks to the simplicity of their design, they do not have to be technically sophisticated. This solution is used in small alternators (e.g. bicycle dynamos). The disadvantage of excitation using permanent magnets is that it cannot be regulated.

A excitation field capable of being regulated can be generated by electromagnets. An electromagnet consists of an iron core and a winding (excitation winding), through which an excitation current is passed. The number of windings determines, together with the magnitude of the excitation current, the magnetic field strength. The magnetizable iron core in the electromagnet conducts the magnetic field generated by the coil.

By altering the excitation current, it is possible to adjust the magnetic field and with it the magnitude of the induced voltage.

When an external power source (e.g. a battery) supplies the excitation current, this is referred to as external excitation. When the excitation current is tapped directly from the generated alternator current inside the alternator, this is referred to as self-excitation.

Alternator design

The essential components of an alternator are the three- or multi-phase armature windings and the excitation system (Figure 2). Because the design of the armature-winding system is more complex than that of the excitation system and the currents generated in the armature winding are much greater than the excitation current, the armature windings are accommodated in the stationary stator. The magnet poles with the excitation winding are located on the rotating part, the rotor. The rotor's magnetic field is generated as soon as an excitation current passes through this winding. The alternator must feature a high number of pole pairs so that it can generate a high induced voltage already at low speeds. A high number of poles generates a high response per rotation and thus a high induced voltage. This is the prerequisite for a high alternator output.

By applying the claw-pole principle, the magnetic field of a single field coil can be split in such a way that the required 12 to 16 poles or 6 to 8 pole pairs are created (Figure 2 and Figure 3).

In the alternator, the armature contains three of more identical windings (phases) which are spatially offset in relation to one another (Figure 2). Because of the spatial offset of the windings, the sinusoidal alternating voltages generated in them are likewise out of phase with one another (temporally offset, Figure 4). The resulting alternating current is called three-phase current.

Rectification of alternating voltage

The alternating voltage generated by the alternator must be rectified, since direct current is required to supply the battery and the electronics in the vehicle electrical system.

Two power diodes are connected to each phase – one diode on the positive side which is usually connected via the

Figure 2: Basic design of a claw-pole alternator with collector rings
1 Rotor,
2 Exciter winding,
3 Armature winding,
4 Stator,
5 Slip ring,
6 Brushes,
7 Rectifier diodes.
B+ Battery positive terminal,
B– Battery negative terminal.

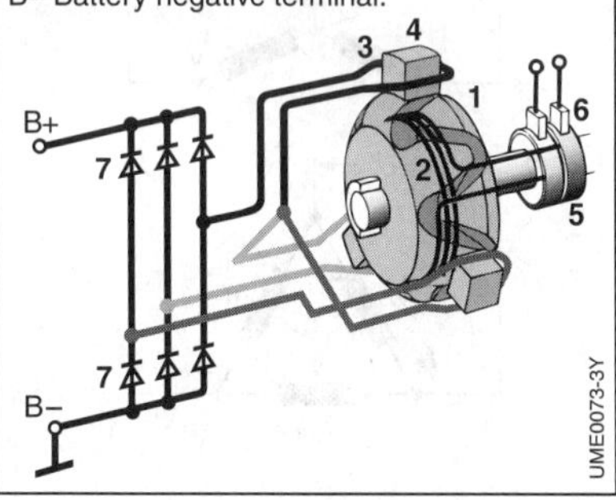

Figure 3: Components of a 12-pole claw-pole rotor
1 A-side (drive-end) claw pole,
2 Field coil,
3 B-side claw pole,
4 Chamfer on claw-pole-finger edge,
5 Alternator shaft.

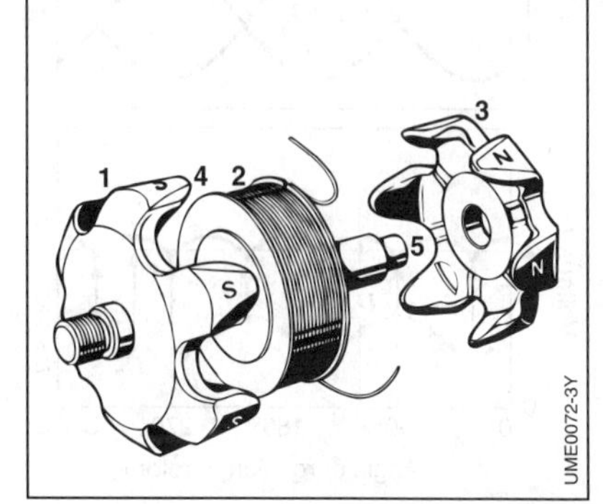

positive heat sink of the rectifier to the B+ stud and one diode on the negative side which is connected via the negative heat sink of the rectifier to the alternator casing (B– terminal) (Figure 2). The alternator casing is also electrically connected to ground via the alternator mounting points. The positive half-waves are conducted by the diodes on the positive side (B+ stud), the negative half-waves by the diodes on the negative side (vehicle ground). This principle is called full-wave rectification (Figure 4).

Figure 4: Three-phase current rectification
a) Three-phase alternating voltage,
b) Alternator voltage formed by the envelopes of the positive and negative half-waves,
c) Rectified alternator voltage.
U_P Phase voltage,
U_G Voltage at rectifier (negative not to ground),
U_{G-} Alternator direct voltage (negative to ground),
$U_{G,rms}$ Effective value of direct voltage.
u, v, w Strands.

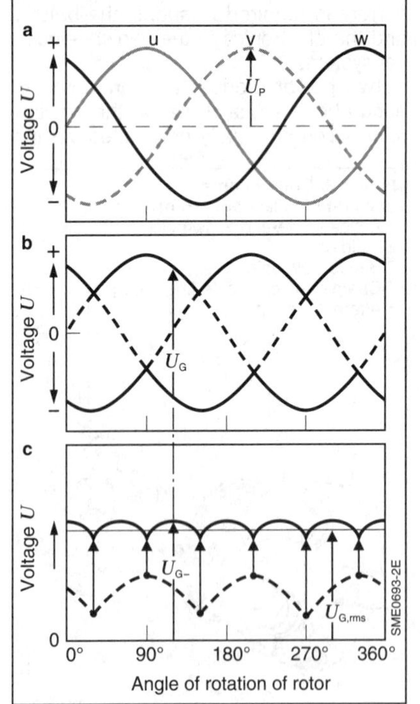

In the case of alternators where the stator winding is star-connected (Figure 5a), two additional diodes (therefore also called "supplementary diodes") rectify the positive and negative half-wave of the star point. By rectifying the third harmonic in the phase voltages, these supplementary diodes can increase the alternator output current at 6,000 rpm by up to 10%. Alternator efficiency is thus significantly improved in the upper speed range. In the lower speed range, the amplitude of the third harmonic is below the vehicle system voltage such that the supplementary diodes cannot deliver any current. Supplementary diodes are rarely used in modern-day alternators.

The direct current which the alternator delivers under electrical load via terminals B+ and B– to the vehicle electrical system is not smooth, but slightly rippled. This ripple is further smoothed by the battery connected in parallel with the alternator and, if necessary, by capacitors in the vehicle electrical system.

Instead of high-blocking-capability power diodes, modern-day motor-vehicle alter-

Figure 5: Connection types of three-phase stator windings (three-phase windings)
a) Star connection,
b) Delta connection.
U Alternator voltage,
U_{Str} Phase voltage,
I Alternator current,
I_{Str} Phase current.
u, v, w Strands.

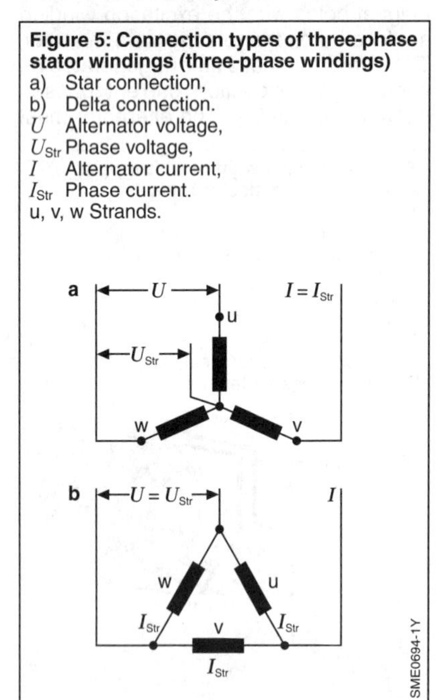

nators use Zener diodes to rectify the alternating voltage. Zener diodes limit the high-energy voltage peaks to levels which are harmless for the alternator and the regulator (load-dump protection). In addition, Zener diodes can be used to provide remote protection for other voltage-sensitive equipment in the vehicle electrical system. When using a 14 V alternator, the response voltage of a rectifier fitted with Zener diodes ranges from 25 to 30 V.

Alternator circuits

In three-phase alternators, six current leads would be needed to transmit electrical energy in the case of non-connected windings. This number can be reduced to three by linking the three circuits. The circuits are linked in a star connection (Figure 5a) or in a delta connection (Figure 5b).

In the case of a star connection, the ends of the three winding phases are interconnected at a single point, the star point. Without a neutral conductor the sum total of the three currents to the star point is zero at every moment.

Reverse-current block

The rectifier diodes in the alternator serve not only to rectify the alternator voltage, but also to prevent the battery from discharging via the stator windings. If the engine is stopped or if the engine revs are so low (e.g. starting speed) that the alternator is not yet excited, without diodes a battery current would flow through the stator winding. The diodes are polarized with regard to the battery voltage in the reverse direction so as to prevent the flow of any significant battery discharge current. Current can therefore only flow from the alternator to the battery.

Generally, alternators are not provided with polarity-reversal protection. Reversal of battery polarity (e.g. mixing up the battery poles when using an external battery to start the vehicle) can lead to destruction of the alternator diodes and poses a threat to the semiconductor components of other devices.

Figure 6: Alternator
A Regulator, B Alternator, C Vehicle electrical system.
1 Free-wheeling diode, 2 Excitation winding, 3 Armature windings, 4 Rectifier diodes,
5 Evaluation (monitoring) circuits,
6 Relay (switches equipment which is to be switched on only when alternator is active),
7 Alternator indicator lamp, 8 Ignition switch, 9 Battery, 10 Equipment.
DF Dynamo field, DFM (DF monitoring),
L Lamp connection,
B+ Battery positive, B– Battery negative,
15 Terminal 15.

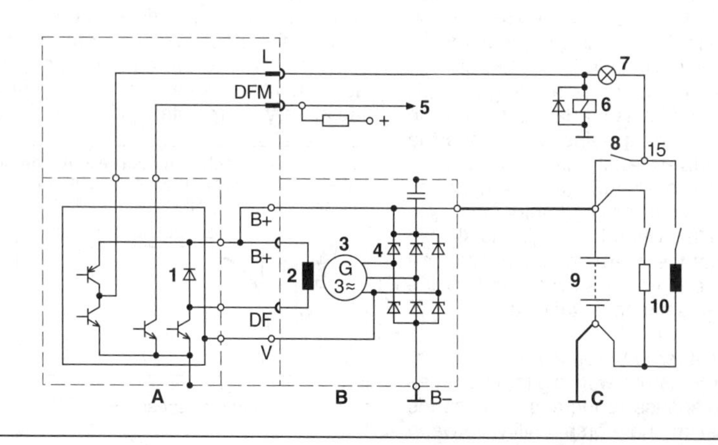

Voltage regulation

At constant excitation current, the alternator voltage is dependent on the alternator speed and load. The function of voltage regulation is to keep the alternator voltage – and thus also the vehicle system voltage – constant over the full speed range of the vehicle engine, irrespective of electrical load. To this end, the voltage regulator regulates the magnitude of the excitation current by cyclical actuation (activation and deactivation of the voltage at the excitation winding). The regulator adjusts the on/off ratio (ratio of ON time to OFF time) as a function of the voltage generated in the alternator. In the periods in which the regulator deactivates the voltage, the excitation current, driven by the high inductance of the field coil, flows on via the free-wheeling diode connected in parallel with the excitation winding (Figure 6). In this way, the regulator keeps the vehicle system voltage constant and prevents the battery from being overcharged or discharged during vehicle operation.

Regulators

Voltage regulators were formerly produced using discrete components. Today they incorporate hybrid or monolithic circuits (Figure 6). When monolithic technology is applied, the control and regulator IC, the power transistor, and the free-wheeling diode are all accommodated on a single chip.

Multifunction regulators perform special functions in addition to the task of voltage regulation described above. The load-response (LR) function is worthy of special mention. A distinction is made between LR while the engine is running and LR when the engine is started. LR while the engine is running helps to improve the running and exhaust-gas characteristics of the internal-combustion engine by means of a limited rate of output increase over time. LR when the engine is started means that the alternator is inactive for a specific period after starting so that the engine can settle to a stable idle.

Intelligent alternator regulation

Voltage regulators with digital interfaces are the response to increased demands for greater mutual compatibility between engine-management and alternator-regulation systems. Various interfaces (e.g. bit-synchronous interface or LIN interface) are used here depending on the vehicle manufacturer.

The control voltage can be adjusted via the regulator interface. If a low control voltage is set, this results in low or no alternator output. The power input and the torque input at the alternator shaft decrease to low levels. On the other hand, the alternator voltage can be stepped up via the interface so that the alternator charges the battery and supplies the electrical equipment in the vehicle electrical system.

Because the alternator converts mechanical energy into electrical energy, the power input at the alternator shaft increases as the electrical power output increases. The internal-combustion engine needs more fuel to cover the alternator's power requirement. However, if the alternator voltage is stepped up in overrun phases, the energy tapped by the alternator at the crankshaft does not cost any additional fuel, because overrun fuel cutoff stops the fuel supply. When the driver presses the accelerator pedal, the engine ECU steps down the alternator voltage again with the aid of the communication interface to the regulator in order to relieve the strain on the engine by the alternator torque. This process is also known as intelligent alternator regulation and recuperation.

Interface regulators allow a fine-tuning of load-response functions to engine operating state, optimization of torque patterns to reduce fuel consumption, and adjustment of charging voltage to improve the battery charge state.

Figure 7: Characteristic curve at maximum alternator current at constant voltage
n_L Idle speed, n_{max} High idle speed.

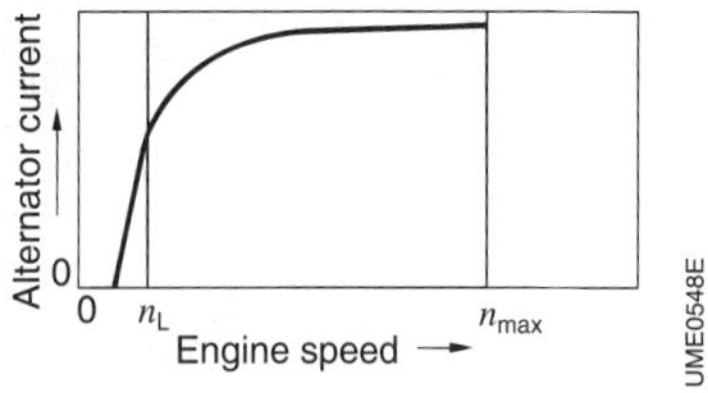

Alternator current
0
0
n_L
n_{max}
Engine speed
UME0548E

Alternator characteristics
The alternator can only deliver current to the vehicle electrical system when the induced voltage U_{ind} in the stator windings is greater than the sum total of the two diode forward voltages and the vehicle system voltage applied to B+ and B–. Because the induced voltage

$$U_{ind} \sim \frac{d\Phi}{dt}$$

is dependent on the frequency of the change in flux, i.e. on the rotor speed, there is a speed range between zero and the self-excitation speed in which the alternator delivers no current (Figure 7). Thereafter, the alternator current rises sharply under full-load conditions as the speed increases. At higher speeds the voltage drop at the magnetizing reactance X_h (Figure 8) and the leakage reactance X_σ becomes so great that, in spite of an increasing induced voltage, the current fails to rise significantly. The magnetizing reactance is obtained from that part of the coil inductance which generates the profitable induced voltage and whose flux takes the envisaged paths in the rotor and the stator. The leakage reactance forms that part of the flux which is not linked to the rotor – i.e. leakage fields which short, for example, in the end winding or directly via the groove.

Figure 8: Simplified symbol of a phase with rectification
B+ Battery positive terminal,
B– Battery negative terminal,
D Rectifier diodes,
I_1 Alternator current,
R_1 Ohmic resistance,
X_σ Leakage reactance,
X_h Magnetizing reactance,
U_{ind} Induced voltage,
U_{Gen} Alternator voltage.

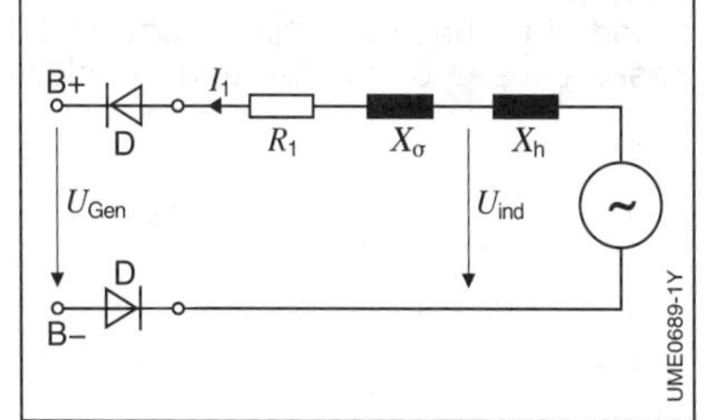

Operating conditions

Rotational speed

The utilization of an alternator (generable energy per kilogram of mass) increases as the speed increases. It is therefore necessary to aim for as high a transmission ratio as possible between the crankshaft and the alternator. However, it is important to ensure that at maximum engine speed the maximum permissible alternator speed is not exceeded. Widening of the claw poles and the service life of the ball bearings used determine the maximum permissible speed of motor-vehicle alternators. Typical values for the maximum speed of passenger-car alternators range between 18,000 rpm and 22,000 rpm. The transmission ratio for passenger cars is between 1:2.2 and 1:3, and for commercial vehicles up to 1:5.

Cooling

The losses arising during conversion from mechanical to electrical energy results in the alternator components being heated. The surrounding air in the engine compartment is used to cool the majority of the motor-vehicle alternators used today. If the surrounding air is not sufficient to cool the alternator, fresh-air induction from cooler areas or liquid cooling are suitable measures for cooling the components adequately.

Engine vibrations

The alternator can, depending on the mounting conditions and the vibration characteristic of the engine, be exposed to vibration accelerations of 500–800 m/s^2. The mountings and the components of the alternator are thereby subjected to high forces. Critical natural frequencies in the alternator assembly must therefore be avoided at all costs.

Engine-compartment climate

The alternator is exposed to splash water, dirt, oil and fuel mist, and where necessary road salt (grit). The alternator must be protected against corrosion so as to prevent the formation of any creepage paths between live parts.

Acoustic behavior

The stringent demands placed on the noise emissions of modern-day vehicles and the superior smooth running of modern-day internal-combustion engines require quiet alternators. As well as chamfers on the claw-pole fingers which reduce the hard chopping of the magnetic flux at finger edges, there are other possibilities for reducing the noise magnetically generated by motor-vehicle alternators. The following may be used: firstly, five phases in a pentagram connection (Figure 9), and secondly, two three-phase systems which electrically offset by 30°.

Figure 9: Pentagram-connected five-phase stator winding with connected bridge rectifier
B+ Battery positive terminal,
B– Battery negative terminal.

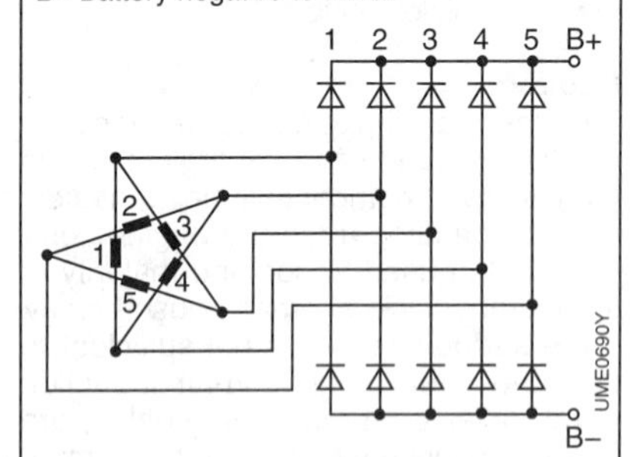

Figure 10: Loss distribution of a 220 A alternator
1 Alternator power output,
2 Iron losses,
3 Copper losses in the stator,
4 Rectifier losses,
5 Friction losses,
6 Copper losses in the excitation winding (rotor).

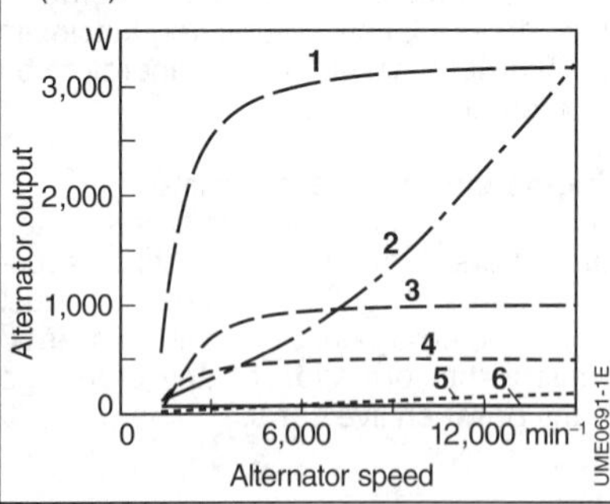

Efficiency

The ratio between the power which is supplied to the unit and the actual power output is known as the efficiency. Losses are an unavoidable byproduct of all processes in which mechanical or kinetic energy is converted into electrical energy. The losses in the claw-pole alternator are categorized as follows (Figure 10).

Losses in the claw-pole alternator

Copper losses in the stator *and excitation windings*
The ohmic losses in the stator windings and the rotor winding are called copper losses. They are proportional to the square of the current.

Iron losses in the laminated stator *core*
Iron losses result from hysteresis and eddy currents produced by alternating magnetic fields in the iron of the stator.

Eddy-current losses on the claw-pole surface
Eddy-current losses on the claw-pole surface are caused by fluctuations in flux brought about by the stator slotting.

Rectifier losses
Rectifier losses are caused by the voltage drop at the diodes. Rectifier losses can be reduced and thus efficiency improved by using semiconductors with a lower voltage drop, e.g. by using high-efficiency diodes (HEDs).

Mechanical losses
Mechanical losses include frictional loss in the roller bearings and at the sliding contacts, air resistance encountered by the rotor and, above all, fan resistance which rises dramatically as speed increases.

Added to these are the aerodynamic losses caused by the fan and the claw poles.

Efficiency optimization

In regular automotive operation, the alternator operates in the part-load range. Efficiency at medium speeds is then 70–80%. The use of a larger (and heavier) alternator allows it to operate in a more favorable part-load efficiency band for the same electrical load. The efficiency gains provided by the larger alternator more than compensate for losses in fuel economy associated with greater weight. However, the higher mass moment of inertia in the belt drive system must be taken into consideration.

Figure 11: Compact-diode alternator
1 Collector-ring end shield,
2 Rectifier heat sink,
3 Power diode,
4 Excitation diode,
5 Drive end shield with mounting flanges,
6 Belt pulley,
7 External fan,
8 Stator,
9 Claw-pole rotor,
10 Transistor regulator.

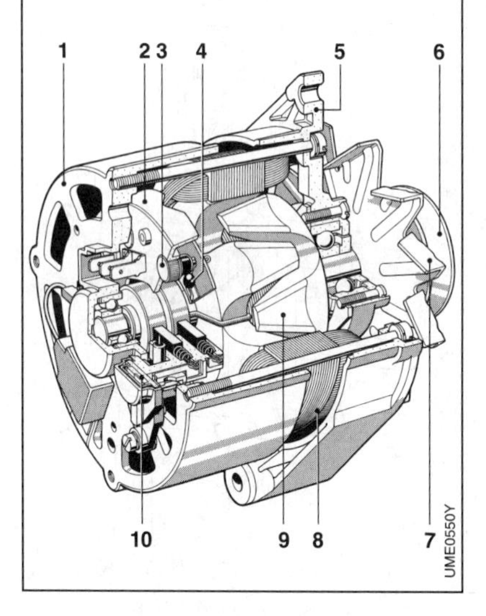

Types of claw-pole alternator

Claw-pole alternators have completely replaced the direct-current generators previously used as standard in motor vehicles. Based on equal outputs for both concepts, a claw-pole alternator weighs 50% less, and is also less expensive to manufacture. Large-scale application (at the beginning of the 1960s) only became feasible with the availability of compact, powerful, inexpensive, and reliable silicone diodes.

Compact-diode alternators

The classic alternator design is characterized by the large external fan that provides single-flow axial ventilation (compact-diode design, Figure 11). Because the rectifier, regulator and brush/collector-ring system are arranged inside the end shield, the shaft inside the collector rings must be relatively thick to be able to transmit the belt-drive forces to the external ball bearing. The collector rings therefore have a large diameter, only permitting a limited brush service life.

Compact alternators

Modern-day passenger-car alternators (Figure 12) are cooled by double-flow ventilation provided by two internal fans. Cooling air is drawn from the surrounding air in the axial plane, and exits the alternator radially in the vicinity of the stator winding heads, at the drive and collector-ring end shields. The major advantages of the compact alternator are:
- High utilization thanks to high maximum speed
- Low aerodynamic noise thanks to small fan diameters
- Low magnetic noise
- Long carbon-brush life thanks to small collector-ring diameter.

Alternators with windingless rotors

The windingless-rotor alternator is a special design variant of the claw-pole alternator where only the claw poles rotate, while the excitation winding remains stationary. Instead of being connected directly to the shaft, one of the pole wheels is held in place by the opposite pole wheel via a nonmagnetic intermediate ring. The magnetic flux must cross two additional air gaps beyond the normal working gap. With this design, the regulator supplies directly to the excitation winding; sliding contacts are not required. This arrangement obviates the wear factor represented by the brush/collector-ring system, making it possible to design alternators for a much longer service life. They are therefore suitable for use in, for example, construction machinery or railroad applications. The units weigh somewhat more than claw-pole alternators with collector rings due to the fact that additional iron and copper is required to conduct the magnetic flux through two additional air gaps.

The windingless rotor is also used in liquid-cooled alternators (Figure 13), among other applications. In the case of this alternator, engine coolant flows around the complete jacket and rear of the alternator casing. The electronic components are mounted on the drive-end shield.

Figure 13: Liquid-cooled alternator with windingless rotor
1 Belt pulley, 2 Rectifier, 3 Regulator, 4 Drive end shield, 5 Alternator casing, 6 Coolant jacket, 7 Jacket casing for engine mounting, 8 Stationary excitation winding, 9 Stator core, 10 Stator winding, 11 Windingless rotor, 12 Nonmagnetic intermediate ring, 13 Conductive element.

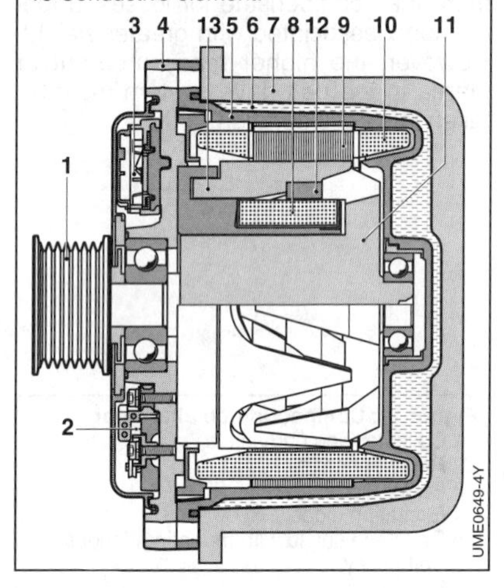

Figure 12: Bosch EL Series (Efficiency Line) compact alternator
1 Belt pulley,
2 A-side (drive-end) ball bearing,
3 A-side fan,
4 Stator winding heads,
5 Laminated stator core,
6 Rotor winding (excitation winding),
7 B-side fan,
8 B-side end shield,
9 Protective cap,
10 Brush holder,
11 Carbon brush,
12 Collector ring,
13 A-side end shield,
14 A-side claw pole,
15 B-side claw pole,
16 B-side ball bearing,
17 Negative heat sink of rectifier.
18 Positive heat sink of rectifier.

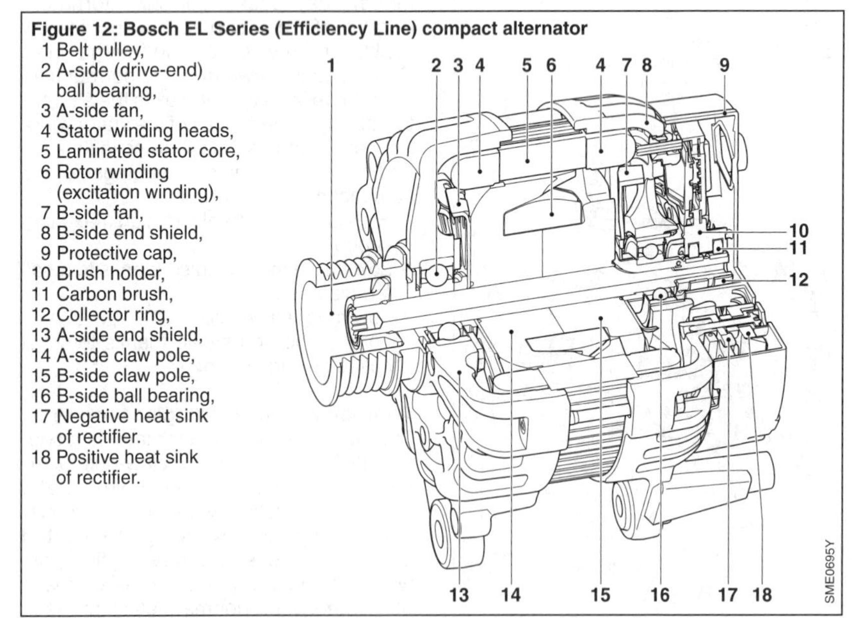

Starting systems

Starter

Requirements

Internal-combustion engines in motor vehicles require start assistance to start. Starting systems consist of the following assemblies:
- DC motor (starter)
- Switchgear and control units
- Battery
- Wiring.

The starter speed, which is much higher than the engine speed, is matched to the engine speed by means of a suitable gear ratio (between 1/10 and 1/20) arranged between the starter pinion and the engine-flywheel ring gear. A small starter is capable of achieving the required rotational speed for sustained operation of the engine (gasoline engines, between 60 and 100 rpm; diesel engines, between 80 and 200 rpm). Compression and decompression in the cylinders means that the torque required to turn the engine over fluctuates considerably, as a result of which the momentary engine speed also fluctuates considerably. Figure 1 shows a typical graph for engine speed and starter-motor current on cold start.

Figure 1: Graph of engine speed and starter current on cold start

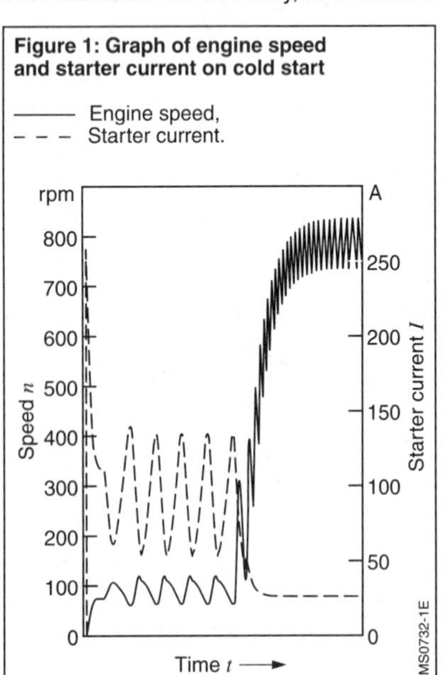

The starter motor itself must satisfy the following technical requirements:
- Readiness to function at any time
- Sufficient starting power at different temperatures
- High service life
- Robust design
- Low weight and low dimensions
- Maintenance-free operation.

Design factors

To achieve the air/fuel (A/F) mixture necessary for sustained operation of gasoline engines and the auto-ignition temperature for diesel engines, the starter must drive the internal-combustion engine at a minimum speed (cranking speed). The cranking speed largely depends on the characteristics of the internal-combustion engine (engine type, engine swept volume, number of cylinders, compression, bearing friction, engine oil, fuel-management system, additional loads driven by the engine), and the ambient temperature.

In general, starting torque and starting rotational speed require a gradual increase in starting power as temperatures decline. However, the power supplied by a starter battery falls as temperatures drop because its internal resistance increases. This opposing relationship of electrical load requirements and available power means that the least favorable operating conditions which a starting system must be capable of dealing with is a cold start.

Because of the high current draw of a starter motor, the voltage drop occurring on the supply lead significantly influences the performance characteristics of the starter motor.

Classification

Auto starting systems are equipped with a rated output of up to 2.5 kW at a rated voltage of 12 V. These can start gasoline engines with up to approximately 7 liters engine displacement and diesel engines with up to approximately 3 liters.

Starters may be classified by the following criteria according to their technical type:
- Type of power transfer: Direct starter or reduction-gear starter
- Type of magnetic field generation in electric motor: permanent magnet or electrically excited starter
- Type of engagement: sliding gear, Bendix or pre-engaged-drive starter.

Permanently excited pre-engaged-drive starters with reduction gear are predominantly used in the automobile these days. These combine the benefit of a high starting power with a compact size.

Starter design and operation
A starter (Figure 2) essentially consists of the electric motor, the engagement system and, from a starter performance of approximately 1 kW, a reduction gear.

During the start, the starter pinion engages with the ring gear by means of an engagement relay. The starter motor is either coupled to the starter pinion directly or by means of a gear set, which reduces the rotational speed of the DC motor. The starter pinion drives the internal-combustion engine via the engine-flywheel ring gear until the engine can run at sustained operation. After the engine starts, it can acceleration quickly to high rotational speeds. After only several ignitions, the engine accelerates so powerfully that the starter can no longer match its speed. The internal-combustion engine "overruns" the starter and would then accelerate the armature to extremely high speeds if the overrunning clutch installed between the pinion and armature did not cancel the non-positive lock. As soon as the driver releases the ignition key, the starter relay

Figure 2: R70 reduction-gear starter
1 Drive shaft, 2 Stop ring, 3 Starter pinion, 4 Roller-type clutch, 5 Meshing spring, 6 Engagement lever, 7 Engagement relay, 8 Hold-in winding, 9 Pull-in winding, 10 Return spring, 11 Contact bridge, 12 Contact, 13 Electrical connection, 14 Commutator end shield, 15 Commutator, 16 Brush holder, 17 Armature, 18 Magnets, 19 Terminal housing, 20 Planetary gear train.

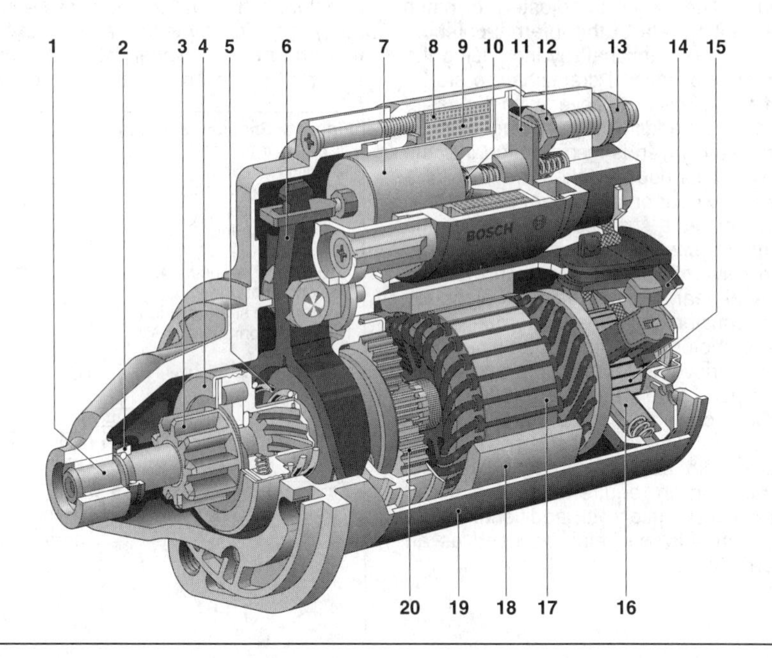

drops out. The starter current is interrupted and the demeshing spring disengages the pinion from the ring gear with the help of the helical spline.

Electric motor
The electric motor is a conventional DC motor. The predominant design for a starter motor is the 6-pole design. The magnetic materials available today have allowed the development of starters which are resistant to demagnetization, and have a highly effective magnetic flux to deliver a high starting power. As the magnetic field is generated by a permanent magnet, and the retroactive effect of the armature magnetic field is very minor, excitation is practically constant over the entire operating range.

Reduction gear
The objective of starter development is to minimize the weight and dimensions of the starter by reducing the volume of the electric motor etc. In order to achieve the same starting power at the same time, a higher armature speed is required to compensate for the existing lower armature torque. The torque is adjusted to match the rotation rate of the internal-combustion engine crankshaft by increasing the overall gear ratio of crankshaft to starter armature. This is achieved with the assistance of an additional gear stage (reduction gear), which is incorporated into the starter. This usually takes the form of a planetary gear on passenger car starters. It comprises a sun gear mounted on the armature shaft, a planetary-gear carrier normally with planet gears and a fixed internal gear. The spur-toothed planetary gear transfers the armature torque via the starter motor drive shaft to the pinion free from transverse forces. The high armature speed here (15,000 – 25,000 rpm) is stepped down at a ratio of $i \approx 3 – 6$.

The standard is a plastic internal gear made from fiber-reinforced polyamide. Depending on requirements, a sintered steel internal gear with additional damping elements may also be used as an alternative.

Pinion-engaging systems
The pinion-engaging system assembly ensures that the pinion meshes with the ring gear. It consists of a pinion, a roller-type overrunning clutch, a meshing spring, and an engagement relay.

Engagement relay
The engagement relay used in the starter (Figure 3) consists of a switch housing, relay armature, magnetic core, contact carrier (contact plate, contact spring), a pull-in winding and a hold-in winding, return spring and a switch cover with integrated terminal. The engagement relay has to fulfill two functions:

- It advances the drive pinion over the engagement lever into the engine's ring gear and
- switches the starter motor's primary electric circuit by closing and opening the contact bridge.

In order for the relay armature to ensure the movement of the pinion over the engagement travel, a relay current of approximately 30 A is required to generate the necessary magnetic force. If the relay armature is fully pulled in (air gap equals zero), significantly less magnetic excitation and hence a lower relay current is required (approximately 8 A) to hold the

Figure 3: Engagement relay
1 Relay armature,
2 Pull-in winding,
3 Holding-in winding,
4 Magnetic core,
5 Contact spring,
6 Contacts,
7 Electrical connection,
8 Contact bridge,
9 Armature shaft (divided),
10 Return spring.

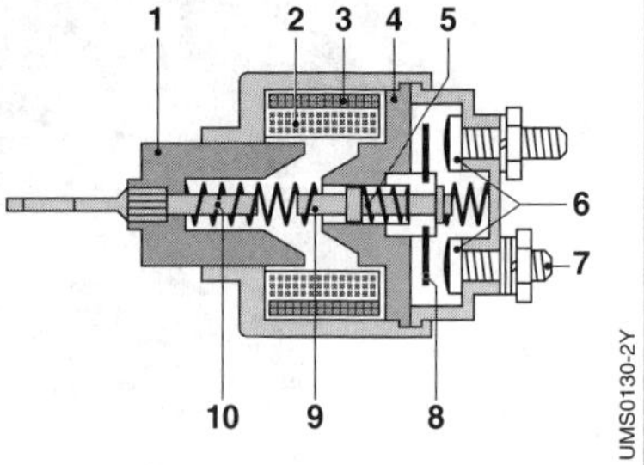

1 2 3 4 5 6 7 8 9 10
UMS0130-2Y

relay armature in the end position. The winding is divided into a pull-in winding and a hold-in winding, above all in order to limit heating of the relay winding. Electrically, the two windings are connected in parallel (Figure 4), which means that the magnetic excitations of the two windings add together. The start of the two windings are both applied to terminal 50 of the engagement relay. The winding end of the pull-in winding is connected via the starter armature, that of the hold-in winding is connected directly to the ground potential.

When a voltage is applied to terminal 50 (ignition switch in start position) of the engagement relay, the relay armature is pulled axially into the housing by the magnetic force generated by the pull-in winding and hold-in winding. This movement pushes the pinion forwards by the engagement lever towards the ring gear. The contact bridge is closed and the main starter current is switched on only once the relay armature is almost completely pulled in. This prevents the starter motor beginning to rotate before the starter pinion has engaged in the ring gear. As the two winding ends for the pull-in winding are now connected to positive, current flows only in the hold-in winding. The lower magnetic force of the hold-in winding is sufficient to hold the relay armature securely until the ignition/starting switch is opened.

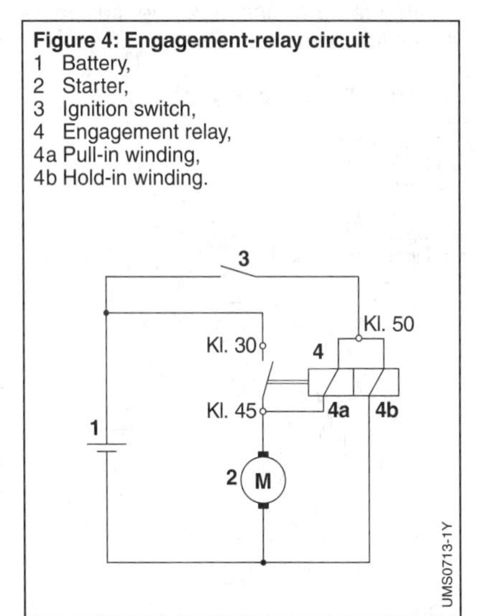

Figure 4: Engagement-relay circuit
1 Battery,
2 Starter,
3 Ignition switch,
4 Engagement relay,
4a Pull-in winding,
4b Hold-in winding.

Pre-engaged-drive starter

The pre-engaged-drive starter has become a worldwide standard for the passenger car, as secure operation over the entire operating range is guaranteed with this engagement principle. On pre-engaged drive starter motors, the engagement travel is made up of a lever-travel phase and a helical-travel phase. The engagement relay armature pushes the pinion toward the ring gear via the engagement lever (lever travel). As the starter-motor current is not yet switched on, the starter pinion does not yet rotate. If a pinion tooth meshes directly with a tooth gap on the ring gear (tooth-gap positioning) when the pinion meets the ring gear, the pinion engages as far as the movement of the relay allows.

If the pinion tooth strikes a tooth on the ring gear (tooth-tooth positioning) when the pinion meets the ring gear – this occurs approximately 80 % of the time – the relay armature tensions the meshing spring via the engagement lever, as the pinion can travel no further axially.

Once the pinion travel limit generated by the engagement relay has been reached, the contact bridge of the relay armature closes and the starter armature begins to rotate. In the case of tooth-gap positioning, the rotating electric motor screws the pinion fully into the ring gear via the helical spline (helical travel). The helical spline creating the helical travel also ensures that the pinion can only transfer the full torque of the starter motor once the pinion has reached the stop (end of helical travel). This ensures that the teeth of the pinion and ring gears are not mechanically overloaded.

Starting from a tooth-tooth position, the electric motor turns the pinion in front of the ring gear until a pinion tooth finds a tooth gap in the ring gear. The pretensioned meshing spring then pushes the pinion and overrunning clutch forward. The rotating electric motor screws the pinion fully into the ring gear via the helical spline.

When the relay winding is de-energized, the armature return spring pushes the relay armature – and the pinion and overrunning clutch via the engagement lever – back into the rest position. The overrunning torque caused by the friction of the overrunning clutch generates an axial force in conjunction with the helical spline, which supports the pinion's demeshing process.

The meshing spring considerably reduces the wear in the pinion-ring gear area by limiting the axial force, thus increasing the service life and reliability of the system.

Sliding-gear starters

Sliding-gear starters are used to start large internal-combustion engines. Because the requirements on service life are generally significantly higher for commercial vehicles, the engagement process usually takes place in two stages in order to protect the pinion and ring gear. The first stage starts the engagement of the pinion in the ring gear, while in the second stage, the pinion is softly meshed in the ring gear. The pinion may be rotated to resolve tooth-on-tooth positions by mechanical or electrical means.

Inertia-drive starters

The inertia drive is the simplest engagement principle, which is used above all on low power engines (e.g. lawn mower engines).

When the starter is switched on, the unloaded armature begins to rotate freely. The pinion and overrunning clutch do not yet rotate due to their mass moment of inertia, but are pushed along the helical spline. When the pinion meshes with the ring gear, the overrunning clutch begins to transmit the armature torque to the ring gear via the pinion. The starter then begins to crank the internal-combustion engine.

When the internal-combustion engine overruns, the overrunning clutch releases the non-positive connection. The overrunning torque caused by the friction of the overrunning clutch generates an axial force in conjunction with the helical spline, which disengages the pinion from the ring gear. This demeshing operation is assisted by the demeshing spring.

The engagement relay's only task in Bendix-type starters is to switch the starter-motor current. Thus it does not have to be mounted on the starter, but may be fitted at any position in the vehicle or in the drive system. The absence of the meshing spring increases the wear on the pinion and ring gear and thus reduces the service life and reliability of the starting system.

Overrunning clutch

In all starter designs, the drive torque is transmitted by an overrunning clutch. This overrunning clutch is installed between the starter and the pinion. Its purpose is to drive the pinion while the starter is cranking the internal-combustion engine, and then release the connection between the pinion and drive shaft as soon as the internal-combustion engine is turning faster than the starter.

Roller-type overrunning clutches are normally used on the starter designs being considered here. The "roller-type overrunning clutch" assembly comprises driver with clutch shell, roller race, rollers, springs, pinion, pinion shaft with helical spline and end cap. The roller-type overrunning clutch pushes individual sprung rollers into wedge-shaped pockets (Figure 5).

When the starter armature shaft is driven, the cylindrical rollers are clamped in the constricting roller race and create a non-positive connection between the internal shaft and the driver.

When overrunning takes place, the rollers are released against the force of the compression springs and move into the expanding section of the race. The clamping non-positive force is practically completely canceled. The spring-loaded rollers create a friction torque, known as the clutch overrunning torque.

Figure 5: Roller-type clutch
1 Pinion,
2 Clutch shell,
3 Roller race,
4 Roller,
5 Pinion shaft,
6 Spring.
a Direction of rotation.

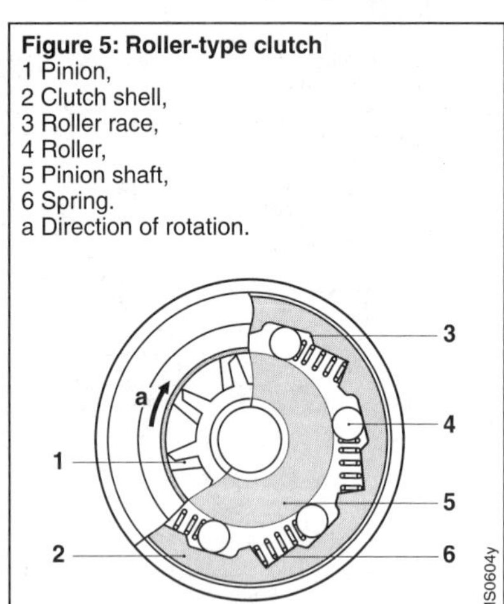

Triggering the starter

Conventional control

During conventional starts, the driver connects the battery voltage (ignition key in starting position) to the starter relay. The relay current (approx. 30 A for passenger cars to approx. 70 A for commercial vehicles) generates power in the relay. This pushes the pinion toward the engine-flywheel ring gear and activates the starter primary current (200–1,000 A for passenger cars, approx. 2,000 A for commercial vehicles).

The starter is switched off at the ignition switch, which opens and interrupts the starter-relay voltage.

Automatic starting systems

The high demands on vehicles with regard to convenience, safety, quality, and low noise levels have resulted in an increase in the use of automatic starting systems. An automatic starting system differs from a conventional one by virtue of additional components (Figure 6): one or more ballast relays, as well as hardware and software components (e.g. an engine ECU) to control the starting sequence.

The driver then no longer directly controls the starter relays current; instead the ignition key is used to send a request signal to the control unit, which then performs a series of checks before initiating the starting sequence. The safety check comprises a wide range of options and can verify any of the following: Is the driver authorized to start the vehicle (theft-deterrence feature)? Is the internal-combustion engine stationary (prevents the pinion from meshing with the moving ring gear)? Is the state of charge (in relation to engine temperature) sufficient to carry out the start? For automatic transmissions, is the selector lever in neutral or, for on manually shifted transmissions, is the clutch disengaged?

When the check has been completed successfully, the control unit initiates the start. On starting, the starting system compares the engine speed with a sustained operation speed of the engine (which may also depend on the engine temperature). Once the engine reaches sustained operation speed, the ECU switches the starter off. This always achieves the shortest possible starting time, reduces the noise levels, and lessens starter wear.

This process can also be used as the basis to implement start-stop operation. Here, the internal-combustion engine is switched off when the vehicle is not in motion, e.g. at traffic lights, and restarts automatically when required. A start-stop system requires a higher-level control system to implement the switch off and restart strategy. A start-stop system requires an electrical energy management system that incorporates battery-charge detection. Measures may also be required to stabilize the vehicle electrical system during the starting phase to avoid unacceptable voltage drops. The control equipment and starting system must therefore be matched. The control units have to fulfill their function at a significantly lowered supply voltage.

At the same time, the internal-combustion engine must also be optimized for quick-start response. What is needed is a starter that incorporates service-life prolonging features in order to meet service-life requirements and also guarantee quicker, low-noise starts. The design of the pinion and ring-gear geometry must be optimized to reduce wear and noise emissions.

Figure 6: Automatic starting system (circuit diagram)
1 Starting signal from driver,
2 Ballast relay,
3 ECU,
4 Park-neutral position or clutch signal,
5 Starter.

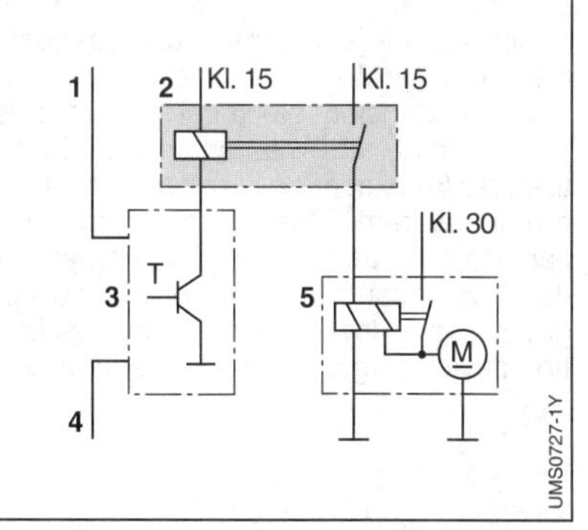

Actuators

Overview

Function

Actuators (final-control elements) form the interface between the electronic signal processor (data processing) and the actual process (mechanical motion). They convert the low-power signals conveying the positioning information into operating signals of an energy level adequate for process control. Signal converters are combined with amplifier elements to exploit the physical transformation principles governing the interrelationships between various forms of energy (electrical – mechanical – fluid – thermal).

Classification

This type of energy conversion represents one option for classifying electromechanical actuators. The energy emanating from the source is transformed into magnetic or electrical field energy, or converted to thermal energy. The individual force-generation principle is determined by these forms of energy, and is based on either field forces or certain specific material characteristics.

Magnetostrictive materials make it possible to design actuators for applications in the micropositioning range. This category also includes piezoelectric actuators, which are built according to a multilayer design similar to ceramic capacitors, and are actuators for high-speed fuel injectors. Thermal actuators depend exclusively on the exploitation of characteristics of specific materials.

Actuators in a motor vehicle are mostly electromagnetomechanical converters and, by extension, electrical servomotors, translational, and rotational solenoid actuators. An exception is the pyrotechnical airbag system. The solenoid actuators can themselves be the servo element, or they can assume a control function by governing a downstream force-amplification device (e.g. mechanical-hydraulic).

Electrodynamic and electromagnetic converters

Force generation in the magnetic field

The distinction between the electrodynamic and the electromagnetic actuator principles stems from the manner in which forces are generated in the magnetic field. Common to both principles is the magnetic circuit with soft-magnetic material and the coil for excitation of the magnetic field. A major difference lies in the force which can be extracted from the unit under technically feasible conditions. Under identical conditions, the force produced by application of the electromagnetic principle is greater by a factor of 40. The electrical time constant for this type of actuator is comparable to the mechanical time constants. Both force-generation principles are applied in linear and rotary drive mechanisms.

Electrodynamic principle

The electrodynamic principle is based on the force exerted on moving charges and charged conductors within a magnetic field (Lorentz force, Figure 1a). A field coil or a permanent magnet generates a constant magnetic field. The electrical energy destined for conversion is applied to the moving armature coil (plunger or immersion coil). A high degree of actuator precision is achieved by designing the armature coil with low mass and low inductance. The two energy storage units (one on the fixed and one on the moving component) produce two active force directions via current-direction reversal in the armature and field coils.

The secondary field produced by the armature current flows in an open magnetic circuit, thereby diminishing the effects of saturation. Approximately speaking, the force (torque) exerted by an electrodynamic actuator over its setting range is proportional to current and independent of travel.

Figure 1: Electrodynamic and electromagnetic converters
a) Electrodynamic principle: force on moving charge $F = B \cdot I \cdot l \cdot \sin \alpha$ (Lorentz force),
b) Electromagnetic principle: force between pole faces $F = B^2 \cdot A/2\mu 0$.

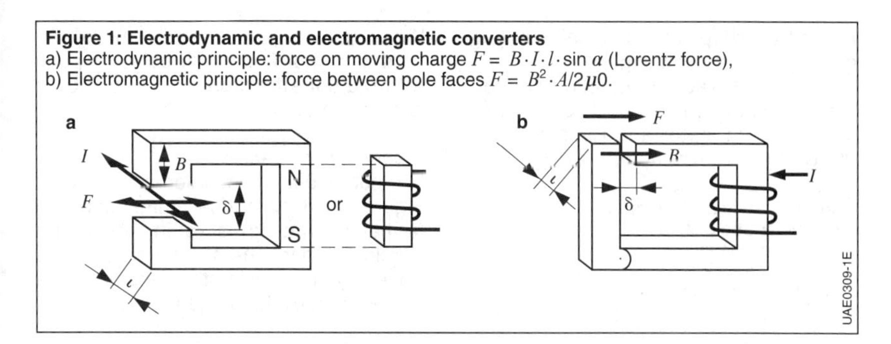

Electromagnetic principle
The electromagnetic principle (Figure 1b) exploits the mutual attraction displayed by soft ferrous materials in a magnetic field. The electromagnetic actuator is equipped with only one coil, which generates both the field energy and the energy to be transformed. In accordance with the operating principles, the field coil is equipped with an iron core to provide higher inductance. However, as the force is proportional to the square of the density of the magnetic flux, the unit is operative in only a single force-transfer direction. The electromagnetic actuator thus requires a return element (such as a mechanical spring or a magnetic return mechanism).

Dynamic response
The dynamic response of an electromechanical actuator, i.e. the activation and deactivation operations, is defined by the equation of mechanical motion, the differential equation of electrical circuits and Maxwell's equations of dynamics. The current- and position-dependent force follows from Maxwell's equations.

The most basic electrical circuit consists of an inductance with an ohmic resistor. One means of enhancing the dynamic response is through over-excitation at the instant of activation, while deactivation can be accelerated by a Zener diode. In each case, increasing the dynamic response of the electric circuit involves additional expenditure and increased losses in the actuator's triggering electronics.

Field diffusion is a delay effect which is difficult to influence in actuators with high dynamic response. Rapid switching operations are accompanied by high-frequency field fluctuations in the soft-magnetic material of the actuator's magnetic circuit. These fluctuations, in turn, induce eddy currents, which counteract their cause (build-up and decay of the magnetic field). The resultant delay in the build-up and reduction of forces can only be reduced by selecting appropriate materials with low electric conductivity and permeability.

Design
Design selection is essentially determined by operating conditions (e.g. installation space, required force/travel curve, and dynamic response).

Electromagnetic actuators
A typical form for translational electromagnetic actuators is the switching solenoid (Figure 2) with a force/travel curve which falls as a function of the square of positioning travel (Figure 3). The precise shape of the characteristic curve is deter-

Figure 2: Switching solenoid
1 Armature,
2 Coil,
3 Magnetic yoke.
F Force,
s Travel.

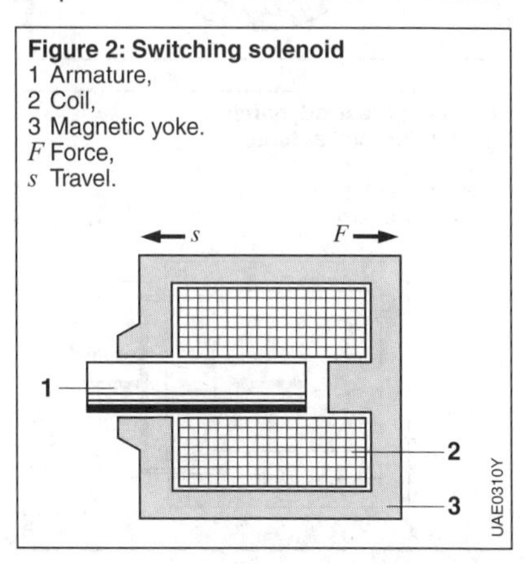

mined by the type of working air gap (e.g. conical or solenoid plunger).

Rotational electromagnetic actuators are characterized by a defined pole arrangement in stator and rotor (e.g. single-winding rotary actuator, Figure 4).

Figure 3: Switching solenoid (characteristic curves)
1 Solenoid plunger,
2 Conical armature,
3 Cylindrical armature.

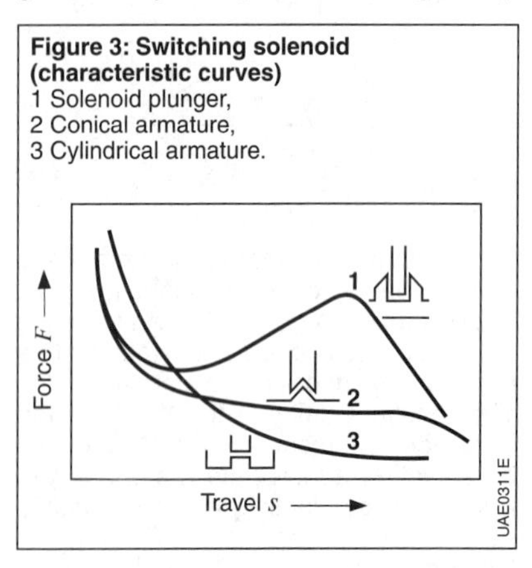

Figure 4: Electromagnetic single-winding rotary actuator
1 Coil,
2 Stator,
3 Rotor.

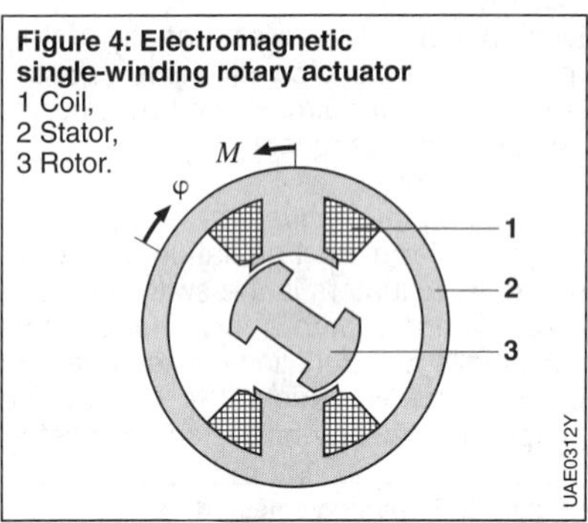

Figure 5: Electrodynamic immersion-coil actuator
1 Immersion coil,
2 Permanent magnet,
3 Magnetic yoke.

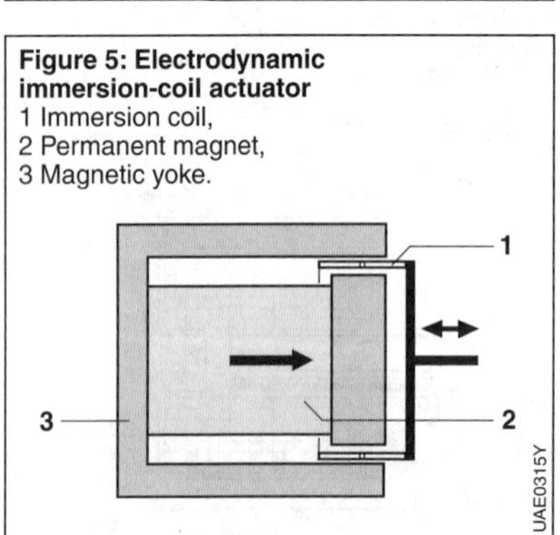

When current is applied to one of the coils, the rotor and stator poles respond with mutual attraction, and in doing so generate a torque.

Electrodynamic actuators

In a pot magnet (immersion-coil actuator, Figure 5), a cylindrical immersion coil (armature winding) is set in motion in a working air gap. The adjustment range is limited by the axial length of the armature winding and by the air gap.

Application

Electromechanical actuators are direct-action control elements, and without an intermediate ratio-conversion mechanism, they convert the energy of the electrical control signal into a mechanical positioning factor or work. Typical applications include positioning of flaps, sleeves, and valves. The described actuators are final-control elements without internal return mechanisms, i.e. without a stable operating point. They are only capable of carrying out positioning operations from a stable initial position (operating point) when a counterforce is applied (e.g. return spring and electrical control).

A solenoid plunger provides a stable static operating point when its force/travel curve is superimposed on the characteristic response of a return spring. A variation of the coil current in the solenoid shifts the operating point. Simple positioning is achieved by controlling the current. However, particular attention must be paid here to the nonlinearity of the force-current characteristic and the positioning system's sensitivity to interference factors (e.g. mechanical friction, pneumatic, and hydraulic forces). The temperature sensitivity of the coil resistance results in positioning errors, making corrective current control necessary. A high-precision positioning system with good dynamic response must incorporate a position sensor and a controller.

Piezo actuators

Physical principle

The essential functioning principle of piezo actuators, the indirect piezoelectric effect, is based on the direct piezoelectric effect discovered in 1880 by the Curie brothers on tourmaline crystals, which is characterized by the conversion of a mechanical deformation of the crystal into a voltage on the crystal surface proportional to it.

The reversibility of this principle is referred to as the indirect piezoelectric effect. The application of a voltage to a piezoelectric material results in a rapid directional deformation (deflection) of the material of a magnitude of a few µm which can be used as an actuating movement.

Essential to the occurrence of the piezoelectric effect are electric dipoles in the elementary cells of the material, which as a result of reciprocal-action processes form larger connected areas of the same orientation which by analogy to magnetism are termed domains (Figure 6a). When an electric field is applied across a privileged direction, the predominant share of these domains can be aligned in the field direction (domain folding), where the electric dipoles in the domains are simultaneously extended (lattice expansion, Figure 6b). The material there demonstrates a macroscopic linear deformation during the polarization process.

Even after the deactivation of the field and the associated relaxation of the dipoles, the domains remain in this aligned (polarized) state (Figure 6c) so that the piezoelectric material can again and again be extended reversibly across the privileged direction when an electric field is applied.

Materials

Piezoelectric materials are a subcategory of dielectrics; in other words, they are generally electrically non-conducting, nonmetallic materials without freely mobile charge carriers. As well as the first piezoelectric monocrystals analyzed, such as tourmaline and quartz, a large number of polycrystalline piezoelectric materials have since become familiar, including many ceramic materials.

Because of the only small specific linear deformations of piezoelectric crystals, only materials which have a particularly effective capacity to convert electrical energy into mechanical energy are relevant to the technical application of the piezoelectric effect. This property can be described by the coupling factor k, which describes the ratio of the mechanical energy stored in the piezo actuator to the total energy supplied ($0 < k < 1$).

The PZT ceramics (Plumbum Zirconate Titanate) based on the oxidic mixed-crystal system of lead zirconate titanate, which demonstrate high coupling factors through specific adjustment of the Ti-Zr ratio and the use of suitable doping agents, are of particular importance to piezo actuators.

Body

The basic design of a piezo actuator consists of a polarized piezoelectric solid in which the electric field needed to extend the actuator can be generated by means of surface external contacting. In order to achieve a technically relevant deflection with this bulk piezo actuator (Figure 7a), as a rule large material thicknesses and very high activation voltages in the two-digit kV range are required.

In a multilayer piezo actuator (Figure 7b), the piezoelectric solid is divided into many parallel-connected active layers by inner electrodes led out alternately to the posi-

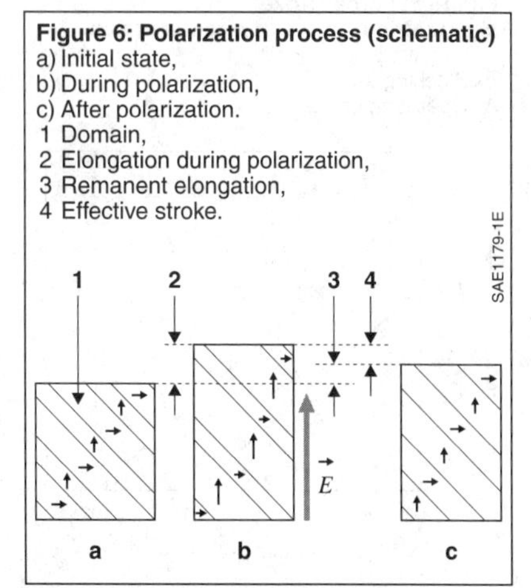

Figure 6: Polarization process (schematic)
a) Initial state,
b) During polarization,
c) After polarization.
1 Domain,
2 Elongation during polarization,
3 Remanent elongation,
4 Effective stroke.

tive and negative outer electrodes. This significantly reduces the required activation voltage with virtually the same deflection of the overall combination. Multilayer piezo actuators can thus be operated in safety-relevant fields of application with moderate voltages (U < 200 V), e.g. in fuel-injection systems of motor vehicles.

Because of their sensitivity to mechanical overstressing and harmful environmental influences (e.g. conductive or corrosive media), ceramic piezo actuators are often installed in a housing under defined initial stress. This ensures that the piezoceramic structure is not subjected to any tensile stresses or unwanted interactions with media even during dynamic actuator operation.

Figure 7: Design of piezo actuators
a) Bulk piezo actuator,
b) Multilayer piezo actuator.
1 Outer electrodes,
2 Inner electrodes,
3 Piezoelectric material.

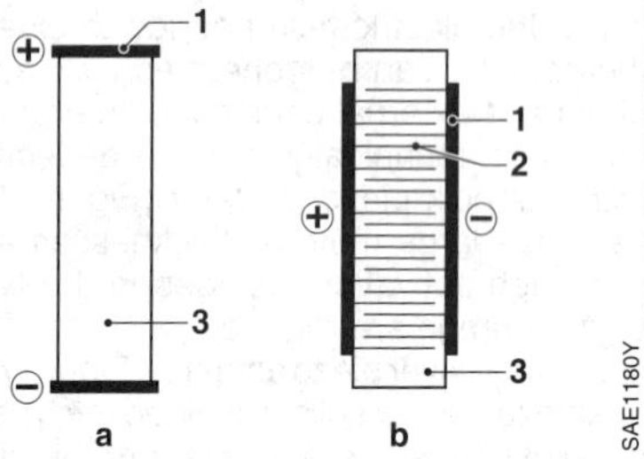

Figure 8: Force/stroke characteristic curve of piezo actuator
x_0 Idle stroke,
F_B Blocking force,
A Operating point.

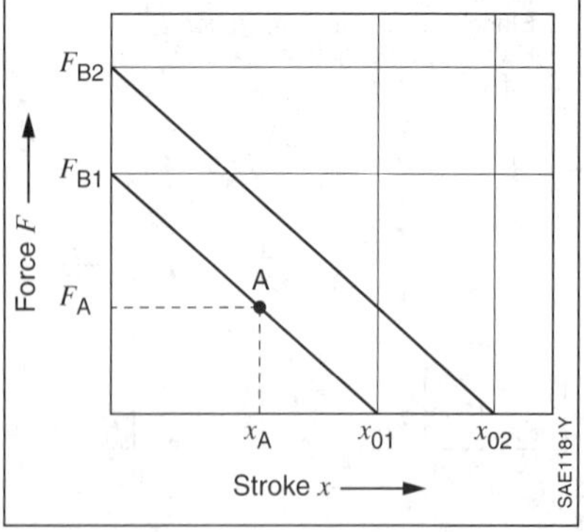

Energy capability

Like all electromechanical actuators, the piezo actuator is also a converter which converts the electrical energy supplied to it into mechanical energy and thus provides the energy capability required for an actuating movement. The effective energy capability is dependent on the efficiency of the actuator, i.e. on the extent of the electrical and mechanical losses incurred during the conversion process.

Depending on the field of application, the operating point of a piezo actuator can lie between its maximum force (blocking force) and its maximum elongation capability (idle stroke). The basic connection here is that the maximum switching force that can be transferred by the piezo actuator at this operating point is dependent on its deflection permitted at this point in the system. This behavior is described by the actuator's force/stroke characteristic curve (Figure 8). Because of the proportionality between activation voltage and deflection of the piezo actuator, but also its force, this characteristic curve is a direct function of the activation voltage. The maximum force and stroke values that can be achieved by increasing the same are limited in the upward direction by the specific electric strength of the material used.

If a higher actuator force is required for a specific application, this can be achieved without changing the piezoelectric material only by the fact that the active cross-section of the actuator is increased under the influence of the electric field. Increasing the actuator stroke, on the other hand, requires an increase in the number of active layers in the actuator. The active cross-section and the number of active layers are thus the essential design features for configuring a multilayer piezo actuator to suit the particular application.

Fluid-mechanical actuators

Hydraulic and pneumatic actuators utilize similar principles for the conversion and regulation of energy. The crucial difference is in the medium used. Hydraulic actuators operate with virtually incompressible fluids, usually oil, in a pressure range up to approximately 30 MPa. Pressures around 200 MPa are achieved for diesel-fuel injectors. Pneumatic actuators operate with compressible gases, usually air, in a pressure range around 1 MPa. For vacuum actuators, the pressure range is 0.05 MPa.

In most applications, fluid-mechanical actuator drives are in the form of hydrostatic energy converters. These operate according to the displacement principle, converting the pressure energy of the fluid medium into mechanical work and vice versa.

In contrast, hydrodynamic converters operate by converting flow energy (kinetic or velocity energy of the moving fluids) into mechanical work (example: hydrodynamic clutch).

Losses during energy conversion stem from leakage and friction. Fluid-thermal losses are caused by flow resistance, in which throttle action transforms the hydraulic energy into heat. A portion of this heat is dissipated into the environment, and some of it is absorbed and carried away by the fluid medium. The following applies:

$$Q_{\text{heat}} = Q_1 p_1 - Q_2 p_2 .$$

In the case of incompressible fluids:

$$Q_{\text{heat}} = Q_1 (p_1 - p_2) .$$

The flow develops into turbulence at restrictions. The flow rate of the fluid is then largely independent of viscosity. On the other hand, viscosity does play a role in the case of laminar flow in narrow pipes and apertures.

Fluid-mechanical amplifiers (Figure 9) control the conversion of energy from fluid to mechanical state. The regulating mechanism must be designed for control with only a very small proportion of the energy required for the ultimate positioning operation.

Switching valves open and close the orifice governing the flow to and from a fluid-mechanical energy converter (Figure 10). Provided that the control-element opens sufficiently, the throttling losses remain negligible. Pulse-width-modulated opening and closing can be applied to achieve quasi-continuous control of the fluid/mechanical energy conversion process with virtually no losses. In practice, however, pressure fluctuations and mechanical contact between the valve elements result in undesirable vibration and noise.

Figure 9: Fluid-mechanical actuators

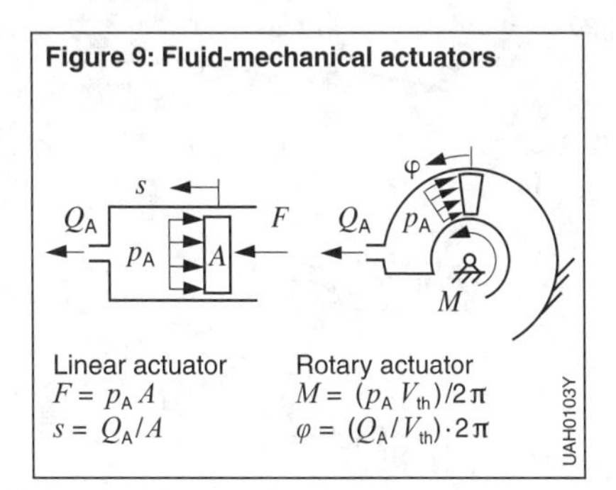

Figure 10: Energy control with 2/2-way switching valves

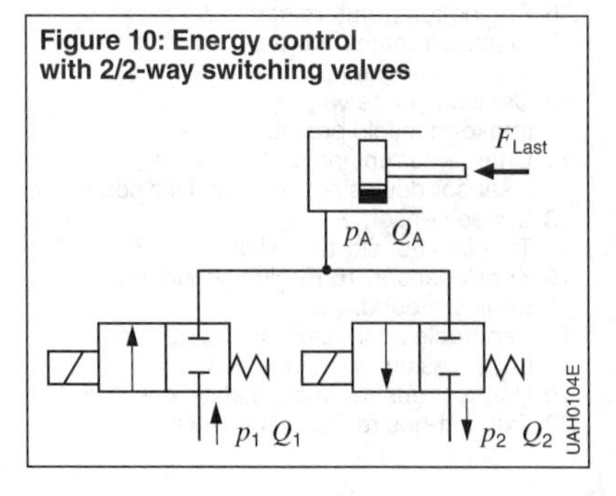

Wiring harnesses and plug-in connections

Wiring harnesses

Requirements

The purpose of the wiring harness is to distribute power and signals within a motor vehicle. A wiring harness in the present day, mid-class passenger car with average equipment has approximately 750 different lines, their length totaling around 1,500 meters (Table 1). In recent years, the number of contact points has practically doubled due to the continuous rise in functions in the motor vehicle. A distinction is made between the engine-compartment and body wiring harness. The latter is subject to less demanding temperature, vibration, media and tightness requirements.

Wiring harnesses have considerable influence on the costs and quality of a vehicle. The following points must be taken into consideration in wiring harness development:

- Leak-tightness
- Electromagnetic compatibility
- Temperatures
- Protection of the wires against damage
- Wire routing
- Ventilation of the wiring harness.

It is therefore necessary to involve wiring harness experts as early as in the system definition stage. Figure 1 shows a wiring harness that was developed as a special intake-module wiring harness. Thanks to the optimization of routing and security in conjunction with engine and wiring harness development, it was possible to achieve an advancement of quality as well as to yield cost and weight advantages.

Dimensioning and selection of materials

The most important tasks for the wiring harness developer are:

- Dimensioning of line cross-sections
- Material selection
- Selection of suitable plug-in connections
- Routing of lines under consideration of ambient temperature, engine vibrations, acceleration and EMC
- Consideration of the environment in which the wiring harness is routed (topology, assembly stages in vehicle manufacture and equipment on the assembly line).

Line cross-sections

Line cross-sections are defined based on permissible voltage drops. The lower cross-section limit is determined by the line strength. Convention has it that no lines with a cross-section of less than 0.5 mm^2 are used. With additional measures (e.g. supports, protective tubes, tension relief), even a cross-section of 0.35 mm^2 may be permissible.

Figure 1: Wiring harness (example)

Connector for
1 Ignition coil module, 2 Channel deactivation,
3 Fuel injectors, 4 Throttle device,
5 Oil-pressure switch,
6 Engine temperature sensor,
7 Intake-air temperature sensor,
8 Camshaft sensor,
9 Canister-purge valve,
10 Intake-manifold pressure sensor,
11 Charge current indicator lamp,
12 λ sensor downstream of catalytic converter,
13 Speed sensor,
14 Terminal 50, starter switch,
15 Knock sensor, 16 Engine control unit,
17 Engine ground,
18 Separable connector for engine and transmission wiring harness,
19 λ sensor upstream of catalytic converter,
20 Exhaust-gas recirculation valve.

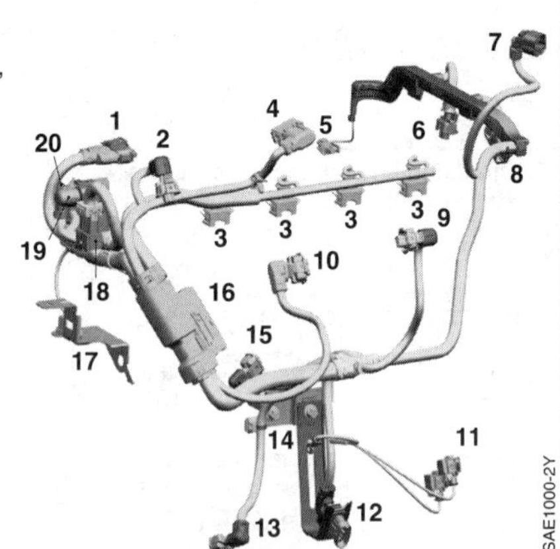

Table 1: Complexity of wiring harnesses (typical values)

	Small vehicle	Medium class	Luxury class
Number of plug connectors	70	120	250
Number of contact points	700	1,500	3,000
Number of wires	350	750	1,500
Total length of the wiring in meters	700	1,500	3,200

Materials

Copper is usually used as the conductive material. The insulation materials for the lines are defined by the temperature to which they are exposed. It is necessary to use materials that are suitable for the high temperatures of continuous operation. Here, the ambient temperature must be taken into consideration as much as the heating caused by the flow of current. The materials used are thermoplastics (e.g. PE, PA, PVC), fluoropolymers (e.g. ETFE, FEP) and elastomers (e.g. CSM, SIR).

If the lines are not routed past particularly hot parts (e.g. exhaust pipe, exhaust gas recirculation) in the engine topology, one of the criteria for the selection of the insulation material and the cable cross-section could be the derating curve of the contact with its associated line. The derating curve represents the relationship between current, the temperature increase that it causes, and the ambient temperature of the plug-in connection. Normally, the heat generated in the contacts can only be carried away along the lines themselves. It should also be noted that the change in temperatures results in a change in the modulus of elasticity of the contact material (metal relaxation). It is possible to influence the relationships described by means of larger line cross-sections and the use of suitable contact types and more noble surfaces (e.g. gold, silver) and thus higher limit temperatures. For highly fluctuating current intensities, it is often useful to measure the contact temperature.

Plug-in connections and contacts

The type of plug-in connections and contacts used depends on various factors:

- Current strength
- Ambient temperatures
- Vibration load
- Resistance to substances
- Installation space.

Line routing and EMC measures

Lines should be routed in such a way as to prevent damage and line breaks. This is achieved by means of fasteners and supports. Vibration loads on contacts and plug-in connections are reduced by fastening the wiring harness as close to the plug as possible and at the same level as the vibration where possible. The line routing must be determined in close cooperation with the engine and vehicle developers.

To avoid EMC problems, it is recommended to route sensitive lines and lines with steep current flanks separately. Shielded lines are not straightforward to produce and are therefore expensive. They also need to be grounded. The twisting of lines is a more cost-favorable and effective measure.

Line protection

Lines need to be protected against chafing and against making contact with sharp edges and hot surfaces. Adhesive tapes are used for this purpose. The level of protection is determined by the interval and winding density. Corrugated tubing (material savings from corrugation) with the necessary connecting pieces are often used as line protection. However, tape fixing is still an essential means of preventing movement of individual lines inside the corrugated tube. Optimal protection is offered by cable ducts.

Wiring harnesses are at risk from rodents. A remedy may be provided by rodent-resistant, extruded plastic tubing.

Plug-in connections

Function and requirements

The high integration density of electronics in the motor vehicle places high demands on a car's plug-in connections. Not only do they carry high currents (e.g. activation of ignition coils), they also carry analog signal currents with low voltage and current intensity (e.g. signal voltage of the engine temperature sensor). Throughout the service life of the vehicle, the plug-in connections must ensure the reliable transmission of signals between control units and to the sensors whilst maintaining tolerances.

The increasing demands of emission-control legislation and active vehicle safety are forcing the ever more precise transmission of signals through the contacts of the plug-in connections. A large number of parameters must be taken into consideration (Figure 2) in the design, arrangement and testing of the plug-in connections.

The most common cause of failure of a plug-in connection is wearing of the contact caused by vibrations or temperature change. The wear promotes oxidation. This results in an increase in ohmic resistance – the contact may, for example, be subjected to thermal overload. The contact part may be heated beyond the melting point of the copper alloy. In the case of highly resistant signal contacts, the vehicle controller often detects a plausibility error by comparison with other signals; the controller then enters fault mode. These problem areas in the plug-in connection are diagnosed by the on-board diagnosis (OBD) required by emission-control legislation. However, it is difficult to diagnose the error in the service workshops because this defect is displayed as being a component failure. It is only possible to diagnose the faulty contact indirectly.

For the assembly of the plug-in connection, there are various functional elements on the plug housing intended to ensure that the cables with their crimped contacts can be joined to the plug-in connection reliably and defect-free. Modern plug-in connections have a joining force of less than 100 N so that the assembly operative is able to reliably join the connector to the component or control-unit interface. The risk of plug-in connections being connected to the interface incorrectly increases with higher connecting forces. The plug would come loose during vehicle operation.

Figure 2: Parameters for the design of plug-in connections

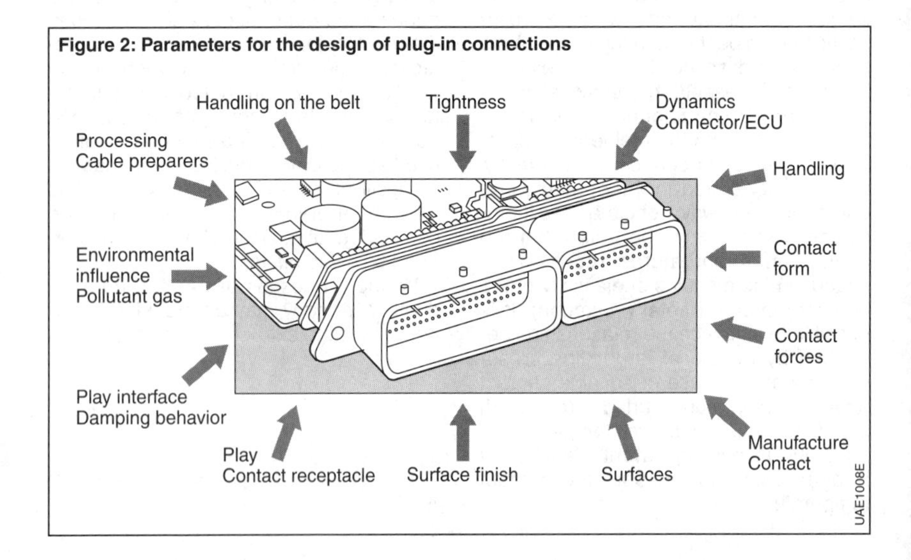

Design and types

Plug-in connections have different areas of application (Table 2). These are characterized by the number of pins and ambient conditions. There are different classes of plug-in connection: hard engine attachment, soft engine attachment, and body attachment. Another difference is the temperature class of the installation location.

Table 2: Uses of plug-in connections

	No. of pins	Special features	Applications
Low-pin-count	1 to 10	No joining force support	Sensors and actuators (many different requirements)
High-pin-count	10 to 150	Joining force support by slide, lever, modules	Control units (several, similar requirements)
Special plugs	any	e.g. integrated electronics	Special applications (individual, agreed requirements)

High-pin-count plug-in connections

High-pin-count plug-in connections are used for all control units in the vehicle. They differ in their number of pins and the geometry of the pins (Table 3). Figure 3 shows a typical design of a high-pin-count plug-in connection. The complete connector is sealed against the male ECU connector concerned by a circumferential radial seal in the connector casing. This, together with three sealing lips, ensures a reliable seal against the control unit sealing collar.

The contacts are protected against the ingress of humidity along the cable by a flat seal, through which the contacts are inserted and the line crimped to them. A silica-gel mat or silica mat is used for this purpose. Larger contacts and lines may also be sealed using a single-core seal (see "Low-pin-count plug-in connections").

Table 3: Number of pins and pin geometry

		Pin thickness in mm	
		0.6	0.8
Pin width in mm	0.6	x	
	1.2	x	
	1.5	x	
	2.8		x
	4.8		x

Figure 3: High-pin-count plug-in connection
a) View, b) Cross-section.
1 Pressure plate, 2 Flat seal, 3 Radial seal, 4 Slide pin (secondary lock), 5 Contact carrier, 6 Contact, 7 Lever, 8 Slide mechanism.

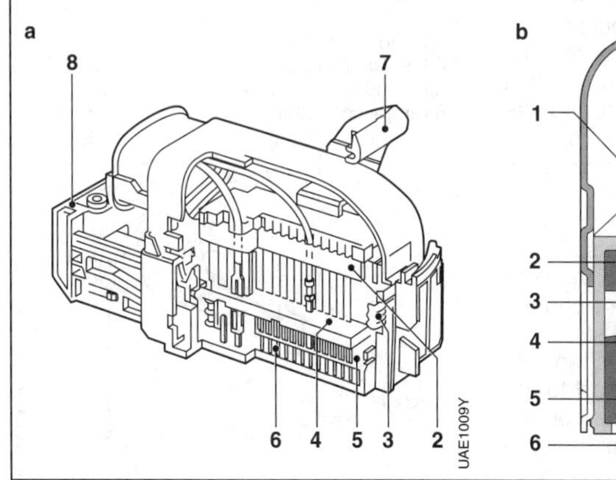

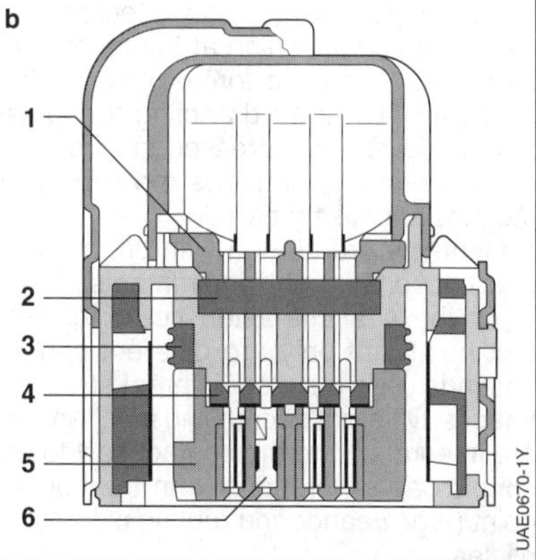

When the plug is assembled, the contact with the line attached is inserted through the flat seal that is already in the plug. The contact slides home into its position in the contact holder. The contact latches on its own by a locking spring that engages in an undercut in the plastic housing of the plug. Once all contacts are in their final position, a slide pin is inserted to provide a second contact safeguard, or secondary lock. This is an additional security measure and increases the retaining force of the contact in the plug-in connection. In addition, the sliding movement is a means of checking that the contacts are in the correct position. The operating force of the plug-in connection is reduced by a lever and a slider mechanism.

Low-pin-count plug-in connections
Low-pin-count plug-in connections (Figure 4) are used for actuators (e.g. fuel injectors) and sensors. Their design is similar in principle to that of a high-pin-count plug-in connection. The operating force of the plug-in connection is not usually supported.

The connection between a low-pin-count plug contact system and the interface is sealed with a radial seal. Inside the plastic housing, however, the lines are sealed with single-core seals secured to the contact.

Contact systems
Two-part contact systems are used in the motor vehicle (Figure 5). The inner part – the live part – is pressed from a high-quality copper alloy. It is protected by a steel overspring, which at the same time increases the contact forces of the contact by means of an inwardly acting spring element. A catch arm pressed out from the steel overspring engages the contact in the plastic housing part.

Contacts are coated with tin, silver or gold, depending on requirements. To improve the wear characteristics of the contact point, not only are different contact coatings used but also different structural shapes. Different decoupling mechanisms are integrated into the contact part to decouple cable vibrations from the contact point (e.g. meandering routing of the supply lead).

The cables are crimped onto the contact. The crimp geometry on the contact must be adapted to the cable concerned. Pliers or fully automatic, process-monitored crimping presses with contact-specific tools are available for the crimping process.

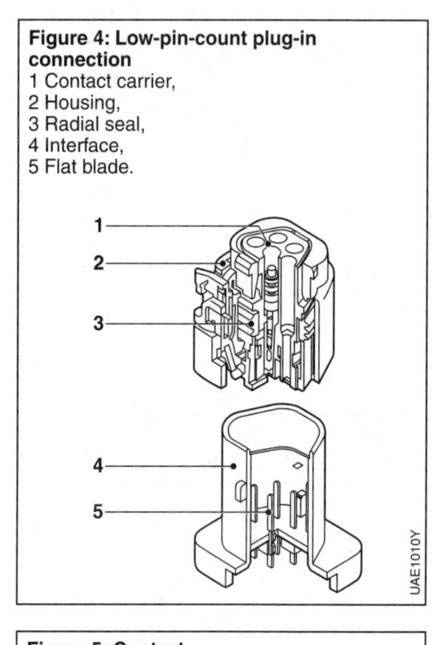

Figure 4: Low-pin-count plug-in connection
1 Contact carrier,
2 Housing,
3 Radial seal,
4 Interface,
5 Flat blade.

Figure 5: Contact
1 Steel overspring,
2 Single core,
3 Conductor crimp,
4 Insulation crimp,
5 Meander,
6 Single-core seal.

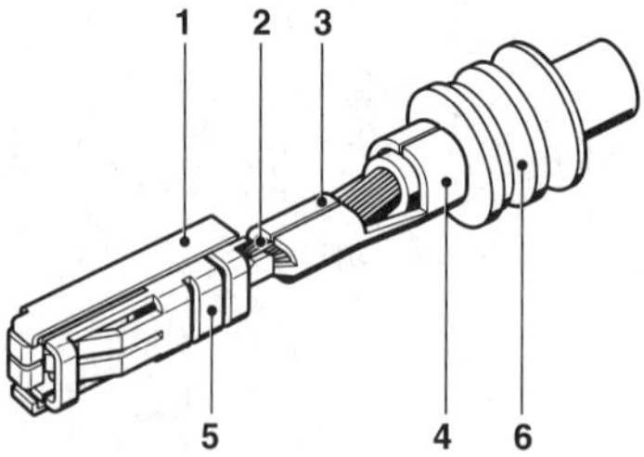

Electromagnetic compatibility (EMC)

Requirements

Electromagnetic compatibility (EMC) is the general term for the target state in which electrotechnical devices and state-of-the-art technologies (e.g. wireless systems) do not have unintended mutual interference from electrical, magnetic or electromagnetic effects.

Particularly in vehicle electronics, the importance of electromagnetic compatibility is increasing due to the higher amount of electrical equipment and the use of new technologies, such as electric and hybrid drives with new energy accumulators (e.g. high-performance batteries and fuel cells) on the one hand and mobile communication systems (e.g. phone, navigation and Internet) on the other. This leads to a rising level of complexity in today's automobiles (Figure 1).

Electronic power train systems installed in the vehicle (e.g. engine and transmission control, electric drive), safety systems (e.g. antilock braking system, Electronic Stability Program, and airbag), comfort and convenience electronics (e.g. climate control, electrical adjusting devices) and mobile communication systems (e.g. radio, navigation, Internet) are arranged next to each other in close proximity. The associated high density and number of fast-switching, high-performance electronic components on the one hand and the requirements of today's communication on the other pose a great and, in part, new challenge to electromagnetic compatibility (Figure 2).

Figure 2: Electromagnetic compatibility in motor vehicles

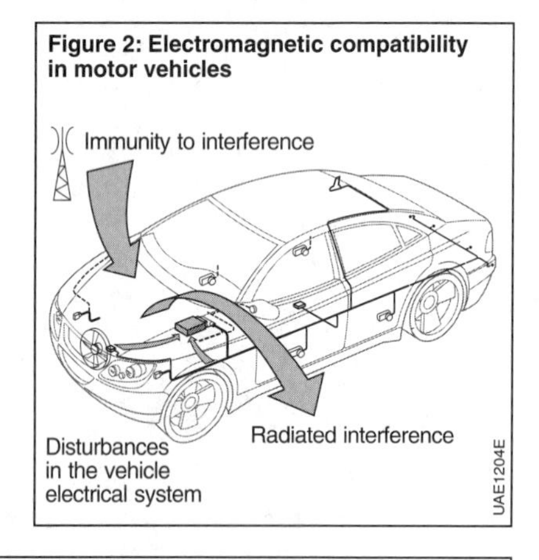

Figure 1: Electronic systems and wiring harness in a modern motor vehicle

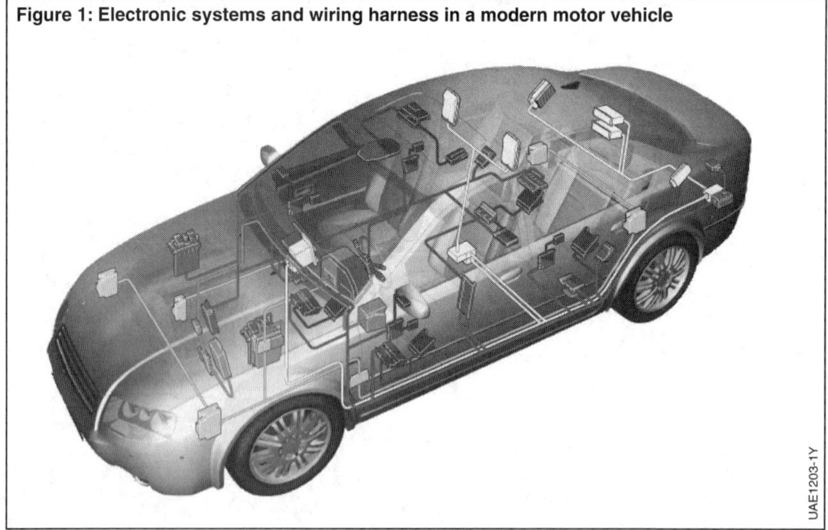

Interference emission and interference immunity

Sources of interference in the DC vehicle electrical system

Ripple in vehicle electrical system

In a conventional vehicle with combustion engine, the alternator supplies the vehicle electrical system with rectified three-phase current. Although the current is smoothed by the vehicle battery, a residual ripple remains. In addition, the energy requirements of the electrical and electronic systems influence the DC voltage supply.

The amplitude of the ripple in the vehicle electrical system depends on the load on the vehicle electrical system and the wiring. Its frequency changes according to alternator speed and the consumer behavior. The fundamental oscillation is in the kilohertz range. It can penetrate the vehicle sound systems – either directly (conductive) or inductive – where the ripple is heard as a howl in the loudspeaker system.

Pulses in the vehicle electrical system

Voltage pulses are generated on the supply lines when electrical equipment is switched on and off. They are routed to adjacent systems directly through the power supply (through conductive coupling, Figure 3a) and indirectly by coupling from connecting lines (through inductive and capacitive coupling, Figure 3b and Figure 3c). These unwanted interference pulses can cause anything from malfunction to total destruction of the adjacent systems. The signal shapes and amplitudes acting as disturbance values depend on the configuration of the vehicle electrical system, such as the ground concept, the position of the wiring harnesses and the individual lines in the wiring harness.

The wide variety of pulses that occur in the vehicle are classified into typical pulse shapes. Major parameters are pulse amplitude, signal rise and fall times as well as internal resistance of the pulse source. A suitable selection of permitted values for interference emission from the sources of interference and the required values for interference immunity ensures that there is no unintended behavior of the electronic systems without unnecessarily high effort for interference suppression. This can be done using various concepts. For example, a higher interference emission for the electric actuators and motors can be allowed than for typical pulse sources if the electronic components, such as control units and sensors, are designed with corresponding interference resistance.

Figure 3: Conductive, capacitive and inductive coupling
a) Conductive coupling of interference signals,
b) Capacitive coupling of interference signals,
c) Inductive coupling of interference signals.
$\underline{Z}_i$ Internal resistance (complex size),
$\underline{Z}_a$ Terminating resistance (complex size),
$\underline{u}_1$, $\underline{u}_2$ Voltage sources (complex size),
$\underline{u}_s$ Interference voltage (complex size),
R_E Input resistance,
C_E Input capacitance,
$C_{1,2}$ Capacitance between both conductors,
L_1, L_2 Inductance of the conductors,
M Inductive coupling.

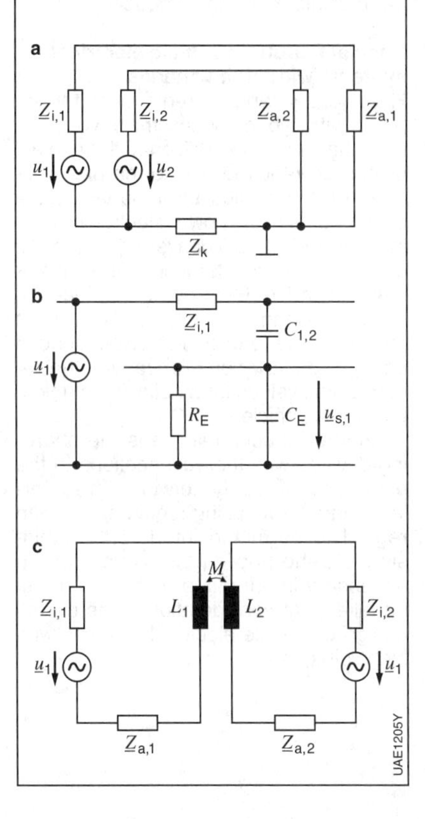

Potentially susceptible devices

Electronic control units and sensors are potentially susceptible devices for interference signals that enter the vehicle electrical system from the outside (Figure 2). These interference signals come from the neighboring systems in the vehicle. Malfunctions occur when the system loses the ability to distinguish between interference signals and useful signals. The possibility of taking effective measures depends on the characteristics of useful and interference signals.

The control unit is unable to distinguish between useful and interference signals if the characteristic of an interference signal is similar to that of a useful signal – for instance, when a pulse-shaped interference signal is at the same frequency as the signal from a wheel-speed sensor. Critical frequencies are those in the range of useful signal frequencies ($f_S \approx f_N$) and a low multiple of a signal frequency.

Special feature in vehicle electrical systems with high voltages

High voltage and currents in motor vehicles with hybrid and electric drives pose new challenges. In addition to the contact protection required due to the high voltages, EMC is particularly important. In the components of the power electronics (e.g. inverter and DC/DC converter) and in the electrical machine, large interference signals can arise. Only by using shielding concepts along with suitable filter circuits can these interference signals be reduced enough to allow compliance with the required limit values for mobile communication in the vehicle.

Another special feature is the charging devices with their connections to the public power supply network. This interface, and the resulting requirements with regard to permitted interference emission into the public power grid and the necessary interference protection, must be taken into consideration in the overall design of vehicle electrical systems with high voltages.

High-frequency vibrations and oscillations in the vehicle electrical system

In addition to the low-frequency ripple in vehicle electrical systems and the pulses in these systems, high-frequency vibrations and oscillations are caused in many electromechanical and electronic components by switching operations. Examples include commutation operations in brush motors or in electronically commutated electric motors, activation of output stages or circuits in digital electronics (such as the CPU core of an electronic control unit). These oscillations can spread – more or less attenuated – into the vehicle electrical system via the connected lines, particularly the power supply lines, and can get into the systems for communications electronics via the power supply lines or the capacitive and inductive coupling in the wiring harness.

Depending on whether the measured spectrum of radio interference voltage has a more or less continuous curve or consists of discrete lines, we refer to it as broad-band interference or narrow-band interference. Broad-band interference is caused, for example, by electric motors in wiper drives, control systems, fans, fuel pumps or alternators, but also by certain electronic components. Narrow-band interference is caused, for example, by electronic control units with microprocessors. This allocation depends on the used bandwidth of the radio service under consideration or the bandwidth used during the measurement compared to the characteristics of the interference signal. The various interference signals also have various effects on analog and digital radio systems.

Disturbances can travel through the electric lines in the wiring harness to the supply and signal connections and into the connecting cables of the antennas. They can also be received by the antennas directly and, just like the desired useful signals, they can get into the receiver circuits of the mobile communication devices. These high-frequency disturbances can interfere with vehicle communication systems, because they are often found in the frequency and amplitude range of the useful signal. Narrow-band interference is

Figure 4: Time signal and spectrum for a trapezoidal pulse
a) Dependency on time,
b) Dependency on frequency.
T Period duration,
T_r Rise time (from 10% to 90%),
T_f Release time (from 90% to 10%),
T_i Pulse duration,
A_0 Amplitude,
$\hat{u}$ Pulse amplitude,
f_0 Basic frequency of the time signal,
f_{n-1} Lines,
f_{min} Periodic minimums,
f_g Fundamental frequencies,
k Pulse duty factor,
m Number of lines between minimums,
H Envelope,
Dec Decade.

$$f_0 = \frac{1}{T}, \quad k = \frac{T}{T_i},$$
$$f_{n-1} = \frac{n}{T} = n f_0, \quad f_{min} = \frac{n}{T_i},$$
$$f_{g1} = \frac{1}{\pi T_i}, \quad f_{g2} = \frac{1}{\pi T_r},$$
$$m = \frac{T}{T_i}, \quad A_0 = 2 \hat{u} T_i.$$

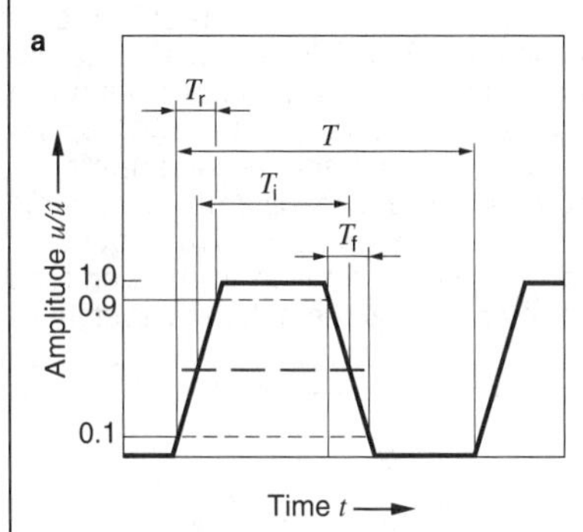

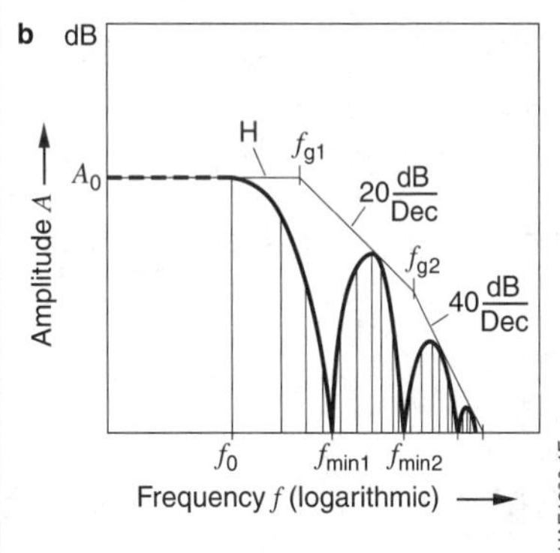

particularly critical, because it has signal characteristics that are very similar to the spectrum of transmitters (Figure 4).

Electrostatic discharge

The subject of potential danger to components and electronic circuits from electrostatic discharge (ESD) also belongs to the field of EMC. The task here is to protect components and equipment from interference or damage by static discharge from humans or from machinery (exposed components) during operation, production and maintenance. This involves both adopting appropriate methods for handling equipment and designing equipment in such a way that the voltages (up to several thousand volts) produced by electrostatic discharges are reduced to acceptable levels.

Immunity to interference from electromagnetic fields

Sources of interference and potentially susceptible devices

During operation, vehicles and their electronic equipment are exposed to various electromagnetic waves, for example from radio, television and wireless transmitters, which are operated at a fixed location, or even from within the same or a neighboring vehicle. Electronic circuits must not experience interference from the electromagnetic fields of the transmitters and the resulting, unwanted voltages and currents.

The large number of electric lines in the wiring harness as well as internal structures in devices (such as printed circuit boards in electronic control units or design technology and joining and bonding techniques in actuators) function as an antenna structure. Depending on their geometric dimensions and the electromagnetic wave frequencies, they receive transmitter signals and conduct them to the semiconductor components. Unmodulated and modulated high-frequency signals can be demodulated at the *pn* transitions in the semiconductor components. This can lead to level shifts caused by the direct component or to the superposition of transient interference signals as a result of demodulated LF components of the interference signal. The carrier frequency $f_{S,HF}$ is usually much higher than the useful-signal frequencies f_N. LF components of the interference signal are particularly critical if its frequency $f_{S,NF}$ is in the range of the useful-signal frequencies f_N. Interference signals with frequencies far lower than those of the useful signals can lead to disturbances due to intermodulations.

Electronic components must be designed in such a way that their function is not impaired by the externally generated interference signals in the electronic circuits.

EMC-oriented development

The importance of EMC-oriented development in automotive electronics is increasing at all levels (see V model). Nowadays, safeguarding of EMC in vehicles influences the design of semiconductor components and of modules (including suitable design technology and the joining and bonding techniques of integrated circuits for components and systems) as well as the configuration of the entire vehicle. Thus EMC-oriented design is an integral part of a state-of-the-art development process at all levels. Today, retroactively suppressing interference is usually no longer possible or, at the very least, extremely time consuming and costly.

EMC requirements analysis

When development of a new electronic system for a motor vehicle begins, necessary requirements must first be analyzed and codified in the customer specifications (Figure 5). In addition, the vehicle manufacturer compiles EMC requirements documents, which take into account legal requirements and expectations of vehicle customers. These EMC specifications contain the requirements for vehicles and the resulting requirements for electric systems and components that are being installed in the vehicles. From these requirements, the system or

Figure 5: Requirements analysis and EMC development based on the V model

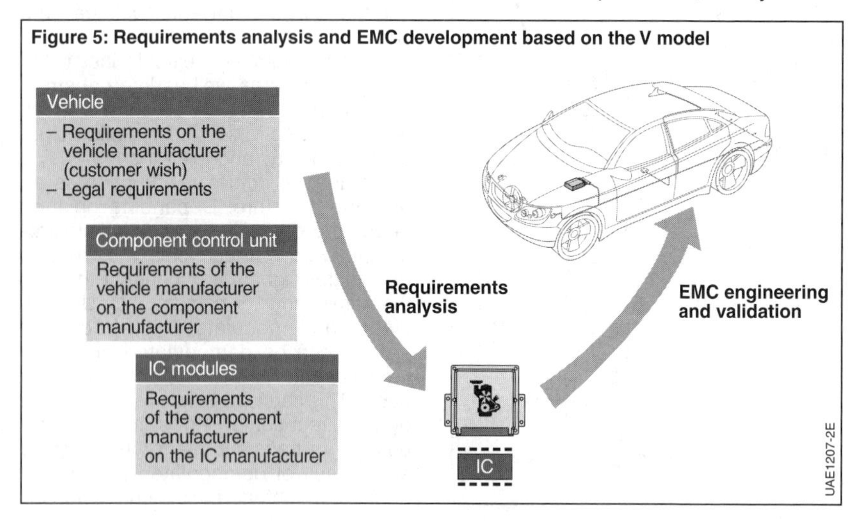

component manufacturer derives its own requirements for components and specifies requirements for design elements, electronic wiring and semiconductor components. Finally, IC and component manufacturers derive from this the requirements for their own products.

EMC development and verification

During the development stages, these requirements must be defined immediately at all levels in the customer specifications (Figure 5, see also V-model) and taken in to account early on during the vehicle design stage (e.g. the wire harness configuration and its position in the vehicle, power supply and ground concept, and the installing point of electronic components). The requirements derived from the vehicle must be taken into account, keeping in mind the system and component design (e.g. system topology, circuit design, housing design, circuit board layout, design and joining and bonding techniques). Similarly, the component manufacturer takes into account the EMC requirements during the IC and filter design stages. During the development process, the effectiveness of individual development steps is verified, in which various prototypes are metrologically assessed or variant studies using numerical simulation methods are implemented.

EMC validation

During the closing stages of EMC development, the requirements set out in the functional specifications and resulting customer specifications are validated in order to prove their worth. This validation is typically carried out using standardized test methods in accordance with legal mandates, standards and EMC specifications for component and vehicle manufacturers. The measurements to be carried out, operating statuses used and limit values to be observed are written down in test plans and the results are documented in a qualification report.

EMC measuring techniques

EMC measuring techniques are an important tool with regard to EMC-compliant development. Through the use of suitable test methods, the effectiveness of design standards like a selection of suitable semiconductor components, circuit design, circuit board layout, design and housing design are drafted and checked. On the other hand, the EMC test methods used in assessing are conducive to complying with EMC and legal requirements with regard to the release of components and vehicles.

A wide variety of test methods is used for testing interference immunity and interference emission. Depending on the methods used to assess interference phenomena, they can be roughly divided into methods operating in the time domain (pulse generators, oscilloscopes), and those operating in the frequency domain (sine-wave generators, test receivers, spectrum analyzers).

In measuring technology, interference signals are specified as relative quantities of interference emission in dB (decibels). The values for interference suppression (pulse amplitude, transmitter field strengths) are typically specified directly (Table 1).

Table 1: Measured values

Physical quantity	Reference variable	Unit	Calculations
	Interference emission		
Voltage L_U	1 µV	dB (µV)	$L_U = 20 \lg (U/1\ \mu V)$
Current L_I	1 µA	dB (µA)	$L_I = 20 \lg (I/1\ \mu A)$
Field strength L_E	1 µV/m	dB (µV/m)	$L_E = 20 \lg (E/1\ \mu V/m)$
power L_P	1 mW	dB (mW)	$L_P = 10 \lg (P/1\ mW)$
	Interference immunity		
Voltage U	–	V	–
Current I	–	A	–
Field strength E Field strength H	– –	V/m A/m	– –
Power P	–	W	–

EMC test methods

EMC test methods are described in relevant standards and divided into methods for the entire vehicle, for the components and systems (e.g. control units, sensors and actuators) and for integrated circuits (IC) and modules.

IC test methods

During the metrological assessment of integrated circuits (IC), the methods used are designed such that only the component itself is assessed instead of the combination of components with a peripheral circuit and larger conductor structures. The goal is to find out more about the EMC peformance of an IC, independent of its various applications, e.g. in order to compare different ICs of the same type. The standardized test methods for this are divided into conducted and irradiated test methods for interference emission (IEC 61967, see Table 3), interference immunity with respect to electromagnetic fields (IEC 62132, see Table 3) and test methods for sending and influencing pulses (IEC 62215, see Table 3) and ESD (electrostatic discharges). Figure 6 shows an example of a test circuit measuring conducted disturbances on individual IC pins.

Component measurement methods

In assessing the devices in the laboratory, conducted test methods involving radiation are used. Test objects are always operated under standardized conditions. The power is supplied via vehicle line impedance stabilization networks (LISN), which simulate a uniform wiring harness. The measurement configuration is typically completed on a laboratory table with a ground plane. The test object is connected to measurement peripherals, which simulate its functioning in a realistic manner. In order to ensure isolation from its surroundings, the test setup is run in a shielded room.

Conducted disturbances

High-frequency interference voltages at the supply lines are capacitively decoupled in the Line Impedance Stabilisation Network, and interference currents are measured using suitable current measurement coils (CISPR 25, see Table 2).

Figure 6: IC measurements, line-bound interference emission (IEC 61967-4)

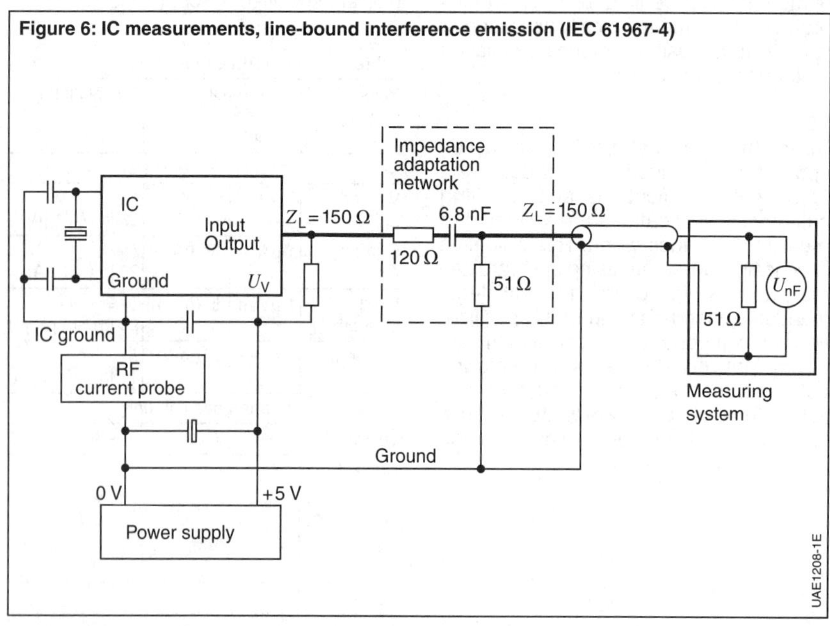

Immunity to pulsed interference is tested using special pulse generators which produce standardized test impulses in accordance with ISO 7637-2, Table 2. The coupling of pulsed interference on signal and control lines is reproduced using capacitive coupling clamps in accordance with ISO 7637-3. As with high-frequency interference emission measurements, the interference emission of pulses is measured in the standardized test setup through the use of relevant switches and oscilloscopes.

High-frequency interference coupling and interference emission

Interference immunity test

Coupling of electromagnetic waves for component measurements is achieved through the use of TEM waveguides like the stripline and the TEM cell (TEM, transverse electromagnetic mode) or the power coupling procedure BCI (Bulk Current Injection) and irradiation with antennas.

The principal measurement configuration always consists of a coupler for high frequencies in a shielded measuring room, a test object with a wiring harness, measurement peripherals, and devices that produce high frequencies and recording and processing measured values (Figure 7).

In the case of a stripline, (ISO 11452-5, Table 2) the wiring harness is arranged in line with the direction of propagation of the electromagnetic wave between a strip conductor and a base plate. When a TEM cell is used (ISO 11452-3), the test object and a section of the wiring harness are arranged at a right angle to the propagation direction of the electromagnetic waves. The BCI method (ISO 11452-4) involves superimposing an RF current (radio frequency) on the wiring harness by means of a current clamp.

When the test setup is irradiated, the transmitting antennas are set up at various locations. Through this process, both the coupling of electromagnetic fields in the wiring harness and the coupling in the test object are reproduced.

These and further test methods, such as one for testing low-frequency magnetic fields (ISO11452-8) and test methods for reproducing the interference coupling of mobile transmitters within close range (ISO 11452-9), are described in various portions of the ISO 11452 (Table 2).

Figure 7: Basic design of interference immunity tests

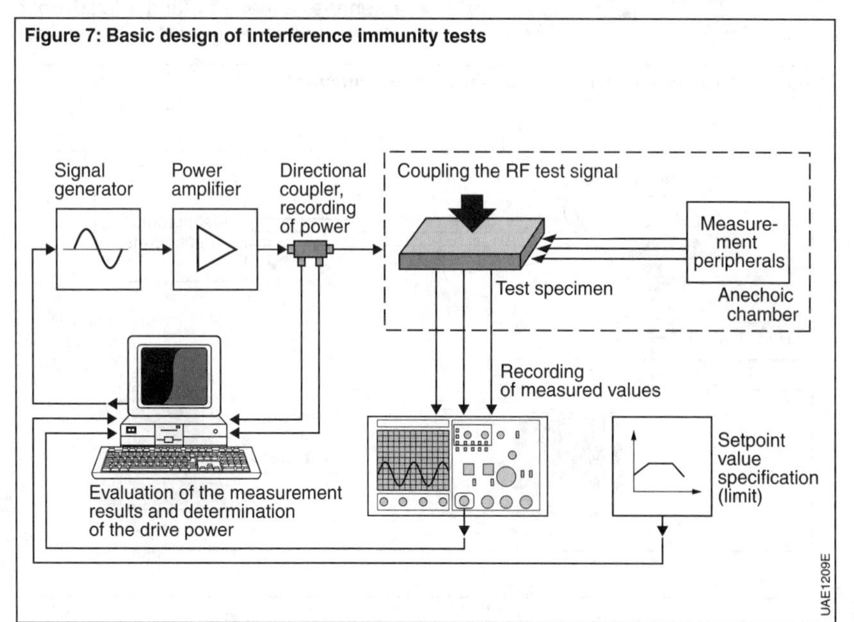

Measuring interference emission
The measuring principles for the immunity test can essentially be used to measure interference emission as well (CISPR 25). The TEM waveguides, current clamps and antennas function as receiving elements for the interference emitted by the test objects. For measuring the interference, the test receivers are connected directly to the receiving measuring instruments (Figure 8).

To decouple from the environment – that is, to ensure that only the interference produced by the test specimen is actually measured during interference emission measurements – and to minimize the emission of high-frequency signals in the environment when interference immunity tests are in progress, the testing for high-frequency disturbance values is carried out in electromagnetically shielded rooms. In order to avoid reflections and room resonance, shielded rooms are lined with high frequency absorbers when radiated signals are used.

Test methods for electrostatic discharges
When assessing the interference immunity of ESD, special high voltage impulse generators are used. In these generators, the ESD impulse is reproduced by charging a smoothing capacitor and a targeted discharge via a discharge resistor. The capacity of the smoothing capacitor and the discharge resistor determine the energy output and the pulse shape. With a suitable discharge probe, the ESD pulse is applied to the coupling point, for example a control unit pin, either by a targeted rollover or after contact via a discharge switch in the alternator.

Vehicle testing procedures
Measuring interference immunity
The immunity of electronic systems to electromagnetic fields radiated by high-power transmitters is tested on vehicle level in a special anechoic chambers (Figure 9). Here, high electrical and magnetic field strengths can be generated, and the whole vehicle can be exposed to them (ISO 11451, Table 2).

An interference emission of the entire vehicle is carried out with external antennas, either in standardized open area test sites or in anechoic chambers

Figure 8: Basic design of interference emission measurement

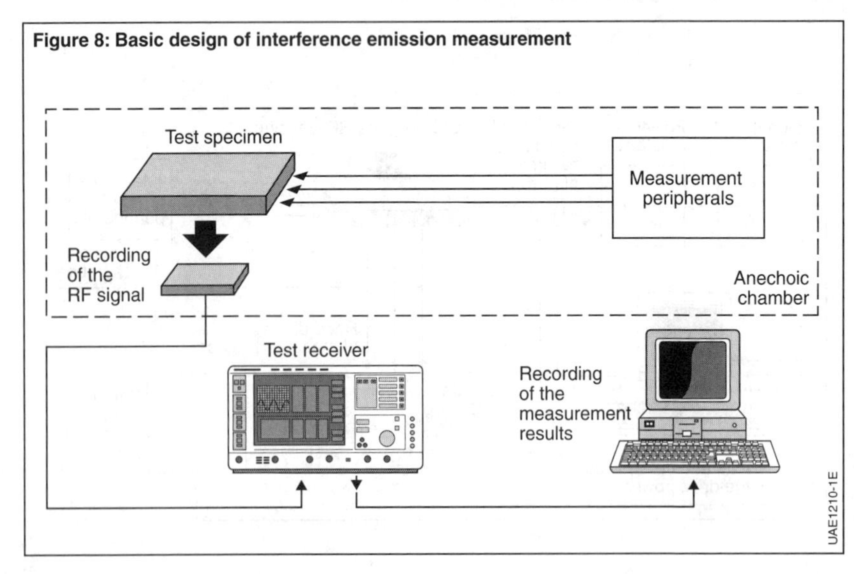

(IEC/CISPR 12, Table 2). The interference effects of the vehicle electrics and electronics on radio reception in the vehicle are measured using highly sensitive test receivers or spectrum analyzers. As far as possible, the original vehicle antenna is left in place, and measurements are taken at the receiver input terminal inside the vehicle in order to collect accurate data. With the help of a suitable test circuit, the impedance of the test receiver is adjusted to the input impedance of the receiver unit. The necessary limit values and measurement parameters like bandwidth and high frequency detectors are derived from the operating parameters of the radio services (IEC/CISPR 25).

Selection of EMC tests

The EMC tests carried out on an electric or electronic device depend on the component's area of application and its internal design. Simple electromechanical devices, which do not contain electronic components, do not have to be tested for interference immunity to electromagnetic fields. On the other hand, a variety of individual tests must be specified and implemented in EMC test plans for components containing electronic parts.

Figure 9: Interference immunity measurement in an EMC anechoic chamber for vehicles

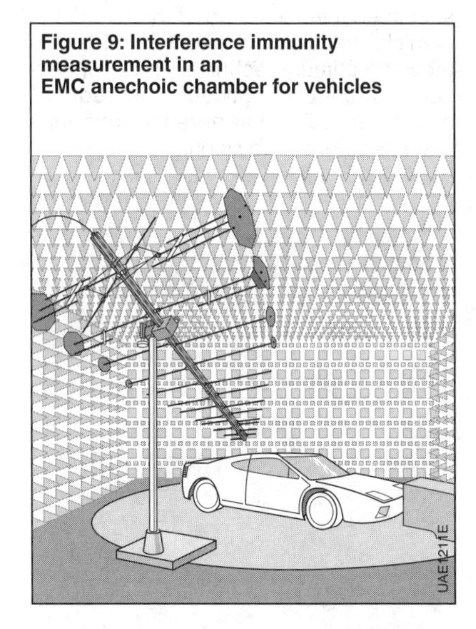

EMC simulation

Nowadays, the application of numerical calculation methods is a method for the EMC development process equivalent to EMC measuring technology. Different calculation methods are employed depending on the problem at hand. What is critical in each case is accurate modeling of the sources of interference, potentially susceptible devices and the coupling path.

If the conducted coupling is dominant or the design of the electrical circuit (for example, filter structures) is to be examined, circuit simulation programs are usually used. If the field-coupled influence is dominant or geometric configurations (such as a metallic structure for shielding the electromagnetic fields or antenna geometry) are to be examined, various methods are used to calculate electromagnetic fields.

Thus the user has to select the method to be used depending on the problem to be solved. The EMC simulation can be used both when developing and assessing individual components, for designing components and systems, and for examining the entire vehicle, for example, to determine the optimum location for an antenna.

A significant advantage of the EMC simulation is that a wide variety of variants (for example, using different components or various geometric configurations) can be examined and compared without having to build and measure a design model for each one.

Legal requirements and standards

EMC type approval

The form of electromagnetic compatibility for motor vehicles is prescribed by law such that, in addition to other requirements (for example, for brakes, light, exhaust gas), requirements concerning interference immunity to electromagnetic fields and regarding the maximum permissible interference emission (interference suppression) also have to be fulfilled for the type approval of vehicles. The currently applicable directive UNECE R10, Revision 4 from 2012 [1] is based on a directive from 1972, which merely contained requirements regarding interference suppression for the vehicle. Since then, this directive has been repeatedly adapted and supplemented to account for technical progress.

The current edition takes into account not only the requirements for conventional motor vehicles, but also requirements for electric and hybrid vehicles with built-in charging devices. Since these vehicles come into contact with the public power grid, there must be assurance of appropriate interference immunity to line-bound interference from the power grid and a corresponding limit to the interference from the vehicle into the power grid.

Two different paths can be taken for the type approval of vehicles. Usually, vehicle manufacturers apply for type approval via type test for the entire vehicle. However, it is also possible to obtain a type approval for electrical and electronic submodules. In this case, in addition to the limit values for interference immunity to electromagnetic fields and the interference radiation, requirements regarding the pulsed interference in the vehicle electrical system also have to be fulfilled.

In addition to regulations governing the type approval procedure for vehicles or components, UNECE R10 also specifies test methods and maximum limits. The test methods are based on the applicable international standards (Table 2); the limit values are explicitly specified in the directive.

The defined limit values represent minimum requirements. In practice, just barely complying with the limit values specified is frequently insufficient to ensure interference-free reception or mobile communication in the vehicle. Depending on their vehicle concepts, therefore, the motor vehicle manufacturers specify in their own customer specifications both a raised requirement for interference immunity and a lower interference emission level for protecting radio reception. For the development of electric and electronic components, therefore, these requirements have to be taken into account from the start by consultation between the vehicle manufacturers and suppliers of vehicle electrical systems.

Standards

The individual, different test methods are described in the international ISO and IEC/CISPR standards according to Table 2 for vehicles and their electric/electronic components and systems. The IEC standards according to Table 3 treat the measuring methods for integrated circuits and semiconductor components. Usually the standards do not specify fixed limit values, but limit value classes. Depending on the individual vehicle concepts, this should enable the users of the standards (vehicle manufacturers and suppliers) to provide optimum technical and economical coordination between the requirements regarding interference immunity and interference emission.

Table 2: International standards for motor vehicles and components

Interference immunity

Designation	Title
Road vehicles – Electrical disturbances from conduction and coupling	
ISO 7637-1	Part 1: Definitions and general considerations
ISO 7637-2	Part 2: Electrical transient conduction along supply lines only
ISO 7637-3	Part 3: Vehicles with nominal 12 V or 24 V supply voltage – Electrical transient transmission by capacitive and inductive coupling via lines other than supply lines
Road vehicles – Vehicle test methods for electrical disturbances from narrowband radiated electromagnetic energy	
ISO 11451-1	Part 1: General principles and terminology
ISO 11451-2	Part 2: Off-vehicle radiation sources
ISO 11451-3	Part 3: Onboard transmitter simulation
ISO 11451-4	Part 4: Bulk current injection (BCI)
Road vehicles – Component test methods for electrical disturbances from narrowband radiated electromagnetic energy	
ISO 11452-1	Part 1: General principles and terminology
ISO 11452-2	Part 2: Absorber-lined shielded enclosure
ISO 11452-3	Part 3: Transverse electromagnetic mode (TEM) cell
ISO 11452-4	Part 4: Bulk current injection (BCI)
ISO 11452-5	Part 5: Stripline
ISO 11452-7	Part 7: Direct radio frequency (RF) power injection
ISO 11452-8	Part 8: Immunity to magnetic fields
ISO 11452-9	Part 9: Portable transmitters
ISO 11452-10	Part 10: Immunity to conducted disturbances in the extended audio frequency range
ISO 11452-11	Part 11: Reverberation chamber
ESD – Electrostatic discharge	
ISO 10605	Road vehicles – Test methods for electrical disturbances from electrostatic discharge

Interference emission

Designation	Title
Vehicles, boats and internal combustion engines	
IEC/CISPR 12	Radio disturbance characteristics – Limits and methods of measurement for the protection of off-board receivers
IEC/CISPR 25	Limits and methods of measurement for the protection of on-board receivers

Table 3: International standards for semiconductor components (IC)

Pulses

Designation	Title
Integrated circuits – Measurement of impulse immunity	
IEC/TS 62215-2	Part 2: Synchronous transient injection method
IEC 62215-3	Part 3: Non-synchronous transient injection method

Interference immunity

Designation	Title
Integrated circuits – Measurement of electromagnetic immunity	
IEC 62132-1	150 kHz to 1 GHz – Part 1: General conditions and definitions
IEC 62132-2	Part 2: Measurement of radiated immunity – TEM cell and wideband TEM cell method
IEC 62132-3	150 kHz to 1 GHz – Part 3: Bulk current injection (BCI) method
IEC 62132-4	150 kHz to 1 GHz – Part 4: Direct RF power injection method
IEC 62132-5	150 kHz to 1 GHz – Part 5: Workbench Faraday cage method
IEC 62132-6	150 kHz to 1 GHz – Part 6: Local injection horn antenna (LIHA) method
IEC 62132-8	Part 8: Measurement of radiated immunity – IC stripline method
IEC/TS 62132-9	Part 9: Measurement of radiated immunity – Surface scan method

Interference emission

Designation	Title
Integrated circuits – Measurement of electromagnetic emissions	
IEC 61967-1	150 kHz to 1 GHz – Part 1: General conditions and definitions
IEC/TR 61967-1-1	Part 1: General conditions and definitions – Near-field scan data exchange format
IEC 61967-2	150 kHz to 1 GHz – Part 2: Measurement of radiated emissions – TEM cell and wideband TEM cell method
IEC/TS 61967-3	150 kHz to 1 GHz – Part 3: Measurement of radiated emissions – Surface scan method
IEC 61967-4	150 kHz to 1 GHz – Part 4: Measurement of conducted emissions; 1 Ohm/150 Ohm direct coupling method
IEC/TR 61967-4-1	150 kHz to 1 GHz – Part 4-1: Measurement of conducted emissions – 1 Ohm/150 Ohm direct coupling method – Application guidance to IEC 61967-4
IEC 61967-5	150 kHz to 1 GHz – Part 5: Measurement of conducted emissions; Workbench Faraday cage method
IEC 61967-6	150 kHz to 1 GHz – Part 6: Measurement of conducted emissions – Magnetic probe method
IEC 61967-8	Part 8: Measurement of radiated emissions – IC stripline method

Additional references

[1] UNECE R10, Revision 4: Uniform provisions concerning the approval of vehicles with regard to electromagnetic compatibility

Symbols and circuit diagrams

Circuit symbols

Symbols and circuit diagrams

The electrical systems in motor vehicles contain a large number of electric and electronic devices for controlling and managing the engine, and for safety and comfort and convenience systems. An overview of the complex vehicle electrical system circuits is only possible with meaningful symbols and circuit diagrams. Circuit diagrams such as schematic diagrams and terminal diagrams help in troubleshooting, simplify the installation of additional devices, and facilitate fault-free connection when retrofitting or altering the electrical equipment in motor vehicles.

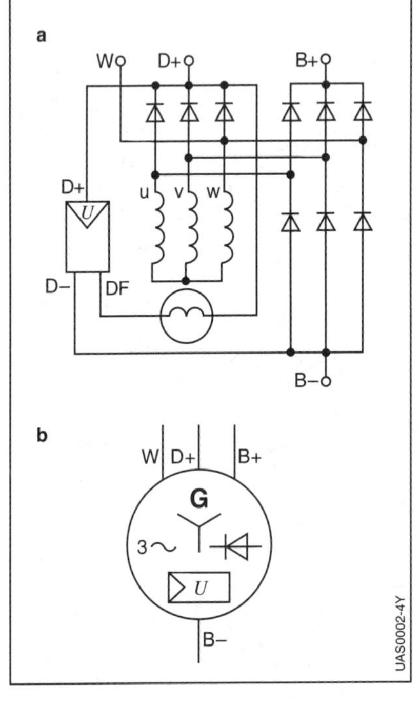

Figure 1: Circuit diagram and circuit symbol of an alternator with regulator
The circuit symbols contains, in addition to the symbol for the alternator/generator G, the symbols for the three windings (phases), the star connection, the diodes, and the regulator.

a) With internal circuitry,
b) Circuit symbol.

Circuit symbols

Standards

The circuit symbols shown in Table 1 are a selection of standardized circuit symbols which are suitable for automotive electrics. But for a few exceptions, they correspond to the standards of the International Electrotechnical Commission (IEC).

The European Standard EN 60617 (Graphical symbols for diagrams, [1]) corresponds to the International Standard IEC 60617. It exists in three official versions (German, English and French). The standard contains symbol elements, identifiers and above all circuit symbols for the following areas: General applications (Part 2), Conductors and connecting devices (Part 3), Basic passive components (Part 4), Semiconductors and electron tubes (Part 5), Production and conversion of electrical energy (Part 6), Switchgear, controlgear and protective devices (Part 7), Measuring instruments, lamps and signalling devices (Part 8), Telecommunications: Switching and peripheral equipment (Part 9), Telecommunications: Transmission (Part 10), Architectural and topographical installation plans and diagrams (Part 11), Binary logic elements (Part 12) and Analogue elements (Part 13).

Requirements

Circuit symbols are the smallest elements in a circuit diagram and the simplified graphical representation of an electrical device or a part thereof. The circuit symbols show the operating concept of a device and represent in circuit diagrams the functional correlations of a technical sequence. Circuit symbols do not take into consideration the shape and dimensions of the device and the position of the connections on the device. A detached representation in the schematic diagram is possible by abstraction only.

A circuit symbol should have the following properties: It should be easily remembered, easily comprehensible, uncomplicated in its graphic representation, and clearly within a classification group.

Circuit symbols consist of circuit-symbol elements and qualifying symbols. The following are examples of qualifying symbols: letters, numbers, symbols, mathematical signs and symbols, unit symbols, and characteristic curves.

If a circuit diagram becomes too elaborate due to the representation of the internal circuitry of a device (Figure 1a) or if not all the details of the circuit are needed to identify the function of the device, the circuit diagram for this special device can be replaced by a single circuit symbol (without internal circuitry) (Figure 1b).

In the case of integrated circuits which demonstrate a high degree of economy of space (this is synonymous with large scale integration of functions in a component), a simplified circuit representation is preferred.

Representation

The circuit symbols are shown without the effect of a physical quantity, i.e. in a de-energized and mechanically non-actuated state. An operating state of a circuit symbol that deviates from this standard representation (basic status) is denoted by an adjacent double arrow (Figure 2).

Circuit symbols and connecting lines (representing electric lines and mechanical linkages) have the same line width.

To avoid unnecessary kinks and crossings of the connecting lines, it is possible to rotate circuit symbols in 90° increments or show a mirror image of them, provided this does not alter their meaning. The direction of continuing lines can be freely selected. Exceptions are the circuit symbols for resistors (connection symbols are only permitted here on the narrow sides) and connections for electromechanical drives (connection symbols are only permitted here on the wide sides, Figure 3).

Junctions are represented both with and without a dot. Where there are crossings without a dot, there is no electrical connection. Connecting points on devices are for the most part not specifically represented. Connecting point, plug, jack or screwed connections are identified by a circuit symbols only at the points required for installation and removal. Other connecting points are identified as standard by dots.

Contact elements with a common drive are identified in an assembled representation in such a way that on actuation they follow a direction of motion which is established by the mechanical linkage (– – –) (Figure 4).

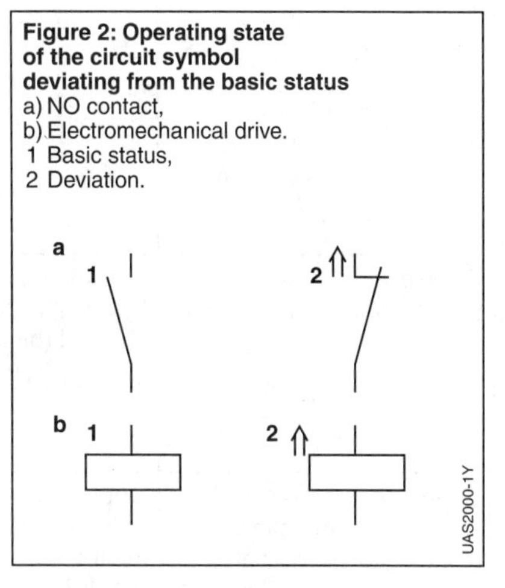

Figure 2: Operating state of the circuit symbol deviating from the basic status
a) NO contact,
b) Electromechanical drive.
1 Basic status,
2 Deviation.

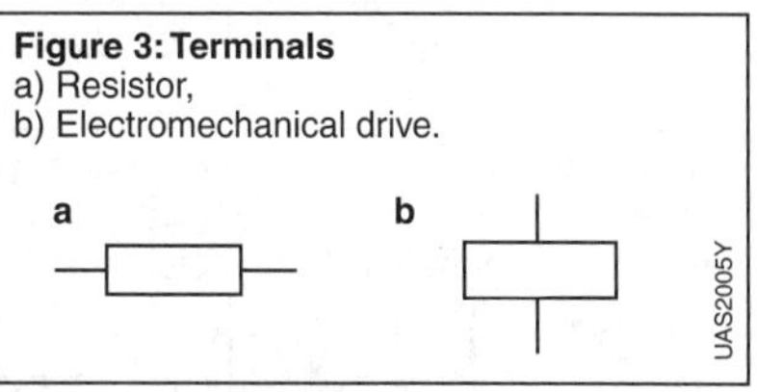

Figure 3: Terminals
a) Resistor,
b) Electromechanical drive.

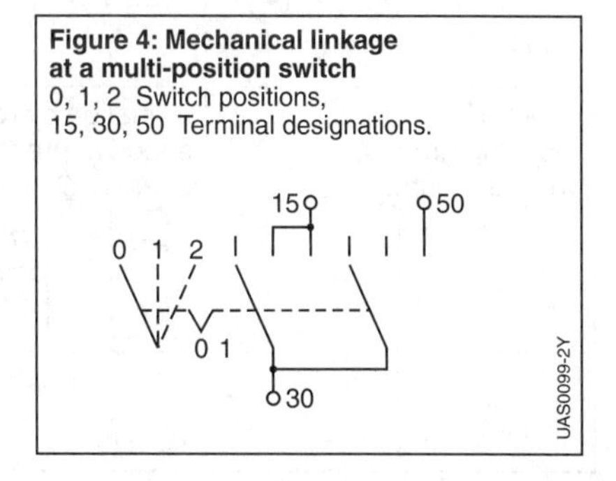

Figure 4: Mechanical linkage at a multi-position switch
0, 1, 2 Switch positions,
15, 30, 50 Terminal designations.

Table 1: Selection of circuit symbols as per EN 60617

Connections

Line; line intersection, without/with a connection

Line; line intersection, without/with a connection

Mechanical linkage; electrical conductor (laid at later stage)

Crossovers (without/with connections)

Connection, general; separable connection (if indication necessary)

Plug connection; socket; plug; three-plug connection

Ground (housing ground, vehicle ground)

Mechanical function

Switch positions (home position: solid line)

0 1 2

0 1 2

Activation is manual, by sensor (cam), thermal (bimetal)

Detent; non-automatic/automatic return in direction of arrow (button)

Actuator, general (mech., pneum., hydraul.); piston actuation

Actuation by rotational speed n, pressure p, quantity Q, time t, temp. $t°$

n p Q t $t°$

Variability/adjustability, not intrinsic (external), general

Variability/adjustability, intrinsic, caused by applied physical variable, linear/non-linear

Variability/adjustability, general

Switches

Pushbutton switch, NO/NC contact

Detent switch, NO/NC contact

Changeover contact, non-bridging/bridging

UAS1245E

Switches	Various components	
Two-way normally open contact with three positions (e.g. turn-signal indicator switch) 	Actuators with one winding 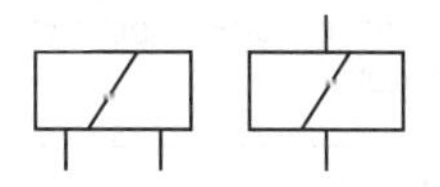	Resistor 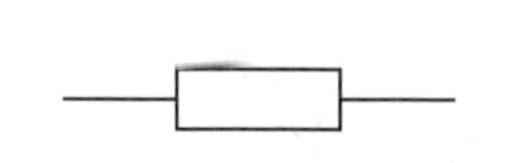
NO/NC contact	Actuator with two windings acting in same direction 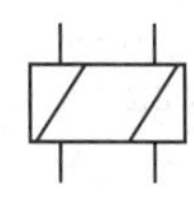	Potentiometer (with three connections) 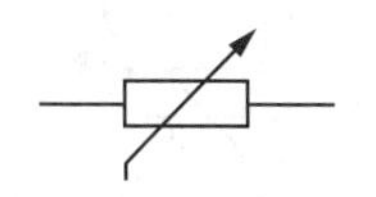
Contact with two makes 	Actuator with two windings acting in opposing directions 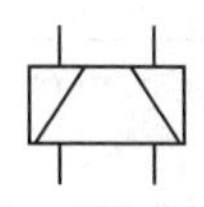	Heating resistor, glow plug, flame plug, heated window 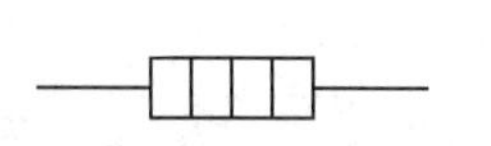
Multiple-position switch 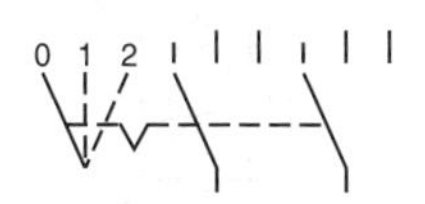	Electrothermal actuator, thermal relay 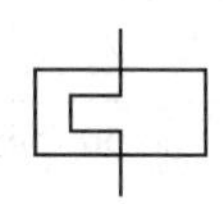	Antenna 
Cam-operated switch (e.g. contact breaker) 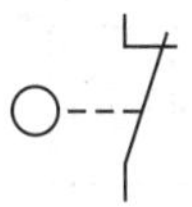	Electrothermal actuator, linear solenoid 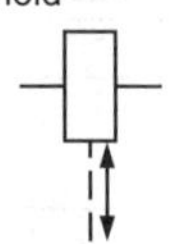	Fuse 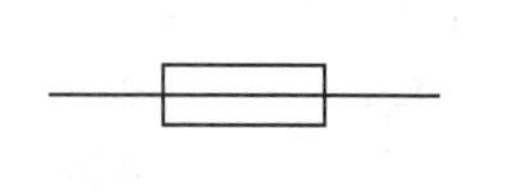
Thermostatic switch 	Solenoid valve, closed 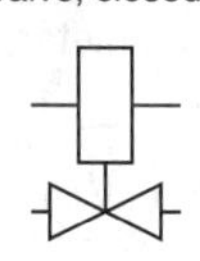	Permanent magnets 
Release/trip device 	Relay (actuator and switch), example: non-delayed-break NC contact and delayed-break NO contact 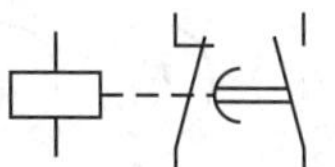	Winding, inductive 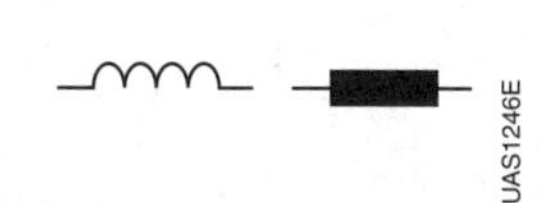

UAS1246E

Various components

PTC resistor

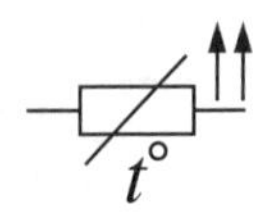

NTC resistor

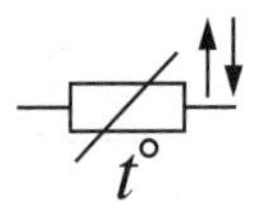

Diode, general, current in direction of triangle tip

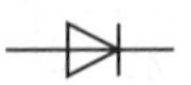

PNP transistor
NPN transistor

E = Emitter (arrow points in direction of flow)
C = Collector, positive
B = Base (horizontal), negative

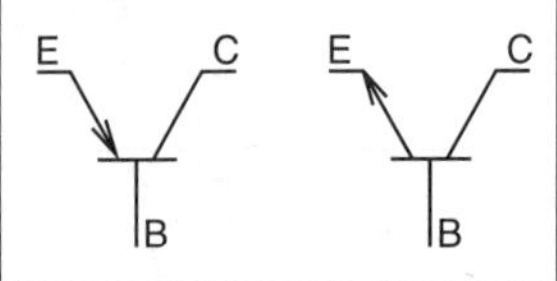

Light-emitting diode (LED)

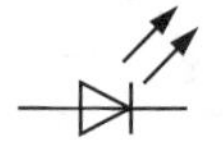

Hall generator

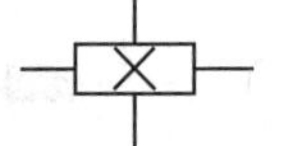

Devices in motor vehicle

Dotted/dashed line used to delineate or group together associated circuit sections

Shielded device, dashed line connected to ground

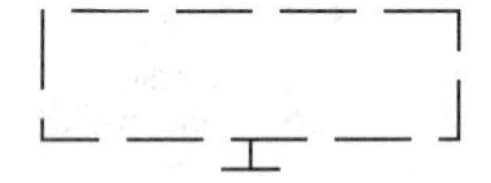

Regulator, general

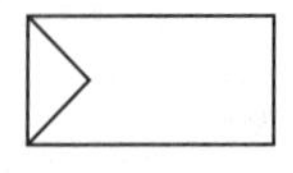

Electronic control units (ECUs)

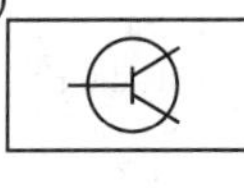

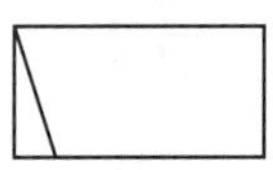

Indicating instrument, general; voltmeter; clock

Rotational-speed indicator; temperature indicator; linear-speed indicator

Battery

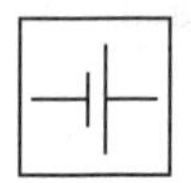

Plug-and-socket connection

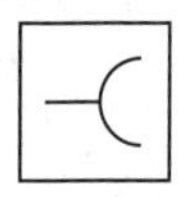

Light, headlamp

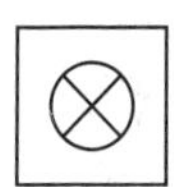

Horn, fanfare horn

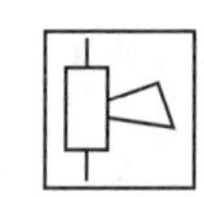

Heated rear window (general heating resistor)

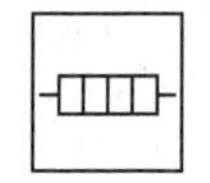

Switch, general, without indicator lamp

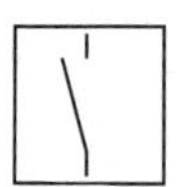

Switch, general, with indicator lamp

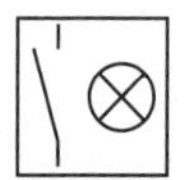

UAS1247E

Devices in motor vehicle

Pressure switch

Relay, general

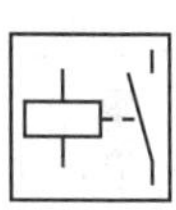

Solenoid valve, fuel injector, cold-start valve

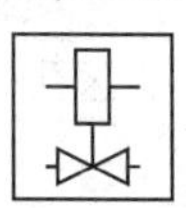

Thermo-time switch

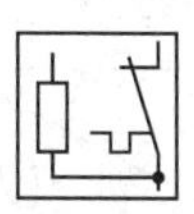

Throttle-valve switch

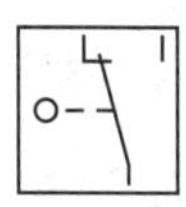

Rotary actuator

Auxiliary-air valve with electrothermal actuator

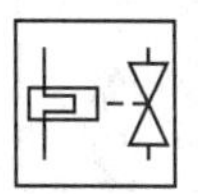

Spark plug

Ignition coil

Ignition distributor, general

Voltage regulator

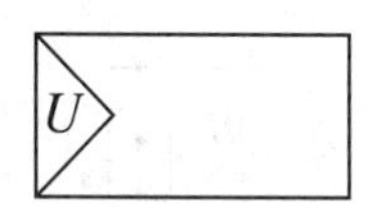

Alternator with regulator (without/with internal circuitry)

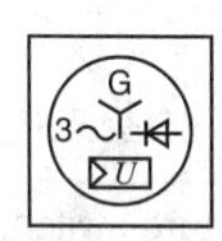

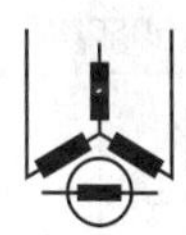

Electric fuel pump, engine drive for hydraulic pump

Motor with blower, fan

Starter motor with engagement relay (without/with internal circuitry)

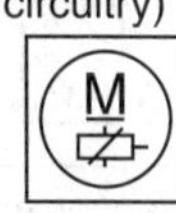

Wiper motor (one/two wiper speeds)

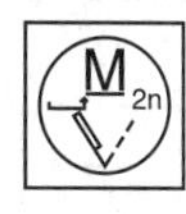

Intermittent-wiper relay

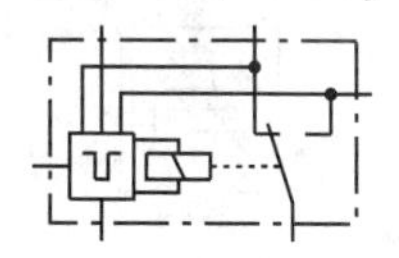

Car radio

UAS1248E

Devices in motor vehicle

Loudspeaker

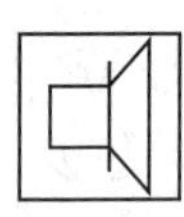

Piezoelectric sensor

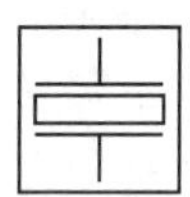

Linear-speed sensor

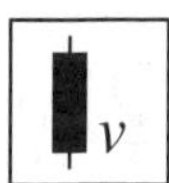

Voltage stabilizer, stabilizer

Resistance sensor

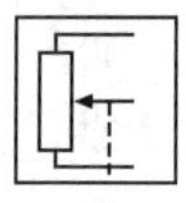

ABS rotational-speed sensor

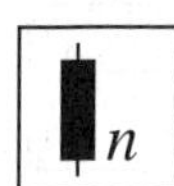

Inductive sensor, controlled with reference mark

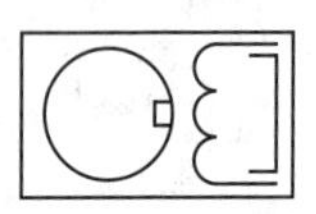

Air-flow sensor

Hall sensor

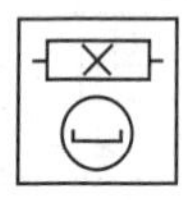

Turn-signal flasher, pulse generator, intermittent relay

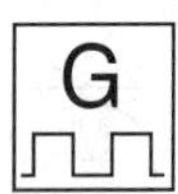

Air-mass meter

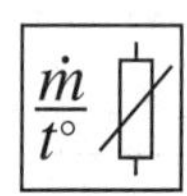

Converter (rate, voltage)

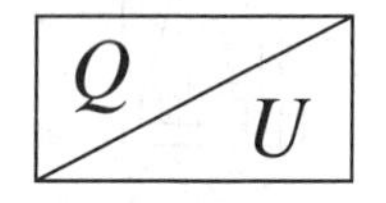

Lambda oxygen sensor (unheated/heated)

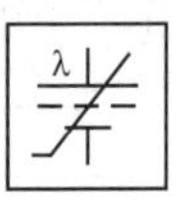

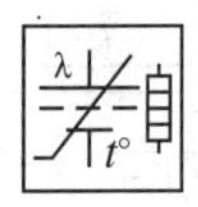

Flow-quantity sensor, fuel-level sensor

Temperature switch, temperature sensor

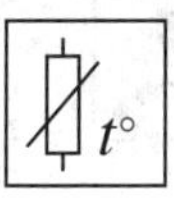

Inductive sensor

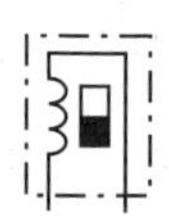

Instrument-cluster device (dashboard)

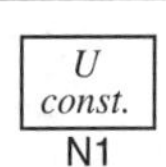
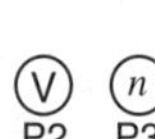
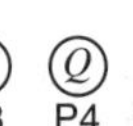

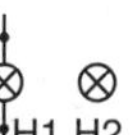

Circuit diagrams

Circuit diagrams are idealized representations of electrical devices, rendered in the form of symbols. Such diagrams also include illustrations and simplified design drawings as needed. A circuit diagram illustrates the relationship between the various devices and shows how they are connected to each other. A circuit diagram may be supplemented by tables, graphs, or descriptions. The type of circuit diagram actually used is determined by its particular purpose (e.g. illustrating the operation of a system), and by the way in which the circuit is represented (Figure 5).

For a circuit diagram to be "legible", it must satisfy the following requirements:

- It must comply with the requirements of the appropriate standards, and deviations must be explained.
- Current paths should be arranged so that current flow or mechanical action takes place from left to right or from top to bottom.

In automotive electrics, block diagrams with single inputs and outputs and with internal circuitry omitted provide a swift overview of the function of a system or a device. The schematic diagram in different methods of representation (layout of the circuit symbols) is the detailed representation of a circuit to identify the function and to carry out repairs. The terminal diagram (with terminal locations of the devices) is used by the Aftersales Service when replacing or retrofitting devices.

According to the method of representation, a distinction is made between

- single- or multi-line representation, and (according to the layout of the circuit symbols)
- assembled, semi-assembled, detached and topographical representation, which can be combined in one and the same circuit diagram.

Figure 5: Classification of the circuit diagrams (in accordance with EN 81346, Part 1 [2])

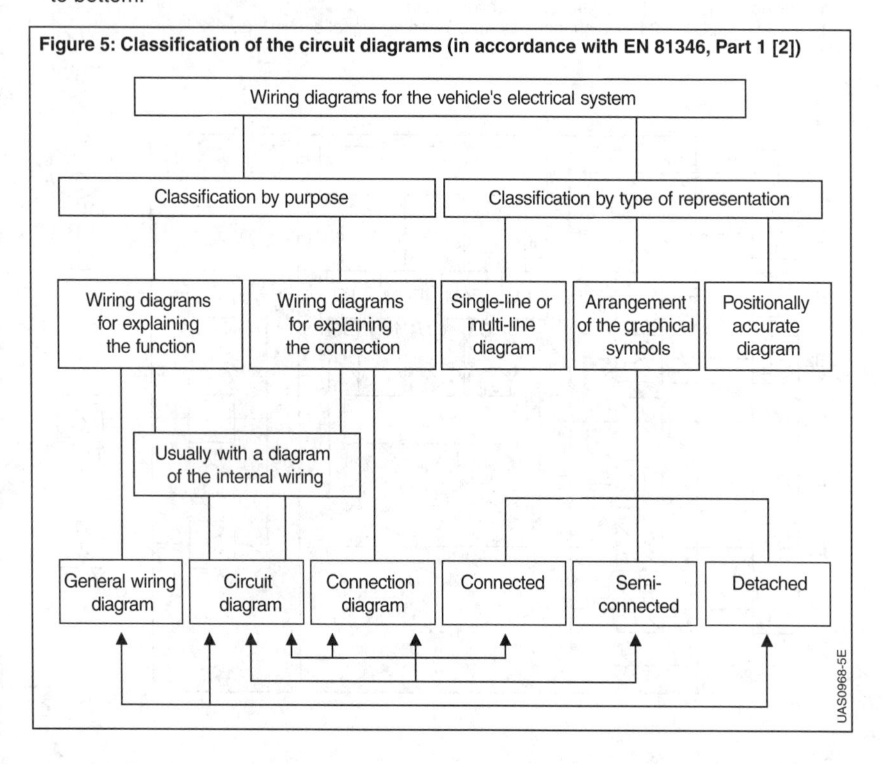

Block diagram

The block diagram is the simplified representation of a circuit, where only the essential parts are taken into consideration (Figure 6). It is intended to provide a quick overview of the task, design, structure and function of an electrical system or a part thereof, and to serve as a guide for more detailed circuit documentation (schematic diagram).

The devices are depicted by squares, rectangles and circles with qualifying symbols inside them similar to EN 60617, Part 2. The lines are for the most part drawn as single lines.

Schematic diagram

A schematic diagram is the detailed representation of a circuit. By clearly depicting individual current paths, it also indicates how an electrical circuit operates. In a schematic diagram, the clear presentation of the circuit's operation, which makes the diagram easy to read, must not be interfered with by the presentation of the individual circuit components and their spatial relationships. Figure 7 shows the schematic diagram of a starter motor in assembled and detached representations.

Figure 6: Block diagram of a Motronic ECU as an example
A1 ECU,
B1 Sensor for engine speed, B2 Sensor for reference mark,
B3 Sensor for air mass, B4 Sensor for intake-air temperature,
B5 Sensor for engine temperature, B6 Throttle-valve switch,
D1 Computing unit (CPU), D2 Address bus, D3 Main memory (RAM),
D4 Program data memory (ROM), D5 Input – Output,
D6 Data bus, D7 Microcomputer,
G1 Battery, K1 Pump relay, M1 Electric fuel pump,
N1 to N3 Driver stages,
S1 Ignition/starter switch, S2 Map changeover switch,
T1 Ignition coil, U1, U2 Pulse shapers, U3 to U6 Analog-digital converters,
Y1 Fuel injector.

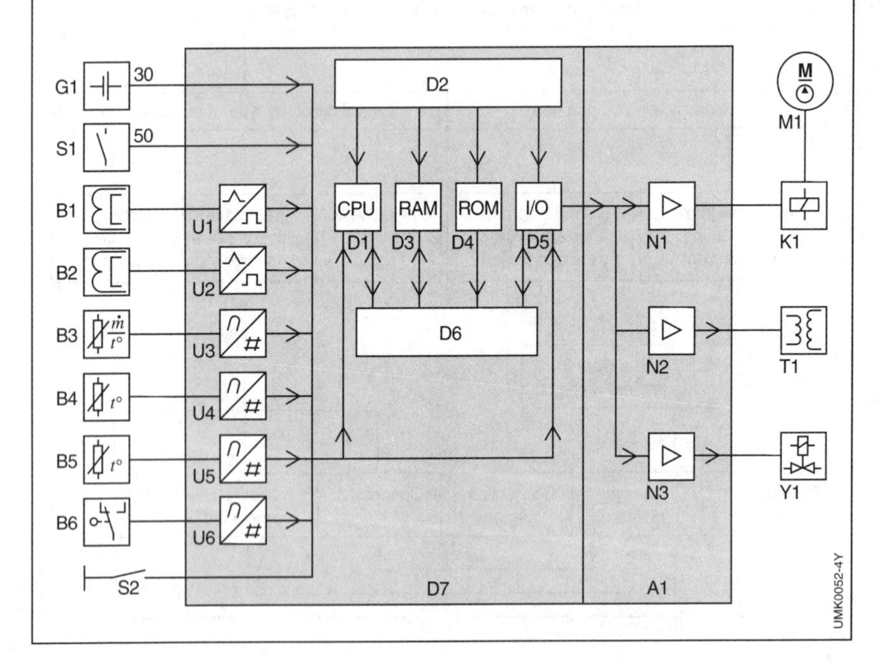

The schematic diagram must contain:
- Circuit
- Device identification (EN 81346, Part 2, [2]), and
- Pin designation and terminal designation (DIN 72552, [3]).

The schematic diagram can contain:
- A complete representation with internal circuitry in order to facilitate testing, fault localization, maintenance and replacement (retrofitting).
- Reference designations to better locate circuit symbols and destinations, particularly in a detached representation.

Representation of the circuit
Mostly multi-line representation is used in the schematic diagram. For the layout of the circuit symbols, there are, in accordance with EN 81346 Part 1, the following methods of representation which can be combined in the same circuit diagram.

Assembled representation
All the parts of a device are directly represented together assembled and connected to each other by a double slash or discontinuous connecting lines to identify the mechanical linkage. This representation can be used for simple, not very com-

Figure 7: Schematic diagram of a KB-type starter motor for parallel operation using two methods of representation
a) Assembled representation,
b) Detached representation.
K1 Control relay, K2 Engagement relay, holding winding and pull-in winding,
M1 Starter motor with series and shunt windings.
30, 30f, 31, 50b Terminal designations.

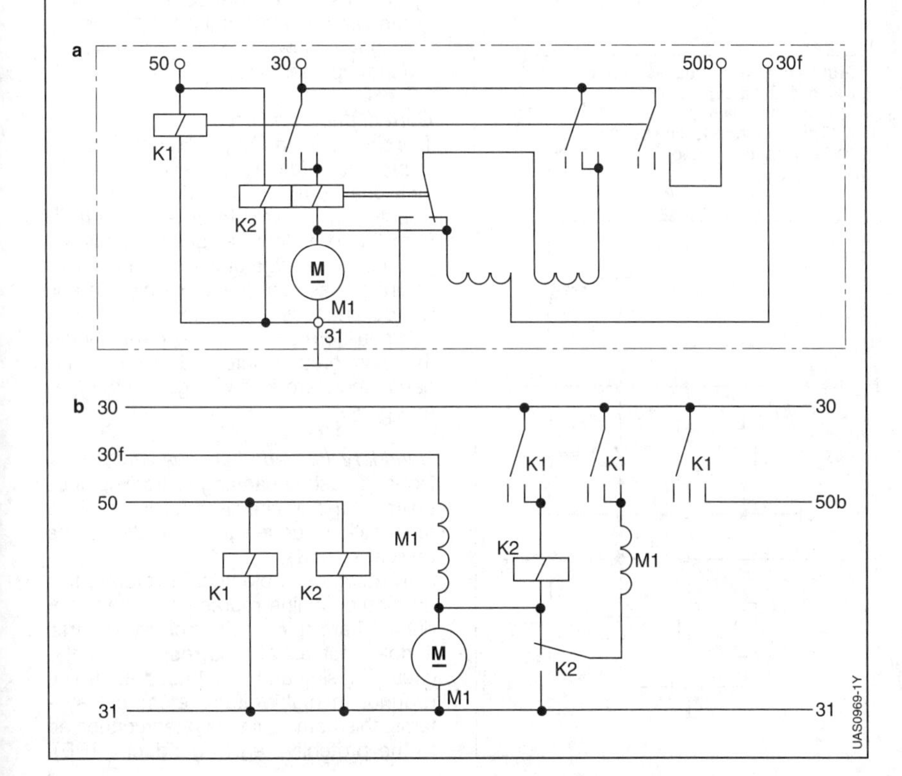

prehensive circuits without a loss of clarity (Figure 7a).

Detached representation

Circuit symbols of parts of electrical devices are represented separately and arranged in such a way that each current path can be traced as easily as possible (Figure 7b). The spatial association of individual devices or their parts is not taken into consideration. The priority is for the layout of the current paths to be as linear, clear and crossing-free as possible. The main purpose of this representation is to identify the function of a circuit.

The association of the individual parts must be identified using an identification system in accordance with EN 81346, Part 2. The identification associated with a device is located on each individual, separately represented circuit symbol for that device. Devices shown detached must be indicated once complete and assembled at a point in the circuit diagram if this is necessary to understand the circuit.

Topographical representation

In this representation, the position of the circuit symbol entirely or partially corresponds to the spatial position within the device or part.

Ground representation

For most motor vehicles, the single-conductor system, in which ground (metal parts of the vehicle) acts as the return conductor, is preferred for its simplicity. If a perfectly conducting connection of the individual ground parts is not guaranteed or if voltages in excess of 42 V are involved, the return conductor is also laid insulated from ground.

All the ground symbols represented in a circuit are electrically connected to each other via device or vehicle ground.

All the devices which contain a ground symbol must be connected to vehicle ground and be electrically conductive.

Figure 8 shows different possibilities of ground representation.

Current paths and lines

The circuits are arranged in such a way as to produce and clear representation. The individual current paths, with the effective direction preferably from left to right or from top to bottom, should generally run parallel to the edge of the circuit diagram as linearly as possible, crossing-free and without changes of direction.

When there is a bundling of parallel lines, these are grouped into units of three lines each, and there follows a gap to the next group.

Boundary lines and framings

Dot-and-dash separating or framing lines delimit parts of circuits in order to show the functional or design association of the devices or parts.

In automotive electrical systems, this dot-and-dash line represents a non-conductive framing of devices or circuit parts; it does not always correspond to the circuit housing and is not used as device ground. In high-voltage electrical systems, this framing line is often connected to the protective earth conductor (PE),

Figure 8: Ground representation
a) Individual ground symbols,
b) Ground continuity,
c) With ground collecting point.
31 Terminal designation.

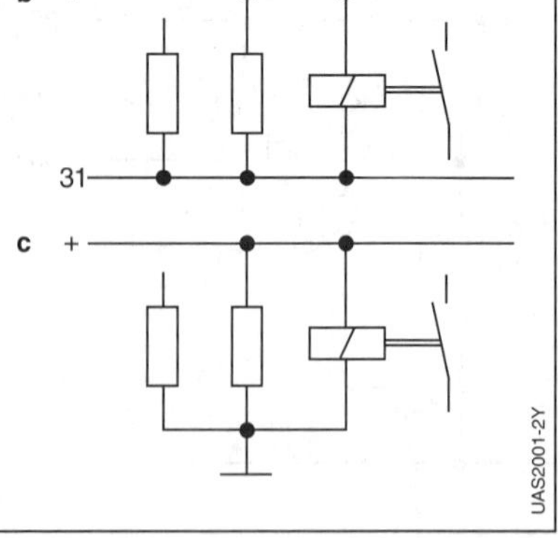

UAS2001-2Y

which is also indicated by a dot-and-dash line.

Break points, identifier, destination designation
Connecting lines (lines and mechanical linkages) which run over a longer distance of the schematic diagram can be interrupted in order to improve clarity. Only the beginning and the end of the connecting line are represented. The association of these break points must be clearly identifiable. The identifier and the destination designation are used for this purpose (Figure 9).

The identifier at associated break points matches. Terminal designations in accordance with DIN 72552 (Figure 9a), the specification of the operating concept and specifications in the form of alphanumeric characters serve as the identifier.

The destination designation is enclosed in parentheses so that it is not confused with the identifier; it consists of the section number of the destination (Figure 9b).

Section designation
The section designation indicated at the top edge of the diagram (formerly called the current path) is used to locate circuit parts. This designation can take three different forms:

- consecutive numbers at the same intervals from left to right (Figure 10a),
- reference to the contents of the circuit sections (Figure 10b),
- or a combination of the two (Figure 10c).

Labeling
Devices, parts or circuit symbols are identified in circuit diagrams with a letter and a number in accordance with EN 81346, Part 2. This identification is placed on the left or underneath the circuit symbol.

The qualifying symbol specified in the standard for the type of devices can be omitted if this does not give rise to ambiguity.

In the case of nested devices, a device is a component part of another, e.g. starter M1 with built-in engagement relay K6. The device designation is then: – M1 – K6.
Designation of associated circuit symbols in a detached representation: Each individual, separately represented circuit symbol of a device is given the designation common to the device.

Figure 9: Identification of break points
a) By terminal designation, e.g. term. 15,
b) By destination designation, e.g. in section 8 and 2.

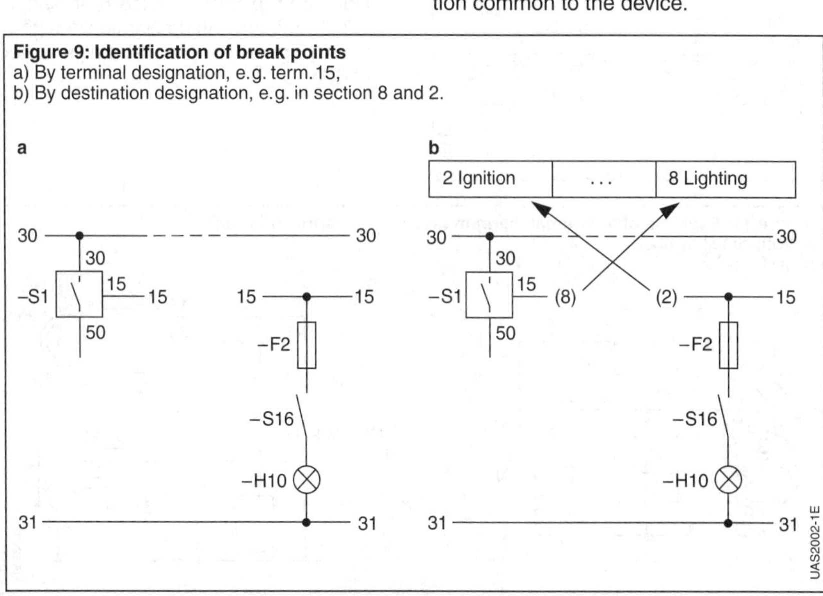

Pin designations (e.g. in accordance with DIN 72552) must be written outside the circuit symbol, and in the case of framing lines preferably outside the framing.

Where the current paths run horizontally: The specifications assigned to the individual circuit symbols are written under the relevant circuit symbols. The terminal designation is located directly outside the actual circuit symbols above the connecting line.

Where the current paths run vertically: The specifications assigned to the individual circuit symbols are written on the left next to the relevant circuit symbols. The terminal designation is located directly outside the actual circuit symbol, on the right if the format is horizontal and on the left next to the connecting line if the format is vertical.

Terminal diagram

The terminal diagram shows the terminal locations of electrical devices and the connected outer and – if necessary – inner conductive connections (lines).

Representation

The individual devices are represented by squares, rectangles, circles and circuit symbols or even pictorially represented, and can be topographically arranged. Circle, dot, plug-in connection or only the led out line are used as terminal points.

The following methods of representation are customary in automotive electrics:

- Assembled, circuit symbols corresponding to EN 60617 (Figure 11a).
- Assembled, pictorial device representation (Figure 11b).
- Detached, device representation with circuit symbols, terminals with destination designations (Figure 12a); lines can be color coded.
- Detached, pictorial device representation, terminals with destination designations (Figure 12b); lines can be color coded.

Figure 10: Possible section designations
a) With consecutive numbers,
b) With references to the sections,
c) With a combination of a and b.

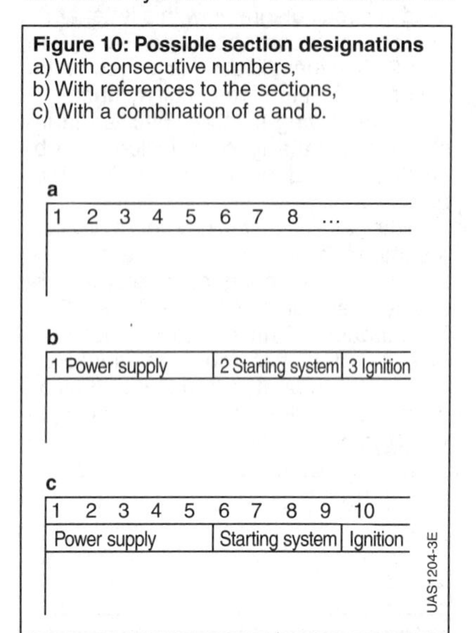

Figure 11: Example of a terminal diagram (assembled representation)
a) With circuit symbols,
b) With devices.

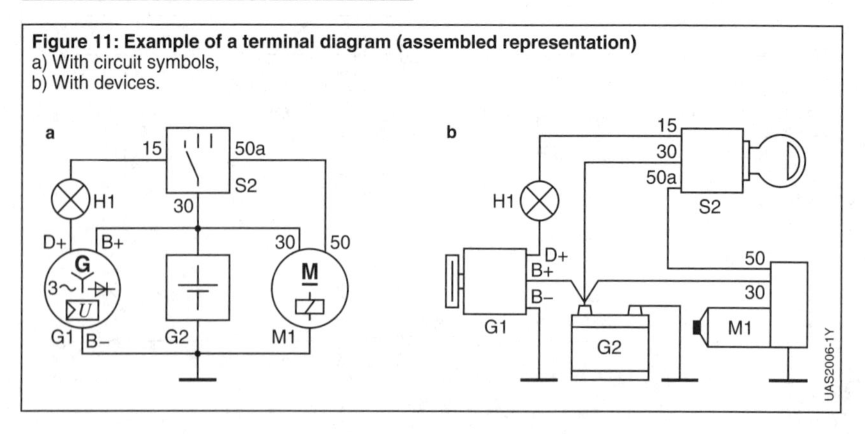

Labeling

The devices are labeled in accordance with EN 81346, Part 2. Terminals and plug-in connections are identified with the terminal designations present on the device (Figure 11).

In the detached method of representation, the continuous connecting lines between the individual devices are omitted. All outgoing lines from a device are given a destination designation (EN 81346, Part 2), consisting of the identification of the destination device and its terminal designation, and – if necessary – the line color is also given (Figure 13).

Figure 13: Device designation (example: alternator)
a) Device designation
(code letter and code number),
b) Terminal designation on device,
c) Device to ground,
d) Destination reference
(code letter and code number/terminal designation/wire color).

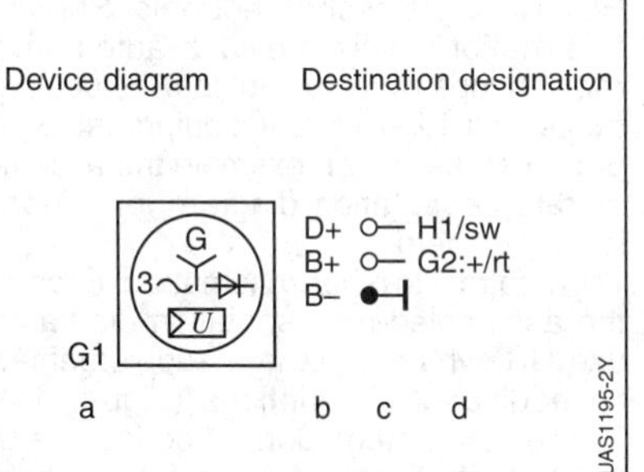

Figure 12: Terminal diagram (detached representation)
a) With circuit symbols and destination designations,
b) With devices and destination designations.
G1 Alternator with regulator, G2 Battery, H1 Alternator indicator lamp,
M1 Starter motor, S2 Ignition/starter switch,
XX Device ground to vehicle ground, YY Terminal for ground connection,
15, 30, 50, 50a Line potential, e.g. terminal 15.

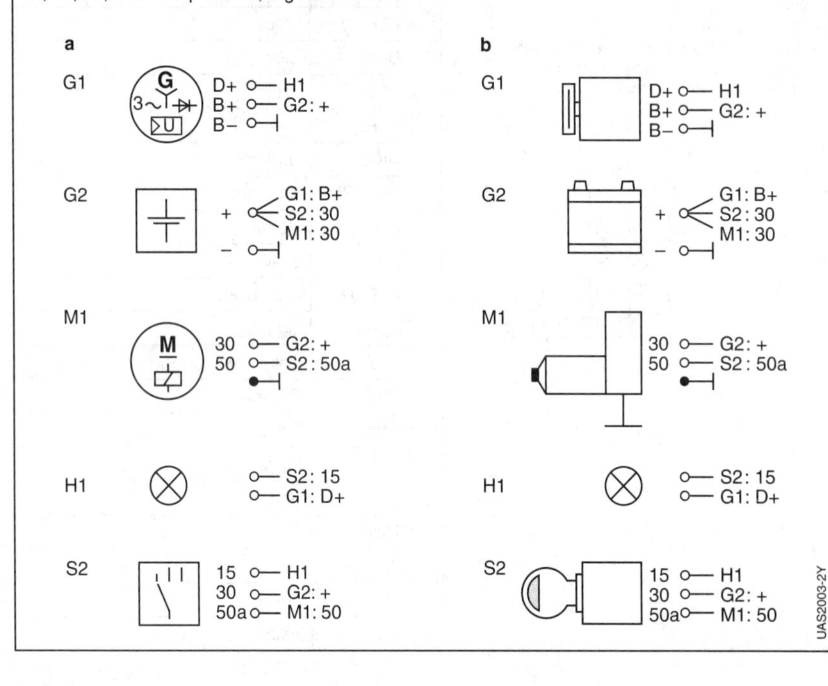

Assembled-representation diagram
For troubleshooting on complex and extensively networked systems with self-diagnosis function, Bosch has developed system-specific schematic diagrams. Bosch makes assembled-representation diagrams for further systems in a great number of motor vehicles available in ESI[tronic] (Bosch Electronic Service Information). This provides automotive repair shops with a useful tool for locating faults or wiring retrofit equipment. Figure 15 shows as an example the assembled-representation diagram for a door locking system.

In contrast to other schematic diagrams, the assembled-representation diagrams use US symbols that are supplemented by additional descriptions (Figure 14). These include component codes – e.g. "A28" (theft-deterrence system) –, which are explained in Table 2 and an explanation of the wire colors (Table 3). Both tables can be called up in ESI[tronic].

Figure 14: Additional descriptions on assembled-representation diagrams
1 Wire color code,
2 Connector number,
3 Pin number (a dashed line between pins indicates that all pins belong to the same connector).

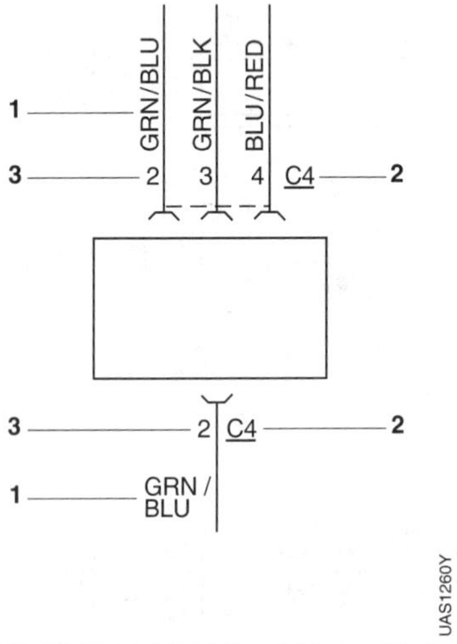

UAS1260Y

Table 2: Explanation of component codes

Position	Designation
A1865	Electrically adjustable seat system
A28	Theft-deterrence system
A750	Fuse box, relay box
F53	Fuse C
F70	Fuse A
M334	Supply pump
S1178	Warning-buzzer switch
Y157	Vacuum actuator
Y360	Actuator, door, front, right
Y361	Actuator, door, front, left
Y364	Actuator, door, rear, right
Y365	Actuator, door, rear, left
Y366	Actuator, filler cap
Y367	Actuator, lock, luggage compartment, trunk lid, tailgate

Table 3: Explanation of wire colors

Position	Designation
BLK	Black
BLU	Blue
BRN	Brown
CLR	Clear
DK BLU	Dark blue
DK GRN	Dark green
GRN	Green
GRY	Gray
LT BLU	Light blue
LT GRN	Light green
NCA	Color not known
ORG	Orange
PNK	Pink
PPL	Purple
RED	Red
TAN	Tan
VIO	Violet
WHT	White
YEL	Yellow

Figure 15: Assembled-representation diagram of a door locking system (example)

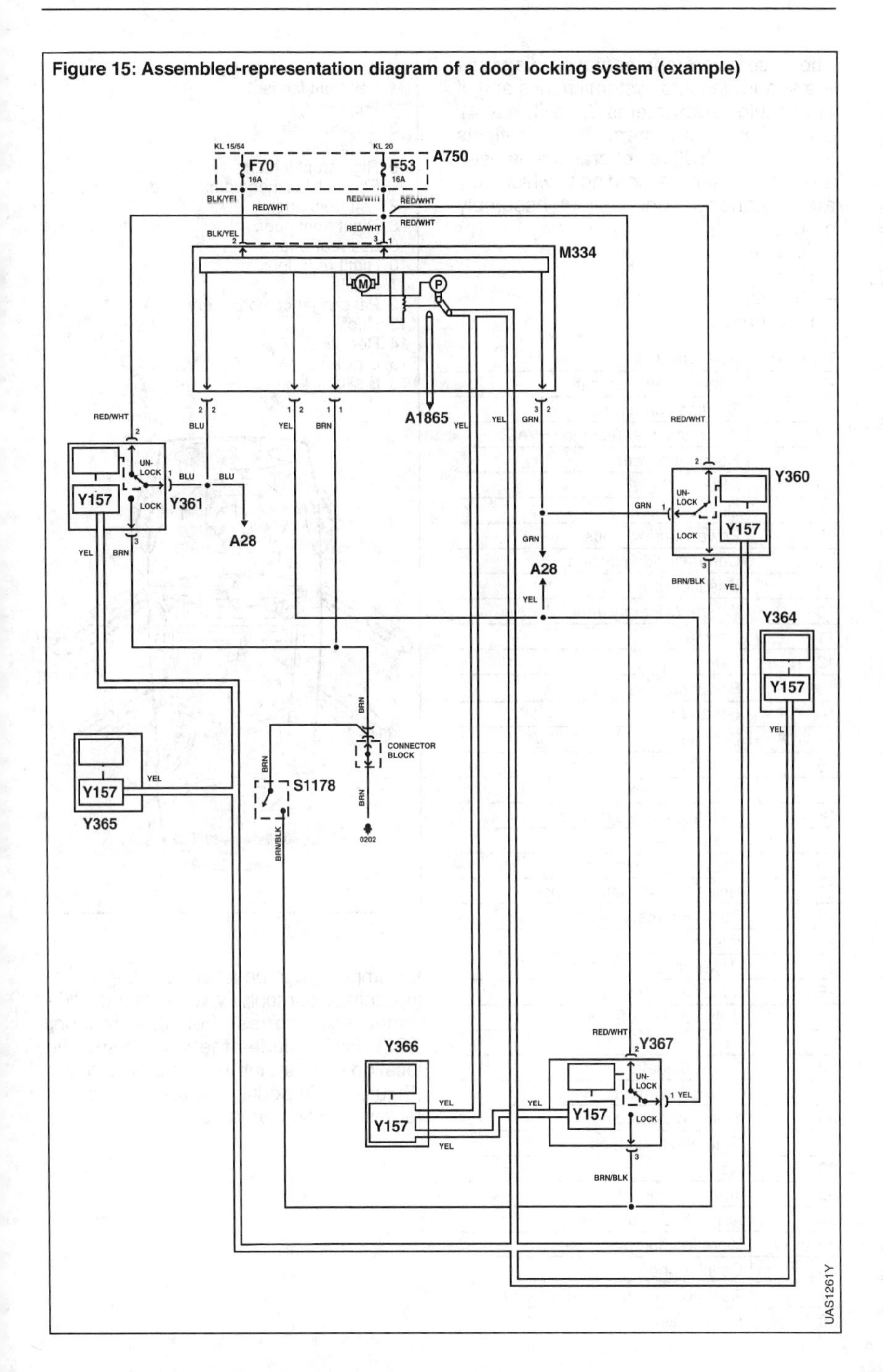

The assembled-representation diagrams are subdivided into system circuits and, if applicable, subsystems (see Table 4). Classification of system circuits reflects the standard ESI[tronic] practice as used for other systems, according to which they are assigned to one of four assembly groups:

- Engine
- Body
- Chassis/suspension
- Drivetrain.

Table 4: System circuits

1	Engine management
2	Starting/charging
3	Air conditioner/heating (HVAC)
4	Radiator blower
5	ABS
6	Cruise control
7	Power-window units
8	Central locking system
9	Dashboard
10	Washer/wiper system
11	Headlamps
12	Exterior lights
13	Power supply
14	Grounding
15	Data line
16	Shift lock
17	Theft deterrence
18	Passive safety systems
19	Power antenna
20	Warning system
21	Heated windshield/mirrors
22	Supplementary safety systems
23	Interior lights
24	Power steering
25	Mirror adjuster
26	Soft-top controls
27	Horn
28	Trunk, tailgate
29	Seat adjustment
30	Electronic damping
31	Cigarette lighter, socket
32	Navigation
33	Transmission
34	Active body components
35	Vibration damping
36	Cellular phone
37	Automotive sound system, Hi-Fi
38	Immobilizer

Figure 16: Grounding points
1 Left front fender,
2 Front end,
3 Engine,
4 Bulkhead,
5 Right front fender,
6 Footwell bulkhead/dashboard,
7 Left front door,
8 Right front door,
9 Left rear door,
10 Right rear door,
11 A pillars,
12 Passenger compartment,
13 Roof,
14 Rear end,
15 C pillars,
16 B pillars.

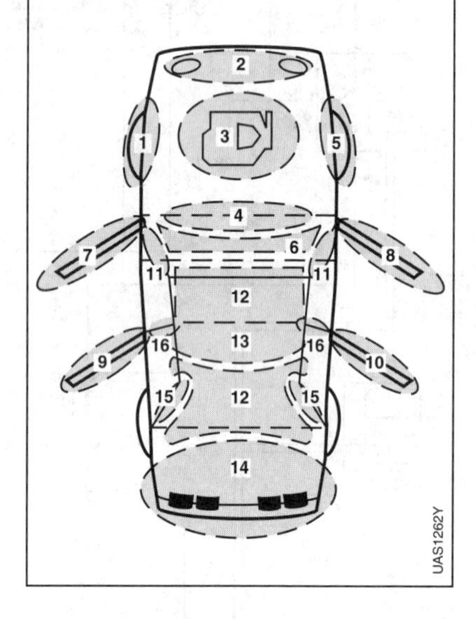

It is important to be aware of the grounding points, particularly when fitting additional accessories. For this reason, ESI[tronic] includes the vehicle-specific location diagram for the grounding points (Figure 16) in addition to the assembled-representation diagrams.

Designations for electrical devices

The designations in accordance with EN 81346 Part 2 (Table 5) are used to provide a clear, internationally comprehensible identification of systems, parts, etc. which are represented by circuit symbols in a circuit diagram. They appear next to the circuit symbols and consist of a series of defined qualifying symbols, letters and numbers (Figure 17).

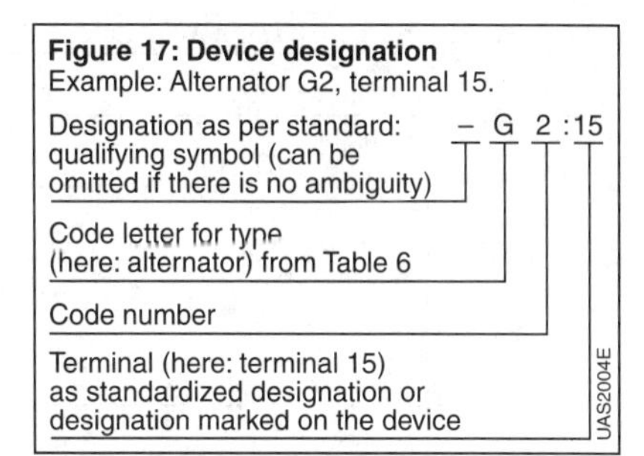

Example: Alternator G2, terminal 15.
Designation as per standard:
– G 2 :15
qualifying symbol (can be omitted if there is no ambiguity)
Code letter for type (here: alternator) from Table 6
Code number
Terminal (here: terminal 15) as standardized designation or designation marked on the device
UAS2004E

Figure 17: Device designation

Table 5: Code letters for designating electrical devices as per EN 81346-2

Code letter	Type	Examples
A	System, assembly, component group	ABS ECU, automotive sound system, car radiotelephone, car phone, theft-deterrence system, device assembly, control device, ECU, cruise control
B	Converter of non-electrical into electrical quantities or vice versa	Reference-mark transmitter, pressure switch, fanfare horn, horn, λ oxygen sensor, loudspeaker, air-flow sensor, microphone, oil-pressure switch, sensors of all types, ignition trigger
C	Capacitor	Capacitors of all types
D	Binary element, memory	On-board computer, digital equipment, integrated circuit, pulse counter, magnetic tape recorder
E	Assorted devices and equipment	Heater, air-conditioning system, lamp, headlamp, spark plug, ignition distributor
F	Protective device	Trigger (bimetal), polarity protection device, fuse, current protection circuit
G	Power supply, alternator	Battery, alternator, charger
H	Check device, signaling unit, signaling device	Acoustic signaling unit, function lamp, turn-signal indicator, turn-signal lamp, brake-pad indicator, stop lamp, high-beam indicator, alternator charge indicator, indicator lamp, signaling unit, oil-pressure indicator, optical signaling unit, signal lamp, warning buzzer
K	Relay, contactor	Battery relay, turn-signal flasher, turn-signal relay, engagement relay, starting relay, hazard warning light flasher
L	Inductance	Inductance coil, coil, winding
M	Engine type	Blower motor, fan motor, pump motor for ABS, TCS and ESP hydraulic modulators, windshield-washer and windshield-wiper motors, starter motor, servomotor
N	Regulator, amplifier	Regulator (electronic or electromechanical), voltage stabilizer

Table 5: Code letters for designating electrical devices (continued)

Code letter	Type	Examples
P	Tester	Ammeter, diagnostic socket, rev counter, pressure indication, tachograph, measuring point, check point, speedometer
R	Resistance	Sheathed-element glow plug, flame glow plug, heating resistor, NTC resistor, PTC resistor, potentiometer, regulating resistor, series resistor
S	Switch	Switches and buttons of all types, ignition contact breaker
T	Transformer	Ignition coil, ignition transformer
U	Modulator, converter	DC converter
V	Semiconductor, tube	Darlington, diode, electron tube, rectifier, semiconductors of all types, variable capacitance diode, transistor, thyristor, Zener diode
W	Transmission path, line, antenna	Vehicle antenna, shield part, shielded line, lines of all types, line bundle, ground (continuity) line
X	Terminal, plug, plug-in connection	Terminal stud, electrical terminals of all kinds, spark-plug connector, terminal, terminal strip, electrical cable coupler, line connector, plug, socket, plug connector, (multiple) plug-in connection, distributor connector
Y	Electrically actuated mechanical equipment	Permanent magnet, (solenoid) fuel injector, electromagnetic clutch, electromagnetic brake, electric air slider, electric fuel pump, electromagnet, electric starting valve, transmission control, linear solenoid, kickdown solenoid valve, headlight leveling control, level control valve, switching valve, starting valve, door locking, central locking system, auxiliary-air slider
Z	Electric filter	Screening unit, suppression filter, filter chain, timer

Terminal designations

The purpose of the terminal-designation system for automotive electrical systems specified by the standard (DIN 72552) is to enable the most accurate connection of wires to all the various devices, above all when making repairs and installing spare parts. The terminal designations (Table 7) do not identify the wires because devices with different terminal designations can be connected at the two ends of each wire. For this reason, they need not be written on the wires.

In addition to the terminal designations listed, designations according to DIN VDE standards may also be used on electrical machines. Multiple connectors, for which the number of terminal designations as per DIN 72552 no longer suffice, are numbered by consecutive numbers or letters whose function assignment is not specified by standards.

Table 7: Terminal designations according to DIN 72552

Terminal	Definition
	Ignition coil
1	Low voltage
4	High voltage
4a	From ignition coil I, terminal 4
4b	From ignition coil II, terminal 4
15	Switched positive after battery (ignition-switch output)
15a	Output at the series resistor to the ignition coil and to the starter
	Glow-plug and starter switch
17	Start
19	Preglow
	Battery
30	Line from battery positive terminal (direct)
30a	Battery changeover 12/24 V Line from battery II positive terminal
31	Return wire from battery Negative or ground (direct)
31 b	Return wire to battery Negative or ground via switch or relay (switched negative)
	Battery changeover relay 12/24 V
31a	Return line to battery II negative
31c	Return line to battery I negative
	Electric motors
32	Return line [1]
33	Main terminal [1]
33a	Self-parking
33b	Shunt field
33f	for second reduced-rpm operation
33g	for third reduced-rpm operation
33h	for fourth reduced-rpm operation
33L	Rotation to left (counterclockwise)
33R	Rotation to right (clockwise)
	Starter
45	Separate starter-motor relay, output; starter, input (primary current)
	Dual starters, parallel activation Starting relay for pinion-engagement current
45a	Starter I output, Starter I and II input
45b	Starter II output
48	Terminal on starter and start repeating relay (monitoring the starting process)
	Turn-signal flasher (pulse generator)
49	Input
49a	Output
49b	Output to second flasher circuit
49c	Output to third flasher circuit
	Starter
50	Starter control (direct)
	Battery changeover relay
50a	Output for starter control
	Starter control
50b	In parallel operation of two starter motors with sequence control
	Starting relay for sequence control of engagement current in parallel operation of two starter motors
50c	Input at starting relay for starter I
50d	Input at starting relay for starter II
	Start-locking relay
50e	Input
50f	Output
	Start repeating relay
50g	Input
50h	Output

[1]) Polarity reversal terminal 32/33 possible

Table 7: Terminal designations according to DIN 72552 (continued)

Terminal	Definition
	Wiper motors
53	Wiper motor, input (+)
53a	Wiper (+), self-parking
53b	Wiper (shunt winding)
53c	Electric windshield-washer pump
53e	Wiper (brake winding)
53i	Wiper motor with permanent magnet and third brush for higher speed)
	Lighting technology
55	Fog lamps
56	Headlamps
56a	High beam with indicator lamp
56b	Low beam (dipped beam)
56d	Headlamp-flasher contact
57a	Parking lamp
57L	Parking lamp, left
57R	Parking lamp, right
58	Side-marker, tail, license-plate and instrument lamps
58L	left
58R	right
	Alternators and voltage regulators
61	Alternator charge indicator
B+	Battery positive terminal
B–	Battery negative terminal
D+	Alternator positive terminal
D–	Alternator negative terminal
DF	Alternator field winding
DF1	Alternator field winding 1
DF2	Alternator field winding 2
U, V, W	Three-phase terminals
	Audio systems
75	Radio, cigarette lighter
76	Loudspeaker
	Switches
	NC contact/changeover contact
81	Input
81a	1st output, NC side
81b	2nd output, NC side
	NO contact
82	Input
82a	1st output
82b	2nd output
82z	1st input
82y	2nd input
	Multiple-position switch
83	Input
83a	Output, position 1
83b	Output, position 2
83L	Output, position left
83R	Output, position right
	Current relay
84	Input, output, relay contact
84a	Output, drive
84b	Output, relay contact
	Switching relay
85	Output, drive (end of winding negative or ground)
86	Input, drive (start of winding)
86a	Start of winding / 1st winding
86b	Winding tap / 2nd winding
	Relay contact for NC contact and changeover contact:
87	Input
87a	1st output (NC side)
87b	2nd output
87c	3rd output
87z	1st input
87y	2nd input
87x	3rd input
	Relay contact for NO contact:
88	Input
	Relay contact for NO contact and changeover contact (NO contact side):
88a	1st output
88b	2nd output
88c	3rd output
	Relay contact for NO contact:
88z	1st input
88y	2nd input
88x	3rd input
	Turn-signal lamp (turn-signal flasher)
C	1st indicator light
C0	Main terminal for check circuits separate from flasher
C2	2nd indicator lamp
C3	3rd indicator lamp (e.g for dual trailer operation)
L	Left-side turn-signal lamp
R	Right-side turn-signal lamp

References
[1] EN 60617: Graphical symbols for diagrams.
Part 2: Symbol elements, qualifying symbols and other symbols having general application.
Part 3: Conductors and connecting devices.
Part 4: Basic passive components.
Part 5: Semiconductors and electron tubes.
Part 6: Production and conversion of electrical energy.
Part 7: Switchgear, controlgear and protective devices.
Part 8: Measuring instruments, lamps and signalling devices.
Part 9: Telecommunications: Switching and peripheral equipment.
Part 10: Telecommunications: Transmission.
Part 11: Architectural and topographical installation plans and diagrams.
Part 12: Binary logic elements.
Part 13: Analogue elements.
[2] EN 81346: Industrial systems, installations and equipment and industrial products – Structuring principles and reference designations.
Part 1: Basic rules.
Part 2: Classification of objects and codes for classes.
[3] DIN 72552: Terminal markings for motor vehicles;
Part 1: Scope, principles, requirements.
Part 2: Codes.
Part 3: Examples for application on circuit diagrams.
Part 4: Summary.

Electronic control unit (ECU)

Functions

Digital technology with programmable controllers has revolutionized automotive engineering. A whole range of functions can be realized with this technology. Many influencing factors are taken into account in the control of the systems in the vehicle. Engine management, for example, performs the entire open- and closed-loop control of the engine (e.g. ignition and fuel injection) and of many assemblies in its periphery (e.g. exhaust-gas turbocharging, exhaust-gas recirculation). Without these possibilities it would not be possible to adhere to current emission limits and low consumption figures under high engine-performance conditions.

Microcontrollers with high computing-capacity requirements are required for electronic open-loop control of systems. The electronic components are accommodated in an electronic control unit (ECU). The number of ECUs has increased sharply in the last few years; more and more systems are being electronically controlled. Further examples are the Electronic Stability Program (ESP), the vehicle power supply control unit, the A/C control unit, and the door control unit.

Compared with the engine ECU the range of functions and the derived performance e.g. with regard to the door control unit, which controls the power-window drives, are low.

All ECUs operate according to the same principle. The ECU detects signals from sensors and control elements, evaluates them, and activates actuators (Figure 1).

Requirements

Operating conditions

High demands are made of ECUs. They are exposed to different loads, depending on their installation location. Thus, an engine ECU installed in the engine compartment is subject to influences that are quite different from the door control unit installed in the passenger compartment. ECUs may be exposed to the following loads:

- ambient temperatures from –40 °C to +125 °C,
- temperature changes,
- vibrational accelerations up to 20 g,
- moisture influences,
- operating fluids such as oil, fuel and brake fluid.

The following additional demands are made of ECUs:

- operational reliability in the event of voltage fluctuations in the vehicle electrical system (e.g. voltage drop when cold starting),
- power dissipation up to 70 W,
- electromagnetic compatibility (resistance to electromagnetic radiation and low radiation of high-frequency interference signals).

Range of functions and data-processing speed

The performance of the electronic components used in an ECU is increasing continuously and facilitates ever more complex control algorithms. This applies particularly to the engine-management system, which must take ever more stringent emission limits into consideration. In response to new requirements the range of functions and thus the demand for memory capacity are increasing continuously, while the trend is toward ever-decreasing external device dimensions. Development is therefore heading towards greater functional integration and to miniaturization of both electronic and mechanical components such as plug-in connections.

The increasing range of functions is accompanied by an increase in data-processing speed. All functions must be executed within a specified timeframe. Added to this is that fact that many program sections must be executed in real time so that the processes in the ECU keep pace with the physical processes (real-time capability).

All this has in the last few years resulted in the use of ever more powerful microcontrollers with increasing computing speed and greater memory resources. At the beginning of the 1990s 32 kByte for the program memory, 8 kByte for the data memory and 12 MHz clock frequency were sufficient for an engine-management system. Present-day engine ECUs now require 4 MByte program memory, 128 kByte data memory and a clock frequency of 270 MHz. And development continues apace, with numerous microcontrollers being used in ECUs in the future (multicore computers).

ECU components

The ECU is the control device of an electronic system. It detects the operating conditions via sensors, processes them and activates actuators. Signal processing takes place in the ECU's computer core.

Computer core

Microprocessor

The microprocessor represents the integration of a computer's central processing unit on a single chip. Microprocessor design ensures that units can be programmed to meet the varied requirements associated with specific operating conditions. There are two different main groups of processor. Earlier PCs (personal computers) used CISC processors (CISC: Complex Instruction Set Computing). A very large amount of special commands which are executed in a specific number of clock pulses is implemented in these processors.

Figure 1: Signal processing in the ECU

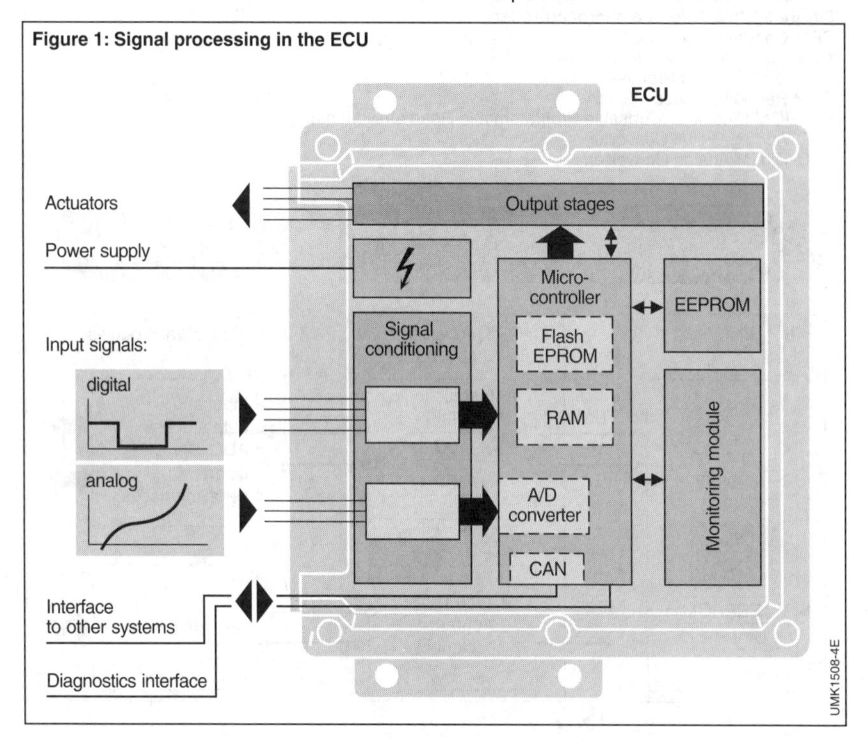

RISC processors (RISC, Reduced Instruction Set Computing) are usually used in ECUs, i.e. also in vehicle ECUs. The advantage of RISC processors lies in the fact that the commands usually only require one clock pulse and can be implemented by the simpler and faster commands of higher-performance systems.

Today's RISC processors with more than 300 commands have more commands than older CISC processors, therefore the more accurate difference between CISC and RISC is the mean number of clock pulses for a command and no longer the number of commands.

A microprocessor cannot usefully operate by itself; it always acts as part of a microcomputer.

Microcomputer

A microcomputer consists of a microprocessor serving as the central processing unit (CPU). The microprocessor contains the controller and the arithmetic-logic unit (Figure 2). The arithmetic-logic unit performs arithmetic and logic operations and operations involving digital signal processing, while the controller retrieves the commands and data from the memory. Pipeline stages, which preprocess the commands (e.g. retrieval of instructions and data, writing of data), are used to achieve higher clock frequencies. The arithmetic-logic unit is integrated in the pipeline.

The program memory (ROM, Read Only Memory; PROM, Programmable Read Only Memory; EPROM, Erasable Programmable Read Only Memory, or Flash EPROM) contains the non-volatile working program (user program) (see Program and data memory). The data are still available after switching on again even when there is no supply voltage. In current applications the program memory is usually a flash EPROM.

The data being run at any given time are stored in the data memory. These

Figure 2: Structure of a microcontroller
CPU Central Processing Unit,
ALU Arithmetic-Logic Unit,
DSP Digital Signal Processor,
RAM Random Access Memory,
EEPROM Electrically Erasable Programmable Read Only Memory,
ADC Analog-Digital Converter,
DAC Digital-Analog Converter.

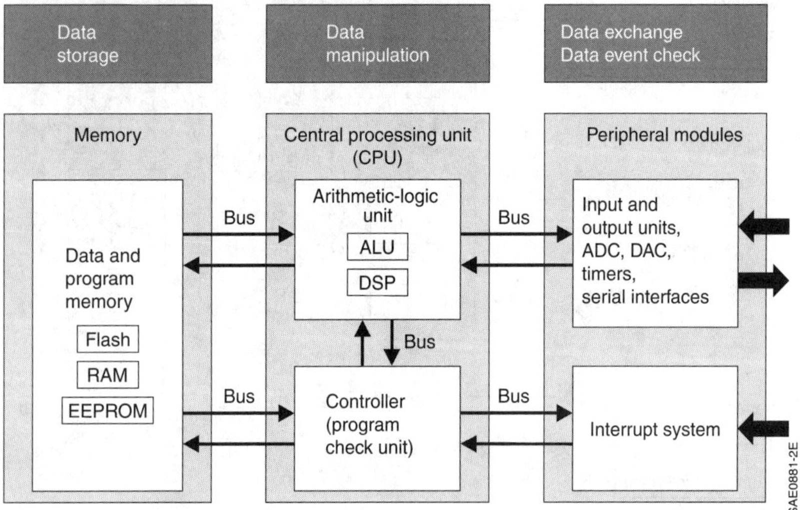

Data storage
Data manipulation
Data exchange
Data event check
Memory
Data and program memory
Flash
RAM
EEPROM
Central processing unit (CPU)
Arithmetic-logic unit
ALU
DSP
Controller (program check unit)
Peripheral modules
Input and output units, ADC, DAC, timers, serial interfaces
Interrupt system
Bus
SAE0881-2E

data change, and are stored in a RAM (Random Access Memory).

A cache as a high-speed buffer memory for program and data memories (in series production in Bosch engine-management systems since 2006) can be integrated on the microcomputer. If the required commands and data were always sent over buses and there were no cache, the high data rates (which the processor needs for working) and the relatively slow access to the memories would drastically reduce CPU speed and there would be no use in deploying a high-speed processor.

The bus system links the individual elements in a microcomputer. A clock generator makes sure that all the operations in the microcomputer take place within a specified timeframe.

Logic circuits are chips that carry out special tasks, such as for interrupts (interrupt) or the reset logic (to reset the microprocessor). Input and outputs units (I/O) are required to connect with the peripherals. The peripherals include, for example, inputs for the crankshaft and camshaft signals, operating switches, analog inputs, and power drivers such as H-bridges or switches.

Microcontroller

In-car applications are subject to the requirement that the sequence of the ECU program keep pace with the physical sequences. Thus, system must respond within the shortest possible time to, for example, changes in input signals. The systems must therefore have "real-time capability". The function of the timer unit is to encapsulate the real time. There are two basic principles here – capture and compare.

The capture function for input signals assigns a time or angle stamp (angle as reference for example to the crankshaft) to the events at the input and stores the capture values in registers. Thus the processor does not have to process the signal further immediately; it can first close the running software routine and then process the capture values.

The compare function for output signals generates an event (e.g. switching of the ignition coil at the moment of ignition) at the output pin at a specific time or angle. These values are stored beforehand in registers.

Microcomputers which have real-time capability are called microcontrollers. Nowadays they are typically integrated on a chip and are not, as in the past, designed as discrete components on the printed-circuit board. This is possible due to the fact that highly complex systems can be integrated on a chip thanks to modern technologies.

For future applications it will no longer be sufficient to have just one CPU, since the clock frequency cannot be increased in any way the user likes and this also results in very high current inputs. For this reason, microcontrollers will come in variants containing more than one CPU (more than five CPUs on one chip for Bosch engine-management systems, to be produced in series from 2015) – the same trend as with PC systems.

The microcontrollers which are used in the automotive industry differ from standard microcontrollers on various points.

- Temperature range: Junction temperature from –40 to 165 °C.
- Failure rate: Less than 1 ppm (one failure per one million components) – compared with 100 to 200 ppm for PCs.
- Service life: This is current specified at 40,000 hours in active operation (this corresponds in the case of a PC that is switched on for eight hours a day to an operating period of approx. 14 years).
- Significantly longer availability: Whereas no new microcontrollers have to be subsequently delivered for cellular phones and PCs after a production cycle of 1 – 3 years, complex projects in the automotive field start series production only after 3 – 5 years development, after which they remain in series production for a period of 15 years and are then subsequently delivered for a further 15 years.

Semiconductor memories

Memory principle

Data storage comprises the following operations: recording (writing), permanent storage (data storage in the narrow sense), retrieval, and output (reading) of information. Figure 3 provides an overview of the different memory types. The memory operates by exploiting physical properties that facilitate unambiguous production and recognition of two opposed states (binary information). In semiconductor memories the states produced are either "conductive" and "nonconductive" or "charged" and "discharged". A technology in frequent use today is the flash memory. Memories operating according to this principle are electrically programmable and erasable.

New memory types will also be used in the future. The FRAM (Ferroelectric Random Access Memory) with the ferroelectric memory principle, the MRAM (Magnetoresistive Random Access Memory) with the magnetic effect as the memory principle, and the PCM (Phase Change Memory) utilizes as its memory effect the change of state of a material from crystalline to amorphous and the associated change in resistance. MRAMs and PCMs have the potential to supersede the flash memories (non-volatile but slow) and RAMs (fast but volatile) used today, since they retain their data after the power supply is switched off and also have very short access times (not to be introduced in engine management prior to 2020).

Program and data memory

The microcontroller requires a program – the software – for the calculations. It is stored in a non-volatile program memory. The CPU reads out the values, interprets them as commands, and executes these commands in order.

Variant-specific data (individual data, characteristic curves and program maps) are additionally stored in this memory. The software can be adapted to different vehicle variants (e.g. different ignition maps) with these data.

The software requires a read/write memory to store variable data (variables) such as e.g. computing values.

Flash memory

The flash EPROM has superseded the conventional EPROM erasable with UV light (Erasable Programmable Read Only Memory) as the program memory. It is electrically erasable so that the ECU can be reprogrammed in the service workshop without it having to be opened. The ECU is connected via a serial interface to the reprogramming station.

RAM

All the variable data calculated in the software are stored in the RAM (Random Access Memory, i.e. in a read/write memory). The memory capacity of the RAM integrated in the microcontroller is not enough for complex application such that an additional RAM chip is required. It is connected via the address bus and the data bus to the microcontroller.

The RAM loses the entire data stock (volatile memory) when the ECU is disconnected from the supply voltage. The RAM is permanently supplied with voltage to prevent adaptation values learned during vehicle operation from being lost when the ignition is switched off. These values are however lost when the battery is disconnected.

EEPROM

Data that change in vehicle operation and which must not be lost when the battery is disconnected (e.g. important adaptation values, codes for the immobilizer) must be stored in a non-volatile memory (e.g. EEPROM).

It is also possible to use separately erasable areas of the flash EPROM as a non-volatile data memory for these data. An EEPROM is emulated in the flash (in separate areas – outside the program and data memory) for this purpose. The current data in each case are written to the end of a table, which gradually fills the blocks of the flash area. The old data can then be completely erased with the block. This ensures that the correct data are in the memory even when the supply voltage is disconnected.

Monitoring module

A monitoring module is required for safety-relevant systems. This is effected by a logic circuit in the voltage regulator (e.g. for engine-management systems) or in a separate integrated circuit (e.g. for transmission management). The monitoring module can also be effected by a separate computer.

The microcontroller and the monitoring module use a "question-and-answer game" to monitor each other. When a fault is detected, the ECU of both is rendered safe independently of each other – in the case of engine management e.g. by the deactivation of torque-relevant output stages.

Voltage regulator

The electronic components in the ECU need a stable voltage of 5 V. Some applications also require further voltage values, e.g. 3.3 V. The voltage regulator maintains the battery voltage, which can be 6 to 16 V depending on the battery state and load, at these constant values. Suppressor circuits suppress high interference voltages from the vehicle electrical system.

The voltage regulator contains an enabling logic which ensures that the voltages run up under defined conditions and then enable resets. This ensures that the microcontroller runs up under defined conditions after the supply voltage is switched on.

Figure 3: Overview of semiconductor memories

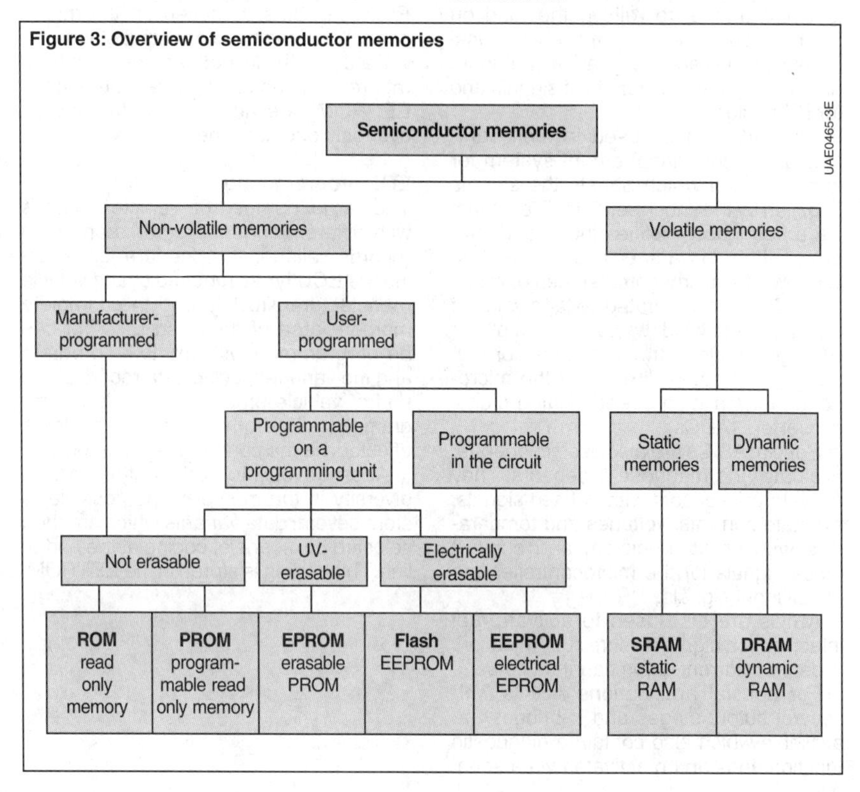

ASIC chips

ASIC chips are application-specific integrated circuits. They are designed and produced in accordance with the specification of ECU development. Integrating a large number of functions in an IC reduces the space requirement, reduces manufacturing costs and increases operational reliability.

Thus, for example, the power supply for the ECU and several power output stages with the associated diagnostic circuits can be integrated on one chip. An example of this is the "U-chip", which features on one IC a power supply, the monitoring module, serial interface drivers and output stages, and an H-bridge with diagnostics.

Another function of an ASIC module is, for example, to relieve the load on the microcontroller or to make additional hardware available (e.g. additional input and output channels for static signals and for PWM signals).

An ASIC can be used in the spark-ignition engine-management system for knock control which boosts the signals supplied by the knock sensors, filters them via a bandpass, rectifies them, and integrates them via a specified crank-angle window. The microcontroller then only has to read out the integrated voltage value at the end of the window. However, in newer ECU generations the knock-sensor signals are detected directly by the microcontroller via high-speed analog-digital converters (ADCs).

Special ASICs are also used to operate, evaluate and diagnose λ sensors. They activate the sensors with defined signals, evaluate currents, voltages and temperature with great precision, and convert these signals for the microcontroller into digital input signals.

ASICs are also used to activate fuel injectors for high-pressure fuel injection. A defined current curve can thus be set.

For special applications ASICs with several output stages and H-bridges are available which also contain a diagnostic function and can be activated via a serial bus.

Communication

The peripheral components communicate with the microcontroller via the address bus and the data bus. The microcontroller outputs via the address bus for example the RAM address whose memory content is to be read. The RAM chip then places the data on the data bus, on which they can be read by the microcontroller.

To save on pins on the components, the addresses and the data are output in the multiplex process. Address bus and data bus use the same lines. First the address is applied to the bus, then the data are transmitted.

Communication with the ASICs is effected via SPI (Serial Peripheral Interface) or via MSC (Micro Second Channel). Ports can thus be saved on the microcontroller in the case of output stages.

Data which do not require a high bit rate (e.g. fault-memory data stored in the EEPROM) are also serially transferred over only one data line.

EOL programming

The large number of vehicle variants with different control programs and data records calls for a procedure for reducing the ECU types required by the vehicle manufacturer. To this end the complete memory area of the flash EPROM can be programmed with the ECU program and the variant-specific data record at the end of vehicle production with EOL programming (End of Line). Data transfer is effected via the communication interface.

Alternatively, to reduce the variant diversity in the memory, it is possible to store several data variants which are then selected by means of coding at the end of line. This coding is stored in the EEPROM.

Data processing

Input signals

Control elements and sensors form the interface between the vehicle and the ECU. The electrical signals are supplied to the ECU via the wiring harness and connectors over different interfaces.

Analog interface

Physical quantities such as the engine temperature or the intake-manifold pressure are forwarded by sensors as analog measured values to the ECU. These input signals can assume any voltage value within a certain range. They are converted in the microcontroller by an analog-digital converter (ADC) into a digital value. The typical resolution of the integrated converters is 10 bits. These 1,024 stages deliver for a reference voltage of 5 V a typical resolution of approx. 5 mV.

In future sensors which detect analog values will increasingly incorporate additional electronics which digitize the analog value locally and output the measured value via a digital interface. One of these interfaces is PSI5, on which data can be transferred bidirectionally (see PSI5).

Digital interface

Digital input signals have just two states "High" (logic 1) and "Low" (logic 0). They can be evaluated directly by the ECU. Examples of digital signals are switching signals of control elements, speed signals from a Hall-effect sensor (for measuring wheel speed or engine speed), or static signals from a Hall-effect sensor (e.g. position sensor in the transmission).

Signal conditioning

The input signals are limited by suppressor circuits to permissible voltage levels. Filters remove most of the superimposed noise from the useful signals, which are then amplified if necessary to the microcontroller's permissible input voltage (0–5 V).

Special ACICs for signal conditioning are used (see ASIC chips) for some sensors, such as λ sensors.

Signal processing

The ECU is the control center for the functional sequences of an electronic system. The control algorithms run in the microcontroller. The input signals provided by the sensors and the interfaces to other systems for example via the CAN bus serve as input variables. They are checked for plausibility again in the microcontroller. The algorithms are processed with the aid of the ECU program (software). The ECU program is stored in the read-only memory (e.g. flash EPROM), the calculated intermediate values are stored in the data memory (RAM). The software calculates the output signals to activate the actuators.

Output signals

The microcontroller uses the output signals to activate output stages, which delivery sufficient power for direct connection of the actuators. Several output stages can be integrated in a single ASIC. Consumers with high currents to be switched (e.g. engine fans) are switched via relays or with output stages directly on the assembly.

Signal types

Switching signals

Switching signals are used to switch actuators on and off, depending on the current operating point (e.g. fuel injectors, engine fans).

PWM signals

Digital output signals can be output as a PWM signal (pulse width modulation) with constant frequency and variable on-time. Actuators such as the exhaust-gas recirculation valve, the throttle valve or the charge-pressure actuator can, depending on the clock ratio, be moved into any operating positions.

Application-specific output stages
Some actuators require special current and voltage curves for operation. Examples of such are the activation of high-pressure fuel injectors for gasoline direct injection (see High-pressure fuel injector) or for common-rail injectors (see injectors). The actuators are activated by complex output-stage modules with integrated control logic and analog signal conditioning. These modules deliver a high voltage during switch-on, with the high making current producing a fast breaking operation. The system is then regulated back to a lower holding current.

Switching variants
There are different switching variants for activating the actuators. Low-side output stages control inductive and ohmic loads which are connected to the battery voltage (e.g. valves, relays, ignition coils). High-side output stages switch consumers which are connected to ground. DC motors are activated via bridge-type output stages; here, both connections of the motor are switched by the ECU. This makes it possible to reverse the direction of rotation (e.g. power-window drives, activation of the throttle valve).

Self-protection function
The output stages are protected against short-circuits to ground, against short-circuits to battery voltage, and against being destroyed due to electrical or thermal overload. These malfunctions and interrupted lines are detected as faults by the output-stage IC and signaled to the microcontroller via a serial interface.

Communication interfaces
The ECU has one or more communication interfaces for communicating with other electronic systems. Thus, for example, the driving speed can be sent by the Electronic Stability Program (ESP) to all the systems that need this signal (e.g. the instrument cluster for displaying the driving speed). The signals therefore only have to be calculated one in the ECU network. Data transfer is effected via bus systems (see Automotive networking, buses in the vehicle).

Automotive software engineering

Motivation

Aim of development

Any development has the aim of creating a new function or improving an existing function of the vehicle. Such functions bring about an added value for the user of the vehicle (e.g. occupants, mechanics in the workshop or haulage contractors), compliance with statutory requirements, simplified maintenance, or improved development or manufacturing efficiency. Technical implementation can be mechanical, hydraulic, electrical or electronic. Often it is combinations of these technologies, where electronics increasingly plays a key role in the realization of many innovations in a motor vehicle. Thanks to the use of electrics, electronics and software – the logical core of systems – "intelligent" functions of the drive, the chassis and the rest of the vehicle are cost-effectively realized.

Virtually all vehicle functions in all vehicle classes are now electronically controlled or monitored. Ongoing advances in the technology and performance of electronics hardware allow the realization of many new, more powerful functions through the use of software. The various electronic systems are being increasingly cross-linked in the vehicle. Efforts are pursuing cross-linking vehicles to each other and with the environment (e.g. Internet).

Software requirements

The specific requirements of the software are highly varied. Many systems for the engine and for driving safety must operate with "real-time capability", i.e. the response of the control operation must keep pace with the physical process. When very fast physical processes are being controlled, such as for example engine management or driving-dynamics control, the calculation must therefore be made very quickly. Even the requirements with regard to reliability are high in many fields. This is particularly true of safety-relevant functions. The software and the electronics are monitored by complex diagnostics.

The software is used across the many variants of a vehicle or even across model series. It must then be adaptable to the corresponding destination system. It contains calibration parameters and program maps for this purpose. These amount to several 10,000 per vehicle. These manipulated variables are dependent on each other in many ways. In addition, a functionality is distributed beyond multiple systems or control units more and more frequently.

The software is for the most part developed specially for the corresponding applications and is integrated into the entire system. It is called "embedded software". The many functions are often developed and further developed over a long period of time in many locations around the world. Because spare parts must remain available even after the production of a particular vehicle has come to an end, the electronics in the vehicle must have a relatively long life cycle of up to 30 years.

For cost reasons, ECUs often contain microcontrollers with limited computing power and limited memory space. This requires optimization measures in software development to reduce the hardware resources required.

The characteristic features of software differ depending on the field of application. While the software for the drivetrain is extensive, the emphasis is on real-time performance in the chassis applications. In safety and comfort and convenience applications, the emphasis is on efficiency, i.e. the consumption of resources, and in multimedia applications large volumes of data must be processed in a short period of time.

The complexity resulting from these requirements and features must be economically mastered in a cooperation of development between vehicle manufacturers and suppliers.

Design of software in motor vehicles

The software in a motor vehicle is made up of many components. A distinction is essentially made, as is the case with a PC, between the "perceptible functions" of the software, the application software and platform software partly dependent on the hardware (Figure 1). The interaction between all the functions is defined in the architecture. Different views can be adopted here. The static view describes hierarchically the function groups, the signals and the distribution of the resources. On the other hand, the functional view describes the signal progression through the different functions. The dynamic, i.e. time-dependent view, considers the time response in the execution of the various tasks. Standards have already been introduced at an early stage to ensure the interaction between and further development of the individual components. The most important standards are explained in the following.

Important standards for software in motor vehicles

Bodies/committees

The "Association for Standardization of Automation and Measuring Systems" (ASAM) is a standardization body in the automotive industry for data models, interfaces and syntax specifications [5]. The ASAM developed different standards for connecting an ECU to a computer or a data-entry terminal. The standard ASAM-MCD1 (MCD stands for Measurement, Calibration and Diagnosis) supports different data transport protocols for this purpose. By using ASAM-MCD2 specifications, the binary data in the ECU can be addressed and the relevant quantities can at the same time be displayed in connected tools as physical values and processed. The standard ASAM-MCD3 also permits an automation of such processes, for example for the automatic calibration of data records. Further ASAM standards cover, for example, the exchange of functional descriptions and data.

Figure 1: Main component parts of the static software architecture for microcontrollers and standardized software components

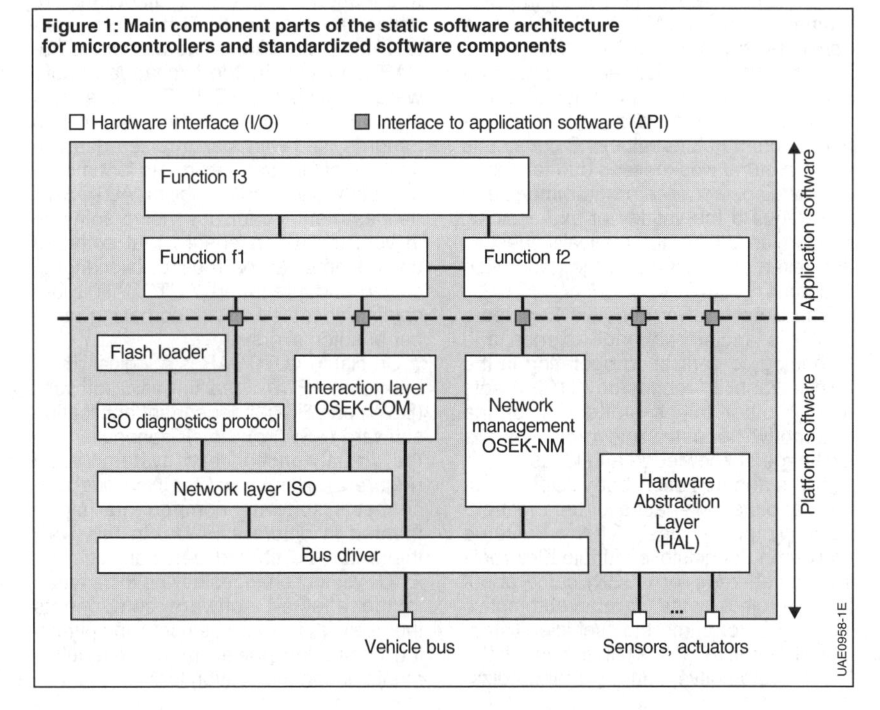

The FlexRay Consortium developed the specification for the FlexRay field bus for open and closed-loop control in the automobile field. Thanks to high transfer rates with predefined bus arbitration and a fault-tolerant design it is particularly suitable for use in active safety systems and in the drivetrain (see FlexRay).

The International Electrotechnical Commission (IEC) is an international standardization body in the fields of electrical engineering and electronics [6]. The IEC offers three evaluation systems with which conformity to the international standards can be verified. The IEC works in close cooperation with the International Organization for Standardization (ISO) [7], the International Telecommunication Union (ITU) and numerous standardization bodies (including the Institute of Electrical and Electronics Engineers, IEEE) [8].

The Motor Industry Software Reliability Association (MISRA) is an organization in the automotive industry that devises rules for the reliable development and application of software in vehicle systems [9]. The most widely known is the programming standard MISRA-C, which was developed by the MISRA. It prescribes programming rules for reliable programming in the programming language C. The purpose of this standard is to avoid runtime errors due to unreliable C constructs and structural weaknesses due to misunderstandings between programmers, and to safeguard the validity of expressions. Many rules can be automatically checked and taken into account in the generation of codes.

The Society of Automotive Engineers (SAE) is an international science and technology organization operating in the field of mobility technology [10]. It sets among other things standards for the automotive industry and promotes the exchange of knowledge and ideas.

The standardization body "Open Systems and their Interfaces for the Electronics in Motor Vehicles" (Offene Systeme und deren Schnittstellen für die Elektronik im Kraftfahrzeug, or OSEK) came about from a project by the German automotive industry. Later came the "Vehicle Distributed Executive" (VDX) initiative of the French automotive industry. Standardizations of basic-software components are established under the term OSEK/VDX in the following fields [11]:
- Communication (data exchange within and between ECUs)
- Operating system (real-time execution of ECU software and basic services for other OSEK/VDX modules)
- Network management (configuration and monitoring).

The Japan Automotive Software Platform and Architecture (JasPar) is an initiative for reducing costs and developing technology in automotive electronics. It encourages Japanese companies to develop jointly non-competition-relevant technologies such as network solutions, service functions and basic software. JasPar works in close cooperation with AUTOSAR and FlexRay.

AUTOSAR

AUTOSAR (Automotive Open System Architecture) [12] is a development partnership of vehicle manufacturers, ECU manufacturers and manufacturers of development tools, ECU basic software and microcontrollers. The aim of AUTOSAR is to simplify the exchange of software on different ECUs. To this end, a standardized software architecture has been devised with standardized description and configuration formats for embedded software in automobiles. AUTOSAR defines methods for describing software in vehicles which ensure that software components can be reused, exchanged, scaled and integrated. AUTOSAR is becoming accepted by an increasing number of automakers.

Crucial to AUTOSAR is a logical distribution into ECU-specific basic software (BSW) and ECU-independent application software (ASW) and their connection via the virtual function bus system (VFB) (Figure 2). This virtual function bus also connects software components implemented in different ECUs. In this way, these can be shifted between different ECUs without changes having to be made in the affected software components themselves. This can be useful in optimizing computing power, memory requirements or communication load.

The functional software components (SWC) are strictly separated from each other and from the basic software. They typically contain specific control algorithms executed at run time, the "runnable entities". They communicate via the AUTOSAR interface with other functions and the ECU interfaces. These interfaces (API) are defined in SWC XML descriptions.

The run-time environment (RTE) provides the communication services between the functional software components and the corresponding basic software on the ECU. The RTE is tailored to the specific ECU and the application. It can be generated to a large extent automatically from the interface requirements.

The basic software contains the ECU-specific program parts, such as the communication interfaces, diagnostics and memory management. The basic software also contains the service layer. This software combines software components for general service functions (SRV), communication (COM) and the operating system partly dependent on the ECU used (OS) [4, 11]. The latter is based on the OSEK/VDX OS. In this field, the resources of the ECU are subdivided and managed so as to arrive at optimal network support, memory management, diagnostics, etc.

The hardware used is encapsulated in two layers based on each other. Abstraction of the microprocessor (Micro-Controller Abstraction Layer, MCAL) with direct access to the interface modules of the ECU continues in a further layer (ECU Abstraction). "Complex Device Drivers" (CDD) facilitate direct access to microcontroller resources for applications with special functionality and timing requirements. They are also an integral part of the basic software such that the application software can be developed independently of the hardware, even when the services of the complex device drivers are required.

In addition to the ECU architecture, the development methods are also standardized in part by AUTOSAR. This relates, above all, to the structure and the dependencies of the different work products (e.g. files). These are needed in order to generate from the different software component descriptions executable programs for the respective ECUs.

Figure 2: AUTOSAR architecture
ECU electronic control unit,
ASW application software,
SWC software components,
VFB virtual function bus,
RTE run-time environment,
BSW basic software,
OS operating system.

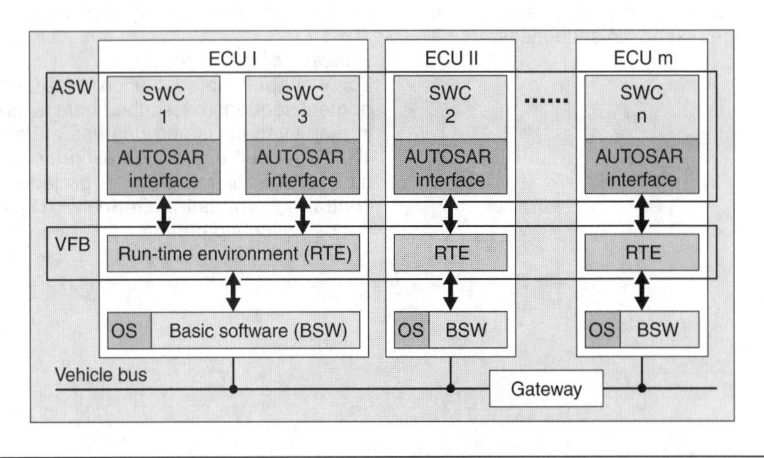

Diagnostic standards
Vehicle-specific systems for diagnostics for vehicle development and production and for workshops prove to be high-maintenance, cost-intensive and inflexible. They tie the manufacturers to suppliers and prevent simple data communication in the case of cross-company cooperations. Some diagnostic standards have therefore been developed (e.g. [13] and [14]).

The Automotive Electronics working group (ASAM-AE) [5] has devised three specifications for data-based (i.e. software-based) vehicle diagnostics which are published as international standards in the ISO 22900 standard group [15]:
- Interface between run-time environment and communication hardware (MCD-1D and PDU-API, ISO 22900).
- ODX standard for exchange of diagnostic data, e.g. for supplying the workshop tester with data (MCD-2D, ISO 22901, [16]).
- Object-oriented programming interface (MCD-3D, ISO 22900) for diagnostic applications, such as e.g. prompted troubleshooting.

The MCD-1D standard takes into consideration existing standard tools such as, for example, devices for flashing ECUs.

Requirements with regard to an exchange format, the "Open Test Sequence Exchange Format" (OTX), are currently being defined within the framework of an ISO draft in order to create, use and exchange diagnostic sequences.

The development process

The focal point of software development is the depiction of the logical system architecture in a concrete software system with all programs and data. Here, the entire processor-controlled system of the vehicle is considered. Particular importance is attached to a clear separation of specification, design and implementation. Specification of the software functions is performed on the physical level, while design and implementation of programs and data are geared towards the specific microcontroller.

In order to satisfy the above-mentioned requirements in the development of software in automobiles, the defined sequences (processes) are, in addition to the technology and the tools, an integral part of development.

Process description models
Numerous more or less complicated models are used to describe the workflows in software development. They serve to make the sequences transparent, to compare them, to identify problem areas, and to verify conformity in accordance with defined standards. They were, however, not originally conceived to improve directly the quality of the software itself, to increase efficiency or to eliminate systematic faults in the sequences. Process description models are therefore only partly suitable. The common V model will be described here by way of example.

Principle of the V model
The V-shaped representation of the development sequence described here is used in many variations and degrees of detail. The "V model" of the federation for planning and implementing IT projects of central government in Germany [17] will not be described here.

The V model divides the process steps associated directly with development along a V, where the x-axis depicts the development progress and the y-axis the depth, i.e. the degree of detail, of the corresponding process step (Figure 3). A process step can be described by the required input variables, the procedure, the methods, the roles, the tools, the quality criteria, and the output variables. The process steps defined on the left arm are verified on the right arm. These steps can also run several times or be divided.

In the extended V model, accompanying processes such as, for example, request, change, project and quality management, can be considered.

Process evaluation models

Process evaluation models provide, in addition to a pure description of the tasks and sequences, information on the maturity and the quality of the processes. Work steps can thus be compared, evaluated and certified. In this way, it is also possible to identify process gaps which impact, for example, on the product quality or the costs. However, the information on the quality of the processes does not provide a complete picture of the quality of the products themselves here either. The three most important process evaluation models will be described here.

ISO 9000 and ISO/TS 16949

The process-oriented EN ISO 9000 standard series [18] and following stipulates the requirements with regard to a quality management system. The focus here is on the interactions and the interfaces. The original emphasis was on production and the customer interfaces.

The ISO/TS 16949 technical specification [19] was devised in the North American and European automotive industries and standardizes requirements relating to quality management systems. The aim of the standard is to improve effectively system and process quality in order to increase customer satisfaction, to identify failures and risks in the production process and in the supply chain, to eliminate their causes, and to check the effectiveness of the corrective and preventive ac-

Figure 3: Extended V model

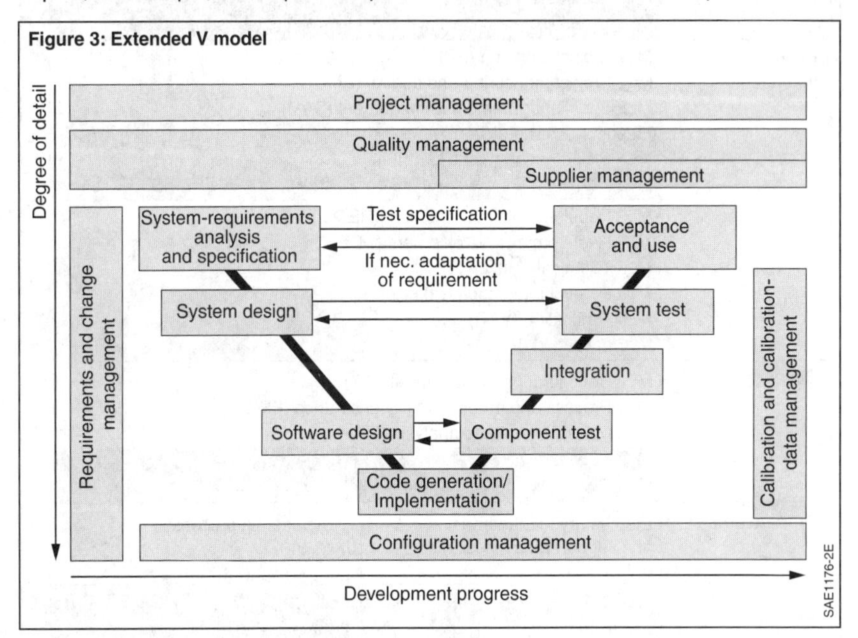

tions taken. The core of the specification is not the discovery, but rather the avoidance of failures.

Compliance with ISO 9000 and ISO/TS 16949 can be verified by certification.

CMMI

"Capability Maturity Model Integration" (CMMI) is a model for evaluating and systematically improving development organizations and their processes [20] which was originally developed by the Software Engineering Institute (SEI). It describes a collection of requirements with regard to the processes and their dependencies (Figure 4). CMMI offers a framework, the implementation of which calls for a business-oriented interpretation and organization of content. It describes what has to be done. The organization must shape the "how" appropriately. The content of CMMI is based on essential industry "best practices". CMMI provides a procedure for process improvement created in the longer term from the path of organization development through to learning organization.

CMMI has much in common in terms of content with ISO 9000 and ISO TS 16949. In this context, CMMI has a higher degree of detail, while ISO 9000 and ISO TS 16949 cover a wider range of application.

CMMI distinguishes between five maturity levels (ML) for an organization unit (Figs. 4, 5). Different process areas are considered, depending on the maturity

Figure 4: CMMI process overview
ML Maturity level.

Category	Process areas
Process management	ML3 Organizational process focus (OPF)
	ML3 Organizational process definition (OPD)
	ML3 Organizational training (OT)
	ML4 Organizational process performance (OPP)
	ML5 Organizational innovation & deployment (OID)
Project management	ML2 Project planning (PP)
	ML2 Project monitoring and control (PMC)
	ML2 Supplier agreement management (SAM)
	ML3 Integrated project management (IPM)
	ML3 Risk management (RSKM)
	ML3 Quantitative project management (IPM)
Engineering	ML2 Requirements management (REQM)
	ML3 Requirements development (RD)
	ML3 Technical solution (TS)
	ML3 Product integration (PI)
	ML3 Verification (VER)
	ML3 Validation (VAL)
Support processes	ML2 Configuration management (CM)
	ML2 Process and product quality assurance (PPQA)
	ML2 Measurement and analysis (MA)
	ML5 Cause analysis and rectification (CAR)
	ML3 Decision analysis and resolution (DAR)

Process areas		
	Elementary management and support processes (ML2)	Engineering processes (ML3)
	Organization processes and further management processes (ML3)	"High-maturity" processes (ML4 and ML5)

SAE1177-2E

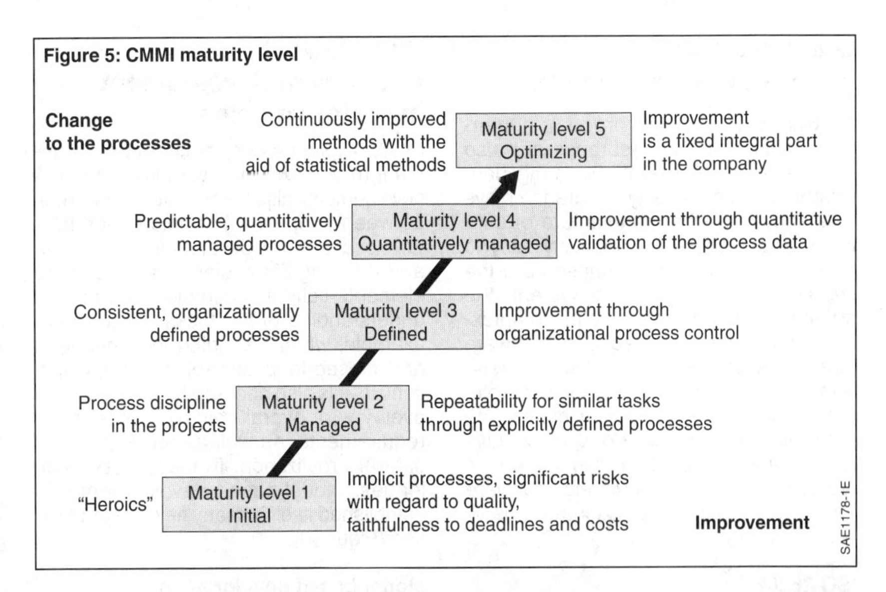

Figure 5: CMMI maturity level

level. A maturity level is reached when all the corresponding process areas are mastered and verified by an assessment. To reach a higher maturity level, the process areas of the maturity levels underneath must also be verified again.

CMMI is used as an improvement and evaluation model for development organizations and offers good support for organization-wide process optimization and auditing of suppliers.

Automotive SPICE

SPICE stands for "Software Process Improvement and Capability Determination". Automotive SPICE [21] is a motor-vehicle-specific variant of the international ISO/IEC 15504 standard (processes in the life cycle of software, [22]). It is a model for project-specific evaluation of software development processes and concentrates, like CMMI, on the requirements for systematic development. The model content of Automotive SPICE and of CMMI is therefore very similar. Automotive SPICE concentrates more on the requirements level and affords less scope for the business-oriented interpretation of requirements. Automotive SPICE focuses (currently) only on the software and only on individual projects. In contrast, the ranges of application of CMMI go further, comprising development activities and services of all kinds and their direction through the organization. Automotive SPICE is used by motor-vehicle manufacturers as an evaluation model for software projects of suppliers.

The process assessments are carried out using the two-dimensional reference and assessment model. The "process dimension" serves to identify and select the processes to be analyzed in the assessment, the "maturity-level dimension" on the other hand serves to determine and evaluate their respective capability. The maturity-level dimension consists of the six maturity-level stages "incomplete", "performed", "controlled", "established", "predictable" and "optimizing".

Quality assurance in software development

As with any technical product, numerous tools for quality assurance are also used on software. In contrast to mechanics and electrics, quality assurance plays a lesser role in the manufacture of software, since software is relatively easy to reproduce. Important emphases are the overall functionality of the system, the quality standards, mastery of complexity, and application. Because the software in a motor vehicle also includes safety-relevant systems such as, for example, driving-dynamics and driver-assistance systems, the verifiability of quality plays an important role. Even the economic representation of the aimed-for software quality, especially in complex systems, is very important.

ISO 26262

Based on the IEC 61508 standard [23], the ISO 26262 standard [24] has been introduced by the automotive industry for drafting safety-related electric and electronic systems in motor vehicles. This comprises requirements relating to both the product and the development process, and thus comprises conception, planning, development, implementation, startup, maintenance, modification, shutdown and uninstallation both of the safety-relevant system itself and of the safety-related (risk-reducing) systems. The standard designates these phases in their entirety as the "entire safety life cycle". The products are subdivided into Safety Integrity Level SIL 1 through SIL 4 (ISO 61508) and Automotive SIL, ASIL A through ASIL D (ISO 26262). SIL 1 and ASIL A are the lowest, SIL 4 and ASIL D the highest safety integrity level.

Workflows of software development in motor vehicles

Interdisciplinary cooperation in development (e.g. between drive and electronics development), distributed development (e.g. between supplier and vehicle manufacturer or at different development locations) and the long life cycles of the software elements call for a common and holistic understanding of the tasks involved. For example, when designing the open-loop and closed-loop control functions of a vehicle, it is also necessary to maintain an overview of overall reliability and safety requirements, as well as the aspects of software realization. To meet these challenges, model-based development has established itself in many fields of application (Figure 6).

Model-based development

Model-based development distinguishes between two areas. The logical system architecture comprises and describes the virtual area of the models, while the technical system architecture contains the real ECUs and vehicles. The "logical system architecture" is gray in the figures and the "technical system architecture" is white. This procedure is described in terms of open-loop and closed-loop control functions, but is also suitable for general realization of functions – for example, for monitoring and diagnostic functions.

A graphics-based function model considers all system components and can be used as the basis for a general understanding. In software development, the use of model-based development methods, with notations such as block diagrams or finite state machines, is replacing written software specifications to an increasing extent. This method of modeling software functions has still other advantages. If the specification model is formally described, i.e. unambiguously and without leeway for interpretation, as a mathematical function, the specification can be executed on a computer in a simulation, and tested quickly and realistically in the vehicle itself by means of "rapid control prototyping". Inconsistencies can also be uncovered more easily.

Methods of automated code generation can be used to implement the specified function models as software components for ECUs. The function models, therefore, need to contain additional software design information. These may include optimization measures, depending on the product characteristics required by the electronic system. Automated code generation also ensures that the code has consistent quality characteristics.

In the next step, virtual environment models, which are supplemented if necessary by real components such as injectors, simulate the environment of the ECUs, facilitating "In-the-loop tests" in the laboratory. Compared to test-bench and road tests, this increases flexibility and enhances test depth, thereby making test cases easier to reproduce.

The calibration of the software functions in the electronic system must consider the vehicle-specific settings, for example, parameters stored in the form of characteristic values, characteristic curves, and characteristic maps of these functions. In many cases, this matching does not take place until a later stage in the development process, frequently directly in the vehicle while the systems are running. The trend, however, is increasingly towards an earlier preliminary data feed; i.e. starting in the early development phases the most realistic calibration data possible is determined using models or empirical values. Because of the multitude of calibration variables and the mutual dependence, the calibration requires suitable procedures and tools, because in the end the quality of the calibration engineering, i.e. the exact adaptation of the software to the vehicle, decides the extent to which the potential of the software is utilized.

Efforts to shift development steps ahead are increasing, i.e. in the realm of simulated (virtual) environment, to detect faults early and save on expensive test hardware and experimental vehicles

Figure 6: Development steps in the model-based development of software functions

Step 1: Modeling and simulation of the software functions for the ECU and of the vehicle, driver and the environment on the computer (software functions are all the functional program parts of the software).

Step 2: Rapid control prototyping of the software functions in the real vehicle,

Step 3: Implementation of the software functions in the real vehicle,

Step 4: Integration and testing of the ECUs with In-the-Loop test systems, laboratory assemblies and test benches,

Step 5: Testing and calibration of the software functions and of the ECUs in the vehicle.

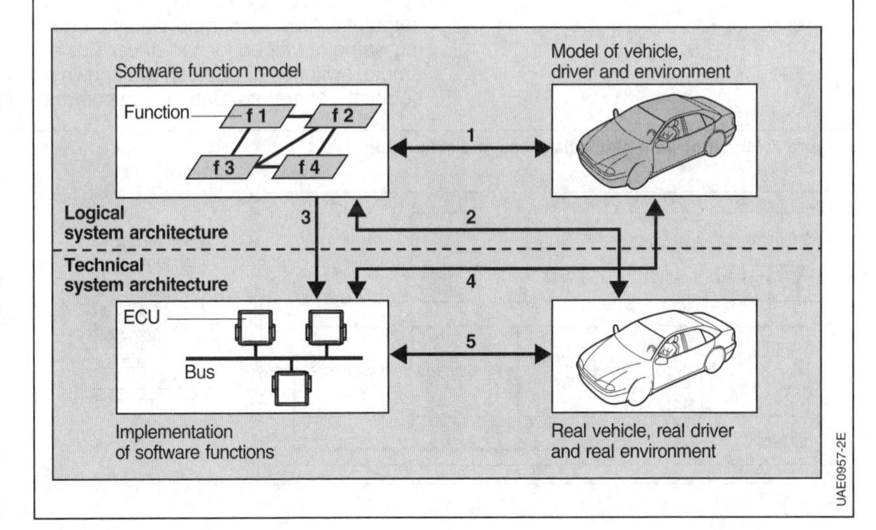

(front loading). High-performance simulation tools and virtual ECU environments make this possible.

Function and ECU networks

This procedure can also be applied to the development of function networks and ECU networks. In these cases, however, additional aspects are introduced, such as:
- combinations of modeled, virtual, and functions already realized in ECU code, and
- combinations of modeled, virtual, and realized mechanical components and hardware.

It is useful, therefore, to distinguish between functions at an abstract level, and technical realization at a more concrete level. The concept of a separate abstract and concrete approach can be applied to all vehicle components, the driver, and the environment.

Modeling and simulation of software functions

Modeling

Block diagrams should be used where possible to model open-loop and closed-loop control systems. These diagrams use blocks to represent the response of components, and arrows represent the flow of signals between the blocks (Figure 7). As most of the systems are multivariable systems, all signals are generally in vector form. These are classified into:
- measurement or feedback variables y,
- open- or closed-loop control output variables u^*,
- reference or setpoint variables w,
- driver setpoint values w^*,
- open- or closed-loop control variables y^*,
- manipulated variables u,
- disturbance values z.

The blocks are classified into:
- open-or closed-loop control,
- models of actuators,
- system model,
- models of setpoint generators and sensors, and
- model of the driver and the environment.

The driver can influence the functions of the open- or closed-loop controller by defining setpoint values. All components for entering setpoint values for the driver (for example, switches or pedals), are known as setpoint generators. Sensors, in contrast,

Figure 7: Modeling with block diagrams and simulation

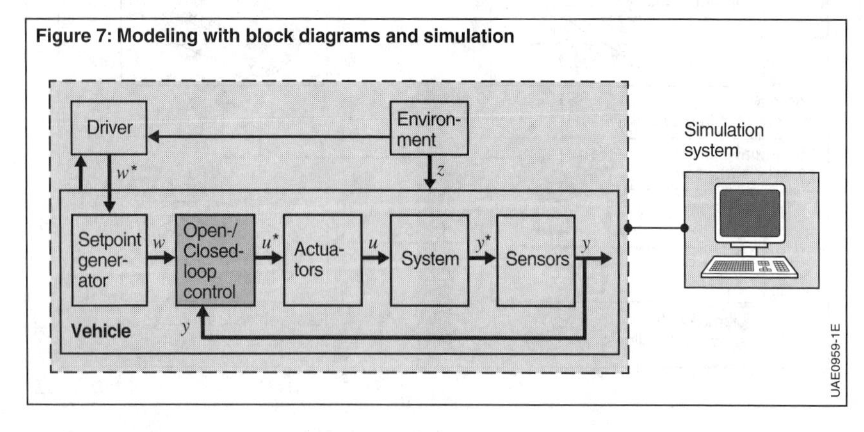

record signals from the plant. A model of this type can be executed on a simulation system (for example, a PC), thus allowing it to be analyzed in more detail.

Rapid control prototyping of software functions

Rapid control prototyping in this context includes all methods for the early implementation of specifications of open-loop and closed-loop control functions in the real vehicle. To allow this, the modeled open- or closed-loop control functions must be implemented in the test. Experimental systems can be used as an implementation platform for the software parts of the open- and closed-loop control functions (Figure 8).

The experimental systems are connected to the setpoint generators, sensors and actuators, as well as the other vehicle ECUs that belong to the overall system. The interfaces to the real vehicle mean that the software functions implemented in the experimental system – and in the ECU – take into account real-time requirements.

Real-time computer systems with considerably higher computing power are usually deployed as ECUs in experimental systems. PCs are increasingly being used as the processor core for this task. This allows the model of a software function to be automatically converted from a specification into an implementable model using a rapid control prototyping tool that is subject to standardized rules. The specified behavior can then be modeled as accurately as possible.

Experimental systems with a modular structure can be configured specifically for the application, for example, in terms of the required interfaces for input and output signals. The overall system is designed for deployment in the vehicle, and is operated using a computer, e.g. a PC. This allows the testing of software function specifications directly in the vehicle at an early stage. The specifications can then be changed as required.

When using experimental systems, there is a choice between bypass and fullpass applications.

Bypass or fullpass applications

Bypass applications are mainly used if only a few software functions are under development, and an ECU with tried-and-true basic functionality is available – for example, from a previous project.

Bypass applications are also suitable if the sensor and actuator functions of an ECU are very complex, and supporting them requires significant effort on the part of the experimental system (for example, in the case of engine ECUs).

Figure 8: Rapid control prototyping for software functions in the real vehicle

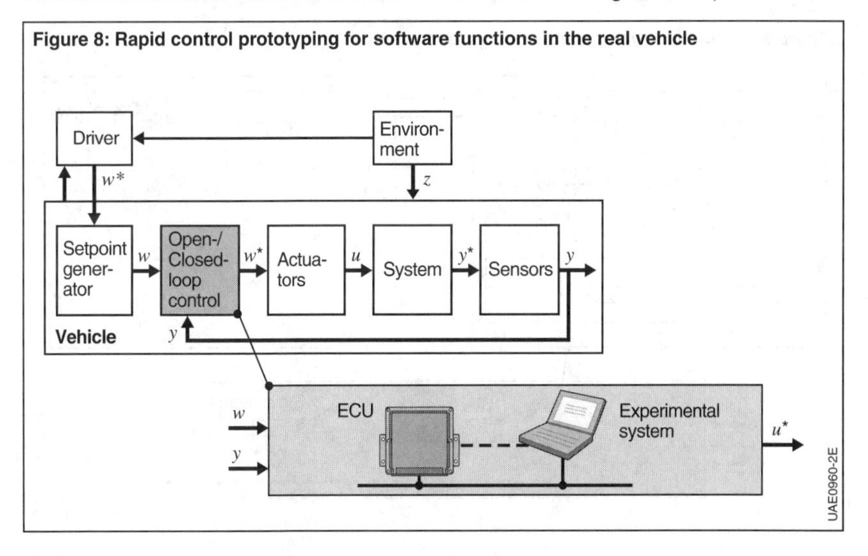

Fullpass applications are more suitable if an ECU of this type is not available, if additional setpoint generators, sensors, and actuators also need to be tested, and if the scope of the ECU is of manageable complexity.

It is also possible to combine a bypass of individual software parts and a fullpass of the entire software. This has the advantage of increased flexibility.

Bypass applications

Bypass development is suitable for early testing of an additional or modified software function of an ECU in the vehicle. The new or modified software function is defined by a model and run on the experimental system. This requires an ECU that can run the basic functionality of the software system, support all the necessary desired-value generators, sensors, and actuators, and provide a bypass interface to the experimental system. The new or modified software function is developed using a rapid control prototyping tool. It is then run on the experimental system (Figure 9).

This approach is also suitable for further developments to existing ECU functions. In this case, the existing functions in the ECU are still used, but they are modified to the extent that the input values are sent via the bypass interface, and the output values from the newly developed bypass function are used. The required software modifications to the ECU are called the bypass hooks. These can also be used in previously compiled software with modern development tools. For the essential synchronization of functional computing between the ECU and the experimental system, a procedure is normally adopted in which the ECU triggers computation of the bypass function on the experimental system via a control flow interface. The ECU monitors the output values of the bypass function for plausibility.

The bypass can also be effected via the vehicle bus (e.g. CAN). Even direct access to the CPU of the ECU via the microcontroller interfaces is possible by means of an emulator probe.

Fullpass calibration

If a completely new function is to be tested in the vehicle, and an ECU with a bypass interface is not available, the test can be carried out using a "fullpass" development. In this case, the experimental system must support all the setpoint generators, sensors, and actuator interfaces for the function. The real-time behavior of the function must also be defined and guaranteed by the experimental system (Figure 10). In general, this is performed by a real-time operating system on the fullpass computer.

Figure 9: Prototype development with bypass system

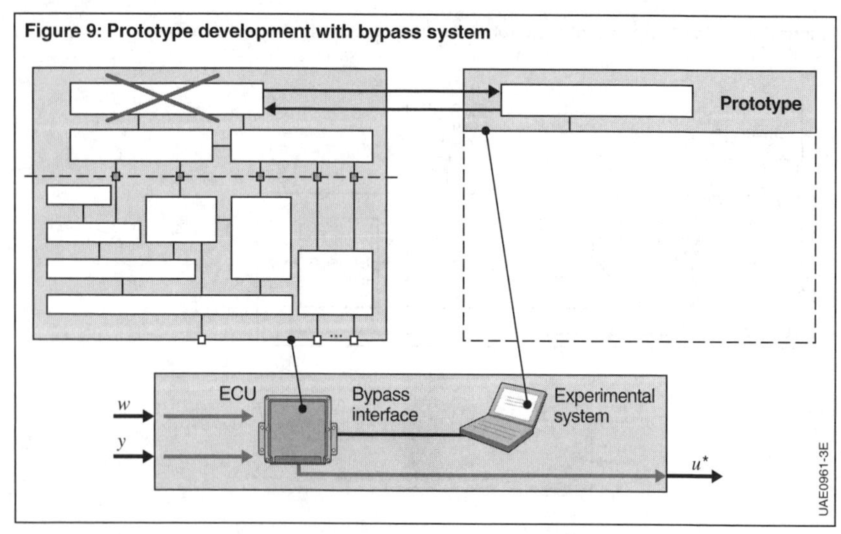

Virtual prototyping
In the case of complex systems, it is advantageous to test functions as early as possible. One possibility is offered by virtual prototyping. Here the prototype is tested on a virtual environment model. The operating system of the later ECU (e.g. RTA) runs on the experimental system. This enables the time response of the later software to be considered as well.

Figure 10: Prototype development with fullpass system

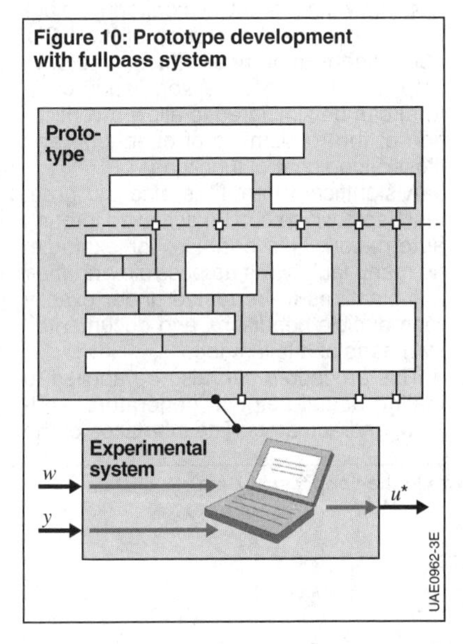

Design and implementation of software functions

Based on the specification of data, the functional behavior and the real-time behavior of a software function, all the technical details of the ECU network, the implemented microcontroller, and the software architecture must be taken into account at the design stage. The final implementation of the software functions can then be defined and executed on the basis of software components (Figure 11).

In addition to decisions regarding the design and behavior of a software function, that takes the time and discrete-related functions of the microcontrollers into consideration, an implementation includes design decisions regarding real-time behavior, distribution and integration of microcontrollers and ECUs, and reliability and safety requirements of the electronic systems. All requirements for electronic systems and vehicles from the production and service viewpoints must also be taken into account (for example, monitoring and diagnostics concepts, the parameterization of software functions, or software updates for ECUs in the field).

Generation of the code and of the accompanying data (e.g. data for documentation, variant management or preliminary data feed of the calibration) is often performed automatically in accordance with established standards.

Figure 11: Realization of open-loop and closed-loop control functions using an ECU network

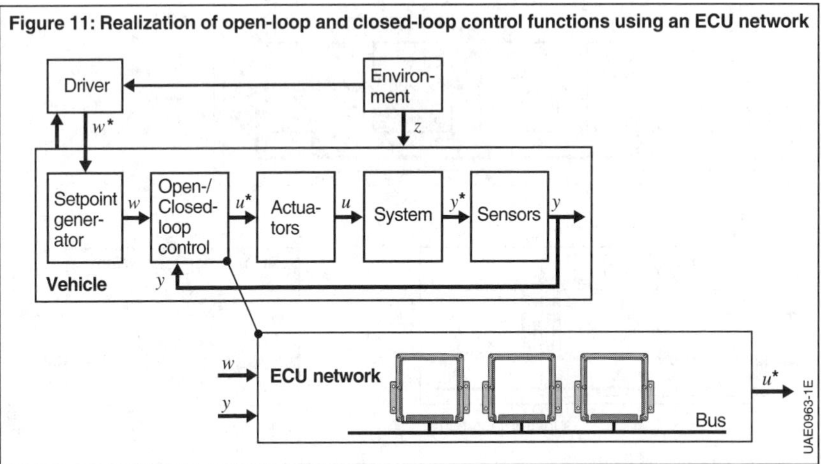

Integration and testing of software and ECUs

Requirements

Prototype vehicles are often only available in limited numbers. It means that a component supplier often does not have a complete or up-to-date integration and test environment for the components supplied. Restrictions in the test environment can sometimes limit the possible test steps. For this reason, the environment models often serve as the basis for test systems and test benches in the integration and test phase.

Component integration is a synchronization point for all the individual component developments involved. The integration test, system test, and acceptance test cannot be carried out until all the components are available. For ECUs, this means that the software functions can only be tested when all the components in a vehicle system are available (ECUs, setpoint generators, sensors, actuators, and system). The use of In-the-Loop test systems in the laboratory allows early verifications on ECUs in a virtual test environment in the absence of actual peripheral components (Figure 12).

This allows tests to be performed and automated under reproducible laboratory conditions with a high level of flexibility. In contrast to tests on the test bench or in the real vehicle, the full, unrestricted range of operating states can be tested (for example, an engine ECU can be tested at all load and rpm situations). Vehicle wear and failure situations are easy to simulate, and allow testing of the monitoring, diagnostic, and safety functions of the ECU. Component tolerances (for example, in setpoint generators, sensors, and actuators) can be simulated to allow a verification of the robustness of open-loop and closed-loop control functions.

A significant benefit is also the great test depth which can be achieved through automation. This enables, for example, as many fault types as possible and their combinations to be tested under exactly reproducible conditions and documented by means of a failure log.

This procedure can also be applied to testing actual setpoint generators, sensors, and actuators. The interfaces of the

Figure 12: Integration and testing of ECUs with an In-the-Loop test system

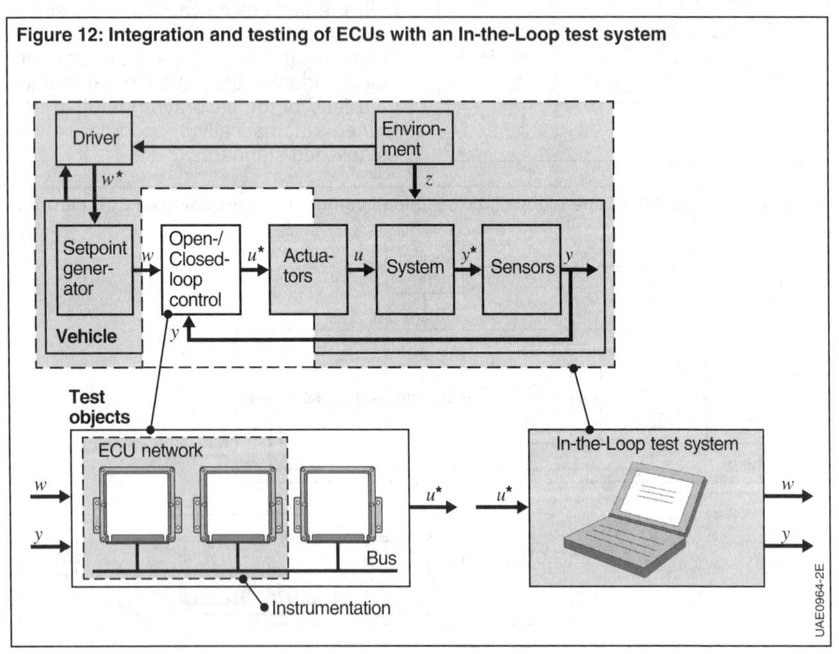

test system need to be modified accordingly. Any intermediate steps can also be incorporated into the procedure.

A structure as illustrated in Figure 12 depicts the ECUs in the form of a black box. The behavior of the ECU functions can only be assessed on the basis of the input and output signals w, y and u^*. This "black-box view" is sufficient for simple software functions. But testing more complicated functions requires the integration of a measurement procedure for internal ECU intermediate variables. This type of measuring technique is also known as instrumentation. The testing of diagnostic functions also requires access to the fault memory via the ECU diagnostics interface, and this requires the integration of a measurement and diagnostics system.

In-the-loop test systems

In-the-Loop is a test procedure in which an embedded electronic system is connected via interfaces to a real (e.g. sensors, actuators) or virtual environment (mathematical models). The reaction of the system is analyzed and played back to the system. Depending on the type of test specimens, a distinction is made between the following test systems:

- In the case of Model-in-the-Loop (MiL), the function model of the software is tested. The model runs on a development computer.
- In the case of Software-in-the-Loop (SiL) the software code is tested. It runs on a development computer.
- In the case of Function-in-the-Loop (FiL), the software code is also tested. However, in contrast to SiL, this runs on the destination hardware. Coupling between the software and the environment model is performed by means of hooks and an emulator probe.
- In the case of Hardware-in-the-Loop (HiL), the complete ECU is tested by means of the input and output interfaces (I/O). Combinations of FiL and HiL are also used. PCs are increasingly being used as simulation computers.

In-the-Loop test systems can be used to validate and further develop software and hardware.

Figure 13: Integration and testing of ECUs in the real vehicle

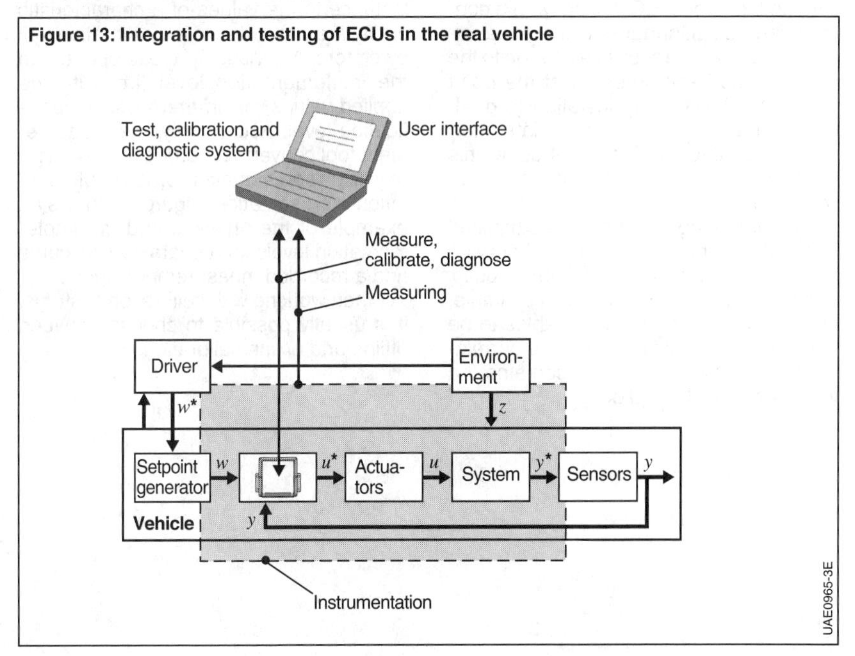

Calibration of software functions

Procedure

Each electronically controlled vehicle system can only then develop its capability when it is optimally adapted to the respective vehicle type. In order for software functions to be used in as many vehicle variants as possible, these contain variable parameters. Adapting these parameters to the corresponding vehicle variant and for every operating condition (e.g. cold running, extreme heat or altitude) is known as calibration. To reach the desired level of functionality in the complete vehicle, a multitude of characteristic values, characteristic curves and program maps – the calibration data – must be applied.

Most changes to the vehicle require a change to the calibration. An example here: The λ oxygen sensor in the exhaust-gas system measures the residual oxygen in the exhaust gas. The fuel quantity actually injected can be determined on the basis of this signal. In this way, it is possible to adapt the activation parameters in the engine ECU exactly on a continuous basis. Changes to the exhaust-gas system which result in a change to the exhaust-gas back pressure at the point where the λ sensor is installed (e.g. altered exhaust manifold, particulate filter) must be adapted. If this is not done, this will result in a deterioration in consumption and emissions.

Variant management can be simplified in development, production and service by separating program and data statuses. The program status contains, for example, all the information on the variables to be applied and their limit values and correlations, while the data status contains the variables actually applied.

Calibration takes place in the laboratory on engine and vehicle test benches during vehicle tests and under real environmental conditions on test tracks. In addition to a measurement and diagnostic system, a calibration system is often required to calibrate internal ECU parameters (such as characteristic curves and program maps). On completion of calibration, the determined data are comprehensively checked. These values are then stored in the read-only memory (EPROM or flash) of the series ECU.

The parameter values must be variable during calibration. A calibration system, therefore, consists of one or more ECUs with a suitable interface to a measurement and calibration tool (Figure 14). In addition to their deployment in vehicles, measurement, calibration and diagnostic systems can also be used in in-the-loop test systems and on test benches. It is being applied more and more often in a virtual environment. Here, high-performance tools assist in finding the trade-off (optimal compromise of the calibration data).

Changes to the parameter values, for instance to the values of a characteristic curve, are supported in the calibration tool by editors. Alternatively, these operate on the implementation level (i.e. with the applied values) or on the physical specification level. Accordingly, the measurement tool converts recorded values into a physical representation or an implementation representation. Figure 14 shows an example of the physical and the implementation levels for a characteristic curve and a recorded measurement signal.

When working with calibration systems, it is usually possible to choose between offline and online calibration.

Offline calibration

In offline calibration, the execution of open-loop control, closed-loop control, and monitoring functions, i.e. of the "drive program", are interrupted to modify or adjust parameter values. Offline application therefore leads to many restrictions. In particular when deployed on test benches and during in-vehicle testing, it always causes an interruption to the test-bench or road test.

Online calibration

In online calibration, the parameter values can be adjusted while the microcontrollers are executing the software functions. It is, therefore, possible to adjust parameter values while executing open-loop control, closed-loop control, and monitoring functions at the same time, and hence during regular use at the test bench or in vehicle.

Online calibration places higher demands on the stability of the open-loop control, closed-loop control, and monitoring functions, since the driving schedule must remain stable during all adjustment procedures if exceptions occur, for example, if the distribution of interpolation points in characteristic curves fails to rise monotonically for short periods of time. Online calibration is suitable for long-term modification of less dynamic parameters (for example, tuning of engine control functions on the engine test bench).

For applying more dynamic functions or safety-critical functions (for example, setting the software functions of an ABS system for braking maneuvers during a test drive), the settings are not adjusted during actual braking maneuvers. In this case, online calibration can still save time by avoiding any interruptions in the program execution, thus reducing the interval between two test drives.

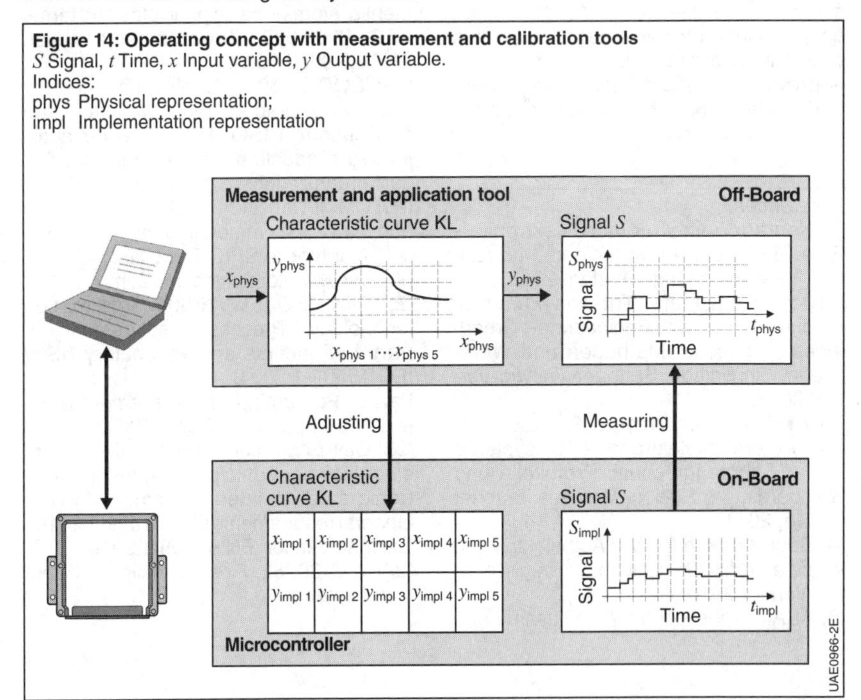

Figure 14: Operating concept with measurement and calibration tools
S Signal, t Time, x Input variable, y Output variable.
Indices:
phys Physical representation;
impl Implementation representation

Outlook

New vehicle functions and technologies will continue to lead to an increase in the scope of software in motor vehicles – even in the low price segment. In the process, the number of ECUs stays the same in many vehicles. As a consequence, the scope of functions of some ECUs is continuing to increase. In all, the system limits between the different ECUs are disappearing more and more, such as, for example, in the management of a hybrid drive. The electronics in motor vehicles are thus becoming ever more complex – including beyond an individual vehicle. The economic mastery of complexity is therefore certainly a key element in future automotive software engineering and the associated development environments.

Another trend is the increasing virtualization of development. Virtualization involves integrating parts of later development steps, such as testing and calibration, into the earlier development phases such that the function models can be optimally tested in-the-loop and fed with data on a preliminary basis early on. In this way, failures are detected earlier, support is provided for distributed development and bottlenecks shortly before series startup are avoided.

References

[1] Konrad Reif: Automobilelektronik – Eine Einführung für Ingenieure, 4th Edition, Vieweg+ Teubner Verlag, 2012.
[2] Jörg Schäuffele, Thomas Zurawka: Automotive Software Engineering – Grundlagen, Prozesse, Methoden und Werkzeuge, 5th Edition, Springer-Vieweg-Verlag, 2013.
[3] Werner Zimmermann, Ralf Schmidgall: Automobilelektronik – Bussysteme in der Fahrzeugtechnik: Protokolle und Standards, 4th Edition, Vieweg+ Teubner Verlag, 2011.
[4] Robert Bosch GmbH: Autoelektrik und Autoelektronik, 6th Edition, Vieweg+Teubner Verlag, 2010.
[5] ASAM website: http://www.asam.net/.
[6] IEC website: http://www.iec.ch/.
[7] ISO website: http://www.iso.org/.
[8] IEEE website: http://www.ieee.org/portal/site.
[9] MISRA website: http://www.misra.org.uk/.
[10] SAE website: http://www.sae.org/.
[11] OSEK VDX Portal website: http://www.osek-vdx.org/.
[12] AUTOSAR website: http://www.autosar.org/.
[13] ISO 14230: Road Vehicles – Diagnostic Systems – Keyword Protocol 2000, 1999.
[14] ISO 15765: Road Vehicles – Diagnostic Systems – Diagnostics on CAN, 2000.
[15] ISO 22900-1: Road vehicles – Modular vehicle communication interface (MVCI) – Part 1: Hardware design requirements.
[16] ISO 22901-1: Road vehicles – Open diagnostic data exchange (ODX) – Part 1: Data model specification.
[17] V model of IABG website: http://www.v-modell.iabg.de/.
[18] DIN EN ISO 9000: Quality management systems – basic principles and terms (ISO 9000:2005); Three-language version EN ISO 9000:2005.
[19] ISO/TS 16949: Quality management systems – Particular requirements for the application of ISO 9001:2008 for automotive production and relevant service part organizations.
[20] CMMI website: http://www.sei.cmu.edu/cmmi/.
[21] Automotive SPICE website: http://www.automotivespice.com/.
[22] DIN ISO/IEC 15504: Information technology – Process assessment –
Part 1: Concepts and vocabulary (ISO/IEC 15504-1:2004).
Part 2: Performing an assessment (ISO/IEC 15504-2:2003 + Cor. 1:2004).
[23] DIN EN 61508: Functional safety of electrical/electronic/programmable electronic safety-related systems – Part 1: General requirements (IEC 61508-1:2010); German version EN 61508-1:2010.
[24] ISO 26262: Road vehicles – Functional safety.

Automotive networking

Bus systems

Networks for data communication, also known as bus systems or protocols, are widely used in today's motor vehicles. Various components such as sensors, actuators or ECUs – the "nodes" – are connected to each other via a single channel (Figure 1). A wealth of data, also known as messages, telegrams, packets or frames, is exchanged via this channel. For example, the driving speed is determined in the Electronic Stability Program (ESP) and transferred to all the other ECUs which are networked as nodes in the bus system.

Advantages of bus systems
Compared with conventional cabling, in which the transmitter and receiver of information are each connected by separate lines, bus systems offer significant advantages:
- The costs of materials for the cables are lower (which offsets the higher costs for the electronics).
- The space required and the weight of the cabling are lower.
- The number of plugs susceptible to faults is lower, thus fewer failures occur in total.
- Data can be distributed to various receivers; for example, the signals of a sensor can be used by several systems.
- All the systems in the vehicle which are connected via the bus can be reached from one access point. This provides for simpler diagnostics and the configuration of all the ECUs at the end of the line.
- The performance of calculations can be distributed to different ECUs.
- Analog sensor signals must be digitized for data processing. The sensor signals can be conditioned directly in the sensor, the information is then disseminated via the bus.

Requirements of buses
General requirements
In order to be able to be used in vehicles, buses must satisfy typical requirements. A bus must use a transmission process which has control over the propagation times, attenuations and reflections of the signals for cables up to 40 m long with which typically all the parts of the vehicle are reached. A network must be able to connect several dozen users in the process.

A bus must be able to withstand the rough ambient conditions in the vehicle with regard to temperature, vibrations and electromagnetic interference.

Because vehicles are produced in high numbers, even small savings potentials in the cost of the bus hardware of the individual vehicle must be utilized in the design of the bus. It is equally necessary for there to be several competing manufacturers of bus components.

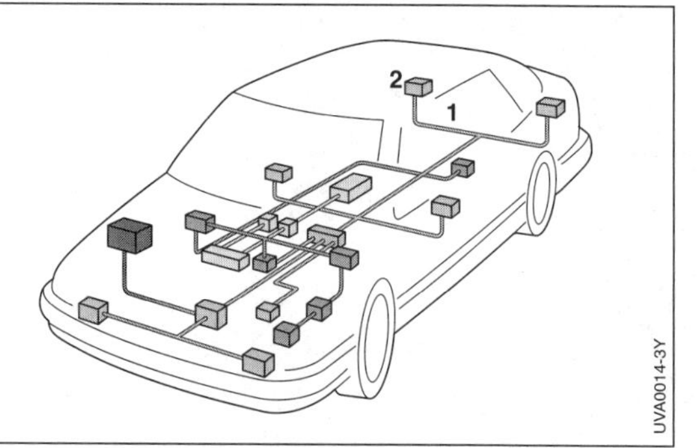

Figure 1: Automotive networking
Schematic representation of nodes and data lines.
1 Data line,
2 Node (network user).

There is a large number of equipment variants of a vehicle type which must be covered by a basic configuration of the bus. The additional installation of an optional extra must not affect the other vehicle systems.

The performance of the bus must be clearly defined in a public standard in order to have a benchmark against which components can be verified. This ensures that verified components of different suppliers function together in a single network.

Special requirements
Different systems in the motor vehicle have different requirements and therefore necessitate the use of different buses. The required data transfer rate very much depends on the application. It ranges from a few bit/s for switching the lights, through engine management with a few 100 kbit/s, right through to video applications with several Mbit/s.

For systems in which a failure or a delayed execution of a function is safety-relevant (e.g. airbag, electric power steering), a maximum duration for data transfer (latency) must be guaranteed in all cases. When the time response of the bus is defined and reproducible at all times, this is referred to as determinism. In this way, transfer times for messages, in particular, are already known in the design phase.

For safety systems such as electric power steering it must be verified that everything has been done in terms of state-of-the-art technology to avoid flaws/faults in design and to ensure correct functioning under all permissible boundary conditions.

Transfer malfunctions which inevitably occur during operation must be detected and dealt with for the relevant application.

Technical principles

Components of a bus

In order for a network to function as planned, calculations must be carried out constantly in every network user. This is performed for the most part by a special piece of hardware (Figure 2), the "communication controller". This removes the load from the actual computer (host), which can also perform this function – although less efficiently. The communication controller can be implemented as a separate semiconductor component; in the case of many microcontrollers, however, communication controllers for some buses are already integrated.

A further component, the bus driver or transceiver, converts the signal which it receives from the communication controller into the physical signal on the bus line (e.g. voltage level). It converts data received on the bus line and forwards them to the communication controller.

OSI reference model

The ISO (International Standardization Organization) developed the OSI reference model (Open Systems Interconnection), which is often used as the basis for describing and comparing communications protocols. In this model, the function of a data-communication system is

Figure 2: Hardware of a network user

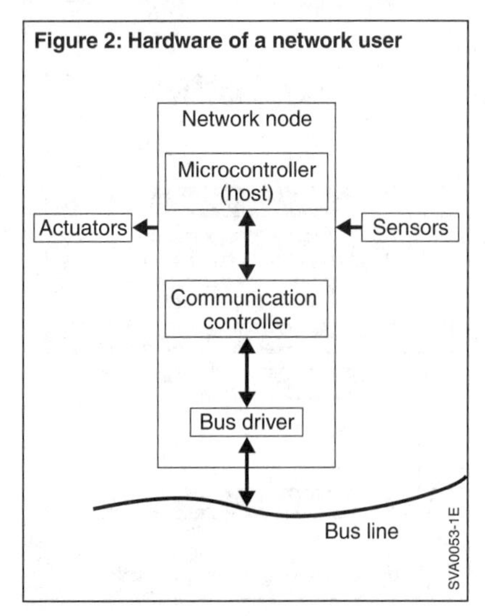

broken down into different hierarchical layers which each use the functions made available by a different layer. The OSI reference model is an aid to structuring conceptually the function of a communication system. However, finding efficient solutions is not necessarily supported by observing this model.

Seven layers are defined (Figure 3), but many protocols are only defined on one part thereof. The upper layers in the OSI reference model (application layer, presentation layer and session layer) are for the most part not served by motor-vehicle buses.

Physical layer
The physical layer defines the physical properties of the transmission medium (e.g. voltage level or shape of plugs). In motor vehicles for the most part, electromagnetic signals in the kHz to MHz range are used on special cables and optical fibers. Radio of different frequencies or the joint use of cables already provided with electrical power for supply purposes are technologies which have not yet been tested in large batches.

Figure 3: OSI reference model

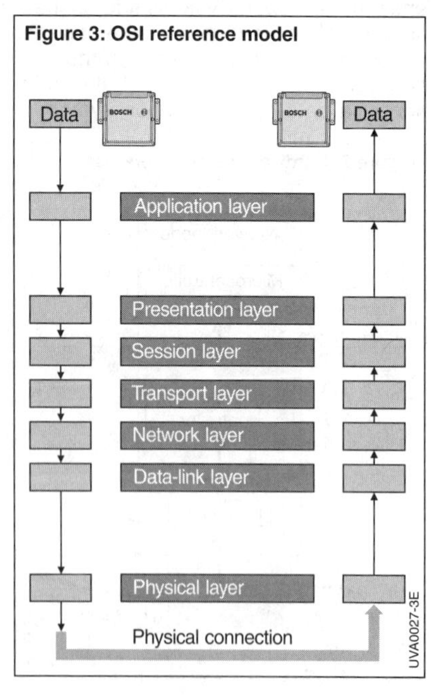

The most widely used method is, because of the low costs involved, transmission by cable, particularly as a voltage difference on twisted two-wire cables, with or without shielding against radiation, or on single-wire cables with a voltage reference to ground.

Optical fibers made from plastic or glass fibers (for the infrared spectral range) are used above all wherever high data rates are needed. They are insensitive to electromagnetic radiation (e.g. by the ignition system), but are costly to lay and their resistance to aging has not yet been adequately assured.

The physical possibilities of the medium also limit the types of encoding of a bit. With optical media, there are the two states "light" and "no light", which offer encoding by means of amplitude (brightness).

High-frequency electrical voltage signals on a cable offer different possibilities for representing bits. The simplest and usual method used in motor-vehicle buses is encoding in which each bit is assigned a voltage value (amplitude modulation) which is applied for the entire bit duration (NRZ, Non Return to Zero). In this frequency range, open cable ends must, if necessary, be terminated by terminating resistors in order to prevent reflections from disrupting the signal shape.

Data-link layer
The data-link layer effects the correct transport of data between neighboring nodes. Data bits are arranged into frames. By adding further bits such as checksums or numberings, it is possible to detect or even correct errors that occur during transfer. Alternatively, an error can be rectified by requesting a new transfer.

Network layer
If not every network user is directly connected with every other, a route must be found for the data which passes through intermediate stations. The search for a route is also called routing and includes different characteristics of the transmission paths (particularly bandwidth) and the intermediate stations, also called routers.

Transport layer
The transport layer's functions include disassembling large data packets and reassembling them at the receiver, which the parts potentially reach via different routes at different times, or if transmission errors occur ensuring that a packet is transmitted again. The transport layer also serves to hide the properties of the transmission path from the application and to offer services independently.

Access methods
Data are usually transmitted as fixed-structure packets (frames), which, as well as the actual content, also contain control and test data such as source, destination, priority and check bits to protect against corruption.

A central function of a network to which several users have access at the same time is the administration of the token in order to avoid conflicts. This is known in buses as arbitration. It should be borne in mind here how urgently a message has to be transported. Different mechanisms are circulated.

Time control
With time-controlled buses, conflict-free access is possible whereby each node receives permanently allocated times (slots, columns) for its emissions. These recurring times are combined into cycles, and cycles combined into communication matrices (Figure 4).

The messages (reference messages, Sync Frame) themselves are used to synchronize the clocks in the nodes. If there are numerous timing nodes (time master) in the network, the clocks are synchronized on the based of the measured

Figure 4: Communication matrix of a time-controlled bus (example: TTCAN)

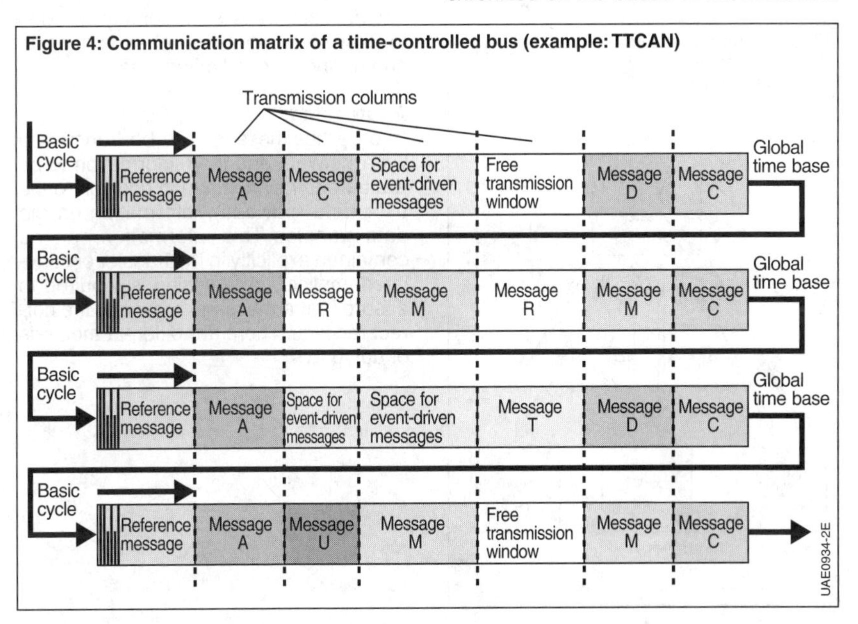

reception times with the expected receptions times. The strict periodicity (determinism) makes it possible to quickly detect when a message fails to materialize, which is required particularly in safety systems such as electric-motor-operated steering.

In addition, the time-controlled approach promotes the possibility that several subsystems are developed independently of each other, because in this way the composability of the subsystems into predictable behavior of the overall system is supported. When each subsystem has its independent time slice, two nodes cannot disrupt each other by simultaneously occupying the bus.

Master-slave

A designated node (master) grants one of the other nodes (slave) limited access (Figure 5). In this way, it determines the communication frequency by interrogating its subordinate nodes. A slave responds only when it is addressed by the master. However, some master-slave protocols permit a slave to register with the master in order to send a message.

Figure 5: Master-slave method
A Master inquiry,
B Slave response.

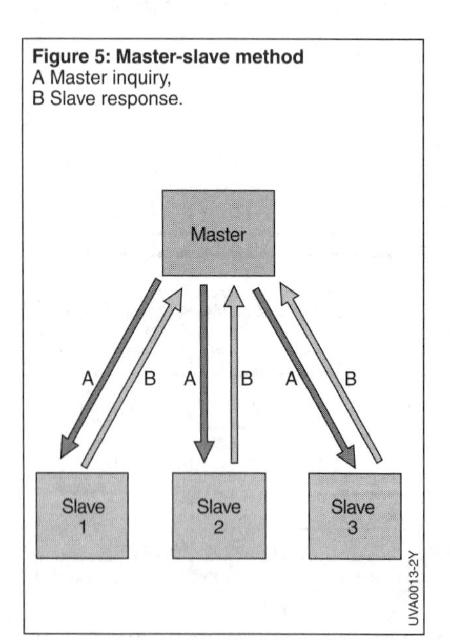

Multi-master

In a multi-master network, various nodes can access the bus automatically and send a message if the bus appears free. In other words, each node can perform the master role and all the nodes can equally start a message transfer. However, this also means that methods for identifying and dealing with access conflicts must be in place. This can be done, for example, by way of a decision phase with prioritization or a delayed resending. The use of priority control prevents a bus conflict when several nodes want to occupy the bus at the same time. The network node which has a high priority or which would like to transmit a high-priority message asserts itself and sends its message first. When the line is free again, all the nodes which have to send messages – particularly the node that is waiting – start with a renewed attempt.

The multi-master architecture has a positive effect on system availability because communication is not controlled by any single node, the failure of which would result in a total communication failure.

Token passing

The right of access is temporarily awarded to a node, which then passes it on to another node for a limited time.

Addressing

In order for messages to be transferred via a network and their information to be evaluated, they contain, in addition to the user data (payload), information on the data transfer. This information can be contained explicitly in the transfer or stipulated implicitly. Addressing is required to ensure that a message arrives at the correct receiver. There are different methods of doing this.

User-oriented method
Data are exchanged here on the basis of node addresses (Figure 6a). The message transmitted by the sender contains, as well as the data to be transferred, the address of the destination node. All the receivers compare the transmitted receiver address with their own and only those receivers with the correct address evaluate the message. Messages can be directed to individual addressees, to groups (multicast) or to all the nodes in the network (broadcast).

Most conventional communication systems (e.g. Ethernet) operate in accordance with the principle of user addressing.

Message-oriented method
With this method, it is not the receiver nodes but the messages themselves which are addressed (Figure 6b). A message is identified according to its content by a message identifier which was established in advance for this information type. With this method, the sender does not require any knowledge of the destination of the message, since each receiver node decides for itself whether to process the message. Several nodes can also accept and evaluate the message.

Figure 6: Addressing types
a) User-oriented method,
b) Message-oriented method.
Adr i, $i = 1 \ldots n$, address of nodes.
Id i, $i = 1 \ldots n$, message identifiers.

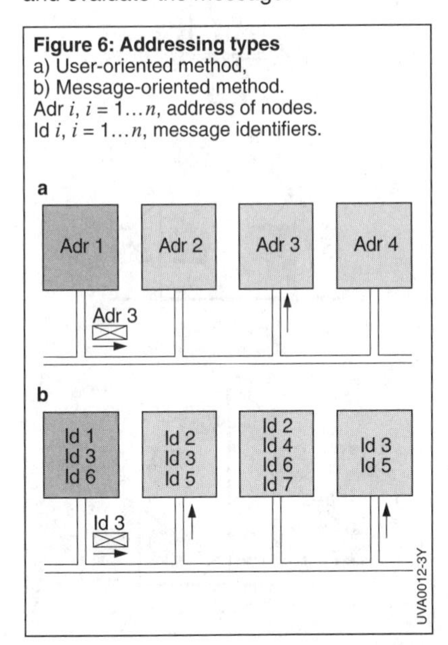

Most bus systems in the automotive field (e.g. CAN, FlexRay) operated according to the message-oriented method.

Transfer-oriented method
Even transfer features can be used to identify a message. If a message is always sent in a defined time slot, it can be identified using this item. For safeguarding purposes, this addressing is also combined in many cases with a message- or user-oriented addressing.

Network topology
A network topology refers to the structure of network nodes and connections. This shows which nodes are connected to each other, but not the details, such as the length of the connection for example. For different applications of communication networks, there are different requirements of the topology used.

Bus topology
This network topology is also referred to as a linear bus. The core element is a single cable, to which all the nodes are connected by way of short connecting cables (Figure 7a). In this topology, it is very easy to expand the network by additional users. Messages are sent out from the individual bus users and distributed to the entire bus.

If a node fails, the data expected from this node are not available to the other nodes in the network, but the remaining nodes can continue to exchange messages. However, a network with a bus topology will fail completely if the central cable has a fault (e.g. in the event of a cable break).

Star topology
The star topology consists of a central node (repeater, hub, star), to which all the other nodes are connected by way of individual connections (Figure 7b). A network with this topology can therefore be easily expanded if free connections are available at the central element.

Data are exchanged via the individual node connections with the central star, where a distinction is made between active and passive stars. The active star contains a computer which processes and forwards the data. The network's performance is significantly determined by the performance of this computer. The central node, however, does not necessarily have to have special control intelligence. A passive star only joins the bus lines of the network users.

If one network user fails or a connecting cable to the central node is faulty, the remaining network continues to be operational. If, on the other hand, the central node fails, then the entire network is rendered inoperational.

Ring topology

In the ring topology, each node is connected to its two neighbors. This creates a closed ring (Figure 7c). In a ring, data are transferred in only one direction from one station to the next. The data are checked in each case after they are received. If they are not intended for this station, they are renewed (repeater function), amplified, and sent forward to the next station. The data to be transferred are forwarded in the ring, i.e. from one station to the next, until they reach their intended destination or arrive at the starting point again. When a message has passed through the complete ring, reception is acknowledged by all the nodes. Once one station in a single ring fails, data transfer is interrupted and the network fails completely.

Rings can also be structured in the shape of a double ring in which data are transferred in both directions. In this topology, the failure of one station or of a connection between two stations can be managed, since all the data continue to be transferred to all the operational stations in the ring.

Daisy-chain topology

The daisy-chain topology looks like the ring topology when a connection is removed. Here, the first component is connected directly to data-processing equipment (e.g. a computer). The subsequent components are connected to their preceding components (series-connection principle), thereby creating a chain. Messages thus pass through several nodes until the destination is reached.

Figure 7: Network topologies
a) Bus topology, b) Star topology,
c) Ring topology, d) Mesh topology,
e) Star-bus topology,
f) Star-ring topology.

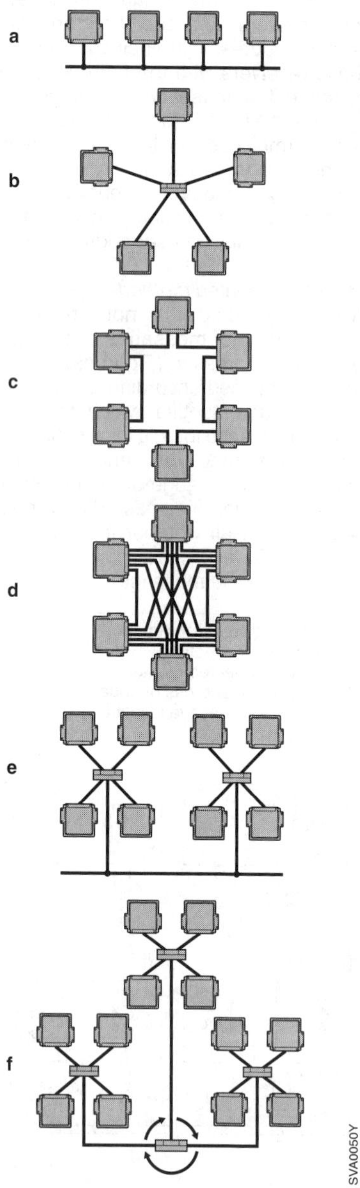

Mesh topology
In a mesh topology, each node is connected to one or more further nodes (Figure 7d). In the event of the failure of a node or a connection, there are diversions along which the data can be routed. This network is therefore characterized by high failure tolerance. However, the costs of networking and transporting the messages are high.

Hybrid topologies
In hybrid topologies, different network topologies are connected. The following combinations are possible, for example:
- Star-bus topology: The hubs of various star networks are connected to each other as a linear bus (Figure 7e).
- Star-ring topology: The hubs of various star networks are connected to the main hub (Figure 7f). In this main hub, the hubs of the star networks are connected in a ring shape.

Transfer reliability
An important feature of buses, particularly if they are to be used in safety-relevant systems, is their ability to identify faults and, if necessary, to maintain limited operation. This includes the possibility of detecting the corruption of data. Electromagnetic radiation from the surroundings (e.g. from the ignition coil) on the cables may cause some bits to assume incorrect values at the receiver. These incidents can be identified by transferring check information in addition to the data. The simplest case is an additional parity bit, which is used to detect whether the correct number of ones is even or uneven. A generalization of this is the “Cyclic Redundancy Check” (CRC), with which different qualities of protection can be obtained using a selectable number of check bits.

One way of ensuring limited operation is to bypass failed lines and prevent blockage of the bus by faulty nodes.

It is ensured that an item of data sent out by a node, e.g. the vehicle speed, is available either to all the addressees in the network or to none (consistency). The situation where two nodes have different images of the speed value therefore cannot arise. For this purpose, a node signals that reception has failed or refrains from acknowledging correct reception (acknowledge). The other nodes then reject the data already correctly received by them.

Buses in motor vehicles

For a quick overview, buses are frequently classified according to the scheme in Table 1.

Table 1: Classification of bus systems

Class A	
Data transfer rates	Low data rates (up to 10 kbit/s)
Applications	Networking of actuators and sensors
Represented by	LIN, PSI5
Class B	
Data transfer rates	Medium data rates (up to 125 kbit/s)
Applications	Complex mechanisms for error handling Networking of control units in the comfort area
Represented by	Low-speed CAN (CAN-B)
Class C	
Data transfer rates	High data rates (up to 1 Mbit/s)
Applications	Real-time requirements Networking of control units in the drive and chassis area
Represented by	High-speed CAN (CAN-C)
Class C+	
Data transfer rates	Very high data rates (up to 10 Mbit/s)
Applications	Real-time requirements Networking of control units in the drive and chassis area
Represented by	FlexRay
Class D	
Data transfer rates	Very high data rates (from 10 Mbit/s)
Applications	Networking of control units in the areas of telematics and multimedia
Represented by	MOST, Ethernet

CAN

Overview

The CAN bus (Controller Area Network) has established itself as the standard since its first series introduction in motor vehicles in 1991. But it is also often used in automation technology. The major features are:

- Priority-controlled message transmission with non-destructive arbitration.
- Low costs through the use of a low-cost twisted two-wire cable and use of a simple protocol with low computing-power demand.
- A data-transfer rate up to 1 Mbit/s for the high-speed CAN (CAN-C) and up to 125 kbit/s for the low-speed CAN (CAN-B, lower expenditure for the hardware).
- High reliability of data transfer through recognition and signaling of sporadic faults and permanent faults and through network-wide consistency via Acknowledge.
- The multi-master principle.
- High availability by locating failed stations.
- Standardization in accordance with ISO 11898 [1].

Transfer system

Logic bus states and encoding

To communicate, the CAN bus uses the two states "dominant" and "recessive", with which the information bits are transmitted. The dominant state represents "0", the recessive "1". The NRZ process (Non Return to Zero) is used as encoding for transmission, in which a zero state is not always returned to between two equal transfer states and, therefore, the time interval between two edges which is required for synchronization can become too great.

For the most part a two-wire cable with, depending on the ambient conditions, a non-twisted or twisted pair is used. The two bus lines are called CAN_H and CAN_L (Figure 1).

The two-wire cable facilitates a symmetrical data transfer, in which the bits are transferred via both bus lines using different voltages. This reduces sensitivity to common-mode interference, since interference affects both lines and can be filtered out by creating the difference (Figure 2).

The single-wire cable represents a way of lowering manufacturing costs by saving on the second cable. However, a common ground connection which assumes the function of the second cable must be available to all the bus users for this purpose. The single-wire version of the CAN bus is therefore only possible for a communication system with limited spatial reach. Data transfer on the single-wire cable is more susceptible to interference radiation, since it is not possible to filter out interference pulses as on the two-wire cable. A higher level signal on the bus line is therefore required. This in turn has a negative effect on the interference radiation. The edge steepness of the bus signals must therefore be reduced compared with the two-wire cable. This is associated with a lower data-transfer rate. For this reason, the single-wire cable is only used for the low-speed CAN in the area of body and comfort and convenience electronics. The single-wire solution is not described in the CAN specification.

There are also so-called fault tolerant transceivers, which communicate via a two-wire cable, but if a cable breaks continue operating as a single-wire system.

Voltage levels

High-speed and low-speed CANs use different voltage levels to transfer dominant and recessive states. The voltage levels of the low-speed CAN are shown in Figure 1a, those of the high-speed CAN in Figure 1b.

The high-speed CAN uses, in the recessive state on both lines, a nominal voltage of 2.5 V. In the dominant state, nominal voltages of 3.5 V and 1.5 V are applied to CAN_H and CAN_L respectively. In the low-speed CAN in the recessive state, a voltage of 0 V (max. 0.3 V) is applied to CAN_H and a voltage of 5 V (min. 4.7 V)

Figure 1: Voltage level of CAN data transfer
a) Low-speed CAN (CAN-B),
b) High-speed-CAN (CAN-C).
CAN_H CAN High level,
CAN_L CAN Low level.

a
Nominal potential
V
5
3.6
1.4
0
CAN_L
CAN_H
recessive dominant recessive
Level states
b
Nominal potential
V
3.5
2.5
1.5
0
CAN_H
CAN_L
recessive dominant recessive
Level states
SVC0018-2E

Figure 2: Filtering out of interference on the CAN bus
a) Signal level of CAN lines with interference on both lines,
b) Difference signal.
1 Interference pulse,
2 Difference signal.
CAN_H CAN High level,
CAN_L CAN Low level.

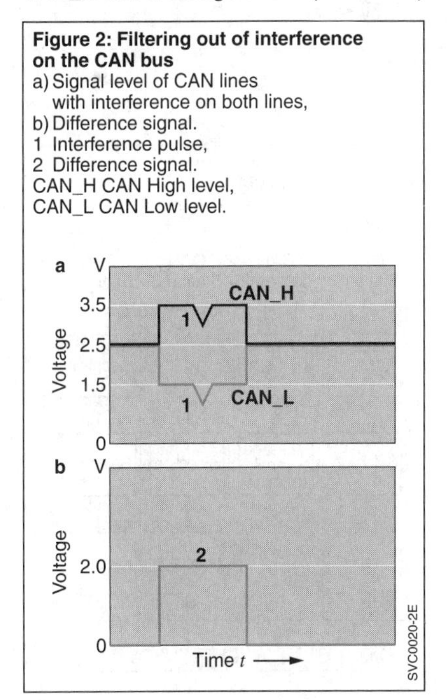

is applied to CAN_L. In the dominant state, the voltage at CAN_H is min. 3.6 V and at CAN_L max. 1.4 V.

Limits
It is important for the arbitration method in the case of CAN that all the nodes in the network see the bits of the message identification (frame identifier) simultaneously so that a node – while it is still transmitting a bit – sees whether another node is also transmitting. Delays result from the signal propagation time on the data bus and the processing times in the transceiver. The maximum permissible transfer rate is thus dependent on the overall length of the bus. ISO specifies 1 Mbit/s for 40 m. For longer lines, the possible transfer rate is roughly inversely proportional to the line length. Networks with a 1 km reach can be operated with 40 kbit/s.

CAN protocol

Bus configuration
CAN operates according to the multi-master principle, in which a linear bus topology connects several nodes of equal priority rating.

Content-based addressing
CAN uses message-based addressing. This involves assigning a fixed identifier to each message. The identifier (ID) classifies the content of the message (e.g. engine speed). Each station processes only messages whose identifiers are stored in the list of messages to be accepted. This is called acceptance checking (Figure 3). Thus, the CAN requires no station addresses for data transmission, and the nodes are not involved in managing the system configuration. This facilitates adaptation to variations in equipment levels.

Logic bus states
The CAN protocol is based on two logic states: The bits are either "recessive" (logic 1) or "dominant" (logic 0). When at least one station transmits a dominant bit, it overwrites the recessive bits simultaneously sent by other stations.

Bus arbitration and priority assignments
Each station can begin transmitting a message as soon as the bus is unoccupied. When several stations start to transmit simultaneously, a "wired-and arbitration scheme" becomes active to resolve the resulting bus-access conflicts. The arbitration scheme ensures that the domi-

Figure 3: Addressing and acceptance checking
Station 2 sends, stations 1 and 4 receive the data.

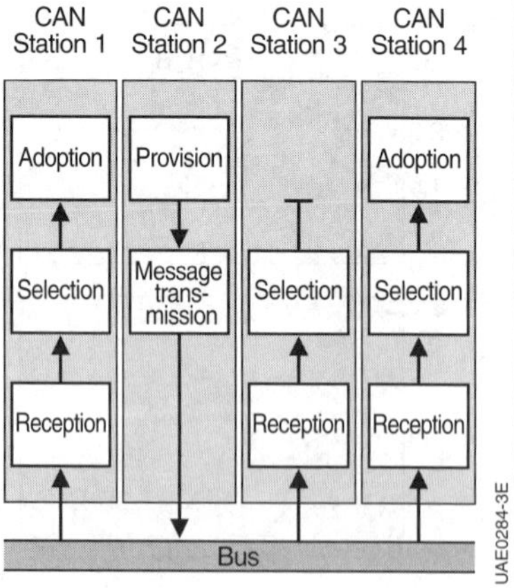

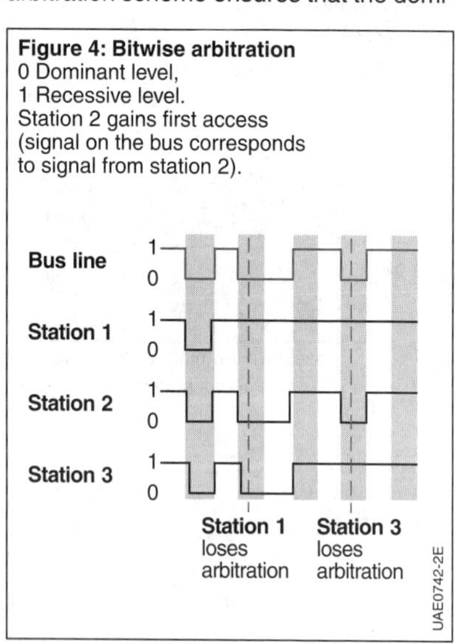

Figure 4: Bitwise arbitration
0 Dominant level,
1 Recessive level.
Station 2 gains first access (signal on the bus corresponds to signal from station 2).

nant bits transmitted by a station overwrite the recessive bits of other stations (Figure 4). Each station gives bit by bit – the highest significant bit first – the identifier of its message on the bus. During this arbitration phase (selection phase) each transmitting station compares the applied bus level with the actual level. Each station that transmits a recessive bit but observes a dominant bit loses arbitration. The station with the lowest identifier – i.e. with the highest priority – is assigned first access on the bus without having to repeat the message (non-destructive access control). Transmitters respond to failure to gain bus access by automatically switching to receive mode; they then repeat the transmission attempt as soon as the bus is free again.

Data frame and message format

The CAN supports two different message formats, which primarily differ with respect to the length of their identifiers. The standard format has 11 bits, while the extended format has 29 bits. Thus, the data frame to be transmitted contains a maximum of 130 bits (in the standard format) and 150 bits (in the extended format). This ensures minimum waiting time until the next transmission (which could be urgent). The data frame for the standard format consists of seven consecutive fields (Figure 5). The "Start of Frame" indicates the beginning of a message and synchronizes all nodes.

Figure 5: Data frame

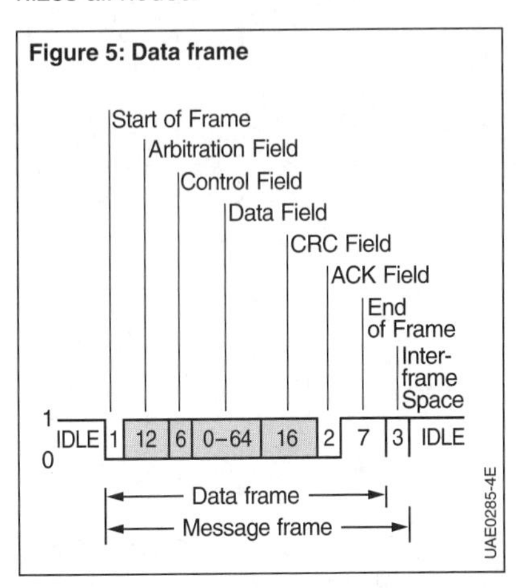

The "Arbitration Field" consists of the message's identifier and an additional control bit. While this field is being transmitted, the transmitter accompanies the transmission of each bit with a check to ensure that no higher-priority message is being transmitted (which would cancel the access authorization). The control bit determines whether the message is classified under "data frame" or "remote frame".

The "Control Field" contains the code indicating the number of data bytes in the "Data Field".

The "Data Field" has an information content of between 0 and 8 bytes. A message of data length 0 can be used to synchronize distributed processes.

The "CRC Field" (Cyclic Redundancy Check) contains a checksum for detecting possible transmission interference.

The "Ack Field" contains the acknowledgement signals with which the receivers indicate the error-free receipt of messages.

The "End of Frame" marks the end of the message.

Then comes the "Interframe Space" to provide a separation from the next data frame.

Transmitter initiative

The transmitter will usually initiate a data transmission by sending a data frame. However, the receiver can also request data from the transmitter. This involves the receiver sending off a "Remote Frame". The "Data Frame" and the corresponding "Remote Frame" have the same identifier. They are distinguished by the bit that follows the identifier.

Error detection
The CAN incorporates a number of monitoring features for detecting errors. These include:
- 15-bit CRC: Each receiver compares the CRC sequence which it receives with the calculated sequence.
- Monitoring: Each transmitter reads its own transmitted message from the bus and compares each transmitted and scanned bit.
- Bit stuffing: Between "Start of Frame" and the end of the "CRC Field", each "Data Frame" or "Remote Frame" may contain a maximum of five consecutive bits of the same polarity. The transmitter follows up a sequence of five bits of the same polarity by inserting a bit of the opposite polarity in the bitstream. The receivers eliminate these bits again as the messages arrive.
- Frame check: The CAN protocol contains several bit fields with a fixed format for verification by all stations.

Error handling
When a CAN controller detects an error, it aborts the current transmission by sending an error flag. An error flag consists of six dominant bits; it functions by deliberately violating the stuffing convention and the formats.

Fault confinement with local failure
Defective stations can severely impair the ability to process bus traffic. Therefore, the CAN controllers incorporate mechanisms which can distinguish between intermittent and permanent errors, and local station failures. This process is based on statistical evaluation of error conditions.

Implementations
The semiconductor manufacturers offer different implementations of CAN controllers which differ primarily in the extent to which they can store and manage messages. In this way, the host computer can be relieved of protocol-specific activities. A standard categorization is Basic CAN controllers, which have only a few message memories, and Full CAN controllers, in which there is room for all the messages necessary for an ECU.

Standardization
The CAN has been standardized for data exchange in motor vehicles; for applications with a low transfer rate up to 125 kbit/s as ISO 11898-3 [1] and for applications with a high transfer rate over 125 kbit/s as ISO 11898-2 [1] and SAE J 1939 (truck and bus, [2]).

Time-triggered CAN
The extension of the CAN protocol to include the capability of operating in time-triggered mode is called "Time-Triggered CAN" (TTCAN). It is fully configurable with regard to the proportion of time-triggered to event-driven communication components, and is therefore fully compatible with CAN networks. TTCAN is standardized as ISO 11898-4 [1].

CAN with flexible data rate
CAN-FD extends CAN by a second bit rate and a wider data field. Unlike the nominal bit rate previously applicable to CAN, the data bit rate is not limited to 1 Mbit/s. The data bit rate acts exclusively on the data field within the data frame, whereas the nominal bit rate still acts on the control data.

In addition, CAN-FD increases the data field from 8 to up to 64 bytes. This also necessitates adaptations to the checksum so as not to compromise transmission reliability. CAN-FD is standardized as ISO 11898-7 [1].

FlexRay

Overview

FlexRay is a bus which was designed for control engineering in the automotive field. Particular attention was given in development to its suitability for use in active safety systems without a mechanical fallback level (X-by-Wire), where determinism and fault tolerance are required. Essentially, thanks to the high transfer rate of up to 20 Mbit/s for non-redundant transmissions, its use in the field of audio transmission or for highly compressed video transmission is also conceivable. The major features are:

- Time-triggered transmission with guaranteed latency
- Possibility of event-driven transmission of information with prioritization
- Transmission of information via one or two channels
- High transfer rate of up to 10 Mbit/s, with parallel transfer via two channels up to 20 Mbit/s
- Structure as a linear bus, in a star configuration or as a mixed form.

FlexRay is the first automotive communication standard to have been created in a consortium of vehicle manufacturers, suppliers and semiconductor manufacturers. It contains elements from TTCAN, Byteflight and other technologies. The specifications published by the FlexRay Consortium are today (only) available as ISO standard 17458 [3].

Figure 6: Bus states and voltages for FlexRay

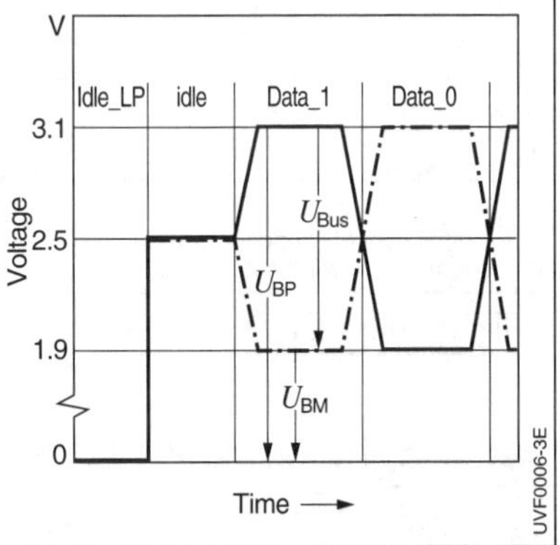

Transmission media

The transmission medium used in a FlexRay system is a twisted-pair two-wire cable where both shielded and non-shielded cables can be used. Each FlexRay channel consists of two strands, Bus-Plus (BP) and Bus-Minus (BM). FlexRay uses NRZ (Non Return to Zero) for encoding.

The bus state is identified by measuring the voltage difference between Bus-Plus and Bus-Minus. Data transfer is thus less sensitive to external electromagnetic influences, since these act equally on both strands and cancel each other out in the difference.

When different voltages are applied to the two strands of a channel, four bus states can be assumed; these bus states are referred to as Idle_LP (LP, Low Power), Idle, Data_0, and Data_1 (Figure 6). Idle_LP is the state in which a low voltage between −200 mV and 200 mV (to ground) is applied to Bus-Plus and Bus-Minus. In the Idle state, a voltage of 2.5 V with a maximum difference of 30 mV is applied to BP and BM. In order to place the channel in the Data_0 state, at least one transmitting node must apply a negative differential voltage of −600 mV to the channel, for Data_1,600 mV, based on a mid-level of 2.5 mV.

Topologies

FlexRay networks can be structured as bus topologies and star topologies. Two stars can be cascaded – when signal delays are taken into consideration in the stars. Topologies in which several buses are connected to a star are also possible.

Because both channels of a FlexRay system can be structured independently of each other, it is possible for different topologies to be used for both channels. For example, one channel can be structured as an active star topology, the other as a bus topology.

On account of the frequencies, which can be ten times those of CAN, it is necessary when designing a FlexRay network to ensure in all topologies in particular that parameters such as line length and terminating resistors are selected in such a way that signal distortions remain in the permissible range.

Bus access, time control

In order to achieve determinism, i.e. the guarantee of a maximum duration for the transmission of a message, communication is performed in the FlexRay bus on a time-controlled basis in cycles of constant duration. Each cycle first features a static segment which is divided into time slots of equal length (Figure 7). Each time slot is permanently assigned a maximum of one node which is permitted to transmit at this time.

This is followed by a dynamic segment in which bus access is regulated by the priority of messages. The split between the static and dynamic segments is freely configurable, but cannot be altered during operation. The same applies to the lengths of the time slots, which are configurable, but must remain constant during operation.

The "Symbol Window" can optionally be defined as the third element in the cycle. This can be used to transmit a single symbol. Symbols are provided to wake a network and to test functionalities.

Synchronization

Each network node requires its own time generator, which it utilizes to decide the time for transmitting and to determine the duration of the bits. The internal time generators of several nodes may deviate from each other on account of temperature and voltage fluctuations and manufacturing tolerances. In a bus system such as FlexRay, which controls bus access via time slots, it is therefore necessary to ensure that the deviation of the clocks from one another remains within a permissible range through regular corrections. To this end, some nodes assume the role of time generators, to which the other nodes regularly synchronize their internal clocks. The procedure adapts both the zero points (offset) of the clocks and their rate. It can continue to operate even if individual nodes fail. To be able to make corrections, each cycle ends with a short phase (NIT, Network Idle Time) in which the zero point of the cycle can be shifted.

Thanks to this procedure a "global time" is provided in all the nodes; this time is given in macroticks. The synchronization mechanism causes the length of one macrotick to be the same on average in all the nodes.

When the network is switched on, a common conception of time on the part of all the nodes must first be established. The start-up process, which takes a little time, serves this purpose. Likewise, a node which intends to synchronize itself to a running network needs a period of time to be taken into consideration.

Figure 7: Time control for FlexRay (example)
A1 Node A transmits message 1,
A2 Node A transmits message 2,
MTS Media Test Symbol.

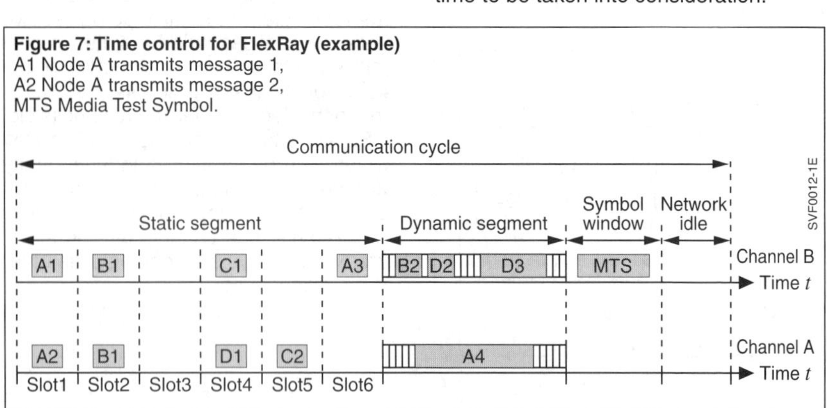

Arbitration in the dynamic segment

Messages can be given different priorities in the dynamic segment. The duration until a message is transmitted cannot be guaranteed for this. The priority is established by the frame ID, which may be allocated only once in the network. The messages are transmitted in the sequence of their frame ID. To this end, each node runs a counter (slot ID), which is increased when a message is received. If the slot ID assumes the value of the frame ID of a message ready in this node, it is transmitted. If the length of the dynamic segment is not sufficient for all the messages, the transmission process must be shifted to a later cycle.

The data frames in the dynamic segment can have different lengths. The limits of the dynamic slots on the two channels are independent of each other. Thus, there may be messages with different slot IDs on the channels at one time.

Data frame

FlexRay uses in both the static and dynamic segments the same data-frame format, which can be divided into the three sections of Header, Payload and Trailer (Figure 8).

Figure 8: Data frame

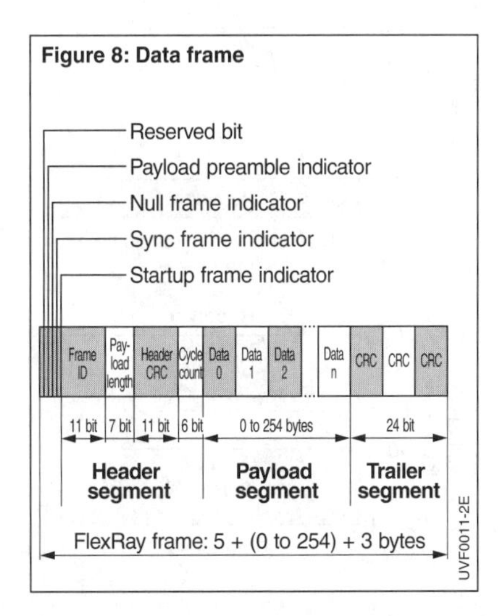

Header

The header comprises:

- The reserved bit for future protocol changes.
- The payload preamble indicator, which identifies whether the payload contains a network management vector.
- The null frame indicator, which identifies that the data have not been updated since the last cycle.
- The sync frame indicator, which signals that this data frame is to be used to synchronize the system.
- The startup frame indicator, which identifies that this frame is used in the network's start-up phase.
- The frame ID; this corresponds to the number of the slot in which the data frame is transmitted.
- The payload length which contains the size of the user data. For all the slots in the static segment this field always contains the same value. Data frames in the dynamic segment can have different lengths.
- The header CRC, which provides this part of the data frame with extra protection, because it is crucial to the time response.
- The cycle count; the number of the cycle in which the transmitting network node is located is transmitted in this field.

Payload

The user data which are processed further by the host are transmitted in the payload segment. For data frames in the static segment the first payload bytes can optionally be declared to the network management vector. The controllers or all the vectors received in the cycle and make them accessible to the host. For data frames in the dynamic segment the first payload bytes can optionally be declared to a 16-bit message ID. In both cases, subsequent handling is left to the software.

The user data have a maximum length of 254 bytes, which are transmitted in 2-byte words.

Trailer
The trailer contains a 24-bit checksum (frame CRC), which acts on the entire data frame.

Generating a frame bitstream
Before a node can transmit a data frame with the host's data, the data frame is converted into a "bitstream". To this end, the data frame is first deconstructed into individual bytes. A Transmission Start Sequence (TSS) of configurable bit length is inserted at the start of the data frame, followed by a bit Frame Start Sequence (FSS). Then an extended byte sequence is generated from the bytes of the data frame, where a two bit Byte Start Sequence (BSS) is inserted ahead of each frame byte.

To conclude the bitstream, a two bit Frame End Sequence (FES) is appended to the bitstream.

In the event that the data frame is in the dynamic segment, it is possible to append to the bitstream a further Dynamic Trailing Sequence (DTS) of configurable bit length. This prevents another node from starting its transmission via the channel early.

Operating modes
FlexRay can be placed in a mode in which the nodes need only minimal power and in which all the operations of the encoding and decoding process are stopped, but can be woken up by a signal on the bus line. Here, the bus driver is still capable of detecting special signals on the bus and then also of activating its host by means of a corresponding signal. Each node can transmit a wake-up.

LIN

Overview
The LIN bus (Local Interconnect Network) was designed to cover the communication requirements for Class A systems (see Table 1) with the most cost-effective hardware possible in the node. Typical applications are the door module with door lock, the power-window units, door-mirror adjustment, and air-conditioning system (transmission of signals from the control element, activation of the fresh-air blower).

The current LIN specification can be accessed on the LIN Consortium website [4].

Important features of the LIN bus are:
- Single-master and multi-slave concepts
- Low hardware costs on account of data transfer via non-shielded single-wire cable
- Self-synchronization of the slaves even without a quartz oscillator
- Communication in the form of very short messages
- Transfer rate max. 20 kbit/s
- Bus length up to 40 m and up to 16 nodes.

Figure 9: Voltage level on data line of LIN bus

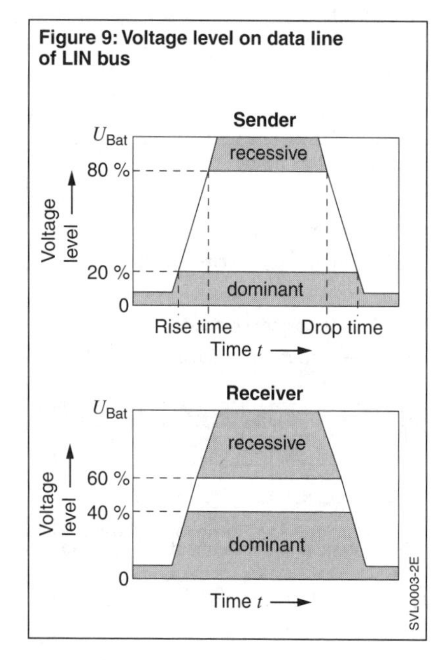

Transfer system

The LIN bus is designed as a non-shielded single-wire cable. The bus level can assume two logic states. The dominant level corresponds to a voltage of approximately 0 V (ground) and represents logic 0. The recessive level corresponds to the battery voltage U_{batt} and represents the logic 1 state.

Because of the different design variations of the circuitry, the levels may be subject to differences. The definition of tolerances for transmitting and receiving in the field of the recessive and dominant levels ensures a stable data transfer. The tolerance bands are wider at the receiving end (Figure 9) so that valid signals can also be received in spite of interference radiation.

The transfer rate of the LIN bus is limited to 20 kbit/s. This is a compromise between the demand for high edge steepness to synchronize the slaves easily on the one hand and the demand for low edge steepness to improve EMC performance on the other. Recommended transfer rates are 2,400 bit/s, 9,600 bit/s and 19,200 Bit/s. The minimum permissible value for the transfer rate is 1 kbit/s.

The maximum number of nodes is not stipulated in the LIN specification. Theoretically, it is limited by the number of available content-related message identifiers. Line and node capacities and edge steepnesses limit the combination of length and node number of a LIN network; a maximum of 16 nodes is recommended.

The bus users are usually arranged in a linear bus topology; however, this topology is not explicitly prescribed.

Figure 10: LIN frame
SB Synch Break,
SF Synch Field,
IF Ident Field,
DF Data Field,
CS Checksum.

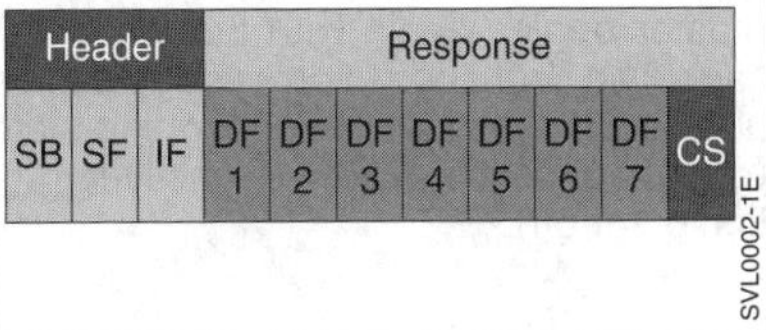

Header
Response
SB SF IF DF 1 DF 2 DF 3 DF 4 DF 5 DF 6 DF 7 CS
SVL0002-1E

Bus access

In the LIN bus, access is provided on the basis of the master-slave access method. The network features a master, which initiates each message. The slave has the opportunity to respond. The messages are exchanged between the master and one, several or all the slaves.

The following relationships are possible during communication between master and slave:

- Message with slave response: The master transmits a message to one or more slaves and asks for data (e.g. switch states of measured values).
- Message with master instruction: The master issues control instructions to a slave (e.g. switching on a servomotor).
- Message for initialization: The master initiates a communication between two slaves.

LIN protocol

Data frame

The information on the LIN bus is embedded in a defined data frame (LIN frame) (Figure 10). A message initiated by the master begins with a header. The message field (response) contains different information depending on the type of message. If the master transmits control instructions for a slave, it describes the message field with the data to be utilized by the slave. In the event of a data request, the addressed slave describes the message field with the data requested by the master.

Header

The header is made up of the synchronization break (Synch Break), the synchronization field (Synch Field), and the identifier field (Ident Field).

Synchronization
A synchronization takes place at the start of each data frame to ensure a consistent data transfer between master and slaves. First the start of a data frame is clearly identified by the Synch Break. It consists of at least 13 consecutive dominant levels and one recessive level.

After the Synch Break, the master transmits the Synch Field, consisting of the bit sequence 01010101. The slaves thus have the opportunity to adapt themselves to the master's time base. The clock pulse of the master should not deviate from the nominal value by more than ±0.5%. The clock pulse of the slaves may deviate prior to synchronization by up to ±15% if the synchronization achieves a deviation of max. ±2% up to the end of the message. The slaves can thus be designed without an expensive quartz oscillator, for example with a cost-effective RC circuit.

Identifier
The third byte in the header is used as the LIN identifier. Similarly to the CAN bus, content-based addressing is used – the identifier therefore gives information about the content of a message. All the nodes connected to the bus decide on the basis of this information whether they intend to receive and process the message or ignore it (acceptance filtering).

Six of the eight bits of the identifier field determine the identifier itself, from which 64 possible identifiers (ID) are obtained. They have the following meanings:

- ID = 0–59: Transmission of signals.
- ID = 60: Master request for commands and diagnostics.
- ID = 61: Slave response to ID 60.
- ID = 62: Reserved for manufacturer-specific communication.
- ID = 63: Reserved for future expansions of the protocol.

Of the 64 possible messages, 32 may contain only two data bytes, 16 four data bytes, and the remaining 16 eight data bytes.

The last two bits in the Ident Field contain two checksums, with which the identifier is protected against transmission errors and resulting incorrect message allocations.

Data field
Transmission of the actual data begins after the master node has transmitted the header. The slaves identify from the transmitted identifier whether they are addressed and, if necessary, transmit back the response in the data field.

Several signals can be packed into a data frame. Here, each signal has exactly one generator, i.e. it is always described by the same network node. During operation it is not permitted to change the signal allocation to another generator, as would be possible in other time-controlled networks.

The data in the slave response are safeguarded by a checksum (CS).

LIN description file
The configuration of the LIN bus, i.e. the specification of network users, signals and data frames, is performed in the LIN description file. The LIN specification provides for a suitable configuration language for this purpose.

From the LIN description file, tools automatically generate program sections which are used to implement the master and slave functions in the ECUs located on the bus. The LIN description file thus serves to configure the entire LIN network. It is a common interface between the vehicle manufacturer and the suppliers of the master and slave modules.

Message scheduling
The scheduling table in the LIN description file determines the order and time frame in which the messages are transmitted. Frequently needed information is transmitted from time to time. When the table has been worked through, the master begins again with the first message. The sequence of processing can be altered depending on the operating state (e.g. diagnostics active or inactive, ignition on or off).

Thus, the transmission frame of each message is known. The deterministic performance is guaranteed by the fact that all the transmissions are initiated by the master in the case of master-slave access control.

Network management
The nodes of a LIN network can be placed in sleep mode in order to minimize closed-circuit current. Sleep mode can be achieved in two ways. The master transmits the "Go to Sleep" command with the reserved identifier 60, or the slaves automatically go into sleep mode if there has been no data transfer on the bus for an extended period of time (four seconds). Both the master and the slaves can wake up the network again. The wake-up signal must be transmitted for this purpose. This consists of a data byte with the number 128 as content. After a break of 4 to 64 bit times (wake-up delimiter), all the nodes must be initialized and be able to respond to the master.

Ethernet

Overview

The term Ethernet refers to a family of buses in which the addressing, the format of the messages and access control are identical (laid down in IEEE 802, [5]). Ethernet and the Internet Protocol (IP) were developed for data communication between computers or peripherals which are locally separate and where, during operation, reconfigurations of the network through the addition of new users or the failure of users can occur. The Ethernet buses are identified by the following important features:

- The transfer rate is in the range of 10 Mbit/s up to 10 Gbit/s.
- Data transfer is possible via assorted media such as coaxial cable, twisted two-wire cable, radio or glass fiber.
- The technology involved is standardized and very widely used.
- Simple insertion and removal of nodes is possible.
- The time response in the case of real-time applications is not guaranteed.

Ethernet is used in series-production vehicles; e.g. in the BMW 7 Series, where it is used to input vehicle data at the end of production.

Transfer system

The Ethernet versions differ in terms of transfer rate, physical design of the channel, and encoding. Coaxial cables, twisted two-wire cables with one or more core pairs, optical fibers, radio paths, or even power-supply cables are specified as channels. Encoding differs accordingly.

Originally, coaxial cables in a bus topology were used as the medium. Here, the transceivers of the nodes were connected either directly or with T-pieces to the cable. Today, twisted two-wire cables are widely used. The transfer rates have been increased from initially 10 Mbit/s via the Fast Ethernet with 100 Mbit/s and the Gigabit Ethernet with 1,000 Mbit/s up to 10 Gbit/s.

Topology

The size of a network is limited by the fact that the signal propagation time between two nodes influences the arbitration process. This can be bypassed by subdividing into segments, which are connected by way of special components – hubs and switches. A hub functions as an amplifier which re-establishes the ideal signal shape of a bit if it has been corrupted by interference or dispersion on the transmission medium. A switch checks entire packets for correctness with regard to the checksum and directs packets without collision to another output if the destination address can be reached by this route. To this end, it must have the option of storing messages temporarily. As well as the costs involved in the hardware, a disadvantage when such elements are used is that the data stream is delayed. For this purpose, however, nodes with different data rates can be connected.

Today, networks are usually designed in such a way that each node is connected to the output of a switch, i.e. there is no direct connection between nodes. Switches themselves are in turn connected via a higher-level switch in such a way that a tree-shaped structure is created.

Figure 11: Frame format of Ethernet protocol

Preamble	Destination address	Source address	Length, type	Data, filler bytes	CRC
8 bytes	6 bytes	6 bytes	2 bytes	46–1,500 bytes	4 bytes

SVF0051-1E

Ethernet protocol

Bus access

To transmit, a node checks whether there are signals on the bus. It starts to transmit when it deems the line to be clear. Because of the signal propagation time between two nodes, the situation may arise where two nodes deem the bus to be clear and start to transmit virtually simultaneously. The data frames transmitted in the process are destroyed. The nodes identify this, abort their transmission, and wait a certain amount of time – different for each node – until they begin with a new transmission attempt. This destruction of data frames reduces the effective transfer rate to a tolerable extent provided bus utilization is not too high.

This arbitration process limits the length of messages and the propagation time, i.e. the reach. There are no priorities among the messages. A maximum duration for transmission can, therefore, not be guaranteed.

Each node adopts from all the messages those which contain its own as the destination address for further processing.

Data frame

Figure 11 shows the slightly simplified structure of a data frame. The preamble is a periodic bit sequence (101010–1011) and thereby generates a signal to synchronize the receiver. Messages contain the address of their source and their destination. Each network card has a unique address. The receiving nodes compare the destination address with their own card address and accept the data frame if they match. With multicast and broadcast addresses, several receivers can also be addressed.

PSI5

Overview

The Peripheral Sensor Interface 5 (PSI5) is a digital interface published by the PSI Consortium ([6]) for sensor applications in motor vehicles and can be assigned by the application to Class A (see Table 1). The PSI5 is based on already existing interfaces for peripheral airbag sensors, but has been developed as an open standard which can be used and implemented at no extra charge. The technical characteristics mentioned in the following, the low implementation expenditure and the low additional costs compared with an analog sensor connection make the PSI5 an attractive option for sensor applications in motor vehicles too.

Transfer system

The PSI5 is a two-wire current interface in which the same lines are used to power the sensors and for Manchester-encoded data transfer. The bus master in the ECU modulates a voltage to the sensor for this purpose. Data transfers from the sensor to the ECU are effected by current modulation of the supply lines. In this way, high EMC robustness and low electromagnetic radiation are achieved. A wide range of supply currents for the sensors can be supported.

The different PSI5 operating modes define the topology and parameters of communication between ECU and sensors (Figure 12):

- Communication modes: The asynchronous mode can be used for a unidirectional point-to-point connection. In the three synchronous bus modes (parallel, universal or daisy-chain cabling) several sensors can communicate on a time-controlled basis with the bus master bidirectionally and using TDMA processes.
- Data-word width: The PSI5 supports a variable data-word width of 8, 10, 16, 20 or 24 bits.
- Error detection: This can be effected either with a parity bit with even parity or with three CRC checksum bits.
- Cycle times: These are specified in µs.
- Number of time slots per cycle.
- Data-transfer rate: As standard 125 kbit/s or optionally 189 kbit/s.

For example, the operating mode "PSI5-P10P-500/3L" denotes a parallel synchronous mode with ten bits per data word and parity bit for error detection. The data are transmitted every 500 µs with three time slots per cycle and a low bit rate.

Figure 12: Designation of PSI5 operating modes
A Asynchronous mode,
P Parallel, synchronous mode,
U Universal cabling,
D Daisy-chain cabling,
P Parity bit,
CRC Cyclic redundancy check,
L Low (low bit rate),
H High (high bit rate).

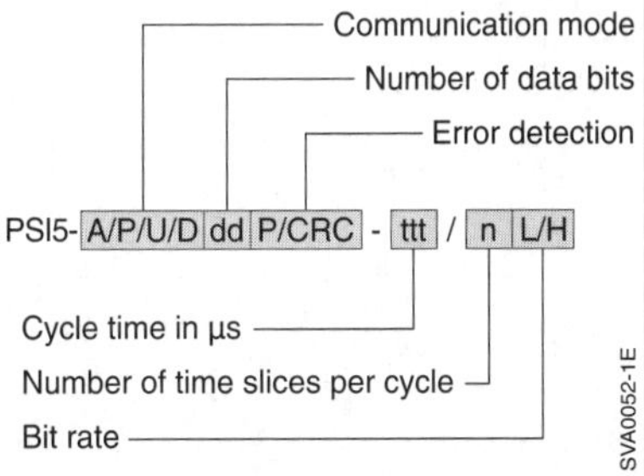

Communication mode
Number of data bits
Error detection
PSI5- A/P/U/D dd P/CRC - ttt / n L/H
Cycle time in µs
Number of time slices per cycle
Bit rate
SVA0052-1E

Figure 13: Synchronization signal of PSI5
T Trigger point.

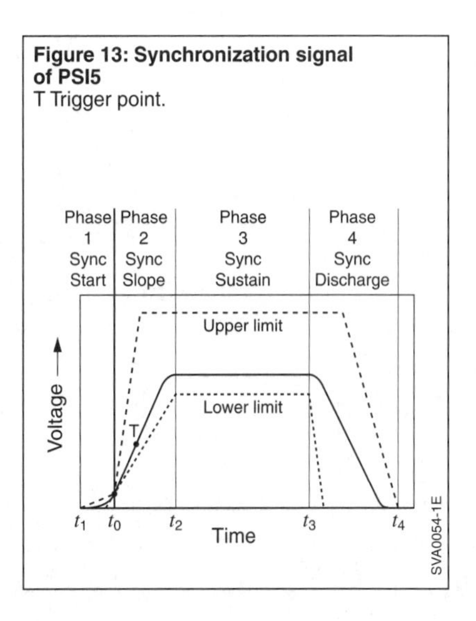

During communication from sensor to ECU, a "low level" is represented by a normal (non-oscillating) current input of the sensor. A "high level" is generated by an increased current sink of the sensor (typically 26 mA). This current modulation is detected by the receiver in the ECU.

Each PSI5 data packet consists of N bits, in which there are in each case two start bits and one parity bit (or three CRC bits) and $N-3$ (or $N-5$) data bits. The data bits are transmitted with the least significant bit (LSB) first. Error detection with a parity bit is recommended for eight or ten bits, three CRC bits for longer data words.

In a PSI5 message, the data and value ranges have different meanings. One range serves to transmit the sensor output signals (≈94 %), one range the status and error messages (≈3 %), and one range the initialization data (≈3 %).

After each starting or undervoltage reset, the sensor conducts an internal initialization, after which the sensor is in an executable mode.

Whereas communication by the sensor with the ECU is effected with current signals, voltage modulation of the supply lines is used during communication by the ECU with the sensors. A logic 1 is represented by the synchronization signal, a logic 0 by the absence of the expected synchronization signal within the designated time slot. The synchronization signal consists of four voltage phases (Figure 13):

- Sync Start (nom. 3 µs, < 0.5 V),
- Sync Slope (nom. 7 µs, > 0.5 V),
- Sync Sustain (nom. 9 µs, > 3.5 V), and
- Sync Discharge (nom. 19 µs, < 3.5 V).

MOST

Overview

Application

The MOST bus (Media Oriented Systems Transport) was specially developed for the networking of multimedia applications in motor vehicles (infotainment bus). Alongside the classical entertainment functions such as radio receivers and CD players, infotainment systems also provide video functions (DVD and TV), navigation functions, and access to mobile communication and information. The MOST bus supports the logical networking of up to 64 devices and provides a fixed and reserved transmission bandwidth. MOST defines the protocol, the hardware, the software, and the system layers. MOST is jointly developed and standardized by automobile manufacturers and suppliers within the MOST Cooperation [7]. With a data rate of more than 10 Mbit/s, the MOST bus is a Class D bus system (see Table 1).

For data transfer, the MOST bus supports the following transmission channels:
- Control channel to transport control commands
- Multimedia channel (synchronous channel) for the transmission of audio and video data
- Package data channel (asynchronous channel), for example to transmit configuration data for a navigation system and to update software in control units.

Requirements

The transmission of multimedia data – both audio and video data – requires a high data rate and also synchronization of the data transfer between source and sink as well as between a number of sinks.

Transfer system

Physical layer

The MOST standard specifies both optical and electrical technologies of the physical layer (transmission layer). The optical transmission layer is widespread and currently uses fiber-optic cables (polymer optical fibers, POF) made of polymethyl methacrylate as its transfer medium. These have a 1 mm core diameter and are used in combination with LEDs and silicon photo diodes as receivers (see Optical fibers/waveguides).

The outstanding feature of MOST 50 is its suitability for the electrical transmission of data. This enables data transfer across unshielded, twisted copper cables (UTP, unshielded twisted pairs). Whereas MOST 25 technology has continued to develop in Europe for a many years and has become established in the Korean market, the Japanese market in particular prefers MOST 50, the second generation of the multimedia standard.

The identification number, for example in the case of MOST 25, stands for a transfer rate of approx. 25 Mbit/s. The exact data rate depends on the sampling rate that the system uses. With a sampling rate of 44.1 kHz, the MOST frame (data frame) is transferred 44,100 times per second; a frame length of 512 bits results in a data rate of 22.58 Mbit/s. For MOST 50, the same sampling rate results in double the data rate, as the frame is 1,024 bits long. Higher data rates of 150 MBit/s (MOST 150) are currently also available.

Special features of MOST 150

In addition to the higher bandwidth of 150 Mbit/s, MOST 150 contains an isochronous transport mechanism to transfer compressed data of HD videos efficiently. MPEG transport streams (MPEG, Moving Picture Experts Group) are transported directly here. With a corresponding MPEG4-based video codec, it is possible to transfer resolutions of up to 1,080 p (1,080 screen lines), as delivered by BluRay players for example. Alongside this, MOST 150 provides an Ethernet channel for the efficient transmission of IP package data (IP = internet protocol).

In contrast to the MAMAC protocol (MOST Asynchronous Medium Access Control) used in the case of MOST 25, the Ethernet channel is able to transfer Ethernet frames. The Ethernet channel transfers unmodified Ethernet data blocks, which means that software stacks and applications from the fields of consumer electronics and IT can be integrated seamlessly in vehicles with much shorter innovation cycles. TCP/IP stacks or protocols that use TCP/IP (TCP = Transmission Control Protocol) can thus communicate via MOST 150 without changes.

The MOST Network Interface Controller (NIC) is a hardware controller that is responsible for control of the physical layer and implements important transfer mechanisms.

Protocol

Data transfer

Data transfer on the MOST bus is organized in data frames that are created by the timing master with a fixed data rate and passed on by the devices in the ring.

Data frames

The timing master usually creates data frames with a clock rate of 44.1 kHz, more rarely also 48 kHz. The size of the data frames thus determines the bus speed of a MOST system. In the case of MOST 25, the size of a data frame is 512 bits (Figure 14). The synchronous and asynchronous areas of MOST 25 jointly use 60 bytes of the data frame. The division between the synchronous channels and the asynchronous channel is determined by the value of the boundary descriptors with a resolution of 4 bytes. The synchronous area must have at least 24 bytes (six stereo channels). This means that between 24 and 60 bytes are permitted for the synchronous area and between zero and 36 bytes are permitted for the asynchronous area. The preamble is used for synchronization; the parity bit is used to recognize bit errors.

Transmission of control messages

The control channel is used to signal device statuses and for the messages required for system administration. So that the control channel does not occupy too much bandwidth per frame, it has been distributed to 16 frames grouped to form a block. Each frame transports two bytes of the channel (Figure 14). In order to ensure recognition of the block start, the preamble of the first frame of a block bears a special bit pattern. In the case of MOST 25, the control channel has a gross bandwidth of 705.6 kbit/s.

Transmission of multimedia data

The synchronous channels are used for realtime communication of audio and video data, whereby the data interchange is controlled via the corresponding control commands on the control channel. A synchronous channel can be assigned a certain bandwidth, which occurs with a resolution of one byte of a data frame. A stereo audio channel with a resolution of 16 bits, for example, requires four bytes. In the case of MOST 25, depending on the value of the boundary descriptor, a maximum of 60 bytes are available for synchronous channels; this corresponds to 15 stereo audio channels.

Transmission of package data

Data are transferred packet-wise on the asynchronous channel. It is therefore suitable for the transmission of information that has no fixed data rate but requires high data rates at short notice. Examples are the transmission of track information of an MP3 player or a software update.

Figure 14: MOST frame structure

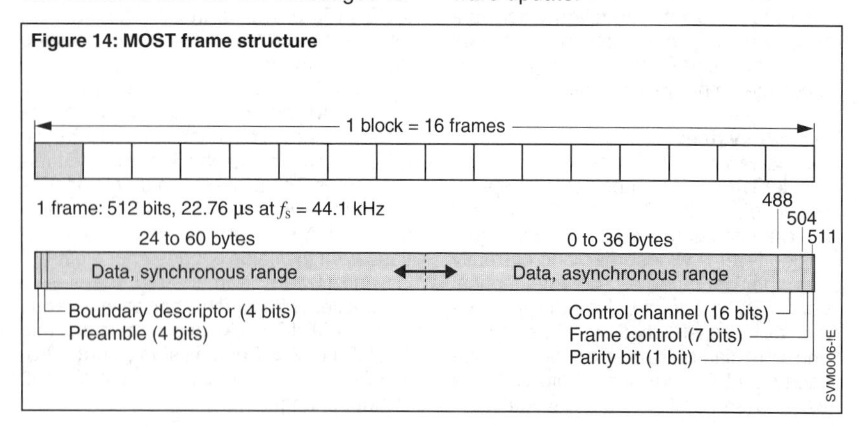

In the case of MOST25, the asynchronous channel has a gross bandwidth of up to 12.7 MBit/s and it currently supports two modes: a slower 48-byte mode in which 48 bytes are available in each package for the net data transmission as well as a 1,014-byte mode that is more complex to implement. In order to ensure reliable transmission and flow control for the typically large asynchronous channel data volumes, an additional transport protocol (Data Link Protocol) is usually deployed; this is implemented in a driver layer at a higher level. This is either the MOST High Protocol (MHP), which has been specially developed for the MOST, or the common TCP/IP protocol that is placed on a corresponding adaptation layer called the MOST Asynchronous Medium Access Control (MAMAC).

Topology
MOST is organized in a ring structure (Figure 15). This is a point-to-multipoint data flow system (i.e. the streaming data have one source and a number of sinks) and all the devices therefore share a joint system frequency that they acquire from the data flow. The devices are thus in phase and can transfer all data synchronously. This makes mechanisms for signal buffering and signal processing superfluous. A certain device acts as "timing master" and generates the data frames used for data transfer and to which the other devices synchronize.

Figure 15: Ring structure of the MOST bus

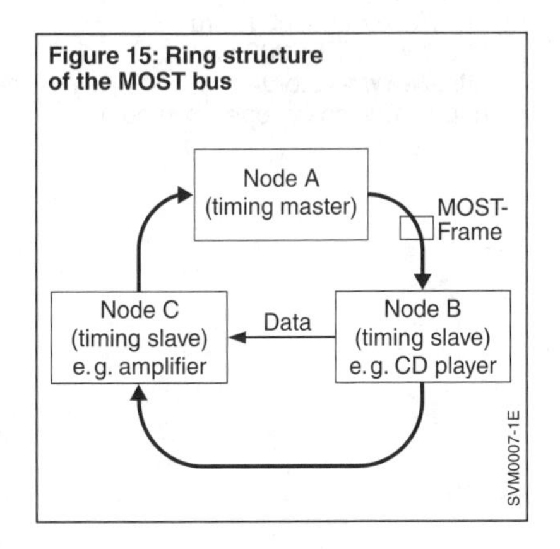

Addressing
The devices are addressed on the MOST bus via a 16-bit address. Various types of addressing are available: Logical and physical addressing as well as group addressing for simultaneous addressing of a defined group of control units.

Administration functions
The MOST standard defines the management mechanisms (network and connection masters) necessary for the operation of a MOST system. These mechanisms are described below.

Network master
The network master is implemented by a marked device in a MOST system and is responsible for the configuration of the system. In current systems, the network master is usually implemented using the head unit (i.e. control panel) of the infotainment system. This device is often also the timing master at the same time. The other devices of the MOST system are referred to as network slaves in this context.

Connection master
The connection master manages the synchronous connections that exist in a MOST system at any particular point in time.

MOST application layer
For the transfer of control commands, status information, and events, the MOST standard defines a corresponding protocol on the application level. This protocol enables triggering of a certain function of an application interface (i.e. of an FBlock = function block) that is provided by any device within the MOST system.

The protocol for MOST control messages includes the following elements of a control message:
- The address of a device in the MOST system (DeviceID)
- An identifier for an FBlock (FBlockID) implemented by this device and its instance in the MOST system (InstID)
- The identifier for the function called up within the FBlock (FunctionID)
- The type of an operation (OpType).

Function block

A function block (FBlock) defines the interface of a certain application or of a system service. The sinks and sources for multimedia data are each assigned to an FBlock that provides the corresponding functions for their administration. An FBlock can therefore have a number of sources and sinks enumerated via a source and sink number.

An FBlock has functions that deliver information regarding the number and type of sources and sinks that it provides (SyncDataInfo, SourceInfo and SinkInfo). Over and above this, each FBlock with a source has an "Allocate" function that it uses to request a synchronous channel and connect the source to that channel. Accordingly, an FBlock with a sink has a "Connect" function to connect the sink to a certain synchronous channel and a "DisConnect" function to break off this connection.

An FBlock is addressed by means of an 8-bit FBlockID, which specifies the type of the FBlock, and an additional 8-bit InstID.

Function classes

To standardize the way in which functions are defined, the MOST standard specifies a series of function classes for properties. These determine which properties the function has and which operations are permissible.

Applications

As well as defining the lower layers necessary for data transfer, the MOST standard defines the interfaces for typical applications from the area of vehicle infotainment systems, e.g. a CD changer, amplifier or radio tuner.

The FBlocks defined by the MOST Cooperation are summarized in a function catalog.

Standardization

The MOST standard is maintained by the MOST Cooperation, which also publishes the corresponding specifications. The specifications are available on the homepage of the MOST Cooperation [7].

The MOST Corporation was founded in 1998 by BMW, Daimler, Becker Radio, and OASIS Silicon Systems with the aim of standardizing MOST technology.

References

[1] ISO 11898: Road vehicles – Controller area network (CAN).
Part 1: Data link layer and physical signalling.
Part 2: High-speed medium access unit.
Part 3: Low-speed, fault-tolerant, medium-dependent interface.
Part 4: Time-triggered communication.
Part 5: High-speed medium access unit with low-power mode.
Part 6: High-speed medium access unit with selective wake-up functionality.
[2] SAE J 1939: Serial Control and Communications Heavy Duty Vehicle Network.
[3] ISO 17458: Road vehicles – FlexRay communications system.
Part 1: General information and use case definition.
Part 2: Data link layer specification.
Part 3: Data link layer conformance test specification.
Part 4: Electrical physical layer specification.
Part 5: Electrical physical layer conformance test specification.
[4] http://www.lin-subbus.org/.
[5] http://www. IEEE802.org/.
[6] http://www.psi5.org/.
[7] http://www.mostcooperation.com.

Architecture of electronic systems

Overview

Owing to the increasing demand for safety, comfort, entertainment, and environmental protection, electronic systems in motor vehicles realize an increasing number of functions and are characterized by a high level of networking and complexity. The latest processes, methods and tools of system architecture are required to keep on top of this structure in the future too.

History

Over the many decades of automobile history, there has been a manageable number of electrical systems in motor vehicles: ignition, lighting, windshield wipers, horn, fuel gage, various indicator lamps, and a vehicle radio. Semiconductors were used initially – except in vehicle radios – only for rectification (direct-current generator replaced by alternator from approximately 1963) and then later for electronic control (transistorized ignition from 1965).

Certain in-vehicle functions were realizable with electromechanical means or with discrete electronic components either not at all or only with disproportionately high complexity. Thus, for example, the first electronic antilock braking system (ABS) had already been developed in 1970, but was never ready for series production on account of its size, weight and cost. By the mid-1970s the development of integrated circuits for a broad range of applications had also reached and revolutionized automotive engineering.

One of the first instances of the networking of electronic systems came about during the development of the traction control system (TCS). This networking was initially realized by purely mechanical means. The throttle valve in the air-intake system of the internal-combustion engine was fitted with a device which could be activated directly by the traction control system. It was not discernible to the engine management whether the driver or the traction control system was moving the throttle valve.

The next stage involved the realization of an electronic connection to the engine control unit via a PWM interface (pulse-width modulation) to improve dynamic response. This could be used to transfer the signal to the engine control unit for reducing the drive torque. This was then implemented in the form of an air-supply throttling, an injection blank-out or an ignition-timing advance.

Figure 1: ECU interconnection in a modern mid-size vehicle

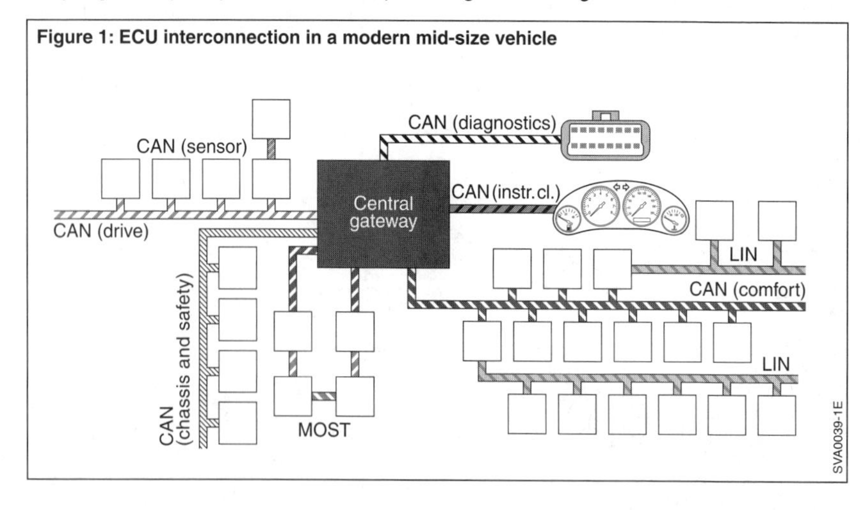

On account of the ever more stringent exhaust-emission regulations the possibilities represented up until that point of coupling between the traction control system and the engine management were no longer sufficient. What was now required was to transmit to the engine management how a reduction of the drive torque requested by the traction control system is effected in the air, fuel or ignition path. It was therefore necessary to come up with a more powerful interface via which a desired torque and a dynamic-response request could be transmitted from the traction control system to the engine management. By contrast, the actual torque, the engine speed and the current setting reserve were to be transmitted to the TCS control unit. It proved complex and expensive in terms of the number of cables required to transfer these different data via discrete and, for example, pulse-width-modulated interfaces. The CAN bus system (Controller Area Network) was introduced in 1991 as an alternative to discrete cabling. In this way, the foundations for the modern networking of systems in motor vehicles were laid.

Technology of the present day

In today's vehicles, virtually all the ECUs are networked directly or indirectly (e.g. via gateways) with each other (Figure 1). Gateways enable data to be exchanged between communication systems or across vehicle borders (connection to radio systems and the Internet). Networking goes to some extent so far that 60 or more ECUs communicate via several CAN buses and further communication systems, such as FlexRay, MOST (Media Oriented Systems Transport) or LIN (Local Interconnect Network), with each other. Thus, for example, the ESP control unit (Electronic Stability Program) supplies the network with the information on the vehicle speed. The vehicle radio can use this information, for example, to adapt the volume to the vehicle speed in each case.

Because of the powerful networking between the ECUs, a good many new performance features can even be achieved completely without additional hardware, i.e. purely by means of data communication and software. One example of this is the opening of the side windows through longer actuation of the radio remote control for the central-locking system. Thus, for example, the vehicle can be uniformly ventilated in the summer months when the doors are opened. The power-window units and the central-locking system exchange the necessary information for this purpose. The software runs either on the ECU for the central-locking system or the ECU for the power-window units. In many vehicles the two systems share a common ECU so that new software-based performance features can be integrated even more easily.

Figure 2: Comparison of decentralized control with centralized control
a) Decentralized control, b) Centralized control.
1 Mirror, 2 Door ECU,
3 Electric power-window unit,
4 Operating element,
5 Mirror, 6 Central ECU.

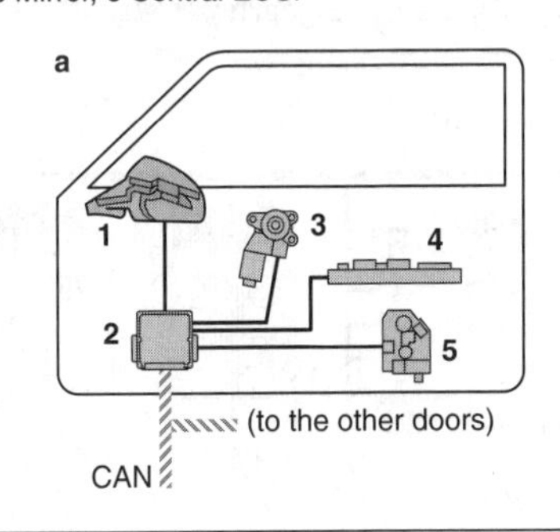

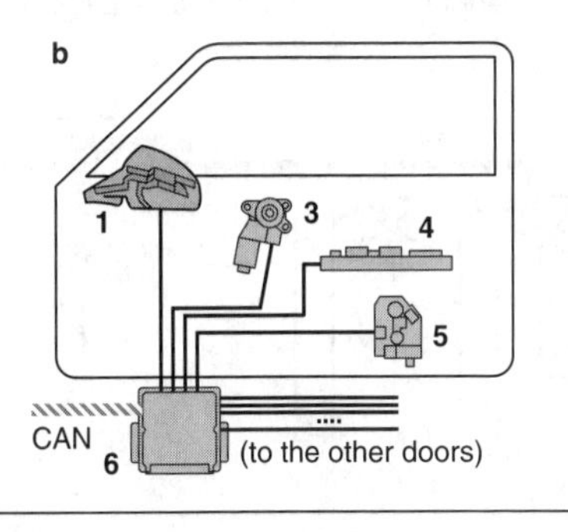

This demonstrates a trend which is initially encountered in body electronics: The integration of individual ECUs to form central ECUs (Figure 2). These central ECUs are connected with the sensors and actuator either via discrete, analog cables or via buses. The latter reduce significantly the number of pins in the ECU plug and thereby also reduce the cabling costs. Sensors and actuators connected via buses are also known as "intelligent" sensors and actuators. These must for the purpose of bus connection have electronic circuitry, which in many cases also contains the sensor-signal conditioning or actuator driver functions. At the same time, however, the use of electronic circuitry gives rise to higher costs in the sensors or actuators. Minimizing the overall costs of comprising the electronics and wiring-harness costs thus represents an important task when defining networking concepts.

Thus, for example, the logic circuit for the finger-protection function of the power-window units is located in many a design variation directly in the ECU on the power-window motor. The activation signal for normal operation, e.g. the mentioned window opening by radio remote control, is transmitted via a LIN bus from a central ECU of the body electronics (BCM, Body Computer Module). A client-server architecture is referred to in this respect.

Development trends

The above-mentioned centralization and the use of intelligent sensors and actuators in the field of body electronics have become established in other function areas in the vehicle (domains) (driver information, driving dynamics and safety) and will continue to expand in the coming vehicle generations. In addition to the combination of functions of different ECUs in a single ECU, domain master computers are used for this purpose to assume central functions (Figure 3). The ECUs of the intelligent sensors and actuators distributed in the vehicle are dependent on these master computers (BCM, IHU, etc.). Functions which require a high degree of networking of information control commands are predominantly reproduced on these central computers in the software. A standard software architecture is required to enable these functions also to run on different ECU platforms and thus to be reused. This is to be achieved by means of the AUTOSAR Initiative (see "AUTOSAR" section).

The domain master computers are networked with each other via a powerful "backbone" – in Figure 3 configured as CAN and FlexRay and connected via the central gateway. Central network access for diagnostics and software downloading is also connected via the gateway here.

Figure 3: Possible scenario for a future luxury-class vehicle
CAN Controller Area Network, CGW Central Gateway,
BCM Body Computer Module, IHU Integrated Head Unit,
VDU Vehicle Dynamics Unit, PSM Passive Safety Manager,
EPM Engine & Powertrain Manager, WLAN Wireless Local Area Network,
LIN Local Interconnected Network, MOST Media Oriented Systems Transport,
PSI Peripheral Sensor Interface, LVDS Low Voltage Differential Signaling.

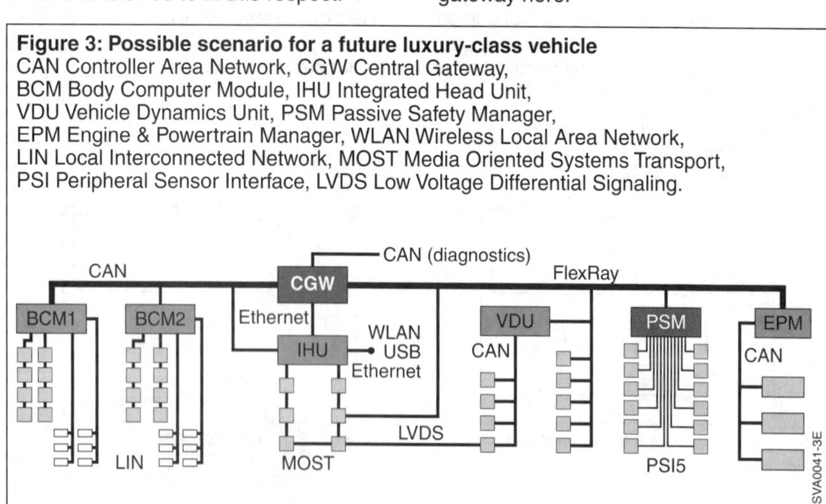

Architecture methods of electronic systems

Architecture

As the amount of electronics and networking in the vehicle increases, so too the demand for powerful development processes and their description methods for the architecture of electrical and electronic systems increases.

The term "architecture" generally refers to the art of building. In the construction industry, the architect designs a building by drawing up plans for the different views and contractual work based on the client's wishes and boundary conditions. A plan abstracts the reality with regard to a particular aspect (e.g. geometric conditions or electric cabling). The building can finally be erected on the basis of the plans of all the necessary aspects.

When carried over to a motor vehicle, this is referred to as the "E/E architecture". "E/E" denotes the electrical and electronic aspects of the motor vehicle. The "plans" of the E/E architect are referred to in the following with the general term of "model".

Figure 4: Models of E/E architecture

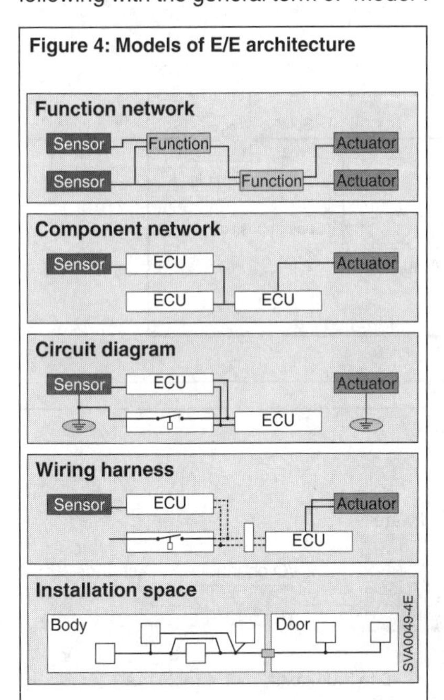

Automobile manufacturers and their suppliers have different views on how many models of which type are needed to describe completely the electrical and electronic systems in the vehicle. The models presented in the following have proven successful in practice and are a necessary framework for describing the E/E scope.

The term architecture is often used in the literature and in publications to denote the models themselves. Here, a clear distinction is made between work operation (architecture development) and presentation of the result (model).

Models of E/E architecture

The models of E/E architecture reflect the results of the different integration aspects of the electronic systems in the vehicle (Figure 4). These aspects are usually dealt with simultaneously because both the geometry (the "body structure") and new systems are addressed in the concept phase. In the course of vehicle development, the situation may for instance arise where an electronic system in the chosen technology does not fit into the available space. Compromises must be found in this case.

Function network

Function model

Function models are the preliminary stage of concrete technical systems. They describe the transfer elements which are needed to realize the required performance features without going into their concrete technology. For the example of superimposed steering, this means a breakdown into transfer elements such as

- variable steering ratio,
- stabilization control,
- vehicle model,
- actuator,
- vehicle,
- driver.

The function models (Figure 5) are usually created as a signal-flow diagram in accordance with DIN 19226 ([5]).

Component network

Technology model

The technology model describes what technical realization is used for the specified transfer elements without already combining these into modules, such as electronic control units (ECUs). "Technology blocks" are created.

Thus, signal filtering can be realized with discrete components by means of a digital circuit or filter software on a microcontroller. Even a controller function can be executed with a discrete electronic circuit or a microcontroller. Voltage stabilization can be achieved by either a smoothing capacitor or a DC/DC converter.

The decision as to which technical realization to opt for is influenced on the one hand by the function and on the other hand by the costs. Before the technology blocks are combined into modules in the form of ECUs, the first step is to look for synergy with the further technology blocks to be integrated. A technological active chain is created (Figure 6). If, for example, a specific sensor technology is available for an active-chain link whose signal is required by another active chain, this will also be used. This occurs even if this sensor is overspecified for the additional user, i.e. there are lower requirements, for example, of the available signal range or of the accuracy.

It is nevertheless important to store the original requirement in a database as this synergy may no longer be present in another vehicle.

The automotive industry usually uses the nomenclature according to DIN EN 60617 ([1]) to describe the hardware.

Figure 5: Signal-flow diagram with standard elements on example of superimposed steering
δ denotes the steering angles,
$\dot{\psi}$ the yaw rate.

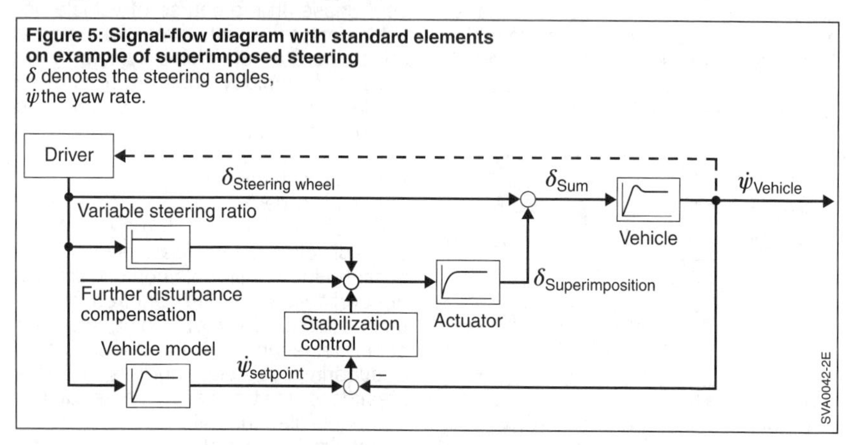

Figure 6: Example of a technological active chain

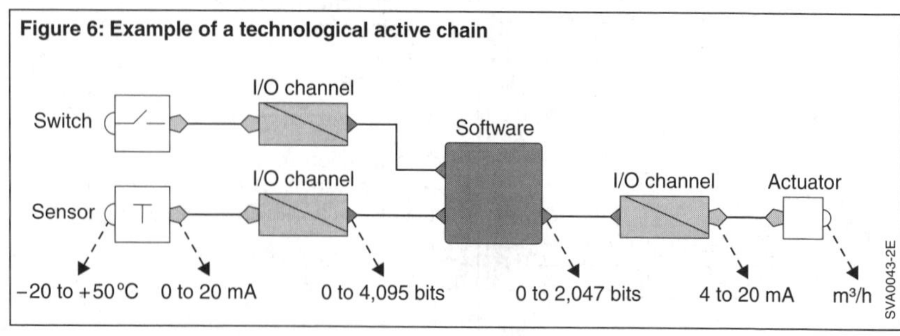

Node model
The links of the technological active chains are combined into groups at different locations, the "nodes". Here, strict adherence to the optimum cost of integrating the technology blocks is maintained. Thus, attempts are made, for example, to integrate the software parts of several technological active chains on a common microcontroller. Sensor signals are used repeatedly where possible and actuators used commonly where possible.

But the history shows that there are remarkable synergies even in the mechanical field, such as, for example, vacuum supply of a pneumatic brake booster by the intake port of the spark-ignition engine.

ECU hardware model
This model represents the structure of the electronic hardware of an individual ECU. It is created by allocating specific electronic components from the technological active chains to an electronic module in a node. An ECU is therefore, generally speaking, a collecting point for electronic components of different systems, an "integration platform".

Software for controlling different systems from different sources (automobile manufacturers or their suppliers) is integrated on the microcontrollers located in the ECU. By networking the ECUs, it is possible to facilitate complex distributed functions, which utilize sensors and actuators from different installation locations in the vehicle.

During development, initially customary circuit diagrams will be used for the electrical or electronic parts of the ECU. Then the ECU mechanics and the design and connection technology will be established. The model is confined to a very rough representation in the early concept phase.

ECU software model
There are from classic information technology (for PC systems) some recognized methods for software-architecture development with associated models (e.g. Product Line Approach). However, no standard for architecture development has developed up to now in the automotive industry. Through the introduction of AUTOSAR the software architecture in the motor vehicle is currently being directly defined.

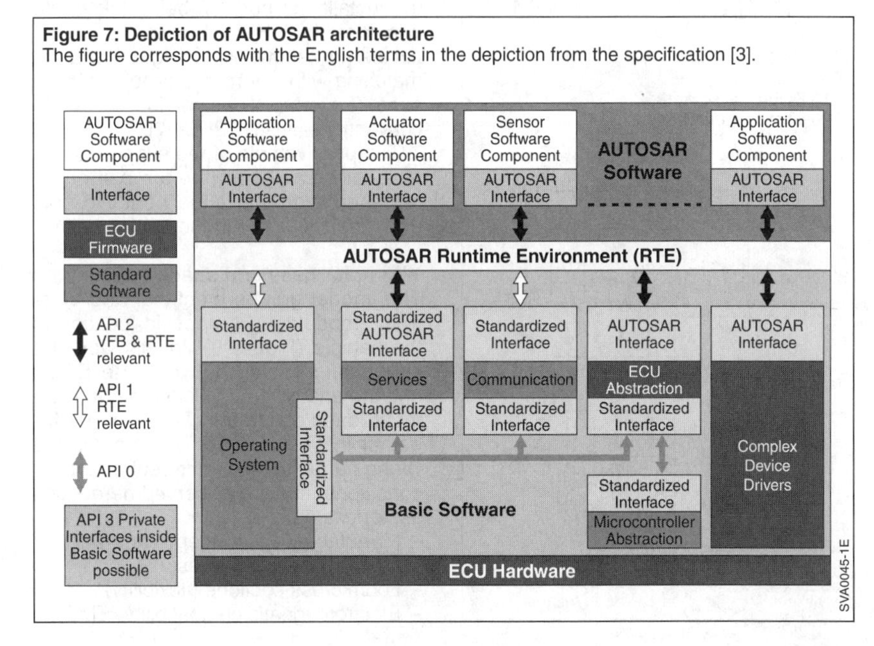

Figure 7: Depiction of AUTOSAR architecture
The figure corresponds with the English terms in the depiction from the specification [3].

The AUTOSAR standard defines the structuring of software close to hardware level and its interface to the application functions and establishes interfaces between application functions (Figure 7). AUTOSAR additionally defines standardized exchange formats which are supported by popular modeling tools.

Typically a distinction is made between basic and application software. Blocks of basic software are, for example, device-driver software, communication software, operating system, and hardware abstraction.

Network model of communication
Because all the technology blocks of a vehicle have been allocated to ECUs in the previous steps, a network of these ECUs is now in place with their communication relationships. The network model of communication represents all the ECUs in the motor vehicle which have bus communication and are thus directly or indirectly networked with each other.

Figure 8: Circuit diagram using example of a vehicle radio
15, 30, 31 Terminal designations,
A2 Vehicle radio,
W1 Vehicle antenna,
F Fuse,
B11, B12 Speakers,
P6 Timer,
X18 Diagnosis socket,
1–8 Section identification.

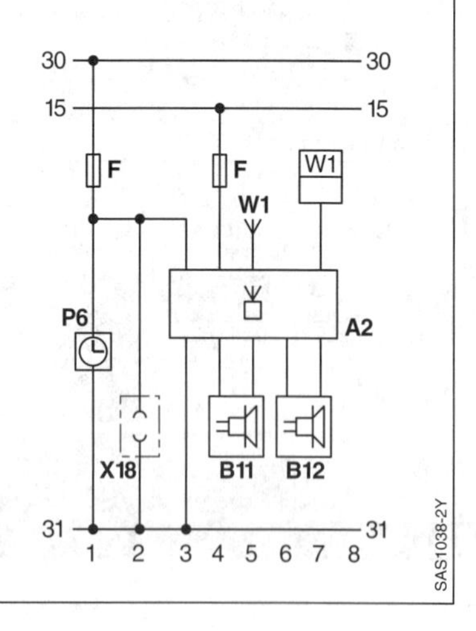

Each signal which is exchanged between two or more ECUs is assigned to a suitable bus system. AUTOSAR defines for this purpose standardized exchange formats which enable bus communication to be described. The AUTOSAR exchange format contains from Release 3.0 the ASAM standard FIBEX [4].

Circuit diagram

Network model of energy supply
The allocation of the technology blocks to ECUs and sensor and actuator modules has also given rise to a network of electrical loads/consumers which requires a suitable energy supply. On the one hand, it is important to fuse individual electric circuits so that a short-circuit does not affect the entire network. On the other hand, not all circuits should be supplied with electrical energy in each operating state. The principle of "terminals" was therefore introduced for this purpose. Thus, for example, terminal 15 is only supplied with electrical energy when the ignition is switched on.

The electrical circuit diagram (Figure 8) shows the electrical networking and fusing of the individual modules without the installation position being taken into consideration. Here the colors of the cables (not shown in the figure) and the matching with a terminal or fuse can be seen. The terminal designations follow the conventions in DIN 72552 [2].

The positive pole of the supply voltage is usually featured in the top half of the representation while the negative pole (ground) is featured in the bottom half.

Wiring harness and space

This model groups electrical and electronic modules in a specific location in the vehicle (Figure 9). In this way, the connecting cables between the ECUs and the energy-supply leads of the electrical loads/consumers are brought together in cable looms. This creates the wiring harness. Many different boundary conditions must be observed here, such as, for example:

- Manufacturing concept (one-part or multi-part wiring harness)
- Loom cross-sections (flexibility)
- Electromagnetic compatibility (EMC)

- Heat dissipation
- Weight
- Costs (e.g. of copper)
- Structure of the wiring harness in the vehicle.

The structure describes the possible routing paths in the body, such as for example the H structure, which consists of two main connections from the front end to the rear end of the vehicle and a cross cable (crossing) from the left to the right side of the vehicle.

Two-dimensional models are usually sufficient in the concept phase of a vehicle; detailed three-dimensional models are used in the later development phase.

E/E development process

The E/E development process links the individual draft stages with each other on a logic and time basis and provides quality criteria at the beginning and the end of a draft stage.

Figure 9: Example of a two-dimensional space model
RL rear left,
RR rear right,
FL front left,
FR front right,
FM front middle,
IL inside left,
IR inside right.

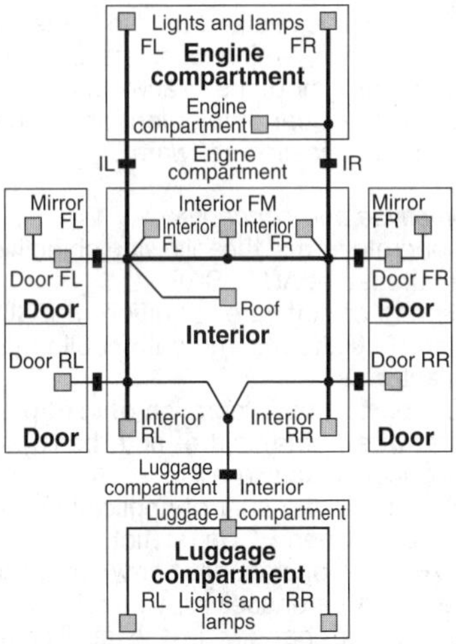

Because E/E architecture for automotive applications is still a young discipline, the processes at automobile manufacturers and their suppliers still differ greatly. This relates both to the number and sequence of the draft stages and to the quality criteria.

Requirement management

The requirements decisively determine the decisions of the E/E architect. It is advisable to distinguish between functional and non-functional requirements.Functional requirements refer to the desired performance features when the vehicle is being used.Non-functional requirements refer to the technical solution and are therefore also known as draft restrictions.

Such a restriction can be, for example, the space available in the center console for installing ECUs. Another restriction can be the maximum permissible heat dissipation in a location which influences the power electronics positioned there. Thus, for example, the audio amplifier in vehicles is frequently installed in the luggage-compartment areas since the heat in the cockpit area cannot be adequately dissipated.

Once the documentation of the functional and non-functional requirements has been completed, the actual development of the E/E architectures begins.

Development of E/E architectures

The development of the E/E architecture can follow two paths: The bottom-up approach, which starts out from existing components, and the top-down approach, which features the implementation of all the previously described modeling steps starting out from the functional and non-functional requirements.

The bottom-up approach involves – during the creation of the E/E architecture starting out from the functionality of existing components – these components being additionally supplemented by function and communication aspects and the corresponding modeling steps being carried out. This approach is typically chosen for creating E/E architectures of follow-up generations of existing vehicle platforms.

The top-down approach focuses on the function complexity and is typically chosen for creating E/E architectures of new vehicle platforms.

The use of E/E concept tools allows an exchange of data with development partners for electronic components or for the wiring harness.

Evaluation of models

The following must be observed for all approaches: During the transition from one model hierarchy to the next (e.g. from the function model to the technology model), a list of evaluation criteria (e.g. reuse or testability) is compared with a portfolio of specimen solutions (e.g. bus technologies). Evaluation of the specimen solutions using the criteria allows a solution to take shape on the basis of the purely functional requirements and irrefutable boundary conditions ("MUST criteria"). This procedure is also known as QFD (Quality Function Deployment).

An alternative procedure consists in comparing a reference solution (e.g. the previous networking model) using the evaluation criteria with alternative solutions. This does indeed deliver fast results, but possibly not the global optimum.

Because the evaluation criteria are generally weighted differently by the automobile manufacturers, the electronic systems of the vehicles sometimes differ considerably from each other.

E/E development tools

Ideally a tool is used for architecture modeling which can represent and network with each other the different models and model levels of the E/E architecture work. Fully connected documentation is created in this way. This enables the different disciplines involved in the development process to come on board at the relevant points. It should furthermore be possible to record the modeling properties numerically in order to be able to provide them with an evaluation.

Different E/E development tools have in the meantime come onto the market which enable architecture modeling to be performed on a tool-assisted basis. An important point is the standardization of the models and their data formats. Only this facilitates competition between the tool manufacturers and opens up to the different disciplines involved the opportunity of joining the process at different points.

With regard to a future seamless transition into the system and ECU configuration with the aid of AUTOSAR technologies, the E/E development tool is required to support AUTOSAR exchange formats such as "System Description" and "SW Component Description". AUTOSAR will therefore be discussed in more detail in the following.

AUTOSAR

The AUTOSAR Partnership (Automotive Open Systems Architecture) was founded in July 2003 by automobile manufacturers and their suppliers – Bosch among them. Its global objective is the joint development of an open software architecture for future automotive applications. The Partnership's objectives include the standardization of a fundamental ECU infrastructure (basic software), exchange formats and functional interfaces. These are intended to replace the previous company-specific individual solutions. Model-based concepts and methods are used to keep on top of the continuous rise in complexity brought about by new functions. The demands for quality and reliability are fulfilled by the multiple use of proven standards.

On the basis of standardized infrastructure software, which consists primarily of standard modules, each automobile manufacturer can implement its specific content (application software).

Objectives and concepts

In concrete terms, the following objectives are pursued at AUTOSAR:

- Reuse of software for different ECUs, vehicle platforms and automobile manufacturers.
- Support in the integration of third-party software in the field of both basic and application software.
- Support in the shift of application software between ECUs (static).
- Exchange of standard hardware blocks (e.g. CAN transceiver) without this resulting in changes in the application software.

The software architecture defined by AUTOSAR (Figure 7) supports a clear separation between the basic software and the application software. This is achieved by several abstraction levels in the basic software from hardware drivers through to complex infrastructure services and the AUTOSAR Runtime Environment (RTE).

Because the interfaces of most basic-software modules within these layers are standardized, standard hardware blocks and the associated drivers can be replaced without the application software incurring changes.

Conversely, application software which limits itself to the use of these standard interfaces can during development be inserted more easily into an ECU or even shifted to another ECU. It is thus possible, for example, for the same application software of a vehicle-speed controller depending on the vehicle platform to run on the engine control unit, the transmission control unit or another ECU without the application software having to be changed. This naturally requires the ECU and the networking to be sufficiently powerful.

Standardization of the basic software and its configurability depending on the requirements of the application software enable basic-software modules to be reused for different vehicle platforms and automobile manufacturers. This increases the quality of the software since there are no product-specific changes, reduces the development costs through reuse and forms a stable basis for the constantly increasing complexity and networking of the application functions.

The AUTOSAR Partnership is also developing model-based concepts for early validation of the system draft. A check is made on the basis of a formalized description as to whether application-software interfaces are consistent with respect to each other without the application software having to be in place as a full program.

Summary and outlook

As a result of the increasing scope and the increasing networking of electronic systems, it is necessary to use suitable processes, methods and tools in E/E architecture development. E/E architecture development has crystallized as an independent task in the automotive industry which has a decisive influence in the development of new vehicles. If the procedures described in this publication are consistently applied and developed further, electronics in motor vehicles will also be manageable in the future and will continue to make significant contributions to improving traffic flow, traffic safety, driving comfort, and economical fuel utilization.

References

[1] DIN EN 60617: Graphical symbols for diagrams.
[2] DIN 72552: Terminal markings for motor vehicles.
[3] AUTOSAR Release 3.1, Rev. 2.
[4] AUTOSAR Release 4.0: ASAM Fibex - Field Bus Exchange Format, Version 3.1.
[5] DIN 19226: Control technology.

Automotive sensors

Basic principles

Function

Sensors convert a physical or a chemical (usually non-electrical) quantity Φ into an electrical quantity E. This is often also done via further, non-electrical intermediate stages.

Figure 1: Types of characteristic curve
S Output signal,
X Measured variable.
a) Continuous, linear,
b) Continuous, nonlinear,
c) Discontinuous, multistage,
d) Discontinuous, dual-stage with hysteresis.

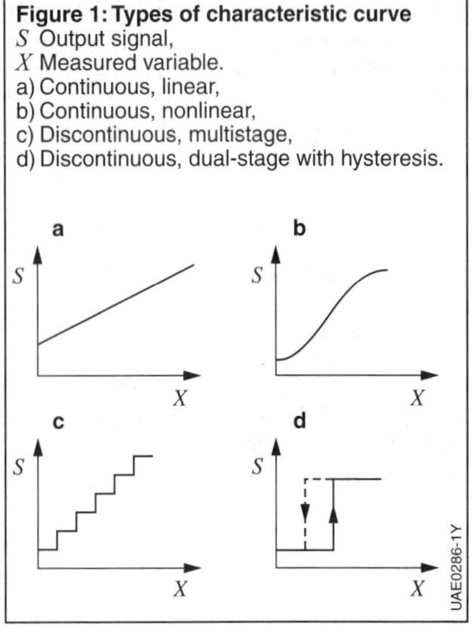

Figure 2: Signal shapes (examples)
a) Frequency f,
b) Pulse duration T_p,
U Output signal,
t Time.

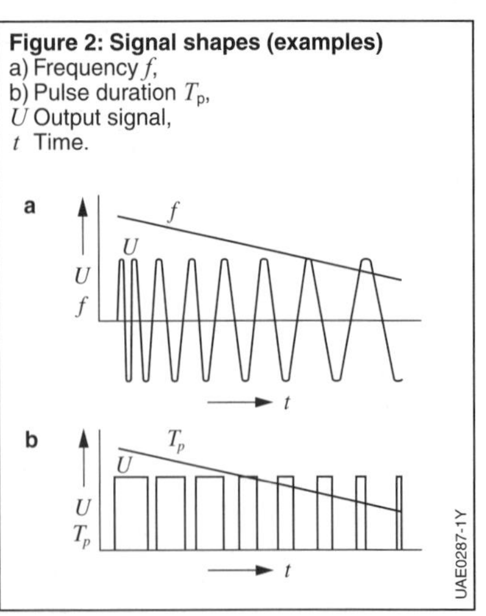

Sensor classification

Sensors can be classified according to very different standpoints. With regard to their use in motor vehicles, they can be categorized as follows.

Assignment and application

- Functional sensors (e.g. pressure sensor, temperature sensor, air-mass sensor), primarily for open- and closed-loop control functions.
- Sensors for safety (passenger protection: airbag and Electronic Stability Program, ESP) and protection (theft deterrence).
- Sensors for vehicle monitoring (onboard diagnostics, consumption and wear quantities) and for providing information to driver and passengers.

Type of characteristic curve

- Continuous linear characteristic curves (Figure 1a) are used particularly for control functions across a broad measurement range. Linear characteristic curves furthermore have the advantage that they are easy to check and adjust.
- Continuous nonlinear characteristic curves (Figures 1b) are often used for closed-loop control of a measured variable within a very narrow measurement range (e.g. control of the air/fuel ratio at $\lambda = 1$, control of the compression level).
- Discontinuous dual-stage characteristic curves, with hysteresis in some cases (Figure 1d) are used to monitor limit values where an easy remedy is possible when they are reached. If remedy is more difficult, early warning can be effected at an earlier stage by means of multiple stages (Figure 1c).

Type of output signal

Analog output signal (Figure 2a):

- current or voltage, or corresponding amplitude
- frequency or period duration
- pulse duration or pulse duty factor.

Discrete output signal (Figure 2b):
- dual stage (binary encoded)
- multistage – irregular graduation (analog encoded)
- multistage – equidistant (analog or digital encoded).

Continuous signals are constantly available at the sensor output, discontinuous signals are available only at discrete times. A bit-serially output signal is inevitably discontinuous.

Automotive applications
In their function as peripheral elements, sensors and actuators form the interface between the vehicle with its complex drive, braking, chassis, suspension, and body functions (including guidance and navigation functions), and the usually digital-electronic control unit (ECU) as the processing unit (Figure 3). An adapter circuit is generally used to convert the sensor's signals into the standardized form (measuring chain, measured-data registration system) required by the ECU. In addition, system operation can be influenced by sensor information from other processing elements for example via bus systems and/or by driver-operated switches.

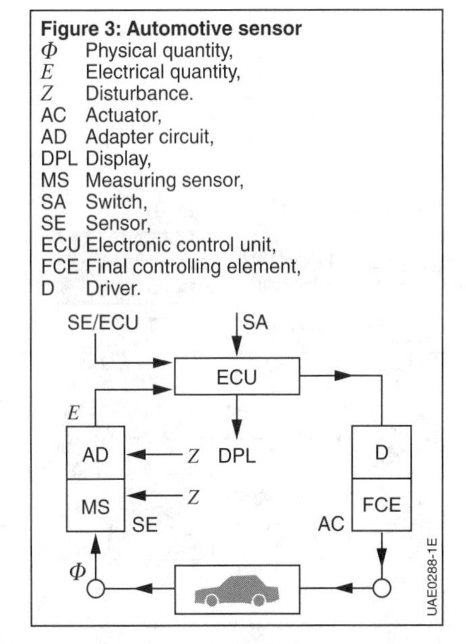

Figure 3: Automotive sensor
Φ Physical quantity,
E Electrical quantity,
Z Disturbance.
AC Actuator,
AD Adapter circuit,
DPL Display,
MS Measuring sensor,
SA Switch,
SE Sensor,
ECU Electronic control unit,
FCE Final controlling element,
D Driver.

Main requirements
Operating conditions
Automotive sensors are sometimes exposed to extreme conditions in their installation locations. These are mechanical stresses (e.g. vibrations and impacts), climatic influences (e.g. low and extremely high temperatures, moisture), chemical influences (e.g. splash water, salt spray, fuel, engine oil, battery acid), and electromagnetic influences (e.g. high-frequency interference, mains-borne interference pulses, overvoltage). The degree of stress to which a sensor is subjected is determined by the operating conditions at its installation location.

Reliability
According to application and technical requirements, automotive sensors are assigned to one of three reliability classes:
- Steering, brakes, passenger protection
- Engine, drivetrain, suspension, tires
- Comfort and convenience, information, diagnosis, theft deterrence

Size
The constantly increasing number of electronic systems in vehicles on the one hand and the ever more compact shape of vehicles on the other hand force engineers to adopt extremely small designs. In addition, the pressure of increased fuel economy necessitates a consistent reduction of vehicle weight. Various miniaturization concepts are employed to achieve compact sensor dimensions:
- Substrate and hybrid technologies (e.g. for strain-dependent resistors, thermistors, and magnetoresistors)
- Semiconductor technology (e.g. for Hall-effect crankshaft sensors)
- Surface and bulk micromechanics (e.g. for silicon pressure, acceleration, and yaw sensors)
- Microsystem technology (integration of micromechanical or microoptical components with microelectronic circuits in a complex system).

Manufacturing costs
Electronic systems in vehicles contain up to 150 sensors. This abundance of sensors forces engineers to radically reduce costs compared with other areas of use.

Extensively automated manufacturing processes working at great efficiency levels are used. In other words, each process step is always performed simultaneously for a larger number of sensors. The manufacture of semiconductor sensors typically involves the simultaneous integration of 100 to 1,000 sensors on one silicon wafer. The automotive industry's huge demand for sensors has set new standards here.

Accuracy requirements
The accuracy requirements for automotive sensors are lower than for sensors for the process industry for example. The permissible deviations are generally greater than 1 % of the upper range value of the measurement range. The inevitable influences of aging must also be taken into consideration here.

However, ever more sophisticated and complex systems call for higher levels of accuracy. This can be achieved to a certain extent by reducing the manufacturing tolerances and by improving the adjustment and compensation technologies. "Integrated sensors" provide for a significant advance here.

Integrated sensors
Systems range from hybrid and monolithic integrated sensors and electronic signal-processing circuits at the measuring point, through to complex digital circuits, such as analog-digital converters and microcomputers, for complete utilization of the sensor's inherent accuracy ("intelligent" sensors, Figure 4). These systems offer the following benefits and options:
- Reduction of load on the ECU
- Uniform, flexible, bus-compatible interface
- Multiple application of sensors
- Multisensor designs
- Use of small measuring effects and high-frequency measuring effects (local amplification and demodulation)
- Correction of sensor deviations at the measuring point, and common calibration and compensation of sensor and circuit, are simplified and improved by storage of the individual correction information in semiconductor memory chips (e.g. PROM).

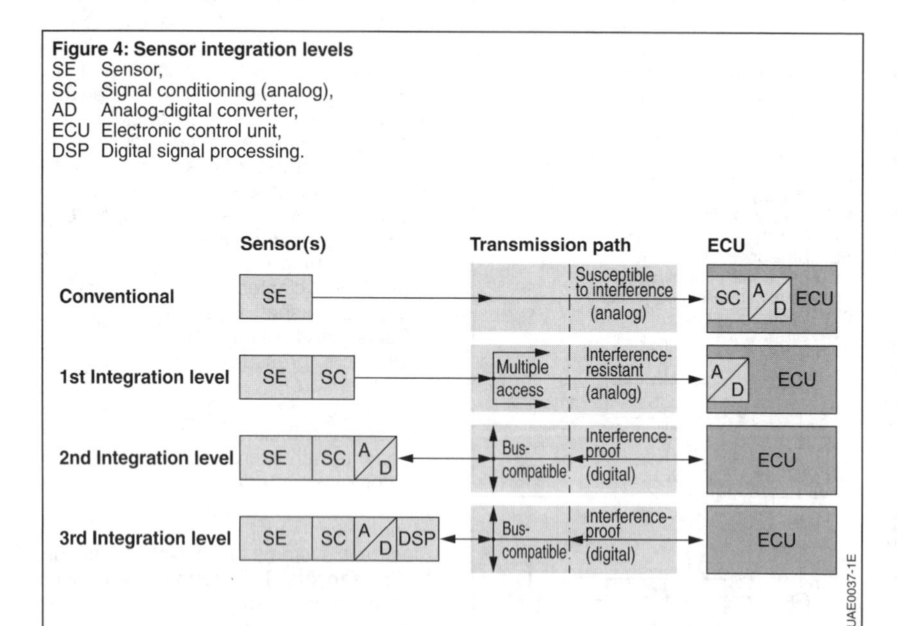

Figure 4: Sensor integration levels
SE Sensor,
SC Signal conditioning (analog),
AD Analog-digital converter,
ECU Electronic control unit,
DSP Digital signal processing.

Micromechanics

Applications

Micromechanics is defined as the application of semiconductor technology in the production of mechanical components from semiconductor materials (usually silicon). This type of application exploits both the semiconducting and the mechanical properties of silicon. The first micromechanical silicon pressure sensors were installed in motor vehicles at the beginning of the 1980s. Typical mechanical dimensions can extend into the micrometer range.

The mechanical properties of silicon (e.g. strength, hardness and modulus of elasticity, see Table 1) can be compared to those of steel. However, silicon is significantly lighter and has greater thermal conductivity than steel. Single-crystal silicon wafers are used with almost perfect mechanical properties. Hysteresis and creepage are negligible. Due to the brittleness of the single-crystal material, the stress-strain curve has no plastic range; the material ruptures when the elastic range is exceeded.

Two methods of manufacturing micromechanical structures in silicon have established themselves: volume micromechanics (VMM) and surface micromechanics (SMM). Both methods use the standard procedures of microelectronics (e.g. epitaxial growth, oxidation, diffusion, and photolithography) together with some additional special procedures [1].

Volume micromechanics

The silicon wafer material is processed at the required depth using anisotropic (alkaline) etching with or without an electrochemical etching stop. From the rear, the material is removed from inside the silicon layer, where there is no etching mask resting on top (Figure 1). Using this method, very small membranes can be produced with typical thicknesses of between 5 and 50 µm, as well as openings, beams and webs as are needed for instance in pressure and acceleration sensors.

The problem when etching with alkaline media lies in the fact that the walls run inwards at an angle. In order to be able to effect high-precision etching at depth with vertical walls, it was necessary to develop a new process. Here Bosch made a breakthrough with the development of the DRIE process (Deep Reactive Ion Etching), which has in the meantime come to be known generally in the industry as the "Bosch process". This involves the creation in a special gas-phase reactor of alternating gases and accompanying conditions for an etching

Table 1: Mechanical properties of silicon

Quantity	Unit	Silicon	Steel (max.)	Stainless steel
Tensile load	10^5 N/cm^2	7.0	4.2	2.1
Knoop hardness	kg/mm^2	850	1,500	660
Modulus of elasticity	10^7 N/cm^2	1.9	2.1	2.0
Density	g/cm^{3}cm^3	2.3	7.9	7.9
Thermal conductivity	W/cm · K	1.57	0.97	0.33
Thermal expansion	10^{-6}/K10^{-6}/K	2.3	12.0	17.3

Figure 1: Electrochemical etching
a) Isotropic (in acid etching media), b) Anisotropic (in alkaline etching media).
1 Etching mask (e.g. oxide or nitride), 2 Silicon.

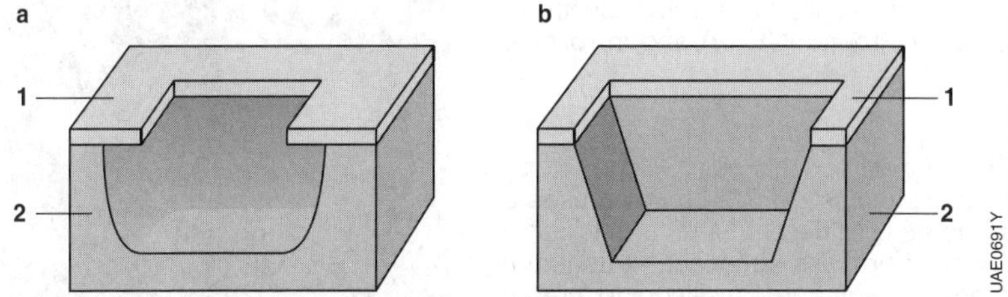

stage and then a passivation stage. This etching and the subsequent passivation of the walls are set such that very precise vertical walls are created. This process can be used for both volume and surface micromechanics.

Surface micromechanics
In contrast to volume micromechanics, surface micromechanics merely uses the silicon wafer as the substrate. Moving structures are usually formed from polycrystalline silicon layers which, similarly to a manufacturing process for integrated circuits, are deposited on the surface of the silicon by epitaxial growth.

When a surface-micromechanical component is made, a sacrificial layer of silicon oxide is first applied on the wafer and structured with standard semiconductor processes, i.e. partly removed again specifically (Figure 2a). An approx. 10 μm thick polysilicon layer (epipoly layer) is then applied at high temperatures in an epitaxial reactor (Figure 2b) and whose desired structure is anisotropically, i.e. vertically, etched with the aid of a lacquer mask (deep etching or trenching, Figure 2c). The vertical side walls are obtained with the Bosch process by alternating etching and passivation cycles. After an etching cycle, the etched side-wall section is provided during passivation with a polymer as protection so that it is not attacked during the subsequent etching. Vertical side walls with a high representation accuracy are created in this way. In the last process stage (Figure 2d) the sacrificial layer beneath the polysilicon layer is removed with gaseous hydrogen fluoride in order to expose the structures (Figure 3).

Among other things, surface micromechanics is used in the manufacture of capacitive acceleration sensors for airbag systems and in the manufacture of yaw sensors for applications in the Electronic Stability Program (ESP) and in rollover sensing.

APSM process
The "Advanced Porous Silicon Membrane" process (APSM process) uses a completely different surface-micromechanical process sequence. It uses the properties of porous silicon to create

Figure 2: Process stages in surface micromechanics
a) Depositing and structuring the sacrificial layer,
b) Depositing polysilicon,
c) Structuring polysilicon by deep etching,
d) Removing the sacrificial layer and so producing freely moving structures on the surface.
1 Silicon,
2 Oxide layer (sacrificial layer),
3 Polysilicon layer ("epipoly").

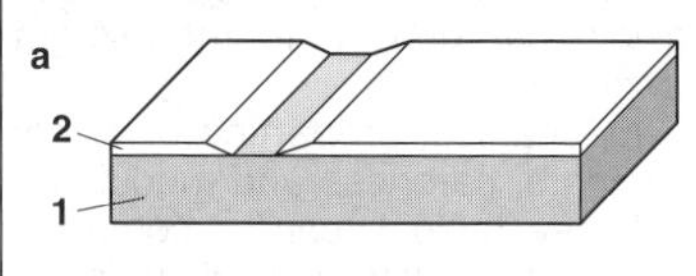

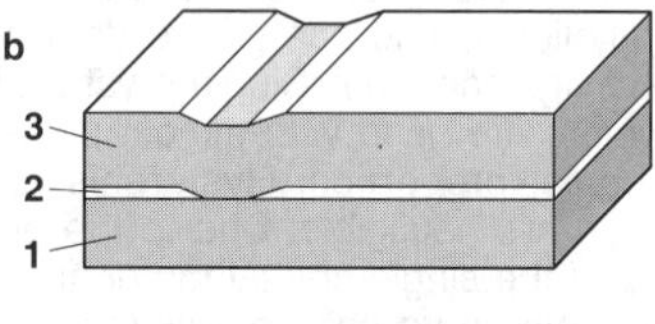

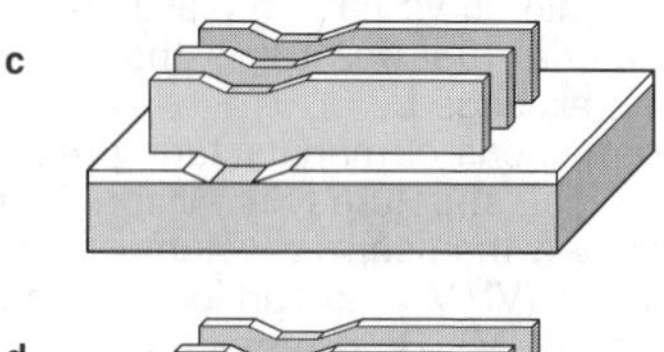

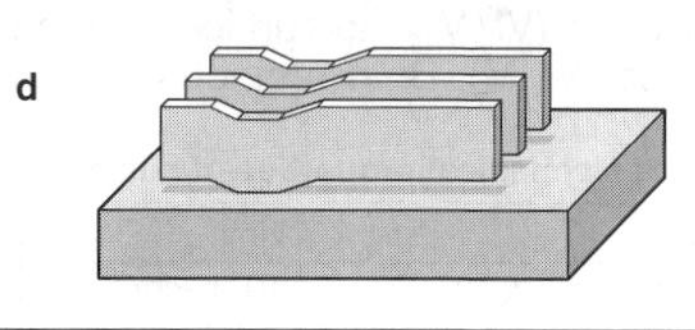

Figure 3: Structure of a surface-micromechanical sensor
Recording with scanning electron microscope.
1 Fixed electrode,
2 Gap,
3 Spring electrode.

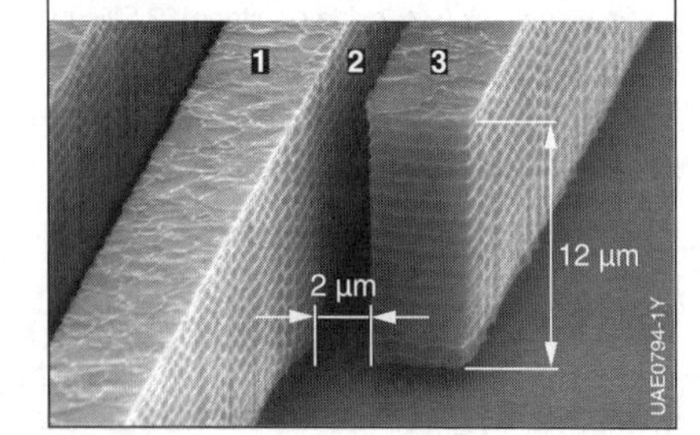

beneath a monocrystalline membrane an exactly defined cavity in which a vacuum is enclosed.

The core of the APSM process is porous silicon. This can be manufactured selectively and on a locally limited basis in p-doped silicon in an electrochemical anodizing process in hydrofluoric acid. In this process some of the silicon is dissolved out of the crystal to leave a porous, sponge-like silicon shell, the "porous silicon".

Porous silicon in turn can be rearranged at high temperatures. The silicon shell dissolves and forms under suitable conditions a thin membrane on the surface. A cavity forms below this membrane (Figure 4). This thin membrane can be enlarged to the target thickness by means of epitaxial growth.

The monocrystalline epitaxial layer is used for example as the diaphragm of a pressure sensor. Evaluating-circuit elements in the epitaxial layer outside the diaphragm provide for a high-precision, small and inexpensive pressure sensor.

Modern pressure sensors, e.g. barometric pressure sensors for engine-management systems, are manufactured using the APSM process.

Wafer bonding

In addition to structuring the silicon, joining two wafers represents another essential task in micromechanical production engineering. Joining technology is required, for example, in order to hermetically seal a reference-vacuum chamber (e.g. for pressure sensors), to protect sensitive structures by applying caps (e.g. in acceleration and yaw sensors, Figure 5), or to join the silicon wafer with intermediate layers which minimize the thermal and mechanical stresses (e.g. glass base on pressure sensors).

In the case of anodic bonding, a Pyrex-glass wafer is joined to a silicon wafer at a voltage of some 100 V and a temperature of approx. 400 °C (Figure 6). A strong electrostatic attraction and an electrochemical reaction (anodic oxidation) result in a permanent hermetic bond between the glass and the silicon.

In the case of seal-glass bonding, two silicon wafers are contacted by way of a glass-solder layer applied in a screen-printing process at approx. 400 °C and under the exertion of pressure. The glass solder melts at this temperature and produces a hermetically sealed bond with the silicon.

References for Sensors – Basic principles

[1] U. Hilleringmann: Mikrosystemtechnik – Prozessschritte, Technologien, Anwendungen. B.G. Teubner-Verlag, Wiesbaden 2006.

Figure 4: Creating exact vacuum caverns in silicon with the APSM process
1 Silicon,
2 Cavity, created from porous silicon by the APSM process,
3 Membrane,
4 Evaluating circuit.

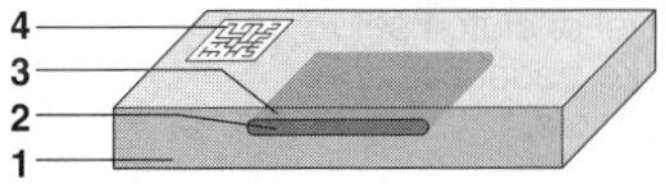

4
3
2
1
SAE1150Y

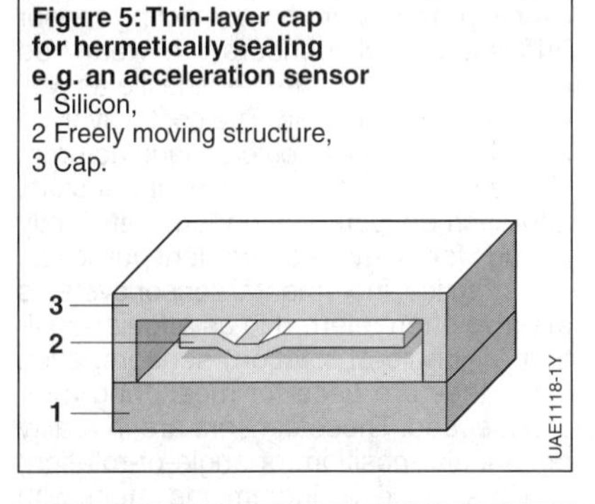

Figure 5: Thin-layer cap for hermetically sealing e.g. an acceleration sensor
1 Silicon,
2 Freely moving structure,
3 Cap.

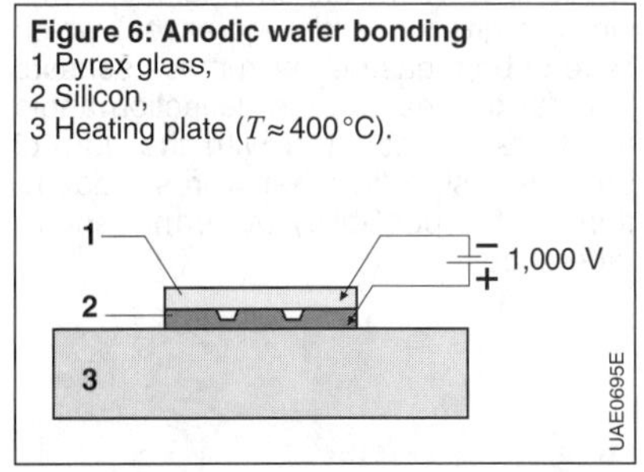

Figure 6: Anodic wafer bonding
1 Pyrex glass,
2 Silicon,
3 Heating plate ($T \approx 400$ °C).

Position and angular-position sensors

Measured variables

These sensors record one- or multidimensional displacement and angle positions (translational and rotatory variables) of the most varied types and ranges. Examples of such measured variables are:

- Accelerator-pedal position for recording the torque request for engine management (driver command)
- Throttle-valve position for controlling the throttle valve
- Fuel-tank level
- Steering-wheel angle
- Position of the transmission selector lever for electronic transmission control
- Seat position
- Mirror position
- Control-rack position in the diesel in-line fuel-injection pump
- Travel of clutch servo unit
- Brake-pedal position
- Tilt angle.

In this sector of applications, activities have long since been directed at changing over to proximity or non-contacting sensor principles. Such sensors are wear-free and thus have a longer service life as well as being more reliable. The costs involved though, often force vehicle manufacturers to retain the "wiper-type" sensor principle, and such sensors still perform efficiently enough for many measurement purposes.

In practice, incremental sensor systems are also often referred to as angular-position (or angle-of-rotation) sensors, even when they are used for measuring rotational speed. These sensors are in reality not angular-position (or angle-of-rotation) sensors. Since the increments (steps with which a given quantity increases) which have to be measured with these sensors in order to measure the deflection angle must first be counted with the correct preceding sign (in other words, added), there is the possibility of permanent interference.

Wiper potentiometer

Measurement principle

The wiper potentiometer – usually designed as an angular-position sensor – measures travel by exploiting the proportional relationship between the length of a wire or film resistor (conductor track made of cermet or "conductive plastic") and its electrical resistance (Figure 1). At present, this is the lowest-priced travel and angular-position sensor.

To protect against overloading, supply voltage is usually applied to the measurement track through series resistors R_V. These resistors can also be used for calibration of zero point and curve slope. The shape of the contour across the width of the measurement track (including that of individual sections) influences the shape of the characteristic curve.

The standard wiper connection is furnished by a contact track consisting of the same material mounted on a low-resistance substrate. Wear and measurement distortions can be avoided by reducing the current at the pick-off (I_A < 1 mA) and sealing the unit against dust and liquid. Also essential to minimizing wear is to have an optimum friction pairing of wiper

Figure 1: Wiper potentiometer
1 Wiper,
2 Potentiometer track (measurement track),
3 Contact track.
U_0 Supply voltage,
U_A Measurement voltage,
I_A Wiper current,
R_0 Resistance of measurement track,
R_α Resistance of part of measurement track,
R_V Series resistors, R_S Protective resistor,
α Measurement angle.

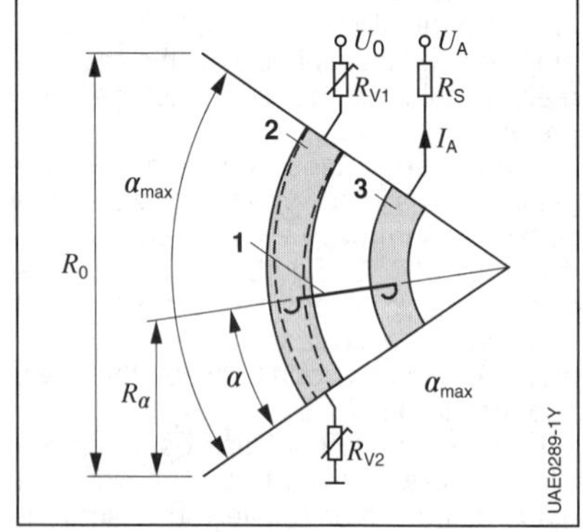

and potentiometer track; here wipers can have a spoon or scraper shape and have a single or multiple design, even in the shape of a broom.

The wiper potentiometer together with further resistors form a voltage divider (Figure 2). The picked-off voltage is a measure of the wiper position.

Advantages of the potentiometer are, for example, the low costs, the simple and clear design, the large measuring effect (measuring stroke corresponds to the supply voltage), the wide operating-temperature range up to 250 °C, and the high degree of accuracy (better than 1 % of the upper range value of the measurement range).

Disadvantages of the potentiometer are, for example, mechanical wear caused by abrasion, measuring errors caused by abraded remnants, wiper lift-off in the event of vibrations, and the limited miniaturization capability.

Wiper-potentiometer applications

Throttle-valve angle sensor

Mounted on the throttle valve in a gasoline engines is a throttle-valve sensor (Figure 3), which signals the position of the throttle valve to the engine-management system. The throttle-valve angle sensor has a redundant design for diagnostic purposes (Figure 2). This makes available two independent signals which can be checked for plausibility against each other. Redundant measurement enables faults to be detected.

Figure 2: Electrical circuit of throttle-valve sensor
1 Throttle valve,
2 Throttle-valve sensor.
U_A Measurement voltages,
U_V Operating voltage,
R_1, R_2 Potentiometer tracks 1 and 2,
R_3, R_4 Trimming resistors,
R_5, R_6 Protective resistors.

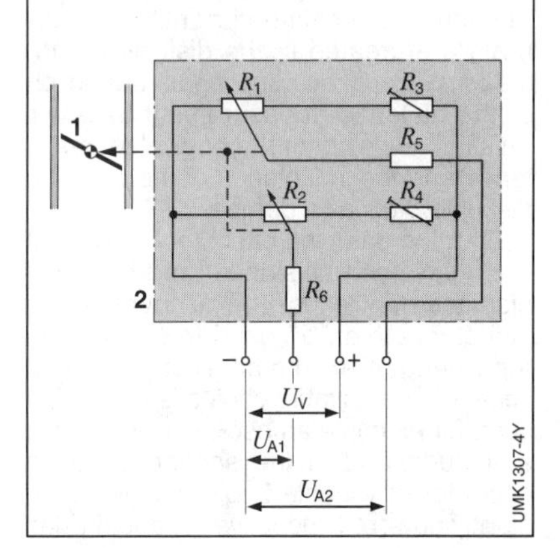

Accelerator-pedal sensor

The accelerator-pedal (pedal-travel) sensor senses the travel or the angle of the pressed accelerator pedal. The engine-management system interprets this value as a torque request from the driver. The accelerator-pedal sensor is integrated together with the accelerator pedal in the accelerator-pedal module. With these units that are ready for installation there is no need for adjustment on the vehicle.

With the potentiometric sensor design the measurement signal is generated by a potentiometer. The engine ECU converts the measured voltage using the stored sensor curve into the covered pedal travel or into the accelerator pedal's angular position.

Like the throttle-valve sensor the accelerator-pedal sensor has a redundant design. One sensor version operates with a second potentiometer that always delivers half the voltage of the first potentiometer at all operating points. Two independent

Figure 3: Throttle-valve sensor
1 Throttle-valve shaft,
2 Potentiometer track 1,
3 Potentiometer track 2,
4 Wiper arm with wipers,
5 Electrical connection.

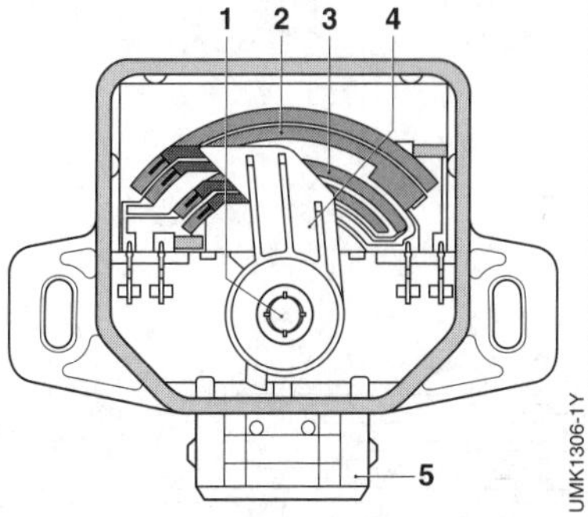

signals are thus available for the purpose of fault detection. Sensor versions with low-idle switches were also customary in the past; the state of these sensors had to be plausible with the measurement signal of the potentiometer.

A further switch for detecting a kick-down can be integrated in the accelerator-pedal module of vehicles with automatic transmissions. Alternatively, this signal can be derived from the rate of change of the potentiometer voltage or generated when a threshold value is exceeded.

Fuel-level sensor

The level in the fuel tank is transmitted by the float via the float lever to the potentiometer (Figure 4). Firstly, the prepared measured value is transmitted for display in the instrument cluster; secondly, the fuel level is used to measure fuel consumption (e.g. for displaying the range calculated by the on-board computer).

Figure 4: Potentiometric fuel-level sensor
1 Electrical connections,
2 Wiper spring,
3 Contact rivet,
4 Resistor board,
5 Bearing pin,
6 Twin contact,
7 Float lever,
8 Float,
9 Base of fuel tank.

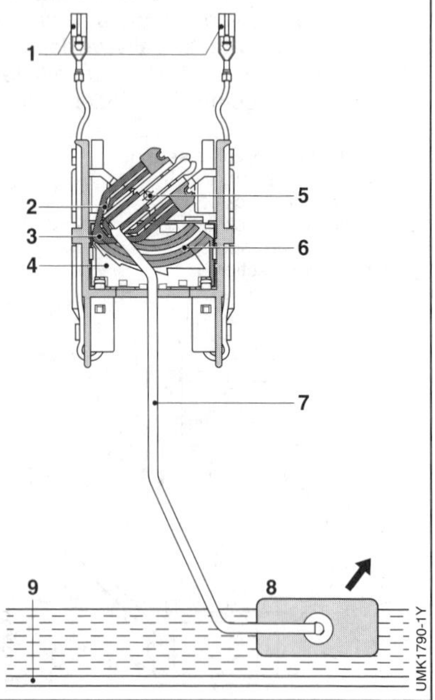

Magnetically inductive sensors

Of all the sensors which perform non-contacting position measurement, magnetic sensors are particularly non-susceptible to interference and robust. This applies in particular to those principles relying on alternating current, in other words magnetically inductive principles. Compared with a micromechanical sensor though, the coil configuration needed here requires far more space. This means therefore that there is no favorable possibility of redundant (parallel measurement) design.

Measurement principle, short-circuiting-ring sensors

Of all the established principles of magnetically inductive sensors it is mainly short-circuiting-ring sensors that have up to now been used by Bosch in motor vehicles. These sensors sense for example the control-rack position in diesel in-line fuel-injection pumps [1]. However, these are no longer used in a macromechanical design form for new developments.

Measurement principle, eddy-current sensors

When an electrically conducting disk (e.g. made of aluminum or copper) approaches a (usually air-core) coil supplied with high-frequency alternating current, eddy currents are created in the disk by the increasing magnetic coupling as a function of the distance (measurement travel s). This influences both the equivalent resistance and the inductance of the coil. The disk thus acts as a damper disk.

Both the damping effect (on account of the equivalent resistance) and the field-displacement effect (on account of the inductance) can be used to convert the measuring effect into an electrical output voltage. The suitable choice is in the first case for example an oscillator of variable amplitude and in the second case an oscillator of variable frequency or a constantly powered, inductive voltage divider.

A high operating frequency is required due to the low coil inductance. This necessitates a direct assignment of electronic circuitry to the sensor.

Application: Position sensor for transmission control

In automatic transmissions position sensors sense the position of final controlling elements (e.g. selector-lever shaft, selector valve or parking-lock cylinder). Because of the complex requirements from the different transmission topologies and the space and functional requirements different physical measurement principles (Hall, AMR, GMR and eddy-current principles) and types (linear and rotatory sensing) are used.

In the eddy-current sensor the rotor rotates with the selector-lever shaft (Figure 5). A return coil is mounted on the rotor. The fixed sensor board contains redundant transmitter and receiver coils with the associated evaluation electronics. The transmitter coil induces in the return coil an eddy current opposing its cause whose magnetic field induces a voltage in the receiver coil. The geometries of the return coil and the receiver coils are matched to each other in such a way that continuously variable rotor positions can be sensed. The rotatory position sensor thus determines the selector-lever positions P, R, N, D, 4, 3, and 2.

Figure 5: Position sensor using eddy-current principle
a) Sensor board with protective circuit,
b) Rotor.
1 Selector-lever shaft,
2 Sensor board,
3 Redundant transmitter and receiver coils,
4 Redundant evaluation electronics,
5 Rotor with return coil.

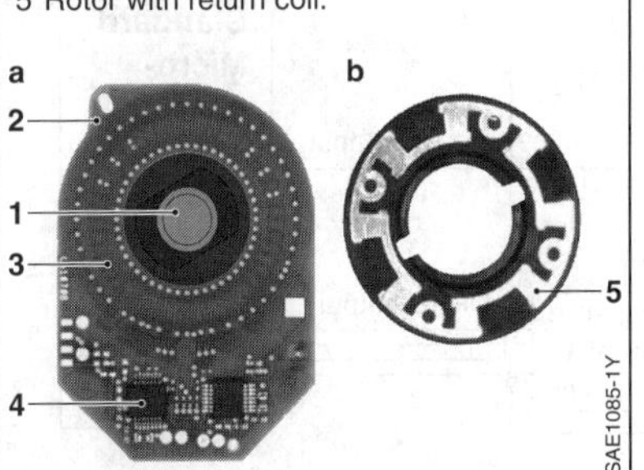

For safety reasons the sensor generates two independent, opposite signals which are checked for plausibility against each other.

Sensors with rotating alternating fields

Two- or multipole alternating-field structures can be arranged either in a circle or linearly with coils energized with alternating current of angular frequency Ω or coil-like configurations (such as meander conductor-track structures) (Figure 6). Compared with a usually fixed set of receiver coils, which have the same pole pitch, these pole structures with fixed pole pitch can be displaced by the movement of the system to be measured – whether it be rotatory or translational in nature. Here, the amplitudes of the receiver signals (U_1, U_2, U_3, etc.) change sinusoidally with the movement. If the receiver coils are offset against each other by a certain part of the pole pitch T (e.g. $T/4$ or $T/3$), the sine shape is phase-displaced in each case by a corresponding angle (e.g. by 90° or 120°). After rectification, the angle

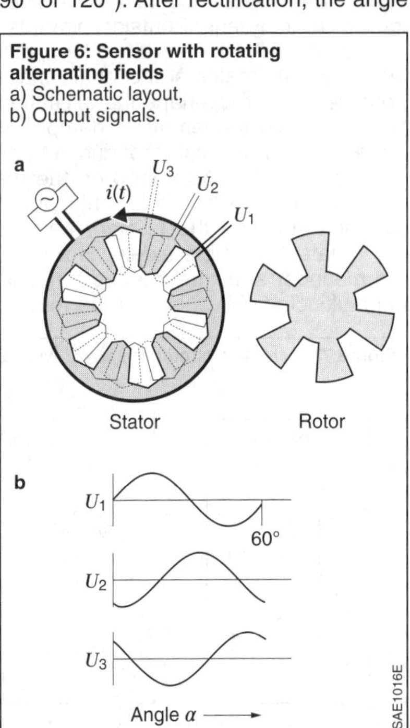

Figure 6: Sensor with rotating alternating fields
a) Schematic layout,
b) Output signals.

of rotation to be measured can be calculated with great accuracy from these voltages. This is the mode of operation of the sensors referred to in classical metrology as synchro, resolver, or inductosyn processes, and preferably configured as angular-position encoders.

The angular-position sensor shown in Figure 6 most resembles the inductosyn process. This is a sensor with a six-figure pole structure ($n = 6$), which in electrical terms converts an angle of rotation of $\varphi = 60°$ into a phase displacement of signal amplitudes of $\alpha = 360°$. All the necessary conductor-track structures are at least in the case of the fixed part (stator) situated on multi-layer board material. The rotor part can also if necessary be designed as a stamping, either self-supported or mounted on plastic carriers (by hot stamping).

Situated on the stator is a circular conductor-track loop which, independently of the angle of rotation, induces at an operating frequency of 20 MHz an eddy current in a self-contained meander loop of roughly equal outside diameter on the rotor. Like the excitation loop, this eddy current creates a secondary magnetic field which overlaps the excitation field such that it attempts to damp it. If there were only a circular conductor track congruent to the stator loop on the rotor instead of the meander, this would simply extinguish the primary field to the greatest possible extent. However, the meander structure creates a resulting multipole field, which can be rotated with the rotor and whose total flux is also virtually zero.

This multipole alternating field is sensed by concentric receiver coils (meanders) of identical shape likewise located on the stator. These coils or meanders are displaced within a pole pitch (of e.g. 60°) by 1/3 in each case, i.e. electrically in their signal amplitude by 120° in each case (Figure 6b). However, the receiver coils stretch over all the n pole pairs (series connection) and utilize the sum total of all the pole fields.

Figure 7 shows the receiver coils star-connected. For the purpose of determining the electrical phase angle α (or the mechanical angle of rotation), the coil signals are routed to an ASIC, which performs the necessary rectification, selection, and ratio formation. The ASIC receives the necessary digital control signals from a nearby microcontroller. A different version of an ASIC is, however, also able to operate the sensor completely independently (stand-alone). The ASICs also permit an end-of-line adjustment of the mechanical and electrical tolerances in production.

For applications with increased safety requirements, it is also possible to design a redundant system with two isolated signal paths and two ASICs. The sensor principle can also be configured very advantageously in an "open" form as a travel sensor. The sensor can therefore be used at many points in a motor vehicle (e.g. recording the throttle-valve angle in the throttle device, lever position on auto-

Figure 7: Protective circuit of sensor with rotating alternating fields

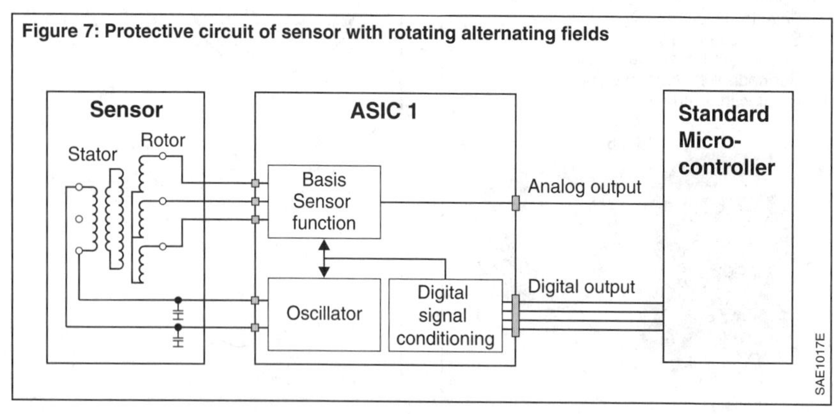

matic transmissions, headlight position for headlight leveling control).

Magnetostatic sensors: Overview
Magnetostatic sensors measure a DC magnetic field. In contrast to the magnetically inductive sensors with coils, they are far more suitable for miniaturization and can be manufactured at reasonable cost using microsystem techniques. Since DC magnetic fields can easily penetrate through housing walls made of plastic but also of non-ferromagnetic metal, magnetostatic sensors have the advantage that the sensitive, generally fixed part can be well encapsulated and protected in respect of the rotatory part – usually a permanent magnet or a soft-magnetic conductive element – and in respect of the environment. Above all, galvanomagnetic effects (Hall and Gaussian effects) and magnetoresistive effects (AMR and GMR) are used.

Hall effect
The Hall effect occurs in thin semiconductor chips. If such a current-carrying chip is permeated vertically by a magnetic induction B, the charge carriers are deflected by the Lorentz force vertically to the field and to the current I by the angle α from their otherwise straight path. It is thus possible to pick off a voltage U_H proportional to the field B and to the current I transversally to the current direction between two opposing marginal points of the chip (Hall voltage, Figure 8):

$$U_H = \frac{R_H I B}{d}$$

where
R_H Hall coefficient,
d Chip thickness.

The coefficient R_H crucial to the measurement sensitivity of the chip is only comparatively small in the case of silicon. However, because the chip thickness d can be made extremely thin using diffusion techniques, the Hall voltage U_H attains a technically acceptable value again.

When silicon is used as the basic material, a signal-conditioning circuit can be integrated on the chip. Sensors using these principles are thus cheap to produce.

In terms of measurement sensitivity, temperature coefficient, and temperature range, however, silicon is by no means the best semiconductor material for Hall sensors. III-V semiconductors such as gallium arsenide or indium antimonide, for example, demonstrate better properties.

Gaussian effect
In addition to the transversally directed Hall effect, semiconductor chips are also subject to a longitudinal resistance effect, otherwise known as the Gaussian effect. Independently of the field direction the series resistance increases according to a roughly parabolic characteristic curve. Elements which utilize this effect are called magnetoresistors (Siemens trade name) and are manufactured from a III-V semiconductor, crystalline indium antimonide (InSb).

Figure 8: Hall effect
B Magnetic flux density,
I Supply current,
U Supply voltage,
U_H Hall voltage,
I_H Hall current.

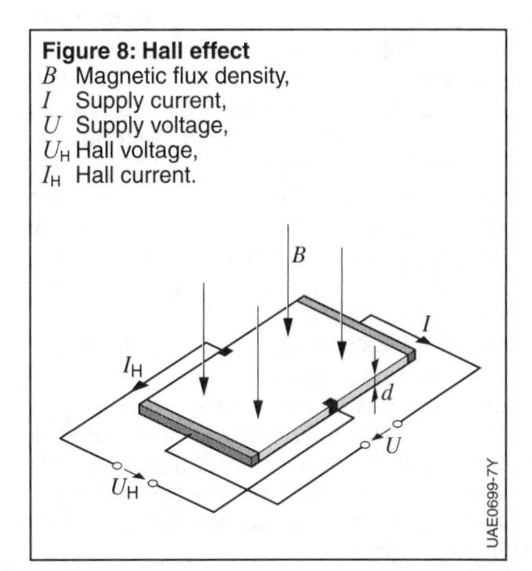

Hall sensors

Hall switch

In the most simple case, in the Hall switch the Hall voltage is applied to an electronic threshold circuit integrated in the sensor (Schmitt trigger), which then outputs a digital signal. If the magnetic induction B applied at the sensor is below a given minimum threshold level, the Schmitt trigger's output value corresponds to a logical 0 (release status); if it exceeds a given upper threshold, the output value corresponds to a logical 1 (operate status). Since this behavior is guaranteed across the complete operating-temperature range and for all sensors of a given type, the two threshold values are relatively far apart (approx. 50 mT). In other words, it takes a considerable induction jump ΔB to trigger the Hall switch.

Such Hall sensors manufactured for the most part in bipolar technology are indeed very economical, but are at best only good for switch operation (e.g. Hall vane switches for triggering ignition in earlier ignition systems, digital steering-angle sensor). Hall vane switches are too inaccurate for recording analog variables.

Hall sensors using the spinning-current principle

A disadvantage of a simple Hall sensor is its simultaneous sensitivity to mechanical stresses (piezoelectric effect), which are inevitable due to packaging and result in an unfavorable offset temperature coefficient. This disadvantage has been overcome by the application of the spinning-current principle (Figure 9), combined with the transition to CMOS technology. The piezoelectric effect does arise here too, but is compensated by time averaging of the signal, since it occurs in the event of very fast, electronically controlled switching (rotation) of the electrodes with a different polarity sign.

If the desired aim is to avoid the expense of complex electronics to switch the electrodes, it is also possible to integrate in close proximity several Hall sensors (two, four or eight) with correspondingly different alignment of the current path, and to add up their signals in the form of an averaging.

Only through these measures was it possible for Hall ICs to become well suited to analog sensor applications. These measures though barely resulted in a reduction of the sometimes considerable effects of temperature on the measurement sensitivity.

Such Hall ICs are suitable above all for the measurement of small travel distances, in which they register the fluctuating field strength of a permanent magnet as it approaches (used for example in the force sensor for sensing front-passenger weight). Up to then, similarly good results could only be achieved by the use of individual Hall elements of, for example, III-V semiconductors (e.g. GaAs) with a hybrid downstream amplifier.

Figure 9: Hall sensor using the spinning-current principle

a) Rotation phase φ_1,
b) Rotation phase φ_2, = $\varphi_1 + 45°$.
1 Semiconductor chip,
2 Active electrode,
3 Passive electrode.
I Supply current
U_H Hall voltage.

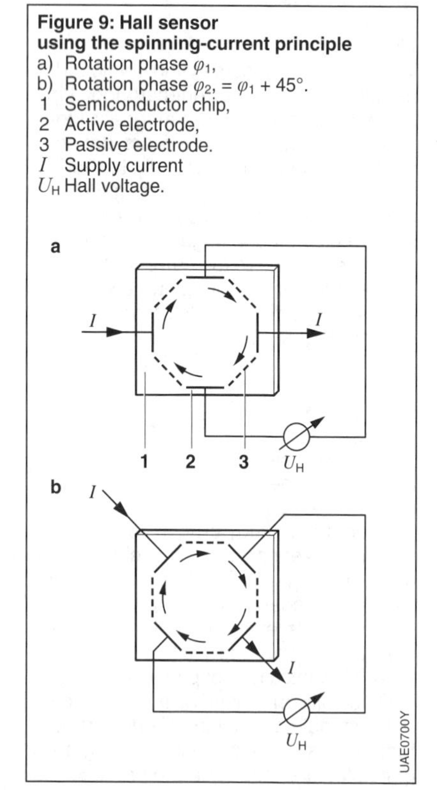

Applications of Hall sensors

Position sensors for transmission control

In this type of position sensor for transmission control for linear position sensing four Hall switches are arranged on a printed-circuit board in such a way that they sense the magnetic encoding of a linearly displaceable multipolar permanent magnet (Figure 10). The magnet carriage is coupled to the linearly actuated selector valve (hydraulic valve in the transmission control plate) or the parking-lock cylinder.

The position sensor senses the positions of the selector valve (P, R, N, D, 4, 3, 2) and the intermediate ranges and outputs these in the form of a 4-bit code to the transmission-control system. For safety reasons the encoding of the position is configured as a single step. In other words, two bit changes are always required until a new position is detected. The sequence of these bit changes corresponds to a Gray code.

Axle sensors

Axle sensors sense the body's inclination angle that changes as a result of vehicle load or during braking and acceleration. The range of the headlamps can be adapted to the conditions with this information (automatic headlamp leveling control).

Vehicle inclination is measured by angle-of-rotation sensors (axle sensors). These are mounted to the body at the front and rear. Compression is measured by a rotary lever which is connected via a pushrod to the vehicle axle. The vehicle inclination is obtained from the difference between the sensor signals on the front and rear axles.

The axle-sensor stator (Figure 11) incorporates a Hall element located in the ring magnet's homogeneous field. The magnetic field induces in the Hall element a Hall voltage that is proportional to the magnetic field strength. During compression the axle-sensor shaft rotates with the ring magnet, and thus the magnetic field is also altered by the Hall element. The Hall element delivers a signal that is dependent on the angle of rotation of the shaft and thus on the compression travel.

Accelerator-pedal sensors

Using a rotatable magnetic ring ("movable magnet"), together with a number of fixed

Figure 10: Encoding of position sensor for transmission control
a) Magnetic encoding,
b) Position ranges.
1 Moving carriage,
2 Fixed position of Hall elements.

a
Magnetic north pole
Magnetic south pole
1
2
4-bit code

b
P R N D 4 3 2
Gearshift and intermediate ranges
Transition range
SAE1084-1E

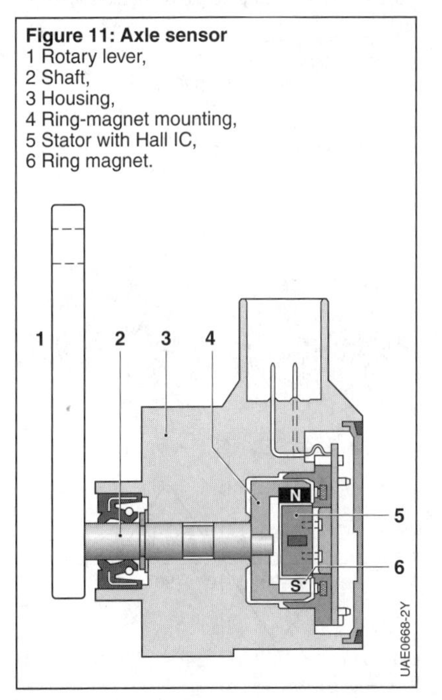

Figure 11: Axle sensor
1 Rotary lever,
2 Shaft,
3 Housing,
4 Ring-magnet mounting,
5 Stator with Hall IC,
6 Ring magnet.

soft-magnetic conductive elements, a linear output signal can be generated for a larger angular range without conversion being necessary (Figure 12). Here, the movable magnet's bipolar field is directed through a Hall sensor located between semicircular conductive elements. The effective magnetic flux through the Hall-effect sensor is dependent on the angle of rotation φ. The angular range that can be sensed is 180° here.

The Hall angular-position sensor with a measuring range of approximately 90° pictured in Figure 13a is derived from the basic "movable magnet" principle. The magnetic flux from a practically semicircular permanent-magnet disc is returned to the magnet through a pole-shoe, two additional conductive elements each of which contains a Hall sensor in its magnetic path, and the shaft which is also ferromagnetic. Depending upon the angular setting, the flux is led to a greater or lesser degree through the two conductive elements, in the magnetic path of which a Hall sensor is situated. Using this principle, it is possible to achieve a practically linear characteristic curve (Figure 13b).

The simplified version pictured in Figure 14 does without soft magnetic conductive elements. In this version, the magnet moves around the Hall sensor in a circular arc (Figure 14a). Only a relatively small section of the resulting sinusoidal characteristic curve features good linearity. If the Hall sensor is located slightly outside the center of the circular arc, the characteristic curve increasingly deviates from the sinusoidal, and now features a short measuring range of almost 90°, and a longer measuring range of something more than 180° with good linearity. A great disadvantage though is the low level of shielding against external fields, as well as the remaining dependence on the geometric tolerances of the

Figure 12: Angular-position sensor using the movable-magnet principle
a) Sensor in initial position,
b) Excursion through angle of rotation φ,
c) Output signal.
1 Magnetic yoke,
2 Stator (soft iron),
3 Rotor (permanent magnet),
4 Air gap,
5 Hall sensor.
φ Angle of rotation.

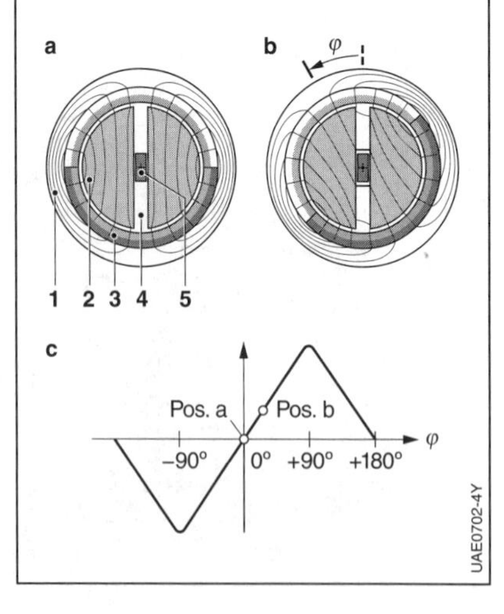

Figure 13: Hall angular-position sensor with linear characteristic curve up to 90° using the movable-magnet principle
a) Layout,
b) Characteristic curve with working range A.
1 Rotor disk (permanent magnet),
2 Pole shoe,
3 Conductive element,
4 Air gap,
5 Hall sensor, 6 Shaft.
φ Angle of rotation.

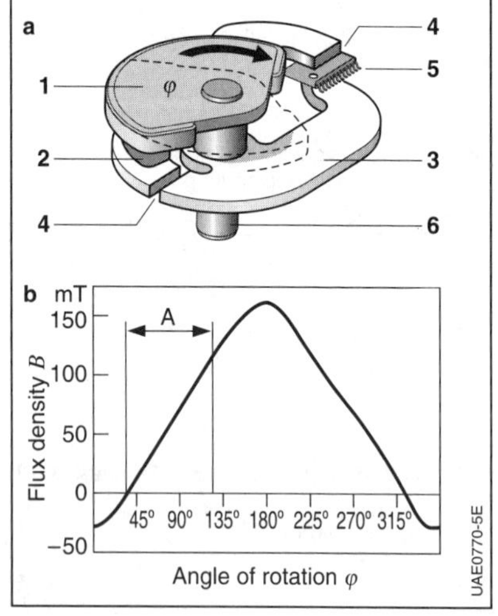

magnetic circuit, and the intensity fluctuations of the magnetic flux in the permanent magnet as a function of temperature and age.

In the Hall angular-position sensor pictured in Figure 15 it is not the field strength but rather the magnetic-field direction that is evaluated. The magnetic-field lines are sensed by four Hall elements lying in one plane and radially arranged in the x- and y-directions. The sensor is located between two magnets to generate a homogeneous magnetic field. The sensor signal is formed from the measurement signals of the Hall elements (sine and cosine signals) in an evaluation circuit using the arctan function.

Like the potentiometric accelerator-pedal sensor these systems also contain two measuring elements in order to obtain two redundant voltage signals for diagnostic purposes.

Figure 14: Hall angular-position sensor with linear characteristic curve up to 180° using the movable-magnet principle
a) Principle (Hall IC in this depiction shifted from center point),
b) Characteristic curve.
1 Curve for arrangement with Hall IC at center point,
2 Curve for arrangement with Hall IC outside center point,
3 Hall IC,
4 Magnet.

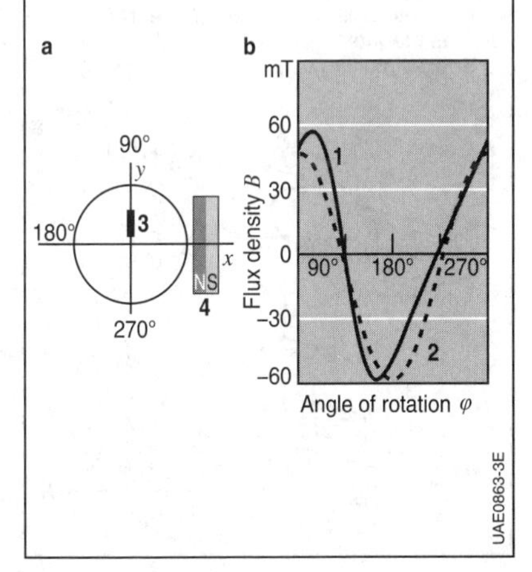

Magnetoresistive sensors

In contrast to the Hall sensors, the optimum chip shape for a magnetoresistor tends to be shorter and squatter, and represents a very low resistance. In order to arrive at technically applicable resistance values in the kΩ range, it is therefore necessary to connect a large number of these chips in series. This problem is solved elegantly by adding microscopically fine, highly conductive nickel-antimonide needles to the semiconductor crystal. These are located obliquely to the direction of current flow. A further measure is to apply meander techniques to the semiconductor resistor.

The dependence of the resistance on the magnetic flux density B follows a square-law function up to inductances of approximately 0.3 T, and above this point

Figure 15: Hall angular-position sensor with fourfold Hall sensor with evaluation of magnetic-field direction for a measurement range over 360°
a) Layout,
b) Measurement principle,
c) Measurement signals.
1 IC with Hall elements,
2 Magnet (the opposite magnet is not visible here),
3 Conductive element,
4 Hall elements for sensing the x component of B,
5 Hall elements for sensing the y component of B.

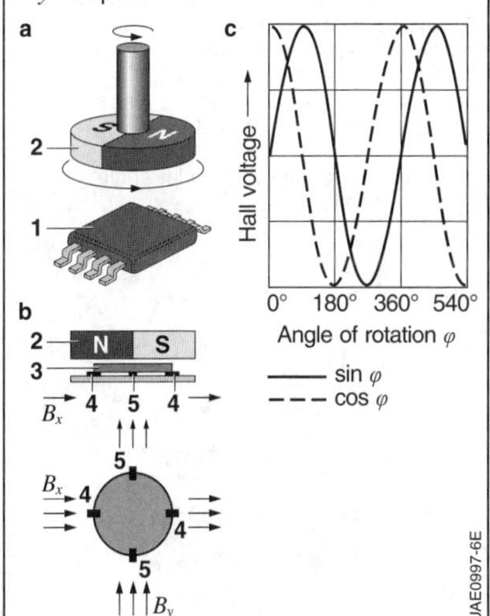

it is increasingly linear. There is no upper limit to the control range, and dynamic response in technical applications can – as is the case with the Hall sensor – be regarded as free from lag.

Since the temperature sensitivity has a pronounced effect on the resistance of magnetoresistors (approximately 50% reduction for 100 K), they are usually delivered only in the dual-configuration form in voltage-divider circuits (differential magnetoresistors). For the particular application, each of the two resistor sections must then be magnetically triggered (as far as possible with opposite polarities). Notwithstanding the high temperature coefficient of the individual resistors, the voltage-divider circuit guarantees good stability of the working-point (that point at which both resistor sections have the same value).

In order to achieve good measurement sensitivity, it is best to operate the magnetoresistors at a magnetic working point between 0.1 and 0.3 T. Generally, the required magnetic bias is supplied by a small permanent magnet (Figure 16) the effects of which can be increased by using a small magnetic return plate. The magnetoresistor features pronounced temperature sensitivity so that it is used almost exclusively in incremental angle-of-rotation and displacement sensors, or in binary limit-value sensors (with switching characteristic). A magnetoresistive increment sensor is used to measure start-of-injection adjustment in diesel distributor injection pumps.

The magnetoresistor's main advantage is its high signal level which is usually in the volts range. This means that amplification is unnecessary, as well as the local electronic circuitry and the associated protective measures which would otherwise be needed. Furthermore, in their role as passive, resistive components, they are highly insensitive to electromagnetic interference and, as a result of their high bias voltage, practically immune against external magnetic fields.

AMR sensors

Anisotropic magnetoresistive effect

NiFe layers only roughly 30 to 50 nm thick display electromagnetically anisotropic properties, i.e. their electrical resistance changes under the influence of a magnetic field. Resistance structures of this type are therefore called AMR sensors (anisotropic magnetoresistive sensors). The metal alloy generally used is also called permalloy.

Types

On an elongated resistance strip, as shown in Figure 17a, a small, spontaneous magnetization M_S occurs in the longi-

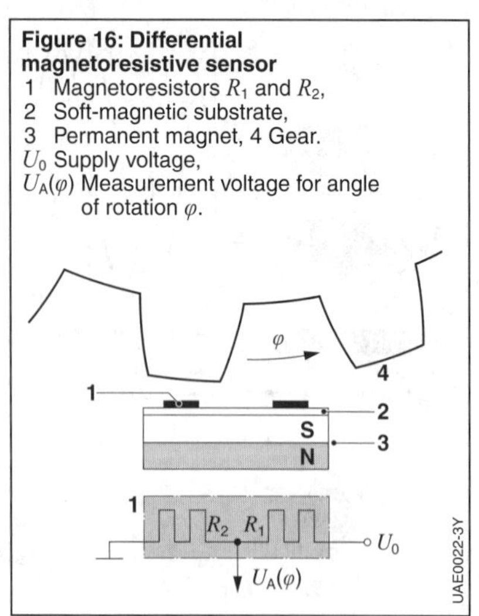

Figure 16: Differential magnetoresistive sensor
1 Magnetoresistors R_1 and R_2,
2 Soft-magnetic substrate,
3 Permanent magnet, 4 Gear.
U_0 Supply voltage,
$U_A(\varphi)$ Measurement voltage for angle of rotation φ.

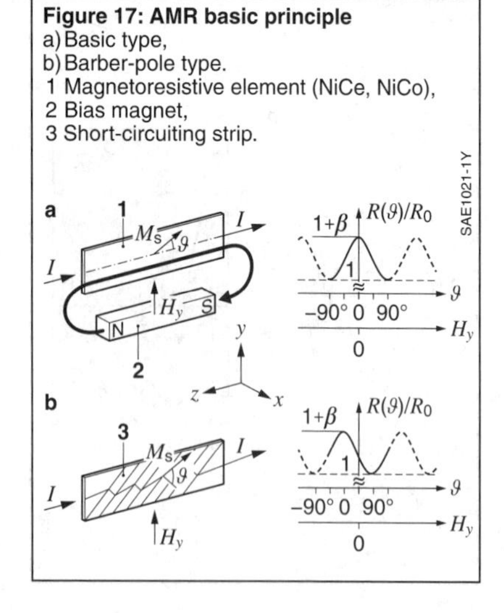

Figure 17: AMR basic principle
a) Basic type,
b) Barber-pole type.
1 Magnetoresistive element (NiCe, NiCo),
2 Bias magnet,
3 Short-circuiting strip.

tudinal direction of the conductor track even without the application of an external control field (form anisotropy). In order to give this magnetization a clearly defined direction – theoretically, it could be in the other direction – AMR sensors are often, as depicted, provided with weak bias magnets. In this state, the series resistance has its greatest value $R_{\parallel}$. If the magnetization vector is rotated under the influence of an additional external field H_y through the angle ϑ, the series resistance decreases gradually until it assumes at $\vartheta = 90°$ a minimum value $R_{\perp}$. Here, the resistance is only dependent on the angle ϑ contained by the resulting magnetization M_S and the current I; it has a roughly cosinusoidal characteristic as a function of ϑ:

$$R = R_0(1 + \beta \cos^2 \vartheta).$$

Thus the following relationships are obtained for the maximum and minimum values:

$$R_{\parallel} = R_0(1 + \beta) \text{ and } R_{\perp} = R_0.$$

Figure 18: Pseudo-Hall sensor
a) Full basic form,
b) Modified form with hollowed-out surface,
c) Replacement circuit diagram for b).
B Magnetic flux density,
U_H Hall voltage,
I Supply current,
φ Angle of rotation,
R_H Resistors of AMR elements.

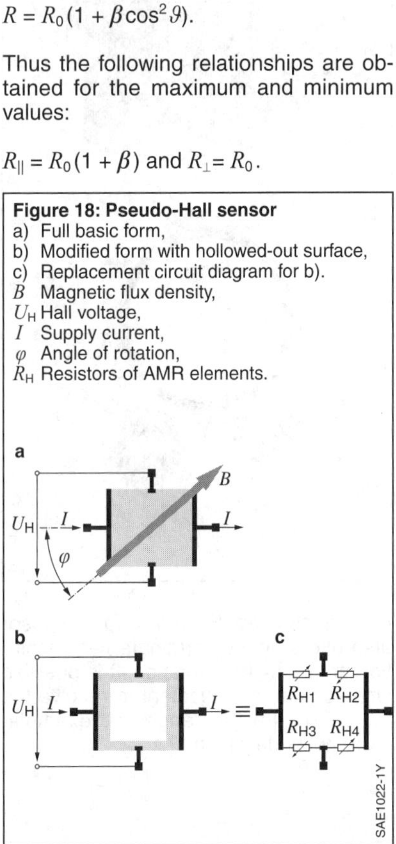

SAE1022-1Y

The coefficient β denotes the maximum possible resistance variation, and is approximately 3%. If the external field is much stronger than the spontaneous magnetization, and this is usually the case when control magnets are used, the effective angle is almost completely a function of the direction of the external field. The field strength is now irrelevant, in other words, the sensor is now operating in the "saturated state".

Highly conductive short-circuiting strips (for instance of gold) on the AMR film force the current to flow at an angle of below 45° to the spontaneous magnetization (longitudinal direction) without the application of an external field. In this version – also referred to as the "barber-pole" sensor – the sensor curve shifts by 45° compared to that of the simple resistor (Figure 17b). This means, therefore, that even when the external field strength $H_y = 0$, the curve is at the point of maximum measurement sensitivity (reversal point). The "striping of two resistors in opposite directions" means that they change their resistances in opposite directions under the effects of the same field. In other words, one of the resistances increases while the other drops.

In addition to the simple, two-pole AMR elements, there are also pseudo-Hall sensors, for instance square NiFe thin-film structures, which, like the normal Hall sensors already described, have four terminals: two for the current path and two transverse ones for picking off a Hall voltage (Figure 18a). Unlike a normal Hall sensor, however, a pseudo-Hall sensor is sensitive to magnetic fields in the film plane and not vertically to it. Also, a pseudo-Hall sensor does not demonstrate a proportional characteristic curve; instead, it has a sinusoidal curve with very high faithfulness of shape to the sine, which is in no way dependent on the strength of the control field and the temperature. For a field parallel to the current path, the output voltage disappears in order then to describe a sinusoidal half-period when the field is rotated up to the angle $\varphi = 90°$. The sinusoidal voltage therefore results with amplitude $\hat{u}_H$ as follows:

$$U_H = \hat{u}_H \sin 2\varphi.$$

If the external control field is rotated once through φ = 360°, the output voltage therefore follows two full sinusoidal periods. The amplitude $\hat{u}_H$ is, however, very much dependent on the temperature and the air gap between the sensor and the control magnet; it decreases as the temperature and the air gap increase.

The measurement sensitivity of this so-called pseudo-Hall element can be considerably increased (without excessively falsifying the sinusoidal shape) by "hollowing out" the element from the inside so that only the "frame" remains (Figure 18b). This modification converts the pseudo-Hall sensor to a full bridge consisting of four AMR resistors (Figure 18c). Even when the bridge resistors are meander-shaped, provided a given minimum conductor width is not dropped below, this still has negligible effect on the signal's sinusoidal shape.

Applications

Steering-angle sensor

The function of driving-dynamics systems is to keep the vehicle on the desired course specified by the driver with targeted brake interventions. The steering-wheel angle must be known for this purpose. Potentiometers, optical code sensing and magnetic measurement principles are used for measurement.

A steering-wheel-angle sensor mounted on the steering shaft senses the steering-wheel position. Several rotations of the steering shaft can be measured with a dual configuration of pseudo-Hall angle-of-rotation sensors (each for 180 °). The two associated permanent magnets are rotated via a high-ratio gear train (Figure 19). Because the two smaller output gears, which carry the control magnets, differ by one tooth, their mutual phase angle (difference of angles of rotation: $\Psi - \Theta$) is a clear measure of the absolute angular position of the steering shaft. The system is designed in such a way that this phase difference does not exceed 360° for a total of four rotations of the steering shaft, and thereby the clearness of the measurement is assured. Each individual sensor also offers an indeterminate fine resolution of the angle of rotation. It is possible with such an arrangement to resolve the full steering-angle range, for example, more accurately than 1°.

Figure 19: AMR steering-angle sensor
a) Layout,
b) Angle relations.
1 Steering shaft,
2 AMR measuring cells,
3 Gear with m teeth,
4 Evaluation electronics,
5 Magnets,
6 Gear with $n > m$ teeth,
7 Gear with $m + 1$ teeth.

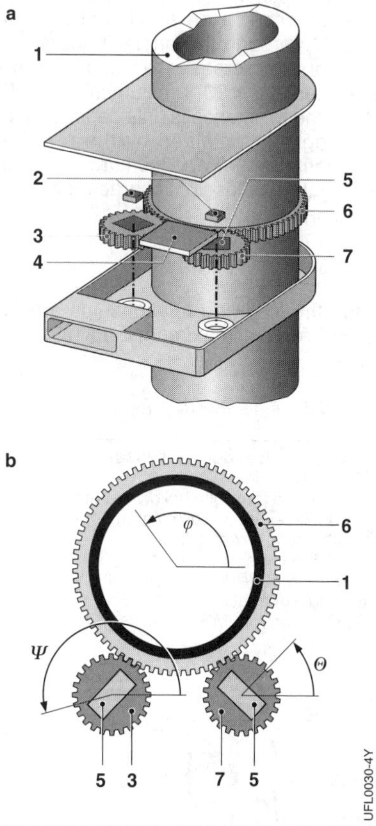

a
1
2
5
6
3
4
7

b
φ
6
1
Ψ
Θ
5 3
7 5
UFL0030-4Y

GMR sensors

Giant magnetoresistive effect

GMR-sensor technology (giant magnetoresistive sensor) is used in automotive applications for angular-position and rpm sensing. The main advantages of GMR sensors over AMR sensors are the natural clearness range within 360° for angular-position sensing and the higher magnetic-field sensitivity for rpm sensing.

GMR layer structures consist of antiferromagnetic, ferromagnetic and nonmagnetic function layers (Figure 20). The individual layer thicknesses in both systems are in the range of 1 to 5 nm, and therefore only comprise a few atom layers. For angular-position sensing the required reference magnetization is generated by the fact that the direction of magnetization of one of the ferromagnetic layers (PL) is pinned by the interaction with a neighboring antiferromagnetic layer (AF). This is therefore also referred to as a "pinned layer". On the other hand, the magnetization of the second ferromagnetic layer (FL) magnetically isolated to a large extent via a non-magnetic interlayer (NML) can be freely rotated with the external magnetic field. This is accordingly referred to as a "free layer".

The resistance changes with a cosinusoidal dependence on the angle between the external field direction and the reference direction. Crucial to the accuracy of the angular-position measurement is the stability of the reference magnetization against the effect of the external field. This stability is greatly increased by using an additional artificial antiferromagnet (SAF).

Figure 20: GMR layer structure
FL Free layer,
NML Non-magnetic interlayer,
RL Reference layer,
PL Pinned layer,
AF Antiferromagnet,
SAF Artificial antiferromagnet.

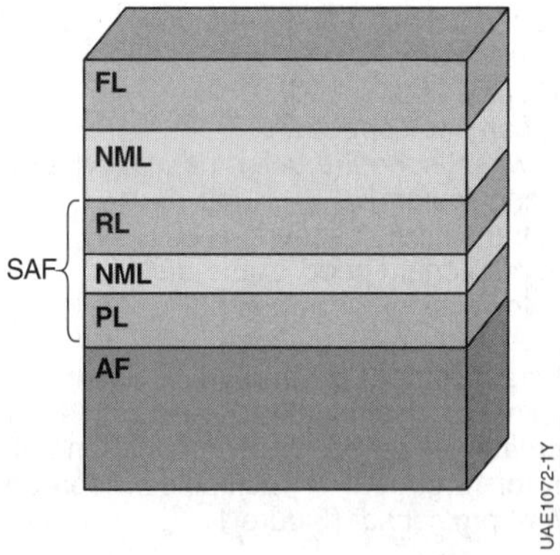

Applications of GMR sensors

Steering-wheel-angle sensor

The mechanical structure and the operating principle of this GMR sensor is equivalent to those of the steering-wheel-angle sensor with AMR elements. In view of the greater sensitivity in comparison with the AMR effect the GMR sensor can operate with weaker magnets and larger air gaps. This results in costs benefits with regard to materials and design. The 360° angular-position measurement range of a single GMR element (180° is typical for AMR) enables smaller gears to be used. This in turn means that it takes up less space.

References

[1] K. Reif (Editor): Klassische Diesel-Einspritzsysteme – Bosch Fachinformation Automobil. 1st Ed., Vieweg+Teubner, 2012.
[2] K. Reif (Editor): Sensoren im Kraftfahrzeug – Bosch Fachinformation Automobil. Vieweg+Teubner, 2010.

Rpm sensors

Measured variables

rpm sensors measure the time required for an angle to be covered during a rotational motion. From this the number of revolutions per unit of time can be determined. In a motor vehicle this usually involves relative measured variables which occur between two components. Examples include:

- crankshaft speed,
- camshaft speed,
- wheel speed (e.g. for the antilock braking system),
- transmission speed,
- rotational speed of the diesel distributor-type injection pump.

Measurement principles

rpm sensing usually occurs with an incremental sensor system, consisting of a rotor (e.g. gear or multipole wheel) and the rpm sensor (Figure 1).

Passive rpm sensors

Conventional inductive sensors used in the past were based on the inductive measuring effect. A sensor of this type consists of a permanent magnet and a soft magnetic pole pin surrounded by an induction coil. The pole pin is opposite a ferromagnetic gear. The distance between the gear and the pole pin changes as the gear turns. The time change in the magnetic flux that this causes gives rise to an induced voltage in the coil.

The measuring effect of the inductive sensor is relatively high, eliminating the need for local electronics. It is therefore called a passive sensor. However, the signal amplitude is rotational-speed-dependent. The sensor is therefore unsuitable for the lowest rotational speeds; it permits only a comparatively small air-gap tolerance between pole pin and gear and for the most part is not able to distinguish air-gap fluctuations (chatter) from rotational-speed pulses.

Active rpm sensors

Active rpm sensors operate in accordance with the magnetostatic principle. The amplitude of the output signal is not dependent on the rotational speed. rpm sensing is thus possible even at very low rotational speeds (quasistatic rpm sensing).

Hall sensor

A voltage U_H proportional to the magnetic field (Hall voltage) can be picked off transversely to the current direction on a current-carrying wafer which is permeated vertically by a magnetic induction B (see Hall effect). In the sensor arrangement with a ferromagnetic gear (pulse wheel) the magnetic field is generated by a permanent magnet (Figure 1a). Located between the magnet and the pulse wheel is the Hall-sensor element. The magnetic flux which permeates this element depends on whether the sensor is opposite a tooth or a gap. This results in a Hall voltage that is equivalent to the progression of the teeth. This rotational-speed infor-

Figure 1: Hall sensors
a) Sensor arrangement with passive rotor (ferromagnetic gear),
b) Sensor arrangement with active rotor (multipole wheel).
1 Increment rotor,
2 Hall IC,
3 Permanent magnet,
4 Multipole wheel,
5 Housing.
ψ Rotational speed.

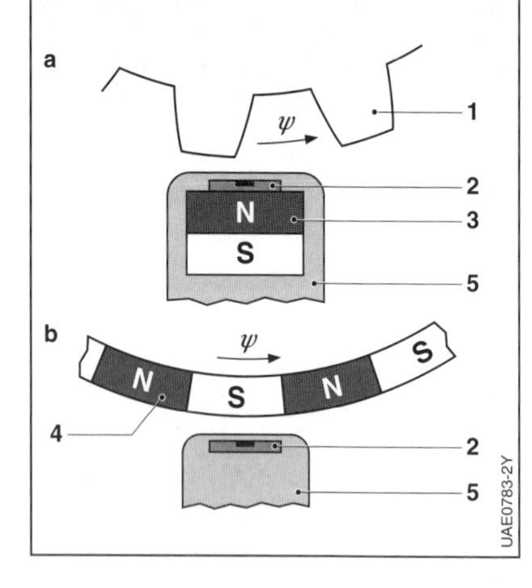

mation is conditioned and amplified, and then transmitted as injected current in the form of square-wave signals. Typical values are 7 mA for the low level and 14 mA for the high level. The current is converted into a signal voltage by a measuring shunt in the ECU.

In a differential Hall sensor there are two Hall-sensor elements situated between the magnet and the pulse wheel. Determining the difference between the signals from the two sensor elements reduces magnetic interference signals and improves the signal-to-noise ratio.

Multipole wheels are also used instead of the ferromagnetic pulse wheel. Here, a magnetizable plastic is applied to a non-magnetically metallic substrate and alternately magnetized. These north and south poles assume the function of the teeth on a pulse wheel (Figure 1b).

AMR sensor

The electrical resistance of magnetoresistive material (AMR, Anisotropic Magneto Resistive) is anisotropic, i.e. it is dependent on the direction of the magnetic field to which it is exposed. This property is exploited in the AMR sensor. The sensor is located between a magnet and the pulse wheel. The magnetic field lines change their direction when the pulse wheel turns. This gives rise to a sinusoidal voltage which is amplified in an evaluation circuit in the sensor and converted into a square-wave signal.

Figure 2: Different sensor shapes
a) Fork shape (vane principle),
b) Rod shape (proximity principle).
d_L Air gap.

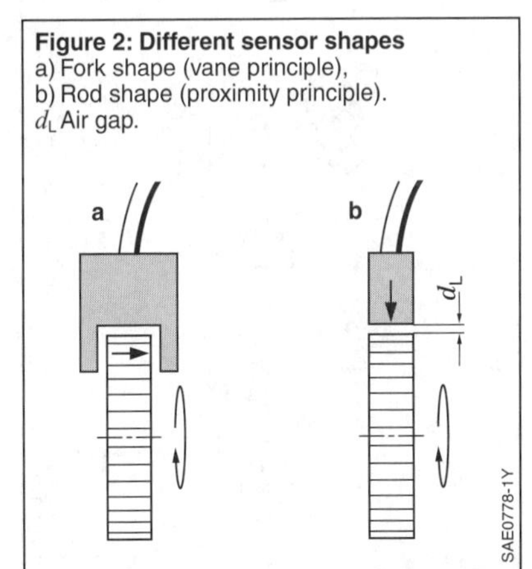

GMR sensor

The use of GMR technology (Giant Magneto Resistance) constitutes a further development of active sensors. Their greater sensitivity than AMR sensors means that larger air gaps are possible, which in turn opens up the possibility of use in difficult areas of application. The greater sensitivity also means lower noise at the signal edges.

Sensor shapes

Different sensor shapes are used (Figure 2): the rod shape, the fork shape, and the inner and outer ring shape. Due to its ease of mounting and simplicity, the rod sensor is the most widespread sensor shape. The rod sensor is located near the rotor, the teeth of which approach it and pass by in close proximity. However, it also has the lowest measurement sensitivity. The fork-shaped sensor, which is insensitive to axial and radial play, is permissible in some cases, and is also in service in the field. This sensor must be roughly aligned to the rotor when installed. The sensor type in which the sensor surrounds the rotor shaft in the form of a ring is practically no longer used.

Rotor shapes

In the incremental detection of relative yaw rate, depending upon the number and size of the scanned peripheral rotor markings, a distinction is made between the following types of sensor (Figure 3):

- Simple sensor, with only a single scanned marking per revolution, so that only the average rotational speed can be registered

Figure 3: Recording of relative rotational speed
a) Increment sensor,
b) Segment sensor,
c) Simple rpm sensor.

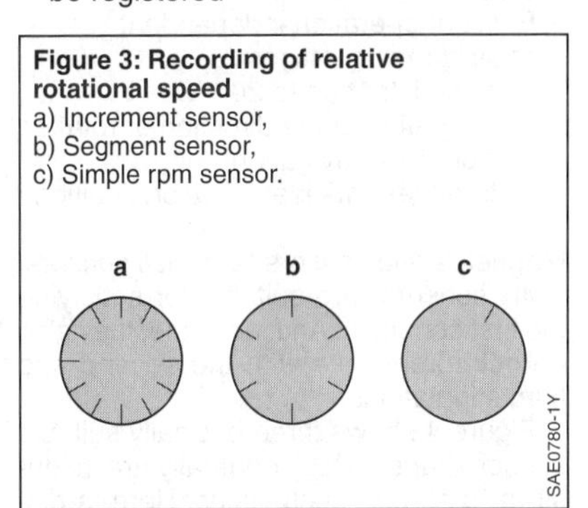

- Segment sensor, with only a small number of scanned peripheral segments (for instance, equivalent to the number of engine cylinders)
- Increment sensor with closely spaced peripheral markings. Up to a certain point, this form of sensor permits instantaneous speed to be measured at points on the circumference, and thus the registration of very fine angular divisions.

The rotor is crucially important to measuring rotational speed; however, it is usually part of the vehicle manufacturer's delivery specification, while the actual sensor comes from the supplier. The rotors used almost exclusively up to now have been magnetically passive types, which therefore consist of soft-magnetic material, usually iron. They are cheaper than hard magnetic pole wheels and are easier to handle, since they are not magnetized, and there is no danger of mutual demagnetization during storage. As a rule, presuming the same increment width and output signal, the pole wheel's intrinsic magnetism (a pole wheel is defined as a magnetically active rotor) permits a considerably larger air gap.

Demands made of rpm sensors

The following demands are made of rpm sensors:

- Static detection (i.e. speed approaching zero, extremely low engine-starting and wheel speeds)
- Large air gaps (non-adjusted mounting on air gap greater than zero)
- Small size
- Efficient operation independent of air-gap fluctuations
- Heat-resistant up to 200 °C
- Identification of the direction of rotation (optional for navigation)
- Reference-mark identification (ignition).

Magnetostatic sensors (e.g. Hall sensors, AMR sensors) are suitable for satisfying the first condition. And, as a rule, they also permit compliance with the second and third stipulations.

Figure 4 shows three basically suitable sensor shapes which generally are insensitive to air-gap fluctuations. Here, a distinction must be made between sensors which sense radially and those which sense tangentially (see also Figure 5). This means that independent of the air gap, magnetostatic sensors are always able to differentiate between the north and south poles of a magnetically active pole wheel. In the case of magnetically passive rotors, the sign of the output signal is then no longer independent of the air gap when they register the tangential-field strength (here though, the fact that the air gap is often enlarged due to the rotor is a disadvantage). Here, though, the fact that the air gap between the rotor and the permanent magnet is often enlarged due to the sensor is a disadvantage. However, types frequently used as well are radially measuring differential-field or gradient sensors, which essentially only register the gradient of the radial field component,

Figure 4: Sensor arrangements which are insensitive to air-gap fluctuations
a) Radial-field sensor with pole wheel,
b) Tangential sensor,
c) Differential sensor with gear.
The sensor measures that part of the magnetic field in a certain direction in space. The arrow marks this direction.

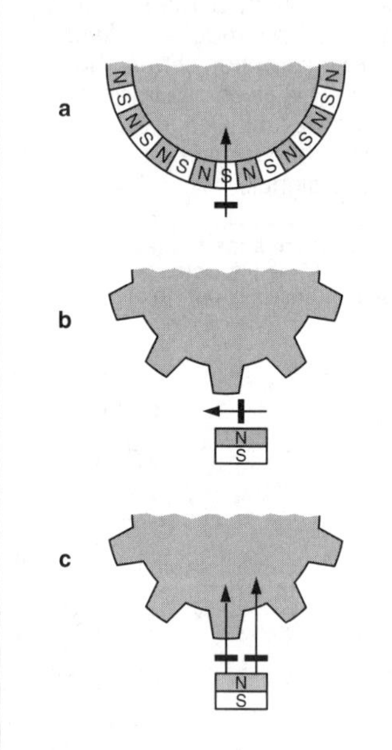

which changes with regard to its sign not with the air gap, but only with the angle of rotation.

In the case of wheel-speed sensors, at the least the sensor tip, because of its proximity to the brake, should be able to withstand higher temperatures.

Sensor arrangements

Gradient sensors

Gradient sensors (e.g. based on differential Hall sensors) incorporate a permanent magnet on which the pole surface facing the gear is homogenized with a thin ferromagnetic wafer. Two sensor elements are located on each element's sensor tip, at a distance of roughly one half a tooth interval. Thus, one of the elements is always opposite a gap between teeth when the other is opposite to a tooth. The sensor measures the difference in field intensity at two adjacent locations on the circumference. The output signal is roughly proportional to the diversion of field strength as a function of the angle at the circumference; polarity is therefore independent of the air gap.

Tangential sensors

Tangential sensors differ from gradient sensors by their reaction to variations in polarity and intensity in the components of a magnetic field located tangentially to the periphery of the rotor. Design options include AMR thin-film technology (barber pole) or single permalloy resistors featuring full- or half-bridge circuits [1]. In contrast to gradient sensors, they need not be matched to the particular tooth pitch of the rotor and can be designed to sense at a given point. Local amplification is necessary, even though their measuring effect is 1 to 2 orders of magnitude larger than that of the silicon-Hall sensors.

In the case of a bearing-integrated crankshaft speed sensor (Simmer shaft-seal module), the AMR thin-film sensor is mounted together with an evaluation IC on a common lead frame. For the purposes of space saving and temperature protection, the evaluation IC is bent at an angle of 90° and also located further away from the sensor tip.

Figure 5: AMR rpm sensor as tangential-field sensor
1 Multipole ring,
2 Measuring cell,
3 Sensor housing.

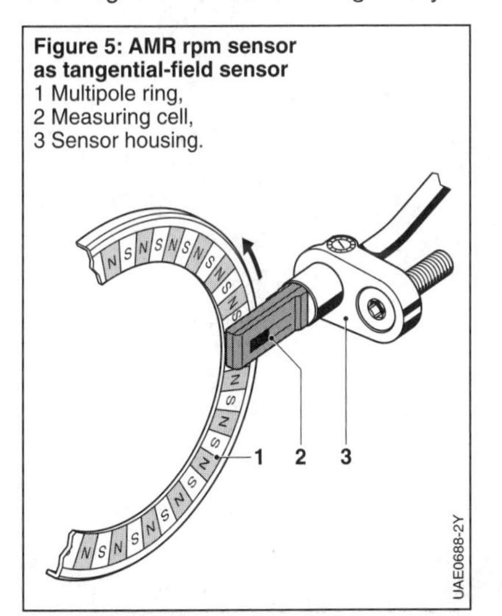

Applications

Crankshaft-speed sensor

Crankshaft-speed sensors are used to measure the engine speed (crankshaft speed). In addition, the engine-management system requires the position of the crankshaft (position of the pistons) to be able for example to activate ignition coils and fuel injectors at the correct crank angle. The pulse wheel mounted on the crankshaft exhibits a tooth gap for this purpose. With a tooth pitch of 6° and two missing teeth the pulse wheel has 58 teeth. The rpm sensor delivers a square-wave signal (tooth signal) that corresponds to the teeth on the pulse wheel.

The microcontroller in the engine ECU records the value of a running timer at the falling edges of the signal and forms the tooth spacing in terms of time from the difference between two edges. At the first tooth after the gap the measured time is very much greater than the preceding time; at the second tooth after the gap the time is again very much less. This tooth is defined as the reference mark. The pulse wheel is mounted so that this position corresponds to a defined angle opposite the top dead center of cylinder no. 1.

At uniform rotational speed the time between the edges around the tooth gap is three times as long as between two normal edges. Even in the event of greatly fluctuating rotational speed when the engine is being started, the reference mark can be reliably inferred on account of the measured differences of the tooth spacing.

Camshaft-speed sensor
The camshaft is stepped down by a ratio of 1:2 compared with the crankshaft. It indicates whether a piston moving to top dead center is in the compression cycle or the exhaust cycle. For this reason a camshaft-speed sensor is required in addition to the crankshaft-speed sensor.

Simple pulse wheels lead to different sensor-signal levels only in the area of the reference mark. More complex pulse wheels with segments of different lengths enable the engine position to be detected more quickly when starting and thus provides for rapid starting.

Wheel-speed sensor
Wheel-speed sensors sense the rotational speed of the wheels. This information is required by for example the Electronic Stability Program (ESP) to calculate driving speed and slip. The navigation system requires the wheel-speed signals to calculate the distance traveled when there is no GPS reception (e.g. in tunnels).

The pulse wheel is permanently connected to the wheel hub. Both steel pulse wheels and multipole rings can be used. This sensor has fewer teeth because the pulse wheel is smaller in diameter than that on an engine rpm sensor.

The sensor element with the signal amplifier and signal conditioning is integrated in an IC, hermetically sealed with plastic, and seated in the sensor head.

Digital signal conditioning enables encoded additional information to be transmitted via a pulse-width-modulated output signal. One example is direction-of-rotation detection, which is required for Hill Hold Control that prevents the vehicle from rolling back when starting. The direction of rotation is also used for vehicle navigation to detect backing up (reversing).

Speed sensor for transmission control
Transmission-speed sensors sense shaft speeds in automatic transmissions. Standard geometries are generally not possible due to the compact designs of transmissions. Specific sensor versions are therefore required for each transmission. Evaluation circuits with varying degrees of evaluation-algorithm complexity are used to cover the full spectrum of functional requirements.

References for rpm sensors
[1] K. Reif (Editor): Sensoren im Kraftfahrzeug. 2nd Ed., Springer-Vieweg, 2012.

Oscillation gyrometers

Measured variables

Oscillation gyrometers measure the absolute yaw rate along the vehicle vertical axis, the vehicle transverse axis (pitch axis), and the vehicle longitudinal axis (roll rate). The signals generated here are used in numerous systems such as for example the Electronic Stability Program (ESP), the rollover protection system, vehicle navigation, and damping control.

Measurement principle

Oscillation gyrometers are similar in principle to mechanical gyroscopes and for measurement purposes utilize the Coriolis acceleration a_C that occurs during rotary motion in conjunction with an impressed oscillating motion (Figure 1). The acceleration is calculated from the cross product of velocity and yaw rate:

(1) $\vec{a}_C = \vec{a}_x = 2\vec{v}_y \times \vec{\Omega}_z$.

The velocity v is produced by the sinusoidal driving of the inert mass of the sensor along its y-axis at frequency ω. Thus:

(2) $v_y = \hat{v}_y \sin(\omega t)$.

Figure 1: Creation of Coriolis acceleration
When a mass point m moves in the y-direction at velocity v_y and the system simultaneously rotates about the vertical axis z with yaw rate Ω_z, the mass point experiences a Coriolis acceleration a_C in the x-direction.

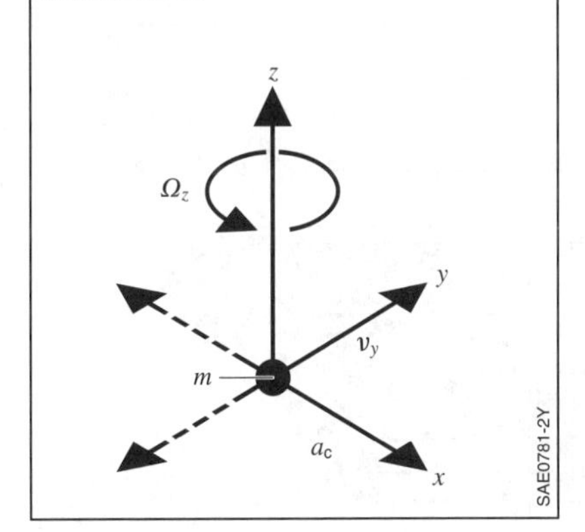

With a constant yaw rate Ω about the z-axis a Coriolis acceleration is produced along the x-axis. It has the same frequency and phase as the drive velocity v, and the associated amplitude is

(3) $\hat{a}_C = 2\,\hat{v}_y \Omega_z$.

The Coriolis force is one of the so-called pseudo forces and can only be measured in the rotating reference system (like the passenger car here).

The signal evaluation of the sensor demodulates the sinusoidal Coriolis signal of the micromechanical sensor ($a_C \propto \sin(\omega t)$) and in this way determines the yaw rate Ω. In the process, unwanted acceleration from the outside (e.g. bodywork acceleration) is removed.

The utilized measuring effects not only are very small, but also require complex signal conditioning. Signal conditioning directly at the sensor is therefore necessary.

Applications

Micromechanical yaw-rate sensors

A seismic mass in the sensor must be induced to vibrate in order to generate a Coriolis force to measure a yaw rate. In a micromechanical yaw-rate sensor (Figure 2) this is achieved in a permanent-magnet field by utilizing the Lorentz force (electrodynamic drive).

This sensor is based on a formulation of bulk micromechanics: Two thick oscillating elements fashioned from the wafer by means of bulk micromechanics oscillate in push-pull mode at their resonant frequency (< 2 kHz), which is determined by their mass and their coupling-spring stiffness. Each of them is provided with a capacitive, surface-micromechanical acceleration sensor (SMM) which measures Coriolis acceleration.

For drive purposes a sinusoidally modulated current is passed through a simple printed conductor on the relevant oscillating element. This generates within the permanent-magnetic field operating perpendicularly to the chip surface a Lorentz force which starts the oscillating element moving.

If this sensor is rotated at yaw rate Ω about the vertical axis, the drive in the oscillation direction results in a Coriolis acceleration which is measured by the acceleration sensors.

The different physical natures of the drive and sensor systems prevent unwanted crosstalk between the two parts. In order to suppress external acceleration (common-mode signal), the two opposing sensor signals are subtracted from each other. Summation, however, can also be used to measure the external acceleration).

The high-precision micromechanical construction helps to suppress the effects of high oscillatory acceleration which is several factors of 10 higher than the low-level Coriolis acceleration (cross sensitivity far below 40 dB). The drive and measuring systems are mechanically and electrically isolated in rigorous terms here.

Surface-micromechanical yaw-rate sensors

Micromechanical yaw-rate sensors can also be manufactured entirely using SMM technology (surface micromechanics). Here the oscillating element is electrostatically driven by capacitors. In this case the isolation of the drive and detection systems is not longer strictly separated from each other since both systems are capacitive in design.

A centrally mounted rotary oscillator is electrostatically driven to oscillate using comb structures (Figure 3).

When the sensor is subjected to a rotation Ω, the Coriolis force induces a simultaneous "out-of-plane" tilting motion. Its amplitude is proportional to the yaw rate and is detected capacitively by electrodes located under the rotary oscillator. To prevent this motion from being excessively damped, it is essential to operate the sensor in a vacuum.

The smaller chip size and the simpler manufacturing process do indeed reduce the cost of such a sensor, but the reduction in size also diminishes the already slight measuring effect and thus the attainable accuracy. It places higher demands on the electronics.

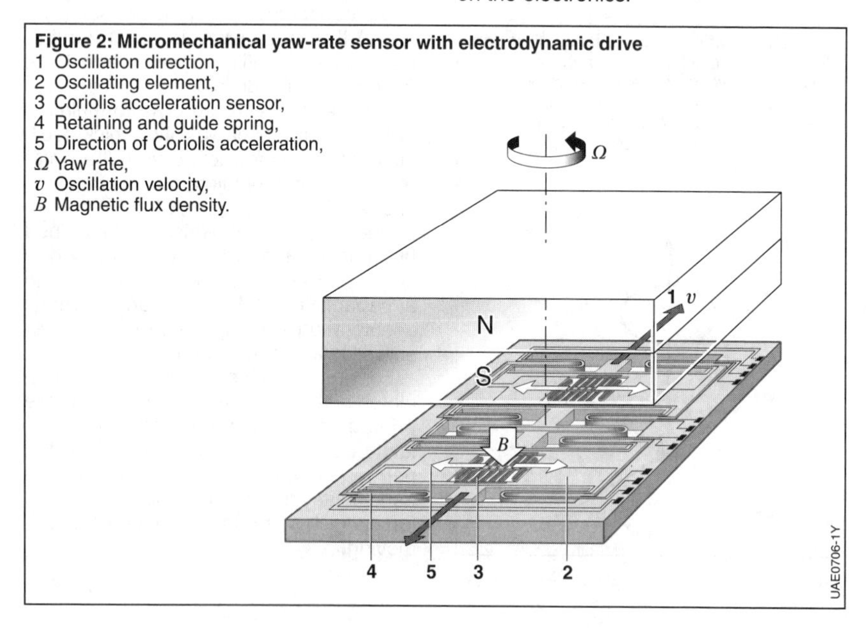

Figure 2: Micromechanical yaw-rate sensor with electrodynamic drive
1 Oscillation direction,
2 Oscillating element,
3 Coriolis acceleration sensor,
4 Retaining and guide spring,
5 Direction of Coriolis acceleration,
Ω Yaw rate,
v Oscillation velocity,
B Magnetic flux density.

Figure 3: Surface-micromechanical yaw-rate sensor with electrostatic drive
1 Comb structure, 2 Rotary oscillator, 3 Measurement axis.
C_{Drv} Capacitance of drive electrodes (drive by application of a sinusoidal voltage),
C_{Det} Capacitance of rotary-oscillation tap (measurement of Coriolis force),
C_{Drv_Det} Capacitance of drive tap (measurement of drive oscillation),
F_C Coriolis force,
v Oscillation velocity,
Ω Yaw rate to be measured (Ω = const $\cdot \Delta C_{Det}$).

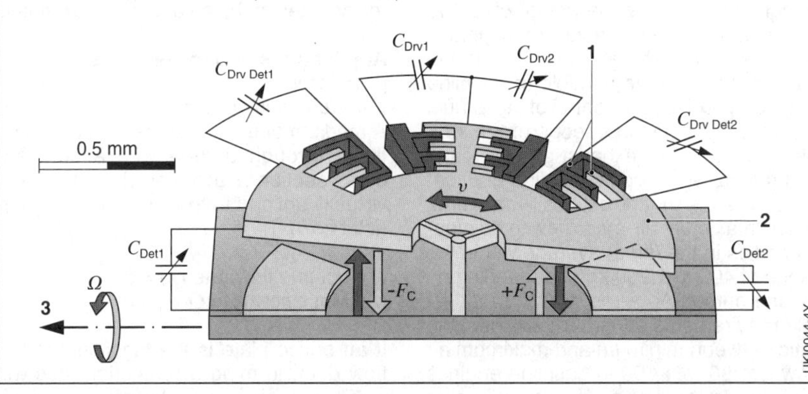

The influence of external accelerations is suppressed in SMM yaw-rate sensors by a special micromechanical design. Here the MEMS element (MEMS, Micro-Electro-Mechanical Systems) exhibits two seismic masses. Their deflection is effected by Coriolis forces in opposite directions, but by external accelerations in the same direction. A differential evaluation of the two oscillators then frees the signal of the unwanted disturbance values.

Further developments of yaw-rate sensors

Ongoing development of systems in the field of active and passive safety means that increased demands are placed on the signal quality and robustness of the yaw-rate sensor. These systems require not only the measurement of the yaw rate, but also the detection of pitching and rolling motion.

For example, an ESP system is based among other things on measurement of the yaw rate Ω_z and on measurement of the lateral acceleration a_y. ACC systems (Adaptive Cruise Control) utilize sensor signals of the yaw rate Ω_z, the lateral acceleration a_y, and the linear acceleration a_x. Rollover protection systems on the other hand use sensor information about the roll axis Ω_x, the vertical acceleration a_z, and also the lateral acceleration a_y.

New sensor generations of micromechanical sensor elements are used here. These sensors measure the yaw rate about either the x-axis or the z-axis; many of the these sensors also incorporate sensing elements for detecting accelerations in the x-, y- or z-direction.

Measurement of the yaw rate is based here on the principle of the surface-micromechanical yaw-rate sensor, while accelerations are detected with capacitive surface-micromechanical acceleration sensors.

Flowmeters

Measured variables

The amount of air supplied to the engine must be measured in order to control the combustion process. The mass ratios are the main factors in the chemical process of combustion, thus the actual objective is to measure the mass flow of the intake and charge air. The air-mass flow is the most important load variable in gasoline engines; precise pilot control of the air/fuel ratio presupposes knowledge of the supplied air mass. The exhaust-gas recirculation rate is regulated in gasoline and diesel engines by means of the air-mass flow.

The maximum air-mass flow to be measured lies in the (time) average within the range of 400 to 1,200 kg/h, depending on engine output. As a result of the modest idle requirements of modern engines, the ratio between minimum and maximum air flow is 1:90 to 1:100 in gasoline engines and due to the higher idle requirements 1:20 to 1:40 in diesel engines. Because of stringent exhaust-emission and consumption requirements, levels of accuracy of 1 to 2% from the measured value must be achieved. Referred to the measurement range, this can mean an – unusually high for motor vehicles – measurement accuracy of 10^{-4}.

The air is not drawn in continuously by the engine, but rather in time with the opening of the intake valves. For this reason, the air flow pulses strongly – particularly with the throttle valve wide open – at the measuring point (Figure 1). The measuring point is always located in the intake tract between the air filter and the throttle valve. Due to resonances of the intake manifold, pulsation of the intake manifold – above all in 4-cylinder engines in which the induction or charge phases do not overlap – is so strong that even brief reverse flows occur. These must be correctly detected by an accurate flowmeter.

Figure 1: Qualitative characteristic of intake-air pulsation in a 4-cylinder gasoline engine
Operating point: n = 3,000 rpm, full load.
Q_L Mean air flow rate.

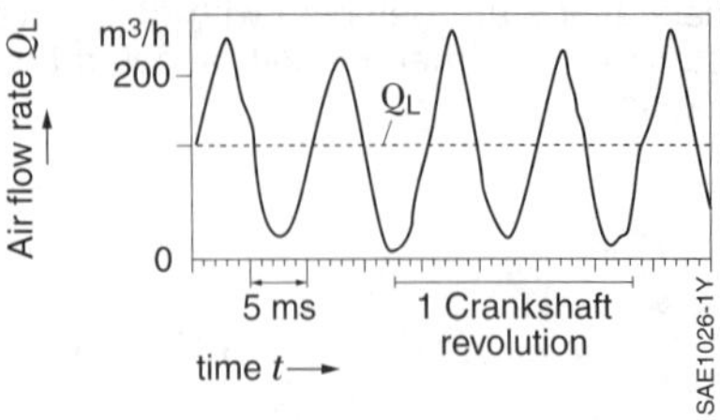

Air flow rate Q_L →
m³/h
200
0
Q_L
5 ms
1 Crankshaft revolution
time t →
SAE1026-1Y

Applications and measurement principles

Pitot-tube flowmeters

A medium of uniform density at all points flows through a tube with a constant cross-section A at a velocity v which is virtually uniform in the tube cross-section (inlet flow). Thus:

(1) Volume flow rate $Q_V = v \cdot A$,
(2) Mass flow rate $Q_M = \rho \cdot v \cdot A$.

If an orifice plate is then installed in the flow duct, forming a restriction, this will result in a pressure differential (dynamic pressure) Δp in accordance with Bernoulli's Law. Thus:

(3) $\Delta p = \text{const} \cdot \rho \cdot v^2 = \text{const} \cdot Q_V \cdot Q_M$,

where the constant is dependent on the tube and the orifice-plate cross-sections. This pressure difference can be measured directly with a differential-pressure sensor or as a force acting on a sensor plate. Pivoting, variable-position sensor flaps (Figure 2) leave a variable section of the flow cross-section unobstructed, with the size of the free diameter being dependent on the flow rate. The sensor plate is pressed as the flow increases against a mostly constant counterforce. A potentiometer monitors the characteristic flap positions for the respective flow rates.

Measurement errors can occur in cases where the sensor-plate mechanical inertia prevents it from keeping pace with a rapidly pulsating air current (full-load condition at high engine speeds).

If the density ρ changes due to temperature fluctuations or the altitude, the measurement signal changes with $\sqrt{\rho}$. An air-temperature sensor and a barometric pressure sensor must be used to compensate for this.

The higher sensitivity of the output signal required for a large measurement range with small air masses is obtained from the mechanical and electrical design of the air-flow sensor used in for example the earlier L-Jetronic gasoline injection system. Air-flow sensors for KE-Jetronic are designed for a linear characteristic.

This method of flow measurement has not been used in new developments for some time now.

Hot-wire air-mass meters

Hot-wire air-mass meters operate without mechanically moving parts. The hot wire is made from for example platinum whose electrical resistance R increases with temperature. The wire heats up when a current I_H is passed through it. It is cooled when it is in an air flow. The cooling process reduces the resistance of the wire and increases the electric current to produce an equilibrium between the input power P_{el} and the power P_V output by the flow:

$$(4)\ P_{el} = I_H^2\, R = P_V = c_1\, \lambda\, \Delta T.$$

The dimensions of the hot wire and its air flow assume the constant c_1. The heat conduction λ is approximately proportional to the root of the mass flow ($\sqrt{Q_M}$). Furthermore, if thermal convection is taken into consideration with the medium at rest (no air flow) with the coefficient c_2, the following equation is obtained for the heating current I_H:

$$(5)\ I_H = c_1 \cdot \sqrt{\left(\sqrt{Q_M} + c_2\right)} \cdot \sqrt{\frac{\Delta T}{R}}\ .$$

The closed-loop control circuit in the sensor housing (Figure 3, the controller is shown as a box) maintains the heater element (platinum hot wire) at a constant overtemperature in relation to the air temperature. The air temperature is taken into consideration with the compensation resistor R_K, whose resistance value is dependent on the air temperature. The current required for heating provides an extremely precise – albeit nonlinear – index of air-mass flow rate. The associated ECU generally converts the signal into linear form and performs other signal-evaluation tasks, with the air-mass meter signal being sensed in millisecond intervals. Due to its closed-loop design, this type of flowmeter can also monitor fast flow-rate variations in the millisecond range since, because of the constant overtemperature of the heater element, its heat content does not have to be changed by means of time-consuming heat transfers.

However, the flow direction is not recognized with this method, and for this reason

Figure 2: Pitot-tube flowmeter
1 Sensor flap,
2 Compensation flap,
3 Damping volume.
Q Flow.

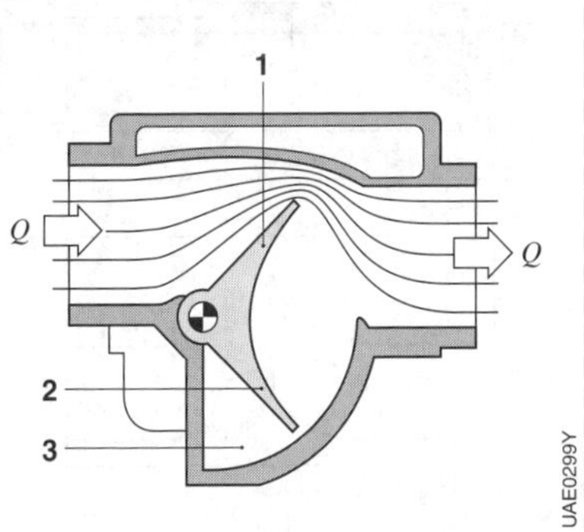

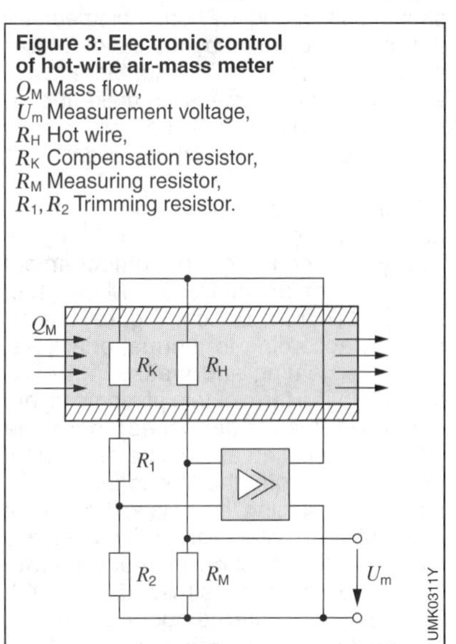

Figure 3: Electronic control of hot-wire air-mass meter
Q_M Mass flow,
U_m Measurement voltage,
R_H Hot wire,
R_K Compensation resistor,
R_M Measuring resistor,
R_1, R_2 Trimming resistor.

these sensors sometimes exhibit substantial deviations when strong pulsations occur in the intake manifold.

To ensure stable and reliable performance throughout an extended service life, the system must burn off all accumulated deposits from the hot wire's surface at approx. 1,000 °C after each phase of active operation (when the ignition is switched off).

The hot-wire air-mass meter was used in earlier engine-management systems (Motronic) and has since been replaced by the hot-film air-mass meter (HFM).

Hot-film air-mass meters using thick-film technology

The first hot-film air-mass meter still manufactured using thick-film technology (HFM2 from Bosch) operates according to the same principle as the hot-wire air-mass meter, but combines all the measuring elements and the control electronics on a single substrate. The flat-design heating resistor is located on the back of the base wafer, with the corresponding temperature sensor on the front. The greater thermal inertia of the ceramic wafer results in somewhat greater response lag than that associated with the hot-wire air-mass meter. The compensation resistor and the heater element are thermally decoupled by means of a laser cut in the ceramic substrate. More favorable air-flow characteristics make it possible to dispense with the hot-wire meter's burn-off decontamination process.

Micromechanical hot-film air-mass meters

Extremely compact micromechanical hot-film air-mass meters (HFM5, HFM6, and HFM7 from Bosch, Figure 4) also operate according to thermal principles. Here the heating and measuring resistors are in the form of thin platinum layers sputtered (vapor-deposited) onto the silicon chip acting as substrate. Thermal decoupling from the mounting is obtained by installing this chip in the area of the heating resistor H on a micromechanically thinned section of the substrate (similar to a pressure-sensor diaphragm). The adjacent heater-temperature sensor S_H and the air-temperature sensor S_L (on the thick edge of the silicon chip) maintain the heating resistor H at a constant overtemperature. This method differs from earlier techniques in dispensing with the heating current as an output signal. Instead, the signal is derived from the temperature difference in the diaphragm detected by the two temperature sensors S_1 and S_2. Temperature sensors are located in the flow path upstream and downstream from the heating resistor H. Without air inflow the temperature profile is the same on both sides of the heating zone; for the temperature at the measuring points $T_1 = T_2$ applies (Figure 4).

If, on the other hand, air flows through the sensor measuring cell, the area upstream of the heater element is cooled by the cold air and a lower temperature is measured at S_1. Downstream of the heater element the air flowing past is heated by the heater element; a higher temperature is therefore measured at S_2.

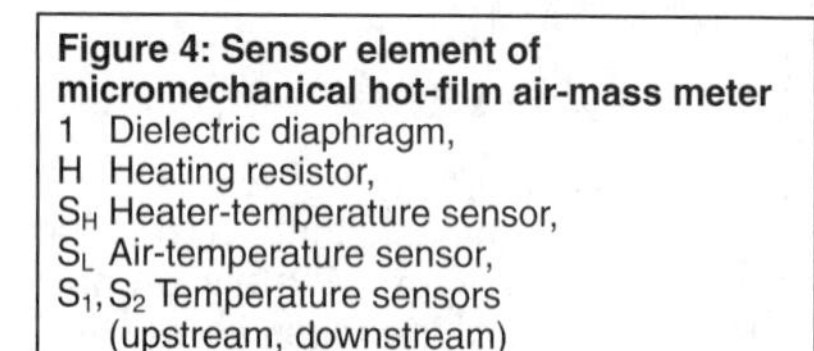

Figure 4: Sensor element of micromechanical hot-film air-mass meter
1 Dielectric diaphragm,
H Heating resistor,
S_H Heater-temperature sensor,
S_L Air-temperature sensor,
S_1, S_2 Temperature sensors (upstream, downstream)
Q_M Air-mass flow,
s Measurement point,
T Temperature.

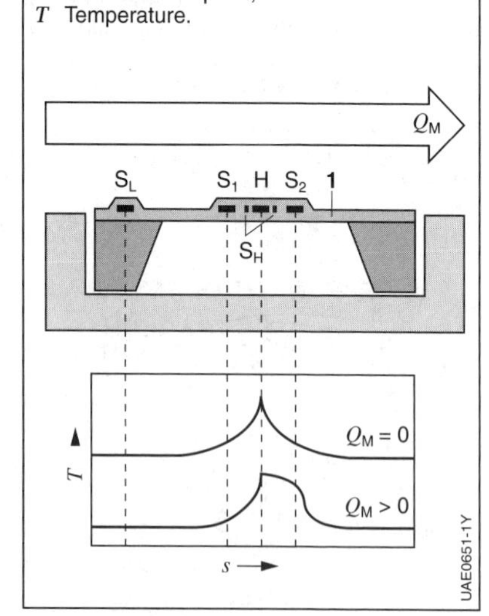

The temperature difference is, irrespective of the absolute temperature of the air flowing past, a measure of the mass of the air flow. Although (as with the earlier process) the response pattern remains nonlinear, the fact that the initial value also indicates the flow direction represents an improvement over the former method using the heating current (Figure 6).

In the HFM5 the evaluation electronics integrated in the sensor convert the temperature difference measured at S_1 and S_2 with the aid of an analog circuit into an analog voltage signal between 0 V and 5 V. In the HFM6 and HFM7 the signal is processed with the aid of digital electronics to achieve greater accuracy and a frequency signal is generated. The HFM7 is a further-developed version of the HFM6 with improved electronics and the possibility of outputting an analog signal as an alternative to the frequency signal. This voltage or the frequency is converted in the engine ECU via the sensor curve stored in the program into an air-mass flow.

The hot-film air-mass meter projects with its housing into a measuring tube (Figure 5). A flow rectifier mounted on the measuring tube (e.g. a wire grill) ensures a uniform flow in the measuring tube. The measuring tube is installed after the air filter in the intake tract.

Because of the small size of the measuring element, this flowmeter is a partial flowmeter, which detects only a specific and very small part of the total flow. The constancy and reproducibility of this division factor have a direct impact on the accuracy of the sensor. A calibration establishes the interrelationship between the air mass Q_M flowing through the measuring tube and the measurement signal generated from the partial air flow.

The inlet and outlet to the micromechanical measuring element are designed and optimized in such a way that heavier particles, such as dust particles and liquid droplets, do not approach the measuring element directly, but are rather diverted ahead of it. As well as the changed form of signal processing the HFM6 and HFM7 feature a modified measuring passage in order to provide improved protection against contamination directly upstream of the sensor element.

The HFM7 is the current Bosch air-mass sensor that is used in engine-management systems for gasoline and diesel engines.

Figure 5: Section through hot-film air-mass meter
1 Electrical connections,
2 Measuring-tube or air-filter housing wall,
3 Evaluation electronics (hybrid circuit),
4 Sensor measuring cell,
5 Sensor housing,
6 Partial-flow measuring passage,
7 Outlet, partial air flow Q_M,
8 Inlet, partial air flow Q_M.

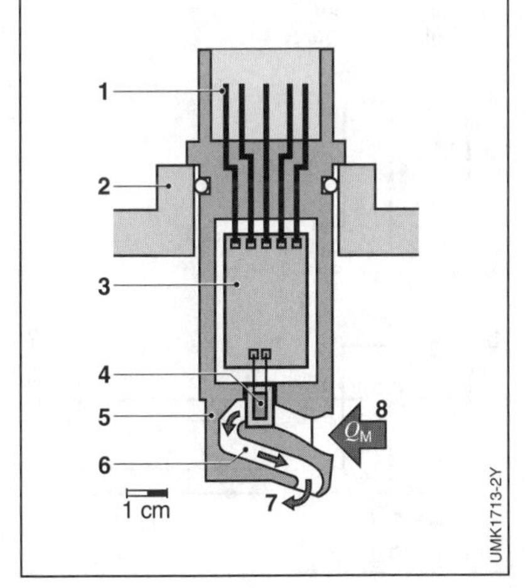

Figure 6: Characteristic curve of hot-film air-mass meter

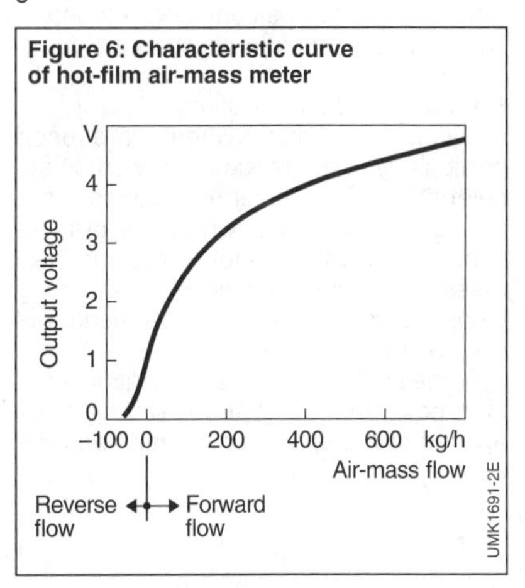

Acceleration and vibration sensors

Measured variables

The acceleration values a to be measured in motor vehicles are frequently expressed as a multiple of gravitational acceleration g ($1\,g \approx 9.81\ \mathrm{m/s^2}$). Acceleration and vibration sensors are used in the following applications:

- Triggering restraint systems such as e.g. airbag and seat-belt pretensioner (35...100 g).
- Side-impact and upfront sensing (100...500 g).
- Rollover detection (3...7 g).
- Sensing acceleration by the vehicle for the antilock braking system (ABS) and the Electronic Stability Program (ESP) (0.8...1.8 g).
- Evaluating bodywork acceleration for chassis control systems: body acceleration (1...2 g), axle and damper (10...20 g).
- Detecting changes in vehicle inclination for anti-theft alarm systems (approx. 1 g).
- Knock control in gasoline engines (measurement range up to 40 g).

Measurement principles

Acceleration sensors measure the force exerted by an acceleration a on an inert mass m:

(1) $F = m\,a$.

There are both position- and (mechanically) stress-measuring systems.

Position-measuring systems

Position-measuring systems are used particularly in the range of very small accelerations. All acceleration sensors are spring-bound right down to the gravitational pendulum. In other words, the inert mass is elastically connected to the body whose acceleration a is to be measured (Figure 1a).

In the static case the acceleration force is in equilibrium with the restoring force applied to the spring which has been deflected by x:

(2) $F = m\,a = c\,x$
where c = spring constant.

The resulting excursion is converted by means of a suitable measurement method (e.g. piezoelectric, capacitive, piezoresistive or thermal) into an electrical signal.

The system's measurement sensitivity S is therefore:

(3) $S = \frac{x}{a} = \frac{m}{c}$.

In other words, a large mass together with low spring stiffness (or constant) result in high measurement sensitivity.

In the dynamic case a damping force and an inertial force must be taken into account as well as the spring force. The damping force is proportional to the velocity $\dot{x}$ and is described with the damping coefficient p. The inertial force is proportional to the acceleration $\ddot{x}$. The oscillatory (resonant) system is described by the following equation:

(4) $F = m\,a = c\,x + p\,\dot{x} + m\,\ddot{x}$.

Figure 1: Displacement-measuring acceleration sensors
a) Excursion-measuring acceleration sensor,
b) Position-controlled acceleration sensor.
a Measurement acceleration,
x System excursion,
F_M Measuring force (inertial force on mass m),
F_K Compensation force,
I_A Output current,
U_A Output voltage.

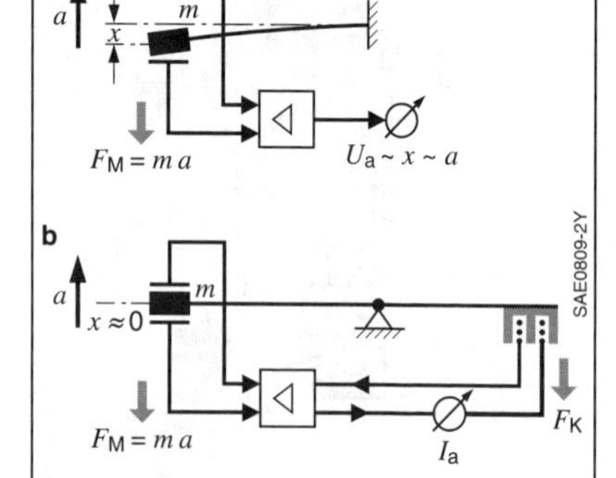

If negligible friction ($p \approx 0$) is presumed, the system resonates at the natural frequency ω_0:

(5) $\omega_0 = \sqrt{\frac{c}{m}}$.

Thus, according to equation (3), the measurement sensitivity S is directly linked to the resonant frequency ω_0 in the following manner:

(6) $S\,\omega_0^2 = 1$.

In other words, it can be expected that sensitivity is reduced to a factor of ¼ when the resonant frequency is increased by a factor of 2. It is only below their resonant frequency that such spring-mass systems display adequate proportionality between measured variable and excursion.

Position-controlled sensor
Position-measuring systems also permit the use of the compensation method in which the system excursion caused by acceleration is compensated for by an equivalent restoring force (Figure 1b). In practical terms the system then ideally always operates very close to its zero point (high linearity, minimum cross sensitivity, high temperature stability). These position-controlled systems also have, on account of their control, higher stiffness and cutoff frequency than excursion systems of the same kind. Any lack of mechanical damping can be electronically generated here.

System behavior
In non-position-controlled systems the damping and resonance of the systems play key roles in system behavior. The application of Lehr's damping ratio D has proved effective in describing this behavior:

(7) $D = \frac{p}{m} \cdot \frac{1}{2\omega_0}$.

Equation (4) can then be formulated as

(8) $\frac{F}{m} = \ddot{x} + 2\,\omega_0 D\,\dot{x} + \omega_0^2 x$.

Using the dimensionless variable D permits a simple comparison to be made of different oscillatory systems: transient response and resonant response are defined to a large extent by this damping ratio.

With very low damping ($D \rightarrow 0$) the system exhibits a resonance sharpness at ω_0. This is often unwelcome in that on the one hand it can lead to the system being destroyed and on the other hand it results in a nonuniform transfer function ($G >> 1$).

As damping factors increase, the system's resonance sharpness continues to decrease. While no more resonance sharpness is obtained under periodic excitation for damping factors $D = \sqrt{2}/2$ (Figure 2), any oscillating transient condition in the case of a step excitation disappears for factors $D > 1$.

In order to achieve as uniform a system response as possible in the region $G \approx 1$ over a wide frequency range, it is important when designing the system to define its damping precisely. Ideally this is influenced for example only slightly by

Figure 2: Amplitude-resonance curve
G Transfer function,
D Damping,
ω Angular frequency,
ω_0 Resonant frequency,
Ω normalized angular frequency.

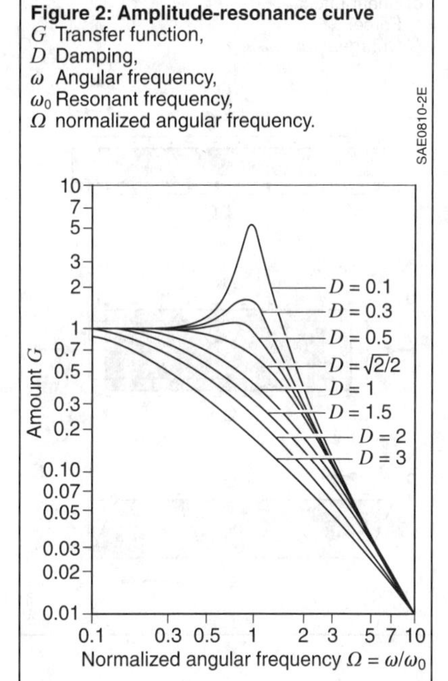

fluctuations in temperature. In practice sensors are therefore often designed in the range $D = 0.5...1.0$.

Mechanical stress-measuring systems
The piezoelectric effect can also be utilized to measure an acceleration. In this case the exertion of force by the external acceleration causes a mechanical stress in the piezoelectric material. These piezoelectric materials generate a charge Q under the influence of a force F on its surfaces, which are fitted with electrodes (Figure 3). This charge is proportional to the mechanical stress generated by the force.

For use as a sensor the generated charges are discharged via the external resistor of a measuring circuit or via an internal resistor of the sensor. The sensor therefore can measure dynamically, but not statically. The typical cutoff frequency is above 1 Hz.

The piezoelectric effect is utilized for example in piezoceramic bimorphous spring elements and in piezoelectric knock sensors.

Thermal acceleration sensors
These sensors generate a confined heated gas zone (Figure 4). The gas in this zone has a lower density than the surrounding cooler gas. During an acceleration the low-density gas zone moves within the surrounding cooler gas. The resulting asymmetry can be sensed by temperature sensors which are interconnected in a bridge circuit. The bridge voltage constitutes the acceleration signal.

This principle is used in numerous applications such as for example rollover sensing, the Electronic Stability Program (ESP), and even in smart phones.

Figure 3: Piezoelectric effect
a) Longitudinal effect,
b) Transversal effect,
c) Shear effect.
F Force,
Q Charge.

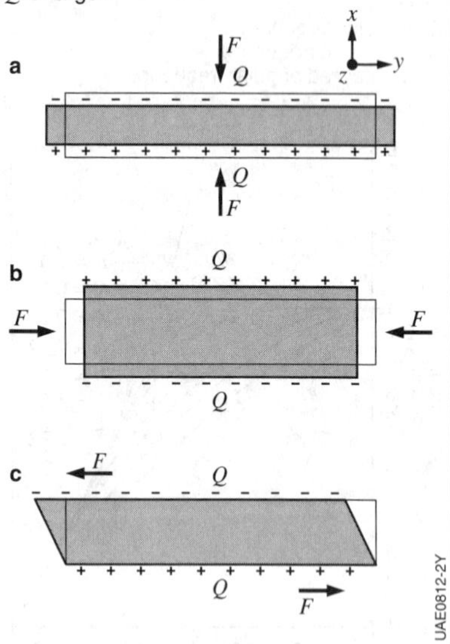

a

b

c

UAE0812-2Y

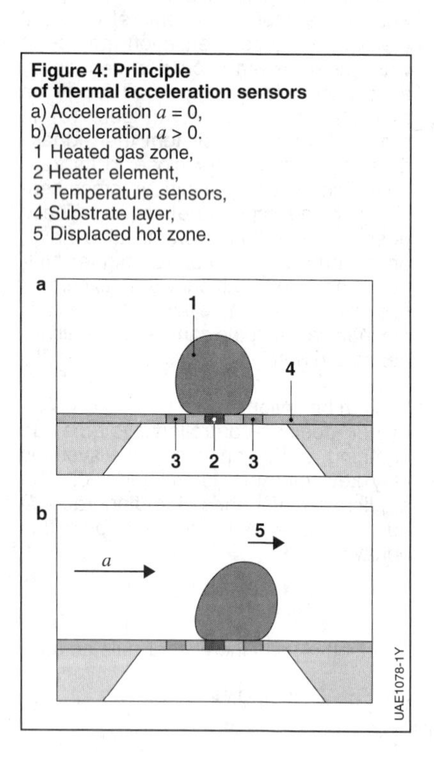

Figure 4: Principle of thermal acceleration sensors
a) Acceleration $a = 0$,
b) Acceleration $a > 0$.
1 Heated gas zone,
2 Heater element,
3 Temperature sensors,
4 Substrate layer,
5 Displaced hot zone.

Applications

Piezoelectric acceleration sensors

Piezoelectric bimorphous spring elements or two-layer piezoceramics (Figure 5) are used for restraint-system sensors for triggering seat bolt tighteners, airbags, and rollover bars. Their intrinsic mass causes them to deflect under acceleration to provide a dynamic (not DC response pattern) signal with excellent processing characteristics (typical cutoff frequency: 10 Hz). The sensor element is located in a sealed housing shared with the initial signal-amplification stage. It is sometimes encased in gel for physical protection. It consists of two oppositely polarized piezoelectric layers which are bonded to each other (bimorphous). An acceleration acting on it causes a mechanical tensile stress ($\varepsilon > 0$) in one layer and a compressive stress ($\varepsilon < 0$) in the other layer. The metallic layers on the upper and lower sides of the spring element serve as electrodes via which the resulting voltage is tapped.

This structure is packed together with the evaluation electronics in a hermetically sealed housing. The electronic circuit consists of an impedance transformer and an adjustable amplifier with specified filter characteristic.

The sensor's actuating principle can also be inverted: The sensor can be checked with an additional actuator electrode (on-board diagnosis).

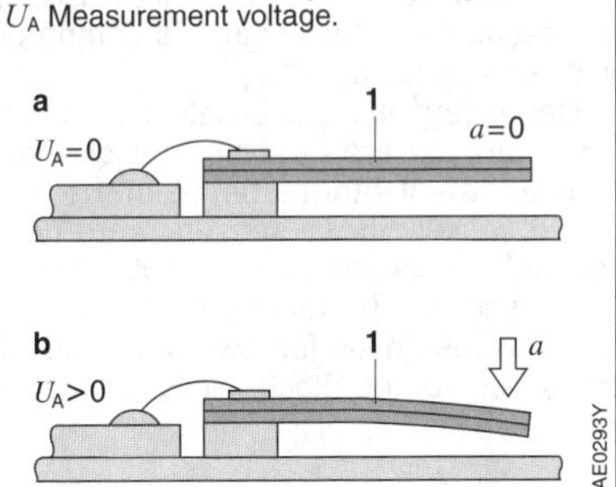

Figure 5: Piezoelectric sensor
a) At rest,
b) Subject to acceleration a.
1 Piezoceramic bimorphous spring element.
U_A Measurement voltage.

Micromechanical bulk silicon acceleration sensors

The first generation of micromechanical sensors relied on anisotropic and selective etching techniques to fabricate the required spring-mass system from the full silicon wafer (bulk silicon micromechanics) and produce the spring profile (Figure 6).

Capacitive taps have proven especially effective for the high-precision measurement of this seismic-mass deflection. This design entails the use of supplementary silicon or glass wafers with counter-electrodes above and below the spring-held seismic mass. This leads to a 3-layer structure, whereby the wafers and their counter-electrodes also provide overload protection. This layout corresponds to a series connection of two capacitors ($C_{1\text{-}M}$ and $C_{2\text{-}M}$). Alternating voltages are supplied at the connections C_1 and C_2, the superposition of these voltages is tapped at C_M, i.e. at the seismic mass. In the rest state the capacitances $C_{1\text{-}M}$ and $C_{2\text{-}M}$ are ideally equal. When an acceleration a acts in the measurement direction, the silicon center wafer is deflected as the seismic

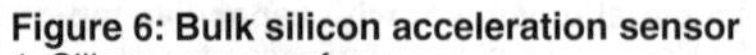

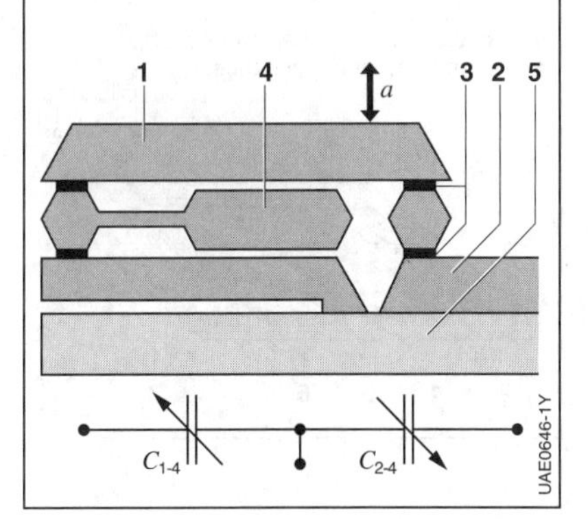

Figure 6: Bulk silicon acceleration sensor
1 Silicon upper wafer,
2 Silicon lower wafer,
3 Silicon oxide,
4 Silicon center wafer M (seismic mass),
5 Glass substrate.
a Acceleration,
C Measurement capacitances.

mass. The change in distance between the upper and lower wafers gives rise to a change in capacitance in the capacitors $C_{1\text{-M}}$ and $C_{2\text{-M}}$, and thus a difference C that is proportional to the applied acceleration. This in turn changes the electrical signal at C_M, which is amplified and filtered in the evaluation electronics.

Filling the sensor's hermetically sealed oscillatory system with a precisely metered charge of air leads to a very space-saving, inexpensive form of damping which also exhibits low temperature sensitivity.

This type of sensor is mainly used for low-level accelerations ($< 2\ g$) and relies upon a two-chip concept (sensor chip + CMOS processing chip).

Surface-micromechanical acceleration sensors

Surface-micromechanical sensors (SMM) have more compact dimensions than bulk silicon sensors. In SMM technology an additive process is employed to construct the spring-mass system on the surface of the silicon wafer (Figure 7). In the sensor core the seismic mass with its comb-shaped electrodes is connected by spring elements to anchor points of the silicon oxide. The chip features fixed, also comb-shaped electrodes on both sides of these moving electrodes. The fixed and moving electrode fingers create individual capacitances which are connected in parallel. This produces an effective total capacitance of 300 fF up to 1 pF. Two rows of electrode fingers connected in parallel produce two useful capacitances (C_1-C_M and C_2-C_M), which change in opposite directions when the seismic mass is deflected. The spring-mass system is subjected during an acceleration to a deflection which behaves linearly to the acceleration via the spring restoring force. An electrical output signal that is linearly dependent on the acceleration can be obtained by evaluation of this differential capacitor.

Figure 7: Surface-micromechanical acceleration sensor with capacitive tap

1 Spring-mounted seismic mass with electrodes,
2 Spring,
3 Fixed electrodes with capacitance C_1,
4 Printed aluminum conductor (for self-test function),
5 Bond pad,
6 Fixed electrodes with capacitance C_2,
7 Silicon oxide.
a Acceleration in sensing direction,
C_M Measurement capacitance.

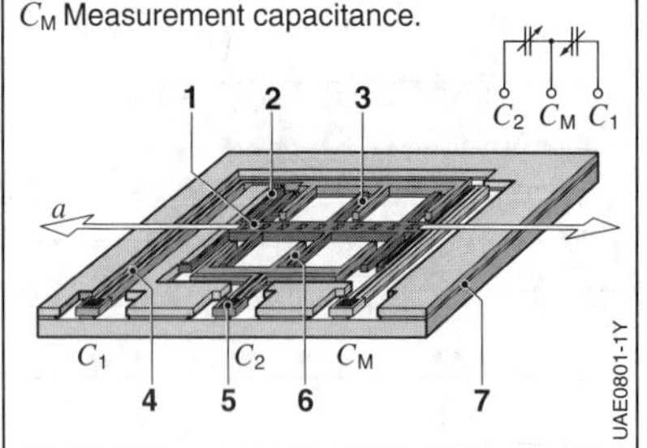

Because of the low capacitance of typically 1 pF the evaluation electronics are therefore integrated together with the sensor on the same chip, or very closely connected to it on the same substrate or leadframe. The raw signal is amplified, filtered, and conditioned for the output interface. Analog voltages, pulse-width-modulated signals, SPI protocols (Serial Peripheral Interface) or PSI5 protocols (current interfaces, see PSI5) are usual for this purpose.

The mechanical and electrical signal paths of the sensor can be checked with a self-test function. The sensor structure is deflected by means of an electrostatic force (simulation of an acceleration) and the resulting sensor signal is compared with a setpoint value.

Depending on the applications, the sensor elements are designed for different measurement ranges between 1 g and 500 g. Initially these sensors were used for high acceleration rates (50...100 g, for passenger-protection systems), but are now also used for lower acceleration rates (e.g. for the Electronic Stability Program, ESP).

Piezoelectric knock sensor

The knock sensor is in principle a vibration sensor. It senses structure-borne sound vibrations that occur as "knocking" in a gasoline engine during uncontrolled combustions (Figure 7). The structure-borne sound vibrations must be able to be introduced from the measuring location on the engine block undamped and without resonance into the knock sensor. A suitable measuring point and a fixed screw connection are required for this purpose.

The knock sensor is screwed to the engine block (Figure 8). A seismic mass exerts as a result of its inertia pressure forces in time with the exciting vibrations on an annular piezoceramic element. These forces induce a charge transfer within the ceramic element. This generates between the upper and lower sides of the ceramic element a voltage that is tapped via contact disks. The knock signal generated in the sensor is forwarded to the engine ECU, where it is conditioned and evaluated.

When knocking noises are detected the engine ECU retards the ignition timing and thus counteracts any further knocking (see Knock control). The vibration frequency is typically 5 to 25 kHz.

Figure 8: Knock sensor (design and mounting)
1 Annular piezoceramic element,
2 Seismic mass with pressure forces F,
3 Housing,
4 Screw,
5 Contact,
6 Electrical connection,
7 Engine block.
V Vibration.

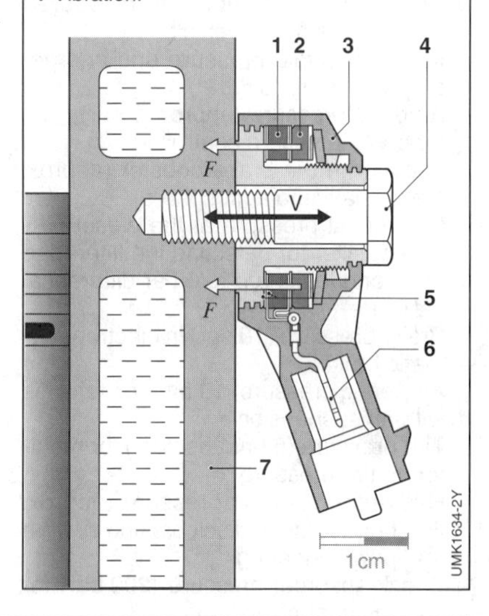

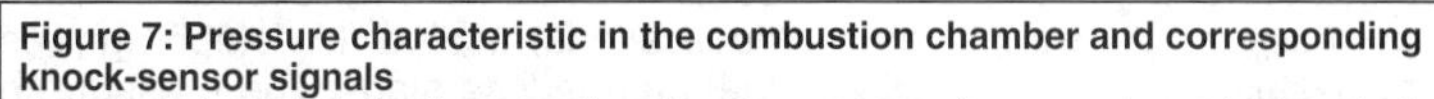

Figure 7: Pressure characteristic in the combustion chamber and corresponding knock-sensor signals
a) Typical characteristic of combustion-chamber pressure (measured on a test engine),
b) Bandpass-filtered signal of combustion-chamber pressure,
c) Structure-borne noise signal sensed by the knock sensor.

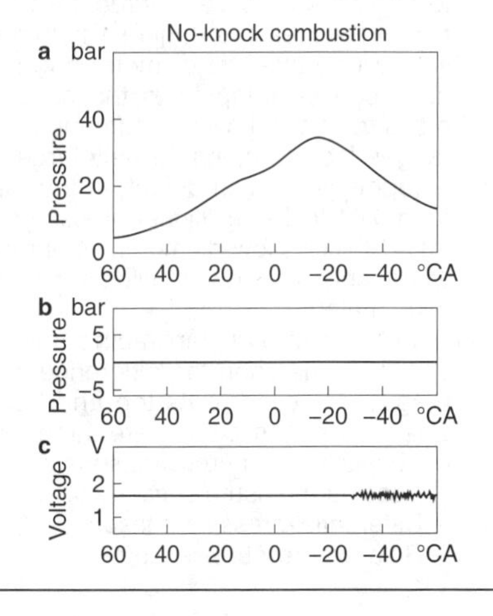

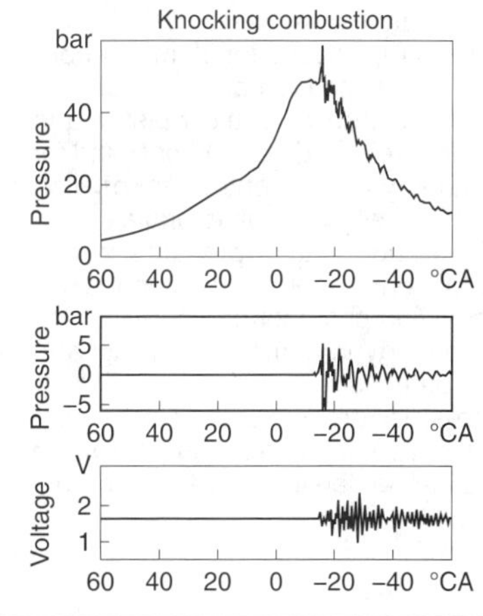

Pressure sensors

Measured variables

The measured variable pressure is defined as a nondirectional force acting in all directions which occurs in gases and liquids. It still propagates very well in gelatinous substances and soft sealing compounds. The most important measured pressures in a motor vehicle are:

- Intake-manifold pressure and charge-air pressure (1 to 6 bar)
- Ambient pressure (approx. 1 bar), e.g. for boost-pressure control
- Vacuum in the brake booster (approx. 1 bar relative to atmosphere)
- Differential pressure at the diesel particulate filter for detecting the filter-load state and leaks (up to 1 bar differential pressure)
- Brake pressure (10 bar) in electropneumatic brakes
- Air-spring pressure (16 bar) for vehicles with air suspension
- Tire pressure (5 bar absolute) for monitoring tire pressure
- Hydraulic reservoir pressure (approx. 200 bar) for the antilock braking system and power steering
- Shock-absorber pressure (200 bar) for chassis-control systems
- Refrigerant pressure (35 bar) in air-conditioning systems
- Modulation pressure (35 bar) in automatic transmissions
- Fluid pressure for clutch actuation in double-clutch transmissions (20 bar)
- Engine-oil pressure for demand-controlled oil pumps (10 bar)
- Brake pressure in brake master and wheel-brake cylinders (200 bar) and for automatic yaw-moment compensation in electronically controlled brakes
- Positive and vacuum pressure in fuel tank (0.5 bar) for on-board diagnosis
- Combustion-chamber pressure (100 bar, dynamic) for misfiring and knock detection
- Pump-side pressure in diesel distributor-type injection pumps (up to 1,000 bar, dynamic) for electronic governing
- Fuel pressure in the low-pressure circuit for demand-controlled fuel pumps (diesel and gasoline; 10 bar)
- Rail pressure in LPG and CNG systems (4 to 16 bar)
- Rail pressure for diesel common rail (up to 2,700 bar)
- Rail pressure for gasoline direct injection (up to 280 bar)
- Oil pressure for taking into account the engine load in the service display.

Measurement principles

Dynamically and statically acting pickups or sensors are used to measure pressures. Dynamically acting pressure sensors include for example microphones, which are insensitive to static pressures and are only used to measure pressure pulsations in gaseous or liquid media. Since up to now, practically only static pressure sensors have been used in automobiles, these will be dealt with in more detail here. Static pressure measurement is effected directly by the deformation of a diaphragm.

Diaphragm-type sensors

Design

The most common method used for pressure measurement (also in automotive applications) uses a thin diaphragm as the intermediate stage. The pressure to be measured is first of all applied to one side of this diaphragm so that this bends to a greater or lesser degree as a function of the pressure (Figure 1a). Within a very wide range, its diameter and thickness can be adapted to the particular pressure range. Low-pressure measuring ranges lead to comparatively large diaphragms which can deform in the range from 0.01 to 1 mm. Higher pressures demand thicker, low-diameter diaphragms which only deform very slightly by a few micrometers.

To be precise, a diaphragm's deformation depends upon the difference in the pressure applied to its top and bottom sides. Accordingly, there are three different basic types of pressure sensor:

- Absolute-pressure sensors
- Differential-pressure sensors
- Relative-pressure sensors.

Relative-pressure sensors measure the differential pressure relative to the ambient pressure.

Voltage-measuring methods dominate in all pressure ranges and here in practical terms only strain-gage technology is used.

Strain-gage technology

The strains on the diaphragm when a diaphragm-type sensor is subjected to deformation are sensed by strain gages (strain-gage resistors) which are applied to the diaphragm (e.g. diffused or vapor-deposited) (Figure 1b). Under the influence of mechanical stresses their electrical resistance changes on account of the piezoresistive effect. The K factor (gage factor) specifies the relative change of the strain-gage resistor referred to the relative change of its length. For the resistors diffused into the monocrystalline silicon in micromechanical pressure sensors this factor is especially high, typically $K \approx 100$.

Figure 1: Semiconductor absolute-pressure sensor
a) Cutaway view,
b) Wheatstone-bridge circuit.
1 Diaphragm,
2 Silicon chip,
3 Reference vacuum,
4 Glass (Pyrex),
p Measured pressure,
U_0 Supply voltage,
U_M Measurement voltage,
R_1 Strain-gage resistor (compressed),
R_2 Strain-gage resistor (stretched).

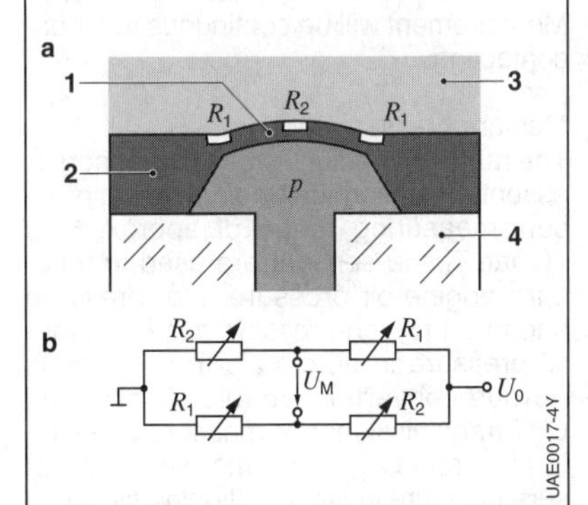

The resistors are interconnected to form a Wheatstone bridge (Figure 1b). The diaphragm is deformed to varying degrees depending on the effective pressure; in the process two resistors are stretched and two resistors are compressed. This alters their electrical resistance and thus also the measurement voltage (bridge voltage). The measured voltage is a measure of the pressure. This circuit produces a higher measurement voltage than is the case when a single resistor is evaluated. The Wheatstone bridge thus provides for a highly sensitive sensor.

Applications

Micromechanical pressure sensors

Micromechanical pressure sensors are used for pressure ranges of less than 6 bar (in low-pressure sensors) and for pressure ranges of less than 70 bar (in medium-pressure sensors).

The measuring cell of the micromechanical pressure sensor consists of a silicon chip in which a thin diaphragm is micromechanically etched (Figure 1). Four resistors arranged in a Wheatstone-bridge circuit are diffused on the diaphragm (see above). The bridge voltage is a measure of the pressure on the diaphragm.

The bridge signal must still be linearized and amplified. Temperature effects must also be compensated. This is performed either in a circuit integrated on the measuring chip or in a separate ASIC (application-specific integrated circuit). The output signal can be transferred either in the form of an analog voltage (0 to 5 V) or in digital form, e.g. via the SENT interface.

The advantage of digital transfer protocols over analog signals is that interface tolerances (contact and cable resistances) play no role and that as a rule additional information (fault codes, temperature signals, etc.) can still be transferred.

In the case of sensors for very aggressive or liquid measured media (e.g. for fuel-pressure and charge-air pressure sensor) “reverse assembly” is often used in which the measured pressure is conducted to an electronically passive cavity recessed into the side of the sensor

Figure 2: Absolute-pressure sensor
a) Reverse assembly (reference vacuum on structure side),
b) Simplified assembly with protective gel,
c) "Outer packaging" for intake-manifold installation with integrated air-temperature sensor.
1,3 Electrical connections with glass-enclosed lead-in,
2 Reference-vacuum chamber,
4 Measuring cell (chip) with evaluation electronics,
5 Glass base,
6 Cap,
7 Supply for measured pressure p,
8 Protective gel,
9 Gel frame,
10 Ceramic hybrid,
11 Cavity with reference-vacuum chamber,
12 Bonded connection,
13 Manifold wall,
14 Housing,
15 Sealing ring,
16 Temperature sensor,
17 Electrical connection,
18 Housing cover,
19 Measuring cell.
p Measured pressure.

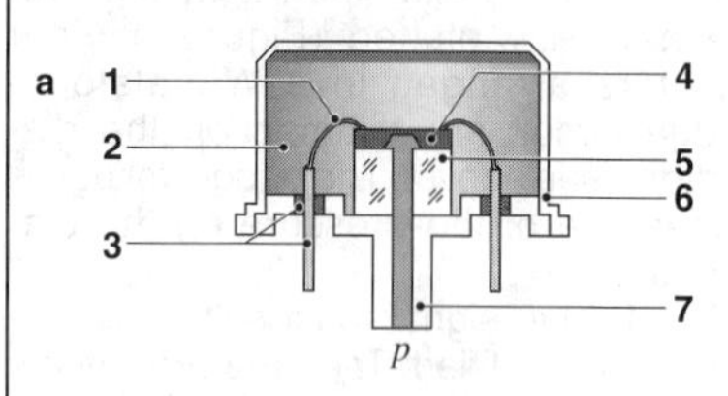

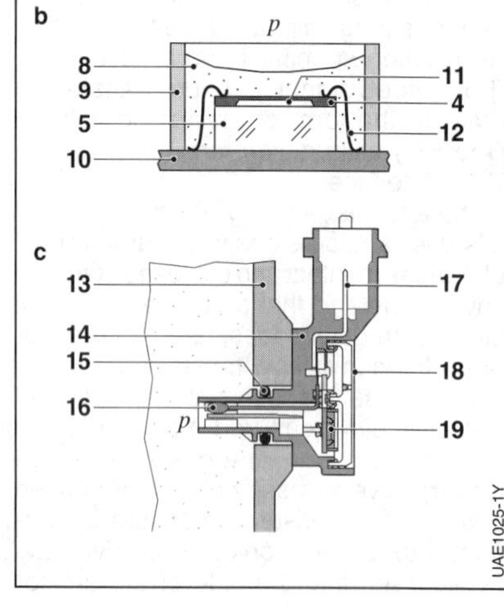

chip (Figure 2a). For maximum protection, the – much more sensitive – side of the chip with the measuring resistors, evaluation electronics and contacts is enclosed in a reference-vacuum chamber located between the housing's base and the soldered metal cap.

However, a cheaper arrangement is to seat the silicon chip with etched diaphragm and four strain-gage resistors as a measuring cell on a glass base. The reference-vacuum chamber is located in a cavity between the diaphragm and the glass base (Figure 2b). Alternatively, measuring chips made of porous silicon with an enclosed reference-vacuum chamber can be used for absolute-pressure sensors. The silicon chip is subjected to pressure from the side on which the measuring resistors and the evaluation electronics are located. This sensitive side of the chip is protected by a suitable gel against the pressure medium. The compact design of integrated single-chip sensors means that they are suitable for mounting directly on the intake manifold (Figure 2c).

Low-pressure sensors

This design is used in virtually all low-pressure sensors (maximum pressure less than 6 bar) such as for example charge-air pressure sensors and barometric sensors. For differential-pressure sensors (e.g. for diesel particulate filters or brake boosters) both sides of the diaphragm are subjected to the pressure medium.

These sensors will also be available for use in tire-pressure monitoring systems. Measurement will be continuous and non-contacting.

Medium-pressure sensors

The range of medium-pressure sensors is essentially described by a maximum pressure-measuring range of approx. 6 to 70 bar. These sensors are used to measure engine-oil pressure, fuel pressure ahead of the high-pressure pump, hydraulic pressure in automatic transmissions (torque-converter, double-clutch, and CVT transmissions), to measure pressure in LPG and CNG systems, and to measure pressure in air-conditioning systems.

The wide variety of applications means that there is a broad range of mechanical and electrical interfaces. The sensors are screwed into lines or into the crankcase and transmission case. A metallic taper seal has proven effective, but O-rings and sealing rings are also used. The design of the medium-pressure sensor very much follows that of the high-pressure sensor, with the metal diaphragm being replaced by a silicon-micromechanics-technology sensor element. In this case the sensor elements used in low-pressure sensors have been adapted by optimized etching methods for higher pressures and increased bursting-pressure strength. Here too the measurement signal of a Wheatstone-bridge circuit is evaluated. The sensor signal is conditioned and issued as an analog or a digital output signal. This is done either integrated in a silicon chip (1-chip technology) or by measuring element and evaluation ASIC (2-chip technology).

Combined sensors for pressure and temperature are used in some applications (e.g. in the engine oil-level sensor). For this purpose an NTC is integrated in the threaded fitting and the thermistor of the NTC is evaluated in the ECU by suitable circuits. The SENT protocol offers the opportunity to issue the pressure and temperature signals in a joint output signal. This dispenses with the need for a separate line for the temperature signal.

High-pressure sensors with metal diaphragm

Diaphragms made of high-quality spring steel are used at very high pressures as have to be measured for example in the rail of the common-rail system or gasoline direct injection for closed-loop control (Figure 4). Further applications include braking systems such as ABS and ESP, and also industrial hydraulics.

Four thin-film strain gages are applied to the diaphragm; the measurement principle is the same as for micromechanical pressure sensors. The steel diaphragm insulates the measured medium, differs from silicon in retaining a yield range for enhanced burst resistance, and is easy to install in metallic housings.

Insulated sputtered (vapor-deposited) metallic thin-film strain gages ($K \approx 2$) and even polysilicon strain gages ($K \approx 40$) offer permanently high sensor accuracy. Amplification, calibration, and compensation elements can be combined in a single ASIC, which is then integrated together with the required EMC protection on a small carrier in the sensor housing.

Figure 4: High-pressure sensor with metal diaphragm, measuring element
1 SiN_x passivation,
2 Gold contact,
3 Polysilicon strain-gage resistor,
4 SiO_2 insulation,
5 Steel diaphragm.
p Measured pressure.

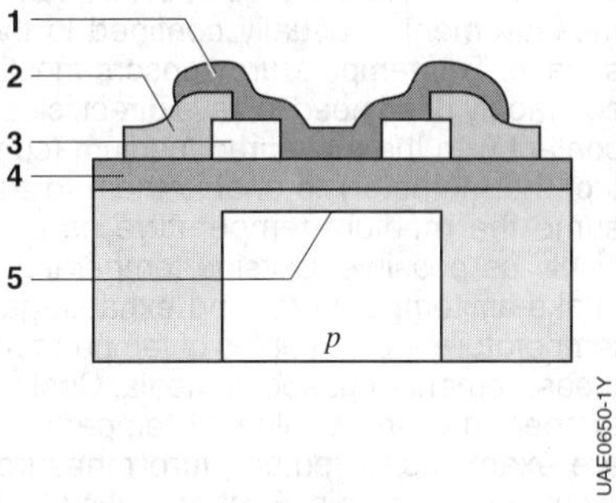

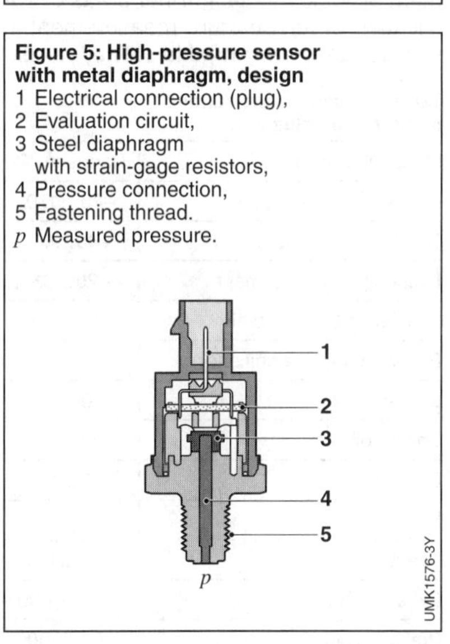

Figure 5: High-pressure sensor with metal diaphragm, design
1 Electrical connection (plug),
2 Evaluation circuit,
3 Steel diaphragm with strain-gage resistors,
4 Pressure connection,
5 Fastening thread.
p Measured pressure.

Temperature sensors

Measured variables

Temperature is defined as a nondirectional quantity which characterizes the energy state of a given medium, and which can be a function of time and location.

(1) $T = T(x, y, z, t)$,
x, y, z Space coordinates
t Time
T measured according to the Celsius or Kelvin scale.

In the case of gaseous and liquid measured media, it is generally possible to measure temperature at all points without any problems; in the case of solid bodies, measurement is usually confined to the surface. The temperature sensors mostly commonly used need to have direct, close contact with the measured medium (contact thermometer) to enable them to assume the medium temperature as precisely as possible. Outside temperature, intake-air temperature, and exhaust-gas temperature are examples of temperature measurement of gaseous media. Coolant temperature and engine-oil temperature are examples of temperature measurement of liquid media. Further areas of application of temperature measurement in vehicles are shown in Table 1. They can differ greatly in terms of required measuring accuracy and permissible measuring time.

For special cases non-contact temperature sensors are also used which determine the temperature of a body or a medium on the basis of the (infrared) heat radiation emitted by it (radiation thermometer, pyrometer, thermal imaging camera). These temperature sensors are used in night-vision systems (see Far-infrared systems) and for pedestrian recognition with infrared cameras.

At many locations temperature is also measured as a secondary variable in order that it can be compensated in cases where temperature variations trigger faults or act as an undesirable influencing variable – for example in sensors for other physical variables whose measured value is to depend as little as possible on the temperature.

Table 1: Temperature ranges in motor vehicles

Measuring point	Range in °C
Intake and charge air	–40…170
Outside atmosphere	–40…60
Passenger compartment	–20…85
Ventilation & heating air	–20…60
Evaporator (air conditioner)	–10…50
Coolant	–40…130
Engine oil	–40…170
Battery	–40…100
Fuel	–40…120
Tire air	–40…120
Exhaust emissions	100…1,000
Brake calipers	–40…2,000

Measurement principle

Temperature measurements in motor vehicles are conducted almost entirely by exploiting the sensitivity to temperature variation found in electrical resistance materials with a positive temperature coefficient (PTC, NTC thermistor) or with a negative temperature coefficient (NTC, PTC thermistor) as contact thermometers. Depending on the temperature, the sensor exhibits a certain resistance

Figure 1: Temperature curve with limits of variation of an NTC thermistor

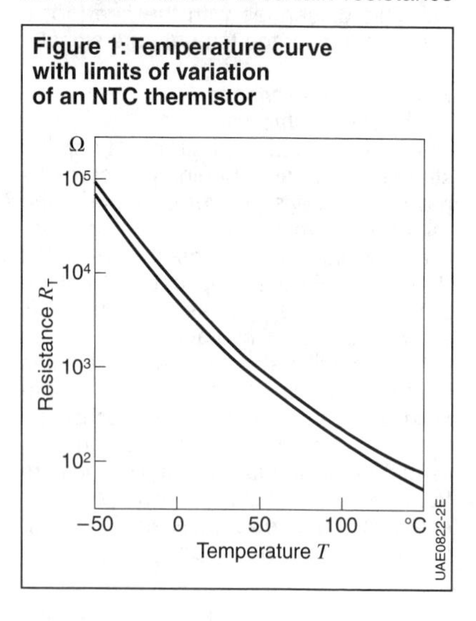

value (Figure 1). Conversion of the resistance variation into an analog voltage is performed predominantly by adding a temperature-neutral or inversely sensitive resistance to a voltage divider (Figure 2a) or by supplying an injected current (Figure 2b). For the voltage-divider circuit the following is obtained

$$U_A(T) = U_0 \frac{R(T)}{R(T)+R_V}.$$

With the injected current I_0 the measurement voltage is obtained at

$$U_A(T) = I_0 R(T).$$

The voltage measured at the measuring resistor is thus temperature-dependent. This analog voltage is digitized in the ECU by an analog-digital converter and then assigned via a characteristic curve to a temperature value.

Alternatively, signal processing can also be performed by an evaluation circuit integrated in the sensor; the temperature value is then transferred via a digital interface (see PSI5).

Sensor types

Sintered-ceramic resistors (NTC)

Semiconducting sintered-ceramic resistors made of heavy-metal oxides and oxidized mixed crystals (sintered in pearl or plate form) display an exponentially falling temperature curve (Figure 1). It can be described to good approximation by the following exponential equation:

$$R(T) = R_0\, e^{B\left(\frac{1}{T} - \frac{1}{T_0}\right)},$$

where
$R_0 = R(T_0)$,
B = 2,000...5,000 K,
T Temperature in K.

The resistance value can vary up to five powers of ten, for example typically from some 100 kiloohms to some 10 ohms. High thermal sensitivity means that applications are restricted to a "window" of approximately 200 K; however, this range can be defined within a latitude of −40 °C to approximately 850 °C by selecting an appropriate NTC. Close tolerances of ±0.5 K at a selectable reference point are possible for example by means of read-out, but this is also reflected in the cost.

Figure 2: Resistance-measurement methods
a) Voltage-divider circuit,
b) Measurement with injected current.
U_0 Supply voltage,
I_0 Injected current,
R_V Voltage-divider resistor,
$R(T)$ Measuring resistor,
$U_A(T)$ Measurement voltage.

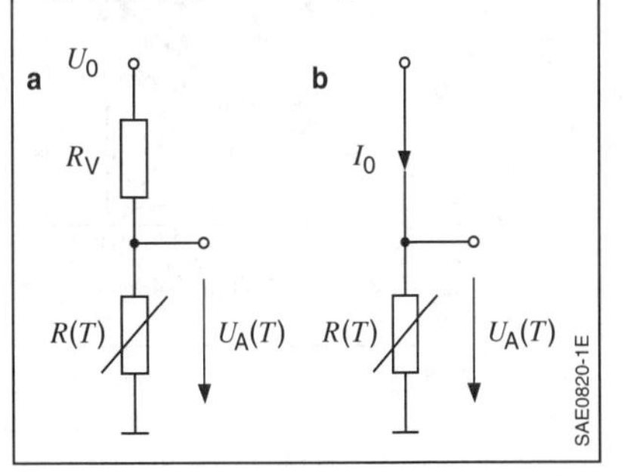

Thin-film metallic resistors (PTC)

Thin-film metallic resistors, integrated on a single substrate wafer (e.g. by vapor-depositing or sputtering) with two supplementary, temperature-neutral trimming resistors (Figure 3a), are characterized by extreme precision, as they can be manufactured and then "trimmed" with lasers to maintain exact response-curve tolerances over long periods of time (Figure 3b). Here the values of the trimming resistor connected in series to the measuring resistor and of the trimming resistor connected in parallel are changed.

The use of layer technology makes it possible to adapt the substrate (ceramic, glass, plastic foil) and the upper layers (plastic molding or paint, sealed foil, glass and ceramic coatings) to the respective application, and thus provide protection against the monitored medium.

Although metallic layers are less sensitive to thermal variations than the ceramic-oxide semiconductor sensors, both linearity and reproducibility are better. The following formulation is used to provide the computational description:

$$R(T) = R_0\,(1+\alpha\,\Delta T + \beta\,\Delta T^2+\ldots)$$

where
$\Delta T = T - T_0$,
T_0 = 20 °C (reference temperature),
α linear temperature coefficient,
β quadratic temperature coefficient.

The coefficient β is small for metals, but not entirely negligible. Therefore in sensors of this type the measurement sensitivity is characterized with a mean temperature coefficient, "TC 100". It corresponds to the mean characteristic-curve gradient between 0 °C and 100 °C (Figure 4) and is obtained from the following relationship:

$$\text{TC}\,100 = \frac{R(100\,°\text{C}) - R(0\,°\text{C})}{R(0\,°\text{C}) \cdot 100\ \text{K}} = \alpha_{100}\,.$$

Figure 3: Metal-film temperature resistor
a) Design,
b) Curve trimming.
1 Auxiliary contacts, 2 Bridge.
R_{Ni} Nickel-film resistor,
$R(T)$ Resistor referred to temperature T,
R_P, R_S Trimming resistors (parallel and serial).

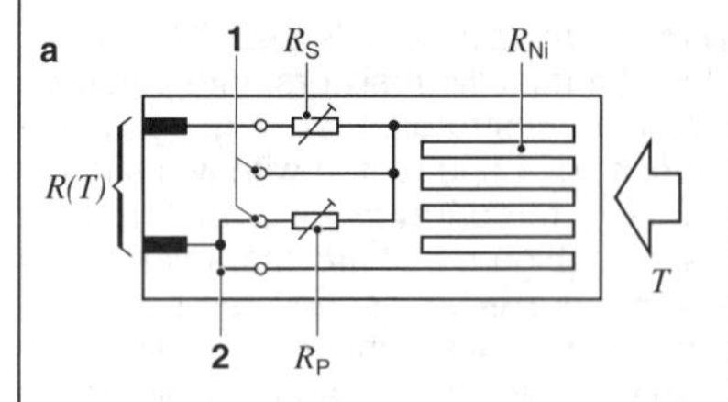

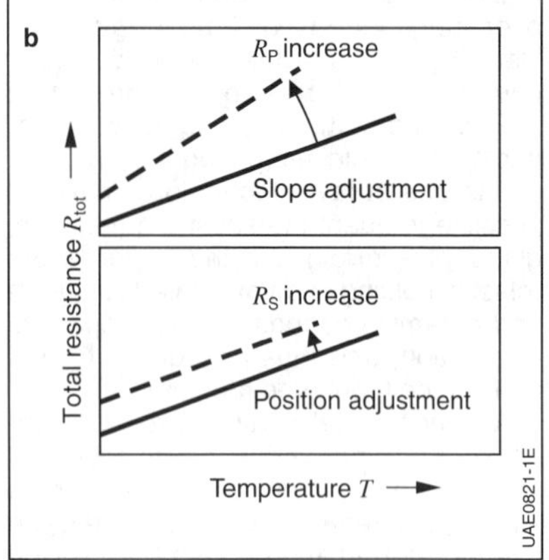

Table 2 shows the mean temperature coefficients for some metals.

Metal-film resistors are generally manufactured and marketed with a basic value (resistance value at 20 °C) of 100 Ω or 1,000 Ω (e.g. under the designation Pt 100 or Pt 1000) in different tolerance classes. Platinum resistors (Pt) do have the lowest temperature coefficient, but are very accurate and resistant to aging.

In addition to compensating elements, further active and passive circuit elements for the thin-film sensor can basically be integrated on the carrier chip. This facilitates initial, advantageous signal conditioning and conversion at the measuring point.

Thick-film resistors (PTC and NTC)
Thick-film resistors consist of a carrier chip (e.g. thin ceramic wafer) on which the resistor material is applied as paste (e.g.

Table 2: Mean temperature coefficient of some metals

Sensor material	TC 100	Measurement range
Nickel (Ni)	$6.18 \cdot 10^{-3}$/K	–60...250 °C
Copper (Cu), value dependent on composition	$3.8 \cdot 10^{-3}$/K up to $4.3 \cdot 10^{-3}$/K	–50...200 °C
Platinum (Pt) (DIN EN 60751, [1])	$3.85 \cdot 10^{-3}$/K	–200...850 °C

Figure 4: Definition of mean temperature coefficient TC 100 = α_{100}

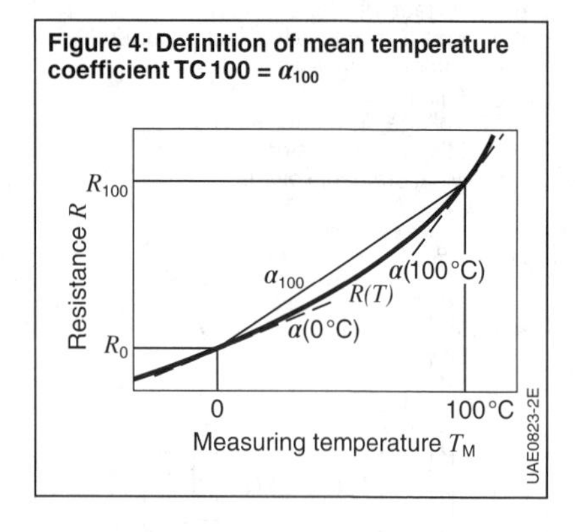

metal oxide) using thick-film technology and sintered at 800 °C to the substrate.

Thick-film resistors with their higher resistivity (low surface-area requirement) and positive and negative temperature coefficients are generally employed as temperature sensors for compensation purposes. They have nonlinear response characteristics (without, however, the extreme variations of the massive NTC resistor) and can for example be laser-trimmed. The measurement effect can be enhanced by using NTC and PTC materials to form voltage-divider circuits.

Monocrystalline silicon semiconductor resistors (PTC)

When monocrystalline semiconductor materials, such as silicon, are used to manufacture the temperature sensor, it is possible to integrate additional active and passive circuitry on the sensor chip. This facilitates initial signal conditioning at the measuring point. Due to the closer tolerancing which is possible, the measuring element is manufactured according to the "spreading-resistance" principle. The current flows through the measuring resistor and through a surface-point contact before arriving at the silicon bulk material. It then proceeds, widely distributed, to a counter-electrode covering the base of the sensor chip (Figure 5).

As well as the highly reproducible material constants, the high current density after the contact point (high precision achieved through photolithographic manufacture) almost exclusively determines the sensor's resistance value. To eliminate polarity sensitivity the sensors are mostly series-connected in a pair in opposite alignment (double-hole version, Figure 5). The bottom electrode can then be designed as a metallic temperature contact (with no electrical function).

Measurement sensitivity is practically double that of a platinum resistor ($TC100 = 7.73 \cdot 10^{-3}$/K). However, the temperature-response curve is less linear than that of a metallic sensor. The measuring range is limited to approximately 150 °C by the material's intrinsic conductivity.

Applications in motor vehicles

These temperature sensors are used in many motor-vehicle applications. Here are some examples:

- Coolant-temperature sensor
- Fuel-temperature sensor
- Engine-oil temperature sensor
- Intake-air temperature sensor
- Charge-air temperature sensor
- Outside-temperature sensor
- Interior-temperature sensor
- Exhaust-gas temperature sensor, etc.

References

[1] DIN EN 60751: Industrial platinum resistance thermometers and platinum temperature sensors.

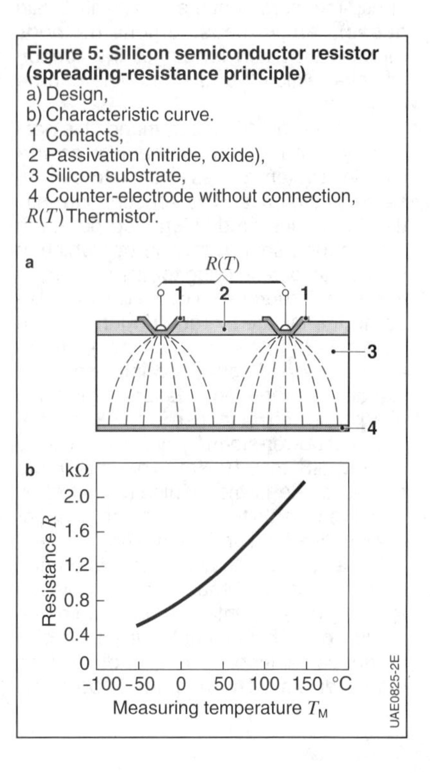

Figure 5: Silicon semiconductor resistor (spreading-resistance principle)
a) Design,
b) Characteristic curve.
1 Contacts,
2 Passivation (nitride, oxide),
3 Silicon substrate,
4 Counter-electrode without connection,
$R(T)$ Thermistor.

Torque sensor

Measured variables and application

Torque measurement is utilized in a variety of applications in motor vehicles. Torque sensors are used for example to sense the steering torque initiated by the driver.

Electromechanical power-steering systems are increasingly being used in vehicles. The essential benefits are simple installation and start-up in the vehicle, energy saving, and the suitability of these systems in the vehicle's ECU network for assistance systems to increase comfort and safety.

Measurement principle

When it comes to torque measurement, there are two different methods: angle and stress measurement. In contrast to stress-measurement methods (with strain gages), angle-measurement methods require a certain length of the torsion bar over which the torsion angle can be picked off.

To sense the driver command, it is necessary in an electromechanical power-steering system to measure the torque initiated by the driver. The sensors currently used in series production incorporate in the steering shaft a torsion bar which in the course of a steering torque applied by the driver undergoes a defined torsion linear to the initiated torque (Figure 1). The torsion is in turn measured by appropriate means and converted into electrical signals. The required measurement range of a torque sensor for use in an electromechanical power-steering system is usually around ±8 to ±10 Nm. The maximum torsion angle is mechanically limited by carrier elements to protect the torsion bar against overload and destruction.

To be able to measure the torsion and thus the applied torque, a magnetoresistive sensor is mounted on one side of the torsion bar which samples the field of a magnetic multipole wheel mounted on the other side. The number of poles on this wheel is chosen so that the sensor outputs a clear signal within its maximum measurement range to provide at all times a clear indication of the applied torque. The magnetoresistive sensor used supplies over the measurement range two signals which describe a sine signal and a cosine signal when shown against the torsion-bar torsion angle. The torsion angle and thus the torque are calculated in an ECU with the aid of an arc-tangent function.

Because the two signals are always permanently assigned over the defined measurement range, in the event of a deviation from these signals sensor faults can be detected and the necessary substitute measures initiated.

Figure 1: Torque sensor
a) Sensor module,
b) Measurement principle.
1 Torsion bar (torsion range internal),
2 Input shaft (from steering wheel),
3 Volute spring for electrical contacting,
4 Sensor module with magnetoresistive sensor chip and signal amplification,
5 Steering shaft,
6 Magnetic multipole wheel.

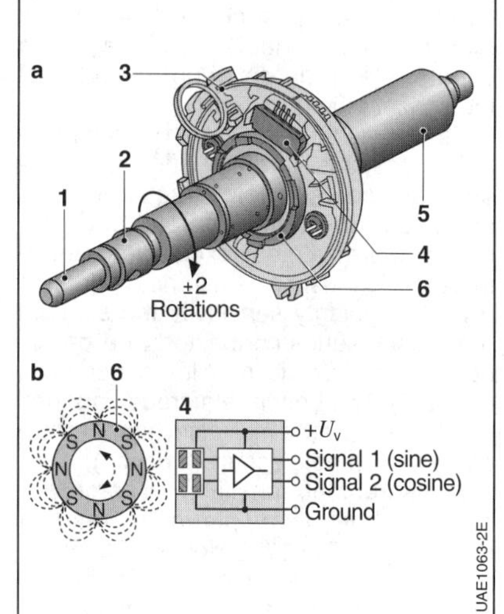

A volute spring with the required number of contacts is used to contact the sensor over the torsion range of approx. ±2 steering-wheel turns. This volute spring serves to deliver the supply voltage and transmit the measured values.

Figure 2: Measurement principle of iBolt force sensor
a) Conditions for $F_G < 850$ N (inside the measurement range),
b) Conditions for $F_G > 850$ N (outside the measurement range).
1 Swinging arm, 2 Air gap, 3 Sleeve,
4 Seat rail,
5 Double bending beam,
6 Magnet,
7 Hall IC.
F_G Weight,
F_R Bearing force.
The dashed lines indicate the flow of force in the force sensor.

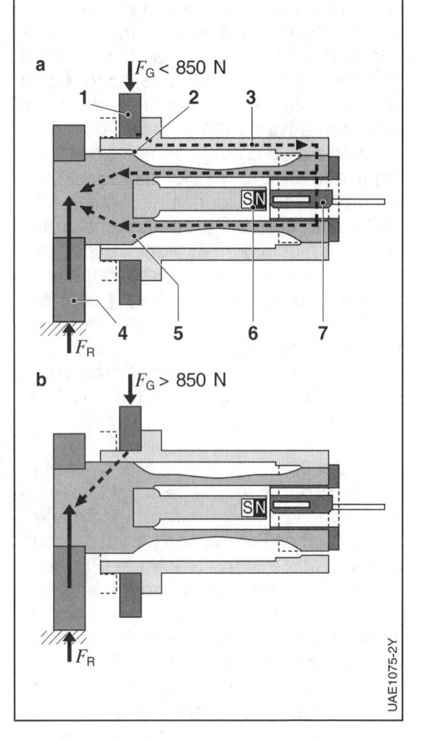

Force sensor

Measured variables and application

One application of force measurement in motor vehicles is the sensing of the front passenger's weight. Classifying the front passenger by measuring their weight enables the airbag to be deactivated when a small child is sitting in the seat.

Measurement principle

The operating principle of the iBolt sensor is based on measuring the deflection of a bending beam in response to the front passenger's weight. The extent of this deflection is detected by measuring the magnetic field strength with a Hall sensor (Figure 2a).

The sensor is designed in such a way that preferably the vertical component of the front passenger's weights causes a deflection of the bending beam. The magnet and the Hall IC are arranged in the sensor in such a way that the static magnetic field that penetrates the Hall IC produces an electrical signal linear to the deflection of the bending beam. The special design of the sensor prevents a horizontal deflection of the Hall IC in respect of the magnet in order to minimize the influence of transverse forces and torques. In addition, the maximum stress in the bending beam is limited by a mechanical overload stop (Figure 2b). This protects the sensor particularly in overload situations in the event of a crash.

The force generated by the front passenger's weight is directed from the upper seat structure via the sleeve into the bending beam (Figure 2a). The force is then directed from the bending beam into the lower seat structure. The bending beam is designed as a double bending beam as this has an S-shaped deformation line. Here the two vertical connecting points of the double bending beam remain vertical for the entire deflection range. This ensures a linear and parallel movement of the Hall IC in respect of the magnet, producing a linear output signal.

Gas and concentration sensors

λ sensors

Measurement principle

λ sensors measure the oxygen content in the exhaust gas. They are used to regulate the air/fuel ratio in motor vehicles. The name is derived from the excess-air factor λ. This factor indicates the ratio of the current air volume to the theoretical air volume which is required for a full combustion of the fuel. It cannot be determined directly in the exhaust gas, but only indirectly via the oxygen volume present in the exhaust gas or required to fully convert combustible components. λ sensors consist of platinum electrodes which are mounted on a ceramic solid electrolyte that conducts oxygen ions (e.g. ZrO_2).

The signal from all the λ sensors is based on electrochemical processes with the involvement of oxygen (see Electrochemistry). The platinum electrodes used catalyze the reaction of remnants of oxidizable components in the exhaust gas (CO, H_2 and hydrocarbons $C_xH_yO_z$) with exhaust-gas oxygen. λ sensors consequently measure not the real oxygen content in the exhaust gas, but instead that content which corresponds to the chemical equilibrium of the exhaust gas.

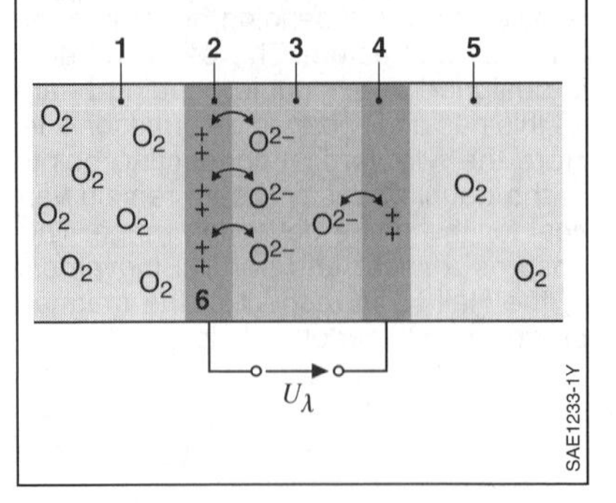

Figure 1: Nernst cell
1 Reference gas, 2 Anode,
3 Solid electrolyte of Y-doped ZrO_2,
4 Cathode,
5 Exhaust gas,
6 Residual charges on the anode.
O^{2-} Oxygen ion,
U_λ Sensor voltage (Nernst voltage).

Applications

λ sensors are used in gasoline engines during the regulation of a stoichiometric mixture (λ = 1) to deliver the lowest-pollutant exhaust gas possible. Optimum impact by the three-way catalytic converter is guaranteed in this range.

Two-point λ sensors (step-change or switching-type sensors) indicate whether the mixture is rich ($\lambda < 1$, excess fuel) or lean ($\lambda > 1$, excess air). With these sensors the oxygen partial pressure of stoichiometric air/fuel mixtures can be measured very accurately due to the steep part of the sensor characteristic curve in this environment. Outside this environment, however, the characteristic curve is very flat (Figure 2).

Only the large measurement range of wide-band λ sensors (from λ = 0.6 to clean air) allow their use in systems with direct injection in stratified-charge operation and in diesel engines. λ control that can be achieved with wide-band λ sensors results in significant system and emission advantages over two-point control with a step-change sensor and further applications such as, for example, more accurate monitoring of the catalytic converter and controlled component protection. The high signal dynamic response of wide-band λ sensors with response times of less than 100 ms improves λ control.

Operating concept

All the exhaust-gas sensors described in the following consist of two modules: Nernst cell and pump cells.

Nernst cell

The incorporation and removal of oxygen ions in the lattice of the solid electrolyte is dependent on the oxygen partial pressure on the surface of the electrode (Figure 1). Thus at low partial pressure more oxygen ions leave than enter. The vacant spaces in the lattice are reoccupied by advancing oxygen ions. An electric field is generated on account of the resulting charge separation at different oxygen partial pressures at the two electrodes. The electric field forces push advancing oxygen ions back and an equilibrium develops at the so-called Nernst voltage.

Pump cell

When a voltage is applied that is less or greater than the developing Nernst voltage, this state of equilibrium can be changed and oxygen ions actively transported through the ceramic. This creates between the electrodes a current, borne by oxygen ions. Crucial to the direction and intensity is the difference between the applied voltage (pump voltage U_P) and the developing Nernst voltage. This process is called electrochemical pumping.

Two-point λ sensors

Design

Two-point λ sensors indicate whether the mixture is rich ($\lambda < 1$, excess fuel) or lean ($\lambda > 1$, excess air). With these sensors the oxygen partial pressure of virtually stoichiometric air/fuel mixtures can be measured very accurately due to the steep part of the characteristic curve in this environment and the lowest-pollutant exhaust gas possible achieved by regulating the fuel quantity.

Figure 2: Characteristic curve of a two-point λ sensor at different sensor-element temperatures

U_λ Sensor voltage,
$p_A(O_2)$ Oxygen partial pressure in the exhaust gas,
λ Excess-air factor.
1 U_λ at 500 °C,
2 U_λ at 700 °C,
3 U_λ at 900 °C,
4 $p_A(O_2)$ at 500 °C,
5 $p_A(O_2)$ at 700 °C,
6 $p_A(O_2)$ at 900 °C.

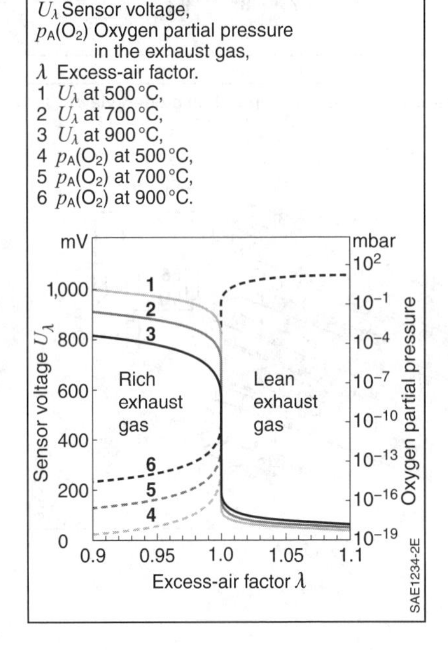

Operating concept

The operating concept is based on the principle of a Nernst cell (Figure 1). The useful signal is the Nernst voltage U_λ that develops between an electrode exposed to the exhaust gas and an electrode exposed to the reference gas. The characteristic curve is very steep at $\lambda = 1$ (Figure 2). In lean mixtures the Nernst voltage increases linearly with the temperature. In rich mixtures, on the other hand, the influence of temperature on the oxygen partial pressure dominates in the state of equilibrium. The higher the temperature, the higher this oxygen partial pressure.

The establishment of equilibrium at the exhaust-gas electrode is also the cause of very small deviations of the λ jump from the exact value. The exhaust-gas electrode is covered with a porous protective ceramic layer to protect against contamination and to encourage the establishment of equilibrium by limiting the number of arriving gas particles. Hydrogen and oxygen diffuse through the porous protective layer and are converted on the electrode. More oxygen must be available on the protective layer to ensure that the faster-diffusing hydrogen is fully converted on the electrode; in all there must be a slightly lean mixture in the exhaust gas. The characteristic curve is therefore shifted in the lean direction. This "λ shift" is electronically compensated during control.

To deliver the signal a reference gas is required that is separated (gas-tight) from the exhaust gas by the ZrO_2 ceramic. Figure 3 depicts the structure of a planar sensor element with reference-air passage. Ambient air is used as reference gas in this type. Figure 4 shows the element in the sensor housing. The exhaust-gas and reference-gas sides are separated from each other (gas-tight) by means of a packing seal. The reference-gas side in the housing is permanently supplied with reference air via the supply leads.

Systems with a "pumped" reference are increasingly being used as an alternative to reference air. Pumping refers here to the active transportation of oxygen in the ZrO_2 ceramic by injecting a current, in which the current is set so low that it does not interfere with the actual measurement. The reference electrode itself is connected via a tighter output in the element to the reference-gas chamber. This results in a build-up of excess oxygen pressure at the reference electrode. This system offers additional protection against unwanted gas components penetrating into the reference-gas chamber.

Robustness

The ceramic sensor element is protected by a protective tube ahead of the direct exhaust-gas flow (Figure 4). This contains openings through which only a small amount of the exhaust gas is directed to the sensor element. It prevents heavy thermal stresses from the exhaust-gas flow and at the same time provides mechanical protection for the ceramic element.

The housing must meet stringent temperature requirements, meaning that high-quality materials must be used. Temperatures in excess of 1,000 °C can be recorded in the exhaust gas, as well as 700 °C at the hexagon head and up to 280 °C at the cable outlet. For this reason, only ceramic and metallic materials are used in the hot area of the sensor.

Most two-point λ sensors are also fitted with a heater element (Figure 3). This quickly heats (FLO, fast light-off) the sensor element to operating temperature and provides for fast control readiness.

Figure 4: Two-point λ sensor, sensor element in the housing
1 Protective tube,
2 Sensor element,
3 Hexagon head,
4 Reference gas,
5 Electrical supply lead,
6 Exhaust-gas side,
7 Packing seal,
8 Support ceramic,
9 Contacting.

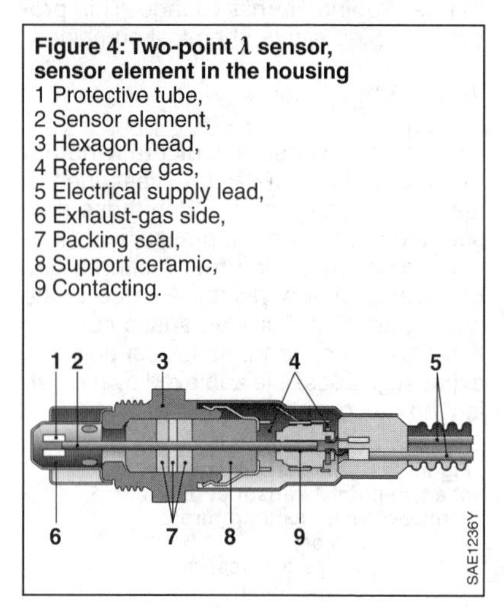

Figure 3: Structure of a planar two-point λ sensor with protective circuit (exploded view)
The vertical lines symbolize conductive connections.
1 Exhaust gas,
2 Porous protective layer,
3 Outer electrode (platinum electrode),
4 ZrO_2 ceramic with Nernst cell,
5 Reference electrode (platinum electrode),
6 Al_2O_3 insulation layer,
7 Heater element,
8 Reference air.
U_λ Sensor voltage,
U_H Heating voltage.

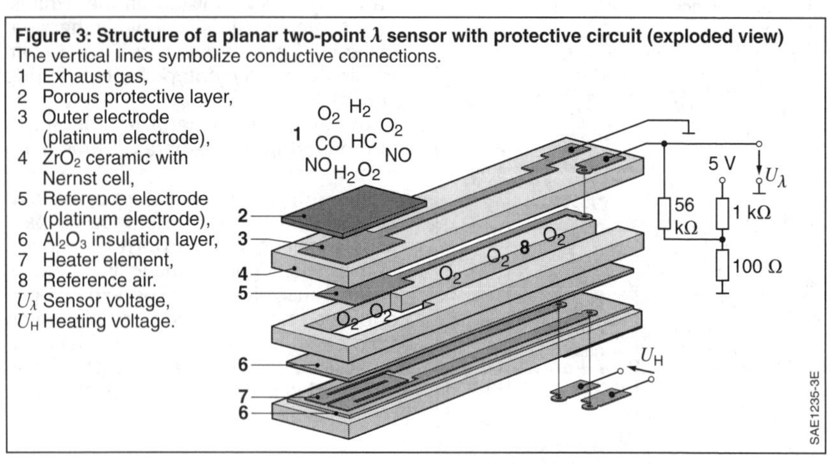

In practice the λ sensor after engine starting is switched on only after a delay. Water that is produced as a combustion product and condenses again in the cold exhaust-gas system is transported by the exhaust gas and can reach the sensor element. If a droplet of this water touches a hot sensor element, it vaporizes immediately and abstracts a large amount of heat locally from the sensor element. The heavy mechanical stresses that occur as a result of thermal shock can cause the ceramic sensor element to break. For this reason, the sensor is often switched on only after the exhaust-gas system has been sufficiently heated. In more recent developments the ceramic elements are enclosed in a further porous, ceramic layer which significantly increases robustness against thermal shock. When a water droplet touches the elements, it spreads in the porous layer. Local cooling is spread over a wider area and mechanical stresses are reduced.

Protective circuit

Figure 3 shows the protective circuit of a two-point λ sensor. Because the sensor when cold cannot generate a signal due to the absence of conductivity of the ZrO_2 ceramic, it is connected via a resistor to a voltage divider. In the cold state the sensor signal is therefore 450 mV, the value of a stoichiometrically combusted gas ($\lambda = 1$). As the temperature increases the sensor is able to develop the Nernst voltage. After approximately 10 s the sensor is at sufficiently high temperature to indicate externally specified lean/rich changes. It is then possible to switch to closed-loop control in the vehicle.

Types

Two-point λ sensors come in different types. The sensor elements can be shaped in the form of a finger with a separate heater element or as a planar element with an integrated heater element which is manufactured using film technology (Figure 3).

Wide-band λ sensor

Design and function

With the two-point λ sensor the oxygen partial pressure of stoichiometric air/fuel mixtures can be measured very accurately in the steep part of the characteristic curve. However, where there is excess air ($\lambda > 1$) or excess fuel ($\lambda < 1$), the characteristic curve is very flat (Figure 2).

The large measurement range of wide-band λ sensors ($0.6 < \lambda < \infty$) enables them to be used in systems with direct injection and stratified-charge operation and in diesel engines. A continuous control concept that can be achieved with a wide-band λ sensor delivers significant system advantages, such as for example controlled component protection. The high signal dynamic response of wide-band λ sensors ($t_{63} < 100$ ms) improves exhaust emissions to the point of low-emission vehicles, which necessarily calls for measures such as individual-cylinder closed-loop control.

Design and operating principle

The Nernst principle is reversed to wide the measurement range. When a voltage greater than the Nernst voltage is applied to the measuring cell, oxygen is transported as oxygen ions through the pump-cell ceramic to the internal pump and discharged there as free, molecular oxygen into the reference-air chamber. The resulting pump voltage is obtained from the difference between pump voltage (U_P) and the developing Nernst voltage.

The inflow of oxygen molecules is limited in order to establish a linear connection with the externally applied oxygen concentration. To this end the admission of exhaust-gas is limited by a porous, ceramic structure with specifically set pore radii, the so-called diffusion barrier.

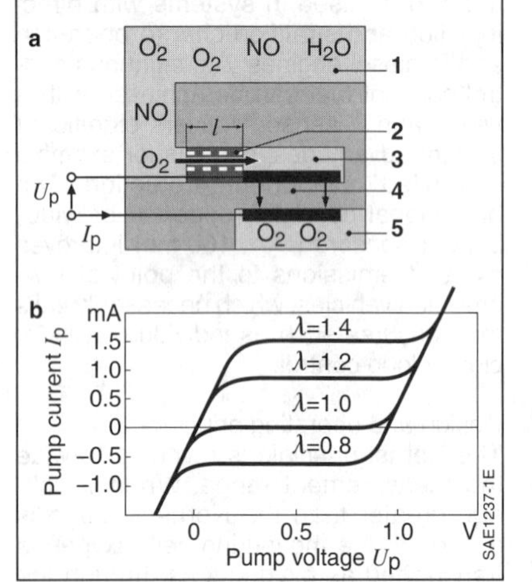

Figure 5: Once-cell λ sensor in lean exhaust gas
a) Cross-section,
b) Characteristic curves.
l Length of diffusion barrier.
1 Lean exhaust gas,
2 Diffusion barrier,
3 Cavity,
4 Pump cell,
5 Reference air.
The arrows in the pump cell indicate the pumping direction.

The pump current I_P that flows at sufficient pump voltage is, on account of the law of diffusion, directly proportional to the partial pressure in the exhaust gas:

$$\frac{I_P}{4F} = I_M = \frac{A\,D(T)}{R\,T\,l}\left(p_A(O_2) - p_H(O_2)\right).$$

Here $p(O_2)$ is the oxygen partial pressure in the exhaust gas ($p_A(O_2)$) or in the cavity ($p_H(O_2)$), T the temperature, $D(T)$ the temperature-dependent diffusion constant of the diffusion barrier, A its cross-sectional area, and l its length (Figure 5). F is the Faraday constant (F = 96485.3365 C/mol), R the general gas constant (R = 8.3144621 J/mol·K).

If rich exhaust gas is present, the created Nernst voltage of approx. 1,000 mV counteracts the applied pump voltage such that the resulting negative voltage pumps oxygen in the opposite direction into the cavity and thus the linear shape of the characteristic curve is extended into the rich range. The oxygen is obtained for this purpose from the reduction of water and CO_2 on the outer electrode.

The drawback of this simple form of wide-band λ sensor is that the fixed pump voltage must be sufficient in rich exhaust gas to pump oxygen into the cavity and in

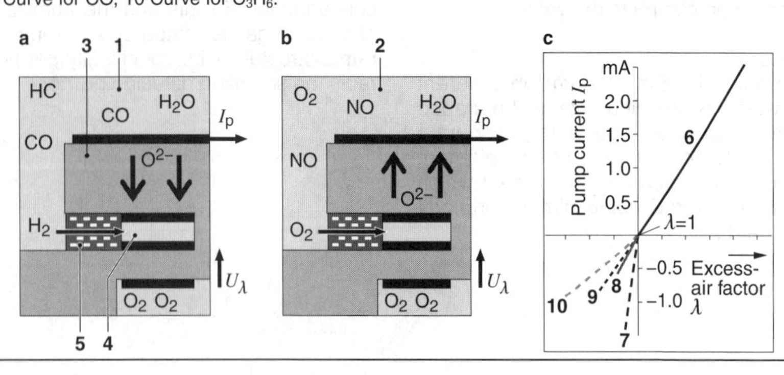

Figure 6: Two-cell λ sensor in rich and lean exhaust gas
a) and b) Cross-section,
c) Sensor characteristic curves.
Depending on the polarity of pump current I_p predominantly reducing exhaust-gas constituents (section a) or oxygen (section b) diffuse through the diffusion barrier.
For λ < 1 the characteristic curve (section c) is dependent on the exhaust-gas composition, here the characteristic curves of individual exhaust-gas constituents are plotted.
1 Rich exhaust gas, 2 Lean exhaust gas, 3 Pump cell, 4 Nernst cell, 5 Diffusion barrier,
6 Curve for O_2, 7 Curve for H_2, 8 Curve for CH_4,
9 Curve for CO, 10 Curve for C_3H_6.

lean exhaust gas to pump oxygen out of it. The internal resistance of the pump cell should therefore be very low. In addition, the measurement range in the rich range is limited by the molecule inflow in the reference passage. An oxygen deficiency in the reference caused by a reference-air contamination or over-range cannot be clearly distinguished. In the case of exhaust-gas exchanges the dynamic response of the one-cell sensor is limited by recharging of the electrode capacities when the pump voltage is changed.

To eliminate these drawbacks, the Nernst cell familiar from the step-change sensor is combined with the oxygen pump cell described above. In the case of rich exhaust gas (Figure 6a), in response to reversing the voltage at the outer pump electrode, oxygen is generated from H_2O and CO_2, transported inwards through the ceramic, and discharged again in the cavity. There the oxygen reacts with the diffusing, reducing exhaust-gas components. The created inert reaction products H_2O and CO_2 diffused through the diffusion barrier in an outward direction.

The oxygen partial pressure in the cavity is measured by the Nernst cell also connected to the cavity and the pump voltage is corrected by a controller (Figure 7) to a specified reference voltage (e.g. 450 mV). There is then an oxygen partial pressure of 10^{-2} Pa in the cavity.

Because the diffusion limit current increases with the sensor temperature, the temperature must be kept as constant as possible. The highly temperature-dependent resistance of the Nernst cell is measured for this purpose. The operating electronics regulate it by means of the heater element operated with a pulse-width-modulated voltage.

Where the exhaust gas is rich the different diffusion coefficients, which are functions of the masses of the gas molecules, become noticeable through different sensitivities (Figure 6). The signal is therefore used in the ECU for a characteristic curve applied for the relevant gas composition.

The sensor's diffusion limit current and thus the sensitivity are very much dependent on the geometry of the diffusion barrier. It is necessary to adjust the pump current in order to achieve the high accuracy requirement in the manufacturing tolerance. In some sensor types this is done by a hybrid resistor in the sensor plug which acts together with the measuring resistor as current divider. Alternatively, the diffusion limit current can be adjusted already in the sensor-element production process by means of specific openings so that there is no need to make an adjustment at the plug. For subsequent calibration of the sensor in the vehicle the oxygen concentration of the air can be measured in overrun conditions and thus the characteristic curve corrected in the ECU.

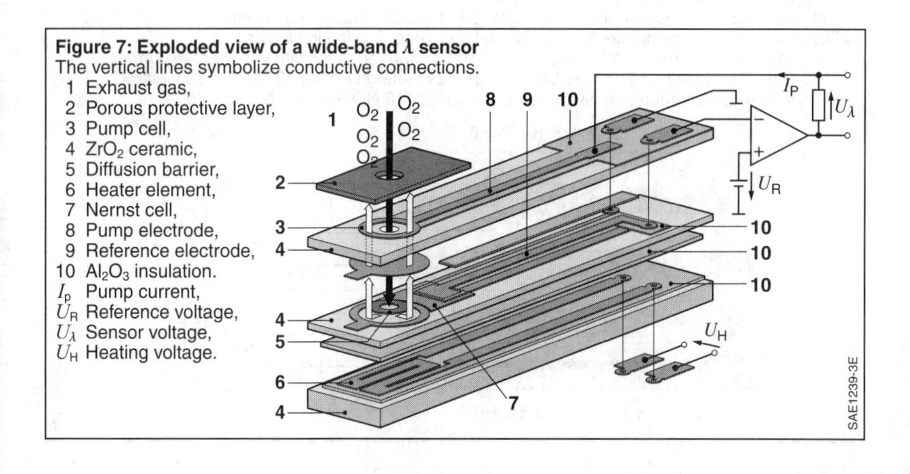

Figure 7: Exploded view of a wide-band λ sensor

The vertical lines symbolize conductive connections.

1 Exhaust gas,
2 Porous protective layer,
3 Pump cell,
4 ZrO_2 ceramic,
5 Diffusion barrier,
6 Heater element,
7 Nernst cell,
8 Pump electrode,
9 Reference electrode,
10 Al_2O_3 insulation.
I_p Pump current,
U_R Reference voltage,
U_λ Sensor voltage,
U_H Heating voltage.

NO_x sensor

Application

NO_x sensors are used in the denitrification systems of diesel and gasoline engines. In systems with diesel engines they are installed upstream and downstream of SCR catalytic converters (Selective Catalytic Reduction) and downstream of NO_x storage catalytic converters (NSC, NO_x storage catalysts). In systems with gasoline engines they are used downstream of the NO_x storage catalytic converter.

In these positions the NO_x sensors determine the nitrogen-oxide and oxygen concentrations in the exhaust gas and downstream of SCR catalytic converters also the ammonia concentration as a cumulative signal. In this way the engine-management system receives the current residual concentration of nitrogen oxides, provides for exact metering, and detects any faults in the exhaust system.

Nitrogen oxides are stored as nitrates in NO_x storage catalytic converters. In short rich phases the catalytic converter is regenerated, whereby the nitrates are reduced with the aid of carbon monoxide and hydrogen to nitrogen.

Design and operating principle

The NO_x sensor in Figure 8 is a planar three-cell limit-current sensor. A Nernst concentration cell and two modified oxygen pump cells (oxygen pump cell and NO_x cell), as are already familiar from wide-band λ sensors, form the overall sensor system.

The sensor element consists of several oxygen-conducting, ceramic solid-electrolyte layers insulated against each other, on which six electrodes are mounted. The sensor is fitted with an integrated heater element which heats the ceramic to an operating temperature of 600 °C to 800 °C.

The outer pump electrode exposed to the exhaust gas and the inner pump electrode in the first cavity, which is separated from the exhaust gas by a diffusion barrier, form the oxygen pump cell.

Figure 9: Characteristic curve of the nitrogen-oxide signal

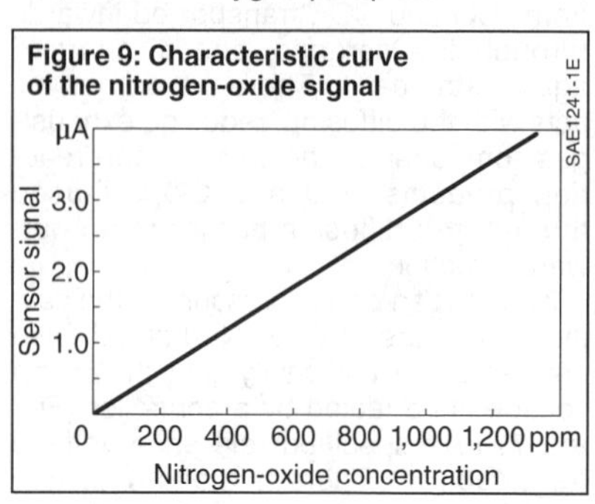

Figure 8: Cross-section of a NO_x sensor

A Oxygen pump cell, B Nernst cell, C NO_x cell.

1 Diffusion barrier 1, 2 Outer pump electrode, 3 Porous aluminum-oxide layer, 4 First cavity, 5 Inner pump electrode, 6 Nernst electrode, 7 Diffusion barrier 2, 8 Second cavity, 9 Common return conductor, 10 Oxygen regulator with voltage transformer, 11 NO current amplifier and voltage transformer, 12 Heater element, 13 Reference electrode, 14 Reference-gas chamber, 15 NO_x counter-electrode, 16 NO_x pump electrode.

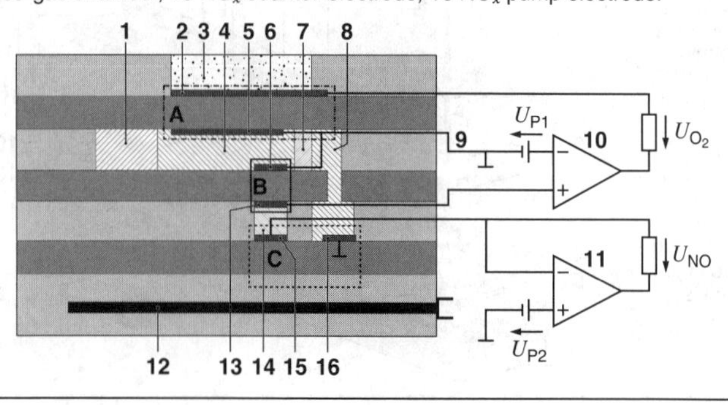

The first cavity contains the Nernst electrode, while the reference-gas chamber contains the reference electrode. Together they form the Nernst cell. These are the functional components which are identical to those of wide-band λ sensors.

In addition, there is a third cell: the NO_x pump electrode and its counter-electrode. The former is situated in a second cavity, which is separated from the first by a further diffusion barrier, the latter is located in the reference-gas chamber. All the electrodes in the first and second cavities have a common return conductor.

The inner pump electrode is, unlike the inner pump electrode in wide-band λ sensors, greatly limited in its catalytic activity by the alloying of platinum with gold. The applied pump voltage U_{P1} is only enough to split (dissociate) oxygen molecules. NO is only minimally dissociated at the adjusted pump voltage and passes the first cavity with low losses. NO_2 as a strong oxidizing agent is converted on the inner pump electrode directly into NO. Ammonia reacts to this in the presence of oxygen and at temperatures of 650 °C into NO and water.

Because of the higher voltage at the NO pump electrode and its catalytically improved activity due to the admixture of rhodium, NO is completely dissociated at this electrode and the oxygen is pumped away through the solid electrolyte.

Electronics

In contrast to the other ceramic exhaust-gas sensors, the NO_x sensor is equipped with evaluation electronics (SCU, sensor control unit). It delivers via CAN bus the oxygen signal, the NO_x signal, and in each case the status of these signals.

Incorporated in the evaluation electronics are a microcontroller, an ASIC (Application-Specific Integrated Circuit) for operating the oxygen pump cell, and a high-precision instrument amplifier for the very small NO signal currents.

Characteristic curves

The oxygen signal is 3.7 mA for air. The oxygen characteristic curve is identical to that familiar from wide-band λ sensors (Figure 6c). The NO_x characteristic curve is shown in Figure 9.

Particulate sensor

The emission regulations in the USA and in Europe necessitate the use of diesel particulate filters (DPF) in diesel vehicles. Particulate sensors in addition to the differential pressure sensor are required to comply with future on-board diagnosis legislation (OBD), which places more stringent demands on the monitoring of the functional capability of these particulate filters. The particulate sensors measure the soot emissions downstream of the particulate filter.

Figure 10: Exploded view of the particulate sensor

1 Exhaust gas,
2 Interdigital structure with two comb-shaped electrodes,
3 Ceramic,
4 Insulation,
5 Heater element,
6 Platinum meander.
U Direct voltage, e.g. 60 V,
I_S Sensor current,
U_H Heating voltage.

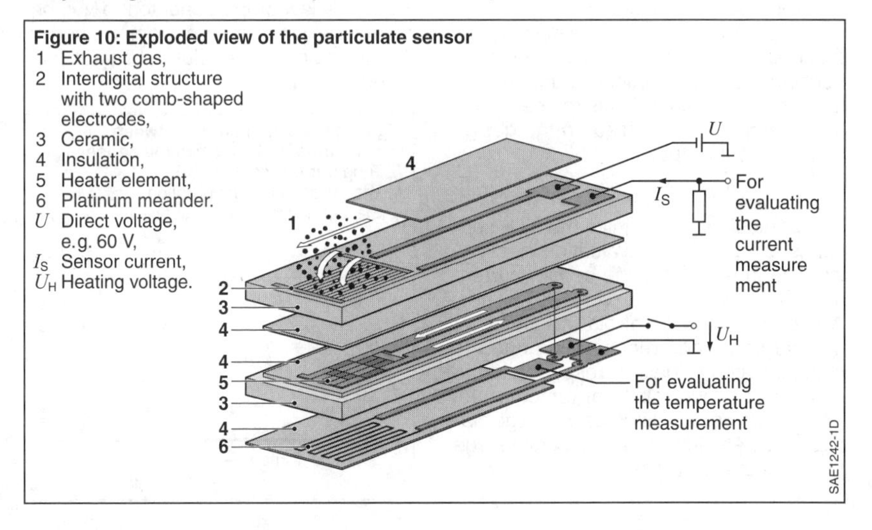

Design
The sensor element consists of an interdigital structure with two comb-shaped platinum electrodes on a ceramic substrate with an integrated heater element, as is familiar from the λ sensor (Figure 10). A platinum meander serves to measure the temperature of the sensor element.

Operating principle
The particulate sensor is a resistive sensor. A direct voltage of for example 60 V is applied at the interdigital structure with initially very high electrical resistance. As a result of the field forces soot particles from the exhaust gas collect on the interdigital electrodes and form increasingly conductive soot paths between the two comb-shaped structures. This produces a monotonously increasing current between the electrodes (Figure 11). A predefined current threshold value is reached after a certain collection time. This triggers regeneration, in which the initial state is re-established when the sensor element is heated and the soot is burned off at temperatures above 600 °C.

The time between the start of measurement and the start of regeneration is defined as the trigger time. An integrated temperature-measurement meander is used to control the temperature to ensure that a controlled regeneration of the sensor element takes place.

The particulate-sensor element is, like that of the λ sensor, installed in a sensor housing. Because particulate deposits on the sensor element are desired here, particular emphasis is placed on the design of the protective tube.

Electronics
The particulate sensor, like the NO_x sensor, is equipped with evaluation electronics. The latter supply the sensor current and the signal status via a CAN bus. A microcontroller controls the timing of measurement and regeneration and the compensation of data in response to the measured sensor-element temperature. The electronics also include a voltage stabilizer, a CAN driver, and the output stage for the heater element.

Hydrogen sensors
The use of fuel cells in motor vehicles necessitates the incorporation of hydrogen sensors. This sensor system has two functions – safety monitoring with leak detection (measurement range 0 to 4 % hydrogen content in air) and process control with adjustment of operating conditions (measurement range 0 to 100 %).

The most common measurement principles are described in descending frequency of use. For the most part different methods are combined to reduce cross-sensitivity to other gases or to increase sensitivity and widen the measurement range [3].

Electrochemical measurement principle
In the case of amperometric measurement a constant voltage is applied to an electrolyte that conducts only protons (e.g. sulfonated tetrafluoroethylene polymer). The electrolyte is covered by a diffusion barrier which is very selective in admitting only hydrogen. These sensors thus have long-term stability. At the electrode hydrogen splits into protons, which flow as pump current through the electrolyte. The electrons flow through the electrodes, driven by the applied voltage, via the outer electric circuit to the other side of the electrolyte. Here they recombine with oxygen on the electrode to form water. The current intensity is according to Faraday's law proportional to the number of protons, which in turn is determined by the diffusion by the diffusion barrier (see λ sensor).

Figure 11: Current curve between the electrodes of a particulate sensor
I_A Trigger threshold,
t_A Trigger point for regeneration,
t_B Start of next measurement cycle.

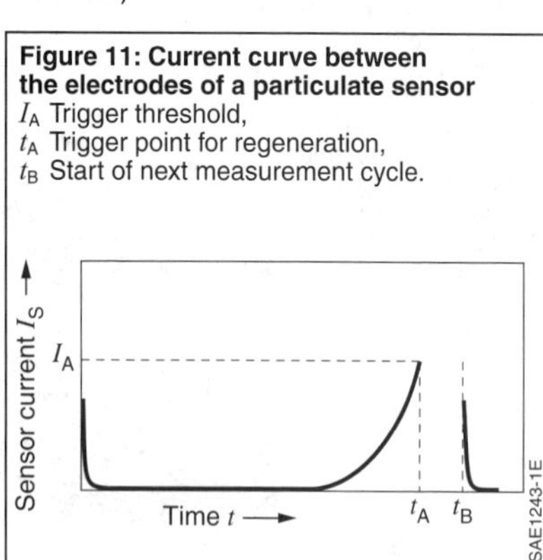

Potentiometric measurement methods utilize the potential difference between two electrodes. One electrode with a platinum or palladium catalyst is situated in the hydrogen environment, the other electrode is situated for example in air. According to the Nernst equation a potential difference develops (see Fuel cell). The potentiometric sensors are cross-sensitive to carbon monoxide that covers the catalyst. The age quickly and have a high drift. The measurement range is between 100 ppm and 100%.

Resistance-based measurement principle

Palladium and semiconductor metal oxides (e.g. SnO_2) exhibit a change in resistance after the adsorption of hydrogen. The semiconductor metal-oxide sensor (Figure 12) consists of a gas-filtering diaphragm which should only admit hydrogen. The sensitive layer absorbs the hydrogen and induces the change in resistance, which is measured by a Wheatstone bridge.

A MOSFET (see Field-effect transistor) can also be used with a palladium gate as a variant. These alters the current between source and drain proportionally to the hydrogen absorbed at the gate. In a Schottky-diode arrangement (see Schottky diode) the breakdown voltage decreases when hydrogen is adsorbed.

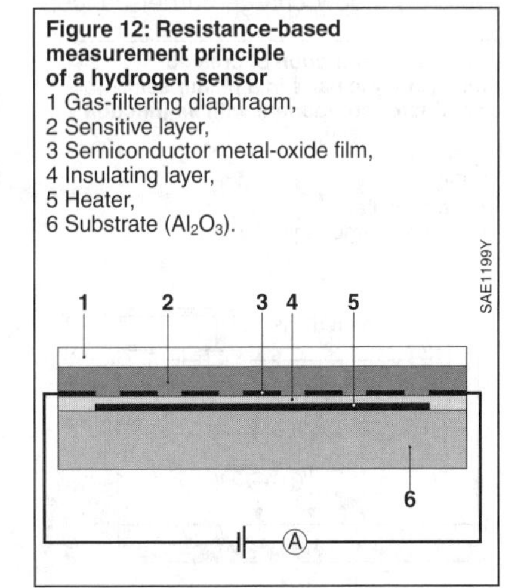

Figure 12: Resistance-based measurement principle of a hydrogen sensor
1 Gas-filtering diaphragm,
2 Sensitive layer,
3 Semiconductor metal-oxide film,
4 Insulating layer,
5 Heater,
6 Substrate (Al_2O_3).

The measurement range for the resistance-based measurement principles is between 10 ppm and 2%.

Catalytic measurement principle

Heat is generated when adsorbed gases oxidize on the surface of a catalyst. In a pellistor a wire consisting of catalyst material (e.g. platinum or palladium) exposed to the hydrogen is heated and as a result undergoes a change in resistance dependent on the hydrogen concentration, while the temperature and resistance of a reference wire remain unchanged. The resistors are integral parts of a Wheatstone-bridge circuit, with which the change in resistance is measured and from this the hydrogen concentration is inferred.

In the thermoelectric method the voltage build-up due to the temperature difference (Seebeck effect) is utilized.

The catalytic methods are cross-sensitive to other oxidizable gases and require a minimum quantity of oxygen (5 to 10%) as the oxidation partner; the measurement range therefore is between 1% and 90 to 95% hydrogen concentration.

References for gas and concentration sensors

[1] Nernstgleichung: Zeitschrift für physikalische Chemie, IV. Volume Book 1, Verlag Wilhelm Engelmann, 1889, Published by W. Ostwald, J.H. Van't Hoff, W. Nernst: Die elektromotorische Wirksamkeit der Ionen.

[2] T. Baunach, K. Schänzlin, L. Diehl. Sauberes Abgas durch Keramiksensoren. Physik Journal 5 (2006) No. 5.

[3] T. Hübert et al.: Hydrogen sensors – A review, Sensors and Actuators B 157 (2011) 329 – 352.

Optoelectronic sensors

Internal photoelectric effect

The internal photoelectric effect forms the basis of optoelectronic sensor elements. Light can be seen as a stream of individual light quanta (photons). The energy E_{Ph} of a photon is only dependent on its frequency f or its wavelength λ:

$$(1)\quad E_{Ph} = hf = \frac{hc}{\lambda}$$

where
h Planck's action quantum,
c Speed of light.

When photons strike atoms, they can each release an electron from the outer electron shell if the energy is sufficient. The energy required for release corresponds to the difference between the energy level E_V of the valence band of the atom and the level E_L of the conduction band, i.e. the band gap E_g.

$$(2)\quad E_g = E_L - E_V .$$

To release an electron, the photon energy E_{Ph} must therefore be greater than the band gap E_g. In a pure semiconductor, charge-carrier pairs (electrons and holes) are created by the absorption of photons. The band gap to be overcome of, for example, silicon is $E_g = 1.12$ eV at room temperature. Without special measures the charge-carrier pairs created recombine already after a short time. However, the radiation created in the process is not in the visible range in the case of silicon.

In highly doped semiconductors, the above-mentioned intrinsic photoelectric effect is supplemented by the extrinsic photoelectric effect. Because the energy gap to be overcome is significantly lower in such extrinsic sensors, they are also suitable for radiating a larger wavelength (infrared range).

No further release occurs for energies $E_{Ph} < E_g$. According to equation (1), this corresponds in the case of silicon to a limit wavelength of $\lambda_g = 1.1$ µm (near infrared). Light with larger wavelengths or lower frequency is not longer absorbed; silicon becomes transparent here.

Light-sensitive sensor elements

Photoresistors

Incident light causes charge-carrier pairs to be created in a sensor configured as a resistor (LDR, light-dependent resistor); these charge-carrier pairs increase the conductance G. They recombine after a short time (milliseconds range); but nevertheless the charge-carrier concentration increases in stationary equilibrium with the illuminance E according to the following law:

$$(3)\quad G = \text{const}\cdot E^{\gamma} \quad \text{where } \gamma = 0.7 \text{ to } 1.$$

The light-sensitive materials usually used are cadmium sulfide CdS ($E_g = 1.8$ eV; $\lambda_g = 0.7$ µm) and cadmium selenide CdSe ($E_g = 1.5$ eV; $\lambda_g = 0.8$ µm) on ceramic substrates.

Semiconductor pn junctions

There is essentially no difference between a photocell, a photodiode, and a phototransistor. They all use the photocurrent or no-load voltage in illuminated pn semiconductor contacts as the measuring effect. The elements differ, however, in the way they operate.

Charge carriers which are created by the internal photoeffect in the depletion layer of a pn semiconductor contact (Figure 1) are immediately accelerated by the electric field in the space-charge region with low charge-carrier concen-

Figure 1: Separation of created electron-hole pairs in a planar semiconductor component with pn junction
1 Optical coating,
2 Contact,
3 SiO_2,
4 Metal contact,
5 Space-charge region.

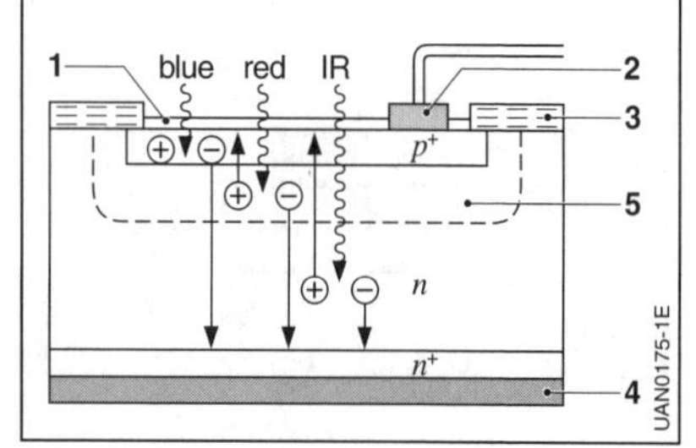

tration, as a result of which the charge carriers are immediately split after they have been created (drift current). Their recombination is practically prevented and photosensitivity is significantly increased.

Photocells

Photocells are operated without external bias and can be operated both at no load (photovoltaic effect) and in a short-circuit. They have accordingly a low background noise and thus a high detection capability.

The characteristic curves applicable to these operating modes (Figure 2) can be easily derived as special cases from a diode polarized in the forward direction with the voltage U with the thermally conditioned off-state saturation current I_S and the photocurrent I_{ph} also flowing in reverse direction:

(4) $I = I_S\, e^{\frac{eU}{kT}} - I_S - I_{ph}$

where
e Elementary charge,
k Boltzmann's constant,
T Absolute temperature.

Special cases:

(5) $U = 0$ (short-circuit)
$\rightarrow I = I_K = -I_{ph}$;

(6) $I = 0$ (no load):

$$\rightarrow U = U_L = \frac{kT}{e} \cdot \ln\left(\frac{I_{ph}}{I_S} + 1\right).$$

Photocells are usually configured with a very large irradiation-sensitive surface, and provide accordingly relatively high photovoltaic current also (e.g. I_{ph} = 250 µA at E = 1,000 lx). Their time constant is proportionally high and is typically approximately 20 ms.

Photodiodes and phototransistors

Photodiodes are operated with constant bias U_S in the reverse direction, where the photocurrent flowing as reverse current is linearly dependent on the illuminance E (Figure 3). The space-charge region is increased in size by the applied

Figure 3: Characteristic curves for a photodiode for constant illuminance E

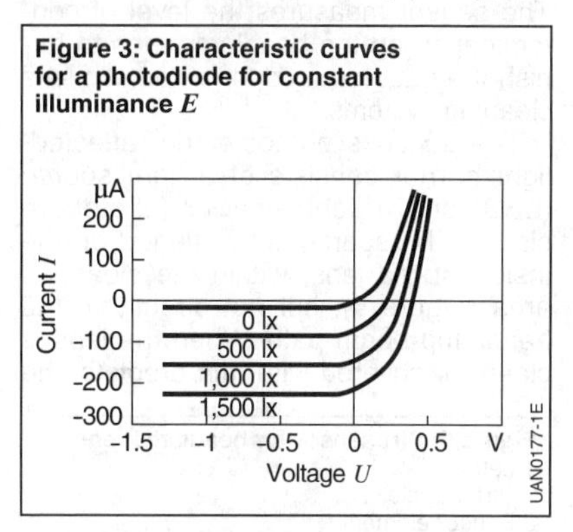

Figure 2: Characteristic curves for a photocell as a function of illuminance E
a) Short-circuit current I_K,
b) Open-circuit voltage U_L.

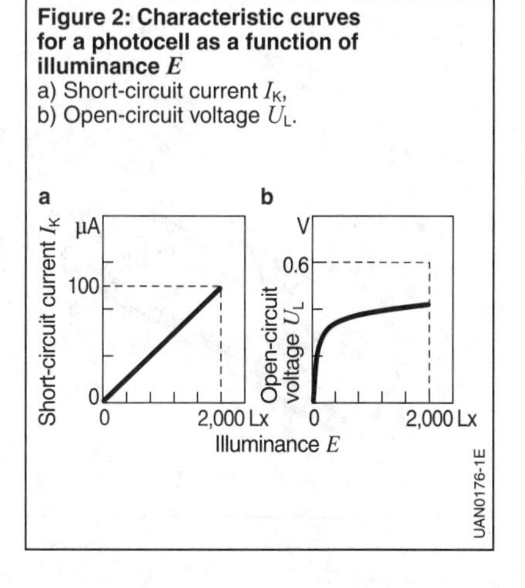

Figure 4: Characteristic curves for a phototransistor for constant illuminance E

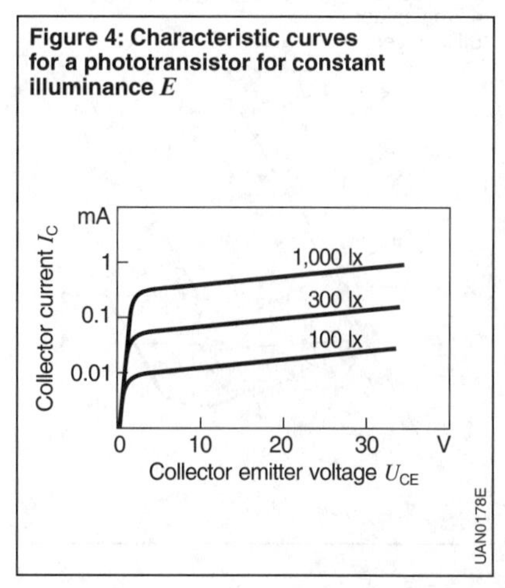

reverse voltage. This reduces the junction capacitance in such a way that the cutoff frequency of such a photodiode is typically a few MHz.

In the case of the phototransistor shown in Figure 4 (*npn* type), the collector-base diode polarized in the reverse direction acts as a photodiode. In this way, the collector, like every transistor, supplies a photocurrent higher by the current-amplification factor B ($\approx$ 100 to 500) (corresponding to base current). However, the higher sensitivity comes at the expense of a slightly worse frequency dynamic and a slightly worse thermal characteristic.

Application

Dirt sensor

The sensor measures the level of contamination on the headlamp lens to furnish the data required for automatic lens cleaning systems.

The sensor's photoelectric reflected-light barrier consists of a light source (LED) and a light receiver (phototransistor). The source is positioned on the inside of the lens, within the cleansed area (Figure 5), but not directly in the headlamp beam path. When the lens is clean, or covered with rain droplets, the infrared measurement beam emitted by the unit passes through the lens without being obstructed. Only a minuscule part is reflected back to the light receiver. However, if it encounters dirt particles on the outer surface of the lens, it is reflected back to the receiver at an intensity that is proportional to the degree of contamination and automatically activates the headlamp washer unit once a defined level is reached.

Rain sensor

The rain sensor recognizes rain droplets on the windshield, so that the windshield wipers can be triggered automatically. The driver, however, can still use the manual controls.

The sensor consists of an optical transmission and reception path (similar to the dirt sensor). However, the light is directed toward the windshield at an angle. A dry outer surface reflects all the light (total reflection) back to the receiver which is also mounted at an angle (Figure 6). When water droplets are present on the outer surface, a substantial amount of the light is refracted outward, thus weakening the return signal. The windshield wiper also responds to dirt once the activation threshold is exceeded.

Figure 5: Dirt sensor for headlamp lens
1 Lens,
2 Dirt particle,
3 Sensor element,
4 Transmitter,
5 Receiver.

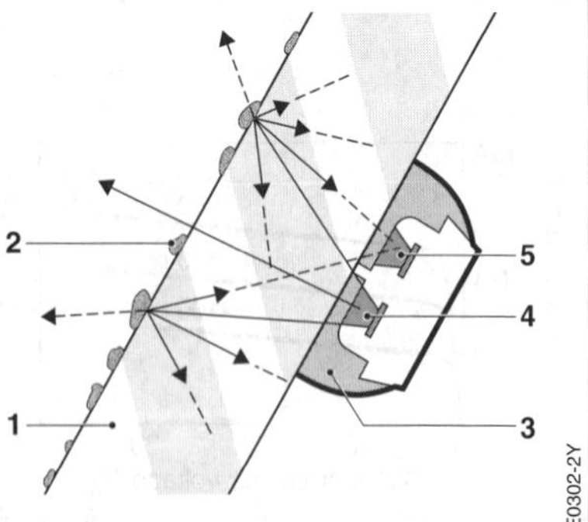

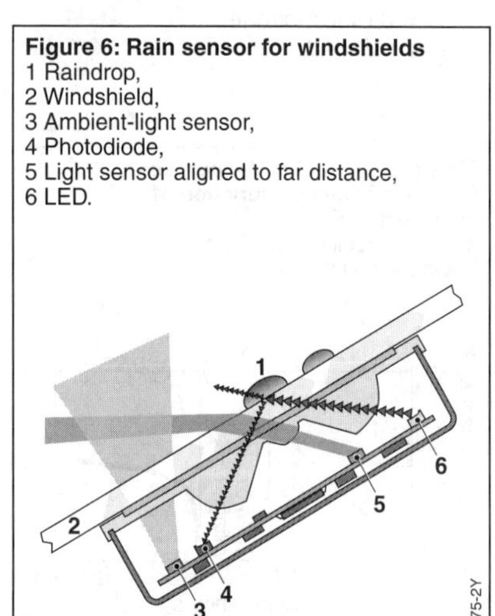

Figure 6: Rain sensor for windshields
1 Raindrop,
2 Windshield,
3 Ambient-light sensor,
4 Photodiode,
5 Light sensor aligned to far distance,
6 LED.

Ultrasonic sensor

Application

Today's parking-aid and maneuvering systems (see Parking-aid systems) use ultrasound-technology ultra-short-range sensors with a detection range of up to 4.5 m. They are integrated in the bumpers of motor vehicles and serve to calculate distances to obstacles and to monitor the vehicle's surroundings when parking and maneuvering. Sound or even visual alerts are issued to the driver as the vehicle approaches an obstacle.

Sensors with ranges of up to around 4.5 m also permit the use of a parking assistant, which either issues instructions to the driver for optimum parking or steers the vehicle into the parking space while the driver only needs to take care of longitudinal guidance of the vehicle.

Ultrasonic-sensor design

The ultrasonic sensor (Figure 1) is comprised of a plastic case with integrated plug-in connection, an ultrasonic transducer (aluminum pot with diaphragm onto the inside of which has been glued a piezoceramic element), and a printed-circuit board with transmit and evaluation electronics. They are electrically connected to the control unit by three wires, two of which supply the power. The third, bidirectional wire is responsible for activating the transmit function and returning the evaluated received signal to the ECU.

Ultrasonic-sensor operating principle

The ultrasonic sensor receives a digital transmit pulse from the ECU. The electronic circuit then induces the aluminum diaphragm to oscillate with square-wave pulses of approx. 300 µs duration at resonant frequency (around 48 kHz) with the result that ultrasonic pulses are emitted. The diaphragm, which has meanwhile returned to rest, is made to vibrate again by the sound reflected back from an obstacle. No reception is possible during the relaxation time of approx. 900 µs. These renewed oscillations are outputted by the piezoceramic element as analog electrical signals and amplified and converted into a digital signal by the sensor electronics (Figure 2).

Ultrasonic sensors usually have for the described application a selective emission characteristic with a wide horizontal sensing range (for sensing as many objects as possible) and a narrow vertical sensing range (for avoiding earth reflections).

Figure 1: Cutaway view of an ultrasonic sensor
1 Piezoceramic element,
2 Decoupling ring,
3 Plastic case with plug connector,
4 PCB with transmit and evaluation electronics,
5 Transformer,
6 Bonded wire,
7 Aluminum diaphragm.

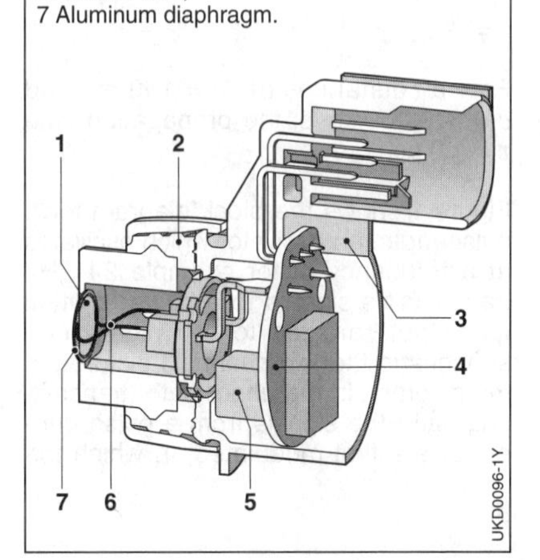

Figure 2: Block diagram of ultrasonic sensor
1 Bidirectional wire to the ECU.

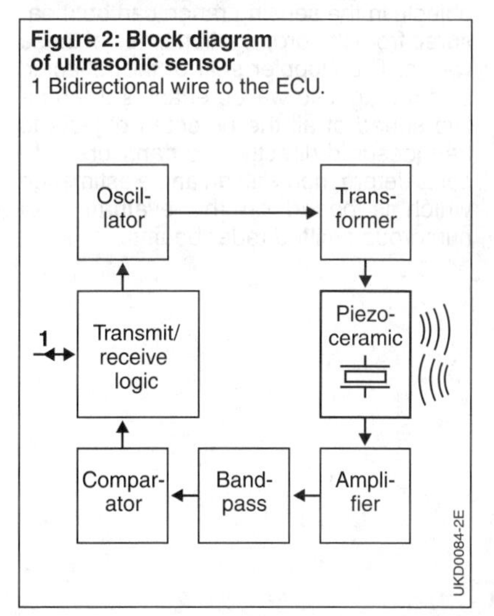

Radar sensors

Application

Radar technology has since 1999 been used with increasing regularity in motor vehicles and supports drivers in their driving tasks. The radar sensor detects objects in front of the driver's vehicle (e.g. other vehicles, pedestrians, roadside structures) and measures their distances, relative speeds and transverse offset to the driver's vehicle. The ACC system (ACC, Adaptive Cruise Control) uses these measurements, adapts the vehicle's driving speed to the traffic in front, and thereby relieves the driver in their driving tasks (see Adaptive cruise control).

Thanks to its excellent properties in terms of fast and precise measurement of distance, relative speed and transverse offset, the radar sensor is particularly well suited for use in active and passive safety functions. Further examples of where radar technology is used are predictive emergency braking systems (PEBS) and pre-crash detection.

Radar principle

The electromagnetic waves emitted by the radar device are rebounded against metal surfaces or other reflecting materials and are picked up again by the radar's receiving section. The distance to the objects in the sensing range can be measured from the propagation times of these waves. The Doppler shift of the reflected electromagnetic waves enables the relative speed of all the detected objects to be measured directly. The transverse offset is determined with an angle estimation which is based on the evaluation of numerous emitted radar beams.

Test methods

The signals received are compared in time or frequency with the transmitted signals. The methods used differ considerably particularly in the way in which the signals are compared. The transmitted waves are modulated to enable a received signal to be uniquely assigned to a transmitted signal. The most common modulations are pulse modulation in which 0.5 to 30 ns pulses are generated, corresponding to a wavelength of 0.15 to 10 m, and frequency modulation in which, during transmission, the instantaneous frequency of the waves is varied over time.

For all radar methods, the distance measurement is based on the direct or indirect propagation-time measurement between when the radar signal is transmitted and when the echo signal echo is received.

Pulse modulation

In the case of a pulse-modulated signal, the propagation time τ (time difference) between transmitted pulse and received pulse is measured. The received wave packet must be demodulated in order to extract the required information. Using the speed of light, the distance to the vehicle traveling in front can be calculated from this time differential. With direct reflection, this is defined by twice the distance d to the reflector divided by the speed of light c:

$$\tau = \frac{2d}{c} \,.$$

For a distance of $d = 150$ m and $c \approx 300{,}000$ km/s, the propagation time $\tau \approx 1.0$ µs.

Figure 1 shows the block diagram for a pulse radar. An oscillator which oscillates at a frequency of for example 24 GHz transmits its signals to a power splitter. Its outputs are fed to two high-speed switches in the two channels pictured in the diagram. In the upper path (transmitting path) the signals from a pulse generator are first modulated, in which the square-wave pulses are converted into a suitable shape to activate and transmit the carrier signal. The modulated signal is then output to the high-speed switch (high-frequency modulation switch). The signals pass from this assembly to the transmitting antenna.

In the lower parallel path (receiving path), to determine the propagation time, a variable delay generates reference signals which feed to a high-speed switch in the receiving path. The received echo signal is coherently mixed with the output signal from the oscillator in order to identify frequency changes in the received echo signal. Coherence in this context means that the phase of the transmitted pulse remains is retained in the reference signal. This procedure presupposes a phase-stable oscillator. The frequency change is determined by the Doppler filter and determines the relative speed.

A radiated peak power of 20 dBm EIRP (power level with the reference quantity 1 mW, Equivalent Isotropic Radiated Power) produces a measurement distance of 20 to 50 m, depending on the size and reflection properties of the object in question and the sensitivity of the receiving path. The minimum measurement distance is typically 0.25 cm.

FMCW modulation

Figure 2 shows the block diagram for an FMCW radar (Frequency Modulated Continuous Wave). A 77 GHz VCO (voltage-controlled oscillator) supplies the transmitting antenna. The receiving antenna transmits the signal reflected by the object to the reception mixers, which mix the received signal with the current transmitted signal of the VCO and thus transform to low frequencies in the range 0 to 500 kHz. The signals are amplified, digitized, and to determine the frequencies subjected to a high-speed Fourier analysis effected in software (see Fourier transform).

The operating principle of frequency generation is explained in the following. The frequency of the 77 GHz VCO is compared continually by PLL closed-loop control (phase-locked loop) with a stable quartz-based reference oscillator and regulated to a prespecified setpoint value. The PLL is altered during a measurement in such a way that for the transmit frequency f_S a linearly rising frequency ramp is created over time – followed by a linearly falling frequency ramp (Figure 3). The mean transmit frequency is f_0. The signal reflected and received from a vehicle driving in front f_E is delayed according to the propagation time, i.e. it has in the rising ramp a lower frequency and in the falling ramp a frequency f_E higher by the same amount Δf_{FMCW}. The frequency difference Δf_{FMCW} is a direct measure of the distance d, dependent on the gradient s of the ramp:

$$\Delta f_{FMCW} = |f_S - f_E| = \frac{2s}{c} \cdot d\,.$$

Figure 1: Block diagram of pulse radar
1 Pulse generator,
2 Modulator for pulse shaping,
3 High-speed switch,
4 Transmitting antenna,
5 24 GHz oscillator,
6 Power splitter,
7 Variable time delay,
8 Modulator for pulse shaping,
9 High-speed switch,
10 Mixer,
11 Receiving antenna,
12 Doppler filter,
13 Output.

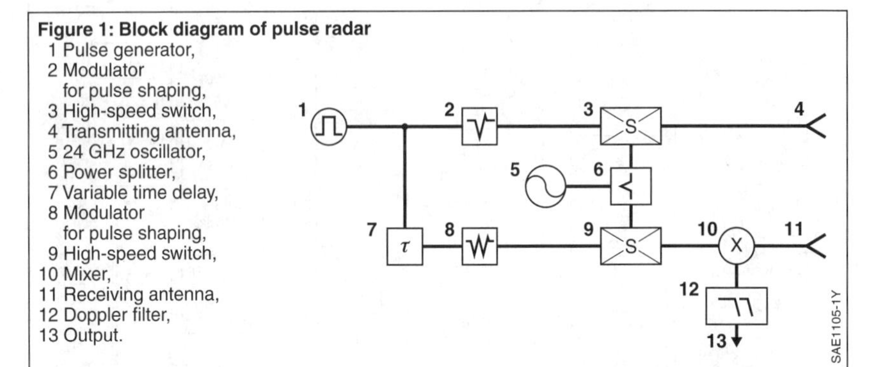

If there is additionally a relative speed Δv to the vehicle traveling in front, the receive frequency f_E is, on account of the Doppler effect, increased (when approaching) or decreased (when the distance increases) in both the rising and falling ramps by a specific, proportional amount Δf_D:

$$\Delta f_D = \frac{2 f_0}{c} \cdot \Delta v \quad \text{(approximation for } v << c\text{)}.$$

In other words, there are two different difference frequencies Δf_1 and Δf_2. For the rising ramp:

$$|\Delta f_1| = |f_S - f_E| = \Delta f_{FMCW} - \Delta f_D$$
$$= \frac{2}{c} \cdot (s\,d - f_0 \Delta v).$$

For the falling ramp:

$$|\Delta f_2| = |f_S - f_E| = \Delta f_{FMCW} + \Delta f_D$$
$$= \frac{2}{c} \cdot (s\,d + f_0 \Delta v).$$

Their addition produces the distance d, and their subtraction the relative speed Δv of the objects:

$$d = \frac{c}{4s} \cdot (\Delta f_2 + \Delta f_1),$$

$$\Delta v = \frac{c}{4 f_0} \cdot (\Delta f_2 - \Delta f_1).$$

Antenna system and determining the angle

The antenna system has not only to transmit and receive the high-frequency signals, but also to estimate the transverse offset of the objects. This is required

Figure 3: Distance and speed measurement with a linear FMCW radar
1 Transmit frequency f_S,
2 Receive frequency f_E without relative speed,
3 Receive frequency f_E with relative speed.
Δf_{FMCW} Frequency difference between transmitted and received radar signals,
Δf_1 Frequency difference on the rising ramp with relative speed,
Δf_2 Frequency difference on the falling ramp with relative speed,
Δf_D Change of receive frequency with relative speed.

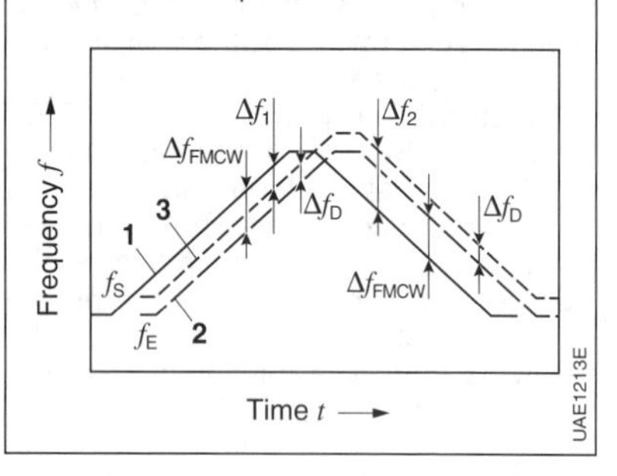

Figure 2: Block diagram of a 4-channel FMCW radar
SiGe MMIC: mm-wave-IC, integrated circuit (silicon-germanium technology).

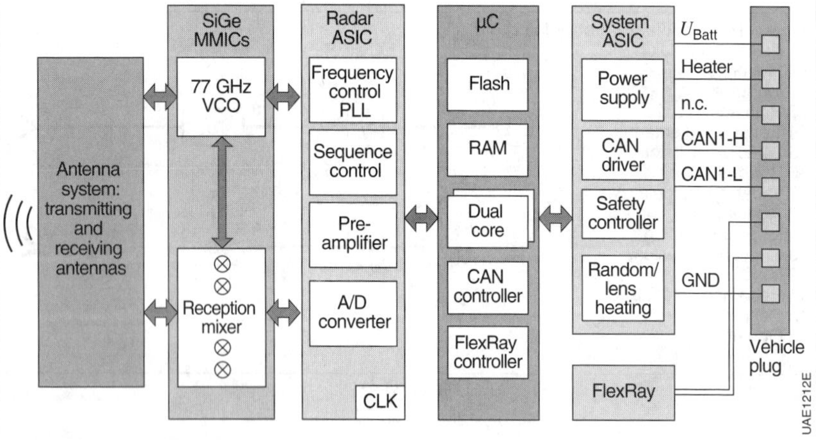

to assign the objects to lanes. Radar systems determine the relative position by estimating the angle under which the receiving antenna locates an object. At least two, or better still more than two receiving antennas are required for this purpose. This can be done by directing a single radar beam (scanning) and also by several parallel, overlapping radar beams.

The gains of the complex amplitudes (phases and amounts) which are measured for an object in neighboring beams allow a conclusion to be drawn about the relative reception angle to the radar-sensor axis. In practice four radar beams are frequently used, achieving an angle accuracy of up to 0.1° and an angle separation of up to 4°.

The are many ways of technically realizing the antenna system. The two most common variants are the lens-antenna system and the patch-array-antenna system.

Lens-antenna system
The lens-antenna system consists of a number of single patches (typically a metallic rectangle seated on a p.c.b. suitable for high frequencies) which are arranged in the focal plane of a dielectric plastic lens. The lens concentrates the radar beams of the single patches horizontally and vertically, thereby increasing the range of the radar sensor. The lateral offset of the single patches creates an angle offset of the radar beams and thus fan-shaped transmit and receive characteristics, which are used to determine the positioning field and to determine the angle.

The lens-antenna system is usually a monostatic system, i.e. the transmitting patches also serve as the receiving patches such that directional couplers are used in the reception mixers to separate the received signals from the transmitted signals.

Patch-array-antenna system
The patch-array-antenna system, in contrast to the lens-antenna system, as a rule has a bistatic design. The transmitting antenna in this system consists of a large number of single patches arranged in rows and columns which are wired in such a way that superposition produces a concentrated radar beam along similar lines to concentration by a lens.

The receiving antenna typically consists of a number of single patches arranged in columns which are electrically separated from each other and arranged next to each other. The offset of the individual reception columns generates a phase shift between the received signals of a detected object which is used to determine the angle. A reception column can in an individual case also consist of a number of wired columns.

Variants of radar sensors
In the past a distinction was made between short-range radars (SRR, usually in the 24 GHz frequency band) and long-range radars (LRR, in the 76 GHz frequency band). In the meantime the applications are merging and sensors are increasingly covering both ranges such that mid-range radars (MRR, in the 76 to 81 GHz Band) will dominate the market, supplemented by a long-range radar. Short-range radars have a typical range of 20 to 50 m and a beam angle of up to 160°. Long-range radars have ranges of 250 m and beam angles of up to 30°; mid-range radars have ranges and beam angles in between.

Vehicles in the future will have up to five radar sensors all round the outer skin, placing a virtually continuous envelope around the vehicle and providing the opportunity for further assistance functions to make driving safer and more comfortable.

Lidar sensors

Applications

Lidar sensors (Light Detection and Ranging) have been used for adaptive cruise control applications (ACC) in Japan and occasionally in the USA since around 2000. Since 2008 the multi-beam lidar has opened up the volume segment for automatic emergency-braking functions. In 2015 a multi-level laser scanner for traffic-jam assistance and NCAP pedestrian protection (New Car Assessment Program) is to go into series production. By 2020 a rapidly increasing number of lidar sensors for highly automated driving functions is expected, driven by the need for a redundant sensor configuration with different physical measurement principles where possible.

Measurement principle

Today's lidar sensors typically achieve ranges of 150 to 250 m with a distance accuracy of 5 to 15 cm. The angle resolution is typically 0.25 to 1°. The measurement rate is for the most part between 10 and 25 Hz. Simple, non-scanning lidar sensors with one or a few fixed beams supply individual measuring points of the surrounding area (distance and intensity). Complex lidar sensors have a number of beams which can be specifically deflected to scan the surrounding area. Such high-resolution lidar sensors achieve several thousand measuring points per measuring cycle. With the available intensity information the measured data are comparable with video images.

Unlike most radar sensors, a lidar sensor does not measure the object speed directly. Instead, it is calculated by differentiating the distance signal, which results in a certain delay and reduced signal quality. On the other hand, the good lateral resolution of a scanning lidar is far superior to the typical radar sensors used today.

Lidar sensors operate in the close infrared range between 800 and 1,000 nm, or to increase eye safety around 1,550 nm. Lidar beams can occasionally be signifi-

Figure 1: Lidar block diagrams for pulse lidar and CW lidar
a) Pulse lidar,
b) CW lidar (continuous wave).
1 Transmitter diode,
2 Receiver diode.
DSP Digital signal processor,
TDC Time-to-digital converter,
A/D Analog-digital converter,
µC Microcontroller.

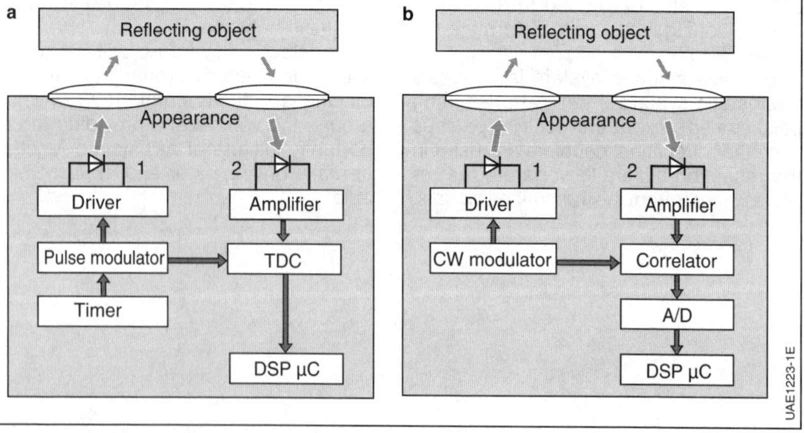

cantly damped by fog and conditions of poor visibility, in particular spray. This can reduce the measurement range accordingly. They are therefore less well suited than radar sensors for safety applications.

Infrared radiation is modulated in its intensity, not in its frequency. The block diagram for a lidar sensor is shown in Figure 1. A lidar sensor emits modulated infrared radiation, which is reflected by an object and received by one or more photodiodes in the sensor. Forms of modulation can be: square waves, sinusoidal oscillations or pulses. The modulator transmits the modulation information to the receiver. In this way, the received signal can be compared with the transmitted signal in order to determine either the phase difference of the signals or their propagation time, and from this to calculate the distance to the object.

The signal-to-noise ratio is very much dependent on the type of modulation, the best results being achieved with pulse modulation. Pulse modulation is therefore the method used in practice for lidar sensors with long ranges.

There are currently three fundamental principles for measuring light propagation time:

Direct pulse-propagation-time method

With the direct pulse-propagation-time method a very strong, short pulse of light (typically up to 75 W pulse height, 2 to 10 ns pulse duration, pulse repetition rate up to 150 kHz) is emitted. In the received light the pulse reflected by the object is detected using different methods (e.g. scanning and correlation analysis or edge detection by threshold-value comparison) with regard to chronological position and signal shape (time-to-digital converter, TDC). The distance is calculated from the chronological position relative to the emitted pulse via the speed of light. The reflectivity can be inferred from the signal shape. But a classification can also be made as to whether the reflection comes from a hard target (object) or a soft target (atmospheric interference, e.g. fog). Sometimes the signal shape is also used to improve the distance estimation (e.g. walk-error compensation). To adapt to the large dynamic range required, some receiver circuits operate primarily in saturation, with the result that only edge detection is possible. Other methods perform a dynamic gain adjustment, e.g. increasingly with rising propagation time. Newer methods work with statistical methods, e.g. threshold-value comparisons within the noise of ambient light (during the continuous event measurement).

Lateral and vertical resolution is achieved by a multibeam configuration, by mechanical scanning or by special reception arrays.

Multibeam lidars have a very limited resolution. The use of such lidars is only sensible where rough volume elements are to be checked for the presence of an object (e.g. during an emergency-braking function for areas in front of the vehicle for which only collision mitigation can still be achieved).

Mechanical scanning has the advantage of a very fine angle resolution when only one or a few transmit and receive elements are used. Beam deflection was until recently achieved either by a rotating mirror (prism mirror) or by directing the optical element of the transmitter or receiver. Now solutions are being used more frequently in which transmitter, receiver and optical elements are located on a rotating plate. Energy transmission to the plate is effected via sliding contacts or by non-contacting means.

MEMS micro-mirrors (Micro-Electro-Mechanical Systems) are used as deflectors for miniaturization purposes. This trend could gain acceptance at least for deflecting the transmitted beam.

In a newer form reception arrays (focal-plane arrays, FPA) with up to 256×256 pixels are used in which each pixel performs a distance measurement by scanning the received pulse. On the transmission side either a single, very strong pulse is emitted (flash lidar) or light pulses are deflected by means of a scanning device. The maturity of these technologies is still low today.

In a further special form, range gated viewing or range scanning, a very short (lasting a few nanoseconds) gate window of the receiver is shifted incrementally in time and for each increment merely the intensity of the received light of the gate window is measured. The advantage is simple conversion in the receiver. The disadvantage is that a new laser pulse is required for each distance increment.

Indirect propagation-time methods
With the indirect propagation-time method with phase displacement an amplitude-modulated light signal (in theory for the most part assumed to be sinusoidal, up to a few 10 W CW output) is emitted and correlated with the reflected light in the detector. The light propagation time and thus the distance can be inferred directly from the phase displacement. In order to determine the phase displacement, the signal amplitude, the phase relation, and the signal offset must be determined from the received signal. To determine the three unknown quantities, four measured values are recorded and the solution is ascertained from the overdetermined system. The measured values are recorded in four measurement phases corresponding to the phase displacements 0°, 90°, 180°, and 270°. Over a period of a few milliseconds charge carriers converted from photons are distributed – controlled by the amplitude of the transmission modulation – to two charge zones (charge-carrier swing). After all the partial measurements have been completed the distance to the object can be inferred from the charge conditions. Because of this mixer comparable with the radar in the pixel imagers structured in this way are also called photonic mixing devices (PMD). In order to prevent a saturation of the charge zones for example by strong background light, it is possible to remove equal components in the charge-carrier swing (SBI, suppression of background illumination).

In a form involving very many PMD pixels such lidar sensors are also called range imagers because they combine characteristics of both lidar technology and camera technology. They can be considered to be video sensors which have the additional ability to measure the distance to the nearest object with each camera pixel. If the problems that still exist can be surmounted, PMD technology must be considered a serious alternative to other sensors in the short and medium ranges.

Indirect propagation-time method with high-speed exposure control
With the indirect propagation-time method with high-speed exposure control a long, if possible square-wave pulse of high amplitude is emitted (several 100 ns pulse duration, up to 100 W pulse height). The receive pulse is integrated in the receiver in two exposure sections. The first exposure section corresponds to the transmission duration so that as the distance to the object increases an ever-decreasing proportion of the reflected light still shines into the gate window and consequently can be integrated. If the object distance is above the distance of the corresponding transmission duration (upper distance limit), no more light can be integrated. This first partial measurement is ambiguous with regard to object reflectivity and object distance. Clarity can be achieved with the second exposure section. This corresponds to twice the transmission duration so that the entire reflected pulse is incident and thus contains a clear indication of the entire quantity of light reflected by the object (dimension of the reflectivity for this distance). Both partial measurements contain a further offset as a result of the background light. The background light is therefore determined in a third partial measurement without pulse lighting. This technology, in view of its integrating character, currently functions only in bright ambient light.

There are currently two forms of implementation. In this first variant a high-speed-switching standard camera imager with very high resolution is used to perform the three partial measurements. The distance images are calculated in a downstream stage on a computer. In the second variant the measurement principle is implemented directly on the CMOS imager. However, because of the smaller space factor (space requirement for implementing the measurement principle), only lower resolutions can be achieved.

Video sensors

Application

Images contain the greatest amount of information for human beings. Consequently, the obvious approach within the context of developing driver-assistance systems is to record images, to extract the relevant details from such images, and to identify dangerous situations by means of image-processing methods.

The first stage has already seen the introduction onto the market of video-based functions such as, for example, night-vision systems, lane departure warning, and road-sign recognition. In a second stage, functions which (above all by the interaction of several sensors) intervene in the vehicle dynamics by acting on the brakes, steering and throttle open up new, effective perspectives for the sustained avoidance of accidents and mitigation of the consequences of accidents.

Two different tasks are performed in an automobile system within this context. When the requirement is to generate a particularly high-contrast, brilliant image, as is needed in night-vision systems, image editing is performed. The edited image is then output directly in a display. The second task involves extracting specific image content by using special algorithms (image processing, e.g. road-sign recognition). The information thereby acquired can then be used to issue an alert to the driver in the display or to generate a vehicle intervention by way of actuators.

Basic principles of photosensing

When photons shine into a semiconductor, electron-hole pairs are generated. These in turn generate an electric field, recombine and generate a photoelectric current. Here, the quantity "quantum efficiency η" describes how many electron-hole pairs are generated by one photon.

Virtually all the photons which penetrate into the semiconductor are converted into electric charges. There is however a strong spectral dependence of the mean absorption length in which this photoconversion occurs. Short-wave light is principally absorbed on the surface of the semiconductor, while long-wave light penetrates deep into the semiconductor. Images with a high content of red and infrared wavelength ranges (e.g. systems for night-vision improvement) therefore demonstrate much less contrast than images which are recorded in the short-wave spectral range. It is therefore essential for night-vision display systems to process images in the interests of improved contrast with brilliant images. For consumer applications optical filters are frequently incorporated in front of the camera in order to cut off the infrared content of the spectrum.

The photoelectric current is proportional over many factors of ten (greater than ten) to the incident luminous flux and is linear over a wide dynamic range. This is what makes semiconductor photosensors such an attractive prospect for numerous consumer and measurement applications.

The two most important photosensitive semiconductor structures are the photodiode (Figure 1) and the metal-oxide semiconductor capacitor (MOS capacitor, Figure 2), as is used in CCD sensors. Both these semiconductor structures are manufactured with standard semiconductor processes.

The photodiode consists of a combination of semiconductor materials of differing conductive properties. An electric field exists in the space-charge region at the junction between the two semiconductor materials. At the same time this space-charge region demonstrates a certain capacitance which behaves in inverse

Figure 1: Photodiode
1 Contact,
2 n^+ silicon substrate,
3 p silicon substrate,
4 Space-charge region.

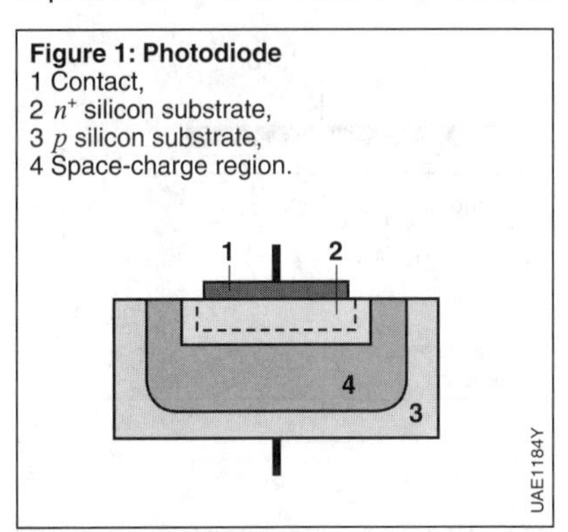

proportion to its thickness. Photodiodes typically work by being charged to a certain potential and then being exposed to light. The photoelectrically generated charges now spread to the entire space-charge region and are stored in the photodiode capacitor. The residual voltage is measured once the photodiode has been exposed to light. The difference between this voltage and the reset voltage is a measure of the quantity of incident light.

The MOS capacitor (Figure 2) consists of semiconductor material which is covered by a thin oxide layer. The oxide layer carries a metallic conductor layer. If a positive voltage is applied to the metal electrode of the MOS element, a space-charge region of stationary positive charges is created underneath the insulating oxide layer. In the event of incident light through the transparent, insulated electrode (front-side exposure) or through the substrate (rear-side exposure), the photoelectrically generated electrons collect in this area without being able to recombine or flow off.

A typical value for the capacitance of a photodiode and of an MOS capacitor is 0.1 fF/µm ([1], [2]).

Figure 2: MOS capacitor working as an integrating photosensor
1 Electrode,
2 Space-charge region,
3 Silicon oxide,
4 *p* silicon.
A Electron,
B Hole.

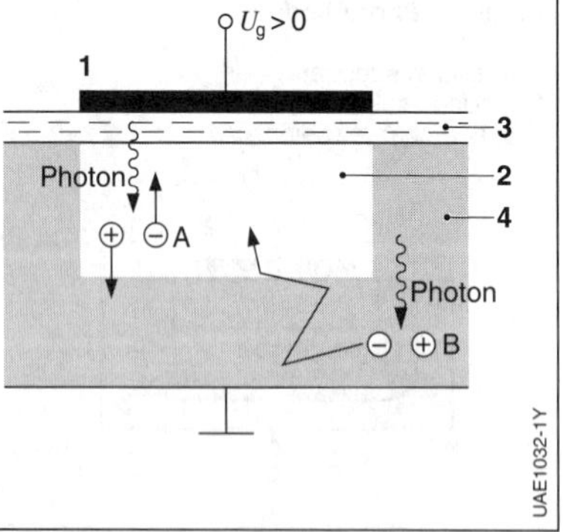

UAE1032-1Y

CCD imaging sensors

For the purpose of manufacturing imaging sensors (imagers), many photodiodes or MOS capacitors are interconnected into "arrays" with numerous pixels. While the output signals of photodiodes correspond to the instantaneous value of the luminous flux (illuminance), the following two structures are integrating in character. Their signal corresponds to the total number of photons which have penetrated the sensor during the exposure time. Such sensors are required primarily to manufacture line or uniplanar sensor arrays according to the CCD principle (charge-coupled devices).

In the case of these *pn* photodiodes, only a small part of the *pn* junction is radiation-sensitive on account of a vapor-deposited screen. The photoelectrically generated charges spread, however, to the entire space-charge region and are accumulated there. When a MOSFET switch is closed, they can flow off to a commonly used signal line (video output). The switch is controlled by a clock generator via a shift register (Figure 3). The charges flowing serially via the video line are a measure of the radiation dose of the photodiodes activated in each case.

To be able to shift the measurement charge laterally after exposure, further electrodes are as shown in Figure 4 arranged next to the exposable zone or the collector electrode; these electrodes are at zero potential during the integration phase. If the potential of a side transfer electrode is then increased to a positive

Figure 3: Line arrangement of photodiodes with serial output line

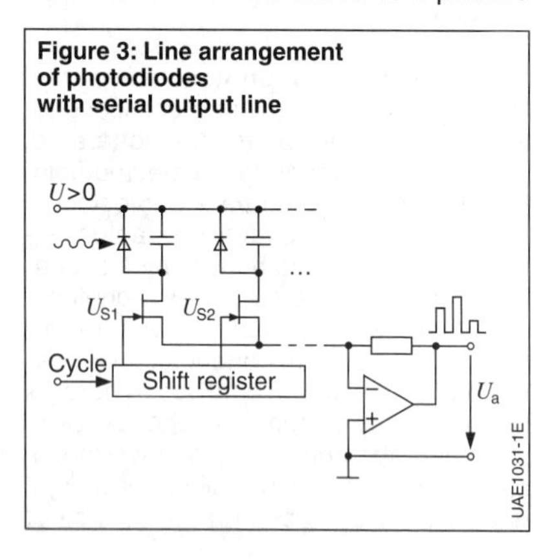

value while the collector-electrode potential is simultaneously reduced, the charge can be shifted to a neighboring MOS element which is shielded by a screen against the incidence of light.

This principle of charge transfer forms the basis of charge-coupled devices (CCDs). According to this principle, analog charges can be shifted or transferred over many stations until they are finally converted at the end of the transfer chain by means of a charge amplifier, for example, into a voltage signal, which can be supplied to a fast A/D converter.

This method of charge transfer, which can also be viewed as a kind of analog shift register, facilitates the simple setup of long line-shaped multiple structures, but also of matrix-shaped structures. One single element of these structures is also called a pixel (contraction of picture element). As things stand today, the maximum possible number of pixels for line sensors is roughly 6,000, while that for matrix sensors is roughly 5,000 × 5,000, i.e. approximately 25 million. Today's imaging sensors in motor vehicles manage with under one million pixels. For more sophisticated automotive applications, however, a much higher number of pixels would be desirable. In consumer applications cameras utilize imaging sensors with over ten million pixels.

The size of the pixels which receive their light from a conventional imaging lens today lies typically in the range of 5 to 20 µm edge length. The chip surface of the sensor is thus in the cm² range. If the intention is to further reduce the individual pixels in size in order to increase the resolution or also to reduce chip costs, it is important to bear in mind that this will also mean a reduction in the number of incident photons per pixel. Thus, limits are placed on sensible reduction by the unavoidable noise process; the increased pixel resolution is then impaired by an increased noise level.

Even the charge that can be absorbed by the individual, integrating cells is limited. If this limit is exceeded, the charge can "spill over" into neighboring cells. This is also called the "blooming effect", which in principle limits the dynamic light/dark response of the CCD technology. Even with additional antiblooming protective measures, this dynamic response can barely be increased over a value of roughly 50 dB without additional aids such as a variable screen and exposure time.

Today, CCD imaging sensors are the most widely used semiconductor-based imaging-sensor technology. However, the limited dynamic light/dark response, the relatively high power requirement compared with other technologies at three different operating voltages, and the limited temperature range have restricted their wider use in automobiles.

Figure 4: MOS capacitor with rear-side lighting and transfer electrodes for charge transfer
a) Collection of charge carriers by incident light,
b) Transfer of charge carriers.
1 Space-charge region,
2 Silicon oxide.
$0 < U_{2t} < U_{2s}$.

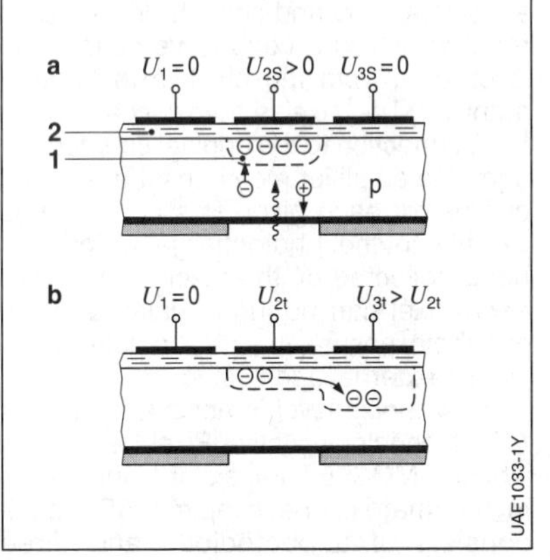

CMOS imaging sensors

Today, CMOS imaging sensors represent a more forward-looking solution than CCD sensors, and are already widely used in many applications. The term "CMOS sensor" may be confusing here; this is because CMOS technology denotes a special semiconductor technology. CCD technology, on the other hand, does not need this, but also contains MOS structures. CMOS sensors essentially differ from their CCD counterparts not only in the manufacturing technology, but also in a range of features:

- The pixels are no longer read out serially; instead – similarly to a memory cell in a RAM or a pixel in an LCD flatscreen display – they are arranged in a matrix structure and can be individually activated. Active electronics are also integrated for each pixel for this purpose (APS, Active Pixel Sensor).
- Integrating photodiodes are not used; instead, those which are to a large extent not dependent on the exposure time are used.
- The brightness values are not proportionally converted into electrical signals; instead, they are logarithmized before being read out. They therefore have a similar characteristic to the human eye. Only through this measure is it possible to extend the dynamic light/dark response to more than 100 dB without additional measures.
- CMOS imaging sensors are not realized using standard CMOS technology. Instead, CMOS technology optimized to the photoelectric element is used which, because of the much lower power requirement than CCD sensors, allows further activation and evaluation electronics to be integrated on the imaging-sensor chip. Because the access time to the individual pixels is in the range of a few 10 ns, somewhat higher image frequencies are possible with CMOS sensors – particularly when use is made of the possibility of reading out subimages only (subframing), which is not possible with CCD sensors.

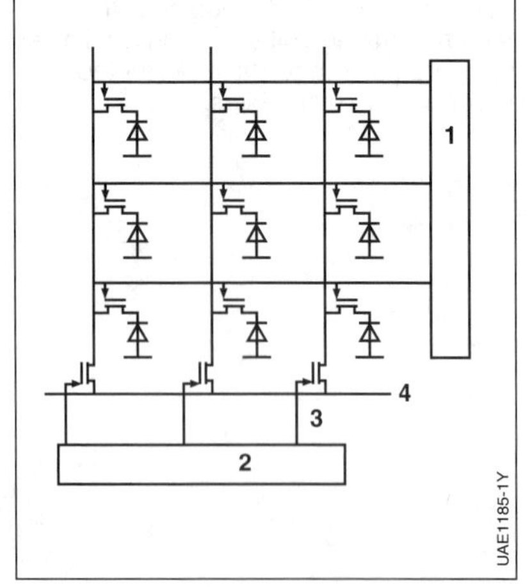

Figure 5: CMOS imaging sensor
Photodiode array
with one photodiode and one activation transistor per pixel
1 Line addressing,
2 Column addressing,
3 Column selection,
4 Column signal line.

Figure 5 shows an extract from a CMOS-imager structure. The individual pixel consists of a photodiode and a MOSFET (MOS field-effect transistor) as the switching element. Each individual pixel can be activated and read out via the matrix structure.

All photodiodes are charged to a bias countervoltage of around 5 V. The individual pixels are discharged to specific voltages under the influence of incident light. A pixel is read out by activating the associated line and column drivers, as a result of which a conductive connection is created from the pixel to the output amplifier. The pixel is then recharged via this connection to the original countervoltage. The amplifier measures the required charge for each pixel. This corresponds exactly to the photocharge which has been collected by the pixel. In this way, each pixel can be individually read out and the exposure time can be determined via the external addressing circuit.

A low noise level is achieved with this APS technology (Active Pixel Sensor), in which a MOSFET transistor is integrated in the matrix. The simplest APS pixel consists of a photodiode and three

MOSFETs. Figure 6 is a schematic representation of the structure of an HDRC pixel (High Dynamic Range CMOS technology). The light-sensitive element of this version of a CMOS sensor is a photodiode polarized in the blocking direction which is connected in series to the PMOS transistor (M1) operated below its opening voltage. The diode current proportional to the illuminance must also flow through the blocked transistor. Its gate-source voltage UGS is dependent in a very wide range virtually ideally logarithmically on the drain current flowing through (photocurrent). The other two transistors M2 and M3 serve to decouple the signal, which is supplied via a multiplexer to a fast 10-bit A/D converter.

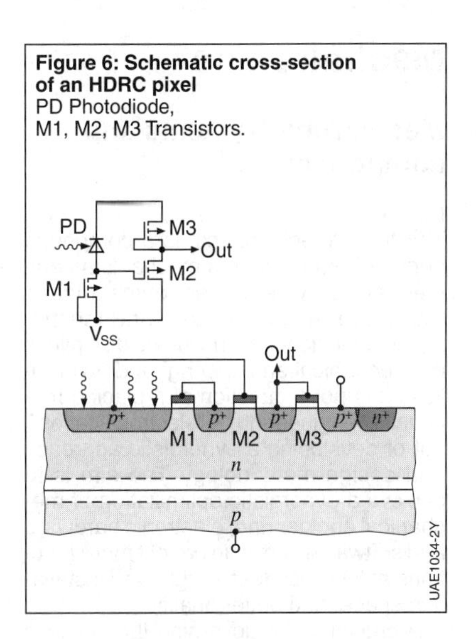

Figure 6: Schematic cross-section of an HDRC pixel
PD Photodiode,
M1, M2, M3 Transistors.

References
[1] B. Jaehne: (Ed.) Handbook of Computer Vision and Applications, Vol. 1. Academic Press, 1999.
[2] P.M. Knoll: Ch. 7.8 "Video Sensors" in: J. Marek, H.-P. Trah, Y. Suzuki, I. Yokomori: Sensors for Automotive Technology. Wiley-VCH, Weinheim, 2003.

Mechatronics

Mechatronic systems and components

Definition

The term "mechatronics" is a compound derived from the words mechanisms and electronics, where electronics means "hardware" and "software", and mechanisms is the generic term for the disciplines of "mechanical engineering" and "hydraulics". It is not a question of replacing mechanical engineering by "electronification", but of developing a synergistic approach and design methodology. The aim is to achieve a synergistic optimization of mechanical engineering, electronic hardware and software in order to project more functions at low cost, less weight and installation space, and better quality.

A crucial factor governing the success of a mechatronic approach to solving problems is to regard the two previously separate disciplines as a single entity.

Application

Mechatronic systems and components are now used in practically all aspects of the automobile today, starting with engine management and fuel injection for gasoline and diesel engines, transmission-shift control, and electrical and thermal energy management, through to a wide variety of braking and vehicle-dynamic systems. It even includes communication and information systems, with many different requirements when it comes to operability. Besides systems and components, mechatronics are also playing an increasingly vital role in the field of micromechanics.

Examples at system level

A general trend is emerging in the advanced development of systems for fully automated vehicle handling and guidance: In future, mechanical systems will be replaced to an increasing extent by "x-by-wire" systems. One system that has been in existence for quite some time is "drive-by-wire", i.e. electronic throttle control. "Brake-by-wire" is replacing the mechanical and hydraulic link between the brake pedal and the wheel brake. Sensors detect the driver's brake command and transmit this information to an electronic control unit. The unit then generates the required braking effect at the wheels by means of actuators.

A possible option for implementing "brake-by-wire" is the electrohydraulic braking system (SBC, Sensotronic Brake Control). When the brake pedal is pressed or when the Electronic Stability Program (ESP) intervenes in the brake system to stabilize the vehicle, the SBC ECU calculates the required braking pressures for each of the wheels. Since the unit calculates the required braking pressures separately for each wheel and detects the actual values separately, it can also regulate the braking pressure to each wheel via the wheel-pressure modulators. The four pressure modulators each consist of an inlet and an outlet valve controlled by electronic output stages which together produce a finely metered pressure regulation.

In the common-rail system, pressure generation and fuel injection are separated from each other. A high-pressure accumulator, i.e. the common rail, stores constantly the fuel pressure required for each of the engine's operating states. A solenoid-valve-controlled injector with integrated nozzle assumes the function of injecting the fuel directly into the combustion chamber of each cylinder. The engine electronics constantly request data on accelerator-pedal position, rotational speed, operating temperature, fresh-air intake flow, and rail pressure in order to optimize the control of fuel metering as a function of the operating conditions.

Examples at component level

Fuel injectors are crucial components in determining the future potential of diesel-engine technology. Common-rail injectors are an excellent example of the fact that an extremely high degree of functionality and, ultimately, customer utility can only be achieved by controlling all the physical domains (electrodynamics, mechanical engineering, fluid dynamics) to which these components are subjected.

In-vehicle CD drives are exposed to particularly tough conditions. Besides wide temperature ranges, they must withstand extreme vibrations that have a critical impact on such precision-engineered systems. The drives are usually equipped with a spring-damper system to isolate the playback unit from vibrations that occur when the vehicle is moving. Any considerations to reduce the weight and installation space of CD drives immediately raise questions concerning these spring-damper systems. If the damper system is eliminated from a CD drive, the main focus is on designing a mechanical system with zero clearances and producing additional reinforcement for the focus and tracking controllers at high frequencies. Only by considering both measures from a mechatronic viewpoint is it possible to achieve an optimized vibration-proof solution for an automotive environment. Besides weight savings of about 15%, the installation height has also been reduced by about 20%.

The new mechatronic approach for electrically driven coolant motors is based on brushless, electronically commutated DC motors. Initially, they are more expensive (motor with electronics) than previous motors equipped with brushes. However, the overall optimization approach has a positive tradeoff: Brushless DC motors can be used as "wet rotors" with a much simpler design. This reduces the number of single parts by roughly 60%. Taking the aggregate view, the sturdier design has double the service life, almost half the weight, about 40% less the overall length, while maintaining costs at a comparable level.

Examples in the field of micromechanics
Another area of applications for mechatronics is the field of micromechanical sensors, with noteworthy examples such as hot-film air-mass meters and yaw-rate sensors.

The design of microsystems also requires an interdisciplinary approach owing to the close interaction between the subsystems involving individual disciplines such as mechanical engineering, electrostatics, fluid dynamics (where necessary), and electronics.

Development methodology

Simulation

The special challenges that designers face when developing mechatronic systems are the ever shorter development times and the increasing complexity of the systems. At the same time it is vital to ensure that the developments will result in useful products.

Complex mechatronic systems consist of a large number of components from a wide range of physical domains: hydraulics, mechanical engineering, and electronics. The interaction between these domains is a decisive factor governing the function and performance of the overall system. Simulation models are required to review key design decisions, especially in the early development stages when there is no prototype available.

Basic issues can often be clarified by producing relatively simple models of the components. If more detail is required, more refined component models are needed. The detailed models focus mainly on a specific physical domain.

As a result, there are detailed hydraulic models of common-rail injectors. They are simulated using special programs whose algorithms are matched precisely to the hydraulic systems. For example, the requirements here would be to take cavitation phenomena into consideration.

Detailed models are also needed to design the power electronics for activating the injectors. Again, this involves the use of simulation tools which must be developed specifically to design electronic circuits.

Tools that are specially designed for this specific part of the overall system are also required to develop and simulate the control-unit software which controls the high-pressure pump and power electronics using signals from the sensors.

As the components in the overall system interact with each other, it is not sufficient to consider specific detailed models of the components in isolation. The optimum solution is also to take into account the models of other system components. In most cases, these components can be portrayed by much simpler models. For example, a system simulation focused on hydraulics only requires a simple model of the power electronics.

The application of various domain-specific simulation tools during the design of mechatronic systems is only efficient if there is some sort of support for exchanging models and parameters between the simulation tools. The direct exchange of models is highly problematic due to the specific languages used for describing the models of each of the tools.

However, an analysis of the typical components in mechatronic systems shows that they can be composed of a few simple elements specific to the domains. These standard elements include, for example:
– in hydraulics: restrictor, valve, or pipe,
– in electronics: resistor, capacitor, or transistor,
– in mechanical engineering: mass with friction, transmission, or clutch (accordingly for micromechanics).

The preferable solution is that these elements should be stored in a central standard model library (Figure 1) that is also decentrally accessible to product development. The kernel of the standard model library is a documentation of all the standard elements. For each element, this comprises:
– a textual description of the physical behavior,
– the physical equations, parameters (e.g. conductivity or permeability), and state variables (e.g. current, voltage, magnetic flux, pressure),
– a description of the associated interfaces.

Figure 1: Model library for a micromechanical yaw-rate sensor

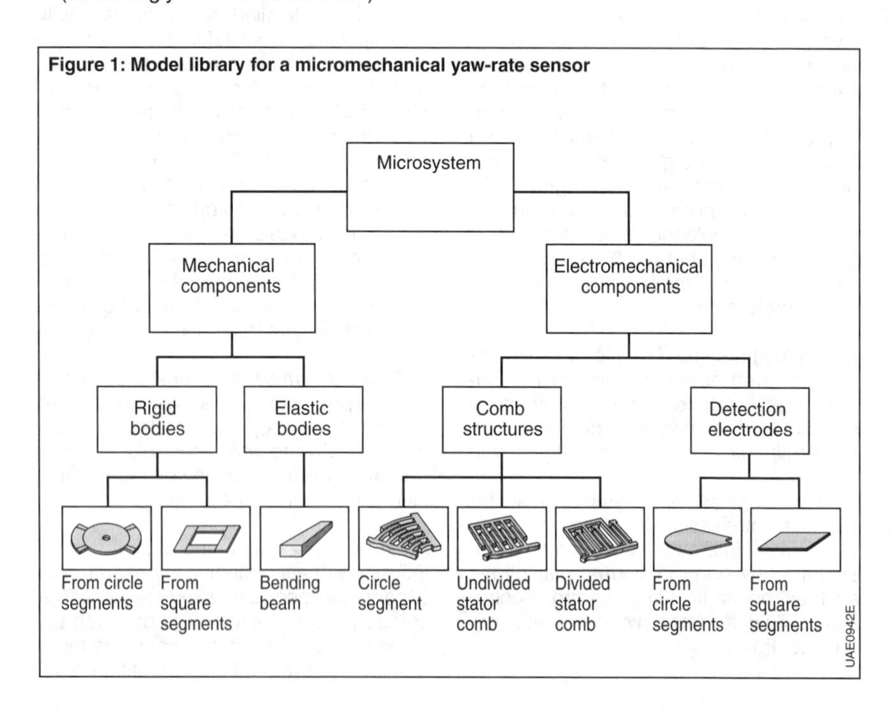

V model

The "V model" contains relationships between the various stages of product development, from the requirements definition and development, implementation, and test, through to system deployment (Figure 2). A project passes through three "top-down" levels during the development stage:

- customer-specific functions,
- system, and
- components.

A requirements specification ("what") must first be produced at each level in the form of specifications (Figure 3). This is then used to produce the design specifications based on design decisions (the actual creative engineering output). The performance specifications describe "how" a requirement can be met. The performance specs form the basis for a model description which allows a review (i.e. validation) of the correctness of each design stage together with previously defined test cases. This procedure passes through each of three stages, and, depending on the technologies applied, for each of the associated domains (mechanical engineering, hydraulics, fluid dynamics, electrics and electronics, software).

Recursions at each of the design levels shorten the development stages significantly. Simulations, rapid prototyping, and simultaneous engineering are tools that allow rapid verification, and they create the conditions for shortening product cycles.

Figure 2: General overview of V model

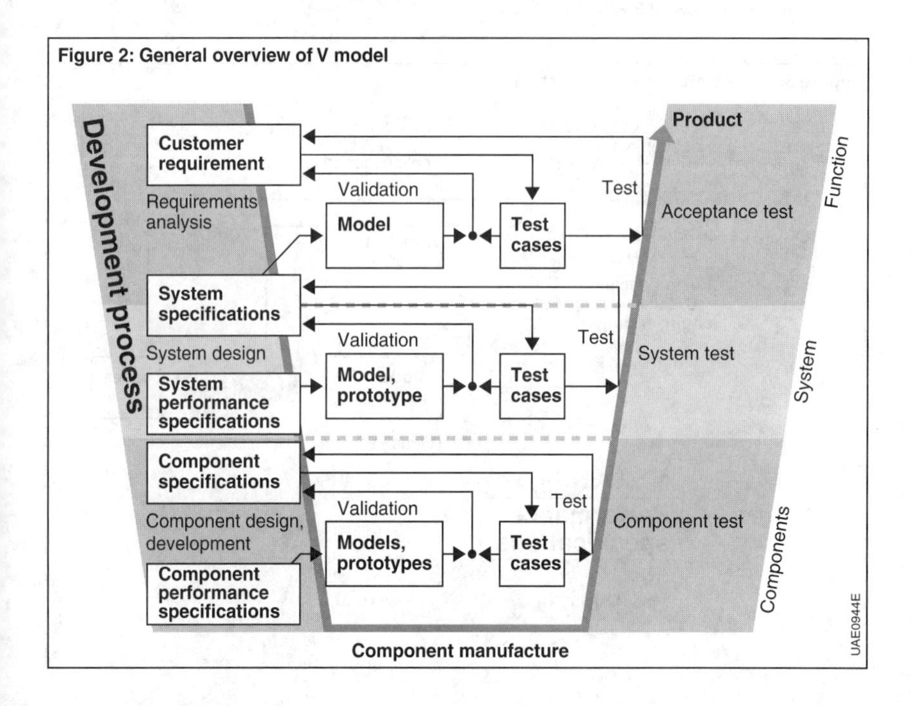

Outlook

The driving force behind mechatronics is the continuous progress in microelectronics. Mechatronics benefits from computer technology in the form of ever more powerful integrated computers in standard applications. Accordingly, there is a huge potential for further increases in safety and comfort in motor vehicles, accompanied by further reductions in pollutant emissions and fuel consumption. On the other hand, engineers face new challenges in mastering the new technologies for these systems.

Even in the event of a fault, future "X-by-wire" systems must continue to be capable of fulfilling a prescribed functionality without reverting to a mechanical or hydraulic fallback level. The condition for their implementation is a high-reliability and high-availability mechatronic architecture which requires a "simple" proof of safety. This affects both single components as well as energy and signal transmissions.

Besides "x-by-wire" systems, driver-assistance systems and their related man-machine interfaces are another field in which significant progress can be achieved for users and automotive manufacturers by the systematic implementation of mechatronic systems.

The design approaches of mechatronic systems should strive toward continuity in several aspects:
- Vertical: "Top-down" from system simulation, with the objective of overall optimization, through to finite-element simulation to achieve a detailed understanding, and "bottom-up" design engineering from component testing through to system testing.
- Horizontal: "Simultaneous engineering" across several disciplines in order to deal with all product-related aspects at the same time.
- Across corporate boundaries: The concept of a "virtual sample" is approaching gradually step by step.

Another challenge is training in order to further an interdisciplinary mindset and develop suitable DE processes and forms of organization and communication.

Figure 3: Recursion methodology at one level

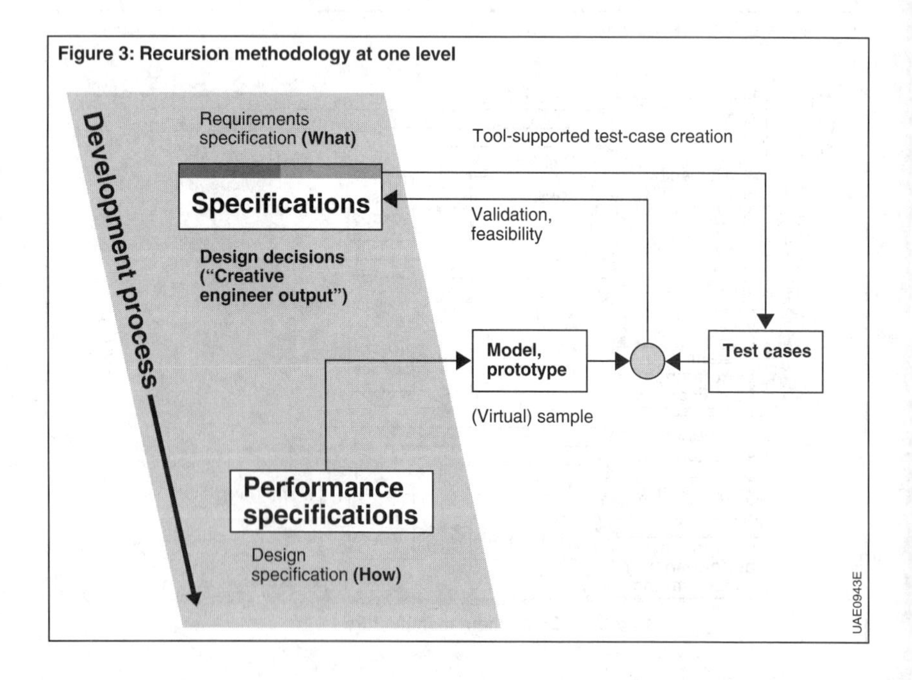

Passenger-compartment climate control

Climate-control requirements

The system for controlling the climate in the passenger compartment of automobiles and commercial vehicles must fulfill many functions. Firstly, it must create a comfortable climate for all the occupants. To this end, cooled or heated air (depending on the outside temperature) is fed into the passenger compartment so as to meet the physiological needs of the occupants. A further function of climate control is to keep the passenger compartment free of unpleasant odors and pollutants.

Climate control also has the safety-related function of demisting and de-icing the windshield and side windows. Statutory requirements that vary from country to country must be met.

The comfort requirements and thus the functions of the A/C unit normally increase with the vehicle category. Regionally different climate-control requirements must also be taken into consideration. In Europe, for instance, vehicles with a system of climate control which is as unobtrusive as possible are preferred, whereas in the USA a cool, noticeable air current is desired.

Design and operating principle of the A/C unit

Air intake

The A/C unit is installed out of sight of the occupants under the vehicle's instrument panel. Fresh air is usually drawn in at the base of the windshield (Figure 1). Rainwater and snow are removed en route to the A/C unit. The outside air is drawn in by a radial fan and delivered through an opening in the bulkhead (partition between engine and passenger compartments). There is a further possibility of drawing in passenger-compartment air (recirculated air) through a second air inlet. The proportions of fresh air and recirculated air from the drawn-in air are controlled by flaps.

Fresh air is needed to supply oxygen to the passenger compartment. In recirculated-air mode, unpleasant odors and pollutants are prevented from entering the passenger compartment. In addition, power consumption is significantly reduced in recirculated-air mode, since the air to be temperature-regulated does not have to be heated and cooled so much in heating and cooling modes respectively. However, cold and thus dry fresh air must sometimes be used to prevent the windows from misting over due to condensation of damp passenger-compartment air.

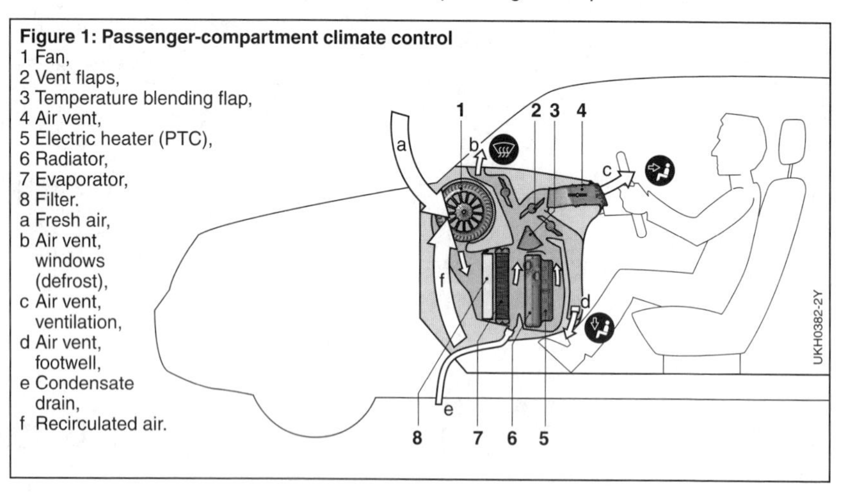

Figure 1: Passenger-compartment climate control
1 Fan,
2 Vent flaps,
3 Temperature blending flap,
4 Air vent,
5 Electric heater (PTC),
6 Radiator,
7 Evaporator,
8 Filter.
a Fresh air,
b Air vent, windows (defrost),
c Air vent, ventilation,
d Air vent, footwell,
e Condensate drain,
f Recirculated air.

Air cleaning
To clean the air, a sheet of filter fleece is pleated, i.e. folded, and preferably installed in the shape of a rectangular filter element before and after the fan in the A/C unit. The filter element is mainly positioned in such a way as to filter not only the fresh air, but also the recirculated air. The filter element has a limited retention capacity and must therefore be replaced. An activated-carbon layer can also be applied to the filter fleece (hybrid filter) in order to filter out particulates and pollutants.

Air cooling
The cleaned air is then cooled on the surface of an evaporator, which is now usually made entirely of aluminum. The air dissipates its energy proportionally to the evaporating refrigerant which is routed through the evaporator.

Because cold air can absorb less moisture than warm air, water is separated out as it cools and routed downwards along the evaporator. The water collects in the housing section under the evaporator, the condensate drip pan, and is discharged via the condensate outlet from the A/C unit to the atmosphere.

Air temperature regulation and distribution
After the evaporator the desired air temperature is set by a blending-flap system. To this end, the air is directed entirely or in part through a radiator (through which engine coolant flows) or – in pure cooling mode – past the radiator directly to the air vents. The air is continuously distributed to the two channels in any proportion by one or more temperature blending flaps. The cold and warm air currents are blended to suit demand in the downstream blending chamber. The air currents are blended in such a way as to cater for the comfort sensation of "cool head and warm feet". For this purpose air is blown out at different temperatures at the three main "Footwell", "Ventilation" (air vents in the instrument panel) and "Windows" vent levels, generating temperature stratification of up to 15 K.

The volume of air blowing out of the individual vents is continuously adjustable to any proportion by hand or by motor-driven vent flaps.

In further expansion stages of the A/C unit additional vents can be provided for the rear passengers or temperature regulation of the passenger compartment can be set differently for different seats. The first expansion stage is two-zone temperature regulation. The interior temperatures on the driver and front-passenger sides can be set independently of one another. Three-zone climate control caters for the rear-seat area as an additional climate zone, while four-zone climate control provides for individual climate control for driver, front passenger and two passengers in the back.

A new feature is individual adaptation of the footwell temperature, or "variable stratification". In this case, the outlet temperature in the footwell can be increased while the basic setting remains unaltered (for example for people with feet which are sensitive to cold).

Booster heater
In modern optimized-consumption internal-combustion engines the waste heat available in the engine coolant is reduced, and thus with it the heat supply in the heating circuit. In this case, an electric PTC booster heater is installed in the air passage after the radiator. By using PTC ceramic heating elements (PTC, Positive Temperature Coefficient), the booster heater automatically reduces its heat output at a defined limit temperature, since the resistance of the PTC elements increases abruptly. The PTC booster heater is therefore intrinsically safe.

The utilization of further heat sources, such as heat recovery from exhaust gas for example, is currently being developed.

Air routing
The temperature-regulated air is directed through plastic blow-channels to the relevant vents. Window and footwell outlets are not adjustable. The blow direction (nozzles) and the air volume can be individually adjusted at the air vents.

Supplementary A/C units

In luxury-class vehicles supplementary A/C units are sometimes use for the rear passengers. These are designed either, like the main A/C units, as heater-coolers or as pure coolers, and are usually supplied with recirculated air. Supplementary A/C units are installed, for example, in the center console, above the rear wheel arch, in the spare-wheel recess, or behind the rear seat bench.

Alternative A/C-unit concepts

The A/C unit described above sets the desired temperature by blending cold and warm air and represents the standard design used today. Alternative A/C-unit concepts may be used in which the air after the evaporator is routed directly through the radiator. A valve can be used to steplessly adjust the water flow rate through the radiator and thus the desired level to which the air flowing through is heated. The blending-flap system for blending cold and warm air is omitted.

Because of the thermal inertia of the water-filled aluminum radiator, desired changes to the air outlet temperature can only be completed slowly. To ensure that the passenger compartment is quickly cooled with the maximum air mass, these concepts normally utilize an additional air-side bypass past the radiator.

The advantage of this concept is the slightly more compact design of the A/C unit and the possibility of shutting off the radiator on the water side and thus avoid the unwanted situation of the air being heated in cooling mode by hot components in the A/C unit. The disadvantage is the expense of integrating a water valve.

Regulation of climate control

The climate of the passenger compartment is controlled by making inputs at the A/C operating unit, by adjusting the air vents and – in more luxurious vehicles – by means of various sensors. The variations range from direct mechanical activation of the A/C unit via Bowden cables and flexible shafts to fully automatic, processor-controlled air conditioning, where only the desired temperature is set.

The variables that can be influenced are blow-out temperature, air flow rate, air moisture, and air distribution to the different vents. The secondary variables are interior temperature including temperature stratification, air flow, acoustics at each seat, and misting status of the windows.

Most people find the following configurations pleasant: in summer mode cold air from the vents in the instrument panel, in winter mode warm air from the footwell and defrost vents (for keeping the windows clear), and in transition mode, heated air in variable proportions from the defrost, footwell and ventilation vents in the passenger compartment. Additional influencing factors can play a part in automatic air conditioning. The status and intensity of the Sun can be determined by solar sensors and included in the control operation. Air-quality sensors detect pollutants and odors in the ambient air and prevent them from entering the passenger compartment by closing the fresh-air inlets. Moisture sensors for the windshield enable action to be taken to prevent imminent window misting. The desired and actual passenger-compartment temperatures are compared by of temperature sensors.

Climate-control systems

Heating circuit

The waste heat from the engine is a source of energy which can be utilized without great expenditure to heat the passenger compartment. The hot coolant flows through the radiator in the A/C unit. The air passing through the radiator is heated.

In order to maintain the interior temperature in winter after a journey when the engine is switched off, it is possible to use an electric water pump, which forwards the coolant and makes the heat stored in the cooling system usable.

Refrigerant circuit

The function of the refrigerant circuit is to absorb the thermal energy of the air to be cooled in the evaporator and dissipate it at another point, outside the passenger compartment, to atmosphere. This is effected by using the cold-loss process in a closed refrigerant circuit (Figure 2). The main components of this circuit are the evaporator with expansion valve in the A/C unit, the condenser at the front of the vehicle, usually mounted directly in front of the coolant radiator, and the compressor secured to the engine block and driven by the engine.

The components are connected by metallic pipes which can contain flexible sections for decoupling. The compressor draws the gaseous refrigerant from the evaporator and compresses it, i.e. the pressure and temperature levels are greatly increased. In the condenser the hot refrigerant dissipates heat to the ambient air flowing through the condenser. This cooling condenses the vapor-state refrigerant. If the air stream is not enough to cool the refrigerant sufficiently, the radiator fan increases the air flow rate through the condenser. The refrigerant – now in liquid form and under high pressure – is atomized in the expansion valve and directed to the evaporator. The pressure level drops suddenly, as a result of which the refrigerant is highly cooled as it evaporates, and absorbs the heat of the air to be cooled in the A/C unit.

The evaporated refrigerant exits the evaporator in gaseous form and passes the expansion valve again, which, on the basis of the pressure and the temperature of the exiting refrigerant, alters the throttling cross-section on the inlet side so as to inject into the evaporator just that amount of refrigerant that can evaporate in the current operating state.

The cold gaseous refrigerant is drawn in again by the compressor and the circuit begins afresh.

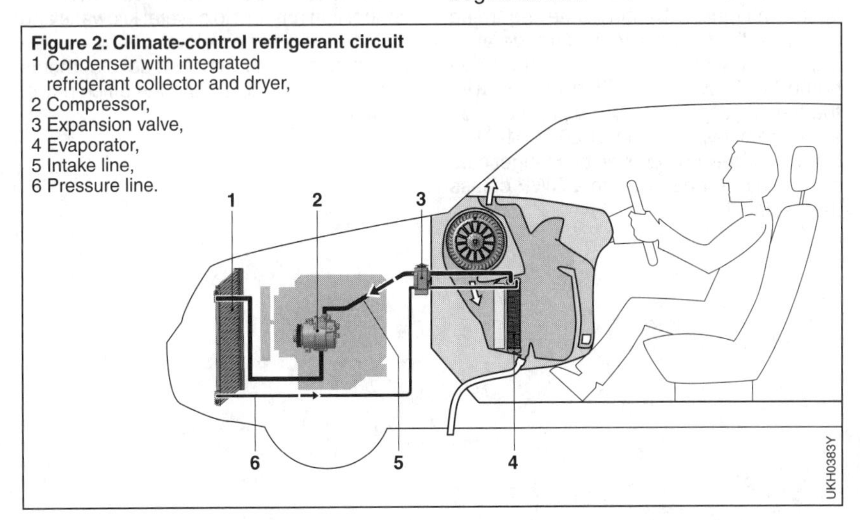

Figure 2: Climate-control refrigerant circuit
1 Condenser with integrated refrigerant collector and dryer,
2 Compressor,
3 Expansion valve,
4 Evaporator,
5 Intake line,
6 Pressure line.

In order to prevent ice from forming on the air side of the evaporator, which would reduce the evaporator air cross-section, the minimum air temperature must not drop below 0 °C. To measure the air temperature after the evaporator, either an air-temperature sensor is used which regulates the compressor accordingly if the temperature drops below a fixed level, or the minimum intake pressure and thus the evaporator temperature are limited in the compressor. The refrigerant output can be adjusted between zero kilowatts (refrigerant does not flow through the evaporator) and a maximum output of approx. 8 kilowatts (for maximum refrigerant quantity and air flow rate).

Refrigerant
The fluorocarbon 1,1,1,2-Tetrafluorethane, which goes by the trade name R134a, is used as the refrigerant for vehicles with homologation before 2011-01-01 or vehicles where an existing type approval has been expanded.

Because of this refrigerant's GWP (Global Warming Potential) of approx. 1,400 and the statutory provision which will come into force in Europe in 2011, which prescribes a refrigerant with a GWP of less than 150 for new vehicle models, an alternative refrigerant must be used here. There is currently still a lively debate regarding applicability of the alternative refrigerants. 2,3,3,3-Tetrafluoropropene with the trade name R-1234yf and carbon dioxide with the trade name R-744 are being pursued as alternatives. Starting at 2017-01-01 in Europe, all newly approved passenger cars must use a refrigerant with a GWP of less than 150.

Climate control for hybrid and electric vehicles

In vehicles with an engine start/stop function, it is necessary to keep the interior climate constant when the engine is switched off. For heating, an electric water pump can be used to utilize the residual heat from the engine circuit. For cooling, evaporators have been developed with a cold accumulator on the basis of a material which can store cold by means of a phase change and dissipate it again when stopped at traffic lights with the engine switched off. The interior climate can thus be kept at a pleasant level for up to two minutes.

In vehicles which run partly or entirely on electricity, a compressor is used which is electrically powered. Two solutions are available for heating. Either the water required for the radiator is electrically heated and supplied to the circuit by an electric pump, or the radiator is replaced by an electric heating element, the high-voltage PTC. Electric heater and electric compressor also open up the possibility of independent climate control. When the battery is sufficiently charged or connected to a charging station, the vehicle can be both preheated and precooled.

Because the power required for climate control can amount to several kilowatts and thus the range of an electric vehicle is reduced significantly, current developments are focused on increasing the efficiency of climate control.

Auxiliary heater systems

Function and designs

Auxiliary heater systems use the fuel carried in the vehicle's tank to generate heat. Air heaters (Figure 4) are used to heat the passenger compartment directly. They transfer the heat generated during combustion to the cabin air. Coolant heaters (Figures 5, 6, 7) introduce heat into the cooling water and are thus suitable both for preheating engines and for heating the passenger compartment via the vehicle's heat exchanger with blower.

Air heaters are particularly efficient due to the direct transfer of heat to the air and are suitable for long periods of heating. Coolant heaters use the cooling circuit and the vehicle's existing air ducts and are therefore particularly suitable for covering the heat deficit of low-consumption drives and for preheating the passenger compartments and engines of passenger cars.

Further advantages of auxiliary heater systems are:

- Ice-free windows thanks to preheating
- Optimized heating comfort
- Use of a truck sleeping cabin even in winter
- Less wear and lower emissions when starting with a preheated engine (coolant heater only)
- Operating temperature of the catalytic converter reached more quickly (coolant heater only).

Of increasing importance is their use in passenger cars with deficient heat output from waste heat caused by increased drivetrain efficiency by means of electrification. Particularly in purely battery-powered vehicles fuel-powered heaters, on account of their great efficiency in converting chemically bound energy into heat, ensures that the range is not limited in the case of the storage capacity of the traction batteries being specified.

Air heaters

Application

Air heaters are the most common type of auxiliary heater in trucks and buses. Their main advantages are low costs, faster installation, and lower power and fuel consumption. A dramatic increase in their use is being witnessed in the USA, since legislation prohibits the running of the engine to provide heat during breaks in driving and the heaters can heat the cabs with clearly lower fuel consumption.

Operating principle

Air heaters (Figure 4) operate independently of the vehicle's own heat balance. A combustion-air blower draws the air required for combustion from the surrounding atmosphere, and blows it into the burner. An electric metering pump delivers diesel or gasoline as fuel via an evaporator to the burner, where it mixes with combustion air, is ignited by a sheathed-element glow plug and burns.

The heating-air blower draws in the fresh air to be heated, blows it through a heat exchanger, and into the vehicle cabin.

In cold countries a heat output of approx. 4 kW is required. In temperate climates units with approx. 2 kW are sufficient.

An important safety aspect is that combustion air and exhaust gas from the cabin air must be kept entirely separate. This prevents exhaust gas from entering the cabin.

The heat output is regulated by varying the quantity of fuel, combustion air and fresh air. If the temperature measured by the room-temperature sensors deviates from what is set at the operating element, the output of the heater is adapted until the desired temperature is reached once again. The combination sensor registers unacceptable temperatures through defects or malfunctions that occur (e.g. overheating due to blocked heating vents) and switches off in good time if necessary.

Installation

Air heaters for commercial vehicles can usually be fitted directly inside the cab. In trucks, for instance, the preferred locations are in the co-driver's footwell, on the rear cab wall, under the bunk bed, on the outside of the cab wall, or in a stowage compartment. The exhaust pipe always runs under the floor (either into the wheel arch, or out of the rear cab wall).

Most sensor units in the fuel tanks of passenger cars and trucks already have a connection available for supplying fuel to the air heater. If necessary, an extra tank sensor unit is fitted. A fuel reserve for the vehicle's engine must be retained.

Coolant heaters

Application

Coolant heaters up to approx. 5 kW heating capacity are preferred for use in passenger cars. They use the cooling circuit and already existing heat exchangers to heat the engine and interiors and are thus easy to retrofit. Heaters up to approx. 12 kW are used in trucks to preheat the engine. The large body surface area and high interior volume of city and tour buses mean that they require heating capacities of up to 35 kW.

The main types are auxiliary and booster heaters for passenger cars. These heaters are integrated directly in the cooling-circuit feed line between the engine and the heat exchanger of the passenger cell (Figure 5). They utilize the existing devices: vehicle heat exchanger with blower, air flaps and vents.

Quick-starting heaters are used to meet high demands with regard to heating-up speed; these reach their full heat output level after approx. 30 s. The required acceleration of fuel vaporization is achieved by heating the vaporizing apparatus electrically during the starting sequence.

Auxiliary heaters for preheating engine and passenger compartment before starting driving

Heat is generated through combustion of fuel along similar lines to air heaters. The generated heat is transferred to the cooling water flowing between the water jacket and heat exchanger (Figure 7) of the heater.

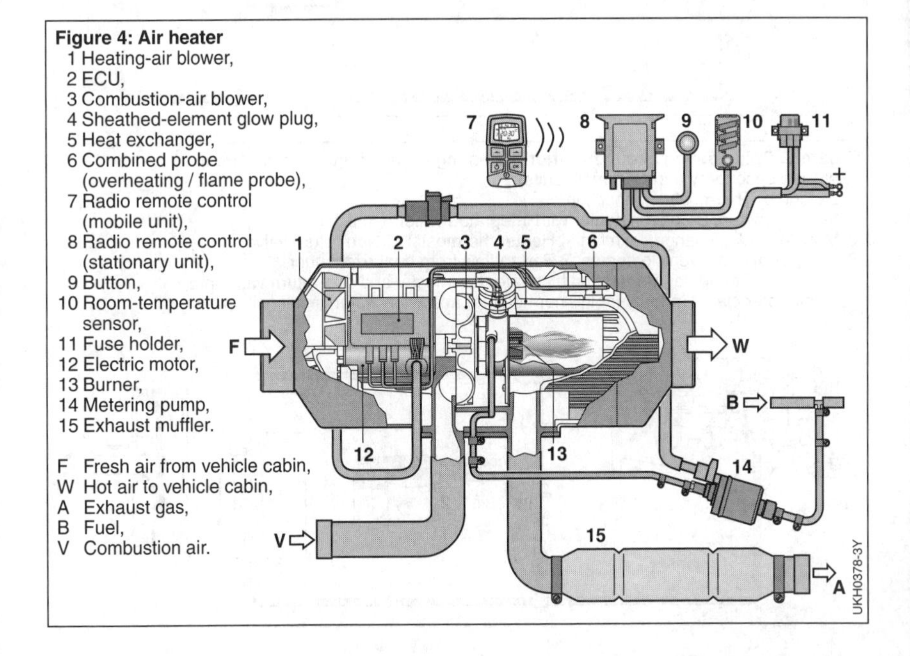

Figure 4: Air heater

1 Heating-air blower,
2 ECU,
3 Combustion-air blower,
4 Sheathed-element glow plug,
5 Heat exchanger,
6 Combined probe (overheating / flame probe),
7 Radio remote control (mobile unit),
8 Radio remote control (stationary unit),
9 Button,
10 Room-temperature sensor,
11 Fuse holder,
12 Electric motor,
13 Burner,
14 Metering pump,
15 Exhaust muffler.

F Fresh air from vehicle cabin,
W Hot air to vehicle cabin,
A Exhaust gas,
B Fuel,
V Combustion air.

The simplest method of integrating them in the vehicle's cooling system is to connect them in series between the engine and the heat exchanger in the primary circuit (Figure 5). A disadvantage with engines over approx. 2.5 l is that engine heating makes heating the passenger compartment much slower. In such cases, the passenger cell can be heated as a matter of priority by a secondary circuit (Figure 6); the engine is heated with this secondary flow only after the thermostat opens when a high coolant temperature is reached. The non-return valve prevents the circuit from being bypassed via the thermostat.

The water pump (Figure 7) in auxiliary heaters pumps engine coolant to the system's heat exchanger, where it is heated up. On the one hand, the heated water preheats the vehicle engine; on the other hand, the vehicle's own heat exchanger (Figures 5 and 6) dissipates the heat. The heated air is blown controllably through the existing ventilation system into the vehicle cabin. Operation is controlled for manual immediate operation and for programing switch-on times and heating duration by a timer, by radio remote control, or even by cellular phone or landline telephone.

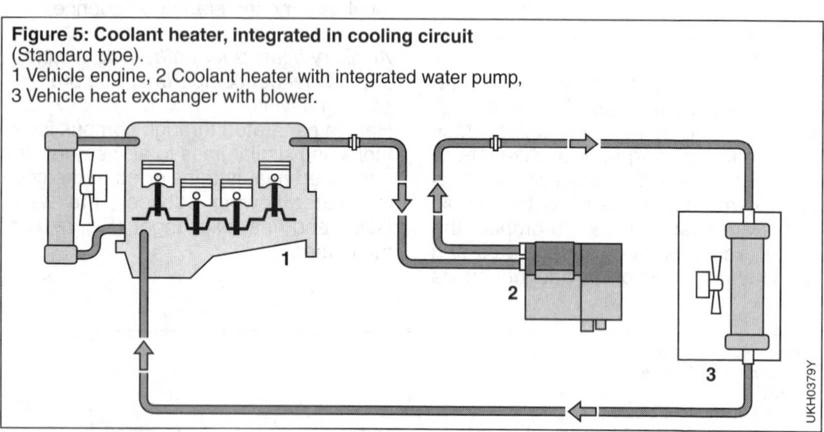

Figure 5: Coolant heater, integrated in cooling circuit
(Standard type).
1 Vehicle engine, 2 Coolant heater with integrated water pump,
3 Vehicle heat exchanger with blower.

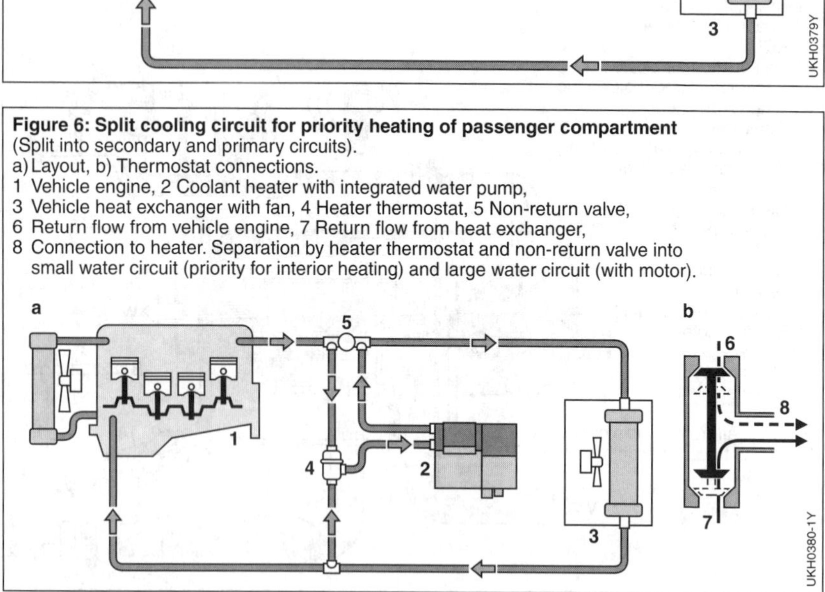

Figure 6: Split cooling circuit for priority heating of passenger compartment
(Split into secondary and primary circuits).
a) Layout, b) Thermostat connections.
1 Vehicle engine, 2 Coolant heater with integrated water pump,
3 Vehicle heat exchanger with fan, 4 Heater thermostat, 5 Non-return valve,
6 Return flow from vehicle engine, 7 Return flow from heat exchanger,
8 Connection to heater. Separation by heater thermostat and non-return valve into small water circuit (priority for interior heating) and large water circuit (with motor).

Booster heaters for covering heat deficits while driving

Efficiency-optimized drivetrains, such as highly turbocharged diesel engines with small displacements and hybrid drives, generate on account of their efficiency insufficient waste heat to heat the passenger compartment. Booster heaters are used to make up this heat deficit. They operate only at outside temperatures below +5 °C and when the engine is running. They therefore do not require their own water pump.

Booster heaters can be upgraded to an auxiliary heater by retrofitting a water pump and a control element with ECU.

When used in an electric vehicle, the fuel-powered heater can cover the entire heat demand.

Installation

To avoid heat losses through long cooling-water hoses, coolant heaters are usually installed in the engine compartment. The confined installation conditions necessitate a very compact construction. Installation is simplified by integrating all the components required to generate and deliver heat in the heater unit (Figure 7).

Regulations

All auxiliary heaters and fuel-operated booster heaters have EU type approval in line with Directive 2001/56/EC. If a heater with this type approval is retrofitted in line with the manufacturer's installation instructions, no inspection/approval by an expert or inspection body is prescribed.

The installation of auxiliary heaters in vehicles used for international transport of hazardous goods is governed by the European ADR agreement (Accord européen relatif au transport international des marchandises Dangereuses par Route). The heater must be switched off for safety reasons before the vehicle enters a hazardous area (e.g. refinery or filling station). The heating also switches off automatically as soon as the vehicle engine is shut down or an ancillary (e.g. auxiliary drive for discharge pump) is switched on.

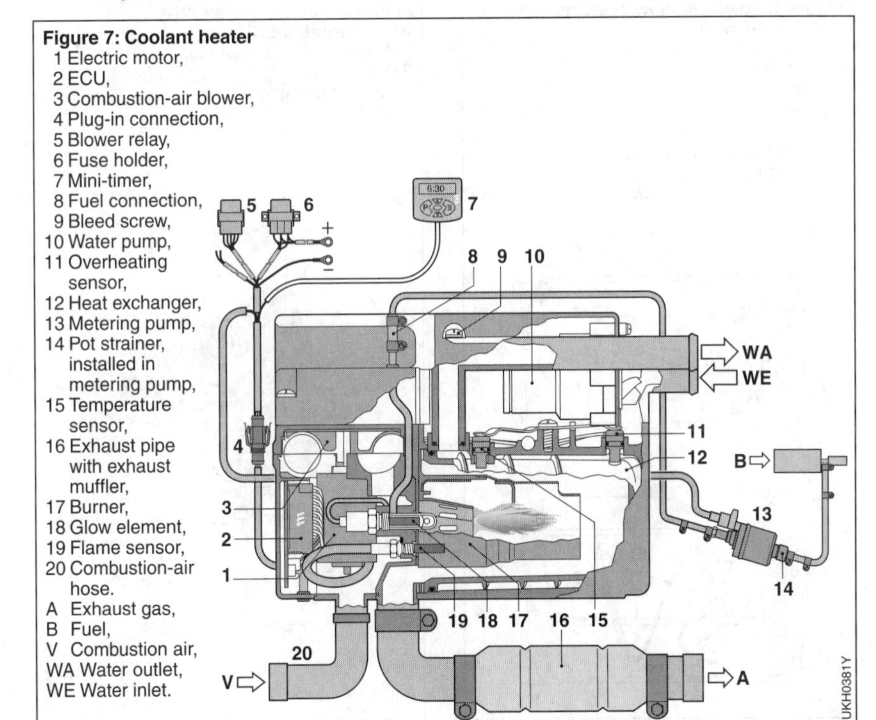

Figure 7: Coolant heater
1 Electric motor,
2 ECU,
3 Combustion-air blower,
4 Plug-in connection,
5 Blower relay,
6 Fuse holder,
7 Mini-timer,
8 Fuel connection,
9 Bleed screw,
10 Water pump,
11 Overheating sensor,
12 Heat exchanger,
13 Metering pump,
14 Pot strainer, installed in metering pump,
15 Temperature sensor,
16 Exhaust pipe with exhaust muffler,
17 Burner,
18 Glow element,
19 Flame sensor,
20 Combustion-air hose.
A Exhaust gas,
B Fuel,
V Combustion air,
WA Water outlet,
WE Water inlet.

Comfort and convenience systems in the door and roof areas

Power-window systems

Power windows have mechanisms that are driven by electric motors. There are two types of design in use (Figure 1):

- Arm window lift: The window-lift-drive pinion engages with a quadrant gear, which is connected to a rod linkage. This system is highly efficient. On the down side, the system requires a large amount of space for fitting and is heavy.
- Flexible cable window lift: The window-lift drive turns a cable reel, which operates a flexible cable mechanism. Systems with two guide rails are used for the most part in front doors. Systems with one guide rail are preferred in rear doors. A crucial advantage of these designs is their excellent ability to guide the window glass.

Power-window drive

The power-window drive consists of a DC motor with downstream reduction gearing (Figure 2). For comfort and safety reasons, many power-window drives are equipped with an ECU. The gearing is designed as a helical bevel gear to create the necessary self-inhibiting effect which prevents the window from opening inadvertently of its own accord or being opened by force. Special tribological measures and the magnetic-circuit configuration of the DC motor further improve the drive's self-inhibiting properties. Space limitations inside the door make narrow (flat) drive units imperative.

Dampers integrated in the gearing mechanism provide good damping characteristics at the window-glass end positions.

Figure 1: Power-window systems
a) Arm window lift,
b) Flexible cable window lift.
1 Power-window drive,
2 Guide rail,
3 Driver,
4 Rod linkage,
5 Drive cable (Bowden cable),
6 Quadrant gear.

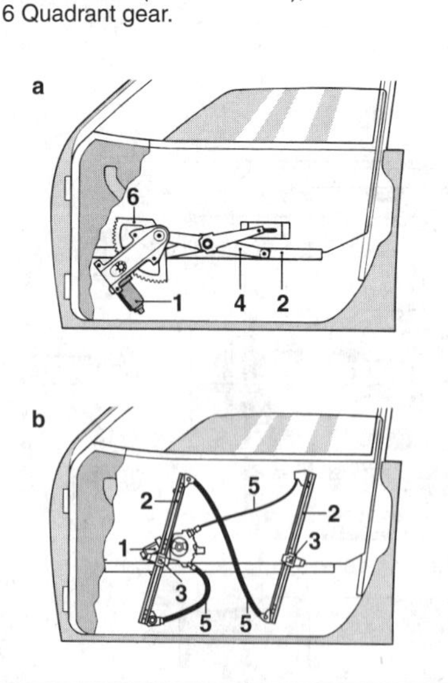

Figure 2: Power-window drive with integrated ECU
1 Power-window drive motor,
2 Worm,
3 Helical bevel gear,
4 ECU.

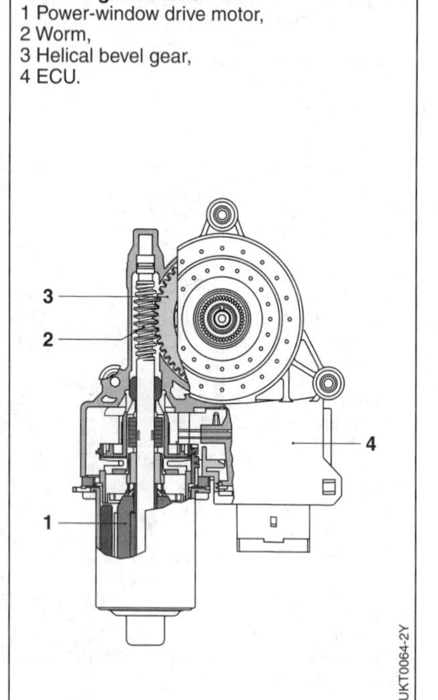

Electronic control unit (ECU)

Two different operating modes are used. In manual mode the power-window drive is operated using a switch which is pressed over the full extent of window travel. In automatic mode the window glass is opened or closed by a short pressing of the switch.

A force-limitation device is required by law to avoid injuries in the course of automating closing. In Germany, paragraph 30 of the StVZO (Road Traffic Licensing Regulations) stipulates that the force-limitation device must be effective when the window is moving upwards within a range of 200 to 4 mm from the upper edge of the window aperture. The power-window drive includes integral Hall sensors to monitor motor speed during operation (Figure 3). If a drop in speed is detected, the DC motor's direction of rotation is immediately reversed. The window closing force, measured for a spring rate of 10 N/mm, must not exceed 100 N. To be able to close the window safely, the force-limitation device is automatically deactivated before the window glass enters the door seal. The window position is monitored over its entire range of movement.

Depending on the vehicle topology, the electronic control may be accommodated in a central ECU or decentrally in the door, preferably directly in the power-window drive. Decentralized electronics can be networked via a LIN-bus interface. The advantages of such a solution are fault diagnosis of the electronics and reduction of the amount of wiring.

Figure 3: Power-window control unit featuring electronic force-limitation device
1 Microcomputer,
2 Relay output stage,
3 Control commands,
4 Networking via CAN,
5 Hall sensors.

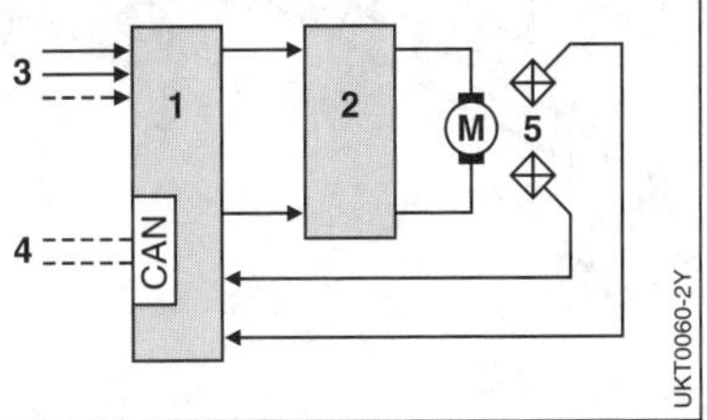

Sunroof systems

Three different types of sunroof systems may be used:

- The simple sliding/tilting sunroof: A panel (frequently made of glass) can be tilted or slid to the rear underneath the roof skin.
- The large sunroof or panoramic sunroof: This system utilizes up to three sliding-sunroof drives with different adjustment options (e.g. adjusting a glass panel in the front and rear areas, or with a fixed glass panel and an electrically adjustable roller sun blind).
- The spoiler sunroof: Here the glass panel is tilted and then moved to the rear over the roof skin (hence the term "spoiler").

Power-sunroof drives

Sunroofs and roller blinds are primarily adjusted by means of torsion- and pressure-resistant control cables or plastic toothed straps with mechatronic drives. These are located in the roof between the windshield and the sliding sunroof or at the rear of the vehicle.

In the event of an electric-adjustment failure, the sunroofs can be closed by means of a hand crank provided on the drive.

The drive comprises a DC motor with worm gearing and an ECU. Control is provided by means of exact position pulses. It can, in addition to exact positioning, provide a force-limitation facility in order to prevent injuries when the sunroof is being closed.

The drive is actuated via external signal inputs using pushbuttons, switches, and analog or digital preselector switches.

Control can be integrated in the vehicle's bus system (CAN, LIN) and diagnostic information can be output. Electronic control offers the opportunity to easily implement a wide range of functionalities and comfort and convenience functions (e.g. preselectable position control, closing via remote control).

Comfort and convenience functions in the passenger compartment

Electrical seat adjustment

Electrical seat adjustment both ensures greater operating convenience than mechanical operation and is preferred on account of its lower space requirement or various adjustment levels in the case of poor accessibility.

Up to ten motors control the following functions: seat-cushion height adjustment, front and rear, seat fore/aft adjustment, seat-cushion depth adjustment, backrest-tilt adjustment, lumbar support adjustment (height and curvature), shoulder-support tilt angle (top third of backrest), adjustment of the backrest and seat-cushion width, head-restraint height adjustment.

Direct drive

A common seat-bottom frame includes four motors which drive the height-adjustment gearing and a combined fore/aft and height-adjustment gearing (Figure 1). On simple seats, it is not possible to adjust the seat cushion depth.

Universal drive

The universal drive consists of three identical motors with four height and two fore/aft adjustment gear sets. The gear assemblies are driven by the motors via flexible shafts. This is a highly flexible system and can be installed on any seat design. The drive units are becoming smaller and smaller on the one hand. However, on the other hand, the flexible shafts generate additional costs and are susceptible to noise, which means that the universal drive is being replaced by the direct drive to an increasing degree.

Integrated concept

Modern seats (especially for sports cars) do not merely affix the lap belt to the seat frame, they also attach the shoulder strap – together with its height adjustment, inertia reel and seat-belt pretensioner mechanism – to the backrest. This type of seat design ensures optimized seat-belt positioning for a wide range of different occupant sizes and for all adjustable seat positions. It makes an important contribution to occupant safety. The seat frame must be reinforced for this type of design, while both the gearset components and their connections to the frame must be strengthened.

Pneumatic and mechanical adjustment possibilities for the seat type are offered as an additional comfort option. These are also available in a dynamic-driving version and can be combined with a massage function.

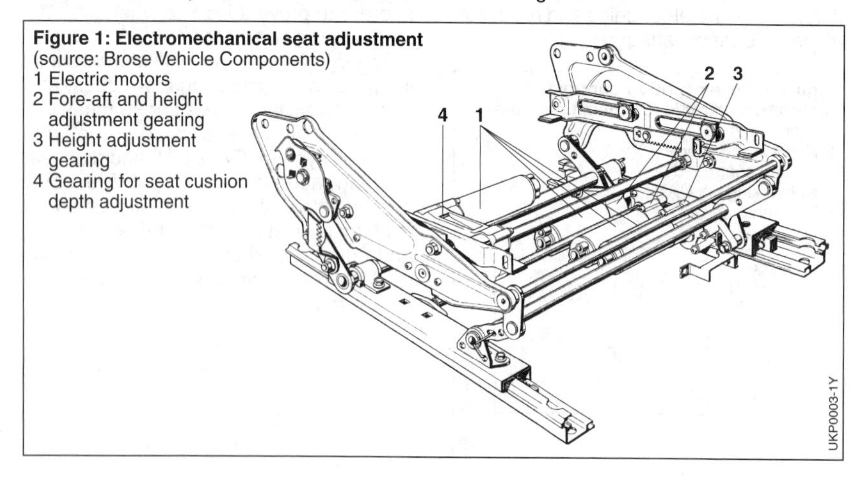

Figure 1: Electromechanical seat adjustment
(source: Brose Vehicle Components)
1 Electric motors
2 Fore-aft and height adjustment gearing
3 Height adjustment gearing
4 Gearing for seat cushion depth adjustment

Display and control

Interaction channels

Drivers have to process a constantly increasing stream of information originating from their own and other vehicles, from the road, and from telecommunication equipment. All this information must be conveyed to drivers on suitable display and indicating equipment that complies with ergonomic requirements.

Visual channel – seeing

A human being discerns their environment predominantly through their sense of sight (Figure 1). Other road users, their position, their suspected behavior, the lane, and objects in the road area are detected with the human being's sight apparatus and their subsequent highly powerful image processing and interpretation, selected and evaluated by further structures in the brain with regard to their development and relevance.

Even the infrastructure in road traffic calls above all on the visual channel: Road signs convey regulations and directions, markings delimit lanes from one another, direction indicators indicate a change in driving direction, stop lamps warn against vehicles that are slowing down. The visual channel is therefore extremely important to the driving environment. This applies to conscious vision, whereby the driver turns their eyes specifically to objects and focuses on them, but also to peripheral vision, which is crucial to positioning the vehicle within the lane. For this reason, additional glances at displays in the vehicle during interaction with driver-information and driver-assistance systems or their monitoring must be carefully assessed as to possible effects on road safety.

Acoustic channel – speaking and hearing

For communication with other road users, particularly the display and signaling of danger, the acoustic channel is used by human beings and by driver-assistance systems. In the driver's own vehicle this channel is also used to input commands via speech input systems and to output warnings and information from the driver-assistance system to the driver with acoustic signals and speech output.

The input of voice commands does not require the driver to turn their eyes, but does take up the driver's attention. Hearing does not require the driver to turn their eyes, but can convey spatial and complex information (e.g. description of the situa-

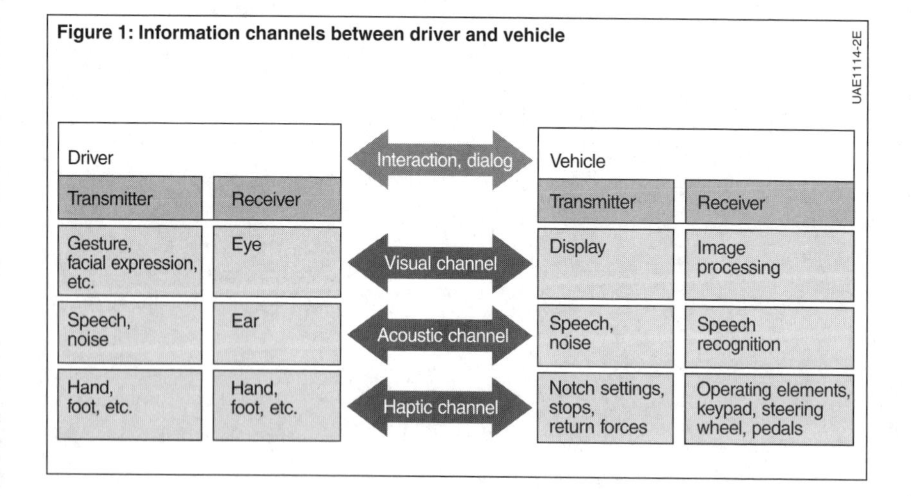

Figure 1: Information channels between driver and vehicle

tion at intersections) poorly. Even drivers who do not have an adequate sense of hearing must be able to communicate with the driver-assistance system.

Haptic channel – operation and feeling

The haptic channel provides feedback to the driver in the case of all motor operations when switches are operated and during steering and braking. Warning the driver by tightening their seat belt briefly and by vibrating the seat in the event of an imminent departure from the present lane has already been introduced in series production. A further possibility for arousing the driver's attention is the vibration of the steering wheel. An increased resistance on the accelerator pedal can support a driving-speed recommendation and be used as an indicator for proper steering behavior to stay in the lane. A torque that the driver can feel can be exerted on the steering wheel to suggest an evasive maneuver.

Even the kinesthetic channel, which helps the driver to perceive instances of acceleration while driving, is already in use in production vehicles to get the driver's attention, for instance by way of a short braking jolt.

The haptic channel is however continuously required for steering and cruise control, and additional manual operations (e.g. operation of the cellular phone's keypad) may impair lateral guidance by distracting the driver.

Instrumentation

Information and communication areas

In the vehicle there are four information and communication areas, which place different demands on the properties of the respective display or indicator: the instrument cluster, the windshield, the center console, and the rear cabin.

The supply of available information and the necessary, advisable or desirable information for the respective occupant determine which communication areas are used in the case. Dynamic information (e.g. driving speed) and monitoring information (e.g. fuel level), to which the driver should respond, is displayed in the instrument cluster as close as possible to the driver's primary field of vision.

A head-up display (HUD), which reflects the information on the windshield, is ideally suited to engaging the driver's attention (e.g. in the case of warnings from a driver-assistance system or for route directions). The head-up display is also suitable for displaying the speed in a digital form. A supplementary feature is the output of acoustic signals or speech.

Status information or more extensive operation dialogs (e.g. for vehicle navigation) are preferably shown in the central display in the center console. The operating units are usually assigned to this central display for more compact vehicles. For more spacious vehicles with a high-mounted central display (see Figures 2c and 2d), the preferred place to arrange the control units is in the primary reach area in the form of a rotary/push control element on the transmission hump.

Information of an entertainment nature is featured in the vehicle rear compartment, away from the primary field of vision. This is also the ideal location for a mobile office. The backrest of the front-passenger seat is a suitable installation location for the monitor and operator unit of a laptop computer.

"Dual View" or "Split View" displays for the center console have additionally been available since 2009; these displays can show different information to driver and passenger so that the passenger can, for example, watch videos without distracting the driver.

Instrument clusters

Earlier single instruments for the optical output of information (e.g. driving speed, engine rpm, fuel level, and engine temperature) were initially superseded by more cost-effective, well-lit and antireflective instrument clusters (combination of several information units in one housing). The passage of time, with the continual increase in information, saw the creation of the modern instrument cluster in the existing space available, with several needle instruments and numerous indicator lamps (Figure 2a). This first generation of instrument cluster was dominated by the eddy-current measuring instrument for the speedometer, the moving-coil measuring instrument for the rev counter and the hot-wire instrument for the fuel gauge.

Figure 2: Driver information area (development stages)
a) Needle instrument,
b) Needle instrument with TN LCD and separate AMLCD in center console,
c) Needle instrument with integrated (D)STN LCD or AMLCD and AMLCD-technology screen in center-console area,
d) Freely programmable instrument with two AMLCD components.
For designations, see Display types.

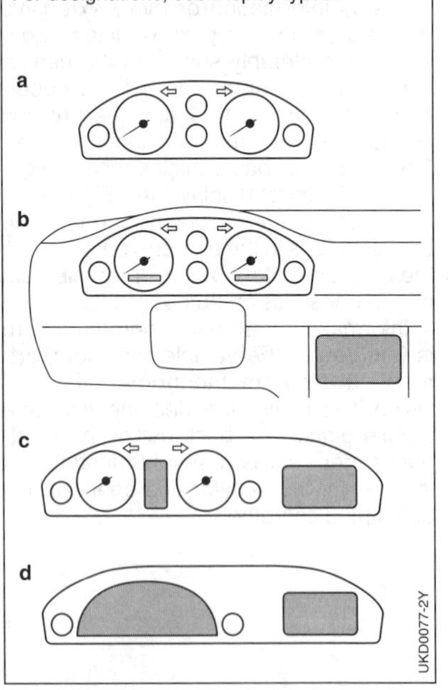

Measuring instruments

Even today most instruments still operate with mechanical needles and dial faces (Figures 2b and 2c). However, slim-fitting stepping motors are for the most part currently being used as measuring instruments. Due to a compact magnetic circuit and (mostly) two-stage gearing with power consumption of only approx. 100 mW, these motors permit swift and highly accurate needle positioning (Figure 3).

Digital displays

The digital instruments fitted up to the 1990s (e.g. for indicating driving speed) displayed information using vacuum-fluorescence display (VFD) technology and, later, liquid-crystal (LCD) technology, but they have now largely disappeared.

Figure 3: Instrument cluster (design)
1 Indicator lamp,
2 Printed-circuit board,
3 Stepping motor,
4 Reflector,
5 View cover,
6 Needle, 7 LED,
8 Dial face,
9 Optical waveguide,
10 LCD.

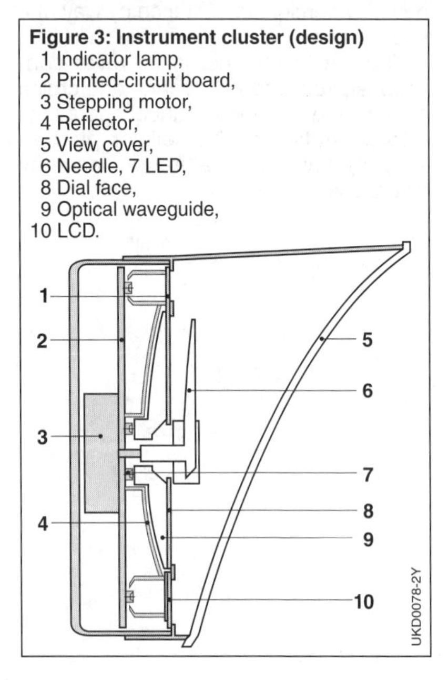

Instead, conventional analog needle instruments are used in combination with displays. At the same time, there is an increase in the size, resolution and color representation of the displays.

Lighting
To illuminate instrument clusters, backlighting technology has, on account of its attractive appearance, gained acceptance to a large extent over the frontlighting technology used previously. Bulbs have been replaced by long-lasting light-emitting diodes (LEDs). LEDs are suitable both as warning lamps and as backlighting for scales, displays and needles (using plastic optical waveguides if necessary).

The efficient yellow, orange, and red InAlGaP technology LEDs are now in widespread use. The more recent InGaN technology has produced significant efficiency improvements for the colors green, blue, and white. Here, the color white is obtained by combining a blue LED chip with an orange-emitting luminescent material (yttrium-aluminum garnet).

Color LCDs, on account of the very low transmission (typically approx. 6 %), had initially required cold cathode fluorescent lamps (CCFLs) for the backlighting in order to achieve a good contrast in daylight. These have since been replaced by white LEDs, which are becoming ever more efficient.

Graphics modules in the instrument area
Fitting vehicles with a driver's airbag and power-assisted steering as standard has resulted in a reduction in the view through the top half of the steering wheel. At the same time the amount of information that has to be displayed in the installation space available has increased. This creates the need for additional display modules with graphics capabilities and display areas that can show any information flexibly and in prioritized form. This tendency results in instrumentation featuring a classical needle instrument, but supplemented by a graphics display.

The graphical modules in the instrument cluster mainly enable an indication of driver-related functions such as service intervals, check functions covering the vehicle's operating state, as well as vehicle diagnostics as needed by a workshop. They can also show route-direction information from the navigation system (no digitized map excerpts, instead only route-direction symbols such as arrows as turnoff instructions or Intersection symbols). The original monochrome units are now being superseded on higher-specification vehicles by color displays (usually TFT – Thin Film Transistor – screens) which can be read more quickly and easily because of their color resolution.

TFT displays have also been in use to represent analog instruments since 2005 (Figure 2d). One example is the instrumentation in the Mercedes S-Class, in which the speed indication is emulated using an LCD screen. During changeover to the night-vision function the camera's conditioned video image appears, while the engine rpm indication is shown in bar form under the video image. For cost reasons, however, this technology is only gradually replacing conventional displays.

Central display and operating unit in the center-console area
With the advent of navigation and driver-information systems, screens and keyboards became widespread initially in the center console. Such systems combine all the additional information from function units and information components in a central display and operating unit, also sometimes including information from driver-assistance systems such as parking and maneuvering aids as of late. The components are interconnected in a network and are capable of interactive communication.

Initially the central screen was positioned in the center console, because there was sufficient installation space there. The vehicle manufacturers have since switched to the ergonomically more favorable solution and install the screen in the center-console area at the height of the instrument cluster (Figure 2c). A wealth of information is shown on this central screen, e.g. navigation, communication (telephone, SMS, Internet), audio (radio, digital sound recording media), video and TV when the vehicle is stationary, climate control, and settings (date, time).

The important issue for all visual displays is that they can be easily read within the driver's primary field of vision or its immediate vicinity so as to ensure that the driver does not have to look away from the road for long periods. Positioning this terminal, which is of universal use to driver and passenger, in the upper center-console area is effective and necessary from both ergonomic and technical standpoints. The optical information appears in a graphics display. The demands placed by TV reproduction and the navigation system on the video and map display determine its resolution and color reproduction.

For the central display monitor with an integrated information system, there has been a change from a 4:3 aspect ratio to a wider format with a 16:9 aspect ratio, which allows additional route-direction symbols to be displayed as well as the map. There are even displays with an 8:3 aspect ratio to enable to images to be displayed next to each other at the same time.

Figure 4: Functioning principle of a liquid crystal display (nematic cell)
1 Polarizer,
2 Glass,
3 Orientation and insulation layer,
4 Electrode,
5 Polarizer.
a Segment area.

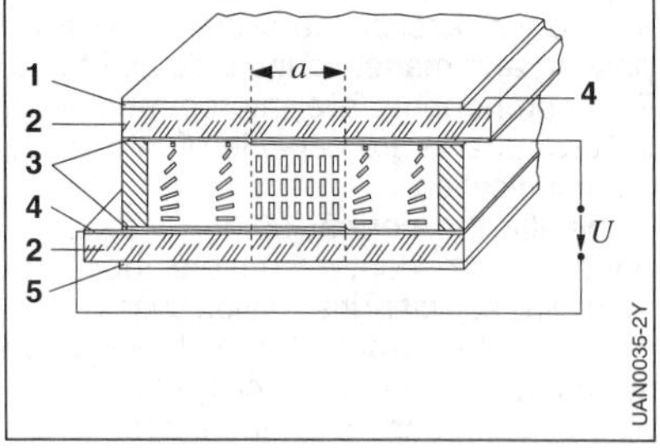

UAN0035-2Y

Display types

Liquid crystal display

The liquid crystal display (LCD) is a passive display element, as it does not emit any light itself. The liquid crystal substance is held between two glass plates (Figure 4). In the display area the glass plates are covered with a transparent, conductive layer, to which a voltage is applied. Between the layers an electric field is generated, subjecting the liquid crystal molecules to a reorientation.

TN LCD

TN LCD (Twisted Nematic Liquid Crystal Display) is the most commonly used form of display. The term stems from the twisted arrangement of the elongated liquid-crystal molecules between the locating glass plates with transparent electrodes. The operating range is −40 to 110 °C. The switching times are slightly longer at low temperatures due to the viscous properties of the liquid crystal material.

An additional orientation layer inside the LCD causes uniform alignment of the liquid crystal molecules at the boundary surfaces. When installing the cell, the glass substrate is arranged to create a 90° twist from the molecules in the de-energized state. This causes the polarization plane of the light passing through the cell to rotate.

TN LCDs can be operated in positive contrast (dark characters on a light background) or negative contrast (light characters on a dark background). In the positive-contrast cell the outer polarizers are arranged in such a way that their polarization directions are vertical to each other. The cell is then transparent in the de-energized state. In the area of the opposing electrodes, the liquid crystal molecules are aligned in the direction of the electric field by applying voltage. Rotation of the plane of polarization is now suppressed and the display area becomes opaque.

The TN LCD is usually operated in negative contrast in the vehicle since this corresponds to the familiar appearance of printed dials. In this negative-contrast cell the polarizers are arranged parallel to each other. The cell is then opaque in the de-energized state and becomes transparent when voltage is applied.

Positive-contrast cells are suitable for front- and backlighting operation, negative-contrast cells essentially require backlighting.

Separately controlled segment areas can be used to portray numbers, letters, and symbols. LCD picture elements arranged in matrix form and individually activated by thin-film transistors (TFT) form the basis of TFT flatscreen displays.

TN technology is suitable not only for smaller display modules but also for larger display areas in modular, or even full-size, LCD instrument clusters.

Dot matrix displays with graphics capabilities are needed to display infinitely variable information. They are activated by line scanning and therefore require multiplex characteristics. The multiplex properties of TN LCDs are limited in motor vehicles (temperature dependence).

STN LCD and DSTN LCD

For higher multiplex rates, STN (Super Twisted Nematic) and DSTN technology (Double-layer STN) are possible with medium resolution. The molecule structure of these displays is more heavily twisted within the cell than in a conventional TN display. STN LCDs allow only monochrome displays.

Due to this limitation and the minimal cost advantage compared to AMLCD technology described below, these displays no longer carry much importance.

AMLCD

The task of the visually sophisticated and rapidly changing display of complex information in the area of the instrument cluster and the center console can only be performed effectively by an active matrix liquid crystal display (AMLCD) which has high-resolution liquid crystal monitors with video capabilities. The pixels are addressed with thin-film film transistors (TFT LCD, Thin-Film Transistor LCD). Display monitors with diagonals of 3.5 to 10 inches in the center-console area and an extended temperature range (–25 to 95 °C) are available for motor vehicles. Sizes of 10 inches and more are used for programmable instrument clusters (series production in the 2005 Mercedes S class).

TFT LCDs consist of the "active" glass substrate and the opposing plate with the color-filter structures. The active substrate accommodates the pixel electrodes made from tin-indium oxide, the metallic row and column conductors, and semiconductor structures. At each intersecting point of the row and column conductors, there is a thin-film field-effect transistor, which is etched in several masking steps from a previously applied sequence of layers. A capacitor is generated at each pixel as well.

The opposite glass plate accommodates the color filters and a "black matrix" structure for covering the metallic row and column conductors, which improves display contrast. These structures are applied to the glass in a sequence of photolithographic processes. A continuous counter-electrode is applied on top of them for all the pixels. The color filters are applied either in the form of continuous strips (good reproduction of graphics information) or as mosaic filters (especially suitable for video pictures).

Organic light-emitting diodes

The luminescence excitation for organic, electroluminescent display units (OLED, organic light emitting diode) occurs, as in light-emitting diodes (LED), through injected charge carriers in a layer of organic materials. Three processes have to take place within the system of coatings for electroluminescence to occur: Injection, transportation of both types of charge carriers (electrons, holes) through the layer as well as their recombination. In molecules, the electron orbitals of the outer shells of the atoms involved overlap each other. This can lead to the formation of molecular orbitals with two electron states. The absorption of electromagnetic radiation can also raise electrons to higher energy levels and cause recombinations into one molecule during transport through molecular layers. These excited states (excitons) can go through both radiative and non-radiative decay.

Depending on the number of layers used in an OLED structure, systems differ between single, bilayer and trilayer systems. In real-world use, bilayer systems are dominant due to the low quantum efficiency in single-layer systems.

A heterostructure (Figure 5) has a barrier for electrons and holes between the two organic layers. This results in areas of increased charge-carrier concentrations near the boundary surface. This causes an increased probability of recombination for the charge carriers that cross the barrier. In these systems, the functions of the electron and hole transportation is separated. One of the layers also assumes the task of recombination and light emission. The barriers for the charge carriers significantly increases the internal quantum efficiency.

The organic layers are embedded between two electrodes. Figure 6 shows the structure of an OLED picture element.

Glass coated with indium tin oxide (ITO) is typically used as the anode. This glass lets the generated light through. The cathode consists of a suitable metal.

Figure 5: Band diagram of a OLED bilayer system
1 Radiative recombination inside the layer
2 Injection (thermal or by tunneling)

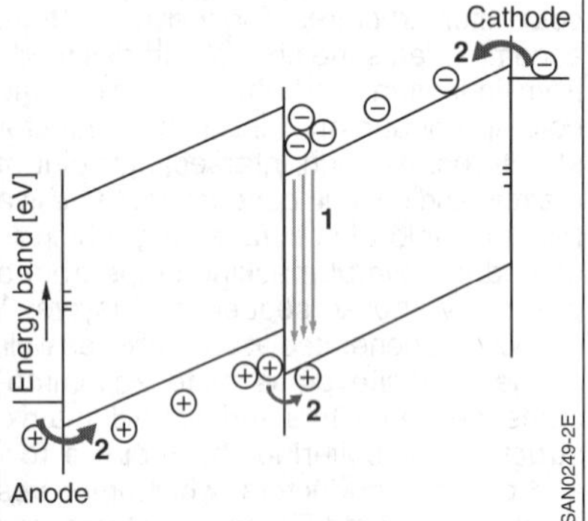

Figure 6: Structure of an OLED picture element
1 Metal cathode (Al, In)
2 Polymer or monomer layer system
3 Transparent anode
4 Carrier made of glass or flexible plastic foil (PET)

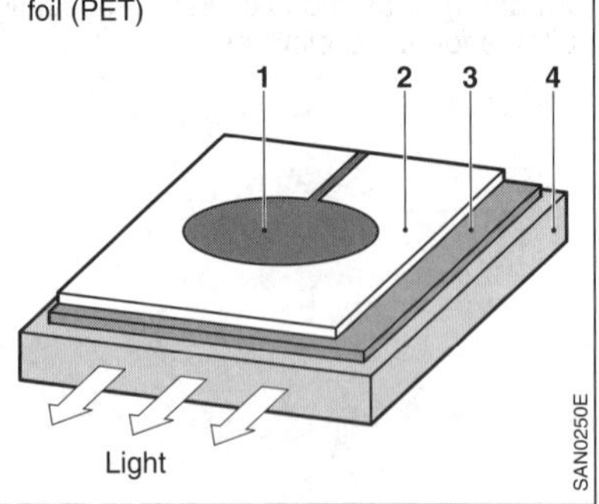

Figure 7: OLED matrix display
1 Substrate
2 Anode strips
3 Organic layer system
4 Cathode strips
5 Pixels

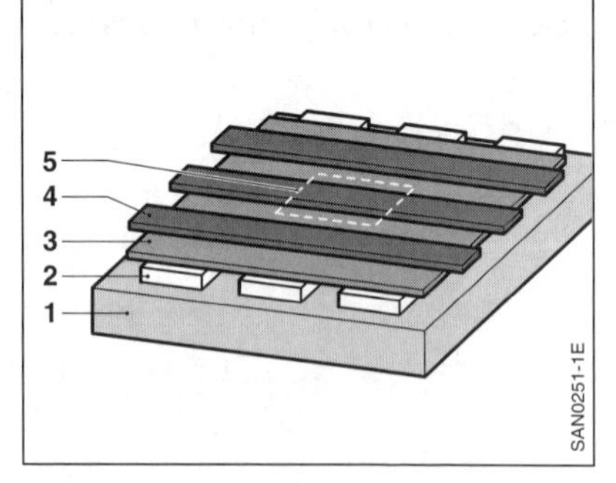

The organic material is located between them. Charge carriers are injected into the material using the electric contact points by applying a voltage. The metal cathode injects electrons; the anode absorbs them (defect electrons, holes). The charge carriers are moved toward each other by the polymer layer. When charge carriers of different charges meet (electrons and holes), this causes the radiative recombination as described above. The radiative recombination described above occurs when charge carriers of different charges meet (electrons and holes).

Figure 7 shows the structure of an OLED matrix display. The organic layer is located between a matrix of row and column electrodes arranged perpendicularly to each other. Multi-color reproduction is achieved by OLED materials in a dot matrix that emit different colors.

OLED displays have already been established in the consumer area (e.g. mobile phones) for years. Due to the high requirements for resistance to climatic conditions, OLEDs have only been used for small, monochrome, alphanumeric and simple graphic displays to date, e.g. for displaying air conditioner functions in Mercedes vehicles. The option to construct OLEDs with curved surfaces using plastic substrates could attract interest in the future from the standpoint of providing good anti-reflection conditions in a vehicle.

Figure 8: Head-up display (principle)
1 Virtual image
2 Reflection in the windshield
3 Display module: liquid crystal display (LCD) with backlighting
4 Optical system

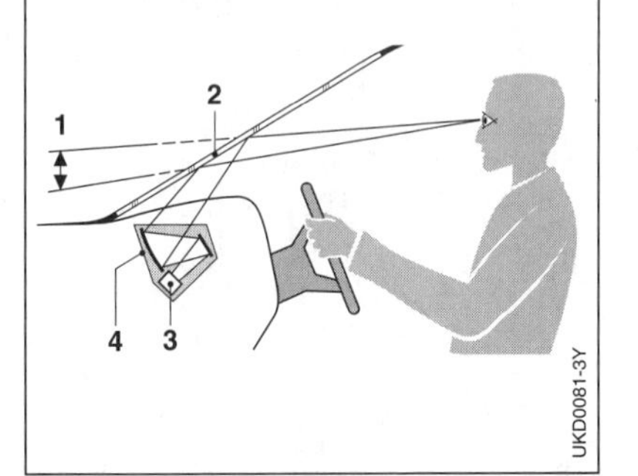

Head-up display

Conventional instrument clusters have a viewing distance of 0.8 to 1.2 m. In order to read information in the area of the instrument cluster, the driver must adjust his or her eyes from long distance (observing the road scene) to the short viewing distance for the instrument. This accustomization process usually takes 0.3 to 0.5 seconds.

Head-up displays (HUDs) have been used in military aviation applications since 1950. Such displays for motor-vehicle applications have been available in a simple design, usually as a digital speedometer, for many years in Japan and in the USA as an optional extra; in the meantime some European manufacturers have started to offer them in their vehicles.

The image in the head-up display is projected through the windshield into the driver's primary field of vision. The optical system generates a virtual image at such a viewing distance that the human eye can remain adjusted to long distance. Head-up displays do not require the driver to divert their eyes from the road – thus allowing the driver to register at all times safety-critical developments in the driving situation. There is no need to glance at the speedometer if the speed or other important information is likewise shown in the head-up display.

Design

A typical head-up display (Figure 8) features an activated display module for generating the image, a lighting facility, an optical imaging unit, and a combiner, which reflects the image to the driver's eyes.

In a motor vehicle it is usually the windshield which acts as the combiner. To avoid double images – caused by reflections on the inner and outer boundary surfaces – the windshield (or more exactly: the plastic foil of the safety glass) is slightly wedge-shaped in design. From the driver's perspective the two images created at the boundary surfaces then coincide.

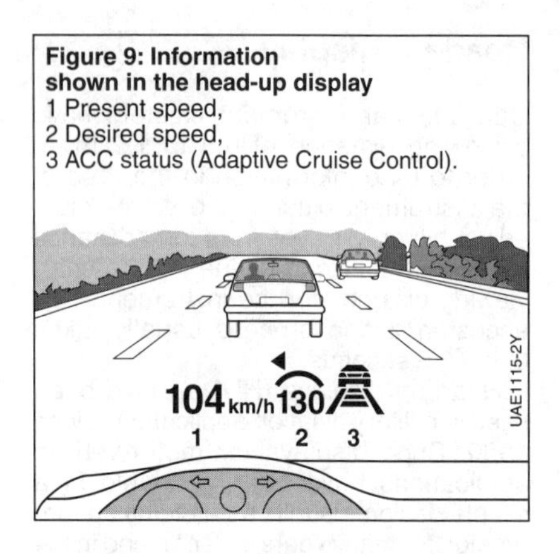

Figure 9: Information shown in the head-up display
1 Present speed,
2 Desired speed,
3 ACC status (Adaptive Cruise Control).

A real image is created in the display module. This module can be a display or a back-projection onto a diffusing surface, e.g. with a scanning laser beam. This real image is only roughly 20×40 mm in size, but the luminance required for good readability in daylight is more than 50,000 cd/m², i.e. 100 times more than a conventional display.

This image is reflected through the windshield into the driver's eyes. For the driver this virtual image of the driving situation in front of the vehicle is superimposed. Optical elements (lenses, concentrating reflectors) are generally inserted into the beam path to increase the virtual distance of the image.

Particularly high-contrast segment TN LCDs can be used as the display for monochrome head-up displays with modest levels of information content. More sophisticated, multicolor displays use polysilicon-technology AMLCDs.

Contact-analogous head-up displays are being developed which, for example, project the warning of an obstacle below the line of sight and at a virtual distance below which the driver would also see the obstacle.

Display of information by head-up displays

The virtual image should not cover and obscure the road scene ahead; it is therefore displayed in an area with low information content, in other words "floating above the hood" (Figure 9). In order to prevent the driver from being overwhelmed with stimuli in his or her primary field of vision, the head-up display should not be overloaded with information, and is therefore not a practical substitute for the conventional instrument cluster. It is, however, particularly well suited for displaying navigation notices (arrows) and safety-related information such as warnings or information from the ACC system (Adaptive Cruise Control, see Adaptive cruise control system).

According to a press release, the number of head-up displays is to increase worldwide from 1.2 million currently to around 9 million by 2020 [1].

References

[1] Automotive electronics 8/9, 2013.

Radio and TV reception in motor vehicles

Wireless signal transmission

Wireless transmission technology permits the simultaneous broadcasting of information to large populations. It is also of major significance to mobile radio reception as used in motor vehicles. At present, the importance of digital transmission methods is growing. The wireless section of the transmission chain also uses analog signal transmission, so the two technologies are basically identical.

Radio and TV broadcasting

Wireless radio and TV broadcasting is primarily used for terrestrial transmissions. In the case of analog radio broadcasting, the audio signal is modulated on the high-frequency signal. In the receiver, the received high-frequency signal is then converted to the baseband frequency and demodulated. The final signal obtained in this way is identical to the useful signal.

In telecommunications, the propagation of an electromagnetic wave is used to transmit information whereby the amplitude, the phase or the frequency of this oscillation is altered depending on the information to be transmitted. The customary frequencies range from a few kilohertz up to 100 GHz. Some of the commonly used frequency ranges are set out in Table 1. The use of frequency bands is subject to legal control (Telecommunications Act for Germany, 2004). Each country's frequency allocation plan is based on international agreements that are set out in Article S5 of the Radio Regulations of the ITU (International Telecommunication Union, [1]).

Information transmission using highfrequency waves

The variation in a high-frequency signal used to transmit a useful signal from a transmitter to a receiver is called modulation. The modulated high-frequency signal is emitted within a precisely determined, narrowly defined frequency band by an antenna. The receiver selects precisely that frequency band from the large number of frequencies received by the antenna. In this way, wave propagation between transmitter and receiver is a link within the signal transmission chain.

For example, when it comes to the transmission of an audio signal, the useful signal in contrast to the high-frequency carrier signal is made up of various frequencies ranging up to a maximum of 20 kHz. The high-frequency carrier is modulated with this low-frequency signal. A transmitter antenna beams out the carrier wave.

The maximum distance at which the signal can still be received and the recep-

Table 1: Overview of some frequency bands in radio and TV broadcasting

Wavebands	Wavelengths λ in m	Frequencies f in MHz	Examples
Long wave (LW)	≈ 2,000 – ≈ 1,000	0.148 – 0.283	Analog radio broadcasting
Medium wave (MW)	≈ 1,000 – ≈ 100	0.526 – 1.606	Digital broadcasting (DRM)
Short wave (SW)	≈ 100 – ≈ 10	3.950 – 26.10	
Very high frequency (VHF)	≈ 10 – ≈ 1	30 – 300	
Band 1		47 – 68	TV
Band 2		87.5 – 108	Digital radio broadcasting (DAB),
Band 3		174 – 223	Digital TV broadcasting (DVB-T)
Ultra high frequency (UHF)	≈ 1 – ≈ 0.1	300 – 3,000	
Band 4		470 – 582	TV
Band 5		610 – 790	Digital TV broadcasting (DVB-T) Digital TV broadcasting (DVB-H)
L-band		1,453 – 1,491	Digital radio broadcasting (DAB)
Super high frequency (SHF)	≈ 0.1 – ≈ 0.01	3,000 – 30,000	
		10,700 – 12,750	Digital TV broadcasting (DVB-S)

tion quality depend on the frequency, among other things. Short-wave and long-wave have very long ranges, in some cases intercontinental, whereas the reception range of VHF transmissions is hardly any furthor than the line of sight.

The receiver station demodulates the signal. The resulting low-frequency electrical oscillation is then converted by a loudspeaker into acoustic oscillations.

Amplitude modulation
Amplitude modulation (AM) is the change in amplitude A_H of the high-frequency oscillation at frequency f_H in synchronization with the low-frequency oscillation (A_N, f_N) (Figure 1).

Amplitude modulation is used in the short-wave, medium-wave and long-wave bands, for example.

Frequency modulation
Frequency modulation (FM) is the change in frequency f_H of the high-frequency oscillation in synchronization with the low-frequency oscillation (Figure 2).

Frequency modulation is used for FM radio and the sound channel of analog TV transmissions, for example. The transmissions of frequency-modulated signals are impaired by amplitude-modulation interference (e.g. caused by the ignition systems of spark-ignition engines) to a lesser degree than the transmissions of amplitude-modulation transmitters.

Digital modulation processes
In the case of digital modulation processes, the amplitude or the frequency of the carrier is discretely altered. One or more bits can thus be allocated to each of these carrier states so as to enable digital information to be transmitted.

Reception problems
VHF signals have virtually a linear propagation. As a result, a car radio may lose the signal from a VHF transmitter that is only 30 km away if there is high ground between the car and the transmitter. On the other hand, reception may be unimpaired at a location further away if there is a clear "line of sight" between the car and the transmitter. Such "radio shadows" are thus often covered by a fill-in transmitter.

Signals may be reflected off the sides of valleys or high-rise buildings. The reflected signals then arrive at the receiving antenna with a time delay and are superimposed on the signals received directly from the transmitter. This produces what is known as "multi-channel reception". It causes multipath interference and results in the deterioration of sound quality in radio reception.

Propagation of electromagnetic waves is impaired by conductors within the transmitter's radiation field (e.g. steel masts or power lines) and even nearby forests, houses, or locations in deep valleys. The characteristics of wave propagation are important when it comes to suppressing

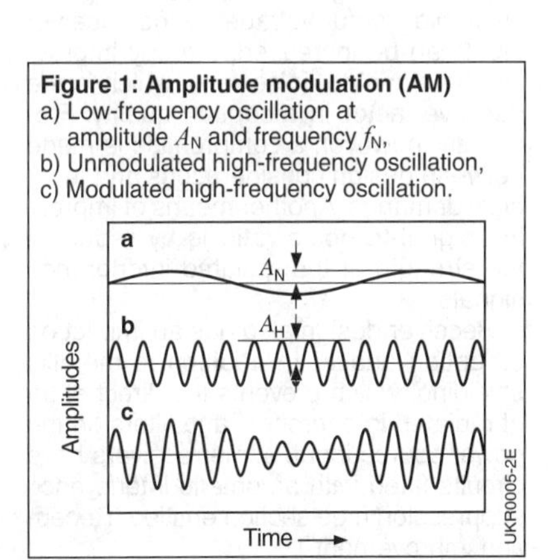

Figure 1: Amplitude modulation (AM)
a) Low-frequency oscillation at amplitude A_N and frequency f_N,
b) Unmodulated high-frequency oscillation,
c) Modulated high-frequency oscillation.

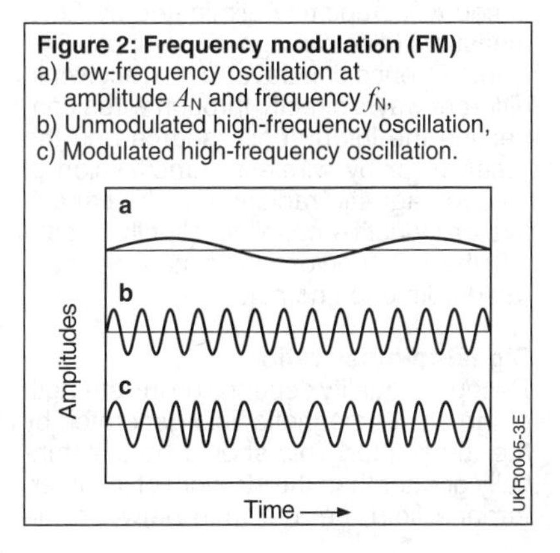

Figure 2: Frequency modulation (FM)
a) Low-frequency oscillation at amplitude A_N and frequency f_N,
b) Unmodulated high-frequency oscillation,
c) Modulated high-frequency oscillation.

interference in motor vehicles effectively. Interference-free reception is impossible if the signals received from the transmitter are too weak. Thus, reception of a previously perfectly interference-free signal will break down suddenly when a car enters a tunnel, for example. This can be explained by the shielding effect of the reinforced-concrete tunnel walls which reduce the useful field strength of the transmitter signal to which the radio is tuned. At the same time, the interference-field strength remains the same. Under certain conditions, it may not be possible at all to continue to receive the radio-station signal. Similar phenomena can also be experienced when traveling in mountains, for example.

Radio interference

Radio interference is caused by undesirable high-frequency waves which are directed together with the desired signal to the receiver. It occurs wherever electrical currents are suddenly interrupted or switched on. High-frequency interference waves are thus produced, for example, by the ignition of a spark-ignition engine, the actuation of a switch, or the switching operations on the commutator of an electric motor. Such rapid changes of current generate high-frequency waves which interfere with radio reception by receivers located close by. The effect of the interference depends on the steepness of the signal pulse and its amplitude, among other things.

Radio interference caused by such steeply rising current pulses can be reduced or completely eliminated by EMC measures (electromagnetic compatibility).

Interference can travel to the receiver in different ways: directly through wires connecting the interference source and the receiver, or by wireless transmission of electromagnetic radiation, or by capacitive or inductive coupling. Strictly speaking, the last three options cannot be separated from one another.

Signal-to-noise ratio

Reception quality depends on the strength of the electromagnetic field generated by the transmitter. This should be substantially greater than the strength of the interference field, i.e. the ratio between the strength of the transmitter signal and the strength of the interference field – the signal-to-noise ratio – should be as large as possible.

A receiver close to the interference source receives not only the useful signal from the desired transmitter, but also the undesired interference signal, if it is transmitted at the same frequency. Nevertheless, good-quality reception is still possible provided that the field strength of the desired transmitter signal at the point of reception is very high in comparison with the strength of the electromagnetic field generated by the interference source. The useful field strength of the signal from the transmitter depends on transmitter power, transmitter frequency, the distance between the transmitter and the receiver, and the propagation characteristics of the electromagnetic waves. In the case of medium-wave and long-wave signals, the field strength of the transmitter signal can be weakened by difficult topology to such an extent that the signals of even powerful transmitters have low useful field strengths at some reception locations. VHF signals may be subject to heavy fluctuations in useful field strength under certain conditions. Receivers in motor vehicles may also suffer from relatively low useful voltages at the receiver input due to the short effective height of the antenna. Accordingly, the opportunities for improving the signal-to-noise ratio in the receiver are very limited.

By optimizing antenna positioning, the available useful voltage at the receiver input can be increased, thereby improving the signal-to-noise ratio, which is the decisive factor in reception quality. Frequently, however, a compromise is made between design considerations and technical demands. Another means of improving signal-to-noise ratio is by reducing the strength of the radiated interference signals.

Receiver design also has an impact on reception quality. In addition to metallic shielding, which prevents the direct entry of radiated interference, and filters on the power source input, some receivers have circuits fitted with automatic interference suppression (see section entitled "Reception improvement").

Radio tuners

Radio tuners in motor vehicles are frequently called car radios or automotive sound systems. However, this term refers not only to a radio tuner but also to devices with a large number of integrated information and entertainment functions. These include, for instance, the analysis of supplementary information (e.g. traffic news), players for reproducing from storage media (e.g. CDs and SD cards), as well as integrated radio communication interfaces, cellular phones and other devices.

In the past few years, conventional analog transmission technology has developed new systems. For this reason, modern car tuners are capable of receiving a wide variety of radio broadcast systems worldwide. As well as conventional radio broadcasting, these systems include among others DAB (Digital Audio Broadcasting), DRM (Digital Radio Mondiale) and SDARS (Satellite Digital Audio Radio Services).

The conventional radio tuner is designed to receive analog FM and AM modulation and has an analog signal path from the antenna to the audio signal. Modern car tuners with the highest received power, on the other hand, process signals digitally. The IF signal (intermediate frequency) supplied by the tuner is digitized by an analog-digital converter, and then processed. In standards covering digital modulation essentially only demodulation in the digital part of the signal path changes.

Conventional tuners

Signal processing

The antenna, which is predominantly designed as a rod or window antenna, picks up the electromagnetic signal emitted by the transmitter. This consists of different channels with a fixed frequency spacing. The high-frequency alternating voltage generated at the base of the antenna is routed to the receiver and processed there.

The conventional tuner for analog radio reception basically contains two signal paths: one for processing amplitude-modulated signals and one for processing frequency-modulated signals. These are usually divided into the blocks described in the following (Figure 3).

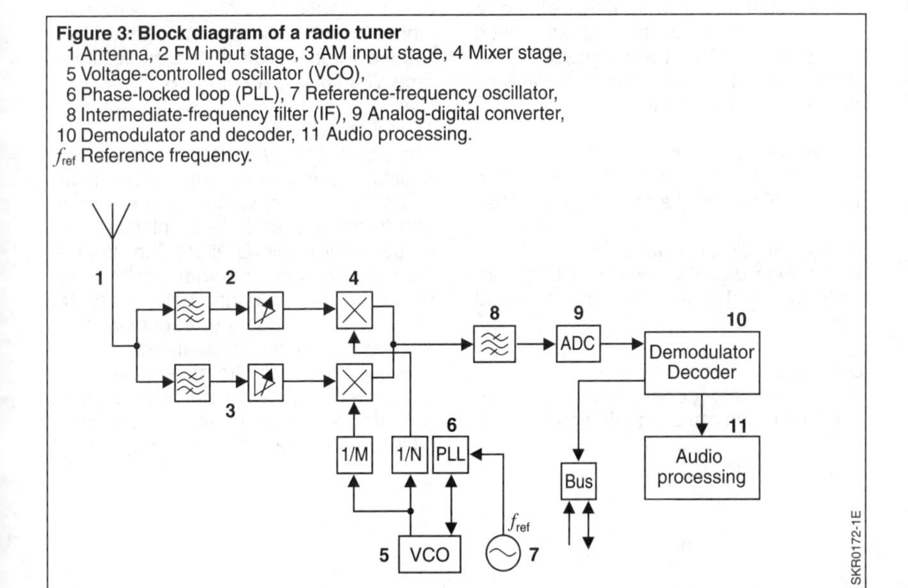

Figure 3: Block diagram of a radio tuner
1 Antenna, 2 FM input stage, 3 AM input stage, 4 Mixer stage,
5 Voltage-controlled oscillator (VCO),
6 Phase-locked loop (PLL), 7 Reference-frequency oscillator,
8 Intermediate-frequency filter (IF), 9 Analog-digital converter,
10 Demodulator and decoder, 11 Audio processing.
f_{ref} Reference frequency.

AM input stage
A bandpass filter limits the amplitude-modulated signal in the LW, MW and SW bands, and the resulting signal is amplified in the following stage with low noise.

FM input stage
The frequency-modulated VHF signal is received by a separate input stage. The input filter is either tuned to the frequency to be received or configured to the entire received band. The received-signal level is then adjusted to the desired input level for the following mixer stage with low noise by an automatically controlled amplifier.

Voltage-controlled oscillator
The oscillator (VCO, voltage-controlled oscillator), whose frequency is controlled by a phase-locked loop (PLL), generates a high-frequency oscillation which is subdivided. The input signal is converted with the aid of this signal into a constant intermediate frequency (IF) in the mixer stage. A quartz-stabilized signal serves as reference frequency.

Mixer stage
The mixer stage converts the input signal into a constant intermediate frequency. Different mixer stages are frequently used for receiving FM and AM signals. On the other hand, the principle of frequency conversion is the same.

IF filter and amplifier
The IF signal recovered in this way is then fed to an IF filter and a controlled amplifier.

Analog-digital converter (ADC)
The analog-digital converter (ADC) converts the analog IF signal into a digital signal.

Demodulator
The demodulator generates the digital audio signal from the digital IF signal.

Decoder
Additional information, such as RDS data (Radio Data System), is decoded by the decoder and passed on to a processor for processing.

Audio processing
After demodulation, the audio signal may be adapted, for example, to the vehicle conditions and the listener's preferences. This can be achieved by using the appropriate controls to adjust tone and volume or to vary the level between front and rear or between left and right.

Digital receivers
A digital receiver (ADR, advanced digital receiver) is a highly integrated receiver module whose input signal is the analog or digital IF signal. The analog signal is converted into a digital signal, the digital signal is processed on the digital level. This technology allows a type of signal processing that would not be possible using analog technology. It generates, for example, IF filters which provide for exceptionally good harmonic-distortion levels, and whose bandwidth is variable and can be adapted to the reception conditions. In addition, there is a larger number of other means of processing the received signal in such a way as to substantially reduce interference in the audio signal (see SHARX, DDA and DDS).

Digital equalizer
The digital equalizer (DEQ) consists of a multiband parametric equalizer which provides separate adjustment features for the mid-frequency, and the amplification/attenuation of individual filters. The suppression of unwanted resonances thus obtained optimizes sound quality inside the vehicle. The frequency response of the speakers can also be linearized.

Preset equalizer filters can also be implemented in some devices. These can be activated according to music genre or vehicle type (e.g. jazz or pop, van or sedan).

Digital Sound Adjustment
"Digital Sound Adjustment" (DSA) is a system which automatically analyzes and corrects frequency response in in-car systems. A microphone and a digital signal processor (DSP) are used to pick up and analyze a test signal generated by the speakers. The optimum sound curve for the vehicle is then set on the equalizer.

Dynamic Noise Covering
While the vehicle is in motion, the "Dynamic Noise Covering" (DNC) function uses a microphone to constantly detect and analyze the vehicle-noise spectrum that is blocking out the audio signal and impairing the perceived sound quality. The function uses selective amplification or selective dynamic compression (reduction of dynamic response between minimum and maximum values) of the interference frequencies to maintain optimum reproduction of sound quality, regardless of driving noise.

Reception quality
Analog radio broadcasting is primarily used for terrestrial transmissions. The transmission path is not always ideal, with the result that reception quality may be impaired, depending on the transmitter/receiver constellation and environmental conditions. In the case of VHF reception, reception locations may be critical because of the transmission-path problems described below.

Fading
Fading is caused by fluctuations in signal reception level due to obstacles in the signal path, such as tunnels, high-rise buildings, or mountains.

Multipath reception
Multipath reception is caused by the reflection of signals off buildings, trees or water. This can easily lead to a substantial dip in reception field strength, and even total signal loss. The differences in the strength of the received signal field strength occur within a few centimeters of each other. Such fluctuation has a particularly detrimental effect on mobile receivers, such as car tuners.

Adjacent-channel interference
Adjacent-channel interference occurs when there is another channel received with a high field strength close to the channel received.

High-level signal interference
High-level signal interference occurs close to the transmitter at high field strength. The receiver protects its input by reducing the field strength. This has the effect of attenuating the weaker signals from other transmitters, i.e. they become quieter.

Overmodulation
Some transmitters increase the modulation level in order to achieve a greater range or higher loudness level. The disadvantages of this process are greater distortion factor and greater susceptibility to multipath interference.

Ignition interference
High-frequency interference sources (such as the ignition of a spark-ignition engine, the operation of a switch, or switching operations on the commutators of electric motors) produce reception interference.

Reception improvement
A large number of functions for improving received power are implemented in modern automotive sound systems. An overview of the most important functions is provided in the following.

Radio Data System
Radio Data System (RDS) is a digital data transmission system for FM radio. This format is standardized throughout Europe. It provides the tuner with extra information in addition to the desired audio signal by using alternative received frequencies with the same modulation. This enables, for example, the tuner to switch continuously to the frequency with the least interference. Table 2 provides an overview of the information transmitted.

Table 2: RDS codes

Code	Information transmitted
PS	Name of the station received
AF	List of alternative frequencies on which the station also broadcasts
PI	Identifies the broadcast station
TP, TA	Traffic-program station identification, traffic-announcement identification
PTY	Identification of the station type
EON	Signals traffic announcements on a parallel station
TMC	Standardized traffic information
CT	Time (for synchronizing the clock in the vehicle)
RT	Text transmission (e.g. title of music track)

Digital Directional Antenna
The "Digital Directional Antenna" (DDA) system developed by Bosch uses the signals from two antennas to calculate a synthetic antenna with a new directional characteristic. This permits suppression of interference caused by multipath reception as shown in Figure 4.

Digital Diversity System
The reception characteristics of FM radio are heavily dependent on location. A Digital Diversity System (DDS) has a number of antennas at its disposal so that it can switch between them in order to enhance

Figure 4: Optimized antenna signal
1 Direct signal,
2 Reflection,
3 Directional diagram.

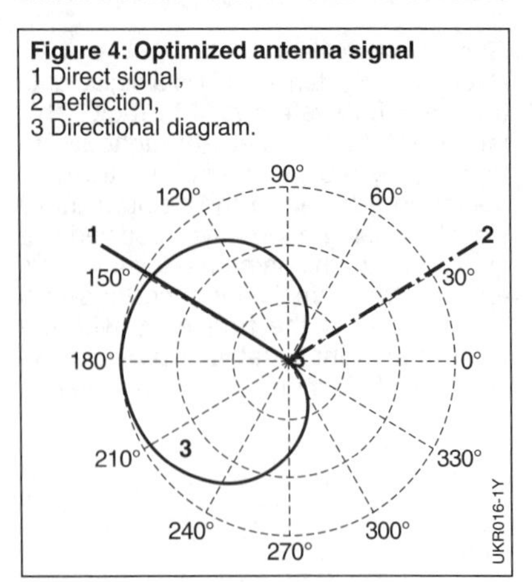

reception. The Digital Diversity System integrated in the digital receiver uses exactly the same signal for its switching strategy and the signal is available as the audio signal after demodulation.

High-cut
Interference caused by such effects as fading and multipath reception has a greater effect on higher audio frequencies. For this reason, modern automotive sound systems have a facility for detecting such interference and reducing the level of the audio signal at higher audio frequencies when there is interference.

SHARX
SHARX is a function which automatically adjusts the bandwidth of the intermediate-frequency filter for FM reception to suit reception conditions. If different stations transmit at frequencies that are very close to one another, this function significantly increases the clarity of separation by reducing the bandwidth and allows virtually interference-free reception. If there are no adjacent channels, the bandwidth can be increased and thus harmonic distortion reduced.

Automatic interference suppression
A further measure for improving reception is automatic interference suppression, which suppresses interference signals generated by interference sources from both the present vehicle and other vehicles. To this end, the demodulated signal, which contains interference pulses, as well as the useful signal, is blanked for that moment of interference and the gap created is filled.

References
[1] Radio Regulations. International Telecommunication Union (ITU).

Traffic telematics

Transmission paths

Traffic telematics refers to systems that transmit traffic-related information to and from vehicles and, generally, which analyze this information automatically. Both unidirectional broadcasting (radio) and bidirectional mobile communications connections are available as transmission paths. Analog and digital radio provides only the path into the vehicle, with the information received identical for all receivers. Mobile communications connections on the other hand enable messages to be specifically received by way of individual requests and thus information to be exchanged between a vehicle and appropriate service providers in both directions.

The transmission paths also differ in the bandwidth limiting the information volume that is available and the transmission costs. The available bandwidth in mobile communications has increased by several orders of magnitude with each generation in recent years through the developments of GPRS (General Packet Radio Service), UMTS (Universal Mobile Telecommunication System), and finally LTE (Long Term Evolution). While data transmission by radio does not involve additional costs for receiving sound radio, costs are generally incurred for the information volumes on the mobile communications channel in proportion to the data volume.

In the future communication with "roadside units" in accordance with the standard WLAN 802.11p (Wireless Local Area Network) will serve as a further transmission path providing a high bandwidth with no additional transmission costs.

Standardization

The standardization of message content is an important precondition to enable the evaluation of information from various sources by different on-board terminals.

RDS/TMC standard
The widespread RDS/TMC standard (Radio Data System/Traffic Message Channel) for FM radio comprises content regarding the type of traffic disruptions (e.g. traffic jam, full road block), the causes (e.g. accident, icy roads), the expected duration, as well as identification of the road sections concerned. A numeric coding for traffic junctions, expressway sections and geographic regions already exists in many countries. However, it limits the messages for the most part to the major roads (expressways and highways).

TPEG standard
It is now also possible, via the TPEG standard (Transport Protocol Experts Group), to transmit traffic forecasts and recommended alternative routes. A limitation by numeric, predefined junctions and road sections within the road network, as used in TMC, is no longer necessary when the AGORA-C standard is used. This standard allows messages for any road to be coded without transmitters and receivers having to use identical-version reference tables. However, because of the bandwidth required, transmission is only possible via digital channels.

Information recording

The benefit of traffic telematics is dependent on the quality of the messages and how up-to-date they are. There are different data sources for recording information about traffic flow. “Historical” data are additionally used so as also to offer – independently of current messages – the best possible route planning.

Local information
The first data sources for automatically determining the traffic situation were induction loops integrated in the road surfaces which can record the average speed and number of vehicles at individual road points and transmit the data to an evaluating control center. This form of recording was supplemented by the introduction of sensors on expressway bridges which also collect this information.

Area-wide information
In order to receive area-wide information instead of local measurements, newer approaches resort to floating-car data or floating-phone data. The basic idea is to infer traffic situations from the movement data of many cars or mobile phones.

Floating-car principle
In the case of the floating-car principle, the in-car navigation device transmits the car’s position and speed by mobile communications to a control center, which then calculates the current traffic situation by statistically analyzing these data.

Floating-phone principle
In the case of the floating-phone principle, the movement patterns of mobile phones are analyzed. This principle exploits the fact that every mobile phone is constantly transmitting information on the current reception situation to its base station. Characteristic patterns can be derived from the changed messages which are accompanied by the mobile communication device’s change of location. These patterns are then used to infer the position of the device. Statistical processes are used first to evaluate whether the devices in question are moving devices in vehicles and then to infer the traffic situation from them.

Benefits of the floating processes
The two floating processes are used not only to detect the traffic disruption itself, but also to determine and forward reference values for the delay caused by the disruption.

The movement profiles of the vehicles can also be collected, assigned a time, and compressed. This practice produces so-called hydrographs which reproduce for individual road sections the average driving speeds depending on the day of the week and the time of the day. In this way, recurring traffic disruptions such as slow-moving traffic and regular rush-hour traffic jams are depicted. The vehicle-navigation system can use these hydrographs as “historical” data as a function of the current time to specify a route and journey time adapted to the expected situation.

Dynamic route guidance
Through the standardized coding of traffic messages, it is possible also to display these messages in the vehicle in comprehensible text form for the driver. Furthermore, the computers in route-guidance systems can determined on the basis of the standardized coding whether a better alternative route exists in the event of disruptions. The relevant messages are filtered out from those available using the vehicle location and, if necessary, its movement along a route. If the associated amount of time lost is communicated as well as the actual disruption, this time lost can also be included in the route computation. In the event of alternative route guidance, the driver is instructed that the route has been recalculated on account of traffic announcements received. Further route recommendations will follow in accordance with the new route (see Vehicle navigation).

Furthermore, subscription traffic services not only transmit the more serious traffic jams and disruptions, but also indicate for all higher road categories the existence of restricted driving speeds identifiable from the floating data. Thus, for example, the traffic situation can be made transparent by a color-coded, high-detail display of the traffic flows in the vehicle navigation's map display even when route guidance is inactive.

Vehicle-vehicle communication

Vehicle-vehicle communication is currently being tested as the logical next step in traffic telematics. The ultimate aim is to further increase road safety and to reduce economic losses due to traffic disruptions. A system of direct information transfer between vehicles can be used to send warnings about, for example, broken-down vehicles, approaching emergency vehicles, the location of the tail end of a jam, right through to relevant individual maneuvers such as hard braking. This gives the driver in the receiving vehicle time to react in an suitable way to the present situation.

The standard WLAN 802.11p (Wireless Local Area Network) is being tested and standardized as the basis for direct vehicle-vehicle communication. The basic principle behind the warning functions is the timely and correct communication of precise location information between the transmitting vehicle and a receiving vehicle. This involves both the continuous transmission of movement messages and the additional transmission of an extra warning message when disruptions and events are detected. The process of regularly repeating or even passing on received messages to other vehicles ensures that all the affected vehicles can be reached.

To prepare for market launch, engineers are currently testing the functionality and efficacy of the systems under everyday conditions in a variety of field tests conducted in for example Germany and the USA.

Driver-assistance systems

Introduction – driver assistance

Driver-assistance systems in modern vehicles can basically be divided into the categories comfort/convenience systems and safety systems. Comfort/convenience systems relieve the driver in performing monotonous, repetitive driving tasks. Typical examples are automatic turn-signal resetting after turning or Adaptive Cruise Control (ACC). Safety systems, on the other hand, are intended to support the driver in critical driving situations so that an accident is avoided or its consequences are lessened. Typical examples are the Electronic Stability Program (ESP) and airbags. As the specified system categories increasingly interact in the face of the increasing level of system networking in vehicles and driver support by comfort/convenience systems is intended to provide increased safety (avoidance of critical situations already in advance), there is increasingly a flowing transition between comfort/convenience and safety systems. For further details, refer to the descriptions in the section "Comfort/convenience and safety systems".

Critical driving situations

Driver-assistance systems aim to make the vehicle capable of perceiving its surroundings, interpret them, identify critical situations, and assist the driver in performing driving maneuvers. The aim is to identify critical situations early, i.e. already before they occur, and anticipate and defuse them, or at best to prevent accidents during critical situations or at worst to minimize as much as possible the consequences of an accident.

In critical driving situations, just fractions of a second are often decisive in determining whether an accident occurs or not. According to the study in [1], approximately 60 % of rear-end collisions and approximately 30 % of frontal impacts could have been avoided if the driver had reacted only half a second earlier. Every second accident at an intersection could be prevented by faster and correct driver reaction.

At the end of the 1980s, when the possibility of highly efficient and partially automated road traffic was presented as part of the EU "Prometheus" project, the electronic components for this task were not in existence. The highly sensitive sensors and powerful microcomputers now available have brought such vehicles driven by highly automated means a step nearer

Figure 1: Operating areas of passenger-car safety functions in accidents involving personal injury by accident type
(Analysis of GIDAS accident database [2]).

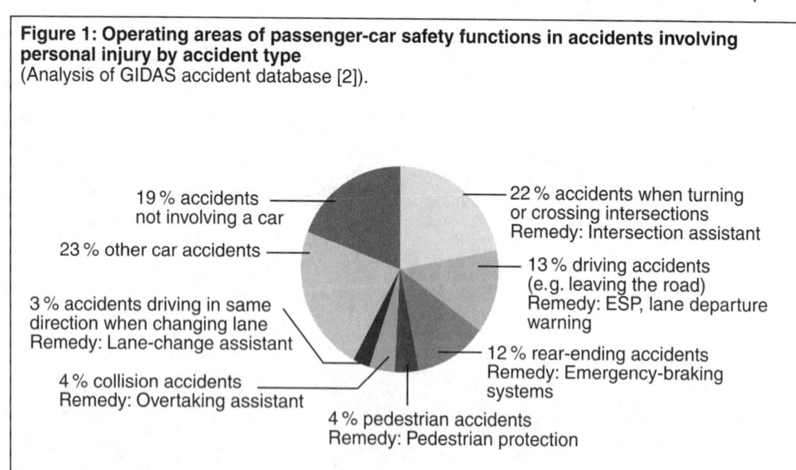

to realization. The first driver-assistance systems to provide partially automated vehicle guidance in specific traffic situations will be available on the market in a few years (e.g. bottleneck assistant, roadworks assistant). Here the environment sensors scan the vehicle's surroundings and detect relevant objects. The driver-assistance systems follow a cascade of intervention and generate warnings or directly perform the maneuvers required (automatic steering or braking interventions). All this is done at those crucial fractions of a second faster, since a computer-controlled system can essentially react faster than a person.

Accident situation and actions

Accident research plays a significant role in the development of driver-assistance functions. Accident research at Bosch provides support among others in the design and development of new vehicle-safety functions while taking into account current traffic-accident occurrences. Assessments are conducted as to the effectiveness of these systems – i.e. of the potential for accident avoidance – and their effects on future accident occurrences.

To assess the effectiveness of passenger-car driver-assistance functions, an analysis of accidents involving personal injury in Germany was conducted (as at 2011: 306,266 accidents involving personal injury) on the basis of the GIDAS accident database (German In-Depth Accident Study [2]) (Figure 1).

Around 22% of all accidents involving personal injury which are caused by a passenger car occur when turning or crossing. Here different intersection-assistant functions will be able in future to reduce accidents.

Around one eighth of all accidents involving personal injury (13%) can be put down to driver error; in many such cases accidents can already be avoided or the consequences of the accidents lessened with the Electronic Stability Program (ESP) and lane departure warning.

A further 12% of all accidents involving personal injury arise when one vehicle rams into another vehicle driving in the same direction. Collision-warning systems can counteract these accidents in an initial stage (e.g. Adaptive Cruise Control). In a further stage collision-avoidance systems prevent the accident by actively intervening in the driving dynamics, for example through brake intervention with the emergency-braking system (Automatic Emergency Braking, AEB).

Accidents involving pedestrian demonstrate a high degree of complexity. The first emergency-braking pedestrian-protection functions are already available on the market. These take effect in accidents where pedestrians are involved (4% of all accidents). Research is currently being conducted into extending these pedestrian-protection systems to further reduce the consequences of accidents by means of automatic avoidance maneuvers.

A further 4% are collision accidents which usually occur during overtaking and can be positively influenced by overtaking-assistance systems. A further reduction in accidents (3% share) can be expected when the driver is supported when changing lanes by an assistance system.

In spite of the already high traffic-safety standards, according to analysis of the GIDAS accident database 42% of accidents involving personal injury still do not involve the operation of a vehicle-safety system – accident research is helping here to developing further vehicle-safety functions.

Applications

Driver-assistance systems have a wide range of applications. They can be divided into active systems which intervene in the driving dynamics and into passive system with no intervention in the driving dynamics.

As already mentioned, a further distinction is made between comfort/convenience systems to reduce the burden on the driver (driver support) with the long-term goal of fully automated driving and safety systems with the aim of accident avoidance or mitigation of accident consequences.

Comfort/convenience and safety systems

Figure 2 shows according to the selected system the range of driver-assistance systems and functions.

Passive safety functions

Passive safety functions (Figure 2, bottom left quadrant) contain the actions for mitigating the consequences of accidents such as precrash actions (airbag deployment) and the functions for passive pedestrian protection (specific design of the front of the vehicle to mitigate the consequences of collisions with pedestrians).

Driver support

Systems for reducing the burden on the driver (driver support) without active vehicle intervention (Figure 2, bottom right quadrant) are the preliminary stage to fully automatic vehicle guidance. These systems give the driver tips on vehicle guidance. The parking assistant with short-range sensors (ultrasound sensors) aids the driver in looking for a parking space and actually parking, while specific infrared video sensors can be used to good effect to improve driver vision during nighttime driving. In the case of the lane-departure warning system, a video camera uses the road markings to detect the direction of the lane ahead of the vehicle and warns the driver if he/she changes lane without activating a turn signal. The warning can be issued acoustically through the car radio's speakers or mechanically in the form of steering-wheel vibrations.

Automated vehicle guidance

The automated vehicle-guidance systems (Figure 2, top right quadrant) include lane-keeping support, which prevents the vehicle from leaving the lane by means of specific intervention in the lateral guidance. Lane-keeping support is therefore a

Figure 2: Systematic classification of driver-assistance systems (with selected example functions)
Pre-fill: Pre-fill the brake.

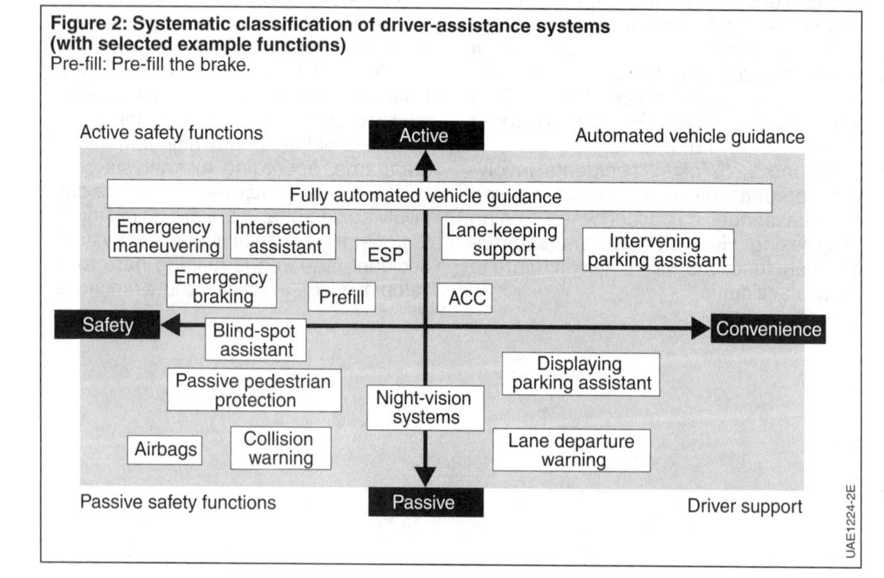

further development of the lane-departure warning system. Adaptive Cruise Control (ACC) is also one of the automated vehicle-guidance systems. A further development of this system relieves the burden on the driver in slow-moving congested traffic – first by braking the vehicle to a complete stop, and then by moving it forward again at low speed (ACC Stop & Go). The traffic-jam assistant as a further development also permits lateral guidance at low speeds. Functions designed to provide automated longitudinal and lateral guidance on expressways/interstates (expressway pilot) are currently in development. The next conceivable development step is full longitudinal and lateral guidance in the urban range at medium speeds (City ACC). Such a system already comes close to the goal of highly automated vehicle guidance.

Active safety functions
The active safety functions (Figure 2, top left quadrant) concern all the active emergency measures for avoiding or mitigating the damage caused by accidents. In the event of a collision danger that can no longer be averted by the driver, these systems in their maximum configuration enables computer-aided maneuvers to be carried out. Collision prevention or mitigation of the consequences is achieved by automated braking or steering here. The intermediate steps consist of a brake prefill when a danger is detected, short hard braking (kinesthetic warning signal to the driver) or a steering pulse.

Standard architecture for driver-assistance systems

Driver-assistance systems can be subdivided in accordance with a distributed standard architecture (Figure 3) into the submodules sensor technology, sensor-data fusion (with the aim of creating an environment model), situation analysis (with the aim of understanding a traffic situation and its further development), function (with the aim of action planning), actuator technology, and human-machine interaction.

The sensor layer consists of various sensors defined and designed in response to the functional requirements in type and parameterization. In the sensor-data fusion module the environment is modeled in accordance with the function specification (non-moving environment, function-relevant moving objects) on the basis of the sensor measured values.

In the following situation analysis the criticality with regard to function-relevant scenarios is checked on the basis of environment modeling. Thus, the position of the object and the actual position of the vehicle are predicted into the future by way of example for an emergency-braking function and the risk of collision is evaluated.

The criticality measurements derived in this way serve as the criterion in the following function module for a decision on whether to deploy the function. The function module typically uses a finite state machine which as well as thresholds for the criticality measurements also checks whether the function-specific system limits are adhered to (e.g. in the event of the vehicle being subject to high yaw rates an emergency-braking function would not be triggered for the situation where vehicles are traveling in the same direction). In the event of a critical situation the function module runs through a cascade of intervention. This ranges, depending on the situation and function, from early driver notification and warning (visual, acoustic, haptic) to an automated steering or braking intervention to avoid a collision or mitigate the consequences of an accident.

Figure 3: Standard architecture for driver-assistance systems

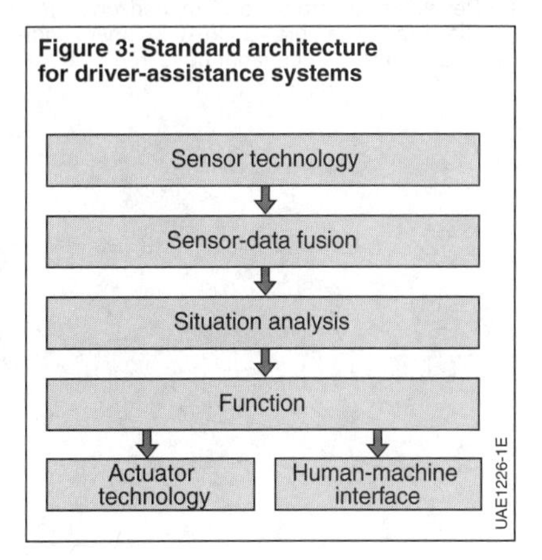

The corresponding actuators and the human-machine interface are now activated in accordance with the function-specific warning cascade.

Typical approaches for the environment-sensor system, sensor-data fusion and function are described in more detail in the following on the basis of the standard architecture introduced.

Environment-sensor system for electronic all-around visibility

Using "electronic all-around visibility", numerous driver-assistance systems are achievable – both for passive and for active intervention purposes. Figure 4 shows the detection ranges of the current all-around visibility sensors.

Long range

Mainly long-range radar (LRR) sensors are used for long-range applications. Today's sensors have an operating frequency of 76.5 GHz and a range of approximately 250 m. Lidar sensors are also used (e.g. for adaptive cruise control) primarily in Japan. These sensors operating in the close infrared range have ranges of between 150 and 250 m.

Medium range

In the medium range mid-range radar (MRR) sensors with an operating frequency of 24 GHz can meet the exacting demands with regard to range and angle resolution. Since 2005 video sensors have also been used in the medium range. They increasingly play a central role in driver-assistance systems as they enable numerous driver-assistance functions to be implemented due to the large information content of image data (among others thanks to stereo video, optical flow, object classification, see Computer vision) and a very good cost-benefit ratio. For example, infrared cameras in night-vision systems support the driver by detecting situations in the dark.

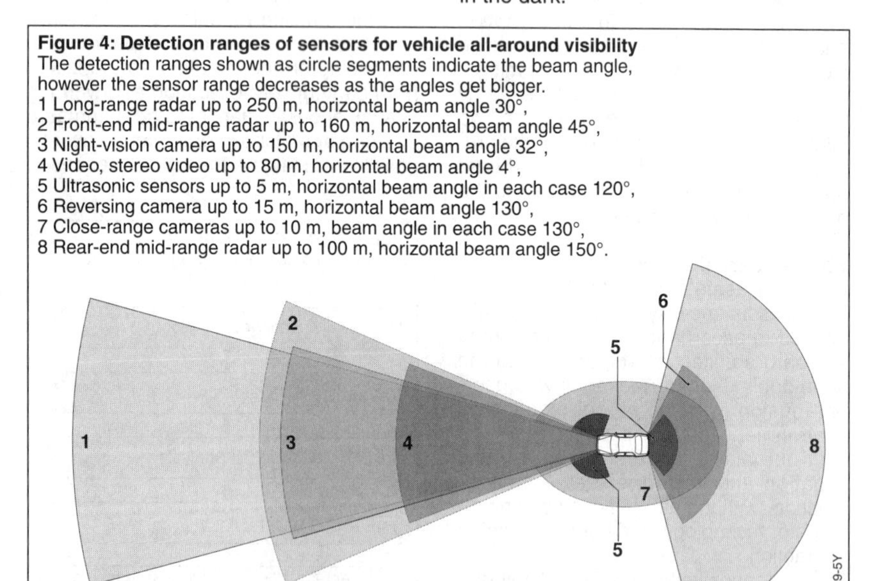

Figure 4: Detection ranges of sensors for vehicle all-around visibility
The detection ranges shown as circle segments indicate the beam angle, however the sensor range decreases as the angles get bigger.
1 Long-range radar up to 250 m, horizontal beam angle 30°,
2 Front-end mid-range radar up to 160 m, horizontal beam angle 45°,
3 Night-vision camera up to 150 m, horizontal beam angle 32°,
4 Video, stereo video up to 80 m, horizontal beam angle 4°,
5 Ultrasonic sensors up to 5 m, horizontal beam angle in each case 120°,
6 Reversing camera up to 15 m, horizontal beam angle 130°,
7 Close-range cameras up to 10 m, beam angle in each case 130°,
8 Rear-end mid-range radar up to 100 m, horizontal beam angle 150°.

Close range
Ultrasonic sensors with higher scanning ranges than previously can form a "virtual safety shield" around the vehicle within certain limits (range smaller than 20 m). This shield can be used to implement a number of functions. The signals from objects inside this safety shield serves as the database of various safety and comfort/convenience systems. Even the driver's "blind spot" can be monitored by these sensors. Since 2008 multi-beam lidar sensors have been used to monitor the range up to 8 m ahead of the front of the vehicle for automatic emergency-braking functions and have been rapidly taking hold in the volume segment.

Ultra-close range
The parking assistant monitors the ultra-close range with the aid of ultrasound technology. Today's sensors have a range of up to 5 m. These sensors are therefore suitable for a variety of parking functions. Rear-end video sensors (in their simplest form) can assist the ultrasound-based ParkPilot system during parking maneuvers.

Sensor-data fusion
In the event of long-range functional requirements a number of sensors (based on different sensor principles) can be used with an overlapping field of vision in order among other things to increase the field of vision and improve the accuracy and reliability of object detection. This procedure is known as sensor-data fusion.

Thus, the camera mounted at the front end can, depending on its functionality, also be used to extend the measured values of the long-range radar. This enables not only the distance to an object to be measured, but also the object itself to be classified. The object class enables the system to draw conclusions about the position and nature of future object movement and thereby helps to increase the reliability of the function. Connecting the video system to the long-range radar creates synergy effects. This significantly increases the beam angle of the ACC system or lane-keeping support and makes object detection even faster and more reliable.

Future driver-assistance systems will incorporate a large number of environment sensors. Because the sensor data typically arrive at different frequencies and asynchronously and are subject to different time delays, a common time allocation must first be established in the fusion module. This is achieved typically using ring memory structures and time stamps. The vehicle's surroundings are modeled on the basis of the sensor data. Grid approaches are increasingly being used to model the non-moving (stationary) surroundings [3]. Moving (dynamic) objects are typically modeled as objects with various function-relevant attributes (such as position, expansion, speed) and time-stabilized by a Kalman filter. Hybrid approaches which combine object models and stationary grids are still in the research stage [4].

Function module
The environment sensors and their functional processing will now be explained in more detail for selected examples in the following.

The front-end video camera can be used for various assistance functions, such as for example lane detection and road-sign recognition. The "lane detection system" can identify the lane boundaries and the lane direction ahead. If the vehicle is about to move out of its lane unintentionally, the system alerts the driver (see Lane-departure warning system). In a further expansion stage the vehicle – on the basis of lane detection – is moved back to its lane by an active steering intervention (see Lane-keeping support). In combination with ACC and a traffic-jam assistant a powerful system is created to reduce the burden on the driver in stop-and-go traffic.

As from 2015 high-resolution multi-level laser scanners with a range of 80 to 150 m are to be used to monitor the central and side front area for traffic-jam assistance functions and for NCAP pedestrian protection (European New Car Assessment Program). Miniaturized laser scanners are designed to provide high-resolution environment sensing at the vehicle sides and used for lane-changing assistance and maneuvering assistance. Their range starts at 5 cm in the ultra-close range and extends to over 40 m.

Another function that makes use of the information from the video sensor is road-sign recognition. This system is capable of recognizing and interpreting road signs (e.g. speed limit or no-overtaking signs). A video camera films the surrounding area and transmits the video signal to the image-processing computer. While the vehicle is moving, the system searches for objects which on the basis of their external shape could be road signs (object detection). When such an object is found, it is tracked until it is close enough to be picked up by the video camera (object classification). Road-sign recognition presupposes that the designs to be recognized have been taught beforehand into the image-processing computer.

The speed limit recognized and interpreted as such by the image-processing computer (Figure 5) is adopted by the instrument cluster and displayed as a symbol in the graphic display. If the driver fails to observe this speed limit, the system can issue an additional acoustic or haptic warning. Reliable road-sign recognition is now possible at speeds of up to 160 km/h and also in conditions of rain and road spray.

Recognized road signs are increasingly being used in a variety of assistance functions. For example, navigation maps can be updated with recognized road signs and the adaptive cruise control system can in a further development of this func-

Figure 5: Road-sign recognition
a) Road image,
b) Symbols in instrument-cluster graphic display.

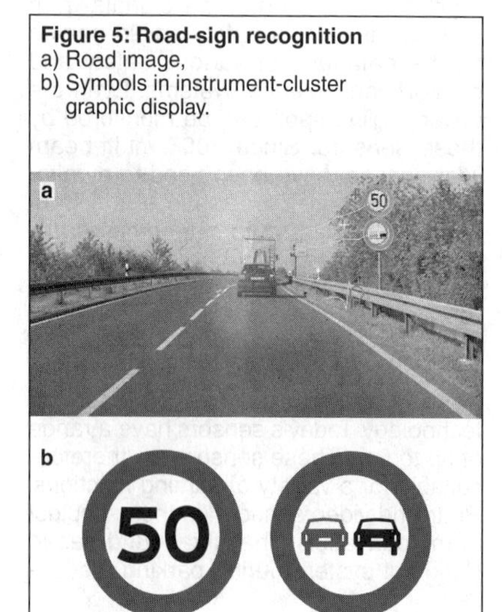

Table 1: Systematic classification of driver-assistance functions

Function name	Warning (W) Longitudinal guidance (Lng) Lateral guidance (Lat)	City (C) Ordinary road (O) Expressway/ interstate (E)	Year of market launch	Typical sensor technology
Adaptive cruise control (ACC)	Lng	C, O, E	1998	Radar Lidar (occasionally)
Lane departure warning	W	C, E	2001	Mono video
Lane-keeping system	W, Lat	C, E	2003	Mono video
Roadworks assistant	W, Lng, Lat	E	Open	Mono video Stereo video Radar Ultrasound (side)
Maneuvering assistant	W, Lat	C	Open	Stereo video Mono video (side)
Bottleneck assistant	W, Lat	C, O	Open	Stereo video
Traffic-jam assistant	W, Lng, Lat	C, E	2013	Stereo video
Intersection assistant	W, Lng	C, O	2013	Radar (side)
Emergency-braking system (driving in same direction)	W, Lng	C, O, E	2006	Radar Stereo video
Active pedestrian protection	W, Lng	C, O	2008	Stereo video Mono video

tion maintain the vehicle's speed at the identified speed limit.

A rear-end camera offers greater benefit if the detected objects can be interpreted by image-processing software and the driver is alerted in critical situations. This is the case for example if the driver while backing up has failed to see a pedestrian crossing the road.

Video technology was first used in information-providing driver-assistance systems. However, advanced methods of image processing (already available on the market) demonstrate the potential of these sensors. Thus, today's video sensors use different approaches to provide three-dimensional measurement of the surroundings, opening up the possibility of new driver-assistance functions.

Table 1 uses delimiting criteria to provide an overview of the driver-assistance functions discussed here.

References

[1] K. Enke: Possibilities for Improving Safety within the Driver Vehicle Environment Loop; 7th International Technical Conference on Experimental Safety Vehicle, Paris (1979).
[2] http://www.bast.de/nn_42740/DE/Aufgaben/abteilung-f/referat-f2/gidas/gidas.html.
[3] C. Coue et al.: Bayesian Occupancy Filtering for Multitarget Tracking: an Automotive Application. The International Journal of Robotics Research 25(1): 19–30, 2006
[4] J. Effertz: Autonome Fahrzeugführung in urbaner Umgebung durch Kombination objekt- und kartenbasierter Umfeldmodelle. Institut für Regelungstechnik, TU Braunschweig, Dissertation, 2009.

Computer vision

The computer vision activity field encompasses concepts and methods for extracting information about the surrounding area made up of a static background (e.g. roadway, buildings) and dynamic objects (e.g. pedestrians, vehicles) and scanned by one or more cameras. The three-dimensional environment changing over time is projected into a sequence of two-dimensional image signals through image acquisition, meaning that reconstruction of the environment poses a complex inverse problem [1].

While people appear to be able to handle complex dynamic scenarios without difficulty, machine vision, and consistent interpretation of the image sequence data in particular, also represents a challenging, interdisciplinary activity field after over five decades of intensive research. A large portion of the basic elementary principles behind the image signal processing had already been set in the 1970s and 1980s. This includes operators for filtering image signals, extracting simple geometric structures and matching image signals.

A crucial turning point in the area of environment reconstruction was reached in the 1990s with the development of robust methods for projective reconstruction based on uncalibrated camera systems. The 2000s were characterized by extensive progress in the area of machine learning and the learning-based methods involving detection and classification of objects based on training examples.

Video-based driver-assistance systems such as automatic light control, lane departure warning systems and traffic sign recognition are already available today in a wide variety of vehicles. Future systems will take over complex, safety-related functions and advanced comfort functions and in doing so, will be one step closer to the long-term objective of automated driving.

Camera model

Mathematical model

The mathematical model of a distortion-free camera that delivers a perspective map of the 3D environment on a 2D image plane, is described in full using a pinhole camera as shown in Figure 1 [2] with:

$$m = c_x + f s_x \frac{X}{Z} = c_x + \alpha_x x \text{ and}$$

$$n = c_y + f s_y \frac{Y}{Z} = c_y + \alpha_y y$$

where $\alpha_x = f s_x$, $\alpha_y = f s_y$, $x = X/Z$, $y = Y/Z$. For an image with $M \times N$ pixels, the whole numbers parts of the image coordinates (m,n) assume values in the range $0 \leq m \leq M-1$ and $0 \leq n \leq N-1$.

The projection center $\boldsymbol{C}$ provides the origin for the camera coordinate system (X,Y,Z). The Z axis forms the optical axis of the camera that runs through the optical center and is perpendicular to the image plane. (c_x,c_y) defines the main point of the image sensor, i.e. the intersection of the optical axis with the image plane. (s_x, s_y) are scaling factors with the unit [pel/m] and correspond to the reciprocal values of the pixel pitch in the x and y direction (q_x, q_y) up to the dimensionless pseudo-unit [pel]. The (α_x, α_y) parameters of the camera model represent the scaled focal length f (camera constant) in pixel units.

By implementing uniform image coordinates, the camera model can be compactly written as a linear transformation:

$$\begin{pmatrix} m \\ n \\ 1 \end{pmatrix} = \boldsymbol{KX} \cong \begin{pmatrix} \alpha_x & 0 & c_x \\ 0 & \alpha_y & c_y \\ 0 & 0 & 1 \end{pmatrix} \begin{pmatrix} X \\ Y \\ Z \end{pmatrix}$$

The transformation is conveyed by the calibration matrix $\boldsymbol{K}$. In this context, the symbol $\cong$ means equality up to a scalar factor $\lambda \neq 0$.

The inverse transformation $\boldsymbol{K}^{-1}$ yields the standardized camera coordinates:

$$\begin{pmatrix} x \\ y \\ 1 \end{pmatrix} = \boldsymbol{K}^{-1} \begin{pmatrix} m \\ n \\ 1 \end{pmatrix} = \begin{pmatrix} X/Z \\ Y/Z \\ 1 \end{pmatrix} \cong \begin{pmatrix} X \\ Y \\ Z \end{pmatrix}$$

These result from the scaling with spatial depth Z from the camera coordinates $\boldsymbol{X} = (X,Y,Z)$.

The relationship between the camera coordinate system (X,Y,Z) and the world coordinate system (X',Y',Z') is determined by a Euclidean transformation $(\boldsymbol{R},\boldsymbol{t})$, whereby $\boldsymbol{R}$ represents a (3×3) rotation matrix and $\boldsymbol{t}$ represents a (3×1) a translation vector (Figure 2).

Subsequently, the transformation from world to camera coordinates is then

$$\boldsymbol{X} = \boldsymbol{R}(\boldsymbol{X}' - \boldsymbol{t}') = \boldsymbol{R}\boldsymbol{X}' + \boldsymbol{t},$$

where $\boldsymbol{t}'$ describes the position vector of the camera center in world coordinates. Accordingly, $\boldsymbol{t} = -\boldsymbol{R}\boldsymbol{t}'$ represents the origin of the world coordinate system in camera coordinates.

These definitions enable the projection of a point in world coordinates to be written to the corresponding image coordinate system as a linear transformation

$$\begin{pmatrix} m \\ n \\ 1 \end{pmatrix} = \boldsymbol{K}(\boldsymbol{R}\boldsymbol{X}' + \boldsymbol{t}) \cong \boldsymbol{K}(\boldsymbol{R}|\boldsymbol{t})\begin{pmatrix} \boldsymbol{X}' \\ 1 \end{pmatrix}$$

where

$$\boldsymbol{P} = \boldsymbol{K}(\boldsymbol{R}|\boldsymbol{t})$$

Figure 2: Camera coordinate system (X,Y,Z) and world coordinate system (X',Y',Z') as well as Euclidean transformation (R,t)

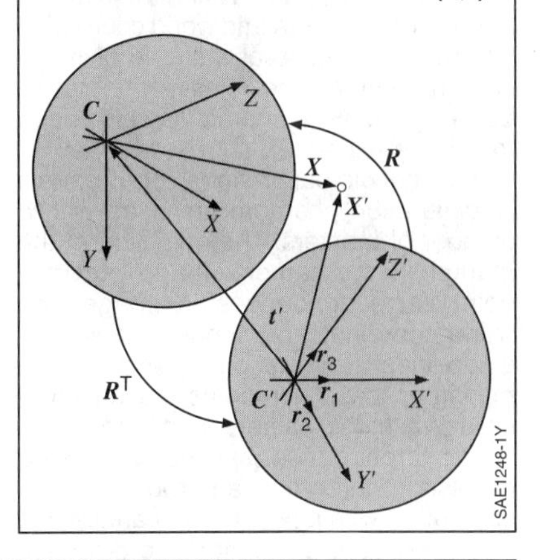

Figure 1: Pinhole camera model
Model with physical and virtual image sensor for defining image coordinates (m,n), camera coordinate system (X,Y,Z) and the parameters for the calibration matrix $\boldsymbol{K}$.
$\boldsymbol{C}$ projection center, f focal length (distance $\boldsymbol{C}$ to the main level of the image sensor), s_x, s_y scaling factors, q_x, q_y pixel pitch, c_x, c_y main point of the image sensor, $\boldsymbol{X}$ pixel.

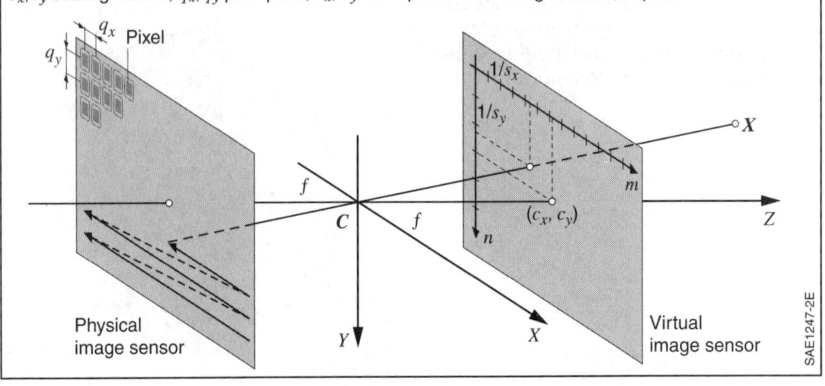

refers to the camera matrix, which has ten degrees of freedom (four intrinsic parameters, three degrees of freedom for rotation, three degrees of freedom for translation) in the case under consideration here and fully describes the transformation geometry of an ideal, distortion-free camera.

Camera calibration

The geometric calibration of a camera includes estimating the intrinsic parameters of the camera, i.e. determining the elements of the calibration matrix $\boldsymbol{K}$, estimating the parameters of optical distortion and, if necessary, the determining the relative orientation $\boldsymbol{R}$ and translation $\boldsymbol{t}$ between the camera and world coordinate system. In the case of a stereo or multi-camera system, the relative orientation between the two cameras may also need to be determined [2].

For intrinsic calibration, target panels or calibration phantom are used with a number of markers. The detection of the marker in the camera image provides point correspondences in image and world coordinates as input variables for a non-linear optimization problem for estimating optical distortion as well as intrinsic and extrinsic camera parameters.

The target panel depicted in Figure 3 consists of three plates perpendicular to each other with a number of markers and makes it possible to determine distortion parameters as well as the elements of the calibration matrix from a single image. Another option for camera calibration is to use a simple, level calibration phantom recorded under different orientations and positions, so that a large number of point correspondences are available to determine the calibration parameters.

Image processing

Image preprocessing

The photographic sequence provided by the camera generally has to be preprocessed for adaptation to the subsequent image preprocessing steps. This includes aspects such as filter operations for noise reduction or even smoothing and down-sampling image data. Methods for the dynamic compression of image data are used to achieve a compact representation for processing and storing the image data, especially for HDR (high dynamic range) image sequences. A representation of the image data on multiple scaling levels in the form of a resolution pyramid is frequently useful to enable a largely scale-invariant postprocessing with limited calculating complexity [3].

Feature extraction

Suitable features are frequently extracted from an image signal for efficient representation and processing of an image sequence. Corners, lines, circles and ellipses provide simple examples that frequently correspond to semantic structures in the environment. This makes it possible, for example, to find lane markings using an edge detector, which is based on evaluation of the image signal gradient

$$\nabla I = \begin{pmatrix} I_x \\ I_y \end{pmatrix}.$$

Figure 3: Calibration body for camera calibration

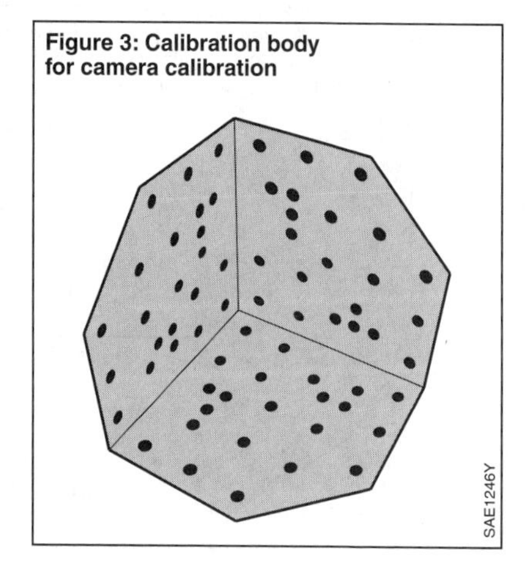

Here I represents the luminance component of the image signal, while I_x and I_y represent the derivatives of I in the x and y direction of the image plane.

Circles, ellipses and complex geometric structures that possess an sufficiently compact parametric description can also be detected with elementary methods on the basis of image signal gradients. One obvious application is detecting traffic signs, which feature simple geometry and are designed for visual detection in regard to their reflective qualities.

Corner-like structures represent another important class of image features, since (in contrast to line-like structures) they enable point-by-point localization and the formation of correspondences, such as are required for determining the optical flow in an image sequence.

Pattern recognition and object classification

A frequently applied method for pattern or object recognition is based on comparing image regions with predefined texture patterns. If these are limited to the binary texture patterns shown in Figure 4, for example, the integral image

$$S(n,m) = \sum_{n'=0}^{n} \sum_{m'=0}^{m} I(n',m')$$

Figure 4: Elementary image texture pattern for designing efficient object classification methods

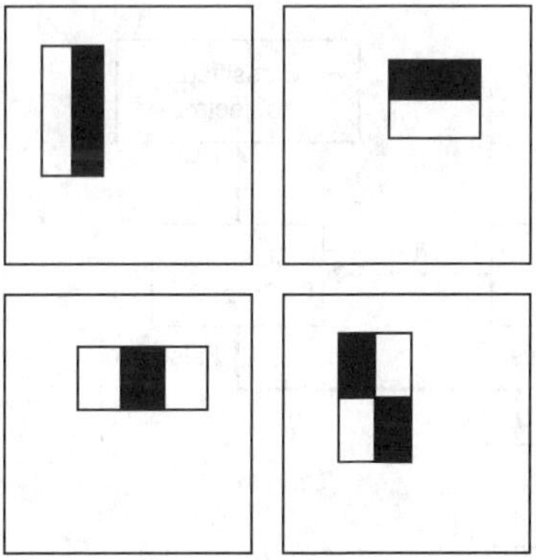

SAE1250Y

with a luminance value of $I(n',m')$ at position (n',m') results in an efficient implementation with constant computing complexity regardless of the size of the texture pattern. Selecting the difference of the gray values summed by the respective image regions bounded by rectangles as the gap dimension makes it possible to determine the necessary subtotals immediately by referring to the integral image.

Figure 5 demonstrates this principle using the determination of the grayscale value total within the image region with the corner points P_i as an example. This region is determined by

$$S_R = S(P_1) + S(P_4) - (S(P_2) + S(P_3)).$$

Consequently, the evaluation of the texture patterns as shown in Figure 4 only requires six (two adjoining rectangles), eight (three rectangles) or nine (four rectangles) times accessing the pre-calculated integral image S.

Figure 5: The sum of grayscale values in the highlighted image region can be efficiently determined by accessing the integral image S four times at the positions P_i

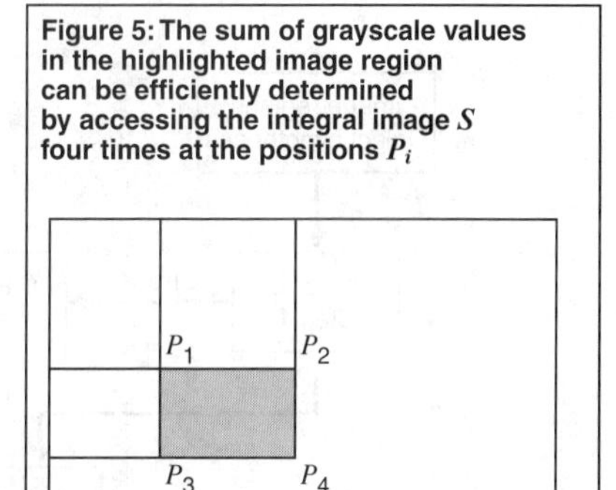

For object classification, classifiers are determined as part of a training process based on example images. The classifiers include increasingly more complex combinations of presented simple texture patterns for object description (Figure 6).

The training process is controlled such that the individual classification levels reach a very high object detection probability, which is associated with a relatively high probability of an erroneous detection by necessity. Very high quality in the overall classifier can be achieved, however, through the cascade structure of the process provided the existence of stochastic independent classification levels. Let us use a ten-level classification cascade with detection probability $P_{d,i}$ = 0.995 and probability of erroneous detection $P_{f,i}$ = 0.4 for all sublevels as an example. This yields an object classifier with

$$P_d = \prod_{i=1}^{N} P_{d,i} = 0.995^{10} = 0.95 \text{ and}$$

$$P_f = \prod_{i=1}^{N} P_{f,i} = 0.4^{10} = 10^{-4}.$$

Although extensive theories on pattern recognition, classification and statistical learning methods already exist, the design engineering of efficient, robust processes for solving object classification problems represents an active research area in real-world applications.

Stereo video

A stereo video camera has two separate optical paths with synchronized image recorders. This enables a spatial 3D reconstruction of the recorded environment based on simple point-by-point triangulation. The measuring principle is depicted in Figure 7 [2].

The projection centers ($\boldsymbol{C}$,$\boldsymbol{C}'$) of the stereo camera system span a stereo base with base width b. The projections of the camera centers on the other respective image plane are referred to as epipoles ($\boldsymbol{e}$,$\boldsymbol{e}'$). The camera centers and the reference point $\boldsymbol{X}$ span a epipolar plane where the intersections with the image planes (Π, Π') form the epipolar lines ($\boldsymbol{l}$,$\boldsymbol{l}'$). Obviously, pixel $\boldsymbol{x}$ corresponding to pixel $\boldsymbol{x}'$ must lie on epipolar line $\boldsymbol{l}'$. This means that determining the corresponding pixels of a stereo camera system represents a one-dimensional search problem that can be efficiently implemented accordingly.

A particularly simple geometry results if the optical axes of the stereo camera system are aligned parallel to each other and curve vertically to the spanned base. Here, epipolar lines and image lines overlap, so that correspondences can be

Figure 6: Object classifier with N stages

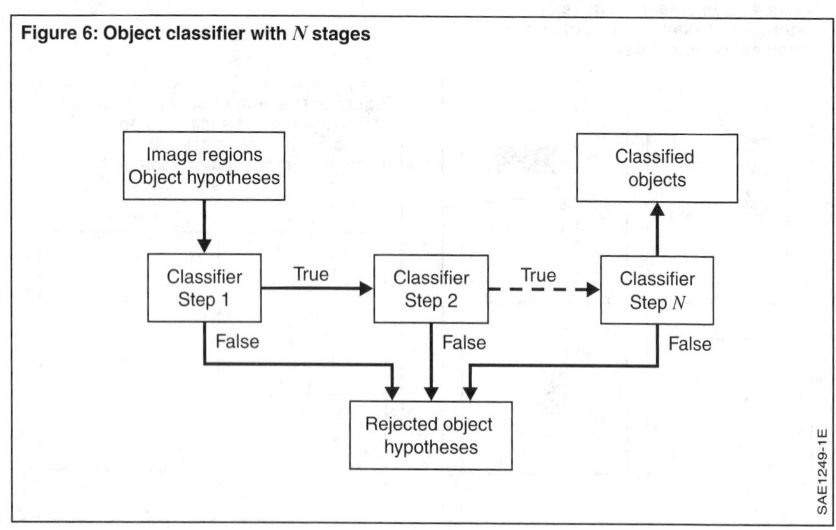

Figure 7: Stereo camera system for explaining the epipolar geometry and one-dimensional point correspondence searching along epipolar lines

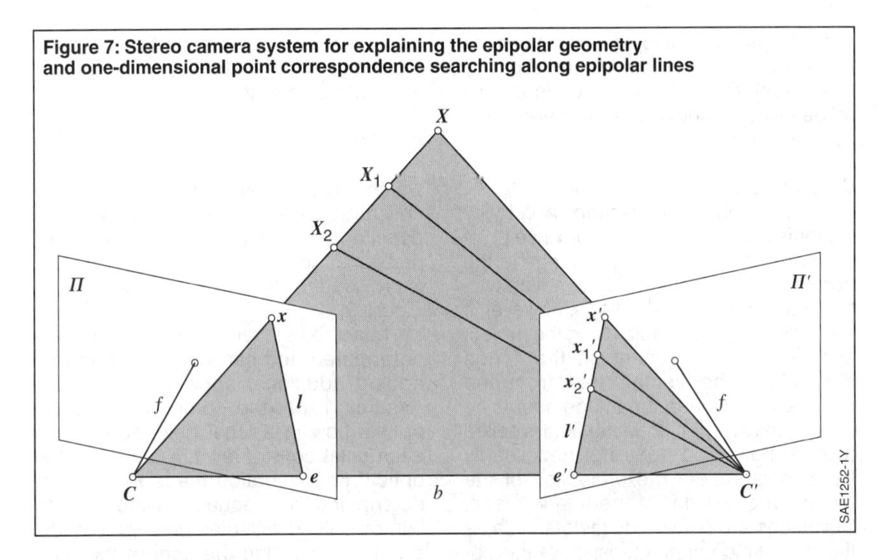

sought using the pixels of a single scan line. The difference

$$d = x - x'$$

means disparity of the corresponding pixels and, in accordance with Figure 8, directly determines the depth

$$Z = \frac{bf}{d}$$

Figure 8: Determining the depth Z of the reference point X from disparity d between corresponding pixels x and x'

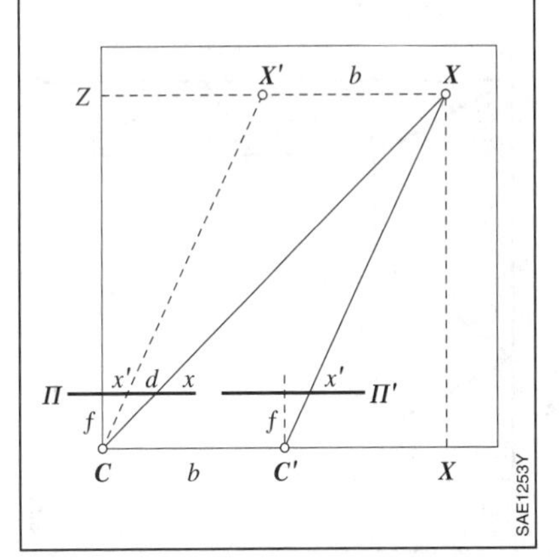

of the reference point $\boldsymbol{X}$ due to the similarity of the triangle $\Delta(\boldsymbol{C},\boldsymbol{x},\boldsymbol{x}')$ and $\Delta(\boldsymbol{C},\boldsymbol{X},\boldsymbol{X}')$. The 3D coordinates (X,Y,Z) of the observed reference point $\boldsymbol{X}$ are thus completely determined by the disparity d and the coordinates $(x = f\,X/Z,\ y = f\,Y/Z)$ of the pixel $\boldsymbol{x}$.

The relatively accuracy of the depth estimate is

$$\frac{|\Delta Z|}{Z} = \frac{|\Delta d|}{d},$$

so that a disparity of Δd = 10 pel is required assuming a disparity estimate of d = 0.5 pel to achieve a relative depth estimation error of 5 %. The absolute error in the depth estimate is

$$|\Delta Z| = \frac{|\Delta d|}{bf} Z^2$$

and results in a base width of b = 0.25 m and a horizontal opening angle of approximately 50° at a range of Z = 32 m for a single measurement for a camera system with M = 1,280 pixels per scan line, as is in the observed example. By averaging several individual measurements, accuracy and range can be further increased on object planes.

Stereo video systems enable robust implementation of safety-related assistant functions in a vehicle with interaction in the longitudinal and lateral control of the vehicle, such as building and bottleneck assistance and automatic emergency brake systems. This is due to its largely model-free method of operation and direct 3D measurement.

Optical flow

At a given observation time, a velocity vector is assigned to each reference point of the environment via the change in the position vector over time. This motion vector field is thus clearly defined, however, it is not directly observable using the projection of the environment on the image sequence. The vector field from the change over time in the image sequence in relation to a specific vector field is called "optical flow" and does not necessarily have to represent the projection of the motion vector field on the image plane. Deviations are caused by factors such as the reflectance characteristics of the observed objects based on the direction of incident light, changes in lighting over time, moving shadows or superimposed movements in the case of transparent objects. If these phenomena are not dominant, then the optical flow supplies a consistent estimate of the projected motion vector field, forming an essential component for detecting and interpreting environment scenarios changing over time.

For designing estimation procedures to determine the optical flow from the observed image sequence, it is often assumed that the luminance component of the pixels along the movement trajectories is constant, i.e.:

$$I(x+\delta x, y+\delta y, t+\delta t) = I(x, y, t).$$

The luminance or the grayscale value information of the image sequence is thus considered a constant that, after linearization for small time intervals where

$$u = dx = \dot{x}\,dt, \; v = dy = \dot{y}\,dt,$$

yields the optical flow equation

$$I_x u + I_y v + I_t = 0$$

for determining $\boldsymbol{u} = (u,v)^{\mathrm{T}}$. The spatial and chronological derivations that occur are replaced by corresponding difference quotients during implementation. Using the image signal gradient introduced previously, the condition equation is

$$\nabla I\, \boldsymbol{u}^{\mathrm{T}} = -I_t,$$

which means that components of the optical flow can be specified only in the direction of the grayscale gradient. This problem, known as the aperture problem, is provided to illustrate the correlations in Figure 9.

Thus, the estimation problem is underdetermined and can only be solved by adding additional ancillary conditions. Assuming that the components of the optical flow in a local neighborhood of each pixel considered are constant, the optical flow equation leads to an overdetermined linear equation system, which can be solved with elementary methods, e.g. by minimizing the sum of the quadratic residue.

Figure 9: Aperture problems in estimating the motion vector field (solid lines) from the optical flow

The optical flow is not determined perpendicular to the direction of the grayscale gradient. If only one image section with two vertical or parallel lines is being considered, it is impossible to determine whether the lines are moving vertically with respect to their alignment or at an angle. All plotted vectors satisfy the optical flow equation, so that the motion vector field cannot be completely determined using local methods.
Solid lines: Actual direction of motion of the object.
Dashed lines: Possible direction of motion.

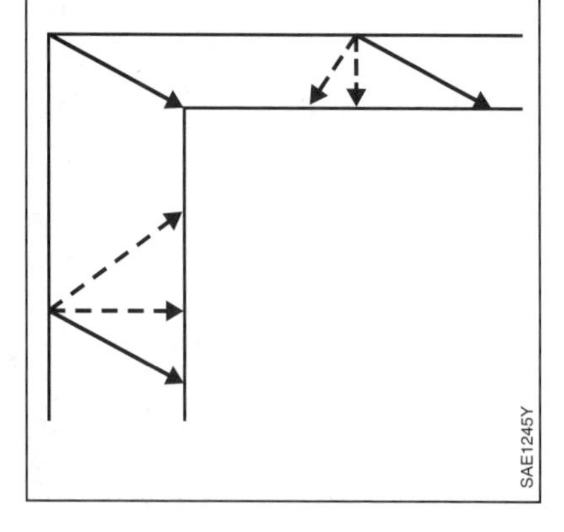

Modern methods for estimating optical flow avoid the linearization of the optical flow equation and ascribe the estimation problem to the minimization of a suitably selected energy function as

$$E(u,v) = E_d + \alpha E_\delta\,.$$

For example,

$$E_d = \sum_{x,y} \psi(I(x+u,\, y+v,\, t+T) - I(x,y,t))$$

can be selected for the data term here, where ψ represents a robust weighting function, which is often selected as

$$\psi(w) = \sqrt{w^2 + \varepsilon^2}\,.$$

The smoothing term

$$E_s = \sum_{x,y} \psi(|\nabla u|^2 + |\nabla v|^2)$$

is based on the assumption of a piecewise smooth vector field and leads to a regularization of the estimation task, where the degree of regularization is controlled via the universal parameter α. Efficient numeric methods are available for solving this and other refined variation problems for determining optical flow. However, the robust estimation of the motion vector field in the case of difficult ancillary conditions, as they occur in video-based assistant functions in motor vehicles, continue to pose a challenging research and development task.

A significant application of optical flow is estimating movement of the camera itself, which enables 3D reconstruction of the environment. Optical flow, however, can also be used directly for segmenting image content, for example, to detect pedestrians crossing the street or merging vehicles.

3D reconstruction from image sequences

The basic principle of 3D reconstruction of the environment from the chronological image sequence of a camera follows from the camera arrangement shown previously in Figure 7. Now, however, this is not regarded as a chronologically synchronized camera pair in a stereo camera system. Instead, it describes the position and orientation of a moving camera at two points in time. Triangulation, and therefore a 3D measurement of the reference point P, is also possible in this case, provided this is a fixed location during the time interval in question and the camera's own movement. However, only the static environment can be reconstructed now using this method. Since the movement of the camera itself provides the basis required for 3D reconstruction, the method is also designated as "Structure from Motion" (SfM). The related conceptualization "motion stereo" is also frequently used [2].

The core task of the SfM method is to estimate the camera's own movement with sufficient accuracy. If, as in Figure 7, you choose the camera's coordinate system with the origin $\boldsymbol{C}$ as the reference coordinate system, then the following relation is true for the corresponding camera model:

$$\boldsymbol{P} = \boldsymbol{K}(\boldsymbol{I}|\boldsymbol{O}),$$
$$\boldsymbol{P}' = \boldsymbol{K}(\boldsymbol{R}|\boldsymbol{t})$$

with the unit vector $\boldsymbol{I}$ and the zero vector $\boldsymbol{O}$. For the camera system's epipoles, therefore,

$$\boldsymbol{e} = \boldsymbol{P}\boldsymbol{C}' \cong \boldsymbol{K}\boldsymbol{R}^{\mathrm{T}}\boldsymbol{t},$$
$$\boldsymbol{e}' = \boldsymbol{P}'\boldsymbol{C} \cong \boldsymbol{K}\boldsymbol{t}.$$

If the epipoles are known, then the epipolar lines, which define the one-dimensional search area for efficient determination of corresponding pixels, are completely determined:

$$\boldsymbol{l} = \boldsymbol{e} \times \boldsymbol{x} \cong [\boldsymbol{K}\boldsymbol{R}^{\mathrm{T}}\boldsymbol{t}]_{\times}\boldsymbol{x},$$
$$\boldsymbol{l}' = \boldsymbol{e}' \times \boldsymbol{x}' \cong [\boldsymbol{K}\boldsymbol{t}]_{\times}\boldsymbol{x}',$$

whereby the vector operator $[\boldsymbol{a}]_\times$ designates the skew symmetric (3×3) matrix

$$[\boldsymbol{a}]_\times = \begin{pmatrix} 0 & -a_3 & a_2 \\ a_3 & 0 & -a_1 \\ -a_2 & a_1 & 0 \end{pmatrix}$$

These correlations describe the epipolar geometry of the camera configuration and lead directly to the fundamental matrix

$$\boldsymbol{F} = \boldsymbol{K}^{-\mathrm{T}} [\boldsymbol{t}]_\times \boldsymbol{R}\boldsymbol{K}^{-1},$$

which describes the projection

$$F\colon \boldsymbol{x} \rightarrow \boldsymbol{l}' = \boldsymbol{F}\boldsymbol{x}$$

of the pixel $\boldsymbol{x}$ from the first camera onto the epipolar line $\boldsymbol{l}'$ of the second camera. The pixel $\boldsymbol{x}'$ lies on the epipolar line $\boldsymbol{l}'$, resulting in the epipolar condition

$$\boldsymbol{x}'\boldsymbol{F}\boldsymbol{x} = 0.$$

If at least eight pixel correspondences are known, this condition leads to a uniform linear system of equations for determining the F matrix. If this is known, the cameras $\boldsymbol{P}, \boldsymbol{P}'$ and the static reference points $\boldsymbol{X}_i$ are determined except for an unknown projective transformation $\boldsymbol{H}$ (homography), that is, a projective reconstruction of the environment is already possible based only on the pixel correspondences $(\boldsymbol{x}, \boldsymbol{x}')$ of two uncalibrated cameras. This enables far-reaching statements to be made about the observed environment, since incidence relations (intersections of points, lines, surfaces) in particular as well as straight lines are preserved, because these variables represent invariants of the projective geometry. By applying previous knowledge about the environment (such as parallelism of lines, angle magnitudes, and lengths), the number of degrees of freedom of the homography $\boldsymbol{H}$ can be reduced incrementally, so that affine and, finally, even metric reconstructions of the environment are made possible.

If the calibration matrix $\boldsymbol{K}$ is known, then, when using normalized camera coordinates $(\boldsymbol{K}^{-1}\boldsymbol{x}, \boldsymbol{K}^{-1}\boldsymbol{x}')$, the F matrix changes into the essential matrix

$$\boldsymbol{E} = [\boldsymbol{t}]_\times \boldsymbol{R}$$

which possesses only five degrees of freedom (three for rotation and two for normalized translation) and can be determined accordingly based on five point correspondences using non-linear methods. If the E matrix is known, the environment can be metrically reconstructed up to a global scaling factor. If the vehicle has a driver assist camera, global scaling can be determined from the odometry data or the installation height of the camera above ground.

References

[1] D.A. Forsyth, J. Ponce: Computer Vision: A Modern Approach. 2nd Edition, Englewood Cliffs, NJ, Prentice Hall, 2011.
[2] R.I. Hartley, A. Zisserman: Multiple View Geometry in Computer Vision. 2nd Edition, Cambridge University Press, 2004.
[3] B. Jähne: Digitale Bildverarbeitung und Bildgewinnung. 7th ed., Springer-Verlag, 2012.

Vehicle navigation

Navigation systems

Determination of the position of a vehicle is based primarily on using the GPS satellite positioning system (Global Positioning System). The navigation system compares the determined position with a digital map and uses this map to calculate the route to the specified destination.

Vehicles can be equipped at the factory (original equipment) with permanently installed navigation devices (Figure 1). Portable navigation devices are also in widespread use. Compared with portable devices, permanently installed navigation devices in the vehicle provide better positioning quality and thus better route-guidance quality, since additional sensors for distance and direction signals (wheel-speed and yaw sensors) can be evaluated and the antenna can be mounted in a more favorable place for satellite reception. As original equipment, networking with other components is also common, i.e. integration in the operating concept of the vehicle is possible. Voice outputs can be issued via the audio system and muted for telephone calls. The route-guidance information can be displayed in the instrument cluster or in the head-up display and thus in the driver's primary field of vision.

Functions of navigation

Positioning

Satellite positioning system GPS

GPS is based on a network of 24 US military satellites which can be used globally for positioning (Figure 2). They circle the Earth in six different orbits at an altitude of approximately 20,000 km in twelve-hourly intervals. They are distributed in such a way that from every point on the Earth always at least four (for the most part up to eight) are visible over the horizon.

The satellites transmit special position, identification and time signals 50 times per second on a carrier frequency of 1.57542 GHz. The satellites each have on board two cesium and two rubidium atomic clocks which vary by less than 20 to 30 ns for the high-precision determination of the transmission time.

Position determination

The signals from the satellites arrive at the vehicle at different times on account of the different propagation times. The position of the receiver is calculated using the trilateration process. When the signals from at least three satellites arrive, the navigation device can calculate its own geographical position in two dimensions (terrestrial longitude and latitude). There is precisely one point which satisfies the distance conditions (signal propagation times).

Figure 1: Navigation system

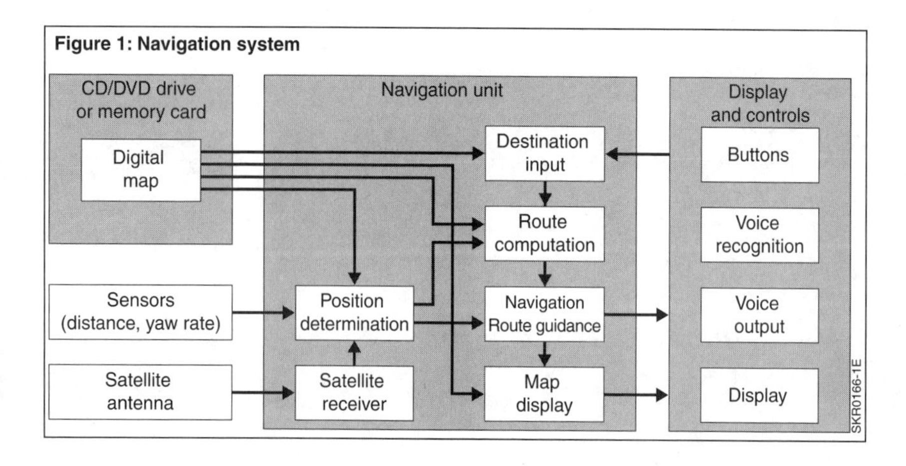

When the signals from at least four satellites arrive, this makes possible a three-dimensional position calculation (with altitude). Figure 3 shows this process in simplified form in only two dimensions.

Accuracy
The achievable accuracy is dependent on the position of the satellites that can be received relative to the vehicle. The greater the solid angle of the satellites to the vehicle, the better the possibility of position determination. The achievable accuracy is in the plane approximately 3 to 5 m, and for the determination of height approximately 10 to 20 m.

In deep urban canyons, satellite signals can only be received if the satellites are arranged to a large extent in a line, i.e. in the direction of the street/road. However, the solid angle formed by the satellites is then very small and position determination inaccurate.

Errors in position determination may arise as a result of the reflection of the satellite signals, on metal-coated building fronts, for example.

Travel-direction determination
The direction of travel can be quickly sensed from the differences in the receive frequency of the satellites which are brought about by the Doppler effect. When a car drives towards a satellite, the GPS receiver of the navigation devices sees a higher frequency than the transmit frequency. It receives a lower frequency from satellites behind. This effect is sufficiently great from a driving speed of approximately 30 km/h to determine the direction.

Dead reckoning
Dead reckoning ensures position determination even if no GPS signals can be received, e.g. in tunnels. It adds cyclically recorded distance elements vectorially by magnitude and direction. The speedometer signal transmitted via the CAN bus is used to measure the distance. Changes in direction are recorded by a yaw sensor. In this way, the direction of travel is determined starting out from an absolute direction which was calculated with the GPS signals received last via the Doppler effect.

Map matching
The procedure known as map matching continuously compares the located position with the route on the digital map. In this way, the exact vehicle position can also be shown on the map, even if the located position is inaccurate (caused for example by the lack of a GPS signal and errors in dead reckoning). Route-guidance driving recommendations can thus be output at the best possible location. In addition, sensor errors and cumulative dead-reckoning errors can be compensated.

Figure 2: GPS satellite positioning system
1 to 24 satellites for determining the position of a vehicle.

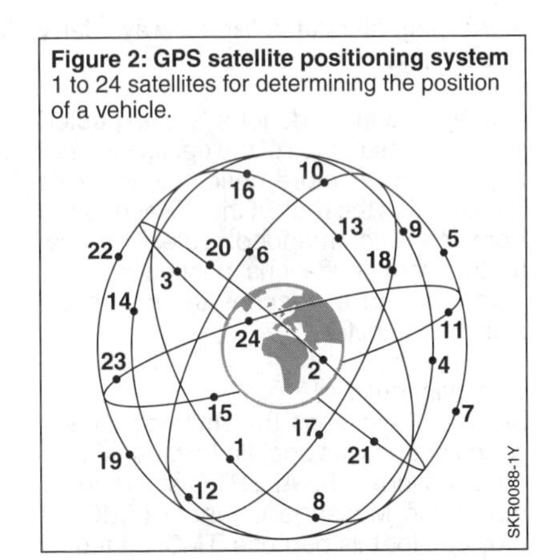

Figure 3: Position determination with GPS
(Simplified two-dimensional view). Where the positions of the satellites are known, the possible reception locations for the measured signal propagation times t_1 and t_2 are on two circles around the satellites. The point of intersection A on the Earth is the sought-after location.

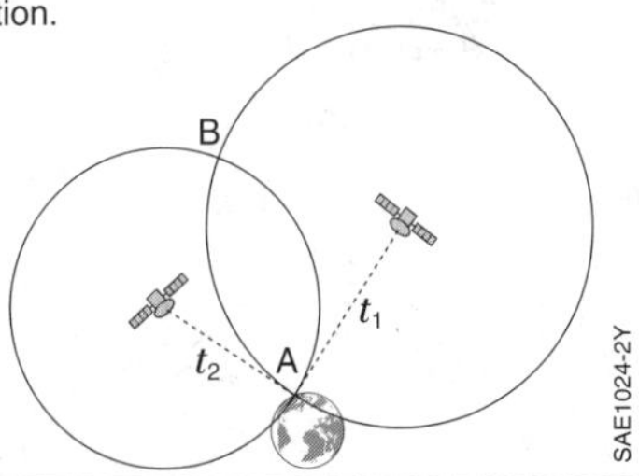

B
t_1
t_2
A
SAE1024-2Y

Destination entry

The destination is entered, for example, via the navigation-device panel with buttons, directly via the touchscreen, or by voice input. The user enters all the necessary details from a menu or in response to prompting by acoustically output instructions.

The digital map contains directories so that a destination can be entered as an address. Lists of all known place names are required for this purpose. In turn, all locations are allocated lists containing the names of the stored street names. To pinpoint a destination even further, the user can then also select road/street intersections, or the number of a building.

The destination can be quickly entered via the destination memory, with which already entered destinations can be called up again (e.g. the last destinations used or the destinations saved as favorites).

For airports, train stations, filling stations, multi-storey parking garages, etc. there are thematic directories in which these destinations (POI, Points of Interest) are listed. These directories make it possible, for example, to locate a nearby filling station or multi-storey parking garage (Figure 4).

In many navigation systems, destinations can also be marked in the map display (e.g. via the touchscreen).

Figure 4: Example of a map display with POI

Figure 5: Dynamic route guidance
1 Original main route, 2 Congestion,
3 Alternative route estimated by driver,
4 Best alternative route calculated by dynamic route guidance.
A Interstate/expressway,
B National highway,
L Secondary road.

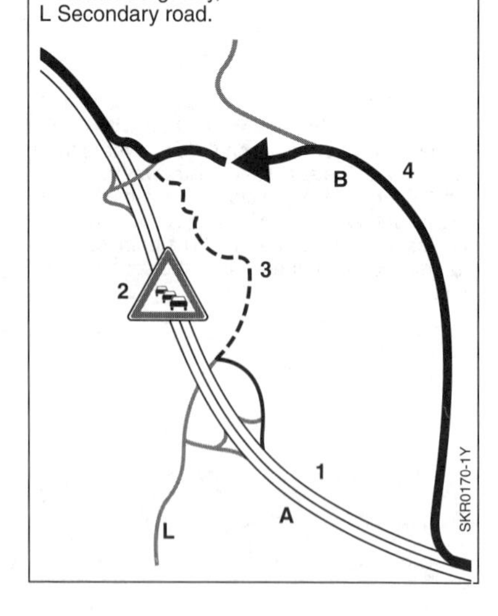

Route computation

Standard computation

Starting from the current location, the navigation device calculates a route leading to the entered destination. The way in which the route is calculated can be adapted to suit the driver's preferences. He or she can stipulate different options such as, for example, optimizing the route

- based on driving time,
- based on an economic mean of driving time and driving distance,
- based on an estimated minimum fuel consumption,
- avoiding interstates/expressways, ferry crossings, or toll roads.

Route recommendations are expected within a few seconds of the destination being entered. A more critical scenario is recomputing the route if the driver departs from the recommended route. The new route recommendations must be supplied before the driver reaches the next intersection or junction.

Dynamic routing

Many radio stations transmit traffic messages not only as spoken text, but also in encoded form. The ALERT-C standard for the Traffic Message Channel (TMC) is provided for this purpose. TMC content is transmitted via the Radio Data System (RDS) of FM radio.

The encoded messages contain, among other things, the location of a holdup, its extent, the actual length, and the reason for it. A navigation system can receive such an encoded message and determine whether a holdup is situated on a planned route. If this is the case, the route is recalculated, where the held-up section is evaluated with a longer driving time. A new route bypassing the holdup may be obtained (Figure 5). Further route recommendations will follow in accordance with the new route.

The TMC codes required for this are limited to interstates/expressways and major highways.

Route guidance

Route recommendations

Route guidance is performed by comparing the present vehicle position with the computed route. Route recommendations are mainly reproduced acoustically (voice output). This allows the driver to follow the directions without being distracted from the task of driving. Graphics, preferably positioned in the primary field of vision (e.g. in the instrument cluster), support the driver in understanding the directions. These graphics range from simple arrow symbols (Figure 6) through to the display of a map excerpt optimally adapted in size (Figure 4).

Figure 6: Pictogram as example of route guidance

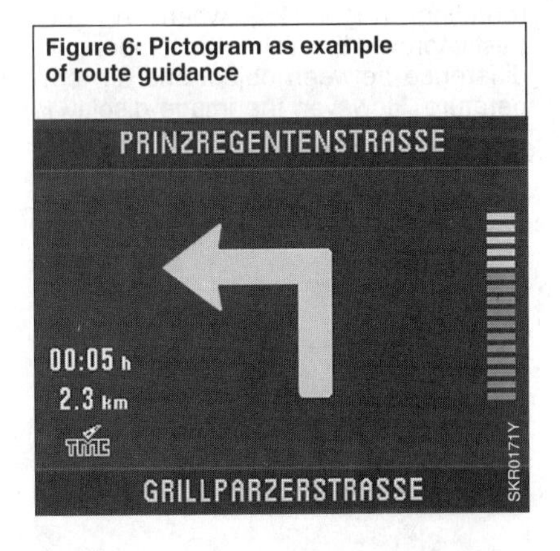

Digital map

Map display

Depending on the system, the map is represented on color displays using scales as of approx. 1:2,000 in 2D, perspective 2D (pseudo 3D), or real 3D. This is helpful for obtaining a general overview of the route in the immediate locality or over a wider area. Additional information such as, for example, bodies of water, railroads and forests provide added orientation.

Digitization

High-precision official maps and satellite and aerial photographs provide the basis for digitizing the data. If the copies are insufficient or not up to date, on-site measurements/surveys are carried out. Digitization is performed manually from map material or satellite and aerial images. Names and classification of objects (e.g. streets, bodies of water, borders) are then integrated into the database.

Specially equipped vehicles drive along roads to record additional traffic-relevant attributes (e.g. one-way streets, passage restrictions, overpasses and underpasses, no turning off at intersections), after which the initial-digitization data are checked on site. The results from these recording vehicles are incorporated into the database and are used to produce digital maps.

Data memory

Because their storage capacity is at least seven times larger, DVDs have now superseded CDs as the storage medium for digital maps. On the other hand, new systems are mainly being supplied with hard disks or SD cards, where the latter clearly dominate in portable systems. These writable storage media open up the additional possibility of the navigation systems being able to adapt to their users' preferences.

Night-vision systems

Applications

Darkness leads to two significant restrictions to a driver's sense of sight. Firstly, despite modern lighting technologies such as halogen or xenon headlamps, only a very limited illuminated field of vision is available. For example, the range of a typical low-beam headlamp is just 40 to 50 m. The visibility range can be increased to 120 to 150 m by using high-beam headlamps, this can only rarely be used during night driving, however, due to the tendency to cause glare for oncoming traffic.

Secondly, the capability to recognize objects in the dark by their color and contrast – which differ from those during daylight – is often greatly reduced. Pedestrians wearing dark clothing, for example, are frequently difficult to recognize even when they are within the range of the low beam headlamp. Night-vision systems help in this respect to increase road safety.

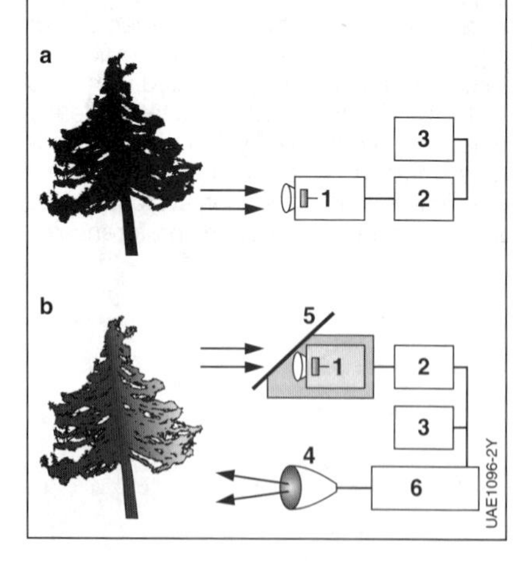

Figure 1: Comparison of far-infrared and near-infrared systems
a) Far-infrared system (7...12 µm),
b) Near-infrared system (780...980 nm).
1 Camera,
2 ECU,
3 Display,
4 Infrared headlight,
5 Windshield,
6 Vehicle.

Far-infrared systems

Operating principle

Night vision systems based on far infrared (FIR) have been used in military applications for years. They were used for the first time in motor-vehicle applications in 2000 in the USA. They use a thermal imaging camera to detect thermal radiation in the wavelength range of 7–2 µm radiated by objects (Figure 1a). These are referred to as passive systems; they do not need any additional radiation sources to illuminate objects.

The pyroelectric thermal imaging camera, or microbolometer camera, is only sensitive in the wavelength range of 7–12 µm. Because the glass in the windshield is not transparent for these wavelengths and the camera must be located on the outside of the vehicle.

Thermal imaging cameras use germanium lenses, or tex-glass with high amounts germanium and a corresponding transmission band. Currently available cameras have a resolution up to VGA resolution (video graphics array, 640 × 480 pixels). The camera signals are processed in an ECU. The generated video signal is output to a display located in the instrument cluster where the thermal image can be viewed.

Image display

Hot objects are shown in the image as light contours against dark (cold) surroundings (Figure 2a), where the contrast improves the higher the temperature difference between object and air temperature. However, the image display is rather unusual for the viewer since the appearance does not correspond to that of a normal reflected image.

Near-infrared systems

Operating principle

Near-infrared systems (NIR) are based on infrared radiation near the visible spectrum (near infrared). However, objects do not emit any radiation in this wavelength range. The area ahead of the vehicle is illuminated with infrared headlamps (Figure 1b). They are referred to as active systems for this reason. An infrared-sensitive camera records the scene, i.e. the infrared radiation reflected by the objects. The image signal is transferred to an ECU that in turn forwards the processed image to the display (e.g. in the instrument cluster). The driver sees a current image of the road situation (Figure 2b).

NIR-based systems were first introduced in 2003 in a simple version in terms of image display and with conventional CCD camera technology in Japan. In 2005, an NIR night-vision system with significantly improved performance was introduced in the Mercedes S-Class (Night Vision).

Figure 2: Comparison of the appearance of far-infrared and near-infrared systems
a) Far-infrared system,
b) Near-infrared system.

Functioning principle

Near-infrared night-vision systems exploit the different spectral sensitivities of the human eye and electronic silicon-based images. The spectral sensitivity of the human eye is described by the $V(\lambda)$ curve (Figure 3) and comprises the wavelength range from 380 nm (violet) to 780 nm (red). The maximum sensitivity lies in the range of 550 nm (green). Compared to this, the spectral sensitivity of an imager reaches considerably further into the range of high wavelengths, only ending at around 1,000 nm.

Figure 3: Spectral sensitivity
1 Spectral sensitivity curve $V(\lambda)$ of the human eye (curve for photopic vision),
2 Spectral sensitivity of video camera,
3 Spectral transmission of filter.

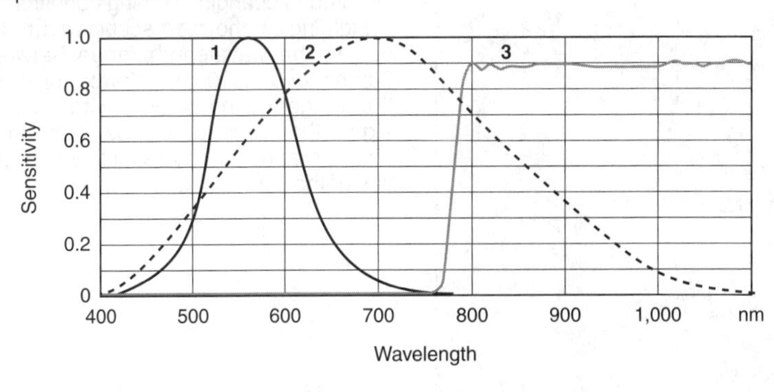

Infrared headlamps
Halogen lamps, like those commonly used for automobile headlamps, have a infrared radiation content. This ranges from the limit of the visible spectrum (380–780 nm) up to wavelengths beyond 2,000 nm, with a maximum between 900 nm and 1,000 nm. The upper limit of the useful wavelength when a video camera is used is 1,100 nm, the sensitivity limit of silicon.

Conventional halogen high-beam headlamp models are used as infrared headlamps for this reason. In halogen high-beam headlamp models, visible light is blocked by an additional filter in the light path without significantly reducing the usable infrared part (Figure 3). The emitted radiation cannot be perceived by the human eye.

The infrared headlamps that are used have a similar range and spatial characteristics to those of conventional high-beam headlamps, which means that the visual range achieved is comparable but without negatively dazzling other road users (Figure 4).

Suitable selection of the filter characteristic curve takes competing requirements into account. The spectral sensitivity of the imager decreases as the wavelength rises. In order to exploit its sensitive range, the filter edge for low wavelengths should lie near to the limit for visible light. On the other hand, in order to prevent non-permitted red impression for front headlamps, the filter edge should be shifted more towards higher wavelengths. Both requirements can be met by carefully selecting the transmission in the visible range, the position of the filter edge, and the edge steepness.

The infrared headlamps are either integrated into the headlamp module or mounted as external modules e.g. in the front apron of the vehicle.

Figure 4: Range of vision of night-vision system
1 Diagram (display),
2 Low beam,
3 Infrared high beam,
4 Range of vision of video camera.

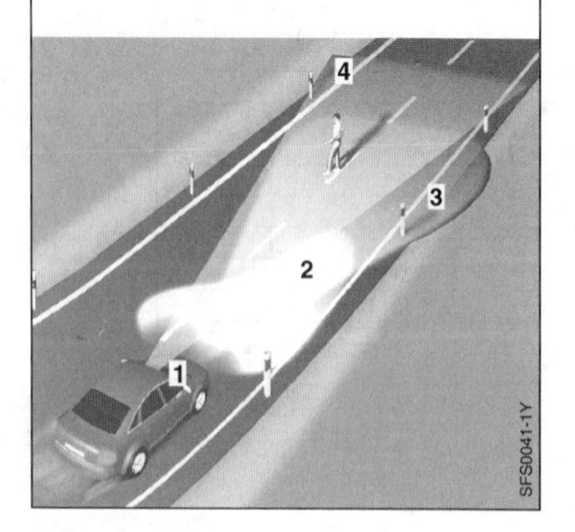

Imagers
The extreme lighting conditions – darkness and very high intensity dynamics, e.g. due to headlamps of oncoming traffic – result in very high requirements for the projection properties of the camera lens, as well as for the imager. The sensitivity to darkness of the imager determines the visual range the system can achieve; the extent of dynamics is determined to a great extent by the resistance to dazzling.

The CCD chip (Charge Coupled Device) features very high sensitivity to darkness, but as a rule it does not achieve the required intensity dynamics without additional measures. The CMOS chip (Complementary Metal Oxide Semiconductor) has lower sensitivity to darkness, but doe achieve an intensity dynamic of over 100 dB. There needs to be compensation for every pixel of the fixed pattern noise of this technology. Adjustable, nonlinear characteristic curves mean that these imagers can be very easily regulated and adapted to changing lighting conditions.

Lighting up the road scene with radiation in the wavelength range between 700 nm and 1,000 nm (infrared) provides the imager of the camera with a useful signal that depends on the illumination strength and spectral reflective capabilities of the illuminated scene.

Activation strategy
One major aspect of the system is the automatic activation of the infrared headlamps. They emit light of high intensity that the human cannot perceive or can only perceive to a minor degree. In order to prevent eye damage due to unnoticed exposure at short distances, the infrared headlamps are only activated automatically at a certain driving speed as an addition to the low beam that is already on. If the speed falls below the threshold, it is automatically deactivated. This excludes occurrence of the critical combinations of long exposure and spatial proximity to the emission source. In addition, the infrared headlamps only light up if the adjacent low beam is also switched on. Through this, the blinking reflex is triggered when the headlamp is looked at, as a result of which the exposure of the eye is greatly reduced.

1st generation night vision systems
The first system on the American market showed an image from a FIR camera in a head-up display (HUD). This was closely followed by NIW systems on the Japanese market, which also used a head-up display.

Due to the risk of distracting the driver caused by moving images in the primary field of view, such systems could not be implemented throughout the European market. In these initial systems, the processed image from the camera (in both far and near infrared systems) was displayed on a graphics-capable monitor in the instrument cluster or in the center-console area. When positioning the monitor in the vehicle, it is important to ensure that the monitor is as close as possible to the windshield and not too far removed from the driver's normal line of sight so that the driver does not have to divert his/her attention from the road and traffic for too long. Ergonomic tests featured the best results in relationship to read-off time and distraction in an arrangement with the monitor in the instrument cluster.

2nd generation night vision systems
New image-processing procedures enable pedestrians to be classified by reference to their typical outlines (head and shoulder area). This makes it possible to mark the detected pedestrian on the monitor and to improve the driver's recognition of a dangerous condition. This is particularly important in far-infrared systems, because the image on account of its strange appearance is more difficult to interpret than a near-infrared image. This feature is important insofar as the majority of fatal accidents involving pedestrians happens at night. In this way, such systems can help to significantly reduce nighttime accidents involving pedestrians. These systems with pedestrian recognition are now fitted as standard in both technologies (near- and far-infrared systems) in German vehicles.

An additional system version completely omits the image display. The new Mercedes S class features a pixel light, which makes it possible to create a light cone of almost any shape. The image processing system can then light up a detected pedestrian with a flash of light. This directs the driver's view to the pedestrian and warns the pedestrian at the same time.

Future prospects
For vehicles that do not feature an expensive pixel light, there is the option to display a warning in a head-up display. Displaying a warning symbol on the windshield at the position where the object will appear for the driver later on represents an effective display format with minimal distractions. The same applies for an optical warning on the base of the windshield that displays the direction of the obstacle. These display formats would be a cost-effective alternative for lower classes of vehicles.

Parking and maneuvering systems

Applications

On virtually all motor vehicles, the bodies have been designed and developed in such a way as to achieve the lowest possible drag-coefficient values in order to reduce fuel consumption. Generally speaking, this trend has resulted in a gentle wedge shape which greatly restricts the driver's rear view when maneuvering. Obstacles are difficult to recognize. This also applies to obstacles beside the vehicle and objects in the driver's blind spot.

Ultrasonic parking aids were initially developed in the past to improve the overview around the vehicle. Thanks to new algorithms and the inclusion of video technology, these systems have undergone further development to assist maneuvering.

A distinction is made between passive and active systems based on the system configuration. Passive systems warn or inform the driver about hazardous situations, while active systems intervene and steer the vehicle, such as into a parking space.

Ultrasonic parking aid

Parking aids with ultrasonic sensors support parking maneuvers. They monitor an area of approx. 20 cm to 250 cm behind and, if necessary, in front of the vehicle. Obstacles are detected and their distance to the vehicle is indicated by visual or acoustic signals.

Numerous vehicle manufacturers offer parking aids as optional equipment; in some vehicles they are now fitted as standard equipment. Systems which can also be adapted to older vehicles are offered for retrofitting.

System

The parking aid essentially consists of ultrasonic sensors (see Ultrasonic sensor), the electronic control unit (ECU) and warning elements. The protection area is determined by the range and number of sensors and by their emission characteristic.

As a rule, vehicles with purely rear-end protection have four ultrasonic sensors in the rear bumper; some large vehicles, e.g. off-road vehicles (SUVs), use six sensors. Front-end protection is provided by four to six additional ultrasonic sensors in the

Figure 1: Scanning range of the parking system with front and rear-end protection
1 Parked vehicle,
2 Parking vehicle,
3 Ultrasonic sensor at the front-end,
4 Ultrasonic sensors at the rear-end.

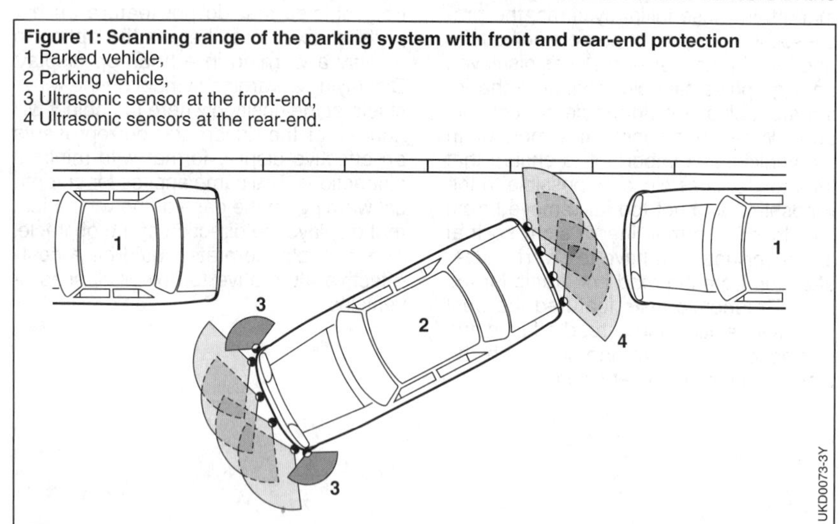

front bumper (Figure 1). For vehicle integration, the installation angle and spacing of the sensors are determined specific to the vehicle. This data is taken into account in the control unit's calculation algorithms. Specifically adapted mounting brackets secure the sensors in their respective positions in the bumper (Figure 2).

The system is activated automatically when reverse gear is engaged. In the case of systems with front-end protection, the system is activated automatically when the vehicle speed falls below a threshold of approx. 15 km/h (9.3 mph). During operation, the self-test function ensures continuous monitoring of all system components.

Ranging

Echo sounding method

Following the echo depth-sounding process (Figure 3), the sensors transmit ultrasonic pulses and measure the time taken for the echo pulses to be reflected back from obstacles. The distance l between the transceiver probe and the nearest obstacle is derived from the propagation time t_e of the first echo pulses and the velocity of sound c according to the following equation:

$$l = 0.5\, t_e\, c$$

Figure 2: Principle of mounting ultrasonic sensor in bumper
1 Sensor,
2 Decoupling ring,
3 Installation housing,
4 Bumper.

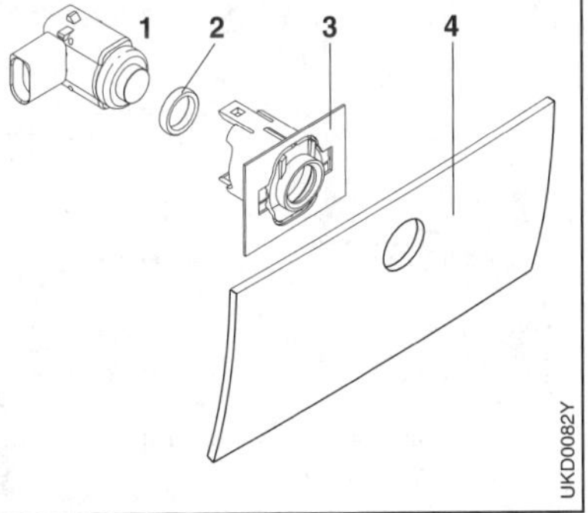

where
t_e propagation time of ultrasonic signal,
c velocity of sound in air
($c \approx 340$ m/s).

Figure 3: Principle of the distance measurement with ultrasound (echo sounding method)
a) Structure: An ultrasonic pulse is output by the emitter, reflected by the obstacle and picked up by the receiver.
b) Signal characteristic.

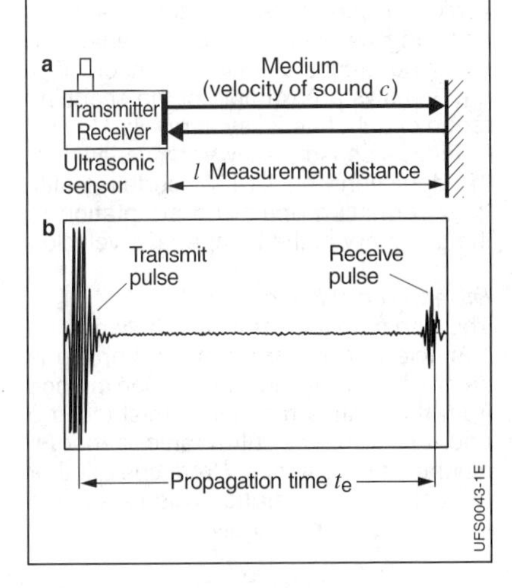

Figure 4: Antenna emission diagram of an ultrasonic sensor
1 Horizontal
2 Vertical

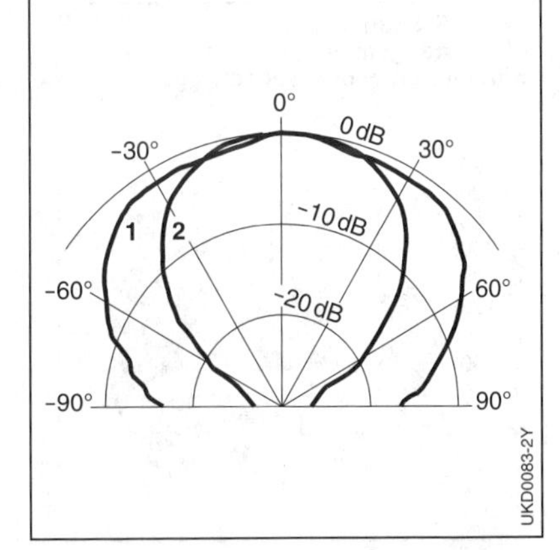

Detection characteristics

To enable the system to cover as large a spatial area as possible, the detection characteristics must meet special requirements. In the horizontal direction, a wide detection angle is desirable; in the vertical direction, on the other hand, a compromise is required. In order to avoid interference from ground reflections, the detection angle must not be too wide. However, existing obstacles must be reliably detected. Figure 4 shows the emission characteristic in the horizontal and vertical planes. A nearly seamless scanning range is produced by the overlapping scanning ranges of multiple sensors. Figure 5 shows an example of the scanning range of a 4-channel system.

Today's sensors allow for a variation of the detection angle within certain limits, thus permitting optimized adaptation to the geometry of the bumper and vehicle.

Distance calculation

The geometrical distance a between an obstacle and the front of the vehicle is determined using the triangulation method from the results measured (distances b and c) from two ultrasonic sensors, mounted at a distance d from one another (Figure 6). The distance a is calculated from the following equation:

$$a = \sqrt{c^2 - \frac{(d^2 + c^2 - b^2)^2}{4d^2}} \ .$$

Figure 5: Scanning range of an arrangement with four ultrasonic sensors

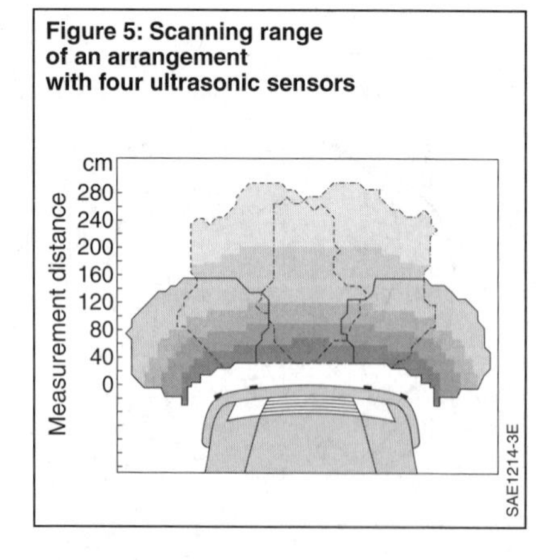

Electronic control unit (ECU)

The ECU contains a voltage stabilizer for the sensors, a microcontroller, and all the interface circuits needed to adapt the different input and output signals. The software assumes the following functions:
- Activating the sensors and receiving the echo
- Evaluating the propagation time and calculating the distance to the obstacle
- Activating the warning elements
- Evaluating the input signals from the vehicle (e.g. signal for engaged reverse gear)
- Monitoring the system components including fault storage
- Providing the diagnosis function.

Warning elements

The warning elements display the distance from an obstacle. Their design is specific to the vehicle, and they usually provide for a combination of acoustic signal and optical display. Both LEDs and LCDs are currently used for optical displays.

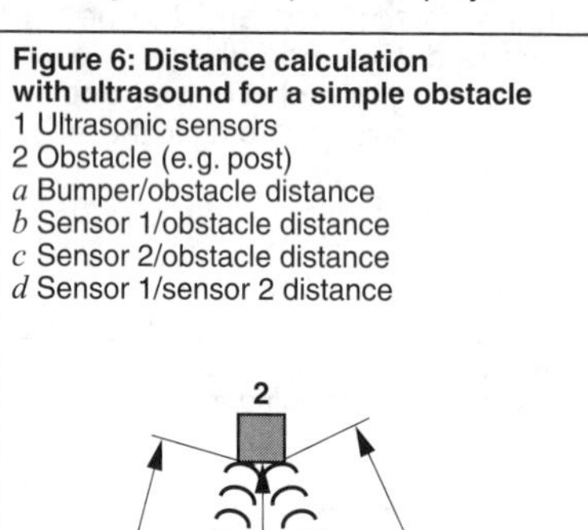

Figure 6: Distance calculation with ultrasound for a simple obstacle
1 Ultrasonic sensors
2 Obstacle (e.g. post)
a Bumper/obstacle distance
b Sensor 1/obstacle distance
c Sensor 2/obstacle distance
d Sensor 1/sensor 2 distance

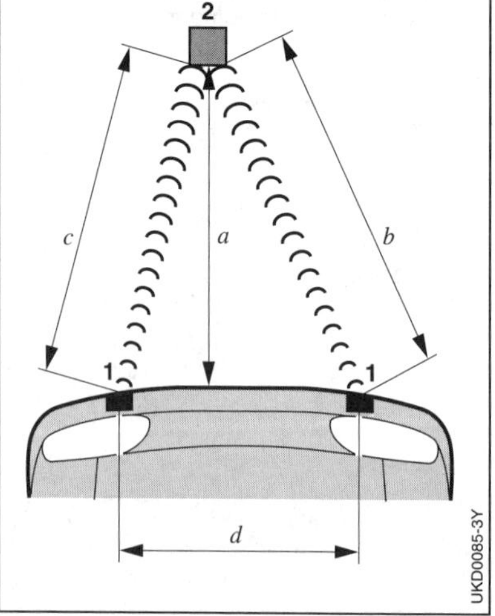

In the case of vehicles with a display monitor in the center-console area or in the instrument cluster, the vehicle can be shown together with the obstacles detected by the system, e.g. from a bird's eye view. This generally improves the driver's view of the situation.

Figure 7: Display of the parking assistant of the multifunctional display in an instrument cluster
a) Display of the information-providing parking assistant
b) Display of the steering parking assistant

SAE1215-1E

a

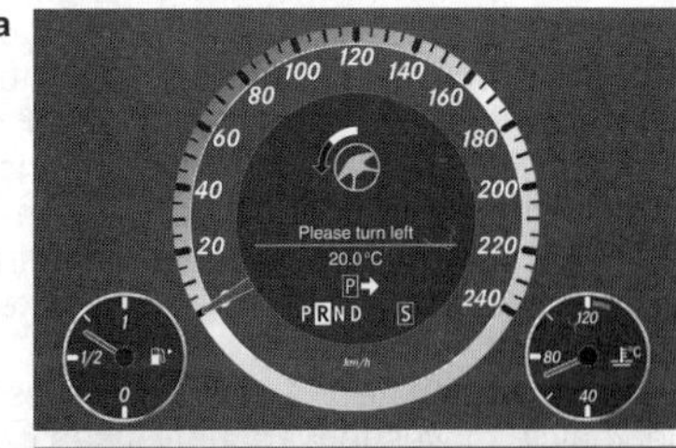

b

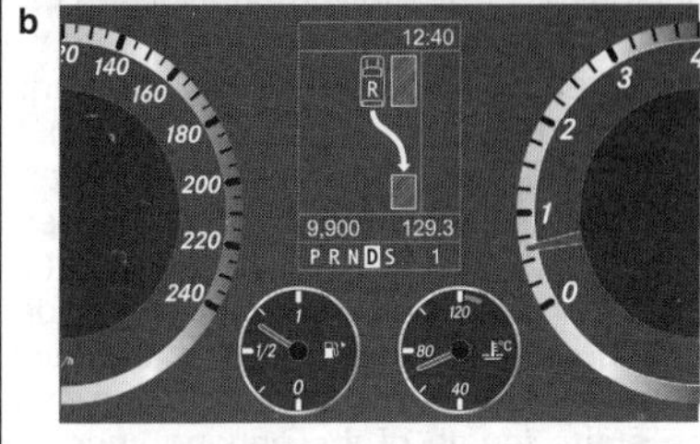

Ultrasonic parking assistant

The parking assistant is based on the ultrasonic parking aid and is designed as an evolution in stages. Each stage is a self contained function.

All ultrasonic systems shown in the following are already in series.

Parallel parking

Information-providing parking-aid assistant

After the system is activated by the driver, an ultrasonic sensor mounted on each side of the vehicle measures the length and depth of the parking gap as the vehicle drives past the gap (Figure 8). The length is obtained by evaluating the ESP speed signal (ESP = Electronic Stability Program). The parking-aid assistant then issues the driver a signal as to whether the parking gap is long enough. If there is an obstacle in the parking space, this is rejected. This requires the above-mentioned sensors with a range of up to approximately 4.5 m.

Once the parking space has been measured, the determined geometry of the environment can be used to ascertain the optimal trajectory for the parking process. The parking assistant requires the signals from the steering-angle sensor (from the Electronic Stability Program, ESP) for this purpose. During the parking process, the system can now give the driver a recom-

Figure 8: Parking assistant
1 Parked vehicle, 2 Parking vehicle, 3 Parking space boundary (e.g. curb), 4 Side-detecting ultrasonic sensor, 5 Ultrasonic sensor in the rear area.
a Measured depth of the parking space, l Length of the parking space.

UKD0090-5Y

mendation regarding how he or she should optimally turn the steering wheel in order to park in the space as smoothly as possible. During the parking operation, the trajectory is continuously recalculated and shown in a display. Figure 7a shows an example of a driving recommendation from the system.

Steering parking-aid assistant
The next evolutionary stage is a system with electric steering activation. A prerequisite is that the vehicle must have electrically activated power steering.

As before, the driver has to inform the system of his or her desire to park. After the system measures the parking space, the driver is notified whether the length of the parking space is sufficient. This is accomplished by a corresponding graphical display in the instrument cluster (Figure 7b). Now the driver simply needs to put vehicle in reverse and control the vehicle's linear motion (by braking and accelerating) – the parking assistant takes over the steering. When the parking process is done, the driver is notified accordingly.

One-step and multi-step parking
In the case of both the information-providing and the steering parking assistant, the system decides whether to park in one step or multiple steps, based on the length of the parking space. One-step parking is always selected if the parking space is long enough. If the parking space is short, the parking assistant decides on multi-step parking, based on the length of the space and the vehicle's possible steering radius. Then the system prompts the driver to shift gears accordingly.

Perpendicular parking
Semi-automated parking is not limited to the parallel parking functions described, but also permits parking in spaces perpendicular to the direction of travel (Cross-Parking Assist, Figure 9).

As with parallel parking, the length of the parking space is measured as you drive by it. To do this, the corner of the vehicle on the left (marked as point 1 in the figure) is taken as the initial reference point for the parking maneuver. If there is also a vehicle to the right of the parking space, its corner (point 2) is also taken into account.

Before beginning the parking process, the vehicle moves forward until it can calculate a parking trajectory. Then, after reverse gear is engaged, the vehicle parks along the trajectory, which is continuously recalculated and corrected if necessary. The signals from the rear ultrasonic sensors are continuously factored into the parking maneuver. Due to the limited detection range of the ultrasonic sensors, the driver is responsible for making sure the entire length of the parking space is free of obstacles.

Figure 9: Perpendicular parking
a) Detection corner point 1,
b) Preparation for parking,
c) Parking in the parking space.
1 Parking vehicle,
2 Corner point 1 (parked vehicle on left),
3 Corner point 2 (parked vehicle on right).

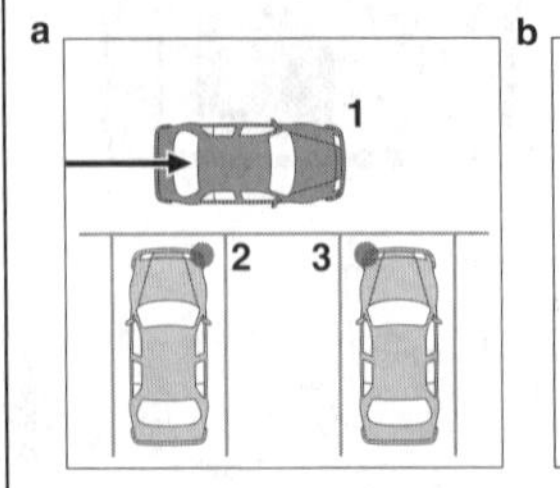

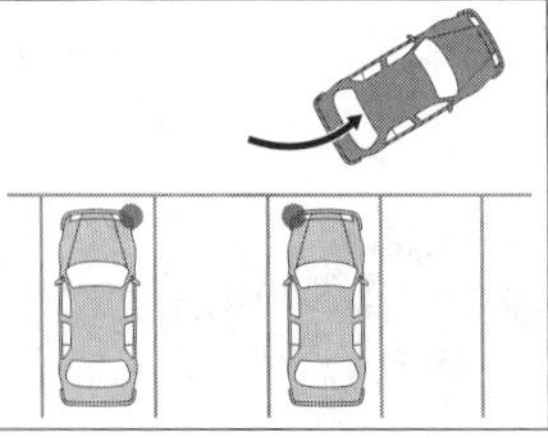

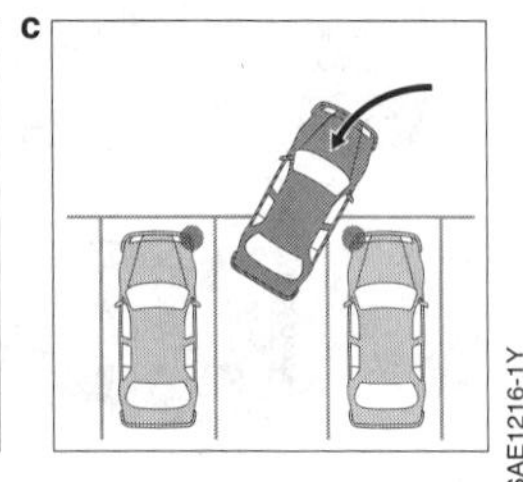

As with the previously described parking assistants for parallel parking, for this function there are also information-providing and – if the vehicle is equipped with electric power steering – steering system versions.

Assistant for leaving a parking space
In principle, the parking aid can also be used for leaving parallel parking spaces. To do so, the system first moves the vehicle into an appropriate starting position, so that the maneuver can be done in one step. The driver is responsible for paying attention to the traffic and accelerating or braking at the right moment. Simply touching the steering wheel turns off the function.

Maneuvering assistant
Traces of paint on parking garage pillars and entrances testify to frequent collisions with vehicles. The system for warning about objects next to the vehicle helps prevent that.

The two ultrasonic sensors mounted on the front of the vehicle detect permanent obstacles at a slow driving speed (up to 30 km/h, 18.6 mph) (Figure 10). The maneuvering assistant saves their position, so that this is known even after the obstacles are no longer in the scanning range of the sensors. The side distance to the saved obstacles is continuously calculated using the steering angle sensor signal. If a risk of collision is detected, the driver receives an audible warning (side distance warning). A bird's-eye view of the situation can also be shown on the central display. This system was introduced to serial production in 2013.

Blind-spot detection
Blind-spot detection (side view assist) makes use of two more ultrasonic sensors installed in the rear end at a horizontal angle of about 45° to the direction of travel, in addition to the two ultrasonic sensors already installed in the front end at a 90° angle to the direction of travel. All four sensors have an increased range of about 4.5 m. The front sensors are used to suppress warnings of oncoming vehicles and stationary objects. On roads with only two lanes, therefore, the system does not issue a warning about oncoming traffic or vehicles parked on the side of the road. The rear sensors detect objects in the vehicle's blind spot (Figure 11). Dangerous situations are identified from the position of detected vehicles and their movement relative to the driver's vehicle, and the driver is alerted. This system was introduced to serial production in 2010.

Figure 10: Maneuvering assistant
1 Vehicle driving into a parking garage,
2 Parked vehicles, 3 Obstruction.

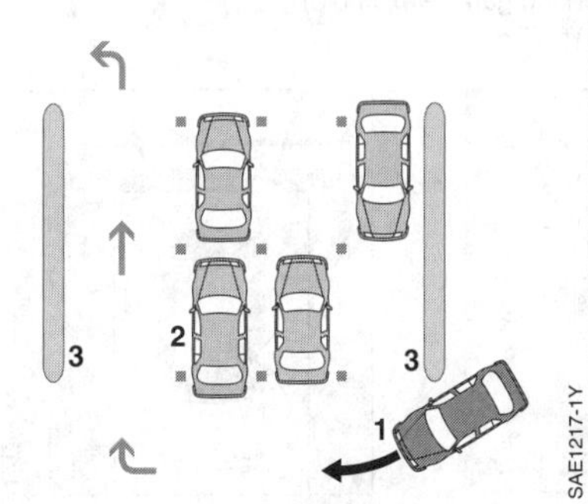

Figure 11: Blind spot detection
1 Vehicle pulling out,
2 Vehicle in the scanning range of vehicle 1,
3 Scanning range of vehicle 1.

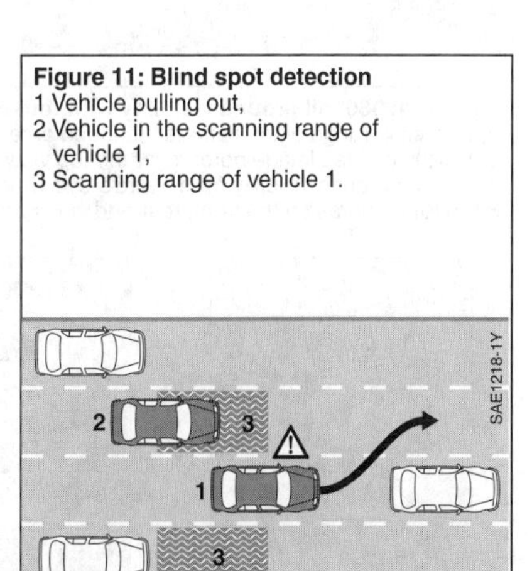

Video systems

Rearview camera

Rearview cameras were first introduced in Japan, but did not become prevalent in Europe due to the heavily distorted picture of the super-wide-angle lens. The breakthrough in Europe didn't come until there were systems to correct the image distortion.

Today's systems have a wide-angle camera installed in the rear end of the vehicle, usually in the recess of the handle to the trunk. The camera image appears in the center console display when driving in reverse. Additional lines to help estimate the distance and display the predicted lane make it easier for the driver to maneuver (Figure 12).

Figure 12: Driving assistance with rearview camera
Video display with additional information displayed. The distance information is calculated from the optical parameters of the single video camera.
1 Trajectory without turning the steering wheel,
2 Calculated trajectory calculated by the current steering angle.

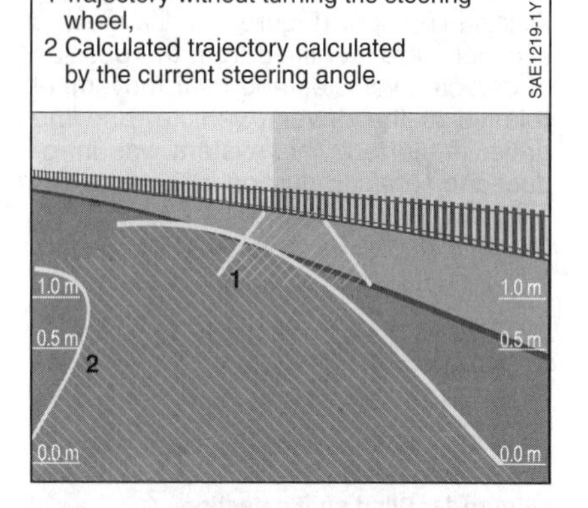

SAE1219-1Y

Figure 13: Driving assistance with front camera
View of the main road when driving out of a private driveway.

SAE1220-1Y

Today, the camera is intended only to provide information and does not provide the driver with any warnings. In the future, the image processing method may enable warnings to be issued.

Front-end camera

Visibility all around the front of the vehicle is also provided by a video camera with a 180° wide-angle lens. It also provides a view of traffic from the sides, for example, when pulling out of a driveway (Figure 13). Since the driver is seated farther back in the vehicle, the view from there is not as good as from the camera installed in the front of the vehicle.

360° all-around visibility

The whole area around the vehicle can be detected using four super-wide-angle cameras installed around the vehicle (top-

Figure 14: 360° all-around visibility with four cameras
a) Detection range of the cameras (shaded areas indicate where they overlap)
b) Strictly defined image borders for the top view image
c) 360° all-around visibility from a bird's-eye view (parking garage example)
K1 to K4 cameras on the front, rear and sides of the vehicle.

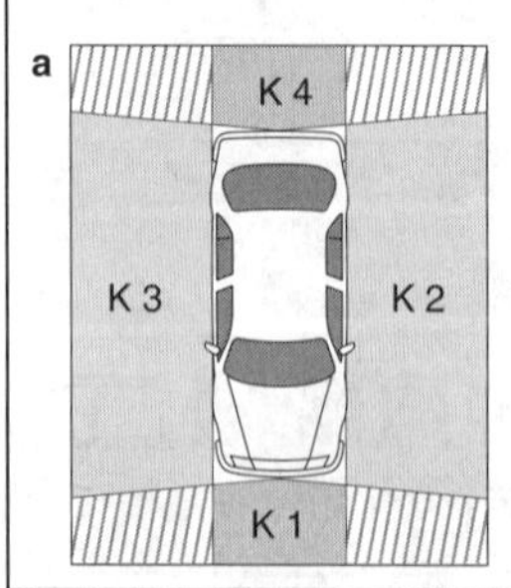

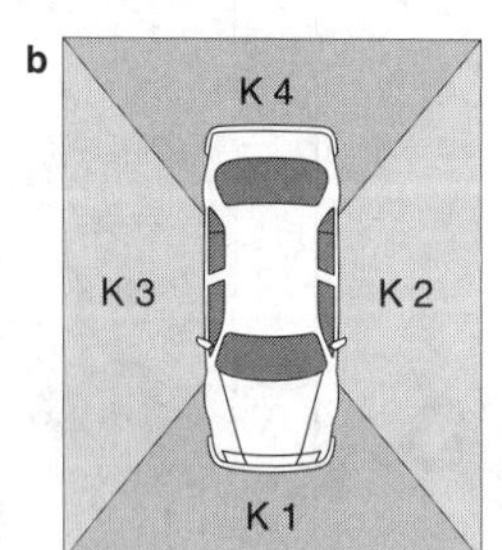

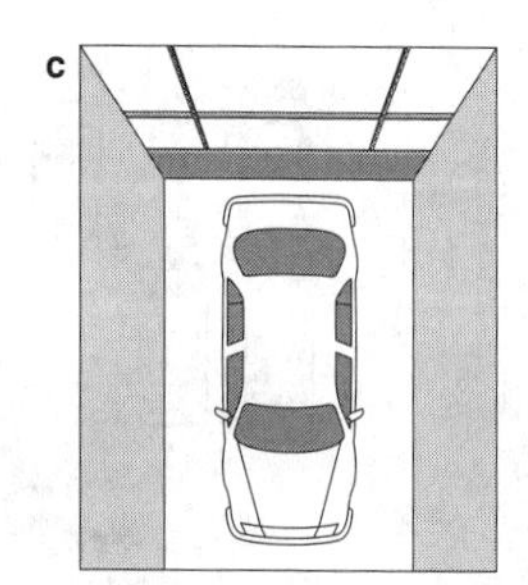

SAE1221-1Y

view system, Figure 14). The fields of view of the individual cameras overlap in the shaded areas. The four individual images are converted into one overall image using the stitching method. This makes it possible to provide the driver with a bird's-eye view of his or her own vehicle in the display.

Fusion of video and ultrasound technology

Further improvements can be achieved through data fusion of the ultrasound system's signals with the video camera's signals. For example, in addition to the lane information, distance information from the ultrasound system can be displayed in the image from the mono video camera as well, which cannot be used for measuring distance (Figure 15). By doing so, the driver receives even more meaningful information about the area by the rear of the vehicle.

Further development

With intelligent image processing algorithms (see Computer vision), even more functions and options become available, such as detection of parking lines, curbs, flat obstacles, and pedestrians at close range. The market launch is anticipated by about 2015.

The combination of intelligent camera systems with ultrasonic sensors form the basis for automated parking and maneuvering assistance systems. The following systems are in development.

Figure 15: Video image with distance information of the ultrasonic sensor
1 Obstacle (e.g. wall)
2 Distance assistance lines on the 50 cm grid, calculated from measured values of the ultrasonic sensors
3 Trajectory at maximum steering angle.

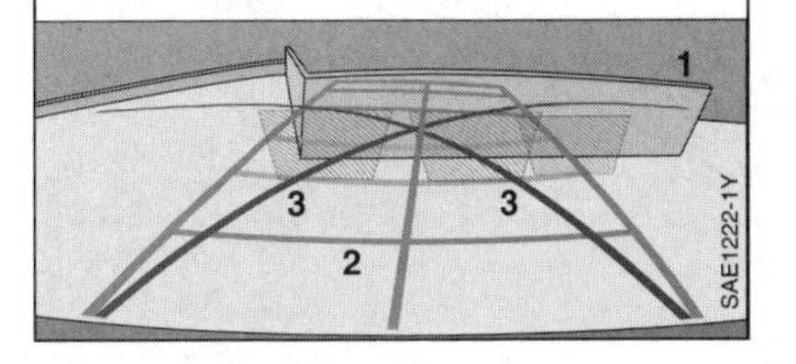

Parking maneuver assistant

Further development of the steering parking assistant has the vehicle now controlling linear motion as well. To initiate the parking maneuver, the driver needs only to put the gear in reverse; the system handles the rest. Series introduction: 2013/2014.

Remote-controlled parking assistant

In principle it is possible to park without having a driver in the vehicle. The driver can park the vehicle from outside by activating a control element using a remote control, an intelligent radio-operated key or, in the future, even using a smartphone. When the control element is released, the vehicle stops immediately. Series introduction: approx. 2015.

Valet parking assistant

With this function, a parking maneuver is manually programmed into the vehicle once for a specific parking space, such as in an underground parking garage. Afterwards the driver can park the vehicle by stepping out at the garage entrance and activating a control element; then the vehicle drives to the programmed parking space in a fully automated process. The same control element can be used to call the vehicle out of the garage. Series introduction: approx. 2020.

Autonomous parking

A vehicle that parks automatically was presented as part of the "Urban Challenge", a competition for automatic driving in an urban environment. This means that drivers park their vehicles at the entrance of the parking lot and the vehicle looks for an available parking spot itself and parks there.

References

[1] http://www.urbanchallenge.com/

Adaptive Cruise Control

Function

Like the basic cruise-control system that has been available as a standard feature for quite some time, Adaptive Cruise Control (ACC, also known as Tempomat) can be categorized as a driver-assistance system. The vehicle-speed controller regulates the driving speed to the desired speed set by the driver using the operating unit of the cruise control. In addition to this function, adaptive cruise control senses the distance and the relative speed for the vehicle in front and uses this information with additional data from the driver's own vehicle (e.g. steering angle, yaw rate) to regulate the time gap between the vehicles. It thus adapts the speed to the vehicle ahead and maintains a safe distance. The adaptive cruise control is equipped with a long-range radar to detect vehicles that are traveling in front in the same lane or also obstacles that are moving in the sensor's sensing range, and necessitate that the vehicle be braked (Figure 1).

Design and function

Adaptive cruise control is a convenience system that relieves the driver of routine tasks, but not of his/her responsibility to maintain control of the vehicle. This is why the driver can override this function at any time through intervention or deactivation (e.g. by operating the accelerator or brake pedal).

Ranging sensor

Current ACC systems for the most part have a radar sensor which operates in a frequency range of 76 to 77 GHz (see Radar sensors). The radar beams emitted by the radar sensor are reflected by vehicles ahead and analyzed with regard to propagation time, Doppler shift, and amplitude ratio. These values are used to calculate the distance, speed, and angle position relative to the vehicles ahead. The evaluation and control electronics (radar-sensor check unit) are integrated in the sensor housing. They receive and transmit data via a CAN data bus from or to other electronic control units that influence the engine torque and brakes (Figure 2).

Figure 1: Adaptive cruise control for a passenger car
1 Engine control unit,
2 Radar-sensor control unit,
3 Active brake intervention via Electronic Stability Program (ESP),
4 Control and display unit,
5 Engine intervention (on a spark-ignition engine with electronically activated throttle valve),
6 Sensors.

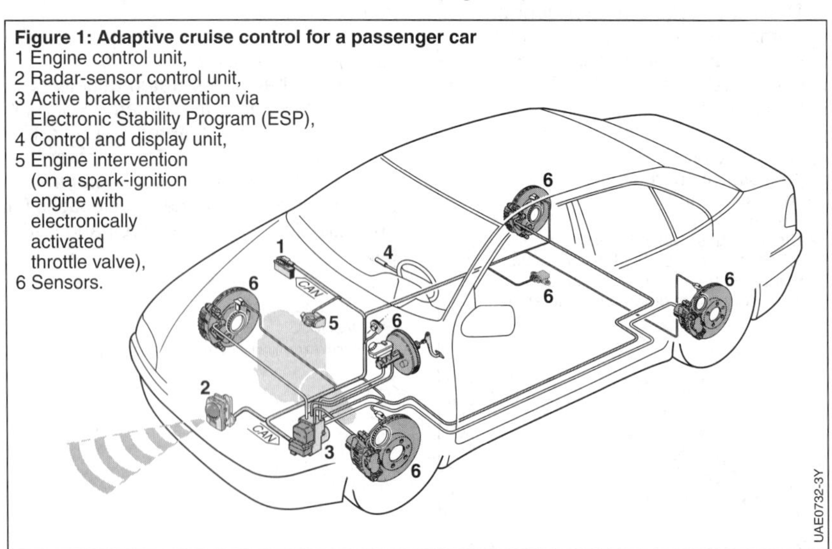

There are also ACC systems that work with laser beams in the infrared range (see Lidar, Light Detection and Ranging). The functioning principle is similar, whereby the use of an optical beam means that restrictions compared to the radar systems must be accepted in poor weather conditions (fog, rain, snow) (see e.g. [5]).

Course setting
To ensure reliable ACC operation no matter what the situation – e.g. also on curves/bends – it is essential that the vehicles traveling ahead can be allocated to the correct lane(s). For this purpose, the information from the Electronic Stability Program (ESP) sensor system is evaluated with regard to the ACC-equipped vehicle's own cornering status (yaw rate, steering angle, wheel speeds and lateral acceleration).

Setting options
The driver inputs the required speed and the required time gap; the time gap available to the driver ranges from 1 to 2 s. The time gap to the vehicle in front is calculated from the radar signals and compared with the required time gap specified by the driver. If the time gap is shorter than the required gap, the ACC system responds in a manner appropriate to the traffic situation by initially reducing engine torque, and only if necessary by automatically braking the vehicle. If the required time gap is exceeded, the vehicle accelerates until either the speed of the vehicle ahead or the desired speed set by the driver is reached.

Engine intervention
Cruise control is run via the electronic engine-performance control system (e.g. Motronic, electronic diesel engine control). This system allows the vehicle to be accelerated to the required speed or, if an obstacle appears, to be decelerated by reducing the drive torque.

Brake intervention
If the deceleration caused by engine control intervention is insufficient, the vehicle must be braked. To achieve this, a passenger car requires the Electronic Stability Program (ESP), which can initiate brake intervention. For commercial vehicles, an electronic braking system (EBS) is sufficient. Usually, this also involves the available retarder or engine brake for wear-free deceleration.

Due to the configuration of ACC as a comfort system, the deceleration calculated by the controller is limited nowadays

Figure 2: Basic structure of ACC system

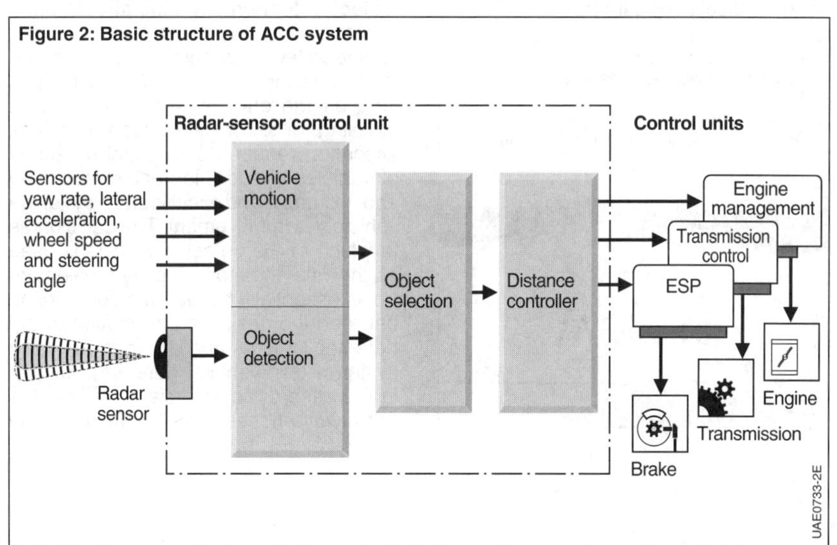

to approx. 2 to 3 m/s^2. Should this be insufficient due to the current traffic situation (e.g. vehicles in front braking sharply), a visual and audible signal is issued to the driver indicating that he or she must assume control. The driver must then initiate the corresponding braking deceleration via the standard brake. The ACC does not include safety functions such as emergency braking.

If necessary, when the ACC is activated, the stabilizing ABS systems (antilock braking system) or ESP (Electronic Stability Program) become are activated as usual. Depending on the parameterization of ACC, stabilizing interventions from ABS, TCS or ESP cause ACC to shut off.

Display

The driver must be provided with at least the following information:

- Display of the desired speed
- Indication of the switch-on status
- Indication of the required time gap selected by the driver
- Indication of the follow-up mode, which informs the driver as to whether the system is controlling the distance to a detected target object or not.

The information can be displayed, for example, in the instrument cluster or in a head-up display (Figure 3).

Figure 3: Information shown in the head-up display
1 Present speed,
2 Desired speed,
3 ACC status (Adaptive Cruise Control).

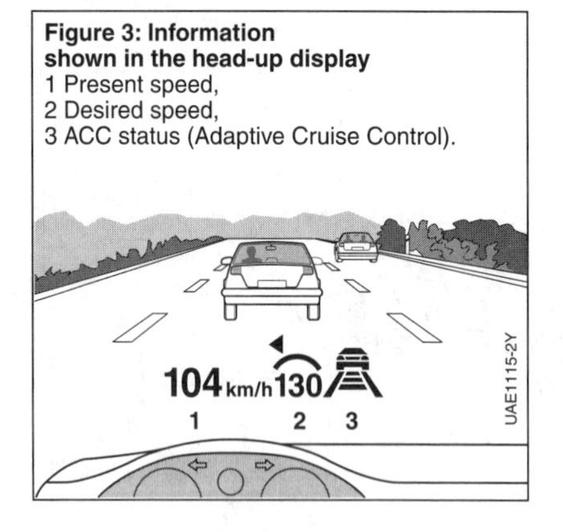

Control algorithms

Control modules

As a general principle, the control system for cars as well as for trucks consists of three control modules.

Cruise control
If the radar sensor has not detected any vehicles in front, the system maintains the vehicle's speed at the cruising speed set by the driver.

Tracking control
The radar sensor has detected vehicles in front. Control essentially maintains the time gap to the nearest vehicle at a constant setting.

Cornering control
When negotiating tight bends, the radar sensor can "lose sight" of the vehicle in front because of the limited width of its field of vision. Until the vehicle comes in sight of the radar again, or until the system is switched to normal cruise control, special measures come into effect. Depending on the vehicle manufacturer, the speed, for example, is kept constant, current lateral acceleration is adapted or ACC is switched off.

Object detection and lane allocation

The central task of the radar sensor and its integrated electronics is to detect objects and allocate them either to the same lane as the one on which the truck is traveling, or to a different lane. This lane allocation requires both exact detection of vehicles ahead (high angle resolution and accuracy) and exact knowledge of the vehicle's own movement. The decision as to which of the detected objects is used as the reference for adaptive cruise control is essentially based on a comparison between the positions and motion of the detected objects and the motion of the system's own vehicle. Here, in particular in the case of commercial vehicles, it is not always necessarily the next vehicle ahead that is selected. Under certain circumstance, it makes better sense to use the next vehicle but one, e.g. when a passenger car filters in from the slip lane and continues rapid acceleration to change quickly to the fast lane.

Electronic structure
Apart from the data transmitted by sensors (Figure 2), ACC requires additional data from the engine, retarder (for commercial vehicles), transmission and ESP control units, and which are transmitted via the CAN data bus. Conversely, these control units convert the accelerations requested by ACC into drive and braking torques. The coordination of the actuators (e.g. distribution of the required braking torque to the available brakes) can take place both in the ESP control unit and in a guidance computer or in the ACC control unit itself.

Adjustment
The radar sensor is fitted in the frontal area of the vehicle. Its radar lobe scans are aligned relative to the vehicle longitudinal axis. This is done by means of the adjustment screws in the mounting area of the sensor. If it is moved out of alignment by physical force, e.g. deformation of the mounting due to accident damage or any other effect, realignment must be carried out. Small degrees of misalignment are automatically corrected by the permanently active alignment routines implemented in the software. If manual realignment is required, this is indicated to the driver.

Area of application and functional expansions

Use in passenger cars
Use of ACC is linked to the availability of the Electronic Stability Program (ESP). This is a requirement for active brake intervention without action on the part of the driver.

Use in commercial vehicles
Brake intervention
For brake intervention, the standard brake system is only activated when the available retarder (or engine brake) is no longer able to decelerate the vehicle in the manner requested by ACC. Brake wear caused by ACC is therefore at the same level as the wear caused by a driver with an anticipatory driving style.

Alongside these ACC systems, there are also ACC systems for commercial vehicles that only influence the engine torque or engine torque and retarder. This restricts the range of application for these systems to significantly lower possible deceleration, i.e. the driver must intervene manually more frequently.

Requirements
As a general principle, the same type of ACC system can be used for buses, trucks and tractor units. The different requirements with regard to drive and braking systems, manual, semi-automatic or fully automatic transmissions are also covered. Only with the configuration of the system are there different requirements in buses, in particular with regard to comfort.

Some of the general conditions for a commercial vehicle ACC differ significantly to those for a passenger car ACC:

- Regulation of braking and acceleration must take account of the widely variable parameters of payload and engine size.
- Overtaking and merging maneuvers follow a dynamic pattern that is slower on trucks than on cars, and therefore result in different control requirements and settings.

- The control dynamics must be capable of handling a situation where several vehicles equipped with ACC are driving in a convoy (trucks drive in a line at nearly the same speed).
- A degree of simplification is provided by the fact that trucks have a more limited speed range than cars.
- The commercial deployment of trucks means that the focus during configuration is placed more on economy than on sports and comfort. Fuel consumption and wear must be at least as good with ACC as with an average driver.

ACCplus

First generation ACC systems had limitations in their range of functions resulting from the limited functionality of sensors and actuators. The limited object-sensing range and the limited horizontal resolution capacity only permitted operation above 30 km/h. Therefore ACC could not initially be used down to low speeds and at a standstill. The ACCplus system (in series production since 2009) has a further-developed radar sensor with wider angular coverage (±15°) and improved detection properties, and thus permits braking to a standstill and restarting by driver intervention. Pressing the accelerator pedal within a predetermined time limit is sufficient to reactivate the system.

Thanks to its greater reliability in destination selection and its even better ability to detect objects at close range, ACC can now also be used in traffic-jam situations.

ACC with low-speed following

During low-speed following (ACC Stop & GO, ACC LSF, Low Speed Following), the data from the long-range radar sensor is combined with the data from the medium or short-range sensors (short-range radar sensors or ultrasonic sensors) in order to provide an exact measurement of objects in front of the entire vehicle.

It operates in a speed range of 0 to 200 km/h, brakes the vehicle to a standstill, and automatically restarts within a predetermined time limit.

Sensor-data fusion with a video camera

Object measurement and object classification can be performed by means of sensor-data fusion with a video camera. This enables robust control over the vehicle with regard to stationary objects.

By combining with video sensor technology, it will be possible for low-speed following to perform full linear guidance at all speed ranges and also in urban traffic in the future (FSR, full speed range). ACC has also provided the basis for developing assistance systems which automatically intervene in critical driving situations to avoid accidents or to mitigate the consequences of accidents. Further developments are geared towards automatic evasive maneuvering by automatic steering intervention.

Current ongoing developments

Electronic horizon

The quality of the ACC function is being improved based on digital maps. The navigation system determines the position of the vehicle and estimates the expected path that the vehicle will travel (most probable path, MPP). A wide variety of predictive information is provided along this path by accessing digital navigation maps (electronic horizon). Examples of this include the road class, the geometric shape of the road, expressway exits and on ramps, speed limits and the gradient of the road. Manufacturer-specific protocols or the ADASIS standard [4] using CAN bus are used to provide the electronic horizon to the ECU.

The ACC system can access this information to improve its function. For example, the predictive course of the road can be used to more reliably assign oncoming radar specifically to a lane (reducing adjacent lane interference). The acceleration of the vehicle can be prevented when approaching an expressway exit if an oncoming radar object is lost when switching to the deceleration lane. The set speed can be reduced automatically before coming to a very tight curve.

In the future, the increased enhancement of the digital map with data and improvement in data quality will enable innovative assistance functions in combination with the ACC system, including predictive, fully automated specification of the set speed. The driver no longer has to enter the intended speed manually; the ACC system determines it automatically by communicating with the navigation system.

This allows the ACC system to implement predictive driving strategies in combination with the engine control unit, which minimize CO_2 emissions. If the ACC "sees" a speed limit or curve in advance via the electronic horizon, the vehicle can roll or coast to it with optimal fuel consumption. The gradient of the road is also taken into account when initiating the rolling or coasting procedure.

Networking with the cloud

The higher the ACC function's degree of automation, the greater the requirements are on the timeliness, precision and reliability of the data in the electronic horizon. To increase the data quality of the digital map, the vehicle itself will contribute to recording street environment data using its sensors. The function: Information such as traffic signs or the cornering speed are recorded by the vehicle's sensors, supplemented with data from the navigation system (e.g. position) and sent to the server (cloud) via mobile communications. In the central server, the incoming signals from many vehicles are processed, compiled and prepared for retrieval. The server information can be requested by vehicles with navigation systems. This ensures that the most recent information is always available. The incoming data from the server is applied to the local navigation map as new knowledge and provided to the ACC system immediately via the electronic horizon.

References

[1] H. Winner: Handbuch Fahrerassistenzsysteme. 2nd Ed., Vieweg+Teubner Verlag, 2012.
[2] H. Wallentowitz, K. Reif: Handbuch Kraftfahrzeugelektronik. 2nd Ed., Vieweg+Teubner Verlag, 2011.
[3] H.-H. Braess, U. Seiffert: Handbuch Kraftfahrzeugtechnik. 6th Ed., Vieweg+Teubner Verlag, 2012.
[4] C. Ress et al.: ADASIS PROTOCOL FOR ADVANCED IN-VEHICLE APPLICATIONS, ADASIS-Forum (http://www.ertico.com/assets/pdf/ADASISv2-ITS-NY-Paper-Finalv4.pdf).
[5] K. Reif: Automobilelektronik. 4th Ed., Vieweg Teubner, 2012.

Lane assistance

Drifting from the lane unintentionally frequently leads to serious accidents. It is caused for the most part by the driver being distracted or tired (e.g. microsleep). Lane departure warning (LDW) is designed to help avoid such accidents by detecting the lane markings ahead and warning the driver if there is a danger of a lane marking being crossed without the turn signal having been activated. The lane-keeping support system (LKS) also actively intervenes in vehicle guidance.

Assistance systems that afford the driver further lateral-guidance support on expressways/interstates with roadworks (roadworks assistant) or in urban traffic (bottleneck/constriction assistant) are currently in development.

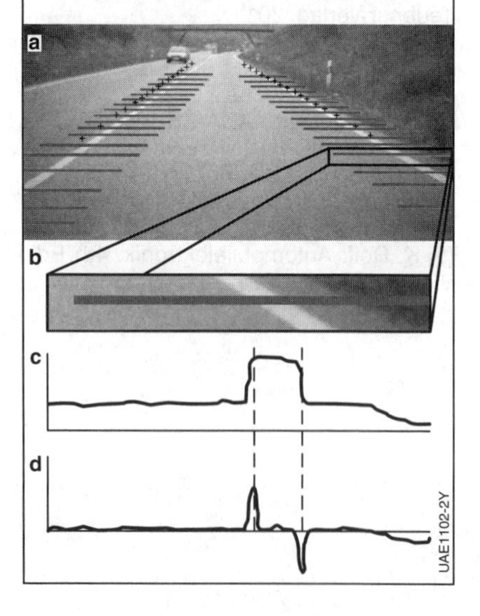

Figure 1: Principle of lane detection
a) Camera image with search lines,
b) Detailed view with search line,
c) Luminance signal (high level for light lane marking),
d) Edge information by high-pass filtering of luminance signal (peaks at light/dark transitions).

Lane departure warning

Lane detection

Lane-departure warning systems use video cameras to detect the lanes ahead of the vehicle and can have both mono and stereo designs. In good weather conditions and with good lane markings, the range is in the region of up to 100 m.

Figure 1 shows the principle of video-based lane detection. The image-processing system searches for lane markings by analyzing differences in contrast between the road surface and the lane markings. Figure 1a shows the camera image with search lines, Figure 1b shows a detailed view. The crosses in Figure 1a mark the course of the lane which have been calculated by the image-processing computer. In order to detect a line, the luminance signal (brightness) inside the search line is analyzed (Figure 1c). The limits of the lane markings are detected by high-pass filtering (Figure 1d).

Driver warning

A warning for the driver can be derived on the basis of the detected lanes in the event of the vehicle drifting out of lane. Different forms of warning are conceivable here. The first systems used an acoustic warning from the vehicle speaker in the form of a warning sound or "rumblestrip rattling" (stereo sound can additionally convey a direction). However, the systems that have become the most widely used since then feature warning the driver via the haptic sensory channel by vibrating the steering wheel or subjecting the steering to a slight counter-torque. The advantage of issuing the warning through the steering wheel lies in the driver directly associating the danger with the steering.

Lane-keeping support

Supporting interventions

Lane-keeping support is a functional enhancement of lane departure warning in that it not only warns the driver against drifting unintentionally from the lane, but also actively supports him in maintaining course.

If the vehicle drifts too far from the lane, lane-keeping support makes a course correction, either by means of an active steering intervention or an asymmetrical braking intervention. In this way, the system supports the driver in keeping his vehicle on course and at the same time helps to increase road safety by avoiding accidents caused by drifting unintentionally from the lane.

However, lane-keeping support itself cannot assume full lateral guidance of a vehicle, but is instead intended to support the driver in maintaining course. The system does not relieve the driver of his responsibility to follow and react to the traffic conditions attentively and to drive the vehicle consciously and deliberately. A system intervention can be overridden by the driver at any time.

To prevent misuse, the lane-keeping support system is fitted with a device which monitors whether the driver has both hands on the steering wheel and is actively steering the vehicle. If the driver is driving with no hands, an instruction is promptly issued to the driver to take control of the vehicle and the assistance afforded by lane-keeping support is terminated.

Applications

Lane-keeping support is designed primarily for use on expressways/interstates, well developed highways, and main city streets.

Like lane departure warning, lane-keeping support is dependent on detecting lanes ahead of the vehicle. Short-term failures in lane detection, caused for example by hidden or faded lane markings, are compensated for by the system. However, lane-keeping support is deactivated if no lane information is received for an extended period of time.

Roadworks assistant

Roadworks assistant and bottleneck/constriction assistant are further developments of lane-keeping support. In addition to lane markings, these systems also detect raised objects and obstacles and take them into consideration in lateral guidance.

Applications

The roadworks assistant supports the driver in negotiating roadworks on expressways/interstates and thus extends the lane-keeping support system to include this special scenario.

Types of support

The type and the extent of support provided can vary greatly. The system can run through all the stages of a cascade of intervention, from the visual, acoustic or haptic warning right up to interventions in vehicle guidance.

An intervention in vehicle guidance can involve interventions both in the lateral vehicle dynamics (steering torque or asymmetrical braking intervention) and in the longitudinal dynamics (deceleration or acceleration).

Detection of lane restrictions

Because roadworks are more complex in their structure than the typical expressway/interstate environment, the roadworks assistant detects roadworks-specific lane restrictions (traffic beacons, cones, safety barriers) in addition to the regular lane markings.

The sensor functions are typically performed by a front-end stereo video camera and side-mounted ultrasonic sensors. These sensors enable the roadworks assistant to distinguish between static objects (e.g. roadside structures) and dynamic objects (e.g. vehicles). In this way, the system can take the necessary collision-avoidance action in the event of lane constrictions at roadworks caused by traffic in neighboring lanes. This in turn affords the driver support to suit the situation as he drives through expressway roadworks and as a result offers the best support possible for maintaining a sufficient safety distance.

Bottleneck/constriction assistant

Applications

The bottleneck/constriction assistant extends the roadworks assistant with regard to the typical speed range in urban areas (0 to 60 km/h) and with regard to the supported radii of bends. Collision-avoiding steering assistance is offered in urban bottlenecks/constrictions accordingly.

Functionally negotiable bottlenecks/constrictions can also be caused by objects traveling in the same direction or by oncoming objects. The bottleneck/constriction assistant ensures that as the vehicle passes through bottlenecks/constrictions a sufficient safety distance to side obstacles is maintained and that if there is insufficient space to pass through an early warning is issued to the driver.

The bottleneck/constriction assistant is currently in the research stage and is part of the publicly sponsored UR:BAN project [1].

Sensor technology

A functional requirements analysis reveals that a forward-mounted stereo video camera and alternatively side-mounted ultrasonic sensors or monocular video sensors must be used. The stationary environment and moving objects (those traveling in the same direction and oncoming objects) must be represented within the framework of environment modeling. An occupancy grid presents itself for the stationary environment. Dynamic objects are depicted by way of object models with various function-relevant attributes. The situation analysis must make an assessment of the collision risk in accordance with the planned trajectory of the vehicle and the objects, and incorporating the stationary environment. Criticality measures which are used in the function module to detect critical situations where action is required are typically calculated for this purpose.

The function module checks that the system limits are observed (e.g. whether the maximum supported bend is exceeded) and activates where necessary steering assistance or issues an early warning to the driver. Figure 2 visualizes some central calculation results by the bottleneck/constriction assistant in a test scenario with a passable bottleneck/constriction while using typical test equipment (safety curbs, cones, balloon car). The rectangles represent detected collision-relevant objects. The gray area (in the graphic on the left of the pictured boundary line) codes the available road width in the bottleneck/constriction. The stripes on the level show the predicted corridor for the vehicle.

References
[1] www.urban-online.org

Figure 2: Bottleneck/constriction assistant – visualization of calculation results
1 Inflatable obstacle (balloon car),
2 Bottleneck/constriction,
3 Collision-relevant objects (e.g. cones, safety curbs),
4 Available road width,
5 Predicted corridor.

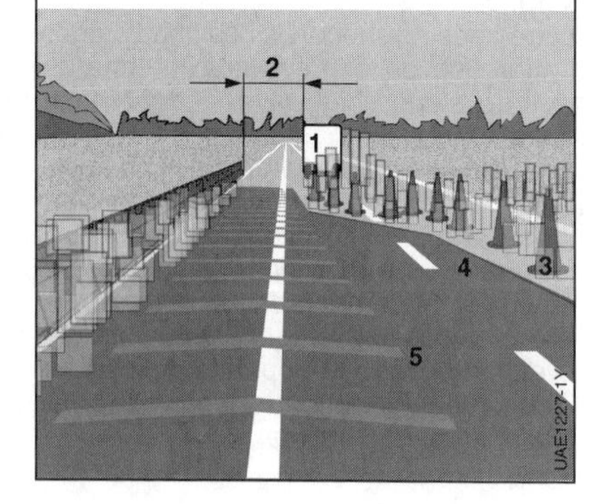

Emergency-braking systems

Emergency brake assist and automatic emergency braking

A significant number of road accidents are collisions between vehicles traveling in the same direction, i.e. rear-end collisions. Driver-assistance systems which can avoid such accidents or at least mitigate their consequences have been in series production for some years now. A simple feature is the brake assistant (see Brake-assistance functions). Systems that use sensor technology to observe and interpret the vehicle's surroundings have a particularly high potential [1]. This enables such a system to make the necessary anticipatory intervention in vehicle guidance in order for example to avoid an accident with emergency braking.

Environment detection

To be able to avoid rear-end collisions, it is necessary to observe the vehicle traveling in front and its movement. If the system detects that a collision is imminent, it can take appropriate action, for example by issuing a warning to the driver or by performing emergency braking.

Various sensors are suitable for observing the vehicle environment. Their range and sensing range constitute an important function-relevant feature. The required look-ahead distance of the sensors is derived above all from the speed range in which the function is to be active. Radar sensors are used predominantly used for long ranges which are required at high driving speeds. Mono and stereo video sensors typically achieve medium look-ahead distances. Lidar sensors can be used for both medium distances and close range.

Several sensors are frequently combined for a automatic emergency-braking system. This offers two advantages: Firstly, the strengths of the individual measurement concepts complement each other, providing for a more accurate measurement of vehicles traveling in front. Secondly, system reliability is enhanced in that possible faults in individual sensors can be detected. In particular, unjustified system responses can be avoided.

Figure 1: Braking-response model of Emergency Brake Assist and Automatic Emergency Braking for an unobservant driver who fails to brake adequately or does not brake at all

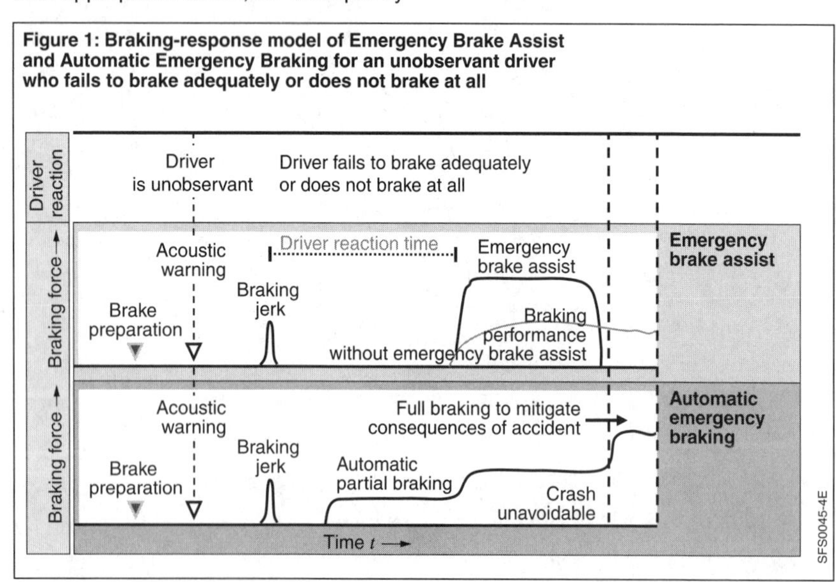

Response model

Different system reactions can be initiated if a collision is imminent and this is detected by the environment-sensor system. The time sequence of an exemplary system response is illustrated in Figure 1. The following system-response stages exists in standard systems:

Brake preparation

If the probability of an imminent collision increases, the brake system can be prepared (pre-fill) in such a way that any subsequent braking (by the driver or by the system) can be effected more rapidly. This is done by pre-filling the brake system.

Warning

The driver can be warned in the event of a heightened collision risk. A warning is suitable to force the driver into a tried-and-tested response, especially if he has not yet recognized the danger. An accident can thus be avoided by the driver himself if the advance-warning time is long enough.

However, when configuring a warning function, it is important not to set the advance-warning time at too long; in such a case the warning would frequently be issued already in easy-to-control, non-critical situations and would thus be perceived to be unjustified.

Braking jerk

The braking jerk constitutes a further warning escalation stage. Here the brake system is automatically activated for a brief period and then released again immediately. This has two advantages: Firstly, the driver is warned by haptic means of the danger; secondly, the vehicle speed is already slightly reduced.

Partial braking

Partial braking works along similar lines to the braking jerk. Here a slight deceleration is automatically set, but unlike the braking jerk is not released again immediately.

Brake assistant

Frequently the driver will react to the critical situation by initiating a braking operation that is too weak to avoid the accident. In this event the system can boost the braking force requested by the driver so as to bring the vehicle to a stop in good time before the obstacle and avoid the accident.

Full braking

If the system responses described above are not sufficient or if the driver fails to react appropriately to avoid the imminent collision, a full-braking operation can be automatically initiated at the last moment. This action will frequently prevent an accident entirely; but at least will greatly reduce the impact speed. Different algorithms can be used to predict a collision and subsequently make a decision as to initiating full braking.

Regulations and standards

As from 2014 automatic emergency braking when traveling in the same direction will become part of the Euro NCAP assessment (European New Car Assessment Program). Emergency-braking systems are to be assessed in standardized tests and the results will be reflected in the number of NCAP stars.

Available systems

Various systems are currently available which can significantly reduce speed in critical situations and frequently avoid the collision entirely. However, an automatic emergency-braking system can involve the additional risk of a rear-end impact and in the event of unjustified deployments give rise to acceptance problems. It is therefore essential to ensure that no unjustified emergency-braking operations occur.

References

[1] A. Georgi, M. Zimmermann, T. Lich, L. Blank, N. Kickler, R. Marchthaler: New approach of accident benefit analysis for rear end collision avoidance and mitigation systems. In Proceedings of 21st International Technical Conference on the Enhanced Safety of Vehicles (2009).

Intersection assistant

Motivation

Cause of accidents

Around 22% of all accidents involving injury in Germany occur when crossing intersections and turning [1]. Driver-assistance systems which are designed to prevent this common type of accident are called intersection assistants. This covers safety functions with the aim of avoiding accidents or mitigating their severity. To achieve this, the driver is alerted at an early stage to an imminent accident or an automatic braking of the vehicle is performed.

Environment sensing

Other road users are detected with radar sensors, video cameras or lidar sensors and their distance, speed and direction of movement are determined. The information is processed by the microcomputer of an ECU and the situation development – such as an imminent collision for example – is predicted.

Critical and non-critical situations

It is particularly difficult to distinguish between critical and non-critical situations in intersection scenarios as this would require the turning behavior of the vehicles involved to be determined several seconds in advance. This is almost impossible, especially when the vehicles involved are being driven in a sporty style. For this reason, an imminent accident can only be unequivocally detected just before the collision. Because unjustified braking interventions must be avoided without fail, accident prevention is not always possible in all critical situations. But even if the system cannot avoid an accident, it can mitigate the consequences of the accident through an automatic braking intervention since the impact speed is reduced and the point of impact favorably shifted under certain circumstances.

A distinction is made between the intersection-assistance systems described in the following.

Intersection-assistance systems

Left-turn assistant

Causes of accidents

The left-turn assistant is designed to prevent accidents which happen when turning left in the face of oncoming traffic (Figure 1). The accident is usually caused by the driver incorrectly judging the speed of the oncoming vehicle – the accident perpetrator turns although there is no longer enough time to complete the maneuver.

An imminent left-turn maneuver by the driver can be detected by an evaluation of turn signal, accelerator pedal and steering angle.

System response

If the turn cannot be made safely due to the approaching oncoming traffic, the driver is warned in an initial stage. If there is an acute risk of collision, automatic braking or start prevention is finally initiated.

To avoid the situation where the vehicle in the event of an incorrect response cannot be moved in a potentially dangerous situation, the driver can override the start prevention by means of a kickdown.

Figure 1: Situation when turning left with oncoming traffic
1 Turning vehicle,
2 Oncoming vehicle,
3 Sensing range of sensor system (e.g. mid-range radar).

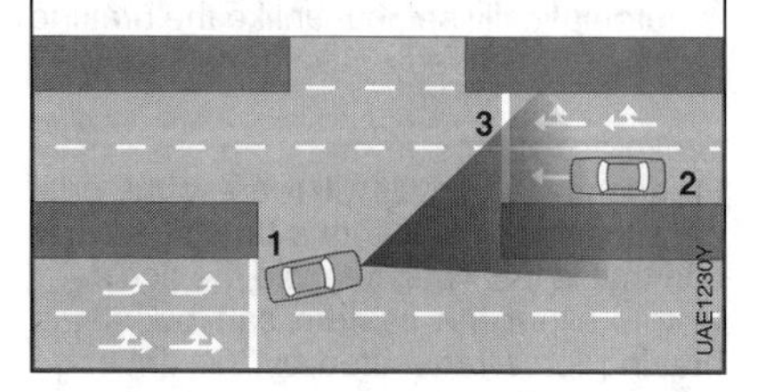

Cross-traffic assistant

Causes of accidents

The cross-traffic assistant is designed to prevent accidents which happen when crossing intersections and turning in the face of vehicles that are themselves crossing (Figure 2). Because passengers involved in side collisions are much less adequately protected than in head-on collisions due to the lack of crumple zones, the risk of injury is particularly high.

Obscurations/obstructions are frequent causes of accidents as well as distracted and inattentive drivers.

Environment sensing

As with the left-turn assistant, crossing vehicles are detected by environment sensors. Compared to the left-turn assistant, however, the field of vision to be covered is much wider (≥180°), meaning that generally several sensors which are mounted at the front corners of the vehicle must be used.

System response

If the system detects a crossing vehicle approaching, the driver is alerted in an initial stage to the presence of a potentially dangerous traffic situation (e.g. visually by way of a neutral indication in the headup display). If the situation escalates further, the initial alert is followed by a warning (conspicuous visual or acoustic indications) and finally by an automatic braking intervention.

Figure 2: Situation with cross-traffic
1 Vehicle driving straight ahead,
2 Crossing vehicle,
3 Sensing range of sensor system (e.g. mid-range radar, utilizing sensors with widened beam angle).

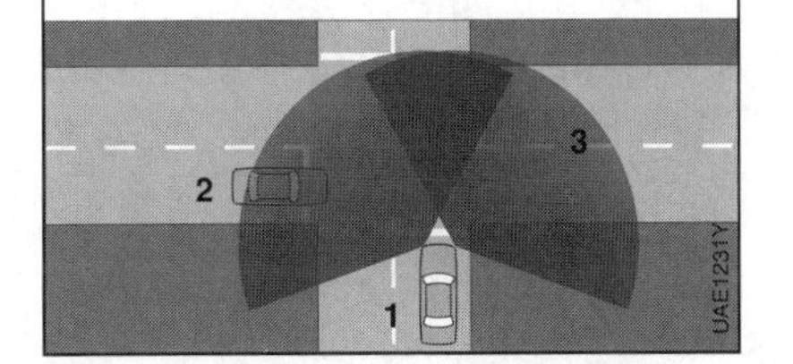

Outlook

Because obscurations / obstructions caused for example by parked vehicles are frequently encountered in intersection situations, presumably car-to-car communication (C2C) will also play an important role in the future. Here the vehicles exchange information by radio signal on their position, speed and steering angle. The position is determined beforehand by satellite positioning.

Traffic-light and stop-sign assistants

Causes of accidents

Unlike the left-turn and cross-traffic assistants, the traffic-light and stop-sign assistants are designed not to prevent an imminent collision with other road users, but to prevent the driver from accidentally running a red light or a stop sign. This is intended to prevent accidents which happen as a result of these traffic violations.

Environment sensing

A video camera equipped with image-recognition algorithms scans the vehicle environment for traffic lights and road signs (see Road-sign recognition), whereupon their status and their meaning are interpreted. To reduce the risk of erroneous detections, digital maps can be used in which the positions of road signs are marked.

An alternative is offered by car-to-infrastructure communication (C2I). Here, the traffic light for example communicates its status by radio signal to vehicles equipped with special receivers in the surrounding area.

System response

If the driver is on the point of not stopping at a red light or a stop sign, a warning is issued first. If the driver then fails to react appropriately, the vehicle is automatically braked.

References

[1] GIDAS database 2011.

Active pedestrian protection

Motivation
The protection of weaker road users, particularly pedestrians, has increasingly taken center stage in recent years. Initially this increased focus resulted in the improvement of passive safety systems for pedestrian protection (e.g. specific design of the vehicle front end to mitigate the consequences of collisions with pedestrians). Now, increasingly, active systems – i.e. systems which clearly act against potential collision – are also going into series production.

Active pedestrian-protection systems react similarly to emergency-braking systems for vehicles traveling in the same direction. However, the environment-sensing requirements are different; this can be identified in an analysis of the frequently occurring types of accident.

Causes of accidents
The majority of pedestrian accidents involve persons who cross the road while the vehicle is driving straight ahead. The pedestrian steps out from the side in front of the car, often from a concealed position (i.e. the pedestrian is initially hidden behind another object).

Another significant number of pedestrian accidents happen at intersections, particularly when vehicles are turning.

Comparison with emergency-braking systems for vehicles traveling in the same direction

System responses
As already mentioned, the system responses for active pedestrian protection are comparable with emergency braking for vehicles traveling in the same direction. The system responds to an imminent collision with a pedestrian with a warning to the driver together with partial and full braking. As pedestrians frequently step out unexpectedly from a concealed position into the road, often only part of the response cascade, i.e. only the partial and full braking for example, is activated.

Environment detection
Pedestrians usually step out from the side into the road. Therefore, when compared with emergency-braking systems for vehicles traveling in the same direction, a wider area in front of the vehicle must be monitored by the sensor system. To make the correct decision on deployment, the sensor system must be able to measure precisely the lateral velocity of pedestrians in particular. Because pedestrians frequently step out from a concealed position into the road, sensors must also be able to detect the pedestrians very quickly. This is the only way of ensuring a punctual system response (e.g. emergency braking).

Various sensor configurations for active pedestrian protection are in series production, often in the form of sensor-data fusion. Sensors used are mono video cameras, stereo video cameras or different radar sensors.

Situation analysis
Pedestrians can very abruptly stop in their tracks, start to run or change direction. These different movements must be taken into consideration in the system so as to avoid unjustified system interventions (e.g. emergency braking).

Assessment standards
There are still no standardized assessment standards for active pedestrian-protection systems. They are expected to be taken into consideration in the Euro NCAP assessment (European New Car Assessment Program) as from 2016. Here the most important types of accident are recreated in a test bay with pedestrian dummies and the system response is evaluated.

High-beam assistant

Function

The high-beam assistant automates the process of switching between low beam and high beam. It detects vehicles driving in front and oncoming vehicles and automatically switches the high beam on or off to suit the traffic situation. This significantly increases safety and comfort when driving at night, mainly on highways and expressways.

Motivation

Statistics show that people are exposed to a significant risk of being involved in a serious accident when driving at night. Tests further show that the average driver typically engages the high beam between 5% and 19% of his journeys at night and drives on low beam for the rest of the time, even though the option exists to engage the high beam. Some drivers do not engage the high beam at all and continuously drive on low beam. A system with high-beam assistant would activate the high beam in more than 30% to 40% of the distance driven at night (more than twice as often as a typical driver activating manually), thereby improving significantly the illumination of the road compared with the low beam. The figures from the statistics are very much dependent on the traffic density – the lower the density, the greater the use of the high beam. It has also been established that fatigue and lack of concentration causes drivers not to manually turn off the high beam reliably and thus to dazzle other road users.

An automatic system offers a dramatic improvement for all road users. The driver of a vehicle with a high-beam assistant experiences a noticeable reduction of the burden and, thanks to automatic activation, a more frequent improvement in illumination. Automatic deactivation of the high beam ensures that drivers of oncoming vehicles are not dazzled.

System design and function

The system consists of a light-sensitive sensor or a camera, combined with an ECU which controls the headlamp. The camera is either integrated in the rear-view mirror or installed in a standalone sensor assembly on the windshield. The camera detects and distinguishes between vehicles and stationary street lights. The system detects oncoming vehicles up to a distance of 600 m and tail lamps of vehicles driving in front up to a distance of 400 m. The positions of detected objects are transmitted to the headlamp ECU, with the data being transmitted via the CAN bus used in the vehicle. The headlamp ECU calculates from the position data whether an object is in the headlamp's beam range and dazzling can occur. The high beam is automatically deactivated if the sensor system detects an object that can be dazzled. If the sensor system does not detect an object, the high beam is automatically activated. The high beam is deactivated even in the presence of street lights for example in urban areas which the system detects and can distinguish from other objects.

Further developments

In the further development of the high-beam assistance function the camera-controlled "glare-free high beam" will further establish itself in vehicles in the years to come. In response to image analysis of the camera in the face of oncoming traffic or traffic in front, the high beam in these systems is only deactivated in the areas or shaded by screens in which other road users are located. New optical systems are used in the headlamp for this purpose. Most of the high beam can thus remain switched on and contribute to the driver's safety and comfort without dazzling other road users.

Future of driver assistance

Long-term goal – autonomous driving

Driving a car can be fun, but can also be tiring or even dangerous. Driver-assistance systems support drivers in difficult driving situations or relieve them of tiresome and monotonous driving tasks.

A trend that has been observed in recent years is the increasing degree of automation of the systems on the market. This trend will continue in the future [1], [2].

Degree of automation
The German Federal Highway Research Institute (BASt) divides driving functions based on the degree of automation into the following categories [3]:

Assisted
The driver permanently performs either longitudinal or lateral guidance. The remaining driving task is performed within certain limits by the assistance system. The driver must constantly monitor the system and be ready at all times to take complete control of vehicle guidance/handling.

Partially automated
The system performs longitudinal and lateral guidance for specific situations or for a limited period of time. The driver must constantly monitor the system and be ready at all times to take complete control of vehicle guidance/handling.

Highly automated
This degree of automation equates to "Partially automated", but in this case the driver does not have to permanently monitor the system. If the system limits are reached, the driver can be prompted to take control within a certain period of time.

Fully automated
The vehicle performs all the driving tasks for a specific application without the driver having to monitor the system. If the system limits are reached, the system can still establish a safe state without the driver having to intervene.

Autonomous driving
The logical extension of this development is a system that can permanently perform all the driving tasks without the driver having to monitor the system. The driver becomes a passenger and can use the journey time for other activities, such as for example reading, working or even sleeping.

Further applications would include the vehicle independently looking for a parking space (see "Parking and maneuvering systems"), driving to the garage or taking the children to school without a driver actually having to be in the vehicle. Last but not least, it is hoped to reduce the number of accident fatalities in that a computer-controlled system can essentially react faster and more reliably than a person.

This results in an extension of the list published in [3]. Opinions differ as to whether and when autonomous driving will catch on.

Obstacles to automated driving

The idea of driverless driving has been pursued for several decades in numerous research projects [4]. As early as 1986 Professor Dickmanns (University of the Federal Armed Forces, Munich) introduced a "robot vehicle" which was able to drive autonomously on expressways/interstates at speeds of up to 96 km/h. In 1995 Professor Dickmanns' team drove another vehicle from Munich to Copenhagen and back (covering a distance of 1,758 km) with automatic longitudinal and lateral guidance. The system was available for 95% of the journey time.

In 2004 the US Department of Defense organized the DARPA Grand Challenge (DARPA, Defense Advanced Research Projects Agency), the first competition over a longer distance for autonomous vehicles.

The DARPA Urban Challenge organized in 2007 involved the autonomous vehicles having to perform a range of tasks in quasi-urban traffic in a former airforce base.

As the various research projects have shown, today it is perfectly possible to build an autonomous or at least fully automated prototype vehicle. But before such vehicles can be put into series production, some technical and non-technical obstacles have to be overcome.

At the beginning of 2013 Robert Bosch GmbH announced it would be joining various vehicle manufacturers and other suppliers in working on the technologies that are required for automated driving.

Technical obstacles

The technical aspects involved are, from a functional standpoint, essentially environment detection, understanding of the situation, and the action planning derived from them.

Environment detection

With regard to environment detection, all aspects of the vehicle's surroundings as are a matter of course to a human driver must be sensed and detected. These aspects include detecting other road users and detecting the road, obstacles that cannot be driven over, and the traffic infrastructure (e.g. road markings, road signs and traffic lights).

Understanding of the situation

In the final analysis, the gestures of other road users and traffic police officers must also be interpreted. Both the reliable measurement of all these aspects with close-to-production sensor technology and the processing of the data supplied by these environment sensors with close-to-production ECUs are currently stretched to their limits.

Action planning

A further functional challenge is to understand the situation regarding the traffic conditions and to derive appropriate actions (action planning) in real time. Traffic situations are often highly complex, particularly in urban environments, due to the fact that a multitude of road users, traffic regulations and special situations has to be taken into account.

Outlook

With the inexorable advances in development (e.g. increasing computing capacity) it appears that these functional obstacles can be surmounted in the foreseeable future.

As for a further technical obstacle, the validation of functional reliability, there is as yet no solution close to hand: A fully automated vehicle naturally may not jeopardize its occupants or other road users in any situation through failures/faults. The failure rate will therefore have to satisfy extremely stringent requirements for a series-production launch. To ensure this, many millions of kilometers would have to be completed in test drives, which is currently not economically feasible.

Legal obstacles

Finally, there are also legal (non-technical) obstacles that need to be removed. It is unclear for instance who is liable for the damage/injury caused by a fully automated vehicle when the driver no longer has responsibility for vehicle handling.

Aside from the question of liability, highly and fully automated driving functions in public traffic are currently not legal in many countries. According to the Vienna Convention on Road Traffic of 1968, which is legally binding in Germany and in many other countries: "Every driver shall at all times be able to control his vehicle or to guide his animals" and "Every driver of a vehicle shall in all circumstances have his vehicle under control so as to be able to exercise due and proper care and to be at all times in a position to perform all manoeuvres required of him".

This means that the driver may not relinquish his responsibility for vehicle guidance and control and must at least monitor the automated driving function.

But there are also countries where this Convention is not valid. It has not been signed or ratified by the USA, for example. However, that country is covered by other regulations and provisions such as, for example, the Geneva Convention on Road Traffic of 1949. Overall, it can be said that the legal situation worldwide is very inconsistent.

Stages on the road to autonomous driving

In view of the obstacles mentioned, it is sensible not to introduce autonomous driving (some time) all at once, but instead to proceed in stages. In other words, to put into series production initially systems with a lower degree of automation – for example partially automated systems – or alternatively to restrict the situations considered by the system accordingly (see road map, Figure 1). Sensible restrictions or situations which can be kept under control more easily are for example

- expressway/interstate situations, due to the more simply structured and less complex environment,
- traffic-jam, parking and maneuvering situations, since they typically occur in the low-speed range, the sensors involved only need to have a short range, and the vehicle can be brought to a stop very quickly and with only a minimal stopping distance.

Combinations are naturally also conceivable, such as the traffic-jam pilot described in the next section.

Future assisting driving functions

Assisting driving functions such as adaptive cruise control (ACC), lane-keeping support and parking assistant have been established on the market for some years now. Comparatively new are partially automated functions such as the traffic-jam assistant. Here the system performs lateral and longitudinal guidance completely in traffic-jam situations. At low speeds the driver does not have to keep his hands on the steering wheel, but does have to monitor the system and intervene if necessary. This system is also expected to be further developed in the near future for traveling at higher speeds.

The expressway/interstate assistant performs longitudinal and lateral guidance at typical expressway/interstate speeds and also performs lane changes independently. But ultimate responsibility remains with the driver. He must monitor the system so that he can intervene at any time.

One of the first fully automated functions that is expected is the traffic-jam pilot, which assumes full responsibility for vehicle handling/guidance in traffic jams. To realize the follow-on expressway/interstate pilot, it is necessary to provide the emergency stop subfunction, which in an emergency can bring the vehicle to a safe stop on the hard shoulder (emergency lane). Unlike the expressway/interstate assistant, the expressway/interstate pilot assumes responsibility for vehicle handling/guidance. The driver can perform secondary activities, but must where necessary be able to take over the driving task from the system within a maximum on ten seconds.

On top of the expressway/interstate functions, the parking and maneuvering functions are also subject to ever-increasing automation. The first functions expected are those that drive the vehicle under fully automated control into a parking space. The valet-parking function also searches for a free parking space and performs the parking maneuvers in parking garages or parking lots. The driver can leave his vehicle in the entrance area and then pick it up again later at reception (see "Parking and maneuvering systems").

Because traffic situations are at their most complex in urban environments, automated driving in towns and cities is something that is still long off in the future.

Long-term goal

The logical extension of development is a fully automated vehicle that can take the passengers from door to door like a taxi.

References

[1] H. Winner, S. Hakuli, G. Wolf: Handbuch Fahrerassistenzsysteme – Grundlagen, Komponenten und Systeme für aktive Sicherheit und Komfort. 2nd Ed., Vieweg+Teubner, 2012.

[2] M. Buehler, K. Iagnemma, S. Singh: The DARPA Urban Challenge: Autonomous Vehicles in City Traffic (Springer Tracts in Advanced Robotics). Springer-Verlag, 2010.

[3] Bundesanstalt für Straßenwesen: Rechtsfolgen zunehmender Fahrzeugautomatisierung (2012). http://www.bast.de/nn_42254/DE/Publikationen/Berichte/unterreihe-f/2013-2012/f83.html.

[4] http://www.autonomes-fahren.de/geschichte-des-autonomen-fahrens/.

Figure 1: Road map of future driver-assistance functions

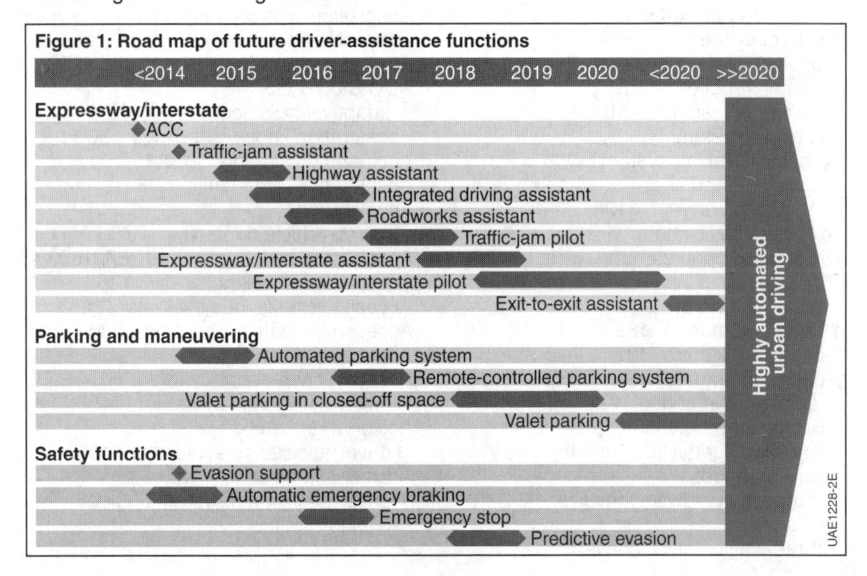

Index of technical terms

Symbols

A

C

D

E

F

G

H

I

J

K

L

M

N

O

Open cycles, thermal engines 399

P

Q

R

S

T

U

V

W

X

Y

Z

Abbreviations

A:

AAS: Atomic Absorption Spectroscopy
ABS: Antilock Braking System
ABS: Acrylnitrile-Butadien-Styrene
AC: Accessory Control
AC: Alternating Current
ACC: Adaptive Cruise Control
ACEA: Association des Constructeurs Européens de l'Automobile (Association of European Automobile Manufacturers)
ADASIS: Advanced Driver Assistance Systems Interface Specifications
ADC: Analog-Digital Converter
ADR: Accord européen relatif au transport international des marchandises Dangereuses par Route (European Agreement concerning the International Carriage of Dangerous Goods by Road)
ADR: Advanced Digital Receiver
ADR: Australian Design Rule
AE: Automotive Electronics
AEB: Automatic Emergency Braking
AFC: Anti-Friction-Coating
AFS: Adaptive Frontlighting System
AGI: Air Gap Insulated
AGM: Absorbent Glass Mat
AGR: Exhaust-gas recirculation (EGR)
AHP: Accelerator Heel Point
AI: Articulation Index
ALB: Automatic load-sensitive braking-force metering (German: Automatische lastabhängige Bremskraftregelung)
ALU: Arithmetic-Logic Unit
AM: Amplitude Modulation
AMLCD: Active Message Liquid Crystal Display
AMR: Anisotropic Magnetoresistive
ANC: Active Noise Control
API: Application Programming Interface
API: American Petroleum Institute
APS: Active Pixel Sensor
APSM: Advanced Porous Silicon Membrane
AS: Air System
ASAM: Association for Standardization of Automation and Measuring Systems
ASF: Audi Space Frame (aluminum frame structure of supporting integrated sheet panels from Audi)
ASIC: Application Specific Integrated Circuit
ASM: Asynchronous Machine
ASM: Trailer control module (German: Anhängersteuermodul)
ASR: Traction control system (German: Antriebsschlupfregelung)
ASS: Anti-Snow System
ASTM: American Society of Testing and Materials
ASU: Automatic interference suppression (German: Automatische Störunterdrückung)
ASVP: Air Saturated Vapour Pressure
ASW: Application Software
AT: Advanced Technology
ATF: Automatic-Transmission Fluid
ATL: Exhaust-gas turbochargers (German: Abgasturbolader)
AUTOSAR: Automotive Open Systems Architecture

B:

BBA: Service-brake system (German: Betriebsbremsanlage)
BCD: Bipolar, CMOS, DMOS
BCI: Bulk Current Injection
BCM: Body Computer Module
BCU: Battery Control Unit
BDC: Bottom Dead Center
BDE: Gasoline direct injection (German: Benzin-Direkteinspritzung)
BEM: Boundary-Element Method
BIP: Beginning of Injection Period
BIPM: Bureau International des Poids et Mesures (International Bureau of Weights and Measures)
BM: Bus-Minus

BMD: Bag Mini Diluter
BMS: Battery Management System
BP: Bus-Plus
BSS: Byte-Start-Sequence
BSW: Basic Software
BtL: Biomass to Liquid
BWC: Butan Working Capacity
BZ: Fuel cell (German: Brennstoffzelle)
BZE: Battery-status recognition (German: Batteriezustandserkennung)

C:

C2C: Car-to-Car communication
C2I: Car-to-Infrastructure communication
CAD: Computer-Aided Design
CAE: Computer-Aided Engineering
CAFE: Corporate Average Fuel Economy
CAL: Computer-Aided Lighting
CAN: Controller Area Network
CARB: California Air Resource Board
CBS: Combined Brake System
CCD: Charge-Coupled Device
CCFL: Cold-Cathode Fluorescence Lamp
CCMC: Comité des Constructeurs d'automobiles du Marché Commun (CCMC manufacturers committee)
CD: Compact Disc
CDD: Complex Device Drivers
CDPF: Catalyzed Diesel Particulate Filter
CE: Coordination Engine
CEMEP: European Committee of Manufacturers of Electrical Machines and Power Electronics
CEN: Comité Européen de Normalisation (European Committee for Standardization)
CF: Carbon Fibers
CFD: Computational Fluid Dynamics
CFPP: Cold Filter Plugging Point
CFR: Cooperative Fuel Research
CFR: Code of Federal Regulations of the United States
CFV: Critical Flow Venturi
CGI: Compacted Graphite Iron
CGPM: Conférence Générale des Poids et Mésures (General Conference on Weights and Measures)
CGW: Central Gateway
CI: Cetane Index
CISC: Complex-Instruction Sot Computing
CISPR: Comité International Spécial des Perturbations Radioélectriques (International Special Committee for Radio Interference)
CLD: Chemiluminescence Detector
CMMI: Capability Maturity Model Integration
CMOS: Complementary Metal Oxide Semiconductor
CMRR: Common Mode Rejection Ratio
CMVSS: Canadian Motor Vehicle Safety Standard
CN: Cetane Number
CNG: Compressed Natural Gas
CO: Communication
COD: Conversion of Olefins to Distillates
COM: Communication
COP: Conformity of Production
CP: Circular Pitch
CPC: Condensation Particulate Counter
Cpsi: Channels per square inch
CPU: Central Processing Unit
CR: Common Rail
CR: Chloroprene Rubber
CRC: Cyclic Redundancy Check
CRS: Common-Rail-System
CRT: Continuously Regenerating Trap
CS: Checksum
CSC: Cell Supervisory Circuit
CtL: Coal to Liquid
CTS: Conti Tire System
CVD: Chemical Vapor Deposition
CVS: Constant Volume Sampling
CVT: Continuously Variable Transmission

D:

DAB: Digital Audio Broadcasting
DAC: Digital-Analog Converter
DAPRA: Defense Advanced Research Projects Agency
DC: Direct Current
DCT: Dynamic Coupling Torque
DCT-C: Dynamic Coupling Torque at Center
DDA: Digital Directional Antenna
DDS: Digital Diversity System
DEF: Diesel Exhaust Fluid
DEQ: Digital Equalizer
DF: Dynamo Field
DFM: Dynamo-Field Monitor
DFV: Vapor/liquid ratio (German: Dampf-Flüssigkeits-Verhältnis)
DI: Direct Injection
DIN: German Institute for Standardization
DIS: Draft International Standard
DK: Dielectric constant
DLC: Diamond-like Carbon
DMOS: Double-diffused Metal-Oxide Semiconductor
DMS: Differential Mobility Spectrometer
DMS: Strain Gages (Strain-gage resistor) (German: Dehnmessstreifen)
DNC: Dynamic Noise Covering
DOC: Diesel Oxidation Catalyst
DOHC: Double Overhead Camshaft
DOT: Department of Transportation
DPF: Diesel Particulate Filter
DRIE: Deep Reactive Ion Etching
DRM: Digital Radio Mondial
DRO: Dielectric Resonance Oscillator
DS: Diagnostic System
DSA: Digital Sound Adjustment
DSM: Diagnostic System Management
DSP: Digital Signal Processor
DSTN-LCD: Double-layer Twisted Nematic-Liquid Crystal Display (German: Doppelschicht Twisted Nematic-Liquid Crystal Display)
DTCO: Digital Tachograph
DTS: Dynamic-Trailing-Sequence
DVB: Digital Video Broadcasting
DVBE: Dry Vapour Pressure Equivalent
DVB-C: Digital Video Broadcasting Cable
DVB-H: Digital Video Broadcasting Handhelds
DVB-S: Digital Video Broadcasting Satellite
DVB-T: Digital Video Broadcasting Terrestrial
DVD: Digital Versatile Disc
DVR: Compression set
DWT: Dynamic Wheel Torque Distribution

E:

E/E: Electrics/Electronics
EAC: Electronic Air Control
EB: Electrical steel sheet and strip (German: Elektroblech und -band)
EBA: Emergency Brake Assist
EBS: Electronic Braking System
EBS: Electronic Battery Sensor
EBV: Electronic braking-force distribution (German: Elektronische Bremskraftverteilung)
EC: European Community
EC: Electronic Commutated
EC: Exhaust Closes
ECE: Economic Commission for Europe
ECU: Electronic Control Unit
ECVT: Electrical Continuously Variable Transmission
EDC: Electronic Diesel Control
EEC: European Economic Community
EEM: Electrical Energy Management (Vehicle electrical system)
EEPROM: Electrically Erasable Programmable Read Only Memory
EEV: Enhanced Environmentally-Friendly Vehicle
EFB: Enhanced Flooded Battery
EFF: Designations of efficiency classes for electrical machines (German: Bezeichnung von Effizienzklassen für elektrische Maschinen)

EFU: Activation spark suppression (German: Einschaltfunkenunterdrückung)
EG: European Community (German: Europäische Gemeinschaft)
EG: Self-steering gradient (German: Eigenlenkgradient)
EGS: Electronic transmission control (German: Elektronische Getriebesteuerung)
EH: Extended Hump
EHB: Electrohydraulic Brake
EHD: Elastohydrodynamics
EIPR: Equivalent Isotropic Radiated Power
EIR: Emissions Information Report
EIS: Electrochemical Impedance Spectroscopy
EKP: Electric fuel pump (German: Elektrokraftstoffpumpe)
ELPI: Electrical Low Pressure Impactor
ELR: European Load Response
EMC: Electromagnetic Compatibility
EMD: Engine Manufacturer Diagnostics
EMP: Electromechanical Parking Brake
EN: European standard
EO: Exhaust Opens
EOBD: European On-Board Diagnostics
EOL: End of Line
EP: Extreme Pressure
EP: Epoxy
EPA: Environment Protection Agency
EPDM: Ethylene Propylene Diene Monomer
EPM: Engine & Powertrain Manager
EPROM: Erasable Programmable Read Only Memory
ES: Exhaust System
ESC: European Steady-State Cycle
ESD: Electrostatic Discharge
ESI: Electronic Service Information
ESP: Electronic Stability Program
ESS: Energy-Saving System
ET: Rim offset (German: Einpresstiefe)
ETBE: Ethyl Tertiary Butyl Ether
ETC: Electronic Throttle Control
ETC: European Transient Cycle
ETK: Emulator probe (German: Emulator-Tastkopf)
ETN: European Type Number
ETRTO: European Tyre and Rim Technical Organization
EU: European Union
euATL: Electrically assisted exhaust-gas turbocharger
EUC: Enhanced Understeering Control
EUDC: Extra Urban Driving Cycle
EUV: Electric changeover valve (German: Elektroumschaltventil)
EV: Fuel injector (German: Einspritzventil)
EV: Electric Vehicle
EVAP: Evaporation
EWIR: Emissions Warranty Information Report

F:

FA: Front Axle
FAEE: Fatty Acid Ethyl Ester
FAME: Fatty Acid Methyl Ester
FAS: Driver-assistance system (German: Fahrerassistenzsystem)
FBA: Parking-brake system (German: Feststellbremsanlage)
FCHV: Fuel Cell Hybrid Vehicle
FC-REX: Fuel Cell Range Extender
FEA: Finite-Element Analysis
FEM: Finite-Element Method
FES: Frame-End-Sequence
FET: Field-Effect Transistor
FFV: Flexible Fuel Vehicle
FGR: Vehicle-speed controller (German: Fahrgeschwindigkeitsregler)
FH: Flat Hump
FID: Flame Ionization Detector
FiL: Function in the Loop
FIR: Far Infrared
FIR: Field Information Report
FIS: Driver-information system (German: Fahrerinformationssystem)
FKM: Forschungskuratorium Maschinenbau
FLO: Fast-Light-Off

FM: Frequency Modulation
FMCW: Frequency Modulated Continuous Wave
FMEA: Failure Mode and Effects Analysis
FMVSS: Federal Motor Vehicle Safety Standard
FPA: Focall-Plane-Array
FRAM: Ferroelectric RAM
FS: Fuel System
FSN: Filter Smoke Number
FSR: Full Speed Range
FSS: Frame-Start-Sequence
FST: Federal Smoke Test
FTA: Fault Tree Analysis
FTIR: Fourier Transform Infrared spectroscopy
FTP: Federal Test Procedure
FW: Front Wheel

G:

GC: Gas-Chromatography Column
GCC: Global Chassis Control
GF: Glass Fibers
GIDAS: German In-Depth Accident Study
GJL: Gray cast iron with lamellar graphite
GJS: Gray cast iron with spheroid graphite
GLP: Glow Plug
GMA: Yaw-moment build-up delay (German: Giermomentaufbau-verzögerung)
GMR: Giant Magnetoresistive
GTDC: Gas-exchange TDC (Top dead center in the exhaust cycle)
GPRS: General Packet Radio Service
GPS: Global Positioning System
GSM: Global System for Mobile Communication
GSY: Speed symbol (tires) (German: Geschwindigkeits-symbol)
GtL: Gas to Liquid
GWP: Global Warming Potential
GZS: Glow control unit (German: Glühzeitsteuergerät)

H:

HBA: Secondary-brake system (German: Hilfsbremsanlage)
HC: Hydrocarbon
HCCI: Homogeneous Charge Compression Ignition
HCI: Hydro Carbon Injection
HD: Heavy Duty
HDDTC: Heavy-Duty Diesel Transient Cycle
HDEV: High-pressure injector (German: Hochdruck-Einspritzventil)
HDG: Light/dark boundary (German: Hell-Dunkel-Grenze)
HDP: High-pressure pump (German: Hochdruckpumpe)
HED: High-Efficiency Diodes
HEV: Hybrid Electric Vehicle
HF: High Frequency
HFM: Hot-film air-mass meters
HFRR: High-Frequency Reciprocating Rig
HiL: Hardware in the Loop
HLDT: Heavy Light-Duty Truck
HLM: Hot-wire air-mass meters
HMM: Homogeneous lean
HNBR: Hydrogenated Nitrile Butadiene Rubber
HNS: Homogeneous Numerically Calculated Surface
HSLA: High Strength Low Alloy
HSV: High-pressure Switching Valve
HTD: High Torque Drive
HTHS: High Temperature High Shear
HUD: Head-Up Display
HV: High Voltage
HVA: Hydraulic valve-clearance compensation

I:

IC: Intake Closes
IC: Integrated Circuit
ICC: Integrated Chassis Control
ICP-OES: Inductively Coupled Plasma Optical Emission Spectrometry
ICM: Integrated Chassis Management
ID: Identifier
IDI: Indirect Injection
IEC: International Electrotechnical Commission

IEEE: Institute of Electrical and Electronics Engineers
IF: Intermediate Frequency
IGBT: Insulated Gate Bipolar Transistor
IGES: Initial Graphics Exchange Specification
IHU: Integrated Head Unit
IHU: Internal high-pressure forming (German: Innenhochdruck-umformung)
IIR: Isobutene-isoprene rubber
ILSAC: International Lubricants Standardization and Approval Committee
I/M: Inspection and Maintenance
IMC: Integrated Magnetic Concentrator
IO: Intake Opens
IP: Ignition Point
IP: Internet Protocol
IR: Infrared
IR: Individual control (ABS) (German: Individualregelung)
IRM: Individual control, modified (ABS) (German: Individual-regelung modifiziert)
IS: Ignition System
ISM: Industrial, Scientific and Medical
ISO: International Organization for Standardization
IT: Information Technology
ITDC: Ignition Top Dead Center
ITO: Indium Tin Oxide
ITSEC: Information Technology Security Evaluation Criteria
ITU: International Telecommunications Union
IUMPR: In-Use Monitor Performance Ratio
IUPAC: International Union of Pure and Applied Chemistry

J:

JAMA: Japan Automobile Manufacturers Association
JASPAR: Japan Automotive Software Platform Architecture
JFET: Junction-gate Field-Effect Transistor

K:

KAMA: Korea Automobile Manufacturers Association
KBA: Federal Road Transport Office (German: Kraftfahrt-Bundesamt)
Kfz: Motor vehicle (German: Kraftfahrzeug)
KOM: Motor bus (German: Kraftomnibus)
KP: Critical point (German: Kritischer Punkt)
KTL: Cathodic deposition (German: Kathodische Tauchlackierung)
KTS: Diagnostics Hardware small testers (German: Diagnostics Hardware Kleintester)
KW: Short waves (German: Kurzwellen)
KW: Crankshaft (German: Kurbelwelle)

L:

LAC: Load Adaptive Control
LAN: Local Area Network
LCD: Liquid Crystal Display
LCV: Light Commercial Vehicles
LDR: Light-Dependent Resistor
LDT: Light-Duty Truck
LDV: Light-Duty Vehicle
LDW: Lane Daparture Warning
LED: Light-Emitting Diode
LEV: Low-Emission Vehicle
LI: Load Index (tires)
Lidar: Light Detection and Ranging
LIN: Local Interconnect Network
LKS: Lane Keeping Support
Lkw: Truck (German: Lastkraftwagen)
LLDT: Light Light-Duty Truck
LLK: Intercooling (charge-air cooling) (German: Ladeluftkühlung)
LMM: Air-flow sensor (German: Luftmengenmesser)
LNG: Liquefied Natural Gas
LP: Low Power
LPG: Liquefied Petroleum Gas
LR: Load Response
LRF: Load Response Driving (German: Load Response Fahrt)
LRR: Long-Range Radar
LRS: Load Response Start
LSF: Low Speed Following

LSG: Laminated Safety Glass
LSI: Large Scale Integration
LSU: Lambda oxygen sensor (universal or wideband sensor) (German: Lambda-Sonde)
LTE: Long Term Evolution
LVDS: Low Voltage Differential Signing
LW: Long Waves (German: Langwellen)
LWL: Optical fibers (German: Lichtwellenleiter)
LWS: Steering-angle sensor (German: Lenkwinkelsensor)

M:

M+S: "Mud and Snow" winter tires
MAG: Active-gas metal-arc (German: Metall-Aktivgas)
MAMAC: MOST Asynchronous Medium Access Control
MB: Mercedes Benz
MCAL: Micro-Controller Abstraction Layer
MCD: Measurement, Calibration and Diagnosis
MDPV: Medium-Duty Passenger Vehicle
MDV: Medium-Duty Vehicle
MEMS: Micro-Electro-Mechanical Systems
MIG: Metal Inert Gas
MIL: Malfunction Indicator Lamp
MIL: Military
MiL: Model in the Loop
MIM: Metal Injection Molding
MISRA: Motor Industry Software Reliability Association
MKS: Multibody system (German: Mehr-Körper-System)
ML: Maturity Level
MMT: Methylcyclopentadienyl Mangan Tricarbonyl
MNEFZ: Modified New European Driving Cycle (German: Modifizierter Neuer Europäischer Fahrzyklus)
MO: Monitoring
MOS: Metal-Oxide Semiconductor
MOSFET: MOS field-effect transistor
MOST: Media-Oriented Systems Transport
MOZ: Motor octane number (German: Motor-Oktanzahl)
MP: Momentary Pole
MPEG: Moving Picture Experts Group
mpg: Miles per gallon
MPP: Most Propable Path
MPT: Multi-Purpose Tire
MRAM: Magnetoresistive RAM
MSG: Gas-shielded metal-arc (German: Metall-Schutzgas)
MSI: Medium-Scale Integration
MSI: Multi-Spark Ignition
MSR: Engine drag-torque control (German: Motorschlepp-momentregelung)
MTBE: Methyl Tertiary Butyl Ether
MTTF: Mean Time To Failure
MW: Medium Waves (German: Mittelwellen)

N:

NAFTA: North American Free Trade Agreement
NBR: Nitrile Butadiene Rubber
NCAP: New Car Assessment Programme
NDIR: Non-Dispersive Infrared (analyzer)
NEDC: New European Driving Cycle
Nfz: Commercial Vehicle (German: Nutzfahrzeug)
Nht: Nitriding depth (German: Nitrierhärtetiefe)
NHTSA: National Highway Traffic Safety Administration
NIC: Network Interface-Controller
NiMH: Nickel-Metal-Hydride
NIR: Near-Infrared
NIT: Network Idle Time
Nkw: Commercial Vehicles (German: Nutzkraftwagen)
NLGI: National Lubricating Grease Institute
NMHC: Non-Methane Hydrocarbon
NMOG: Non-Methane Organic Gases
NMOS: N-channel MOS Transistor
NR: Natural Rubber
NRZ: Non Return to Zero
NSC: NO_x Storage Catalyst

NSM: Shunt-wound machine (German: Nebenschluss-maschine)
NTC: Negative Temperature Coefficient
NTE: Not To Exceed
NVH: Noise Vibration Harshness
NYCC: New York City Cycle

O:

OBD: On-Board Diagnostics
OBD: On-Board Diagnostics (system)
OC: Occupant Classification
OD: Operating Data
ODB: Offset Deformable Barrier Crash
OEM: Original Equipment Manufacturer
OHC: Overhead Camshaft
OHV: Overhead Valves
OLED: Organic Light Emitting Diode
OMM: Surface micromechanics (German: Oberflächenmikro-mechanik)
OP: Operational amplifiers
OPV: Operational amplifiers
OS: Operating System
OSEK: Open systems and their interfaces for electronics in motor vehicles
OSI: Open Systems Interconnection
OTDC: Overlap Top Dead Center
OTX: Open Test Sequence Exchange Format

P:

PA: Polyamide
PAO: Poly-α-Olefin
PAS: Peripheral Acceleration Sensor
PAX: Pneu Accrochage, X = synonym for Michelin radial tire technology
PBT: Polybutylene Terephthalate
PC: Personal Computer
PC: Polycarbonate
PCB: Printed Circuit Board
PCM: Phase Change Memory
PCW: Predictive Collision Warning
PDE: Unit injector system (German: Pumpe-Düse-Einheit)
PDP: Positive Displacement Pump
PDU: Protocol Data Unit
PE: Polyethylene
PEBS: Predictive Emergency Braking System
PEEK: Polyetheretherketone
PEM: Polymer Electrolyte Membrane
PEMS: Portable Emission Measurement System
PES: Poly-Ellipsoid System (headlamp)
PET: Polyethylene Terephthalate
PF: Phenol Formaldehyde
pH: Potentia Hydrogenii
PHV: Plug-in Hybrid Electric Vehicle
PK: Profile code
Pkw: Passenger cars
PLN: Pump-Line Nozzle (Unit Pump System) (German: Pumpe-Leitung-Düse)
PLL: Phase-Locked Loop
PM: Particulate Matter
PMD: Paramagnetic Detector
PMD: Photonic Mixing Device
PMM: Polymethylmethacrylate
PMOS: P-channel MOS Transistor
POF: Polymer Optical Fibers
POI: Points of Interest
POM: Polyoximethylene
PP: Polypropylene
PPS: Polyphenylene Sulfide
PPS: Peripheral Pressure Sensor
PROM: Programmable Read Only Memory
PS: Polystyrene
PSD: Power Spectral Density
PSI: Peripheral Sensor Interface
PSM: Passive Safety Manager
PSOC: Partial State of Charge
PSRR: Power Supply Rejection Ratio
PTB: Physikalisch-Technische Bundesanstalt
PTC: Positive Temperature Coefficient
PTFE: Polytetrafluoroethylene (teflon)
PUR: Polyurethane
PVB: Polyvinyl Butyral
PVC: Polyvinyl Chloride
PVD: Physical Vapor Deposition
PVDF: Polyvinylidene Fluoride
PWM: Pulse Width Modulation
PZEV: Partial Zero-Emission Vehicle
PZT: Plumbum Zirconate Titanate

Q:

QFD: Quality Function Deployment
QM: Quality Management
QVGA: Quarter Video Graphics Array

R:

RA: Rear Axle
RAM: Random Access Memory
RAMSIS: Computer-aided anthropological-mathematical system for passenger simulation (German: Rechnerunterstütztes anthropologisch-mathematisches System zur Insassensimulation)
RDS: Radio Data System
RF: Radio Frequency
RFA: X-ray fluorescence analysis (German: Röntgenfluoreszenzanalyse)
RGM: Reliability-Growth Management
RIM: Reaction-Injection Moulding
RISC: Reduced Instruction-Set Computing
RME: Rapeseed Methyl Ester (alternative fuel)
RMQ: Root Mean Quad
RMS: Root Mean Square
RMV: Rollover Mitigation Function
ROM: Read Only Memory
RON: Research Octane Number
RR: Rolling Resistance
RREG: Council Directive of the European Community (today: European Union)
RRIM: Reinforced Reaction-Injection Moulding
RSM: Series-wound machine (German: Reihenschlussmaschine)
RTA: Real Time Architect
RTE: Run-Time Environment
RW: Rear Wheel
RWAL: Rear Wheel Anti Lock Brake System

S:

SAE: Society of Automotive Engineers
SBC: Sensotronic Brake Control
SBI: Suppression of Background Illumination
SBR: Styrene Butadiene Rubber
SC: System Control
SCR: Selective Catalytic Reduction
SCU: Sensor Control Unit
SD: Secure Digital
SD: System Document
SDARS: Satellite Digital Audio Radio Service
SEI: Software Engineering Institute
SEL: Sound Exposure Level
SFTP: Supplement Federal Test Procedure
SG: Float-angle gradient
SHED: Sealed Housing for Evaporative Determination
SHF: Super High Frequency
SI: Système International (International System of Units)
SI: Speed Index, speed symbol (tires)
SiL: Software in the Loop
SIL: Safety Integrity Level
SIS: Service Information System
SL: Select-Low
SLI: Starting, Lighting, Ignition
SM: Synchronous Machine
SMD: Surface-Mounted Device
SMK: Inertia weight class (German: Schwungmassenklasse)
SMPS: Scanning Mobility Particle Sizer
SMT: Surface-Mount Technology
SOC: State of Charge (battery)
SOF: State of Function (battery)
SOH: State of Health (battery)
SPC: Statistic Process Control
SPICE: Software Process Improvement and Capability Determination
SR: Slew Rate
SRET: Scanning Reference-Electrode Techniques
SRR: Short-Range Radar
SSI: Small-Scale Integration
STEP: Standard for the Exchange of Product Model Data

STN-LCD: Super Twisted Nematic-Liquid Crystal Display
StVZO: Straßenverkehrszulassungs-ordnung (Road Traffic Licens-ing Regulations, for Germany)
SULEV: Super Ultra-Low-Emission Vehicle
SUV: Sport Utility Vehicle
SV: Switchover Valve
SWC: Software Component

T:

TA: Type Approval
TC: Transient Cycle
TCP: Transmission Control Protocol
TD: Torque Demand
TDC: Time to Digital Converter
TDC: Top Dead Center
TDMA: Time Division Multiplex Access
TEM: Transversal Electromagnetic
TFT: Thin-Film Transistor
TLEV: Transitional Low-Emission Vehicle
TMC: Traffic Message Channel
TN-LCD: Twisted Nematic-Liquid Crystal Display
TOF: Time of Flight
TP: Triple Point
TPA: Thermoplastic copolyamides
TPC: Thermoplastic copolyester elastomers
TPE: Thermoplastic Elastomers
TPO: Thermoplastic elastomers on an olefin basis
TPU: Thermoplastic elastomers on a urethane basis
TPV: Cross-linked thermoplastic elastomers on an olefin basis
TPEG: Transport Protocol Experts Group
TPMS: Tire Pressure Monitoring System
TRX: Tension Répartie (distributed tension)
TS: Technical Specification
TS: Torque Structure
TSG: Single-pane Toughened Safety Glass
TSM: Trailer Sway Mitigation
TSS: Transmission-Start-Sequence
TTCAN: Time Triggered CAN
TTL: Transistor-Transistor Logic (Bipolar intogratod digitalcircuit)
TV: Television

U:

UDC: Urban Driving Cycle
UDDS: Urban Dynamometer Driving Schedule
UFOME: Used Frying Oil Methyl Ester
UHF: Ultra-High Frequency
UIS: Unit Injector System
ULEV: Ultra-Low-Emission Vehicle
ULSD: Ultra Low Sulfur Diesel
ULSI: Ultra Large Scale Integration
UMTS: Universal Mobile Telecommunication System
UN: United Nations
UNECE: United Nations Economic Commission for Europe
UPS: Unit Pump System (Pump-Line-Nozzle)
UPS: Ultraviolet Photo Spectroscopy
USA: United States of America
UTP: Unshielded Twisted Pair
UV: Ultraviolet
UWB: Ultra Wide Band

V:

VCI: Volatile Corrosion Inhibitor
VCO: Voltage-Controlled Oscillator
VCR: Variable Compression Ratio
VDA: Association of German Automobile Manufacturers (German: Verband der Automobilindustrie)
VDA-FS: Association of German Automobile Manufacturers – Surface Interface (German: Verband der Automobilindustrie – Flächenschnittstelle)

VDE: Association for Electrical, Electronic and Information Technologies (German: Verband der Elektrotechnik Elektronik Informationstechnik e.V.)
VDI: Association of German Engineers (German: Verein Deutscher Ingenieure)
VDM: Vehicle Dynamics Management
VDU: Vehicle Dynamics Unit
VDX: Vehicle Distributed Executive
VE: Distributor-type injection pump (German: Verteilereinspritzpumpe)
VFB: Virtual Function Bus
VFD: Vacuum Fluorescence Design (Vacuum-fluorescence display technology)
VGA: Video Graphics Array
VHAD: Vehicle Headlamp Aiming Device (Adjusters for front headlamps)
VHD: Vertical Hall Device
VHF: Very High Frequency
VHF: VHF transmissions
VI: Viscosity Index
VLI: Vapor-Lock-Index
VLSI: Very Large Scale Integration
VMM: Volume Micromechanics
VOL: Visual Optical Aim Left
VOR: Visual Optical Aim Right
VPI: Vapor-Phase Inhibitor
VR: Distributor-type injection pump, radial-piston pump (German: Verteilereinspritzpumpe, Radialkolbenpumpe)
VRLA: Valve-Regulated Lead-Acid
VTG: Variable Turbine Geometry (exhaust-gas turbocharger)
VVT: Variable Valve Timing

W:

WdK: Economic Association of the German Rubber Industry (German: Wirtschaftsverband der deutschen Kautschukindustrie)
WHSC: World Harmonized Stationary Cycle
WHTC: World Harmonized Transient Cycle
WLAN: Wireless Local Area Network
WLTP: World Harmonized Light Duty Test Procedure
WNTE: Worldwide Harmonized Not To Exceed
WWH: Worldwide Harmonized
WWW: World Wide Web

X:

XML: Extensible Markup Language
XPS: X-ray Photoelectron Spectroscopy

Z:

ZEV: Zero-Emission Vehicle
zGm: Gross weight rating (German: Zulässige Gesamtmasse)
ZME: Metering unit (German: Zumesseinheit)

Symbols:

3PMSF: Three Peak Mountain Snow Flake (Snow-flake symbol)

Handwritten notes

Handwritten notes

Handwritten notes